Table 3
Derived SI units obtained by combining base units and units with special names

Quantity	Units	Quantity	Units
Acceleration	m/s^2	Molar entropy	$J/mol \cdot K$
Angular acceleration	rad/s^2	Molar heat capacity	$J/mol \cdot K$
Angular velocity	rad/s	Moment of force	$N \cdot m$
Area	m^2	Permeability	H/m
Concentration	mol/m^3	Permittivity	F/m
Current density	$A/m2$	Radiance	$W/m^2 \cdot sr$
Density, mass	kg/m^3	Radiant intensity	W/sr
Electric charge density	C/m^3	Specific heat capacity	$J/kg \cdot K$
Electric field strength	V/m	Specific energy	J/kg
Electric flux density	$C/m2$	Specific entropy	$J/kg \cdot K$
Energy density	J/m^3	Specific volume	m^3/kg
Entropy	J/K	Surface tension	N/m
Heat capacity	J/K	Thermal conductivity	$W/m \cdot K$
Heat flux density	W/m^2	Velocity	m/s
Irradiance	W/m^2	Viscosity, dynamic	$Pa \cdot s$
Luminance	cd/m^2	Viscosity, kinematic	m^2/s
Magnetic field strength	A/m	Volume	m^3
Molar energy	J/mol	Wavelength	m

Table 4
SI prefixes

Multiplication factor	Prefix[a]	Symbol
$1\ 000\ 000\ 000\ 000 = 10^{12}$	tera	T
$1\ 000\ 000\ 000 = 10^{9}$	giga	G
$1\ 000\ 000 = 10^{6}$	mega	M
$1\ 000 = 10^{3}$	kilo	k
$100 = 10^{2}$	hectot	h
$10 = 10^{1}$	deka[b]	da
$0.1 = 10^{-1}$	deci[b]	d
$0.01 = 10^{-2}$	centi[b]	c
$0.001 = 10^{-3}$	milli	m
$0.000\ 001 = 10^{-6}$	micro	μ
$0.000\ 000\ 001 = 10^{-9}$	nano	n
$0.000\ 000\ 000\ 001 = 10^{-12}$	pico	p
$0.000\ 000\ 000\ 000\ 001 = 10^{-15}$	femto	f
$0.000\ 000\ 000\ 000\ 000\ 001 = 10^{-18}$	atto	a

[a] The first syllable of every prefix is accented so that the prefix will retain its identity. Thus, the preferred pronunciation of kilometer places the accent on the first syllable, not the second.

[b] The use of these prefixes should be avoided, except for the measurement of areas and volumes and for the nontechnical use of centimeter, as for body and clothing measurements.

Student

Please enter my subscription for **Engineering News-Record**.

6 months ☐ $29.50 (Domestic)

Name

Address

City State Zip

☐ Payment enclosed ☐ Bill me later

McGraw_Hill CONSTRUCTION ENR
5EN2DMHE

Friendly

Please enter my subscription for **Architectural Record**.

6 months ☐ $19.50 (Domestic)

Name

Address

City State Zip

☐ Payment enclosed ☐ Bill me later

McGraw_Hill CONSTRUCTION Architectural Record 5AR2DMHE

Savings

Please enter my subscription for **Aviation Week & Space Technology**.

6 months ☐ $29.95 (Domestic)

Name

Address

City State Zip

☐ Payment enclosed ☐ Bill me later

AVIATION WEEK'S
AVIATION WEEK & SPACE TECHNOLOGY
www.AviationNow.com/awst

CAW34EDU

Wastewater Engineering

The McGraw-Hill Series in Civil and Environmental Engineering

Wastewater Engineering

Treatment and Reuse

Fourth Edition

Metcalf & Eddy, Inc.

Revised by

George Tchobanoglous
Professor Emeritus of Civil and Environmental Engineering
University of California, Davis

Franklin L. Burton
Consulting Engineer
Los Altos, California

H. David Stensel
Professor of Civil and Environmental Engineering
University of Washington, Seattle

Boston   Burr Ridge, IL   Dubuque, IA   Madison, WI   New York
San Francisco   St. Louis   Bangkok   Bogotá   Caracas   Kuala Lumpur
Lisbon   London   Madrid   Mexico City   Milan   Montreal   New Delhi
Santiago   Seoul   Singapore   Sydney   Taipei   Toronto

McGraw-Hill Higher Education

A Division of The McGraw-Hill Companies

WASTEWATER ENGINEERING, TREATMENT AND REUSE
FOURTH EDITION

Published by McGraw-Hill, a business unit of The McGraw-Hill Companies, Inc., 1221 Avenue of the Americas, New York, NY 10020. Copyright © 2003, 1991, 1979, 1972 by The McGraw-Hill Companies, Inc. All rights reserved. No part of this publication may be reproduced or distributed in any form or by any means, or stored in a database or retrieval system, without the prior written consent of The McGraw-Hill Companies, Inc., including, but not limited to, in any network or other electronic storage or transmission, or broadcast for distance learning.

Some ancillaries, including electronic and print components, may not be available to customers outside the United States.

This book is printed on acid-free paper.

International   3 4 5 6 7 8 9 0 DOC/DOC 0 9 8 7 6 5 4
Domestic   3 4 5 6 7 8 9 0 DOC/DOC 0 9 8 7 6 5 4

ISBN 0–07–041878–0
ISBN 0–07–112250–8 (ISE)

General manager: *Thomas E. Casson*
Publisher: *Elizabeth A. Jones*
Sponsoring editor: *Suzanne Jeans*
Developmental editor: *Amy W. Hill*
Executive marketing manager: *John Wannemacher*
Senior project manager: *Kay J. Brimeyer*
Production supervisor: *Sherry L. Kane*
Coordinator of freelance design: *Rick D. Noel*
Cover designer: *Gino Cieslik*
Background image: *Courtesy of George Tchobanoglous*
Inset image: *Deer Island Wastewater Treatment Facilities, Boston, Massachusetts Kevin Kirwin/RVA/MWRA*
Senior photo research coordinator: *Lori Hancock*
Lead supplement producer: *Audrey A. Reiter*
Media technology senior producer: *Phillip Meek*
Compositor: *Lachina Publishing Services*
Typeface: *10/12 Times Roman*
Printer: *R. R. Donnelley & Sons Company, Crawfordsville, IN*

Photographs: All of the photographs for this textbook were taken by George Tchobanoglous, unless otherwise noted.

Library of Congress Cataloging-in-Publication Data

Wastewater engineering : treatment and reuse / Metcalf & Eddy, Inc. — 4th ed. / revised by George Tchobanoglous, Franklin L. Burton, H. David Stensel.
      p.   cm. — (McGraw-Hill series in civil and environmental engineering)
      Includes bibliographical references and indexes.
      ISBN 0–07–041878–0
      1. Sewerage. 2. Sewage disposal. 3. Water reuse. I. Metcalf & Eddy. II. Tchobanoglous, George. III. Burton, Franklin L. (Franklin Louis), 1927–. IV. Stensel, H. David. V. Series.

TD645 .W295 2003
628.3—dc21
                                                        2001053724
                                                        CIP

INTERNATIONAL EDITION ISBN 0–07–112250–8
Copyright © 2003. Exclusive rights by The McGraw-Hill Companies, Inc., for manufacture and export. This book cannot be re-exported from the country to which it is sold by McGraw-Hill. The International Edition is not available in North America.

www.mhhe.com

George Tchobanoglous is a professor emeritus of environmental engineering in the Department of Civil and Environmental Engineering at the University of California at Davis. He received a B.S. degree in civil engineering from the University of the Pacific, an M.S. degree in sanitary engineering from the University of California at Berkeley, and a Ph.D. in environmental engineering from Stanford University. His principal research interests are in the areas of wastewater treatment, wastewater filtration, UV disinfection, aquatic wastewater management systems, solid waste management, and wastewater management for small systems. He has authored or coauthored over 350 technical publications, including 12 textbooks and two reference works. The textbooks are used in more than 200 colleges and universities throughout the United States. The textbooks and reference books are also used extensively by practicing engineers both in the United States and abroad. Professor Tchobanoglous serves nationally and internationally as consultant to both governmental agencies and private concerns. An active member of numerous professional societies, he is a past president of the Association of Environmental Engineering and Science Professors. He is a registered civil engineer in California.

Franklin L. Burton spent 30 years with Metcalf & Eddy serving as vice president and chief engineer in their western regional office in Palo Alto, CA. He received a B.S. degree in mechanical engineering from Lehigh University and an M.S. degree in civil engineering from the University of Michigan. He has been involved in the planning, design, and technical review of over 50 wastewater treatment plants. He was the coauthor of the third edition of this textbook. He retired from Metcalf & Eddy in 1986 and is in private practice in Los Altos, CA, specializing in treatment technology evaluation, energy management, facilities design review, and value engineering. He is a registered civil and mechanical engineer in California and is a life member of the American Society of Civil Engineers, Water Environment Federation, and the American Water Works Association.

H. David Stensel is a professor of civil and environmental engineering at the University of Washington, Seattle, WA. Prior to his academic positions, he spent 10 years in practice developing and applying industrial and municipal wastewater treatment processes. He received a B.S. degree in civil engineering from Union College, Schenectady, NY, and M.E. and Ph.D. degrees in environmental engineering from Cornell University. His principal research interests are in the areas of wastewater treatment, biological nutrient removal, sludge processing methods, biodegradation of hazardous substances, and stormwater treatment. He has authored or coauthored over 100 technical publications and a textbook on biological nutrient removal. He has received the ASCE Rudolf Hering Medal and twice received the Water Environment Federation Harrison

Prescott Eddy Medal for his research contributions. He is a member of numerous professional societies, and has served as chair of the ASCE Environmental Engineering Division, treasurer of the American Association of Environmental Engineering Professors, and associate editor of the *Water Environment Federation Research Journal.* He is a registered professional engineer and a diplomate in the American Academy of Environmental Engineers.

Contents

Appendixes

Indexes

Preface

During the past 12 years since the publication of the third edition of this textbook, the number of new developments and changes that have occurred in the field of wastewater engineering has been dramatic, especially with respect to

(1) the characterization of the constituents found in wastewater, both in terms of the range of constituents and the detection limits;
(2) a greater fundamental understanding of the mechanisms of biological wastewater treatment;
(3) the application of advanced treatment methods for the removal of specific constituents;
(4) the increased emphasis on the management of the biosolids resulting from the treatment of wastewater; and
(5) the issuance of more comprehensive and restrictive permit requirements for the discharge and reuse of treated wastewater.

The fourth edition of this textbook has been prepared to address the significant new developments and changes that have occurred in the field and to correct other issues with the third edition to make the fourth edition more user friendly. For example, the theory and practice chapters, separated in the third edition, are now combined in the fourth edition to provide subject continuity and eliminate redundancy. Because of the importance of biological wastewater treatment, four separate chapters have been devoted to this subject. The chapter on advanced wastewater treatment has been expanded to include processes that are increasingly required to meet more stringent discharge requirements. A new chapter on disinfection has been added to deal with recent developments in the field. The chapter on reclamation and reuse has been revised completely and much new material has been added. Because of the importance of biosolids management, an entire chapter is devoted to this subject. The issues of process design and performance to meet more stringent permit requirements, including the upgrading of existing treatment plants, are considered in Chapter 15.

IMPORTANT FEATURES OF THIS BOOK

Following the practice in the third edition, more than 100 new example problems have been worked out in detail to enhance the readers' understanding of the basic concepts presented in the text. Wherever possible, spreadsheet solutions are presented. To aid in the planning, analysis, and design of wastewater management systems, design data and information are summarized and presented in more than 300 tables, most of which are new. To illustrate the principles and facilities involved in the field of wastewater management, over 570 illustrations, graphs, diagrams, and photographs are included. To help the readers of this textbook hone their analytical skills and mastery of the material, problems and discussion topics are included at the end of each chapter. Selected references are also provided for each chapter.

The International System (SI) of Units is used in the fourth edition. The use of SI units is consistent with teaching practice in most U.S. universities and in many countries throughout the world. In general, dual sets of units (i.e., SI and U.S. customary) have been used for the data tables. Where the use of double units was not possible, conversion factors are included as a footnote to the table.

To further increase the utility of this textbook, several appendixes have been included. Conversion factors from International System (SI) of Units to U.S. Customary Units and the reverse are presented in Appendixes A–1 and A–2, respectively. Conversion factors commonly used for the analysis and design of wastewater management systems are presented in Appendix A–3. Abbreviations for SI and U.S. Customary units are presented in Appendixes A–4 and A–5, respectively. Physical characteristics of air and selected gases and water are presented in Appendixes B and C, respectively. Dissolved oxygen concentrations in water as a function of temperature are presented in Appendix D. Tables of most probable numbers (MPN) are presented in Appendix E, carbonate equilibrium is considered in Appendix F, and Moody diagrams for the analysis of flow in pipes are presented in Appendix G.

USE OF THIS BOOK

Enough material is presented in this textbook to support a variety of courses for one or two semesters or three quarters at either the undergraduate or graduate level. The specific topics to be covered will depend on the time available and the course objectives. Suggested course outlines follow.

For a one-semester introductory course on wastewater treatment, the following material is suggested:

Topic	Chapter	Sections
Introduction to wastewater treatment	1	All
Wastewater characteristics	2	All
Wastewater flowrates and constituent loadings	3	All
Introduction to process analysis	4	All
Physical unit operations	5	5–1 to 5–8
Chemical unit operations	6	6–1, 6–2,
Introduction to biological treatment of wastewater	7	All
Disinfection	12	12–1 to 12–5, 12–9
Water reuse	13	13–1 to 13–2
Biosolids management	14	All
Introduction to treatment plant performance	15	15–1 to 15–3

For a two-semester course on wastewater treatment, the following material is suggested:

Topic	Chapter	Sections
Introduction to wastewater treatment	1	All
Wastewater characteristics	2	All
Wastewater flowrates and constituent loadings	3	All
Introduction to process analysis	4	All
Introduction to treatment plant performance	15	15–1 to 15–3
Physical unit operations	5	All
Chemical unit operations	6	All
Introduction to biological treatment of wastewater	7	All
Suspended growth biological treatment processes	8	All
Attached growth and combined biological treatment processes	9	9–1 to 9–5
Anaerobic suspended and attached growth treatment processes	10	10–1, 10–2, 10–4
Disinfection	12	All
Water reuse	13	All
Biosolids management	14	All
Process control and upgrading treatment plant performance	15	15–3 to 15–7

For a one-semester course on biological wastewater treatment, the following material is suggested:

Topic	Chapter	Sections
Introduction to wastewater treatment	1	All
Wastewater characteristics	2	All
Introduction to process analysis	4	All
Introduction to treatment plant performance	15	15–1 to 15–3
Introduction to biological treatment of wastewater	7	All
Suspended growth biological treatment processes	8	All
Attached growth and combined biological treatment processes	9	All
Anaerobic suspended and attached growth treatment processes	10	All
Anaerobic and aerobic digestion and composting	14	14–9 to 14–11

For a one-semester course on wastewater reclamation and reuse, the following material is suggested:

Topic	Chapter	Sections
Introduction to wastewater treatment	1	All
Wastewater characteristics	2	All
Introduction to water reclamation and reuse	13	13–1
Introduction to risk assessment	13	13–3
Introduction to treatment plant performance	15	15–1 to 15–3
Advanced wastewater treatment (optional)	11	11–6
Disinfection	12	12–1 to 12–5, 12–7 to 12–9
Water reclamation technologies	13	13–4
Storage of reclaimed water	13	13–5
Reuse of reclaimed water	13	13–6 to 13–9
Planning consideration for reclamation and reuse	13	10

For a one-semester course on physical and chemical unit operations and processes, the following material is suggested. It should be noted that material listed below could be supplemented with additional examples from water treatment.

Topic	Chapter	Sections
Introduction to process analysis	4, 15	All
Introduction to treatment plant performance	15	15–1 to 15–3
Introduction to physical unit operations		
Mixing and flocculation	5	5–4
Sedimentation	5	5–5, 5–7, 5–8
Gas transfer	5	5–11 to 5–12
Filtration (conventional depth filtration)	11	11–3, 11–4
Membrane filtration	11	11–6
Adsorption	11	11–7
Gas stripping	5, 11	5–13, 11–8
UV disinfection	12	12–9
Introduction to chemical unit processes		6–2
Coagulation	6	6–2
Chemical precipitation	6	6–3 to 6–5
Ion exchange	11	11–9
Water stabilization	6	6–7
Chemical oxidation (conventional)	6	6–6
Advanced oxidation processes	11	11–9

ACKNOWLEDGMENTS

A book of this magnitude could not have been written without the assistance of numerous individuals. First and foremost, Mr. Harold Leverenz, a doctoral candidate at the University of California at Davis, provided exceptional assistance. He read and commented on all of the drafts, checked the problems, and prepared many of the new figures for this text. In addition, he helped review the page proofs. His devotion to the task of making this book student-friendly was beyond the call of duty.

Other individuals who contributed, arranged in alphabetical order, are: Mr. Mike Anderson of Nolte and Associates, reviewed portions of the text and worked several design examples; Professor Takashi Asano of the University of California at Davis, the 2001 Stockholm Water Prize Laureate, revised Chap. 13, which he had contributed to the third edition; Dr. Keith Bourgeous of Carollo Engineers contributed and reviewed portions of Chap. 11; Mr. Max Burchett of Whitley Burchett & Associates in numerous discussions over many years has contributed valuable insights on the application of theory to practice; Ing. Ermanno Cacciari of Austep Environmental Protection, Milan, Italy, contributed to the section on anaerobic sludge blanket processes in Chap. 10; Dr. Robert Cooper of BioVir laboratories reviewed and provided valuable input for Chap. 2; Professor John C. Crittenden of Michigan Technological University reviewed Chap. 4 and the section on carbon adsorption in Chap. 11; Dr. Alex Ekster, of Ekster and Associates, contributed the sections on process control optimization in Chap. 15; Dr. Robert Emerick of Ecologic Engineers, contributed the section on UV disinfection in Chap. 11; Mr. William Hartnett of Montgomery/Watson contributed writeups on program logic controllers and piping and instrument diagrams in Chap. 15; Dr. Tim Haug of the City of Los Angeles reviewed and provided valuable insight on Chaps. 7 through 10 and 15; Professor David Jenkins of the University of California at Berkeley, provided photomicrographs of filamentous microorganisms; Ms. Sarah Mayhew printed most of the photographs; Ms. Margie Nellor of the County Sanitation Districts of Los Angeles County provided a photograph of the Rio Hondo spreading basins, Professor Kara Nelson of the University of California at Berkeley reviewed Chap. 2; Mr. Andrew Salveson of Whitley Burchett & Associates, reviewed Chaps. 2 and 4 and provided data for Chap. 11; Professor Edward D. Schroeder of the University of California at Davis reviewed an early draft of Chap. 7 and provided valuable organizational and technical comments and guidance, and as a colleague of 30 years, Professor Schroeder has contributed significantly to the education of the senior author; Dr. Richard E. Speece of Vanderbilt University reviewed the section on anaerobic digestion in Chap. 14 and provided valuable insight; Mr. Jeff Sollar of EOA, contributed to the probabilistic analysis of multiple processes in Chap. 13; Dr. Rhodes Trussell of Montgomery/Watson in numerous discussions offered valuable insights on disinfection, flocculation, and mixing processes; and Mr. Mike Wehner of the Orange County Water District provided a photograph of the water spreading basins in Orange County. The collective efforts of these individuals is invaluable.

Reviewers, selected by the publisher, who were asked to assess and comment on the prospectus for the fourth edition included Professor James J. Bisogni Jr. of Cornell University, Professor Alan R. Bowers of Vanderbilt University, Professor Jeff Kuo of California State University at Fullerton, Professor Bruce Logan of Pennsylvania State University, Professor John A. Olofsson of the University of Alaska at Anchorage, and

Professor Tian C. Zhang of the University of Nebraska at Lincoln. Their input, early on, helped guide the development of the final format of the fourth edition. Their contributions are acknowledged gratefully.

The reviewers, selected by the publisher, who read the entire manuscript of the fourth edition, were Professor Syed A. Hashsham of Michigan State University; Professor Robert Lang of California Polytechnic University at San Luis Obispo; Professor John T. Novak of Virginia Polytechnic Institute and State University; and Professor Robert M. Sykes of Ohio State University. They provided valuable and timely comments that improved the content, organization, and readability of this textbook. Their contributions were significant and are acknowledged gratefully.

The assistance of the staff of Metcalf & Eddy in the preparation of this text is also acknowledged. The efforts of Mr. James Anderson were especially important in making this book possible and in managing the resources made available by Metcalf & Eddy to the authors. Mr. Jonathan Doane organized the team of staff reviewers and provided liaison between the authors and the reviewers. Valuable comments were provided that reflect current design practice in both the United States and overseas.

The McGraw-Hill staff was also critical to the production of this textbook. Mr. Eric Munson, formerly of McGraw-Hill, was instrumental in the early development of this project. Ms. Amy Hill, Developmental Editor, served as the overall project manager. Her organizational skills and tireless efforts on our behalf have made this book a reality. Her sunny personality was also a great help. Ms. Kay J. Brimeyer served as production coordinator and helped keep all of the loose ends together, while maintaining her sense of humor. Ms. Susan Sexton served as the technical editor. The publishers were Mr. Tom Casson and Ms. Betsy Jones.

Finally to Rosemary Tchobanoglous and Nancy Burton who suffered, supported, and encouraged us through the writing of this textbook, we are eternally grateful. Support for Dave Stensel by Carleen Clark and Pat Halikas was especially helpful.

George Tchobanoglous *Davis, CA*   **Franklin L. Burton** *Los Altos, CA*   **H. David Stensel** *Seattle, WA*

Foreword

Almost 90 years have passed since Metcalf & Eddy first published *American Sewage Practice,* the legendary three-volume treatise that established design standards for sewerage facilities. Subsequently, the three volumes were combined into a single text, *Sewerage and Sewage Disposal,* in 1922 and a second edition was published in 1930 for class use in engineering schools. In 1972, a new version of the textbook was published, *Wastewater Engineering: Collection, Treatment, and Disposal,* followed in 1979 by a second edition, *Wastewater Engineering: Treatment, Disposal, Reuse.* A companion textbook, *Wastewater Engineering: Collection and Pumping of Wastewater,* was also published in 1981. The most recent publication was the third edition of *Wastewater Engineering* in 1991. Even though the wastewater practice has continued to evolve and grow during this period, no time period in the past can equal the last decade in terms of technological development.

In addition, the awareness of environmental issues among the U.S and world communities has reached a level not seen before. This active awareness is driving our industry to achieve levels of performance far beyond those envisioned even as recently as the last decade. This fourth edition of *Wastewater Engineering* incorporates these concepts as an essential part of a central theme. As a result, the fourth edition has been designed with a forward-looking focus. For example, emerging fields of biological process modeling and genetic engineering are addressed with some predictions on where these concepts may fit into future wastewater engineering activities.

Since the third edition was published in 1991, much focus in the wastewater practice has turned to nutrient removal, with particular emphasis on biological nutrient removal (BNR). Research in this field is being carried on worldwide, and new discoveries that challenge some of the conventional theory continue to be made. The chapters in the fourth edition that pertain to biological processes and BNR are therefore essentially new. They deal with both research results and their applications to wastewater engineering design.

Pressure for environmental compliance today is greater than ever. Regulatory requirements are, as always, present and forceful. Support from the community for environment-related programs is becoming a stronger driving force than ever before. *Stakeholders,* as they are often referred to, are quite demanding, well organized, and informed. They challenge wastewater engineers to stretch the performance of existing infrastructure through applied research programs. This concept, referred to by Metcalf & Eddy as "infrastretching," represents one of the most significant challenges to the practice of wastewater engineering in the new century.

ACKNOWLEDGMENTS

Metcalf & Eddy, Inc. has been privileged to have as our principal authors Dr. George Tchobanoglous of the University of California at Davis; Franklin L. Burton, a retired vice president of Metcalf & Eddy; and Dr. David Stensel of the University of Washington.

The principal authors were responsible for scope of the textbook, writing, editing, and coordination with the reviewers and the publisher. Two of our principal authors have enjoyed a long association with the textbook. The third principal author, Dr. Stensel, has joined the team for the fourth edition. Dr. Tchobanoglous was the principal author for the first three editions of this text and for the first two editions of the companion volume, *Wastewater Engineering: Collection and Pumping of Wastewater.* Mr. Burton was technical reviewer for the second edition of this text and the companion text and a principal author for the third edition.

The staff of Metcalf & Eddy played a significant role in the preparation of this edition. We acknowledge the contributions of the following members of the staff for their valuable efforts in review of manuscripts:

Kevin L. Anderson	Jonathan W. Doane
David P. Bova	Robert Gay
John G. Chalas	Charles E. Pound

We would also like to offer our sincere gratitude to Dr. Roger T. Haug for his review of the chapters that cover biological wastewater treatment processes. Dr. Haug's comments were very helpful with respect to achieving one of our goals: the presentation of wastewater engineering in a forward-looking fashion.

Special appreciation is extended to Jonathan W. Doane for his efforts in managing Metcalf & Eddy's efforts and to John G. Chalas, the former director of technology of Metcalf & Eddy, now retired, who served as special advisor.

An effort such as this fourth edition could not be successful without the professional encouragement and support of corporate management. We gratefully acknowledge the contributions of Robert H. Fisher, chairman and CEO of Metcalf & Eddy, Inc., and John Somerville, president of Metcalf & Eddy. We would also like to acknowledge the vision of our parent company, AECOM Technologies, Inc., acting through Richard Newman, chairman, and Raymond Holdsworth, president, for the commitment they made to complete this effort and to pave the way for future editions.

Metcalf & Eddy would like to make special note of the passing in 2001 of Harry L. Kinsel, a former president of Metcalf & Eddy. Harry Kinsel was known for his professionalism and commitment to education. Metcalf & Eddy established an internal competition for technical papers in 1973 in Harry Kinsel's honor and this competition will continue as an inspiration to Metcalf & Eddy staff and the industry.

Finally, I would like to personally express my appreciation to the entire team of authors, editors, and reviewers for their tireless effort, and the senior management of Metcalf & Eddy, Inc., and AECOM Technologies, Inc., for the opportunity to direct this effort for Metcalf & Eddy.

James Anderson
Senior Vice President
Chief Engineer
Metcalf & Eddy, Inc.

1

Wastewater Engineering: An Overview

Every community produces both liquid and solid wastes and air emissions. The liquid waste—wastewater—is essentially the water supply of the community after it has been used in a variety of applications (see Fig. 1–1). From the standpoint of sources of generation, wastewater may be defined as a combination of the liquid or water-carried wastes removed from residences, institutions, and commercial and industrial establishments, together with such groundwater, surface water, and stormwater as may be present.

When untreated wastewater accumulates and is allowed to go septic, the decomposition of the organic matter it contains will lead to nuisance conditions including the production of malodorous gases. In addition, untreated wastewater contains numerous

Figure 1-1

Schematic diagram of a wastewater management infrastructure.

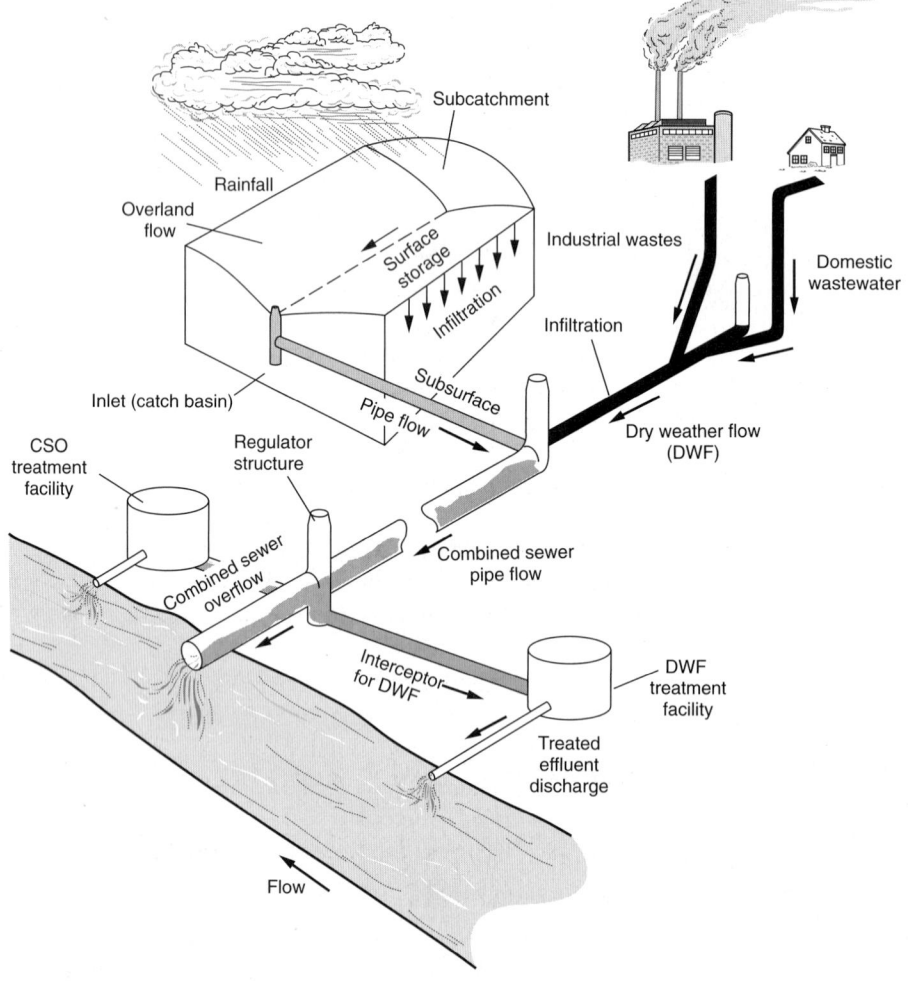

pathogenic microorganisms that dwell in the human intestinal tract. Wastewater also contains nutrients, which can stimulate the growth of aquatic plants, and may contain toxic compounds or compounds that potentially may be mutagenic or carcinogenic. For these reasons, the immediate and nuisance-free removal of wastewater from its sources of generation, followed by treatment, reuse, or dispersal into the environment is necessary to protect public health and the environment.

Wastewater engineering is that branch of environmental engineering in which the basic principles of science and engineering are applied to solving the issues associated with the treatment and reuse of wastewater. The ultimate goal of wastewater engineering is the protection of public health in a manner commensurate with environmental, economic, social, and political concerns. To protect public health and the environment, it is necessary to have knowledge of (1) constituents of concern in wastewater, (2) impacts of these constituents when wastewater is dispersed into the environment, (3) the transformation and long-term fate of these constituents in treatment processes, (4) treatment methods that can be used to remove or modify the constituents found in wastewater, and (5) methods for beneficial use or disposal of solids generated by the treatment systems.

To provide an initial perspective on the field of wastewater engineering, common terminology is first defined followed by (1) a discussion of the issues that need to be addressed in the planning and design of wastewater management systems and (2) the current status and new directions in wastewater engineering.

1-1 TERMINOLOGY

In the literature, and in governmental regulations, a variety of terms have been used for individual constituents of concern in wastewater. The terminology used commonly for key concepts and terms in the field of wastewater management is summarized in Table 1-1. In some cases, confusion and undue negative perceptions arise with the use of the terms *contaminants, impurities,* and *pollutants,* which are often used interchangeably. To avoid confusion, the term *constituent* is used in this text in place of these terms to refer to an individual compound or element, such as ammonia nitrogen. The term *characteristic* is used to refer to a group of constituents, such as physical or biological characteristics.

The term "sludge" has been used for many years to signify the residuals produced in wastewater treatment. In 1994, the Water Environment Federation adopted a policy defining "biosolids" as a primarily organic, solid wastewater treatment product that can be recycled beneficially. In this policy, "solids" are defined as the residuals that are derived from the treatment of wastewater. Solids that have been treated to the point at which they are suitable for beneficial use are termed "biosolids." In this text, the terms of solids and biosolids are used extensively, but "sludge" continues to be used, especially in cases where untreated solid material and chemical residuals are referenced.

1-2 IMPACT OF REGULATIONS ON WASTEWATER ENGINEERING

From about 1900 to the early 1970s, treatment objectives were concerned primarily with (1) the removal of colloidal, suspended, and floatable material, (2) the treatment of biodegradable organics, and (3) the elimination of pathogenic organisms. Implementation in the United States of the Federal Water Pollution Control Act Amendments of 1972 (Public Law 92-500), also known as the Clean Water Act (CWA), stimulated substantial changes in wastewater treatment to achieve the objectives of "fishable and swimmable" waters. Unfortunately, these objectives were not uniformly met.

From the early 1970s to about 1980, wastewater treatment objectives were based primarily on aesthetic and environmental concerns. The earlier objectives involving the reduction of biological oxygen demand (BOD), total suspended solids (TSS), and pathogenic organisms continued but at higher levels. Removal of nutrients, such as nitrogen and phosphorus, also began to be addressed, particularly in some of the inland streams and lakes, and estuaries and bays such as Chesapeake Bay and Long Island Sound. Major programs were undertaken by both state and federal agencies to achieve more effective and widespread treatment of wastewater to improve the quality of the surface waters. These programs were based, in part, on (1) an increased understanding of the environmental effects caused by wastewater discharges; (2) a greater appreciation of the adverse long-term effects caused by the discharge of some of the specific constituents

Table 1–1

Terminology commonly used in the field of wastewater engineering[a]

Term	Definition
Biosolids	Primarily an organic, semisolid wastewater product that remains after solids are stabilized biologically or chemically and are suitable for beneficial use
Class A biosolids[b]	Biosolids in which the pathogens (including enteric viruses, pathogenic bacteria, and viable helminth ova) are reduced below current detectable levels
Class B biosolids[b]	Biosolids in which the pathogens are reduced to levels that are unlikely to pose a threat to public health and the environment under specific use conditions. Class B biosolids cannot be sold or given away in bags or other containers or applied on lawns or home gardens
Characteristics (wastewater)	General classes of wastewater constituents such as physical, chemical, biological, and biochemical
Composition	The makeup of wastewater, including the physical, chemical, and biological constituents
Constituents[c]	Individual components, elements, or biological entities such as suspended solids or ammonia nitrogen
Contaminants	Constituents added to the water supply through use
Disinfection	Reduction of disease-causing microorganisms by physical or chemical means
Effluent	The liquid discharged from a processing step
Impurities	Constituents added to the water supply through use
Nonpoint sources	Sources of pollution that originate from multiple sources over a relatively large area
Nutrient	An element that is essential for the growth of plants and animals. Nutrients in wastewater, usually nitrogen and phosphorus, may cause unwanted algal and plant growths in lakes and streams
Parameter	A measurable factor such as temperature
Point sources	Pollutional loads discharged at a specific location from pipes, outfalls, and conveyance methods from either municipal wastewater treatment plants or industrial waste treatment facilities
Pollutants	Constituents added to the water supply through use
Reclamation	Treatment of wastewater for subsequent reuse application or the act of reusing treated wastewater
Recycling	The reuse of treated wastewater and biosolids for beneficial purposes
Repurification	Treatment of wastewater to a level suitable for a variety of applications including indirect or direct potable reuse
Reuse	Beneficial use of reclaimed or repurified wastewater or stabilized biosolids
Sludge	Solids removed from wastewater during treatment. Solids that are treated further are termed biosolids
Solids	Material removed from wastewater by gravity separation (by clarifiers, thickeners, and lagoons) and is the solid residue from dewatering operations

[a] Adapted, in part, from Crites and Tchobanoglous (1998).
[b] U.S. EPA (1997b).
[c] To avoid confusion, the term "constituents" is used in this text in place of contaminants, impurities, and pollutants.

found in wastewater; and (3) the development of national concern for the protection of the environment. As a result of these programs, significant improvements have been made in the quality of the surface waters.

Since 1980, the water-quality improvement objectives of the 1970s have continued, but the emphasis has shifted to the definition and removal of constituents that may cause long-term health effects and environmental impacts. Health and environmental concerns are discussed in more detail in the following section. Consequently, while the early treatment objectives remain valid today, the required degree of treatment has increased significantly, and additional treatment objectives and goals have been added. Therefore, treatment objectives must go hand in hand with the water quality objectives or standards established by the federal, state, and regional regulatory authorities. Important federal regulations that have brought about changes in the planning and design of wastewater treatment facilities in the United States are summarized in Table 1–2. It is interesting to note that the clean air acts of 1970 and 1990 have had a significant impact on industrial and municipal wastewater programs, primarily through the implementation of treatment facilities for the control of emissions.

Table 1–2
Summary of significant U.S. federal regulations that affect wastewater management

Regulation	Description
Clean Water Act (CWA) (Federal Water Pollution Control Act Amendments of 1972)	Establishes the National Pollution Discharge Elimination System (NPDES), a permitting program based on uniform technological minimum standards for each discharger
Water Quality Act of 1987 (WQA) (Amendment of the CWA)	Strengthens federal water quality regulations by providing changes in permitting and adds substantial penalties for permit violations. Amends solids control program by emphasizing identification and regulation of toxic pollutants in sewage sludge
40 CFR Part 503 (1993) (Sewage Sludge Regulations)	Regulates the use and disposal of biosolids from wastewater treatment plants. Limitations are established for items such as contaminants (mainly metals), pathogen content, and vector attraction
National Combined Sewer Overflow (CSO) Policy (1994)	Coordinates planning, selection, design, and implementation of CSO management practices and controls to meet requirements of CWA. Nine minimum controls and development of long-term CSO control plans are required to be implemented immediately
Clean Air Act of 1970 and 1990 Amendments	Establishes limitations for specific air pollutants and institutes prevention of significant deterioration in air quality. Maximum achievable control technology is required for any of 189 listed chemicals from "major sources," i.e., plants emitting at least 60 kg/d
40 CFR Part 60	Establishes air emission limits for sludge incinerators with capacities larger than 1000 kg/d (2200 lb/d) dry basis
Total maximum daily load (TDML) (2000) Section 303(d) of the CWA	Requires states to develop prioritized lists of polluted or threatened water bodies and to establish the maximum amount of pollutant (TMDL) that a water body can receive and still meet water quality standards

Pursuant to Section 304(d) of Public Law 92-500 (see Table 1–2), the U.S. Environmental Protection Agency (U.S. EPA) published its definition of minimum standards for secondary treatment. This definition, originally issued in 1973, was amended in 1985 to allow additional flexibility in applying the percent removal requirements of pollutants to treatment facilities serving separate sewer systems. The definition of secondary treatment is reported in Table 1–3 and includes three major effluent parameters: 5-day BOD, TSS, and pH. The substitution of 5-day carbonaceous BOD ($CBOD_5$) for BOD_5 may be made at the option of the permitting authority. These standards provided the basis for the design and operation of most treatment plants. Special interpretations of the definition of secondary treatment are permitted for publicly owned treatment works (1) served by combined sewer systems, (2) using waste stabilization ponds and trickling filters, (3) receiving industrial flows, or (4) receiving less concentrated influent wastewater from separate sewers. The secondary treatment regulations were amended further in 1989 to clarify the percent removal requirements during dry periods for treatment facilities served by combined sewers.

In 1987, Congress enacted the Water Quality Act of 1987 (WQA), the first major revision of the Clean Water Act. Important provisions of the WQA were: (1) strengthening federal water quality regulations by providing changes in permitting and adding substantial penalties for permit violations, (2) significantly amending the CWA's formal sludge control program by emphasizing the identification and regulation of toxic pollutants in sludge, (3) providing funding for state and U.S. EPA studies for defining nonpoint and toxic sources of pollution, (4) establishing new deadlines for compliance including priorities and permit requirements for stormwater, and (5) a phase-out of the construction grants program as a method of financing publicly owned treatment works (POTW).

Table 1–3
Minimum national standards for secondary treatment[a, b]

Characteristic of discharge	Unit of measurement	Average 30-day concentration[c]	Average 7-day concentration[c]
BOD_5	mg/L	30[d]	45
Total suspended solids	mg/L	30[d]	45
Hydrogen-ion concentration	pH units	Within the range of 6.0 to 9.0 at all times[e]	
$CBOD_5$[f]	mg/L	25	40

[a] *Federal Register* (1988, 1989).
[b] Present standards allow stabilization ponds and trickling filters to have higher 30-day average concentrations (45 mg/L) and 7-day average concentrations (65 mg/L) of BOD/suspended solids performance levels as long as the water quality of the receiving water is not adversely affected. Exceptions are also permitted for combined sewers, certain industrial categories, and less concentrated wastewater from separate sewers. For precise requirements of exceptions, *Federal Register* (1988) should be consulted.
[c] Not to be exceeded.
[d] Average removal shall not be less than 85 percent.
[e] Only enforced if caused by industrial wastewater or by in-plant inorganic chemical addition.
[f] May be substituted for BOD_5 at the option of the permitting authority.

Recent regulations that affect wastewater facilities design include those for the treatment, disposal, and beneficial use of biosolids (40 CFR Part 503). In the biosolids regulation promulgated in 1993, national standards were set for pathogen and heavy metal content and for the safe handling and use of biosolids. The standards are designed to protect human health and the environment where biosolids are applied beneficially to land. The rule also promotes the development of a "clean sludge" (U.S. EPA, 1999).

The total maximum daily load (TMDL) program was promulgated in 2000 but is not scheduled to be in effect until 2002. The TMDL rule is designed to protect ambient water quality. A TMDL represents the maximum amount of a pollutant that a water body can receive and still meet water quality standards. A TMDL is the sum of (1) the individual waste-load allocations for point sources, (2) load allocations for nonpoint sources, (3) natural background levels, and (4) a margin of safety (U.S. EPA, 2000). To implement the rule, a comprehensive watershed-based water quality management program must be undertaken to find and control nonpoint sources in addition to conventional point source discharges. With implementation of the TMDL rule, the focus on water quality shifts from technology-based controls to preservation of ambient water quality. The end result is an integrated planning approach that transcends jurisdictional boundaries and forces different sectors, such as agriculture, water and wastewater utilities, and urban runoff managers, to cooperate. Implementation of the TMDL rule will vary depending on the specific water quality objectives established for each watershed and, in some cases, will require the installation of advanced levels of treatment.

1–3 HEALTH AND ENVIRONMENTAL CONCERNS IN WASTEWATER MANAGEMENT

As research into the characteristics of wastewater has become more extensive, and as the techniques for analyzing specific constituents and their potential health and environmental effects have become more comprehensive, the body of scientific knowledge has expanded significantly. Many of the new treatment methods being developed are designed to deal with health and environmental concerns associated with findings of recent research. However, the advancement in treatment technology effectiveness has not kept pace with the enhanced constituent detection capability. Pollutants can be detected at lower concentrations than can be attained by available treatment technology. Therefore, careful assessment of health and environment effects and community concerns about these effects becomes increasingly important in wastewater management. The need to establish a dialogue with the community is important to assure that health and environmental issues are being addressed.

Water quality issues arise when increasing amounts of treated wastewater are discharged to water bodies that are eventually used as water supplies. The waters of the Mississippi River and many rivers in the eastern United States are used for municipal and industrial water supplies and as repositories for the resulting treated wastewater. In southern California, a semiarid region, increasing amounts of reclaimed wastewater are being used or are planned to be used for groundwater recharge to augment existing potable water supplies. Significant questions remain about the testing and levels of treatment necessary to protect human health where the commingling of highly treated wastewater with drinking water sources results in indirect potable reuse. Some professionals object in principle to the indirect reuse of treated wastewater for potable purposes; others express concern that current techniques are inadequate for detecting all microbial and chemical contaminants of health significance (Crook et al., 1999). Among the latter concerns are (1) the lack of sufficient information regarding the health risks posed by some microbial pathogens and chemical constituents in wastewater, (2) the nature of unknown or unidentified chemical constituents and potential pathogens, and (3) the effectiveness of treatment processes for their removal. Defining risks to public health based on sound science is an ongoing challenge.

Because new and more sensitive methods for detecting chemicals are available and methods have been developed that better determine biological effects, constituents that were undetected previously are now of concern (see Fig. 1–2). Examples of such chemical constituents found in both surface and groundwaters include: n-nitrosodimethylamine (NDMA), a principal ingredient in rocket fuel, methyl tertiary butyl ether (MTBE), a highly soluble gasoline additive, medically active substances including endocrine disruptors, pesticides, industrial chemicals, and phenolic compounds commonly found in nonionic surfactants. Endocrine-disrupting chemicals are a special health concern as they can mimic hormones produced in vertebrate animals by causing an exaggerated response, or they can block the effects of a hormone on the body (Trussell, 2000). These chemicals can cause problems with development, behavior, and reproduction in a variety of species. Increases in testicular, prostate, and breast cancers have been blamed on endocrine-disruptive chemicals (Roefer et al., 2000). Although treatment of these chemicals is not currently a mission of municipal wastewater treatment, wastewater treatment facilities may have to be designed to deal with these chemicals in the future.

Other health concerns relate to: (1) the release of volatile organic compounds (VOCs) and toxic air contaminants (TACs) from collection and treatment facilities, (2) chlorine disinfection, and (3) disinfection byproducts (DBPs). Odors are one of the most serious environmental concerns to the public. New techniques for odor measurement are used to quantify the development and movement of odors that may emanate from wastewater facilities, and special efforts are being made to design facilities that minimize the development of odors, contain them effectively, and provide proper treatment for their destruction (see Fig. 1–3).

Figure 1–2

Atomic adsorption spectrometer used for the detection of metals. Photo was taken in wastewater treatment plant laboratory. The use of such analytical instruments is now commonplace at wastewater treatment plants.

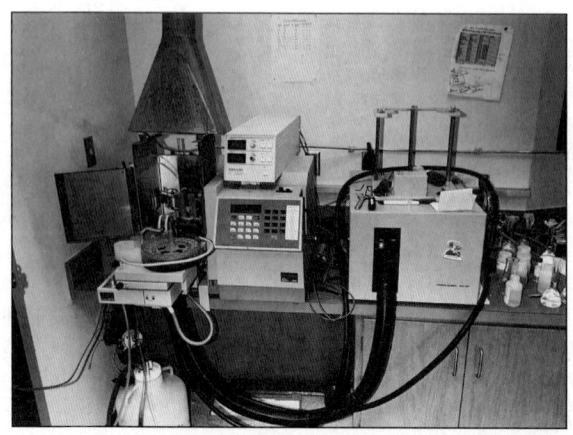

Figure 1–3

Covered treatment plant facilities for the control of odor emissions.

Many industrial wastes contain VOCs that may be flammable, toxic, and odorous, and may be contributors to photochemical smog and tropospheric ozone. Provisions of the Clean Air Act and local air quality management regulations are directed toward (1) minimizing VOC releases at the source, (2) containing wastewater and their VOC emissions (i.e., by adding enclosures), treating wastewater for VOC removal, and collecting and treating vapor emissions from wastewater. Many VOCs, classified as TACs, are discharged to the ambient atmosphere and transported to downwind receptors. Some air management districts are enforcing regulations based on excess cancer risks for lifetime exposures to chemicals such as benzene, trichloroethylene, chloroform, and methylene chloride (Card and Corsi, 1992). Strategies for controlling VOCs at wastewater treatment plants are reviewed in Chap. 5.

Effluents containing chlorine residuals are toxic to aquatic life, and, increasingly, provisions to eliminate chlorine residuals are being instituted. Other important health issues relate to the reduction of disinfection byproducts (DBPs) that are potential carcinogens and are formed when chlorine reacts with organic matter. To achieve higher and more consistent microorganism inactivation levels, improved performance of disinfection systems must be addressed. In many communities, the issues of safety in the transporting, storing, and handling of chlorine are also being examined.

1–4 WASTEWATER CHARACTERISTICS

Prior to about 1940, most municipal wastewater was generated from domestic sources. After 1940, as industrial development in the United States grew significantly, increasing amounts of industrial wastewater have been and continue to be discharged to municipal collection systems. The amounts of heavy metals and synthesized organic compounds generated by industrial activities have increased, and some 10,000 new organic compounds are added each year. Many of these compounds are now found in the wastewater from most municipalities and communities.

As technological changes take place in manufacturing, changes also occur in the compounds discharged and the resulting wastewater characteristics. Numerous compounds generated from industrial processes are difficult and costly to treat by conventional wastewater treatment processes. Therefore, effective industrial pretreatment becomes an essential part of an overall water quality management program. Enforcement of an industrial pretreatment program is a daunting task, and some of the regulated pollutants still escape to the municipal wastewater collection system and must be treated. In the future with the objective of pollution prevention, every effort should be made by industrial dischargers to assess the environmental impacts of any new compounds that may enter the wastewater stream before being approved for use. If a compound cannot be treated effectively with existing technology, it should not be used.

Improved Analytical Techniques

Great strides in analytical techniques have been made with the development of new and more sophisticated instrumentation. While most constituent concentrations are reported in milligrams per liter (mg/L), measurements in micrograms per liter (μg/L) and nanograms per liter (ng/L) are now common. As detection methods become more sensitive and a broader range of compounds are monitored in water supplies, more contaminants that affect humans and the environment will be found. Many trace compounds and microorganisms, such as *Giardia lamblia* and *Cryptosporidium parvum,* have been identified that potentially may cause adverse health effects. Increased analytical sophistication also allows the scientist and engineer to gain greater knowledge of the behavior of wastewater constituents and how they affect process performance and effluent quality.

Importance of Improved Wastewater Characterization

Because of changing wastewater characteristics and the imposition of stricter limits on wastewater discharges and biosolids that are used beneficially, greater emphasis is being placed on wastewater characterization. Because process modeling is widely used in the design and optimization of biological treatment processes (e.g., activated sludge), thorough characterization of wastewater, particularly wastewaters containing industrial waste, is increasingly important. Process modeling for activated sludge as it is currently conceived requires experimental assessment of kinetic and stoichiometric constants. Fractionization of organic nitrogen, chemical oxygen demand (COD), and total organic carbon into soluble and particulate constituents is now used to optimize the performance of both existing and proposed new biological treatment plants designed to achieve nutrient removal. Techniques from the microbiological sciences, such as RNA and DNA typing, are being used to identify the active mass in biological treatment processes. Because an understanding of the nature of wastewater is fundamental to the design and operation of wastewater collection, treatment, and reuse facilities, a detailed discussion of wastewater constituents is provided in Chap. 2.

1-5 WASTEWATER TREATMENT

Wastewater collected from municipalities and communities must ultimately be returned to receiving waters or to the land or reused. The complex question facing the design engineer and public health officials is: What levels of treatment must be achieved in a given application—beyond those prescribed by discharge permits—to ensure protection of public health and the environment? The answer to this question requires detailed analyses of local conditions and needs, application of scientific knowledge and engineering judgment based on past experience, and consideration of federal, state, and local regulations. In some cases, a detailed risk assessment may be required. An overview of wastewater treatment is provided in this section. The reuse and disposal of biosolids, vexing problems for some communities, are discussed in the following section.

Treatment Methods

Methods of treatment in which the application of physical forces predominate are known as *unit operations*. Methods of treatment in which the removal of contaminants is brought about by chemical or biological reactions are known as *unit processes*. At the present time, unit operations and processes are grouped together to provide various levels of treatment known as preliminary, primary, advanced primary, secondary (without or with nutrient removal), and advanced (or tertiary) treatment (see Table 1–4). In preliminary treatment, gross solids such as large objects, rags, and grit are removed that may damage equipment. In primary treatment, a physical operation, usually sedimentation, is used to remove the floating and settleable materials found in wastewater (see Fig. 1–4). For advanced primary treatment, chemicals are added to enhance the removal of suspended solids and, to a lesser extent, dissolved solids. In secondary treatment, biological and chemical processes are used to remove most of the organic matter. In advanced treatment, additional combinations of unit operations and processes are used to remove residual suspended solids and other constituents that are not reduced significantly by conventional secondary treatment. A listing of unit operations and processes used for

Table 1–4
Levels of wastewater treatment[a]

Treatment level	Description
Preliminary	Removal of wastewater constituents such as rags, sticks, floatables, grit, and grease that may cause maintenance or operational problems with the treatment operations, processes, and ancillary systems
Primary	Removal of a portion of the suspended solids and organic matter from the wastewater
Advanced primary	Enhanced removal of suspended solids and organic matter from the wastewater. Typically accomplished by chemical addition or filtration
Secondary	Removal of biodegradable organic matter (in solution or suspension) and suspended solids. Disinfection is also typically included in the definition of conventional secondary treatment
Secondary with nutrient removal	Removal of biodegradable organics, suspended solids, and nutrients (nitrogen, phosphorus, or both nitrogen and phosphorus)
Tertiary	Removal of residual suspended solids (after secondary treatment), usually by granular medium filtration or microscreens. Disinfection is also typically a part of tertiary treatment. Nutrient removal is often included in this definition
Advanced	Removal of dissolved and suspended materials remaining after normal biological treatment when required for various water reuse applications

[a] Adapted, in part, from Crites and Tchobanoglous (1998).

Figure 1–4

Typical primary sedimentation tanks used to remove floating and settleable material from wastewater.

the removal of major constituents found in wastewater and addressed in this text is presented in Table 1–5.

About 20 years ago, biological nutrient removal (BNR)—for the removal of nitrogen and phosphorus—was viewed as an innovative process for advanced wastewater treatment. Because of the extensive research into the mechanisms of BNR, the advantages of its use, and the number of BNR systems that have been placed into operation, nutrient removal, for all practical purposes, has become a part of conventional wastewater treatment. When compared to chemical treatment methods, BNR uses less chemical, reduces the production of waste solids, and has lower energy consumption. Because of the importance of BNR in wastewater treatment, BNR is integrated into the discussion of theory, application, and design of biological treatment systems.

Land treatment processes, commonly termed "natural systems," combine physical, chemical, and biological treatment mechanisms and produce water with quality similar to or better than that from advanced wastewater treatment. Natural systems are not covered in this text as they are used mainly with small treatment systems; descriptions may be found in the predecessor edition of this text (Metcalf & Eddy, 1991) and in Crites and Tchobanoglous (1998) and Crites et al. (2000).

Current Status

Up until the late 1980s, conventional secondary treatment was the most common method of treatment for the removal of BOD and TSS. In the United States, nutrient removal was used in special circumstances, such as in the Great Lakes area, Florida, and the Chesapeake Bay, where sensitive nutrient-related water quality conditions were identified. Because of nutrient enrichment that has led to eutrophication and water quality degradation (due in part to point source discharges), nutrient removal processes have evolved and now are used extensively in other areas as well.

As a result of implementation of the Federal Water Pollution Control Act Amendments, significant data have been obtained on the numbers and types of wastewater facilities used and needed in accomplishing the goals of the program. Surveys are conducted by U.S. EPA to track these data, and the results of the 1996 Needs Assessment Survey (U.S. EPA, 1997a) are reported in Tables 1–6 and 1–7. The number and types of

Table 1–5
Unit operations and processes used to remove constituents found in wastewater

Constituent	Unit operation or process	See Chap.
Suspended solids	Screening	5
	Grit removal	5
	Sedimentation	5
	High-rate clarification	5
	Flotation	5
	Chemical precipitation	6
	Depth filtration	11
	Surface filtration	11
Biodegradable organics	Aerobic suspended growth variations	8, 14
	Aerobic attached growth variations	9
	Anaerobic suspended growth variations	10, 14
	Anaerobic attached growth variations	10
	Lagoon variations	8
	Physical-chemical systems	6, 11
	Chemical oxidation	6
	Advanced oxidation	11
	Membrane filtration	8, 11
Nutrients		
Nitrogen	Chemical oxidation (breakpoint chlorination)	6
	Suspended-growth nitrification and denitrification variations	8
	Fixed-film nitrification and denitrification variations	9
	Air stripping	11
	Ion exchange	11
Phosphorus	Chemical treatment	6
	Biological phosphorus removal	8, 9
Nitrogen and phosphorus	Biological nutrient removal variations	8, 9
Pathogens	Chlorine compounds	12
	Chlorine dioxide	12
	Ozone	12
	Ultraviolet (UV) radiation	12
Colloidal and dissolved solids	Membranes	11
	Chemical treatment	11
	Carbon adsorption	11
	Ion exchange	11
Volatile organic compounds	Air stripping	5, 11
	Carbon adsorption	11
	Advanced oxidation	11
Odors	Chemical scrubbers	15
	Carbon adsorption	11, 15
	Biofilters	15
	Compost filters	15

facilities needed in the future (~20 yr) are also shown in Table 1–7. These data are useful in forming an overall view of the current status of wastewater treatment in the United States.

The municipal wastewater treatment enterprise is composed of over 16,000 plants that are used to treat a total flow of about 1400 cubic meters per second (m^3/s) [32,000 million gallons per day (Mgal/d)]. Approximately 92 percent of the total existing flow is handled by plants having a capacity of 0.044 m^3/s [1 million gallons per day (Mgal/d)] and larger. Nearly one-half of the present design capacity is situated in plants

Table 1–6
Number of U.S. wastewater treatment facilities by flow range (1996)[a]

| Flow ranges | | Number of facilities | Total existing flowrate | |
Mgal/d	m³/s		Mgal/d	m³/s
0.000–0.100	0.000–0.00438	6,444	287	12.57
0.101–1.000	0.0044–0.0438	6,476	2,323	101.78
1.001–10.000	0.044–0.438	2,573	7,780	340.87
10.001–100.00	0.44–4.38	446	11,666	511.12
>100.00	>4.38	47	10,119	443.34
Other[b]		38	—	—
Total		16,204	32,175	1,409.68

[a] Adapted from U.S. EPA (1997a).
[b] Flow data unknown.

Table 1–7
Number of U.S. wastewater treatment facilities by design capacity in 1996 and in the future when needs are met[a]

| Level of treatment | Existing facilities | | | Future facilities (when needs are met) | | |
	Number of facilities	Mgal/d	m³/s	Number of facilities	Mgal/d	m³/s
Less than secondary	176	3,054	133.80	61	601	26.33
Secondary	9,388	17,734	776.98	9,738	17,795	779.65
Greater than secondary[b]	4,428	20,016	876.96	6,135	28,588	1,252.53
No discharge[c]	2,032	1,421	62.26	2,369	1,803	78.99
Total	16,024	42,225	1,850.00	18,303	48,787	2,137.50

[a] Adapted from U.S. EPA (1997a).
[b] Treatment plants that meet effluent standards higher than those given in Table 1–3.
[c] Plants that do not discharge to a water body and use some form of land application.

providing greater than secondary treatment. Thus, the basic material presented in this text is directed toward the design of plants larger than 0.044 m³/s (1 Mgal/d) with the consideration that many new designs will provide treatment greater than secondary.

In the last 10 years, many plants have been designed using BNR. Effluent filtration has also been installed where the removal of residual suspended solids is required. Filtration is especially effective in improving the effectiveness of disinfection, especially for ultraviolet (UV) disinfection systems, because (1) the removal of larger particles of suspended solids that harbor bacteria enhances the reduction in coliform bacteria and (2) the reduction of turbidity improves the transmittance of UV light. Effluent reuse systems, except for many that are used for agricultural irrigation, almost always employ filtration.

New Directions and Concerns

New directions and concerns in wastewater treatment are evident in various specific areas of wastewater treatment. The changing nature of the wastewater to be treated, emerging health and environmental concerns, the problem of industrial wastes, and the impact of new regulations, all of which have been discussed previously, are among the most important. Further, other important concerns include: (1) aging infrastructure, (2) new methods of process analysis and control, (3) treatment plant performance and reliability, (4) wastewater disinfection, (5) combined sewer overflows, (6) impacts of stormwater and sanitary overflows and nonpoint sources of pollution, (7) separate treatment of return flows, (8) odor control (see Fig. 1–5) and the control of VOC emissions, and (9) retrofitting and upgrading wastewater treatment plants.

Aging Infrastructure. Some of the problems that have to be addressed in the United States deal with renewal of the aging wastewater collection infrastructure and upgrading of treatment plants. Issues include repair and replacement of leaking and undersized sewers, control and treatment of overflows from sanitary and combined collection systems, control of nonpoint discharges, and upgrading treatment systems to achieve higher removal levels of specific constituents. Upgrading and retrofitting treatment plants is addressed later in this section.

Figure 1–5

Facilities used for chemical treatment of odors from treatment facilities.

Portions of the collection systems, particularly those in the older cities in the eastern and midwestern United States, are older than the treatment plants. Sewers constructed of brick and vitrified clay with mortar joints, for example, are still used to carry sanitary wastewater and stormwater. Because of the age of the pipes and ancillary structures, the types of materials and methods of construction, and lack of repair, leakage is common. Leakage is in the form of both infiltration and inflow where water enters the collection system, and exfiltration where water leaves the pipe. In the former case, extraneous water has to be collected and treated, and oftentimes may overflow before treatment, especially during wet weather. In the latter case, exfiltration causes untreated wastewater to enter the groundwater and/or migrate to nearby surface water bodies. It is interesting to note that while the standards for treatment have increased significantly, comparatively little or no attention has been focused on the discharge of untreated wastewater from sewers through exfiltration. In the future, however, leaking sewers are expected to become a major concern and will require correction.

Process Analysis and Control.   Because of the changing characteristics of the wastewater (discussed above), studies of wastewater treatability are increasing, especially with reference to the treatment of specific constituents. Such studies are especially important where new treatment processes are being considered. Therefore, the engineer must understand the general approach and methodology involved in: (1) assessing the treatability of a wastewater (domestic or industrial), (2) conducting laboratory and pilot plant studies, and (3) translating experimental data into design parameters.

Computational fluid dynamics (CFD), computer-based computational methods for solving the fundamental equations of fluid dynamics (i.e., continuity, momentum, and energy), is now being used to improve and optimize the hydraulic performance of wastewater treatment facilities. Applications of CFD include the design of new systems or the optimization of systems such as vortex separators, mixing tanks, sedimentation tanks, dissolved-air flotation units, and chlorine contact tanks to reduce or eliminate dead zones and short circuiting. Improved UV disinfection systems are being designed using CFD. One of the main advantages of CFD is simulating a range of operating conditions to evaluate performance before designs and operating changes are finalized. Another advantage is that dynamic models can be integrated with the process control system to optimize ongoing operation.

Treatment Process Performance and Reliability.   Important factors in process selection and design are treatment plant performance and reliability in meeting permit requirements. In most discharge permits, effluent constituent requirements, based on 7-day and 30-day average concentrations, are specified (see Table 1–3). Because wastewater treatment effluent quality is variable because of varying organic loads, changing environmental conditions, and new industrial discharges, it is necessary to design the treatment system to produce effluent concentrations equal to or less than the limits prescribed by the discharge permit. Reliability is especially important where critical water quality parameters have to be maintained such as in reuse applications. On-line monitoring of critical parameters such as total organic carbon (TOC), transmissivity, turbidity, and dissolved oxygen is necessary for building a database and for improving process control. Chlorine residual monitoring is useful for dosage control, and pH monitoring assists in controlling nitrification systems.

Treatment plant reliability can be defined as the probability that a system can meet established performance criteria consistently over extended periods of time. Two components of reliability, the inherent reliability of the process and mechanical reliability, are discussed in Chap. 15. As improved microbiological techniques are developed, it will be possible to optimize the disinfection process.

The need to conserve energy and resources is fundamental to all aspects of wastewater collection, treatment, and reuse. Operation and maintenance costs are extremely important to operating agencies because these costs are funded totally with local moneys. Detailed energy analyses and audits are important parts of treatment plant design and operation as significant savings can be realized by selecting energy-efficient processes and equipment. Large amounts of electricity are used for aeration that is needed for biological treatment. Typically, about one-half of the entire plant electricity usage is for aeration. In the design of wastewater treatment plants power use can be minimized by paying more careful attention to plant siting, selecting energy-efficient equipment, and designing facilities to recover energy for in-plant use. Energy management in treatment plant design and operation is also considered in Chap. 15.

Wastewater Disinfection.   Changes in regulations and the development of new technologies have affected the design of disinfection systems. Gene probes are now being used to identify where specific groups of organisms are found in treated secondary effluent (i.e., in suspension or particle-associated). Historically, chlorine has been the disinfectant of choice for wastewater. With the increasing number of permits requiring low or nondetectable amounts of chlorine residual in treated effluents, dechlorination facilities have had to be added, or chlorination systems have been replaced by alternative disinfection systems such as ultraviolet (UV) radiation (see Fig. 1–6). Concerns about chemical safety have also affected design considerations of chlorination and dechlorination systems. Improvements that have been made in UV lamp and ballast design within the past 10 years have improved significantly the performance and reliability of UV disinfection systems. Effective guidelines have also been developed for the application and design of UV systems (NWRI, 2000). Capital and operating costs have also been lowered. It is anticipated that the application of UV for treated drinking water and for stormwater will continue to increase in the future. Because UV produces essentially no troublesome byproducts and is also effective in the reduction of NDMA and other related compounds, its use for disinfection is further enhanced as compared to chlorine compounds.

Combined Sewer Overflows (CSOs), Sanitary Sewer Overflows (SSOs), and Nonpoint Sources.   Overflows from combined sewer and sanitary sewer collection systems have been recognized as difficult problems requiring solution, especially for many of the older cities in the United States. The problem has become more critical as greater development changes the amount and characteristics of stormwater runoff and increases the channelization of runoff into storm, combined, and sanitary collection systems. Combined systems carry a mixture of wastewater and stormwater runoff and, when the capacity of the interceptors is reached, overflows occur to the receiving waters. Large overflows can impact receiving water quality and can prevent attainment of mandated standards. Recreational beach closings and shellfish

Figure 1-6

UV lamps used for the disinfection of wastewater.

bed closures have been attributed to CSOs (Lape and Dwyer, 1994). Federal regulations for CSOs are still under development and have not been issued at the time of writing this text (2001).

A combination of factors has resulted in the release of untreated wastewater from parts of sanitary collection systems. These releases are termed sanitary system overflows (SSOs). The SSOs may be caused by (1) the entrance of excessive amounts of stormwater, (2) blockages, or (3) structural, mechanical, or electrical failures. Many overflows result from aging collection systems that have not received adequate upgrades, maintenance, and repair. The U.S. EPA has estimated that at least 40,000 overflows per year occur from sanitary collection systems. The untreated wastewater from these overflows represents threats to public health and the environment. The U.S. EPA is proposing to clarify and expand permit requirements for municipal sanitary collection systems under the Clean Water Act that will result in reducing the frequency and occurrence of SSOs (U.S. EPA 2001). At the time of writing this text (2001) the proposed regulations are under review. The U.S. EPA estimates that nearly $45 billion is required for constructing facilities for controlling CSOs and SSOs in the United States (U.S. EPA, 1997a).

The effects of pollution from nonpoint sources are growing concerns as evidenced by the outbreak of gastrointestinal illness in Milwaukee traced to the oocysts of *Cryptosporidium parvum,* and the occurrence of *Pfiesteria piscicida* in the waters of Maryland and North Carolina. *Pfiesteria* is a form of algae that is very toxic to fish life. Runoff from pastures and feedlots has been attributed as a potential factor that triggers the effects of these microorganisms.

The extent of the measures that will be needed to control nonpoint sources is not known at this time of writing this text (2001). When studies for assessing TMDLs are completed (estimated to be in 2008), the remedial measures for controlling nonpoint sources may require financial resources rivaling those for CSO and SSO correction.

Treatment of Return Flows. Perhaps one of the significant future developments in wastewater treatment will be the provision of separate facilities for treating return flows from biosolids and other processing facilities. Treatment of return flows will be especially important where low levels of nitrogen are to be achieved in the treated effluent. Separate treatment facilities may include (1) steam stripping for removal of ammonia from biosolids return flows, now typically routed to the plant headworks; (2) high-rate sedimentation for removing fine and difficult-to-settle colloidal material that also shields bacteria from disinfection; (3) flotation and high-rate sedimentation for treating filter backwash water to reduce solids loading on the liquid treatment process; and (4) soluble heavy metals removal by chemical precipitation to meet more stringent discharge requirements. The specific treatment system used will depend on the constituents that will impact the wastewater treatment process.

Control of Odors and VOC Emissions. The control of odors and in particular the control of hydrogen sulfide generation is of concern in collection systems and at treatment facilities. The release of hydrogen sulfide to the atmosphere above sewers and at treatment plant headworks has occurred in a number of locations. The release of excess hydrogen sulfide has led to the accelerated corrosion of concrete sewers, headworks structures, and equipment, and to the release of odors. The control of odors is of increasing environmental concern as residential and commercial development continues to approach existing treatment plant locations. Odor control facilities including covers for process units, special ventilation equipment, and treatment of odorous gases need to be integrated with treatment plant design. Control of hydrogen sulfide is also fundamental to maintaining system reliability.

The presence of VOCs and VTOCs in wastewater has also necessitated the covering of treatment plant headworks and primary treatment facilities and the installation of special facilities to treat the compounds before they are released. In some cases, improved industrial pretreatment has been employed to eliminate these compounds.

Retrofitting and Upgrading Wastewater Treatment Plants. Large numbers of wastewater treatment plants were constructed in the United States during the 1970s and 1980s when large sums of federal money were available for implementation of the CWA. Much of the equipment, now over 20 years old, is reaching the end of its useful life and will need to be replaced. Process changes to improve performance, meet stricter permit requirements, and increase capacity will also be needed. For these reasons, significant future efforts in the planning and design of wastewater treatment plants in the United States will be directed to modifying, improving, and expanding existing treatment facilities. Fewer completely new treatment plants will be constructed. In developing countries, opportunities for designing and building completely new facilities may be somewhat greater. Upgrading and retrofitting treatment plants is addressed in Chap. 15.

Future Trends in Wastewater Treatment

In the U.S. EPA Needs Assessment Survey, the total treatment plant design capacity is projected to increase by about 15 percent over the next 20 to 30 years (see Table 1–7). During this period, the U.S. EPA estimates that approximately 2,300 new plants may have to be built, most of which will be providing a level of treatment greater than secondary. The design capacity of plants providing greater than secondary treatment is expected to increase by 40 percent in the future (U.S. EPA, 1997). Thus, it is clear that the future trends in wastewater treatment plant design will be for facilities providing higher levels of treatment.

Some of the innovative treatment methods being utilized in new and upgraded treatment facilities include vortex separators, high rate clarification, membrane bioreactors, pressure-driven membrane filtration (ultrafiltration and reverse osmosis—see Fig. 1–7), and ultraviolet radiation (low-pressure, low- and high-intensity UV lamps, and medium-pressure, high-intensity UV lamps). Some of the new technologies, especially those developed in Europe, are more compact and are particularly well suited for plants where available space for expansion is limited.

In recent years, numerous proprietary wastewater treatment processes have been developed that offer potential savings in construction and operation. This trend will likely continue, particularly where alternative treatment systems are evaluated or facilities are privatized. Privatization is generally defined as a public-private partnership in which the private partner arranges the financing, design, building, and operation of the treatment facilities. In some cases, the private partner may own the facilities. The reasons for privatization, however, go well beyond the possibility of installing proprietary processes. In the United States, the need for private financing appears to be the principal rationale for privatization; the need to preserve local control appears to be the leading pragmatic rationale against privatization (Dreese and Beecher, 1997).

1–6 WASTEWATER RECLAMATION AND REUSE

In many locations where the available supply of fresh water has become inadequate to meet water needs, it is clear that the once-used water collected from communities and municipalities must be viewed not as a waste to be disposed of but as a resource that

Figure 1-7

Reverse osmosis membrane system used for the removal of residual suspended solids remaining after conventional secondary treatment.

must be reused. The concept of reuse is becoming accepted more widely as other parts of the country experience water shortages. The use of dual water systems, such as now used in St. Petersburg in Florida and Rancho Viejo in California, is expected to increase in the future. In both locations, treated effluent is used for landscape watering and other nonpotable uses. Satellite reclamation systems such as those used in the Los Angeles basin, where wastewater flows are mined (withdrawn from collection systems) for local treatment and reuse, are examples where transportation and treatment costs of reclaimed water can be reduced significantly. Because water reuse is expected to become of even greater importance in the future, reuse applications are considered in Chap. 13.

Current Status

Most of the reuse of wastewater occurs in the arid and semiarid western and southwestern states of the United States; however, the number of reuse projects is increasing in the south especially in Florida and South Carolina. Because of health and safety concerns, water reuse applications are mostly restricted to nonpotable uses such as landscape and agricultural irrigation. In a report by the National Research Council (1998), it was concluded that indirect potable reuse of reclaimed water (introducing reclaimed water to augment a potable water source before treatment) is viable. The report also stated that direct potable reuse (introducing reclaimed water directly into a water distribution system) was not practicable. Because of the concerns about potential health effects associated with the reclaimed water reuse, plans are proceeding slowly about expanding reuse beyond agricultural and landscape irrigation, groundwater recharge for repelling saltwater intrusion, and nonpotable industrial uses (e.g., boiler water and cooling water).

New Directions and Concerns

Many of the concerns mentioned in the National Research Council (NRC, 1998) report regarding potential microbial and chemical contamination of water supplies also apply to water sources that receive incidental or unplanned wastewater discharges. A number of communities use water sources that contain a significant wastewater component. Even though these sources, after treatment, meet current drinking water standards, the growing knowledge of the potential impacts of new trace contaminants raises concern. Conventional technologies for both water and wastewater treatment may be incapable of reducing the levels of trace contaminants below where they are not considered as a potential threat to public health. Therefore, new technologies that offer significantly improved levels of treatment or constituent reduction need to be tested and evaluated. Where indirect potable reuse is considered, risk assessment also becomes an important component of a water reuse investigation. Risk assessment is addressed in Chap. 13.

Future Trends in Technology

Technologies that are suitable for water reuse applications include membranes (pressure-driven, electrically driven, and membrane bioreactors), carbon adsorption, advanced oxidation, ion exchange, and air stripping. Membranes are most significant developments as new products are now available for a number of treatment applications. Membranes had been limited previously to desalination, but they are being tested increasingly for wastewater applications to produce high-quality treated effluent suitable for reclamation. Increased levels of contaminant removal not only enhance the product for reuse but also lessen health risks. As indirect potable reuse intensifies to augment existing water supplies, membranes are expected to be one of the predominant treatment technologies. Advanced wastewater treatment technologies are discussed in Chap. 11, and water reuse is considered in Chap. 13.

1–7 BIOSOLIDS AND RESIDUALS MANAGEMENT

The management of the solids and concentrated contaminants removed by treatment has been and continues to be one of the most difficult and expensive problems in the field of wastewater engineering. Wastewater solids are organic products that can be used beneficially after stabilization by processes such as anaerobic digestion and composting. With the advent of regulations that encourage biosolids use, significant efforts have been directed to producing a "clean sludge" (Class A biosolids—see definition in Table 1–1) that meets heavy metals and pathogen requirements and is suitable for land application. Regulations for Class B biosolids call for reduced density in pathogenic bacteria and enteric viruses, but not to the levels of Class A biosolids. Further, the application of Class B biosolids to land is strictly regulated, and distribution for home use is prohibited (see Table 1–1).

Other treatment plant residuals such as grit and screenings have to be rendered suitable for disposal, customarily in landfills. Landfills usually require some form of dewatering to limit moisture content. With the increased use of membranes, especially in wastewater reuse applications, a new type of residual, brine concentrate, requires further processing and disposal. Solar evaporation ponds and discharge to a saltwater environment are only viable in communities where suitable and environmental geographic conditions prevail; brine concentration and residuals solidification are generally too complex and costly to implement.

Current Status

Treatment technologies for solids processing have focused on traditional methods such as thickening, stabilization, dewatering, and drying. Evolution in the technologies has not occurred as rapidly as in liquid treatment processes, but some significant improvements have occurred. Centrifuges that produce a sludge cake with higher solids content, egg-shaped digesters that improve operation, and dryers that minimize water content are just a few examples of products that have come into use in recent years. These developments are largely driven by the need to produce biosolids that are clean, have less volume, and can be used beneficially.

Landfills still continue to be used extensively for the disposal of treatment plant solids, either in sludge-only monofills or with municipal solid waste. The number and capacity of landfills, however, have been reduced, and new landfill locations that meet public and regulatory acceptance and economic requirements are increasingly difficult to find. Incineration of solids by large municipalities continues to be practiced, but incineration operation and emission control are subject to greater regulatory restrictions and adverse public scrutiny. Alternatives to landfills and incineration include land application of liquid or dried biosolids and composting for distribution and marketing. Land application of biosolids is used extensively to reclaim marginal land for productive uses and to utilize nutrient content in the biosolids. Composting, although a more expensive alternative, is a means of stabilizing and distributing biosolids for use as a soil amendment. Alkaline stabilization of biosolids for land application is also used but to a lesser extent.

New Directions and Concerns

Over the last 30 years, the principal focus in wastewater engineering has been on improving the quality of treated effluent through the construction of secondary and advanced wastewater treatment plants. With improved treatment methods, higher levels of treatment must be provided not only for conventional wastewater constituents but also for the removal of specific compounds such as nutrients and heavy metals. A byproduct of these efforts has been the increased generation of solids and biosolids per person served by a municipal wastewater system. In many cases, the increase in solids production clearly taxes the capacity of existing solids processing and disposal methods.

In addition to the shear volume of solids that has to be handled and processed, management options continue to be reduced through stricter regulations. Limitations that affect options are: (1) landfill sites are becoming more difficult to find and have permitted, (2) air emissions from incinerators are more closely regulated, and (3) new requirements for the land application of biosolids have been instituted. In large urban areas, haul distances to landfill or land application sites have significantly affected the cost of solids processing and disposal. Few new incinerators are being planned because of difficulties in finding suitable sites and obtaining permits. Emission control regulations of the Clean Air Act also require the installation of complex and expensive pollution control equipment.

More communities are looking toward (1) producing Class A biosolids to improve beneficial reuse opportunities or (2) implementing a form of volume reduction, thus lessening the requirements for disposal. The issue—"are Class A biosolids clean enough?"—will be of ongoing concern to the public. The continuing search for better methods of solids processing, disposal, and reuse will remain as one of the highest priorities in the future. Additionally, developing meaningful dialogue with the public about health and environmental effects will continue to be very important.

Future Trends in Biosolids Processing

New solids processing systems have not been developed as rapidly as liquid unit operations and processes. Anaerobic digestion remains the principal process for the stabilization of solids. Egg-shaped digesters, developed in Europe for anaerobic digestion, are being used more extensively in the United States because of advantages of easier operation, lower operation and maintenance costs, and, in some cases, increased volatile solids destruction (which also increases the production of reusable methane gas) (see Fig. 1–8). Other developments in anaerobic and aerobic digestion include temperature-phased anaerobic digestion and autothermal aerobic digestion (ATAD), another process developed in Europe. These processes offer advantages of improved volatile solids destruction and the production of stabilized biosolids that meet Class A requirements.

High solids centrifuges and heat dryers are expected to be used more extensively. High solids centrifuges extract a greater percentage of the water in liquid sludge, thus providing a dryer cake. Improved dewatering not only reduces the volume of solids

Figure 1-8

Egg-shaped digesters used for the anaerobic treatment of biosolids.

requiring further processing and disposal, but allows composting or subsequent drying to be performed more efficiently. Heat drying provides further volume reduction and improves the quality of the product for potential commercial marketing. Each of the newer methods of biosolids processing is described in Chap. 14.

REFERENCES

Boyd, J. (2000) "Unleashing the Clean Water Act, the Promise and Challenge of the TMDL Approach to Water Quality," *Resources*, Issue 139.

Card, T. R., and R. L. Corsi (1992) "A Flexible Fate Model for VOCs in Wastewater," *Water Environment & Technology*, vol. 4, no. 3.

Crites, R. W., S. C. Reed, and R. K. Bastian (2000) *Land Treatment Systems for Municipal and Industrial Wastes*, McGraw-Hill, New York.

Crites, R., and G. Tchobanoglous (1998) *Small and Decentralized Wastewater Management Systems*, McGraw-Hill, New York.

Crook, J., J. A. MacDonald, and R. R. Trussel (1999) "Potable Use of Reclaimed Water," *Journal American Water Works Association*, vol. 91, no. 8.

Curren, M. D. (1999) "Total Maximum Daily Loads," *Environmental Protection*, vol. 10, no. 11.

Dreese, G. R., and J. A. Beecher (1997) "To Privatize or Not to Privatize," *Water Environment & Technology*, vol. 9, no. 1, Water Environment Federation, Alexandria, VA.

Federal Register (1988) 40 CFR Part 133, Secondary Treatment Regulation.

Federal Register (1989) 40 CFR Part 133, Amendments to the Secondary Treatment Regulations: Percent Removal Requirements During Dry Weather Periods for Treatment Works Served by Combined Sewers.

Federal Register (1993) 40 CFR Parts 257 and 503, Standards for the Disposal of Sewage Sludge.

Lape, J. L., and T. J. Dwyer (1994) A New Policy on CSOs, *Water Environment & Technology*, vol. 6, no. 6.

Metcalf & Eddy, Inc. (1991) *Wastewater Engineering: Treatment, Disposal and Reuse*, 3d ed., McGraw-Hill, New York.

National Research Council (1998) *Issues in Potable Reuse—The Viability of Augmenting Drinking Water Supplies with Reclaimed Water*, National Academy Press, Washington, DC.

NWRI and AWWA (2000) *Ultraviolet Disinfection Guidelines for Drinking Water and Wastewater Reclamation*, NWRI-00-03, National Water Research Institute and American Water Works Association Research Foundation, Fountain Valley, CA.

Roefer, P., S. Snyder, R. E. Zegers, D. J. Rexing, and J. L. Frank (2000) "Endocrine-Disrupting Chemicals in Source Water," *Journal American Water Works Association*, vol. 92, no. 8.

Trussell, R. R. (2001) "Endrocrine Disruptors and the Water Industry," *Journal American Water Works Association*, vol. 93, no. 2.

U.S. EPA (1997*a*) *1996 Clean Water Needs Survey Report to Congress*, U.S. Environmental Protection Agency, EPA 832-R-97-003, Washington, DC.

U.S. EPA (1997*b*) *Control of Pathogens and Vectors in Sewage Sludge*, U.S. Environmental Protection Agency, EPA 625-R-92-013 (revised), Washington, DC.

U.S. EPA (2000) *Total Maximum Daily Load (TMDL)*, U.S. Environmental Protection Agency, EPA 841-F-00-009, Washington, DC.

U.S. EPA (2001) *Proposed Rule to Protect Communities from Overflowing Sewers*, U.S. Environmental Protection Agency, EPA 833-01-F-001, Washington, DC.

2

Constituents in Wastewater

An understanding of the nature of wastewater is essential in the design and operation of collection, treatment, and reuse facilities, and in the engineering management of environmental quality. To promote this understanding, the information in this chapter is presented in nine sections dealing with (1) an introduction to the constituents found in wastewater, (2) sampling and analytical procedures, (3) physical characteristics, (4) inorganic nonmetallic constituents, (5) metallic constituents, (6) aggregate organic constituents, (7) individual organic constituents and compounds, (8) biological characteristics, and (9) toxicity tests. The material in this chapter has been organized in a manner similar to that used in Standard Methods (1998), the standard reference work for the characterization of wastewater in the field of environmental engineering.

2–1 WASTEWATER CONSTITUENTS

The physical, chemical, and biological constituents found in wastewater and the constituents of concern in wastewater are introduced briefly in the following discussion.

Constituents Found in Wastewater

Wastewater is characterized in terms of its physical, chemical, and biological composition. The principal physical properties and the chemical and biological constituents of wastewater, and their sources, are reported in Table 2–1. It should be noted that many of the physical properties and chemical and biological characteristics listed in Table 2–1 are interrelated. For example, temperature, a physical property, affects both the amounts of gases dissolved in the wastewater and the biological activity in the wastewater.

Constituents of Concern in Wastewater Treatment

The important constituents of concern in wastewater treatment are listed in Table 2–2. Secondary treatment standards for wastewater are concerned with the removal of biodegradable organics, total suspended solids, and pathogens. Many of the more stringent standards that have been developed recently deal with the removal of nutrients, heavy metals, and priority pollutants. When wastewater is to be reused, standards normally include additional requirements for the removal of refractory organics, heavy metals, and in some cases, dissolved inorganic solids.

2–2 SAMPLING AND ANALYTICAL PROCEDURES

Proper sampling and analytical techniques are of fundamental importance in the characterization of wastewater. Sampling techniques, the methods of analysis, the units of measurement for chemical constituents, and some useful concepts from chemistry are considered below.

Sampling

Sampling programs are undertaken for a variety of reasons such as to obtain (1) routine operating data on overall plant performance, (2) data that can be used to document the performance of a given treatment operation or process, (3) data that can be used to implement proposed new programs, and (4) data needed for reporting regulatory compliance. To meet the goals of the sampling program, the data collected must be:

1. *Representative.* The data must represent the wastewater or environment being sampled.
2. *Reproducible.* The data obtained must be reproducible by others following the same sampling and analytical protocols.
3. *Defensible.* Documentation must be available to validate the sampling procedures. The data must have a known degree of accuracy and precision.
4. *Useful.* The data can be used to meet the objectives of the monitoring plan (Pepper et al., 1996).

Because the data from the analysis of the samples will ultimately serve as a basis for implementing wastewater management facilities and programs, the techniques used in a wastewater sampling program must be such that representative samples are obtained.

Table 2-1
Common analyses used to assess the constituents found in wastewater[a]

Test[b]	Abbreviation/ definition	Use or significance of test results
Physical characteristics		
Total solids	TS	
Total volatile solids	TVS	
Total fixed solids	TFS	
Total suspended solids	TSS	
Volatile suspended solids	VSS	To assess the reuse potential of a wastewater and to determine the most suitable type of operations and processes for its treatment
Fixed suspended solids	FSS	
Total dissolved solids	TDS (TS − TSS)	
Volatile dissolved solids	VDS	
Total fixed dissolved solids	FDS	
Settleable solids		To determine those solids that will settle by gravity in a specified time period
Particle size distribution	PSD	To assess the performance of treatment processes
Turbidity	NTU[c]	Used to assess the quality of treated wastewater
Color	Light brown, gray, black	To assess the condition of wastewater (fresh or septic)
Transmittance	% T	Used to assess the suitability of treated effluent for UV disinfection
Odor	TON[d]	To determine if odors will be a problem
Temperature	°C or °F	Important in the design and operation of biological processes in treatment facilities
Density	ρ	
Conductivity	EC	Used to assess the suitability of treated effluent for agricultural applications
Inorganic chemical characteristics		
Free ammonia	NH_4^+	
Organic nitrogen	Org N	
Total Kjeldahl nitrogen	TKN (Org N + NH_4^+-N)	
Nitrites	NO_2^-	
Nitrates	NO_3^-	Used as a measure of the nutrients present and the degree of decomposition in the wastewater; the oxidized forms can be taken as a measure of the degree of oxidation
Total nitrogen	TN	
Inorganic phosphorus	Inorg P	
Total phosphorus	TP	
Organic phosphorus	Org P	

(continued)

| **Table 2–1** (Continued)

Test[b]	Abbreviation/ definition	Use or significance of test results
Inorganic chemical characteristics (continued)		
pH	$pH = -\log [H+]$	A measure of the acidity or basicity of an aqueous solution
Alkalinity	$\Sigma\ HCO_3^- + CO_3^{-2}$ $+ OH^- - H^+$	A measure of the buffering capacity of the wastewater
Chloride	Cl^-	To assess the suitability of wastewater for agricultural reuse
Sulfate	SO_4^{-2}	To assess the potential for the formation of odors and may impact the treatability of the waste sludge
Metals	As, Cd, Ca, Cr, Co, Cu, Pb, Mg, Hg, Mo, Ni, Se, Na, Zn	To assess the suitability of the wastewater for reuse and for toxicity effects in treatment. Trace amounts of metals are important in biological treatment
Specific inorganic elements and compounds		To assess presence or absence of a specific constituent
Various gases	O_2, CO_2, NH_3, H_2S, CH_4	The presence or absence of specific gases
Organic chemical characteristics		
Five-day carbonaceous biochemical oxygen demand	$CBOD_5$	A measure of the amount of oxygen required to stabilize a waste biologically
Ultimate carbonaceous biochemical oxygen demand	UBOD (also BOD_u, BOD_L)	A measure of the amount of oxygen required to stabilize a waste biologically
Nitrogenous oxygen demand	NOD	A measure of the amount of oxygen required to oxidize biologically the nitrogen in the wastewater to nitrate
Chemical oxygen demand	COD	Often used as a substitute for the BOD test
Total organic carbon	TOC	Often used as a substitute for the BOD test
Specific organic compounds and classes of compounds	MBAS[e], CTAS[f]	To determine presence of specific organic compounds and to assess whether special design measures will be needed for removal
Biological characteristics		
Coliform organisms	MPN (most probable number)	To assess presence of pathogenic bacteria and effectiveness of disinfection process
Specific microorganisms	Bacteria, protozoa, helminths, viruses	To assess presence of specific organisms in connection with plant operation and for reuse
Toxicity	TU_a and TU_c	Toxic unit acute, Toxic unit chronic

[a] Adapted, in part, from Crites and Tchobanoglous (1998).
[b] Details on the various tests may be found in Standard Methods (1998).
[c] NTU = nephelometric turbidity unit
[d] TON = threshold odor number
[e] MBAS = methylene blue active substances
[f] CTAS = cobalt thiocyanate active substances

Table 2–2
Principal constituents of concern in wastewater treatment

Constituent	Reason for importance
Suspended solids	Suspended solids can lead to the development of sludge deposits and anaerobic conditions when untreated wastewater is discharged in the aquatic environment
Biodegradable organics	Composed principally of proteins, carbohydrates, and fats, biodegradable organics are measured most commonly in terms of BOD (biochemical oxygen demand) and COD (chemical oxygen demand). If discharged untreated to the environment, their biological stabilization can lead to the depletion of natural oxygen resources and to the development of septic conditions
Pathogens	Communicable diseases can be transmitted by the pathogenic organisms that may be present in wastewater
Nutrients	Both nitrogen and phosphorus, along with carbon, are essential nutrients for growth. When discharged to the aquatic environment, these nutrients can lead to the growth of undesirable aquatic life. When discharged in excessive amounts on land, they can also lead to the pollution of groundwater
Priority pollutants	Organic and inorganic compounds selected on the basis of their known or suspected carcinogenicity, mutagenicity, teratogenicity, or high acute toxicity. Many of these compounds are found in wastewater
Refractory organics	These organics tend to resist conventional methods of wastewater treatment. Typical examples include surfactants, phenols, and agricultural pesticides
Heavy metals	Heavy metals are usually added to wastewater from commercial and industrial activities and may have to be removed if the wastewater is to be reused
Dissolved inorganics	Inorganic constituents such as calcium, sodium, and sulfate are added to the original domestic water supply as a result of water use and may have to be removed if the wastewater is to be reused

There are no universal procedures for sampling; sampling programs must be tailored individually to fit each situation (see Fig. 2–1). Special procedures are necessary to handle sampling problems that arise when wastes vary considerably in composition.

Before a sampling program is undertaken, a detailed sampling protocol must be developed along with a quality assurance project plan (QAPP) (known previously as quality assurance/quality control, QA/QC). As a minimum, the following items must be specified in the QAPP (Pepper et al., 1996). Additional details on the subject of sampling may be found in Standard Methods (1998).

1. *Sampling plan.* Number of sampling locations, number (see homework problem 2–5) and type of samples, time intervals (e.g., real-time and/or time-delayed samples).
2. *Sample types and size.* Catch or grab samples, composite samples, or integrated samples, separate samples for different analyses (e.g., for metals). Sample size (i.e., volume) required.

Figure 2-1

Collection of samples for analysis: (*a*) collection of an effluent sample from a pilot plant treatment unit and (*b*) view of an uncapped monitoring well equipped with sampling outlets for four different well depths. Samples are collected from each depth to monitor a groundwater injection system.

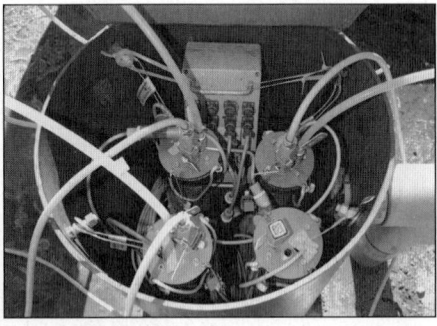

(b)

(a)

3. *Sample labeling and chain of custody.* Sample labels, sample seals, field log book, chain of custody record, sample analysis request sheets, sample delivery to the laboratory, receipt and logging of sample, and assignment of sample for analysis.
4. *Sampling methods.* Specific techniques and equipment to be used (e.g., manual, automatic, or sorbent sampling).
5. *Sampling storage and preservation.* Type of containers (e.g., glass or plastic), preservation methods, maximum allowable holding times.
6. *Sample constituents.* A list of the parameters to be measured.
7. *Analytical methods.* A list of the field and laboratory test methods and procedures to be used, and the detection limits for the individual methods.

If the physical, chemical, and/or biological integrity of the samples is not maintained during interim periods between sample collection and sample analysis, a carefully performed sampling program will become worthless. Considerable research on the problem of sample preservation has failed to perfect a universal treatment or method, or to formulate a set of fixed rules applicable to samples of all types. Prompt analysis is undoubtedly the most positive assurance against error due to sample deterioration. When analytical and testing conditions dictate a lag between collection and analysis, such as when a 24-h composite sample is collected, provisions must be made for preserving samples (see Fig. 2–2). Current methods of sample preservation for the analysis of properties subject to deterioration must be used (Standard Methods, 1998). Probable errors due to deterioration of the sample should be noted in reporting analytical data.

Figure 2-2

Typical automatic composite sampler used to collect and refrigerate samples over a 24-h period.

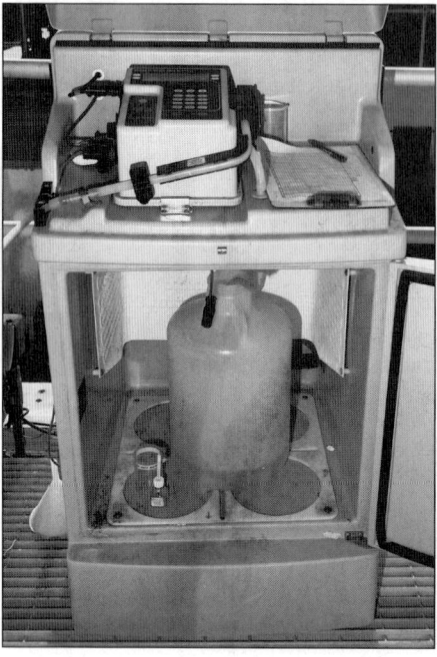

Methods of Analysis

The analyses used to characterize wastewater vary from precise quantitative chemical determinations to the more qualitative biological and physical determinations. The quantitative methods of analysis are either gravimetric, volumetric, or physicochemical. In the physicochemical methods, properties other than mass or volume are measured. Instrumental methods of analysis such as turbidimetry, colorimetry, potentiometry, polarography, adsorption spectrometry, fluorometry, spectroscopy, and nuclear radiation are representative of the physicochemical analyses. Details concerning the various analyses may be found in Standard Methods (1998), the accepted reference that details the conduct of water and wastewater analyses.

Regardless of the method of analysis used, the detection level must be specified. Several detection limits are defined and are listed below in order of increasing levels (Standard Methods, 1998).

1. *Instrumental detection level (IDL).* Constituent concentration that produces a signal greater than five times the signal/noise ratio of the instrument. The IDL is, in many respects, similar to "critical level" and "criterion of detection." The latter level is stated as 1.645 times the standard deviation s of blank analyses.
2. *Lower level of detection (LLD).* Constituent concentration in reagent water that produces a signal ($2 \times 1.645s$) above the mean of blank analyses, where s is the standard deviation. The value at $2 \times 1.645s$ sets both Type I and Type II errors at 5 percent. Type I and Type II errors are used to test the null hypothesis (Larson

and Farber, 2000). Other names for this level are "detection level" and "level of detection" (LOD).

3.   *Method detection level (MDL).* Constituent concentration that, when processed through the complete method, produces a signal with a 99 percent probability that it is different from the blank. For seven replicates of the sample, the mean must be $3.14s$ above the blank, where s is the standard deviation of the seven replicates. Compute MDL from replicate measurements one to five times the actual MDL. The MDL will be larger than the LLD because of the few replications and the sample processing steps, and may vary with constituent and matrix.

4.   *Level of quantification (LOQ).* Constituent concentration that produces a signal sufficiently greater than the blank that can be detected within specified levels by good laboratories during routine operating conditions. Typically it is the concentration that produces a signal $10s$ above the reagent blank signal.

Units of Measurement for Physical and Chemical Parameters

The results of the analysis of wastewater samples are expressed in terms of physical and chemical units of measurement. The most common units for these measurements are reported in Table 2–3. Measurements of chemical parameters are usually expressed in the physical unit of milligrams per liter (mg/L) or grams per cubic meter (g/m^3). The concentration of trace constituents is usually expressed as micrograms per liter ($\mu g/L$) or nanograms per liter (ng/L).

As noted in Table 2–3, the concentration can also be expressed as parts per million (ppm), which is a mass-to-mass ratio. The relationship between mg/L and ppm is:

$$ppm = \frac{mg/L}{\text{specific gravity of fluid}} \qquad (2\text{--}1)$$

For dilute systems, such as those encountered in natural waters and wastewater, in which one liter of sample weighs approximately one kilogram, the units of mg/L or g/m^3 are interchangeable with ppm. The terms "parts per billion" (ppb) and "parts per trillion" (ppt) are used interchangeably with $\mu g/L$ and ng/L, respectively. Dissolved gases, considered to be chemical constituents, are expressed in parts per million by volume (ppm_v), $\mu g/m^3$, or mg/L. Conversion of gas concentrations between ppm_v and $\mu g/m^3$ is given by Eq. (2–44), the universal gas law. Gases that evolve as byproducts of wastewater treatment, such as carbon dioxide and methane (from anaerobic decomposition), are measured in terms of L or m^3 (ft^3). Parameters such as temperature, odor, hydrogen ion, and biological organisms are expressed in other units, as explained below.

Useful Chemical Relationships

Other useful relationships from chemistry used in the analysis and evaluation of wastewater test results and in the design of treatment facilities include mole fraction, electroneutrality, chemical equilibrium, activity coefficient, ionic strength, and solubility product.

Mole Fraction.   The ratio of the number of moles of a given solute to the total number of moles of all components in solution is defined as the *mole fraction*. Along with

Table 2–3
Units commonly used to express analytical results

Basis	Application	Unit[a]
Physical analyses:		
Density	$\dfrac{\text{Mass of solution}}{\text{Unit of volume}}$	$\dfrac{kg}{m^3}$
Percent by volume	$\dfrac{\text{Volume of solution} \times 100}{\text{Total volume of solution}}$	% (by vol)
Percent by mass	$\dfrac{\text{Mass of solution} \times 100}{\text{Combined mass of solute + solvent}}$	% (by mass)
Volume ratio	$\dfrac{\text{Milliliters}}{\text{Liter}}$	$\dfrac{mL}{L}$
Mass per unit volume	$\dfrac{\text{Picograms}}{\text{Liter of solution}}$	$\dfrac{pg}{L}$
	$\dfrac{\text{Nanograms}}{\text{Liter of solution}}$	$\dfrac{ng}{L}$
	$\dfrac{\text{Micrograms}}{\text{Liter of solution}}$	$\dfrac{\mu g}{L}$
	$\dfrac{\text{Milligrams}}{\text{Liter of solution}}$	$\dfrac{mg}{L}$
	$\dfrac{\text{Grams}}{\text{Cubic meter of solution}}$	$\dfrac{g}{m^3}$
Mass ratio	$\dfrac{\text{Milligrams}}{10^9 \text{ milligrams}}$	ppb
	$\dfrac{\text{Milligrams}}{10^6 \text{ milligrams}}$	ppm
Chemical analyses:		
Molality	$\dfrac{\text{Moles of solute}}{1000 \text{ grams solvent}}$	$\dfrac{mol}{kg}$
Molarity	$\dfrac{\text{Moles of solute}}{\text{Liter of solution}}$	$\dfrac{mol}{L}$
Normality	$\dfrac{\text{Equivalents of solute}}{\text{Liter of solution}}$	$\dfrac{eq}{L}$
	$\dfrac{\text{Milliequivalents of solute}}{\text{Liter of solution}}$	$\dfrac{meq}{L}$

[a]Note: $1 \text{ g} = 10^3 \text{ mg} = 10^6 \text{ }\mu g = 10^9 \text{ ng} = 10^{12} \text{ pg}$
mg/L = g/m³
ppm = parts per million (by mass), ppb = parts per billion (by mass), ppt = parts per trillion (by mass)

its importance in solution chemistry, the mole fraction is of importance in the mass transfer of gases into and out of liquids. In equation form,

$$x_B = \frac{n_B}{n_A + n_B + n_C + \cdots + n_N}$$ (2-2)

where x_B = mole fraction of solute B
$\quad n_B$ = number of moles of solute B
$\quad n_A$ = number of moles of solute A
$\quad n_C$ = number of moles of solute C
$\quad n_N$ = number of moles of solute N

The application of Eq. (2-2) is illustrated in Example 2-1.

EXAMPLE 2-1 **Determination of Mole Fraction** Determine the mole fraction of oxygen in water if the concentration of dissolved oxygen is 10.0 mg/L.

Solution

1. Determine the mole fraction of oxygen using Eq. (2-2) written as follows:

$$x_{O_2} = \frac{n_{O_2}}{n_{O_2} + n_W}$$

a. Determine the moles of oxygen.

$$n_{O_2} = \frac{(10 \text{ mg/L})}{(32 \times 10^3 \text{ mg/mole O}_2)} = 3.125 \times 10^{-4} \text{ mole/L}$$

b. Determine the moles of water.

$$n_W = \frac{(1000 \text{ g/L})}{(18 \text{ g/mole of water})} = 55.556 \text{ mole/L}$$

c. The mole fraction of oxygen is:

$$x_{O_2} = \frac{3.125 \times 10^{-4}}{3.125 \times 10^{-4} + 55.556} = 5.62 \times 10^{-6}$$

Electroneutrality. The principle of *electroneutrality* requires that the sum of the positive ions (cations) must equal the sum of negative ions (anions) in solution, thus:

$$\Sigma \text{ cations} = \Sigma \text{ anions}$$ (2-3)

where cations = positively charged species in solution expressed in terms of equivalent weight per liter, eq/L or milliequvalent weight per liter, meq/L
anions = negatively charged species in solution, eq/L or meq/L

The equivalent weight of a compound is defined as:

$$\text{Equivalent weight, g/eq} = \frac{\text{molecular weight, g}}{Z}$$ (2-4)

where Z = (1) the absolute value of the ion charge, (2) the number of H^+ or OH^- ions a species can react with or yield in an acid-base reaction, or (3) the absolute value of the change in valence occurring in an oxidation reduction reaction (Sawyer et al., 1994).

Equation (2–3) can be used to check the accuracy of chemical analyses by taking into account the percentage difference defined as follows (Standard Methods, 1998):

$$\text{Percent difference} = 100 \times \left(\frac{\Sigma \text{ cations} - \Sigma \text{ anions}}{\Sigma \text{ cations} + \Sigma \text{ anions}} \right) \qquad (2\text{–}5)$$

The acceptance criteria are as given below.

Σ Anions, meq/L	Acceptable difference
0–3.0	±0.2 meq/L
3.0–10.0	±2%
10–800	5%

The application of Eqs. (2–3) and (2–5) is illustrated in Example 2–2.

EXAMPLE 2–2 **Checking the Accuracy of Analytical Measurements** The following analysis has been completed on a filtered effluent, from an extended aeration wastewater-treatment plant, that is to be used for landscape watering. Check the accuracy of the analysis to determine if the analysis is sufficiently accurate, based on the criteria given above.

Cation	Conc., mg/L	Anion	Conc., mg/L
Ca^{2+}	82.2	HCO_3^-	220.0
Mg^{2+}	17.9	SO_4^{2-}	98.3
Na^+	46.4	Cl^-	78.0
K^+	15.5	NO_3^-	25.6

Solution

1. Prepare a cation-anion balance.

Cation	Conc., mg/L	mg/meq[a]	meq/L	Anion	Conc., mg/L	mg/meq[a]	meq/L
Ca^{2+}	82.2	20.04[b]	4.10	HCO_3^-	220	61.02	3.61
Mg^{2+}	17.9	12.15	1.47	SO_4^{2-}	98.3	48.03	2.05
Na^+	46.4	23.00	2.02	Cl^-	78.0	35.45	2.20
K^+	15.5	39.10	0.40	NO_3^-	25.6	62.01	0.41
		Σ cations	7.99			Σ anions	8.27

[a] Eq weight = molecular weight in grams/Z
[b] For calcium, eq wt = 40.08/2 = 20.04 g/eq or 20.04 mg/meq

2. Check the accuracy of the cation-anion balance using Eq. (2–5).

$$\text{Percent difference} = 100 \times \left(\frac{\Sigma \text{ cations} - \Sigma \text{ anions}}{\Sigma \text{ cations} + \Sigma \text{ anions}} \right)$$

$$\text{Percent difference} = 100 \times \left(\frac{7.99 - 8.27}{7.99 + 8.27} \right) = -1.72\%$$

For a total anion concentration between 3 and 10 meq/L, the acceptable difference must be equal to or less than 2 percent (see table given above); thus, the analysis is of sufficient accuracy.

Comment   If the cation-anion balance is not of sufficient accuracy, the problem may be analytical or a constituent of significant concentration may be missing.

Chemical Equilibrium.   A reversible chemical reaction in which reactants A and B combine to yield products C and D may be written as:

$$aA + bB \leftrightarrow cC + dD \tag{2-6}$$

Where the *stoichiometry* coefficients a, b, c, and d correspond to the number of moles of constituents A, B, C, and D, respectively. The *stoichiometry* of reaction refers to the definition of the quantities of chemical compounds involved in a reaction (e.g., a of A, b of B, etc.). When the chemical species come to a state of equilibrium, as governed by the law of mass action, the numerical value of the ratio of the products over the reactants is known as the *equilibrium constant K* and is written as:

$$\frac{[C]^c [D]^d}{[A]^a [B]^b} = K \tag{2-7}$$

For a given reaction, the value of the equilibrium constant will change with temperature and the ionic strength of the solution. It should also be noted that in Eq. (2–7) it is assumed that that activity of the individual ions is equal to 1.

Brackets are used in Eq. (2–7) to denote molar concentrations. The use of molal concentrations (see Table 2–3) is more correct theoretically, but for dilute solutions encountered in wastewater applications molar concentrations are used. Molal concentrations must be used for brine solutions and seawater. To account for nonideal conditions encountered due to ion-ion interactions, a new concentration term called "activity" is used. The activity of an ion is defined as follows.

$$a_i = \gamma [C_i] \tag{2-8}$$

where a_i = activity of ith ion, mol/L
$\qquad \gamma$ = activity coefficient for the ith ion
$\qquad C_i$ = concentration of ith ion in solution, mol/L

If Eq. (2–7) is written in terms of activity and activity coefficients rather than concentrations, the resulting expression is

$$\frac{[a_C]^c [a_D]^d}{[a_A]^a [a_B]^b} = \frac{[\gamma_C C]^c [\gamma_D D]^d}{[\gamma_A A]^a [\gamma_B B]^b} = K \tag{2-9}$$

Activity Coefficient. The activity coefficient can be estimated using the following expression, derived from the Debye-Huckel theory, and known as the Guntelberg approximation [see Snoeyink and Jenkins (1998); McMurry and Fay (1998) for discussion].

$$\log \gamma = -\frac{0.5(Z_i)^2 \sqrt{I}}{1 + \sqrt{I}} \tag{2-10}$$

where Z_i = charge on ith ionic species
I = ionic strength

The above relationship is valid for solutions with an ionic strength that does not exceed 0.1 M. A number of other similar relationships will be found in the literature including those proposed by Davies (1962). The equation proposed by Davies, which can be used for higher ionic strengths (>0.1 M), is essentially the same as Eq. (2–10) with the exception that a factor of $0.3I$ is subtracted from the right-hand side of Eq. (2–10). The Davies equation is used in homework problem 2–3. Computation of the activity coefficient is illustrated in Example 2–3 following the discussion of ionic strength and solubility.

Ionic Strength. The ionic strength of a solution is a measure of the concentration of dissolved chemical constituents. The ionic strength of a solution can be estimated using the following expression:

$$I = \frac{1}{2} \Sigma C_i Z_i^2 \tag{2-11}$$

where I = ionic strength
C_i = concentration of the ith species, mole/L
Z_i = valence (or oxidation) number of the ith species [see Eq. (2–4)]

The ionic strength can also be estimated based on the total dissolved solids concentration using the following expression:

$$I = 2.5 \times 10^{-5} \times \text{TDS} \tag{2-12}$$

where TDS = total dissolved solids, mg/L or g/m^3
Equation (2–12) is often used to estimate the ionic strength of treated wastewater in groundwater recharge applications (see Chap. 13).

Solubility Product. The equilibrium constant for a reaction involving a precipitate and its constituent ions is known as the *solubility product*. For example, the reaction for calcium carbonate (CaCO$_3$) is

$$\text{CaCO}_3 \leftrightarrow \text{Ca}^{2+} + \text{CO}_3^{2-} \tag{2-13}$$

Because the activity of the solid phase is usually taken as 1, the solubility product is written as:

$$[\text{Ca}^{2+}]\,[\text{CO}_3^{2-}] = K_{sp} \tag{2-14}$$

where K_{sp} = solubility product constant.

It is important to note that the value of the equilibrium constant will change with temperature of the solution. Written in terms of activity coefficients, Eq. (2–14) becomes

$$\gamma_{Ca^{2+}}[Ca^{2+}]\gamma_{CO_3^{2-}}[CO_3^{2-}] = K_{sp} \tag{2–15}$$

The application of Eq. (2–15) is illustrated in Example 2–3.

EXAMPLE 2–3 **Determine the Activity Coefficients and Solubility of Calcium Carbonate** Determine the activity coefficients for the mono and divalent ions in the wastewater given in Example 2–2. Using the value of the activity coefficient for a divalent ion, estimate the equilibrium concentration of calcium in solution needed to satisfy the solubility product for calcium carbonate ($CaCO_3$) at 25°C. The value of the solubility product constant K_{sp} at 25°C is 5×10^{-9}.

Solution

1. Determine the ionic strength of the wastewater using Eq. (2–11).
   a. Prepare a computation table to determine the summation term in Eq. (2–10) using the data from Example 2–2.

Ion	Conc. C, mg/L	$C \times 10^3$, mole/L	Z^2	$CZ^2 \times 10^3$
Ca^{2+}	82.2	2.051	4	8.404
Mg^{2+}	17.9	0.736	4	2.944
Na^+	46.4	2.017	1	2.017
K^+	15.5	0.396	1	0.397
HCO_3^-	220	3.607	1	3.607
SO_4^{2-}	98.3	1.024	4	4.096
Cl^-	78.0	2.200	1	2.200
NO_3^-	25.6	0.413	1	0.413
Sum				23.876

   b. Determine the ionic strength for the concentration C.

   $$I = \frac{1}{2}\Sigma C_i Z_i^2 = \frac{1}{2}(23.876 \times 10^{-3}) = 11.938 \times 10^{-3}$$

2. Determine the activity coefficients for Ca^{2+} and CO_3^{2-}. Because both species have a valence (charge) of 2, the activity of each will be the same.
   a. For monovalent ions

   $$\log \gamma = -\frac{0.5(Z_i)^2\sqrt{I}}{1 + \sqrt{I}} = \frac{0.5(1)^2\sqrt{11.938 \times 10^{-3}}}{1 + \sqrt{11.938 \times 10^{-3}}} = 0.0492$$

   $$\gamma = 0.893$$

b. For divalent ions

$$\log \gamma = -\frac{0.5(Z_i)^2\sqrt{I}}{1 + \sqrt{I}} = -\frac{0.5(2)^2\sqrt{11.938 \times 10^{-3}}}{1 + \sqrt{11.938 \times 10^{-3}}} = -0.1968$$

$$\gamma = 0.636$$

3. Determine the minimum solubility of calcium using Eq. (2–15).
   a. Because the molar concentrations of calcium and carbonate ions are the same, Eq (2–15) can be written as follows:

   $$\gamma^2[C]^2 = K_{sp}$$

   b. Solve for the concentration C.

   $$C = \sqrt{\frac{K_{sp}}{\gamma^2}} = \sqrt{\frac{5 \times 10^{-9}}{(0.636)^2}} = 11.1 \times 10^{-5} \text{ mole/L}$$

   c. Convert the molar concentration of calcium carbonate to mg/L.

   $$Ca = 11.1 \times 10^{-5} \text{ mole/L} \times 40,000 \text{ mg/mole} = 4.45 \text{ mg/L}$$

Comment The computed value represents the minimum concentration of calcium that would be required in solution to be in equilibrium with solid calcium carbonate.

2–3 PHYSICAL CHARACTERISTICS

The most important physical characteristic of wastewater is its total solids content, which is composed of floating matter, settleable matter, colloidal matter, and matter in solution. Other important physical characteristics include particle size distribution; turbidity; color; transmittance; temperature; conductivity; and density, specific gravity, and specific weight. Odor, sometimes considered a physical factor, is considered in the following section.

Solids

Wastewater contains a variety of solid materials varying from rags to colloidal material. In the characterization of wastewater, coarse materials are usually removed before the sample is analyzed for solids. The various solids classifications are identified in Table 2–4. The interrelationship between the various solids fractions found in wastewater is illustrated graphically on Fig. 2–3. The standard test for settleable solids consists of placing a wastewater sample in a 1-liter Imhoff cone (see Fig. 2–5) and noting the volume of solids in millimeters that settle after a specified time period (1 h). Typically, about 60 percent of the suspended solids in a municipal wastewater are settleable. Total solids (TS) are obtained by evaporating a sample of wastewater to dryness and measuring the mass of the residue. As shown on Fig. 2–3, a filtration step is used to separate the total suspended solids (TSS) from the total dissolved solids (TDS). The apparatus used to determine TSS is shown on Fig. 2–4. The principal types of materials that comprise the filterable and nonfilterable solids in wastewater and their approximate size range are reported on Fig. 2–6.

Total Suspended Solids. Because a filter is used to separate the TSS from the TDS, the TSS test is somewhat arbitrary, depending on the pore size of the filter paper used for the test. Filters with nominal pore sizes varying from 0.45 μm to about 2.0 μm have been used for the TSS test (see Fig. 2–7). More TSS will be measured as the pore size of the filter used is reduced. Thus, it is important to note the pore size of the filter paper used, when comparing reported TSS values. It is also important to note that the TSS test itself has no fundamental significance. The principal reasons that the test lacks a fundamental basis are as follows. First, the measured values of TSS are dependent on

Table 2–4
Definitions for solids found in wastewater[a]

Test[b]	Description
Total solids (TS)	The residue remaining after a wastewater sample has been evaporated and dried at a specified temperature (103 to 105°C)
Total volatile solids (TVS)	Those solids that can be volatilized and burned off when the TS are ignited (500 ± 50°C)
Total fixed solids (TFS)	The residue that remains after TS are ignited (500 ± 50°C)
Total suspended solids (TSS)	Portion of the TS retained on a filter (see Fig. 2–4) with a specified pore size, measured after being dried at a specified temperature (105°C). The filter used most commonly for the determination of TSS is the Whatman glass fiber filter, which has a nominal pore size of about 1.58 μm
Volatile suspended solids (VSS)	Those solids that can be volatilized and burned off when the TSS are ignited (500 ± 50°C)
Fixed suspended solids (FSS)	The residue that remains after TSS are ignited (500 ± 50°C)
Total dissolved solids (TDS) (TS − TSS)	Those solids that pass through the filter, and are then evaporated and dried at specified temperature. It should be noted that what is measured as TDS is comprised of colloidal and dissolved solids. Colloids are typically in the size range from 0.001 to 1 μm
Total volatile dissolved solids (VDS)	Those solids that can be volatilized and burned off when the TDS are ignited (500 ± 50°C)
Fixed dissolved solids (FDS)	The residue that remains after TDS are ignited (500 ± 50°C)
Settleable solids	Suspended solids, expressed as milliliters per liter, that will settle out of suspension within a specified period of time

[a] Adapted from Standard Methods (1998).
[b] With the exception of settleable solids, all solids values are expressed in mg/L.

Figure 2-3

Interrelationships of solids found in water and wastewater. In much of the water quality literature, the solids passing through the filter are called dissolved solids. (Tchobanoglous and Schroeder, 1985.)

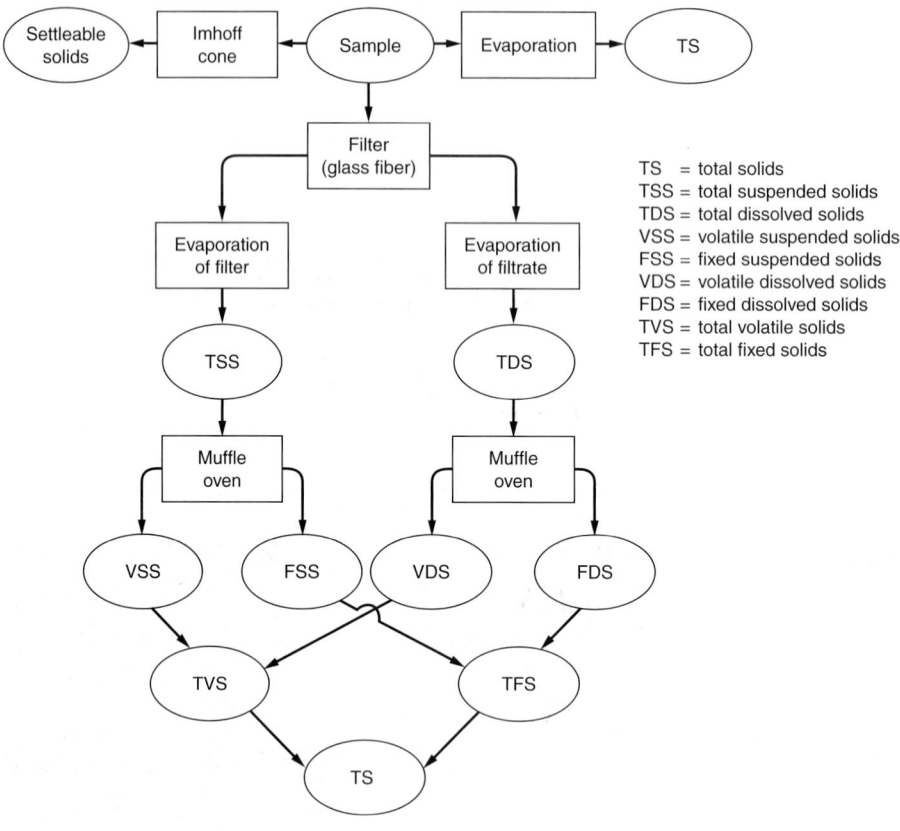

TS = total solids
TSS = total suspended solids
TDS = total dissolved solids
VSS = volatile suspended solids
FSS = fixed suspended solids
VDS = volatile dissolved solids
FDS = fixed dissolved solids
TVS = total volatile solids
TFS = total fixed solids

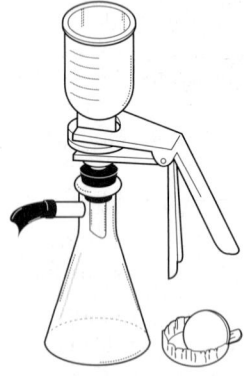

Figure 2-4

Apparatus used for the determination of total suspended solids. After wastewater sample has been filtered, the preweighted filter paper is placed in an aluminum dish for drying before weighing.

the type and pore size of the filter paper used in the analysis. Second, depending on the sample size used for the determination of TSS, autofiltration, where the suspended solids that have been intercepted by the filter also serve as a filter, can occur. Autofiltration will cause an apparent increase in the measured TSS value over the actual value. Third, depending on the characteristics of the particulate matter, small particles may be removed by adsorption to material already retained by the filter. Fourth, TSS is a lumped parameter, because the number and size distribution of the particles that comprise the measured value is unknown. Nevertheless, TSS test results are used routinely to assess the performance of conventional treatment processes and the need for effluent filtration in reuse applications. Finally, TSS is one of the two universally used effluent standards (along with BOD) by which the performance of treatment plants is judged for regulatory control purposes. The analysis of laboratory data is illustrated in Example 2–4.

Total Dissolved Solids. By definition, the solids contained in the filtrate that passes through a filter with a nominal pore size of 2.0 μm or less are classified as dissolved (Standard Methods, 1998). Yet it is known that wastewater contains a high fraction of colloidal solids. The size of colloidal particles in wastewater is typically in the range from 0.01 to 1.0 μm. It should be noted that some researchers have classified the

Figure 2–5

Imhoff cone used to determine settleable solids in wastewater. Solids that accumulate in the bottom of the cone after 60 min are reported as mL/L.

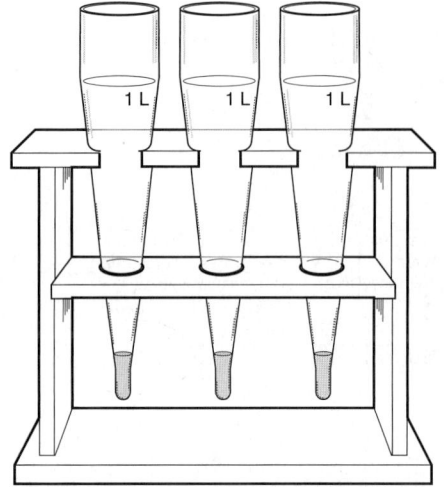

Figure 2–6

Micrographs of two laboratory filters used for the measurement of suspended solids in wastewater. (a) Polycarbonate membrane filter with a nominal pore size of 1.0 μm and (b) glass fiber filter with a nominal pore size of 1.2 μm.

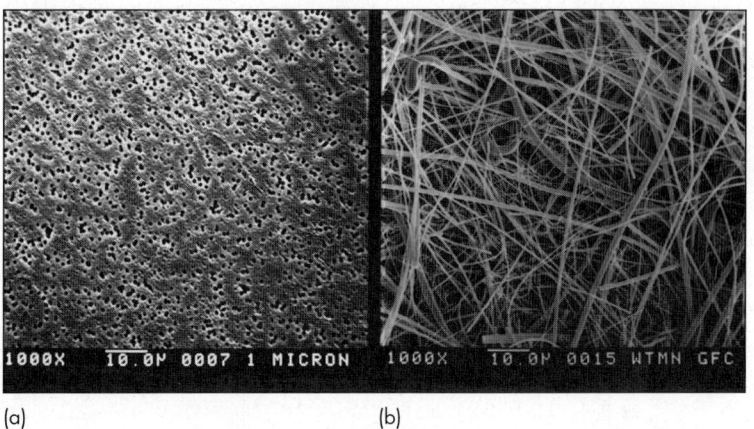

(a)   (b)

size range for colloidal particles as varying from 0.001 to 1.0 μm, others from 0.003 to 1.0 μm. The size range for colloidal particles considered in this text is from 0.01 to 1.0 μm. The number of colloidal particles in untreated wastewater and after primary sedimentation is typically in the range from 10^8 to 10^{12}/mL. The fact that the distinction between colloidal particles and truly dissolved material has not been made routinely has led to confusion in the analysis of treatment plant performance and in the design of treatment processes.

Volatile and Fixed Solids. Material that can be volatilized and burned off when ignited at 500 ± 50°C is classified as volatile. In general, volatile solids (VS) are presumed to be organic matter, although some organic matter will not burn and some inorganic solids break down at high temperatures. Fixed solids (FS) comprise the residue that remains after a sample has been ignited. Thus, TS, TSS, and TDS are comprised of both fixed solids and volatile solids. The ratio of the VS to FS is often used to characterize the wastewater with respect to amount of organic matter present.

Figure 2-7

Size ranges of organic contaminants in wastewater and size separation and measurement techniques used for their quantification.

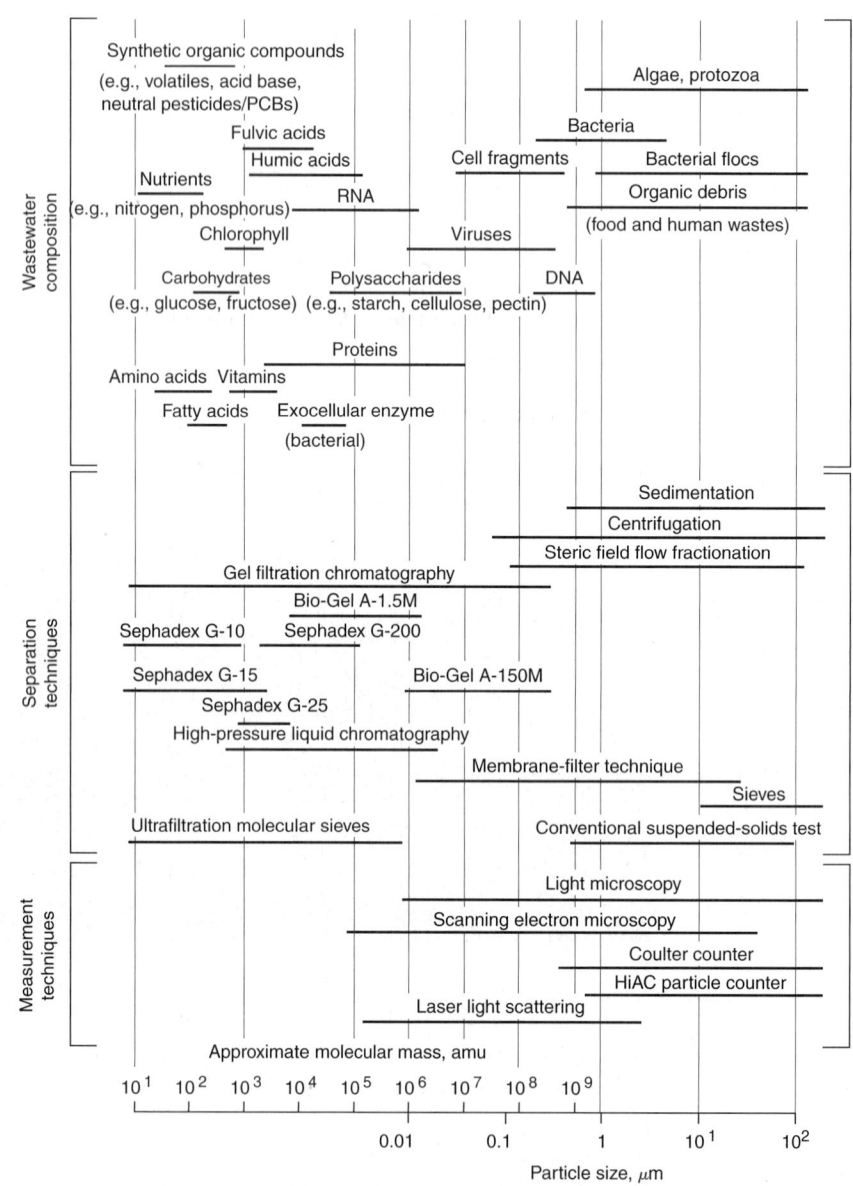

EXAMPLE 2–4 Analysis of Solids Data The following test results were obtained for a wastewater sample taken at the headworks to a wastewater-treatment plant. All of the tests were performed using a sample size of 50 mL. Determine the concentration of total solids, total volatile solids, suspended solids, volatile suspended solids, total dissolved solids, and total volatile dissolved solids. The samples used in the solids analyses were all either evaporated, dried, or ignited to constant weight.

Tare mass of evaporating dish = 53.5433 g

Mass of evaporating dish plus residue after evaporation at 105°C = 53.5794 g

Mass of evaporating dish plus residue after ignition at 550°C = 53.5625 g

Tare mass of Whatman GF/C filter after drying at 105°C = 1.5433 g

Mass of Whatman GF/C filter and residue after drying at 105°C = 1.5554 g

Mass of Whatman GF/C filter and residue after ignition at 550°C = 1.5476 g

Solution

1. Determine total solids.

$$TS = \frac{\left(\begin{array}{c}\text{mass of evaporating}\\\text{dish plus residue, g}\end{array}\right) - \left(\begin{array}{c}\text{mass of evaporating}\\\text{dish, g}\end{array}\right)}{\text{sample size, L}}$$

$$TS = \frac{[(53.5794 - 53.5433)\text{ g}](10^3\text{ mg/g})}{0.050\text{ L}} = 722\text{ mg/L}$$

2. Determine total volatile solids.

$$TVS = \frac{\left(\begin{array}{c}\text{mass of evaporating}\\\text{dish plus residue, g}\end{array}\right) - \left(\begin{array}{c}\text{mass of evaporating dish}\\\text{plus residue after ignition, g}\end{array}\right)}{\text{sample size, L}}$$

$$TVS = \frac{[(53.5794 - 53.5625)\text{ g}](10^3\text{ mg/g})}{0.050\text{ L}} = 338\text{ mg/L}$$

3. Determine the total suspended solids.

$$TSS = \frac{\left(\begin{array}{c}\text{residue on filter}\\\text{after drying, g}\end{array}\right) - \left(\begin{array}{c}\text{tare mass of filter}\\\text{after drying, g}\end{array}\right)}{\text{sample size, L}}$$

$$TSS = \frac{[(1.5554 - 1.5433)\text{ g}](10^3\text{ mg/g})}{0.050\text{ L}} = 242\text{ mg/L}$$

4. Determine the volatile suspended solids.

$$VSS = \frac{\left(\begin{array}{c}\text{residue on filter}\\\text{after drying, g}\end{array}\right) - \left(\begin{array}{c}\text{residue on filter}\\\text{after ignition, g}\end{array}\right)}{\text{sample size, L}}$$

$$VSS = \frac{[(1.5554 - 1.5476)\text{ g}](10^3\text{ mg/g})}{0.050\text{ L}} = 156\text{ mg/L}$$

5. Determine the total dissolved solids.

   TDS = TS − TSS = 722 − 242 = 480 mg/L

6. Determine the volatile dissolved solids.

   VDS = TVS − VSS = 338 − 156 = 182 mg/L

Particle Size Distribution

As noted above, TSS is a lumped parameter. In an effort to understand more about the nature of the particles that comprise the TSS in wastewater, measurement of particle size is undertaken and an analysis of the distribution of particle sizes is conducted (Tchobanoglous, 1995). Information on particle size is of importance in assessing the effectiveness of treatment processes (e.g., secondary sedimentation, effluent filtration, and effluent disinfection). Because the effectiveness of both chlorine and UV disinfection is dependent on particle size, the determination of particle size has become more important, especially with the move toward greater effluent reuse in the western United States.

Information on the size of the biodegradable organic particles is significant from a treatment standpoint, as the biological conversion rate of these particles is dependent on size (see discussion in Sec. 2–6, which deals with biochemical oxygen demand). Methods that have been used to determine particle size are summarized in Table 2–5. As reported in Table 2–5, the methods can be divided into two general categories: (1) methods based on observation and measurement and (2) methods based on separation and analysis techniques. The methods used most commonly to study and quantify the particles in wastewater are: (1) serial filtration, (2) electronic particle counting, and (3) direct microscopic observation.

Serial Filtration. In the serial filtration method, a wastewater sample is passed sequentially through a series of membrane filters (see Fig. 2–8) with circular openings of known diameter (typically 12, 8, 5, 3, 1, and 0.1 μm), and the amount of suspended solids retained in each filter is measured. Typical results from such a measurement are reported in Table 2–6. What is interesting to note in Table 2–6 is the amount of colloidal material found between 0.1 and 1.0 μm. If a 0.1-μm filter had been used to determine TSS for the treated effluent at Monterey instead of a filter with a nominal pore size equal to or greater than 1.0 μm (2.0 μm as specified in Standard Methods for the TSS test), more than 20 mg/L of additional TSS would have been measured. Although some information is gained on the size and distribution of the particles in the wastewater sample, little information is gained on the nature of the individual particles. This method is useful in assessing the effectiveness of treatment methods (e.g., microfiltration) for the removal of residual TSS.

Electronic Particle Size Counting. In electronic particle size counting, particles in wastewater are counted by diluting a sample and then passing the diluted sample through a calibrated orifice or past laser beams. As the particles pass through the orifice,

Table 2–5
Analytical techniques applicable to particle size analysis of wastewater contaminants[a]

Technique	Typical size range, μm
Observation and measurement	
Microscopy	
Light	0.2–>100
Transmission electron	0.2–>100
Scanning electron	0.002–50
Image analysis	0.2–>100
Particle counters	
Conductivity difference	0.2–>100
Equivalent light scattering	0.005–>100
Light blockage	0.2–>100
Separation and analysis	
Centrifugation	0.08–>100
Field flow fractionation	0.09–>100
Gel filtration chromatography	< 0.0001–>100
Sedimentation	0.05–>100
Membrane filtration (see Chap. 11)	0.0001–1

[a] Adapted from Levine et al. (1985).

Figure 2–8

Definition sketch for the determination of the particle size distribution (by mass) using serial filtration with membrane filters. *(From Crites and Tchobanoglous, 1998.)*

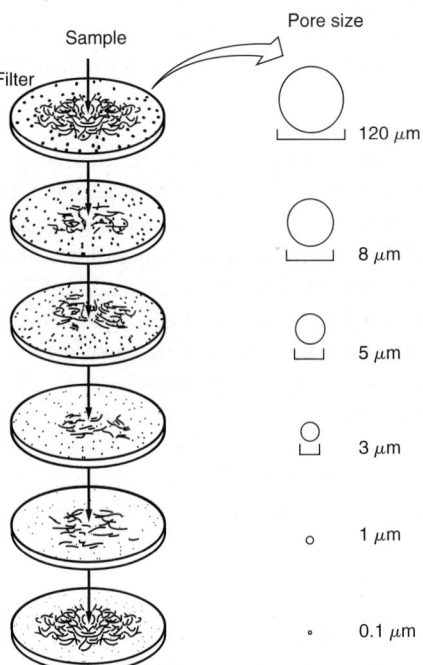

Pore size

120 μm

8 μm

5 μm

3 μm

1 μm

0.1 μm

Table 2–6
Typical data on the distribution of filterable solids and filtrate turbidity in treated (effluent) wastewater obtained by serial filtration

Sample location and (date)	Total, TSS[a]	Initial sample turbidity, NTU	Mass retained in indicated size range, μm, and turbidity of filtrate passing through filter with the smallest pore size					
			>0.1 <1.0	>1.0 <3.0	>3.0 <5.0	>5.0 <8.0	>8.0 <12.0	>12.0
Monterey[b] (12/08/93)								
TSS, mg/L	39.9		22.2	3.3	1.8	2.4	1.3	8.9
Turbidity, NTU		9.0	0.56	3.9	5.2	5.7	6.2	6.7
Monterey[b] (01/24/94)								
TSS, mg/L	38.3		21.3	8.7	4.3	0.8	0.4	2.8
Turbidity, NTU		10.0	1.4	4.6	6.4	9.8	9.8	9.8
UCD[c] (8/17/97)								
TSS, mg/L	4.67		0.55	0.53	0.24	0.20	0.24	2.91
Turbidity, NTU		0.58	0.14	0.25	0.35	0.41	0.42	0.50
UCD[d] (8/17/97)								
TSS, mg/L	1.47		0.45	0.33	0.30	0.14	0.10	0.15
Turbidity, NTU		0.45	0.14	0.31	0.33	0.38	0.40	0.41

[a] Total TSS retained on a polycarbonate membrane filter with a pore size of 0.1 μm.
[b] Secondary effluent, Monterey, CA (courtesy of Jaques, 1994).
[c] Secondary effluent, University of California, Davis, CA.
[d] Tertiary filtered secondary effluent, University of California, Davis, CA.

the conductivity of the fluid changes, owing to the presence of the particle. The change in conductivity is correlated to the size of an equivalent sphere. In a similar fashion, as a particle passes by a laser beam, it reduces the intensity of the laser because of light scattering. The reduced intensity is correlated to the diameter of the particle. The particles that are counted are grouped into particle size ranges (e.g., 0.5 to 2, 2 to 5, 5 to 20 μm, etc.). In turn, the volume fraction corresponding to each particle size range can be computed.

Typical effluent volume fraction data from two activated sludge treatment plants are reported on Fig. 2–9. As shown, the particle size data for small particles are the same for both treatment plants. However, the particle size data for the large particles are quite different, owing primarily to the design and operation of the secondary clarifiers (see discussion in Chaps. 5 and 8). Particle size information, such as that shown on Fig. 2–9, is useful in assessing the performance of secondary sedimentation facilities, effluent filtration, and the potential for chlorine and ultraviolet radiation disinfection.

Microscopic Observation. Particles in wastewater can also be enumerated microscopically by placing a small sample in a particle counting chamber and counting the individual particles. To aid in differentiating different types of particles, various types of stains

Figure 2–9

Volume fraction of particle sizes found in the effluent from two activated-sludge plants with clarifiers having different side water depths.

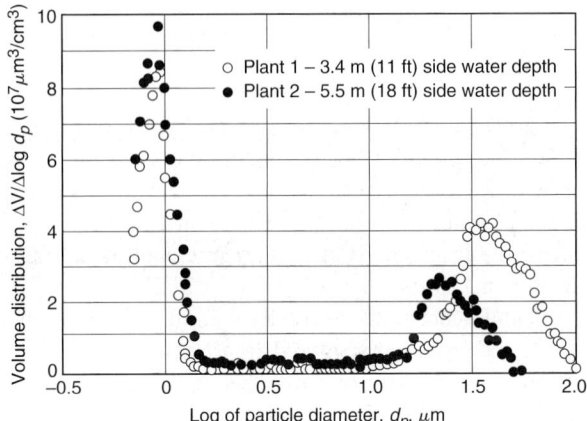

can be used. In general, microscopic counting of particles is impractical on a routine basis, given the number of particles per milliliter of wastewater. Nevertheless, this method can be used to qualitatively assess the nature and size of the particles in wastewater.

A quantitative assessment of wastewater particles can be obtained with a microscope by means of a process called optical imaging. A small sample of wastewater is placed on a microscope slide. The images of the wastewater particles are collected with a video camera attached to a microscope and transmitted to a computer where various measurements of the wastewater particles can be assessed. The types of measurements that can be obtained are dependent on the computer software but typically include the mean, minimum, and maximum diameter, the aspect ratio (length to width ratio), the circumference, the surface area, the volume, and the centroid of various particles. Particle imaging greatly reduces the time required to measure various characteristics of wastewater particles, but the cost of the software and equipment is often prohibitive for many small laboratories.

Turbidity

Turbidity, a measure of the light-transmitting properties of water, is another test used to indicate the quality of waste discharges and natural waters with respect to colloidal and residual suspended matter. The measurement of turbidity is based on comparison of the intensity of light scattered by a sample to the light scattered by a reference suspension under the same conditions (Standard Methods, 1998). Formazin suspensions are used as the primary reference standard. The results of turbidity measurements are reported as nephelometric turbidity units (NTU). Colloidal matter will scatter or absorb light and thus prevent its transmission. It should be noted that the presence of air bubbles in the fluid will cause erroneous turbidity readings. In general, there is no relationship between turbidity and the concentration of total suspended solids in untreated wastewater. There is, however, a reasonable relationship between turbidity and total suspended solids for the settled and filtered secondary effluent from the activated sludge process. The general form of the relationship is as follows:

$$\text{TSS, mg/L} \approx (\text{TSS}_f)\,(T) \tag{2–16}$$

where TSS = total suspended solids, mg/L

> TSS_f = factor used to convert turbidity readings to total suspended solids, (mg/L TSS)/NTU
>
> T = turbidity, NTU

The specific value of the conversion factor will vary for each treatment plant, depending primarily on the operation of the biological treatment process. The conversion factors for settled secondary effluent and for secondary effluent filtered with a granular-medium depth filter will typically vary from 2.3 to 2.4 and 1.3 to 1.6, respectively.

One of the problems with the measurement of turbidity (especially low values in filtered effluent) is the high degree of variability observed, depending on the light source (incandescent light versus light-emitting diodes) and the method of measurement (reflected versus transmitted light). Another problem often encountered is the light-absorbing properties of the suspended material. For example, the turbidity of a solution of lampblack will essentially be equal to zero. As a result, it is almost impossible to compare turbidity values reported in the literature. However, turbidity readings at a given facility can be used for process control. Some on-line turbidity meters used to monitor the performance of microfiltration units are affected by the air used to clean the membranes.

Color

Historically, the term "condition" was used along with composition and concentration to describe wastewater. Condition refers to the age of the wastewater, which is determined qualitatively by its color and odor. Fresh wastewater is usually a light brownish-gray color. However, as the travel time in the collection system increases, and more anaerobic conditions develop, the color of the wastewater changes sequentially from gray to dark gray, and ultimately to black. When the color of the wastewater is black, the wastewater is often described as septic. Some industrial wastewaters may also add color to domestic wastewater. In most cases, the gray, dark gray, and black color of the wastewater is due to the formation of metallic sulfides, which form as the sulfide produced under anaerobic conditions reacts with the metals in the wastewater.

Absorption/Transmittance

The absorbance of a solution is a measure of the amount of light, of a specified wavelength, that is absorbed by the constituents in a solution. Absorbance, measured using a spectrophotometer and a fixed path length (usually 1.0 cm), is given by the following relationship:

$$A = \log\left(\frac{I_o}{I}\right) \tag{2–17}$$

where A = absorbance, absorbance units/centimeter, a.u./cm

I_o = initial detector reading for the blank (i.e., distilled water) after passing through a solution of known depth

I = final detector reading after passing through solution containing constituents of interest

Absorbance is measured with a spectrophotometer using a specified wavelength, typically 254 nm. Typical absorbance values for various wastewaters at 254 nm are:

1. Primary = 0.55 to 0.30/cm
2. Secondary = 0.35 to 0.15/cm
3. Nitrified secondary = 0.25 to 0.10/cm
4. Filtered secondary = 0.25 to 0.10/cm
5. Microfiltration effluent = 0.10 to 0.04/cm
6. Reverse osmosis effluent = 0.05 to 0.01/cm

The transmittance of a solution is defined as

$$\text{Transmittance } T, \% = \left(\frac{I}{I_o}\right) \times 100 \tag{2-18}$$

Thus, the transmittance can also be derived from absorption measurements using the following relationship:

$$T = 10^{-(\text{a.u./cm})} \tag{2-19}$$

The term percent transmittance, commonly used in the literature, is

$$T, \% = 10^{-(\text{a.u./cm})} \times 100 \tag{2-20}$$

The extreme values of A and T are (Delahay, 1957)

For a perfectly transparent solution $A = 0$, $T = 1$

For a perfectly opaque solution $A \to \infty$, $T = 0$

Typical transmittance values for treated wastewater from several activated-sludge biological treatment plants and two lagoon systems are presented on Fig. 2–10. Percent transmittance is affected by all substances in wastewater that can absorb or scatter light. Unfiltered and filtered transmittance are measured in wastewater in connection with the evaluation and design of UV disinfection systems (see Chap. 12).

The principal wastewater characteristics that affect the percent transmission include selected inorganic compounds (e.g., copper, iron, etc.), organic compounds (e.g., organic dyes, humic substances, and conjugated ring compounds such as benzene and toluene), and TSS. Of the inorganic compounds which affect transmittance, iron is considered to be the most important with respect to UV absorbance because dissolved iron can absorb

Figure 2–10

Transmittance measured at various wavelengths for activated-sludge effluents and lagoon effluents.

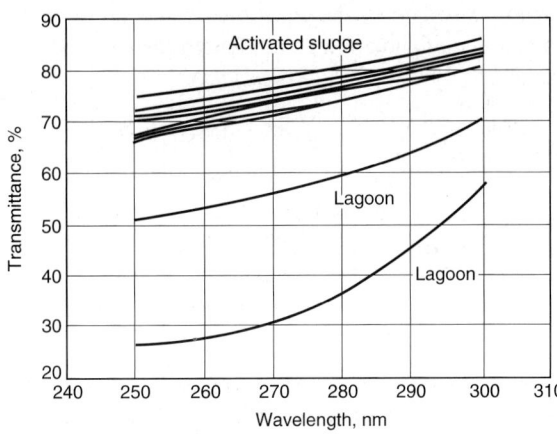

UV light directly and because iron will adsorb onto suspended solids, bacterial clumps, and other organic compounds. The sorbed iron can prevent the UV light from penetrating the particle and inactivating organisms that may be embedded within the particle. Where iron salts are added in the treatment process, dosage control is extremely important when UV disinfection is to be used. Organic constituents, identified as being absorbers of UV light, are compounds with six conjugated carbons or a five- or six-member conjugated ring. The reduction in transmittance observed during storm events is often ascribed to the presence of humic substances from stormwater flows.

Temperature

The temperature of wastewater is commonly higher than that of the local water supply, because of the addition of warm water from households and industrial activities. As the specific heat of water is much greater than that of air, the observed wastewater temperatures are higher than the local air temperatures during most of the year and are lower only during the hottest summer months. Depending on the geographic location, the mean annual temperature of wastewater in the United States varies from about 3 to 27°C (37 to 81°F); 15.6°C (60°F) is a representative value. Temperatures as high as 30 to 35°C (84 to 98°F) have been reported for countries in Africa and the Middle East. The variation that can be expected in influent wastewater temperatures is illustrated on Fig. 2–11. Depending on the location and time of year, the effluent temperatures can be either higher or lower than the corresponding influent values.

Effects of Temperature.   The temperature of water is a very important parameter because of its effect on chemical reactions and reaction rates, aquatic life, and the suitability of the water for beneficial uses. Increased temperature, for example, can cause a change in the species of fish that can exist in the receiving water body. Industrial establishments that use surface water for cooling-water purposes are particularly concerned with the temperature of the intake water.

In addition, oxygen is less soluble in warm water than in cold water. The increase in the rate of biochemical reactions that accompanies an increase in temperature, combined with the decrease in the quantity of oxygen present in surface waters, can often cause serious depletions in dissolved oxygen concentrations in the summer months. When sig-

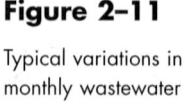

Figure 2–11

Typical variations in monthly wastewater temperatures.

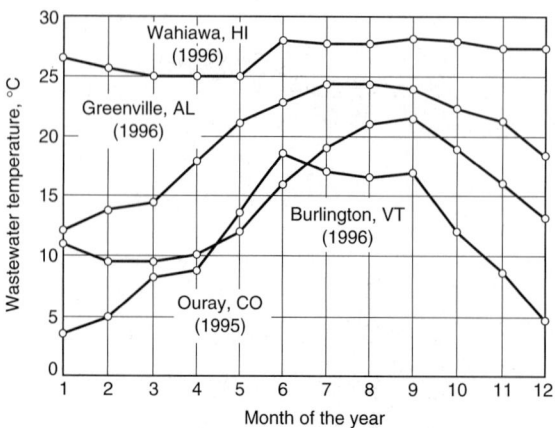

nificantly large quantities of heated water are discharged to natural receiving waters, these effects are magnified. It should also be realized that a sudden change in temperature can result in a high rate of mortality of aquatic life. Moreover, abnormally high temperatures can foster the growth of undesirable water plants and wastewater fungus.

Optimum Temperatures for Biological Activity. Optimum temperatures for bacterial activity are in the range from 25 to 35°C. Aerobic digestion and nitrification stops when the temperature rises to 50°C. When the temperature drops to about 15°C, methane-producing bacteria become quite inactive, and at about 5°C, the autotrophic-nitrifying bacteria practically cease functioning. At 2°C, even the chemoheterotrophic bacteria acting on carbonaceous material become essentially dormant. The effects of temperature on the performance of biological treatment processes are considered in greater detail in Chaps. 7 and 8.

Estimation of Temperature Effects on Reaction Rates. Equilibrium constants, solubility product constants, and specific reaction-rate constants are all dependent on temperature. The temperature dependence of the rate and equilibrium constants can be estimated using the van't Hoff-Arrhenius relationship.

$$\frac{d(\ln k)}{dT} = \frac{E}{RT^2} \tag{2-21}$$

where k = reaction rate constant
T = temperature, K = 273.15 + °C
E = a constant characteristic of the reaction (e.g., activation energy), J/mol
R = ideal gas constant, 8.314 J/mol·K (1.99 cal/mol·K)

Integration of Eq. (2–21) between the limits T_1 and T_2 gives

$$\ln \frac{k_2}{k_1} = \frac{E(T_2 - T_1)}{RT_1T_2} = \frac{E}{RT_1T_2}(T_2 - T_1) \tag{2-22}$$

With k_1 known for a given temperature and with E known, k_2 can be estimated using Eq. (2–22). The activation energy E can be calculated by determining k values at two different temperatures. The application of the van't Hoff-Arrhenius temperature relationship will recur throughout this and the following chapters.

Because most wastewater-treatment operations and processes are carried out at or near the ambient temperature, the quantity $E/(RT_1T_2)$ in Eq. (2–22) may be assumed to be a constant for all practical purposes. If the value of the quantity $E/(RT_1T_2)$ is designated by C, then Eq. (2–22) can be rewritten as

$$\ln \frac{k_2}{k_1} = C(T_2 - T_1) \tag{2-23}$$

$$\frac{k_2}{k_1} = e^{C(T_2-T_1)} \tag{2-24}$$

Replacing e^C in Eq. (2–23) with a temperature coefficient θ yields

$$\frac{k_2}{k_1} = \theta^{(T_2-T_1)} \tag{2-25}$$

which is commonly used in environmental engineering to adjust the value of the operative rate constant to reflect the effect of temperature. An alternative form of the temperature correction equation may be obtained by expanding Eq. (2–21) as a series and dropping all but the first two terms. It should be noted, however, that although the value of θ is assumed to be constant, it will often vary considerably with temperature. Therefore, caution must be used in selecting appropriate values for θ for different temperature ranges. Typical values for various operations and processes for different temperature ranges are given, where available, in the sections in which the individual topics are discussed.

Conductivity

The electrical conductivity (EC) of a water is a measure of the ability of a solution to conduct an electrical current. Because the electrical current is transported by the ions in solution, the conductivity increases as the concentration of ions increases. In effect, the measured EC value is used as a surrogate measure of total dissolved solids (TDS) concentration. At present, the EC of a water is one of the important parameters used to determine the suitability of a water for irrigation. The salinity of treated wastewater to be used for irrigation is estimated by measuring its electrical conductivity.

The electrical conductivity in SI units is expressed as millisiemens per meter (mS/m) and in micromhos per centimeter (μmho/cm) in U.S. customary units. It should be noted that 1 mS/m is equivalent to 10 μmho/cm. Equation (2–26) can be used to estimate the TDS of a water sample based on the measured EC value (Standard Methods, 1998).

$$\text{TDS (mg/L)} \cong \text{EC (dS/m or } \mu\text{mho/cm)} \times (0.55 - 0.70) \tag{2–26}$$

The above relationship does not necessarily apply to raw wastewater or high-strength industrial wastewater. The above relationship can also be used to check the acceptability of chemical analyses (see Standard Methods, 1998).

The electrical conductivity can also be used to estimate the ionic strength of a solution using the following relationship (Russell, 1976):

$$I = 1.6 \times 10^{-5} \times \text{EC (dS/m or } \mu\text{mho/cm)} \tag{2–27}$$

Equation (2–26) is used to estimate the ionic strength of treated wastewater in groundwater recharge applications (see Chap. 13).

Density, Specific Gravity, and Specific Weight

The density of wastewater ρ_w is defined as its mass per unit volume expressed as g/L or kg/m^3 in SI units and as lb$_m$/ft^3 in U.S. customary units. Density is an important physical characteristic of wastewater because of the potential for the formation of density currents in sedimentation tanks, chlorine contact tanks, and other treatment units. The density of domestic wastewater, which does not contain significant amounts of industrial waste, is essentially the same as that of water at the same temperature (see Appendix C).

In some cases the specific gravity of the wastewater s_w is used in place of the density. The specific gravity is defined as

$$s_w = \frac{\rho_w}{\rho_o} \tag{2–28}$$

where ρ_w = density of wastewater

ρ_o = density of water

Both the density and specific gravity of wastewater are temperature-dependent and will vary with the concentration of total solids in the wastewater.

The specific weight of a fluid γ is its weight per unit volume. In SI units the specific weight is expressed as kN/m^3 and as lb_f/ft^3 in U.S. customary units. The relationship between γ, ρ, and the acceleration due to gravity g is $\gamma = \rho g$. At normal temperatures γ is about 9.81 kN/m^3 (62.4 lb_f/ft^3). Values for both density and specific weight as a function of temperature in both SI and U.S. customary units are given in Appendix C.

2–4 INORGANIC NONMETALLIC CONSTITUENTS

The chemical constituents of wastewater are typically classified as inorganic and organic. Inorganic chemical constituents of concern include nutrients, nonmetallic constituents, metals, and gases. Organic constituents of interest in wastewater are classified as aggregate and individual. Aggregate organic constituents are comprised of a number of individual compounds that cannot be distinguished separately. Both aggregate and individual organic constituents are of great significance in the treatment, disposal, and reuse of wastewater. Aggregate organic constituents and individual organic compounds are considered in Secs. 2–6 and 2–7, respectively. Inorganic nonmetallic chemical constituents are considered in this section. Metallic constituents are considered in Sec. 2–5.

The sources of inorganic nonmetallic and metallic constituents in wastewater derive from the background levels in the water supply and from the additions resulting from domestic use, from the addition of highly mineralized water from private wells and groundwater, and from industrial use. Domestic and industrial water softeners also contribute significantly to the increase in mineral content and, in some areas, may represent the major source. Occasionally, water added from private wells and groundwater infiltration will (because of its high quality) serve to dilute the mineral concentration in the wastewater. Because concentrations of various inorganic constituents can greatly affect the beneficial uses made of the waters, the constituents in each wastewater must be considered separately. Inorganic nonmetallic constituents considered in this section include pH, nitrogen, phosphorus, alkalinity, chlorides, sulfur, other inorganic constituents, gases, and odors. Because the concentration of the species of most chemical constituents is dependent on the hydrogen-ion concentration in solution, pH is considered first in the following discussion.

pH

The hydrogen-ion concentration is an important quality parameter of both natural waters and wastewaters. The usual means of expressing the hydrogen-ion concentration is as pH, which is defined as the negative logarithm of the hydrogen-ion concentration.

$$pH = -\log_{10} [H^+] \tag{2-29}$$

The concentration range suitable for the existence of most biological life is quite narrow and critical (typically 6 to 9). Wastewater with an extreme concentration of hydrogen ion is difficult to treat by biological means, and if the concentration is not altered before discharge, the wastewater effluent may alter the concentration in the natural waters. For treated effluents discharged to the environment the allowable pH range usually varies from 6.5 to 8.5.

The hydrogen-ion concentration in water is connected closely with the extent to which water molecules dissociate. Water will dissociate into hydrogen and hydroxyl ions as follows:

$$H_2O \leftrightarrow H^+ + OH^- \tag{2-30}$$

Applying the law of mass action [Eq. (2–7)] to Eq. (2–30) yields

$$\frac{[H^+][OH^-]}{H_2O} = K \tag{2-31}$$

where the brackets indicate concentration of the constituents in moles per liter. Because the concentration of water in a dilute aqueous system is essentially constant, this concentration can be incorporated into the equilibrium constant K to give

$$[H^+][OH^-] = K_w \tag{2-32}$$

K_w is known as the ionization constant or ion product of water and is approximately equal to 1×10^{-14} at a temperature of 25°C. Equation (2–32) can be used to calculate the hydroxyl-ion concentration when the hydrogen-ion concentration is known, and vice versa.

With pOH, which is defined as the negative logarithm of the hydroxyl-ion concentration, it can be seen from Eq. (2–32) that, for water at 25°C,

$$pH + pOH = 14 \tag{2-33}$$

The pH of aqueous systems typically is measured with a pH meter (see Fig. 2–12). Various pH papers and indicator solutions that change color at definite pH values are also used. The pH is determined by comparing the color of the paper or solution to a series of color standards.

Figure 2–12

Typical meter used for the measurement of pH and specific ion concentrations.

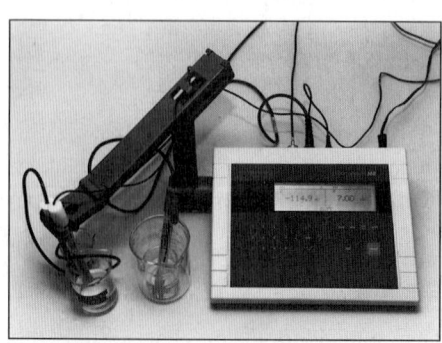

Chlorides

Chloride is a constituent of concern in wastewater as it can impact the final reuse applications of treated wastewater. Chlorides in natural water result from the leaching of chloride-containing rocks and soils with which the water comes in contact, and in coastal areas from saltwater intrusion. In addition, agricultural, industrial, and domestic wastewaters discharged to surface waters are a source of chlorides.

Human excreta, for example, contain about 6 g of chlorides per person per day. In areas where the hardness of water is high, home regeneration type water softeners will also add large quantities of chlorides. Because conventional methods of waste treatment do not remove chloride to any significant extent, higher than usual chloride concentrations can be taken as an indication that a body of water is being used for waste disposal. Infiltration of groundwater into sewers adjacent to saltwater is also a potential source of high chlorides as well as sulfates.

Alkalinity

Alkalinity in wastewater results from the presence of the hydroxides [OH^-], carbonates [CO_3^{2-}], and bicarbonates [HCO_3^-] of elements such as calcium, magnesium, sodium, potassium, and ammonia. Of these, calcium and magnesium bicarbonates are most common. Borates, silicates, phosphates, and similar compounds can also contribute to the alkalinity. The alkalinity in wastewater helps to resist changes in pH caused by the addition of acids. Wastewater is normally alkaline, receiving its alkalinity from the water supply, the groundwater, and the materials added during domestic use. The concentration of alkalinity in wastewater is important where chemical and biological treatment is to be used (see Chaps. 6 and 7, respectively), in biological nutrient removal (see Chap. 8), and where ammonia is to be removed by air stripping (see Chap. 11).

Alkalinity is determined by titrating against a standard acid; the results are expressed in terms of calcium carbonate, mg/L as $CaCO_3$. For most practical purposes alkalinity can be defined in terms of molar quantities, as

$$\text{Alk, eq/m}^3 = \text{meq/L} = [HCO_3^-] + 2\,[CO_3^{2-}] + [OH^-] - [H^+] \tag{2–34}$$

The corresponding expression in terms of equivalents is

$$\text{Alk, eq/m}^3 = (HCO_3^-) + (CO_3^{2-}) + (OH^-) - (H^+) \tag{2–35}$$

In practice, alkalinity is expressed in terms of calcium carbonate. To convert from meq/L to mg/L as $CaCO_3$, it is helpful to remember that

$$\text{Milliequivalent mass of } CaCO_3 = \frac{100 \text{ mg/mmole}}{2 \text{ meq/mmole}} \tag{2–36}$$

$$= 50 \text{ mg/meq}$$

Thus 3 meq/L of alkalinity would be expressed as 150 mg/L as $CaCO_3$.

$$\text{Alkalinity, Alk as } CaCO_3 = \frac{3.0 \text{ meq}}{L} \times \frac{50 \text{ mg } CaCO_3}{\text{meq } CaCO_3}$$

$$= 150 \text{ mg/L as } CaCO_3$$

Nitrogen

The elements nitrogen and phosphorus, essential to the growth of microorganisms, plants, and animals, are known as nutrients or biostimulants. Trace quantities of other elements, such as iron, are also needed for biological growth, but nitrogen and phosphorus are, in most cases, the major nutrients of importance. Because nitrogen is an essential building block in the synthesis of protein, nitrogen data will be required to evaluate the treatability of wastewater by biological processes. Insufficient nitrogen can necessitate the addition of nitrogen to make the waste treatable. Nutrient requirements for biological waste treatment are discussed in Chaps. 7 and 8. Where control of algal growths in the receiving water is necessary, removal or reduction of nitrogen in wastewater prior to discharge may be desirable.

Sources of Nitrogen.   The principal sources of nitrogen compounds are (1) the nitrogenous compounds of plant and animal origin, (2) sodium nitrate, and (3) atmospheric nitrogen. Ammonia derived from the distillation of bituminous coal is an example of nitrogen obtained from decayed plant material. Sodium nitrate ($NaNO_3$) is found principally in mineral deposits in Chile and in the manure found in seabird rookeries. The production of nitrogen from the atmosphere is termed *fixation*. Because fixation is a biologically mediated process and because $NaNO_3$ deposits are relatively scarce, most sources of nitrogen in soil/groundwater are of biological origin.

Forms of Nitrogen.   The chemistry of nitrogen is complex, because of the several oxidation states that nitrogen can assume and the fact that changes in the oxidation state can be brought about by living organisms. To complicate matters further, the oxidation state changes brought about by bacteria can be either positive or negative depending upon whether aerobic or anaerobic conditions prevail. The oxidation states of nitrogen are summarized below (Sawyer et al., 1994).

$$-\,\text{III} \quad 0 \quad \text{I} \quad \text{II} \quad \text{III} \quad \text{IV} \quad \text{V}$$
$$NH_3 - N_2 - N_2O - NO - N_2O_3 - NO_2 - N_2O_5 \tag{2-37}$$

The most common and important forms of nitrogen in wastewater and their corresponding oxidation state in the water/soil environment are ammonia (NH_3, $-$III), ammonium (NH_4^+, $-$III), nitrogen gas (N_2, 0), nitrite ion (NO_2^-, $+$III), and nitrate ion (NO_3^-, $+$V). The oxidation state of nitrogen in most organic compounds is $-$III.

Total nitrogen, as reported in Table 2–7, is comprised of organic nitrogen, ammonia, nitrite, and nitrate. The organic fraction consists of a complex mixture of compounds including amino acids, amino sugars, and proteins (polymers of amino acids). The compounds that comprise the organic fraction can be soluble or particulate. The nitrogen in these compounds is readily converted to ammonium through the action of microorganisms in the aquatic or soil environment. Urea, readily converted to ammonium carbonate, is seldom found in untreated municipal wastewaters.

Organic nitrogen is determined analytically using the Kjeldahl method. The aqueous sample is first boiled to drive off the ammonia, and then it is digested. During digestion the organic nitrogen is converted to ammonium through the action of heat and acid. Total Kjeldahl nitrogen (TKN) is determined in the same manner as organic nitrogen, except that the ammonia is not driven off before the digestion step. Total Kjeldahl nitrogen is therefore the total of the organic and ammonia nitrogen.

Table 2–7
Definition of the various terms used to define various nitrogen species

Form of nitrogen	Abbrev.	Definition
Ammonia gas	NH_3	NH_3
Ammonium ion	NH_4^+	NH_4^+
Total ammonia nitrogen	TAN[a]	$NH_3 + NH_4^+$
Nitrite	NO_2^-	NO_2^-
Nitrate	NO_3^-	NO_3^-
Total inorganic nitrogen	TIN[a]	$NH_3 + NH_4^+ + NO_2^- + NO_3^-$
Total Kjeldahl nitrogen	TKN[a]	Organic N $+ NH_3 + NH_4^+$
Organic nitrogen	Organic N[a]	$TKN - (NH_3 + NH_4^+)$
Total nitrogen	TN[a]	Organic N $+ NH_3 + NH_4^+ + NO_2^- + NO_3^-$

[a] All species expressed as N.

As biological nutrient removal has become more common, information on the various organic nitrogen fractions has become important. The principal fractions are particulate and soluble. In biological treatment studies, the particulate and soluble fractions of organic nitrogen are fractionated further to assess wastewater treatability (see discussion in Sec. 8–2 in Chap. 8). Fractions that have been used include (1) free ammonia, (2) biodegradable soluble organic nitrogen, (3) biodegradable particulate organic carbon, (4) nonbiodegradable soluble organic nitrogen, and (5) nonbiodegradable particulate organic nitrogen. Unfortunately, there is little standardization on the definition of soluble versus particulate organic nitrogen (see discussion of under Solids in Sec. 2–3). Where filtration is the technique used to fractionate the sample, the relative distribution between soluble and particulate organic nitrogen will vary depending on the pore size of the filter used. In many cases, colloidal organic nitrogen has been classified as soluble or dissolved. The lack of standardized definition will also affect other aggregate constituents (i.e., chemical oxygen demand and total organic carbon).

Ammonia nitrogen exists in aqueous solution as either the ammonium ion (NH_4^+) or ammonia gas (NH_3), depending on the pH of the solution, in accordance with the following equilibrium reaction:

$$NH_4^+ \leftrightarrow NH_3 + H^+ \tag{2–38}$$

Applying the law of mass action [Eq. (2–7)] to Eq. (2–38) and, assuming the activity of water is 1, yields

$$\frac{[NH_3][H^+]}{[NH_4^+]} = K_a \tag{2–39}$$

where K_a = acid ionization (dissociation) constant = $10^{-9.25}$ or 5.62×10^{-10}

Because the distribution of the ammonia species is a function of the pH, the percentage ammonia can be determined using the following relationship.

$$NH_3, \% = \frac{[NH_3] \times 100}{[NH_3] + [NH_4^+]} = \frac{100}{1 + [NH_4^+]/[NH_3]} = \frac{100}{1 + [H^+]/K_a} \tag{2–40}$$

Figure 2-13

Distribution of ammonia (NH$_3$) and ammonium ion (NH$_4^+$) as a function of pH.

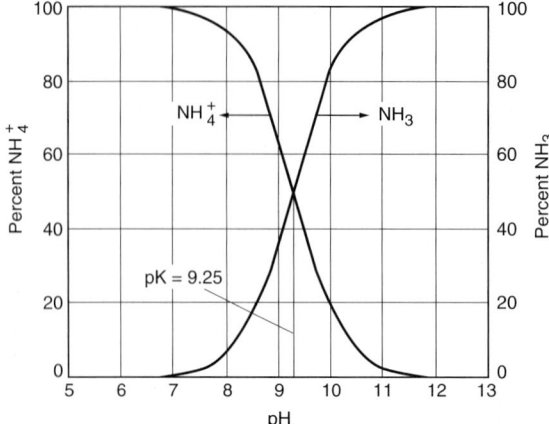

Using Eq. (2–40) the distribution of the ammonia species as a function of pH is shown on Fig. 2–13. At pH levels above 7, the equilibrium is displaced to the right; at levels below pH 7, the ammonium ion is predominant. Ammonia is determined by raising the pH, distilling off the ammonia with the steam produced when the sample is boiled, and condensing the steam that absorbs the gaseous ammonia. The measurement is made colorimetrically, titrimetrically, or with specific-ion electrodes.

Nitrite nitrogen, determined colorimetrically, is relatively unstable and is easily oxidized to the nitrate form. It is an indicator of past pollution in the process of stabilization and seldom exceeds 1 mg/L in wastewater or 0.1 mg/L in surface waters or groundwaters. Although present in low concentrations, nitrite can be very important in wastewater or water pollution studies because it is extremely toxic to most fish and other aquatic species. Nitrites present in wastewater effluents are oxidized by chlorine and thus increase the chlorine dosage requirements and the cost of disinfection.

Nitrate nitrogen is the most oxidized form of nitrogen found in wastewaters. Where secondary effluent is to be reclaimed for groundwater recharge, the nitrate concentration is important. The U.S. EPA primary drinking water standards (U.S. EPA, 1977) limit nitrogen to 45 mg/L as NO$_3^-$, because of its serious and occasionally fatal effects on infants. Nitrates may vary in concentration from 0 to 20 mg/L as N in wastewater effluents. Assuming complete nitrification has taken place, the typical range found in treated effluents is from 15 to 20 mg/L as N. The nitrate concentration is typically determined by colorimetric methods or with specific-ion electrodes.

Nitrogen Pathways in Nature. The various forms of nitrogen that are present in nature and the pathways by which the forms are changed in an aquatic environment are depicted on Fig. 2–14. The nitrogen present in fresh wastewater is primarily combined in proteinaceous matter and urea. Decomposition by bacteria readily changes the organic form to ammonia. The age of wastewater is indicated by the relative amount of ammonia that is present. In an aerobic environment, bacteria can oxidize the ammonia nitrogen to nitrites and nitrates. The predominance of nitrate nitrogen in wastewater indicates that the waste has been stabilized with respect to oxygen demand. Nitrates, however, can be used by plants and animals to form protein. Death and decomposition

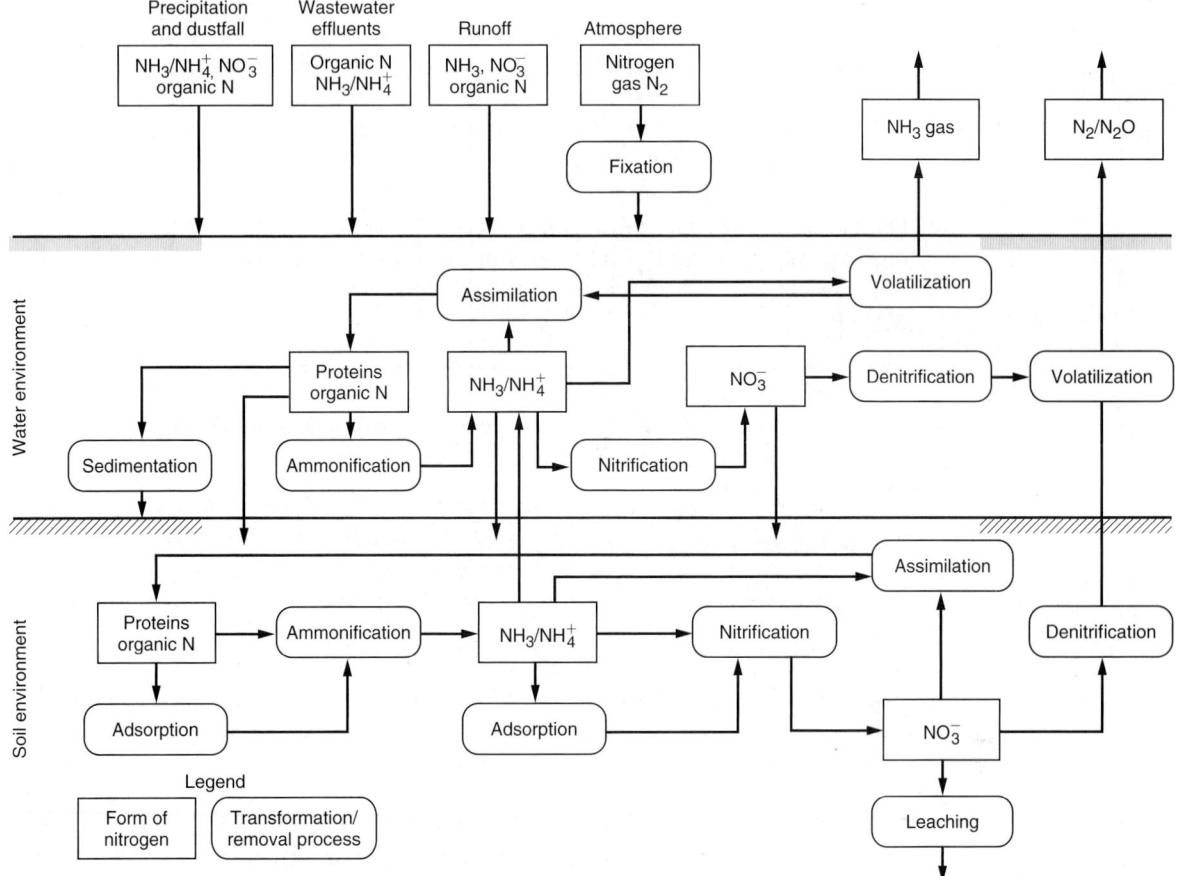

Figure 2–14

Generalized nitrogen cycle in the aquatic and soil environment.

of the plant and animal protein by bacteria again yields ammonia. Thus, if nitrogen in the form of nitrates can be reused to make protein by algae and other plants, it may be necessary to remove or reduce the nitrogen that is present to prevent these growths.

Phosphorus

Phosphorus is also essential to the growth of algae and other biological organisms. Because of noxious algal blooms that occur in surface waters, there is presently much interest in controlling the amount of phosphorus compounds that enter surface waters in domestic and industrial waste discharges and natural runoff. Municipal wastewaters, for example, may contain from 4 to 16 mg/L of phosphorus as P.

The usual forms of phosphorus that are found in aqueous solutions include the orthophosphate, polyphosphate, and organic phosphate. The orthophosphates, for example, PO_4^{3-}, HPO_4^{2-}, $H_2PO_4^-$, H_3PO_4, are available for biological metabolism without further breakdown. The polyphosphates include those molecules with two or more phosphorus atoms, oxygen atoms, and, in some cases, hydrogen atoms combined in a complex molecule. Polyphosphates undergo hydrolysis in aqueous solutions and revert to the orthophosphate forms; however, this hydrolysis is usually quite slow. The organically bound phosphorus is usually of minor importance in most domestic wastes, but it can be an important constituent of industrial wastes and wastewater sludges.

Orthophosphate can be determined by directly adding a substance such as ammonium molybdate which will form a colored complex with the phosphate. The polyphosphates and organic phosphates must be converted to orthophosphates using an acid digestion step before they can be determined in a similar manner.

Sulfur

The sulfate ion occurs naturally in most water supplies and is present in wastewater as well. Sulfur is required in the synthesis of proteins and is released in their degradation. Sulfate is reduced biologically under anaerobic conditions to sulfide which, in turn, can combine with hydrogen to form hydrogen sulfide (H_2S). The following generalized reactions are typical.

$$\text{Organic matter} + SO_4^{2-} \xrightarrow{\text{bacteria}} S^{2-} + H_2O + CO_2 \tag{2-41}$$

$$S^{2-} + 2H^+ \rightarrow H_2S \tag{2-42}$$

If lactic acid is used as the precursor organic compound, the reduction of sulfate to sulfide occurs as follows:

$$\underset{\text{lactic acid}}{2CH_3CH(OH)COOH} + \underset{\text{sulfate}}{SO_4^{2-}} \xrightarrow{\text{bacteria}} \underset{\text{acetate}}{2CH_3COOH} + \underset{\text{sulfide ion}}{S^{2-}} + 2H_2O + 2CO_2 \tag{2-43}$$

Hydrogen sulfide gas, which will diffuse into the headspace above the wastewater in sewers that are not flowing full, tends to collect at the crown of the pipe. The accumulated H_2S can then be oxidized biologically to sulfuric acid, which is corrosive to concrete sewer pipes. This corrosive effect, known as "crown rot," can seriously threaten the structural integrity of the sewer pipe (ASCE, 1989; U.S. EPA, 1985e).

Sulfates are reduced to sulfides in sludge digesters and may upset the biological process if the sulfide concentration exceeds 200 mg/L. Fortunately, such concentrations are rare. The H_2S gas, which is evolved and mixed with the wastewater gas ($CH_4 + CO_2$), is corrosive to the gas piping and, if burned in gas engines, the products of combustion can damage the engine and severely corrode exhaust gas heat recovery equipment, especially if allowed to cool below the dew point.

Gases

Gases commonly found in untreated wastewater include nitrogen (N_2), oxygen (O_2), carbon dioxide (CO_2), hydrogen sulfide (H_2S), ammonia (NH_3), and methane (CH_4). The first three are common gases of the atmosphere and will be found in all waters exposed to air. The latter three are derived from the decomposition of the organic matter present in wastewater and are of concern with respect to worker health and safety. Although not found in untreated wastewater, other gases with which the environmental engineer must be familiar include chlorine (Cl_2) and ozone (O_3) (for disinfection and odor control), and the oxides of sulfur and nitrogen (in combustion processes). The following discussion is limited to those gases that are of interest in untreated wastewater. Under most circumstances, the ammonia in untreated wastewater will be present as the ammonium ion (see "Nitrogen"). However, before discussing the individual gases it will be useful to review the ideal gas law and to consider the solubility of gases in water and Henry's law as applied to the gases of interest.

Solubility of Gases in Water. The actual quantity of a gas that can be present in solution is governed by (1) the solubility of the gas as defined by Henry's law, (2) the partial pressure of the gas in the atmosphere, (3) the temperature, and (4) the concentration of the impurities in the water (e.g., salinity, suspended solids, etc.).

The Ideal Gas Law. The ideal gas law, derived from a consideration of Boyle's law (volume of a gas is inversely proportional to pressure at constant temperature) and Charles' law (volume of a gas is directly proportional to temperature at constant pressure) is

$$PV = nRT \qquad (2\text{--}44)$$

where P = absolute pressure, atm
V = volume occupied by the gas, L, m^3
n = moles of gas, mole
R = universal gas law constant, 0.082057 atm·L/mole·K
 = 0.000082057 atm·m^3/mole·K
T = temperature, K (273.15 + °C)

Using the universal gas law, it can be shown that the volume of gas occupied by one mole of a gas at standard temperature [0°C, (32°F)] and pressure (1.0 atm) is equal to 22.414 L.

$$V = \frac{nRT}{P}$$

$$V = \frac{(1 \text{ mole})(0.082057 \text{ atm·L/mole·K})\left[(273.15 + 0)\text{K}\right]}{1.0 \text{ atm}} = 22.414 \text{ L}$$

The following relationship, based on the ideal gas law, is used to convert between gas concentrations expressed in ppm$_v$ and $\mu g/m^3$:

$$\mu g/m^3 = \frac{(\text{concentration, ppm}_v)(\text{mw, g/mole of gas})(10^6 \mu g/g)}{(22.414 \times 10^{-3} \, m^3/\text{mole of gas})} \qquad (2\text{--}45)$$

The application of Eq. (2–45) is illustrated in Example 2–5.

EXAMPLE 2–5   **Conversion of Gas Concentration Units**   The off gas from a wastewater force main (i.e., pressure sewer) was found to contain 9 ppm$_v$ (by volume) of hydrogen sulfide (H_2S). Determine the concentration in $\mu g/m^3$ and in mg/L at standard conditions (0°C, 101.325 kPa).

Solution

1.   Compute the concentration in $\mu g/L$ using Eq. (2–45). The molecular weight of $H_2S = 34.08$ $[2(1.01) + 32.06]$

$$9 \text{ ppm}_v = \left(\frac{9 \text{ m}^3}{10^6 \text{ m}^3}\right)\left(\frac{(34.08 \text{ g/mole H}_2\text{S})}{(22.4 \times 10^{-3} \text{ m}^3/\text{mole of H}_2\text{S})}\right)\left(\frac{10^6 \mu g}{g}\right) = 13,693 \ \mu g/m^3$$

2.   The concentration in mg/L is

$$13,693 \ \mu g/m^3 = \left(\frac{13,693 \ \mu g}{m^3}\right)\left(\frac{1 \text{ mg}}{10^3 \ \mu g}\right)\left(\frac{1 \text{ m}^3}{10^3 \text{ L}}\right) = 0.0137 \text{ mg/L}$$

Comment   If gas measurements, expressed in $\mu g/L$, are made at other than standard conditions, the concentration must be corrected to standard conditions, using the ideal gas law, before converting to ppm.

Henry's Law for Dissolved Gases.   The equilibrium or saturation concentration of gas dissolved in a liquid is a function of the type of gas and the partial pressure of the gas in contact with the liquid. The relationship between the mole fraction of the gas in the atmosphere above the liquid and the mole fraction of the gas in the liquid is given by the following form of Henry's law:

$$p_g = \frac{H}{P_T} x_g \qquad\qquad (2\text{–}46)$$

where p_g = mole fraction of gas in air, mole gas /mole of air

H = Henry's law constant, $\dfrac{\text{atm (mole gas/mole air)}}{\text{(mole gas/mole water)}}$

P_T = total pressure, usually 1.0 atm

x_g = mole fraction of gas in water, mole gas/mole water

$\qquad = \dfrac{\text{mole gas}(n_g)}{\text{mole gas }(n_g) + \text{mole water }(n_w)}$

In Eq. (2–46) it is helpful to remember that the mole fraction of a gas corresponds to the partial pressure of the gas. If the partial pressure of the gas is used, Eq. (2–46) is written as follows:

$$P_g = Hx_g \qquad\qquad (2\text{–}47)$$

where P_g = partial pressure of gas, atm

$\qquad$ other terms are as defined above.

In practice and in the literature, the Henry's law constant is often reported as atm, with the mole fraction being implied. Henry's law constant is a function of the type of

Table 2–8
Henry's law constants at 20°C, unitless Henry's law constants at 20°C, and temperature-dependent coefficients[a]

Parameter	Henry's constant, atm	Henry's constant, unitless	Temperature coefficients	
			A	**B**
Air	66,400	49.68	557.60	6.724
Ammonia	0.75	5.61×10^{-4}	1887.12	6.315
Carbon dioxide	1420	1.06	1012.40	6.606
Carbon monoxide	53,600	40.11	554.52	6.621
Chlorine	579	0.43	875.69	5.75
Chlorine dioxide	1500	1.12	1041.77	6.73
Hydrogen	68,300	51.10	187.04	5.473
Hydrogen sulfide	483	0.36	884.94	5.703
Methane	37,600	28.13	675.74	6.880
Nitrogen	80,400	60.16	537.62	6.739
Oxygen	41,100	30.75	595.27	6.644
Ozone	5,300	3.97	1268.24	8.05
Sulfur dioxide	36	2.69×10^{-2}	1207.85	5.68

[a]Adapted in part from Montgomery (1985), Cornwell (1990), and Hand et al. (1998).

gas, temperature, and nature of the liquid. Values of Henry's law constants for various gases in water at 20°C are given in Table 2–8. It is important to note that reported values of the Henry's law constant found in the literature will vary depending on the date of the reference and the specific method used to estimate the constant. Use of the data in Table 2–8 is illustrated in Example 2–6.

EXAMPLE 2–6 **Saturation Concentration of Oxygen in Water**   What is the saturation of oxygen in water in contact with dry air at 1 atm and 20°C?

Solution

1. Dry air contains about 21 percent oxygen by volume (see Appendix D). Therefore, $p_g = 0.21$ mole O_2/mole air.
2. Determine x_g.
   a. From Table 2–8, at 20°C, Henry's constant is

$$H = 4.11 \times 10^4 \frac{\text{atm (mole gas/mole air)}}{\text{(mole gas/mole water)}}$$

b. Using Eq (2–46), the value of x_g is

$$x_g = \frac{P_T}{H} p_g$$

$$= \frac{1.0 \text{ atm}}{\left[4.11 \times 10^4 \dfrac{\text{atm (mole gas/mole air)}}{\text{(mole gas/mole water)}} \right]} (0.21 \text{ mole gas/mole air})$$

$$= 5.11 \times 10^{-6} \text{ mole gas/mole water}$$

3. One liter of water contains 1000 g/(18 g/mole) = 55.6 mole, thus

$$\frac{n_g}{n_g + n_w} = 5.11 \times 10^{-6}$$

$$\frac{n_g}{n_g + 55.6} = 5.11 \times 10^{-6}$$

Because the number of moles of dissolved gas in a liter of water is much less than the number of moles of water,

$$n_g + 55.6 \approx 55.6$$

and $n_g \approx (55.6) \, 5.11 \times 10^{-6}$

$$n_g \approx 2.84 \times 10^{-4} \text{ mole } O_2/\text{L}$$

4. Determine the saturation concentration of oxygen.

$$C_s \approx \frac{(2.84 \times 10^{-4} \text{ mole } O_2/\text{L}) (32 \text{ g/mole } O_2)}{(1 \text{ g}/10^3 \text{ mg})}$$

$$\approx 9.09 \text{ mg/L}$$

Comment   The computed value (9.09 mg/L) compares well with the value given in Appendix D (9.08 mg/L). It should be noted that the values for the Henry's law constant given in Table 2–8 will vary depending on the source and the method used to derive them.

The change in the Henry's law constant with temperature can be estimated using the following modified form of the van't Hoff-Arrhenius relationship [see Eq. (2–20)].

$$\log_{10} H = \frac{-A}{T} + B \tag{2–48}$$

where H = Henry's law constant at temperature T, atm
  A = empirical constant that takes into account the enthalpy change in water due to the dissolution of a component in water and the universal gas law constant
  T = temperature, K = 273.15 + °C
  B = empirical constant

Values of A and B for various gases of interest in wastewater treatment are presented in Table 2–8. It should be noted that the values given in Table 2–8 for A and B are approximate and will vary depending on the source and the method used to derive them.

Unitless Form of Henry's Law. In the literature, the unitless form of Henry's law is often used to compute the solubility of trace gases in water or wastewater. The unitless form is usually written as

$$\frac{C_g}{C_s} = H_u \tag{2–49}$$

where C_g = concentration of constituent in gas phase, $\mu g/m^3$, mg/L
C_s = saturation concentration of constituent in liquid, $\mu g/m^3$, mg/L
H_u = Henry's law constant, unitless

The unitless form is obtained by noting that at 1.0 atm pressure and 0°C, the volume occupied by 1.0 mole of air is 22.414 L. At other temperatures, 1.0 mole of air is equal to $0.082\, T$ L of air, where T is temperature in Kelvin, K $(273.15 + °C)$. Using these conversions, the unitless form of Henry's law is (Cornwell, 1990)

$$H_u = \left[H\,\frac{\text{atm (mole gas/mole air)}}{\text{(mole gas/mole water)}}\right]\left(\frac{\text{mole air}}{0.082\,T\,L}\right)\left(\frac{L}{55.6\text{ mole water}}\right)$$

$$H_u = \left(\frac{H}{4.559\,T}\right) \tag{2–50}$$

For example, at 20°C, H_u equals

$$H_u = \left[\frac{H}{4.559(273.15 + 20)}\right] = H \times (7.49 \times 10^{-4})\text{ at }20°C$$

If atmospheric conditions prevail and Henry's constant is expressed in terms of atm·m³/mole (another form of Henry's law used commonly in the literature), the unitless form of Henry's law is obtained as follows:

$$H_u = \frac{H}{RT} \tag{2–51}$$

where H_u = Henry's law constant, unitless as used in Eq. (2–49)
H = Henry's law constant values expressed in atm·m³/mole
R = universal gas law constant, 0.000082057 atm·m³/mole·K
T = temperature, K $= 273.15 + °C$

Dissolved Oxygen. Dissolved oxygen is required for the respiration of aerobic microorganisms as well as all other aerobic life forms. However, oxygen is only slightly soluble in water. The actual quantity of oxygen (other gases too) that can be present in solution is governed by (1) the solubility of the gas, (2) the partial pressure of the gas in the atmosphere, (3) the temperature, and (4) the concentration of the impurities in the water (e.g., salinity, suspended solids, etc.). The interrelationship of these variables is delineated in Chap. 6 and is illustrated in Appendix D, where the effect of temperature and salinity on dissolved oxygen concentration is presented.

Because the rate of biochemical reactions that use oxygen increases with increasing temperature, dissolved oxygen levels tend to be more critical in the summer months. The problem is compounded in summer months because stream flows are usually lower, and thus the total quantity of oxygen available is also lower. The presence of dissolved oxygen in wastewater is desirable because it prevents the formation of noxious odors. The role of oxygen in wastewater treatment is discussed in Chaps. 7, 8, and 9.

Hydrogen Sulfide.   Hydrogen sulfide is formed, as mentioned previously, from the anaerobic decomposition of organic matter containing sulfur or from the reduction of mineral sulfites and sulfates. It is not formed in the presence of an abundant supply of oxygen. This gas is a colorless, inflammable compound having the characteristic odor of rotten eggs. Hydrogen sulfide is also toxic, and great care must be taken in its presence. High concentrations can overwhelm olfactory glands, resulting in a loss of smell. This loss of smell can lead to a false sense of security that is very dangerous. The blackening of wastewater and sludge usually results from the formation of hydrogen sulfide that has combined with the iron present to form ferrous sulfide (FeS). Various other metallic sulfides are also formed. Although hydrogen sulfide is the most important gas formed from the standpoint of odors, other volatile compounds such as indol, skatol, and mercaptans, which may also be formed during anaerobic decomposition, may cause odors far more offensive than that of hydrogen sulfide.

Methane.   The principal byproduct from the anaerobic decomposition of the organic matter in wastewater is methane gas (see Chaps. 10 and 14). Methane is a colorless, odorless, combustible hydrocarbon of high fuel value. Normally, large quantities are not encountered in untreated wastewater because even small amounts of oxygen tend to be toxic to the organisms responsible for the production of methane. Occasionally, however, as a result of anaerobic decay in accumulated bottom deposits, methane has been produced. Because methane is highly combustible and the explosion hazard is high, access ports (manholes) and sewer junctions or junction chambers where there is an opportunity for gas to collect should be ventilated with a portable blower during and before the time required for operating personnel to work in them for inspection, renewals, or repairs. In treatment plants, methane is produced from the anaerobic treatment process used to stabilize wastewater sludges (see Chap. 14). In treatment plants where methane is produced, notices should be posted about the plant warning of explosion hazards, and plant employees should be instructed in safety measures to be maintained while working in and about the structures where gas may be present.

Odors

Odors in domestic wastewater usually are caused by gases produced by the decomposition of organic matter or by substances added to the wastewater. Fresh wastewater has a distinctive, somewhat disagreeable odor, which is less objectionable than the odor of wastewater which has undergone anaerobic (devoid of oxygen) decomposition. The most characteristic odor of stale or septic wastewater is that of hydrogen sulfide, which, as discussed previously, is produced by anaerobic microorganisms that reduce sulfate to sulfide. Industrial wastewater may contain either odorous compounds or compounds that produce odors during the process of wastewater treatment.

Odors have been rated as the foremost concern of the public relative to the implementation of wastewater-treatment facilities. Within the past few years, the control of odors has become a major consideration in the design and operation of wastewater-collection, treatment, and disposal facilities, especially with respect to the public acceptance of these facilities. In many areas, projects have been rejected because of the concern over the potential for odors. In view of the importance of odors in the field of wastewater management, it is appropriate to consider the effects they produce, how they are detected, and their characterization and measurement.

Effects of Odors. The importance of odors at low concentrations in human terms is related primarily to the psychological stress they produce rather than to the harm they do to the body. Offensive odors can cause poor appetite for food, lowered water consumption, impaired respiration, nausea and vomiting, and mental perturbation. In extreme situations, offensive odors can lead to the deterioration of personal and community pride, interfere with human relations, discourage capital investment, lower socioeconomic status, and deter growth. Also, some odorous compounds (e.g., H_2S) are toxic at elevated concentrations. These problems can result in a decline in market and rental property values, tax revenues, payrolls, and sales.

Detection of Odors. The malodorous compounds responsible for producing psychological stress in humans are detected by the olfactory system, but the precise mechanism involved is at present not well understood. Since 1870, more than 30 theories have been proposed to explain olfaction. One of the difficulties in developing a universal theory has been the inadequate explanation of why compounds with similar structures may have different odors and why compounds with very different structures may have similar odors. At present, there appears to be some general agreement that the odor of a molecule must be related to the molecule as a whole.

Over the years, a number of attempts have been made to classify odors in a systematic fashion. The major categories of offensive odors and the compounds involved are listed in Table 2–9. All these compounds may be found or may develop in domestic wastewater, depending on local conditions. The odor thresholds for specific malodorous compounds associated with untreated wastewater are listed in Table 2–10.

Odor Characterization and Measurement. It has been suggested that four independent factors are required for the complete characterization of an odor: intensity, character, hedonics, and detectability (see Table 2–11). To date, detectability is the only factor that has been used in the development of statutory regulations for nuisance odors. As shown on Fig. 2–15, odor can be measured by sensory methods, and specific odorant concentrations can be measured by instrumental methods. It has been shown that, under carefully controlled conditions, the sensory (organoleptic) measurement of odors by the human olfactory system can provide meaningful and reliable information. Therefore, the sensory method is often used to measure the odors emanating from wastewater-treatment facilities. The availability of a direct reading meter for hydrogen sulfide (see Fig. 2–16) which can be used to detect concentrations as low as 1 ppb is a significant development.

Table 2–9
Major categories of odorous compounds associated with untreated wastewater

Odorous compound	Chemical formula	Odor quality
Amines	CH_3NH_2, $(CH_3)_3NH$	Fishy
Ammonia	NH_3	Ammoniacal
Diamines	$NH_2(CH_2)_4NH_2$, $NH_2(CH_2)_5NH_2$	Decayed flesh
Hydrogen sulfide	H_2S	Rotten eggs
Mercaptans (e.g., methyl and ethyl)	CH_3SH, $CH_3(CH_2)SH$	Decayed cabbage
Mercaptans (e.g., T = butyl and crotyl)	$(CH_3)_3CSH$, $CH_3(CH_2)_3SH$	Skunk
Organic sulfides	$(CH_3)_2S$, $(C_6H_5)_2S$	Rotten cabbage
Skatole	C_9H_9N	Fecal matter

Table 2–10
Odor thresholds of odorous compounds associated with untreated wastewater[a]

Odorous compound	Chemical formula	Molecular weight	Odor threshold, ppm_v[b]	Characteristic odor
Ammonia	NH_3	17.0	46.8	Ammoniacal, pungent
Chlorine	Cl_2	71.0	0.314	Pungent, suffocating
Crotyl mercaptan	$CH_3-CH=CH-CH_2-SH$	90.19	0.000029	Skunklike
Dimethyl sulfide	CH_3-S-CH_3	62	0.0001	Decayed vegetables
Diphenyl sulfide	$(C_6H_5)_2S$	186	0.0047	Unpleasant
Ethyl mercaptan	CH_3CH_2-SH	62	0.00019	Decayed cabbage
Hydrogen sulfide	H_2S	34	0.00047	Rotten eggs
Indole	C_8H_6NH	117	0.0001	Fecal, nauseating
Methyl amine	CH_3NH_2	31	21.0	Putrid, fishy
Methyl mercaptan	CH_3SH	48	0.0021	Decayed cabbage
Skatole	C_9H_9N	131	0.019	Fecal, nauseating
Sulfur dioxide	SO_2	64.07	0.009	Pungent, irritating
Thiocresol	$CH_3-C_6H_4-SH$	124	0.000062	Skunk, rancid

[a] Adapted from Patterson et al. (1984) and U.S. EPA (1985e).
[b] Parts per million by volume.

In the sensory method, human subjects (often a panel of subjects) are exposed to odors that have been diluted with odor-free air, and the number of dilutions required to reduce an odor to its minimum detectable threshold odor concentration (MDTOC) is noted. The detectable odor concentration is reported as the dilutions to the MDTOC, commonly called D/T (dilutions to threshold). Thus, if four volumes of diluted air must be added to one unit volume of sampled air to reduce the odorant to its MDTOC, the

Table 2–11
Factors that must be considered for the complete characterization of an odor

Factor	Description
Character	Relates to the mental associations made by the subject in sensing the odor. Determination can be quite subjective
Detectability	The number of dilutions required to reduce an odor to its minimum detectable threshold odor concentration (MDTOC)
Hedonics	The relative pleasantness or unpleasantness of the odor sensed by the subject
Intensity	The perceived strength of the odor. Usually measured by the butanol olfactometer or calculated from the D/T (dilutions to threshold ratio) when the relationship is established

Figure 2–15

Classification of methods used to detect odors.

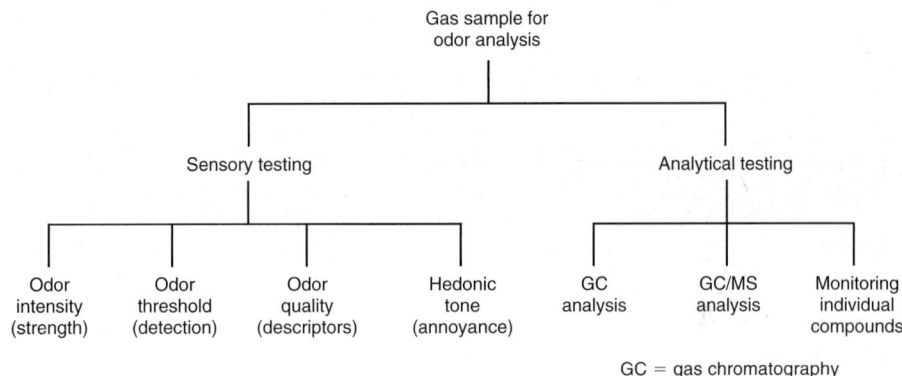

Figure 2–16

Portable H_2S meter used for field odor studies. (*From Arizona Instrument Corporation, Jerome Instrument Division.*)

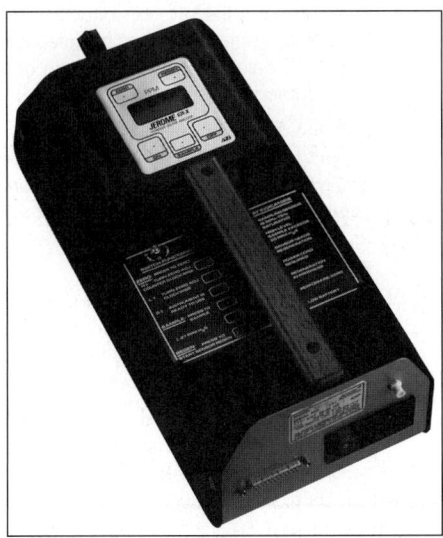

odor concentration would be reported as four dilutions to MDTOC. Other terminology commonly used to measure odor strength is ED_{50}. The ED_{50} value represents the number of times an odorous air sample must be diluted before the average person (50 percentile) can barely detect an odor in the diluted sample. Details of the test procedure are provided in ASTM (1979). However, the sensory determination of this minimum threshold concentration can be subject to a number of errors. Adaptation and cross adaptation, synergism, subjectivity, and sample modification (see Table 2–12) are the principal errors. To avoid errors in sample modification during storage in sample collection containers, direct-reading olfactometers have been developed to measure odors at their source without using sampling containers.

The threshold odor of a water or wastewater sample is determined by diluting the sample with odor-free water. The "threshold odor number" (TON) corresponds to the greatest dilution of the sample with odor-free water at which an odor is just perceptible. The recommended sample size is 200 mL. The numerical value of the TON is determined as follows:

$$TON = \frac{A + B}{A} \qquad\qquad (2\text{–}52)$$

where TON = threshold odor number
  A = mL of sample
  B = mL of odor-free water

Table 2–12
Types of errors in the sensory detection of odors[a]

Description	Type of error
Adaptation and cross adaptation	When exposed continually to a background concentration of an odor, the subject is unable to detect the presence of that odor at low concentrations. When removed from the background odor concentration, the subject's olfactory system will recover quickly. Ultimately, subjects with an adapted olfactory system will be unable to detect the presence of an odor to which their system has adapted
Sample modification	Both the concentration and composition of odorous gases and vapors can be modified in sample-collection containers and in odor-detection devices. To minimize problems associated with sample modification, the period of odor containment should be minimized or eliminated, and minimum contact should be allowed with any reactive surfaces
Subjectivity	When the subject has knowledge of the presence of an odor, random error can be introduced in sensory measurements. Often, knowledge of the odor may be inferred from other sensory signals such as sound, sight, or touch
Synergism	When more than one odorant is present in a sample, it has been observed that it is possible for a subject to exhibit increased sensitivity to a given odor because of the presence of another odor

[a] Wilson (1975).

The odor emanating from the liquid sample is determined as discussed above with human subjects (often a panel of subjects). Details for this procedure may be found in Standard Methods (1998). The application of Eq. (2–52) is illustrated in Example 2–7.

EXAMPLE 2–7 **Determination of Threshold Odor Number** A 25-mL sample of treated wastewater requires 175 mL of distilled water to reduce the odor to a level that is just perceptible. What is the threshold odor number (TON)?

Solution

1. Determine the TON using Eq. (2–52).

$$TON = \frac{A + B}{A} = \frac{25 \text{ mL} + 175 \text{ mL}}{25 \text{ mL}} = 8.0$$

Comment The value of the threshold number will depend on the nature of the odor compound. Also, depending on the odor compound, different persons will respond differently in assessing when an odor is just perceptible.

With regard to the instrumental measurement of odors, air-dilution olfactometry provides a reproducible method for measuring threshold odor concentrations. Equipment used to analyze odors includes (1) the triangle olfactometer, (2) the butanol wheel, and (3) the scentometer. The triangle olfactometer enables the operator to introduce the sample at different concentrations at six different cups (see Fig. 2–17). At each cup, two ports contain purified air, and one port contains diluted sample. Each odor panel member (usually six) then sniffs each of the three ports and selects the port which he or she believes to contain the sample. The butanol wheel is a device used to measure the intensity of the odor against a butanol scale containing various concentrations of butanol. A scentometer (see Fig. 2–18) is a hand-held device in which malodorous air passes through graduated orifices and is mixed with air that has been purified by passing through activated-carbon beds. The dilution ratios are determined by the ratio of the size of the malodorous to purified inlets. The scentometer is very useful in the field for making odor determinations over a large area surrounding a treatment plant. Often a mobile odor laboratory, which contains several types of olfactory and analytical equipment in a single van-type vehicle, is used for field sites.

It is often desirable to know the specific compounds responsible for odor. Although gas chromatography has been used successfully for this purpose, it has not been used as successfully in the detection and quantification of odors derived from wastewater-collection, treatment, and disposal facilities. Equipment developed and found useful in the chemical analysis of odors is the triple-stage quadrupole mass spectrometer. The spectrometer can be used as a conventional mass spectrometer to produce simple mass spectra or as a triple-stage quadrupole to produce collesionally activated disassociation spectra. The former operating mode provides the masses of molecular or parent ions present in samples, while the latter provides positive identification of compounds. Types of compounds that can be identified include ammonia, amino acids, and volatile organic compounds.

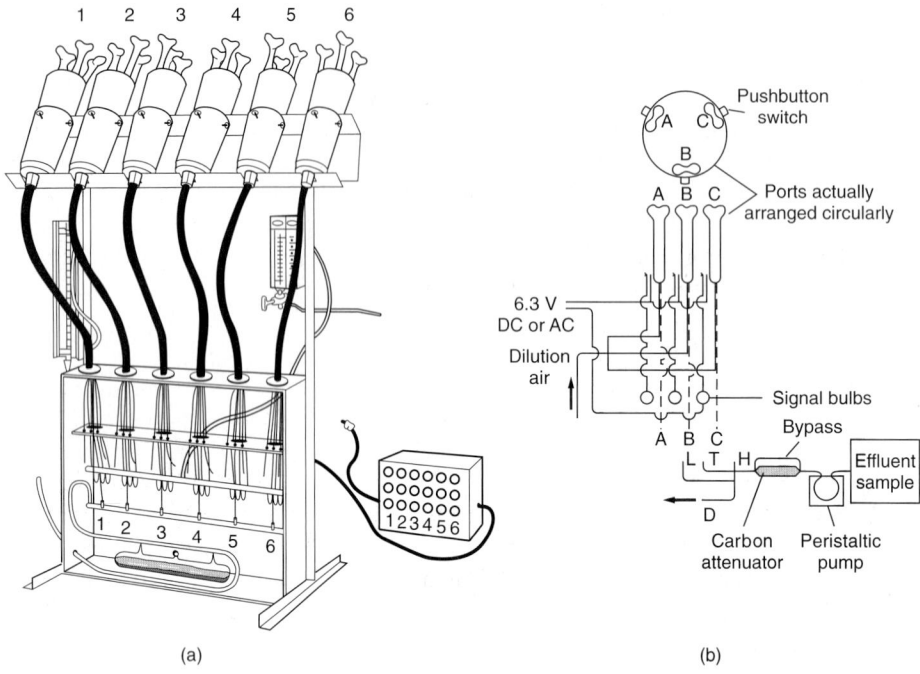

(a)   (b)

Figure 2–17

Dynamic forced-choice triangle olfactometer: (a) schematic and (b) flow diagram.

Figure 2–18

Scentometer used for field studies of odors: (a) schematic and (b) front view looking at nosepieces (125 × 150 × 67 mm, from Barnebey & Sutcliffe Corp.).

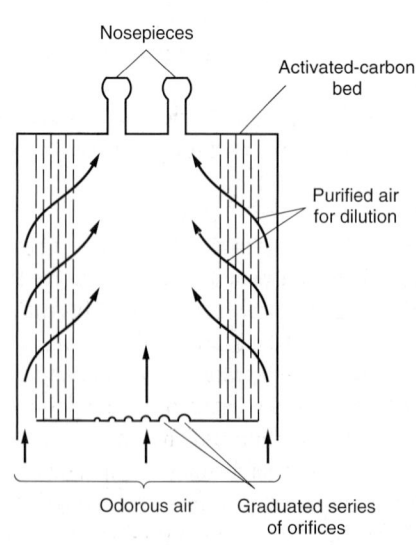

(a)

(b)

2–5 METALLIC CONSTITUENTS

Trace quantities of many metals, such as cadmium (Cd), chromium (Cr), copper (Cu), iron (Fe), lead (Pb), manganese (Mn), mercury (Hg), nickel (Ni), and zinc (Zn) are important constituents of most waters. Many of these metals are also classified as priority pollutants. However, most of these metals are necessary for growth of biological life, and absence of sufficient quantities of them could limit growth of algae, for example. The presence of any of these metals in excessive quantities will interfere with many beneficial uses of the water because of their toxicity; therefore, it is frequently desirable to measure and control the concentrations of these substances.

Importance of Metals

Metals of importance in the treatment, reuse, and disposal of treated effluents and biosolids are summarized in Table 2–13. All living organisms require varying amounts (macro and micro) of metallic elements, such as iron, chromium, copper, zinc, and cobalt, for proper growth. Although macro and micro amounts of metals are required for proper growth, the same metals can be toxic when present in elevated concentrations. As more use is made of treated wastewater effluent for irrigation and landscape watering, a variety of metals must be determined to assess any adverse effects that may occur. Calcium, magnesium, and sodium are of importance in determining the sodium adsorption ratio (SAR), which is used to assess the suitability of treated effluent for agricultural use (see Chap. 13). Where composted sludge is applied in agricultural applications, arsenic, cadmium, copper, lead, mercury, molybdenum, nickel, selenium, and zinc must be determined.

Sources of Metals

The sources of trace metals in wastewater include the discharges from residential dwellings, groundwater infiltration, and commercial and industrial discharges. Many of the sources of heavy metals are identified in Table 2–14. For example, cadmium, chromates, lead, and mercury are often present in industrial wastes. These are found particularly in metal-plating wastes and should be removed by pretreatment at the site of the industry rather than be mixed with the municipal wastewater. Fluoride, a toxic anion, is found commonly in wastewater from electronics manufacturing facilities.

Sampling and Methods of Analysis

Methods for determining the concentrations of these substances vary in complexity according to the interfering substances that may be present (Standard Methods, 1998). Metals are determined typically by flame atomic absorption, electrothermal atomic absorption, inductively coupled plasma, or IPC/mass spectrometry. Various classes of metals are defined as: (1) *dissolved metals* are those metals present in unacidified samples that pass through a 0.45-μm membrane filter, (2) *suspended metals* are those metals present in unacidified samples that are retained on a 0.45-μm membrane filter, (3) *total metals* is the total of the dissolved and suspended metals or the concentration of metals determined on an unfiltered sample after digestion, and (4) *acid extractable metals* are those metals in solution after an unfiltered sample is treated with a hot dilute mineral acid (Standard Methods, 1998).

Table 2–13
Metals of importance in wastewater management[a]

Metal	Symbol	Nutrients necessary for biological growth		Concentration threshold of inhibitory effect on heterotrophic organisms, mg/L	Used to determine SAR for land application of effluent	Used to determine if biosolids are suitable for land application
		Macro	Micro[b]			
Arsenic	As			0.05		✔
Cadmium	Cd			1.0		✔
Calcium	Ca	✔			✔	
Chromium	Cr		✔	10[c], 1[d]		
Cobalt	Co		✔			
Copper	Cu		✔	1.0		✔
Iron	Fe	✔				
Lead	Pb		✔	0.1		✔
Magnesium	Mg	✔	✔		✔	
Manganese	Mn		✔			
Mercury	Hg			0.1		✔
Molybdenum	Mo		✔			✔
Nickel	Ni		✔	1.0		✔
Potassium	K	✔				
Selenium	Se		✔			✔
Sodium	Na	✔			✔	
Tungsten	W		✔			
Vanadium	V		✔			
Zinc	Zn		✔	1.0		✔

[a] From Crites and Tchobanoglous (1998).
[b] Often identified as trace elements needed for biological growth.
[c] Total chromium.
[d] Hexavalent chromium.

Typical Effluent Discharge Limits for Metals

Increasingly, metallic constituents in effluent discharges and in biosolids are being regulated. Typical discharge requirements for metals and other toxic constituents are reported in Table 2–15. In addition to complying with existing U.S. EPA requirements, many states have adopted more restrictive standards to protect specific beneficial uses.

Table 2–14
Typical waste compounds produced by commercial, industrial, and agricultural activities that have been classified as priority pollutants

Name	Formula	Use	Concern
Arsenic	As	Alloying additive for metals, especially lead and copper as shot, battery grids, cable sheaths, boiler tubes. High-purity (semiconductor) grade	Carcinogen and mutagen. *Long-term*—sometimes can cause fatigue and loss of energy; dermatitis
Barium	Ba	Getter alloys in vacuum tubes, deoxidizer for copper, Frary's metal, lubricant for anode rotors in x-ray tubes, spark-plug alloys	Flammable at room temperature in powder form. *Long-term*—Increased blood pressure and nerve block
Cadmium	Cd	Electrodeposited and dipped coatings on metals, bearing and low-melting alloys, brazing alloys, fire protection system, nickel-cadmium storage batteries power transmission wire, TV phosphors, basis of pigments used in ceramic glazes, machinery enamels, fungicide, photography and lithography, selenium rectifiers, electrodes for cadmium-vapor lamps and photoelectric cells	Flammable in powder form. Toxic by inhalation of dust or fume. A carcinogen. Soluble compounds of cadmium are highly toxic. *Long-term*—concentrates in the liver, kidneys, pancreas, and thyroid; hypertension suspected effect
Chromium	Cr	Alloying and plating element on metal and plastic substrates for corrosion resistance, chromium-containing and stainless steels, protective coating for automotive and equipment accessories, nuclear and high-temperature research, constituent of inorganic pigments	Hexavalent chromium compounds are carcinogenic and corrosive on tissue. *Long-term*—skin sensitization and kidney damage
Lead	Pb	Storage batteries, gasoline additive, cable covering, ammunition, piping, tank linings, solder and fusible alloys, vibration damping in heavy construction, foil, babbitt and other bearing alloys	Toxic by ingestion or inhalation of dust or fumes. *Long-term*—brain and kidney damage; birth defects
Mercury	Hg	Amalgams, catalyst electrical apparatus, cathodes for production of chlorine and caustic soda, instruments, mercury vapor lamps, mirror coating, arc lamps, boilers	Highly toxic by skin absorption and inhalation of fume or vapor. *Long-term*—toxic to central nervous system, may cause birth defects
Selenium	Se	Electronics, xerographic plates, TV cameras, photocells, magnetic computer cores, solar batteries (rectifiers, relays), ceramics (colorant for glass), steel and copper, rubber accelerator, catalyst, trace element in animal feeds	*Long-term*—red staining of fingers, teeth, and hair; general weakness; depression; irritation of nose and mouth
Silver	Ag	Manufacture of silver nitrate, silver bromide, photochemicals; lining vats and other equipment for chemical reaction vessels, water distillation, etc.; mirrors, electric conductors, silver plating electronic equipment; sterilant, water purification, surgical cements, hydration and oxidation catalyst, special batteries, solar cells, reflectors for solar towers, low-temperature brazing alloys, table cutlery, jewelry, dental, medical, and scientific equipment, electrical contacts, bearing metal, magnet windings, dental amalgams, colloidal silver used as a nucleating agent in photography and medicine, often combined with protein	Toxic metal. *Long-term*—permanent gray discoloration of skin, eyes, and mucous membranes

Table 2–15
Typical discharge limits for toxic constituents found in secondary effluent

Constituent	Units	Average value[a] Daily	Average value[a] Monthly
Arsenic	μg/L	20	
Cadmium	μg/L	1.1	
Chromium	μg/L	11	
Copper	μg/L	4.9	
Lead[b]	μg/L	5.6	
Mercury	μg/L	2.1	0.012
Nickel[b]	μg/L	7.1	
Selenium[b]	μg/L	5.0	
Silver	μg/L	2.3	
Zinc[b]	μg/L	58	
Dieldrin[c]	μg/L	0.0019	0.00014
Lindane	μg/L	0.16	0.063
Tributylin	μg/L	0.01	0.005
PAHs[d,e]	μg/L	0.049	

[a] Limits apply to the average concentration of all samples collected during the averaging period (daily—24-h period; monthly—calendar month).

[b] Effluent limitation may be met as a 4-d average. If compliance is to be determined based on a 4-d average, then concentrations of four 24-h composite samples must be reported as well as the average of four.

[c] Compliance will be based on the practical quantification level (PQL), 0.07 μg/L.

[d] PAHs = polynuclear aromatic hydrocarbons.

[e] Compliance will be based on the practical quantification level (PQL) for each PAH, 4 μg/L.

Source: Bay Area Regional Water Quality Control Board, Oakland, CA.

2–6 AGGREGATE ORGANIC CONSTITUENTS

Organic compounds are normally composed of a combination of carbon, hydrogen, and oxygen, together with nitrogen in some cases. The organic matter in wastewater typically consists of proteins (40 to 60 percent), carbohydrates (25 to 50 percent), and oils and fats (8 to 12 percent). Urea, the major constituent of urine, is another important organic compound contributing to fresh wastewater. Because urea decomposes rapidly, it is seldom found in other than very fresh wastewater. Along with the proteins, carbohydrates, fats and oils, and urea, wastewater typically contains small quantities of a very large number of different synthetic organic molecules, with structures ranging from simple to extremely complex.

Over the years, a number of different analyses have been developed to determine the organic content of wastewaters. In general, the analyses may be classified into those used to measure aggregate organic matter comprising a number of organic constituents with similar characteristics that cannot be distingushed separately, and those analyses used to quantify individual organic compounds (Standard Methods, 1998). Aggregate organic compounds are considered in this section. Individual organic compounds are considered in the following section.

Measurement of Organic Content

In general, the analyses used to measure aggregate organic material may be divided into those used to measure gross concentrations of organic matter greater than about 1.0 mg/L and those used to measure trace concentrations in the range of 10^{-12} to 10^{0} mg/L. Laboratory methods commonly used today to measure gross amounts of organic matter (typically greater than 1 mg/L) in wastewater include: (1) biochemical oxygen demand (BOD), (2) chemical oxygen demand (COD), and (3) total organic carbon (TOC). Complementing these laboratory tests is the theoretical oxygen demand (ThOD), which is determined from the chemical formula of the organic matter.

Other methods used in the past included: (1) total, albuminoid, organic, and ammonia nitrogen, and (2) oxygen consumed. These determinations, with the exception of albuminoid nitrogen and oxygen consumed, are still included in complete wastewater analyses. Their significance, however, has changed. Whereas formerly they were used almost exclusively to indicate organic matter, they are now used to determine the availability of nitrogen to sustain biological activity in industrial waste-treatment processes and to determine whether undesirable algal growths will occur in receiving waters.

Trace organics in the range of 10^{-12} to 10^{-3} mg/L are determined using instrumental methods including gas chromotography and mass spectroscopy. Within the past 10 years, the sensitivity of the methods used for the detection of trace organic compounds has improved significantly, and detection of concentrations in the range of 10^{-9} mg/L is now almost a routine matter.

Biochemical Oxygen Demand (BOD)

The most widely used parameter of organic pollution applied to both wastewater and surface water is the 5-day BOD (BOD_5). This determination involves the measurement of the dissolved oxygen used by microorganisms in the biochemical oxidation of organic matter. Despite the widespread use of the BOD test, it has a number of limitations, as discussed later in this section. It is hoped that, through the continued efforts of workers in the field, one of the other measures of organic content, or perhaps a new measure, will ultimately be used in its place. Why, then, if the test suffers from serious limitations, is further space devoted to it in this text? The reason is that BOD test results are now used (1) to determine the approximate quantity of oxygen that will be required to biologically stabilize the organic matter present, (2) to determine the size of waste-treatment facilities, (3) to measure the efficiency of some treatment processes, and (4) to determine compliance with wastewater discharge permits. Because it is likely that the BOD test will continue to be used for some time, it is important to know the details of the test and its limitations.

Basis for BOD Test. If sufficient oxygen is available, the aerobic biological decomposition of an organic waste will continue until all of the waste is consumed. Three more or less distinct activities occur. First, a portion of the waste is oxidized to end products to obtain energy for cell maintenance and the synthesis of new cell tissue.

Simultaneously, some of the waste is converted into new cell tissue using part of the energy released during oxidation. Finally, when the organic matter is used up, the new cells begin to consume their own cell tissue to obtain energy for cell maintenance. This third process is called endogenous respiration. Using the term COHNS (which represents the elements carbon, oxygen, hydrogen, nitrogen, and sulfur) to represent the organic waste and the term $C_5H_7NO_2$ [first proposed by Hoover and Porges (1952)] to represent cell tissue, the three processes are defined by the following generalized chemical reactions:

Oxidation:

$$COHNS + O_2 + bacteria \rightarrow CO_2 + H_2O + NH_3 + other\ end\ products + energy$$

$$(2-53)$$

Synthesis:

$$COHNS + O_2 + bacteria + energy \rightarrow \underset{\text{New cell tissue}}{C_5H_7NO_2}$$

$$(2-54)$$

Endogenous respiration:

$$C_5H_7NO_2 + 5O_2 \rightarrow 5CO_2 + NH_3 + 2H_2O$$

$$(2-55)$$

If only the oxidation of the organic carbon that is present in the waste is considered, the ultimate BOD is the oxygen required to complete the three reactions given above. This oxygen demand is known as the ultimate carbonaceous or first-stage BOD, and is usually denoted as UBOD.

BOD Test Procedure. In the standard BOD test (see Fig. 2–19a), a small sample of the wastewater to be tested is placed in a BOD bottle (volume = 300 mL). The bottle is then filled with dilution water saturated in oxygen and containing the nutrients required for biological growth. To ensure that meaningful results are obtained, the sample must be suitably diluted with a specially prepared dilution water so that adequate nutrients and oxygen will be available during the incubation period. Normally, several dilutions are prepared to cover the complete range of possible values. The ranges of BOD that can be measured with various dilutions based on percentage mixtures and direct pipetting are reported in Table 2–16. Before the bottle is stoppered, the oxygen concentration in the bottle is measured (see Fig. 2–20).

After the bottle is incubated for 5 days at 20°C, the dissolved oxygen concentration is measured again. The BOD of the sample is the difference in the dissolved oxygen concentration values, expressed in milligrams per liter, divided by the decimal fraction of sample used. The computed BOD value is known as the 5-day, 20°C biochemical oxygen demand. When testing waters with low concentrations of microorganisms, a seeded BOD test is conducted (see Fig. 2–19b). The organisms contained in the effluent from primary sedimentation facilities are used commonly as the seed for the BOD test. Seed organisms can also be obtained commercially. When the sample contains a large population of microorganisms (e.g., untreated wastewater), seeding is not necessary.

The standard incubation period is usually 5 days at 20°C, but other lengths of time and temperatures can be used. Longer time periods (typically 7 days), which correspond to work schedules, are often used, especially in small plants where the laboratory

Figure 2–19

Procedure for setting up BOD test bottles: (a) with unseeded dilution water and (b) with seeded dilution water. (Tchobanoglous and Schroeder, 1985.)

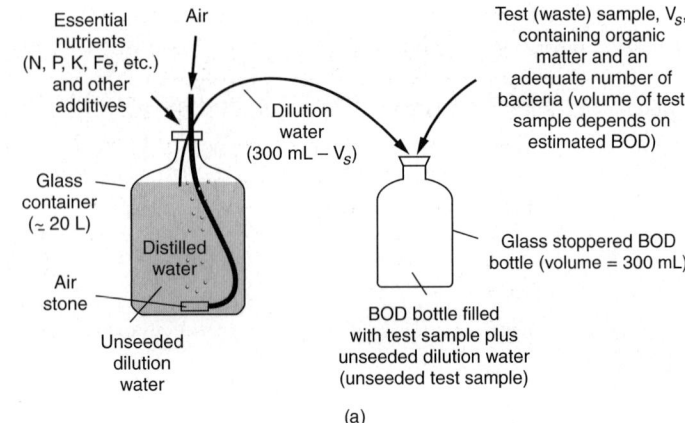

(a)

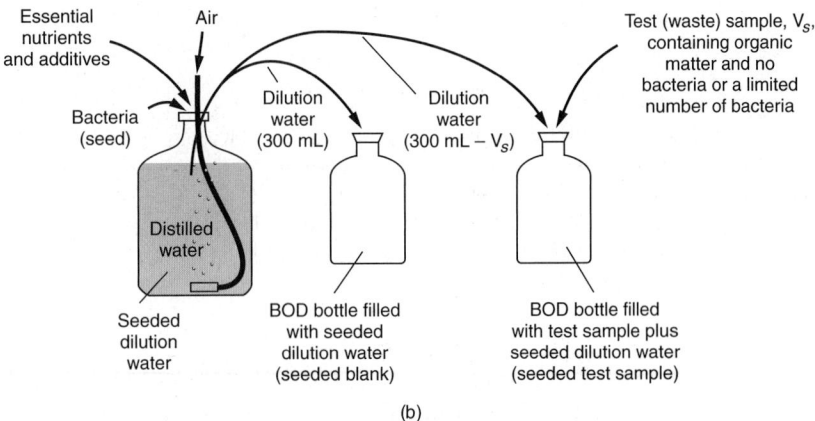

(b)

Figure 2–20

Measurement of oxygen in BOD bottle with a DO probe equipped with a stirring mechanism.

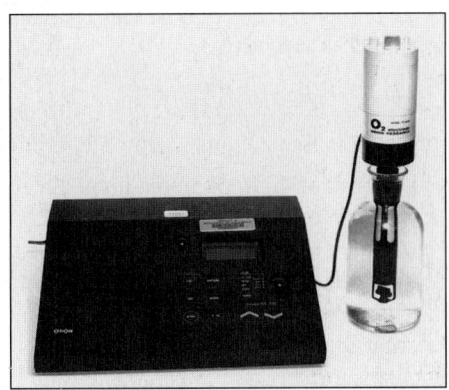

staff is not available on the weekends. The temperature, however, should be constant throughout the test. The 20°C temperature used is an average value for slow-moving streams in temperate climates and is easily duplicated in an incubator. Different results would be obtained at different temperatures, because biochemical reaction rates are

Table 2-16
Measurable BOD using various dilutions of samples

By using percent mixtures		By direct pipetting into 300-mL bottles	
% mixture	Range of BOD, mg/L	mL	Range of BOD, mg/L
0.01	20,000–70,000	0.02	30,000–105,000
0.02	10,000–35,000	0.05	12,000–42,000
0.05	4,000–14,000	0.10	6,000–21,000
0.1	2,000–7,000	0.20	3,000–10,500
0.2	1,000–3,500	0.50	1,200–4,200
0.5	400–1,400	1.0	600–2,100
1.0	200–700	2.0	300–1,050
2.0	100–350	5.0	120–420
5.0	40–140	10.0	60–210
10.0	20–70	20.0	30–105
20.0	10–35	50.0	12–42
50.0	4–14	100.0	6–21
100.0	0–7	300.0	0–7

temperature-dependent. After incubation, the dissolved oxygen of the sample is measured (see Fig. 2–20) and the BOD is calculated using Eq. (2–56) or (2–57).

When the dilution water is not seeded:

$$\text{BOD, mg/L} = \frac{D_1 - D_2}{P} \qquad (2\text{–}56)$$

When the dilution water is seeded:

$$\text{BOD, mg/L} = \frac{(D_1 - D_2) - (B_1 - B_2)f}{P} \qquad (2\text{–}57)$$

where D_1 = dissolved oxygen of diluted sample immediately after preparation, mg/L
D_2 = dissolved oxygen of diluted sample after 5-day incubation at 20°C, mg/L
B_1 = dissolved oxygen of seed control before incubation, mg/L
B_2 = dissolved oxygen of seed control after incubation, mg/L
f = fraction of seeded dilution water volume in sample to volume of seeded dilution water in seed control
P = fraction of wastewater sample volume to total combined volume

Biochemical oxidation theoretically takes an infinite time to go to completion because the rate of oxidation is assumed to be proportional to the amount of organic matter remaining. Within a 20-day period, the oxidation of the carbonaceous organic matter is about 95 to 99 percent complete, and in the 5-day period used for the BOD test, oxidation is from 60 to 70 percent complete.

EXAMPLE 2–8 **Determination of BOD from Laboratory Data** The following information is available for a seeded 5-day BOD test conducted on a wastewater sample. Fifteen mL of the waste sample was added directly into a 300-mL BOD incubation bottle. The initial DO of the diluted sample was 8.8 mg/L and the final DO after 5 days was 1.9 mg/L. The corresponding initial and final DO of the seeded dilution water was 9.1 and 7.9, respectively. What is the 5-day BOD (BOD_5) of the wastewater sample?

Solution

1. Determine the 5-day BOD using Eq. (2–57).

$$BOD_5, mg/L = \frac{(D_1 - D_2) - (B_1 - B_2)f}{P}$$

$$f = [(300 - 15)/300] = 0.95$$

$$P = 15/300 = 0.05$$

$$BOD_5, mg/L = \frac{(8.8 - 1.9) - (9.1 - 7.9)0.95}{0.05} = 115.2 \ mg/L$$

Modeling of BOD Reaction. The rate of BOD oxidation ("exertion") is modeled based on the assumption that the amount of organic material remaining at any time t is governed by a first-order function (see Chap. 4), as given below.

$$\frac{dBOD_r}{dt} = -k_1 BOD_r \qquad (2\text{–}58)$$

Integrating between the limits of UBOD and BOD_t and $t = 0$ and $t = t$ yields

$$BOD_r = UBOD(e^{-k_1 t}) \qquad (2\text{–}59)$$

where BOD_r = amount of waste remaining at time t (days) expressed in oxygen equivalents, mg/L
k_1 = first-order reaction rate constant, 1/d
UBOD = total or ultimate carbonaceous BOD, mg/L
t = time, d

Thus the BOD exerted up to time t is given by

$$BOD_t = UBOD - BOD_r = UBOD - UBOD(e^{-k_1 t}) = UBOD(1 - e^{-k_1 t}) \qquad (2\text{–}60)$$

Equation (2–60) is the standard expression used to define the BOD for wastewater. The basis for this equation is discussed in Sec. 4–3 in conjunction with the analysis of a batch reactor. It should be noted that in the literature dealing with the characterization of wastewater, the terms "L" and "BOD_u" are often used to denote ultimate carbonaceous BOD (UBOD).

The value of k_1 for untreated wastewater is generally about 0.12 to 0.46 d^{-1} (base e), with a typical value of about 0.23 d^{-1}. The range of k_1 values for effluents from biological treatment processes is from 0.12 to 0.23 d^{-1}. For a given wastewater, the value of k_1 at 20°C can be determined experimentally by observing the variation with time of the dissolved oxygen in a series of incubated samples. If k_1 at 20°C is equal to 0.23 d^{-1}, the 5-day oxygen demand is about 68 percent of the ultimate first-stage demand. Occasionally, the first-order reaction rate constant will be expressed in log (base 10) units. The relationship between k_1 (base e) and K_1 (base 10) is as follows:

$$K_1(\text{base 10}) = \frac{k_1(\text{base } e)}{2.303} \tag{2-61}$$

As discussed above, the temperature at which the BOD of a wastewater sample is determined is usually 20°C. It is possible, however, to determine the reaction constant k at a temperature other than 20°C using the following relationship developed in the discussion on temperature in Sec. 2–2.

$$k_{1_T} = k_{1_{20}} \theta^{T-20} \tag{2-25}$$

The value of θ has been found to vary from 1.056 in the temperature range between 20 and 30°C to 1.135 in the temperature range between 4 and 20°C (Schroepfer et al., 1964). A value of θ often quoted in the literature is 1.047 (Phelps, 1944), but it has been observed that this value does not apply at cold temperatures (e.g., below 20°C). Equation (2–60), along with Eq. (2–25), makes it possible to convert test results from different time periods and temperatures to the standard 5-day 20°C test, as illustrated in Example 2–9.

EXAMPLE 2-9 Calculation of BOD Determine the 1-day BOD and ultimate first-stage BOD for a wastewater whose 5-day 20°C BOD is 200 mg/L. The reaction constant k (base e) = 0.23 d^{-1}. What would have been the 5-day BOD if the test had been conducted at 25°C?

Solution

1. Determine the ultimate carbonaceous BOD.

$$\text{BOD}_5 = \text{UBOD} - \text{BOD}_r = \text{UBOD}(1 - e^{-k_1 t})$$

$$200 = \text{UBOD}(1 - e^{-5 \times 0.23}) = \text{UBOD}(1 - 0.316)$$

$$\text{UBOD} = 293 \text{ mg/L}$$

2. Determine the 1-day BOD.

$$\text{BOD}_t = \text{UBOD}(1 - e^{-k_1 t})$$

$$\text{BOD}_1 = 293(1 - e^{-0.23 \times 1}) = 293(1 - 0.795) = 60.1 \text{ mg/L}$$

3. Determine the 5-day BOD at 25°C.

$$k_{1_T} = k_{1_{20}}(1.047)^{T-20}$$

$$k_{1_{25}} = 0.23(1.047)^{25-20} = 0.29 \text{ d}^{-1}$$

$$\text{BOD}_5 = \text{UBOD}(1 - e^{-k_1 t}) = 293(1 - e^{-0.29 \times 5}) = 224 \text{ mg/L}$$

Figure 2–21

Effect of the rate constant k_1 on BOD (for a unit UBOD value).

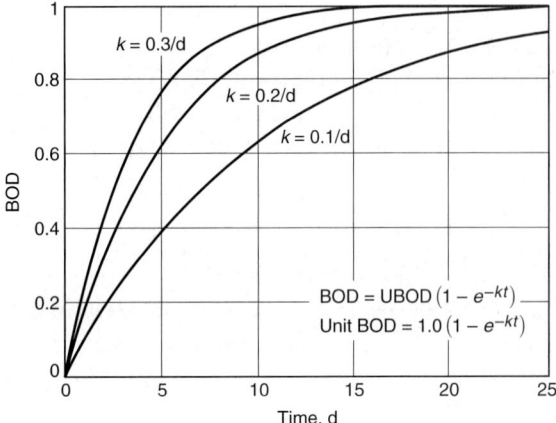

For polluted water and wastewater, a typical value of k_1 (base e at 20°C) is 0.23 d^{-1} (K_1, base 10, = 0.10 d^{-1}). The value of reaction rate constant varies significantly, however, with the type of waste. The range may be from 0.05 to 0.3 d^{-1} (base e) or more. For the same ultimate BOD, the oxygen uptake will vary with time and with different reaction rate constant values (see Fig. 2–21).

Nitrification in the BOD Test. Noncarbonaceous matter, such as ammonia, is produced during the hydrolysis of proteins. It is now known that a number of bacteria are capable of oxidizing ammonia to nitrite and subsequently to nitrate. The generalized reactions are as follows:

Conversion of ammonia to nitrite (as typified by *Nitrosomonas*):

$$NH_3 + 3/2O_2 \rightarrow HNO_2 + H_2O \tag{2–62}$$

Conversion of nitrite to nitrate (as typified by *Nitrobacter*):

$$HNO_2 + 1/2O_2 \rightarrow HNO_3 \tag{2–63}$$

Overall conversion of ammonia to nitrate:

$$NH_3 + 2O_2 \rightarrow HNO_3 + H_2O \tag{2–64}$$

The oxygen demand associated with the oxidation of ammonia to nitrate is called the nitrogenous biochemical oxygen demand (NBOD). The normal exertion of the oxygen demand in a BOD test for a domestic wastewater is shown on Fig. 2–22. Because the reproductive rate of the nitrifying bacteria is slow, it normally takes from 6 to 10 days for them to reach significant numbers to exert a measurable oxygen demand. However, if a sufficient number of nitrifying bacteria is present initially, the interference caused by nitrification can be significant.

When nitrification occurs in the BOD test, erroneous interpretations of treatment operating data are possible. For example, assume the effluent BOD from a biological treatment process is 20 mg/L without nitrification and 40 mg/L with nitrification. If the influent BOD to the treatment process is 200 mg/L, then the corresponding BOD removal efficiency would be reported as 90 and 80 percent without and with nitrification, respectively. Thus, if nitrification is occurring but is not suspected, it might be concluded

Figure 2–22

Definition sketch for
the exertion of the
carbonaceous and
nitrogenous biochemical
oxygen demand in a
waste sample.

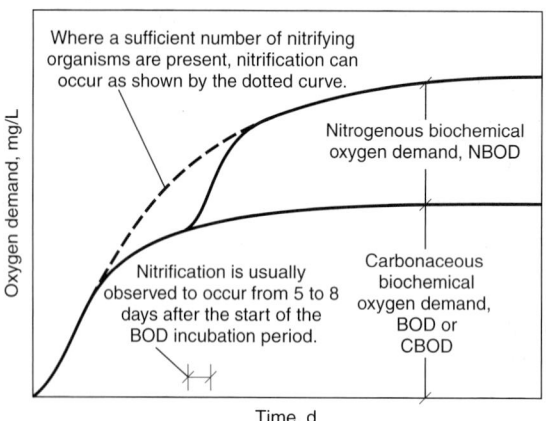

Carbonaceous Biochemical Oxygen Demand. When nitrification occurs, the measured BOD value will be higher than the true value due to the oxidation of carbonaceous material (see Fig. 2–22). If a given percentage of carbonaceous biochemical oxygen demand (CBOD) removal must be achieved to meet regulatory permit limits, early nitrification can pose a serious problem. The effects of nitrification can be overcome either by using various chemicals to suppress the nitrification reactions, or by treating the sample to eliminate the nitrifying organisms (Young, 1973). Pasteurization and chlorination/dechlorination are two methods that have also been used to suppress the nitrifying organisms.

When the nitrification reaction is suppressed, the resulting BOD is known as the carbonaceous biochemical oxygen demand (CBOD). In effect, the CBOD is a measure of the oxygen demand exerted by the oxidizable carbon in the sample. The CBOD test, in which the nitrification reaction is suppressed chemically, should be used only on samples that contain small amounts of organic carbon (e.g., treated effluent). Large errors will occur in the measured BOD values (up to 20 percent) when the CBOD test is used on wastewater containing significant amounts of organic matter such as untreated wastewater (Albertson, 1995).

Analysis of BOD Data. The value of k is needed if the BOD_5 is to be used to obtain UBOD, the ultimate or 20-day BOD. The usual procedure followed when these values are unknown is to determine k_1 and UBOD from a series of BOD measurements. There are several ways of determining k_1 and UBOD from the results of a series of BOD measurements, including the method of least squares, the method of moments (Moore et al., 1950), the daily-difference method (Tsivoglou, 1958), the rapid-ratio method (Sheehy, 1960), the Thomas method (Thomas, 1950), and the Fujimoto method (Fujimoto, 1961). The least-squares method and the Fujimoto method are illustrated in the following discussion.

The least-squares method involves fitting a curve through a set of data points, so that the sum of the squares of the residuals (the difference between the observed value and the value of the fitted curve) must be a minimum. Using this method, a variety of different types of curves can be fitted through a set of data points. For example, for a time series of BOD measurements on the same sample, the following equation may be written for each of the various n data points:

$$\frac{dy}{dt}\bigg|_{t=n} = k_1(\text{UBOD} - y) \tag{2-65}$$

where k_1 = first-order reaction rate constant
y = BOD at $t = n$

In the above equation both k_1 and UBOD are unknown. If it is assumed that dy/dt represents the value of the slope of the curve to be fitted through all the data points for a given k_1 and UBOD value, then because of experimental error, the two sides of Eq. (2–65) will not be equal but will differ by an amount R. Rewriting Eq. (2–65) in terms of R for the general case yields

$$R = k_1(\text{UBOD} - y) - \frac{dy}{dt} \tag{2-66}$$

Simplifying and using the notation y' for dy/dt gives

$$R = k_1\text{UBOD} - k_1 y - y' \tag{2-67}$$

Substituting a for k_1 UBOD and $-b$ for k_1 gives

$$R = a + by - y' \tag{2-68}$$

Now, if the sum of the squares of the residuals R is to be a minimum, the following equations must hold:

$$\frac{\partial}{\partial a} \Sigma R^2 = \Sigma 2R \frac{\partial R}{\partial a} \tag{2-69}$$

$$\frac{\partial}{\partial b} \Sigma R^2 = \Sigma 2R \frac{\partial R}{\partial b} \tag{2-70}$$

If the indicated operations in Eqs. (2–69) and (2–70) are carried out using the value of the residual R defined by Eq. (2–68), the following set of equations result:

$$na + b\Sigma y - \Sigma y' = 0 \tag{2-71}$$

$$a\Sigma y + b\Sigma y^2 - \Sigma yy' = 0 \tag{2-72}$$

where n = number of data points
$a = -b\text{UBOD}$
$b = -k_1(\text{base } e)$
$\text{UBOD} = -a/b$
$y = y_t$, mg/L
$y' = \dfrac{y_{n+1} - y_{n-1}}{2\Delta t}$

Application of the least-squares method in the analysis of BOD data is illustrated in Example 2–9 following the discussion of the Fujimoto method.

In the Fujimoto method an arithmetic plot is prepared of BOD_{t+1} versus BOD_t (Fujimoto, 1961). The value at the intersection of the plot with a line of slope 1 corresponds to the ultimate BOD. After the UBOD has been determined, the rate constant is determined using Eq. (2–59) and one of the BOD values. The application of the Fujimoto method is illustrated in Example 2–10.

EXAMPLE 2–10 **Calculation of BOD Constants Using the Least-Squares and the Fujimoto Methods** Compute k_1 and UBOD using the least-squares and Fujimoto methods for the following BOD data reported for a stream receiving some treated effluent:

t, d	2	4	6	8	10
y, mg/L	11	18	22	24	26

Solution

1. Determine k_1 and UBOD using the method of least squares.
    a. Set up a computation table and perform the indicated steps.

Time, d	y	y²	y'	yy'
0	0			
2	11	121	4.50	49.5
4	18	324	2.75	49.5
6	22	484	1.50	33.0
8	24	576	1.00	24.0
10	26ᵃ			
Sums	75	1505	9.75	156.0

ᵃ Value not included in total.

    b. The slope y' is computed as follows:

$$\frac{dy}{dt} = y' = \frac{y_{n+1} - y_{n-1}}{2\Delta t}$$

    c. Substituting the values computed in Step 1 in Eqs. (2–71) and (2–72), solve for a and b.

$$4a + 75b - 9.75 = 0$$

$$75a + 1505b - 156.0 = 0$$

$$a = 7.5 \text{ and } b = -0.271$$

    d. Determine the values of k_1 and UBOD.

$$k_1 = -b = -(-0.271) = 0.271 \text{ (base } e)$$

$$\text{UBOD} = -\frac{a}{b} = -\frac{7.5}{(-0.271)} = 27.7$$

2. Determine k_1 and UBOD using the Fujimoto method.
   a. Prepare an arithmetic plot of BOD_{t+1} versus BOD_t and on the same plot draw a line with a slope of 1. The value at the intersection of the two lines (BOD = 27.8 mg/L) corresponds to the ultimate BOD.

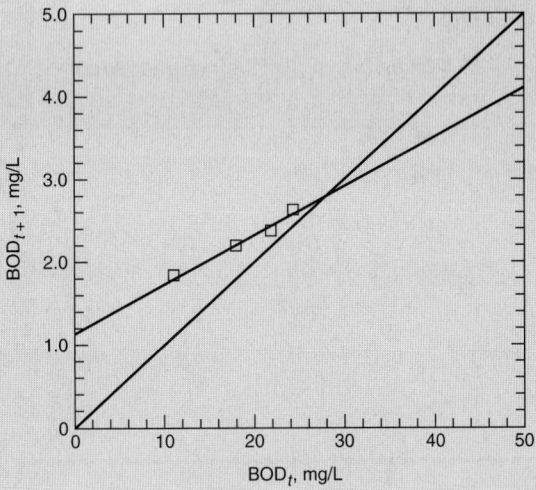

   b. Determine the k_1 value using Eq. (2–59).

$$y_6 = UBOD - BOD_6 = UBOD(1 - e^{-6k_1})$$

$$22 = UBOD - BOD_6 = 27.8(1 - e^{-6k_1})$$

$$k_1 = 0.261 \text{ d}^{-1}$$

Respirometric Determination of BOD. Determination of the BOD value and the corresponding rate constant k_1 can be accomplished more effectively in the laboratory using an instrumented large-volume (1.0-L) electrolysis cell respirometer. An electrolysis cell may also be used to obtain a continuous BOD (Young and Baumann, 1976a, 1976b). Within the cell, oxygen pressure over the sample is maintained constant by continuously replacing the oxygen used by the micoorganisms. Oxygen replacement is accomplished by means of an electrolysis reaction in which oxygen is produced in response to changes in the pressure. The BOD readings are determined by noting the length of time that the oxygen was generated and correlating it to the amount of oxygen produced by the electrolysis reaction. The modern electrolysis cell respirometer has replaced the Gilson and Warburg respirometer, used previously, in which the oxygen consumed was calculated from pressure drop measurements made with a manometer (Tchobanoglous and Burton, 1991). The principal advantages of the electrolysis cell over the Gilson and Warburg respirometers are that (1) the use of a large (1-L) sample minimizes the errors of grab sampling and pipetting in dilutions, and (2) the value of the BOD

Figure 2-23

Commercial electrolytic respirometer apparatus. (Courtesy Challenge Environmental Systems, Inc.)

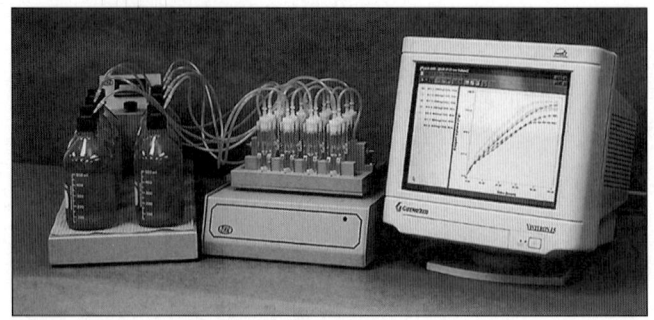

Table 2-17

Effect of the size of the biodegradable particles found in wastewater on observed BOD reaction rates[a]

Fraction	Size range, μm	k (base 10), d^{-1}
Settleable	>100	0.08
Supracolloidal	1–100	0.09
Colloidal	0.1–1.0	0.22
Soluble	<0.1	0.39

[a] Adapted from Balmat (1957).

is available directly. A typical example of a commercially available electrolytic respirometer with multiple electrolysis cells is also shown on Fig. 2–23.

Effect of Particle Size on BOD Reaction Rates. If a separation and analysis technique, such as membrane filtration (see Figs. 2–4 and 2–8), is used to quantify the size distribution of the solids in the influent wastewater, the various size fractions can be correlated to observed oxygen (BOD) uptake rates, determined using a respirometer. As reported in Table 2–17, the observed BOD reaction rate coefficients are affected significantly by the size of the particles in wastewater. Based on the data given in Table 2–17, it is clear that the treatment of a wastewater can be effected by modifying the particle size distribution. Further, wastewaters with significantly different particle size distributions will respond differently, depending on the method of treatment (e.g., in constructed wetlands).

Limitations in the BOD Test. The limitations of the BOD test are as follows: (1) a high concentration of active, acclimated seed bacteria is required; (2) pretreatment is needed when dealing with toxic wastes, and the effects of nitrifying organisms must be reduced; (3) only the biodegradable organics are measured; (4) the test does not have stoichiometric validity after the soluble organic matter present in solution has been used (see Fig. 2–24); and (5) the relatively long period of time required to obtain test results (originally chosen to minimize variance). Of the above, perhaps the most serious limitation is that the 5-day period may or may not correspond to the point where the solu-

Figure 2–24

Functional analysis of the BOD test: (a) Interrelationship of organic waste, bacterial mass (cell tissue, total organic waste, and oxygen consumed in BOD test) and (b) idealized representation of the BOD test. *(Tchobanoglous and Schroeder, 1985.)*

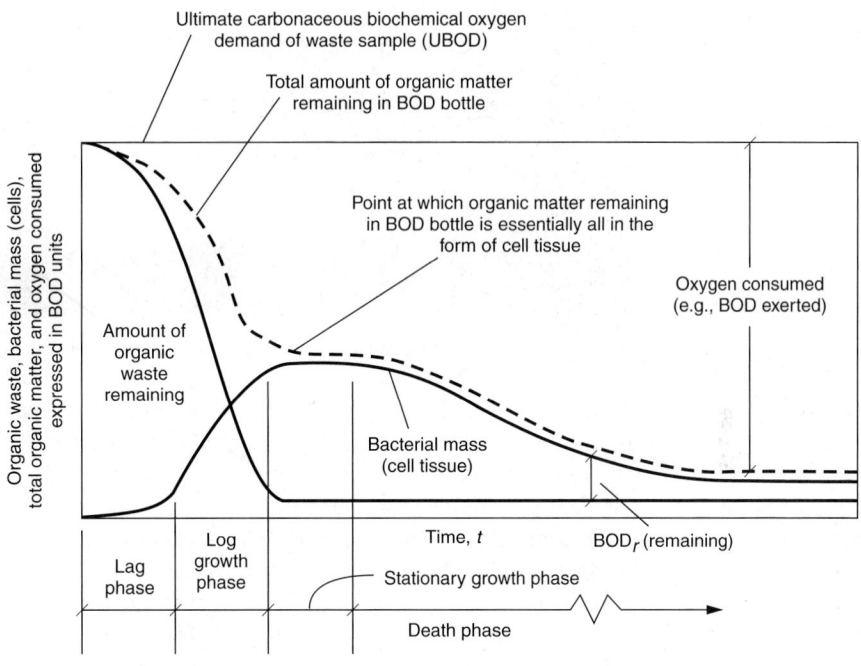

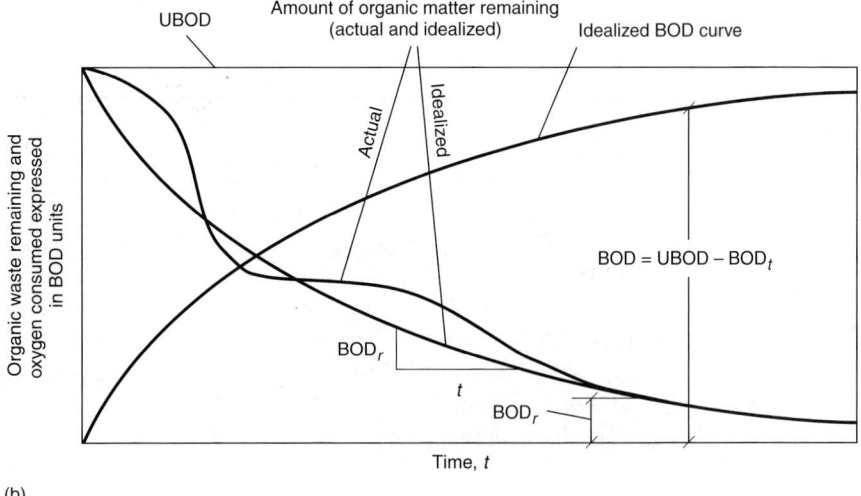

ble organic matter that is present has been used. The lack of stoichiometric validity at all times reduces the usefulness of the test results.

Total and Soluble Chemical Oxygen Demand (COD and SCOD)

The COD test is used to measure the oxygen equivalent of the organic material in wastewater that can be oxidized chemically using dichromate in an acid solution, as illustrated in the following equation, when the organic nitrogen is in the reduced state (oxidation number = −3) (Sawyer et al., 1994).

$$C_nH_aO_bN_c + dCr_2O_7^{2-} + (8d + c)H^+ \rightarrow$$

$$nCO_2 + \frac{a + 8d - 3c}{2} H_2O + cNH_4^+ + 2d\,Cr^{3+} \quad (2\text{-}73)$$

where $d = \dfrac{2n}{3} + \dfrac{a}{6} - \dfrac{b}{3} - \dfrac{c}{2}$

Although it would be expected that the value of the ultimate carbonaceous BOD would be as high as the COD, this is seldom the case. Some of the reasons for the observed differences are as follows: (1) many organic substances which are difficult to oxidize biologically, such as lignin, can be oxidized chemically, (2) inorganic substances that are oxidized by the dichromate increase the apparent organic content of the sample, (3) certain organic substances may be toxic to the microorganisms used in the BOD test, and (4) high COD values may occur because of the presence of inorganic substances with which the dichromate can react. From an operational standpoint, one of the main advantages of the COD test is that it can be completed in about 2.5 h, compared to 5 or more days for the BOD test. To reduce the time further, a rapid COD test that takes only about 15 min has been developed.

As new methods of biological treatment have been developed, especially with respect to biological nutrient removal, it has become more important to fractionate the COD. The principal fractions are particulate and soluble COD. In biological treatment studies, the particulate and soluble fractions are fractionated further to assess wastewater treatability (see discussion in Chap. 8, Sec. 8–2). Fractions that have been used include: (1) readily biodegradable soluble COD, (2) slowly biodegradable colloidal and particulate (enmeshed) COD, (3) nonbiodegradable soluble COD, and (4) nonbiodegradable colloidal and particulate COD. The readily biodegradable soluble COD is often fractionated further into complex COD that can be fermented to volatile fatty acids (VFAs) and short chain VFAs (see Fig. 8–31 in Chap. 8). Unfortunately, as noted previously, there is little standardization on the definition of soluble versus particulate COD. Where filtration is the technique used to fractionate the sample, the relative distribution between soluble and particulate COD will vary greatly depending on the pore size of the filter. An alternative method used to determine the soluble COD involves precipitation of the suspended solids and a portion of the colloidal material. The COD of the clarified liquid corresponds to the soluble COD.

Total and Dissolved Organic Carbon (TOC and DTOC)

The TOC test, done instrumentally, is used to determine the total organic carbon in an aqueous sample. The test methods for TOC utilize heat and oxygen, ultraviolet radiation, chemical oxidants, or some combination of these methods to convert organic carbon to carbon dioxide which is measured with an infrared analyzer or by other means. The TOC of a wastewater can be used as a measure of its pollutional characteristics, and in some cases it has been possible to relate TOC to BOD and COD values. The TOC test is also gaining in favor because it takes only 5 to 10 min to complete. If a valid relationship can be established between results obtained with the TOC test and the results of the BOD test for a given wastewater, use of the TOC test for process control is recommended.

More recently, a continuous on-line TOC analyzer has been developed, in conjunction with the space program, that can be used to detect TOC concentrations in the ppb (parts per billion) range. Such instruments are currently being used to detect the residual TOC in the treated effluent from microfiltration and reverse osmosis (RO) treatment units. Continuous TOC measurements may be used to monitor the performance of the full-scale RO units, to be used in conjunction with repurification projects in which repurified effluent is proposed to be blended with other waters.

Along with COD, it has also become more important to fractionate the TOC. The principal fractions are particulate TOC and dissolved (soluble) DTOC. As with COD, the particulate and soluble TOC fractions are fractionated further to assess treatability. It should be noted that the pore size of the filter paper recommended in Standard Methods (1998) for differentiating between dissolved and particulate TOC is 0.45 μm, in contrast to the pore size (2.0 μm or less) used to define TSS and TDS. Again, because of the pore size of the filter paper used, the colloidal material that passes through the filter will be classified as dissolved. Because of the interest in the chemical constituents that make up the DTOC, advanced methods of analysis have been developed to quantify the constituent groupings as illustrated on Fig. 2–25.

UV-Absorbing Organic Constituents

A number of organic compounds are found in wastewater, including humic substances, lignin, tannin, and various aromatic compounds, that strongly absorb ultraviolet (UV) radiation. As a result, UV absorption has been used as a surrogate measure for the organic compounds cited above. The UV wavelengths at which adsorption is determined are typically in the range from 200 to 400 nm, with the value of 254 nm being reported most commonly. The results of UV absorption measurements are reported in units of cm^{-1}, along with the pH and the UV wavelength (e.g., UV^{pH}_λ where λ is the UV wavelength). This method has proved useful in assessing the aggregate presence of UV absorbing compounds in wastewater, although interfering compounds can render the test invalid.

The results of UV absorption measurements at a wavelength of 254 nm are also correlated to the amount of dissolved organic carbon (DOC) present in a sample which

Figure 2–25

Procedure for the characterization of the organic fractions that comprise the TOC (From Jerry Leenheer, U.S. Geologic Survey.)

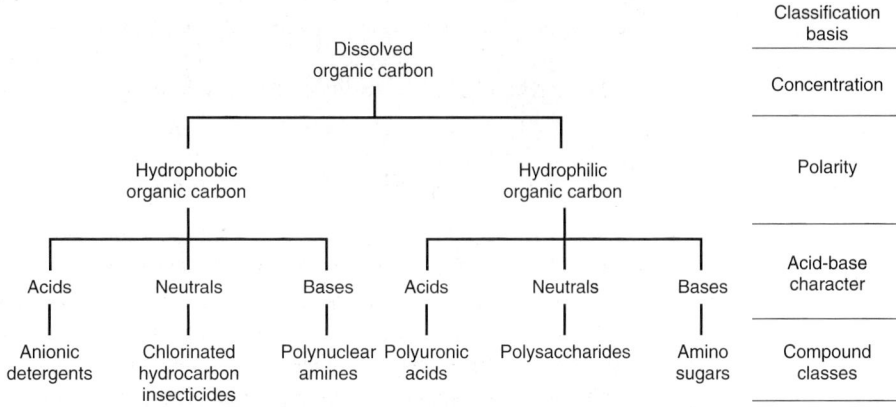

has been filtered through a filter with a pore size of 0.45 μm. The results are reported as the *specific ultraviolet adsorption* (SUVA) per mg/L of DOC. It should be noted that although the UV measurement is correlated to the DOC, SUVA is in fact a measure of the nature of the carbon in the sample being analyzed, more specifically the extent to which the carbon is aromatic. Thus, the SUVA test is used most commonly to distinguish between different water samples. The SUVA test has also been used to assess the potential for the formation of trihalomethanes (THMs) (see Sec. 2–7).

Theoretical Oxygen Demand (ThOD)

Organic matter of animal or vegetable origin in wastewater is generally a combination of carbon, hydrogen, oxygen, and nitrogen. The principal groups of these elements present in wastewater are, as previously noted, carbohydrates, proteins, oils and grease, and products of their decomposition. The biological decomposition of the substances is discussed in Chap. 7. If the chemical formula of the organic matter is known, the ThOD may be computed as illustrated in Example 2–11.

EXAMPLE 2–11 **Calculation of ThOD** Determine the ThOD for glycine ($CH_2(NH_2)COOH$) using the following assumptions:

1. In the first step, the organic carbon and nitrogen are converted to carbon dioxide (CO_2) and ammonia (NH_3), respectively.
2. In the second and third steps, the ammonia is oxidized sequentially to nitrite and nitrate.
3. The ThOD is the sum of the oxygen required for all three steps.

Solution

1. Write a balanced reaction for the carbonaceous oxygen demand.

$$CH_2(NH_2)COOH + 3/2O_2 \rightarrow NH_3 + 2CO_2 + H_2O$$

2. Write balanced reactions for the nitrogenous oxygen demand.
   a. $NH_3 + 3/2O_2 \rightarrow HNO_2 + H_2O$
   b. $\underline{HNO_2 + 1/2O_2 \rightarrow HNO_3}$
   $NH_3 + 2O_2 \rightarrow HNO_3 + H_2O$
3. Determine the ThOD.

$$ThOD = (3/2 + 4/2)\ \text{mol}\ O_2/\text{mol glycine}$$

$$= 7/2\ \text{mol}\ O_2/\text{mol glycine} \times 32\ \text{g/mol}\ O_2$$

$$= 112\ \text{g}\ O_2/\text{mol glycine}$$

Interrelationships between BOD, COD, and TOC

Typical values for the ratio of BOD/COD for untreated municipal wastewater are in the range from 0.3 to 0.8 (see Table 2–18). If the BOD/COD ratio for untreated wastewater is 0.5 or greater, the waste is considered to be easily treatable by biological means. If the ratio is below about 0.3, either the waste may have some toxic components or accli-

Type of wastewater	BOD/COD	BOD/TOC
Untreated	0.3–0.8	1.2–2.0
After primary settling	0.4–0.6	0.8–1.2
Final effluent	0.1–0.3[a]	0.2–0.5[b]

Table 2–18
Comparison of ratios of various parameters used to characterize wastewater

[a] CBOD/COD.
[b] CBOD/TOC.

mated microorganisms may be required in its stabilization. The corresponding BOD/TOC ratio for untreated wastewater varies from 1.2 to 2.0. In using these ratios it important to remember that they will change significantly with the degree of treatment the waste has undergone, as reported in Table 2–18. The theoretical basis for these ratios is explored in the following example.

EXAMPLE 2–12 Determination of BOD/COD, BOD/TOC, and TOC/COD Ratios Determine the theoretical BOD/COD, BOD/TOC, and TOC/COD ratios for the following compound $C_5H_7NO_2$. Assume the value of the BOD first-order reaction rate constant is 0.23/d (base e) (0.10/d base 10).

Solution

1. Determine the COD of the compound using Eq. (2–55).

$$C_5H_7NO_2 + 5O_2 \rightarrow 5CO_2 + NH_3 + 2H_2O$$
mw = 113   mw = 160

COD = 160/113 = 1.42 mg O_2/mg $C_5H_7NO_2$

2. Determine the BOD of the compound.

$$\frac{BOD}{UBOD} = (1 - e^{-k_1 t}) = (1 - e^{-0.23 \times 5}) = 1 - 0.32 = 0.68$$

BOD = 0.68 × 1.42 mg O_2/mg $C_5H_7NO_2$ = 0.97 mg BOD/mg $C_5H_7NO_2$

3. Determine the TOC of the compound.

TOC = (5 × 12)/113 = 0.53 mg TOC/mg $C_5H_7NO_2$

4. Determine BOD/COD, BOD/TOC, and TOC/COD ratios.

$$\frac{BOD}{COD} = \frac{0.68 \times 1.42}{1.42} = 0.68$$

$$\frac{BOD}{TOC} = \frac{0.68 \times 1.42}{0.53} = 1.82$$

$$\frac{TOC}{COD} = \frac{0.53}{1.42} = 0.37$$

Oil and Grease

The term *oil and grease,* as commonly used, includes the fats, oils, waxes, and other related constituents found in wastewater. The term *fats, oil, and grease* (FOG) used previously in the literature has been replaced by the term *oil and grease*. The oil and grease content of a wastewater is determined by extraction of the waste sample with trichlorotrifluoroethane (oil and grease are soluble in trichlorotrifluoroethane). Other extractable substances include mineral oils, such as kerosene and lubricating and road oils. Oil and grease are quite similar chemically; they are compounds (esters) of alcohol or glycerol (glycerin) with fatty acids. The glycerides of fatty acids that are liquid at ordinary temperatures are called oils, and those that are solids are called grease (or fats).

If grease is not removed before discharge of treated wastewater, it can interfere with the biological life in the surface waters and create unsightly films. The thickness of oil required to form a translucent film on the surface of a water body is about 0.0003048 mm (0.0000120 in), as given below.

Appearance	Film thickness		Quantity spread	
	in	mm	gal/mi^2	L/ha
Barely visible	0.0000015	0.0000381	25	0.365
Silvery sheen	0.0000030	0.0000762	50	0.731
First trace of color	0.0000060	0.0001524	100	1.461
Bright bands of color	0.0000120	0.0003048	200	2.922
Colors begin to dull	0.0000400	0.0010160	666	9.731
Colors are much darker	0.0000800	0.0020320	1332	19.463

Source: Eldridge (1942).

Fats and oils are contributed to domestic wastewater in butter, lard, margarine, and vegetable fats and oils. Fats are also commonly found in meats, in the germinal area of cereals, in seeds, in nuts, and in certain fruits. The low solubility of fats and oils reduces their rate of microbial degradation. Mineral acids attack them, however, resulting in the formation of glycerin and fatty acid. In the presence of alkalies, such as sodium hydroxide, glycerin is liberated, and alkali salts of the fatty acids are formed. These alkali salts are known as soaps. Common soaps are made by saponification of fats with sodium hydroxide. They are soluble in water, but in the presence of hardness constituents, the sodium salts are changed to calcium and magnesium salts of the fatty acids, or so-called mineral soaps. These are insoluble and are precipitated.

Kerosene, lubricating, and road oils are derived from petroleum and coal tar and contain essentially carbon and hydrogen. These oils sometimes reach the sewers in considerable volume from shops, garages, and streets. For the most part, they float on the wastewater, although a portion is carried into the sludge on settling solids. To an even greater extent than fats, oils, and soaps, the mineral oils tend to coat surfaces. The particles interfere with biological action and cause maintenance problems.

Surfactants

Surfactants, or surface-active agents, are large organic molecules that are slightly soluble in water and cause foaming in wastewater treatment plants and in the surface waters into which the waste effluent is discharged. Surfactants are most commonly composed of a strongly hydrophobic group combined with a strongly hydrophilic group. Typically, the hydrophobic group is a hydrocarbon radical (R) made up of 10 to 20 carbon atoms. Two types of hydrophobic groups are used: those that will and those that will not ionize in water. Anionic surfactants are negatively charged [e.g., $(RSO_3N)^-Na^+$]; whereas cationic surfactants are positively charged [e.g., $(RMe_3N)^+Cl^-$]. Nonionizing (nonionic) surfactants commonly contain a polyoxyethylene hydrophilic group ($ROCH_2CH_2OCH_2CH_2 \ldots OCH_2CH_2OH$, often abbreviated Re_n, where n is the average number of $-OCH_2CH_2-$ units in the hydrophilic group). Hybrids of these types also exist. In the United States, ionic surfactants amount to about two-thirds of the total surfactants used and nonionics to about one-third (Standard Methods, 1998).

Surfactants tend to collect at the air-water interface with the hydrophilic in the water and the hydrophobic group in the air. During aeration of wastewater, these compounds collect on the surface of the air bubbles and thus create a very stable foam. Before 1965, the type of surfactant present in synthetic detergents, called alkyl-benzene-sulfonate (ABS), was especially troublesome because it resisted breakdown by biological means. As a result of legislation in 1965, ABS has been replaced in detergents by linear-alkyl-sulfonate (LAS), which is biodegradable. Because surfactants come primarily from synthetic detergents, the foaming problem has been greatly reduced. It should be noted that so-called "hard" synthetic detergents are still used extensively in many foreign countries.

Two tests are now used to determine the presence of surfactants in water and wastewater. The MBAS (methylene blue active substances) test is used for anionic surfactants. The determination of surfactants is accomplished by measuring the color change in a standard solution of methylene blue dye. Nonionic surfactants are measured using the CTAS (cobalt thiocyanate active substances) test. Nonionic surfactants will react with the CTAS to produce a cobalt-containing product which can be extracted into a organic liquid and then measured. It should be noted that the CTAS method requires sublimation to remove nonionic surfactants and ion exchange to remove the cationic and anionic surfactants (Standard Methods, 1998).

2–7 INDIVIDUAL ORGANIC COMPOUNDS

Individual organic compounds are determined to assess the presence of priority pollutants identified by the U.S. Environmental Protection Agency (U.S. EPA) and a number of new emerging compounds of concern. Priority pollutants (both inorganic and organic) have been and are continuing to be selected on the basis of their known or suspected carcinogenicity, mutagenicity, teratogenicity, or high acute toxicity. As the techniques used to identify specific compounds continue to improve, a number of other organic compounds have been detected in public water supplies and in treated wastewater effluents.

Priority Pollutants

The Environmental Protection Agency has identified approximately 129 priority pollutants in 65 classes to be regulated by categorical discharge standards (*Federal Register*, 1981). Priority pollutants (both inorganic and organic) were selected on the basis of their known or suspected carcinogenicity, mutagenicity, teratogenicity, or high acute toxicity. Many of the organic priority pollutants are also classified as volatile organic compounds (VOCs).

Two types of standards are used to control pollutant discharges to publicly owned treatment works (POTWs). The first, "prohibited discharge standards," applies to all commercial and industrial establishments which discharge to POTWs. Prohibited standards restrict the discharge of pollutants that may create a fire or explosion hazard in sewers or treatment works, are corrosive (pH < 5.0), obstruct flow, upset treatment processes, or increase the temperature of the wastewater entering the plant to above 40°C. "Categorical standards" apply to industrial and commercial discharges in 25 industrial categories ("categorical industries"), and are intended to restrict the discharge of the 129 priority pollutants. It is anticipated that this list will continue to be expanded in the future.

Analysis of Individual Organic Compounds

The analytical methods used to determine individual organic compounds require the use of sophisticated instrumentation capable of measuring trace concentrations in the range of 10^{-12} to 10^{-3} mg/L. Gas chromatographic (GC) and high-performance liquid chromatographic (HPLC) methods are most commonly used to detect individual organic compounds. Different types of detectors are used with each method, depending on the nature of the compound being analyzed. Typical detectors used in conjunction with gas chromatography include electrolytic conductivity, electron capture (ECD), flame ionization (FID), photoionization (PID), and mass spectrometer (GCMS). Typical detectors for high-performance liquid chromatography include photodiode array (PDAD) and post column reactor (PCR). It should also be noted that many of the individual organic constituents can be determined by two or more of the above methods (Standard Methods, 1998).

Over 180 individual organic compounds can be determined using one or more of the methods cited above. The principal categories containing the individual organic compounds are reported in Table 2–19. As instrumental methods of analysis have improved, the detection limits for these compounds have become increasingly small, typically below 10 ng/L. The specific organic compounds that are analyzed will depend on the application. For example, for indirect reuse applications scans of disinfection byproducts may be required where chlorine is used for disinfection.

Volatile Organic Compounds (VOCs)

Organic compounds that have a boiling point less than or equal to 100°C and/or a vapor pressure greater than 1 mm Hg at 25°C are generally considered to be volatile organic compounds (VOCs). For example, vinyl chloride, which has a boiling point of -13.9°C and a vapor pressure of 2548 mm Hg at 20°C, is an example of an extremely volatile organic compound. Volatile organic compounds are of great concern because (1) once such compounds are in the vapor state they are much more mobile and therefore more likely to be released to the environment, (2) the presence of some of these compounds in the atmosphere may pose a significant public health risk, and (3) they contribute to a general increase in reactive hydrocarbons in the atmosphere, which can lead to the formation of photochemical oxidants. The release of these compounds in sewers and at treatment plants, especially at the headworks, is of particular concern with respect to the health of collection system and treatment plant workers. The physical phenomena involved in the release and control of VOCs are considered in more detail in Chap. 5.

Table 2-19
Typical classes of organic compounds whose members are identified as individual compounds

Name	Occurrence/source	Concern
Volatile organic compounds	Found in ground- and surface waters	Potential for tetratogenesis or carcinogenesis in humans
1,2-Dibromoethane (EDB) and 1,2-dibromo-3-chloropropane (DBCP)	Found in groundwater supplies, especially where these compounds have been used as fumigants	Detrimental effects on human health
Trihalomethanes (THMs)	Found in most chlorinated water supplies	Disinfection byproduct. Potential human carcinogen
Chlorinated organic solvents	Found in raw water supplies resulting from industrial contamination	Potential human carcinogen
Haloacetic acids (HAAs)	Formed from the chlorination of natural organic matter (humic and fulvic acids)	Disinfection byproduct. Potential human carcinogen
Trichlorophenol	Formed from the chlorination of natural organic matter (humic and fulvic acids)	Disinfection byproduct. Dichloroacetic acid and trichloroacetic acid are animal carcinogens
Aldehydes	Formed from the application of ozone to water containing organic matter	Disinfection byproduct
Extractable base/neutral and acids	Many semivolatile compounds including polynuclear aromatic hydrocarbons, phthalates, phenolics, organochlorine pesticides, and PCBs	Many of the listed compounds are toxic or carcinogenic
Phenols	Generally traceable to industrial discharges or landfills	Impart a taste to water at low levels. May have detrimental impact on human health at higher levels
Polychlorinated biphenyls (PCB)	Found in water supplies contaminated by transformer oils	These compounds are toxic, bioaccumulative, and extremely stable in water
Polynuclear aromatic hydrocarbons (PAHs)	Byproducts of petroleum processing or combustion	Many compounds in this group are highly carcinogenic at relatively low levels
Carbamate pesticides	Found in water supplies contaminated by pesticides	
Organochlorine pesticides	Found in water supplies contaminated by pesticides	Many compounds in this group are bioaccumulative, and relatively stable, as well as toxic or carcinogenic
Acidic herbicide compounds	Used for weed control, these compounds are found in aquatic systems	
Glyphosphate herbicide	Broad spectrum nonselective postemergence herbicide. Water supplies can become contaminated through runoff and spray drift	

Disinfection Byproducts

It has been found that when chlorine is added to water containing organic matter a variety of organic compounds containing chlorine are formed. Collectively, these compounds, along with others, are known as disinfection byproducts (DBPs). Although generally present in low concentrations, they are of concern because many of them are known or suspected potential human carcinogens. Typical classes of compounds include trihalomethanes (THMs), haloacetic acids (HAAs), trichlorophenol, and aldehydes.

More recently, N-nitrosodimethylamine (NDMA) has been found in the effluent from wastewater-treatment plants. The reason for concern over this compound is that as a group of compounds nitrosamines are among the most powerful carcinogens known (Snyder, 1995). These compounds are also known to be strongly carcinogenic to various fish species at low concentrations. The U.S. EPA action limit for NDMA is 2 parts per trillion. Based on the results of recent studies, NDMA appears to be formed during the chlorination process. In treated effluent, the nitrite ion can react with hydrochloric acid, present as a result of the use of chlorine for disinfection, to form nitrous acid. In turn, nitrous acid can react with dimethylamine to form NDMA (Hill, 1988). The compound dimethylamine is common in wastewater and surface waters, and is found in urine, feces, algae, and plant tissues. Dimethylamine is also part of some polymers used for water treatment (such as polydiallyl dimethylamine) and for ion-exchange resins. The formation of NDMA under basic and alkaline conditions has been reported by Wainwright (1986).

Because of the concern over the formation of DBPs and NDMA, considerable attention has been focused over the past five years on the use of ultraviolet (UV) disinfection as a possible replacement for chlorine. In addition, considerable attention has been focused on the modifications to conventional treatment processes to improve the treatment of these compounds and to advanced treatment processes for the removal of these substances. The use of UV radiation for disinfection and the destruction of NDMA is considered in Chaps. 11 and 12.

Pesticides and Agricultural Chemicals

Pesticides, herbicides, and other agricultural chemicals are toxic to many organisms and therefore can be significant contaminants of surface waters. These chemicals are not common constituents of domestic wastewater but result primarily from surface runoff from agricultural, vacant, and park lands. Concentrations of these chemicals can result in fish kills, in contamination of the flesh of fish that decreases their value as a source of food, and in impairment of water supplies.

Emerging Organic Compounds

In addition to the compounds discussed above, for which requirements have been established, a variety of new (emerging) compounds have been identified in many of the nation's water supplies and in treated wastewater effluents. The compounds in question are derived, in large part, from (1) veterinary and human antibiotics, (2) human prescription and nonprescription drugs, (3) industrial and household wastewater products, and (4) sex and steroidal hormones. Typical examples of the types of compounds involved are reported in Table 2–20. As more becomes known about the health impacts of these compounds, it is anticipated that discharge limits may be developed for a number of

Table 2–20
Emerging organic compounds found in the stream waters[a]

Veterinary and human antibiotics

Carbadox	Norfloxacin	Sulfamethazine
Chlortetracycline	Oxytetracycline	Sulfamethiazole
Ciprofloxacin	Roxarsone	Sulfamethoxazole
Doxycycline	Roxithromycin	Sulfathiazole
Enrofloxacin	Sarafloxacin	Tetracycline
Erythromycin	Spectinomycin	Trimethoprim
Erythromycin-H_2O	Sulfachlorpyridazine	Tylosin
Ivermectin	Sulfadimethoxine	Virginiamycin
Lincomycin	Sulfamerazine	

Human prescription and nonprescription drugs

Acetaminophen (antipyretic)	Gemfibrozil (lipotropic agent)
Amoxicillin (antibiotic)	Ibuprofen (anti-inflammatory)
Caffeine (stimulant)	Metformin (antidiabetic agent)
Cimetidine (antacid)	Paraxanthine (caffeine metabolite)
Cotinine (nicotine metabolite)	Paroxetine (paxil metabolite)
Dehydronifedipine (antianginal)	Ranitidine (antacid)
Digoxigenin (digoxin metabolite)	Salbutamol (antiasthmatic)
Diltiazem (antihypertensive)	Sulfamethoxazole (antibiotic)
Enalaprillat (antihypertensive)	Trimethoprim (antibiotic)
Fluoxetine (antidepressant)	

Industrial and household wastewater products

Acetophenone (fragrance)	Cholesterol (fecal indicator)
Anthracene (PAH)	Cis-chlordane (pesticide)
Benzo(a)pyrene (PAH)	Coarbaryl (pesticide)
2,6-di-tert-para-benzoquinone (antioxidant)	Codeine (analgesic)
5-methyl 1 H benzotriazole (antioxidant)	3b-coprostanol (carnivore fecal indicator)
Bisphenol A (used in polymers)	Cotinine (nicotine metabolite)
Bis(2-ethylhexyl)phthalate (plasticizer)	Para-cresol (wood preservative)
2,6-di-tert-butylphenol (antioxidant)	Diazinon (pesticide)
Butylated hydroxyanisole (BHA)	1,4-dichlorobenzene (fumigant)
Butylated hydroxytoluene (BHT) (antioxidant)	Dieldrin (pesticide)
Caffeine (stimulant)	Diethylphthalate (plasticizer)
Chlorpyrifos (pesticide)	N,N-diethyltoluamide (DEET) (insecticide)

(continued)

| **Table 2–20** (*Continued*) |

Industrial and household wastewater products (continued)	
Ethanol, 2-butoxy-, phosphate (plasticizer)	Phenol (disinfectant)
Fluoranthene (PAH)	Para-nonylphenol-total (detergent metabolite)
Lindane (pesticide)	Phthalic anhydride (used in plastics)
Methyl parathion (pesticide)	Pyrene (PAH)
Naphthalene (PAH)	Stigmastanol (plant sterol)
NPEO1-total (detergent metabolite)	Tetrachloroethylene (solvent)
NPEO2-total (detergent metabolite)	Triclosan (antimicrobial disinfectant)
OPEO1	Tri(2-chloroethyl)phosphate (fire retardant)
OPEO2	Tri(dichlorisopropyl)phosphate (fire retardant)
Phenanthrene (PAH)	Triphenyl phosphate (plasticizer)

Sex and steroidal hormones		
Cis-androsterone	17a-estradiol	Mestranol
3b-coprostanol	17b-estradiol	19-norethisterone
Cholesterol	Estriol	Progesterone
Equilenin	Estrone	Testosterone
Equilin	17a-ethynylestradiol	

[a]From U.S. GS (2000).

these compounds. Given that over 30 million organic compounds are known to exist, it is clear that the list of emerging compounds will continue to grow as analytical techniques continue to improve.

2–8   BIOLOGICAL CHARACTERISTICS

The biological characteristics of wastewater are of fundamental importance in the control of diseases caused by pathogenic organisms of human origin, and because of the extensive and fundamental role played by bacteria and other microorganisms in the decomposition and stabilization of organic matter, both in nature and in wastewater-treatment plants. The purpose of this section is to introduce (1) the microorganisms found in surface waters and wastewater, (2) the pathogenic microorganisms associated with human disease, (3) the use of indicator organisms, (4) the methods and techniques used for the enumeration of bacteria, (5) the method of enumerating viruses, and (6) a brief discussion of newly recognized or emerging organisms. The organisms responsible for the treatment of wastewater are considered further in Chap. 7.

Microorganisms Found in Surface Waters and Wastewater

Organisms found in surface water and wastewater include bacteria, fungi, algae, protozoa, plants and animals, and viruses. Bacteria, fungi, algae, protozoa, and viruses can only be observed microscopically. The general classification of these organisms and a general description of the organisms found in wastewater are considered in the following discussion. The growth and metabolic and environmental requirements of microorganisms are considered in detail in Chap. 7.

General Classification.   Living single-cell microorganisms that can only be seen with a microscope are responsible for the activity in biological wastewater treatment. The basic functional and structural unit of all living matter is the cell. Living organisms are divided into either prokaryote or eukaryote cells as a function of their genetic information and cell complexity. The prokaryotes have the simplest cell structure and include bacteria, blue-green algae (cyanobacter), and archaea. The archaea are separated from bacteria due to their DNA composition and unique cellular chemistry, such as differences in the cell wall and ribosome structure. Many archaea are bacteria that can grow under extreme conditions of temperature and salinity, and also include methanogenic methane-producing bacteria, important in anaerobic treatment processes. In contrast to the prokaryotes, the eukaryotes are much more complex and contain plants and animals and single-celled organisms of importance in wastewater treatment including protozoa, fungi, and green algae.

Some of the key differences between prokaryote and eukaryote microorganisms are summarized in Table 2–21, and a schematic of their cell structures is shown on Fig. 2–26. The prokaryote organisms are generally much smaller compared to eukaryote organisms. The absence of a nuclear membrane to contain the cell DNA is also a distinguishing feature of the prokaryote organisms. The eukaryotic organisms have much more complex internal structures. These include the endoplasmic reticulum, which is a distinct organelle that contains the sites of ribosomes with internal membranes. The golgi bodies are also distinct membrane structures and contain sites for the secretion of enzymes and other macromolecules. The mitochondrion is a complex internal membrane structure where respiration occurs for eukaryotic cells, and is lacking in prokaryotic cells. While the prokaryotes can have photosynthetic pigments, they do not contain chloroplasts, which are used in photosynthesis by green algae. Additional information on prokaryote cell structure and composition and the role and importance of DNA and ribonucleic acid (RNA) is discussed in Chap. 7.

Viruses are obligate *intracellular parasites* that require the machinery of a host cell to support their growth. Although viruses contain the genetic information (either DNA or RNA) needed to replicate themselves, they are unable to reproduce outside of a host cell. Viruses are composed of a nucleic acid core (RNA or DNA) surrounded by an outer coat of protein and glycoprotein. Viruses are classified separately according to the host infected. Bacteriophage, as the name implies, are viruses that infect bacteria.

General Description.   A general description of the microorganisms found in wastewater is given in Table 2–22 using the terminology introduced in the previous paragraphs. Data on the shape, resistant form, and size of the microorganisms found in wastewater are presented in Table 2–23. Information on the size of the microorganisms, especially the resistant form, is needed to determine the type of treatment that will be required to treat and/or remove them.

Table 2-21
Comparison of prokaryote and eukaryote cells[a]

Cell characteristic	Prokaryote	Eukaryote (Eukarya)
Phylogenetic group	Bacteria, blue-green algae (cyanobacter), archaea	Single cell: algae, fungi, protozoan; multicell: plants, animals
Size[b]	Small, 0.2–3.0 μm	2–100 μm for single-cell organisms
Cell wall	Composed of peptidoglycan (bacteria), other polysaccharides, protein, glycoprotein (archaea)	Absent in animals and most protozoan; present in plants, algae, fungi: usually polysaccharide
Nuclear structure:		
Nuclear membrane	Absent	Present
DNA	Single molecular, plasmids	Several chromosomes
Internal membranes	Simple, limited	Complex, endoplasmic reticulum, golgi, mitochondria; several present
Membrane organelles	Absent	Several present
Photosynthetic pigments	In internal membranes; chloroplasts absent	In chloroplasts
Respiratory system	Part of cytoplasmic membrane	Mitochondria

[a] Adapted from Ingraham and Ingraham (1995), Madigan et al. (2000), and Stanier et al. (1986).
[b] For additional size information see Table 2–23.

Figure 2-26

Typical structure of microorganism cells: (a) prokaryotic and (b) eukaryotic.

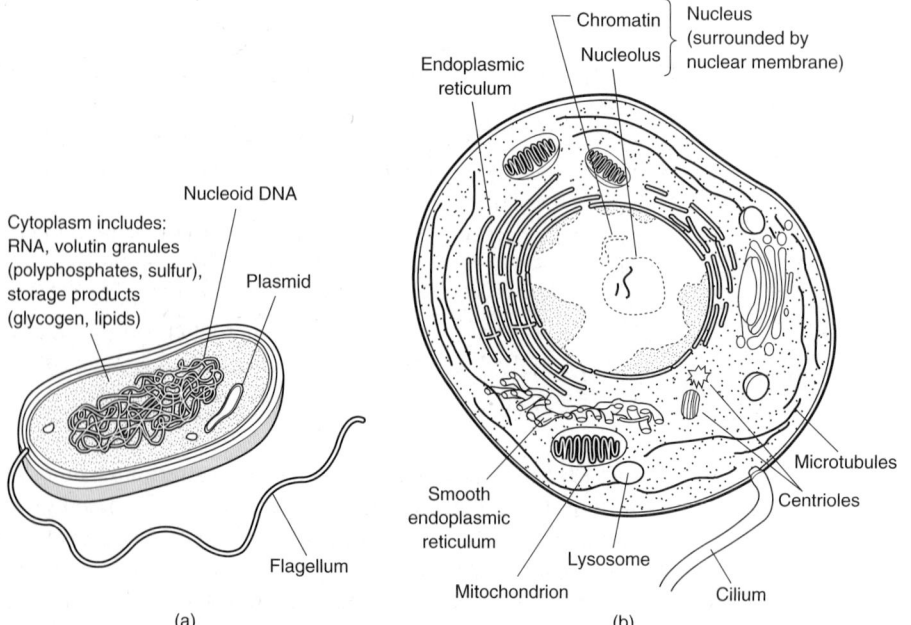

Table 2–22
Typical descriptions of the microorganisms found in natural waters, wastewater, and wastewater-treatment processes

Organism	Description
Bacteria	Bacteria are single-cell prokaryotic organisms. The interior of the cell contains a colloidal suspension of proteins, carbohydrates, and other complex organic compounds, called the cytoplasm. The cytoplasmic area contains ribonucleic acid (RNA), whose major role is in the synthesis of proteins. Also within the cytoplasm is deoxyribonucleic acid (DNA). DNA contains all the information necessary for the reproduction of all the cell components and may be considered to be the blueprint of the cell. Their usual mode of reproduction is by binary fission, although some species reproduce sexually or by budding
Archaea	Similar to bacteria in size and basic cell components. Their cell wall, cell material, and RNA composition are different. Important in anaerobic processes and also found under extreme conditions of temperature and chemical composition
Fungi/yeast	Fungi are multicellular, nonphotosynthetic, heterotrophic eukaryotes. Most fungi are either strict or facultative aerobes which reproduce sexually or asexually, by fission, budding, or spore formation. Molds, or "true fungi," produce microscopic units (hyphae), which collectively form a filamentous mass called the mycelium. Yeasts are fungi that cannot form a mycelium and are therefore unicellular. Fungi have the ability to grow under low-moisture, low-nitrogen conditions and can tolerate an environment with a relatively low pH. The ability of the fungi to survive under low-pH and nitrogen-limiting conditions, coupled with their ability to degrade cellulose, makes them very important in the composting of sludge
Protozoa	Protozoa are motile, microscopic eukaryotes that are usually single cells. The majority of protozoa are aerobic heterotrophs, some are aerotolerant anaerobes, and a few are anaerobic. Protozoa are generally an order of magnitude larger than bacteria and often consume bacteria as an energy source. In effect, the protozoa act as polishers of the effluents from biological waste-treatment processes by consuming bacteria and particulate organic matter
Rotifers	Rotifers are aerobic heterotrophic animal eukaryotes. The name is derived from the fact that they have two sets of rotating cilia on their head which are used for motility and capturing food. Rotifers are very effective in consuming dispersed and flocculated bacteria and small particles of organic matter. Their presence in an effluent indicates a highly efficient aerobic biological purification process
Algae	Algae are unicellular or multicellular, autotrophic, photosynthetic eukaryotes. They are of importance in biological treatment processes. In wastewater-treatment lagoons, the ability of algae to produce oxygen by photosynthesis is vital to the ecology of the water environment. The blue-green alga cyanobacter is a prokaryotic organism
Viruses	Viruses are composed of a nucleic acid core (either DNA or RNA) surrounded by an outer shell of protein called a capsid. Viruses are obligate intracellular parasites that multiply only within a host cell, where they redirect the cell's biochemical system to reproduce themselves. Viruses can also exist in an extracellular state in which the virus particle (known as a viron) is metabolically inert. Bacteriophages are viruses that infect bacteria as the host; they have not been implicated in human infections

An important feature of microorganisms is their ability to form resistant forms. For example, selected species of bacteria can form endospores (formed within the cell), the structure of which is extremely complex. The endospore which contains all of the information necessary for reproduction is coated with several layers of proteins. Endospores are extremely resistant to heat and disinfecting chemicals. It has been speculated that endospores may remain dormant for decades, and perhaps even centuries. A spore can

Table 2-23

Typical data on the shape, size, and resistant forms of classes of microorganisms and selected species found in wastewater[a]

Microorganism	Shape	Size, μm[b]	Envionmentally resistant form
Bacteria:			
Bacilli	Rod	$0.3–1.5\ D \times 1–10\ L$	Endospores or dormant cells
Bacillus (E. coli)	Rod	$0.6–1.2\ D \times 2–3\ L$	Endospores or dormant cells
Cocci	Spherical	0.5–4	Endospores or dormant cells
Spirilla	Spiral	$0.6–2\ D \times 20–50\ L$	Endospores or dormant cells
Vibrio	Rod, curved	$0.4–2\ D \times 1–10\ L$	Endospores or dormant cells
Protozoa:			
Cryptosporidium[c]			
Oocysts	Spherical	3–6	Oocysts
Sporozoite	Teardrop	$1–3\ W \times 6–8\ L$	
Entamoeba histolytica			
Cysts	Spherical	10–15	Cysts
Trophozoite	Semispherical	10–20	
Giardia lamblia[d]			
Cysts	Ovid	$6–8\ W \times 8–14\ L$	Cysts
Trophozoite	Pear or kite	$6–8\ W \times 12–16\ L$	
Helminths:			
Ancylostoma duodenale (hookworm) eggs	Elliptical or egg	$36–40\ W \times 55–70\ L$	Filariform larva
Ascaris lumbricoides (roundworm) eggs	Lemon or egg	$35–50\ W \times 45–70\ L$	Embryonated egg
Trichuris trichiura (whipworm) eggs	Elliptical or egg	$20–24\ W \times 50–55\ L$	Embryonated egg
Viruses:			
MS2	Spherical	0.022–0.026	Virion
Enterovirus	Spherical	0.020–0.030	Virion
Norwalk	Spherical	0.020–0.035	Virion
Polio	Spherical	0.025–0.030	Virion
Rotavirus	Spherical	0.070–0.080	Virion

[a] From Crites and Tchobanoglous (1998).
[b] D = diameter, L = length, and W = width.
[c] Member of the phylum Apicomplexa.
[d] Member of the phylum Sarcomastigophora, order Diplomonadida.

become viable in a suitable environment in a three-step process: activation, germination, and outgrowth (Madigan et al., 2000). The resistant forms in protozoans are known as cysts or oocysts. Resistant forms in helminths are eggs and oocysts.

The term *parasite* is used to describe an organism that lives at the expense of another. Parasites that live on the surface of a host organism are *ectoparasites*. Parasites that live internally within the host are known as *endoparasites* (Roberts and Janovy, 1996).

Pathogenic Organisms

Pathogenic organisms found in wastewater may be excreted by human beings and animals who are infected with disease or who are carriers of a particular infectious disease. The pathogenic organisms found in wastewater can be classified into four broad categories, bacteria, protozoa, helminths, and viruses. The principal pathogenic organisms found in untreated wastewater are reported in Table 2–24, along with the diseases and disease symptoms associated with each pathogen. Bacterial pathogenic organisms of human origin typically cause diseases of the gastrointestinal tract, such as typhoid and paratyphoid fever, dysentery, diarrhea, and cholera. Because these organisms are highly infectious, they are responsible for many thousands of deaths each year in areas with poor sanitation, especially in the tropics. It has been estimated that up to 4.5 billion people are or have been infected with some parasite (Madigan et al., 2000). Typical data on the quantity of selected pathogenic organisms found in wastewater and the corresponding concentration needed for an infectious dose are reported in Table 2–25.

Bacteria.   Many types of harmless bacteria colonize the human intestinal tract and are routinely shed in the feces. Because pathogenic bacteria are present in the feces of infected individuals, domestic wastewater contains a wide variety and concentration range of nonpathogenic and pathogenic bacteria. One of the most common bacterial pathogens found in domestic wastewater is the genus *Salmonella*. The *Salmonella* group contains a wide variety of species that can cause disease in humans and animals. Typhoid fever, caused by *Salmonella typhi*, is the most severe and serious. The most common disease associated with *Salmonella* is food poisoning identified as salmonellosis. *Shigella*, a less common genus of bacteria, is responsible for an intestinal disease known as bacillary dysentery or shigellosis. Waterborne outbreaks of shigellosis have been reported from recreational swimming areas and where wastewater has contaminated wells used for drinking water (Crook, 1998; Maier et al., 2000).

Other bacteria isolated from raw wastewater include *Vibrio, Mycobacterium, Clostridium, Leptospira,* and *Yersinia* species. *Vibrio cholerae* is the disease agent for cholera, which is not common in the United States but is still prevalent in other parts of the world. Humans are the only known hosts, and the most frequent mode of transmission is through water. *Mycobacterium tuberculosis* has been found in municipal wastewater, and outbreaks have been reported among persons swimming in water contaminated with wastewater (Crook, 1998; Maier et al., 2000).

Waterborne gastroenteritis of unknown cause is frequently reported, with the suspected agent being bacterial. One potential source of this disease is certain gram-negative bacteria normally considered to be nonpathogenic. These include the enteropathogenic *Escherichia coli* and certain strains of *Pseudomonas*, which may affect the newborn and have been implicated in gastrointestinal disease outbreaks. *Campylobacter jejuni* has

Table 2–24

Infectious agents potentially present in untreated domestic wastewater[a]

Organism	Disease	Remarks/symptoms
Bacteria:		
Campylobacter jejuni	Gastroenteritis	Diarrhea
Escherichia coli (enteropathogenic)	Gastroenteritis	Diarrhea
Legionella pneumophila	Legionnaires' disease	Malaise, myalgia, fever, headache, respiratory illness
Leptospira (spp.)	Leptospirosis	Jaundice, fever (Weil's disease)
Salmonella ($\approx$2100 serotypes)	Salmonellosis	Food poisoning
Salmonella typhi	Typhoid fever	High fever, diarrhea, ulceration of small intestine
Shigella (4 spp.)	Shigellosis	Bacillary dysentery
Vibrio cholerae	Cholera	Extremely heavy diarrhea, dehydration
Yersinia enterocolitica	Yersiniosis	Diarrhea
Protozoa:		
Balantidium coli	Balantidiasis	Diarrhea, dysentery
Cryptosporidium parvum	Cryptosporidiosis	Diarrhea
Cyclospora cayetanensis	Cyclosporasis	Severe diarrhea, stomach cramps, nausea, and vomiting lasting for extended periods
Entamoeba histolytica	Amebiasis (amoebic dysentery)	Prolonged diarrhea with bleeding, abscesses of the liver and small intestine
Giardia lamblia	Giardiasis	Mild to severe diarrhea, nausea, indigestion
Helminths:[b]		
Ascaris lumbricoides	Ascariasis	Roundworm infestation
Enterobius vermicularis	Enterobiasis	Pinworm
Fasciola hepatica	Fascioliasis	Sheep liver fluke
Hymenolepis nana	Hymenolepiasis	Dwarf tapeworm
Taenia saginata	Taeniasis	Beef tapeworm
T. solium	Taeniasis	Pork tapeworm
Trichuris trichiura	Trichuriasis	Whipworm
Viruses:		
Adenovirus (31 types)	Respiratory disease	
Enteroviruses (more than 100 types, e.g., polio, echo, and coxsackie viruses)	Gastroenteritis, heart anomalies, meningitis	
Hepatitis A virus	Infectious hepatitis	Jaundice, fever
Norwalk agent	Gastroenteritis	Vomiting
Parvovirus (2 types)	Gastroenteritis	
Rotavirus	Gastroenteritis	

[a] Adapted from Feachem et al. (1983), Madigan et al. (2000), and Crook (1998).

[b] The helminths listed are those with a worldwide distribution.

Table 2-25
Microorganism concentrations found in untreated wastewater and the corresponding infectious dose[a]

Organism	Concentration in raw wastewater,[b] MPN/100 mL	Infectious dose, number of organisms[c]
Bacteria:		
Bacterioides	10^7–10^{10}	
Coliform, total	10^7–10^9	
Coliform, fecal[d]	10^6–10^8	10^6–10^{10}
Clostridium perfringens	10^3–10^5	1–10^{10}
Enterococci	10^4–10^5	
Fecal streptococci	10^4–10^7	
Pseudomonas aeruginosa	10^3–10^6	
Shigella	10^0–10^3	10–20
Salmonella	10^2–10^4	10^1–10^8
Protozoa:		
Cryptosporidium parvum oocysts	10^1–10^3	1–10
Entamoeba histolytica cysts	10^{-1}–10^1	10–20
Giardia lamblia cysts	10^3–10^4	<20
Helminth:		
Ova	10^1–10^3	
Ascaris lumbricoides	10^{-2}–10^0	1–10
Viruses:		
Enteric virus	10^3–10^4	1–10
Coliphage	10^3–10^4	

[a] Adapted in part from Crook (1998) and Feachem et al. (1983).
[b] Value will vary with portion of population shedding at any given time.
[c] Infectious dose will vary with serotype or strain of organism, and the individual's general health.
[d] *Escherichia coli* (enteropathogenic).

been identified as the cause of a form of bacterial diarrhea in humans. While it has been well established that this organism causes disease in animals, it has also been implicated as the etiologic agent in human waterborne disease outbreaks (Crook, 1998).

Protozoa. Of the disease causing organisms reported in Table 2–24, the protozoans *Cryptosporidium parvum, Cyclospora,* and *Giardia lamblia* (see Fig. 2–27) are of great concern because of their significant impact on individuals with compromised immune systems, including very young children, the elderly, persons undergoing treatment for cancer, and individuals with acquired immune deficiency syndrome (AIDS). The life cycle of *Cryptosporidium parvum* and *Giardia lamblia* is illustrated on Fig. 2–28. As shown, infection is caused by the ingestion of water contaminated with oocysts and cysts. It is also important to note that numerous nonhuman sources of

Figure 2–27

Defintion sketch for (a) *Giardia lamblia* cyst and trophozoite and (b) *Cryptosporidium parvum* oocyst and sporozoite.

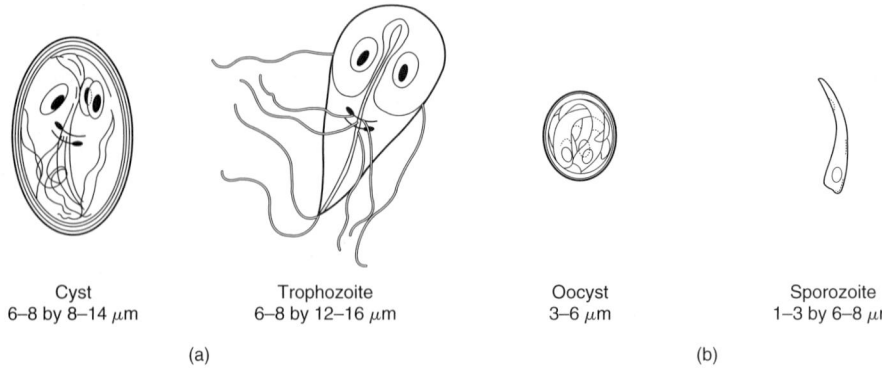

Cyst	Trophozoite	Oocyst	Sporozoite
6–8 by 8–14 μm	6–8 by 12–16 μm	3–6 μm	1–3 by 6–8 μm

(a)   (b)

Figure 2–28

Life cycle of *Cryptosporidium parvum* and *Giardia lamblia*.

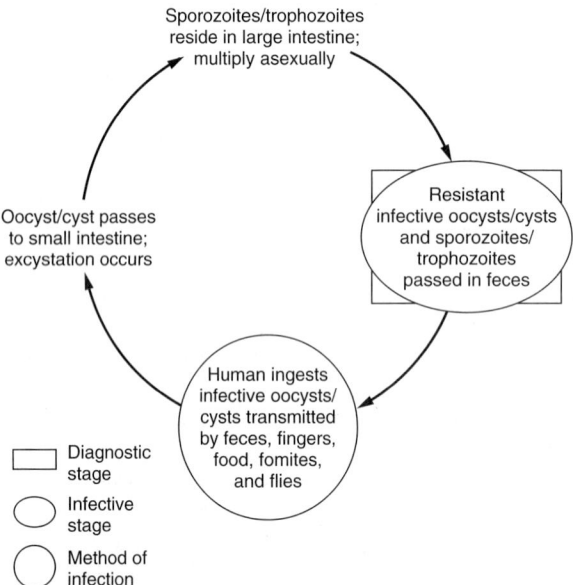

Cryptosporidium parvum and *Giardia lamblia* are present in the environment. Further, not all of the oocysts and cysts that are present are viable in terms of their ability to cause disease. To determine the potential risk from these microorganisms, infectivity studies must be conducted.

Pathogenic protozoan disease outbreaks have been significant, highlighted by the 1993 outbreak of cryptosporidiosis in Milwaukee in which 400,000 persons became ill and outbreaks of cyclosporiasis in ten states. As noted in Table 2–24, these protozoan organisms may cause symptoms which can include severe diarrhea, stomach cramps, nausea, and vomiting lasting for extended periods. Despite intensive trials in humans and animals, no effective treatment has been found for cryptosporidiosis (Roberts and Janovy, 1996). The oocysts of *Cryptosporidium parvum* and the cysts of *Giardia lamblia* are the most resistant forms (see Table 2–23). These organisms are of particular concern because they are found in almost all wastewaters, and because conventional disinfection techniques using chlorine have not proved to be effective in their inactivation or destruction. However, in recent studies, it has been found that UV disinfection is extremely effective in the inactivation of oocysts of *Cryptosporidium parvum* and the cysts of *Giardia lamblia*.

Helminths. The term *helminths* is used to describe worms collectively. In the United States, as a result of the improvements in the provision of sanitation and wastewater treatment facilities and in food-handling practices, the prevalence of helminth infections has decreased dramatically over the last century. Nevertheless, owing to increased levels of immigration to the United States of persons from countries where worms are endemic, the transmission of helminths by wastewater and particularly by biosolids remains a concern. In fact, the eggs of worms are found in wastewater throughout the United States. In particular, small nonparasitic nematodes are universally present, even in finished drinking water at the tap (Cooper, 2001). Worldwide, worms are one of the principal causative agents of human disease. It is estimated that the number of human infections caused by helminths collectively is on the order of 4.5 billion (Roberts and Janovy, 1996).

Most of the helminths fall into three major phyla: Nematoda (roundworms), Platyhelminthes (flatworms), and Annelida (segmented worms). Most human infections are associated with nematodes and flatworms, while the segmented worms are primarily ectoparasitic, such as leaches. The phylum Nematoda collectively represents one of the most abundant animal groups on earth, most of which are harmless to humans. Included among its members are the large roundworm *Ascaris lumbricoides,* the whipworm *Trichuris trichiura,* the hookworms *Necator americanus* and *Ancylostoma duodenale,* and the threadworm *Strongyloides stercoralis. Ascaris lumbricoides* is considered to be the most prevalent parasitic infection worldwide with over one and a half billion persons infected (Crompton, 1999; Maier et al., 2000; Roberts and Janovy, 1996). It has been estimated that there are on the order of 4 million cases in the United States (Khuroo, 1996).

The phylum platyhelminthes includes the tapeworms *Taenia saginata* (beef tapeworm) and *Taenia solium* (pork tapeworm) and *Schistosoma* species. *Taenia saginata,* transmitted primarily by infected beef products, is the most common tapeworm found in humans. The trematodes *Schistosoma mansoni, S. haematobium,* and *S. japonicum,* also known as blood flukes, are medically important members of the trematoda class. More than 200 million infections are ascribed to these worms worldwide. It is estimated that more than 400,000 infected individuals, most of whom were infected outside of the United States, live in the United States (West and Olds, 1992).

The human infective stage of helminths varies; in some species it is either the adult organism or larvae, while in other species it is the eggs, but it is primarily the eggs that are present in wastewater. Helminth eggs, which range in size from about 10 μm to more than 100 μm, can be removed by many commonly used wastewater-treatment processes such as sedimentation, filtration, and stabilization ponds. However, some helminth eggs are extremely resistant to environmental stresses and may survive usual wastewater and sludge disinfection procedures. Chlorine disinfection and mesophilic anaerobic digestion, for example, are not effective at inactivating many helminth eggs. In a recent study, it has been found that the eggs of *Ascaris* can survive for up to 10 years in the sediments of oxidation ponds (Nelson, 2001). The long survival times of

Ascaris and other worm eggs is of particular importance in the management of biosolids.

Viruses. More than 100 different types of enteric viruses capable of producing infection or disease are excreted by humans. Enteric viruses multiply in the intestinal tract and are released in the fecal matter of infected persons. From the standpoint of health, the most important human enteric viruses are the enteroviruses (polio, echo, and coxsackie), Norwalk viruses, rotaviruses, reoviruses, caliciviruses, adenoviruses, and hepatitis A virus. Of the viruses that cause diarrheal disease, only the Norwalk virus and rotavirus have been shown to be major waterborne pathogens. The reoviruses and adenoviruses, known to cause respiratory illness, gastroenteritis, and eye infections, have been isolated from wastewater. There is no evidence that the human immunodeficiency virus (HIV), the pathogen that causes the acquired immunodeficiency syndrome (AIDS), can be transmitted via the waterborne route (Crook, 1998; Madigan et al., 2000; Maier et al., 2000; Rose and Gerba, 1991). The biology of viruses is delineated in Voyles (1993).

Survival of Pathogenic Organisms. Of great concern in the management of disease-causing organisms is the survival of these organisms in the environment. Typical data on the survival of microorganisms in the environment are presented in Table 2–26. Although the data given in Table 2–26 can be used as a rough guide, numerous

Table 2–26
Typical pathogen survival times at 20–30°C in various environments[a]

Pathogen	Survival time, days		
	Fresh water and wastewater	Crops	Soil
Bacteria:			
Fecal coliforms[b]	<60 but usually <30	<30 but usually <15	<120 but usually <50
Salmonella spp.[b]	<60 but usually <30	<30 but usually <15	<120 but usually <50
Shigella[b]	<30 but usually <10	<10 but usually <5	<120 but usually <50
Vibrio cholerae[c]	<30 but usually <10	<5 but usually <2	<120 but usually <50
Protozoa:			
E. histolytica cysts	<30 but usually <15	<10 but usually <2	<20 but usually <10
Helminths:			
A. lumbricoides eggs	Many months	<60 but usually <30	<Many months
Viruses:[b]			
Enteroviruses[d]	<120 but usually <50	<60 but usually <15	<100 but usually <20

[a] Adapted from Feachem et al. (1983).
[b] In seawater, viral survival is less, and bacterial survival is very much less than in fresh water.
[c] *V. cholerae* survival in aqueous environments is a subject of current uncertainty.
[d] Includes polio, echo, and coxsackie viruses.

U.S. EPA (1984) *Environmental Regulations and Technology, Use and Disposal of Municipal Wastewater Sludge,* EPA/625/10-84-003, U.S. Environmental Protection Agency.

U.S. EPA (1985) *Seminar Publication Composting of Municipal Wastewater Sludges,* U.S. Environmental Protection Agency, EPA/625/4-85/014.

U.S. EPA (1987*a*) *Design Information Report—Design, Operational, and Cost Considerations for Vacuum Assisted Sludge Dewatering Bed Systems,* U.S. Environmental Protection Agency.

U.S. EPA (1987*b*) *Design Information Report—Sidestreams in Wastewater Treatment Plants,* U.S. Environmental Protection Agency, *Journal Water Pollution Control Federation,* 59:54, Water Environment Federation, Alexandria, VA.

U.S. EPA (1987*c*) *Design Manual, Dewatering Municipal Wastewater Sludges,* U.S. Environmental Protection Agency, EPA/625/1-87/014.

U.S. EPA (1987*d*) *Innovations in Sludge Drying Beds, A Practical Technology,* U.S. Environmental Protection Agency.

U.S. EPA (1989) *Summary Report, In-Vessel Composting of Municipal Wastewater Sludge,* EPA/625/8-89/016.

U.S. EPA (1992) *Control of Pathogens and Vector Attraction in Sewage Sludge,* U.S. Environmental Protection Agency, EPA/625/R-92/013, Office of Research and Development, Washington, D.C.

U.S. EPA (1995) *Process Design Manual—Land Application of Sewage Sludge and Domestic Septage,* EPA/625/R-95/001, Center for Environmental Research Information, U.S. Environmental Protection Agency.

Vesilind, P. A., and J. Martel (1990) "Freezing of Water and Wastewater Sludges," American Society of Civil Engineers, *Journal of the Environmental Engineering Division,* May issue.

WEF (1980) *Sludge Thickening,* Manual of Practice no. FD-1, Water Environment Federation, Alexandria, VA.

WEF (1983) *Sludge Dewatering,* Manual of Practice no. 20, Water Environment Federation, Alexandria, VA.

WEF (1987*a*) *Anaerobic Digestion,* Manual of Practice no. 16, 2d ed., Water Environment Federation, Alexandria, VA.

WEF (1987*b*) "Anaerobic Digester Mixing Systems," *Journal Water Pollution Control Federation,* vol. 59, p. 162, Water Environment Federation, Alexandria, VA.

WEF (1988) *Sludge Conditioning,* Manual of Practice no. FD-14 Water Environment Federation, Alexandria, VA.

WEF (1992) *Sludge Incineration,* Manual of Practice no. OM-11 Water Environment Federation, Alexandria, VA.

WEF (1995*b*) *Biosolids Composting,* Water Environment Federation, Alexandria, VA.

WEF (1995*a*) *Wastewater Residuals Stabilization,* Manual of Practice no. FD-9, Water Environment Federation, Alexandria, VA.

WEF (1996) *Operation of Wastewater Treatment Plants,* 5th ed., Manual of Practice No. 11, 5th ed., vol. 3, Chaps. 27–33, Water Environment Federation, Alexandria, VA.

WEF (1998) *Design of Wastewater Treatment Plants,* 4th ed., Manual of Practice no. 8, vol. 3, Chaps. 17–24, Water Environment Federation, Alexandria, VA.

Wegner, G. (1992) "The Benefits of Biosolids from a Farmer's Perspective." *Proceedings, The Future Direction of Municipal Sludge (Biosolids) Management,* WEF Specialty conference, pp. 39–44, Portland, OR.

Wilson, T. E., and N. A. Dichtl (2000) "Two-Phase Anaerobic Digestion: An Update on the AG Process," *Proceedings of WEFTEC 2000,* Water Environment Federation, Alexandria, VA.

exceptions have been reported in the literature. Additional data on the effect of temperature on the survival of microorganisms are given in Chap 13.

Use of Indicator Organisms

Because the numbers of pathogenic organisms present in wastes and polluted waters are usually few and difficult to isolate and identify, microorganisms, which are more numerous and more easily tested for, are commonly used as surrogate (i.e., an indicator) organisms for the target pathogen(s). The general features of an ideal indicator organism and the use of bacterial and other indicators are considered briefly in the following discussion.

Characteristics of an Ideal Indicator Organism. An ideal organism should have the following characteristics (adapted from Cooper, 2001; Maier et al., 2000):

1. The indicator organism must be present when fecal contamination is present.
2. The numbers of indicator organisms present should be equal to or greater than those of the target pathogenic organism (e.g., pathogenic viruses).
3. The indicator organism must exhibit the same greater survival characteristics in the environment as the target pathogen organism for which it is a surrogate.
4. The indicator organism must not reproduce outside of the host organism (i.e., the culturing procedure itself should not produce a serious health threat to laboratory workers).
5. The isolation and quantification of the indicator organism must be faster than that of the target pathogen (i.e, the procedure must be cheaper and easier for cultivating the indicator organisms than for the target pathogen).
6. The organism should be a member of the intestinal microflora of warmblooded animals.

Some authors have stated the first characteristic as "The indicator organism must be present when the target pathogen is present." Unfortunately, the target pathogen(s) may not be present during the entire year, because the shedding of pathogenic organisms is not uniform throughout the year. Thus, it is important that the indicator organism be present when fecal contamination is present, if public health is to be protected. To date, no ideal indicator organism has been found.

Bacterial Indicators. The intestinal tract of humans contains a large population of rod-shaped bacteria known collectively as coliform bacteria. Each person discharges from 100 to 400 billion coliform bacteria per day, in addition to other kinds of bacteria. Thus, the presence of coliform bacteria in environmental samples has, over the years, been taken as an indication that pathogenic organisms associated with feces (e.g., viruses) may also be present. The absence of coliform bacteria is taken as an indication that the water is free from disease-producing organisms. Microorganisms that have been proposed for use as indicators of fecal contamination are summarized in Table 2–27. Indicator organisms that have been used to establish performance criteria for various water uses are reported in Table 2–28.

The coliform bacteria include a number of genera and species of bacteria that have common biochemical and morphological attributes. Typically, these organisms are gram-negative (a staining procedure, non-spore-forming rod-shaped organisms (see Fig. 2–29) that ferment lactose in 24 to 48 h at $35 \pm 0.5°C$ (BioVir, 2001). The term

Table 2–27
Specific organisms that have been used or proposed for use as indicators of fecal contamination

Indicator organism	Characteristics
Total coliform bacteria	Species of gram-negative rods that may ferment lactose with gas production (or produce a distinctive colony within 24 ± 2 h to 48 ± 3 h incubation on a suitable medium) at $35 \pm 0.5°C$. There are strains that do not conform to the definition. The total coliform group includes four genera in the Enterobacteriaceae family. These are *Escherichia, Citrobacter, Enterobacter,* and *Klebsiella.* Of the group, the *Escherichia* genus (*E. coli* species) appears to be most representative of fecal contamination
Fecal coliform bacteria	A fecal coliform bacteria group was established based on the ability to produce gas (or colonies) at an elevated incubation temperature ($44.5 \pm 0.2°C$ for 24 ± 2 h)
Klebsiella	The total coliform population includes the genera *Klebsiella.* The thermotolerant *Klebsiella* are also included in the fecal coliform group. This group is cultured at $35 \pm 0.5°C$ for 24 ± 2 h
E. coli	The *E. coli* is one of the coliform bacteria population and is more representative of fecal sources than other coliform genera
Bacteroides	Bacteroides, an anaerobic organism, has been proposed as a human specific indicator
Fecal streptococci	This group had been used in conjunction with fecal coliforms to determine the source of recent fecal contamination (man or farm animals). Several strains appear to be ubiquitous and cannot be distinguished from the true fecal streptococci under usual analytical procedures, which detract from their use as an indicator organism
Enterococci	Two strains of fecal streptococci, *S. faecalis* and *S. faecium,* are the most human-specific members of the fecal streptococcus group. By eliminating the other strains through the analytical procedures, the two strains known as enterococci can be isolated and enumerated. The enterococci are generally found in lower numbers than other indicator organisms; however, they exhibit better survival in seawater
Clostridium perfringens	This organism is a spore-forming anaerobic persistent bacteria, and the characteristics make it a desirable indicator where disinfection is employed, where pollution may have occurred in the past, or where the interval before analysis is protracted
P. aeruginosa and *A. hydrophila*	These organisms may be present in domestic wastewater in large numbers. Both can be considered aquatic organisms and can be recovered in water in the absence of immediate sources of fecal pollution

gram-negative refers to a staining procedure used to differentiate groups of organisms. The organism *Escherichia coli* (*E. coli*), found in the feces of warmblooded animals, has historically been the target organism tested for with the *total coliform test.* Unfortunately, it was found early on that the coliform test was not specific for *E. coli* and that a variety of other coliform organisms were included in the test results. For example,

Table 2–28
Indicator organisms used in establishing performance criteria for various water uses

Water use	Indicator organism
Drinking water	Total coliform
Freshwater recreation	Fecal coliform
	E. coli
	Enterococci
Saltwater recreation	Fecal coliform
	Total coliform
	Enterococci
Shellfish-growing areas	Total coliform
	Fecal coliform
Agricultural irrigation (for reclaimed water)	Total coliform
Wastewater effluent disinfection	Total coliform
	Fecal coliform
	MS2 coliphage

Figure 2–29

Micrograph of a pure culture of E. coli.

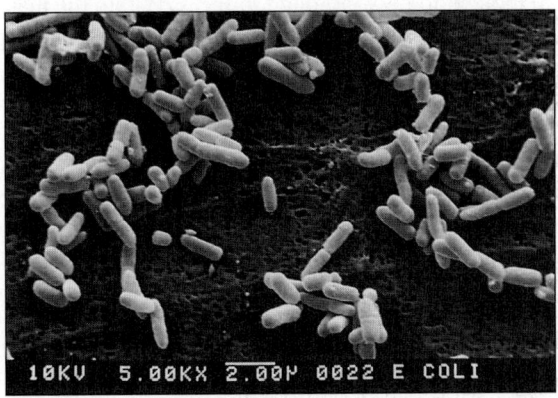

10KV 5.00KX 2.00µ 0022 E COLI

some of the coliform organisms of the genus *Escherichia* can grow in soil. Thus, the presence of coliforms does not always mean contamination with human wastes. In recent years, tests have been developed that distinguish among total coliforms, fecal coliforms, and *E. coli*, and all three are being reported in the literature.

One of the most recent and important developments in the evolution of the coliform test is the ability to identify and quantify *E. coli* through the use of elevated temperatures and a specific growth medium (MUG). When *E. coli* are present, they are able to cleave the fluorogenic substrate MUG (4-methylumbelliferyl-β-D–glucuronide) from the growth medium, because they possess a specific enzyme (β-glucuronidase). The presence of a bright blue fluorescence is taken as a positive response for *E. coli*. In turn, the occurrence of *E. coli* is taken as a specific indicator of fecal contamination and the possible presence of enteric pathogens (Standard Methods, 1998).

Other Indicator Organisms.   While total and fecal coliform organisms and *E. coli* may be present, it has not been demonstrated that they are indicators of the presence of enteric viruses and protozoa. Further, concerns for newly emerging pathogenic organisms which may arise from nonhuman reservoirs (e.g., pathogenic *E. coli, Cryptosporidium parvum,* and *Giardia lamblia*) have led to the questioning of the use of indicators that arise primarily from fecal inputs. In a recently completed study, it was concluded that coliform bacteria are adequate indicators for the potential presence of pathogenic bacteria and viruses, but are inadequate as an indicator of the presence of waterborne protozoa. It was also found that waterborne disease outbreaks have occurred in water systems that have not violated their microbial water quality standards (Craun et al., 1997).

Given the limitations in using coliform organisms as indicators of potential contamination by wastewater, attention has now focused on the use of bacteriophages as an indicator organism and more specifically as indicators of enteric viruses. Bacteriophages are viruses that can infect prokaryotic cells. There are six major families of bacteriophages, five of which are DNA-based and one of which is RNA-based. Of the five DNA-based bacteriophages, three are double-stranded and two are single-stranded. Bacteriophages that infect *E. coli* are known as coliphages. Coliphages that attach directly to the cell wall are known as *somatic.* Coliphages that infect only male strains of *E. coli* (possess pili) are known as male-specific (F+) coliphages. Male-specific phages are thought to only be found in feces. Within the male-specific family there are four serotypes. Groups II and III are primarily of human origin whereas groups I and IV are of animal origin, with the exception of pigs, which may harbor groups II and III. Analytically, somatic coliphages are determined using *E. coli* as the host organism whereas F+ coliphages are determined using *E. coli* with pilli. Interest in using coliphages as indicators of enterovirses is based on the fact that the phages are approximately the same size as pathogenic viruses of interest (e.g., polio), are of fecal origin, and are always present in raw municipal wastewater. Coliphages have been used extensively in disinfection studies (see Sec. 12–8 in Chap. 12).

Enumeration and Identification of Bacteria

Individual bacteria are typically enumerated by one of four methods: (1) direct microscopic count, (2) pour and spread plate counts, (3) membrane filtration, and (4) multiple tube fermentation. Colonies of bacteria are often enumerated but not identified using the heterotrophic plate count (HPC) method. In addition, a number of staining and fluorescent methods have been developed for the identification of specific bacteria. These tests are considered in the following discussion.

Direct Counts.   Direct counts can be obtained by microscopically using a Petroff-Hauser counting chamber (see Fig. 2–30). Counting cells, as shown on Fig. 2–30, are designed so that each square in the counting chamber corresponds to a given volume (depth is known). Because it is difficult to differentiate between live and dead cells, the measured counts are total counts. Acridine orange (vital) stain is often used to differentiate between live and dead cells, but questions have been raised about the accuracy of this method of identification. Another technique for obtaining direct counts is with an electronic particle counter in which a sample containing bacteria is passed though an

Figure 2–30

Schematic of Petroff-
Hauser counting chamber
for bacteria. *(From Crites
and Tchobanoglous,
1998.)*

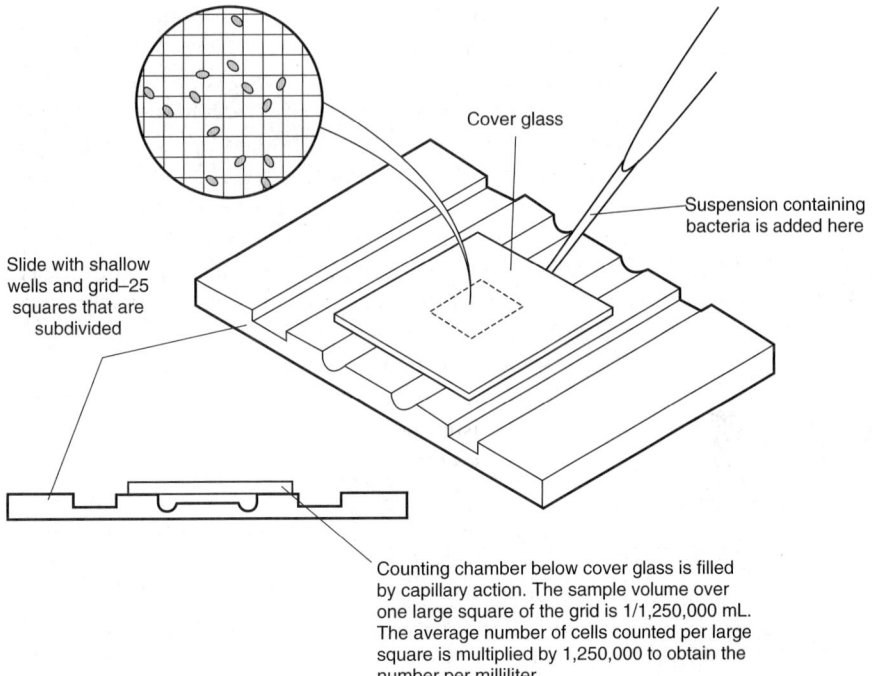

Cover glass

Suspension containing
bacteria is added here

Slide with shallow
wells and grid–25
squares that are
subdivided

Counting chamber below cover glass is filled
by capillary action. The sample volume over
one large square of the grid is 1/1,250,000 mL.
The average number of cells counted per large
square is multiplied by 1,250,000 to obtain the
number per milliliter.

orifice. As each bacteria passes through the orifice, the electrical conductivity of the fluid in the orifice decreases. The number of times the conductivity is reduced and the values to which it is reduced are correlated to the number of bacteria. Unfortunately the electronic particle counter cannot differentiate between bacteria (whether live or dead) and inert particles, both of which are counted as particles. Also, orifice clogging is a serious problem with this method.

Pour and Spread Plate Method.   The pour and spread plate count methods are used to culture, identify, and enumerate bacteria. In the pour plate method (see Fig. 2–31a), a sample of wastewater to be tested is diluted serially. A small amount of each dilution is then mixed with a warmed liquid (agar) culture medium, poured into a culture dish, allowed to solidify, and incubated under controlled conditions. The separate distinct bacterial colonies formed on the plates after incubation are counted, and the results reported as colony-forming units (cfu) per unit volume of sample (typically cfu/mL). In the past, it was assumed that each colony developed from a single bacterium, but use of the term cfu does not assume that one bacteria formed each colony. The total number of bacteria are determined using the appropriate dilutions. In the spread plate method (see Fig. 2–31b), a small amount of the diluted wastewater is placed and spread on the surface of a prepared culture dish containing a suitable solid medium.

Membrane-Filter Technique.   In the membrane-filter (MF) technique (see Fig. 2–32) a known volume of water sample is passed through a membrane filter that has a small pore size (typically 0.45 μm). Bacteria are retained on the filter because they are

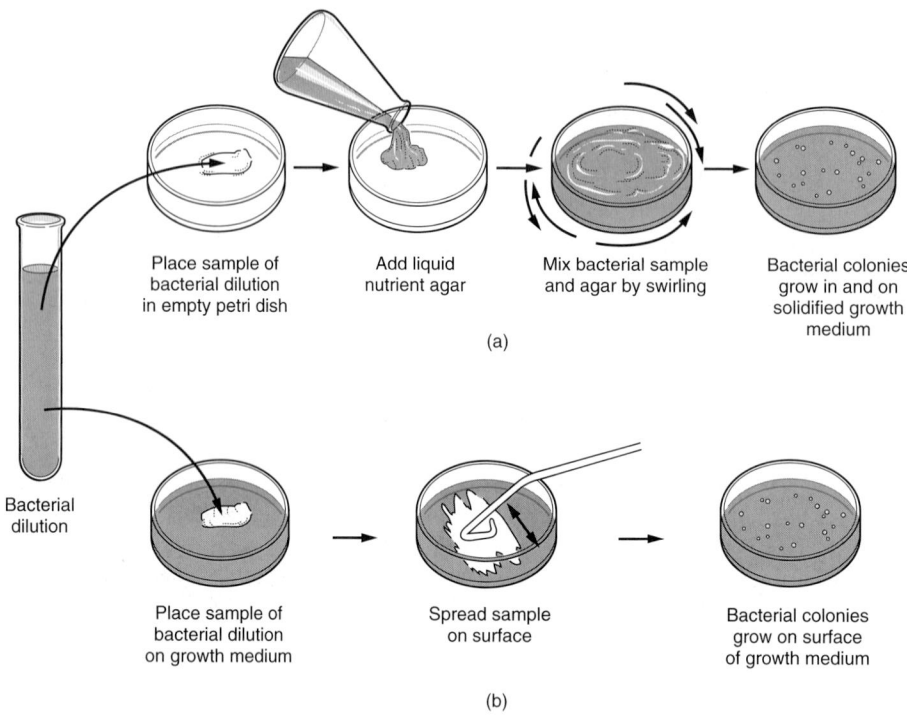

(a)

(b)

Figure 2–31

Schematic of plate culture methods used for the enumeration of bacteria: (a) pour plate and (b) spread plate. (*From Crites and Tchobanoglous, 1998.*)

Figure 2–32

Membrane filter apparatus used to test for bacteria in relatively clean waters. After centering the membrane filter on the filter support, the funnel top is attached and the water sample to be tested is poured into the funnel. To aid in the filtration process, a vacuum line is attached to the base of the filter apparatus. After the sample has been filtered, the membrane filter is placed in a petri dish containing a culture medium for bacterial analysis.

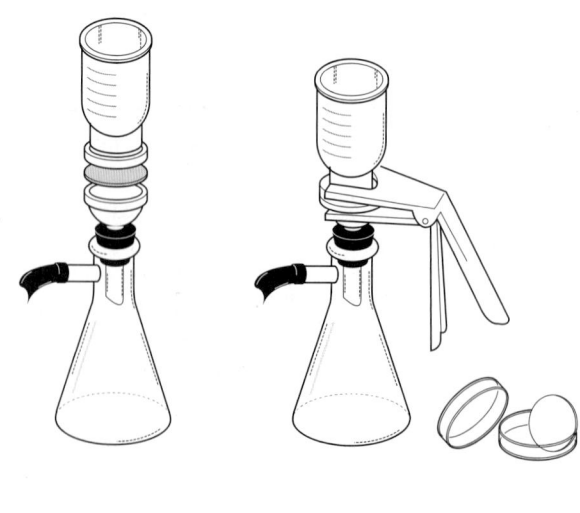

larger than the size of the pores of the membrane filter. The membrane filter containing the bacteria is then placed, right side up, in contact with an agar that contains the nutrients necessary for the growth of the specific target bacteria. After incubation, the colonies formed on the surface of the filter can be counted and the concentration in the original water sample determined. The membrane filter technique has the advantage of being faster than the MPN procedure and of giving a direct count of the number of organisms (e.g., coliform organisms). In the environmental field, the MF method is used for coliform and fecal streptococcus enumeration as opposed to direct and pour and spread plate methods. All of these methods are subject to limitations in interpretation (Standard Methods, 1998).

Multiple-Tube Fermentation. The multiple-tube fermentation technique is based on the principle of dilution to extinction as illustrated on Fig. 2–33. Concentrations of total coliform bacteria are most often reported as the *most probable number per 100 mL* (MPN/100 mL). The MPN is based on the application of the Poisson distribution for extreme values to the analysis of the number of positive and negative results obtained when testing multiple portions of equal volume and in portions constituting a geometric series. It is emphasized that the MPN is not the absolute concentration of organisms that are present, but only a statistical estimate of that concentration. The

Figure 2–33

Schematic illustration of the methods used to obtain bacterial counts: (a) multiple tube fermentation technique using a liquid medium and (b) use of a solid medium.

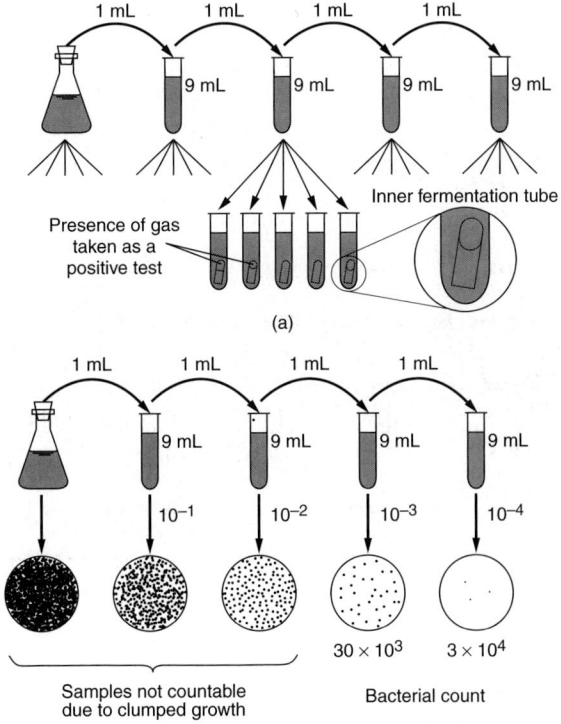

complete multiple-tube fermentation procedure for total coliform involves three test phases identified as the presumptive, confirmed, and completed test. A similar procedure is available for the fecal coliform group as well as for other bacterial groups (Standard Methods, 1998).

The MPN can be determined using the Poisson distribution directly, MPN tables derived from the Poisson distribution, or the Thomas equation.

The joint probability (based on the Poisson distribution) of obtaining a given result from a series of three dilutions is given by Eq. (2–74). It should be noted that Eq. (2–74) can be expanded to account for any number of serial dilutions.

$$y = \frac{1}{a}\left[(1 - e^{-n_1\lambda})^{p_1}(e^{-n_1\lambda})^{q_1}\right]\left[(1 - e^{-n_2\lambda})^{p_2}(e^{-n_2\lambda})^{q_2}\right]\left[(1 - e^{-n_3\lambda})^{p_3}(e^{-n_3\lambda})^{q_3}\right] \quad (2-74)$$

where y = probability of occurrence of a given result
a = constant for a given set of conditions
n_1, n_2, n_3 = sample size in each dilution, mL
λ = coliform density, number/mL
p_1, p_2, p_3 = number of positive tubes in each sample dilution
q_1, q_2, q_3 = number of negative tubes in each sample dilution

When the Poisson equation or MPN tables are not available, the Thomas equation (Thomas, 1942) can be used to estimate the MPN.

$$\text{MPN}/100 \text{ mL} = \frac{\text{number of positive tubes} \times 100}{\sqrt{\left(\begin{array}{c}\text{mL of sample in}\\\text{negative tubes}\end{array}\right) \times \left(\begin{array}{c}\text{mL of sample in}\\\text{all tubes}\end{array}\right)}} \quad (2-75)$$

In applying the Thomas equation to situations in which some of the dilutions have all five tubes positive, the count of positive tubes should begin with the highest dilution in which at least one negative result has occurred. The application of the Thomas equation is illustrated in Example 2–13.

EXAMPLE 2–13 Calculation of MPN Using Multiple-Tube Fermentation Test Results   The results of a coliform analysis using the multiple-tube fermentation test for the effluent from an intermittent sand filter (see Chap. 11) are as given below. Using these data, determine the coliform density (MPN/100 mL) using the Poisson equation, the Thomas equation, and the MPN tables given in Appendix E.

Size of portion, mL	Number positive	Number negative
1.0	4	1
0.1	3	2
0.01	2	3
0.001	0	5

Solution

1. Determine the MPN using the Poisson equation [Eq. (2–74)]. Substitute the appropriate values for n, p, and q and solve the Poisson equation by successive trials.

   $$n_1 = 1.0 \qquad p_1 = 4 \qquad q_1 = 1$$
   $$n_2 = 0.1 \qquad p_2 = 3 \qquad q_2 = 2$$
   $$n_3 = 0.01 \qquad p_3 = 2 \qquad q_3 = 3$$
   $$n_4 = 0.001 \qquad p_4 = 0 \qquad q_4 = 5$$

   a. Substitute the coefficient values in Eq. (2–74) and determine ya values for selected values of λ.

$$y = \frac{1}{a}\left[(1 - e^{-1.0\lambda})^4 (e^{-1.0\lambda})^1\right]\left[(1 - e^{-0.1\lambda})^3 (e^{-0.1\lambda})^2\right]\left[(1 - e^{-0.01\lambda})^2 (e^{-0.01\lambda})^3\right]\left[(1 - e^{-0.001\lambda})^0 (e^{-0.001\lambda})^5\right]$$

λ	ya
3.80	3.6754×10^{-7}
3.84	3.6773×10^{-7}
3.85	3.6774×10^{-7}
3.86	3.6773×10^{-7}
3.90	3.6755×10^{-7}

   b. The maximum value of ya occurs for a λ value of 3.85 organisms per milliliter. Thus the MPN/100 mL is

   $$\text{MPN/100 mL} = 100 \times 3.85 = 385$$

2. Determine the MPN using the Thomas equation [Eq. (2–75)].
   a. Number of positive tubes $(4 + 3 + 2) = 9$
   b. mL of sample in negative tubes $= [(1 \times 1.0) + (2 \times 0.1) + (3 \times 0.01) + (5 \times 0.001)] = 1.235$
   c. mL of sample in all tubes $= [(5 \times 1.0) + (5 \times 0.1) + (5 \times 0.01) + (5 \times 0.001)] = 5.555$

   $$\text{MPN/100 mL} = \frac{9 \times 100}{\sqrt{(1.235) \times (5.555)}} = 344/100 \text{ mL}$$

3. From Appendix G, eliminating the portion with no positive tubes, as outlined, the MPN/100 mL is 390.

Comment It should be noted that MPN tables were developed for use before the advent of the small hand-held scientific calculator as a means of computing the results from a multiple-tube fermentation test. With the use of the scientific calculator, the results from all of the serial dilutions can be considered.

Presence-Absence Test. The presence-absence (P-A) test for coliform organisms is a modification of the multiple-tube fermentation technique described above. The test is intended for use for samples collected from water distribution systems or water-treatment plants. Rather than using multiple dilutions, a single 100-mL sample is tested for the P-A of coliform organisms using lauryl sulfate tryptose lactose broth as used in the MPN test. Coliform organisms are present if a distinct yellow color forms, indicating that lactate fermentation has occurred in the sample. The test is based on the rationale that no organisms should be present in 100 mL. It has been used in wastewater for highly treated samples.

In addition to the modified MPN test, several commercial enzymatic assays have been developed that can be used to detect both total coliform bacteria as well as *E. coli*. In the enzymatic assays, wastewater samples are added to bottles or MPN tubes containing powdered ingredients comprised of salts and specific enzyme substrates that serve as the sole carbon source. Samples containing coliform organisms turn yellow, and samples containing *E. coli* will fluoresce when exposed to long-wave UV illumination. Other tests are also available for detecting *E. coli* (Standard Methods, 1998).

Heterotrophic Plate Count. The heterotrophic plate count (HPC) is a procedure for estimating the number of live heterotrophic bacteria in wastewater samples. The HPC is often used to evaluate the performance of treatment processes and regrowth in effluent distribution systems in reuse applications. The HPC can be determined using the (1) pour plate method, (2) spread plate method, or (3) membrane filter method as described above. In the HPC test, colonies of bacteria, which may be derived from pairs, chains, clusters, or single cells, are measured. The results are reported as colony-forming units per milliliter (CFU/mL). Test details may be found in Standard Methods, 1998.

Identification of Specific Bacteria and Protozoa. Over the years, a variety of techniques have been developed for the identification of specific bacteria including growth-dependent methods, the use of fluorescent antibodies, and the use of nucleic acid probes (BioVir, 1997; Madigan et al., 2000). A brief description of the latter two methods is given because of their use in the study of biological treatment and disinfection processes. The availability of the techniques described below has made it possible to study specific reactions and organisms, and will ultimately help to further our understanding of the microbial interactions which occur in natural systems.

In the fluorescence method, an antibody (a soluble protein, also known as immunoglobulin, produced by a single B cell clone), is tagged with a fluorescent dye such as rhodamine B or fluorescein. The covalent attachment of the dye to the antibody does not affect the specificity of the antibody. Once the tagged antibody has found the organism in question and becomes attached to the surface (see Fig. 2–34), the sample can be examined using fluorescence microscopy. Organisms to which the antibodies attach will glow when exposed to the fluorescent light of the microscope. This method of analysis is commonly used for bacteria, and is also the method of choice for the identification of *Cryptosporidium parvum* and *Giardia lamblia*.

A nucleic acid probe, as used in microbial studies, is a molecule having a strong interaction only with a genomic sequence unique to the targeted organisms in question, and possessing a means for detection once a probe-target interaction has been achieved (Keller and Manak, 1989). The genetic importance of DNA and RNA is discussed in

Figure 2–34

Schematic of the fluorescence method used for the enumeration of bacteria. *(From Crites and Tchobanoglous, 1998.)*

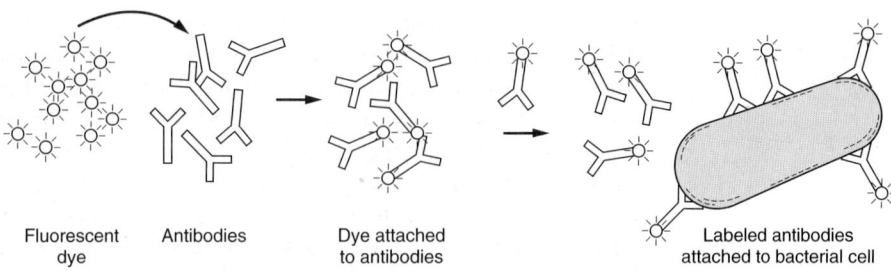

Fluorescent dye   Antibodies   Dye attached to antibodies   Labeled antibodies attached to bacterial cell

Figure 2–35

Schematic of the DNA probe method used for the enumeration of bacteria. *(From Crites and Tchobanoglous, 1998.)*

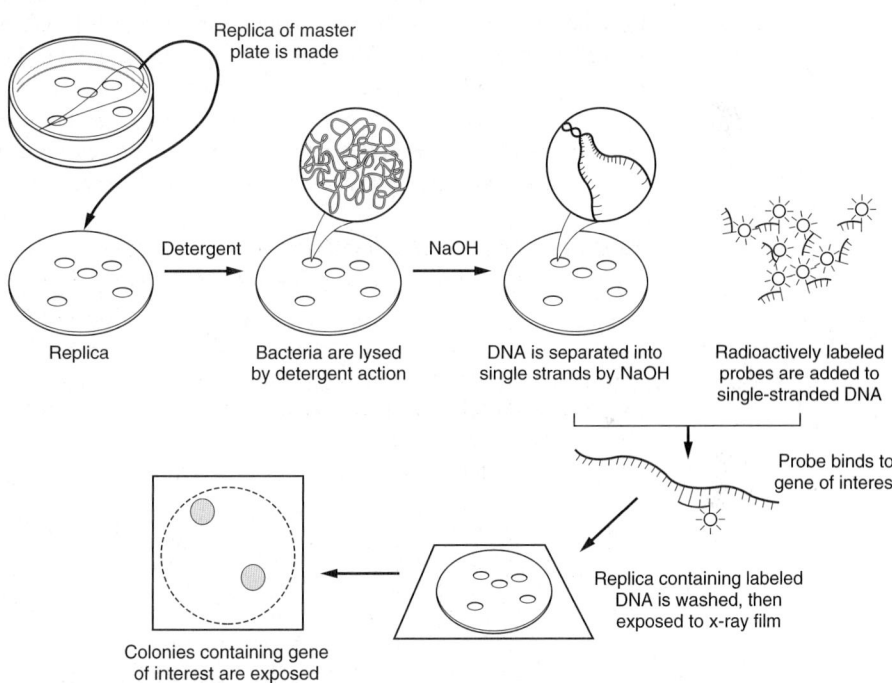

Replica of master plate is made

Replica   Detergent   Bacteria are lysed by detergent action   NaOH   DNA is separated into single strands by NaOH   Radioactively labeled probes are added to single-stranded DNA

Probe binds to gene of interest

Replica containing labeled DNA is washed, then exposed to x-ray film

Colonies containing gene of interest are exposed

greater detail in Sec. 7–2 in Chap. 7. In the nucleic acid probe method, a nucleic acid probe is synthesized that is complementary to either a DNA or RNA sequence unique to the targeted organism or organisms (see Fig. 2–35). The probe is then labeled with either a radioisotope, a fluorescent dye, or an enzyme. The use of fluorescent or enzyme labels has greatly enhanced the use of probes, and has reduced the use of radioactive probes and handling problems associated with the management of radioactive materials. The next step in the nucleic acid probe technique is to hybridize the probe to the nucleic acid in a bacterial cell. With a nucleic acid probe specific to DNA, the hybridization step generally involves filtering a solution of lysed organisms onto a polycarbonate filter and then soaking the filter in a liquid medium that contains the labeled nucleic acid probe. This process is commonly referred to as dot blot hybridization. With a nucleic acid probe specific to RNA, either dot blot hybridization or a method called in situ hybridization can be used.

In situ hybridization or fluorescent in situ hybridization (FISH) involves adding the nucleic acid probe to a solution of cells with permeablized cell walls (see Fig. 2–36). The final step in the nucleic acid probe approach is to detect the hybridized probe which is dependent on the method used to hybridize the probe. When dot blot hybridization is used, the hybridized probe is typically detected by placing the filter on a sheet of film sensitive to the type of label put on the nucleic acid probe. When FISH hybridization is used, cells containing the hybridized probe are generally detected using fluorescent microscopy. In situ hybridization has the advantage over dot blot hybridization of allowing the location of cells in the native habitat to be visualized (Loge, 1997).

Nucleic acid probes, with their ability to identify specific nucleic acid sequences, have become very important in the identification of pathogenic organisms. Nucleic acid probes have also been used to study biological treatment processes and the disinfection of wastewater following biological wastewater treatment. With the accelerated development of nucleic acid probes, it is certain that greater use will be made of this technique in developing a better understanding of the biological processes used to treat wastewater. Additional details may be found in Madigan et al. (2000).

Enumeration and Identification of Viruses

Virus particles, which vary in size from 0.02 to 0.08 μm (see Table 2–23), are too small to see with a light microscope. While an electron or transmission scanning microscope can be used, the process of preparing the sample for examination is costly and time-consuming, and most commercial laboratories do not have an electron scanning microscope. The methods used to determine enteric viruses and bacteriophage are as follows.

Enteric Viruses. There are more than 100 viral entities associated with human feces collectively known as enteric viruses. The health significance of these agents in humans ranges from poliomyelitis (polio), hepatitis, and gastroenteritis, to innocuous infections. These viruses are extremely small particles ranging from 20 to 80 nanometers (nm) in diameter. In comparison a human red blood cell averages 7600 nm in diameter. Each virus contains a single type of nucleic acid, either RNA or DNA, which is enclosed by a protein "shell" called a capsid. Replication of viruses can only take place in a living host cell. The virus is capable of using its genetic information to commandeer the host cell's machinery to produce more virus particles. In the case of enteric viruses, these particles (virons) are released into the host's feces and subsequently find their way into the wastewater environment. While these viruses can be visualized by transmission electron microscopy, the cost and small sample size makes this technology impractical for determining their presence in environmental samples.

In treated domestic wastewater, a secondary effluent will frequently have culturable virus concentration of less than 10 per liter. The numbers may be greater but, based upon the virus assay methods presently available, these are the order of magnitude commonly observed. Because of the low number of viruses expected in water and wastewater and restrictions in viral assay procedures, virus concentration methods are used. Large sample volumes, as large as 1000 gallons in the case of drinking water and one or more gallons in wastewater, are passed through filters in such a manner that viruses present are absorbed to the filter medium. Most often these are 10-inch spun-glass filters that are electropositive in charge and through which the sample water is

Figure 2–36

Schematic of the RNA probe method used for the in situ identification of bacteria. *(From Crites and Tchobanoglous, 1998.)*

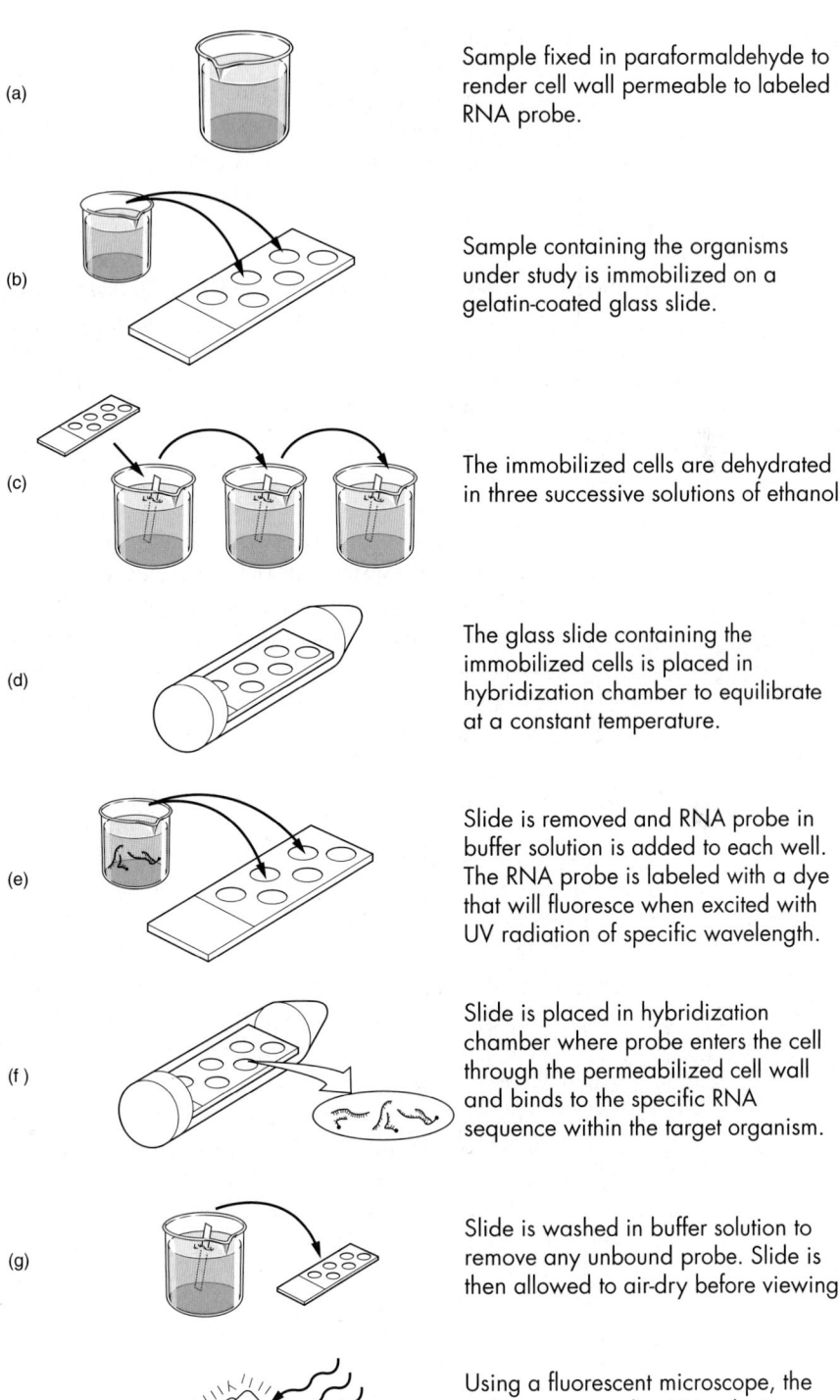

(a) Sample fixed in paraformaldehyde to render cell wall permeable to labeled RNA probe.

(b) Sample containing the organisms under study is immobilized on a gelatin-coated glass slide.

(c) The immobilized cells are dehydrated in three successive solutions of ethanol.

(d) The glass slide containing the immobilized cells is placed in hybridization chamber to equilibrate at a constant temperature.

(e) Slide is removed and RNA probe in buffer solution is added to each well. The RNA probe is labeled with a dye that will fluoresce when excited with UV radiation of specific wavelength.

(f) Slide is placed in hybridization chamber where probe enters the cell through the permeabilized cell wall and binds to the specific RNA sequence within the target organism.

(g) Slide is washed in buffer solution to remove any unbound probe. Slide is then allowed to air-dry before viewing.

(h) Using a fluorescent microscope, the cells are exposed to UV radiation, which causes them to fluoresce.

Figure 2-37

Schematic of the technique used for the enumeration of viruses.

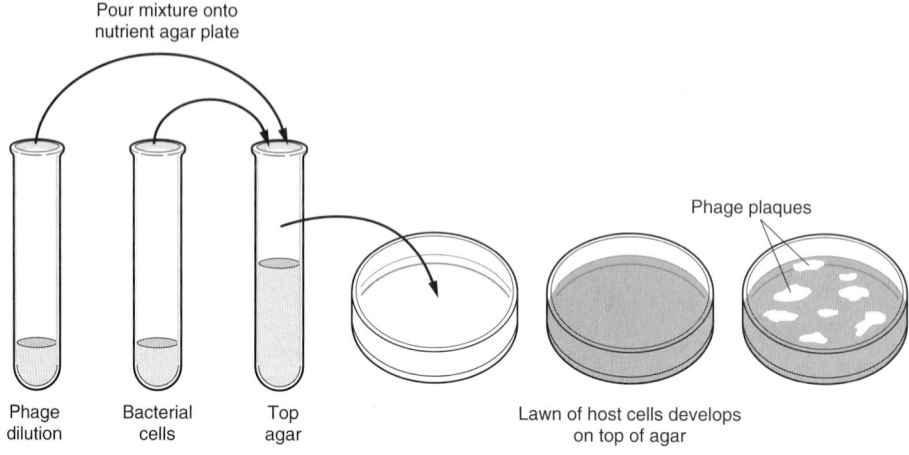

passed at a rate of approximately 1 gal/min. The absorbed viruses are then eluted from the concentrating filter using as small a volume as possible, and this concentrate is assayed for the presence of culturable viruses. In some cases, such as in raw wastewater, the filtration step may be bypassed and virus from smaller sample volumes concentrated by a process combining flocculation, centrifugation, and elution from the consolidated flock. Laboratory studies have shown that the efficiency of virus recovery using these methods can be as low as 10 percent.

Virus assays are performed in the laboratory by inoculating the sample concentrate onto monolayers of cultured cells (a process commonly referred to as tissue culture), the most common of which is a buffalo green monkey kidney (GBMK) cell line. If viruses are present, they will grow and destroy the host cells in 10 to 14 days. One of the most common methods for the enumeration of viruses is the plaque assay (see Fig. 2–37). In this method the inoculated tissue culture cells are overlain with agar to localize the released viruses. After suitable incubation, any culturable virus present in the sample concentrate will begin to destroy the infected cells. The destroyed cells appear as a hole or plaque in the cell monolayer. Each plaque (plaque-forming unit, or PFU) is the result of the presence of a single or a clump of viruses. The plaques are counted and their numbers are equal to the PFU per volume of inoculum. Using the appropriate adjustments, the number of PFU per volume of water sampled can be calculated.

A most probable number (MPN) method may also be employed in which the destruction of the tissue culture cell sheet, called cytopathic effect (CPE), takes place without the agar overlay. Dilutions of the sample are inoculated into separate cell culture flasks, the flasks incubated and observed for the presence or absence of CPE. The most likely concentration of viruses in the original sample is calculated based upon the distribution of positive and negative CPE associated with the dilutions used.

Bacteriophage. Viruses that infect bacterial cells are called bacteriophage. There are a variety of these viruses called coliphage which infect many subspecies of *Escherichia coli*. These phages, as opposed to human enteric viruses, are consistently present in wastewater in relatively large numbers. Their source is the feces of humans and animals. A variety of coliphages are called "male-specific" because they infect the bacterium via the pili (small appendages on the bacterium's surface) and bacteria with these appendages are called "male." Certain of these male-specific (also called F+) phages are the same shape and size as the smallest enteroviruses and, as with many enteroviruses, contain single-stranded RNA.

In the laboratory, these viruses are detected by the formation of plaques on "lawns" of susceptible *E. coli*. These lawns are formed by mixing the various dilution of phage sample with large numbers of host bacteria in a semisoft agar medium that will support bacterial growth. After incubation the bacteria will grow to large numbers, forming a "lawn." As the phages develop, they will lyse infected bacteria, forming plaques in the "lawn." The number of plaques is recorded as PFU per volume of sample. Phage can also be enumerated by an MPN method in which host bacteria and dilutions of phage sample are mixed in tubes containing liquid nutrient media and incubated. At the end of the incubation period the presence or absence of phage is verified by placing a small amount of liquid from each tube on a newly seeded agar "lawn" of host bacteria. After incubation the appearance of a clear spot in the lawn is evidence of a positive tube. The MPN is calculated as described previously.

These procedures are easier and less costly than the use of tissue culture as employed in the detection of animal viruses. Assay results are usually available within 24 to 48 hours. This group of viruses is gaining importance as surrogates for animal viruses when evaluating virus-removing capability of treatment processes. This is so because of their relative abundance in wastewater; their ease of detection; the low cost of the assay; and growing evidence that these viruses, in comparison with human enteric viruses, show an equal or greater resistance to environmental factors, including disinfection.

Polymerase Chain Reaction (PCR)

Polymerase chain reaction (PCR) is a technique, developed within the past 15 years, that has been used for the rapid detection of potential pathogenic microorganisms. The technique involves the amplification of the DNA of the genome of the microorganisms being tested by using a complementary cell fragment known as a *primer*. The primer triggers a reaction that results in the production of many millions of copies of the microorganism DNA. The basis for the technique is the repetitive enzymatic synthesis of DNA and the fact that amplification only takes place if the specific nucleic acid of the microorganism is present. To identify viruses that only contain RNA, the RNA must first be converted to DNA with a reverse transcriptase enzyme. The procedure for identifying RNA viruses in known as RT-PCR.

It should be noted, however, that there are some problems associated with the use of this technique, including (1) the small sample volumes that can be assayed, (2) inhibition by interfering constituents in environmental samples, (3) the fact that the test is, at present, not quantitative (i.e., the number of organisms is not enumerated), and (4) inability of the test to differentiate between viable, inactivated, or dead microorganisms. The inability to differentiate between viable and inactivated microorganisms (e.g., organisms inactivated as a result of exposure to UV radiation) limits the use of this technique for some environmental studies (e.g., disinfection). At present this is the only technique that can be used to detect certain viruses (such as the Norwalk virus) for which culture techniques are not now available. With the rapid improvements that have been made in the test procedure since it was first introduced, it appears that many of the current limitations of the test will be overcome. Some of the limitations can be overcome now by combining the PCR technique with one or more of the other available techniques, but the time and effort involved is considerable.

Development of Microorganisms Typing Techniques

The development of techniques that can be used for the identification (typing) of specific microorganisms is another area of microbial research that is developing rapidly. For example, the ability to differentiate between fecal coliform organisms of animal, avian, and human origin would make it possible to trace the sources of the organisms and the transmission routes for specific pathogens. The principal techniques that are now used to identify specific microorganisms are discussed in Sec. 7–2 in Chap. 7.

New and Reemerging Microorganisms

Within the past 5 years there has been a disturbing increase in the number of disease outbreaks in the United States and in many other parts of the world, especially in light of the fact it was thought that a number of endemic contagious diseases had been controlled or eliminated (only smallpox to date) (Levins et al., 1994). The bacteria *Legionella pneumophila,* the causative agent in Legionnaires' disease, found in wastewater and reclaimed wastewater, is an example of a disease-causing organism that has only recently been identified (Levins et al., 1994). The recent high incidence of tuberculosis reported in Africa is an example of the reemergence of a disease that was thought to be under control or essentially eliminated. The significance of the identification of new disease organisms, disease outbreaks, and the reemergence of old diseases is that the concern for public health must remain the primary objective of wastewater management.

2–9 TOXICITY TESTS

Toxicity tests are used to:

1. Assess the suitability of environmental conditions for aquatic life
2. Establish acceptable receiving water concentrations for conventional parameters (such as DO, pH, temperature, salinity, or turbidity)
3. Study the effects of water quality parameters on wastewater toxicity
4. Assess the toxicity of wastewater to one or more freshwater, estuarine, or marine test organisms
5. Establish relative sensitivity of a group of standard aquatic organisms to effluent as well as standard toxicants
6. Assess the degree of wastewater treatment needed to meet water pollution control requirements
7. Determine the effectiveness of wastewater-treatment methods
8. Establish permissible effluent discharge rates
9. Determine compliance with federal and state water quality standards and water quality criteria associated with National Pollution Discharge Elimination System (NPDES) permits (Standard Methods, 1998)

Such tests provide results that are useful in protecting human health, aquatic biota, and the environment from impacts caused by the release of constituents found in wastewater into surface waters. Toxicity identification, in which the constituents or compounds responsible for the observed toxicity are delineated, is another important aspect of toxicity assessment.

During the past several decades, pollution control measures were focused primarily on conventional pollutants (such as oxygen-demanding materials, suspended solids, etc.) which were identified as causing water quality degradation. In the past 10 years, increased attention has been focused on the control of toxic substances, expecially those contained in wastewater-treatment plant discharges. The early requirements for monitoring and regulating toxic discharges were on a "chemical-specific" basis. The chemical-specific approach has many shortcomings, including the inability to identify synergistic effects or the bioavailability of the toxin. The more contemporary whole-effluent, or toxicity-based, approach to toxicity control involves the use of toxicity tests to measure the toxicity of treated wastewater discharges. The whole-effluent test procedure is used to determine the aggregate toxicity of unaltered effluent discharged into receiving waters; toxicity is the only parameter measured.

The national policy prohibiting the discharge of toxic pollutants in toxic amounts is documented in Section 101(a) (3) of the federal Clean Water Act. Because it is not economically feasible to determine the specific toxicity of each of the thousands of potentially toxic substances in complex effluents, whole-effluent toxicity testing using aquatic organisms is a direct, cost-effective means of determining effluent toxicity. Whole-effluent toxicity testing involves the introduction of appropriate bioassay organisms into test aquariums (see Fig. 2–38) containing various concentrations of the effluent in question and observing their responses. Toxicity terminology, general test procedures, the evaluation of test results, the application of the test results, and the means of identifying classes of toxic compounds are described in the following discussion.

Toxicity Terminology

Terms commonly encountered when considering the conduct of toxicity tests and the analysis, interpretation, and application of test results are summarized in Table 2–29. Because the terms reported in Table 2–29 are subject to change as new and improved

Figure 2–38

Typical setup used to conduct whole-effluent toxicity tests using fish where mortality is the test endpoint.

Table 2–29
Terms used in evaluating the effects of contaminants on aquatic test organisms[a,b]

Term	Description
Acute toxicity	Exposure that will result in significant response shortly after exposure (typically a response is observed within 48 or 96 h)
Chronic toxicity	Exposure that will result in sublethal response over a long term, often 1/10 of the life span or more
Chronic value (ChV)	Geometric mean of the NOEC and LOEC from partial and full cycle tests and early-life-stages tests
Cumulative toxicity	Effects on an organism caused by successive exposures
Dose	Amount of a constituent that enters the test organism
Effective concentration (EC)	Constituent concentration estimated to cause a specified effect in a specified time period (e.g., 96-h EC_{50})
Exposure time	Time period during which a test organism is exposed to a test constituent
Inhibiting concentration (IC)	Constituent concentration estimated to cause a specified percentage inhibition or impairment in a qualitative function
In vitro (in glass or test tube)	Tests conducted in glass petri dishes or test tubes
In vivo (in living organism)	Toxicity tests conducted using the whole organism
Lethal concentration (LC)	Constituent concentration estimated to produce death in a specified number of test organisms in a specified time period (e.g., 96-h LC_{50})
Lowest-observed-effect concentration (LOEC)	Lowest constituent concentration in which the measured values are statistically different from the control
Maximum-allowable-toxicant concentration (MATC)	Constituent concentration that may be present in receiving water without causing significant harm to productivity or other uses
Median tolerance limit (TL_m)	An older term used to denote the constituent concentration at which at least 50 percent of the test organisms survive for a specified period of time. Use of the term "median tolerance limit" has been superseded by the terms median lethal concentration (LC_{50}) and median effective concentration (EC_{50})
No-observed-effect concentration (NOEC)	Highest constituent concentration at which the measured effects are no different from the control
Sublethal toxicity	Exposure that will damage organism, but not cause death
Toxicity	Potential for a test constituent to cause adverse effects on living organisms

[a]Adapted from Hughes (1996) and Standard Methods (1998).
[b]It should be noted that the terms given in this table apply only to aquatic organisms and are, for the most part, distinct from the terms used for animals and humans.

methods of toxicity testing are developed, it is imperative that the latest version of Standard Methods and related U.S. EPA protocols be reviewed before undertaking any toxicity testing.

Toxicity Testing

Toxicity tests are classified according to (1) duration: short-term, intermediate, and/or long-term; (2) method of adding test solutions: static, recirculation, renewal, or flow-through; (3) type of test: in vitro (i.e., tests in petri dishes or test tubes) or in vivo (i.e., toxicity tests using the whole organism); and (4) purpose: NPDES permit requirements, mixing zone determinations, etc. In vitro toxicity testing has been validated widely in recent years. Even though organisms vary in sensitivity to effluent toxicity, the U.S. EPA has documented that (1) toxicity of effluents correlated well with toxicity measurements in the receiving waters when effluent dilution was measured; and (2) predictions of impacts from both effluent and receiving water toxicity tests compare favorably with ecological community responses in the receiving waters. The U.S. EPA has conducted nationwide tests with freshwater, estuarine, and marine ecosystems. Methods include both acute and chronic exposures. Typical short-term chronic toxicity test methods are reported in Table 2–30. Detailed contemporary testing and analysis protocols are sum-

Table 2–30
Typical examples of short-term chronic toxicity test methods using various freshwater and marine/estuarine aquatic species[a]

Species/common name	Test duration	Test endpoints
Freshwater species		
Cladoceran (*Ceriodaphnia dubia*)	Approximately 7 d (until 60 percent of control have 3 broods)	Survival, reproduction
Fathead minnow (*Pimephales promelas*)	7 d 9 d	Larval growth, survival Embryo-larval survival, percent hatch, percent abnormality
Freshwater algae (*Selenastrum capricomutum*)	4 d	Growth
Marine/estuarine species		
Sea urchin (*Arbacia punctulata*)	1.5 h	Fertilization
Red macroalgae (*Champia parvula*)	7–9 d	Cystocarp production (fertilization)
Mysid (*Mysidopsis bahia*)	7 d	Growth, survival, fecundity
Sheepshead minnow (*Caprinodon variegatus*)	7 d 7–9 d	Larval growth, survival Embryo-larval survival, percent hatch, percent abnormality
Inland silverside (*Menidia beryijina*)	7 d	Larval growth, survival

[a] From U.S. EPA (1988, 1989).

marized in Standard Methods (1998) and in U.S. EPA publications (U.S. EPA, 1985a, b, c, d).

Analysis of Toxicity Test Results

Methods used to analyze both short-term (acute) and long-term (chronic) toxicity data are considered in the following discussion.

Acute Toxicity Data.   The median lethal concentration (LC_{50}) when mortality is the test endpoint, or median effective concentration (EC_{50}) when a sublethal effect (e.g., immobilization, fatigue in swimming, "avoidance") is the endpoint, are typically used to define acute toxicity (Stephen, 1982). A typical bioassay setup using fish where mortality is the test endpoint is shown on Fig. 2–38. A fish swimming chamber is used to assess sublethal effects. A fish is placed in a chamber where the flow-through velocity can be increased until the fish is swept out of the chamber. The washout velocity for fish exposed to a specific compound can be compared to the washout velocity for the control fish.

Because the LC_{50} value is the median value, it is important to provide some information on the variability of the test population. The LC_{50} values can be determined graphically or analytically using the Spearman Karber, moving average, binomial, and probit methods. The 95 percent confidence limits are usually specified. Most standard statistical packages available for desktop computers include a probit analysis program. Determination of LC_{50} values, both graphically and by means of probit analysis, is illustrated in Example 2–14. Typically, LC_{50} values are computed based on survival at both 48- and 96-h exposures.

EXAMPLE 2–14   **Analysis of Acute Toxicity Data**   Determine graphically and by probit analysis the 48- and 96-h LC_{50} values in percent by volume for the following toxicity test data obtained using flathead minnows:

Concentration of waste, % by volume	No. of test animals	No. of test animals dead after[a]	
		48 h	**96 h**
60	20	16 (80)	20 (100)
40	20	12 (60)	18 (90)
20	20	8 (40)	16 (80)
10	20	4 (20)	12 (60)
5	20	0 (0)	6 (30)
2	20	0 (0)	2 (10)

[a] Percentage values are given in parentheses.

Solution

1. Plot the concentration of wastewater in percent by volume (log scale) against the test animals that have died in percent (probability scale).

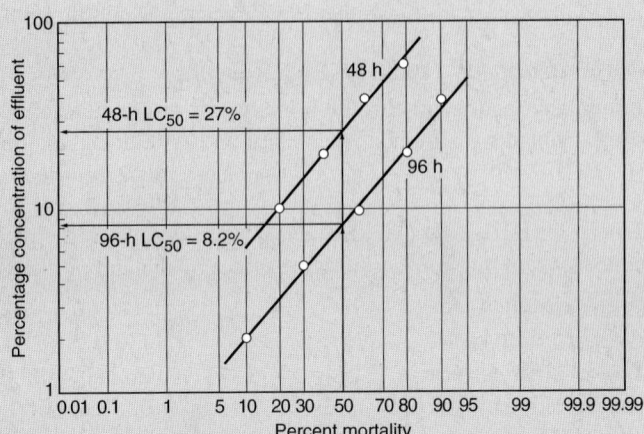

2. Fit a line to the data points by eye, giving most consideration to the points lying between 16 and 84% mortality, which corresponds to approximately one standard deviation.

3. Find the wastewater concentration causing 50% mortality. The estimated LC_{50} values are:
   a. 48-h LC_{50} = 27.0%
   b. 96-h LC_{50} = 8.2%

4. Compare the results obtained with a probit analysis to the values determined in Step 3. The probit analysis results are as follows.
   a. 48-h LC_{50} = 27.6%, 95% confidence limits 21.0 and 37.8%
   b. 96-h LC_{50} = 8.1%, 95% confidence limits 5.8 and 10.9%

Comment Although the LC_{50} values obtained using the graphical analysis approach are approximate, they are quite close to the values obtained using the probit analysis approach and serve as a good check. To obtain confidence limits, a probit or similar analysis must be performed.

Chronic Toxicity Data. Results of chronic toxicity tests often are analyzed statistically to determine the lowest-observed-effect concentration (LOEC), the no-observed-effect concentration (NOEC), and the chronic value (ChV). Generally, statistical significance is assumed to be at the $p = 0.05$ level. The chronic value (ChV) is calculated as the geometric mean of the LOEC and the NOEC.

Chronic toxicity limits may be specified with either NOEC or ChV as the endpoint. The term maximum acceptable toxicant concentration (MATC) often is used interchangeably with the chronic value. Similar to acute toxicity data, lethal concentration (LC) or effective concentration (EC) values can be used with chronic toxicity data to describe chronic toxicity tolerance levels. Recently, the concept of the inhibiting concentration (IC) has been introduced to characterize effects in chronic tests. A variety of nonparametric and parametric statistical methods are available to determine NOECs and LOECs and LCs, ECs, and ICs (Standard Methods, 1998).

Application of Toxicity Test Results

In applying acute and chronic toxicity test results, the toxic units (TU) approach has been adopted by a number of federal and state agencies. In the toxic units approach (U.S. EPA, 1985), a TU concentration is established for the protection of aquatic life.

Toxic Unit Acute (TU$_a$).   The TU$_a$ is defined as the reciprocal of the wastewater concentration that caused the acute effect by the end of the exposure period.

$$TU_a = 100/LC_{50} \tag{2-76}$$

Toxic Unit Chronic (TU$_c$).   The TU$_c$ is defined as the reciprocal of the effluent concentration at which the measured effects, by the end of the chronic exposure period, are no different from the control.

$$TU_c = 100/NOEC \tag{2-77}$$

where NOEC = no-observed-effect concentration

Depending on the use to be made of the toxicity test results, a variety of different numerical values have been used for TU$_a$ and TU$_c$, as a basis for assessing the suitability of a given effluent for discharge to the environment. For example, to protect against acute toxicity it has been suggested that the MATC should be less than $0.3 \times TU_c$. Because the limiting values vary from location to location, current regulatory standards must be reviewed in applying toxicity results. The application of toxicity test results is illustrated in Example 2–15. Some typical effluent toxicity requirements are summarized in Table 2–31.

EXAMPLE 2–15   **Application of Toxicity Test Results**   A critical initial dilution of 100:1 is achieved for a treated effluent discharged to marine receiving waters. Toxicity tests were conducted with the wastewater-treatment plant effluent using three marine species. Based on the toxicity test results, it was found that *Champia parvula* was the most sensitive species acute endpoint (2.59 percent effluent) as measured by the EC$_{50}$, and also the most sensitive species chronic endpoint (1.0 percent) as measured by the NOEC. For protection of the aquatic environment, the acute and chronic toxicity requirements have been set at 10 TU$_a$ and 1.0 Tu$_c$, respectively.

Results of acute toxicity tests

Species	Control exposure, h	Survival, %	Percent effluent LC$_{50}$ or EC$_{50}$[a]	NOEC
Mysidopis bahia	96	100	18.66	10.0
Cyprinodon variegatus	96	100	>100	50.0
Champia parvula	48/168	100	2.59	12.25

[a]EC50 results based on reduction of cystocarp production.

Results of chronic toxicity tests

Species	Control exposure, days	Survival, %	Percent effluent NOEC	Percent effluent LOEC
Mysidopis bahia	7	82	6.0	10.0
Cyprinodon variegatus	7	98.8	15.0	>15.0
Champia parvula	7	100	1.0	2.25

Solution

1. Check compliance with acute toxicity requirements.
   a. Based on data for the most sensitive species tested, the number of acute toxic units (TU_a), based on Eq. (2–76), is

   $$TU_a = 100/ LC_{50} = 100/2.59 = 38.6$$

   b. Following an initial dilution of 100, the TU_a value is

   $$TU_a/100 = 38.6/100 = 3.86 \ TU_{ad} \text{ (after dilution)}$$

   Because the TU_a value after dilution (3.86 TU_{ad}) is less than 10 TU_a, the acute toxicity requirement has been met.

2. Check compliance chronic toxicity requirements.
   a. Based on data for the most sensitive species tested, the number of chronic toxic units (TU_c), based on Eq. (2–77), is

   $$TU_c = 100/NOEC = 100/1.0 = 100$$

   b. Following an initial dilution of 100, the TU_c value is

   $$TU_c = 100/100 = 1.0 \ TU_{cd} \text{ (after dilution)}$$

   Because the TU_c value after dilution (1.0 TU_{cd}) is equal to 1.0 TU_c, the chronic toxicity requirement has been met.

In summary, there are a number of advantages to the use of whole-effluent toxicity testing. In this approach, the bioavailability of the toxics is measured and the effects of any synergistic interactions are also considered. Because the aggregate toxicity of all components of the wastewater effluent is determined, the toxic effect can be limited by limiting only one parameter, the effluent toxicity. Because contemporary receiving water management strategies are based on site-specific water quality criteria, toxicity testing facilitates comparison of effluent toxicity with site-specific water quality criteria designed to protect representative, sensitive species and yet allow for establishment of discharge limitations that will protect aquatic environments.

Identification of Toxicity Components

Toxicity testing can also be used to determine the source of toxicity, an especially important test for industrial discharges. For example, is the toxicity caused by the suspended

Table 2–31
Description of separation techniques that can be used to fractionate a wastewater sample[a]

Separation technique	Description
Filtration for suspended solids	Filtration is generally performed first to determine whether the toxicity is related to the soluble or insoluble phase of the sample. Typically, 1-μm glass fiber filters that have been prewashed with ultrapure water are used. The insoluble phase should be resuspended in control water to ensure that filtration, not adsorption on the filter medium, removed the toxicity
Filtration for colloidal solids	A 0.1-μm filter should be used to determine if the colloidal fraction is responsible for the toxicity
Ion exchange	Inorganic toxicity can be studied by using cationic and anionic exchange resins to remove potentially toxic inorganic compounds or ions
Molecular weight classification	Evaluating the molecular weight distribution of the influent, and the toxicity of each molecular weight range, can often narrow the list of suspected contaminants
Biodegradability test	Controlled biological treatment of effluent samples in the lab can result in almost compete oxidation of the biodegradable portion of organics. Bioassay analysis can then quantify the toxicity associated with the nonbiodegradable components, as well as the reduction in toxicity attainable by biological treatment
Oxidant reduction	Residual chemical oxidants carried over from a process (e.g., chlorine and chloramines used for disinfection, or ozone and hydrogen peroxide used in sludge conditioning) can be toxic to most organisms. A simple batch reduction of these oxidants at various concentrations, using an agent such as sodium thiosulfate, can be used to assess the toxicity of any remaining oxidants
Metal chelation	The toxicity of the sum of all cationic metals (with the exception of mercury) can be determined by chelation of samples, using varying concentrations of ethylenediaminetetraacetic acid (EDTA) and evaluating the change in toxicity
Air stripping	Batch air stripping at acid, neutral, and basic pH can remove essentially all volatile organics. At a basic pH, ammonia is removed as well. Thus, if both volatile organics and ammonia are suspected toxicants, an alternative ammonia removal technique, such as a zeolite exchange, should be used. (Note that ammonia is toxic in the nonionized form, so ammonia toxicity is very pH-dependent.)
Resin adsorption and solvent extraction	Specific nonpolar organics can sometimes be identified as toxics, using a resin-adsorption/solvent-extraction process. A sample is adsorbed on a long-chain organic resin, the organics are reextracted from the resin with a solvent (e.g., methanol), and the toxicity of the sample is determined using a bioassay test procedure

[a] Adapted from Eckenfelder (2000).

Figure 2–39

Separation techniques that can be used to fractionate a wastewater sample (see also Fig. 2–25). *(Adapted from Eckenfelder, 2000.)*

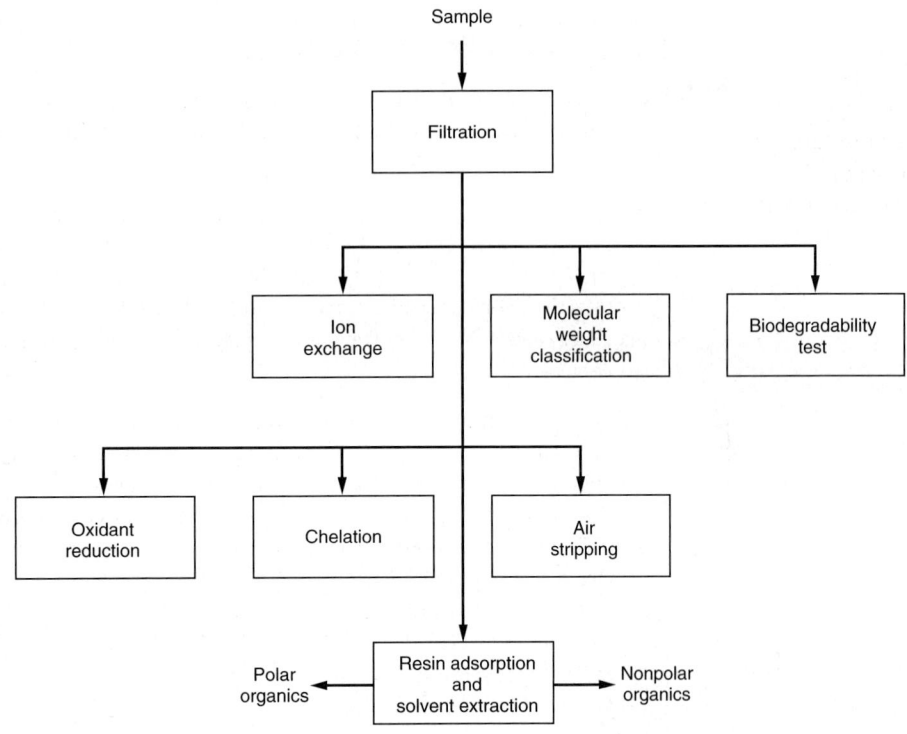

solids, the colloidal solids, the long- or short-chain dissolved organic constituents, or the dissolved inorganic constituents? To determine the source(s) of the toxicity, each of the potential sources must be isolated from the other constituents in the sample and tested for toxicity. The various methods that can be used to fractionate a wastewater sample are illustrated graphically on Fig. 2–39 and described in Table 2–31.

PROBLEMS AND DISCUSSION TOPICS

2–1   The following test results were obtained for four water samples. Check the accuracy of the analytical measurements for one of the waters (to be selected by the instructor). Do you suspect that a constituent of significance has been neglected? If so, is it a cation or anion?

Cation	Concentration, mg/L				Anion	Concentration, mg/L			
	A	**B**	**C**	**D**		**A**	**B**	**C**	**D**
Ca^{2+}	121.3	76.0	190.2	93.8	HCO_3^-	280	128.2	260.0	167.4
Mg^{2+}	36.2	27.2	84.1	28.0	SO_4^{2-}	116	240.0	64.0	134.0
Na^+	8.1	22.9	75.2	13.7	Cl^-	61	37.2	440.4	92.5
K^+	12	18.7	5.1	30.2	NO_3^-	15.6	2.0	35.1	—
Fe^{2+}	—	2.1	0.2	—	CO_3^{2-}	—	—	30	—

2-2 Determine the activity coefficients and activity for the constituents for one of the waters in Problem 2–1 (to be selected by instructor).

2-3 Compare the values obtained for the activity coefficients obtained using Eq. (2–10) and the equation proposed by Davies (1962) as given below for one of the waters given in Problem 2–1 (to be selected by instructor).

$$\log \gamma = -0.51 (z_i)^2 \left(\frac{\sqrt{I}}{1 + \sqrt{I}} - 0.3\,I \right) \text{ at } 25°C$$

2-4 Estimate the TDS for one of the water samples in Problem 2–1 (to be selected by instructor) using Eq. [2–12] and by summing the mass of the individual ionic species. How do the computed values compare?

2-5 Obtain the most recent edition of Standard Methods and develop a sampling plan including sample size, type of container, type of preservatives, if used, the time before analysis, and any other requirements to determine the concentration of the following constituents (to be selected by instructor). Wastewater samples are to be collected at the headworks and at the effluent discharge point from a wastewater-treatment plant.
a. Organic carbon (BOD, TOC, and COD)
b. Particulate material (TS, TSS, TDS, TVS, VSS, VDS, and turbidity)
c. Nutrients (TN, TKN, NO_3^-, NO_2^-, NH_4^+, TP, Org P, and Inorg P)
d. Indicator organisms (coliform and coliphages)
e. Metals (Cd, Cu, Hg, and Zn)

2-6 A sample of primary effluent has an expected BOD_5 of 280 mg/L ± 150 mg/L. Determine the number of tests and the volume of sample, seeded dilution water, and unseeded dilution water that would be required for each test to measure the BOD_5 of the sample.

2-7 The following test results were obtained for four different wastewater samples. The size of the sample was 100 mL. Determine the concentration of total and volatile solids, expressed as mg/L, for one of the samples (to be selected by instructor).

		Weight, g			
Item	**Unit**	**A**	**B**	**C**	**D**
Sample size	mL	90	100	120	200
Tare mass of evaporating dish	g	22.6435	22.6445	22.6550	22.6445
Mass of evaporating dish plus residue after evaporation at 105°C	g	22.6783	22.6832	22.6995	22.6667
Mass of evaporating dish plus residue after ignition at 550°C	g	22.6768	22.6795	22.6832	22.6433

2-8 The total suspended solids for a series of wastewater samples was found to be 175, 197, 113, and 247 mg/L, respectively. If the following test results were obtained, determine the size of sample (to be selected by instructor) used in the analysis.

Item	Weight, g			
	A	**B**	**C**	**D**
Tare mass of glass fiber filter	1.5244	1.5243	1.5241	1.5246
Mass of glass fiber filter plus residue after drying at 105°C	1.5953	1.5379	1.5449	1.5667

2-9 The following test results were obtained for a wastewater sample taken at an industrial facility. All of the tests were performed using a sample size of 100 mL. Determine the concentration of total solids, total volatile solids, total suspended solids, and dissolved solids for one of the samples (to be selected by instructor).

Item	Weight, g			
	A	**B**	**C**	**D**
Tare mass of evaporating dish	54.6422	54.6423	54.6424	54.6423
Mass of evaporating dish plus residue after evaporation at 105°C	54.7022	54.7173	54.7224	54.7148
Mass of evaporating dish plus residue after ignition at 550°C	54.6722	54.6893	54.6801	54.6818
Tare mass of Whatman GF/C filter	1.5348	1.5347	1.5347	1.5346
Mass of Whatman GF/C filter plus residue after drying at 105°C	1.5553	1.5586	1.5622	1.5571
Mass of Whatman GF/C filter plus residue after ignition at 550°C	1.5453	1.5454	1.5471	1.5418

2-10 The following test results were obtained for a wastewater sample taken at the headworks to a wastewater-treatment plant. All of the tests were performed using a sample size of 50 mL. Determine the concentration of total solids, total volatile solids, total suspended solids, volatile suspended solids, and dissolved solids for one of the samples (to be selected by instructor).

Item	Weight, g			
	A	**B**	**C**	**D**
Tare mass of evaporating dish	53.5435	53.5434	53.5436	53.5433
Mass of evaporating dish plus residue after evaporation at 105°C	53.5765	53.5693	53.5725	53.5793
Mass of evaporating dish plus residue after ignition at 550°C	53.5750	53.5652	53.5695	53.5772
Tare mass of Whatman GF/C filter	1.5433	1.5435	1.5436	1.5434
Mass of Whatman GF/C filter plus residue after drying at 105°C	1.5653	1.5705	1.5735	1.5625
Mass of Whatman GF/C filter plus residue after ignition at 550°C	1.5502	1.5575	1.5631	1.5531

2-11 The following data were obtained from a serial filtration test of a settled effluent after biological treatment. Prepare a plot of one of the samples (to be selected by instructor). How great an error in the measurement of the total suspended and colloidal solids would occur if a filter paper with a nominal pore size of 1.2 μm was used to determine the total suspended solids, as compared to using a filter with a nominal pore size of 0.1 μm filter.

Nominal pore size, μm	Weight, mg/L			
	A	**B**	**C**	**D**
12	20.2	29.4	22.5	25.1
8	8.8	11.5	8.0	15.1
5	4.1	3.5	4.9	2.2
3	7.5	5.1	11.6	8.9
1	15.1	13.5	21.2	25.0
0.1	9.9	15.1	24.9	17.5

2-12 Estimate the alkalinity, expressed as mg/L as $CaCO_3$, for one of the water samples in Problem 2–1 (to be selected by instructor).

2-13 What is the molar mass of a gas at 20°C assuming the gas has a density of 0.68 g/L at standard temperature and pressure (STP).

2-14 Compare the saturation concentrations of O_2, N_2, and CO_2 between San Francisco (sea level) and Taos, NM (elevation 2150 m), Denver, CO (elevation 1600 m), or La Paz, Bolivia (elevation 4270 m) (city to be selected by instructor).

2-15 Determine the equilibrium concentration of oxygen in the liquid phase in a covered pure oxygen treatment process subject to 1, 1.5, 2.0, 2.5, and 3.0 atms of pressure (pressure to be selected by instructor). Assume the composition of the gas above the liquid is 84 percent O_2, 12 percent N_2, and 4 percent CO_2.

2-16 Using Henry's law, determine the saturation concentration of O_2, N_2, or CO_2 (gas to be selected by instructor) in water at 0, 10, 20, 30, 40, and 50°C. Construct a plot of O_2, N_2, or CO_2 saturation concentration as a function of temperature.

2-17 The local Air Pollution Control District has threatened to fine and penalize the local wastewater-management agency, your client, because of frequently recurring odor complaints from residents who live downwind of the plant. The plant manager, a full-time employee at the treatment plant, claims that no problem exists. He proves his point by consistently finding less than 5 dilutions to MDTOC at the plant boundary using a hand-held sniff dilution olfactometer as employed by the local Air Pollution Control District. You, however, live downwind of the plant and have frequently detected odors from it. Why do these differences exist? How would you resolve them objectively?

2-18 You have been asked to review an odor-control system that has apparently failed to control odors from a sludge-dewatering building adequately. The wastewater-management agency, your client, claims the system has failed to perform according to specifications. The engineering contractor who installed the system claims that the specifications were not adequate.

In your investigation you find that the agency employed a reputable odor consultant to develop the odor-control-system specifications. The consultant used the ASTM Panel Method for odor measurement, using evacuated glass cylinders for sample collection. Several measurements were made, and the maximum observed value was doubled to develop the control-system specifications. In this way a 90 percent odor-removal requirement was established to meet the desired final odor-emission limit of 2.8×10^4 odor units per minute (the product of airflow in m^3/min and number of dilutions to MDTOC).

Using a direct-reading olfactometer, you find that the control system removes 99 percent of the odor, and that at a rate of 10^6 odor units per minute the final odor emission is 10^6 odor units per minute. What reasons might explain your findings? How would you resolve the problem?

2–19 You have been asked by a wastewater-management agency to review the adequacy of their odor-control program. What would be your major considerations in making such a review?

2–20 In a BOD determination, 6 mL of wastewater are mixed with 294 mL of diluting water containing 9.1 mg/L of dissolved oxygen. After a 5-d incubation at 20°C, the dissolved oxygen content of the mixture is 2.8 mg/L. Calculate the BOD of the wastewater. Assume that the initial dissolved oxygen of wastewater is zero.

2–21 Solve Problem 2–20 for one of the following conditions (to be selected by instructor):
a. sample size, 8 mL, oxygen in dilution water, 9.0 mg/L, oxygen in mixture after 7-d incubation at 20°C, 1.8 mg/L
b. sample size, 6 mL, oxygen in dilution water, 9.2 mg/L, oxygen in mixture after 6-d incubation at 20°C, 1.65 mg/L
c. sample size, 6 mL, oxygen in dilution water, 8.9 mg/L, oxygen in mixture after 4-d incubation at 20°C, 1.5 mg/L
d. sample size, 10 mL, oxygen in dilution water, 9.15 mg/L, oxygen in mixture after 5-d incubation at 20°C, 1.42 mg/L

2–22 Determine the UBOD and BOD_5 (in mg/L) of a mixture of 150 mg/L glutamic acid ($C_5H_{10}N_2O_3$) and 150 mg/L glucose ($C_6H_{12}O_6$). Assume the value of the BOD_5 first-order reaction rate constant is 0.23/d (base e).

2–23 The BOD_5 of a waste sample was found to be 40.0 mg/L. The initial oxygen concentration of the BOD dilution water was equal to 9 mg/L, the DO concentration measured after incubation was equal to 2.74 mg/L, and the size of sample used was equal to 40 mL. If the volume of the BOD bottle used was equal to 300 mL, estimate the initial DO concentration in the waste sample.

2–24 Solve Problem 2–23 for one of the following conditions (to be selected by instructor):
a. sample size, 20 mL, BOD of sample 78 mg/L, oxygen in dilution water, 9.15 mg/L, oxygen in mixture after 5-d incubation at 20°C, 3.83 mg/L
b. sample size, 10 mL, BOD of sample 175 mg/L, oxygen in dilution water, 9.1 mg/L, oxygen in mixture after 5-d incubation at 20°C, 2.97 mg/L
c. sample size, 5 mL, BOD of sample 275 mg/L, oxygen in dilution water, 9.1 mg/L, oxygen in mixture after 5-d incubation at 20°C, 4.47 mg/L
d. sample size, 2 mL, BOD of sample 750 mg/L, oxygen in dilution water, 9.1 mg/L, oxygen in mixture after 5-d incubation at 20°C, 4.04 mg/L

2–25 What size of sample expressed as a percent is required if the 5-d BOD is 120, 230, 333, or 400 mg/L (to be selected by instructor) and the total oxygen consumed in the BOD bottle is limited to 2 mg/L.

2-26 A wastewater sample is diluted by a factor of 10 using seeded dilution water. If the following results are obtained, determine the 5-d BOD.

	Dissolved oxygen, mg/L	
Time, d	Diluted sample	Seeded sample
0	8.65	8.75
1	5.35	8.60
2	5.12	8.56
3	4.45	8.50
4	3.75	8.48
5	2.80	8.44
6	2.20	8.37

2-27 Using the data from Problem 2–26, determine the 2-, 4- , and 6-d BOD.

2-28 The 5-d 20°C BOD of a wastewater is 185 mg/L. What will be the ultimate BOD (UBOD)? What will be the 10-d demand? If the bottle had been incubated at 33°C, what would the 5-d BOD have been? $k = 0.23$ d^{-1}.

2-29 The 5-d BOD at 20°C is equal to 350 mg/L for three different samples, but the 20°C k values are equal to 0.25 d^{-1}, 0.35 d^{-1}, and 0.46 d^{-1}. Determine the ultimate BOD (UBOD) of each sample.

2-30 The BOD value of a wastewater was measured at 2 and 8 d, 1 and 9 d, 2 and 7 d, and 3 and 10 d (days to be selected by instructor) and found to be 125 and 225 mg/L, respectively. Determine the 5-d value using the first-order rate model.

2-31 The following BOD results were obtained on a series of untreated wastewater samples at 20°C. Determine for one of the samples (to be selected by instructor) the reaction rate constant k and the ultimate BOD (UBOD) using both the least-squares and the Fujimoto methods.

Time, t	BOD, mg/L				
d	A	B	C	D	E
0	0	0	0	0	0
1	65	52	170	20	110
2	109	82	235	32	160
3	138	108	275	40	193
4	158	123	300	42	220
5	178	137	322	46	241
6	190	149	335	48	258
7	200	158	342	49	270
8	205	165	349	50	280
9	210	170	351	50	288
10	212	175	354		292
11		178	358		293
12		180	359		294

2-32   Given the following results, determined for a wastewater sample at 20°C the ultimate carbonaceous oxygen demand, the ultimate nitrogenous oxygen demand (NOD), the carbonaceous BOD
reaction-rate constant (k), and the nitrogenous NOD reaction-rate constant (k_n). Determine
$k(\theta = 1.05)$ and $k_n(\theta = 1.08)$ at 25°C.

	BOD, mg/L			
Time, d	A	B	C	D
0	0	0	0	0
2	18	30	45	36
4	26	43	75	58
6	30	52	95	70
8	33	58	114	80
10	56	60	135	90
12	69	90	144	98
14	77	104	149	102
16	82	114	151	145
18	84	120	152	170
20	87	125	152	182
25	90	135	170	210
30	91.5	142	239	222
35	92.5	147	260	233
40	93	148	268	239
45	94	149	271	240
50	94.5	150	272	241

2-33   Compute the carbonaceous and nitrogenous oxygen demand of a waste represented by the formula $C_9N_2H_6O_2$, $C_6N_2H_4O$, $C_{12}N_4H_6O_2$, or $C_{10}N_2H_8O_2$ (formula to be selected by instructor).
Assume N is converted to NH_3 in the first step.

2-34   Determine the carbonaceous and nitrogenous oxygen demand in mg/L for a l-L solution containing 300, 350, 333, or 450 mg (value to be selected by instructor) of acetic acid (CH_3COOH).

2-35   The following data have been obtained for five different wastewater samples. Estimate the total
quantity of oxygen in mg/L that must be furnished to stabilize completely one of the wastewaters
(to be selected by instructor). What is the corresponding COD and the ThOD for the wastewater?

		Wastewater				
Item	Unit	A	B	C	D	E
BOD	mg/L	400	375	225	185	325
k (base e)	d^{-1}	0.29	0.23	0.027	0.025	0.023
NH$_3$	mg/L	80	65	75	67	83

2-36 An industrial wastewater is known to contain only stearic acid ($C_{18}H_{36}O_2$), glycine ($C_2H_5O_2N$), and glucose ($C_6H_{12}O_6$). The results of a laboratory analysis for four different samples are as follows. For one these samples (to be selected by instructor), determine the concentration of each of the three constituents in mg/L.

a. organic nitrogen = 11 mg/L, organic carbon = 130 mg/L, and COD = 425 mg/L. Determine the concentration of each of the three constituents in mg/L.

b. organic nitrogen = 13 mg/L, organic carbon = 109 mg/L, and COD = 440 mg/L. Determine the concentration of each of the three constituents in mg/L.

c. organic nitrogen = 9 mg/L, organic carbon = 123 mg/L, and COD = 625 mg/L. Determine the concentration of each of the three constituents in mg/L.

d. organic nitrogen = 12 mg/L, organic carbon = 143 mg/L, and COD = 425 mg/L. Determine the concentration of each of the three constituents in mg/L.

2-37 How many mg/L of $Cr_2O_7^{2-}$ are consumed if the COD of a wastewater sample is found to 450, 325, 533, or 622 mg/L (value to be selected by instructor)?

2-38 Given the absorption calibration data for methylene blue dye as measured with a spectrophotometer, determine the concentration of methylene blue at absorbance measurements of 0.366, 0.812, 2.010, and 2.310.

Concentration, mL/L	Absorbance at 280 nm
0.1	0.065
0.25	0.147
0.5	0.288
1	0.578
2	1.182
3	1.994
4	2.503
5	2.507
6	2.519
7	2.519

2-39 Bacteria have equivalent diameters of 2×10^{-6} μm and densities of approximately 1 kg/L. Under optimal conditions, bacteria can divide every 30 min. Determine the mass of bacteria that would accumulate in 72 h under continuing optimal growth conditions. Can this occur? Explain.

2-40 If the bacteria found in feces have an average volume of 2.0 μm³, determine the concentration of suspended solids that would be represented by a bacterial density equal to 10^8 organisms/mL. Assume the density of the bacteria is 1.005 kg/L.

2-41 Derive, from fundamental considerations, an expression that can be used to compute the MPN based on a single sample comprised of 5 fermentation tubes.

2-42 Use the Poisson formula to compute the most probable number (MPN) of coliform organisms per 100 mL based on the following test results (sample one column only). Positive test results are indicated by (+), and negative results by (−).

Sample volume, mL	A (+)	A (−)	B (+)	B (−)	C (+)	C (−)	D (+)	D (−)
100	5	0	5	0	—	—	5	0
10	5	0	5	0	—	—	5	0
1	4	1	4	1	5	0	4	1
0.1	3	2	2	3	2	3	4	1
0.01	1	4	—	—	0	5	0	5

2–43 The results of a presumptive coliform test were 2 of 5 tubes of 10-mL portions positive. What is the MPN per 100 mL?

2–44 The results of a presumptive coliform test were 4 of 5 tubes of 10-mL portions positive. What is the MPN per 100 mL?

2–45 Six weekly effluent samples have been analyzed for bacterial content using the standard confirmed test. Determine the coliform density, expressed as MPN, for three of the weekly samples (to be selected by instructor) using the Poisson equation (Eq. 2–74). Check the answers obtained using the standard MPN tables and with the Thomas equation (Eq. 2–75).

Size of portion, mL	Sample number 1	2	3	4	5	6
100.0			5/5	5/5	5/5	5/5
10.0		4/5	4/5	5/5	5/5	5/5
1.0	4/5	5/5	5/5	5/5	5/5	5/5
0.1	3/5	3/5	3/5	2/5	1/5	5/5
0.01	1/5	2/5	2/5	3/5	2/5	5/5
0.001					0/5	1/5

2–46 Using the data given in Problem 2–45 determine the coliform density, expressed as MPN, for the fourth, fifth, and sixth weekly samples using the Poisson equation (Eq. 2–74). Check the answers obtained using the standard MPN tables and with the Thomas equation (Eq. 2–75).

2–47 Discuss the advantages and disadvantages of using the fecal coliform and fecal streptococci tests to indicate bacteriological contamination. Cite a minimum of three references from the literature.

2–48 Determine graphically the 48- and 96-hr LC_{50} values in percent by volume for the following toxicity test data below.

Concentration of waste, % by volume	Number of test animals	Number of test animals dead after 48 hr	96 hr
80	20	17	20
60	20	13	20

(continued)

(Continued)

Concentration of waste, % by volume	Number of test animals	Number of test animals dead after	
		48 hr	96 hr
40	20	10	15
20	20	6	13
10	20	3	9
5	20	1	4
2	20	0	2

2–49 Determine graphically the 48- and 96-hr LC_{50} values in percent by volume for the following toxicity test data below.

Concentration of waste, % by volume	Number of test animals	Number of test animals dead after		
		24 h	48 h	96 h
12	20	0	2	8
10	20	0	5	10
8	20	1	8	13
6	20	3	11	16
4	20	6	16	20
2	20	14	20	20

REFERENCES

Albertson, O. E. (1995) Is $CBOD_5$ Test Viable for Raw and Settled Wastewater? *Journal of Environmental Engineering Division*, American Society of Civil Engineers, vol. 121, no. 7, pp. 515–520.

ASCE (1989) Sulfide in Wastewater Collection and Treatment Systems, Manual of Practice No. 69, *American Society of Civil Engineers*, New York.

ASTM (1979) Standard Practice for the Determination of Odor and Taste Thresholds by the Forced-Choice Ascending Concentration Series Method of Limits, *American Society for Testing and Materials*, E679, Philadelphia, PA.

Balmat, J. L. (1957) Biochemical Oxidation of Various Particulate Fractions of Sewage, *Sewage and Industrial Wastes,* vol. 29, no. 7.

BioVir Laboratories, Inc. (2001) Literature on Enteric Virus, *Cryptosporidium,* and *Giardia,* Benicia, CA.

Cooper, R. C. (2001) Personal Communication, BioVir Laboratories, Benicia, CA.

Cornwell, D. A. (1990) Air Stripping and Aeration, in F. W. Pontius (ed.) *Water Quality and Treatment: A Handbook of Community Water Supplies,* 4th ed., McGraw-Hill, Inc., New York.

Craun, G. F., P. S. Berger, and R. L. Calderon (1997) Coliform Bacteria and Waterborne Disease Outbreaks, *Journal American Water Works Association,* vol. 89, no. 3.

Crites, R., and G. Tchobanoglous (1998) *Small and Decentralized Wastewater Management Systems,* McGraw-Hill, New York.

Crompton, D. W. T. (1999) How much human helminthias is there in the world? *Journal of Parasitology,* vol. 85, pp. 379–403.

Crook, J. (1998) Chapter 7 Water Reclamation and Reuse Criteria, in T. Asano (ed.) *Wastewater Reclamation and Reuse,* Technomic Publishing Co., Ltd. Lancaster, PA.

Davies, C. W. (1962) *Ion Association,* Butterworth and Company, Ltd., London.

Delahay, P. (1957) *Instrumental Analysis,* The Macmillan Company, New York, NY.

Eckenfelder, W. W., Jr. (2000) *Industrial Water Pollution Control,* 3rd ed., McGraw Hill, Boston, MA.

Eldrige, E. F. (1942) *Industrial Waste Treatment Practice,* McGraw-Hill Book Company, Inc., New York, NY.

Feachem, R. G., D. J. Bradley, H. Garelick, and D. D. Mara (1983) *Sanitation and Disease: Health Aspects of Excreta and Wastewater Management,* Published for the World Bank by John Wiley & Sons, New York.

Federal Register (January 8, 1981) 46 CRF, 2264.

Fujimoto, Y. (1961) "Graphical Use of First-Stage BOD Equation," *Journal Water Pollution Control Federation,* vol. 36, no. 1, p. 69.

Hand, D. W., D. R. Hokanson, and J. C. Crittenden (1999) "Chapter 6 Air Stripping and Aeration," in R. D. Letterman, ed., *Water Quality and Treatment: A Handbook of Community Water Supplies,* 5th ed., American Water Works Association, McGraw-Hill, New York, NY.

Hill, M. J. (1988) *Nitrosomines, Toxicology and Microbiology,* Ellis Horwood Publishing, England.

Hoover, S. R., and N. Porges (1952) "Assimilation of Dairy Wastes by Activated Sludge II: The Equations of Synthesis and Oxygen Utilization," *Sewage and Industrial Wastes,* vol. 24.

Hughes, W. W. (1996) *Essentials of Environmental Toxicology,* Taylor & Francis, Philadelphia, PA.

Ingraham, J. L., and C. A. Ingraham (1995) *Introduction to Microbiology,* Wadsworth Publishing Company, Belmont, CA.

Jaques, R. S. (1994) Personal communication, Monterey Regional Water Pollution Control Agency, Monterey, CA.

Keller, G. H., and M. L. Manak (1989) *DNA Probes,* Stockton Press, New York.

Khuroo, M. S. (1996) *Ascariasis, Gasteroenteriology Clinics of North America,* vol. 25, no. 3, pp. 553–577.

Larson, R., and B. Farber (2000) *Elementary Statistics: Picturing the World,* Prentice-Hall Inc., Upper Saddle River, NJ.

Levine, A. D., G. Tchobanoglous, and T. Asano (1985) Characterization of the Size Distribution of Contaminants in Wastewater: Treatment and Reuse Implications, *Journal Water Pollution Control Federation,* vol. 57, no. 7, pp. 205–216.

Levine, A. D., G. Tchobanoglous, and T. Asano (1991) Size Distributions of Particulate Contaminants in Wastewater and Their Impact on Treatability, *Water Research,* vol. 25, no. 8, pp. 911–922.

Levins, R., T. Awerbuch, U. Brinkrnann, I. Eckardt, P. Epstein, N. Makhoul, C. Albuquerque de Possas, C. Puccia, A. Spielman, and M. F. Wilson (1994) The Emergence of New Diseases: Lessons learned from the emergence of new diseases and the resurgence of old ones may help us prepare for future epidemics, *American Scientist,* vol. 82, pp. 53–60.

Loge, F. (1997) Personal communication, Department of Civil and Environmental Engineering, University of California at Davis, Davis, CA.

Madigan, M. T., J. M. Martinko, and J. Parker (2000) *Brock Biology of Microorganisms,* 9th ed, Prentice-Hall, Upper Saddle River, NJ.

Maier, R. M., I. L. Pepper, and C. P. Gerba (2000) *Environmental Microbiology,* Academic Press, A Harcourt Science and Technology Company, San Diego, CA.

McMurry, J., and R. C. Fay (1998) *Chemistry,* 2nd ed., Prentice-Hall, Upper Saddle River, NJ.

Montgomery, J. M., Consulting Engineers, Inc. (1985) *Water Treatment Principles and Design,* A Wiley-Interscience Publication, John Wiley & Sons, New York.

Moore, E. W., H. A. Thomas, and W. B. Snow (1950) Simplified Method for Analysis of BOD Data, *Sewage & Industrial Wastes,* vol. 22, no. 10, 1950.

Nelson, K. (2001) Personal communication, Department of Civil and Environmental Engineering, University of California at Davis, Davis, CA.

Patterson, R. G., R. C. Jain, and S. Robinson (1984) "Odor Controls for Sewage Treatment Facilities," Presented at the 77th Annual Meeting of the Air Pollution Control Association, San Francisco, CA.

Pepper, I. L., C. P. Gerba, and M. L. Brusseau (eds.) (1996) *Pollution Science,* Academic Press, San Diego, CA.

Perry, R. H., D. W. Green, and J. O. Maloney (1984) *Perry's Chemical Engineers' Handbook,* 6th ed., McGraw-Hill Book Company, New York.

Phelps, E. B. (1944) *Stream Sanitation,* John Wiley & Sons, New York.

Roberts, L. S., and J. Janovy, Jr. (1996) *Foundations of Parasitology,* 5th ed., WCB, Wm. C. Brown Publishers, Dubuque, IA.

Rose, J. B., and C. P. Gerba (1991) "Assessing Potential Health Risks from Viruses and Parasites in Reclaimed Water in Arizona and Florida, U.S.A.," Water Science Technology, vol. 23, pp. 2091–2098.

Russell, L. L. (1976) "Chemical Aspects of Groundwater Recharge with Wastewaters," Ph.D Thesis, University of California, Berkeley, CA.

Sawyer, C. N., P. L. McCarty, and G. F. Parkin (1994) *Chemistry for Environmental Engineering,* 4th ed., McGraw-Hill, Inc., New York, NY.

Schroepfer, G. J., M. L. Robins, and R. H. Susag (1964) The Research Program on the Mississippi River in the Vicinity of Minneapolis and St. Paul, *Advances in Water Pollution Research,* vol.1, Pergamon, London.

Sheehy, J. P. (1960) Rapid Methods for Solving Monomolecular Equations, *Journal Water Pollution Control Federation,* vol. 32, no. 6.

Snoeyink, V. L., and D. Jenkins (1988) *Water Chemistry,* 2nd ed., John Wiley & Sons, New York.

Snyder, C. H. (1995) *The Extraordinary Chemistry of Ordinary Things,* 2nd ed., John Wiley & Sons, Inc., New York.

Standard Method (1998) *Standard Methods for the Examination of Water and Waste Water,* 20th ed., American Public Health Association, Washington, D.C.

Stanier, R.Y., J. L. Ingraham, M. L. Wheelis, and P. R. Painter (1986) *The Microbial World,* 5th ed., Prentice-Hall, Englewood Cliffs, NJ.

Stephen, C. E. (1982) Methods for Calculating an LC 50, in F. L. Mayer and J. L. Hamelink (eds.), *Aquatic Toxicology and Hazard Evaluation,* ASTM STP 634, American Society for Testing and Materials, pp. 65–84, Philadelphia, PA.

Tchobanoglous, G. (1995) Particle-Size Characterization: The Next Frontier, *Journal of Environmental Engineering Division,* ASCE, vol. 121, no. 12.

Tchobanoglous, G., and F. L. Burton (1991) *Wastewater Engineering, Treatment, Disposal, Reuse,* 3rd ed., McGraw-Hill Inc., New York.

Tchobanoglous, G., and E. D. Schroeder (1985) *Water Quality: Characteristics, Modeling, Modification,* Addison-Wesley Publishing Company, Reading, MA.

Thomas, H. A., Jr. (1942) Bacterial Densities from Fermentation Tube Tests, *Journal American Water Works Association,* vol. 34, no. 4, p. 572.

Thomas, H. A., Jr. (1950) Graphical Determination of BOD Curve Constants, *Water & Sewage Works,* vol. 97, p.123.

Thompson, D. E. (1997) Personal communication, Department of Civil and Environmental Engineering, University of California at Davis, Davis, CA.

Tsivoglou, E. C. (1958) "Oxygen Relationships in Streams," Robert A. Taft Sanitary Engineering Center, Technical Report W-58-2, Cincinnati, OH.

U.S. EPA (1977, effective date) *National Interim Primary Drinking Water Regulations,* Federal Register, 40(248), December 24, 1975, Section 59566–59588, Washington, DC (note the NIPDWRs have been amended four times since being issued and numerous rules have been added).

U.S. EPA (1977b, effective date) *National Interim Primary Drinking Water Regulations,* EPA-570/9-76-003, U.S. Environmental Protection Agency, Washington, DC.

U.S. EPA (1985a) *Methods for Measuring the Acute Toxicity of Effluents to Freshwater and Marine Organisms,* U.S. EPA Environmental Monitoring and Support Laboratory, EPA-600/4-85/013, U.S. Environmental Protection Agency, Cincinnati, OH.

U.S. EPA (1985b) *Technical Support Document for Water Quality-Based Toxics Control,* U.S. EPA Office of Water, U.S. Environmental Protection Agency, Washington, DC.

U.S. EPA (1985c) *Short Term Methods for Estimating Chronic Toxicity of Effluents and Receiving Waters to Freshwater Organisms,* EPA-660/4-85/014, U.S. Environmental Protection Agency, Washington, DC.

U.S. EPA (1985d) *User's Guide to the Conduct and Interpretation of Complex Effluent Toxicity Tests at Estuarine/Marine Sites,* EPA-600/X-86/224, U.S. Environmental Protection Agency, Washington, DC.

U.S. EPA (1985e) *Odor Control and Corrosion Control in Sanitary Sewerage Systems and Treatment Plants,* Design Manaual, EPA-625/1-85-018, U.S. Environmental Protection Agency, Washington, DC.

U.S. EPA (1988) *Short Term Methods for Estimating the Chronic Toxicity of Effluents and Receiving Waters to Marine and Estuarine Organisms,* EPA-600/4-88/028, U.S. Environmental Protection Agency, Washington, DC.

U.S. EPA (1989) *Short Term Methods for Estimating Chronic Toxicity of Effluents and Receiving Waters to Freshwater Organisms,* EPA-660/2nd ed., U.S. Environmental Protection Agency, Washington, DC.

U.S. GS (2000) "National Reconnaissance of Emerging Contaminants in the Nation's Stream Waters," U.S. Geological Survey, http://toxics.usgs.gov/regional/contaminants.html.

Voyles, B. A. (1993) *The Biology of Viruses,* Mosby, St. Louis, MO.

Wainwright, T. (1986), The Chemistry of Nitrosamine Formation: Relevance to Malting and Brewing, *Journal Institute of Brewing,* vol. 92, pp. 49–64.

WEF (1995) *Odor Control in Wastewater Treatment Plants,* WEF Manual of Practice No. 22, ASCE Manuals and Reports on Engineering Practice No. 82, Water Environment Federation, Alexandria, VA.

West, P. M., and F. R. Olds (1992) Chinical Schistosomiasis, *R.I. Medical Journal,* vol. 75, p. 179.

Wilson, G. (1975) "Odors: Their Detection and Measurement," EUTEK, Process Development and Engineering, Sacramento, CA.

Wood, D. K., and G. Tchobanoglous (1975) Trace Elements in Biological Waste Treatment, *Journal Water Pollution Control Association.* vol. 47, no. 7.

Young, J. C. (1973) Chemical Methods for Nitrification Control, *Jounal Water Pollution Control Association,* vol. 45, no. 4.

Young, J. C., and E. R. Baumann (1976a) The Electrolytic Respirometer—I: Factors Affecting Oxygen Uptake Measurements, *Water Research,* vol. 10, no. 11.

Young, J. C., and E. R. Baumann (1976b) The Electrolytic Respirometer—II: Use in Water Pollution Control Plant Laboratories, *Water Research,* vol. 10, no. 12.

3

Analysis and Selection of Wastewater Flowrates and Constituent Loadings

Determining wastewater flowrates and constituent mass loadings is a fundamental step in initiating the conceptual process design of wastewater treatment facilities. Reliable data for existing and projected flowrates affect the hydraulic characteristics, sizing, and operational considerations of the treatment system components. Constituent mass loading, the product of constituent concentration and flowrate, is necessary to determine the capacity and operational characteristics of the treatment facilities and ancillary equipment to ensure that treatment objectives are met.

Important factors and issues, typical to most planning and design projects, addressed in this chapter include: (1) components of wastewater flows; (2) wastewater sources and flowrates; (3) analysis of flowrate data; (4) analysis of constituent mass loading data; (5) statistical analysis of flowrates, constituent concentrations, and mass loadings; and (6) selection of design flowrates and mass loadings.

3–1 COMPONENTS OF WASTEWATER FLOWS

The components that make up the wastewater flow from a community depend on the type of collection system used and may include:

1. *Domestic (also called sanitary) wastewater.* Wastewater discharged from residences and from commercial, institutional, and similar facilities.
2. *Industrial wastewater.* Wastewater in which industrial wastes predominate.
3. *Infiltration/inflow (I/I).* Water that enters the collection system through indirect and direct means. Infiltration is extraneous water that enters the collection system through leaking joints, cracks and breaks, or porous walls. Inflow is stormwater that enters the collection system from storm drain connections (catch basins), roof leaders, foundation and basement drains, or through access port (manhole) covers.
4. *Stormwater.* Runoff resulting from rainfall and snowmelt.

Three types of collection systems are used for the removal of wastewater and stormwater: sanitary collection systems, storm collection systems, and combined collection systems. Where separate collection systems are used for the collection of wastewater (sanitary collection systems) and stormwater (storm collection systems), wastewater flows in sanitary collection systems consist of three major components: (1) domestic wastewater, (2) industrial wastewater, and (3) infiltration/inflow. Where only one collection system (combined) is used, wastewater flows consist of these three components plus stormwater. In both cases, the percentage of the wastewater components will vary with local conditions and the time of the year.

3–2 WASTEWATER SOURCES AND FLOWRATES

Data that can be used to estimate average wastewater flowrates from various domestic, commercial, institutional, and industrial sources and the infiltration/inflow contribution are presented in this section. Variations in the flowrates that must be established before collection systems and treatment facilities are designed are also discussed.

Domestic Wastewater Sources and Flowrates

The principal sources of domestic wastewater in a community are the residential areas and commercial districts. Other important sources include institutional and recreational facilities. For areas now served with collection systems, wastewater flowrates are commonly determined from existing records or by direct field measurements (see Fig. 3–1). For new developments, wastewater flowrates are derived from an analysis of population data and estimates of per capita wastewater flowrates from similar communities.

Water consumption records may also be used for estimating flowrates. These records are especially useful in other parts of the world where water use for landscape irrigation is limited and 90 percent or more of the water used becomes wastewater. In the United States, on the average about 60 to 90 percent of the per capita water consumption becomes wastewater. The higher percentages apply to the northern states during cold weather; the lower percentages are applicable to the semiarid region of the southwestern United States where landscape irrigation is used extensively. When water consumption records are used for estimating wastewater flowrates, the amount of water consumed for purposes such as landscape irrigation (that is not discharged to the collection system), leakage from water mains and service pipes, or product water that is used by manufacturing establishments must be evaluated carefully.

Residential Areas. For many residential areas, wastewater flowrates are commonly determined on population and the average per capita contribution of wastewater.

Figure 3–1

Flowrate measurement using a Parshall flume: (a) Parshall flume equipped with a sonic indicator to determine the depth of flow that is correlated to the flowrate and (b) Parshall flume equipped with a float indicator to determine the depth of flow.

(a)   (b)

Table 3–1
Typical wastewater flowrates from urban residential sources in the United States[a]

Household size, no. of persons	Flowrate, gal/capita·d		Flowrate, L/capita·d	
	Range	Typical	Range	Typical
1	75–130	97	285–490	365
2	63–81	76	225–385	288
3	54–70	66	194–335	250
4	41–71	53	155–268	200
5	40–68	51	150–260	193
6	39–67	50	147–253	189
7	37–64	48	140–244	182
8	36–62	46	135–233	174

[a] Adapted in part from AWWARF (1999).

For residential areas where large residential development is planned, it is often advisable to develop flowrates on the basis of land-use areas and anticipated population densities. Where possible, these rates should be based on actual flow data from selected similar communities, preferably in the same locale. In the past, the preparation of population projections for use in estimating wastewater flowrates was often the responsibility of the engineer, but today population projection data are usually available from local, regional, and state planning agencies.

Wastewater flowrates can vary depending on the quantity and quality of the water supply; rate structure; and economic, social, and other characteristics of the community. Data on ranges and typical flowrate values are given in Table 3–1 for residential sources in the United States. Beginning in recent years, greater attention is now being given to water conservation and the installation of water-conserving devices and appliances. Reduced household water use changes not only the quantity of wastewater generated but, as discussed later in this chapter, the characteristics of wastewater as well.

Commercial Districts. Depending on the function and activity, unit flowrates for commercial facilities can vary widely. Because of the wide variations that have been observed, every effort should be made to obtain records from actual or similar facilities. If no other records are available, estimates for selected commercial sources, based on function or persons served, may be made using the data presented in Table 3–2. In the past, commercial wastewater flowrates were often based on existing or anticipated future development or comparative data. Flowrates were generally expressed in terms of quantity of flow per unit area [i.e., m^3/ha·d (gal/ac·d)]. Typical unit-flowrate allowances for commercial developments normally range from 7.5 to 14 m^3/ha·d (800 to 1500 gal/ac·d). The latter approach can be used to check the values obtained from existing records or estimates made using Table 3–2.

Table 3–2
Typical wastewater flowrates from commercial sources in the United States[a]

Source	Unit	Flowrate, gal/unit·d		Flowrate, L/unit·d	
		Range	Typical	Range	Typical
Airport	Passenger	3–5	4	11–19	15
Apartment	Bedroom	100–150	120	380–570	450
Automobile service station	Vehicle served	8–15	10	30–57	40
	Employee	9–15	13	34–57	50
Bar/cocktail lounge	Seat	12–25	20	45–95	80
	Employee	10–16	13	38–60	50
Boarding house	Person	25–65	45	95–250	170
Conference center	Person	6–10	8	40–60	30
Department store	Toilet room	350–600	400	1300–2300	1500
	Employee	8–15	10	30–57	40
Hotel	Guest	65–75	70	150–230	190
	Employee	8–15	10	30–57	40
Industrial building (sanitary waste only)	Employee	15–35	20	57–130	75
Laundry (self-service)	Machine	400–550	450	1500–2100	1700
	Customer	45–55	50	170–210	190
Mobile home park	Unit	125–150	140	470–570	530
Motel (with kitchen)	Guest	55–90	60	210–340	230
Motel (without kitchen)	Guest	50–75	55	190–290	210
Office	Employee	7–16	13	26–60	50
Public lavatory	User	3–5	4	11–19	15
Restaurant:					
Conventional	Customer	7–10	8	26–40	35
With bar/ cocktail lounge	Customer	9–12	10	34–45	40
Shopping center	Employee	7–13	10	26–50	40
	Parking space	1–3	2	4–11	8
Theater (Indoor)	Seat	2–4	3	8–15	10

[a] Adapted from Metcalf & Eddy (1991), Salvato (1992), and Crites and Tchobanoglous (1998).

Table 3-3
Typical wastewater flowrates from institutional sources in the United States[a]

Source	Unit	Flowrate, gal/unit·d		Flowrate, L/unit·d	
		Range	Typical	Range	Typical
Assembly hall	Guest	3–5	4	11–19	15
Hospital	Bed	175–400	250	660–1500	1000
	Employee	5–15	10	20–60	40
Institutions other than hospitals	Bed	75–125	100	280–470	380
	Employee	5–15	10	20–60	40
Prison	Inmate	80–150	120	300–570	450
	Employee	5–15	10	20–60	40
School, day:					
With cafeteria, gym, and showers	Student	15–30	25	60–120	100
With cafeteria only	Student	10–20	15	40–80	60
School, boarding	Student	75–100	85	280–380	320

[a]Adapted from Metcalf & Eddy (1991), Salvato (1992), and Crites and Tchobanoglous (1998).

Institutional Facilities. Typical flowrates from some institutional facilities are shown in Table 3–3. Again, it is stressed that flowrates vary with the region, climate, and type of facility. The actual records of institutions are the best sources of flow data for design purposes.

Recreational Facilities. Wastewater flowrates from many recreational facilities are subject to seasonal variations. Typical data on wastewater flowrates from recreational facilities are presented in Table 3–4.

Strategies for Reducing Interior Water Use and Wastewater Flowrates

Because of the importance of conserving both resources and energy, various means for reducing wastewater flowrates and pollutant loadings from domestic sources are available. The reduction of wastewater flowrates from domestic sources results directly from the reduction in interior water use. Therefore, the terms *interior water use* and *domestic wastewater flowrates* are used interchangeably. Representative water use rates for various devices and appliances are reported in Table 3–5. Information on the relative distribution of water use within a residence is reported in Table 3–6. Devices and appliances that can be used to reduce interior domestic water use and wastewater flows are described in Table 3–7.

Table 3-4

Typical wastewater flowrates from recreational facilities in the United States[a]

Facility	Unit	Flowrate, gal/unit·d		Flowrate, L/unit·d	
		Range	Typical	Range	Typical
Apartment, resort	Person	50–70	60	190–260	230
Cabin, resort	Person	8–50	40	30–190	150
Cafeteria	Customer	2–4	3	8–15	10
	Employee	8–12	10	30–45	40
Camp:					
With toilets only	Person	15–30	25	55–110	95
With central toilet and bath facilities	Person	35–50	45	130–190	170
Day	Person	15–20	15	55–76	60
Cottages, (seasonal with private bath)	Person	40–60	50	150–230	190
Country club	Member present	20–40	25	75–150	100
	Employee	10–15	13	38–57	50
Dining hall	Meal served	4–10	7	15–40	25
Dormitory, bunkhouse	Person	20–50	40	75–190	150
Fairground	Visitor	1–3	2	4–12	8
Picnic park with flush toilets	Visitor	5–10	5	19–38	19
Recreational vehicle park:					
With individual connection	Vehicle	75–150	100	280–570	380
With comfort station	Vehicle	40–50	45	150–190	170
Roadside rest areas	Person	3–5	4	10–19	15
Swimming pool	Customer	5–12	10	19–45	40
	Employee	8–12	10	30–45	40
Vacation home	Person	25–60	50	90–230	190
Visitor center	Visitor	3–5	4	10–19	15

[a] Adapted from Metcalf & Eddy (1991), Salvato (1992), and Crites and Tchobanoglous (1998).

Another method of achieving flow reduction that has been adopted by a number of communities is to restrict the use of appliances, such as automatic dishwashers and kitchen food-waste grinders (i.e., garbage disposal units), that tend to increase water consumption. The use of one or more of the flow-reduction devices is specified for all new residential dwellings in many communities; in others, the use of waste food grinders has been limited in new housing developments. Further, many individuals concerned about conservation have installed such devices on their own, as a means of

Table 3–5
Typical rates of water use for various devices and appliances in the United States[a]

Device or appliance	U.S. Customary Units		SI Units	
	Units	Range	Units	Range
Automatic home-type washing machine:				
Top loading	gal/load	34–57	L/load	130–216
Front loading	gal/load	12–15	L/load	45–60
Automatic home-type dishwasher	gal/load	9.5–15.5	L/load	36–60
Bathtub	gal/use	30	L/use	114
Kitchen food-waste grinder	gal/d	1–2	L/d	4–8
Shower	gal/min·use	2.5–3	L/min·use	9–11
Toilet, tank, conservation type	gal/use	1.6–3.5	L/use	6–13
Toilet, tank type, standard	gal/use	4–6	L/use	15–23
Washbasin	gal/min·use	2–3	L/min·use	8–11

[a] Adapted from Metcalf & Eddy (1991), Salvato (1992), and Crites and Tchobanoglous (1998).

Table 3–6
Typical distribution of residential interior water use in the United States[a,b]

Use	Percent of total	
	Range	Typical
Shower	12–20	16.8
Bath	1–3	1.7
Faucet	12–18	15.7
Dishwashing	1–2	1.4
Clothes washing	12–28	21.8
Other domestic	0–9	2.2
Toilet flushing	23–31	26.7
Leakage	5–22	13.7
Total		**100**

[a] Adapted from AWWARF (1999).
[b] Mean indoor water use = 260 L/capita·d (69 gal/capita·d).

reducing water consumption. New designs in front-loading clothes washers also offer significant reductions in water use, on the order of 50 to 75 percent of older models. A comparison of residential interior water use (and resulting per capita wastewater flows), from a recent survey, is given in Table 3–8 for homes without and with water-conserving fixtures (AWWA, 1998). The potential saving of employing selected water-efficient devices is illustrated in Example 3–1.

Table 3–7
Flow-reduction
devices and
appliances

Device/appliance	Description and/or application
Faucet aerators	Increases the rinsing power of water by adding air and concentrating flow, thus reducing the amount of wash water used
Flow-limiting showerheads	Restricts and concentrates water passage by means of orifices that limit and divert shower flow for optimum use by the bather
Low-flush toilets	Reduces the amount of water per flush
Pressure-reducing valve	Maintains home water pressure at a lower level than that of the water distribution system. Decreases the probability of leaks and dripping faucets
Pressurized shower	Water and compressed air are mixed together. Impact provides the sensation of conventional shower
Retrofit kits for bathroom	Kits may consist of shower flow restrictors, toilet dams or fixture displacement bags, and toilet leak-detector tablets
Toilet dam	A partition in the toilet tank that reduces the amount of water per flush
Toilet leak detectors	Tablets that dissolve in the toilet tank and release dye to indicate leakage of the flush valve
Vacuum toilet	A vacuum along with a small amount of water is used to remove solids from toilet
Water-efficient dishwasher	Reduces the amount of water used to wash dishes
Water-efficient clothes washer	Reduces the amount of water used to wash clothes. New front-loading machines have been developed that not only use less water but are more energy-efficient

Table 3–8
Typical comparisons of interior water use without and with water-conservation practices and devices in the United States[a]

Use	Flow, gal/capita·d		Flow, L/capita·d	
	Without water conservation	With water conservation	Without water conservation	With water conservation
Bathing	1.3	1.3	5	5
Showers	13.2	11.1	50	42
Dishwashing	1.0	1.0	4	4
Clothes washing	16.8	11.8	64	45
Faucets	11.4	11.1	43	42
Toilets	19.3	9.3	73	35
Leaks	9.4	4.7	36	18
Other domestic use	1.6	1.6	6	6
Totals	74	51.9	281	197

[a] Adapted from AWWA (1998).

EXAMPLE 3–1 Determine Water Savings by Employing Water-Efficient Appliances A new subdivision of 2000 homes is planned, and a condition of the building permit is to determine the savings in water consumption (and wastewater flows) if the following water-efficient appliances are used: front-loading washing machines, ultra-low flush toilets, and ultra-low-flow showerheads. Use 3.5 residents per home and values for devices and appliances from Table 3–8.

The estimated water use and percentage savings are illustrated in the following table.

Appliance/device	No. of Residents	Unit water use, L/capita·d		Water use, L/day	
		No conservation	With conservation	No conservation	With conservation
Clothes washing	7000	64	45	448,000	315,000
Toilets	7000	73	35	511,000	245,000
Showers	7000	50	42	350,000	294,000
Total				1,309,000	854,000
Percent savings					34.8

Comment Three of the largest water-using appliances and devices utilized in the home are those described in this example. Interior water use and the generation of wastewater can be reduced significantly with the community-wide installation of water-efficient appliances and devices, thus reducing the flows that will have to be handled by the collection system and treatment plant. Where high infiltration rates occur within the collection system, it is difficult or impossible to assess the beneficial effects of using water conservation devices.

Water Use in Developing Countries

The typical flowrates and use patterns presented in Tables 3–1 through 3–6 and Table 3–8 are based on water use and wastewater flowrate data from communities and facilities in the United States. Many developed countries have flowrates in similar ranges. Water use and, consequently, wastewater-generation rates in developing countries, however, are significantly lower. In some cases, the water supply is only available for limited periods of the day. Water-use data from some developing countries are given in Table 3–9.

Sources and Rates of Industrial (Nondomestic) Wastewater Flows

Nondomestic wastewater flowrates from industrial sources vary with the type and size of the facility, the degree of water reuse, and the onsite wastewater-treatment methods, if any. Extremely high peak flowrates may be reduced by the use of onsite detention tanks and equalization basins. Typical design values for estimating the flows from industrial areas that have no or little wet-process-type industries are 7.5 to 14 m³/ha·d (1000 to 1500 gal/ac·d) for light industrial developments and 14 to 28 m³/ha·d (1500 to

Table 3–9

Water consumption in developing countries and areas[a]

Country/Area	Per capita water consumption	
	Gal/d	L/d
China	21	80
Africa	4–9	15–35
Southeast Asia	8–19	30–70
Western Pacific	8–24	30–90
Eastern Mediterranean	11–23	40–85
Algeria, Morocco, Turkey	5–17	20–65
Latin America and Caribbean	19–51	70–190
World average	9–24	35–90

[a] Adapted, in part, from Salvato (1992).

3000 gal/ac·d) for medium industrial developments. For industries without internal water recycling or reuse programs, it can be assumed that about 85 to 95 percent of the water used in the various operations and processes will become wastewater. For large industries with internal water-reuse programs, separate estimates based on actual water consumption records must be made. Average domestic (sanitary) wastewater contributed from industrial facilities may vary from 30 to 95 L/capita·d (8 to 25 gal/capita·d).

Infiltration/Inflow

Extraneous flows in collection systems, described as infiltration and inflow, are illustrated on Fig. 3–2 and are defined as follows:

Infiltration. Water entering a collection system from a variety of entry points including service connections and from the ground through such means as defective pipes, pipe joints, connections, or access port (manhole) walls.

Steady inflow. Water discharged from cellar and foundation drains, cooling-water discharges, and drains from springs and swampy areas. This type of inflow is steady and is identified and measured along with infiltration.

Direct inflow. Those types of inflow that have a direct stormwater runoff connection to the sanitary collection system and cause an almost immediate increase in wastewater flowrates. Possible sources are roof leaders, yard and areaway drains, access port covers, cross connections from storm drains and catch basins, and combined systems.

Total inflow. The sum of the direct inflow at any point in the system plus any flow discharged from the system upstream through overflows, pumping station bypasses, and the like.

Delayed inflow. Stormwater that may require several days or more to drain through the collection system. Delayed inflow can include the discharge of sump pumps from cellar drainage as well as the slowed entry of surface water through access ports (manholes) in ponded areas.

Figure 3–2

Graphic identification of infiltration/inflow.

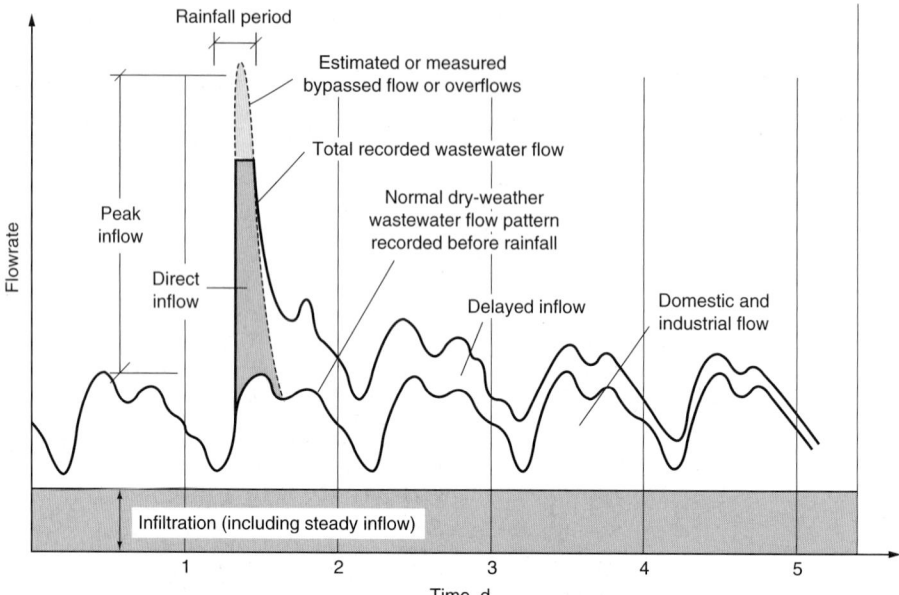

The initial impetus in the United States for defining and identifying infiltration/inflow was the Federal Water Pollution Control Act Amendments of 1972. As a condition of receiving a federal grant for the design and construction of wastewater-treatment facilities, grant applicants had to demonstrate that their wastewater-collection systems were not subject to excessive infiltration/inflow. By correcting infiltration/inflow problems and "tightening" the collection system, benefits to the community include (1) reducing wastewater backups and overflows in the collection system, (2) increasing the efficiency of operation of wastewater-treatment facilities, and (3) improving the utilization of collection system hydraulic capacity for wastewater requiring treatment instead of for infiltration/inflow. Because an understanding of the effects of infiltration/inflow is important in determining treatment-plant flowrates, a discussion of excessive infiltration/inflow is included in this section.

Infiltration into Collection Systems. One portion of the rainfall in a given area runs quickly into the stormwater systems or other drainage channels; another portion evaporates or is absorbed by vegetation; and the remainder percolates into the ground, becoming groundwater. The proportion of the rainfall that percolates into the ground depends on the character of the surface and soil formation and on the rate and distribution of the precipitation. Any reduction in permeability, such as that due to buildings, pavements, or frost, decreases the opportunity for precipitation to become groundwater and increases the surface runoff correspondingly. The amount of groundwater flowing from a given area may vary from a negligible amount for a highly impervious district or a district with a dense subsoil to 25 or 30 percent of the rainfall for a semipervious district with a sandy subsoil permitting rapid passage of water. The percolation of water through the ground from rivers or other bodies of water sometimes has considerable effect on the groundwater table, which rises and falls continually.

The presence of high groundwater results in leakage into the collection systems and in an increase in the quantity of wastewater and the expense of disposing of it. The amount of flow that can enter a collection system from groundwater, or infiltration, may range from 0.01 to 1.0 m³/d·mm·km (100 to 10,000 gal/d·in-mi) or more. The number of millimeter-kilometers (inch-miles) in a wastewater-collection system is the sum of the products of sewer diameters, in millimeters (inches), times the lengths, in kilometers (miles), of sewers of corresponding diameters.

Infiltration may also be estimated based on the area served by the collection system and may range from 0.2 to 28 m³/ha·d (20 to 3000 gal/ac·d) (Metcalf & Eddy, 1981). The variation in the amount of infiltration encompasses a wide range because the lot sizes may vary in area, which in turn affects the length and extent of the collection system network. During heavy rains, when there may be leakage through access port covers, or inflow as well as infiltration, the rate may exceed 500 m³/ha·d (50,000 gal/ac·d).

Infiltration/inflow is a variable part of the wastewater, depending on the quality of the material and workmanship in constructing the collection systems and building connections, the character of the maintenance, and the elevation of the groundwater compared with that of the collection system. The rate and quantity of infiltration depend on the length of the collection system, the area served, the soil and topographic conditions, and, to a certain extent, the population density (which affects the number and total length of house connections). Although the elevation of the water table varies with the quantity of rain and melting snow percolating into the ground, the leakage through defective joints, porous concrete, and cracks has been large enough, in some cases, to lower the groundwater table to the level of the collection system.

Most of the piping systems built during the first half of the 20th century were laid with cement mortar joints or hot-poured bituminous compound joints. Access ports were almost always constructed of brick masonry. Deterioration of pipe joints, pipe-to-access port joints, and the waterproofing of brickwork has resulted in a high potential for infiltration into these old sewers. The use of high-quality pipe with dense walls, precast access port sections, and joints sealed with rubber or synthetic gaskets is standard practice in modern collection-system design. The use of these improved materials has greatly reduced infiltration into and exfiltration from newly constructed collection systems, and infiltration rates with time are expected to be much slower than with older sewers.

Inflow into Collection Systems.   As described previously, the type of inflow that causes a "steady flow" cannot be identified separately and so is included in the measured infiltration. The direct inflow can cause an almost immediate increase in flowrates in sanitary systems. The effects of inflow on peak flowrates that must be handled by a wastewater treatment plant are shown in Example 3–2.

EXAMPLE 3–2   Determine Infiltration/Inflow from Wastewater Flow Records   A large city has measured high flowrates during the wet season of the year. The flow during the dry period of the year, when rainfall is rare and groundwater infiltration is negligible, averages 120,000 m³/d (31.7 Mgal/d). During the wet period when groundwater levels are elevated, the flowrate averaged 230,000 m³/d (60.8 Mgal/d) excluding those days during and following any significant rainfall events. During a recent storm, hourly flowrates were recorded during the peak flow period, as well as several days following the storm. The flow plots are shown on the accompanying figure. Compute the infiltration and inflow and determine if the infiltration is excessive. Excessive infiltration is defined by the local regulatory agency as rates over 0.75 m³/d·mm-km (8000 gal/d·in-mi) of collection system. The composite diameter-length of the collection system is 270,000 mm-km (6600 in-mi).

Solution

1. Determine the infiltration and inflow components during the wet season.
   a. Because infiltration is low during dry periods, high groundwater infiltration is computed as wet weather average flow minus base (dry weather) flow:

   Infiltration = (230,000 − 120,000) m³/d
   $$= 110,000 \text{ m}^3/\text{d}$$

   b. The maximum hourly inflow is determined graphically from the accompanying figure as the difference between the peak hourly wet weather flow during the storm and the comparable flowrate on the preceding day. The maximum flowrate at hour 35 is 606,000 m³/d, and the flow at hour 11 is 340,000 m³/d. In this case, the maximum inflow is

   Inflow = (606,000 − 340,000) m³/d
   $$= 266,000 \text{ m}^3/\text{d}$$

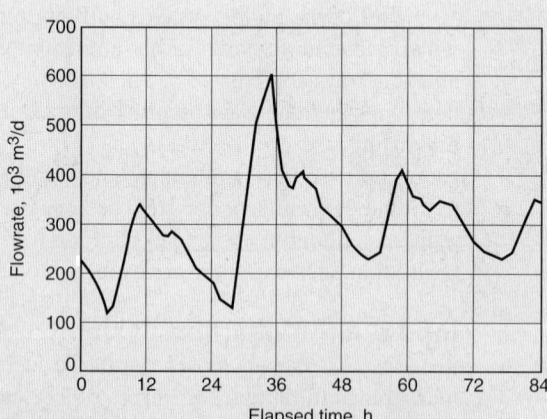

2. Determine if the infiltration is excessive.
   a. Calculate the infiltration by dividing the calculated flowrate by the composite diameter-length of the collection system.

   $$\text{Infiltation} = \frac{(110,000 \text{ m}^3 \cdot \text{d})}{(270,000 \text{ mm-km})} = 0.407 \text{ m}^3/\text{d·mm-km}$$

   b. Using the regulatory agency criterion of 0.75 m³/d·mm-km, the infiltration is not excessive.

Comment   In this example, the peak flow during the storm period was 4.7 times the average dry weather flow. As discussed later in this chapter, the peak flow factor is high for a system of this size. Because inflow represents over 50 percent of the peak flow and requires oversizing of the hydraulic capacity of the treatment plant, methods of inflow reduction should be investigated to decrease the hydraulic load on the collection system and treatment facilities.

Exfiltration from Collection Systems

Collection systems that have high infiltration rates and are in need of rehabilitation also may exhibit high exfiltration. When exfiltration occurs, untreated wastewater leaks out of pipe joints and service connections. If the piping and joints are in poor condition, significant quantities of wastewater may seep into the ground, travel through the gravel bedding of the piping system, or even surface in extreme cases. Seepage of untreated wastewater into the ground near shallow wells can result in pollution of the water supply. Well contamination has occurred in urban areas such as Los Angeles, California, where collection systems are within 300 m (1000 ft) of water wells. Exfiltration in collection systems near surface water bodies can also contribute to ongoing high coliform counts in those water bodies that may be difficult to correct. Reduction of inflow/filtration in collection systems may serve to limit exfiltration and remove potential threats to water supplies and public health. The potential effects of exfiltration on surface water quality are illustrated in Example 3–3.

EXAMPLE 3–3   **Determine the Pollution Contributing Effects of Exfiltration on Nearby Water Body**   Untreated wastewater from a damaged collection system pipeline leaks into a nearby lake. The leakage is estimated to be 10,000 L/d. What increase in concentration of organisms can be expected in the lake if the initial coliform count is 10^7 organisms/100 mL and the wastewater is diluted by 1000 to 1? Assume the number of coliform organisms in the dilution water is zero.

Solution

1. To solve this problem, a concentration balance (see Chap. 4 for additional details for mass and concentration balances) must be made as follows:

Total number of organisms in mixture	=	Total number of organisms in leakage	+	Total number of organisms in leakage

$$Q_M C_M = Q_L C_L + Q_{DW} C_{DW}$$

Where Q_M = volume of the mixture

C_M = concentration of organisms/100 mL in the mixture

Q_L = volume of leakage

C_L = concentration of organisms/100 mL in the leakage
Q_{DW} = volume of the dilution water
C_{DW} = concentration of organisms/100 mL in the dilution water

2.   Substitute for the various quantities and solve for the number of organisms in the mixture.

$$[10^4 \, \text{L} + 10^3 \, (10^4 \, \text{L})]\left(\frac{C_M}{100 \, \text{mL}}\right) = 10^4 \, \text{L}\left(\frac{10^7}{100 \, \text{mL}}\right) + 10^3(10^4 \, \text{L})\left(\frac{0}{100 \, \text{mL}}\right)$$

$$(10^4 \, \text{L} + 10^7 \, \text{L})\left(\frac{C_M}{100 \, \text{mL}}\right) = 10^4 \, \text{L}\left(\frac{10^7}{100 \, \text{mL}}\right)$$

$$C_M \approx \frac{10^4}{100 \, \text{mL}}$$

Comment Because untreated wastewater contains a high concentration of coliform organisms, the receiving water, even under the assumed condition that the exfiltration is well mixed, can contain high concentration levels. Such high coliform counts would be taken as indicators that a potential health hazard may exist. Therefore, in this example, because the exfiltration from the collection system into the lake is significant, receiving water-quality objectives will be difficult to maintain.

Combined System Flowrates

Flow in the combined system is composed mainly of rainfall runoff and wastewater. Flow enters the combined system continuously during both dry and wet weather from the contributing wastewater sources (see Fig. 1–1 in Chap. 1). This flow may include domestic, commercial, and industrial wastewater and infiltration. During a rainfall event, the amount of storm flow is normally much larger than the dry-weather wastewater flow, and the observed flows during wet weather can mask completely the dry weather flow patterns.

As flow proceeds through the combined system to the interceptor, it is modified by hydraulic routing effects as well as any surcharged conditions within the system (surcharging results when the pipeline capacity is exceeded). When the collection system capacity is exceeded, a portion of the flow may be discharged directly into a receiving body through overflows, either intentionally or accidentally, or routed to a special combined sewer overflow (CSO) treatment facility. In some cases where the combined system is undersized, flooding or surcharging may occur at various upstream locations within the system. Either condition (untreated overflow to receiving waters or flooding) is undesirable and most likely will result in a violation of the discharge permit and/or public health regulations.

The effects of combined system flowrates are illustrated on Fig. 3–3. The catchment hydrograph (flow versus time), as illustrated on Fig. 3–3b, closely resembles that

Figure 3–3

Flow variations in a
combined collection
system during wet
weather; (a) rainfall
hyetograph, (b) typical
catchment flowrate, and
(c) observed treatment
plant flowrate.

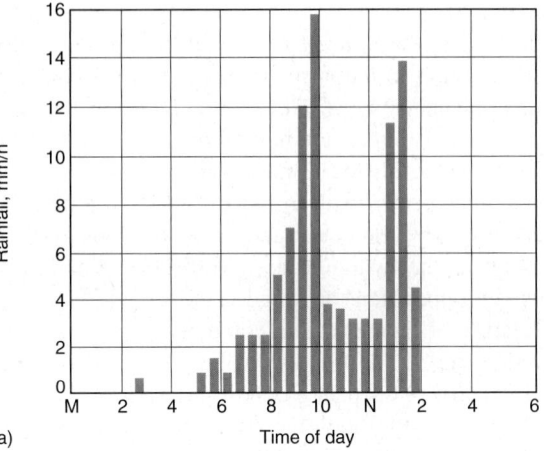

(a)

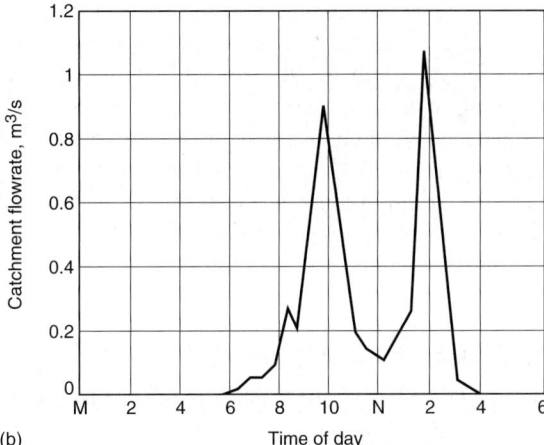

(b)

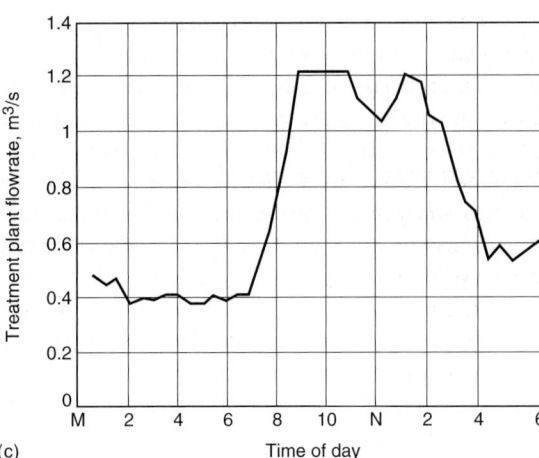

(c)

of the variations in rainfall intensity. The short response time between the rainfall event and the increase in the flowrate can be taken as an indication of a short travel time for flow from all points in the upstream combined system. In contrast, the hydrograph at the treatment plant (see Fig. 3–3c) shows less distinct flow peaks and a lag time of several hours for flows to return to normal dry-weather levels following rainfall cessation. The higher flows at this location are due to the larger contributing combined system, and the smoothed peaks result from loss of flow through overflows and hydraulic routing effects. The peak flowrates and accompanying mass loadings, however, must be accounted for in the hydraulic design of the treatment plant and in the selection of appropriate unit operations and processes.

Calculation of flowrates in a combined system is a complicated and challenging task. The first step in the process involves quantifying wastewater, rainfall runoff, and other sources of flow such as groundwater infiltration. These sources of flow are then combined and routed through the various components of the system. Finally, the volumes of flow exiting the system through CSO outlets, entering the downstream treatment facility, or being transported to other points in the system are determined.

Owing to the complexity of combined systems, it is normally necessary to use computer models to simulate the complete combined system including dry-weather wastewater flows, hydraulic routing through the piping system, discharges through the outlets, and the flow through the interceptor and to the treatment plant. Computer models that are in the public domain and are suitable for the assessment of combined systems include the Storm Water Management Model (SWMM) and the Corps of Engineers STORM and Hydrologic Simulation Model—Fortran (HSPF). Summaries of these programs are found in Nayyer (1992).

3–3 STATISTICAL ANALYSIS OF FLOWRATES, CONSTITUENT CONCENTRATIONS, AND MASS LOADINGS

The statistical analysis of wastewater flowrate and constituent concentration data involves the determination of statistical parameters used to quantify a series of measurements. Commonly used statistical parameters and graphical techniques for the analysis of wastewater management data are reviewed below.

Common Statistical Parameters

Commonly used statistical measures include the mean, median, mode, standard deviation, and coefficient of variation, based on the assumption that the data are distributed normally. Although the terms just cited are the most commonly used statistical measures, two additional statistical measures are needed to quantify the nature of a given distribution. The two additional measures are the coefficient of skewness and the coefficient of kurtosis. If a distribution is highly skewed, as determined by the coefficient of skewness, normal statistics cannot be used. For most wastewater data that are skewed, it has been found that the log of the value is normally distributed. Where the log of the values is normally distributed, the distribution is said to be log-normal. The common statistical measures used for the analysis of wastewater management data [Eqs. (3–1) through (3–9)] are summarized in Table 3–10.

Table 3–10
Statistical parameters used for the analysis of wastewater management data[a]

Parameter	Definition
Mean value $$\bar{x} = \frac{\Sigma f_i x_i}{n} \qquad (3\text{–}1)$$	$\bar{x}$ = mean value f_i = frequency (for ungrouped data $f_i = 1$) x_i = midpoint of the *i*th data range (for ungrouped data x_i = the *i*th observation) n = number of observations (Note $\Sigma f_i = n$)
Standard deviation $$s = \sqrt{\frac{\Sigma f_i (x_i - \bar{x})^2}{n-1}} \qquad (3\text{–}2)$$	s = standard deviation C_v = coefficient of variation, percent α_3 = coefficient of skewness
Coefficient of variation $$C_v = \frac{100s}{\bar{x}} \qquad (3\text{–}3)$$	α_4 = coefficient of kurtosis M_g = geometric mean s_g = geometric standard deviation $\log x_g = \log x_i - \log M_g$
Coefficient of skewness $$\alpha_3 = \frac{\Sigma f_i (x_i - \bar{x})^3/(n-1)}{s^3} \qquad (3\text{–}4)$$	**Median value** If a series of observations are arranged in order of increasing value, the middlemost observation, or the arithmetic mean of the two middlemost observations, in a series is known as the median
Coefficient of kurtosis $$\alpha_4 = \frac{\Sigma f_i (x_i - \bar{x})^4/(n-1)}{s^4} \qquad (3\text{–}5)$$	**Mode** The value occurring with the greatest frequency in a set of observations is known as the mode. If a continuous graph of the frequency distribution is drawn, the mode is the value of the high point, or hump, of the curve. In a symmetrical set of observations, the mean, median, and mode will be the same value
Geometric mean $$\log M_g = \frac{\Sigma f_i (\log x_i)}{n} \qquad (3\text{–}6)$$	**Coefficient of skewness** When a frequency distribution is asymmetrical, it is usually defined as being a skewed distribution
Geometric standard deviation $$\log s_g = \sqrt{\frac{\Sigma f_i (\log^2 x_g)}{n-1}} \qquad (3\text{–}7)$$	**Coefficient of kurtosis** Used to define the peakedness of the distribution. The value of the kurtosis for a normal distribution is 3. A peaked curve will have a value greater than 3, whereas a flatter curve will have a value less than 3
Using probability paper $$s = P_{84.1} - \bar{x} \quad \text{or} \quad P_{15.9} + \bar{x} \qquad (3\text{–}8)$$ $$s_g = \frac{P_{84.1}}{M_g} = \frac{M_g}{P_{15.9}} \qquad (3\text{–}9)$$	

[a] Adapted from Metcalf & Eddy (1991) and Crites and Tchobanoglous (1998).

Graphical Analysis of Data

Graphical analysis of wastewater management data is used to determine the nature of the distribution. For most practical purposes, the type of the distribution can be determined by plotting the data on both arithmetic- and logarithmic-probability (log-probability) paper and noting whether the data can be fitted with a straight line. The three steps involved in the use of arithmetic and log-probability paper are as follows.

1. Arrange the measurements in a data set in order of increasing magnitude and assign a rank serial number.
2. Compute a corresponding plotting position for each data point using Eq. (3–10).

$$\text{Plotting position } (\%) = \left(\frac{m}{n+1}\right) \times 100 \tag{3-10a}$$

where m = rank serial number
$\quad\quad n$ = number of observations

The term $(n + 1)$ is used to correct for a small sample bias. The plotting position represents the percent or frequency of observations that are equal to or less than the indicated value. Another expression often used to define the plotting position is known as Blom's transformation:

$$\text{Plotting position, } \% = \frac{m - 3/8}{n + 1/4} \times 100 \tag{3-10b}$$

3. Plot the data on arithmetic- and log-probability paper. The probability scale is labeled "Percent of values equal to or less than the indicated value."

If the data, plotted on arithmetic-probability paper, can be fit with a straight line, then the data are assumed to be normally distributed. Significant departure from a straight line can be taken as an indication of skewness. If the data are skewed, log-probability paper can be used. The implication here is that the logarithm of the observed values is normally distributed. On log-probability paper, the straight line of best fit passes through the geometric mean M_g and through the intersection of $M_g \times s_g$ at a value of 84.1 percent and M_g/s_g at a value of 15.9 percent. The geometric standard deviation s_g can be determined using Eq. (3–9) given in Table 3–10. The use of arithmetic- and logarithmic-probability paper is illustrated in Example 3–4.

EXAMPLE 3–4 **Statistical Analysis of Wastewater Constituent Concentration Data** Determine the appropriate statistical parameters for the following set of effluent data from a wastewater-treatment plant, collected over a 24-month period.

Month	Value, g/m³	
	TSS	COD
1	13.50	15.00
2	25.90	11.25
3	28.75	35.35
4	10.75	13.60
5	12.50	15.30
6	9.85	15.75
7	13.90	16.80

(continued)

(*Continued*)

Month	Value, g/m³	
	TSS	**COD**
8	15.10	15.20
9	23.40	18.75
10	21.90	37.50
11	23.70	27.00
12	18.00	23.30
13	37.00	46.60
14	30.10	36.25
15	21.25	30.00
16	23.50	25.75
17	16.75	17.90
18	8.35	11.35
19	18.10	25.20
20	9.25	16.10
21	9.90	16.75
22	8.75	15.80
23	15.50	19.50
24	7.60	9.40

Solution

1. Determine the nature of the distribution by plotting the data on arithmetic- and log-probability paper.

   a. Determine the plotting position using Eq. (3–10a) where n equals 24.

Number	Plotting position, %[a]	Value, g/m³	
		TSS	**COD**
1	4	7.60	9.40
2	8	8.35	11.25
3	12	8.75	11.35
4	16	9.25	13.60
5	20	9.85	15.00
6	24	9.90	15.20
7	28	10.75	15.30
8	32	12.50	15.75
9	36	13.50	15.80
10	40	13.90	16.10
11	44	15.10	16.75

(*continued*)

(Continued)

Number	Plotting position, %[a]	Value, g/m³ TSS	COD
12	48	15.50	16.80
13	52	16.75	17.90
14	56	18.00	18.75
15	60	18.10	19.50
16	64	21.25	23.30
17	68	21.90	25.20
18	72	23.40	25.75
19	76	23.50	27.00
20	80	23.70	30.00
21	84	25.90	35.35
22	88	28.75	36.25
23	92	30.10	37.50
24	96	37.00	46.60

[a] Plotting position (%) $= \left(\dfrac{m}{n+1}\right) \times 100.$

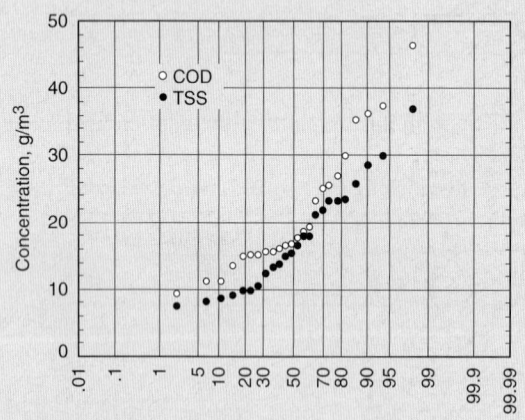

Percentage of values equal or less than indicated value

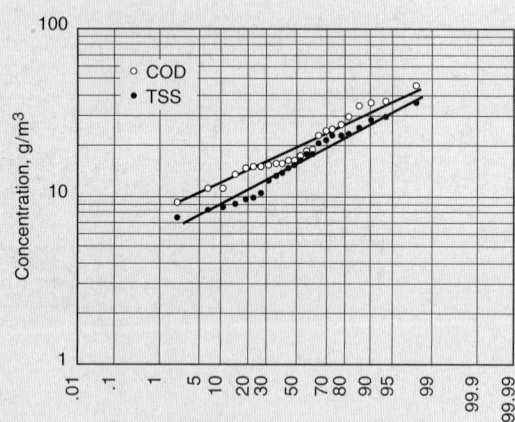

Percentage of values equal or less than indicated value

   b. Plot the above data on both arithmetic- and log-probability paper. As shown in the accompanying plot, both the TSS and COD data are log-normal.

2. Determine the geometric mean for TSS and COD and the corresponding geometric standard deviation using Eq. (3–9).

$$s_g = \frac{P_{84.1}}{M_g} = \frac{M_g}{P_{15.9}}$$

Constituent	M_g	s_g
TSS	17	1.5
COD	20	1.47

EXAMPLE 3–5   **Statistical Analysis of Wastewater Flowrate Data**   Using the following weekly flowrate data obtained from an industrial discharger for a calendar quarter of operation, determine the statistical characteristics and predict the maximum weekly flowrate that will occur during a full year's operation.

Week no.	Flowrate, m³/wk	Week no.	Flowrate, m³/wk
1	2900	8	3675
2	3040	9	3810
3	3540	10	3450
4	3360	11	3265
5	3770	12	3180
6	4080	13	3135
7	4015		

Solution

1. Plot the flowrate data using the log-probability method.
    a. Set up a data analysis table with three columns as described below.
        i. In column 1, enter the rank serial number starting with number 1.
        ii. In column 2, arrange the flowrate data in ascending order.
        iii. In column 3, enter the probability plotting position.

Rank serial no., m	Flowrate, m³/wk	Plotting position,[a] %
1	2900	7.1
2	3040	14.3
3	3135	21.4
4	3180	28.6
5	3265	35.7
6	3360	42.9
7	3450	50.0
8	3540	57.1
9	3675	64.3
10	3770	71.4

(continued)

(Continued)

Rank serial no., m	Flowrate, m³/wk	Plotting position,ᵃ %
11	3810	78.6
12	4015	85.7
13	4080	92.9

ᵃPlotting position = $[m/(n + 1)]100$.

b. Plot the weekly flowrates expressed in m³/wk versus the plotting position. The resulting plots are presented below. Because the data fall on a straight line on both plots, the flowrate data can be described adequately by either distribution. This fact can be taken as indication that the distribution is not skewed significantly and that normal statistics can be applied.

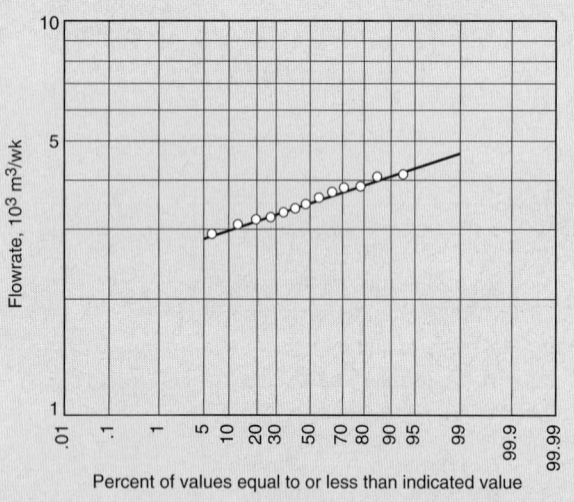

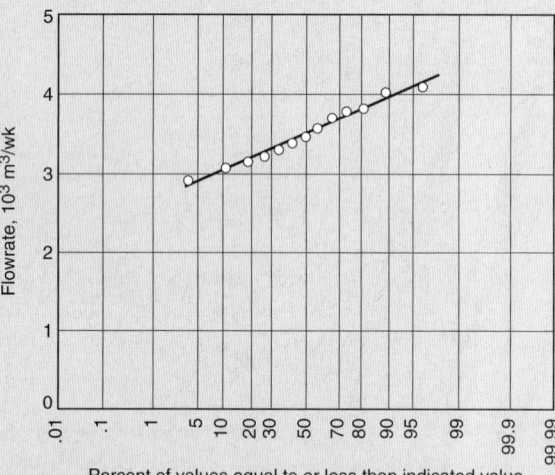

2. Determine the statistical characteristics of the flowrate data.
   a. Set up a data analysis table to obtain the quantities needed to determine the statistical characteristics.

Flowrate, m³/wk	$(x - \bar{x})$	$(x - \bar{x})^2$	$(x - \bar{x})^3$ 10^{-6}	$(x - \bar{x})^4$ 10^{-9}
2,900	−578	334,084	−193	11,161
3,040	−438	191,844	−84	3,680
3,135	−343	117,649	−40	1,384
3,180	−298	88,804	−26	789

(continued)

(Continued)

Flowrate, m³/wk	$(x - \bar{x})$	$(x - \bar{x})^2$	$(x - \bar{x})^3$ 10^{-6}	$(x - \bar{x})^4$ 10^{-9}
3,265	−213	45,369	−9.6	206
3,360	−118	13,924	−1.6	19.4
3,450	−28	784	−0.02	0.06
3,540	62	3,844	0.24	1.48
3,675	197	38,809	7.6	151
3,770	292	85,264	25	727
3,810	332	110,224	37	1,215
4,015	537	288,369	155	8,316
4,080	602	362,404	218	13,134
45,220		1,681,372	88.62	40,784

b. Determine the statistical characteristics using the parameters given in Table 3–10.

 i. Mean

$$\bar{x} = \frac{\Sigma x}{n}$$

$$\bar{x} = \frac{45,220}{13} = 3478 \text{ m}^3/\text{wk}$$

 ii. Median (the middlemost value)

Median $= 3450$ m³/wk (see data table above)

 iii. Mode

Mode $= 3(\text{Med}) - 2(\bar{x}) = 3(3450) - 2(3478) = 3394$ m³/wk

 iv. Standard deviation

$$s = \sqrt{\frac{\Sigma(x - \bar{x})^2}{n - 1}}$$

$$s = \sqrt{\frac{1,681,372}{12}} = 374.3 \text{ m}^3/\text{wk}$$

 v. Coefficient of variation

$$C_V = \frac{100\,s}{\bar{x}}$$

$$C_V = \frac{100(374.3)}{3478} = 10.8\%$$

vi. Coefficient of skewness

$$\alpha_3 = \frac{[\Sigma(x - \bar{x})^3/(n - 1)]}{s^3}$$

$$\alpha_3 = \frac{(88.62 \times 10^6/12)}{(374.3)^3} = 0.141$$

vii. Coefficient of kurtosis

$$\alpha_4 = \frac{[\Sigma(x - \bar{x})^4/(n - 1)]}{s^4}$$

$$\alpha_4 = \frac{(40,784 \times 10^9/12)}{(374.3)^4} = 1.73$$

Reviewing the statistical characteristics, it can be seen that the distribution is somewhat skewed ($\alpha_3 = 0.141$ versus 0 for a normal distribution) and is considerably flatter than a normal distribution would be ($\alpha_4 = 1.73$ versus 3.0 for a normal distribution).

3. Determine the probable annual maximum weekly flowrate.
   a. Determine the probability factor

$$\text{Peak week} = \frac{m}{n + 1} = \frac{52}{52 + 1} = 0.981$$

   b. Determine the flowrate from the figure given in Step 1b at the 98.1 percentile.

   Peak weekly flowrate = 4500 m³/wk

Comment   The statistical analysis of data is important in establishing the design conditions for wastewater-treatment plants. The application of statistical analysis to the selection of design flowrates and mass loadings is considered in the following section.

3–4 ANALYSIS OF WASTEWATER FLOWRATE DATA

Because the hydraulic design of both collection and treatment facilities is affected by variations in wastewater flowrates, the flowrate characteristics have to be analyzed carefully from existing records. In cases where only flowrate data in the collection system is available, it must be recognized that the flowrates may differ somewhat from the flowrate entering the treatment plant because of the flow-dampening effect of the sewer system. Peak hourly flowrates may also be attenuated by the available storage capacity in the sewer system.

Definition of Terms

Before considering the variations in flowrates and constituent concentrations, it will be helpful to define some terminology that is used commonly to quantify the variations that are observed. The principal terms used to describe these observed variations are defined in Table 3–11. As will be discussed in Sec. 3–6, these terms are also of importance in the selection and sizing of individual unit treatment processes and operations.

Variations in Wastewater Flowrates

Wastewater flowrates vary during the time of day, day of the week, season of the year, or depending upon the nature of the dischargers to the collection system. Short-term, seasonal, and industrial variations in wastewater flowrates are briefly discussed here.

Short-Term Variations. The variations in wastewater flows observed at treatment plants tend to follow a diurnal pattern, as shown on Fig. 3–4. Minimum flows occur during the early morning hours when water consumption is lowest and when the base flow consists of infiltration and small quantities of sanitary wastewater. The first peak flow generally occurs in the late morning when wastewater from the peak morning water use reaches the treatment plant. A second peak flow generally occurs in the

Table 3–11
Terminology used to quantify observed variations in flowrate and constituent concentrations[a]

Item	Description
Average dry-weather flow (ADWF)	The average of the daily flows sustained during dry-weather periods with limited infiltration
Average wet-weather flow (AWWF)	The average of the daily flows sustained during wet-weather periods when infiltration is a factor
Average annual daily flow	The average flowrate occurring over a 24–h period based on annual flowrate data
Instantaneous peak	Highest record flowrate occurring for a period consistent with the recording equipment. In many situations the recorded peak flow may be considerably below the actual peak flow because of metering and recording equipment limitations
Peak hour	The average of the peak flows sustained for a period of 1 hour in the record examined (usually based on 10-min increments)
Maximum day	The average of the peak flows sustained for a period of 1 day in the record examined (the duration of the peak flows may vary)
Maximum month	The average of the maximum daily flows sustained for a period of 1 month in the record examined
Minimum hour	The average of the minimum flows sustained for the period of 1 hour in the record examined (usually based on 10-min increments)
Minimum day	The average of the minimum flows sustained for the period of a day in the record examined (usually for the period from 2 A.M. to 6 A.M.)
Minimum month	The average of the minimum daily flows sustained for the period of 1 month in the record examined
Sustained flow (and load)	The value (flowrate or mass loading) sustained or exceeded for a given period of time (e.g., 1 hour, 1 day, or 1 month)

[a]Adapted in part from Crites and Tchobanoglous (1998).

Figure 3–4

Typical hourly variations in domestic wastewater flowrates.

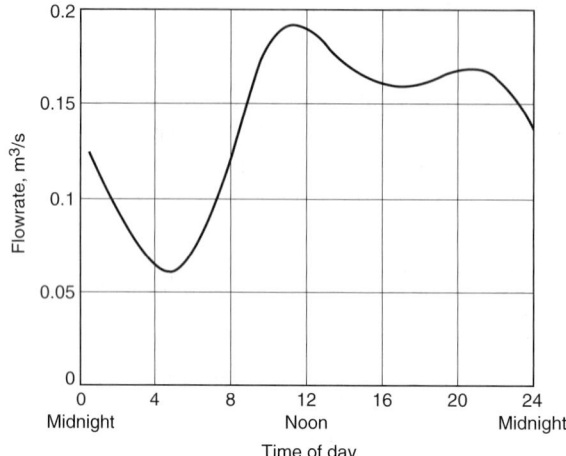

early evening between 7 and 9 P.M. The time of occurrence and the amplitude of the peak flowrates vary with the size of the community and the length of the collection system. As the community size increases, the variations between the high and low flows decrease due to (1) the increased storage in the collection system of large communities that tends to equalize flowrates and (2) changes in the economic and social makeup of the community. When extraneous flows are minimal, wastewater discharge curves resemble water consumption curves, but with a lag of several hours.

Seasonal Variations.   Seasonal variations in domestic wastewater flows are commonly observed at resort areas, in small communities with college campuses, and in communities that have seasonal commercial and industrial activities. The magnitude of the variations to be expected depends on both the size of the community and the seasonal activity.

Industrial Variations.   Industrial wastewater discharges are difficult to predict. Many manufacturing facilities generate relatively constant flowrates during production, but the flowrates change markedly during cleanup and shutdown. While internal process changes may lead to reduced discharge rates, plant expansion and increased production may lead to increased wastewater generation. Where joint municipal and industrial treatment facilities are to be constructed, special attention should be given to industrial flowrate projections, whether they are prepared by the industry or jointly with the city's staff or engineering consultant. Industrial discharges are most troublesome in smaller wastewater-treatment plants where there is limited capacity to absorb shock loadings.

Wastewater Flowrate Factors

Quantifying the variations in flowrates is important in the design and operation of wastewater-treatment plants. One of the measures used in determining the "peak" or maximum flows is the peaking factor. Peaking factors can be developed based on maximum hour, maximum day, maximum month, or other time periods. The peaking factor is particularly useful in estimating the maximum hydraulic conditions that might occur and have to be accommodated. Peaking factors can also be applied to mass loadings. Peaking factors are applied most frequently to determine the peak hourly flowrate. Because it is difficult to compare numerical peak flow values from different wastewater-treatment plants, peak hourly flowrate values are normalized by dividing by the long-term average flowrate. The resultant ratio, known as a peaking factor, is defined as follows:

$$\text{Sustained peaking factor, PF} = \frac{\text{peak flowrate (e.g., hourly, daily)}}{\text{average long-term flowrate}} \qquad (3\text{--}11)$$

The most common method of determining the peaking factor is from the analysis of flowrate data. Where flowrate records are available, at least 3 years of data should be analyzed to define the peak to average day peaking factor.

3–5 ANALYSIS OF CONSTITUENT MASS LOADING DATA

The analysis of wastewater data involves the determination of the flowrate and mass loading variations. The analysis may involve determining the concentrations of specific constituents, mass loadings (flowrate times concentration), or sustained mass loadings, loadings that occur over a defined period of time.

From the standpoint of treatment processes, one of the most serious deficiencies results when the design of a treatment plant is based on average flowrates and average BOD and TSS loadings, with little or no recognition of peak conditions. In many communities, peak influent flowrates and BOD and TSS loadings can reach two or more times average values. It must also be emphasized that, in nearly all cases, peak flowrates and BOD and TSS mass-loading rates do not occur at the same time. Analysis of current records is the best method of arriving at appropriate peak and sustained mass loadings.

The principal factors responsible for loading variations are (1) the established habits of community residents, which cause short-term (hourly, daily, and weekly) variations; (2) seasonal conditions, which usually cause longer-term variations; and (3) industrial activities, which cause both long- and short-term variations.

Wastewater Constituent Concentrations

The physical, chemical, and biological characteristics of wastewater vary throughout the day. An adequate determination of the waste characteristics will result only if the sample tested is representative. Typically, composite samples made up of portions of samples collected at regular intervals during a day are used (see Fig. 3–5). The amount of liquid used from each sample is proportional to the rate of flow at the time the sample was collected. Adequate characterization of wastewater is of fundamental importance in the design of treatment and disposal processes.

Quantity of Waste Discharged by Individuals in the United States.

Typical data on the total quantities of waste discharged per person per day (dry weight basis) from individual residences are reported in Table 3–12. The data presented in Table 3–12 have been gathered from numerous sources (primarily the United States) by

Figure 3-5

Samplers used to collect wastewater samples for analysis: (a) refrigerated unit used to collect daily composite samples and (b) portable sampler used to collect individual hourly samples throughout a day at different locations. The individual samples are composited to obtain flow-weighted mass loadings.

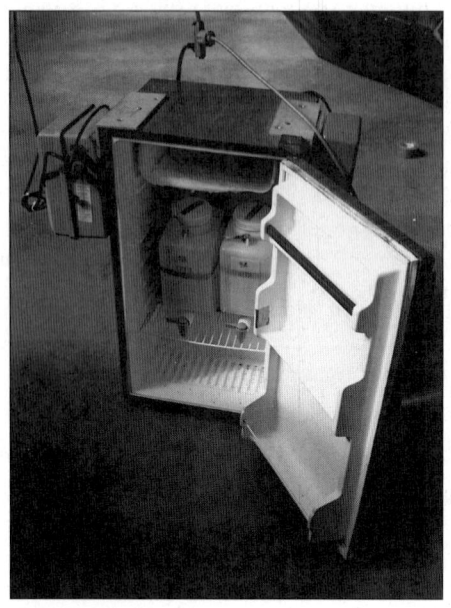

(a)

(b)

Table 3-12
Quantity of waste discharged by individuals on a dry weight basis[a]

Constituent (1)	Range (2)	Value, lb/capita·d		Range (5)	Value, g/capita·d	
		Typical without ground-up kitchen waste (3)	Typical with ground-up kitchen waste (4)		Typical without ground-up kitchen waste (6)	Typical with ground-up kitchen waste (7)
BOD$_5$	0.11–0.26	0.180	0.220	50–120	80	100
COD	0.30–0.65	0.420	0.480	110–295	190	220
TSS	0.13–0.33	0.200	0.250	60–150	90	110
NH$_3$ as N	0.011–0.026	0.017	0.019	5–12	7.6	8.4
Organic N as N	0.009–0.022	0.012	0.013	4–10	5.4	5.9
TKN[b] as N	0.020–0.048	0.029	0.032	9–21.7	13	14.3
Organic P as P	0.002–0.004	0.0026	0.0028	0.9–1.8	1.2	1.3
Inorganic P as P	0.004–0.006	0.0044	0.0048	1.8–2.7	2.0	2.2
Total P as P	0.006–0.010	0.0070	0.0076	2.7–4.5	3.2	3.5
Oil and grease	0.022–0.088	0.0661	0.075	10–40	30	34

[a] Adapted from Crites and Tchobanoglous (1998).
[b] TKN is total Kjeldahl nitrogen.

Crites and Tchobanoglous, 1998. The total number of pathogenic organisms discharged will depend on whether an individual is ill and is shedding pathogens. If one or more members of a family are ill and shedding pathogens, the number of measured organisms can increase by several orders of magnitude. The corresponding constituent concentrations, assuming the quantities of waste given in Table 3–12 were diluted in 190 and 460 L (50 and 120 gal) of water, are reported in Table 3–13. As noted in Table 3–13, the typical mass amounts of waste discharged (column 3) are based on the assumption that 25 percent of the homes are equipped with kitchen food-waste grinders. The concentrations are based on flowrates ranging from 190 L/capita·d (50 gal/capita·d) to 460 L/capita·d (120 gal/capita·d). The lower flowrate is typical for rural areas; the higher flowrate is typical for urban areas and is recognized by the U.S. EPA as not containing excessive inflow/infiltration (*Federal Register*, 1989).

Quantity of Waste Discharged by Individuals in Countries Outside of the United States.
The amounts of waste discharged by individuals in other countries can vary significantly due to cultural and socioeconomic differences. Constituent data for nine other countries as compared to the United States are reported in Table 3–14. Water use by individuals also differs significantly in other countries; in most cases the quantities used are significantly less. Consequently, the strength of the wastewater may be much higher than found in U.S. practice and can significantly affect wastewater treatability. In some cases, the wastewater composition may be high in organic content but low in alkalinity and thus cannot be nitrified fully. Concentrations of TSS and BOD for another culture compared to the United States are illustrated in Example 3–6.

Table 3–13
Typical unit loading factors and expected wastewater constituent concentrations from individual residences in the United States[a]

Constituent	Unit	Typical value[b]	Concentration, mg/L	
			Volume, L/capita·d (gal/capita·d)	
			190 (50)	460 (120)
BOD$_5$	g/capita·d	85	450	187
COD	g/capita·d	198	1,050	436
TSS	g/capita·d	95	503	209
NH$_3$ as N	g/capita·d	7.8	41.2	17.2
Organic N as N	g/capita·d	5.5	29.1	12.1
TKN as N	g/capita·d	13.3	70.4	29.3
Organic P as P	g/capita·d	1.23	6.5	2.7
Inorganic P as P	g/capita·d	2.05	10.8	4.5
Total P as P	g/capita·d	3.28	17.3	7.2
Oil and grease	g/capita·d	31	164	68

[a] Adapted from Crites and Tchobanoglous (1998).
[b] Data from Table 3–12, Columns 6 and 7, assuming 25 percent of the homes have kitchen waste-food grinders.

Table 3-14
Typical wastewater constituent data for various countries[a]

Country/ constituent	BOD, g/capita·d	TSS, g/capita·d	TKN, g/capita·d	NH₃-N, g/capita·d	Total P, g/capita·d
Brazil	55–68	55–68	8–14	ND	0.6–1
Denmark	55–68	82–96	14–19	ND	1.5–2
Egypt	27–41	41–68	8–14	ND	0.4–0.6
Germany	55–68	82–96	11–16	ND	1.2–1.6
Greece	55–60	ND	ND	8–10	1.2–1.5
India	27–41	ND	ND	ND	ND
Italy	49–60	55–82	8–14	ND	0.6–1
Japan	40–45	ND	1–3	ND	0.15–0.4
Palestine[b]	32–68	52–72	4–7	3–5	0.4–0.7
Sweden	68–82	82–96	11–16	ND	0.8–1.2
Turkey	27–50	41–68	8–14	9–11	0.4–2
Uganda	55–68	41–55	8–14	ND	0.4–0.6
United States[c]	50–120	60–150	9–22	5–12	2.7–4.5

[a] Adapted from Henze et al. (1997), Ozturk et al. (1992), Andreadakis (1992), and Nashashibi and van Duijl (1995).
[b] West Bank and Gaza Strip.
[c] From Table 3–11.

EXAMPLE 3-6 **Estimate Waste Constituent Concentration** Using data from Table 3–14, determine the BOD, TSS, and ammonia nitrogen concentrations for the West Bank and the Gaza Strip assuming the water supply is intermittent and the wastewater flowrate is 60 L/capita·d.

Solution

1. From Table 3–14, use the following average constituent contributions:
   a. BOD = 50 g/capita·d
   b. TSS = 60 g/capita·d
   c. NH₃-N = 3 g/capita·d
2. Compute BOD concentration

$$\text{BOD} = \left(\frac{50 \text{ g/capita·d}}{60 \text{ L/capita·d}}\right)\left(\frac{10^3 \text{ L}}{\text{m}^3}\right) = 833 \text{ g/m}^3$$

3. Compute TSS concentration

$$\text{TSS} = \left(\frac{60 \text{ g/capita·d}}{60 \text{ L/capita·d}}\right)\left(\frac{10^3 \text{ L}}{\text{m}^3}\right) = 1000 \text{ g/m}^3$$

4. Compute the NH_3-N concentration

$$NH_3\text{-}N = \left[\frac{(3 \text{ g/capita·d})}{(60 \text{ L/capita·d})}\right]\left(\frac{10^3 \text{ L}}{m^3}\right) = 50 \text{ g/m}^3$$

Comment   In many parts of the world where water usage is low, constituent concentrations for BOD and TSS may range up to 1000 g/m^3 (mg/L). In the above example, the concentrations of BOD and TSS are nearly 2 to 4 times the BOD and TSS concentrations typically found in the United States (see Table 3–13). Ammonia nitrogen (NH_3-N) concentrations may also be higher, perhaps on the order of two times or more of those in the United States.

Composition of Wastewater in Collection Systems.   Typical data on the composition of untreated domestic wastewater as found in wastewater-collection systems (in the United States) are reported in Table 3–15. The data presented in Table 3–15 for medium-strength wastewater are based on an average flow of 460 L/capita·d (120 gal/capita·d) and include constituents added by commercial, institutional, and industrial sources. Typical concentrations for low-strength and high-strength wastewater, which reflect different amounts of infiltration, are also given. Because there is no "typical" wastewater, it must be emphasized that the typical data presented in Table 3–15 *should only be used as a guide.* The constituent concentrations presented in Table 3–13, developed from the waste amounts given in Table 3–12, can also be compared to the values given in Table 3–15. It is interesting to note that the values given in Table 3–13 correspond closely to the values given in Table 3–15 for typical wastewater.

Mineral Increase Resulting from Water Use.   Data on the increase in the mineral content of wastewater resulting from water use, and the variation of the increase within a collection system, are especially important in evaluating the reuse potential of wastewater. Typical data on the incremental increase in mineral content that can be expected in municipal wastewater resulting from domestic use are reported in Table 3–16. Increases in the mineral content of wastewater may be due in part to addition of highly mineralized water from private wells and groundwater and from industrial use. Domestic and industrial water softeners also contribute significantly to the increase in mineral content and, in some areas, may represent the major source. Occasionally, water added from private wells and groundwater infiltration will (because of its high quality) serve to dilute the mineral concentration in wastewater.

Variations in Constituent Concentrations

Several types of constituent concentration variations can occur depending upon the characteristics of the contributors to the wastewater-collection system. Types of variations are discussed below.

Short-Term Variation in Constituent Values.   Constituent concentration variations may change significantly during the course of a day. An example of typical

Table 3–15
Typical composition of untreated domestic wastewater

Contaminants	Unit	Concentration[a]		
		Low strength	Medium strength	High strength
Solids, total (TS)	mg/L	390	720	1230
Dissolved, total (TDS)	mg/L	270	500	860
Fixed	mg/L	160	300	520
Volatile	mg/L	110	200	340
Suspended solids, total (TSS)	mg/L	120	210	400
Fixed	mg/L	25	50	85
Volatile	mg/L	95	160	315
Settleable solids	mL/L	5	10	20
Biochemical oxygen demand, 5–d, 20°C (BOD, 20°C)	mg/L	110	190	350
Total organic carbon (TOC)	mg/L	80	140	260
Chemical oxygen demand (COD)	mg/L	250	430	800
Nitrogen (total as N)	mg/L	20	40	70
Organic	mg/L	8	15	25
Free ammonia	mg/L	12	25	45
Nitrites	mg/L	0	0	0
Nitrates	mg/L	0	0	0
Phosphorus (total as P)	mg/L	4	7	12
Organic	mg/L	1	2	4
Inorganic	mg/L	3	5	8
Chlorides[b]	mg/L	30	50	90
Sulfate[b]	mg/L	20	30	50
Oil and grease	mg/L	50	90	100
Volatile organic compounds (VOCs)	μg/L	<100	100–400	>400
Total coliform	No./100 mL	10^6–10^8	10^7–10^9	10^7–10^{10}
Fecal coliform	No./100 mL	10^3–10^5	10^4–10^6	10^5–10^8
Cryptosporidum oocysts	No./100 mL	10^{-1}–10^0	10^{-1}–10^1	10^{-1}–10^2
Giardia lamblia cysts	No./100 mL	10^{-1}–10^1	10^{-1}–10^2	10^{-1}–10^3

[a] Low strength is based on an approximate wastewater flowrate of 750 L/capita·d (200 gal/capita·d).
Medium strength is based on an approximate wastewater flowrate of 460 L/capita·d (120 gal/capita·d).
High strength is based on an approximate wastewater flowrate of 240 L/capita·d (60 gal/capita·d).
[b] Values should be increased by amount of constituent present in domestic water supply.
Note: mg/L = g/m³.

Table 3-16
Typical mineral
increase from
domestic water use

Constituent	Increment range, mg/L[a,b]
Anions:	
Bicarbonate (HCO₃)	50–100
Carbonate (CO₃)	0–10
Chloride (Cl)	20–50
Sulfate (SO₄)	15–30
Cations:	
Calcium (Ca)	6–16
Magnesium (Mg)	4–10
Potassium (K)	7–15
Sodium (Na)	40–70[c]
Other constituents:	
Aluminum (Al)	0.1–0.2
Boron (B)	0.1–0.2
Fluoride (F)	0.2–0.4
Manganese (Mn)	0.2–0.4
Silica (SiO₂)	2–10
Total alkalinity (as CaCO₃)	60–120
Total dissolved solids (TDS)	150–380

[a] Based on 460 L/capita·d (120 gal/capita·d).
[b] Values do not include commercial and industrial additions.
[c] Excluding the addition from domestic water softeners.
Note: mg/L = g/m³.

variations in domestic wastewater strength is shown on Fig. 3–6. The BOD variation generally follows the flow. The peak BOD (organic matter) concentration often occurs in the evening.

Seasonal Variation in Constituent Values. For domestic flow only, and neglecting the effects of infiltration, the unit (per capita) loadings and the strength of the wastewater from most seasonal sources, such as resorts, will remain about the same on a daily basis throughout the year even though the total flowrate varies. The total mass of BOD and TSS of the wastewater, however, will increase directly with the population served.

Infiltration/inflow, as discussed earlier in this chapter, is another source of water flow into the collection system. In most cases, the presence of this extraneous water tends to decrease the concentrations of BOD and TSS, depending on the characteristics of the water entering the sewer. In some cases, concentrations of some inorganic constituents may actually increase where the groundwater contains high levels of dissolved constituents.

Figure 3–6

Typical hourly variations in flow and strength of domestic wastewater.

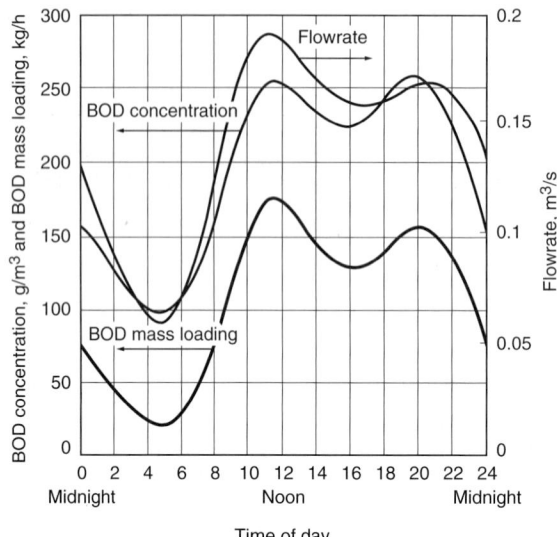

Variations in Industrial Wastewater. The composition of wastewater from industrial operations varies widely depending on the function and activity of the particular industry. Examples of the constituent concentration variability are illustrated in Table 3–17. From these examples, it can be observed that flow values and water quality measurements may vary by several orders of magnitude over a period of a year. Because of this variation, it is often difficult to define "typical operating conditions" for industrial activities.

The concentrations of both BOD and TSS in industrial wastewater can vary significantly throughout the day. For example, the BOD and TSS concentrations contributed from vegetable-processing facilities during the noon wash-up period may far exceed those contributed during working hours. Problems with high short-term loadings most commonly occur in small treatment plants that have limited reserve capacity to handle "shock loadings." The seasonal impact of industrial wastes such as canneries can cause both the flow and BOD loadings to increase from two to five times average conditions.

If industrial wastes are to be discharged to the collection system for treatment in a municipal wastewater facility, it will be necessary to characterize the wastes adequately to identify the ranges in constituent concentrations and mass loadings. Such characterization is also needed to determine if pretreatment is required before the waste is permitted to be discharged into the collection system. If pretreatment is needed, the effluent from the pretreatment facilities must also be characterized. Further, any proposed future process changes should also be assessed to determine what effects they might have on the wastes to be discharged. Where data are not available, every effort should be made to obtain information from similar facilities. With sufficient characterization of the wastewater from industrial discharges, suitable pretreatment facilities can be provided and plant upsets can be avoided.

Table 3–17
Range of effluent constituent concentrations for two industrial activities

Constituent	Unit	Wool textile mill[a]		Tomato cannery[b]	
		Annual average	Daily maximum	Peak season[c]	Off season[d]
Flowrate	m³/d	–	–	4164–22,300	1140–6400
pH	–	5.92[e]	–	7.2–8.0	7.2–8.0
BOD	mg/L	90.7	169	460–1100	29–56
COD	mg/L	529	1240	–	–
SS	mL/L	–	–	6–80	0.5–2.2
TSS	mg/L	93.4	860	270–760	69–120
TDS	mg/L	–	–	480–640	360–520
Nitrate-N	mg/L	–	–	0.4–5.6	2.2–0.1
Ammonia-N	mg/L	8.1	54	–	–
Phosphorus	mg/L	–	–	1.5–7.4	0.3–3.9
Sulfate	mg/L	–	–	23–15	9.9–7.1
DO	mg/L	–	–	0.9–3.8	1.6–9.8
Oil and grease	mg/L	27.4	45.2	–	–
Temperature	°C	–	–	18–23	13–19

[a] Yohe and Rich (1995).
[b] Adapted from Crites and Tchobanoglous (1998).
[c] Peak season is from early July to late September, when fresh-harvest tomatoes are canned. Treatment consists of screening and brief sedimentation.
[d] Off season is from November to June, when canned tomatoes are remanufactured into tomato paste, tomato sauce, and other tomato products (e.g., salsa, ketchup, spaghetti sauce). Treatment typically consists of screening, aeration, and sedimentation.
[e] Median value.

Note: m³/d $\times$ 0.264 x 10^{-3} = Mgal/d.

Variations in Constituent Values in Combined Collection Systems.
Flowrates, constituent concentrations, and mass loads emanating from combined collection systems can vary widely. Typical factors influencing the characteristics of wastewater from combined collection systems are shown in Table 3–18. Example variations of BOD, TSS, and fecal coliform measured in a combined system are shown on Fig. 3–7, during and after a storm event. As shown, the BOD and fecal coliform bacteria concentrations are low during the storm when runoff flows are high. After the storm, when runoff subsides and the flow consists primarily of wastewater, concentrations rise significantly. When this rise occurs, it can be concluded that the BOD and fecal coliform concentrations in the stormwater are significantly lower than in the wastewater component. Unlike BOD and fecal coliform bacteria, TSS concentrations rise slightly during the storm, and remain unchanged after the storm, indicating that TSS concentrations from stormwater runoff and wastewater are similar. The slight rise in the TSS

Table 3–18
Typical factors influencing the characteristics of combined wastewater

Parameter	Quantity-related factors	Quality-related factors
Precipitation	Rainfall depth and volume Storm intensity Storm duration	Regional atmospheric quality
Wastewater sources	Flowrate and variability Type of contributing sources (residential, commercial, etc.)	Type of contributing sources
Drainage basin characteristics	Size, time of concentration Land-use type Impervious area Soil characteristics Runoff control practices	Pollutant buildup and wash-off Watershed management practices
Sewer system, interceptor design and condition	Pipe size, slope, and shape Quantity of infiltration Surcharging or backwater conditions Type of flow regulation or diversion Capacity reduction from sediment buildup	Chemical and biological transformations Quality of infiltration Sediment load resuspended from collection system

Figure 3–7

Typical variations of flowrate, BOD, TSS, and fecal coliform in a combined collection system during a storm event.

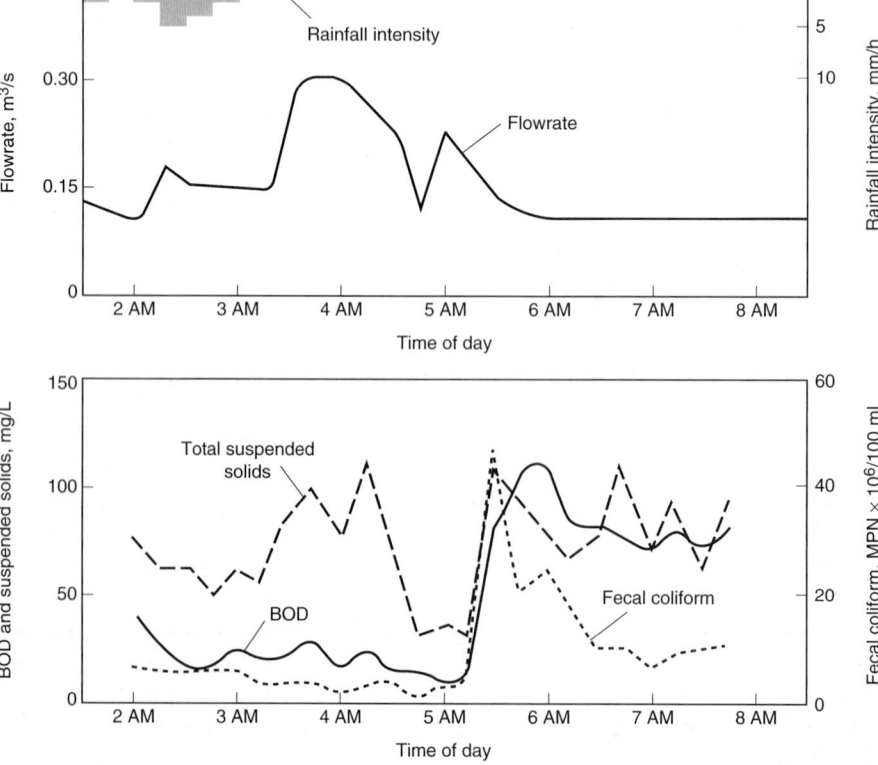

concentration during the peak flow may be due to a phenomenon common to many combined sewer systems known as the "first flush." The first flush has often been observed following the initial phase of a rainfall event in which much of the accumulated surface contaminants are washed into the combined system. In combined collection systems, the increased flows may be capable of resuspending material deposited previously during low-flow periods. Together, the resuspended material and contaminants washed off surfaces result in high contaminant concentrations. Factors known to contribute to the magnitude and frequency of the first-flush effect include combined sewer slopes; street and catch basin cleaning frequency and design; rainfall intensity and duration; and surface buildup of debris and contaminants.

Wastewater from combined collection systems usually contains more inorganic matter than wastewater from sanitary collection systems because of the larger quantities of storm drainage that enter the combined sewer system. Characteristics of combined wastewater and stormwater as compared to municipal wastewater are shown in Table 3–19.

Table 3-19
Comparison of characteristics of combined wastewater with other sources

Parameter	Unit	Range of parameter concentrations			
		Rainfall[a]	Stormwater runoff[b]	Combined wastewater[c]	Municipal wastewater[d]
Total suspended solids, TSS	mg/L	<1	67–101	270–550	120–370
Biochemical oxygen demand, BOD	mg/L	1–13	8–10	60–220	120–380
Chemical oxygen demand, COD	mg/L	9–16	40–73	260–480	260–900
Fecal coliform bacteria	MPN/100 mL		10^3–10^4	10^5–10^6	10^5–10^7
Nitrogen (as N):					
Total Kjeldahl nitrogen	mg/L		0.43–1.00	4–17	20–45
Nitrate	mg/L	0.05–1.0	0.48–0.91		0
Phosphorus (total as P)	mg/L	0.02–0.15	0.67–1.66	1.2–2.8	4–12
Metals:					
Copper, Cu	μg/L		27–33		
Lead, Pb	μg/L	30–70	30–144	140–600	
Zinc, Zn	μg/L		135–226		

[a] Adapted from Huber (1984).
[b] Adapted from U.S. EPA (1983).
[c] Adapted from Metcalf & Eddy (1977).
[d] From Table 3–15.

Flow-Weighted Constituent Concentrations

Flow-weighted constituent concentrations are obtained by multiplying the flow (typically hourly values over a 24-h period) times the corresponding constituent concentration, summing the results, and dividing by the summation of the flows as given by Eq. (3–12).

$$C_w = \frac{\sum\limits_{i=1}^{n} C_i Q_i}{\sum\limits_{i=1}^{n} Q_i} \qquad (3\text{--}12)$$

where C_w = flow-weighted average concentration of the constituent

n = number of observations

C_i = average concentration of the constituent during ith time period

Q_i = average flowrate during ith time period

Whenever possible, flow-weighted constituent concentrations should be used because they are a more accurate representation of the actual wastewater strength that must be treated. Determination of the simple arithmetic average and flow-weighted constituent concentrations is illustrated in Example 3–7.

EXAMPLE 3–7 Calculation of Flow-Weighted BOD and TSS Concentrations Compute the flow-weighted BOD and TSS values using the data provided on Fig. 3–6 for a community of about 5000 persons. Compare the flow-weighted values to the simple arithmetic averages. What is the significance of the difference?

Solution

1. Create a spreadsheet for calculating the flow-weighted values. Divide the BOD, TSS, and flow curves into 24 one-hour periods. Enter the time intervals (e.g., 12 to 1 A.M.) in column 1.

2. For each time period, calculate the average BOD value during the interval. For example, the average BOD value during the first interval (12 to 1 A.M.) is

    Value at beginning of interval = 155 g/m³

    Value at end of interval = 150 g/m³

    $$\text{Average BOD} = \frac{155 + 150}{2} = 152.5 \text{ g/m}^3$$

    The average BOD values for each successive time interval are entered in column 2.

3. Enter the average values for TSS and flow in columns 3 and 4, respectively.

4. For each time period, multiply the average BOD value (column 2) times the average flowrate (column 4), and enter the results in column 5.

5. For each time period, multiply the average TSS (column 3) value by the average flowrate (column 4), and enter the results in column 6.

6. Calculate the sum and simple arithmetic average for columns 2 through 6.
7. Divide the sum of the values in columns 5 and 6 (BOD × flow and TSS × flow, respectively) by the sum of the values in column 4 (flow) to obtain the flow-weighted average for BOD and TSS. The resulting values are given in the last two lines of the spreadsheet.

Time interval (1)	BOD, g/m³ (2)	TSS, g/m³ (3)	Flowrate Q, m³/s (4)	BOD × Q, kg/d (5) = (2) × (4) × 86.4ᵃ	TSS × Q, kg/d (6) = (3) × (4) × 86.4ᵃ
12–1 A.M.	152.5	157.5	0.12	1,581	1,633
1–2	135.0	140.0	0.10	1,166	1,210
2–3	115.0	122.5	0.08	795	847
3–4	102.5	112.5	0.0625	554	608
4–5	92.5	102.5	0.056	448	496
5–6	97.5	92.5	0.059	497	472
6–7	115.0	102.5	0.0775	770	686
7–8	142.5	140.0	0.105	1,293	1,270
8–9	182.5	195.0	0.1475	2,326	2,485
9–10	226.5	244.0	0.1835	3,591	3,868
10–11	251.5	269.0	0.1895	4,118	4,404
11–12	256.5	279.0	0.1965	4,355	4,737
12–1 P.M.	246.5	270.0	0.185	3,940	4,316
1–2	232.0	250.0	0.1725	3,458	3,726
2–3	222.0	237.5	0.165	3,165	3,386
3–4	214.5	226.0	0.161	2,984	3,144
4–5	214.0	212.5	0.157	2,903	2,883
5–6	219.5	202.5	0.158	2,996	2,764
6–7	232.0	199.0	0.162	3,247	2,785
7–8	262.5	209.0	0.1675	3,799	3,025
8–9	290.0	218.0	0.173	4,335	3,258
9–10	265.0	206.5	0.1675	3,855	2,988
10–11	209.5	192.5	0.1575	2,851	2,620
11–12	174.0	174.0	0.1425	2,142	2,142
Totals	4651.0	4554.5	3.3455	61,148	59,753
Average values	193.8	189.8	0.1393		
Flow-weighted concentration values				211.5	206.7

ᵃConversion factor: g/m³ × m³/s × 86,400 s/d × 1 kg/10³ g = 86.4 × C × Q = kg/d.

Comment When comparing the computation of a simple average to a flow-weighted value, the differences can be significant. In this example, if simple averages were used, the BOD loading would have been understated by 17.7 mg/L (9.1 percent), and the TSS loading by 16.9 mg/L (8.9 percent). If simple averages had been used in establishing process loading values in this case, the treatment facilities would be underdesigned by 9 percent.

Calculation of Mass Loadings

Constituent mass loadings are usually expressed in kilograms per day and may be computed using Eq. (3–13a) when the flowrate is expressed in cubic meters per day, or Eq. (3–13b) when the flowrate is expressed in million gallons per day. Note that in the SI system of units, the concentration expressed in milligrams per liter is equivalent to grams per cubic meter.

$$\text{Mass loading, kg/d} = \frac{(\text{concentration, g/m}^3)(\text{flowrate, m}^3/\text{d})}{(10^3 \text{g/kg})} \qquad (3\text{–}13a)$$

$$\text{Mass loading, lb/d} = (\text{concentration, mg/L})(\text{flowrate, Mgal/d})\left[\frac{8.34 \text{ lb}}{\text{Mgal} \cdot (\text{mg/L})}\right]$$

$$(3\text{–}13b)$$

To design treatment processes to function properly under varying loading conditions, data must be available for the sustained peak mass loadings of constituents that are to be expected. In the past, such information has seldom been available. When the data are not available, curves similar to those shown on Fig. 3–8 can be used. The curves for BOD, TSS, TKN (total Kjeldahl nitrogen), NH_3 (ammonia), and phosphorus were derived from an analysis of the records of over 50 treatment plants throughout the country. It should be noted that significant variations will be observed from plant to plant, depending on the size of the system, the percentage of combined wastewater, the size and slope of the interceptors, and the types of wastewater contributors.

The procedure used to develop the mass loading curves shown on Fig. 3–10 is as follows. First, the average mass loading is determined for the period of record. Second, the records are reviewed for the highest and lowest sustained one-day mass loading. These values are divided by the average mass loading and the numbers are plotted. Third, the same procedures are followed for two consecutive days, three consecutive days, etc., sustained loadings until ratio values are found for the period of interest (usually 10 to 30 days).

The daily mass loading rates for the various plants were developed using hourly data and the following expression:

$$\text{Daily mass loading, kg/d} = \sum_{i=1}^{24} \frac{(\text{concentration, g/m}^3)(\text{flowrate, m}^3/\text{d})}{(10^3 \text{ g/kg})} \qquad (3\text{–}14a)$$

Figure 3–8

Typical information on the ratio of averaged peak and low-constituent mass loadings to average mass loadings for: (a) BOD, (b) TSS, and (c) nitrogen and phosphorus.

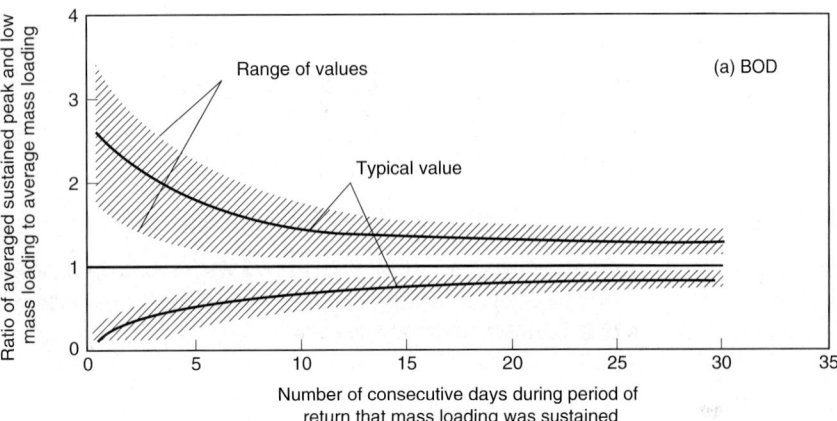

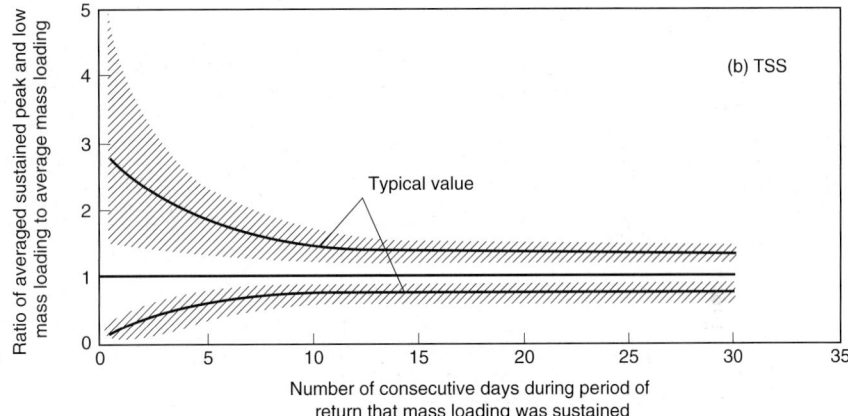

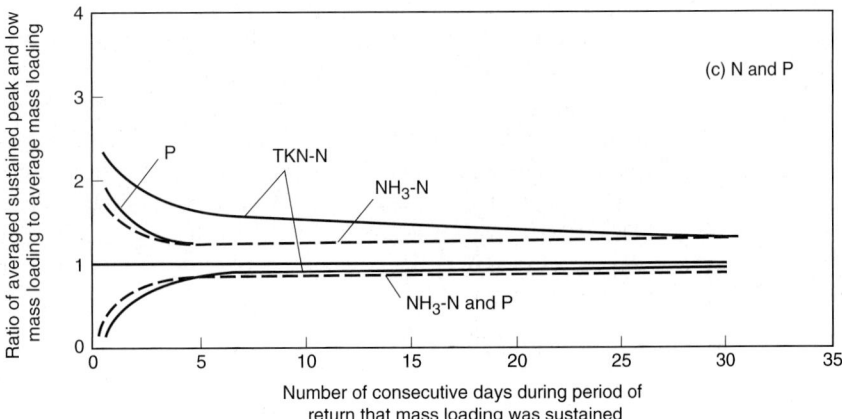

Daily mass loading, lb/d

$$= \sum_{i=1}^{24} (\text{concentration, mg/L})(\text{flowrate, Mgal/d}) \left[\frac{8.34 \text{ lb}}{\text{Mgal} \cdot (\text{mg/L})} \right] \quad (3\text{-}14b)$$

The development of a sustained peak mass loading curve is illustrated in Example 3–8.

EXAMPLE 3–8   **Development of BOD Sustained Mass Loading Values**   Develop a sustained BOD peak mass loading curve for a treatment plant with a design flowrate of 1 m³/s (22.8 Mgal/d). Assume that the long-term daily average BOD concentration is 200 g/m³.

Solution

1.   Compute the daily mass loading value for BOD.

     Daily BOD mass loading, kg/d

$$= \frac{(200 \text{ g/m}^3)(1 \text{ m}^3/\text{s})(86,400 \text{ s/d})}{(10^3 \text{ g/kg})} = 17,280 \text{ kg/d}$$

2.   Set up a computation table for the development of the necessary information for the peak sustained BOD mass loading curve (see following table).
3.   Obtain peaking factors for the sustained peak BOD loading rate from Fig. 3–8a, and determine the sustained mass loading rates for various time periods [see table, cols. (1), (2), and (3)].
4.   Develop data for the sustained mass loading curve and prepare a plot of the resulting data (see following figure).

Length of sustained peak, d (1)	Peaking factor[a] (2)	Peak BOD mass loading, kg/d (3)	Total mass loading, kg[b] (4)
1	2.4	41,472	41,472
2	2.1	36,288	72,576
3	1.9	32,832	98,496
4	1.8	31,104	124,416
5	1.7	29,376	146,880
10	1.4	24,192	241,920
15	1.3	22,464	336,960
20	1.25	21,600	432,000
30	1.21	19,872	596,160
365	1.0	17,280	

[a] From Fig. 3–8a.
[b] Col. 1 × col. 3 = col. 4.

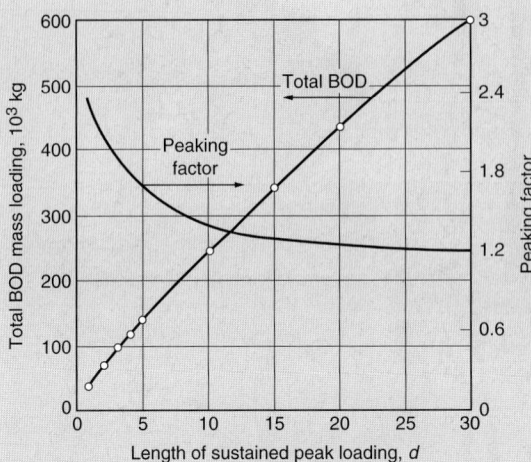

Comment The interpretation of the curve plotted for this example is as follows. If the sustained peak loading period were to last for 10 days, the total amount of BOD that would be received at a treatment facility during the 10-day period would be 241,695 kg. The corresponding amounts for sustained peak periods of 1 and 2 days would be 41,401 and 72,451 kg, respectively. Computations for an example of this type can be facilitated by using a personal computer spreadsheet program.

Effect of Mass Loading Variability on Treatment Plant Performance

During the course of a day, the mass loading that is received by the treatment plant can vary widely. An example of a diurnal mass loading curve is illustrated on Fig. 3–9. The variation in loading rates and the compounding effects during the high-flow and concentration periods is illustrated clearly on Fig. 3–9. The variations are more pronounced in small collection systems where the collection system storage capacity does not provide a significant dampening effect. The impact of these load variations is seen most dramatically in the effects on biological treatment operating conditions. The maximum hourly BOD loading may vary as much as 3 to 4 times the minimum hourly BOD load in a 24-h period. Over longer periods of time, the mass loadings can also vary widely (see Fig. 3–10). These types of variations have to be accounted for in the design of the biological treatment system. In extreme cases, flow equalization may be required.

3–6 SELECTION OF DESIGN FLOWRATES AND MASS LOADINGS

The rated capacity of wastewater-treatment plants is normally based on the average annual daily flowrate at the design year plus an allowance for future growth. As a practical matter, however, wastewater-treatment plants have to be designed to meet a number of conditions that are influenced by flowrates, wastewater characteristics, and constituent

Figure 3–9

Illustration of diurnal wastewater flow, BOD, and mass loading variability.

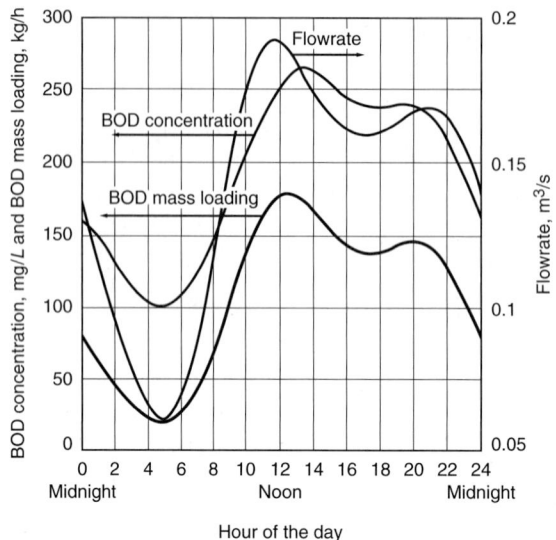

Figure 3–10

Example variations of TSS and BOD concentrations and mass loadings over a monthly period.

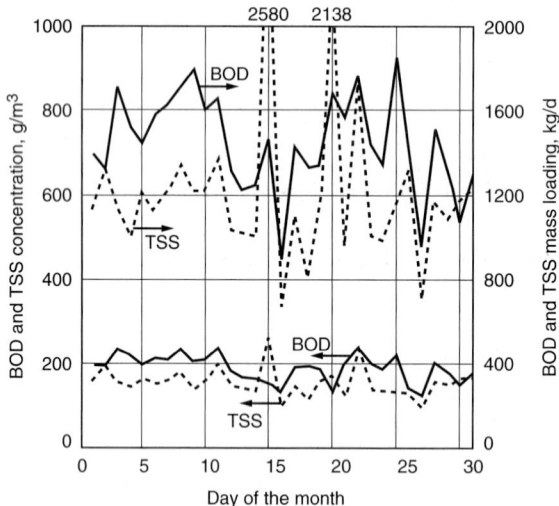

concentrations, and a combination of both (mass loading). Conditions that must be considered include peak and minimum hydraulic flowrates and the maximum, minimum, and sustained process constituent mass loading rates. Additionally, periods of initial operation and low flows and loads must be taken into consideration in design. The importance of wastewater flowrates and mass loadings in process design and operation is considered in this section.

Typical flowrate and mass loading factors that are important in the design and operation of wastewater-treatment facilities are described in Table 3–20. The overall

Table 3–20
Typical flowrate and mass loading factors used for the design and operation of wastewater-treatment facilities[a]

Factor	Purpose for design and operation
Flowrate	
Average daily flow	Development of flowrate ratios and for estimating pumping and chemical costs
Minimum hour	Sizing turndown of pumping facilities and determining low range of plant flowmeter
Minimum day	Sizing of influent channels to control solids deposition; sizing effluent recycle requirements for trickling filters
Minimum month	Selection of minimum number of operating units required during low-flow periods; scheduling shutdown for maintenance
Peak hour	Sizing of pumping facilities and conduits; sizing of physical unit operations: grit chambers, sedimentation tanks, and filters; sizing chlorine contact tanks. Also important in developing process control strategies for managing high flows
Maximum day	Sizing of equalization basins, chlorine contact tanks, sludge pumping system
Maximum month	Record keeping and reporting; sizing of chemical storage facilities
Mass loading	
Minimum month	Process turndown requirements
Minimum day	Sizing of trickling filter recycle rates
Maximum day	Sizing of selected process units
Maximum month	Sizing of sludge storage facilities; sizing of composting requirements
15-day maximum	Sizing anaerobic and aerobic digesters
Sustained loading	Sizing of selected process units and ancillary process equipment

[a] Adapted in part from Metcalf & Eddy (1991) and Crites and Tchobanoglous (1998).

objective of wastewater treatment is to provide a wastewater-treatment system that is capable of coping with a wide range of probable wastewater conditions while complying with the overall performance requirements.

Design Flowrates

The development and forecasting of flowrates is necessary to determine the design capacity as well as the hydraulic requirements of the treatment system. Flowrates need to be developed both for the initial period of operation and for the future (design) period. Consideration of the flowrates during the early years of operation is often overlooked, and oversizing of equipment and inefficient operation can result. The focus of the following discussion is on the development of various design flowrates.

Rationale for the Selection of Flowrates. The rationale for selecting flowrates is based on hydraulic and process considerations. As stated, the process units and hydraulic conduits must be sized to accommodate the anticipated peak flowrates that will pass through the treatment plant. Provisions have to be made to ensure bypassing of wastewater does not occur either in the collection system or at the treatment plant. Many of the process units are designed based on detention time or overflow rate (flowrate per unit of surface area) to achieve the desired removal rates of BOD and TSS. Because the performance of these units can be affected significantly by varying flowrate conditions and mass loadings, minimum and peak flowrates must be considered in design.

Forecasting Flowrates. In determining the design flowrate, elements to be considered are (1) the existing base flows; (2) estimated future flows for residential, commercial, institutional, and industrial sources; and (3) nonexcessive infiltration/inflow. Existing base flows equal actual metered flowrates minus excessive infiltration/inflow (defined as infiltration/inflow that can be controlled by cost-effective improvements to the collection system).

A yardstick by which total dry-weather base flow can be measured is 460 L/capita·d (120 gal/capita·d), established by the U.S. EPA as a historical average where infiltration is not excessive. The base flow includes 270 L/capita·d (70 gal/capita·d) for domestic flows, 40 L/capita·d (10 gal/capita·d) for commercial and small industrial flows, and 150 L/capita·d (40 gal/capita·d) for infiltration (*Federal Register,* 1989).

A useful technique in forecasting flowrates is probability analysis, discussed earlier in this chapter. Where flowrate data are available, preferably for at least 2 years, future flowrates for design can be predicted with a reasonable certainty. An example of a probability analysis of flowrates, as well as BOD and TSS concentrations and mass loadings, is shown on Fig. 3–11. The probability analysis can be used to estimate occurrences of peak flows and loads, and to establish a basis for selecting design flows and loads. For example, a maximum 1-day occurrence can be determined based on a 99.7 percent probability; the value will not be equaled or exceeded in the time period analyzed. A probability value, such as the 95 percentile, can also be established for forecasting the design loadings to meet permit requirements.

Minimum Flowrates. As noted in Table 3–20, low flowrates are also of concern in treatment-plant design, particularly during the initial years of operation when the plant is operating well below the design capacity, and in designing pumping stations. In cases where very low nighttime flow is expected, provisions for recycling treated effluent may have to be included to sustain the process (e.g., biological treatment processes such as trickling filters and to maintain optimal flowrates through ultraviolet disinfection systems). In the absence of measured flowrate data, minimum daily flowrates may be assumed to range from 30 to 70 percent of average flowrates for medium- to large-size communities, respectively (WEF, 1998).

Sustained Flowrates. Sustained flowrates are those that are equaled or exceeded for a specified number of consecutive days based on annual operating data. Data for sustained flowrates may be used in sizing equalization basins and other plant

Figure 3–11

Typical probability plots for flowrate, BOD, and TSS.

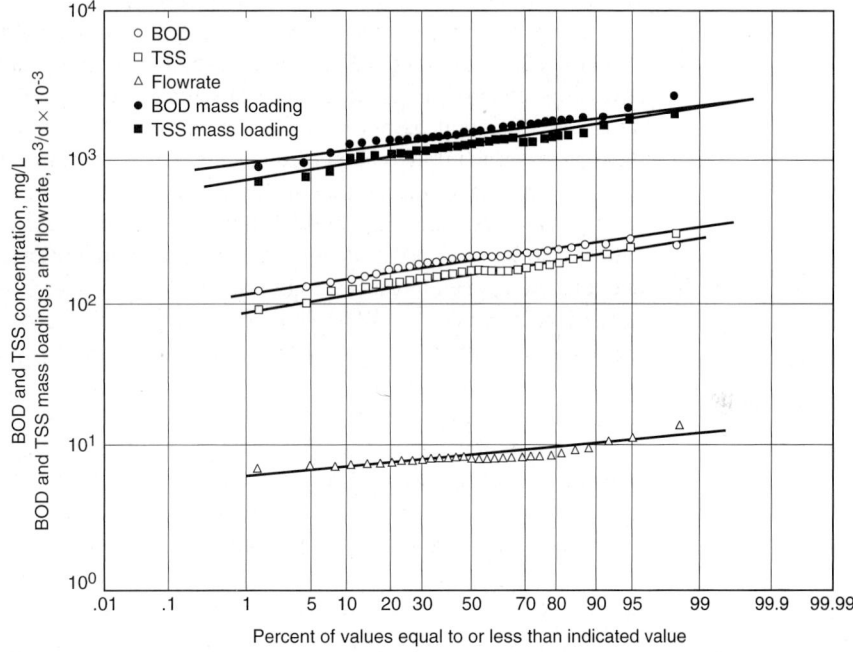

Figure 3–12

Typical ratios of averaged sustained peak and low daily flowrates to average annual daily flowrates for time periods up to 30 days.

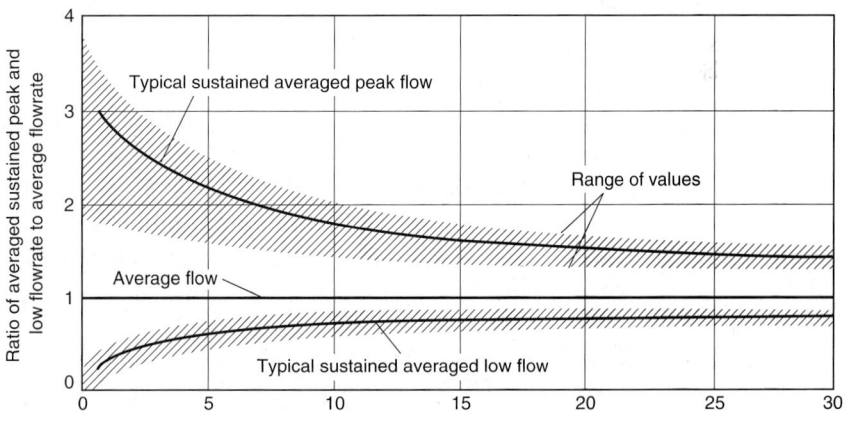

hydraulic components. An example plot of sustained and low flowrates is shown on Fig. 3–12. When developing plots similar to Fig. 3–12, the longest available period should be used.

Peak Flowrate Factors. The flowrate peaking factors (the ratio of peak flowrate to average flowrate) most frequently used in design are those for peak hour and maximum day (see Table 3–20). Peak hourly flowrates are used to size the hydraulic conveyance system and other facilities such as sedimentation tanks and chlorine contact tanks where little volume is available for flow dampening. Other peaking factors such as maximum week or maximum month may be used for treatment facilities such as pond systems that have long detention times or for sizing solids and biosolids processing facilities that also have long detention times or ample storage. Peaking factors may be developed from flowrate records or based on published curves or data from similar communities.

The most common method of determining peaking factors is from the analysis of flowrate data (see Fig. 3–1). Where flowrate records are available, at least 2 years of data should be analyzed to develop the peak to average flowrate factors. These factors may then be applied to estimated future average flowrates, adjusted for any anticipated future special conditions. Where commercial, institutional, or industrial wastewaters are expected to make up a significant portion of the average flowrates (say 25 percent or more of all flows, exclusive of infiltration), peaking factors for the various categories of flow should be estimated separately. Peak flows from each category most probably will not occur simultaneously; therefore, some adjustment may have to be made to the total peak flow to prevent overestimating the peaking conditions. If possible, peaking factors for industrial wastewater should be estimated on the basis of average water use, number of shifts worked, and pertinent details of plant operations.

If flow measurement records are inadequate to establish peaking factors, published information may be used. Many sources for peaking factor data are available including state agencies, cities and special districts that provide wastewater collection and treatment services, and professional publications from organizations such as the Water Environment Federation and the American Society of Civil Engineers (see WEF, 1998). An example peaking factor curve is given on Fig. 3–13, and it may be used for estimating peak hourly flowrates from domestic sources. The curve given on Fig. 3–13 was developed from analyses of the records of numerous communities throughout the United States. The curve is based on average residential flowrates, exclusive of infiltration/inflow, and includes small amounts of commercial flows and industrial wastes.

In developing factors for peak hourly flowrates, the characteristics of the collection system serving the wastewater-treatment plant must be considered carefully. Improve-

Figure 3–13

Peaking factor curve (ratio of peak hourly to average daily flow).

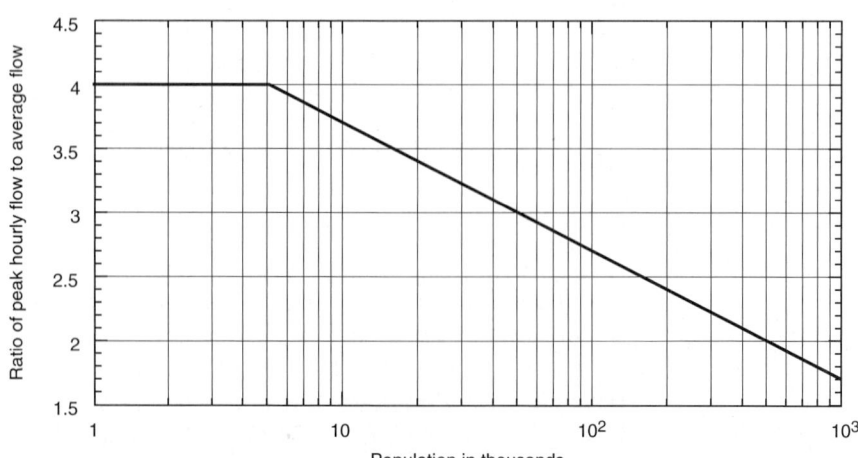

ments to or rehabilitation of the collection system may also increase or decrease the peaking factors. For pumped flows where reliable metering data are not available, factors to be considered include:

- Interviews with operators regarding observations of operating conditions
- Review of pumping records (historical data on number of pumps in service and running time, if available)
- Operating speed of pumps
- Condition of pumps from maintenance records (unit output will be lower if impellers are worn)

Field testing at pumping stations can also be performed to measure the combined output of a simulated historical high-flow event. Assistance in performing pump tests is often available from the local energy service provider.

Where flow to the treatment plant is by gravity, the peak flowrate can be estimated based on the following:

- Capacity of the influent sewers
- Investigation of upstream access ports (i.e., manholes) to determine if a high watermark is visible
- Interviews with operating staff and review of any documented field records

Forecasting design flowrates, including the use of peaking factors, is illustrated in Example 3–9.

EXAMPLE 3–9 **Forecasting Design Flowrates** A residential community with a current population of 15,000 is planning to expand its wastewater-treatment plant. In 20 years, the population is estimated to increase to 25,000 residents and 2000 day students are expected to attend a proposed junior college. A new industry will also move in and contribute an average flowrate of 840 m³/d and a peak flowrate of 1260 m³/d. The plant will operate 8 h/d and will shut down one day per week. The present average daily wastewater flowrate is 6500 m³/d and the infiltration/inflow has been determined to be nonexcessive. Infiltration is estimated to be 100 L/capita·d at average flow and 150 L/capita·d at peak flow. Residential water use in the new homes is expected to be 10 percent less than the existing residences because of the installation of water-saving appliances and fixtures. Compute the future average, peak, and minimum design flowrates. For peak residential flowrates, use a peaking factor of 2.75 and assume the ratio of minimum to average flowrate is 0.35.

Solution

1. Compute the present and future per capita wastewater flowrates
   a. For present conditions, compute the average domestic flowrate excluding infiltration
      i. Compute infiltration

$$\text{Infiltration} = 15{,}000 \text{ persons} \times 100 \text{ L/capita·d} \times \frac{1 \text{ m}^3}{1000 \text{ L}} = 1500 \text{ m}^3/\text{d}$$

ii. Compute average domestic flowrate

$$\text{Domestic flow, m}^3/\text{d} = \text{total average flow} - \text{infiltration}$$

$$= 6500 - 1500 = 5000 \text{ m}^3/\text{d}$$

    b. Compute present per capita flowrate by dividing the existing domestic flowrate by the present population

$$\text{Per capita flowrate} = \frac{(5000 \text{ m}^3/\text{d})}{15,000 \text{ persons}} = 0.33 \text{ m}^3/\text{capita·d}$$

    c. For future conditions, reduce existing per capita flowrate by 10%

Future per capita flowrate $= 0.33 \times 0.9 = 0.297 \text{ m}^3/\text{capita·d}$

		Flowrate, m³/d
2.	Compute future average flowrate	
a.	Existing residents =	5,000
b.	Future residents = 10,000 × 0.297 m³/capita·d =	2,970
c.	Day students (assume 95 L/capita·d from Table 3–3) = 2000 × 0.095 m³/capita·d	190
	Subtotal residential	8,160
d.	Industrial flowrate =	840
e.	Infiltration = (25,000)(100 L/capita·d)(1 m³/10³ L) =	2,500
f.	Total future average flowrate =	11,500
3.	Compute future peak flowrate	
a.	Residential peak flowrate = 8160 × 2.75 =	22,440
b.	Industrial peak flowrate =	1,260
c.	Infiltration = (25,000)(150 L/capita·d)(1 m³/10³ L) =	3,750
d.	Total future peak flowrate =	27,450
4.	Compute the minimum flowrate	
a.	Residential: As indicated on Fig. 3–4, the low flowrate usually occurs in the early morning hours. The future minimum flowrate, excluding college day students = 0.35 × (5000 + 2970) =	2,780
b.	Industrial (facilities are shut down at night)=	0
c.	Infiltration = (25,000)(100 L/capita·d)(1 m³/10³ L) =	2,500
d.	Total minimum flowrate =	5,280

Comment In this example, infiltration/inflow contributes nearly 50 percent of the minimum flowrate and over 20 percent of the average flowrate; an illustration of the influence of extraneous flows on treatment plant design. If wastewater-flow records are not adequate or are unavailable, future average daily flow may be calculated based on the future population and unit wastewater flowrates, similar to those given in Tables 3–1 to 3–5. Appropriate adjustments should be made in the calculations to account for any special conditions such as flow reduction, infiltration/inflow allowances, and industrial flows. When peak flowrates for more than one flow component are calculated, some adjustment in the total peak flowrate should be made if the peaks from the components do not occur simultaneously.

Upstream Control of Peak Flowrates. Planning wastewater facilities to handle peak flowrates may involve other considerations including (1) improvements to the collection system to reduce peak flow related to infiltration/inflow (I/I), (2) installation of flow-equalization facilities to provide storage either in the collection system or at the treatment plant. Other alternatives for peak flowrate control, namely, provision for flow splitting and sidestream treatment facilities, are discussed in Chaps. 5 and 15.

Improvement to the collection system may involve a lengthy and costly process and may not have an immediate effect on significantly reducing peak flowrates. In some cases, the amount of flow reduction resulting from collection system rehabilitation has been less than anticipated, particularly if infiltration is a significant component of I/I. In unusual circumstances, the flowrates have actually increased after completion of the collection system improvement program. Therefore, safety factors should be considered when estimating possible peak flowrate reduction resulting from collection system improvements.

Flow equalization can be an effective measure in reducing peak flows. Benefits derived by upstream flow equalization include (1) reduced hydraulic loading on already overtaxed collection facilities, (2) reduced collection system overflows (and reduced public health threats), and (3) reduced peak hydraulic loading of the treatment plant. Equalization depends on available volume and may be of limited value in extreme peak flow conditions. Siting of equalization facilities in the collection system is often difficult because of limited available space at locations that are compatible with the system hydraulics. Operation and maintenance may also be difficult to manage, particularly in remote areas. Ease of operation, maintenance, and control and environmental factors are major reasons that many equalization facilities are located at treatment plants. The analysis for sizing flow equalization facilities is presented in Chap. 5.

Design Mass Loadings

The importance of mass loading in the design of wastewater-treatment facilities is identified in Table 3–20. For example, the sizing of aeration facilities and the amounts of solids and biosolids produced are directly related to the mass of BOD that must be processed. Further, the performance of the preliminary and primary treatment facilities has to be taken into account as ineffective operation of these facilities can result in the transfer of greater organic loads to the biological treatment system. Peak process loading rates are also important in sizing the process units and their support systems so that treatment plant performance objectives can be achieved consistently and reliably. The performance of primary sedimentation tanks is discussed in Chap. 5.

PROBLEMS AND DISCUSSION TOPICS

3-1 A community is rapidly approaching the design capacity of its wastewater-treatment plant. As an alternative to expanding the plant, a water conservation program has been proposed. If the current average residential unit flowrate is 320 L/capita·d and the proposed water conservation rate reduction is 25, 35, 40 percent (to be selected by instructor), is the proposal reasonable? State your justification.

3-2   The new commercial development is being considered, and four developers have submitted competing proposals. An Environmental Assessment Report is being prepared to evaluate the wastewater flows from each of the competing proposals. Estimate the average daily and peak wastewater flowrates from one of the proposed developments (to be selected by instructor). The proposals consist of the following elements:

Type of facility	Units	Developer A	B	C	D
Hotel	Number of guest rooms	120	80	60	250
	Number of employees	25	16	14	40
Department stores	Number of toilet rooms	8	12	16	
	Number of employees	40	60	80	
Self-service laundry	Number of machines		20	16	18
Restaurant, no bar	Number of seats	125	100	100	50
Restaurant, with bar	Number of seats	100	125	75	80
Theater (indoor)	Number of seats	500	400		350

3-3   Estimate the average and maximum flowrates from one of the following recreational areas (to be selected by instructor) that consists of the following facilities:

Type of facility	Units	Area A	B	C	D
Visitor center	No. of visitors	250	300	400	500
Motel with kitchen	No. of guests		60	100	60
Resort cabins	No. of guests		100	40	
Cottages	No. of guests	60		60	120
Campground (toilets only)	No. of persons	140	120		200
Recreational vehicle park	Number of individual connection	40	50	20	50
Self-service laundry	Number of machines	8	10	6	10
Shopping center	Number of employees	10	15	15	20
	Number of parking spaces	30	40	40	60
Automobile service station	Number of vehicles	80	120	160	200
Restaurant with bar	Number of customers	200	300	400	500

State all of your assumptions clearly.

3-4   A college dormitory complex is planning to institute a water conservation program to reduce its wastewater flows. The complex consists of four dormitories each with the following characteristics:

Characteristics	Unit	Dormitory A	B	C	D
Wastewater flowrate	L/d	125,000	105,000	140,000	160,000
Number of beds	No	300	250	300	350
Existing fixture flowrates					
Toilets,	L/use	9	10	11	11
Showerheads,	L/min·use	18	20	23	23
Faucets,	L/capita·d	10	8	9	11
Flowrate reduction required	%	15	20	25	25

Using the data given above, develop a water conservation program that will accomplish the flowrate reduction goal for one of the facilities (to be selected by instructor).

3–5 Compute the flow-weighted BOD and TSS concentrations from one of the following flowrate regimes (to be selected by instructor).

Time	BOD, mg/L	TSS, mg/L	Flowrate A, m³/d	Flowrate B, m³/d	Flowrate C, m³/d
0200	130	150	8,000	7,200	10,000
0400	110	135	6,000	6,400	8,400
0600	160	150	9,400	9,800	13,600
0800	220	205	12,800	13,500	19,200
1000	230	210	13,000	13,800	19,500
1200	245	220	14,400	14,500	21,800
1400	225	210	12,000	12,500	18,500
1600	220	200	9,600	10,000	14,800
1800	210	205	11,000	10,500	15,000
2000	200	210	8,000	8,500	11,500
2200	180	185	9,000	8,200	12,600
2400	160	175	8,400	7,700	11,600

3–6 The data in the table below consist of population values for a city, and average monthly influent flow values at the city wastewater-treatment plant, from 1996 through 1999. The city is located in an area of high groundwater. Using these data answer the following questions.

a. What is the nature of the distribution of the monthly flowrate values? Use the plotting position method described in Sec. 3–3 (see Eq. 3–10) to plot the monthly influent values versus the corresponding probabilities for each year on arithmetic- and log-probability paper, and check for linearity.

b. What is the average annual flow, average dry-weather flow (ADWF), and average wet-weather flow (AWWF) for each year? If the data are arithmetically distributed, use the arithmetic mean; if the data are log-normally distributed, use the geometric mean (see Table 3–10).

Assume the dry season occurs from June to October, and the wet season occurs from November to May.

c. What is the per capita flow contribution from commercial and light industrial activities? Assume that the residential contribution to the dry-weather flow is 260 L/capita·d, and that commercial and light industrial activities make up the remaining flow.

d. What is the per capita flow contribution from infiltration and inflow for each year? Assume that the difference between the wet- and dry-weather flows is due to infiltration and inflow.

Community population and average monthly influent flowrate data for the period from 1996 through 1999

Year	1996	1997	1998	1999
Population	8,690	9,400	11,030	12,280
Month	Influent flowrate, m³/d			
January	8,800	13,900	8,300	10,000
February	6,200	9,900	11,800	18,400
March	6,800	8,100	9,400	13,000
April	4,000	4,200	6,500	5,000
May	4,000	5,700	5,300	7,600
June	3,600	3,600	4,800	4,600
July	2,400	2,600	3,300	3,800
August	2,000	1,500	3,800	3,100
September	2,800	2,000	2,800	2,200
October	3,200	4,800	4,400	4,400
November	4,800	3,200	6,000	6,500
December	5,200	6,700	7,300	8,600

3-7 Consider the city described in Problem 3–6. Assume that the build-out population for the community is 16,000, and that the residential wastewater flowrate will be 300 L/capita·d. The commercial flowrate in 1999 (1000 m³/d) is 80 percent of what it will be at build-out. Due to high infiltration and inflow (I/I) rates, a sewer repair program will be implemented. The I/I contribution will be either 500, 400, 300, or 200 L/cap·d (to be selected by instructor), depending on the degree of repair achieved. Estimate the ADWF, AWWF, and the average annual flow that will be received at the community treatment plant at build out. Justify the use of the AWWF as the nominal design capacity for the treatment plant.

3-8 Consider the treatment plant for the city described in Problem 3–6. Use the plotting position method described in Sec. 3–3 (see Eq. 3–10) to plot the monthly influent values versus the corresponding probabilities for each year. Determine the nature of the distribution by plotting the values on arithmetic- and log-probability paper and checking for linearity. If the average annual flow to the treatment facility at build-out is estimated to be 8000 m³/d, what will the peak monthly flow be? (Hint: the average annual flow occurs at the 50 percent line on the probability graph. Use the slope of the wettest year to pass a line through the 50 percent line at 8000 m³/d, and read the flow value for the highest month from the graph.)

3-9 Use the following influent data for a wastewater-treatment plant to conduct the statistical analyses described in Problem 3–6, steps a, b, c, and d. If the data in the table below are not distributed either arithmetic-normally or log-normally, suggest a method for determining the required parameters.

Community population and average monthly influent flowrate data for the period from 1996 through 1999

Year	1996	1997	1998	1999
Population	17,040	17,210	17,380	17,630
Month	Influent flowrate, m^3/d			
January	8,800	7,760	9,360	7,600
February	9,440	7,280	7,920	7,840
March	8,640	7,200	8,800	7,680
April	7,840	6,960	8,080	7,440
May	7,440	6,800	7,680	7,280
June	7,200	6,880	7,520	7,360
July	7,120	6,960	7,280	7,200
August	7,040	6,720	7,200	7,280
September	6,880	6,880	7,040	7,200
October	6,960	6,800	7,280	7,440
November	7,120	7,120	7,360	7,680
December	7,360	7,600	7,680	8,000

3-10 A treatment plant is planned to be installed in a recreational resort that contains a developed campground for 200 persons, lodges and cabins for 100 persons, and resort apartments for 150 persons. Assume that persons staying in lodges use the dining hall for 3 meals per day and that a 50-seat cafeteria with 4 employees and an estimated 200 customers per day has been constructed. Daily attendance at the visitor center is 500 persons. Other facilities include a 10-machine laundromat, a 20-seat cocktail lounge, and three gas stations at 1100 L/d per station. Determine the average wastewater flowrate in L/d using the unit flows assuming the housing facilities are at maximum capacity.

3-11 Obtain an annual report or one year of flow and BOD data from your local wastewater-treatment facility. From these records, prepare probability plots for the flowrate and mass loadings. Determine the values at 50 and 95 percent probability.

3-12 From the flowrate data obtained for Problem 3–11, determine the mean and standard deviation.

3-13 The wastewater-treatment plant has been experiencing high wastewater flowrates during the wet weather months. The average monthly flows for four flow regimes are reported below (one regime to be selected by instructor). The rapid increase in flows during the winter months is due mainly to increased infiltration/inflow. Infiltration is estimated to be 67 percent of the excess flow. The lengths of the collection system pipelines that need to be repaired are also listed below. The average repair cost is $200,000/km and the repair will be effective in reducing the infiltration by 30 percent. How many years from now will it take to pay back the cost of the sewer repair program based on the annual savings in treatment cost, assuming the future annual flowrates are equal to those in the table below? The current cost of treatment is $1.00/m^3$ and the future cost of treatment is estimated to escalate at 6 percent per year. Assume the sewer repair will be complete in three years for Areas A and B, and four years for Areas C and D.

| | Average monthly flowrate, m³/d | | | |
| | Flowrate regime | | | |
Month	A	B	C	D
January	293,000	410,000	460,000	470,000
February	328,000	459,000	440,000	485,000
March	279,000	391,000	515,000	560,000
April	212,000	296,000	333,000	400,000
May	146,000	204,000	230,000	300,000
June	108,000	151,000	170,000	225,000
July	95,000	133,000	150,000	200,000
August	89,000	125,000	141,000	188,000
September	93,000	130,000	140,000	165,000
October	111,000	155,000	167,000	192,000
November	132,000	185,000	200,000	240,000
December	154,000	215,000	225,000	215,000
Length of collection system, km	300	400	450	600

3-14 Nine months of flow records have been collected for a new wastewater-treatment plant. Inspection of the records indicates that weekend flowrates tend to be higher than weekday flowrates. The weekday and weekend flowrates have been averaged and arranged in ascending order as given below. Develop arithmetic- and log-probability plots for both the weekday and weekend data for either Flowrate regime A or Flowrate regime B (to be selected by instructor) and comment on the skewness of the data. Determine the mean and 95 percentile values for each set of flow data, and determine the probable one-day maximum flowrate. Discuss the significance of the data analysis.

| | Flowrate regime A | | Flowrate regime B | |
| | Weekday average flowrate, m³/d × 10³ | Weekend average flowrate, m³/d × 10³ | Weekday average flowrate, m³/d × 10³ | Weekend average flowrate, m³/d × 10³ |
Number				
1	39.7	42.8	55.7	56.4
2	40.5	43.1	56.1	57.5
3	40.9	43.5	56.6	58.1

(continued)

(*Continued*)

Number	Flowrate regime A		Flowrate regime B	
	Weekday average flowrate, m³/d × 10³	**Weekend average flowrate, m³/d × 10³**	**Weekday average flowrate, m³/d × 10³**	**Weekend average flowrate, m³/d × 10³**
4	41.3	43.9	57.2	58.6
5	42.0	44.3	57.7	59.5
6	42.1	44.7	58.2	60.6
7	42.2	45.0	58.5	60.8
8	42.4	45.4	59.1	61.1
9	42.9	45.8	59.6	61.8
10	43.5	46.2	60.1	62.6
11	43.9	46.6	60.7	63.2
12	44.3	46.7	60.8	63.8
13	44.7	46.9	61.0	64.4
14	45.0	47.7	62.1	64.8
15	45.4	47.9	62.3	65.4
16	45.6	48.8	63.5	65.8
17	45.7	49.2	64.0	66.1
18	46.0	50.0	65.1	66.5
19	46.4	50.3	65.5	66.8
20	46.9	51.1	66.5	67.9
21	47.7	51.5	67.0	69.6
22	48.4	53.0	69.0	70.7
23	48.8	53.4	69.5	71.2
24	49.0	53.7	69.9	71.5
25	49.2	54.9	71.5	72.2
26	49.6	55.3	72.0	72.4
27	50.5	56.0	72.9	73.7
28	51.1	56.8	73.9	74.6
29	52.2	57.2	74.5	76.2
30	53.0	58.3	75.9	77.5
31	53.2	59.1	76.9	78.6
32	54.3	60.6	78.9	80.7
33	55.3	60.9	79.2	82.8
34	56.0	61.7	80.3	85.0
35	60.6	62.1	80.8	88.4
36	62.5	63.6	82.8	91.1

3-15 Land use in a new development is given in the following first table. A new school being built will have 1500 students. The average flowrate is 75 L/student, and the peaking factor (ratio of peak flow to average flow) is 4.0. Average flowrate allowances and the peaking factors for the other developments are shown in the second table. Determine the peak wastewater flowrate from one of the areas (to be selected by instructor).

Type of development	Area A, ha	Area B, ha	Area C, ha	Area D, ha
Residential	125	150	150	160
Commercial	11	10	15	16
School	4	4	4	4
Industrial	6	8	20	10

Type of development	Average flowrate, m³/ha·d	Peaking factor
Residential	40	3.0
Commercial	20	2.0
Industrial	30	2.5

REFERENCES

Andreadakis, A. D. (1992) "The Use of a Water Quality Model for the Evaluation of the Impact of Marine Sewage Disposal on the Evoilos Gulf, Greece." *Water Science and Technology,* vol. 25, 165–172.

AWWA (1998) *1998 Residential Water Use Survey,* American Water Works Association, Denver, CO.

AWWARF (1999) *Residential End Uses of Water,* American Water Works Association Research Foundation, Denver, CO.

Crites, R. W., and G. Tchobanoglous (1998) *Small and Decentralized Wastewater Management Systems,* McGraw-Hill, New York.

Federal Register (1989) Amendment to the Secondary Treatment Regulations: Percent Removal Requirements During Dry Weather Periods for Treatment Works Served by Combined Sewers, 40 CFR Part 133.

Geyer, J. C., and Lentz, J. J. (1962) *Evaluation of Sanitary Sewer System Designs,* Johns Hopkins University School of Engineering.

Henze, M., P. Harremoes, J. Jansen, and E. Arvin (1997) *Biological and Chemical Processes,* 2nd ed., Springer-Verlag Berlin Heidelberg, p. 27.

Huber, W.C. (1984) *Storm Water Management Model, User's Manual,* Version III, Report to U.S. Environmental Protection Agency, Project No. CR-805664.

Metcalf & Eddy, Inc. (1977) *Urban Stormwater Management and Technology, Update and User's Guide,* Report to U.S. Environmental Protection Agency, Report no. EPA 600/8-77-014.

Metcalf & Eddy, Inc. (1981) *Wastewater Engineering: Collection and Pumping of Wastewater,* McGraw-Hill, New York.

Metcalf & Eddy, Inc. (1991) *Wastewater Engineering: Treatment, Disposal, and Reuse,* 3d ed., McGraw-Hill, New York.

Nashashibi, M., and L. A. van Duijl (1995) "Wastewater Characteristics in Palestine," *Water Science and Technology,* vol. 32, pp. 65–75.

Nayyer, M. L. (1992) *Piping Handbook,* 6th ed., McGraw-Hill, New York.

Ozturk, I., T. Zambal, A. Smasunlu, and E. Göknel (1992) "Environmental Impact Evaluation of Istanbul Wastewater Treatment and Marine Disposal System," *Water Science and Technology,* vol. 25, 85–92.

Salvato, J. A. (1992) *Environmental Engineering and Sanitation,* 4th ed., Wiley Interscience Publishers, New York.

U.S. DHUD (1984) *Residential Water Conservation Projects, Summary Report,* U.S. Department of Housing and Urban Development, Washington, D.C.

U.S. DHUD (1984) *Water Saved by Low-Flush Toilets and Low-Flow Shower Heads,* U.S. Department of Housing and Urban Development, Washington, D.C.

U.S. DHUD (1984) *Survey of Water Fixture Use,* U.S. Department of Housing and Urban Development, Washington, D.C.

U.S. EPA (1983) *Results of the Nationwide Urban Runoff Program,* vol. 1, Final Report, NTIS PB84-185552, U.S. Environmental Protection Agency.

WEF (1998) *Design of Municipal Wastewater Treatment Plants, Manual of Practice* no. 8, 4th ed., vol. 1, Water Environment Federation, Alexandria, VA.

Yohe, T. M., and J. E. Rich (1995) "Textile Mill Builds for Today, Plans for Tomorrow," *Industrial Wastewater,* Water Environment Federation, pp. 30–35, May/June.

4 Introduction to Process Analysis and Selection

The constituents of concern found in wastewater are removed by physical, chemical, and biological methods. The individual methods usually are classified as physical unit operations, chemical unit processes, and biological unit processes. Treatment methods in which the application of physical forces predominate are known as *physical unit operations.* Examples of physical unit operations include screening, mixing, sedimentation, gas transfer, filtration, and adsorption. Treatment methods in which the removal or conversion of constituents is brought about by the addition of chemicals or by other chemical reactions are known as *chemical unit processes.* Examples of chemical unit processes include disinfection, oxidation, and precipitation. Treatment methods in which the removal of constituents is brought about by biological activity are known as *biological unit processes.* Biological treatment is used primarily to remove the biodegradable organic constituents and nutrients in wastewater. Examples of biological treatment processes include the activated-sludge and trickling-filter processes. Unit operations and processes occur in a variety of combinations in treatment flow diagrams.

From practical observations, the rates at which physical, chemical, and biological reactions and conversions occur are important, as they will affect the size of the treatment facilities that must be provided (see Fig. 4–1). The rate at which reactions and conversions occur, and the degree of their completion, is generally a function of the constituents involved, the temperature, and the type of reactor (i.e., container or tank in which the reactions take place). Hence, both the effects of temperature and the type of

Figure 4-1

Overview of Harford County, MD, Sod Run biological nutrient removal (BNR) wastewater-treatment plant. The capacity of the plant is 76,000 m³/d (20 Mgal/d).

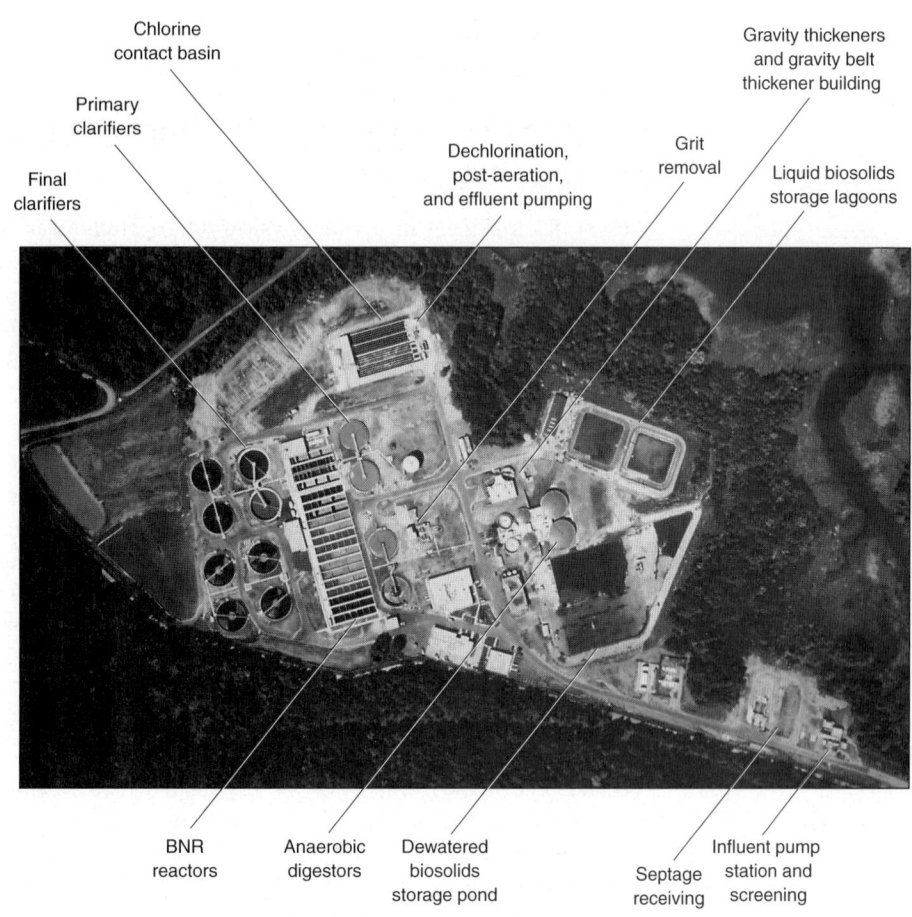

reactor employed are important in the selection of treatment processes. In addition, a variety of environmental and other physical constraints must be considered in process selection. The fundamental basis for the analysis of the physical, chemical, and biological unit operations and processes used for wastewater treatment is the *materials mass balance principle* in which an accounting of mass is made before and after reactions and conversions have taken place.

Therefore, the purpose of this chapter is to introduce and discuss (1) the types of reactors used for wastewater treatment; (2) the preparation of mass balances to determine process performance; (3) modeling ideal flow in reactors; (4) the analysis of reactor hydraulics using tracers; (5) modeling nonideal flow in reactors; (6) reactions, reaction rates, and reaction rate coefficients; (7) modeling treatment kinetics, which involves the coupling of reactors and reaction rates; (8) treatment processes involving mass transfer; and (9) important factors involved in process analysis and selection. The information in this chapter is intended to serve as an introduction to the subject of process analysis, and to provide a basis for the analysis of the unit operations and processes that will be presented in subsequent chapters. By dealing with the basic concepts first, it will be possible to apply them (without repeating the details) in the remaining chapters. Processes involving physical removal such as sedimentation and filtration are considered in Chaps. 5 and 11, respectively.

4–1 REACTORS USED FOR THE TREATMENT OF WASTEWATER

Wastewater treatment involving physical unit operations and chemical and biological unit processes is carried out in vessels or tanks commonly known as "reactors." The types of reactors that are available and their applications are introduced in this section. The analysis of the hydraulic characteristics and the performance of reactors used for wastewater treatment is considered in Secs. 4–3, 4–4, and 4–5, following the discussion of the mass-balance principle.

Types of Reactors

The principal types of reactors used for the treatment of wastewater, illustrated on Fig. 4–2, are (1) the batch reactor, (2) the complete-mix reactor [also known as the continuous-flow stirred-tank reactor (CFSTR) in the chemical engineering literature], (3) the plug-flow reactor (also known as a tubular-flow reactor), (4) complete-mix reactors in series, (5) the packed-bed reactor, and (6) the fluidized-bed reactor. Brief descriptions of these reactors are presented below.

Batch Reactor. In the batch reactor (see Fig. 4–2a), flow is neither entering nor leaving the reactor (i.e., flow enters, is treated, and then is discharged, and the cycle repeats). The liquid contents of the reactor are mixed completely. For example, the BOD test discussed in Chap. 2 is carried out in a batch reactor (i.e., BOD bottle as shown on Fig. 2–19), although it should be noted that the contents are not mixed completely during the incubation period. Batch reactors are often used to blend chemicals or to dilute concentrated chemicals.

Complete-Mix Reactor. In the complete-mix reactor (see Fig. 4–2b), it is assumed that complete mixing occurs instantaneously and uniformly throughout the reactor as fluid particles enter the reactor. Fluid particles leave the reactor in proportion to their statistical population. Complete mixing can be accomplished in round or square reactors if the contents of the reactor are uniformly and continuously redistributed. The actual time required to achieve completely mixed conditions will depend on the reactor geometry and the power input.

Plug-Flow Reactor. Fluid particles pass through the reactor with little or no longitudinal mixing and exit from the reactor in the same sequence in which they entered. The particles retain their identity and remain in the reactor for a time equal to the theoretical detention time. This type of flow is approximated in long open tanks with a high length-to-width ratio in which longitudinal dispersion is minimal or absent (see Fig. 4–2c) or closed tubular reactors (e.g., pipelines, see Fig. 4–2d).

Complete-Mix Reactors in Series. The series of complete-mix reactors (see Fig. 4–2e) is used to model the flow regime that exists between the ideal hydraulic flow

Figure 4–2

Definition sketch for the different types of reactors used for wastewater treatment: (a) batch reactor, (b) complete-mix reactor, (c) plug-flow open reactor, (d) plug-flow closed reactor also known as a tubular reactor, (e) complete-mix reactors in series, (f) packed-bed reactor, (g) packed-bed upflow reactor, and (h) expanded-bed upflow reactor.

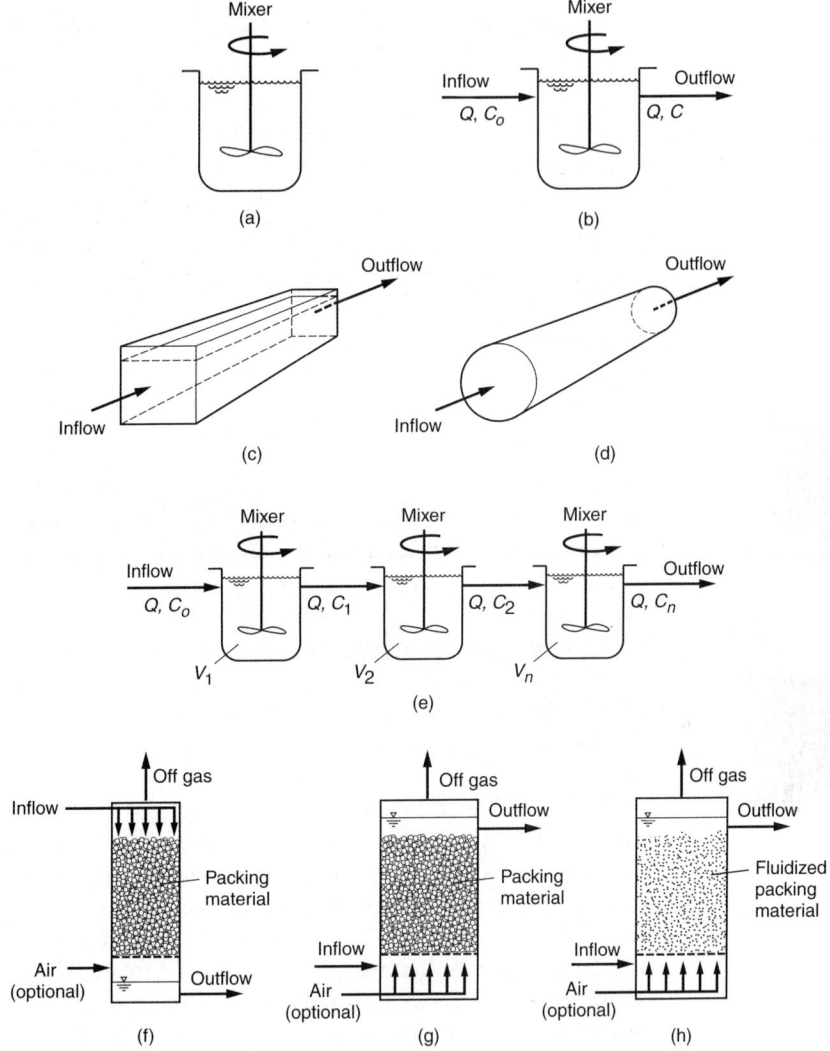

patterns corresponding to the complete-mix and plug-flow reactors. If the series is composed of one reactor, the complete-mix regime prevails. If the series consists of an infinite number of reactors in series, the plug-flow regime prevails.

Packed-Bed Reactors. The packed-bed reactor is filled with some type of packing material, such as rock, slag, ceramic, or, now more commonly, plastic. With respect to flow, the packed-bed reactor can be operated in either the downflow or upflow mode. Dosing can be continuous or intermittent (e.g., trickling filter). The packing material in packed-bed reactors can be continuous (see Fig. 4–2f) or arranged in multiple stages, with flow from one stage to another. A packed-bed upflow anaerobic (without oxygen) reactor is shown on Fig. 4–2g.

Table 4–1
Principal applications of reactor types used for wastewater treatment

Type of reactor	Application in wastewater treatment
Batch	Activated-sludge biological treatment in a sequence batch reactor, mixing of concentrated solutions into working solutions
Complete-mix	Aerated lagoons, aerobic sludge digestion
Complete-mix with recycle	Activated-sludge biological treatment
Plug-flow	Chlorine contact basin, natural treatment systems
Plug-flow with recycle	Activated-sludge biological treatment, aquatic treatment systems
Complete-mix reactors in series	Lagoon treatment systems, used to simulate nonideal flow in plug-flow reactors
Packed-bed	Nonsubmerged and submerged trickling-filter biological treatment units, depth filtration, natural treatment systems, air stripping
Fluidized-bed	Fluidized-bed reactors for aerobic and anaerobic biological treatment, upflow sludge blanket reactors, air stripping

Fluidized-Bed Reactor.   The fluidized-bed reactor is similar to the packed-bed reactor in many respects, but the packing material is expanded by the upward movement of fluid (air or water) through the bed (see Fig. 4–2h). The expanded porosity of the fluidized-bed packing material can be varied by controlling the flowrate of the fluid.

Application of Reactors

The principal applications of reactor types used for wastewater treatment are reported in Table 4–1. Operational factors that must be considered in the selection of the type of reactor or reactors to be used in the treatment process include (1) the nature of the wastewater to be treated, (2) the nature of the reaction (i.e., homogeneous or heterogeneous), (3) the reaction kinetics governing the treatment process, (4) the process performance requirements, and (5) local environmental conditions. Homogeneous and heterogeneous reactions and reaction kinetics are discussed in Sec. 4–6. In practice, the construction costs and operation and maintenance costs also affect reactor selection. Because the relative importance of these factors varies with each application, each factor should be considered separately when the type of reactor is to be selected. Reactor selection is considered further in Sec. 4–9.

Hydraulic Characteristics of Reactors

Complete-mix and plug-flow reactors are the two reactor types used most commonly in the field of wastewater treatment. The hydraulic flow characteristics of complete-mix and plug-flow reactors can be described as varying from ideal and nonideal, depending on the relationship of the incoming flow to outgoing flow. Ideal and nonideal flow are described in the following discussion. The mathematical modeling of the flow in ideal and nonideal reactors is presented in Secs. 4–3 and 4–5, respectively, following a discussion of the mass-balance analysis and tracer studies.

Ideal Flow in Complete-Mix and Plug-Flow Reactors. The ideal hydraulic flow characteristics of complete-mix and plug-flow reactors are illustrated on Fig. 4–3 in which dye tracer response curves are presented for pulse (slug-dose) and

Figure 4–3

Output tracer response curves from reactors subject to pulse and step inputs of a tracer: (a) complete-mix reactor and (b) plug-flow reactor.

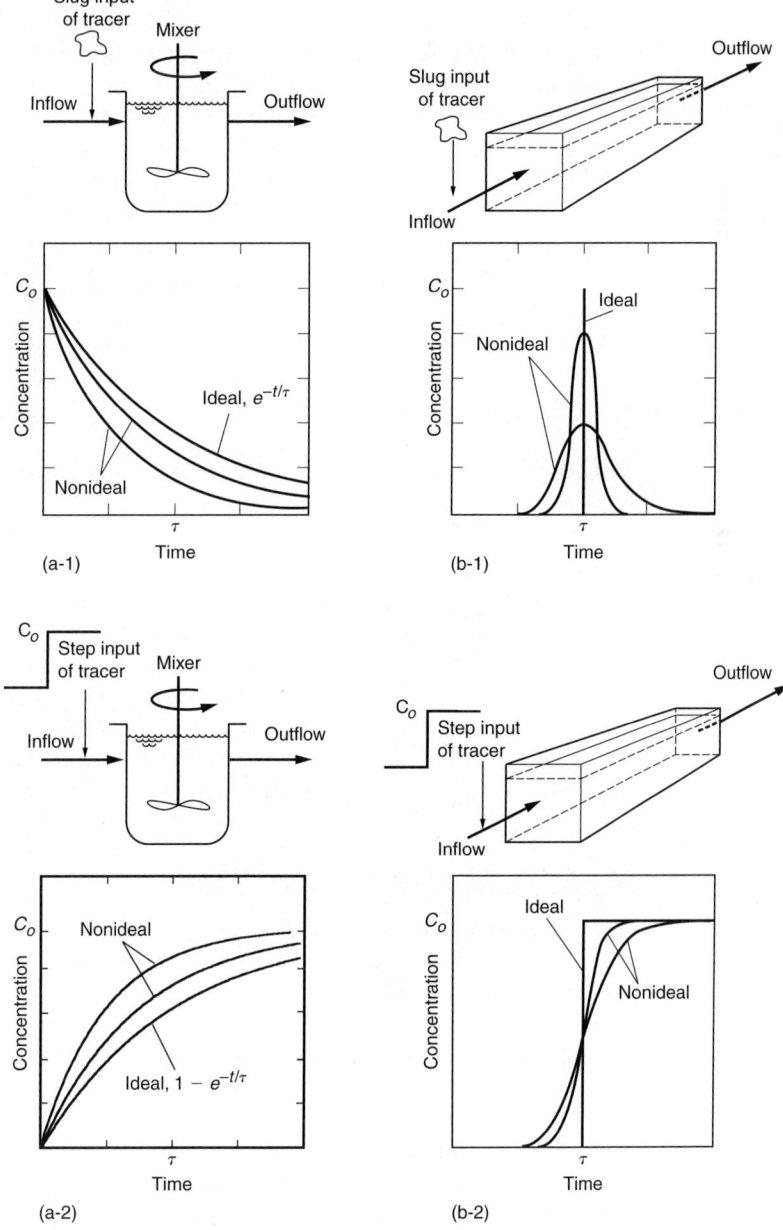

step inputs (continuous injection). On Fig. 4–3, t is the actual time and τ is equal to the theoretical hydraulic detention time defined as follows:

$$\tau = V/Q \tag{4-1}$$

where τ = hydraulic detention time, T
 V = volume of the reactor, L^3
 Q = volumetric flowrate, L^3T^{-1}

    If a pulse (slug) input of a conservative (i.e., nonreactive) tracer is injected and dispersed instantaneously in an ideal-flow complete-mix reactor, with a continuous inflow of clear water, the output tracer concentration would appear as shown on Fig. 4–3(a-1). If a continuous step input of a conservative tracer at concentration C_o is injected into the inlet of an ideal complete-mix reactor, initially filled with clear water, the appearance of the tracer at the outlet would occur as shown on Fig. 4–3(a-2).

    In the case of an ideal plug-flow reactor, the reactor is initially filled with clear water before being subjected to a pulse or a step input of tracer. If an observer were positioned at the outlet of the reactor, the appearance of the tracer in the effluent for a pulse input, distributed uniformly across the reactor cross section, would occur as shown on Fig. 4–3(b-1). If a continuous step input of a tracer were injected into such a reactor at an initial concentration C_o, the tracer would appear in the effluent as shown on Fig. 4–3(b-2).

Nonideal Flow in Complete-Mix and Plug-Flow Reactors. In practice, the flow in complete-mix and plug-flow reactors is seldom ideal. For example, when a reactor is designed, how is the flow to be introduced to satisfy the theoretical requirement of instantaneous and complete dispersion? In practice, there is always some deviation from ideal conditions, and it is the precautions taken to minimize these effects that are important. Nonideal flow occurs when a portion of the flow that enters the reactor during a given time period arrives at the outlet before the bulk of the flow that entered the reactor during the same time period arrives. Nonideal flow is illustrated on Fig. 4–3a and 4–3b. The causes of nonideal flow are considered in Sec. 4–4. The important issue with nonideal flow is that a portion of the flow will not remain in the reactor as long as may be required for a biological or chemical reaction to go to completion.

4–2 MASS-BALANCE ANALYSIS

The fundamental approach used to study the hydraulic flow characteristics of reactors and to delineate the changes that take place when a reaction is occurring in a reactor (e.g., a container), or in some definable portion of a body of liquid, is the mass-balance analysis. The basic concepts involved in such an analysis are described in this section. The application of the mass-balance approach to problem solving will be illustrated in the sections that follow and throughout this text.

The Mass-Balance Principle

The mass-balance analysis is based on the principle that mass is neither created nor destroyed, but the form of the mass can be altered (e.g., liquid to a gas). The mass-balance analysis affords a convenient way of defining what occurs within treatment reactors as a

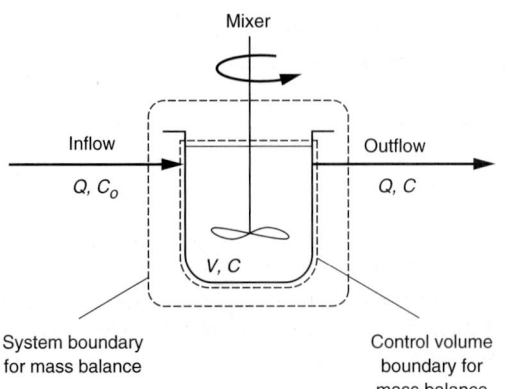

Figure 4–4

Definition sketch for the application of materials mass-balance analysis for a complete-mix reactor with inflow and outflow. The presence of a mixer is used to represent symbolically the fact that the contents of the reactor are mixed completely. The photo is of a typical complete-mix activated sludge reactor used for the biological treatment of wastewater.

function of time. To illustrate the basic concepts involved in the preparation of a mass-balance analysis, consider the reactor shown on Fig. 4–4. The *system boundary* is drawn to identify all of the liquid and constituent flows into and out of the system. The *control volume* is used to identify the actual volume in which change is occurring. In most cases, the system and control volume boundaries will coincide. For a given reactant, the general mass-balance analysis is given by

1. General word statement:

$$
\begin{array}{cccc}
\text{Rate of accumulation} & \text{rate of flow of} & \text{rate of flow of} & \text{rate of generation} \\
\text{of reactant within} = & \text{reactant into the} - & \text{reactant out of the} + & \text{of reactant within} \\
\text{the system boundary} & \text{system boundary} & \text{system boundary} & \text{the system boundary} \\
(1) & (2) & (3) & (4)
\end{array}
$$

(4–2)

2. The corresponding simplified word statement is

$$
\begin{array}{cccc}
\text{Accumulation} = & \text{inflow} - & \text{outflow} + & \text{generation} \\
(1) & (2) & (3) & (4)
\end{array}
$$
(4–3)

The mass balance is made up of the four terms cited above. Depending on the flow regime or treatment process, one or more of the terms can be equal to zero. For example, in a batch reactor in which there is no inflow or outflow the second and third terms will be equal to zero. In the analysis of the hydraulic characteristics of reactors, considered in the following sections, as well as in the analysis of separation processes discussed in Sec. 4–8, the fourth term, the rate of generation, will be equal to zero. In Eq. (4–2), a positive sign is used for the rate-of-generation term because the necessary sign for the operative process is part of the rate expression (e.g., $r_c = -kC$ for a decrease in the reactant or $r_c = +kC$ for an increase in the reactant).

Preparation of Mass Balances

In preparing mass balances it is helpful if the following steps are followed, especially as the techniques involved are being mastered.

1. Prepare a simplified schematic or flow diagram of the system or process for which the mass balance is to be prepared.
2. Draw a system or control volume boundary to define the limits over which the mass balance is to be applied. Proper selection of the system or control volume boundary is extremely important because, in many situations, it may be possible to simplify the mass-balance computations.
3. List all of the pertinent data and assumptions that will be used in the preparation of the materials balance on the schematic or flow diagram.
4. List all of the rate expressions for the biological or chemical reactions that occur within the control volume.
5. Select a convenient basis on which the numerical calculations will be based.

It is recommended that the above steps be followed routinely, to avoid the errors that are often made in the preparation of mass-balance analyses.

Application of the Mass-Balance Analysis

To illustrate the application of the mass-balance analysis, consider the complete-mix reactor shown on Fig. 4–4. First, the control volume boundary must be established so that all the flows of mass into and out of the system can be identified. On Fig. 4–4a, the control volume boundary is shown by the inner dashed line.

To apply a mass-balance analysis to the liquid contents of the reactor shown on Fig. 4–4, it will be assumed that:

1. The volumetric flowrate into and out of the control volume is constant.
2. The liquid within the control volume is not subject to evaporation (constant volume).
3. The liquid within the control volume is mixed completely.
4. A chemical reaction involving a reactant A is occurring within the reactor.
5. The rate of change in the concentration of the reactant A that is occurring within the control volume is governed by a first-order reaction ($r_c = -kC$).

Using the above assumptions, the mass balance can be formulated as follows:

1. Simplified word statement:

$$\text{Accumulation} = \text{inflow} - \text{outflow} + \text{generation} \tag{4-4}$$

2. Symbolic representation (refer to Fig. 4–4):

$$\frac{dC}{dt} V = QC_o - QC + r_c V \tag{4-5}$$

Substituting $-kC$ for r_c yields

$$\frac{dC}{dt} V = QC_o - QC + (-kC)V \tag{4-6}$$

where dC/dt = rate of change of reactant concentration within the control volume, $ML^{-3}T^{-1}$

$\qquad V$ = volume contained within control volume, L^3

$\qquad Q$ = volumetric flowrate into and out of control volume, L^3T^{-1}

$\qquad C_o$ = concentration of reactant entering the control volume, ML^{-3}

$\qquad C$ = concentration of reactant leaving the control volume, ML^{-3}

$\qquad r_c$ = first-order reaction, $(-kC)$, $ML^{-3}T^{-1}$

$\qquad k$ = first-order reaction rate coefficient, T^{-1}

Before attempting to solve any mass-balance expression, a unit check should always be made to assure that units of the individual quantities are consistent. If the following units are substituted into Eq. (4–6)

$V = m^3$

$dC/dt = g/m^3 \cdot s$

$Q = m^3/s$

$C_o, C = g/m^3$

$k = 1/s$

the resulting unit check yields

$$\frac{dC}{dt}V = QC_o - QC + (-kC)V$$

$(g/m^3 \cdot s)\, m^3 = m^3/s \,(g/m^3) - m^3/s \,(g/m^3) + (-1/s)(g/m^3)\, m^3$

$g/s = g/s - g/s - g/s$ (units are consistent)

The analytical procedures that are adopted for the solution of mass-balance equations usually are governed by (1) the nature of the rate expression, (2) the type of reactor under consideration, (3) the mathematical form of the final materials-balance expression (i.e., ordinary or partial differential equation), and (4) the corresponding boundary conditions. The mass balance for a plug-flow reactor, as will be illustrated in the following section, results in a partial differential equation. A variety of solution procedures for mass balances in both ordinary and partial differential equation formats are presented in the following sections.

Steady-State Simplification

Fortunately, in most applications in the field of wastewater treatment, the solution of mass-balance equations, such as the one given by Eq. (4–6), can be simplified by noting that the steady-state (i.e., long-term) concentration is of principal concern. If it is assumed that only the steady-state effluent concentration is desired, then Eq. (4–6) can be simplified by noting that, under steady-state conditions, the rate accumulation is zero ($dC/dt = 0$). Thus, Eq. (4–6) can be written as

$$0 = QC_o - QC - r_cV \qquad (4–7)$$

When solved for r_c, Eq. (4–7) yields the following expression:

$$r_c = \frac{Q}{V}(C - C_o) \qquad (4–8)$$

The solution to the expression given by Eq. (4–8) will depend on the nature of the rate expression (e.g., zero-, first-, or second-order; see Sec. 4–6 for discussion of rate expressions).

4-3 **MODELING IDEAL FLOW IN REACTORS**

Modeling of the hydraulic characteristics of reactors is important because the results can be used to determine the actual amount of time a given volume of water will remain in the reactor and its average age. In turn, the average ages can be related to the degree of treatment achieved, based on the applicable kinetics. The coupling of reactor hydraulic characteristics and reaction rates to determine treatment process performance is considered in Sec. 4–7. Comparison of actual hydraulic characteristics of a reactor, measured using tracers, to the expected theoretical response can be used to assess the degree to which the design ideal has been achieved. The complete-mix and plug-flow reactors, as noted previously, are the reactor types used most commonly in the field of wastewater treatment. The mathematical analysis of ideal flow in complete-mix and plug-flow reactors is considered below. The modeling of nonideal flow is considered in Sec. 4–5.

Ideal Flow in Complete-Mix Reactor

If a pulse input of a conservative (i.e., nonreactive) tracer is injected into an ideal complete-mix reactor, with a continuous inflow of clear water, the output tracer concentration would appear as shown on Fig. 4–3(a-1). If a continuous step input of a conservative tracer at concentration C_o is injected into the inlet of an ideal complete-mix reactor, initially filled with clear water, the appearance of the tracer at the outlet would occur as shown on Fig. 4–3(a-2).

Analytically, using the mass-balance approach introduced in Sec. 4–2, the effluent tracer concentration as a function of time for a pulse (slug) input of tracer, which is mixed instantaneously and is purged with clear water [see Fig. 4–3(a-1)], can be determined by writing a mass balance around the reactor.

1. General word statement:

| Rate of accumulation of tracer within the reactor | = | rate of flow of tracer into the reactor | − | rate of flow of tracer out of the reactor | (4–9) |

2. Simplified word statement:

   Accumulation = inflow − outflow   (4–10)

3. Symbolic representation [refer to Fig. 4–3(a-1)]:

$$\frac{dC}{dt} V = QC_o - QC \qquad (4\text{–}11)$$

Rewriting Eq. (4–11) and simplifying by noting that $C_o = 0$ yields

$$\frac{dC}{dt} = -\frac{Q}{V} C \qquad (4\text{–}12)$$

Integrating between the limits of $C = C_o$ to $C = C$, and $t = 0$ to $t = t$ yields

$$\int_{C_o}^{c} \frac{dC}{C} = -\frac{Q}{V} \int_{0}^{t} dt \qquad (4\text{–}13)$$

The resulting expression after integration is

$$C = C_o e^{-t(Q/V)} = C_o e^{-t/\tau} = C_o e^{-\theta} \qquad (4\text{–}14)$$

where C = concentration of the tracer in the reactor at time t, ML^{-3}
$\quad\quad\ C_o$ = initial concentration of the tracer in the reactor, ML^{-3}
$\quad\quad\ t$ = time, T
$\quad\quad\ Q$ = volumetric flowrate, L^3T^{-1}
$\quad\quad\ V$ = reactor volume, L^3
$\quad\quad\ \tau$ = theoretical detention time V/Q, T
$\quad\quad\ \theta$ = normalized detention time t/τ, unitless

The corresponding response for a continuous step input of tracer [see Fig. 4–3(a-2)] which is mixed instantaneously is given by

$$C = C_o(1-e^{-t(Q/V)}) = C_o(1-e^{-t/\tau}) = C_o(1-e^{-\theta}) \qquad (4\text{–}15)$$

It will be noted that Eq. (4–15) has the same form as the BOD equation given previously in Chap. 2 [see Eq. (2-60)].

Ideal Flow in Plug-Flow Reactor

In the case of an ideal plug-flow reactor, the reactor is initially filled with clear water before being subjected to a pulse or a step input of tracer. If an observer were positioned at the outlet of the reactor, the appearance of the tracer in the effluent for a pulse input, distributed uniformly across the reactor cross section, would occur as shown on Fig. 4–3(b-1). If a continuous step input of a tracer were injected into such a reactor at an initial concentration C_o, the tracer would appear in the effluent as shown on Fig. 4–3(b-2). Under ideal plug-flow conditions t, the measured detention time should be the same as τ, the theoretical detention time (V/Q).

To verify the form of the plot given on Fig. 4–3(b-2), it will be instructive to prepare a materials balance for an ideal plug-flow reactor (no axial dispersion) in which the concentration C of a nonreactive tracer is distributed uniformly across the cross-sectional area of the control volume. The materials balance for a nonreactive tracer for the differential volume element shown on Fig. 4–5e can be written as follows:

1. General word statement:

    Rate of accumulation of   rate of flow of tracer   rate of flow of tracer
    tracer within differential = into differential − out of differential   (4–16)
    volume element   volume element   volume element

2. Simplified word statement:

    Accumulation = inflow − outflow   (4–17)

3. Symbolic representation (refer to Fig. 4–5e)

    $$\frac{\partial C}{\partial t} \Delta V = QC|_x - QC|_{x+\Delta x} \qquad (4\text{–}18)$$

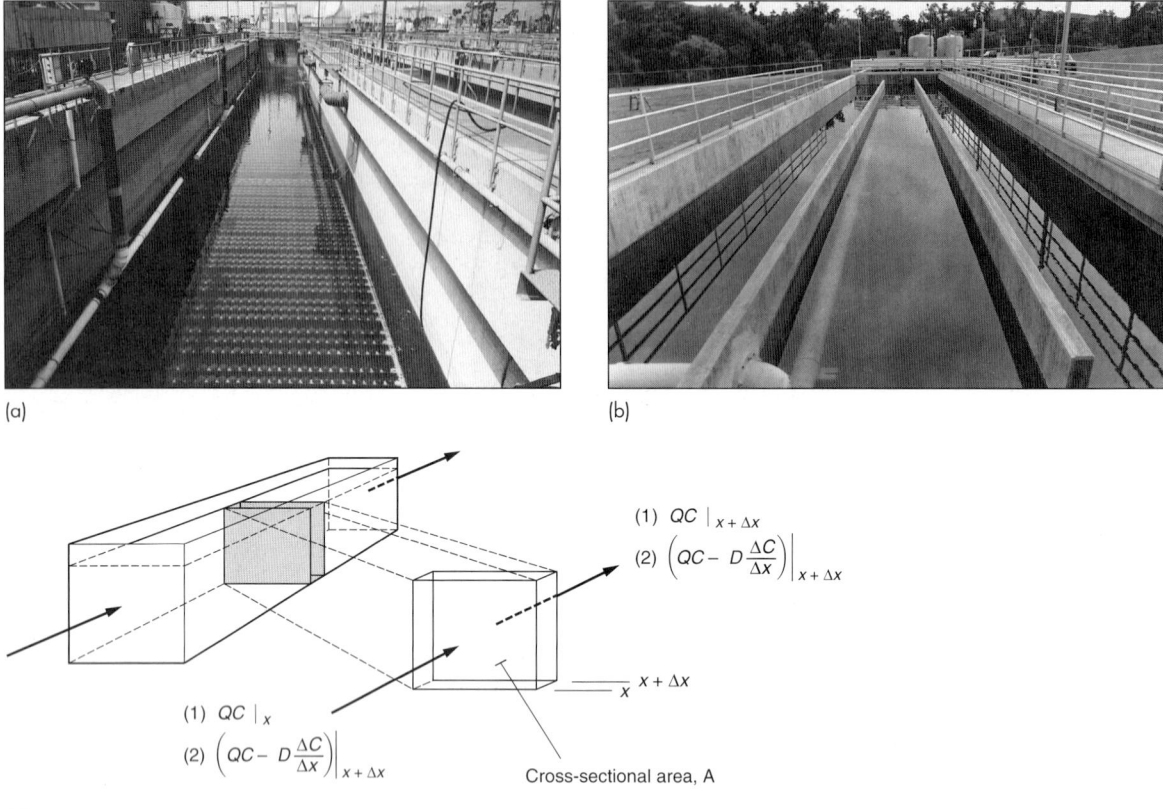

(a)
(b)

(1) $QC \mid_{x + \Delta x}$

(2) $\left(QC - D \dfrac{\Delta C}{\Delta x} \right)\Big|_{x + \Delta x}$

$x + \Delta x$

(1) $QC \mid_{x}$

(2) $\left(QC - D \dfrac{\Delta C}{\Delta x} \right)\Big|_{x + \Delta x}$

Cross-sectional area, A

(c)

Figure 4–5

Views of plug-flow reactors and definition sketch: (a) empty plug-flow activated sludge process reactor, (b) plug-flow chlorine contact basin with long narrow channels, and (c) definition sketch for the hydraulic analysis of a plug-flow reactor with (1) advection only and (2) with advection and axial dispersion.

where $\partial C/\partial t$ = constituent concentration, ML^{-3}, (g/m^3)

ΔV = differential volume element, L^3, (m^3)

t = time T, (s)

Q = volumetric flow rate, $L^3 T^{-1}$, (m^3/s)

x = some point along the reactor length L, (m)

Δx = differential distance L, (m)

The change in concentration with time term ($\partial C/\partial t$) is written as partial differential because the concentration is also changing with distance (i.e., the change in concentration is a function of both time and distance).

Substituting the differential form for the terms $QC\big|_x$ and $QC\big|_{x + \Delta x}$ in Eq. (4–18) results in

$$\frac{\partial C}{\partial t} \Delta V = QC - Q\left(C + \frac{\Delta C}{\Delta x} \Delta x \right)$$

(4–19)

Substituting $A\Delta x$ for ΔV, where A is the cross-sectional area in the x direction, and simplifying yields

$$\frac{\partial C}{\partial t} A\Delta x = -Q\frac{\Delta C}{\Delta x}\Delta x \tag{4-20}$$

where $\Delta C/\Delta t$ = change in constituent concentration C with time, $ML^{-3}T^{-1}$, (g/m^3)

Dividing by A and Δx yields

$$\frac{\partial C}{\partial t} = -\frac{Q}{A}\frac{\Delta C}{\Delta x} \tag{4-21}$$

Taking the limit as Δx approaches zero yields

$$\frac{\partial C}{\partial t} = -\frac{Q}{A}\frac{\partial C}{\partial x} = -v\frac{\partial C}{\partial x} \tag{4-22}$$

where v = velocity of flow, LT^{-1}, (m/s)

Because both sides of the equation are the same (note $\partial t = \partial x/v$), except for the minus sign, the only way that the equation can be satisfied is if the change in concentration with distance is equal to zero. Thus, the effluent concentration must be equal to the influent concentration, which is consistent with the depiction on Fig. 4–3(b-2).

4–4 ANALYSIS OF NONIDEAL FLOW IN REACTORS USING TRACERS

Although many of the practical aspects of process and reactor design are discussed in detail in the chapters that follow, the analysis of reactor hydraulic performance using tracers is considered in this section. The discussion of nonideal flow in this section will serve as an introduction to the modeling of nonideal flow considered in the following section. Attention is called to this subject because often it is neglected or not considered properly. Because of a lack of appreciation for the hydraulics of reactors, many of the treatment plants that have been built do not perform hydraulically as designed. Included in this section is a discussion of (1) the factors leading to nonideal flow in reactors, (2) the need for tracer analysis, (3) the conduct of tracer tests, (4) the types of tracers used, (5) the analysis of tracer response curves, and (6) the practical interpretation of tracer measurements.

Factors Leading to Nonideal Flow in Reactors

As noted previously, nonideal flow is often defined as short circuiting that occurs when a portion of the flow that enters the reactor during a given time period arrives at the outlet before the bulk of the flow that entered the reactor during the same time period arrives. Factors leading to nonideal flow in reactors include:

1. Temperature differences. In complete-mix and plug-flow reactors, nonideal flow (short circuiting) can be caused by density currents due to temperature differences. When the water entering the reactor is colder or warmer than the water in the tank, a portion of the water can travel to the outlet along the bottom of or across the top of the reactor without mixing completely (see Fig. 4–6a).

Figure 4–6

Definition sketch for short circuiting caused by (a) density currents caused by temperature differences, (b) wind circulation patterns, (c) inadequate mixing, (d) fluid advection and dispersion.

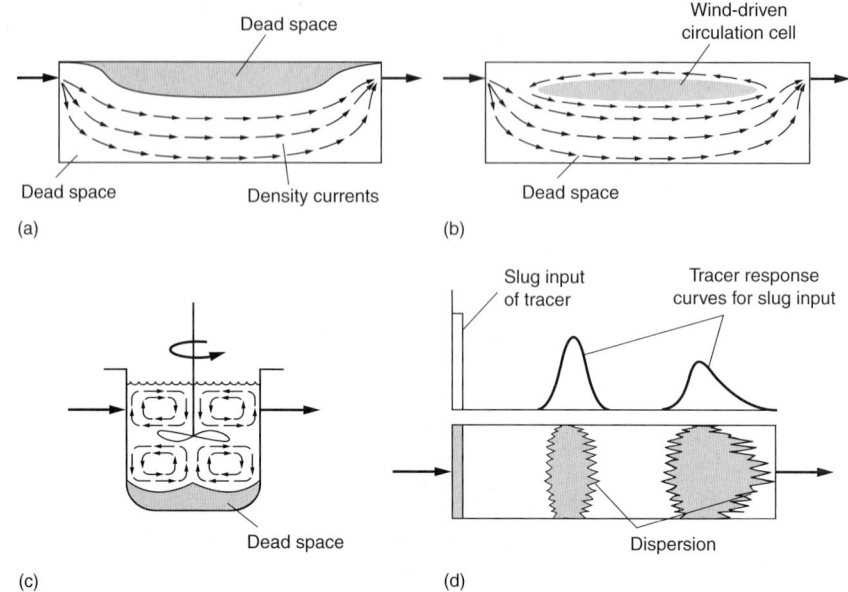

2. **Wind-driven circulation patterns.** In shallow reactors, wind-circulation patterns can be set up that will transport a portion of the incoming water to the outlet in a fraction of the actual detention time (see Fig. 4–6*b*).

3. **Inadequate mixing.** Without sufficient energy input, portions of the reactor contents may not mix with the incoming water (see Fig. 4–6*c*).

4. **Poor design.** Depending on the design of the inlet and outlet of the reactor relative to the reactor aspect ratio, dead zones may develop within the reactor that will not mix with the incoming water (see Fig. 4–6*d*).

5. **Axial dispersion in plug-flow reactors.** In plug-flow reactors the forward movement of the tracer is due to advection and dispersion. *Advection* is the term used to describe the movement of dissolved or colloidal material with the current velocity. *Dispersion* is the term used to describe the axial and longitudinal transport of material brought about by velocity differences, turbulent eddies, and molecular diffusion. The distinction between molecular diffusion, turbulent diffusion, and dispersion is considered in the subsequent discussion dealing with "Modeling Nonideal Flow In Reactors." In a tubular plug-flow reactor (e.g., a pipeline), the early arrival of the tracer at the outlet can be reasoned partially by remembering that the velocity distribution in the pipeline will be parabolic.

Ultimately, the inefficient use of the reactor volume due to short circuiting resulting from temperature differences, the presence of dead zones resulting from poor design, inadequate mixing, and dispersion (see Fig. 4–6) can result in reduced treatment performance. The subject of short circuiting in a series of complete-mix reactors was examined extensively in early papers by MacMullin and Weber (1935); Fitch (1956); and Morrill (1932). Morrill examined the effects of short circuiting on the performance of sedimentation tanks.

Need for Tracer Analysis

One of the more important practical considerations involved in reactor design is how to achieve the ideal conditions postulated in the analysis of their performance. The use of dyes and tracers for measuring the residence time distribution curves is one of the simplest and most successful methods now used to assess the hydraulic performance of full-scale reactors. Important applications of tracer studies include (1) the assessment of short circuiting in sedimentation tanks and biological reactors, (2) the assessment of the contact time in chlorine contact basins, (3) the assessment of the hydraulic approach conditions in UV reactors, and (4) the assessment of flow patterns in constructed wetlands and other natural treatment systems. Tracer studies are also of critical importance in assessing the degree of success that has been achieved with corrective measures.

Types of Tracers

Over the years, a number of tracers have been used to evaluate the hydraulic performance of reactors. Important characteristics for a tracer include (adapted in part from Denbigh and Turner, 1965):

- The tracer should not affect the flow (should have essentially the same density as water when diluted).
- The tracer must be conservative so that a mass balance can be performed.
- It must be possible to inject the tracer over a short time period.
- The tracer should be able to be analyzed conveniently.
- The molecular diffusivity of the tracer should be low.
- The tracer should not be absorbed on or react with the exposed reactor surfaces.
- The tracer should not be absorbed on or react with the particles in wastewater.

Dyes and chemicals that have been used successfully in tracer studies include congo red, fluorescein, fluorosilicic acid (H_2SiF_6), hexafluoride gas (SF_6), lithium chloride (LiCl), Pontacyl Brilliant Pink B, potassium, potassium permanganate, rhodamine WT, and sodium chloride (NaCl). Pontacyl Brilliant Pink B (the acid form of rhodamine WT) is especially useful in the conduct of dispersion studies because it is not readily adsorbed onto surfaces. Because fluorescein, rhodamine WT, and Pontacyl Brilliant Pink B can be detected at very low concentrations using a fluorometer (see Fig. 4–7), they are the dye tracers used most commonly in the evaluation of wastewater-treatment facilities. Lithium chloride is commonly used for the study of natural systems. Sodium chloride, used commonly in the past, has a tendency to form density currents unless mixed. Hexafluoride gas (SF_6) is used most commonly for tracing the movement of groundwater.

Conduct of Tracer Tests

In tracer studies, typically a tracer (i.e., a dye, most commonly) is introduced into the influent end of the reactor or basin to be studied. The time of its arrival at the effluent end is determined by collecting a series of grab samples for a given period of time or by measuring the arrival of a tracer using instrumental methods (see Fig. 4–8). The method used to introduce the tracer will control the type of response observed at the downstream end. Two types of dye input are used, the choice depending on the influent

Figure 4–7

Schematic of setup used to conduct tracer studies of plug-flow reactors: (a) slug of tracer added to flow and (b) continuous input of tracer added to flow. Tracer response curve is measured continuously (see Fig. 4–8).

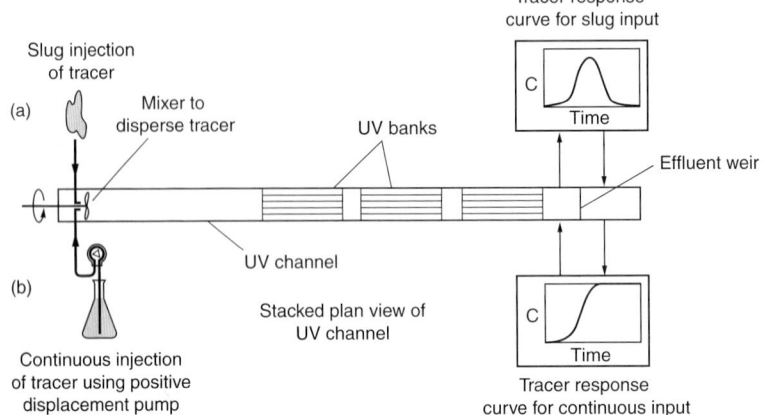

Figure 4–8

Field measurement of concentration versus time curve for a UV disinfection system: (a) fluorometer used for the continuous measurement of concentration versus time curve and (b) overview of UV disinfection system.

(a)

(b)

and effluent configurations. The first method involves the injection of a quantity of dye (sometimes referred to as a pulse or slug of dye) over a short period of time. Initial mixing is usually accomplished with a static mixer or an auxiliary mixer. With the slug injection method it is important to keep the initial mixing time short relative to the detention time of the reactor being measured. The measured output is as described on Fig. 4–3(a-1 and b-1). In the second method, a continuous step input of dye is intro-

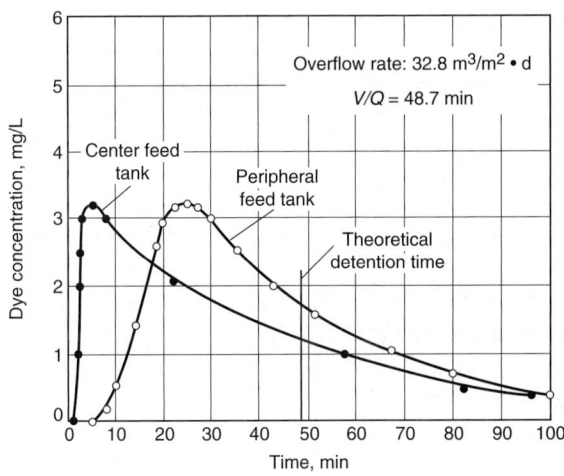
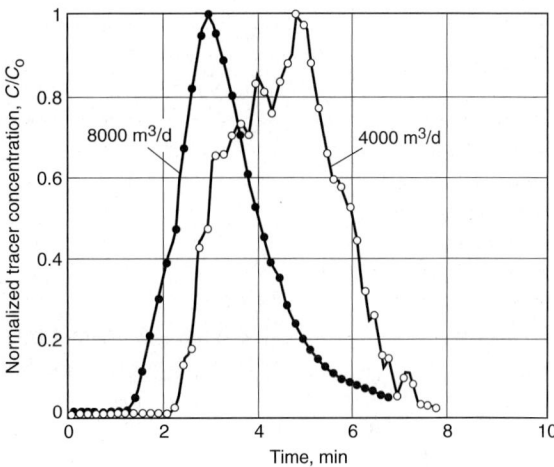

Figure 4–9

Typical tracer response curves: two different types of circular clarifiers (adapted from Dague and Baumann, 1961) and open channel UV disinfection system. *(Courtesy Andy Salveson, Whitley Burchett & Associates.)*

duced until the effluent concentration matches the influent concentration. The measured response is as shown on Fig. 4–3(*a*-2 and *b*-2). Another response curve can be measured after the dye injection has ceased and the dye in the reactor is flushed out.

Analysis of Tracer Response Curves

Tracers of various types are used commonly to assess the hydraulic performance of reactors used for wastewater treatment. Typical examples of tracer response curves are shown on Fig. 4–9. Tracer response curves, measured using a short-term and continuous injection of tracer, are known as C (concentration versus time) and F (fraction of tracer remaining in the reactor versus time) curves, respectively. The fraction remaining is based on the volume of water displaced from the reactor by the step input of tracer. The analysis of tracer response curves is considered in detail in the following discussion. The terms used to characterize tracer response curves, the analysis of concentration versus time, and the development of residence time distribution (RTD) curves are described below. The material to be discussed below will be used in the subsequent analysis of the hydraulic characteristics of nonideal reactors.

Terms Used to Characterize Tracer Response Curves. Because of the complexity of the hydraulic response of full-scale reactors, tracer response curves are used to analyze the hydraulics of reactors. Over the years, a number of numerical values have been used to characterize output tracer curves (see Table 4–2). As reported in Table 4–2, τ, as given previously, is the theoretical detention (residence) time defined as V/Q, where V is the volume and Q is the flowrate in consistent units. As noted in Table 4–2, various other symbols have been used in the literature for the theoretical detention time.

Table 4-2
Various terms used to describe the hydraulic performance of reactors used for wastewater treatment[a]

Term	Definition
τ, θ, θ_h[b]	Theoretical hydraulic residence time (V, volume/Q, flowrate)
t_i	Time at which tracer first appears
t_p	Time at which the peak concentration of the tracer is observed (mode)
t_g	Mean time to reach centroid of the RTD curve
t_{10}, t_{50}, t_{90}	Time at which 10, 50, and 90 percent of the tracer had passed through the reactor
t_{90}/t_{10}	Morrill dispersion index, MDI
$1/MDI$	Volumetric efficiency as defined by Morrill (1932)
t_i/τ	Index of short circuiting. In an ideal plug-flow reactor, the ratio is 1, and approaches zero with increased mixing
t_p/τ	Index of modal retention time. Ratio will approach 1 in a plug-flow reactor, and 0 in a complete-mix reactor. For values of the ratio greater than or less than 1.0, the flow distribution in the reactor is not uniform
t_g/τ	Index of average retention time. A value of 1 would indicate that full use is being made of the volume. A value of the ratio greater than or less than 1.0 indicates the flow distribution is not uniform
t_{50}/τ	Index of mean retention time. The ratio t_{50}/τ to is a measure of the skew of the RTD curve. In an effective plug-flow reactor, the RTD curve is very similar to a normal or Gaussian distribution (U.S. EPA, 1986). A value of t_{50}/τ of less than 1.0 corresponds to an RTD curve that is skewed to the left. Similarly, for values greater than 1.0 the RTD curve is skewed to the right

[a] Adapted from Morrill (1932), Fair and Geyer (1954), and U.S. EPA (1986).
[b] The symbols θ and θ_h have also been used for the theoretical hydraulic residence time.

The relationships of the terms in Table 4–2 and a typical tracer response curve are illustrated on Fig. 4–10.

Concentration versus Time Tracer Response Curves. As noted previously, tracer response curves measured using a short-term and continuous injection of tracer are known as C curves (concentration versus time). To characterize C curves, such as those shown on Fig. 4–9, the mean value is given by the centroid of the distribution. For C curves, the theoretical mean residence time is determined as follows.

$$\bar{t}_c = \frac{\displaystyle\int_0^\infty tC(t)dt}{\displaystyle\int_0^\infty C(t)dt} \tag{4-23}$$

where $\bar{t}_c$ = mean residence time derived from tracer curve, T
t = time, T
$C(t)$ = tracer concentration at time t, ML^{-3}

Figure 4-10

Definition sketch for the parameters used in the analysis of concentration versus time tracer response curves.

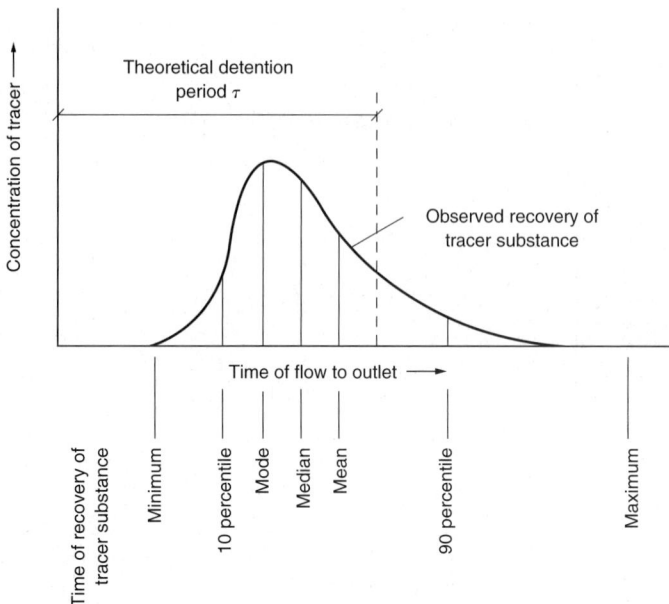

The variance σ_c^2 used to define the spread of the distribution is defined as

$$\sigma_c^2 = \frac{\displaystyle\int_0^\infty (t - \bar{t})^2 C(t)dt}{\displaystyle\int_0^\infty C(t)dt} = \frac{\displaystyle\int_0^\infty t^2 C(t)dt}{\displaystyle\int_0^\infty C(t)dt} - (\bar{t}_c)^2 \tag{4-24}$$

It will be recognized that the integral term in the denominator in Eqs. (4–23) and (4–24) corresponds to the area under the concentration versus time curve. It will also be recognized that the mean and variance are equal to the *first* and *second moments* of the distribution about the *y* axis.

If the concentration versus time tracer response curve is defined by a series of discrete time step measurements, the theoretical mean residence time is typically approximated as

$$\bar{t}_{\Delta c} \approx \frac{\Sigma t_i C_i \Delta t_i}{\Sigma C_i \Delta t_i} \tag{4-25}$$

where $\bar{t}_{\Delta c}$ = mean detention time based on discrete time step measurements, T
t_i = time at ith measurement, T
C_i = concentration at ith measurement, ML^{-3}
Δt_i = time increment about C_i, T

Figure 4–11

Concentration versus time tracer response curve for a long narrow chlorine contact basin.

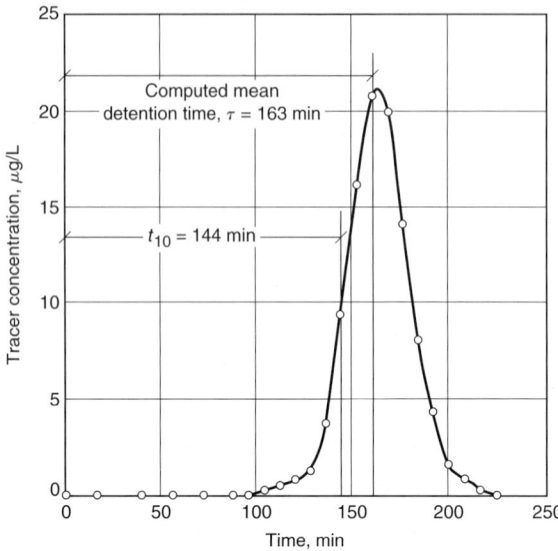

The variance for a concentration versus time tracer response curve, defined by a series of discrete time step measurements, is defined as

$$\sigma_{\Delta c}^2 \approx \frac{\Sigma t_i^2 C_i \Delta t_i}{\Sigma C_i \Delta t_i} - (\bar{t}_{\Delta c})^2 \tag{4–26}$$

where $\sigma_{\Delta}^2 c$ = variance based on discrete time measurements, T^2

A typical concentration versus time tracer curve for a chlorine contact basin is presented on Fig. 4–11. Additional details on the physical features for the chlorine contact basin used to develop Fig. 4–11 are presented in Example 4–1. The application of Eqs. (4–25) and (4–26) is illustrated in Example 4–1. Additional details on the analysis of tracer response curves may be found in Levenspiel (1972).

EXAMPLE 4–1

Determination of the Mean Residence Time and Variance for a Plug-Flow Chlorine Contact Basin The following data were obtained from a dye tracer test to evaluate the hydraulic performance of a long chlorine contact basin, such as shown on Fig. 4–5. The chlorine contact system is designed to provide a minimum contact time of 2 h at a flowrate of 300,000 m³/d (80 Mgal/d). The flowrate at the time when the tracer test was conducted was 240,000 m³/d. The chlorine contact system consists of four chlorine contact tanks. Each of the four chlorine contact tanks is baffled into 13 individual contact channels. Each channel is 36.6 m (120 ft) long, 3.0 m (10 ft) wide, and has a normal water depth of 4.9 m (16 ft). Using these data and the tracer test data, determine the mean residence time and variance for the tracer response curve.

Time, min	Concentration, μg/L	Time, min	Concentration, μg/L
0	0.0	144	9.333
16	0.0	152	16.167
40	0.0	160	20.778
56	0.0	168	19.944
72	0.0	176	14.111
88	0.0	184	8.056
96	0.056	192	4.333
104	0.333	200	1.556
112	0.556	208	0.889
120	0.833	216	0.278
128	1.278	224	0.000
136	3.722		

Solution

1. Determine the mean residence time and variance for the tracer response curve using Eqs. (4–25) and (4–26).

   a. Set up computation table to solve Eqs. (4–25) and (4–26). In setting up the computation table given below, the Δt value was omitted as it appears in both the numerator and the denominator of Eqs. (4–25) and (4–26).

Time t, min	Conc. C, μg/L	$t \times C$	$t^2 \times C$
88	0.000	0.000	0
96	0.056	5.338	512.41
104	0.333	34.663	3,604.97
112	0.556	62.227	6,969.45
120	0.833	99.996	11,999.52
128	1.278	163.558	20,935.48
136	3.722	506.219	68,845.81
144	9.333	1,343.995	193,535.31
152	16.167	2,457.384	373,522.37
160	20.778	3,324.480	531,916.80
168	19.944	3,350.592	562,899.46
176	14.111	2,483.536	437,102.34
184	8.056	1,482.230	272,730.39
192	4.333	831.994	159,742.77
200	1.556	311.120	62,224.00
208	0.889	184.891	38,457.37
216	0.278	60.005	12,961.04
224	0.000	0.000	
Total	102.222	16,702.229	2,757,959.48

b. Determine the mean residence time and variance

$$\bar{t}_{\Delta c} = \frac{\Sigma t_i C_i \Delta t_i}{\Sigma C_i \Delta t_i} = \frac{16,702.23}{102.22} = 163.4 \text{ min} = 2.7 \text{ h}$$

c. Determine the variance

$$\sigma_{\Delta c}^2 = \frac{\Sigma t_i^2 C_i \Delta t_i}{\Sigma C_i \Delta t_i} - (\bar{t}_{\Delta c})^2 = \frac{2,757,959.48}{102.22} - (163.4)^2 = 280.5 \text{ min}^2$$

$$\sigma_{\Delta c} = 16.7 \text{ min}$$

Comment The mean and variance determined in this analysis will be used in Example 4–4 to determine the dispersion number and the coefficient of dispersion.

Residence Time Distribution (RTD) Curves. To standardize the analysis of output concentration versus time curves for a pulse input, such as those shown on Fig. 4–9, the output concentration measurements are often normalized by dividing the measured concentration values by an appropriate function such that the area under the normalized curve is equal to 1. The normalized curves are known, more formally, as *residence time distribution (RTD) curves* (see Fig. 4–12). When a pulse addition of tracer is used, the area under the normalized curve is known as an *E* curve (also known

Figure 4–12

Normalized residence time distribution curves. The curve on the bottom, known as the exit age curve, is identified as the *E* curve. The curve on the top, known as the cumulative residence time curve, is identified as the *F* curve.

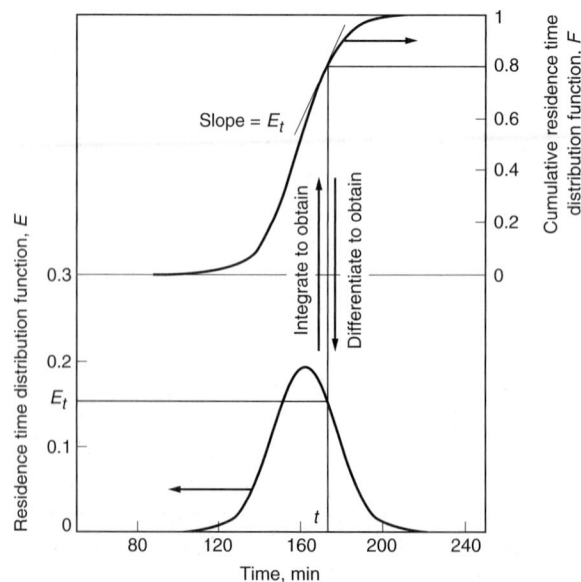

as the exit age curve). The most important characteristic of an E curve is that the area under the curve is equal to 1, as defined by the following integral:

$$\int_0^\infty E(t)dt = 1 \qquad (4\text{--}27)$$

where $E(t)$ is the residence time distribution function. The $E(t)$ value is related to the $C(t)$ value as follows:

$$E(t) = \frac{C(t)}{\int_0^\infty C(t)dt} \qquad (4\text{--}28)$$

As in Eqs. (4–23) and (4–24), the integral term in the denominator in Eq. (4–28) corresponds to the area under the concentration versus time curve. Applying Eq. (4–28) to the expression, obtained previously, for the complete-mix reactor, the exit age curve $E(t)$ for a complete-mix reactor is obtained as follows:

$$E(t) = \frac{C(t)}{\int_0^\infty C(t)dt} = \frac{C_o e^{-t/\tau}}{\int_0^\infty C_o e^{-t/\tau}\,dt} = \frac{e^{-t/\tau}}{\tau} \qquad (4\text{--}29)$$

and the corresponding value based on normalized time, $\theta = t/\tau$, is

$$E(\theta) = \tau E(t) = e^{-\theta} \qquad (4\text{--}30)$$

The mean residence time for the $E(t)$ curve, given by Eq. (4–30), can be derived by applying Eq. (4–23); the resulting expression is given by

$$t_m = \frac{\int_0^\infty tE(t)dt}{\int_0^\infty E(t)dt} = \int_0^\infty tE(t)dt \qquad (4\text{--}31)$$

In a similar manner when a step input is used, the normalized concentration curve is known as the cumulative residence time distribution curve and is designated as the F curve. The F curve is defined as

$$F(t) = \int_0^t E(t)dt \qquad (4\text{--}32)$$

where $F(t)$ is the cumulative residence time distribution function. As shown on Fig. 4–12, the $F(t)$ curve is the integral of the $E(t)$ curve while the $E(t)$ curve is the derivative of the $F(t)$ curve. In effect, $F(t)$ represents the amount of tracer that has been in the reactor for less than the time t. The development of E and F RTD curves is illustrated in Example 4–2. Additional details on the analysis of E and F curves may be found in Denbigh and Turner (1965), Fogler (1999), Levenspiel (1972), and in the chemical engineering literature.

EXAMPLE 4-2 Development of Residence Time Distribution (RTD) Curves from Concentration versus Time Tracer Curves Use the concentration versus time tracer data given in Example 4–1 to develop E and F residence time distribution (RTD) curves for the chlorine contact basin. Using the E curve, compute the residence time and compare to the value obtained in Example 4–1. Plot the resulting E and F curves.

Solution

1. Using the values for time and concentration from Example 4–1, set up a computation table to calculate $E(t)$ values. The computation table is shown below. The $E(t)$ values are calculated by finding the sum of the $C \times \Delta t$ values, which corresponds to the area under the C (concentration) curve, and dividing the original concentrations by the sum as illustrated below.

   Area under C curve $= \Sigma C \Delta t$

   and

   $$E(t) = \frac{C}{\Sigma C \Delta t}$$

   a. For each time interval, multiply the concentration by the time step ($\Delta t = 8$ min) and obtain the sum of the multiplied values. As shown in the computation table, the sum (which is the approximate area under the curve) is 817.778 $(\mu g/L) \cdot min$.

   b. Calculate the $E(t)$ values by dividing the original concentration values by the area under the curve.

   c. Confirm that the $E(t)$ values are correct by multiplying each one by the time step, and calculating the sum. According to Eq. (4–28) written in differential form, the sum should be 1.00.

   $$\Sigma E(t)\Delta t = \Sigma\left(\frac{C\Delta t}{\Sigma C\Delta t}\right) = \Sigma\left(\frac{C}{\Sigma C}\right) = 1$$

Time t, min	Conc. C, $\mu g/L$	$C \times \Delta t$ $\mu g/L \cdot min$	$E(t)$, min^{-1}	$E(t) \times \Delta t$, unitless	$t \times E(t) \times \Delta t$, min
88	0.000	0.000	0.00000	0.00000	0.000
96	0.056	0.445	0.00007	0.00054	0.077
104	0.333	2.666	0.00041	0.00326	0.333
112	0.556	4.445	0.00068	0.00544	0.627
120	0.833	6.666	0.00102	0.00815	0.960
128	1.278	10.222	0.00156	0.01250	1.638
136	3.722	29.778	0.00455	0.03641	5.005

(continued)

(Continued)

Time t, min	Conc. C, μg/L	$C \times \Delta t$ μg/L·min	$E(t)$, min^{-1}	$E(t) \times \Delta t$, unitless	$t \times E(t) \times \Delta t$, min
144	9.333	74.666	0.01141	0.09130	13.133
152	16.167	129.336	0.01977	0.15816	24.077
160	20.778	166.224	0.02541	0.20326	32.512
168	19.944	159.552	0.02439	0.19510	32.794
176	14.111	112.888	0.01726	0.13804	24.358
184	8.056	64.445	0.00985	0.07880	14.573
192	4.333	34.666	0.00530	0.04239	8.141
200	1.556	12.445	0.00190	0.01522	3.040
208	0.889	7.111	0.00109	0.00870	1.830
216	0.278	2.222	0.00034	0.00272	0.518
224	0.000	0.000	0.00000	0.00000	0.000
Total	102.222	817.778	—	1.00000	163.616

2. Determine the mean residence time using the following summation form of Eq. (4–28).

$$\bar{t} = \Sigma(t)E(t)\Delta t$$

The required computation is presented in the final column of the computation table given above. As shown, the computed mean residence time (163.6 min) is essentially the same as the value (163.4 min) determined in Example 4–1.
3. Develop the F RTD curve. The values for plotting the F curve are obtained by summing cumulatively the $E(t)\Delta t$ values to obtain the coordinates of the F curve.

Time t, min	$E(t)$, min^{-1}	$E(t) \times \Delta t$, unitless	Cumulative total F, unitless
88	0.00000	0.00000	0.0000
96	0.00007	0.00054	0.0005
104	0.00041	0.00326	0.0038
112	0.00068	0.00544	0.0092
120	0.00102	0.00815	0.0174
128	0.00156	0.01250	0.0299
136	0.00455	0.03641	0.0663
144	0.01141	0.09130	0.1576
152	0.01977	0.15816	0.3158

(continued)

(*Continued*)

Time t, min	$E(t)$, min^{-1}	$E(t) \times \Delta t$, unitless	Cumulative total F, unitless
160	0.02541	0.20326	0.5191
168	0.02439	0.19510	0.7142
176	0.01726	0.13804	0.8522
184	0.00985	0.07880	0.9310
192	0.00530	0.04239	0.9734
200	0.00190	0.01522	0.9886
208	0.00109	0.00870	0.9973
216	0.00034	0.00272	1.0000
224	0.00000	0.00000	1.0000
Total		1.00000	

4.   Plot the resulting E and F curves. The plot of the E (column 3) and F (column 4) curves using the values determined above is shown below.

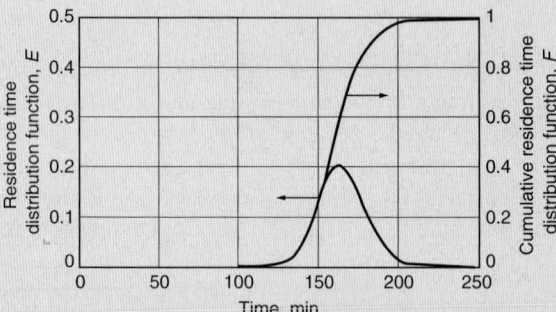

Comment   The use of normalized RTD curves to obtain coefficients of dispersion is included in the discussion of the Hydraulic Characteristics of Reactors. Additional details may be found in Denbigh and Turner (1965) and Levenspiel (1972).

Practical Interpretation of Tracer Measurements

In 1932, based on his studies of sedimentation basins, Morrill (1932) suggested that the ratio of the 90 percentile to the 10 percentile value from the cumulative tracer curve could be used as a measure of the dispersion index, and that 1 over the dispersion index is a measure of the volumetric efficiency. The dispersion index as proposed by Morrill is given by

$$\text{Morrill dispersion index, MDI} = \frac{P_{90}}{P_{10}} \qquad (4\text{–}33)$$

where P_{90} = 90 percentile value from log-probability plot
P_{10} = 10 percentile value from log-probability plot

The percentile values are obtained from a log-probability plot of the time (log scale) versus the cumulative percentage of the total tracer which has passed out of the basin (on probability scale). The value of the MDI for an ideal plug-flow reactor is 1.0 and about 22 for a complete-mix reactor. A plug-flow reactor with an MDI value of 2.0 or less is considered by the U.S. EPA to be an effective plug-flow reactor (U.S. EPA, 1986). The volumetric efficiency is given by the following expression:

$$\text{Volumetric efficiency, \%} = \frac{1}{\text{MDI}} \times 100 \qquad (4\text{–}34)$$

The determination of the Morrill dispersion index and the volumetric efficiency is illustrated in Example 4–3.

EXAMPLE 4–3 **Determination of the Morrill Dispersion Index and Volumetric Efficiency for a Plug-Flow Chlorine Contact Basin** Determine the MDI and the corresponding basin volumetric efficiency for the chlorine contact basin given in Example 4–1.

Solution

1. Determine the MDI and the volumetric efficiency using the tracer response data given in Example 4–1.
   a. Set up a computation table to obtain cumulative concentration and percentage values.

Time t, min	Conc. C, ppb	Cumulative concentration	Cumulative percentage
88	0.000		
96	0.056	0.05	0.05
104	0.333	0.39[a]	0.38[b]
112	0.556	0.94	0.92
120	0.833	1.78	1.74
128	1.278	3.06	2.99
136	3.722	6.78	6.63
144	9.333	16.11	15.75
152	16.167	32.28	31.58

(continued)

(Continued)

Time t, min	Conc. C, ppb	Cumulative concentration	Cumulative percentage
160	20.778	53.06	51.91
168	19.944	73.00	71.41
176	14.111	87.11	85.22
184	8.056	95.17	93.10
192	4.333	99.50	97.34
200	1.556		
208	0.889		
216	0.278		
224	0.000		
Total	102.222		

[a]$0.056 + 0.333 = 0.39.$
[b]$(0.39/102.222) \times 100 = 0.38.$

b. Plot the cumulative concentration as a percentage on log-probability paper. The cumulative percentage is plotted on the probability scale and the time is plotted on the log scale. The required plot is shown below.

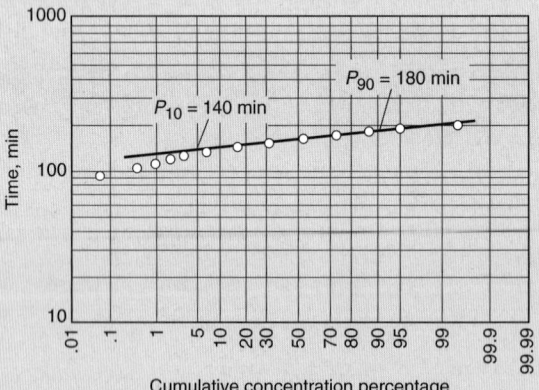

c. Determine the Morrill dispersion index from the plot prepared in the previous step.

$$\text{Morrill dispersion index, MDI} = \frac{P_{90}}{P_{10}} = \frac{180}{140} = 1.29$$

d. Determine the volumetric efficiency.

$$\text{Volumetric efficiency, \%} = \frac{1}{\text{MDI}} \times 100 = \frac{1}{1.29} \times 100 = 78\%$$

Comment The MDI value (1.29) for the chlorine contact basin is very low. As noted above, an MDI value below 2.0 has been established as an effective design (U.S. EPA, 1986). Similarly the volumetric efficiency is very high, signifying essentially plug flow with small amounts of axial dispersion.

4-5 MODELING NONIDEAL FLOW IN REACTORS

The hydraulic characteristics of nonideal reactors can be modeled by taking dispersion into consideration. For example, if dispersion becomes infinite, the plug-flow reactor with axial dispersion is equivalent to a complete-mix reactor. However, before considering nonideal flow in reactors it will be helpful to examine the distinction between the coefficient of molecular diffusion, turbulent diffusion, and dispersion as applied to the analysis of reactors used for wastewater treatment. Both the plug-flow reactor with axial dispersion and complete-mix reactors in series are considered in the following discussion.

The Distinction between Molecular Diffusion, Turbulent Diffusion, and Dispersion

Under quiescent flow conditions (i.e., no flow), the mass transfer of material is brought about by "molecular diffusion." The transfer of mass by molecular diffusion in stationary systems can be represented as a function of the concentration gradient as follows:

$$r = -D_m \frac{\partial C}{\partial x} \tag{4-35}$$

where r = rate of mass transfer per unit area per unit time, $ML^{-2}T^{-1}$
D_m = coefficient of molecular diffusion in the x direction, L^2T^{-1}
C = concentration of constituent being transferred, ML^{-3}
x = distance, L

In the chemical engineering literature the symbol J is used to denote mass transfer in concentration units whereas the symbol N, as used in Eq. (4–73), is used to denote the transfer of mass expressed as moles. The negative sign in Eq. (4–35) is used to denote the fact that diffusion takes place in the direction of decreasing concentration (Shaw, 1966). Equation (4–35) is known as Fick's first law of diffusion. Fick derived the first and second laws of diffusion [Eqs. (4–35) and (4–140)] in the 1850s by direct analogy to the equations used to describe the conduction of heat in solids as proposed by Fourier (Crank, 1957). Determination of numerical values for the coefficient of molecular diffusion is illustrated in Sec. 4–8.

Under turbulent conditions (e.g., mixing) without flow, the transfer of mass is brought about by microscale turbulence known as "turbulent or eddy diffusion." For this case, the molecular diffusion term D_m in Eq. (4–35) is replaced by the turbulent diffusion term D_e or E.

Table 4–3

Typical range of values for molecular diffusion, turbulent diffusion, and dispersion[a]

Item	Symbol	Range of values, cm²/s
Molecular diffusion	D_m	10^{-8}–10^{-4}
Turbulent or eddy diffusion	D_e, E	10^{-4}–10^{-2}
Dispersion	D	10^2–10^6

[a] Adapted from Schnoor (1996), Shaw (1966), and Thibodeaux (1996).

Under turbulent-flow conditions, the longitudinal spreading of a tracer is caused by dispersion, in which case the molecular diffusion term D_m is replaced by the "coefficient of dispersion" D. The coefficient of dispersion is an inclusive term which includes the effects of dispersion caused by advection (i.e., transport caused by differences in the fluid velocity), turbulent eddies (i.e., eddy diffusion), and molecular diffusion. While the magnitude of the molecular diffusion depends primarily on the chemical and fluid properties, turbulent or eddy diffusion and dispersion depend primarily on the flow regime.

In general, the following relationship holds between the numerical values of the various terms described above.

$$D >> D_e >> D_m \tag{4-36}$$

Typical observed ranges for the coefficient of molecular diffusion, turbulent or eddy diffusion, or dispersion are reported in Table 4–3. In all cases, it is important to remember that regardless of whether the coefficient of molecular diffusion, turbulent or eddy diffusion, or dispersion is operative, the driving force for mass transfer is the concentration gradient.

Plug-Flow Reactor with Axial Dispersion

In the following analysis only the one-dimensional problem is considered. However, it should be noted that all dispersion problems are three-dimensional, with the dispersion coefficient varying with direction and the degree of turbulence. Using the relationship given above [Eq. (4–35)] and referring to Fig. 4–5c, given previously, the one-dimensional materials mass balance for the transport of a conservative dye tracer by advection and dispersion is

Accumulation = inflow − outflow

$$\frac{\partial C}{\partial t} A\Delta x = \left(vAC - AD\frac{\Delta C}{\Delta x} \right)\bigg|_x - \left(vAC - AD\frac{\Delta C}{\Delta x} \right)\bigg|_{x+\Delta x} \tag{4-37}$$

where $\partial C/\partial t$ = change in concentration with time, $ML^{-3}T^{-1}$, (g/m³·s)

A = cross-sectional area in x direction, L^2, (m²)

Δx = differential distance, L, m

v = average velocity in x direction, LT^{-1}, (m/s)

C = constituent concentration, ML^{-3}, (g/m³)

D = coefficient of axial dispersion, L^2T^{-1}, (m²/s)

In Eq. (4–37) the term vAC represents the transport of mass due to advection and the term $AD(\Delta C/\Delta x)$ represents the transport brought about by dispersion. Taking the limit of Eq. (4–37) as Δx approaches zero results in the following two expressions:

$$\frac{\partial C}{\partial t} = -D\frac{\partial^2 C}{\partial x^2} - v_x \frac{\partial C}{\partial x} \tag{4–38}$$

$$\frac{\partial C}{\partial t} = -D\frac{\partial^2 C}{\partial x^2} - \frac{\partial C}{\partial t} \tag{4–39}$$

In Eq. (4–39) the hydraulic detention time ∂t has been substituted for the term $\partial x/v_x$. Equation (4–39) has been solved for small amounts of axial dispersion (see below). The solution for a unit pulse input leading to symmetrical output tracer response curves for small amounts of axial dispersion is given by Levenspiel (1972).

$$C_\theta = \frac{1}{2\sqrt{\pi(D/uL)}} \exp\left[-\frac{(1-\theta)^2}{4(D/uL)}\right] \tag{4–40}$$

where C_θ = normalized tracer response C/C_o, unitless
θ = normalized time t/τ, unitless
t = time T, (s)
τ = theoretical detention time V/Q, T, (s)
D = coefficient of axial dispersion $L^2 T^{-1}$, (m^2/s)
u = fluid velocity LT^{-1}, (m/s)
L = characteristic length L, (m)

The solution given by Eq. (4–40) is for what is known as a *closed system* in which it is assumed that there is no dispersion upstream or downstream of the boundaries of the reactor (e.g., a reactor with inlet and outlet weirs). Nevertheless, for small amounts of dispersion, Eq. (4–40) can be used to approximate the performance of an open or closed reactor, regardless of the boundary conditions.

It will be noted that Eq. (4–40) has the same general form as the equation for the normal probability distribution. Thus, the corresponding mean and variance are

$$\bar{\theta} = \frac{\bar{t}_c}{\tau} = 1 \tag{4–41}$$

$$\sigma_\theta^2 = \frac{\sigma_c^2}{\tau^2} = 2\frac{D}{uL} \tag{4–42}$$

where $\bar{\theta}$ = normalized mean detention time $\bar{t}_c/\tau$, unitless
$\bar{t}_c$ = mean detention (or residence) time derived from C curve [see Eq. (4–26)], T, h, min
τ = theoretical detention (or residence) time T, (s)
σ_θ^2 = variance of normalized tracer response C curve T^2, (s^2)
σ_c^2 = variance derived from C curve [see Eq. (4–24)] T^2, (s^2)

Defining the exact extent of axial dispersion is difficult. To provide an estimate of dispersion, the following unitless dispersion number has been defined:

$$d = \frac{D}{uL} = \frac{Dt}{L^2} \tag{4–43}$$

Figure 4–13

Typical concentration versus time tracer response curves for a plug-flow reactor with small amounts of axial dispersion subject to a pulse (slug) input of tracer.

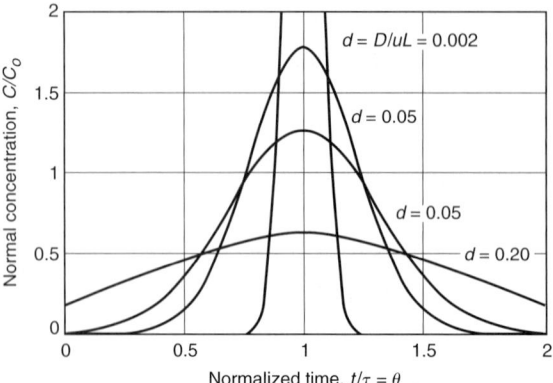

where d = dispersion number, unitless

D = coefficient of axial dispersion L^2T^{-1}, (m²/s)

u = fluid velocity LT^{-1}, (m/s)

L = characteristic length L, (m)

t = travel time (L/u), T, (s)

Normalized effluent concentration versus time curves, obtained using Eq. (4–40), for a plug-flow reactor with limited axial dispersion for various values of the dispersion number are shown on Fig. 4–13.

When dispersion is large, the output curve becomes increasingly nonsymmetrical, and the problem becomes sensitive to the boundary conditions. In environmental problems, a wide variety of entrance and exit conditions are encountered, but most can be considered approximately open; that is, the flow characteristics do not change greatly as the boundaries are crossed. The solution to Eq. (4–39) for a unit pulse input in an *open system* with larger amounts of dispersion is as follows (Fogler, 1999; Levenspiel, 1972):

$$C_\theta = \frac{1}{2\sqrt{\pi\theta(D/uL)}} \exp\left[-\frac{(1-\theta)^2}{4\theta(D/uL)} \right]$$ (4–44)

The corresponding mean and variance are

$$\bar{\theta} = \frac{\bar{t_c}}{\tau} = 1 + 2\frac{D}{uL}$$ (4–45)

$$\sigma_\theta^2 = \frac{\sigma_c^2}{\tau^2} = 2\frac{D}{uL} + 8\left(\frac{D}{uL}\right)^2$$ (4–46)

where the terms are as defined previously for Eqs. (4–41), (4–42), and (4–43).

The mean as given by Eq. (4–45) is greater than the hydraulic detention time because of the forward movement of the tracer due to dispersion. Effluent concentration versus time curves, obtained using Eq. (4–44), for a plug-flow reactor with significant axial dispersion for various dispersion factors are shown on Fig. 4–14. A more

Figure 4-14

Typical concentration versus time tracer response curves for a plug-flow reactor with large amounts of axial dispersion subject to a pulse (slug) input of tracer.

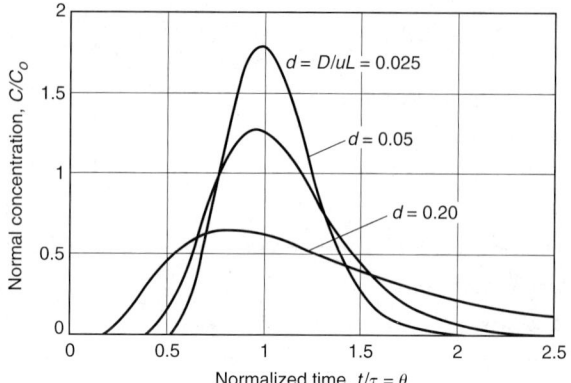

detailed discussion of closed and open reactors may be found in Fogler (1999) and Levenspiel (1972).

In the literature, the inverse of Eq (4–43), as given below, is often identified as the Peclet number of longitudinal dispersion (Kramer and Westererp, 1963).

$$P_e = \frac{uL}{D} \qquad (4\text{–}47)$$

In effect, the Peclet number represents the ratio of the mass transport brought about by advection and dispersion. If the Peclet number is significantly greater than 1, advection is the dominant factor in mass transport. If the Peclet number is significantly less than 1, dispersion is the dominant factor in mass transport. Although beyond the scope of this presentation, it can also be shown that the number of complete-mix reactors in series required to simulate a plug-flow reactor with axial dispersion is approximately equal to the Peclet number divided by 2. Thus, for a dispersion factor of 0.025, the Peclet number is equal to 40 and the corresponding number of reactors in series needed to simulate the dispersion in a plug-flow reactor is equal to 20. This relationship will be illustrated in the following discussion dealing with complete-mix reactors in series. The Peclet number is also used to define transverse diffusion in packed-bed plug-flow reactors (Denbigh and Turner, 1965).

For practical purposes, the following dispersion values can be used to assess the degree of axial dispersion in wastewater-treatment facilities.

No dispersion $d = 0$ (ideal plug-flow)
Low dispersion $d = \, < 0.05$
Moderate dispersion $d = 0.05$ to 0.25
High dispersion $d = \, > 0.25$
                       $d \rightarrow \infty$ (complete-mix)

Typical dispersion numbers determined for actual treatment facilities are given in Table 4–4. The considerable range in the dispersion numbers reported in Table 4–4 for

Table 4–4
Typical dispersion numbers for various wastewater treatment facilities[a]

Treatment facility	Range of values for dispersion number
Rectangular sedimentation tanks	0.2–2.0
Activated-sludge aeration reactors	
Long plug-flow	0.1–1.0
Complete-mix	3.0–4.0+
Oxidation ditch activated-sludge process	3.0–4.0+
Waste-stabilization ponds	
Single ponds	1.0–4.0+
Multiple cells in series	0.1–1.0
Mechanically aerated lagoons	
Long rectangular shaped	1.0–4.0+
Square shaped	3.0–4.0+
Chlorine contact basins	0.02–0.004

[a]Adapted from Arceivala (1998).

individual and similar treatment facilities is due, most often, to one or more of the following factors (Arceivala, 1983):

1. The scale of the mixing phenomenon
2. Geometry (i.e., aspect ratio) of the unit
3. Power input per unit volume (i.e., mechanical and pneumatic)
4. Type and disposition of the inlets and outlets of the treatment units
5. Inflow velocity and its fluctuations
6. Density and temperature differences between the inflow and the contents of the reactor

Because of the wide range of dispersion numbers that can result for individual treatment processes, special attention must be devoted to the factors cited above in the design of treatment facilities. Determination of the dispersion number and the coefficient of dispersion using the tracer response curves is illustrated in Example 4–4.

Evaluation of the coefficient of axial dispersion D for existing facilities is done experimentally using the results of tracer tests, as discussed previously. Because the systems encountered in wastewater treatment are large, experimental work is often difficult and expensive, and is, unfortunately, usually after the fact. To take into account axial dispersion in the design of treatment facilities both scaled models and empirical relationships have been developed for a variety of treatment units including oxidation ponds (Polprasert and Bhattarai, 1985) and chlorine contact basins (Montgomery, 1985). An approximate value of D for water for large Reynolds numbers is (Davies, 1972)

$$D = 1.01\nu N_R^{0.875}$$

(4–48)

where D = coefficient of dispersion, L^2T^{-1}, (m²/s)

$\quad\quad\quad v$ = kinematic viscosity, L^2T^{-1}, m²/s (see Appendix C)

$\quad\quad N_R$ = Reynolds number, unitless

$\quad\quad\quad\quad$ = $4vR/\nu$

$\quad\quad\quad v$ = velocity in open channel, LT^{-1}, (m/s)

$\quad\quad\quad R$ = hydraulic radius = area/wetted perimeter, L, (m)

Values for N_R found in open channel flow in wastewater-treatment plants are typically in the range from 10^3 to 10^4. The corresponding values for D range from 0.0004 to 0.003 m²/s (4 to 30 cm²/s).

EXAMPLE 4–4 **Determination of the Dispersion Number and the Coefficient of Dispersion from Concentration versus Time Tracer Response Curves** Use the concentration versus time tracer data given in Example 4–1 to determine the dispersion number and the coefficient of dispersion for the chlorine contact basin described in Example 4–1. Compare the value of the coefficient of dispersion computed, using the tracer data, to the value computed using Eq. (4–48). Assume the following data are applicable:

1. The flowrate at the time when the tracer test was conducted = 240,000 m³/d
2. Number of chlorine contact basin = 4
3. Number of channels per contact basin = 13
4. Channel dimensions
   a. Length = 36.6 m
   b. Width = 3.0 m
   c. Depth = 4.9 m
5. Temperature = 20°C

Solution

1. From Example 4–1, the mean value and variance for the tracer response C curve are:
   a. Mean, $t_{\Delta c}$ = 2.7 h
   b. Variance, $\sigma^2_{\Delta c}$ = 280.5 min²
2. Determine the theoretical detention time for the chlorine contact basin using the given data.

$$\tau = \frac{4 \times 13 \times (36.6 \text{ m} \times 3.0 \text{ m} \times 4.9 \text{ m})}{(240,000 \text{ m}^3/\text{d})}$$

$$= 0.1166 \text{ d} = 2.8 \text{ h} = 167.9 \text{ min}$$

3. Determine the mean detention time using Eq. (4–41). Use the approximate value of $\bar{t}_{\Delta c}$ for $\bar{t}_c$.

$$\bar{\theta}_\Delta = \frac{\bar{t}_{\Delta c}}{\tau} \approx \frac{2.7}{2.8} \approx 0.96 \approx 1.0$$

4. Determine the dispersion number using Eq. (4–42). Use the approximate value of $\sigma_{\Delta c}^2$ for σ_c^2.

$$\sigma_{\Delta \theta}^2 = \frac{\sigma_{\Delta c}^2}{\tau^2} \approx 2\frac{D}{uL} = 2d$$

$$d \approx \frac{1}{2}\frac{\sigma_{\Delta c}^2}{\tau^2} \approx \frac{1}{2}\frac{280.5 \text{ min}^2}{(167.9 \text{ min})^2} = 0.00498$$

5. Using Eq. (4–40), determine the coefficient of dispersion

$$D = d \times u \times L$$

$$u = \frac{Q}{A} = \frac{(240,000 \text{ m}^3/\text{d})}{(4 \times 13 \times 3.0 \text{ m} \times 4.9 \text{ m})} = 314.0 \text{ m/d} = 0.00363 \text{ m/s}$$

$$D = (0.00498)(0.00363 \text{ m/s})(36.6 \text{ m}) = 6.62 \times 10^{-4} \text{ m}^2/\text{s}$$

6. Compare the value of the coefficient of dispersion computed in Step 5 with the value computed using Eq. (4–48).

$$D = 1.01 \, \nu N_R^{0.875}$$

a. Compute the Reynolds number

$$N_R = 4\nu R/\nu$$

$$\nu = 1.002 \times 10^{-6} \text{ m}^2/\text{s}$$

$$N_R = \frac{(4)(0.000363 \text{ m/s})[(3.0 \text{ m} \times 4.9 \text{ m})/(2 \times 4.9 \text{ m} + 3.0 \text{ m})]}{(1.002 \times 10^{-6} \text{ m}^2/\text{s})} = 1664$$

b. Compute the coefficient of dispersion

$$D = 1.002 \times 10^{-6}(1664)^{0.875} = 6.60 \times 10^{-4} \text{ m}^2/\text{s}$$

Comment   Based on the computed value of the dispersion number (0.00498), the chlorine contact basin would be classified as having low dispersion (i.e., $d = < 0.05$). The coefficients of dispersion determined using the results of the tracer study and Eq. (4–48) are remarkably close, given the nature of such measurements.

Complete-Mix Reactors in Series

When varying amounts of axial dispersion are encountered, the flow is sometimes identified as *arbitrary flow*. The output from a plug-flow reactor with axial dispersion (arbitrary flow) is often modeled using a number of complete-mix reactors in series, as outlined below. In some situations, the use of a series of complete-mix reactors may have certain advantages with respect to treatment. To understand the hydraulic characteristics of reactors in series (see Figs. 4–2e and 4–15), assume that a pulse input (i.e., a slug) of tracer is injected into the first reactor in a series of equally sized reactors so that

Figure 4-15

Definition sketch for the analysis of complete-mix reactors in series.

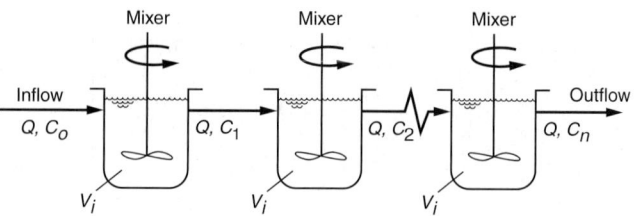

the resulting instantaneous concentration of tracer in the first reactor is C_o. The total volume of all the reactors is V, the volume of an individual reactor is V_i, and the hydraulic residence time V_i/Q is τ_i. The effluent concentration from the first reactor as given by Eq. (4–14) is

$$C_1 = C_o e^{-1/\tau_i} \tag{4–49}$$

Writing a materials balance for the second reactor results in the following:

Accumulation = inflow − outflow

$$\frac{dC_2}{dt} + \frac{1}{\tau_i} C_1 - \frac{1}{\tau_i} C_2 \tag{4–50}$$

Rearranging the terms in Eq. (4–47) and substituting Eq. (4–46) for C_1 results in

$$\frac{dC_2}{dt} + \frac{C_2}{\tau_i} = \frac{C_o}{\tau_i} e^{-n\theta} \tag{4–51}$$

The general non-steady-state solution for Eq. (4–51) is obtained by first noting that Eq. (4–51) has the form of the standard first-order linear differential equation. The solution procedure outlined in Eqs. (4–52) through (4–59) involves the use of an integrating factor. It should be noted that Eq. (4–51) can also be solved using the separation of variables method. The solution to Eq. (4–51) is included here because these types of equations are encountered frequently in the field of environmental engineering and in this text. The first step in the solution is to rewrite Eq. (4–51) in the form

$$C_2' + \frac{C_2}{\tau_i} = \frac{C_o}{\tau_i} e^{-t/\tau_i} \tag{4–52}$$

where C_2' is used to denote the derivative dC_2/dt. In the next step, both sides of the expression are multiplied by the integrating factor $e^{\beta t}$, where $\beta = (1/\tau_i)$.

$$e^{\beta t}(C_2' + \beta C_2) = (\beta C_o e^{-\beta t}) = \beta C_o \tag{4–53}$$

The left-hand side of the above expression can be written as a differential as follows:

$$(C_2 e^{\beta t})' = \beta C_o \tag{4–54}$$

The differential sign is removed by integrating the above expression

$$C_2 e^{\beta t} = \beta C_o \int dt \tag{4–55}$$

Integration of Eq. (4–55) yields

$$C_2 e^{\beta t} = \beta C_o t + K \text{ (constant of integration)} \tag{4–56}$$

Figure 4-16

Effluent concentration curves for each of four complete-mix reactors in series subject to a slug input of tracer into the first reactor of the series.

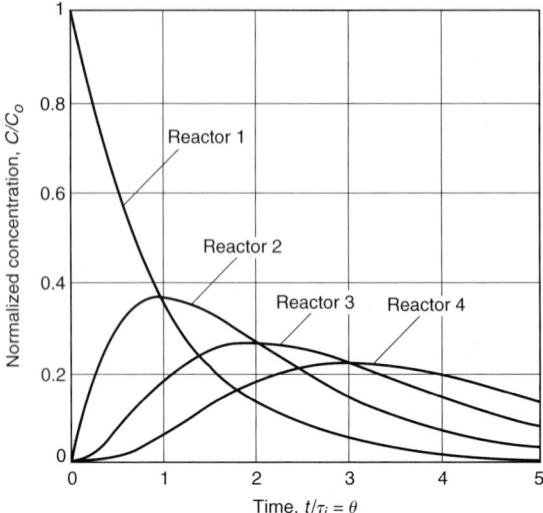

Dividing by $e\beta t$ yields

$$C_2 = \beta C_o t e^{-\beta t} + K e^{-\beta t} \qquad (4\text{-}57)$$

But when $t = 0$, $C_2 = 0$ and K is equal to 0. Thus,

$$C_2 = C_o \frac{t}{\tau_i} e^{-t/\tau_i} \qquad (4\text{-}58)$$

Following the same solution procedure, the generalized expression for the effluent concentration for the ith reactor is

$$C_i = \frac{C_o}{(i-1)!} \left(\frac{t}{\tau_i}\right)^{i-1} e^{-t/\tau_i} \qquad (4\text{-}59)$$

The effluent concentration from each of four complete-mix reactors in series is shown on Fig. 4–16.

The generalized RTD for n reactors in series, obtained using Eq. (4–59), is

$$E(t) = \frac{t^{n-1}}{(n-1)\tau_i^n} e^{-t/\tau_i} \qquad (4\text{-}60)$$

Fraction Remaining. Equation (4–59) can also be used to obtain the fraction of tracer remaining in a series of complete-mix reactors at any time t. The fraction of tracer remaining, F, at time t, is equal to

$$F = \frac{C_1 + C_2 + \cdots + C_n}{C_o} \qquad (4\text{-}61)$$

Figure 4–17

Fraction of tracer remaining in a system comprised of reactors in series.

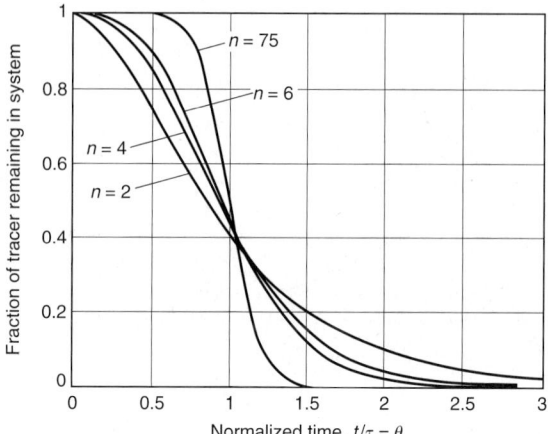

Using Eq. (4–59) to obtain the individual effluent concentrations, the fraction remaining in four equal-sized reactors in series is given by:

$$F_{4R} = \frac{C_o e^{-4t/\tau_i} + C_o(4t/\tau_i)e^{-4t/\tau_i} + (C_o/2)(4t/\tau_i)^2 e^{-4t/\tau_i} + (C_o/6)(4t/\tau_i)^3 e^{-4t/\tau_i}}{C_o}$$

$$F_{4R} = \left[1 + (4t/\tau_i) + \frac{(4t/\tau_i)^2}{2} + \frac{(4t/\tau_i)^3}{6} \right] e^{-4t/\tau_i} \qquad (4\text{–}62)$$

The fraction of a tracer remaining in a series of 2, 4, 6, and 75 complete-mix reactors in series is given on Fig. 4–17.

Comparison of Nonideal Plug-Flow Reactor and Complete-Mix Reactors in Series.

In many cases it will be useful to model the performance of plug-flow reactors with axial dispersion, as discussed previously, with a series of complete-mix reactors in series. To obtain the normalized residence time distribution curve for n reactors in series, Eq. (4–60) can be written as follows by noting that the total volume is nV_i and $\tau_i = \tau/n$:

$$E(\theta) = \frac{n}{(n-1)!} (n\theta)^{n-1} e^{-n\theta} \qquad (4\text{–}63)$$

In effect, in Eq. (4–63) it is assumed that the same amount of tracer is always added to the first reactor in series. Effluent residence time distribution curves, obtained using Eq. (4–63), for 1, 2, 4, 6, and 75 reactors in series are shown on Fig. 4–18. It is interesting to note that a model comprised of four complete-mix reactors in series is often used to describe the hydraulic characteristics of constructed wetlands. On Fig. 4–18, the concentration increases as the number of reactors in series increases because the same amount of tracer is used regardless of the number of reactors in series. It should also be

Figure 4–18

Effluent tracer concentration curves for reactors in series, subject to a slug input of tracer into the first reactor of the series. Concentration values greater than 1 occur because the same amount of tracer is placed in the first reactor in each series of reactors.

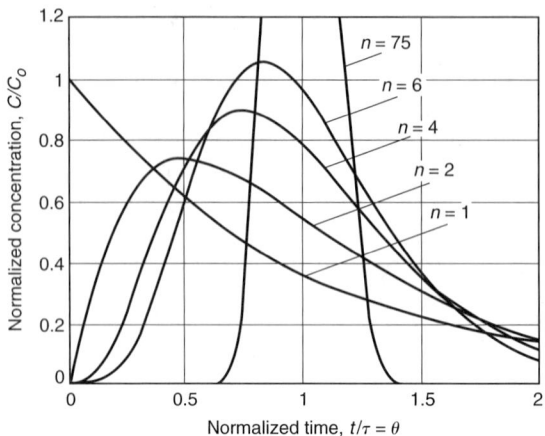

Figure 4–19

Comparison of effluent response curves for a plug-flow reactor with a dispersion factor of 0.05 and reactor systems comprised of six, eight, and ten reactors in series subject to a pulse input of tracer.

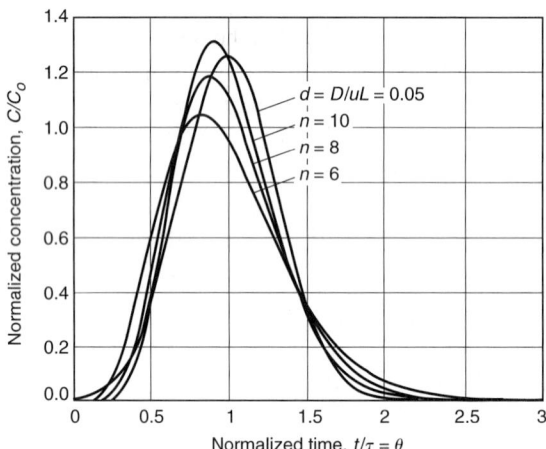

noted that the F curves shown on Fig. 4–17 can also be obtained by integrating the E curves given on Fig. 4–18.

A comparison of the residence time distribution curves obtained for a plug-flow reactor, with a dispersion number of 0.05, to the residence time distribution curves obtained for six, eight, and ten complete-mix reactors in series is shown on Fig. 4–19. As shown, all three of the complete-mix reactors in series can be used, for practical purposes, to simulate a plug-flow reactor with a dispersion factor of 0.05. As noted previously, the number of reactors in series needed to simulate a plug-flow reactor with dispersion is approximately equal to the Peclet number divided by 2. Thus, for a dispersion factor of 0.05, the Peclet number is equal to 20 and the corresponding number of reactors in series needed to simulate the dispersion in a plug-flow reactor is equal to 10. As shown on Fig. 4–19, the response curves computed using ten reactors in series and Eq. (4–44) with a dispersion number $d = 0.05$ are essentially the same.

4–6 REACTIONS, REACTION RATES, AND REACTION RATE COEFFICIENTS

From the standpoint of process selection and design, the controlling stoichiometry and the rate of the reaction are of principal concern. The number of moles of a substance entering into a reaction and the number of moles of the substances produced are defined by the stoichiometry of a reaction. The *stoichiometry* of reaction refers to the definition of the quantities of chemical compounds involved in a reaction. The rate at which a substance disappears or is formed in any given stoichiometric reaction is defined as the rate of reaction. These and other related topics are discussed in this section. The rate expressions discussed in this section will be integrated with the hydraulic characteristics of the reactors, discussed previously, to define treatment kinetics.

Types of Reactions

The two principal types of reactions that occur in wastewater treatment are classified as homogeneous and heterogeneous (nonhomogeneous).

Homogeneous Reactions. In homogeneous reactions, the reactants are distributed uniformly throughout the fluid so that the potential for reaction at any point within the fluid is the same. Homogeneous reactions are usually carried out in the batch, complete-mix, and plug-flow reactors (see Figs. 4–2a, b, c, and d). Homogeneous reactions may be either irreversible or reversible.

Examples of irreversible reactions are

a. Simple reactions

$$A \rightarrow B \tag{4-64}$$

$$A + A \rightarrow C \tag{4-65}$$

$$aA + bB \rightarrow C \tag{4-66}$$

b. Parallel reactions

$$A + B \rightarrow C \tag{4-67}$$

$$A + B \rightarrow D \tag{4-68}$$

c. Consecutive reactions

$$A + B \rightarrow C \tag{4-69}$$

$$A + C \rightarrow D \tag{4-70}$$

Examples of reversible reactions are

$$A \leftrightarrow B \tag{4-71}$$

$$A + B \leftrightarrow C + D \tag{4-72}$$

As will be discussed subsequently, for both irreversible and reversible reactions, the rate of reaction will be an important consideration in the design of the treatment facilities in which these reactions will be carried out. Special attention must be given to the design of mixing facilities, especially for reactions that are rapid.

Heterogeneous Reactions. Heterogeneous reactions occur between one or more constituents that can be identified with specific sites, such as those on an ion-exchange resin in which one or more ions is replaced by another ion.

Reactions that require the presence of a solid-phase catalyst are also classified as heterogeneous. Heterogeneous reactions are usually carried out in packed and fluidized-bed reactors (see Fig. 4–2f, g, and h). These reactions are more difficult to study because a number of interrelated steps may be involved. The typical sequence of these steps, as quoted from Smith (1981), is as follows:

1. Transport of reactants from the bulk fluid to the fluid-solid interface (external surface of catalyst particle)
2. Intraparticle transport of reactants into the catalyst particle (if it is porous)
3. Adsorption of reactants at interior sites of the catalyst particle
4. Chemical reaction of adsorbed reactants to adsorbed products (surface reaction)
5. Desorption of adsorbed products
6. Transport of products from the interior sites to the outer surface of the catalyst particle

Rate of Reaction

The rate of reaction is the term used to describe the change (decrease or increase) in the number of moles of a reactive substance per unit volume per unit time (for homogeneous reactions), or per unit surface area or mass per unit time (for heterogeneous reactions) (Denbigh and Turner, 1965).

For homogeneous reactions, the rate of reaction r is given by

$$r = \frac{1}{V}\frac{d[N]}{dt} = \frac{\text{moles}}{(\text{volume})(\text{time})} \tag{4–73}$$

If N is replaced by the term VC, where V is the volume and C is the concentration, Eq. (4–73) becomes

$$r = \frac{1}{V}\frac{d(VC)}{dt} = \frac{1}{V}\frac{VdC + CdV}{dt} \tag{4–74}$$

If the volume remains constant (i.e., isothermal conditions, no evaporation), Eq. (4–74) reduces to

$$r = \pm\frac{dC}{dt} \tag{4–75}$$

where the plus sign indicates an increase or accumulation of the substance, and the minus sign indicates a decrease of the substance.

For heterogeneous reactions where S is the surface area, the corresponding expression is

$$r = \frac{1}{S}\frac{d[N]}{dt} = \frac{\text{moles}}{(\text{area})(\text{time})} \tag{4–76}$$

For reactions involving two or more reactants with unequal stoichiometric coefficients, the rate expressed in terms of one reactant will not be the same as the rate for the other reactants. For example, for the reaction

$$aA + bB \rightarrow cC + dD \qquad (4\text{--}77)$$

the concentration changes for the various reactants are given by

$$-\frac{1}{a}\frac{d[A]}{dt} = -\frac{1}{b}\frac{d[B]}{dt} = \frac{1}{c}\frac{d[C]}{dt} = \frac{1}{d}\frac{d[D]}{dt} \qquad (4\text{--}78)$$

Thus, for reactions in which the stoichiometric coefficients are not equal, the rate of reaction is given by

$$r = \frac{1}{c_i}\frac{d[C_i]}{dt} \qquad (4\text{--}79)$$

where the coefficient term $(1/c_i)$ is negative for reactants and positive for products.

The rate at which a reaction proceeds is an important consideration in wastewater treatment. For example, in many cases the reaction usually takes too long to go to completion. In such cases, treatment processes are designed on the basis of the rate at which the reaction proceeds rather than the equilibrium position of the reaction. Often, quantities of chemicals in excess of the stoichiometric or exact reacting amount may be used to accomplish the treatment step in a shorter period of time by driving the reaction to completion.

Reaction Order

The rate at which reactions occur usually is determined by measuring the concentration of either a reactant or a product as the reaction proceeds to completion. The measured results are then compared to the corresponding results obtained from various standard rate equations by which the reaction under study is expected to proceed.

The order of a reaction with respect to a specified compound is equal to the stoichiometric coefficient for that compound. For example, in the following reaction, the reaction order for compound A is a, for compound B is b, and so on.

$$aA + bB + \cdots \rightarrow pP + qQ + \cdots \qquad (4\text{--}80)$$

If the rate is experimentally found to be proportional to the first power of the concentration of A (i.e., $a = 1$), then the reaction is said to be first-order with respect to A.

When the mechanism of reaction is not known, the reaction rate for Eq. (4–80) may be approximated with the following expression:

$$r = kC_A^a C_B^b C_C^c \cdots C_P^p = kC_A^n \qquad (4\text{--}81)$$

where a and b are the reaction orders with respect to reactants A and B, and n is the overall reaction order ($n = a + b + \cdots + p$). The sum of the exponents to which the concentration(s) are raised is known as the *order* of the reaction. The following discussion of other types of rate expressions illustrates the relationships that are used to identify reactions of various orders.

Types of Rate Expressions

Other types of rate expressions that have been used to describe the conversion of waste-water constituents in treatment processes and the fate of constituents released in the environment include the following:

$$r = \pm k \qquad \text{(zero-order)} \qquad (4\text{--}82)$$

$$r = \pm kC \qquad \text{(first-order)} \qquad (4\text{--}83)$$

$$r = \pm k(C - C_s) \qquad \text{(first-order)} \qquad (4\text{--}84)$$

$$r = \pm kC^2 \qquad \text{(second-order)} \qquad (4\text{--}85)$$

$$r = \pm kC_A C_B \qquad \text{(second-order)} \qquad (4\text{--}86)$$

$$r = \pm \frac{kC}{K + C} \qquad \text{(saturation or mixed-order)} \qquad (4\text{--}87)$$

$$r = \pm \frac{kC}{(1 + r_t t)^n} \qquad \text{(first-order retarded)} \qquad (4\text{--}88)$$

Zero-order reactions [Eq. (4–82)] proceed at a rate independent of the concentration of the reactants, whereas first-order reactions [Eqs. (4–83) and (4–84)] proceed at a rate directly proportional to the concentration of one of the reactants. For example, the first-order reaction ($r_c = -kC$) is used to model the exertion of BOD and bacterial decay, as discussed in Chap. 2. Second-order reactions [Eq. (4–85)] proceed at a rate proportional to the second power of a single reactant. Although Eq. (4–86) is second-order overall, it is first-order with respect to C_A and C_B, individually. Equation (4–87), known as a saturation type of equation, is illustrated on Fig. 4–20. As shown on Fig. 4–20, when the concentration C is large, the rate of reaction is zero-order, and when the concentration is low, the rate of reaction is first-order.

Figure 4–20

Rate of reaction versus concentration for a saturation-type expression. Beyond a concentration of about 20 mg/L the rate of reaction is essentially zero-order.

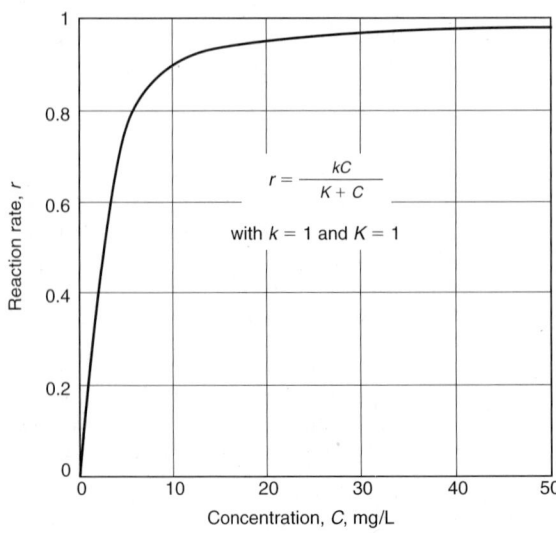

$$r = \frac{kC}{K + C}$$

with $k = 1$ and $K = 1$

Reaction rate, r

Concentration, C, mg/L

The rate expression given by Eq. (4–88) is known as a retarded first-order rate expression, because the rate constant changes with distance or time or with the degree of treatment which, in turn, can be related to distance or time. The term r_t in the denominator is the retardation factor. In wastewater-treatment applications, the exponent n in Eq. (4–88) is related to the particle size distribution. For example, if all of the particles are the same size and composition, the value of the exponent n is equal to 1 and the retardation factor r_t is equal to zero. The retarded rate expression is also applied to the removal of organic matter from mixtures where the biodegradability of the individual constituents comprising the organic matter is different (Tchobanoglous et al., 2000). Application of the retarded first-order rate expression is considered further in Sec. 4–7.

Rate Expressions Used in Environmental Modeling

The physical, chemical, and biological processes that control the fate of the constituents dispersed to the environment are numerous and varied. Important constituent transformations and removal processes (i.e., fate processes) operative in the environment, along with the constituents affected, are reported in Table 4–5. The various processes listed in Table 4–5 will be referred to throughout this text. For example, bacterial conversion was considered previously in Chap. 2 in the analysis of the BOD reaction, and will be considered in greater detail in the chapters dealing with biological treatment. Because all of the processes summarized in Table 4–5 are rate-dependent, representative rate expressions used to model these processes are presented in Table 4–6. The important thing to note about Table 4–6 is the variety of different rate expressions that have been used to model constituent transformation and removal processes. The application of various rate expressions will be illustrated in this chapter. Additional details on the application of the rate expressions in environmental engineering modeling may be found in Hemond and Fechner (1994), Orlob (1983), Schnoor (1996), Thibodeaux (1996), and Thomann and Mueller (1987).

Effects of Temperature on Reaction Rate Coefficients

The temperature dependence of the specific reaction rate constants is important because of the need to adjust for other temperatures. The temperature dependence of the rate constant is given by the van't Hoff–Arrhenius relationship presented previously in Chap. 2.

$$\ln \frac{k_2}{k_1} = \frac{E(T_2 - T_1)}{RT_1T_2} = \frac{E}{RT_1T_2}(T_2 - T_1) \tag{2-21}$$

where k_2 = reaction rate constant at temperature T_2, (1/s, 1/min, etc.)
   k_1 = reaction rate constant at temperature T_1, (1/s, 1/min, etc.)
   T = temperature, K = 273.15 + °C
   E = a constant characteristic of the reaction (e.g., activation energy), J/mol
   R = ideal gas constant, 8.314 J/mol·K (1.99 cal/mol·K)

As noted previously in the discussion of temperature in Chap. 2, most wastewater-treatment operations and processes are carried out over a relatively narrow temperature range. Given that the temperature differences are not great, the following relationship, developed from the van't Hoff–Arrhenius relationship (see discussion of temperature in

Table 4–5

Constituent transformation and removal processes (i.e., fate processes) in the environment

Process	Comments	Constituents affected
Adsorption/ desorption	Many chemical constituents tend to attach or sorb onto solids. The implication for wastewater discharges is that a substantial fraction of some toxic chemicals is associated with the suspended solids in the effluent. Adsorption combined with solids settling results in the removal from the water column of constituents that might not otherwise decay	Metals, trace organics, NH_4^+, PO_4^{3-}
Algal synthesis	The synthesis of algal cell tissue using the nutrients found in wastewater	NH_4^+, NO_3^-, PO_4^{3-}, pH, etc.
Bacterial conversion	Bacterial conversion (both aerobic and anaerobic) is the most important process in the transformation of constituents released to the environment. The exertion of BOD and NOD is the most common example of bacterial conversion encountered in water-quality management. The depletion of oxygen in the aerobic conversion of organic wastes is also known as deoxygenation. Solids discharged with treated wastewater are partly organic. Upon settling to the bottom, they decompose bacterially either anaerobically or aerobically, depending on local conditions. The bacterial transformation of toxic organic compounds is also of great significance	BOD, nitrification, denitrification, sulfate reduction, anaerobic fermentation (in bottom sediments), conversion of priority organic pollutants, etc.
Chemical reactions	Important chemical reactions that occur in the environment include hydrolysis, photochemical, and oxidation-reduction reactions. Hydrolysis reactions occur between contaminants and water	Chemical disinfection, decomposition of organic compounds, specific ion exchange, element substitution
Filtration	Removal of suspended and colloidal solids by straining (mechanical and chance contact), sedimentation, interception, impaction, and adsorption	TSS, colloidal particles
Flocculation	Flocculation is the term used to describe the aggregation of smaller particles into larger particles that can be removed by sedimentation and filtration. Flocculation is brought about by Brownian motion, differential velocity gradients, and differential settling in which large particles overtake smaller particles and form larger particles	Colloidal and small particles
Gas absorption/ desorption	The process whereby a gas is taken up by a liquid is known as absorption. For example, when the dissolved oxygen concentration in a body of water with a free surface is below the saturation concentration in the water, a net transfer of oxygen occurs from the atmosphere to the water. The rate of transfer (mass per unit time per unit surface area) is proportional to the amount by which the dissolved oxygen is below saturation. The addition of oxygen to water is also known as reaeration. Desorption occurs when the concentration of the gas in the liquid exceeds the saturation value, and there is a transfer from the liquid to the atmosphere	O_2, CO_2, CH_4, NH_3, H_2S

(continued)

Table 4-5	(Continued)	
Process	**Comments**	**Constituents affected**
Natural decay	In nature, contaminants will decay for a variety of reasons, including mortality in the case of bacteria and photooxidation for certain organic constituents. Natural and radioactive decay usually follow first-order kinetics	Plants, animals, algae, fungi, protozoa, eubacteria (most bacteria), archaebacteria, viruses, radioactive substances, plant mass
Photochemical reactions	Solar radiation is known to trigger a number of chemical reactions. Radiation in the near-ultraviolet (UV) and visible range is known to cause the breakdown of a variety of organic compounds	Oxidation of inorganic and organic compounds
Photosynthesis/ respiration	During the day, algal cells in water bodies will produce oxygen by means of photosynthesis. Dissolved oxygen concentrations as high as 30 to 40 mg/L have been measured. During the evening hours algal respiration will consume oxygen. Where heavy growths of algae are present, oxygen depletion has been observed during the evening hours	Algae, duckweed, submerged macrophytes, NH_4^+, PO_4^{3-}, pH, etc.
Sedimentation	The suspended solids discharged with treated wastewater ultimately settle to the bottom of the receiving water body. This settling is enhanced by flocculation and hindered by ambient turbulence. In rivers and coastal areas, turbulence is often sufficient to distribute the suspended solids over the entire water depth	TSS
Sediment oxygen demand	The residual solids discharged with treated wastewater will, in time, settle to the bottom of streams and rivers. Because the particles are partly organic, they can be decomposed anaerobically as well as aerobically, depending on conditions. Algae which settle to the bottom will also be decomposed, but much more slowly. The oxygen consumed in the aerobic decomposition of material in the sediment represents another dissolved oxygen demand in the water body	O_2, particulate BOD
Volatilization	Volatilization is the process whereby liquids and solids vaporize and escape to the atmosphere. Organic compounds that readily volatilize are known as VOCs (volatile organic compounds). The physics of this phenomenon are very similar to gas absorption, except that the net flux is out of the water surface	VOCs, NH_3, CH_4, H_2S, other gases

Chap. 2), can be used to adjust the value of the operative rate constant to reflect the effect of temperature:

$$\frac{k_2}{k_1} = \theta^{(T_2 - T_1)} \qquad (2\text{-}25)$$

Typical values for θ vary from about 1.020 to 1.10 for some biological treatment systems.

Table 4–6

Typical rate expressions for selected processes given in Table 4–5[a]

Process	Rate expression	Comments
Bacterial conversion	$r_c = -kC$	r_c = rate of conversion, M/L^3T
		k = first-order reaction rate coefficient, $1/T$
		C = concentration of organic material remaining, M/L^3
Chemical reactions	$r_c = \pm kC^n$	r_c = rate of conversion, M/L^3T
		k = reaction rate coefficient, $(M/L^3)^{n-1}/T$
		C = concentration of constituent, $(M/L^3)^n$
		n = reaction order (e.g., for second-order $n = 2$)
Gas absorption/ desorption	$r_{ab} = k_{ab}\dfrac{A}{V}(C_s - C)$	r_{ab} = rate of absorption, M/L^3T
		r_{de} = rate of desorption, M/L^3T
		k_{ab} = coefficient of absorption, L/T
	$r_{de} = k_{de}\dfrac{A}{V}(C - C_s)$	k_{de} = coefficient of desorption, L/T
		A = area, L^2
		V = volume, L^3
		C_s = saturation concentration of constituent in liquid, M/L^3 [see Eq. (2–49)]
		C = concentration of constituent in liquid, M/L^3
Natural decay	$r_d = -k_d N$	r_d = rate of decay, no./T
		k_d = first-order reaction rate coefficient, $1/T$
		N = amount of organisms remaining, no.
Sedimentation	$r_s = \dfrac{v_s}{H}(SS)$	r_s = rate of sedimentation, $1/T$
		v_s = settling velocity, L/T
		H = depth, L
		SS = settleable solids, L^3/L^3
Volatilization	$r_v = -k_v(C - C_s)$	r_v = rate of volatilization per unit time per unit volume, M/L^3T
		k_v = volatilization constant, $1/T$
		C = concentration of constituent in liquid, M/L^3
		C_s = saturation concentration of constituent in liquid, M/L^3 [see Eq. (2–49)]

[a]Adapted in part from Ambrose et al. (1988), Tchobanoglous and Burton (1991).

Analysis of Reaction Rate Coefficients

Typically, reaction rate coefficients are determined using the results obtained from batch experiments (i.e., no inflow or outflow), from continuous-flow experiments, and from pilot- and field-scale experiments. Using the data from batch experiments, the

Table 4–7

Integration and differential methods used to determine reaction rate coefficients

Rate expression		Method used to determine the reaction rate coefficient
Integration method		
Zero-order reaction $$r_c = \frac{dC}{dt} = -k$$	Integrated form $$C - C_o = -kt$$	Graphically, by plotting C versus t (see Fig. 4–21a)
First-order reaction $$r_c = \frac{dC}{dt} = -kC$$	$$\ln \frac{C}{C_o} = kt$$	Graphically, by plotting $-\ln(C/C_o)$ versus t (see Fig. 4–21b)
Second-order reaction $$r_c = \frac{dC}{dt} = -kC^2$$	$$\frac{1}{C} - \frac{1}{C_o} = kt$$	Graphically, by plotting $1/C$ versus t (see Fig. 4–21c)
Saturation reaction $$r_c = \frac{dC}{dt} = -\frac{kC}{K+C}$$	$$kt = K\ln\frac{C_o}{C_t} + (C_o - C_t)$$	Graphically, by plotting $1/t \ln(C_o/C_t)$ versus $(C_o - C_t)/t$ (see Fig. 4–21d)
Differential method		
$$r_c = \frac{dC}{dt} = -kC^n$$		Analytically, by solving for n $$n = \frac{\log(-d[C_1]/dt) - \log(-d[C_2]/dt)}{\log[C_1] - \log[C_2]}$$ Once the order of the reaction is known, the reaction rate coefficient can be determined by substitution

coefficients can be determined using a variety of methods including (1) the method of integration and (2) the differential method (see Table 4–7).

As summarized in Table 4–7, the method of integration involves the substitution of the measured data on the amount of reactant remaining at various times into the integrated form of the rate expression. Plots of the integrated forms of the reaction rate expressions used to determine the reaction rate coefficients are shown on Fig. 4–21. In the differential method, where the order of the reaction is unknown, the concentrations remaining at two different times are used to solve the differential form of the rate expression for the order of the reaction. Once the reaction order is known, the reaction rate coefficient is determined by substitution using the test data. The application of these two methods is illustrated in Example 4–5.

Figure 4–21

Graphical analysis for the determination of reaction order and reaction rate coefficients; (a) zero-order reaction, (b) first-order reaction, (c) second-order reaction, and (d) saturation-type reaction.

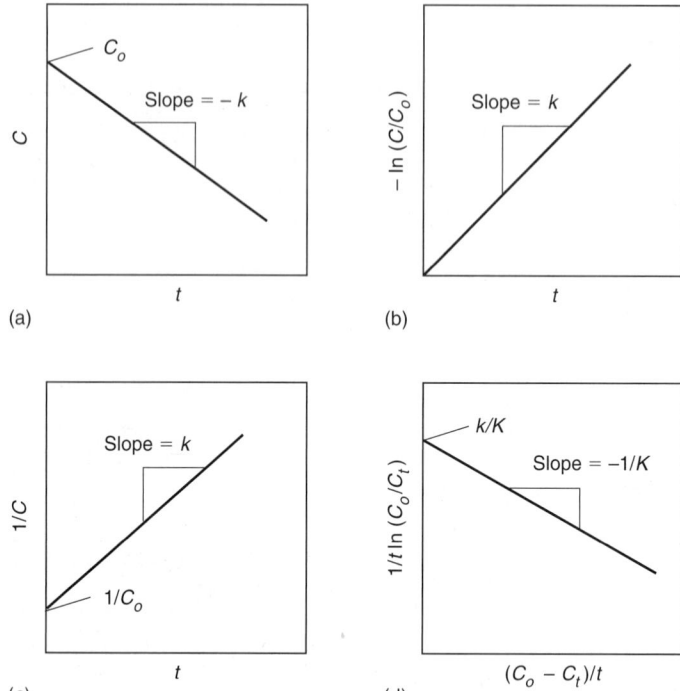

EXAMPLE 4–5 **Determination of the Reaction Order and the Reaction Rate Coefficient**
Given the following set of data obtained using a batch reactor (see Fig. 4–2a), determine the order of the reaction and the reaction rate coefficient using the integration and differential methods.

Time, d	Concentration, [C], mole/L
0	250
1	70
2	42
3	30
4	23
5	18
6	16
7	13
8	12

Solution—Part 1

1. Determine the reaction order and the reaction rate constant using the integration method. Develop the data needed to plot the experimental data functionally, assuming the reaction is either first- or second-order.

Time, d	$[C]$, mole/L	$-\log [C/C_o]$	$1/[C]$
0	250	0	0.004
1	70	0.553	0.014
2	42	0.775	0.024
3	30	0.921	0.033
4	23	1.036	0.044
5	18	1.143	0.056
6	16	1.194	0.063
7	13	1.284	0.077
8	12	1.319	0.083

2. To determine whether the reaction is first- or second-order, plot $\log [C/C_o]$ and $1/[C]$ versus t as shown below. Because the plot of $1/C$ versus t is a straight line, the reaction is second-order with respect to the concentration C.

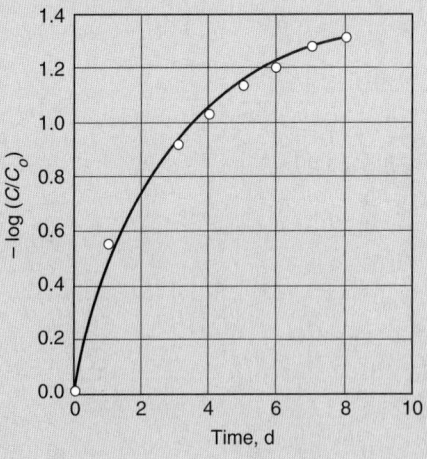 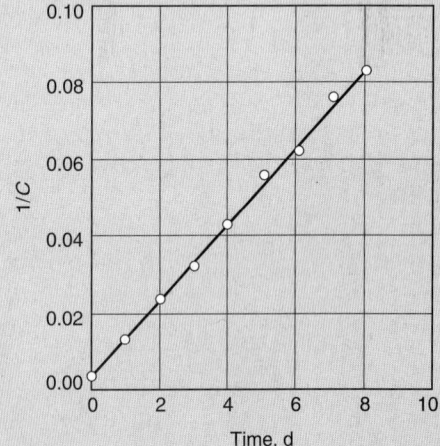

3. Determine the reaction rate coefficient.

   Slope = k

   The slope from the plot $= \dfrac{0.084 - 0.024}{8d - 2d} = 0.010/\text{mole·}d$

   $k = 0.010/\text{mole·}d$

Solution—Part 2

1. Determine the reaction order and the reaction rate constant using the differential method.

$$n = \frac{\log(-d[C_1]/dt) - \log(-d[C_2]/dt)}{\log[C_1] - \log[C_2]}$$

a. Use the experimental data obtained at day 2 and 5.

Time, d	[C], mole/L	$\dfrac{C_{t+1} - C_{t-1}}{2}$	$\approx \dfrac{dC_t}{dt}$
0	250		
1	70		
2	42	(30 − 70)/2	−20.0
3	30		
4	23		
5	18	(16 − 23)/2	−3.5
6	16		
7	13		
8	12		

b. Substitute and solve for n.

$$n = \frac{\log(20.0) - \log(3.5)}{\log(42.0) - \log(18.0)} = 2.06 \qquad \text{use } n = 2$$

c. The reaction is second-order
d. The reaction rate constant is

$$\frac{1}{C} - \frac{1}{C_o} = kt$$

$$\frac{1}{42} - \frac{1}{250} = k(2)$$

$$k = 0.0103/\text{mole·}d \quad \text{use } k = 0.010/\text{mole·}d$$

In the applications described above, the initial concentration of a constituent is generally known. However, in the conventional BOD test, described previously in Chap. 2, both UBOD and k_1 are unknown. To determine these values, the usual procedure is to run a series of BOD measurements with time. Using these measurements, the UBOD and k_1 values can be determined using a number of methods including the method of least squares, the method of moments, the daily-difference method, the rapid-ratio method, the Thomas method, and the Fujimoto method, as discussed in Sec. 2–6. The application of the least squares and the Fujimoto methods is illustrated in Example 2–9 in Chap. 2.

4-7 MODELING TREATMENT PROCESS KINETICS

In wastewater treatment, the chemical and biological reactions that are needed to bring about the treatment of wastewater are carried out in the reactors described previously in Sec. 4–1. Treatment process kinetics involves the coupling of reactors and reaction rates to determine treatment process performance. In this section, the focus is on modeling the reactions that occur in the reactors used for wastewater treatment. The reactors considered include (1) batch, (2) complete-mix, (3) complete-mix reactors in series, (4) ideal plug-flow, (5) ideal plug-flow with retarded reaction rate, and (6) plug-flow with axial dispersion.

Batch Reactor with Reaction

The derivation of the materials-balance equation for a batch reactor (see Fig. 4–2a) for a reactive constituent is written as follows:

Accumulation = inflow − outflow + generation

$$\frac{dC}{dt}V = QC_o - QC + r_cV \tag{4-89}$$

Because $Q = 0$ the resulting equation for a batch reactor is

$$\frac{dC}{dt} = r_C \tag{4-90}$$

Before proceeding further, it will be instructive to explore the difference between the rate of change term that appears as part of the accumulation term and the rate of generation or utilization or decay term. In general, these terms are not equal, except in the special case of a batch reactor in which there is no inflow or outflow from the control volume. The key point to remember is that when flow is not occurring, the concentration per unit volume is changing according to the applicable rate expression. On the other hand, when flow is occurring, the concentration in the reactor is also being modified by the inflow and outflow from the reactor.

If the rate of reaction is defined as first-order (i.e., $r_c = -kC$), integrating between the limits $C = C_o$ and $C = C$ and $t = 0$ and $t = t$ yields

$$\int_{C=C_o}^{C=C} \frac{dC}{C} = -k \int_{t=0}^{t=t} dt = kt \tag{4-91}$$

The resulting expression is

$$\frac{C}{C_o} = e^{-kt} \tag{4-92}$$

Equation (4–92) is the same as the BOD equation [Eq. (2–59)] considered previously in Chap. 2.

Complete-Mix Reactor with Reaction

The general form of the mass-balance equation for a complete-mix reactor as shown on Figs. 4–2a and 4–4, in which the liquid in the reactor is mixed completely, is given below.

Accumulation = inflow − outflow + generation

$$\frac{dC}{dt}V = QC_o - QC + r_cV \tag{4-93}$$

Assuming first-order removal kinetics ($r_c = -kC$), Eq. (4–93) can be rearranged and written as follows:

$$C' + \beta C = \frac{Q}{V}C_o \tag{4-94}$$

where $C' = dC/dt$
$\beta = k + Q/V$

To solve Eq. (4–94) both sides of the expression are multiplied by the integrating factor $e^{\beta t}$:

$$e^{\beta t}(C' + \beta C) = \frac{Q}{V}C_o e^{\beta t} \tag{4-95}$$

The left-hand side of the above expression can be written as a differential as follows:

$$(Ce^{\beta t})' = \frac{Q}{V}C_o e^{\beta t} \tag{4-96}$$

The differential sign is removed by integrating the above expression

$$Ce^{\beta t} = \frac{Q}{V}C_o \int^{e^{\beta t}} dt \tag{4-97}$$

Integration of Eq. (4–97) yields

$$Ce^{\beta t} = \frac{Q}{V}\frac{C_o}{\beta}e^{\beta t} + K \tag{4-98}$$

Dividing by $e^{\beta t}$ yields

$$C = \frac{Q}{V}\frac{C_o}{\beta} + Ke^{-\beta t} \tag{4-99}$$

But when $t = 0$, $C = C_o$ and K is equal to

$$K = C_o - \frac{Q}{V}\frac{C_o}{\beta} \tag{4-100}$$

Substituting for K in Eq. (4–99) and simplifying yields the following expression, which is the non-steady-state solution of Eq. (4–93):

$$C = \frac{Q}{V}\frac{C_o}{\beta}(1 - e^{-\beta t}) + C_o e^{-\beta t} \tag{4-101}$$

The solution to Eq. (4–93) under steady-state conditions [i.e., the rate accumulation term is equal to zero $(dC/dt = 0)$] is given below.

$$C = \frac{C_o}{[1 + k(V/Q)]} = \frac{C_o}{[1 + k\tau]} \tag{4-102}$$

It should also be noted that when $t \to \infty$, Eq. (4–101) becomes the same as Eq. (4–102).

Complete-Mix Reactors in Series with Reaction

When complete-mix reactors are used in series, the steady-state solution is of concern as it is used for design. Two approaches are presented for the analysis of reactors in series: (1) analytical and (2) graphical. The graphical approach also applies to cascades of reactors, as discussed in Sec. 4–5 when dealing with mass-transfer equilibria.

Analytical Solution. The steady-state form of the mass balance for the second reactor of the two-reactor system (see Fig. 4–15), is given by

Accumulation = inflow − outflow + generation

$$\frac{dC_2}{dt}\frac{V}{2} = 0 = QC_1 - QC_2 + r_c\frac{V}{2} \tag{4-103}$$

Assuming first-order removal kinetics ($r_c = -kC_2$), Eq. (4–103) can be rearranged and solved for C_2, yielding

$$C_2 = \frac{C_1}{[1 + (kV/2Q)]} \tag{4-104}$$

But from Eq. (4–102), the value of C_1 is equal to

$$C_1 = \frac{C_o}{[1 + (kV/2Q)]} \tag{4-105}$$

Combining the above two expressions yields

$$C_2 = \frac{C_o}{[1 + (kV/2Q)]^2} \tag{4-106}$$

For n reactors in series the corresponding expression is

$$C_n = \frac{C_o}{[1 + (kV/nQ)]^n} = \frac{C_o}{[1 + (k/n\tau)]^n} \tag{4-107}$$

For example, consider three 1000 m^3 complete-mix reactors in series with a flowrate of 100 m^3/d and first-order kinetics with a reaction rate coefficient value of $k = 0.1$/d. Using Eq. (4–107), the effluent concentration from the third reactor, assuming the starting concentration was 100 mg/L, is

$$C_3 = \frac{C_o}{[1 + (kV/3Q)]^3} = \frac{100}{\left\{1 + \left[\dfrac{(0.1/\text{d})(3000\ \text{m}^3)}{(3 \times 100\ \text{m}^3/\text{d})}\right]\right\}^3} = 12.5\ \text{mg/L}$$

Solving Eq. (4–107) for the detention time yields

$$t_o = \frac{V}{Q} = \left[\frac{1}{(C_n/C_o)^{1/n}} - 1 \right]\left(\frac{n}{k}\right) \quad \text{or} \quad t_o = \left[\left(\frac{C_o}{C_n}\right)^{1/n} - 1 \right]\left(\frac{n}{k}\right) \tag{4–108}$$

Graphical Solution. The graphical solution for 3 (or for n) reactors in series is obtained as follows. For a single reactor, Eq. (4–93) can be written as follows:

Accumulation = inflow − outflow + generation

$$0 = QC_o - QC_1 - r_c V \tag{4–109}$$

The first step in developing a graphical solution is to draw a graph of r_c versus C (see Fig. 4–22). To plot r_c versus C, Eq. (4–109) is now rewritten as follows:

$$r_c = -\frac{Q}{V}(C_1 - C_o) = -\frac{1}{\tau}(C_1 - C_o) \tag{4–110}$$

The above equation can be represented graphically by a straight line drawn from the point $r_c = 0$ and $C = 100$ mg/L with a slope of $-1/\tau$. The line drawn will intersect the graph of r_c versus C at $r_c = 5.0$ and $C_1 = 50$ mg/L as shown on Fig. 4–22. The value $C_1 = 50$ mg/L is the solution of Eq. (4–110) for a single reactor for the stated conditions used to derive the analytical solution, presented previously. If the procedure is repeated for a second and a third reactor, the final effluent concentration from the third reactor is found to be 12.5 mg/L, which is the same as the analytical solution determined above. The graphical approach is especially useful in solving phase-separation processes as described in Sec. 4–8. To use the graphical procedure, the reaction rate coefficient must be a function of a single variable (e.g., C). Additional details on the graphical solution of design equations may be found in Eldridge and Piret (1950) and Smith (1981).

Figure 4–22

Graphical analysis used to determine the effluent concentration from a series of complete-mix reactors.

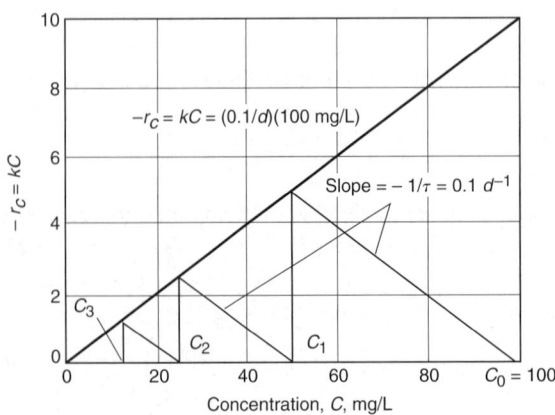

EXAMPLE 4–6 **Analysis of Reactors in Series Using Both an Analytical and Graphical Approach** Two 1000 m³ complete-mix reactors are to be used in series with a flowrate of 500 m³/d and second-order kinetics with a k value of 0.01/mg·d. Determine the effluent concentration from the second reactor assuming the starting concentration is 100 mg/L.

Solution

1. Determine the effluent concentration from the series of two complete-mix reactors analytically.
   a. At steady state, the mass balance for the first complete-mix reactor is

   $$0 = QC_o - QC_1 - kC_1^2V$$

   Substituting the given values and solving for C_1 yields

   $$0 = \frac{(500 \text{ m}^3/\text{d})}{1000 \text{ m}^3}(100 \text{ mg/L}) - \frac{(500 \text{ m}^3/\text{d})}{1000 \text{ m}^3}C_1 - (0.01/\text{mg·d})C_1^2$$

   $$C_1 = 50 \text{ mg/L}$$

   b. At steady state, the mass balance for the second complete-mix reactor is

   $$0 = QC_1 - QC_2 - kC_2^2V$$

   Substituting the given values and solving for C_2 yields

   $$0 = \frac{(500 \text{ m}^3/\text{d})}{1000 \text{ m}^3}(50 \text{ mg/L}) - \frac{(500 \text{ m}^3/\text{d})}{1000 \text{ m}^3}C_2 - (0.01/\text{mg·d})C_2^2$$

   $$C_2 = 30 \text{ mg/L}$$

2. Determine the effluent concentration from the series of two complete-mix reactors graphically.
   a. Prepare a plot of r_c versus C as shown below.

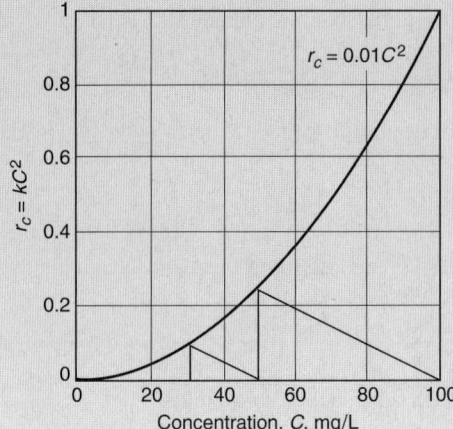

b. Linearize the mass-balance equation from Step 1a

$$r_c = -\frac{Q}{V}(C_1 - C_o)$$

c. On the plot prepared above draw a straight line from the point $r_c = 0$ and $C = 100$ mg/L with a slope of $-Q/V$ equal to $-0.5/d$ [$-(500\ m^3/d)/1000\ m^3$]. The line drawn intersects the graph of r_c versus C at $r_c = 0.25$ and $C_1 = 50$ mg/L. Repeating the above procedure the final effluent concentration from the second reactor is found to be 30 mg/L, which is the same as the analytical solution determined in Step 1.

Ideal Plug-Flow Reactor with Reaction

The derivation of the materials-balance equation for an ideal plug-flow reactor, in which the concentration C of the constituent is uniformly distributed across the cross-sectional area of the control volume and there is no longitudinal dispersion, can be illustrated by considering the differential volume element shown on Fig. 4–5. For the differential volume element ΔV shown on Fig. 4–5, the materials balance on a reactive constituent C is written as follows:

Accumulation = inflow − outflow + generation

$$\frac{\partial C}{\partial t}\Delta V = QC|_x - QC|_{x+\Delta x} + r_c\Delta V \tag{4-111}$$

Accumulation = inflow − outflow + generation

where $\partial C/\partial t$ = change in average concentration with time, $ML^{-3}T^{-1}$ (g/m³·s)
C = constituent concentration, ML^{-3} (g/m³)
ΔV = differential volume element, L^3 (m³)
Q = volumetric flow rate, L^3T^{-1} (m³/s)
r_c = reaction rate for constituent C, $ML^{-3}T^{-1}$, (g/m³·s)

Substituting the differential form for the term $QC|_{x+\Delta x}$ in Eq. (4–111) results in

$$\frac{\partial C}{\partial t}\Delta V = QC - Q\left(C + \frac{\Delta C}{\Delta x}\Delta x\right) + r_c\Delta V \tag{4-112}$$

Substituting $A\Delta x$ for ΔV and dividing by A and Δx yields

$$\frac{\partial C}{\partial t} = -\frac{Q}{A}\frac{\Delta C}{\Delta x} + r_C \tag{4-113}$$

Taking the limit as Δx approaches zero yields

$$\frac{\partial C}{\partial t} = -\frac{Q}{A}\frac{\partial C}{\partial x} + r_C \tag{4-114}$$

If steady-state conditions are assumed ($\partial C/\partial t = 0$) and the rate of reaction is defined as $r_c = -kC^n$, integrating between the limits $C = C_o$ and $C = C$ and $x = 0$ and $x = L$ yields

$$\int_{C_o}^{C} \frac{dC}{C^n} = -k\frac{A}{Q}\int_{0}^{L} dx = -k\frac{AL}{Q} = -k\frac{V}{Q} = -k\tau \qquad (4\text{--}115)$$

Equation (4–115) is the steady-state solution to the materials-balance equation for a plug-flow reactor without dispersion. If it is assumed that n is equal to 1, Eq. (4–115) becomes

$$\frac{C}{C_o} = e^{-k\tau} \qquad (4\text{--}116)$$

which is equivalent to Eq. (4–92), derived previously for the batch reactor.

Comparison of Complete-Mix and Plug-Flow Reactors with Reaction

The combined effect of reactor type (e.g., complete-mix versus plug-flow) and kinetics is also of interest. The total volume required for various removal efficiencies for first-order kinetics, using 1, 2, 4, 6, 8, or 10 reactors in series, is reported in Table 4–8 and shown graphically on Fig. 4–23. The corresponding volume required for a plug-flow reactor is also reported in Table 4–8. As shown in Table 4–8, as the number of reactors in series is increased, the total reactor volume required approaches that of a plug-flow reactor. A comparison of reactor types for second-order kinetics is examined in Example 4–7.

It should be noted, however, that for zero-order kinetics the volume of the two reactors will be the same. It is also important to note that biological processes do not obey the results presented in Table 4–8 (i.e., plug-flow is more efficient than complete-mix) because biological processes are modeled using BOD and COD, which includes

Table 4–8
Required reactor volumes expressed in terms of Q/k for complete-mix reactors in series and plug-flow reactors for various removal efficiencies for first-order kinetics[a]

No. of reactors in series	K values where V = K(Q/k)			
	85% removal efficiency	90% removal efficiency	95% removal efficiency	98% removal efficiency
1	5.67	9.00	19.00	49.00
2	3.16	4.32	6.94	12.14
4	2.43	3.11	4.46	6.64
6	2.23	2.81	3.89	5.52
8	2.14	2.67	3.63	5.05
10	2.09	2.59	3.49	4.79
Plug flow	1.90	2.30	3.00	3.91

[a] Volume of individual reactors equals value in table divided by the number of reactors in series.

Figure 4-23

Definition sketch for the total volume required versus the number of complete-mix reactors in series for various removal efficiencies. The K value on the vertical axis is multiplied by the flowrate and divided by the reaction coefficient to obtain the total volume required. The volume of an individual reactor is equal to the total volume divided by the number of reactors in series.

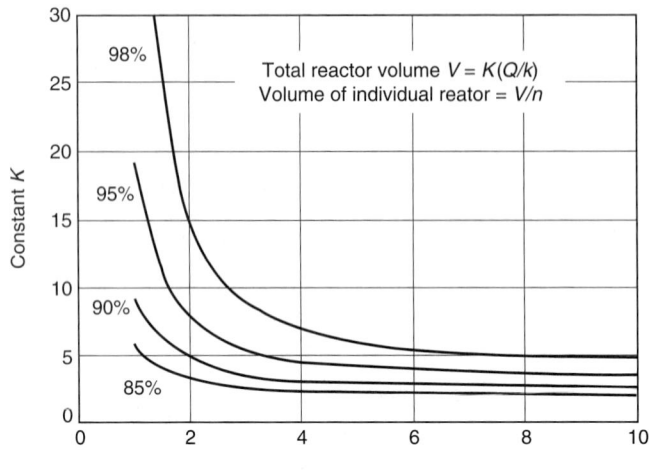

microbial products in addition to any residual substrate. As a result, the volumes required for the two reactors will be the same. The use of a plug-flow reactor, or mixed cells in series, is often favored to help control the growth of filamentous organisms (see discussion in Chap. 7).

EXAMPLE 4-7 **Comparison of Required Reactor Volumes for Second-Order Kinetics**
Assuming that second-order kinetics apply ($r_c = -kC^2$), compare the required volume of a complete-mix reactor to the volume of a plug-flow reactor to achieve a 90 percent reduction in the concentration ($C_o = 1$ and $C_e = 0.1$).

Solution

1. Compute the required volume for a complete-mix reactor in terms of Q/k.
   a. At steady state, a mass balance for a complete-mix reactor yields

   $$0 = QC_o - QC_e - kC_e^2 V$$

   b. Simplify and substitute the given data.

   $$V = \frac{Q}{k}\left(\frac{C_o - C_e}{C_e^2}\right) = \frac{Q}{k}\frac{1 - 0.1}{(0.1)^2} = 90\frac{Q}{k}$$

2. Compute the required volume for a plug-flow reactor in terms of Q/k.
   a. At steady state, a mass balance for a plug-flow reactor yields

   $$0 = -Q\frac{dC}{dx}dx + A\,dx(-kC^2)$$

b. The integrated form of the steady-state equation is

$$V = \frac{Q}{k}\int_{C_o}^{C_e}\frac{dC}{C^2} = \frac{Q}{k}\frac{1}{C}\bigg|_{C_o}^{C_e} = \frac{Q}{k}\left(\frac{1}{C_e} - \frac{1}{C_o}\right)$$

c. Substituting the given concentration values yields

$$V = \frac{Q}{k}\left(\frac{1}{0.1} - \frac{1}{1}\right) = \frac{9Q}{k}$$

3. Determine the volume ratio.

$$\frac{V_{CMR}}{V_{PFR}} = \frac{(90\,Q/k)}{(9\,Q/k)} = 10$$

Ideal Plug-Flow Reactor with Retarded Reaction

A retarded rate expression is used when the rate constant is changing with distance or time. To illustrate the concept, consider the removal of total suspended solids in an overland flow-treatment system. As shown on Fig. 4–24, the removal rate constant for the original particle size distribution is k_1. If it is assumed that the largest particle size will be removed after the wastewater has passed a unit distance, then the new removal rate constant for the remaining particle size distribution will be k_2. Based on actual observations, it has been found that k_1 is greater than k_2, and that k_2 will be greater than k_3, and so on. It is interesting to note that the same analysis can be applied to the settling of the particulate matter found in wastewater. The same argument can also be made for the removal of BOD comprised of several different organic compounds, each having a different reaction rate coefficient (Tchobanoglous et al., 2000).

Figure 4–24

Definition sketch to illustrate the change that can occur in the removal rate coefficient with treatment for an influent wastewater with a particle size distribution (Tchobanoglous, 1969).

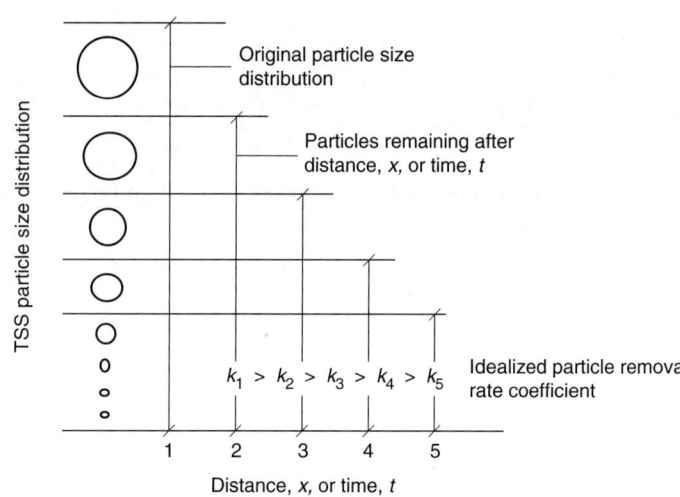

The expression for a plug-flow reactor in which a retarded first-order reaction is occurring can be written as follows:

$$\int_{C_o}^{C} \frac{dC}{C} = -\int_0^t \frac{k}{(1 + r_t t)^n} dt \tag{4-117}$$

The integrated forms of Eq. (4–114) for $n = 1$ (all of the particles are of the same size and composition) and $n \neq 1$ are as given below

$$C = C_o \exp\left[-\frac{k}{r_t} \ln (1 + r_t t)\right] \qquad \text{(for } n = 1) \tag{4-118}$$

$$C = C_o \exp\left\{-\frac{k}{r_t(n-1)}\left[1 - \frac{1}{(1 + r_t t)^{n-1}}\right]\right\} \qquad \text{(for } n \neq 1) \tag{4-119}$$

EXAMPLE 4–8   **Analysis of Impact of Reaction Rate Retardation.**   Determine the effluent concentration from a treatment system, designed as an ideal plug-flow reactor, assuming the removal of the constituent in question can be described by a retarded first-order equation. Assume the exponent in the retardation term is equal to 1.5. If the value of the retardation coefficient is 0.2, compare the effluent concentration with and without retardation after 2 days. Determine the retarded effluent concentration if the value of n is equal to 1.0.

Solution

1.   Determine the effluent concentration using Eq. (4–119) for the given conditions.

$$C = C_o \exp\left[-\frac{k}{r_t(n-1)}\left(1 - \frac{1}{(1 + r_t t)^{n-1}}\right)\right]$$

where $C_o = 1.0$ (assumed)
$\quad\quad k = 0.3/\text{d}$
$\quad\quad r_t = 0.8/\text{d}$
$\quad\quad n = 1.5$ (unitless)
$\quad\quad t = 2.0$ d

$$C = 1.0 \exp\left[-\frac{0.3}{0.8(1.5 - 1)}\left(1 - \frac{1}{(1 + 0.8 \times 2.0)^{1.5-1}}\right)\right] = 0.75$$

2.   Determine the effluent concentration without retardation using Eq. (4–116).

$$\frac{C}{C_o} = e^{-kt}$$

where $C_o = 1.0$ (assumed)
$\quad\quad k = 0.3/\text{d}$
$\quad\quad t = 2.0$ d

$$C_{(\text{unretarded})} = 1.0\, e^{-0.3 \times 2.0} = 0.55$$

3. Determine the effluent concentration for a wastewater with an n value of 1.0 using Eq. (4–118).

$$C = C_o \exp \left[-\frac{k}{r_t} \ln (1 + r_t t) \right]$$

where $C_o = 1.0$ (assumed)
$k = 0.3/d$
$r_t = 0.8/d$
$t = 2.0 \text{ d}$

$$C = 1.0 \exp \left[-\frac{0.3}{0.8} \ln (1 + 0.8 \times 2.0) \right] = 0.69$$

Comment From the above computations it can be seen that the effect of retardation can be significant for a first-order reaction. If the value of n were equal to 1, the corresponding effluent concentration would be equal to 0.54. In this example, the retardation coefficient has the greatest impact. The impact of retardation is not as great for second-order reactions. By assuming that an unretarded removal rate coefficient applies throughout, the performance of the treatment system would be overestimated.

Plug-Flow Reactor with Axial Dispersion and Reaction

In most full-scale plug-flow reactors, the flow usually is nonideal because of entrance and exit flow disturbances and axial dispersion. Depending on the magnitude of these effects, the ideal effluent-tracer curves may look like the curves shown on Fig. 4–25. Using first-order removal kinetics, Wehner and Wilhelm (1958) developed a solution for a plug-flow reactor with dispersion numbers varying from complete-mix ($d = \infty$) to ideal plug-flow ($d = 0$). The equation developed by Wehner and Wilhelm is as follows:

$$\frac{C}{C_o} = \frac{4a \, \exp(1/2d)}{(1 + a)^2 \exp(a/2d) - (1 - a)^2 \exp(-a/2d)} \tag{4-120}$$

Figure 4–25

Theoretical and generalized nonideal response curves for a reactor with axial dispersion.

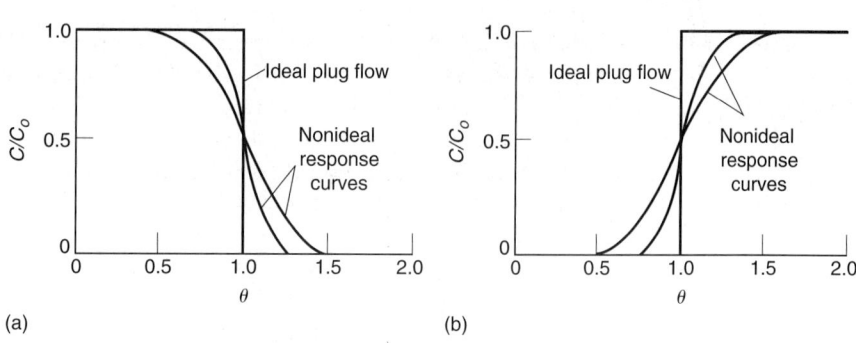

(a) (b)

Figure 4-26

Values of $k\tau$ in the Wehner and Wilhelm equation [Eq. (4–100)] versus percent remaining for various dispersion factors and first-order kinetics. (After Thirumurthi, 1969.)

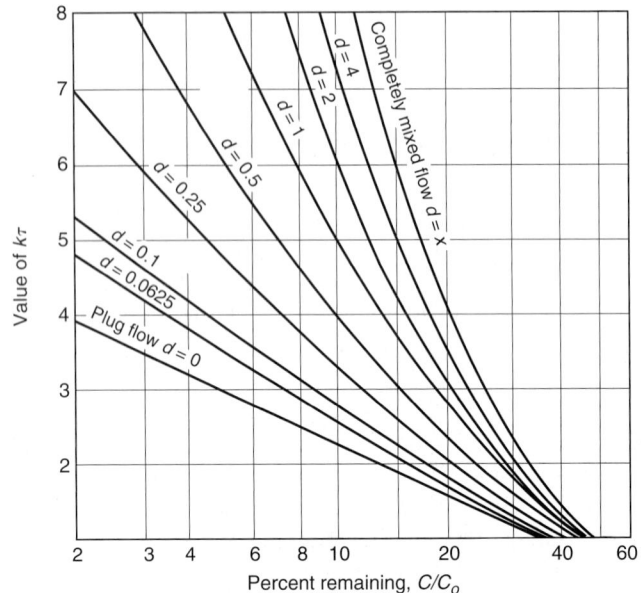

where C = effluent concentration, ML^{-3}
C_o = influent concentration, ML^{-3}
$a = \sqrt{1 + 4k\tau d}$
d = dispersion factor = D/uL [see Eq. (4–40)]
k = first-order reaction constant, T^{-1}, (1/h)
τ = hydraulic detention time V/Q, T, (h)

To facilitate the use of Eq. (4–120) for the design of treatment processes such as stabilization ponds and natural systems, Thirumurthi (1969) developed Fig. 4–26, in which the term $k\tau$ is plotted against C/C_o for dispersion factors varying from zero for an ideal plug-flow reactor to infinity for a complete-mix reactor. The application of Fig. 4–26 is illustrated in Example 4–9.

EXAMPLE 4–9   **Comparison of the Performance of a Treatment Process Occurring in a Plug-Flow Reactor without and with Axial Dispersion**   A treatment process reactor was designed assuming ideal plug flow with a first-order BOD removal rate constant of 0.5/d at 20°C and a detention time of 5 d. Once in operation, a considerable amount of axial dispersion was observed in the reactor. What effect will the observed axial dispersion have on the performance of the treatment process? The dispersion factor for the reactor d has been estimated to be about 0.5. Determine how much longer the detention time must be for a reactor with a dispersion factor of 0.5 to achieve the same degree of treatment as expected initially with the ideal plug-flow reactor.

Solution

1. Estimate the percentage removal for an ideal plug-flow reactor using Eq. (4–116).
   a. The BOD remaining is

   $$\frac{C}{C_o} = e^{-k\tau}$$

   $$\frac{C}{C_o} = e^{-0.5\times5} = 0.082 = 8.2\%$$

   b. The percentage removal is

   Percentage removal $100 - 8.2 = 91.8\%$

2. Determine the percentage removal for the reactor using Fig. 4–26.
   a. The value of $k\tau$ equals

   $$k\tau = (0.5/d \times 5\ d) = 2.5$$

   b. The percent remaining from Fig. 4–26 is equal to

   $$C/C_o = 0.20 = 20\%$$

   Percentage removal $100 - 20 = 80.0\%$

3. Determine the required detention time to achieve 91.8 percent removal.
   a. The value of $k\tau$ from Fig. 4–26 for a C/C_o value of 8.2% is 4.6.
   b. The required detention time is

   $$k\tau = 4.6$$

   $$\tau = 4.6/0.5 = 9.2\ d$$

Comment Clearly, axial dispersion can affect the predicted performance of a treatment process designed to function as an ideal plug-flow reactor. Because of axial dispersion and temperature effects, the actual performance of the treatment process will generally be less than expected.

Other Reactor Flow Regimes and Reactor Combinations

In the previous discussions of complete-mix and plug-flow reactors, a single-pass straight-through flow pattern has been used for the purpose of analysis. In practice, other flow regimes and reactor combinations are also used. Some of the more common alternative flow regimes are shown schematically on Fig. 4–27. The flow regime shown on Fig. 4–27a is used to achieve intermediate levels of treatment by blending various amounts of treated and untreated wastewater. The flow regime used on Fig. 4–27b is often adopted to achieve greater process control and will be considered specifically in Chaps. 7 and 8, which deal with biological wastewater treatment. The flow regime shown on Fig. 4–27c is used to reduce the loading applied to the process. On Fig. 4–27d,

Figure 4–27

Flow regimes commonly used in the treatment of wastewater. (a) Direct input with bypass flow (plug-flow or complete-mix reactor), (b) direct input with recycle flow (plug-flow or complete-mix reactor), (c) stepped input with recycle (recycle flow mixed with influent, recycle type 1), and (d) stepped input with recycle (recycle flow introduced at influent end of reactor, recycle type 2).

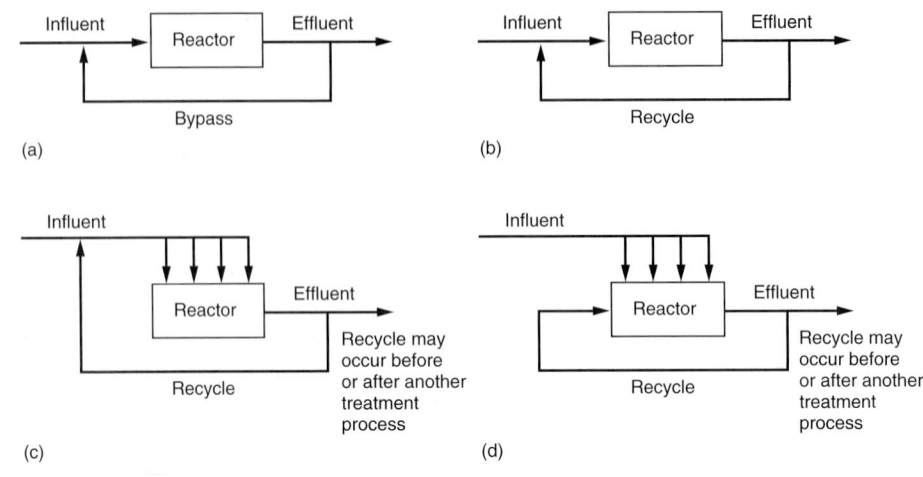

Figure 4–28

Hybrid reactor systems. (a) Plug-flow reactor followed by complete-mix reactor and (b) complete-mix reactor followed by plug-flow reactor.

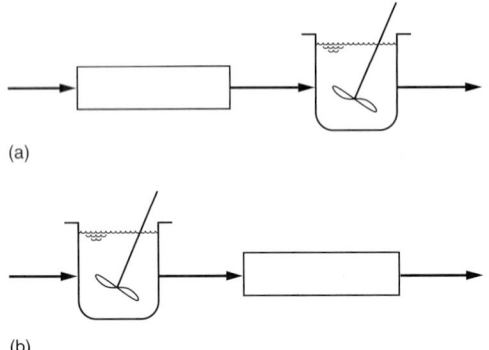

the return flow is not mixed with the influent but is introduced at the entrance of the reactor to achieve greater initial dilution of the wastewater to be treated. Each of these hydraulic regimes is considered further in the following chapters.

Among the numerous types of reactor combinations that are possible and that have been used, two combinations using a plug-flow reactor and a continuous-flow stirred-tank reactor are shown on Fig. 4–28. In the arrangement shown on Fig. 4–28a, complete mixing takes place second; in the arrangement shown on Fig. 4–28b, it occurs first. If no reaction takes place and the reactors are used only to equalize temperature, for example, the result will be identical. If a reaction is occurring, however, the product yields of the two reactor systems can be different. The use of such hybrid reactor systems will depend on the specific product requirements. Additional details on the analysis of such processes may be found in Denbigh and Turner (1965), Kramer and Westererp (1963), and Levenspiel (1972).

4–8 TREATMENT PROCESSES INVOLVING MASS TRANSFER

In the previous section, the removal of constituents was accomplished using chemical and/or biological conversion reactions. In this section, a group of separation operations based on the transfer of material from one homogeneous phase to another is considered. The operations are designed to reduce the concentration of a given component in one stream and to increase the concentration in another stream. Unlike mechanical separations, the driving force for the transfer of material is a pressure or concentration gradient (McCabe and Smith, 1976). The separation operations are sometimes identified as equilibrium phase separations or equilibrium contact separations, because the transfer of a component will cease when equilibrium conditions prevail (Treybal, 1980).

Common operations and processes in wastewater treatment involving mass transfer are identified in Table 4–9. The most important mass transfer operations in wastewater treatment involve (1) the transfer of material across gas-liquid interfaces as in aeration and in the removal of unwanted gaseous constituents found in wastewater by air stripping, and (2) the removal of unwanted constituents from wastewater by adsorption onto solid surfaces such as activated carbon and ion exchange. To introduce the concepts involved in mass transfer, the basic principle of mass transfer is reviewed followed by a consideration of gas-liquid and liquid-solid mass transfer operations.

Basic Principle of Mass Transfer

The transfer of mass is represented by Eq. (4–35), given previously, and repeated here for easy reference.

$$r = -D_m \frac{\partial C}{\partial x} \tag{4–35}$$

Table 4–9

Principal applications of mass transfer operations and processes in wastewater treatment[a]

Type of reactor	Phase equilibria	Application
Absorption	Gas → liquid	Addition of gases to water (e.g., O_2, O_3, CO_2, Cl_2, SO_2), NH_3 scrubbing in acid
Adsorption	Gas → solid	Removal of organics with activated carbon
	Liquid → solid	Removal of organics with activated carbon, dechlorination
Desorption	Solid → liquid	Sediment scrubbing
	Solid → gas	Reactivation of spent activated carbon
Drying (evaporation)	Liquid → gas	Drying of sludges
Gas stripping (also known as desorption)	Liquid → gas	Removal of gases (e.g., CO_2, O_2, H_2S, NH_3, volatile organic compounds, NH_3 from digester supernatant)
Ion exchange	Liquid → solid	Selective removal of chemical constituents, demineralization

[a]Adapted from Crittenden (1999), McCabe and Smith (1976), and Montgomery (1985).

where r = rate of mass transfer per unit area per unit time, $ML^{-2}T^{-1}$
D_m = coefficient of molecular diffusion in the x direction, L^2T^{-1}
C = concentration of constituent being transferred, ML^{-3}
x = distance, L

The coefficient of molecular diffusion is related to the frictional coefficient of a particle by Stokes-Einstein's law of diffusion. For spherical particles the coefficient of diffusion is given by the following expression (Shaw, 1966):

$$D = \frac{kT}{6\pi\eta r_p} = \frac{RT}{6\pi\eta r_p N} \tag{4-121}$$

where D = coefficient of diffusion, m^2/s
k = Boltzmann constant 1.3805×10^{-23} J/K
T = temperature, K = $273.15 + °C$
R = universal gas law constant, 8.3145 J/mol·K
η = dynamic viscosity, N·s/m^2
r_p = radius of particle, m
N = Avogadro's number, 6.02×10^{23} molecules/g-mol

The terms in the denominator in Eq. (4–121) correspond to the coefficient of friction for a particle as defined by Stokes' law. The coefficient of diffusion for a particle with a radius of 10^{-7} m (0.01 μm), which corresponds to the size of the smallest bacteria, for the following conditions is

$T = 20°C$

$\eta = 1.002 \times 10^{-3}$ N·s/m^2

$$D = \frac{RT}{6\pi\eta r_p A} = \frac{(8.3145 \text{ J/mol·K})[(273.15 + 20)\text{K}]}{6(3.14)(1.002 \times 10^{-3}\text{N·s/m}^2)(10^{-7}\text{m})(6.02 \times 10^{23}/\text{mol})}$$

$$= 21.43 \times 10^{-13} \text{ m}^2/\text{s} = 2.143 \times 10^{-8} \text{ cm}^2/\text{s}$$

From the above computation it is easy to see that as the particles get smaller the coefficient of molecular diffusion increases. Depending on the fluid regime, the coefficient of molecular diffusion in Eq. (4–121) will be replaced by the turbulent coefficient of diffusion or the coefficient of dispersion. Determination of the turbulent coefficient of diffusion and the coefficient of dispersion was considered previously in Sec. 4.5.

Typically, the concentration gradient arises from the difference in the equilibrium concentrations in the two streams (phases). In practice, the approach to equilibrium is brought about in a number of contact stages so that the concentration of the material being transferred changes progressively from one stream to another. Thus, two features common to all equilibrium contact operations are the attainment of, or approach to, equilibrium conditions and the provision of contact stages, which may occur in a continuous packed bed or in separated discrete stages.

Gas-Liquid Mass Transfer

Over the past 50 years a number of mass transfer theories have been proposed to explain the mechanism of gas transfer across gas-liquid interfaces. The simplest and most commonly used is the two-film theory proposed by Lewis and Whitman (1924). The penetration model proposed by Higbie (1935) and the surface-renewal model proposed by Danckwerts (1951) are more theoretical and take into account more of the physical phenomena involved. The two-film theory remains popular because, in more than 95 percent of the situations encountered, the results obtained are essentially the same as those obtained with the more complex theories. Even in the 5 percent where there is some disagreement between the two-film theory and other theories, it is not clear which approach is more correct. For these reasons, the two-film theory will be described in the following discussion.

The Two-Film Theory. The two-film theory is based on a physical model in which two films exist at the gas-liquid interface, as shown on Fig. 4–29. Two conditions are shown on Fig. 4–29: (*a*) "absorption," in which a gas is transferred from the gas phase to the liquid phase, and (*b*) "desorption," in which a gas is transferred out of the liquid phase into the gas phase. The two films, one liquid and one gas, provide the resistance to the passage of gas molecules between the bulk-liquid and the bulk-gaseous phases. It is very important to note that in the application of the two-film theory it is assumed that the concentration and partial pressure in both the bulk-liquid and bulk-gas phase are uniform (i.e., mixed completely).

Figure 4–29

Definition sketch for the two-film theory of gas transfer: (*a*) absorption and (*b*) desorption.

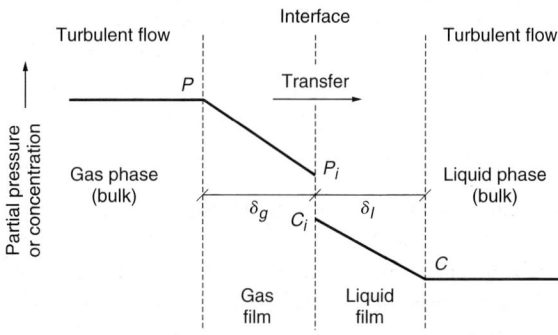

(a)

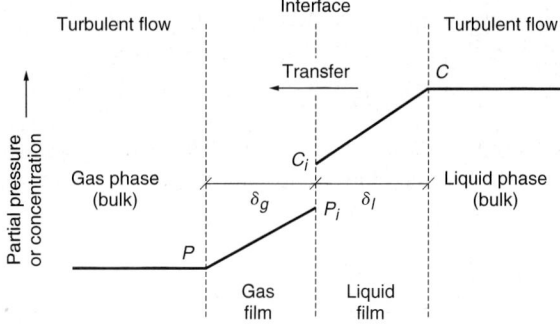

(b)

Under steady-state conditions, the rate of mass transfer of a gas through the gas film must be equal to the rate transfer through the liquid film. Using Fick's first law [Eq. (4–35)], the mass flux for each phase for absorption (gas addition) is written as follows:

$$r = k_G(P_G - P_i) = k_L(C_i - C_L) \tag{4-122}$$

where r = rate of mass transferred per unit area per unit time
  k_G = gas film mass transfer coefficient
  P_G = partial pressure of constituent A in the bulk of the gas phase
  P_i = partial pressure of constituent A at the interface in equilibrium with concentration C_i of constituent A in liquid
  k_L = liquid film mass transfer coefficient
  C_i = concentration of constituent A at the interface in equilibrium with partial pressure P_i of constituent A in the gas
  C_L = concentration of constituent A in the bulk liquid phase

It should be noted that the gas and liquid film mass transfer coefficients depend on the conditions at the interface. The terms $(P_G - P_i)$ and $(C_i - C_L)$ represent the driving force causing transfer in the gas and liquid phase, respectively. If the terms $(P_G - P_i)$ and $(C_i - C_L)$ are divided by their respective film thickness values (δ_G and δ_L), the driving force can be expressed in terms of unit thickness. Thus, the degree of mass transport can be enhanced by reducing the thickness of the film, depending on which is the controlling film.

However, because it is difficult to measure the values of k_G and k_L at the interface, it is common to use overall coefficients K_G and K_L, depending on whether the resistance to mass transfer is on the gas or liquid side. If it is assumed that all of the resistance to mass transfer is caused by the liquid film, then the rate of mass transfer can be defined as follows in terms of the overall liquid mass transfer coefficient:

$$r = K_L(C_s - C_L) \tag{4-123}$$

where r = rate of mass transferred per unit area per unit time
  K_L = overall liquid mass transfer coefficient
  C_s = concentration of constituent A at the interface in equilibrium with the partial pressure of constituent A in the bulk gas phase
  C_L = concentration of constituent A in the bulk liquid phase

If the two expressions given by Eqs (4–123) and (4–122) above are equated, the following relationship can be derived between the overall liquid mass transfer coefficient and the gas and liquid film coefficients:

$$r = K_L(C_s - C_L) = k_G(P_G - P_i) = k_L(C_i - C_L) \tag{4-124}$$

Because it was assumed that all of the resistance to mass transfer is caused by the liquid film, the following relationships, based on Henry's law (see Chap. 2), must apply at the interface:

$$P_G = HC_s \quad \text{and} \quad P_i = HC_i$$

It will now be noted that the overall driving force $(C_s - C_L)$ in Eq. (4–124) can be written as

$$(C_s - C_L) = (C_s - C_i) + (C_i - C_L) \tag{4-125}$$

Substituting for P_G and P_i in Eq. (4–145) and combining Eq. (4–125) and Eq. (4–124), the following relationship is obtained where the liquid film controls the mass transfer:

$$\frac{r}{K_L} = \frac{r}{k_L} + \frac{r}{Hk_G} \qquad \text{or} \qquad \frac{1}{K_L} = \frac{1}{k_L} + \frac{1}{Hk_G} \tag{4-126}$$

In a similar manner it can be shown that the following relationship holds if the transfer of mass is controlled by the gas film:

$$\frac{1}{K_G} = \frac{1}{k_G} + \frac{H}{k_L} \tag{4-127}$$

The relationship between the overall liquid and gas phase transfer coefficients is

$$\frac{1}{K_L} = \frac{1}{K_G H} \tag{4-128}$$

It should be noted that in Eqs. (4–126) and (4–127) the overall transfer coefficients include the resistance to mass transfer offered by both the gas and liquid phases. The fact that the overall resistance to mass transfer is the sum of the gas and liquid phase resistances was first demonstrated by Lewis and Whitman (1924). Referring to Eq. (4–126), it is interesting to note that if the Henry's constant is large, then the liquid phase resistance will typically control the mass transfer process. For the transfer of gas molecules from the gas phase to the liquid phase, slightly soluble gases (e.g., O_2, N_2, and CO_2 in water) encounter the primary resistance to transfer from the liquid film, and very soluble gases (e.g., NH_3 in water) encounter the primary resistance to transfer from the gaseous film. Gases of intermediate solubility (e.g., H_2S in water) encounter significant resistance from both films.

To estimate the flux of a slightly soluble gas from the gas to the liquid phase (liquid film controls transfer rate), Eq. (4–123) can be approximated by substituting C_t for C_L as follows:

$$r = K_L(C_s - C_t) \tag{4-129}$$

where r = rate of mass transferred per unit area per unit time, $ML^{-2}T^{-1}$
K_L = overall liquid mass transfer coefficient, LT^{-1}
C_t = concentration in liquid bulk phase at time t, ML^{-3}
C_s = concentration in equilibrium with gas as given by Henry's law, ML^{-3}

The corresponding rate of mass transfer per unit volume per unit time is obtained by multiplying Eq. (4–129) by the area A and dividing by the volume V.

$$r_v = K_L \frac{A}{V}(C_s - C_t) = K_L a(C_s - C_t) \tag{4-130}$$

where r_v = rate of mass transfer per unit volume per unit time, $ML^{-3}T^{-1}$
$K_L a$ = volumetric mass transfer coefficient, T^{-1}
A = area through which mass is transferred, L^2
V = volume in which constituent concentration is increasing, L^3
a = interfacial area for mass transfer per unit volume, A/V, L^{-1}

The term K_La, known as the volumetric mass transfer coefficient, depends on water quality and the type of aeration equipment and is unique for each situation. Numerical values for K_La are usually determined experimentally (see Chap. 5, Sec. 5–11). Equation (4–127) is the basic relationship used in solving problems involving the addition of oxygen to water as in aeration, the removal of volatile organics from wastewater by bubbling air through the wastewater, and for the stripping of dissolved constituents such as ammonia from digested supernatant.

Absorption of Gases.   The application of the gas-liquid mass transfer relationship developed above will be illustrated by considering the absorption of a gas in a liquid (see Fig. 4–30a). Consider, for example, a storage basin open to the atmosphere with surface area A and depth h. If the concentration of dissolved oxygen in the basin is initially undersaturation, how long would it take for the oxygen concentration to increase by a given amount? The approach to this mass transfer problem can be outlined as follows.

First, a mass balance is written for the open basin as follows:

1.   General word statement:

| Rate of accumulation of a gas within the system boundary | = | rate of flow of a gas into the system boundary | − | rate of flow of a gas out of the system boundary | + | amount of gas absorbed through system boundary | (4–131) |

2.   Simplified word statement:

$$\text{Accumulation} = \text{inflow} - \text{outflow} + \text{increase due to absorption} \qquad (4\text{–}132)$$

3.   Symbolic representation at equilibrium (refer to Fig. 4–29a):

$$\frac{dC}{dt}(V) = 0 - 0 + r_v V \qquad (4\text{–}133)$$

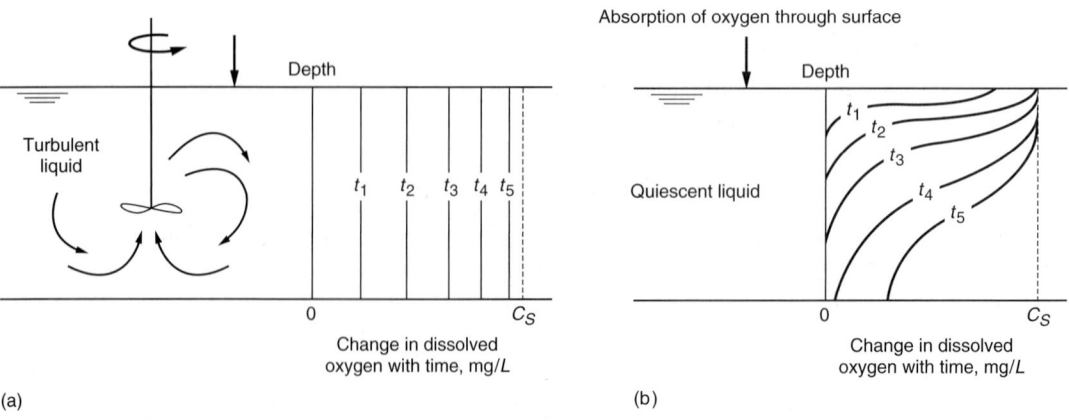

(a)   (b)

Figure 4–30

Definition sketch for the absorption of a gas: (a) under turbulent conditions where the concentration of gas in the gaseous and liquid phases is uniform and (b) under quiescent conditions. *(Adapted from Tchobanoglous and Schroeder, 1985.)*

where dC/dt = change in concentration with time, $ML^{-3}T^{-1}$, $(g/m^3 \cdot s)$

V = volume in which constituent concentration is increasing, L^3, (m^3)

r_v = mass of constituent transferred per unit volume per unit time, $ML^{-3}T^{-1}$, $(g/m^3 \cdot s)$

Using Eq. (4–130) to describe the mass transfer through the surface of the basin, Eq. (4–133) can be written as follows, which is the same as Eq. (4–130):

$$\frac{dC}{dt} = K_L a(C_s - C_t) \qquad (4\text{–}134)$$

Integrating Eq. (4–134) between the limits of $C = C_o$ and $C = C_t$ and $t = 0$ and $t = t$, where C_o is the initial concentration and C_t is the concentration at some time t

$$\int_{C_o}^{C_t} \frac{dC}{C_s - C_t} = K_L a \int_0^t dt \qquad (4\text{–}135)$$

yields

$$\frac{C_s - C_t}{C_s - C_o} = e^{-(K_L a)t} \qquad (4\text{–}136)$$

In Eq. (4–136), the term $(C_s - C_t)$ represents, as noted above, the degree of undersaturation at any time t and the term $(C_s - C_o)$ represents the initial degree of undersaturation. The application of Eq. (4–136) is illustrated in Example 4–10.

EXAMPLE 4–10 **Time Required to Absorb a Gas** Dechlorinated secondary effluent is placed in a storage basin until needed for reuse. If the initial dissolved oxygen concentration is 1.5 mg/L, estimate the time required for the dissolved oxygen concentration to increase to 8.5 mg/L due to surface reaeration, assuming the water in the storage basin is circulated and not stagnant. Assume the K_L value for oxygen is equal to 0.03 m/h. The surface area of the storage basin is 400 m^2 and the depth is 3 m. Assume the temperature is 20°C and that the saturation value estimated in Example 2–6 can be used.

Solution

1. The time required for the concentration of oxygen to be increased from 1.5 to 8.5 can be estimated using Eq. (4–136).

$$\frac{C_s - C_t}{C_s - C_o} = e^{-(K_L a)t}$$

a. The oxygen saturation value from Example 2–6 is 9.09 mg/L.
b. Solve for the term $(K_L a)t$.

$$\ln\left(\frac{9.09 - 8.5}{9.09 - 1.5}\right) = -2.55 = -(K_L a)t$$

2. The time required is
   a. The value for a, the interfacial area for mass transfer per unit volume

   $$a = A/V = 400 \text{ m}^2/10 \text{ m}^3 = 40 \text{ m}^2/\text{m}^3$$

   b. Solve for t.

   $$t = 2.55/[(0.03 \text{ m/h})(40 \text{ m}^2/\text{m}^3)] = 2.12 \text{ h}$$

Comment The importance of the surface area exposed to the atmosphere is illustrated in this example. The larger the surface area relative to the depth, the greater the rate of oxygen transfer. In biological wastewater treatment either many small gas bubbles, released at the bottom of a reactor, are used to transfer oxygen to the active biomass or small droplets of water containing the active biomass are sprayed into the atmosphere to maximize the rate of oxygen transfer.

Absorption of Gases under Quiescent Conditions. The transfer of a slightly soluble gas into a liquid under quiescent conditions (see Fig. 4–30b) occurs as a result of molecular diffusion. Applying the materials-balance approach to the control volume shown on Fig. 4–31, the quiescent transfer of a gas across the open surface can be modeled as follows:

First, a mass balance is written for the open basin as follows:

1. General word statement:

| Rate of accumulation of a gas within the system boundary | = | rate of diffusion of gas into the system boundary | − | rate of diffusion of gas out of the system boundary | (4-137) |

Figure 4-31

Definition sketch for the absorption of a gas under quiescent conditions due to molecular diffusion.

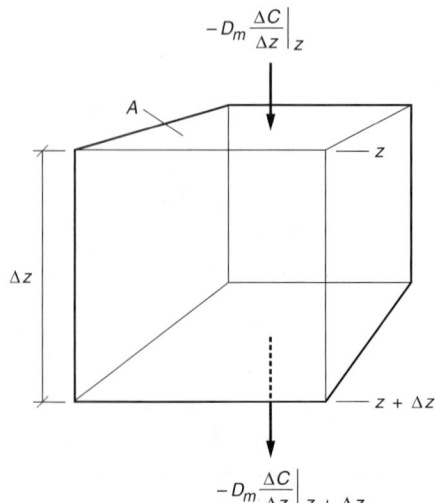

$$-D_m \frac{\Delta C}{\Delta z}\Big|_z$$

$$-D_m \frac{\Delta C}{\Delta z}\Big|_{z + \Delta z}$$

2. Simplified word statement:

$$\text{Accumulation} = \text{inflow} - \text{outflow} \qquad (4\text{–}138)$$

3. Symbolic representation at equilibrium (refer to Fig. 4–31):

$$\frac{\partial C}{\partial t}(A\Delta z) = -D_m A \left.\frac{\Delta C}{\Delta z}\right|_z + D_m A \left.\frac{\Delta C}{\Delta z}\right|_{z+\Delta z} \qquad (4\text{–}139)$$

where $\partial C/\partial t$ = change in concentration per unit time, $ML^{-3}T^{-1}$, (g/m³·s)
$\quad A$ = surface area through which mass is transferred, L^2, (m²)
$\quad \Delta z$ = distance in z direction, L, (m)
$\quad D_m$ = coefficient of molecular diffusion, L^2T^{-1}, (m²/s)
$\quad \Delta C/\Delta z$ = change in concentration with distance, $ML^{-3}L^{-1}$, (g/m³·s)

Taking the limit as Δz approaches zero yields

$$\frac{\partial C}{\partial t} = D_m A \frac{\partial^2 C}{\partial z^2} \qquad (4\text{–}140)$$

Equation (4–140) is also known as Fick's second law of diffusion (Crank, 1957). Typical values for the coefficient of molecular diffusion for gases of low solubility are reported in Table 4–10. Solutions to Eq. (4–140) for a variety of boundary conditions may be found in Carlslaw and Jaeger (1947), Crank (1957), Danckwertz (1970), and Thibodeaux (1996).

Desorption (Removal) of Gases. The application of the gas-liquid mass transfer relationship for the removal of a gas from a liquid will be illustrated by considering the volatilization of a constituent from a liquid. The same approach used for the addition of a gas will be followed, but noting that for the removal of a gas from a liquid Eq. (4–134) is written as follows:

$$\frac{dC}{dt} = -K_L a(C_t - C_s) \qquad (4\text{–}141)$$

where the term $(C_t - C_s)$ represents the degree of supersaturation at any time t. If Eq. (4–141) is integrated between the limits of $C = C_s$ and $C = C_t$ and $t = 0$ and $t = t$, the

Table 4–10
Approximate coefficients of molecular diffusion and coefficients of diffusion for gases of low solubility in water at 20°C[a]

Gas	Coefficient of molecular, diffusion, cm²/h	Gas transfer coefficient, cm/h	Estimated film thickness, cm
Oxygen, O_2	6.7×10^{-2}	$32.3 \times 1.018^{T-20}$	$\sim 2 \times 10^{-3}$
Nitrogen, N_2	6.4×10^{-2}	$34.0 \times 1.019^{T-20}$	$\sim 2 \times 10^{-3}$
Carbon dioxide, CO_2	$\sim 6.5 \times 10^{-2}$		$\sim 2 \times 10^{-3}$
Air		$32.1 \times 1.019^{T-20}$	$\sim 2 \times 10^{-3}$

[a]Adapted from Adeney and Becker (1919), Becker (1924).

integrated form of Eq. (4–141), corresponding to Eq. (4–136), for the volatilization of a gas from a supersatured liquid is given by

$$\frac{C_t - C_s}{C_o - C_s} = e^{-(K_t a)t} \tag{4-142}$$

In Eq. (4–142), the term $(C_o - C_s)$ represents the initial degree of supersaturation. The application of Eq. (4–141) is illustrated in Example 4–11.

EXAMPLE 4–11 **Time Required for a Gas to Volatilize from a Liquid** A quantity of benzene was spilled accidentally into a treated wastewater-storage basin. Estimate the time required for the concentration of benzene to drop by 50 percent from the initial concentration due to volatilization. Assume the $K_L a$ value for benzene is 0.144/h.

Solution

1. If it is assumed that the concentration of the specific volatile chemical is not common to the atmosphere, then $C_s \sim 0$, and Eq. (4–142) can be written as follows:

    $$\frac{C_t}{C_o} = e^{-(K_L a)t}$$

2. Knowing the value of $K_L a$, the time for the concentration to dissipate to one half of the initial concentration can be determined by rewriting the equation developed in Step 1 as

    $$\frac{0.5\,C_o}{1.0\,C_o} = e^{-(K_L a)t_{1/2}}$$

    Solving for $t_{1/2}$ yields

    $$t_{1/2} = \frac{0.69}{K_L a}$$

3. Using a $K_L a$ value of 0.144 h for benzene, the time for 50 percent of the initial concentration to dissipate is

    $$t_{1/2} = \frac{0.69 \times 2}{(0.144/\text{h})} = 9.6\ \text{h}$$

The application of gas-liquid mass transfer is considered further in the discussion of the aeration in Chap. 5, in the discussion of the aeration system in Chap. 9, in the discussion of gas stripping in Chap.11, and in the discussion of the control of volatile organic compounds (VOCs) in Chap 15.

Liquid-Solid Mass Transfer

In the discussion of gas-liquid mass transfer, it was found that mass could be transferred from either phase to the other phase. In liquid-solid mass transfer operations, constituents from the liquid phase are transferred (adsorbed) to a solid phase. Adsorption and ion exchange mass transfer processes are introduced in the following discussion. The application of liquid-solid mass transfer is considered further in the discussion of adsorption and ion exchange in Chap. 11.

Adsorption. The process of accumulating substances that are in solution on a suitable interface is termed *adsorption*. The *adsorbate* is the substance that is being removed from liquid phase at the interface. The *adsorbent* is the solid, liquid, or gas phase onto which the adsorbate accumulates. The adsorption process takes place in three steps: macrotransport, microtransport, and sorption. Macrotransport involves the movement of the adsorbate (e.g., organic matter) through the water to the liquid/solid interface by advection and diffusion. Microtransport involves the diffusion of the organic material through the macropore system of the solid adsorbent, such as activated carbon, to the adsorption sites in the micropores and the solid adsorbent (see Fig. 4–32). Although adsorption also occurs on the surface of the solid adsorbent and in the macropores and mesopores, the surface area of these parts of most solid adsorbents is so extremely small compared with the surface area of the micropores that the amount of material adsorbed there is usually considered negligible.

Two important characteristics of the solid adsorbent are (1) its extremely large surface area to volume ratio and (2) its preferential affinity for certain constituents in the liquid phase. The purpose of the following discussion is to introduce the basic concept of adsorption as a liquid-solid mass transfer operation. Adsorption, in particular carbon adsorption, is considered in detail in Chap. 11. Additional details on activated carbon adsorption may be found in Snoeyink and Summers (1999) and Sontheimer et al. (1988).

As noted above, adsorption is the process of accumulating substances that are in solution on a suitable interface. Granular or powdered activated carbon (GAC or PAC)

Figure 4–32

Macropore, mesopore, micropore, and submicropore adsorption sites on activated carbon.

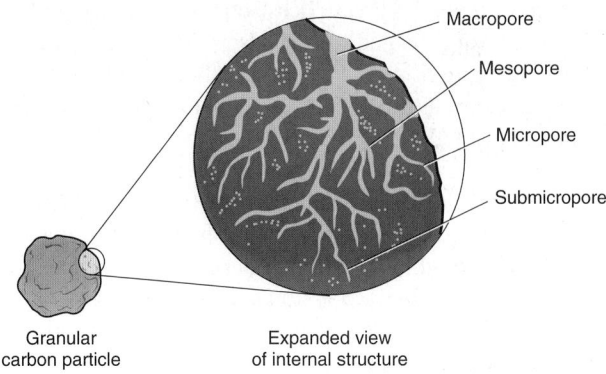

Figure 4–33

Typical plot of a
Freundlich adsorption
isotherm.

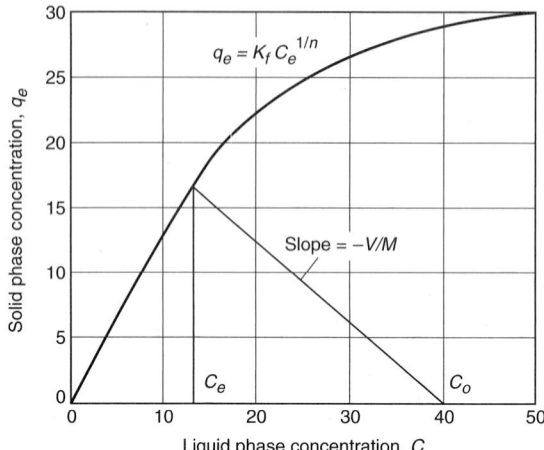

is used most commonly for the removal of selected constituents from wastewater. The accumulation of material is described by what is known as an adsorption isotherm, which is used to define the mass of material adsorbed per unit mass of adsorbing material. A common adsorption isotherm is the Freundlich isotherm. Derived from empirical considerations, the Freundlich isotherm is defined as follows:

$$q_e = \frac{x}{m} = K_f C_e^{1/n} \tag{4-143}$$

where q_e = mass of material adsorbed (x) per unit mass of adsorbent (m) at
equilibrium
K_f = Freundlich capacity factor
C_e = equilibrium concentration of adsorbate in liquid phase after adsorption,
mg/L
$1/n$ = Freundlich intensity parameter

A plot of a typical Freundlich isotherm is shown on Fig. 4–33. It should be noted that in the case of activated carbon adsorption, the expression governing the mass transfer is the adsorption isotherm. The coefficients in the Freundlich isotherm are determined by linearizing Eq. (4–143) and plotting the experimental results on loglog paper. Determination of the Freundlich isotherm coefficients is illustrated in Sec. 11–7 in Chap. 11.

The removal of a constituent in solution with powdered activated carbon (PAC) can be used to illustrate the adsorption process. If a mass balance is performed for a batch reactor into which a quantity of powdered activated carbon has been added (see Fig. 4–34), the resulting expression at equilibrium, at the completion of the mass transfer process, is given by

Figure 4–34

Definition sketch for mass-balance analysis of carbon adsorption.

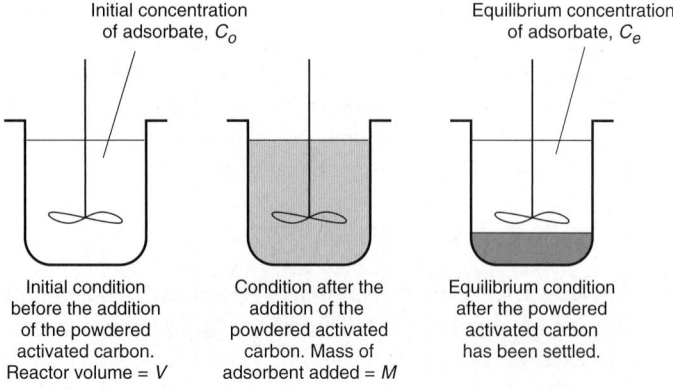

Initial concentration of adsorbate, C_o

Equilibrium concentration of adsorbate, C_e

Initial condition before the addition of the powdered activated carbon. Reactor volume = V

Condition after the addition of the powdered activated carbon. Mass of adsorbent added = M

Equilibrium condition after the powdered activated carbon has been settled.

1. General word statement:

$$\begin{array}{c} \text{Amount of reactant} \\ \text{adsorbed within the} \\ \text{system boundary} \end{array} = \begin{array}{c} \text{initial amount of} \\ \text{reactant within the} \\ \text{system boundary} \end{array} - \begin{array}{c} \text{final amount of} \\ \text{reactant within the} \\ \text{system boundary} \end{array} \quad (4\text{-}144)$$

2. Simplified word statement:

$$\begin{array}{c} \text{Amount} \\ \text{adsorbed} \end{array} = \begin{array}{c} \text{initial amount} \\ \text{of absorbate} \\ \text{present} \end{array} - \begin{array}{c} \text{final amount} \\ \text{of absorbate} \\ \text{present} \end{array} \quad (4\text{-}145)$$

3. Symbolic representation at equilibrium (refer to Fig. 4–34):

$$q_e M = VC_o - VC_e \qquad (4\text{-}146)$$

where q_e = adsorbent phase concentration after equilibrium, mg adsorbate/g adsorbent [see Eq. (4–143)]
M = mass of adsorbent, g
V = volume of liquid in the reactor, L
C_o = initial concentration of adsorbate, mg/L
C_e = final equilibrium concentration of adsorbate after absorption has occurred, mg/L

Equation (4–146) can be written as follows:

$$q_e = -\frac{V}{M}(C_e - C_o) \qquad (4\text{-}147)$$

Because the form of Eq. (4–147) is the same as Eq. (4–110), used previously to develop graphical solutions for reactor design equations, the same graphical approach can be applied to obtain the graphical solution of Eq. (4–147). If a straight line is drawn on the plot of the Freundlich isotherm, presented previously on Fig. 4–33, from the point $q_e = 0$ and $C = C_o$ mg/L with a slope of $-V/M$, as given by the above equation, the intersection of the line and plot of the adsorption isotherm corresponds to the final equilibrium concentration of organic compound that can be achieved with an activated carbon dose of V/M. The straight line on Fig. 4–33 is known as the *operating line* for the equilibrium adsorption of an organic constituent from solution (Sontheimer et al., 1988).

EXAMPLE 4–12   **Activated Carbon Required to Treat a Wastewater**   As a result of effluent chlorination, the amount of chloroform formed was found to be 0.12 mg/L. How much powdered activated carbon will be required to treat an effluent flowrate of 4000 m^3/d to reduce the chloroform concentration to 0.05 mg/L? The Freundlich adsorption isotherm coefficients for chloroform are: $K_f = 2.6$ and $1/n = 0.73$.

Solution

1. Combine Eqs. (4–143) and (4–147) to obtain an expression for V/M as follows:

$$q_e = \frac{x}{m} = K_f C_e^{1/n}$$

$$q_e = -\frac{V}{M}(C_e - C_o)$$

$$-\frac{V}{M} = \frac{K_f C_e^{1/n}}{(C_e - C_o)}$$

2. Substitute the isotherm coefficients and solve for M/V:

$$-\frac{V}{M} = \frac{K_f C_e^{1/n}}{(C_e - C_o)} = \frac{2.6(0.05)^{0.73}}{0.05 - 0.12} = -4.17 \text{ L/g}$$

$$M/V = 1/4.17 = 0.24 \text{ g/L}$$

3. Determine the amount of carbon required to treat 4000 m^3/d:

$$\text{PAC required} = \frac{(0.24 \text{ g/L})(4000 \text{ m}^3/\text{d})(10^3 \text{ L/m}^3)}{(10^3 \text{ g/kg})} = 960 \text{ kg/d}$$

Comment   Given the amount of PAC required to treat the effluent to reduce the residual chloroform to 0.05 mg/L, carbon adsorption is a poor choice for the removal of residual chloroform. Carbon adsorption is examined in greater detail in Sec. 11–7 in Chap. 11.

Ion Exchange.   Ion exchange is a mass transfer process in which ions of a given species are displaced from an insoluble exchange material by ions of a different species in solution. In water treatment, ion exchange is used most commonly to soften hard water through the removal of multivalent cations. In wastewater treatment, the principal uses are for (1) the removal of nitrogen and phosphorus, (2) the removal of heavy metals, and (3) the removal of total dissolved solids (i.e., demineralization) for reuse applications.

The reactions for the removal of sodium (Na^+) and calcium (Ca^{2+}) ions from water using a strong acid synthetic cationic exchange resin R, and the regeneration of the exhausted resins with hydrochloric acid (HCl) and sodium chloride (NaCl) are as follows:

Reaction:

$$R^-H^+ + Na^+ \rightarrow R^-Na^+ + H^+ \tag{4–148}$$

$$2\,R^-Na^+ + Ca^{2+} \rightarrow R_2^-Ca^{2+} + 2Na^+ \tag{4–149}$$

Regeneration:

$$R^-Na^+ + HCl \rightarrow R^-H^+ + NaCl \tag{4–150}$$

$$R_2^-Ca^{2+} + 2NaCl \rightarrow 2R^-Na^+ + CaCl_2 \tag{4–151}$$

As illustrated in the reaction given in Eq. (4–148), the Na^+ ion in solution is exchanged for the H^+ ion on the resin. Similarly, in Eq. (4–149) the Ca^{2+} in solution is exchanged for two Na^+ ions on the resin. To recover the capacity of the resin, the process is reversed during regeneration. The subject of ion exchange is considered in detail in Chap. 11, Sec. 11–9.

4–9 INTRODUCTION TO PROCESS SELECTION

Process selection involves the detailed evaluation of the various factors that must be considered when evaluating unit operations and processes and other treatment methods to meet current and future treatment objectives. The purpose of process analysis is to select the most suitable unit operations and processes and the optimum operational criteria. The purpose of this section is to introduce the important factors that must be considered in process selection and to consider the basis for process design.

Important Factors in Process Selection

The most important factors that must be evaluated in process analysis and selection are identified in Table 4–11. Each factor is important in its own right, but some factors require additional attention and explanation. The first factor, "process applicability," stands out above all others and reflects directly upon the skill and experience of the design engineer. Many resources are available to the designer to determine applicability, including past experience in similar types of projects. Available resources include performance data from operating installations, published information in technical journals, manuals of practice published by the Water Environment Federation, process design manuals published by EPA, and the results of pilot-plant studies. Where the applicability of a process to a given situation is unknown or uncertain, pilot-plant studies must be conducted to determine performance capabilities and to obtain design data upon which a full-scale design can be based.

Table 4-11
Important factors that must be considered when evaluating and selecting unit operations and processes

Factor	Comment
1. Process applicability	The applicability of a process is evaluated on the basis of past experience, data from full-scale plants, published data, and from pilot-plant studies. If new or unusual conditions are encountered, pilot-plant studies are essential
2. Applicable flow range	The process should be matched to the expected range of flowrates. For example, stabilization ponds are not suitable for extremely large flowrates in highly populated areas
3. Applicable flow variation	Most unit operations and processes have to be designed to operate over a wide range of flowrates. Most processes work best at a relatively constant flowrate. If the flow variation is too great, flow equalization may be necessary
4. Influent wastewater characteristics	The characteristics of the influent wastewater affect the types of processes to be used (e.g., chemical or biological) and the requirements for their proper operation
5. Inhibiting and unaffected constituents	What constituents are present and may be inhibitory to the treatment processes? What constituents are not affected during treatment?
6. Climatic constraints	Temperature affects the rate of reaction of most chemical and biological processes. Temperature may also affect the physical operation of the facilities. Warm temperatures may accelerate odor generation and also limit atmospheric dispersion
7. Process sizing based on reaction kinetics or process loading criteria	Reactor sizing is based on the governing reaction kinetics and kinetic coefficients. If kinetic expressions are not available, process loading criteria are used. Data for kinetic expressions and process loading criteria usually are derived from experience, published literature, and the results of pilot-plant studies
8. Process sizing based on mass transfer rates or process loading criteria	Reactor sizing is based on mass transfer coefficients. If mass transfer rates are not available, process loading criteria are used. Data for mass transfer coefficients and process loading criteria usually are derived from experience, published literature, and the results of pilot-plant studies
9. Performance	Performance is usually measured in terms of effluent quality and its variability, which must be consistent with the effluent discharge requirements
10. Treatment residuals	The types and amounts of solid, liquid, and gaseous residuals produced must be known or estimated. Often, pilot-plant studies are used to identify and quantify residuals
11. Sludge processing	Are there any constraints that would make sludge processing and disposal infeasible or expensive? How might recycle loads from sludge processing affect the liquid unit operations or processes? The selection of the sludge processing system should go hand in hand with the selection of the liquid treatment system
12. Environmental constraints	Environmental factors, such as prevailing winds and wind directions and proximity to residential areas, may restrict or affect the use of certain processes, especially where odors may be produced. Noise and traffic may affect selection of a plant site. Receiving waters may have special limitations, requiring the removal of specific constituents such as nutrients
13. Chemical requirements	What resources and what amounts must be committed for a long period of time for the successful operation of the unit operation or process? What effects might the addition of chemicals have on the characteristics of the treatment residuals and the cost of treatment?
14. Energy requirements	The energy requirements, as well as probable future energy cost, must be known if cost-effective treatment systems are to be designed

(continued)

Table 4-11 (Continued)	
Factor	**Comment**
15. Other resource requirements	What, if any, additional resources must be committed to the successful implementation of the proposed treatment system using the unit operation or process being considered?
16. Personnel requirements	How many people and what levels of skills are needed to operate the unit operation or process? Are these skills readily available? How much training will be required?
17. Operating and maintenance requirements	What special operating or maintenance requirements will need to be provided? What spare parts will be required and what will be their availability and cost?
18. Ancillary processes	What support processes are required? How do they affect the effluent quality, especially when they become inoperative?
19. Reliability	What is the long-term reliability of the unit operation or process being considered? Is the operation or process easily upset? Can it stand periodic shock loadings? If so, how do such occurrences affect the quality of the effluent?
20. Complexity	How complex is the process to operate under routine or emergency conditions? What levels of training must the operators have to operate the process?
21. Compatibility	Can the unit operation or process be used successfully with existing facilities? Can plant expansion be accomplished easily?
22. Adaptability	Can the process be modified to meet future treatment requirements?
23. Economic life-cycle analysis	Cost evaluation must consider initial capital cost and long-term operating and maintenance costs. The plant with lowest initial capital cost may not be the most effective with respect to operating and maintenance costs. The nature of the available funding will also affect the choice of process
24. Land availability	Is there sufficient space to accommodate not only the facilities currently being considered but possible future expansion? How much of a buffer zone is available to provide landscaping to minimize visual and other impacts?

Some of the factors (items 2 through 6) listed in Table 4–11 have been discussed previously in Chaps. 2 and 3. The following discussion will deal briefly with process design based on reaction kinetics, mass transfer, and the use of loading criteria, the subjects considered in this chapter. As part of the discussion, the conduct of bench and pilot-plant studies is considered along with process variability. The other factors in Table 4–11 will be discussed throughout the remainder of the book. They are identified here to indicate the diverse nature of the information that must be available to make a proper evaluation of unit operations and processes used for the treatment of wastewater.

Process Selection Based on Reaction Kinetics

In process selection and sizing based on reaction kinetics, particular emphasis is placed on defining the nature of the reactions occurring within the process, the appropriate values of the kinetic coefficients, and the selection of the reactor type.

Nature of the Kinetic Reactions. The nature of the reactions occurring within a process must be known to apply the reaction kinetics approach to design. Selection of reaction rate expressions for the process that is to be designed is typically based on (1) information obtained from the literature, (2) experience with the design and operation of similar systems, or (3) data derived from pilot-plant studies. For example, it is of critical importance to know if the reaction is zero-, first-, retarded first-, or second-order, or if the reaction is a saturation type. Clearly, as demonstrated in this chapter, the order of the reaction will have a significant effect on the type and size of the reactor. The total volume required for various removal efficiencies for first-order kinetics, using 1, 2, 4, 6, 8, or 10 complete-mix reactors in series, was compared to the volume required for a plug-flow reactor previously in Table 4–8. As indicated in Table 4–8, greater reactor volume is required for complete-mix reactors to achieve the same removal efficiencies as plug-flow reactors. It should be noted, however, that for zero-order kinetics the volume of the two reactors will be the same.

Selection of Appropriate Kinetic Rate Coefficients. Selection of appropriate kinetic rate coefficients for the process that is to be designed is also based on (1) information obtained from the literature, (2) experience with the design and operation of similar systems, or (3) data derived from pilot-plant studies. In cases where significantly different wastewater characteristics occur or new applications of existing technology or new processes are being considered, pilot-plant testing is recommended. The various rate expressions that have been developed for biological waste treatment, based on the method of analysis presented in this chapter, are considered in Chaps. 8 through 10.

Selection of Reactor Types. Operational factors that must be considered in the type of reactor or reactors to be used in the treatment process include (1) the nature of the wastewater to be treated, (2) the nature of the reaction kinetics governing the treatment process, (3) special process requirements, and (4) local environmental conditions. As noted previously, for biological treatment with the activated sludge process there is no difference in the size of the reactor required (i.e., $V_{complete-mix} = V_{plug-flow}$). For example, a complete-mix reactor might be selected over a plug-flow reactor, because of its dilution capacity, if the influent wastewater is known to contain toxic constituents that cannot be removed by pretreatment. Alternatively, a plug-flow or multistage reactor might be selected over a complete-mix reactor to control the growth of filamentous microorganisms. In practice, the construction costs and operation and maintenance costs also affect reactor selection.

Process Selection Based on Mass Transfer

In addition to process selection based on reaction kinetics and loading criteria, a number of treatment processes will be based on mass transfer considerations, as introduced in this chapter. The principal operations in wastewater treatment involving mass transfer are aeration, especially the addition of oxygen to water; the drying of biosolids and sludge; the removal of volatile organics from wastewater; the stripping of dissolved constituents such as ammonia from digested supernatant; and the exchange of dissolved constituents as in ion exchange. Fortunately, there is a considerable body of literature on these subjects as well as a vast amount of practical experience. Additional details on these subjects are presented in the subsequent chapters.

Process Design Based on Loading Criteria

If appropriate reaction rate expressions and/or mass transfer coefficients cannot be developed, generalized loading criteria are frequently used. Early design loading criteria for activated sludge biological treatment systems were based on aeration tank capacity [e.g., kg of $BOD/m^3 \cdot d$ (lb $BOD/10^3$ $ft^3 \cdot d$)]. For example, if a process that is loaded at 10 kg/m^3 produces an acceptable effluent and one loaded at 20 $kg/10^3$ m^3 does not, the successful experience tends to be repeated. Unfortunately, records often are not well maintained, and the limits of such loading criteria are seldom defined. Examples of loading criteria are presented in the design chapters for unit operations and processes. It should be noted that with the new activated-sludge biological treatment process variations and new aeration equipment, the use of loading factors should be avoided.

Bench Tests and Pilot-Plant Studies

Where the applicability of a process for a given situation is unknown, but the potential benefits of using the process are significant, bench-scale or pilot-scale tests must be conducted. The purpose of conducting pilot-plant studies is to establish the suitability of the process in the treatment of a specific wastewater under specific environmental conditions and to obtain the necessary data on which to base a full-scale design. Factors that should be considered in planning pilot-plant studies for wastewater treatment are presented in Table 4–12. The relative importance of the factors presented in Table 4–12 will depend on the specific application and the reasons for conducting the testing program. For example, testing of UV disinfection systems is typically done: (1) to verify manufacturers' performance claims, (2) to quantify effects of effluent water quality constituents on UV performance, (3) to assess the effect(s) of system and reactor hydraulics on UV performance, (4) to assess the effect(s) of effluent filtration on UV performance, and (5) to investigate photoreactivation and impacts.

Bench-scale tests are conducted in the laboratory with small quantities of the wastewater in question. Pilot-scale tests are typically conducted with flows that are 5 to 10 percent of the design flows (see Fig. 4–35). In some instances full-scale tests have been conducted, especially where scale-up issues are complex and computational methods are too complex. A discussion of scale-up issues may be found in Johnstone and Thring (1957) and Schmidtke and Smith (1983).

Reliability Considerations in Process Selection

Important factors in process selection and design are treatment plant performance and reliability in meeting permit requirements. In most permits, effluent constituent requirements based on 7-day and 30-day average concentrations are specified. Because wastewater treatment effluent quality is variable for a number of reasons (varying organic loads, changing environmental conditions, etc.), it is necessary to ensure that the treatment system is designed to produce effluent concentrations equal to or less than the permit limits.

Two approaches in process selection and design are (1) the use of arbitrary safety factors, and (2) statistical analysis of treatment plant performance to determine a functional relationship between effluent quality and the probable frequency of occurrence. The latter approach, termed the "reliability concept," is preferred because it can be used

Table 4–12
Considerations in
setting up pilot-plant
testing programs

Item	Consideration
Reasons for conducting pilot testing	Test new process
	Simulation of another process
	Predict process performance
	Document process performance
	Optimize system design
	Satisfy regulatory agency requirements
	Satisfy legal requirements
Pilot-plant size	Bench- or laboratory-scale model
	Pilot-scale tests
	Full- (prototype) scale tests
Nonphysical design factors	Available time, money, and labor
	Degree of innovation and motivation involved
	Quality of water or wastewater
	Location of facilities
	Complexity of process
	Similar testing experience
	Dependent and independent variables
Physical design factors	Scale-up factors
	Size of prototype
	Flow variations expected
	Facilities and equipment required and setup
	Materials of construction
Design of pilot testing program	Dependent variables including ranges
	Independent variables including ranges
	Time required
	Test facilities
	Test protocols
	Statistical design of data acquisition program

to provide a consistent basis for analysis of uncertainty and a rational basis for the analysis of performance and reliability. Treatment plant reliability can be defined as the probability that a system can meet established performance criteria consistently over extended periods of time. Thus, reliability is comprised of two components: inherent reliability and mechanical reliability. These subjects are considered further in Chap. 15.

Figure 4–35

Typical pilot plants used to develop design criteria: (*a*) activated sludge, (*b*) effluent filtration with depth filter, (*c*) effluent filtration with cloth surface filter, and (*d*) effluent filtration with membrane microfilter.

(a)

(b)

(c)

(d)

PROBLEMS AND DISCUSSION TOPICS

4-1   A water storage tank receives a constant feed rate of 0.2 m^3/s and the demand varies according to the relationship 0.2[1 − cos (πt/43,200)] m^3/s. The tank is cylindrical with a cross-sectional area of 1000 m^2. If the depth at $t = 0$ is 5 m, plot the water depth as a function of time.

4-2   Solve Problem 4–1 assuming the feed rate is 0.33 m^3/s and that the storage tank is a square with a cross-sectional area of 1600 m^2.

4-3   A large tank having a floor area of 1000 m² and a sidewall depth of 10 m is used as an equalization reservoir. Flow out of the basin is 0.3 m³/s, while flow into the basin is 0.30[1 + cos
($\pi t/43,200$)] m³/s. Plot the hourly values of water depth versus time, assuming $h = h_o = 5$ m at
$t = 0$.

4-4   Solve Problem 4–3 assuming the feed rate is 0.35[1 + cos ($\pi t/43,000$)] m³/s and that the floor area
for the storage tank is 2000 m². Plot the hourly values of water depth versus time, assuming $h =
h_o = 2.0$ m at $t = 0$.

4-5   Wastewater is being pumped into a 4.2 m diameter tank at the rate of 0.5 m³/min. At the same time,
water leaves the tank at a rate that is dependent on the height of the liquid in the tank. The relationship governing the flow from the tank is $q = [2.1$ (m²/min) × h(m)]. If the tank was initially
empty, develop a relationship that can be used to define the height of the liquid in the tank as a
function of time. What is the steady-state height of the liquid in the tank?

4-6   Solve Problem 4–5 assuming the feed rate is 0.75 m³/min and the tank outflow is $q = [2.7$
(m²/min) × h(m)].

4-7   The following data were obtained from dye tracer studies of five different chlorine contact basins.
Using these data, determine the mean detention time and the corresponding variance for one of
the basins (to be selected by instructor).

Time, min	Concentration, ppb Basin number				
	A	B	C	D	E
0	0.0	0.0	0.0	0.0	0.0
6	0.0	0.0	0.0	0.0	0.0
12	3.5	0.1	0.1	0.0	0.0
18	7.6	2.1	2.1	0.0	0.7
24	7.8	7.5	10.0	0.3	4.0
30	6.9	10.1	12.0	1.8	9.0
36	5.9	10.2	10.2	4.5	12.5
42	4.8	9.7	8.0	8.0	11.5
48	3.8	8.1	6.0	11.0	8.8
54	3.0	6.0	4.3	11.0	5.5
60	2.4	4.4	3.0	9.0	3.0
66	1.9	3.0	2.1	4.3	1.8
72	1.5	1.9	1.5	2.0	0.8
78	1.0	1.0	1.0	1.0	0.4
84	0.6	0.4	0.5	0.2	0.1
90	0.3	0.1	0.1	0.0	0.0
96	0.1	0.0	0.0	0.0	0.0
102	0.0	0.0	0.0	0.0	0.0

4-8   Using the tracer data given in Problem 4–7, develop residence time distribution curves (E and F) for one of the chlorine contact basins (to be selected by instructor).

4-9   Using the tracer data given in Problem 4–7, determine the Morrill dispersion index and the volumetric efficiency for one of the chlorine contact basins (to be selected by instructor).

4-10   Using the tracer data given in Problem 4–7, determine the dispersion number, the Peclet number, and the number of complete-mix reactors in series that would be needed to simulate the tracer curve for one of the chlorine contact basins (to be selected by instructor). Also, plot the tracer curve and the simulated curve.

4-11   When a slug of tracer is added to a series of identical complete-mix reactors, a maximum value of $C(\theta = t/\tau_i)$ occurs at $t > 0$ for all reactors except the first. Determine the time, t, when the peak concentration of tracer occurs in reactor number 5 in a reactor system comprised of 6 reactors in series.

4-12   Solve Problem 4–11 for the ith reactor in a reactor system comprised of any number of identical complete-mix reactors in series.

4-13   The following data were obtained for four different reactants for the reaction $A \rightarrow B + C$. Determine the order of the reaction for one reaction (to be selected by instructor) and the value of the reaction rate constant k.

Time, t, min	Concentration, mg/L			
	A	**B**	**C**	**D**
0	90	1.9	240	113
10	72	1.55	150	80
20	57	1.31	110	56
40	36	0.99	70	28
60	23	0.8	51	14

4-14   A bimolecular reaction $A + B \rightarrow P$ is 10 percent complete in 10 min. If the initial concentration of A and B is equal to 1.0 mole/L, determine the reaction rate constant and how long it will take for the reaction to be 90 percent complete.

4-15   A bimolecular reaction $A + B \rightarrow P$ is 8 percent complete in 12 min. If the initial concentration of A and B is equal to 1.33 mole/L, determine the reaction rate constant and how long it will take for the reaction to be 96 percent complete.

4-16   The following values have been obtained for the rate constant for the reaction $A + B \rightarrow P$. Using these data, determine the activation energy E and the value of the rate constant at 15°C.

$k_{25°C} = 1.5 \times 10^{-2}$ L/mole·min

$k_{45°C} = 4.5 \times 10^{-2}$ L/ mole·min

4-17   Solve Problem 4–16 for the following rate constant values:

$k_{20°C} = 1.25 \times 10^{-2}$ L/mole·min

$k_{35°C} = 3.55 \times 10^{-2}$ L/ mole·min

4-18   An aqueous reaction is being studied in a laboratory-sized complete-mix reactor with a volume of 5 L. The stoichiometry of the reaction is $A \rightarrow 2R$, and reactant A is introduced into the reactor at a concentration of 1 mole/L. From the results given in the following table, find the rate expression for this reaction. Assume steady-state flow.

Run	Feed rate, cm³/s	Temperature, °C	Concentration R in effluent, mole/L
1	2	13	1.8
2	15	13	1.5
3	15	84	1.8

4-19   The rate of reaction for an enzyme-catalyzed substrate in a batch reactor can be described by the following relationship.

$$r_C = -\frac{kC}{K + C}$$

where k = maximum reaction rate, mg/L·min
     C = substrate concentration, mg/L
     K = constant, mg/L

Using this rate expression, derive an equation that can be used to predict the reduction of substrate concentration with time in a batch reactor. If k equals 40 mg/L·min and $K = 100$ mg/L, determine the time required to decrease the substrate concentration from 1000 to 100 mg/L.

4-20   Solve problem 4–19 for the following values: k equals 28 mg/L·min and $K = 116$ mg/L.

4-21   A wastewater is to be treated in a complete-mix reactor. Assuming that the reaction is irreversible and first-order ($r = -kC$) with a reaction rate coefficient equal to 0.15 d, determine the flowrate that can be treated if the reactor has a volume of 20 m³ and 98 percent treatment efficiency is required. What volume would be required to treat the flowrate determined above if the required treatment efficiency is 92 percent?

4-22   For first-order removal kinetics, demonstrate that the maximum treatment efficiency in a series of complete-mix reactors occurs when all the reactors are the same size.

4-23   Determine the number of completely mixed chlorine contact chambers each having a detention time of 30 min that would be required in a series arrangement to reduce the bacterial count of a treated effluent from 10^6 to 14.5 organisms/mL if the first-order removal rate constant is equal to 6.1 h⁻¹. If a plug-flow chlorine contact chamber were used with the same detention time as the series of completely mixed chambers, what would the bacterial count be after treatment?

4-24   The concentration of UBOD in a river entering the first of two lakes that are connected in a series is equal to 20 mg/L. If the first-order BOD reaction rate constant (k) equals 0.35 d⁻¹ and complete mixing occurs in each lake, what is the concentration of UBOD at the outlet of each lake? The flow in the river is equal to 4000 m³/d and the volume of the first and second lake is 20,000 and 12,000 m³, respectively. Assume steady-state conditions.

4-25 In Prob. 4–24, if the length of the river connection between the two lakes is equal to 3 km and the velocity in the river is equal to 0.4 m/s, determine the concentration of the UBOD in the effluent from the second lake.

4-26 Plot the ratio of required tank volume for a plug-flow reactor to that of a continuous-flow stirred-tank reactor (VPFR/VCMR) versus the fraction of the original substrate that is converted for the following reaction rates.

$r = -k$

$r = -kC^{0.5}$

$r = -kC$

$r = -kC^2$

What is the value of the required volume ratio for each of these rates when $C = 0.25$ mg/L and $C_o = 1.0$ mg/L?

4-27 Solve problem 4–26 for the following values: $C = 0.17$ mg/L and $C_o = 1.25$ mg/L.

4-28 If second-order reaction kinetics are applicable ($r = -kC^2$), determine the effluent concentration for each of the reactor systems shown on Fig. 4–28. To simplify the computations, assume that the following data apply:

$k = 1.0$ m^3/kg·d

$Q = 1.0$ m^3/d

$V_{PFR} = 1.0$ m^3

$V_{CMR} = 1.0$ m^3

$C_o = 1.0$ kg/m^3

Explain your results. What would happen if first- or zero-order kinetics are applicable?

4-29 A portion of the outflow, αQ, from an ideal plug-flow reactor is recycled around the reactor where $\alpha \geq 0$. Assume that the rate of conversion can be defined as $r_c = -kC$.
  a. Sketch the generalized curve of conversion versus the recycle ratio.
  b. Sketch a family of curves showing the effect of the recycle ratio α on the longitudinal concentration gradient.
  c. If a continuous-flow stirred-tank reactor were substituted for the plug-flow reactor, what effect would the recycle have on conversion?

4-30 Determine the effect of recycle on the performance of a complete-mix reactor for first- and second-order reactions.

4-31 Derive an expression that can be used to compute the effluent concentration from an overland flow treatment system, designed as an ideal plug-flow reactor, assuming the removal of the constituent in question can be described by a retarded second-order equation. Assume the exponent n in the retardation term is equal to one. If the value of the retardation coefficient is 0.2, compare the effluent concentration with and without retardation.

4-32 For zero-order removal kinetics demonstrate that the treatment efficiency in a series of complete-mix reactors (use three) is the same as that for a plug-flow reactor for the same total volume.

4-33   A treated effluent flows in a 0.3 m deep open channel for a distance of 1000 m. Estimate the concentration of dissolved oxygen in the effluent at the end of the channel, assuming ideal plug-flow if the average flow velocity in the channel is 0.33 m/s and that the mass transfer coefficient, K, for oxygen is 0.05 m/h. Assume the initial concentration of dissolved oxygen is 0.5 mg/L and the temperature is 20°C.

4-34   Solve Problem 4–33 for a 0.45 m deep open channel, an average flow velocity of 0.25 m/s, and an oxygen mass transfer coefficient value of 0.06 m/h.

4-35   A treated effluent flows in a 0.25 m deep open channel for a distance of 1500 m at a velocity of 0.25 m/h. If the initial concentration of dissolved oxygen is 0.8 mg/L and the concentration at the end of the channel is 2.9 mg/L, estimate the value of the mass transfer coefficient, K, for oxygen at 20°C.

4-36   Solve Problem 4–35 if the initial concentration of dissolved oxygen is 0.6 mg/L and the concentration at the end of the channel is 2.7.

4-37   How much powdered activated carbon will be required to treat an effluent flowrate of 6500 m³/d to reduce the bromoform concentration from 0.09 to 0.025 mg/L? The Freundlich adsorption isotherm coefficients for bromoform are: $K_f = 19.6$ and $1/n = 0.52$ (see Table 11–32 in Chap. 11).

4-38   Solve Problem 4–37 for a toluene. Assume the initial concentration is 0.067 mg/L and that the final concentration is to be 0.01 mg/L. The Freundlich adsorption isotherm coefficients for toluene are: $K_f = 26.1$ and $1/n = 0.44$ (see Table 11–32 in Chap. 11).

4-39   Wastewater containing 0.8 g/m³ of total pesticide percolates through a soil treatment system. Adsorption has been determined to follow a Langmuir isotherm with the following coefficients: $a = 1.0$ and $b = 0.003$ m³/g. Determine the mass of pesticide that will potentially accumulate in the soil per 1.0 m² of cross-sectional and the effective depth is 4 m. The dry density of the soil is approximately 2400 kg/m³. What is the effluent concentration and the volume of wastewater that can be applied assuming all of the available adsorption capacity will be utilized?

4-40   Solve Problem 4–39 for a water containing 0.75 mg/L g/m³ of total pesticide. Assume the Langmuir isotherm coefficients are: $a = 0.9$ and $b = 0.0025$ m³/g.

REFERENCES

Adeney, W. E., and H. G. Becker (1919) "The Determination of the Rate of Solution of Atmospheric Nitrogen and Oxygen by Water," *Philosophical Magazine,* vol. 38, p. 317.

Ambrose, R. B., Jr., J. P. Connolly, E. Southerland, T. O. Barnwell, Jr., and J. L. Schnoor (1988) Waste Allocation Simulation Models, *Journal of the Water Pollution Control Federation,* vol. 60, no. 9, pp. 1646–1655.

Arceivala, S. J. (1983) "Discussion on Hydraulics Modeling for Wastewater Stabilization Ponds," *Journal of the Environmental Engineering Division,* American Society of Civil Engineers, vol. 109, no. 1. p. 265.

Arceivala, S. J. (1998) *Wastewater Treatment for Pollution Control,* 2nd ed., Tata McGraw-Hill Publishing Company Limited, New Delhi, India.

Becker, H. G. (1924) "Mechanism of Absorption of Moderately Soluble Gases in Water," *Industrial Engineering Chemistry,* vol. 16, p. 1666.

Carslaw, H. S., and J. C. Jaeger (1947) *Conduction of Heat in Solids,* Oxford University Press, London.

Cornwell, D. A. (1990) "Air Stripping and Aeration," in F. W. Pontius (ed.) *Water Quality and Treatment: A Handbook of Community Water Supplies,* 4th ed., American Water Works Association, McGraw-Hill, Inc., New York.

Coulson, J. M., and J. F. Richardson (1968) *Chemical Engineering, Volume Two, Unit Operations,* Pergamon Press, Oxford, England.

Crank, J. (1957) *The Mathematics of Diffusion,* Oxford University Press, London.

Crites, R., and G. Tchobanoglous (1998) *Small and Decentralized Wastewater Management Systems,* McGraw-Hill Book Company, New York.

Crittenden, J. C. (1999) Course Notes, CE 456—Hazardous Waste Treatment and Residuals Processing, Michigan Technological University, Houghton, MI.

Dague, R. R., and E. R. Baumann (1961) "Hydraulics of Circular Settling Tanks Determined by Models," Paper presented at the 1961 Annual meeting, Iowa Water Pollution Control Association, Lake Okoboji, IA.

Danckwerts, P. V. (1951) "Significance of Liquid Film Coefficients in Gas Absorption," *Journal Industrial Engineering Chemistry,* vol. 43, p. 1460, 1951.

Danckwerts, P. V. (1970) *Gas-Liquid Reactions,* McGraw-Hill Book Company, New York.

Davies, J. T. (1972) *Turbulence Phenomena,* Academic Press, New York.

Denbigh, K. G., and J. C. R. Turner (1965) *Chemical Reactor Theory,* 2nd ed., Cambridge University Press, England.

Eldridge, J. M., and E. L. Piret (1950) "Chemical Engineering Progress," vol. 46, p. 290.

Fair, G. M., and J. C. Geyer (1954) *Water Supply and Waste-Water Disposal,* John Wiley and Sons, Inc. New York.

Fair, G. M., J. C. Geyer, and D. A. Okun (1988) *Water and Wastewater Engineering,* vol. 2, Wiley, New York.

Fitch, E. B. (1956) "Effect of Flow Path Effect on Sedimentation," *Sewage and Industrial Wastes,* vol. 28, no. 1, p. 1.

Fogler, H. S. (1999) *Elements of Chemical Reaction Engineering,* 3rd ed., Prentice Hall, Inc., Upper Saddle River, NJ.

Hemond, H. F., and E. J. Fechner (1994) *Chemical Fate and Transport in the Environment,* Academic Press, Inc., San Diego, CA.

Higbie, R. (1935) "The Rate of Absorption of Pure Gas into a Still Liquid During Short Periods of Exposure," *Transactions American Institute of Industrial Chemistry,* vol. 31, p. 365.

Johnstone, R. E., and M. W. Thring (1957) *Pilot Plants, Models, and Scale-up Methods in Chemical Engineering,* McGraw-Hill Book Company, New York.

Kramer, H., and K. R. Westererp (1963) *Elements of Chemical Reactor Design and Operation,* Academic Press, Inc., New York.

Levenspiel, O. (1972) *Chemical Reaction Engineering,* 2nd ed., Wiley, New York.

Lewis, W. K., and W. C. Whitman (1924) "Principles of Gas Adsorption," *Journal Industrial and Engineering Chemistry,* vol. 16, pp. 1215–1220.

MacMullin, R. B., and M. Weber, Jr. (1935) "The Theory of Short-Circuiting in Continuous Mixing Vessels in Series and Kinetics of Chemical Reactions in Such Systems," *Journal American Institute of Chemical Engineers,* vol. 31, pp. 409–458.

McCabe, W. L., and J. C. Smith (1976) *Unit Operations of Chemical Engineering,* 3rd ed., McGraw-Hill Book Company, New York.

Montgomery, J. M., Consulting Engineers, Inc. (1985) *Water Treatment Principles and Design,* A Wiley-Interscience Publication, John Wiley & Sons, New York.

Morrill, A. B. (1932) "Sedimentation Basin Research and Design," *Journal American Water Works Association,* vol. 24, p. 1442.

Orlob, G. T. (ed.) (1983) *Mathematical Modeling of Water Quality: Streams, Lakes, and Reservoirs,* A Wiley-Interscience Publication, John Wiley & Sons, Chichester, England.

Polprasert C., and K. K. Bhattarai (1985) "Dispersion Model for Waste Stabilization Ponds," *Journal of the Environmental Engineering Division,* American Society of Civil Engineers, vol. 11. EE1, pp. 45–58.

Schmidtke, N. W., and D. W. Smith (1983) *Scale-Up of Water and Wastewater Treatment Processes,* Butterworth Publishers, Boston, MA.

Schnoor, J. L. (1996) *Environmental Modeling,* A Wiley-Interscience Publication, John Wiley & Sons, Inc., New York.

Shaw, D. J. (1966) *Introduction to Colloid and Surface Chemistry,* Butterworth, London.

Smith, J. M. (1981) *Chemical Engineering Kinetics,* 3rd ed., McGraw-Hill, New York.

Snoeyink, V. L., and R. S. Summers (1999) "Chapter 13: Adsorption of Organic Compounds," in R. D. Letterman, ed., *Water Quality and Treatment: A Handbook of Community Water Supplies,* 5th ed., American Water Works Association, McGraw-Hill, Inc., New York.

Sontheimer, H., J. C. Crittenden, and R. S. Summers (1988) *Activated Carbon for Water Treatment,* 2nd ed. in English, DVGW-Forschungsstelle, Engler-Bunte-Institut, Universitat Karlsruhe, Germany.

Tchobanoglous, G. (1969) A Study of the Filtration of Treated Sewage Effluent, Ph.D. dissertation, Stanford University, Stanford, CA.

Tchobanoglous, G., and F. L. Burton (1991) *Wastewater Engineering. Treatment, Disposal, Reuse,* 3rd ed., McGraw-Hill Inc., New York.

Tchobanoglous, G., and E. D. Schroeder (1985) *Water Quality: Characteristics, Modeling, Modification,* Addison-Wesley Publishing Company, Reading, MA.

Tchobanoglous, G., R. Crites, B. Gearheart, and W. Reed (2000) "A Review of Treatment Kinetics for Constructed Wetlands," in Conference Proceedings *Disinfection 2000: Disinfection of Wastes in the New Millennium,* New Orleans, LA, Water Environment Federation, Alexandria, VA.

Thibodeaux, L. J. (1996) *Chemodynamics: Environmental Movement of Chemicals in Air, Water, and Soil,* 2nd ed., John Wiley & Sons, New York.

Thirumurthi, D. (1969) "Design of Waste Stabilization Ponds," *Journal Sanitary Engineering Division,* American Society of Civil Engineers, vol. 95, no. SA2.

Thomann, R. V., and J. A. Mueller (1987) *Principles of Surface Water Quality Modeling and Control,* Harper & Row, Publishers, Inc., New York.

Treybal, R. E. (1980) *Mass Transfer Operations,* 3rd ed., Chemical Engineering Series, McGraw-Hill, New York.

U.S. EPA (1986) *Design Manual, Municipal Wastewater Disinfection,* U.S. Environmental Protection Agency, EPA/625/1-86/021, Cincinnati, OH.

Wehner, J. F., and R. F. Wilhelm (1958) "Boundary Conditions of Flow Reactor," *Chemical Engineering Science,* vol. 6, p. 89.

5

Physical Unit Operations

Operations used for the treatment of wastewater in which change is brought about by means of or through the application of physical forces are known as *physical unit operations*. Because physical unit operations were derived originally from observations of the physical world, they were the first treatment methods to be used. Today, physical unit operations, as shown on Fig. 5–1, are a major part of most wastewater-treatment systems.

The unit operations most commonly used in wastewater treatment include (1) screening, (2) coarse solids reduction (comminution, maceration, and screenings grinding), (3) flow equalization, (4) mixing and flocculation, (5) grit removal, (6) sedimentation, (7) high-rate clarification, (8) accelerated gravity separation (vortex separators), (9) flotation, (10) oxygen transfer, (11) aeration, and (12) volatilization and stripping of volatile organic compounds (VOCs). The principal applications of these operations and treatment devices used are summarized in Table 5–1. Physical unit operations that apply to advanced wastewater-treatment systems such as packed-bed filters, membrane separation systems, and ammonia stripping are discussed separately in Chap. 11. Unit operations associated with the processing of solids and biosolids (sludge) are covered in Chap. 14.

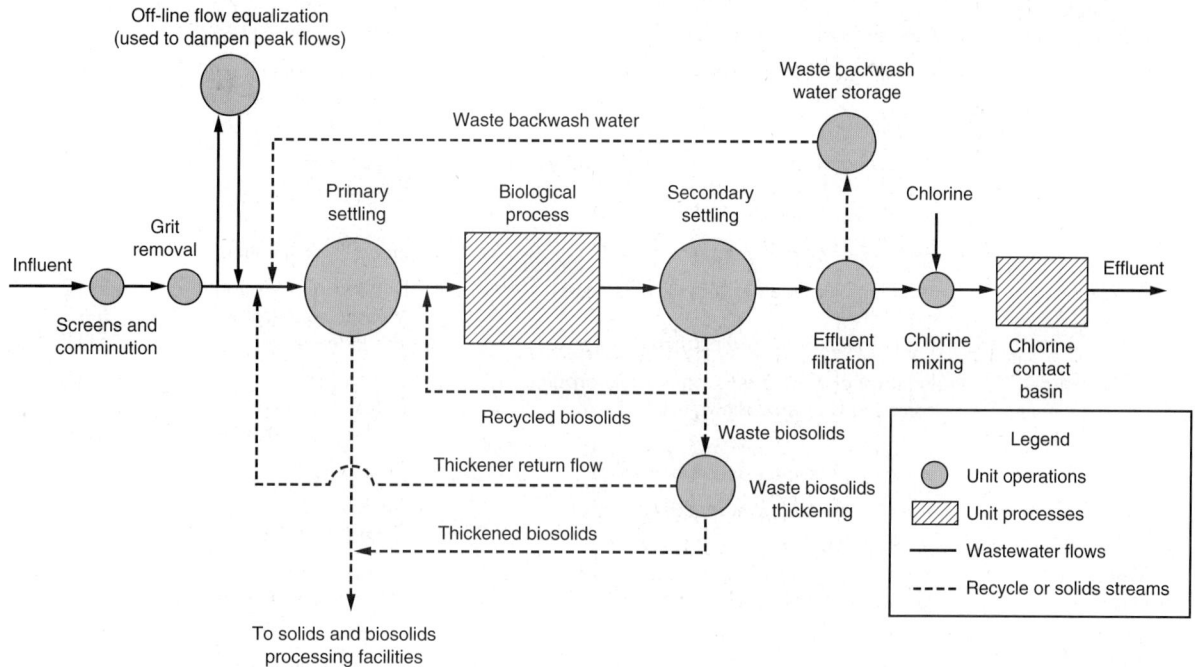

Figure 5–1

Location of physical unit operations in a wastewater-treatment plant flow diagram.

Table 5–1

Typical physical unit operations used for wastewater treatment

Operation	Application	Device	See section
Screening, coarse	Removal of coarse solids such as sticks, rags, and other debris in untreated wastewater by interception (surface straining)	Bar rack	5–1
Screening, fine	Removal of small particles	Fine screen	5–1
Screening, micro	Removal of fine solids, floatable matter, and algae	Microscreen	5–1, 11–5
Comminution	In-stream grinding of coarse solids to reduce size	Comminutor	5–2
Grinding/ maceration	Grinding of solids removed by bar racks Side-stream grinding of coarse solids	Screenings grinder Macerator	5–2 5–2
Flow equalization	Temporary storage of flow to equalize flowrates and mass loadings of BOD and suspended solids	Equalization tank	5–3
Mixing	Blending chemicals with wastewater and for homogenizing and maintaining solids in suspension	Rapid mixer	5–4
Flocculation	Promoting the aggregation of small particles into larger particles to enhance their removal by gravity sedimentation	Flocculator	5–4
Accelerated sedimentation	Removal of grit Removal of grit and coarse solids	Grit chamber Vortex separator	5–6 5–9
Sedimentation	Removal of settleable solids	Primary clarifier High-rate clarifier	5–7 5–8
	Thickening of solids and biosolids	Gravity thickener	14–6
Flotation	Removal of finely divided suspended solids and particles with densities close to that of water; also thickens biosolids	Dissolved-air flotation (DAF)	5–10 14–6
	Removal of oil and grease	Induced-air flotation	5–10
Aeration	Addition of oxygen to biological process	Diffused-air aeration Mechanical aerator	5–12 5–12
	Postaeration of treated effluent	Cascade aerator	5–12
VOC control	Removal of volatile and semivolatile organic compounds from wastewaters	Gas stripper Diffused-air and mechanical aeration	5–13 5–12, 5–13
Depth filtration	Removal of residual suspended solids	Depth filters	11–4
Surface filtration	Removal of residual suspended solids	Discfilter® Cloth-Media Disk Filter®	11–5
Membrane filtration	Removal of suspended and colloidal solids and dissolved organic and inorganic matter	Microfiltration, ultrafiltration, nanofiltration, and reverse osmosis	11–6
Air stripping	Removal of ammonia, hydrogen sulfide, and other gases from wastewater and digester supernatant	Packed tower	11–8

5–1 SCREENING

The first unit operation generally encountered in wastewater-treatment plants is screening. A screen is a device with openings, generally of uniform size, that is used to retain solids found in the influent wastewater to the treatment plant or in combined wastewater-collection systems subject to overflows, especially from stormwater. The principal role of screening is to remove coarse materials from the flow stream that could (1) damage subsequent process equipment, (2) reduce overall treatment process reliability and effectiveness, or (3) contaminate waterways. Fine screens are sometimes used in place of or following coarse screens where greater removals of solids are required to (1) protect process equipment or (2) eliminate materials that may inhibit the beneficial reuse of biosolids.

All aspects of screenings removal, transport, and disposal must be considered in the application of screening devices, including (1) the degree of screenings removal required because of potential effects on downsteam processes, (2) health and safety of the operators as screenings contain pathogenic organisms and attract insects, (3) odor potential, and (4) requirements for handling, transport, and disposal, i.e., removal of organics (by washing) and reduced water content (by pressing), and (5) disposal options. Thus, an integrated approach is required to achieve effective screenings management.

Classification of Screens

The types of screening devices commonly used in wastewater treatment are shown on Fig. 5–2. Two general types of screens, coarse screens and fine screens, are used in preliminary treatment of wastewater. Coarse screens have clear openings ranging from 6 to 150 mm (0.25 to 6 in); fine screens have clear openings less than 6 mm (0.25 in). Microscreens, which generally have screen openings less than 50 μm, are used principally in removing fine solids from treated effluents.

The screening element may consist of parallel bars, rods or wires, grating, wire mesh, or perforated plate, and the openings may be of any shape but generally are circular or rectangular slots. A screen composed of parallel bars or rods is often called a "bar rack" or a coarse screen and is used for the removal of coarse solids. Fine screens are devices consisting of perforated plates, wedgewire elements, and wire cloth that have smaller openings. The materials removed by these devices are known as *screenings*.

Figure 5–2

Definition sketch for types of screens used in wastewater treatment.

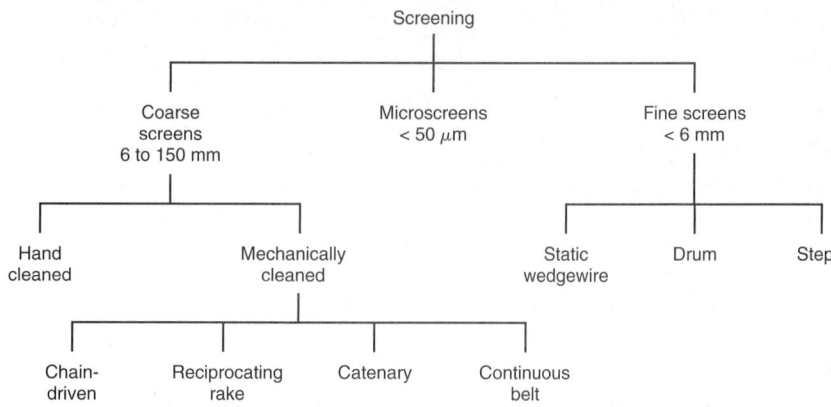

Coarse Screens (Bar Racks)

In wastewater treatment, coarse screens are used to protect pumps, valves, pipelines, and other appurtenances from damage or clogging by rags and large objects. Industrial waste-treatment plants may or may not need them, depending on the character of the wastes. According to the method used to clean them, coarse screens are designated as either hand-cleaned or mechanically cleaned.

Hand-Cleaned Coarse Screens. Hand-cleaned coarse screens are used frequently ahead of pumps in small wastewater pumping stations and sometimes used at the headworks of small- to medium-sized wastewater-treatment plants. Often they are used for standby screening in bypass channels for service during high-flow periods, when mechanically cleaned screens are being repaired, or in the event of a power failure. Normally, mechanically cleaned screens are provided in lieu of hand-cleaned screens to minimize manual labor required to clean the screens and to reduce flooding due to clogging.

Where used, the length of the hand-cleaned bar rack should not exceed the distance that can be conveniently raked by hand, approximately 3 m (10 ft). The screen bars are welded to spacing bars located at the rear face, out of the way of the tines of the rake. A perforated drainage plate should be provided at the top of the rack where the rakings may be stored temporarily for drainage.

The screen channel should be designed to prevent the accumulation of grit and other heavy materials in the channel ahead of the screen and following it. The channel floor should be level or should slope downward through the screen without pockets to trap solids. Fillets may be desirable at the base of the sidewalls. The channel preferably should have a straight approach, perpendicular to the bar screen, to promote uniform distribution of screenable solids throughout the flow and on the screen. Typical design information for hand-cleaned bar screens is provided in Table 5–2.

Table 5–2
Typical design information for manually and mechanically cleaned bar racks

Parameter	U.S. customary units			SI units		
		Cleaning method			Cleaning method	
	Unit	Manual	Mechanical	Unit	Manual	Mechanical
Bar size						
Width	in	0.2–0.6	0.2–0.6	mm	5–15	5–15
Depth	in	1.0–1.5	1.0–1.5	mm	25–38	25–38
Clear spacing between bars	in	1.0–2.0	0.6–3.0	mm	25–50	15–75
Slope from vertical	°	30–45	0–30	°	30–45	0–30
Approach velocity						
Maximum	ft/s	1.0–2.0	2.0–3.25	m/s	0.3–0.6	0.6–1.0
Minimum	ft/s		1.0–1.6	m/s		0.3–0.5
Allowable headloss	in	6	6–24	mm	150	150–600

Mechanically Cleaned Bar Screens. The design of mechanically cleaned bar screens has evolved over the years to reduce the operating and maintenance problems and to improve the screenings removal capabilities. Many of the newer designs include extensive use of corrosion-resistant materials including stainless steel and plastics. Mechanically cleaned bar screens are divided into four principal types: (1) chain-driven, (2) reciprocating rake, (3) catenary, and (4) continuous belt. Cable-driven bar screens were used extensively in the past but largely have been replaced in wastewater applications by the other types of screens. Typical design information for mechanically cleaned is also included in Table 5–2. Examples of the different types of mechanically cleaned bar screens are shown on Fig. 5–3 and the advantages and disadvantages of each type are presented in Table 5–3.

Chain-Driven Screens Chain-driven mechanically cleaned bar screens can be divided into categories based on whether the screen is raked to clean from the front (upstream) side or the back (downstream) side and whether the rakes return to the bottom of the

Figure 5–3

Typical mechanically cleaned coarse screens: (a) front-cleaned, front-return chain-driven, (b) reciprocating rake, (c) catenary, and (d) continuous belt.

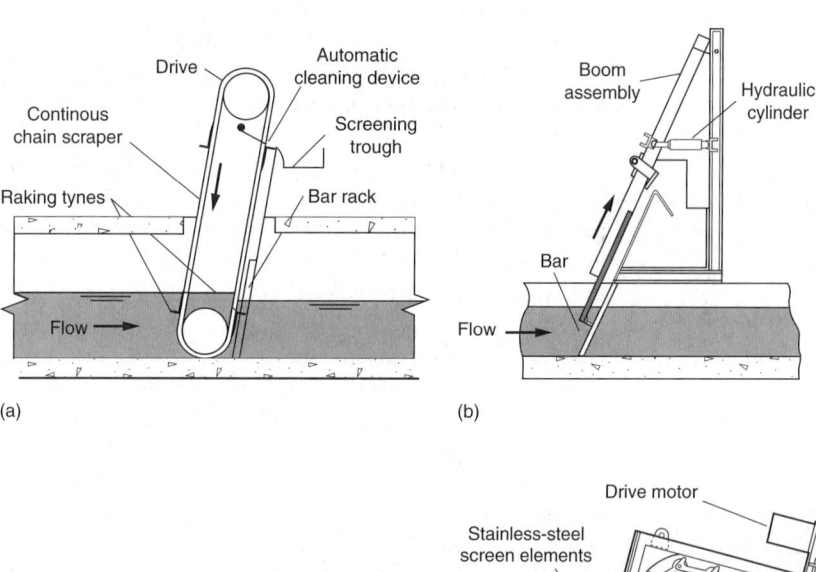

(a)

(b)

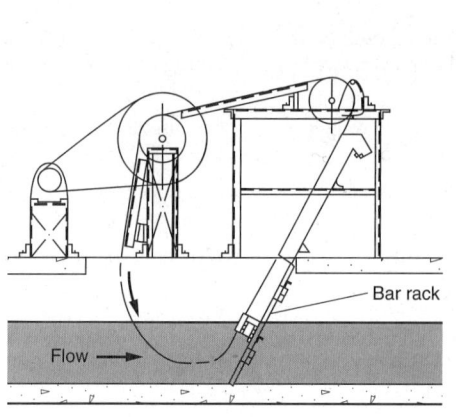

(c)

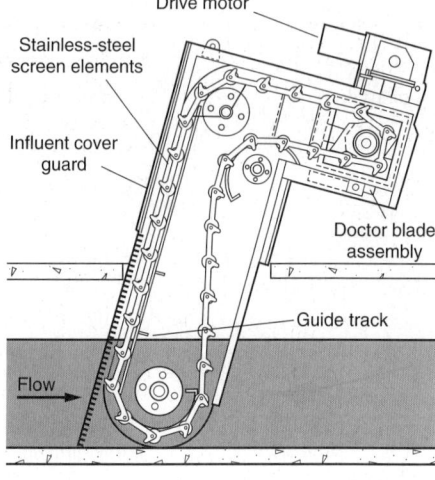

(d)

Table 5–3

Advantages and disadvantages of various types of bar screens

Type of screen	Advantages	Disadvantages
Chain-driven screen		
Front clean/ back return	Multiple cleaning elements (short cleaning cycle)	Unit has submerged moving parts that require channel dewatering for maintenance
	Used for heavy-duty applications	Less efficient screenings removal, i.e., carryover of residual screenings to screened wastewater channel
Front clean/ front return	Multiple cleaning elements (short cleaning cycle)	Unit has submerged moving parts that require channel dewatering for maintenance
	Very little screenings carryover	Submerged moving parts (chains, sprockets, and shafts) are subject to fouling
		Heavy objects may cause rake to jam
Back clean/ back return	Multiple cleaning elements (short cleaning cycle)	Unit has submerged moving parts that require channel dewatering for maintenance
	Submerged moving parts (chains, sprockets, and shafts) are protected by bar rack	Long rake teeth are susceptible to breakage
		Some susceptibility to screenings carryover
Reciprocating rake	No submerged moving parts; maintenance and repairs can be done above operating floor	Unaccounted for high channel water level can submerge rake motor and cause motor burnout
	Can handle large objects (bricks, tires, etc.)	Requires more headroom than other screens
	Effective raking of screenings and efficient discharge of screenings	Long cycle time; raking capacity may be limiting
	Relatively low operating and maintenance costs	Grit accumulation in front of bar may impede rake movement
	Stainless-steel construction reduces corrosion	Relatively high cost due to stainless-steel construction
	High flow capacity	
Catenary	Sprockets are not submerged; most maintenance can be done above the operating floor	Because design relies on weight of chain for engagement of rakes with bars, chains are very heavy and difficult to handle
	Required headroom is relatively low	Because of the angle of inclination of the screen (45 to 75°), screen has a large footprint
	Multiple cleaning elements (short cleaning cycle)	Misalignment and warpage can occur when rakes are jammed
	Can handle large objects	May emit odors because of open design
	Very little screenings carryover	
Continuous belt	Most maintenance can be done above operating floor	Overhaul or replacement of the screening elements is a time-consuming and expensive operation
	Unit is difficult to jam	

bar screen from the front or back. Each type has its advantages and disadvantages, although the general mode of operation is similar. In general, front cleaned, front return screens (see Fig. 5–3a) are more efficient in terms of retaining captured solids, but they are less rugged and are susceptible to jamming by solids that collect at the base of the rake. Front cleaned, front return screens are seldom used for plants serving combined sewers where large objects can jam the rakes. In front cleaned, back return screens, the cleaning rakes return to the bottom of the bar screen on the downstream side of the screen, pass under the bottom of the screen, and clean the bar screen as the rake rises. The potential for jamming is minimized, but a hinged plate, which is also subject to jamming, is required to seal the pocket under the screen.

In back cleaned screens, the bars protect the rake from damage by the debris. However, a back cleaned screen is more susceptible to solids carryover to the downstream side, particularly as rake wipers wear out. The bar rack of the back cleaned, back return screens is less rugged than the other types because the top of the rack is unsupported so the rake tines can pass through. Most of the chain-operated screens share the disadvantage of submerged sprockets that require frequent operator attention and are difficult to maintain. Additional disadvantages include the adjustment and repair of the heavy chains, and the need to dewater the channels for inspection and repair of submerged parts.

Reciprocating Rake (Climber) Screen The reciprocating-rake-type bar screen (see Fig. 5–3b) imitates the movements of a person raking the screen. The rake moves to the base of the screen, engages the bars, and pulls the screenings to the top of the screen where they are removed. Most screen designs utilize a cogwheel drive mechanism for the rake. The drive motors are either submersible electric or hydraulic type. A major advantage is that all parts requiring maintenance are above the waterline and can be easily inspected and maintained without dewatering the channel. The front cleaned, front return feature minimizes solids carryover. The screen uses only one rake instead of multiple rakes that are used with other types of screens. As a result, the reciprocating rake screen may have limited capacity in handling heavy screenings loads, particularly in deep channels where a long "reach" is necessary. The high overhead clearance required to accommodate the rake mechanism can limit its use in retrofit applications.

Catenary Screen A catenary screen is a type of front cleaned, front return chain-driven screen, but it has no submerged sprockets. In the catenary screen (see Fig. 5–3c), the rake is held against the rack by the weight of the chain. If heavy objects become jammed in the bars, the rakes pass over them instead of jamming. The screen, however, has a relatively large "footprint" and thus requires greater space for installation.

Continuous Belt Screen The continuous belt screen is a relatively new development for use in screening applications in the United States. It is a continuous, self-cleaning screening belt that removes fine and coarse solids (see Fig. 5–3d). A large number of screening elements (rakes) are attached to the drive chains; the number of screening elements depends on the depth of the screen channel. Because the screen openings can range from 0.5 to 30 mm (0.02 to 1.18 in), it can be used as either a coarse or a fine screen. Hooks protruding from the belt elements are provided to capture large solids such as cans, sticks, and rags. The screen has no submerged sprocket.

Design of Coarse Screen Installations. Considerations in the design of screening installations include (1) location; (2) approach velocity; (3) clear openings between bars or mesh size; (4) headloss through the screens; (5) screenings handling, processing, and disposal; and (6) controls.

Because the purpose of coarse screens is to remove large objects that may damage or clog downstream equipment, in nearly all cases, they should be installed ahead of the grit chambers. If grit chambers are placed before screens, rags and other stringy material could foul the grit chamber collector mechanisms, wrap around air piping, and settle with the grit. If grit is pumped, further fouling or clogging of the pumps will likely occur.

In hand-cleaned installations, it is essential that the velocity of approach be limited to approximately 0.45 m/s (1.5 ft/s) at average flow to provide adequate screen area for accumulation of screenings between raking operations. Additional area to limit the velocity may be obtained by widening the channel at the screen and by placing the screen at a flatter angle to increase the submerged area. As screenings accumulate, partially plugging the screen, the upstream head will increase, submerging new areas for the flow to pass through. The structural design of the screen should be adequate to prevent collapse if it becomes plugged completely.

For most mechanically cleaned coarse screen installations, two or more units should be installed so that one unit may be taken out of service for maintenance. Slide gates or recesses in the channel walls for the insertion of stop logs should be provided ahead of, and behind, each screen so that the unit can be dewatered for screen maintenance and repair. If only one unit is installed, it is absolutely essential that a bypass channel with a manually cleaned bar screen be provided for emergency use. Sometimes the manually cleaned bar screen is arranged as an overflow device if the mechanical screen should become inoperative, especially during unattended hours. Flow through the bypass channel normally would be prevented by a closed slide or sluice gate. The screen channel should be designed to prevent the settling and accumulation of grit and other heavy materials. An approach velocity of at least 0.4 m/s (1.25 ft/s) is recommended to minimize solids deposition in the channel. To prevent the pass-through of debris at peak flowrates, the velocity through the bar screen should not exceed 0.9 m/s (3 ft/s).

The velocity through the bar screen can be controlled by installation of a downstream head control device such as a Parshall flume, or, for screens located upstream of a pumping station, by controlling the wetwell operating levels. If the channel velocities are controlled by wetwell levels, lower velocities can be tolerated provided flushing velocities occur during normal operating conditions.

Headloss through mechanically cleaned coarse screens is typically limited to about 150 mm (6 in) by operational controls. The raking mechanisms are operated normally based on differential headloss through the screen or by a time clock. For time clock operation, a cycle length of approximately 15 min is recommended; however, either a high-water or high-differential contact should be provided that will place the screen in continuous operation when needed.

Hydraulic losses through bar screens are a function of approach velocity and the velocity through the bars. The headloss through coarse screens can be estimated using the following equation:

$$h_L = \frac{1}{C}\left(\frac{V^2 - v^2}{2g}\right) \tag{5-1}$$

where h_L = headloss, m

C = an empirical discharge coefficient to account for turbulence and eddy losses, typically 0.7 for a clean screen and 0.6 for a clogged screen

V = velocity of flow through the openings of the bar screen, m/s

v = approach velocity in upstream channel, m/s

g = acceleration due to gravity, 9.81 m/s²

The headloss calculated using Eq. (5–1) applies only when the bars are clean. Headloss increases with the degree of clogging. The buildup of headloss can be estimated by assuming that a part of the open space in the upper portion of the bars in the flow path is clogged. The use of Eq. (5–1) is illustrated in Example 5–1.

EXAMPLE 5–1 **Headloss Buildup in Coarse Screens** Determine the buildup of headloss through a bar screen when 50 percent of the flow area is blocked off due to the accumulation of coarse solids. Assume the following conditions apply:

Approach velocity = 0.6 m/s
Velocity through clean bar screen = 0.9 m/s
Open area for flow through clean bar screen = 0.19 m²
Headloss coefficient for a clean bar screen = 0.7

Solution

1. Compute the clean water headloss through bar screen using Eq. (5–1).

$$h_L = \frac{1}{C}\left(\frac{V^2 - v^2}{2g}\right)$$

$$h_L = \frac{1}{0.7}\left[\frac{(0.9\text{ m/s})^2 - (0.6\text{ m/s})^2}{2(9.81\text{ m/s}^2)}\right] = 0.033\text{ m}$$

2. Estimate the headloss through the clogged bar screen (reducing the screen area by 50 percent results in a doubling of the velocity).
   The velocity through the clogged bar screen is

$$V_c = 0.9\text{ m/s} \times 2 = 1.8\text{ m/s}$$

Assuming the flow coefficient for the clogged bar screen is approximately 0.6, the estimated headloss is

$$h_L = \frac{1}{0.6}\left[\frac{(1.8\text{ m/s})^2 - (0.6\text{ m/s})^2}{2(9.81\text{ m/s}^2)}\right] = 0.24\text{ m}$$

Comment Where mechanically cleaned coarse screens are used, the cleaning mechanism typically is actuated by the buildup of headloss. Headloss is determined by measuring the water level before and after the screen. In some cases, the screen is cleaned at predetermined time intervals, as well as at a maximum head differential.

Although most screens use rectangular bars, optional shapes, i.e., "teardrop" and trapezoidal, are available. For the optional shapes, the wider width dimension is located on the upstream side of the bar rack to make it easier to dislodge materials trapped between the bars. The alternative shapes also reduce headloss through the rack.

Screenings from the rake mechanism are usually discharged directly into a hopper or container or into a screenings press. For installations with multiple units, the screenings may be discharged onto a conveyor or into a pneumatic ejector system and transported to a common screenings storage hopper. As an alternative, screenings grinders may be used to grind and shred the screenings. Ground screenings are then returned to the wastewater; however, ground screenings may adversely affect operation and maintenance of downstream equipment such as clogging weir openings on sedimentation tanks or wrapping around air diffusers. Application of screenings grinders is discussed in Sec. 5–3.

Fine Screens

The applications for fine screens range over a broad spectrum; uses include preliminary treatment (following coarse bar screens), primary treatment (as a substitute for primary clarifiers), and treatment of combined sewer overflows. Fine screens can also be used to remove solids from primary effluent that could cause clogging problems in trickling filters.

Screens for Preliminary and Primary Treatment. Fine screens used for preliminary treatment are of the (1) static (fixed), (2) rotary drum, or (3) step type. Typically, the openings vary from 0.2 to 6 mm (0.01 to 0.25 in). Examples of fine screens are illustrated on Fig. 5–4, descriptive information is provided in Table 5–4, and addi-

(a)

(b)

Figure 5–4

Typical fine screens:
(a) static wedgewire,
(b) drum, and (c) step. In step screens, screenings are moved up the screen by means of movable and fixed vertical plates.

(c)

Table 5–4
Description of screening devices used in wastewater treatment

| Type of screening device | Screening surface | | | | Application | See fig. |
| | Size classification | Size range | | Screen medium | | |
		in	mm[a]			
Inclined (fixed)	Medium	0.01–0.1	0.25–2.5	Stainless-steel wedgewire screen	Primary treatment	5–4a
Drum (rotary)	Coarse	0.1–0.2	2.5–5	Stainless-steel wedgewire screen	Preliminary treatment	5–4b
	Medium	0.01–0.1	0.25–2.5	Stainless-steel wedgewire screen	Primary treatment	
	Fine		6–35 μm	Stainless-steel and polyester screen cloths	Removal of residual secondary suspended solids	5–6
Horizontal reciprocating	Medium	0.06–0.17	1.6–4	Stainless-steel bars	Combined sewer overflows/ stormwater	5–5a
Tangential	Fine	0.0475	1200 μm	Stainless-steel mesh	Combined sewer overflows	5–5b

[a] Unless otherwise noted.

Table 5–5
Typical data on the removal of BOD and TSS with fine screens used to replace primary sedimentation[a]

| Type of screen | Size of openings | | Percent removed | |
	in	mm	BOD	TSS
Fixed parabolic	0.0625	1.6	5–20	5–30
Rotary drum	0.01	0.25	25–50	25–45

[a] The actual removal achieved will depend on the nature of the wastewater-collection system and the wastewater travel time.

tional information is given below. In many cases, application of fine screens is limited to plants where headloss through the screens is not a problem.

Fine screens may be used to replace primary treatment at small wastewater-treatment plants, up to 0.13 to m³/s (3 Mgal/d) in design capacity. Typical removal rates of BOD and TSS are reported in Table 5–5. Stainless-steel mesh or special wedge-shaped bars are used as the screening medium. Provision is made for the continuous removal of the collected solids, supplemented by water sprays to keep the screening medium clean. Headloss through the screens may range from about 0.8 to 1.4 m (2.5 to 4.5 ft).

Static Wedgewire Screens Static wedgewire screens (see Fig. 5–4a) customarily have 0.2 to 1.2 mm (0.01 to 0.06 in) clear openings and are designed for flowrates of about 400 to 1200 L/m²·min (10 to 30 gal/ft²·min) of screen area. Headloss ranges from 1.2 to 2 m (4 to 7 ft). The wedgewire medium consists of small, stainless-steel wedge-shaped bars with the flat part of the wedge facing the flow. Appreciable floor area is required for installation and the screens must be cleaned once or twice daily with high-pressure hot water, steam, or degreaser to remove grease buildup. Static wedgewire screens are generally applicable to smaller plants or for industrial installations.

Drum Screens For the drum-type screen (see Fig. 5–4b), the screening or straining medium is mounted on a cylinder that rotates in a flow channel. The construction varies, principally with regard to the direction of flow through the screening medium. The wastewater flows either into one end of the drum and outward through the screen with the solids collection on the interior surface, or into the top of the unit and passing through to the interior with solids collection on the exterior. Internally fed screens are applicable for flow ranges of 0.03 to 0.8 m³/s (0.7 to 19 Mgal/d) per screen, while externally fed screens are applicable for flowrates less than 0.13 m³/s (3 Mgal/d) (Laughlin and Roming, 1993). Drum screens are available in various sizes, from 0.9 to 2 m (3 to 6.6 ft) in diameter and from 1.2 to 4 m (4 to 13.3 ft) in length.

Step Screens Step screens, although widely used in Europe, are a relatively new technology in fine screening in the United States. The design consists of two step-shaped sets of thin vertical plates, one fixed and one movable (see Fig. 5–4c). The fixed and movable step plates alternate across the width of an open channel and together form a single screen face. The movable plates rotate in a vertical motion. Through this motion, solids captured on the screen face are automatically lifted up to the next fixed step landing, and are eventually transported to the top of the screen where they are discharged to a collection hopper. The circular pattern of the moving plates provides a self-cleaning feature for each step. Normal ranges of openings between the screen plates are 3 to 6 mm (0.12 or 0.24 in); however, openings as small as 1 mm (0.04 in) are available. Solids trapped on the screen also create a "filter mat" that enhances solids removal performance. In addition to wastewater screening, step screens can be used for removal of solids from septage, primary sludge, or digested biosolids.

Fine Screens for Combined Sewer Overflows. Screens have also been developed specifically for the removal of floatable and other solids from combined sewer overflows. Two basic types are used: horizontal reciprocating screens and tangential flow screens. The horizontal reciprocating screen is a rigid, weir-mounted screen configured of narrow stainless-steel bars that run the length of the device (see Fig. 5–5a). The screening bars are parallel to the normal direction of flow and are designed in continuous runs with no intermediate supports to collect solids. As the water in the screen channel rises, wastewater begins to pass through the openings in the screen bars. Solids are trapped on the screen, and, as the level continues to rise in the channel due to the entrapped solids on the screen, a hydraulically driven rake assembly is automatically activated to remove the accumulated solids from the screen. The rake carriage travels back and forth across the screen, combing the entrapped solids. The combing tines of the rake assembly carry the solids to one end of the screen for disposal either

Figure 5–5

Devices used for the screening of combined sewer overflows: (a) horizontal type in which the cutting mechanism travels back and forth horizontally along the bars (Photo courtesy of Parkson) and (b) tangential flow in which the flow passes through a cylindrical screen leaving the solids to accumulate in a sump. (Adapted from CDS Technologies Catalog.)

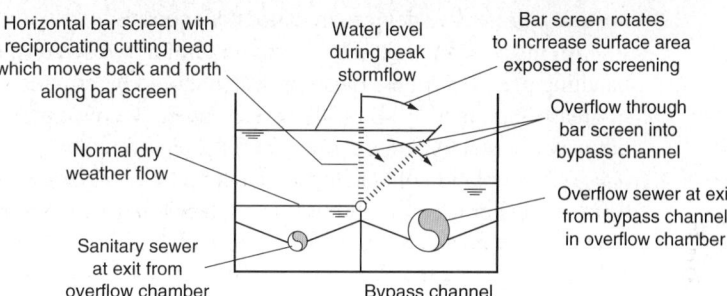

(a)

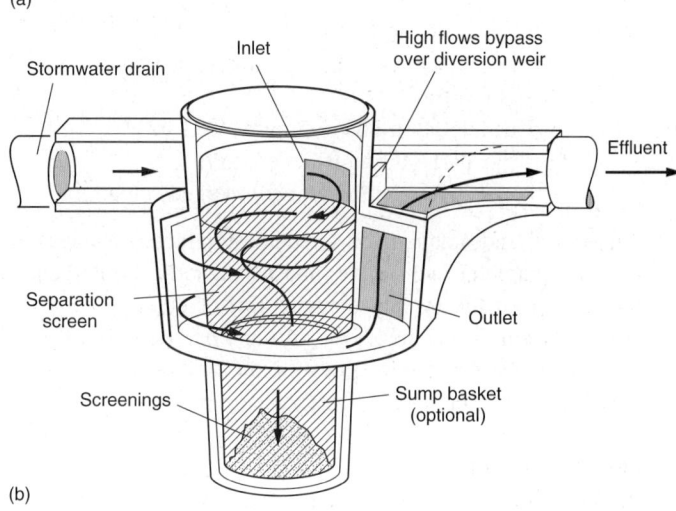

(b)

into the wastewater channel that carries flow to the wastewater-treatment plant or to a solids-collection pit.

In the tangential flow screen, the technology relies on the natural motion of water to screen and trap solids. The separation process is effected using a fine-mesh cylindrical screen and requires no moving parts (see Fig. 5–5b). As the wastewater flows into the separation chamber, a circular motion is generated that is designed to allow water to pass through the cylindrical screen while forcing solids to swirl toward the center of the chamber. The swirling water is regulated so that the tangential flow around the chamber is greater than the radial force attempting to push the solids outward. Thus, the accumulation of solids on the screen is minimized. The solids settle into a central sump where they can be removed. Floatables are trapped in the separation chamber until the flow stops, when they can be removed (Williams, 2000).

Design of Fine-Screen Installations. Mechanically cleaned coarse screens should precede some types of fine screens. Newer designs of internally fed rotary screens that use wedgewire instead of screen fabric are structurally more rugged. These designs can handle coarse solids that are transported through wastewater pumps; thus upstream protective devices may not be required.

An installation should have a minimum of two screens, each with the capability of handling peak flowrates. Flushing water should be provided nearby so that the buildup of grease and other solids on the screen can be removed periodically. In colder climates, hot water or steam is more effective for grease removal.

The calculation of headloss through fine screens differs from that of coarse screens. The clear water headloss through fine screens may be obtained from manufacturers' rating tables, or calculated using Eq. (5–2):

$$h_L = \frac{1}{2g}\left(\frac{Q}{CA}\right)^2 \qquad\qquad (5\text{–}2)$$

where h_L = headloss, m
$\quad\quad C$ = coefficient of discharge for the screen (a typical value for a clean screen is 0.60)
$\quad\quad g$ = acceleration due to gravity, 9.81 m/s^2
$\quad\quad Q$ = discharge through screen, m^3/s
$\quad\quad A$ = effective open area of submerged screen, m^2

Values of C and A depend on screen design factors, such as the size and milling of slots, the wire diameter and weave, and particularly the percent of open area, and must be obtained from the screen manufacturer or determined experimentally. The important determination is the headloss during operation; headloss depends on the size and amount of solids in the wastewater, the size of the apertures, and the method and frequency of cleaning.

Microscreens

Microscreening involves the use of variable low-speed (up to 4 r/min), continuously backwashed, rotating-drum screens operating under gravity-flow conditions (see Fig. 5–6). The filtering fabrics have openings of 10 to 35 μm and are fitted on the drum periphery. The wastewater enters the open end of the drum and flows outward through

Figure 5–6

Microscreens used in wastewater treatment as a replacement for primary treatment: (a) disk type with stainless-steel fabric and (b) drum type with wedgewire screen. The size of the openings on both screens is 250 μm (0.010 in).

(b)

(a)

the rotating-drum screening cloth. The collected solids are backwashed by high-pressure jets into a trough located within the drum at the highest point of the drum. (See also Sec. 11–5 for cloth medium surface filters used in advanced wastewater treatment.) The principal applications for microscreens are to remove suspended solids from secondary effluent and from stabilization-pond effluent.

Typical suspended-solids removal achieved with microscreens ranges from 10 to 80 percent, with an average of 55 percent. Problems encountered with microscreens include incomplete solids removal and inability to handle solids fluctuations. Reducing the rotating speed of the drum and less frequent flushing of the screen have resulted in increased removal efficiencies but reduced capacity.

The functional design of a microscreen involves (1) characterizing the suspended solids with respect to the concentration and degree of flocculation, (2) selecting design parameters that will not only assure sufficient capacity to meet maximum hydraulic loadings with critical solids characteristics but also meet operating performance requirements over the expected range of hydraulic and solids loadings, and (3) providing backwash and cleaning facilities to maintain the capacity of the screen. Typical design information for microscreens is presented in Table 5–6. Because of the variable performance of microscreens, pilot-plant studies are recommended, especially if the units are to be used to remove solids from stabilization-pond effluent, which may contain significant amounts of algae.

Screenings Characteristics and Quantities

Screenings are the material retained on bar racks and screens. The smaller the screen opening, the greater will be the quantity of collected screenings. While no precise definition of screenable material exists, and no recognized method of measuring quantities of screenings is available, screenings exhibit some common properties.

Table 5–6

Typical design information for microscreens used for screening secondary settled effluent[a]

Item	Typical value		Remarks
	U.S. customary units	**SI units**	
Screen size	20–35 μm	20–35 μm	Stainless-steel or polyester screen cloths are available in sizes ranging from 15 to 60 μm
Hydraulic loading rate	75–150 gal/ft²·min	3–6 m³/m²·min	Based on submerged surface area of drum
Headloss through screen	3–6 in	75–150 mm	Bypass should be provided when headloss exceeds 200 mm (8 in)
Drum submergence	70–75% of height; 60–70% of area	70–75% of height; 60–70% of area	Varies depending on screen design
Drum diameter	8–16 ft	2.5–5 m	3 m (10 ft) is most commonly used size; smaller sizes increase backwash requirements
Drum speed	15 ft/min at 3 in headloss	4.5 m/min at 75 mm headloss	Maximum rotational speed is limited to 45 m/min (150 ft/min)
Backwash requirements	2% of throughput at 50 lb$_f$/in²; 5% of throughput at 15 lb$_f$/in²	2% of throughput at 350 kPa; 5% of throughput at 100 kPa	

[a] Adapted in part from Tchobanoglous, 1988.

Screenings Retained on Coarse Screens. Coarse screenings, collected on coarse screens of about 12 mm (0.5 in) or greater spacing, consist of debris such as rocks, branches, pieces of lumber, leaves, paper, tree roots, plastics, and rags. Organic matter can collect as well. The accumulation of oil and grease can be a serious problem, especially in cold climates. The quantity and characteristics of screenings collected for disposal vary, depending on the type of bar screen, the size of the bar screen opening, the type of sewer system, and the geographic location. Typical data on the characteristics and quantities of coarse screenings to be expected at wastewater-treatment plants served by conventional gravity sewers are reported in Table 5–7.

Combined storm and sanitary collection systems may produce volumes of screenings several times the amounts produced by separate systems. The quantities of screenings have also been observed to vary widely, ranging from large quantities during the "first flush" to diminishing amounts as the wet weather flows persist. The quantities of screenings removed from combined sewer flows are reported to range from 3.5 to 84 L/1000 m³ of flow (0.5 to 11.3 ft³/Mgal) (WEF, 1998b).

Screenings Retained on Fine Screens. Fine screenings consist of materials that are retained on screens with openings less than 6 mm (0.25 in). The materials retained on fine screens include small rags, paper, plastic materials of various types, razor blades, grit, undecomposed food waste, feces, etc. Compared to coarse screen-

Table 5-7

Typical information on the characteristics and quantities of screenings removed from wastewater with coarse screens

Size of opening between bars, mm	Moisture content, %	Specific weight, kg/m³	Volume of screenings			
			ft³/Mgal		L/1000 m³	
			Range	Typical	Range	Typical
12.5	60–90	700–1100	5–10	7	37–74	50
25	50–80	600–1000	2–5	3	15–37	22
37.5	50–80	600–1000	1–2	1.5	7–15	11
50	50–80	600–1000	0.5–1.5	0.8	4–11	6

Note: mm × 0.3937 = in
kg/m³ × 8.3492 = lb/1000 gal

Table 5-8

Typical information on the characteristics and quantities of screenings removed from wastewater with fine bar and rotary-drum screens

Operation	Size of opening, mm	Moisture content, %	Specific weight, kg/m³	Volume of screenings			
				ft³/Mgal		L/1000 m³	
				Range	Typical	Range	Typical
Fine bar screens	12.5	80–90	900–1100	6–15	10	44–110	75
Rotary drum[a]	6.25	80–90	900–1100	4–8	6	30–60	45

[a] Following coarse screening.
Note: mm × 0.3937 = in
kg/m³ × 8.3492 = lb/1000 gal

ings, the specific weight of the fine screenings is slightly lower and the moisture content is slightly higher (see Table 5–8). Because putrescible matter, including fecal material, is contained within screenings, they must be handled and disposed of properly. Fine screenings contain substantial grease and scum, which require similar care, especially if odors are to be avoided.

Screenings Handling, Processing, and Disposal. In mechanically cleaned screen installations, screenings are discharged from the screening unit directly into a screenings grinder, a pneumatic ejector, or a container for disposal; or onto a conveyor for transport to a screenings compactor or collection hopper. Belt conveyors and pneumatic ejectors are generally the primary means of mechanically transporting screenings. Belt conveyors offer the advantages of simplicity of operation, low maintenance, freedom from clogging, and low cost. Belt conveyors give off odors and may have to be provided

Figure 5-7

Typical device used for compacting screenings.

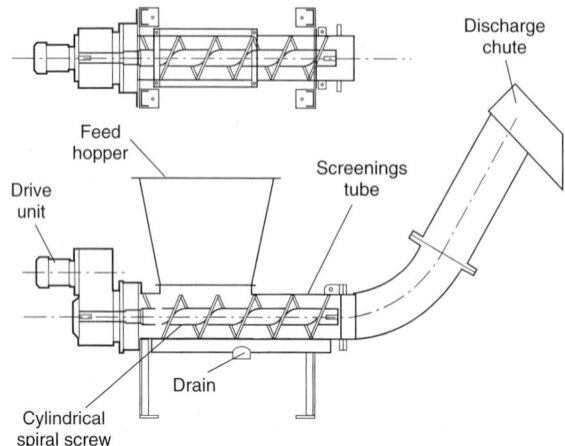

with covers. Pneumatic ejectors are less odorous and typically require less space; however, they are subject to clogging if large objects are present in the screenings.

Screenings compactors can be used to dewater and reduce the volume of screenings (see Fig. 5–7). Such devices, including hydraulic ram and screw compactors, receive screenings directly from the bar screens and are capable of transporting the compacted screenings to a receiving hopper. Compactors can reduce the water content of the screenings by up to 50 percent and the volume by up to 75 percent. As with pneumatic ejectors, large objects can cause jamming, but automatic controls can sense jams, automatically reverse the mechanism, and actuate alarms and shut down equipment.

Means of disposal of screenings include (1) removal by hauling to disposal areas (landfill) including codisposal with municipal solid wastes, (2) disposal by burial on the plant site (small installations only), (3) incineration either alone or in combination with sludge and grit (large installations only), and (4) discharge to grinders or macerators where they are ground and returned to the wastewater. The first method of disposal is most commonly used. In some states, screenings are required to be lime stabilized for the control of pathogenic organisms before disposal in landfills. Grinding the screenings and returning them to the wastewater flow shares many of the disadvantages cited under comminution, as discussed in the following section.

5-2 COARSE SOLIDS REDUCTION

As an alternative to coarse bar screens or fine screens, comminutors and macerators can be used to intercept coarse solids and grind or shred them in the screen channel. High-speed grinders are used in conjunction with mechanically cleaned screens to grind and shred screenings that are removed from the wastewater. The solids are cut up into a smaller, more uniform size for return to the flow stream for subsequent removal by downstream treatment operations and processes. Comminutors, macerators, and grinders can theoretically eliminate the messy and offensive task of screenings handling and disposal. The use of comminutors and macerators is particularly advantageous in a pumping station to protect the pumps against clogging by rags and large objects and to eliminate the need to handle and dispose of screenings. They are particularly useful in cold climates where collected screenings are subject to freezing.

There is a wide divergence of views, however, on the suitability of using devices that grind and shred screenings at wastewater-treatment plants. One school of thought maintains that once coarse solids have been removed from wastewater, they should not be returned, regardless of the form. The other school of thought maintains that once cut up, the solids are more easily handled in the downstream processes. Shredded solids often present downstream problems, particularly with rags and plastic bags, as they tend to form ropelike strands. Rag and plastic strands can have a number of adverse impacts, such as clogging pump impellers, sludge pipelines, and heat exchangers, and accumulating on air diffusers and clarifier mechanisms. Plastics and other nonbiodegradable material may also adversely affect the quality of biosolids that are to be beneficially reused.

Approaches to using comminutors, macerators, and grinders are applicable in many retrofit situations. Examples of retrofit applications include plants where a spare channel has been provided for the future installation of a duplicate unit or in very deep influent pumping stations where the removal of screenings may be too difficult or costly to achieve. Alternative approaches may also be possible, such as using chopper pumps at pumping stations (see Sec. 14–4) or installing grinders ahead of sludge pumps.

Comminutors

Comminutors are used most commonly in small wastewater-treatment plants, less than 0.2 m³/s (5 Mgal/d). Comminutors are installed in a wastewater flow channel to screen and shred material to sizes from 6 to 20 mm (0.25 to 0.77 in) without removing the shredded solids from the flow stream. A typical comminutor uses a stationary horizontal screen to intercept the flow (see Fig. 5–8) and a rotating or oscillating arm that

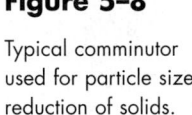

Figure 5–8

Typical comminutor used for particle size reduction of solids.

contains cutting teeth to mesh with the screen. The cutting teeth and the shear bars cut coarse material. The small sheared particles pass through the screen and into the downstream channel. Comminutors may create a string of material, namely, rags, that can collect on downstream treatment equipment. Because of operating problems and high maintenance with comminutors, newer installations generally use a screen or a macerator described below.

Macerators

Macerators are slow-speed grinders that typically consist of two sets of counterrotating assemblies with blades (see Fig. 5–9a). The assemblies are mounted vertically in the flow channel. The blades or teeth on the rotating assemblies have a close tolerance that effectively chops material as it passes through the unit. The chopping action reduces the potential for producing ropes of rags or plastic that can collect on downstream equipment. Macerators can be used in pipeline installations to shred solids, particularly ahead of wastewater and sludge pumps, or in channels at smaller wastewater-treatment plants. Sizes for pipeline applications typically range from 100 to 400 mm (4 to 16 in) in diameter. Grinders for solids (sludge) applications are discussed in Chap. 14.

Another type of macerator used in channel applications is a moving, linked screen that allows wastewater to pass through the screen while diverting screenings to a

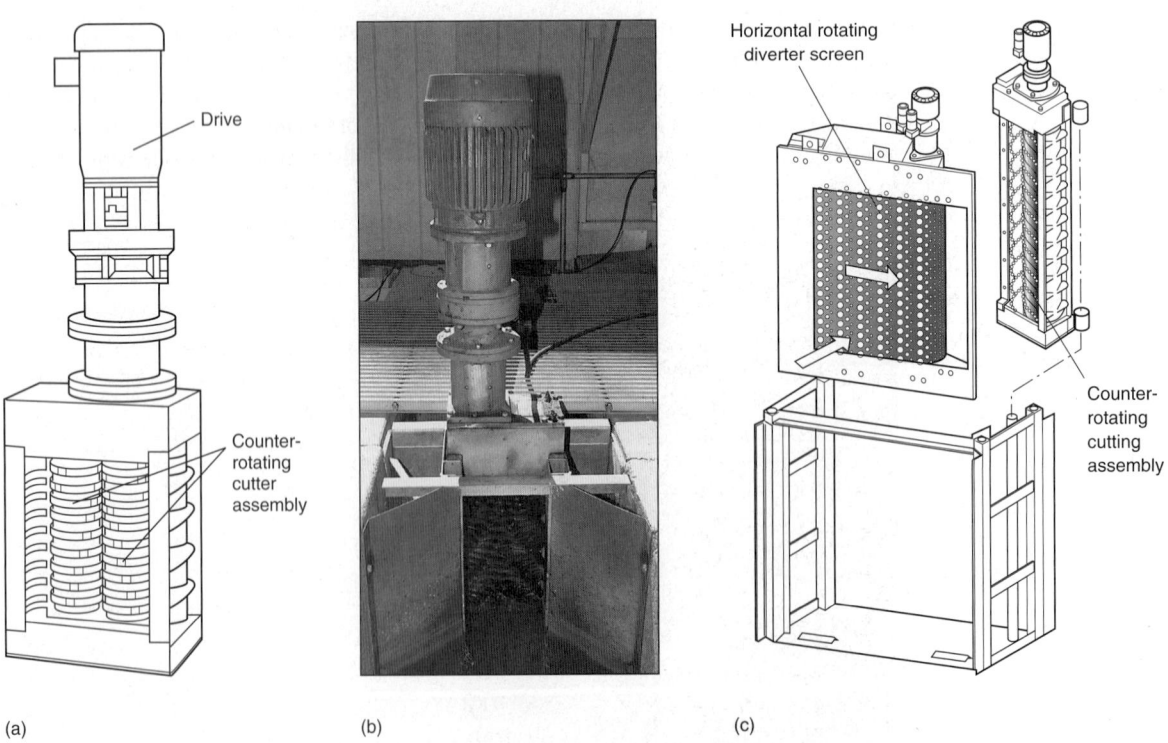

(a) (b) (c)

Figure 5–9

Typical macerators: (a) schematic of in-channel type slow-speed grinder/macerator, (b) view of a macerator mounted in an open channel, and (c) schematic of linked-screen macerator.

grinder located at one side of the channel (see Fig. 5–9c). Standard sizes of this device are available for use in large channels ranging from widths of 750 to 1800 mm (30 to 72 in) and depths of 750 to 2500 mm (30 to 100 in). The headloss is lower than that of the units with counterrotating blades shown on Fig. 5–9a.

Grinders

High-speed grinders, typically referred to as hammermills, receive screened materials from bar screens. The materials are pulverized by a high-speed rotating assembly that cuts the materials passing through the unit. The cutting or knife blades force screenings through a stationary grid or louver that encloses the rotating assembly. Washwater is typically used to keep the unit clean and to help transport materials back to the waste-water stream. Discharge from the grinder can be located either upstream or downstream of the bar screen.

Design Considerations

Comminuting and macerating devices may be preceded by grit chambers to prolong the life of the equipment and to reduce the wear on the cutting surfaces. Comminutors should be constructed with a bypass arrangement so that a manual bar screen is used in case flowrates exceed the capacity of the comminutor or when there is a power or mechanical failure. Stop gates and provisions for dewatering the channel should also be included to facilitate maintenance. Headloss through a comminutor usually ranges from 0.1 to 0.3 m (4 to 12 in), and can approach 0.9 m (3 ft) in large units at maximum flowrates.

In cases where a comminutor or macerator precedes grit chambers, the cutting teeth are subject to high wear and require frequent sharpening or replacement. Units that use cutting mechanisms ahead of the screen grid should be provided with rock traps in the channel upstream of the comminutor to collect material that could jam the cutting blade.

Because these units are complete in themselves, no detailed design is necessary. Manufacturers' data and rating tables for these units should be consulted for recommended channel dimensions, capacity ranges, headloss, upstream and downstream submergence, and power requirements. Because manufacturers' capacity ratings are usually based on clean water, the ratings should be decreased by approximately 80 percent to account for partial clogging of the screen.

5–3 FLOW EQUALIZATION

The variations of influent wastewater flowrate and characteristics at wastewater-treatment facilities were discussed in Chap. 3. Flow equalization is a method used to overcome the operational problems caused by flowrate variations, to improve the performance of the downstream processes, and to reduce the size and cost of downstream treatment facilities.

Description/Application

Flow equalization simply is the damping of flowrate variations to achieve a constant or nearly constant flowrate and can be applied in a number of different situations, depending on the characteristics of the collection system. The principal applications are for the

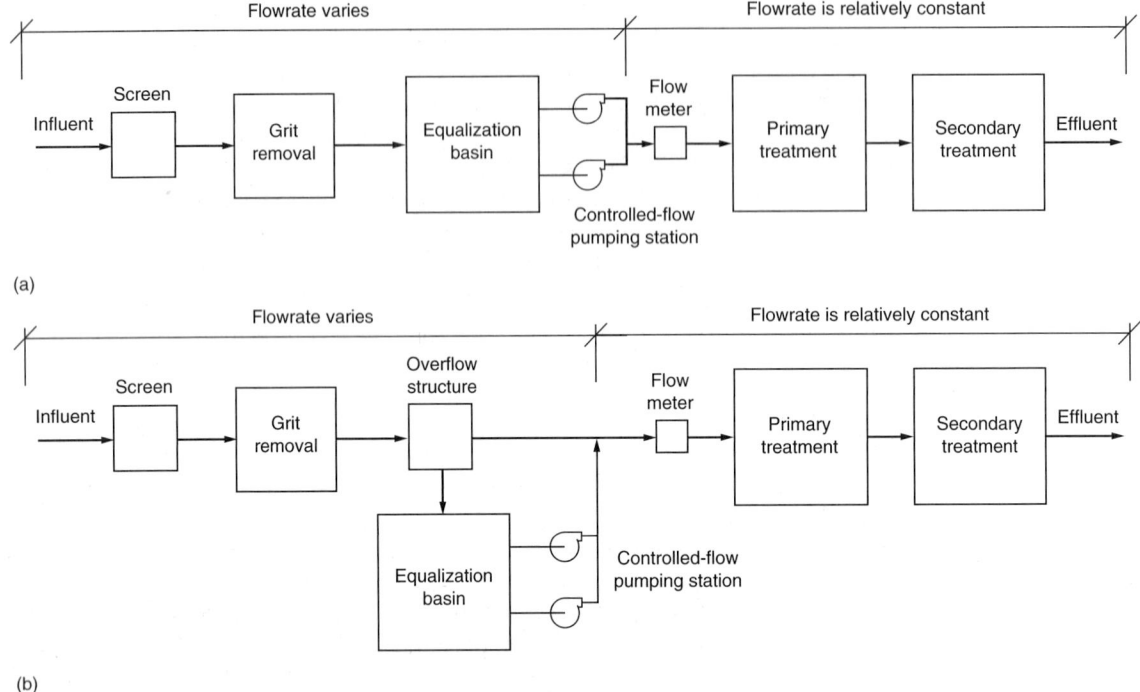

(a)

(b)

Figure 5–10

Typical wastewater-treatment plant flow diagram incorporating flow equalization: (a) in-line equalization and (b) off-line equalization. Flow equalization can be applied after grit removal, after primary sedimentation, and after secondary treatment where advanced treatment is used.

equalization of (1) dry-weather flows to reduce peak flows and loads, (2) wet-weather flows in sanitary collection systems experiencing inflow and infiltration, or (3) combined stormwater and sanitary system flows.

The application of flow equalization in wastewater treatment is illustrated in the two flow diagrams given on Fig. 5–10. In the in-line arrangement (Fig. 5–10a), all of the flow passes through the equalization basin. This arrangement can be used to achieve a considerable amount of constituent concentration and flowrate damping. In the off-line arrangement (Fig. 5–10b), only the flow above some predetermined flow limit is diverted into the equalization basin. Although pumping requirements are minimized in this arrangement, the amount of constituent concentration damping is considerably reduced. Off-line equalization is sometimes used to capture the "first flush" from combined collection systems.

The principal benefits that are cited as deriving from application of flow equalization are: (1) biological treatment is enhanced, because shock loadings are eliminated or can be minimized, inhibiting substances can be diluted, and pH can be stabilized; (2) the effluent quality and thickening performance of secondary sedimentation tanks following biological treatment is improved through improved consistency in solids loading; (3) effluent filtration surface area requirements are reduced, filter performance is improved, and more uniform filter-backwash cycles are possible by lower hydraulic loading; and (4) in chemical treatment, damping of mass loading improves chemical feed control and process reliability. Apart from improving the performance of most treatment operations and processes, flow equalization is an attractive option for upgrading the performance of overloaded treatment plants. Disadvantages of flow equalization include (1) relatively large land areas or sites are needed, (2) equalization facilities may have to be covered for odor control near residential areas, (3) additional operation and maintenance is required, and (4) capital cost is increased.

Design Considerations

The design of flow equalization facilities is concerned with the following questions:

1. Where in the treatment process flowsheet should the equalization facilities be located?
2. What type of equalization flowsheet should be used, in-line or off-line?
3. What is the required basin volume?
4. What are the features that should be incorporated into design?
5. How can the deposition of solids and potential odors be controlled?

Location of Equalization Facilities. The best location for equalization facilities must be determined for each system. Because the optimum location will vary with the characteristics of the collection system and the wastewater to be handled, land requirements and availability, and the type of treatment required, detailed studies should be performed for several locations throughout the system. Where equalization facilities are considered for location adjacent to the wastewater-treatment plant, it is necessary to evaluate how they could be integrated into the treatment process flowsheet. In some cases, equalization after primary treatment and before biological treatment may be appropriate. Equalization after primary treatment causes fewer problems with solids deposits and scum accumulation. If flow-equalization systems are to be located ahead of primary settling and biological systems, the design must provide for sufficient mixing to prevent solids deposition and concentration variations, and aeration to prevent odor problems.

In-Line or Off-Line Equalization. As shown on Fig. 5–10, it is possible to achieve considerable damping of constituent mass loadings to the downstream processes with in-line equalization, but only slight damping is achieved with off-line equalization. The analysis of the effect of in-line equalization on the constituent mass loading is illustrated in Example 5–2.

Volume Requirements for the Equalization Basin. The volume required for flowrate equalization is determined by using an inflow cumulative volume diagram in which the cumulative inflow volume is plotted versus the time of day. The average daily flowrate, also plotted on the same diagram, is the straight line drawn from the origin to the endpoint of the diagram. Diagrams for two typical flowrate patterns are shown on Fig. 5–11.

To determine the required volume, a line parallel to the coordinate axis, defined by the average daily flowrate, is drawn tangent to the mass inflow curve. The required

Figure 5–11

Schematic mass diagrams for the determination of the required equalization basin storage volume for two typical flowrate patterns.

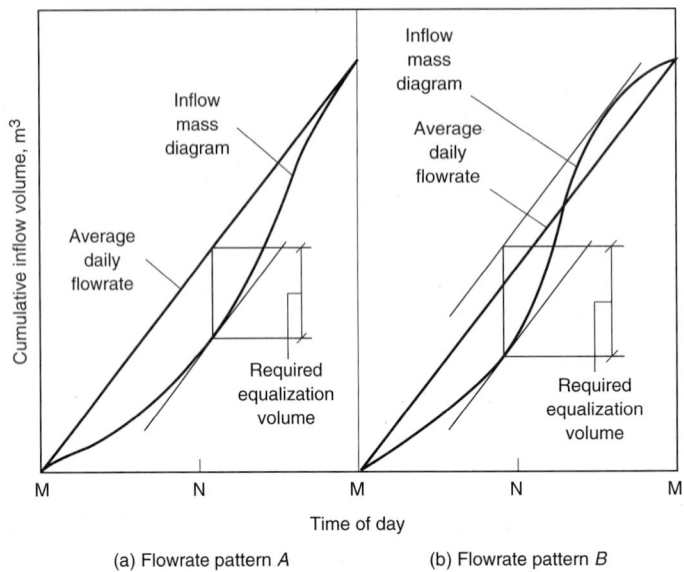

(a) Flowrate pattern A   (b) Flowrate pattern B

volume is then equal to the vertical distance from the point of tangency to the straight line representing the average flowrate (see Fig. 5–11a). If the inflow mass curve goes above the line representing the average flowrate (see Fig. 5–11b), the inflow mass diagram must be bounded with two lines that are parallel to the average flowrate line and tangent to extremities of the inflow mass diagram. The required volume is then equal to the vertical distance between the two lines. The determination of the required volume for equalization is also illustrated in Example 5–2. The procedure is exactly the same as if the average hourly volume were subtracted from the volume flow occurring each hour, and the resulting cumulative volumes were plotted. In this case, the low and high points of the curve would be determined using a horizontal line.

The physical interpretation of the diagrams shown on Fig. 5–11 is as follows. At the low point of tangency (flowrate pattern A) the storage basin is empty. Beyond this point, the basin begins to fill because the slope of the inflow mass diagram is greater than that of the average daily flowrate. The basin continues to fill until it becomes full at midnight. For flowrate pattern B, the basin is filled at the upper point of tangency.

In practice, the volume of the equalization basin will be larger than that theoretically determined to account for the following factors:

1. Continuous operation of aeration and mixing equipment will not allow complete drawdown, although special structures can be built.
2. Volume must be provided to accommodate the concentrated plant recycle streams that are expected, if such flows are returned to the equalization basin (a practice that is not recommended unless the basin is covered because of the potential to create odors).
3. Some contingency should be provided for unforeseen changes in diurnal flow.

Although no fixed value can be given, the additional volume will vary from 10 to 20 percent of the theoretical value, depending on the specific conditions.

EXAMPLE 5–2 Determination of Flowrate Equalization Volume Requirements and Effects on BOD Mass Loading For the flowrate and BOD concentration data (derived from Fig. 3–5) given in the following table, determine (1) the in-line storage volume required to equalize the flowrate, and (2) the effect of flow equalization on the BOD mass loading rate.

Time period	Given data Average flowrate during time period, m³/s	Given data Average BOD concentration during time period, mg/L	Derived data Cumulative volume of flow at end of time period, m³	Derived data BOD mass loading during time period, kg/h
M–1	0.275	150	990	149
1–2	0.220	115	1,782	91
2–3	0.165	75	2,376	45
3–4	0.130	50	2,844	23
4–5	0.105	45	3,222	17
5–6	0.100	60	3,582	22
6–7	0.120	90	4,014	39
7–8	0.205	130	4,752	96
8–9	0.355	175	6,030	223
9–10	0.410	200	7,506	295
10–11	0.425	215	9,036	329
11–N	0.430	220	10,584	341
N–1	0.425	220	12,114	337
1–2	0.405	210	13,572	306
2–3	0.385	200	14,958	277
3–4	0.350	190	16,218	239
4–5	0.325	180	17,388	211
5–6	0.325	170	18,558	199
6–7	0.330	175	19,746	208
7–8	0.365	210	21,060	276
8–9	0.400	280	22,500	403
9–10	0.400	305	23,940	439
10–11	0.380	245	25,308	335
11–M	0.345	180	26,550	224
Average	0.307			213

Note: m³/s × 35.3147 = ft³/s
m³ × 35.3147 = ft³
mg/L = g/m³

Solution

1. Determine the volume of the basin required for the flow equalization.
   a. The first step is to develop a cumulative volume curve of the wastewater flowrate expressed in cubic meters. The cumulative volume curve is obtained

by converting the average flowrate (q^i) during each hourly period to cubic meters, using the following expression, and then cumulatively by summing the hourly values to obtain the cumulative flow volume.

Volume, m^3 = (q_i, m^3/s)(3600 s/h)(1.0 h)

For example, for the first three time periods shown in the data table, the corresponding hourly volumes are as follows:

    For the time period M–1:

$$V_{M-1} = (0.275 \text{ m}^3/\text{s})(3600 \text{ s/h})(1.0 \text{ h}) = 990 \text{ m}^3$$

For the time period 1–2:

$$V_{1-2} = (0.220 \text{ m}^3/\text{s})(3600 \text{ s/h})(1.0 \text{ h}) = 792 \text{ m}^3$$

The cumulative flow, expressed in m^3, at the end of each time period is determined as follows:

    At the end of the first time period M–1:

$$V_1 = 990 \text{ m}^3$$

At the end of the second time period 1–2:

$$V_2 = 990 + 792 = 1782 \text{ m}^3$$

The cumulative flows for all the hourly time periods are computed in a similar manner (see derived data in data table).

b. The second step is to prepare a plot of the cumulative flow volume, as shown in the following diagram. As will be noted, the slope of the line drawn from the origin to the endpoint of the inflow mass diagram represents the average flowrate for the day, which in this case is equal to 0.307 m^3/s.

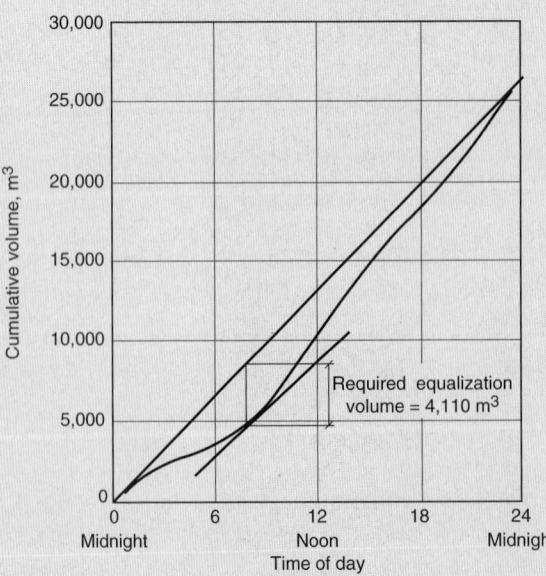

c. The third step is to determine the required storage volume. The required storage volume is determined by drawing a line parallel to the average flowrate tangent to the low point of the inflow mass diagram. The required volume is represented by the vertical distance from the point of tangency to the straight line representing the average flowrate. Thus, the required volume is equal to

Volume of equalization basin, $V = 4110 \text{ m}^3$ (145,100 ft³)

2. Determine the effect of the equalization basin on the BOD mass loading rate. Although there are alternative computation methods, perhaps the simplest way is to perform the necessary computations starting with the time period when the equalization basin is empty. Because the equalization basin is empty at about 8:30 A.M., the necessary computations will be performed starting with the 8–9 time period.

a. The first step is to compute the liquid volume in the equalization basin at the end of each time period. The volume required is obtained by subtracting the equalized hourly flowrate expressed as a volume from the inflow flowrate also expressed as a volume. The volume corresponding to the equalized flowrate for a period of 1 h is $0.307 \text{ m}^3/\text{s} \times 3600 \text{ s/h} = 1106 \text{ m}^3$. Using this value, the volume in storage is computed using the following expression:

$$V_{sc} = V_{sp} + V_{ic} - V_{oc}$$

where V_{sc} = volume in the equalization basin at the end of current time period
V_{sp} = volume in the equalization basin at the end of previous time period
V_{ic} = volume of inflow during the current time period
V_{oc} = volume of outflow during the current time period

Thus, using the values in the original data table, the volume in the equalization basin for the time period 8–9 is as follows:

$$V_{sc} = 0 + 1278 \text{ m}^3 - 1106 \text{ m}^3 = 172 \text{ m}^3$$

For time period 9–10:

$$V_{sc} = 172 \text{ m}^3 + 1476 \text{ m}^3 - 1106 \text{ m}^3 = 542 \text{ m}^3$$

The volume in storage at the end of each time period has been computed in a similar way (see the following computation table).

Time period	Volume of flow during time period, m³	Volume in storage at end of time period, m³	Average BOD concentration during time period, mg/L	Equalized BOD concentration during time period, mg/L	Equalized BOD mass loading during time period, kg/h
8–9	1278	172	175	175	193
9–10	1476	542	200	197	218
10–11	1530	966	215	210	232
11–N	1548	1408	220	216	239

(continued)

(*Continued*)

Time period	Volume of flow during time period, m³	Volume in storage at end of time period, m³	Average BOD concentration during time period, mg/L	Equalized BOD concentration during time period, mg/L	Equalized BOD mass loading during time period, kg/h
N–1	1530	1832	220	218	241
1–2	1458	2184	210	214	237
2–3	1386	2464	200	209	231
3–4	1260	2618	190	203	224
4–5	1170	2680	180	196	217
5–6	1170	2746	170	188	208
6–7	1188	2828	175	184	203
7–8	1314	3036	210	192	212
8–9	1440	3370	280	220	243
9–10	1440	3704	305	245	271
10–11	1368	3966	245	245	271
11–M	1242	4102	180	230	254
M–1	990	3986	150	214	237
1–2	792	3972	115	196	217
2–3	594	3160	75	179	198
3–4	468	2522	50	162	179
4–5	378	1794	45	147	162
5–6	360	1048	60	132	146
6–7	432	374	90	119	132
7–8	738	0	130	126	139
Average					213

Note: m³ × 35.3147 = ft³
kg × 2.2046 = lb
g/m³ = mg/L

b. The second step is to compute the average concentration leaving the storage basin. Using the following expression, which is based on the assumption that the contents of the equalization basin are mixed completely, the average concentration leaving the storage basin is

$$X_{oc} \frac{(V_{ic})(X_{ic}) + (V_{sp})(X_{sp})}{V_{ic} + V_{sp}}$$

where X_{oc} = average concentration of BOD in the outflow from the storage basin during the current time period, g/m³ (mg/L)

V_{ic} = volume of wastewater inflow during the current period, m³

X_{ic} = average concentration of BOD in the inflow wastewater volume, g/m³

V_{sp} = volume of wastewater on storage basin at the end of the previous time period, m³

X_{sp} = concentration of BOD in wastewater in storage basin at the end of the previous time period, g/m³

Using the data given in column 2 of the above computation table, the effluent concentration is computed as follows:

For the time period 8–9:

$$X_{oc} = \frac{(1278 \text{ m}^3)(175 \text{ g/m}^3) + (0)(0)}{1278 \text{ m}^3} = 175 \text{ g/m}^3$$

For the time period 9–10:

$$X_{oc} = \frac{(1476 \text{ m}^3)(200 \text{ g/m}^3) + (172 \text{ m}^3)(175 \text{ g/m}^3)}{(1476 + 172) \text{ m}^3} = 197 \text{ g/m}^3$$

All the concentration values computed in a similar manner are reported in the above computation table.

c. The third step is to compute the hourly mass loading rate using the following expression:

$$\text{Mass loading rate, kg/h} = \frac{(X_{oc}, \text{ g/m}^3)(q_i, \text{ m}^3/\text{s})(3600 \text{ s/h})}{(10^3 \text{ g/kg})}$$

For example, for the time period 8–9, the mass loading rate is

$$\frac{(175 \text{ g/m}^3)(0.307 \text{ m}^3/\text{s})(3600 \text{ s/h})}{(10^3 \text{ g/kg})} = 193 \text{ kg/h}$$

All hourly values are summarized in the computation table. The corresponding values without flow equalization are reported in the original data table.

d. The effect of flow equalization can be shown best graphically by plotting the hourly unequalized and equalized BOD mass loading (see the following plot). The following flowrate ratios, derived from the data presented in the table given in the problem statement and the computation table prepared in Step 2a, are also helpful in assessing the benefits derived from flow equalization:

Ratio	BOD mass loading	
	Unequalized	**Equalized**
$\dfrac{\text{Peak}}{\text{Average}}$	$\dfrac{439}{213} = 2.06$	$\dfrac{271}{213} = 1.27$
$\dfrac{\text{Minimum}}{\text{Average}}$	$\dfrac{17}{213} = 0.08$	$\dfrac{132}{213} = 0.62$
$\dfrac{\text{Peak}}{\text{Minimum}}$	$\dfrac{439}{17} = 25.82$	$\dfrac{271}{132} = 2.05$

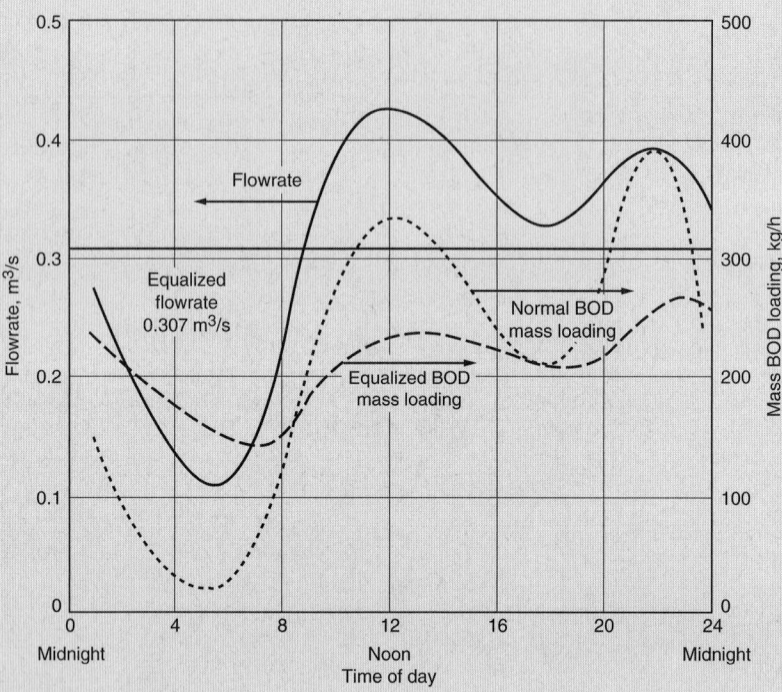

Comment Where in-line equalization basins are used, additional damping of the BOD mass loading rate can be obtained by increasing the volume of the basins. Although the flow to a treatment plant was equalized in this example, flow equalization would be used, more realistically, in locations with high infiltration/inflow or peak stormwater flows.

Basin Configuration and Construction. In equalization basin design, the principal factors that must be considered are (1) basin geometry; (2) basin construction including cleaning, access, and safety; (3) mixing and air requirements; (4) operational appurtenances; and (5) pump and pump control systems.

Basin Geometry The importance of basin geometry varies somewhat, depending on whether in-line or off-line equalization is used. If in-line equalization is used to dampen both the flow and the mass loadings, it is important to use a geometry that allows the basin to function as a continuous-flow stirred-tank reactor insofar as possible. Therefore, elongated designs should be avoided, and the inlet and outlet configurations should be arranged to minimize short circuiting. Discharging the influent near the mixing equipment usually minimizes short circuiting. If the geometry of the basins is controlled by the available land area and an elongated geometry must be used, it may be necessary to use multiple inlets and outlets. Provisions should be included in the basin design for access by cleaning equipment such as front-end loaders. Multiple compartments are also desirable to reduce cleaning costs and for odor control.

Basin Construction New basins may be of earthen, concrete, or steel construction; earthen basins are generally the least expensive. Depending on local conditions, the interior side slopes may vary between 3:1 and 2:1. A section through a typical earthen basin is shown on Fig. 5–12. In most installations, a liner is required to prevent groundwater contamination. Basin depths will vary depending on land availability, groundwater level, and topography. If a liner is used in areas of high groundwater, the effects of hydraulic uplift on the liner must be considered. The freeboard required depends on the surface area of the basin and local wind conditions. If a floating aerator is used to provide mixing and prevent septicity and odor formation, a minimum operating level is needed to protect the aerator. Typically, the minimum water depth can vary from 1.5 to 2 m (5 to 6 ft). With floating aerators, a concrete pad should be provided below the aerators to minimize erosion. To prevent wind-induced erosion in the upper portions of the basin, it may be necessary to protect the slopes with riprap, soil cement, or a partial concrete layer. Fencing should also be provided to prevent public access to the basins.

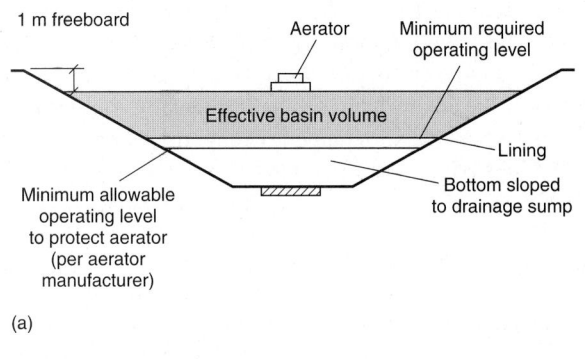

(a)

(b)

Figure 5–12

Typical open type flow equalization basins: (a) typical section through a lined earthen basin, (b) shallow concrete basin, and (c) deep concrete basin.

(c)

In areas of high groundwater, drainage facilities should be provided to prevent embankment failure. To further ensure a stable embankment, the tops of the dikes should be of adequate width. The use of an adequate dike width will facilitate the use of mechanical equipment for maintenance and will also reduce construction costs, especially where mechanical compaction equipment is used.

Mixing and Air Requirements The proper operation of both in-line and off-line equalization basins generally requires proper mixing and aeration. Mixing equipment should be sized to blend the contents of the tank and to prevent deposition of solids in the basin. To minimize mixing requirements, grit-removal facilities should precede equalization basins where possible. Mixing requirements for blending a medium-strength municipal wastewater (see Table 3–15), having a suspended solids concentration of approximately 210 mg/L, range from 0.004 to 0.008 kW/m^3 (0.02 to 0.04 hp/10^3 gal) of storage. Aeration is required to prevent the wastewater from becoming septic and odorous. To maintain aerobic conditions, air should be supplied at a rate of 0.01 to 0.015 m^3/m^3·min (1.25 to 2.0 ft^3/10^3 gal·min). In equalization basins that follow primary sedimentation and have short detention times (less than 2 h), aeration may not be required.

Where mechanical aerators are used, baffling may be necessary to ensure proper mixing, particularly with a circular tank configuration. To protect the aerators in the event of excessive level drawdown, low-level shutoff controls should be provided. Because it may be necessary to dewater the equalization basins periodically, the aerators should be equipped with legs or draft tubes that allow them to come to rest on the bottom of the basin without damage. Various types of diffused air systems may also be used for mixing and aeration including static tube, jet, and aspirating aerators (see Sec. 5–12).

Operational Appurtenances Among the appurtenances that should be included in the design of equalization basins are (1) facilities for flushing any solids and grease that may tend to accumulate on the basin walls; (2) a high-water takeoff for the removal of floating material and foam; (3) water sprays to prevent the accumulation of foam on the sides of the basin and to aid in scum removal; and (4) separate odor control facilities (see Chap. 15) where covered equalization basins must be used. Solids removed from equalization basins should be returned to the head of the plant for processing.

Pumps and Pump Control Because flow equalization imposes an additional head requirement within the treatment plant, pumping facilities are frequently required. Pumping may precede or follow equalization, but pumping into the basin is generally preferred for reliability of treatment operation. In some cases, pumping of both basin influent and equalized flows will be required.

An automatically controlled flow-regulating device will be required where gravity discharge from the basin is used. Where basin effluent pumps are used, instrumentation should be provided to control the preselected equalization rate. Regardless of the discharge method used, a flow-measuring device should be provided on the outlet of the basin to monitor the equalized flow.

5–4 MIXING AND FLOCCULATION

Mixing is an important unit operation in many phases of wastewater treatment including (1) mixing of one substance completely with another, (2) blending of miscible liquids, (3) flocculation of wastewater particles, (4) continuous mixing of liquid suspensions, and (5) heat transfer. Most mixing operations in wastewater can be classified as continuous-rapid (less than 30 s) or continuous (i.e., ongoing). Continuous rapid mixing and continuous mixing are considered in this section. Each unit operation is described separately below followed by a description of the types of mixers and flocculation devices and an analysis of the energy requirements for these operations.

Continuous Rapid Mixing in Wastewater Treatment

Continuous rapid mixing is used, most often, where one substance is to be mixed with another. The principal applications of continuous rapid mixing are in (1) the blending of chemicals with wastewater (e.g., the addition of alum or iron salts prior to flocculation and settling or for dispersing chlorine and hypochlorite into wastewater for disinfection), (2) the blending of miscible liquids, and (3) the addition of chemicals to sludge and biosolids to improve their dewatering characteristics. Typical examples of the types of mixers used in wastewater-treatment facilities for rapid mixing are reported in Table 5–9.

Continuous Mixing in Wastewater Treatment

Continuous mixing is used where the contents of a reactor or holding tank or basin must be kept in suspension such as in equalization basins, flocculation basins, suspended-growth biological treatment processes, aerated lagoons, and aerobic digesters. The two applications considered in the following discussion are flocculation and the maintenance of material in suspension.

Flocculation in Wastewater Treatment. The purpose of wastewater flocculation is to form aggregates or flocs from finely divided particles and from chemically destabilized particles. Flocculation is a transport step that brings about the collisions between the destabilized particles needed to form larger particles that can be removed readily by settling or filtration. Although not used routinely, flocculation of wastewater by mechanical or air agitation may be considered for (1) increasing removal of suspended solids and BOD in primary settling facilities, (2) conditioning wastewater containing certain industrial wastes, (3) improving performance of secondary settling tanks following the activated-sludge process, and (4) as a pretreatment step for the filtration of secondary effluent. When used, flocculation can be accomplished in separate tanks or basins specifically designed for the purpose, in in-line facilities such as in the conduits and pipes connecting the treatment units, or in combination with flocculator-clarifiers.

Flocculation typically follows rapid mixing where chemicals have been added to destabilize the particles. The destabilization of particles resulting from the addition of chemicals is defined as "coagulation" and is considered in Chap. 6. There are two types of flocculation: (1) microflocculation and (2) macroflocculation. The distinction between these two types of flocculation is based on the particle sizes involved.

Microflocculation (also known as *perikinetic flocculation*) is the term used to refer to the aggregation of particles brought about by the random thermal motion of fluid molecules. The random thermal motion of fluid molecules is also known as *Brownian motion* or *movement* (see Fig. 5–13a). Microflocculation is significant for particles that are in the size range from 0.001 to about 1 μm.

Table 5–9

Typical mixing times and applications for different mixing and flocculation devices used in wastewater-treatment facilities

Mixing device	Typical mixing times, s	Applications/remarks
Mixing and blending devices		
Static in-line mixers	<1	Used for chemicals requiring instantaneous mixing such as alum (Al^{3+}), ferric chloride (Fe^{3+}), cationic polymer, chlorine (Cl_2)
In-line mixers	<1	Used for chemicals requiring instantaneous mixing such as alum (Al^{3+}), ferric chloride (Fe^{3+}), cationic polymer, chlorine (Cl_2)
High-speed induction mixers	<1	Used for chemicals requiring instantaneous mixing such as alum (Al^{3+}), ferric chloride (Fe^{3+}), cationic polymer, chlorine (Cl_2)
Pressurized water jets	<1	Used in water-treatment practice and for reclaimed water applications
Turbine and propeller mixers	2–20	Used in back mix reactors for the mixing of alum in sweep floc applications. Actual time depends on the configuration of the vessel in which mixing is taking place. Mixing of chemicals in solution feed tanks
Pumps	<1	Chemicals to be mixed are introduced in the suction intake of the pump
Other hydraulic mixing devices	1–10	Hydraulic jumps, weirs, Parshall flumes, etc.
Flocculation devices		
Static mixers	600–1800	Used for flocculation of coagulated colloidal particles
Paddle mixers	600–1800	Used for flocculation of coagulated colloidal particles
Turbine mixers	600–1800	Used for flocculation of coagulated colloidal particles
Continuous mixing		
Mechanical aerators	Continuous	Used to provide oxygen and to maintain mixed liquor suspended solids in suspension in suspended-growth biological treatment processes
Pneumatic mixing	Continuous	Used to provide oxygen and to maintain mixed liquor suspended solids in suspension in suspended-growth biological treatment processes

Macroflocculation (also known as *orthokinetic flocculation*) is the term used to refer to the aggregation of particles greater than 1 or 2 μm. *Macroflocculation* can be brought about by (1) induced velocity gradients and (2) differential settling.

Particles can be brought together (i.e., flocculated) by inducing velocity gradients in a fluid containing the particles to be flocculated. As illustrated on Fig. 5–13b, faster-moving particles will overtake slower-moving particles in a velocity field. If the parti-

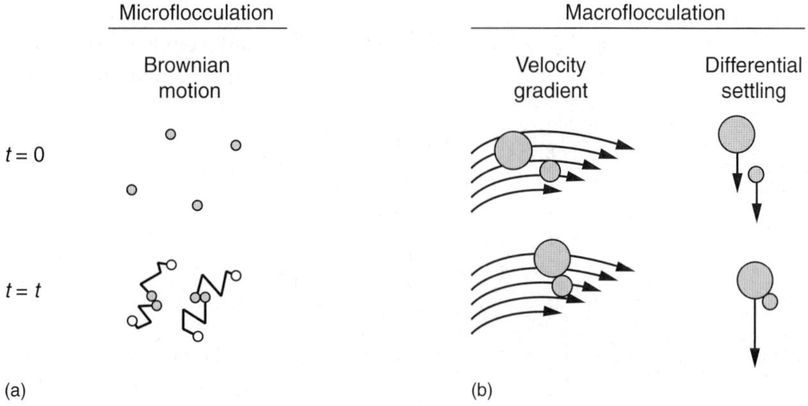

Figure 5–13

Schematic illustration of the two types of flocculation: (a) microflocculation (due to Brownian motion, also known as perikinetic flocculation) and (b) macroflocculation (also known as orthokinetic flocculation) due to (i) fluid shear and (ii) differential settling. (*Adapted from Pankow, 1991; Logan, 2000.*)

Microflocculation

Macroflocculation

Brownian motion

Velocity gradient

Differential settling

$t = 0$

$t = t$

(a)

(b)

cles that collide stick together, a larger particle will be formed that will be easier to remove by gravity separation.

In macroflocculation by differential settling (see Fig. 5–13b), large particles overtake smaller particles during gravity settling. When the two particles collide and stick together, a larger particle is formed that settles at a rate that is greater than that of the larger particle before the two particles collided. Flocculent settling in the absence of induced velocity gradients is considered in the following section.

It should be noted that flocculation brought about by induced velocity gradients is ineffectual until the colloidal particles reach a size of 1 or 2 μm through contacts produced by Brownian motion. For example, macroflocculation cannot be used to aggregate viruses, which are 0.1 μm in size or smaller, until they are microflocculated or adsorbed or enmeshed in larger flocs or particles. Typical examples of the types of mixers used in wastewater-treatment facilities for flocculation are reported in Table 5–9.

Maintaining Material in Suspension. Continuous mixing operations are used in biological treatment processes such as the activated-sludge process to maintain the mixed liquor suspended solids in suspension. In biological treatment systems the mixing device is also used to provide the oxygen needed for the process. Thus, the aeration equipment must be able to provide the oxygen needed for the process and must be able to deliver the energy needed to maintain mixed conditions within the reactor. Both mechanical aerators and dissolved aeration devices are used. In both aerobic and anaerobic digestion, mixing is used to homogenize the contents of the digester to accelerate the biological conversion process, and to distribute uniformly the heat generated from biological conversion reactions.

Energy Dissipation in Mixing and Flocculation

Mixing with an impeller in a reactor or mixing chamber causes two actions to occur: circulation and shearing of the fluid. The power input per unit volume of liquid can be used as a rough measure of mixing effectiveness, based on the reasoning that more input power creates greater turbulence, and greater turbulence leads to better mixing.

Camp and Stein (1943) studied the establishment and effect of velocity gradients in coagulation tanks of various types and developed the following equations for use in the design and operation of systems with mechanical mixing devices (e.g., paddles).

$$G = \sqrt{\frac{P}{\mu V}} \qquad (5\text{--}3)$$

where G = average velocity gradient, T^{-1}, 1/s
$\quad\quad P$ = power requirement, W
$\quad\quad \mu$ = dynamic viscosity, N·s/m^2
$\quad\quad V$ = flocculator volume, m^3

In Eq. (5–3), it is important to note that the velocity gradient G is a measure of the average velocity gradient in the fluid. High G values will be observed near the blades of the mechanical mixing device, while significantly lower values will be observed at some distance from the blades of the mixing device.

As shown, the value of G depends on the power input, the viscosity of the fluid, and the volume of the basin. Multiplying both sides of Eq. (5–3) by the detention time $\tau = V/Q$ yields

$$G\tau = \frac{V}{Q}\sqrt{\frac{P}{\mu V}} = \frac{1}{Q}\sqrt{\frac{PV}{\mu}} \qquad (5\text{--}4)$$

where τ = detention time, s
$\quad\quad Q$ = flowrate, m^3/s

Typical values that have been used for G for various mixing operations are reported in Table 5–10. The power required for various types of mixers is considered in the following discussion. The use of Eq. (5–3) is illustrated in Example 5–3.

Table 5–10

Typical detention time and velocity gradient G values for mixing and flocculation in wastewater[a]

Process	Range of values	
	Detention time	G value, s^{-1}
Mixing		
Typical rapid mixing operations in wastewater treatment	5–30 s	500–1500
Rapid mixing for effective initial contact and dispersion of chemicals	<1 s	1500–6000
Rapid mixing of chemicals in contact filtration processes	<1 s	2500–7500
Flocculation		
Typical flocculation processes used in wastewater treatment	30–60 min	50–100
Flocculation in direct filtration processes	2–10 min	25–150
Flocculation in contact filtration processes	2–5 min	25–200

[a]The limitations associated with the use of the velocity gradient concept, as discussed in the text, must be considered in applying the G values given in this table.

EXAMPLE 5-3 Power Requirement to Develop Velocity Gradients   Determine the theoretical power requirement to achieve a G value of 100/s in a tank with a volume of 2800 m³ ($\sim 10^5$ ft³). Assume that the water temperature is 15°C. What is the corresponding value when the water temperature is 5°C?

Solution

1. Determine the theoretical power requirement at 15°C using Eq. (5–3) rearranged as follows.

    $P = G^2 \mu V$

    μ at 15°C = 1.139×10^{-3} N·s/m² (see Appendix C)

    $P = (100/\text{s})^2 (1.139 \times 10^{-3} \text{ N·s/m}^2)(2800 \text{ m}^3)$

      = 31,892 W

      = 31.9 kW

2. Determine the theoretical power requirement at 5°C

    μ at 5°C = 1.518×10^{-3} N·s/m² (see Appendix C)

    $P = (100/\text{s})^2 (1.518 \times 10^{-3} \text{ N·s/m}^2)(2800 \text{ m}^3)$

      = 42,504 W

      = 42.5 kW

While the use of the velocity gradient G has been popular in the water and wastewater field, it should be noted that use of the velocity gradient concept does not apply to microflocculation and, as will be discussed later, cannot be used as design parameter for some types of mixing devices. Insight into the reason that the velocity gradient G is not an effective measure for microflocculation can be gained by considering the following relationship developed by Kolmogoroff (1941) to describe the size of eddies formed as result of power input to a fluid (Davies, 1972).

$$l_K = \left(\frac{v^3}{P_M} \right)^{1/4} \tag{5-5}$$

where l_K = Kolmogoroff microscale length, m
   v = kinematic viscosity, m²/s
   P_M = power per unit mass, W/kg, [(kg·m²/s³)/kg]
     = $G^2 v$

Substituting $G^2 v$ in Eq. (5–5) yields

$$l_K = \left(\frac{v^2}{G^2} \right)^{1/4} \tag{5-6}$$

Equation (5–6) can now be used to estimate the smallest eddy that can be produced for a given average G value. For example, if $G = 1000/\text{s}$, and $v = 1.003 \times 10^{-6}$ m²/s at 20°C, the corresponding value of the microscale length is 31.7 μm. Thus, particles smaller than 31.7 μm will not be affected. In fact, if the G value were increased to 10,000/s, the corresponding microscale length is 10.0 μm. Based on this analysis, it is clear that if particles smaller than 1 to 10 μm are to be removed they must first be destabilized and allowed to undergo microflocculation caused by Brownian motion. It can be argued that effective mixing is critical in keeping the particles in suspension so that collisions can occur by Brownian motion (Han and Lawler, 1992). Further, because the G value is an average value, the effectiveness of the mixing, which will depend on the pumping characteristics of the mixer and the geometry of the mixing basin, must be evaluated carefully.

Timescale in Mixing

The timescale for mixing is an important consideration in the design of mixing facilities and operations. For example, if the reaction rate between the substance being mixed into a liquid and the liquid is rapid, the time of mixing is extremely important. For slowly reacting substances, the time of mixing is not as critical. The rationale for mixing times for various chemicals is considered in Chap. 6. Typical mixing times may be found in Table 6–19 in Chap. 6. As reported in Table 6–19, recommended mixing times for coagulants such as alum or iron salts and for dispersing chlorine and hypochlorite into solution are less than 1 s. Typical mixing times achievable for various mixing devices are presented in Table 5–9. It should be noted that achieving extremely short mixing times becomes increasingly difficult as the flowrate increases. In some applications, it may be preferable to use multiple mixing devices to achieve optimal mixing times.

Types of Mixers Used for Rapid Mixing in Wastewater Treatment

Many types of mixing devices are available, depending on the application and the timescale required for mixing (see Table 5–9). The principal devices used for rapid mixing in wastewater-treatment applications include static in-line mixers, in-line mixers, high-speed induction mixers, pressurized water jets, and propeller and turbine mixers. Mixing can also be accomplished in pumps and with the aid of hydraulic devices such as hydraulic jumps, Parshall flumes, or weirs. Although hydraulic mixing can sometimes be highly efficient, the principal problem is that the energy input varies with the flowrate, and incomplete and ineffective mixing can occur at low flowrates.

Static Mixers.   Static in-line mixers contain internal vanes or orifice plates that bring about sudden changes in the velocity patterns as well as momentum reversals. Static mixers are principally identified by their lack of moving parts. Typical examples include in-line static mixers that contain elements that bring about sudden changes in the velocity patterns as well as momentum reversals (see Fig. 5–14a) and mixers that contain orifice plates and nozzles (see Fig. 5–14b). Static in-line mixers are used most commonly for mixing of chemicals with wastewater. In-line mixers are available in sizes varying from about 12 mm to 3 m × 3 m open channels. Low-pressure-drop round, square, and rectangular in-line static mixers have been developed for chlorine mixing in open channels and tunnels for flowrates varying from 0.22 to over 8.76 m³/s (5 to over 200 Mgal/d) (Carlson, 2000).

For static in-line mixers with vanes, the longer the mixing elements, the better the mixing; however, the pressure loss increases. It should also be noted that the shear rate and the scale (i.e., size) of the turbulent eddies formed in static mixers with vanes are more limited in range as compared to the wide range of values obtained with mechanical mixers. Mixing also occurs in a plug-flow regime in static in-line mixers. Mixing times in static mixers are quite short, typically less than 1 s. The actual mixing time will vary with the length of the mixer, which depends on the number of mixing elements used, and the

Figure 5-14

Typical mixers used in wastewater treatment for rapid mixing: (a) in-line static mixer with internal vanes, (b) in-line static mixer with orifice for mixing dilute chemicals, (c) in-line mixer, (d) in-line mixer with internal mixer, (e) high-speed induction mixer, (f) pressurized water jet mixer with reactor tube.

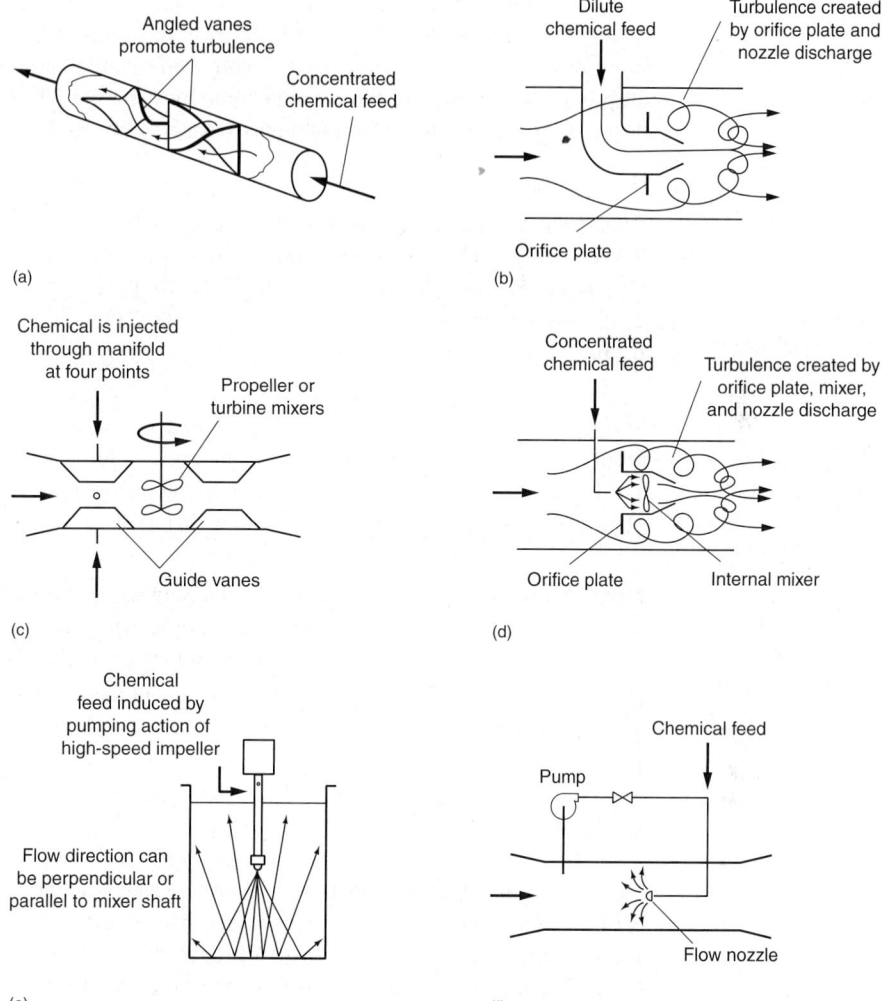

(a)

Angled vanes promote turbulence

Concentrated chemical feed

(b)

Dilute chemical feed

Turbulence created by orifice plate and nozzle discharge

Orifice plate

(c)

Chemical is injected through manifold at four points

Propeller or turbine mixers

Guide vanes

(d)

Concentrated chemical feed

Turbulence created by orifice plate, mixer, and nozzle discharge

Orifice plate   Internal mixer

(e)

Chemical feed induced by pumping action of high-speed impeller

Flow direction can be perpendicular or parallel to mixer shaft

(f)

Chemical feed

Pump

Flow nozzle

internal volume occupied by the mixing element. Thus, because the nature of the mixing that occurs in static mixers is quite different from that of mechanical mixers, use of the velocity gradient concept [see Eq. (5–3)] is inappropriate for static mixers.

For static mixers, the degree of mixing is related to the headloss (i.e., pressure drop) through the mixer. The headloss through the mixer can be estimated using the following expression:

$$h \approx k \left(\frac{v^2}{2g} \right) \approx K_{SM} v^2 \qquad (5\text{-}7)$$

where h = headloss dissipated as liquid passes through mixing device, m

k = empirical coefficient characteristic of the mixing

v = approach velocity, m/s

K_{SM} = overall coefficient for mixing device, s²/m

g = acceleration due to gravity, 9.81 m/s²

Typical values for K_{SM} vary from about 1.0 to 4.0, with a value of 2.5 being typical for the mixers used in wastewater. However, because the specific geometry of the internal mixing vanes used in such mixers varies with each manufacturer, the headloss or pressure-drop curves provided by the manufacturer must be used for estimation purposes. The power dissipated by static mixing devices can be computed using the following equation:

$$P = \gamma Q h \qquad (5\text{--}8)$$

where P = power dissipated, kW
$\quad\ \ \gamma$ = specific weight of water, kN/m³
$\quad\ \ Q$ = flowrate, m³/s

In-line Mixers. In-line mixers are similar to static mixers but contain a rotating mixing element to enhance the mixing process. Typical examples of in-line mixers are illustrated on Fig. 5–14c and d. In the in-line mixer shown on Fig. 5–14c, the power required for mixing is supplied by an external source. For the mixer shown on Fig. 5–14d, the power for mixing is supplied by the energy dissipation caused by the orifice plate and by the power input to the propeller mixer.

High-Speed Induction Mixer. The high-speed induction mixer is an efficient mixing device for a variety of chemicals. A proprietary device, shown on Fig. 5–14e for chlorine mixing, consists of a motor-driven open propeller that creates a vacuum in the chamber directly above the propeller. The vacuum created by the impeller induces the chemical to be mixed directly from the storage container without the need for dilution water. The high operating speed of the impeller (3450 r/min) provides a thorough mixing of the chemical that is being added to the water by the high velocity of the fluid leaving the impeller of the mixing device.

Pressurized Water Jets. Pressurized water jet mixers, such as illustrated on 5.14f, can also be used to mix chemicals. An important design feature of pressurized water jet mixers is that the velocity of the jet containing the chemical to be mixed must be sufficient to achieve mixing in all parts of the pipeline (Chao and Stone, 1979; Pratte and Baines, 1967). As shown on Fig. 5–14f, a reactor tube has been added to achieve effective mixing. With pressurized water jet mixers, the power for mixing is provided by an external source (i.e., the solution feed pump).

Turbine and Propeller Mixers. Turbine and propeller mixers are used commonly in wastewater-treatment processes for mixing and blending of chemicals, for keeping material in suspension, and for aeration. Turbine or propeller mixers are usually constructed with a vertical shaft driven by a speed reducer and electric motor. Two types of impellers are used for mixing: (1) radial-flow impellers and (2) axial-flow impellers. Radial-flow impellers generally have flat or curved blades located parallel to the axis of the shaft. The vertical flat-blade turbine impeller is a typical example of a radial-flow impeller. Axial-flow impellers make an angle of less than 90° with the drive shaft. Axial-flow impellers are further classified as variable pitch–constant angle of attack and constant pitch–variable angle of attack. Propellers and hydrofoils are typical examples. Propeller mixers may be provided with more than one set of propeller blades on a shaft. Typical impellers used for mixing are shown on Fig. 5–15, and information on the different types of impellers is presented in Table 5–11.

Rapid mixing in wastewater-treatment processes usually occurs in the regime of turbulent flow in which inertial forces predominate. As a general rule, the higher the velocity and the greater the turbulence, the more efficient the mixing. On the basis of inertial and viscous forces, the following mathematical relationships can be used to estimate the power requirements for mixing and the pumping capacity of the mixer.

Figure 5–15

Typical impellers used for mixing in wastewater-treatment facilities: (a) disk-type radial-flow impeller, (b) axial-flow pitched (typically 45°) blade impeller, (c) axial-flow hydrofoil-type impeller, and (d) propeller mixer. Note: The flat blade radial-flow turbine mixer looks like the axial-flow impeller (b) with the exception that the blades are set parallel to the axis of the shaft.

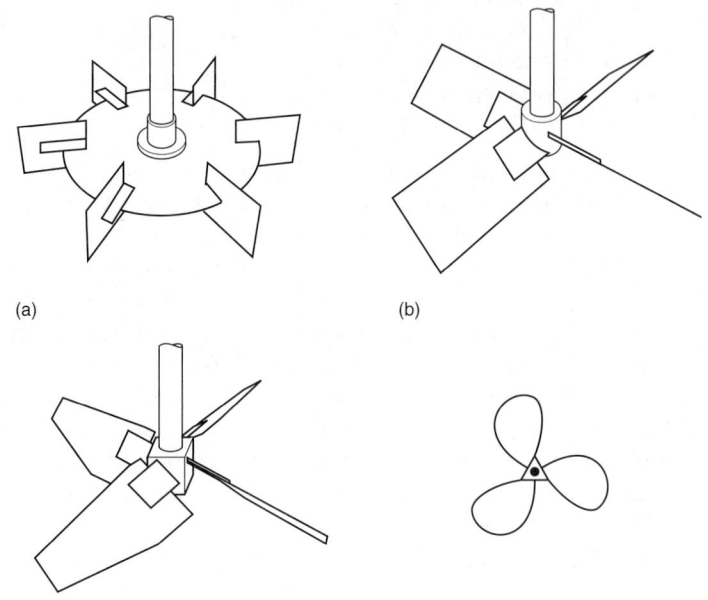

Table 5–11
Typical types of mixing impellers used in wastewater treatment[a]

Type of impeller	Flow	Shear	Pumping capacity	Applications
Vertical flat blade turbine (VFBT)	Radial	High	Low	Vertical-flow flash mixing, suspension of solids, gas dispersion
Disk turbine	Radial	High	Low	Mixing, gas dispersion
Surface impeller	Radial	High	Moderate	Gas transfer
Pitched-blade turbine (45 or 32° PBT)	Axial	Moderate	Moderate	Horizontal flash mixing, suspension of solids
Low-shear hydrofoil (LS)	Axial	Low	High	Horizontal-flow flash mixing, suspension of solids, blending, flocculation
Propeller	Axial	Very low	High	Horizontal-flow flash mixing, suspension of solids, blending, flocculation

[a] Adapted, in part, from Philadelphia Mixer Catalog.

Power for mixing

$$P = N_P \rho n^3 D^5 \qquad (5\text{--}9)$$

Pumping capacity

$$Q_i = N_Q n D^3 \qquad (5\text{--}10)$$

where P = power input, W (kg·m²/s³)
$\quad N_P$ = power number for impeller, unitless
$\quad \rho$ = density, kg/m³
$\quad n$ = revolutions per second, r/s
$\quad D$ = diameter of impeller, m
$\quad Q_i$ = pump discharge, m³/s
$\quad N_Q$ = flow number for impeller, unitless

Typical values for N_P and N_Q are presented in Table 5–12 for various types of impellers. It is important to consult manufacturers' catalogs for the appropriate values for N_P and N_Q for a specific piece of equipment. The values of N_P and N_Q must be adjusted for viscosity, blade characteristics, and the number of impellers on a single shaft.

Based on Eqs. (5–9) and (5–10), for a constant input of power as the impeller size is increased, more power is expended on flow and less on turbulence or shear. Thus, a small impeller operating at a high rotational speed will produce greater fluid shear and less pumping, whereas a large impeller operating at a slow speed will result in a high pumping capacity and low fluid shear. Mixers with small impellers operating at high speeds are best for dispersing gases or small amounts of chemicals in wastewater. Mixers with slow-moving impellers are best for blending two fluid streams, or for flocculation.

Typically, Eq. (5–9) applies if the Reynolds number is in the turbulent range (greater than 10,000). For intermediate values of the Reynolds number, manufacturers' catalogs should be consulted. The Reynolds number is given by

$$N_R = \frac{D^2 n \rho}{\mu} \qquad (5\text{--}11)$$

Table 5–12
Typical power and flow numbers for various impellers[a]

Type of impeller	Power number, N_P	Flow number, N_Q	Pumping capacity
Vertical flat-blade turbine (VFBT)	3.5–4.0	0.84–0.86	Low
Disk turbine			Low
Pitched-blade turbine (45° PBT)	1.6	0.84–0.86	
Pitched-blade turbine (32° PBT)	1.1	0.84–0.86	Moderate
Low-shear hydrofoil (LS, 3-blade)	0.30	0.50	High
Low-shear hydrofoil (LS, 4-blade)	0.60	0.55	High
Propeller			High

[a] Adapted from Philadelphia Mixer Catalog.

where D = diameter of impeller, m

$\quad n$ = rotational speed, r/s

$\quad \rho$ = mass density of fluid, kg/m^3

$\quad \mu$ = dynamic viscosity of fluid, N·s/m^2

Note: $N = $ kg·m/s^2

Mixers are selected on the basis of laboratory or pilot-plant tests or similar data provided by manufacturers. No satisfactory method exists for scaling up from an agitator of one design to a unit of a different design. Geometrical similarity should be preserved, and the power input per unit volume should be kept the same.

Where propeller or turbine mixers are used, it is imperative that vortexing or mass swirling of the liquid be eliminated. Vortexing, in which the liquid to be mixed rotates with the impeller, causes a reduction in the difference between the fluid velocity and the impeller velocity and thus decreases the effectiveness of mixing. If the mixing vessel is fairly small, vortexing can be prevented by mounting the impellers off-center or at an angle with the vertical, or by having them enter the side of the basin at an angle. In circular and rectangular tanks the usual method used to limit vortexing is to install four or more vertical baffles extending approximately one-tenth the diameter out from the wall (see Fig. 5–16). These baffles effectively break up the mass rotary motion and promote vertical mixing.

Typical design criteria for turbine and propeller mixers are given in Table 5–13. As reported in Table 5–13, important design considerations include (1) the velocity gradient G, subject to the caveats discussed previously, (2) the rotational speed, and (3) the ratio of the impeller diameter to the equivalent tank diameter. The rotation speed will vary considerably, depending on whether the flow through the mixer is horizontal or vertical (see Fig. 5–17).

Types of Mixers Used for Flocculation in Wastewater Treatment

The principal types of mixers used for flocculation can be classified as (1) static mixers, (2) paddle mixers, and (3) turbine and propeller mixers. Each of these types of mixers is considered briefly in the following discussion.

Static Mixers. In the most common type of static mixer, the liquid to be treated is subjected to a series of flow reversals in which the direction of flow is changed. Static

Figure 5–16

Definition sketch for the placement of baffles to limit vortexing in mixing tanks and reactors.

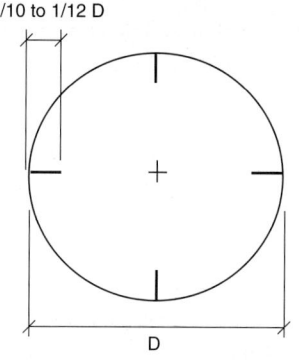
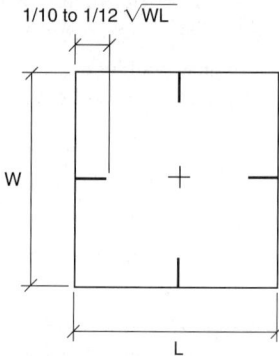

Table 5-13

Typical design parameters for mixing operations[a]

Parameter	Symbol	Unit	Value
Horizontal-flow mixing			
Velocity gradient	G	1/s	500–2500
Rotational speed	n	r/min	40–125
Ratio impeller diameter to equivalent tank diameter[b]	D/T_e	unitless	0.25–0.40
Vertical-flow mixing			
Velocity gradient	G	1/s	500–2500
Rotational speed	n	r/min	25–45
Ratio impeller diameter to equivalent tank diameter	D/T_e	unitless	0.40–0.60

[a] Adapted from Philadelphia Mixer Catalog.

[b] $T_e \approx 1.13\sqrt{L \times W}$ where L = length and W = width.

Figure 5-17

Definition sketch for mixing: (a) horizontal flow through a mixing tank equipped with an axial-flow impeller mixer and (b) vertical flow through a mixing tank equipped with a radial-flow impeller mixer.

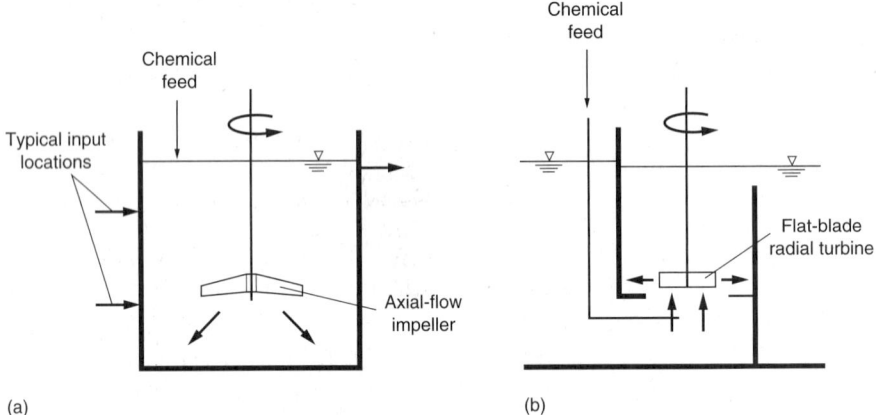

(a)   (b)

mixers can be comprised of over and under narrow flow channels, such as shown on Fig. 5–18a, or the narrow flow channels can be laid out horizontally. The headloss caused by frictional resistance offered by the flow channels and the flow reversals provides the energy for flocculation. In some designs, the channel spacing is varied to provide a decreasing energy gradient so that the large floc particles formed toward the end of the flocculation basin will not be broken apart.

Paddle Mixers. Paddle mixers are used as flocculation devices when coagulants, such as aluminum or ferric sulfate, and coagulant aids, such as polyelectrolytes and lime, are added to wastewater or solids (sludge). Paddle flocculators consist of a series of appropriately spaced paddles mounted on either a horizontal or vertical shaft. Flocculation is promoted by gentle mixing brought about by the slow-moving paddles, which, as shown on Fig. 5–18b, rotate the liquid and promote mixing. Increased particle contact promotes floc growth, but, if the mixing is too vigorous, the increased shear

Figure 5–18

Typical mixers used for flocculation in wastewater-treatment facilities: (a) over and under baffled reactor, (b) paddle mixed in baffled tank, and (c) turbine mixer in a baffled tank.

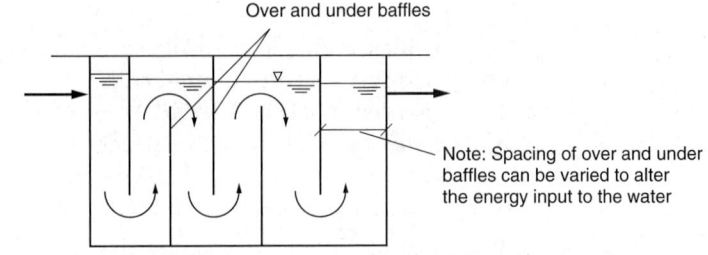

Over and under baffles

Note: Spacing of over and under baffles can be varied to alter the energy input to the water

(a)

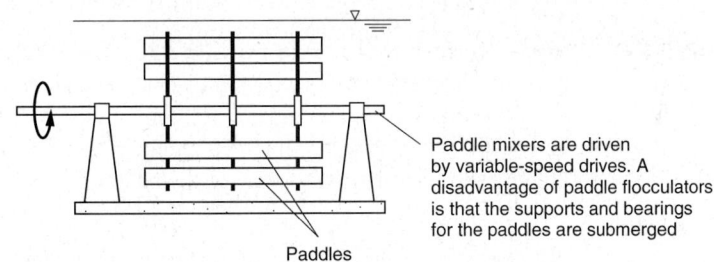

Paddle mixers are driven by variable-speed drives. A disadvantage of paddle flocculators is that the supports and bearings for the paddles are submerged

Paddles

(b)

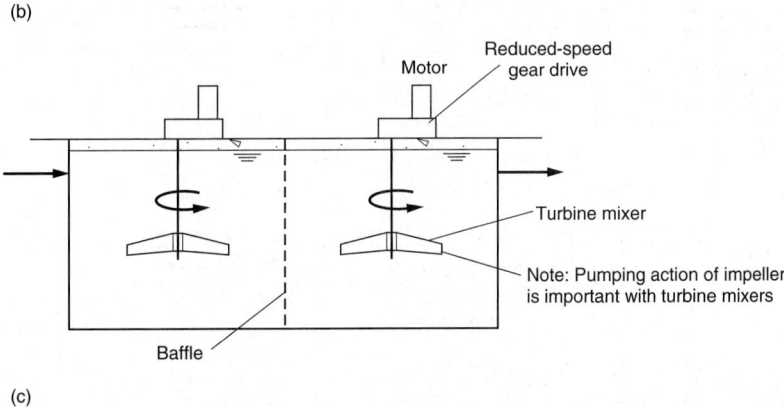

Motor

Reduced-speed gear drive

Turbine mixer

Note: Pumping action of impeller is important with turbine mixers

Baffle

(c)

forces will break up the floc into smaller particles. Agitation should be controlled carefully so that the floc particles will be of suitable size and will settle readily. Variable-speed drives are often used to regulate the paddle speed. There has been a movement away from paddle flocculators to the use of turbine flocculators because of the maintenance problems associated with paddle flocculators.

Power in a mechanical paddle system can be related to the drag force on the paddles as follows.

$$F_D = \frac{C_D A \rho v_p^2}{2} \tag{5–12}$$

$$P = F_D v_p = \frac{C_D A \rho v_p^3}{2} \tag{5–13}$$

where F_D = drag force, N

$\qquad C_D$ = coefficient of drag of paddle moving perpendicular to fluid

$\qquad A$ = cross-sectional area of paddles, m^2

$\qquad \rho$ = mass density of fluid, kg/m^3

$\qquad v_p$ = relative velocity of paddles with respect to the fluid, m/s, usually assumed to be 0.6 to 0.75 times the paddle-tip speed

$\qquad P$ = power requirement, W (kg·m^2/s^3)

The application of Eq. (5–13) is illustrated in Example 5–4.

EXAMPLE 5–4 **Power Requirements and Paddle Area for a Wastewater Flocculator** Determine the theoretical power requirement and the paddle area required to achieve a G value of 50/s in a tank with a volume of 3000 m^3. Assume that the water temperature is 15°C, the coefficient of drag C_D for rectangular paddles is 1.8, the paddle-tip velocity v is 0.6 m/s, and the relative velocity of the paddles v_p is 0.75 v.

Solution

1. Determine the theoretical power requirement using Eq. (5–3), rearranged as shown below.

μ at 15°C = 1.139 × 10^{-3} N·s/m^2 (see Appendix C)

$P = G^2 \mu V$

$\quad = (50/\text{s})^2 \, (1.139 \times 10^{-3} \, \text{N·s/m}^2) \, (3000 \, \text{m}^3)$

$\quad = 8543 \, \text{kg·m}^2/\text{s}^3 = 8543 \, \text{W}$

$\quad = 8.54 \, \text{kW}$

2. Determine the required paddle area using Eq. (5–13).

ρ = 999.1 kg/m^3 (see Appendix C)

$$A = \frac{2P}{C_D \rho v_p^3}$$

$$= \frac{2(8543 \, \text{kg/m}^2 \cdot \text{s}^3)}{1.8 \, (999.1 \, \text{kg/m}^3)(0.75 \times 0.6 \, \text{m/s})^3}$$

$$= 104.3 \, \text{m}^2$$

Turbine- and Propeller-Type Flocculators. The rotating element of turbine- and propeller-type flocculators consists of three or four blades attached to a vertical shaft (see Fig. 5–18c). The flocculator is driven with an external gear reduction system powered by a variable-speed drive. The blades of the propeller may be rectangular in shape or have the shape of a hydrofoil. Blades shaped as hydrofoils are used to limit the amount of floc shearing while at the same time providing the velocity gradients and pumping capacity needed for mixing. Typical design requirements for flocculation are presented in Table 5–14. In sizing turbine- or propeller-type flocculators, both the

Table 5–14
Typical design parameters for flocculation facilities[a]

Parameter	Symbol	Unit	Value
Velocity gradient	G	1/s	100–500
Rotational speed	n	r/min	10–30
Ratio length to width	L/W	unitless	$1 \le L/W \le 1.25$
Ratio impeller diameter to equivalent tank diameter[b]	D/T_e	unitless	0.35–0.45
Ratio height to equivalent tank diameter[b]	H/T_e	unitless	0.9–1.1
Tip speed			
Flat-blade turbine	TS	m/s	0.6–1.5
Pitch-blade turbine (45 or 32°)	TS	m/s	1.8–2.4
Low-shear propeller (3 or 4 blade)	TS	m/s	2–2.7
Superficial velocity[c]	SV	m/min	1–2

[a] Adapted from Philadelphia Mixer Catalog.
[b] $T_e = 1.13\sqrt{L \times W}$ where L = length and W = width in m.
[c] $SV = Q/A$ where Q is the pumping rate and A is the basin cross-sectional area.

power and the pumping requirements must be considered. In addition, the tip speed and the superficial velocity must also be considered. The required power and the pumping capacity can be estimated using Eqs. (5–9) and (5–10), presented previously. In wastewater flocculation, if variable-speed drives are provided for the flocculators (considering the minimum G required to keep particles in suspension), the mixer speed can be adjusted to optimize flocculation and energy use.

Types of Mixers Used for Continuous Mixing in Wastewater Treatment

Continuous mixing operations are used in biological treatment processes such as the activated-sludge process to maintain the mixed liquor suspended solids uniformly mixed state. In biological treatment systems the mixing device is also used to provide the oxygen needed for the process. Thus, the aeration equipment must be able to provide the oxygen needed for the process and the energy needed to maintain mixed conditions within the reactor. Both mechanical aerators and dissolved aeration devices are used. Diffused air is often used to fulfill both the mixing and oxygen requirements. Alternatively, mechanical turbine-aerator mixers may be used. Horizontal, submersible propeller mixers are often used to maintain channel velocities in oxidation ditches, mix the contents of anoxic reactors (see Fig. 5–19), and aid in the destratification of reclaimed water storage reservoirs.

Pneumatic Mixing. In pneumatic mixing, a gas (usually air or oxygen) is injected into the bottom of mixing or activated-sludge tanks, and the turbulence caused by the rising gas bubbles serves to mix the fluid contents of the tank. In aeration, soft

bubbles are formed with an average diameter of 5 mm while the air flow is about 10 percent of the liquid flow. The velocity gradients due to bubble formation range from a $G_{avg} < 200$ s^{-1} to $G_{max} = 8200$ s^{-1} (Masschelein, 1992).

Where air flocculation is employed, the air supply system should be adjustable so that the flocculation energy level can be varied throughout the tank. When air is injected in mixing or flocculation tanks or channels, the power dissipated by the rising air bubbles can be estimated with the following equation:

$$P = p_a V_a \ln \frac{p_c}{p_a}$$ (5-14)

where P = power dissipated, kW

p_a = atmospheric pressure, kN/m^2

V_a = volume of air at atmospheric pressure, m^3/s

p_c = air pressure at the point of discharge, kN/m^2

Equation (5-14) is derived from a consideration of the work done when the volume of air released under compressed conditions expands isothermally. If the flow of air at atmospheric pressure is expressed in terms of m^3/min (ft^3/min) and the pressure is expressed in term of meters (feet) of water, Eq. (5-14) can be written as follows:

$$p_c = KQ_a \ln \left(\frac{h + 10.33}{10.33} \right) \quad \text{S.I. units}$$ (5-15a)

$$p_c = KQ_a \ln \left(\frac{h + 33.9}{33.9} \right) \quad \text{U.S. customary units}$$ (5-15b)

where K = constant = 1.689 (35.28 in U.S. customary units)

Q_a = air flow rate at atmospheric pressure, m^3/min (ft^3/min)

h = air pressure at the point of discharge expressed in meters of water, m (ft)

The velocity gradient G achieved in pneumatic mixing is obtained by substituting P from Eq. (5–15) into Eq. (5–3).

Mechanical Aerators and Mixers.   The principal types of mechanical aerators used for continuous mixing are high-speed surface aerators and slow-speed surface aerators. These devices are discussed in Sec. 5–13, which deals with aeration, and in Chap. 8. Typical power requirements for mixing with mechanical aerators range from 20 to 40 $kW/10^3$ m^3 (0.75 to 1.50 $hp/10^3$ ft^3), depending on the type of mixer and the geometry of the tank, lagoon, or basin.

New Developments in Mixing Technology

New analytical tools that are now being applied to the analysis of and design of mixing devices include (1) computational fluid dynamics (CFD), (2) digital particle image velocimetry (DPIV), (3) laser doppler anemometry (LDA), and (4) laser-induced fluorescence (LIF). Computational fluid dynamics is used to model the fluid flow patterns in mixing devices and for scale-up analysis. In respect to fluid flow, both two- and three-dimensional models are now available. Digital particle image velocimetry is used to understand fluid movement in mixing devices. The movement of neutrally buoyant fluorescent particles is photographed using laser beam illumination. Laser doppler anemometry is used to study turbulence and to obtain data on the mean velocity at a given location in the mixing chamber. To evaluate the mean velocity, two laser beams are focused so that the beams intersect. As a particle passes through the intersection of the beams, light is reflected. The wavelength of the reflected light is a function of the particle velocity. Laser-induced fluorescence is used to measure the mixedness of solutions. Dyes such as rhodimine and other materials will fluoresce when struck by laser light of a given wavelength. The scattering of light is measured to assess the degree of mixedness. This technique is being used to study the diffusion and mixing of a substance by assessing the coefficient of variation of the mixed solution and to evaluate blending times (Chemineer, Inc., 2000).

5–5 GRAVITY SEPARATION THEORY

The removal of suspended and colloidal materials from wastewater by gravity separation is one of the most widely used unit operations in wastewater treatment. A summary of gravitational phenomena is presented in Table 5–15. Sedimentation is the term applied to the separation of suspended particles that are heavier than water, by gravitational settling. The terms *sedimentation* and *settling* are used interchangeably. A sedimentation basin may also be referred to as a sedimentation tank, clarifier, settling basin, or settling tank. Accelerated gravity settling involves the removal of particles in suspension by gravity settling in an accelerated flow field. The fundamentals of gravity separation are introduced in this section. The design of facilities for the removal of grit and TSS are considered in Secs. 5–6 and 5–7, respectively.

Table 5-15
Types of gravitational phenomena utilized in wastewater treatment

Type of separation phenomenon	Description	Application/occurrence
Discrete particle settling	Refers to the settling of particles in a suspension of low solids concentration by gravity in a constant acceleration field. Particles settle as individual entities, and there is no significant interaction with neighboring particles	Removal of grit and sand particles from wastewater
Flocculent settling	Refers to a rather dilute suspension of particles that coalesce, or flocculate, during the settling operation. By coalescing, the particles increase in mass and settle at a faster rate	Removal of a portion of the TSS in untreated wastewater in primary settling facilities, and in upper portions of secondary settling facilities. Also removes chemical floc in settling tanks
Ballasted flocculent settling	Refers to the addition of an inert ballasting agent and a polymer to a partially flocculated suspension to promote rapid settling and improved solids reduction. A portion of the recovered ballasting agent is recycled to the process	Removal of a portion of the TSS in untreated wastewater, wastewater from combined systems, and industrial wastewater. Also reduces BOD and phosphorus
Hindered settling (also called zone settling)	Refers to suspensions of intermediate concentration, in which interparticle forces are sufficient to hinder the settling of neighboring particles. The particles tend to remain in fixed positions with respect to each other, and the mass of particles settles as a unit. A solids-liquid interface develops at the top of the settling mass	Occurs in secondary settling facilities used in conjunction with biological treatment facilities
Compression settling	Refers to settling in which the particles are of such concentration that a structure is formed, and further settling can occur only by compression of the structure. Compression takes place from the weight of the particles, which are constantly being added to the structure by sedimentation from the supernatant liquid	Usually occurs in the lower layers of a deep solids or biosolids mass, such as in the bottom of deep secondary settling facilities and in solids-thickening facilities
Accelerated gravity settling	Removal of particles in suspension by gravity settling in an acceleration field	Removal of grit and sand particles from wastewater
Flotation	Removal of particles in suspension that are lighter than water by air or gas flotation	Removal of greases and oils, light material that floats, thickening of solids suspensions

Description

Sedimentation is used for the removal of grit, TSS in primary settling basins, biological floc removal in the activated-sludge settling basin, and chemical floc removal when the chemical coagulation process is used. Sedimentation is also used for solids concentration in sludge thickeners. In most cases, the primary purpose is to produce a clarified effluent, but it is also necessary to produce sludge with a solids concentration that can be handled and treated easily.

On the basis of the concentration and the tendency of particles to interact, four types of gravitational settling can occur: (1) discrete particle, (2) flocculent, (3) hindered (also called *zone*), and (4) compression. Because of the fundamental importance of the separation processes in the treatment of wastewater, the analysis of each type of separation process is discussed separately. In addition, tube settlers, used to enhance the performance of sedimentation facilities, are also described. Other gravitational separation processes include high-rate clarification, accelerated gravity settling, and flotation and are discussed in subsequent sections.

Particle Settling Theory

The settling of discrete, nonflocculating particles can be analyzed by means of the classic laws of sedimentation formed by Newton and Stokes. Newton's law yields the terminal particle velocity by equating the gravitational force of the particle to the frictional resistance, or drag. The gravitational force is given by

$$F_G = (\rho_p - \rho_w)gV_p \tag{5-16}$$

where F_G = gravitational force, MLT^{-2} (kg·m/s^2)
ρ_p = density of particle, ML^{-3} (kg/m^3)
ρ_w = density of water, ML^{-3} (kg/m^3)
g = acceleration due to gravity, LT^{-2} (9.81 m/s^2)
V_p = volume of particle, L^3 (m^3)

The frictional drag force depends on the particle velocity, fluid density, fluid viscosity, particle diameter, and the drag coefficient C_d (dimensionless), and is given by Eq. (5–17).

$$F_d = \frac{C_d A_p \rho_w v_p^2}{2} \tag{5-17}$$

where F_d = frictional drag force, MLT^{-2} (kg·m/s^2)
C_d = drag coefficient (unitless)
A_p = cross-sectional or projected area of particles in direction of flow, L^2 (m^2)
v_p = particle settling velocity, LT^{-1} (m/s)

Equating the gravitational force to the frictional drag force for spherical particles yields Newton's law:

$$v_{p(t)} = \sqrt{\frac{4g}{3C_d}\left(\frac{\rho_p - \rho_w}{\rho_w}\right)d_p} \approx \sqrt{\frac{4g}{3C_d}(sg_p - 1)d_p} \tag{5-18}$$

where $v_{p\,(t)}$ = terminal velocity of particle, LT^{-1} (m/s)
d_p = diameter of particle, L (m)
sg_p = specific gravity of the particle

The coefficient of drag C_d takes on different values depending on whether the flow regime surrounding the particle is laminar or turbulent. The drag coefficient for various particles is shown on Fig. 5–20 as a function of the Reynolds number. As shown on Fig. 5–20, there are three more or less distinct regions, depending on the Reynolds number: laminar ($N_R < 1$), transitional ($N_R = 1$ to 2000), and turbulent ($N_R > 2000$). Although

Figure 5–20

Coefficient of drag as a function of Reynolds number.

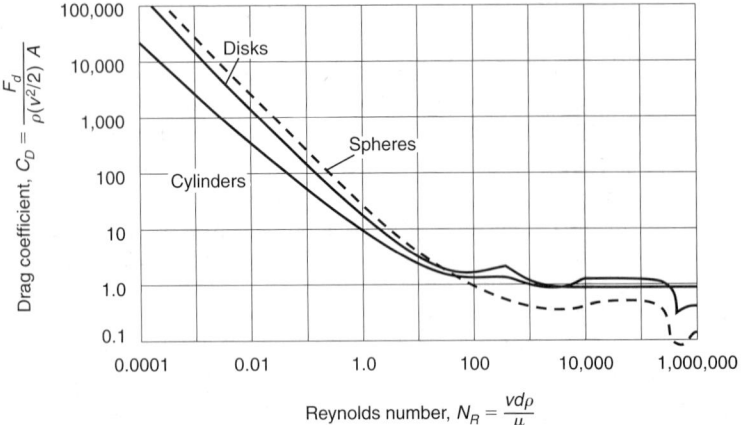

particle shape affects the value of the drag coefficient, for particles that are approximately spherical, the curve on Fig. 5–20 is approximated by the following equation (upper limit of $N_R = 10^4$):

$$C_d = \frac{24}{N_R} + \frac{3}{\sqrt{N_R}} + 0.34 \tag{5-19}$$

The Reynolds number N_R for settling particles is defined as

$$N_R = \frac{v_p d_p \rho_w}{\mu} = \frac{v_p d_p}{\nu} \tag{5-20}$$

where μ = dynamic viscosity, MTL^{-2} (N·s/m^2)
ν = kinematic viscosity, L^{2}T^{-1} (m^2/s)

Other terms are as defined above.

Equation (5–18) must be modified for nonspherical particles. An application that has been proposed is to rewrite Eq. (5–18) as follows (Gregory et al., 1999):

$$v_{p(t)} = \sqrt{\frac{4g}{3C_d\phi}\left(\frac{\rho_p - \rho_w}{\rho_w}\right)d_p} \approx \sqrt{\frac{4g}{3C_d\phi}(sg_p - 1)d_p} \tag{5-21}$$

where ϕ is a shape factor and the other terms are as defined previously. The value of the shape factor is 1.0 for spheres, 2.0 for sand grains, and up to and greater than 20 for fractal floc. The shape factor is especially important in wastewater treatment where few, if any, particles are spherical. The shape factor must also be accounted for in computing N_R. The application of Eq. (5–21) will be considered in subsequent discussions of flocculent and ballasted flocculent settling.

Settling in the Laminar Region. For Reynolds numbers less than about 1.0, viscosity is the predominant force governing the settling process, and the first term in Eq. (5–19) predominates. Assuming spherical particles, substitution of the first term of the drag coefficient equation [Eq. (5–19)] into Eq. (5–18) yields Stokes' law:

$$v_p = \frac{g(\rho_p - \rho_w)d_p^2}{18\mu} \approx \frac{g(sg_p - 1)d_p^2}{18\nu} \qquad (5\text{--}22)$$

Terms are as defined previously.

For laminar-flow conditions, Stokes found the drag force to be

$$F_D = 3\pi\mu v_p d_p \qquad (5\text{--}23)$$

Stokes' law [Eq. (5–22)] can also be derived by equating the drag force found by Stokes to the effective weight of the particle [Eq. (5–16)].

Settling in the Transition Region. In the transition region, the complete form of the drag equation [Eq. (5–16)] must be used to determine the settling velocity, as illustrated in Example 5–5. Because of the nature of the drag equation, finding the settling velocity is an iterative process. As an aid in visualizing settling in the transition region, Fig. 5–21 has been prepared, which covers the laminar and the transition region for particle sizes of interest in environmental engineering.

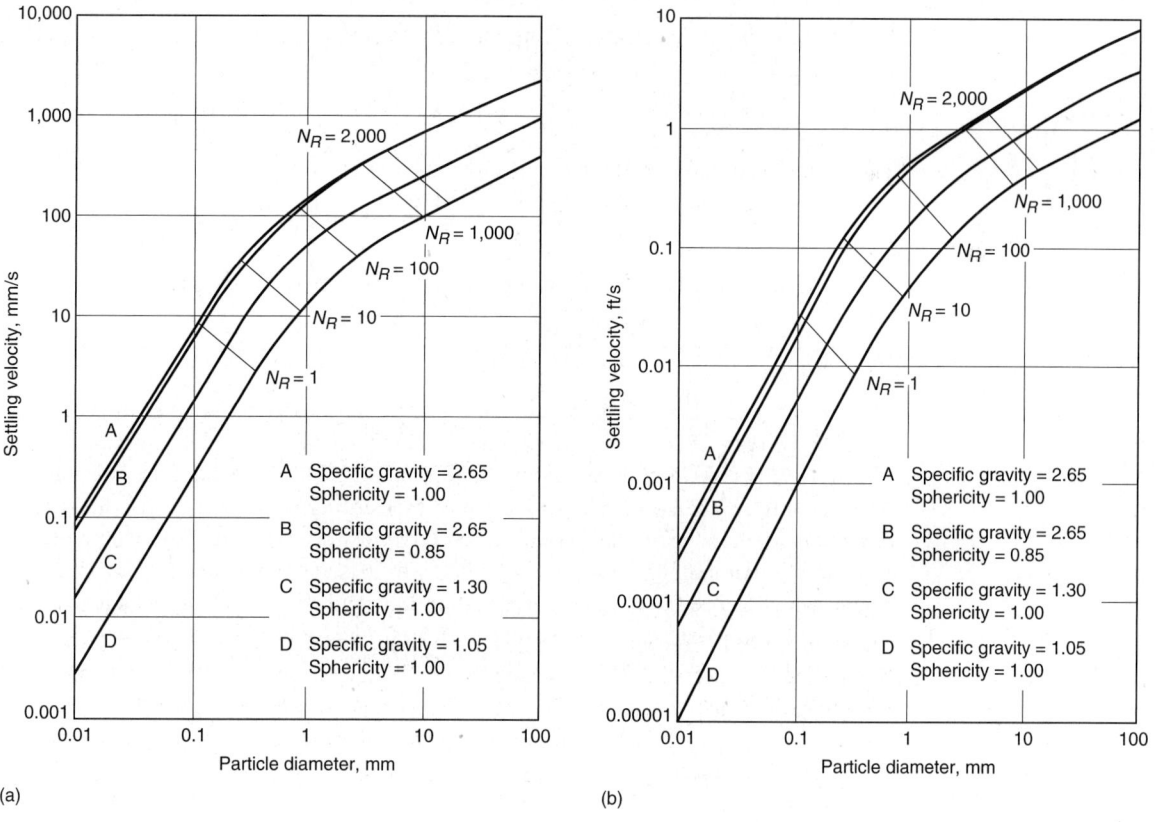

(a) (b)

Figure 5–21

Settling velocities for various particle sizes under varying conditions at 20°C: (a) settling velocity in m/s versus particle size in mm, (b) settling velocity in ft/s versus particle size in mm. *(Crites and Tchobanoglous, 1998.)*

Settling in the Turbulent Region. In the turbulent region, inertial forces are predominant, and the effect of the first two terms in the drag coefficient equation [Eq. (5–19)] is reduced. For settling in the turbulent region, a value of 0.4 is used for the coefficient of drag. If a value of 0.4 is substituted into Eq. (5–21) for C_d, the resulting equation is

$$v_p = \sqrt{3.33g\left(\frac{\rho_p - \rho_w}{\rho_w}\right)d_p} \approx \sqrt{3.33g(sg_p - 1)d_p} \tag{5–24}$$

The use of Eqs. (5–18) through (5–22) is illustrated in Example 5–5.

EXAMPLE 5–5 **Determination of Particle Terminal Settling Velocity** Determine the terminal settling velocity for a sand particle with an average diameter of 0.5 mm (0.00164 ft), a shape factor of 0.85, and a specific gravity of 2.65, settling in water at 20°C (68°F). At this temperature, the kinematic viscosity value given in Appendix C is 1.003×10^{-6} m²/s (1.091×10^{-5} ft²/s).

Solution

1. Determine the terminal settling velocity for the particle using Stokes' law [(Eq. (5–22)].

$$v_p = \frac{g(sg_p - 1)d_p^2}{18\nu}$$

$$= \frac{(9.81 \text{ m/s}^2)(2.65 - 1)(0.5 \times 10^{-3} \text{ m})^2}{18(1.003 \times 10^{-6} \text{ m}^2/\text{s})} = 0.224 \text{ m/s}$$

2. Check the Reynolds number [Eq. (5–20)] (include the shape factor ϕ).

$$N_R = \frac{\phi v_p d_p}{\nu} = \frac{0.85(0.224 \text{ m/s})(0.5 \times 10^{-3} \text{ m})}{(1.003 \times 10^{-6} \text{ m}^2/\text{s})} = 94.9$$

The use of Stokes' law is not appropriate for Reynolds number > 1.0. Therefore, Newton's law [Eq. (5–18)] must be used to determine the settling velocity in the transition region (see Fig. 5–21). The drag coefficient term in Newton's equation is dependent on the Reynolds number, which is a function of the settling velocity. Because the settling velocity is not known, an initial settling velocity must be assumed. The assumed velocity is used to compute the Reynolds number, which is used to determine the drag coefficient, which is used in the Newton equation to calculate the settling velocity. A solution is achieved when the initial assumed settling velocity is approximately equal to the settling velocity resulting from Newton's equation. The solution process is iterative, as illustrated below.

3. For the first assumed settling velocity, use the Stokes' law settling velocity calculated above. Using the resulting Reynolds number, also determined previously, compute the drag coefficient.

$$C_d = \frac{24}{N_R} + \frac{3}{\sqrt{N_R}} + 0.34 = \frac{24}{94.9} + \frac{3}{\sqrt{94.9}} + 0.34 = 0.901$$

4. Use the drag coefficient in Newton's equation to determine the particle settling velocity.

$$v_p = \sqrt{\frac{4g(sg-1)d}{3C_d}} = \sqrt{\frac{4(9.81 \text{ m/s}^2)(2.65-1)(0.5 \times 10^{-3} \text{ m})}{3 \times 0.901}} = 0.109 \text{ m/s}$$

Because the initial assumed settling velocity (0.224 m/s) does not equal the settling velocity calculated by Newton's equation (0.109 m/s), a second iteration is necessary.

5. For the second iteration, assume a settling velocity value of 0.09 m/s, and calculate the Reynolds number. Use the Reynolds number to determine the drag coefficient, and use the drag coefficient in Newton's equation to find the settling velocity.

$$N_R = \frac{0.85(0.09 \text{ m/s})(0.5 \times 10^{-3} \text{ m})}{(1.003 \times 10^{-6} \text{ m}^2/\text{s})} = 38.1$$

$$C_d = \frac{24}{38.1} + \frac{3}{\sqrt{38.1}} + 0.34 = 1.456$$

$$v_p = \sqrt{\frac{4(9.81 \text{ m/s}^2)(2.65-1)(0.5 \times 10^{-3} \text{ m})}{3 \times 1.456}} = 0.086 \text{ m/s}$$

Although the assumed settling velocity (0.09 m/s) and the calculated settling velocity (0.086 m/s) still do not agree, they are approaching closure. Successive iterations to calculate the actual settling velocity will be done as a homework problem.

Because the settling velocity used to compute the Reynolds number agrees with the settling velocity value from Newton's equation, the solution approach has been confirmed.

Discrete Particle Settling

In the design of sedimentation basins, the usual procedure is to select a particle with a terminal velocity v_c and to design the basin so that all particles that have a terminal velocity equal to or greater than v_c will be removed. The rate at which clarified water is produced is equal to

$$Q = Av_c \qquad (5\text{–}25)$$

where Q = flowrate, L^3T^{-1} (m³/s)

A = surface of the sedimentation basin, L^2 (m²)

v_c = particle settling velocity, LT^{-1} (m/s)

Rearranging Eq. (5–25) yields

$$v_c = \frac{Q}{A} = \text{overflow rate, } LT^{-1} \text{ (m}^3/\text{m}^2 \cdot \text{d)}$$

Thus, the critical velocity is equivalent to the overflow rate or surface loading rate. A common basis of design for discrete particle settling recognizes that the flow capacity is independent of the depth.

For continuous-flow sedimentation, the length of the basin and the time a unit volume of water is in the basin (detention time) should be such that all particles with the design velocity v_c will settle to the bottom of the tank. The design velocity, detention time, and basin depth are related as follows:

$$v_c = \frac{\text{depth}}{\text{detention time}} \tag{5-26}$$

In actual practice, design factors must be adjusted to allow for the effects of inlet and outlet turbulence, short circuiting, sludge storage, and velocity gradients due to the operation of sludge-removal equipment. These design factors are discussed in Sec. 5–7. In the above discussion ideal settling conditions have been assumed.

Idealized discrete particle settling in three different types of settling basins is illustrated on Fig. 5–22. Particles that have a velocity of fall less than v_c will not all be removed during the time provided for settling. Assuming that the particles of various sizes are uniformly distributed over the entire depth of the basin at the inlet, it can be seen from an analysis of the particle trajectory on Fig. 5–23 that particles with a settling velocity less than v_c will be removed in the ratio

$$X_r = \frac{v_p}{v_c} \tag{5-27}$$

where X_r is the fraction of the particles with settling velocity v_p that are removed.

In most suspensions encountered in wastewater treatment, a large gradation of particle sizes will be found. To determine the efficiency of removal for a given settling time, it is necessary to consider the entire range of settling velocities present in the system. The settling velocities of the particles can be obtained by use of a settling column test. The particle settling data are used to construct a velocity settling curve as shown on Fig. 5–24.

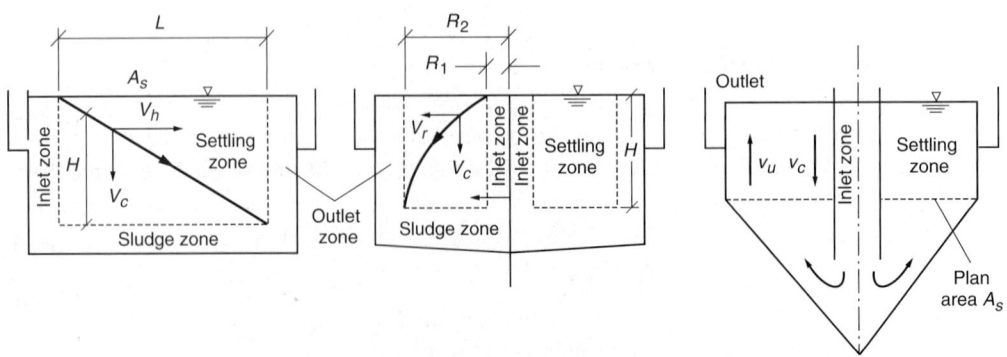

Figure 5-22

Definition sketch for the idealized settling of discrete particles in three different types of settling basins: (a) rectangular, (b) circular, and (c) upflow. *(Crites and Tchobanoglous, 1998.)*

Figure 5–23

Definition sketch for the analysis of ideal discrete particle settling.

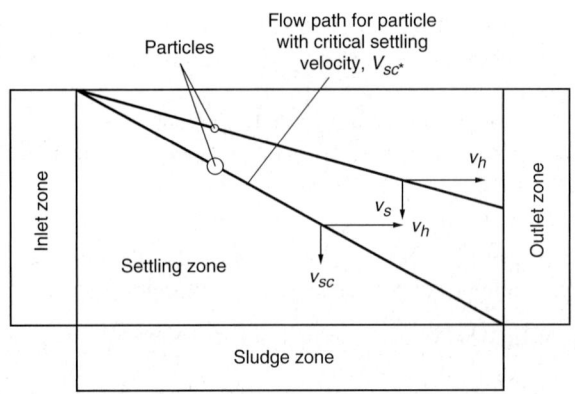

Figure 5–24

Definition sketch for the analysis of flocculent settling.

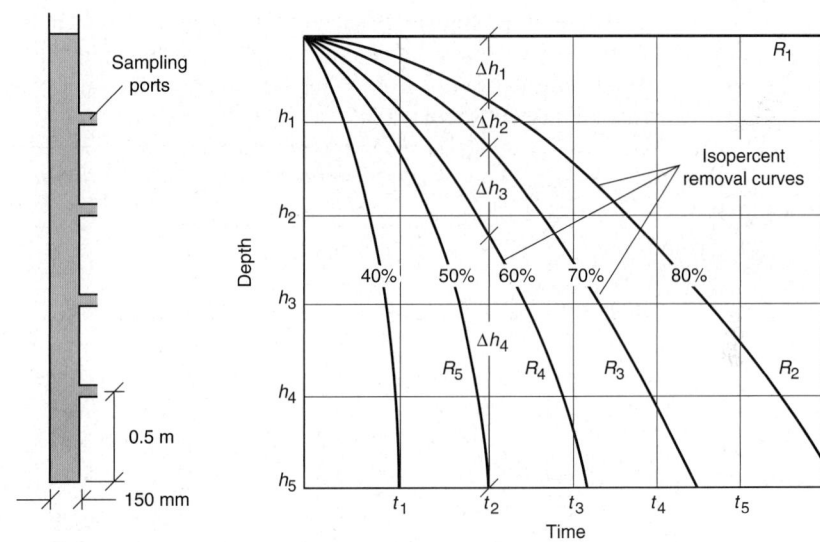

For a given clarification rate Q where

$$Q = v_c A \tag{5–28}$$

only those particles with a velocity greater than v_c will be completely removed. The remaining particles will be removed in the ratio v_p/v_c. The total fraction of particles removed for a continuous distribution is given by Eq. (5–29).

$$\text{Fraction removed} = (1 - X_c) + \int_0^{x_c} \frac{v_p}{v_c}\, dx \tag{5–29}$$

where $1 - X_c$ = fraction of particles with velocity v_p greater than v_c

$\int_0^{x_c} \dfrac{v_p}{v_c}\, dx$ = fraction of particles removed with v_p less than v_c

For discrete particles within a given settling velocity range, the following expression may be used

$$\text{Total fraction removed} = \frac{\sum\limits_{i=1}^{n} \dfrac{v_{n_i}}{v_c}(n_i)}{\sum\limits_{i=1}^{n} n_i} \tag{5-30}$$

where v_n = average velocity of particles in the ith velocity range
n_i = number of particles in the ith velocity range

The use of Eq. (5–30) is illustrated in Example 5–6.

EXAMPLE 5-6 **Calculation of Removal Efficiency for a Primary Sedimentation Basin**
Determine the removal efficiency for a sedimentation basin with a critical overflow velocity of 2 m³/m²·h in treating a wastewater containing particles whose settling velocities are distributed as given in the table below. Plot the particle histogram for the influent and effluent wastewater.

Settling velocity, m/h	Number of particles per liter × 10⁻⁵
0.0–0.5	30
0.5–1.0	50
1.0–1.5	90
1.5–2.0	110
2.0–2.5	100
2.5–3.0	70
3.0–3.5	30
3.5–4.0	20
Total	500

Solution

1. Create a table for calculating the percentage removal for each particle size. Enter the particle settling velocity ranges in column (1).

Settling velocity range, m/h (1)	Average settling velocity, m/h (2)	Number of particles in influent, × 10⁻⁵ (3)	Fraction of particles removed (4)	Number of particles removed, × 10⁻⁵ (5)	Particles remaining in effluent, × 10⁻⁵ (6)
0.0–0.5	0.25	30	0.125	3.75	26.25
0.5–1.0	0.75	50	0.375	18.75	31.25

(continued)

(Continued)

Settling velocity range, m/h (1)	Average settling velocity, m/h (2)	Number of particles in influent, × 10⁻⁵ (3)	Fraction of particles removed (4)	Number of particles removed, × 10⁻⁵ (5)	Particles remaining in effluent, × 10⁻⁵ (6)
1.0–1.5	1.25	90	0.625	56.25	33.75
1.5–2.0	1.75	110	0.875	96.0	14.0
2.0–2.5	2.25	100	1.000	100.0	0.0
2.5–3.0	2.75	70	1.000	70.0	0.0
3.0–3.5	3.25	30	1.000	30.0	0.0
3.5–4.0	3.75	20	1.000	20.0	0.0
Total		500		394.75	105.25

2. Calculate the average particle settling velocity for each velocity range by taking the average of the range limits, and enter the values in column (2). For the first velocity range, the average settling velocity is (0.0 + 0.5)/2 = 0.25 m/h.

3. Enter the number of influent particles for each velocity range in column (3).

4. Calculate the removal fraction for each velocity range by dividing the average settling velocity by the critical overflow velocity (2.0 m/h), and enter the result in column (4). For the first velocity range

$$\text{Fraction removed} = \frac{v_{n_i}}{v_c} = \frac{0.25}{2.0} = 0.125$$

Where the result is greater than 1.0, enter a value of 1.0, because all of the particles are removed.

5. Determine the number of particles removed by multiplying the number of influent particles by the percent removal [column (3) × column (4)]. Enter the values in column (5).

6. Calculate the particles remaining by subtracting the particles removed from the number of influent particles [column (3) − column (5)]. Enter the result in column (6).

7. Compute the removal efficiency by calculating the sum of particles removed and dividing the sum by the total number of particles in the influent.

$$\text{Total percent removed} = \frac{\sum_{i=1}^{n} \frac{v_{n_i}}{v_c} (n_i)}{\sum_{i=1}^{n} n_i} \times 100$$

$$= \frac{394.75 \times 10^{-5}}{500 \times 10^{-5}} \times 100 = 78.95\%$$

8. Plot the particle histogram for the influent and effluent wastewater.

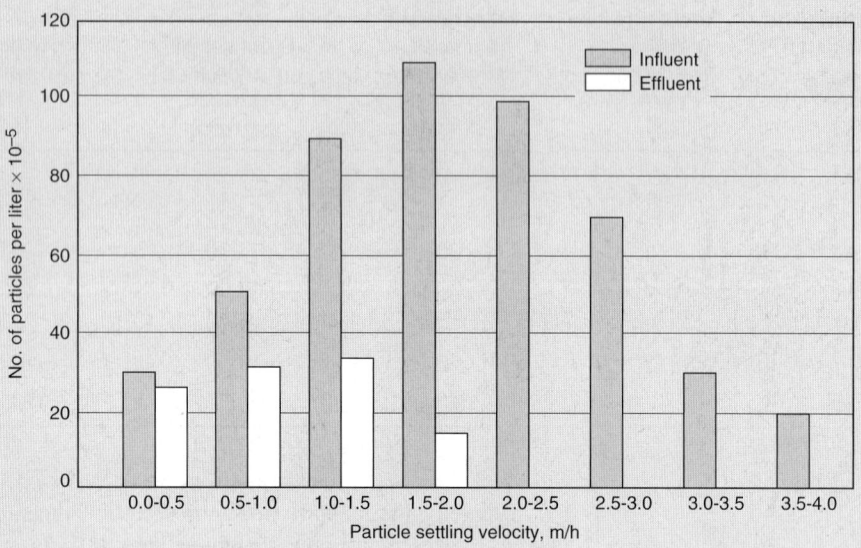

Flocculent Particle Settling

Particles in relatively dilute solutions will not act as discrete particles but will coalesce during sedimentation. As coalescence or flocculation occurs, the mass of the particle increases, and it settles faster. The extent to which flocculation occurs is dependent on the opportunity for contact, which varies with overflow rate, depth of the basin, velocity gradients in the system, concentration of particles, and range of particle sizes. The effects of these variables can be determined only by sedimentation tests.

The settling characteristics of a suspension of flocculent particles can be obtained by using a settling column test. Such a column can be of any diameter but should be equal in height to the depth of the proposed tank. The solution containing the suspended matter should be introduced into the column in such a way that a uniform distribution of particle sizes occurs from top to bottom. Care should be taken to ensure that a uniform temperature is maintained throughout the test to eliminate convection currents. Settling should take place under quiescent conditions. The duration of the test should be equivalent to the settling time in the proposed tank.

At the conclusion of the settling time, the settled matter that has accumulated at the bottom of the column is drawn off, the remaining liquid is mixed, and the TSS of the liquid is measured. The TSS of the liquid is then compared to the sample TSS before settling to obtain the percent removal.

The more traditional method of determining settling characteristics of a suspension is to use a column similar to the one described above but with sampling ports inserted at approximately 0.5 m (1.5 ft) intervals. At various time intervals, samples are withdrawn from the ports and analyzed for suspended solids. The percent removal is computed for each sample analyzed and is plotted as a number against time and depth, as elevations are plotted on a survey grid. Curves of equal percent removal are drawn as shown on Fig. 5–24. From the curves shown on Fig. 5–24, the overflow rate for various settling is determined by noting the value where the curve intersects the x axis. The settling velocity v_c is

$$v_c = \frac{H}{t_c} \qquad (5\text{--}31)$$

where H = height of the settling column, L (m)
$\quad\quad t_c$ = time required for a given degree of removal to be achieved, T (min)

The fraction of particles removed is given by

$$R, \% = \sum_{h=1}^{n} \left(\frac{\Delta h_n}{H} \right) \left(\frac{R_n + R_{n+1}}{2} \right) \qquad (5\text{--}32)$$

where R = TSS removal, %
$\quad\quad n$ = number of equal percent removal curve
$\quad\quad \Delta h_n$ = distance between curves of equal percent removal, L (m)
$\quad\quad H$ = total height of settling column, L (m)
$\quad\quad R_n$ = equal percent removal curve number n
$\quad\quad R_{n+1}$ = equal percent removal curve number $n + 1$

The advantage of the more traditional method is that it is possible to obtain removal data at various depths of settling. The removal percentage obtained using the curve given on Fig. 5–24 is illustrated in Example 5–7.

EXAMPLE 5–7 **Removal of Flocculent Suspended Solids** Using the results of the settling test shown on Fig. 5–24, determine the overall removal of solids if the detention time is t_2 and the depth is h_5. Also demonstrate that the same result is obtained when the solids are measured after settling has occurred.

Solution

1. Determine the percent removal.

    Percent removal

    $$= \frac{\Delta h_1}{h_5} \times \frac{R_1 + R_2}{2} + \frac{\Delta h_2}{h_5} \times \frac{R_2 + R_3}{2} + \frac{\Delta h_3}{h_5} \times \frac{R_3 + R_4}{2} + \frac{\Delta h_4}{h_5} \times \frac{R_4 + R_5}{2}$$

2. For the curves shown on Fig. 5–24, a total removal for quiescent settling is 65.7 percent. The computations follow.

$$\frac{\Delta h_n}{h_5} \times \frac{R_n + R_{n+1}}{2} = \text{percent removal}$$

$0.20 \times \dfrac{100 + 80}{2} =$	18.00
$0.11 \times \dfrac{80 + 70}{2} =$	8.25
$0.15 \times \dfrac{70 + 60}{2} =$	9.75
$0.54 \times \dfrac{60 + 50}{2} =$	29.70
$\overline{1.00}$	$\overline{65.70}$

3. Determine the percent removal if the liquid had been mixed and the solids were measured.
    a. Assume the initial solids concentration is equal to 100 and that at the end of the settling period the concentration of the solids at the top of the column is equal to zero.
    b. Set up a computation table and determine the remaining solids after settling.

$$\Delta h \times \frac{TSS_n \times TSS_{n+1}}{2} = \text{average TSS}$$

$0.20 \times \dfrac{0 + 20}{2} =$	2.00
$0.11 \times \dfrac{20 + 20}{2} =$	2.75
$0.15 \times \dfrac{30 + 40}{2} =$	5.25
$0.54 \times \dfrac{40 + 50}{2} =$	24.30
	$\overline{34.30}$

The percent removal is $R_t = 100 - 34.30 = 65.70$

Comment To account for the less than optimum conditions encountered in the field, the design settling velocity or overflow rate obtained from column studies often is multiplied by a factor of 0.65 to 0.85, and the detention times are multiplied by a factor of 1.25 to 1.5.

Inclined Plate and Tube Settling

Inclined plate and tube settlers are shallow settling devices consisting of stacked offset trays or bundles of small plastic tubes of various geometries (see Fig. 5–25a) that are used to enhance the settling characteristics of sedimentation basins. They are based on

Figure 5–25

Plate and tube settlers:
(a) module of inclined
tubes, (b) tubes installed
in a rectangular
sedimentation tank,
(c) operation, and
(d) definition sketch for
the analysis of settling
in a tube settler.

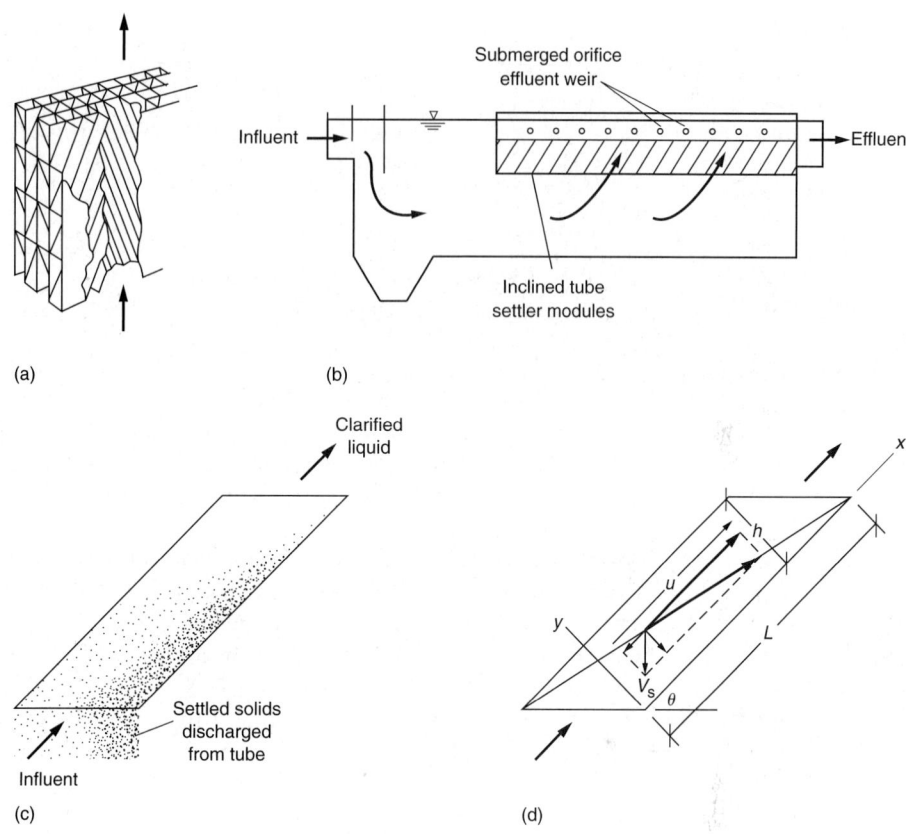

(a)

(b)

(c)

(d)

the theory that settling depends on the settling area rather than detention time. Although they are used predominantly in water-treatment applications, plate and tube settlers are used in wastewater-treatment for primary, secondary, and tertiary sedimentation. In primary sedimentation applications, however, fine screening should be provided ahead of the settling operation to prevent plugging of the plates or tubes.

To be self-cleaning, plate or tube settlers are usually set at an angle between 45 and 60° above the horizontal. When the angle is increased above 60°, the efficiency decreases. If the plates and tubes are inclined at angles less than 45°, solids will tend to accumulate within the plates or tubes. Nominal spacing between plates is 50 mm (2 in), with an inclined length of 1 to 2 m (3 to 6 ft). To control biological growths and the production of odors (the principal problems encountered with their use), the accumulated solids must be flushed out periodically (usually with a high-pressure water). The need for flushing poses a problem with the use of plate and tube settlers when the characteristics of the solids to be removed vary from day to day.

The main objective in inclined settler development has been to obtain settling efficiencies close to theoretical limits. Attention must be given to providing equal flow distribution to each settler, producing good flow distribution within each settler, and collecting settled solids while preventing resuspension.

Figure 5–26

Alternative flow patterns through tube settlers: (a) countercurrent with respect to the movement of solids, (b) cocurrent with the respect to the movement of solids, and (c) cross-flow. (AWWA, 1999.)

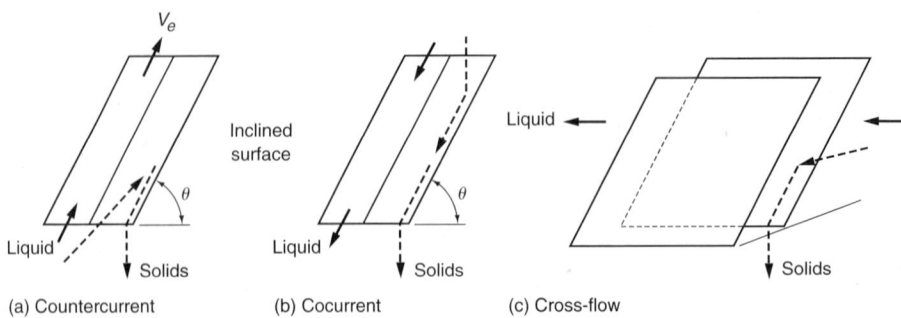

(a) Countercurrent   (b) Cocurrent   (c) Cross-flow

Inclined settling systems are generally constructed for use in one of three ways with respect to the direction of liquid flow relative to the direction of particle settlement: (1) countercurrent, (2) cocurrent, and (3) cross-flow. The flow patterns are shown schematically on Fig. 5–26.

Countercurrent Settling. With countercurrent flow, wastewater suspension in the basin passes upward through the plate or tube modules and exits from the basin above the modules (see Fig. 5–25a). The solids that settle out within the plates or tubes move by gravity countercurrently downward and out of the modules to the basin bottom (see Fig. 5–25c). Tube settlers are mostly used in the countercurrent mode.

In countercurrent settling, the time t for a particle to settle the vertical distance between two parallel inclined surfaces is (AWWA, 1999):

$$t = \frac{w}{v \cos \theta} \tag{5-33}$$

where w = perpendicular distance between surfaces, L (m)
v = settling velocity, LT^{-1} (m/s)
θ = angle of the surface inclination from the horizontal

The length of surface L_p needed to provide this time, if the liquid velocity between the surfaces is v_θ, is

$$L_p = \frac{w(v_\theta - v \sin \theta)}{v \cos \theta} \tag{5-34}$$

By rearranging this equation, all particles with a settling velocity v and greater are removed if

$$v \geq \frac{v_\theta w}{L_p \cos \theta - w \sin \theta} \tag{5-35}$$

When many plates or tubes are used

$$v_\theta = \frac{Q}{Nwb} \tag{5-36}$$

where Q = flowrate, L^3T^{-1} (m³/s)
N = number of channels made by $N+1$ plates or tubes

b = dimension of the surface at right angles to w and Q, L (m)

w = perpendicular distance between surfaces, L (m)

v_θ = settling velocity, LT^{-1} (m/s)

A proprietary settler, the Lamella® Gravity Settler, manufactured by the Parkson Corporation, is based on countercurrent settling with modifications (see Fig. 5–27). The feed stream is introduced into the settler by means of a feed duct to the feed box, which is a bottomless channel between plate sections. The flow is directed downward toward individual side-entry plate slots. The feed is distributed across the width of the plates and flows upward under laminar flow conditions. The plates are inclined 55° from the horizontal. The solids settle on the plates and the clean supernatant exits the plates through orifice holes. The orifice holes are placed immediately above each plate and are sized to induce a calculated pressure drop to ensure the feed is hydraulically distributed

Figure 5–27

Example of a Lamella plate settler. *(Courtesy Parkson Corporation.)*

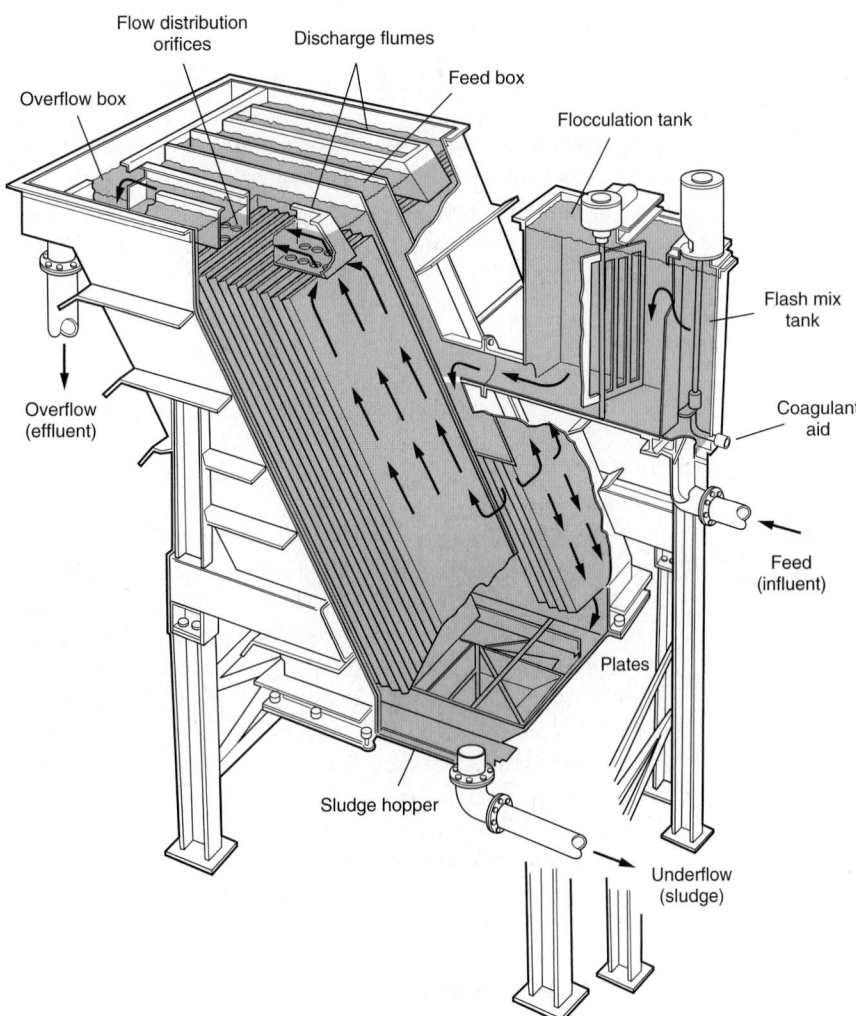

equally among the plates. The solids slide down the plates into a collection hopper. Further thickening of the solids occurs in the hopper due to compression in the quiescent zone made possible by feeding the plates from the side rather than from the bottom. Plate packs can also be retrofitted into existing clarifiers to improve performance.

Cocurrent Settling. In cocurrent settling, the solids suspension is introduced above the inclined surfaces and the flow is down through the tubes or plates (see Fig. 5–26b). The time for a particle to settle the vertical distance between two surfaces is the same as for countercurrent settling. The length of surface needed, L_p, however, has to be based on downward and not upward liquid flow, as follows:

$$L_p = w \frac{(v_\theta - v \sin \theta)}{v \cos \theta} \tag{5–37}$$

Consequently, the condition for removal of particles is given by

$$v \geq \frac{v_\theta w}{L_p \cos \theta - w \sin \theta} \tag{5–38}$$

Cross-Flow Settling. In cross-flow settling, the liquid flow is horizontal and does not interact with the vertical settling velocity (see Fig. 5–26c). The length of the surface L_p is determined by

$$L_p = \frac{w v_\theta}{v \cos \theta} \tag{5–39}$$

and

$$v \geq \frac{v_\theta w}{L_p \cos \theta} \tag{5–40}$$

Hindered (Zone) Settling

In systems that contain a high concentration of suspended solids, both hindered or zone settling and compression settling usually occur in addition to discrete (free) and flocculent settling. The settling phenomenon that occurs when a concentrated suspension, initially of uniform concentration throughout, is placed in a graduated cylinder, is illustrated on Fig. 5–28. Because of the high concentration of particles, the liquid tends to move up through the interstices of the contacting particles. As a result, the contacting particles tend to settle as a zone or "blanket," maintaining the same relative position with respect to each other. The phenomenon is known as hindered settling. As the particles settle, a relatively clear layer of water is produced above the particles in the settling region. The scattered, relatively light particles remaining usually settle as discrete or flocculent particles, as discussed previously. In most cases, an identifiable interface develops between the upper region and the hindered settling region on Fig. 5–28. The rate of settling in the hindered settling region is a function of the concentration of solids and their characteristics.

As settling continues, a compressed layer of particles begins to form on the bottom of the cylinder in the compression settling region. The particles apparently form a struc-

Figure 5-28

Definition sketch for hindered (zone) settling: (a) settling column in which the suspension is transitioning through various phases of settling and (b) the corresponding interface settling curve.

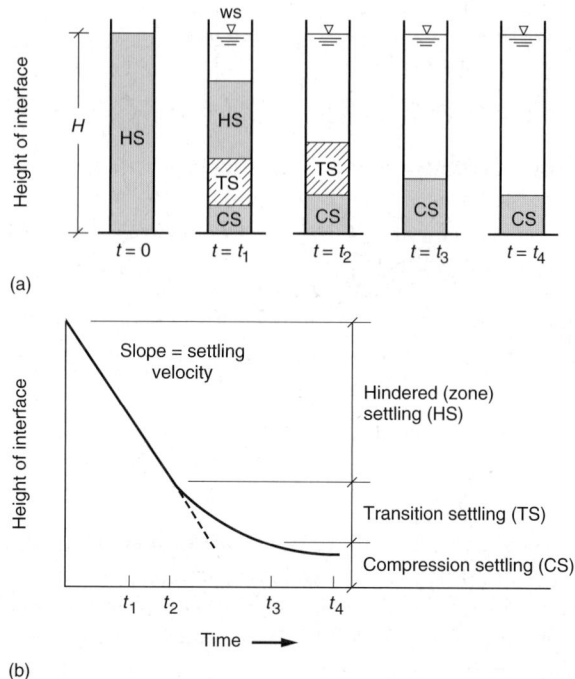

(a)

(b)

ture in which there is close physical contact between the particles. As the compression layer forms, regions containing successively lower concentrations of solids than those in the compression region extend upward in the cylinder. Thus, in actuality the hindered settling region contains a gradation in solids concentration from that found at the interface of the settling region to that found in the compression settling region.

Because of the variability encountered, settling tests are usually required to determine the settling characteristics of suspensions where hindered and compression settling are important considerations. On the basis of data derived from column settling tests, two different design approaches can be used to obtain the required area for the settling/thickening facilities. In the first approach, the data derived from one or more batch settling tests are used. In the second approach, known as the solids flux method, data from a series of settling tests conducted at different solids concentrations are used. Both methods are described in the following discussion. The solids flux method is considered further in Sec. 8-8 in Chap. 8. It should be noted that both methods have been used where existing plants are to be expanded or modified. These methods are, however, seldom used in the design of small treatment plants.

Area Requirement Based on Single-Batch Test Results. For purposes of design, the final overflow rate selected should be based on a consideration of the following factors: (1) the area needed for clarification, (2) the area needed for thickening, and (3) the rate of sludge withdrawal. Column settling tests, as previously described, can be used to determine the area needed for the free settling region directly. However, because the area required for thickening is usually greater than the area required for the settling, the rate of free settling rarely is the controlling factor. In the case of the activated-sludge

Figure 5–29

Graphical analysis of hindered (zone) interface settling curves.

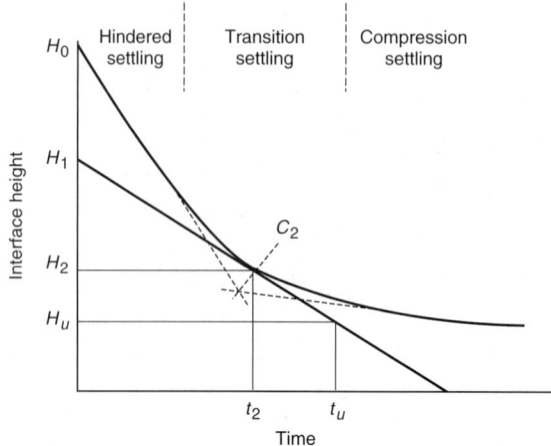

process where stray, light fluffy floc particles may be present, it is conceivable that the free flocculent settling velocity of these particles could control the design.

The area requirement for thickening is determined according to a method developed by Talmadge and Fitch (1955). A column of height H_0 is filled with a suspension of solids of uniform concentration C_0. The position of the interface as time elapses and the suspension settles is given on Fig. 5–29. The rate at which the interface subsides is then equal to the slope of the curve at that point in time. According to the procedure, the area required for thickening is given by Eq. (5–41):

$$A = \frac{Qt_u}{H_0} \tag{5-41}$$

where A = area required for sludge thickening, L^2 (m²)
  Q = flowrate into tank, L^3T^{-1} (m³/s)
  H_0 = initial height of interface in column, L (m)
  t_u = time to reach desired underflow concentration, T (s)

The critical concentration controlling the solids handling capability of the tank occurs at a height H_2 where the concentration is C_2. This point is determined by extending the tangents to the hindered settling and compression regions of the subsidence curve to the point of intersection and bisecting the angle thus formed, as shown on Fig. 5–29. The time t_u can be determined as follows:

1. Construct a horizontal line at the depth H_u that corresponds to the depth at which the solids are at the desired underflow concentration C_u. The value of H_u is determined using the following expression:

    $$H_u = \frac{C_0 H_0}{C_u} \tag{5-42}$$

2. Construct a tangent to the settling curve at the point indicated by C_2.
3. Construct a vertical line from the point of intersection of the two lines drawn in steps 1 and 2 to the time axis to determine the value of t_u.

With this value of t_u, the area required for the thickening is computed using Eq. (5–41). The area required for clarification is then determined. The larger of the two areas is the controlling value. Application of this procedure is illustrated in Example 5–8.

EXAMPLE 5–8   **Sizing an Activated-Sludge Settling Tank**   The settling curve shown in the following diagram was obtained for an activated sludge with an initial solids concentration C_0 of 3000 mg/L. The initial height of the interface in the settling column was at 0.75 m (2.5 ft). Determine the area required to yield a thickened solids concentration, C_u of 12,000 mg/L with a total flow of 3800 m³/d (1 Mgal/d). Determine also the solids loading (kg/m²·d) and the overflow rate (m³/m²·d).

Solution

1. Determine the area required for thickening using Eq. (5–42).
    a. Determine the value of H_u

    $$H_u = \frac{C_0 H_0}{C_u}$$

    $$= \frac{(3000 \text{ mg/L})(0.75 \text{ m})}{(12,000 \text{ mg/L})} = 0.188 \text{ m}$$

    On the following settling curve, a horizontal line is constructed at $H_u = 0.188$ m. A tangent is constructed to the settling curve at C_2, the midpoint of the region between hindered and compression settling. Bisecting the angle formed where the two tangents meet determines point C_2. The intersection of the tangent at C_2 and the line $H_u = 0.188$ m determines t_u. Thus $t_u = 47$ min, and the required area is

    $$A = \frac{Q t_u}{H_0} = \left[\frac{(3800 \text{ m}^3/\text{d})}{(24 \text{ h/d})(60 \text{ min/h})} \right]\left(\frac{47 \text{ min}}{0.75 \text{ m}} \right) = 165 \text{ m}^2$$

2. Determine the area required for clarification.

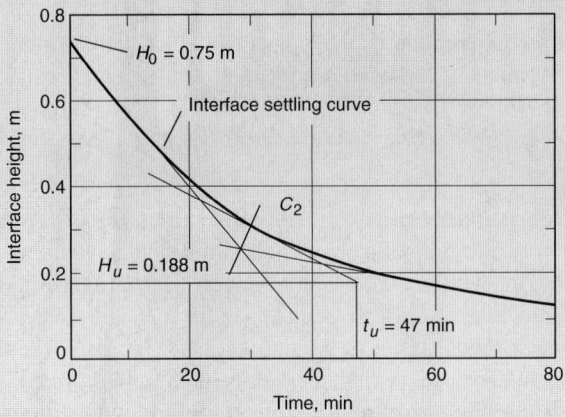

a.   Determine the interface subsidence velocity v. The subsidence velocity is determined by computing the slope of the tangent drawn from the initial portion of the interface settling curve. The computed velocity represents the unhindered settling rate of the sludge.

$$v = \left(\frac{0.75 \text{ m} - 0.3 \text{ m}}{29.5 \text{ m}}\right)\left(\frac{60 \text{ min}}{\text{h}}\right) = 0.92 \text{ m/h}$$

b.   Determine the clarification rate. Because the clarification rate is proportional to the liquid volume above the critical sludge zone, it may be computed as follows:

$$Q = 3800 \text{ m}^3/\text{d}\left(\frac{0.75 \text{ m} - 0.188 \text{ m}}{0.75 \text{ m}}\right) = 2847 \text{ m}^3/\text{d}$$

c.   Determine the area required for clarification. The required area is obtained by dividing the clarification rate by the settling velocity.

$$A = \frac{Q_c}{v} = \frac{(2847 \text{ m}^3/\text{d})}{(24 \text{ h/d})(0.91 \text{ m/h})} = 129 \text{ m}^2$$

3.   The controlling area is the thickening area (165 m²) because it exceeds the area required for clarification (129 m²).

4.   Determine the solids loading. The solids loading is computed as follows:

$$\text{Solids, kg/d} = \frac{(3800 \text{ m}^3/\text{d})(3000 \text{ g/m}^3)}{(10^3 \text{ g/kg})} = 11{,}400 \text{ kg/d}$$

$$\text{Solids loading} = \frac{(11{,}400 \text{ kg/d})}{165 \text{ m}^2} = 69.1 \text{ kg/m}^2\text{·d}$$

5.   Determine the hydraulic loading rate.

$$\text{Hydraulic loading rate} = \frac{(3800 \text{ m}^3/\text{d})}{165 \text{ m}^2} = 23.0 \text{ m}^3/\text{m}^2\text{·d}$$

Comment An alternative approach for sizing the secondary clarifiers using the initial settling velocity of the sludge is given in Sec. 8–8 in Chap. 8.

Area Requirements Based on Solids Flux Analysis.   An alternative method of determining the area required for hindered settling is based on an analysis of the solids (mass) flux (Coe and Clevenger, 1916). In the solids flux method of analysis it is assumed that a settling basin is operating at steady state. Within the tank, the downward flux of solids is brought about by gravity (hindered) settling and by bulk transport due to the underflow that is being pumped out and recycled. The solids flux method of analysis is used to assess the performance of existing facilities and to obtain information for the design of new facilities to treat the same wastewater. Application of the solids flux method of analysis is illustrated in Sec. 8–7 in Chap. 8, and additional information may be found in the following references: Dick and Ewing (1967); Dick and Young (1972); Keinath (1985); Wahlberg and Keinath (1988); and Yoshika et al. (1957).

Compression Settling

The volume required for the sludge in the compression region can also be determined by settling tests. The rate of consolidation has been found to be proportional to the difference in the depth at time t and the depth to which the sludge will settle after a long period of time. The long-term consolidation can be modeled as a first-order decay function, as given by Eq. (5–43).

$$H_t - H_\infty = (H_2 - H_\infty)e^{-i(t-t_2)} \tag{5–43}$$

where H_t = sludge height at time t, L
$\quad H_\infty$ = sludge depth after long settling period, on the order of 24 h, L
$\quad H_2$ = sludge height at time t_2, L
$\quad i$ = constant for a given suspension

Stirring serves to compact solids in the compression region by breaking up the floc and permitting water to escape. Rakes are often used on sedimentation equipment to manipulate the solids and thus produce better compaction.

Gravity Separation in an Accelerated Flow Field

Sedimentation, as described previously, occurs under the force of gravity in a constant acceleration field. The removal of settleable particles can also be accomplished by taking advantage of a changing acceleration field. A number of devices that take advantage of both gravitational and centrifugal forces and induced velocities have been developed for the removal of grit from wastewater. The principles involved are illustrated on Fig. 5–30. In appearance, the separator looks like a large diameter cylinder with a conical bottom. Wastewater, from which grit is to be separated, is introduced tangentially near the top and exits through the opening in the top of the unit. The liquid is removed at the top. Grit is removed through an opening in the bottom of the unit.

Because the top of the separator is enclosed, the rotating flow creates a free vortex within the separator. The most important characteristic of a free vortex is that the product of the tangential velocity times the radius is a constant:

$$Vr = \text{constant} \tag{5–44}$$

where V = tangential velocity, LT^{-1} (m/s)
$\quad r$ = radius, L (m)

The significance of Eq. (5–44) can be illustrated by the following example. Assume the tangential velocity in a separator with a 1.5 m (5 ft) radius is 0.9 m/s (3 ft/s). The product of the velocity times the radius at the outer edge of the separator is equal to 1.35 m^2/s (15 ft^2/s). If the discharge port has a radius of 0.9 m (1 ft), then the tangential velocity at the entrance to the discharge port is 4.5 m/s (15 ft/s). The centrifugal force experienced by a particle within this flow field is equal to the square of the velocity divided by the radius. Because the centrifugal force is also proportional to the inverse of the radius, a fivefold decrease in the radius results in a 125-fold increase in the centrifugal force.

Figure 5-30

Accelerated gravity separator: (a) outline sketch and (b) definition sketch. (From Eutek.)

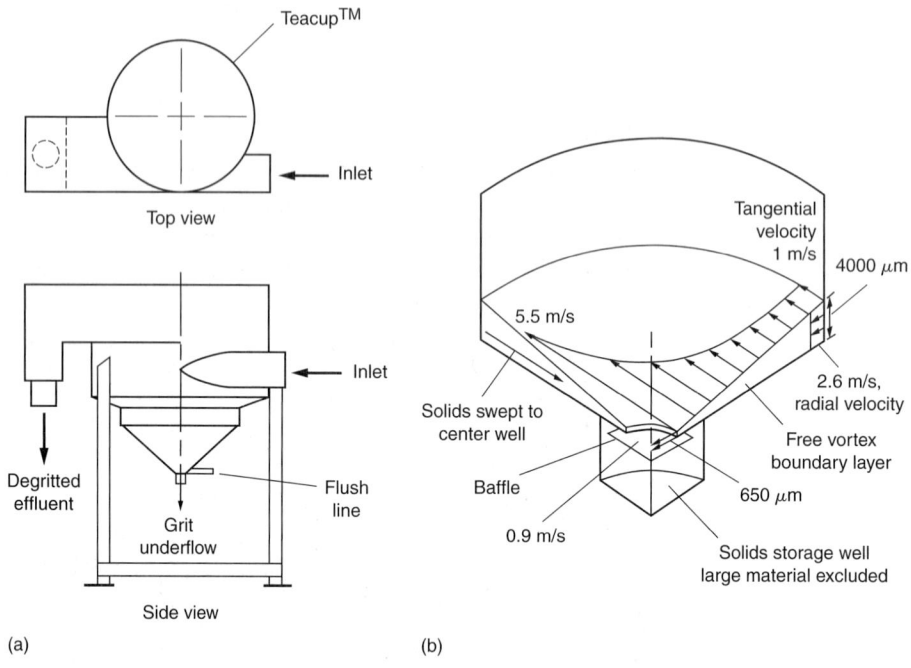

Because of the high centrifugal forces near the discharge port, some of the particles, depending on their size, density, and drag, are retained within the body of the free vortex near the center of the separator, while other particles are swept out of the unit. Grit and sand particles will be retained while organic particles are discharged from the unit. Organic particles having the same settling velocity as sand will typically be from four to eight times as large. The corresponding drag forces for these organic particles will be from 16 to 64 times as great. As a result, the organic particles tend to move with the fluid and are transported out of the separator. The particles held in the free vortex ultimately settle to the bottom of the unit under the force of gravity. Organic particles that sometimes settle usually consist of oil and grease attached to grit or sand particles.

5-6 GRIT REMOVAL

Removal of grit from wastewater may be accomplished in grit chambers or by the centrifugal separation of solids. Grit chambers are designed to remove grit, consisting of sand, gravel, cinders, or other heavy solid materials that have subsiding velocities or specific gravities substantially greater than those of the organic putrescible solids in wastewater. Grit chambers are most commonly located after the bar screens and before the primary sedimentation tanks. Primary sedimentation tanks function for the removal of the heavy organic solids. In some installations, grit chambers precede the screening facilities. Generally, the installation of screening facilities ahead of the grit chambers makes the operation and maintenance of the grit removal facilities easier.

Locating grit chambers ahead of wastewater pumps, when it is desirable to do so, would normally involve placing them at considerable depth at added expense. It is therefore usually deemed more economical to pump the wastewater, including the grit, to grit chambers located at a convenient position ahead of the treatment plant units, recognizing that the pumps may require greater maintenance.

Types of Grit Chambers

Grit chambers are provided to (1) protect moving mechanical equipment from abrasion and accompanying abnormal wear; (2) reduce formation of heavy deposits in pipelines, channels, and conduits; and (3) reduce the frequency of digester cleaning caused by excessive accumulations of grit. The removal of grit is essential ahead of centrifuges, heat exchangers, and high-pressure diaphragm pumps.

There are three general types of grit chambers: horizontal-flow, of either a rectangular or a square configuration; aerated; or vortex type. In the horizontal-flow type, the flow passes through the chamber in a horizontal direction and the straight-line velocity of flow is controlled by the dimensions of the unit, an influent distribution gate, and a weir at the effluent end. The aerated type consists of a spiral-flow aeration tank where the spiral velocity is induced and controlled by the tank dimensions and quantity of air supplied to the unit. The vortex type consists of a cylindrical tank in which the flow enters tangentially creating a vortex flow pattern; centrifugal and gravitational forces cause the grit to separate. Design of grit chambers is commonly based on the removal of grit particles having a specific gravity of 2.65 and a wastewater temperature of 15.5°C (60°F). However, analysis of grit-removal data indicates the specific gravity ranges from 1.3 to 2.7 (WPCF, 1985).

Horizontal-Flow Grit Chambers

Rectangular and square horizontal-flow grit chambers have been used for many years. Their use, however, in new installations has been limited in favor of aerated and vortex-type grit chambers. Representative design data for horizontal-flow grit chambers are presented in Table 5–16.

Table 5–16
Typical design information for horizontal-flow grit chambers

	U.S. customary units			SI units		
	Unit	Range	Typical	Unit	Range	Typical
Detention time	s	45–90	60	s	45–90	60
Horizontal velocity	ft/s	0.8–1.3	1.0	m/s	0.25–0.4	0.3
Settling velocity for removal of:						
0.21 mm (65-mesh) material	ft/min[a]	3.2–4.2	3.8	m/min[a]	1.0–1.3	1.15
0.15 mm (65-mesh) material	ft/min[a]	2.0–3.0	2.5	m/min[a]	0.6–0.9	0.75
Headloss in a control section as percent of depth in channel	%	30–40	36[b]	%	30–40	36[b]
Added length allowance for inlet and outlet turbulence	%	25–50	30	%	25–50	30

[a] If the specific gravity of the grit is significantly less than 2.65, lower velocities should be used.
[b] For Parshall flume control.

Rectangular Horizontal-Flow Grit Chambers. The oldest type of grit chamber used is the rectangular horizontal-flow, velocity-controlled type. These units were designed to maintain a velocity as close to 0.3 m/s (1.0 ft/s) as practical and to provide sufficient time for grit particles to settle to the bottom of the channel. The design velocity will carry most organic particles through the chamber and will tend to resuspend any organic particles that settle but will permit the heavier grit to settle out.

The basis of design of rectangular horizontal-flow grit chambers is that, under the most adverse conditions, the lightest particle of grit will reach the bed of the channel prior to its outlet end. Normally, grit chambers are designed to remove all grit particles that will be retained on a 0.21-mm-diameter (65-mesh) screen, although many chambers have been designed to remove grit particles retained on a 0.15-mm-diameter (100-mesh) screen. The length of channel will be governed by the depth required by the settling velocity and the control section, and the cross-sectional area will be governed by the rate of flow and by the number of channels. Allowance should be made for inlet and outlet turbulence.

Grit removal from horizontal-flow grit chambers is accomplished usually by a conveyor with scrapers, buckets, or plows. Screw conveyors or bucket elevators are used to elevate the removed grit for washing or disposal. In small plants, grit chambers are sometimes cleaned manually.

Square Horizontal-Flow Grit Chambers. Square horizontal-flow grit chambers, such as those shown on Fig. 5–31, have been in use for over 60 years. Influent to the units is distributed over the cross section of the tank by a series of vanes or gates, and the distributed wastewater flows in straight lines across the tank and overflows a weir in a free discharge. Where square grit chambers are used, it is generally advisable to use at least two units. These types of grit chambers are designed on the basis of overflow rates that are dependent on particle size and the temperature of the wastewater. They are nominally designed to remove 95 percent of the 0.15-mm-diameter (100-mesh) particles at peak flow. A typical set of design curves is shown on Fig. 5–32.

In square grit chambers, the solids are removed by a rotating raking mechanism to a sump at the side of the tank. Settled grit may be moved up an incline by a reciprocating rake mechanism (see Fig. 5–31) or grit may be pumped from the tank through a cyclone degritter to separate the remaining organic material and concentrate grit. The concentrated grit then may be washed again in a classifier using a submerged reciprocating rake or an inclined-screw conveyor. By either method, organic solids are separated from the grit and flow back into the basin, resulting in a cleaner, dryer grit.

Aerated Grit Chambers

In aerated grit chambers, air is introduced along one side of a rectangular tank to create a spiral flow pattern perpendicular to the flow through the tank. The heavier grit particles that have higher settling velocities settle to the bottom of the tank. Lighter, principally organic, particles remain in suspension and pass through the tank. The velocity of roll or agitation governs the size of particles of a given specific gravity that will be removed. If the velocity is too great, grit will be carried out of the chamber; if it is too small, organic material will be removed with the grit. Fortunately, the quantity of air is easily adjusted. With proper adjustment, almost 100 percent removal will be obtained,

Figure 5–31

Typical square horizontal-flow grit chamber:
(a) schematic, and
(b) photo of empty basin.
The two rakes are used to move settled grit to the periphery for removal.

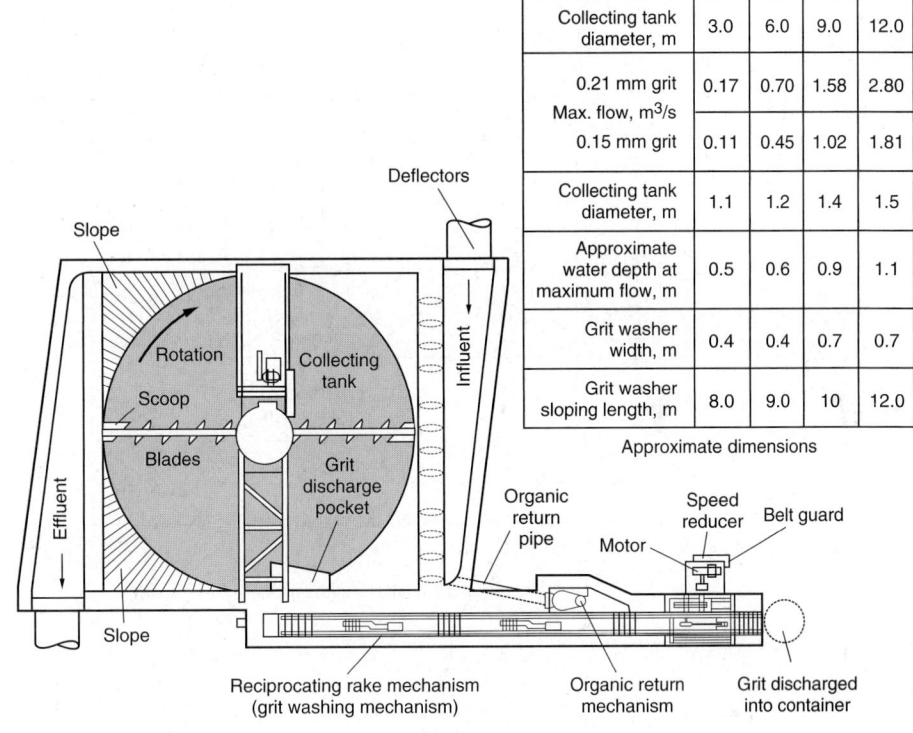

Collecting tank diameter, m	3.0	6.0	9.0	12.0
0.21 mm grit Max. flow, m³/s	0.17	0.70	1.58	2.80
0.15 mm grit	0.11	0.45	1.02	1.81
Collecting tank diameter, m	1.1	1.2	1.4	1.5
Approximate water depth at maximum flow, m	0.5	0.6	0.9	1.1
Grit washer width, m	0.4	0.4	0.7	0.7
Grit washer sloping length, m	8.0	9.0	10	12.0

Approximate dimensions

Note: m × 3.2808 = ft; m³/s × 22.8245 = Mgal/d; mm × 0.03937 = in.

(a)

(b)

and the grit will be well washed. (Grit that is not well washed and contains organic matter is an odor nuisance and attracts insects.)

Aerated grit chambers are nominally designed to remove 0.21-mm-diameter (65-mesh) or larger, with 2- to 5-minute detention periods at the peak hourly rate of flow. The cross section of the tank is similar to that provided for spiral circulation in activated-sludge aeration tanks, except that a grit hopper about 0.9 m (3 ft) deep with steeply

Figure 5–32

Area required for settling grit particles with a specific gravity of 2.65 in wastewater at indicated temperatures.

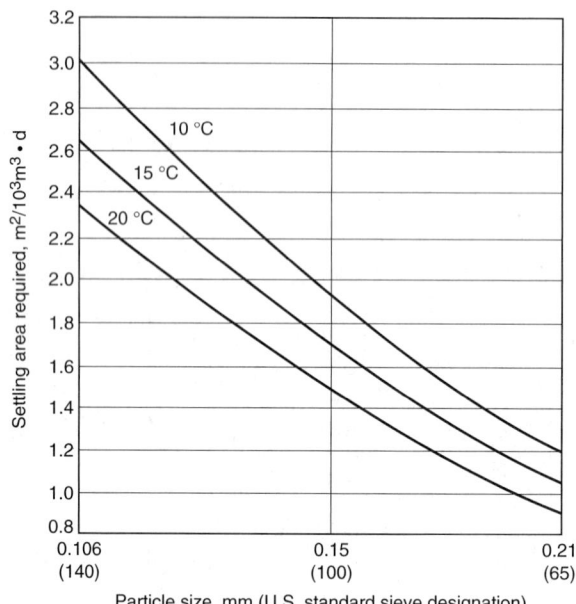

sloping sides is located along one side of the tank under the air diffusers (see Fig. 5–33). The air diffusers are located about 0.45 to 0.6 m (1.5 to 2 ft) above the normal plane of the bottom. Influent and effluent baffles are used frequently for hydraulic control and improved grit-removal effectiveness. Basic design data for aerated grit chambers are presented in Table 5–17. The design of aerated grit chambers is illustrated in Example 5–9.

Wastewater will move through the tank in a spiral path (see Fig. 5–34) and will make two to three passes across the bottom of the tank at maximum flow and more passes at lesser flows. Wastewater should be introduced in the direction of the roll. To determine the required headloss through the chamber, the expansion in volume caused by the air must be considered.

For grit removal, aerated grit chambers are often provided with grab buckets traveling on monorails and centered over the grit collection and storage trough (see Fig. 5–35). An added advantage of a grab bucket grit-removal system is that dropping the grit from the bucket through the tank contents can further wash grit. Other installations are equipped with chain-and-bucket conveyors, running the full length of the storage troughs, which move the grit to one end of the trough and elevate it above the wastewater level in a continuous operation. Screw conveyors, tubular conveyors, jet pumps, and airlifts have also been used. Grit-removal equipment for aerated grit chambers is subject to the same wear as experienced in the horizontal-flow units. For large installations, traveling-bridge grit collectors, as shown on Fig. 5–36, can be used. Grit pumps are immersed in the grit chambers and travel the entire length, pumping grit into a stationary grit trough collection system. The pumps can operate continuously or they can be programmed to run on cycles based on time or flow.

Figure 5–33

Typical section through an aerated grit chamber.

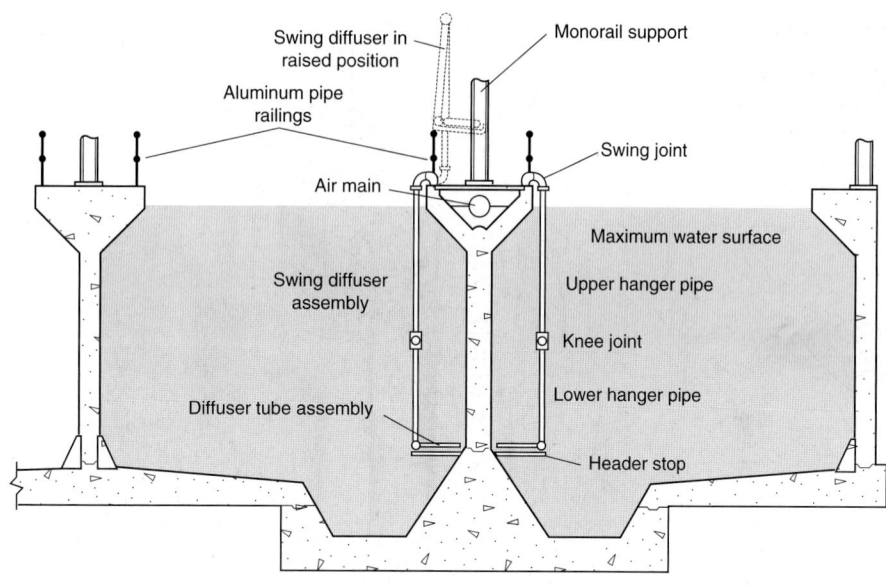

Table 5–17

Typical design information for aerated grit chambers

Item	U.S. customary units			SI units		
	Unit	Range	Typical	Unit	Range	Typical
Detention time at peak flowrate	min	2–5	3	min	2–5	3
Dimensions:						
Depth	ft	7–16		m	2–5	
Length	ft	25–65		m	7.5–20	
Width	ft	8–23		m	2.5–7	
Width-depth ratio	Ratio	1:1 to 5:1	1.5:1	Ratio	1:1 to 5:1	1.5:1
Length-width ratio	Ratio	3:1 to 5:1	4:1	Ratio	3:1 to 5:1	4:1
Air supply per unit of length	$ft^3/ft \cdot min$	3–8		$m^3/m \cdot min$	0.2–0.5	
Grit quantities	$ft^3/Mgal$	0.5–27	2	$m^3/10^3 \, m^3$	0.004–0.20	0.015

In areas where industrial wastewater is discharged to the collection system, the release of volatile organic compounds (VOCs) by the air agitation in aerated grit chambers needs to be considered. The release of significant amounts of VOCs can be a health risk to the treatment plant operators. Where release of VOCs is an important consideration, covers may be required or nonaerated-type grit chambers used.

Figure 5–34

Helical flow pattern in an aerated grit chamber.

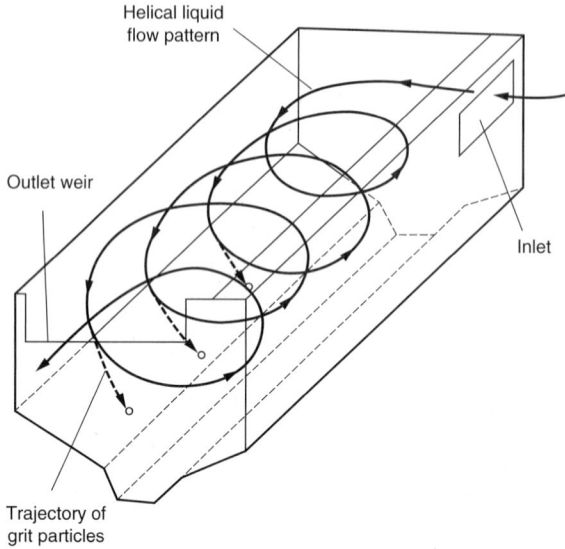

Helical liquid flow pattern

Outlet weir

Inlet

Trajectory of grit particles

Figure 5–35

Grab bucket used to remove grit from aerated grit chamber.

EXAMPLE 5-9   Design of Aerated Grit Chamber   Design an aerated grit chamber for the treatment of municipal wastewater. The average flowrate is 0.5 m³/s, and the peaking factor curve given on Fig. 3–13 is applicable.

Figure 5–36

Aerated grit chamber with traveling-bridge-type grit removal system. Pumps are mounted on the traveling bridge for removal of grit from the grit hopper. The diffusers create the helical flow pattern as shown in Fig. 5–34.

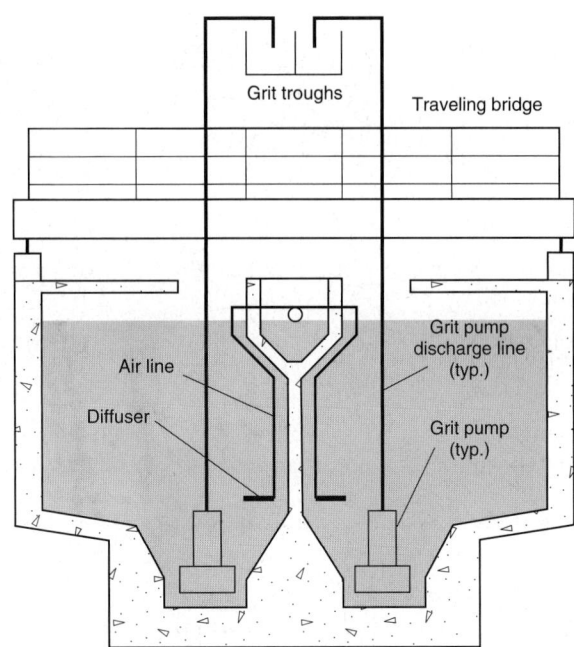

Grit troughs

Traveling bridge

Grit pump discharge line (typ.)

Air line

Diffuser

Grit pump (typ.)

Solution:

1. Establish the peak flowrate for design. Assume that the aerated grit chamber will be designed for the 1-day sustained peak flowrate. From Fig. 3–11, the peaking factor is found to be 2.75, and the peak design flowrate is

   Peak flowrate = 0.5 m³/s × 2.75 = 1.38 m³/s

2. Determine the grit chamber volume. Because it will be necessary to drain the chamber periodically for routine maintenance, use two chambers. Assume the average detention time at the peak flowrate is 3 min.

   $$\text{Grit chamber volume, m}^3 \text{ (each)} = (1/2)(1.38 \text{ m}^3/\text{s}) \times 3 \text{ min} \times 60 \text{ s/min}$$
   $$= 124.2 \text{ m}^3$$

3. Determine the dimensions of each grit chamber. Use a width-to-depth ratio of 1.2:1 and assume that the depth is 3 m.
   a. Width = 1.2 (3 m) = 3.6 m
   b. $\text{Length} = \dfrac{\text{volume}}{\text{width} \times \text{depth}} = \dfrac{124.2 \text{ m}^3}{3 \text{ m} \times 3.6 \text{ m}} = 11.5 \text{ m}$

4. Determine the detention time in each grit chamber at average flow.

   $$\text{Detention time} = \frac{124.2 \text{ m}^3}{(0.25 \text{ m}^3/\text{s})} = 496.8 \text{ s} \left(\frac{1 \text{ min}}{60 \text{ s}} \right) = 8.28 \text{ min}$$

5. Determine the air supply requirement. Assume that 0.3 m^3/min·m of length will be adequate.

Air required (length basis) = 11.5 m × 0.3 m^3/min·m

= 3.45 m^3/min for each grit chamber

Total air supply required = 3.45 × 2 = 6.9 m^3/min

6. Estimate the quantity of grit at peak flow. Assume a value of 0.05 m^3/10^3 m^3 at peak flow.

Volume grit = [(1.38 m^3/s)(86,400 s/d)(0.05 m^3/10^3 m^3)]

= 5.96 m^3/d

Comment   In designing aerated grit chambers, methods of regulating the air flow rate should be provided to control grit-removal rates and improve the cleanliness of the grit.

Vortex-Type Grit Chambers

Grit is also removed in devices that use a vortex flow pattern. Two types of devices are shown on Fig. 5–37. In one type, illustrated on Fig. 5–37a, wastewater enters and exits tangentially. The rotating turbine maintains constant flow velocity, and its adjustable-pitch blades promote separation of organics from the grit. The action of the propeller produces a toroidal flow path for grit particles. The grit settles by gravity into the hopper in one revolution of the basin's contents. Solids are removed from the hopper by a grit pump or an airlift pump. Grit removed by a grit pump can be discharged to a hydro-clone for removal of the remaining organic material. Grit removed by an airlift may be dewatered on a wedgewire screen (see Screening, Sec. 5–1, for description). Typical design data are presented in Table 5–18. If more than two units are installed, special arrangements for flow splitting are required.

In the second type, illustrated on Fig. 5–37b, a vortex is generated by the flow entering tangentially at the top of the unit. Effluent exits the center of the top of the unit from a rotating cylinder, or "eye" of the fluid. Centrifugal and gravitational forces within this cylinder minimize the release of particles with densities greater than water. Grit settles by gravity to the bottom of the unit, while organics, including those separated from grit particles by centrifugal forces, exit principally with the effluent. Organics remaining with the settled grit are separated as the grit particles move along the unit floor. Headloss in the unit is a function of the size particle to be removed and increases significantly for very fine particles. Vortex grit-removal units are sized to handle peak flowrates up to 0.3 m^3/s (7 Mgal/d) per unit. Grit is removed from the unit by a cleated belt conveyor. Because of its overall height, this type of grit system requires a deep basement, or a lift station if it is installed above grade.

Solids (Sludge) Degritting

Where grit chambers are not used and the grit is allowed to settle in the primary settling tanks, grit removal is accomplished by pumping dilute quantities of primary

Figure 5–37

Vortex-type grit
chambers: (a) Pista
(Courtesy Smith &
Loveless) and (b) Teacup.
(Courtesy Eutek.)

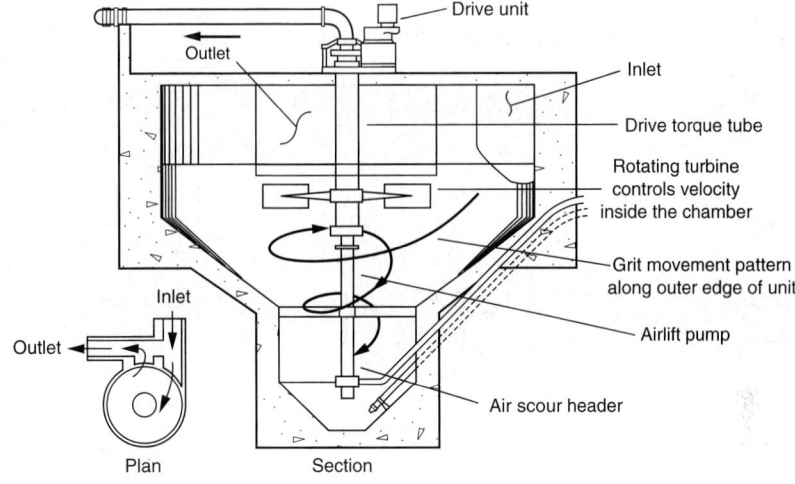

(a)

(b)

sludge to a cyclone degritter. The cyclone degritter acts as a centrifugal separator in
which the heavy particles of grit and solids are separated by the action of a vortex and
discharged separately from the lighter particles and the bulk of the liquid. The principal advantage of cyclone degritting is the elimination of the cost of building, operating, and maintaining grit chambers. The disadvantages are (1) pumping of dilute quantities of solids usually requires solids thickeners, and (2) pumping of grit with the
liquid primary solids increases the cost of operating and maintaining solids collectors
and the primary sludge pumps.

Table 5-18

Typical design information for vortex-type grit chambers

Item	U.S. customary units			SI units		
	Unit	Range	Typical	Unit	Range	Typical
Detention time at average flowrate	s	20–30	30	s	20–30	30
Diameter						
Upper chamber	ft	4.0–24.0		m	1.2–7.2	
Lower chamber	ft	3.0–6.0		m	0.9–1.8	
Height	ft	9.0–16.0		m	2.7–4.8	
Removal rates						
0.30 mm (50 mesh)	%	92–98	95+	%	92–98	95+
0.24 mm (70 mesh)	%	80–90	85+	%	80–90	85+
0.15 mm (100 mesh)	%	60–70	65+	%	60–70	65+

Grit Characteristics, Quantities, Processing, and Disposal

Grit consists of sand, gravel, cinders, or other heavy materials that have specific gravities or settling velocities considerably greater than those of organic particles. In addition to these materials, grit includes eggshells, bone chips, seeds, coffee grounds, and large organic particles. The characteristics, quantities, processing, and disposal of grit are discussed below.

Characteristics of Grit. Generally, what is removed as grit is predominantly inert and relatively dry material. However, grit composition can be highly variable, with moisture content ranging from 13 to 65 percent, and volatile content from 1 to 56 percent. The specific gravity of clean grit particles reaches 2.7 for inerts but can be as low as 1.3 when substantial organic material is agglomerated with inerts. A bulk density of 1600 kg/m^3 (100 lb/ft^3) is commonly used for grit. Often, enough organics are present in the grit so that it quickly putrefies if not properly handled after removal from the wastewater. Grit particles 0.2 mm and larger have been cited as the cause of most downstream problems.

The actual size distribution of retained grit exhibits variation due to differences in collection system characteristics, as well as variations in grit-removal efficiency. Generally, most grit particles are retained on a 0.15-mm (100-mesh) sieve, reaching nearly 100 percent retention in some instances; however, grit can be much finer. In the southeastern United States, where fine sand known as "sugar sand" constitutes a portion of the grit, less than 60 percent of the grit was retained on a 0.15-mm (100-mesh) screen in some cases.

The character of grit normally collected in grit chambers and from cyclone degritters varies widely from what might be normally considered as clean grit, to grit that includes a large proportion of putrescible organic material. Unwashed grit may contain

Table 5–19
Comparison of quantities of grit removed from wastewater from separate and combined collection systems[a]

Type of system	Ratio of maximum day to average day	Average grit quantity	
		ft³/Mgal	m³/1000 m³
Separate	1.5 to 3:1	0.53–5	0.004–0.037
Combined	3 to 15:1	0.53–24	0.004–0.18

[a] WEF (1998b).

50 percent or more of organic material, has a distinctly disagreeable odor, and, unless promptly disposed of, may attract insects and rodents.

Quantities of Grit. The quantities of grit will vary greatly from one location to another, depending on the type of sewer system, the characteristics of the drainage area, the condition of the sewers, the frequency of street sanding to counteract icing conditions, the types of industrial wastes, the number of household garbage grinders served, and in areas with sandy soils. Typical values for aerated grit chambers are reported in Table 5–17. A comparison of the quantities of grit removed from separate and combined sewer systems is shown in Table 5–19.

It is difficult to interpret grit-removal data because grit itself is poorly characterized and almost no data exist on relative removal efficiencies. The information on grit characteristics derives from what has been removed as grit. Sieve analyses are not normally performed on grit chamber influents and effluents. For these reasons, the efficiencies of grit-removal systems cannot be compared.

Grit Separation and Washing. Grit separators and grit washers may accomplish removal of a major part of the organic material contained in grit. When some of the heavier organic matter remains with the grit, grit washers are commonly used to provide a second stage of volatile solids separation. Examples of grit separation and washing units are shown on Fig. 5–38.

Two principal types of grit washers are available. One type relies on an inclined submerged rake that provides the necessary agitation for separation of the grit from the organic materials and, at the same time, raises the washed grit to a point of discharge above water level (similar to the inclined ramp shown on Fig. 5–31). Another type of grit washer (see Fig. 5–38) uses an inclined screw and moves the grit up the ramp. Both types can be equipped with water sprays to assist in the cleansing action. Hydroclone separators are often installed at the inlet to the grit washer to improve grit separation and organics removal.

Disposal of Grit. The most common method of grit disposal is transport to a landfill. In some large plants, grit is incinerated with solids. As with screenings, some states require grit to be lime stabilized before disposal in a landfill. Disposal in all cases should be done in conformance with the appropriate environmental regulations.

In larger plants where trucks are used to transfer grit, elevated grit storage facilities may be provided with bottom-loading gates. Difficulties experienced in getting the grit to flow freely from the storage hoppers have been minimized by using steep slopes on

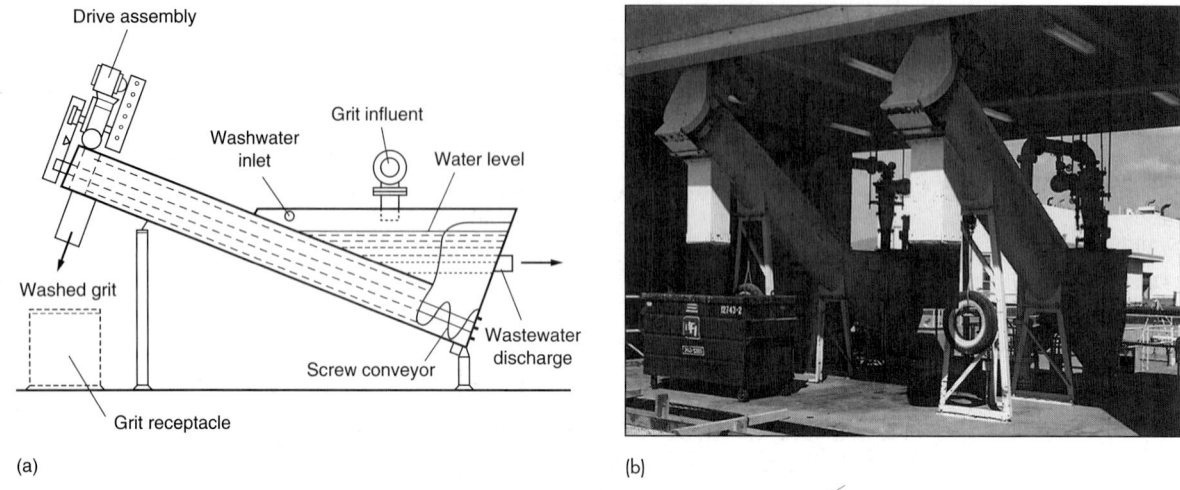

(a)   (b)

Figure 5-38

Example of grit separation and washing unit: (a) schematic and (b) view of typical installation.

the storage hoppers, by applying air beneath the grit, and by the use of hopper vibrators. Drainage facilities for collection and disposal of drippings from the bottom-loading gates are desirable. Grab buckets operating on a monorail system may also be used to load trucks directly from the grit chambers.

Pneumatic conveyors are sometimes used to convey grit short distances. Advantages of pneumatic conveying include (1) no elevated storage hoppers are required, and (2) attendant odor problems associated with storage are eliminated. The principal disadvantage is the considerable wear on piping, especially at bends.

5-7 PRIMARY SEDIMENTATION

The objective of treatment by sedimentation is to remove readily settleable solids and floating material and thus reduce the suspended solids content. Primary sedimentation is used as a preliminary step in the further processing of the wastewater. Efficiently designed and operated primary sedimentation tanks should remove from 50 to 70 percent of the suspended solids and from 25 to 40 percent of the BOD.

Sedimentation tanks have also been used as stormwater retention tanks, which are designed to provide a moderate detention period (10 to 30 min) for overflows from either combined sewers or storm sewers. The purpose of sedimentation is to remove a substantial portion of the organic solids that otherwise would be discharged directly to the receiving waters. Sedimentation tanks have also been used to provide detention periods sufficient for effective disinfection of such overflows.

The purpose of this section is to describe the various types of sedimentation facilities, to consider their performance, and to review important design considerations. Sedimentation tanks used for secondary treatment are considered in Chap. 8.

Description

Almost all treatment plants use mechanically cleaned sedimentation tanks of standardized circular or rectangular design (see Fig. 5–39). The selection of the type of sedimentation unit for a given application is governed by the size of the installation, by rules and regulations of local control authorities, by local site conditions, and by the experience and judgment of the engineer. Two or more tanks should be provided so that the process may remain in operation while one tank is out of service for maintenance and repair work. At large plants, the number of tanks is determined largely by size limitations. Typical design information and dimensions for rectangular and circular sedimentation tanks used for primary treatment are presented in Tables 5–20 and 5–21.

Rectangular Tanks.   Rectangular sedimentation tanks may use either chain-and-flight solids collectors or traveling-bridge-type collectors. A rectangular tank that uses a chain-and-flight-type collector is shown on Fig. 5–40. Equipment for settled solids removal generally consists of a pair of endless conveyor chains, manufactured of alloy steel, cast iron, or thermoplastic. Attached to the chains at approximately 3-m (10-ft) intervals are scraper flights made of wood or fiberglass, extending the full width of the tank or bay. The solids settling in the tank are scraped to solids hoppers in small tanks and to transverse troughs in large tanks. The transverse troughs are equipped with collecting mechanisms (cross collectors), usually either chain-and-flight or screw-type collectors, which convey solids to one or more collection hoppers. In very long units (over 50 m), two collection mechanisms can be used to scrape solids to collection points near the middle of the tank length. Where possible, it is desirable to locate solids pumping facilities close to the collection hoppers.

Rectangular tanks may also be cleaned by a bridge-type mechanism that travels up and down the tank on rubber wheels or on rails supported on the sidewalls. One or more scraper blades are suspended from the bridge. Some of the bridge mechanisms are designed so that the scraper blades can be lifted clear of the solids blanket on the return travel.

Figure 5–39

Aerial view of wastewater-treatment plant with circular primary and secondary sedimentation basins. *(Photo courtesy F. Wayne Hill Water Resources Center, Gwinnett County, GA.)*

Table 5–20
Typical design information for primary sedimentation tanks[a]

Item	U.S. customary units			SI units		
	Unit	Range	Typical	Unit	Range	Typical
Primary sedimentation tanks followed by secondary treatment						
Detention time	h	1.5–2.5	2.0	h	1.5–2.5	2.0
Overflow rate						
Average flow	gal/ft²·d	800–1200	1000	m³/m²·d	30–50	40
Peak hourly flow	gal/ft²·d	2000–3000	2500	m³/m²·d	80–120	100
Weir loading	gal/ft·d	10,000–40,000	20,000	m³/m·d	125–500	250
Primary settling with waste activated-sludge return						
Detention time	h	1.5–2.5	2.0	h	1.5–2.5	2.0
Overflow rate						
Average flow	gal/ft²·d	600–800	700	m³/m²·d	24–32	28
Peak hourly flow	gal/ft²·d	1200–1700	1500	m³/m²·d	48–70	60
Weir loading	gal/ft·d	10,000–40,000	20,000	m³/m·d	125–500	250

[a]Comparable data for secondary clarifiers are presented in Chap. 8.

Table 5–21
Typical dimensional data for rectangular and circular sedimentation tanks used for primary treatment of wastewater

Item	U.S. customary units			SI units		
	Unit	Range	Typical	Unit	Range	Typical
Rectangular:						
Depth	ft	10–16	14	m	3–4.9	4.3
Length	ft	50–300	80–130	m	15–90	24–40
Width[a]	ft	10–80	16–32	m	3–24	4.9–9.8
Flight speed	ft/min	2–4	3	m/min	0.6–1.2	0.9
Circular:						
Depth	ft	10–16	14	m	3–4.9	4.3
Diameter	ft	10–200	40–150	m	3–60	12–45
Bottom slope	in/ft	3/4–2/ft	1.0/ft	mm/mm	1/16–1/6	1/12
Flight speed	r/min	0.02–0.05	0.03	r/min	0.02–0.05	0.03

[a]If widths of rectangular mechanically cleaned tanks are greater than 6 m (20 ft), multiple bays with individual cleaning equipment may be used, thus permitting tank widths up to 24 m (80 ft) or more.

Figure 5–40

Typical rectangular primary sedimentation tank: (a) plan and (b) section.

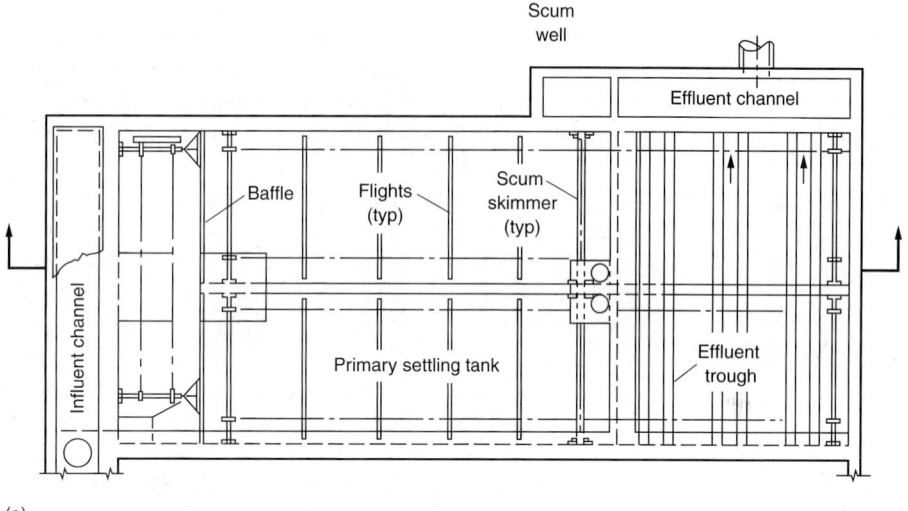

(a)

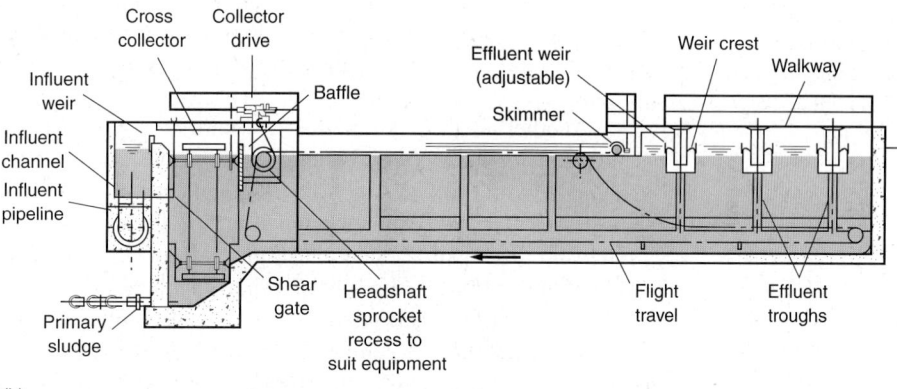

(b)

Where cross collectors are not provided, multiple solids hoppers must be installed. Solids hoppers have operating difficulties, notably solids accumulation on the slopes and in the corners and arching over the solids drawoff piping. Wastewater may also be drawn through the solids hopper, bypassing some of the accumulated solids, resulting in a "rathole" effect. A cross collector is more advisable, except possibly in small plants, because a more uniform and concentrated solids can be withdrawn and many of the problems associated with solids hoppers can be eliminated.

Because flow distribution in rectangular tanks is critical, one of the following inlet designs is used: (1) full-width inlet channels with inlet weirs, (2) inlet channels with submerged ports or orifices, (3) or inlet channels with wide gates and slotted baffles. Inlet weirs, while effective in spreading flow across the tank width, introduce a vertical velocity component into the solids hopper that may resuspend the solids particles. Inlet ports can provide good distribution across the tank width if the velocities are maintained in the 3 to 9 m/min (10 to 30 ft/min) range. Inlet baffles are effective in reducing the high initial velocities and distribute flow over the widest possible cross-sectional area. Where full-width baffles are used, they should extend from 150 mm (6 in) below the surface to 300 mm (12 in) below the entrance opening.

For installations of multiple rectangular tanks, below-grade pipe and equipment galleries can be constructed integrally with the tank structure and along the influent end. The galleries are used to house the sludge pumps and sludge drawoff piping. The galleries also provide access to the equipment for operation and maintenance. Galleries can also be connected to service tunnels for access to other plant units.

Scum is usually collected at the effluent end of rectangular tanks with the flights returning at the liquid surface. The scum is moved by the flights to a point where it is trapped by baffles before removal. Water sprays can also move the scum. The scum can be scraped manually up an inclined apron, or it can be removed hydraulically or mechanically, and for scum removal a number of means have been developed. For small installations, the most common scum drawoff facility consists of a horizontal, slotted pipe that can be rotated by a lever or a screw. Except when drawing scum, the open slot is above the normal tank water level. When drawing scum, the pipe is rotated so that the open slot is submerged just below the water level, permitting the scum accumulation to flow into the pipe. Use of this equipment results in a relatively large volume of scum liquor.

Another method for removing scum is by a transverse rotating helical wiper attached to a shaft. Scum is removed from the water surface and moved over a short inclined apron for discharge to a cross-collecting scum trough. The scum may then be flushed to a scum ejector or hopper ahead of a scum pump. Another method of scum removal consists of a chain-and-flight type of collector that collects the scum at one side of the tank and scrapes it up a short incline for deposit in scum hoppers, whence it can be pumped to disposal units. Scum is also collected by special scum rakes in rectangular tanks that are equipped with the carriage or bridge type of sedimentation tank equipment. In installations where appreciable amounts of scum are collected, the scum hoppers are usually equipped with mixers to provide a homogeneous mixture prior to pumping. Scum is usually disposed of with the solids and biosolids produced at the plant; however, separate scum disposal is used at many plants.

Multiple rectangular tanks require less land area than multiple circular tanks and find application where site space is at a premium. Rectangular tanks also lend themselves to nesting with preaeration tanks and aeration tanks in activated-sludge plants, thus permitting common wall construction and reducing construction costs. They are also used generally where tank roofs or covers are required.

Circular Tanks. In circular tanks the flow pattern is radial (as opposed to horizontal in rectangular tanks). To achieve a radial flow pattern, the wastewater to be settled can be introduced in the center or around the periphery of the tank, as shown on Fig. 5–41. Both flow configurations have proved to be satisfactory generally, although the center-feed type is more commonly used, especially for primary treatment. In the center-feed design (see Fig. 5–41a), the wastewater is transported to the center of the tank in a pipe suspended from the bridge, or encased in concrete beneath the tank floor.

Figure 5–41

Typical circular primary sedimentation tanks:
(a) center feed and
(b) peripheral feed.
(Crites and
Tchobanoglous, 1998.)

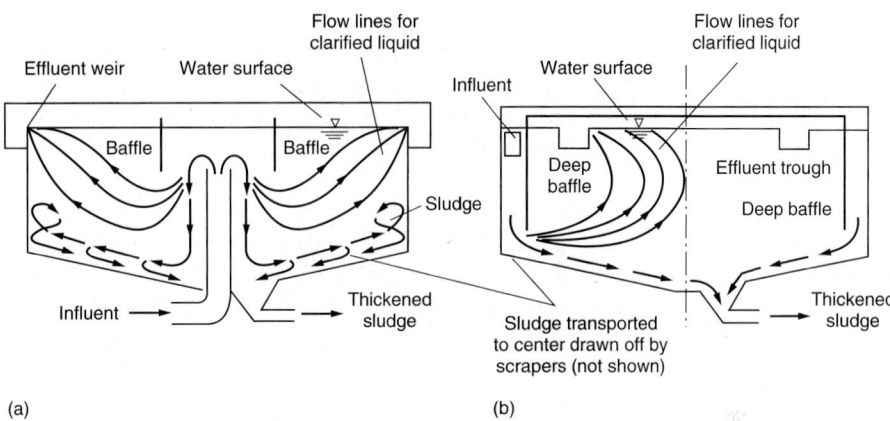

(a) (b)

Figure 5–42

Typical circular primary sedimentation tank.

At the center of the tank, the wastewater enters a circular well designed to distribute the flow equally in all directions (see Fig. 5–42). The center well has a diameter typically between 15 and 20 percent of the total tank diameter and ranges from 1 to 2.5 m (3 to 8 ft) in depth and should have a tangential energy-dissipating inlet within the feedwell.

The energy-dissipating device (see Fig. 5–43) functions to collect influent from the center column and discharge it tangentially into the upper 0.5 to 0.7 m of the feedwell. The discharge ports are sized to produce a velocity of ≤ 0.75 m/s at maximum flow and 0.30 to 0.45 m/s at average flow. The feedwell should be sized so that the maximum downward velocity does not exceed 0.75 m/s. The depth of the feedwell should extend about 1 meter below the energy-dissipating inlet ports (Randall et al., 1992).

In the peripheral-feed design (see Fig. 5–41b), a suspended circular baffle forms an annular space into which the inlet wastewater is discharged in a tangential direction. The wastewater flows spirally around the tank and underneath the baffle, and the clarified liquid is skimmed off over weirs on both sides of a centrally located weir trough. Grease and scum are confined to the surface of the annular space. Peripheral feed tanks are used generally for secondary clarification.

Circular tanks 3.6 to 9 m (12 to 30 ft) in diameter have the solids-removal equipment supported on beams spanning the tank. Tanks 10.5 m (35 ft) in diameter and larger have a central pier that supports the mechanism and is reached by a walkway or bridge.

Figure 5–43

Typical energy-
dissipating and flow
distribution inlet for a
center-feed sedimentation
tank. The inner ring is
used to create a
tangential flow pattern.
(Randall, et al., 1992.)

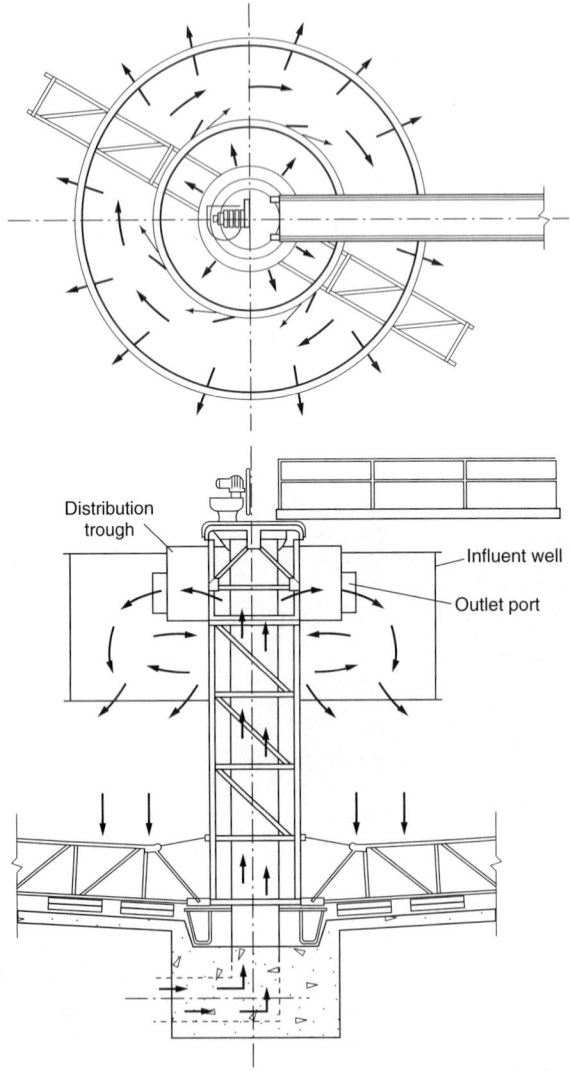

The bottom of the tank is sloped at about 1 in 12 (vertical:horizontal) to form an inverted cone, and the solids are scraped to a relatively small hopper located near the center of the tank.

Multiple tanks are customarily arranged in groups of two or four. The flow is divided among the tanks by a flow-split structure, commonly located between the tanks. Solids are usually withdrawn by sludge pumps for discharge to the solids processing and disposal units.

Combination Flocculator-Clarifier.   Combination flocculator-clarifiers are often used in water treatment and sometimes used for wastewater treatment, especially in cases where enhanced settling, such as for industrial wastewater treatment or for

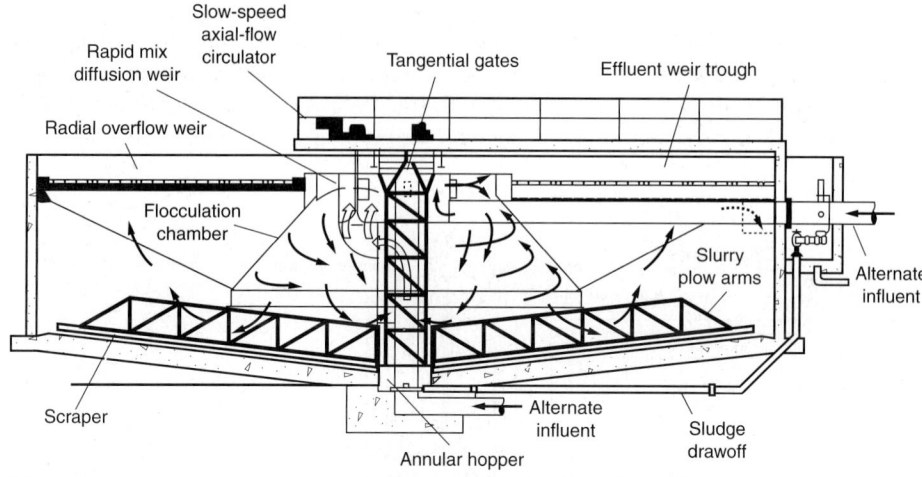

(a)

Figure 5–44

Typical flocculator-clarifier: (a) schematic and (b) view of empty tank. In some designs, turbine or propeller mixers are included in the flocculation chamber.

(b)

biosolids concentration, is required. Inorganic chemicals or polymers can be added to improve flocculation. Circular clarifiers are ideally suited for incorporation of an inner, cylindrical flocculation compartment (see Fig. 5–44). Wastewater enters through a center shaft or well and flows into the flocculation compartment, which is generally equipped with a paddle-type or low-speed mixer. The gentle stirring causes flocculent particles to form. From the flocculation compartment, flow then enters the clarification zone by passing down and radially outward. Settled solids and scum are collected in the same way as in a conventional clarifier.

Stacked (Multilevel) Clarifiers. Stacked clarifiers originated in Japan in the 1960s where limited land area is available for the construction of wastewater-treatment facilities. Since that time, stacked clarifiers have been used in the United States, the most notable installation of which is at the Deer Island Wastewater Treatment Plant constructed in Boston harbor. Design of these types of clarifiers recognizes the importance of settling area to settling efficiency. Operation of stacked rectangular clarifiers is similar to conventional rectangular clarifiers in terms of influent and effluent flow patterns and solids collection and removal. The stacked clarifiers are actually two (or more) tanks, one located above the other, operating on a common water surface (see Fig. 5–45). Each clarifier is fed independently, resulting in parallel flow through the lower and upper tanks. Settled solids are collected from each tank with chain and flight solids collectors, discharging to a common hopper. In addition to saving space, advantages claimed for stacked clarifiers include less piping and pumping requirements.

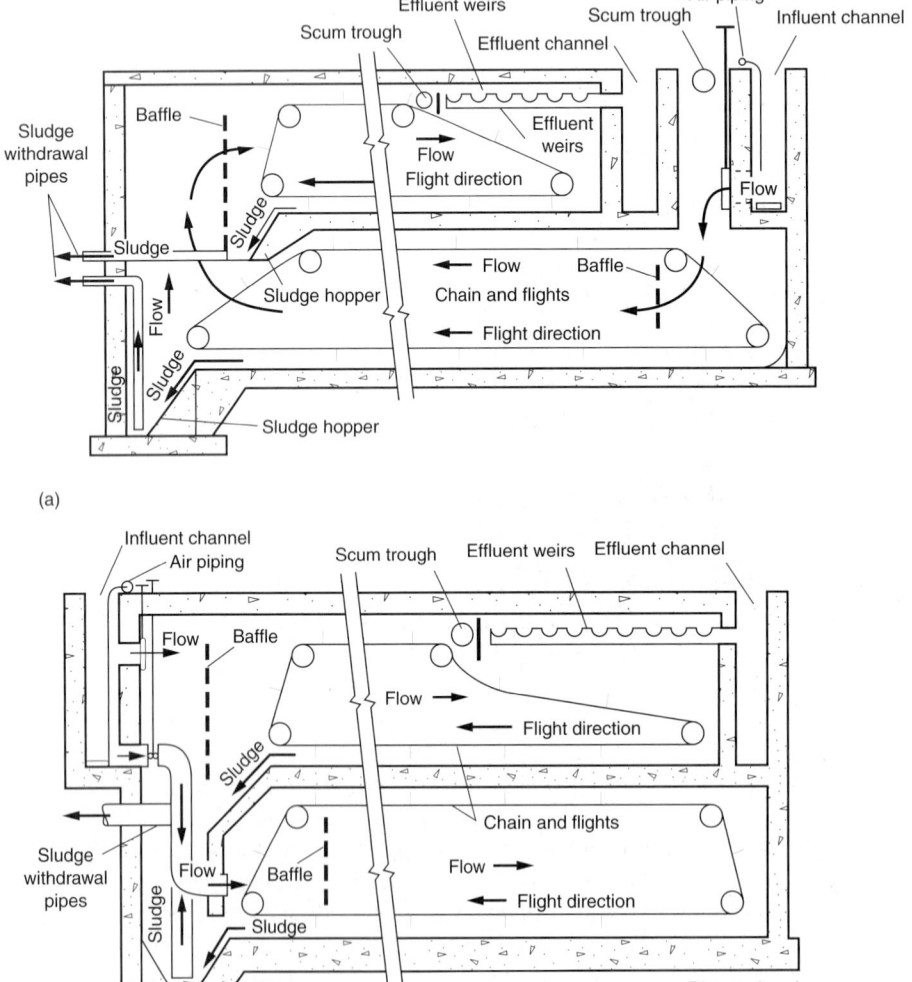

(a)

(b)

Figure 5–45

Typical section through a stacked clarifier: (a) series flow and (b) parallel flow type used at the Deer Island Wastewater Treatment Plant, Boston, MA. Note: In the parallel flow type, the upper effluent weirs serve both the upper and lower clarifiers. Channels for the discharge of effluent from the lower to the upper clarifier are located on either side of the sludge collection mechanism in the upper clarifier.

Because the facilities are more compact and have less exposed surface area, better control of odors and volatile organic compound emissions is possible. Disadvantages include higher construction cost than conventional clarifiers and more complex structural design. Design criteria for stacked clarifiers, as regards overflow and weir rates, are similar to conventional primary and secondary clarifiers.

Sedimentation Tank Performance

The efficiency of sedimentation basins with respect to the removal of BOD and TSS is reduced by (1) eddy currents formed by the inertia of the incoming fluid, (2) wind-induced circulation cells formed in uncovered tanks, (3) thermal convection currents, (4) cold or warm water causing the formation of density currents that move along the bottom of the basin and warm water rising and flowing across the top of the tank, and (5) thermal stratification in hot arid climates (Fair and Geyer, 1954). Factors that affect performance are considered in the following discussion.

BOD and TSS Removal. Typical performance data for the removal of BOD and TSS in primary sedimentation tanks, as a function of the detention time and constituent concentration, are presented on Fig. 5–46. The curves shown on Fig. 5–46 are derived from observations of the performance of actual sedimentation tanks. The curvilinear relationships in the figure can be modeled as rectangular hyperbolas using the following relationship (Crites and Tchobanoglous, 1998).

$$R = \frac{t}{a + bt} \tag{5-45}$$

where R = expected removal efficiency
t = nominal detention time T
a, b = empirical constants

Figure 5–46

Typical BOD and TSS removal in primary sedimentation tanks. (Greeley, 1938.)

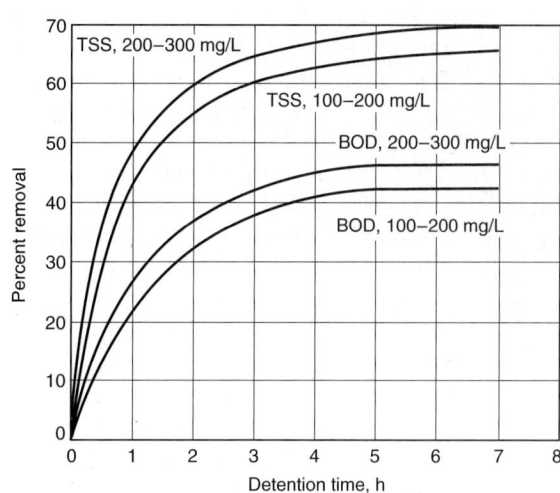

Typical values for the empirical constants in Eq. (5–45) at 20°C are as follows:

Item	a	b
BOD	0.018	0.020
TSS	0.0075	0.014

A fact that is often overlooked in sedimentation tank performance is the change in the wastewater characteristics that occurs through the sedimentation process. Larger, more slowly biodegradable suspended solids settle first, leaving a more volatile fraction in suspension that remains in the primary tank effluent. The strict use of removal curves, such as those given on Fig. 5–46, does not account for the transformation in wastewater characteristics that actually occurs. Where possible for domestic wastewater, primary tank influent and effluent should be characterized to determine concentration and composition of the constituents. Such characterization is important when determining the organic loading required to be treated by the succeeding biological treatment units. Further discussion on the effects of wastewater characterization on biological processes is contained in Chaps. 7 and 8.

Short Circuiting and Hydraulic Stability.   In an ideal sedimentation basin (see Fig. 5–47a), a given block of entering water should remain in the basin for the full detention time. Unfortunately, in practice sedimentation basins are seldom ideal and considerable short circuiting will be observed for one or more of the reasons cited above. To determine if short circuiting exists and to what extent, tracer studies, as discussed in Sec. 4–4 in Chap. 4, should be performed. Time-concentration curves should be developed for analysis. If in the repeated tests the time-concentration curves are similar, then the basin is stable. If the time-concentration curves (also known as residence time distribution (RTD) curves and discussed in Sec. 4–4) are not repeatable, the basin is unstable and the performance of the basin will be erratic (Fair and Geyer, 1954). The method of influent flow distribution, as discussed above, will also affect short circuiting.

Temperature Effects.   Temperature effects can be significant in sedimentation basins. It has been shown that a 1° Celsius temperature differential between the incoming wastewater and the wastewater in the sedimentation tank will cause a density current to form (see Figs. 5–47b and c). The impact of the temperature effects on performance will depend on the material being removed and its characteristics.

Wind Effects.   Wind blowing across the top of open sedimentation basins can cause circulation cells to form (see Fig. 5–47d). When circulation cells form, the effective volumetric capacity of the basin is reduced. As with temperature effects, the impact of the reduced volume on performance will depend on the material being removed and its characteristics.

Design Considerations

If all solids in wastewater were discrete particles of uniform size, uniform density, uniform specific gravity, and uniform shape, the removal efficiency of these solids would be dependent on the surface area of the tank and time of detention. The depth of the tank would have little influence, provided that horizontal velocities would be maintained

Figure 5-47

Typical flow patterns observed in rectangular sedimentation tanks: (a) ideal flow, (b) effect of density flow or thermal stratification (water in tank is warmer than influent), (c) effect of thermal stratification (water in tank is colder than influent), and (d) formation of wind-driven circulation cell. (Crites and Tchobanoglous, 1998.)

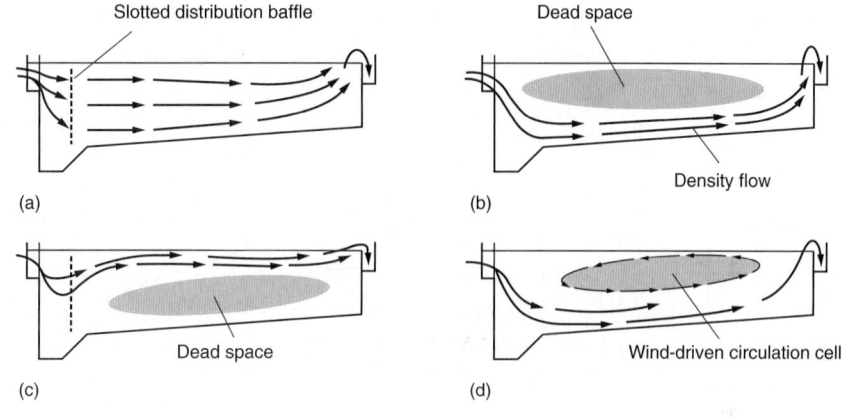

below the scouring velocity. However, the solids in most wastewaters are not of such regular character but are heterogeneous in nature, and the conditions under which they are present range from total dispersion to complete flocculation. Design parameters for sedimentation are considered below. Typical design data for sedimentation tanks are presented in Tables 5–17 and 5–18. Additional details on the analysis and design of sedimentation tanks may be found in WPCF, 1985. A design procedure is illustrated in Example 5–10.

Detention Time. The bulk of the finely divided solids reaching primary sedimentation tanks is incompletely flocculated but is susceptible to flocculation. Flocculation is aided by eddying motion of the fluid within the tanks and proceeds through the coalescence of fine particles, at a rate that is a function of their concentration and of the natural ability of the particles to coalesce upon collision. As a general rule, coalescence of a suspension of solids becomes more complete as time elapses, thus, detention time is a consideration in the design of sedimentation tanks. The mechanics of flocculation are such, however, that as the time of sedimentation increases, less and less coalescence of remaining particles occurs.

Normally, primary sedimentation tanks are designed to provide 1.5 to 2.5 h of detention based on the average rate of wastewater flow. Tanks that provide shorter detention periods (0.5 to 1 h), with less removal of suspended solids, are sometimes used for preliminary treatment ahead of biological treatment units. In cold climates, increases in water viscosity at lower temperatures retard particle settling in clarifiers and reduce performance at wastewater temperatures below 20°C (68°F). A curve showing the increase in detention time necessary to equal the detention time at 20°C is presented on Fig. 5–48 (WPCF, 1985). For wastewater having a temperature of 10°C, for example, the detention period is 1.38 times that required at 20°C to achieve the same efficiency. Thus, in cold climates, safety factors should be considered in clarifier design to ensure adequate performance.

Surface Loading Rates. Sedimentation tanks are normally designed on the basis of a surface loading rate (commonly termed "overflow rate") expressed as cubic

Figure 5–48

Curve of the increase in detention time required at cooler temperatures to achieve the same sedimentation performance as achieved at 20°C.

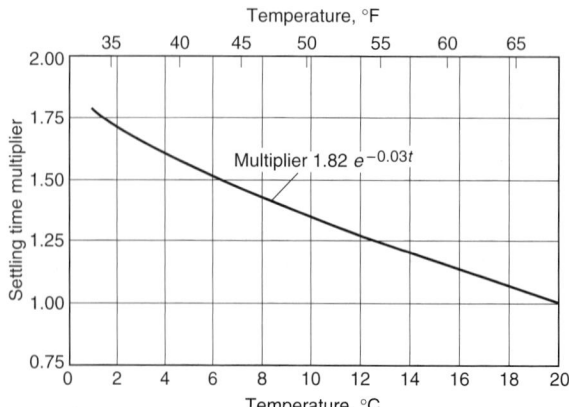

meters per square meter of surface area per day, m³/m²·d (gallons per square foot of surface area per day, gal/ft²·d). The selection of a suitable loading rate depends on the type of suspension to be separated. Typical values for various suspensions are reported in Table 5–20. Designs for municipal plants must also meet the approval of state regulatory agencies, many of which have adopted standards for surface loading rates that must be followed. When the area of the tank has been established, the detention period in the tank is governed by water depth. Overflow rates in current use result in nominal detention periods of 2.0 to 2.5 h, based on average design flow.

The effect of the surface loading rate and detention time on suspended solids removal varies widely depending on the character of the wastewater, proportion of settleable solids, concentration of solids, and other factors. It should be emphasized that overflow rates must be set low enough to ensure satisfactory performance at peak rates of flow, which may vary from over 3 times the average flow in small plants to 2 times the average flow in large plants (see discussion of peak flowrates in Chap. 3).

Weir Loading Rates.   In general, weir loading rates have little effect on the efficiency of primary sedimentation tanks and should not be considered when reviewing the appropriateness of clarifier design. For general information purposes only, typical weir loading rates are given in Table 5–20. The placement of weirs and baffles in secondary sedimentation applications is discussed in Sec. 8–8 of Chap. 8.

Scour Velocity.   To avoid the resuspension (scouring) of settled particles, horizontal velocities through the tank should be kept sufficiently low. Using the results from studies by Shields (1936), Camp (1946) developed the following equation for the critical velocity.

$$V_H = \left[\frac{8k(s-1)gd}{f} \right]^{1/2} \tag{5-46}$$

where V_H = horizontal velocity that will just produce scour, LT^{-1} (m/s)

k = constant that depends on type of material being scoured (unitless)

s = specific gravity of particles

g = acceleration due to gravity, LT^{-2} (9.81 m/s^2)

d = diameter of particles, L

f = Darcy-Weisbach friction factor (unitless)

Typical values of k are 0.04 for unigranular sand and 0.06 for more sticky, interlocking matter. The term f (the Darcy-Weisbach friction factor) depends on the characteristics of the surface over which flow is taking place and the Reynolds number. Typical values of f are 0.02 to 0.03. Either SI or U.S. customary units may be used in Eq. (5–46), so long as they are consistent, because k and f are dimensionless.

EXAMPLE 5–10 **Design of a Primary Sedimentation Basin** The average flowrate at a small municipal wastewater-treatment plant is 20,000 m^3/d. The highest observed peak daily flowrate is 50,000 m^3/d. Design rectangular primary clarifiers with a channel width of 6 m (20 ft). Use a minimum of two clarifiers. Calculate the scour velocity, to determine if settled material will become resuspended. Estimate the BOD and TSS removal at average and peak flow. Use an overflow rate of 40 m^3/m^2·d at average flow (see Table 5–20) and a side water depth of 4 m (13.1 ft).

Solution

1. Calculate the required surface area. For average flow conditions, the required area is

$$A = \frac{Q}{OR} = \frac{(20{,}000 \text{ m}^3/\text{d})}{(40 \text{ m}^3/\text{m}^2\text{·d})} = 500 \text{ m}^2$$

2. Determine the tank length.

$$L = \frac{A}{W} = \frac{500 \text{ m}^2}{2 \times 6 \text{ m}} = 41.7 \text{ m}$$

However, for the sake of convenience, the surface dimensions will be rounded to 6 m by 42 m.

3. Compute the detention time and overflow rate at average flow. Using the assumed sidewater depth of 4 m,

Tank volume = 4 m × 2(42 m × 6 m) = 2016 m^2

$$\text{Overflow rate} = \frac{Q}{A} = \frac{(20{,}000 \text{ m}^3/\text{d})}{2(6 \text{ m} \times 42 \text{ m})} = 39.7 \text{ m}^3/\text{m}^2 \cdot \text{d}$$

$$\text{Detention time} = \frac{\text{Vol.}}{Q} = \frac{(2016 \text{ m}^3)(24 \text{ h}/\text{d})}{(20{,}000 \text{ m}^3/\text{d})} = 2.42 \text{ h}$$

4. Determine the detention time and overflow rate at peak flow.

$$\text{Overflow rate} = \frac{Q}{A} = \frac{(50{,}000 \text{ m}^3/\text{d})}{2(6 \text{ m} \times 42 \text{ m})} = 99.2 \text{ m}^3/\text{m}^2 \cdot \text{d}$$

$$\text{Detention time} = \frac{\text{Vol.}}{Q} = \frac{(2016 \text{ m}^3)(24 \text{ h}/\text{d})}{(50{,}000 \text{ m}^3/\text{d})} = 0.97 \text{ h}$$

5. Calculate the scour velocity [Eq. (5–46)], using the following values:

Cohesion constant	$k = 0.05$
Specific gravity	$s = 1.25$
Acceleration due to gravity	$g = 9.81 \text{ m/s}^2$
Diameter of particles	$d = 100 \ \mu\text{m} = 100 \times 10^{-6} \text{ m}$
Darcy-Weisbach friction factor	$f = 0.025$

$$V_H = \left[\frac{8k(s-1)gd}{f} \right] = \left[\frac{(8)(0.05)(0.25)(9.81)(100 \times 10^{-6})}{0.025} \right]^{1/2} = 0.063 \text{ m/s}$$

6. Compare the scour velocity calculated in the previous step to the peak flow horizontal velocity (the peak flow divided by the cross-sectional area through which the flow passes).

    The peak flow horizontal velocity through the settling tank is

$$V = \frac{Q}{A_x} = \left[\frac{(50{,}000 \text{ m}^3/\text{d})}{2(6 \text{ m} \times 4 \text{ m})} \right] \left[\frac{1}{(24 \text{ h}/\text{d})(3600 \text{ s}/\text{h})} \right] = 0.012 \text{ m/s}$$

    The horizontal velocity value, even at peak flow, is substantially less than the scour velocity. Therefore, settled matter should not be resuspended.

7. Use Eq. (5–45) and the accompanying coefficients to estimate the removal rates for BOD and TSS at average and peak flow.

    a. At average flow:

$$\text{BOD removal} = \frac{t}{a + bt} = \frac{2.42}{0.018 + (0.020)(2.42)} = 36\%$$

$$\text{TSS removal} = \frac{t}{a + bt} = \frac{2.42}{0.0075 + (0.014)(2.42)} = 58\%$$

    b. At peak flow:

$$\text{BOD removal} = \frac{t}{a + bt} = \frac{0.97}{0.018 + (0.020)(0.97)} = 26\%$$

$$\text{TSS removal} = \frac{t}{a + bt} = \frac{0.97}{0.0075 + (0.014)(0.97)} = 46\%$$

Table 5–22
Typical values of specific gravity and solids concentration of solids and scum removed from primary sedimentation tanks

Type of solids (sludge)	Specific gravity	Solids concentration, %[a] Range	Typical
Primary only:			
Medium-strength wastewater	1.03	4–12	6
From combined sewer system	1.05	4–12	6.5
Primary and waste-activated sludge	1.03	2–6	3
Primary and trickling filter humus sludge	1.03	4–10	5
Scum	0.95	[b]	—

[a] Percent dry solids.
[b] Range is highly variable.

Characteristics and Quantities of Solids (Sludge) and Scum

Typical values of specific gravity and solids concentration of solids (sludge) and scum removed from primary sedimentation tanks are presented in Table 5–22. Scum consists of a variety of floatable materials, and solids concentrations vary widely.

In primary sedimentation tanks used in activated-sludge plants, provision may be required for handling the excess activated sludge that may be discharged into the influent of the primary tanks for settlement and consolidation with the primary sludge. For treatment plants where waste-activated sludge is returned to the primary sedimentation tanks, the primary sedimentation tanks should include provisions for light flocculent solids of 98 to 99.5 percent moisture and for concentrations ranging from 1500 to 10,000 mg/L in the influent mixed liquor.

The volume of solids produced in primary settling tanks must be known or estimated so that these tanks and subsequent solids pumping, processing, and disposal facilities can be properly designed. The solids volume will depend on (1) the characteristics of the untreated wastewater, including strength and freshness; (2) the period of sedimentation and the degree of purification to be effected in the tanks; (3) the condition of the deposited solids, including specific gravity, water content, and changes in volume under the influence of tank depth or mechanical solids-removal devices; and (4) the period between solids-removal operations. Additional information on the characteristics and quantities of solids produced during primary sedimentation and other treatment operations and processes is provided in Chap. 14.

5–8 HIGH-RATE CLARIFICATION

High-rate clarification employs physical/chemical treatment and utilizes special flocculation and sedimentation systems to achieve rapid settling. The essential elements of high-rate clarification are enhanced particle settling and the use of inclined plate or tube settlers. Advantages of high-rate clarification are (1) units are compact and thus reduce space requirements, (2) start-up times are rapid (usually less than 30 min) to achieve

Figure 5–49

Schematic of microsand ballasted floc particles. Polymer layer is used to absorb chemical flocs onto sand grains. (Adapted from Krüger Catalog.)

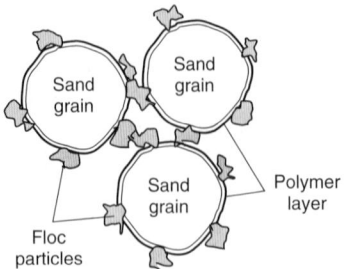

peak efficiency, and (3) a highly clarified effluent is produced. Enhanced particle flocculation and high-rate clarification applications are discussed in this section. Inclined plate and tube settlers were discussed previously in Sec. 5–5.

Enhanced Particle Flocculation

Enhanced particle flocculation has been used in Europe for more than 15 years but has only been introduced relatively recently in the United States. In its most basic form, enhanced particle flocculation involves the addition of an inert ballasting agent (usually silica sand or recycled chemically conditioned sludge) and a polymer to a coagulated and partially flocculated suspension. The polymer appears to coat the ballasting particles and forms the "glue" that binds the chemical floc to the ballasted particles (see Fig. 5–49). After contact with the ballasting agent, the mixture is stirred gently in a maturation tank that allows the floc particles to grow. The particles grow as the larger, faster-settling particles overtake and collide with slower-settling particles (see Fig. 5–13b). The velocity gradient G for flocculation is important as a high gradient will cause a breakdown in the floc particles, and insufficient agitation will inhibit floc formation. Velocity gradients for enhanced particle settling of wastewater generally range from 200 to 400 s^{-1}.

Analysis of Ballasted Particle Flocculation and Settling

The settling velocity of the ballasted particle is increased, when compared to an unballasted floc particle, by (1) increasing the density of the particle, (2) decreasing the coefficient of drag and increasing the Reynolds number, and (3) decreasing the shape factor through the formation of more dense spherical-shaped particles [see (Eq. 5–18)]. The ballasted floc particles appear to be more spherical than the floc particles alone. In effect, ballasted flocculent particles settle with a velocity closer to that of a discrete particle than that of flocculent particles that have very high shape factors. The comparative settling velocities of ballasted particles compared to other particles are illustrated in Example 5–11.

EXAMPLE 5–11   Calculation of Settling Velocities of Ballasted Floc and Other Particles
Determine the settling velocities for ballasted floc, spherical, and wastewater particles having the characteristics tabulated below.

	Particle type		
Parameter	**Ballasted floc**	**Spherical**	**Wastewater particle**
Average diameter, μm	200	150	500
Particle specific gravity	2.6	2.65	1.0035
Shape factor	2.5	1	18
Wastewater specific gravity	1.003	1.003	1.003

Solution   Using Eq. (5–22) and the same computation procedure as Example 5–5, calculate the particle settling velocity for a ballasted floc particle and compare to spherical and wastewater floc particles.

1. Determine the terminal settling velocity for the ballasted floc particle.

$$v_p = \frac{g(sg_p - 1)d_p^2}{18\nu}$$

$$= \frac{(9.81 \text{ m/s}^2)(2.6 - 1)(200 \times 10^{-6} \text{ m})}{18(1.003 \times 10^{-6} \text{ m}^2/\text{s})}$$

$$= 0.0348 \text{ m/s}$$

$$(d_p = 200 \ \mu\text{m} = 200 \times 10^{-6} \text{ m})$$

2. Check the Reynolds number [Eq. (5–20)] (include the shape factor ϕ).

$$N_R = \frac{\phi v_p d_p}{\nu} = \frac{(2.5)(0.0348 \text{ m/s})(200 \times 10^{-6} \text{ m})}{(1.003 \times 10^{-6} \text{ m}^2\text{s})} = 17.3$$

Because the Reynolds number > 1.0, Newton's law [Eq. (5–18)] must be used to determine the settling velocity in the transition region (see Fig. 5–21). Follow the iterative procedure described in Example 5–5.

3. For the first assumed settling velocity, use the Stokes' law settling velocity calculated above. Using the resulting Reynolds number, also determined previously, compute the drag coefficient.

$$C_d = \frac{24}{N_R} + \frac{3}{\sqrt{N_R}} + 0.34 = \frac{24}{17.3} + \frac{3}{\sqrt{17.3}} + 0.34 = 2.445$$

4. Use the drag coefficient in Newton's equation to determine the particle settling velocity in Eq. (5–21), which incorporates the shape factor.

$$v_p = \sqrt{\frac{4g(sg - 1)d}{3C_d\phi}} = \sqrt{\frac{4(9.81 \text{ m/s}^2)(2.6 - 1)(200 \times 10^{-6} \text{ m})}{3 \times 2.445 \times 2.5}}$$

$$= 0.026 \text{ m/s}$$

Because the initial assumed settling velocity (0.345 m/s) does not equal the Newton's equation settling velocity (0.026 m/s), additional iterations are necessary.

5. For a following iteration, assume a settling velocity value of 0.022 m/s, and calculate the Reynolds number (the assumed value is based on several trial iterations for various velocities). Use the Reynolds number to determine the drag coefficient, and use the drag coefficient in Newton's equation to find the settling velocity.

$$N_R = \frac{(2.5)(0.022 \text{ m/s})(200 \times 10^{-6} \text{ m})}{(1.003 \times 10^{-6} \text{ m}^2/\text{s})} = 10.97$$

$$C_d = \frac{24}{10.97} + \frac{3}{\sqrt{10.97}} + 0.34 = 3.43$$

$$v_p = \sqrt{\frac{4(9.81 \text{ m/s}^2)(2.6 - 1)(200 \times 10^{-6} \text{ m})}{3 \times 3.43 \times 1}} = 0.022 \text{ m/s}$$

The assumed and calculated settling velocities (0.022 m/s) are in agreement.

6. Using the same computational procedure, calculate the settling velocities for the spherical and wastewater particles. The results of the calculations for the three particle settling velocities are summarized below:

Ballasted floc particle = 0.022 m/s = 79 m/h

Spherical particle = 0.0164 m/s = 59 m/h

Wastewater particle = 0.00038 m/s = 1.4 m/h

Comment Although the particle settling velocities vary widely because of the particle characteristics (specific gravity, shape, and size), as illustrated in this example, by forming a ballasted floc particle the settling velocity can be enhanced significantly. Increasing the size and density of the wastewater particles by various means of ballasting is one of the premises of high-rate clarification. Because an iterative process is needed to calculate settling velocities in the transition zone, a spreadsheet program can expedite the computation process by allowing several assumed velocities to be tried to effect closure with the computed velocity.

Process Application

Three basic types of process are used for high-rate clarification: (1) ballasted flocculation with lamella plate clarification, (2) three-stage flocculation with lamella plate clarification, and (3) dense-solids flocculation/clarification with lamella plate clarification. Each of these processes can operate at high overflow rates that allow significant reduction in the physical size of the sedimentation units. A summary of the principal features of each process is presented in Table 5–23. Applications for high-rate clarification include (1) providing advanced primary treatment, (2) treating wet-weather flows and combined sewer overflows, (3) treating waste filter backwash water, and (4) treating return flows from solids-processing facilities. Ranges of overflow rates and BOD and TSS removals for treating wet-weather flows (domestic wastewater plus infiltration/

inflow) are reported in Table 5–24 (Sawey, 1998). The processes are illustrated on Fig. 5–50 and are discussed below.

Ballasted Flocculation. Ballasted flocculation employs a proprietary process, shown on Fig. 5–50a, in which a flocculation aid and a ballasting agent (typically a silica microsand) are used to form dense microfloc particles. The resulting floc particles are thus "ballasted" and settle rapidly. The treatment system consists of three compartments

Table 5–23
Summary of features of high-rate clarification processes

Process	Features
Microsand ballasted flocculation and clarification	• Microsand provides nuclei for floc formation • Floc is dense and settles rapidly • Lamella clarification, when used, provides high-rate settling in a small tank volume
Chemical addition, multistage flocculation, and lamella clarification	• Three-stage flocculation enhances floc formation • Lamella clarification provides high-rate settling in a small tank volume
Two-stage flocculation with chemically conditioned recycled sludge followed by lamella clarification	• Settled sludge solids are recycled to accelerate floc formation • Dense floc is formed that settles rapidly • Lamella clarification provides high-rate settling in a small tank volume

Table 5–24
Ranges of overflow rates and BOD and TSS removals from high-rate clarification processes treating wet-weather flows[a]

Parameter/Process	Ballasted flocculation	Lamella plate clarification	Dense sludge
Overflow rates			
Low, m³/m²·d (gal/ft²·min)	1200–2900 (20–50)	880 (15)	2300 (40)
Medium, m³/m²·d (gal/ft²·min)	1800–3500 (30–60)	1200 (20)	2900 (50)
High, m³/m²·d (gal/ft²·min)	2300–4100 (40–70)	1800 (30)	3500 (60)
BOD removals, %			
At low overflow rates	35–50	45–55	25–35
At medium overflow rates	40–60	35–40	40–50
At high overflow rates	30–60	35–40	50–60
TSS removals, %			
At low overflow rates	70–90	60–70	80–90
At medium overflow rates	40–80	65–75	70–80
At high overflow rates	30–80	40–50	70–80

[a] Adapted from EPRI (1999).

Figure 5-50

High-rate clarification processes: (a) ballasted flocculation, (b) lamella plate clarification, and (c) dense-sludge.

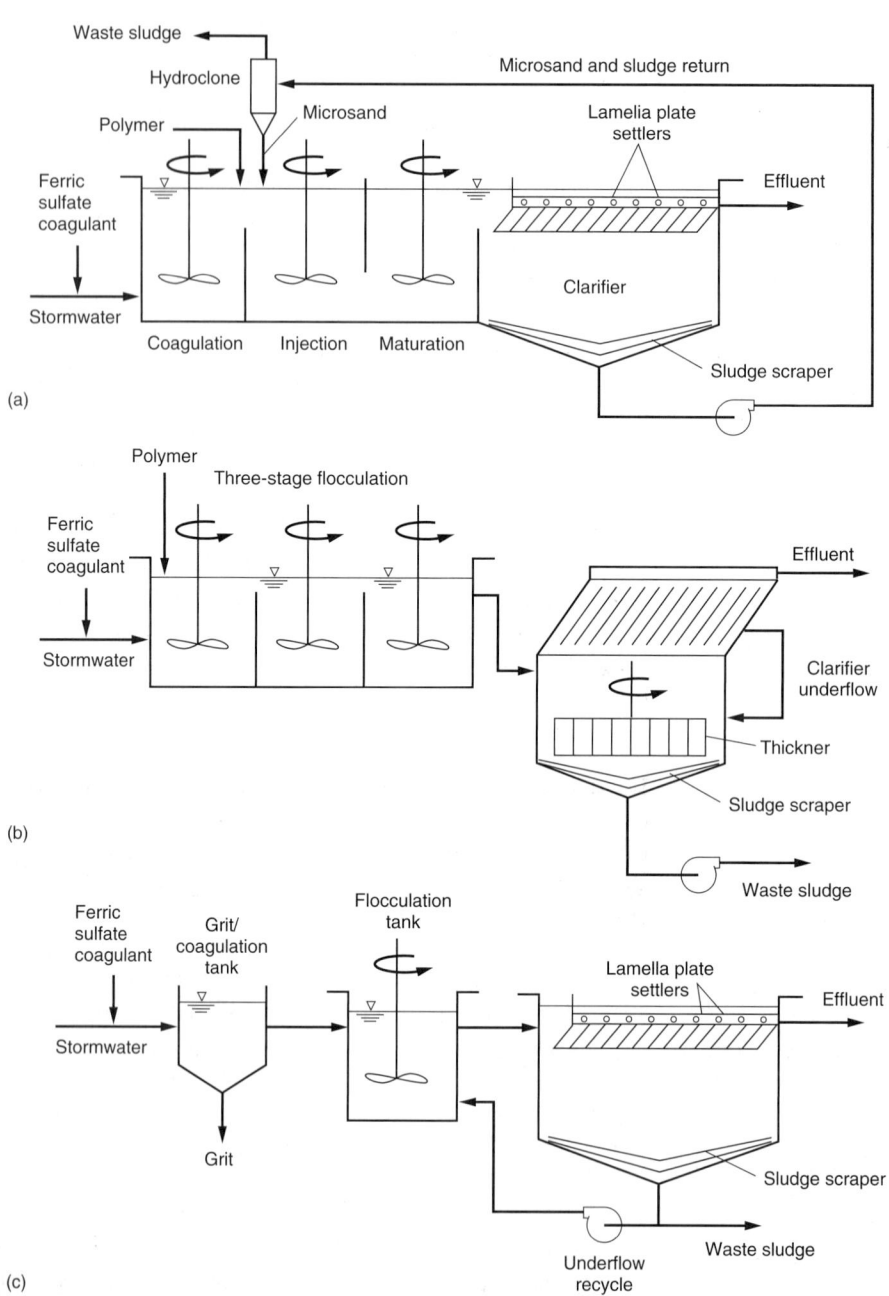

or zones: a mixing zone, maturation zone, and settling zone. Depending on the manufacturer of the process equipment, separate, serial compartments can be used to perform the process functions, or the functions can be combined in a single vessel. Either lamella plate settling or conventional gravity clarification can be used.

Typically, screened wastewater is introduced to the ballasted flocculation reactor where a chemical coagulant (typically an iron salt) is injected to destabilize the solids. The wastewater then enters a mixing zone where microsand and polymer are injected to maximize the efficiency of flocculation and enhance settling of suspended solids. In the mixing zone, the polymer acts as a bonding agent for adhering the destabilized solids to the microsand. The maturation zone follows and is used to keep the solids in suspension while floc particles continue to develop and grow. Once developed, the ballasted floc particles settle rapidly to the bottom of the clarifier. Sand and floc particles removed from the clarified water are pumped to a cyclone separator (hydroclone) for separation of the sand. The separated sand is returned to the injection tank, and solids from the hydroclone are sent to the biosolids-processing facilities. The microsand usually ranges in size from 100 to 150 μm for treating wastewater and combined wastewater flows and has a specific gravity greater than 2.6 to enhance settling.

Lamella Plate Clarification. Lamella plate clarification uses chemical addition followed by three-stage flocculation and a lamella plate clarifier (see Fig. 5–50b). Coagulant and polymer are injected into the influent wastewater prior to entrance into the flocculation zone. When chemically conditioned wastewater passes through each of the three flocculation zones, the mixing energy gradient is decreased as the wastewater proceeds from one stage to the next. The chemically conditioned/flocculated wastewater then passes to the lamella clarifier for solids separation. A portion of the clarifier underflow can be recycled to the influent of the process to enhance settling, or the entire underflow can be sent to a thickening tank and the solids-processing facilities.

Dense-Sludge Process. The dense-sludge system is a proprietary process and differs from ballasted flocculation in that recycled chemically conditioned solids are used to form microfloc particles with the incoming wastewater instead of microsand. As shown on Fig. 5–50c, the influent wastewater enters an air-mixing zone where grit separation occurs and coagulant (usually ferric sulfate) is injected. After mixing, the wastewater flows into the first stage of a two-stage flocculation tank where polymer is added together with chemically conditioned, recirculated solids. Recirculated solids accelerate the flocculation process and ensure the formation of dense, homogeneous floc particles. In the second stage of flocculation, grease and scum begin separating and are removed. Flow from the flocculation tank enters a presettling zone and then passes into a lamella plate settler. Most of the suspended flocculated solids are separated directly in the presettling zone; the residual flocculated particles are removed in the lamella settler. A portion of the settled solids is recirculated, and the remainder is sent to the solids processing and disposal system.

5–9 LARGE-SCALE SWIRL AND VORTEX SEPARATORS FOR COMBINED WASTEWATER AND STORMWATER

Solids-separation devices such as swirl concentrators and vortex separators have been used in Europe and, to a lesser extent, in the United States for the treatment of combined sewer overflows (CSOs) and stormwater. These devices are compact solids-separation units with no moving parts. A typical vortex-type CSO solids-separation unit is illustrated on Fig. 5–51. Operation of vortex separators is based on the movement of

Figure 5-51

Typical vortex separator used for solids removal from combined sewer overflows. *(Pisano, et al., 1990.)*

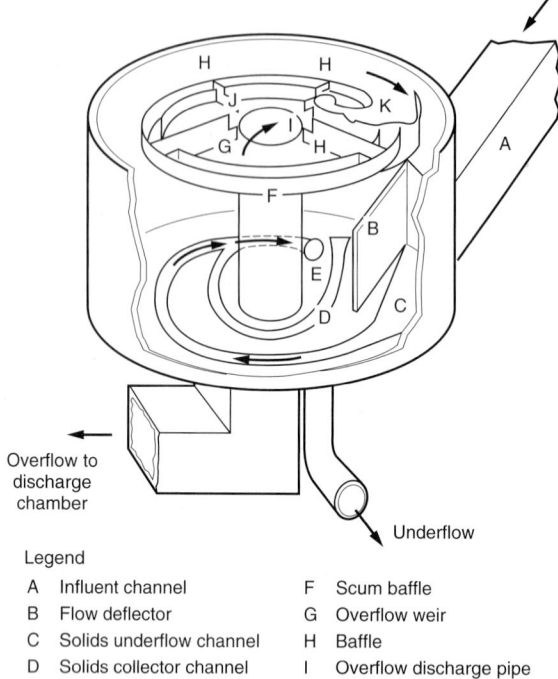

Overflow to discharge chamber

Underflow

Legend

A	Influent channel	F	Scum baffle
B	Flow deflector	G	Overflow weir
C	Solids underflow channel	H	Baffle
D	Solids collector channel	I	Overflow discharge pipe
E	Underflow discharge pipe	J	Scum trap plate
		K	Scum trap

particles within the unit. Water velocity moves the particles in a swirling action around the separator, additional flow currents move the particles toward the vortex, gravity pulls particles down, and a sweeping action moves heavier particles across the sloping floor toward the central drain.

During wet weather, the outflow from the unit is throttled, causing the unit to fill and to self-induce a swirling vortexlike flow regime. In the device shown on Fig. 5–51, secondary flow currents rapidly separate settleable grit and floatable matter. Concentrated foul matter is intercepted for treatment while the cleaner, treated flow discharges to receiving waters. The device is intended to operate under extremely high flow regimes.

A device more recently developed and termed the continuous deflection separator (CDS) differs from the more traditional vortex separator in that it utilizes a filtration mechanism for solids separation and does not rely on secondary flow currents induced by the vortex action. The CDS system (pictured previously on Fig. 5–5b) involves a single flow path and has one outlet point while other types of vortex separators discharge flows at the top and bottom of the units. The flow conditions within the CDS separation chamber have a different velocity profile. The surface velocity increases with increasing distance from the center of the separation chamber of the CDS unit, the reverse of that normally observed in conventional vortex separators. Solids separation is enhanced by a large expanded stainless-steel plate that acts as a filter screen with an outer volute outlet passage. The perforations in the separation screen are typically elongated in shape and are aligned with the longer axis in the vertical direction. The separation

Figure 5–52

Particle capture from a Continuous Deflection Separator (CDS) vortex separator. *(Wong et al., 1997.)*

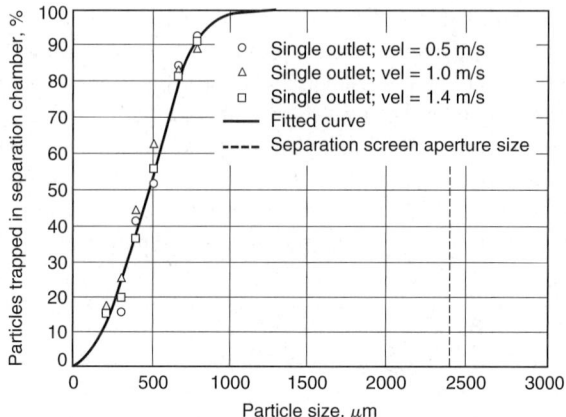

screen is installed so that the leading edge of each perforation extends into the flow stream of the containment chamber. Perforations in the screen can range from 1200 to 4700 μm (0.0475 to 0.185 in).

The CDS device is most appropriately used to capture the "first flush" and set up to divert all flows up to a threshold discharge. In tests conducted by Wong (1997), greater than 90 percent capture was reported for solids as small as 900 microns (see Fig. 5–52). Headloss through the separation unit varies depending on flowrate and screen openings.

5–10 FLOTATION

Flotation is a unit operation used to separate solid or liquid particles from a liquid phase. Separation is brought about by introducing fine gas (usually air) bubbles into the liquid phase. The bubbles attach to the particulate matter, and the buoyant force of the combined particle and gas bubbles is great enough to cause the particle to rise to the surface. Particles that have a higher density than the liquid can thus be made to rise. The rising of particles with lower density than the liquid can also be facilitated (e.g., oil suspension in water).

In wastewater treatment, flotation is used principally to remove suspended matter and to concentrate biosolids (see Chap. 14). The principal advantages of flotation over sedimentation are that very small or light particles that settle slowly can be removed more completely and in a shorter time. Once the particles have been floated to the surface, they can be collected by a skimming operation.

Description

The present practice of flotation as applied to wastewater treatment is confined to the use of air as the flotation agent. Air bubbles are added or caused to form by (1) injection of air while the liquid is under pressure, followed by release of the pressure (dissolved-air flotation), and (2) aeration at atmospheric pressure (dispersed-air flotation). In these systems, the degree of removal can be enhanced through the use of various chemical additives. In municipal wastewater treatment, dissolved-air flotation is frequently used, especially for thickening of waste biosolids.

Dissolved-Air Flotation. In dissolved-air flotation (DAF) systems, air is dissolved in the wastewater under a pressure of several atmospheres, followed by release of the pressure to the atmospheric level (see Fig. 5–53). In small pressure systems, the entire flow may be pressurized by means of a pump to 275 to 350 kPa (40 to 50 lb/in² gage) with compressed air added at the pump suction (see Fig. 5–53a). The entire flow is held in a retention tank under pressure for several minutes to allow time for the air to dissolve. It is then admitted through a pressure-reducing valve to the flotation tank where the air comes out of solution in very fine bubbles.

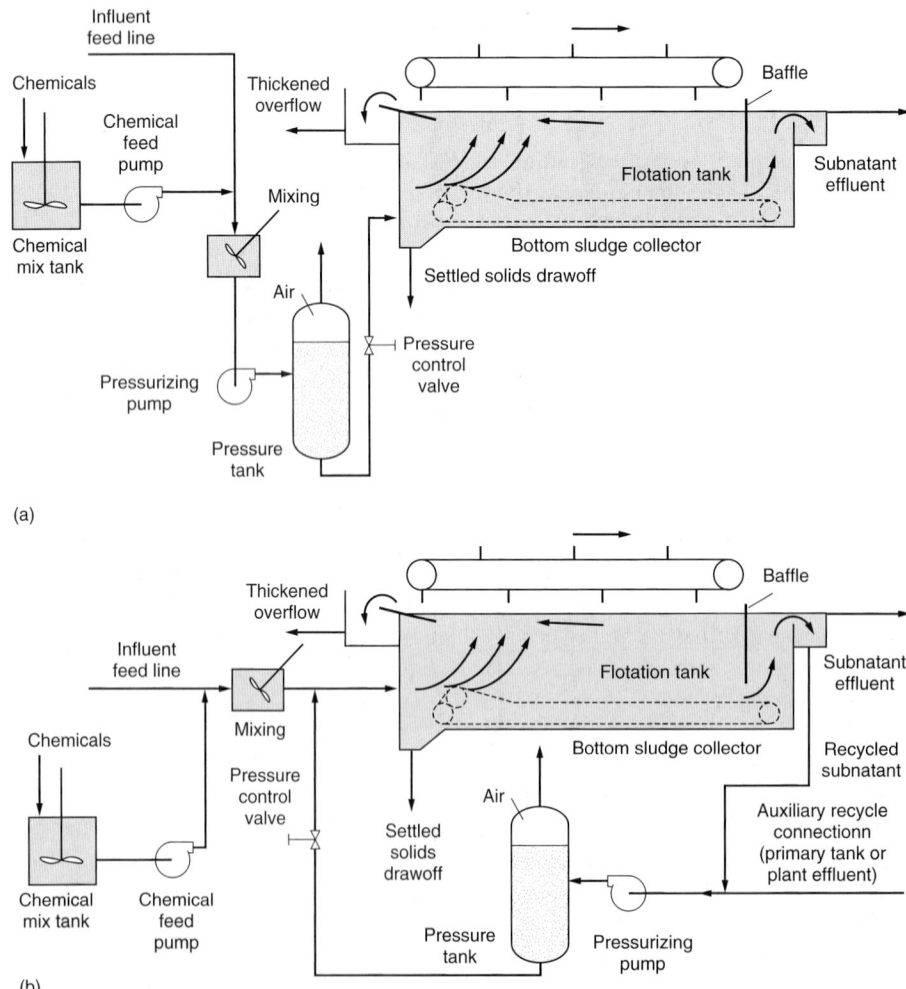

(a)

(b)

Figure 5–53

Schematic of dissolved-air flotation systems: (a) without recycle in which the entire flow is passed through the pressurizing tank and (b) with recycle in which only the recycle flow is pressurized. The pressurized flow is mixed with the influent before being released into the flotation tank.

In the larger units, a portion of the DAF effluent (15 to 120 percent) is recycled, pressurized, and semisaturated with air (Fig. 5–53b). The recycled flow is mixed with the unpressurized main stream just before admission to the flotation tank, with the result that the air comes out of solution in contact with particulate matter at the entrance to the tank. Pressure types of units have been used mainly for the treatment of industrial wastes and for the concentration of solids.

Dispersed-Air Flotation. Dispersed-air (sometimes referred to as induced-air) flotation is seldom used in municipal wastewater treatment, but it is used in industrial applications for the removal of emulsified oil and suspended solids from high-volume waste or process waters. In dispersed-air flotation systems, air bubbles are formed by introducing the gas phase directly into the liquid phase through a revolving impeller. The spinning impeller acts as a pump, forcing fluid through disperser openings and creating a vacuum in the standpipe (see Fig. 5–54). The vacuum pulls air (or gas) into the standpipe and thoroughly mixes it with the liquid. As the gas/liquid mixture travels through the disperser, a mixing force is created that causes the gas to form very fine bubbles. The liquid moves through a series of cells before leaving the unit. Oil particles and suspended solids attach to the bubbles as they rise to the surface. The oil and suspended solids gather in dense froth at the surface and are removed by skimming paddles. The advantages of a dispersed-air flotation system are (1) compact size, (2) lower capital cost, and (3) capacity to remove relatively free oil and suspended solids. The disadvantages of induced-air flotation include higher connected power requirements than the pressurized system, performance is dependent on strict hydraulic control, and less flocculation flexibility. The quantities of float skimmings are significantly higher

Figure 5–54

Dispersed-air flotation unit. Air is induced and dispersed into the liquid by pumping action of the inductors. *(Courtesy Eimco.)*

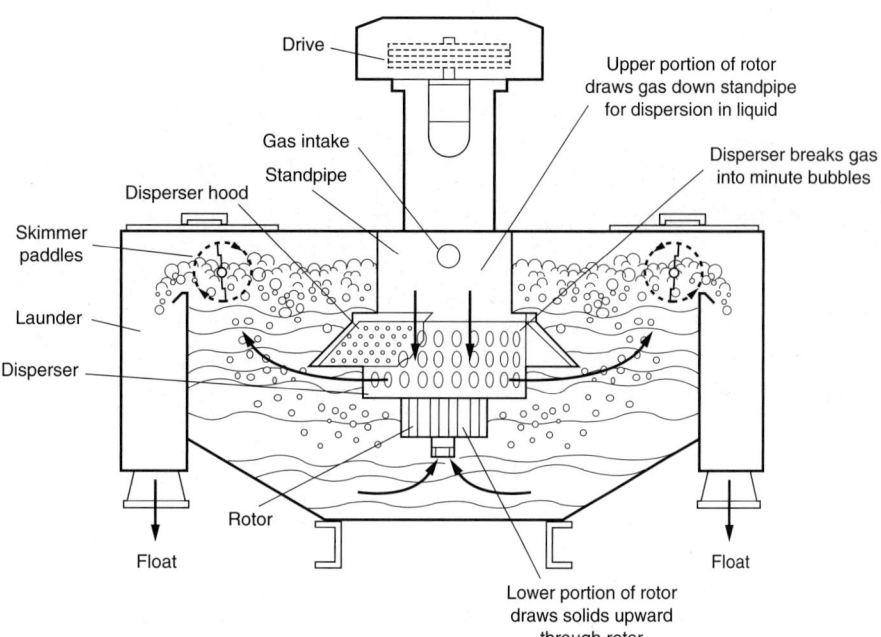

than the pressurized unit: 3 to 7 percent of the incoming flow as compared to less than 1 percent for dissolved-air systems (Eckenfelder, 2000).

Chemical Additives. Chemicals are commonly used to aid the flotation process. These chemicals, for the most part, function to create a surface or a structure that can easily absorb or entrap air bubbles. Inorganic chemicals, such as the aluminum and ferric salts and activated silica, can be used to bind the particulate matter together and, in so doing, create a structure that can easily entrap air bubbles. Various organic polymers can be used to change the nature of either the air-liquid interface or the solid-liquid interface, or both. These compounds usually collect on the interface to bring about the desired changes.

Design Considerations for Dissolved-Air Flotation Systems

Because flotation is very dependent on the type of surface of the particulate matter, laboratory and pilot-plant tests should be performed to yield the necessary design criteria. Factors that must be considered in the design of flotation units include the concentration of particulate matter, quantity of air used, the particle-rise velocity, and the solids loading rate. In the following analysis, dissolved-air flotation is discussed because it is the method most commonly used.

The performance of a dissolved-air flotation system depends primarily on the ratio of the volume of air to the mass of solids (A/S) required to achieve a given degree of clarification. The ratio will vary with each type of suspension and must be determined experimentally using a laboratory flotation cell. A typical laboratory flotation cell is shown on Fig. 5–55. Procedures for conducting the necessary tests may be found in Higbie (1935) and WEF (1998c). Typical A/S ratios encountered in the thickening of solids and biosolids in wastewater-treatment plants vary from about 0.005 to 0.060.

The relationship between the A/S ratio and the solubility of air, the operating pressure, and the concentration of solids for a system in which all the flow is pressurized is given in Eq. (5–47).

$$\frac{A}{S} = \frac{1.3\, s_a (fP - 1)}{S_a} \qquad (5\text{–}47)$$

Figure 5–55

Schematic of dissolved-air flotation test apparatus.

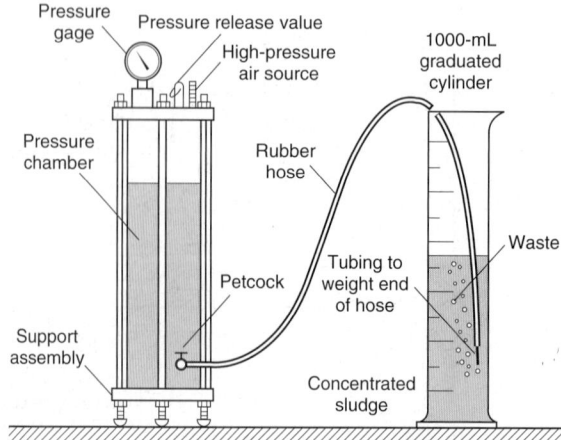

where A/S = air to solids ratio, mL (air)/mg(solids)
   s_a = air solubility, mL/L
   f = fraction of air dissolved at pressure P, usually 0.5
   P = pressure, atm

$$= \frac{p + 101.35}{101.35} \text{ (SI units)}$$

$$= \frac{p + 14.7}{14.7} \text{ (U.S. customary units)}$$

   p = gage pressure, kPa (lb/in^2 gage)
   S_a = influent suspended solids, g/m^3 (mg/L)

Temp., °C	0	10	20	30
s_a, mL/L	29.2	22.8	18.7	15.7

The corresponding equation for a system with only pressurized recycle is

$$\frac{A}{S} = \frac{1.3 \, s_a(fP - 1)R}{S_a Q} \qquad (5\text{–}48)$$

where R = pressurized recycle, m^3/d (Mgal/d)
   Q = mixed-liquor flow, m^3/d (Mgal/d)

In both of the foregoing equations, the numerator represents the weight of air and the denominator the weight of the solids. The factor 1.3 is the weight in milligrams of 1 mL of air and the term (-1) within the brackets accounts for the fact that the system is to be operated at atmospheric conditions. The use of these equations is illustrated in Example 5–12. Additional information about the use of flotation for treating oily wastewaters can be found in Eckenfelder (2000).

The required area of the thickener is determined from a consideration of the rise velocity of the solids, 8 to 160 L/m^2·min (0.2 to 4.0 gal/min·ft^2), depending on the solids concentration, degree of thickening to be achieved, and the solids loading rate (see Table 14–20).

EXAMPLE 5–12   **Flotation Thickening of Activated-Sludge Mixed Liquor**   Design a flotation thickener without and with pressurized recycle to thicken the solids in activated-sludge mixed liquor from 0.3 to about 4 percent. Assume that the following conditions apply:

1. Optimum A/S ratio = 0.008 mL/mg
2. Temperature = 20°C
3. Air solubility = 18.7 mL/L
4. Recycle-system pressure = 275 kPa
5. Fraction of saturation = 0.5
6. Surface-loading rate = 8 L/m^2·min
7. Sludge flowrate = 400 m^3/d

Solution
(without Recycle)

1. Compute the required pressure using Eq. (5–47).

$$\frac{A}{S} = \frac{1.3\, s_a(fP - 1)}{S_a}$$

$$0.008\ \text{mL/mg} = \frac{1.3\ (18.7\ \text{mL/L})(0.5P - 1)}{(3000\ \text{mg/L})}$$

$$0.5\,P = 0.99 + 1$$

$$P = 3.98\ \text{atm} = \frac{p + 101.35}{101.35}$$

$$p = 302\ \text{kPa}\ (43.8\ \text{lb}_f/\text{in}^2\text{gage})$$

2. Determine the required surface area.

$$A = \frac{(400\ \text{m}^3/\text{d})(10^3\ \text{L/m}^3)}{(8\ \text{L/m}^2\!\cdot\!\text{min})(1440\ \text{min/d})} = 34.7\ \text{m}^2$$

3. Check the solids loading rate.

$$\text{kg/m}^2 \cdot \text{d} = \frac{(400\ \text{m}^3/\text{d})(3000\ \text{g/m}^3)}{(34.7\ \text{m}^2)(10^3\ \text{g/kg})} = 34.6\ \text{kg/m}^2 \cdot \text{d}$$

Solution
(with Recycle)

1. Determine pressure in atmospheres.

$$p = \frac{275 + 101.35}{101.35} = 3.73\ \text{atm}$$

2. Determine the required recycle rate using Eq. (5–48).

$$\frac{A}{S} = \frac{1.3 s_a(fP - 1)R}{S_a Q}$$

$$0.008\ \text{mL/mg} = \frac{1.3\ (18.7\ \text{mL/L})[0.5\ (3.73) - 1]R}{(3000\ \text{mg/L})(400\ \text{m}^3/\text{d})}$$

$$R = 461.9\ \text{m}^3/\text{d}$$

3. Determine the required surface area.

$$A = \frac{(461.9\ \text{m}^3/\text{d})(10^3\ \text{L/m}^3)}{(8\ \text{L/m}^2\!\cdot\!\text{min})(1440\ \text{min/d})} = 40.1\ \text{m}^2$$

Comment Alternatively, the recycle flowrate could have been set and the pressure determined. In an actual design, the costs associated with the recycle pumping, pressurizing systems, and tank construction can be evaluated to find the most economical combination.

5–11 **OXYGEN TRANSFER**

Oxygen transfer, the process by which oxygen is transferred from the gaseous to the liquid phase, is a vital part of a number of wastewater-treatment processes. The functioning of aerobic processes, such as activated sludge, biological filtration, and aerobic digestion, depends on the availability of sufficient quantities of oxygen. The application of oxygen transfer is covered in this section, and the types of aeration systems are presented in Sec. 5–12. The fundamental theory of liquid-gas mass transfer is discussed in Chap. 4, Sec. 4–8.

Description

The most common application of oxygen transfer is in the biological treatment of wastewater. Because of the low solubility of oxygen and the consequent low rate of oxygen transfer, sufficient oxygen to meet the requirements of aerobic waste treatment does not enter water through normal surface air-water interfaces. To transfer the large quantities of oxygen that are needed, additional interfaces must be formed. Either air or oxygen can be introduced into the liquid, or the liquid in the form of droplets can be exposed to the atmosphere.

Oxygen can be supplied by means of air or pure-oxygen bubbles introduced to the water to create additional gas-water interfaces. In wastewater-treatment plants, submerged-bubble aeration is most frequently accomplished by dispersing air bubbles in the liquid at depths up to 10 m (~30 ft); depths up to 30 m (~100 ft) have been used in some European designs. Hydraulic shear devices may also be used to create small bubbles by impinging a flow of liquid at an orifice to break up the air bubbles into smaller sizes. Turbine mixers may be used to disperse air bubbles introduced below the center of the turbine; they are designed both to mix the liquid in the basin and to expose it to the atmosphere in the form of small liquid droplets.

Evaluation of Oxygen Transfer Coefficient

For a given volume of water being aerated, aeration devices are evaluated on the basis of the quantity of oxygen transferred per unit of air introduced to the water for equivalent conditions (temperature and chemical composition of the water, depth at which the air is introduced, etc.). The evaluation of the oxygen transfer coefficient in clean water and wastewater is considered in the following discussion.

Oxygen Transfer in Clean Water. The accepted procedure for determining the overall oxygen transfer coefficient in clean water, as detailed in ASCE (1992), may be outlined as follows. The accepted test method involves the removal of dissolved oxygen (DO) from a known volume of water by the addition of sodium sulfite followed by reoxygenation to near the saturation level. The DO of the water volume is monitored during the reaeration period by measuring DO concentrations at several different points selected to best represent the contents of the tank. The minimum number of points, their distribution, and range of DO measurements made at each determination point are specified in the procedure (ASCE, 1992).

The data obtained at each determination point are then analyzed by a simplified mass transfer model (see Sec. 4-8 in Chap. 4):

$$\frac{C_s - C_t}{C_s - C_o} = e^{-(K_L a)t} \tag{4-136}$$

where $K_L a$ = overall liquid film coefficient
C_t = concentration in liquid bulk phase at time t, mg/L
C_s = concentration in equilibrium with gas as given by Henry's law
C_o = initial concentration

Equation (4–136) is used to estimate the apparent volumetric mass transfer coefficient $K_L a$ and the equilibrium concentration C_x^*, obtained as the aeration period approaches infinity. The term C_x^* is substituted for the term C_s in Eq. (4–136). A nonlinear regression analysis is employed to fit Eq. (4–136) to the DO profile measured at each determination point during reoxygenation test period. In this way, estimates of $K_L a$ and C_x^* are obtained at each determination point. These estimates are adjusted to standard conditions and the standard oxygen transfer rate (mass of oxygen dissolved per unit time at a hypothetical concentration of zero DO) is obtained as the average of the products of the adjusted point $K_L a$ values, the corresponding adjusted point C_x^* values, and the tank volume (ASCE, 1992).

Oxygen Transfer in Wastewater. In an activated-sludge system, the $K_L a$ value can be determined by considering the uptake of oxygen by microorganisms. Typically, oxygen is maintained at a level of 1 to 3 mg/L and the microorganisms use the oxygen as rapidly as it is supplied. In equation form,

$$\frac{dC}{dt} = K_L a(C_s - C) - r_M \tag{5-49}$$

where C = concentration of oxygen in solution
r_M = rate of oxygen used by the microorganisms

Typical values of r_M vary from 2 to 7 g/d per gram of mixed-liquor volatile suspended solids (MLVSS). If the oxygen level is maintained at a constant level, dC/dt is zero and

$$r_M = K_L a (C_s - C) \tag{5-50}$$

C in this case is constant also.

Values of r_M can be determined in a laboratory by using a respirometer. In this case, $K_L a$ can easily be determined as follows:

$$K_L a = \left(\frac{r_M}{C_s - C}\right) \tag{5-51}$$

Prediction of oxygen transfer rates in aeration systems is nearly always based on an oxygen rate model. The overall oxygen mass transfer coefficient $K_L a$ is usually determined in test or full-scale facilities. If pilot-scale facilities are used to determine $K_L a$ values, scale-up must be considered. The mass transfer coefficient $K_L a$ is also a function of temperature, intensity of mixing (and hence of the type of aeration device used and the geometry of the mixing chamber), and constituents in the water (Tchobanoglous and Schroeder, 1985). The effects of temperature, mixing intensity, tank geometry, and wastewater characteristics and the application of correction factors are discussed below. Determination of $K_L a$ is illustrated in Example 5–13.

Effect of Temperature on Oxygen Transfer. Temperature effects are treated in the same manner as they were treated in establishing the BOD rate coefficient (i.e., by using an exponential function to approximate the van't Hoff–Arrhenius relationship):

$$K_L a_{(T)} = K_L a_{(20°C)} \, \theta^{T-20} \tag{5-52}$$

where $K_L a_{(T)}$ = oxygen mass transfer coefficient at temperature T, s^{-1}
$K_L a_{(20°C)}$ = oxygen mass transfer coefficient at 20°C, s^{-1}

Reported values for θ vary with the test conditions. Typical θ values are in the range of 1.015 to 1.040. The θ value of 1.024 is typical for both diffused and mechanical aeration devices.

EXAMPLE 5–13 **Determination of $K_L a$ Value** The following data have been obtained from a surface aeration test. Using the data, determine the $K_L a$ value at 20°C using a linear regression analysis. The temperature of the water was 15°C.

Time, min	DO conc., mg/L
4	0.8
7	1.8
10	3.3
13	4.5
16	5.5
19	5.2
22	7.3

Solution

1. To analyze the field data, rewrite Eq. (4–136) in a linear form.

$$\log (C_s - C_t) = \log (C_s - C_o) - \frac{K_L a}{2.303} t$$

2. Determine $C_s - C_t$, and plot $C_s - C_t$ versus t on semilog paper.
   a. $C_{s(15°C)} = 10.07$ (see Appendix E)

Time, min	$C_s - C_t$, mg/L
4	9.27
7	8.27
10	6.77
13	5.57
16	4.57
19	4.87
22	2.77

b. Plot $C_s - C_t$ versus t. See the following plot:

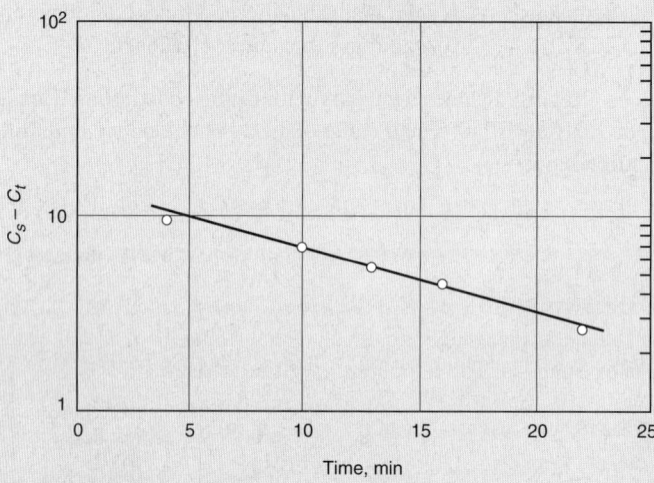

3. Determine the value of $K_L a$ at 20°C.

a. From the plot, the value of $K_L a$ at 15°C is

$$K_L a = 2.303 \frac{\log C_{t_1} - \log C_{t_2}}{t_2 - t_1} (60)$$

$$K_L a = 2.303 \left(\frac{\log 8.35 - \log 2.85}{22 - 7} \right) (60)$$

$$K_L a = 4.39 \text{ h}^{-1}$$

b. The approximate value of $K_L a$ at 20°C is

$$K_L a_{20} = (1.91)\, 1.024^{20-15}$$

$$= 4.94 \text{ h}^{-1}$$

Comment The value of $K_L a$ determined in this example is approximate because a linear regression analysis was used. To obtain a more accurate value of $K_L a$, the nonlinear method outlined in ASCE (1992) should be used.

Effects of Mixing Intensity and Tank Geometry. Effects of mixing intensity and tank geometry are difficult to deal with on a theoretical basis but must be considered in the design process because aeration devices are often chosen on the basis of efficiency. Efficiency is strongly related to the K_La value associated with a given aeration unit. In most cases an aeration device is rated for a range of operating conditions using tap water having a low TDS concentration. A correction factor α is used to estimate the K_La value in the actual system:

$$\alpha = \frac{K_La \ (\text{wastewater})}{K_La \ (\text{tap water})} \tag{5–53}$$

where α is the correction factor. Values of α vary with the type of aeration device, the basin geometry, the degree of mixing, and the wastewater characteristics. Values of α vary from about 0.3 to 1.2. Typical values for diffused and mechanical aeration equipment, discussed in the following section, are in the range of 0.4 to 0.8 and 0.6 to 1.2, respectively. If the basin geometry in which the aeration device is to be used is significantly different from that used to test the device, great care must be exercised in selecting an appropriate α value.

Effects of Wastewater Characteristics. The correction factor β is used to correct the test system oxygen transfer rate for differences in oxygen solubility due to constituents in the water such as salts, particulates, and surface-active substances:

$$\beta = \frac{C_s \ (\text{wastewater})}{C_s \ (\text{clean water})} \tag{5–54}$$

Values of β vary from about 0.7 to 0.98. A β value of 0.95 is commonly used for wastewater. Because the determination of β is within the capability of most wastewater-treatment plant laboratories, experimental verification of assumed values is recommended.

Application of Correction Factors. The actual amount of oxygen required must be obtained by applying factors to a standard oxygen requirement that reflect the effects of salinity-surface tension (beta factor), temperature, elevation, diffused depth (for diffused aeration systems), the desired oxygen operating level, and the effects of mixing intensity and basin configuration. The interrelationship of these factors is given by the following expression:

$$\text{AOTR} = \text{SOTR} \left(\frac{\beta C_{\bar{s},T,H} - C_L}{C_{s,20}} \right) (1.024^{T-20})(\alpha)(F) \tag{5–55}$$

where AOTR = actual oxygen transfer rate under field conditions, kg O_2/h
　　　　SOTR = standard oxygen transfer rate in tap water at 20°C, and zero dissolved oxygen, kg O_2/h
　　　　　　β = salinity–surface tension correction factor, typically 0.95 to 0.98, see Eq. (5–54)
　　　$C_{\bar{s},T,H}$ = average dissolved oxygen saturation concentration in clean water in aeration tank at temperature T and altitude H, mg/L

$$= (C_{s,T,H}) \frac{1}{2} \left(\frac{P_d}{P_{atm,H}} + \frac{O_t}{21} \right) \text{ (Note: For surface aerators, } C_{\bar{s},T,H} = C_{s,T,H})$$

The term in the brackets when multiplied by one-half represents the average pressure at mid depth and accounts for the loss of oxygen to biological uptake. If the biological uptake is not considered, then the following expression can be used:

$$= (C_{s,T,H})\left(\frac{P_{atm,H} + P_{w,\,mid\,depth}}{P_{atm,H}}\right),$$

$C_{s,T,H}$ = oxygen saturation concentration in clean water at temperature T and altitude H (see Appendix D), mg/L

P_d = pressure at the depth of air release, kPa

$P_{atm,H}$ = atmospheric pressure at altitude H (see Appendix B), kPa

$P_{w,\,mid\,depth}$ = pressure at mid depth, above point of air release, due to water column

O_t = percent oxygen concentration leaving tank, usually 18 to 20 percent

C_L = operating oxygen concentration, mg/L

$C_{s,20}$ = dissolved oxygen saturation concentration in clean water at 20°C and 1 atm, mg/L

T = operating temperature, °C

α = oxygen transfer correction factor for waste [(Eq. 5–53)]

F = fouling factor, typically 0.65 to 0.9

Note that the AOTR and SOTR values given above can also be expressed as transfer efficiencies. The fouling factor F is used to account for both internal and external fouling of air diffusers (see Sec. 5–12). Internal fouling is caused by impurities in the compressed air, whereas external fouling is caused by the formation of biological slimes and inorganic precipitants. The oxygen necessary for the biological process can be supplied by using air or pure oxygen. Three methods of introducing oxygen to the contents of the aeration tank used commonly are (1) mechanical aeration, (2) injection of diffused air, and (3) injection of high-purity oxygen.

5–12 AERATION SYSTEMS

There are several types of aeration systems used for wastewater treatment. The systems used depend on the function to be performed, type and geometry of the reactor, and cost to install and operate the system. In this section, the commonly used types of aeration systems and devices are described.

Types of Aeration Systems

The various types of aeration systems used and their applications are described in Table 5–25. The principal types, diffused-air systems, mechanical aeration, and high-purity oxygen systems, are discussed in the following paragraphs. Postaeration, which is a special application for aeration, is also discussed in the latter part of this section.

Diffused-Air Aeration

The two basic methods of aerating wastewater are (1) to introduce air or pure oxygen into the wastewater with submerged diffusers or other aeration devices or (2) to agitate the wastewater mechanically so as to promote solution of air from the atmosphere. A diffused-air system consists of diffusers that are submerged in the wastewater, header

Table 5–25
Description of commonly used devices for wastewater aeration

Classification	Description	Use or application
Submerged:		
Diffused air		
Fine-bubble (fine-pore) system	Bubbles generated with ceramic, plastic, or flexible membranes (domes, tubes, disks, plates, or panel configurations)	All types of activated-sludge processes
Coarse-bubble (nonporous) system	Bubbles generated with orifices, injectors and nozzles, or shear plates	All types of activated-sludge processes, channel and grit chamber aeration, and aerobic digestion
Sparger turbine	Low-speed turbine and compressed-air injection	All types of activated-sludge processes and aerobic digestion
Static tube mixer	Short tubes with internal baffles designed to retain air injected at bottom of tube in contact with liquid	Aerated lagoons and activated-sludge processes
Jet	Compressed air injected into mixed liquor as it is pumped under pressure through jet device	All types of activated-sludge processes, equalization tank mixing and aeration, and deep tank aeration
Surface:		
Low-speed turbine aerator	Large-diameter turbine used to expose liquid droplets to the atmosphere	Conventional activated-sludge processes, aerated lagoons, and aerobic digestion
High-speed floating aerator	Small-diameter propeller used to expose liquid droplets to the atmosphere	Aerated lagoons and aerobic digestion
Aspirating	Inclined propeller assembly	Aerated lagoons
Rotor-brush or rotating-disk assembly	Blades or disks mounted on a horizontal central shaft are rotated through the liquid. Oxygen is induced into the liquid by the splashing action of the rotor and by exposure of liquid droplets to the atmosphere	Oxidation ditch, channel aeration, and aerated lagoons
Cascade	Wastewater flows over a series of steps in sheet flow	Postaeration

pipes, air mains, and the blowers and appurtenances through which the air passes. The following discussion covers the selection of diffusers, the design of blowers, and air piping design.

Diffusers. In the past, the various diffusion devices have been classified as either fine bubble or coarse bubble, with the connotation that fine bubbles were more efficient in transferring oxygen. The definition of terms and the demarcation between fine and coarse bubbles, however, have not been clear, but they continue to be used. The current preference is to categorize the diffused aeration systems by the physical characteristics of the equipment. Three categories are defined: (1) porous or fine-pore diffusers, (2) nonporous diffusers, and (3) other diffusion devices such as jet aerators, aspirating aerators, and U-tube aerators. The various types of diffused-air devices are described in Table 5–26.

Table 5-26
Description of commonly used air diffusion devices[a]

Type of diffuser or device	Transfer efficiency	Description	See fig. no.
Porous			
Disk	High	Rigid ceramic disks mounted on air-distribution pipes near the tank floor	5–56a, c
Dome	High	Dome-shaped ceramic diffusers mounted on air-distribution pipes near the tank floor	5–56b
Membrane	High	Flexible porous membrane supported on disk mounted on an air-distribution grid	5–56d
Panel	Very high	Rectangular panel with a flexible plastic perforated membrane	5–58
Nonporous			
Fixed orifice			
Orifice	Low	Devices usually constructed of molded plastic and mounted on air-distribution pipes	5–59a
Slotted tube	Low	Stainless-steel tubing containing perforations and slots to provide a wide band of diffused air	5–59b
Static tube	Low	Stationary vertical tube mounted on basin bottom and functions like an air-lift pump	5–60a

[a] Adapted in part from WEF (1998b).

Porous Diffusers Porous diffusers are made in many shapes, the most common being domes, disks, and membranes (see Fig. 5–56a, b, c, and d). Tubes are also used. Plates were once the most popular but are costly to install and difficult to maintain. Porous domes, disks, and membranes have largely supplanted plates in newer installations. Domes, disks, or tube diffusers are mounted on or screwed into air manifolds, which may run the length of the tank close to the bottom and along one or two sides, or short manifold headers may be mounted on movable drop pipes on one side of the tank. Dome and disk diffusers may also be installed in a grid pattern on the bottom of the aeration tank to provide uniform aeration throughout the tank (see Fig. 5–57).

Numerous materials have been used in the manufacture of porous diffusers. These materials generally fall into the categories of rigid ceramic and plastic materials and flexible plastic, rubber, or cloth sheaths. The ceramic materials consist of rounded or irregular-shaped mineral particles bonded together to produce a network of interconnecting passageways through which compressed air flows. As the air emerges from the surface pores, pore size, surface tension, and air flowrate interact to produce the bubble size. Porous plastic materials are newer developments. Similar to the ceramic materials, the plastics contain a number of interconnecting channels or pores through which the compressed air can pass. Thin, flexible sheaths made from soft plastic or synthetic rub-

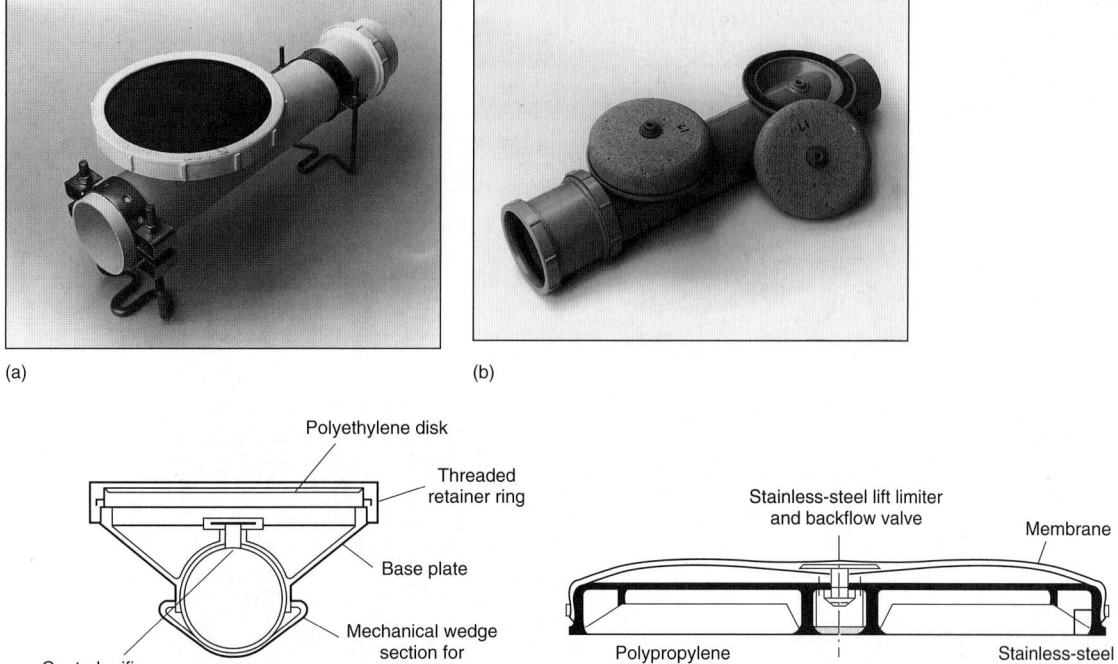

(a)   (b)

Polyethylene disk

Threaded
retainer ring

Stainless-steel lift limiter
and backflow valve

Membrane

Base plate

Mechanical wedge
section for
attaching base

Polypropylene
support disk

Threaded
connection

Stainless-steel
clamping ring

Control orifice
and check valve

(c)   (d)

Figure 5-56

Typical porous air diffusers: (a) aluminum oxide disk, (b) ceramic dome, (c) polyethylene disk, and (d) perforated membrane.

ber have also been developed and adapted to disks and tubes. Air passages are created by punching minute holes in the sheath material. When the air is turned on, the sheath expands and each slot acts as a variable aperture opening; the higher the air flowrate, the greater the opening.

Rectangular panels that use a flexible polyurethane sheet (see Fig. 5–58) are also used in activated-sludge aeration. The panels are constructed with a stainless-steel frame and are placed on or close to the bottom of the tank and anchored. Advantages cited for aeration panels are (1) ultra-fine bubbles are produced that significantly improve oxygen transfer and system energy efficiency, (2) large areas of the tank floor can be covered, which facilitates mixing and oxygen transfer, and (3) foulants can be dislodged by "bumping," i.e., increasing the airflow to flex the membrane. Disadvantages are (1) the panel is a proprietary design and thus lacks competitive bidding, (2) the membrane has a higher headloss, which may affect blower performance in retrofit applications, and (3) increased blower air filtration is required to prevent internal fouling.

With all porous diffusers, it is essential that the air supplied be clean and free of dust particles that might clog the diffusers. Air filters, often consisting of viscous-impingement and dry-barrier types, are commonly used. Precoated bag filters and electrostatic filters have also been used. The filters should be installed on the blower inlet.

Figure 5-57

Plug-flow aeration tank
equipped with dome
aeration devices.

Figure 5-58

Ultra-fine pore
membrane aeration panels:
(a) schematic (*Courtesy
Parkson Corp.*) and (b) panels
placed in bottom of an
activated-sludge reactor.

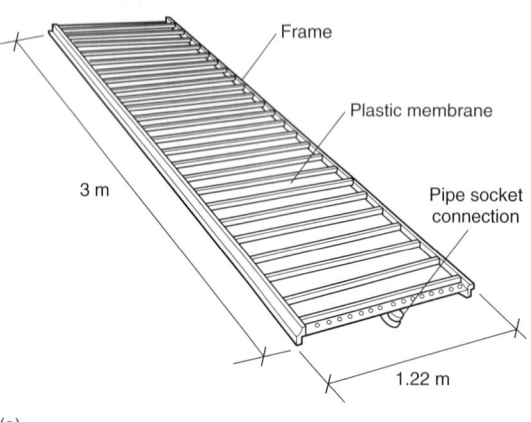

(a)

Frame

Plastic membrane

3 m

Pipe socket
connection

1.22 m

(b)

Nonporous Diffusers Several types of nonporous diffusers are available (see Fig. 5–59a and b). Nonporous diffusers produce larger bubbles than porous diffusers and consequently have lower aeration efficiency; but the advantages of lower cost, less maintenance, and the absence of stringent air-purity requirements may offset the lower oxygen transfer efficiency and energy cost. Typical system layouts for orifice diffusers closely parallel the layouts for porous dome and disk diffusers; however, single- and dual-roll spiral patterns using narrow- or wide-band diffuser placement are the most common. Applications for orifice and tube diffusers include aerated grit chambers, channel aeration, flocculation basin mixing, aerobic digestion, and industrial waste treatment (WEF, 1998b).

Figure 5–59

Nonporous diffusers used for the transfer of oxygen: (a) orifice and (b) tube.

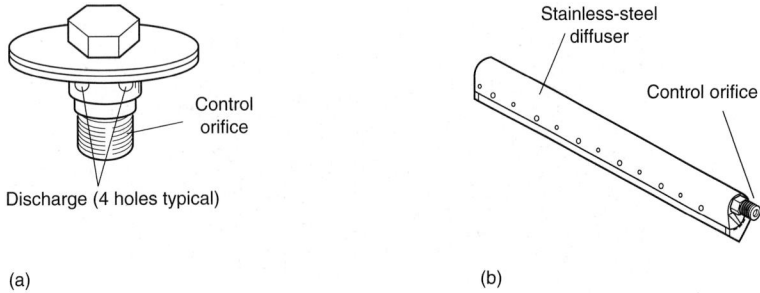

Figure 5–60

Other devices used for the transfer of oxygen: (a) static tube mixer where air is introduced at the base of the aerator that contains mixing elements, (b) jet reactor in which pressurized air and liquid are combined in a mixing chamber (As the jet is emitted, the surrounding liquid is entrained to enhance oxygen transfer.), (c) jet aerator in a manifold arrangement, and (d) aspirating aerator.

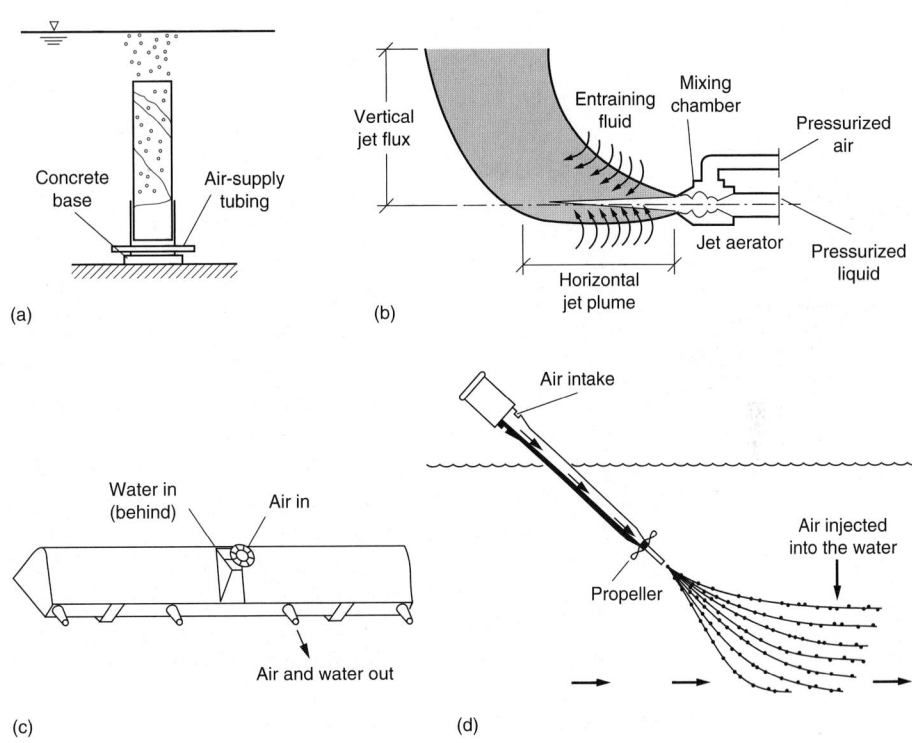

In the static tube aerator (see Fig. 5–60a), air is introduced at the bottom of a circular tube that can vary in height from 0.5 to 1.25 m (1.5 to 4.0 ft). Internally, the tubes are fitted with alternately placed deflection plates to increase the contact of the air with the wastewater. Mixing is accomplished because the tube aerator acts as an airlift pump. Static tubes are normally installed in a grid-type floor coverage pattern.

Other Air-Diffusion Devices Jet aeration (see Fig. 5–60b and c) combines liquid pumping with air diffusion. The pumping system recirculates liquid in the aeration basin, ejecting it with compressed air through a nozzle assembly. This system is particularly suited for deep (>8 m) tanks. Aspirating aeration (Fig. 5–60d) consists of a

Figure 5–61

U-tube aerator.

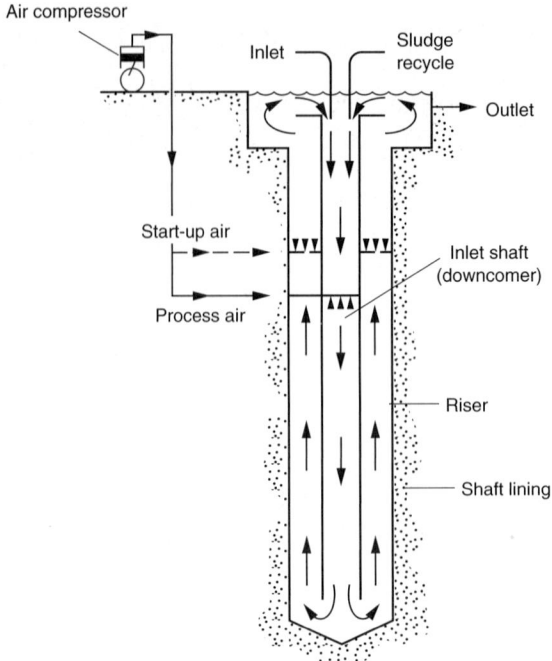

motor-driven aspirator pump. The pump draws air in through a hollow tube and injects it underwater where both high velocity and propeller action create turbulence and diffuse the air bubbles. The aspirating device can be mounted on a fixed structure or on pontoons. U-tube aeration consists of a deep shaft that is divided into two zones (see Fig. 5–61). Air is added to the influent wastewater in the downcomer under high pressure; the mixture travels to the bottom of the tube and then back to the surface. The great depth to which the air-water mixture is subjected results in high oxygen transfer efficiencies because the high pressure forces all the oxygen into solution. U-tube aeration has particular application for high-strength wastes.

Diffuser Performance.   The efficiency of oxygen transfer depends on many factors, including the type, size, and shape of the diffuser; the air flowrate; the depth of submersion; tank geometry including the header and diffuser location; and wastewater characteristics. Aeration devices are conventionally evaluated in clean water and the results adjusted to process operating conditions through widely used conversion factors. Typical clean water transfer efficiencies and air flowrates for various diffused-air devices are reported in Table 5–27. Typically, the standard oxygen transfer efficiency (SOTE) increases with depth; the transfer efficiencies in Table 5–27 are shown for the 4.5-m (15-ft) depth, the most common depth of submergence. Data on the variation of SOTE with water depth for various diffuser types can be found in WPCF (1988). The variation of oxygen transfer efficiencies with the type of diffuser and diffuser arrangement are also illustrated in Table 5–27. Additional data on the effects of diffuser arrangement on transfer efficiency are reported in U.S. EPA (1989).

Table 5-27
Typical information on the clean water oxygen transfer efficiency of various air diffuser systems[a]

Diffuser type and placement	Air flowrate/diffuser ft³/min	Air flowrate/diffuser m³/min	SOTE (%) at 4.5 m (15 ft) submergence[b]
Ceramic disks—grid	0.4–3.4	0.01–0.1	25–35
Ceramic domes—grid	0.5–2.5	0.015–0.07	27–37
Ceramic plates—grid	2.0–5.0[c]	0.6–1.5[d]	26–33
Rigid porous plastic tubes			
Grid	2.4–4.0	0.07–0.11	28–32
Dual spiral roll	3.0–11.0	0.08–0.3	17–28
Single spiral roll	2.0–12.0		13–25
Nonrigid porous plastic tubes			
Grid	1.0–7.0	0.03–0.2	26–36
Single spiral roll	2.0–7.0	0.06–0.2	19–37
Perforated membrane tubes			
Grid	1.0–4.0	0.03–0.11	22–29
Quarter points	2.0–6.0	0.6–0.17	19–24
Single spiral roll	2.0–6.0	0.6–0.17	15–19
Perforated membrane panels	N/A	N/A	38–43[e]
Jet aeration			
Side header	54–300	1.5–8.5	15–24
Nonporous diffusers			
Dual spiral roll	3.3–10	0.1–0.28	12–13
Midwidth	4.2–45	0.12–1.25	10–13
Single spiral roll	10–35	0.28–1.0	9–12

[a] Adapted in part from WPCF (1988) and U.S. EPA (1989).
[b] SOTE = standard oxygen transfer efficiency. Standard conditions: tap water 20°C (68°F); at 101.3 kN/m² (14.7 lb$_f$/in²); and initial dissolved oxygen = 0 mg/L.
[c] Units are ft³/ft² of diffuser·min.
[d] Units are m³/m² of diffuser·min.
[e] Personal communication, Parkson Corporation.
N/A = not applicable.

Oxygen transfer efficiency (OTE) of porous diffusers may also decrease with use due to internal clogging or exterior fouling. Internal clogging may be due to impurities in the compressed air that have not been removed by the air filters. External fouling may be due to the formation of biological slimes or inorganic precipitants. The effect of fouling on OTE is described by the term F. The rate at which F decreases with time is designated f_F, which is expressed as the decimal fraction of OTE lost per unit time. The rate of fouling

Figure 5-62

Aeration hood used to measure oxygen transfer rates in a biological wastewater treatment reactor.

will depend on the operating conditions, changes in wastewater characteristics, and the time in service. The fouling rates are important in determining the loss of OTE and the expected frequency of diffuser cleaning. Fouling and the rate of fouling can be estimated by (1) conducting full-scale OTE tests over a period of time, (2) monitoring aeration system efficiency (see Fig. 5–62), and (3) conducting OTE tests of fouled and new diffusers.

Factors commonly used to convert the oxygen transfer required for clean water to wastewater are the alpha, beta, and theta factors as described in Sec. 5–11. The alpha factor, the ratio of the K_La of wastewater to the K_La of clean water, is especially important because alpha factor varies with the physical features of the diffuser system, the geometry of the reactor, and the characteristics of the wastewater. Wastewater constituents may affect porous diffuser oxygen transfer efficiencies to a greater extent than other aeration devices, resulting in lower alpha factors. The presence of constituents such as detergents, dissolved solids, and suspended solids can affect the bubble shape and size and result in diminished oxygen transfer capability. Values of alpha varying from 0.4 to 0.9 have been reported for fine-bubble diffuser systems (Hwang and Stenstrom, 1985). Therefore, considerable care must be exercised in the selection of the appropriate alpha factors.

Another measure of the performance of porous diffusers is the combination of the alpha and fouling factors, designated by the term αF. In a number of in-process studies, the values of αF have ranged widely, from 0.11 to 0.79 with a mean of < 0.5, and were significantly lower than anticipated (U.S. EPA, 1989). The variability of αF was found to be site-specific, and demonstrated the need for the designer to investigate and evaluate carefully the environmental factors that may affect porous diffuser performance in selecting an appropriate α or αF factor.

Because the amount of air used per kilogram (pound) of BOD removed varies greatly from one plant to another, and there is risk in comparing the air use at different plants, not only because of the factors mentioned above but also because of different loading rates, control criteria, and operating procedures. Extra-high air flowrates applied along one side of a tank reduce the efficiency of oxygen transfer and may even reduce the net oxygen transfer by increasing circulating velocities. The result is a shorter residence time of air bubbles as well as larger bubbles with less transfer surface.

Figure 5–63

Typical blowers used for diffused-air aeration: (a) centrifugal and (b) rotary-lobe positive displacement.

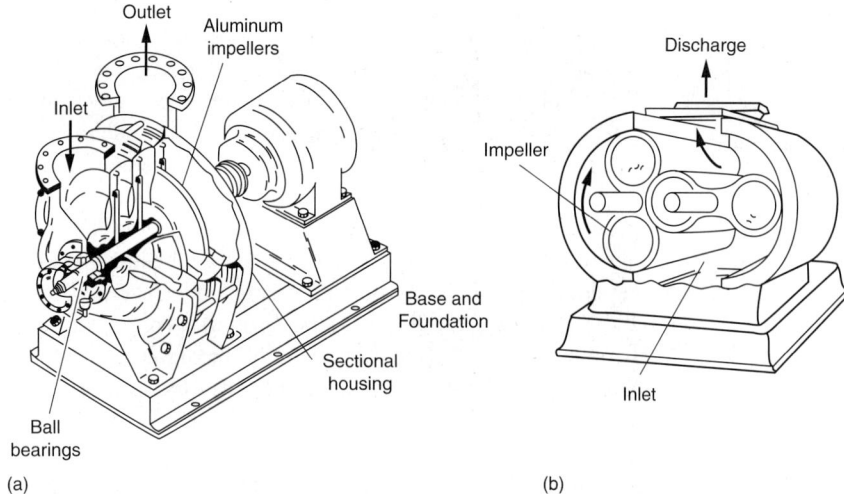

Methods of cleaning porous diffusers may consist of refiring of ceramic plates, high-pressure water sprays, brushing, or chemical treatment with acid or caustic baths. Additional details on cleaning methods may be found in U.S. EPA (1989).

Blowers. There are three types of blowers commonly used for aeration: centrifugal, rotary lobe positive displacement, and inlet guide vane-variable diffuser. Centrifugal blowers (see Fig. 5–63a) are almost universally used where the unit capacity is greater than 425 m³/min (15,000 ft³/min) of free air. Rated discharge pressures range normally from 48 to 62 kN/m² (7 to 9 lb$_f$/in²). Centrifugal blowers have operating characteristics similar to a low-specific-speed centrifugal pump. The discharge pressure rises from shutoff to a maximum at about 50 percent of capacity and then drops off. The operating point of the blower is determined, similar to a centrifugal pump, by the intersection of the head-capacity curve and the system curve.

In wastewater-treatment plants, the blowers must supply a wide range of airflows with a relatively narrow pressure range under varied environmental conditions. A blower usually can only meet one particular set of operating conditions efficiently. Because it is necessary to meet a wide range of airflows and pressures at a wastewater-treatment plant, provisions have to be included in the blower system design to regulate or turn down the blowers. Methods to achieve regulation or turndown are (1) flow blowoff or bypassing, (2) inlet throttling, (3) adjustable discharge diffuser, (4) variable-speed driver, and (5) parallel operation of multiple units. Inlet throttling and an adjustable discharge diffuser are applicable only to centrifugal blowers; variable-speed drivers are more commonly used on positive-displacement blowers. Flow blowoff and bypassing is also an effective method of controlling surging of a centrifugal blower, a phenomenon that occurs when the blower operates alternately at zero capacity and full capacity, resulting in vibration and overheating. Surging occurs when the blower operates in a low volumetric range.

For higher discharge pressure applications [> 55 kN/m² (8 lb$_f$/in²)] and for capacities smaller than 425 m³/min (15,000 ft³/min) of free air per unit, rotary-lobe positive-displacement blowers are commonly used (see Fig. 5–63b). The positive-displacement blower is a machine of constant capacity with variable pressure. The units cannot be throttled, but capacity control can be obtained by the use of multiple units or a variable-speed drive. Rugged inlet and discharge silencers are essential.

A relatively new blower design, the inlet guide vane-variable diffuser that was developed in Europe, mitigates some of the problems and considerations associated with standard centrifugal and positive-displacement aeration blowers. The design is based on a single-stage centrifugal operation that incorporates actuators to position the inlet guide vane and variable diffusers to vary blower flowrate and optimize efficiency. The blowers are especially well suited to applications with medium to high fluctuations in inlet temperature, discharge pressure, and flowrate. Blower capacities range from 85 to 1700 m³/min (3000 to 60,000 ft³/min) at pressures up to 170 kN/m² (25 lb_f/in²). Turndown rates of up to 40 percent of maximum capacity are possible without significant reduction in operating efficiency over the range of operation. Principal disadvantages are high initial cost and a sophisticated computer control system to ensure efficient operation.

The performance curve for a centrifugal blower is a plot of pressure versus inlet air volume and resembles the performance curve for a centrifugal pump. The performance curve typically is a falling-head curve where the pressure decreases as the inlet volume increases. Blowers are rated at standard air conditions, defined as a temperature of 20°C (68°F), a pressure of 760 mm Hg (14.7 lb_f/in²), and a relative humidity of 36 percent. Standard air has a specific weight of 1.20 kg/m³ (0.0750 lb/ft³). The air density affects the performance of a centrifugal blower; any change in the inlet air temperature and barometric pressure will change the density of the compressed air. The greater the gas density, the higher the pressure will rise. As a result, greater power is needed for the compression process (see Fig. 5–64). (Typical values for the specific weight of ambient air are presented in Appendix B). Blowers must be selected to have adequate capacity for a hot summer day, and be provided with a driver with adequate power for the coldest winter weather. The power requirement for adiabatic compression is given in Eq. (5–56).

$$P_w = \frac{wRT_1}{29.7\, n\, e}\left[\left(\frac{p_2}{p_1}\right)^{0.283} - 1\right] \text{SI units}$$ (5–56a)

Figure 5–64

Characteristic curves for centrifugal blower at various inlet air temperatures.

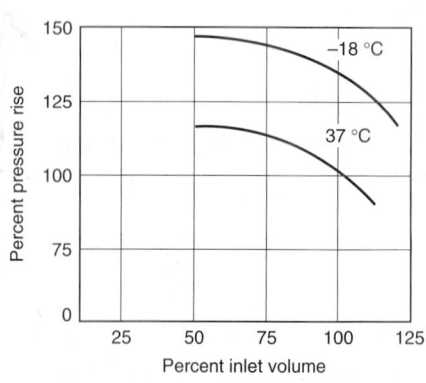

(a)

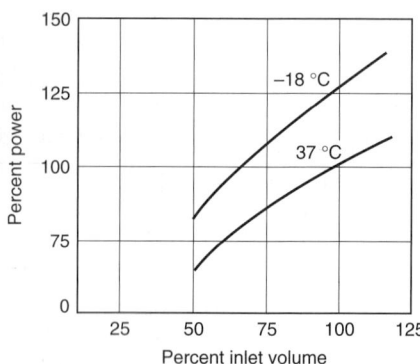

(b)

$$P_w = \frac{wRT_1}{550\,n\,e}\left[\left(\frac{p_2}{p_1}\right)^{0.283} - 1\right] \text{(U.S. customary units)} \qquad (5\text{-}56b)$$

where P_w = power requirement of each blower, kW (hp)

  w = weight of flow of air, kg/s (lb/s)

  R = engineering gas constant for air, 8.314 kJ/k mol K (SI units)

    53.3 ft·lb/(lb air)·°R (U.S. customary units)

  T_1 = absolute inlet temperature, K (°R)

  p_1 = absolute inlet pressure, atm (lb$_f$/in^2)

  p_2 = absolute outlet pressure, atm (lb$_f$/in^2)

  n = $(k - 1)/k$ = 0.283 for air

  k = 1.395 for air

  29.7 = constant for SI units conversion

  550 = ft·lb/s·hp

  e = efficiency (usual range for compressors is 0.70 to 0.90)

Air Piping. Air piping consists of mains, valves, meters, and other fittings that transport compressed air from the blowers to the air diffusers. Because the pressures are low [less than 70 kN/m^2 (10 lb$_f$/in^2)], lightweight piping can be used.

The piping should be sized so that losses in air headers and diffuser manifolds are small in comparison to the losses in the diffusers. Typically, if headlosses in the air piping between the last flow-split device and the farthest diffuser are less than 10 percent of the headloss across the diffusers, good air distribution through the aeration basin can be maintained. Valves and control orifices are an important consideration in piping design (WEF, 1998b). Typical velocities in air piping are given in Table 5–28.

Friction losses in air piping can be calculated using the Darcy-Weisbach equation written in the following form:

$$h_L = f\,\frac{L}{D}\,h_i \qquad (5\text{-}57)$$

where h_L = friction loss, mm (in) of water

  f = dimensionless friction factor obtained from Moody diagram (see Appendix G) based on relative roughness. It is recommended that f be increased by at least 10 percent to allow for an increase in friction factor as the pipe ages.

Table 5–28
Typical air velocities in aeration header pipes

| Pipe diameter | | Velocity[a] | |
in	mm	ft/min	m/min
1–3	75–225	1200–1800	360–540
4–10	100–250	1800–3000	540–900
12–24	300–600	2700–4000	800–1200
30–60	750–1500	3800–6500	1100–2000

[a] At standard conditions.

L = equivalent length of pipe, m (ft)

D = pipe diameter, m (ft)

h_i = velocity head of air, mm (in) of water

The friction factor for steel pipes carrying air can be approximated by Eq. (5–58) (McGhee, 1991).

$$f = \frac{0.029 \, (D)^{0.027}}{Q^{0.148}} \tag{5–58}$$

where Q = airflow, m³/min under prevailing pressure and temperature conditions

D = pipe diameter, m

Headloss in a straight pipe can be computed by substituting Eq. (5–58) in Eq. (5–57), which yields

$$h_L = 9.82 \times 10^{-8} \left(\frac{fLTQ^2}{PD} \right) \tag{5–59}$$

where P = air discharge pressure, atm

T = temperature in pipe, K [from Eq. (5–60)]

$$T = T_o \, (P/P_o)^{0.283} \tag{5–60}$$

where T_o = ambient air temperature, K (maximum summer air temperature)

P_o = ambient barometric temperature, atm

Losses in elbows, tees, valves, etc., can be computed as a fraction of velocity head using headloss coefficient K values given in the companion volume to this text (Metcalf & Eddy, 1981) or in standard hydraulic texts. Minor losses can also be computed as equivalent lengths of straight pipe, as follows:

$$L = 55.4 \, CD^{1.2} \tag{5–61}$$

where L = equivalent length of pipe, L (m)

D = pipe diameter, L (m)

C = resistance factor (see Table 5–29)

Table 5–29
Resistance factors for fittings in aeration piping systems

Fitting	C factor
Long-radius ell or run of standard tee	0.33
Medium-radius ell or run of tee reduced by 25 percent	0.42
Standard ell or run of tee reduced 50 percent	0.67
Tee through side outlet	1.33
Gate valves	0.25
Globe valve	2.00
Angle valve	0.90

Table 5–30
Typical headlosses through air filters, blower silences, and check valves[a]

Device	Headloss	
	mm	**in**
Air filter	13–76	0.5–3
Silencer		
Centrifugal blower	13–38	0.5–1.5
Positive-displacement blower	152–216	6–8.5
Check valve	20–203	0.8–8

[a] Adapted from Qasim (1999).

Meter losses can be estimated as a fraction of the differential velocity head across the meter, depending on the type of meter. Losses in air filters, blower silencers, and check valves should be obtained from equipment manufacturers, but approximate values given in Table 5–30 can be used as a guide (Qasim, 1999).

The discharge pressure at the blowers will be the sum of the above losses, the depth of water over the air diffusers, and the loss through the diffusers.

Because of the high temperature of the air discharged by blowers [60 to 80°C (140 to 180°F)], condensation in the air piping is not a problem, except where piping is submerged in the wastewater. It is essential, however, that provisions be made for pipe expansion and contraction. Where porous diffusers are used, pipes must be made of nonscaling materials or must be lined with material that will not corrode. Pipe materials are often stainless steel, fiberglass, or plastics suitable for higher temperatures. Other materials used include mild steel or cast iron with external coatings (e.g., coal-tar epoxy or vinyl). Interior surfaces include cement lining or coal tar or vinyl coatings.

Mechanical Aerators

Mechanical aerators are commonly divided into two groups based on major design and operating features: aerators with vertical axis and aerators with horizontal axis. Both groups are further subdivided into surface and submerged aerators. In surface aerators, oxygen is entrained from the atmosphere; in submerged aerators, oxygen is entrained from the atmosphere and, for some types, from air or pure oxygen introduced in the tank bottom. In either case, the pumping or agitating action of the aerators helps to keep the contents of the aeration tank or basin mixed. In the following discussion, the various types of aerators will be described, along with aerator performance and the energy requirement for mixing.

Surface Mechanical Aerators with Vertical Axis. Surface mechanical aerators with a vertical axis are designed to induce either updraft or downdraft flows through a pumping action (see Fig. 5–65). Surface aerators consist of submerged or partially submerged impellers that are attached to motors mounted on floats or on fixed structures. The impellers are fabricated from steel, cast iron, noncorrosive alloys, and fiberglass-reinforced plastic and are used to agitate the wastewater vigorously, entraining air in the wastewater and causing a rapid change in the air-water interface to facilitate solution of the air. Surface aerators may be classified according to the type of

Figure 5–65

Typical mechanical aerators: (a) schematic low-speed surface aerator, (b) low-speed surface aerator mounted on fixed platform, (c) schematic high-speed surface aerator, and (d) view of high-speed surface aerator in aeration test basin.

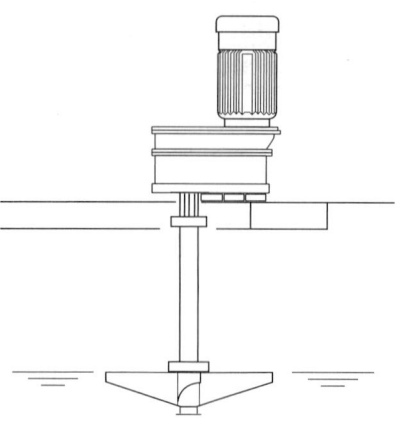

(a)

(b)

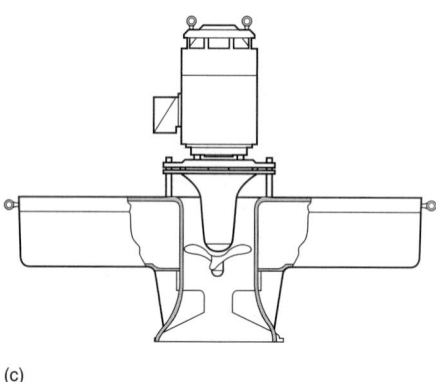

(c)

(d)

impeller used: centrifugal, radial-axial, or axial; or the speed of rotation of the impeller: low and high speed. Centrifugal impellers belong to the low-speed category; the axial-flow impeller type aerators operate at high speed. In low-speed aerators, the impeller is driven through a reduction gear by an electric motor (see Fig. 5–65a). The motor and gearbox are usually mounted on a platform that is supported either by piers extending to the bottom of the tank or by beams that span the tank. Low-speed aerators may also be mounted on floats. In high-speed aerators, the impeller is coupled directly to the rotating element of the electric motor (see Fig. 5–65c). High-speed aerators are almost always mounted on floats. These units were originally developed for use in ponds or lagoons where the water surface elevation fluctuates, or where a rigid support would be impractical. Surface aerators may be obtained in sizes from 0.75 to 100 kW (1 to150 hp).

Submerged Mechanical Aerators with Vertical Axis. Most surface mechanical aerators are upflow types that rely on violent agitation of the surface and air entrainment for their efficiency. With submerged mechanical aerators, however, air or

Figure 5–66

Typical submerged mechanical aerators: (a) turbine type with supplementary air or oxygen feed introduced below the turbine, and (b) draft tube turbine aerator equipped with an air sparger. (Adapted from Philadelphia Mixer Catalog.)

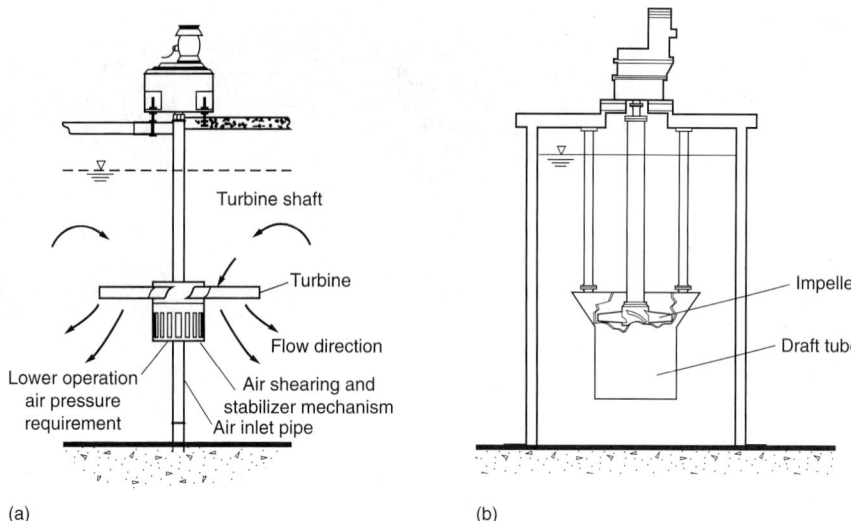

pure oxygen may also be introduced by diffusion into the wastewater beneath the impeller or downflow of radial aerators (see Fig. 5–66a). The impeller is used to disperse the air bubbles and mix the tank contents. A draft tube may be used with either upflow or downflow models to control the flow pattern of the circulating liquid within the aeration tank (see Fig. 5–66b). The draft tube is a cylinder, usually with flared ends, mounted concentrically with the impeller. The length of the draft tube depends upon the aerator manufacturer. Submerged mechanical aerators may be obtained in sizes from 0.75 to 100 kW (1 to 150 hp).

Mechanical Aerators with Horizontal Axis. Mechanical aerators with horizontal axis are divided into two groups: surface and submerged aerators. The surface aerator is patterned after the original Kessener brush aerator, a device used to provide both aeration and circulation in oxidation ditches. The brush-type aerator had a horizontal cylinder with bristles mounted just above the water surface. The bristles were submerged in the water and the cylinder was rotated rapidly by an electric motor drive, spraying wastewater across the tank, promoting circulation, and entraining air in the wastewater. Angle steel, steel of other shapes, or plastic bars or blades are now used instead of bristles. A typical horizontal-axis surface aerator is shown on Fig. 5–67.

Submerged horizontal-axis aerators are similar in principle to surface aerators except disks or paddles attached to rotating shafts are used to agitate the water. The disk aerator has been used in numerous applications for channel and oxidation ditch aeration. The disks are submerged in the wastewater for approximately one-eighth to three-eighths of the diameter and enter the water in a continuous, nonpulsating manner. Recesses in the disks introduce entrapped air beneath the surface as the disk turns. Spacing of the disks can vary depending on the oxygen and mixing requirements of the process. Typical power requirements are reported as 0.1 to 0.75 kW/disk (0.15 to 1.00 hp/disk) (WPCF, 1988).

Figure 5–67

Horizontal-axis aerators: (a) rotary brush (also known as a Kessener brush), and (b) disk aerators.

(a)

(b)

Table 5–31
Typical ranges of oxygen transfer capabilities for various types of mechanical aerators[a]

Aeration system	Transfer rate, lb O$_2$/hp·h		Transfer rate, kg O$_2$/kW·h	
	Standard[b]	Field[c]	Standard	Field
Surface low-speed	2.5–3.5	1.2–2.4	1.5–2.1	0.7–1.5
Surface low-speed with draft tube	2.0–4.6	1.2–2.1	1.2–2.8	0.7–1.3
Surface high-speed	1.8–2.3	1.2–2.0	1.1–1.4	0.7–1.2
Submerged turbine with draft tube	2.0–3.3	1.2–1.8[d]	1.2–2.0	0.6–1.1
Submerged turbine	1.8–3.5		1.1–2.1	
Submerged turbine with sparger	2.0–3.3	1.2–1.8[d]	1.2–2.0	0.7–1.0
Horizontal rotor	1.5–3.6	0.8–1.8	1.5–2.1	0.5–1.1

[a] Derived in part from WPCF (1988) and WEF (1998b).
[b] Standard conditions: tap water 20°C (68°F); at 100 kN/m^2 (14.7 lb$_f$/in^2); and initial dissolved oxygen = 0 mg/L
[c] Field conditions: wastewater, 15°C (59°F); altitude 150 m (500 ft), $\alpha = 0.85$, $\beta = 0.9$; operating dissolved oxygen = 2 mg/L
[d] Based on research results, it appears that α values may be lower than 0.85; reported ranges vary from 0.3 to 1.1, WEF (1998b).

Aerator Performance. Mechanical aerators are rated in terms of their oxygen transfer rate expressed as kilograms of oxygen per kilowatt-hour (pounds of oxygen per horsepower-hour) at standard conditions. Standard conditions exist when the temperature is 20°C, the dissolved oxygen is 0.0 mg/L, and the test liquid is tap water. Testing and rating are normally done under non-steady-state conditions using fresh water, deaerated with sodium sulfite. Commercial-size surface aerators range in efficiency from 1.20 to 2.4 kg O$_2$/kW·h (2 to 4 lb O$_2$/hp·h). Oxygen transfer data for various types of mechanical aerators are reported in Table 5–31. Efficiency claims for aerator performance should be accepted by the design engineer only when they are supported by actual test data for the actual model and size of aerator under consideration. For design purposes, the standard performance data must be adjusted to reflect anticipated field conditions by using the following equation. The term within the brackets represents the correction factor.

$$N = N_o \left(\frac{\beta C_{\text{walt}} - C_L}{9.17} \right) 1.024^{T-20} \alpha \tag{5-62}$$

where N = kg O_2/kW·h (lb O_2/hp·h) transferred under field conditions
N_o = kg O_2/kW·h (lb O_2/hp·h) transferred in water at 20°C, and zero dissolved oxygen
β = salinity–surface tension correction factor, usually 1
C_{walt} = oxygen saturation concentration for tap water at given temperature and altitude (see Appendix D), mg/L and Fig. 5–68
C_L = operating oxygen concentration, mg/L
T = temperature, °C
α = oxygen transfer correction factor for waste (see Table 5–32)

The application of this equation is illustrated in Chap. 8, Sec. 8–7, which deals with the design of aerated lagoons.

Figure 5–68

Oxygen solubility correction factor versus elevation.

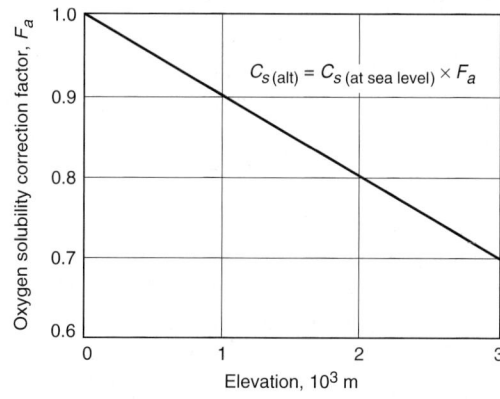

$$C_{s\,(\text{alt})} = C_{s\,(\text{at sea level})} \times F_a$$

Table 5–32

Typical values of alpha factors for low-speed surface aerators and selected wastewater types[a]

Wastewater type	BOD$_5$, mg/L		α factor[b]	
	Influent	Effluent	Influent	Effluent
Municipal wastewater	180	3	0.82	0.98
Pulp and paper	187	50	0.68	0.77
Kraft paper	150–300	37–48	0.48–0.68	0.7–1.1
Bleached paper	250	30	0.83–1.98	0.86–1.0
Pharmaceutical plant	4500	380	1.65–2.15	0.75–0.83
Synthetic fiber plant	5400	585	1.88–3.25	1.04–2.65

[a] WPCF (1988).
[b] Research suggests that α values may be lower and more variable than values listed in table.

Energy Requirement for Mixing in Aeration Systems

As with diffused-air systems, the size and shape of the aeration tank are very important if good mixing is to be achieved. Aeration tanks may be square or rectangular and may contain one or more aerators. The depth and width of the aeration tanks for mechanical surface aerators are dependent on aerator size, and typical values are given in Table 5–33. Depths up to 11 m (35 ft) have been used with submerged-draft tube mixers.

In diffused-air systems, the air requirement to ensure good mixing varies from 20 to 30 $m^3/10^3$ m^3·min (20 to 30 $ft^3/10^3$ ft^3·min) of tank volume, for a spiral-roll aeration pattern. For a grid system of aeration in which the diffusers are installed uniformly along the aeration basin bottom, mixing rates of 10 to 15 $m^3/10^3$ m^3·min (10 to 15 $ft^3/$ 10^3 ft^3·min) have been suggested (WPCF, 1988). Typical power requirements for maintaining a completely mixed flow regime with mechanical aerators vary from 20 to 40 $kW/10^3$ m^3 (0.75 to 1.50 $hp/10^3$ ft^3), depending on the type and design of the aerator, the nature and concentration of the suspended solids, the temperature, and the geometry of the aeration tank, lagoon, or basin. In the design of aerated lagoons for the treatment of domestic wastewater, it is extremely important that the mixing power requirement be checked because, in most instances, it will be the controlling factor.

Generation and Dissolution of High-Purity Oxygen

After the quantity of oxygen required is determined, it is necessary, where high-purity oxygen is to be used, to specify the type of oxygen generator that will best serve the needs of the plant. There are two basic oxygen generator designs: (1) a pressure swing adsorption (PSA) system for smaller and more common plant sizes [less than 150,000 m^3/d (40 Mgal/d)], and (2) the traditional cryogenic air-separation process for large applications. Liquid oxygen can also be trucked in and stored onsite in a liquid form.

Pressure-Swing Adsorption.   The pressure-swing adsorption system uses a multibed adsorption process to provide a continuous flow of oxygen gas (U.S. EPA, 1974). A schematic diagram of the four-bed system is shown on Fig. 5–69a. The operating principle of the pressure-swing adsorption generator is that the oxygen is separated from the feed air by adsorption at high pressure, and the adsorbent is regenerated by "blowdown" to low pressure. The process operates on a repeated cycle having two

Table 5–33
Typical aeration tank dimensions for mechanical surface aerators

Aerator size		Tank dimensions			
		U.S. customary units		SI units	
hp	kW	Depth, ft	Width, ft	Depth, m	Width, m
10	7.5	10–12	30–40	3–3.5	9–12
20	15	12–14	35–50	3.5–4	10–15
30	22.5	13–15	40–60	4–4.5	12–18
40	30	12–17	45–65	3.5–5	14–20
50	37.5	15–18	45–75	4.5–5.5	14–23
75	55	15–20	50–85	4.5–6	15–26
100	75	15–20	60–90	4.5–6	18–27

Figure 5–69

Schematic diagrams for the generation of oxygen used in the high-purity oxygen-activated sludge process: (a) pressure-swing adsorption and (b) cryogenic air separation.

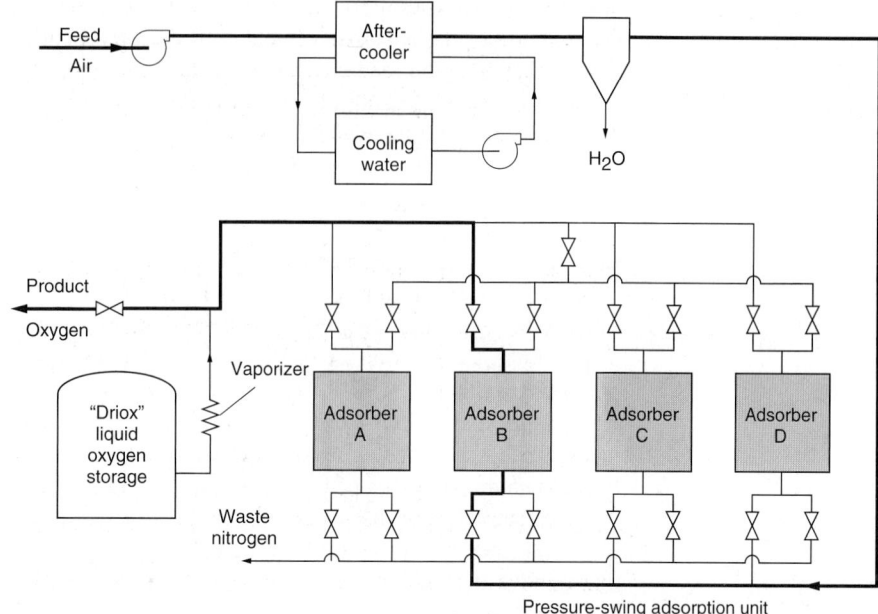

(a)

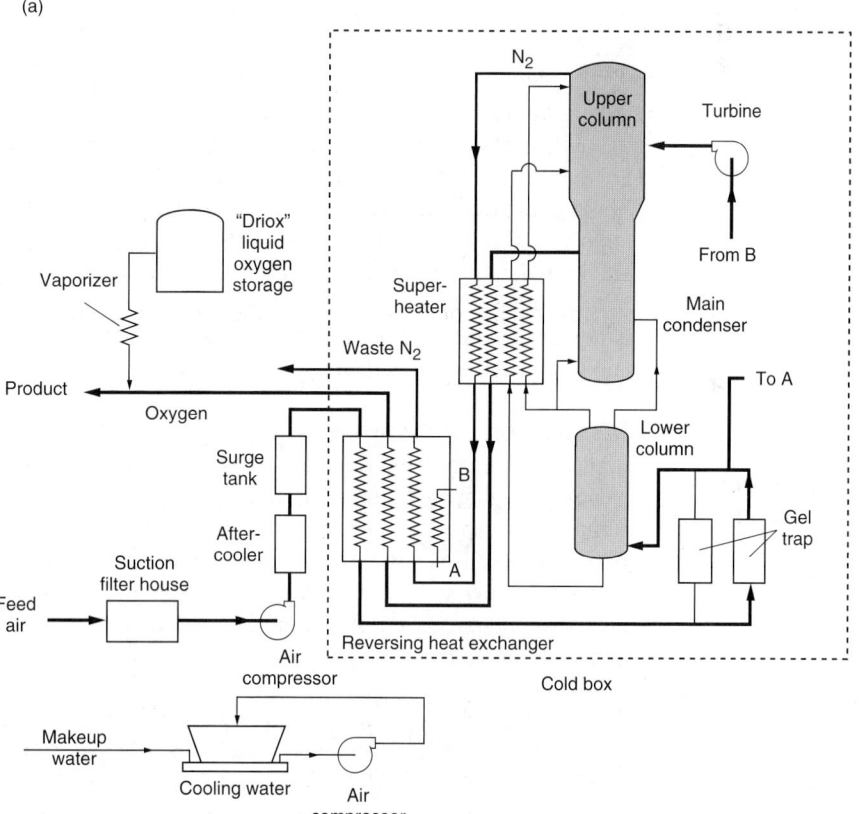

(b)

basic steps, adsorption and regeneration. During the adsorption step, feed air flows through one of the adsorber vessels until the adsorbent is partially loaded with impurity. At that time the feed-air flow is switched to another adsorber, and the first adsorber is regenerated. During regeneration, the impurities are cleaned from the adsorbent so that the bed will be available again for the adsorption step. Regeneration is carried out by depressurizing to atmospheric pressure, purging with some of the oxygen, and repressurizing back to the pressure of the feed air.

Cryogenic Air Separation. The cryogenic air separation process involves the liquefaction of air, followed by fractional distillation to separate it into its components (mainly nitrogen and oxygen). A schematic diagram of this process is shown on Fig. 5–69b. First, the entering air is filtered and compressed, and then it is fed to the reversing heat exchangers, which perform the dual function of cooling and removing the water vapor and carbon dioxide by freezing these mixtures out into the exchanger surfaces. Periodically switching or reversing the feed air and the waste nitrogen streams through identical passes of the exchangers to regenerate their water vapor and carbon dioxide removal capacity accomplishes this process.

Next, the air is processed through "cold and gel traps," which are adsorbent beds that remove the final traces of carbon dioxide as well as most hydrocarbons from the feed air. The processed air is then divided into two streams. The first stream is fed directly to the lower column of the distillation unit. The second stream is returned to the reversing heat exchangers and partially warmed to provide the required temperature difference across the exchanger. This stream is then passed through an expansion turbine and fed into the upper column of the distillation unit. An oxygen-rich liquid exits from the bottom of the lower column and the liquid nitrogen exits from the top. Both streams are then subcooled and transferred to the upper column. In this column, the descending liquid phase becomes progressively richer in oxygen, and the liquid that subsequently collects in the condenser reboiler is the oxygen product stream. This oxygen is recirculated continually through an adsorption trap to remove all possible residual traces of hydrocarbons. The waste nitrogen exits from the top portion of the upper column and is heat exchanged along with the oxygen product to recover all available refrigeration and to regenerate the reversing heat exchangers.

Dissolution of Commercial Oxygen. Oxygen is very insoluble in water—even pure oxygen—and requires special considerations to ensure high absorption efficiency. Oxygen dissolution equipment designed for air only optimizes energy consumption because the air is free and efficient oxygen absorption is not relevant. However, because of the cost of commercial oxygen, the facilities used for its dissolution must be designed to both efficiently absorb the commercial oxygen as well as minimize the unit energy consumption. These requirements rule out the more common aeration equipment alternatives.

Dissolution Time. A key feature that must be incorporated into a commercial oxygen dissolution system is oxygen retention time. To optimize the absorption of pure oxygen it has been found that a detention time of about 100 s is required. Further, two-phase flow must be maintained to avoid the coalescence of the oxygen bubbles to maintain absorption efficiency. Unfortunately, some pure oxygen dissolution systems consume as much energy to dissolve a ton of pure oxygen as standard surface aerators consume in dissolving a ton of oxygen from air.

Speece Cone (Downflow Bubble Contactor). A system that incorporates prolonged oxygen bubble contact time and high rates of oxygen transfer is a cone-shaped chamber, downflow bubble contact aerator termed the Speece cone (see Fig. 5–70a). Water enters the chamber at the apex with a velocity of approximately 3 m/s (10 ft/s). This inlet velocity provides the energy to maintain a two-phase bubble swarm in the cone, ensuring a very high bubble/water interface, resulting in a proportionately high gas transfer rate. The expanding horizontal cross section of the cone reduces the downward flow velocity of the water to less than 0.3 m/s (1 ft/s). Because the bubbles have a nominal buoyant velocity of about 0.3 m/s, if the downflow velocity of the water is reduced below the buoyant velocity of the bubbles, they will remain indefinitely in the cone, thus satisfying the required bubble residence time. The water, however, has a residence time of about 10 s, reflecting a relatively small volume of reactor cone. This system incorporates the desired features of relatively small size, high rate of oxygen transfer, and more than adequate bubble residence time (Speece and Tchobanoglous, 1990).

U-Tube Contactor. Another oxygen transfer system that incorporates desirable features for efficient dissolution of commercial oxygen with low unit energy consumption is the U-tube (see Fig. 5–70b). At a depth of 30 m (100 ft) and a throughput velocity of 2.5 m/s (8 ft/s), the residence time is 25 s. Because a contact time of 25 s is low, off-gas recycle back through the system can be used to increase the contact time to about 100 s where efficient absorption occurs. The energy requirements are low because the bubble/water mixture is pumped through a filled U-tube pipe that is hydrostatically pressurized by its vertical configuration. Use of the U-tube enhances gas transfer significantly.

Figure 5–70

Pure oxygen dissolution systems: (a) Speece cone (downflow bubble contactor) and (b) U-tube contactor. (Speece and Tchobanoglous, 1990.)

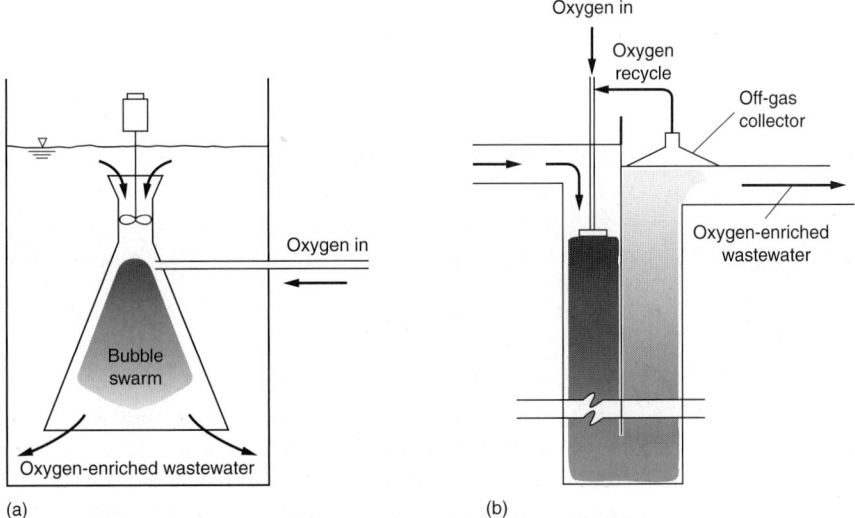

(a)   (b)

Energy consumption is about 54 kWh/Mg O_2 (60 kWh/ton) while producing an effluent dissolved oxygen of 60 mg/L (Speece and Tchobanoglous, 1990).

Conventional Diffused Aeration. Conventional diffused aeration or surface aerators must operate in a closed headspace to absorb commercial oxygen efficiently. A concrete cover usually is placed over the aeration tank to enclose the headspace.

Postaeration

The requirement for postaeration systems has developed in recent years with the introduction of effluent standards and permits that include high dissolved oxygen levels (5 to 8 mg/L). Dissolved oxygen levels have become standard for discharge to water-quality-limited stream sections and to effluent dominated waters. The regulatory intent is to ensure that low dissolved oxygen levels in the treated effluent do not cause immediate depression after mixture with the waters of the receiving stream. To meet postaeration requirements, three methods are most commonly used: (1) cascade aeration, (2) mechanical aeration, and (3) diffused air. The Speece cone, described above, can also be used for postaeration and for reaerating reclaimed water-storage reservoirs.

Cascade Aeration. If site constraints and hydraulic conditions permit gravity flow, the least costly method to raise dissolved oxygen levels is with the use of cascade aeration. Cascade aeration consists of using the available discharge head to create turbulence as the wastewater falls in a thin film over a series of concrete steps. Performance depends on the initial dissolved oxygen level, required discharge dissolved oxygen, and wastewater temperature. Typical design information is given in Table 5–34. Where the cascade aeration facility joins the chlorine contact basin, the postaeration structure may be made equal to the chlorine contact basin width for ease of construction.

The most commonly used method for determining required cascade height is based on the following equations developed by Barrett (1960).

$$H = \frac{R - 1}{0.361 \, ab \, (1 + 0.046 \times T)} \qquad \text{SI units} \qquad (5\text{–}63a)$$

Table 5–34
Typical design information for a cascade-type postaeration system

Parameter	U.S. customary units			SI units		
	Units	Range	Typical	Units	Range	Typical
Hydraulic loading rate at average design flow	gal/ft of width·d	100,000–500,000	240,000	m³/m of width·d	1240–6200	3000
Step dimensions						
Height	in	6–12	8	mm	150–300	200
Length	in	12–24	18	mm	300–600	450
Cascade height	ft	6–16		m	2–5	

$$H = \frac{R - 1}{0.11 \, ab \, (1 + 0.046 \times T)} \qquad \text{U.S. customary units} \qquad (5\text{-}63b)$$

where R = deficit ratio $= \dfrac{C_s - C_o}{C_s - C}$

$\quad C_s$ = dissolved oxygen saturation concentration of the wastewater at temperature T, mg/L

$\quad C_o$ = dissolved oxygen concentration of the postaeration influent, mg/L

$\quad C$ = required final dissolved oxygen level after postaeration, mg/L

$\quad a$ = water-quality parameter equal to 0.8 for a wastewater-treatment plant effluent

$\quad b$ = weir geometry parameter for a weir, $b = 1.0$; for steps, $b = 1.1$; for step weir, $b = 1.3$

$\quad T$ = water temperature, °C

$\quad H$ = height through which water falls, m (ft)

A key element is the proper selection of the critical wastewater temperature, which affects the dissolved oxygen saturation concentration C_s. The effect of temperature is illustrated in Example 5-14.

EXAMPLE 5-14 **Calculation of Cascade Aeration Height** Calculate the height of a cascade aeration system for a wastewater-treatment plant in a warm climate where the wastewater temperature averages 20°C in the winter and 25°C in the summer. The dissolved oxygen in the influent to the postaeration system C_o is 1.0 mg/L and the required final dissolved oxygen concentration C is 6.0 mg/L.

Solution

1. Determine the dissolved oxygen saturation concentration C_s at the wastewater temperatures.
   a. From Appendix D, the dissolved oxygen solubilities are:

   for 20°C = 9.08 mg/L; for 25°C = 8.24 mg/L.

2. Calculate the cascade height for $T = 20$°C using Eq. (5-63a).
   a. Calculate the deficit ratio

   $$R = \text{deficit ratio} = \frac{C_s - C_o}{C_s - C} = \frac{9.08 - 1.0}{9.08 - 6.0} = 2.62$$

   b. Calculate the cascade height, assuming steps.

   $$H = \frac{R - 1}{0.361 \, ab \, (1 + 0.046T)}$$

   $$H = \frac{2.62 - 1}{0.361 \, (0.8)(1.1)(1 + 0.046 \times 20)}$$

   $$H = 2.66 \text{ m } (8.73 \text{ ft})$$

3. Calculate the cascade height for $T = 25°C$.
   a. Calculate the deficit ratio

   $$R = \frac{8.24 - 1.0}{8.24 - 6.0} = 3.23$$

   b. Calculate the cascade height, assuming steps and using the same computation procedure as in 2b above.

   $$H = 3.12 \text{ m (9.99 ft)}$$

Comment The increased wastewater temperature increases the dissolved oxygen deficit ratio and therefore affects the height of the cascade. The maximum wastewater temperatures should be checked to ensure the cascade height is not underdesigned.

Mechanical Aeration. Two major types of mechanical aeration equipment are commonly used for postaeration systems: low-speed surface aerators and submerged turbine aerators. Low-speed surface aerators are preferred because they are usually the most economical, except where high oxygen transfer rates are required. For high oxygen transfer rates, submerged turbine units are preferred. Most installations consist of two or more aerators in rectangular basins. Detention times for postaeration using either mechanical or diffused-air aeration are usually 10 to 20 min at peak flowrates.

Diffused-Air Aeration. In larger treatment plants, diffused aeration systems may be more appropriate. Coarse- or fine-bubble diffusers may be used. Depending on the depth of submergence, transfer efficiencies of 5 to 8 percent may be attained with nonporous diffusers, and 15 to 25 percent with fine-pore diffusers. After secondary treatment, the alpha factors should range from 0.85 to 0.95 for nonporous systems, and from 0.70 to 0.85 for fine-pore systems.

EXAMPLE 5–15 **Derivation of Equation for Estimating Diffused-Air Requirements for Post-aeration** Develop an expression that can be used to estimate the diffused-air requirement for the postaeration of effluent following chlorination. Assume that aeration will be accomplished in a plug-flow reactor.

Solution

1. Write an equation for the oxygen-solution rate. The appropriate expression, similar to Eq. (5–50), is

   $$r_M = \frac{dm}{dt} = K_T' (C_s - C)$$

   where K_T' = overall mass transfer coefficient for the given conditions
   $$K_T' = K_{20}' \times (1.024)^{T-20}$$

2. Write an expression for the oxygen-transfer efficiency. The efficiency may be defined as

$$E = \frac{(dm/dt)_{20°C,C=0}}{M}$$

where

E = oxygen transfer efficiency
$(dm/dt)_{20°C,C=0}$ = oxygen-solution rate at 20°C and zero dissolved oxygen
M = mass rate at which oxygen is introduced

3. Develop a differential expression for the mass rate at which oxygen is introduced. The mass rate at which oxygen is introduced is given by

$$M = \frac{1}{E}\left(\frac{dm}{dt}\right)_{20°C,C=0}$$

$$= \frac{1}{E}\left(\frac{dm}{dt}\right)_T \frac{(dm/dt)_{20°C,C=0}}{(dm/dt)_T}$$

Substituting for $(dm/dt)_{20, C=0}$ and $(dm/dt)_T$ yields

$$M = \frac{1}{E}\left(\frac{dm}{dt}\right)_T \frac{(C_s)_{20°C}}{(C_s - C)_T(1.024)^{T-20}}$$

If the expression is applied to an infinitesimal transverse segment of the tank and $Q\,dC$ is substituted for dm/dt [note that $V(dC/dt) = dm/dt$ and $Q = V/dt$], then the differential form of the above expression can be rewritten as

$$dm = \frac{Q(C_s)_{20°C}}{E(1.024)^{T-2}}\left(\frac{dC}{C_s - C}\right)_T$$

4. Derive the integrated form of the differential expression that was derived in step 3. The integrated form of the equation can be obtained by integrating the expression from the inlet of the tank where $C = C_i$ to the outlet of the tank where $C = C_o$:

$$\int_0^M dM = \frac{Q(C_s)_{20°C}}{E(1.024)^{T-20}}\int_{C_i}^{C_o}\frac{dc}{C_s - C}$$

$$M = \frac{Q(C_s)_{20°C}}{E(1.024)^{T-20}}\left(\ln\frac{C_s - C_i}{C_s - C_o}\right)_T$$

5. Rewrite the equation derived in step 4 in a more practical format. Note that the density of air is 1.23 kg/m³ and air contains about 23 percent oxygen by weight. Using these values, the rate of oxygen input, in terms of the equivalent air flowrate expressed in m³/s, is equal to

$$Q_a = 3.53 \times 10^{-3}\frac{Q(C_s)_{20°C}}{E(1.024)^{T-20}}\left(\ln\frac{C_s - C}{C_s - C_o}\right)$$

where Q_a = required air flowrate, m³/s
  Q = wastewater flowrate, m³/s
  C_s = saturation concentration of oxygen at 20°C, g/m³

Comment   The value of Q_a is usually multiplied by a factor of 1.1 to account for the fact that the saturation value of oxygen in wastewater is about 95 percent of that in distilled water, and to account for the difference in the transfer rates.

Aeration Using the Speece Cone. In postaeration with a Speece cone, a sidestream of the main flow is oxygenated and then reblended with the main flow before discharge. In open reservoirs used for storing reclaimed water, problems occur due to temperature stratification, low dissolved oxygen, and release of odors, principally hydrogen sulfide (see Sec. 13–4). The Speece cone, which has a very high oxygen transfer rate and is shown on Fig. 5–70a, is ideally suited for reservoir aeration and temperature destratification. For reservoir aeration applications, compressed air can be used instead of high-purity oxygen.

5–13 REMOVAL OF VOLATILE ORGANIC COMPOUNDS (VOCs) BY AERATION

At some wastewater-treatment facilities, volatile organic compounds (VOCs) such as trichloroethylene (TCE) and 1,2-dibromo-3-chloropropane (DBCP) have been detected in wastewater. The uncontrolled release of such compounds that now occurs in wastewater-collection systems and wastewater-treatment plants is an area of concern. The mechanisms governing the release of these compounds, the locations where the release of these compounds is most prevalent, and the methods of controlling the discharge of these compounds to the atmosphere are discussed in this section.

Emission of VOCs

The principal mechanisms governing the release of VOCs in wastewater collection and treatment facilities are (1) volatilization and (2) gas stripping. These mechanisms and the principal locations where VOCs are released are considered in the following discussion.

Volatilization. The release of VOCs from wastewater surfaces to the atmosphere is termed volatilization. Volatile organic compounds are released because they partition between the gas and water phase until equilibrium concentrations are reached (Roberts et al., 1984). The mass transfer (movement) of a constituent between these two phases is a function of the constituent concentration in each phase relative to the equilibrium concentration. Thus, the transfer of a constituent between phases is greatest when the concentration in one of the phases is far from equilibrium. Because the concentration of VOCs in the atmosphere is extremely low, the transfer of VOCs usually occurs from wastewater to the atmosphere.

Gas Stripping. Gas stripping of VOCs occurs when a gas (usually air) is temporarily entrained in wastewater or is introduced purposefully to achieve a treatment objective. When gas is introduced into a wastewater, VOCs are transferred from the wastewater to the gas. The forces governing the transfer between phases are the same as described above. For this reason, gas (air) stripping is most effective when contaminated wastewater is exposed to contaminant-free air. In wastewater treatment, air stripping occurs most commonly in aerated grit chambers, aerated biological treatment processes, and aerated transfer channels. Specially designed facilities (e.g., stripping towers) for gas stripping are considered in Sec. 11–8 in Chap 11.

Locations Where VOCs Are Emitted. The principal locations where VOCs are emitted from wastewater collection and treatment facilities are summarized in Table 5–35. The degree of VOC removal at a given location will depend on local conditions. Mass transfer is considered in the following section.

Mass Transfer Rates for VOCs

The mass transfer of VOCs can, for practical purposes, be modeled using the following equation (Roberts et al., 1984, and Thibodeaux, 1979):

$$r_{VOC} = -K_L a_{VOC}(C - C_s) \tag{5-64}$$

Where r_{VOC} = rate of VOC mass transfer, $\mu g/m^3 \cdot h$
$(K_L a)_{VOC}$ = overall VOC mass transfer coefficient, 1/h
C = concentration of VOC in liquid, $\mu g/m^3$
C_s = saturation concentration of VOC in liquid, $\mu g/m^3$

Due to chemical handling and analytical requirements, measuring $K_L a_{VOC}$ is much more difficult than measuring $K_L a_{O_2}$. Therefore, a practical approach is to relate the $K_L a_{VOC}$ to the $K_L a_{O_2}$. The following equation relates the mass transfer coefficients as a function of the VOC and O_2 diffusion coefficients in water:

$$K_L a_{VOC} = K_L a_{O_2} \left(\frac{D_{VOC}}{D_{O_2}} \right)^n \tag{5-65}$$

where $K_L a_{VOC}$ = system mass transfer coefficient, T^{-1} (1/h)
$K_L a_{O_2}$ = system oxygen mass transfer coefficient, T^{-1} (1/h)
D_{VOC} = diffusion coefficient of VOC in water, $L^2 T^{-1}$ (cm^2/s)
D_{O_2} = diffusion coefficient of oxygen in water, $L^2 T^{-1}$ (cm^2/s)
n = coefficient

Diffusion coefficient values for different compounds can be obtained from Schwarzenbach et al. (1993) or in other handbooks, and some variation in literature values will be found. Experimental investigations of $K_L a_{VOC}$ versus $K_L a_{O_2}$ have shown that Eq. (5–65) can be generally applicable, and that the value for n varies depending on whether the gas/liquid transfer is accomplished by surface aeration, diffused aeration, or a packed column air stripper, and the power intensity of the gas transfer device (Roberts and Dandliker, 1983; Matter-Muller et al.; 1981, Hsieh et al., 1993; Libra, 1993; and Bielefeldt and Stensel, 1999). For practical power intensities of less than 100 W/m^3 a reasonable value of n is 0.50 for packed columns and mechanical aeration and 1.0 for diffused aeration. For higher power intensities the work of Hsieh et al. (1993) should be consulted. The $K_L a_{VOC}$ was also found to be essentially the same in wastewater as in tap water (Bielefeldt and Stensel, 1999).

Table 5-35
Sources, methods of release, and control of VOCs from wastewater facilities

Source	Method of release	Suggested control strategies
Domestic, commercial, and industrial discharges	Discharge of small amounts of VOCs in liquid wastes	Institute active source control program to limit the discharge of VOCs to municipal sewers
Wastewater sewers	Volatilization from the surface enhanced by flow-induced turbulence	Seal existing manholes. Eliminate the use of structures that create turbulence and enhance volatilization
Sewer appurtenances	Volatilization due to turbulence at junctions, etc. Volatilization and air stripping at drop manholes and junction chambers	Isolate and cover existing appurtenances
Pump stations	Volatilization and air stripping at influent wetwell inlets	Vent gases from wetwell to VOC treatment unit. Use variable-speed pumps to reduce size of wetwell
Bar racks	Volatilization due to turbulence	Cover units. Reduce headloss through bar racks
Comminutors	Volatilization due to turbulence	Cover units. Use in-line enclosed comminutors
Parshall flume	Volatilization due to turbulence	Cover units. Use alternative measuring device
Grit chamber	Volatilization due to turbulence in horizontal-flow grit chambers. Volatilization and air stripping in aerated grit chambers. Volatilization in vortex-type grip chambers	Cover aerated and vortex-type grit chambers. Reduce turbulence in horizontal-flow grit chambers; cover if necessary
Equalization basins	Volatilization from surface enhanced by local turbulence. Air stripping where diffused air is used	Cover units. Use submerged mixers. Reduce air flow
Primary and secondary sedimentation tanks	Volatilization from surface. Volatilization and air stripping at overflow weirs, in effluent channel, and at other discharge points	Cover tanks. Replace overflow weirs with drops with submerged launders
Biological treatment	Air stripping in diffused-air activated sludge. Volatilization in activated-sludge processes with surface aerators. Volatilization from surface enhanced by local turbulence. Volatilization from trickling filters	Cover units. In activated-sludge systems, use submerged mixers and reduce aeration rate
Transfer channels	Volatilization from surface enhanced by local turbulence. Volatilization and air stripping in aerated transfer channels	Use enclosed transfer channels
Digester gas	Uncontrolled release of digester gas. Discharge of incompletely combusted or incinerated digester gas	Controlled thermal incineration, combustion, or flaring of digester gas

Mass Transfer of VOCs from Surface and Diffused-Air Aeration Processes

The amount of VOCs released from a complete-mix reactor used for the activated-sludge process will depend on the method of aeration (e.g., surface aeration or diffused aeration).

Complete-Mix Reactor with Surface Aeration. A materials balance for the stripping of a VOC written around a complete-mix reactor is as follows, assuming no other removal mechanisms for the VOC compound such as biodegradation or solids sorption are applicable.

1. General word statement:

| Rate of accumulation of VOC within the system boundary | = | Rate of flow of VOC into the system boundary | − | Rate of flow of VOC out of the system boundary | + | Amount of VOC removed through system boundary by stripping | (5–66) |

2. Simplified word statement:

   Accumulation = inflow − outflow + decrease due to stripping   (5–67)

3. Symbolic representation:

$$\frac{dC}{dt} V = QC_i - QC_e + r_{VOC}V \tag{5–68}$$

   where dC/dt = rate of change in VOC concentration in reactor
   V = volume of complete mix reactor, L^3 (m³)
   Q = liquid flowrate, $L^3 T^{-1}$ (m³/s)
   C_i = concentration of VOC in influent to reactor, ML^{-3} ($\mu g/m^3$)
   C_e = concentration of VOC in effluent from reactor, ML^{-3} ($\mu g/m^3$)
   r_{VOC} = rate of VOC mass transfer, $ML^{-3}T^{-1}$ ($\mu g/m^3 \cdot h$)

Substituting for r_{VOC} from Eq. (5–64) and τ for V/Q yields

$$\frac{dC}{dt} = \frac{C_i - C_e}{\tau} + [- (K_L a)_{VOC}(C_e - C_s)] \tag{5–69}$$

If steady-state conditions are assumed and it is further assumed that C_s is equal to zero, then the amount of VOC that can be removed by surface aeration is given by the following expression:

$$1 - \frac{C_e}{C_i} = 1 - [1 + (K_L a)\tau]^{-1} \tag{5–70}$$

If a significant amount of the VOC is adsorbed or biodegraded, the results obtained with the above equation will be overestimated. The above analysis can also be used to estimate the release of VOCs at weirs and drops by assuming the time period is about 30 s.

Complete Mix Reactor with Diffused-Air Aeration. The corresponding expression to Eq. (5–67) for a complete-mix reactor with diffused-air aeration is developed by a mass balance on the VOC compound. At steady state the VOC in equals the VOC out, and the corresponding mass balance is

Inflow in	=	outflow	+	outflow
liquid stream		liquid stream		in exit gas

$$QC_i = QC_e + Q_g C_{g,e} \tag{5-71}$$

where Q = liquid flowrate, $L^3 T^{-1}$ (m³/s)
  C_i = VOC concentration in influent, ML^{-3} ($\mu g/m^3$)
  C_e = VOC concentration in effluent, ML^{-3} ($\mu g/m^3$)
  Q_g = gas flowrate, $L^3 T^{-1}$ (m³/s)
  $C_{g,e}$ = VOC concentration in exit gas, ML^{-3} ($\mu g/m^3$)

The general expression for the removal of VOC by gas sparging through a liquid is (Bielefeldt and Stensel, 1999)

$$Q_g C_{g,e} = Q_g H_u C_e (1 - e^{-\phi}) \tag{5-72}$$

where ϕ = VOC saturation parameter defined as

$$\phi = \frac{(K_L a)_{VOC} V}{H_u Q_g} \tag{5-73}$$

Eq. (5–72) can be rearranged as follows:

$$Q(C_I - C_e) = H_u C_e (1 - e^{-\phi}) \tag{5-74}$$

Solving Eq. (5–74) for C_e/C_i yields

$$\frac{C_e}{C_i} = \left[1 + \frac{Q_g}{Q} H_u (1 - e^{-\phi}) \right]^{-1} \tag{5-75}$$

and the fraction removed is given by

$$1 - \frac{C_e}{C_i} = 1 - \left[1 + \frac{Q_g}{Q} H_u (1 - e^{-\phi}) \right]^{-1} \tag{5-76}$$

The application of the above equations is illustrated in Example 5–16.

EXAMPLE 5-16 **Stripping of Benzene in the Activated-Sludge Process** Determine the amount of benzene that can be stripped in a complete-mix activated-sludge reactor equipped with a diffused-air aeration system. Assume the following conditions apply:

1. Wastewater flowrate = 4000 m³/d
2. Aeration tank volume = 1000 m³
3. Depth of aeration tank = 6 m
4. Air flowrate = 50 m³/min at standard conditions
5. Oxygen mass transfer rate = 6.2/h
6. Influent concentration of benzene = 100 $\mu g/m^3$
7. $H = 5.49 \times 10^{-3}$ m³·atm/mol (see Table 5–36)

Table 5-36
Physical properties of selected volatile and semivolatile organic compounds[a],[b]

Compounds	mw	mp, °C	bp, °C	vp, mmHg	vd	sg	Sol., mg/L	C_s, g/m³	H, m³·atm/mole	log K_{ow}
Benzene	78.11	5.5	80.1	76	2.77	0.8786	1,780	319	5.49×10^{-3}	2.1206
Chlorobenzene	112.56	−45	132	8.8	3.88	1.1066	500	54	3.70×10^{-3}	2.18–3.79
o-Dichlorobenzene	147.01	18	180.5	1.60	5.07	1.036	150	N/A	1.7×10^{-3}	3.3997
Ethylbenzene	106.17	−94.97	136.2	7	3.66	0.867	152	40	8.43×10^{-3}	3.13
1,2-Dibromoethane	187.87	9.8	131.3	10.25	0.105	2.18	2,699	93.61	6.29×10^{-4}	N/A
1,1-Dichloroethane	98.96	−97.4	57.3	297	3.42	1.176	7,840	160.93	5.1×10^{-3}	N/A
1,2-Dichloroethane	98.96	−35.4	83.5	61	3.4	1.25	8,690	350	1.14×10^{-3}	1.4502
1,1,2,2-Tetrachloroethane	167.85	−36	146.2	14.74	5.79	1.595	2,800	13.10	4.2×10^{-4}	2.389
1,1,1-Trichloroethane	133.41	−32	74	100	4.63	1.35	4,400	715.9	3.6×10^{-3}	2.17
1,1,2-Trichloroethane	133.4	−36.5	133.8	19	N/A	N/A	4,400	13.89	7.69×10^{-4}	N/A
Chloroethene	62.5	−153	−13.9	2548	2.15	0.912	6,000	8521	6.4×10^{-2}	N/A
1,1-Dichloroethene	96.94	−122.1	31.9	500	3.3	1.21	5,000	2640	1.51×10^{-2}	N/A
c-1,2-Dichloroethene	96.95	−80.5	60.3	200	3.34	1.284	800	104.39	4.08×10^{-3}	N/A
t-1,2-Dichloroethene	96.95	−50	48	269	3.34	1.26	6,300	1428	4.05×10^{-3}	N/A
Tetrachloroethene	165.83	−22.5	121	15.6	N/A	1.63	160	126	2.85×10^{-2}	2.5289
Trichloroethene	131.5	−87	86.7	60	4.54	1.46	1,100	415	1.17×10^{-2}	2.4200
Bromodichloromethane	163.8	−57.1	90	N/A	N/A	1.971	N/A	N/A	2.12×10^{-3}	N/A
Chlorodibromomethane	208.29	<−20	120	50	N/A	2.451	N/A	N/A	8.4×10^{-4}	N/A
Dichloromethane	84.93	−97	39.8	349	2.93	1.327	20,000	1702	3.04×10^{-3}	N/A
Tetrachloromethane	153.82	−23	76.7	90	5.3	1.59	800	754	2.86×10^{-2}	2.7300
Tribromomethane	252.77	8.3	149	5.6	8.7	2.89	3,130	7.62	5.84×10^{-4}	N/A
Trichloromethane	119.38	−64	62	160	4.12	1.49	7,840	1027	3.10×10^{-3}	1.8998
1,2-Dichloropropane	112.99	−100.5	96.4	41.2	3.5	1.156	2,600	25.49	2.75×10^{-3}	N/A
2,3-Dichloropropene	110.98	−81.7	94	135	3.8	1.211	insol.	110	N/A	N/A
t-1,3-Dichloropropene	110.97	N/A	112	99.6	N/A	1.224	515	110	N/A	N/A
Toluene	92.1	−95.1	110.8	22	3.14	0.867	515	110	6.44×10^{-3}	2.2095

[a] Data were adapted from Lang et al., 1987.

[b] All values are reported at 20°C.

Note: mw = molecular weight, mp = melting point, bp = boiling point, vd = vapor density, sg = specific gravity, Sol = solubility, C_s = saturation concentration in air, H = Henry's law constant, log K_{ow} = logarithm of the octanol-water partition coefficient.

8. $n = 1.0$
9. Temperature $= 20°C$
10. Oxygen diffusivity $= 2.11 \times 10^{-5}$ cm²/s
11. Benzene diffusivity $= 0.96 \times 10^{-5}$ cm²/s

Solution

1. Determine the quantity of air referenced to the middepth of the aeration tank, which represents the depth for an average bubble size

$$Q_g = (50 \text{ m}^3/\text{min}) \times \frac{10.33}{10.33 + 3} = 38.7 \text{ m}^3/\text{min}$$

2. Determine the air/liquid ratio

$$Q = \frac{(4000 \text{ m}^3/\text{d})}{(1440 \text{ min/d})} = 2.78 \text{ m}^3/\text{min}$$

$$\frac{Q_g}{Q} = \frac{(38.7 \text{ m}^3/\text{min})}{(2.78 \text{ m}^3/\text{min})} = 13.9$$

3. Estimate the mass transfer coefficient for benzene using Eq. (5–65).

$$K_L a_{\text{VOC}} = K_L a_{O_2}\left(\frac{D_{\text{VOC}}}{D_{O_2}}\right)^n = (6.2/\text{h})\left[\frac{(0.96 \times 10^{-5} \text{ cm}^2/\text{s})}{(2.11 \times 10^{-5} \text{ cm}^2/\text{s})}\right]^{1.0}$$

$$K_L a_{\text{VOC}} = 2.82/\text{h} = 0.047/\text{min}$$

4. Determine the dimensionless value of Henry's constant using Eq. (2–51).

$$H_u = \frac{H}{RT}$$

$$H_u = \frac{(0.00549 \text{ m}^3\cdot\text{atm/mole})}{(0.000082057 \text{ atm}\cdot\text{m}^3/\text{mole}\cdot\text{K})[(273.15 + 20)\text{K}]} = 0.228$$

5. Determine the saturation parameter ϕ using Eq. (5–73).

$$\phi = \frac{(K_L a)_{\text{VOC}} V}{H_u Q_g}$$

$$\phi = \frac{(0.047/\text{min} \times 1000 \text{ m}^3)}{(0.228 \times 38.7 \text{ m}^3/\text{min})} = 5.33$$

6. Determine the fraction of benzene removed from the liquid phase using Eq. (5–76).

$$1 - \frac{C_e}{C_i} = 1 - \left[1 + \frac{Q_g}{Q}(H_u)(1 - e^{-\phi})\right]^{-1}$$

$$1 - \frac{C_e}{C_i} = 1 - [1 + 13.9\,(0.228)(1 - e^{-5.33})]^{-1}$$

$$1 - \frac{C_e}{C_i} = 1 - 0.24 = 0.76$$

Comment   The computations presented in this example problem are based on the assumption that the concentration of benzene in the influent is not being reduced by adsorption or biological degradation.

Control Strategies for VOCs

Volatilization and gas stripping are, as noted previously, the principal means by which VOCs are released from wastewater-treatment facilities. In general, it can be shown that the release of VOCs from open surfaces is quite low compared to the release of VOCs at points of liquid turbulence and by gas stripping. Thus, the principal strategies for controlling the release of VOCs, as reported in Table 5–35, are (1) source control, (2) elimination of points of turbulence, and (3) the covering of various treatment facilities. Three serious problems associated with the covering of treatment facilities are (1) treatment of the off-gases containing VOCs, (2) corrosion of mechanical parts, and (3) provision for confined space entry for personnel for equipment maintenance. The treatment of off-gases is considered here.

Treatment of Off-Gases.   The off-gases containing VOCs from covered treatment facilities will have to be treated before they can be discharged to the atmosphere. Options for the off-gas treatment include (see Fig. 5–71): (1) vapor-phase adsorption on granular activated carbon or other VOC selective resins, (2) thermal incineration, (3) catalytic incineration, (4) combustion in a flare, (5) biofiltration, and (6) combustion in a boiler or process heater (U.S. EPA, 1986; WEF, 1997). The application of these processes will depend primarily on the volume of air to be treated and the types and concentrations of the VOCs contained in the airstream. The first four of these off-gas treatment processes are considered in greater detail in the following discussion. Biofiltration is discussed in Chap. 15 under Sec. 15–3, Odor Management. Use of a boiler or heater is done only where a combustion process is included as part of the plant facilities.

Vapor-Phase Adsorption   Adsorption is the process whereby hydrocarbons and other compounds are adsorbed selectively on the surface of such materials as activated carbon, silica gel, or alumina. Of the available adsorbents, activated carbon is used most

Figure 5–71

Options for treating off-gases containing VOCs. *(Adapted from Eckenfelder, 2000.)*

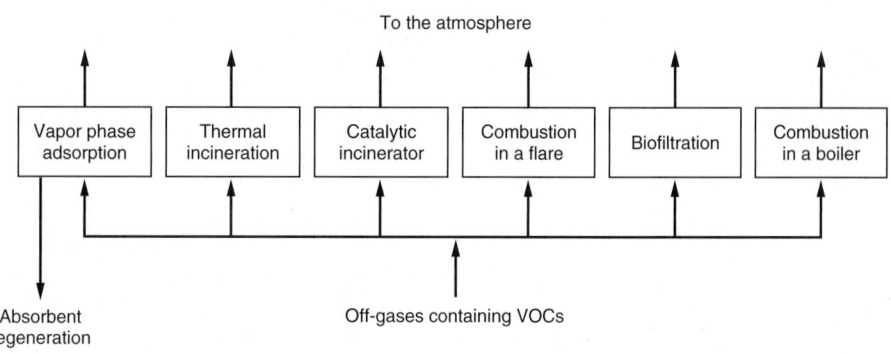

widely. The adsorption capacity of an adsorbent for a given VOC is often represented by adsorption isotherms that relate the amount of VOC adsorbed (adsorbate) to the equilibrium pressure (or concentration) at constant temperature. Typically, the adsorption capacity increases with the molecular weight of the VOC being adsorbed. In addition, unsaturated compounds are generally more completely adsorbed than saturated compounds, and cyclical compounds are more easily adsorbed than linearly structured materials. Also, the adsorption capacity is enhanced by lower operating temperatures and higher concentrations. VOCs characterized by low vapor pressures are more easily adsorbed than those with high vapor pressures (U.S. EPA, 1986).

Carbon adsorption is usually carried out as a batch operation, involving multiple beds (see Fig. 5–72). The two main steps in the adsorption operation include adsorption and regeneration, usually performed in sequence. For control of continuous emission streams, at least one bed remains on-line in the adsorption mode while the other is being regenerated. In a typical batch operation, the off-gas containing VOCs is passed through the carbon bed where the VOCs are adsorbed on the bed surface. As the adsorption capacity of the bed is approached, traces of VOCs appear in the exit stream, indicating that the breakthrough point of the bed has been attained. The off-gas is then directed to a parallel bed containing regenerated adsorbent, and the process continued. Concurrently, the saturated bed is regenerated by the passage of hot air (see Fig. 5–72, Mode A), hot inert gases (see Fig. 5–72, Mode B), low-pressure steam, or a combination of vacuum and hot gas. Because adsorption is a reversible process, the VOCs adsorbed on the bed can be desorbed by supplying heat (equivalent to the amount of heat released during adsorption). Small residual amounts of VOCs are always left on the carbon bed, because complete desorption is technically difficult to achieve and economically

Figure 5–72

Gas phase carbon adsorption and regeneration system for the treatment of VOCs in off-gas.

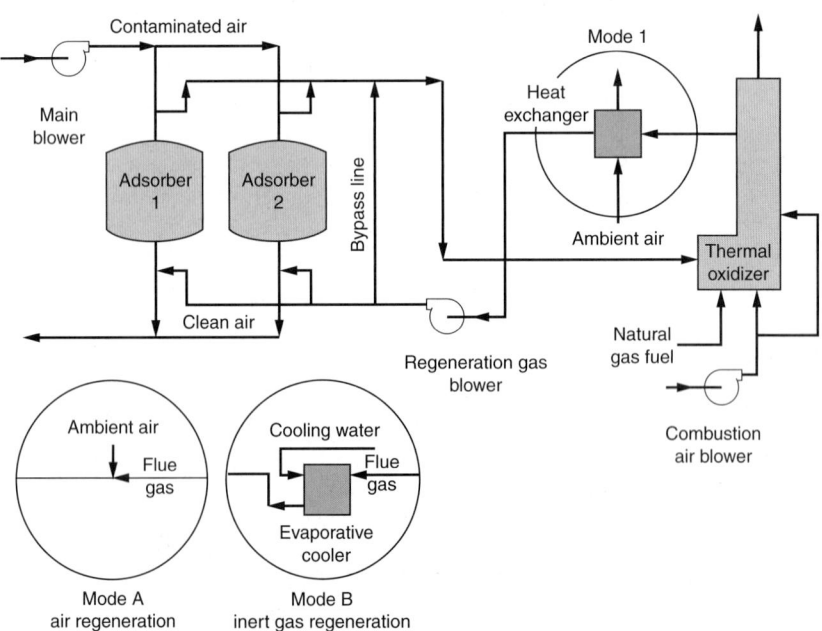

impractical. Regeneration with hot air and a hot inert gas is considered in the following discussion.

Hot air regeneration is used when the VOCs are either nonflammable or have a high ignition temperature and thus do not pose a risk of carbon fires. A portion of the hot flue gas in the oxidizer is mixed with ambient air to cool the gas to below 180°C (350°F). The regeneration gas is driven upflow (or countercurrent to adsorption flow) through the GAC adsorber. As the temperature of the carbon bed rises, the desorbed organics are transferred to the regeneration gas stream. The regeneration gas containing the desorbed VOCs is sent directly to the thermal oxidizer where the VOCs are destroyed. After the bed has been maintained at the desired regeneration temperature for a sufficient period of time, regeneration is ended. The bed is then cooled to approximately ambient temperature by shutting off the hot regeneration gas and continuing to pass ambient air through the carbon bed. The regeneration and cooling times are predetermined based on the amount of carbon in the adsorber and the expected loading on the carbon (U.S. EPA, 1986).

Where the VOCs contained in the off-gas include compounds such as ketones and aldehydes that may pose fire risks at elevated temperatures in the presence of oxygen, inert gas regeneration is used. A relatively inert gas can be obtained by passing a portion of the hot flue gas from the thermal oxidizer through an evaporative cooler. It is possible, therefore, to keep the oxygen concentration in the regeneration gas at a low of 2 to 5 percent by volume. The desorbed VOCs are transferred along with the regeneration gas to the thermal oxidizer. A controlled amount of secondary air is added to the oxidizer. The addition of air ensures complete combustion of the VOCs but limits the excess oxygen level in the oxidizer to an acceptable range (e.g., 2 to 5 percent by volume). Regeneration is complete when the carbon bed has reached the necessary temperature for a given period of time, and VOCs are no longer being desorbed from the bed. Cooling of the bed is accomplished by increasing the water flowrate to the evaporative cooler and reducing the regeneration gas temperature to between 105 and 120°C (220 and 250°F).

Thermal Incineration Thermal incineration (see Fig. 5–73) is used to oxidize VOCs at high temperatures. The most important variables to consider in thermal incinerator design are the combustion temperature and residence time because these design variables determine the VOC destruction efficiency of the incinerator. Further, at a given

Figure 5–73

Schematic diagram of a thermal incinerator system for VOCs in off-gas released from treatment facilities.

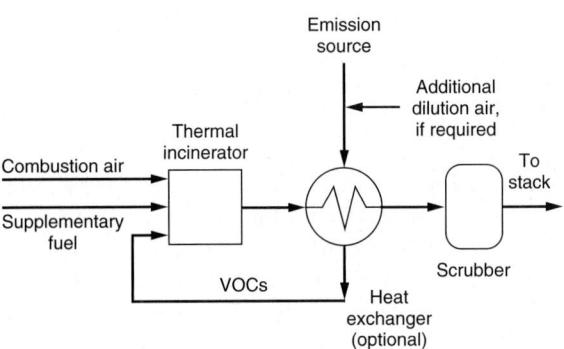

Figure 5–74

Schematic diagram of a catalytic incinerator system for VOCs in off-gas released from treatment facilities.

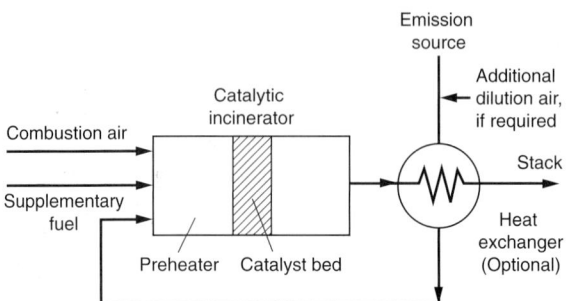

combustion temperature and residence time, destruction efficiency is also affected by the degree of turbulence, or mixing of the emission stream and hot combustion gases, in the incinerator. In addition, halogenated organics are more difficult to oxidize than unsubstituted organics; hence, the presence of halogenated compounds in the emission stream requires higher temperature and longer residence times for complete oxidation. When emission streams treated by thermal incineration are dilute (i.e., low heat content), supplementary fuel is required to maintain the desired combustion temperatures. Supplementary fuel requirements may be reduced by recovering the energy contained in the hot flue gases from the incinerator.

Catalytic Incineration In catalytic incineration (see Fig. 5–74), VOCs in an emission stream are oxidized with the help of a catalyst. A catalyst is a substance that accelerates the rate of a reaction at a given temperature without being appreciably changed during the reaction. Catalysts typically used for VOC incineration include platinum and palladium; other formulations are also used, including metal oxides for emission streams containing chlorinated compounds. The catalyst bed (or matrix) in the incinerator is generally a metal mesh-mat, ceramic honeycomb, or other ceramic matrix structure designed to maximize catalyst surface area. The catalysts may also be in the form of spheres or pellets. Before passing through the catalyst bed, the emission stream is preheated, if necessary, in a natural gas-fired preheater (U.S. EPA, 1986).

The performance of a catalytic incinerator is affected by several factors including (1) operating temperature, (2) space velocity (reciprocal of residence time), (3) VOC composition and concentration, (4) catalyst properties, and (5) presence of catalyst poisons or inhibitors in the emission stream. In catalytic incinerator design, the important variables are the operating temperature at the catalyst bed inlet and the space velocity. The operating temperature for a particular destruction efficiency is dependent on the concentration and composition of the VOC in the emission stream and the type of catalyst used (U.S. EPA, 1986).

Combustion in a Flare Flares, commonly used for disposal of waste digester gas, can be used to destroy most VOCs found in off-gas streams. Flares can be designed and operated to handle fluctuations in emission VOC content, inerts content, and flowrate. Several different types of flares are available including steam-assisted, air-assisted, and pressure-head flares. Steam-assisted flares are employed in cases where large volumes of waste gases are released. Air-assisted flares are generally used for moderate off-gas gas flows. Pressure-head flares are used for small gas flows.

Disposal of Off-Gases.   Another method for the ultimate disposal of treated off-gases is to discharge them through high stacks. Stacks as high as 30 to 40 m (100 to 130 ft) have been constructed for dispersal of off-gases into the atmosphere. The dispersion lowers the concentration of any residuals left in the gas stream by dilution with the ambient air. With the passage of the Clean Air Act in 1990, discharge of untreated off-gases has been phased out.

PROBLEMS AND DISCUSSION TOPICS

5-1   A bar screen is inclined at a 60° angle from the horizontal. The bars have a diameter of 20 mm and a clear spacing of 25 mm. Determine the headloss when the bars are cleaned and the velocity approaching the screen is 1 m/s. Is this a very realistic computation in terms of what actually happens at a treatment plant?

5-2   Design an aerated grit chamber for a plant with an average flowrate of 15,000 m^3/d and a peak hourly flowrate of 40,000 m^3/d. Determine the amount of air required and the pressure at the discharge of the blowers. Allow a 250-mm loss in the diffusers, and add the submergence plus 30 percent for loss in piping and valves. Determine the power required using an appropriate blower formula. Use a blower efficiency of 70 percent. Determine the monthly power bill, assuming a motor efficiency of 90 percent and a power cost of $0.08/kWh.

5-3   Design an aerated grit chamber installation for an average wastewater flowrate of 0.3 m^3/s and a peak flowrate of 1.0 m^3/s. The average depth is 3 m, the width-to-depth ratio is 1.5:1, and the detention time at peak flow is 3.5 min. The aeration rate is 0.4 m^3/min per m of tank length. Determine the dimensions of the grit chambers and the total air required.

5-4   Discuss the advantages and disadvantages of aerated grit chambers versus vortex-type grit chambers. Cite two references.

5-5   Visit your local treatment plant and review the grit and screenings operations. What methods do they use and what problems do they have? How might their operations be improved as compared to alternative methods described in this chapter?

5-6   Using the information given in the data table presented in Example 5–2, determine (*a*) the off-line storage volume needed to equalize the flowrate and (*b*) the effect of flow equalization on the BOD mass loading rate. How does the BOD mass loading-rate curve determined in this problem compare with the curve shown in step 2d? In your estimation, does the difference in the mass loading rate justify the cost of the larger basin required for in-line storage?

5-7   Using the information given in the data table presented in Example 5–2, estimate the in-line volume required to reduce the variation in the BOD mass loading rate between the maximum and minimum from the existing ratio of 25.8:1 to a peak value of 5:1.

5-8   The contents of a tank are to be mixed with a turbine impeller that has six flat blades. The diameter of the impeller is 3 m, and the impeller is installed 1.25 m above the bottom of a 6 m deep tank. If the temperature is 30°C and the impeller is rotated at 30 r/min, what will be the power consumption? Find the Reynolds number using Eq. (5–7).

5-9 It is desired to flash-mix some chemicals with incoming wastewater that is to be treated. Mixing is to be accomplished using a flat-paddle mixer 500 mm in diameter having six blades. If the temperature of the coming water is 10°C and the mixing chamber power number is 1.70, determine:
(a) The speed of rotation when the Reynolds number is approximately 100,000
(b) Why it is desirable to have a high Reynolds number in most mixing operations
(c) The required mixer motor size, assuming an efficiency factor of 20 percent.

5-10 Assuming that a given flocculation process can be defined by a first-order reaction ($r_N = -kN$), complete the following table assuming the process is occurring in a plug-flow reactor with a detention time of 10 min. What would the value be after 5 min if a batch reactor were used instead, assuming the rate constant is the same?

Time, t	0	5	10
Particles, No./unit volume	10	(?)	3

5-11 If the steady-state effluent from a continuous-flow stirred-tank reactor used as a flocculator contained 3 particles/unit volume, determine the concentration of particles in the effluent 5 min after the process started before steady-state conditions are reached. Assume that the influent contains 10 particles/unit volume, the detention time in the CFSTR is equal to 10 min, and that the first-order kinetics apply ($r_N = -kN$).

5-12 An air flocculation system is to be designed. If a G value of 60 s^{-1} is to be used, estimate the air flowrate that will be necessary for a 200 m^3 flocculation chamber. Assume the depth of the flocculation basin is to be 4 m and that the air is released 0.5 m above the tank bottom.

5-13 Determine the required air flowrate to accomplish the flocculation operation in Example 5–4 pneumatically. Assume the air will be released at a depth of 3 m.

5-14 Derive Stokes' law by equating Eq. (5–23) to the particle mass.

5-15 Determine the settling velocity in m/s of a sand particle with a specific gravity of 2.65 and a diameter of 1 mm. Assume that the Reynolds number is 275.

5-16 Prepare a spreadsheet for computing the settling velocity of a spherical particle for the data given in Example 5–5. What are the values of Reynolds number and coefficient of drag necessary to have the assumed velocities agree with the calculated velocities?

5-17 Determine the removal efficiency for a sedimentation basin with a critical velocity V_o of 2 m/h in treating a wastewater containing particles whose settling velocities are distributed as given in the table below. Plot the particle histogram for the influent and effluent wastewater.

Velocity, m/h	Number of particles
0.0–0.5	20
0.5–1.0	40
1.0–1.5	80
1.5–2.0	120
2.0–2.5	100
2.5–3.0	70
3.0–3.5	20
3.5–4.0	10

5-18 The rate of flow through an ideal clarifier is 8000 m^3/d, the detention time is 1 h, the depth is 3 m. If a full-length movable horizontal tray is set 1 m below the surface of the water, determine the percent removal of particles having a settling velocity of 1 m/h. Could the removal efficiency of the clarifier be improved by moving the tray? If so, where should the tray be located and what would be the maximum removal efficiency? What effect would moving the tray have if the particle settling velocity were equal to 0.3 m/h?

5-19 In Example 5–5, the assumed and calculated particle settling velocities did not reach closure. Continue the calculation iterations to find the actual settling velocity.

5-20 For a flocculent suspension, determine the removal efficiency for a basin 3 m deep with an overflow rate V_o equal to 3 m/h using the laboratory settling data presented in the following table.

Time, min	Percent suspended solids removed at indicated depth (m)				
	0.5	**1.0**	**1.5**	**2**	**2.5**
20	61				
30	71	63	55		
40	81	72	63	61	57
50	90	81	73	67	63
60	....	90	80	74	68
70	....	....	86	80	75
80	....	....	....	86	81

5-21 The curve following was obtained from a settling test in a 3-m cylinder. The initial solids concentration was 3600 mg/L. Determine the thickener area required for a concentration C_u of 12,000 mg/L with a flow of 1500 m^3/d.

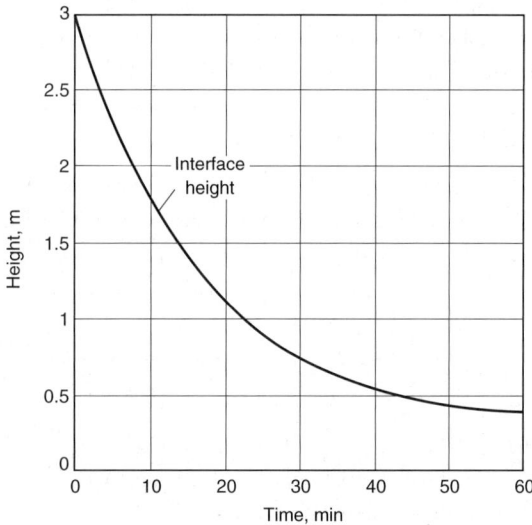

5-22 Design a circular radial-flow sedimentation tank for a town with a projected population of 45,000. Assume that the wastewater flow is 400 L/capita·d. Design for 2 h detention at the average flow.

Determine the tank depth and diameter to produce an overflow rate of 36 m³/m²·d for average flow. Assume standard tank dimensions to fit mechanisms made in diameters with increments of 1.0 m and in depth increments of 0.5 m.

5-23 A rectangular settling tank has an overflow rate of 30 m³/m²·d and dimensions of 2.75 m deep by 6 m wide by 15 m long. Determine whether or not particles with a diameter of 0.1 mm and a specific gravity of 2.5 will be scoured from the bottom. Use $f = 0.03$ and $k = 0.04$.

5-24 Determine the percentage increase in the hydraulic and solids-loading rates of the primary settling facilities of a treatment plant when 200 m³/d of settled waste activated sludge containing 2000 mg/L of suspended solids is discharged to the existing primary facilities for thickening. The average plant flowrate is 20,000 m³/d and the influent suspended-solids concentration is about 350 mg/L. The design overflow rate for the primary settling tanks without the waste solids is 32 m³/m²·d and the detention time is 2.8 h. Do you believe that the added incremental loadings will affect the performance of the primary settling facilities? Document the basis for your answer.

5-25 Prepare a table and compare the data from a minimum of five references with regard to the following primary sedimentation tank design parameters: (1) detention time (with and without preaeration); (2) expected BOD removal; (3) expected suspended-solids removal; (4) mean horizontal velocity; (5) surface loading rate in m³/m²·d; (6) effluent weir overflow rate per unit length; (7) size of organic particle removed; (8) length-to-width ratio (rectangular tanks); (9) average depth. List all references.

5-26 A medium-size treatment plant is being designed, and circular and rectangular primary sedimentation tanks are being considered. What factors should be considered in the evaluation and selection of the type of tank? List the advantages and disadvantages for each type. Cite at least three recent references (since 1980).

5-27 Contrast dissolved-air flotation with sedimentation discussing the following parameters:
(a) Detention time
(b) Surface-loading rate
(c) Power input
(d) Efficiency
(e) Most favorable application for each type

5-28 The following data were obtained from a test program designed to evaluate a new diffused-air aeration system. Using these data, determine the value of $K_L a$ at 20°C and the equilibrium dissolved-oxygen concentration in the test tank. The test program was conducted using tap water at a temperature of 24°C.

C, mg/L	1.5	2.7	3.9	4.8	6.0	7.0	8.2
dC/dt, mg/L·h	8.4	7.5	5.3	4.9	4.2	2.8	2.0

5-29 If the volume of the test tank used to evaluate the aeration system in Prob. 5–28 were equal to 100 m³ and the air flowrate were equal to 2 m³/min, determine the maximum oxygen-transfer efficiency at 20°C and 1.0 atmosphere.

5-30 Using the equation developed in Example 5–15, estimate the air flowrate in m³/min required to increase the oxygen content of chlorinated effluent from zero to 4 mg/L. The effluent flowrate is equal to 20,000 m³/d. Assume that the transfer efficiency is 6 percent and the temperature is 15°C. What is the air requirement when the temperature is 25°C?

5–31 Alternative diffused-air aeration devices are being considered for installation with a submergence of 4.5 m in an aeration tank at an activated-sludge treatment plant. The oxygen required for biological treatment is 7000 kg/d. Determine the standard oxygen transfer rate and theoretical air requirements for ceramic dome diffusers installed in a grid pattern as compared to nonporous diffusers installed for a dual-spiral roll. The wastewater temperature is 20°C and the α factors are 0.64 for ceramic domes and 0.75 for the nonporous diffusers, respectively.

5–32 Using the data in Prob. 5–31 and a wintertime wastewater temperature of 10°C, determine the theoretical air requirements for cold-weather operation. How will equipment selection be affected in meeting both summer and winter operation?

REFERENCES

Amirtharajah, A., M. C. Clark, and R. R. Trussell (1991) *Mixing in Coagulation and Flocculation,* American Water Works Association Research Foundation, Denver, CO.

ASCE (1989) "Sulfide in Wastewater Collection and Treatment Systems," ASCE Manuals and Reports on Engineering Practice No. 69, American Society of Civil Engineers, Reston, VA.

ASCE (1992) *ASCE Standard—Measurement of Oxygen Transfer in Clean Water,* ANSI/ASCE 2-91, 2d ed., Reston, VA.

AWWA (1999) *Water Quality and Treatment,* 5th ed., American Water Works Association, Denver, CO.

Barrett, M. J. (1960) "Aeration Studies of Four Weir Systems," *Water and Wastes Engineering,* vol. 64, no. 9.

Bielefeldt, A. R., and H. D. Stensel (1999) "Treating VOC-Contaminated Gases in Activated Sludge: Mechanistic Model to Evaluate Design and Performance," *Environmental Science and Technology,* vol. 33, pp. 3234–3240.

Braley, B.G. (1990) "Wastewater Treatment Reaches New Levels," *Journal Water Pollution Control Federation,* vol. 2, no. 2, pp. 11–13.

Camp, T. R. (1946) "Sedimentation and the Design of Settling Tanks," *Trans. ASCE,* vol. 111.

Camp, T. R., and P. C. Stein (1943) "Velocity Gradients and Internal Work in Fluid Motion," *Journal Boston Society of Civil Engineers,* vol. 30, p. 209.

Carlson, R. F. (2000) "Static Mixers for Chlorine Flash Mixing: Square and Rectangular Channel Mixers," Presented at the 2000 WEF Specialty Conference Disinfection 2000: Disinfection of Waste in the New Millennium, New Orleans, LA, sponsored by the Water Environment Federation.

Chao, J. I., and B. G. Stone (1979) "Initial Mixing by Jet Injection Blending," *Journal Environment Engineering Division,* American Society of Civil Engineers, vol. 106, no. 10, pp. 570–573.

Chemineer, Inc. (2000) Notes on Mixing Technology, Chemineer, Inc., a unit of Robbins & Myers, North Andover, MA.

Coe, H. S., and G. H. Clevenger (1916) "Determining Thickener Unit Areas," *Transactions American Institute of Mechanical Engineers,* vol. 55, no. 3.

Crites, R., and G. Tchobanoglous (1998) *Small and Decentralized Wastewater Management Systems,* McGraw-Hill, New York.

Danckwertz, P. V. (1951) "Significance of Liquid Film Coefficients in Gas Absorption," *Journal Industrial Engineering Chemistry,* vol. 43, p. 1460.

Davies, J. T. (1972) *Turbulence Phenomena,* Academic Press, New York.

Dick, R. I., and B. B. Ewing (1967) "Evaluation of Activated Sludge Thickening Theories," *Journal Sanitary Engineering Division, American Society of Civil Engineers,* vol. 93, no. SA-4.

Dick, R. I., and K. W. Young (1972) "Analysis of Thickening Performance of Final Settling Tanks," *Proceedings of the 27th Industrial Waste Conference,* Purdue University, Engineering Extension Series 141, Purdue, IN.

Eckenfelder, W. W., Jr. (2000) *Industrial Water Pollution Control,* McGraw-Hill, New York.

EPRI (1999) "High-Rate Clarification for the Treatment of Wet Weather Flows," TechCommentary, Electric Power Research Institute, St. Louis, MO.

Fair, G. M., and J. C. Geyer (1954) *Water Supply and Waste-Water Disposal,* Wiley, New York.

Greeley, S. A. (1938) "Sedimentation and Digestion in the United States," in L. Pearse (ed.) *Modern Sewer Disposal: Anniversary Book of the Federation of Sewage Works Associations,* Lancaster Press, Inc., New York.

Gregory, R., T. F. Zabel, and J. K. Edzwald (1999) *Water Quality and Treatment,* 5th ed., Chap. 7, American Water Works Association, Denver, CO.

Han, M., and D. F. Lawler (1992) "The (Relative) Insignificance of G in Flocculation," *Journal American Water Works Association,* vol. 84, no. 10, p. 79.

Higbie, R. (1935) "The Rate of Absorption of Pure Gas into a Still Liquid during Short Periods of Exposure," *Transaction American Institute of Chemical Engineers,* vol. 31, p. 365.

Holland, F. A., and Chapman, F. S. (1966) *Liquid Mixing and Processing in Stirred Tanks,* Reinhold Publishing Corp., London.

Hsieh, C., K. S. Ro, and M. K. Stenstrom (1993) "Estimating Emissions of 20 VOCs II, Diffused Aeration," *Journal Environmental Engineering,* vol. 119, no. 6, pp 1099–1118.

Hwang, H. J., and M. K. Stenstrom (1985) "Evaluation of Fine-Bubble Alpha Factors in Near Full-Scale Equipment," *Journal Water Pollution Control Federation,* vol. 57, no. 12.

Keinath, T. M. (1985) "Operational Dynamics and Control of Secondary Clarifiers," *Journal Water Pollution Control Federation,* vol. 57, no. 7, p. 770.

Kolmogoroff, A. N. (1941) *Compt. Rend. Acad. Sci., U.R.F.F.* 30, 301, vol. 32, p. 16, USSR.

Lang, R. J., T. A. Herrera, D. P. Y. Chang, G. Tchobanoglous, and R. G. Spicher (1987) "Trace Organic Constituents in Landfill Gas," Prepared for the California Waste Management Board, Department of Civil Engineering, University of California—Davis, Davis, CA.

Laughlin, J. E., and W. C. Roming (1993) "Design of Rotary Fine Screen Facilities in Wastewater Treatment," *Public Works,* April.

Lewis, W. K., and W. C. Whitman (1924) "Principles of Gas Adsorption," *Industrial Engineering Chemistry,* vol. 16, pp. 1215, 1924.

Libra, J. A. (1993) "Stripping of Organic Compounds in an Aerated Stirred Tank Reactor," *Fortschc.-ber. VD1* Rhhe 15, Nr. 102, VDI-Verlag, Düsseldorf.

Logan, B. E. (2000) *Environmental Transport Processes,* Wiley, New York.

Masschelein, W. J. (1992) *Unit Processes in Drinking Water,* Marcel Dekker, New York.

Matter-Muller, C., W. Gujer, and W. Giger (1981) "Transfer of Volatile Substances from Water to the Atmosphere," *Water Research,* 15, pp. 1271–1279.

McGhee, T. J. (1991) *Water Supply and Sewerage,* 6th ed., McGraw-Hill, New York.

Metcalf & Eddy, Inc. (1981) *Wastewater Engineering: Collection and Pumping of Wastewater,* McGraw-Hill, New York.

Morrill, A. B. (1932) "Sedimentation Basin Research and Design," *Journal American Water Works Association,* vol. 24, p. 1442.

Pankow, J. F. (1991) *Aquatic Chemistry Concepts,* Lewis Publishers, Chelsea, MI.

Pisano, W. C., N. Thibault, and G. Forber (1990) "The Vortex Solids Separator," *Water Environment & Technology,* vol. 2, no. 5, pp. 64–71.

Pratte, R. D., and W. D. Baines (1967) "Profiles of the Round Turbulent Jet in a Cross Flow," *Journal Hydraulic Division,* American Society of Civil Engineers, HY6, vol. 93, no. 11, pp. 53–64.

Qasim, S. (1999) *Wastewater Treatment Plants, Planning, Design, and Operation,* 2d ed., Technomic Publishing Co., Lancaster, PA.

Randall, C. W., J. L. Barnard, and H. D. Stensel (1992) *Design and Retrofit of Wastewater Treatment Plants for Biological Nutrient Removal,* Technomic Publishing Co., Lancaster, PA.

Roberts, P. V., and P. G. Dandliker (1983) "Mass Transfer of Volatile Organic Contaminants from Aqueous Solution to the Atmosphere During Surface Aeration," *Environmental Science and Technology,* 17, pp. 484–489.

Roberts, P. V., C. Munz, P. G. Dandliker, and C. Matter-Muller (1984) "Volatilization of Organic Pollutants in Wastewater Treatment-Model Studies," EPA-600/S2-84-047.

Rushton, J. H. (1952) "Mixing of Liquids in Chemical Processing," *Industrial Engineering Chemistry,* vol. 44, no. 12.

Sawey, R. (1998) "Physical-Chemical Processes Make Treatment of Peak Flows Affordable," *Water Environment & Technology,* vol. 10, no. 9.

Schwarzenbach, R. P., P. M. Gschwend, and D. M. Imboden (1993) *Environmental Organic Chemistry,* Wiley, New York.

Shields, A. (1936) *Application of Similitude Mechanics and Turbulence Research to Bed-Load Movement,* Mitt. der Pruess, Versuchsanstalt für Wasserbau und Schiffbau, No. 26, Berlin.

Speece, R. E., and G. Tchobanoglous (1990) "Commercial Oxygen Utilization in Water Quality Management," *Water Environment & Technology,* vol. 2, no. 7.

Talmadge, W. P., and E. B. Fitch (1955) "Determining Thickener Unit Areas," *Industrial Engineering Chemistry.,* vol. 47, no. 1.

Tchobanoglous, G. (1988) "Filtration of Treated Wastewater Effluents," presented at the 61st Annual Conference of the Water Pollution Control Federation, Dallas, TX.

Tchobanoglous, G., and E. D. Schroeder (1985) *Water Quality: Characteristics, Modeling, Modification,* Addison-Wesley Publishing Company, Reading, MA.

Thibodeaux, L. J. (1979) *Chemodynamics: Environmental Movement of Chemicals in Air, Water, and Soil,* Wiley, New York.

U.S. EPA (1974) *Oxygen Activated Sludge in Wastewater Treatment Systems: Design Criteria and Operating Experience,* U.S. Environmental Protection Agency, re. ed., Technology Transfer Publication.

U.S. EPA (1986) *Handbook: Control Technologies for Hazardous Air Pollutants,* U.S. Environmental Protection Agency: EPA/625/6-86/014.

U.S. EPA (1987) *Design and Operational Considerations—Preliminary Treatment,* U.S. Environmental Protection Agency: EPA 430/09-87-007.

U.S. EPA (1989) *Design Manual Fine Pore Aeration Systems,* EPA/625/1-89/023, U.S. Environmental Protection Agency, Cincinnati, OH.

U. S. EPA (1993) *Manual, Combined Sewer Overflow Control,* EPA-625/R-93/007 U.S. Environmental Protection Agency, Cincinnati, OH.

Wahlberg, E. J., and T. M. Keinath (1988) "Development of Settling Flux Curves Using SVI," *Journal Water Pollution Control Federation,* vol. 60, p. 2095.

WEF (1994) *Preliminary Treatment for Wastewater Facilities,* WEF Manual of Practice OM-2, Water Environment Federation, Alexandria, VA.

WEF (1997) *Biofiltration: Controlling Air Emissions through Innovative Technology,* Project 92-VOC-1, Water Environment Federation, Alexandria, VA.

WEF (1998*a*) *Design of Municipal Wastewater Treatment Plants,* vol. 1: Chaps. 1–8, WEF Manual of Practice No. 8, ASCE Manual and Report on Engineering Practice No. 76, Water Environment Federation, Alexandria, VA.

WEF (1998*b*) *Design of Municipal Wastewater Treatment Plants,* vol. 2: Chaps. 9–16, WEF Manual of Practice No. 8, ASCE Manual and Report on Engineering Practice No. 76, Water Environment Federation, Alexandria, VA.

WEF (1998*c*) *Design of Wastewater Treatment Plants,* 4th ed., vol. 3, Chaps. 17–24, WEF Manual of Practice No. 8, ASCE Manual and Report on Engineering Practice No. 76, Water Environment Federation, Alexandria, VA.

Williams, R. (2000) "Louisville Tests New Treatment to Meet EPA CSO Guidelines," *Public Works,* February.

Wong, T. H. F. (1997) "Continuous Deflection Separation: Its Mechanism and Applications," *Proceedings of WEFTEC '97,* Chicago, IL, vol. 2, pp. 703–714.

WPCF (1985) *Clarifier Design,* WPCF Manual of Practice FD-10, Water Pollution Control Federation, Alexandria, VA.

WPCF (1988) *Aeration,* Manual of Practice FD-13, Water Pollution Control Federation, Alexandria, VA.

Yoshika, N., et al. (1957) "Continuous Thickening of Homogeneous Flocculated Slurries," *Kagaku Kogaku,* vol. 26 (also in *Journal of Chemical Engineering of Japan,* vol. 21, Tokyo, 1957).

6

Chemical Unit Processes

Those processes used for the treatment of wastewater in which change is brought about by means of or through chemical reactions are known as *chemical unit processes.* In the field of wastewater treatment, chemical unit processes usually are used in conjunction with the physical unit operations discussed in Chap. 5, and the biological unit processes to be discussed in Chaps. 7 through 10, to meet specific treatment objectives.

The purpose in this chapter is to present and discuss: (1) the role of chemical unit processes in wastewater treatment; (2) some fundamentals of chemical coagulation; (3) the precipitation reactions that occur when various chemicals are added to improve the performance of wastewater-treatment facilities; (4) the chemical reactions involved in the precipitation of phosphorus from wastewater; (5) the precipitation of heavy metals and dissolved inorganic substances; (6) chemical oxidation; (7) chemical neutralization, scale control, and stabilization; and (8) chemical storage, feeding, piping, and control systems. Advanced chemical oxidation and ion exchange are considered in Chap. 11. Adsorption, sometimes classified as a chemical unit process, but which, more correctly, should be classified as a physical unit operation, is considered in Chap. 11. Chemical disinfection, an extremely important and much-used chemical unit process, is considered separately in Chap. 12.

6-1   ROLE OF CHEMICAL UNIT PROCESSES IN WASTEWATER TREATMENT

The principal chemical unit processes used for wastewater treatment include (1) chemical coagulation, (2) chemical precipitation, (3) chemical disinfection, (4) chemical oxidation, (5) advanced oxidation processes, (6) ion exchange, and (7) chemical neutralization, scale control, and stabilization. The applications and the limitations involved in the use of these processes are considered in the following discussion.

Application of Chemical Unit Processes

Applications of chemical unit processes for the management and treatment of wastewater are reported in Table 6–1. Chemical processes, in conjunction with various physical operations, have been developed for the complete secondary treatment of untreated (raw) wastewater, including the removal of either nitrogen or phosphorus or both. Chemical processes have also been developed to remove phosphorus by chemical precipitation, and are designed to be used in conjunction with biological treatment. Other

Table 6–1

Applications of chemical unit processes in wastewater treatment

Process	Application	See section or chapter
Advanced oxidation processes	Removal of refractory organic compounds	Chap. 11
Chemical coagulation	The chemical destabilization of particles in wastewater to bring about their aggregation during perikinetic and orthokinetic flocculation	6–2
Chemical disinfection	Disinfection with chlorine, chlorine compounds, bromine, and ozone	Chap. 12
	Control of slime growths in sewers	Chap. 12
	Control of odors	Chap. 15
Chemical neutralization	Control of pH	6–7
Chemical oxidation	Removal of BOD, grease, etc.	6–6
	Removal of ammonia (NH_4^+)	6–6
	Destruction of microorganisms	Chap. 12
	Control of odors in sewers, pump stations, and treatment plants	Chap. 15
	Removal of resistant organic compounds	6–6, Chap. 11
Chemical precipitation	Enhancement removal of total suspended solids and BOD in primary sedimentation facilities	6–3
	Removal of phosphorus	6–4
	Removal of heavy metals	6–5
	Physical-chemical treatment	6–3
	Corrosion control in sewers due to H_2S	
Chemical scale control	Control of scaling due to calcium carbonate and related compounds	6–7
Chemical stabilization	Stabilization of treated effluents	6–7
Ion exchange	Removal of ammonia (NH_4^+), heavy metals, total dissolved solids	Chap. 11
	Removal of organic compounds	Chap. 11

Figure 6-1

Typical lime clarification facilities following secondary treatment used as pretreatment step for advanced treatment of wastewater using reverse osmosis. Lime storage is in the silo shown behind the building that is used to house the lime slaking facilities and the reverse osmosis units, used for advanced treatment. Granular medium-depth filters are shown to the right of the lime clarifier in the foreground.

chemical processes have been developed for the removal of heavy metals and for specific organic compounds and for the advanced treatment of wastewater. Currently the most important applications of chemical unit processes in wastewater treatment are for (1) the disinfection of wastewater, (2) the precipitation of phosphorus, and (3) the coagulation of particulate matter found in wastewater at various stages in the treatment process (see Fig. 6–1).

Considerations in the Use of Chemical Unit Processes

In considering the application of the chemical unit processes to be discussed in this chapter, it is important to remember that one of the inherent disadvantages associated with most chemical unit processes, as compared with the physical unit operations, is that they are additive processes (i.e., something is added to the wastewater to achieve the removal of something else). As a result, there is usually a net increase in the dissolved constituents in the wastewater. For example, where chemicals are added to enhance the removal efficiency of particulate sedimentation, the total dissolved solids (TDS) concentration of the wastewater is always increased. Similarly, when chlorine is added to wastewater, the TDS of the effluent is increased. If the treated wastewater is to be reused, the increase in dissolved constituents can be a significant factor. This additive aspect is in contrast to the physical unit operations and the biological unit processes, which may be described as being subtractive, in that wastewater constituents are removed from the wastewater. A significant disadvantage of chemical precipitation processes is the handling, treatment, and disposal of the large volumes of sludge that is produced. Another disadvantage of chemical unit processes is that the cost of most chemicals is related to the cost of energy. As a result, the end user has little control over chemical costs.

6-2 FUNDAMENTALS OF CHEMICAL COAGULATION

Colloidal particles found in wastewater typically have a net negative surface charge. The size of colloids (about 0.01 to 1 μm) is such that the attractive body forces between particles are considerably less than the repelling forces of the electrical charge. Under these stable conditions, Brownian motion keeps the particles in suspension. Brownian motion (i.e., random movement) is brought about by the constant thermal bombardment of the colloidal particles by the relatively small water molecules that surround them. Coagulation is the process of destabilizing colloidal particles so that particle growth can occur as a result of particle collisions. The theory of chemical coagulation reactions is very complex. The simplified reactions used in this and other textbooks to describe coagulation and chemical precipitation processes can only be considered approximations as the reactions may not necessarily proceed as indicated.

Coagulation reactions are often incomplete, and numerous side reactions with other substances in wastewater may take place depending on the characteristics of the wastewater which will vary throughout the day as well as seasonally. To introduce the subject of chemical coagulation the following topics are discussed in this section: (1) basic definitions for coagulation and flocculation, (2) the nature of particles in wastewater, (3) the development and measurement of surface charge, (4) consideration of particle-particle interaction, (5) particle destabilization with potential determining ions and electrolytes, (6) particle destabilization and aggregation with polyelectrolytes, and (7) particle destabilization and removal with hydrolyzed metal ions. The following discussion is meant to serve as an introduction to the nature of the phenomena and processes involved in the coagulation process.

Basic Definitions

The term "chemical coagulation" as used in this text includes all of the reactions and mechanisms involved in the chemical destabilization of particles and in the formation of larger particles through perikinetic flocculation (aggregation of particles in the size range from 0.01 to 1 μm). *Coagulant* and *flocculent* are terms that will also be encountered in the literature on coagulation. In general, a *coagulant* is the chemical that is added to destabilize the colloidal particles in wastewater so that floc formation can result. A *flocculent* is a chemical, typically organic, added to enhance the flocculation process. Typical coagulants and flocculants include natural and synthetic organic polymers, metal salts such as alum or ferric sulfate, and prehydrolized metal salts such as polyaluminum chloride (PACl) and polyiron chloride (PICl). Flocculants, especially organic polymers, are also used to enhance the performance of granular medium filters and in the dewatering of digested biosolids. In these applications, the flocculant chemicals are often identified as filter aids.

The term "flocculation" is used to describe the process whereby the size of particles increases as a result of particle collisions. As noted in Chap. 5, there are two types of flocculation: (1) *microflocculation* (also known as *perikinetic flocculation*), in which particle aggregation is brought about by the random thermal motion of fluid molecules known as Brownian motion or movement and (2) *macroflocculation* (also known as *orthokinetic flocculation*), in which particle aggregation is brought about by inducing velocity gradients and mixing in the fluid containing the particles to be flocculated. Another form of macroflocculation is brought about by differential settling in which large particles overtake small particles to form larger particles. The purpose of flocculation is to produce particles, by means of aggregation, that can be removed by inexpensive particle-separation procedures such as gravity sedimentation and filtration. Again as noted in Chap. 5, macroflocculation is ineffectual until the colloidal particles reach a size of 1 to 10 μm through contacts produced by Brownian motion and gentle mixing.

Nature of Particles in Wastewater

The particles in wastewater may, for practical purposes, be classified as suspended and colloidal. Suspended particles are generally larger than 1.0 μm and can be removed by gravity sedimentation. In practice, the distinction between colloidal and suspended particles is blurred because the particles removed by gravity settling will depend on the design of the sedimentation facilities. Because colloidal particles cannot be removed by sedimentation in a reasonable period of time, chemical methods (i.e., the use of chemical coagulants and flocculant aids) must be used to help bring about the removal of these particles.

To understand the role that chemical coagulants and flocculant aids play in bringing about the removal of colloidal particles, it is important to understand the characteristics of the colloidal particles found in wastewater. Important factors that contribute to the characteristics of colloidal particles in wastewater include (1) particle size and number, (2) particle shape and flexibility, (3) surface properties including electrical characteristics, (4) particle-particle interactions, and (5) particle-solvent interactions (Shaw, 1966). Particle size, particle shape and flexibility, and particle-solvent interactions are considered below. Because of their importance, the development and measurement of surface charge and particle-particle interactions are considered separately.

Particle Size and Number. The size of colloidal particles in wastewater considered in this text is typically in the range from 0.01 to 1.0 μm. As noted in Chap. 2, some researchers have classified the size range for colloidal particles as varying from 0.001 to 1 μm. The number of colloidal particles in untreated wastewater and after primary sedimentation is typically in the range from 10^6 to 10^{12}/mL. It is important to note that the number of colloidal particles will vary depending on the location where the sample is taken within a treatment plant. The number of particles, as will be discussed later, is of importance with respect to the method to be used for their removal.

Particle Shape and Flexibility. Particle shapes found in wastewater can be described as spherical, semispherical, ellipsoids of various shapes (e.g., prolate and oblate), rods of various length and diameter (e.g., *E. coli*), disk and disklike, strings of various lengths, and random coils. Large organic molecules are often found in the form of coils which may be compressed, uncoiled, or almost linear. The shape of some larger floc particles is often described as fractal. The particle shape will vary depending on the location within the treatment process that is being evaluated. The shape of the particles will affect the electrical properties, the particle-particle interactions, and particle-solvent interactions. Because of the many shapes of particles encountered in wastewater, the theoretical treatment of particle-particle interactions is an approximation at best.

Particle-Solvent Interactions. There are three general types of colloidal particles in liquids: hydrophobic or "water-hating," hydrophilic or "water-loving," and association colloids. The first two types are based on the attraction of the particle surface for water. Hydrophobic particles have relatively little attraction for water; while hydrophilic particles have a great attraction for water. It should be noted, however, that water can interact to some extent even with hydrophobic particles. Some water molecules will generally adsorb on the typical hydrophobic surface, but the reaction between water and hydrophilic colloids occurs to a much greater extent. The third type of colloid is known as an association colloidal, typically made up of surface-active agents such as soaps, synthetic detergents, and dyestuffs which form organized aggregates known as micelles.

Development and Measurement of Surface Charge

An important factor in the stability of colloids is the presence of a surface charge. It develops in a number of different ways, depending on the chemical composition of the medium (wastewater in this case) and the nature of the colloid. Surface charge develops most commonly through (1) isomorphous replacement, (2) structural imperfections, (3) preferential adsorption, and (4) ionization, as defined below. Regardless of how it develops, the surface charge, which promotes stability, must be overcome if these particles are to be aggregated (flocculated) into larger particles with enough mass to settle easily.

Isomorphous Replacement.   Charge development through isomorphous replacement occurs in clay and other soil particles, in which ions in the lattice structure are replaced with ions from solution (e.g., the replacement of Si^{4+} with Al^{3+}).

Structural Imperfections.   In clay and similar particles, charge development can occur because of broken bonds on the crystal edge and imperfections in the formation of the crystal.

Preferential Adsorption.   When oil droplets, gas bubbles, or other chemically inert substances are dispersed in water, they will acquire a negative charge through the preferential adsorption of anions (particularly hydroxyl ions).

Ionization.   In the case of substances such as proteins or microorganisms, surface charge is acquired through the ionization of carboxyl and amino groups (Shaw, 1966). This ionization can be represented as follows, where R represents the bulk of the solid (Fair et al., 1968):

$$R_{NH_2}^{COO^-} \qquad R_{NH_3^+}^{COOH} \qquad R_{NH_2^+}^{COO^-} \qquad\qquad (6-1)$$

$$\text{at high pH} \qquad \text{at low pH} \qquad \begin{array}{c}\text{at isoelectric}\\ \text{point}\end{array}$$

The Electrical Double Layer.   When the colloid or particle surface becomes charged, some ions of the opposite charge (known as counterions) become attached to the surface (see Fig. 6–2). They are held there through electrostatic and van der Waals forces of attraction strongly enough to overcome thermal agitation. Surrounding this fixed layer of ions is a diffuse layer of ions, which is prevented from forming a compact double layer by thermal agitation, as illustrated schematically on Fig. 6–2. The electrical double layer consists of a compact layer (Stern) in which the potential drops from ψ_o to ψ_s, and a diffuse layer in which the potential drops from ψ_s to 0 in the bulk solution.

Figure 6–2

Stern model of electrical double layer. *(Shaw, 1966.)*

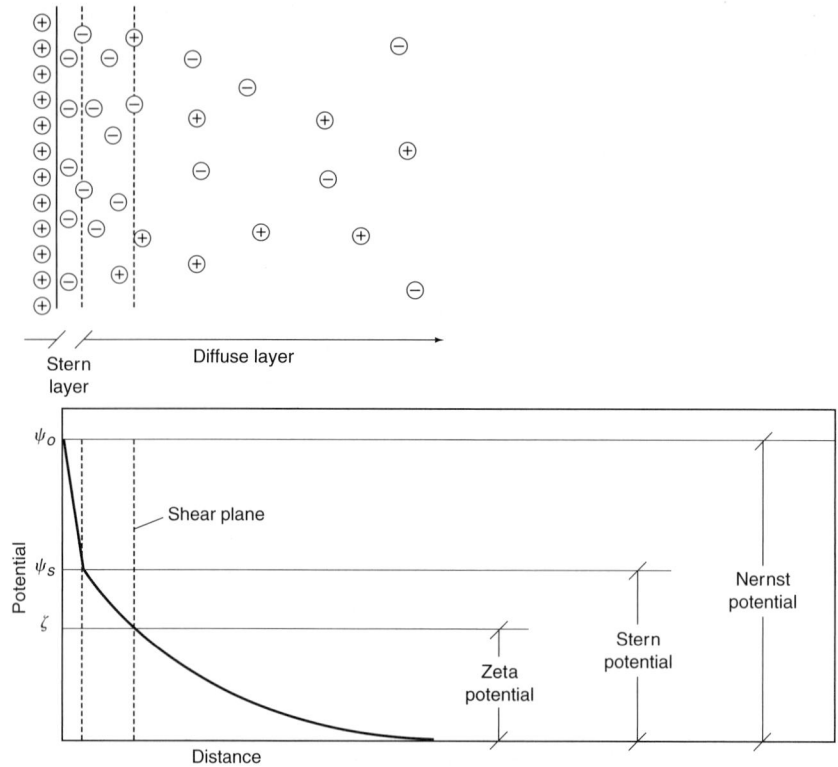

Measurement of Surface Potential. If a particle is placed in an electrolyte solution, and an electric current is passed through the solution, the particle, depending on its surface charge, will be attracted to one or the other of the electrodes, dragging with it a cloud of ions. The potential at the surface of the cloud (called the surface of shear) is sometimes measured in wastewater-treatment operations. The measured value is often called the zeta potential. Theoretically, however, the zeta potential should correspond to the potential measured at the surface enclosing the fixed layer of ions attached to the particle, as shown on Fig. 6–2. The use of the measured zeta potential value is limited because it will vary with the nature of the solution components.

Particle-Particle Interactions

Particle-particle interactions are extremely important in bringing about aggregation by means of Brownian motion. The theory that has been developed to describe particle-particle interactions is based on the consideration of interaction between two charged flat plates and between two charged spheres (Deryagin and Landau, 1941; Verwey and Overbeek, 1948). As neither of these developments is directly applicable to the particles found in wastewater, as described above, the analysis for two charged flat plates will be used for illustrative purposes. The interaction between two plates is illustrated on Fig. 6–3. As shown on Fig. 6–3, the two principal forces involved are the forces of repulsion, due to the electrical properties of the charged plates, and the van der Waals

Figure 6–3

Definition sketch for particle-particle interactions based on the repulsion due to particle surface charge and van der Waals forces of attraction.

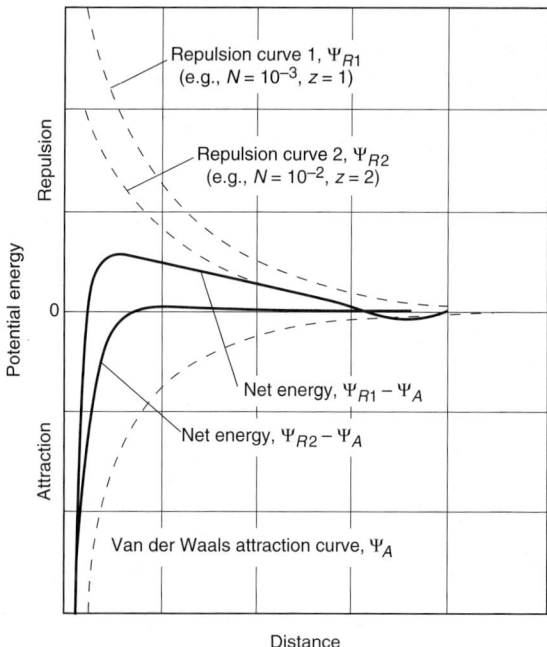

forces of attraction. It should be noted that the van der Waals forces of attraction do not come into play until the two plates are brought together in close proximity to each other.

The net total energy shown by the solid lines on Fig. 6–3 is the difference between the forces of repulsion and attraction. The two conditions, with respect to the forces of repulsion, are illustrated on Fig. 6–3. As shown for condition 1, the forces of attraction will predominate at short and long distances. The net energy curve for condition 1 contains a repulsive maximum that must be overcome if the particles, represented as the two plates, are to be held together by the van der Waals force of attraction. In condition 2, there is no energy barrier to overcome. Clearly, if colloidal particles are to be removed by microflocculation, the repulsive force must be reduced. Although floc particles can form at long distances as shown by the net energy curve for condition 1, the net force holding these particles together is weak and the floc particles that are formed can be ruptured easily.

Particle Destabilization with Potential-Determining Ions and Electrolytes

To bring about particle aggregation through microflocculation, steps must be taken to reduce particle charge or to overcome the effect of this charge. The effect of the charge can be overcome by (1) the addition of potential-determining ions, which will be taken up by or will react with the colloid surface to lessen the surface charge and (2) the addition of electrolytes, which have the effect of reducing the thickness of the diffuse electric layer and, thereby, reduce the zeta potential.

Use of Potential-Determining Ions. The addition of potential-determining ions to promote coagulation can be illustrated by the addition of strong acids or bases

Figure 6–4

Definition sketch for the effects of the addition of counterions and electrolytes to solutions containing charged colloidal particles.

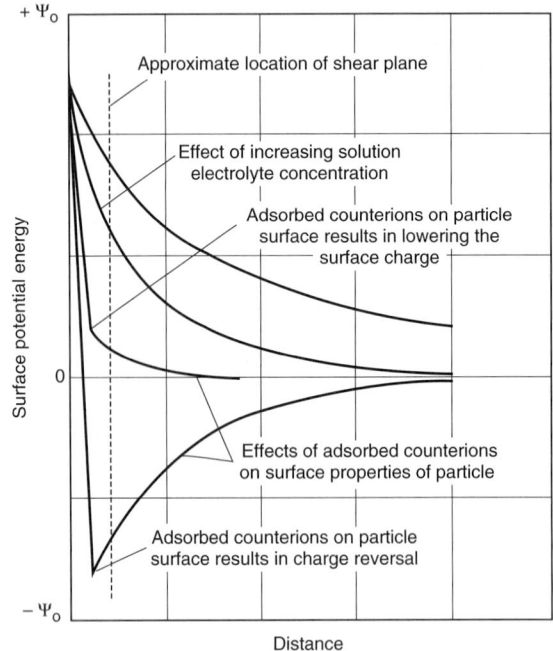

to reduce the charge of metal oxides or hydroxides to near zero so that coagulation can occur. The effect of adding potential-determining ions in a solution containing charged particles is illustrated on Fig. 6–4. The magnitude of the effect will depend on the concentration of potential-determining ions added. The following ratios, known as the Shultz-Hardy rule, can be used to assess the effectiveness of potential-determining or counterions:

$$1 : \frac{1}{2^6} : \frac{1}{3^6} \qquad \text{or} \qquad 100 : 1.6 : 0.13 \tag{6–2}$$

It is interesting to note that depending on the concentration and nature of the counterions added, it is possible to reverse the charge of the double layer and develop a new stable particle.

The effect of adding counterions to a solution containing charged particles is illustrated on Fig. 6–5. The upper curve on Fig. 6–5 represents the surface charge of the particle as a function of the concentration of counterions added. The lines designated kT represent the thermal kinetic energy of the particle. The lower diagram is a plot of the turbidity that would result if the particles that have been destabilized and have undergone microflocculation were removed by settling. As shown, when the surface charge (either positive or negative) is greater than the thermal kinetic energy of the particles, the particles will not flocculate and the original turbidity is observed.

Additional details on the use of counterions may be found in Shaw (1966). The use of potential determining ions is not feasible in either water or wastewater treatment because of the massive concentration of ions that must be added to bring about sufficient compression of the electrical double layer to effect perikinetic flocculation.

Figure 6–5

Definition sketch for the reversal of particle surface charge due to the addition of counterions.

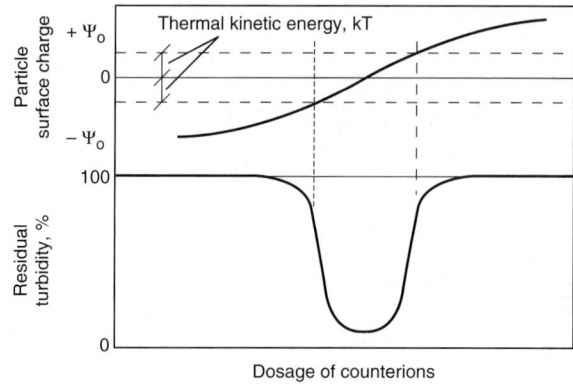

Use of Electrolytes. Electrolytes can also be added to coagulate colloidal suspensions. Increased concentration of a given electrolyte will cause a decrease in zeta potential and a corresponding decrease in repulsive forces as illustrated in condition 2 on Fig. 6–3 and on Fig. 6–4. The concentration of an electrolyte that is needed to destabilize a colloidal suspension is known as the *critical coagulation concentration* (CCC). Increasing the concentration of an indifferent electrolyte will not result in the restabilization of the colloidal particles. As with the addition of potential-determining ions, the use of electrolytes is also not feasible in wastewater treatment. As discussed subsequently, a change in the particle charge will occur when chemicals are added to adjust the pH of the wastewater to optimize the performance of hydrolyzed metal ions used as coagulants.

Particle Destabilization and Aggregation with Polyelectrolytes

Polyelectrolytes may be divided into two categories: natural and synthetic. Important natural polyelectrolytes include polymers of biological origin and those derived from starch products such as cellulose derivatives and alginates. Synthetic polyelectrolytes consist of simple monomers that are polymerized into high-molecular-weight substances. Depending on whether their charge, when placed in water, is negative, positive, or neutral, these polyelectrolytes are classified as anionic, cationic, and nonionic, respectively. The action of polyelectrolytes may be divided into the following three general categories.

Charge Neutralization. In the first category, polyelectrolytes act as coagulants that neutralize or lower the charge of the wastewater particles. Because wastewater particles normally are charged negatively, cationic polyelectrolytes are used for this purpose. In this application, the cationic polyelectrolytes are considered to be primary coagulants. To effect charge neutralization, the polyelectrolyte must be adsorbed to the particle. Because of the large number of particles found in wastewater, the mixing intensity must be sufficient to bring about the adsorption of the polymer onto the colloidal particles. With inadequate mixing, the polymer will eventually fold back on itself and its effectiveness in reducing the surface charge will be diminished. Further, if the number of colloidal particles is limited, it will be difficult to remove them with low polyelectrolyte dosages.

Figure 6-6

Definition sketch for
interparticle bridging
with organic polymers.

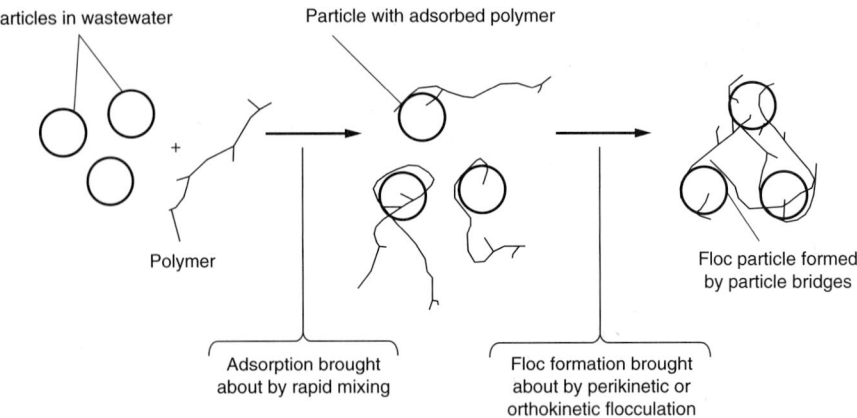

Particles in wastewater

Particle with adsorbed polymer

Polymer

Floc particle formed
by particle bridges

Adsorption brought
about by rapid mixing

Floc formation brought
about by perikinetic or
orthokinetic flocculation

Polymer Bridge Formation.   The second mode of action of polyelectrolytes is interparticle bridging (see Fig. 6–6). In this case, polymers that are anionic and non-ionic (usually anionic to a slight extent when placed in water) become attached at a number of adsorption sites to the surface of the particles found in the wastewater. A bridge is formed when two or more particles become adsorbed along the length of the polymer. Bridged particles become intertwined with other bridged particles during the flocculation process. The size of the resulting three-dimensional particles grows until they can be removed easily by sedimentation. Where particle removal is to be achieved by the formation of particle-polymer bridges, the initial mixing of the polymer and the wastewater containing the particles to be removed must be accomplished in a matter of seconds. Instantaneous initial mixing is usually not required as the polymers are already formed, which is not the case with the polymers formed by metal salts (see discussion of hydrolyzed metal ions given below). As noted above, the mixing intensity must be sufficient to bring about the adsorption of the polymer onto the colloidal particles. If inadequate mixing is provided, the polymer will eventually fold back on itself, in which case it is not possible to form polymer bridges.

Charge Neutralization and Polymer Bridge Formation.   The third type of polyelectrolyte action may be classified as a charge neutralization and bridging phenomenon, which results from using cationic polyelectrolytes of extremely high molecular weight. Besides lowering the surface charge on the particle, these polyelectrolytes also form particle bridges as described above.

Particle Destabilization and Removal with Hydrolyzed Metal Ions

In contrast with the aggregation brought about by the addition of chemicals that act as counterions, electrolytes, and polymers, aggregation brought about by the addition of alum or ferric sulfate is a more complex process. To understand particle destabilization and the removals achieved with hydrolyzed metal ions, it will be instructive to consider first the formation of metal ion hydrolysis products. Operating ranges for action of metal salts and the importance of initial mixing are also considered in light of the formation of these particles.

Formation of Hydrolysis Products. In the past, it was thought that free Al^{3+} and Fe^{3+} were responsible for the effects observed during particle aggregation; it is now known, however, that their hydrolysis products are responsible (Stumm and Morgan, 1962; Stumm and O'Melia, 1968). Although the effect of these hydrolysis products is only now appreciated, it is interesting to note that their chemistry was first elucidated in the early 1900s by Pfeiffer (1902–1907), Bjerrum (1906–1920), and Werner (1907) (Thomas, 1934). In the early 1900s, Pfeiffer proposed that the hydrolysis of trivalent metal salts, such as chromium, aluminum, and iron, could be represented as

$$
\begin{bmatrix}
H_2O & & OH_2 \\
H_2O{-}Me{-}OH_2 \\
H_2O & & OH_2
\end{bmatrix}^{3+}
\leftrightarrow
\begin{bmatrix}
H_2O & & OH \\
H_2O{-}Me{-}OH_2 \\
H_2O & & OH_2
\end{bmatrix}^{2+}
+ H^+
\tag{6-3}
$$

with the extent of the dissociation depending on the anion associated with the metal and on the physical and chemical characteristics of the solution. Further, it was proposed that, upon the addition of sufficient base, the dissociation can proceed to produce a negative ion (Thomas, 1934), such as

$$
\begin{bmatrix}
H_2O & & OH \\
H_2O{-}Me{-}OH \\
HO & & OH
\end{bmatrix}^{-}
\tag{6-4}
$$

It should be noted that the complex compounds given in Eqs. (6–3) and (6–4) are known as *coordination compounds,* which are defined as a central metal ion (or atom) attached to a group of surrounding molecules or ions by coordinate covalent bonds. The surrounding molecules or ions are known as *ligands,* and the atoms attached directly to the metal ion are called ligand donor atoms (McMurry and Fay, 1998). Ligand compounds of interest in wastewater treatment include carbonate (CO_3^{2-}), chloride (Cl^-), hydroxide (OH^-), ammonia (NH_3), and water (H_2O).

In addition, a number of the coordination compounds are also *amphoteric* in that they can exist both in strong acids and in strong bases (McMurry and Fay, 1998). For example, aluminum hydroxide behaves as follows in acidic and basic solutions:

In acid: $Al(OH)_3(s) + 6H_3O^+(aq) \leftrightarrow Al^{3+}(aq) + 6H_2O$ $\hfill$ (6-5)

In base: $Al(OH)_3(s) + OH^-(aq) \leftrightarrow Al(OH)_4^-(aq)$ $\hfill$ (6-6)

As shown in Eq. (6–5), $Al(OH)_3$ will dissolve in the presence of excess acid to form aqueous Al^{3+}. In the presence of excess hydroxide, $Al(OH)_3$ will dissolve to form the aluminate ion, $Al(OH)_4^-$. The acid and base properties of the hydroxides and the nature of the covalent bonds will depend on the position of the element on the periodic table. Further, it should be noted some basic hydroxides will dissolve in strong acid, but not in a strong base (McMurry and Fay, 1998).

Over the past 50 years, it has been observed that the intermediate hydrolysis reactions of Al(III) are much more complex than would be predicted on the basis of a model in which a base is added to the solution. At the present time the complete chemistry for the formation of hydrolysis reactions and products is not well understood (Letterman et al., 1999). A hypothetical model [see Eq. (6–7)], proposed by Stumm (Fair et al., 1968) for Al(III), is useful for the purpose of illustrating the complex reactions involved. A number of alternative formation sequences have also been proposed (Letterman, 1991).

$$[Al(H_2O)_6]^{3+} \xrightarrow{\text{OH}^-} [Al(OH)(H_2O)_5]^{2+} \xrightarrow{\text{OH}^-} [Al(OH)_2(H_2O)_4]^+$$

mononuclear species mononuclear species mononuclear species

$$\text{OH}^-$$
$$\text{OH}^-$$

$$[Al_6(OH)_{15}]^{3+} \text{ (aq)} \quad \text{or} \quad [Al_8(OH)_{20}]^{4+} \text{ (aq)} \xrightarrow{\text{OH}^-}$$

polynuclear species polynuclear species

$$[Al(OH)_3(H_2O)_3](s) \xrightarrow{\text{OH}^-} [Al(OH)_4(H_2O)_2]^- \qquad (6\text{--}7)$$

mononuclear species precipitate mononuclear species aluminate ion

Before the reaction proceeds to the point where a negative aluminate ion is produced, polymerization as depicted in the following formula will usually take place (Thomas, 1934).

$$2Me(H_2O)_5OH^{2+} \leftrightarrow [(H_2O)_4Me \underset{OH}{\overset{OH}{<}} Me(H_2O)_4]^{4+} + 2H_2O \qquad (6\text{--}8)$$

As illustrated by Eqs. (6–7) and (6–8), the possible combinations of the various hydrolysis products are endless, and their enumeration is not the purpose here. What is important, however, is the realization that one or more of the hydrolysis products and/or polymers may be responsible for the observed action of aluminum or iron.

Further, because the hydrolysis reactions follow a stepwise process, the effectiveness of aluminum and iron will vary with time. For example, an alum slurry that has been prepared and stored will behave differently from a freshly prepared solution when it is added to a wastewater. For a more detailed review of the chemistry involved, the excellent articles on this subject by Stumm and Morgan (1962) and Stumm and O'Melia (1968) are recommended. Additional details on the chemistry of aluminum and iron may be found in Benefield et al. (1982), Morel and Hering (1993), Pankow (1991), Snoeyink and Jenkins (1980), Sawyer et al. (1994), and Stumm and Morgan (1981).

Action of Hydrolyzed Metal Ions. The action of hydrolyzed metal ions in bringing about the destabilization and removal of colloidal particles may be divided into the following three categories:

1. Adsorption and charge neutralization
2. Adsorption and interparticle bridging
3. Enmeshment in sweep floc

Adsorption and charge neutralization involves the adsorption of mononuclear and polynuclear metal hydrolysis species [see Eq. (6–5)] on the colloidal particles found in wastewater. It should be noted that it is also possible to get charge reversal with metal salts, as described previously with the addition of counterions (see Fig. 6–5). Adsorption and interparticle bridging involves the adsorption of polynuclear metal hydrolysis species and polymer species [see Eqs. (6–5) and (6–6)] which, in turn, will ultimately form particle-polymer bridges, as described previously. As the coagulant requirement for adsorption and charge neutralization is satisfied, metal hydroxide precipitates and soluble metal hydrolysis products will form as defined by Eq. (6–5). If a sufficient concentration of metal salt is added, large amounts of metal hydroxide floc will form. Following macroflocculation, large floc particles will be formed that will settle readily. In turn, as these floc particles settle, they sweep through the water containing colloidal particles. The colloidal particles that become enmeshed in the floc will thus be removed from the wastewater. In most wastewater applications, the sweep floc mode of operation is used most commonly where particles are to be removed by sedimentation.

The sequence of reactions and events that occur in the coagulation and removal of particles can be illustrated pictorially as shown on Fig. 6–7. In zone 1, sufficient coagulant has not been added to destabilize the colloidal particles, even though some reduction in surface charge may occur due to the presence of Fe^{3+} and some mononuclear hydrolysis species. In zone 2, the colloidal particles have been destabilized by the adsorption of mono- and polynuclear hydrolysis species, and, if allowed to flocculate and settle, the residual turbidity would be lowered as shown. In zone 3, as more coagulant is added, the surface charge of the particles has reversed due to the continued adsorption of mono- and polynuclear hydrolysis species (see Fig. 6–5). As the colloidal particles are now positively charged, they cannot be removed by perikinetic flocculation. As more coagulant is added, zone 4 is reached, where large amounts of hydroxide floc will form. As the floc particles settle, the colloidal particles will be removed by the sweep action of the settling floc particles, and the residual turbidity will be lowered as shown. The coagulant dosage required to reach any of the zones will depend on the nature of the colloidal particles and the pH and temperature of the wastewater. Specific constituents (e.g., organic matter) will also have an effect on the coagulant dose.

It is also very important to note that the example reaction sequence given by Eq. (6–5) and the coagulation process illustrated on Fig. 6–7 are time-dependent. For example, if it is desired to destabilize the colloidal particles in wastewater with mono- and polynuclear species, then *rapid* and *intense* initial mixing of the metal salt and the

Figure 6–7

Definition sketch for the effects of the continued addition of a coagulant (e.g., alum) on the destabilization and flocculation of colloidal particles.

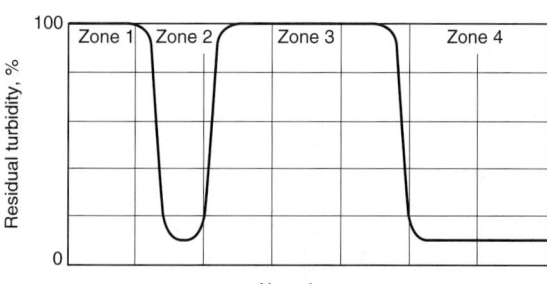

wastewater containing the particles to be destabilized is of critical importance. If the reaction is allowed to proceed to the formation of metal hydroxide floc, it will be difficult to contact the chemical and the particles. As discussed below, it has been estimated that the formation of the mono- and polynuclear and polymer hydroxide species occurs in a fraction of a second.

Solubility of Metal Salts.   To further appreciate the action of the hydrolyzed metal ions, it will be useful to consider the solubility of the metal salts. The solubility of the various alum [Al(III)] and iron [Fe(III)] species is illustrated on Fig. 6–8a and b, respectively, in which the log molar concentrations have been plotted versus pH. In preparing these diagrams, only the mononuclear species for alum and iron have been plotted. The various mononuclear species for alum and iron are given in Table 6–2, along with the corresponding range of acid solubility products reported in the literature. The formation of some of the mononuclear species is also illustrated in Eq. (6–7). It should be noted that Hayden and Rubin (1974) compared experimental and predicted

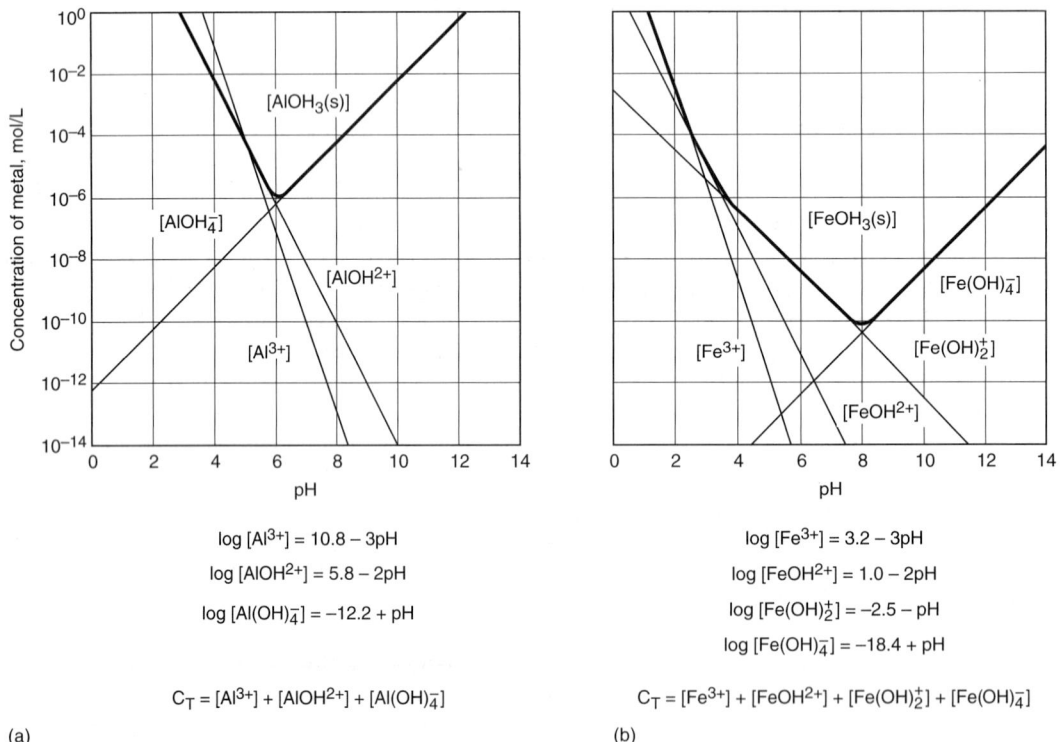

$$\log [Al^{3+}] = 10.8 - 3pH$$
$$\log [AlOH^{2+}] = 5.8 - 2pH$$
$$\log [Al(OH)_4^-] = -12.2 + pH$$

$$C_T = [Al^{3+}] + [AlOH^{2+}] + [Al(OH)_4^-]$$

(a)

$$\log [Fe^{3+}] = 3.2 - 3pH$$
$$\log [FeOH^{2+}] = 1.0 - 2pH$$
$$\log [Fe(OH)_2^+] = -2.5 - pH$$
$$\log [Fe(OH)_4^-] = -18.4 + pH$$

$$C_T = [Fe^{3+}] + [FeOH^{2+}] + [Fe(OH)_2^+] + [Fe(OH)_4^-]$$

(b)

Figure 6–8

Solubility diagram for alum [Al(III)] and iron [Fe(III)]. It should be noted that only the mononuclear species have been plotted. The polynuclear species are extremely dependent on the chemistry of the wastewater. The mononuclear species Al(OH)$_2^+$ has not been included in the development of Fig. 6–8a. Further, because of the wide variation in the solubility and formation constants for the various metal hydroxides, the curves presented in this figure should only be used as a reference guide.

Table 6–2
Reactions and associated equilibrium constants for aluminum and iron species in equilibrium with amorphorous aluminum hydroxide and ferric hydroxide[a]

Reaction	Equilibrium constant	Acid equilibrium constants Range[a]	Used for Fig. 6–8
Aluminum, Al(III)			
$Al(OH)_{3(s)} + 3H^+ = Al^{3+} + 3H_2O$	K_{s0}	9.0–10.8	10.8
$Al(OH)_{3(s)} + 2H^+ = AlOH^{2+} + 2H_2O$	K_{s1}	4.0–5.8	5.8
$Al(OH)_{3(s)} + H^+ = Al(OH)_2^+ + H_2O$[b]	K_{s2}	1.5	1.5
$Al(OH)_{3(s)} = Al(OH)_3$	K_{s3}	−4.2	−4.2
$Al(OH)_{3(s)} + H_2O = AlOH_4^- + H^+$	K_{s4}	−7.7–(−12.5)	−12.2
Species not considered: $Al_2(OH)_2^{4+}$;			
$Al_8(OH)_{20}^{4+}$; $Al_{13}O_4(OH)_{24}^{7+}$; $Al_{14}(OH)_{32}^{10+}$			
Iron, Fe(III)			
$Fe(OH)_{3(s)} + 3H^+ = Fe^{3+} + 3H_2O$	K_{s0}	3.2–4.891	3.2
$Fe(OH)_{3(s)} + 2H^+ = FeOH^{2+} + 2H_2O$	K_{s1}	0.91–2.701	1.0
$Fe(OH)_{3(s)} + H^+ = Fe(OH)_2^+ + H_2O$	K_{s2}	−0.779–(−2.5)	−2.5
$Fe(OH)_{3(s)} = Fe(OH)_3$	K_{s3}	−8.709–(−12.0)	−12.0
$Fe(OH)_{3(s)} + H_2O = FeOH_4^- + H^+$	K_{s4}	−16.709–(−19)	−18.4
Species not considered: $Fe_2(OH)_2^{4+}$; $Fe_3(OH)_4^{5+}$			

[a] Abstracted from Benefield et al. (1982), McMurry and Fay (1998), Morel and Hering (1993), Pankow (1991), Snoeyink and Jenkins (1980), Sawyer et al. (1994), and Stumm and Morgan (1981).
[b] Hayden and Rubin (1974) compared experimental and predicted values and concluded that $Al(OH)_2^+$ is not an important mononuclear species.

values and concluded that $Al(OH)_2^+$ is not an important mononuclear species. Accordingly, mononuclear species $Al(OH)_2^+$ has not been included in the development of Fig. 6–8a. The solid lines trace the approximate total concentration of residual soluble alum (see Fig. 6–8a) and iron (see Fig. 6–8b) after precipitation. Aluminum hydroxide and ferric hydroxide are precipitated within the shaded areas, and polynuclear and polymeric species are formed outside at higher and lower pH values. The region within the square boxes is where most precipitation operations are conducted when these coagulants are used in a sweep floc mode of operation. As shown, the operating region for alum precipitation is from a pH range of 5 to about 7, with minimum solubility occurring at a pH of 6.0, and from about 7 to 9 for iron precipitation, with minimum solubility occurring at a pH of 8.0. A more thorough development of the equations used to develop Fig. 6–8 may be found in Benefield et al. (1982), Benjamin (2001), Morel and Hering (1993), Pankow (1991), Snoeyink and Jenkins (1980), Sawyer et al. (1994), and Stumm and Morgan (1981).

Figure 6–9

Typical operating ranges
for alum coagulation.
*(Adapted from
Amirtharajah and
Mills, 1982.)*

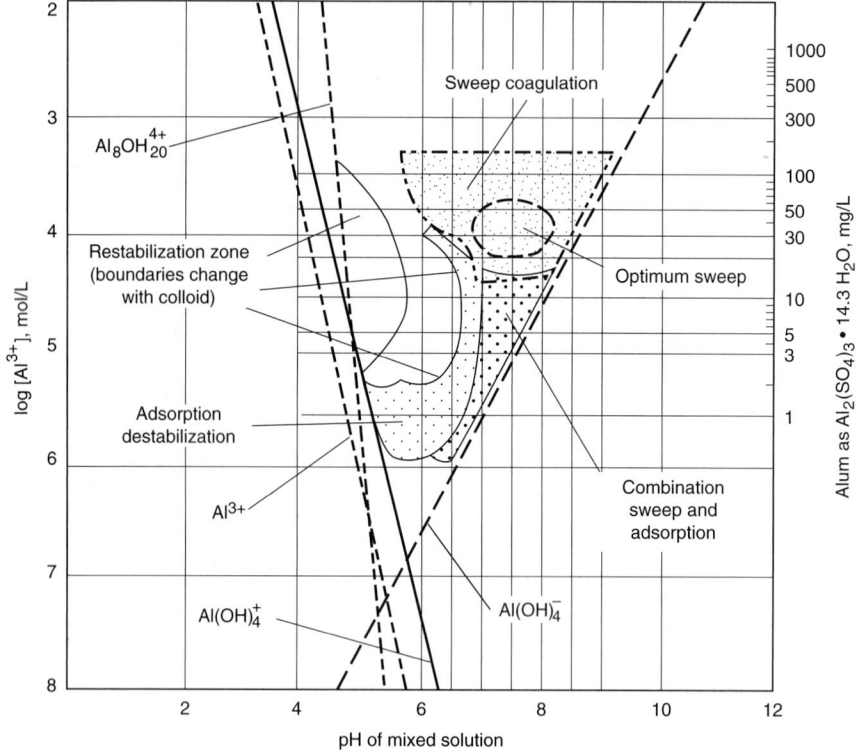

Figure 6–9

Typical operating ranges for alum coagulation. (Adapted from Amirtharajah and Mills, 1982.)

Operating Regions for Action of Metal Salts. Because the chemistry of the various reactions is so complex, there is no complete theory to explain the action of hydrolyzed metal ions. To quantify qualitatively the application of alum as a function of pH, taking into account the action of alum as described above, Amirtharajah and Mills (1982) developed the diagram shown on Fig. 6–9. Although Fig. 6–9 was developed for water treatment applications, it has been found to apply reasonably well to most wastewater applications, with minor variations. As shown on Fig. 6–9, the approximate regions in which the different phenomena associated with particle removal in conventional sedimentation and filtration processes are operative are plotted as a function of the alum dose and the pH of the treated effluent after alum has been added. For example, optimum particle removal by sweep floc occurs in the pH range of 7 to 8 with an alum dose of 20 to 60 mg/L. Generally, for many wastewater effluents that have high pH values (e.g., 7.3 to 8.5), low alum dosages in the range of 5 to 10 mg/L will not be effective. With proper pH control it is possible to operate with extremely low alum dosages. Because the characteristics of wastewater will vary from treatment plant to treatment plant, bench-scale and pilot-plant tests must be conducted to establish the appropriate chemical dosages.

Importance of Initial Chemical Mixing with Metal Salts. Perhaps the least appreciated fact about chemical addition of metal salts is the importance of the rapid initial mixing of the chemicals with the wastewater to be treated. In a 1967 article, Hudson and Wolfner (1967) noted that "coagulants hydrolyze and begin to poly-merize in a fraction of second after being added to water." Hahn and Stumm (1968) studied the coagulation of silica dispersions with Al(III). They reported that the time required for the formation of mono- and polynuclear hydroxide species appears to be extremely short, on the order of 10^{-3} s. The time of formation for the polymer species was on the order of 10^{-2} s. Further, they found that the rate-limiting step in the coagulation process was the time required for the colloidal transport step brought about by Brownian motion (i.e., perikinetic flocculation) which was estimated to be on the order of 1.5 to 3.3 $\times$ 10^{-3} s. The importance of initial and rapid mixing is also discussed by Amirtharajah and Mills (1982) and Vrale and Jorden (1971). Clearly, based on the literature and actual field evaluations, the instantaneous rapid and intense mixing of metal salts is of critical importance, especially where the metal salts are to be used as coagulants to lower the surface charge of the colloidal particles. It should be noted that although achieving extremely low mixing times in large treatment plants is often difficult, low mixing times can be achieved by using multiple mixers. Typical mixing times for various chemicals are reported in Table 6–19 in Sec. 6–8.

6–3 CHEMICAL PRECIPITATION FOR IMPROVED PLANT PERFORMANCE

Chemical precipitation, as noted previously, involves the addition of chemicals to alter the physical state of dissolved and suspended solids and facilitate their removal by sedimentation. In the past, chemical precipitation was often used to enhance the degree of TSS and BOD removal: (1) where there were seasonal variations in the concentration of the wastewater (such as in cannery wastewater), (2) where an intermediate degree of treatment was required, and (3) as an aid to the sedimentation process. Since about 1970, the need to provide more complete removal of the organic compounds and nutrients (nitrogen and phosphorus) contained in wastewater has brought about renewed interest in chemical precipitation. In current practice, chemical precipitation is used (1) as a means of improving the performance of primary settling facilities, (2) as a basic step in the independent physical-chemical treatment of wastewater, (3) for the removal of phosphorus, and (4) for the removal of heavy metals. The first two applications are considered in the following discussion. Phosphorus and heavy metals removal is considered in Secs. 6–4 and 6–5, respectively.

Aside from the determination of the required chemical dosages, the principal design considerations related to the use of chemical precipitation involve the analysis and design of the necessary sludge processing facilities, and the selection and design of the chemical storage, feeding, piping, and control systems. Chemical storage, feeding, piping, and control systems are considered in Sec. 6–8.

Chemical Reactions in Wastewater Precipitation Applications

Over the years a number of different substances have been used as precipitants. The degree of clarification obtained depends on the quantity of chemicals used and the care with which the process is controlled. It is possible by chemical precipitation to obtain a clear effluent, substantially free from matter in suspension or in the colloidal state. The chemicals added to wastewater interact with substances that are either normally present in the wastewater or added for this purpose. The most common chemicals are listed in

Table 6–3

Inorganic chemicals used most commonly for coagulation and precipitation processes in wastewater treatment

Chemical	Formula	Molecular weight	Equivalent weight	Availability Form	Percent
Alum	$Al_2(SO_4)_3 \cdot 18H_2O^a$	666.5		Liquid	8.5 (Al_2O_3)
				Lump	17 (Al_2O_3)
	$Al_2(SO_4)_3 \cdot 14H_2O^a$	594.4	114	Liquid	8.5 (Al_2O_3)
				Lump	17 (Al_2O_3)
Aluminum chloride	$AlCl_3$	133.3	44	Liquid	
Calcium hydroxide (lime)	$Ca(OH)_2$	56.1 as CaO	40	Lump	63–73 as CaO
				Powder	85–99
				Slurry	15–20
Ferric chloride	$FeCl_3$	162.2	91	Liquid	20 (Fe)
				Lump	20 (Fe)
Ferric sulfate	$Fe_2(SO_4)_3$	400	51.5	Granular	18.5 (Fe)
Ferrous sulfate (copperas)	$FeSO_4 \cdot 7H_2O$	278.1	139	Granular	20 (Fe)
Sodium aluminate	$Na_2Al_2O_4$	163.9	100	Flake	46 (Al_2O_3)

^{a} Number of bound water molecules will typically vary from 14 to 18.

Table 6–3. Information on the handling, storage, and feeding of these compounds may be found in Sec. 6–8. The reactions involved with (1) alum, (2) lime, (3) ferrous sulfate (copperas) and lime, (4) ferric chloride, (5) ferric chloride and lime, and (6) ferric sulfate and lime are considered in the following discussion (Metcalf & Eddy, 1935).

Alum. When alum is added to wastewater containing calcium and magnesium bicarbonate alkalinity, a precipitate of aluminum hydroxide will form. The overall reaction that occurs when alum is added to water may be illustrated as follows:

$$\underset{\substack{\text{Calcium} \\ \text{bicarbonate} \\ \text{(soluble)}}}{\overset{3 \times 100 \text{ (as CaCO}_3)}{3Ca(HCO_3)_2}} + \underset{\substack{\text{Aluminum} \\ \text{sulfate} \\ \text{(soluble)}}}{\overset{666.5}{Al_2(SO_4)_3 \cdot 18H_2O}} \leftrightarrow$$

$$\underset{\substack{\text{Aluminum} \\ \text{hydroxide} \\ \text{(insoluble)}}}{\overset{2 \times 78}{2Al(OH)_3}} + \underset{\substack{\text{Calcium} \\ \text{sulfate} \\ \text{(soluble)}}}{\overset{3 \times 136}{3CaSO_4}} + \underset{\substack{\text{Carbon} \\ \text{dioxide} \\ \text{(soluble)}}}{\overset{6 \times 44}{6CO_2}} + \overset{18 \times 18}{18H_2O} \qquad (6\text{–}9)$$

The numbers above the chemical formulas are the combining molecular weights of the different substances and, therefore, denote the quantity of each one involved. The precipitation reaction given above also occurs with the addition of aluminum chloride ($AlCl_3$). The insoluble aluminum hydroxide is a gelatinous floc that settles slowly through the wastewater, sweeping out suspended material and producing other changes. The reaction is exactly analogous when magnesium bicarbonate is substituted for the calcium salt.

Because alkalinity in Eq. (6–9) is reported in terms of calcium carbonate ($CaCO_3$), the molecular weight of which is 100, the quantity of alkalinity required to react with 10 mg/L of alum is

$$(10.0 \text{ mg/L}) \left[\frac{3(100 \text{ g/mole})}{(666.5 \text{ g/mole})} \right] = 4.5 \text{ mg/L}$$

If less than this amount of alkalinity is available, it must be added. Lime is commonly used for this purpose when necessary, but it is seldom required in the treatment of wastewater.

Lime. When lime alone is added as a precipitant, the principles of clarification are explained by the following reactions for the carbonic acid [Eq. (6–10)] and the alkalinity [Eq. (6–11)]:

44 (as CO_2) 56 (as CaO)   100   2 × 18

$$H_2CO_3 + Ca(OH)_2 \leftrightarrow CaCO_3 + 2H_2O \qquad (6\text{–}10)$$

Carbonic   Calcium   Calcium
  acid   hydroxide   carbonate
(soluble)   (slightly   (somewhat
             soluble)   soluble)

100 (as $CaCO_3$) 56 (as CaO)   2 × 100   2 × 18

$$Ca(HCO_3)_2 + Ca(OH)_2 \leftrightarrow 2CaCO_3 + 2H_2O \qquad (6\text{–}11)$$

 Calcium   Calcium   Calcium
bicarbonate   hydroxide   carbonate
(soluble)   (slightly   (somewhat
              soluble)   soluble)

A sufficient quantity of lime must therefore be added to combine with all the free carbonic acid and with the carbonic acid of the bicarbonates (half-bound carbonic acid) to produce calcium carbonate. Much more lime is generally required when it is used alone than when sulfate of iron is also used (see the following discussion). Where industrial wastes introduce mineral acids or acid salts into the wastewater, these must be neutralized before precipitation can take place.

Ferrous Sulfate and Lime. In most cases, ferrous sulfate cannot be used alone as a precipitant because lime must be added at the same time to form a precipitate. When ferrous sulfate alone is added to wastewater, the following reactions occur:

      278   100 (as $CaCO_3$)   178   136   7 × 18

$$FeSO_4 \cdot 7H_2O + Ca(HCO_3)_2 \leftrightarrow Fe(HCO_3)_2 + CaSO_4 + 7H_2O \qquad (6\text{–}12)$$

  Ferrous   Calcium   Ferrous   Calcium
  sulfate   carbonate   bicarbonate   sulfate
(soluble)   (soluble)   (soluble)   (soluble)

$$\underset{\text{Ferrous}\atop\text{bicarbonate}\atop\text{(soluble)}}{\overset{278}{Fe(HCO_3)_2}} \rightarrow \underset{\text{Ferrous}\atop\text{hydroxide}\atop\text{(very slightly}\atop\text{soluble)}}{\overset{4 \times 89.9}{Fe(OH)_2}} + \underset{\text{Carbon}\atop\text{dioxide}\atop\text{(soluble)}}{CO_2} \qquad (6\text{-}13)$$

If sufficient alkalinity is not available, lime is often added in excess in conjunction with ferrous sulfate. The resulting reaction is

$$\underset{\text{Ferrous}\atop\text{bicarbonate}\atop\text{(soluble)}}{\overset{178}{Fe(HCO_3)_2}} + \underset{\text{Calcium}\atop\text{hydroxide}\atop\text{(slightly}\atop\text{soluble)}}{\overset{2 \times 56 \text{ (as CaO)}}{2Ca(OH)_2}} \leftrightarrow \underset{\text{Ferrous}\atop\text{hydroxide}\atop\text{(very slightly}\atop\text{soluble)}}{\overset{89.9}{Fe(OH)_2}} + \underset{\text{Calcium}\atop\text{carbonate}\atop\text{(somewhat}\atop\text{soluble)}}{\overset{2 \times 100}{2CaCO_3}} + \overset{2 \times 18}{2H_2O} \qquad (6\text{-}14)$$

The ferrous hydroxide can be oxidized to ferric hydroxide, the final form desired, by oxygen dissolved in the wastewater. The reaction is:

$$\underset{\text{Ferrous}\atop\text{hydroxide}\atop\text{(very slightly}\atop\text{soluble)}}{\overset{89.9}{Fe(OH)_2}} + \underset{\text{Oxygen}\atop\text{(soluble)}}{\overset{1/4 \times 32}{1/4 O_2}} + \overset{1/2 \times 18}{1/2\,H_2O} \leftrightarrow \underset{\text{Ferric}\atop\text{hydroxide}\atop\text{(insoluble)}}{\overset{106.9}{Fe(OH)_3}} \qquad (6\text{-}15)$$

The insoluble ferric hydroxide is formed as a bulky, gelatinous floc similar to the alum floc. The alkalinity required for a 10 mg/L dosage of ferrous sulfate [see Eq. (6–12)] is

$$(10.0 \text{ mg/L})\left[\frac{(100 \text{ g/mole})}{(278 \text{ g/mole})}\right] = 3.6 \text{ mg/L}$$

The lime required [see Eq. (6–14)] is

$$(10.0 \text{ mg/L})\left[\frac{2(56 \text{ g/mole})}{(278 \text{ g/mole})}\right] = 4.0 \text{ mg/L}$$

The oxygen required [see Eq. (6–15)] is

$$(10.0 \text{ mg/L})\left[\frac{(32 \text{ g/mole})}{4(278 \text{ g/mole})}\right] = 0.29 \text{ mg/L}$$

Because the formation of ferric hydroxide is dependent on the presence of dissolved oxygen, the reaction given in Eq. (6–15) cannot be completed in most wastewaters, and, as a result, ferrous sulfate is not used commonly in wastewater.

Ferric Chloride.   Because of the many problems associated with the use of ferrous sulfate, ferric chloride is the iron salt used most commonly in precipitation applications. When ferric chloride is added to wastewater, the following reactions take place.

$$\underset{\text{Ferric}\atop\text{chloride}\atop\text{(soluble)}}{\overset{2 \times 162.2}{2FeCl_3}} + \underset{\text{Calcium}\atop\text{bicarbonate}\atop\text{(soluble)}}{\overset{3 \times 100 \text{ (as CaCO_3)}}{3Ca(HCO_3)_2}} \leftrightarrow \underset{\text{Ferric}\atop\text{hydroxide}\atop\text{(insoluble)}}{\overset{2 \times 106.9}{2Fe(OH)_3}} + \underset{\text{Calcium}\atop\text{sulfate}\atop\text{(soluble)}}{3CaCl_2} + \underset{\text{Carbon}\atop\text{dioxide}\atop\text{(soluble)}}{6CO_2} \qquad (6\text{-}16)$$

Ferric Chloride and Lime. If lime is added to supplement the natural alkalinity of the wastewater, the following reaction can be assumed to occur:

$$\underset{\substack{\text{Ferric} \\ \text{chloride} \\ \text{(soluble)}}}{\underset{2 \times 162.2}{2FeCl_3}} + \underset{\substack{\text{Calcium} \\ \text{hydroxide} \\ \text{(slightly} \\ \text{soluble)}}}{\underset{3 \times 56 \text{ (as CaO)}}{3Ca(OH)_2}} \leftrightarrow \underset{\substack{\text{Ferric} \\ \text{hydroxide} \\ \text{(insoluble)}}}{\underset{2 \times 106.9}{2Fe(OH)_3}} + \underset{\substack{\text{Calcium} \\ \text{chloride} \\ \text{(soluble)}}}{\underset{3 \times 111}{3CaCl_2}} \qquad (6\text{-}17)$$

Ferric Sulfate and Lime. The overall reaction that occurs when ferric sulfate and lime are added to wastewater may be represented as follows:

$$\underset{\substack{\text{Ferric} \\ \text{sulfate} \\ \text{(soluble)}}}{\underset{399.9}{Fe_2(SO_4)_3}} + \underset{\substack{\text{Calcium} \\ \text{hydroxide} \\ \text{(slightly} \\ \text{soluble)}}}{\underset{3 \times 56 \text{ (as CaO)}}{3Ca(OH)_2}} \leftrightarrow \underset{\substack{\text{Ferric} \\ \text{hydroxide} \\ \text{(insoluble)}}}{\underset{2 \times 106.9}{2Fe(OH)_3}} + \underset{\substack{\text{Calcium} \\ \text{sulfate} \\ \text{(soluble)}}}{\underset{3 \times 136}{3CaSO_4}} \qquad (6\text{-}18)$$

Enhanced Removal of Suspended Solids in Primary Sedimentation

The degree of clarification obtained when chemicals are added to untreated wastewater depends on the quantity of chemicals used, mixing times, and the care with which the process is monitored and controlled. With chemical precipitation, it is possible to remove 80 to 90 percent of the total suspended solids (TSS) including some colloidal particles, 50 to 80 percent of the BOD, and 80 to 90 percent of the bacteria. Comparable removal values for well-designed and well-operated primary sedimentation tanks without the addition of chemicals are 50 to 70 percent of the TSS, 25 to 40 percent of the BOD, and 25 to 75 percent of the bacteria. Because of the variable characteristics of wastewater, the required chemical dosages should be determined from bench- or pilot-scale tests. Recommended surface loading rates for various chemical suspensions to be used in the design of the sedimentation facilities are given in Table 6–4.

Table 6–4
Recommended surface loading rates for sedimentation tanks for various chemical suspensions

	Overflow rate			
	gal/ft²·d		m³/m²·d	
Suspension	Typical range	Peak flow	Typical range	Peak flow
Alum floc[a]	700–1400	1200	30–70	80
Iron floc[a]	700–1400	1200	30–70	80
Lime floc[a]	750–1500	1500	35–80	90
Untreated wastewater	600–1200	1200	30–70	80

[a] Mixed with the settleable suspended solids in the untreated wastewater and colloidal or other suspended solids swept out by the floc.
Note: m³/m²·d × 24.5424 = gal/ft²·d

Independent Physical-Chemical Treatment

In some localities, industrial wastes have rendered municipal wastewater difficult to treat by biological means. In such situations, physical-chemical treatment may be an alternative approach. This method of treatment has met with limited success because of its lack of consistency in meeting discharge requirements, high costs for chemicals, handling and disposal of the great volumes of sludge resulting from the addition of chemicals, and numerous operating problems. Based on typical performance results of full-scale plants using activated carbon, the activated-carbon columns removed only 50 to 60 percent of the applied total BOD, and the plants did not meet consistently the effluent standards for secondary treatment. In some instances, substantial process modifications have been required to reduce the operating problems and meet performance requirements, or the process has been replaced by biological treatment. Because of these reasons, new applications of physical-chemical treatment for municipal wastewater are rare. Physical-chemical treatment is used more extensively for the treatment of industrial wastewater. Depending on the treatment objectives, the required chemical dosages and application rates should be determined from bench- or pilot-scale tests.

A flow diagram for the physical-chemical treatment of untreated wastewater is presented on Fig. 6–10. As shown, after first-stage precipitation and pH adjustment by recarbonation (if required), the wastewater is passed through a granular-medium filter to remove any residual floc and then through carbon columns to remove dissolved organic compounds. The filter is shown as optional, but its use is recommended to reduce the blinding and headloss buildup in the carbon columns. The treated effluent from the carbon column is usually chlorinated before discharge to the receiving waters.

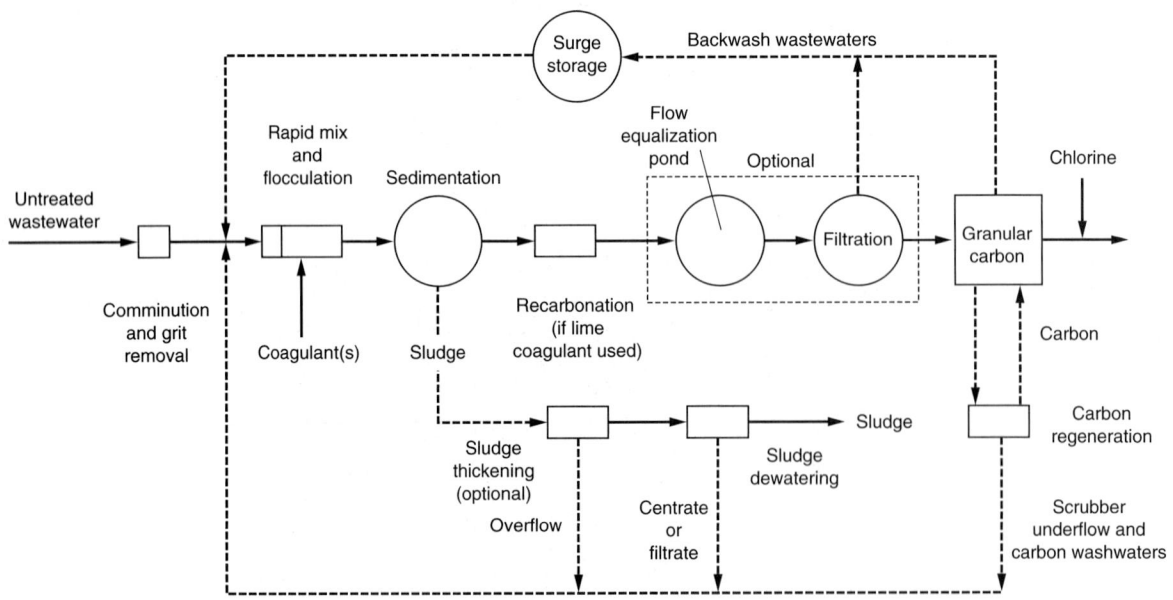

Figure 6–10

Typical flow diagram of an independent physical-chemical treatment plant.

Estimation of Sludge Quantities from Chemical Precipitation

The handling and disposal of the sludge resulting from chemical precipitation is one of the greatest difficulties associated with chemical treatment. Sludge is produced in great volume from most chemical precipitation operations, often reaching 0.5 percent of the volume of wastewater treated when lime is used. The computational procedures involved in estimating the quantity of sludge resulting from chemical precipitation with ferric chloride and lime are illustrated in Example 6–1.

EXAMPLE 6–1   **Estimation of Sludge Volume from Chemical Precipitation of Untreated Wastewater**   Estimate the mass and volume of sludge produced from untreated wastewater without and with the use of ferric chloride for the enhanced removal of TSS. Also estimate the amount of lime required for the specified ferric chloride dose. Assume that 60 percent of the TSS is removed in the primary settling tank without the addition of chemicals, and that the addition of ferric chloride results in an increased removal of TSS to 85 percent. Also, assume that the following data apply to this situation:

1.	Wastewater flowrate, m^3/d	1000
2.	Wastewater TSS, mg/L	220
3.	Wastewater alkalinity as $CaCO_3$, mg/L	136
4.	Ferric chloride ($FeCl_3$) added, $kg/1000 \ m^3$	40
5.	Raw sludge properties	
	Specific gravity	1.03
	Moisture content, %	94
6.	Chemical sludge properties (from Chap. 15)	
	Specific gravity	1.05
	Moisture content, %	92.5

Solution

1. Compute the mass of TSS removed without and with chemicals.
   a. Determine the mass of TSS removed without chemicals.

   $$M_{TSS} = \frac{0.6(220 \ g/m^3)(1000 \ m^3/d)}{(10^3 \ g/kg)} = 132.0 \ kg/d$$

   b. Determine the mass of TSS removed with chemicals.

   $$M_{TSS} = \frac{0.85(220 \ g/m^3)(1000 \ m^3/d)}{(10^3 \ g/kg)} = 187.0 \ kg/d$$

2. Using Eq. (6–16), determine the mass of ferric hydroxide [$Fe(OH)_3$] produced from the addition of 40 kg/1000 m^3 of ferrous sulfate ($FeCl_3$).

   $$Fe(OH)_3 \ formed = 40 \times \left(\frac{2 \times 106.9}{2 \times 162.2} \right) = 26.4 \ kg/1000 \ m^3$$

3. Using Eq. (6–17), determine the mass of lime required to convert the ferric chloride to ferric hydroxide $Fe(OH)_3$.

$$\text{Lime required} = 40 \times \left(\frac{3 \times 56}{2 \times 162.2} \right) = 20.7 \text{ kg/1000 m}^3$$

Because there is sufficient natural alkalinity no lime addition will be required.

4. Determine the total amount of sludge on a dry basis resulting from chemical precipitation.

Total dry solids $= 187 + 26.4 = 213.4$ kg/1000 m^3

5. Determine the total volume of sludge resulting from chemical precipitation, assuming that the sludge has a specific gravity of 1.05 and a moisture content of 92.5 percent.

$$V_s = \frac{(213.4 \text{ kg/d})}{(1.05)(1000 \text{ kg/m}^3)(0.075)} = 2.71 \text{ m}^3/\text{d}$$

6. Determine the total volume of sludge without chemical precipitation, assuming that the sludge has a specific gravity of 1.03 and a moisture content of 94 percent.

$$V_s = \frac{(132 \text{ kg/d})}{(1.03)(1000 \text{ kg/m}^3)(0.06)} = 2.1 \text{ m}^3/\text{d}$$

7. Prepare a summary table of sludge masses and volumes without and with chemical precipitation.

	Sludge	
Treatment	**Mass, kg/d**	**Volume, m³/d**
Without chemical precipitation	132.0	2.13
With chemical precipitation	213.4	2.71

Comment The magnitude of the sludge disposal problem when chemicals are used is evident from a review of the data presented in the summary table given in Step 7. Even larger volumes of sludge are produced when lime is used as the primary precipitant (see Example 6–3).

6–4 CHEMICAL PRECIPITATION FOR PHOSPHORUS REMOVAL

The removal of phosphorus from wastewater involves the incorporation of phosphate into TSS and the subsequent removal of those solids. Phosphorus can be incorporated into either biological solids (e.g., microorganisms) or chemical precipitates. The fundamentals of biological phosphorus removal are considered in Chap. 8. The removal of phosphorus in chemical precipitates is introduced in this section. The topics to be considered include (1) the chemistry of phosphate precipitation, (2) strategies for phosphorus removal, (3) phosphorus removal using metal salts and polymers, and (4) phosphorus removal using lime.

Chemistry of Phosphate Precipitation

The chemical precipitation of phosphorus is brought about by the addition of the salts of multivalent metal ions that form precipitates of sparingly soluble phosphates. The multivalent metal ions used most commonly are calcium [Ca(II)], aluminum [Al(III)], and iron [Fe(III)]. Polymers have been used effectively in conjunction with alum and lime as flocculant aids. Because the chemistry of phosphate precipitation with calcium is quite different than with aluminum and iron, the two different types of precipitation are considered separately in the following discussion.

Phosphate Precipitation with Calcium. Calcium is usually added in the form of lime $Ca(OH)_2$. From the equations presented previously, it will be noted that when lime is added to water it reacts with the natural bicarbonate alkalinity to precipitate $CaCO_3$. As the pH value of the wastewater increases beyond about 10, excess calcium ions will then react with the phosphate, as shown in Eq. (6–19), to precipitate hydroxylapatite $Ca_{10}(PO_4)_6(OH)_2$.

$$10Ca^{2+} + 6PO_4^{3-} + 2OH^- \leftrightarrow Ca_{10}(PO_4)_6(OH)_2 \qquad (6\text{–}19)$$
$$\text{Hydroxylapatite}$$

Because of the reaction of lime with the alkalinity of the wastewater, the quantity of lime required will, in general, be independent of the amount of phosphate present and will depend primarily on the alkalinity of the wastewater (see Fig. 6–11). The quantity of lime required to precipitate the phosphorus in wastewater is typically about 1.4 to 1.5 times the total alkalinity expressed as $CaCO_3$. Because a high pH value is required to precipitate phosphate, coprecipitation is usually not feasible. When lime is added to raw

Figure 6–11

Lime dosage required to raise the pH to 11 as a function of untreated wastewater alkalinity.

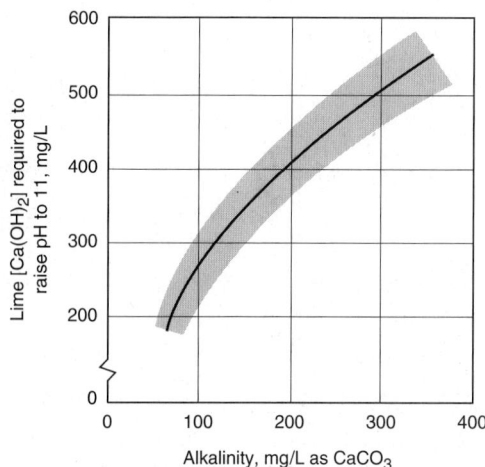

wastewater or to secondary effluent, pH adjustment is usually required before subsequent treatment or disposal. Recarbonation with carbon dioxide (CO_2) is used to lower the pH value.

Phosphate Precipitation with Aluminum and Iron.
The basic reactions involved in the precipitation of phosphorus with aluminum and iron are as follows.

Phosphate precipitation with aluminum:

$$Al^{3+} + H_nPO_4^{3-n} \leftrightarrow AlPO_4 + nH^+ \tag{6-20}$$

Phosphate precipitation with iron:

$$Fe^{3+} + H_nPO_4^{3-n} \leftrightarrow FePO_4 + nH^+ \tag{6-21}$$

In the case of alum and iron, 1 mole will precipitate 1 mole of phosphate; however, these reactions are deceptively simple and must be considered in light of the many competing reactions and their associated equilibrium constants, and the effects of alkalinity, pH, trace elements, and ligands found in wastewater. Because of the many competing reactions, Eqs. (6–20) and (6–21) cannot be used to estimate the required chemical dosages directly. Therefore, dosages are generally established on the basis of bench-scale tests and occasionally by full-scale tests, especially if polymers are used. For example, for equimolar initial concentrations of Al(III), Fe(III), and phosphate, the total concentration of soluble phosphate in equilibrium with both insoluble $AlPO_4$ and $FePO_4$ is shown on Fig. 6–12. The solid lines trace the concentration of residual soluble phosphate after precipitation. Pure metal phosphates are precipitated within the shaded area, and mixed complex polynuclear species are formed outside toward higher and lower pH values.

Figure 6–12

Concentration of aluminum and ferric phosphate in equilibrium with soluble phosphorus: (a) Al(III)-phosphate; (b) Fe(III)-phosphate.

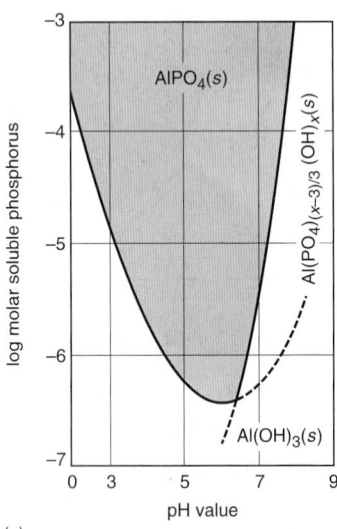

(a)

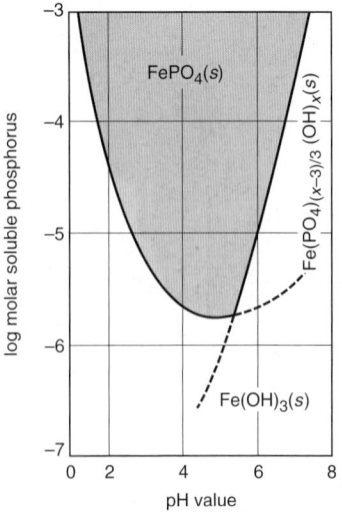

(b)

EXAMPLE 6–2 **Determination of Alum Dosage for Phosphorus Removal** Determine the amount of liquid alum required to precipitate phosphorus in a wastewater that contains 8 mg P/L. Also determine the required alum storage capacity if a 30-d supply is to be stored at the treatment facility. Based on laboratory testing, 1.5 mole of Al will be required per mole of P. The flowrate is 12,000 m^3/d. The following data are for the liquid alum supply.

1. Formula for liquid alum $Al_2(SO_4)_3 \cdot 18H_2O$
2. Alum strength = 48 percent
3. Density of liquid alum solution = 1.2 kg/L

Solution

1. Determine the weight of aluminum (Al) available per liter of liquid alum.
   a. The weight of alum per liter is

   Alum/L = (0.48)(1.2 kg/L) = 0.576 kg/L

   b. The weight of aluminum per liter is

   Molecular weight of alum = 666.5 (see Table 6–2)

   Aluminum/L = (0.58 kg/L)(2 × 26.98/666.5) = 0.0466 kg/L

2. Determine the weight of Al required per unit weight of P.
   a. Theoretical dosage = 1.0 mole Al per 1.0 mole P [see Eq. (6–20)]
   b. Aluminum required = 1.0 kg × (mw Al/mw P)
      = 1.0 kg × (26.98/30.97) = 0.87 kg Al/kg P

3. Determine the amount of alum solution required per kg P.

$$\text{Alum dose} = 1.5 \times \left(\frac{0.87 \text{ kg Al}}{1.0 \text{ kg P}} \right)\left(\frac{\text{L alum solution}}{0.0466 \text{ kg}} \right)$$

   = 28.0 L alum solution/kg P

4. Determine the amount of alum solution required per day.

$$\text{Alum} = \frac{(12{,}000 \text{ m}^3/\text{d})(8 \text{ g P/m}^3)(28.0 \text{ L alum/kg P})}{(10^3 \text{ g/kg})}$$

   = 2688 L alum solution/d

5. Determine the required alum solution storage capacity based on average flow.

   Storage capacity = (2688 L alum solution/d)(30 d)

   = 80,640 L = 80.6 m^3

Strategies for Phosphorus Removal

The precipitation of phosphorus from wastewater can occur in a number of different locations within a process flow diagram (see Fig. 6–13). The general locations where

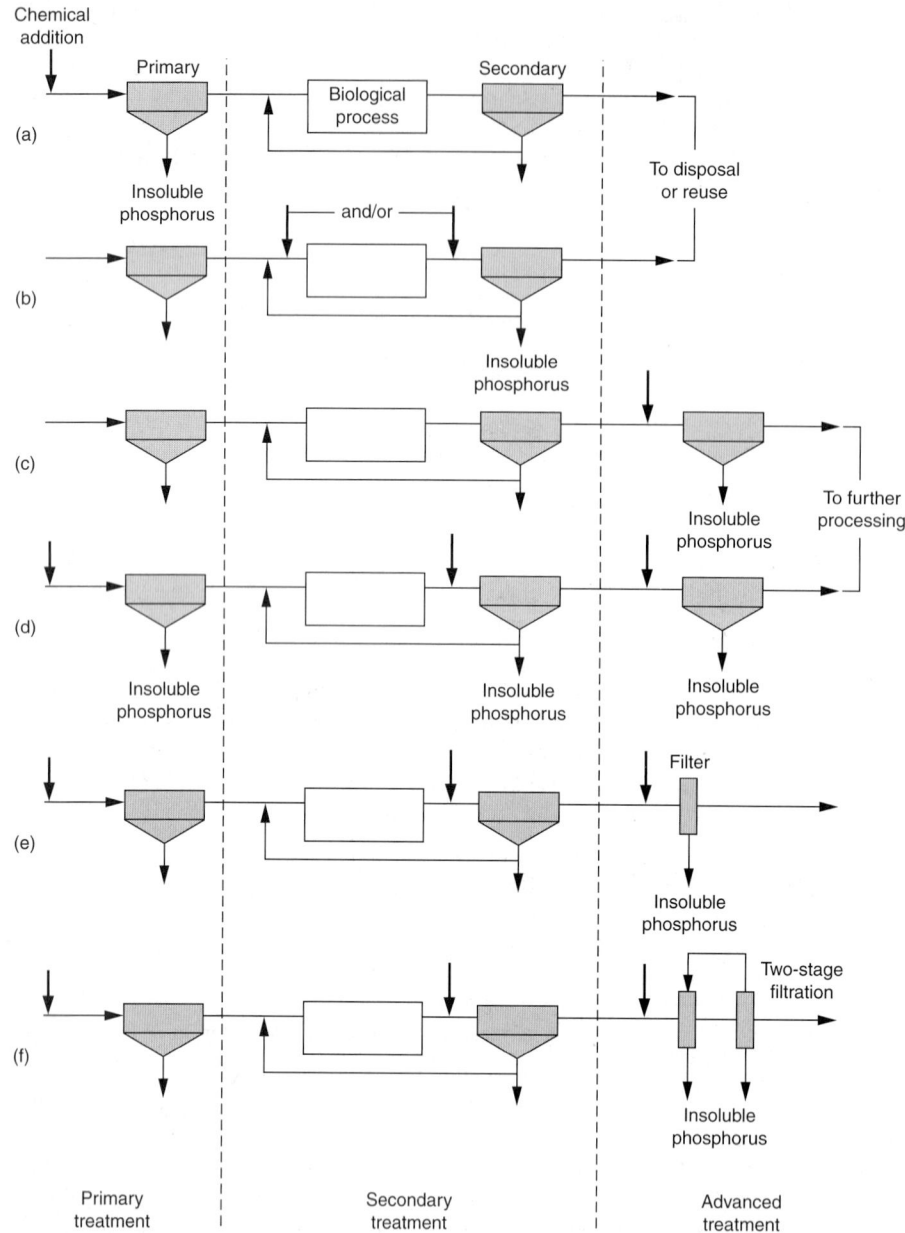

Figure 6-13

Alternative points of chemical addition for phosphorus removal: (a) before primary sedimentation, (b) before and/or following biological treatment, (c) following secondary treatment, and (d–f) at several locations in a process (known as "split treatment").

Table 6–5
Factors affecting the choice of chemical for phosphorus removal[a]

1. Influent phosphorus level
2. Wastewater suspended solids
3. Alkalinity
4. Chemical cost (including transportation)
5. Reliability of chemical supply
6. Sludge handling facilities
7. Ultimate disposal methods
8. Compatibility with other treatment processes

[a] Adapted in part from Kugelman (1976).

phosphorus can be removed may be classified as (1) pre-precipitation, (2) coprecipitation, and (3) postprecipitation (Sedlak, 1991). Factors affecting the choice of chemical to use for phosphorus removal are reported in Table 6–5.

Pre-precipitation. The addition of chemicals to raw wastewater for the precipitation of phosphorus in primary sedimentation facilities is termed "pre-precipitation." The precipitated phosphate is removed with the primary sludge.

Coprecipitation. The addition of chemicals to form precipitates that are removed along with waste biological sludge is defined as "coprecipitation." Chemicals can be added to (1) the effluent from primary sedimentation facilities, (2) the mixed liquor (in the activated-sludge process), or (3) the effluent from a biological treatment process before secondary sedimentation.

Postprecipitation. Postprecipitation involves the addition of chemicals to the effluent from secondary sedimentation facilities and the subsequent removal of chemical precipitates. In this process, the chemical precipitates are usually removed in separate sedimentation facilities or in effluent filters (see Fig. 6–13).

Phosphorus Removal Using Metal Salts and Polymers

As noted above, iron or aluminum salts can be added at a variety of different points in the treatment process (see Fig. 6–13), but because polyphosphates and organic phosphorus are less easily removed than orthophosphorus, adding aluminum or iron salts after secondary treatment (where organic phosphorus and polyphosphorus are transformed into orthophosphorus) usually results in the best removal. Some additional nitrogen removal occurs because of better settling, but essentially no ammonia is removed unless chemical additions to primary treatment reduce BOD loadings to the point where nitrification can occur. A number of the important features of adding metal salts and polymers at different points in the treatment process are discussed in this section.

Metal Salt Addition to Primary Sedimentation Tanks. When aluminum or iron salts are added to untreated wastewater, they react with the soluble orthophosphate to produce a precipitate. Organic phosphorus and polyphosphate are removed by more complex reactions and by adsorption onto floc particles. The insolubilized

Table 6–6

Typical alum dosage requirements for various levels of phosphorus removal[a]

Phosphorus reduction, %	Mole ratio, Al:P	
	Range	Typical
75	1.25:1–1.5:1	1.4:1
85	1.6:1–1.9:1	1.7:1
95	2.1:1–2.6:1	2.3:1

[a]Developed in part from U.S. EPA (1976).

Figure 6–14

Soluble phosphorus removal by ferric chloride addition.

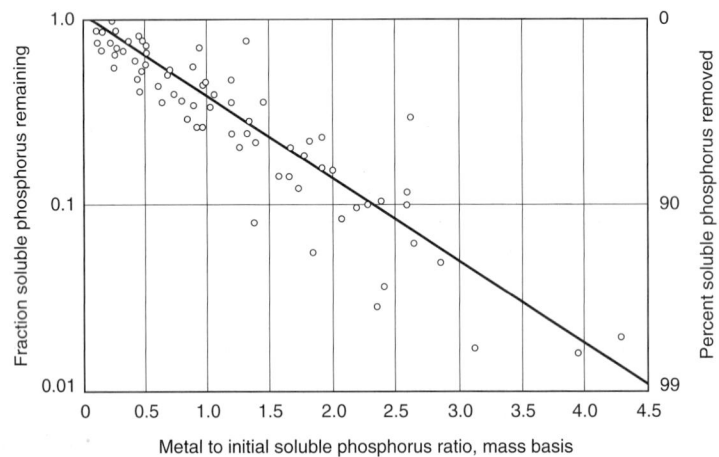

phosphorus, as well as considerable quantities of BOD and TSS, are removed from the system as primary sludge. Adequate initial mixing and flocculation are necessary upstream of primary facilities, whether separate basins are provided or existing facilities are modified to provide these functions. Polymer addition may be required to aid in settling. In low-alkalinity waters, the addition of a base is sometimes necessary to keep pH in the 5 to 7 range. Alum generally is applied in a molar ratio in the range of a 1.4 to 2.5 mole Al/mole P (see Table 6–6). Molar ratios for ferric chloride are shown on Fig. 6–14. The exact application rate is determined by onsite testing, and varies with the characteristics of the wastewater and the desired phosphorus removal.

Metal Salt Addition to Secondary Treatment. Metal salts can be added to the untreated wastewater, in the activated-sludge aeration tank, or the final clarifier influent channel. In trickling filter systems, the salts are added to the untreated wastewater or to the filter effluent. Multipoint additions have also been used. Phosphorus is removed from the liquid phase through a combination of precipitation, adsorption, exchange, and agglomeration, and removed from the process with either the primary or secondary sludges, or both. Theoretically, the minimum solubility of $AlPO_4$ occurs at about pH 6.3, and that of $FePO_4$ occurs at about pH 5.3; however, practical applications have yielded good phosphorus removal anywhere in the range of pH 6.5 to 7.0, which is compatible with most biological treatment processes.

The use of ferrous salts is limited because they produce low phosphorus levels only at high pH values. In low-alkalinity waters, either sodium aluminate and alum or ferric plus lime, or both, can be used to maintain the pH higher than 5.5. Improved settling and lower effluent BOD result from chemical addition, particularly if polymer is also added to the final clarifier. Dosages generally fall in the range of a 1 to 3 metal ion-phosphorus molar ratio.

Metal Salt and Polymer Addition to Secondary Clarifiers. In certain cases, such as trickling filtration and extended aeration activated-sludge processes, solids may not flocculate and settle well in the secondary clarifier. This settling problem may become acute in plants that are overloaded. The addition of aluminum or iron salts will cause the precipitation of metallic hydroxides or phosphates, or both. Aluminum and iron salts, along with certain organic polymers, can also be used to coagulate colloidal particles and to improve removals on filters. The resultant coagulated colloids and precipitates will settle readily in the secondary clarifier, reducing the TSS in the effluent and effecting phosphorus removal. Dosages of aluminum and iron salts usually fall in the range of 1 to 3 metal ion/phosphorus on a molar ratio basis if the residual phosphorus in the secondary effluent is greater than 0.5 mg/L. To achieve phosphorus levels below 0.5 mg/L, significantly higher metal salt dosages and filtration will be required.

Polymers may be added (1) to the mixing zone of a highly mixed or internally recirculated clarifier, (2) preceding a static or dynamic mixer, or (3) to an aerated channel. Although mixing times of 10 to 30 seconds have been used for polymers, shorter mixing times are favored (typically less than 10 s, see Table 6–19). Polymers should not be subjected to insufficient or excessive mixing, as noted previously, because the process efficiency will diminish, resulting in poor settling and thickening characteristics.

Phosphorus Removal Using Lime

The use of lime for phosphorus removal is declining because of (1) the substantial increase in the mass of sludge to be handled compared to metal salts and (2) the operation and maintenance problems associated with the handling, storage, and feeding of lime (U.S. EPA, 1987). When lime is used, the principal variables controlling the dosage are the degree of removal required and the alkalinity of the wastewater. The operating dosage must usually be determined by onsite testing. Lime has been used customarily either as a precipitant in the primary sedimentation tanks or following secondary treatment clarification.

Although lime recalcination lowers chemical costs, it is a feasible alternative only for large plants. Where a lime recovery system is required for a cost-effective operation, it includes a thermal regeneration facility, which converts the calcium carbonate in the sludge to lime by heating to 980°C (1800°F). The carbon dioxide from this process or other onsite stack gas (containing 10 to 15 percent carbon dioxide) is generally used as the source of recarbonation for pH adjustment of the wastewater.

Lime Addition to Primary Sedimentation Tanks. Both low and high lime treatment can be used to precipitate a portion of the phosphorus (usually about 65 to 80 percent). When lime is used, both the calcium and the hydroxide react with the orthophosphorus to form an insoluble hydroxyapatite $[Ca_5(OH)(PO_4)_3]$. A residual phosphorus level of 1.0 mg/L can be achieved with the addition of effluent filtration facilities to which chemicals can be added. In the high lime system, sufficient lime is added to raise the pH to about 11 (see Fig. 6–11). After precipitation, the effluent must be recarbonated before biological treatment. In activated-sludge systems, the pH of the primary effluent should not exceed 9.5 or 10; higher pH values can result in biological process upsets. In the trickling filter process, the carbon dioxide generated during treatment is usually sufficient to lower the pH without recarbonation. The dosage for low lime treatment is usually in the range of 75 to 250 mg/L as $Ca(OH)_2$ at pH values of 8.5 to 9.5. In low lime systems, however, the conditions required for precipitation are more specialized; the Ca^{2+}/Mg^{2+} mole ratio is $\leq 5/1$ (Sedlak, 1991).

Lime Addition Following Secondary Treatment. Lime can be added to the waste stream after biological treatment to reduce the level of phosphorus and TSS. Single-stage process and two-stage process flow diagrams for lime addition are shown on Fig. 6–15. On Fig. 6–15a, a single-stage lime precipitation process is used for the treatment of secondary effluent. In the first-stage clarifier of the two-stage process shown on Fig. 6–15b, sufficient lime is added to raise the pH above 11 to precipitate the soluble phosphorus as basic calcium phosphate (apatite). The calcium carbonate precipitate formed in the process acts as a coagulant for TSS removal. An example of a large lime precipitation unit is shown on Fig. 6–16. The excess soluble calcium is removed in the second-stage clarifier as a calcium carbonate precipitate by adding carbon dioxide gas to reduce the pH to about 10. Generally, there is a second injection of carbon dioxide to the second-stage effluent to reduce the formation of scale. To remove the residual levels of TSS and phosphorus, the secondary clarifier effluent is passed through a multimedia filter or a membrane filter. Care should be taken to limit excess calcium in the filter feed to ensure cementing of the filter media will not occur.

Phosphorus Removal with Effluent Filtration

Depending on the quality of the settled secondary effluent, chemical addition has been used to improve the performance of effluent filters. Chemical addition has also been used to achieve specific treatment objectives including the removal of specific contaminants such as phosphorus, metal ions, and humic substances. The removal of phosphorus by chemical addition to the contact filtration process is used in many parts of the country to remove phosphorus from wastewater treatment plant effluents which are discharged to sensitive water bodies. A two-stage filtration process (see Fig. 11–17 and discussion in Chap. 11) has proved to be very effective for the removal of phosphorus. Based on the performance data from full-scale installations, phosphorus levels equal to or less than 0.02 mg/L have been achieved in the filtered effluent. To achieve such low levels of phosphorus removal, the backwash water from the second filter which contains small particles and residual coagulant is recycled to the first filter to improve floc formation within the first-stage filter and the influent to waste ratio.

Comparison of Chemical Phosphorus Removal Processes

The advantages and disadvantages of the removal of phosphorus by the addition of chemicals at various points in a treatment system are summarized in Table 6–7. It is recommended that each alternative point of application be evaluated carefully.

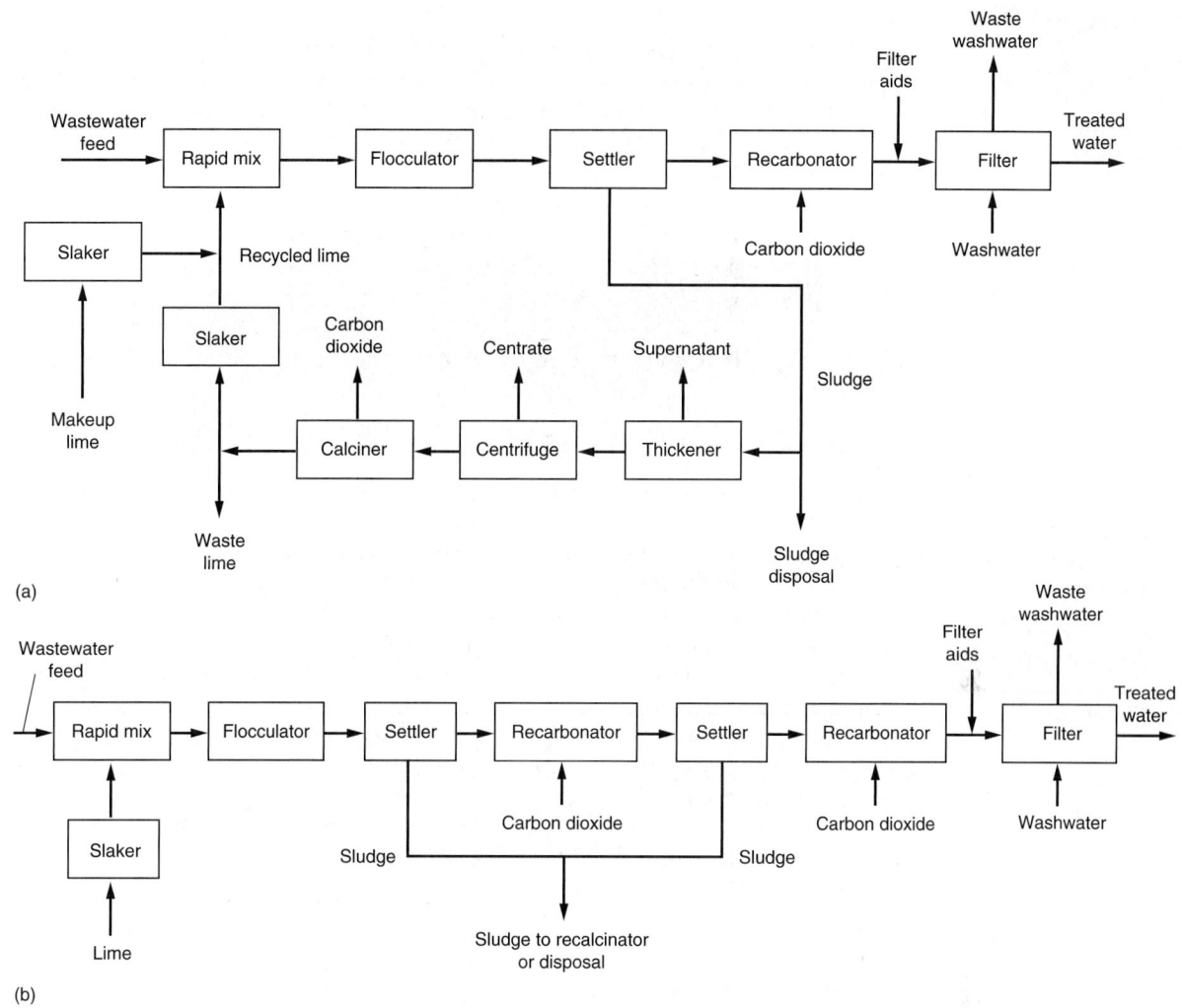

(a)

(b)

Figure 6–15

Typical lime treatment process flow diagrams for phosphorus removal: (a) single-stage system, and (b) two-stage system.

Estimation of Sludge Quantities from Phosphorus Precipitation

The additional BOD and TSS removals afforded by chemical addition to primary treatment may also solve overloading problems on downstream biological systems, or may allow seasonal or year-round nitrification, depending on biological system designs. The BOD removal in the primary sedimentation operation is on the order of 50 to 60 percent

Figure 6-16

Large reactor clarifier used for the lime precipitation of secondary effluent. The settled effluent is treated further by microfiltration before reuse in industrial applications.

Table 6-7

Advantages and disadvantages of chemical addition in various sections of a treatment plant for phosphorus removal[a]

Level of treatment	Advantages	Disadvantages
Primary	Applicable to most plants; increased BOD and suspended solids removal; lowest degree of metal leakage; lime recovery demonstrated	Least efficient use of metal; polymer may be required for flocculation; sludge more difficult to dewater than primary sludge
Secondary	Lowest cost; lower chemical dosage than primary; improved stability of activated sludge; polymer not required	Overdose of metal may cause low pH toxicity; with low-alkalinity wastewaters, a pH control system may be necessary; cannot use lime because of excessive pH; inert solids added to activated-sludge mixed liquor, reducing the percentage of volatile solids
Advanced— precipitation	Lowest phosphorus effluent; most efficient metal use; lime recovery demonstrated	Highest capital cost; highest metal leakage
Advanced—single- and two-stage filtration	Low cost can be combined with the removal of residual suspended solids	Length of filter run may be reduced with single-stage filtration. Additional expense with two-stage filtration process

[a] Adapted from U.S. EPA (1976).

at a pH of 9.5. The amount of primary sludge will also increase significantly. A summary of the pertinent reactions required for determining the quantity of sludge when using alum, iron, or lime for the precipitation of phosphorus is given in Table 6–8. The computational procedures involved in estimating the quantity of sludge resulting from the chemical precipitation of phosphorus with lime are illustrated in Example 6–3.

Table 6–8

Summary of pertinent reactions required to determine quantities of sludge produced during the precipitation of phosphorus with lime, alum, and iron

Reaction	Equation	Chemical species in sludge
Lime, $CaCO_3$		
1. $10Ca^{2+} + 6PO_4^{3-} + 2OH^- \leftrightarrow Ca_{10}(PO_4)_6(OH)_2$	(6–19)	$Ca_{10}(PO_4)_6(OH)_2$
2. $Mg^{2+} + 2OH^- \leftrightarrow Mg(OH)_2$	(6–22)	$Mg(OH)_2$
3. $Ca^{2+} + CO_3^{2-} \leftrightarrow CaCO_3$	(6–23)	$CaCO_3$
Alum, Al (III)		
1. $Al^{3-} + PO_4^{3-} \leftrightarrow AlPO_4$	(6–24)	APO_4
2. $2Al^{3-} + 3OH^- \leftrightarrow Al(OH)_3$	(6–25)	$Al(OH)_3$
Iron, Fe (III)		
1. $Fe^{3+} + PO_4^{3-} \leftrightarrow FePO_4$	(6–26)	$FePO_4$
2. $Fe^{3+} + 3OH^- \leftrightarrow Fe(OH)_3$	(6–27)	$Fe(OH)_3$

EXAMPLE 6–3 **Estimation of Sludge Volume from the Chemical Precipitation of Phosphorus with Lime in a Primary Sedimentation Tank** Estimate the mass and volume of sludge produced in a primary sedimentation tank from the precipitation of phosphorus with lime. Assume that 60 percent of the TSS is removed without the addition of lime, and that the addition of 400 mg/L of $Ca(OH)_2$ results in an increased removal of TSS to 85 percent. Assume the following data apply:

1.	Wastewater flowrate, m^3/d	1000
2.	Wastewater TSS, mg/L	220
3.	Wastewater volatile TSS, mg/L	150
4.	Wastewater PO_4^{3-} as P, mg/L	10
5.	Wastewater total hardness as $CaCO_3$, mg/L	241.3
6.	Wastewater Ca^{2+}, mg/L	80
7.	Wastewater Mg^{2+}, mg/L	10
8.	Effluent PO_4^{3-} as P, mg/L	0.5
9.	Effluent Ca^{2+}, mg/L	60
10.	Effluent Mg^{2+}, mg/L	0
11.	Chemical sludge properties	
	Specific gravity	1.07
	Moisture content, %	92.5

Solution

1. Compute the mass and volume of solids removed without chemicals, assuming that the sludge contains 94 percent moisture and has a specific gravity of 1.03.
   a. Determine the mass of TSS removed.

$$M_{TSS} = \frac{0.6(220 \text{ g/m}^3)(1000 \text{ m}^3/\text{d})}{(10^3 \text{ g/kg})} = 132 \text{ kg/d}$$

   b. Determine the volume of sludge produced.

$$V_s = \frac{(132 \text{ kg/d})}{(1.03)(1000 \text{ kg/m}^3)(0.06)} = 2.14 \text{ m}^3/\text{d}$$

2. Using the equations summarized in Table 6–8, determine the mass of $Ca_5(PO_4)_3OH$, $Mg(OH)_2$, and $CaCO_3$ produced from the addition of 400 mg/L of lime.
   a. Determine the mass of $Ca_5(PO_4)_3OH$ formed.
      i. Determine the moles of P removed.

$$\text{mole P removed} = \frac{(10 - 0.5) \text{ mg/L}}{(30.97 \text{ g/mole})(10^3 \text{ mg/g})}$$

$$= 0.307 \times 10^{-3} \text{ mole/L}$$

      ii. Determine the moles of $Ca_5(PO_4)_3OH$ formed.

$$\text{mole } Ca_5(PO_4)_3OH \text{ formed} = 1/3 \times 0.307 \times 10^{-3} \text{ mole/L}$$

$$= 0.102 \times 10^{-3} \text{ mole/L}$$

      iii. Determine the mass of $Ca_5(PO_4)_3OH$ formed.

$$\text{Mass } Ca_5 (PO_4)_3OH = 0.102 \times 10^{-3} \text{ mole/L} \times 502 \text{ g/mole} \times 10^3 \text{ mg/g}$$

$$= 51.3 \text{ mg/L}$$

   b. Determine the mass of $Mg(OH)_2$ formed.
      i. Determine the moles of Mg^{2+} removed.

$$\text{mole } Mg^{2+} \text{ removed} = \frac{(10 \text{ mg/L})}{(24.31 \text{ g/mole})(10^3 \text{ mg/g})}$$

$$= 0.411 \times 10^{-3} \text{ mole/L}$$

      ii. Determine the mass of $Mg(OH)_2$ formed.

$$\text{mole } Mg(OH)_2 = 0.411 \times 10^{-3} \text{ mole/L} \times 58.3 \text{ g/mole} \times 10^3 \text{ mg/g}$$

$$= 24.0 \text{ mg/L}$$

   c. Determine the mass of $CaCO_3$ formed.
      i. Determine the mass of Ca^{2+} in $Ca_5(PO_4)_3(OH)$.

$$\text{Mass } Ca^{2+} \text{ in } Ca_5(PO_4)_3(OH)$$

$$= 5(40 \text{ g/mole})(0.102 \times 10^{-3} \text{ mole/L})(10^3 \text{ mg/g})$$

$$= 20.4 \text{ mg/L}$$

ii. Determine the mass of Ca^{2+} added in the original dosage.

$$\text{Mass } Ca^{2+} \text{ in } Ca(OH)_2 = \frac{(40 \text{ g/mole})(400 \text{ mg/L})}{(74 \text{ g/mole})}$$

$$= 216.2 \text{ mg/L}$$

iii. Determine the mass of Ca present as $CaCO_3$.

$$Ca^{2+} \text{ in } CaCO_3 = \text{Ca in } Ca(HO)_2 + Ca^{2+} \text{ in influent wastewater} -$$

$$Ca^{2+} \text{ in } Ca_5(PO_4)_3OH - Ca^{2+} \text{ in effluent wastewater}$$

$$= 216.2 + 80 - 20.4 - 60$$

$$= 215.8 \text{ mg/L}$$

iv. Determine the mass of $CaCO_3$.

$$\text{Mass } CaCO_3 = \frac{(100 \text{ g/mole})(215.8 \text{ mg/L})}{(40 \text{ g/mole})}$$

$$= 540 \text{ mg/L}$$

3. Determine the total mass of solids removed as a result of the lime dosage.
   a. TSS in wastewater

   $$M_{TSS} = \frac{0.85(220 \text{ g/m}^3)(1000 \text{ m}^3/\text{d})}{(10^3 \text{ g/kg})} = 187 \text{ kg/d}$$

   b. Chemical solids

   $$M_{Ca_5(PO_4)_3OH} = \frac{(51.2 \text{ g/m}^3)(1000 \text{ m}^3/\text{d})}{(10^3 \text{ g/kg})} = 51.2 \text{ kg/d}$$

   $$M_{Mg(OH)_2} = \frac{(24 \text{ g/m}^3)(1000 \text{ m}^3/\text{d})}{(10^3 \text{ g/kg})} = 24.0 \text{ kg/d}$$

   $$M_{CaCO_3} = \frac{(540 \text{ g/m}^3)(1000 \text{ m}^3/\text{d})}{(10^3 \text{ g/kg})} = 540 \text{ kg/d}$$

   c. Total mass of solids removed

   $$M_T = (187 + 51.3 + 24 + 540) \text{ kg/d}$$

   $$= 802.3 \text{ kg/d}$$

4. Determine the total volume of sludge resulting from chemical precipitation, assuming that the sludge has a specific gravity of 1.07 and a moisture content of 92.5 percent (see Chap. 14).

$$V_s = \frac{(802.3 \text{ kg/d})}{(1.07)(1000 \text{ kg/m}^3)(0.075)} = 10.0 \text{ m}^3/\text{d}$$

5. Prepare a summary table of sludge masses and volumes without and with chemical precipitation.

Treatment	Sludge	
	Mass, kg/d	Volume, m³/d
Without chemical precipitation	132.0	2.14
With chemical precipitation	802.3	10.0

Comment The magnitude of the sludge disposal problem associated with high lime treatment for phosphorus removal as compared to biological phosphorus removal is illustrated in this example.

6-5 CHEMICAL PRECIPITATION FOR REMOVAL OF HEAVY METALS AND DISSOLVED INORGANIC SUBSTANCES

The technologies available for the removal of heavy metals from wastewater include chemical precipitation, carbon adsorption, ion exchange, and reverse osmosis. Of these technologies, chemical precipitation is most commonly employed for most of the metals. Common precipitants include hydroxide (OH) and sulfide (S^{2-}). Carbonate (CO_3^{2-}) has also been used in some special cases. Metal may be removed separately or coprecipitated with phosphorus.

Precipitation Reactions

Metals of interest include arsenic (As), barium (Ba), cadmium (Cd), copper (Cu), mercury (Hg), nickel (Ni), selenium (Se), and zinc (Zn). Most of these metals can be precipitated as hydroxides or sulfides. Solubility products for free metal concentrations in equilibrium with hydroxide and sulfide precipitates are reported in Table 6–9. In wastewater treatment facilities, metals are precipitated most commonly as metal hydroxides through the addition of lime or caustic to a pH of minimum solubility. However, several of these compounds are, as discussed previously, amphoteric (i.e., capable of either accepting or donating a proton) and exhibit a point of minimum solubility. The pH value at minimum solubility varies with the metal in question as illustrated on Fig. 6–17 for hydroxide precipitation. The solid line on Fig. 6–17 represents the total metal in solution in equilibrium with the precipitate. The curves were developed based on the mononuclear hydroxide species using the same procedures as illustrated on Fig. 6–8 for Al^{3+} and Fe^{3+}. It is important to remember that the location of the minimum solubility will vary depending on the constituents in the wastewater. The curves given on Fig. 6–17 are useful in establishing the pH ranges for testing. Metals can also be precipitated as the sulfides as illustrated on Fig. 6–18. The minimum effluent concentration levels that

Table 6–9
Solubility products for free metal ion concentrations in equilibrium with hydroxides and sulfides[a,b]

Disinfectant	Half reaction	pK_{sp}
Cadmium hydroxide	$Cd(OH)_2 \leftrightarrow Cd^{2+} + 2OH^-$	13.93
Cadmium sulfide	$CdS \leftrightarrow Cd^{2+} + S^{2-}$	28
Chromium hydroxide	$Cr(OH)_3 \leftrightarrow Cr^{3+} + 3OH^-$	30.2
Copper hydroxide	$Cu(OH)_2 \leftrightarrow Cu^{2+} + 2OH^-$	19.66
Copper sulfide	$CuS \leftrightarrow Cu^{2+} + S^{2-}$	35.2
Iron (II) hydroxide	$Fe(OH)_2 \leftrightarrow Fe^{2+} + 2OH^-$	14.66
Iron (II) sulfide	$FeS \leftrightarrow Fe^{2+} + S^{2-}$	17.2
Lead hydroxide	$Pb(OH)_2 \leftrightarrow Pb^{2+} + 2OH^-$	14.93
Lead sulfide	$PbS \leftrightarrow Pb^{2+} + S^{2-}$	28.15
Mercury hydroxide	$Hg(OH)_2 \leftrightarrow Hg^{2+} + 2OH^-$	23
Mercury sulfide	$HgS \leftrightarrow Hg^{2+} + S^{2-}$	52
Nickel hydroxide	$Ni(OH)_2 \leftrightarrow Ni^{2+} + 2OH^-$	15
Nickel sulfide	$NiS \leftrightarrow Ni^{2+} + S^{2-}$	24
Silver hydroxide	$AgOH \leftrightarrow Ag^+ + OH^-$	14.93
Silver sulfide	$(Ag)_2S \leftrightarrow 2Ag^+ + S^{2-}$	28.15
Zinc hydroxide	$Zn(OH)_2 \leftrightarrow Zn^{2+} + 2OH^-$	16.7
Zinc sulfide	$ZnS \leftrightarrow Zn^{2+} + S^{2-}$	22.8

[a] Adapted from Bard (1966).
[b] To obtain the complete solubility of a metal, all of the complex species must be considered such as reported in Table 6–2 for aluminum and iron.

Figure 6–17

Residual soluble metal concentration as a function of pH for the precipitation of metals as hydroxides. Because of the wide variation in the solubility and formation constants for the various metal hydroxides, the curves presented in this figure should only be used as a reference guide (see also Table 6–10).

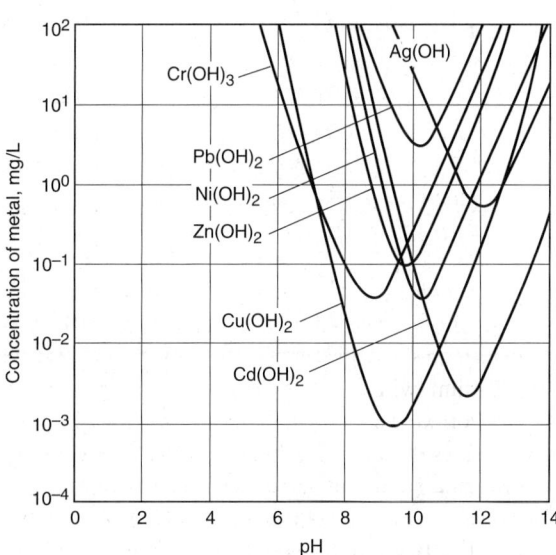

Figure 6–18

Residual soluble metal concentration as a function of pH for the precipitation of metals as sulfides. Because of the wide variation in the solubility and formation constants for the various metal sulfides, the curves presented in this figure should only be used as a reference guide (see also Table 6–10).

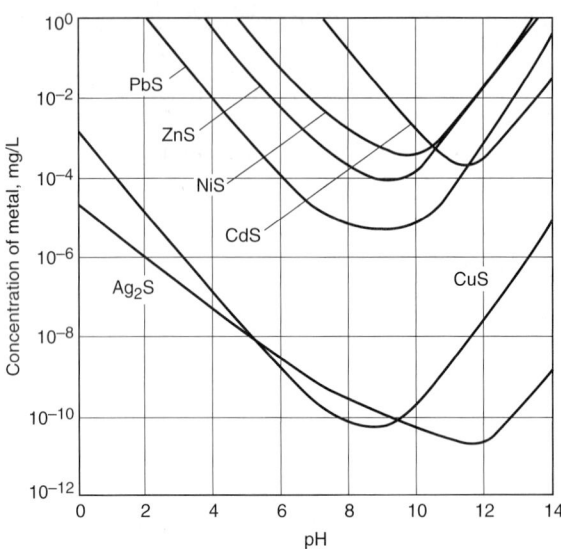

Table 6–10

Practical effluent concentration levels achievable in heavy metals removal by precipitation[a]

Metal	Achievable effluent concentration, mg/L	Type of precipitation and technology
Arsenic	0.05	Sulfide precipitation with filtration
	0.005	Ferric hydroxide coprecipitation
Barium	0.5	Sulfate precipitation
Cadmium	0.05	Hydroxide precipitation at pH 10–11
	0.05	Coprecipitation with ferric hydroxide
	0.008	Sulfide precipitation
Copper	0.02–0.07	Hydroxide precipitation
	0.01–0.02	Sulfide precipitation
Mercury	0.01–0.02	Sulfide precipitation
	0.001–0.01	Alum coprecipitation
	0.0005–0.005	Ferric hydroxide coprecipitation
	0.001–0.005	Ion exchange
Nickel	0.12	Hydroxide precipitation at pH 10
Selenium	0.05	Sulfide precipitation
Zinc	0.1	Hydroxide precipitation at pH 11

[a]From Eckenfelder (2000).

can be achieved in the chemical precipitation of heavy metals are reported in Table 6–10. In practice, the minimum achievable residual metal concentrations will also depend on the nature and concentration of the organic matter in the wastewater as well as the temperature. Because of the many uncertainties associated with the precipitation of metals, laboratory bench-scale or pilot-plant testing should be conducted.

Coprecipitation with Phosphorus

As discussed previously, precipitation of phosphorus in wastewater is usually accomplished by the addition of coagulants, such as alum, lime or iron salts, and polyelectrolytes. Coincidentally with the addition of these chemicals for the removal of phosphorus is the removal that occurs of various inorganic ions, principally some of the heavy metals. Where both industrial and domestic wastes are treated together, it may be necessary to add chemicals to the primary settling facilities, especially if onsite pretreatment measures prove to be ineffective. When chemical precipitation is used, anaerobic digestion for sludge stabilization may not be possible because of the toxicity of the precipitated heavy metals. As noted previously, one of the disadvantages of chemical precipitation is that it usually results in a net increase in the total dissolved solids of the wastewater that is being treated.

6–6 CHEMICAL OXIDATION

Chemical oxidation in wastewater treatment typically involves the use of oxidizing agents such as ozone (O_3), hydrogen peroxide (H_2O_2), permanganate (MnO_4), chloride dioxide (ClO_2), chlorine (Cl_2) or (HOCl), and oxygen (O_2), to bring about change in the chemical composition of a compound or a group of compounds. Included in the following discussion is an introduction of the fundamental concepts involved in chemical oxidation, an overview of the uses of chemical oxidation in wastewater treatment, and a discussion of the use of chemical oxidation for the reduction of BOD and COD, the oxidation of ammonia, and oxidation of nonbiodegradable organic compounds. Advanced oxidation process (AOPs) in which the free hydroxyl radical (HO°) is used as a strong oxidant to destroy specific organic constituents and compounds that cannot be oxidized by conventional oxidants such as ozone and chlorine are discussed in Chap. 11, which deals with advanced treatment methods.

Fundamentals of Chemical Oxidation

The purpose of the following discussion is to introduce the basic concepts involved in chemical oxidation reactions. The topics to be discussed include (1) oxidation-reduction reactions, (2) half reaction potentials, (3) reaction potentials, (4) equilibrium constants for redox equations, and (5) rate of oxidation-reduction reactions.

Oxidation-Reduction Reactions. Oxidation-reduction reactions (known as redox equations) take place between an *oxidizing agent* and a *reducing agent*. In oxidation-reduction reactions both electrons are exchanged as are the oxidation states of the constituents involved in the reaction. While an oxidizing agent causes the oxidation to occur, it is reduced in the process. Similarly, a reducing agent that causes a reduction to occur is oxidized in the process. For example, consider the following reduction:

$$Cu^{2+} + Zn \leftrightarrow Cu + Zn^{2+} \tag{6–28}$$

In the above reaction copper (Cu) changes from a +2 to zero oxidation state and the zinc (Zn) changes from a zero to a +2 state. Because of the electron gain or loss, oxidation-reduction reactions can be separated into two half reactions. The oxidation half reaction involves the loss of electrons while the reduction half reaction involves the gain of electrons. The two half reactions that comprise Eq. (6–28) are as follows:

$$Zn - 2e^- \leftrightarrow Zn^{2+} \quad \text{(oxidation)} \tag{6-29}$$

$$Cu^{2+} + 2e^- \leftrightarrow Cu \quad \text{(reduction)} \tag{6-30}$$

Referring to the above equations, there is a two-electron change.

Half-Reaction Potentials. Because of the almost infinite number of possible reactions, there are no summary tables of equilibrium constants for oxidation-reduction reactions. What is done instead is the chemical and thermodynamic characteristics of the half reactions, such as those given by Eqs. (6–29) and (6–30), are determined and tabulated so that any combination of reactions can be studied. Half reactions for disinfection processes are given in Table 6–11 and other representative half reactions are given in Table 6–12. Of the many properties that can be used to characterize oxidation-reduction reactions, the electrical potential (i.e., voltage) or emf of the half reaction is used most commonly. Thus, every half reaction involving an oxidation or reduction has a *standard potential $E°$* associated with it. The potentials for the half reactions given by Eqs. (6–29) and (6–30) are as follows:

$$Cu^{2+} + 2e^- \leftrightarrow Cu \quad E° = 0.34 \text{ volt} \tag{6-31}$$

$$Zn + 2e^- \leftrightarrow Zn^{2+} \quad E° = -0.763 \text{ volt} \tag{6-32}$$

The potentials for a number of half reactions are given in Table 6–12. The half-reaction potential is a measure of the tendency of a reaction to proceed to the right. Half reactions with large positive potential, $E°$, tend to proceed to the right as written. Conversely, half reactions with large negative potential, $E°$, tend to proceed to the left.

Table 6–11
Standard electrode potentials for oxidation half reactions for chemical disinfectants[a]

Disinfectant	Half reaction	Oxidation potential,[b] V
Ozone	$O_3 + 2H^+ + 2e^- \leftrightarrow O_2 + H_2O$	+2.07
Hydrogen peroxide	$H_2O_2 + 2H^+ + 2e^- \rightarrow 2H_2O$	+1.78
Permanganate	$MnO_4^- + 4H^+ + 3e^- \leftrightarrow MnO_2 + 2H_2O$	+1.67
Chlorine dioxide	$ClO_2 + e^- \leftrightarrow ClO_2^-$	+1.50
Hypochlorous acid	$HOCl + H^+ + 2e^- \leftrightarrow Cl^- + H_2O$	+1.49
Hypoiodous acid	$HOI + H^+ + e^- \leftrightarrow 1/2I_2 + H_2O$	+1.45
Chlorine gas	$Cl_2 + 2e^- \leftrightarrow 2Cl^-$	+1.36
Oxygen	$O_2 + 4H^+ + 4e^- \leftrightarrow 2H_2O$	+1.23
Bromine	$Br_2 + 2e^- \leftrightarrow 2Br^-$	+1.09
Hypochlorite	$OCl^- + H_2O + 2e^- \leftrightarrow Cl^- + 2OH^-$	+0.90
Chlorite	$ClO_2^- + 2H_2O + 4e^- \leftrightarrow Cl^- + 4OH^-$	+0.76
Iodine	$I_2 + 2e^- \leftrightarrow 2I^-$	+0.54

[a] Derived in part form Bard (1966) and White (1999).
[b] Reported values will vary depending on source.

Table 6–12
Selected standard electrode potentials for oxidation-reduction half reactions[a]

Half reaction	Oxidation potential,[b] V
$Li^+ + e^- \rightarrow Li$	−3.03
$K^+ + e^- \rightarrow K$	−2.92
$Ba^{2+} + 2e^- \rightarrow Ba$	−2.90
$Ca^{2+} + 2e^- \rightarrow Ca$	−2.87
$Na^+ + e^- \rightarrow Na$	−2.71
$Mg(OH)_2 + 2e^- \rightarrow Mg + 2OH^-$	−2.69
$Mg^{2+} + 2e^- \rightarrow Mg$	−2.37
$Al^{3+} + 3e^- \rightarrow Al$	−1.66
$MnO_4^- + 8H^+ + 5e^- \rightarrow Mn^{2+} + 4H_2O$	−1.51
$Mn^{2+} + 2e^- \rightarrow Mn$	−1.18
$2H_2O + 2e^- \rightarrow H_2 + 2OH^-$	−0.828
$Zn^{2+} + 2e^- \rightarrow Zn$	−0.763
$Fe^{2+} + 2e^- \rightarrow Fe$	−0.440
$Cd^{2+} + 2e^- \rightarrow Cd$	−0.40
$Ni^{2+} + 2e^- \rightarrow Ni$	−0.250
$S + 2H^+ + 2e^- \rightarrow H_2S$	−0.14
$Pb^{2+} + 2e^- \rightarrow Pb$	−0.126
$2H^+ + 2e^- \rightarrow H_2$	0.000
$Cu^{2+} + e^- \rightarrow Cu^+$	+0.15
$N_2 + 4H^+ + 3e^- \rightarrow NH_4^+$	+0.27
$Cu^{2+} + 2e^- \rightarrow Cu$	+0.34
$I_2 + 2e^- \rightarrow 2I^-$	+0.54
$O_2 + 2H^+ + 2e^- \rightarrow H_2O_2$	+0.68
$Fe^{3+} + e^- \rightarrow Fe^{2+}$	+0.771
$Ag^+ + e^- \rightarrow Ag$	+0.799
$ClO^- + H_2O + 2e^- \rightarrow Cl^- + 2OH^-$	+0.90
$Br_2(aq) + 2e^- \rightarrow 2Br^-$	+1.09
$O_2 + 4H^+ + 4e^- \rightarrow 2H_2O$	+1.229
$Cl_2(g) + 2e^- \rightarrow 2Cl^-$	+1.360
$H_2O_2 + 2H^+ + 2e^- \rightarrow 2H_2O$	+1.776
$O_3 + 2H^+ + 2e^- \leftrightarrow O_2 + H_2O$	+2.07
$F_2 + 2H^+ + 2e^- \rightarrow 2HF$	+2.87

[a] Adapted in part from Bard (1966) and Benefield et al. (1982).
[b] Reported values will vary depending on source.

Reaction Potentials.   The half-reaction potentials, discussed above, can be used to predict whether a reaction comprised of two half reactions will proceed as written. The tendency of a reaction to proceed is obtained by determining the $E^\circ_{reaction}$ for the entire reaction as given by the following expression.

$$E^\circ_{reaction} = E^\circ_{reduction} - E^\circ_{oxidation} \qquad (6\text{--}33)$$

where $E^\circ_{reaction}$ = potential of the overall reaction
$\qquad E^\circ_{reduction}$ = potential of the reduction half reaction
$\qquad E^\circ_{oxidation}$ = potential of the oxidation half reaction

For example, for the reaction between copper and zinc [see Eq. (6–28)] the $E^\circ_{reaction}$ of the reaction is determined as follows:

$$E^\circ_{reaction} = E^\circ_{Cu^{2+},Cu} - E^\circ_{Zn^{2+},Zn} \qquad (6\text{--}34)$$

$$E^\circ_{reaction} = 0.34 - (-0.763) = +1.103 \text{ volts} \qquad (6\text{--}35)$$

The positive value for the $E^\circ_{reaction}$ is taken as an indication that the reaction will proceed as written. The magnitude of the value, as will be illustrated subsequently, can be taken as a measure of the extent to which the reaction as written will proceed. For example, if Eq. (6–28) had been written as follows:

$$Cu + Zn^{2+} \leftrightarrow Cu^{2+} + Zn \qquad (6\text{--}36)$$

The corresponding $E^\circ_{reaction}$ for this reaction is

$$E^\circ_{reaction} = E^\circ_{Zn^{2+},Zn} - E^\circ_{Cu^{2+},Cu} \qquad (6\text{--}37)$$

$$E^\circ_{reaction} = (-0.763) - 0.34 = -1.103 \text{ volts} \qquad (6\text{--}38)$$

Because the $E^\circ_{reaction}$ for the reaction is negative, the reaction will proceed in the opposite direction from what is written.

EXAMPLE 6–4   **Determination of Reaction Potential**   Determine whether hydrogen sulfide (H_2S) can be oxidized with hydrogen peroxide (H_2O_2). The pertinent half reactions from Table 6–12 are as follows.

$$H_2S \leftrightarrow S + 2H^+ + 2e^- \qquad E^\circ = -0.14$$
$$H_2O_2 + 2H^+ + 2e^- \leftrightarrow 2H_2O \qquad E^\circ = +1.776$$

Solution

1.   Determine the overall reaction by adding the two half reactions.

$$H_2S \leftrightarrow S + 2H^+ + 2e^-$$
$$\underline{H_2O_2 + 2H^+ + 2e^- \leftrightarrow 2H_2O}$$
$$H_2S + H_2O_2 \leftrightarrow S + 2H_2O$$

2. Determine the $E^\circ_{\text{reaction}}$ for overall reaction.

$$E^\circ_{\text{reaction}} = E^\circ_{H_2O_2{}^{2+},H_2O} - E^\circ_{s^{2-},s}$$

$$E^\circ_{\text{reaction}} = (1.78) - (-0.14) = +1.92 \text{ volts}$$

Because the $E^\circ_{\text{reaction}}$ for the reaction is positive, the reaction will proceed as written.

Equilibrium Constants for Redox Equations. The equilibrium constant for oxidation reduction reactions is calculated using the Nernst equation as defined below.

$$\ln K = \frac{nFE^\circ_{\text{reaction}}}{RT} \tag{6-39a}$$

$$\log K = \frac{nFE^\circ_{\text{reaction}}}{2.303 \, RT} \tag{6-39b}$$

where K = equilibrium constant
$\quad n$ = number of electrons exchanged in the overall reaction
$\quad F$ = Faraday's constant
$\quad\quad$ = 96,485 a·s/g eq = 96,485 C/g eq $\quad$ (*Note:* C = coulomb)
$\quad E^\circ_{\text{reaction}}$ = reaction potential
$\quad\quad R$ = universal gas constant
$\quad\quad$ = 8.3144 J (abs)/mole·K
$\quad\quad T$ = temperature, K (273.15 + °C)

For example, at 25°C

$$\log K = \frac{n \,(96{,}485 \text{ C/g eq}) \, E^\circ_{\text{reaction}}}{(2.303)(8.3144 \text{ J/mole·K})[(273.15 + 25)K]} = \frac{nE^\circ_{\text{reaction}}}{0.0592}$$

The application of this equation is illustrated in Example 6–5.

EXAMPLE 6-5 **Determination of Equilibrium Constant for Oxidation-Reduction Reaction**
Determine the equilibrium constant for the following oxidation-reduction reactions:

$$Cu^{2+} + Zn \leftrightarrow Cu + Zn^{2+}$$

$$H_2S + H_2O_2 \leftrightarrow S + 2H_2O$$

Solution

1. Determine the equilibrium constant for the following equation using Eq. (6–39):

$$Cu^{2+} + Zn \leftrightarrow Cu + Zn^{2+}$$

As computed above, the $E°_{\text{reaction}}$ for the reaction is $+1.1$ volts, and the number of electrons exchanged is 2. Using this information, the K value is determined as follows:

$$\log K = \frac{nE°_{\text{reaction}}}{0.0592} = \frac{2(1.10)}{0.0592} = 37.2$$

$$K = 1.58 \times 10^{37} = \frac{[Zn^{2+}]}{[Cu^{2+}]}$$

2. Determine the equilibrium constant for the following equation using Eq. (6–39):

$$H_2S + H_2O_2 \Leftrightarrow S + 2H_2O$$

From Example 6–4, the $E°_{\text{reaction}}$ for the above reactions is $+1.92$ volts. The value of the equilibrium constant is

$$\log K = \frac{nE°_{\text{reaction}}}{0.0592} = \frac{2(1.92)}{0.0592} = 64.9$$

$$K = 7.94 \times 10^{64} = \frac{[S]}{[H_2S][H_2O_2]}$$

Rate of Oxidation-Reduction Reactions. As noted previously, the half-reaction potentials can be used to predict whether a reaction will proceed as written. Unfortunately, the reaction potential provides no information about the rate at which the reaction will proceed. Chemical oxidation reactions often require the presence of one or more catalysts for the reaction to proceed or to increase the rate of reaction. Transition metal cations, enzymes, pH adjustment, and a variety of proprietary substances have been used as catalysts.

Applications

Some of the more important applications of chemical oxidation in wastewater management are summarized in Table 6–13. In the past, chemical oxidation was used most commonly to (1) reduce the concentration of residual organics, (2) control odors, (3) remove ammonia, and (4) reduce the bacterial and viral content of wastewaters. Chemical oxidation is especially effective for the elimination of odorous compounds (e.g., oxidation of sulfides and mercaptans). Because of its importance, chemical disinfection is considered separately in Chap. 12. In addition to the applications reported in Table 6–13, chemical oxidation is now commonly used to (1) improve the treatability of nonbiodegradable (refractory) organic compounds, (2) eliminate the inhibitory effects of certain organic and inorganic compounds to microbial growth, and (3) reduce or eliminate the toxicity of certain organic and inorganic compounds to microbial growth and aquatic flora. The chemical oxidation of BOD and COD, ammonia, and

Table 6-13

Typical applications of chemical oxidation in wastewater collection, treatment, and disposal

Application	Chemicals used[a]	Remarks
Collection		
Slime-growth control	Cl_2, H_2O_2	Control of fungi and slime-producing bacteria
Corrosion control (H_2S)	Cl_2, H_2O_2, O_3	Control brought about by oxidation of H_2S
Odor control	Cl_2, H_2O_2, O_3	Especially in pumping stations and long, flat sewers
Treatment		
Grease removal	Cl_2	Added before preaeration
BOD reduction	Cl_2, O_3	Oxidation of organic substances
Ferrous sulfate oxidation	Cl_2[b]	Production of ferric sulfate and ferric chloride
Filter-ponding control	Cl_2	Maintaining residual at filter nozzles
Filter-fly control	Cl_2	Maintaining residual at filter nozzles during fly season
Sludge-bulking control	Cl_2, H_2O_2, O_3	Temporary control measure
Control of filamentous microorganisms	Cl_2	Dilute chlorine solution sprayed on foam caused by filamentous organisms
Digester supernatant oxidation	Cl_2	
Digester foaming control	Cl_2	
Ammonia oxidation	Cl_2	Conversion of ammonia to nitrogen gas
Odor contol	Cl_2, H_2O_2, O_3	
Oxidation of refractory organic compounds	O_3	
Dispersal		
Bacterial reduction	Cl_2, H_2O_2, O_3	Plant effluent, overflows, and stormwater
Odor control	Cl_2, H_2O_2, O_3	

[a] Cl_2 = chlorine, H_2O_2 = hydrogen peroxide, O_3 = ozone.
[b] $6FeSO_4 \cdot 7H_2O + 3Cl_2 \rightarrow 2FeCl_3 + 2Fe_2(SO_4)_3 + 42H_2O$.

refractory organic compounds is considered in this section. Advanced oxidation processes are considered in Chap. 11.

Chemical Oxidation of BOD and COD

The overall reaction for the oxidation of organic molecules comprising BOD, for example, with chlorine, ozone, and hydrogen peroxide, can be represented as follows:

$$
\begin{array}{ccccc}
\text{Organic} & Cl_2, O_3 & \text{Intermediate} & Cl, O_3 & \text{Simple} \\
\text{molecule} & \rightarrow\rightarrow\rightarrow & \text{oxygenated} & \rightarrow\rightarrow\rightarrow & \text{end products} \\
\text{(e.g., BOD)} & H_2O_2 & \text{molecules} & H_2O_2 & \text{(e.g., } CO_2, H_2O, \text{ etc.)}
\end{array}
\qquad (6\text{-}40)
$$

Multiple arrows in the direction of the reaction are used to signify that a number of steps are involved in the overall reaction sequence. The use of oxidizing agents such as oxygen, chlorine, ozone, and hydrogen peroxide is termed "simple oxidation." In general the overall reaction rates are usually too slow to be applicable generally for wastewater treatment (SES, 1994). Advanced oxidation processes (AOPs), which typically involve the use of the hydroxyl radical for the oxidation of complex organic molecules, are considered in Chap. 11.

Chemical Oxidation of Nonbiodegradable Organic Compounds.

Typical chemical dosages for both chlorine and ozone for the oxidation of the organics in wastewater are reported in Table 6–14. As shown in Table 6–14, the dosages increase with the degree of treatment, which is reasonable when it is considered that the organic compounds that remain after biological treatment are typically composed of low-molecular-weight polar organic compounds and complex organic compounds built around the benzene ring structure. Because of the complexities associated with composition of wastewater, chemical dosages for the removal of refractory organic compounds cannot be derived from the chemical stoichiometry, assuming that it is known. Pilot-plant studies must be conducted when either chlorine, chlorine dioxide, or ozone is to be used for the oxidation of refractory organics to assess both the efficacy and required dosages. The application of AOPs involving more than one oxidizing agent (e.g., ozone/hydrogen peroxide) is considered in Chap. 11.

Chemical Oxidation of Ammonia

The chemical process in which chlorine is used to oxidize the ammonia nitrogen in solution to nitrogen gas and other stable compounds is known as breakpoint chlorination. Perhaps the most important advantage of this process is that, with proper control, all the ammonia nitrogen in the wastewater can be oxidized. However, because the process has a number of disadvantages including the buildup of acid (HCl) which will react with the alkalinity, the buildup of total dissolved solids, and the formation of unwanted chloro-organic compounds, ammonia oxidation is seldom used today.

Table 6–14
Typical chemical dosages for the oxidation of organics in wastewater[a]

Chemical	Use	Dosage, kg/kg destroyed	
		Range	Typical
Chlorine	BOD reduction		
	Settled wastewater	0.5–2.5	1.75
	Secondary effluent	1.0–3.0	2.0
Ozone	COD reduction		
	Settled wastewater	2.0–4.0	3.0
	Secondary effluent	3.0–8.0	6.0

[a]Derived in part from White (1999).

Analysis. Although the theory of breakpoint chlorination is described in greater detail in Sec. 12–3 in Chap. 12, it is, nevertheless, instructive to assess whether the reaction is feasible using the oxidation-reduction concepts presented above. The pertinent half reactions are as follows:

$$HOCl + H^+ + 2e^- \leftrightarrow Cl^- + H_2O \qquad E° = +1.49 \text{ volts} \qquad (6\text{--}41)$$

$$N_2 + 8H^+ + 6e^- \leftrightarrow 2NH_4^+ \qquad E° = +0.27 \text{ volt} \qquad (6\text{--}42)$$

Rewrite Eq. (6–42) as a reduction

$$2NH_4^+ \leftrightarrow N_2 + 8H^+ + 6e^- \qquad E° = -0.27 \text{ volt} \qquad (6\text{--}43)$$

Combining Eqs. (6–43) and (6–41) yields

$$2NH_4^+ \leftrightarrow N_2 + 8H^+ + 6e^-$$

$$\frac{(3)HOCl + (3)H^+ + (3)2e^- \leftrightarrow (3)Cl^- + (3)H_2O}{3HOCl + 2NH_4^+ \leftrightarrow N_2 + 3HCl + 8H^+ + 3H_2O} \qquad (6\text{--}44)$$

Determine the $E°_{reaction}$ for overall reaction

$$E°_{reaction} = E°_{HOCl,Cl^-} - E°_{NH_4^+,N_2}$$

$$E°_{reaction} = (1.49) - (-0.27) = +1.96 \text{ volts}$$

Because the $E°_{reaction}$ for the reaction is positive, the reaction will proceed as written. The stoichiometric mass ratio of chlorine as Cl_2 to ammonia as N, as computed using Eq. (6–44), is 7.6:1. In practice, the ratio has been found to vary from 8:1 to 10:1.

From both laboratory studies and full-scale testing, it has been found that the optimum operating pH range for breakpoint chlorination is between 6 and 7. If breakpoint chlorination is accomplished outside this range, it has been observed that the chlorine dosage required to reach the breakpoint increases significantly and that the rate of reaction is slower. Temperature does not appear to have a major effect on the process in the ranges normally encountered in wastewater treatment.

Application. The breakpoint chlorination process can be used for the removal of ammonia nitrogen from treatment-plant effluents, either alone or in combination with other processes. To avoid the large chlorine dosages required when used alone, breakpoint chlorination can be used following biological nitrification to achieve low levels of ammonia in the effluent.

To optimize the performance of this process and to minimize equipment and facility costs, flow equalization is usually required. Also, because of the potential toxicity problems that may develop if chlorinated compounds are discharged to the environment (see Sec. 12-3, Chap. 12), it is usually necessary to dechlorinate the effluent. The use of the breakpoint chlorination process for seasonal nitrogen control is considered in Example 6–6.

EXAMPLE 6–6 **Analysis of Breakpoint Chlorination Process Used for Seasonal Control of Nitrogen** Estimate the daily required chlorine dosage and the resulting buildup of total dissolved solids when breakpoint chlorination is used for the seasonal control of nitrogen. Assume that the following data apply to this problem:

1. Plant flowrate = 3800 m^3/d
2. Effluent characteristics after conventional treatment
   a. BOD = 20 g/m^3 (mg/L)
   b. TSS = 25 g/m^3
   c. $NH_3 - N$ concentration = 23 g/m^3
3. Required effluent $NH_3 - N$ concentration = 1.0 g/m^3

Solution

1. Estimate the required Cl_2 dosage. Assume that the required mass ratio of chlorine to ammonia is 9:1.

$$\text{kg } Cl_2/d = \frac{(3800 \text{ m}^3/\text{d}) \left[(23 - 1) \text{ g/m}^3 \right] (9.0)}{(10^3 \text{ g/kg})} = 752.4 \text{ kg/d}$$

2. Determine the increment of total dissolved solids added to the wastewater. Using the data reported in Table 12-10 in Chap. 12, the total dissolved solids increase per mg/L of ammonia consumed is equal to 6.2

Total dissolved solids increment = 6.2 (23 − 1) g/m^3 = 136 g/m^3

Comment In this example, it was assumed that the acid produced from the breakpoint reaction would not require the addition of a neutralizing agent such as NaOH (sodium hydroxide). If the addition of NaOH were required, the total dissolved solids increase would have been significantly large. Although breakpoint chlorination can be used to control nitrogen, it may be counterproductive if in the process the treated effluent is rendered unusable for reuse applications because of the buildup of total dissolved solids.

6–7 CHEMICAL NEUTRALIZATION, SCALE CONTROL, AND STABILIZATION

The removal of excess acidity or alkalinity by treatment with a chemical of the opposite composition is termed *neutralization*. In general, all treated wastewaters with excessively low or high pH will require neutralization before they can be dispersed to the environment. Scaling control is required for nanofiltration and reverse osmosis treatment to control the formation of scale, which can severely impact performance. Chemical stabilization is often required for highly treated wastewaters to control their aggressiveness with respect to corrosion. These subjects are considered briefly below.

pH Adjustment

In a variety of wastewater-treatment operations and processes, there is often a need for pH adjustment. Because a number of chemicals are available that can be used, the choice will depend on the suitability of a given chemical for a particular application and prevailing economics. General information on the chemicals used most commonly for pH adjustment is given in Table 6–15. Wastewater that is acidic can be neutralized with any number of basic chemicals, as reported in Table 6–15. Sodium hydroxide (NaOH, also known as caustic soda) and sodium carbonate, although somewhat expensive, are convenient and are used widely by small plants or for treatment where small quantities are adequate. Lime, which is cheaper but somewhat less convenient, is the most widely used chemical. Lime can be purchased as quicklime or slaked hydrated lime, high-calcium or

Table 6–15
Chemicals used most commonly for the control of pH (neutralization)[a]

Chemical	Formula	Molecular weight	Equivalent weight	Availability Form	Availability Percent
		Chemicals used to raise pH			
Calcium carbonate	$CaCO_3$	100.0	50.0	Powder granules	96 to 99
Calcium hydroxide (lime)	$Ca(OH)_2$	74.1	37.1	Powder granules	82 to 95
Calcium oxide	CaO	56.1	28.0	Lump, pebble, ground	90 to 98
Dolomitic hydrated lime	$[Ca(OH)_2]_{0.6}$ $[Mg(OH)_2]_{0.4}$	67.8	33.8	Powder	58 to 65
Dolomitic quicklime	$(CaO)_{0.6}(MgO)_{0.4}$	49.8	24.8	Lump, pebble ground	55–58 CaO
Magnesium hydroxide	$Mg(OH)_2$	58.3	29.2	Powder	
Magnesium oxide	MgO	40.3	20.2	Powder, granules	99
Sodium bicarbonate	$NaHCO_3$	84.0	84.0	Powder, granules	99
Sodium carbonate (soda ash)	Na_2CO_3	106.0	53.0	Powder	99.2
Sodium hydroxide (caustic soda)	$NaOH$	40.0	40.0	Solid flake, ground flake, liquid	98
		Chemicals used to lower pH			
Carbonic acid	H_2CO_3	62.0	31.0	Gas (CO_2)	
Hydrochloric acid	HCl	36.5	36.5	Liquid	27.9, 31.45, 35.2
Sulfuric acid	H_2SO_4	98.1	49.0	Liquid	77.7 (60° Be) 93.2 (66° Be)

[a] Adapted in part from Eckenfelder (2000).

dolomitic lime, and in several physical forms. Limestone and dolomitic limestone are cheaper but less convenient to use and slower in reaction rate. Because they can become coated in certain waste-treatment applications, their use is limited. Calcium and magnesium chemicals often form sludges that require disposal.

Alkaline wastes are less of a problem than acid wastes but nevertheless often require treatment. If acidic waste streams are not available or are not adequate to neutralize alkaline wastes, sulfuric acid is commonly employed. In some treatment plants, carbon dioxide (CO_2) in the form of flue gas has been used to neutralize alkaline wastewaters, as illustrated by the following reactions:

$$2OH^- + CO_2 \rightarrow CO_3^{2-} + H_2O \tag{6-45}$$

$$CO_3^{2-} + CO_2 + H_2O \rightarrow 2HCO_3^- \tag{6-46}$$

Based on the chlorine dose used for disinfection, the pH of the disinfected effluent will be lower than that allowed for reuse applications and for dispersal to the environment. In such cases, neutralization is controlled by automatic instruments using a feedback loop, and the final effluent pH is recorded. Depending on the sensitivity of the environment, two-stage neutralization may be required. The reagent chemicals can be fed automatically, in the form of solutions, slurries, or dry materials. If the reaction rate is slow, instrumentation and control design must take this factor into account.

Analysis of Scaling Potential

With the increasing use that is being made of nanofiltration, reverse osmosis, and electrodialysis in wastewater reuse applications, adjustment of the scaling characteristics of the effluent to be treated is important to avoid calcium carbonate and sulfate scale formation. Depending on the recovery rate, the concentration of salts can increase by a factor of up to 10 within the treatment module. When such a salt concentration increase occurs, it is often possible to exceed the solubility product of calcium carbonate and other scale-forming compounds. The formation of scale within the treatment module will cause a deterioration in the performance, ultimately leading to the failure of the membrane module.

The tendency to develop calcium carbonate ($CaCO_3$) scale during the advanced treatment of treated effluent can be approximated by calculating the Langelier saturation index (LSI) of the concentrate stream (Langelier, 1946).

$$LSI = pH - pH_s \tag{6-47}$$

where pH = measured pH in concentrate stream water sample
pH_s = saturation pH for calcium carbonate

The scaling criteria for the Langelier saturation index are:

LSI > 0   Water is supersaturated with respect to calcium carbonate ($CaCO_3$) and scaling may occur.

LSI < 0   Water is undersaturated with respect to calcium carbonate. Undersaturated water has a tendency to remove existing calcium carbonate protective coatings in pipelines and equipment.

LSI = 0   Water is considered to be neutral (i.e., neither scale-forming nor scale-removing).

It should be noted that undersaturated water is also sometimes referred to as *corrosive*, but use of the term corrosive is incorrect, as the LSI applies only to the presence or absence of a calcium carbonate scale.

An alternative index known as the *stability index* was proposed by Ryzner (1944) and is used in a number of industrial applications. The Ryzner index (RI) is given by the following expression:

$$RI = 2\,pH_s - pH \qquad (6\text{–}48)$$

The scaling criteria for the Ryzner index are as follows:

RI < 5.5	Heavy scale will form
5.5 < RI < 6.2	Scale will form
6.2 < RI < 6.8	No difficulties
6.8 < RI < 8.5	Water is aggressive
RI > 8.5	Water is very aggressive

The saturation pH_s can be computed using the following expression:

$$pH_s = -\log\left(\frac{K_{a2}\gamma_{Ca^{2+}}[Ca^{2+}]\gamma_{HCO_3^-}[HCO_3^-]}{K_{sp}}\right) \qquad (6\text{–}49)$$

where K_{a2} = equilibrium constant for the dissociation of bicarbonate
$\gamma_{Ca^{2+}}$ = activity coefficient for calcium
Ca^{2+} = concentration of calcium, mole
$\gamma_{HCO_3^-}$ = activity coefficient for bicarbonate
HCO_3^- = concentration of bicarbonate, mole
K_{sp} = solubility product constant for the dissociation of calcium carbonate

The activity coefficient can be estimated using Eq. (2–9) given in Chap. 2.

$$\log\gamma = \frac{0.5\,(Z_i)^2\sqrt{I}}{1+\sqrt{I}} \qquad (2\text{–}9)$$

where Z = charge on ionic species
I = ionic strength

The ionic strength can be estimated using Eq. (2–11) given in Chap. 2.

$$I = 2.5 \times 10^{-5} \times TDS \qquad (2\text{–}11)$$

The saturation pH_s for calcium carbonate ($CaCO_3$) solubility in the pH range from 6.5 to 9.0 is given by

$$pH_s = pK_{a2} - pK_{sp} + p[Ca^{2+}] + p[HCO_3^-] - \log\gamma_{Ca^{2+}} - \log\gamma_{HCO_3^-} \qquad (6\text{–}50)$$

where pK_{a2} = negative logarithm of the equilibrium constant for the dissociation of bicarbonate
pK_{sp} = negative logarithm of solubility product constant for the dissociation of calcium carbonate
pCa^{2+} = negative logarithm of calcium concentration
$pHCO_3^-$ = negative logarithm of bicarbonate concentration

Values of K_1, K_2, and K_{sp} for the carbonate system are given in Table 6–16 as a function of temperature. The application of these equations is illustrated in Example 6–7. The diagram presented on Fig. 6–19 can also be used to estimate LSI for concentrated solutions as illustrated in the accompanying computation.

Table 6–16
Carbonate equilibrium constants as function of temperature[a]

Temperature, °C	Equilibrium constant[b]		
	$K_{a1} \times 10^7$	$K_{a2} \times 10^{11}$	$K_{sp} \times 10^9$
5	3.020	2.754	8.128
10	3.467	3.236	7.080
15	3.802	3.715	6.02
20	4.169	4.169	5.248
25	4.467	4.477	4.571
40	5.012	6.026	3.090

[a] Adapted from Snoeyink and Jenkins (1980).
[b] The reported values have been multiplied by the indicated exponents. Thus, the value K_{a2} at 20°C is equal to 4.169×10^{-11}.

Example:

Temp = 20°C
pH = 8.19
Ca^{2+} = 800 mg/L as $CaCO_3$
HCO_3 = 774.8 mg/L as $CaCO_3$
TDS = 7853.6 mg/L (use 5000 mg/L)
pCa = 2.10
$pHCO_3$ = 1.81
$K = pK_{a2} - pK_{sp} = 2.37$
$pH_s = pCa^{2+} + pHCO_3^- + K$
$pH_s = 2.10 + 1.81 + 2.37 = 6.28$
LSI = 8.19 – 6.28 = 1.91

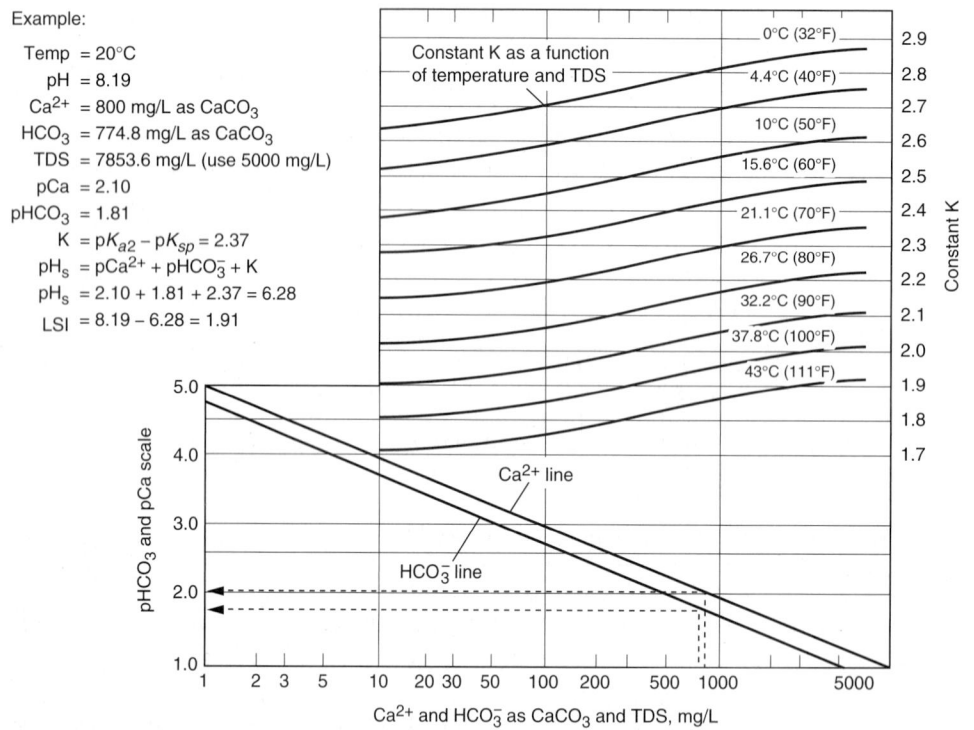

Figure 6–19
Chart for determining the value of the Langelier saturation index. *(Adapted from Du Pont Company, 1992.)*

EXAMPLE 6–7 **Analysis of Scaling Potential** Estimate the scaling potential at 20°C using both the Langelier and Ryzner indexes for a treated wastewater with the following chemical characteristics.

	Concentration	
Constituent	g/m³	mole/L
Ca^{2+}	5	0.125×10^{-3}
HCO_3^-	10	0.164×10^{-3}
TDS	20	
pH	7.7	

Solution

1. Determine the ionic strength of the treated water using Eq. (2–11).

   $$I = 2.5 \times 10^{-5} \times TDS$$

   $$I = 2.5 \times 10^{-5} \times 20 = 50 \times 10^{-5}$$

2. Determine the activity coefficients for calcium and bicarbonate using Eq. (2–10).
   a. For calcium

   $$\log \gamma_{Ca^{2+}} = -\frac{0.5(Z_i)^2 \sqrt{I}}{1 + \sqrt{I}} = -\frac{0.5(2)^2 \sqrt{50 \times 10^{-5}}}{1 + \sqrt{50 \times 10^{-5}}} = -0.0437$$

   $$\gamma_{Ca^{2+}} = 0.903$$

   b. For bicarbonate

   $$\log \gamma_{HCO_3^-} = -\frac{0.5(Z_i)^2 \sqrt{I}}{1 + \sqrt{I}} = -\frac{0.5(1)^2 \sqrt{50 \times 10^{-5}}}{1 + \sqrt{50 \times 10^{-5}}} = -0.0109$$

   $$\gamma_{HCO_3^-} = 0.975$$

3. Determine the saturation pH_s using Eq. (6–49).

   $$pH_s = -\log\left(\frac{K_{a2}\gamma_{Ca^{2+}}[Ca^{2+}]\gamma_{HCO_3^-}[HCO_3^-]}{K_{sp}}\right)$$

   $$pH_s = -\log\left(\frac{(4.17 \times 10^{-11})(0.903)(0.125 \times 10^{-3})(0.975)(0.164 \times 10^{-3})}{5.25 \times 10^{-9}}\right)$$

   $$pH_s = -\log(1.43 \times 10^{-10}) = 9.84$$

4. Determine the Langelier and Ryzner indexes.
   a. Langelier saturation index

   $$LSI = pH - pH_s = 7.7 - 9.84 = -2.14$$

   $LSI < 0$ (water is undersaturated with respect to calcium carbonate)

b. Ryzner Index

$$RI = 2pH_s - pH = 2(9.84) - 7.7 = 11.98$$

$$(RI = 11.98) > 8.5 \text{ (water is very aggressive)}$$

Comment   Although both indexes are used, the Langelier index is used most commonly in the water and wastewater field whereas the Ryzner index is used most commonly in industrial applications.

Scaling Control

Usually, $CaCO_3$ scale control can be achieved using one or more of the following methods:

- Acidifying to reduce pH and alkalinity
- Reducing calcium concentration by ion exchange or lime softening
- Adding a scale inhibitor chemical (antiscalant) to increase the apparent solubility of $CaCO_3$ in the concentrate stream
- Lowering the product recovery rate

Because it is not possible to predict a priori the value of pH in water treated with reverse osmosis, it is usually necessary to conduct pilot-scale studies using the same modules that will be used in the full-scale installation.

Stabilization

Wastewater effluent that is demineralized with reverse osmosis will generally require pH and calcium carbonate adjustment (stabilization) to prevent metallic corrosion, due to the contact of the demineralized water with metallic pipes and equipment. Corrosion occurs because material from the solid is removed (solubilized) to satisfy the various solubility products. Demineralized water typically is stabilized by adding lime to adjust the LSI, using the procedure outlined above.

6-8 CHEMICAL STORAGE, FEEDING, PIPING, AND CONTROL SYSTEMS

The design of chemical precipitation operations involves not only the sizing of the various unit operations and processes but also the necessary appurtenances. Because of the corrosive nature of many of the chemicals used and the different forms in which they are available, special attention must be given to the design of chemical storage, feeding, piping, and control systems. The following discussion is intended to serve as an introduction to this subject.

In domestic wastewater-treatment systems, the chemicals employed can be in a solid, liquid, or gaseous form. Coagulants in the dry solid form generally are converted to solution or slurry form prior to introduction into the wastewater. Coagulants in the liquid form are usually delivered to the plant in a concentrated form and have to be

Figure 6–20

Classification of chemical-feed systems.

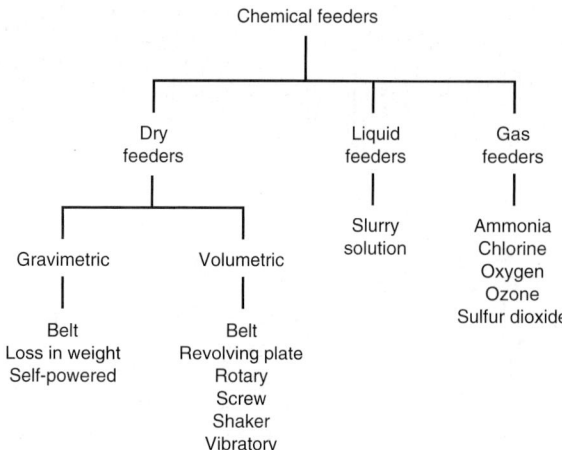

diluted prior to introduction into the wastewater. Chemicals in the gas form (generally stored as a liquid), typically used for disinfection purposes, are either dissolved in water before injection or are injected directly into the wastewater. The types of chemical-feed systems used for these chemicals are termed *dry, liquid* (also known as *wet*), or *gas* feed. The various types of feeders are classified on Fig. 6–20. Chemical feeders are generally designed to be (1) proportioning, feeding chemical in proportion to the influent wastewater flowrate, and (2) constant feed, designed to deliver chemical at a fixed rate regardless of the influent flowrate.

Chemical Storage and Handling

General information on the handling, storage, and feeding requirements for various chemicals is presented in Table 6–17. The specific storage facilities required will depend on the form in which the chemical is available locally. For small treatment plants the available forms are usually limited. A typical storage facility for chemicals for a small treatment facility is shown on Fig. 6–21.

Dry Chemical-Feed Systems

A dry chemical-feed system typically consists of a storage hopper, dry chemical feeder, a dissolving tank, and a pumped or gravity distribution system (see Figs. 6–22 and 6–23). The units are sized according to the volume of wastewater, treatment rate, and optimum length of time for chemical feeding and dissolving. Hoppers used with powdered chemicals that are compressible and can form an arch such as lime are equipped with positive agitators and a dust-collection system. As shown on Fig. 6–20, dry chemical feeders are of either the volumetric or the gravimetric type. In the volumetric type the volume of the dry chemical fed is measured whereas in the gravimetric type the weight of chemical fed is measured. A brief description of the chemical feeders used most commonly is presented in Table 6–18, and the chemical feeders are illustrated on Figs. 6–24 and 6–25. Of the dry chemical-feed systems, the loss in weight (see Fig. 6–25*b*) feed system is used most commonly in wastewater-treatment operations.

With a dry feed system, the dissolving operation is critical. The capacity of the dissolving tank is based on the detention time, which is directly related to the wettability

Table 6-17

Handling, storage, and feeding requirements for various chemicals used in wastewater treatment[a]

Chemical	Formula	Shipping form	Feeding form	Feeding type	Accessory equipment	Handling materials
Chemicals used for precipitation						
Aluminum sulfate	$Al_2(SO_4)_3 \cdot 18H_2O$	Lump, ground, or powdered solutions	Liquid	Metering pump	Slurry tank, slaker	Iron, steel
Aluminum chloride	$AlCl_3$	Liquid	Liquid	Metering pump	Storage tank	Hastelloy B. plastic
Calcium hydroxide (lime)	$Ca(OH)_2$	Bags, barrels, and bulk	Liquid	Metering pump	Slurry tank	Iron, steel
Ferric chloride	$FeCl_3$	Bags, carboys, and bulk	Liquid	Metering pump	Slurry tank	Iron, steel
Ferric sulfate	$Fe_2(SO_4)_3$	Bags, barrels, and bulk	Liquid	Metering pump	Slurry tank	Iron, steel
Copperas	$FeSO_4 \cdot 7H_2O$		Liquid	Metering pump	Slurry tank	Iron, steel
Chemicals used for neutalization						
Calcium carbonate	$CaCO_3$	Bags, drums, or bulk	Slurry, dry slurry in fixed beds	Metering pump Volumetric pump	Slurry tank	Iron, steel
Calcium oxide	CaO	Bags (22.5 kg), barrels, or bulk	Dry or slurry, slaked to $Ca(OH)_2$	Metering pump	Slurry tank, slaker	Iron, steel, plastic, rubber hose
Sodium bicarbonate	$NaHCO_3$	Bags or drums	Dry or slurry	Metering pump	Dissolving tank	Iron, steel, rubber, plastic
Sodium carbonate	Na_2CO_3	Bags (45.5 kg), bulk	Dry or slurry	Metering pump	Dissolving tank	Iron, steel, rubber, plastic
Sodium hydroxide	$NaOH$	Drum (45.5, 204.5, 367.5 kg)	Dry or slurry	Metering pump	Solution tank	Iron, steel, plastic, rubber hose
Carbonic acid	H_2CO_3		Gas (CO_2)	Gas feeder		
Hydrochloric acid	HCl	Barrels, drums, bulk	Liquid	Metering pump	Dilution tank	Hastelloy A, plastic, rubber[b]
Sulfuric acid	H_2SO_4	Carboys, drums, and bulk	Liquid	Metering pump		Iron, steel, plastic, glass

[a] Adapted in part from Eckenfelder (2000).

Figure 6–21

Typical chemical storage facilities: (a) outdoor facility at a small wastewater-treatment plant. The chemical storage tanks are located within a containment structure in case of a chemical spill, and (b) chemical storage tanks located indoors in a containment area.

(a)

(b)

Figure 6–22

Schematic of typical dry chemical-feed system.

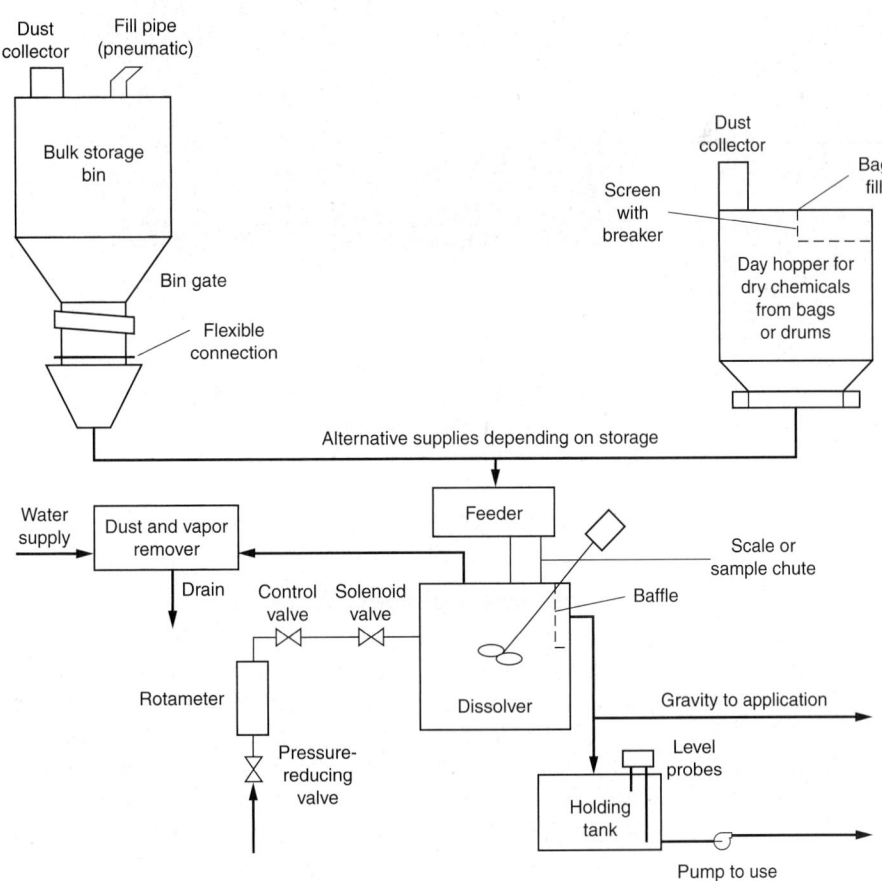

Figure 6-23

Typical dry chemical-feed system. The two chemical storage hoppers are coupled through flexible connectors to individual chemical feeders and dissolvers as illustrated schematically on Fig. 6–22.

or rate of solution of the chemical. When the water supply is controlled for the purpose of forming a constant-strength solution, mechanical mixers should be used. Depending on the flow pattern within the mixing tank, it may be necessary to add baffles for effective mixing. In smaller mixing tanks, the mixer can be set at an angle to avoid the use of baffles. Solutions or slurries are often stored after dissolving and discharged to the application point at metered rates by chemical-feed pumps.

Liquid Chemical-Feed Systems

Liquid chemical-feed systems typically include a solution storage tank, transfer pump, day tank for diluting the concentrated solution, and chemical-feed pump for distribution to the application point (see Figs. 6–26 and 6–27). In general, liquid feed systems provide for better initial contact and dispersion of the chemical and the wastewater. In systems where the liquid chemical does not require dilution, the chemical-feed pumps draw liquid directly from the solution storage tank. The storage tank is sized based upon the stability of the chemical, feed rate requirements, delivery constraints (cost, size of tank truck, etc.), and availability of the supply. Solution feed pumps are usually of the positive-displacement type for accurate metering of the chemical feed.

Type of feeder	Description
Volumetric	
Conveyor belt	Consists of a belt located below a hopper. The feed rate is adjusted by varying the speed of the belt (see Fig. 6–24a)
Revolving plate	Consists of a rotating plate below the storage hopper (see Fig. 6–24b). As the plate is rotated, material to be fed is drawn from the hopper. The amount of material feed is controlled by the rate of rotation
Rotary	Consists of rotating shaft with vanes that form pockets (see Fig. 6–24c). The amount of material feed is controlled by the rate of rotation
Shaker	Consists of a shaker pan mounted below a storage hopper (see Fig. 6–24d). As the pan oscillates, the material to be fed is moved forward and dropped into the feed chute
Screw	Consists of a variable-pitch screw mounted below a feed hopper (see Fig. 6–24e). The amount of material feed is controlled by the rate of rotation of the screw
Vibratory	Consists of vibrating pan or chute positioned below a chemical storage hopper. The pan or chute, which vibrates back and forth by the oscillating electromagnetic driver, delivers the material to be fed forward. The amount of material feed can be controlled by adjusting the rate of oscillation (see Fig. 6–24f)
Gravimetric	
Belt	Consists of volumetric feeder that transfers the material to be fed from the feed hopper to the weigh belt. The signal generated from the weigh belt is used to control the volumetric feeder (see Fig. 6–25a)
Loss in weight	Consists of a feed hopper mounted on scale and a chemical feeder (see Fig. 6–25b). The chemical feed rate can be controlled with a screw or vibratory feeder. The feed rate is controlled by the loss in weight measured by the scale
Self-powered	Consists of a counterbalanced control gate mounted below a storage hopper (see Fig. 6–25c). As shown in Fig. 6–25c, the weight of the material in the hopper is counterbalanced by the setting on the beam balance. The rate at which material is fed is controlled by the impact pan. Although not very accurate, this device does not require any power source

Table 6–18
Typical characteristics of chemical feeders[a]

[a] Adapted from Liptak (1974).

Gas Chemical-Feed Systems

Chemicals that are used as a gas include ammonia, chlorine, oxygen, ozone, and sulfur dioxide. Gas feed systems are used mostly for feeding chemicals used for disinfection and dechlorination. Chlorine, the most commonly used chemical for disinfection, is supplied in a liquid form within the storage container and evaporates continuously as the gas is drawn from the headspace above the liquid in the storage container. Feed systems for disinfection chemicals are illustrated in Chap. 12, which deals with disinfection.

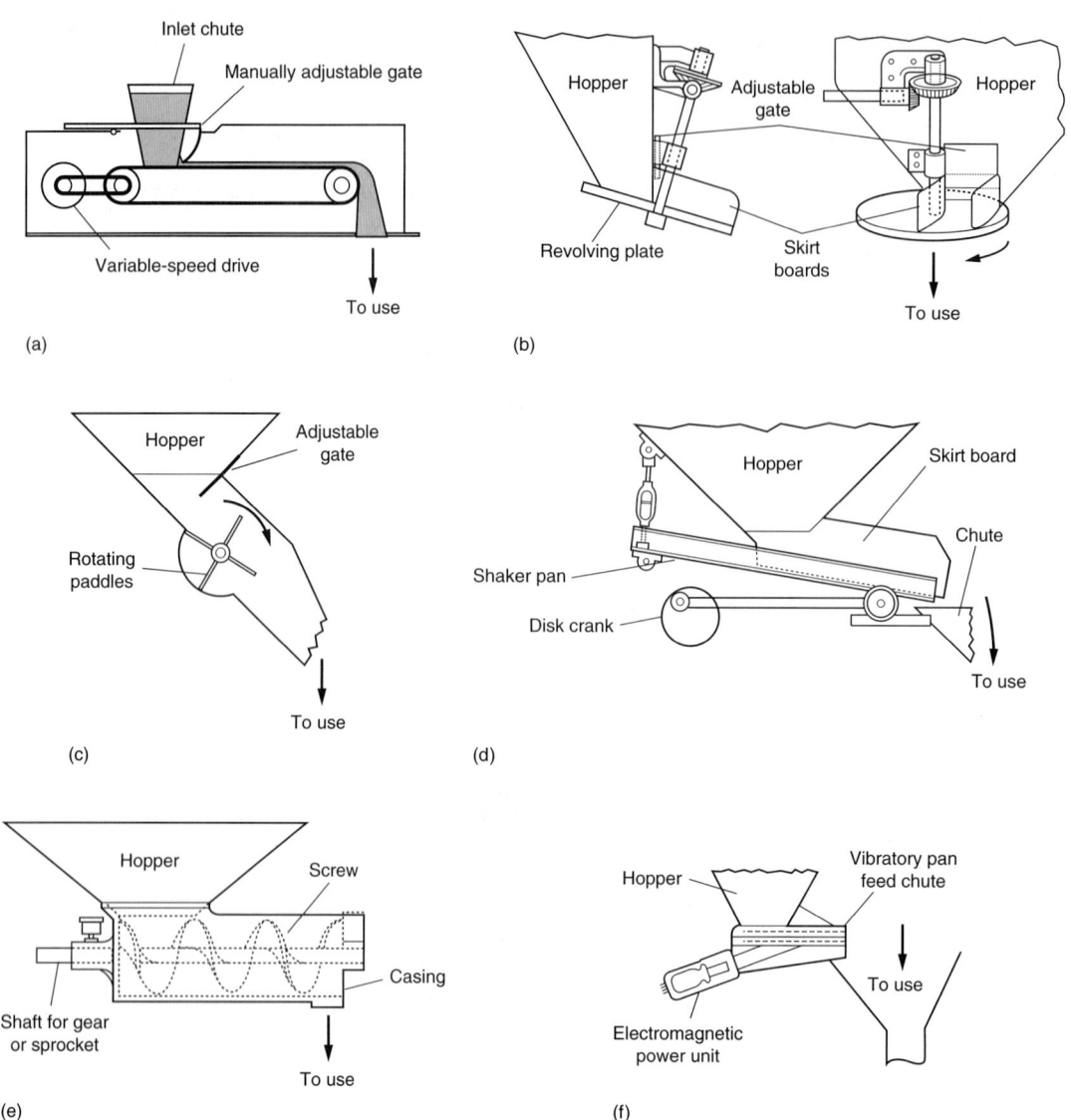

Figure 6–24

Typical volumetric chemical feeders: (a) conveyor belt, (b) revolving plate, (c) rotary, (d) shaker, (e) screw, and (f) vibratory. (Adapted from Liptak, 1974 and Acrison, Inc.)

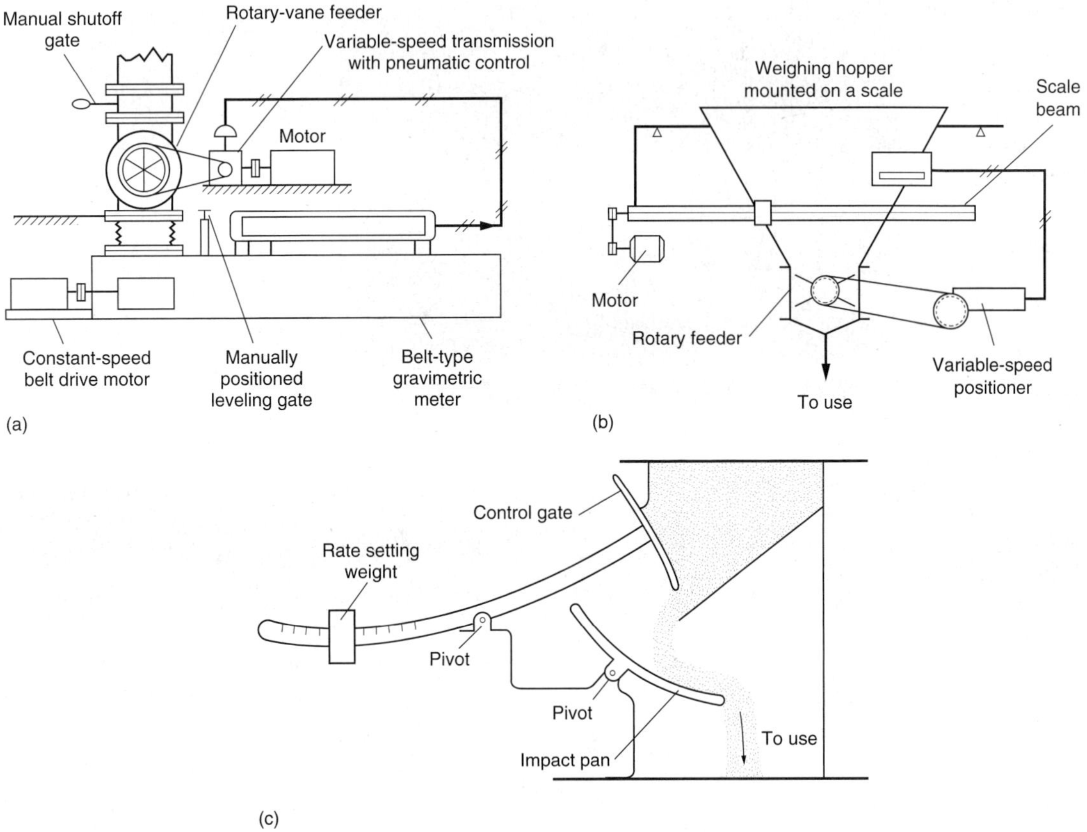

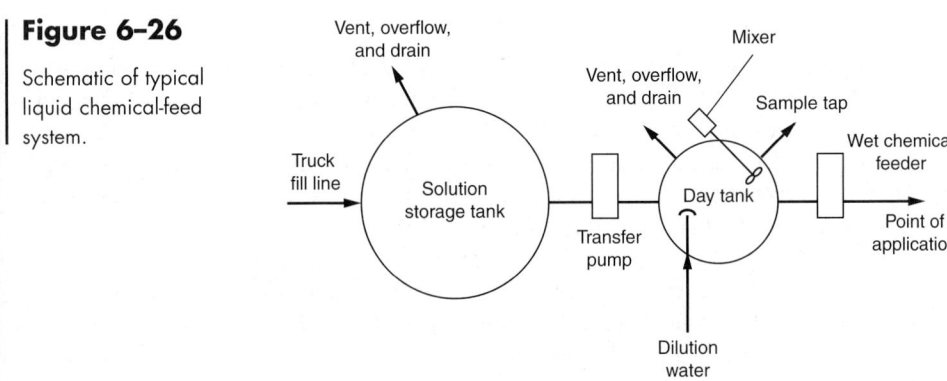

Figure 6–25

Typical gravimetric chemical feeders: (a) belt, (b) loss in weight, and (c) self-powered. *(Adapted from Liptak, 1974 and Acrison, Inc.)*

Figure 6–26

Schematic of typical liquid chemical-feed system.

Figure 6–27

Typical outdoor liquid chemical-feed systems. The chemical feeders are covered to limit deterioration due to exposure to the elements.

Table 6–19

Typical mixing times for various chemicals used in wastewater-treatment facilities

Chemical	Applications	Recommended mixing times, s
Alum, Al^{3+}, Ferric chloride, Fe^{3+}	Coagulation of colloidal particles	<1
Alum, Al^{3+}, Ferric chloride, Fe^{3+}	Sweep floc precipitation	1–10
Lime $Ca(OH)_2$	Chemical precipitation	10–30
Chlorine, Cl_2	Chemical disinfection	<1
Chloramine, NH_2Cl	Chemical disinfection	5–10
Cationic polymers	Destabilization of colloidal particles	<1
Anionic polymers	Particle bridging	1–10
Polymers, nonionic	Filter aids	1–10

Initial Chemical Mixing

Perhaps the least appreciated fact about chemical addition is the importance of both the initial and uniform mixing of the chemical with the wastewater to be treated. The optimal time for mixing can, as discussed in this section, vary from a fraction of a second to several seconds or more. Because of the difficulties in achieving extremely low mixing times in large treatment plants with a single mixing device, the use of multiple mixing devices is recommended. The particular mixing device selected for a given application must be based on a consideration of the reaction times and operative mechanisms for the chemicals that are being used. Typical mixing times for the chemicals used in wastewater-treatment facilities are reported in Table 6–19. Typical devices used for mixing chemicals in wastewater-treatment plants are discussed in Chap. 5 and throughout this textbook. Approximate mixing times achievable with various mixing devices may be found in Table 5–9 in Chap. 5.

PROBLEMS AND DISCUSSION TOPICS

6-1 To aid sedimentation in the primary settling tank, 25, 40, 60, or 75 g/m³ (value to be selected by instructor) of ferrous sulfate (FeSO₄·7H₂O) is added to the wastewater. Determine the minimum alkalinity required to react initially with the ferrous sulfate. How many grams of lime should be added as CaO to react with Fe(HCO₃)₂ and the dissolved oxygen in the wastewater to form insoluble Fe(OH)₃?

6-2 Copperas (FeSO₄·7H₂O) is to be added at a rate of 20, 25, 30, or 40 kg/1000 m³ (value to be selected by instructor) to a wastewater to improve the efficiency of an existing primary sedimentation tank. Assuming sufficient alkalinity is present as Ca (HCO₃)₂ determine:
   a. How many kg of lime should be added as CaO to complete the reaction,
   b. The concentration of oxygen needed in the wastewater to oxidize the ferrous hydroxide that is formed,
   c. The mass of sludge that will result per 1000 m³ of wastewater, and
   d. The amount (kg) of alum needed to obtain the same quantity of sludge as in Part (c), assuming Al(OH)₃ is the precipitate that is formed.

6-3 Assume that 40, 45, 50, and 55 kg (value to be selected by instructor) of (a) alum (mol wt 666.5) and (b) ferrous sulfate and lime as Ca(OH)₂ is added per 4000 m³ of wastewater. Also assume that all insoluble and very slightly soluble products of the reactions, with the exception of 15 g/m³ CaCO₃, are precipitated as sludge. How many kg of sludge/1000 m³ will result in each case?

6-4 Raw wastewater is to be treated chemically for the removal of total suspended-solids and phosphorus through coagulation, flocculation, and sedimentation. The wastewater characteristics are as follows: Q = 0.75 m³/s; orthophosphorus = 10 g/m³ as P; alkalinity = 200 g/m³ expressed as CaCO₃ [essentially all due to the presence of Ca(HCO₃)₂]; total TSS = 220 g/m³.
   a. Determine the sludge production in kg dry wt/d and m³/d under the following conditions: (1) alum [Al₂(SO₄)₃·14.3 H₂O] dosage of 120, 130, 140, or 150 g/m³ (to be selected by instructor); (2) 100 percent removal of orthophosphorus as insoluble AlPO₄; (3) 95 percent removal of original TSS; (4) all alum not required for reaction with phosphate reacts with alkalinity to form Al(OH)₃, which is 100 percent removed; (5) wet sludge has a water content of 93 percent and a specific gravity of 1.04.
   b. Determine the sludge production in a kg dry wt/d and m³/d under the following conditions: (1) Lime [Ca(OH)₂] dosage of 450 g/m³ to give pH of approximately 11.2; (2) 100 percent removal of orthophosphorus as insoluble hydroxylapatite [Ca₁₀(PO₄)₆(OH)₂]; (3) 95 percent removal of original TSS; (4) added lime (i) reacts with phosphate, (ii) reacts with all alkalinity to form CaCO₃, 20 g/m³ of CaCO₃ is soluble and remains in solution and the rest is 100 percent removed, and (iii) remainder stays in solution; (5) wet sludge has a water content of 92 percent and a specific gravity of 1.05.
   c. Determine the net increase in calcium hardness in g/m³ as CaCO₃ for the treatment specified in part b.

6-5 Verify the correctness of the plots given on Fig. 6–8, using the equations given below the figures.

6-6 Obtain equilibrium values from the literature and verify one of the solubility curves given on Fig. 6–17.

6-7 Obtain equilibrium values from the literature and verify one of the solubility curves given on Fig. 6–18.

6-8 Using the potentials for the following two half reactions given in Table 6–12, estimate the equilibrium constant for the ionization of water.

$$2H_2O + 2e^- \rightarrow H_2 + 2OH^-$$

$$2H^+ + 2e^- \rightarrow H_2$$

6-9 Using the half reactions given in Table 6–12, evaluate the feasibility of the following reaction.

$$2Fe^{2+} + 2H^+ + H_2O_2 \leftrightarrow 2Fe^{3+} + 2H_2O$$

6-10 Using the half reactions given in Table 6–12, evaluate the feasibility of the following reaction.

$$2Fe^{2+} + Cl_2 \leftrightarrow 2Fe^{3+} + 2Cl^-$$

6-11 Using the half reactions given in Table 6–12, evaluate the feasibility of the following reaction.

$$H_2S + Cl_2 \leftrightarrow S + 2HCl$$

6-12 Using the half reactions given in Table 6–12, evaluate the feasibility of the following reaction.

$$H_2S + O_3 \leftrightarrow S + O_2 + H_2O$$

6-13 Estimate the scaling potential of one the treated wastewater samples (to be selected by instructor) using both the Langelier and Ryzner indexes. Use a temperature of 20°C.

		Wastewater sample			
Constituent	**Unit**	**A**[a]	**B**	**C**	**D**
Ca^{2+}	mg/L as $CaCO_3$	5	12	245	15
HCO_3^-	mg/L as $CaCO_3$	7	9	200	16
TDS	mg/L	30	275	600	500
pH	unitless	6.5	8.0	6.9	6.5

[a]Typical values are for runoff from snow melt (adapted from Benefield et al. (1982).

6-14 Estimate the scaling potential of one the treated wastewater samples (to be selected by instructor) using both the Langelier and Ryzner indexes. The measured pH values for the four samples are A = 7.2, B = 6.9, C = 7.3, and D = 6.8.

	Concentration, mg/L					Concentration, mg/L			
Cation	**A**	**B**	**C**	**D**	**Anion**	**A**	**B**	**C**	**D**
Ca^{2+}	121.3	64.0	42.1	44.0	HCO_3^-	280	96.0	158.7	91.5
Mg^{2+}	36.2	15.1	14.6	25.2	SO_4^{2-}	116	80	48.0	57.6
Na^+	8.1	20.5	46.0	4.6	Cl^-	61	17.3	63.8	17.7
K^+	12	10.0	11.7		NO_3^-	15.6	5		
					H_2CO_3				8.8
					CO_3^{2-}			12.0	

6–15 Review the literature concerning the Langelier (1946) and Ryzner (1944) indexes and discuss the difference in approach to scaling used in the development of these two indexes.

REFERENCES

Amirtharajah, A., and K. M. Mills (1982) "Rapid Mix Design for Mechanisms of Alum Coagulation," *Journal American Water Works Association,* vol. 74. p. 210.

Bard, A. J. (1966) *Chemical Equilibrium,* Harper & Row, New York.

Benefield, L. D., J. F. Judkins, Jr., and B. L. Weand (1982) *Process Chemistry for Water and Wastewater Treatment,* Prentice-Hall, Englewood Cliffs, NJ.

Benjamin, M. M. (2001) *Water Chemistry,* McGraw-Hill, New York.

Deryagin, B. V., and L. D. Landau (1941) "Theory of Stability of Strongly Charged Lyophobic Soles and Coalescence of Strongly Charged Particles in Solutions of Electrolytes," *Acta Phys.-Chim. USSR,* vol. 14, p. 633.

Du Pont Company (1992) *PEM Permasep Products Engineering Manual,* Wilmington, DE.

Eckenfelder, W. W., Jr. (2000) *Industrial Water Pollution Control,* 3d ed., McGraw-Hill, Boston, MA.

Fair, G. M., J. C. Geyer, and D. A. Okun (1968) *Water and Wastewater Engineering,* vol. 2, Wiley, New York.

Hahn, H. H., and W. Stumm (1968) "Kinetics of Coagulation with Hydrolyzed AL(III)," *Journal of Colloidal and Interface Science,* vol. 28, no. 1, pp. 134–144.

Hayden, P. L., and A. J. Rubin (1974) "Systematic Investigation of the Hydrolysis and Precipitation of Aluminum (III)," in A. J. Rubin (ed.), *Aqueous-Environmental Chemistry of Metals,* Ann Arbor Science, Ann Arbor, MI.

Hudson, H. E., and J. P. Wolfner (1967) "Design of Mixing and Flocculating Basins," *Journal American Water Works Association,* vol. 59, no. 10, pp. 1257–1267.

Kugelman, I. J. (1976) "Status of Advanced Waste Treatment," in H. W. Gehm and J. I. Bregman (eds.), *Handbook of Water Resources and Pollution Control,* Van Nostrand, New York.

Langelier, W. F. (1946) "The Analytical Control of Anti-Corrosion Water Treatment," *Journal American Water Works Association,* vol. 28, no. 10, p. 1500.

Letterman, R. D. (1991) *Filtration Strategies to Meet the Surface Water Treatment Rule,* American Water Works Association, Denver, CO.

Letterman, R. D., A. Amirtharajah, and C. R. O'Melia (1999) "Coagulation and Flocculation," in R. D. Letterman (ed.), *Water Quality and Treatment: A Handbook of Community Water Supplies,* 5th ed., American Water Works Association, McGraw-Hill, New York.

Liptak, B. G. (ed.) (1974) *Environmental Engineers' Handbook,* vol. 1, *Water Pollution,* Chilton Book Company, Radnor, PA.

McMurry, J., and R. C. Fay (1998) *Chemistry,* 2d ed., Prentice-Hall, Upper Saddle River, NJ.

Metcalf, L., and H. P. Eddy (1935) *American Sewerage Practice,* vol. III, 3d ed., McGraw-Hill, New York.

Morel, F. M. M., and J. G. Hering (1993) *Principles and Applications of Aquatic Chemistry,* A Wiley-Interscience Publication, New York.

Pankow, J. F. (1991) *Aquatic Chemistry Concepts,* Lewis Publishers, Chelsea, MI.

Ryzner, J. W. (1944) "A New Index for Determining Amount of Calcium Carbonate Formed by a Water," *Journal American Water Works Association,* vol. 36, no. 4, p. 472.

Sawyer, C. N., P. L. McCarty, and G. F. Parkin (1994) *Chemistry for Environmental Engineering,* 4th ed., McGraw-Hill, New York.

Sedlak, R. I. (1991) *Phosphorus and Nitrogen Removal from Municipal Wastewater: Principles and Practice,* 2d ed., The Soap and Detergent Association, Lewis Publishers, New York.

SES (1994) *The UV/Oxidation Handbook,* Solarchem Environmental Systems, Markham, Ontario, Canada.

Shaw, D. J (1966) *Introduction to Colloid and Surface Chemistry,* Butterworth, London.

Snoeyink, V. L., and D. Jenkins (1980) *Water Chemistry,* Wiley, New York.

Stumm, W., and J. J. Morgan (1962) "Chemical Aspects of Coagulation," *Journal American Water Works Association,* vol. 54, no. 8, pp. 971–994.

Stumm, W., and J. J. Morgan (1981) *Aquatic Chemistry,* 2d. ed., Wiley-Interscience, New York.

Stumm, W., and C. R. O'Melia (1968) "Stoichiometry of Coagulation," *Journal American Water Works Association,* vol. 60, p. 514.

Thomas, A. W. (1934) *Colloid Chemistry,* McGraw-Hill, New York.

U.S. EPA (1973) *Physical-Chemical Wastewater Treatment Plant Design,* U.S. Environmental Protection Agency, Technology Transfer Seminar Publication, U.S. Environmental Protection Agency.

U.S. EPA (1976) *Process Design Manual for Phosphorus Removal,* Office of Technology Transfer, U.S. Environmental Protection Agency, Washington, D.C.

U.S. EPA (1987) *Phosphorus Removal Design Manual,* EPA/625/1-87/001, U.S. Environmental Protection Agency, Washington, D.C.

Verwey, E. J. W., and J. Th. G. Overbeek (1948) *Theory of the Stability of Lyophobic Colloids,* Elsevier, Amsterdam.

Vrale, L., and R. M. Jorden (1971) "Rapid Mixing in Water Treatment," *Journal American Water Works Association,* vol. 63, no. 1, pp. 52–58.

White, G. C. (1999) *Handbook of Chlorination and Alternative Disinfectants,* 4th. ed., A Wiley-Interscience Publication, John Wiley & Sons, Inc., New York.

7

Fundamentals of Biological Treatment

With proper analysis and environmental control, almost all wastewaters containing biodegradable constituents can be treated biologically. Therefore, it is essential that the environmental engineer understand the characteristics of each biological process to ensure that the proper environment is produced and controlled effectively. The principal purposes of this chapter are (1) to provide fundamental background information on the microorganisms used to treat wastewater and (2) to consider the application of biological process fundamentals for the biological treatment of wastewater. The information presented in this chapter provides the necessary background material needed for the design of biological treatment processes discussed in Chaps. 8 through 10. For ease of computation, constituent concentrations in this chapter and in Chaps. 8, 9, and 10 are expressed in g/m^3 instead of mg/L because flowrate is given in units of m^3/s or m^3/d.

The fundamentals of biological treatment introduced in the first seven sections of this chapter include (1) an overview of biological wastewater treatment, (2) the composition and classification of the microorganisms used for wastewater treatment, (3) an introduction to important aspects of microbial metabolism, (4) bacterial growth and energetics, (5) microbial growth kinetics, (6) modeling suspended growth treatment processes, and (7) modeling attached-growth treatment processes. Following the presentation of fundamentals, the remaining seven sections deal with an introduction to the general classes of biological processes used for the treatment of wastewater. The topics covered include (1) aerobic oxidation, (2) biological nitrification, (3) biological denitrification, (4) biological phosphorus removal, (5) anaerobic oxidation, (6) biological removal of toxic and recalcitrant organic compounds, and (7) biological removal of heavy metals.

7-1 OVERVIEW OF BIOLOGICAL WASTEWATER TREATMENT

The objectives of biological treatment, some useful definitions, the role of microorganisms in the biological treatment of wastewater, and biological processes used for wastewater treatment are introduced in this section to provide a perspective for the material to be presented in this chapter.

Objectives of Biological Treatment

The overall objectives of the biological treatment of domestic wastewater are to (1) transform (i.e., oxidize) dissolved and particulate biodegradable constituents into acceptable end products, (2) capture and incorporate suspended and nonsettleable colloidal solids into a biological floc or biofilm, (3) transform or remove nutrients, such as nitrogen and phosphorus, and (4) in some cases, remove specific trace organic constituents and compounds. For industrial wastewater, the objective is to remove or reduce the concentration of organic and inorganic compounds. Because some of the constituents and compounds found in industrial wastewater are toxic to microorganisms, pretreatment may be required before the industrial wastewater can be discharged to a municipal collection system. For agricultural irrigation return wastewater, the objective is to remove nutrients, specifically nitrogen and phosphorus, that are capable of stimulating the growth of aquatic plants. Schematic flow diagrams of various treatment processes for domestic wastewater incorporating biological processes are shown on Fig. 7–1.

Some Useful Definitions

Common terms used in the field of biological wastewater treatment and their definitions are presented in Table 7–1. All of the terms presented in Table 7–1 are discussed in greater detail in the remainder of this chapter and in Chaps. 8, 9, and 10, which follow. The first five entries in Table 7–1 refer to the metabolic function of the processes. As reported, the principal processes used for the biological treatment of wastewater can be classified with respect to their metabolic function as aerobic processes, anaerobic processes, anoxic processes, facultative processes, and combined processes. Terminology used to describe the types of treatment processes is presented in the second group of entries in Table 7–1. The principal processes used for wastewater treatment are classified as suspended-growth, attached-growth, or combinations thereof. The third group of terms are descriptors for the different types of treatment functions.

Role of Microorganisms in Wastewater Treatment

The removal of dissolved and particulate carbonaceous BOD and the stabilization of organic matter found in wastewater is accomplished biologically using a variety of microorganisms, principally bacteria. Microorganisms are used to oxidize (i.e., convert) the dissolved and particulate carbonaceous organic matter into simple end products and additional biomass, as represented by the following equation for the aerobic biological oxidation of organic matter.

$$\nu_1 \text{ (organic material)} + \nu_2 O_2 + \nu_3 NH_3 + \nu_4 PO_4^{3-} \xrightarrow{\text{microorganisms}}$$

$$\nu_5 \text{(new cells)} + \nu_6 CO_2 + \nu_7 H_2O \quad (7\text{–}1)$$

where ν_i = the stoichiometric coefficient, as defined previously in Sec. 4–6 in Chap. 4.

In Eq. (7–1), oxygen (O_2), ammonia (NH_3), and phosphate (PO_4^{3-}) are used to represent the nutrients needed for the conversion of the organic matter to simple end products [i.e., carbon dioxide (CO_2) and water]. The term shown over the directional arrow is used to denote the fact that microorganisms are needed to carry out the oxidation process. The term *new cells* is used to represent the biomass produced as a result of the oxidation of the organic matter. Microorganisms are also used to remove nitrogen and

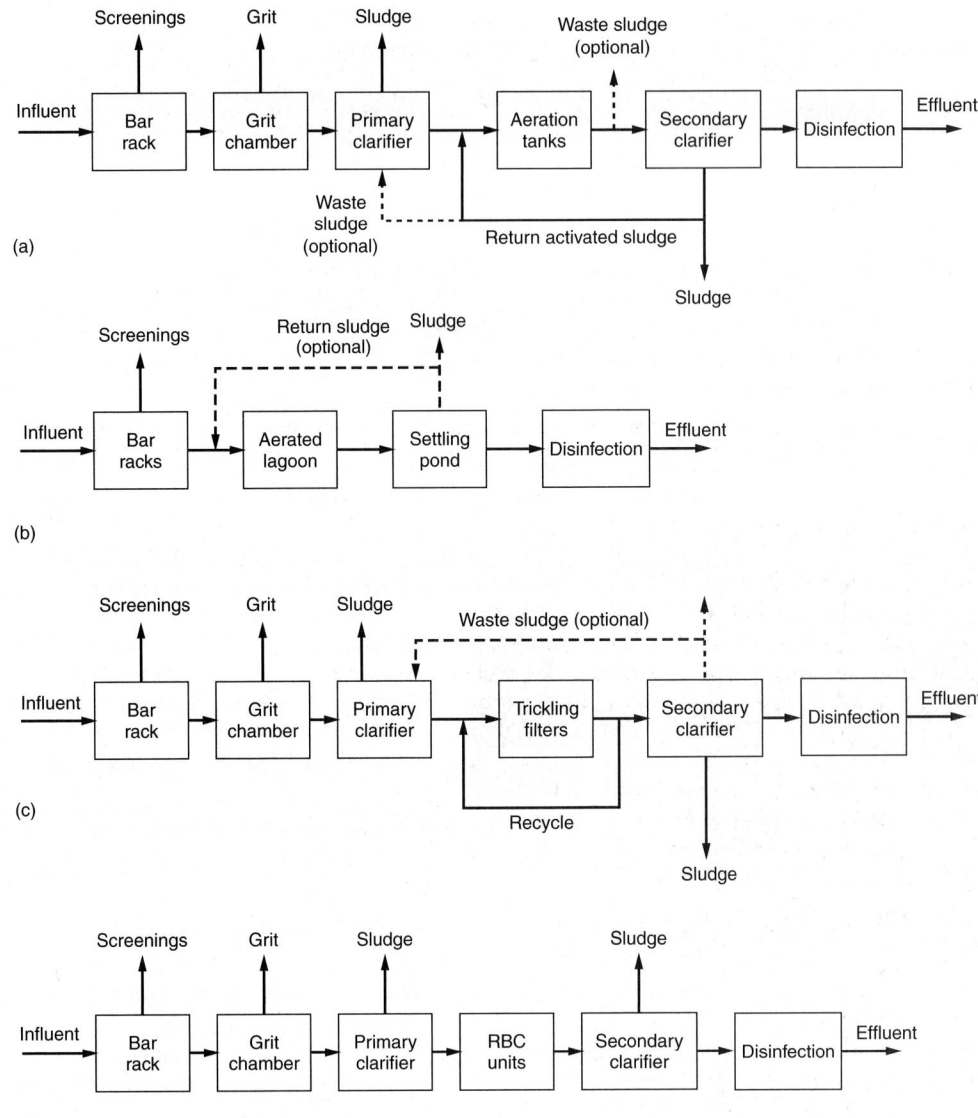

(a)

(b)

(c)

(d)

Figure 7-1

Typical (simplified) flow diagrams for biological processes used for wastewater treatment: (a) activated-sludge process, (b) aerated lagoons, (c) trickling filters, and (d) rotating biological contactors.

phosphorus in wastewater treatment processes. Specific bacteria are capable of oxidizing ammonia (nitrification) to nitrite and nitrate, while other bacteria can reduce the oxidized nitrogen to gaseous nitrogen. For phosphorus removal, biological processes are configured to encourage the growth of bacteria with the ability to take up and store large amounts of inorganic phosphorus.

Table 7–1

Definitions of common terminology used for biological wastewater treatment[a]

Term	Definition
Metabolic function	
Aerobic (oxic) processes	Biological treatment processes that occur in the presence of oxygen
Anaerobic processes	Biological treatment processes that occur in the absence of oxygen
Anoxic processes	The process by which nitrate nitrogen is converted biologically to nitrogen gas in the absence of oxygen. This process is also known as denitrification
Facultative processes	Biological treatment processes in which the organisms can function in the presence or absence of molecular oxygen
Combined aerobic/ anoxic/anaerobic processes	Various combinations of aerobic, anoxic, and anaerobic processes grouped together to achieve a specific treatment objective
Treatment processes	
Suspended-growth processes	Biological treatment processes in which the microorganisms responsible for the conversion of the organic matter or other constituents in the wastewater to gases and cell tissue are maintained in suspension within the liquid
Attached-growth processes	Biological treatment processes in which the microorganisms responsible for the conversion of the organic matter or other constituents in the wastewater to gases and cell tissue are attached to some inert medium, such as rocks, slag, or specially designed ceramic or plastic materials. Attached-growth treatment processes are also known as fixed-film processes
Combined processes	Term used to describe combined processes (e.g., combined suspended and attached growth processes)
Lagoon processes	A generic term applied to treatment processes that take place in ponds or lagoons with various aspect ratios and depths
Treatment functions	
Biological nutrient removal	The term applied to the removal of nitrogen and phosphorus in biological treatment processes
Biological phosphorus removal	The term applied to the biological removal of phosphorus by accumulation in biomass and subsequent solids separation
Carbonaceous BOD removal	Biological conversion of the carbonaceous organic matter in wastewater to cell tissue and various gaseous end products. In the conversion, it is assumed that the nitrogen present in the various compounds is converted to ammonia
Nitrification	The two-step biological process by which ammonia is converted first to nitrite and then to nitrate
Denitrification	The biological process by which nitrate is reduced to nitrogen and other gaseous end products
Stabilization	The biological process by which the organic matter in the sludges produced from the primary settling and biological treatment of wastewater is stabilized, usually by conversion to gases and cell tissue. Depending on whether this stabilization is carried out under aerobic or anaerobic conditions, the process is known as aerobic or anaerobic digestion
Substrate	The term used to denote the organic matter or nutrients that are converted during biological treatment or that may be limiting in biological treatment. For example, the carbonaceous organic matter in wastewater is referred to as the substrate that is converted during biological treatment

[a] Adapted from Crites and Tchobanoglous (1998).

Because the biomass has a specific gravity slightly greater than that of water, the biomass can be removed from the treated liquid by gravity settling. It is important to note that unless the biomass produced from the organic matter is removed on a periodic basis, complete treatment has not been accomplished because the biomass, which itself is organic, will be measured as BOD in the effluent. Without the removal of biomass from the treated liquid, the only treatment achieved is that associated with the bacterial oxidation of a portion of the organic matter originally present.

Types of Biological Processes for Wastewater Treatment

The principal biological processes used for wastewater treatment can be divided into two main categories: *suspended growth* and *attached growth* (or *biofilm*) processes (see Table 7–1). Typical process applications for suspended and attached growth biological treatment processes are given in Table 7–2, along with other treatment processes. The successful design and operation of the processes listed in Table 7–2 require an understanding of the types of microorganisms involved, the specific reactions that they perform, the environmental factors that affect their performance, their nutritional needs, and their reaction kinetics. These subjects are considered in the sections that follow.

Suspended Growth Processes.   In suspended growth processes, the microorganisms responsible for treatment are maintained in liquid suspension by appropriate mixing methods. Many suspended growth processes used in municipal and industrial wastewater treatment are operated with a positive dissolved oxygen concentration (aerobic), but applications exist where suspended growth anaerobic (no oxygen present) reactors are used, such as for high organic concentration industrial wastewaters and organic sludges. The most common suspended growth process used for municipal wastewater treatment is the activated-sludge process shown on Fig. 7–2 and discussed below.

The activated-sludge process was developed around 1913 at the Lawrence Experiment Station in Massachusetts by Clark and Gage (Metcalf and Eddy, 1930), and by Ardern and Lockett (1914) at the Manchester Sewage Works in Manchester, England. The activated-sludge process was so named because it involved the production of an activated mass of microorganisms capable of stabilizing a waste under aerobic conditions. In the aeration tank, contact time is provided for mixing and aerating influent wastewater with the microbial suspension, generally referred to as the mixed liquor suspended solids (MLSS) or mixed liquor volatile suspended solids (MLVSS). Mechanical equipment is used to provide the mixing and transfer of oxygen into the process. The mixed liquor then flows to a clarifier where the microbial suspension is settled and thickened. The settled biomass, described as *activated sludge* because of the presence of active microorganisms, is returned to the aeration tank to continue biodegradation of the influent organic material. A portion of the thickened solids is removed daily or periodically as the process produces excess biomass that would accumulate along with the nonbiodegradable solids contained in the influent wastewater. If the accumulated solids are not removed, they will eventually find their way to the system effluent.

An important feature of the activated-sludge process is the formation of floc particles, ranging in size from 50 to 200 μm, which can be removed by gravity settling, leaving a relatively clear liquid as the treated effluent. Typically, greater than 99 percent of the suspended solids can be removed in the clarification step. As will be discussed in

Table 7–2
Major biological treatment processes used for wastewater treatment.

Type	Common name	Use[a]
Aerobic processes		
Suspended growth	Activated-sludge process(es)	Carbonaceous BOD removal, nitrification
	Aerated lagoons	Carbonaceous BOD removal, nitrification
	Aerobic digestion	Stabilization, carbonaceous BOD removal
Attached growth	Trickling filters	Carbonaceous BOD removal, nitrification
	Rotating biological contactors	Carbonaceous BOD removal, nitrification
	Packed-bed reactors	Carbonaceous BOD removal, nitrification
Hybrid (combined) suspended and attached growth processes	Trickling filter/activated sludge	Carbonaceous BOD removal, nitrification
Anoxic processes		
Suspended growth	Suspended-growth denitrification	Denitrification
Attached growth	Attached-growth denitrification	Denitrification
Anaerobic processes		
Suspended growth	Anaerobic contact processes	Carbonaceous BOD removal
	Anaerobic digestion	Stabilization, solids destruction, pathogen kill
Attached growth	Anaerobic packed and fluidized bed	Carbonaceous BOD removal, waste stabilization, denitrification
Sludge blanket	Upflow anaerobic sludge blanket	Carbonaceous BOD removal, especially high-strength wastes
Hybrid	Upflow sludge blanket/attached growth	Carbonaceous BOD removal
Combined aerobic, anoxic, and anaerobic processes		
Suspended growth	Single- or multistage processes, various proprietary processes	Carbonaceous BOD removal, nitrification, denitrification, and phosphorus removal
Hybrid	Single- or multistage processes with packing for attached growth	Carbonaceous BOD removal, nitrification, denitrification, and phosphorus removal
Lagoon processes		
Aerobic lagoons	Aerobic lagoons	Carbonaceous BOD removal
Maturation (tertiary) lagoons	Maturation (tertiary) lagoons	Carbonaceous BOD removal, nitrification
Facultative lagoons	Facultative lagoons	Carbonaceous BOD removal
Anaerobic lagoons	Anaerobic lagoons	Carbonaceous BOD removal, waste stabilization

[a] Adapted from Tchobanoglous and Schroeder (1985).

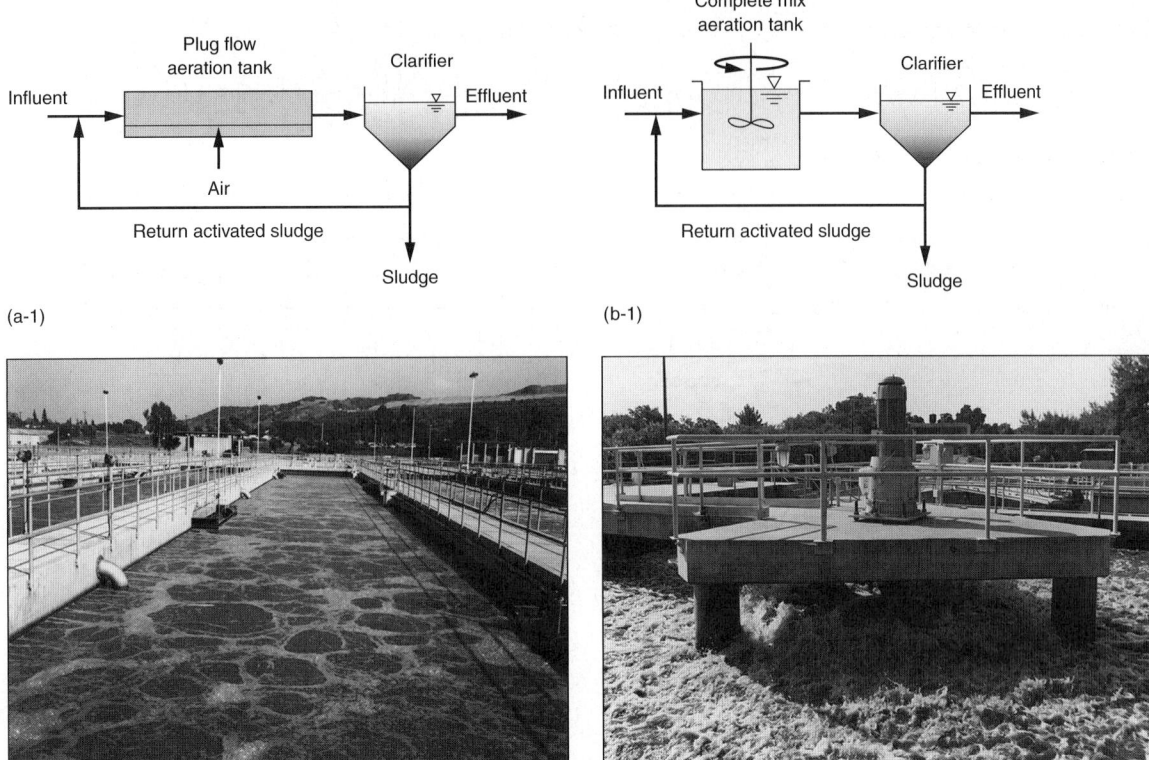

Figure 7–2

Suspended growth biological treatment process (a-1) schematic and (a-2) view of plug-flow activated-sludge process and (b-1) schematic and (b-2) view of complete-mix activated-sludge process.

Chap. 8, the characteristics and thickening properties of the flocculent particles will affect the clarifier design and performance.

Attached Growth Processes. In attached growth processes, the microorganisms responsible for the conversion of organic material or nutrients are attached to an inert packing material. The organic material and nutrients are removed from the wastewater flowing past the attached growth also known as a biofilm. Packing materials used in attached growth processes include rock, gravel, slag, sand, redwood, and a wide range of plastic and other synthetic materials. Attached growth processes can also be operated as aerobic or anaerobic processes. The packing can be submerged completely in liquid or not submerged, with air or gas space above the biofilm liquid layer.

The most common aerobic attached growth process used is the trickling filter in which wastewater is distributed over the top area of a vessel containing nonsubmerged packing material (see Fig. 7–3). Historically, rock was used most commonly as the packing material for trickling filters, with typical depths ranging from 1.25 to 2 m (4 to

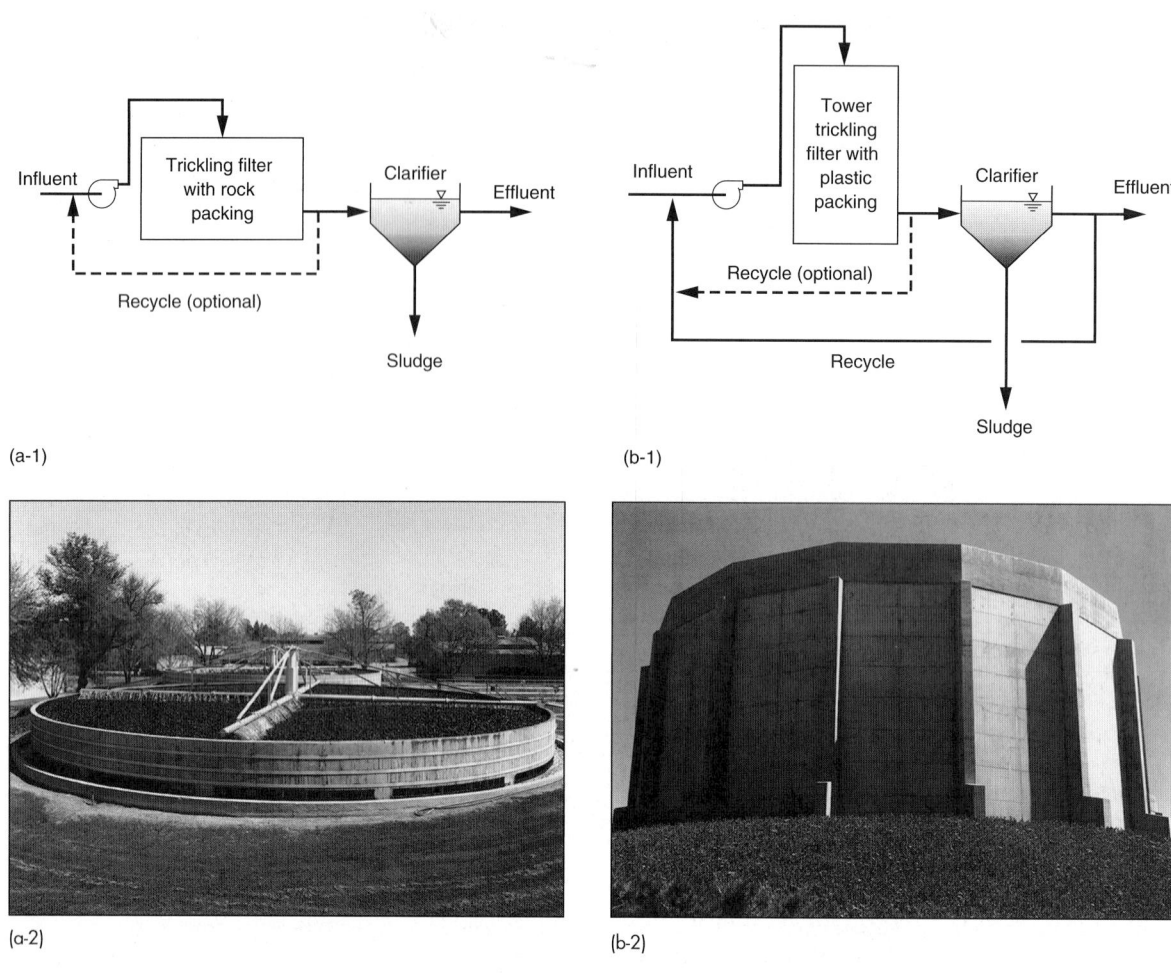

(a-1)

(b-1)

(a-2)

(b-2)

Figure 7–3

Attached growth biological treatment process (a-1) schematic and (a-2) view of tricking filter with rock packing and (b-1) schematic and (b-2) view of tower tricking filter with plastic packing (see Fig. 7–15b). The tower filter is 10 m high and 50 m in diameter.

6 ft). Most modern trickling filters vary in height from 5 to 10 m (16 to 33 ft) and are filled with a plastic packing material for biofilm attachment. The plastic packing material is designed such that about 90 to 95 percent of the volume in the tower consists of void space. Air circulation in the void space, by either natural draft or blowers, provides oxygen for the microorganisms growing as an attached biofilm. Influent wastewater is distributed over the packing and flows as a nonuniform liquid film over the attached biofilm. Excess biomass sloughs from the attached growth periodically and clarification is required for liquid/solids separation to provide an effluent with an acceptable suspended solids concentration. The solids are collected at the bottom of the clarifier and removed for waste-sludge processing.

7–2 COMPOSITION AND CLASSIFICATION OF MICROORGANISMS

Biological processes for wastewater treatment consist of mixed communities with a wide variety of microorganisms, including bacteria, protozoa, fungi, rotifers, and possibly algae. The basic characteristics and important roles of these organisms have been described in Chap. 2. In some cases, biological treatment goals can only be accomplished by the presence of a specific microbial species. To provide a basic understanding of the nature of microorganisms, the topics introduced in this section are: (1) cell components, (2) cell composition, (3) environmental factors that affect microbial activity, and (4) methods used to identify and classify microorganisms. The focus here is mainly on prokaryotes (see Sec. 2–8 in Chap. 2), because of their major role in biological wastewater treatment.

Cell Components

The important components of the prokaryotic cell and their functions are illustrated on Fig. 7–4a and described in Table 7–3; the eukaryotic cell is illustrated on Fig. 7–4b. Key components that relate to the cell's genetic information and specific enzymes produced, which determine the capability of the microorganism in wastewater treatment, are deoxyribose nucleic acid (DNA) and the ribosomes. Ribosomes are the sites of protein synthesis, which are necessary for enzyme production, and the DNA provides the genetic information used to determine the protein structure synthesized. To understand how DNA codes for the cell proteins, the DNA structure and nucleotide sequence and the structure and role of ribose nucleic acid (RNA) are reviewed.

Figure 7–4

Typical internal structure of cells: (a) prokaryotic and (b) eukaryotic.

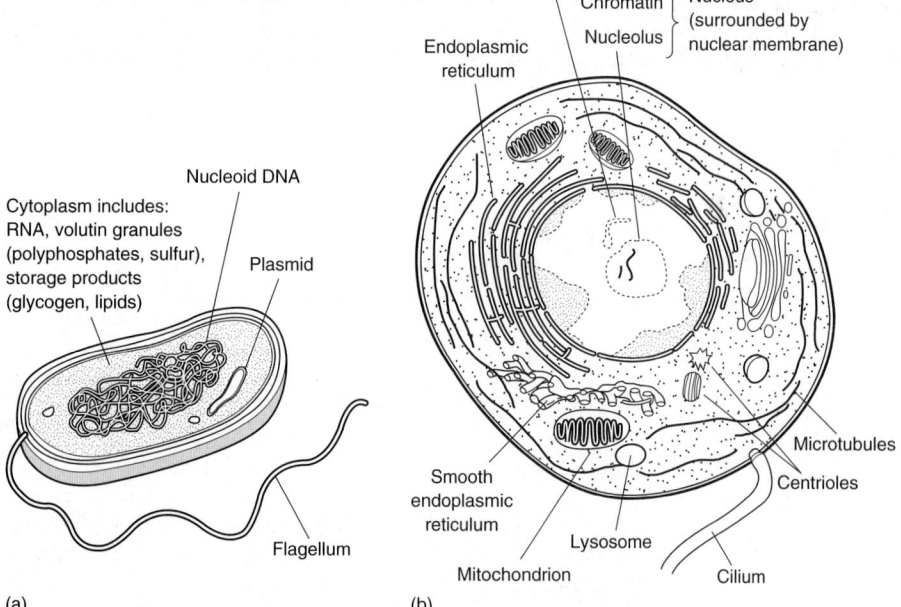

Table 7–3
Description of bacteria cell components

Cell component	Function
Cell wall	Provides strength to maintain the cell shape and protects the cell membrane. Some bacteria can produce a sticky polysaccharide layer outside the cell wall, called a capsule or slime layer
Cell membrane	Controls the passage of dissolved organics and nutrients into the cell and the waste materials and metabolic by-products out of the cell
Cytoplasm	Contains the material within the cell to carry out cell functions and includes water, nutrients, enzymes, ribosomes, and small organic molecules
Cytoplasmic inclusions	Contains storage material that can provide carbon, nutrients, or energy. These may be carbohydrate deposits, such as polyhydroxybutyrate (PHB) or glycogen, polyphosphates, lipids, and sulfur granules
Deoxyribonucleic acid (DNA)	A double-stranded helix-shaped molecule that contains genetic information that determines the nature of the cell protein and enzymes that are produced
Plasmid DNA	Small circular DNA molecules that can also provide genetic characteristics for the bacteria
Ribosomes	Particles in the cytoplasm that are composed of ribonucleic acid (RNA) and protein and are the sites where proteins are produced
Flagella	Protein hairlike structures that extend from the cytoplasm membrane several bacteria lengths out from the cell and provide mobility by rotating at high speeds
Fimbriae and pili	Short protein hairlike structures (pili is longer) that enable bacteria to stick to surfaces. Pili also enable bacteria to attach to each other

Figure 7–5

Nucleotide structure of deoxyribose nucleic acid (DNA) and ribose nucleic acid (RNA).

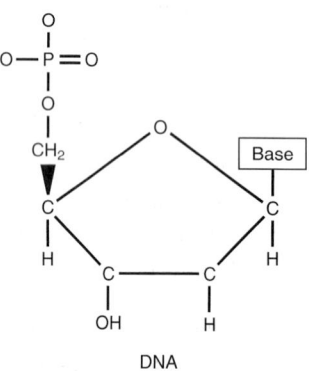

The nucleic acids, DNA and RNA, are composed of a series of nucleotides. Each nucleotide consists of a five-carbon sugar compound, a nitrogen base, and a phosphate molecule (see Fig. 7–5). To form the nucleotide chain of DNA or RNA, the phosphate group bonds to the third carbon (clockwise from the oxygen bond) of the sugar molecule. The nitrogen bases for DNA can be one of four pyrimidine or purine compounds:

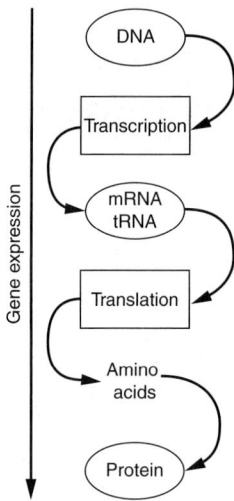

Figure 7–6

Gene expression leads to the formation of proteins by transcription of a segment of the DNA genetic code and translation in the ribosome via messenger RNA (mRNA) and transfer RNA (tRNA) to form a series of amino acids, which form polypeptides and finally protein.

cytosine (C), thymine (T), adenine (A), and guanine (G). For RNA, four nitrogen bases are also involved, including A, C, and G, and thymine is replaced with uracil (U). The DNA is a double-strand helix structure with bonding between the nitrogen bases of each strand. Base bonding is very specific with bonds only between G and C, and A and T. The RNA is a single strand of a nucleotide sequence of various combinations of A, C, G, and U. The sequence of nucleotides in DNA contains the necessary genetic codes for the cell, which determine the specific proteins and enzymes that the bacteria can produce. The number of nucleotides in DNA is very high, and the size of the DNA molecule is described in terms of the number of thousands of nucleotide bases (kilobase) per molecule. The bacterium *Escherichia coli* has 4.7 million nucleotides in each DNA strand or 4700 kilobase pairs.

Gene expression as illustrated on Fig. 7–6 involves the transcription and translation of a segment (gene) of the DNA to form a specific protein. The first step in the process is transcription, in which a small segment of the DNA is unraveled to form a single strand that is used to form a single strand on RNA by complementing base pairing of the nitrogen bases in the DNA nucleotides. For example, adenine (A) in the DNA strand pairs with uracil (U), and guanine (G) with cytosine (C) in the mRNA. The order of the nucleotides in the mRNA will determine the order of amino acids that form polypeptides and the protein structure produced. Translation of the mRNA occurs in the ribosome by tRNA. Each of the nucleotides on the mRNA in series is matched by complementary base pairing with the tRNA in the ribosome, and for each match, another segment of the tRNA, containing three nucleotides, selects a specific amino acid. The three nucleotide sequences in the mRNA are called codons. In essence, each codon selects for a specific amino acid, and there is more than one codon for each of the possible 21 amino acids that can be found in living cells. Thus, the length and nucleotide sequence expressed in the DNA represents a gene that determines what specific protein is formed. Because protein is an essential component of cellular enzymes, the DNA gene composition then determines the microbial cell functions and degradative capabilities. Additional details on gene expression may be found in Madigan et al. (2000).

Cell enzymes, consisting of protein and a cofactor such as a metal ion (e.g., zinc, iron, copper, manganese, or nickel), determine the metabolic capability of microorganisms in wastewater treatment. Enzymes are large organic molecules with molecular weights in the range of 10,000 to 1,000,000. Enzymes catalyze biological reactions necessary for cell functions, such as hydrolysis, oxidation-reduction reactions, and cell synthesis reactions. Cells may also produce enzymes for activity outside the cell wall (extracellular enzymes). An example of the function of extracellular enzymes is the hydrolysis of particulates and large molecules so that the material can be transported across the cell membrane for use by the cell. Enzymes can also be *constitutive* or *inducible.* Constitutive enzymes are produced continuously by the cell, while inducible enzymes are produced in response to the presence of a particular compound. The rate of enzyme activity is affected by temperature and pH.

Cell Composition

To support microbial growth in biological systems, appropriate nutrients must be available. Reviewing the composition of a typical microbial cell will provide a basis for understanding the nutrients needed for growth. Prokaryotes are composed of about 80 percent water and 20 percent dry material, of which 90 percent is organic and 10 percent

Table 7–4
Typical composition of bacteria cells[a]

Constituent or element	Percent of dry weight
Major cellular material	
Protein	55.0
Polysaccharide	5.0
Lipid	9.1
DNA	3.1
RNA	20.5
Other (sugars, amino acids)	6.3
Inorganic ions	1.0
As cell elements	
Carbon	50.0
Oxygen	22.0
Nitrogen	12.0
Hydrogen	9.0
Phosphorus	2.0
Sulfur	1.0
Potassium	1.0
Sodium	1.0
Calcium	0.5
Magnesium	0.5
Chloride	0.5
Iron	0.2
Other trace elements	0.3

[a] Adapted from Madigan et al. (1997).

is inorganic. Typical values for the composition of prokaryote cells are reported in Table 7–4. The most widely used empirical formula for the organic fraction of cells is $C_5H_7O_2N$ first proposed by Hoover and Porges (1952). About 53 percent by weight of the organic fraction is carbon. The formulation $C_{60}H_{87}O_{23}N_{12}P$ can be used when phosphorus is also considered. It should be noted that both formulations are approximations and may vary with time and species, but they are used for practical purposes. Nitrogen and phosphorus are considered macronutrients because they are required in comparatively large amounts. Prokaryotes also require trace amounts of metallic ions, or micronutrients, such as zinc, manganese, copper, molybdenum, iron, and cobalt. Because all of these elements and compounds must be derived from the environment, a shortage of any of these substances would limit and, in some cases, alter growth.

Environmental Factors

Environmental conditions of temperature and pH have an important effect on the selection, survival, and growth of microorganisms. In general, optimal growth for a particu-

Table 7–5
Temperature
classification
of biological
processes

Type	Temperature range, °C	Optimum range, °C
Psychrophilic	10–30	12–18
Mesophilic	20–50	25–40
Thermophilic	35–75	55–65

lar microorganism occurs within a fairly narrow range of temperature and pH, although most microorganisms can survive within much broader limits. Temperatures below the optimum typically have a more significant effect on growth rate than temperatures above the optimum; it has been observed that growth rates double with approximately every 10°C increase in temperature until the optimum temperature is reached. According to the temperature range in which they function best, bacteria may be classified as *psychrophilic*, *mesophilic*, or *thermophilic*. Typical temperature ranges for microorganisms in each of these categories are presented in Table 7–5. For a more detailed discussion of the organisms in the various temperature ranges, see Madigan et al. (1997, 2000).

The pH of the environment is also a key factor in the growth of organisms. Most bacteria cannot tolerate pH levels above 9.5 or below 4.0. Generally, the optimum pH for bacterial growth lies between 6.5 and 7.5. Different archaea are able to grow at thermophilic and ultrathermophilic (60 to 80°C) temperatures, extremely low pH, and high salinity.

Microorganism Identification and Classification

With the use of more sophisticated biological treatment processes and developments in the use of molecular tools for microbial applications, environmental engineers have expanded their interests from the general functionality of microorganisms to understanding the presence and role of specific bacterial species in biological treatment systems. Environmental engineers have also shown how certain conditions in engineered systems may be controlled to create selective pressures that will favor the growth of preferred microorganisms. Thus, the identification of microorganisms and their metabolic characteristics is very important, and is described here along with molecular methods used in biological processes.

In the past, the methods used to identify bacteria relied on physical taxonomic (morphologic) and metabolic characteristics (phenotypic analysis). With modern tools from molecular biology, bacterial identification is now based on cellular genetic information. The basic taxonomic unit in the identification of bacteria is the species, which represents a collection of similar strains of bacteria that exhibit characteristics significantly different from other groups of bacteria. Species that share one or more major properties are grouped and this collection is termed a genus (plural genera). All bacteria are given a genus and species name. The genus name is capitalized and placed before the species name and can be abbreviated ahead of the species name after first identified. The genus and species names are italicized. For example, the genus *Bacillus* contains several species including *B. subtilis*, *B. cerus*, and *B. stearothermophilis*, based on different morphological, physiological, and ecological traits (Madigan et al., 1997, 2000).

Taxonomic Classification. Conventional taxonomic methods used to identify a bacterium rely on physical properties of the bacteria and metabolic characteristics. To apply this approach, a pure culture must first be isolated. The culture may be isolated by serial dilution and growth in selective growth media. The cells are harvested and grown as pure culture using sterilization techniques to prevent contamination. In some cases, isolation of a species is not possible, which may be due to the need for synergistic growth with other species or the lack of a specific growth factor. Historically, the types of tests that are used to characterize a pure culture include (1) microscopic observations, to determine morphology (size and shape); (2) gram staining, to determine if the bacteria cell wall will absorb crystal violet dye; (3) the type of electron acceptor (i.e., oxygen, CO_2, etc.) used in oxidation-reduction reactions; (4) the type of carbon source used for cell growth; (5) the ability to use various nitrogen and sulfur sources; (6) nutritional needs; (7) cell wall chemistry; (8) cell characteristics including pigments, segments, cellular inclusions, and storage products; (9) resistance to antibiotics; and (10) environmental effects of temperature and pH. An alternative to taxonomic classification is a newer method, termed phylogeny.

Phylogenetic Classification.   In the late 1970s, microbiologists began to use tools that allowed them to study microorganisms at the molecular level, and to observe genetic relationships related to the evolutionary history of living cells. The characterization of microorganisms based on genetic information and evolutionary location in time is termed *phylogeny,* which is the more current method of identification and classification. To identify accurately the microbes and determine the true evolutionary relationships between species, the choice of cellular genetic material is critical. The genetic code for ribosomal RNA was chosen as the evolutionary chronometer for cell identification because the code (1) is of evolutionary significance, (2) is well conserved across broad phylogenetic distances, and (3) contains a large number of nucleotide sequences so that similarity in sequences between two organisms indicates a phylogenetic relationship.

Ribosomal RNA can be separated into two components, 30 S (Svedberg units) and 50 S, based on different centrifugal forces in ultracentrifugation. The 30 S units consist of 16S rRNA, containing about 1500 nucleotides, and 21 proteins. While the 16S rRNA can be extracted from cells for nucleotide sequencing using molecular techniques, more recent developments have led to using a section of DNA that contains the rRNA genes (the DNA that encodes the 16S rRNA). This method includes genome DNA extraction from the cell material, followed by a polymerase chain reaction (PCR) procedure that uses DNA primers and a DNA polymerase enzyme to reproduce and amplify artificially the DNA material by a factor of 10^6 or more from the small amount of material extracted from the cell. The amplified 16S rRNA gene is then subjected to sequencing to determine its nucleotide sequence. The sequencing result is compared to the ribosome sequences available in a data base to determine the identity of the organism and its phylogenetic relationship to known organisms.

Molecular phylogeny involves a systematic organization and classification of microorganisms based on their genetic traits. The phylogenetic tree of life with distinct kingdoms as determined from rRNA sequencing is shown on Fig. 7–7. Cellular life is divided into three basic domains, two composed of prokaryotic cells (Archaea and Bacteria), as indicated earlier, and the third composed of Eukarya cells. For the bacteria, some were recognized based on similar morphology and physiology, but most contain a mixture of phenotypic properties. In addition to having different rRNA sequencing properties, the archaea have a number of different phenotypic properties compared to bacte-

Figure 7–7

Phylogenetic tree of life. *(Adapted from Madigan et al., 2000.)*

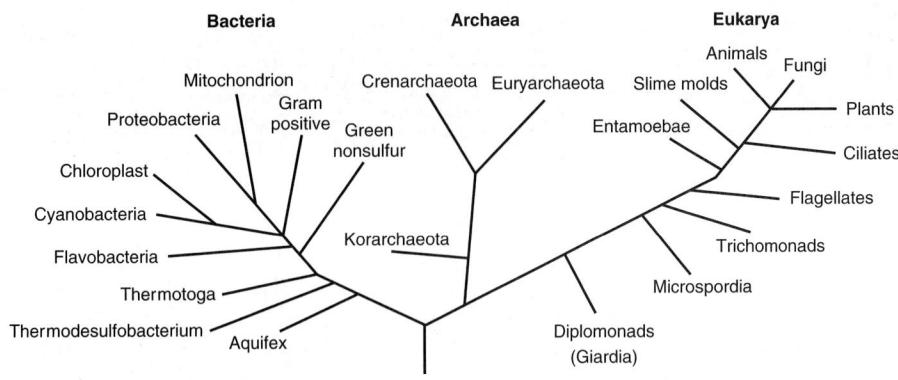

ria. These differences are found in cell wall composition, cell membrane lipid chemistry, RNA polymerase composition, and protein synthesis mechanisms in the ribosome.

Use of Molecular Tools

Besides the ability to identify and classify specific bacteria, molecular tools can be used to provide information previously unavailable about active microbial communities, and to study water or wastewater treatment plant effluents for specific pathogenic organisms. Molecular techniques are evolving continuously as a result of the interest in microbial cell research. These techniques were introduced and described in Sec. 2–8 in Chap. 2. Two significant techniques involve the use of oligonucleotide probes and restrictive fragment length polymorphism (RFLP). An oligonucleotide or nucleic acid probe takes advantage of the fact that unique nucleic acid sequences exist in rRNA or in DNA for a particular microorganism. In RFLP, DNA is extracted from cells in a mixed microbial community and analyzed in a way that provides a genetic fingerprint of the community.

The oligonucleotide probe is constructed so that it will form a double-stranded nucleic acid by complementary base pairing with a single-strand nucleic acid representing the target genetic information in the microorganism. The number of base pairs may be as low as 15 to 18 to over 100. The complementary base pairing of two single-stranded nucleic acids is termed hybridization. The probe is labeled so that it can be detected by fluorescent or radioactive techniques. With the more common fluorescent techniques, the probe contains a reagent that produces a color product or fluorescence, which can be detected with x-ray film. Reagents used to tag the nucleic acid probe include digoxigenin, biotin, and fluorescein. Radioactive labeling of the probe is used less often and is done by using a radio-labeled substance such as ^{32}P, which is incorporated into the phosphate structure of the nucleotide. Autoradiography is used to detect the radio-labeled hybridized probe, which involves the use of a scintillation counter or photographic film. Probes can be used to analyze DNA or RNA extracted from cells or can be transferred to react with rRNA and fluoresce within the cell (see Fig. 7–8). The latter is referred to as fluorescence in situ hybridization (FISH) (Maier et al., 2000).

Figure 7-8

Dot blot technique to probe DNA genetic information.

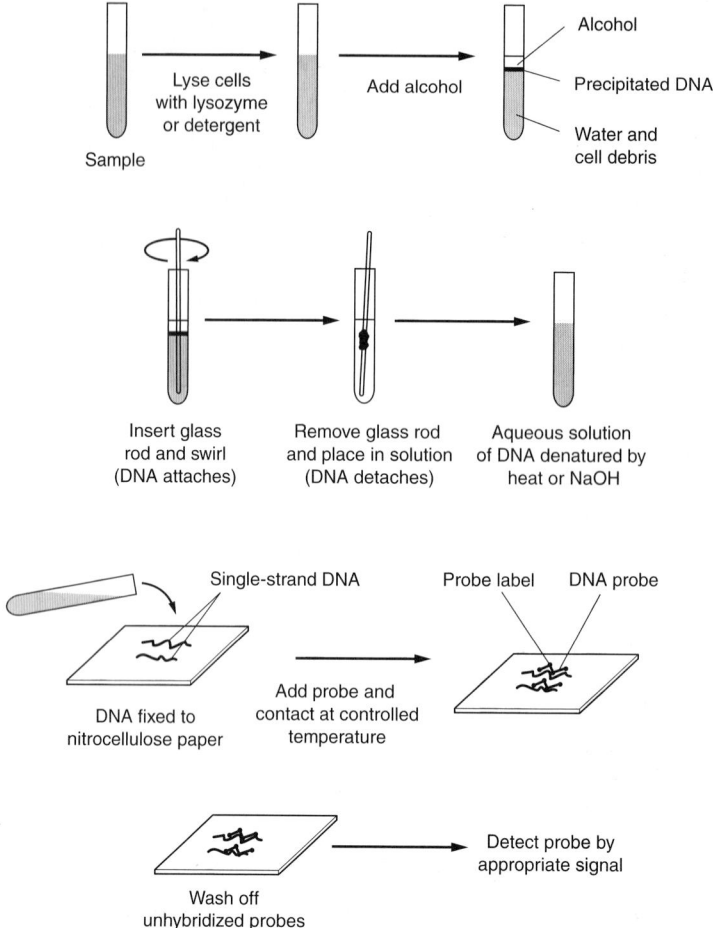

Wagner et al. (1995) demonstrated the use of FISH to identify dense clusters (up to 3000 cells) of ammonia-oxidizing (*Nitrosomonas*) bacteria in several activated-sludge plant samples. The probe, termed the NEU probe, contained 18 nucleotides. A probe used for the study of *E. coli* developed to study UV disinfection was illustrated previously on Fig. 2–36 in Chap. 2. One of the distinct advantages of FISH is that the distribution of a species in its environment can be observed. Also, FISH techniques have been employed with more than one nucleic acid probe added to the sample. Using this technique, a number of bacterial species or strains can be identified and their relative distribution within the matrix can be observed.

In most applications RFLP is not used to identify a particular organism, but to characterize a microbial community. In RFLP, restriction enzymes are used to cut DNA extracted from the microbial cells into various fragments. Different organisms will yield different DNA fragment lengths and different nucleotide sequences within these fragments. From an analysis of the DNA fragments, it is possible to obtain information about the diversity of the microbial community, and the identification of a particular organism responsible for the type of microbial degradation observed.

Table 7-6

Classification of microorganisms by electron donor, electron acceptor, sources of cell carbon, and end products

Type of bacteria	Common reaction name	Carbon source	Electron donor (substrate oxidized)	Electron acceptor	Products
Aerobic heterotrophic	Aerobic oxidation	Organic compounds	Organic compounds	O_2	CO_2, H_2O
Aerobic autotrophic	Nitrification	CO_2	NH_3^-, NO_2^-	O_2	NO_2^-, NO_3^-
	Iron oxidation	CO_2	Fe(II)	O_2	Ferric Iron Fe(III)
	Sulfur oxidation	CO_2	H_2S, $S°$, $S_2O_3^{2-}$	O_2	SO_4^{2-}
Facultative heterotrophic	Denitrification anoxic reaction	Organic compounds	Organic compounds	NO_2^-, NO_3^-	N_2, CO_2, H_2O
Anaerobic heterotrophic	Acid fermentation	Organic compounds	Organic compounds	Organic compounds	Volatile fatty acids (VFAs) (acetate, propionate, butyrate)
	Iron reduction	Organic compounds	Organic compounds	Fe(III)	Fe(II), CO_2, H_2O
	Sulfate reduction	Organic compounds	Organic compounds	SO_4	H_2S, CO_2, H_2O
	Methanogenesis	Organic compounds	Volatile fatty acids (VFAs)	CO_2	Methane

7–3 INTRODUCTION TO MICROBIAL METABOLISM

Basic to the design of a biological treatment process, or to the selection of the type of biological process to be used, is an understanding of the biochemical activities of microorganisms. The classification of microorganisms by sources of cell carbon, electron donor, electron acceptor, and end products is summarized in Table 7–6. Different microorganisms can use a wide range of electron acceptors, including oxygen, nitrite, nitrate, iron (III), sulfate, organic compounds, and carbon dioxide. Schematic representations of some common types of bacterial metabolism are given on Fig. 7–9 showing the energy production and cell carbon sources. The two major topics considered in this section are (1) the general nutritional requirements of the microorganisms commonly encountered in wastewater treatment, and (2) the nature of microbial metabolism based on the need for molecular oxygen.

Carbon and Energy Sources for Microbial Growth

To continue to reproduce and function properly, an organism must have sources of energy, carbon for the synthesis of new cellular material, and inorganic elements (nutrients) such as nitrogen, phosphorus, sulfur, potassium, calcium, and magnesium. Organic nutrients (growth factors) may also be required for cell synthesis. Carbon and

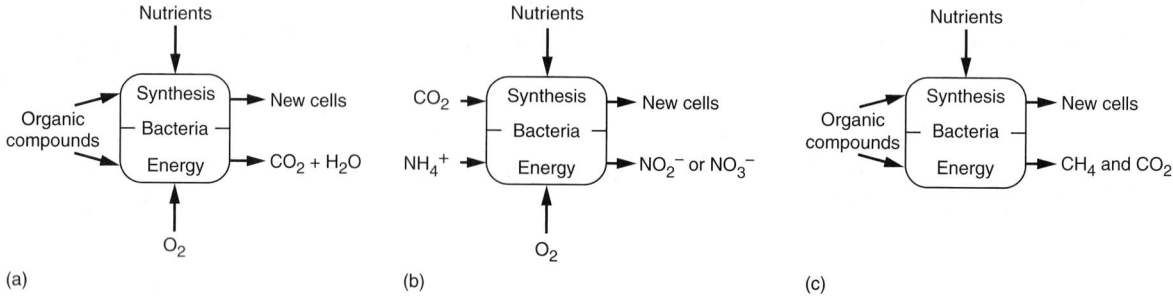

Figure 7–9

Examples of bacteria metabolism: (*a*) aerobic, heterotrophic, (*b*) aerobic, autotrophic, (*c*) anaerobic, heterotrophic.

energy sources, usually referred to as *substrates,* and nutrient and growth factor requirements for various types of organisms are considered in the following discussion.

Carbon Sources.   Microorganisms obtain their carbon for cell growth from either organic matter or carbon dioxide. Organisms that use organic carbon for the formation of new biomass are called *heterotrophs,* while organisms that derive cell carbon from carbon dioxide are called *autotrophs.* The conversion of carbon dioxide to cellular carbon compounds requires a reductive process, which requires a net input of energy. Autotrophic organisms must therefore spend more of their energy for synthesis than do heterotrophs, resulting in generally lower yields of cell mass and growth rates.

Energy Sources.   The energy needed for cell synthesis may be supplied by light or by a chemical oxidation reaction. Bacteria can oxidize organic or inorganic compounds to gain energy. Those organisms that are able to use light as an energy source are called *phototrophs.* Phototrophic organisms may be either heterotrophic (certain sulfur-reducing bacteria) or autotrophic (algae and photosynthetic bacteria). Organisms that derive their energy from chemical reactions are known as *chemotrophs.* As with the phototrophs, chemotrophs may be either heterotrophic (protozoa, fungi, and most bacteria) or autotrophic (i.e., nitrifying bacteria). *Chemoautotrophs* obtain energy from the oxidation of reduced *inorganic* compounds, such as ammonia, nitrite, ferrous iron, and sulfide. *Chemoheterotrophs* usually derive their energy from the oxidation of *organic* compounds.

The energy-producing chemical reactions by chemotrophs are oxidation-reduction reactions that involve the transfer of electrons from an electron donor to an electron acceptor. The electron donor is oxidized and the electron acceptor is reduced. The electron donors and acceptors can be either organic or inorganic compounds, depending on the microorganism. The electron acceptor may be available within the cell during metabolism (endogenous) or it may be obtained from outside the cell (i.e., dissolved oxygen) (exogenous). Organisms that generate energy by enzyme-mediated electron transport to an *external* electron acceptor are said to have a *respiratory metabolism.* The use of an internal electron acceptor is termed *fermentative metabolism* and is a less efficient energy-yielding process than respiration. Heterotrophic organisms that are strictly fermentative are characterized by lower growth rates and cell yields than respiratory heterotrophs.

When oxygen is used for the electron acceptor the reaction is termed *aerobic,* and reactions involving other electron acceptors are considered anaerobic. The term *anoxic* is used to distinguish the use of nitrite or nitrate for electron acceptors from the others under anaerobic conditions. Under anoxic conditions nitrite or nitrate reduction to gaseous nitrogen occurs, and this reaction is also referred to as biological denitrification. Organisms that can only meet their energy needs with oxygen are called *obligate aerobic* microorganisms. Some bacteria can use oxygen or nitrate/nitrite as electron acceptors when oxygen is not available. These bacteria are called *facultative aerobic* bacteria.

Organisms that generate energy by fermentation and that can exist only in an environment that is devoid of oxygen are *obligate anaerobes. Facultative anaerobes* have the ability to grow in either the presence or absence of molecular oxygen and fall into two subgroups, based on their metabolic abilities. True facultative anaerobes can shift from fermentative to aerobic respiratory metabolism, depending upon the presence or absence of molecular oxygen. *Aerotolerant anaerobes* have a strictly fermentative metabolism but are relatively insensitive to the presence of molecular oxygen.

Nutrient and Growth Factor Requirements

Nutrients, rather than carbon or energy sources, may at times be the limiting material for microbial cell synthesis and growth. The principal inorganic nutrients needed by microorganisms are N, S, P, K, Mg, Ca, Fe, Na, and Cl. Minor nutrients of importance include Zn, Mn, Mo, Se, Co, Cu, and Ni (Madigan et al., 2000). Required organic nutrients, known as growth factors, are compounds needed by an organism as precursors or constituents of organic cell material, which cannot be synthesized from other carbon sources. Although growth factor requirements differ from one organism to another, the major growth factors fall into the following three classes: (1) amino acids, (2) nitrogen bases (i.e., purines and pyrimidines), and (3) vitamins.

For municipal wastewater treatment sufficient nutrients are generally present, but for industrial wastewaters nutrients may need to be added to the biological treatment processes. The lack of sufficient nitrogen and phosphorus is common especially in the treatment of food-processing wastewaters or wastewaters high in organic content. Using the formula $C_{12}H_{87}O_{23}N_{12}P$ (given previously) for the composition of cell biomass, about 12.2 g of nitrogen and 2.3 g of phosphorus are needed per 100 g of cell biomass.

7–4 BACTERIAL GROWTH AND ENERGETICS

In the description of microbial metabolism it was noted that as microorganisms consume substrate and carry out oxidation-reduction reactions, growth occurs by the production of additional cells. Thus, in wastewater treatment applications biomass is produced continuously as the substrate in the wastewater is consumed and biodegraded. Topics considered in this section include (1) bacterial reproduction, (2) bacterial growth patterns in a batch reactor, (3) bacterial growth and biomass yield, (4) methods used to measure biomass growth, (5) estimating cell yield and oxygen requirements from stoichiometry, (6) estimating cell yield from bioenergetics, and (7) observed versus synthesis yield. The material presented in this section will serve as a basis for the sections that follow and the material presented in Chaps. 8, 9, and 10.

Bacterial Reproduction

Bacteria can reproduce, as noted in Chap. 2, by binary fission, by asexual mode, or by budding. Generally, they reproduce by binary fission, in which the original cell becomes two new organisms. The time required for each division, which is termed the generation time, can vary from days to less than 20 min. For example, if the generation time is 30 min, one bacterium would yield 16,777,216 (i.e., 2^{24}) bacteria after a period of 12 h. Assuming spherical-shaped bacteria with a 1 μm diameter and specific gravity of 1.0, the weight of 1 cell is approximately 5.0×10^{-13} g. In 12 h the bacteria mass would be about 8.4×10^{-6} g or 8.4 μg; thus the number of cells is quite large compared to the mass. This rapid change in biomass with time is a hypothetical example, for in biological treatment systems bacteria would not continue to divide indefinitely because of environmental limitations, such as substrate and nutrient availability.

Bacterial Growth Patterns in a Batch Reactor

Bacterial growth in a batch reactor is characterized by identifiable phases as illustrated on Fig. 7–10. The curves shown on Fig. 7–10 represent what occurs in a batch reactor in which, at time zero, substrate and nutrients are present in excess and only a very small population of biomass exists. As substrate is consumed, four distinct growth phases develop sequentially.

1. The lag phase. Upon addition of the biomass, the lag phase represents the time required for the organisms to acclimate to their new environment before significant cell division and biomass production occur. During the lag phase enzyme induction may be occurring and/or the cells may be acclimating to changes in salinity, pH, or temperature. The apparent extent of the lag phase may also be affected by the ability to measure the low biomass concentration during the initial batch phase.

2. The exponential-growth phase. During the exponential-growth phase, bacterial cells are multiplying at their maximum rate, as there is no limitation due to substrate or nutrients. The biomass growth curve increases exponentially during this period. With unlimited substrate and nutrients the only factor that affects the rate of exponential growth is temperature.

Figure 7–10

Batch process biomass growth phases with changes in substrate and biomass versus time.

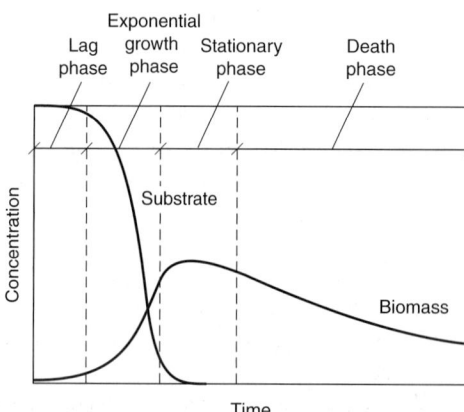

Time

3. The stationary phase. During this phase, the biomass concentration remains relatively constant with time. In this phase, bacterial growth is no longer exponential and the amount of growth is offset by the death of cells.

4. The death phase. In the death phase, the substrate has been depleted so that no growth is occurring, and the change in biomass concentration is due to cell death. An exponential decline in the biomass concentration is often observed as an approximate constant fraction of the biomass remaining that is lost each day.

Bacterial Growth and Biomass Yield

In biological treatment processes, cell growth occurs concurrent with the oxidation of organic or inorganic compounds, as described above. The ratio of the amount of biomass produced to the amount of substrate consumed (g biomass/g substrate) is defined as the *biomass yield,* and typically is defined relative to the electron donor used.

$$\text{Biomass yield } Y = \frac{\text{g biomass produced}}{\text{g substrate utilized (i.e., consumed)}} \tag{7-2}$$

For example, for aerobic heterotrophic reactions with organic substrates, the yield is expressed as g biomass/g organic substrate; for nitrification the yield is expressed as g biomass/g NH_4-N oxidized; and for the anaerobic degradation of volatile fatty acids (VFAs) to produce methane, the yield is expressed as g biomass/g VFAs used. Where specific compounds are measured and known such as ammonia, the yield is quantified relative to the amount of compound used. For aerobic or anaerobic treatment of municipal and industrial wastewater containing a large number of organic compounds, the yield is based on a measurable parameter reflecting the overall organic compound consumption, such as COD or BOD. Thus, the yield would be g biomass/g COD removed or g biomass/g BOD removed.

Measuring Biomass Growth

Because biomass is mostly organic material, an increase in biomass can be measured by volatile suspended solids (VSS) or particulate COD (total COD minus soluble COD). Other parameters that are used to indicate biomass growth are protein content, DNA, and ATP, a cellular compound involved in energy transfer. Of these growth measurement parameters, VSS is the parameter used most commonly to follow biomass growth in full-scale biological wastewater treatment systems because its measurement is simple and minimal time is required for analysis. It should be noted that the VSS measured includes other particulate organic matter in addition to biomass. Most wastewaters contain some amount of nonbiodegradable VSS and possibly influent VSS that may be slowly degraded in the biological reactor. These solids are included with biomass in the VSS measurement. Nevertheless the VSS measurement is used as an apparent indicator of biomass production and also provides a useful measurement of reactor solids in general.

For laboratory research on biological treatment processes, growth parameters that can be related to true microbial mass are often used. Of these, protein is the most popular growth parameter due to the relative ease of measurement and the fact that about 50 percent of biomass dry weight is protein. Both ATP and DNA have also been used, especially where the reactor solids contain proteins and other solids that are not associated with biomass. Where very low biomass concentrations are involved, turbidity measurements may be used to provide a rapid and simple means of observing cell growth. Bacterial cell counts have also been used to enumerate the biomass population. A portion of a diluted sample is applied to an agar growth plate, and after incubation, the number of colonies formed are counted and used to determine the number of bacterial cells in the culture. It should be noted, however, that not all bacteria are culturable.

Estimating Biomass Yield and Oxygen Requirements from Stoichiometry

As given by Eq. (7–1), a definite stoichiometric relationship exists between the substrate removed, the amount of oxygen consumed during aerobic heterotrophic biodegradation, and the observed biomass yield. The most common approach used to define the fate of the substrate is to prepare a COD mass balance. The COD is used because the substrate concentration in the wastewater can be defined in terms of its oxygen equivalence, which can be accounted for by being conserved in the biomass or oxidized. In general, the exact stoichiometry involved in the biological oxidation of a mixture of wastewater compounds is never known. However, for the purpose of illustration, assume organic matter can be represented as $C_6H_{12}O_6$ (glucose) and new cells can be represented as $C_5H_7NO_2$ (Hoover and Porges, 1952). Thus, neglecting nutrients other than nitrogen, Eq. (7–1) can be written as

$$3C_6H_{12}O_6 + 8O_2 + 2NH_3 \rightarrow 2C_5H_7NO_2 + 8CO_2 + 14H_2O \qquad (7\text{–}3)$$
$$\quad\;\; 3(180) \qquad 8(32) \quad 2(17) \qquad\quad 2(113)$$

As given by the above equation, the substrate used (glucose in this case) is divided between that found in new cells and that oxidized. The yield based on the glucose consumed can be obtained as follows:

$$Y = \frac{\Delta(C_5H_7NO_2)}{\Delta(C_6H_{12}O_6)} = \frac{2(113 \text{ g/mole})}{3(180 \text{ g/mole})}$$

$$= 0.42 \text{ g cells/g glucose used}$$

In practice, COD and VSS are used to represent the organic matter and the new cells, respectively. To express the yield on a COD basis, the COD of glucose must be determined. The COD of glucose can be determined by writing a balanced stoichiometric reaction for the oxidation of glucose to carbon dioxide as follows:

$$C_6H_{12}O_6 + 6O_2 \rightarrow 6CO_2 + 6H_2O \qquad (7\text{–}4)$$
$$\quad (180) \qquad 6(32)$$

The COD of glucose is

$$COD = \frac{\Delta(O_2)}{\Delta(C_6H_{12}O_6)} = \frac{6(32 \text{ g/mole})}{(180 \text{ g/mole})} = 1.07 \text{ g } O_2/\text{g glucose}$$

The theoretical yield expressed in terms of COD, accounting for the portion of the substrate converted to new cells, is

$$Y = \frac{\Delta(C_5H_7NO_2)}{\Delta(C_6H_{12}O_6 \text{ as COD})} = \frac{2(113 \text{ g/mole})}{3(180 \text{ g/mole})(1.07 \text{ g COD/g glucose})}$$

$$= 0.39 \text{ g cells/g COD used}$$

It should be noted that the actual observed yield in a biological treatment process will be less than the value given above, because a portion of the substrate incorporated into the cell mass will be oxidized with time by the bacteria to obtain energy for cell maintenance.

The quantity of oxygen utilized can be accounted for by considering (1) the oxygen used for substrate oxidation to CO_2 and H_2O, (2) the COD of the biomass, and (3) the COD of any substrate not degraded. Based on the formula $C_5H_7NO_2$, the oxygen equivalent of the biomass (typically measured as VSS) is approximately 1.42 g COD/g biomass VSS, as given below.

$$C_5H_7NO_2 + 5O_2 \rightarrow 5CO_2 + NH_3 + 2H_2O \qquad (7-5)$$
$$\underset{(113)}{} \quad \underset{5(32)}{}$$

The COD of cell tissue is

$$COD = \frac{\Delta(O_2)}{\Delta(C_5H_7NO_2)} = \frac{5(32 \text{ g/mole})}{(113 \text{ g/mole})} = 1.42 \text{ g } O_2/\text{g cells}$$

Based on the above relationships, the oxygen consumed per unit of COD utilized for the reaction given by Eq. (7–3) can be determined from a mass balance on COD as follows:

$$\text{COD utilized} = \text{COD cells} + \text{COD of oxidized substrate} \qquad (7-6)$$

The COD of the oxidized substrate is equal to oxygen consumed; thus

$$\text{Oxygen consumed} = \text{COD utilized} - \text{COD cells} \qquad (7-7)$$

$$= \left(\frac{1.07 \text{ g } O_2}{\text{g glucose}} \right) \left(3 \text{ mole} \times \frac{180 \text{ g glucose}}{\text{mole}} \right)$$

$$- \left(\frac{1.42 \text{ g } O_2}{\text{g cells}} \right) \left(2 \text{ moles} \times \frac{113 \text{ g cells}}{\text{mole}} \right)$$

$$= 577.8 \text{ g } O_2 - 320.9 \text{ g } O_2 = 256.9 \text{ g } O_2$$

Thus, the oxygen consumed per unit of COD used is

$$\frac{\text{Oxygen consumed}}{\text{Glucose as COD}} = \frac{256.9 \text{ g } O_2}{3 \text{ mole } (1.07 \text{ g COD/g glucose})(180 \text{ g glucose/mole})}$$

$$= 0.44 \text{ g } O_2/\text{g COD used}$$

The amount of oxygen required based on the COD balance as given above is in agreement with the oxygen use based on the stoichiometry as defined by Eq. (7–3) in which 8 moles of oxygen are required for 3 moles of glucose.

$$\frac{\text{Oxygen used}}{\text{Glucose as COD}} = \frac{8(32 \text{ g } O_2/\text{mole})}{3(180 \text{ g/mole}) (1.07 \text{ g COD/g glucose})}$$

$$= 0.44 \text{ g } O_2/\text{g COD used}$$

The relationship of the observed biomass yield to the oxygen consumed for substrate oxidation by aerobic heterotrophic biomass based on typical measurements made at wastewater-treatment plants is illustrated in Example 7–1.

EXAMPLE 7–1 **Observed Biomass Yield and Oxygen Consumption** The aerobic complete-mix biological treatment process without recycle, as shown below, receives wastewater with a biodegradable soluble COD (bsCOD) concentration of 500 g/m^3. The flowrate is 1000 m^3/d and the reactor effluent bsCOD and VSS concentrations are 10 and 200 g/m^3, respectively. Based on these data:

1. What is the observed yield in g VSS/g COD removed?
2. What is the amount of oxygen used in g O_2/g COD removed and in g/d?

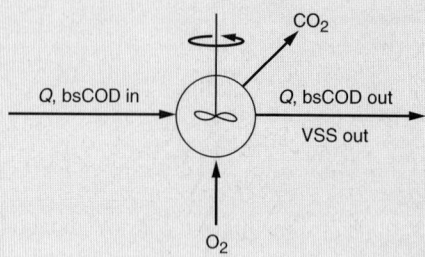

Solution

1. Determine the observed yield. Assume the following general reaction is applicable:

$$\text{Organic matter} + O_2 + \text{nutrients} \rightarrow C_5H_7NO_2 + CO_2 + H_2O$$
    $\quad$ 500 g COD/m^3 $\qquad\qquad\qquad\qquad\qquad$ 200 g VSS/m^3

   a. The g VSS/d produced is

   g VSS/d = 200 g/m^3 (1000 m^3/d) = 200,000 g VSS/d

   b. The g bsCOD/d removed is

   g COD /d = (500 − 10) g COD/m^3 (1000 m^3/d)

   $\qquad\qquad\quad$ = 490,000 g COD/d

   c. The observed yield is

   $$Y_{obs} = \frac{(200{,}000 \text{ g VSS/d})}{(490{,}000 \text{ g COD/d})} = 0.41 \text{ g VSS/g COD removed}$$

2. Determine the amount of oxygen used per g bsCOD removed.
   a. Prepare a steady-state COD mass balance around the reactor.

   Accumulation = inflow − outflow + conversion

   0 = COD_{in} − COD_{out} − oxygen used (expressed as COD)

   Oxygen used = COD_{in} − COD_{out}

   COD_{in} = 500 g COD/m^3 (1000 m^3/d) = 500,000 g COD/d

   COD_{out} = bsCOD_{out} + biomass COD_{out}

   bsCOD_{out} = 10 g/m^3 (1000 m^3/d) = 10,000 g COD/d

$$\text{Biomass COD}_{\text{out}} = 200{,}000 \text{ g VSS/d } (1.42 \text{ g COD/g VSS})$$

$$= 284{,}000 \text{ g COD/d}$$

$$\text{Total COD}_{\text{out}} = 10{,}000 \text{ g/d} + 284{,}000 \text{ g/d} = 294{,}000 \text{ g COD/d}$$

b. The oxygen used is

$$\text{Oxygen used} = 500{,}000 \text{ g COD/d} - 294{,}000 \text{ g COD/d}$$

$$= 206{,}000 \text{ g COD/d} = 206{,}000 \text{ g O}_2\text{/d}$$

c. Amount of oxygen used per unit COD removed is

$$\text{Oxygen/COD} = (206{,}000 \text{ g/d})/(490{,}000 \text{ g/d}) = 0.42 \text{ g O}_2\text{/g COD}$$

Comment   *Note:* The general COD balance that accounts for cell production and COD oxidation is:

g COD cells + g COD oxidized = g COD removed

$(0.41 \text{ g VSS/g COD})(1.42 \text{ g O}_2\text{/g VSS}) + 0.42 \text{ g O}_2\text{/g COD} = 1.0 \text{ g O}_2\text{/g COD}$

Estimating Biomass Yield from Bioenergetics

Most cell yield values are based on measurements from laboratory reactors, pilot plants, or full-scale systems. However, an approach that has been developed to estimate cell yield, based on bioenergetics, involves the application of thermodynamic principles to biological reactions. An introduction to bioenergetics and the application of bioenergetics to estimate the biomass yield for different types of biological reactions is provided in this section (McCarty 1971, 1975).

Chemical reactions, which involve changes in energy, can be described thermodynamically by a change in the free energy $G°$, known as the *Gibbs free energy*. The change in energy due to the reaction is termed $\Delta G°$. The superscript is used to indicate that the free energy values were obtained at standard conditions of pH = 7.0 and 25°C. The net Gibbs free energy, positive or negative, can be evaluated for reactants and products based on standard free energy values available for the half reactions. Half reactions describe the transfer of 1 mole of electron in oxidation-reduction and synthesis reactions. Free energy changes for various half reactions are listed in Table 7–7. Reactions that result in a negative change in the free energy release energy are called *exergonic* reactions. These reactions will proceed spontaneously in the direction shown. However, if the free energy change results in a positive value, the reaction is termed *endergonic*, and such a reaction will not occur spontaneously. Reactions with positive free energy values require energy to proceed in the direction indicated.

The basis of the analysis of free energy changes for reactions is that in oxidation reduction reactions one compound loses electrons (electron donor) and the other compound gains electrons (electron acceptor) (see discussion in Sec. 6–6 in Chap. 6). Half reactions that show the moles of compound used as an electron acceptor and electron donor per mole of electron (e^-) transferred along with the change in free energy are

Table 7–7
Half reactions for biological systems[a]

Reaction number	Half reaction	ΔG° (W),[b] kJ per electron equivalent
	Reactions for bacterial cell synthesis (R_{cs})	
	Ammonia as nitrogen source:	
1.	$\frac{1}{5} CO_2 + \frac{1}{20} HCO_3^- + \frac{1}{20} NH_4^+ + H^+ + e^- = \frac{1}{20} C_5H_7O_2N + \frac{9}{20} H_2O$	
	Nitrate as nitrogen source:	
2.	$\frac{1}{28} NO_3^- + \frac{5}{28} CO_2 + \frac{29}{28} H^+ + e^- = \frac{1}{28} C_5H_7O_2N + \frac{11}{28} H_2O$	
	Reactions for electron acceptors (R_a)	
	Nitrite:	
3.	$\frac{1}{3} NO_2^- + \frac{4}{3} H^+ + e^- = \frac{1}{6} N_2 + \frac{2}{3} H_2O$	−93.23
	Oxygen:	
4.	$\frac{1}{4} O_2 + H^+ + e^- = \frac{1}{2} H_2O$	−78.14
	Nitrate:	
5.	$\frac{1}{5} NO_3^- + \frac{6}{5} H^+ + e^- = \frac{1}{10} N_2 + \frac{3}{5} H_2O$	−71.67
	Sulfite:	
6.	$\frac{1}{6} SO_3^{2-} + \frac{5}{4} H^+ + e^- = \frac{1}{12} H_2S + \frac{1}{12} HS^- + \frac{1}{2} H_2O$	13.60
	Sulfate:	
7.	$\frac{1}{8} SO_4^{2-} + \frac{19}{16} H^+ + e^- = \frac{1}{16} H_2S + \frac{1}{16} HS^- + \frac{1}{2} H_2O$	21.27
	Carbon dioxide (methane fermentation):	
8.	$\frac{1}{8} CO_2 + H^+ + e^- = \frac{1}{8} CH_4 + \frac{1}{4} H_2O$	24.11
	Reactions for electron donors (R_d)	
	Organic donors (heterotrophic reactions)	
	Domestic wastewater:	
9.	$\frac{9}{50} CO_2 + \frac{1}{50} NH_4^+ + \frac{1}{50} HCO_3^- + H^+ + e^- = \frac{1}{50} C_{10}H_{19}O_3N + \frac{9}{25} H_2O$	31.80
	Protein (amino acids, proteins, nitrogenous organics):	
10.	$\frac{8}{33} CO_2 + \frac{2}{33} NH_4^+ + \frac{31}{33} H^+ + e^- = \frac{1}{66} C_{16}H_{24}O_5N_4 + \frac{27}{66} H_2O$	32.22
	Formate:	
11.	$\frac{1}{2} HCO_3^- + H^+ + e^- = \frac{1}{2} HCOO^- + \frac{1}{2} H_2O$	48.07

(continued)

| **Table 7-7** (Continued)

Reaction number	Half reaction		ΔG° (W),[b] kJ per electron equivalent
	Glucose:		
12.	$\frac{1}{4}CO_2 + H^+ + e^-$	$= \frac{1}{24}C_6H_{12}O_6 + \frac{1}{4}H_2O$	41.96
	Carbohydrate (cellulose, starch, sugars):		
13.	$\frac{1}{4}CO_2 + H^+ + e^-$	$= \frac{1}{4}CH_2O + \frac{1}{4}H_2O$	41.84
	Methanol:		
14.	$\frac{1}{6}CO_2 + H^+ + e^-$	$= \frac{1}{6}CH_3OH + \frac{1}{6}H_2O$	37.51
	Pyruvate:		
15.	$\frac{1}{5}CO_2 + \frac{1}{10}HCO_3^- + H^+ + e^-$	$= \frac{1}{10}CH_3COCOO^- + \frac{2}{5}H_2O$	35.78
	Ethanol:		
16.	$\frac{1}{6}CO_2 + H^+ + e^-$	$= \frac{1}{12}CH_3CH_2OH + \frac{1}{4}H_2O$	31.79
	Propionate:		
17.	$\frac{1}{7}CO_2 + \frac{1}{14}HCO_3^- + H^+ + e^-$	$= \frac{1}{14}CH_3CH_2COO^- + \frac{5}{14}H_2O$	27.91
	Acetate:		
18.	$\frac{1}{8}CO_2 + \frac{1}{8}HCO_3^- + H^+ + e^-$	$= \frac{1}{8}CH_3COO^- + \frac{3}{8}H_2O$	27.68
	Grease (fats and oils):		
19.	$\frac{4}{23}CO_2 + H^+ + e^-$	$= \frac{1}{46}C_8H_{16}O + \frac{15}{46}H_2O$	27.61
	Inorganic donors (autotrophic reactions):		
20.	$Fe^{3+} + e^-$	$= Fe^{2+}$	−74.40
21.	$\frac{1}{2}NO_3^- + H^+ + e^-$	$= \frac{1}{2}NO_2^- + \frac{1}{2}H_2O$	−40.15
22.	$\frac{1}{8}NO_3^- + \frac{5}{4}H^+ + e^-$	$= \frac{1}{8}NH_4^+ + \frac{3}{8}H_2O$	−34.50
23.	$\frac{1}{6}NO_2^- + \frac{4}{3}H^+ + e^-$	$= \frac{1}{6}NH_4^+ + \frac{1}{3}H_2O$	−32.62
24.	$\frac{1}{6}SO_4^{2-} + \frac{4}{3}H^+ + e^-$	$= \frac{1}{6}S + \frac{2}{3}H_2O$	19.48
25.	$\frac{1}{8}SO_4^{2-} + \frac{19}{16}H^+ + e^-$	$= \frac{1}{16}H_2S + \frac{1}{16}HS^- + \frac{1}{2}H_2O$	21.28
26.	$\frac{1}{4}SO_4^{2-} + \frac{5}{4}H^+ + e^-$	$= \frac{1}{8}S_2O_3^{2-} + \frac{5}{8}H_2O$	21.30
27.	$\frac{1}{6}N_2 + \frac{4}{3}H^+ + e^-$	$= \frac{1}{3}NH_4^+$	27.47
28.	$H^+ + e^-$	$= \frac{1}{2}H_2$	40.46
29.	$\frac{1}{2}SO_4^{2-} + H^+ + e^-$	$= SO_3^{2-} + H_2O$	44.33

[a] Adapted from McCarty (1975) and Sawyer et al. (1994).
[b] Reactants and products at unit activity except $[H^+] = 10^{-7}$.

used to develop energy balances in bioenergetic analyses. Determination of the free energy change resulting from the oxidation of hydrogen by oxygen is illustrated in Example 7–2.

EXAMPLE 7–2 **Free Energy Change from Hydrogen Oxidation by Molecular Oxygen**
Determine the free energy change resulting from the oxidation of hydrogen by molecular oxygen.

Solution

1. Identify the electron donor and acceptor
    Electron donor: hydrogen
    Electron acceptor: oxygen
2. Determine the change in free energy. From Table 7–7, the free energy change values for the half reactions are:

		$G°$, kJ/mole e^-
No. 28	$H_2 \rightarrow 2H^+ + 2e^-$	-40.46
No. 4	$\frac{1}{2}O_2 + 2H^+ + 2e^- \rightarrow H_2O$	-78.14
Overall	$H_2 + \frac{1}{2}O_2 \rightarrow H_2O$ $\Delta G = -118.60$	

Because the ΔG value is negative for this oxidation reduction reaction, energy is released and the overall reaction will proceed as written.

Exergonic reactions are catalyzed by enzymes within the microbial cell, making energy available to support cell growth. Only a portion (40 to 80 percent) of the energy produced is captured by the bacteria, while the rest escapes as heat. McCarty (1971) assumed 60 percent energy capture efficiency, but the exact amount varies. The energy that is not captured or released can result in an elevated temperature in the surrounding liquid, where high biomass concentrations exist and high reaction rates are occurring. An example is autothermal aerobic digestion in which liquid temperatures are increased from 20°C to as high as 60°C due to biological oxidation and energy release. The volatile solids concentration in autothermal aerobic digesters may be in the range of 20 to 40 g/L.

The key steps in bioenergetics analysis are to (1) identify the electron donor (substrate oxidized) and electron acceptor, (2) determine the energy produced from the bacteria oxidation reduction reaction, (3) determine the amount of energy needed for converting the growth carbon source into cell matter, and (4) calculate the cell yield based on a balance between energy produced and energy needed for cell yield. The energy production step was illustrated above for oxidation of hydrogen with oxygen as the electron acceptor.

The amount of energy required for cell synthesis depends on the specific carbon and nitrogen compounds used for growth. The bioenergetics analysis discussed here is for heterotrophic bacteria. A different procedure is used for the case of autotrophic bacteria, and additional details can be found in McCarty (1971, 1975) and Rittman and McCarty (2001). For heterotrophic bacteria, many carbon sources may be used for growth with different energetic effects. The analysis assumes pyruvate as an intermediate organic compound in cell synthesis, and energy will be either produced or consumed depending on the free energy of the organic compound relative to pyruvate. Pyruvate was selected by McCarty (1971) because it is at the end of the glycolysis pathway and just ahead of the Kreb cycle. When CO_2 is used for the carbon source, as for autotrophic bacteria, a considerable amount of energy is needed to incorporate CO_2 into cell mass. If nitrogen is not available in the form of ammonia, additional energy is needed to convert the nitrogen source to ammonia. The energy required for cell synthesis is estimated as follows, with pyruvate as the organic intermediate for cellular carbon constituents:

$$\Delta G_s = \frac{\Delta G_P}{K^m} + \Delta G_c + \frac{\Delta G_N}{K} \tag{7–8}$$

where ΔG_s = free energy to convert 1 electron equivalent (e^- eq) of the carbon source to cell material

ΔG_P = free energy to convert 1 e^- eq of the carbon source to the pyruvate intermediate

K = fraction of energy transfer captured

m = $+1$ if ΔG_P is positive and -1 if energy is produced

ΔG_C = free energy to convert 1 e^- eq of pyruvate intermediate to 1 e^- eq of cells

ΔG_N = free energy per e^- eq of cells to reduce nitrogen to ammonia

The value for ΔG_c is $+31.41$ kJ/e^- eq of cells (McCarty, 1971) and the ΔG_N for the following nitrogen sources are $+17.46$, $+13.61$, $+15.85$, and 0.00 kJ/e^- eq of cells for NO_3^-, NO_2^-, N_2, and NH_4^+, respectively. The value for ΔG_P is estimated by using the free energy half reactions to convert the carbon source to the pyruvate intermediate.

The electron donor used for heterotrophic reactions is divided between portions oxidized to produce energy or used in cell synthesis. The energy balance relative to the substrate used is illustrated in the following equation in which the energy made available (on the left side of the equation) equals the energy used for cell growth (right side of the equation).

$$K\Delta G_R\left(\frac{f_e}{f_s}\right) = -\Delta G_S \tag{7–9}$$

$$f_e + f_s = 1.0 \tag{7–10}$$

where K = fraction of energy captured

ΔG_R = energy released from oxidation reduction reactions, kJ/mole e^- transferred

f_e = e^- mole of substrate oxidized per e^- mole of substrate used

f_s = e^- mole of substrate used for cell synthesis per e^- mole of substrate used

ΔG_S = energy used for cell growth, kJ/mole e^- transfer for cell growth

Equations (7–9) and (7–10) are used with half reactions and their free energy values to estimate cell yield by solving for f_e and f_s. The terms f_e and f_s represent the fraction of substrate oxidized or used in cell synthesis, respectively. The substrate is expressed as COD because a mole of COD contains a set quantity of electron moles of oxygen transfer. Thus, the values for f_e and f_s can also be expressed as COD fractions. The oxidation of acetate by heterotrophic bacteria with different acceptors is used in Example 7–3 to illustrate the bioenergetics analysis.

EXAMPLE 7–3 **Estimate Biomass Yield Using Energetics** Compare the cell yield in g cells as COD/g COD used and g cells as VSS/g COD used for acetate utilization by heterotrophic bacteria with oxygen and then carbon dioxide as the electron acceptor, and ammonia as the nitrogen source. Assume 60 percent energy capture efficiency.

Solution Part A—Oxygen as the Electron Acceptor

1. Solve for the energy produced and captured ($K\Delta G_R$) using reaction No. 18 for acetate oxidation and reaction No. 4 for oxygen reduction from Table 7–7.

		kJ/mole e⁻
No. 18	$\frac{1}{8}CH_3COO^- + \frac{3}{8}H_2O \rightarrow \frac{1}{8}CO_2 + \frac{1}{8}HCO_3^- + H^+ + e^-$	−27.68
No. 4	$\frac{1}{4}O_2 + H^+ + e^- \rightarrow \frac{1}{2}H_2O$	−78.14
	$\frac{1}{8}CH_3COO^- + \frac{1}{4}O_2 \rightarrow \frac{1}{8}CO_2 + \frac{1}{8}HCO_3^- + \frac{1}{8}H_2O$	$\Delta G_R = -105.82$

Energy captured by cell:

$$K(\Delta G_R) = 0.60(-105.82) = -63.42 \text{ kJ/mole e}^-$$

2. Solve for the energy needed per electron mole of cell growth (ΔG_S).

$$\Delta G_C = 31.41 \text{ kJ/mole e}^- \text{ cells}$$

$$\Delta G_N = 0$$

ΔG_P [acetate (reaction No. 18) to pyruvate (reaction No. 15)]

		ΔG kJ/mole e⁻
No. 18	$\frac{1}{8}CH_3COO^- + \frac{3}{8}H_2O \rightarrow \frac{1}{8}CO_2 + \frac{1}{8}HCO_3 + e^-$	−27.66
No. 15	$\frac{1}{5}CO_2 + \frac{1}{10}HCO_3^- + H^+ + e^- \rightarrow \frac{1}{10}CH_3COCOO^- + \frac{2}{5}H_2O$	+35.78
	$\frac{1}{8}CH_3COO^- + \frac{3}{40}CO_2 \rightarrow \frac{1}{10}CH_3COCOO^- + \frac{1}{40}HCO_3^- + \frac{1}{40}H_2O$	$\Delta G_P = +8.12$

Because ΔG_P is positive, energy is required and $m = +1$

$$\Delta G_S = \left[\frac{8.12}{(0.6)^{1.0}} + 31.41 + 0 \right] = 44.94 \text{ kJ/mole e}^-$$

3. Determine f_e and f_s using Eq. (7–9).

$$\frac{f_e}{f_s} = \frac{-\Delta G_S}{K\Delta G_R} = \frac{-44.94 \text{ kJ/mole e}^-}{-63.42 \text{ kJ/mole e}^-}$$

$$\frac{f_e}{f_s} = 0.707$$

$$f_e + f_s = 1.0$$

Solve for f_e and f_s:

$$f_e = 0.41$$

$$f_s = 0.59 \frac{\text{g cell COD}}{\text{g COD used}}$$

4. Determine the yield based on COD.

   For biomass ($C_5H_7NO_2$), 1 g cells = 1.42 g COD

   Thus, the yield is

   $$Y = \frac{(0.59 \text{ g COD/g COD})}{(1.42 \text{ g COD/g VSS})} = 0.42 \text{ gVSS/g COD}$$

5. Determine the yield based on BOD assuming a conversion factor of 1.6 g COD/g BOD (see discussion in Sec. 8–2 in Chap. 8).

   Thus, the yield is

   $$Y = \frac{(0.42 \text{ g VSS/g COD})}{(\text{g BOD}/1.6 \text{ g COD})} = 0.67 \text{ g VSS/g BOD}$$

Solution **Part B—Carbon Dioxide as the Electron Acceptor**

1. Solve for energy produced and captured ($K\Delta G_R$) using reaction No. 18 for acetate oxidation and reaction No. 8 for CO_2 reduction to methane from Table 7–7.

		ΔG kJ/mole e$^-$
No. 18	$\frac{1}{8}CH_3COO^- + \frac{3}{8}H_2O \rightarrow \frac{1}{8}CO_2 + \frac{1}{8}HCO_3^- + H^+ + e^-$	-27.66
No. 8	$\frac{1}{8}CO_2 + H^+ + e^- \rightarrow \frac{1}{8}CH_4 + \frac{1}{8}HCO_3^-$	$+24.11$
	$\frac{1}{8}CH_3COO^- + \frac{3}{8}HO_2 \rightarrow \frac{1}{8}CH_4 + \frac{1}{4}HCO_3^-$	$\Delta G_R = -3.57$

   Energy captured by cell:

   $$(K\Delta G_R) = 0.60 \, (-3.57) = 2.142 \text{ kJ/mole e}^-$$

2. Solve for the energy needed per electron mole at cell growth ΔG_S.

   $$\Delta G_C = 31.41 \text{ kJ/mole e}^- \text{ cells}$$

   $$\Delta G_N = 0$$

   $$\Delta G_P \text{ (same as for acetate/O}_2)$$

   $$\Delta G_S = 44.94 \text{ kJ/mole e}^-$$

3. Determine the values of f_e and f_s using Eq. (7–9).

$$\frac{f_e}{f_s} = \frac{-\Delta G_S}{K\Delta G_R} = \frac{-44.94}{-2.142} = 21.0$$

$$f_e + f_s = 1.0$$

Solve for f_e and f_s.

$$f_e = 0.954 \qquad f_s = 0.046 \text{ g cell COD/g COD used}$$

4. Determine the yield based on COD

$$Y = \frac{(0.046 \text{ g cell COD/g COD used})}{(1.42 \text{ g COD/g VSS})} = 0.032 \text{ g VSS/g COD}$$

5. Compare yields for acetate oxidation.

Electron acceptor	Yield, g VSS/g COD	Product
O_2	0.42	CO_2, H_2O
CO_2	0.032	CH_4

Comment Based on bioenergetics calculations, estimated yield values for anaerobic reactions using carbon dioxide as the electron acceptor are much lower as compared to oxygen. The lower yield values are due to the much lower energy production with carbon dioxide as the electron acceptor in lieu of oxygen. The cell synthesis yield values calculated for these electron acceptors are very similar to yield values reported in the literature.

Stoichiometry of Biological Reactions

With the values determined for f_e and f_s, the stoichiometry of the biological reactions can be described according to the following relationship (McCarty, 1971, 1975):

$$R = f_e R_a + f_s R_{cs} - R_d \tag{7-11}$$

where R = overall balanced reaction
$\quad f_e$ = fraction of electron donor used for energy
$\quad R_a$ = half reaction for electron acceptor
$\quad f_s$ = fraction of electron donor used for cell synthesis
$\quad R_{cs}$ = half reaction for synthesis of cell tissue
$\quad R_d$ = half reaction for electron donor
$\quad f_s + f_e = 1$

The minus sign in Eq. (7–11) means that the electron donor equation given in Table 7–7 must be reversed and then added to the other two equations. In the first equation given in Table 7–7, the term $C_5H_7O_2N$ (Hoover and Porges, 1952), is used to represent bacterial cell tissue. Application of Eq. (7–11) is illustrated in Example 7–4.

EXAMPLE 7–4 **Write a Balanced Reaction for the Biological Oxidation of Acetate Using Oxygen** Using Eq. (7–11) and the half reactions given in Table 7–7, write a balanced reaction for the biological oxidation of acetate with oxygen. Use the values for f_e and f_s determined in Example 7–3, Part A (f_e = 0.41 and f_s = 0.59). The COD of acetate is 1.07 g COD/g acetate, computed as illustrated in Eq. (7–4).

Solution

1. Develop the balanced stoichiometric reaction for the oxidation of acetate.

$$R = f_e R_a + f_s R_{cs} - R_d$$

$$R = 0.41 \text{ (No. 4)} + 0.59 \text{ (No. 1)} - \text{No. 18}$$

$$(0.41)\,(\text{No. 4}) = 0.103O_2 + 0.41H^+ + 0.4e^- \rightarrow 0.205H_2O$$

$$(0.59)\,(\text{No. 1}) = 0.118CO_2 + 0.0295HCO_3^- + 0.0295NH_4^+ + 0.59H^+ + 0.59e^-$$
$$\rightarrow 0.0295C_5H_7O_2N + 0.2655H_2O$$

$$-\text{No. 18} = \quad 0.125CH_3COO^- + 0.375H_2O \rightarrow 0.125CO_2 + 0.125HCO_3^- + H^+ + e^-$$

$$R = \quad 0.125CH_3COO^- + 0.0295NH_4^+ + 0.103O_2$$
$$\rightarrow 0.0295C_5H_7O_2N + 0.0955H_2O + 0.095HCO_3^- + 0.007CO_2$$

2. Determine the cell yield from the stoichiometry.
    a. Cells produced from oxidation of acetate

        Cells produced = 0.0295 moles (113 g VSS/mole) = 3.334 g VSS

    b. Acetate utilized for cell production

        Acetate used = 0.125 mole acetate (60 g/mole)(1.07 g COD/g acetate)
        $$= 8.03 \text{ g COD}$$

    c. Determine the cell yield.

        $$Y = \frac{3.334 \text{ g VSS}}{8.03 \text{ g COD}} = 0.42 \text{ g VSS/g COD}$$

    The results are the same as in Example 7–3.

Biomass Synthesis Yields for Different Growth Conditions

In Example 7–3, it was demonstrated that the biomass synthesis yield is related to the energy produced by the electron transfer from the electron donor (acetate) to the electron acceptor (oxygen). From a review of the half reaction $\Delta G°$ values in Table 7–7, it can be seen that the energy production that occurs from the oxidation reduction reactions is reduced as the electron acceptor is changed from oxygen to nitrate to sulfate and to carbon dioxide. Accordingly, a lower cell yield would be predicted theoretically. Observed cell synthesis yields have been in accordance with predicted values using the half reactions given in Table 7–7. Typical synthesis yield coefficients are given in Table 7–8 for common electron donors and acceptors in wastewater treatment.

Table 7-8
Typical bacteria synthesis yield coefficients for common biological reactions in wastewater treatment

Growth Condition	Electron donor	Electron acceptor	Synthesis yield
Aerobic	Organic compound	Oxygen	0.40 g VSS/g COD
Aerobic	Ammonia	Oxygen	0.12 g VSS/g NH_4-N
Anoxic	Organic compound	Nitrate	0.30 g VSS/g COD
Anaerobic	Organic compound	Organic compound	0.06 g VSS/g COD
Anaerobic	Acetate	Carbon dioxide	0.05 g VSS/g COD

Observed versus Synthesis Yield

In the evaluation and modeling of biological treatment systems a distinction is made between the *observed yield* and the *synthesis yield* (or *true yield*). The observed biomass yield is based on the actual measurements of biomass production and substrate consumption and is actually less than the synthesis yield, because of cell loss (Sec. 7–6) concurrent with cell growth. In full-scale wastewater treatment processes the term *solids production* (or *solids yield*) is also used to describe the amount of VSS generated in the treatment process. The term is different from the synthesis biomass yield values because it contains other organic solids from the wastewater that are measured as VSS, but are not biological.

The synthesis yield is the amount of biomass produced immediately upon consumption of the growth substrate or oxidation of the electron donor in the case of autotrophic bacteria. The synthesis yield is seldom measured directly and is often interpreted from evaluating biomass production data for reactors operating under different conditions. Synthesis yield values for bacterial growth are affected by the energy that can be derived from the oxidation reduction reaction, by the growth characteristics of the carbon source, by the nitrogen source, and by environmental factors such as temperature, pH, and osmotic pressure. As illustrated in this section, the synthesis yield can be estimated if the stoichiometry or the amount of energy produced in the oxidation reduction reaction is known.

7–5 MICROBIAL GROWTH KINETICS

The performance of biological processes used for wastewater treatment depends on the dynamics of substrate utilization and microbial growth. Effective design and operation of such systems requires an understanding of the biological reactions occurring and an understanding of the basic principles governing the growth of microorganisms. Further, the need to understand all of the environmental conditions that affect the substrate utilization and microbial growth rate cannot be overemphasized, and it may be necessary to control such conditions as pH and nutrients to provide effective treatment. The purpose of this section is to present an introduction to microbial growth kinetics. The topics considered in this section include (1) microbial growth kinetics terminology, (2) rate of utilization of soluble substrate, (3) other rate expressions for the utilization of soluble substrate, (4) rate of soluble substrate production from biodegradable particulate organic matter, (5) the rate of biomass growth with soluble substrates, (6) kinetic coefficients for substrate utilization and biomass growth, (7) the rate of oxygen uptake,

(8) effects of temperature, (9) total volatile suspended solids and active biomass, and (10) net biomass and observed yield.

Microbial Growth Kinetics Terminology

The kinetics of microbial growth govern the oxidation (i.e., utilization) of substrate and the production of biomass, which contributes to the total suspended solids concentration in a biological reactor. Common terms used to describe the transformations that occur in substrate oxidation and biomass growth as well as in biological treatment processes are presented in Table 7–1. Because municipal and industrial wastewaters contain numerous substrates, the concentration of organic compounds is defined, most commonly, by the *biodegradable COD* (bCOD) or UBOD, both of which are comprised of soluble (dissolved), colloidal, and particulate biodegradable components. Both bCOD and UBOD represent measurable quantities that apply to all of the compounds. In the formulation of kinetic expressions in this chapter *biodegradable soluble COD* (bsCOD) will be used to quantify the fate of biodegradable organic compounds because it easily relates to the stoichiometry of substrate oxidized or used in cell growth [see Eq. (7–7)]. It should be noted that in Sec. 8.2 in Chap. 8, bsCOD is fractionated further into readily and slowly biodegradable components.

The biomass solids in a bioreactor are commonly measured as *total suspended solids* (TSS) and *volatile suspended solids* (VSS). The mixture of solids resulting from combining recycled sludge with influent wastewater in the bioreactor is termed *mixed liquor suspended solids* (MLSS) and *mixed liquor volatile suspended solids* (MLVSS). The solids are comprised of biomass, *nonbiodegradable volatile suspended solids* (nbVSS), and *inert inorganic total suspended solids* (iTSS). The nbVSS is derived from the influent wastewater and is also produced as cell debris from endogenous respiration. The iTSS originates in the influent wastewater. Additional wastewater characterization terminology is considered in Sec. 8–2 in Chap. 8.

Rate of Utilization of Soluble Substrates

In the introduction to this chapter, it was noted that one of the principal concerns in wastewater treatment is the removal of substrate. Stated another way, the goal in biological wastewater treatment is, in most cases, to deplete the electron donor (i.e., organic compounds in aerobic oxidation). For heterotrophic bacteria the electron donors are the organic substances being degraded; for autotrophic nitrifying bacteria it is ammonia or nitrite or other reduced inorganic compounds. The substrate utilization rate in biological systems can be modeled with the following expression for soluble substrates. Because the mass of substrate is decreasing with time due to substrate utilization and Eq. (7–12) is used in substrate mass balances, a negative value is shown.

$$r_{su} = -\frac{kXS}{K_s + S} \qquad (7\text{–}12)$$

where r_{su} = rate of substrate concentration change due to utilization, g/m³·d
k = maximum specific substrate utilization rate, g substrate/g microorganisms·d
X = biomass (microorganism) concentration, g/m³
S = growth-limiting substrate concentration in solution, g/m³

Figure 7–11

Rate of change of substrate utilization versus biodegradable soluble COD concentration based on the saturation-type model [see Eq. (7–12)].

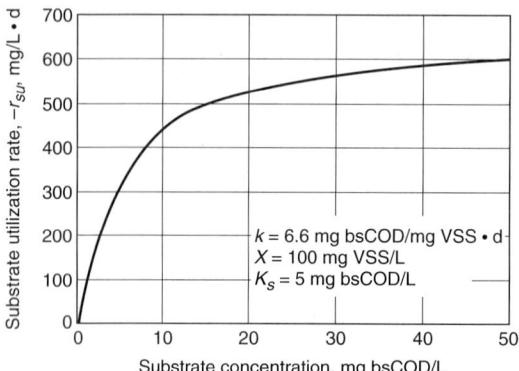

K_s = half-velocity constant, substrate concentration at one-half the maximum specific substrate utilization rate, g/m^3

Equation (7–12) will be recognized as a saturation-type equation, as described previously in Chap. 4. For substrate removal, Eq. (7–12) has been referred to as the Michaelis-Menten equation (Bailey and Ollis, 1986). Equation (7–12) is also of the form proposed by Monod for the specific growth rate of bacteria in which the limiting substrate is available to the microorganisms in a dissolved form (Monod, 1942, 1949). A plot of r_{su} as a utilization rate versus the substrate concentration is shown on Fig. 7–11. As shown on Fig. 7–11, the maximum substrate utilization rate occurs at high substrate concentrations. Further, as the substrate concentration decreases below some critical value, the value of $-r_{su}$ also decreases almost linearly. In practice, biological treatment systems are designed to produce an effluent with extremely low substrate values.

When the substrate is being used at its maximum rate, the bacteria are also growing at their maximum rate. The maximum specific growth rate of the bacteria is thus related to the maximum specific substrate utilization rate as follows.

$$\mu_m = kY \tag{7-13}$$

and

$$k = \frac{\mu_m}{Y} \tag{7-14}$$

where μ_m = maximum specific bacterial growth rate, g new cells/g cells·d
$\quad\quad k$ = maximum specific substrate utilization rate, g/g·d
$\quad\quad Y$ = true yield coefficient, g/g (defined earlier)

Using the definition for the maximum specific substrate utilization rate given by Eq. (7–12), the substrate utilization rate is also reported in the literature as

$$r_{su} = -\frac{\mu_m XS}{Y(K_s + S)} \tag{7-15}$$

Other Rate Expressions for the Utilization of Soluble Substrate

In reviewing kinetic expressions used to describe substrate utilization and biomass growth rates, it is very important to remember that the expressions used to model biological processes are all empirical, based on experimentally determined coefficient values. Besides the substrate limited relationship presented above, other expressions that have been used to describe substrate utilization rates include the following:

$$r_{su} = -k \tag{7-16}$$

$$r_{su} = -kS \tag{7-17}$$

$$r_{su} = -kXS \tag{7-18}$$

$$r_{su} = -kX\frac{S}{S_o} \tag{7-19}$$

The particular rate expression used to define kinetics of substrate utilization depends mainly on the experimental data available to fit the kinetic equations and the application of the kinetic model. In some cases, the first-order model shown by Eq. (7–18) is satisfactory for describing substrate utilization rates when the biological treatment process will be operated at relatively low substrate concentrations. Fundamental in the use of any rate expression is its application in a mass-balance analysis to be discussed in the following section. Also with regard to modeling biological treatment processes, kinetic models should not be applied outside of the range of the conditions used to develop model coefficients.

Rate of Soluble Substrate Production from Biodegradable Particulate Organic Matter

The rate expressions for substrate utilization and biomass growth presented thus far are based on the utilization of soluble substrates. In municipal wastewater treatment only about 20 to 50 percent of the degradable organic material enters as soluble compounds, and for some industrial wastewaters the soluble organic material may be a low to moderate fraction of the total degradable organic substrates. Bacteria cannot consume the particulate substrates directly and employ extracellular enzymes to hydrolyze the particulate organics to soluble substrates, which can be used according to the rate expressions described above. The particulate substrate conversion rate is also a rate-limiting process that is dependent on the particulate substrate and biomass concentrations. A rate expression for particulate substrate conversion is shown as follows (Grady et al., 1999):

$$r_{sc,P} = -\frac{k_P(P/X)X}{(K_X + P/X)} \tag{7-20}$$

where $r_{sc,P}$ = rate of change of particulate substrate concentration due to conversion to soluble substrate, g/m^3·d
k_P = maximum specific particulate conversion rate, g P/g X·d
P = particulate substrate concentration, g/m^3
X = biomass concentration, g/m^3
K_X = half-velocity degradation coefficient, g/g

The particulate degradation concentration is expressed relative to the biomass concentration, because the particulate hydrolysis is related to the relative contact area between the nonsoluble organic material and the biomass. The effect of particulate organic constituents is considered further in Chap. 8.

Rate of Biomass Growth with Soluble Substrates

In Sec. 7–4, the biomass growth rate was shown to be proportional to the substrate utilization rate by the synthesis yield coefficient, and biomass decay was shown to be proportional to the biomass present. Thus, the following relationship between the rate of growth and the rate of substrate utilization is applicable in both batch and continuous culture systems.

$$r_g = -Yr_{su} - k_dX \tag{7-21}$$

$$= Y\frac{kXS}{K_s + S} - k_dX \tag{7-22}$$

where r_g = net biomass production rate, g VSS/m^3·d
$\quad\quad Y$ = synthesis yield coefficient, g VSS/g bsCOD
$\quad\quad k_d$ = endogenous decay coefficient, g VSS/g VSS·d
$\quad\quad$ Other terms are as defined above.

If both sides of Eq. (7–22) are divided by the biomass concentration X, the specific growth rate is defined as follows:

$$\mu = \frac{r_g}{X} = Y\frac{kS}{K_s + S} - k_d \tag{7-23}$$

where μ = specific biomass growth rate, g VSS/g VSS·d

As shown, the specific growth rate corresponds to the change in biomass per day relative to the amount of biomass present, and is a function of the substrate concentration and the endogenous decay coefficient.

The endogenous decay coefficient accounts for the loss in cell mass due to oxidation of internal storage products for energy for cell maintenance, cell death, and predation by organisms higher in the food chain. It should be noted that microorganisms in all growth phases require energy for cell maintenance; however, it appears that the decay coefficient most probably changes with cell age. Usually, these factors are lumped together, and it is assumed that the decrease in cell mass caused by them is proportional to the concentration of organisms present. The decrease in mass is often identified in the literature as the *endogenous decay*. In Eq. (7–21) the coefficient k_d is the *endogenous decay* rate coefficient. An alternative approach used to describe the endogenous decay known as a *lysis-regrowth* model is described in Sec. 8–10 in Chap. 8.

In biological treatment processes, both the substrate utilization and biomass growth rates are controlled by some limiting substrate, as given by Eqs. (7–15) and (7–22). The growth limiting substrate can be any of the essential requirements for cell growth (i.e., electron donor, electron acceptor, or nutrients), but often it is the electron donor that is limiting, as other requirements are usually available in excess. Thus, when the term substrate is used to describe growth kinetics, it generally refers to the electron donor.

Kinetic Coefficients for Substrate Utilization and Biomass Growth

The coefficient values (k, K_s, Y, and k_d) used to predict the rate of substrate utilization and biomass growth can vary as a function of the wastewater source, microbial population, and temperature. Kinetic coefficient values are determined from bench-scale testing or full-scale plant test results. For municipal and industrial wastewater the coeffi-

Table 7–9

Typical kinetic coefficients for the activated-sludge process for the removal of organic matter from domestic wastewater

Coefficient	Unit	Value[a]	
		Range	Typical
k	g bsCOD/g VSS·d	2–10	5
K_s	mg/L BOD	25–100	60
	mg/L bsCOD	10–60	40
Y	mg VSS/mg BOD	0.4–0.8	0.6
	mg VSS/mg bsCOD	0.3–0.6	0.4
k_d	g VSS/g VSS·d	0.06–0.15	0.10

[a] Values reported are for 20°C.

cient values represent the net effect of microbial kinetics on the simultaneous degradation of a variety of different wastewater constituents. Typical kinetic coefficient values are reported in Table 7–9 for the aerobic oxidation of BOD in domestic wastewater. Additional kinetic coefficient values are given in Chaps. 8, 9, and 10.

Rate of Oxygen Uptake

The rate of oxygen uptake is related stoichiometrically to the organic utilization rate and growth rate (see Sec. 7–4). Thus, the oxygen uptake rate can be defined as

$$r_o = -r_{su} - 1.42r_g \qquad (7\text{–}24)$$

where r_o = oxygen uptake rate, g O_2/m³·d
r_{su} = rate of substrate utilization, g bsCOD/m³·d
1.42 = the COD of cell tissue, g bsCOD/g VSS
r_g = rate of biomass growth, g VSS/m³·d

A negative sign is required in front of the term r_{su} in Eq. (7–24), because the rate of substrate utilization as given by Eq. (7–12) is negative (i.e., the substrate concentration decreases with time). The factor 1.42 represents the COD of cell tissue as defined previously by Eq. (7–5).

Effects of Temperature

The temperature dependence of the biological reaction-rate constants is very important in assessing the overall efficiency of a biological treatment process. Temperature not only influences the metabolic activities of the microbial population but also has a profound effect on such factors as gas-transfer rates and the settling characteristics of the biological solids. The effect of temperature on the reaction rate of a biological process is expressed using the same type of relationship developed previously in Chap. 2 [see Eq. (2–25)] and repeated here for ease of reference.

$$k_T = k_{20}\theta^{(T-20)} \qquad (2\text{–}25)$$

where k_T = reaction-rate coefficient at temperature T, °C
k_{20} = reaction-rate coefficient at 20°C
θ = temperature-activity coefficient
T = temperature, °C

Values for θ in biological systems can vary from 1.02 to 1.25. Temperature correction factors for various kinetic coefficients are given in Chap. 8.

Total Volatile Suspended Solids and Active Biomass

The kinetic expressions used to describe biological kinetics and growth are related to the active biomass concentration X in the treatment reactor. In reality the VSS in a reactor consists of more than active biomass, and the fraction of active biomass can vary depending on the wastewater characteristics and operating conditions. The other components that contribute to the VSS concentration are cell debris, following endogenous decay, and non-biodegradable VSS (nbVSS) in the influent wastewater fed to the biological reactor.

During cell death, cell lysis occurs with the release of cellular materials into the liquid for consumption by other bacteria. A portion of the cell mass (cell wall) is not dissolved and remains as nonbiodegradable particulate matter in the system. The remaining nonbiodegradable material is referred to as cell debris and represents about 10 to 15 percent of the original cell weight. Cell debris is also measured as VSS and contributes to the total VSS concentration measured in the reactor mixed liquor. The rate of production of cell debris is directly proportional to the endogenous decay rate.

$$r_{Xd} = f_d(k_d)X \tag{7-25}$$

where r_{Xd} = rate of cell debris production, g VSS/m^3·d
$\quad\quad f_d$ = fraction of biomass that remains as cell debris, $0.10 - 0.15$ g VSS/g VSS

The nbVSS concentration resulting from cell debris is typically a relatively small fraction of the VSS in a bioreactor used to treat municipal and some industrial waste-waters. As noted above, a variable amount of MLVSS that is not biomass originates from the nbVSS in the influent wastewater. For typical untreated municipal wastewaters the nbVSS concentration may be in the range from 60 to 100 mg/L, and following primary treatment may range from 10 to 40 mg/L.

Total Volatile Suspended Solids. The VSS production rate in the aeration tank can be defined as the sum of the biomass production given by Eq. (7–21), the nbVSS production given by Eq. (7–25), and the nbVSS in the influent:

$$r_{X_T,\text{VSS}} = \underbrace{-Yr_{su} - k_dX}_{\substack{\text{net nbVSS} \\ \text{from soluble} \\ \text{bCOD}}} + \underbrace{f_d(k_d)X}_{\substack{\text{nbVSS} \\ \text{from cells}}} + \underbrace{Q\,X_{o,i}/V}_{\substack{\text{nbVSS} \\ \text{in influent}}} \tag{7-26}$$

where $r_{X_T,\text{VSS}}$ = total VSS production rate, g/m^3·d
$\quad\quad Q$ = influent flowrate, m^3/d
$\quad\quad X_{o,i}$ = influent nbVSS concentration, g/m^3
$\quad\quad V$ = volume of reactor, m^3
Other terms are as defined previously.

Active Biomass. From Eq. (7–26), the fraction of active biomass in the mixed-liquor VSS (MLVSS) is the ratio of the sum of the growth and decay term divided by the total MLVSS production:

$$F_{X,\text{act}} = (-Yr_{su} - k_dX)/r_{X_T,\text{VSS}} \tag{7-27}$$

where $F_{X,\text{act}}$ = active fraction of biomass in MLVSS, g/g

Net Biomass Yield and Observed Yield

The term true yield was defined in Sec. 7–4 as the amount of biomass produced during cell synthesis relative to the amount of substrate degraded. In the design and analysis of biological treatment processes, two other yield terms are important: (1) the *net biomass yield* and (2) the *observed solids yield*. The first is used as an estimate of the amount of active microorganisms in the system, and the second as the amount of sludge production.

Net Biomass Yield. The net biomass yield is the ratio of the net biomass growth rate to the substrate utilization rate:

$$Y_{bio} = -r_g/r_{su} \tag{7-28}$$

where Y_{bio} = net biomass yield, g biomass/g substrate used

Observed Yield. The observed yield accounts for the actual solids production that would be measured for the system and is shown as follows:

$$Y_{obs} = -r_{X_T,VSS}/r_{su} \tag{7-29}$$

where Y_{obs} = observed yield, g VSS produced/g substrate removed

EXAMPLE 7–5 Determine Biomass and Solids Yields For an industrial wastewater activated-sludge process, the amount of bsCOD in the influent wastewater is 300 g/m³ and the influent nbVSS concentration is 50 g/m³. The influent flowrate is 1000 m³/d, the biomass concentration is 2000 g/m³, the reactor bsCOD concentration is 15 g/m³, and the reactor volume is 105 m³. If the cell debris fraction f_d is 0.10, determine the net biomass yield, the observed solids yield, and the biomass fraction in the MLVSS. Use the kinetic coefficients given in Table 7–9.

Solution

1. Determine the net biomass yield using Eq. (7–28).

   $$Y_{bio} = -r_g/r_{su}$$

   a. Solve for r_{su} using Eq. (7–12) and the information given in Table 7–9.

   $$r_{su} = -\frac{kXS}{K_S + S}$$

   $$= -\frac{(5/d)(2000 \text{ g/m}^3)(15 \text{ g bsCOD/m}^3)}{(40 + 15) \text{ g/m}^3}$$

   $$= -2727 \text{ g bsCOD/m}^3\cdot\text{d}$$

   b. Determine the net biomass production rate r_g using Eq. (7–21).

   $$r_g = -Yr_{su} - k_dX$$

   $$= -(0.40 \text{ g VSS/g bsCOD})(-2727 \text{ g bsCOD/m}^3\cdot\text{d})$$
   $$- (0.10 \text{ g VSS/g VSS}\cdot\text{d})(2000 \text{ g VSS/m}^3)$$
   $$= 891 \text{ g VSS/m}^3\cdot\text{d}$$

c. Calculate the net biomass yield.

$$Y_{bio} = -r_g/r_{su} = (891 \text{ g VSS/m}^3 \cdot \text{d})/(2727 \text{ g bsCOD/m}^3 \cdot \text{d})$$

$$= 0.33 \text{ g VSS/g bsCOD}$$

2. Determine VSS production rate using Eq. (7–26).

$$r_{X_T, \text{VSS}} = -Yr_{su} - k_d X + f_d(k_d)X + QX_{o,i}/V$$

$$= 891 \text{ g VSS/m}^3 \cdot \text{d}$$

$$+(0.10 \text{ g VSS/g VSS})(0.10 \text{ g VSS/g VSS} \cdot \text{d})(2000 \text{ g VSS/m}^3)$$

$$+ (1000 \text{ m}^3/\text{d})(50 \text{ g VSS/m}^3)/105 \text{ m}^3$$

$$= (891 + 20 + 476) \text{ g VSS/m}^3 \cdot \text{d}$$

$$= 1387 \text{ g VSS/m}^3 \cdot \text{d}$$

3. Calculate the observed solids yield using Eq. (7–29).

$$Y_{obs} = -r_{X_T, \text{VSS}}/r_{su}$$

$$= -(1387 \text{ g VSS/m}^3 \cdot \text{d})/(-2727 \text{ g bsCOD/m}^3 \cdot \text{d})$$

$$= 0.51 \text{ g VSS/g bsCOD}$$

4. Calculate the active biomass fraction in the MLVSS. Using Eq. (7–27),

$$F_{X, \text{act}} = (-Yr_{su} - k_d X)/r_{X_T, \text{VSS}}$$

$$= (891 \text{ g VSS/m}^3 \cdot \text{d})/(1387 \text{ g VSS/m}^3 \cdot \text{d})$$

$$= 0.64$$

Comment   Thus, accounting for the nbVSS in the wastewater influent and cell debris produced, the MLVSS contains 64 percent active biomass.

7–6 MODELING SUSPENDED GROWTH TREATMENT PROCESSES

Before discussing the individual biological processes used for the treatment of wastewater as given in Secs. 7–8 through 7–14, the general application of the kinetics of biological growth and substrate removal will be explained. The purpose here is to illustrate (1) the development of biomass and substrate balances, (2) the prediction of effluent biomass and soluble substrate concentrations, (3) the prediction of the reactor biomass and MLSS/MLVSS concentrations and amount of waste sludge produced daily, and (4) the prediction of the oxygen requirements. Attached growth processes are considered in Sec. 7–7.

Description of Suspended Growth Treatment Processes

The complete-mix reactor with recycle will be considered in the following discussion as a model for suspended growth processes. The schematic flow diagrams shown on Fig. 7–12 include the nomenclature used in the following mass-balance equations. A similar complete-mix reactor may be used in laboratory studies to assess wastewater treatability and to obtain model kinetic coefficients.

All biological treatment reactor designs are based on using mass balances across a defined volume for each specific constituent of interest (i.e., biomass, substrate, etc.). The mass balance includes the flowrates for the mass of the constituent entering and/or leaving the system and appropriate reaction rate terms for the depletion or production of the constituent within the system. The units for a mass balance are usually given in mass per volume per time. For all mass balances a check of the units is recommended to assure that the mass-balance equations are correct.

Biomass Mass Balance

A mass balance for the mass of microorganisms in the complete-mix reactor shown on Fig. 7–12a can be written as follows: As discussed in Chap. 4, the sign of the generation or formation term [net growth in Eq. (7–30)] is always positive. The actual sign for the process is always part of the rate expression. For example, if the rate expression is $r = -kC$, the generation term is negative and a decrease will occur; similarly if $r = +kC$ the generation term is positive and an increase will occur.

1. General word statement:

$$
\begin{array}{c}
\text{Rate of accumulation} \\
\text{of microorganism} \\
\text{within the system} \\
\text{boundary}
\end{array}
=
\begin{array}{c}
\text{rate of flow of} \\
\text{microorganism} \\
\text{into the system} \\
\text{boundary}
\end{array}
-
\begin{array}{c}
\text{rate of flow of} \\
\text{microorganism} \\
\text{out of the system} \\
\text{boundary}
\end{array}
+
\begin{array}{c}
\text{net growth of} \\
\text{microorganism} \\
\text{within the} \\
\text{boundary}
\end{array}
\qquad (7\text{–}30)
$$

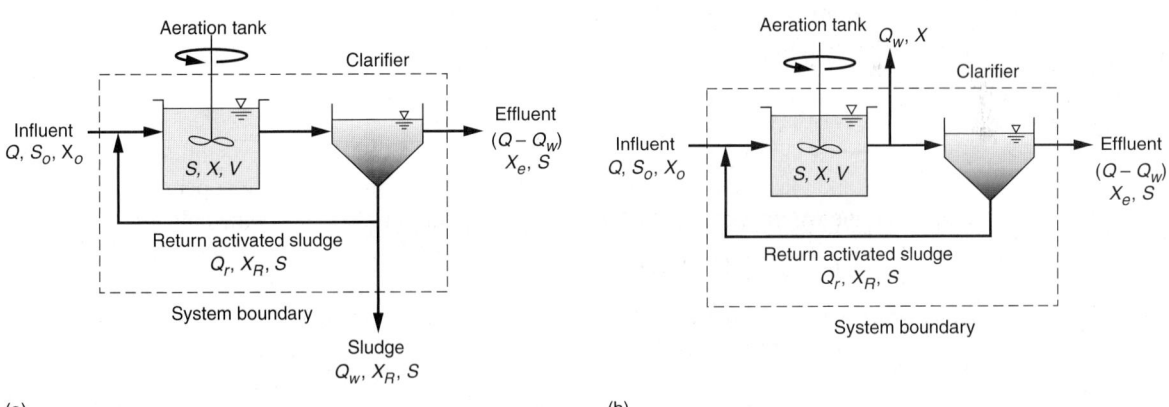

(a)   (b)

Figure 7–12

Schematic diagram of activated-sludge process with model nomenclature: (a) with wasting from the sludge return line and (b) with wasting from the aeration tank.

2. Simplified word statement:

Accumulation = inflow − outflow + net growth   (7–31)

3. Symbolic representation:

$$\frac{dX}{dt}V = QX_o - [(Q - Q_w)X_e - Q_wX_R] + r_gV \tag{7–32}$$

where dX/dt = rate of change of biomass concentration in reactor measured as g VSS/m³·d

    V = reactor volume (i.e., aeration tank), m³

    Q = influent flowrate, m³/d

    X_o = concentration of biomass in influent, g VSS/m³

    Q_w = waste sludge flowrate, m³/d

    X_e = concentration of biomass in effluent, g VSS/m³

    X_R = concentration of biomass in return line from clarifier, g VSS/m³

    r_g = net rate of biomass production, g VSS/m³·d

If it is assumed that the concentration of microorganisms in the influent can be neglected and that steady-state conditions prevail ($dX/dt = 0$), Eq. (7–32) can be simplified to yield

$$(Q - Q_w)X_e + Q_wX_R = r_gV \tag{7–33}$$

If Eq. (7–33) is combined with Eq. (7–21), the result is

$$\frac{(Q - Q_w)X_e + Q_wX_R}{VX} = -Y\frac{r_{su}}{X} - k_d \tag{7–34}$$

where X = concentration of the biomass in the reactor, g/m³

The inverse of the term on the left-hand side of Eq. (7–34) is defined as the average *solids retention time* (SRT) as given below.

$$SRT = \frac{VX}{(Q - Q_w)X_e + Q_wX_R} \tag{7–35}$$

Note that the numerator in Eq. (7–35) represents the total mass of solids in the aeration tank and the denominator corresponds to the amount of solids lost per day in the effluent and by intentional wasting (Q_w). By definition the SRT is the solids in the system divided by the mass of solids removed per day. Using the above definition of SRT, Eq. (7–34) can be written as

$$\frac{1}{SRT} = -Y\frac{r_{su}}{X} - k_d \tag{7–36}$$

The term 1/SRT is also related to μ, the specific biomass growth rate, as given by Eq. (7–37):

$$\frac{1}{SRT} = \mu \tag{7–37}$$

Equation (7–37) is derived from Eq. (7–36) by dividing by X and substituting r_g from Eq. (7–21) for the right-hand side of the equation, and using the definition for μ given in Eq. (7–23).

Thus, for a complete-mix activated-sludge process, the SRT (which can be controlled by solids wasting) is the inverse of the average specific rate that relates to the process biokinetics.

In Eq. (7–36) the term $(-r_{su}/X)$ is known as the *specific substrate utilization rate* U. The specific substrate utilization rate U is calculated as follows:

$$U = \frac{r_{su}}{X} = \frac{Q(S_o - S)}{VX} = \frac{S_o - S}{\tau X} \tag{7–38}$$

where U = specific substrate utilization rate, g BOD or COD/g VSS·d
Q = wastewater flowrate, m³/d
S_o = influent soluble substrate concentration, g BOD or bsCOD/m³
S = effluent soluble substrate concentration, g BOD or bsCOD/m³
V = volume of aeration tank, m³
X = biomass concentration, g/m³
τ = hydraulic detention time, V/Q, d
Other terms as defined above.

Substituting Eq. (7–12) into Eq (7–36) yields

$$\frac{1}{\text{SRT}} = \frac{YkS}{K_s + S} - k_d \tag{7–39}$$

The solids retention time (SRT) is an important design and operating parameter for the activated-sludge process. The SRT is the average time the activated-sludge solids are in the system. Assuming that the solids inventory in the clarifier shown on Fig. 7–12a is negligible compared to that in the aeration tank, the SRT is determined by dividing the mass of solids in the aeration tank by the solids removed daily via the effluent and by wasting for process control. For many activated-sludge processes, where good flocculation occurs and the clarifier is designed properly, the effluent VSS is typically less than 15 g/m³. Where the effluent VSS is low, excess solids must be removed from the system by wasting. Wasting is accomplished most commonly by removing biomass (sludge) from the clarifier underflow recycle line as shown on Fig. 7–12a. Alternatively, wasting can be accomplished from the aeration tank as shown on Fig. 7–12b.

Solving Eq. (7–39) for the effluent dissolved substrate concentration S yields

$$S = \frac{K_s[1 + (k_d)\text{SRT}]}{\text{SRT}(Yk - k_d) - 1} \tag{7–40}$$

It should be noted that in Eq. (7–40), the effluent soluble substrate concentration for a complete-mix activated-sludge process is only a function of the SRT and kinetic coefficients for growth and decay. The effluent substrate concentration is not related to the influent soluble substrate concentration, but as will be shown in other mass balances, the influent concentration affects the biomass concentration.

Substrate Mass Balance

The mass balance for substrate utilization in the aeration tank (see Fig. 7–12a) is

Accumulation = inflow − outflow + generation

$$\frac{dS}{dt} V = QS_o - QS + r_{su}V \tag{7-41}$$

where S_o = influent soluble substrate concentration, g/m^3

Substituting the value for r_{su} [Eq. (7–12)] and assuming steady-state conditions (dS/dt = 0), Eq. (7–41) can be rewritten as

$$S_o - S = \left(\frac{V}{Q}\right)\left(\frac{kXS}{K_s + S}\right) \tag{7-42}$$

The volume of the aeration tank divided by the influent flowrate is τ, the hydraulic retention time.

If Eq. (7–39) is solved for the term $S/(K_s + S)$ and substituted into Eq. (7–42), the following expression is obtained for the biomass concentration in the aeration tank:

$$X = \left(\frac{\text{SRT}}{\tau}\right)\left[\frac{Y(S_o - S)}{1 + (k_d)\text{SRT}}\right] \tag{7-43}$$

As given by Eq. (7–43), the reactor biomass concentration is a function of the system SRT, the aerobic aeration tank τ, the synthesis yield coefficient, the amount of substrate removed ($S_o - S$), and the endogenous decay coefficient.

The same equations can be applied to describe an activated-sludge process with no clarifier and thus no return sludge flow. For the case with no return sludge, all of the solids produced are present in the effluent from the aeration tank and the SRT equals the τ.

$$\text{SRT} = VX/QX = \tau \tag{7-44}$$

The importance of the system SRT in determining the effluent soluble substrate concentration and aeration tank biomass concentration is clear from an examination of Eqs. (7–40) and (7–43). As indicated by the waste sludge flow term (Q_w) in Eq. (7–35), a selected SRT value can be maintained by the amount of solids wasted per day to control the process performance. Similarly, it can be shown that by wasting from the aeration tank, the SRT can be controlled by wasting a given percentage of the aeration tank volume each day.

Mixed Liquor Solids Concentration and Solids Production

The solids production from a biological reactor represents the mass of material that must be removed each day to maintain the process. It is of interest to quantify the solids production in terms of TSS, VSS, and biomass. By definition, the SRT also provides a convenient expression to calculate the total sludge produced daily from the activated-sludge process:

$$P_{X_T,\text{VSS}} = \frac{X_T V}{\text{SRT}} \tag{7-45}$$

where $P_{X_T,\text{VSS}}$ = total solids wasted daily, g VSS/d

$\quad\quad X_T$ = total MLVSS concentration in aeration tank, g VSS/m^3

$\quad\quad V$ = volume of reactor, m^3

$\quad\quad$ SRT = solids retention time, d

Because the 1/SRT in Eq. (7–45) represents the fraction of solids wasted per day and the mixed liquor can be assumed to be a homogeneous mixture of biomass and other solids, Eq. (7–45) can be used to calculate the amount of solids wasted for any of the mixed liquor components. For the amount of biomass wasted per day (P_X), the biomass concentration X can be used in place of X_T in Eq. (7–45).

Mixed Liquor Solids Concentration. The total MLVSS in the aeration tank equals the biomass concentration X plus the nbVSS concentration X_i:

$$X_T = X + X_i \tag{7–46}$$

A mass balance is needed to determine the nbVSS concentration in addition to the active biomass VSS concentration. The MLVSS nbVSS concentration is affected by the amount of nbVSS in the influent wastewater, the amount of nbVSS wasted per day, and the amount of cell debris produced from cell decay. A materials balance on the inert material is as follows:

Accumulation = inflow − outflow + generation

$$(dX_i/dt)V = QX_{o,i} - X_iV/\text{SRT} + r_{X,i}V \tag{7–47}$$

where $X_{o,i}$ = nbVSS concentration in influent, g/m^3

$\quad\quad X_i$ = nbVSS concentration in aeration tank, g/m^3

$\quad\quad r_{X,i}$ = rate of nbVSS production from cell debris, g/m^3·d

At steady-state ($dX_i/dt = 0$) and substituting Eq. (7–25) for $r_{X,i}$ in Eq. (7–47) yields

$$0 = QX_{o,i} - X_iV/\text{SRT} + (f_d)(k_d)XV \tag{7–48}$$

$$X_i = X_{o,i}\,(\text{SRT})/\tau + (f_d)(k_d)X(\text{SRT}) \tag{7–49}$$

Combining Eq. (7–43) and Eq. (7–49) for X and X_i produces the following equation that can be used to determine the total MLVSS concentration:

$$X_T = \left(\frac{\text{SRT}}{\tau}\right)\left[\frac{Y(S_o - S)}{1 + (k_d)\text{SRT}}\right] + (f_d)(k_d)(X)\text{SRT} + \frac{(X_{o,i})\text{SRT}}{\tau} \tag{7–50}$$

$$\underset{\substack{\text{Heterotrophic}\\\text{biomass}}}{(A)} \quad\quad \underset{\text{Cell debris}}{(B)} \quad\quad \underset{\substack{\text{Nonbiodegradable}\\\text{VSS in influent}}}{(C)}$$

Solids Production. By substituting Eq. (7–50) for X_T in Eq. (7–45) and replacing τ with V/Q, the amount of VSS produced and wasted daily is determined as follows:

$$P_{X,\text{VSS}} = \frac{QY(S_o - S)}{1 + (k_d)\text{SRT}} + f_d(k_d)X(V) + QX_{o,i} \tag{7–51}$$

Figure 7-13

Biodegradable soluble COD, biomass, and MLVSS concentrations versus SRT for complete-mix activated-sludge process.

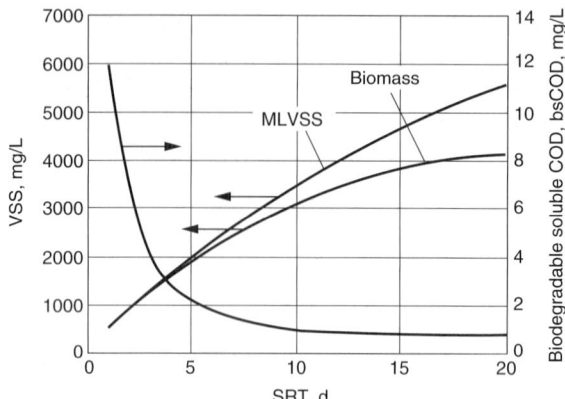

Equation (7–43) is substituted for the biomass concentration (X) in Eq. (7–51) to show the daily VSS production rate in terms of the substrate removed, influent nbVSS, and kinetic coefficients as follows:

$$P_{X,VSS} = \frac{QY(S_o - S)}{1 + (k_d)SRT} + \frac{(f_d)(k_d)YQ(S_o - S)SRT}{1 + (k_d)SRT} + QX_{o,i} \qquad (7\text{–}52)$$

$$\underbrace{\qquad\qquad}_{\substack{(A) \\ \text{Heterotrophic} \\ \text{biomass}}} \quad \underbrace{\qquad\qquad}_{\substack{(B) \\ \text{Cell debris}}} \quad \underbrace{\qquad}_{\substack{(C) \\ \text{Nonbiodegradable} \\ \text{VSS in influent}}}$$

The effect of SRT on the performance of an activated-sludge system for soluble substrate removal is illustrated on Fig. 7–13. In addition to the soluble substrate concentration, the total VSS concentration which includes nbVSS is also shown. As the SRT increases, more biomass decays and thus more cell debris accumulates, so that the difference between MLVSS and biomass VSS concentration increases with SRT. Also illustrated on Fig. 7–13 is the fact that the soluble substrate concentration is very low (<5 mg/L) at SRTs above 2 d. The low substrate concentration is typical of the activated-sludge process when used for the treatment of municipal wastewaters, and illustrates how effectively the organic compounds are degraded in the activated-sludge process. As will be shown in Chap. 8, organic substrate degradation is usually not the major factor in selecting a design SRT value.

The total mass of dry solids wasted per day is based on the TSS, which includes the VSS plus inorganic solids. Inorganic solids are in the influent wastewater (TSS − VSS) and the biomass contains 10 to 15 percent inorganic solids by dry weight. The influent inorganic solids are not soluble, and are assumed captured in the mixed liquor solids and removed in the wasted solids. Equation (7–52) is modified to calculate the solids production in terms of TSS by adding the influent inorganic solids and by calculating the biomass in terms of TSS by assuming a typical biomass VSS/TSS ratio of 0.85. The ratio of VSS/TSS may vary from 0.80 to 0.90.

$$P_{X,\,TSS} = \frac{A}{0.85} + \frac{B}{0.85} + C + Q(TSS_o - VSS_o) \qquad (7\text{–}53)$$

where $P_{X,TSS}$ = net waste activated sludge produced each day, measured in terms of total suspended solids, kg/d

TSS_o = influent wastewater TSS concentration, g/m^3

VSS_o = influent wastewater VSS concentration, g/m^3

The mass of MLVSS and MLSS can be obtained by using Eqs. (7–52) and (7–53), respectively, with Eq. (7–45) as follows:

$$\text{Mass of MLVSS} = (X_{VSS})(V) = (P_{X,VSS})\, SRT \tag{7-54}$$

$$\text{Mass of MLSS} = (X_{TSS})(V) = (P_{X,TSS})\, SRT \tag{7-55}$$

By selecting an appropriate MLSS concentration, the aeration volume can be determined from Eq. (7–55). MLSS concentrations in the range of 2000 to 4000 mg/L may be selected, and must be compatible with the sludge settling characteristics and clarifier design as discussed in Sec. 8–7 in Chap. 8.

The Observed Yield

The observed yield Y_{obs} is based on the amount of solids production measured relative to the substrate removal, and may be calculated in terms of g TSS/g bsCOD or g BOD, or relative to VSS as g VSS/g bsCOD or g BOD. The measured solids production is the sum of the solids in the system effluent flow and the solids intentionally wasted, which equals the term P_X defined in Eqs. (7–45), (7–52), and (7–53). The observed yield for VSS can be calculated by dividing Eq. (7–52) by the substrate removal rate, which is $Q(S_o - S)$:

$$Y_{obs} = \underbrace{\frac{Y}{1 + (k_d)SRT}}_{\substack{(A)\\ \text{Heterotrophic}\\ \text{biomass}}} + \underbrace{\frac{(f_d)(k_d)(Y)SRT}{1 + (k_d)SRT}}_{\substack{(B)\\ \text{Cell debris}}} + \underbrace{\frac{X_{o,i}}{S_o - S}}_{\substack{(C)\\ \text{Nonbiodegradable}\\ \text{VSS in influent}}} \tag{7-56}$$

where Y_{obs} = g VSS/g substrate removed

For wastewaters with no nbVSS in the influent the solids production consists of only active biomass and cell debris, and the observed yield for VSS is as follows:

$$Y_{obs} = \frac{Y}{1 + (k_d)SRT} + \frac{(f_d)(k_d)(Y)SRT}{1 + (k_d)SRT} \tag{7-57}$$

The impact of nonbiodegradable influent VSS in Eq. (7–56) on the observed yield depends on the wastewater characteristics and the type of pretreatment. The effluent substrate concentration is generally very low compared to S_o and the term $X_{o,i}/(S_o - S)$ can be approximated by $X_{o,i}/S_o$, which is the g nbVSS/g influent substrate. For municipal wastewater $X_{o,i}/S_o$ values range from 0.10 to 0.30 g/g with primary treatment and 0.30 to 0.50 without primary treatment.

Oxygen Requirements

The oxygen required for the biodegradation of carbonaceous material is determined from a mass balance using the bCOD concentration of the wastewater treated and the amount of biomass wasted from the system per day. If all of the bCOD were oxidized to CO_2 and H_2O, the oxygen demand would equal the bCOD concentration, but bacteria only oxidize a portion of the bCOD to provide energy and use a portion of the bCOD for cell growth. Oxygen is also consumed for endogenous respiration, and the amount will depend on the system SRT. For a given SRT, a mass balance on the system can be done where the bCOD removal equals the oxygen used plus the biomass VSS remaining (in terms of an oxygen equivalent), as was shown by Eq. (7–7). Thus, for a suspended growth process, the oxygen used is

$$\text{Oxygen used} = \text{bCOD removed} - \text{COD of waste sludge} \qquad (7\text{–}58)$$

$$R_o = Q(S_o - S) - 1.42\, P_{X,bio} \qquad (7\text{–}59)$$

where R_o = oxygen required, kg/d
$P_{X,bio}$ = biomass as VSS wasted per day, kg/d

It is important to note that $P_{X,bio}$ includes active biomass and cell debris derived from cell growth and is thus the sum of terms A and B in Eq. (7–52).

EXAMPLE 7–6 **Design of a Complete-Mix Suspended Growth Process** A complete-mix activated-sludge system with recycle is used to treat municipal wastewater after primary sedimentation. The characteristics of the primary effluent are: flow = 1000 m^3/d, bsCOD = 192 g/m^3, nbVSS = 30 g/m^3, and inert inorganics = 10 g/m^3. The aeration tank MLVSS = 2500 g/m^3. Using these data and the kinetic coefficients given below, design a system with a 6-d SRT and determine the following:

1. What is the effluent bsCOD concentration?
2. What value of τ should be used so that the MLVSS concentration is 2500 g/m^3?
3. What is the daily sludge production in kg/d as VSS and TSS?
4. What is the fraction of biomass in the MLVSS?
5. What is the observed solids yield in g VSS/g bsCOD and g TSS/g bsCOD?
6. What is the oxygen requirement in kg/d?

Kinetic coefficients:

$k = 12.5$ g COD/g VSS·d $\qquad K_s = 10$ g COD/m^3

$Y = 0.40$ g VSS/g COD used $\qquad f_d = 0.15$ g VSS/g VSS

$k_d = 0.10$ g VSS/g VSS·d $\qquad$ Biomass VSS/TSS = 0.85

Solution

1. Determine the effluent bsCOD concentration using Eq. (7–40).

$$S = \frac{K_s[1 + (k_d)\,\text{SRT}]}{\text{SRT}(Yk - k_d) - 1}$$

$$= \frac{(10 \text{ g bsCOD}/m^3)[1 + (0.10 \text{ g VSS/g VSS·d})(6 \text{ d})]}{(6 \text{ d})[(0.40 \text{ g VSS/g COD})(12.5 \text{ g COD/g VSS·d}) - (0.10 \text{ g VSS/g VSS·d})] - 1}$$

$$= 0.56 \text{ g bsCOD}/m^3$$

2. Determine τ for 2500 g/m³ MLVSS concentration.

Solve for τ in Eq. (7–50).

$$X_T = Y(S_o - S)SRT/[1 + (k_d)SRT](\tau) + (f_d)(k_d)(X)SRT + (X_{o,i})SRT/\tau$$

2500 g VSS/m³ = (0.40 g VSS/g COD)[(192 − 0.56) g COD/m³](6 d)/

[(1 + 0.10 g VSS/g VSS·d (6 d)(τ)]

+ (0.15 g VSS/g VSS)(0.10 g VSS/g VSS·d)(X)(6 d)

+ 30 g VSS/m³ (6 d/τ)

2500 = 287.2/τ + 0.09(X) + 180/τ

The biomass concentration X is determined using Eq. (7–43).

$$X = [Y(S_o - S)]SRT/[1 + (k_d)SRT](\tau)$$

$$= \frac{(0.40 \text{ g VSS/g COD})[(192 - 0.56) \text{ g COD/m}^3](6 \text{ d})}{[1 + (0.10 \text{ g VSS/g VSS})(6 \text{ d})](\tau)}$$

$$= (287.2 \text{ g/m}^3 \cdot \text{d})/\tau$$

Substituting for X in the above expression yields:

2500 = 287.2/τ + 180/τ + 25.8/τ = 493/τ

and solving the above expression for τ yields

τ = 0.197 d

Aeration tank volume = τ(Q) = 0.197 d (1000 m³/d) = 197 m³

3. Determine the total sludge production as kg VSS/d using Eq. (7–45).

$$Px_{T,\text{VSS}} = X_T(V)/(\text{SRT})$$

$$= (2500 \text{ g VSS/m}^3)(197 \text{ m}^3)(1 \text{ kg}/10^3 \text{ g})/6 \text{ d} = 82.1 \text{ kg VSS/d}$$

4. Determine the total sludge production as kg TSS/d using Eq. (7–53) and the assumed coefficients.

$$P_{X,\text{TSS}} = \frac{QY(S_o - S)}{1 + (k_d)SRT}\left(\frac{1}{0.85}\right) + \frac{(f_d)(k_d)YQ(S_o - S)SRT}{1 + (k_d)SRT}\left(\frac{1}{0.85}\right) + QX_{o,i} + Q(\text{TSS}_o - \text{VSS}_o)$$

$$= \frac{(1000 \text{ m}^3/\text{d})(0.40 \text{ g VSS/g COD})[(192 - 0.56) \text{ g COD/m}^3]}{[1 + (0.10 \text{ g VSS/g VSS·d})(6 \text{ d})](0.85)}$$

$$+ \frac{(0.15)(0.10)(1000 \text{ m}^3/\text{d})(0.40)[(192 - 0.56)\text{g COD/m}^3](6 \text{ d})}{[1 + (0.10 \text{ g VSS/ g VSS·d})(6 \text{ d})](0.85)}$$

$$+ (1000 \text{ m}^3/\text{d})(30 \text{ g/m}^3) + (1000 \text{ m}^3/\text{d})(10 \text{ g/m}^3)$$

$$= (56.3 + 5.1 + 30 + 10)(10^3 \text{ g/d}) = 101.4 \times 10^3 \text{ g/d} = 101.4 \text{ kg/d}$$

5. Determine the biomass fraction from the values for X and X_T.

$X = (287.2 \text{ g/m}^3\cdot\text{d})/\tau = (287.2 \text{ g/m}^3\cdot\text{d})/0.197 \text{ d} = 1458 \text{ g VSS/m}^3$

Biomass fraction $= X/X_T = 1458/2500 = 0.58$

6. Calculate the observed solids yield, g VSS/g bsCOD removed and g TSS/g bsCOD removed.

Solids wasted/d $= P_{X_T} = 82.2 \text{ kg VSS/d and } 101.4 \text{ kg TSS/d}$

bsCOD removed/d $= Q(S_o - S)$

$$= (1000 \text{ m}^3/\text{d})[(192 - 0.56) \text{ g COD/m}^3](1 \text{ kg}/10^3 \text{ g})$$

$$= 191,440 \text{ g COD/d} = 191.4 \text{ kg/d}$$

As VSS, $Y_{obs} = 82.2/191.4 = 0.43 \text{ g VSS/g bsCOD}$

As TSS, $Y_{obs} = 101.4/191.4 = 0.53 \text{ g TSS/g bsCOD}$

7. Determine the oxygen required using Eq. (7–59).

$R_o = Q(S_o - S) - 1.42 P_{X, \text{bio}}$

$P_{X, \text{bio}} = P_{X_T, \text{VSS}} - P_{\text{nbVSS}}$

$$= 82.2 \text{ kg/d} - (1000 \text{ m}^3/\text{d})(30 \text{ g VSS/m}^3)(1 \text{ kg}/10^3 \text{ g}) = 52.2 \text{ kg/d}$$

$R_o = (1000 \text{ m}^3/\text{d})[(192 - 0.56) \text{ g COD/m}^3](1 \text{ kg}/10^3 \text{ g}) - 1.42 (52.2 \text{ kg VSS/d})$

$$= 117.7 \text{ kg O}_2/\text{d}$$

Comment The same approach can be used to treat wastewater with particulate biodegradable COD by assuming it is equal to bsCOD. For complete-mix suspended growth designs if the SRT is 3 d or more, essentially all of the degradable particulate COD will be converted to bsCOD.

Design and Operating Parameters

In the mass balance for the complete-mix reactor, presented above, the SRT was introduced as the fundamental process parameter that affects the treatment efficiency and general performance for the activated-sludge process. Two other activated-sludge process parameters used for the design and operation of the activated-sludge process, the food to microorganism ratio and the volumetric loading rate, are introduced below.

Food to Microorganism (F/M) Ratio. The F/M ratio is defined as the rate of BOD or COD applied per unit volume of mixed liquor:

$$F/M = \frac{\text{total applied substrate rate}}{\text{total microbial biomass}} = \frac{QS_o}{VX} \tag{7-60}$$

and

$$F/M = \frac{S_o}{\tau X} \tag{7–61}$$

where F/M = food to biomass ratio, g BOD or bsCOD/g VSS·d
      Q = influent wastewater flowrate, m³/d
      S_o = influent BOD or bsCOD concentration, g/m³
      V = aeration tank volume, m³
      X = mixed liquor biomass concentration in the aeration tank, g/m³
      τ = hydraulic retention time of aeration tank, V/Q, d

Specific Substrate Utilization Rate. The F/M ratio can be related to the specific substrate utilization rate U, defined earlier [see Eq. (7–38)], by the process efficiency:

$$U = \frac{(F/M)E}{100} \tag{7–62}$$

where E = BOD or bsCOD process removal efficiency as defined by Eq. (7–63):

$$E, \% = \frac{S_o - S}{S_o} \times 100 \tag{7–63}$$

Substituting Eq. (7–61) for the F/M ratio and the term $[(S_o - S)/S_o](100)$ for the process efficiency yields the specific substrate utilization rate U, as given previously:

$$U = \frac{S_o - S}{\tau X} \tag{7–38}$$

The value of U can also be calculated by dividing $-r_{su}$ in Eq. (7–12) by X.

$$U = \frac{kS}{K_S + S} \tag{7–64}$$

By combining Eqs. (7–64) and (7–39),

$$\frac{1}{SRT} = YU - k_d \tag{7–65}$$

Again, the net specific growth rate μ is also equal to 1/SRT. In turn, as given by Eq. (7–65), the specific growth rate is equal to the synthesis yield times the specific substrate utilization rate minus the endogenous decay rate. By substituting Eq. (7–62) into Eq. (7–65) the SRT can be related to the F/M ratio as follows:

$$\frac{1}{SRT} = Y(F/M)\frac{E}{100} - k_d \tag{7–66}$$

For systems designed for the treatment of municipal wastewater with activated-sludge SRT values in the 20- to 30-d range, the F/M value may range from 0.10 to 0.05 g BOD/g VSS·d, respectively. At SRTs in the range of 5 to 7 d, the value may range from 0.3 to 0.5 g BOD/g VSS·d, respectively.

Organic Volumetric Loading Rate. The organic volumetric loading rate, defined as the amount of BOD or COD applied to the aeration tank volume per day, is

$$L_{org} = \frac{(Q)(S_o)}{(V)(10^3 \text{ g/kg})} \qquad (7\text{–}67)$$

where L_{org} = volumetric organic loading, kg BOD/m³·d
Q = influent wastewater flowrate, m³/d
S_o = influent BOD concentration, g/m³
V = aeration tank volume, m³

Process Performance and Stability

The effects of the kinetics considered above on the performance and stability of the system shown on Fig. 7–14 will now be examined further. It was shown in Eqs. (7–65) and (7–37) that 1/SRT, the net microorganism specific growth rate, and U, the specific substrate utilization ratio, are related directly. For a specified waste, a given biological community, and a particular set of environmental conditions, the kinetic coefficients Y, k, K_s, and k_d are fixed. It is important to note that domestic wastewater may have significant variability in its composition and may not always be treated as a single waste type in evaluating the kinetic coefficients. For given values of the coefficients, the effluent substrate concentration from the reactor is a direct function of the SRT, as shown in Eq. (7–40). Setting the SRT value fixes the values of U and μ and also defines the efficiency of biological waste stabilization. Equation (7–40) for substrate is plotted on Fig. 7–14 for a growth-specified complete-mix system with recycle. As shown, the treatment efficiency and the substrate concentration are related directly to the SRT, and the reactor hydraulics (i.e., complete-mix or plug-flow).

It can also be seen from Fig. 7–14 that there is a certain value of SRT below which waste stabilization does not occur. The critical SRT value is called the minimum solids retention residence time SRT$_{min}$. Physically, SRT$_{min}$ is the residence time at which the cells are washed out or wasted from the system faster than they can reproduce. The min-

Figure 7–14

Effluent substrate concentration and removal efficiency for complete-mix and plug-flow reactors with recycle versus SRT.

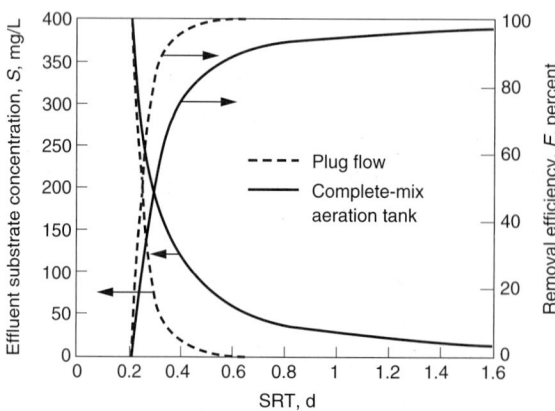

imum SRT can be calculated using Eq. (7–39), in which $S = S_o$. When washout occurs, the influent concentration S_o is equal to the effluent waste concentration S.

$$\frac{1}{\mathrm{SRT_{min}}} = \frac{YkS_o}{K_s + S_o} - k_d \tag{7–68}$$

In many situations encountered in waste treatment, S_o is much greater than K_s so that Eq. (7–68) can be rewritten to yield

$$\frac{1}{\mathrm{SRT_{min}}} \approx Yk - k_d \tag{7–69}$$

or

$$\frac{1}{\mathrm{SRT_{min}}} \approx \mu_m - k_d \tag{7–70}$$

Equations (7–69) and (7–70) can be used to determine the $\mathrm{SRT_{min}}$. Typical kinetic coefficients that can be used to solve for $\mathrm{SRT_{min}}$ for BOD removal systems are given in Table 7–9. Obviously, biological treatment processes should not be designed with SRT values equal to $\mathrm{SRT_{min}}$. To ensure adequate waste treatment, biological treatment processes are usually designed and operated with a design SRT value from 2 to 20 times $\mathrm{SRT_{min}}$. In effect, the ratio of the design SRT ($\mathrm{SRT_{des}}$) to $\mathrm{SRT_{min}}$ can be considered to be a process safety factor SF against system failure (Lawrence and McCarty, 1970).

$$\mathrm{SF} = \frac{\mathrm{SRT_{des}}}{\mathrm{SRT_{min}}} \tag{7–71}$$

Modeling Plug-Flow Reactors

The plug-flow system with cellular recycle, shown schematically on Fig. 7–2a-1 and pictorially on Fig. 7–2a-2, can be used to model certain forms of the activated-sludge process. The distinguishing feature of this recycle system is that the hydraulic regime of the reactor is of a plug-flow nature. In a true plug-flow model, all the particles entering the reactor stay in the reactor an equal amount of time. Some particles may make more passes through the reactor because of recycle, but while they are in the tank, all pass through in the same amount of time.

A kinetic model of the plug-flow system is mathematically difficult, but Lawrence and McCarty (1970) have made two simplifying assumptions that lead to a useful kinetic model of the plug-flow reactor:

1. The concentration of microorganisms in the influent to the reactor is approximately the same as that in the effluent from the reactor. This assumption applies only if $\mathrm{SRT}/\tau > 5$. The resulting average concentration of microorganisms in the reactor is symbolized as $\bar{X}$.

2. The rate of substrate utilization as the waste passes through the reactor is given by the following expression:

$$r_{su} = -\frac{kS\bar{X}}{K_s + S} \tag{7–72}$$

Integrating Eq. (7–72) over the retention time of the wastewater in the tank, substituting Eq. (7–43) for $\overline{X}$, and simplifying, the following expression is obtained:

$$\frac{1}{\text{SRT}} = \frac{Yk(S_o - S)}{(S_o - S) + (1 + \alpha)K_s \ln (S_i/S)} - k_d \tag{7–73}$$

where S_o = influent concentration
S = effluent concentration
S_i = influent concentration to reactor after dilution with recycle flow
$\quad = \dfrac{S_o + \alpha S}{1 + \alpha}$
α = recycle ratio
Other terms are as defined previously.

Equation (7–73) is similar to Eq. (7–36), which is applied to complete-mix systems, with or without recycle. The main difference in the two equations is that in Eq. (7–73), SRT is also a function of the influent waste concentration S_o. A version of Eq. (7–73) in which Eq. (7–43) is not substituted for $\overline{X}$ is shown in Chap. 8 in the design of sequencing batch reactors.

The true plug-flow recycle system is theoretically more efficient in the stabilization of most soluble wastes than in the continuous-flow stirred-tank recycle system. A graphical representation is shown on Fig. 7–14. In actual practice, a true plug-flow regime is essentially impossible to obtain because of longitudinal dispersion caused by aeration and mixing. By dividing the aeration tank into a series of reactors, the process approaches plug-flow kinetics with improved treatment efficiency compared to a complete-mix process. Because of the greater dilution with the influent wastewater, the complete-mix system can handle shock loads better than staged reactors in series. Reactor selection is discussed further in Chap. 8.

7–7 SUBSTRATE REMOVAL IN ATTACHED GROWTH TREATMENT PROCESSES

In an attached growth treatment process, a biofilm consisting of microorganisms, particulate material, and extracellular polymers is attached and covers the support packing material, which may be plastic, rock, or other material (see Fig. 7–15). The growth and substrate utilization kinetics described for the suspended growth process were related to the dissolved substrate concentration in the bulk liquid. For attached growth processes, substrate is consumed within a biofilm. Depending on the growth conditions and the hydrodynamics of the system, the biofilm thickness may range from 100 μm to 10 mm (WEF, 2000). A stagnant liquid layer (diffusion layer) separates the biofilm from the bulk liquid that is flowing over the surface of the biofilm or is mixed outside of the fixed film (see Fig. 7–16a). Substrates, oxygen, and nutrients diffuse across the stagnant liquid layer to the biofilm, and products of biodegradation from the biofilm enter the bulk liquid after diffusion across the stagnant film.

The substrate concentration at the surface of the biofilm, S_s as shown on Fig. 7–17, decreases with biofilm depth as the substrate is consumed and diffuses into the biofilm layers. As a result, the process is said to be *diffusion limited*. The substrate and oxygen

(a)

(b)

Figure 7–15

Typical packing for trickling filters: (a) rock and (b) plastic.

Figure 7–16

Schematic representation of the cross section of a biological slime in a trickling filter: (a) pictorial and (b) idealized.

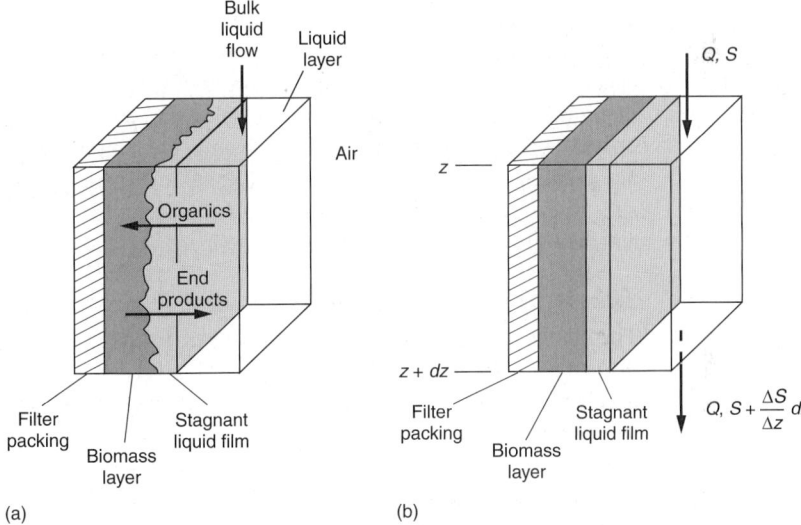

(a)

(b)

concentrations within the film are lower than the bulk liquid concentration and change with biofilm depth and the substrate utilization rate. The overall substrate utilization rate is less than would be predicted based on the bulk liquid substrate concentration.

The total amount of substrate used per unit of biofilm cross-sectional area must diffuse across the stagnant layer. This rate of mass transfer is termed the surface flux and is expressed as mass per unit area per unit time (g/m^2·d). The biofilm layer is not simply a planar surface as depicted on Fig. 7–16b (Costerton et al., 1995). The biofilm layers are in fact very complex nonuniform structures with uneven protrusions much like

Figure 7-17

Definition sketch for the analysis of substrate concentration in the biofilm.

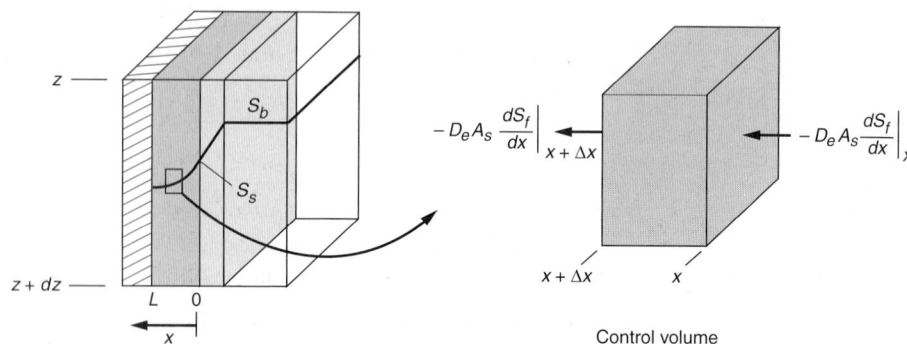

Control volume

peaks and valleys, and are believed to have vertical and horizontal pores through which liquid flows. The biomass can be very dense in biofilms, and may also vary in density and depth. Biofilm VSS concentrations may range from 40 to 100 g/L. Uniform growth across the support packing also does not occur, because of periodic sloughing, as well as the hydrodynamics and media configuration (Hinton and Stensel, 1991).

Mechanistic models have been developed by a number of investigators to describe mass transfer and biological substrate utilization kinetics in biofilms (Williamson and McCarty, 1976; Rittman and McCarty, 1980; Kissel et al., 1984; Saez and Rittman, 1992; Suidan and Wang, 1985; Wanner and Gujer, 1986; and Rittman and McCarty, 2001) and provide useful tools for the evaluation of biofilm processes. However, because of the complexity of attached growth reactors and the inability to define accurately the physical parameters and model coefficients, empirical relationships, based on observed performance, are used for design. The empirical relationships used for design are presented and illustrated in Chap. 8. Fundamental concepts of mass transfer and substrate utilization that can be used to model the behavior of substrate removal in attached growth processes are presented here.

Substrate Flux in Biofilms

The substrate flux across the stagnant layer to the biofilm, a function of the substrate diffusion coefficient and concentration, is given by Eq. (7–74). The negative sign is used because the substrate concentration is decreasing along the stagnant layer and substrate is removed from the bulk liquid.

$$r_{sf} = -D_w \frac{dS}{dx} = -D_w \frac{(S_b - S_s)}{L} \tag{7-74}$$

where r_{sf} = rate of substrate surface flux, g/m²·d
 D_w = diffusion coefficient of substrate in water, m²/d
 dS/dx = substrate concentration gradient, g/m³·m
 S_b = bulk liquid substrate concentration, g/m³
 S_s = substrate concentration at outer layer of biofilm, g/m³
 L = effective thickness of stagnant film, m

The thickness of the stagnant layer will vary with the fluid properties and fluid velocity. Higher velocities result in thinner films with greater substrate flux rates (Grady et al., 1999).

Mass transfer within the biofilm is described by Fick's law for diffusion (see Sec. 4–8 in Chap. 4) in an aqueous solution, with a modification to the diffusivity constant to account for the effect of the biofilm structure on the effective diffusion.

$$r_{bf} = -D_e \frac{\partial S_f}{\partial x} \tag{7-75}$$

where r_{bf} = rate of substrate flux in biofilm due to mass transfer, g/m²·d
D_e = effective diffusivity coefficient in biofilm, m²/d
$\partial S_f/\partial x$ = substrate concentration gradient, g/m³·m

The substrate utilization rate within the biofilm at any point can be defined as a saturation-type reaction (Eq. 7–12) for the substrate concentration (S_f) at that location:

$$r_{su} = -\frac{kS_f X}{K_s + S_f} \tag{7-76}$$

where r_{su} = rate of substrate utilization in biofilm, g/m²·d
S_f = substrate concentration at a point in the biofilm, g/m³

Substrate Mass Balance for Biofilm

A substrate mass balance around a differential element (dx) shown on Fig. 7–17 within the biofilm yields:

$$
\begin{array}{c}
\text{Rate of substrate} \\
\text{accumulation within} \\
\text{differential element}
\end{array}
=
\begin{array}{c}
\text{rate of substrate} \\
\text{flow into} \\
\text{differential element}
\end{array}
-
\begin{array}{c}
\text{rate of substrate} \\
\text{flow out of} \\
\text{differential element}
\end{array}
+
\begin{array}{c}
\text{rate of substrate} \\
\text{utilization in} \\
\text{differential element}
\end{array}
\tag{7-77}
$$

For steady-state conditions, the mass balance is

Accumulation = inflow − outflow + generation

$$0 = -D_e A_s \left.\frac{dS_f}{dx}\right|_x + D_e A_s \left.\frac{dS_f}{dx}\right|_{x+\Delta x} - \Delta x A_s \left(\frac{kS_f X}{K_s + S_f}\right) \tag{7-78}$$

where A_s = biofilm area normal to the substrate flux, m²
Δx = width of differential section, m

Dividing both sides by A_s and dx, and taking the limit as Δx approaches zero yields the following general equation for the change in substrate concentration within the biofilm:

$$D_e \frac{d^2 S_f}{dx^2} - \left(\frac{kS_f X}{K_s + S_f}\right) = 0 \tag{7-79}$$

Solutions to the above equation require two boundary conditions. The first boundary condition is that the substrate flux at the biofilm surface equals the substrate flux through the stagnant film, as given by Eq. (7–74). The second boundary condition is that there is no flux at the packing surface.

$$\left.\frac{dS_f}{dx}\right|_{x=L} = 0 \tag{7-80}$$

Solutions for Eq. (7–79) vary, depending on (1) whether a deep biofilm exists such that the biofilm substrate concentration approaches zero toward the support surface,

(2) whether a shallow film exists such that S_f is a finite value throughout the film, and (3) the relative concentration of S_f compared to K_s. Solution approaches are provided in a number of references, including Williamson and McCarty (1976), Grady et al. (1999), and Rittman and McCarty (2001).

Substrate Flux Limitations

An important implication of diffusion-limited processes is the relationship between the bulk liquid electron donor and electron acceptor concentrations. An assumption in the mechanistic models used is that either the electron donor or electron acceptor (i.e., oxygen or nitrate) is limiting. The substrate limitation may be due to reaction rates within the biofilm or to bulk liquid concentrations and diffusion rates across the stagnant layer. These are referred to by Williamson and McCarty (1976) as substrate and surface flux limitations, respectively. There are situations where the substrate limitation may switch between electron donor and electron acceptor with depth in the biofilm. For the situation where the substrate limitation can switch, numerical analysis techniques must be used to evaluate the biofilm behavior. A simple method that can be used to evaluate whether a surface flux limitation exists has been proposed by Williamson and McCarty (1976). The proposed method can also be used to assess the relative electron acceptor bulk liquid substrate concentrations needed to sustain electron donor utilization within the biofilm.

The effect of surface flux substrate limitation described by Williamson and McCarty (1976) is summarized in the following two equations:

$$\nu_d + \nu_a + \text{growth requirements} \rightarrow \text{end products} + \text{cells} \tag{7-81}$$

$$S_{ba} < \frac{D_{wd}\nu_a mw_a}{D_{wa}\nu_d mw_d} S_{bd} \tag{7-82}$$

where ν_d = molar stoichiometric reaction coefficient for electron donor, moles
ν_a = molar stoichiometric reaction coefficient for electron acceptor, moles
S_{ba} = bulk liquid electron acceptor substrate concentration, mg/L
S_{bd} = bulk liquid electron donor substrate concentration, mg/L
D_{wd} = diffusivity coefficient of electron donor in water, cm²/d
D_{wa} = diffusivity coefficient of electron acceptor in water, cm²/d
mw_a = molecular weight of electron acceptor, g
mw_d = molecular weight of electron donor, g

Nitrification rates in fixed-film systems are often limited by the bulk liquid DO concentration, and the following example uses Eqs. (7–81) and (7–82) to illustrate this important issue regarding fixed-film process applications.

EXAMPLE 7–7   Oxygen Limitation for Nitrification in a Biofilm   For bulk liquid NH₄-N concentrations of 1.0, 2.0, and 3.0 mg/L, respectively, what bulk liquid DO concentration must be present so that the nitrification rate in the biofilm is not limited due to the surface flux rate of oxygen? Assume the following conditions apply:

Electron donor = NH₄-N, $mw_d = 14$

Electron acceptor = oxygen, $mw_a = 32$

NH$_4$-N diffusivity coefficient at 20°C = D_{wd} = 1.6 cm²/d

Oxygen diffusivity coefficient at 20°C = D_{wa} = 2.6 cm²/d

Solution

1. Determine the stoichiometric coefficients from the reaction stoichiometry.

   $$NH_4^+ + 2O_2 \rightarrow NO_3^- + 2H^+ + H_2O$$

   $$\nu_d = 1.0$$

   $$\nu_a = 2.0$$

2. Determine the DO concentration where oxygen is flux-limited using Eq. (7–82).

   $$S_{ba} < \frac{D_{wd}\nu_a mw_a}{D_{wa}\nu_a mw_d} S_{bd} < \frac{(1.6 \text{ cm}^2/\text{d})(2.0)(32 \text{ g/mole})}{(2.6 \text{ cm}^2/\text{d})(1.0)(14 \text{ g/mole})} S_{bd} = 2.8\, S_{bd}$$

   Thus, if S_{ba} is equal to 2.8 (S_{bd}), the nitrification rate is not hindered by the oxygen flux rate through the stagnant layer. Bulk liquid DO concentrations necessary to prevent an oxygen flux limitation for nitrification are summarized in the following table.

Bulk liquid NH$_4$-N concentration, g/m³	Bulk liquid DO concentration, g/m³
1.0	2.8
2.0	5.6
3.0	8.4

Comment   For low bulk liquid NH$_4$-N concentrations, which result in lower nitrification rates in the biofilm, lower DO concentrations can be tolerated.

7–8 AEROBIC BIOLOGICAL OXIDATION

Dating back to the early 1900s, the primary purpose of biological wastewater treatment has been to (1) remove organic constituents and compounds to prevent excessive DO depletion in receiving waters from municipal or industrial point discharges, (2) remove colloidal and suspended solids to avoid the accumulation of solids and the creation of nuisance conditions in receiving waters, and (3) reduce the concentration of pathogenic organisms released to receiving waters. The U.S. EPA secondary treatment regulatory standards, set in 1972 and still in effect, were focused mainly on the removal of BOD and TSS, and require 85 percent removal of each (see Table 1–3 in Chap. 1). Most treatment applications involve the removal of organic constituents and compounds. Because a wide range of constituents and compounds exist in wastewater, the organic content is quantified in terms of biodegradable soluble COD (bsCOD) or BOD. Additional information on the characterization of the organic constituents in wastewater is presented in Sec. 8–2 in Chap. 8.

Process Description

The removal of BOD can be accomplished in a number of aerobic suspended growth or attached (fixed film) growth treatment processes as illustrated on Figs. 7–2 and 7–3, respectively, and described in detail in Chaps. 8 and 9. Both require sufficient contact time between the wastewater and heterotrophic microorganisms, and sufficient oxygen and nutrients. During the initial biological uptake of the organic material, more than half of it is oxidized and the remainder is assimilated as new biomass, which may be further oxidized by endogenous respiration. For both suspended and attached growth processes, the excess biomass produced each day is removed and processed to maintain proper operation and performance. The biomass is separated from the treated effluent by gravity separation, and more recent designs using membrane separation are finding applications.

Microbiology

A wide variety of microorganisms are found in aerobic suspended and attached growth treatment processes used for the removal of organic material. Aerobic heterotrophic bacteria found in these processes are able to produce extracellular biopolymers that result in the formation of biological flocs (or biofilms for attached growth processes) that can be separated from the treated liquid by gravity settling with relatively low concentrations of free bacteria and suspended solids.

Protozoa also play an important role in aerobic biological treatment processes. By consuming free bacteria and colloidal particulates, protozoa aid effluent clarification. Protozoa require a longer SRT than aerobic heterotrophic bacteria, prefer dissolved oxygen concentrations above 1.0 mg/L, and are sensitive to toxic materials. Thus, their presence is a good indicator of a trouble-free stable process operation. Because of their size, protozoa can easily be observed with a light microscope at 100 to 200 magnifications. Rotifers can also be found in activated sludge and in biofilms, as well as nematodes and other multicellular microorganisms. These organisms occur at longer biomass retention times, and their importance has not been well defined.

Aerobic attached growth processes, depending on the biofilm thickness, generally have a much more complex microbial ecology than activated sludge with films containing bacteria, fungi, protozoan, rotifers, and possibly annelid worms, flatworms, and nematodes (WEF, 2000). The nature of biofilms and their microbial composition will be discussed in more detail in Chap. 9.

Depending on process loadings and environmental conditions, a number of nuisance organisms can also develop in the activated-sludge process. The principal problem caused by nuisance organisms is a condition known as *bulking sludge,* in which the biological floc has poor settling characteristics. In the extreme, bulking sludge can result in high effluent suspended solids concentrations and poor treatment performance. Another nuisance condition, foaming, has been related to the development of two bacteria genera *Nocardia* and *Microthrix* (Pitt and Jenkins, 1990), which have hydrophobic cell surfaces and attach to air bubble surfaces, where they stabilize the bubbles to cause foam (see Fig. 7–18). The organisms can be found at high concentrations in the

(a)   (b)

Figure 7-18

Examples of foam caused by *Nocardia* accumulated on the surface of activated-sludge aeration tanks.

foam above the activated-sludge liquid. The above types of nuisance organisms along with others are considered in Chap. 8.

Stoichiometry of Aerobic Biological Oxidation

The stoichiometry for aerobic oxidation was discussed previously but is repeated here for completeness. In aerobic oxidation, the conversion of organic matter is carried out by mixed bacterial cultures in general accordance with the stoichiometry shown below.

Oxidation and synthesis:

$$\underset{\substack{\text{Organic}\\\text{matter}}}{\text{COHNS}} + O_2 + \text{nutrients} \xrightarrow{\text{bacteria}} CO_2 + NH_3 + \underset{\text{new cells}}{C_5H_7NO_2} + \underset{\substack{\text{other}\\\text{end}\\\text{products}}}{} \quad (7\text{--}83)$$

Endogenous respiration:

$$\underset{\substack{\text{Cells}\\ \text{mw} = 113\\ 1}}{C_5H_7NO_2} + 5O_2 \xrightarrow{\text{bacteria}} 5CO_2 + 2H_2O + NH_3 + \text{energy} \quad (7\text{--}84)$$

mw = 113 mw = 160; 160/113 = 1.42
 1   1.42

In Eq. (7–83), COHNS is used to represent the organic matter in wastewater, which serves as the electron donor while the oxygen serves as the electron acceptor. Although the endogenous respiration reaction [Eq. (7–84)] is shown as resulting in relatively simple end products and energy, stable organic end products are also formed. From Eq. (7–5), it was shown that if all of the cells (i.e., the electron donor) were oxidized completely, the UBOD or COD of the cells is equal to 1.42 times the concentration of cells as VSS. At longer SRT values, a greater portion of the cells will be oxidized.

Using the half reactions given in Table 7–7, the stoichiometry for the aerobic oxidation of acetate (the electron donor) can be represented as given below. Assuming ammonia will serve as the nitrogen source for cell tissue, oxygen is the electron acceptor, and f_s for the reaction is 0.59 (see Example 7–4).

$$0.125CH_3COO^- + 0.0295NH_4^+ + 0.103O_2 \rightarrow 0.0295C_5H_7O_2N + 0.0955H_2O$$
$$+ 0.0955HCO_3^- + 0.007CO_2$$

$$(7\text{--}85)$$

Growth Kinetics

The form of the rate expressions for substrate utilization and biomass growth for the heterotrophic oxidation of organic substrates, based on the stoichiometry given above, were presented previously but are repeated below for ease of reference.

$$r_{su} = -\frac{kXS}{K_s + S} \tag{7-12}$$

$$r_g = -Yr_{su} - k_dX \tag{7-21}$$

$$= Y\frac{kXS}{K_s + S} - k_dX \tag{7-22}$$

Both of the above expressions are of the saturation type. As noted previously, these expressions are similar to the saturation equation proposed by Monod (1942) for growth and the Michaelis-Menten equation for substrate utilization (Bailey and Ollis, 1986). Typical k and K_s values at 20°C vary from 8 to 12.0 g COD/g VSS·d and 10 to 40 g bsCOD/m³, respectively. As noted in Sec. 7–3, the K_s value can vary depending on the nature and complexity of the bsCOD components. For easily biodegradable single substrates, K_s values of less than 1.0 mg bsCOD/L have been measured (Bielefeldt and Stensel, 1999).

Applying the above expressions for substrate utilization and biomass growth leads to the development of a series of design parameters including the solids retention time (SRT), the food to microorganism ratio (F/M), and the specific utilization rate (U). These design parameters are applied to the design of a variety of activated-sludge processes in Chap. 8. With the exception of some difficult-to-degrade constituents in industrial wastewaters, the kinetics for aerobic oxidation of organic substrates seldom control the SRT design value for the activated-sludge process. For good floc formation, sufficient time is needed for the biomass in the activated-sludge aeration tank to develop extracellular polymers and a floc structure. More optimal flocculation and TSS removal in clarification occur typically at SRT values greater than 2.5 to 3.0 d at 20°C and 3 to 5 d at 10°C. However, some wastewater-treatment plants in warmer climates operate at SRT values varying from less than 1 to 1.5 d. Excessively long SRTs (>20 d) may lead to floc deterioration with the development of small pinpoint floc particles that produce a more turbid effluent. However, even with pinpoint floc, effluent suspended solids concentrations of less than 30 mg/L are generally achieved. The SRT may be varied in treatment plant operations to find the most optimal settling condition.

Environmental Factors

For carbonaceous removal, pH in the range of 6.0 to 9.0 is tolerable, while optimal performance occurs near a neutral pH. A reactor DO concentration of 2.0 mg/L is commonly used, and at concentrations above 0.50 mg/L there is little effect of the DO concentration on the degradation rate. For industrial wastewaters care must be taken to assure that sufficient nutrients (N and P) are available for the amount of bsCOD to be treated. Heterotrophic bacteria responsible for BOD removal can tolerate higher concentrations of toxic substances as compared to the bacteria responsible for ammonia oxidation or the production of methane.

7–9 BIOLOGICAL NITRIFICATION

Nitrification is the term used to describe the two-step biological process in which ammonia (NH_4-N) is oxidized to nitrite (NO_2-N) and nitrite is oxidized to nitrate (NO_3-N). The need for nitrification in wastewater treatment arises from water quality concerns over (1) the effect of ammonia on receiving water with respect to DO concentrations and fish toxicity, (2) the need to provide nitrogen removal to control eutrophication, and (3) the need to provide nitrogen control for water-reuse applications including groundwater recharge. For reference, the current (2001) drinking water maximum contaminant level (MCL) for nitrate nitrogen is 45 mg/L as nitrate or 10 mg/L as nitrogen. The total concentration of organic and ammonia nitrogen concentration in municipal wastewaters is typically in the range from 25 to 45 mg/L as nitrogen based on a flowrate of 450 L/capita·d (120 gal/capita·d). In many parts of the world with limited water supplies, total nitrogen concentrations in excess of 200 mg/L as N have been measured in domestic wastewater.

Process Description

As with BOD removal, nitrification can be accomplished in both suspended growth and attached growth biological processes. For suspended growth processes, a more common approach is to achieve nitrification along with BOD removal in the same single-sludge process, consisting of an aeration tank, clarifier, and sludge recycle system (see Fig. 7–19a). In cases where there is a significant potential for toxic and inhibitory substances in the wastewater, a two-sludge suspended growth system may be considered (see Fig. 7–19b). The two-sludge system consists of two aeration tanks and two clarifiers in series with the first aeration tank/clarifier unit operated at a short SRT for BOD removal. The BOD and toxic substances are removed in the first unit, so that nitrification can proceed unhindered in the second. A portion of influent wastewater usually has to be bypassed to the second sludge system to provide a sufficient amount of solids for efficient solids flocculation and clarification. Because the bacteria responsible for nitrification grow much more slowly than heterotrophic bacteria, systems designed for nitrification generally have much longer hydraulic and solids retention times than those for systems designed only for BOD removal.

In attached growth systems used for nitrification, most of the BOD must be removed before nitrifying organisms can be established. The heterotrophic bacteria have a higher biomass yield and thus can dominate the surface area of fixed-film systems over nitrifying bacteria. Nitrification is accomplished in an attached growth reactor after BOD removal or in a separate attached growth system designed specifically for nitrification. The design of attached growth biological systems is described in Chap. 9.

Microbiology

Aerobic autotrophic bacteria are responsible for nitrification in activated sludge and biofilm processes. Nitrification, as noted above, is a two-step process involving two

Figure 7–19

Process configuration used for biological nitrification: (a) single-sludge suspended growth system and (b) two-sludge suspended growth system.

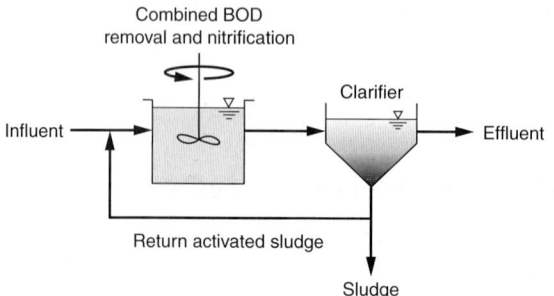

(a)

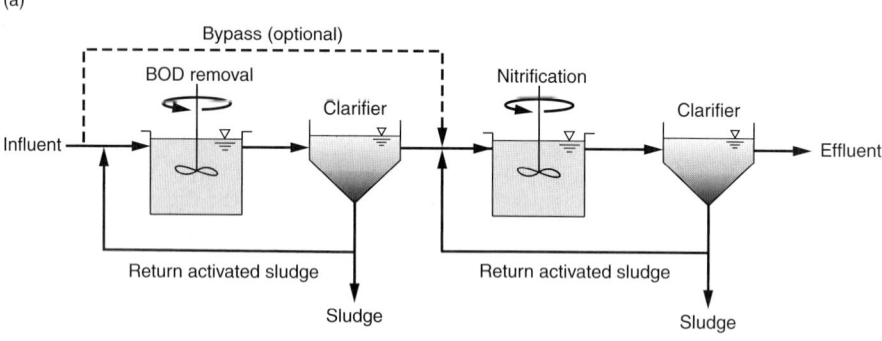

(b)

groups of bacteria. In the first stage, ammonia is oxidized to nitrite by one group of autotrophic bacteria. In the second stage, nitrite is oxidized to nitrate by another group of autotrophic bacteria. It should be noted that the two groups of autotrophic bacteria are distinctly different. Starting with classical experiments on nitrification by Winogradsky (1891), the bacteria genera commonly noted for nitrification in wastewater treatment are the autotrophic bacteria *Nitrosomonas* and *Nitrobacter*, which oxidize ammonia to nitrite and then to nitrate, respectively. Other autotrophic bacteria genera capable of obtaining energy from the oxidation of ammonia to nitrite (prefix with *Nitroso-*) are *Nitrosococcus, Nitrosospira, Nitrosolobus,* and *Nitrosorobrio* (Painter, 1970). It should be noted that during the 1990s, many more autotrophic bacteria were identified as being capable of oxidizing ammonia.

Besides *Nitrobacter*, nitrite can also be oxidized by other autotrophic bacteria (*Nitro-*) genera: *Nitrococcus, Nitrospira, Nitrospina,* and *Nitroeystis*. Using oligonucleotide probes for ammonia-oxidizing bacteria, Wagner et al. (1995) showed that *Nitrosomonas* was common in activated-sludge systems. For nitrite oxidation in activated sludge, Teske et al. (1994) found that *Nitrococcus* was quite prevalent. Whether different growth conditions can select for different genera of nitrifying bacteria or if their nitrification kinetics are significantly different is unknown at present.

Stoichiometry of Biological Nitrification

The energy-yielding two-step oxidation of ammonia to nitrate is as follows:

Nitroso-bacteria:

$$2NH_4^+ + 3O_2 \rightarrow 2NO_2^- + 4H^+ + 2H_2O \qquad (7\text{-}86)$$

Nitro-bacteria:

$$2NO_2^- + O_2 \rightarrow 2NO_3^- \tag{7-87}$$

Total oxidation reaction:

$$NH_4^+ + 2O_2 \rightarrow NO_3^- + 2H^+ + H_2O \tag{7-88}$$

Based on the above total oxidation reaction, the oxygen required for complete oxidation of ammonia is 4.57 g O_2/g N oxidized with 3.43 g O_2/g used for nitrite production and 1.14 g O_2/g NO_2 oxidized. When synthesis is considered, the amount of oxygen required is less than 4.57 g O_2/g N. In addition to oxidation, oxygen is obtained from fixation of carbon dioxide and nitrogen into cell mass.

Neglecting cell tissue, the amount of alkalinity required to carry out the reaction given in Eq. (7–88) can be estimated by writing Eq. (7–88) as follows:

$$NH_4^+ + 2HCO_3^- + 2O_2 \rightarrow NO_3^- + 2CO_2 + 3H_2O \tag{7-89}$$

In the above equation, for each g of ammonia nitrogen (as N) converted, 7.14 g of alkalinity as $CaCO_3$ will be required [2 × (50 g $CaCO_3$/eq)/14].

Along with obtaining energy, a portion of the ammonium ion is assimilated into cell tissue. The biomass synthesis reaction can be represented as follows:

$$4CO_2 + HCO_3^- + NH_4^+ + H_2O \rightarrow C_5H_7O_2N + 5O_2 \tag{7-90}$$

As noted previously, the chemical formula $C_5H_7O_2N$ is used to represent the synthesized bacterial cells.

The half reactions provided in Table 7–7 can be used to create an equation for the overall nitrification reaction. As demonstrated in Example 7–4, half reactions for cell synthesis, oxidation of ammonia to nitrate, and reduction of oxygen to water can be combined to create Eq. (7–91) ($f_s = 0.05$). Due to rounding of the coefficients, the equation does not balance exactly; however, the error introduced by rounding is negligible (Crites and Tchobanoglous, 1998).

$$NH_4^+ + 1.863O_2 + 0.098CO_2 \rightarrow 0.0196C_5H_7NO_2 + 0.98NO_3^-$$
$$+ 0.0941H_2O + 1.98H^+ \tag{7-91}$$

From the above equation it will be noted that for each g of ammonia nitrogen (as N) converted, 4.25 g of O_2 are utilized, 0.16 g of new cells are formed, 7.07 g of alkalinity as $CaCO_3$ are removed, and 0.08 g of inorganic carbon are utilized in the formation of new cells. The oxygen required to oxidize 1.0 g of ammonia nitrogen to nitrate (4.25 g) is less than the theoretical value of 4.57 g computed using Eq. (7–88), because the ammonia for cell synthesis is not considered in Eq. (7–87). Similarly, the alkalinity required for nitrification in Eq. (7–91) (7.07 g/g) is less than the value of 7.14 g calculated without considering the conversion of some of the ammonia to cellular nitrogen [Eq. (7–89)]. It should be recognized that the coefficient values in Eq. 7–91 are dependent upon the value of f_s that is used (see Sec. 7–4). [Note that $f_s = 0.05$ was assumed for Eq. (7–91).] Werzernak and Gannon (1967) found that the actual total oxygen consumption was 4.33 g O_2/g N with 3.22 g O_2/g N used for ammonia oxidation and 1.11 g O_2/g N used for nitrite oxidation.

The wastewater nitrogen concentration, BOD concentration, alkalinity, temperature, and potential for toxic compounds are major issues in the design of biological nitrification processes. Nitrifying bacteria need CO_2 and phosphorus for cell growth, as well as trace elements. With such a low cell yield, the CO_2 in air is adequate and phosphorus is seldom limiting. Trace element concentrations that have been found to stimulate the growth of nitrifying bacteria in pure culture work are: Ca = 0.50, Cu = 0.01, Mg = 0.03, Mo = 0.001, Ni = 0.10, and Zn = 1.0 mg/L (Poduska, 1973).

Growth Kinetics

For nitrification systems operated at temperatures below 28°C ammonia-oxidation kinetics versus nitrite-oxidation kinetics are rate-limiting, so that designs are based on saturation kinetics for ammonia oxidation as given below, assuming excess DO is available.

$$\mu_n = \left(\frac{\mu_{nm} N}{K_n + N} \right) - k_{dn} \qquad (7\text{–}92)$$

where μ_n = specific growth rate of nitrifying bacteria, g new cells/g cells·d
$\quad\quad \mu_{nm}$ = maximum specific growth rate of nitrifying bacteria, g new cells/g cells·d
$\quad\quad N$ = nitrogen concentration, g/m^3
$\quad\quad K_n$ = half-velocity constant, substrate concentration at one-half the maximum specific substrate utilization rate, g/m^3
$\quad\quad k_{dn}$ = endogenous decay coefficient for nitrifying organisms, g VSS/g VSS·d

A wide range of maximum specific growth rates have been reported as a function of temperature (Randall et al., 1992). At 20°C, reported μ_{nm} varies from 0.25 to 0.77 g VSS/g VSS. The wide range of nitrification growth rates may be due to the presence of inhibitory substances in the wastewater and/or variations in experimental techniques and methods of analysis. In any event, the μ_{nm} values for nitrifying organisms are much lower than the corresponding values for heterotrophic organisms, requiring much longer SRT values for nitrifying activated-sludge systems. Typical design SRT values may range from 10 to 20 d at 10°C to 4 to 7 d at 20°C. Above 28°C, both ammonia and nitrite oxidation kinetics should be considered. At elevated temperatures, the relative kinetics of NH_4-N and NO_2-N oxidation change, and NO_2-N will accumulate at lower SRT values.

For fully acclimated complete-mix activated sludge nitrification systems, at temperatures below 25°C with sufficient DO present, the NO_2-N concentration may be less than 0.10 mg/L as compared to NH_4-N concentrations in the range of 0.50 to 1.0 mg/L. However, during the initiation of nitrification, NO_2-N concentrations will be greater than NH_4-N concentrations, as the growth of nitrite-oxidizing bacteria cannot occur until the ammonia-oxidizing bacteria generate nitrite. Under transient conditions, NO_2-N concentrations of 5 to 20 mg/L are possible.

Nitrification rates are affected by the liquid DO concentration in activated sludge (attached growth effects are described in Chap. 9). In contrast to what has been observed for aerobic heterotrophic bacteria degradation of organic compounds, nitrification rates increase up to DO concentrations of 3 to 4 mg/L. To account for the effects of DO, the expression for the specific growth rate [Eq. (7–92)] is modified as follows:

$$\mu_n = \left(\frac{\mu_{nm} N}{K_n + N} \right) \left(\frac{DO}{K_o + DO} \right) - k_{dn} \qquad (7\text{–}93)$$

where DO = dissolved oxygen concentration, g/m^3
$\quad\quad K_o$ = half-saturation coefficient for DO, g/m^3
Other terms as defined previously.

While the above kinetic model and coefficients, based on observed results, can be used to describe nitrification in systems at low to modest organic loadings, the use of these models will generally overpredict nitrification rates in systems with high organic loadings. Stenstrom and Song (1991) have shown experimentally that the effect of DO on nitrification is affected by the activated-sludge floc size and density, and total oxygen demand of the mixed liquor. Nitrifying bacteria are distributed within a floc containing heterotrophic bacteria and other solids, with floc diameters ranging from 100 to 400 μm. Oxygen from the bulk liquid diffuses into floc particles, and bacteria deeper within the floc are exposed to lower DO concentrations. At higher organic loadings, there is a greater substrate concentration in the mixed liquor, which causes a higher oxygen consumption rate within the floc. Therefore, a higher bulk liquid DO concentration is needed to maintain the same internal floc DO concentration and subsequent nitrification rate.

At low DO concentrations (<0.50 mg/L) where nitrification rates are greatly inhibited, the low DO inhibition effect has been shown to be greater for *Nitrobacter* than for *Nitrosomonas*. In such cases, incomplete nitrification will occur with increased NO_2-N concentrations in the effluent. The presence of nitrite in the effluent is particularly troublesome for plants that use chlorination for disinfection, as nitrite is readily oxidized by chlorine requiring 5 g chlorine/g NO_2-N.

Environmental Factors

Nitrification is affected by a number of environmental factors including pH, toxicity, metals, and un-ionized ammonia.

Hydrogen-Ion Concentration (pH). Nitrification is pH-sensitive and rates decline significantly at pH values below 6.8. At pH values near 5.8 to 6.0, the rates may be 10 to 20 percent of the rate at pH 7.0 (U.S. EPA, 1993). Optimal nitrification rates occur at pH values in the 7.5 to 8.0 range. A pH of 7.0 to 7.2 is normally used to maintain reasonable nitrification rates, and for locations with low-alkalinity waters, alkalinity is added at the wastewater-treatment plant to maintain acceptable pH values. The amount of alkalinity added depends on the initial alkalinity concentration and amount of NH_4-N to be oxidized. Alkalinity may be added in the form of lime, soda ash, sodium bicarbonate, or magnesium hydroxide depending on costs and chemical handling issues.

Toxicity. Nitrifying organisms are sensitive to a wide range of organic and inorganic compounds and at concentrations well below those concentrations that would affect aerobic heterotrophic organisms. In many cases, nitrification rates are inhibited even though bacteria continue to grow and oxidize ammonia and nitrite, but at significantly reduced rates. In some cases, toxicity may be sufficient to kill the nitrifying bacteria.

Nitrifiers have been shown to be good indicators of the presence of organic toxic compounds at low concentrations (Blum and Speece, 1991). Toxic organic compounds have been listed by Hockenbury and Grady (1977) and Sharma and Ahlert (1977). Compounds that are toxic include solvent organic chemicals, amines, proteins, tannins, phenolic compounds, alcohols, cyanates, ethers, carbamates, and benzene. Because of the numerous compounds that can inhibit nitrification, it is difficult to pinpoint the source of nitrification toxicity for wastewater plants with inhibition, and extensive sampling of the collection system is normally needed to find the source.

Metals. Metals are also of concern for nitrifiers, and Skinner and Walker (1961) have shown complete inhibition of ammonia oxidation at 0.25 mg/L nickel, 0.25 mg/L chromium, and 0.10 mg/L copper.

Un-ionized Ammonia. Nitrification is also inhibited by un-ionized ammonia (NH_3) or free ammonia, and un-ionized nitrous acid (HNO_2). The inhibition effects are dependent on the total nitrogen species concentration, temperature, and pH. At 20°C and pH 7.0, the NH_4-N concentrations at 100 mg/L and 20 mg/L may initiate inhibition of NH_4-N and NO_2-N oxidation, respectively, and NO_2-N concentrations at 280 mg/L may initiate inhibition of NO_2-N oxidation (U.S. EPA, 1993).

7–10 BIOLOGICAL DENITRIFICATION

The biological reduction of nitrate to nitric oxide, nitrous oxide, and nitrogen gas is termed *denitrification*. Biological denitrification is an integral part of biological nitrogen removal, which involves both nitrification and denitrification. Compared to alternatives of ammonia stripping (see Chap. 11), breakpoint chlorination (see Chap. 12), and ion exchange (see Chap. 11), biological nitrogen removal is generally more cost-effective and used more often. Biological nitrogen removal is used in wastewater treatment where there are concerns for eutrophication, and where groundwater must be protected against elevated NO_3-N concentrations where wastewater-treatment plant effluent is used for groundwater recharge and other reclaimed water applications.

Process Description

Two modes of nitrate removal can occur in biological processes, and these are termed assimilating and dissimilating nitrate reduction (see Fig. 7–20). Assimilating nitrate reduction involves the reduction of nitrate to ammonia for use in cell synthesis. Assimilation occurs when NH_4-N is not available and is independent of DO concentration. On the other hand, dissimilating nitrate reduction or biological denitrification is coupled to the respiratory electron transport chain, and nitrate or nitrite is used as an electron acceptor for the oxidation of a variety of organic or inorganic electron donors.

Two basic flow diagrams for activated-sludge denitrification and the conditions that drive the denitrification reaction rates are illustrated on Fig. 7–21. The first flow diagram (Fig. 7–21a) is for the Modified Ludzak-Ettinger (MLE) process (U.S. EPA, 1993), the most common process used for biological nitrogen removal in municipal wastewater treatment. The process consists of an anoxic tank followed by the aeration tank where nitrification occurs. Nitrate produced in the aeration tank is recycled back to the anoxic tank. Because the organic substrate in the influent wastewater provides the electron donor for oxidation reduction reactions using nitrate, the process is termed *substrate denitrification*. Further, because the anoxic process precedes the aeration tank, the process is known as a *preanoxic denitrification*.

In the second process shown on Fig. 7–21b, denitrification occurs after nitrification and the electron donor source is from endogenous decay. The process illustrated on Fig. 7–21b is generally termed a *postanoxic denitrification* as BOD removal has occurred first and is not available to drive the nitrate reduction reaction. When a postanoxic denitrification process depends solely on endogenous respiration for energy, it has a much slower rate of reaction than for the preanoxic processes using wastewater

Figure 7–20

Nitrogen transformations in biological treatment processes. (*Adapted from Sedlak, 1991.*)

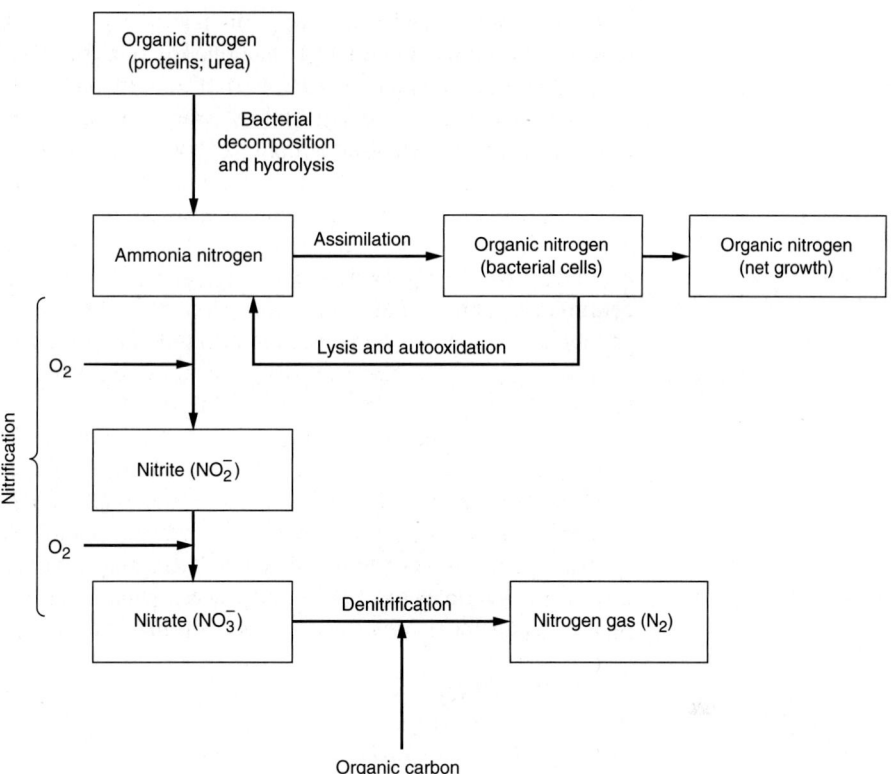

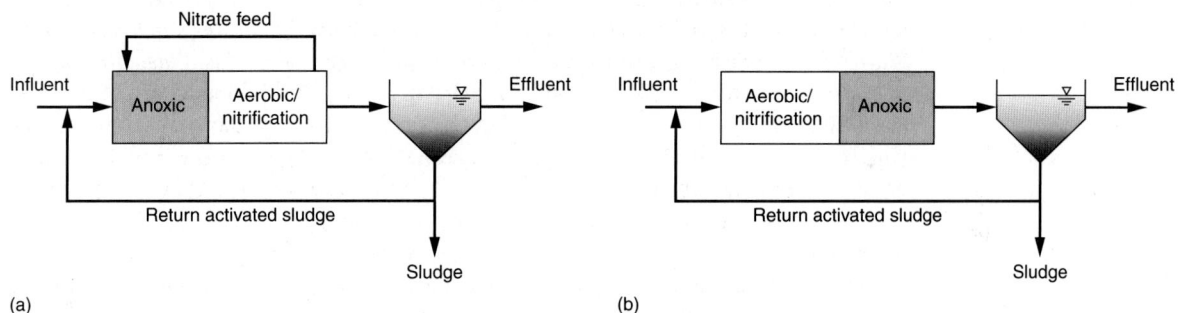

(a)

(b)

Figure 7–21

Types of denitrification processes and the reactors used for their implementation: (*a*) substrate driven (preanoxic denitrification) and (*b*) endogenous driven (postanoxic denitrification).

BOD. Often an exogenous carbon source such as methanol or acetate is added to postanoxic processes to provide sufficient BOD for nitrate reduction and to increase the rate of denitrification. Postanoxic processes include both suspended and attached growth systems. In one attached growth granular-medium filtration process, both nitrate reduction and effluent suspended solids removal occur in the same reactor.

The denitrification preanoxic and postanoxic processes described employ heterotrophic bacteria for nitrate reduction, but other pathways for biological nitrogen removal exist. Ammonia can be converted to nitrogen gas by novel autotrophic bacteria under anaerobic conditions and by heterotrophic-nitrifying bacteria under aerobic conditions. The organisms identified with these reactions are presented in the Microbiology section below. Littleton et al. (2000) looked for the possible influence of these autotrophic-denitrifying bacteria or to heterotrophic nitrifying bacteria for nitrogen removal in aerated full-scale, long-SRT, single-sludge, activated-sludge systems but could not find any significant activity for them. However, the use of novel autotrophic denitrifying bacteria has been demonstrated in a process for the treatment of high ammonia strength anaerobic digestion centrate in a fluidized bed reactor at 30 to 35°C (Strous et al. 1997, Jetten et al. 1999). The fluidized bed reactor is effective in maintaining these relatively slow-growing bacteria. Under anaerobic conditions NH_4^+ is oxidized by NO_2^- to produce nitrogen gas and a small amount of nitrate. About 1.3 moles of NO_2^- are used per mole of NH_4^+ (Strous et al. 1999a). A portion of the centrate stream must be nitrified to nitrite, but no carbon source is needed for nitrogen removal. The process is termed the Anammox process, which stands for anaerobic ammonium oxidation. The oxidation of ammonia to nitrite in anaerobic digester centrate has been done in a process termed the SHARON process, which can be used with the Anammox process (Jetten et al. 1999) and is described in Chap. 8.

Microbiology

A wide range of bacteria has been shown capable of denitrification, but similar microbial capability has not been found in algae or fungi. Bacteria capable of denitrification are both heterotrophic and autotrophic. The heterotrophic organisms include the following genera: *Achromobacter, Acinetobacter, Agrobacterium, Alcaligenes, Arthrobacter, Bacillus, Chromobacterium, Corynebacterium, Flavobacterium, Hypomicrobium, Moraxella, Neisseria, Paracoccus, Propionibacterium, Pseudomonas, Rhizobium, Rhodopseudomonas, Spirillum,* and *Vibrio* (Payne, 1981). In addition, Gayle (1989) lists *Halobacterium* and *Methanomonas. Pseudomonas* species are the most common and widely distributed of all the denitrifiers, and have been shown to use a wide array of organic compounds including hydrogen, methanol, carbohydrates, organic acids, alcohols, benzoates, and other aromatic compounds (Payne, 1981). Most of these bacteria are facultative aerobic organisms with the ability to use oxygen as well as nitrate or nitrite, and some can also carry out fermentation in the absence of nitrate or oxygen. Other autotrophic bacteria that can denitrify use hydrogen and reduced sulfur compounds as electron donors during denitrification. Both groups of organisms can grow heterotrophically if an organic carbon source is present (Gayle, 1989).

Nitrogen removal can also be accomplished by heterotrophic- and autotrophic-nitrifying bacteria under certain conditions, and by a unique bacteria associated with the Anammox process, which was discovered in the mid-1990s. Denitrification can occur under aerobic conditions by heterotrophic nitrifying bacteria (Robertson and Kuenen, 1990, and Patureau et al. 1994), so that simultaneous nitrification and denitrification exist with the conversion of ammonia to gaseous nitrogen products. The heterotrophic bacteria, *Paracoccus pantotropha,* have been studied extensively for simultaneous ammonia oxidation and nitrate reduction. The oxidation of ammonia by heterotrophic bacteria requires energy, which can be obtained by nitrate or nitrite reduction by *P. pan-*

totropha under aerobic conditions. A readily available substrate, such as acetate, is also needed. Due to the need for the carbon substrate, which is in limited supply in aerobic activated-sludge systems, little growth of heterotrophic nitrifiers is expected (van Loosdrecht and Jetten, 1998) in aerobic wastewater treatment systems.

Autotrophic nitrifying bacteria, such as *Nitrosomonas europaea,* can use nitrite to oxidize ammonia, with the production of nitrogen gas, when dissolved oxygen is not present (Bock et al. 1995). With oxygen present, these bacteria oxidize the ammonia with oxygen as the electron acceptor.

Ammonia oxidation with the reduction of nitrite under anaerobic conditions has also been shown at temperatures above 20°C in the Anammox process (Strous et al. 1997). The bacteria in the Anammox process are different than the autotrophic nitrifying bacteria described above, in that it cannot use oxygen for ammonia oxidation (Jetten et al. 1999). The Anammox bacteria could not be isolated and grown in pure culture (Strous et al. 1999b), but an enrichment was obtained by density purification for 16S rRNA extraction and analysis. Phylogenetic analysis showed that the bacteria is in the order *Planctomycetales,* a division with the domain Bacteria. Under anaerobic conditions the ammonia oxidation rate by the Annamox bacteria was shown to be 6 to 10 times faster than that for *N. Europaea* (Jetten et al. 1999).

Stoichiometry of Biological Denitrification

Biological denitrification involves the biological oxidation of many organic substrates in wastewater treatment using nitrate or nitrite as the electron acceptor instead of oxygen. In the absence of DO or under limited DO concentrations, the nitrate reductase enzyme in the electron transport respiratory chain is induced, and helps to transfer hydrogen and electrons to nitrate as the terminal electron acceptor. The nitrate reduction reactions involve the following reduction steps from nitrate to nitrite, to nitric oxide, to nitrous oxide, and to nitrogen gas:

$$NO_3^- \rightarrow NO_2^- \rightarrow NO \rightarrow N_2O \rightarrow N_2 \tag{7–94}$$

In biological nitrogen removal processes, the electron donor is typically one of three sources: (1) the bsCOD in the influent wastewater, (2) the bsCOD produced during endogenous decay, and (3) an exogenous source such as methanol or acetate. The latter has been added in separate treatment units, such as polishing filters, after nitrification where almost no bsCOD remains. Reaction stoichiometry for different electron donors is shown as follows. The term $C_{10}H_{19}O_3N$ is often used to represent the biodegradable organic matter in wastewater (U.S. EPA, 1993).

Wastewater:

$$C_{10}H_{19}O_3N + 10NO_3^- \rightarrow 5N_2 + 10CO_2 + 3H_2O + NH_3 + 10OH^- \tag{7–95}$$

Methanol:

$$5CH_3OH + 6NO_3^- \rightarrow 3N_2 + 5CO_2 + 7H_2O + 6OH^- \tag{7–96}$$

Acetate:

$$5CH_3COOH + 8NO_3^- \rightarrow 4N_2 + 10CO_2 + 6H_2O + 8OH^- \tag{7–97}$$

In all the above heterotrophic denitrification reactions, one equivalent of alkalinity is produced per equivalent of NO_3-N reduced, which equates to 3.57 g of alkalinity (as

$CaCO_3$) production per g of nitrate nitrogen reduced. Recall from nitrification that 7.14 g alkalinity (as $CaCO_3$) was consumed per g of NH_4-N oxidized, so that by denitrification about one-half of the amount destroyed by nitrification can be recovered.

From oxidation reduction half reactions, the oxygen equivalent of using nitrate or nitrite as electron acceptors can be determined. The half reactions per mole e^- transferred are determined from Table 7–7 and are shown as follows:

For oxygen:

$$0.25O_2 + H^+ + e^- \rightarrow 0.5H_2O \tag{7–98}$$

For nitrate:

$$0.20NO_3^- + 1.2H^+ + e^- \rightarrow 0.1N_2 + 0.6H_2O \tag{7–99}$$

For nitrite:

$$0.33NO_2^- + 1.33H^+ + e^- \rightarrow 0.67H_2O + 0.17N_2 \tag{7–100}$$

Comparing the above half reactions for oxygen [Eq. (7–98)] and nitrate [Eq. (7–99)], it should be noted that 0.25 mole of oxygen is equivalent to 0.2 mole of nitrate for electron transfer in oxidation reduction. Thus, the oxygen equivalent of nitrate is (0.25 × 32 g O_2/mole) divided by the nitrate gram equivalent (0.20 × 14 g N/mole) equals 2.86 g O_2/g NO_3-N. The oxygen equivalent is a useful design factor when calculating the total oxygen required for nitrification-denitrification biological treatment systems. Similarly, for nitrite as the electron acceptor, the oxygen equivalent of nitrite is 1.71 g O_2/g NO_2-N.

In biological denitrification, the principal design goal is to reduce nitrate biologically. Thus, an important design parameter for denitrification processes is the amount of bsCOD or BOD needed to provide a sufficient amount of electron donor for nitrate removal. As a general rule, Barth et al. (1968) estimated that 4 g of BOD is needed per g of NO_3 reduced. However, the actual value will depend on the system operating conditions and the type of electron donor used for denitrification. Based on information presented in Sec. 7–6, in which it was shown that the amount of oxygen used per unit of bsCOD was related to the biomass yield, the bsCOD/NO_3-N ratio is similarly related to the system biomass yield. An expression that can be used to determine the bsCOD/NO_3-N ratio can be developed based on the following assumptions and facts (Randall et al., 1992). It should be noted that because of the long SRT values involved in nitrification, the following analysis also applies to bCOD, which includes colloidal and particulate components. From a steady-state COD balance it can be shown that the bsCOD removed is oxidized or accounted for in cell growth.

$$bsCOD_r = bsCOD_{syn} + bsCOD_o \tag{7–101}$$

where $bsCOD_r$ = bsCOD utilized, g bsCOD/d
$bsCOD_{syn}$ = bsCOD incorporated into cell synthesis, g bsCOD/d
$bsCOD_o$ = bsCOD oxidized, g bsCOD/d

For cell synthesis, the $bsCOD_{syn}$ is calculated from the net biomass yield and the ratio of 1.42 g O_2/g VSS. The oxygen equivalent of the biomass is equal to the bsCOD incorporated into biomass.

$$bsCOD_{syn} = 1.42\, Y_n\, bsCOD_r \tag{7–102}$$

where Y_n = net biomass yield, g VSS/g $bsCOD_r$

and

$$Y_n = \frac{Y}{1 + (k_{dn})\text{SRT}} \qquad (7\text{–}103)$$

Thus

$$\text{bsCOD}_r = \text{bsCOD}_o + 1.42 Y_n \, \text{bsCOD}_r \qquad (7\text{–}104)$$

Rearranging yields

$$\text{bsCOD}_o = (1 - 1.42 \, Y_n) \, \text{bsCOD}_r \qquad (7\text{–}105)$$

In Eq. (7–105), bsCOD_o is the COD oxidized and is equal to the oxygen equivalent of the NO_3-N used for bsCOD oxidation. Hence,

$$\text{bsCOD}_o = 2.86 \, \text{NO}_x \qquad (7\text{–}106)$$

where 2.86 = O_2 equivalent of NO_3-N, g O_2/g NO_3-N

NO_x = NO_3-N reduced, g/d

Substituting Eq. (7–105) in Eq. (7–106) yields

$$2.86 \, NO_3 = (1 - 1.42 \, Y_n) \, \text{bsCOD}_r \qquad (7\text{–}107)$$

or

$$\frac{\text{bsCOD}}{NO_3\text{-N}} = \frac{2.86}{1 - 1.42 Y_n} \qquad (7\text{–}108)$$

Thus, g bsCOD/g NO_3-N $= \dfrac{2.86}{1 - 1.42 Y_n}$

Growth Kinetics

For biological denitrification, the biokinetic equations used to describe bacteria growth and substrate utilization are similar to those described previously in this chapter for aerobic heterotrophic bacteria. The rate of soluble substrate utilization is also controlled by the soluble substrate concentration [Eq. (7–40)], with nitrate serving as the electron acceptor in lieu of oxygen. The nitrate concentration controls the substrate utilization kinetics only at very low NO_3-N concentrations, near 0.1 mg/L. Nitrate serves as an electron acceptor in the same way as oxygen from a biokinetics perspective, and thus the nitrate utilization rate (denitrification rate) is proportional to the substrate utilization rate.

There are two general cases where the substrate utilization rate controls the denitrification rate. The first is for anoxic/aerobic processes where the organic substrate (electron donor) is from the influent wastewater fed into the anoxic reactor. The second is for postanoxic denitrification, where nitrate reduction is done after secondary treatment in a reactor receiving another carbon source. Because the BOD is depleted from secondary treatment, an exogenous carbon source is used to drive the nitrate reduction reaction. In most cases methanol is supplied to create the demand for the electron acceptor in suspended growth or attached growth processes. Endogenous respiration creates a demand for nitrate in addition to that caused by substrate utilization and oxidation. This reaction occurs in the mixed liquor of anoxic tanks and is at a much lower rate than the denitrification rate caused by substrate utilization.

In the case where wastewater provides the electron donor, an anoxic reactor receives the influent wastewater and it is followed in the treatment system by an aerobic reactor where nitrification occurs. Heterotrophic bacterial growth occurs in both the anoxic and aerobic zones with nitrate and oxygen consumption, respectively. The mixed liquor biomass concentration can be calculated based on the total amount of BOD removed, but only a portion of that biomass can use both nitrate and oxygen as electron acceptors. The microorganisms in the other portion are strict aerobes and can only use oxygen as the electron acceptor. To apply biokinetic expressions for denitrification, the substrate utilization rate expression is modified to account for the fact that only a portion of the biomass is active in the anoxic zone.

The substrate utilization rate r_{su} expression [Eq. (7–12)] is modified by a term to show a lower utilization rate in the anoxic zone as follows:

$$r_{su} = -\frac{k\, XS\eta}{K_s + S} \tag{7–109}$$

where η = fraction of denitrifying bacteria in the biomass, g VSS/g VSS
k, X, S, K_s are as defined previously

When nitrate is used as an electron acceptor instead of oxygen, the maximum specific substrate utilization rate (k) may be lower than the rate with oxygen as the electron acceptor. In Eq. (7–109), the k value used is the same as that used for oxygen, and any effect of lower kinetic rate is incorporated into the η term. The K_s value has been found to be similar, whether nitrate or oxygen is the electron acceptor (Stensel and Horne, 2000).

The value for η has been found to vary from 0.20 to 0.80 for preanoxic denitrification reactors fed domestic wastewaters (Stensel and Horne, 2000). The activated-sludge configuration, the system SRT, and the fraction of influent BOD removed with nitrate appear to affect the η value. For anoxic/aerobic processes with substantial substrate and nitrate removal in the preanoxic zone η may be close to 0.80. In spite of the fact that only a portion of the mixed liquor biomass can use nitrate, the anoxic reactor volumes used for anoxic/aerobic processes range from only 10 to 30 percent of the total volume (anoxic plus aerobic) for treating domestic wastewater.

For postanoxic suspended growth or attached growth processes the biomass is developed under mainly anoxic conditions and with a selected single organic substrate. In this case the η term is not necessary because the biomass consists of mainly denitrifying bacteria. The biokinetic equations presented previously can then be used with the appropriate kinetic coefficient values (k, K_s, Y, k_d) to design a postanoxic complete-mix suspended growth process. The kinetic coefficient values for growth using methanol have been developed at 10°C and 20°C in laboratory studies (Randall et al., 1992), and a suspended growth design with methanol as the electron donor substrate is illustrated in Chap. 10.

The kinetics for methanol utilization are such that the SRTs required for a denitrification suspended growth process are in the same range as SRTs for aerobic systems designed for BOD removal only, about 3 to 6 d.

Effect of Dissolved Oxygen Concentration. Dissolved oxygen can inhibit nitrate reduction by repressing the nitrate reduction enzyme. In activated-sludge flocs and biofilms, denitrification can proceed in the presence of low bulk liquid DO concentrations. A dissolved oxygen concentration of 0.2 mg/L and above has been reported to inhibit denitrification for a *Pseudomonas* culture (Skerman and MacRae, 1957; Terai and Mori, 1975) and by Dawson and Murphy (1972) for activated-sludge treating domestic wastewater. Nelson and Knowles (1978) reported that denitrification ceased in a highly dispersed growth at a DO concentration of 0.13 mg/L. The effect of nitrate and DO concentration on the biokinetics is accounted for by two correction factors to Eq. (7–109) expressed in the form of two saturation terms as follows:

$$r_{su} = - \left(\frac{kXS}{K_s + S} \right) \left(\frac{NO_3}{K_{s,NO_3} + NO_3} \right) \left(\frac{K'_o}{K'_o + DO} \right) (\eta) \qquad (7\text{–}110)$$

where K'_o = DO inhibition coefficient for nitrate reduction, mg/L
   K_{s,NO_3} = half velocity coefficient for nitrate limited reaction, mg/L
   Other terms are as defined previously.

The value of K'_o is system-specific. Values in the range from 0.1 to 0.2 mg/L have been proposed for K'_o and 0.1 mg/L for K_{s,NO_x} (Barker and Dold, 1997). Assuming a K_o value of 0.1 mg/L the rate of substrate utilization with nitrate as the electron acceptor at DO concentrations of 0.10, 0.20, and 0.50 mg/L would be at 50, 33, and 17 percent of the maximum rate, respectively.

Effect of Simultaneous Nitrification-Denitrification.   In activated-sludge systems, the issue of DO concentration is confounded by the fact that the measured bulk liquid DO concentration does not represent the actual DO concentration within the activated-sludge floc. Under low DO concentration conditions, denitrification can occur in the floc interior, while nitrification is occurring at the floc exterior. Also in activated-sludge tanks operated at low DO concentrations, both aerobic and anaerobic zones exist depending on mixing conditions and distance from the aeration point, so that nitrification and denitrification can be occurring in the same tank. Under these conditions, nitrogen removal that occurs in a single aeration tank is referred to as simultaneous nitrification and denitrification. Although both nitrification and denitrification are occurring at reduced rates as indicated by the DO effects described for both processes, if a sufficient SRT and τ exist, the overall nitrogen removal can be significant. Rittman and Langeland (1985) reported greater than 90 percent nitrogen removal by nitrification and denitrification in an activated-sludge system used to treat municipal wastewater at DO concentrations below 0.50 mg/L and with values of τ greater than 25 hours.

Environmental Factors

Alkalinity is produced in denitrification reactions and the pH is generally elevated, instead of being depressed as in nitrification reactions. In contrast to nitrifying organisms, there has been less concern about pH influences on denitrification rates. No significant effect on the denitrification rate has been reported for pH between 7.0 and 8.0, while Dawson and Murphy (1972) showed a decrease in the denitrification rate as the pH was decreased from 7.0 to 6.0 in batch unacclimated tests.

7–11   BIOLOGICAL PHOSPHORUS REMOVAL

The removal of phosphorus by biological means is known as *biological phosphorus removal*. Phosphorus removal is generally done to control eutrophication because phosphorus is a limiting nutrient in most freshwater systems. Treatment plant effluent discharge limits have ranged from 0.10 to 2.0 mg/L of phosphorus depending on plant location and potential impact on receiving waters. Chemical treatment using alum or iron salts is the most commonly used technology for phosphorus removal (see Sec. 6–4 in Chap. 6), but since the early 1980s success in full-scale plant biological phosphorus removal has encouraged further use of the technology. The principal advantages of biological phosphorus removal are reduced chemical costs and less sludge production as compared to chemical precipitation.

Process Description

In the biological removal of phosphorus, the phosphorus in the influent wastewater is incorporated into cell biomass, which subsequently is removed from the process as a result of sludge wasting. Phosphorus accumulating organisms (PAOs) are encouraged to grow and consume phosphorus in systems that use a reactor configuration that provides PAOs with a competitive advantage over other bacteria. The reactor configuration utilized for phosphorus removal is comprised of an anaerobic tank having a τ value of 0.50 to 1.0 h that is placed ahead of the activated-sludge aeration tank (see Fig. 7–22). The contents of the anaerobic tank are mixed to provide contact with the return activated sludge and influent wastewater. Anaerobic contact tanks have been placed in front of many different types of suspended growth processes (see detailed discussion in Sec. 8–6 in Chap. 8), with aerobic SRT values ranging from 2 to 40 d.

Figure 7–22

Biological phosphorus removal: (a) typical reactor configuration. Photos below flow diagram are of (b) transmission electron microscope images of polyhydroxybutyrate storage and (c) polyphosphate storage granules.

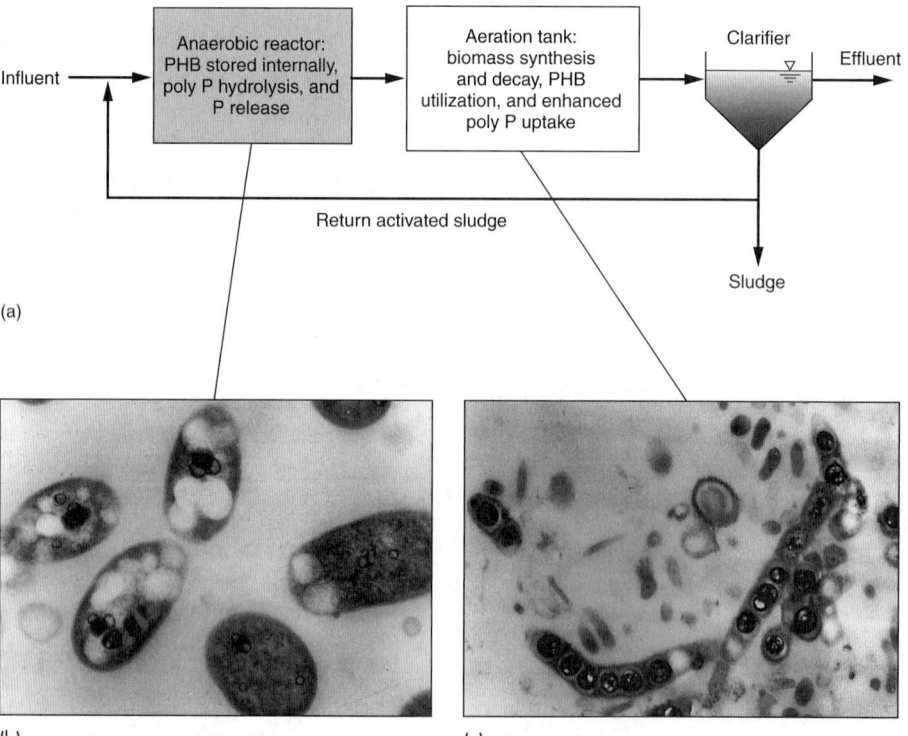

Phosphorus removal in biological systems is based on the following observations (Sedlak, 1991):

1. Numerous bacteria are capable of storing excess amounts of phosphorus as polyphosphates in their cells.
2. Under anaerobic conditions, PAOs will assimilate fermentation products (e.g., volatile fatty acids) into storage products within the cells with the concomitant release of phosphorus from stored polyphosphates.
3. Under aerobic conditions, energy is produced by the oxidation of storage products and polyphosphate storage within the cell increases.

A simplified version of the processes occurring in the anaerobic and aerobic/anoxic reactors or zones is presented below. In many applications for phosphorus removal, an anoxic reactor follows the anaerobic reactor and precedes the aerobic reactor. Most PAOs can use nitrite in place of oxygen to oxidize their stored carbon source. A more comprehensive description of the biochemistry and intracellular transformations can be found in Wentzel et al. (1991).

Processes Occurring in the Anaerobic Zone

- Acetate is produced by fermentation of bsCOD which, as defined earlier, is dissolved degradable organic material that can be assimilated easily by the biomass. Depending on the value of τ for the anaerobic zone, some colloidal and particulate COD is also hydrolyzed and converted to acetate, but the amount is generally small compared to that from the bsCOD conversion.
- Using energy available from stored polyphosphates, the PAOs assimilate acetate and produce intracellular polyhydroxybutyrate (PHB) storage products. Some glycogen contained in the cell is also used. Concurrent with the acetate uptake is the release of orthophosphate ($O-PO_4$), as well as magnesium, potassium, and calcium cations.
- The PHB content in the PAOs increases while the polyphosphate decreases.

Processes Occurring in the Aerobic/Anoxic Zone

- Stored PHB is metabolized, providing energy from oxidation and carbon for new cell growth.
- Some glycogen is produced from PHB metabolism.
- The energy released from PHB oxidation is used to form polyphosphate bonds in cell storage so that soluble orthophosphate ($O-PO_4$) is removed from solution and incorporated into polyphosphates within the bacterial cell. Cell growth also occurs due to PHB utilization and the new biomass with high polyphosphate storage accounts for phosphorus removal.
- As a portion of the biomass is wasted, stored phosphorus is removed from the biotreatment reactor for ultimate disposal with the waste sludge.

The events occurring in the anaerobic and aerobic zones are illustrated graphically on Fig. 7–23.

Microbiology

Phosphorus is important in cellular energy transfer mechanisms via adenosine triphosphate (ATP) and polyphosphates. As energy is produced in oxidation reduction reactions,

Figure 7–23

Fate of soluble BOD and phosphorus in nutrient removal reactor. (Adapted from Sedlak, 1991.)

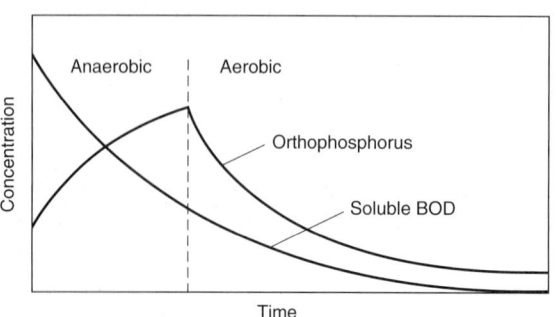

adenosine diphosphate (ADP) is converted to ATP with 7.4 kcal/mole of energy captured in the phosphate bond. As the cell uses energy, ATP is converted to ADP with phosphorus release. For common heterotrophic bacteria in activated-sludge treatment the typical phosphorus composition is 1.5 to 2.0 percent. However, many bacteria are able to store phosphorus in their cells in the form of energy-rich polyphosphates, resulting in phosphorus content as high as 20 to 30 percent by dry weight. The polyphosphates are contained in volutin granules within the cell along with Mg^{2+}, Ca^{2+}, and K^+ cations.

In the anaerobic zone, concentrations of $O\text{-}PO_4$ as high as 40 mg/L can be measured in the liquid, as compared to wastewater influent concentrations of 5 to 8 mg/L. The high concentration of $O\text{-}PO_4$ can be taken as an indication that phosphorus release by the bacteria has occurred in this zone. Also in this zone, significant amounts of poly-β-hydroxybutyrate (PHB) are found stored in bacteria cells, but the PHB concentration declines appreciably in the subsequent anoxic and/or aerobic zones and can be measured and quantified. The $O\text{-}PO_4$ is taken up from solution in the aerobic and anoxic zones, generally leading to very low remaining concentrations. Based on investigations of biological phosphorus removal, it was found that acetate was essential to forming the PHB under anaerobic conditions, which provided a competitive advantage for the PAOs.

The anaerobic zone in the anaerobic/aerobic treatment process is termed a "selector," because it provides conditions that favor the proliferation of the PAOs, by the fact that a portion of the influent bCOD is consumed by the PAOs instead of other heterotrophic bacteria. Because the PAOs prefer low-molecular-weight fermentation product substrates, the preferred food source would not be available without the anaerobic zone that provides for the fermentation of the influent bsCOD to acetate. Because of the polyphosphate storage ability, the PAOs have energy available to assimilate the acetate in the anaerobic zone. Other aerobic heterotrophic bacteria have no such mechanism for acetate uptake, and they are starved while the PAOs assimilate COD in the anaerobic zone. It should also be noted that the PAOs form very dense, good settling floc in the activated sludge, which is an added benefit. In some facilities, the anaerobic/aerobic process sequence has been used because of the sludge settling benefits, even though biological phosphorus removal was not required.

Care must be taken in the handling of the waste sludge from biological phosphorus removal systems. When the sludge is held under anaerobic conditions, phosphorus release will occur. Release of $O\text{-}PO_4$ is possible even without acetate addition as the bacteria use the stored polyphosphate for an energy source. The release of $O\text{-}PO_4$ can also occur after extended contact time in the anaerobic zone of the biological phosphorus treatment system. In that case the released phosphorus may not be taken up in the aerobic zone because the release was not associated with acetate uptake and PHB storage for later oxidation. The release of O-PO$_4$ under these conditions is termed *secondary release* (Barnard, 1984), which can lead to a lower phosphorus removal efficiency for the biological process.

Stoichiometry of Biological Phosphorus Removal

Based on the description of the phosphorus removal mechanism, acetate uptake in the anaerobic zone is critical in determining the amount of PAOs that can be produced and, thus, the amount of phosphorus that can be removed by this pathway. If significant amounts of dissolved oxygen or nitrate enter the anaerobic zone, the acetate can be depleted before it is taken up by the PAOs, and treatment performance will be hindered. Biological phosphorus removal is not used in systems that are designed with nitrification without including a means for denitrification to minimize the amount of nitrate in the return sludge flow to the anaerobic zone.

The amount of phosphorus removed by biological storage can be estimated from the amount of bsCOD that is available in the wastewater influent as most of the bsCOD will be converted to acetate in the short anaerobic hydraulic detention time τ. Methods for determining the amount of bsCOD in the influent will be described in Chap. 8 under wastewater characterization. The following assumptions are used to evaluate the stoichiometry of biological phosphorus removal: (1) 1.06 g acetate/g bsCOD will be produced as most of the COD fermented will be converted to VFAs due to the low cell yield of the fermentation process, (2) a cell yield of 0.30 g VSS/g acetate, and (3) a cell phosphorus content 0.3 g P/g VSS. Using these assumptions, about 10 g of bsCOD will be required to remove 1 g of phosphorus by the biological storage mechanism. Other bCOD removal in the activated-sludge system will result in additional phosphorus removal by normal cell synthesis.

Better performance for biological phosphorus removal systems is achieved when bsCOD or acetate is available at a steady rate. Periods of starvation or low bsCOD concentrations result in changes in the intracellular storage reserves of glycogen, PHB, and polyphosphates and rapidly lead to decreased phosphorus removal efficiency (Stephens and Stensel, 1998). The amount of phosphorus that can be removed from a wastewater is illustrated in Example 7–8.

EXAMPLE 7–8 **Estimating the Amount of Phosphorus Removal** Given the following influent wastewater characteristics and the corresponding biological process information, estimate the effluent phosphorus concentration.

Influent	Concentration, g/m³
COD	300
bCOD	200
bsCOD	50
Phosphorus	6

1. Heterotrophic synthesis yield, $Y = 0.40$ g VSS/g COD
2. Endogenous decay coefficient, $k_d = 0.08$ g VSS/g VSS·d

3. SRT = 5 d
4. Phosphorus content of PAOs = 0.30 g P/g VSS
5. Phosphorus content of other bacteria = 0.02 g P/g VSS
6. Clarifier effluent VSS concentration = 8 g/m^3

Solution

1. Determine phosphorus removed by PAOs due to the fermentation of the 50 g rbsCOD/m^3 in the wastewater influent.
   a. Determine biomass produced using Eq. (7–57) neglecting cell debris

   $$\text{Biomass produced} = \left[\frac{Y}{1 + (k_d)\text{SRT}}\right]\text{bsCOD}$$

   $$= \left\{\frac{(0.40 \text{ g VSS/g COD})}{[1 + (0.08 \text{ g/g·d})(5 \text{ d})]}\right\}(50 \text{ g bsCOD/m}^3) = 14.3 \text{ g VSS/m}^3$$

   b. Determine the phosphorus removed

   P removed = (0.30 g P/g VSS)(14.3 g VSS/m^3) = 4.3 g/m^3

2. Determine phosphorus removed by heterotrophs from the conversion of colloidal and particulate bCOD.
   a. Determine the COD removed

   COD removed = bCOD − bsCOD = 200 − 50 g/m^3 = 150 g/m^3

   $$\text{Biomass produced from bpCOD} = \left[\frac{Y}{1 + (k_d)\text{SRT}}\right]\text{bpCOD}$$

   $$= \left\{\frac{(0.40 \text{ g VSS/g COD})}{[1 + (0.08 \text{ g/g·d})(5 \text{ d})]}\right\}150 \text{ g bpCOD/m}^3 = 42.9 \text{ g VSS/m}^3$$

   b. Determine the phosphorus removed

   P removed = 0.02 g P/g VSS = 0.02(42.9) = 0.86 g/m^3

3. Determine total phosphorus removed and effluent phosphorus concentration.

   Total P removed = 4.3 + 0.86 = 5.16 g/m^3

   Effluent soluble concentration = 6.0 − 5.16 = 0.84 g/m^3

4. Estimate P content of effluent VSS.

   Average P content of effluent VSS

   $$= \frac{(0.30 \text{ g P/g VSS})(14.3 \text{ g/m}^3) + (0.02 \text{ g P/g VSS})(42.9 \text{ g/m}^3)}{[(14.3 + 42.9) \text{ g/m}^3]}$$

   = 0.09 g P/g VSS

   Phosphorus in effluent VSS = 0.09(8 g/m^3) = 0.72 g/m^3

   Total effluent P concentration = 0.84 + 0.72 = 1.56 g/m^3

Growth Kinetics

Biological phosphorus growth kinetics are within the same order of magnitude of other heterotrophic bacteria. Mamais and Jenkins (1992) showed that biological phosphorus removal could be maintained in anaerobic/aerobic systems at SRTs greater than 2.5 d at 20°C. A maximum specific growth rate at 20°C is given as 0.95 g/g·d (Barker and Dold, 1997).

Environmental Factors

System performance is not affected by DO as long as the aerobic zone DO concentration is above 1.0 mg/L. At pH values below 6.5, phosphorus removal efficiency is greatly reduced (Sedlak, 1991).

In biological phosphorus removal systems, sufficient cations associated with polyphosphate storage must also be available. The recommended molar ratios of Mg, K, and Ca to phosphorus are 0.71, 0.50, and 0.25, respectively (Wentzel et al., 1989). Thus, for an influent soluble phosphorus concentration of 10 mg/L, 5.6, 6.3, and 3.2 mg/L of Mg, K, and Ca, respectively, would be required. The relative amounts of these cations associated with phosphate storage are 0.28, 0.26, and 0.09 mole/mole of phosphorus, respectively (Sedlak, 1991). Most municipal wastewaters have sufficient amounts of these inorganic elements, but care must be taken to assure sufficient amounts in industrial applications or laboratory experiments.

7–12 ANAEROBIC FERMENTATION AND OXIDATION

Anaerobic fermentation and oxidation processes are used primarily for the treatment of waste sludge (see Fig. 7–24) and high-strength organic wastes. However, applications for dilute waste streams have also been demonstrated and are becoming more common. Anaerobic fermentation processes are advantageous because of the lower

Figure 7–24

Views of anaerobic digesters: (a) Kuwait City, Kuwait and (b) egg-shaped digester at Okinawa, Japan.

(a)

(b)

biomass yields and because energy, in the form of methane, can be recovered from the biological conversion of organic substrates. Although most fermentation processes are operated in the mesophilic temperature range (30 to 35°C), there is increased interest in thermophilic fermentation alone or before mesophilic fermentation. The latter is termed temperature phased anaerobic digestion (TPAD) and is typically designed with a sludge SRT of 3 to 7 d in the first thermophilic phase at 50 to 60°C and 7 to 15 d in the final mesophilic phase (Han and Dague, 1997). Thermophilic anaerobic digestion processes, considered in Chap. 14, are used to accomplish high pathogen kill to produce *Class A biosolids,* which can be used for unrestricted reuse applications.

For treating high-strength industrial wastewaters, anaerobic treatment has been shown to provide a very cost-effective alternative to aerobic processes with savings in energy, nutrient addition, and reactor volume. Because the effluent quality is not as good as that obtained with aerobic treatment, anaerobic treatment is commonly used as a pretreatment step prior to discharge to a municipal collection system or is followed by an aerobic process. Suspended and attached growth anaerobic treatment process designs for liquid streams are presented in Chap. 10 and anaerobic digester designs for sludge treatment are presented in Chap. 14.

Process Description

Three basic steps are involved in the overall anaerobic oxidation of a waste: (1) hydrolysis, (2) fermentation (also known as acidogenesis), and (3) methanogenesis. The three steps are illustrated schematically on Fig. 7–25. The starting point on the schematic for a particular application depends on the nature of the waste to be processed.

Hydrolysis. The first step for most fermentation processes, in which particulate material is converted to soluble compounds that can then be hydrolyzed further to simple monomers that are used by bacteria that perform fermentation, is termed *hydrolysis.* For some industrial wastewaters, fermentation may be the first step in the anaerobic process.

Fermentation. The second step is *fermentation* (also referred to as *acidogenesis*). In the fermentation process, amino acids, sugars, and some fatty acids are degraded further, as shown on Fig. 7–25. Organic substrates serve as both the electron donors and acceptors. The principal products of fermentation are acetate, hydrogen, CO_2, and propionate and butyrate. The propionate and butyrate are fermented further to also produce hydrogen, CO_2, and acetate. Thus, the final products of fermentation (acetate, hydrogen, and CO_2) are the precursors of methane formation (methanogenesis). The free energy change associated with the conversion of propionate and butyrate to acetate and hydrogen requires that hydrogen be at low concentrations in the system ($H_2 < 10^{-4}$ atm), or the reaction will not proceed (McCarty and Smith, 1986).

Methanogenesis. The third step, *methanogenesis,* is carried out by a group of organisms known collectively as methanogens. Two groups of methanogenic organisms are involved in methane production. One group, termed *aceticlastic methanogens,* split acetate into methane and carbon dioxide. The second group, termed hydrogen-utilizing methanogens, use hydrogen as the electron donor and CO_2 as the electron acceptor to produce methane. Bacteria within anaerobic processes, termed *acetogens,* are also able to use

Theoretical
stages

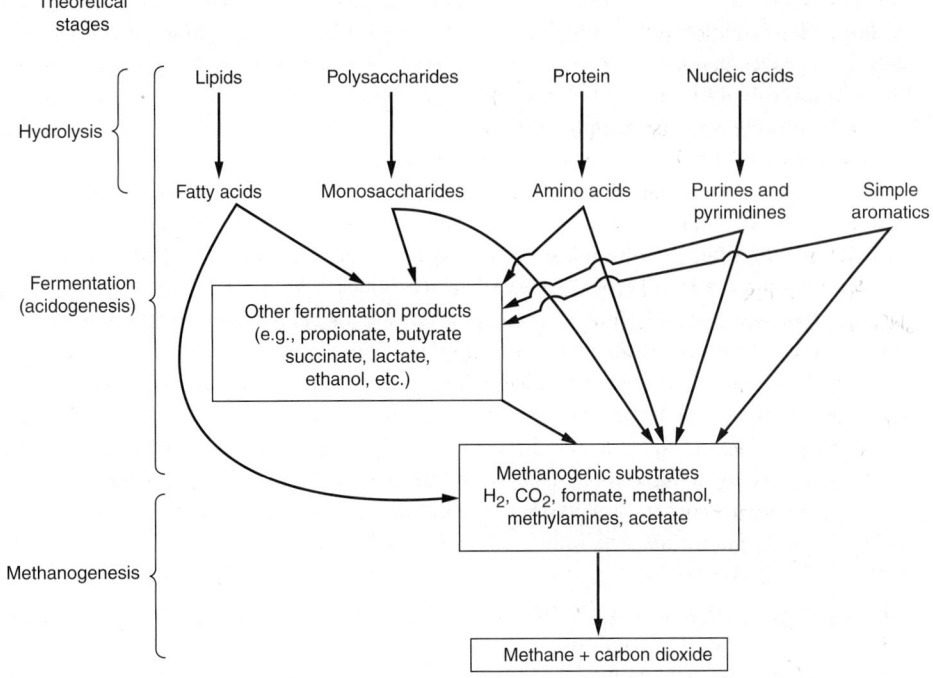

Figure 7–25

Anaerobic process schematic of hydrolysis, fermentation, and methanogenesis. *(Adapted from McCarty and Smith, 1991.)*

Figure 7–26

Carbon and hydrogen flow in anaerobic digestion process. The given percentage values are based on COD. *(Adapted from Jeris and McCarty, 1963 and McCarty, 1981.)*

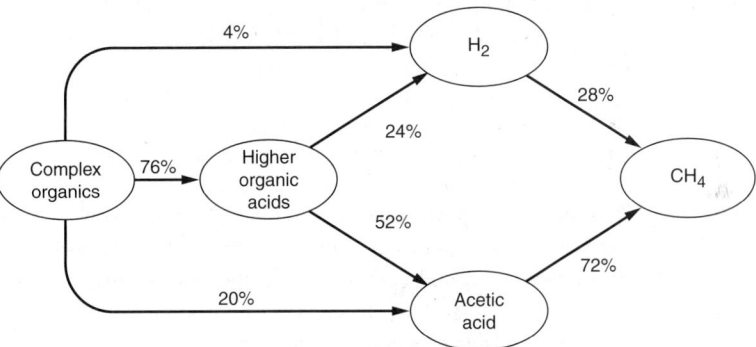

CO_2 to oxidize hydrogen and form acetic acid. However, the acetic acid will be converted to methane, so the impact of this reaction is minor. As shown on Fig. 7–26, about 72 percent of the methane produced in anaerobic digestion is from acetate formation.

Microbiology

The group of nonmethanogenic microorganisms responsible for hydrolysis and fermentation consists of facultative and obligate anaerobic bacteria. Organisms isolated from anaerobic digesters include *Clostridium* spp., *Peptococcus anaerobus, Bifidobac-terium* spp., *Desulphovibrio* spp., *Corynebacterium* spp., *Lactobacillus, Actinomyces, Staphylococcus,* and *Escherichia coli.* Other physiological groups present include those producing proteolytic, lipolytic, ureolytic, or cellulytic enzymes.

The microorganisms responsible for methane production, classified as archaea, are strict obligate anaerobes. Many of the methanogenic organisms identified in anaerobic digesters are similar to those found in the stomachs of ruminant animals and in organic sediments taken from lakes and rivers. The principal genera of microorganisms that have been identified at mesophilic conditions include the rods (*Methanobacterium, Methanobacillus*) and spheres (*Methanococcus, Methanothrix,* and *Methanosarcina*).

Methanosarcina and *Methanothrix* (also termed *Methanosaeta*) are the only organisms able to use acetate to produce methane and carbon dioxide. The other organisms oxidize hydrogen with carbon dioxide as the electron acceptor to produce methane. The acetate-utilizing methanogens were also observed in thermophilic reactors (van Lier, 1996; Zinder and Koch, 1984; and Ahring, 1995). Some species of *Methanosarcina* were inhibited by temperature at 65°C, while others were not, but no inhibition of *Methanothrix* was shown. For hydrogen-utilizing methanogens at temperatures above 60°C, *Methanobacterium* was found to be very abundant.

Syntrophic Relationships in Fermentation.   The methanogens and the acidogens form a syntrophic (mutually beneficial) relationship in which the methanogens convert fermentation end products such as hydrogen, formate, and acetate to methane and carbon dioxide. Because the methanogens are able to maintain an extremely low partial pressure of H_2, the equilibrium of the fermentation reactions is shifted toward the formation of more oxidized end products (e.g., formate and acetate). The utilization of the hydrogen produced by the acidogens and other anaerobes by the methanogens is termed *interspecies hydrogen transfer.* In effect, the methanogenic organisms serve as a hydrogen sink that allows the fermentation reactions to proceed. If process upsets occur and the methanogenic organisms do not utilize the hydrogen produced fast enough, the propionate and butyrate fermentation will be slowed with the accumulation of volatile fatty acids in the anaerobic reactor and a possible reduction in pH.

Nuisance Organisms.   Nuisance organisms in anaerobic operations are the sulfate-reducing bacteria, which can be a problem when the wastewater contains significant concentrations of sulfate. These organisms can reduce sulfate to sulfide, which can be toxic to methanogenic bacteria at high enough concentrations. Where high sulfide concentrations occur, one solution is to add iron at controlled amounts to form iron sulfide precipitate. Sulfate-reducing bacteria, obligate anaerobes of the domain Bacteria, are morphologically diverse, but share the common characteristic of being able to use sulfate as an electron acceptor and are divided into one of two groups depending on whether they produce fatty acids or use acetate. Group I sulfate reducers can use a diverse array of organic compounds as their electron donor, oxidizing them to acetate and reducing sulfate to sulfide. A common genus found in anaerobic biochemical operations is *Desulfovibrio.* Group II sulfate reducers oxidize fatty acids, particularly acetate, to carbon dioxide, while reducing sulfate to sulfide. A bacteria commonly found in this group is in the genus *Desulfobacter.*

Stoichiometry of Anaerobic Fermentation and Oxidation

A limited number of substrates are used by the methanogenic organisms and reactions defined as CO_2, and methyl group type reactions are shown as follows (Madigan et al., 1997), involving the oxidation of hydrogen, formic acid, carbon monoxide, methanol, methylamine, and acetate, respectively.

$$4H_2 + CO_2 \rightarrow CH_4 + 2H_2O \tag{7-111}$$

$$4HCOO^- + 4H^+ \rightarrow CH_4 + 3CO_2 + 2H_2O \tag{7-112}$$

$$4CO + 2H_2O \rightarrow CH_4 + 3CO_2 \tag{7-113}$$

$$4CH_3OH \rightarrow 3CH_4 + CO_2 + 2H_2O \tag{7-114}$$

$$4(CH_3)_3N + H_2O \rightarrow 9CH_4 + 3CO_2 + 6H_2O + 4NH_3 \tag{7-115}$$

$$CH_3COOH \rightarrow CH_4 + CO_2 \tag{7-116}$$

In the reaction for the aceticlastic methanogens as given by Eq. (7–116), the acetate is cleaved to form methane and carbon dioxide.

A COD balance can be used to account for the changes in COD during fermentation. Instead of oxygen accounting for the change in COD, the COD loss in the anaerobic reactor is accounted for by the methane production. By stoichiometry the COD equivalent of methane can be determined. The COD of methane is the amount of oxygen needed to oxidize methane to carbon dioxide and water.

$$CH_4 + 2O_2 \rightarrow CO_2 + 2H_2O \tag{7-117}$$

From the above, the COD per mole of methane is $2(32 \text{ g } O_2/\text{mole}) = 64 \text{ g } O_2/\text{mole}$ CH_4. The volume of methane per mole at standard conditions (0°C and 1 atm) is 22.414 L, so the CH_4 equivalent of COD converted under anaerobic conditions is 22.414/64 = 0.35 L CH_4/g COD.

EXAMPLE 7–9 **Prediction of Methane Gas Production** An anaerobic reactor, operated at 35°C, processes a wastewater stream with a flow of 3000 m^3/d and a bsCOD concentration of 5000 g/m^3. At 95 percent bsCOD removal and a net biomass synthesis yield of 0.04 g VSS/g COD used, what is the amount of methane produced in m^3/d?

Solution

1. Prepare a steady-state mass balance for COD to determine the amount of the influent COD converted to methane.
    a. The required steady-state mass balance is

    $$0 = \begin{array}{c} \text{Influent} \\ \text{COD} \end{array} - \begin{array}{c} \text{portion of} \\ \text{influent COD} \\ \text{in effluent} \end{array} - \begin{array}{c} \text{influent COD} \\ \text{converted to} \\ \text{cell tissue} \end{array} - \begin{array}{c} \text{influent COD} \\ \text{converted to} \\ \text{methane} \end{array}$$

    $$COD_{in} = COD_{eff} + COD_{VSS} + COD_{methane}$$

    b. Determine the values of the individual mass balance terms.

$$COD_{in} = (5000 \text{ g/m}^3)(3000 \text{ m}^3/\text{d}) = 15{,}000{,}000 \text{ g/d}$$

$$COD_{eff} = (1 - 0.95)(5000 \text{ g/m}^3)(3000 \text{ m}^3/\text{d}) = 750{,}000 \text{ g/d}$$

$$COD_{VSS} = (1.42 \text{ g COD/g VSS})(0.04 \text{ g VSS/g COD})(0.95)(15{,}000{,}000 \text{ g/d})$$
$$= 809{,}400 \text{ g/d}$$

    c. Solve for the COD converted to methane.

$$COD_{methane} = 15{,}000{,}000 - 750{,}000 - 809{,}400 = 13{,}440{,}600 \text{ g/d}$$

2. Determine the amount of methane produced at 35°C.
    a. Determine the volume of gas occupied by 1 mole of gas at 35°C.

$$V = \frac{nRT}{P}$$

$$V = \frac{(1 \text{ mole})(0.082057 \text{ atm·L/mole·K})[(273.15 + 35)\text{K}]}{1.0 \text{ atm}}$$

$$= 25.29 \text{ L}$$

    b. The CH_4 equivalent of COD converted under anaerobic conditions is (25.29 L/mole)/(64 g COD/mole CH_4) = 0.40 L CH_4/g COD.

    c. Methane production

$$CH_4 \text{ production} = (13{,}440{,}600 \text{ g COD/d})(0.40 \text{ L } CH_4/\text{g COD})(1 \text{ m}^3/10^3 \text{ L})$$

$$= 5376 \text{ m}^3/\text{d}$$

At 65 percent methane the total gas flow = (5376 m³/d)/0.65

$$= 8271 \text{ m}^3/\text{d}$$

Comment   It is important to determine the volume occupied by the gas at the actual operating temperature.

Growth Kinetics

In anaerobic processes two rate-limiting concepts are important: (1) the hydrolysis conversion rate and (2) the soluble substrate utilization rate for fermentation and methanogenesis. The hydrolysis of colloidal and solid particles does not affect the process operation and stability but does affect the total amount of solids converted. In anaerobic digestion processes used for municipal waste sludges, greater than 30 days detention time is needed to approach full conversion of solids. The soluble substrate utilization kinetics are of great concern to develop a stable anaerobic process.

    Because of the relatively low free energy change for anaerobic reactions, growth yield coefficients are considerably lower than the corresponding values for aerobic oxidation. Typical synthesis yield and endogenous decay coefficients for fermentation and methanogenic anaerobic reactions are $Y = 0.10$ and 0.04 g VSS/g COD and $k_d = 0.04$ and 0.02 g VSS/g VSS·d, respectively.

The process is more stable when the volatile fatty acid (VFA) concentrations approach a minimal level, which can be taken as an indication that a sufficient methanogenic population exists and sufficient time is available to minimize hydrogen and VFA concentrations. The rate-limiting step is the conversion of VFAs by the methanogenic organisms and not the fermentation of soluble substrates by the fermenting bacteria. Thus, the methanogenic growth kinetics are of most interest in anaerobic process designs. Appropriate system SRTs are selected based on kinetics and treatment goals. At 20, 25, and 35°C, the washout or SRT_{min} values for methanogenesis are 7.8, 5.9, and 3.2 d, respectively (Lawrence and McCarty, 1970). Thus, with a factor of safety of 5, design SRT values would be about 40, 30, and 15 d, respectively, for a suspended growth process. Safety factors higher than 5 have been used to provide a more stable process (Parker and Owen, 1986).

Environmental Factors

Anaerobic processes are sensitive to pH and inhibitory substances. A pH value near neutral is preferred and below 6.8 the methanogenic activity is inhibited. Because of the high CO_2 content in the gases developed in anaerobic processes (30 to 35 percent CO_2), a high alkalinity is needed to assure pH near neutrality. An alkalinity concentration in the range of 3000 to 5000 mg/L as $CaCO_3$ is often found. For sludge digestion sufficient alkalinity is produced by the breakdown of protein and amino acids to produce NH_3, which combines with CO_2 and H_2O to form alkalinity as $NH_4(HCO_3)$. For industrial wastewater applications, especially for waste containing mainly carbohydrates, it is necessary to add alkalinity for pH control. Substances inhibitory to anaerobic processes (e.g., NH_3, H_2S, and various other inorganic and organic compounds) are considered in Chap. 10.

7–13 BIOLOGICAL REMOVAL OF TOXIC AND RECALCITRANT ORGANIC COMPOUNDS

Most of the organic compounds in domestic wastewater and some in industrial wastewaters are of natural origin and can be degraded by common bacteria in aerobic or anaerobic processes. However, currently there are over 70,000 synthetic organic chemicals, termed *xenobiotic compounds,* in general use (Schwarzenbach et al., 1993). Unfortunately, some of these organic compounds pose unique problems in wastewater treatment, due to their resistance to biodegradation and potential toxicity to the environment and human health. Organic compounds that are difficult to treat in conventional biological treatment processes are termed *refractory.* In addition, there are naturally occurring substances, such as those found in petroleum products, that are of similar concern. Examples of petroleum compounds and synthetic organic compounds found in different types of wastewater are reported in Table 7–10.

Development of Biological Treatment Methods

Since the early 1970s, information and knowledge related to the biodegradation of toxic and refractory compounds has increased significantly, based on work with specific

Table 7-10
Examples of toxic and recalcitrant organic compounds found in wastewater[a]

Type of waste	Types of organic compounds
Petroleum	Alkanes, alkenes, polyaromatic hydrocarbons, monocyclic aromatics—benzene, toluene, ethylbenzene, xylenes, naphthenes
Nonhalogenated solvents	Alcohols, ketones, esters, ethers, aromatic and aliphatic hydrocarbons, glycols, amines
Halogenated solvents	Chlorinated methanes—methylene chloride, chloroform, carbon tetrachloride; chlorinated ethenes—tetrachloroethene, trichloroethene; chlorinated ethanes—trichloroethane; chlorinated benzenes
Pesticides, including insecticides, herbicides, fungicides	Organochloride compounds, organophosphate compounds, carbamate esters, phenyl ethers, creosotes, chlorinated phenols
Munitions and explosives	Nitroaromatics, trinitrotoluene, nitramines, nitrate esters
Industrial intermediates	Phthalate esters, benzene, phenol, chlorobenzenes, chlorophenols, xylenes
Transformer and hydraulic fluids	Polychlorinated biphenyls
Production products	Dioxin, furans

[a] Adapted from Watts (1997).

industrial wastewaters (i.e., petrochemical, textile, pesticide, pulp and paper, and pharmaceutical industries). In addition, since the 1980s, significant progress has also been made on the biodegradation of organic substances found at hazardous waste sites. Work in both of these fields has expanded knowledge on the capabilities and limitations of biodegradation. With a few exceptions most organic compounds can be biodegraded eventually, but in some cases the rates may be slow, unique environmental conditions may be required (i.e., redox potential, pH, temperature), fungi may be needed instead of prokaryotes, or specific bacteria capable of degrading the xenobiotic compounds may be needed. For example, anaerobic degradation of polychlorinated biphenyls (PCB) occurred using bacteria seed from sediment in the Hudson River where PCB had accumulated over decades, but after 1.5 years of exposure in a laboratory anaerobic digester used to treat municipal wastewater plant sludge, bacteria could not be developed to degrade PCB (Ballapragada et al., 1998).

Importance of Specific Microorganisms. The ability to degrade toxic and recalcitrant compounds will depend primarily on the presence of appropriate microorganism(s) and acclimation time. In some cases, special seed sources are needed to provide the necessary microorganisms. Once the critical microorganism is present, long-term exposure to the organic compound may be needed to induce and sustain the enzymes and bacteria required for degradation. Acclimation times can vary from hours to weeks depending on the microorganism population and organic compound. Melcer et al. (1994) found that a period of 3 weeks was required before complete removal of dichlorobenzene (DCB) occurred in a municipal activated-sludge plant. They noted that intermittent addition of DCB resulted in much lower treatment efficiencies. Without acclimation and no biodegradation, the DCB was removed from the activated-sludge aeration tank by volatilization as described in Chap. 5. Strand et al. (1999) showed that after 4 weeks of constant exposure to dinitrophenol in a laboratory activated-sludge process, seeded from a municipal wastewater plant, dinitrophenol degradation increased from 0 to 98 percent. When dinitrophenol was not added to the process, the ability to degrade dinitrophenol was eventually lost. Thus, it appears that a relatively constant supply of toxic and recalcitrant organic compounds can lead to better biodegradation performance than intermittent additions.

Biodegradation Pathways. The three principal types of degradation pathways that have been observed are (1) the compound serves as a growth substrate; (2) the organic compound provides an electron acceptor; and (3) the organic compound is degraded by cometabolic degradation. In cometabolic degradation, the compound that is degraded is not part of the microorganism's metabolism. Degradation of the compound is brought about by a nonspecific enzyme and provides no benefit to the cell growth. Complete biodegradation of toxic and recalcitrant organic compounds to harmless end products such as CO_2 and H_2O or methane may not always occur, and instead biotransformation to a different organic compound is possible. Care must be taken to determine if the organic compound produced is innocuous, or just as harmful as or more harmful than the initial compound.

Anaerobic Degradation

Many toxic and recalcitrant organic compounds are degraded under anaerobic conditions, with the compound serving as a growth substrate with fermentation and ultimate methane production. Typical examples include nonhalogenated aromatic and aliphatic compounds such as phenol, toluene, alcohols, and ketones. However, most chlorinated organic compounds are not attacked easily under anaerobic conditions and do not serve as growth substrates. Fortuitously, many of these compounds also serve as electron acceptors in anaerobic oxidation reduction reactions. Most of the work and application for anaerobic degradation of chlorinated organic compounds have been related to subsurface contamination of chlorinated solvents at hazardous waste sites (McCarty, 1999).

Examples of chlorinated compounds degraded under anaerobic conditions include tetrachloroethene, trichloroethene, carbon tetrachloride, trichlorobenzene, pentachlorophenol, chlorohydrocarbons, and PCBs. The chlorinated compound serves as the electron acceptor, and hydrogen produced from fermentation reactions provides the main electron donor. Hydrogen replaces chlorine in the molecule, and such reactions have generally been referred to as *anaerobic dehalogenation* or *anaerobic dechlorination*. For example, dechlorination of tetrachloroethene proceeds sequentially with a loss of chlorine in each step via trichloroethene to dichloroethene to vinyl chloride and finally to ethene.

As the number of chlorine molecules on the organic molecule decreases, the reactions tend to be slower and less complete. Dechlorination of tetrachloroethene, trichlorobenzene, and pentachlorophenol has been demonstrated in lab-scale anaerobic digesters (Ballapragada et al., 1998) treating municipal primary and secondary sludge. However, the reaction rates were slow with mono- and dichlorophenol and mono-I dichlorobenzenes remaining. Conversion of tetrachloroethene to vinyl chloride and ethene occurred in the digesters after one year of acclimatization and constant exposure of the chloroethenes.

Aerobic Biodegradation

With proper environmental conditions, seed source, and acclimation time, a wide range of toxic and recalcitrant organic compounds have been found to serve as growth substrates for heterotrophic bacteria. Such compounds include phenol, benzene, toluene, polyaromatic hydrocarbons, pesticides, gasoline, alcohols, ketones, methylene chloride, vinyl chloride, munitions compounds, and chlorinated phenols. However, many chlorinated organic compounds cannot be attacked readily by aerobic heterotrophic bacteria and thus do not serve as growth substrates. Some of the lesser chlorinated compounds, such as dichloromethane, 1,2-dichloroethane, and vinyl chloride can be used as growth substrates by aerobic bacteria. Fortunately, a number of chlorinated organic compounds are degradable by cometabolic degradation. It should be noted that organic compounds that are saturated fully with chlorine are degraded only by anaerobic dechlorination (Stensel and Bielefeldt, 1997).

Chlorinated organic compounds that have been degraded by cometabolic degradation include trichloroethene, dichloroethene, vinyl chloride, chloroform, dichloromethane, and trichloroethane. Cometabolic degradation is possible by bacteria that produce nonspecific mono-oxygenase or dioxygenase enzymes. These enzymes mediate a reaction with oxygen and hydrogen and change the structure of the chlorinated compound. Bacteria that produce oxygenase enzymes oxidize certain substrates that induce the enzyme. Oxygenase-producing bacteria include methanotrophic bacteria that oxidize methane, a number of bacteria that can oxidize phenol or toluene, a number of bacteria that can oxidize propane, and nitrifying bacteria that oxidize ammonia to nitrite.

The reaction of the nonspecific oxygenase enzyme with the organic chlorinated compound typically produces an intermediate compound that is degraded by other aerobic heterotrophic bacteria in the biological consortia. Various reactor designs have been developed to apply this biological process for treatment of contaminated groundwater or vapor extraction gas streams (Lee et al., 2000). While such reactions are possible in municipal and industrial biological wastewater treatment processes, a large amount of the chlorinated organic compounds that may be present are more likely lost from the process by volatilization during aeration, because of their high volatility and the minimal potential for cometabolic bacteria to be present.

Abiotic Losses

Due to concerns about environmental and health effects of toxic and recalcitrant compounds, it is important to understand their fate and transport in biological treatment processes. For many toxic and recalcitrant organic compounds entering biological wastewater treatment processes, nonbiological or abiotic losses may be more significant than biodegradation. Abiotic losses include adsorption of the compound to the mixed-liquor solids in the reactor with subsequent transport out of the system by the waste sludge and volatilization with release of the compound to the surrounding atmosphere.

Losses Due to Adsorption.   For certain compounds, removal by partitioning (i.e., adsorbing) onto the biomass can be more significant than biodegradation or volatilization. To describe solids partitioning, the Freundlich Isotherm model (see Sec. 4–8 in Chap. 4

and Sec. 11–7 in Chap. 11) is modified to a general linear equilibrium relationship ($n = 1$) for adsorption to solids at relatively low liquid organic concentrations:

$$q = K_p S \tag{7–118}$$

where q = g organic adsorbed/g adsorbent
$\quad\quad K_p$ = partition coefficient, L/g
$\quad\quad S$ = concentration of organic compound in liquid, g/L

Absorption of organic compounds in biological treatment processes has been observed to be relatively fast (Melcer et al., 1994), so that Eq. (7–118) can be used to describe the distribution of the compound between the solid biomass and liquid phases as a function of the partition coefficient K_p for the compound. The value for K_p depends on the hydrophobic nature of the compound and the adsorption characteristics of the solids. Solids with high carbon content and greater surface area result in higher K_p values. The following equation by Dobbs et al. (1989) can be used to obtain an estimate of the partition coefficient for wastewater treatment process solids as a function of the compound octanol-water partition coefficient K_{ow}:

$$\log K_p = 0.58 \log K_{ow} + 1.14 \tag{7–119}$$

where K_p = partition coefficient, L/g
$\quad\quad K_{ow}$ = octanol/water partition coefficient (dimensionless)

Octanol-water partition coefficient values have been developed for many organic compounds (Schwarzenbach et al., 1993; LaGrega et al., 2001). Octanol/water partition coefficients for selected compounds are given in Table 5–36 in Chap. 5. The test to determine K_{ow} involves measuring the concentration of the organic compound in an octanol/water mixture after quiescent separation of the octanol layer above the water layer. Greater amounts of more hydrophobic compounds will be found in the octanol layer, and these compounds will have greater K_p values in solids/water mixtures. Ranges of K_p values for various types of compounds are shown in Table 7–11. Based on the information given in Table 7–11, benzopyrene and PCBs are more likely to be found on solids than in the liquid than compounds like benzene and trichloroethene, as

Table 7-11
Comparison of selected estimated partition coefficients (K_p) values for different types of organic compounds[a]

Organic compound	K_p, L/g
Benzene	0.23
Dinitrotoluene	0.29
Trichloroethene	0.33
Dieldrin	1.48
Phenanthrene	5.33
Pentachlorophenol	10.96
Polychlorinated biphenyl	43.87
Benzopyrene	45.15

[a] Adapted from LaGrega et al. (2001).

their K_p values are greater by a factor of about 150. Such high partition coefficients result in very low liquid concentrations, which minimize the amount of the organic compound lost by biodegradation and volatilization.

Using the equilibrium partition coefficient, the amount of organic compound removed by sludge wasting can be estimated from the ratio of the mass of organic compound adsorbed to the mass of solids wasted per day:

$$q = \frac{r_{ad}}{r_{X,w}} \tag{7-120}$$

where q = g organic compound adsorbed/g solids
$\quad r_{ad}$ = rate of organic compound absorbed daily, g/d
$\quad r_{X,w}$ = rate of solids wasted daily, g/d

By substituting Eq. (7-118) for q and solving for r_{ad}, the amount of compound lost daily due to adsorption is

$$r_{ad} = -r_{X,w}K_p S \tag{7-121}$$

At steady state, the amount of solids wasted daily is related to the average SRT value for the activated-sludge system as given by Eq. (7-45):

$$r_{X,w} = \frac{X_T V}{\text{SRT}} \tag{7-122}$$

Substituting Eq. (7-122) into Eq. (7-121) yields the following expression for mass loss due to adsorption:

$$r_{ad} = -\frac{X_T V K_p S}{\text{SRT}} \tag{7-123}$$

Losses Due to Volatilization. The removal of volatile organic compounds due to aeration (volatilization) was discussed previously in Sec. 5-13 in Chap. 5 and is reviewed briefly here.

$$r_{sv} = -K_L a_s(S) \tag{7-124}$$

where r_{sv} = loss due to volatilization, mg/L·d
$\quad K_L a_s = K_L a$ of organic compound, d^{-1}
$\quad S$ = concentration of organic compound in liquid, mg/L

Modeling Biotic and Abiotic Losses

A number of models have been developed and evaluated that account for the fate of recalcitrant organic compounds in biological treatment processes (Melcer et al., 1995; Melcer et al., 1994; Monteith et al., 1995; Parker et al., 1993; Grady et al., 1997; and Lee et al., 1998a). In general, the models contain basic mechanisms and mass balances that account for the mass rate of the organic compound entering the treatment process, and leaving in the liquid effluent, by biodegradation, by volatilization, and by adsorption on waste solids.

In the following discussion, all of these mechanisms as discussed above are combined into a general model to describe the fate of specific compounds in a biological treatment process. The following steady-state mass balance (i.e., accumulation = 0) across a complete-mix activated-sludge process can be prepared to predict the fate of an organic compound subject to biotic and abiotic processes.

	organic	loss of organic	loss of organic	loss of organic	loss of organic
0 =	constituent	− constituent due	− constituent due	− constituent due	− constituent
	in influent	to biodegradation	to sorption	to volatilization	in effluent

$$QS_o = r_{su} + r_{ad} + r_{sv} + QS \qquad (7\text{--}125)$$

where QS_o = mass of compound in wastewater influent, g/d
$\quad r_{su}$ = biodegradation rate, g/d
$\quad r_{ad}$ = solids adsorption rate, g/d
$\quad r_{sv}$ = volatilization rate, g/d
$\quad QS$ = mass of compound in wastewater effluent, g/d

Substituting the appropriate reaction terms for each component of the mass balance yields the following expression:

$$QS_o = \left(\frac{1}{Y}\right)\frac{\mu_m S}{(K_s + S)}(X_s)(V) + \frac{X_T V K_p S}{SRT} + K_L a_s SV + QS \qquad (7\text{--}126)$$

The fate of the compound in the influent wastewater as a function of the solids concentration, liquid concentration, τ, SRT, and rate terms is obtained by dividing Eq. (7–126) by Q:

$$S_o = \left(\frac{1}{Y}\right)\frac{\mu_m S}{(K_s + S)}(X_s)\tau + K_p SX_T\left(\frac{\tau}{SRT}\right) + K_L a_s S(\tau) + S \qquad (7\text{--}127)$$

Note in Eqs. (7–126) and (7–127), X_s is the biomass concentration capable of degrading the specific organic compound, and X_T is the total MLVSS concentration that includes all the biomass grown on various substrates plus the nonbiodegradable VSS. The value of X_s can be calculated as a function of the amount of substrate that is biodegraded and kinetic coefficients, and the system τ, and SRT. The following expression is used to calculate X_s at steady state for a complete-mixed reactor with consideration for losses of the substrate by volatilization and adsorption:

$$X_s = \frac{Y[(S_o - S) - K_p SX_T(\tau/SRT) - K_L a_s S(\tau)]}{k_d(\tau) + (\tau/SRT)} \qquad (7\text{--}128)$$

For most cases where biodegradation is occurring and the loss is not overwhelming due to volatilization and/or solids adsorption, the liquid constituent concentration, based on Eq. (7–40), for a complete-mix reactor at steady state is given by

$$S = \frac{K_s[1 + (k_d)SRT]}{SRT(\mu_m - k_d) - 1} \qquad (7\text{--}129)$$

For the other exceptional cases, Eq. (7–127) and (7–128) can be solved simultaneously.

The approach outlined above can be used to estimate the fate of an organic compound in a complete-mix activated-sludge reactor assuming steady-state conditions, a constant input of the organic constituent, and a fully acclimated culture. Predicting the fate of an organic constituent in an activated-sludge treatment process is illustrated in Example 7–10.

EXAMPLE 7–10 **Predicting the Fate of Benzene in an Activated-Sludge Treatment Process**
A complete-mix activated-sludge system is used to treat domestic wastewater, but receives a wastewater discharge containing benzene. Given the following information on the activated-sludge process design, and biotic and abiotic rate information for benzene, what is the effluent soluble benzene concentration and the relative amounts of benzene lost through biodegradation, sorption to solids, volatilization, and in the liquid effluent?

1. Influent benzene concentration, $S_o = 2.0$ g/m^3
2. System SRT $= 6.0$ d
3. Aeration tank $\tau = 0.25$ d
4. MLVSS concentration, $X_T = 2500$ g/m^3
5. $K_p = 0.234 \times 10^{-3}$ m^3/g
6. $K_L a_s = 3/\text{h} = 72/\text{d}$
7. $\mu_m = 2.0$ g VSS/g VSS·d
8. $K_s = 0.50$ g/m^3
9. $k_d = 0.10$ VSS/g VSS·d
10. $Y = 0.60$ g VSS/g benzene

Solution

1. Determine the liquid benzene concentration using Eq. (7–129).

$$S = \frac{(0.5 \text{ g/m}^3)[1 + (0.10 \text{ g VSS/g VSS·d})(6.0 \text{ d})]}{(6.0 \text{ d})[(2.0 - 0.10)\text{g VSS/g VSS·d}] - 1} = 0.077 \text{ g/m}^3$$

2. Determine concentration of biomass degrading benzene X_s from Eq. (7-128):

$$X_s = \frac{Y[(S_o - S) - K_p S X_T (\tau/\text{SRT}) - K_L a_s S(\tau)]}{k_d(\tau) + (\tau/\text{SRT})}$$

$$X_s = \frac{(0.60 \text{ g/g})\{[(2.0 - 0.077)\text{g/m}^3] - (0.234 \times 10^{-3} \text{ m}^3/\text{g})(0.077 \text{ g/m}^3)}{(2500 \text{ g/m}^3)(0.25 \text{ d/6.0 d}) - (72/\text{d})(0.077 \text{ g/m}^3)(0.25 \text{ d})\}}{(0.10 \text{ g/g·d})(0.25 \text{ d}) + (0.25 \text{ d/6.0 d})}$$

$$= 4.83 \text{ g/m}^3$$

3. Determine the loss of benzene due to biodegradation [1st term in Eq. (7–127)]:

$$\left(\frac{1}{Y}\right)\frac{\mu_m(S)}{(K_s + S)}(X_s)\tau = \left[\frac{1}{(0.6 \text{ g/g})}\right]\left\{\frac{(2.0/\text{d})(0.077 \text{ g/m}^3)}{[(0.5 + 0.077) \text{ g/m}^3]}\right\}(4.83 \text{ g/m}^3)(0.25 \text{ d})$$

$$= 0.537 \text{ g/m}^3$$

4. Determine the loss of benzene due to sorption [2nd term in Eq. (7–127)]:

$$\frac{K_pS(X_T)\tau}{\text{SRT}} = \frac{(0.234 \times 10^{-3} \text{ m}^3/\text{g})(0.077 \text{ g/m}^3)(2500 \text{ g/m}^3)(0.25 \text{ d})}{6.0 \text{ d}}$$

$$= 0.0019 \text{ g/m}^3$$

5. Determine the loss of benzene due to volatilization [3rd term in Eq. (7–127)]:

$$(K_La_s)(S)(\tau) = (72/\text{d})(0.077 \text{ g/m}^3)(0.25 \text{ d}) = 1.386 \text{ g/m}^3$$

6. Summarize the losses due to the various mechanisms.

Pathway	Influent fate, g/m³	Fraction of total
Effluent	0.077	0.039
Biodegradation	0.537	0.268
Sorption	0.002	0.001
Volatilization	1.386	0.692
Total	2.002ᵃ	1.000

ᵃ0.002 is due to round-off error.

a. For benzene, which is a volatile organic compound with a low solids partition coefficient, 69.2 percent of the influent benzene is transferred to the atmosphere, 26.8 percent is biodegraded, 3.9 percent remains dissolved in the liquid effluent, and 0.1 percent is sorbed onto the solids leaving the process.

b. The effect of the aeration system benzene K_La value on the relative amounts biodegraded or stripped to the atmosphere is illustrated in the following summary table:

Parameter	Unit	K_La, h⁻¹			
		1.5	**3.0**	**4.0**	**6.3**
Effluent	g/m³	0.077	0.077	0.077	0.052
Effluent	%	3.9	3.9	3.9	2.6
Biodegraded	%	61.4	26.8	3.7	0.0
Volatilized	%	34.6	69.2	92.3	97.3

7–14 **BIOLOGICAL REMOVAL OF HEAVY METALS**

Metal removal in biological treatment processes is mainly by adsorption and complexation of the metals with the microorganisms. In addition, processes that result in transformations and precipitation of metals are possible. Microorganisms combine with metals and adsorb them to cell surfaces because of interactions between the metal ions and the negatively charged microbial surfaces. Metals may also be complexed by carboxyl groups found in microbial polysaccharides and other polymers, or absorbed by protein materials in the biological cell.

The removal of metals in biological processes has been found to fit adsorption characteristics displayed by the Freundlich isotherm model (see Sec. 11–7 in Chap. 11) (Mullen et al., 1989; Kunz et al., 1976). A significant amount of soluble metal removal has been observed in biological processes, with removals ranging from 50 to 98 percent depending on the initial metal concentration, the biological reactor solids concentrations, and system SRT. In anaerobic processes the reduction of sulfate to hydrogen sulfide can promote the precipitation of metal sulfides. A classic example is the addition of ferric or ferrous chloride to anaerobic digesters to remove sulfide toxicity by the formation of iron sulfide precipitates. The precipitation of heavy metals by hydrogen sulfide is discussed in Sec. 6–5 in Chap. 6.

PROBLEMS AND DISCUSSION TOPICS

7–1 Prepare a recipe for an inorganic medium to be used in a laboratory chemostat to grow 500, 1000, or 1200 mg VSS/d (value to be selected by instructor) of bacteria biomass, assuming that the chemical formula for the biomass can be described as $C_5H_7NO_2$. Determine the concentration of essential inorganic compounds as reported in Table 7–4 for a feed rate of 1 L/d. Assume that phosphorus is added as KH_2PO_4, sulfur as Na_2SO_4, nitrogen as NH_4Cl, and other cations associated with chloride.

7–2 Protein is a major component of bacterial enzymes. List the key cell components involved and the major steps that lead to protein production.

7–3 From the literature (e.g., *Journal of Applied and Environmental Microbiology*) identify the key physiological, metabolic characteristics, and phylogenetic classification of a bacteria that may have a role in biological wastewater treatment or toxic degradation. Cite a minimum of 3 references.

7–4 From the literature, describe an application using molecular biology (e.g., molecular probes or other methods) techniques that can be related to biological wastewater treatment. Cite a minimum of 3 references.

7–5 A 1-L sample contains 20, 28, or 36 g (value to be selected by instructor) of casein ($C_8H_{12}O_3N_2$). If 18 g of bacterial cell tissue ($C_5H_7NO_2$) is synthesized per 50 g of casein consumed, determine the amount of oxygen required to complete the oxidation of casein to end products and cell tissue. The end products of the oxidation are carbon dioxide (CO_2), ammonia (NH_3), and water. Assume that the nitrogen not incorporated in cell-tissue production will be converted to ammonia.

7-6 A complete-mix suspended growth reactor, without a clarifier and recycle, is used to treat a wastewater flow containing only soluble organic substances. The influent BOD and COD are as follows.

Influent	Unit	Wastewater		
		A	**B**	**C**
BOD	mg/L	200	180	220
COD	mg/L	450	450	480

If the effluent dissolved BOD concentration is 2 mg/L, and the effluent volatile suspended solids concentration is 100 mg/L, determine (wastewater to be selected by the instructor): (a) the observed yield in terms of g VSS/g BOD, g VSS/g COD, and g TSS/g BOD, (b) the effluent total sCOD concentration including nonbiodegradable dissolved COD, and (c) the fraction of the influent BOD that is oxidized to CO_2 and H_2O. Assume the biodegradable COD/BOD ratio is 1.6 and 1.42 g O_2 equivalent/g biomass.

7-7 An aerobic complete-mix reactor (no recycle) with a volume of 1000 L receives a 500 L/d wastewater flow and has an effluent soluble COD concentration of 10 mg/L. For one of the wastewaters with the characteristics given below (to be selected by instructor), determine: (a) the τ value for the reactor in days, (b) the oxygen used per day in (g/d), (c) the effluent volatile suspended solids concentration (assume biomass oxygen equivalent of 1.42 g O_2/g VSS), and (d) the observed yield in g VSS/g bsCOD removed.

Item	Unit	Wastewater		
		A	**B**	**C**
Influent sCOD	mg/L	1000	1800	600
Reactor oxygen uptake rate	mg/L·h	10	15	8

7-8 Using the half-reaction free energy values given in Table 7–7, calculate and compare the biomass yields (g VSS/g COD_r) for the degradation of methanol, carbohydrate mixture, or ethanol (constituent to be selected by instructor) with oxygen and nitrate as the electron acceptors. Assume ammonia is available for cell synthesis needs and 1.42 g O_2 equiv/g biomass.

7-9 Nitrate and sulfate are both available in an anaerobic laboratory chemostat with both nitrate-reducing and sulfate-reducing bacteria present. The chemostat is fed continuously a solution containing the electron acceptors in equal amounts, glucose, and a nutrient media. Which biological populations will remain after long-term operation? Explain.

7-10 For Example 7–3, use the half reactions to write a balanced equation of acetate oxidation by methanogenic bacteria.

7-11 For the synthesis yield values given in Table 7–8 for organic compound degradation, what are the respective f_e and f_s values?

7-12 Compare the end products of organic compound degradation under the following conditions and discuss how the bacterial synthesis yields are affected by them: aerobic (oxygen as electron acceptor), fermentation (organic compound as electron acceptor), and methanogenesis (CO_2 as electron acceptor).

7-13 If bacterial cells are of the coccus type with a diameter of 1, 1.2, or 1.5 μm and are 80 percent water with 90 percent of the dry weight as organic, determine (cell diameter to be selected by the instructor): (a) the volume and organic mass of one cell and (b) the number of cells present in one liter of a biomass suspension with a concentration of 100 mg VSS/L.

7-14 For aerobic bacteria with an assumed generation time of 20, 30, or 60 min (time to be selected by instructor), how many bacteria would be present after 12 hours, if 20 cells are present at time zero? Using the bacteria volume and mass from Problem 7–13 for a 1-μm diameter bacteria, what would be the dry weight of the bacteria after 12 hours in mg volatile suspended solids?

7-15 Consider a batch reaction with nitrifying bacteria in a chemostat. The initial concentration of nitrifying bacteria is 10 mg/L and the initial substrate concentration is 40 mg NH_4-N/L. The NH_4-N is oxidized to NO_2-N and the cell yield is 0.12 g VSS/g NH_4-N oxidized. Other kinetic coefficients related to substrate utilization and growth are one of the following selected by instructor:

Coefficient	Unit	Wastewater		
		A	B	C
μ_m	g VSS/g VSS·d	0.50	0.75	0.60
K_s	mg/L	0.50	0.50	0.75
k_d	g VSS/g VSS·d	0.01	0.08	0.04

What is the substrate and biomass concentration at 0.80 days? Plot the substrate and biomass concentration versus time up to 1.0 day. (Hint: one solution approach is to use a spreadsheet to solve for the biomass and substrate concentration at small time increments.)

7-16 Curves A and B represent the Monod kinetics for two different bacteria capable of degrading the same substrate. You are to operate a laboratory continuous flow CMAS reactor without recycle that is inoculated with bacteria A and B. In the first experiment (I) a high SRT is used (10 d or greater) and in the second (II) a very low SRT is used (about 1.1 d). Which bacteria will be dominant in experiments I and II? Explain why.

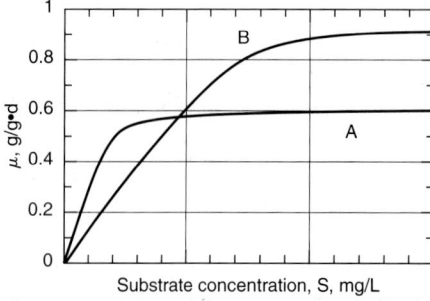

7-17   A complete-mix aerobic reactor without solids recycle is used to treat a wastewater containing 100 mg/L phenol (C_6H_6O) at 20°C. Using the following kinetic coefficients (coefficient set A, B, or C to be selected by instructor) determine: (a) the minimal hydraulic retention time τ in days at which the biomass can be washed out faster than they can grow, (b) the minimum τ value at 10°C, assuming the temperature-activity coefficient θ is 1.07 for k and 1.04 for k_d, (c) the effluent phenol and biomass concentration at a τ value of 4.0 d at 20°C, and (d) the amount of oxygen required in kg/d for a τ value of 4.0 d assuming a flow of 100 m³/d. Plot the phenol and biomass concentration and the amount of oxygen required versus τ in days, for τ from 3.3 to 15 d at 20°C.

		Wastewater		
Coefficient	Unit	A	B	C
k	g phenol/g VSS·d	0.90	0.80	0.90
K_s	mg phenol/L	0.20	0.15	0.18
Y	g VSS/g phenol	0.45	0.45	0.40
k_d	g VSS/g VSS·d	0.10	0.08	0.06

7-18   Laboratory test reactors have been operated at different SRT values at steady state to obtain biological kinetic coefficients for a wastewater with soluble constituents only. The reactors are complete-mix and aerated with clarifiers and solids recycle. The τ value in all cases is 0.167 d and the SRT values were varied for the five tests. The influent and effluent soluble COD and reactor MLVSS concentrations are summarized as follows:

Test no.	SRT, d	S_o, mg COD/L	S, mg COD/L	X, mg VSS/L
1	3.1	400	10.0	3950
2	2.1	400	14.3	2865
3	1.6	400	21.0	2100
4	0.8	400	49.5	1050
5	0.6	400	101.6	660

From these results determine the values for the biokinetic coefficients, k, K_s, μ_m, Y, and k_d. (Note: calculate the solids production at each SRT.)

7-19   The following data were obtained using four bench-scale continuous-flow activated-sludge units to treat a food-processing waste. Using these data, determine Y and k_d.

		Parameter	
Unit	X, g MLVSS/L	r_g, g MLVSS/L·d	U, g BOD$_5$/g MLVSS·d
1	18.81	0.88	0.17
2	7.35	1.19	0.41
3	7.65	1.42	0.40
4	2.89	1.56	1.09

7–20 Using the data given below for three different complete-mix activated-sludge reactors, determine (reactor to be selected by instructor): (a) the system SRT, (b) how much oxygen is required in kg/d if the effluent soluble COD concentration = 5 mg/L, and (c) the oxygen uptake rate, expressed in mg/L·h, at steady state in the aeration tank. Assume 1.42 g COD/g VSS.

Item	Unit	Reactor A	B	C
Aeration tank MLVSS	mg/L	3000	3000	3000
Aeration tank volume	m³	1000	1000	1000
Influent flow rate	m³/d	5000	5000	5000
Waste sludge flow rate	m³/d	59	45	65
Waste sludge VSS concentration	mg/L	8000	8000	8000
Influent soluble COD concentration	mg/L	400	400	400

7–21 A complete-mix activated-sludge process with secondary clarification and sludge recycle is used to treat a dairy wastewater at a flowrate of 1000 m³/d with a degradable influent COD of 3000 mg/L and BOD of 1875 mg/L. The MLSS concentration is 2800, 3300, or 3500 mg/L (MLSS value to be selected by instructor), MLVSS/MLSS ratio is 0.80, effluent TSS concentration is 20 mg/L, τ is 24 hours, recycle MLSS concentration is 10,000 mg/L, and waste sludge flowrate from the recycle line is 85.5 m³/d. Using the given information, determine: (a) the system SRT, the F/M ratio in g BOD/g MLVSS·d, and the volumetric BOD loading (kg/m³·d), (b) the observed yield in terms of g TSS/BOD and g TSS/g COD, and (c) the synthesis yield, assuming that $k_d = 0.10$ g VSS/g VSS·d and $f_d = 0.15$ g VSS/g VSS.

7–22 A conventional activated-sludge plant is operated at SRT values of 8, 10, or 12 d (value to be selected by instructor). The reactor volume is 8000 m³ and the MLSS concentration is 3000 mg/L. Determine (a) the sludge production rate, (b) the sludge wasting flowrate when wasting from the reactor, and (c) the sludge wasting flowrate when wasting from the recycle line. Assume that the concentration of suspended solids in the recycle is equal to 10,000 mg/L, and the solids loss in the secondary clarifier effluent is minor and can be neglected.

7–23 A complete-mix activated-sludge process with a clarifier and sludge recycle receives an influent wastewater flow of 2000 m³/d and influent particulate concentration of 400, 500, or 600 mg VSS/L (value to be selected by instructor) that is entirely biodegradable. The volume of the activated-sludge reactor is 500 m³. The biokinetic coefficients for particulate degradation (Eq. 7–20) are $k_p = 2.2$ g VSS/g biomass·d and $K_x = 0.15$ g VSS/g biomass. The yield and endogenous decay coefficients are 0.50 g biomass/g VSS and 0.10 g VSS/g VSS·d, respectively. Using the given information: (a) develop a steady-state mass balance for particulate removal in the activated-sludge system, (b) develop equations for the aeration tank particulate and biomass concentrations as a function of SRT (assume that the effluent flow contains no degradable particulates, particulates only leave the system via the waste sludge, and soluble COD is negligible), (c) determine the biomass and particulate concentrations in the aeration tank at SRT values of 3, 5, and 10 d, and (d) determine the percent removal of particulates at 3, 5, and 10 d.

7–24 The following sets of kinetic coefficients (to be selected by instructor) are given for the treatment of a municipal wastewater with an influent degradable COD of 300 mg/L and influent nbVSS concentration of 100 mg/L. Using these data and assuming the effluent degradable COD concentration is negligible compared to the amount of COD removed, prepare plots of (a) the observed yield (as g VSS/g COD) removed as a function of SRT and (b) the g oxygen used/g COD removed as a function of SRT. On the plot in Part (a) also show the fraction of the yield from cell debris and influent nbVSS.

		Coefficient set		
Coefficient	Unit	A	B	C
Y	g VSS/g COD	0.40	0.40	0.35
k_d	g VSS/g VSS·d	0.10	0.08	0.12
f_d	g VSS/g VSS	0.10	0.15	0.15

7–25 Design a complete-mix activated sludge process with recycle to treat an industrial wastewater with one of the following characteristics (to be selected by instructor) at peak month conditions.

		Wastewater		
Item	Unit	A	B	C
Flow	m³/d	4000	4300	4000
BOD	mg/L	800	600	1000
nbVSS	mg/L	200	200	200
TKN	mg/L	30	30	40
Total phosphorus	mg/L	8	8	6
Temperature	°C	15	15	15

The relevant biokinetic coefficients and operating conditions are:

$Y = 0.45$ g VSS/g COD   SRT = 10 d
$k_d = 0.10$ g VSS/g VSS·d   Return sludge = 8000 mg TSS/L
$\mu_m = 2.5$ g VSS/g VSS·d   Aeration tank MLSS = 2500 mg/L
$K_s = 20$ mg COD/L   Clarifier effluent TSS = 15 mg/L
$f_d = 0.10$ g VSS/g VSS
bCOD = 1.6 (BOD)

Using the given information and biokinetic coefficients, determine (a) the aeration tank volume (m³), the amount of waste solids produced/day (kg/d), the oxygen requirement (kg/d), the aeration tank oxygen uptake rate (mg/L·h), the effluent soluble BOD concentration, the return sludge recycle ratio for the following design conditions, and the MLVSS to MLSS ratio, and (b) whether supplemental nitrogen or phosphorus is required and, if so, how much in mg/L? Assume the biomass contains 12 percent nitrogen and 2 percent phosphorus on a volatile suspended solids basis. Assume no nitrification occurs.

7–26 For the same industrial wastewater application given in Problem 7–23, powdered activated carbon (PAC) is added to the influent at a dose of 50 mg/L to sorb potential toxic substances. The SRT is still held at 10 d. Determine the MLSS concentration, the MLVSS/MLSS ratio, and the total daily sludge production in kg TSS/d with the PAC addition.

7-27 A complete-mix activated-sludge system receives wastewater with one of the following characteristics (wastewater to be selected by instructor):

		Wastewater		
Item	Unit	A	B	C
Flow	m³/d	6000	6000	6000
Biodegradable COD	mg/L	300	400	500
Influent nbVSS	mg/L	100	100	150

The relevant design criteria are:

Flow = 6000 m³/d
Biodegradable COD = 300 mg/L
Influent nbVSS = 100 mg/L

The following biokinetic coefficients can be assumed:

Y = 0.40 g VSS/g COD
k_d = 0.10 g VSS/g VSS·d
f_d = 0.10 g VSS/g VSS
μ_m = 5.0 g VSS/g VSS·d
K_s = 20 mg COD/L

If the system aeration oxygen transfer capacity is 52 kg O_2/h, what maximum SRT can be used
so that the oxygen requirements can be met by the existing oxygen transfer capacity?

7-28 The kinetics for substrate utilization can be described by a first-order relationship (Eq. 7–18)
($r_{su} = -kSX$). (a) Using the given first-order kinetic relationship instead of the Michaelis-Menten
relationship for substrate utilization, derive a steady-state relationship that can be used to calculate the effluent soluble substrate concentration from a complete-mix suspended growth reactor.
Verify that Eq. (7–43) can be used to determine the biomass (X) concentration. (b) For the following reactor conditions and biokinetic information, determine the SRT needed to provide an
effluent soluble substrate concentration of 1.0 mg/L, and the biomass concentration.

S_o = 500 mg/L COD
τ = 0.25 d
Y = 0.50 g VSS/g COD removed
k_d = 0.06, 0.10, or 0.12 g VSS/g VSS·d (to be selected by instructor)
r_{su} = $-kXS$, where k = 0.504 g/g·d

7-29 Consider a biofilm treating a liquid stream containing acetate and dissolved oxygen with a stagnant liquid layer above the biofilm. (a) Using the stoichiometric relationship developed for the
biological degradation of acetate in Example 7–4, determine the maximum acetate concentration
in the bulk liquid that can be satisfied before the aerobic degradation in the biofilm is limited by
the surface flux rate of oxygen, where the bulk liquid DO concentration is 2.0, 3.0, or 4.0 mg/L
(DO value to be selected by instructor). (b) Compare these results to the results of NH_4-N oxidation in Example 7–7. Why is the bulk liquid NH_4-N concentration so much lower at a DO concentration of 2.0 mg/L?

Given:

Acetate diffusivity coefficient = 0.9 cm²/d
Oxygen diffusivity coefficient = 2.6 cm²/d

7–30   An activated-sludge system treating domestic wastewater is operated at a solids retention time of 10 d with a mixed-liquor temperature of 18°C. For many weeks nitrification has occurred, with an effluent NH_4-N concentration reported at less than 1.0 mg/L. After some time, the nitrification performance declines with effluent NH_4-N concentrations exceeding 10 mg/L. As the city engineer you are requested to investigate the cause of the decline in performance and to make recommendations for actions that will get the discharge quality back in compliance. Describe possible causes for the decline in nitrification efficiency and how you would evaluate the problem.

7–31   Using the half-reactions from Table 7–7 calculate the oxygen equivalent of nitrite (g O_2/g NO_2-N), for biological reaction with nitrite as the electron acceptor instead of oxygen.

7–32   An anoxic suspended growth reactor is operated at an SRT of 5.0, 10.0, or 15.0 d (to be selected by instructor) and acetate is added as the electron donor. Given the following kinetic coefficients for acetate under nitrate reduction conditions, determine how much acetate is needed, in kg/d, to remove 20 mg/L of nitrate in a treatment flowrate of 4000 m^3/d.

Y = 0.3 g VSS/g COD removed
k_d = 0.08 g VSS/g VSS·d
f_d = 0.10 g VSS/g VSS

7–33   Two complete-mix suspended growth laboratory reactors with sludge recycle fed the same synthetic wastewater are operated in parallel at the same aerobic SRT. One reactor has an anaerobic/aerobic sequence to promote biological phosphorus removal and the other is operated only with the aerobic portion. The influent flow contains 100, 200, or 300 mg/L acetate (as selected by instructor) for the organic carbon source. The phosphorus and volatile fraction contents of the two mixed liquors are as follows. The lower VSS/TSS ratio for the biological phosphorus removal reactor accounts for both polyphosphate and associated cation in the storage products.

Reactor	g P/g VSS	g VSS/g TSS
Aerobic only	0.015	0.85
Biological P removal	0.25	0.65

Using the following operating conditions and kinetic coefficients, how much phosphorus is removed from the influent for each system in mg/L, and what are the aerobic reactor MLVSS and MLSS concentrations? (Note: for this problem the coefficients are assumed equal for both types of organisms, but in practice they may be different).

Y = 0.40 g VSS/g COD
k_d = 0.10 g VSS/g VSS·d
SRT = 5 d
τ = 3 h
f_d = 0.10 g VSS/g VSS

7–34   A laboratory reactor is to be operated to study operating conditions that affect biological phosphorus removal. The influent phosphorus concentration will be 10, 20, or 30 mg/L (as selected by instructor). What minimum concentrations of magnesium, potassium, and calcium should be in the influent liquid?

7–35   An anaerobic treatment process is used to treat a flow of 500 m^3/d with an influent soluble COD concentration of 2000, 5000, or 9000 mg/L (value to be selected by instructor). The net biomass yield is 0.04 g VSS/g COD removed and 95 percent soluble COD removal occurs at a temperature of 30°C. Assuming the gas contains 65 percent methane, calculate the total gas flow in m³/d. What is the energy value of the gas produced in kJ/d? (The heat value of methane is 50.1 kJ/g at 30°C.)

7-36 Based on a review of the literature (cite a minimum of two references) explain the importance of the syntrophic relationship between methanogens and acid fermenters in an anaerobic process. What is the effect (increase, decrease, or remain the same) on the gas production rate, percent methane in the gas, volatile fatty acid concentration, and pH if an upset occurs to create an imbalance between fermenters and methanogens.

7-37 Modify Eqs. (7–127), (7–128), and (7–129) based on using first-order kinetics for the substrate removal rate versus the Monod growth kinetic model, where: $r_{su} = -kSX$. The first-order model is often used to describe the biodegradation kinetics of a number of priority pollutants.

7-38 Assume a complete-mix reactor is to be used to treat a wastewater containing a priority pollutant with the following characteristics and other easily degradable organic compounds. The priority pollutant is not very volatile so that losses due to stripping can be ignored. Using the following information determine (a) the fate of the compound in terms of biodegradation losses, and removal in the system effluent and waste sludge, and (b) the values computed in Part (a) if the value for μ_m is 3 times higher.

Design data and coefficients:

System SRT (to be selected by instructor) 5, 10, or 15 d
Reactor MLVSS = 2000 mg/L
Reactor τ = 0.25 d
Compound characteristics and biokinetic coefficients:
Influent concentration = 5.0 mg/L
$K_p = 15 \times 10^{-3}$ m³/g
$\mu_m = 2.0$ g VSS/g VSS·d
$K_s = 0.4$ g/m³
$Y = 0.6$ g VSS/g compound
$k_d = 0.08$ g VSS/g VSS·d

REFERENCES

Ahring, B. K. (1995) "Methanogenesis in Thermophilic Biogas Reactors," *Antonie van Leeuwenhoek,* vol. 67, pp. 91–102, The Netherlands.

Ardern, E., and W. T. Lockett (1914) "Experiments on the Oxidation of Sewage without the Aid of Filters," *Journal Society Chemical Industries,* vol. 33, pp. 523, 1122.

Atkinson, B., I. J. Davies, and S. Y. How (1974) "The Overall Rate of Substrate Uptake by Microbial Films," parts I and II, *Transactions Institute of Chemical Engineers* (British).

Bailey, J. E., and D. F. Ollis (1986) *Biochemical Engineering Fundamentals,* 2d ed., McGraw-Hill, New York.

Ballapragada, B., H. D. Stensel, J. F. Ferguson, V. S. Magar, and J. A. Puhakka (1998) *Toxic Chlorinated Compounds: Fate and Biodegradation in Anaerobic Digestion,* Project 91-TFT-3. Water Environment Research Foundation, Alexandria, VA.

Barker, P. S., and P. L. Dold (1997) "General Model for Biological Nutrient Removal in Activated Sludge Systems: Model Presentation," *Water Environment Research,* vol. 69, no 5, pp. 969–984.

Barnard, J. L. (1984) "Activated Primary Tanks for Phosphate Removal," *Water SA,* vol. 10, p. 3.

Barth, E. F., R. C. Brenner, and R. F. Lewis (1968) "Chemical Biological Control of Nitrogen and Phosphorus in Wastewater Effluent," *Journal Water Pollution Control Federation,* vol. 40, p. 2040.

Bielefeldt, A. R., and H. D. Stensel (1999) "Modeling Competitive Inhibition Effects During Biodegradation of BTEX Mixtures," *Water Research Journal,* vol. 33.

Blum, D. J. W., and R. E. Speece (1991) "A Database of Chemical Toxicity to Environmental Bacteria and Its Use in Interspecies Comparisons and Correlations," *Research Journal, Water Pollution Control Federation,* vol. 63, p. 198.

Bock, E., I. Schmidt, R. Stuven, and D. Zart (1995) "Nitrogen Loss Caused by Denitrifying *Nitrosomonas* Cells Using Ammonium or Hydrogen as Electron Donors and Nitrite as Electron Acceptor," *Archive Microbiology,* vol. 163, pp. 16–20.

Costerton, J. W., Z. Lewandowski, D. E. Caldwell, D. R. Korber, and H. M. Lappin-Scott (1995) "Microbial Biofilms," *Annual Review of Microbiology,* vol. 49, pp. 711–745.

Crites, R., and G. Tchobanoglous (1998) *Small and Decentralized Wastewater Management Systems,* McGraw-Hill, New York.

Dawson, R. N., and K. L. Murphy (1972) "The Temperature Dependency of Biological Denitrification," *Water Research,* vol. 6, p. 71.

Dobbs, R. A., L. Wang, and R. Govind (1989) "Sorption of Toxic Organic Compounds on Wastewater Solids: Correlation with Fundamental Properties," *Environmental Science and Technology,* vol. 23, pp. 1092–1097.

Gayle, B. P. (1989) "Biological Denitrification of Water," *Journal Environmental Engineering,* vol. 115, p. 930.

Grady, C. P. L. Jr., G. T. Daigger, and H. C. Lim (1999) *Biological Wastewater Treatment,* 2d ed., rev. and expanded, Marcel Dekker, New York.

Grady, C. P. L., Jr., B. S. Magbanua, S. Brau, and R. W. Sanders II (1997) "A Simple Technique for Estimating the Contribution of Abiotic Mechanisms to the Removal of SOCs by Completely Mixed Activated Sludge," *Water Environment Research,* vol. 69, p. 1232.

Han, Y., and R. R. Dague (1997) "Laboratory Studies on the Temperature-Phased Anaerobic Digestion of Domestic Primary Sludge," *Water Environment Research*, vol. 69, no. 6, pp. 1139–1143.

Hinton, S. W., and H. D. Stensel (1991) "Experimental Observation of Trickling Filter Hydraulics," *Water Research,* vol. 25, pp. 1389–1398.

Hockenbury, M. R., and C. P. L. Grady, Jr. (1977) "Inhibition of Nitrification—Effects of Selected Organic Compounds," *Journal Water Pollution Control Federation,* vol. 49, p. 768.

Hoover, S. R., and N. Porges (1952) "Assimilation of Dairy Wastes by Activated Sludge II: The Equation of Synthesis and Oxygen Utilization," *Sewage and Industrial Wastes,* vol. 24.

Jeris, J. S., and P. L. McCarty (1963) "Biochemistry of Methane Fermentation," *Proceedings 17th Purdue Industrial Waste Conference,* Lafayette, IN.

Jetten, M. S. M., M. Strous, K. T. van de Pas-Schooen, J. Schalk, U. G. J. M. van Dongen, A. A. van de Graaf, S. Logemann, G. Muyzer, M. C. M. van Loosdrecht, and J. G. Kuenen (1999) "The Anaerobic Oxidation of Ammonium," *FEMS Microbiology Reviews,* vol. 22, pp. 421–437.

Kissel, J. C., P. L. McCarty, and R. L. Street (1984) Numerical Simulation of Mixed-Cultures Biofilm, *Journal of Environmental Engineering,* vol. 110, pp. 393–411.

Kunz, B., J. Gianelli, and H. D. Stensel (1976) "Vanadium Removal From Industrial Wastewater," *Journal Water Pollution Control Federation,* vol. 48, p. 76.

LaGrega, M. D., P. L. Buckingham, and J. C. Evans (2001) *Hazardous Waste Management,* 2nd ed., McGraw-Hill, New York.

Lawrence, A. W., and P. L. McCarty (1970) "A Unified Basis for Biological Treatment Design and Operation," *Journal Sanitary Engineering Division,* American Society of Civil Engineers, vol. 96, no. SA3, 1970.

Lee, S. B., J. P. Patton, S. E. Strand, and H. D. Stensel (1998a) "Sustained Biodegradation of Trichloroethylene in a Suspended Growth Gas Treatment Reactor," *Proceedings, First International Conference on Bioremediation of Chlorinated Solvent and Recalcitrant Compounds,* Monterey, CA.

Lee, K. C., B. E. Rittmann, J. Shi, and D. McAvoy (1998b) "Advanced Steady-State Model for the Fate of Hydrophobic and Volatile Compounds in Activated Sludge Wastewater Treatment," *Water Environment Research,* vol. 70, p. 1118.

Lee, S., H. D. Stensel, and S. E. Strand (2000) "Sustained Degradation of Trichloroethylene in a Suspended Growth Gas Treatment Reactor by an Actinomycetes Enrichment," *Environmental Science and Technology,* vol. 34, pp. 3261–3268.

Littleton, H. X., G. T. Daigger, P. F. Strom, and R. A. Cowan (2000) "Evaluation of Autotrophic Denitrification and Heterotrophic Nitrification in Simultaneous Biological Nutrient Removal Systems," Proceedings of the Research Symposium, 73rd Annual Conference, Water Environment Federation, Anaheim, CA.

Madigan, M. T., J. M. Martinko, and J. Parker (1997) *Brock Biology of Microorganisms,* 8th ed., Prentice-Hall, Upper Saddle River, NJ.

Madigan, M. T., J. M. Martinko, and J. Parker (2000) *Brock Biology of Microorganisms,* 9th ed., Prentice-Hall, Upper Saddle River, NJ.

Maier, R. M., I. L. Pepper, and C. P. Gerba (2000) *Environmental Microbiology,* Academic Press, San Diego, CA.

McCarty, P. L. (1971) "Energetics and Bacterial Growth," in S. D. Faust and J. V. Hunter (eds.), *Organic Compounds in Aquatic Environments,* Marcel Dekker, New York.

McCarty, P. L. (1975) "Stoichiometry of Biological Reactions," *Progress in Water Technology,* vol. 7, pp. 157–172.

McCarty, P. L. (1981) "One Hundred Years of Anaerobic Treatment," In D. E. Hughes et al., ed., *Anaerobic Digestion 1981,* Elsevier Biomedical Press, pp. 3–32, Amsterdam.

McCarty, P. L. (1999) "Chlorinated Organics in Environmental Availability of Chlorinated Organics, Explosives, and Heavy Metals on Soils and Groundwater," in W. C. Anderson, R. C. Loerh, and B. P. Smith (eds.), American Academy of Environmental Engineers, Annapolis.

McCarty, P. L., and D. P. Smith (1986) "Anaerobic Wastewater Treatment," *Environmental Science and Technology,* vol. 20, pp. 1200–1226.

Mamais, D., and D. Jenkins (1992) "The Effects of MCRT and Temperature on Enhanced Biological Phosphorus Removal," *Water Science and Technology,* vol. 26, pp. 955–965.

Melcer, H., P. Steel, and W. K. Bedford (1995) "Removal of Polycyclic Aromatic Hydrocarbons and Heterocyclic Nitrogen Compounds in a Municipal Treatment Plant," *Water Environment Research,* vol. 67, p. 926.

Melcer, H., P. Steel, I. P. Bell, D. I. Thompson, C. M. Yendt, and J. Kemp (1994) "Modeling Volatile Organic Contaminant's Fate in Wastewater Treatment Plants," *Journal of Environmental Engineering,* vol. 120, p. 588.

Metcalf, L., and H. P. Eddy (1930) *Sewerage and Sewage Disposal, A Textbook,* 2d. ed., McGraw-Hill, New York.

Monod, J. (1942) *Recherches sur la croissance des cultures bacteriennes,* Herman et Cie., Paris.

Monod, J. (1949) "The Growth of Bacterial Cultures," *Annual Review of Microbiology,* vol. III, pp. 371–394.

Monteith, H. D., W. I. Parker, I. P. Bell, and H. Melcer (1995) "Modeling the Fate of Pesticides in Municipal Wastewater Treatment," *Water Environment Research,* vol. 67, p. 964.

Mullen, M. D., D. C. Wolf, F. G. Ferris, T. J. Beveridge, C. A. Flemming, and G. W. Bailey (1989) "Bacterial Sorption of Heavy Metals," *Applied and Environmental Microbiology,* Vol. 55, p. 3143–3149.

Nelson, L.M. and R. Knowles (1978) "Effect of Oxygen and Nitrate on Nitrogen Fixation and Denitrification by Azospirillum Brasilense Grown in Continuous Culture," *Canada Journal of Microbiology,* vol. 24, p. 1395.

Painter, H. A. (1970) "A Review of Literature on Inorganic Nitrogen Metabolism I Microorganisms," *Water Research,* vol. 4, p. 393.

Parker, G. F., and W. F. Owen (1986) "Fundamentals of Anaerobic Digestion of Wastewater Sludges," *Journal of Environmental Engineering,* vol. 112, no. 5, pp. 867–920.

Parker, W. I., D. I. Thompson, I. P. Bell, and H. Melcer (1993) "Fate of Volatile Organic Compounds in Municipal Activated Sludge Plants," *Water Environment Research,* vol. 65, p. 58.

Patureau, D., J. Davison, N. Bernet, and R. Moletta (1994) "Denitrification Under Various Aeration Conditions in *Comamonas sp.,* Strain SGLY2," *FEMS Microbiology Ecology,* vol. 14, pp. 71–78.

Payne, W. J. (1981) *Denitrification,* Wiley, New York.

Pitt, P., and D. Jenkins (1990) "Cause and Control of *Nocardia* in Activated Sludge," *Journal Water Pollution Control Federation,* vol. 62, pp. 143–150.

Poduska, R. A. (1973) *A Dynamic Model of Nitrification For the Activated Sludge Process,* Ph.D. Thesis, Clemson University.

Randall, C. W., J. L. Barnard, and H. D. Stensel (1992) *Design and Retrofit of Wastewater Treatment Plants for Biological Nutrient Removal, Volume 5,* Water Quality Management Library, Technomic Publishing Co., Lancaster, PA.

Rittman, B. E., and W. E. Langeland (1985) "Simultaneous Denitrification with Nitrification in Single-Channel Oxidation Ditches," *Journal Water Pollution Control Federation,* vol. 57, p. 300.

Rittman, B. E., and P. L. McCarty (1980) "Model of Steady-State-Biofilm Kinetics," *Biotechnology and Bioengineering,* vol. 22, pp. 2343–2357.

Rittman, B. E., and P. L. McCarty (2001) *Environmental Biotechnology: Principles and Applications,* McGraw-Hill, New York.

Robertson, L. A., and J. G. Kuenen (1990) "Combined Heterotrophic Nitrification and Aerobic Denitrification in *Thiosphaera pantotropha* and other Bacteria," *Antonie Van Leeuwenhoek,* vol. 56, pp. 289–299.

Saez, P. B., and B. E. Rittman (1992) "Accurate Pseudo-Analytical Solution for Steady-State Biofilms," *Biotechnology and Bioengineering,* vol. 39, pp. 790–793.

Sawyer, C. N., P. L. McCarty, and G. F. Parkin (1994) *Chemistry for Environmental Engineering,* 4th ed., McGraw-Hill, Inc., New York.

Schwarzenbach, R. P., P. M. Gschwend, and D. M. Imboden (1993) *Environmental Organic Chemistry,* Wiley, New York.

Sedlak, R. I. (1991) *Phosphorus and Nitrogen Removal from Municipal Wastewater,* 2d ed., The Soap and Detergent Association, Lewis Publishers, New York.

Sharma, B., and R. C. Ahlert (1977) "Nitrification and Nitrogen Removal," *Water Research,* vol. 11, p. 897.

Skerman, V. B. D., and J. C. MacRae (1957) "The Influence of Oxygen Availability on the Degree of Nitrate Reduction by *Psudomonas denitrificans,*" *Canada Journal of Microbiology,* vol. 3, p. 505.

Skinner, F. A., and N. Walker (1961) "Growth of *Nitrosomonas europaea* in Batch and Continuous Culture," *Archives Mikrobiology,* vol. 38, p. 339.

Stensel, H. D., R. C. Loehr, and A.W. Lawrence (1973) "Biological Kinetics of Suspended Growth Denitrification," *Journal Water Pollution Control Federation,* vol. 45, p. 249.

Stensel, H. D., and A. R. Bielefeldt (1997) "Anaerobic and Aerobic Degradation of Chlorinated Aliphatic Compounds," in *Bioremediation of Hazardous Wastes: Principles and Practices.* Kluwer Publishers, San Diego.

Stensel, H. D., and G. Horne (2000) "Evaluation of Denitrification Kinetics at Wastewater Treatment Facilities," *Research Symposium Proceedings, 73rd Annual Water Environment Federation Conference,* Anaheim, CA. October 14–18, 2001.

Stenstrom, M. K., and S. S. Song (1991) "Effects of Oxygen Transport Limitations on Nitrification in the Activated-Sludge Process," *Research Journal, Water Pollution Control Federation,* vol. 63, p. 208.

Stephens, H. L., and H. D. Stensel (1998) "Effect of Operating Conditions on Biological Phosphorus Removal," *Water Environment Research Journal,* vol. 68.

Strand, S. E., G. N. Harem, and H. D. Stensel (1999) "Activated Sludge Yield Reduction Using Chemical Uncouplers," *Water Environment Research,* vol. 71, p. 454–458.

Strous, M., J. A. Fuerst, E. H. M. Kramer, S. Logemann, G. Muyzert, K. T. van de Pas-Schoonen, R. Webb, J. G. Kuenen, and M. S. M. Jetten (1999b) "Missing Lithotroph Identified as New Planctomycete," *Nature,* vol. 400.

Strous, M., J. G. Kuenen, and M. S. M. Jetten (1999a) "Key Physiology of Anaerobic Ammonium Oxidation," *Applied and Environmental Microbiology,* vol. 65, pp. 3248–3250.

Strous, M., E. van Gerven, P. Zheng, J. G. Kuenen, and M. S. M. Jetten (1997) "Ammonium Removal From Concentrated Waste Streams with the Anaerobic Ammonium Oxidation (ANAMMOX) Process in Different Reactor Configurations," *Water Research,* vol. 31, pp. 1955–1962.

Suidan, M. T., and Y. T. Wang (1985) "Unified Analysis of Biofilm Kinetics," *Journal of Environmental Engineering,* vol. 111, pp. 634–646.

Tchobanoglous, G., and E. D. Schroeder (1985) *Water Quality: Characteristics, Modeling, Modification,* Addison-Wesley Publishing Company, Reading, MA.

Terai, H., and T. Mori (1975) "Studies on Phosphorylation Coupled with Denitrification and Aerobic Respiration in *Pseudomonas denitrificans*," *Botany Magazine,* vol. 38, p. 231.

Teske, A, E. Alm, J. M. Regan, S. Toze, B. E. Rittman, and D. A. Stahl (1994) "Evolutionary Relationships Among Ammonia and Nitrite-Oxidizing Bacteria," *Journal Bacteriology,* vol. 176, pp. 6623–6630.

U.S. EPA (1993) *Manual Nitrogen Control,* EPA/625/R-93/010, Office of Research and Development, U.S. Environmental Protection Agency, Washington, DC.

van Lier, J. B. (1996) "Limitations of Thermophilic Anaerobic Wastewater Treatment and the Consequences for Process Design," *Antonie van Leeuwenhoek,* vol. 69, pp. 1–14, The Netherlands.

van Loosdrecht, M. C. M., and M. S. M. Jetten (1998) "Microbiological Conversions in Nitrogen Removal," *International Association of Water Quality 19th Biennial International Conference Preprint Book 1, Nutrient Removal,* pp. 1–8.

Wagner, M., G. Rath, R. Amann, H. P. Koops, and K. H. Schleifer (1995) "In Situ Identification of Ammonia—Oxidizing Bacteria," *System Applied Microbiology,* vol. 18, pp. 251–264.

Wanner, O., and W. Gujer (1986) "A Multispecies Biofilm Model," *Biotechnology and Bioengineering,* vol. 28, pp. 314–328.

Watts, R. J. (1997) *Hazardous Wastes: Sources, Pathways, Receptors,* Wiley, New York.

WEF (2000) *Aerobic Fixed-Growth Reactors,* A special publication prepared by the Aerobic Fixed-Growth Reactors Task Force, Water Environment Federation, Alexandria, VA.

Wentzel, M. C., R. E. Loewenthal, G. A. Ekama, and G. v. R. Marais (1989) "Enhanced Polyphosphate Organism Cultures in Activated Sludge Systems—Part 1: Enhanced Culture Development," *Water,* South African, vol. 14, pp. 81–92.

Wentzel, M. C., L. H. Lotter, G. A. Ekama, R. E. Loewenthal, and G. v. R. Marais (1991) "Evaluation of Biochemical Models for Biological Excess Phosphorus Removal," *Water Science and Technology,* vol. 23, pp. 567–576.

Werzernak, C. T., and J. J. Gannon (1967) "Oxygen-Nitrogen Relationships in Autotrophic Nitrification," *Applied Microbiology,* vol. 15, p. 211.

Williamson, K., and P. L. McCarty (1976) "A Model of Substrate Utilization by Bacterial Films," *Journal Water Pollution Control Federation,* vol. 48, pp. 9–24.

Winogradsky, M. S. (1891) "Recherches Sur Les Organismes de la Nitrification," *Annals Institute Pasteur,* vol. 5, p. 92.

Zinder, S. H., and M. Koch (1984) "Non-Aceticlastic Methanogenesis from Acetate: Acetate Oxidation by a Thermophilic Syntropic Coculture," *Archival Microbiology,* vol. 138, pp. 263–272.

8

Suspended Growth Biological Treatment Processes

The theory of biological wastewater treatment is presented and discussed in detail in Chap. 7. As noted in Chap. 7, biological treatment processes may be classified as aerobic and anaerobic suspended growth, attached growth, and various combinations thereof. The focus of this chapter is on suspended growth treatment processes as exemplified by the activated-sludge process for BOD and nitrification and for nitrogen and phosphorus removal. Attached growth and combined processes are discussed in Chap. 9, and other suspended and attached growth anaerobic processes are considered in Chap. 10. Included in this chapter are (1) general descriptions of the basic activated-sludge processes used to meet a wide range of treatment objectives, (2) a description of key wastewater *characteristics* that affect activated-sludge process design, (3) fundamentals of process analysis and control, (4) processes for BOD removal and nitrification, (5) processes for nitrogen removal, (6) processes for phosphorus removal, (7) selection and design of physical facilities for activated-sludge processes, (8) aerated lagoons, (9) biological treatment with membrane separation, and (10) a review of activated-sludge modeling approaches. Stabilization ponds, i.e., non-aerated lagoons, are not covered in this text, as they are used mainly for small rural communities where sufficient land is available and discharge requirements may not be as stringent as in urban areas. For information about the design of stabilization ponds, see Metcalf & Eddy (1991), Reed et al. (1995), and Crites and Tchobanoglous (1998).

8-1 INTRODUCTION TO THE ACTIVATED-SLUDGE PROCESS

To provide a basis for the process designs presented in the subsequent sections of this chapter, it will be useful to consider (1) a brief summary of the historical development of the activated-sludge process, (2) a description of the basic process, (3) a brief review of the evolution of the activated-sludge process, and (4) an overview of recent process developments.

Historical Development

The activated-sludge process is now used routinely for biological treatment of municipal and industrial wastewaters. The antecedents of the activated-sludge process date back to the early 1880s to the work of Dr. Angus Smith, who investigated the aeration of wastewater in tanks and the hastening of the oxidation of the organic matter. The aeration of wastewater was studied subsequently by a number of investigators, and in 1910 Black and Phelps reported that a considerable reduction in putrescibility could be secured by forcing air into wastewater in basins. In experiments conducted at the Lawrence Experiment Station during 1912 and 1913 by Clark and Gage with aerated wastewater, growths of organisms could be cultivated in bottles and in tanks partially filled with roofing slate spaced about 25 mm (1 in) apart and would greatly increase the degree of purification obtained (Clark and Adams, 1914). The results of the work at the Lawrence Experiment Station were so striking that knowledge of them led Dr. G. J. Fowler of the University of Manchester, England to suggest experiments along similar lines be conducted at the Manchester Sewage Works where Ardern and Lockett carried out valuable research on the subject. During the course of their experiments, Ardern and Lockett found that the sludge played an important part in the results obtained by aeration, as announced in their paper of May 3, 1914 (Ardern and Lockett, 1914). The process was named *activated sludge* by Ardern and Lockett because it involved the production of an activated mass of microorganisms capable of aerobic stabilization of organic material in wastewater (Metcalf & Eddy, 1930).

Description of Basic Process

By definition, the basic activated-sludge treatment process, as illustrated on Fig. 8–1a and b, consists of the following three basic components: (1) a reactor in which the microorganisms responsible for treatment are kept in suspension and aerated; (2) liquid-solids separation, usually in a sedimentation tank; and (3) a recycle system for returning

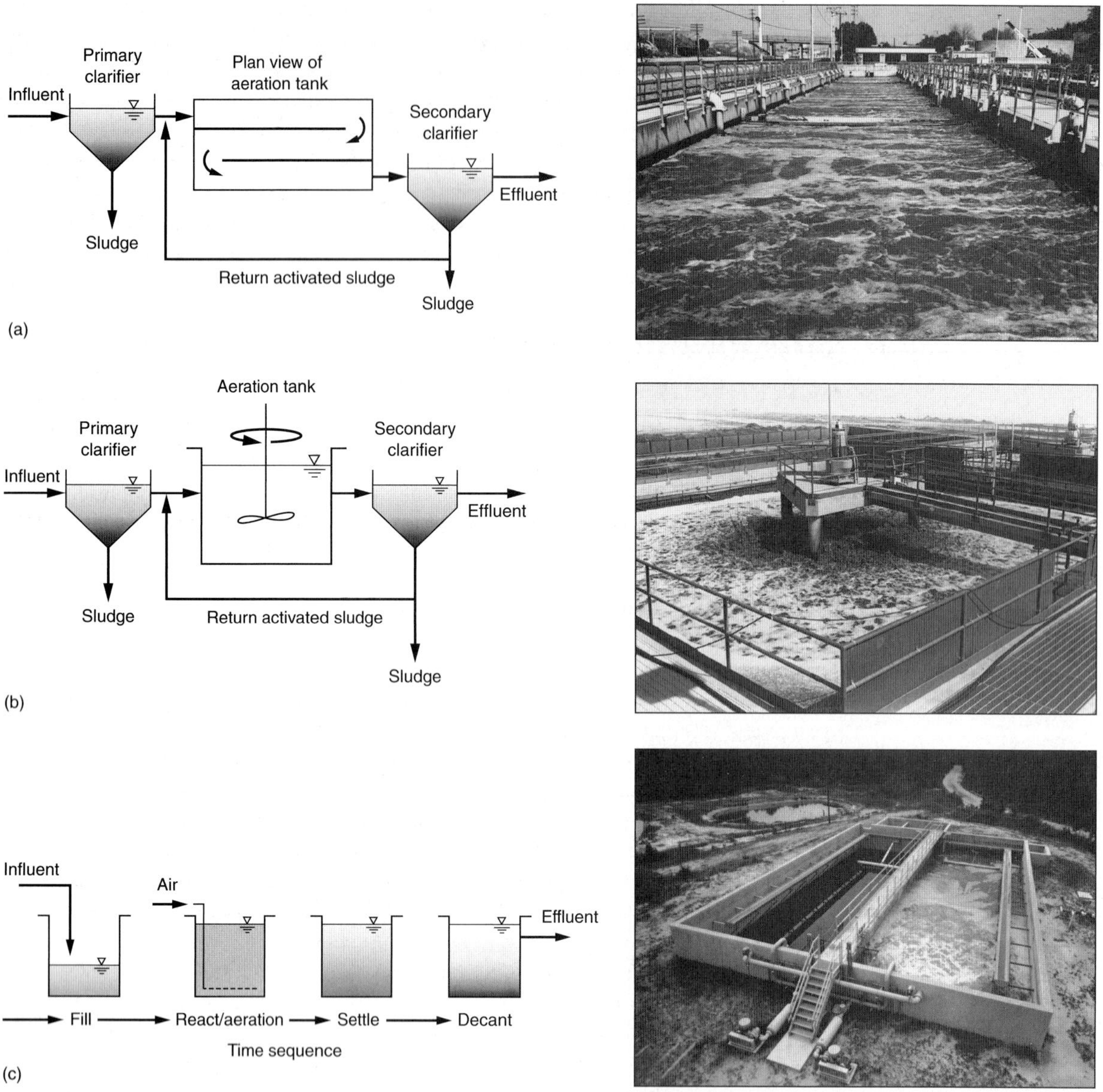

(a)

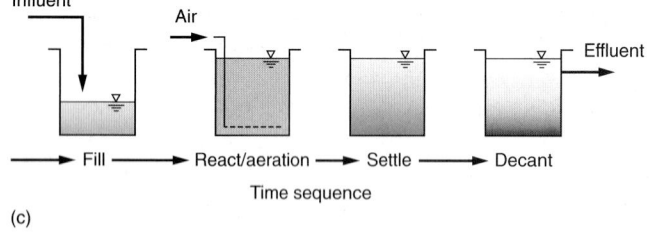

(b)

(c)

Figure 8-1

Typical activated-sludge processes with different types of reactors: (a) schematic flow diagram of plug-flow process and view of plug-flow reactor, (b) schematic flow diagram of complete-mix process and view of complete-mix activated-sludge reactor, and (c) schematic diagram of sequencing batch reactor process and view of sequencing batch reactor. (From H. D. Stensel.)

solids removed from the liquid-solids separation unit back to the reactor. Numerous process configurations have evolved employing these components. An important feature of the activated-sludge process is the formation of flocculent settleable solids that can be removed by gravity settling in sedimentation tanks. In most cases, the activated-sludge process is employed in conjunction with physical and chemical processes that are used for the preliminary and primary treatment of wastewater, and posttreatment, including disinfection and possibly filtration.

Historically, most activated-sludge plants have received wastewaters that were pretreated by primary sedimentation, as shown on Fig. 8–1a and b. Primary sedimentation is most efficient at removing settleable solids, whereas the biological processes are essential for removing soluble, colloidal, and particulate (suspended) organic substances; for biological nitrification and denitrification; and for biological phosphorus removal. For applications such as treating wastewater from smaller-sized communities, primary treatment is often not used as more emphasis is placed on simpler and less operator-intensive treatment methods. Primary treatment is omitted frequently in areas of the world that have hot climates where odor problems from primary tanks and primary sludge can be significant. For these applications, various modifications of conventional activated-sludge processes are used, including sequencing batch reactors, oxidation ditch systems, aerated lagoons, or stabilization ponds.

Evolution of the Activated-Sludge Process

A number of activated-sludge processes and design configurations have evolved since its early conception as a result of (1) engineering innovation in response to the need for higher-quality effluents from wastewater treatment plants; (2) technological advances in equipment, electronics, and process control; (3) increased understanding of microbial processes and fundamentals; and (4) the continual need to reduce capital and operating costs for municipalities and industries. With greater frequency, activated-sludge processes used today may incorporate nitrification, biological nitrogen removal, and/or biological phosphorus removal. These designs employ reactors in series, operated under aerobic, anoxic, and anaerobic conditions, and may use internal recycle pumps and piping. The general types of activated-sludge processes used (i.e., plug flow, complete mix, and sequencing batch reactor) are illustrated on Fig. 8–1.

Since the process came into common use in the early 1920s and up until the late 1970s, the type of activated-sludge process used most commonly was the one in which a plug-flow reactor with large length to width ratios (typically > 10:1) was used (see Fig. 8–1a). In considering the evolution of the activated-sludge process, it is important to note that the discharge of industrial wastes to domestic wastewater collection systems increased in the late 1960s. The use of a plug-flow process became problematic when industrial wastes were introduced because of the toxic effects of some of the discharges. The complete-mix reactor was developed, in part, because the larger volume allowed for greater dilution and thus mitigated the effects of toxic discharges. The more common type of activated-sludge process in the 1970s and early 1980s tended to be single-stage, complete-mix activated-sludge (CMAS) processes (see Fig. 8–1b), as advanced by McKinney (1962). In Europe, the CMAS process has not been adopted generally as ammonia standards have become increasingly stringent. For some nitrification applications, two-stage systems (each stage consisting of an aeration tank and clarifier) were used with the first stage designed for BOD removal, followed by a second stage for nitrification. Other activated-sludge processes that have found application, with their dates of major interest in parentheses, include the oxidation ditch (1950s), contact stabilization (1950s), Krause process (1960s), pure oxygen activated sludge (1970s), Orbal process (1970s), deep shaft aeration (1970s), and sequencing batch reactor process (1980).

With the development of simple inexpensive program logic controllers (PLCs) and the availability of level sensors and automatically operated valves, the sequencing batch reactor (SBR) process (see Fig. 8–1c) became more widely used by the late 1970s, especially for smaller communities and industrial installations with intermittent flows. In recent years, however, SBRs are being used for large cities in some parts of the world. The SBR is a fill-and-draw type of reactor system involving a single complete-mix reactor in which all steps of the activated-sludge process occur. Mixed liquor remains in the reactor during all cycles, thereby eliminating the need for separate sedimentation tanks.

In comparing the plug-flow (Fig. 8–1a) and complete-mix activated-sludge (CMAS) (Fig. 8–1b) processes, the mixing regimes and tank geometry are quite different. In the CMAS process, the mixing of the tank contents is sufficient so that ideally the concentrations of the mixed-liquor constituents, soluble substances (i.e., COD, BOD, NH_4-N), and colloidal and suspended solids do not vary with location in the aeration basin. The plug-flow process involves relatively long, narrow aeration basins, so that the concentration of soluble substances and colloidal and suspended solids varies along the reactor length. Although process configurations employing long, narrow tanks are commonly referred to as plug-flow processes, in reality, true plug flow does not exist. Depending on the type of aeration system, back mixing of the mixed liquor can occur and, depending on the layout of the reactor and the system reaction kinetics, nominal plug flow may be described more appropriately by the series of complete-mix reactors as discussed in Chap. 4.

Activated-sludge process designs before and until the late 1970s generally involved the configurations shown on Fig. 8–1a and b. However, with interest in biological nutrient removal, staged reactor designs consisting of complete-mix reactors in series have been developed (see Fig. 8–2). Some of the stages are not aerated (anaerobic or anoxic stages) and internal recycle flows may be used. For nitrification, a staged aerobic reactor design may also be used to provide more efficient use of the total reactor volume than a single-stage CMAS process. A number of biological nutrient-removal configurations will be shown in subsequent sections of this chapter. Pilot-plant studies are sometimes used to evaluate and optimize biological nutrient-removal processes. A large pilot plant used by the city of New York for the investigation of alternative nitrogen-removal processes is shown on Fig. 8–3.

Recent Process Developments

As noted above, numerous modifications of the activated-sludge process have evolved in the last 10 to 20 years, aimed principally at effective and efficient removal of nitrogen and phosphorus. Nearly all of the various modifications are based on the same fundamental principles of biological treatment as described in Chap. 7. Many of the processes that are being used in full-scale operation are described in Secs. 8–4, 8–5, and 8–6; design examples for the processes most commonly used are also included.

Because of the development of improved membrane design, principally for water treatment applications, membrane technology has found increasing application for

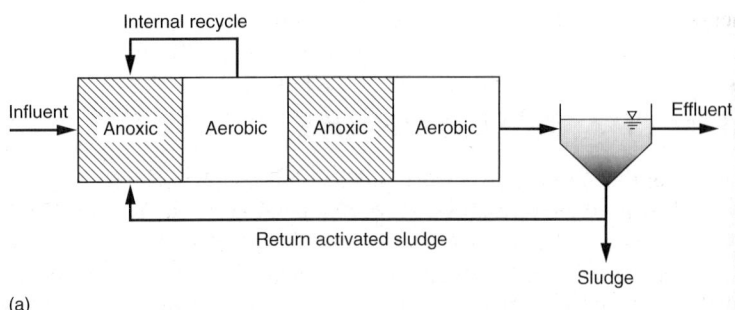

(a)

(b)

Figure 8–2

Bardenpho process with staged reactors for biological nitrogen removal: (a) schematic diagram of staged process and (b) view of a staged Bardenpho treatment plant in Palmetto, FL, the first of its type in the United States. *(From H. D. Stensel.)*

Figure 8–3

Large-scale pilot plant for wastewater treatability investigations. *(Courtesy of the New York City Department of Environmental Protection.)*

enhanced solids separation for water reuse (see Chaps. 11 and 13), and more recently for use in suspended growth reactors for wastewater treatment. Membrane biological reactors (MBRs), which may change the look of wastewater-treatment facilities in the future, are discussed in Sec. 8–9.

Because the design and operation of the activated-sludge process is becoming more complex, computer modeling is an increasingly important tool to incorporate the large number of components and reactions necessary to evaluate activated-sludge performance. The use of simulation models for suspended growth systems is discussed in Sec. 8–10.

8–2 WASTEWATER CHARACTERIZATION

Activated-sludge process design requires determining (1) the aeration basin volume, (2) the amount of sludge production, (3) the amount of oxygen needed, and (4) the effluent concentration of important parameters. To design an activated-sludge treatment process properly, characterization of the wastewater is perhaps the most critical step in the process. For biological nutrient-removal processes, wastewater characterization is essential for predicting performance. Wastewater characterization is an important element in the evaluation of existing facilities for optimizing performance and available treatment capacity. Flow characterization is also important including diurnal, seasonal, and wet-weather flow variations (see Chap. 3). Without comprehensive wastewater characterization, facilities may either be under- or overdesigned, resulting in inadequate or inefficient treatment.

Key Wastewater Constituents for Process Design

Wastewater characteristics of importance in the design of the activated-sludge process can be grouped into the following categories: carbonaceous substrates, nitrogenous compounds, phosphorus compounds, total and volatile suspended solids (TSS and VSS), and alkalinity. Typical wastewater constituents quantified for use in the design of wastewater treatment processes are reported in Table 8–1. While the constituents reported in Table 8–1 are sufficient for some designs, additional information is needed for biological nutrient-removal process designs or to evaluate the capacity of an existing system. A complete listing of wastewater components, with the nomenclature used in this chapter, is presented in Table 8–2. The terms presented in Table 8–2 are introduced, discussed, and applied in the following paragraphs. In the text of this chapter, the units of expression for constituent concentrations are given in mg/L. In the examples,

Table 8–1
Example of typical domestic wastewater characterization parameters and typical values

Component	Concentration, mg/L[a]
COD	430
BOD	190
TSS	210
VSS	160
TKN	40
NH_4-N	25
NO_3-N	0
Total phosphorus	7
Alkalinity	200 (as $CaCO_3$)

[a] Typical medium-strength wastewater in the United States, from Table 3–15.

Table 8–2
Definition of terms used to characterize important wastewater constituents used for the analysis and design of biological wastewater treatment processes

Constituent [a,b]	Definition
BOD	
BOD	Total 5-d biochemical oxygen demand
sBOD	Soluble 5-d biochemical oxygen demand
UBOD	Ultimate biochemical oxygen demand
COD	
COD	Total chemical oxygen demand
bCOD	Biodegradable chemical oxygen demand
pCOD	Particulate chemical oxygen demand
sCOD	Soluble chemical oxygen demand
nbCOD	Nonbiodegradable chemical oxygen demand
rbCOD	Readily biodegradable soluble chemical oxygen demand
bsCOD	Biodegradable soluble chemical oxygen demand
sbCOD	Slowly biodegradable chemical oxygen demand
bpCOD	Biodegradable particulate chemical oxygen demand
nbpCOD	Nonbiodegradable particulate chemical oxygen demand
nbsCOD	Nonbiodegradable soluble chemical oxygen demand
Nitrogen	
TKN	Total Kjeldahl nitrogen
bTKN	Biodegradable total Kjeldahl nitrogen
sTKN	Soluble (filtered) total Kjeldahl nitrogen
ON	Organic nitrogen
bON	Biodegradable organic nitrogen
nbON	Nonbiodegradable organic nitrogen
pON	Particulate organic nitrogen
nbpON	Nonbiodegradable particulate organic nitrogen
sON	Soluble organic nitrogen
nbsON	Nonbiodegradable soluble organic nitrogen
Suspended Solids	
TSS	Total suspended solids
VSS	Volatile suspended solids
nbVSS	Nonbiodegradable volatile suspended solids
iTSS	Inert total suspended solids

[a] *Note:* b = biodegradable; i = inert; n = non; p = particulate; s = soluble.
[b] Measured constituent values, based on the terminology given in this table, will vary depending on the technique used to fractionate a particular constituent.

however, constituent concentrations are expressed as g/m³ (which is equivalent to mg/L) for ease of use in process computations, thus eliminating one unit conversion step.

Carbonaceous Constituents. Carbonaceous constituents measured by BOD or COD analyses are critical to the activated-sludge process design. Higher concentrations of degradable COD or BOD result in (1) a larger aeration basin volume, (2) more oxygen transfer needs, and (3) greater sludge production. While the BOD has been the common parameter to characterize carbonaceous material in wastewater, COD is becoming more common and is the primary carbonaceous parameter used in most current comprehensive computer simulation design models. By using a COD mass balance, the fate of carbonaceous material between the amount oxidized and the amount incorporated into cell mass is followed more easily. The various forms of the COD in wastewater are shown on Fig. 8–4 and defined in Table 8–2.

Unlike BOD, some portion of the COD is not biodegradable, so the COD is divided into *biodegradable* and *nonbiodegradable* concentrations. The next level of interest is how much of the COD in each of these categories is *dissolved* or *soluble*, and how much is *particulate*, comprised of colloidal and suspended solids. The *nonbiodegradable soluble COD* (nbsCOD) will be found in the activated-sludge effluent, and nonbiodegradable particulates will contribute to the total sludge production.

Because the *nonbiodegradable particulate COD* (nbpCOD) is organic material, it will also contribute to the VSS concentration of the wastewater and mixed liquor in the activated-sludge process, and is referred to here as the *nonbiodegradable volatile suspended solids* (nbVSS). The influent wastewater will also contain nonvolatile influent suspended solids that add to the MLSS concentration in the activated-sludge process. These solids are influent *inert TSS* (iTSS) and can be quantified by the difference in influent wastewater TSS and VSS concentrations. For biodegradable COD, understanding the fractions that are measured as soluble, soluble readily biodegradable COD (rbCOD), and particulate is extremely important for activated-sludge process design. The rbCOD portion is quickly assimilated by the biomass, while the particulate and colloidal COD must first be dissolved by extracellular enzymes and are thus assimilated at

Figure 8–4

Fractionation of COD in wastewater. Information on the COD fractions is used in the detailed design of activated-sludge processes.

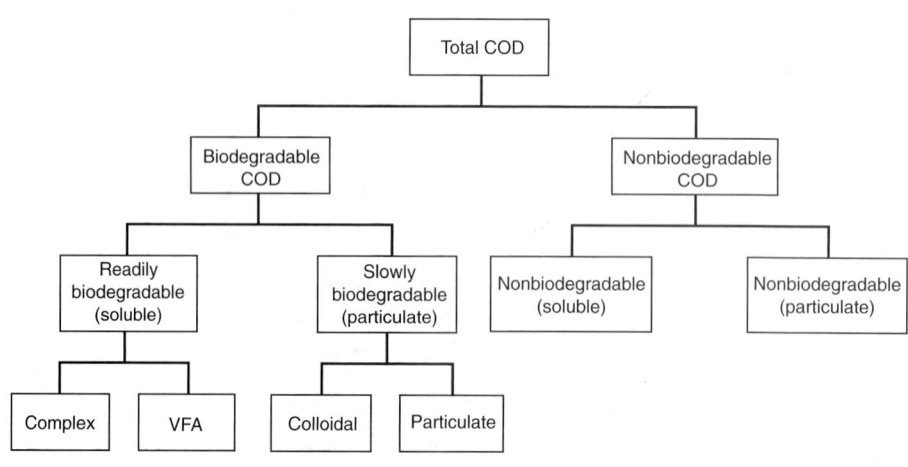

Table 8-3

Biological processes affected by readily biodegradable COD (rbCOD) concentration in influent wastewater

Process	Effect of rbCOD
Activated-sludge aeration	For plug flow or staged aeration zones, there will be a higher oxygen demand toward the front of the tank with a higher fraction of rbCOD in the influent COD
Biological nitrogen removal	For the preanoxic tank, there will be a higher denitrification rate with a higher fraction of rbCOD in the influent COD. Can result in smaller anoxic tank volume
Biological phosphorus removal	Greater influent rbCOD concentration results in a greater amount of biological phosphorus removal
Activated-sludge selector	Higher fraction of rbCOD in influent COD provides more COD for floc-forming bacteria in the selector. Can have a greater impact on improving sludge volume index (SVI)

much slower rates [slowly biodegradable COD (sbCOD) on Fig. 8-4]. Filtration methods, such as a 0.45-μm filter, are used to determine the soluble COD (sCOD), the COD in the filtrate. The sCOD contains rbCOD, a small fraction of the colloidal COD, and nonbiodegradable soluble COD (nbsCOD). For domestic wastewater, the nbsCOD is equal to the soluble COD concentration of an activated-sludge process effluent (sCOD$_e$) when operated at an SRT > 4 d. Special testing techniques, described on page 672, are used to find the biodegradable "true" soluble COD, which is the rbCOD.

The rbCOD fraction of the COD has a direct effect on the activated-sludge biological kinetics and process performance. Process applications where the rbCOD concentration affects the process design and performance are summarized in Table 8-3. For conventional plug-flow or staged aerobic activated-sludge reactors, more oxygen demand will be required toward the front of the aeration tank where there is a greater rbCOD concentration. The rbCOD concentration has a significant effect on the denitrification rate in preanoxic zones in biological nitrogen-removal processes, where it will be consumed before the aeration zone. The greater the amount of rbCOD, the faster will be the nitrate reduction rate. For biological phosphorus removal, the rbCOD can be quickly converted to acetate via fermentation in the anaerobic zone for uptake by the phosphorus-storing bacteria. The rbCOD concentration in the influent wastewater must be known to predict more accurately the performance of biological phosphorus removal.

A further step in the characterization of the influent COD is illustrated on Fig. 8-4. The rbCOD consists of complex soluble COD that can be fermented to volatile fatty acids (VFAs) in the influent wastewater. Wastewaters that are more septic, for example, from collection systems in warm climates with minimal slopes, will contain higher concentrations of VFAs.

BOD test data are necessary to obtain the total biodegradable COD (bCOD). Grady et al. (1999) point out that the bCOD/BOD ratio is greater than the ultimate BOD to BOD ratio (UBOD/BOD), because not all of the bCOD is oxidized in the BOD test. Some of the bCOD is converted into biomass, which can still remain as cell debris and active cells at the end of the long incubation time for the UBOD determination. For domestic wastewater with a measured UBOD/BOD ratio of 1.5, the bCOD/BOD ratio may be 1.6 to 1.7 depending on the biomass yield and cell debris fraction. The bCOD/BOD can be estimated using the following equation, which is based on the fact that the bCOD consumed in the BOD test equals the oxygen consumed (UBOD) plus the oxygen equivalent of the remaining cell debris [bCOD = UBOD + $1.42 f_d (Y_H)$bCOD] after long-term incubation:

$$\frac{bCOD}{BOD} = \frac{UBOD/BOD}{1.0 - 1.42 f_d(Y_H)} \tag{8-1}$$

where f_d = fraction of cell mass remaining as cell debris, g/g
$\quad\quad Y_H$ = synthesis yield coefficient for heterotrophic bacteria, g VSS/g COD used

For example, using values typical of domestic wastewater (UBOD/BOD = 1.5, f_d = 0.15, Y_H = 0.4), the bCOD/BOD ratio is 1.64.

Nitrogenous Constituents. The composition of nitrogen in wastewater is illustrated on Fig. 8–5. The total Kjeldahl nitrogen (TKN) is a measure of the sum of the ammonia and organic nitrogen. About 60 to 70 percent of the influent TKN concentration will be as NH_4-N, which is readily available for bacterial synthesis and nitrification. Organic nitrogen is present in both soluble and particulate forms, and some portion of each of these is nonbiodegradable. The particulate degradable organic nitrogen will be removed more slowly than the soluble degradable organic nitrogen because a hydrolysis reaction is necessary first. The nondegradable organic nitrogen is assumed to be about 6 percent of the nondegradable VSS as COD in the influent wastewater (Grady et al., 1999). The particulate nondegradable nitrogen will be captured in the activated-sludge floc and exit in the waste sludge, but the soluble nondegradable nitrogen will be found in the secondary clarifier effluent. The soluble nondegradable nitrogen contributes to the effluent total nitrogen concentration and is a small fraction of the influent wastewater TKN concentration (<3 percent). The soluble nondegradable organic nitrogen concentration in domestic wastewater typically ranges from 1 to 2 mg/L as N (Parkin and McCarty, 1981). There is also a possibility that some soluble nondegradable organic nitrogen may be produced from endogenous respiration. Further research

Figure 8–5

Fractionation of nitrogen in wastewater. Information on the nitrogen fractions is used in the detailed design of nitrification and denitrification processes.

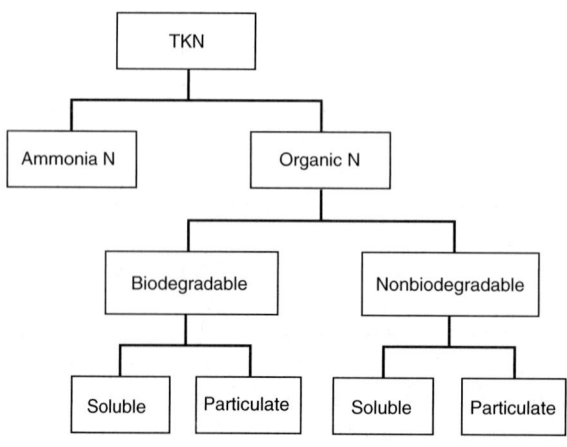

is needed to understand fully the sources and factors affecting nondegradable organic nitrogen.

Alkalinity.   Alkalinity concentration is an important wastewater characteristic that affects the performance of biological nitrification processes. Adequate alkalinity is needed to achieve complete nitrification. In some cases where a wastewater sample is not available, the total alkalinity of the wastewater may be estimated from information on the alkalinity of the potable water plus the alkalinity contributed through domestic use (see Table 3–16).

Measurement Methods for Wastewater Characterization

Special procedures are used to quantify the rbCOD, nbVSS, and soluble organic nitrogen (sON) and nonbiodegradable organic nitrogen (nbON) concentrations in wastewaters. Some of the methods and techniques used to quantify these constituents are discussed below.

Readily Biodegradable COD.   The rbCOD concentration is either determined from a biological response or estimated by a physical separation technique. In the biological response method the oxygen uptake rate (OUR) is followed and recorded with time after mixing the wastewater sample with an acclimated activated-sludge sample. The wastewater may be preaerated so that upon contact with the activated sludge a high DO concentration is present to allow an immediate measurement of the OUR. The wastewater sample and activated sludge are mixed in a batch reactor with separate aeration and mixing. During mixing without aeration the DO concentration declines and is measured with a calibrated DO probe and meter with time to observe the OUR in mg/L·h. When the DO concentration decreases to about 3.0 mg/L, vigorous aeration is applied to elevate the DO concentration to 5 to 6 mg/L, so that another OUR measurement can begin.

An idealized example of the OUR response for a wastewater sample using an activated sludge containing nitrifying bacteria is shown on Fig. 8–6. The OUR versus time can be divided into four areas, which can be used to determine the oxygen consumed for the reaction indicated by the area. Area A is the oxygen used for rbCOD degradation,

Figure 8–6

Idealized oxygen uptake rate (OUR) in aerobic batch test for a mixture of influent wastewater and activated-sludge mixed liquor. Area A represents rbCOD oxygen demand (Barker and Dold, 1997).

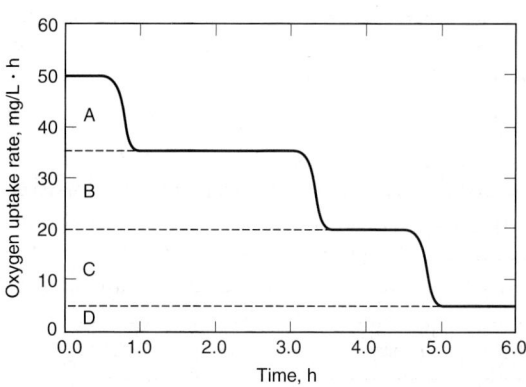

area B for zero-order nitrification, area C for particulate COD degradation, and area D for endogenous decay (Ekama et al., 1986). The rbCOD concentration is based on the amount of oxygen used in area A, and is calculated as follows by accounting for the fact that some of the rbCOD is used for cell synthesis:

$$rbCOD = \frac{O_A}{1 - Y_{H,COD}}\left(\frac{V_{AS} + V_{WW}}{V_{WW}}\right) \tag{8-2}$$

where O_A = oxygen consumed in area A, mg/L

$Y_{H,COD}$ = synthesis yield coefficient for heterotrophic bacteria, g cell COD/g COD used

V_{AS} = volume of activated sludge used in test, mL

V_{WW} = volume of wastewater sample, mL

A value of 0.67 for Y_H is recommended by Ekama et al. (1986).

The volume of activated sludge and wastewater may be selected by successive trials until a response is obtained that provides a satisfactory measurement area for the rbCOD degradation.

A common method used to estimate the rbCOD concentration is the floc/filtration method presented by Mamais et al. (1993) in an attempt to separate the *true* soluble COD in the wastewater sample. The method requires applying the floc/filtration procedure to both the wastewater sample and a secondary effluent sample or a settled supernatant sample after sufficient contact and aeration of the wastewater sample with activated sludge. The soluble COD measured in the secondary effluent sample is the nonbiodegradable soluble COD (nbsCOD) as the rbCOD would be removed by the activated sludge process. The procedure is based on the assumption that suspended solids and colloidal material can be captured effectively and removed by flocculation with a zinc hydroxide precipitate to leave only truly dissolved organic material after filtration. The steps in the method for each sample are as follows: (1) 1 mL of a 100 g/L $ZnSO_4$ solution is added to 100 mL of sample with vigorous mixing for 1 min, (2) the pH is raised to about 10.5 using 6M (molar) NaOH, with 5 to 10 min of gentle mixing for floc formation, (3) the sample is settled for 10 to 20 min and the supernatant is withdrawn and filtered using a 0.45-μm membrane filter, and (4) the filtrate is analyzed for COD concentration. The difference in COD concentration between the wastewater and activated-sludge treated sample is the rbCOD.

The floc/filtration technique may not give the exact results as the rbCOD concentration determination by respirometry, but it provides a reasonable estimate. Floc/filtration is used widely at wastewater-treatment facilities because of its simplicity. As long as the selected procedure is compatible with the design models used to evaluate the activated-sludge process, a useful design approach is possible.

Nonbiodegradable Volatile Suspended Solids.   The wastewater nonbiodegradable volatile suspended solids (nbVSS) concentration can be estimated from analyses for COD, sCOD, BOD, sBOD, and VSS concentration, and by assuming a constant COD/VSS ratio for both biodegradable and nonbiodegradable VSS:

$$nbVSS = \left[1 - \left(\frac{bpCOD}{pCOD}\right)\right]VSS \tag{8-3}$$

$$\frac{bpCOD}{pCOD} = \frac{(bCOD/BOD)(BOD - sBOD)}{COD - sCOD} \tag{8-4}$$

where bpCOD = concentration of biodegradable particulate COD, mg/L
pCOD = concentration of particulate COD, mg/L
sCOD = concentration of soluble COD, mg/L

Care must be taken in sample handling and analyses to obtain reliable nbVSS concentration data. A sufficient number of composite samples must be obtained to assure that the results are representative of the wastewater characteristics. Samples must be well mixed when taken for analyses, and for small sample volumes, the pipettes must have wide openings at the tip to better capture solids. When small sample volumes are used, for example, with the HACH COD analysis, treating the sample first in a high-speed blender is often done. Filtration through 0.45-μm membrane filters is recommended to obtain soluble samples for COD and BOD analyses.

Nitrogen Compounds. For the nitrogen compounds, the soluble organic nitrogen concentration is of interest from the standpoint of its effect on the effluent total nitrogen concentration. A filtered sample from the plant effluent or from a bench-scale treatability reactor can be used to determine the total effluent soluble organic nitrogen concentration by the difference between the TKN concentration of the filtered sample and the effluent NH_4-N concentration. The *nonbiodegradable soluble organic nitrogen* (nbsON) cannot be determined directly, but from a practical standpoint and considering the low concentration of effluent soluble organic nitrogen, the total effluent soluble organic nitrogen concentration usually provides sufficient information.

The *nonbiodegradable particulate organic nitrogen* (nbpON) can be estimated by an analysis of the influent VSS for organic nitrogen and the estimated amount of nbVSS. The fraction of nitrogen in the VSS is as follows:

$$f_N = \frac{(TKN - sON - NH_4 \text{ as } N)}{VSS} \tag{8-5}$$

$$nbpON = f_N(nbVSS) \tag{8-6}$$

where f_N = fraction of organic nitrogen in VSS, g N/g VSS
TKN = total TKN concentration, mg/L
sON = soluble (i.e., filtered) organic nitrogen concentration, g/m^3
nbpON = nonbiodegradable particulate organic nitrogen concentration, g/m^3
Other terms as defined previously.

Summary Tabulation. In summary, the COD and nitrogen wastewater components can be tabulated as follows:

$$COD = bCOD + nbCOD \tag{8-7}$$

$$bCOD = \sim 1.6(BOD) \tag{8-8}$$

$$nbCOD = nbsCOD + nbpCOD \tag{8-9}$$

$$bCOD = sbCOD + rbCOD \tag{8-10}$$

$$TKN = NH_4\text{-}N + ON \tag{8-11}$$

$$ON = bON + nbON \tag{8-12}$$

$$nbON = nbsON + nbpON \tag{8-13}$$

where the terms are as defined in Table 8–2.

The application of the above equations in determining the characteristics of a wastewater is illustrated in Example 8–1.

EXAMPLE 8–1   **Wastewater Characterization Evaluation**   Given the following wastewater characterization results, determine concentrations for the following:

1. bCOD (biodegradable COD)
2. nbpCOD (nonbiodegradable particulate COD)
3. sbCOD (slowly biodegradable COD)
4. nbVSS (nonbiodegradable VSS)
5. iTSS (inert TSS)
6. nbpON (nonbiodegradable particulate organic nitrogen)
7. Total degradable TKN

Influent wastewater characteristics:

Constituent	Concentration, mg/L
BOD	195
sBOD	94
COD	465
sCOD	170
rbCOD	80
TSS	220
VSS	200
TKN	40
NH$_4$-N	26
Alkalinity	200 (as CaCO$_3$)

Activated-sludge effluent:

Constituent	Concentration, mg/L
sCODe	30
sON	1.2

Solution

1. Determine biodegradable (bCOD) using Eq. (8–8).

$$bCOD = \sim 1.6(BOD)$$

$$= 1.6(195 \text{ mg/L}) = 312 \text{ mg/L}$$

2. Determine the nbpCOD.
   a. Determine the nbCOD using Eq. (8–7).

   $$nbCOD = COD - bCOD$$

   $$nbCOD = (465 - 312) \text{ mg/L} = 153 \text{ mg/L}$$

   b. Determine the nbpCOD using Eq. (8–9).

   $$nbpCOD = nbCOD - sCODe$$

   $$= (153 - 30) \text{ mg/L} = 123 \text{ mg/L}$$

3. Determine the sbCOD using Eq. (8–10).

$$sbCOD = bCOD - rbCOD$$

$$= (312 - 80) \text{ mg/L} = 232 \text{ mg/L}$$

4. Determine the nbVSS.
   a. Determine the bpCOD/pCOD ratio using Eq. (8–4).

   $$\frac{bpCOD}{pCOD} = \frac{(bCOD/BOD)(BOD - sBOD)}{COD - sCOD}$$

   $$\frac{bpCOD}{pCOD} = \frac{1.6(195 - 94) \text{ mg/L}}{(465 - 170) \text{ mg/L}} = 0.55$$

   b. Determine the nbVSS using Eq. (8–3).

   $$nbVSS = \left[1 - \left(\frac{bpCOD}{pCOD}\right)\right]VSS$$

   $$nbVSS = (1 - 0.55)(200 \text{ mg/L}) = 90 \text{ mg/L}$$

5. Determine the inert TSS.

$$iTSS = TSS - VSS = (220 - 200) \text{ mg/L} = 20 \text{ mg/L}$$

6. Determine the nbpON.
   a. Determine the organic N content of VSS using Eq. (8–5).

   $$f_N = \frac{(TKN - sON - NH_4\text{-}N)}{VSS}$$

   $$f_N = \frac{(40 - 1.2 - 26) \text{ mg/L}}{(200 \text{ mg/L})} = 0.064$$

b. Determine the nbpON using Eq. (8–6).

$$nbpON = f_N(nbVSS)$$

$$nbpON = 0.064(90 \text{ mg/L}) = 5.8 \text{ mg/L}$$

7. Determine total degradable TKN.

$$bTKN = TKN - nbpON - sON$$

$$= (40 - 5.8 - 1.2) \text{ mg/L}$$

$$= 33 \text{ mg/L}$$

Recycle Flows and Loadings

The impact of recycle flows must also be quantified and included in defining the influent wastewater characteristics to the activated-sludge process. The possible sources of recycle flows include digester supernatant flows (if settling and decanting are practiced in the digestion operation), recycle of centrate or filtrate from solids dewatering equipment, backwash water from effluent filtration processes, and water from odor-control scrubbers. Depending on the source, a significant BOD, TSS, and NH_4-N load may be added to the influent wastewater. The levels of BOD and TSS concentrations possible for various solids processing unit operations are given in Table 14–49 in Chap. 14. Compared to untreated wastewater or primary clarifier effluent, the BOD/VSS ratio is often much lower for recycle streams. In addition, a significant NH_4-N load can be returned to the influent wastewater from anaerobic digestion-related processes. Concentrations of NH_4-N in the range of 1000 to 2000 mg/L are possible in centrate or filtrate from the dewatering of anaerobically digested solids. Thus, the ammonia load from a return flow of about one-half percent of the influent flow can increase the influent TKN load to the activated-sludge process by 10 to 20 percent. The return solids load from effluent polishing filters can be estimated by a mass balance on solids removed across the filtration process, and thus released in the backwash water flow. In all cases, a mass balance for flow and important constituents, such as BOD, TSS/VSS, nitrogen compounds, and phosphorus should be done to account for all contributing flows and loads to the activated-sludge process.

8–3 FUNDAMENTALS OF PROCESS ANALYSIS AND CONTROL

The purpose of this section is to introduce (1) the basic considerations involved in process design, (2) process control measures, (3) operating problems associated with the activated-sludge process, and (4) activated-sludge selector processes. The information presented in this section is applied to the analysis and design of alternative activated-sludge processes in the remainder of this chapter. Many of the equations presented in this chapter were derived previously in Chap. 7 and are summarized in this section for convenient reference.

Process Design Considerations

In the design of the activated-sludge process, consideration must be given to (1) selection of the reactor type, (2) applicable kinetic relationships, (3) solids retention time and loading criteria to be used, (4) sludge production, (5), oxygen requirements and transfer, (6) nutrient requirements, (7) other chemical requirements, (8) settling characteristics of biosolids, (9) use of selectors, and (10) effluent characteristics.

Selection of Reactor Type.   Important factors that must be considered in the selection of reactor types for the activated-sludge process include (1) the effects of reaction kinetics, (2) oxygen transfer requirements, (3) nature of the wastewater, (4) local environmental conditions, (5) presence of toxic or inhibitory substances in the influent wastewater, (6) costs, and (7) expansion to meet future treatment needs. Information on these factors is summarized in Table 8–4.

Kinetic Relationships.   As developed in Chap. 7, kinetic relationships are used to determine biomass growth and substrate utilization, and to define process performance. Important kinetic relationships are summarized in Table 8–5. The derivation of these relationships may be found in Chap. 7.

Selection of Solids Retention Time and Loading Criteria.   Certain design and operating parameters distinguish one activated-sludge process from another. The common parameters used are the solids retention time (SRT), the food to biomass (F/M) ratio (also known as food to microorganism ratio), and the volumetric organic loading rate. While the SRT is the basic design and operating parameter, the F/M ratio and volumetric loading rate provide values that are useful for comparison to historical data and typical observed operating conditions.

Solids Retention Time.   The SRT, in effect, represents the average period of time during which the sludge has remained in the system. As presented previously in Chap. 7, SRT is the most critical parameter for activated-sludge design as SRT affects the treatment process performance, aeration tank volume, sludge production, and oxygen requirements. For BOD removal, SRT values may range from 3 to 5 d, depending on the mixed-liquor temperature. At 18 to 25°C an SRT value close to 3 d is desired where only BOD removal is required and to discourage nitrification and eliminate the associated oxygen demand. To limit nitrification, some activated-sludge plants have been operated at SRT values of 1 d or less. At 10°C, SRT values of 5 to 6 d are common for BOD removal only. Temperature and other factors that affect SRT in various treatment applications are summarized in Table 8–6.

   Because nitrification is temperature-dependent, the design SRT for nitrification must be selected with caution as variable nitrification growth rates have been observed at different sites, presumably due to the presence of inhibitory substances (Barker and Dold, 1997; Fillos et al., 2000). For nitrification design, a safety factor is used to increase the SRT above that calculated based on nitrification kinetics and the required effluent NH_4-N concentration. A factor of safety is used for two reasons: (1) to allow flexibility for operational variations in controlling the SRT, and (2) to provide for additional nitrifying bacteria to handle peak TKN loadings. The influent TKN concentration

Table 8–4
General considerations for the selection of the type of suspended growth reactor

Factor	Description
Effect of reaction kinetics	The two types of reactors used commonly are the complete-mix and the plug-flow reactor. From a practical standpoint, the hydraulic detention times of many of the complete-mix and plug-flow reactors in actual use are about the same. The reason is that the designs for BOD removal are generally governed by an SRT sufficient to assure good settling properties and of a duration longer than that needed for BOD removal. For nitrification, the possible reaction kinetic benefits from using a staged-reactor or plug-flow system may be exploited, provided that the aeration equipment has a high enough oxygen transfer rate in first stage or at the front of a plug-flow tank to satisfy the demand from higher BOD removal and nitrification rates
Oxygen transfer requirements	Historically, in conventional plug-flow aeration systems, sufficient oxygen often could not be supplied at the beginning of the reactor to meet the demand. The inability to supply the needed oxygen led to development of the following modifications to the activated-sludge process: (1) tapered aeration in which an attempt was made to match the air supplied to the oxygen demand, (2) the step-feed process where the incoming wastewater is distributed along the length of the reactor (usually at quarter points), and (3) the complete-mix process where the air supplied uniformly matches or exceeds the oxygen demand. Most of the past oxygen transfer limitations have been overcome by better selection of process operational parameters and improvements in the design and application of aeration equipment
Nature of wastewater	The nature of the wastewater includes the overall characteristics of the wastewater as affected by contributions such as domestic wastewater, industrial discharges, and inflow/infiltration. Alkalinity and pH are important, particularly in the operation of nitrification processes (see Chap. 7). Because low pH values inhibit the growth of nitrifying organisms (and encourage the growth of filamentous organisms), pH adjustments may be required. Industrial waste discharges may also affect the pH in low-alkalinity wastewaters
Local environmental conditions	Temperature is an important environmental condition that affects treatment performance because changes in the wastewater temperature can affect the biological reaction rate. Temperature is especially important in nitrification design as the expected mixed-liquor temperature will affect the design SRT. Precipitation effects and groundwater infiltration are local factors that can affect both flowrates and constituent concentrations. High peak flowrates can cause the washout of solids in biological reactors
Toxic or inhibitory substances	For municipal wastewater treatment systems with a large number of industrial connections, a potential exists for receiving inhibitory substances that can depress biological nitrification rates. Where such potential exists, laboratory treatability studies are recommended to assess nitrification kinetics. If shock loads or toxic discharges are a design consideration, a complete-mix reactor can more easily withstand changing wastewater characteristics because the incoming wastewater is more or less uniformly dispersed with the reactor contents, as compared to a plug-flow reactor. The complete-mix process has been used in a number of installations to mitigate the impacts caused by shock loads and toxic discharges, especially from industrial installations
Cost	Construction and operating costs are very important considerations in selecting the type and size of reactor. Because the associated settling facilities are an integral part of the activated-sludge process, the selection of the reactor and the solids separation facilities must be considered as a unit
Future treatment needs	Potential future treatment needs can have an impact on present process selection. For example, if water reuse is anticipated in the future, the process selection should favor designs that can easily accommodate nitrogen removal and effluent filtration

Table 8–5
Summary of equations used in the analysis of suspended growth processes

Equation	Eq. No.	Definition of terms
$k_T = k_{20}\theta^{(T-20)}$	2-25	DO = dissolved oxygen, ML^{-3}
$r_{su} = -\dfrac{kXS}{K_s + S}$	7-12	F/M = food to microorganism ratio k = maximum rate of substrate utilization, T^{-1}
$\mu_m = kY$	7-13	k_d = endogenous decay coefficient, T^{-1} k_{dn} = endogenous decay coefficient for nitrifying organisms, T^{-1}
$r_{su} = \dfrac{\mu_m XS}{Y(K_s + S)}$	7-15	k_T = reaction rate coefficient at temperature (T) k_{20} = reaction rate coefficient at 20°C
$r_g = Y\dfrac{kXS}{K_s + S} - k_d X$	7-22	K_n = half-velocity constant, ML^{-3} K_o' = oxygen inhibition coefficient, ML^{-3}
$\mu = \dfrac{r_g}{X}$	7-23	K_s = half-velocity constant, ML^{-3} K_{s,NO_3} = half-velocity constant for nitrate limited reaction, ML^{-3}
$SRT = \dfrac{VX}{(Q - Q_w)X_e + Q_w X_R}$	7-35	L_{org} = volumetric organic loading rate, $ML^{-3}T^{-1}$ μ = specific growth rate, T^{-1}
$SRT = \dfrac{1}{\mu}$	7-37	μ_m = maximum specific growth rate, T^{-1} μ_n = specific growth rate for nitrification, T^{-1}
$\dfrac{1}{SRT} = -\dfrac{YkS}{K_s + S} - k_d$	7-39	μ_{nm} = maximum specific growth rate of nitrifying bacteria, T^{-1} N = nitrogen concentration, ML^{-3}
$S = \dfrac{K_s[1 + (k_d)SRT]}{SRT(Yk - k_d) - 1}$	7-40	η = ratio of substrate utilization rate with nitrate versus oxygen as the electron acceptor P_X = solids, MT^{-1}
$X = \left(\dfrac{SRT}{\tau}\right)\left[\dfrac{Y(S_o - S)}{1 + (k_d)SRT}\right]$	7-43	Q = flowrate, L^3T^{-1} Q_w = waste sludge flowrate L^3T^{-1}
$(X_{VSS})(V) = (P_{X,VSS})SRT$	7-54	R_o = oxygen, MT^{-1}
$(X_{TSS})(V) = (P_{X,TSS})SRT$	7-55	r_g = net biomass production rate, $ML^{-3}T^{-1}$
$R_o = Q(S_o - S) - 1.42\,P_{X,bio}$	7-59	r_{su} = soluble substrate utilization rate, $ML^{-3}T^{-1}$ S = concentration of growth-limiting substrate in solution, ML^{-3}
$F/M = \dfrac{QS_o}{VX}$	7-60	SF = safety factor S_o = influent concentration, ML^{-3}
$L_{org} = \dfrac{(Q)(S_o)}{(V)}$	7-67	SRT = solids retention time, T TSS = total suspended solids, M
$SF = SRT_{des}/SRT_{min}$	7-71	τ = hydraulic retention time (V/Q), T θ = temperature activity coefficient
$\mu_n = \left(\dfrac{\mu_{nm}N}{K_n + N}\right)\left(\dfrac{DO}{K_o + DO}\right) - k_{dn}$	7-93	U = substrate utilization rate, T^{-1} V = volume, L^3 VSS = volatile suspended solids, M
$r_{su} = -\left(\dfrac{kXS}{K_s + S}\right)\left(\dfrac{NO_3}{K_{s,NO_3} + NO_3}\right)\left(\dfrac{K_o'}{K_o' + DO}\right)\eta$	7-110	X = biomass concentration, ML^{-3} X_e = concentration of biomass in the effluent, ML^{-3} X_R = concentration of biomass in the return line from clarifier, ML^{-3} Y = biomass yield, M of cell formed per M of substrate consumed

Note: Expressions for units are M = mass, L = length, and T = time.

Table 8–6
Typical minimum SRT ranges for activated-sludge treatment[a]

Treatment goal	SRT range, d	Factors affecting SRT
Removal of soluble BOD in domestic wastewater	1–2	Temperature
Conversion of particulate organics in domestic wastewater	2–4	Temperature
Develop flocculent biomass for treating domestic wastewater	1–3	Temperature
Develop flocculent biomass for treating industrial wastewater	3–5	Temperature/compounds
Provide complete nitrification	3–18	Temperature/compounds
Biological phosphorus removal	2–4	Temperature
Stabilization of activated sludge	20–40	Temperature
Degradation of xenobiotic compounds	5–50	Temperature/specific bacteria/compounds

[a] Adapted from Grady et al. (1999).

and mass loading can vary throughout the day (a peak to average TKN loading of 1.3 to 1.5 is not unusual, depending on plant size) and can also be affected by return flows from digested and dewatered biosolids processing. By increasing the design SRT, the inventory of nitrifying bacteria is increased to meet the NH_4-N concentration at the peak load so that the effluent NH_4-N concentration requirement is achieved.

Typically, the value of the factor of safety is equal to the peak/average TKN load. Because use of the peak/average TKN load is conservative, the NH_4-N concentration during the normal loading period will be lower with the net effect of a composite effluent NH_4-N concentration that is somewhat lower than the design goal. Dynamic simulation models can be used to optimize the design SRT value to meet target effluent NH_4-N concentrations, subject to changing influent flow and TKN concentrations (Barker and Dold, 1997). The steady-state solution approach described in Sec. 8–10 has resulted in reasonable designs.

Food to Microorganism Ratio. A process parameter commonly used to characterize process designs and operating conditions is the food to microorganism (biomass) ratio (*F/M*). Typical values for the BOD *F/M* ratio reported in the literature vary from 0.04 g substrate/g biomass·d for extended aeration processes to 1.0 g/g·d for high rate processes. The BOD *F/M* ratio is usually evaluated for systems that were designed based on SRT to provide a reference point to previous activated-sludge design and operating performance.

Volumetric Organic Loading Rate. The volumetric organic loading rate is defined as the amount of BOD or COD applied to the aeration tank volume per day. Organic loadings, expressed in kg BOD or COD/m^3·d, may vary from 0.3 to more than

3.0. While the mixed-liquor concentration, the *F/M* ratio, and the SRT (which may be considered an operating variable as well as a design parameter) are ignored when such empirical relationships are used, these relationships do have the merit of requiring a minimum aeration tank volume that has proved to be adequate for the treatment of domestic wastewater. These empirical parameters are not adequate for predicting effluent quality, however, when such relationships are used to design facilities for the treatment of wastewater containing industrial wastes or for biological nitrogen- and phosphorus-removal processes. Higher volumetric organic loadings generally result in higher required oxygen transfer rates per unit volume for the aeration system.

Sludge Production. The design of the sludge-handling and disposal/reuse facility depends on the prediction of sludge production for the activated-sludge process. If the sludge-handling facilities are undersized, then the treatment process performance may be compromised. Sludge will accumulate in the activated-sludge process if it cannot be processed fast enough by an undersized sludge-handling facility. Eventually, the sludge inventory capacity of the activated-sludge system will be exceeded and excess solids will exit in the secondary clarifier effluent, potentially violating discharge limits. The sludge production relative to the amount of BOD removed also affects the aeration tank size.

Two methods are used to determine sludge production. The first method is based on an estimate of an observed sludge production yield from published data from similar facilities, and the second is based on the actual activated-sludge process design in which wastewater characterization is done and the various sources of sludge production are considered and accounted for. The first method may be satisfactory for determining an initial activated-sludge process design and an estimated sludge production for a particular activated-sludge process. With the first method, the quantity of sludge produced daily (and thus wasted daily) can be estimated using Eq. (8–14). For a given wastewater, the Y_{obs} value will vary depending on whether the substrate is defined as BOD, bCOD, or COD.

$$P_{X,VSS} = Y_{obs}(Q)(S_o - S)(1 \text{ kg}/10^3 \text{ g}) \tag{8–14}$$

where $P_{X,VSS}$ = net waste activated sludge produced each day, kg VSS/d
Y_{obs} = observed yield, g VSS/g substrate removal
Q = influent flow, m^3/d
S_o = influent substrate concentration, mg/L
S = effluent substrate concentration, mg/L

Observed volatile suspended solids yield values, based on BOD, are illustrated on Fig. 8–7. The observed yield decreases as the SRT is increased due to biomass loss by more endogenous respiration. The yield is lower with increasing temperature as a result of a higher endogenous respiration rate at higher temperature. The yield is higher when no primary treatment is used, as more nbVSS remains in the influent wastewater. The temperature correction value θ for endogenous respiration [see Eq. (2–25) in Table 8–5] is 1.04 between 20 and 30°C, and 1.12 between 10 and 20°C. A θ value of 1.04 has been adopted in this text for the temperature effect on endogenous decay.

With sufficient wastewater characterization, a more accurate prediction of sludge production can be made. The following equation, based on Eq. (7–52) in Chap. 7,

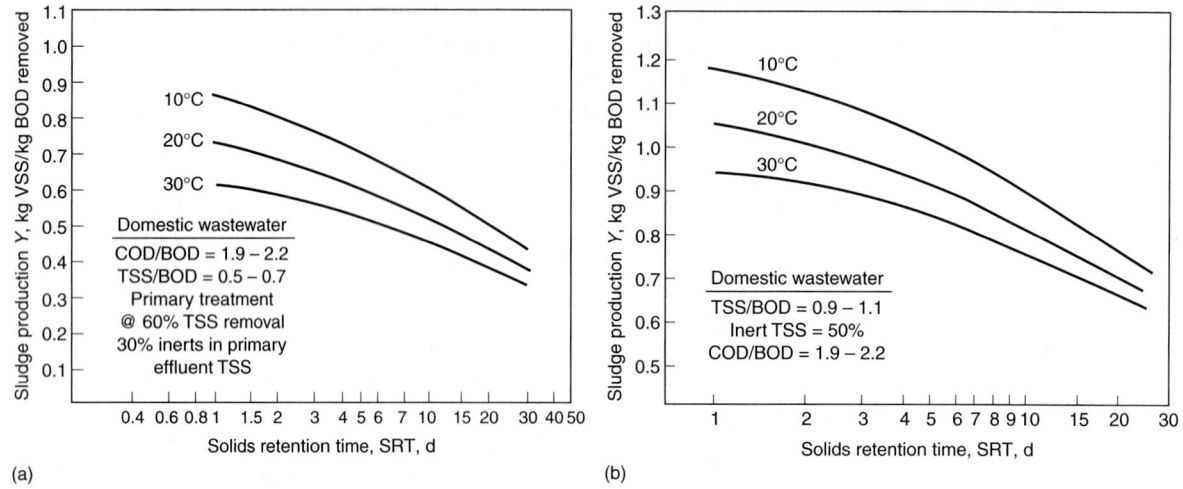

Figure 8–7

Net solids production vs. solids retention time (SRT) and temperature: (a) with primary treatment and (b) without primary treatment.

accounts for the heterotrophic biomass growth, cell debris from endogenous decay, nitrifying bacteria biomass, and nonbiodegradable volatile suspended solids and can be used to estimate sludge production.

$$
P_{X,\text{VSS}} = \underbrace{\frac{QY(S_o - S)(1\ \text{kg}/10^3\ \text{g})}{1 + (k_d)\text{SRT}}}_{\substack{\text{(A)}\\ \text{Heterotrophic}\\ \text{biomass}}} + \underbrace{\frac{(f_d)(k_d)QY(S_o - S)\text{SRT}(1\ \text{kg}/10^3\ \text{g})}{1 + (k_d)\text{SRT}}}_{\substack{\text{(B)}\\ \text{Cell}\\ \text{debris}}}
$$

$$
+ \underbrace{\frac{QY_n(\text{NO}_x)(1\ \text{kg}/10^3\ \text{g})}{1 + (k_{dn})\text{SRT}}}_{\substack{\text{(C)}\\ \text{Nitrifying bacteria}\\ \text{biomass}}} + \underbrace{Q(\text{nbVSS})(1\ \text{kg}/10^3\ \text{g})}_{\substack{\text{(D)}\\ \text{Nonbiodegradable}\\ \text{VSS in influent}}} \qquad (8\text{–}15)
$$

where NO_x = concentration of $\text{NH}_4\text{-N}$ in the influent flow that is nitrified, mg/L
$\quad k_{dn}$ = endogenous decay coefficient for nitrifying organisms, g VSS/g VSS·d
Other terms as defined previously.

The total mass of dry solids wasted/day includes TSS and not just VSS. The TSS includes the VSS plus inorganic solids. Inorganic solids in the influent wastewater $(\text{TSS}_o - \text{VSS}_o)$ contribute to inorganic solids and are an additional solids production term that must be added to Eq. (8–15). The biomass terms in Eq. (8–15) (A, B, and C) contain inorganic solids and the VSS fraction of the total biomass is about 0.85, based on the cell composition given in Table 7–4. Thus, Eq. (8–15) is modified as follows to calculate the solids production in terms of TSS:

$$P_{X,\text{TSS}} = \frac{A}{0.85} + \frac{B}{0.85} + \frac{C}{0.85} + D + Q(\text{TSS}_o - \text{VSS}_o) \qquad (8\text{–}16)$$

$$\underset{\substack{\text{(E)} \\ \text{Inert TSS} \\ \text{in influent}}}{}$$

where TSS_o = influent wastewater TSS concentration, mg/L
$\quad\quad \text{VSS}_o$ = influent wastewater VSS concentration, mg/L

The daily mass of solids in the aeration tanks is determined from the SRT. The daily sludge production can be computed by Eqs. (7–54) and (7–55) in Table 8–5.

By selecting an appropriate MLSS concentration, the aeration volume can be determined using Eq. (7–55) in Table 8–5. Concentrations of MLSS generally in the range of 1200 to 4000 mg/L may be selected, but must be compatible with the sludge settling characteristics and clarifier design as discussed later in this section and in Sec. 8–7.

Oxygen Requirements. The oxygen required for the biodegradation of carbonaceous material is determined from a mass balance using the bCOD concentration of the wastewater treated and the amount of biomass wasted from the system per day. If all of the bCOD were oxidized to CO_2, H_2O, and NH_3, the oxygen demand would equal the bCOD concentration. However, bacteria oxidize a portion of the bCOD to provide energy and use the remaining portion of the bCOD for cell growth. Oxygen is also consumed for endogenous respiration, and the amount will depend on the system SRT. For a given SRT, a mass balance on the system can be done where the bCOD removal equals the oxygen used plus the biomass VSS remaining in terms of an oxygen equivalent. The oxygen requirements for BOD removal without nitrification can be computed using Eq. (7–59) in Table 8–5. As an approximation, for BOD removal only, the oxygen requirement will vary from 0.90 to 1.3 kg O_2/kg BOD removed for SRTs of 5 to 20 d, respectively (WEF, 1998).

When nitrification is included in the process, the total oxygen requirements will include the oxygen required for removal of carbonaceous material plus the oxygen required for ammonia and nitrite oxidation to nitrate (see Sec. 7–9 in Chap. 7) as follows:

$$R_o = Q(S_o - S) - 1.42P_{X,\text{bio}} + 4.33Q(\text{NO}_x) \qquad (8\text{–}17)$$

where R_o = total oxygen required, g/d
$\quad\quad P_{X,\text{bio}}$ = biomass as VSS wasted, g/d [parts A, B, and C of Eq. (8–15)]
$\quad\quad$ Other terms as defined previously.

As shown in Eq. (8–17), NO_x is the amount of TKN oxidized to nitrate. A nitrogen mass balance for the system that accounts for the influent TKN, nitrogen removed for biomass synthesis, and unoxidized effluent nitrogen is done to determine NO_x. Unless a careful wastewater characterization study is done to determine the nonbiodegradable particulate and soluble nitrogen (nbpON and nbsON), these components are ignored. Ignoring these terms results in predicting a slightly higher NO_x concentration (5 to 15 percent) and a more conservative oxygen requirement estimate from Eq. (8–17). The nitrogen mass balance, based on the assumption of 0.12 g N/g biomass ($C_5H_7NO_2$ for biomass), is as follows:

$$\text{Nitrogen} = \text{nitrogen in} - \text{nitrogen in} - \text{nitrogen in}$$
$$\text{oxidized} \quad\quad \text{influent} \quad\quad \text{effluent} \quad\quad \text{cell tissue}$$

$$Q(NO_x) = Q(TKN_o) - \quad QN_e \quad - 0.12\, P_{X,bio}$$

$$NO_x = TKN - N_e - 0.12\, P_{X,bio}/Q \tag{8-18}$$

where NO_x = nitrogen oxidized, mg/L
$\quad\quad TKN_o$ = influent TKN concentration, mg/L
$\quad\quad\quad N_e$ = effluent NH_4-N concentration, mg/L
$\quad\quad$ Other terms as defined previously.

Equation (8–18) can be solved for the NO_x concentration by estimating the effluent NH_4-N concentration from the nitrification process design.

Nutrient Requirements.   If a biological system is to function properly, nutrients must be available in adequate amounts. As discussed in Chaps. 2 and 7, the principal nutrients are nitrogen and phosphorus. Using the formula $C_5H_7NO_2$, for the composition of cell biomass, about 12.4 percent by weight of nitrogen will be required. The phosphorus requirement is usually assumed to be about one-fifth of the nitrogen value. These are typical values, not fixed quantities, because it has been shown that the percentage distribution of nitrogen and phosphorus in cell tissue varies with the system SRT and environmental conditions. The amount of nutrients required can be estimated based on the daily biomass production rate [terms A, B, and C in Eq. (8–15)]. It should be noted that nutrient limitations can occur when the concentrations of nitrogen and phosphorus are in the range of 0.1 to 0.3 mg/L. As a general rule, for SRT values greater than 7 d, about 5 g nitrogen and 1 g phosphorus will be required per 100 g of BOD to provide an excess of nutrients.

Other Chemical Requirements.   In addition to the nutrient requirements, alkalinity is a major chemical requirement needed for nitrification. The amount of alkalinity required for nitrification, taking into account cell growth, is about 7.07 g $CaCO_3$/g NH_4-N [see Eq. (7–91) in Chap. 7]. In addition to the alkalinity required for nitrification, additional alkalinity must be available to maintain the pH in the range from 6.8 to 7.4. Typically the amount of residual alkalinity required to maintain pH near a neutral point (i.e., pH $\sim$ 7) is between 70 and 80 mg/L as $CaCO_3$.

Mixed-Liquor Settling Characteristics.   The settling characteristics of the mixed-liquor solids must be considered when designing the secondary clarifier for liquid-solids separation. Clarifier design must provide adequate clarification of the effluent and solids thickening for the activated-sludge solids. In the design of installations where sludge characteristics are not known, data from other installations must be assumed or experience of the designer with similar suspended growth processes must be utilized. Tests are available (as described below) to evaluate sludge settling or thickening characteristics, and techniques have been developed to apply these fundamental characteristics to clarifier design.

Two commonly used measures developed to quantify the settling characteristics of activated sludge are the sludge volume index (SVI) and the zone settling rate (WEF,

1998). The SVI is the volume of 1 g of sludge after 30 min of settling. The SVI is determined by placing a mixed-liquor sample in a 1- to 2-L cylinder and measuring the settled volume after 30 min and the corresponding sample MLSS concentration. The numerical value is computed using the following expression:

$$SVI = \frac{(\text{settled volume of sludge, mL/L})(10^3 \text{ mg/g})}{(\text{suspended solids, mg/L})} = \frac{mL}{g} \qquad (8\text{-}19)$$

For example, a mixed-liquor sample with a 3000 mg/L TSS concentration that settles to a volume of 300 mL in 30 min in a 1-L cylinder would have an SVI of 100 mL/g. A value of 100 mL/g is considered a good settling sludge (SVI values below 100 are desired). SVI values above 150 are typically associated with filamentous growth (Parker et al., 2001).

Because the SVI test is empirical, it is subject to significant errors. For example, if sludge with a concentration of 10,000 mg/L did not settle at all after 30 min, the SVI value would be 100. To avoid erroneous results and to allow for a meaningful comparison of SVI results for different sludges, the diluted SVI (DSVI) test has been used (Jenkins et al., 1993). In Jenkins's analysis, the sludge sample is diluted with process effluent until the settled volume after 30 min is 250 mL/L or less. The standard SVI test is then followed with this sample.

Many SVI tests at wastewater treatment plants are done in a 2-L settleometer that has a larger diameter than 1- or 2-L graduated cylinders (see Fig. 8–8). To eliminate well effects on solids settling in a small-diameter test apparatus, use of a slow-speed stirring device is encouraged (Wahlberg and Keinath, 1988). The test is called a stirred SVI when a stirring device is used (see Standard Methods, WEF, 1998). The stirred SVI test is used frequently in Europe.

Figure 8–8

Field test for determining sludge volume index (SVI).

The secondary clarifier design has also been related to an expected zone settling velocity (ZSV). The ZSV can also be observed during SVI tests or can be measured in larger, taller cylinders. The ZSV is the settling velocity of the sludge/water interface at the beginning of the sludge settleability test (V_i), and the procedure is described in Standard Methods (WEF, 1998). The surface overflow rate, based on a zone settling velocity, is determined using the following expression:

$$OR = \frac{(V_i)(24)}{SF} \tag{8-20}$$

where OR = surface overflow rate, m³/m²·d
V_i = settling velocity of interface, m/h (m³/m²·h)
24 = conversion factor from m/h to m/d
SF = safety factor, typically 1.75 to 2.5

An advantage of this procedure is that the value for ZSV can be selected to account for the effect of higher MLSS concentrations. At higher MLSS concentrations, the value for V_i will decrease and the resulting clarifier surface area will increase. The V_i value can be estimated using the following equation (Wilson and Lee, 1982; Wilson, 1996):

$$V_i = V_{max} \exp(-K \times 10^{-6})X \tag{8-21}$$

where V_i = settling velocity of interface, m/h
V_{max} = maximum settling velocity of interface, typically 7 m/h
K = constant, typically 600 L/mg for activated-sludge mixed liquor with an SVI of 150
X = average MLSS concentration, mg/L

Correlations between V_i and the MLSS concentration have been developed with SVI as an additional parameter using data from several facilities (Daigger, 1995; Wahlberg, 1995). The results of the correlations are summarized by the following equations:

$$\ln(V_i) = 1.871 - (0.1646 + 0.001586 \text{ SVI}) X_T \tag{8-22}$$

$$\ln(V_i) = 2.028 - (0.1030 + 0.0025555 \text{ DSVI}) X_T \tag{8-23}$$

where DSVI = diluted SVI, mL/g
X_T = MLSS concentration, g/L

Secondary Clarification.   Several approaches are used in the design of secondary clarification facilities. The approach used most commonly is to base the design on a consideration of the surface overflow rate and the solids loading rate. Because steady-state operations seldom occur due to fluctuations in wastewater flowrate, return activated-sludge flowrate, and MLSS concentrations, attention to the occurrence of peak events and use of safety factors [Eq. (8–20)] are important design considerations.

Typical surface overflow rates are given in Table 8–7. Overflow rates are based on wastewater flowrates instead of on the mixed-liquor flowrates because the overflow rate is equivalent to an upward flow velocity. The return sludge flow is drawn off the bottom of the tank and does not contribute to the upward flow velocity. Selection of a surface overflow rate is influenced by the target effluent requirements and the need to provide consistent process performance (see Chap. 15).

Table 8-7

Typical design information for secondary clarifiers for the activated-sludge process[a]

Type of treatment	Overflow rate gal/ft²·d Average	Overflow rate gal/ft²·d Peak	Overflow rate m³/m²·d Average	Overflow rate m³/m²·d Peak	Solids loading lb/ft²·h Average	Solids loading lb/ft²·h Peak	Solids loading kg/m²·h Average	Solids loading kg/m²·h Peak	Depth, m[b]
Settling following air-activated sludge (excluding extended aeration)	400–700	1000–1600	16–28	40–64	0.8–1.2	1.6	4–6	8	3.5–6
Selectors, biological nutrient removal	400–700	1000–1600	16–28	40–64	1.0–1.5	1.8	5–8	9	3.5–6
Settling following oxygen-activated sludge	400–700	1000–1600	16–28	40–64	1.0–1.4	1.8	5–7	9	3.5–6
Settling following extended aeration	200–400	600–800	8–16	24–32	0.2–1.0	1.4	1.0–5	7	3.5–6
Settling for phosphorus removal; effluent concentration, mg/L									
Total P = 2	600–800		24–32						
Total P = 1[c]	400–600		16–24						
Total P = 0.2–0.5[d]	300–500		12–20						

[a] Adapted in part from Kang (1987); WEF (1998).

[b] m × 3.2808 = ft.

[c] Occasional chemical addition required.

[d] Continuous chemical addition required for effluent polishing.

Note: Peak is a 2-h sustained peak.

The solids loading rate on an activated-sludge settling tank may be computed by dividing the total solids applied by the surface area of the tank.

$$\text{SLR} = \frac{(Q + Q_R)(X)}{A} \tag{8-24}$$

where SLR = solids loading rate
Q = influent flowrate, m³/d
Q_R = return activated-sludge flowrate, m³/d
X = MLSS concentration, mg/L
A = clarifier cross-sectional area, m²

The commonly used units for SLR are kilograms per square meter per hour (kg/m²·h) [pounds per square foot per hour (lb/ft²·h)]. If peak flowrates are of short duration, average 24-h values may govern; if peaks are of long duration, peak values should be assumed to govern to prevent the solids from overflowing the tank.

In effect, the solids loading rate represents a characteristic value for the suspension under consideration. In a settling tank of fixed surface area, the effluent quality will deteriorate if the solids loading is increased beyond the characteristic value for the suspension. Typical solids loading values used for the design of biological systems are also given in Table 8–7. Higher rates should not be used for design without extensive experimental work covering all seasons and operating variables.

While the surface overflow rate has been the historical clarifier design parameter, the solids loading rate is considered by some to be the limiting parameter that affects the effluent concentration. Parker et al. (2001) have shown that with a proper hydraulic design and management of solids in the sedimentation tank, the overflow rate has little or no effect on the effluent quality over a wide range of overflow rates, and the design can be based on the solids loading rates. Wahlberg (1995) supports Parker's position and, based on the evaluation of secondary clarifier performance for a number of facilities, found no effect of using surface overflow rates up to 3.4 m/h.

Additional information for the physical design of solids separation facilities is given in Sec. 8–7.

Use of Selectors. Because solids separation is one of the most important aspects of biological wastewater treatment, a biological selector (a small contact tank) is often incorporated in the design to limit the growth of organisms that do not settle well. Selectors are naturally incorporated into the biological nitrogen- and phosphorus-removal processes described. For BOD removal only or BOD removal and nitrification processes, an appropriate selector design can be added before the activated-sludge aeration basin. Selector design concepts and design are discussed at the end of this section.

Effluent Characteristics. The major parameters of interest that determine the effluent quality from biological treatment processes consist of organic compounds, suspended solids, and nutrients as indicated by the following four constituents:

1. Soluble biodegradable organics
    a. Organics that escaped biological treatment

      b. Organics formed as intermediate products in the biological degradation of the waste

      c. Cellular components (result of cell death or lysis)

2. Suspended organic material

      a. Biomass produced during treatment that escaped separation in the final settling tank

      b. Colloidal organic solids in the plant influent that escaped treatment and separation

3. Nitrogen and phosphorus

      a. Contained in biomass in effluent suspended solids

      b. Soluble nitrogen as NH_4-N, NO_3-N, NO_2-N, and organic N

      c. Soluble orthophosphates

4. Nonbiodegradable organics

      a. Those originally present in the influent

      b. Byproducts of biological degradation

In a well-operating activated-sludge process treating domestic wastes with an SRT ≥ 4 d, the soluble carbonaceous BOD of a filtered sample is usually less than 3.0 mg/L. Although use of Eq. (7–40) may result in predicted effluent sBOD concentrations < 1.0 mg/L, the practical sBOD measurement value appears to be in the range of 2 to 4 mg/L. With a proper secondary clarifier design and good settling sludge, the effluent suspended solids may be in the range of 5 to 15 mg/L. Assuming an sBOD of 3 mg/L, a VSS/TSS ratio of 0.85, and an effluent TSS of 10 mg/L, the final effluent BOD concentration, BOD_e, can be estimated as follows:

$$BOD_e = sBOD + \left(\frac{1 \text{ g BOD}}{1.42 \text{ g VSS}} \right) \left(\frac{0.85 \text{ g VSS}}{\text{g TSS}} \right) (\text{TSS, mg/L}) \qquad (8\text{–}25)$$

$$BOD_e = 3 \text{ mg/L} + (0.70)(0.85)(10 \text{ mg/L})$$

$$BOD_e = 8.6 \text{ mg/L}$$

Process Control

To maintain high levels of treatment performance with the activated-sludge process under a wide range of operating conditions, special attention must be given to process control. The principal approaches to process control are (1) maintaining dissolved oxygen levels in the aeration tanks, (2) regulating the amount of return activated sludge (RAS), and (3) controlling the waste-activated sludge (WAS). The parameter used most commonly for controlling the activated-sludge process is SRT. The mixed-liquor suspended solids (MLSS) concentration may also be used as a control parameter. Return activated sludge is important in maintaining the MLSS concentration and controlling the sludge blanket level in the secondary clarifier. The waste activated-sludge flow from the recycle line is selected usually to maintain the desired SRT. Oxygen uptake rates (OURs) are also measured as a means of monitoring and controlling the activated-sludge process. Routine microscopic observations are important for monitoring the microbial characteristics and for early detection of changes that might negatively impact sludge settling and process performance.

Dissolved Oxygen Control. Theoretically, the amount of oxygen that must be transferred in the aeration tanks equals the amount of oxygen required by the microorganisms in the activated-sludge system to oxidize the organic material. In practice, the transfer efficiency of oxygen for gas to liquid is relatively low so that only a small amount of oxygen supplied is used by the microorganisms. When oxygen limits the growth of microorganisms, filamentous organisms may predominate and the settleability and quality of the activated sludge may be poor. In general, the dissolved oxygen concentration in the aeration tank should be maintained at about 1.5 to 2 mg/L in all areas of the aeration tank. Higher DO concentrations (>2.0 mg/L) may improve nitrification rates in reactors with high BOD loads. Values above 4 mg/L do not improve operations significantly, but increase the aeration costs considerably.

Return Activated-Sludge Control. The purpose of the return of activated sludge is to maintain a sufficient concentration of activated sludge in the aeration tank so that the required degree of treatment can be obtained in the time interval desired. The return of activated sludge from the final clarifier to the inlet of the aeration tank is the essential feature of the process. Ample return sludge pump capacity should be provided and is important to prevent the loss of sludge solids in the effluent. The solids form a sludge blanket in the bottom of the clarifier, which can vary in depth with flow and solids loadings variations to the clarifier. At transient peak flows, less time for sludge thickening is available so that the sludge blanket depth increases. Sufficient return sludge pumping capacity is needed, along with sufficient clarifier depth (3.7 to 5.5 m), to maintain the blanket below the effluent weirs. Return sludge pumping rates of 50 to 75 percent of the average design wastewater flowrate are typical, and the design average *capacity* is typically of 100 to 150 percent of the average design flowrate. Return sludge concentrations from secondary clarifiers range typically from 4000 to 12,000 mg/L (WEF, 1998).

Several techniques are used to calculate the desirable return sludge flowrate. Common control strategies for determining the return sludge flowrate are based on maintaining either a target MLSS level in the aeration tanks or a given sludge blanket depth in the final clarifiers. The most commonly used techniques to determine return sludge flowrate are (1) settleability, (2) sludge blanket level control, (3) secondary clarifier mass balance, and (4) aeration tank mass balance.

Settleability. Using the settleability test, the return sludge pumping rate is set so that the flowrate is approximately equal to the percentage ratio of the volume occupied by the settleable solids from the aeration tank effluent to the volume of the clarified liquid (supernatant) after settling for 30 min in a 1000-mL graduated cylinder. This ratio should not be less than 15 percent at any time. For example, if the settleable solids occupied a volume of 275 mL after 30 min of settling, the percentage volume would be equal to 38 percent [(275 mL / 725 mL) × 100]. If the plant flow were 2 m³/s, the return sludge rate should be 0.38 × 2 m³/s = 0.76 m³/s.

Another settleability test method often used to control the rate of return sludge pumping is based on the sludge volume index (SVI). If the SVI is known, then the percentage of return sludge, in terms of the recirculation ratio (Q_r/Q) required to maintain a given percentage of mixed-liquor solids concentration in the aeration tank is

$$Q_r = 100/[(100/P_w \text{SVI}) - 1] \tag{8-26}$$

where Q_r = return activated-sludge flowrate, percent of influent flowrate
   P_w = MLSS expressed as a percentage
   SVI = as defined by Eq. (8–19)

For example, to maintain a mixed-liquor solids concentration of 0.30 percent (3000 mg/L), the percentage of sludge that must be returned when the sludge volume index is 100 is equal to $100/[(100/0.30 \times 100) - 1]$, or 43 percent.

Sludge Blanket Level. With the sludge blanket level control method, an optimum sludge blanket level is maintained in the clarifiers. The optimum level is determined by experience and is a balance between settling depth and sludge storage. The optimum depth of the sludge blanket usually ranges between 0.3 and 0.9 m (1 and 3 ft). The sludge blanket method of control requires considerable operator attention because of the diurnal flow and sludge production variations and changes in the settling characteristics of the sludge. Several methods are available to detect the sludge blanket levels, including withdrawing samples using air-lift pumps, gravity-flow tubes, portable sampling pumps, and core samplers, or using sludge-supernatant interface detectors.

Mass-Balance Analysis. The return sludge pumping rate may also be determined by making a mass-balance analysis around either the settling tank or the aeration tank. The appropriate limits for the two mass-balance analyses are illustrated on Fig. 8–9. Assuming the sludge-blanket level in the settling tank remains constant and that the solids in the effluent from the settling tank are negligible, the mass balance around the settling tank shown on Fig. 8–9a is as follows:

Accumulation = inflow − outflow

$$0 = X(Q + Q_R) - Q_R X_R - Q_w X_R - Q_e X_e \qquad (8\text{--}27)$$

where X = mixed-liquor suspended solids, mg/L
   Q = secondary influent flowrate, m³/s
   Q_R = return sludge flowrate, m³/s
   X_R = return activated-sludge suspended solids, mg/L

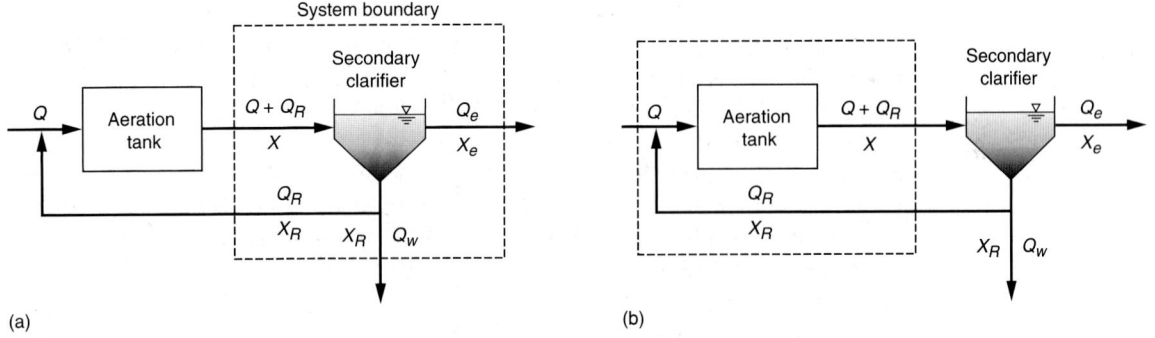

(a)   (b)

Figure 8–9

Definition sketch for suspended solids mass balances for return sludge control: (a) secondary clarifier mass balance and (b) aeration tank mass balance.

Q_w = waste activated-sludge flowrate, m³/s
Q_e = effluent flowrate, m³/s
X_e = effluent suspended solids, mg/L

Assuming X_e is negligible and that $Q_w X_R$ is related to the SRT [Eq. (7–35)], solving Eq. (8–27) for Q_R yields

$$Q_R = \frac{[XQ - (XV/SRT)]}{X_R - X}$$ (8–28)

The recycle ratio ($Q_R/Q = R$) is then

$$R = \frac{1 - (\tau/SRT)}{(X_R/X) - 1}$$ (8–29)

The required RAS pumping rate can also be estimated by performing a mass balance around the aeration tank (see Fig. 8–9b). The solids entering the tank will equal the solids leaving the tank if new cell growth can be considered negligible. Under conditions such as high organic loadings, this assumption may be incorrect. Solids enter the aeration tank in the return sludge and in the influent to the secondary process. However, if the influent solids are negligible compared to the MLSS, the mass balance around the aeration tank results in the following expression:

Accumulation = inflow − outflow

$$0 = X_R Q_R - X(Q + Q_R)$$ (8–30)

Solving for the return activated-sludge ratio R yields

$$Q_R/Q = R = \frac{X}{X_R - X}$$ (8–31)

Sludge Wasting. To maintain a given SRT, the excess activated sludge produced each day must be wasted. The most common practice is to waste sludge from the return sludge line because RAS is more concentrated and requires smaller waste sludge pumps. The waste sludge can be discharged to the primary sedimentation tanks for co-thickening, to thickening tanks, or to other sludge-thickening facilities. An alternative method of wasting sometimes used is withdrawing mixed liquor directly from the aeration tank or the aeration tank effluent pipe where the concentration of solids is uniform. The waste mixed liquor can then be discharged to a sludge-thickening tank or to the primary sedimentation tanks where it mixes and settles with the untreated primary sludge. The actual amount of liquid that must be pumped to achieve process control depends on the method used and the location from which the wasting is to be accomplished. Also, because the solids capture of the sludge processing facilities is not 100 percent, and some solids are returned, the actual wasting rate will be higher than the theoretically determined value. For example, if SRT is used for process control and wasting is from the return sludge line, the wasting rate can be computed by modifying the terms of Eq. (7–32) (see Table 8–5).

$$SRT = \frac{VX}{(Q_w X_R + Q_e X_e)}$$ (8–32)

where V = volume of the reactor, m^3
X = aeration tank mass concentration, mg/L
Q_w = waste sludge flowrate from the return sludge line, m^3/d
X_R = concentration of sludge in the return sludge line, mg/L
Q_e = effluent flowrate from secondary clarifier, m^3/d
X_e = effluent TSS concentration, mg/L

If it is assumed that the concentration of solids in the effluent from the settling tank is low, then Eq. (8–32) reduces to

$$SRT \approx \frac{VX}{Q_w X_R} \tag{8-33}$$

and

$$Q_w \approx \frac{VX}{(X_R)SRT} \tag{8-34}$$

To determine the waste flowrate using Eq. (8–34), the solids concentration in both the aeration tank and the return line must be measured.

If wasting is done from the aeration tank and the solids in the settled effluent are again neglected, then the rate of pumping can be estimated using the following relationship:

$$SRT = \frac{V}{Q_w} \tag{8-35}$$

or

$$Q_w = \frac{V}{SRT} \tag{8-36}$$

where Q_w = waste sludge flowrate from the aeration tank, m^3/d

Thus, the process may be controlled by daily wasting of a quantity of flow equal to the volume of the aeration tank divided by the SRT.

Oxygen Uptake Rates. Microorganisms in the activated-sludge process use oxygen as they consume the substrate. The rate at which they use oxygen, known as the *oxygen uptake rate* (OUR), is a measure of the biological activity and loading on the aeration tank. Values for the OUR are obtained by performing a series of DO measurements over a period of time, and the measured results are conventionally reported as mg O$_2$/L·min or mg O$_2$/L·h.

Oxygen uptake is most valuable for plant operations when combined with VSS data. The combination of OUR with MLVSS yields a value termed the specific oxygen uptake rate (SOUR) or respiration rate. The SOUR is a measure of the amount of oxygen used by microorganisms and is reported as mg O$_2$/g MLVSS·h. It has been shown that the mixed liquor SOUR and the final effluent COD can be correlated, thereby allowing predictions of final effluent quality to be made during transient loading conditions (Huang and Cheng, 1984). Changes in SOUR values may also be used to assess the presence of toxic or inhibitory substances in the influent wastewater.

Microscopic Observations.   Routine microscopic observations provide valuable monitoring information about the condition of the microbial population in the activated-sludge process. Specific information gathered includes changes in floc size and density, the status of filamentous organism growth in the floc, the presence of *Nocardia* bacteria, and the type and abundance of higher life-forms such as protozoans and rotifers. Changes in these characteristics can provide an indication of changes in the wastewater characteristics or of an operational problem. Examples of the changes in predominance of microorganisms versus F/M ratio and SRT are shown on Fig. 8–10. A decrease in the protozoan population may be indicative of DO limitations, operation at a lower SRT, or inhibitory substances in the wastewater. Early detection of filamentous or *Nocardia* growth will allow time for corrective action to be taken to minimize potential problems associated with excessive growth of these organisms. Procedures may be followed to identify the specific type of filamentous organism, which may help identify an operating or design condition that encourages their growth (Jenkins et al., 1993).

Operational Problems

The most common problems encountered in the operation of an activated-sludge plant are bulking sludge, rising sludge, and *Nocardia* foam. Because few plants have escaped these problems, it is appropriate to discuss their nature and methods for their control.

Figure 8–10

Relative predominance of microorganisms versus food to microorganisms (F/M) ratio and solids retention time (SRT). *(Adapted in part from WEF, 1996.)*

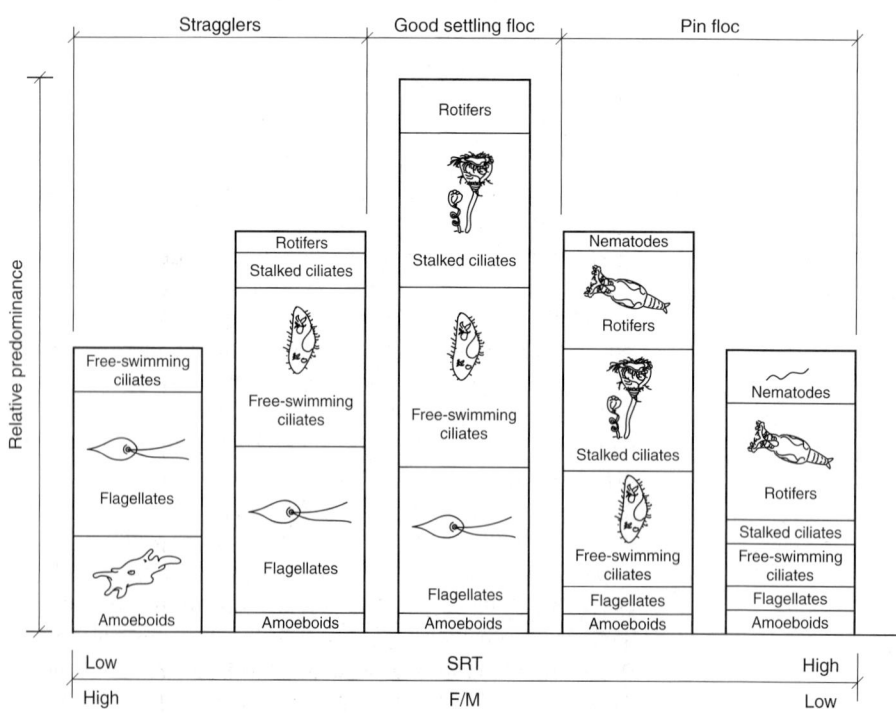

Bulking Sludge. In many cases MLSS with poor settling characteristics has developed into what is known as a *bulking sludge* condition, which defines a condition in the activated-sludge clarifier that can cause high effluent suspended solids and poor treatment performance. In a bulking sludge condition, the MLSS floc does not compact or settle well, and floc particles are discharged in the clarifier effluent. With good settling sludge, sludge levels may be as low as 10 to 30 cm at the bottom of the clarifier. In extreme bulking sludge conditions, the sludge blanket cannot be contained and large quantities of MLSS are carried into the system effluent, potentially resulting in violation of permit requirements, inadequate disinfection, and clogging of effluent filters.

Two principal types of sludge bulking problems have been identified. One type, *filamentous bulking,* is caused by the growth of filamentous organisms or organisms that can grow in a filamentous form under adverse conditions, and is the predominant form of bulking that occurs. The other type of bulking, *viscous bulking,* is caused by an excessive amount of extracellular biopolymer, which produces a sludge with a slimy, jellylike consistency (Wanner, 1994). As the biopolymers are hydrophilic, the activated sludge is highly water-retentive, and this condition is referred to as *hydrous bulking.* The resultant sludge has a low density with low settling velocities and poor compaction. Viscous bulking is usually found with nutrient-limited systems or in a very high loading condition with wastewater having a high amount of rbCOD.

Bulking sludge problems due to the growth of filamentous bacteria are more common. In filamentous growth, bacteria form filaments of single-cell organisms that attach end-to-end, and the filaments normally protrude out of the sludge floc. This structure, in contrast to the preferred dense floc with good settling properties, has an increased surface area to mass ratio, which results in poor settling. On Fig. 8–11, a good settling, dense nonfilamentous floc is contrasted to floc containing filamentous growth. Many types of filamentous bacteria exist, and means have been developed for the identification and classification of filamentous bacteria found commonly in activated-sludge systems (Eikelboom, 2000). The classification system is based on morphology (size and shape of cells, length and shape of filaments), staining responses, and cell inclusions. Common filamentous organisms are summarized in Table 8–8, along with the operating conditions that favor their growth. Sludge bulking can be caused by a variety of factors, including wastewater characteristics, design limitations, and operational issues. Individual items associated with each of these categories are identified in Table 8–9.

Activated-sludge reactor operating conditions (low DO, low F/M, and complete-mix operation) clearly have an effect on the development of filamentous populations. One of the kinetic features of filamentous organisms that relates to these conditions is that they are very competitive at low substrate concentrations whether it be organic substrates, DO, or nutrients. Thus, lightly loaded complete-mix activated-sludge systems or low DO (<0.5 mg/L) operating conditions provide an environment more favorable to filamentous bacteria than to the desired floc-forming bacteria.

Filamentous bacteria such as *Beggiatoa* and *Thiothrix* grow well on hydrogen sulfide and reduced substrates, respectively, that would be found in septic wastewaters (Wanner, 1994). When the influent wastewater contains fermentation products such as volatile fatty acids and reduced sulfur compounds (sulfides and thiosulfate), *Thiothrix* can proliferate. Prechlorination of the wastewaters has been done in some cases to prevent their growth. Besides causing bulking problems in activated-sludge systems, *Beggiatoa* and *Thiothrix*

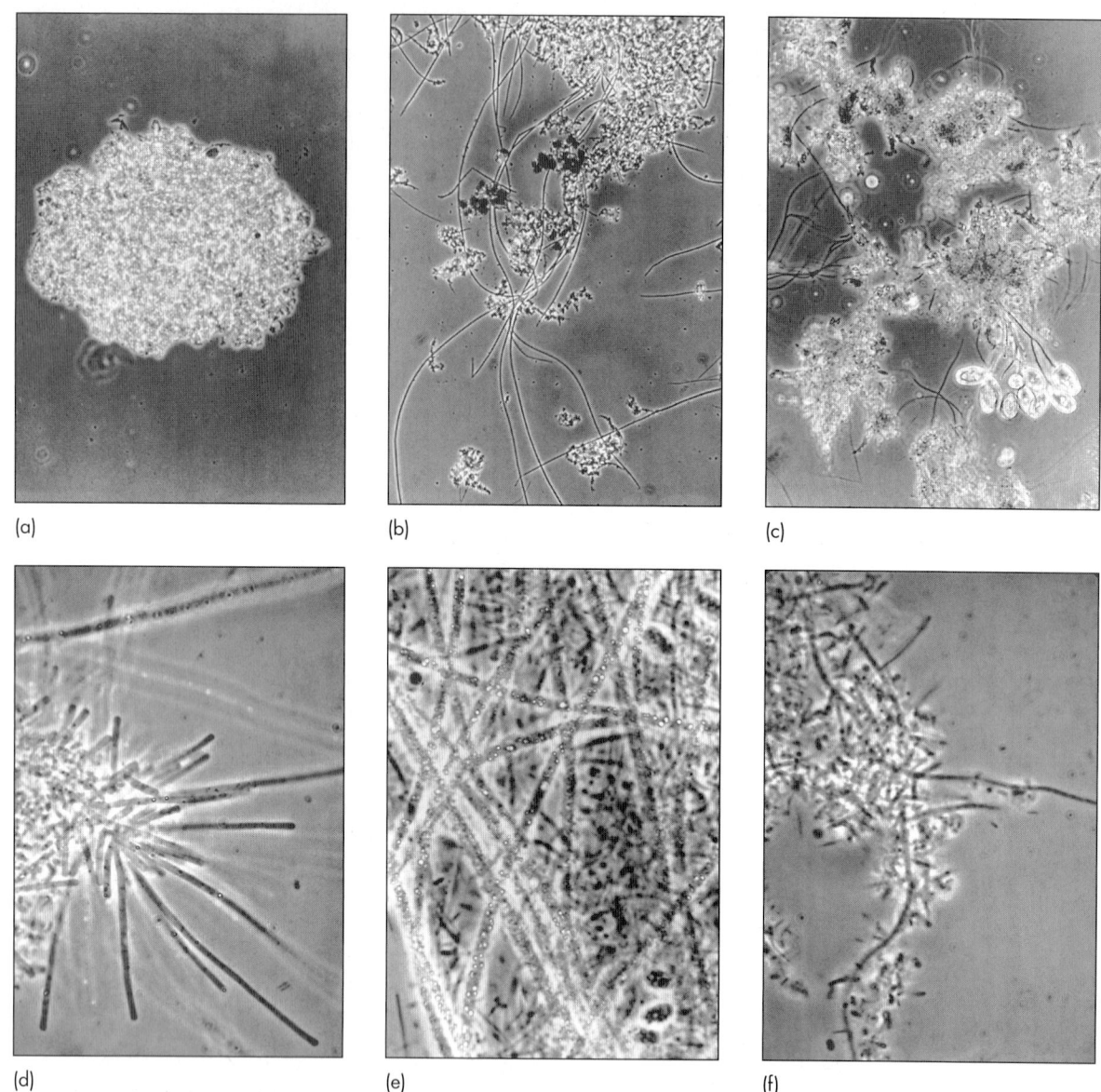

(a)   (b)   (c)

(d)   (e)   (f)

Figure 8–11

Examples of good and poor settling floc particles: (a) nonfilamentous good settling floc, (b) floc particles bridged by filamentous microorganisms, (c) floc particles with limited filamentous microorganisms and secondary form, (d) filaments extending from floc causing poor settling, (e) *Thiothrix* filaments with sulfur granules, and (f) type 1701 filamentous microorganism observed under low dissolved oxygen conditions. *(Courtesy Dr. David Jenkins, University of California, Berkeley.)*

Table 8–8
Filamentous organisms found in activated sludge and associated process conditions[a]

Filament type identified	Cause of filament growth
Sphaerotilus natans, Halsicomenobacter hydrossis, Microthrix parvicella, type 1701	Low dissolved oxygen concentration
M. parvicella, types 0041, 0092, 0675, 1851	Low F/M
H. hydrossis, Nocardia spp., *Nostocoida limicola, S. natans, Thiothrix* spp., types 021N, 0914	Complete-mix reactor conditions
Beggiatoa, Thiothrix spp., types 021N, 0914	Septic wastewater/sulfide available
S. natans, Thiothrix spp., type 021N, possible *H. hydrossis*, types 0041, 0675	Nutrient deficiency
Fungi	Low pH, nutrient deficiency

[a]From Eikelboom (1975).

Table 8–9
Factors that affect sludge bulking

Factor	Description
Wastewater characteristics	Variations in flowrate
	Variations in composition
	pH
	Temperature
	Septicity
	Nutrient content
	Nature of waste components
Design limitations	Limited air supply
	Poor mixing
	Short circuiting (aeration tanks and clarifiers)
	Clarifier design (sludge collection and removal)
	Limited return sludge pumping capacity
Operational issues	Low dissolved oxygen
	Insufficient nutrients
	Low F/M
	Insufficient soluble BOD

can create problems in fixed-film systems, including trickling filters and rotating biological contactors.

In the control of bulking, where a number of variables are possible causes, a checklist of items to investigate is valuable. The following items are recommended: (1) wastewater characteristics, (2) dissolved oxygen content, (3) process loading, (4) return and waste sludge pumping rates, (5) internal plant overloading, and (6) clarifier operation. One of the first steps to be taken when sludge settling characteristics change is to view the mixed liquor under the microscope to determine what type of microbial growth changes or floc structure changes can be related to the development of bulking sludge. A reasonable quality phase-contrast microscope with magnification up to 1000 times (oil immersion) is necessary to view the filamentous bacteria structure and size.

Wastewater Characteristics. The nature of the components found in wastewater or the absence of certain components, such as trace elements, can lead to the development of a bulked sludge (Wood and Tchobanoglous, 1975). If it is known that industrial wastes are being introduced into the system either intermittently or continuously, the quantity of nitrogen and phosphorus in the wastewater should be checked first, because limitations of both or either are known to favor bulking. Nutrient deficiency is a classic problem in the treatment of industrial wastewaters containing high levels of carbonaceous BOD. Wide fluctuations in pH are also known to be detrimental in plants of conventional design. Variations in organic waste loads due to batch-type operations can also lead to bulking and should be checked.

Dissolved Oxygen Concentration. Limited dissolved oxygen has been noted more frequently than any other cause of bulking. If the problem is due to limited oxygen, it can usually be confirmed by operating the aeration equipment at full capacity or by decreasing the system SRT, if possible, to reduce the oxygen demand. The aeration equipment should have adequate capacity to maintain at least 2 mg/L of dissolved oxygen in the aeration tank under normal loading conditions. If 2 mg/L of oxygen cannot be maintained, installation of improvements to the existing aeration system may be required.

Process Loading/Reactor Configuration. The aeration SRT should be checked to make sure that it is within the range of generally accepted values. In many cases, complete-mix systems with long SRTs and subsequent low F/M ratios experience filamentous growths. In such systems, the filamentous organisms are more competitive for substrate. Laboratory research and full-scale investigations have led to activated-sludge design configurations that provide conditions favoring the dominance of *floc-forming* bacteria over filamentous organisms (Jenkins et al., 1993). Reactors in series with various types of environmental conditions, i.e., aerobic, anoxic, and anaerobic, are generally used to augment or replace a complete-mix reactor. The series configurations are called selector processes because they provide conditions that cause selection of floc-forming bacteria in lieu of filamentous organisms as the dominant population. The selector process is discussed in greater detail later in this section.

Internal Plant Overloading. To avoid internal plant overloading, recycle loads should be controlled so they are not returned to the plant flow during times of peak hydraulic and organic loading. Examples of recycle loads are centrate or filtrate from sludge dewatering operations and supernatant from sludge digesters.

Clarifier Operation. The operating characteristics of the clarifier may also affect sludge settling characteristics. Poor settling is often a problem in center-feed circular tanks where sludge is removed from the tank directly under the point where the mixed liquor enters. Sludge may actually be retained in the tank for many hours rather than the desired 30 min and cause localized septic conditions. If this is the case, then the design is at fault, and changes must be made in the inlet feed well and sludge withdrawal equipment.

Temporary Control Measures. In an emergency situation or while the afore-mentioned factors are being investigated, chlorine and hydrogen peroxide may be used to provide temporary help. Chlorination of return sludge has been practiced quite extensively as a means of controlling bulking. A typical design for a low (5 to 10 h) τ system uses 0.002 to 0.008 kg of chlorine per kg MLSS·d (Jenkins et al., 1993). Although chlorination is effective in controlling bulking caused by filamentous growths, it is ineffective when bulking is due to light floc containing bound water. Chlorination normally results in the production of a turbid effluent until such time as the sludge is free of the filamentous forms. Chlorination of a nitrifying sludge will also produce a turbid effluent because of the death of the nitrifying organisms. The use of chlorine also raises issues about the formation of trihalomethanes and other compounds with potential health and environmental effects. Hydrogen peroxide has also been used in the control of filamentous organisms in bulking sludge. Dosage of hydrogen peroxide and treatment time depend on the extent of the filamentous development.

Rising Sludge. Occasionally, sludge that has good settling characteristics will be observed to rise or float to the surface after a relatively short settling period. The most common cause of this phenomenon is denitrification, in which nitrites and nitrates in the wastewater are converted to nitrogen gas. As nitrogen gas is formed in the sludge layer, much of it is trapped in the sludge mass. If enough gas is formed, the sludge mass becomes buoyant and rises or floats to the surface. Rising sludge can be differentiated from bulking sludge by noting the presence of small gas bubbles attached to the floating solids and the presence of more floating sludge on the secondary clarifier surface. Rising sludge is common in short SRT systems, where the temperature encourages the initiation of nitrification, and the mixed liquor is very active due to the low sludge age.

Rising sludge problems may be overcome by (1) increasing the return activated-sludge withdrawal rate from the clarifier to reduce the detention time of the sludge in the clarifier, (2) decreasing the rate of flow of aeration liquor into the offending clarifier if the sludge depth cannot be reduced by increasing the return activated-sludge withdrawal rate, (3) where possible, increasing the speed of the sludge-collecting mechanism in the settling tanks, and (4) decreasing the SRT to bring the activated sludge out of nitrification. For warm climates where it is very difficult to operate at a low enough SRT to limit nitrification, an anoxic/aerobic process is preferred to denitrification to prevent rising sludge and to improve sludge settling characteristics.

***Nocardia* Foam.** Two bacteria genera, *Nocardia* and *Microthrix parvicella,* are associated with extensive foaming in activated-sludge processes. These organisms have hydrophobic cell surfaces and attach to air bubbles, where they stabilize the bubbles to cause foam. The organisms can be found at high concentrations in the foam above the mixed liquor. Both types of bacteria can be identified under microscopic examination. *Nocardia* has a filamentous structure, and the filaments are very short and are contained within the floc particles. *Microthrix parvicella* has thin filaments extending from the floc particles. Foaming on an activated-sludge basin and a microscopic view of *Nocardia* are

(a)

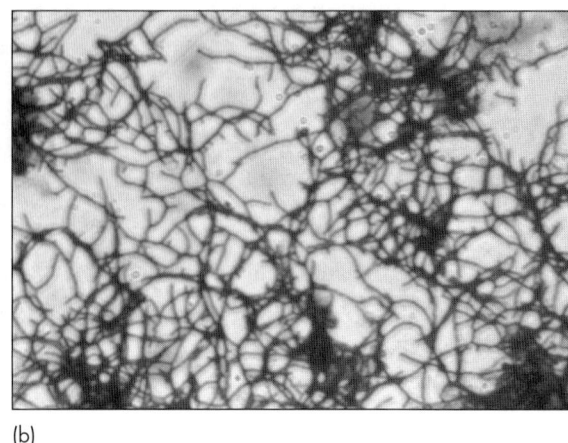

(b)

Figure 8-12

Nocardia foam: (a) example of foam on an aeration tank and (b) microscopic observation of gram-stained *Nocardia* filaments. (Courtesy Dr. David Jenkins, University of California, Berkeley.)

shown on Fig. 8–12. The foam is thick, has a brown color, and can build up in thickness of 0.5 to 1 m.

The foam production can occur with both diffused and mechanical aeration but is more pronounced with diffused aeration and with higher air flowrates. Problems of *Nocardia* foaming in the activated sludge can also lead to foaming in anaerobic and aerobic digesters that receive the waste-activated sludge. *Nocardia* growth is common where surface scum is trapped in either the aeration basin or secondary clarifiers. Aeration basins that are baffled with flow from one cell to the next occurring under the baffles, instead of over the top, encourage *Nocardia* growth and foam collection.

Methods that can be used to control *Nocardia* include (1) avoiding trapping foam in the secondary treatment process, (2) avoiding the recycle of skimmings into the secondary treatment process, and (3) using chlorine spray on the surface of the *Nocardia* foam. The use of a selector design may help to discourage *Nocardia* foaming, but significant foaming has been observed with anoxic/aerobic processes. The addition of a small concentration of cationic polymer has been used with some success for controlling *Nocardia* foaming (Shao et al., 1997). The presence of *Nocardia* has also been associated with the presence of *Nocardia-Microthrix* with fats and edible oils in wastewater. Reducing the oil and grease content from discharges to the collection system from restaurants, truck stops, and meatpacking facilities by effective degreasing processes can help control potential *Nocardia* problems.

Activated-Sludge Selector Processes

In the above discussion, problems caused by nuisance microorganisms in activated sludge were presented, including the effect of filamentous bacteria on sludge settling characteristics and the potential of sludge bulking when the filamentous bacteria are present in high numbers. Prior to the 1970s, filamentous bulking was considered an

Figure 8–13

Typical growth curve for filamentous and nonfilamentous organisms as a function of the substrate concentration.

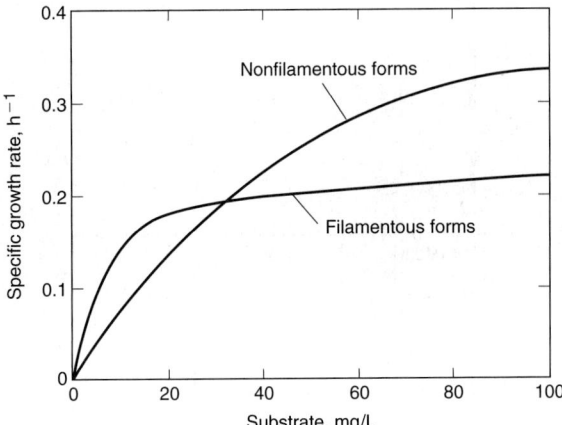

inevitable consequence of activated-sludge treatment, but work by Chudoba et al. (1973) with staged versus complete-mix activated-sludge reactors led to the concept that reactor configuration designs, now termed *selectors,* could be used to control filamentous bulking and improve sludge-settling characteristics. Rensink (1974) has also introduced the relationship between the reactor configuration and bulking control.

The concept of a selector is the use of a specific bioreactor design that favors the growth of floc-forming bacteria instead of filamentous bacteria to provide an activated sludge with better settling and thickening properties. The high substrate concentration in the selector favors the growth of nonfilamentous organisms (see Fig. 8–13). A selector is a small tank (20 to 60 min contact time) or a series of tanks in which the incoming wastewater is mixed with return sludge under aerobic, anoxic, and anaerobic conditions. Various types of selectors are shown on Fig. 8–14. The selector reactor precedes the activated-sludge aeration tank and may be designed as a separate reaction stage for a complete-mix reactor (see Fig. 8–14*a*) or as individual compartments in a plug-flow system (see Fig. 8–14*b* and *c*). Sequencing batch reactors may also be operated to employ the selector concept. The goal in the selector is to have most of the rbCOD consumed by the floc-forming bacteria. Because the particulate degradable COD decomposes at a much slower rate and will be present in the aeration tank, the rbCOD must be utilized for the benefit of the floc-forming bacteria. Selector designs are based on either kinetic or metabolic mechanisms (Albertson, 1987; Jenkins et al., 1993; and Wanner, 1994) and are described in the following paragraphs. The kinetics-based selector designs are called *high F/M selectors,* and the metabolic-based selectors are either anoxic or anaerobic processes.

Kinetics-Based Selector.   Selector designs based on biokinetic mechanisms provide for reactor substrate concentrations that result in faster substrate uptake by the floc-forming bacteria. While filamentous bacteria are more efficient for substrate utilization at low substrate concentrations, the floc-forming bacteria have higher growth rates at high soluble substrate concentrations. A series of reactors at relatively low τ values (minutes) is used to provide high soluble substrate concentrations, in contrast to

Figure 8–14

Typical selector
configurations:
(a) anaerobic/aerobic,
(b) high F/M, and
(c) anoxic selector.

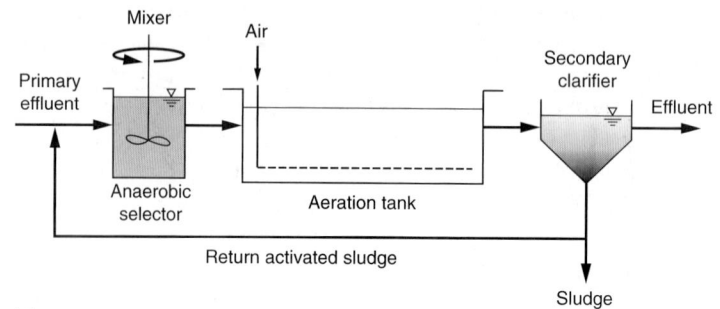

(a)

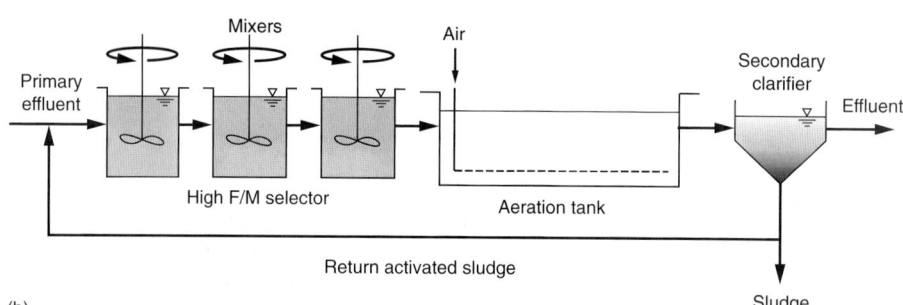

(b)

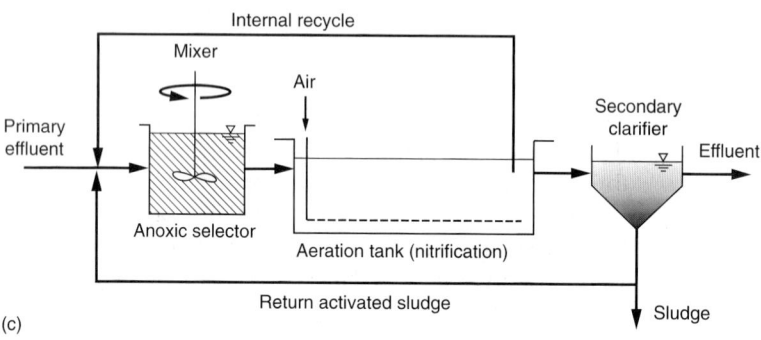

(c)

feeding influent wastewater to aeration tanks with τ values on the order of hours. For
three reactors in series, the following COD F/M ratios, based on the influent flowrate
and COD concentration, are recommended (Jenkins et al., 1993).

- First reactor, 12 g COD/g MLSS·d
- Second reactor, 6 g COD/g MLSS·d
- Third reactor, 3 g COD/g MLSS·d

The F/M ratio is calculated for the first reactor using the volume and MLSS concentration at that reactor and the influent wastewater flowrate and COD concentration. The
F/M value shown for the second reactor includes the volume of the first and second
reactor and the applied loading as the product of the influent flowrate and COD concentration.

Albertson (1987) recommended a similar approach based on a BOD F/M loading of 3 to 5 g BOD/g MLSS in the first reactor, with the second and third reactors being equal to and twice the first reactor volume, respectively. Albertson further notes that if the loading to the first reactor is too high (F/M > 8 g BOD/g MLSS·d), a viscous, nonfilamentous-type bulking can develop. The kinetic concept of a high F/M selector suggests that it be aerobic, and high DO concentrations are needed to maintain an aerobic floc (>6 to 8 mg/L). In many cases, such high DO concentrations are not practical or provided, and the staged selector design (described above) is operated at a low to zero DO concentration so that a metabolic selector mechanism is involved.

A sequencing batch reactor (SBR) can also act as a very effective high F/M selector, depending on the wastewater strength and feeding strategy. For high-strength wastewaters with a relatively large fraction of the SBR volume occupied by the influent wastewater, a high initial F/M ratio can occur. The subsequent reaction by the batch process is equal to that for a plug-flow reactor.

Metabolic-Based Selector. With biological nutrient-removal processes, improved sludge-settling characteristics, and, in many cases, minimal filamentous bacteria growth has been observed. The anoxic or anaerobic metabolic conditions used in these processes favor growth of the floc-forming bacteria. The filamentous bacteria cannot use nitrate or nitrite for an electron acceptor, thus yielding a significant advantage to denitrifying floc-forming bacteria. Similarly, the filamentous bacteria do not store polyphosphates and thus cannot consume acetate in the anaerobic contact zone in biological phosphorus-removal designs, giving an advantage for substrate uptake and growth to the phosphorus-storing bacteria. In some wastewater-treatment facilities (Seattle and San Francisco, for example), an anaerobic selector has been used before the aeration tank in low SRT activated-sludge systems designed for BOD removal, even though phosphorus removal is not required.

Where nitrification is used and phosphorus removal is not required, anoxic selectors (either the staged high F/M gradient or the single-stage designs) have been used. For the high F/M anoxic or anaerobic selectors, the resultant mixed-liquor SVI may be in the range of 65 to 90 mL/g, and for single-tank anoxic selectors, SVI values in the range of 100 to 120 mL/g are more commonly obtained.

The use of selector designs in activated sludge is more common because of the many advantages derived from the minimal investment in a relatively small reactor volume. By improving sludge settling, the activated-sludge treatment capacity may be increased, as higher MLSS concentrations are usually possible. The hydraulic capacity of the secondary clarifiers is also increased.

8–4 PROCESSES FOR BOD REMOVAL AND NITRIFICATION

Important considerations for the design of activated-sludge processes were presented in Sec. 8–3. The purpose of this section is to illustrate in detail the design procedure for three different activated-sludge processes. Following the design examples, detailed information is presented on the alternative processes used for BOD removal and nitrification, typical process design parameters, and process selection considerations. Biological nitrogen and phosphorus removal can be incorporated into most of the processes used for BOD removal and nitrification, but because of additional design factors, nitrogen- and phosphorus-removal processes are covered separately in Secs. 8–5 and 8–6, respectively. Details for the selection and design of physical facilities may be found in Sec. 8–7.

Process Design Considerations

The fundamental principles of wastewater characterization, biological treatment, and process analysis were presented in Chap. 7 and in Secs. 8–2 and 8–3. For wastewater characterization, rbCOD and nbVSS are of greatest significance in process design. For BOD removal and nitrification processes, the rbCOD concentration is important for evaluating the oxygen demand profiles for plug-flow, staged, and batch-fed processes. The effect of nbVSS concentration in the influent will be significant in process sludge production and aeration volume requirements.

In the following paragraphs, three activated-sludge process design examples are provided to demonstrate application of these fundamental principles to BOD removal and nitrification processes. The examples are (1) a single-sludge complete-mix activated-sludge process without and with nitrification, (2) a sequencing batch reactor (SBR) with nitrification, and (3) a staged nitrification process. The types of computations presented are typically used for these three different types of processes. All of the elements shown for the BOD removal and nitrification process design can be applied to a BOD removal-only design by modifying the SRT and removing items that deal with nitrification. The design methodology presented is based fundamentally on using SRT and thus can be applied to the broad range of processes described at the end of this section.

For general design purposes, kinetic expressions used for design were summarized in Table 8–5. Kinetic coefficients for the removal of carbonaceous material (based on bCOD) by heterotrophic bacteria are given in Table 8–10. The μ_m and K_s values given

Table 8–10
Activated-sludge kinetic coefficients for heterotrophic bacteria at 20°C[a]

Coefficient	Unit	Range	Typical value
μ_m	g VSS/g VSS·d	3.0–13.2	6.0
K_s	g bCOD/m³	5.0–40.0	20.0
Y	g VSS/g bCOD	0.30–0.50	0.40
k_d	g VSS/g VSS·d	0.06–0.20	0.12
f_d	Unitless	0.08–0.20	0.15
θ values			
μ_m	Unitless	1.03–1.08	1.07
k_d	Unitless	1.03–1.08	1.04
K_s	Unitless	1.00	1.00

[a]Adapted from Henze et al. (1987a); Barker and Dold (1997); and Grady et al. (1999).

Table 8–11
Activated-sludge nitrification kinetic coefficients at 20°C[a]

Coefficient	Unit	Range	Typical value
μ_{mn}	g VSS/g VSS·d	0.20–0.90	0.75
K_n	g NH_4-N/m^3	0.5–1.0	0.74
Y_n	g VSS/g NH_4-N	0.10–0.15	0.12
k_{dn}	g VSS/g VSS·d	0.05–0.15	0.08
K_o	g/m^3	0.40–0.60	0.50
θ values			
μ_n	Unitless	1.06–1.123	1.07
K_n	Unitless	1.03–1.123	1.053
k_{dn}	Unitless	1.03–1.08	1.04

[a] Adapted from Henze et al. (1987a); Barker and Dold (1997); and Grady et al. (1999).

in Table 8–10 are the default values recommended in the IAWPRC ASM 1 model (Henze et al., 1987a). The kinetic coefficients to be used for the design of nitrification with activated sludge are given in Table 8–11. Because reported nitrification kinetics cover a wide range, bench-scale or in-plant testing should be undertaken to evaluate site-specific nitrification kinetic values. Higher maximum specific growth rates than indicated in Table 8–11 could be found or testing could reveal lower reaction rates and a significant nitrification inhibition problem. Source control measures may be required to mitigate nitrification inhibition. Aeration tank volume requirements and SRT values are directly related to nitrification μ_m values.

Complete-Mix Activated-Sludge Process

A typical complete-mix activated-sludge (CMAS) process is shown on Fig. 8–15. Effluent from the primary sedimentation tank and recycled return activated sludge are introduced typically at several points in the reactor. Because the tank contents are thoroughly mixed, the organic load, oxygen demand, and substrate concentration are uniform throughout the entire aeration tank and the F/M ratio is low. Care should be taken to assure that the CMAS reactor is well mixed and that influent feed and effluent withdrawal points are selected to prevent short-circuiting of untreated or partially treated wastewater. The complete-mix reactor is usually configured in square, rectangular, or round shapes. Tank dimensions depend mainly on the size, type, and mixing pattern of the aeration equipment.

The computational approach used in the design of the activated-sludge process is given in Table 8–12 and makes use of the design equations presented in Chap. 7 and Secs. 8–2 and 8–3. The application of the design approach is presented in Example 8–2. The key design concepts are the selection of the design SRT, the selection of kinetic and stoichiometric coefficients, and application of appropriate mass balances.

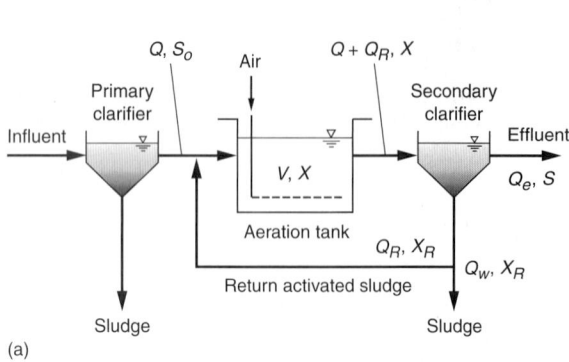

(a)

(b)

Figure 8–15

Complete-mix activated-sludge process: (a) schematic diagram and (b) view of a typical complete-mix reactor.

Table 8–12 Computation approach for the design of the activated-sludge process for BOD removal and nitrification	
	1. Obtain influent wastewater characterization data
	2. Determine the effluent requirements in terms of NH_4-N, TSS, and BOD concentrations
	3. Select an appropriate nitrification safety factor for the design SRT based on expected peak/average TKN loadings. Safety factors may vary from 1.3 to 2.0
	4. Select the minimum DO concentration for the aeration basin mixed liquor. A minimum DO concentration of 2.0 mg/L is recommended for nitrification
	5. Determine the nitrification maximum specific growth rate (μ_m) based on the aeration basin temperature and DO concentration, and determine K_n
	6. Determine the net specific growth rate μ and SRT at this growth rate, to meet the effluent NH_4-N concentration
	7. Obtain the design SRT by applying the safety factor to Step 6
	8. Determine the biomass production
	9. Perform a nitrogen balance to determine NO_x, the concentration of NH_4-N oxidized
	10. Calculate the VSS mass and TSS mass for the aeration basin
	11. Select a design MLSS concentration and determine the aeration basin volume and hydraulic residence time
	12. Determine the overall sludge production and observed yield
	13. Calculate the oxygen demand
	14. Determine if alkalinity addition is needed
	15. Design the secondary clarifier
	16. Design the aeration oxygen transfer system
	17. Summarize the final effluent quality
	18. Prepare a design summary table

EXAMPLE 8-2 **Complete-Mix Activated-Sludge Process Design for BOD Removal Only and for BOD Removal with Nitrification** Design a complete-mix activated-sludge (CMAS) process to treat 22,464 m³/d of primary effluent to (a) meet a BOD_e concentration less than 30 g/m³ and (b) accomplish BOD removal and nitrification with an effluent NH_4-N concentration of 0.50 g/m³ and BOD_e and $TSS_e \le 15$ g/m³. Compare the two design conditions in a summary table. The aeration basin mixed-liquor temperature is 12°C.

Note: Steps 2 through 8 cover condition (a), BOD removal only; and Steps 9 through 20 cover condition (b), BOD removal and nitrification. Clarifier design is discussed in Step 21.

The following wastewater characteristics and design conditions apply:

Wastewater characteristics:

Constituent	Concentration, g/m³
BOD	140
sBOD	70
COD	300
sCOD	132
rbCOD	80
TSS	70
VSS	60
TKN	35
NH_4-N	25
TP	6
Alkalinity	140 as $CaCO_3$
bCOD/BOD ratio	1.6

Note: g/m³ = mg/L.

Design conditions and assumptions:

1. Fine bubble ceramic diffusers with an aeration clean water O_2 transfer efficiency = 35%
2. Liquid depth for the aeration basin = 4.9 m
3. The point of air release for the ceramic diffusers is 0.5 m above the tank bottom
4. DO in aeration basin = 2.0 g/m³
5. Site elevation is 500 m (pressure = 95.6 kPa)
6. Aeration α factor = 0.50 for BOD removal only and 0.65 for nitrification; β = 0.95 for both conditions, and diffuser fouling factor F = 0.90

7. Use kinetic coefficients given in Tables 8–10 and 8–11
8. SRT for BOD removal = 5 d
9. Design MLSS X_{TSS} concentration = 3000 g/m³; values of 2000 to 3000 g/m³ can be considered
10. TKN peak/average factor of safety FS = 1.5

Solution **Part A, BOD Removal without Nitrification**

1. Develop the wastewater characteristics needed for design.
   a. Find bCOD using Eq. (8–8).

   $$bCOD = 1.6(BOD) = 1.6(140 \text{ g/m}^3) = 224 \text{ g/m}^3$$

   b. Find nbCOD using Eq. (8–7).

   $$nbCOD = COD - bCOD = (300 - 224) \text{ g/m}^3 = 76 \text{ g/m}^3$$

   c. Find effluent $sCOD_e$ (assumed to be nonbiodegradable).

   $$sCOD_e = sCOD - 1.6 \text{ sBOD}$$
   $$= (132 \text{ g/m}^3) - (1.6)(70 \text{ g/m}^3) = 20 \text{ g/m}^3$$

   d. Find nbVSS using Eq. (8–4).

   $$nbVSS = (1 - bpCOD/pCOD)VSS$$

   $$\frac{bpCOD}{pCOD} = \frac{(bCOD/BOD)(BOD - sBOD)}{COD - sCOD}$$

   $$\frac{bpCOD}{pCOD} = \frac{1.6(BOD - sBOD)}{COD - sCOD} = \frac{1.6[(140 - 70) \text{ g/m}^3]}{[(300 - 132) \text{ g/m}^3]} = 0.67$$

   $$nbVSS = (1 - 0.67)(60 \text{ g VSS/m}^3) = 20 \text{ g/m}^3$$

   e. Find the iTSS.

   $$iTSS = TSS - VSS$$
   $$= (70 - 60) \text{ g/m}^3 = 10 \text{ g/m}^3$$

2. Design suspended growth system for BOD removal only.
   a. Determine biomass production using parts A and B of Eq. (8–15).

   $$P_{X,vss} = \frac{QY(S_o - S)}{1 + (k_d)SRT} + \frac{(f_d)(k_d)QY(S_o - S)SRT}{1 + (k_d)SRT}$$

   Define input data for above equation.

   $Q = 22{,}464 \text{ m}^3/\text{d}$

   $Y = 0.40 \text{ g VSS/g bCOD}$ (Table 8–10)

   $S_o = 224 \text{ g bCOD/m}^3$ (see Step 1)

Determine S from Eq. (7–40) in Table 8–5. Note: $Yk = \mu_m$

$$S = \frac{K_s[1 + (k_d)SRT]}{SRT(\mu_m - k_d) - 1}$$

Use μ_m and k_d from Table 8–10.

$K_s = 20 \text{ g/m}^3$

$\mu_{m,T} = \mu_m \theta^{T-20}$

$\mu_{m,12°C} = 6.0 \text{ g/g·d}(1.07)^{12-20} = 3.5 \text{ g/g·d}$

$k_{d,T} = k_{20}\theta^{(T-20)}$ Eq. (2–25)

$k_{d,12°C} = (0.12 \text{ g/g·d})(1.04)^{12-20} = 0.088 \text{ g/g·d}$

$$S = \frac{(20 \text{ g/m}^3)[1 + (0.088 \text{ g/g·d})(5 \text{ d})]}{(5 \text{ d})[(3.5 - 0.088) \text{ g/g·d} - 1]} = 1.8 \text{ g bCOD/m}^3$$

b. Substitute the above values in the expression given above and solve for $P_{X,VSS}$

$$P_{X,VSS} = \frac{(22,464 \text{ m}^3/\text{d})(0.40 \text{ g/g})[(224 - 1.8) \text{ g/m}^3](1 \text{ kg}/10^3 \text{ g})}{[1 + (0.088 \text{ g/g·d})(5 \text{ d})]}$$

$$+ \frac{(0.15\text{g/g})(0.088 \text{ g/g·d})(0.40 \text{ g/g})(22,464 \text{ m}^3/\text{d})[(224 - 1.8) \text{ g/m}^3](5 \text{ d})(1 \text{ kg}/10^3 \text{ g})}{[1 + (0.088 \text{ g/g·d})(5 \text{ d})]}$$

$P_{X,VSS} = (1386.5 + 91.5) \text{ kg/d} = 1478 \text{ kg VSS/d}$

3. Determine the mass of VSS and TSS in the aeration basin. The mass of VSS and TSS can be determined using Eqs. (7–54) and (7–55) given in Table 8–5.

Mass $= P_X(SRT)$

a. Determine $P_{X,VSS}$ and $P_{X,TSS}$ using Eqs. (8–15) and (8–16) including parts A, B, and D. Part C $= 0$ because there is no nitrification.

From Eq. (8–15), $P_{X,VSS}$ is:

$P_{X,VSS} = 1478 \text{ kg/d} + Q(\text{nbVSS})$

$= 1478 \text{ kg/d} + (22,464 \text{ m}^3/\text{d})(20 \text{ g/m}^3)(1 \text{ kg}/10^3 \text{ g})$

$= (1478 + 449.3) \text{ kg/d} = 1927.3 \text{ kg/d}$

From Eq. (8–16), $P_{X,TSS}$ is

$P_{X,TSS} = [(1478 \text{ kg/d})/0.85] + (449.3 \text{ kg/d}) + Q(\text{TSS}_o - \text{VSS}_o)$

$= 1738.8 \text{ kg/d} + 449.3 \text{ kg/d} + (22,464 \text{ m}^3/\text{d})(10 \text{ g/m}^3)(1 \text{ kg}/10^3 \text{ g})$

$= 2412.7 \text{ kg/d}$

b. Calculate the mass of VSS and TSS in the aeration basin.
   i. Mass of MLVSS using Eq. (7–54) in Table 8–5

$$(X_{VSS})(V) = (P_{X,VSS})\ SRT$$

$$= (1927.3\ kg/d)(5\ d) = 9637\ kg$$

   ii. Mass of MLSS using Eq. (7–55) in Table 8–5

$$(X_{TSS})(V) = (P_{X,TSS})\ SRT$$

$$= (2412.7\ kg/d)(5.0\ d) = 12,064\ kg$$

4. Select a design MLSS mass concentration and determine the aeration tank volume and detention time using the TSS mass computed in Step 3.
   a. Determine the aeration tank volume using the relationship from Step 3b.

$$(V)(X_{TSS}) = 25,405\ kg$$

   At X_{TSS} = 3000 g/m^3

$$V = \frac{(12,064\ kg)(10^3\ g/kg)}{(3000\ g/m^3)} = 4021\ m^3$$

   b. Determine the aeration tank detention time.
   Use 3 basins at 1340 m^3 each so that one of the basins can be taken off-line for a short period of time when maintenance of the aeration system is necessary.

$$\tau = \frac{V}{Q} = \frac{(4021\ m^3)(24\ h/d)}{(22,464\ m^3/d)} = 4.3\ h$$

   c. Determine MLVSS.

$$\text{Fraction VSS} = \frac{9637\ kg\ VSS}{12,064\ kg TSS} = 0.80$$

$$\text{MLVSS} = 0.80(3000\ g/m^3) = 2400\ g/m^3$$

5. Determine F/M and BOD volumetric loading.
   a. Determine F/M using Eq. (7–60) in Table 8–5.

$$F/M = \frac{QS_o}{XV} = \frac{kg\ BOD}{kg\ MLVSS{\cdot}d}$$

$$= \frac{(22,464\ m^3/d)(140\ g/m^3)}{(2400\ g/m^3)(4021\ m^3)} = 0.33\ g/g{\cdot}d = 0.33\ kg/kg{\cdot}d$$

   b. Determine volumetric BOD loading

$$\text{BOD loading} = \frac{QS_o}{V} = \frac{kg\ BOD}{m^3{\cdot}d}$$

$$= \frac{(22,464\ m^3/d)(140\ g/m^3)}{(4021\ m^3)(10^3\ g/kg)} = 0.78\ kg/m^3{\cdot}d$$

6. Determine the observed yield based on TSS and VSS.
    a. Observed yield based on TSS

    Observed yield = g TSS/g bCOD = kg TSS/kg bCOD

    $P_{X,\,TSS} = 2412.7$ kg/d

    bCOD removed $= Q(S_o - S)$

    $\qquad = (22,464 \text{ m}^3/\text{d})[\,(224 - 1.8)\text{ g/m}^3\,](1 \text{ kg}/10^3 \text{ g})$

    $\qquad = 4991.5$ kg/d

    $$Y_{obs,TSS} = \frac{(2412.7 \text{ kg/d})}{(4991.5 \text{ kg/d})} = \frac{0.48 \text{ kg TSS}}{\text{kg bCOD}} = \frac{0.48 \text{ g TSS}}{\text{g bCOD}}$$

    $$= \left(\frac{0.48 \text{ g TSS}}{\text{g bCOD}}\right)\left(\frac{1.6 \text{ g bCOD}}{\text{g BOD}}\right) = 0.77 \text{ g TSS/g BOD}$$

    b. Observed yield based on VSS

    $Y_{obs,VSS}$: VSS/TSS = 0.80 (see Step 4c)

    $$= \left(\frac{0.48 \text{ g TSS}}{\text{g bCOD}}\right)\left(\frac{0.80 \text{ g VSS}}{\text{g TSS}}\right)$$

    $$= 0.38 \text{ g VSS/g bCOD}$$

    $$= \left(\frac{0.38 \text{ g VSS}}{\text{g bCOD}}\right)\left(\frac{1.6 \text{ g bCOD}}{\text{g BOD}}\right)$$

    $$= 0.61 \text{ g VSS/g BOD}$$

7. Calculate the O_2 demand using Eq. (7–59) in Table 8–5.

    $R_o = Q(S_o - S) - 1.42 P_{X,bio}$

    $\qquad = (22,464 \text{ m}^3/\text{d})[\,(224 - 1.8)\text{ g/m}^3\,](1 \text{ kg}/10^3 \text{ g}) - 1.42(1478 \text{ kg/d})$

    $R_o = 4991.5$ kg/d $- 2098.8$ kg/d

    $\qquad = 2892.7$ kg/d $= 120.5$ kg/h

8. Fine bubble aeration design—determine air flowrate at average design flowrate using Eq. (5–55).

    $$\text{AOTR} = \text{SOTR}\left(\frac{\beta C_{\bar{s},T,H} - C_L}{C_{s,20}}\right)(1.024^{T-20})(\alpha)(F)$$

    where AOTR = actual oxygen transfer rate under field conditions, kg O_2/h
    SOTR = standard oxygen transfer rate in tap water at 20°C and zero dissolved oxygen, kg O_2/h
    Other terms as defined previously in Chap. 5.

a. Determine $C_{\bar{s},T,H}$, the average dissolved oxygen saturation concentration in clean water in aeration tank at temperature T and altitude H, using the following relationship from Eq. (5–55).

$$C_{\bar{s},T,H} = (C_{s,T,H})\frac{1}{2}\left(\frac{P_d}{P_{atm}} + \frac{O_t}{21}\right)$$

i. From Table D–1 (Appendix D), $C_{20} = 9.08$ mg/L and $C_{12} = 10.77$ mg/L

ii. Determine the relative pressure at elevation 500 m to correct the DO concentration for altitude.

From Appendix B

$$\frac{P_b}{P_a} = \exp\left[-\frac{g\,M(z_b - z_a)}{RT}\right]$$

$$= \exp\left\{-\frac{(9.81\text{ m/s}^2)(28.97\text{ kg/kg-mole})[(500 - 0)\text{ m}]}{(8314\text{ kg·m}^2/\text{s}^2\text{·kg-mole·K})[(273.15 + 12)\text{ K}]}\right\} = 0.94$$

The oxygen concentration at 500 m and 12°C is

$$C_{s,T,H} = 10.77\text{ mg/L} \times 0.94 = 10.12\text{ mg/L}$$

iii. Determine atmospheric pressure in m of water at an elevation of 500 m and a temperature of 12°C (see Appendix B and C).

$$P_{atm,H} = \frac{(P_b/P_a)(P_{atm,\,H}\text{ kN/m}^2)}{(\gamma\text{ kN/m}^3)} = \frac{(0.94)(101.325\text{ kN/m}^2)}{(9.802\text{ kN/m}^3)} = 9.72\text{ m}$$

iv. Determine the oxygen concentration assuming the percent oxygen concentration leaving the aeration tank is 19%.

$$C_{\bar{s},T,H} = (C_{s,T,H})\left(\frac{1}{2}\right)\left(\frac{P_{atm,H} + P_{w,\text{Eff depth}}}{P_{atm,H}} + \frac{O_t}{21}\right)$$

$$= 10.12\text{ mg/L}\left(\frac{1}{2}\right)\left[\frac{9.72\text{ m} + (4.9 - 0.5)\text{ m}}{9.72\text{ m}} + \frac{19}{21}\right]$$

$$= 11.93\text{ mg/L}$$

b. Determine the SOTR using $\alpha = 0.50$, $\beta = 0.95$, and diffuser fouling factor $F = 0.9$.

$$\text{SOTR} = \text{AOTR}\left[\frac{C_{s,20}}{\alpha F(\beta C_{\bar{s},T,H} - C_L)}\right](1.024^{20-T})$$

$$= \frac{(120.5\text{ kg/h})(9.08\text{ g/m}^3)(1.024)^{20-12}}{(0.50)(0.9)[(0.95)(11.93\text{ g/m}^3) - 2.0\text{ g/m}^3]} = 315\text{ kg/h}$$

c. Determine the air flowrate.

$$\text{Air flowrate, kg/min} = \frac{(\text{SOTR kg/h})}{[(E)(60\text{ min/h})(\text{kg O}_2/\text{m}^3\text{ air})]}$$

Using the data given in Appendix B, the density of air at 12°C and a pressure of 95.2 kPa (0.94 × 101.325 kPa) is 1.1633 kg/m³. The corresponding amount of oxygen by weight is 0.270 (0.2318 × 1.1633 kg/m³). Thus, the required air flowrate is

$$\text{Air flowrate, m}^3/\text{min} = \frac{(315 \text{ kg/h})}{[(0.35)(60 \text{ min/h})(0.270 \text{ kg O}_2/\text{m}^3 \text{ air})]}$$

$$= 55.5 \text{ m}^3/\text{min}$$

Note: To continue the facilities design for secondary clarifiers for the BOD removal process, move to Step 21. For nitrification design, continue to Step 9.

Part B, BOD Removal and Nitrification

9. Perform the nitrification design following the same steps as for BOD removal except the design SRT must first be determined. Determine the specific growth rate μ_n for the nitrifying organisms using Eq. (7–93). The nitrification rate will control the design because the nitrifying organisms grow more slowly than the heterotrophic organisms that remove organic carbon.

$$\mu_n = \left(\frac{\mu_{n,m}N}{K_n + N}\right)\left(\frac{\text{DO}}{K_o + \text{DO}}\right) - k_{dn}$$

   a. Find $\mu_{n,m}$ at $T = 12°C$.

   $$\mu_{n, m, 12°C} = (0.75 \text{ g/g·d})(1.07)^{12-20} = 0.44 \text{ g/g·d}$$

   b. Find K_n, at $T = 12°C$.

   $$K_{n,12°C} = (0.74 \text{ g/m}^3)(1.053)^{12-20} = 0.49 \text{ g/m}^3$$

   c. Find k_{dn}, at $T = 12°C$.

   $$k_{dn, 12°C} = (0.08 \text{ g/g·d})(1.04)^{12-20} = 0.06 \text{ g/g·d}$$

   d. Substitute the above and given values in Eq. (7–93) and solve for μ_n.

   $$N = 0.50 \text{ g/m}^3, \text{DO} = 2.0 \text{ g/m}^3, K_o = 0.50 \text{ g/m}^3$$

   $$\mu_n = \left\{\frac{(0.44 \text{ g/g·d})(0.50 \text{ g/m}^3)}{[(0.49 + 0.50) \text{ g/m}^3]}\right\}\left\{\frac{(2.0 \text{ g/m}^3)}{[(0.50 + 2.0) \text{ g/m}^3]}\right\} - (0.06 \text{ g/g·d})$$

   $$= 0.12 \text{ g/g·d}$$

10. Determine the theoretical and design SRT.
    a. Find theoretical SRT using Eq. (7–37) in Table 8–5.

    $$\text{SRT} = \frac{1}{\mu_n} = \frac{1}{(0.12 \text{ g/g·d})} = 8.33 \text{ d}$$

    b. Determine the design SRT using Eq. (7–71).

    $$\text{FS} = \text{TKN peak/TKN average} = 1.5$$

Design SRT = (FS)(theoretical SRT)

= 1.5(8.33 d) = 12.5 d

11.   Determine biomass production using Eq. (8–15), parts A, B, and C.

$$P_{X,bio} = \frac{QY(S_o - S)}{1 + (k_d)SRT} + \frac{(f_d)(k_d)Q(Y)(S_o - S)SRT}{1 + (k_d)SRT} + \frac{QY_n(NO_x)}{1 + (k_{dn})SRT}$$

a.   Define input data for the above equation.

$Q = 22{,}464 \text{ m}^3/\text{d}$

$Y = 0.40 \text{ VSS/g bCOD}$

$S_o = 224 \text{ g bCOD/m}^3$ (Step 1)

$k_d = 0.088 \text{ g/g·d}$ (Step 2a)

$\mu_m = 3.5 \text{ g/g·d}$ (Step 2a)

Determine S from Eq. (7–40) in Table 8–5.

$$S = \frac{K_s[1 + (k_d)SRT]}{SRT(\mu_m - k_d) - 1}$$

$$S = \frac{(20 \text{ g/m}^3)[1 + (0.088 \text{ g/g·d})(12.5 \text{ d})]}{(12.5 \text{ d})[(3.5 - 0.088) \text{ g/g·d}] - 1} = 1.0 \text{ g bCOD/m}^3$$

$Y_n = 0.12 \text{ g VSS/g NO}_x$ (Table 8–11)

$k_{dn, 12°C} = 0.06 \text{ g/g·d}$ (Step 9e)

Assume $NO_x \approx 80\%$ (TKN) as nitrogen balance cannot be done yet. The error in assuming that the $NO_x \approx 80\%$ (TKN) is small as the nitrifier VSS yield is a small fraction of total MLVSS concentration.

$NO_x = 0.80(35 \text{ g/m}^3) = 28 \text{ g/m}^3$

b.   Substitute the above values in the expression given above and solve for $P_{X,bio}$.

$$P_{X,bio} = \frac{(22{,}464 \text{ m}^3/\text{d})(0.40 \text{ g/g})[(224 - 1.0)\text{g/m}^3](1 \text{ kg}/10^3 \text{ g})}{[1 + (0.088 \text{ g/g·d})(12.5 \text{ d})]}$$

$$+ \frac{(0.15 \text{ g/g})(0.088 \text{ g/g·d})(0.40 \text{ g/g})(22{,}464 \text{ m}^3/\text{d})[(224 - 1.0) \text{ g/m}^3](12.5 \text{ d})(1 \text{ kg}/10^3 \text{ g})}{[1 + (0.088 \text{ g/g·d})(12.5 \text{ d})]}$$

$$+ \frac{(22{,}464 \text{ m}^3/\text{d})(0.12 \text{ g/g})(28 \text{ g/m}^3)(1 \text{ kg}/10^3 \text{ g})}{[1 + (0.06 \text{ g/g·d})(12.5 \text{ d})]}$$

$P_{X,bio} = 954.2 \text{ kg/d} + 157.4 \text{ kg/d} + 43.1 \text{ kg/d}$

= 1154.7 kg VSS/d

12.   Determine the amount of nitrogen oxidized to nitrate. The amount of nitrogen oxidized to nitrate can be found by performing a nitrogen balance using Eq. (8–18).

$$NO_x = TKN - N_e - 0.12\, P_{X,bio}/Q$$

$$= 35.0 \text{ g/m}^3 - 0.50 \text{ g/m}^3$$

$$- (0.12 \text{ g N/g VSS})(1154.7 \text{ kg VSS/d})(10^3 \text{ g/kg})/(22{,}464 \text{ m}^3/\text{d})$$

$$= (35.0 - 0.50 - 6.2) \text{ g/m}^3 = 28.3 \text{ g/m}^3$$

13. Determine the concentration and mass of VSS and TSS in the aeration basin.

$$\text{Mass} = P_X(\text{SRT})$$

  a. Calculate the concentration of VSS and TSS in the aeration basin
    i. $P_{X,VSS}$, use Eq. (8–15). Parts A, B, and C have already been calculated above as $P_{X,bio}$. Part D must be added to determine $P_{X,VSS}$.

$$P_{X,VSS} = 1154.7 \text{ kg/d} + Q(\text{nbVSS})$$

$$= 1154.7 \text{ kg/d} + (22{,}464 \text{ m}^3/\text{d})(20 \text{ g/m}^3)(1 \text{ kg}/10^3 \text{ g})$$

$$= (1154.7 + 449.3) \text{ kg/d} = 1604.0 \text{ kg/d}$$

    ii. $P_{X,TSS}$, use Eq. (8–16), with the term E added to account for inert influent TSS.

$$P_{X,TSS} = [(1154 \text{ kg/d})/0.85] + (449.3 \text{ kg/d}) + Q(\text{TSS}_o - \text{VSS}_o)$$

$$= 1358.5 \text{ kg/d} + 449.3 \text{ kg/d} + (22{,}464 \text{ m}^3/\text{d})(10 \text{ g/m}^3)(1 \text{ kg}/10^3 \text{ g})$$

$$= 2032.4 \text{ kg/d}$$

  b. Calculate the mass of VSS and TSS in the aeration basin using Eqs. (7–54) and (7–55) in Table 8–5.
    i. Mass of MLVSS

$$(X_{VSS})(V) = (P_{X,VSS})\text{SRT}$$

$$= (1604.0 \text{ kg/d})(12.5 \text{ d}) = 20{,}050 \text{ kg}$$

    ii. Mass of MLSS

$$(X_{TSS})(V) = (P_{X,TSS})\text{SRT}$$

$$= (2032.4 \text{ kg/d})(12.5 \text{ d}) = 25{,}405 \text{ kg}$$

14. Select a design MLSS mass concentration and determine the aeration tank volume and detention time using the TSS mass computed in Step 6.

  a. Determine the aeration tank volume using the relationship from Step 6b.

$$(V)(X_{TSS}) = 25{,}405 \text{ kg}$$

At MLSS $= 3000 \text{ g/m}^3$

$$V = \frac{(25{,}405 \text{ kg})(10^3 \text{ g/kg})}{(3000 \text{ g/m}^3)}$$

$$V = 8468 \text{ m}^3$$

b. Determine the aeration tank detention time.
   Use 3 basins at 2822 m³ each so that 1/3 of the volume can be taken off-line when maintenance of the aeration system is necessary.

$$\tau = \frac{V}{Q} = \frac{(8468 \text{ m}^3)(24 \text{ h/d})}{(22{,}464 \text{ m}^3/\text{d})} = 9.0 \text{ h}$$

c. Determine MLVSS.

$$\text{Fraction VSS} = \frac{20{,}050 \text{ kg VSS}}{25{,}405 \text{ kg TSS}} = 0.79$$

$$\text{MLVSS} = 0.79(3000 \text{ g/m}^3) = 2370 \text{ g/m}^3$$

15. Determine F/M and BOD volumetric loading.
    a. Determine F/M using Eq. (7–60).

$$\text{F/M} = \frac{QS_o}{XV} = \frac{\text{g BOD}}{\text{g MLVSS·d}}$$

$$= \frac{(22{,}464 \text{ m}^3/\text{d})(140 \text{ g/m}^3)}{(2370 \text{ g/m}^3)(8466 \text{ m}^3)} = 0.16 \text{ g/g·d}$$

b. Determine volumetric BOD loading using Eq. (7–61).

$$L_{\text{org}} = \frac{QS_o}{V} = \frac{\text{kg BOD}}{\text{m}^3\text{·d}}$$

$$= \frac{(22{,}464 \text{ m}^3/\text{d})(140 \text{ g/m}^3)}{(10^3 \text{ g/kg})(8466 \text{ m}^3)} = 0.37 \text{ kg/m}^3\text{·d}$$

Note: Both the F/M and L_{org} loading are in the low- to mid-range of the design parameters given later in this chapter in Table 8–16.

16. Determine the observed yield based on TSS and VSS.

Observed yield = g TSS/g bCOD = kg TSS/kg bCOD

$P_{X,\text{TSS}}$ = 2032.4 kg/d

bCOD removed = $Q(S_o - S)$

$$= (22{,}464 \text{ m}^3/\text{d})[(224 - 1.0) \text{ g/m}^3](1 \text{ kg}/10^3 \text{ g})$$

$$= 5009.5 \text{ kg/d}$$

a. Observed yield based on TSS

$$Y_{\text{obs,TSS}} = \frac{(2032.4 \text{ kg/d})}{(5009.5 \text{ kg/d})} = \frac{0.41 \text{ kg TSS}}{\text{kg bCOD}} = \frac{0.41 \text{ g TSS}}{\text{g bCOD}}$$

$$= \left(\frac{0.41 \text{ g TSS}}{\text{g bCOD}}\right)\left(\frac{1.6 \text{ g bCOD}}{\text{g BOD}}\right) = 0.65 \text{ g TSS/g BOD}$$

b. Observed yield based on VSS

$Y_{obs, VSS}$: VSS/TSS = 0.80 (see Step 4c)

$$= \left(\frac{0.41 \text{ g TSS}}{\text{g bCOD}}\right)\left(\frac{0.80 \text{ g VSS}}{\text{g TSS}}\right)$$

$$= 0.328 \text{ g VSS/g bCOD}$$

$$= \left(\frac{0.328 \text{ g VSS}}{\text{g bCOD}}\right)\left(\frac{1.6 \text{ g bCOD}}{\text{g BOD}}\right)$$

$$= 0.52 \text{ g VSS/g BOD}$$

17. Calculate the O_2 demand using Eq. (8–17).

$$R_o = Q(S_o - S) - 1.42\,P_{X,bio} + 4.33\,Q(NO_x)$$

$$= (22,464 \text{ m}^3/\text{d})[(224 - 1.0) \text{ g/m}^3](1 \text{ kg}/10^3 \text{ g}) - 1.42(1154.7 \text{ kg/d})$$

$$+ (4.33 \text{ g } O_2/\text{g N})(22,464 \text{ m}^3/\text{d})(28.3 \text{ g/m}^3)(1 \text{ kg}/10^3 \text{ g})$$

$$R_o = 5009.5 \text{ kg/d} - 1639.7 \text{ kg/d} + 2752.7 \text{ kg/d}$$

$$= 6122.5 \text{ kg/d} = 255.1 \text{ kg/h}$$

18. Fine bubble aeration design—determine air flowrate at average design flowrate (see procedure for Step 8).

a. Determine the SOTR using the values given in the problem statement: $\alpha = 0.65$, $\beta = 0.95$, and $F = 0.9$.

$$\text{SOTR} = \text{AOTR}\left[\frac{C_{s,20}}{\alpha F(\beta C_{\bar{s},T,H} - C)}\right](1.024^{20-T})$$

$$= \frac{(255.1 \text{ kg/h})(9.08 \text{ g/m}^3)(1.024)^{20-12}}{(0.65)(0.9)[(0.95)(11.93 \text{ g/m}^3) - 2.0 \text{ g/m}^3]} = 512.9 \text{ kg/h}$$

b. Determine the air flowrate.

$$\text{Air flowrate, m}^3/\text{min} = \frac{(512.9 \text{ kg/h})}{[(0.35)(60 \text{ min/h})(0.270 \text{ kg } O_2/\text{m}^3 \text{ air})]}$$

$$= 90.5 \text{ m}^3/\text{min}$$

19. Check alkalinity.

a. Prepare an alkalinity balance

Alkalinity to maintain pH ~ 7 = Influent Alk − Alk used + Alk to be added

Influent alkalinity: 140 g/m^3 as $CaCO_3$

Amount of nitrogen converted to nitrate: NO_x = 28.3 g/m^3 (see Step 12)

Alkalinity used for nitrification = (7.14 g $CaCO_3$/g NH_4-N)(28.3 g/m^3)

$$= 202.6 \text{ g/m}^3 \text{ used as } CaCO_3$$

b. Substitute known values and solve for alkalinity needed.
Residual alkalinity concentration needed to maintain pH in the range of 6.8–7.0 = 70 to 80 g/m^3 as $CaCO_3$; select 80 g/m^3

$80 \ g/m^3$ = Influent alk − alk used + alk to be added

$80 \ g/m^3 = 140 \ g/m^3 − 202.6 \ g/m^3$ + alk to be added

Alkalinity needed = 142.6 g/m^3 as $CaCO_3$

$$= (22{,}464 \ m^3/d)(142.6 \ g/m^3)(1 \ kg/10^3 g)$$

$$= 3203 \ kg/d \ as \ CaCO_3$$

c. Determine the alkalinity needed as sodium bicarbonate.
Sodium bicarbonate may be preferred over lime for alkalinity addition due to ease of handling and fewer scaling problems as compared to lime. The amount of $Na(HCO_3)$ needed is as follows:

Equivalent weight of $CaCO_3$ = 50 g/equivalent

Equivalent weight of $Na(HCO_3)$ = 84 g/equivalent

$$Na(HCO_3) \ needed = \frac{(3203 \ kg/d \ CaCO_3)(84 \ g \ NaHCO_3/eq)}{(50 \ g \ CaCO_3/equivalent)}$$

$$= 5380 \ kg/d \ NaHCO_3$$

20. Estimate effluent BOD using Eq. (8–25).

$$BOD = sBODe + \left(\frac{g \ BOD}{1.42 \ g \ VSS}\right)\left(\frac{0.85 \ g \ VSS}{g \ TSS}\right)(TSS, \ g/m^3)$$

Assume sBODe = 3.0 g/m^3

$$TSS = 10 \ g/m^3$$

$$BOD = 3.0 \ g/m^3 + (0.70)(0.85)(10 \ g/m^3)$$

$$= 8.95 \ g/m^3$$

21. Secondary clarifier design (for both BOD removal and BOD removal and nitrification).
a. Define return sludge recycle ratio (see Fig. 8–9):

$Q_r X_r = (Q + Q_r)X$ (assume waste sludge mass is insignificant)

Q_r = RAS flowrate, m^3/d

X_r = return sludge mass concentration, g/m^3

RAS recycle ratio = $Q_r/Q = R$

$R X_r = (1 + R)X$

$$R = \frac{X}{X_r - X}$$

b. Determine size of clarifier.

Assume $X_r = 8000$ g/m^3 (moderate settling/thickening sludge; per Sec. 8–3 range is 4000 to 12,000 mg/L).

$$R = \frac{(3000 \text{ g/m}^3)}{[(8000 - 3000) \text{ g/m}^3]} = 0.60$$

Assume a hydraulic application rate of 22 m^3/m^2·d at average flow for the secondary clarifier (see Table 8–7); the range is 16 to 28 m^3/m^2·d

$$\text{Area} = \frac{(22,464 \text{ m}^3/\text{d})}{(22 \text{ m}^3/\text{m}^2 \cdot \text{d})} = 1021 \text{ m}^3$$

Use 3 clarifiers (1 for each aeration tank)

Area/clarifier $= 312$ m^2

Clarifier diameter $= 20.8$ m; use 20 m

c. Check solids loading.

$$\text{Solids loading} = \frac{\text{kg TSS applied}}{\text{m}^2 \text{ clarifier area·h}}$$

$$\text{Solids loading} = \frac{(Q + Q_r)(\text{MLSS})}{A} = \frac{(1 + R)Q(\text{MLSS})}{A}$$

where A = area of clarifier, m^2 = $(\pi/4)(20 \text{ m})^2 \times 3 = 942$ m^2

$$\text{Solids loading} = \frac{(1 + 0.6)(22,464 \text{ m}^3/\text{d})(3000 \text{ g/m}^3)(1 \text{ kg}/10^3\text{g})}{(942 \text{ m}^2)(24 \text{ h/d})}$$

$$= 4.8 \text{ kg MLSS/m}^2 \cdot \text{h}$$

(within acceptable range of solids loading of 4 to 6 kg/m^2·d given in Table 8–7)

22. Prepare design summary.

Design parameter	Unit	BOD removal only (Part A)	BOD removal and nitrification (Part B)
Average wastewater flow	m^3/d	22,464	22,464
Average BOD load	kg/d	3145	3145
Average TKN load	kg/d	786	786
Aerobic SRT	d	5.0	12.5
Aeration tanks	Number	3	3
Aeration tank volume, ea	m^3	1340	2822
Hydraulic detention time, τ	h	4.3	9.0

(continued)

(Continued)

Design parameter	Unit	BOD removal only (Part A)	BOD removal and nitrification (Part B)
MLSS	g/m³	3000	3000
MLVSS	g/m³	2400	2370
F/M	g/g·d	0.33	0.16
BOD loading	kg BOD/m³·d	0.78	0.37
Sludge production	kg/d	2413	2032
Observed yield	kg TSS/kg bCOD	0.77	0.65
	kg VSS/kg BOD	0.61	0.52
Oxygen required	kg/h	120.5	255.1
Air flowrate at average wastewater flow	sm³/min	55.5	90.5
RAS ratio	Unitless	0.60	0.60
Clarifier hydraulic application rate	m³/m²·d	24	24
Clarifiers	Number	3	3
	Diameter, m	20	20
Alkalinity addition as Na(HCO₃)	kg/d	—	5380
Effluent BOD	g/m³	<30	8.95
TSSe	g/m³	<30	10
Effluent NH₄-N	g/m³	28.8	≤0.5

Comment Examination of the biomass production components for the nitrification design shows that the cell debris and nitrifying bacteria amounts represent a small portion of the total biomass (13.6 and 3.7 percent, respectively). In Step 13, the contribution from nbVSS is significant, 22 percent of the TSS produced. In this example, the design procedure is described for an average wastewater flowrate. In actual design, computations must also be made for peak flow and load conditions.

Sequencing Batch Reactor Process

The sequencing batch reactor (SBR) process utilizes a fill-and-draw reactor with complete mixing during the batch reaction step (after filling) and where the subsequent steps of aeration and clarification occur in the same tank. All SBR systems have five steps in common, which are carried out in sequence as follows: (1) fill, (2) react (aeration), (3) settle (sedimentation/clarification), (4) draw (decant), and (5) idle. Each of these steps is illustrated on Fig. 8–16 and described in Table 8–13. For continuous-flow applications, at least two SBR tanks must be provided so that one tank receives flow

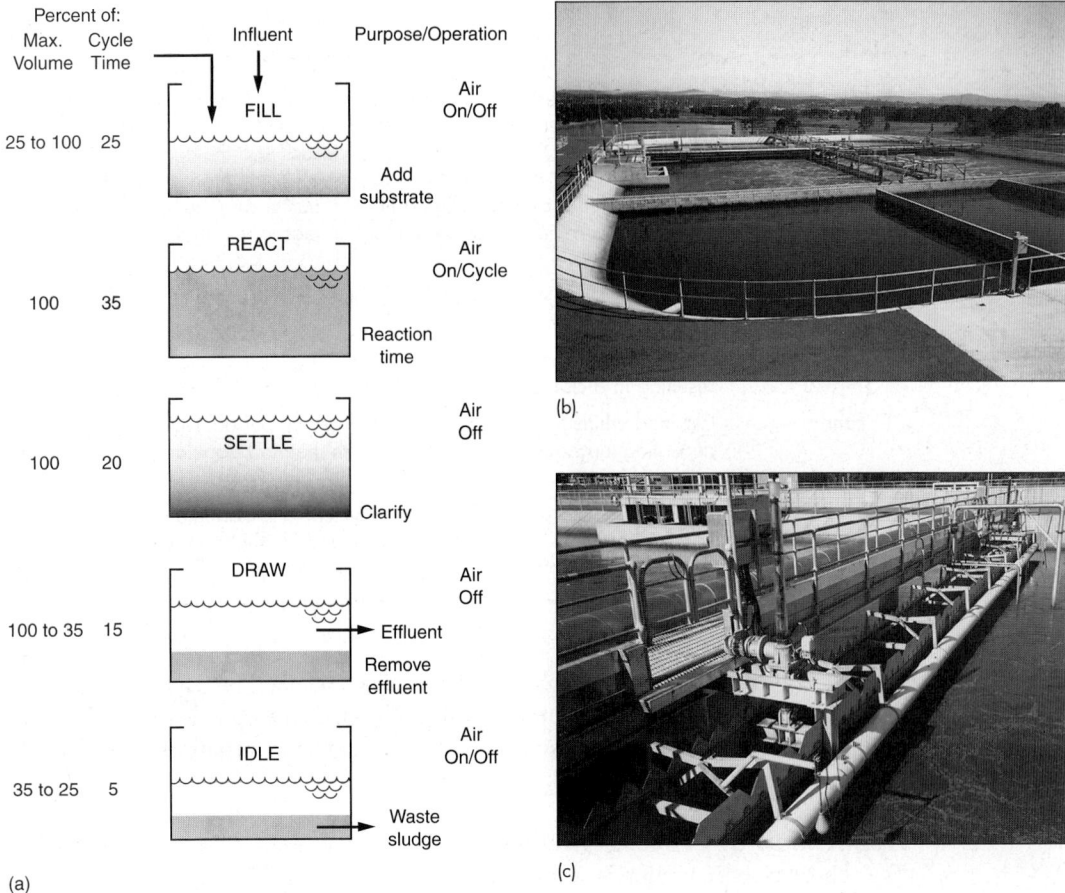

Percent of:			
Max. Volume	Cycle Time		

(a)

(b)

(c)

Figure 8–16

Sequencing batch reactor (SBR) activated-sludge process: (a) schematic diagram and (b) view of a typical SBR reactor and (c) view of movable weir used to decant contents of SBR reactor. Weir is located on the far side of the second dividing wall shown in (b). Photographs were taken in Australia.

while the other completes its treatment cycle. Several process modifications have been made in the times associated with each step to achieve nitrogen and phosphorus removal.

Sludge Wasting in SBRs. Sludge wasting is another important step in the SBR operation that greatly affects performance. Wasting is not included as one of the five basic process steps because there is no set time period within the cycle dedicated to wasting. The amount and frequency of sludge wasting is determined by performance requirements, as with a conventional continuous-flow system. In an SBR operation, sludge wasting usually occurs during the react phase so that a uniform discharge of solids (including fine material and large floc particles) occurs. A unique feature of the SBR system is that there is no need for a return activated-sludge (RAS) system.

Table 8–13
Description of operational steps for the sequencing batch reactor (SBR)

Operational step	Description
Fill	During the fill operation, volume and substrate (raw wastewater or primary effluent) are added to the reactor. The fill process typically allows the liquid level in the reactor to rise from 75% of capacity (at the end of the idle period) to 100%. When two tanks are used, the fill process may last about 50% of the full cycle time. During fill, the reactor may be mixed only or mixed and aerated to promote biological reactions with the influent wastewater
React	During the react period, the biomass consumes the substrate under controlled environmental conditions
Settle	Solids are allowed to separate from the liquid under quiescent conditions, resulting in a clarified supernatant that can be discharged as effluent
Decant	Clarified effluent is removed during the decant period. Many types of decanting mechanisms can be used, with the most popular being floating or adjustable weirs
Idle	An idle period is used in a multitank system to provide time for one reactor to complete its fill phase before switching to another unit. Because idle is not a necessary phase, it is sometimes omitted

Because both aeration and settling occur in the same chamber, no sludge is lost in the react step and none has to be returned to maintain the solids content in the aeration chamber. The SBR process can also be modified to operate in a continuous-flow mode, as discussed later in this chapter.

Application of Process Kinetics. During the react period, batch kinetics apply. The substrate concentration is much higher initially than would be present in a CMAS system, and the substrate decreases gradually as it is consumed by the biomass. The change in substrate concentration with time can be determined by starting with the substrate mass balance in Chap. 7 [Eq. (7–41)] for a continuous-flow complete-mix reactor:

$$\frac{dS}{dt} V = QS_o - QS + r_{su}V \tag{7–41}$$

$$\text{where } r_{su} = -\frac{\mu_m XS}{Y(K_s + S)} \tag{7–15}$$

Other terms as defined previously.

Because $Q = 0$ for the batch reaction, the substrate concentration is

$$\frac{dS}{dt} = -\frac{\mu_m XS}{Y(K_s + S)} \tag{8–37}$$

Integration of Eq. (8–37) with respect to time yields

$$K_s \ln \frac{S_o}{S_t} + (S_o - S_t) = X\left(\frac{\mu_m}{Y}\right)t \tag{8–38}$$

where S_o = initial substrate concentration at $t = 0$, mg/L

t = time, d

S_t = substrate concentration at time t, mg/L

The same kinetic expression applies for nitrification where $X = X_n$, the nitrifying bacteria concentration, $S = N$, the NH_4-N concentration, and the Monod model kinetic coefficients are substituted:

$$K_n \ln \frac{N_o}{N_t} + (N_o - N_t) = X_n \left(\frac{\mu_{mn}}{Y_n} \right) t \qquad (8\text{–}39)$$

where N_o = NH_4-N concentration at $t = 0$, mg/L

N_t = NH_4-N concentration at time t, mg/L

X_n = nitrifying bacteria concentration, mg/L

The maximum specific growth rate for nitrifying bacteria is affected by the DO concentration [Eq. (7–93)], so this effect is included in Eq. (8–40) as follows:

$$K_n \ln \frac{N_o}{N_t} + (N_o - N_t) = X_n \left(\frac{\mu_{mn}}{Y_n} \right) \left(\frac{DO}{K_o + DO} \right) t \qquad (8\text{–}40)$$

The above batch kinetic equations can be used to determine if the react period aeration time selected for SBR design is sufficient to provide the desired amount of degradation. An overall mass balance can be done first, assuming a certain amount of substrate is removed, to determine the biomass concentrations (X and X_n) for use in the equations. The time needed for dissolved BOD removal is relatively short (less than 1 h) due to the batch kinetics for treating domestic wastewater, resulting in a relatively low initial dissolved BOD concentration. For nitrification, SBR aerobic react times may range from 1.0 to 3.0 h (WEF, 1998). It should be noted that the SRTs for the SBR and continuous-flow activated-sludge processes are not comparable. At the same SRT, the SBR may be expected to be more efficient because of its batch kinetics, but the biomass may not be under aeration for a significant period of time so that the effective SRT is lower.

Because of the substrate concentration changes with time, the substrate utilization and oxygen demand rates change, progressing from high to low levels. The aeration system should be designed to reflect the changing requirements in oxygen demand. Additional descriptive material for the SBR process is provided in a later part of this section.

Process Design of SBRs. The following SBR design example requires selecting key design conditions and evaluating the results to determine if the design is appropriate. Because of the many design variables involved in an SBR design, an iterative approach is necessary in which key reactor design conditions are first assumed. A set of different design conditions can be evaluated by use of a spreadsheet analysis to determine the most optimal choice. The key design conditions selected are (1) the fraction of the tank contents removed during decanting and (2) the settle, decant, and aeration times. Because the fill volume equals the decant volume, the fraction of decant volume equals the fraction of the SBR tank volume used for the fill volume per cycle.

The design procedure for the SBR system is presented in Table 8–14 and illustrated in Example 8–3.

Table 8-14
Computation approach for the design of a sequencing batch reactor

1. Obtain influent wastewater characterization data, define effluent requirements, and define safety factors
2. Select the number of SBR tanks
3. Select the react/aeration, settling, and decant times. Determine the fill time and total time per cycle. Determine the number of cycles per day
4. From the total number of cycles per day, determine the fill volume per cycle
5. Select the MLSS concentration and determine the fill volume fraction relative to the total tank volume. Determine the decant depth. Using the computed depths, determine the SBR tank volume
6. Determine the SRT for the SBR process design developed
7. Determine the amount of TKN added that is nitrified
8. Calculate the nitrifier biomass concentration and determine if the aeration time selected is sufficient for the nitrification efficiency needed
9. Adjust the design as needed—additional iterations may be done
10. Determine the decant pumping rate
11. Determine the oxygen required and average transfer rate
12. Determine the amount of sludge production
13. Calculate the F/M and BOD volumetric loading
14. Evaluate alkalinity needs
15. Prepare design summary

EXAMPLE 8-3   Sequencing Batch Reactor Process Design Design a sequencing batch reactor process to treat a domestic wastewater with a flow of 7570 m³/d with the following wastewater characteristics. The reactor mixed-liquor concentration at full volume is 3500 g/m³ and the temperature is 12°C. The required effluent NH_4-N concentration = 0.50 g/m³. Primary treatment is not used.

Wastewater characteristics:

Constituent	Concentration, g/m³
BOD	220
sBOD	80
COD	485
sCOD	160
rbCOD	80
TSS	240

(continued)

(Continued)

Constituent	Concentration, g/m³
VSS	220
TKN	35
NH₄-N	25
TP	6
Alkalinity	200 as CaCO₃
bCOD/BOD ratio	1.6

Note: $g/m^3 = mg/L$.

Design conditions and assumptions:

1. Use 2 tanks
2. Total liquid depth when full = 6 m
3. Decant depth = 30 percent of tank depth
4. SVI = 150 mL/g
5. $NO_x \approx 80$ percent of TKN
6. Use kinetic coefficients in Tables 8–10 and 8–11

Solution

1. Develop wastewater characteristics needed for process design.
   a. Determine bCOD using Eq. (8–8).

   $$bCOD = 1.6(220 \text{ g/m}^3) = 352 \text{ g/m}^3$$

   b. Determine nbVSS concentration using Eq. (8–4).

   $$\frac{bpCOD}{pCOD} = \frac{1.6[(220 - 80) \text{ g/m}^3]}{[(485 - 160) \text{ g/m}^3]} = 0.69$$

   $$nbVSS = (1 - 0.69)(220 \text{ g/m}^3) = 68 \text{ g/m}^3$$

   c. Determine iTSS.

   $$iTSS = TSS_o - VSS_o$$

   $$= (240 - 220) \text{ g/m}^3 = 20 \text{ g/m}^3$$

2. Determine SBR operating cycle
   The total cycle time (T_c) consists of fill (t_F), react/aerate (t_A), settle (t_S), and decant (t_D). An idle time (t_I) can also be added. Thus, the total cycle time $T_c = t_F + t_A + t_S + t_D + t_I$. At least 2 tanks are needed so that when one tank is in the fill period t_F, the following cycles are occuring in the other tank: aeration t_A, settling t_S, and decant t_D cycles. Thus,

   $$t_F = t_A + t_S + t_D$$

Select period times:

Assume: $t_A = 2.0$ h

$\qquad t_S = 0.50$ h

$\qquad t_D = 0.50$ h

$\qquad t_I = 0$

Then $t_F = 2.0 + 0.50 + 0.50 = 3.0$ h for each tank (Note: Some aeration may also be done in the fill period.)

Total cycle time $T_c = t_F + t_A + t_S + t_D = 6.0$ h

$$\text{Number of cycles/tank·d} = \frac{(24 \text{ h/d})}{(6 \text{ h/cycle})} = 4$$

$$\text{Total number of cycles/d} = (2 \text{ tanks}) \left[\frac{(4 \text{ cycles/d})}{\text{tank}} \right]$$

$$= 8 \text{ cycles/d}$$

$$\text{Fill volume/cycle} = \frac{(7570 \text{ m}^3/\text{d})}{(8 \text{ cycles/d})} = 946.3 \text{ m}^3/\text{fill}$$

3. Determine fill fraction per cycle (V_F/V_T) allowed and compare to selected design value of 0.3.

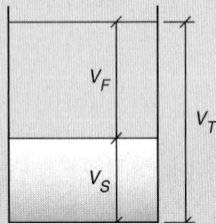

   a. Develop a mass balance based on solids in the reactor.

   Mass of solids at full volume = Mass of settled solids

   $$V_T X = V_S X_S$$

   where V_T = total volume, m^3

   $\qquad X$ = MLSS concentration at full volume, g/m^3

   $\qquad V_S$ = settled volume after decant, m^3

   $\qquad X_S$ = MLSS concentration in settled volume, g/m^3

   b. Solve mass balance and determine the fill fraction/cycle.

   i. Estimate X_s based on an assumed SVI value of 150 mL/g.

   $$X_s = \frac{(10^3 \text{ mg/g})(10^3 \text{ mL/L})}{(\text{SVI, mL/g})} = \frac{(10^3 \text{ mg/g})(10^3 \text{ mL/L})}{(150 \text{ mL/g})}$$

   $$= 6{,}666 \text{ g/m}^3$$

   $$X = 3500 \text{ g/m}^3 \text{ (see problem statement)}$$

ii. Determine the settled fraction.

$$\frac{V_S}{V_T} = \frac{X}{X_S} = \frac{(3500 \text{ g/m}^3)}{(6666 \text{ g/m}^3)} = 0.525$$

Provide 20 percent liquid above the sludge blanket so that solids are not removed by the decanting mechanism.

$$\frac{V_S}{V_T} = 1.2(0.525) = 0.63$$

iii. Determine the fill fraction.

$$V_F + V_S = V_T$$

where V_F = fill volume, m^3

$$\frac{V_F}{V_T} + \frac{V_S}{V_T} = 1.0$$

$$\frac{V_F}{V_T} = 1.0 - 0.63 = 0.37$$

Selected $V_F/V_T = 0.30$ is acceptable

4. Determine the overall hydraulic retention time τ.

Full liquid depth = 6.0 m

Decant depth = 0.3(6.0 m) = 1.80 m

$$V_T = \frac{V_F/\text{tank}}{0.30} = \frac{(946.3 \text{ m}^3/\text{tank})}{0.30} = 3154 \text{ m}^3/\text{tank}$$

$$\text{Overall } \tau = \frac{2 \text{ tanks } (3154 \text{ m}^3/\text{tank})(24 \text{ h/d})}{(7570 \text{ m}^3/\text{d})} = 20 \text{ h}$$

5. Determine the SRT.
    a. Use Eqs. (8–15), (8–16), and (7–58) to obtain a relationship that can be used to solve for $(P_{X,\text{TSS}})$ SRT.

$$(P_{X,\text{TSS}})\text{SRT} = \frac{QY(S_o - S)\text{SRT}}{[1 + (k_d)\text{SRT}](0.85)}$$

$$+ Q(\text{nbVSS})\text{SRT} + \frac{QY_n(\text{NO}_x)\text{SRT}}{[1 + (k_{dn})\text{SRT}](0.85)}$$

$$+ \frac{(f_d)(k_d)Q(Y)(S_o - S)\text{SRT}^2}{[1 + (k_d)\text{SRT}](0.85)} + Q(\text{TSS}_o - \text{VSS}_o)\text{SRT}$$

$$(P_{X,\text{TSS}})\text{SRT} = (V)(X_{\text{MLSS}}) = (3154 \text{ m}^3)(3500 \text{ g/m}^3)$$

$$= 11,039,000 \text{ g}$$

b. Develop input data to solve the above relationship for SRT:

nbVSS = 68 g/m³ (from Step 1b)

Assume $S_o \approx S_o - S$

S_o = bCOD = 352 g/m³ (Step 1a)

Q = (7570 m³/d)/2 tanks = 3785 m³/tank·d

$iTSS_o = TSS_o - VSS_o = (240 - 220)$ g/m³ = 20 g/m³ (Step 1c)

$NO_x = (0.80)(35$ g TKN/m³) = 28 g/m³

Kinetic coefficients from Tables 8–10 and 8–11:

Y = 0.40 g VSS/g bCOD

$k_{d,\,12°C}$ = 0.12 g/g·d $(1.04)^{12-20}$ = 0.088 g/g·d

Y_n = 0.12 g VSS/g NO_x

$k_{dn,\,12°C}$ = 0.08 g/g·d $(1.04)^{12-20}$ = 0.06 g/g·d

f_d = 0.15 g/g

Substituting values and calculations for above equation yields

$$11{,}039{,}000 \text{ g} = \frac{(3785 \text{ m}^3/\text{d})(0.40 \text{ g VSS/g bCOD})(352 \text{ g/m}^3)(\text{SRT})}{[1 + (0.088 \text{ g/g·d})(\text{SRT})](0.85)}$$

$$+ \; (3785 \text{ m}^3/\text{d})(68 \text{ g/m}^3)(\text{SRT})$$

$$+ \; \frac{(3785 \text{ m}^3/\text{d})(0.12 \text{ g/g})(28 \text{ g/m}^3)(\text{SRT})}{[1 + (0.06 \text{ g/g·d})(\text{SRT})](0.85)}$$

$$+ \; \frac{(0.15 \text{ g/g})(0.088 \text{ g/g·d})(0.40 \text{ g VSS/g bCOD})(3785 \text{ m}^3/\text{d})(352 \text{ g/m}^3)(\text{SRT})^2}{[1 + (0.088 \text{ g/g·d})(\text{SRT})](0.85)}$$

$$+ \; (3785 \text{ m}^3/\text{d})(20 \text{ g/m}^3)(\text{SRT})$$

Solve for SRT (use spreadsheet with solver or solve by successive trials)

SRT = 17.0 d

6. Determine MLVSS concentration

a. Solve Eq. (7–54) (SRT = 17.0 d) $(S_o \approx S_o - S)$

$$(P_{X,\text{VSS}})\text{SRT} = V_T \, (X_{\text{MLVSS}})$$

$$(P_{X,\text{VSS}})\text{SRT} = \frac{Q(Y)(S_o - S)\text{SRT}}{1 + (k_d)\text{SRT}} + Q(\text{nbVSS})\text{SRT}$$

$$+ \frac{QY_n(NO_x)\text{SRT}}{1 + (k_{dn})\text{SRT}} + \frac{(f_d)(k_d)(Q)(Y)(S_o - S)(\text{SRT})^2}{1 + (k_d)\text{SRT}}$$

$$= \frac{(3785 \text{ m}^3/\text{d})(0.40 \text{ g VSS/g COD})(352 \text{ g/m}^3)(17.0 \text{ d})}{[1 + (0.088 \text{ g/g·d})(17.0 \text{ d})]} + (3785 \text{ m}^3/\text{d})(68 \text{ g/m}^3)(17.0 \text{ d})$$

$$+ \frac{(3785 \text{ m}^3/\text{d})(0.12 \text{ g VSS/g NO}_x)(28 \text{ g/m}^3)(17.0 \text{ d})}{[1 + (0.06 \text{ g/g·d})(17.0 \text{ d})]}$$

$$+ \frac{(0.15 \text{ g/g})(0.088 \text{ g/g·d})(0.40 \text{ g VSS/g COD})(3785 \text{ m}^3/\text{d})(352 \text{ g/m}^3)(17.0 \text{ d})^2}{[1 + (0.088 \text{ g/g·d})(17.0 \text{ d})]}$$

$$= V_T(X_{\text{MLVSS}})$$

$$V_T(X_{\text{MLVSS}}) = 8{,}926{,}716 (\text{m}^3 \cdot \text{g/m}^3)$$

$$V_T = 3154 \text{ m}^3 \text{ (Step 4)}$$

$$V_T(X_{\text{MLVSS}}) = (3154 \text{ m}^3)(X_{\text{MLVSS}})$$

$$8{,}926{,}716 (\text{m}^3 \cdot \text{g/m}^3) = (3154 \text{ m}^3)(X_{\text{MLVSS}})$$

$$X_{\text{MLVSS}} = 2830 \text{ g/m}^3$$

b. Determine the fraction of MLVSS.

$$\frac{X_{\text{MLVSS}}}{X_{\text{MLSS}}} = \frac{(2830 \text{ g/m}^3)}{(3500 \text{ g/m}^3)} = 0.81$$

7. Determine amount of NH_4-N oxidized (NO_x). Nitrogen balance (Eq. 8–18)

$$NO_x = TKN_o - N_e - 0.12 \, P_{X,\text{bio}}/Q$$

$$P_{X,\text{bio}} = [\text{Items A + B + C in Eq. (8–15)}]$$

$$P_{X,\text{bio}} = \frac{QY(S_o - S)}{1 + (k_d)\text{SRT}} + \frac{QY_n(NO_x)}{1 + (k_{dn})\text{SRT}} + \frac{(f_d)(k_d)QY(S_o - S)\text{SRT}}{1 + (k_d)\text{SRT}}$$

$$= \frac{(3785 \text{ m}^3/\text{d})(0.40 \text{ g VSS/g COD})(352 \text{ g/m}^3)}{[1 + (0.088 \text{ g/g·d})(17.0 \text{ d})]} + \frac{(3785 \text{ m}^3/\text{d})(0.12 \text{ g VSS/g NO}_x)(28 \text{ g/m}^3)}{[1 + (0.06 \text{ g/g·d})(17.0 \text{ d})]}$$

$$+ \frac{(0.15 \text{ g/g})(0.088 \text{ g/g·d})(3785 \text{ m}^3/\text{d})(0.40 \text{ gVSS/g COD})(352 \text{ g/m}^3)(17.0 \text{ d})}{[1 + (0.088 \text{ g/g·d})(17.0 \text{ d})]}$$

$$= 267{,}720 \text{ g/d} = 267.7 \text{ kg/d}$$

$$NO_x = 35.0 - 0.50 - \frac{(0.12)(267.7 \text{ kg/d})(10^3 \text{ g/kg})}{(3785 \text{ m}^3/\text{d})}$$

$$= (35.0 - 0.50 - 8.5) \text{ g/m}^3$$

$$NO_x = 26.0 \text{ g/m}^3$$

8. Check the degree of nitrification to determine whether NH_4-N will be removed to a level of 0.50 g/m^3 in a 2-h aeration period.

a. Determine the amount of oxidizable N available.

$NO_x = 26.0$ g/m$^3 = NH_4$-N in feed flow that can be oxidized

Oxidizable NH_4-N added/cycle:

$V_F(NO_x) = 946.3$ m^3/cycle $(26.0$ g/m$^3)$

$= 24{,}604$ g/fill

NH_4-N remaining before fill $= V_S(N_e)$

$N_e = 0.50$ g/m^3 NH_4-N

$V_S(N_e) = N_e(V - V_F)$

$= (0.50$ g/m$^3)[(3154 - 946.3)$ m$^3]$

$= 1104$ g

Total oxidizable N at beginning of cycle $= (24{,}604 + 1104)$ g $= 25{,}708$ g

$\text{Initial concentration} = \dfrac{25{,}708 \text{ g}}{V_T} = \dfrac{25{,}708 \text{ g}}{3154 \text{ m}^3} = 8.15 \text{ g/m}^3 = N_o$

b. Determine the reaction time.

Using Eq. (8–40), the react time (aeration) after fill to achieve the desired NH_4-N concentration can be calculated. First, the nitrifier concentration must be determined.

$$K_n \ln \frac{N_o}{N_t} + (N_o - N_t) = X_n \left(\frac{\mu_{mn}}{Y_n}\right)\left(\frac{\text{DO}}{K_o + \text{DO}}\right)t$$

i. Nitrifier concentration

$$X_n = \frac{Q(Y_n)(NO_x)\text{SRT}}{[1 + (k_d)\text{SRT}]V}$$

$$= \frac{(3785 \text{ m}^3/\text{d})(0.12 \text{ g VSS/g } NH_4\text{-N})(26.0 \text{ g/m}^3)(17.0 \text{ d})}{[1 + (0.06 \text{ g/g·d})(17.0 \text{ d})](3154 \text{ m}^3)}$$

$= 31.5$ g/m^3

$\mu_{m,12°C} = 0.75$ g/g·d $(1.07)^{12-20} = 0.44$ g/g·d

$K_{n,12°C} = 0.74$ g/m^3 $(1.053)^{12-20} = 0.49$ g/m^3

$K_o = 0.50$ g/m^3

ii. Determine the time for reaction.

Solve for t for $N_o = 8.15$ g/m³, $N_e = 0.50$ g/m³

$$0.49 \ln \left[\frac{(8.15 \text{ g/m}^3)}{(0.5 \text{ g/m}^3)} \right] [(8.15 - 0.5) \text{ g/m}^3]$$

$$= (31.5 \text{ g/m}^3) \left[\frac{(0.44 \text{ g/g·d})}{(0.12 \text{ g/g})} \right] \left(\frac{2.0}{0.5 + 2.0} \right) t$$

$t = 0.113$ days $= 2.7$ h

c. Determine the aeration time.

Required aeration time = 2.7 h

Aeration time selected was 2.0 h; therefore, aeration is required during part of the fill period

Fill time = 3.0 h

At least one half of the fill time should be used for mixing without aeration to provide a selector operation for SVI control.

$$\text{Total aeration time} = \frac{3.0}{2} + 2.0 = 3.5 \text{ h}$$

Thus, there is sufficient time for nitrification. Because the factor of safety for this design is quite low, other iterations can be done to develop a more conservative design, which would result in a larger volume (V_T) and longer SRT. The volume could be increased by making V_F/V_T smaller or by increasing the time for the react period.

9. Determine the decant pumping rate.

Decant volume = fill volume

$V_F = 946.3$ m³

Decant time = 30 min

$$\text{Pumping rate} = \frac{946.3 \text{ m}^3}{30 \text{ min}} = 31.5 \frac{\text{m}^3}{\text{min}}$$

10. Determine oxygen required/tank using Eq. (8–17).

$$R_o = Q (S_o - S) - 1.42 P_{X,\text{bio}} + 4.33 Q(NO_x)$$

$$= (3785 \text{ m}^3/\text{d})(352 \text{ g/m}^3)(1 \text{ kg}/10^3 \text{ g}) - 1.42(267.7 \text{ kg/d})$$

$$+ 4.33(26.0 \text{ g/m}^3)(3785 \text{ m}^3/\text{d})(1 \text{ kg}/10^3 \text{ g})$$

$$R_o = (1332 - 380 + 426) \text{ kg/d} = 1378 \text{ kg/d}$$

Aeration time/tank

Aeration time/cycle = 3.5 h
Number of cycles/d = 4
Total aeration time = 14 h/d

$$\text{Average oxygen transfer rate} = \frac{(1378 \text{ kg } O_2/d)}{(14 \text{ h/d})} = 98.4 \text{ kg/h}$$

Note: The oxygen demand will be higher at the beginning of the aeration period so the aeration system oxygen transfer capacity must be higher than this average transfer rate. The oxygen transfer rate should be multiplied by a factor of 1.5 to 2.0 to provide sufficient oxygen transfer at the beginning of the cycle and to handle peak loads.

11. Determine sludge production using Eq. (7–55). (MLSS = X_{TSS})

$$P_{X,TSS} = \frac{(V)(\text{MLSS})}{\text{SRT}}$$

$$= \frac{(2 \text{ tanks})(3154 \text{ m}^3/\text{tank})(3500 \text{ g/m}^3)(1 \text{ kg}/10^3 \text{ g})}{17.0 \text{ d}}$$

$$= 1299 \text{ kg/d}$$

$$\text{bCOD removed} = (7570 \text{ m}^3/\text{d})(352 \text{ g/m}^3)(1 \text{ kg}/10^3 \text{ g})$$

$$= 2664 \text{ kg/d}$$

$$\text{BOD removed} = \frac{(2664 \text{ kg bCOD/d})}{(1.6 \text{ kg bCOD/kg BOD})} = 1665 \text{ kg/d}$$

$$\text{Observed yield, g TSS/g BOD} = \frac{1299}{1665} = \left(\frac{0.78 \text{ g TSS}}{\text{g BOD}}\right)$$

$$\text{Observed yield, g VSS/g BOD} = \left(\frac{0.78 \text{ g TSS}}{\text{g BOD}}\right)\left(\frac{0.81 \text{ g VSS}}{\text{g TSS}}\right)$$

$$= \frac{0.63 \text{ g VSS}}{\text{g BOD}}$$

$$\text{Observed yield, g TSS/g bCOD} = \frac{1299}{2664} = 0.49 \text{ g TSS/g bCOD}$$

Note: The observed yield is higher than the observed yield in the CMAS design example (Example 8–2) even though the SRT is higher in this design. Because no primary treatment is used, the wastewater has a higher nbVSS concentration. Also note that the endogenous decay coefficient was assumed as a constant regardless of whether the biomass was under aeration or nonaerobic conditions. During a large part of the nonaerated period, nitrates are present, and the endogenous decay coefficient may not be lower. The error introduced by the decay coefficient assumption has little effect on the reliability of the design.

12. Determine F/M and BOD volumetric loading
    a. F/M

$$F/M = \frac{kg\ BOD}{kg\ MLVSS \cdot d} = \frac{QS_o}{XV}$$

$$= \frac{(3785\ m^3/d)(220\ g/m^3)}{(2830\ g/m^3)(3154\ m^3)} = 0.09\ g/g \cdot d$$

    b. BOD volumetric loading

$$L_{org} = \frac{kg\ BOD}{m^3 \cdot d} = \frac{QS_o}{V}$$

$$= \frac{(7570\ m^3/d)(220\ g/m^3)}{(2)(3154\ m^3)(10^3\ g/kg)} = 0.27\ kg/m^3 \cdot d$$

13. Prepare a design summary for average conditions.

Design parameter	Unit	Value
Average flow	m³/d	7570
Average BOD load	kg/d	1665
Average TKN load	kg/d	265
Number of tanks	Number	2
Fill time	h	3.0
React time	h	2.0
Total aeration time	h	3.5
Settle time	h	0.5
Decant time	h	0.5
Cycle time	h	6.0
Total SRT	d	17.0
Tank volume	m³	3154
Fill volume/cycle	m³	946.3
Fill volume/tank volume	Ratio	0.3
Decant depth	m	1.8
Tank depth	m	6.0
MLSS	g/m³	3500
MLVSS	g/m³	2830
F/M	g/g·d	0.09
Volumetric BOD load	kg/m³·d	0.27
Decant pumping rate	m³/min	31.5
Sludge production	kg/d	1299
Observed yield	kg VSS/kg BOD	0.63
	kg TSS/kg bCOD	0.49
Average oxygen required/tank	kg/d	1378
Total aeration time/d-tank	h	14
Average O₂ transfer rate	kg/h	98.4

Staged Activated-Sludge Process

In the conventional plug-flow activated-sludge system, the tank hydraulics and mixing regime may result in two to four effective stages from the standpoint of biological kinetics. Activated-sludge processes can be designed with baffle walls to intentionally create a number of complete-mix activated-sludge zones operating in series (see Fig. 8–17). For the same reactor volume, reactors in series can provide greater treatment efficiency than a single complete-mix reactor, or provide a greater treatment capacity. As a consequence, staged activated-sludge process configurations are used at several full-scale installations.

Oxygen Demand in Staged Designs. The oxygen demand varies in staged complete-mix reactor designs and can be high enough in the first stage to challenge the volumetric oxygen transfer capability of aeration equipment. With high-density fine bubble aeration diffusers, such as membrane aeration panels described in Sec. 5–12 in Chap. 5, oxygen transfer rates of 100 to 150 mg/L·h are possible, with some manufacturers claiming higher rates. The changes in oxygen uptake rates (OURs) in each stage of a four-stage activated-sludge process (defined as a function of oxygen needed for nitrification, rbCOD removal, particulate degradable COD, and endogenous respiration) are depicted on Fig. 8–18. Most of the rbCOD will be consumed in the first stage, and the OUR for pCOD degradation will decrease from stage to stage as a function of the

Figure 8–17

Schematic diagram of a staged activated-sludge process.

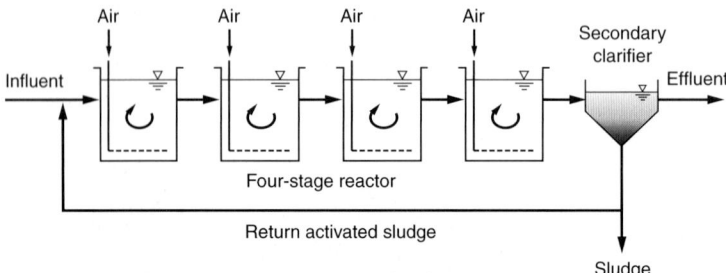

Figure 8–18

Changes in oxygen uptake rates for staged activated-sludge process.

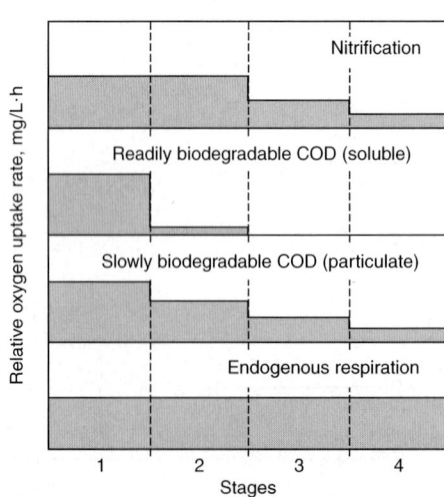

degradation kinetics. Nitrification rates may be at a maximum zero-order kinetic rate for the first one to three stages due to higher NH_4-N concentrations in the early stages. Oxygen demand for endogenous respiration will be relatively constant from stage to stage.

The oxygen demand distribution may be estimated to determine the aeration design for staged processes. The percent of the total oxygen consumption may range from 40, 30, 20, and 10 percent, respectively, for a four-stage system. One design approach that can be used to obtain an estimate of the oxygen demand in a staged system is to calculate the total oxygen demand as would be done for a CMAS process, and then estimate the oxygen demand distribution with consideration to the various components described above. With proper selection of the type and placement of the diffusers and by providing an air supply system with DO control in each portion of the system, the air can be provided where needed. Generally, the approach outlined above is satisfactory because during the life of the process, the oxygen demand will vary across the tank as the load changes.

Use of Simulation Models. The other approach involves the use of simulation models, in which the kinetics and changes in constituent concentrations in each stage are taken into consideration. This approach will typically result in a more optimal design and can be used to assess the real capacity of a given activated-sludge design. The simulation approach involves solving a set of equations in each stage for each constituent, which includes rbCOD, pCOD, NH_4-N, endogenous respiration, and biomass concentration. Models also include phosphorus and the effect of biological phosphorus removal on design and performance. An example of such a comprehensive simulation model is presented at the end of this chapter.

The effect of using a staged system as compared to a single CMAS tank for nitrification is illustrated in Example 8–4.

EXAMPLE 8–4   **Evaluation of Staged Reactors for Nitrification**   Compare the performance of a four-stage activated-sludge system with equal volumes per stage for nitrification to that for a single-stage system. The hydraulic retention time is 8 h for both systems, and the same SRT is used. The single-stage system is designed for an effluent NH_4-N concentration of 0.50 g/m^3 for the following conditions and using kinetic coefficients from Table 8–11.

Design conditions and assumptions:

Item	Unit	Value
Temperature	°C	18
μ_m	g/g·d	0.655
K_n	g NH_4-N /m^3	0.67
Y_n	g VSS/g NH_4-N	0.12
k_{dn}	g/g·d	0.074
Influent NH_4-N	g/m^3	30.0

(continued)

(*Continued*)

Item	Unit	Value
Effluent NH_4-N	g/m^3	0.50
DO	g/m^3	2.0
K_o	g/m^3	0.5
RAS recycle ratio	Unitless	0.5

Note: g/m^3 = mg/L.

Solution

1. Determine the SRT value and the concentration of the nitrifying bacteria for a single-stage system, $\tau = 8$ h $= 0.33$ d, $N = 0.50$ g/m^3.

   a. Solve for the specific growth rate using Eq. (7–93) in Table 8–5.

   $$\mu_n = \left(\frac{\mu_{nm}N}{K_n + N}\right)\left(\frac{DO}{K_o + DO}\right) - k_{dn}$$

   $$= \left\{\frac{(0.655 \text{ g/g·d})(0.5 \text{ g/m}^3)}{[(0.67 + 0.50) \text{ g/m}^3]}\right\}\left\{\frac{(2.0 \text{ g/m}^3)}{[(0.5 + 2.0) \text{ g/m}^3]}\right\} = 0.074 \text{ g/g·d}$$

   b. Solve for SRT using Eq. (7–37).

   $$\text{SRT} = \frac{1}{\mu_n} = \frac{1}{0.15 \text{ g/g·d}} = 6.7 \text{ d}$$

   c. Solve for the concentration of nitrifying bacteria using a modified form of Eq. (7–43).

   $$X_n = \frac{(\text{SRT})Y_n(NO_x)}{\tau[1 + (k_d)\text{SRT}]}$$

   $$= \frac{(6.7 \text{ d})(0.12 \text{ g/g})(30 \text{ g/m}^3)}{(0.33 \text{ d})[1 + (0.074 \text{ g/g·d})(6.7 \text{ d})]} = 48.9 \text{ g/m}^3$$

2. Perform nitrogen mass balances for a four-stage system shown on the following figure using equal volumes per stage. The total volume of the four-stage system is equal to the volume of the CMAS system, τ/stage $= 0.333$ d/4 $= 0.0833$ d/stage.

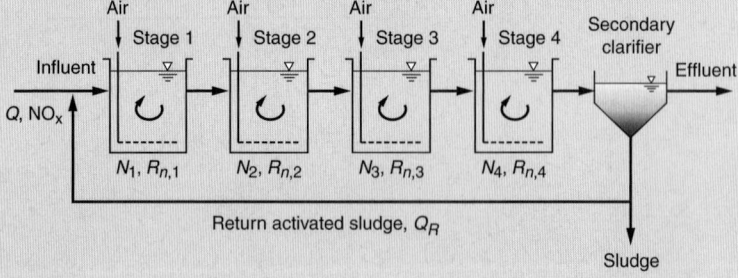

a. For Stage 1

Accumulation = in − out + generation

$$\frac{dN_1}{dt} V = Q(NO_x) + Q_R N_4 - (Q + Q_R)N_1 - R_{n,1}V$$

The rate expression for nitrification, derived from Eq. (7–15), includes a correction for the DO concentration, and is given by

$$R_{n,1} = \left(\frac{\mu_m}{Y_n}\right)\left(\frac{N_1}{K_n + N_1}\right)\left(\frac{DO}{K_o + DO}\right)X_n$$

where Q = wastewater flowrate, m^3/d
NO_x = available influent NH_4-N = 30 g/m^3
Q_R = recycle flowrate from stage 4, m^3/d
Q/Q_R = 0.50
N_4 = NH_4-N concentration for stage 4, g/m^3
N_1 = NH_4-N concentration for stage 1, g/m^3
$R_{n,1}$ = nitrification rate for stage 1, $g/m^3 \cdot d$
X_n = nitrifying bacteria concentration, g/m^3

The nitrifying bacteria concentration is the same as that calculated for the CMAS system because the same amount of NH_4-N is removed and the system is at the same SRT.

At steady state $dN_1/dt = 0$, and

$$NO_x + Q_R/QN_4 - (1 + Q_R/Q)/N_1 - R_{n,1}V/Q = 0$$

$$NO_x + 0.5N_4 = 1.5N_1 + R_{n,1}(\tau)$$

where τ = 0.0833 d, detention time of stage 1
NO_x = 30 g/m^3

b. For Stage 2, use the same procedure as Stage 1.

$$V \frac{dN_2}{dt} = (Q + Q_R)N_1 - (Q + Q_R)N_2 - R_{n,2}V$$

$$1.5N_1 = 1.5N_2 + R_{n,2}(\tau)$$

c. For Stage 3

$$1.5N_2 = 1.5N_3 + R_{n,3}(\tau)$$

d. For Stage 4

$$1.5N_3 = 1.5N_4 + R_{n,4}(\tau)$$

3. $R_{n,i(i=1-4)}$ is a function of the NH_4-N concentration (N) in each stage:
For stage 1,

$$R_{n,1} = \left[\frac{(0.655 \text{ g/g} \cdot \text{d})}{(0.12 \text{ g VSS/g NH}_4\text{-N})}\right]\left\{\frac{N_1}{[(0.67 + N_1) \text{ g/m}^3]}\right\}\left\{\frac{(2.0 \text{ g/m}^3)}{[(0.5 + 2.0) \text{ g/m}^3]}\right\}(48.9 \text{ g/m}^3)$$

4. The above equations for the four stages are solved with a spreadsheet program starting with Stage 1. The value for N_4 is assumed and adjusted in iterative calculations. The following effluent NH_4-N concentrations are computed for each stage:

Stage	NH_4-N concentration, g/m^3
1	12.2
2	4.7
3	0.50
4	0.03

5. Based on the above results, the same amount of nitrification that can be achieved theoretically with staging is 75 percent of the aeration tank volume required for a CMAS design.

Alternative Processes for BOD Removal and Nitrification

Over the last 30 years numerous activated-sludge processes have been developed for the removal of organic material (BOD) and for nitrification. Some of the processes are modifications or variations of basic processes that have evolved to meet different performance objectives. Descriptions and flowsheets are presented in Table 8–15 for representative processes used for BOD removal and nitrification. The processes are grouped according to the basic reactor configuration: plug-flow, complete-mix, and sequentially operated systems. A view of a large wastewater treatment plant using plug-flow reactors is shown on Fig. 8–19.

Figure 8–19

View of plug-flow reactors at Owls Head Wastewater Treatment Plant, New York. *(Courtesy of the New York City Department of Environmental Protection.)*

The processes differ in terms of their aeration configuration, aeration equipment design, solids retention time, operating mode, and ability to remove nitrogen, and some are proprietary. The high-rate aeration, contact stabilization, and high-purity oxygen processes are used primarily for BOD removal only, are designed for relatively short SRTs, and require less space than other processes. Where nitrification is not needed to meet treatment discharge limits, the three processes cited above are particularly attractive for large municipalities where space is limited. The conventional plug-flow, step-feed, and complete-mix processes are used for both BOD removal and nitrification and are applied over a wide range of SRTs, depending on the wastewater temperature and treatment needs. The Kraus process is seldom used, but it is included to show how oxidized nitrogen can be used to help BOD degradation in the first pass of a plug-flow aeration tank.

In contrast to the processes described above, conventional extended aeration, oxidation ditch, Orbal™, and Biolac™ processes represent a different approach to biological wastewater treatment (for the latter three, see Fig. 8–20). The processes employ a

(a)

(b)

Figure 8–20

Views of alternative activated-sludge processes: (a) oxidation ditch with brush aerators, (b) Orbal™ process with disk aerators, and (c) Biolac™ process with contiguous clarifier.

(c)

much simpler system by generally eliminating primary treatment and anaerobic digestion from the overall treatment system. Larger aeration tanks with longer SRTs, usually exceeding 20 d, are used. The process approach is attractive for smaller communities where space is not an issue and less complex operation is preferred. The large aeration tank volume provides good equalization at high flow and loading occurrences, and a high-quality effluent is produced. With the exception of the conventional extended aeration process, the systems are operated usually to promote denitrification in addition to nitrification. The aeration and mixing of the channel-flow processes (oxidation ditch, Orbal™, and CCAS™) require much less energy for mixing than needed for aeration so that aeration equipment design is based on meeting oxygen requirements instead of tank mixing. Less energy is required in comparison to conventional extended aeration processes. In the past, the oxidation ditch and extended aeration processes were thought to need long SRTs to provide well-stabilized biosolids for reuse. However, with stricter regulations governing biosolids stabilization (see Chap. 14), separate aerobic digestion facilities are used to meet the requirements for reuse.

Several sequentially operated activated-sludge processes that do not use separate tanks for liquid-solids separation are also described in Table 8–15. The processes include the sequencing batch reactor, batch decant reactor, and the cycle activated-sludge system. Operation is based usually on long τ and SRT values. The processes are attractive to small communities because of the simplicity of operation and relatively low cost. Sequentially operated processes are also adaptable to nitrogen removal, as discussed in Sec. 8–5.

Process Design Parameters

Typical parameters used for the design and operation of various activated-sludge processes are presented in Table 8–16.

Process Selection Considerations

Selection of an activated-sludge process for BOD removal and nitrification is a function of many considerations including specific site constraints, compatibility with the existing process, compatibility with existing equipment, present and future treatment needs, level of capability of the operating staff, capital costs, and operating costs. Significant features and limitations of the various activated-sludge process alternatives that affect process selection in certain applications are summarized in Table 8–17.

Table 8-15

Description of activated-sludge processes for BOD removal and nitrification

Process	Description
Complete-mix	
(a) Complete-mix activated-sludge (CMAS)	The CMAS process is an application of the flow regime of a continuous-flow stirred-tank reactor. Settled wastewater and recycled activated sludge are introduced typically at several points in the aeration tank. The organic load on the aeration tank, MLSS concentration, and oxygen demand are uniform throughout the tank. An advantage of the CMAS process is the dilution of shock loads that occur in the treatment of industrial wastewaters. The CMAS process is relatively simple to operate but tends to have low organic substrate concentrations (i.e., low F/M ratios) that encourage the growth of filamentous bacteria, causing sludge bulking problems

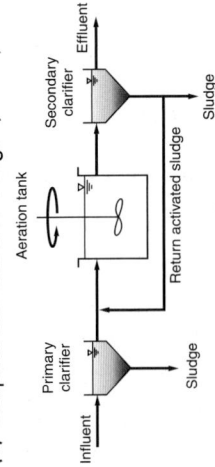

(continued)

Process	Description
Plug-flow	
(b) Conventional plug flow	Settled wastewater and return activated sludge (RAS) enter the front end of the aeration tank and are mixed by diffused air or mechanical aeration. Typically, from 3 to 5 channels (passes) are used. In early designs, air application was generally uniform throughout the tank length; however, low DO concentrations usually occurred in the initial passes of the tank. In modern designs, the aeration system is designed to match the oxygen demand along the length of the tank by tapering the aeration rates, i.e., applying higher rates in the beginning and lower rates near the end of the tank. During the aeration period, adsorption, flocculation, and oxidation of organic matter occur. Activated-sludge solids are separated in a secondary settling tank

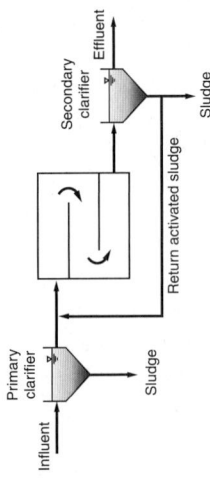

Process	Description
(c) High-rate aeration (see conventional plug flow)	High-rate aeration is a process modification in which low MLSS concentrations are combined with high volumetric BOD loadings. The high-rate system is characterized by short τ, high sludge recycle ratio, high F/M loading, and relatively low MLSS concentration. High-rate systems produce a lesser effluent quality, in terms of BOD and TSS concentration, as compared to conventional plug-flow or complete-mix systems. Because of the high loading used, more care must be taken to keep a stable operation. Adequate mixing and aeration are very important

| Table 8-15 (Continued)

Process

Description

Plug-flow (Continued)
(d) Step feed

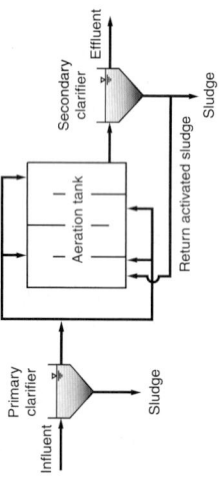

Step feed is a modification of the conventional plug-flow process in which the settled wastewater is introduced at 3 to 4 feed points in the aeration tank to equalize the F/M ratio, thus lowering peak oxygen demand. Generally, three or more parallel channels are used. Flexibility of operation is one of the important features of this process because the apportionment of the wastewater feed can be changed to suit operating conditions. The concentration of MLSS may be as high as 5000 to 9000 mg/L in the first pass, with lower concentrations in subsequent passes as more influent feed is added. The step-feed process has the capability of carrying a higher solids inventory, and thus a higher SRT for the same volume as a conventional plug-flow process. The step-feed process can also be operated in the contact-stabilization mode by feeding only the last pass, and high wet-weather flows can be bypassed to the last pass so that the solids load to the secondary clarifier can be minimized

(e) Contact stabilization

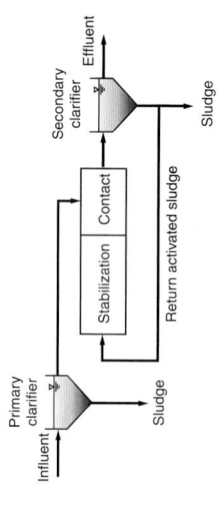

Contact stabilization uses two separate tanks or compartments for the treatment of the wastewater and stabilization of the activated sludge. The stabilized activated sludge is mixed with the influent (either raw or settled) wastewater in a contact zone. The contact zone detention time is relatively short (30 to 60 min), and the MLSS concentration in the contact zone is lower than that in the stabilization zone. Rapid removal of soluble BOD occurs in the contact zone, and colloidal and particulate organics are captured in the activated-sludge floc for degradation later in the stabilization zone. In the stabilization zone, return activated sludge (RAS) is aerated and the detention time is in the order of 1 to 2 h to maintain a sufficient SRT for sludge stabilization. Because the MLSS concentration is so much higher in the stabilization zone, the contact-stabilization process requires much less aeration volume than complete-mix or conventional plug-flow processes for the same SRT. The process was developed for BOD removal, and the short contact time limits the amount of soluble BOD degraded and NH₄-N oxidation

(f) Two-sludge

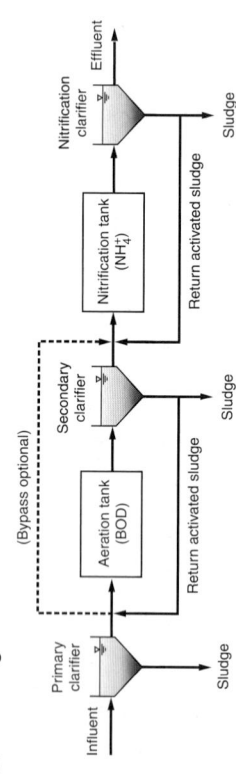

The two-sludge process is a two-stage system using high-rate activated sludge for BOD removal followed by a second stage for nitrification, which is operated at a longer SRT. A portion of the wastewater influent may be bypassed around the first stage to provide BOD and suspended solids for the nitrification process and promote flocculation and solids capture in secondary clarification. The main reason to separate the BOD removal stage from the nitrogen removal stage is to treat toxic substances in the first stage, and thus protect the more sensitive nitrifying bacteria. With better industrial pretreatment programs in place today, BOD removal and nitrification are done more commonly in a single-sludge process

A staged enclosed reactor is used in the high-purity oxygen activated-sludge process (McWhirter, 1978). Three or four stages are generally used and the influent wastewater, RAS, and high-purity oxygen are added to the first stage. The headspace gas and mixed liquor flow concurrently from stage to stage. The oxygen partial pressure in the headspace may range from 40 to 60 percent in the first stage to 20 percent in the last stage. At high oxygen partial pressure, higher volumetric oxygen transfer rates are possible so that pure oxygen systems can have a higher MLSS concentration and operate at a shorter τ and higher volumetric organic loadings than conventional processes. The rate of oxygen addition is about 2 to 3 times greater than that by conventional aeration systems. Onsite oxygen generation equipment is needed to provide the pure oxygen gas for the process, making the process operation more complex than conventional activated-sludge processes. Nitrification ability is limited with the high-purity oxygen processes due to the accumulation of carbon dioxide in the gas headspace, which causes low pH in the mixed liquor (less than 6.5). Major advantages for pure oxygen systems are the reduced space requirement and greatly reduced quantities of off-gas if odor control and VOC control are required

The Kraus process is a variation of the step aeration process used to treat nitrogen-deficient industrial wastewater. Digester supernatant is added as a food source to a portion of the return sludge in a separate aeration tank designed to nitrify. The resulting mixed liquor is then added to the main plug-flow aeration system. Besides providing nitrogen, nitrate is available to serve as an electron acceptor in the event of oxygen limitations

The extended aeration process is similar to the conventional plug-flow process except that it operates in the endogenous respiration phase of the growth curve, which requires a low organic loading and long aeration time. Because of the long SRTs (20 to 30 d) and τ's on the order of 24 h, aeration equipment design is controlled by mixing needs and not oxygen demand. The process is used extensively for preengineered plants for small communities. Generally, primary clarification is not used. Secondary clarifiers are designed at lower hydraulic loading rates than conventional activated-sludge clarifiers to better handle large flowrate variations typical of small communities. Although the biosolids are well stabilized, additional biosolids stabilization is required to permit beneficial reuse (see Chap. 14)

(continued)

(g) High-purity oxygen

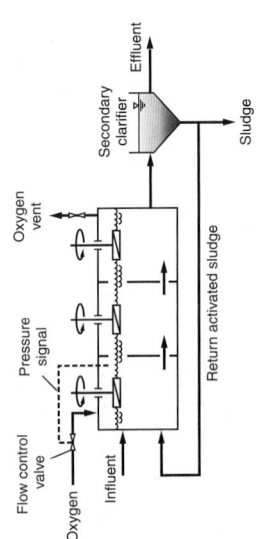

(h) Kraus process

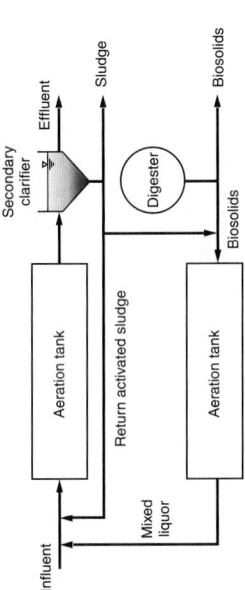

(i) Conventional extended aeration

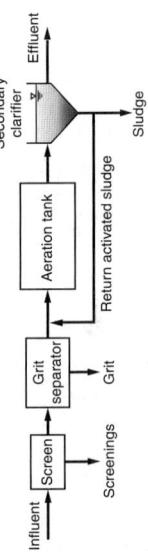

| Table 8-15 (Continued)

Process	Description

Extended aeration

(j) Oxidation ditch

The oxidation ditch consists of a ring- or oval-shaped channel equipped with mechanical aeration and mixing devices. Screened wastewater enters the channel and is combined with the return activated sludge. The tank configuration and aeration and mixing devices promote unidirectional channel flow, so that the energy used for aeration is sufficient to provide mixing in a system with a relatively long hydraulic retention time. The aeration/mixing method used creates a velocity from 0.25–0.30 m/s (0.8–1.0 ft/s) in the channel, which is sufficient to keep the activated sludge in suspension. At these channel velocities, the mixed liquor completes a tank circulation in 5–15 min, and the magnitude of the channel flow is such that it can dilute the influent wastewater flow by a factor of 20–30. As a result, the process kinetics approach that of a complete-mix reactor, but with plug flow along the channels. As the wastewater leaves the aeration zone, the DO concentration decreases and denitrification may occur. Brush-type or surface-type mechanical aerators are used for mixing and aeration (see Sec. 5–12 in Chap. 5). Secondary sedimentation tanks are used for most applications, and in some cases intrachannel clarifiers have been used

(k) Orbal™

The Orbal™ process is a variation of the oxidation ditch and uses a series of concentric channels within the same structure. Wastewater enters the larger outer channel and mixed liquor flows typically toward the center of the structure through at least two more channels before entering an internal clarifier or a distribution box. Disk aerators mounted on a horizontal shaft provide aeration. Channel depths range up to 4.3 m (14 ft). One version of the Orbal design (Bionutre™) limits the aeration rate in the first channel so that both nitrification and denitrification (anoxic condition) occur

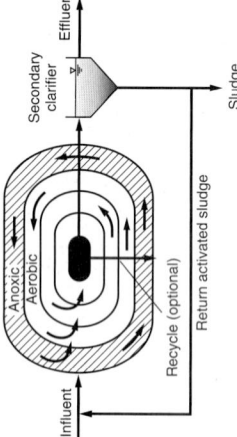

(l) Countercurrent aeration system (CCAS™)

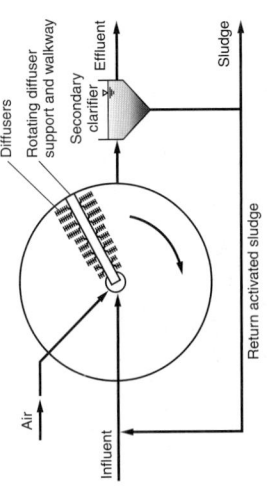

In the CCAS process, a unique aeration system is employed in which air diffusers are mounted at the bottom of a revolving bridge in a circular aeration tank. Because of the circulating motion of the bridge, which is moving faster than the aeration tank contents, fine bubbles are dispersed in a sweeping motion behind the traveling bridge. When the air is turned off, the movement of the diffusers creates enough mixing energy to keep the tank contents in suspension. The process is operated at a DO ranging from 0.7 to 1.0 mg/L. The low DO concentration is sufficient for nitrification at the long SRT, while allowing anoxic conditions to develop to promote denitrification. The system is normally designed with extended aeration SRTs

(m) Biolac™ process

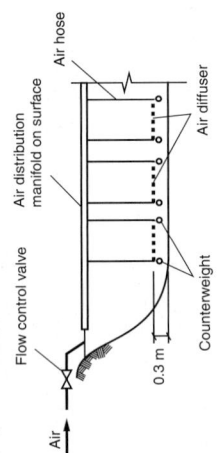

Biolac is a proprietary process that combines long solids retention times with submerged aeration in earthen basins. Fine bubble membrane diffusers are attached to floating aeration chains that are moved across the basin by the air released from the diffusers. Aeration basins are typically 2.4 to 4.6 m (8 to 15 ft) deep. The process can be designed for nitrification since the SRT ranges from 40 to 70 d. The F/M ratio ranges from 0.04 to 0.1 and the MLSS range is from 1500 to 5000 mg/L. A variation of the standard process, known as the "wave oxidation modification," allows biological nitrification and denitrification to occur simultaneously by using timers to cycle the air flowrate to each aeration chain. Either an internal or external clarifier can be used

Sequentially operated systems

(n) Sequencing batch reactor (SBR)

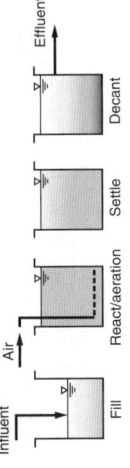

The SBR is a fill-and-draw type of reactor system involving a single complete-mix reactor in which all steps of the activated-sludge process occur. For municipal wastewater treatment with continuous flow, at least 2 basins are used so that one basin is in the fill mode while the other goes through react, solids settling, and effluent withdrawal. An SBR goes through a number of cycles per day; a typical cycle may consist of 3-h fill, 2-h aeration, 0.5-h settle, and 0.5-h for withdrawal of supernatant. An idle step may also be included to provide flexibility at high flows. Mixed liquor remains in the reactor during all cycles, thereby eliminating the need for separate secondary sedimentation tanks. Decanting of supernatant is accomplished by either fixed or floating decanter mechanisms. The τ's for SBRs generally range from 18 to 30 h, based on influent flowrate and tank volume used. Aeration may be accomplished by jet aerators or coarse bubble diffusers with submerged mixers (see Sec. 5–12 in Chap. 5). Separate mixing provides operating flexibility and is useful during the fill period for anoxic operation. Sludge wasting occurs normally during the aeration period

(continued)

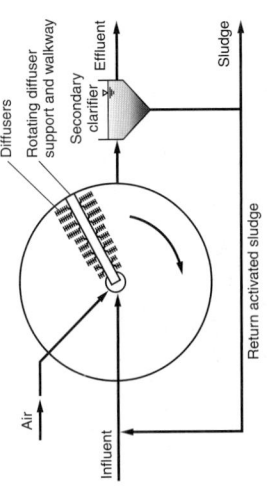

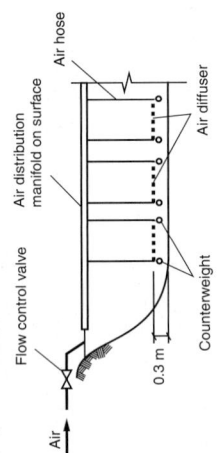

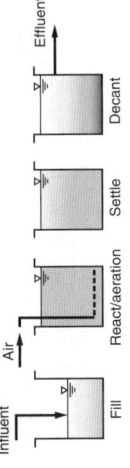

| Table 8-15 (Continued)

Process

(o) Batch decant reactor, intermittent cycle extended aeration system (ICEAS™)

Influent

Decanter floats on water surface

Effluent

Prereact chamber

Main chamber

(p) Cyclic activated-sludge system (CAAS™)

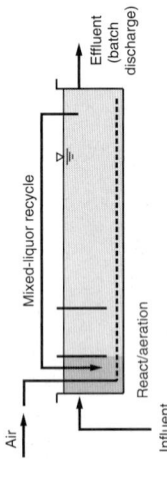

Air

Mixed-liquor recycle

Effluent (batch discharge)

React/aeration

Influent (continuous feed)

Description

The ICEAS process, developed in Australia, is another type of SBR process used for treating flowrates up to 500,000 m^3/d. Influent wastewater is fed continuously through the same cycles of react, settle, and decant as in an SBR. The influent is fed into one side of a baffled chamber (a prereact zone) so the flow does not disturb the mixed liquor during the settling and decant period. Wastewater flows continuously through openings at the bottom of the baffle wall and into the main react zone where BOD and nitrification occur. After aeration and settling, separated liquid is removed by an automated, time-controlled decant mechanism. Sludge wasting is also accomplished during this phase

The CAAS process uses three baffled zones in an approximate volumetric proportion of 1:2:20, and mixed liquor is recycled from Zone 3 to Zone 1. Nitrification occurs because of the long SRTs used. Nitrate reduction is claimed to occur at significant levels in the sludge blanket during the settle and decant periods, as well as in the aeration period by operation at low DO concentrations. As in the ICEAS process, the influent wastewater is fed continuously while the effluent is removed on a batch basis

Table 8-16

Typical design parameters for commonly used activated-sludge processes[a]

Process name	Type of reactor	SRT, d	F/M kg BOD/kg MLVSS·d	Volumetric loading lb BOD/ 1000 ft³·d	kg BOD/ m³·d	MLSS, mg/L	Total τ, h	RAS, % of influent[e]
High-rate aeration	Plug flow	0.5–2	1.5–2.0	75–150	1.2–2.4	200–1000	1.5–3	100–150
Contact stabilization	Plug flow	5–10	0.2–0.6	60–75	1.0–1.3	1000–3000[b] 6000–10000[c]	0.5–1[b] 2–4[c]	50–150
High-purity oxygen	Plug flow	1–4	0.5–1.0	80–200	1.3–3.2	2000–5000	1–3	25–50
Conventional plug flow	Plug flow	3–15	0.2–0.4	20–40	0.3–0.7	1000–3000	4–8	25–75[f]
Step feed	Plug flow	3–15	0.2–0.4	40–60	0.7–1.0	1500–4000	3–5	25–75
Complete mix	CMAS	3–15	0.2–0.6	20–100	0.3–1.6	1500–4000	3–5	25–100[f]
Extended aeration	Plug flow	20–40	0.04–0.10	5–15	0.1–0.3	2000–5000	20–30	50–150
Oxidation ditch	Plug flow	15–30	0.04–0.10	5–15	0.1–0.3	3000–5000	15–30	75–150
Batch decant	Batch	12–25	0.04–0.10	5–15	0.1–0.3	2000–5000[d]	20–40	NA
Sequencing batch reactor	Batch	10–30	0.04–0.10	5–15	0.1–0.3	2000–5000[d]	15–40	NA
Countercurrent aeration system (CCAS™)	Plug flow	10–30	0.04–0.10	5–10	0.1–0.3	2000–4000	15–40	25–75[f]

[a] Adapted from WEF (1998); Crites and Tchobanoglous (1998).

[b] MLSS and detention time in contact basin.

[c] MLSS and detention time in stabilization basin.

[d] Also used at intermediate SRTs.

[e] Based on average flow.

[f] For nitrification, rates may be increased by 25 to 50%.

NA = not applicable.

Table 8–17
Advantages and limitations of activated-sludge processes for BOD removal and nitrification

Process	Advantages	Limitations
Complete mix	Common, proven process	Susceptible to filamentous sludge bulking
	Adaptable to many types of wastewater	
	Large dilution capacity for shock and toxic loads	
	Uniform oxygen demand	
	Design is relatively uncomplicated	
	Suitable for all types of aeration equipment	
Conventional plug flow	Proven process	Design and operation for tapered aeration is more complex
	May achieve a somewhat higher level of ammonia removal than the complete-mix process	May be difficult to match oxygen supply to oxygen demand in first pass
	Adaptable to many operating schemes including step-feed, selector design, and anoxic/aerobic processes	
High rate	Requires less aeration tank volume than conventional plug flow	Less stable operation; produces lower-quality effluent
	Uses less aeration energy	Not suitable for nitrification
		Sludge production is higher
		High peak flows can disrupt operation by washing out MLSS
Contact stabilization	Requires smaller aeration volume	Has little or no nitrification capability
	Handles wet-weather flows without loss of MLSS	Operation somewhat more complex
Step feed	Distributes load to provide more uniform oxygen demand	More complex operation
	Peak wet-weather flows can be bypassed to the last pass to minimize high clarifier solids loading	Flow split is not usually measured or known accurately
	Flexible operation	More complicated design for process and aeration system
	Adaptable to many operating schemes including anoxic/aerobic processes	
Extended aeration	High-quality effluent possible	Aeration energy use is high
	Relatively uncomplicated design and operation	Relatively large aeration tanks
	Capable of treating shock/toxic loads	Adaptable mostly to small plants
	Well-stabilized sludge; low biosolids production	
High-purity oxygen	Requires relatively small aeration tank volume	Limited capability for nitrification
	Emits less VOC and off-gas volume	More complex equipment to install, operate, and maintain
	Generally produces good settling sludge	*Nocardia* foaming
	Operation and DO control are relatively uncomplicated	High peak flows can disrupt operation by washing out MLSS
	Adaptable to many types of wastewater	

(continued)

| **Table 8–17** (*Continued*) | | |

Process	Advantages	Limitations
Oxidation ditch	Highly reliable process; simple operation	Large structure, greater space requirement
	Capable of treating shock/toxic loads without affecting effluent quality	Low F/M bulking is possible
	Economical process for small plants	Some oxidation ditch process modifications are proprietary and license fees may be required
	Uses less energy than extended aeration	Requires more aeration energy than conventional CMAS and plug-flow treatment
	Adaptable to nutrient removal	Plant capacity expansion is more difficult
	High-quality effluent possible	
	Well-stabilized sludge; low biosolids production	
Sequencing batch reactor	Process is simplified; final clarifiers and RAS pumping are not required	Process control more complicated
	Compact facility	High peak flows can disrupt operation unless accounted for in design
	Operation is flexible; nutrient removal can be accomplished by operational changes	Batch discharge may require equalization prior to filtration and disinfection
	Can be operated as a selector process to minimize sludge bulking potential	Higher maintenance skills required for instruments, monitoring devices, and automatic valves
	Quiescent settling enhances solids separation (low effluent SS)	Some designs use less efficient aeration devices
	Applicable for a variety of plant sizes	
Countercurrent aeration	High-quality effluent possible	Fine screening is required to prevent diffuser fouling
	Oxygen transfer efficiencies are higher than conventional aeration systems	Process is proprietary
	Well-stabilized sludge; low biosolids production	Significant downtime of aeration unit for maintenance will affect plant performance
	Process design can be modified to accommodate nutrient removal	Good operator skills required

8–5 PROCESSES FOR BIOLOGICAL NITROGEN REMOVAL

Nitrogen removal is often required before discharging treated wastewater to sensitive water bodies (to prevent eutrophication), or for groundwater recharge or other reuse applications. Nitrogen removal can be either an integral part of the biological treatment system or an add-on process to an existing treatment plant. The purpose of this section, as in the previous section, is to illustrate in detail the design procedure for processes used to remove the nitrogen from wastewater biologically. However, before considering the design examples, it is appropriate to present an overview of the biological nitrogen-removal process and design issues for the anoxic/aerobic process. Following the discussion of design issues, design examples are provided for (1) the anoxic/aerobic process, (2) step-feed anoxic/aerobic process, (3) intermittent aeration, (4) a sequencing batch reactor, and (5) postanoxic denitrification with methanol addition. Descriptions and flow diagrams for several alternative processes for nitrogen removal, typical process design parameters, and process selection considerations are presented following the design examples.

Overview of Biological Nitrogen-Removal Processes

All of the biological nitrogen-removal processes include an aerobic zone in which biological nitrification occurs. Some anoxic volume or time must also be included to provide biological denitrification to complete the objective of total nitrogen removal by both NH_4-N oxidation and NO_3-N and NO_2-N reduction to nitrogen gas. As discussed in Sec. 7–10 in Chap. 7, nitrate reduction requires an electron donor, which can be supplied in the form of influent wastewater BOD, by endogenous respiration, or an external carbon source.

The types of suspended growth biological nitrogen-removal processes can be categorized as (1) single-sludge or (2) two-sludge. The term "single-sludge" means only one solids separation device (normally a secondary clarifier) is used in the process (see Fig. 8–21a, c, and d). The activated-sludge tank may be divided into different zones of anoxic and aerobic conditions and mixed liquor may be pumped from one zone to another (internal recycle), but the liquid-solids separation occurs only once. In the two-sludge system, the most common system consists of an aerobic process (for nitrification) followed by an anoxic process (for denitrification), each with its own clarifier, thus producing two sludges (see Fig. 8–21e). For postanoxic denitrification, an organic substrate, usually methanol, must be added to create a biological demand for the nitrate. Because single-sludge systems are used more often, they are discussed in more detail in the following paragraphs. The two-sludge system is discussed briefly later in this section.

Single-Sludge Biological Nitrogen-Removal Processes

The single-sludge biological nitrogen-removal processes are grouped according to whether the anoxic zone is located before, after, or within the aerobic nitrification zone. These three possibilities, illustrated on Fig. 8–21a, c, and d, are termed (1) preanoxic, where initial contact of the wastewater and return activated sludge is in an anoxic zone; (2) postanoxic, where the anoxic zone follows the aerobic zone; or (3) simultaneous nitrification-denitrification (SNdN) processes where both nitrification and denitrification occur in the same tank. In the preanoxic configuration (see Fig. 8–21a), nitrate produced in the aerobic zone is recycled to the preanoxic compartment. The rate of denitrification is affected by the rbCOD concentration in the influent wastewater, the MLSS concentration, and temperature. Postanoxic designs (Fig. 8–21c) may be operated with or without an exogenous carbon source. Without an exogenous source, postanoxic processes depend on the endogenous respiration of the activated sludge to provide electron donor for nitrate consumption in lieu of oxygen. The denitrification rate is much slower, by a factor of 3 to 8, compared to preanoxic applications that use influent wastewater BOD for the electron donor. A long detention time would be required in this type of postanoxic tank to achieve high nitrate-removal efficiency. Single-tank designs have also been used in which nitrification and denitrification occur simultaneously. The simultaneous nitrification-denitrification (SNdN) applications require DO control or other types of control methods to assure that both nitrification and denitrification occur in a single tank.

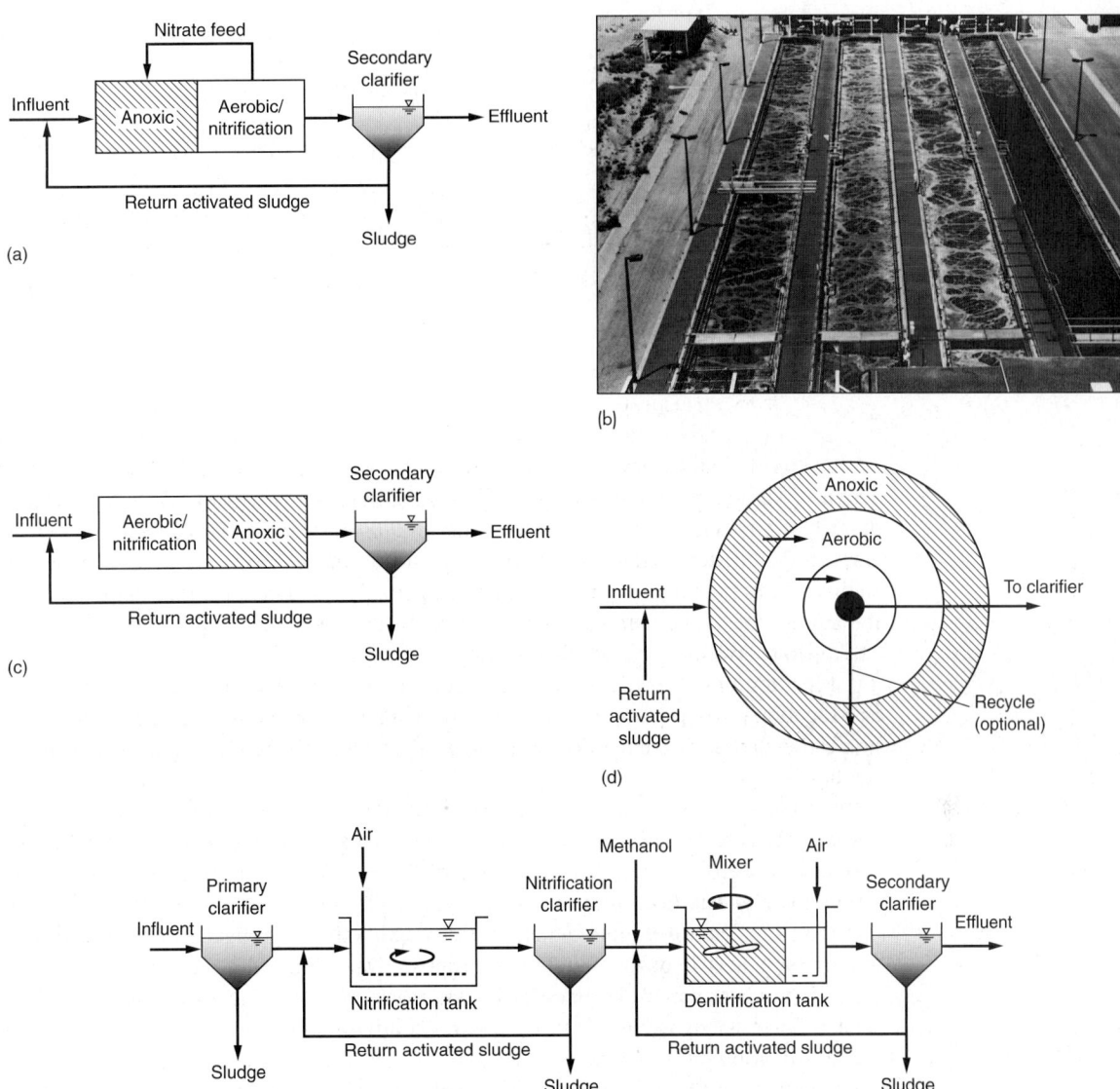

Figure 8–21

Schematic diagrams of four basic nitrogen-removal process configurations and a view of a typical reactor: (a) preanoxic, (b) view of plug-flow activated-sludge reactor used for nitrogen removal (the right-hand channel without aeration is the preanoxic section), (c) postanoxic, (d) simultaneous nitrification/denitrification, and (e) two stage nitrification-denitrification (also known as two-sludge).

Figure 8–22

Diagram of an activated-sludge particle showing aerobic and anoxic zones.

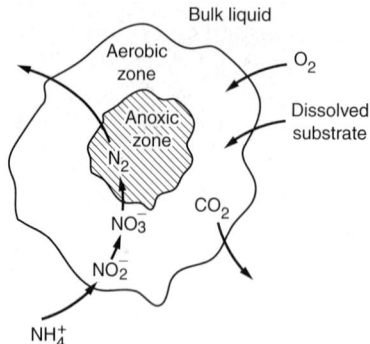

A significant amount of nitrogen removal has been observed in single-sludge activated-sludge systems without distinctive separate anoxic zones. Van Huyssteen et al. (1990) reported significant nitrogen loss in a nitrification aeration basin mixed and aerated by surface mechanical aerators. Significant nitrogen loss can occur in reactors, such as the oxidation ditch or similar process (see Fig. 8–21d) that have long hydraulic retention times. The combination of both nitrification and denitrification may be explained by two possible mechanisms.

First, regions of low DO or zero DO concentration may be present within the basin as a function of the mixing regime. Van Huyssteen et al. (1990) opined that as the mixed liquor traveled away from the surface mechanical aerators, the DO was depleted, creating conditions more favorable for anoxic reactions. Second, activated-sludge floc can contain both aerobic and anoxic zones, as illustrated in a simplified view of a biological floc on Fig. 8–22. Dissolved oxygen and dissolved substrates outside of the floc diffuse into the aerobic zone, and, depending on the DO concentration and concentrations of ammonia and bCOD, oxygen may be depleted at significant rates within the floc so that the DO cannot penetrate the entire floc depth. Nitrate produced by nitrification in the aerobic zone can diffuse into the inner anoxic zone along with substrate so that denitrification occurs within the floc depth. The existence of anoxic zones within the biological floc has been supported by Stenstrom and Song (1991), in which they showed that nitrification rates were related not only to bulk liquid DO concentration but also to the amount of BOD present. At higher soluble BOD concentrations, higher oxygen uptake rates occurred, and lower nitrification rates were observed for the same bulk liquid DO concentration, suggesting that the aerobic zone of the activated-sludge floc was decreased.

Nitrification and denitrification rates should both be at less than optimal levels for simultaneous nitrification/denitrification (SNdN) processes. Only a portion of the biomass is used for each of these reactions. In addition, the nitrification rate is lower due to the low DO concentration, and the denitrification rate is lower due to substrate consumption in the aerobic portion of the floc. However, systems with very long detention times, such as oxidation ditches, have sufficient volume to accommodate lower rates for both nitrification and denitrification.

Of the three basic process configurations, the preanoxic nitrification/denitrification process (Fig. 8–21a) is used most often because of (1) the relative ease of retrofit to existing plants, (2) the benefits of the selector operation for control of bulking sludge, (3) the production of alkalinity before the nitrification step, and (4) the ability to convert an existing biological treatment system to nitrogen removal with relatively short to moderate basin detention times. Most of the nitrogen-removal processes can incorporate biological phosphorus removal, as discussed in Sec. 8–6. Design issues and design examples for commonly used biological nitrogen-removal processes are presented below. The fundamental design concepts that are exemplified can be of use in evaluating other types of suspended growth biological nitrogen-removal processes.

Process Design Considerations

In Sec. 8–4, the process design was based on the fundamental principles presented in Chap. 7 and Secs. 8–2 and 8–3 for BOD removal and nitrification. In this section, an additional treatment step is considered, i.e., denitrification, and information is presented that is specific to the nitrogen-removal process.

Anoxic/Aerobic Reactor Design Considerations. In the anoxic/aerobic process (Fig. 8–21a) nitrate is fed to the anoxic reactor from nitrate in the return activated-sludge flow and by pumping mixed liquor from the aerobic zone. In step-feed anoxic/aerobic processes, nitrate will be fed to the anoxic zone by flow of mixed liquor from a previous nitrification step. The electron donor is provided by the influent wastewater fed to these preanoxic zones. Key design parameters that affect the amount of nitrogen removed are (1) anoxic zone detention time, (2) mixed-liquor volatile suspended solids (MLVSS) concentration, (3) internal recycle rate and return sludge flow, (4) influent BOD or biodegradable COD (bCOD) concentration, (5) the readily biodegradable COD (rbCOD) fraction, and (6) temperature (Randall et al., 1992). The influent rbCOD concentration has a significant effect on the denitrification rate in the anoxic zone. Wastewaters with the same influent bCOD, but with a higher fraction of rbCOD, will undergo higher denitrification rates in the anoxic zone. Anoxic zones have been designed as single-stage or a series of complete mix tanks with equal or different detention times. Typical power requirements for mechanical mixing in the anoxic zone range from 8 to 13 $kW/10^3$ m^3 (0.3 to 0.5 $hp/10^3$ ft^3).

Two design approaches are used to design the anoxic zone volume and to determine the amount of nitrogen removal. One is a desktop design approach that employs mass balances for nitrogen and a commonly used design parameter, the specific denitrification rate (SDNR) in g NO_3-N reduced/g MLVSS·d. More recently, comprehensive mechanistic simulation models have been developed that can relate denitrification rates to fundamental biokinetics, wastewater characteristics, activated-sludge tank volumes and configurations, and SRT.

Most simulation models are based on the ASM1 model developed by a committee under the International Association of Water Pollution Research Control (IAWPRC). The ASM1 model is now referred to as the IWA (International Water Association) model to account for the organization name change. The model is developed around a basic activated-sludge model that can be used to describe biomass growth rates, and follows the fate of degradable COD and nitrogen (in both soluble and particulate forms) for systems that have both aerobic and anoxic treatment zones (Grady et al., 1986).

An ASM2 version of the model incorporates biological phosphorus removal as well as nitrogen removal (Barker and Dold, 1997). In these models, nitrate is consumed by heterotrophic biomass under anoxic conditions during the consumption of either rbCOD or short-chain volatile fatty acids. The rbCOD substrate consumed in the anoxic reactor is from the influent wastewater plus that produced in the reactor by hydrolysis of influent pCOD and released biomass material due to cell lyses. A general description of the ASM1 model is included in Sec. 8–10.

Anoxic Tank Design Using the Specific Denitrification Rate. The desktop design approach is based on using a specific denitrification rate (SDNR), which is the nitrate reduction rate in the anoxic tank normalized to the MLSS concentration. The amount of nitrate removed in the anoxic tank is described by Eq. (8–41).

$$NO_r = (V_{nox})(SDNR)(MLVSS) \tag{8–41}$$

where NO_r = nitrate removed, g/d
$\quad V_{nox}$ = anoxic tank volume, m^3
$\quad$ SDNR = specific denitrification rate, g NO_3-N /g MLVSS·d
$\quad$ MLVSS = mixed liquor volatile suspended solids concentration, mg/L

Values of SDNR observed for preanoxic tanks in full-scale installations have ranged from 0.04 to 0.42 g NO_3-N/g MLVSS·d (Burdick et al., 1982; Henze, 1991; Bradstreet and Johnson, 1994; Reardon et al., 1996; Hong et al., 1997; and Murakami and Babcock, 1998). For postanoxic denitrification without an exogenous carbon source, observed SDNRs have ranged from 0.01 to 0.04 g N/g MLVSS·d.

Based on observed denitrification rates in pilot-plant and full-scale plants, empirical relationships have been developed that relate SDNR to the BOD or COD F/M ratio for the preanoxic tank (U.S. EPA, 1993). One commonly used relationship is described as (Burdick et al., 1982; U.S. EPA, 1993)

$$SDNR = 0.03(F/M) + 0.029 \tag{8–42}$$

where F/M = g BOD applied/g MLVSS·d in the anoxic tank

The above relationship was based on data collected for anoxic/aerobic processes at mixed-liquor temperatures in the range of 20 to 25°C. A conservative SDNR estimate can be made using Eq. (8–42) because the MLVSS in the systems evaluated contained a relatively low active biomass fraction and the wastewater had a relatively low rbCOD concentration. Such empirical relationships as described above are limited in application because the SDNR depends on a number of factors that are site- and design-specific, including the fraction of active biomass in the mixed liquor, rbCOD concentration in the anoxic zone, and temperature. The above factors are affected by the size of the anoxic zone, influent wastewater rbCOD and nbVSS concentrations, and design SRT. Thus, these empirical relationships can provide only a rough estimate of the SDNR.

Design SDNR values at 20°C are presented on Fig. 8–23. The values are generally applicable and can be used for wastewaters with different fractions of rbCOD (rbCOD/bCOD) and inert nonbiodegradable volatile solids. On Fig. 8–23, the F/M_b ratio and $SDNR_b$ values are based only on the active heterotrophic biomass concentration in the mixed liquor, so that the rates can be applicable to many situations regardless of the amount of nondegradable solids in the mixed liquor and the SRT. The F/M_b

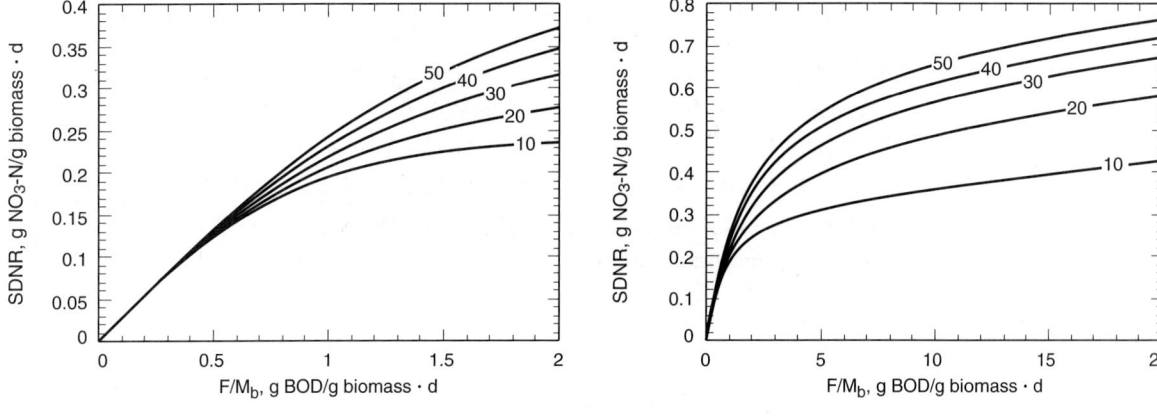

Figure 8–23

Plot of specific denitrification rates (SDNR$_b$) based on biomass concentration at 20°C versus food to biomass (F/M$_b$) ratio for various percentages of rbCOD relative to the biodegradable COD of the influent wastewater.

ratio is defined as a function of the BOD loading to the anoxic volume and active heterotrophic biomass concentration, as follows:

$$F/M_b = \frac{QS_o}{(V_{nox})X_b} \tag{8-43}$$

where F/M_b = BOD F/M ratio based on active biomass concentration, g BOD/g biomass·d

Q = influent flowrate, m³/d

S_o = influent BOD concentration, mg/L

V_{nox} = anoxic volume, m³

X_b = anoxic zone biomass concentration, mg/L

The curves shown on Fig. 8–23 are based on the results of model simulations of biomass, NO$_3$-N, rbCOD, and pbCOD mass balances in the anoxic tank. For lower anoxic τ values, the F/M$_b$ is higher, which resulted in greater rbCOD concentrations in the anoxic zone and thus a higher biological reaction rate and SDNR. Internal recycle rates from the aerobic zone and temperature effects were accounted for. The biokinetic coefficients used in the model simulation are summarized in Table 8–18 and are model parameter default values from the ASM1 model (Grady et al., 1986) along with the rbCOD kinetics under anoxic conditions observed by Stensel and Horne (2000) from testing at different municipal wastewater-treatment facilities. The design procedure using SDNR$_b$ values from Fig. 8–23 requires calculating the active biomass VSS concentration in the mixed liquor using the procedure presented previously in Example 8–2. By using the calculated biomass concentration, the effects of SRT are accounted for in the design procedure. The BOD F/M$_b$ ratio on Fig. 8–23 is also based on the active biomass VSS concentration.

Table 8-18
Biokinetic coefficient values used in model simulation to develop $SDNR_b$ design curves

Kinetic coefficient parameter	Unit	Value
Yield, Y	g VSS/g COD	0.40
Endogenous decay, k_d	g VSS/g biomass·d	0.15
Cell debris, f_d	g VSS/g VSS	0.10
Maximum specific growth rate, μ_m	g VSS/g VSS·d	3.2
Half-velocity, K_s	g/m³	9.0
Particulate hydrolysis maximum specific rate constant, K_h	g VSS/g biomass·d	2.8
Hydrolysis half-velocity constant, K_X	g VSS/g VSS	0.15
COD of biomass	g COD/g VSS	1.42
Fraction of denitrifying bacteria, η	g VSS/g VSS	0.50

For multiple-staged anoxic reactors, the F/M_b ratio [Eq. (8–43)] is calculated using the volume of the stage for which the SDNR is calculated plus the volume of the previous stage(s). At high F/M_b ratios shown on Fig. 8–23, the SDNR reaches a maximum saturated rate, as the rbCOD concentration will be very high in the anoxic reactor. The maximum saturation rate will only be experienced for high F/M_b anoxic selector designs where the first anoxic zones have detention times of less than 10 to 20 min. For long anoxic detention times (in the order of 3 to 6 h), $SDNR_b$ values will be found at the lower F/M_b ratios.

The design procedure requires correcting the SDNR values from Fig. 8–23 for temperature and internal recycle ratios. The temperature correction is made by using the following expression:

$$SDNR_T = SDNR_{20}\theta^{T-20} \qquad (8\text{–}44)$$

where θ = temperature coefficient (1.026)
T = temperature, °C

The SDNR in the preanoxic tank is affected by the internal recycle rate, often defined as an internal recycle (IR) ratio. The IR ratio is the recycle flowrate divided by the influent wastewater flowrate. At higher IR ratios, the influent rbCOD is diluted more in the anoxic reactor by mixed liquor from the aerobic reactor, resulting in a lower denitrification rate. The corrections to the SDNR for designs with internal recycle corrections greater than 1.0 are shown below. The SDNR value can be interpolated from the values listed. If the F/M is less than or equal to 1.0, no correction is required.

$$IR = 2 \qquad SNDR_{adj} = SDNR_{IR1} - 0.0166 \ \ln\,(F/M_b) - 0.0078 \qquad (8\text{–}45)$$

$$IR = 3\text{–}4 \qquad SDNR_{adj} = SDNR_{IR1} - 0.029 \ln\,(F/M_b) - 0.012 \qquad (8\text{–}46)$$

where $SDNR_{adj}$ = SDNR adjusted for the effect of internal recycle
$SDNR_{IR1}$ = SDNR value at internal recycle ratio = 1
F/M_b = BOD F/M ratio based on anoxic zone volume and active biomass concentration, g/g·d

Anoxic/Aerobic Process Design Procedure. The design procedure for an anoxic/aerobic process is presented in Table 8–19. The influent wastewater rbCOD fraction is a critical design parameter, and if unknown, a conservative value in the range of 15 to 25 percent of the total bCOD can be used. After defining the wastewater characteristics, the initial design step requires designing the aerobic reactor for nitrification similar to the procedure illustrated in Sec. 8–4. The aerobic volume is based on using an aerobic SRT for nitrification, and only the aerobic basin volume and mixed liquor are used to compute the sludge wasting for that SRT. The total process SRT will be longer when the mixed liquor and the volume of the anoxic reactor are included. The aeration oxygen requirement will be less than that for the nitrification-only design as nitrate will be used to consume some of the influent bCOD in the preanoxic zone.

A mass balance on nitrogen must be done to determine (1) how much nitrate is produced in the aeration zone, and (2) what the internal recycle ratio must be to meet the desired effluent nitrate concentration. The mass balance accounts for the nitrate produced in the aerobic zone. The amount of nitrate produced in the aerobic zone is based on the influent flowrate and nitrogen concentration, the amount consumed for cell synthesis,

Table 8–19
Computation approach for anoxic/aerobic process design

1. Establish wastewater flowrates and characteristics, including the rbCOD concentration, and effluent requirements
2. Follow the procedure outlined in Table 8–12 for the aerobic zone for a nitrification process with the exception that Steps 13–18 are done after the anoxic reactor design
3. Determine the biomass concentration in the mixed liquor from the nitrification design
4. Determine the internal recycle (IR) ratio, using the NO_x value determined in Step 9 of the nitrification design and desired effluent NO_3-N concentration
5. Calculate the amount of nitrate fed to the anoxic tank. The design is based on the assumption that essentially all of the nitrate fed to the anoxic zone will be reduced. A low nitrate concentration of 0.1 to 0.3 mg/L may remain, depending on the design, because the nitrate limits the denitrification reaction rate at only very low concentrations
6. Select the anoxic volume and configuration; single-stage or multiple-stage reactors
7. Calculate the F/M_b based on the biomass concentration determined for the mixed liquor in the nitrification design
8. Use Fig. 8–23 and appropriate corrections for temperature and IR ratio to obtain the $SDNR_b$ for the anoxic basin(s)
9. Using the $SDNR_b$ biomass concentration, and anoxic volume, compute the amount of nitrate removed in the anoxic basin(s). Compare the amount of removal required to remove all of the nitrate in the recycle streams fed to the anoxic zone
10. Repeat the anoxic zone design steps as necessary to obtain a satisfactory design
11. Calculate the oxygen demand
12. Determine if alkalinity addition is needed
13. Design the secondary clarifier
14. Design the aeration oxygen transfer system
15. Summarize the final effluent quality
16. Prepare a design summary table

and the effluent NH_3-N and soluble organic nitrogen concentrations. As a conservative design approach, all of the influent TKN is assumed to be biodegradable and the effluent soluble organic nitrogen concentration is ignored. The nitrate produced is contained in the total flow leaving the aerobic zone, which includes internal recycle, RAS, and effluent flows. The mass balance is expressed as

$$\begin{array}{ccccc} \text{kg/d of nitrate} & & \text{nitrate} & \text{nitrate in} & \text{nitrate in} \\ \text{produced in} & = & \text{in effluent} + & \text{internal} + & \text{return activated} \\ \text{aerobic zone} & & & \text{recycle} & \text{sludge (RAS)} \end{array}$$

$$Q\,NO_x = N_e(Q + IR\,Q + R\,Q) \qquad (8\text{-}47)$$

$$IR = \frac{NO_x}{N_e} - 1.0 - R \qquad (8\text{-}48)$$

where IR = internal recycle ratio (internal recycle flowrate/influent flowrate)
   R = RAS recycle ratio (RAS flowrate/influent flowrate)
   NO_x = nitrate produced in aeration zone as a concentration relative to influent flow, mg NO_3-N/L
   N_e = effluent NO_3-N concentration, mg/L

The effect of the IR ratio on the effluent NO_3-N concentration for a given amount of nitrate produced (NO_x) and for a RAS recycle ratio of 0.50 is illustrated on Fig. 8–24. A greater IR ratio is needed to produce the same effluent NO_3-N concentration when more NO_x is produced in the aerobic zone. To meet a standard of 10 mg TN/L or less, a design effluent NO_3-N concentration of 5 to 7 mg/L should be used. (Note: Selection of design values based on the variability in wastewater treatment plant performance is discussed in Chap. 15.) An internal recycle ratio in the range of 3 to 4 is typical, but ratios in the range of 2 to 3 are also applied for wastewaters with a lower influent wastewater TKN concentration. Recycle ratios above 4 are generally not warranted, as the incremental removal of NO_3-N is low and more DO is recycled from the aeration zone into the anoxic zone.

The amount of DO fed to the anoxic zone due to the internal recycle flow from the aerobic zone must be minimized because the oxygen will consume rbCOD, leaving less

Figure 8–24

Effect of internal recirculation rate on effluent nitrate concentration (RAS ratio = 0.5) for an anoxic/aerobic process.

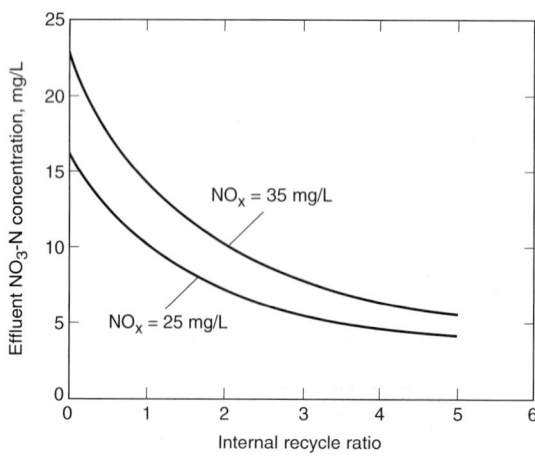

available for NO_3 reduction. In some designs, sections of the aerobic zone are baffled with DO control so that the DO concentration in the recycle can be controlled and minimized. Care must also be taken to ensure that the influent wastewater is not overly aerated when passing through the plant en route to the anoxic tank.

Single-Sludge Simultaneous Nitrification Denitrification (SNdN) Processes. Both high levels of nitrification and denitrification have been reported for oxidation ditch systems operated with low DO concentrations (0.10 to 0.40 mg/L) and with relatively long τ and SRT values (Rittman and Langeland, 1985; Trivedi and Heinen, 2000). The low DO concentration will result in lower nitrification rates, as the activated-sludge floc will be only partially aerobic. Thus, only a portion of the nitrifying bacteria contained in the floc will be active. In addition, as shown in Chap. 7, nitrification rates are lower at low DO concentrations. Denitrification occurs in the anoxic zones established within the floc particles due to oxygen depletion, with the result that simultaneous nitrification and denitrification takes place. The nitrification and denitrification rates are a function of the reaction kinetics, floc size, floc density, floc structure, rbCOD loading, and bulk liquid DO concentration. Because of the complex physical factors, the nitrification and denitrification rates cannot be predicted accurately with present models. Basic modifications to the Monod growth model, however, can be used to estimate the effects of a low DO concentration on nitrification and denitrification rates and system performance.

Nitrification specific growth rates are shown as a function of DO concentration as described in Eq. (7–93) (see also Table 8–5 for definition of terms):

$$\mu_n = \left(\frac{\mu_{nm}N}{K_N + N}\right)\left(\frac{DO}{K_o + DO}\right) - k_{dn} \qquad (7\text{–}93)$$

As the nitrification SRT is the inverse of the specific growth rate μ_n, Eq. (7–93) can be used to calculate required design SRT values as a function of NH_4-N and DO concentrations for a CMAS system. The effect of DO concentration on the design SRT is shown on Fig. 8–25 for an effluent NH_4-N concentration of 1.0 mg/L at 20°C using the kinetic coefficient values for nitrification given in Table 8–11. No safety factor is used for these

Figure 8–25

Effect of DO concentration on SRT required to achieve effluent NH_4-N concentration of 1.0 mg/L at 20°C in a CMAS system based on kinetic coefficients in Table 8–11 and K_o = 0.5 mg/L.

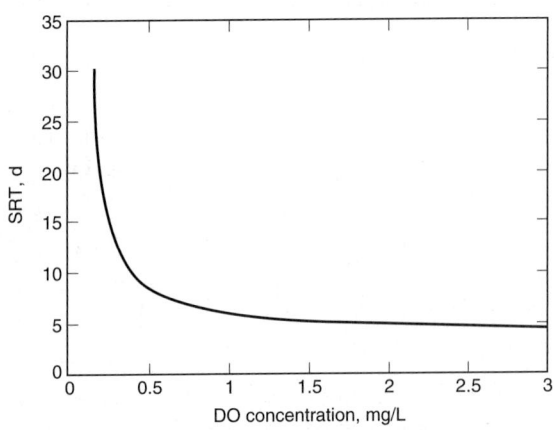

calculations. The actual SRT value used for design would be higher depending on the safety factor selected for a specific system. The nitrification rate at 0.2 mg/L DO concentration is 24 percent of the rate at a 2.0 mg/L concentration based on the calculated SRT values of 19.9 and 4.7 d, respectively.

With many oxidation ditch systems operated at SRT values of 20 to 30 d, the ability to produce complete nitrification at low DO is apparent. The exact nitrification rate at low DO concentrations is site-specific, and determines the K_o value that can be used in Eq. (7–93) to describe the reaction.

The rate of nitrate reduction can be related to the rate of substrate utilization given by Eq. (7–110) (also see Table 8–5 for definition of terms) as follows:

$$r_{su} = -\left(\frac{kXS}{K_s + S}\right)\left(\frac{NO_3}{K_{s,NO_x} + NO_3}\right)\left(\frac{K'_o}{K'_o + DO}\right)(\eta) \tag{7–110}$$

The nitrate utilization rate can be related to the substrate utilization rate shown in Eq. (7–110) by the following procedure that first relates it to an oxygen utilization rate. The oxygen utilization rate (r_o) is derived from Eq.(7–59) by dividing the equation by the reactor volume. The oxygen utilization rate is the difference between the substrate utilization rate and the biomass growth rate on a COD basis as follows:

$$r_o = -r_{su} - 1.42r_g \tag{8–49}$$

Substituting Eq. (7–21) for r_g in Eq. (8–49) yields

$$r_o = -(1 - 1.42Y)r_{su} + 1.42k_dX \tag{8–50}$$

where r_o = oxygen utilization rate, g/m³·d
r_{su} = substrate utilization rate, g/m³·d
r_g = net biomass growth rate, g/m³·d

Using 2.86 g O_2 equivalent/g NO_3-N [from Eqs. (7–98) and (7–99)], the nitrate reduction rate (r_{NO_x}) is

$$r_{NO_x} = r_o/2.86 \tag{8–51}$$

where r_{NO_x} = NO_3-N reduction rate, g/m³·d

By accounting for the fraction of biomass that can use nitrate (η), r_{NO_x} is

$$r_{NO_x} = -\left(\frac{1 - 1.42Y}{2.86}\right)r_{su} + \left(\frac{1.42}{2.86}\right)k_dX\eta \tag{8–52}$$

Substituting Eq. (7–110) into Eq. (8–52) to account for the effect of the nitrate and DO concentration on substrate utilization under denitrifying conditions yields

$$r_{NO_x} = \left(\frac{1 - 1.42Y}{2.86}\right)\left(\frac{kXS}{K_s + S}\right)\left(\frac{NO_3}{K_{s,NO_x} + NO_3}\right)\left(\frac{K'_o}{K'_o + DO}\right)(\eta)$$
$$+ \left(\frac{NO_3}{K_{s,NO_x} + NO_3}\right)\left(\frac{K'_o}{K'_o + DO}\right)\left(\frac{1.42}{2.86}\right)k_dX\eta \tag{8–53}$$

Equation (8–53) is a general expression for the rate of nitrate reduction in an anoxic reactor. The nitrate reduction rate as shown by Eq. (8–53) is a function of the rbCOD,

Figure 8–26

Effect of mixed-liquor DO concentration on maximum denitrification rates (R-NO₃).

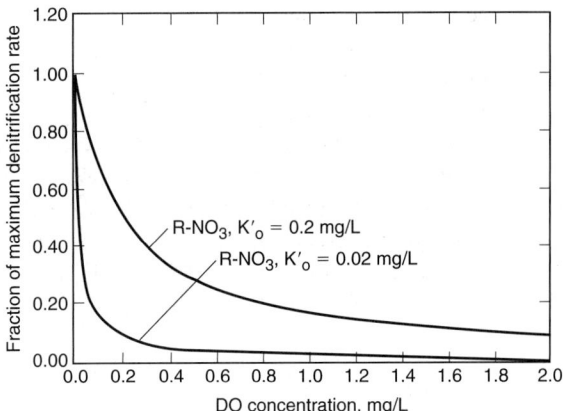

NO₃-N, DO, and biomass concentrations, as well as the various biokinetic coefficients. The DO inhibition coefficient K_o' is difficult to estimate and will be site-specific depending on the floc size and structure. The effect of DO concentration on the denitrification rate is shown on Fig. 8–26 for K_o' values of 0.02 and 0.2 mg/L. At a DO concentration of 0.2 mg/L, the denitrification rate may be 10 to 50 percent of its maximum rate. For long SRT systems with τ values in the range of 18 to 30 h, sufficient time may be available for high nitrate-removal efficiency, even though the rate is somewhat inhibited by having DO present at low concentrations.

Anoxic/Aerobic Process Design

An example of an anoxic/aerobic activated-sludge process using a complete-mix reactor for nitrogen removal is presented in Example 8–5.

EXAMPLE 8–5   **Anoxic/Aerobic Process Design**   Design a preanoxic basin for the CMAS nitrification system process described in Example 8–2 to produce an effluent NO₃-N concentration of 6.0 g/m³. A process schematic flow diagram is shown below. The design condition is based on the following information from Example 8–2.

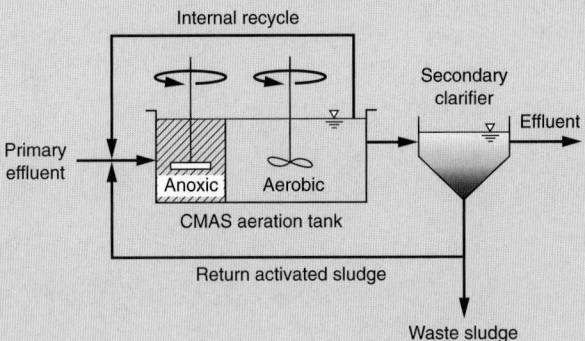

Wastewater characteristics:

Constituent	Concentration, g/m³
BOD	140
bCOD	224
rbCOD	80
NO$_x$	28.3
TP	6
Alkalinity	140 as CaCO$_3$

Design conditions:

Parameter	Unit	Value
Influent flowrate	m³/d	22,464
Temperature	°C	12
MLSS	g/m³	3000
MLVSS	g/m³	2370
Aerobic SRT	d	12.5
Aeration basin volume	m³	8466
Mixing energy	kW/10³ m³	10
RAS ratio	Unitless	0.6
R$_o$	kg/h	255

Note: g/m³ = mg/L.

Assumptions:
1. Nitrate concentration in RAS = 6 g/m³.
2. Use the same coefficients as the nitrification process design.
3. Mixing energy for anoxic reactor = 10 kW/10³ m³.

Solution

1. Determine the active biomass concentration using Eq. (7–43) and substituting V/Q for τ.

$$X_b = \left[\frac{Q(\text{SRT})}{V}\right]\left[\frac{Y(S_o - S)}{1 + (k_d)\text{SRT}}\right]$$

where $S_o - S \approx S_o$

$$X_b = \frac{(22{,}464 \text{ m}^3/\text{d})(12.5 \text{ d})(0.40 \text{ g VSS/g COD})(224 \text{ g bCOD/m}^3)}{[1 + (0.088 \text{ g/g·d})(12.5 \text{ d})](8466 \text{ m}^3)}$$

$$= 1415 \text{ g/m}^3$$

2. Determine the IR ratio using Eq. (8–48).

   Aerobic tank NO_3-N concentration $= N_e = 6.0$ g/m^3

   $$IR = \frac{NO_x}{N_e} - 1.0 - R = \frac{(28.3 \text{ g/m}^3)}{(6 \text{ g/m}^3)} - 1.0 - 0.60 = 3.1$$

3. Determine the amount of NO_3-N fed to the anoxic tank.

   Flowrate to anoxic tank $= IR\, Q + RQ$

   $$= 3.1(22,464 \text{ m}^3/\text{d}) + 0.60(22,464 \text{ m}^3/\text{d})$$

   $$= 83,117 \text{ m}^3/\text{d}$$

   NO_x feed $= (83,117 \text{ m}^3/\text{d})(6.0 \text{ g/m}^3) = 498,700$ g/d

4. Determine the anoxic volume.
   As a first approximation, use a detention time $= 2.5$ h

   $$\tau = \frac{2.5 \text{ h}}{(24 \text{ h/d})} = 0.104 \text{ d}$$

   $V_{nox} = \tau \times Q = 0.104 \text{ d } (22,464 \text{ m}^3/\text{d}) = 2336 \text{ m}^3$

5. Determine F/M$_b$ using Eq. (8–43).

   $$F/M_b = \frac{QS_o}{V_{nox}(X_b)} = \frac{(22,464 \text{ m}^3/\text{d})(140 \text{ g BOD/m}^3)}{(2336 \text{ m}^3)(1415 \text{ g/m}^3)} = 0.95 \text{ g/g·d}$$

6. Determine the SDNR using the curve with an F/M$_b$ range of 0 to 2 given on Fig. 8–23.

   Fraction of rbCOD $=$ rbCOD/bCOD $= (80 \text{ g/m}^3)(224 \text{ g/m}^3) = 0.36 = 36\%$

   From Fig. 8–23, SDNR$_b = 0.22$ g/g·d at 20°C

   Apply temperature correction using Eq. (8–44):

   SDNR$_{12} = 0.22(1.026)^{12-20} = 0.18$ g/g·d

7. Determine the amount of NO_3-N that can be reduced using Eq. (8–41).
   a. Check NO$_r$ based on $\tau = 2.5$ h.

   NO$_r = (V_{nox})$(SDNR)(MLVSS, biomass)

   $$= (2336 \text{ m}^3)(0.18 \text{ g/g·d})(1415 \text{ g/m}^3) = 592,206 \text{ g/d}$$

   Comparing 592,206 g/d versus 498,700 g/d, there is about 20% excess nitrate-removal capacity. Thus, $t = 2.5$ h is acceptable but a lower detention time may be used.

b. Evaluate new value for τ.

Select new τ. If the same SDNR is used, $\tau = 2.5 \text{ h}/1.2 \approx 2.0 \text{ h}$, but SDNR will be higher due to higher F/M ratio for smaller anoxic reactor. Therefore, try $\tau = 1.5 \text{ h}$.

$$V_{nox} = (1.5 \text{ h}/24 \text{ h/d})(22{,}464 \text{ m}^3/\text{d}) = 1404 \text{ m}^3$$

$$F/M_b = \frac{QS_o}{V_{nox}(X_b)} = \frac{(22{,}464 \text{ m}^3/\text{d})(140 \text{ g BOD/m}^3)}{(1404 \text{ m}^3)(1415 \text{ g/m}^3)} = 1.58 \text{ g/g·d}$$

c. Determine new SDNR values.

Using Fig. 8–23, for SDNR at 20°C, rbCOD fraction = 0.36

$$SDNR_{20} = 0.31 \text{ g/g·d}$$

$$SDNR_{12} = 0.31(1.026)^{12-20} = 0.25 \text{ g/g·d}$$

d. Determine the amount of nitrate that can be reduced [Eq. (8–41)].

$$NO_r = (0.25 \text{ g/g·d})(1404 \text{ m}^3)(1415 \text{ g/m}^3) = 501{,}541 \text{ g/d}$$

Capacity ratio = $501{,}541/498{,}700 = 1.01$; therefore, $\tau = 1.5 \text{ h}$ is acceptable.

e. Compare the computed value to conventional observed SDNR values, based on MLSS.

$$SDNR \text{ (MLSS)} = (0.25)(X_b/X_T)$$

$$= (0.25)(1415/3000) = 0.12 \text{ g/g·d}$$

The computed value is in the range of reported SDNR values (0.04 to 0.42 g/g·d).

8. Go to nitrification step in design and determine net oxygen required.

R_o (without denitrification) = 255 kg/h   (See Step 17, Example 8–2)

The amount of oxygen supplied by nitrate reduction is as follows:

$$\text{Oxygen credit} = \left(\frac{2.86 \text{ g O}_2}{\text{g NO}_3\text{-N}}\right)[(28.3 - 6.0) \text{ g/m}^3]\left(\frac{22{,}464 \text{ m}^3}{\text{d}}\right)\left(\frac{1 \text{ kg}}{10^3 \text{ g}}\right)$$

$$= 1433 \text{ kg/d} = 59.7 \text{ kg/h}$$

Net O_2 required = R_o = $(255 - 59.7) \text{ kg/h} = 195.3 \text{ kg/h}$

Note the required aeration rate will decrease in proportion to a lower R_o. The oxygen required can be reduced by 23.4 percent.

9. Check alkalinity.

a. Prepare an alkalinity mass balance.

Alk to be added to maintain pH ~ 7 = Influent Alk $-$ Alk used $+$ Alk produced

   i. Influent alkalinity = 140 g/m^3 as CaCO$_3$

   ii. Alkalinity used = 7.14 (28.3 g NO$_3$-N/m^3) = 202.1 g/m^3

   iii. Alkalinity produced = 3.57 [(28.3 $-$ 6) g/m^3] = 79.6 g/m^3

   iv. Alkalinity needed to maintain neutral pH = 80 g/m^3 as CaCO$_3$

    b. Solve the above expression for Alk to be added.

$$\text{Alk to be added} = (80 - 140 + 202.1 - 79.6) \text{ g/m}^3$$

$$= 62.5 \text{ g/m}^3 \text{ as } CaCO_3$$

$$\text{Mass of alkalinity needed} = (62.5 \text{ g/m}^3)(22{,}464 \text{ m}^3/\text{d})(1 \text{ kg}/10^3 \text{ g})$$

$$= 1404 \text{ kg/d as } CaCO_3$$

    c. Compare to alkalinity needed for nitrification only.
        For the nitrification only design, the alkalinity needed was 3203 kg/d

$$\text{Alkalinity savings} = 3203 - 1404 = 1799 \text{ kg/d}$$

10. Determine anoxic zone mixing energy.

$$\text{Mixing energy} = 10 \text{ kW}/10^3 \text{ m}^3 \text{ (given)}$$

$$\text{Volume} = 1404 \text{ m}^3$$

$$\text{Power} = (1404 \text{ m}^3)(10 \text{ kW}/10^3 \text{ m}^3) = 14 \text{ kW total}$$

11. Prepare summary of anoxic design.

Item	Unit	Value
Effluent NO_3-N	g/m^3	6.0
Internal recycle ratio	Unitless	3.1
RAS recycle ratio	Unitless	0.6
Anoxic volume	m^3	1404
MLSS	g/m^3	3000
Overall SDNR	g NO_3-N/g MLSS·d	0.13
Detention time	h	2.5
Mixing power	kW	14
Alkalinity required	kg/d as $CaCO_3$	1393

Comment   In the above design example, computations are based on an average design condition. In actual design, allowances should be provided for peak flows and loads or a safety factor should be included, as discussed in Chap. 3.

Step-Feed Anoxic/Aerobic Process Design

Step feed for nitrogen removal is similar to the step-feed process described in Sec. 8–4 for BOD and nitrification. For nitrogen removal, wastewater is introduced at several feed points (see Fig. 8–27). In most cases, where a step-feed process is in place for BOD removal and nitrification, it will be relatively easy to upgrade it to a step-feed anoxic/aerobic biological nitrogen-removal process. For such applications the influent

Figure 8–27

Schematic diagram of
a step-feed biological
nitrogen-removal process.

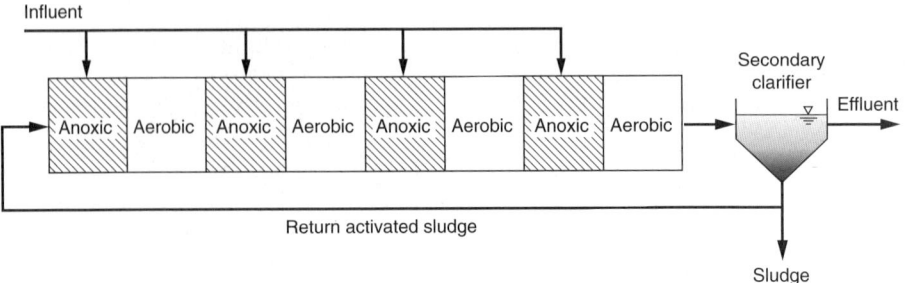

feed points and volumes of the individual channels in the reactor (passes) are already
determined. The tank layout is generally symmetrical and the volume in each pass is
equal. For a new tank design, it is possible to use a nonsymmetrical step-feed design
where the feed split is somewhat equal, but the volume of each pass increases as the
mixed-liquor concentration decreases from the first to last pass. The nonsymmetrical
design approach may utilize the tank volumes more efficiently by using a similar F/M
ratio for each pass.

The variables involved in the design of a step-feed biological nitrogen-removal
process for an existing basin are (1) the flow distribution between passes, (2) the relative split between anoxic and aerobic volumes, and (3) the final pass MLSS concentration. The selection of the final pass MLSS concentration is normally based on using an
acceptable solids loading for the secondary clarifier. As will be shown in the following
design example, the selection of the final pass MLSS concentration, the RAS ratio, the
influent flow split, and wastewater characteristics determine the system SRT. With the
known SRT value, the biomass and nitrifying bacteria concentration in the mixed liquor
can be determined, which can then be used to determine the nitrification and denitrification capacity of the system. The process design procedure involves successive iterations with varying anoxic/aerobic volumes and flow splits evaluated to find the most
satisfactory design. Design of a step-feed biological nitrogen-removal process is presented in Example 8–6.

EXAMPLE 8–6 **Step-Feed Biological Nitrogen-Removal Process Design** Determine the amount
of nitrification and nitrate removal in a four-pass step-feed biological nitrogen-removal
process (see Fig. 8–27), using the same influent flowrate, wastewater characteristics, and
temperature as was used in Example 8–2.

Design conditions and assumptions:

1. The flowrate is 22,464 m³/d
2. The step-feed aeration tank is divided into four equal passes, with equal volumes
   used for the anoxic and aerobic zones
3. The flow split to each pass is 0.10, 0.40, 0.30, and 0.20 of the influent flow, for
   passes 1 through 4, respectively
4. MLSS concentration in the final aerobic zone is 3000 mg/L (same as in Example 8–2)

5. RAS recycle ratio (Q_{RAS}/Q) is 0.6
6. Anoxic volume is 20% of the total reactor volume
7. Total aeration tank volume is 9870 m³ (from Examples 8–2 and 8–5. Anoxic and aeration tank volumes are 1404 and 8466 m³, respectively)
8. Aeration tank DO = 2.0 g/m³
9. Temperature = 12°C
10. Effluent NH_4-N = 0.5 g/m³

Wastewater characteristics:

Constituent	Concentration, g/m³
BOD	140
rbCOD	80
bCOD	224
nbVSS	20
TKN	35
$TSS_o - VSS_o$	10

Note: g/m³ = mg/L.

Kinetic coefficients from Tables 8–10 and 8–11 and adjust for temperature

Heterotrophs:

$Y = 0.40$ g VSS/g bCOD

$k_{d,\,12°C} = 0.12$ g/g·d $(1.04)^{12-20} = 0.088$ g/g·d

$f_d = 0.15$ g/g

Nitrifiers:

$Y_n = 0.12$ g VSS/g NO_x

$k_{dn,\,12°C} = 0.08$ g/g·d $(1.04)^{12-20} = 0.06$ g/g·d

$\mu_{m,\,12°C} = 0.75$ g/g·d $(1.07)^{12-20} = 0.44$ g/g·d

$K_{n,\,12°C} = 0.74$ g/m³ $(1.053)^{12-20} = 0.49$ g/m³

$K_o = 0.50$ g/m³

Solution

1. Determine the aeration and anoxic zone volumes. Anoxic volume is 20% of total volume per problem statement.

V = total volume = 9870 m³

Anoxic volume = 0.20(9870 m³) = 1974 m³

Aerobic volume $= 0.80(9870 \text{ m}^3) = 7896 \text{ m}^3$

Anoxic volume/pass $= (1974 \text{ m}^3)/4 = 493.5 \text{ m}^3$

Aerobic volume/pass $= (7896 \text{ m}^3)/4 = 1974 \text{ m}^3$

2. Determine RAS concentration.

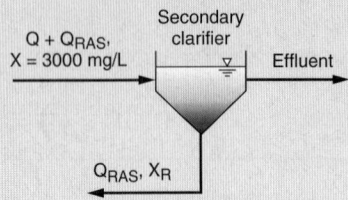

Perform a solids balance (neglect the effluent TSS because it is not significant for the clarifier solids balance).

$Q_{RAS} = 0.6\,Q$ (recycle ratio from problem statement)

The solids balance is

$$(Q + 0.6\,Q)\,3000 \text{ g/m}^3 = 0.6\,Q\,X_R$$

$$X_R = \frac{(Q + 0.6\,Q)(3000 \text{ g/m}^3)}{0.6\,Q}$$

$$X_R = (1.6\,Q/0.6\,Q)(3000 \text{ g/m}^3) = 8000 \text{ g/m}^3$$

3. Determine MLSS concentration in each pass (see figure below)

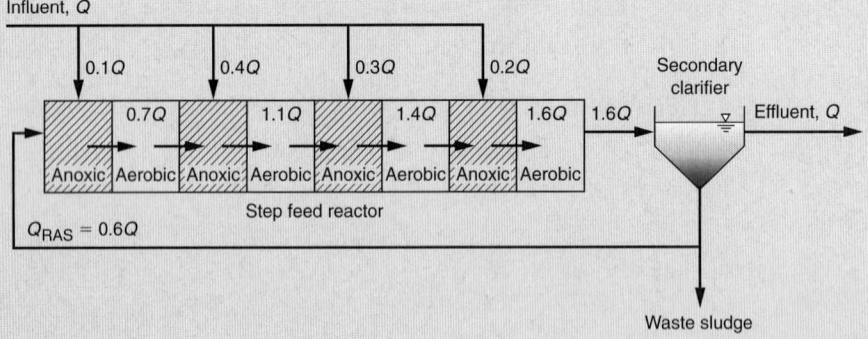

a. Pass 1 mass balance

Solids in = solids out (Note: solids production for a single pass is negligible)

$$0.10\,Q(0) + (Q_{RAS})(8000 \text{ g/m}^3) = (\text{RAS} + 0.1\,Q)X_1$$

$$0.1\,Q(0) + (0.6\,Q)(8000\ \text{g/m}^3) = (0.6\,Q + 0.1\,Q)X_1$$

$$X_1 = (8000\ \text{g/m}^3)(0.6/0.7) = 6860\ \text{g/m}^3$$

b. Pass 2 mass balance

$$(0.7\,Q)X_1 + 0.4\,Q(0) = 1.1\,QX_2$$

$$X_2 = (0.7/1.1)X_1 = (0.7/1.1)6860\ \text{g/m}^3$$

$$= 4365\ \text{g/m}^3$$

c. Pass 3 and 4 are calculated similarly
d. Summary of MLSS concentrations and volumes:

Pass	MLSS, g/m³	Anoxic volume, m³	Aerobic volume, m³
1	6860	493.5	1974
2	4365	493.5	1974
3	3430	493.5	1974
4	3000	493.5	1974

4. Perform solids balance on system and determine aerobic SRT.
a. Solids balance

$$\Sigma X_i V_i = (P_{x,\text{TSS}})(\text{SRT})$$

$$\Sigma X_i V_i = X_1 V_1 + X_2 V_2 + X_3 V_3 + X_4 V_4$$

$$= [(6860 + 4365 + 3430 + 3000)\ \text{g/m}^3](1974\ \text{m}^3)(1\ \text{kg}/10^3\ \text{g})$$

$$= 34{,}851\ \text{kg}$$

b. Apply Eq. (8–16) assuming $NO_x \sim 0.80$ TKN.

MLSS components = biomass + cell debris + nitrifiers + nbVSS
+ inorganic inerts

$$(\text{SRT})(P_{X,\text{TSS}}) = \frac{QY(S_o - S)\text{SRT}}{[1 + (k_d)\text{SRT}](0.85)(10^3\ \text{g/kg})} + \frac{f_d(k_d)QY(S_o - S)(\text{SRT})^2}{[1 + (k_d)\text{SRT}](0.85)(10^3\ \text{g/kg})}$$

$$+ \frac{QY_n(NO_x)\text{SRT}}{[1 + (k_{dn})\text{SRT}](0.85)(10^3\ \text{g/kg})} + \frac{Q(\text{nbVSS})\text{SRT}}{(10^3\ \text{g/kg})}$$

$$+ \frac{Q(\text{TSS}_o - \text{VSS}_o)\text{SRT}}{(10^3\ \text{g/kg})}$$

$$NO_x = 0.80(35\ \text{g/m}^3) = 28\ \text{g/m}^3$$

c. Solve for SRT.
Substituting wastewater and coefficient values yields

$$34{,}851 \text{ kg} = \frac{(22{,}464 \text{ m}^3/\text{d})(0.40 \text{ g/g})(224 \text{ g/m}^3)\text{SRT}}{[1 + (0.088 \text{ g/g·d})\text{SRT}](0.85)(10^3 \text{ g/kg})}$$

$$+ \frac{(0.15 \text{ g/g})(0.088 \text{ g/g·d})(0.40 \text{ g/g})(22{,}464 \text{ m}^3/\text{d})(224 \text{ g/m}^3)(\text{SRT})^2}{[1 + (0.088 \text{ g/g·d})\text{SRT}](0.85)(10^3 \text{ g/kg})}$$

$$+ \frac{(22{,}464 \text{ m}^3/\text{d})(0.12 \text{ g/g})(28 \text{ g/m}^3)(\text{SRT})}{[1 + (0.06 \text{ g/g·d})\text{SRT}](0.85)(10^3 \text{ g/kg})}$$

$$+ \frac{(22{,}464 \text{ m}^3/\text{d})(20 \text{ g/m}^3)\text{SRT}}{(10^3 \text{ g/kg})}$$

$$+ \frac{(22{,}464 \text{ m}^3/\text{d})(10 \text{ g/m}^3)\text{SRT}}{(10^3 \text{ g/kg})}$$

$$34{,}851 \text{ kg} = \frac{2367 \text{ SRT}}{1 + 0.088 \text{ SRT}} + \frac{31.2 \text{ SRT}^2}{1 + 0.088 \text{ SRT}}$$

$$+ \frac{88.8 \text{ SRT}}{1 + 0.06 \text{ SRT}} + 449 \text{ SRT} + 224.6 \text{ SRT}$$

Solve using spreadsheet solver function or by successive iterations, SRT = 19.2 d.

Compare to Example 8–2. In Example 8–2, aerobic volume = 8466 m³ and SRT = 12.5 d. For step feed, aerobic volume = 7896 m³ and SRT = 19.2 d. The lower volume and higher SRT is due to the higher MLSS concentrations in the early passes.

5. Determine the composition of the MLSS and MLVSS using the above solution in Step 4 for MLSS and SRT = 19.2 d. The results of the calculation of the MLVSS components are summarized in the following table:

Item	MLVSS, kg	Fraction of total MLVSS	MLSS, kg
Biomass	14,362	0.53	16,827
Cell debris	3,635	0.13	4,276
Nitrifiers	673	0.0247	792
nbVSS	8,620	0.32	8,620
Inert inorganics			4,312
Total	27,290		34,897

a. Fraction of biomass solids (from above table)
    Thus, MLVSS/MLSS = 27,290/34,897 = 0.78

    Biomass = 0.53(MLVSS)

    Nitrifiers = 0.0247(MLVSS)

b. Nitrogen for nitrifier growth
    With biomass and cell debris calculated, the NO_x for nitrifier growth is calculated as follows:

    Daily biomass + debris production = (14,362 + 3635) kg = 17,997 kg

    Daily wasting = 17,997 kg/19.2 d SRT = 937.3 kg/d

    N used for synthesis = (0.12 g N/g VSS biomass) (937.3 kg/d)

    $$= 112.5 \text{ kg/d}$$

    Based on influent flow the N synthesis is

    $$\text{N synthesis} = \frac{(112.5 \text{ kg/d})(10^3 \text{ g/kg})}{(22,464 \text{ m}^3/\text{d})} = 5.0 \text{ g/m}^3$$

    $$NO_x = \text{TKN} - N_{syn} - (\text{NH}_4\text{-N})_e$$

    $$= (35 - 5.0 - 0.5) \text{ g/m}^3 = 29.5 \text{ g/m}^3$$

c. Nitrifier mass fraction correction

    $$\text{Mass nitrifiers} = \frac{QY_n(NO_x)\text{SRT}}{[1 + (k_{dn})\text{SRT}]}$$

    $$= \frac{(22,464 \text{ m}^3/\text{d})(0.12 \text{ g/g})(29.5 \text{ g/m}^3)(19.2 \text{ d})}{[1 + (0.06 \text{ g/g·d})(19.2 \text{ d})](10^3 \text{ g/kg})} = 709 \text{ kg VSS}$$

    Corrected MLVSS = (27,290 − 673 + 709) kg = 27,326 kg

    Nitrifiers as fraction of MLVSS = 709/27,326 = 0.0259 (as compared to 0.0247)

d. Summary Table
    Based on the above data, prepare a summary table of biomass and nitrifier concentrations in each pass using ratios from above.

Pass	MLSS, g/m³	MLVSS, kg/d	Nitrifiers, g VSS/m³	Biomass, g VSS/ m³
1	6860	5350	140	2823
2	4365	3405	88	1796
3	3430	2675	70	1411
4	3000	2340	61	1234

6. Evaluate nitrification rate in each pass and compare to NH_4-N fed to stage.
   a. Develop equations for nitrification rate.
      The nitrification rate (R_n) in each stage is a function of the specific growth rate from ammonia oxidation and nitrifier concentration as follows:

   $$R_n, \text{g/d} = \left(\frac{\mu_m X_n}{Y_n}\right)V$$

   For growth only related to substrate utilization [Eq. (7–93) in Table 8–5]

   $$\mu_m = \left(\frac{\mu_{n,m}N}{K_n + N}\right)\left(\frac{DO}{K_o + DO}\right)$$

   Substituting Eq. (7–93) into the above

   $$R_n = \left(\frac{\mu_{n,m}}{Y_n}\right)\left(\frac{N}{K_n + N}\right)\left(\frac{DO}{K_o + DO}\right)X_n V$$

   As indicated in the nitrogen mass balance for each pass, the rate of oxidizable nitrogen fed to the pass equals the nitrification rate plus the rate of ammonia nitrogen leaving the pass. The nitrogen entering the pass is related to the influent feed rate of available nitrogen (NO_x) to that pass and the rate of influent nitrogen from a previous pass. Before proceeding with the mass balance, prepare a flowrate summary.
   b. Prepare flowrate summary and mass balances.

Pass	Flowrate from previous reactor	Influent flowrate	Total flowrate
1	0.6 Q	0.1 Q	0.7 Q
2	0.7 Q	0.4 Q	1.1 Q
3	1.1 Q	0.3 Q	1.4 Q
4	1.4 Q	0.2 Q	1.6 Q

The balances for passes 1 and 2 are shown for steady state:

Pass 1 (recycle NH_4-N) = last pass (pass 4) NH_4-N

Rate of influent N = rate of effluent N + nitrification rate

$$RAS \ Q \ N_4 + 0.1 \ Q \ NO_x = 0.7 \ Q \ N_1 + R_{n,1}$$

$$0.6 \ Q \ N_4 + 0.1 \ Q \ NO_x = 0.7 \ Q \ N_1 + R_{n,1}$$

$$R_{n,1} = \left(\frac{\mu_{n,m}}{Y_n}\right)\left(\frac{N_1}{K_n + N_1}\right)\left(\frac{DO}{K_o + DO}\right)X_{n,1}V$$

Pass 2

$$0.7\,Q\,N_1 + 0.4\,Q\,NO_x = 1.1\,Q\,N_2 + R_{n,2}$$

$$R_{n,2} = \left(\frac{\mu_{n,m}}{Y_n}\right)\left(\frac{N_2}{K_n + N_2}\right)\left(\frac{DO}{K_o + DO}\right)X_{n,2}V$$

c. Solve for nitrification rate in each stage.

Thus, N_1, N_2, N_3, and N_4 are solved for each stage using $NO_x = 29.5$ g/m^3 (determined earlier). The value for N_4 is assumed to solve pass 1 and then the final solution is reached by iteration until the value for N_4 equals the value used to solve pass 1. A spreadsheet program with a solver function can assist in the solution. Using the coefficients shown for the design condition, the solution is summarized below.

d. Summary of step-feed solution for NH$_4$-N per pass:

Pass	Nitrifiers, g/m³	Influent flowrate, m³/d	RAS or flow from previous pass, m³/d	NH₄-N, g/m³	R_n, g/d
1	140	2246	13,478	0.05	69,389
2	89	8986	15,724	0.48	254,028
3	70	6739	24,710	0.46	196,129
4	61	4493	31,449	0.31	135,963

Sufficient volume exists to provide an effluent NH$_4$-N concentration of 0.5 g/m^3 or less for each pass. Excess volume is available in pass 1 and pass 4 for an effluent NH$_4$-N concentration of 0.5 g/m^3.

7. Determine the amount of nitrate removal in the anoxic zones and effluent NO$_3$-N concentration.

The amount of nitrate fed to each anoxic zone for passes 2, 3, and 4 equals the nitrification rate (g/d) in the previous pass plus the nitrate not removed in the previous anoxic zones. For the first anoxic zone, the nitrate feed is equal to the effluent NO$_3$-N concentration times the RAS flowrate. For the solution, the first step is to determine the denitrification capacity for each anoxic stage by calculating the F/M$_b$ per pass and using Fig. 8–23 to obtain the SDNR as follows:

a. Calculate F/M$_b$ from Eq. (8–43).

$$F/M_b = \frac{Q\,S_o}{(V_{nox})(X_b)} \quad [\text{from previous computation, } X_b = 0.53\,(\text{MLSS})]$$

b. Obtain SDNR$_b$ from Fig. 8–23.

$$\text{Rate of COD degradability} = \frac{\text{rbCOD}}{\text{bCOD}} = \frac{(80\text{ g/m}^3)}{(224\text{ g/m}^3)} = 0.36\,(\approx 40\%)$$

c. Use Eq. (8–44) for temperature correction.

$$SDNR_{12} = SDNR_{20}(1.026)^{12-20} = SDNR_{20}(0.814)$$

d. NO_3-N removed $= (SDNR_b)(V_{nox})X_b$

Pass	X_b, g/m³	Influent flowrate, m³/d	Anoxic volume, m³	F/M$_b$	SDNR$_{12}$, g/g·d
1	2729	2246	493	0.23ᵃ	0.05ᵇ
2	1736	8986	493	1.47	0.26
3	1364	6739	493	1.40	0.24
4	1193	4493	493	1.07	0.20

Sample calculation for pass 1:

$$^a F/M_b = \frac{(2346 \text{ m}^3/\text{d})(140 \text{ g/m}^3)}{(2729 \text{ g/m}^3)(493 \text{ m}^3)} = 0.23 \text{ g/g·d}$$

^{b}From Fig. 8–23, SDNR $= 0.06$ g/g·d

$SDNR_{12} = 0.06 (1.026)^{12-20} = 0.05$

NO_3-N removed $= (0.05$ g/g·d$)(2729$ g/m³$)(493$ m³$) = 67{,}269$ g/d.

Using the SDNR from the above for each pass, the amount of NO_3-N removed is tabulated below.

Pass	NO$_3$ removal capacity, g/d
1	67,269
2	222,520
3	161,388
4	117,269

8. Nitrate balance and effluent NO_3-N concentration.
    a. Develop equations for nitrate mass balance.
        A nitrate balance for each pass is done to determine how much nitrate remains after the anoxic reactor and the effluent nitrate nitrogen concentration. The nitrogen remaining after each reactor is

NO_3-N influent from RAS or previous pass	−	pass anoxic removal capacity	=	NO_3-N remaining after anoxic reactor

If there is a negative value for NO_3-N remaining because of excess anoxic NO_3-N removal capacity, then a value of zero is assigned [see column (3) in the following table].

The effluent NO_3-N concentration from each pass is then:

$$\frac{NO_3\text{-N remaining}}{\text{after anoxic}} - \frac{NO_3\text{-N produced}}{\text{in the pass}} = \frac{\text{effluent}}{NO_3\text{-N}}$$

b. Prepare table to solve mass balance for each stage.
The NO_3-N balance is illustrated in the following table.

Pass	Total NO_3-N to pass, g/d	Anoxic removal capacity, g/d[a]	NO_3-N remaining after anoxic, g/d[b] (1) − (2)	NO_3-N produced (R_n) in pass, g/d[c]	Effluent NO_3-N, g/d (3) + (4)
	(1)	(2)	(3)	(4)	(5)
1	115,299	67,269	48,030	69,589	117,619
2	117,619	222,520	0	254,028	254,028
3	254,028	161,388	92,640	196,129	288,769
4	288,769	117,269	171,500	135,963	307,483

Effluent NO_3-N = 8.6 g/m³

[a] Source of influent nitrate: for pass 1, RAS; other passes, column (6).
[b] From Step 8.
[c] From Step 6.

c. Determine effluent NO_3-N concentration.
The nitrate fed to pass 1 from RAS is calculated as follows:

$$NO_3\text{-N to pass } 1 = (Q_{RAS})N_e$$

where N_e = effluent NO_3-N concentration, g/m³

The value for N_e is obtained using the effluent NO_3-N shown in column (5) for pass 4:

$$(Q + Q_{RAS})N_e = \text{pass 4 effluent } NO_3\text{-N (g/d)}$$

$$N_e = (307,483 \text{ g/d})/(Q + 0.6\,Q)$$

$$N_e = (307,483 \text{ g/d})/(1.6)(22,464 \text{ m}^3/\text{d}) = 8.6 \text{ g/m}^3$$

Initially, assume an effluent concentration, e.g., 10 g/m³, to calculate the amount of NO_3-N fed to pass 1. Using the spreadsheet, perform successive iterations until the calculated effluent NO_3-N equals the trial.

9. Reevaluate design.
Note that some excess capacity is available and is not realized in the initial passes of the symmetrical step-feed design. Different influent flow splits may be used to reduce the effluent nitrate concentration, the anoxic and aerobic volumes can be changed, the anoxic zone may be staged, and the MLSS concentration may be increased. A spreadsheet model is necessary to evaluate various design changes.

Intermittent Aeration Process Design

Long SRT systems, such as oxidation ditch processes, may employ intermittent aeration to accomplish both nitrification and denitrification in a single tank. During the aeration-off period, the aeration tank operates essentially as an anoxic reactor as nitrate is used in lieu of DO for BOD removal. During the anoxic period, the tank operation is similar to a preanoxic tank because influent BOD is added continuously to drive the denitrification reaction.

Operation of an oxidation ditch using intermittent aeration is shown on Fig. 8–28. Intermittent aeration systems typically are operated with SRT values in the range of 18 to 40 d and hydraulic detention times in excess of 16 h. During the anoxic reaction

Figure 8–28

Operation of a Nitrox™ oxidation ditch process using intermittent aeration: (a) aerobic conditions, (b) anoxic conditions, and (c) variations in ORP, DO, ammonia, and nitrate.

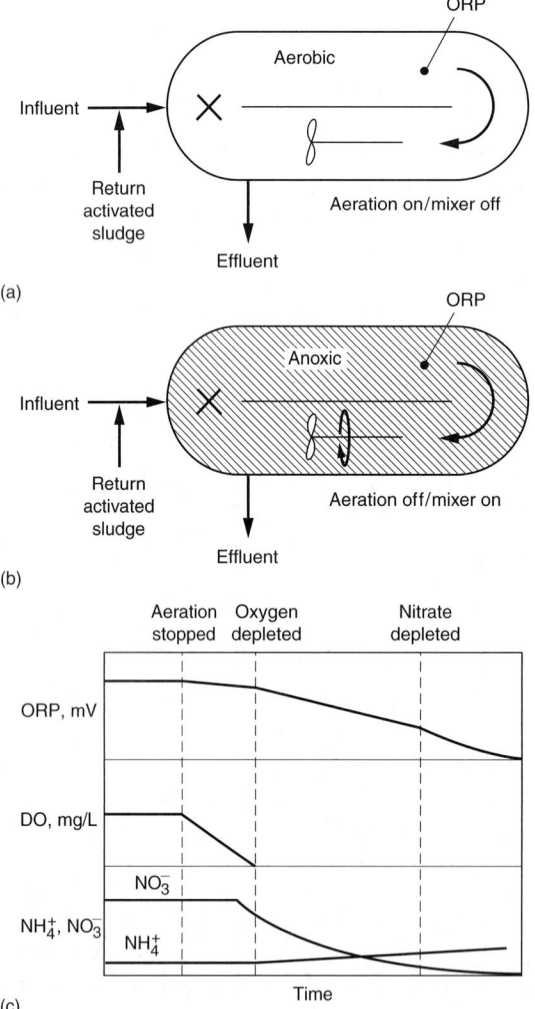

period (Fig. 8–28*b*), aeration is stopped, a submerged mixer is turned on, and nitrate is used as the electron acceptor. The reactor is operated as a complete-mix activated-sludge anoxic process. During the anoxic period, DO and nitrate are depleted and the ammonia concentration increases (see Fig. 8–28*c*). The time for the anoxic and aerobic periods is important in determining the system's treatment performance.

The anoxic/aerobic cycle times may be adjusted manually as part of the system operation to optimize the process performance. Alternatively, a patented process, Nitrox™, uses an oxidation-reduction potential (ORP) measurement to control the intermittent operation. The ORP response during an aeration-off period is shown on Fig. 8–28*c*. As the DO concentrations decline, the ORP value decreases. When the NO_3-N is depleted, a dramatic decline in the ORP value occurs. The ORP decline is called the *ORP knee* and can be identified by calculating the ORP slope with time. In the Nitrox™ process, the ORP values are logged onto a computer, which is programmed to turn on the aeration based on the changing slope of the ORP. The aeration-off periods are selected to occur during different times of the day; the more ideal time is when the influent BOD concentration is high so that nitrate reduction occurs at a faster rate. The process behaves much like an anoxic selector as improved SVIs have been reported (Stensel et al., 1995).

Reported plant performance data for intermittent aeration processes indicate effluent NO_3-N concentrations range from 3.0 to 4.8 mg/L. The processes are also predicted to produce effluent total nitrogen (TN) concentrations <8.0 mg/L (U.S. EPA, 1993). The relatively long τ values used provide sufficient dilution to minimize the effluent NH_4-N concentrations during the OFF period. A sufficiently long SRT is also needed to provide enough nitrification capacity to allow the aeration system to be operated intermittently.

Because of the long SRT and τ, the denitrification kinetics are related to the overall degradation of bCOD, pbCOD, and endogenous decay and are not as strongly influenced by the rbCOD fraction as for the preanoxic denitrification application with the relatively short τ. During the complete-mix anoxic period in an intermittent oxidation ditch operation or in the anoxic zone of a continuously aerated oxidation ditch, the specific denitrification rate is affected by both the endogenous respiration rate and the bCOD in the influent wastewater as continuous feeding occurs. The average specific denitrification rate, which includes these effects, can be estimated by the following equation [based on Stensel (1981) and modified to account for the rate as a function of active heterotrophic biomass]:

$$SDNR_b = \frac{0.175A_n}{(Y_{net})SRT} \tag{8–54}$$

where $SDNR_b$ = specific denitrification rate relative to heterotrophic biomass concentration, g NO_3-N/g biomass·d

A_n = net oxygen utilization coefficient, g O_2/g bCOD removed

Y_{net} = net yield for heterotrophic biomass, g VSS/g bCOD

0.175 = based on 2.86 g O_2 equivalent/g NO_3-N and the assumption that only 50% of heterotrophic biomass can use nitrate in place of oxygen

For a complete-mix activated-sludge reactor, A_n and Y_{net} are determined as follows (Stensel, 1981):

$$A_n = 1.0 - 1.42Y + \frac{1.42(k_d)(Y)\text{SRT}}{1 + (k_d)\text{SRT}}$$ (8–55)

$$Y_{net} = \frac{Y}{1 + (k_d)\text{SRT}}$$ (8–56)

The SDNR_b in Eq. (8–54) is relative to the biomass concentration that is a portion of the MLVSS concentration. Thus, the design procedure used to determine the amount of nitrate removed in the anoxic period during intermittent aeration incorporates some of the elements of the computational steps for anoxic/aerobic design described in Table 8–19. In Step 3 of Table 8–19, the heterotrophic biomass concentration has to be determined. As detailed in Step 9, the biomass concentration, SDNR_b, anoxic volume, and anoxic time are used to determine the amount of nitrate nitrogen removed. Use of the above equations is illustrated in Example 8–7.

EXAMPLE 8-7 **Determine Oxidation Ditch Intermittent Aeration Cycle Time for Nitrogen Removal** Determine what fraction of the time an oxidation ditch system must be operated as an anoxic reactor for an intermittent aeration process to produce an effluent NO_3-N concentration of 7 g/m^3, based on the following design conditions.

Design conditions:

Oxidation ditch volume	= 8700 m^3
SRT	= 25 d
MLSS	= 3500 g/m^3
MLVSS	= 2500 g/m^3
Fraction of biomass	= 0.40 g biomass/g MLVSS
Temperature	= 15°C
Y, k_d	= values given in Table 8–10
Influent flowrate	= 7570 m^3/d
Nitrate concentration in oxidation ditch	= 27 g/m^3 based on influent flowrate

Note: g/m^3 = mg/L.

Solution
1. Determine the SDNR.
   a. Obtain Y and k_d from Table 8–10 and correct k_d for temperature [Eq. (2–25)].

      $Y = 0.4$ g VSS/g bCOD

      $k_{d,20} = 0.12$ g/g·d

$$k_{d,15} = k_{d,20}(1.04)^{15-20}$$
$$= 0.12(1.04)^{15-20} = 0.099 \text{ g/g·d}$$

b. Using Eq. (8–55), determine A_n.

$$A_n = 1.0 - 1.42(0.4) + \frac{1.42(0.099 \text{ g/g·d})(0.40 \text{ g/g})(25 \text{ d})}{[1 + (0.099 \text{ g/g·d})(25 \text{ d})]}$$

$$= 0.84 \text{ g O}_2/\text{g bCOD}$$

c. Using Eq. (8–56) determine Y_{net}.

$$Y_{net} = \frac{Y}{1 + (k_d)\text{SRT}}$$

$$= \frac{0.4}{1 + (0.099 \text{ g/g·d})(25 \text{ d})} = 0.18 \text{ g VSS/g bCOD}$$

d. Using Eq. (8–54), determine SDNR_b.

$$\text{SDNR}_b = \frac{0.175 \, A_n}{(Y_{net}) \, \text{SRT}}$$

$$\text{SDNR}_b = \frac{0.175(0.84 \text{ g O}_2/\text{g bCOD})}{(0.18 \text{ g VSS/g bCOD})(25 \text{ d})} = \frac{0.033 \text{ g NO}_3\text{-N}}{\text{g biomass·d}}$$

2. Determine biomass concentration in the mixed liquor.

$$X_b = (2500 \text{ g/m}^3 \text{ MLVSS})\left(\frac{0.40 \text{ g biomass}}{\text{g MLVSS}}\right)$$

$$= 1000 \text{ g/m}^3 \text{ biomass}$$

3. Determine amount of $\text{NO}_3\text{-N}$ removal needed in g/d.

Nitrate removed concentration $(\text{NO}_r) = (27.0 - 7.0) \text{ g/m}^3 = 20 \text{ g/m}^3$

$\text{NO}_r = (7570 \text{ m}^3/\text{d}) (20 \text{ g/m}^3)$

$$= 151,400 \text{ g/d}$$

4. Determine $\text{NO}_3\text{-N}$ removal rate under anoxic reaction time.

Anoxic $\text{NO}_r = (\text{SDNR}_b)(X_b)(V)$

$$= (0.033 \text{ g NO}_3\text{-N/g biomass·d})(1000 \text{ g/m}^3)(8700 \text{ m}^3)$$

$$= 287,100 \text{ g/d}$$

5. Determine anoxic time needed per day.

$$\text{Anoxic time} = \frac{(151,400 \text{ g/d})(24 \text{ h/d})}{(287,100 \text{ g/d})} = 12.7 \text{ h}$$

Fraction of day $= 12.7 \text{ h}/24 \text{ h} = 0.52$

The computed value is within the range of values reported for the anoxic fraction for cyclic aeration process as given in U.S. EPA (1993). The actual time may be less or the amount of nitrogen removal may be greater due to denitrification within the floc during low DO aeration and in the secondary clarifier. By knowing the approximate amount of time needed for the anoxic conditions, the aerobic SRT available for nitrification can be estimated.

Postanoxic Endogenous Denitrification

Postanoxic denitrification can be done in separate tanks within the same single-sludge system following nitrification, as first proposed by Wuhrmann (1964). The Bardenpho process is a good example of this application. After nitrification, the rbCOD is fully depleted and, depending upon the system SRT, most of the pbCOD is likely to be depleted. Thus, the electron donor that creates the demand for nitrate reduction is mainly from activated-sludge endogenous respiration. Observed SDNRs have ranged from 0.01 to 0.04 g NO_3-N/g MLVSS under endogenous respiration (U.S. EPA, 1993; Stensel et al., 1995). The endogenous oxygen equivalent utilization rate under anoxic conditions has been found to be about 50 percent of that under aerobic conditions (Randall et al., 1992; Wuhrmann, 1964). Based on the cited references, the SDNR under endogenous conditions ($SDNR_b$) can be calculated from the endogenous decay rate as follows:

$$SDNR_b = \frac{1.42 k_d}{2.86}\, \eta = 0.5 (k_d)(\eta) \tag{8-57}$$

where 1.42 = g O_2/g biomass VSS

  2.86 = g O_2 equivalent/g NO_3-N

  η = fraction of biomass that can use NO_3-N in place of O_2 as an electron acceptor

  k_d = biomass endogenous decay coefficient, g VSS/g VSS biomass·d

The values for k_d at 20°C and the temperature correction coefficient are shown in Table 8–10. As discussed previously, the $SDNR_b$ is based on the biomass concentration. As shown in Chap. 7, the fraction of biomass from the MLVSS reaction declines as the SRT increases, so the SDNR value based on the MLVSS concentration would decrease with increasing SRT.

The endogenous respiration SDNR value at 20°C as a function of SRT has also been described by the following empirical relationship (Burdick et al., 1982), in which the SDNR is based on the total MLVSS concentration:

$$SDNR_{MLVSS} = 0.12 (SRT)^{-0.706} \tag{8-58}$$

where $SDNR_{MLVSS}$ = g NO_3-N/g MLVSS·d

A temperature correction value (θ) of 1.08 is recommended for use with Eq. (8–58) (U.S. EPA, 1993).

Sequencing Batch Reactor Process Analysis

In the SBR process and other batch decant processes (see Figs. 8–1c and 8–16b and c) nitrate removal can be accomplished by three methods: (1) nitrate reduction by using a mixed nonaerated fill period, (2) cycling aeration On/Off during the react period, and (3) operating at a low DO concentration to encourage SNdN. Under cyclic aeration conditions, Eq. (8–51) may also be useful to estimate the SDNR because the nitrate-reduction activity is driven mainly by endogenous respiration. Denitrification during a mixed nonaerated fill period provides the most efficient means of nitrate removal and also provides a selector operation to prevent filamentous sludge bulking. Most of the nitrate produced during the previous aerobic cycle remains in the SBR tank because the decant volume is only 20 to 30 percent of the total tank volume. The mass of nitrate remaining after decant can be reduced during the fill period if sufficient BOD and time are available. The following example is used to illustrate how to estimate the amount of nitrate removed during a mixed fill period for an SBR reactor.

EXAMPLE 8–8   **Determine Nitrate Removal During the Fill Period in an SBR Process**   Determine how much nitrate may be removed during a mixed, unaerated fill period in an SBR for the following design conditions. Assume the following design conditions apply.

Design conditions:

Number of tanks	2
Flow/tank, m³/d	3785
Cycle times, h	
Fill	4.0
Aerate	3.0
Settle	0.5
Decant	0.5
Fill volume fraction, V_F/V_T	0.3
Influent BOD, g/m³	200
Influent bCOD, g/m³	320
Influent rbCOD, g/m³	60
Influent TKN, g/m³	35
Effluent NH$_4$-N, g/m³	0.5
SRT, d	20
Temperature, °C	16

Note: g/m³ = mg/L.

Use kinetic coefficients from Table 8–10.

Solution

1.   Determine tank volume.

Cycle time = 4.0 + 3.0 + 0.50 + 0.50 = 8.0 h

$$\text{Cycles/day} = \frac{(24 \text{ h/d})}{(8 \text{ h/cycle})} = 3 \text{ cycles/d}$$

$$\text{Fill volume/cycle} = \frac{(3785 \text{ m}^3/\text{d})}{3 \text{ cycles}} = 1261.7 \text{ m}^3/\text{fill cycle}$$

$$V_F/V_T = 0.30$$

$$V_T = \frac{1261.7 \text{ m}^3}{0.30} = 4206 \text{ m}^3$$

2. Determine nitrate produced (NO_x).
   a. Determine heterotrophic biomass produced.
      Determine heterotrophic biomass production including cell debris to estimate the nitrogen used for synthesis using Eq. (8–15). (Biomass produced from nitrification can be ignored, as it represents a small fraction of the biomass.)

$$P_X = \frac{Q(Y)(S_o - S)}{1 + (k_d)\text{SRT}} + \frac{f_d(k_d)Y(Q)(S_o - S)\text{SRT}}{1 + (k_d)\text{SRT}}$$

   i. Use coefficients from Table 8–10 and adjust k_d for temperature.

$$k_d = 0.12(1.04)^{16-20} = 0.103 \text{ g/g·d}$$

$$Y = 0.40 \text{ g VSS/g bCOD}$$

$$f_d = 0.15$$

   Assume $S_o - S \approx S_o$.
   ii. Substitute and solve for P_X.

$$P_X = \frac{(3785 \text{ m}^3/\text{d})(0.40 \text{ g/g})(320 \text{ g/m}^3)}{[1 + (0.103 \text{ g/g·d})(20 \text{ d})](10^3 \text{ g/kg})}$$

$$+ \frac{(0.15 \text{ g/g})(0.103 \text{ g/g·d})(0.40 \text{ g/g})(3785 \text{ m}^3/\text{d})(320 \text{ g/m}^3)(20 \text{ d})}{[1 + (0.103 \text{ g/g·d})(20 \text{ d})](10^3 \text{ g/kg})}$$

$$= (158.3 + 48.9) \text{ kg/d} = 207.2 \text{ kg/d}$$

   b. Determine N synthesis.

$$NO_x = \text{TKN} - N_{syn} - (NH_4\text{-}N)_e$$

$$N_{syn} = 0.12(P_X) = 0.12(207.2 \text{ kg/d}) = 24.9 \text{ kg/d}$$

$$N_{syn} = \frac{(24.9 \text{ kg/d})(10^3 \text{ g/kg})}{(3785 \text{ m}^3/\text{d})} = 6.6 \text{ g/m}^3$$

   Effluent $NH_4\text{-}N = 0.5 \text{ g/m}^3$

$$NO_x \text{ produced} = (35.0 - 6.6 - 0.5) \text{ g/m}^3 = 27.9 \text{ g/m}^3$$

3. Determine amount of nitrate left in SBR mixed liquor after decant.
    Assume: NO_3-N ≈ 0 before aeration period, which means that all of the nitrate remaining in the SBR in the liquid volume after decanting is removed by denitrification during the mixing and unaerated fill period.
    a. Prepare nitrogen balance.

    $$NO_x = TKN - N_{syn} - (NH_4\text{-}N)_e$$

    $$NO_x = (35.0 - 6.6 - 0.5) \text{ g/m}^3 = 27.9 \text{ g/m}^3$$

    b. Determine NO_x produced per cycle.

    g NO_x produced per cycle $= 27.9 \text{ g/m}^3 (1261.7 \text{ m}^3/\text{fill})$

    $$= 35{,}201 \text{ g } NO_x/\text{fill}$$

    NO_3-N concentration at end of aeration with tank full: $(V = 4206 \text{ m}^3)$

    $$= \frac{35{,}201\text{g}}{4206 \text{ m}^3} = 8.4 \text{ g/m}^3$$

    Effluent NO_3-N $= 8.4 \text{ g/m}^3$ based on all NO_3-N that is produced in the aeration period is removed by denitrification during the unaerated fill period.
    c. Determine nitrate remaining in SBR.

    Volume remaining after decant: $0.70(4206 \text{ m}^3) = 2944 \text{ m}^3$

    NO_3-N present $= 8.4 \text{ g/m}^3 (V_s) = (8.4 \text{ g/m}^3)(2944 \text{ m}^3) = 24{,}729 \text{ g}$

4. Determine $SDNR_b$ in fill period.

    Active biomass concentration (from Step 2aii, biomass $= 158.3 \text{ kg/d}$)

    $$X_b = \frac{(\text{biomass})(\text{SRT})}{V_T} = \frac{(158.3 \text{ kg/d})(20 \text{ d})(10^3 \text{ g/kg})}{4206 \text{ m}^3}$$

    $$= 753 \text{ g/m}^3 \text{ at full volume}$$

    a. Determine F/M_b ratio in fill period.

    Biomass in system $= (753 \text{ g/m}^3)(4206 \text{ m}^3) = 3167 \text{ kg}$

    BOD feed rate $= Q_F S_o$

    $$Q_F = \frac{V_F}{t_F} = (1261.7 \text{ m}^3/4 \text{ h})(24 \text{ h/d})$$

    $$= 7570 \text{ m}^3/\text{d}$$

    $$Q_F S_o = (7570 \text{ m}^3/\text{d})(200 \text{ g/m}^3 \text{ BOD})(1 \text{ kg}/10^3 \text{ g})$$

    $$= 1514 \text{ kg/d}$$

    $$F/M_b = \frac{(1514 \text{ kg/d})}{3167 \text{ kg}} = 0.48 \text{ g/g·d}$$

b. Determine the SDNR$_b$.

Fraction rbCOD $= (60 \text{ g/m}^3)/(320 \text{ g/m}^3) = 0.19$

From Fig. 8–23 (FM$_b$ range of 0 to 2.0) at F/M$_b = 0.48$

SDNR$_b = 0.12$ g/g·d at 20°C

At 16°C, SDNR$_{16} =$ SDNR$_{20}$ θ^{16-20}; $\theta = 1.026$

$$= 0.108 \text{ g/g·d}$$

5. Determine NO$_3$-N removal capacity during the fill period.

NO$_x = $ (SDNR$_b$)(X_b)(V_T) [Note: (X_b)(V_T) $=$ biomass in system]

$$= (0.108 \text{ g/g·d})(753 \text{ g/m}^3)(4206 \text{ m}^3)$$

$$= 342{,}049 \text{ g/d}$$

Fill time $= 4$ h

$$\text{NO}_r \text{ at } 4 \text{ h} = \frac{(342{,}049 \text{ g/d})(4 \text{ h})}{(24 \text{ h/d})} = 57{,}008 \text{ g}$$

NO$_3$-N available from Step 3 $= 24{,}729$ g

Therefore, all of NO$_3$-N can be removed in the fill period.

Note: V_F/V_T controls effluent NO$_3$-N concentration.

Postanoxic Denitrification with an External Carbon Source

The biological nitrogen-removal processes discussed thus far in this text are single-sludge suspended growth processes with the electron donor for denitrification provided by BOD in the wastewater (e.g., preanoxic condition) or from mainly endogenous decay of the activated sludge (postanoxic). Prior to the late 1970s or early 1980s, the principal approach for nitrate removal was to add a process (either suspended growth or attached growth) after nitrification. Nitrification/denitrification using suspended growth processes is also termed the *two-sludge* process. Because only a negligible amount of BOD remains in the nitrified effluent, an external carbon source has to be added to supply the energy for the nitrifying organisms. In this section, postanoxic denitrification using a suspended growth process is considered. Attached growth denitrification is discussed in Chap. 9.

For postanoxic denitrification, nitrified influent is fed to a mixed anoxic tank along with an external carbon source, which is commonly methanol (see Fig. 8–21e). Sufficient detention time and SRT are needed to consume the methanol with NO_3-N as the electron acceptor and to assure good floc settling and thickening characteristics. An SRT of at least 5 d is normally used. The anoxic tank is followed by a short aeration time of 10 to 20 min to release nitrogen gas bubbles from the mixed liquor to promote maximum suspended solids removal in the final clarifier. As in any suspended growth process, wasting of excess sludge is necessary.

The postanoxic suspended growth process can be designed in a similar manner to an activated-sludge process, with the electron donor being the growth substrate. The NO_3-N that must be removed serves as the electron acceptor in lieu of dissolved oxygen. Nitrate limits only the process kinetics at very low concentrations (less than 0.30 mg/L); thus, the process design is based on methanol degradation kinetics. The process is designed on a complete-mix activated-sludge (CMAS) process, using bCOD kinetic expressions from Table 8–5 to describe the methanol consumption and biomass growth. The nitrate consumption is treated in a manner similar to supplying oxygen for an aerobic CMAS process. Because the process and the organic substrate used are so well defined, in contrast to the preanoxic processes using wastewater BOD, the SDNR design approach is not appropriate.

Typical values for the biological kinetic coefficients, based on methanol addition, are summarized in Table 8–20, and a design example using the kinetic coefficients given in Table 8–20 is presented in Example 8–9. The design process procedure is outlined as follows.

1. Determine the amount of NO_3-N to be reduced.
2. Select the anoxic tank SRT.
3. Calculate the anoxic tank residual methanol concentration based on design SRT and kinetic coefficients.
4. Determine the methanol dose based on the amount of nitrate to be removed and the amount of dissolved oxygen in the influent wastewater to be consumed.
5. Calculate the total solids production.
6. Determine the anoxic tank volume based on the solids production and SRT.
7. Select a detention time for the postaeration tank prior to clarification.

Table 8–20

Denitrification kinetic coefficients with methanol as the growth substrate at 10°C and 20°C[a]

Parameter	Unit	Temperature, °C	
		10	20
Synthesis yield, Y	g VSS/g bCOD	0.17	0.18
Endogenous decay, k_d	g VSS/g VSS·d	0.04	0.05
Maximum specific growth rate, μ_m	g VSS/g VSS·d	0.52	1.86
Maximum specific substrate utilization rate, k	g bCOD/g VSS·d	3.1	10.3
Half-velocity, K_s	g/m³	12.6	9.1

[a] Adapted from Stensel et al. (1973).

EXAMPLE 8–9 **Suspended Growth Postanoxic Denitrification Process** Given the following nitrified effluent flow and characteristics, design a postanoxic suspended growth system (see Fig. 8–21e) to achieve an effluent NO_3-N concentration of 2.0 g/m³. Using the given data and assumptions, determine (1) the required methanol dose, (2) the anoxic volume and detention time, and (3) the waste sludge production.

Wastewater characteristics:

Item	Unit	Value
Flowrate	m³/d	4000
NO_3-N	g/m³	25
TSS	g/m³	20
Temperature	°C	15

Note: g/m³ = mg/L.

Design assumptions:

1. Use the following denitrification kinetic coefficients based on the kinetic data given in Table 8–20.

   Design denitrification kinetics at 15°C (interpolated)

   $Y = 0.18$ g VSS/g bCOD used

   $k_d = 0.05$ g VSS/g VSS·d

   $k = 6.7$ g bCOD/g VSS·d

   $K_s = 10.9$ g/m³

2. MLSS = 2000 mg/L.
3. Feed TSS is not significantly degraded (and the BOD from the nitrification reactor is negligible).
4. Methanol exiting the anoxic zone is degraded in the aerobic zone.
5. SRT = 5 d for one anoxic tank.
6. Aeration period following denitrification = 15 min.
7. Clarifier overflow rate = 24 m/d (m³/m²·d).

Solution

1. Determine the residual concentration of methanol in the effluent from the anoxic zone based on process kinetics using Eq. (7–40) in Table 8–5.

   $$S = \frac{K_s[1 + (k_d)SRT]}{SRT(Yk - k_d) - 1}$$

   where S = residual methanol concentration as bCOD

   Substitute known values and solve for S.

   $$S = \frac{(10.9 \text{ g/m}^3)[1 + (0.05 \text{ g/g·d})(5 \text{ d})]}{(5 \text{ d})[(0.18 \text{ g/g})(6.7 \text{ g/g·d}) - 0.05 \text{ g/g·d}] - 1} = 2.9 \text{ g/m}^3 \text{ as bCOD}$$

2. Determine the required methanol dose.
   a. Determine the methanol required for NO_3-N reduction.

   $$NO_3\text{-N reduced} = (25.0 - 2.0)\ g/m^3 = 23.0\ g/m^3$$

   $bCOD/NO_3$-N ratio, Eq. (7–108)

   $$\frac{bCOD}{NO_3\text{-N}} = \frac{2.86}{1 - 1.42 Y_n}$$

   The value of Y_n can be determined using Eq. (7–103).

   $$Y_n = \frac{Y}{[1 + (k_d)\,SRT]} = \frac{(0.18\ g\ VSS/g\ bCOD)}{[1 + (0.05\ g/g{\cdot}d)(5\ d)]} = 0.144\ g\ VSS/g\ bCOD$$

   Thus,

   $$\frac{bCOD}{NO_3\text{-N}} = \frac{2.86}{1 - 1.42(0.144)} = 3.6\ g/g$$

   b. Determine the total methanol required, which is comprised of the amount needed for NO_3-N reduction and the residual bCOD from the anoxic reactor.

   Methanol dose (as COD)

   $$= (3.6\ g/g\ NO_3\text{-N})(23\ g/m^3) + 2.9\ g/m^3$$

   $$= 85.7\ g/m^3\ \text{as COD}$$

   $$\text{Methanol dose} = \frac{(85.7\ g\ COD/m^3)}{(1.5\ g\ COD/g\ CH_3OH)} = 57.1\ g\ CH_3OH/m^3$$

   $$\text{Daily methanol consumption} = (57.1\ g/m^3)(4000\ m^3/d)(1\ kg/10^3\ g)$$
   $$= 228.5\ kg/d$$

3. Determine solids production based on Eqs. (7–52) and (7–53):

   $$P_{X,TSS} = \frac{QY(S_o - S)}{[1 + (k_d)SRT](0.85)} + \frac{f_d(k_d)QY(S_o - S)SRT}{[1 + (k_d)SRT](0.85)} + Q(TSS_o)$$

   Assume $S_o - S \approx S_o$.

   $$P_{X,TSS} = \frac{(4000\ m^3/d)(0.18\ g/g)(85.7\ g/m^3)}{[1 + (0.05\ g/g{\cdot}d)(5\ d)](0.85)}$$

   $$+ \frac{0.15(0.05)(4000\ m^3/d)(0.18\ g/g)(85.7\ g/m^3)(5\ d)}{[1 + (0.05\ g/g{\cdot}d)(5\ d)](0.85)}$$

   $$+ (4000\ m^3/d)(20\ g/m^3)$$

   $$= 58{,}074\ g/d + 2178\ g/d + 80{,}000\ g/d$$

   $$= 140{,}252\ g/d$$

4. Determine the volume of the anoxic reactor using Eq. (7–55).

$$(MLSS)V = (P_{X,TSS})SRT$$

At MLSS = 2000 g/m^3

$$V = \frac{(140{,}252 \text{ g/d})(5 \text{ d})}{(2000 \text{ g/m}^3)} = 350.6 \text{ m}^3$$

$$\tau = \frac{V}{Q} = \frac{(350.6 \text{ m}^3)(24 \text{ h/d})}{(4000 \text{ m}^3/\text{d})} = 2.1 \text{ h}$$

5. Use a detention time in the aeration zone = 15 min = 0.25 h
6. Determine the size of the clarifier.

Use hydraulic application rate = 24 m^3/m^2·d (given)

$$\text{Clarifier area} = \frac{(4000 \text{ m}^3/\text{d})}{(24 \text{ m}^3/\text{m}^2\cdot\text{d})} = 166.7 \text{ m}^2$$

Use two clarifiers:

Area/clarifier = 83.3 m^2

Clarifier diameter = 10.3 m

Round diameter to nearest 0.5 m, thus,

Clarifier diameter = 10.5 m.

Nitrogen Removal in Anaerobic Digestion Recycle Streams

Recycle flows from dewatering of anaerobically digested solids contain high NH$_4$-N concentrations (>1000 mg/L) that can increase wastewater influent load by 15 to 20 percent. The recycle steam is also characterized by relatively high temperature and pH. A sidestream treatment process termed the *Sharon*™ (single-reactor high-activity ammonia removal over nitrite) process has been developed at Delft University of Technology in the Netherlands to remove nitrogen biologically from digester recycle in a relatively short detention time reactor (Hellinga et al., 1998). The process takes advantage of the effect of high temperature on nitrification kinetics that favor more rapid growth of ammonia-oxidizing bacteria over nitrite-oxidizing bacteria. A complete-mix reactor without solids recycle is operated with intermittent aeration for nitrification and denitrification. The BOD in the anaerobic recycle stream is low compared to the NH$_4$-N concentration, so for nitrite reduction, methanol is added during the anoxic period to provide an electron donor. In a bench-scale study, 80 to 85 percent nitrogen removal has been reported for an operation having a temperature of 35°C, 1.5 d SRT, 80 min of aeration, and 40 min of anoxic contacting. The *Sharon*™ process can also be operated without the anoxic step and methanol addition to produce a nitrite-rich recycle stream that can be fed to a preanoxic zone or to the plant headworks.

Alternative Process Configurations for Biological Nitrogen Removal

A variety of activated-sludge process configurations are used to accomplish biological nitrogen removal. Representative process schematics and descriptions for biological nitrogen-removal processes are given in Table 8–21.

Process Design Parameters

Typical parameters used for the design and operation of various biological nitrogen-removal processes are presented in Table 8–22.

Process Selection Considerations

The selection of a specific process for biological nitrogen removal will depend on site-specific conditions, existing processes and equipment, and treatment needs. The advantage and limitations of the processes commonly used and their treatment capability in terms of effluent total nitrogen concentrations are summarized in Table 8–23.

The MLE process is one of the most common methods used for biological nitrogen removal, and can be adapted easily to existing activated-sludge facilities. The amount of nitrate removal is limited by the practical levels of internal recycle to the preanoxic zone, and the process is used more generally to achieve effluent total nitrogen concentrations between 5 and 10 mg/L. Dissolved oxygen control should be used in the zone from which the recycle stream is taken to limit the amount of DO fed to the anoxic zone.

The step-feed process is also applicable for meeting effluent total nitrogen concentrations of less than 10 mg/L. However, it is theoretically possible to achieve lower effluent nitrogen total concentrations (ranging from less than 3 to 5 mg/L) with step-feed BNR using internal recycle, such as in the MLE process, for the last pass of the anoxic-aerobic step-feed process. As in the MLE process, the DO concentration from the aerobic zone must be controlled to minimize dissolved oxygen addition to the anoxic zone. In the case of the step-feed process, more DO control points are required. Influent flow splitting measurement and control are necessary to optimize the step-feed reactor volume for nitrogen removal.

The sequencing batch reactor process provides a high degree of flexibility for nitrogen removal. Mixing during the fill period provides an opportunity for anoxic conditions for nitrate removal. During the aeration react period, the DO concentration may be cycled to provide anoxic operating periods. The batch decant reactor designs are slightly less flexible than the SBR processes because they depend on internal recycle like the MLE for a major portion of the nitrate removal.

Bio-denitro™, Nitrox™, and the oxidation ditch with DO control are all processes with large reactor volumes for nitrogen removal, and represent various methods for optimizing biological nitrogen removal in oxidation ditch systems. Very low effluent total nitrogen concentrations (less than 5 mg/L) have been reported for the Bio-denitro™ process. The Nitrox™ process is generally limited to effluent total nitrogen concentrations of 5 to 8 mg/L. During the aeration "off period" in the Nitrox™ process, ammonia accumulates in the oxidation ditch, resulting in higher effluent NH_4-N concentrations from the process. The effluent NH_4-N and total nitrogen concentration is dependent on the total reactor volume and influent nitrogen concentrations. Higher influent TKN concentrations can result in higher effluent ammonia concentrations.

The Bardenpho process and postanoxic processes with methanol addition have demonstrated the ability to achieve less than 3 mg/L total nitrogen. The second anoxic zone of the Bardenpho process has a very low denitrification rate, resulting in less efficient reactor volume utilization. The addition of methanol to the second anoxic zone reduces the reactor volume requirements.

The SNdN processes, such as the Orbal™ and Sym-Bio™ NADH control process, require large reactor volumes and some operator attention and skill. However, depending on the operating conditions, these processes have been shown to be capable of producing very low effluent total nitrogen concentrations.

Besides removing nitrogen, the preanoxic and SNdN processes have additional advantages over the postanoxic processes. By removing nitrate before or during the nitrification step, the alkalinity produced by denitrification is made available to offset the alkalinity depleted by nitrification. Because 3.57 g of alkalinity (as $CaCO_3$) are produced per g NO_3-N oxidized, and 7.14 g alkalinity (as $CaCO_3$) are used per g NH_4-N oxidized, almost half of the alkalinity used for nitrification can be provided by preanoxic or SNdN processes. The recovery of alkalinity is very important for wastewaters that have low alkalinity. In some applications, alkalinity may have to be added in the form of lime or sodium hydroxide, at significant cost, to maintain an acceptable pH for the nitrification process.

Other advantages of these processes include aeration energy savings and the ability to produce a sludge that settles well. By using nitrate to oxidize influent BOD, the preanoxic and SNdN processes require less oxygen for aeration compared to postanoxic processes. Greater energy savings can be achieved with the SNdN processes as compared to the preanoxic processes, because they are operated at a lower DO concentration. By operating at a lower DO concentration, the DO gradient is larger, which results in more efficient oxygen transfer during aeration. Internal recycle pumping is also eliminated, which accounts for further energy reduction.

The postanoxic suspended growth process with methanol addition lacks the benefits of the preanoxic process, i.e., energy savings, alkalinity production, and filamentous bulking control, and has higher operating cost because of the purchase of methanol. The selection of the postanoxic suspended growth process is driven mainly by the site layout, existing reactor configuration, and equipment considerations.

Table 8-21

Description of suspended growth processes for nitrogen removal

Process	Description
Preanoxic	
(a) Ludzack-Ettinger 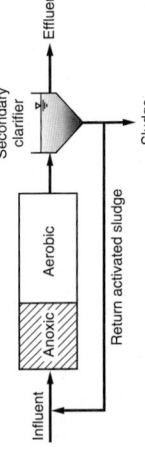	The first concept of a preanoxic BNR was an anoxic-aerobic operating sequence by Ludzack and Ettinger (1962). The influent wastewater was fed to an anoxic zone, which was followed by an aerobic zone. The process relies on the nitrate formed in the aerobic zone being returned via the RAS to the anoxic zone. Because the only nitrate fed to the anoxic zone is that in the RAS, denitrification is limited greatly by the RAS recycle ratio. However, more recently, this process has been used with increased RAS recycle rates to prevent rising sludge in the secondary clarifiers due to denitrification
(b) Modified Ludzack-Ettinger (MLE) 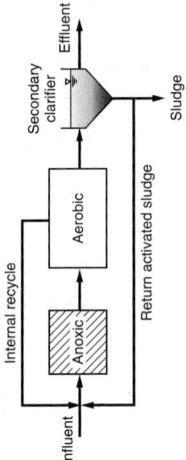	One of the most commonly used BNR processes is the modified Ludzack-Ettinger (MLE) process. Barnard (1973) improved on the original Ludzack-Ettinger design by providing the internal recycle to feed more nitrate to the anoxic zone directly from the aerobic zone. Both the denitrification rate and overall nitrogen-removal efficiency are increased. The internal recycle flow ratio (recycle flowrate divided by influent flowrate) typically ranges from 2 to 4. With sufficient influent BOD and anoxic contact time, these recycle ratios result in an average effluent NO_3-N concentration from 4 to 7 mg/L when treating domestic wastewater. The MLE process is very adaptable to existing activated-sludge facilities and can easily meet a common effluent standard of <10mg/L total nitrogen. A BOD/TKN ratio of 4:1 in the influent wastewater is usually sufficient for effective nitrate reduction by preanoxic processes. Typical anoxic tank detention times for the MLE process range from 2 to 4 h, but when the anoxic zone is divided into 3 to 4 stages in series, denitrification kinetic rates are increased and the total detention time needed may then be 50 to 70 percent of the single-stage design. Another modification of the MLE process, termed "eMLE," uses an additional set of anoxic/aeration basins to enhance nutrient removal

(continued)

| **Table 8-21** (*Continued*)

Process	Description

Preanoxic (*Continued*)

(c) Step feed

Preanoxic zones can also be used in a step-feed BNR process. Because step-feed BNR processes are usually adapted to existing multiple-pass full-scale tanks, symmetrical anoxic/aerobic stages are generally used. However, nonsymmetrical designs with smaller initial anoxic/aerobic stages can take better advantage of the higher MLSS concentration in the early stages, due to less RAS dilution, resulting in greater treatment capacity. A possible influent flow splitting percent distribution for a 4-pass system is 15:35:35:30:20. The final flow portion to the last anoxic/aerobic zone is critical as the nitrate produced in the aerobic zone from that flow will not be reduced, and will thus determine the final effluent NO_3-N concentration. Effluent NO_3-N concentrations of less than 8 mg/L are possible

(d) Sequencing batch reactor (SBR)

The SBR system (see Table 8–15) also employs preanoxic denitrification using BOD in the influent wastewater. Mixing is used during the fill period to contact the mixed liquor with the influent wastewater. For many domestic applications, depending on the wastewater strength, sufficient BOD and fill time are available to remove almost all of the nitrate remaining in the mixed liquor after the settle and decant steps. Some nitrate removal also occurs during the nonaerated settle and decant periods. Separate mixing provides operating flexibility and is useful for anoxic operation during the aeration period, as well as anaerobic or anoxic contacting during the fill period. Mixing without aeration during the fill period is effective in improving sludge-settling properties in addition to nitrogen removal. Many facilities have reported effluent NO_3-N concentrations of less than 5 mg/L.

(e) Bio-denitro™

The Bio-denitro™ process has also been referred to as a phased-isolation oxidation ditch technology. The process was developed in Denmark for nitrogen removal and has been installed in over 75 full-scale facilities producing effluent total nitrogen concentrations of less than 8 mg/L (Stensel and Coleman, 2000). The technology uses at least two oxidation ditches in a series configuration in which the operating sequence of the ditches and operation of the aeration and anoxic zones is varied. Submerged mixers are installed in the ditches so that for some operating phases, the basin is only mixed and not aerated. The basin continues to receive influent wastewater and operates as a preanoxic zone. Similar to the SBR operation, nitrate is available from a previous aerobic nitrification operation. Besides denitrification in the preanoxic zones, nitrate reduction is also possible during the aerobic operation depending on the DO concentration level. A typical duration for phases A, B, C, and D is 1.5, 0.5, 1.5, and 0.5 h, respectively

In the Nitrox™ process, the oxidation ditch operation is switched from an aerobic to an anoxic operating condition by turning off the aeration and operating a submerged mixer to maintain channel velocity. The process depends on the use of oxidation-reduction potential (ORP) control to (1) determine when the nitrate is depleted during the anoxic operation and (2) restart aeration. At selected times, the aerators are turned off and the mixer is turned on. When the nitrate is depleted in the aeration OFF period, the ORP drops dramatically. The ORP data is interpreted by a PC, which starts the aeration. A typical operating condition for the Nitrox™ process is to turn the aerators off at least twice per day, usually in the morning when the load is increasing, and then in the early evening hours (Stensel and Coleman, 2000). The off-time for the nitrate depletion usually lasts 3 to 5 h depending on the plant load and amount of nitrate in the oxidation ditch. Effluent NO_3-N concentrations of less than 8 mg/L and NH_4-N concentrations ranging from 1.0 to 1.5 mg/L have been reported

(f) Nitrox™

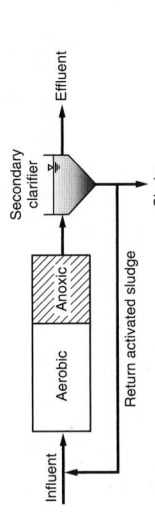

In the single-sludge process (developed by Wuhrmann), nitrogen removal was accomplished in the activated-sludge process by adding a mixed anoxic tank after aerobic nitrification. To achieve high nitrate-removal efficiency, a long detention time is required in the postanoxic tank because the denitrification rate is proportional to the endogenous respiration rate in the mixed liquor

Postanoxic
(g) Single-sludge

Both preanoxic and postanoxic denitrification are incorporated in the Bardenpho process, which was developed and applied at full-scale facilities in South Africa in the mid-1970s, before making its way to the United States in 1978. The detention time of the postanoxic stage is about the same as or larger than that used for the preanoxic zone. In the postanoxic zone, the NO_3-N concentration leaving the aeration zone is typically reduced from about 5 to 7 mg/L to less than 3 mg/L. During pilot-plant testing with higher-strength wastewaters, Barnard (1974) found that biological phosphorus removal occurred as well as nitrogen removal, hence the basis for the process name (the name comes from the first three letters of the inventor's name, Barnard; denitrification; and phosphorus)

(h) Bardenpho (4-stage)

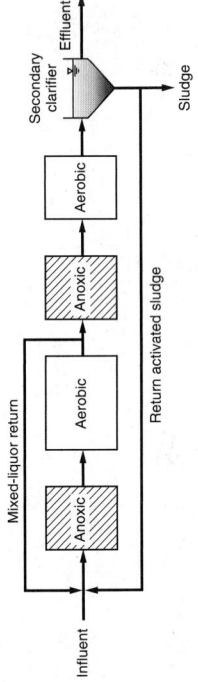

(continued)

| **Table 8-21** (Continued) |

Process	**Description**

Postanoxic (Continued)

(i) Oxidation ditch

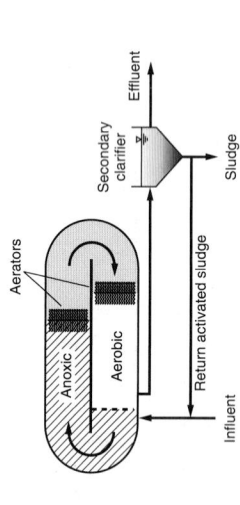

Depending on the aeration design and length of the oxidation ditch channel, anoxic denitrification zones can be established in oxidation ditches to accomplish biological nitrogen removal in a single tank. An aerobic zone exists after the aerator, and as the mixed liquor flows down the channel away from the aerator, the DO concentration decreases due to oxygen uptake by the biomass. At a point where the DO is depleted, an anoxic zone is created in the ditch channel and the nitrate will be used for endogenous respiration activity by the mixed liquor. Most of the readily degradable BOD had been consumed previously in the aerobic zone. Because of the large tank volumes and long SRTs used in oxidation ditch processes, sufficient capacity is available to accommodate nitrification and denitrification zones. DO control is necessary, however, to maintain a sufficient anoxic zone volume to allow for significant nitrogen removal

(j) Two-stage (two-sludge) with an external carbon source

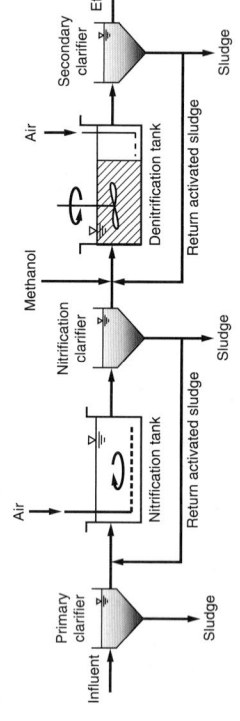

An approach that was most popular in the 1970s and is occasionally used today is a postanoxic design with exogenous external carbon addition, usually methanol. The activated-sludge anoxic zone (1 to 3 h) is mixed and a short aeration time (<30 min) follows to strip nitrogen gas bubbles from the floc and to provide aerobic conditions to improve liquid-solids separation in the clarifier. Methanol is a commonly used substrate because it is more effective than other substrates for denitrification in terms of cost per unit of nitrate removed. Though its absolute cost may be higher than glucose or acetate, methanol is less expensive overall because it has a relatively low biomass yield. The lower yield means that a greater portion of the methanol is oxidized to result in a higher ratio of nitrate used per g of substrate provided. Typical ratios of methanol to nitrate removal are 3.0 to 4.0 g/g depending on the amount of DO in the influent wastewater and the anoxic system SRT. Longer SRT designs have greater amounts of biomass oxidized by endogenous respiration, which consumes nitrate, and thus the influent methanol to nitrate ratio can be lower. The use of methanol requires special storage and monitoring precautions as it is a flammable substance. Because methanol is flammable, other substrates, such as acetate, have been considered. Attached growth denitrification reactors with methanol addition have also been used for postanoxic nitrogen removal (see Chap. 9)

Simultaneous nitrification/denitrification

(k) Low DO oxidation ditch

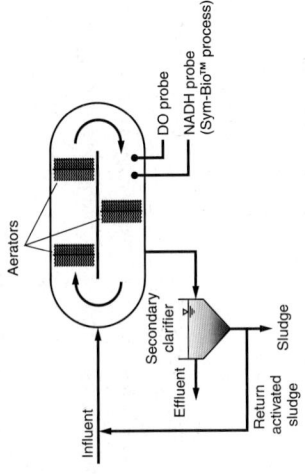

Oxidation ditches that have sufficient volume available are able to accommodate both nitrification and denitrification at lower rates under low DO conditions. An oxidation ditch may be used to maintain DO concentrations below 0.5 mg/L with manual or automated DO control. Where multiple aerators are used, for example, with brush aerators, a low to zero DO is maintained throughout the channel. Besides DO control, other methods may be used, and one such method using NADH measurement is the Sym-Bio™ process.

The Sym-Bio™ process uses both a DO probe and an NADH probe to control simultaneous nitrification/denitrification (Trivedi and Heinen, 2000). The bacteria content of the reduced form of coenzyme nicotinamide adenine dinucleotide (NADH) is measured by the emitted fluorescence after exposed to ultraviolet light emission by the NADH probe. NADH plays a significant role in bacteria reactions that involve the transfer of hydrogen. Under more reduced conditions, the ratio of NADH/NAD will increase in the bacteria cell. The NADH probe can be used to accomplish SNdN by (1) monitoring the changes in NADH in the activated sludge and (2) controlling the operation of activated-sludge systems operated at low to zero DO concentrations. The process involves a series of steps; the DO is lowered and when the NADH/NAD ratio reaches a set value, the DO is increased to a higher level for a predetermined time. Effluent NO_3-N and NH_4-N concentrations of less than 3.0 and 1.0 mg/L, respectively, have been achieved

(continued)

| Table 8–21 (Continued)

Process

Simultaneous nitrification/denitrification (Continued)

(l) Orbal™

Nitrogen removal from digested sludge processing recycle flows

(m) Sharon (single-reactor high-activity ammonia removal over nitrite)

Description

The channels in the Orbal process are operated in series with a zero to low DO (<0.3 mg/L) concentration in the first channel, a 0.5 to 1.5 mg/L DO concentration in the second channel, and a higher DO concentration (2 to 3 mg/L) in the third channel. The first channel receives the influent wastewater and return activated sludge and generally contains about one-half the total basin volume. Volumes of the second and third channels are about one-third and one-sixth of the total basin volumes, respectively. Recycle of mixed liquor from the inner loop to the outer loop allows denitrification of nitrates resulting from nitrification in the inner channels. Variations of the process include operation with and without internal recycle flow (Bionutre™ process) from the third channel to the first channel. An oxygen supply rate of about 50% of the estimated design requirement has been recommended to support SNdN in the first zone

A sidestream treatment process, the Sharon process, has been developed at Delft University of Technology in the Netherlands to remove nitrogen biologically from recycle flows from dewatering of anaerobically digested solids (Hellinga et al. 1998). The recycle stream is also characterized by a relatively high temperature and high pH. The process takes advantage of high-temperature effects on nitrification kinetics, which favors more rapid growth of ammonia-oxidizing bacteria and inhibition of nitrite-oxidizing bacteria. A short residence time (1 to 2 d) is then used to favor mainly nitrite production with an on/off aeration operation that provides an anoxic operating period to reduce nitrite. Methanol is added during the anoxic period to provide an electron donor for nitrite reduction. Alternatively, the nitrite stream can be returned to the headworks to provide an electron acceptor for odor control

Table 8–22
Typical design parameters for commonly used nitrogen-removal processes

Design parameter/ process	SRT, d[a]	MLSS, mg/L	τ, h			RAS, % of influent	Internal recycle, % of influent
			Total	Anoxic zone	Aerobic zone		
MLE	7–20	3000–4000	5–15	1–3	4–12	50–100	100–200
SBR	10–30	3000–5000	20–30	Variable	Variable		
Bardenpho (4-stage)	10–20	3000–4000	8–20	1–3 (1st stage) 2–4 (3rd stage)	4–12 (2nd stage) 0.5–1 (4th stage)	50–100	200–400
Oxidation ditch	20–30	2000–4000	18–30	Variable	Variable	50–100	
Bio-denitro™	20–40	3000–4000	20–30	Variable	Variable	50–100	
Orbal™	10–30	2000–4000	10–20	6–10	3–6 (1st stage) 2–3 (2nd stage)	50–100	Optional

[a] Temperature-dependent.

Table 8–23
Advantages and limitations of nitrogen-removal processes

Process	Advantages	Limitations
Preanoxic— general	Saves energy; BOD is removed before aerobic zone Alkalinity is produced before nitrification Design includes an SVI selector	
MLE	Very adaptable to existing activated-sludge processes 5 to 8 mg/L TN is achievable	Nitrogen-removal capability is a function of internal recycle Potential *Nocardia* growth problem DO control is required before recycle
Step feed	Adaptable to existing step-feed activated-sludge processes With internal recycle in last pass, nitrogen concentrations less than 5 mg/L are possible 5 to 8 mg/L TN is achievable	Nitrogen-removal capability is a function of flow distribution More complex operation than MLE; requires flow split control to optimize operation Potential *Nocardia* growth problem Requires DO control in each aeration zone

(continued)

| **Table 8–23** (Continued)

Process	Advantages	Limitations
Sequencing batch reactor	Process is flexible and easy to operate Mixed-liquor solids cannot be washed out by hydraulic surges because flow equalization is provided Quiescent settling provides low effluent TSS concentration 5 to 8 mg/L TN is achievable	Redundant units are required for operational reliability unless aeration system can be maintained without draining the aeration tank More complex process design Effluent quality depends upon reliable decanting facility May need effluent equalization of batch discharge before filtration and disinfection
Batch decant	5 to 8 mg/L TN is achievable Mixed-liquor solids cannot be washed out by hydraulic surges	Less flexible to operate than SBR Effluent quality depends upon reliable decanting facility
Bio-denitro™	5 to 8 mg/L TN is achievable Large reactor volume is resistant to shock loads	Complex system to operate Two oxidation ditch reactors are required; increases construction cost
Nitrox™	Large reactor volume is resistant to shock loads Easy and economical to upgrade existing oxidation ditch processes Provides SVI control	Nitrogen-removal capability is limited by higher influent TKN concentrations Process is susceptible to ammonia bleed-through Performance is affected by influent variations
Bardenpho (4-stage)	Capable of achieving effluent nitrogen levels less than 3 mg/L	Large reactor volumes required Second anoxic tank has low efficiency
Oxidation ditch	Large reactor volume is resistant to load variations without affecting effluent quality significantly Has good capacity for nitrogen removal; less than 10 mg/L effluent TN is possible	Nitrogen-removal capability is related to skills of operating staff and control methods
Postanoxic with carbon addition	Capable of achieving effluent nitrogen levels less than 3 mg/L May be combined with effluent filtation	Higher operating cost due to purchase of methanol Methanol feed control required
Simultaneous nitrification/ denitrification	Low effluent nitrogen level possible (3 mg/L lower limit) Significant energy savings possible Process may be incorporated into existing facilities without new construction SVI control enhanced Produces alkalinity	Large reactor volume; skilled operation also required Process control system required

8–6 PROCESSES FOR BIOLOGICAL PHOSPHORUS REMOVAL

Over the past 20 years, several biological suspended growth process configurations have been used to accomplish biological phosphorus removal, and they all include the basic steps of an anaerobic zone followed by an aerobic zone. Barnard (1974) was the first to clarify the need for anaerobic contacting between activated sludge and influent wastewater before aerobic degradation to accomplish biological phosphorus removal. Other modifications of the basic process include (1) combining the anaerobic/aerobic sequence with various biological nitrogen-removal designs, (2) recycling mixed liquor to the anaerobic zone from a downstream anoxic zone instead of only from the secondary clarifier underflow, (3) adding volatile fatty acids to the anaerobic zone as either acetate or a liquid stream from a fermentation reactor processing primary clarifier sludge, and (4) using multiple-staged anaerobic and aerobic reactors.

The alternating exposure to anaerobic conditions can be accomplished in the main biological treatment process, or "mainstream," or in the return sludge stream, or "sidestream." Several mainstream biological phosphorus-removal processes and one sidestream process, Phostrip™, are described in this section. Also included in this section are process design considerations, process control, analysis of biological phosphorus-removal performance, design parameters, and process selection considerations. The first "mainstream" biological phosphorus-removal process that was included with biological nitrogen removal is the Bardenpho process at Palmetto, FL, shown on Fig. 8–2b.

Biological Phosphorus-Removal Processes

Three biological phosphorus-removal (BPR) configurations that are more commonly used are shown on Fig. 8–29. Barnard (1975) used the term *Phoredox* to represent any process with an anaerobic/aerobic sequence to promote BPR. Since then, other process names have evolved to designate specific process configurations, such as the A/O™ (anaerobic/aerobic only) or A²O™ (anaerobic/anoxic/aerobic) processes. The A/O™ process is similar to the Phoredox process and was patented and marketed by Air Products and Chemicals, Inc., as well as the A²O™ process.

The main difference between the Phoredox (A/O) process and the A²O processes shown on Fig. 8–29a and b is that nitrification does not occur in the Phoredox (A/O) process. Low operating SRTs are used to prevent the initiation of nitrification. Desirable SRT values range from 2 to 3 d at 20°C and 4 to 5 d at 10°C for biological phosphorus removal to occur without nitrification (Grady et al., 1999). For applications where nitrification is needed to meet discharge requirements, the process must also include biological denitrification to prevent excessive amounts of nitrate from entering the anaerobic reactor by way of the RAS recycle. Heterotrophic bacteria will use nitrate to consume rbCOD in the anaerobic zone, which then leaves less rbCOD available for phosphorus-storing bacteria, thus decreasing the biological phosphorus-removal treatment efficiency.

The A²O and UCT (University of Cape Town) processes are two basic types of mainstream systems used for nitrate removal with BPR. In the A²O process, the return activated-sludge (RAS) recycle, which contains nitrate, is directed to the anaerobic

Figure 8–29

Typical mainstream biological phosphorus-removal processes: (a) Phoredox (A/O), (b) A²O, and (c) University of Cape Town (UCT).

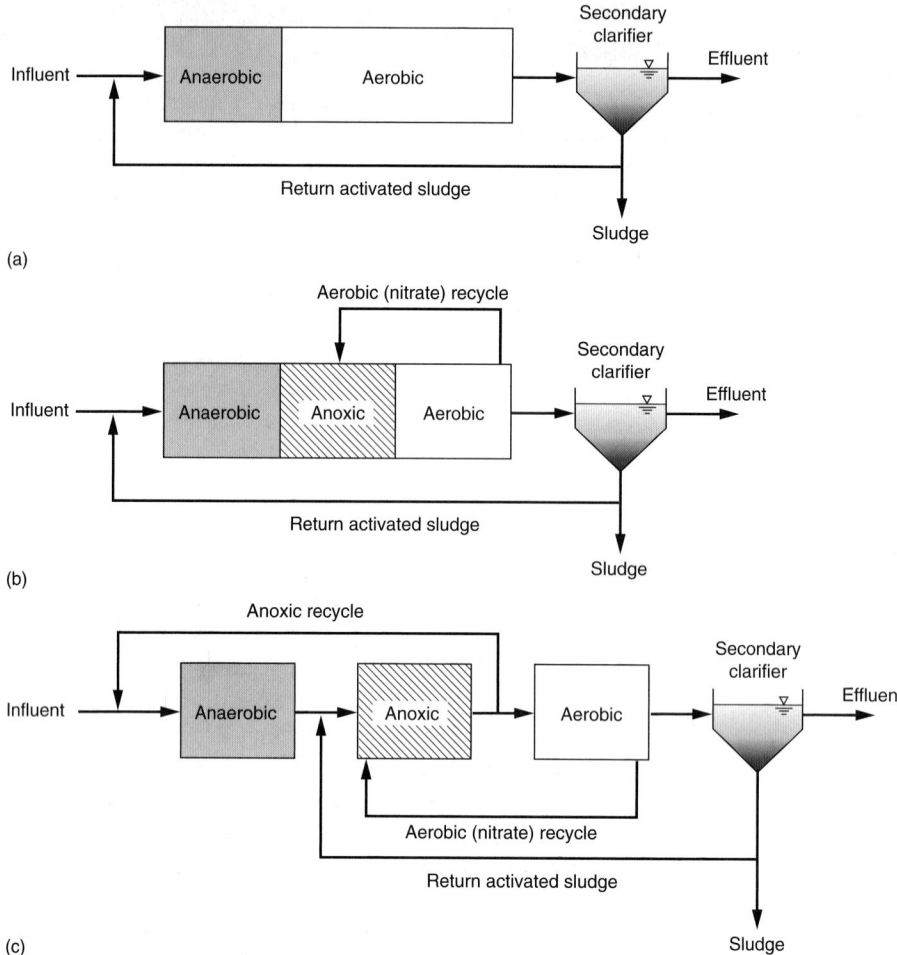

(a)

(b)

(c)

zone (see Fig. 8–29b). In the UCT process (see Fig. 8–29c), the RAS recycle is directed instead to an anoxic zone, and the mixed-liquor recycle to the anaerobic zone is drawn following the anoxic zone where the nitrate concentration is minimal. The UCT and similar processes (see later discussion in this section) are used generally for relatively weak wastewaters where the addition of nitrate would have a significant effect on the BPR performance.

The PhoStrip™ process combines biological and chemical processes for phosphorus removal. A portion of phosphorus-rich return activated sludge is diverted to an anaerobic stripping tank where phosphorus is released in solution (see Fig. 8–30). The phosphorus-rich stripper supernatant is then precipitated with lime and the stripped biomass is returned to the aeration tank. In the PhoStrip process, phosphorus-removal efficiency depends less on the influent rbCOD concentrations than for other biological phosphorus-removal processes (WEF, 1998).

Figure 8–30

PhoStrip sidestream biological treatment process for the removal of phosphorus.

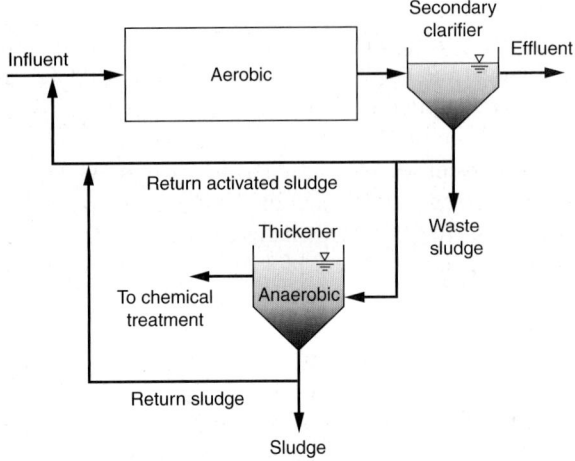

Process Design Considerations

The process design considerations for BPR processes include (1) wastewater characteristics, (2) anaerobic contact time, (3) SRT, (4) waste sludge processing method, and (5) chemical addition capability. The reaction kinetics for BOD removal, nitrification, and denitrification are similar to those discussed in Secs. 8–4 and 8–5.

Wastewater Characteristics.   Wastewater characterization, including rbCOD measurements, is essential to evaluate fully the design and performance of BPR systems. Biological phosphorus removal is initiated in the anaerobic zone where acetate (and propionate) is taken up by phosphorus-storing bacteria and converted to carbon storage products that provide energy and growth in the subsequent anoxic and aerobic zones. The rbCOD is the primary source of volatile fatty acids (VFAs) for the phosphorus-storing bacteria. The conversion of rbCOD to VFAs occurs quickly through fermentation in the anaerobic zone and 7 to 10 mg of acetate results in about 1.0 mg P removal by enhanced phosphorus removal (Wentzel et al., 1989, 1990). The more acetate, the more cell growth, and, thus, more phosphorus removal. Because of the need for organic material for nitrate removal, the amount of rbCOD relative to the amount of TKN in the influent is also an important wastewater parameter.

The diurnal variation in wastewater strength is also an important process consideration. Because the performance of phosphorus-storing bacteria depends on the availability of fermentation substrates, it is important to know if periods of low influent wastewater strength may affect BPR performance. For domestic wastewaters, the influent total BOD and rbCOD concentrations will vary with time over a 24-h period, with lower concentrations in the late evening and early morning hours. For smaller-sized communities, the variations are usually more pronounced and very little rbCOD may be present at certain times. During wet-weather conditions, especially in the winter, BPR may be difficult to achieve due to cold, low strength wastewater that does not readily become anaerobic. Extended periods of reduced rbCOD concentration have been reported to decrease BPR performance for a number of hours after the occurrence of low substrate concentration (Stephens and Stensel, 1998). The impact of continuous acetate feeding at plants where sludge fermentation has been done to produce additional VFAs has shown the benefit of a steady supply of rbCOD for biological phosphorus removal. In parallel modified Bardenpho trains at Kelowna, Canada, one train was fed fermentation liquor and the other train was used as the control. With continuous VFA addition, the effluent soluble phosphorus concentration decreased from 2.5 to 0.3 mg/L (Oldham and Stevens, 1985), and the VFA/P ratio was 6.7 g/g, an amount lower than the estimated 7 to 10 g/g. Based on these results, it appears that continuous acetate addition may provide more efficient biological phosphorus removal.

Anaerobic Contact Time. The role of the anaerobic contact zone has been described in Sec. 7–11. Detention times of 0.25 to 1.0 h are adequate for fermentation of rbCOD. To account for the effect of the MLVSS concentration in the anaerobic contact zone, a 1-d SRT is recommended for the anaerobic contact zone design (Grady et al., 1999). Barnard (1984) cautioned against using too long an anaerobic contact time due to the potential for a *secondary release* of phosphorus, which is phosphorus release not associated with acetate uptake. When secondary release occurs, bacteria have not accumulated polyhydroxybutyrate (PHB) for subsequent oxidation in the aerobic zone. Polyhydroxybutyrate provides energy for phosphorus uptake and storage. From SBR bench-scale studies, Stephens and Stensel (1998) found that secondary phosphorus release occurred for anaerobic contact times in excess of 3.0 h.

Solids Retention Time. Biological nutrient-removal systems with longer SRTs are less efficient for BPR than shorter SRT designs. Observations of the influent BOD to phosphorus removal ratio for a number of plants as a function of their design SRT are summarized in Table 8–24 (Grady et al., 1999). Processes with longer SRT values had less biological phosphorus removal for a given amount of influent BOD. These data were obtained without measurements of the rbCOD concentration in the wastewater, but provide a general trend on the effect of SRT on BPR removal efficiency. Two adverse effects on phosphorus removal efficiency are associated with lightly loaded, long SRT processes. First, because the final amount of phosphorus removed is proportional to the amount of biological phosphorus-storing bacteria wasted, the phosphorus-storing biomass production is lower so that less phosphorus can be removed. Second, at long SRTs the biological phosphorus bacteria are in a more extended endogenous phase, which will deplete more of their intracellular storage products. If the intracellu-

Table 8–24
Summary of observed influent BOD and COD to phosphorus-removal ratios for different BPR processes[a]

Type of BPR process[b]	BOD/P ratio, g BOD/g P	COD/P ratio, g COD/g P	SRT, d
Phoredox, VIP	15–20	26–34	<8.0
A²O, UCT	20–25	34–43	7–15
Bardenpho	>25	>43	15–25

[a] Adapted from Grady et al. (1999).
[b] For process descriptions, see Table 8–25.

lar glycogen is depleted, less efficient acetate uptake and PHB storage will occur in the anaerobic contact zone, thus making the overall BPR process less efficient (Stephens and Stensel, 1998).

Waste Sludge Processing. Because phosphorus is removed in the sludge wasted from BPR processes, consideration must be given to the waste sludge processing methods and the potential to recycle excessive amounts of phosphorus back to the BPR process. In Chap. 7, review of the biological phosphorus-removal mechanism showed that phosphorus is released when the bacteria that contain stored phosphorus are subject to anaerobic conditions. Anaerobic conditions in thickening and/or digestion can thus result in the release of significant amounts of phosphorus. The recycle stream from these processes would, in essence, increase the influent phosphorus concentration that would then require a greater amount of influent rbCOD to produce low effluent phosphorus concentrations.

Thickening of waste sludge by dissolved air flotation, gravity belt thickeners, or rotary-drum thickeners is preferred over gravity thickening of waste sludge to minimize phosphorus release. Phosphorus release would be expected from anaerobic and aerobic digestion processes as well. However, less phosphorus release and recycle have been observed than expected for anaerobic and aerobic digestion evaluations (Randall et al., 1992). Based on bench-scale aerobic digestion studies on sludge from a BPR process at the Little Patuxent wastewater treatment plant (WWTP) in Howard County, MD, only 20 percent P release was observed. At the York River, VA, WWTP, only 27 percent of the phosphorus removed was released in an anaerobic digester. The formation of phosphorus precipitates, such as struvite and brushnite, was credited with keeping phosphorus out of solution. Direct land application of liquid, digested sludge or dewatered raw sludge followed by stabilization such as composting minimizes recycled phosphorus loads.

Chemical Addition Capability. Many BPR facility designs include provisions for phosphorus removal by chemical precipitation in addition to biological removal. Where there are insufficient amounts of rbCOD in the influent wastewater, chemical addition is necessary to provide enough phosphorus removal to meet the effluent discharge concentration needed. Alum or iron salts may be used (see Chap. 6) and may be applied at a number of locations in the liquid stream treatment process. Where effluent filtration is used and the additional amount of phosphorus to be removed is small (less than 2.0 mg/L), chemicals may be added and mixed with the flow before filtration. Chemical addition prior to the secondary clarifier is also possible. Where primary treatment is used, alum or iron salts may be added to remove phosphorus prior to the biological process. The biological process is then depended upon to remove the phosphorus to the low concentration required. Present experience indicates that the phosphorus can be removed biologically to dissolved concentrations as low as 0.20 to 0.30 mg/L, provided sufficient rbCOD is available. Iron salts may be preferred in some cases over alum salts for primary treatment applications, because they have the additional advantage of removing sulfide to help reduce odors.

The choice of the chemical addition point can affect the chemical dosage. When added for polishing to achieve low effluent phosphorus concentrations, the metal salt is added at dosages well above the stoichiometric ratio, which is the theoretical amount needed to form a metal/phosphorus precipitate compound. By adding the metal salt before the biological process, stoichiometric ratios can be used to minimize the chemical dose needed.

Process Control

BPR performance is not just dependent on placing an anaerobic zone in front of the aerobic zone of an activated-sludge process and the amount of rbCOD in the influent wastewater. Process performance is affected by a number of operating conditions including (1) nitrate-removal efficiency in processes in which nitrification occurs, (2) process SRT, (3) control of dissolved oxygen entering the anaerobic zone, (4) phosphorus in recycle streams, and (5) the system effluent suspended solids concentration.

Effect of Dissolved Oxygen and Nitrate in Recycle Flows. Recycle flows to the anaerobic contact zone must be evaluated in terms of their possible effects on BPR, and some recycle streams should be avoided where possible. Filter backwash recycle flows should be sent to the aerobic zone instead of the anaerobic or anoxic zones. Recycle streams with significant concentrations of DO and nitrate can have an adverse impact on process performance. The nitrate concentration in the RAS flow can have a significant effect on the amount of influent rbCOD that is available for BPR. Assuming a synthesis yield of 0.4 g VSS/g rbCOD removed, the amount of rbCOD used by nitrate and oxygen fed to the anaerobic zone can be estimated as follows:

The oxygen used for rbCOD oxidation equals the rbCOD removed minus the COD in the biomass formed:

$$\left(\frac{1.0 \text{ g O}_2}{\text{g rbCOD}} \right) - \left(\frac{1.42 \text{ g O}_2}{\text{g VSS}} \right)\left(\frac{0.4 \text{ g VSS}}{\text{g rbCOD}} \right) = \left(\frac{0.43 \text{ g O}_2}{\text{g rbCOD}} \right) \tag{8–59}$$

or 2.3 g rbCOD used/g DO added

Based on COD used for dissolved oxygen, the COD used for nitrate consumption can be estimated, as the oxygen equivalent of NO_3-N is 2.86 g/g as follows:

$$\frac{\text{g rbCOD}}{\text{g NO}_3\text{-N}} = \left(\frac{2.3 \text{ g rbCOD}}{\text{g O}_2} \right)\left(\frac{2.86 \text{ g O}_2}{\text{g NO}_3\text{-N}} \right) = 6.6 \text{ g rbCOD/g NO}_3\text{-N} \tag{8–60}$$

Based on the above rbCOD/NO_3-N ratio and rbCOD/DO ratio, the impact of DO and nitrate fed into the anaerobic contact zone on the BPR performance can be evaluated. The rbCOD in the influent wastewater added to the anaerobic zone will most likely be removed by bacteria using oxygen and nitrate before it is available for biological phosphorus removal.

Effect of Recycle Streams with Released Phosphorus. As discussed under Process Design Considerations, recycle streams from sludge thickening or digestion processes may contain high phosphorus concentrations. Equalization and control of the recycle flow and phosphorus load with time may help to minimize the impact of the recycled phosphorus on effluent quality. By adding the recycle streams during times of the day when the influent wastewater strength is higher, a better possibility exists of removing recycled phosphorus in the waste sludge. Recycle streams may also be treated separately with chemical addition to minimize the phosphorus load to the liquid treatment process.

Effluent Suspended Solids. The phosphorus content in the mixed-liquor solids is greater than that from the conventional activated-sludge process due to the biological phosphorus storage. The phosphorus content, on a dry solids basis, may be in the range of 3 to 6 percent (Randall et al., 1992). Thus, the total phosphorus concentration in the effluent can be affected significantly by the system effluent TSS concentration. At 3 to 6 percent phosphorus in the solids, the phosphorus contribution in an effluent having a TSS concentration of 10 mg/L would be 0.3 to 0.6 mg/L, values that are significant if the effluent standard is less than 1.0 mg P/L. Fortunately, most BPR processes exhibit good settling characteristics and have secondary clarifier effluent TSS concentrations of 10 mg/L or less. To provide very low effluent phosphorus concentrations, effluent filtration may be required.

Solids Separation Facilities

Operation of the solids separation facilities affects the process design and performance. If chemical addition is necessary for effluent polishing to achieve low phosphorus levels in the effluent, sufficient clarifier capacity is required to handle the additional chemical precipitate. Typical clarifier overflow rates for biological phosphorus removal are also given in Table 8–7.

Methods to Improve Phosphorus-Removal Efficiency in BPR Systems

The performance of BPR systems is very site-specific and depends on the wastewater characteristics and the plant process design and operation. For wastewaters with relatively low influent rbCOD concentrations, effluent soluble phosphorus concentrations may exceed 1.0 to 2.0 mg/L, whereas effluent concentrations below 0.5 or 1.0 mg/L have been shown for higher-strength wastewaters. Methods to improve performance for overall phosphorus removal include the following:

1. Provide supplemental acetate by direct purchase or by primary sludge fermentation.
2. Reduce the process SRT.
3. Add alum or iron salts in primary treatment or for effluent polishing.
4. Reduce the amount of nitrate and/or oxygen entering the anaerobic zone.

Two methods used to provide additional rbCOD for biological phosphorus removal are to import (purchase) an exogenous carbon source (i.e., acetate) or to produce VFAs from fermentation of primary clarifier sludge. Two examples of primary sludge fermentation design are shown on Fig. 8–31; other designs are also possible. As shown on Fig. 8–31a, a fermentation reactor provides residence time and mixing of the primary sludge, and VFAs are released through the primary clarifier for feed to the secondary treatment process anaerobic zone. A deeper depth primary clarifier design (Fig. 8–31b) has also been proposed to provide sufficient holding time for the settled primary sludge for hydrolysis and acid fermentation (Barnard, 1984). The underflow sludge is recycled to release VFAs to the liquid stream. Operating issues of odors, mixing, and accumulation of rags in the fermenters must also be considered. The primary tank sludge fermenters are not heated and SRT values ranging from 3 to 5 d, depending on temperature, are generally used to stay below the point where methanogenic activity can start (Rabinowitz and Oldham, 1985). At SRT values greater than 4 to 5 d, methanogenesis activity can be high enough to consume the VFAs. The VFA production ranges from

Figure 8–31

Examples of fermentation reactors for producing volatile fatty acids (VFAs) used for phosphorus removal.

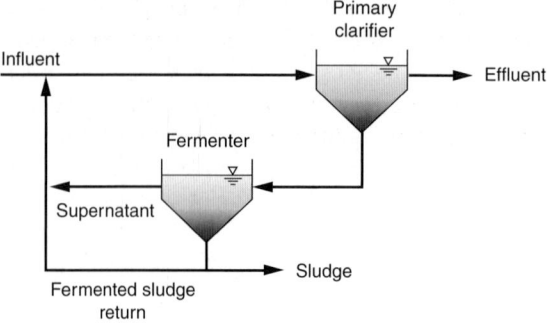

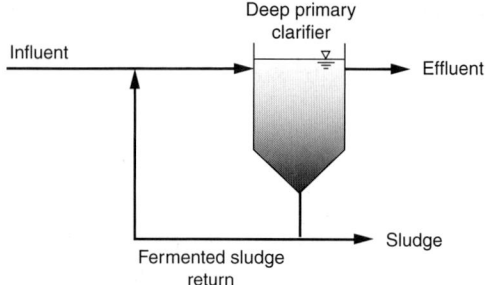

0.1 to 0.2 g VFA/g VSS applied to the fermenters. Fermenters can add an additional 10 to 20 mg/L VFA concentration to the influent wastewater, and lower, more consistent effluent soluble phosphorus concentrations have been observed when fermenters are used.

A good example of process changes that can improve BPR performance has been shown for the Kelowna, British Columbia, plant (Oldham and Stevens, 1985), which was one of the first noted successes with primary sludge fermentation. A two-train modified Bardenpho™ facility is used to provide BPR. The first change was to ferment primary sludge in existing tanks that were operated as sludge thickeners with long holding times. The thickener overflow was fed to the Bardenpho anaerobic zone. The fermenter effluent VFA concentration ranged from 110 to 140 mg/L and, when mixed with the primary effluent, resulted in a VFA concentration of 9 to 10 mg/L. The additional VFA supply to the BPR process decreased the average effluent phosphorus concentration from about 1.5 to 0.5 mg/L. Alum was later added before the secondary clarifiers at a dosage of about 8 mg/L (as alum) to further reduce the effluent phosphorus concentration. With both prefermentation and alum addition, the effluent phosphorus concentration averaged less than 0.20 mg/L for the period from September 1989 to August 1990. During 1993, alum addition was stopped and the last two anoxic-aerobic stages of the Bardenpho™ system were taken out of service, when a higher effluent nitrate concentration was allowed. Removing the latter two stages resulted in a lower SRT for the process, and the average effluent phosphorus concentration decreased to 0.10 mg/L. The other benefit claimed for removing the second anoxic zone was that it had longer than necessary detention times for nitrate removal. The second anoxic zone was suspected of causing undesired secondary phosphorus release.

Biological Phosphorus-Removal Process Performance

The following example is designed to illustrate the methodology used in evaluating the performance of a BPR process. Because biological phosphorus- and nitrogen-removal processes are very complex with many dependent interactions, comprehensive simulation models (Barker and Dold, 1997) are very useful for evaluating biological phosphorus- and nitrogen-removal designs.

EXAMPLE 8–10 **Biological Phosphorus Removal** An A^{2}O™ biological nutrient-removal process receives wastewater with the characteristics shown. The RAS contains 6.0 mg/L NO_3-N and the system is operated at a 7 d SRT. The RAS recycle ratio is 0.5. The anaerobic contact detention time is 0.75 h. Estimate the effluent soluble phosphorus concentration and the phosphorus content of the waste sludge.

Design conditions and assumptions:

1. Wastewater characteristics:

Item	Unit	Value
Flowrate	m³/d	4000
Total BOD	g/m³	140
rbCOD	g/m³	70
nbVSS	g/m³	20
Inorganic inert matter	g/m³	10
TKN	g/m³	35
NO_3-N	g/m³	0
Phosphorus	g/m³	7
Temperature	°C	20

Note: g/m³ = mg/L.

2. 10 g rbCOD/g P is removed by biological phosphorus removal
3. rbCOD/nitrate ratio = 6.6 g rbCOD/g NO_3-N
4. Phosphorus content of heterotrophic biomass = 0.015 g P/g biomass
5. bCOD/BOD ratio = 1.6
6. $NO_x \approx 0.80$ TKN
7. Use coefficients from Tables 8–10 and 8–11

Solution

1. Determine the rbCOD available for biological phosphorus removal using a mass balance at the influent to the reactor.
   a. Develop a mass balance for nitrate

$$Q_{RAS}(NO_3\text{-}N)_{inf} + Q_{RAS}(NO_3\text{-}N)_{RAS} = (Q + Q_{RAS})(NO_3\text{-}N)_{React}$$

where $Q_{RAS} = 0.5Q$
$(NO_3\text{-}N)_{React}$ = nitrate feed to reactor

$$Q(0) + 0.5Q(6 \text{ g/m}^3) = 1.5Q(NO_3\text{-N})_{React}$$

$$(NO_3\text{-N})_{React} = 2 \text{ g}(NO_3\text{-N})/\text{m}^3$$

b. Determine rbCOD available for P removal

$$\text{rbCOD equivalent} = [2 \text{ g}(NO_3\text{-N})/\text{m}^3](6.6 \text{ g rbCOD/g})(NO_3\text{-N})$$

$$= 13.2 \text{ g/m}^3$$

rbCOD available for P removal $= (70 - 13.2) \text{ g/m}^3 = 56.8 \text{ g/m}^3$

2. Phosphorus removed by BPR mechanism

$$\text{Biological P removal} = \frac{(56.8 \text{ g rbCOD/m}^3)}{(10 \text{ g rbCOD/g P})} = 5.7 \text{ g P/m}^3$$

3. Determine phosphorus used for heterotrophic biomass synthesis in addition to phosphorus storage due to BPR.

Biomass production [Eq. (8–15)]: $P_{X,bio} = \dfrac{Q(Y)(S_o - S)}{[1 + (k_d)\text{SRT}]} + \dfrac{Q(Y_n)(NO_x)}{[1 + (k_{dn})\text{SRT}]}$

Assume $S_o \simeq S_o - S$

$\text{bCOD} = 1.6 \text{ BOD} = 1.6(140 \text{ g/m}^3) = 224 \text{ g/m}^3$

$NO_x = 0.80 \text{ TKN} = 0.80(35 \text{ g/m}^3) = 28 \text{ g/m}^3$

$$P_{x,bio} = \frac{(4000 \text{ m}^3/\text{d})(0.4 \text{ g/g})(224 \text{ g/m}^3)}{[1 + (0.12 \text{ g/g·d})(7 \text{ d})]}$$

$$+ \frac{(4000 \text{ m}^3/\text{d})(0.12 \text{ g/g})(28 \text{ g/m}^3)}{[1 + (0.08 \text{ g/g·d})(7 \text{ d})]}$$

$$= (194{,}783 + 8615) \text{ g/d} = 203{,}398 \text{ g/d}$$

P utilized for biomass growth:

P used $= (0.015 \text{ g P/g biomass})(203{,}398 \text{ g/d}) = 3015 \text{ g/d}$

$$\text{In g/m}^3 = \frac{(3015 \text{ g/d})}{(4000 \text{ m}^3/\text{d})} = 0.8 \text{ g/m}^3$$

4. Determine effluent soluble P.

P removed $= (5.7 + 0.8) \text{ g/m}^3 = 6.5 \text{ g/m}^3$

Effluent soluble P $= (7.0 - 6.5) \text{ g/m}^3 = 0.5 \text{ g/m}^3$

5. Determine P content of waste sludge.

Total P in sludge $= (6.5 \text{ g/m}^3)(4000 \text{ m}^3/\text{d})(1 \text{ kg}/10^3 \text{ g}) = 26.0 \text{ kg/d}$

Determine total sludge production using Eqs. (8–15) and (8–16).

$$P_{X,TSS} = \frac{(4000 \text{ m}^3/\text{d})(0.40 \text{ g/g})(224 \text{ g/m}^3)(1 \text{ kg}/10^3 \text{ g})}{[1 + (0.12 \text{ g/g·d})(7 \text{ d})](0.85)}$$

$$+ \frac{(4000 \text{ m}^3)(0.12 \text{ g/g})(28 \text{ g/m}^3)(1 \text{ kg}/10^3 \text{ g})}{[1 + (0.08 \text{ g/g·d})(7 \text{ d})](0.85)}$$

$$+ (20 \text{ g/m}^3)(4000 \text{ m}^3/\text{d})(1 \text{ kg}/10^3 \text{ g})$$

$$+ \frac{(0.15)(0.12 \text{ g/g·d})(0.40 \text{ g/g})(4000 \text{ m}^3/\text{d})(224 \text{ g/m}^3)(7 \text{ d})(1 \text{ kg}/10^3 \text{ g})}{[1 + (0.12 \text{ g/g·d})(7 \text{ d})](0.85)}$$

$$+ (10 \text{ g/m}^3)(4000 \text{ m}^3/\text{d})(1 \text{ kg}/10^3 \text{ g}) + \text{phosphorus stored}$$

$$P_{X,TSS} = (229.2 + 10.1 + 80.0 + 28.9 + 40) \text{ kg/d}$$

$$= 388.2 \text{ kg/d}$$

$$\text{Phosphorus, } \% = \frac{(26.0 \text{ kg/d})}{(388.2 \text{ kg/d})} \times 100 = 6.7$$

Alternative Processes for Biological Phosphorous Removal

Various modifications to the basic Phoredox process are used for both biological phosphorus and nitrogen removal. Most deal with different designs that are intended to minimize the amount of nitrate fed to the anaerobic zone, some involve short SRTs, and some involve using multiple stages for the anaerobic, anoxic, and aerobic zones.

The principal mainstream processes and the Phostrip sidestream process are described in Table 8–25.

Process Design Parameters

Typical parameters used in the design of biological phosphorus-removal systems are presented in Table 8–26.

Process Selection Considerations

The processes described for biological phosphorus removal all incorporate the necessary anaerobic contacting between influent wastewater and activated sludge, followed by an aerobic zone for bio-oxidation of stored PHB and phosphorus uptake by the polyphosphate-storing bacteria. The specific designs selected are governed more by other treatment needs such as BOD or nitrogen removal. The advantages and limitations of the biological phosphorus-removal processes are presented in Table 8–27.

Biological phosphorus-removal efficiency is affected by the overall activated-sludge process as well as the influent wastewater characteristics (Randall et al., 1992). Lower phosphorus-removal efficiency occurs for systems with longer SRTs, more nitrate and/or oxygen input to the anaerobic zone, and less readily biodegradable COD

Table 8–25

Description of suspended growth processes for phosphorus removal

Process	Description
(a) Phoredox (A/O) 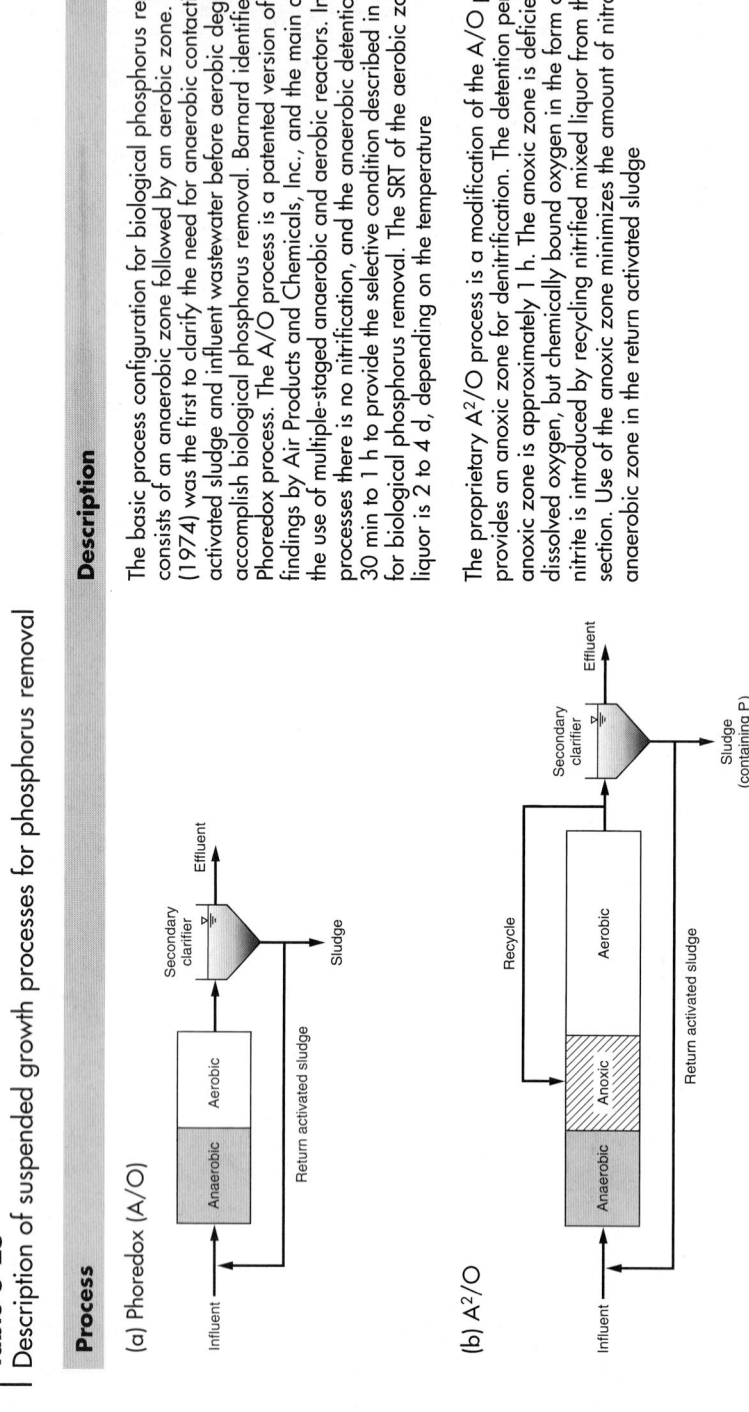	The basic process configuration for biological phosphorus removal consists of an anaerobic zone followed by an aerobic zone. Barnard (1974) was the first to clarify the need for anaerobic contacting between activated sludge and influent wastewater before aerobic degradation to accomplish biological phosphorus removal. Barnard identified it as the Phoredox process. The A/O process is a patented version of Barnard's findings by Air Products and Chemicals, Inc., and the main difference is the use of multiple-staged anaerobic and aerobic reactors. In these processes there is no nitrification, and the anaerobic detention time is 30 min to 1 h to provide the selective condition described in Sec. 7–11 for biological phosphorus removal. The SRT of the aerobic zone mixed liquor is 2 to 4 d, depending on the temperature
(b) A²/O	The proprietary A²/O process is a modification of the A/O process and provides an anoxic zone for denitrification. The detention period in the anoxic zone is approximately 1 h. The anoxic zone is deficient in dissolved oxygen, but chemically bound oxygen in the form of nitrate or nitrite is introduced by recycling nitrified mixed liquor from the aerobic section. Use of the anoxic zone minimizes the amount of nitrate fed to the anaerobic zone in the return activated sludge

The Bardenpho process, described in Table 8–21, can be modified for combined nitrogen and phosphorus removal. The staging sequence and recycle method are different from the A^2/O process. The 5-stage system provides anaerobic, anoxic, and aerobic stages for phosphorus, nitrogen, and carbon removal. A second anoxic stage is provided for additional denitrification using nitrate produced in the aerobic stage as the electron acceptor, and the endogenous organic carbon as the electron donor. The final aerobic stage is used to strip residual nitrogen gas from solution and to minimize the release of phosphorus in the final clarifier. Mixed liquor from the first aerobic zone is recycled to the anoxic zone. The 5-stage process uses a longer SRT (10 to 20 days) than the A^2/O process, and thus increases the carbon oxidation capability

The UCT process stands for the University of Cape Town (South Africa) treatment process where it was developed. The UCT process was developed to minimize the effect of nitrate in weaker wastewaters in entering the anaerobic contact zone. The amount of nitrate in the anaerobic zone is critical to the biological phosphorus-removal efficiency. The UCT process is similar to the A^2/O process with two exceptions. The return activated sludge is recycled to the anoxic stage instead of the aeration stage, and the internal recycle is from the anoxic stage to the anaerobic stage. By returning the activated sludge to the anoxic stage, the introduction of nitrate to the anaerobic stage is eliminated, thereby improving the uptake of phosphorus in the anaerobic stage. The internal recycle feature provides for increased organic utilization in the anaerobic stage. The mixed liquor from the anoxic stage contains substantial soluble BOD but little nitrate. The recycle of the anoxic mixed liquor provides for optimal conditions for fermentation uptake in the anaerobic stage. Because the mixed liquor is at a lower concentration, the anaerobic detention time must be longer than that used in the Phoredox process, and is in the range of 1 to 2 h. The anaerobic recycle rate is typically 2 times the influent flowrate

In the modified UCT process shown on the second diagram, the return activated sludge is directed to an anoxic reactor that does not receive internal nitrate recycle flow. The nitrate is reduced in this tank, and the mixed liquor from the reactor is recycled to the anaerobic tank. The second anoxic tank follows the first anoxic tank and receives internal nitrate recycle flow from the aeration tank to provide the major portion of nitrate removal for the process

(continued)

(c) Modified Bardenpho (5-stage)

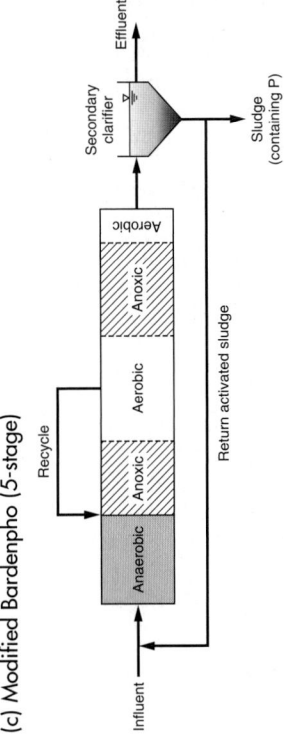

(d) UCT (standard and modified)

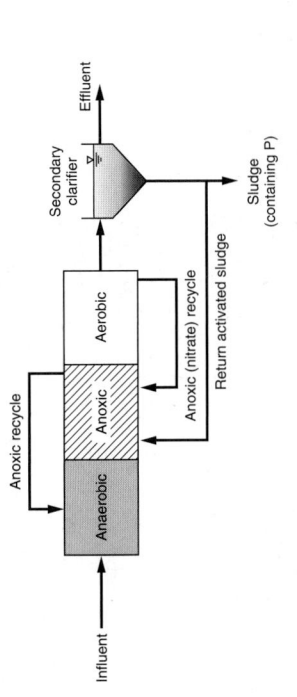

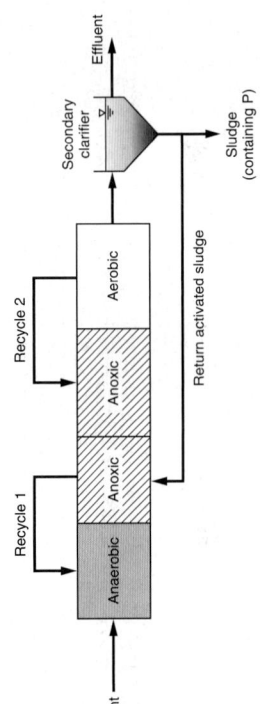

| **Table 8-25** (*Continued*) |

Process

Description

(e) VIP

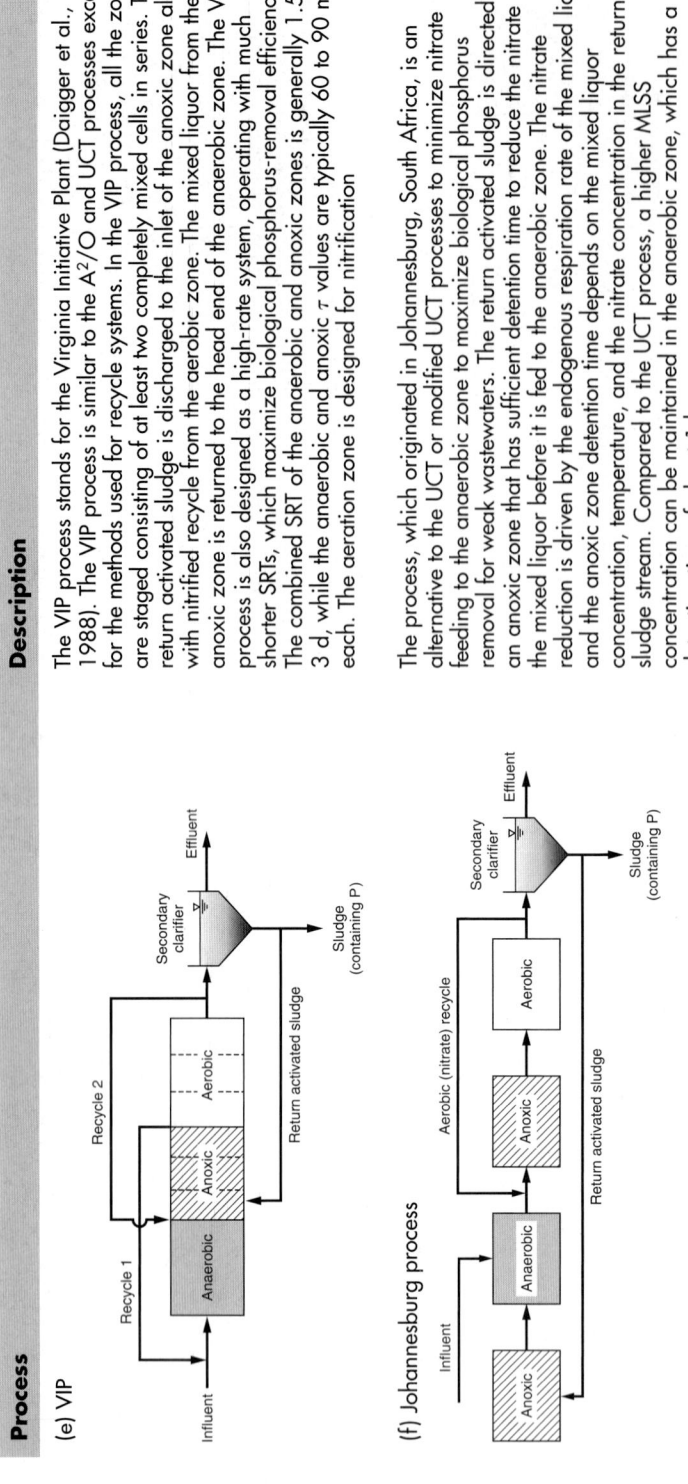

The VIP process stands for the Virginia Initiative Plant [Daigger et al., 1988). The VIP process is similar to the A²/O and UCT processes except for the methods used for recycle systems. In the VIP process, all the zones are staged consisting of at least two completely mixed cells in series. The return activated sludge is discharged to the inlet of the anoxic zone along with nitrified recycle from the aerobic zone. The mixed liquor from the anoxic zone is returned to the head end of the anaerobic zone. The VIP process is also designed as a high-rate system, operating with much shorter SRTs, which maximize biological phosphorus-removal efficiency. The combined SRT of the anaerobic and anoxic zones is generally 1.5 to 3 d, while the anaerobic and anoxic τ values are typically 60 to 90 min each. The aeration zone is designed for nitrification

(f) Johannesburg process

The process, which originated in Johannesburg, South Africa, is an alternative to the UCT or modified UCT processes to minimize nitrate feeding to the anaerobic zone to maximize biological phosphorus removal for weak wastewaters. The return activated sludge is directed to an anoxic zone that has sufficient detention time to reduce the nitrate in the mixed liquor before it is fed to the anaerobic zone. The nitrate reduction is driven by the endogenous respiration rate of the mixed liquor, and the anoxic zone detention time depends on the mixed liquor concentration, temperature, and the nitrate concentration in the return sludge stream. Compared to the UCT process, a higher MLSS concentration can be maintained in the anaerobic zone, which has a detention time of about 1 h

If sufficient nitrate is removed during the SBR operation, an anaerobic reaction period can be developed during and after the SBR fill period. An anoxic operating period is used after a sufficient aerobic time elapses for nitrification and nitrate production. Alternatively cyclic aerobic and anoxic periods can be used during the react period. The nitrate concentration is thus minimized before settling, and little nitrate is available to compete for rbCOD in the fill and initial react period. Thus, anaerobic conditions occur in the fill and initial react period, so that rbCOD uptake and storage by phosphorus-accumulating bacteria can occur instead of rbCOD consumption by nitrate-reducing bacteria

The PhoStrip process is in essence an anaerobic/aerobic process. The anaerobic condition is created by holding RAS long enough in a gravity thickener with residence times generally in the range of 8 to 12 h (Levin et al., 1975). The released phosphorus is elutriated, usually by adding primary effluent or raw wastewater, which also enhances the anaerobic condition. The overflow from the stripper tank is then treated chemically for phosphorus removal and the RAS is directed to the aerobic tank. Lime is usually used as a chemical to precipitate phosphorus from the stripper tank overflow. The lime dose needed to raise the pH for phosphorus removal is a function of the wastewater alkalinity and not the amount of phosphorus present. If alum and ferric salts are used instead, the dose is proportional to the amount of phosphorus released. Even though chemical treatment is used for phosphorus removal, there is some enhanced phosphorus removal in the waste sludge due to the development of phosphorus-storing bacteria. As shown, the phosphorus-removal performance would be hindered by nitrification

(g) SBR with biological phosphorus removal

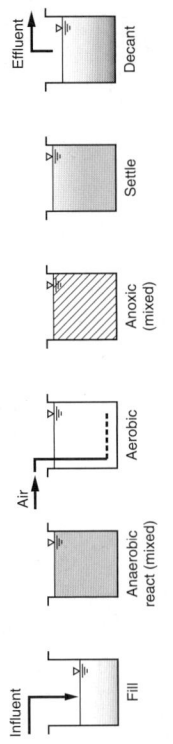

(h) PhoStrip

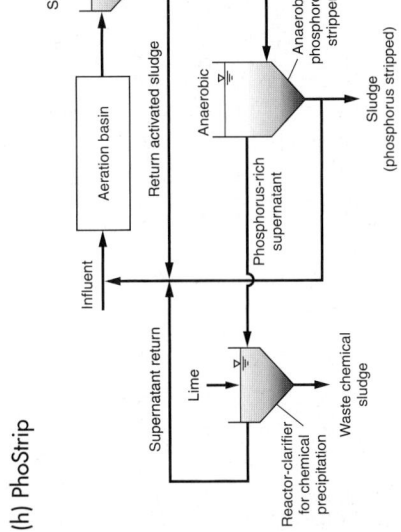

Table 8-26
Typical design parameters for commonly used biological phosphorus-removal processes[a]

Design parameter/process	SRT, d	MLSS, mg/L	τ, h			RAS, % of influent	Internal recycle, % of influent
			Anaerobic zone	Anoxic zone	Aerobic zone		
A/O	2–5	3000–4000	0.5–1.5	—	1–3	25–100	
A²/O	5–25	3000–4000	0.5–1.5	0.5–1	4–8	25–100	100–400
UCT	10–25	3000–4000	1–2	2–4	4–12	80–100	200–400 (anoxic) 100–300 (aerobic)
VIP	5–10	2000–4000	1–2	1–2	4–6	80–100	100–200 (anoxic) 100–300 (aerobic)
Bardenpho (5-stage)	10–20	3000–4000	0.5–1.5	1–3 (1st stage) 2–4 (2nd stage)	4–12 (1st stage) 0.5–1 (2nd stage)	50–100	200–400
PhoStrip	5–20	1000–3000	8–12		4–10	50–100	10–20
SBR	20–40	3000–4000	1.5–3	1–3	2–4		

[a] Adapted from WEF (1998).

Table 8–27

Advantages and limitations of phosphorus-removal processes

Process	Advantages	Limitations
Phoredox (A/O)	Operation is relatively simple when compared to other processes Low BOD/P ratio possible Relatively short hydraulic retention time Produces good settling sludge Good phosphorus removal	Phosphorus removal declines if nitrification occurs Limited process control flexibility is available
A²O	Removes both nitrogen and phosphorus Provides alkalinity for nitrification Produces good settling sludge Operation is relatively simple Saves energy	RAS containing nitrate is recycled to anaerobic zone, thus affecting phosphorus-removal capability Nitrogen removal is limited by internal recycle ratio Needs higher BOD/P ratio than the A/O process
UCT	Nitrate loading on anaerobic zone is reduced, thus increasing phosphorus-removal capability For weaker wastewater, process can achieve improved phosphorus removal Produces good settling sludge Good nitrogen removal	More complex operation Requires additional recycle system
VIP	Nitrate loading on anaerobic zone is reduced, thus increasing phosphorus-removal capability Produces good settling sludge Requires lower BOD/P ratio than UCT	More complex operation Requires additional recycle system More equipment required for staged operation
Bardenpho (5-stage)	Can achieve 3 to 5 mg/L TN in unfiltered effluent Produces good settling sludge	Less efficient phosphorus removal Requires larger tank volumes
SBR	Both nitrogen and phosphorus removal are possible Process is easy to operate Mixed-liquor solids cannot be washed out by hydraulic surges Quiescent settling may produce lower effluent TSS concentration Flexible operation	More complex operation for N and P removal Needs larger volume than SBR for N removal only Effluent quality depends upon reliable decanting facility Design is more complex Skilled maintenance is required More suitable for smaller flowrates
PhoStrip	Can be incorporated easily into existing activated-sludge plants Process is flexible; phosphorus-removal performance is not controlled by BOD/phosphorus ratio Significantly less chemical usage than mainstream chemical precipitation process Can achieve reliable effluent orthophosphate concentrations less than 1 mg/L	Requires lime addition for phosphorus precipitation Requires higher mixed-liquor dissolved oxygen to prevent phosphorus release in final clarifier Additional tank capacity required for stripping Lime scaling may be a maintenance problem

in the influent wastewater. Shorter SRT systems like the VIP or Phoredox processes provide more biological phosphorus removal for the same amount of influent BOD than the modified Bardenpho process, for example. In some applications, other sources of rbCOD are obtained to enhance biological phosphorus removal.

8–7 SELECTION AND DESIGN OF PHYSICAL FACILITIES FOR ACTIVATED-SLUDGE PROCESSES

The physical facilities used in the design of activated-sludge treatment systems are presented and discussed in this section. The subjects discussed include (1) the aeration system, (2) aeration tanks and appurtenances, (3) solids separation, and (4) solids separation facilities. A recently developed technology, membrane separation, is also finding application for biological treatment and solids separation and is discussed in Sec. 8–8 and in Chap. 11.

Aeration System

The aeration system design for the activated-sludge process must be adequate to (1) satisfy the bCOD of the waste, (2) satisfy the endogenous respiration by the biomass, (3) satisfy the oxygen demand for nitrification, (4) provide adequate mixing, and (5) maintain a minimum dissolved oxygen concentration throughout the aeration tank. If the oxygen transfer efficiency of the aeration system is known or can be estimated, the actual air requirements for diffused air aeration or installed power of mechanical surface aerators may be determined. The characteristics of air diffusers and the energy requirements for mixing for diffused air and mechanical aeration systems are discussed in Sec. 5–12 in Chap. 5.

To meet sustained organic loadings at peak conditions discussed in Chap. 3, aeration equipment should be designed with a peaking factor of at least 1.5 to 2.0 times the average BOD load. Aeration equipment should also be sized based on a residual dissolved oxygen (DO) of 2 mg/L in the aeration tank at the average load and 1.0 mg/L at peak load. The Ten States Standards [GLUMRBSS (1978)] require the aeration system to be capable of providing oxygen to meet the diurnal peak oxygen demand or 200 percent of the design average, whichever is larger with a residual DO of 2.0 mg/L. The aeration equipment must be designed with enough flexibility to (1) meet minimum oxygen demands, (2) prevent excessive aeration and save energy, and (3) meet maximum oxygen demands.

Aeration Tanks and Appurtenances

After the activated-sludge process and the aeration system have been selected and a preliminary design has been prepared, the next step is to design the aeration tanks and support facilities. The following discussion covers (1) aeration tanks, (2) flow distribution, and (3) froth control systems. Energy requirements for aeration tank mixing are discussed in Sec. 5–12 in Chap. 5.

Aeration Tanks. Aeration tanks usually are constructed of reinforced concrete and left open to the atmosphere. A sectional view of a typical aeration tank using porous tube diffusers is shown on Fig. 8–32. The rectangular shape permits common-wall con-

Figure 8–32

View of a typical plug-flow activated-sludge reactor with porous diffusers located near the bottom of reactor on left side and shallow coarse bubble diffusers located on right side.

struction for multiple tanks. The total tank capacity required should be determined from the biological process design. For plants in a capacity range of 0.22 to 0.44 m³/s (5 to 10 Mgal/d), at least two tanks should be provided (a minimum of two tanks is preferred for smaller plants as well, for redundancy). In the range of 0.44 to 2.2 m³/s (10 to 50 Mgal/d), four tanks are often provided to allow operational flexibility and ease of maintenance. Large plants, over 2.2 m³/s (50 Mgal/d) in capacity, should contain six or more tanks. Some of the largest plants have from 30 to 40 tanks arranged in several groups or batteries. Although the air bubbles dispersed in the wastewater occupy perhaps 1 percent of the total volume, no allowance is made for this in tank sizing.

If the wastewater is to be aerated with diffused air, the geometry of the tank may significantly affect the aeration efficiency and the amount of mixing obtained. The depth of wastewater in the tank should be between 4.5 and 7.5 m (~15 and 25 ft) to maximize the energy efficiency of the diffuser systems. Freeboard from 0.3 to 0.6 m (1 to 2 ft) above the waterline should be provided. The width of the tank in relation to its depth is important if spiral-flow mixing is used in the plug-flow configuration. The width-to-depth ratio for such tanks may vary from 1.0:1 to 2.2:1, with 1.5:1 being the most common. In large plants, the channels become quite long and sometimes exceed 150 m (~500 ft) per tank. Tanks may consist of one to four channels with round-the-end flow in multiple-channel tanks. The length-to-width ratio of each channel should be at least 5:1. Where complete-mix diffused air systems are used, the length-to-width ratio may be reduced to save construction cost.

For tanks with diffusers on both sides or in a grid or panel pattern, greater widths are permissible. The important point is to restrict the width of the tank so that "dead spots" or zones of inadequate mixing are avoided. The dimensions and proportions of each independent unit should be such as to maintain adequate velocities so that deposition of solids will not occur. In spiral-flow tanks, triangular baffles or fillets may be placed longitudinally in the corners of the channels to eliminate dead spots and to deflect the spiral flow.

Aerator size		Tank depth		Tank width	
hp	kW	ft	m	ft	m
10	7.5	10–12	3–3.6	30–40	9–12
20	15	12–14	3.6–4.2	35–50	10.5–15
30	22.5	13–15	3.9–4.5	40–60	12–18
40	30	12–17	3.6–5.1	45–65	13.5–20
50	37.5	15–18	4.5–5.5	45–75	13.5–23
75	56	15–20	4.5–6	50–85	15–26
100	75	15–20	4.5–6	60–90	18–27

For mechanical aeration systems, the most efficient arrangement is one aerator per tank. Where multiple aerators are installed in the same tank for best efficiency, the length-to-width ratio of the tank should be in even multiples with the aerator centered in a square configuration to avoid interference at the hydraulic boundaries. The width and depth should be sized in accordance with the power rating of the aerator as illustrated in Table 8–28. Two-speed aerators are desirable to provide operating flexibility to cover a wide range of oxygen demand conditions. Freeboard of about 1 to 1.5 m (3.5 to 5 ft) should be provided for mechanical aeration systems.

Individual tanks should have inlet and outlet gates or valves so that they may be removed from service for inspection and repair. The common walls of multiple tanks must therefore be able to withstand the full hydrostatic pressure from either side. Aeration tanks must have adequate foundations to prevent settlement, and, in saturated soil, they must be designed to prevent flotation when the tanks are dewatered. Methods of preventing flotation include thickening the floor slab, installing hold-down piles, or installing hydrostatic pressure relief valves. Drains or sumps for aeration tanks are desirable for dewatering. In large plants where tank dewatering might be more common, it may be desirable to install mud valves in the bottoms of all tanks. The mud valves should be connected to a central dewatering pump or to a plant drain discharging to the wet well of the plant pumping station. Dewatering systems are commonly designed to empty a tank in 12 to 24 h.

Flow Distribution.   For wastewater treatment plants containing multiple units of primary sedimentation basins and aeration tanks, consideration has to be given to equalizing the distribution of flow to the aeration tanks. In many designs, the wastewater from the primary sedimentation basins is collected in a common conduit or channel for transport to the aeration tanks. For efficient use of the aeration tanks, a method of splitting or controlling the flowrate to each of the individual tanks should be used. Methods commonly used are splitter boxes equipped with weirs or control valves or aeration tank influent control gates. Hydraulic balancing of the flow by equalizing the headloss from the primary sedimentation basins to the individual aeration tanks is also practiced. Flow regimes using a form of step feed particularly need a positive means of flow control. Where channels are used for aeration tank influent or effluent transport, they can be equipped with aeration devices to prevent deposition of solids. The air required ranges from 0.2 to 0.5 m³/lin m·min (2 to 5 ft³/lin ft·min) of channel.

Figure 8–33

Typical non-*Nocardia* froth on activated-sludge aeration tank sludge return channels.

Froth Control Systems. Wastewater normally contains soap, detergents, and other surfactants that produce foam when the wastewater is aerated. If the concentration of mixed-liquor suspended solids is high, the foaming tendency is minimized. Large quantities of foam may be produced during startup of the process, when surfactants are present in the wastewater. The foaming action produces a froth that contains sludge solids, grease, and large numbers of wastewater bacteria (see Fig. 8–33). The wind may lift the froth off the tank surface and blow it about, contaminating whatever it touches. The froth, besides being unsightly, is a hazard to those working with it because it is very slippery, even after it collapses. In addition, once the froth has dried, it is difficult to remove.

It is important, therefore, to consider some method for controlling froth formation, particularly in spiral-flow tanks where the froth collects along the side of the tank, aerated channels, and free-fall from weirs. A commonly used system for spiral-roll tanks consists of a series of spray nozzles mounted above the surface in areas where the froth collects. Screened effluent or clear water is sprayed through these nozzles and physically breaks down the froth as it forms. Another approach is to meter a small quantity of antifoaming chemical additive into the spray water.

Nocardia Foam Control. *Nocardia* foam, discussed in Sec. 8–3, is a thick layer of brown, biological foam that forms on the top of aeration tanks and clarifiers. When the *Nocardia* organisms grow in sufficient numbers, they tend to trap air bubbles that subsequently float to the surface and accumulate as scum. When *Nocardia* foam occurs, it should be removed from the system, as it will also cause foaming problems in anaerobic and aerobic digesters. *Nocardia* foam has been controlled by spraying a chlorine solution directly into the foam layer. In some cases, spray nozzles have been installed within a hood located across the width of plug-flow aeration tanks. The addition of a cationic polymer to the activated-sludge process has also been used to control the production of *Nocardia* foam (Shao et al., 1997). Chlorinating the return activated sludge (RAS) for bulking control may not be effective, as it may cause floc breakup and inhibit BOD removal and nitrification.

Solids Separation

The separation of solids in the activated-sludge process is a very important function in order to provide well-clarified effluent and concentrated solids that are returned to the biological treatment system or wasted to the solids processing facilities. The design of secondary clarifiers following suspended growth biological treatment considers two functions: (1) a sufficient time is needed to provide gravity settling of particles and a relatively clear liquid and (2) thickening of the settled solids to provide higher solids concentration in the return activated sludge. The clarification design considerations and parameters that relate to mixed-liquor settling and thickening characteristics were discussed previously in Sec. 8–3.

For solids-thickening considerations, the solids loading rate (kg/m²·d), which is the rate of solids feed relative to the clarifier cross-sectional area, is the primary process parameter. In this section, methods are presented to determine solids loading rates as a function of the sludge-settling properties and clarifier return sludge flowrate. The analysis to determine acceptable loadings requires knowledge of the sludge-thickening characteristics. Thus, the procedures presented in this section are more applicable for the evaluation and optimization of existing systems than the design for new systems. For new systems where data are not available, the design parameters given in Table 8–7 may be used.

Two methods of clarifier analysis are described: solids flux and state point analyses. Both methods are based on the same principles.

Solids Flux Analysis. The area required for thickening of the applied mixed liquor depends on the limiting solids flux that can be transported to the bottom of the sedimentation basin. Because the solids flux varies with the characteristics of the sludge, column settling tests should be conducted to determine the relationship between the sludge concentration and the settling rate. The required area can then be determined using the solids flux analysis procedure described below. The depth of the thickening portion of the sedimentation tank must be sufficient to (1) ensure maintenance of an adequate sludge blanket depth so that unthickened solids are not recycled, and (2) temporarily store excess solids that may be applied.

A method used to determine the area required for hindered settling is based on an analysis of the solids (mass) flux. Data derived from settling tests must be available when applying this method, which is based on an analysis of the mass flux (movement across a boundary) of the solids in the settling basin.

In a settling basin that is operating at steady state, a constant flux of solids is moving downward, as shown on Fig. 8–34. Within the tank, the downward flux of solids is brought about by gravity (hindered) settling and by bulk transport due to the underflow that is being pumped out and recycled. At any point in the tank, the mass flux of solids due to gravity (hindered) settling is

$$SF_g = C_i V_i (1 \text{ kg}/10^3 \text{ g}) \tag{8-61}$$

where SF_g = solids flux due to gravity, kg/m²·h
  C_i = concentration of solids at the point in question, g/m³
  V_i = settling velocity of the solids at concentration C_i, m/h

Figure 8–34

Definition sketch for a settling basin operating at steady state.

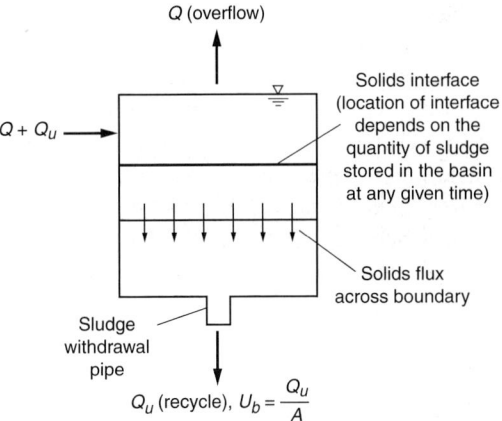

Q (overflow)

$Q + Q_u$ →

Solids interface (location of interface depends on the quantity of sludge stored in the basin at any given time)

Solids flux across boundary

Sludge withdrawal pipe

Q_u (recycle), $U_b = \dfrac{Q_u}{A}$

The mass flux of solids due to the bulk movement of the suspension is

$$SF_u = C_i U_b (1 \text{ kg}/10^3 \text{ g}) = C_i \frac{Q_u}{A} (1 \text{ kg}/10^3 \text{ g}) \qquad (8\text{–}62)$$

where SF_u = solids flux due to underflow, kg/m²·h
$\quad U_b$ = bulk downward velocity, m/h
$\quad Q_u$ = underflow flowrate, m³/h
$\quad A$ = cross-sectional area, m²

The total mass flux SF_t of solids is the sum of previous components and is given by

$$SF_t = SF_g + SF_u \qquad (8\text{–}63)$$

$$SF_t = (C_i V_i + C_i U_b)(1 \text{ kg}/10^3 \text{ g}) \qquad (8\text{–}64)$$

In the above equations, the flux of solids due to gravity (hindered) settling depends on the concentration of solids and the settling characteristics of the solids at that concentration. The procedure used to develop a solids flux curve from column settling test data is illustrated on Fig. 8–35. At low concentrations (below about 1000 mg/L), the movement of solids due to gravity is small, because the settling velocity of the solids is more or less independent of concentration. If the velocity remains essentially the same as the solids concentration increases, the total flux due to gravity starts to increase as the solids concentration starts to increase. At very high solids concentrations, the hindered-settling velocity approaches zero, and the total solids flux due to gravity again becomes extremely low. Thus, it can be concluded that the solids flux due to gravity must pass through a maximum value as the concentration is increased (see Figs. 8–35c and 8–36).

The solids flux due to bulk transport is a linear function of the concentration with slope equal to U_b, the underflow velocity (see Fig. 8–36). The total flux, which is the sum of the gravity and the underflow flux, is also shown on Fig. 8–36. Increasing or decreasing the flowrate of the underflow causes the total-flux curve to shift upward or downward. Because the underflow velocity can be controlled, it is used for process control.

The required cross-sectional area for thickening is determined as follows: As shown on Fig. 8–36, if a horizontal line is drawn tangent to the low point on the total-flux curve, its intersection with the vertical axis represents the limiting solids flux SF_L

Figure 8–35

Procedure for preparing a plot of solids flux due to gravity as a function of solids concentration: (a) hindered settling velocities derived from column settling tests for suspension at different concentrations, (b) plot of hindered settling velocities obtained in step (a) versus corresponding concentration, and (c) plot of completed value of solids flux versus corresponding concentration.

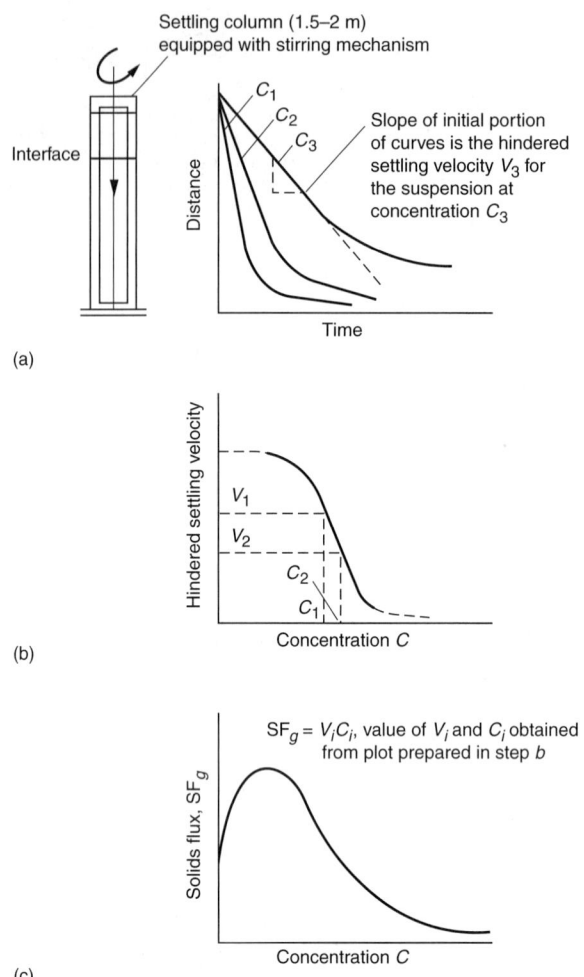

that can be processed in the settling basin. The corresponding underflow concentration is obtained by dropping a vertical line to the X axis from the intersection of the horizontal line and the underflow flux line assuming the gravity flux is negligible at the bottom of the settling basin and the solids are removed by bulk flow. The fact that the gravity flux is negligible at the bottom of the tank can be verified by performing a materials balance around the portion of the settling tank that lies below the depth where the limiting solids flux occurs and comparing the gravity settling velocity of the sludge to the velocity in the sludge withdrawal pipe. If the quantity of solids discharged to the settling basin is greater than the limiting solids-flux value defined on Fig. 8–36, the solids will build up in the settling basin and, if adequate storage capacity is not provided, ultimately overflow at the top. Using the limiting solids-flux value, the required area derived from a materials balance is given by

$$A = \frac{(Q + Q_u)C_o}{\text{SF}_L} \ (1 \text{ kg}/10^3 \text{ g})$$ (8–65)

Figure 8-36

Definition sketch for the analysis of settling data using the solids flux method of analysis.

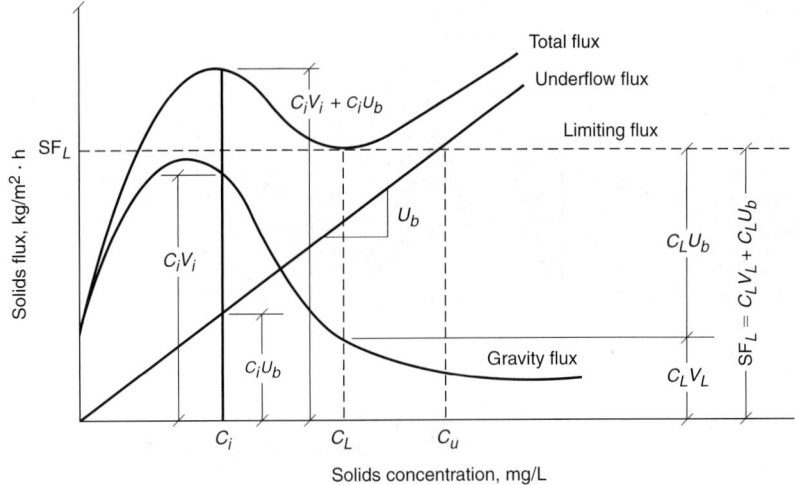

$$A = \frac{(1 + \alpha)QC_o}{SF_L} \; (1 \text{ kg}/10^3 \text{ g}) \tag{8-66}$$

where A = cross-sectional area, m^2
$(Q + Q_u)$ = total volumetric flowrate to settling basin (overflow + underflow), m^3/d
C_o = influent solids concentration, g/m^3
SF_L = limiting solids flux, $kg/m^2 \cdot d$
$\alpha = Q_u/Q$

Referring to Fig. 8-36, if a thicker underflow concentration is required, the slope of the underflow flux line must be reduced. The reduced slope, in turn, will lower the value of the limiting flux and increase the required settling area. In an actual design, the use of several different flowrates for the underflow should be evaluated.

An alternative graphical method of analysis to that presented on Fig. 8-36 for determining the limiting solids flux is shown on Fig. 8-37. The graphical analysis is derived from Eq. (8-64) in which the minimal SF_t is defined where its derivative with respect to C_i equals zero:

$$\frac{\partial SF_t}{\partial C_i} = 0 = V_i - U_b \tag{8-67}$$

Thus, as shown on Fig. 8-37, the solids concentration at the limiting solids flux is C_L and $V_iC_L = U_bC_L$. The value of the limiting flux on the ordinate is obtained by drawing a line tangent to the flux curve passing through the desired underflow and intersecting the ordinate. The geometric relationship of this method to that given on Fig. 8-36 is shown by the line for U_b on Fig. 8-37. The method detailed on Fig. 8-37 is especially useful where the effect of the use of various underflow concentrations on the size of the treatment facilities (aerator and sedimentation basin) is to be evaluated. Application of the solids flux method of analysis is illustrated in Example 8-11.

Figure 8–37

Alternative definition sketch for the analysis of settling data using the solids flux method of analysis.

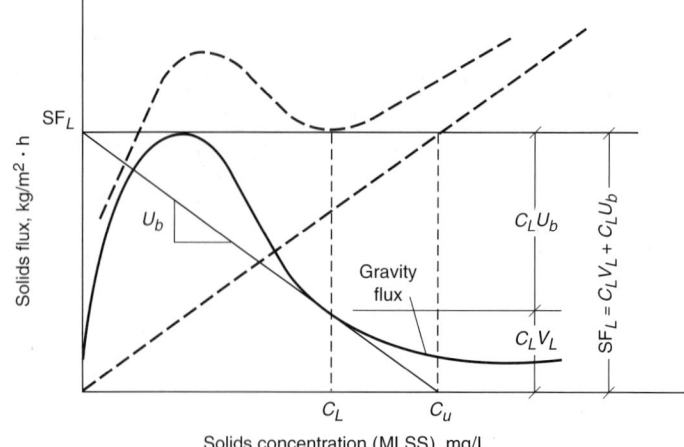

EXAMPLE 8–11 **Application of Solids Flux Analysis** Given the following settling data for a biological sludge, derived from a pure-oxygen activated-sludge pilot plant, estimate the maximum concentration of the aerator mixed-liquor biological suspended solids that can be maintained if the sedimentation tank application rate $Q + Q_r$ has been fixed at 25 m³/m²·d and the sludge recycle rate Q_r is equal to 40 percent. The definition sketch for this problem is shown on Fig. 8–9b on page 691. As shown, settled and thickened biological solids from the sedimentation tank are returned to the aeration tank to maintain the desired level of biological solids in the aerator. Assume that the solids wasting rate Q_w is negligible in this example.

MLSS, g/m³	Initial settling velocity, m/h
2,000	4.27
3,000	3.51
4,000	2.77
5,000	2.13
6,000	1.28
7,000	0.91
8,000	0.67
9,000	0.49
10,000	0.37
15,000	0.15
20,000	0.07
30,000	0.027

Note: g/m³ = mg/L.

Solution

1. Develop the gravity solids-flux curve from the given data and plot the curve.
   a. Set up a computation table to determine the solids-flux values corresponding to the given solids concentrations.

MLSS, g/m^3	Solids flux,[a] kg/m^2·h
2,000	8.54
3,000	10.53
4,000	11.03
5,000	10.65
6,000	7.68
7,000	6.37
8,000	5.36
9,000	4.41
10,000	3.70
15,000	2.25
20,000	1.40
30,000	0.81

[a] Solids flux $= X(g/m^3)V(m/h)$ $(1 \text{ kg}/10^3 \text{ g})$.

   b. Plot the gravity solids-flux curve (see following figure).

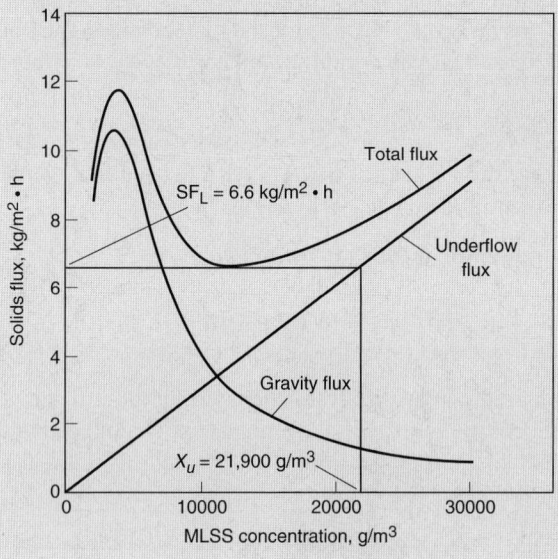

2. Determine the underflow bulk velocity. Referring to the definition sketch, the applied loading on the sedimentation facilities equals $(Q + Q_r)$, which per unit area is equal to 25 $m^3/m^2{\cdot}d$. The underflow velocity U_b is therefore equal to

$$U_b = [0.4Q/(Q + 0.4Q)](25 \text{ m}^3/\text{m}^2{\cdot}\text{d})$$

$$= 7.14 \text{ m/d} = 0.30 \text{ m/h}$$

3. Develop the total-flux curve for the system, and determine the value of the limiting flux and maximum underflow concentration.

    a. Plot the underflow curve on the solids-flux curve using the following relationship:

    $$SF_u = X_i U_b (1 \text{ kg}/10^3 \text{ g})$$

    where X_i = MLSS concentration, g/m^3
    $\qquad\quad U_b$ = bulk underflow velocity, m/h

    For example, at $X_i = 10,000$ g/m^3,

    $$SF_u = (10,000 \text{ g/m}^3)(0.30 \text{ m/h})(1 \text{ kg}/10^3 \text{ g}) = 3.0 \text{ kg/m}^2{\cdot}\text{h}$$

    b. Plot the total solids-flux curve by summing the values of the gravity and underflow solids flux (see solids-flux curve).

    c. From the solids-flux curve, the limiting solids flux is found to be equal to

    $$SF_L = 6.6 \text{ kg/m}^2{\cdot}\text{h}$$

    d. From the solids-flux curve, the maximum underflow solids concentration is equal to 21,900 g/m^3.

4. Estimate the maximum solids concentration that can be maintained in the reactor.

    a. Write a mass balance for the system within the boundary, neglecting the rate of cell growth within the reactor. Let X_o = influent TSS to the aeration tank.

    $$QX_o + Q_R X_R = (Q + Q_R)X$$

    b. Assuming the $X_o = 0$ $(X_o \ll X_R)$ and that $Q_R/Q = 0.4$, solve for the concentration of MLSS in the reactor.

    $$0.4Q\,(21,900 \text{ g/m}^3) = (1 + 0.4)QX$$

    $$X = 6257 \text{ g/m}^3$$

Comment   As shown in the above analysis, the concentration of the return solids will affect the maximum concentration of solids that can be maintained in the aerator. Thus, the sedimentation tank must be considered an integral part of the design of an activated-sludge treatment process.

State Point Analysis. The state point analysis procedure extends the principles of the solids-flux analysis to provide a convenient means to assess different mixed-liquor concentrations and clarifier operating conditions relative to the limiting solids-flux operating condition (Keinath et al., 1977; Keinath, 1985). The state point as shown on Fig. 8–38a is the intersection of the clarifier overflow solids-flux rate and underflow solids-flux rate lines. Thus, the analysis accounts for the actual mixed-liquor concentration, clarifier hydraulic application rate, and return activated-sludge recycle rate, and whether the combination of these operating parameters results in a condition that is within the clarifier solids-flux limitations for a sludge with specific thickening characteristics.

On Fig. 8–38a, the clarifier overflow solids flux is

$$SF_Q = \frac{Q(X)}{A} \tag{8-68}$$

where SF_Q = overflow solids-flux rate, kg/m²·d
  Q = clarifier effluent flowrate, m³/d
  A = clarifier cross-section area, m²
  X = aeration tank MLSS concentration, g/L

The aeration tank MLSS concentration (X) at any point along the overflow solids-flux line is found by constructing a vertical line to the X axis. The underflow operating line represents the negative slope of the clarifier underflow velocity as was also shown on Fig. 8–37. The y intercept of the horizontal line drawn from the point of intersection

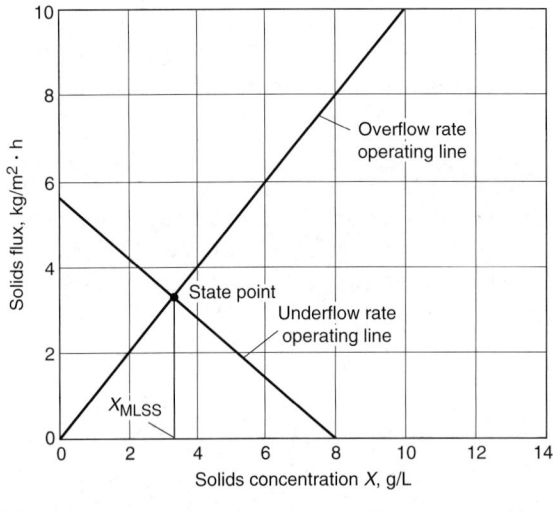

(a)

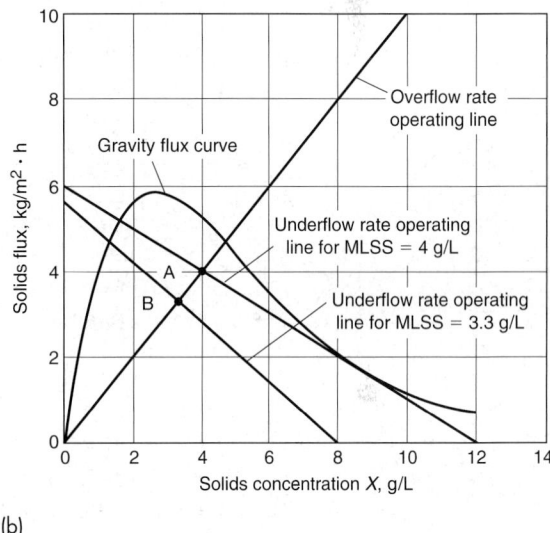

(b)

Figure 8–38

State point analysis for assessing clarifier operating conditions: (a) state point at intersection of overflow rate and underflow rate operating lines and (b) underloaded (B) and critically loaded (A) state points relative to settling flux curve.

of the vertical line and operating line is the total solids flux (SF_X) to the clarifier. Evaluating the slope of the underflow flux rate line shows that U_b equals

$$U_b = \frac{SF_t - SF_Q}{0 - X_{MLSS}} \tag{8-69}$$

$$U_b = \frac{[(Q + Q_R)X_{MLSS}/A]}{- X_{MLSS}} \tag{8-70}$$

$$U_b = -\frac{Q_R}{A} \tag{8-71}$$

The state point and underflow solids-flux line can be compared to the gravity flux curve to determine if the clarifier operation is within its solids-flux limitation (see Fig. 8–38b). The underflow line at state point A is tangent to the gravity flux curve, which, as shown on Fig. 8–37, represents the limiting solids-flux condition. Thus, the clarifier is critically loaded for this underflow velocity and the MLSS concentration at the state point is equal to 4.0 g/L. If the operation is changed to obtain a higher MLSS concentration, and the underflow line crosses the lower limb of the gravity flux curve, the limiting solids flux will be exceeded and the clarifier blanket will rise to the effluent weir. At state point B on Fig. 8–38b, a lower MLSS concentration is used and the underflow line is thus well below the lower limb of the gravity flux curve. An underloaded operation exists relative to the solids loading.

The state point analysis procedure provides a method to evaluate various clarifier overflow rates, and MLSS concentrations give a gravity flux curve that represents the activated-sludge settling properties. The state point analysis technique can be used with settling tests on activated-sludge mixed liquor at an existing facility to determine an optimal MLSS concentration and return sludge recycle ratio for a given influent flow condition. Application of state point analysis is illustrated in Example 8–12.

EXAMPLE 8–12 **Evaluate Secondary Clarifier Operating Conditions Using State Point Analysis** Determine acceptable operating conditions using the following solids-settling test results for one or two clarifiers in operation in an activated-sludge system. The system is to be evaluated at a maximum month design flowrate of 15,070 m³/d and the following design conditions.

Design conditions:

1. Two 20-m-diameter clarifiers are installed.
2. With both clarifiers in operation, the desired MLSS concentration is 3500 mg/L.
3. Evaluate the feasibility of operating the clarifier with underflow concentrations of 10, 12, and 14 g/L and determine the recycle ratios.

4. Determine the MLSS concentration with one clarifier in operation using an underflow solids concentration of 12 g/L.
5. Determine the solids loading to the clarifier for the 12 g/L underflow concentration and MLSS = 3500 g/m³ for (a) two-clarifier operation and (b) one clarifier.
6. The solids settling results are:

Mixed-liquor solids concentration, g/m³	Interfacial settling velocity, m/h
2,000	2.90
3,000	1.90
4,000	1.30
5,000	0.90
6,000	0.60
8,000	0.26
9,000	0.17
10,000	0.12
12,000	0.05
16,000	0.01

Note: g/m³ = mg/L.

Solution

1. Develop gravity flux curve (shown below) using Eq. (8–61), $SF_g = C_i V_i$.

C_i, g/L	V_i, m/h	SF_g, kg/m²·h
2.0	2.90	5.80
3.0	1.90	5.70
4.0	1.30	5.20
5.0	0.90	4.50
6.0	0.60	3.60
8.0	0.26	2.08
9.0	0.17	1.53
10.0	0.12	1.20
12.0	0.05	0.60
16.0	0.01	0.16

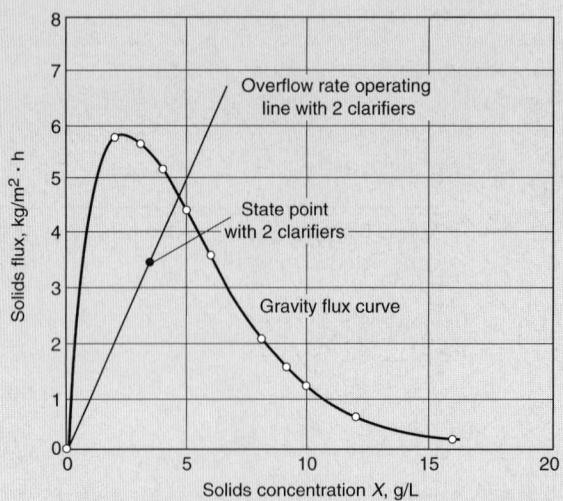

2. Add overflow rate operating line and MLSS concentration state point at 3500 mg/L overflow solids flux using Eq. (8–68), $SF_Q = Q(X)/A$.
    a. Determine the clarifier surface area.

    $$A \text{ (area/clarifier)} = \pi D^2/4 = \pi (20)^2/4 = 314 \text{ m}^2$$

    $$\text{Total area (2 clarifiers)} = 2 \times 314 = 628 \text{ m}^2$$

    b. Determine the overflow rate expressed in terms of X.

    $$SF_Q = \frac{Q(X)}{A} = \frac{(15{,}070 \text{ m}^3/\text{d})(X)}{628 \text{ m}^2} = 23.997 \text{ m/d}(X)$$

    c. Determine the plotting point of an underflow concentration of 3500 mg/L (3.5 g/L).
    Sample calculation: if $X = 3.5 \text{ g/L} = 3.5 \text{ kg/m}^3$,

    $$SF_Q = (23.997 \text{ m/d})(3.5 \text{ kg/m}^3)(1 \text{ d}/24 \text{ h}) = 3.5 \text{ kg/m}^2 \cdot \text{h}$$

3. Evaluate underflow conditions at 10, 12, and 14 g/L (see following figure).
    a. Analysis for an underflow concentration of 14 g/L.
    Draw a line that intercepts 14 g/L on the x axis and passes through the state point. The line intercepts the y axis at 4.67 kg/m²·h; however, this is not a feasible operating condition as the line crosses above the gravity flux curve. For 10 and 12 g/L underflow concentrations, the lines cross below the gravity flux curve; therefore, both concentrations are feasible.

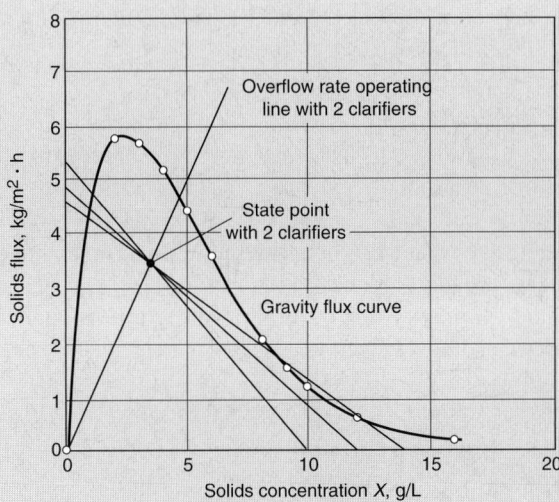

b. Calculate recycle ratio at 10 g/L (10 kg/m³) underflow concentration. Determine the slope of the underflow operating rate curve from the figure. The intercept on the x axis is 5.38 kg/m²·h and the slope is negative and is equal to the recycle velocity, m/h.

$$\text{Operating curve slope} = \frac{[(5.38 - 0) \text{ kg/m}^2\text{·h}]}{[(0 - 10) \text{ g/m}^3]} = -0.538 \text{ m/h}$$

$$\text{Underflow velocity} = -(-0.538 \text{ m/h}) = 0.538 \text{ m/h}$$

$$\text{Clarifier overflow rate} = \frac{(15,070 \text{ m}^3/\text{d})(1 \text{ d/24 h})}{628 \text{ m}^2} = 1.0 \text{ m/h}$$

$$\text{Recycle ratio} = \frac{(0.538 \text{ m/h})}{(1 \text{ m/h})} = 0.538$$

c. Check recycle ratio at 10 g/L underflow concentration using solids balance.

$$X_R Q_R = (Q_R + Q)X$$

$$X_R R = (1 + R)X \text{ where } R = \text{recycle ratio}$$

$$R = \left(\frac{X_R}{X} - 1\right)^{-1} = \left[\frac{(10 \text{ g/L})}{(3.5 \text{ g/L})} - 1\right]^{-1}$$

$$R = 0.538$$

d. Calculate recycle ratio at 12 g/L (12 kg/m³) underflow concentration using the same procedure as above.

$$\text{Underflow velocity} = -\frac{[(4.94 - 0)\ \text{kg/m}^2\cdot\text{h}]}{[(0 - 12)\ \text{g/m}^3]} = -0.41\ \text{m/h}$$

$R = 0.41$

4. Determine what MLSS concentration is possible with one clarifier operating and underflow solids concentration = 12 g/L.

Using the gravity flux curve, draw an overflow operating rate line for one clarifier (see following figure).

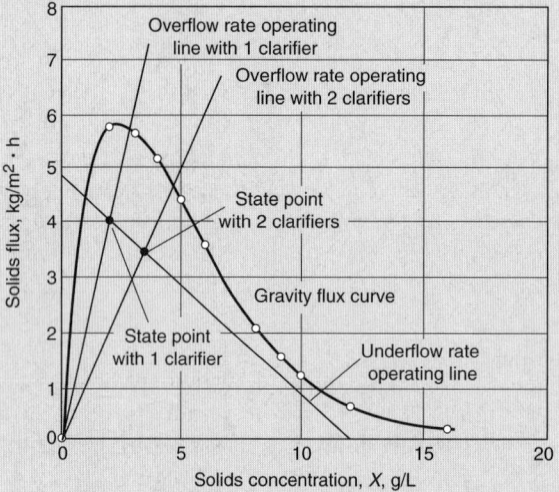

For $A = 314\ \text{m}^3$ and $X = 2$ g/L,

$$SF_Q = \frac{(15{,}070\ \text{m}^3/\text{d})(1\ \text{d}/24\ \text{h})(2\ \text{kg/m}^3)}{314\ \text{m}^2} = 4\ \text{kg/m}^2\cdot\text{h}$$

The MLSS concentration possible is the "state point" where the overflow rate operating line intersects the underflow rate operating line. In the above figure, the state point is approximately 2.1 g/L = 2100 mg/L MLSS.

5. Determine clarifier solids loading
   a. For 2 clarifiers: $A = 628\ \text{m}^2$; MLSS = 3.5 g/L, and $R = 0.41$

   $$\text{Solids loading} = Q(1 + R)(X)/A$$

   $$= \frac{(15{,}070\ \text{m}^3/\text{d})(1 + 0.41)(3.5\ \text{kg/m}^3)}{(628\ \text{m}^2)(24\ \text{h/d})} = 4.93\ \text{kg/m}^2\cdot\text{h}$$

b. For one clarifier: $A = 314 \text{ m}^2$

$$\text{Clarifier overflow rate} = \frac{(15{,}070 \text{ m}^3/\text{d})(1 \text{ d}/24 \text{ h})}{(314 \text{ m}^2)} = 2.0 \text{ m/h}$$

Underflow velocity (from step 3d) = 0.41 m/h

$R = 0.41 \text{ m/h}/2.0 \text{ m/h} = 0.205$

$$\text{Solids loading} = \frac{(15{,}070 \text{ m}^3/\text{d})(1 + 0.205)(2.1 \text{ kg/m}^3)}{(314 \text{ m}^2)(24 \text{ h/d})} = 5.06 \text{ kg/m}^2 \cdot \text{h}$$

Design of Solids Separation Facilities

Solids separation is the final step in the production of a well-clarified, stable effluent low in BOD and suspended solids and, as such, represents a critical link in the operation of an activated-sludge treatment process. Although much of the information presented in Chap. 5 for the design of primary sedimentation tanks is applicable, the presence of a large volume of flocculent solids in the mixed liquor requires that special consideration be given to the design of activated-sludge settling tanks. As mentioned previously, these solids tend to form a sludge blanket in the bottom of the tank that will vary in thickness. The blanket may fill the entire depth of the tank and overflow the weirs at peak flowrates if the return sludge pumping capacity or the size of the settling tank is inadequate. Further, the mixed liquor, on entering the tank, has a tendency to flow as a density current, interfering with the separation of the solids and the thickening of the sludge. To cope successfully with these characteristics, the following factors must be considered in the design of secondary sedimentation tanks: (1) tank types, (2) surface and solids loading rates, (3) sidewater depth, (4) flow distribution, (5) inlet design, (6) weir placement and loading rates, and (7) scum removal.

Tank Types. The most commonly used types of activated-sludge settling tanks are either circular or rectangular (see Fig. 8–39a and b). Square tanks are used on occasion but they are not as effective in retaining separated solids as circular or rectangular tanks. Solids accumulate in the corners of the square tanks and are frequently swept over the weirs by the agitation of the sludge collectors. Circular tanks have been constructed with diameters ranging from 3 to 60 m (10 to 200 ft), although the more common range is from 10 to 40 m (~30 to 140 ft). The tank radius should preferably not exceed five times the sidewater depth.

Two basic types of circular tanks are used for secondary sedimentation: center-feed and rim-feed (see Fig. 5–41 in Chap. 5). Both types use a revolving mechanism to transport and remove the sludge from the bottom of the clarifier. Mechanisms are of two types: those that scrape or plow the sludge to a center hopper similar to the types used in primary sedimentation tanks, and those that remove the sludge directly from the tank bottom through suction orifices that serve the entire bottom of the tank in each revolution. Of the latter, in one type the suction is maintained by reduced static head on the

(a)

(b)

Figure 8–39

Typical secondary clarifiers: (a) circular with peripheral drive mechanism for sludge collection and (b) rectangular.

(a)

(b)

Figure 8–40

Typical circular sludge-collection mechanisms: (a) suction-type and (b) spiral-type scraper used commonly in Europe.

individual suction pipes (Fig. 8–40a). In another patented suction system, sludge is removed through a manifold either hydrostatically or by pumping. Spiral-type scrapers are also used to accelerate movement of settled solids from the tank periphery to the collection sump (Fig. 8–40b).

Rectangular tanks must be proportioned to achieve proper distribution of incoming flow so that horizontal velocities are not excessive. The maximum length of rectangular tanks normally should not exceed 10 times the depth, but lengths up to 90 m (300 ft) have been used successfully in large plants. Where widths of rectangular tanks exceed 6 m (20 ft), multiple sludge collection mechanisms may be used to permit tank widths

Figure 8–41

Typical rectangular
sludge-collection
mechanisms: (a) chain
and flight and
(b) traveling bridge.

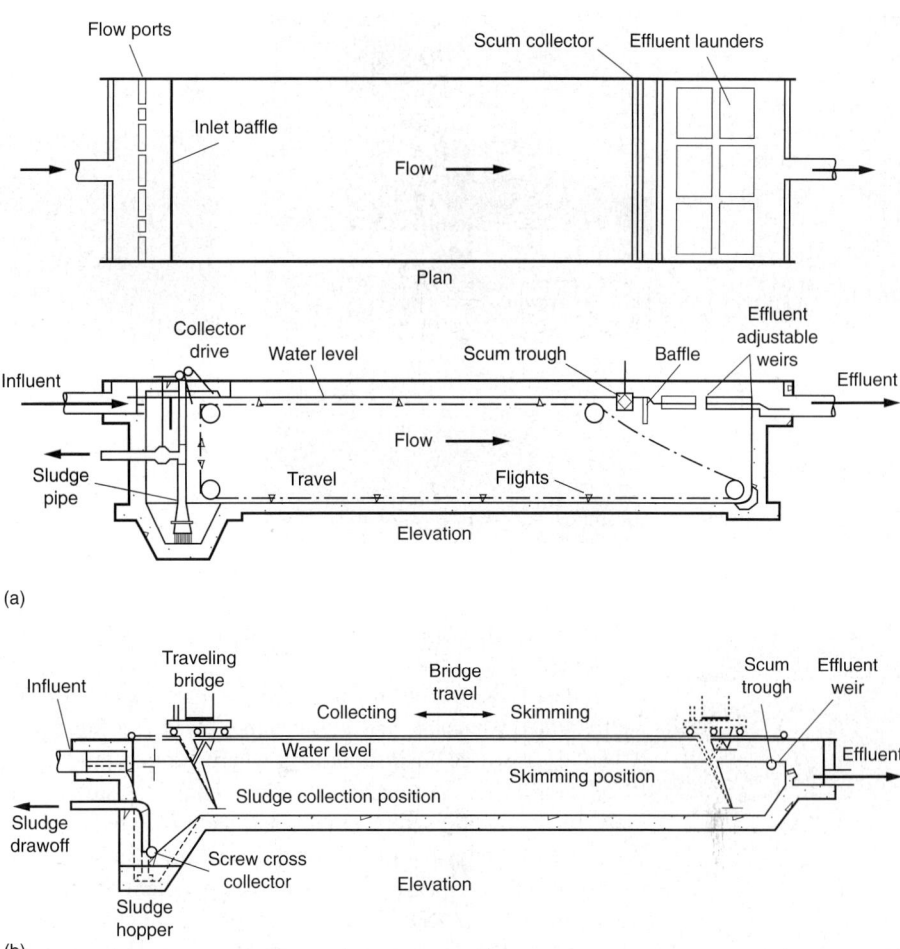

up to 24 m (80 ft). Regardless of tank shape, the sludge collector selected should be able to meet the following operational conditions: (1) the collector should have enough capacity so that when a high sludge-recirculation rate is desired, channeling of the overlying liquid through the sludge will not result, and (2) the mechanism should be sufficiently rugged to transport and remove very dense sludges that could accumulate in the settling tank during periods of mechanical breakdown or power failure.

Two types of sludge collectors are commonly used in rectangular tanks: (1) traveling flights, and (2) traveling bridges (see Figs. 8–41 and 8–42). Traveling flights are similar to those used for the removal of sludge in primary settling tanks. For very long tanks, it is desirable to use two sets of chains and flights in tandem with a central hopper to receive the sludge to minimize the sludge transport distance. Sludge may be collected at the influent or effluent end of the tank. The traveling bridge, which is similar to a traveling overhead crane, travels along the sides of the sedimentation tank or on a support structure if several bridges are used. The bridge serves as the support for the sludge-removal system, which usually consists of a scraper or a suction manifold from

Figure 8–42

Views of sludge-collection mechanisms in rectangular tanks: (a) chain and flight and (b) traveling bridge.

(a)

(b)

which the sludge is pumped. The sludge is discharged to a collection trough that runs the length of the tank.

Other types of settling tanks that are used include stacked clarifiers, tube and plate settlers (see Chap. 5), and intrachannel clarifiers. Stacked clarifiers (see Fig. 5–45 in Chap. 5) are used in installations where limited land area is available for clarifiers. Stacked clarifiers are used at the Deer Island Wastewater Treatment Plant in Boston, MA, for secondary sedimentation and were selected because of limited land area.

The efficiency of conventional or shallow clarifiers may be improved by the installation of tubes or parallel plates to establish laminar flow (see Fig. 5–25 in Chap. 5). Constructed of bundles of tubes or plates set at selected angles (usually 60°) from the horizontal, tube and plate settlers have a very short settling distance and circulation is dampened because of the small size of the tubes. Solids that collect in the tubes or on the plates tend to slide out due to gravitational forces. The major drawback in wastewater treatment is a tendency of the tubes and plates to clog because of the accumulation of biological growths, grease, and small objects that pass through coarse screens.

Intrachannel clarifiers are proprietary devices that have been developed to improve the performance of the oxidation ditch activated-sludge process and to save the cost of constructing external final clarifiers. The concept incorporates a clarifier, configured somewhat similar to a boat, installed in the oxidation ditch channel. These devices permit liquid and solids separation and sludge return to occur within the aeration channel. Sludge wasting is done from either the aeration channel or the intrachannel clarifier. Disadvantages of intrachannel clarifiers include loss of independency of reactors and clarifiers, wasting dilute sludge, loss of RAS chlorination potential for controlling bulking sludge, and lack of metered RAS control (WEF, 1998).

Surface and Solids Loading Rates. When it is necessary to design secondary settling facilities without the benefit of settling tests, published values for surface and solids loading rates are often used. Because of the large amount of solids that may be lost in the effluent if design criteria are exceeded, effluent overflow rates based on peak flow conditions are used commonly. An alternative design basis is to use the average dry-weather flowrate (ADWF) and a corresponding surface loading rate and also check for peak flow and loading conditions. Either condition may govern the design parameters. Typical surface and solids loading rates were given previously in Table 8–7 for secondary clarifier design.

Sidewater Depth. Liquid depth in a secondary clarifier is normally measured at the sidewall in circular tanks and at the effluent end wall for rectangular tanks. The liquid depth is a factor in the effectiveness of suspended solids removal and in the concentration of the return sludge. Other factors such as inlet design, type of sludge-removal equipment, sludge blanket depth, and weir type and location also affect clarifier performance. In recent years, the trend has been toward increasing liquid depths to improve overall performance. Typical sidewater depths are presented in Table 8–7 in Sec. 8–3. Current practice favors a minimum sidewater depth of 3.5 m (~12 ft) for large secondary clarifiers, and depths ranging up to 6 m (20 ft) have been used. The cost of tank construction has to be considered in selecting a sidewater depth, especially in areas of high groundwater levels. Tanks with depths less than 3.5 m (~12 ft) often have difficulty containing the typically low-density activated sludge, and low-density sludge blankets are more easily disturbed by hydraulic fluctuations. Deeper tanks therefore provide greater flexibility of operation and a larger margin of safety when changes in the activated-sludge system occur.

Flow Distribution. Flow imbalance between multiple process units can cause under- or overloading of the individual units and affect overall system performance. In plants where parallel tanks of the same size are used, flow between the tanks should be equalized. In cases where the tanks are not of equal capacity, flows should be distributed in proportion to surface area. Methods of flow distribution to the secondary sedimentation tanks include weirs, flow distribution boxes, flow control valves, hydraulic distribution using hydraulic symmetry, and feed gate or inlet port control (see Fig. 8–43). Effluent weir control, although frequently used to effect flow splitting, is usually ineffective and should be used only where there are two tanks of equal size.

Figure 8–43

Alternative methods of flow splitting: (a) hydraulic symmetry, (b) flow measurement and feedback control, (c) hydraulic split with weirs, and (d) inlet feed gate control.

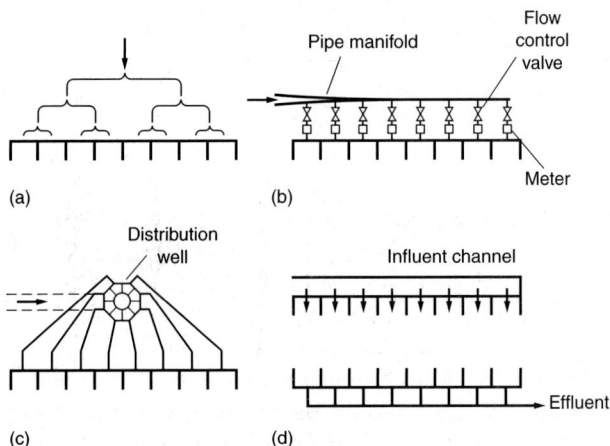

Figure 8–44

Typical secondary clarifier with a flocculating center feed well.

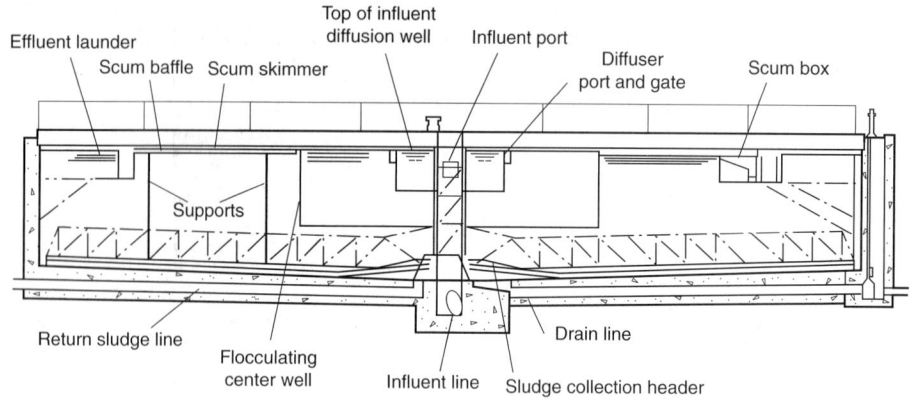

Tank Inlet Design. Poor distribution or jetting of the tank influent can increase the formation of density currents and scouring of settled sludge, resulting in unsatisfactory tank performance. Tank inlets should dissipate influent energy, distribute the flow evenly in horizontal and vertical directions, mitigate density currents, minimize sludge blanket disturbance, and promote flocculation. In circular center-feed tanks, a common design is to use small, solid-skirted, cylindrical baffles to dissipate the influent energy and distribute flow. Studies indicate that a density current waterfall can be created using skirted baffles resulting in poor vertical flow distribution (Crosby and Bender, 1980). Methods to overcome these problems include the use of a large center diffusion well or a flocculating-type clarifier (Fig. 8–44). The large center diffusion well, with a minimum diameter of 25 percent of the tank diameter, provides a greater area for dissipation of the influent energy and distribution of the incoming mixed liquor. The bottom of the feed well should end well above the sludge blanket interface in order to minimize turbulence and resuspension of the solids. Flocculating center-feed clarifiers can incorporate an energy-dissipating inlet (EDI) and means to promote flocculation in the center-feed well (see Fig. 8–45a). Typical flocculation feed wells have diameters of 30 to 35 percent of the tank diameter. An alternative device for dissipating energy, developed by the City of Los Angeles, is shown on Fig. 8–45b. Operationally, the flow is discharged from a centerwell through a series of downward-facing discharge ports. By arranging the discharge ports so they discharge facing each other, the momentum energy is dissipated as the discharge streams impact each other. In rectangular tanks, inlet ports or baffles should be provided to achieve flow distribution. Inlet port velocities are typically 75 to 150 mm/s (15 to 30 ft/min) (WPCF, 1985).

Weir Placement and Loading. When density currents occur in a secondary clarifier, mixed liquor entering the tank flows along the tank bottom until it encounters a countercurrent pattern or an end wall. Unless density currents are considered in the design, solids may be discharged over the effluent weir. Experimental work performed by Anderson (1945) at Chicago on tanks approximately 38 m (126 ft) in diameter indicated that a circular weir trough placed at two-thirds to three-fourths of the radial distance from the center was in the optimum position to intercept well-clarified effluent. With low surface loadings and weir rates, the placement of the weirs in small tanks does

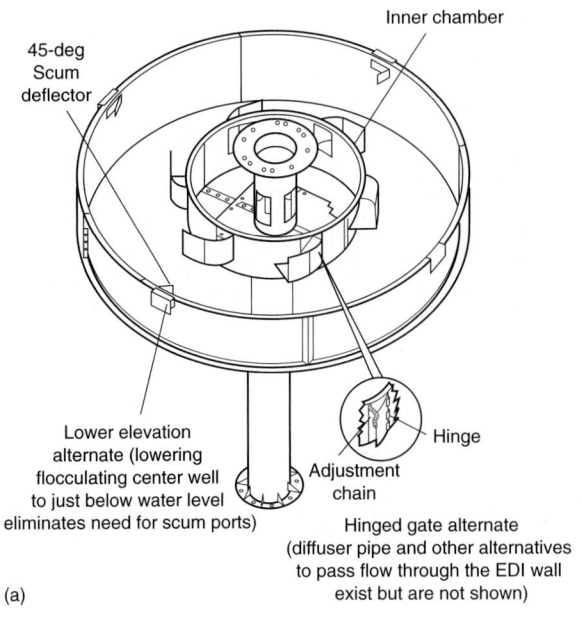

(b)

(a)

Figure 8–45

Energy-dissipating inlet devices used in circular clarifiers: (a) schematic of a center column energy-dissipating inlet and flocculating feed well (WEF, 1998) and (b) view of an energy-dissipating feed well. *(Courtesy City of Los Angeles.)*

not significantly affect the performance of the clarifier. Circular clarifiers are manufactured with overflow weirs located near both the center and the perimeter of the tank. If weirs are located at the tank perimeter or at end walls in rectangular tanks, a baffle should be provided to deflect the density currents toward the center of the tank and away from the effluent weir. Alternative baffle arrangements are shown on Fig. 8–46.

Weir loading rates are used commonly in the design of clarifiers, although they are less critical in clarifier design than hydraulic overflow rates. Weir loading rates used in large tanks should preferably not exceed 375 m³/lin m·d (30,000 gal/lin ft·d) of weir at maximum flow when located away from the upturn zone of the density current, or 250 m³/lin m·d (20,000 gal/lin ft·d) when located within the upturn zone. In small tanks, the weir loading rate should not exceed 125 m³/lin m·d (10,000 gal/lin ft·d) at average flow or 250 m³/lin m·d at maximum flow. The upflow velocity in the immediate vicinity of the weir should be limited to about 3.5 to 7 m/h (~12 to 24 ft/h).

Scum Removal. In many well-operating secondary plants, very little scum is formed in the secondary clarifiers. However, occasions arise when some floating material is present (see "Operating Problems" in Sec. 8–3), necessitating its removal. Where primary settling tanks are not used, skimming of the final tanks is essential. Most designs in recent years provide scum removal for both circular and rectangular secondary clarifiers. Typical scum-removal equipment includes beach and scraper type, rotating pipe-through skimmer, and slotted pipes.

Figure 8–46

Alternative peripheral baffle arrangements: (a) Stamford, (b) unnamed, (c) McKinney (also known as the Lincoln baffle), and (d) interior trough. *(WEF, 1998.)*

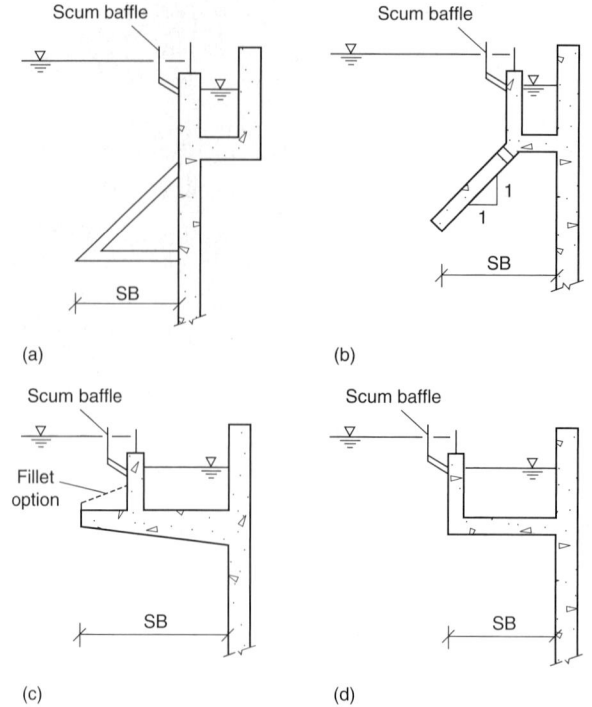

Note: SB will vary from 0.5 to 1.5 m, depending on the diameter of the clarifier

Scum should not be returned to the plant headworks because microorganisms responsible for foaming (typically *Nocardia*) will be recycled, causing foaming problems to persist because of continuous seeding of the unwanted microorganisms. In some plants, scum is discharged to sludge-thickening facilities or is added directly to digester feed streams, as appropriate.

8–8 SUSPENDED GROWTH AERATED LAGOONS

Suspended growth lagoons are relatively shallow earthen basins varying in depth from 2 to 5 m, provided with mechanical aerators on floats or fixed platforms. The mechanical aerators are used to provide oxygen for the biological treatment of wastewater and to keep the biological solids in suspension. In a few cases, diffused aeration has also been used. Suspended growth aerated lagoons are operated on a flow-through basis or with solids recycle. The purpose of this section is to (1) describe the principal types of suspended growth lagoons and (2) present and illustrate the design considerations for the suspended growth flow-through type of lagoon. Lagoons with solids recycle are essentially the same as the activated-sludge process described in detail in Secs. 8–1 and 8–3. The dual-powered flow-through lagoon system, a modification of the conventional flow-through lagoon, is also considered in this section.

Types of Suspended Growth Aerated Lagoons

The principal types of suspended growth lagoon processes, classified based on the manner in which the solids are handled (Arceivala, 1998), are:

1. Facultative partially mixed
2. Aerobic flow through with partial mixing
3. Aerobic with solids recycle and nominal complete mixing

Differences in the manner in which the solids are handled will affect the treatment efficiency, power requirements, hydraulic and solids retention time, sludge disposal, and environmental considerations. The general characteristics of these lagoon systems are summarized in Table 8–29. Each of the lagoon processes is described below.

Table 8–29
Typical characteristics of different types of aerated suspended growth lagoons[a]

Item	Unit	Type of aerated lagoon		
		Facultative	**Aerobic flow-through**	**Aerobic with solids recycling**
TSS	mg/L	50–200	100–400	1500–3000
VSS/TSS	Unitless	50–80	70–80	50–80
Solids retention time, SRT	d	b	3–6[c]	Warm: 10–20
			Typically 5	Moderate: 20–30
				Cold: over 30
Hydraulic retention time, τ	d	4–10	3–6	0.25–2.0
			Typically 5	
Overall BOD removal rate, k	d^{-1}	0.5–0.8[d]	0.5–1.5[d]	e
Temperature coefficient	Unitless	1.04	1.04	1.04
Depth	m	2–5	2–5	2–5
Mixing regime		Partially mixed	Partially mixed	Nominal complete-mix
Minimum power	$kW/10^3 \ m^3$	1–1.25	5.0–8.0	16–20
Sludge		Sludge accumulates internally in the lagoon	Sludge accumulates in external sedimentation facility	Sludge recycled to process from sedimentation tank. Waste sludge is discharged to sludge drying beds
Nitrification		No	Not typically	Likely, especially in warm climates

[a] Adapted from Arceivala (1998).

[b] Because the solids are retained in the lagoon, it is difficult to predict the SRT. It has been estimated that values over 100 d are possible.

[c] Because of incomplete mixing, the actual SRT value can be greater than the τ value.

[d] The removal rate constant for soluble BOD will be considerably higher.

[e] Use kinetic coefficients given in Table 7–9 and the analysis given in Example 8–2 for a complete mix activated sludge process for BOD removal.

Facultative Partially Mixed Lagoon. In the facultative partially mixed lagoon system, the energy input is only sufficient to transfer the amount of oxygen required for biological treatment, but is not sufficient to maintain the solids in suspension. Because the energy input will not maintain the solids in suspension, a portion of the incoming solids will settle along with a portion of the biological solids produced from the conversion of the soluble organic substrate. In time, the settled solids will undergo anaerobic decomposition. The term *facultative* is derived from the observation that the biological conversion, which occurs in the lagoon, is partially aerobic and partially anaerobic. Eventually facultative lagoons must be dewatered and the accumulated solids removed. Because there is essentially no way to manage the events in an aerated facultative lagoon (e.g., wind-driven circulation patterns), the use of facultative lagoons has diminished, especially where discharge limits must be met reliably. Because of the lack of a rational approach to the analysis of facultative lagoons, they are not considered further in this section.

Aerobic Flow-Through Partially Mixed Lagoon. In aerobic flow-through lagoons, the energy input is sufficient to meet the oxygen requirements, but is insufficient to maintain all of the solids in suspension. Operationally, the hydraulic and solids retention times are the same (i.e., $\tau = \text{SRT}$). The solids in the effluent are removed prior to discharge in an external sedimentation facility. The analysis and design of aerobic flow-through lagoons is considered in detail in the balance of this section. A typical aerobic flow-through lagoon is shown on Fig. 8–47.

Aerobic Lagoon with Solids Recycle. The aerobic lagoon with solids recycle is essentially the same as the extended aeration activated-sludge process, with the exception that an earthen (typically lined) basin is used in place of a reinforced-concrete reactor basin. As shown in Table 8–29, the hydraulic detention time, at the high end (i.e., 2 d), is longer than that used in a conventional extended aeration process ($\sim$1 d). It should be noted that the aeration requirement for an aerobic lagoon with recycle must be higher than the values for an aerobic flow-through lagoon to maintain the solids in suspension. The power levels given in Table 8–29 are designed to maintain most of the solids in suspension, hence the term *nominal complete mix* is used. As noted in the intro-

Figure 8–47

View of an aerobic flow-through aerated lagoon with slow-speed surface aerators mounted on floats.

ductory paragraph, because the analysis for the aerobic lagoon with recycle is the same as the activated-sludge process, this process will not be considered further in the following discussion.

Process Design Considerations for Flow-Through Lagoons

Factors that must be considered in the process design of suspended growth flow-through lagoons include (1) BOD removal, (2) effluent characteristics, (3) temperature effects, (4) oxygen requirement, (5) energy requirement for mixing, and (6) solids separation. The first four factors are considered in the following discussion, and their application is illustrated in Example 8–14. The energy required for mixing is discussed in Chap. 5. Solids separation is discussed at the end of this section.

BOD Removal. Because a suspended growth aerobic flow-through lagoon can be considered to be a complete-mix reactor without recycle, the basis of design is SRT, which in this case is equal to the hydraulic retention time τ under ideal flow conditions. The SRT should be selected to assure that (1) the suspended microorganisms will bioflocculate for easy removal by sedimentation and (2) an adequate safety factor is provided. Typical design values of SRT for aerated lagoons used for treating domestic wastes vary from about 3 to 6 d. Once the value of SRT has been selected, the soluble substrate concentration of the effluent can be estimated, and the removal efficiency can then be computed using the equations given earlier in this chapter.

An alternative approach is to assume that the observed BOD removal (either overall, including soluble and suspended solids contribution, or soluble only) can be described in terms of a first-order removal function. The BOD removal is measured between the influent and the lagoon outlet (not the outlet of the sedimentation facilities following the lagoon). The pertinent equation, derived from Eq. (4–102) in Chap. 4, for an ideal single aerated lagoon is

$$\frac{S}{S_o} = \frac{1}{1 + k\tau} \tag{8–72}$$

where S = effluent BOD concentration, g/m^3
S_o = influent BOD concentration, g/m^3
k = overall first-order BOD removal-rate constant, d^{-1}
τ = hydraulic retention time (V/Q), d

Reported overall k values vary from 0.5 to 1.5 d^{-1}. Removal rates for soluble BOD would be higher. Application of this equation is illustrated later in this section. In some cases, to limit short-circuiting in very large lagoons, the total volume required for a flow-through lagoon is distributed between two or three lagoons in series. The appropriate equation for equally sized lagoons in series is Eq. (4–107):

$$C_n = \frac{C_o}{[1 + (kV/nQ)]^n} = \frac{C_o}{[1 + (k/n\tau)]^n} \tag{4–107}$$

where n = number of lagoons in series
V = volume, m^3
Q = flowrate, m^3/s

Effluent Characteristics. The important characteristics of the effluent from an aerated lagoon include the BOD and the TSS concentration. The effluent BOD will be made up of those components previously discussed in connection with the activated-sludge process and occasionally may contain the contribution of small amounts of algae. The solids in the effluent are composed of a portion of the incoming suspended solids, the biological solids produced from waste conversion, and occasionally small amounts of algae. The solids produced from the conversion of soluble organic wastes can be estimated using Eq. (8–15). It should be noted that if the effluent from a suspended growth flow-through aerated lagoon is to meet the minimum standards for secondary treatment (see Chap. 1), settling facilities are needed. Where greater removal of effluent solids is required, slow sand or rock filters are used commonly (Rich, 1999).

Temperature. Because suspended growth aerobic flow-through lagoons are often installed and operated in locations with widely varying climatic conditions, the effects of temperature change must be considered in their design. The two most important effects of temperature are (1) reduced biological activity and treatment efficiency and (2) the formation of ice.

The effect of temperature on biological activity is described in Chap. 7. From a consideration of the influent wastewater temperature, air temperature, surface area of the lagoon, and wastewater flowrate, the resulting temperature in the aerated lagoon can be estimated using the following equation developed by Mancini and Barnhart (1968):

$$T_i - T_w = \frac{(T_w - T_a)fA}{Q} \tag{8–73}$$

where T_i = influent waste temperature, °C
T_w = lagoon water temperature, °C
T_a = ambient air temperature, °C
f = proportionality factor
A = surface area, m²
Q = wastewater flowrate, m³/d

The proportionality factor incorporates the appropriate heat-transfer coefficients and includes the effect of surface area increase due to aeration, wind, and humidity. A typical value for f in the eastern United States is 0.5 in SI units. To compute the lagoon temperature, Eq. (8–73) is rewritten as

$$T_w = \frac{AfT_a + QT_i}{Af + Q} \tag{8–74}$$

Alternatively, if climatological data are available, the average temperature of the lagoon may be determined from a heat budget analysis by assuming the lagoon is completely mixed. It should be noted that surface aerators tend to further cool lagoons in cold weather, but submerged diffused air systems add heat to some extent.

Where icing may be a problem, the effect on the operation of lagoons may be minimized by increasing the depth of the lagoon or by altering the method of operation. The effect of reducing the surface area is illustrated in Example 8–13. As computed, reducing the area by one-half increases the temperature about 2.2°C, which corresponds roughly to about a 20 percent increase in the rate of biological activity. As the depth of the lagoon is increased, maintenance of a completely mixed flow regime becomes difficult. If the depth is increased much beyond 3.7 m, draft tube aerators or diffused aeration must be used.

EXAMPLE 8-13 **Effect of Lagoon Surface Area on Liquid Temperature** Determine the effect of reducing the surface area of an aerated lagoon from 1 ha to 0.5 ha by doubling the depth for the following conditions:

1. Wastewater flowrate Q = 3800 m³/d
2. Wastewater temperature T_i = 15°C
3. Air temperature during coldest month T_a = 2°C
4. Proportionality constant, 0.5

Solution

1. Determine the lagoon water temperature for a surface area of 10,000 m² using Eq. (8–74):

$$T_w = \frac{AfT_a + QT_i}{Af + Q} = \frac{(10{,}000 \text{ m}^2)(0.5)(2) + (3800 \text{ m}^3/\text{d})(15)}{(10{,}000 \text{ m}^2)(0.5) + (3800 \text{ m}^3/\text{d})}$$

$$= 7.6°C$$

2. Determine the lagoon water temperature for a surface of 5000 m².

$$T_w = \frac{AfT_a + QT_i}{Af + Q} = \frac{(5000 \text{ m}^2)(0.5)(2) + (3800 \text{ m}^3/\text{d})(15)}{(5000 \text{ m}^2)(0.5) + (3800 \text{ m}^3/\text{d})}$$

$$= 9.8°C$$

Comment Based on the above computation, the increase in temperature achieved by reducing the surface area and doubling the lagoon depth is probably not justified economically.

In multiple lagoon systems, cold-weather effects can be mitigated by seasonal changes in the method of operation. During the warmer months, the lagoons would be operated in parallel; in the winter, they would be operated in series. In the winter operating mode, the downstream aerators could be turned off and removed, and the lagoon surface is allowed to freeze. In spring when the ice melts, the parallel method of operation is again adopted. With this method of operation it is possible to achieve a 60 to 70 percent removal of BOD even during the coldest winter months. Still another method that can be used to improve performance during the winter months is to recycle a portion of the solids removed by settling.

Oxygen Requirement. The oxygen requirement is computed as previously outlined in Sec. 8–3. Based on operating results obtained from a number of industrial and domestic installations, the amount of oxygen required has been found to vary from 0.7 to 1.4 times the amount of BOD removed. The oxygen requirement is considered in Example 8–14 following the discussion of the mixing requirements.

Mixing Requirements. The power requirements for maintaining the solids in suspension are functions of several variables including (1) the type and design of the aeration system, (2) the concentration and nature of the suspended solids, (3) the temperature of the lagoon contents, and (4) the size and geometry (i.e., aspect ratio) of the lagoon. The following relationship has been found for low-speed mechanical aerators (Rich, 1999):

$$P = 0.004X + 5 \text{ for } X \leq 2000 \text{ mg/L} \tag{8–75}$$

where P = energy input, $\text{kW}/10^3 \text{ m}^3$
X = total suspended solids, mg/L

Other relationships have been proposed by McKinney and Benjes (1965), Fleckseder and Malina (1970), and Murphy and Wilson (1974). The threshold energy input value for the suspension of biosolids is about 1.5 to 1.75 $\text{kW}/10^3 \text{ m}^3$ (7.5 to 8.75 hp/Mgal). The energy requirement to maintain all of the solids in suspension is considerably greater than the threshold value.

As important as the amount of energy required is the placement of the aerators, to achieve effective dispersion of the oxygen and mixing with sufficient overlap of the dispersion and mixing zones. Because so many different types of aerators are used in aerated lagoons (e.g., high and low speed and aspirating aerators, draft tube aerators, static tubes, and perforated tubing) manufacturers' literature should be consulted to determine appropriate aerator spacing. In a report prepared for the U.S. EPA, the recommended maximum spacing between surface aerators should not exceed 75 m (Crown Zellerbach, 1970). In general, a number of smaller aeration devices, spaced more closely, will be more effective than a few larger devices (Rich, 1999). For depths greater than 3.7 m (12 ft), aerators with draft tubes may be considered to prevent solids deposition.

EXAMPLE 8–14 **Design of a Flow-through Aerated Lagoon** Design a flow-through aerated lagoon to treat a wastewater flow of 3800 m^3/d, including the number of surface aerators and their kilowatt rating. The treated liquid is to be held in a settling basin (lagoon) with a 2-d detention time before being discharged. Assume that the following conditions and requirements apply:

1. Influent TSS = 200 g/m^3
2. Influent TSS are not degraded biologically
3. Influent sBOD = 200 g/m^3
4. Effluent sBOD = 30 g/m^3
5. Effluent suspended solids after settling = 20 g/m^3
6. Kinetic coefficients: $Y = 0.65$ g/g, $K_S = 100$ g/m^3, $k = 6.0$ g/g·d, $k_d = 0.07$ g/g·d for $T = 20$ to 25°C
7. Total solids produced are equal to computed volatile suspended solids divided by 0.85
8. First-order observed soluble BOD removal-rate constant $k_{20} = 2.5$ d^{-1} at 20°C. *Note:* The overall BOD removal rate constant, which includes particulate BOD, will be lower (see Table 8–29).

9. Summer air temperature = 30°C
10. Winter air temperature during coldest month = 6°C
11. Wastewater temperature during winter = 16°C
12. Wastewater temperature during summer = 22°C
13. Temperature coefficient, $\theta = 1.06$
14. Aeration constants: $\alpha = 0.85$, $\beta = 1.0$
15. Aerator oxygen transfer rate = 1.8 kg O_2/kWh
16. Elevation = 500 m
17. Oxygen concentration to be maintained in liquid = 1.5 g/m³
18. Lagoon depth = 3.3 m
19. Design SRT = 5 d
20. Power required for mixing = 8 kW/10³/m³

Solution

1. On the basis of an SRT of 5 d, determine the surface area of the lagoon.

   Volume = $Q \times$ SRT = (3800 m³/d)(5 d) = 19,000 m³

   Surface area of lagoon = V/D = 19,000 m³/3.3 m = 5758 m²

2. Estimate the summer and winter liquid temperature using Eq. (8–74).

   Summer:

   $$T_w = \frac{AfT_a + QT_i}{Af + Q} = \frac{(5758 \text{ m}^2)(0.5)(30) + (3800 \text{ m}^3/\text{d})(22)}{(5758 \text{ m}^2)(0.5) + (3800 \text{ m}^3/\text{d})} = 25.4°C$$

   Winter:

   $$T_w = \frac{AfT_a + QT_i}{Af + Q} = \frac{(5758 \text{ m}^2)(0.5)(6) + (3800 \text{ m}^3/\text{d})(16)}{(5758 \text{ m}^2)(0.5) + (3800 \text{ m}^3/\text{d})} = 11.7°C$$

3. Estimate the soluble effluent BOD measured at the lagoon outlet during the summer using Eq. (7–40) from Table 8–5.

   $$S = \frac{K_s[1 + (k_d) \text{ SRT}]}{\text{SRT}(Yk - k_d) - 1}$$

   $$= \frac{(100 \text{ g/m}^3)[1 + (0.07 \text{ g/g·d})(5 \text{ d})]}{(0.5 \text{ d})[(0.65 \text{ g/g})(6.0 \text{ g/g·d}) - (0.07 \text{ g/g·d})] - 1} = 7.4 \text{ g/m}^3$$

   (*Note:* The BOD value in the effluent from the settling lagoon will be essentially the same).

   The effluent sBOD value was computed using kinetic-growth constants derived for the temperature in the range from 20 to 25°C. Thus, during the summer months, the effluent requirement of 20 g/m³ or less will be met easily. Because there is no reliable information on how to correct these constants for the winter temperature of 10°C, an estimate of the effect of temperature can be obtained using the first-order soluble BOD removal-rate constant.

4. Estimate the effluent BOD.
    a. Correct the removal-rate constant for temperature effects using Eq. (2–25) from Table 8–5.

    Summer (25.4°C):

    $$k_{25.4°C} = 2.5(1.06)^{25.4-20} = 3.42/d$$

    Winter (11.7°C):

    $$k_{11.7°C} = 2.5(1.06)^{11.7-20} = 1.54/d$$

    b. Determine the effluent BOD values using Eq. (8–72).

    Summer (25.4°C):

    $$S = \frac{S_o}{[1 + (k)\tau]} = \frac{(200 \text{ g/m}^3)}{[1 + (3.42/d)(5 \text{ d})]} = 11.0 \text{ g/m}^3$$

    Winter (11.7°C):

    $$S = \frac{(200 \text{ g/m}^3)}{[1 + (1.54/d)(5 \text{ d})]} = 22.9 \text{ g/m}^3$$

    Based on the above analysis using removal-rate constants, it appears that the effluent requirement of 30 g/m³ or less will be met during both the summer and winter.

    (*Note:* The foregoing calculations were presented only to illustrate the method. The value of the removal-rate constant must be evaluated for the wastewater in question, in a bench- or pilot-scale test program.)

5. Estimate the concentration of biological solids produced using Eq. (7–43) in Table 8–5. (*Note:* SRT = τ.)

    $$X = \frac{Y(S_o - S)}{[1 + (k_d) \text{ SRT}]} = \frac{(0.65 \text{ g/g})[(200 - 7.4) \text{ g/m}^3]}{[1 + (0.07 \text{ g/g·d})(5 \text{ d})]} = 92.7 \text{ g/m}^3$$

    An approximate estimate of the biological solids produced can be obtained by multiplying the assumed growth yield constant (BOD basis) by the BOD removed.

6. Estimate the suspended solids in the lagoon effluent before settling.

    $$\text{TSS} = 200 \text{ g/m}^3 + \frac{(92.7 \text{ g/m}^3)}{0.85} = 309.1 \text{ g/m}^3$$

    With the extremely low overflow rate provided in a holding basin with a detention time of 2 d, an effluent containing less than 20 g/m³ of suspended solids should be attainable.

7. Estimate the oxygen requirement using Eq. (7–59) in Table 8–5.

    $$R_o = Q(S_o - S) - 1.42 P_{X,bio}$$

a. Determine $P_{X,bio}$, the amount of biological solids wasted per day.

$$P_X = (92.7 \text{ g/m}^3)(3800 \text{ m}^3/\text{d})(1 \text{ kg}/10^3 \text{ g}) = 352.3 \text{ kg/d}$$

b. Assuming the conversion factor for BOD to COD is 1/1.6 (= 0.625), determine the oxygen requirements.

$$R_o = \frac{(3800 \text{ m}^3/\text{d})[(200 - 7.4) \text{ g/m}^3]}{(0.625)(10^3 \text{ g/kg})} - (1.42)(352.3 \text{ kg/d})$$

$$= 670.7 \text{ kg/d}$$

8. Compute the ratio of oxygen required to sBOD removed.

$$\frac{\text{O}_2 \text{ required}}{\text{BOD removed}} = \frac{(670.7 \text{ kg/d})}{(3800 \text{ m}^3/\text{d})[(200 - 7.4) \text{ g/m}^3](1 \text{ kg}/10^3 \text{ g})} = \frac{0.92 \text{ kg O}_2}{\text{kg BOD}}$$

9. Determine the surface aerator power requirements, assuming that the aerators to be used are rated at 1.8 kg O_2/kWh.
    a. Determine the correction factor for surface aerators for summer conditions.
        i. Oxygen saturation concentration at 25.4°C = 8.18 g/m³ (see Appendix E)
        ii. Oxygen saturation concentration at 25.4°C corrected for altitude (see Fig. 5–68)

$$DO = (0.94)(8.18 \text{ g/m}^3) = 7.69 \text{ g/m}^3$$

b. Determine the required AOTR.

$$\text{SOTR} = \text{AOTR}\left[\frac{C_{s,20}}{\alpha(\beta C_{s,T.H} - C)}\right](1.024^{20-T})$$

$$= \frac{(670.7 \text{ kg/d})(9.08 \text{ g/m}^3)(1.024)^{20-25.4}}{(0.85)[(1.0)(7.69 \text{ g/m}^3) - (1.5 \text{ g/m}^3)]} = 1018.3 \text{ kg/d}$$

c. Determine the energy required to supply the needed oxygen.
The amount of O_2 transferred per day per aerator unit is equal to 1.8 kg O_2/kWh. The total power needed to meet the oxygen requirements is

$$\text{Energy} = \frac{(1018.3 \text{ kg/d})}{(1.8 \text{ kg/kWh})(24 \text{ h/d})} = 23.6 \text{ kW}$$

10. Check energy requirement for mixing.
    a. Power requirement

    Lagoon volume = 19,000 m³

    Power required = $(19,000 \text{ m}^3)(8 \text{ kW}/10^3 \text{ m}^3) = 152 \text{ kW}$

    b. Develop aerator options.

Depending upon the lagoon configuration, the following aerator options are possible:

4 — 40 kW aerators
6 — 25 kW aerators
8 — 20 kW aerators

The selection of an aerator option will depend on the geometry of the lagoon site, the type and design of the aerator, and a life cycle cost estimate of the lagoon construction and aerator installation. Special attention should be paid to the zones of mixing and oxygen dispersion (obtained from manufacturers' technical information) to minimize solids deposition and to ensure that aerobic conditions are maintained throughout the lagoon.

Comment In the above example, the energy requirement for mixing is over 6 times the energy requirement to satisfy the oxygen demand. For installations designed to treat domestic wastewater, the energy requirement for mixing is almost always the controlling factor in sizing the aerators. The energy needed to meet the oxygen required is often the controlling factor in sizing the aerators for treating high strength industrial wastes.

Solids Separation. The separation of solids from suspended growth flow through aerated lagoons is accomplished most commonly in a large shallow earthen sedimentation basin (lagoon) designed expressly for the purpose, or in more conventional settling facilities. Where large earthen basins are used, the following requirements must be considered carefully: (1) the detention time must be adequate to achieve the desired degree of suspended solids removal, (2) sufficient volume must be provided for sludge storage, (3) algal growth must be minimized, (4) odors that may develop as a result of the anaerobic decomposition of the accumulated sludge must be controlled, and (5) the need for a lining must be assessed. In some cases, because of local conditions, these requirements may be in conflict with each other.

In most cases, a minimum detention time of 1 d is required to achieve solids separation (Adams and Eckenfelder, 1974). If a 1-d detention time is used, adequate provision must be made for sludge storage so that the accumulated solids will not reduce the actual liquid detention time. Further, if all the solids become deposited in localized patterns, it may be necessary to increase the detention time to counteract the effects of poor hydraulic distribution. Under anaerobic conditions, about 40 to 60 percent of the deposited volatile suspended solids will be degraded each year. Assuming that first-order removal kinetics apply, the following expression can be used to estimate the decay of volatile suspended solids (Adams and Eckenfelder, 1974).

$$W_t = W_o e^{-k_d t} \tag{8-76}$$

where W_t = mass of volatile suspended solids remaining that have not degraded after time t, kg
W_o = mass of solids deposited initially, kg
k_d = decay coefficient, d^{-1} or $year^{-1}$
t = time, d or year

Two problems that are often encountered with the use of settling basins are the growth of algae and the production of odors. Algal growths can usually be controlled by limiting the hydraulic detention time to 2 d or less. If longer detention times must be used, the algal content may be reduced by using either a rock filter or a microstrainer (see Chap. 5). Odors arising from anaerobic decomposition can generally be controlled by maintaining a minimum water depth of 1 m (3 ft). In extremely warm areas, depths up to 2 m (~6 ft) have been needed to eliminate odors, especially those of hydrogen sulfide.

If space for large settling basins is unavailable, conventional settling facilities can be used. To reduce the construction costs associated with conventional concrete and steel settling tanks, lined earthen basins can be used. The design of a large earthen sedimentation basin for an aerated lagoon is illustrated in Example 8–15.

As noted previously, where greater removal of effluent solids is required, it has become common practice to use slow sand or rock filters (Rich, 1999). Rock filters, which may be effective in algae removal, consist of a submerged bed of rocks through which lagoon effluent passes vertically or horizontally. Although rock filters are simple to operate and maintain, effluent quality and dependability of long-term operation are not well documented (U.S. EPA, 1992).

EXAMPLE 8–15 **Design of a Large Earthen Sedimentation Basin (Lagoon) for a Suspended Growth Flow-through Lagoon** Design an earthen sedimentation basin (lagoon) for the suspended growth flow-through lagoon designed in Example 8–14. Assume that the hydraulic detention time is to be 2 d and that the liquid level above the sludge layer at its maximum level of accumulation is 1.25 m. For the purposes of this example, assume that 70 percent of the total solids discharged to the sedimentation basin are volatile and that the effluent from the sedimentation basin will contain 30 g/m³ TSS. Also assume that the sedimentation lagoon is cleaned after 4 years.

Solution

1. Determine the mass of sludge that must be accumulated in the basin each year without anaerobic decomposition.

$$\text{Mass} = Q(\text{TSS}_i - \text{TSS}_e)(365 \text{ d/yr})$$

where TSS_i = total suspended solids in the influent to the sedimentation basin, g/m³

TSS_e = suspended solids in the effluent from the sedimentation basin, g/m³

Q = flowrate, m³/d

a. Compute the total mass of solids added per year.

$$\text{Total solids} = \frac{(3800 \text{ m}^3/\text{d})[(309.1 - 30) \text{ g/m}^3](365 \text{ d/yr})}{(10^3 \text{ g/kg})}$$

$$= 387{,}112 \text{ kg/yr}$$

b. Compute the mass of volatile and fixed solids added per year, assuming that VSS = 0.70 × TSS.

i. Volatile solids:

$$\text{VSS} = 0.70(387{,}112 \text{ kg/yr}) = 270{,}978 \text{ kg/yr}$$

ii. Fixed solids:

$$\text{FSS} = 387{,}112 \text{ kg/yr } (1 - 0.7) = 116{,}134 \text{ kg/yr}$$

2. Determine the amount of sludge that will accumulate at the end of 4 years. Assume that the maximum volatile solids reduction that will occur is equal to 60 percent and that it will occur within 1 year. To simplify the problem, assume that the deposited volatile suspended solids undergo a linear decomposition. Because the volatile solids will decompose to the maximum extent within 1 year, the following relationship can be used to determine the maximum amount of volatile solids available at the end of each year of operation:

$$\text{VSS}_t = [0.7 + 0.4(t - 1)](\text{VSS kg/yr})$$

where $(\text{VSS})_t$ = mass of volatile suspended solids at the end of t yr, kg
$\qquad t$ = time, years

a. Mass of VSS accumulated at the end of 4 yr:

$$\text{VSS}_t = [0.7 + 0.4(4 - 1)] \, (270{,}978 \text{ kg/yr})$$

$$= 514{,}852 \text{ kg}$$

b. Total mass of solids (TSS) accumulated at the end of 4 yr:

$$\text{TSS}_t = 514{,}852 \text{ kg} + (4 \text{ yr})(116{,}134 \text{ kg/yr})$$

$$= 979{,}388 \text{ kg}$$

3. Determine the required liquid volume and the dimensions for the sedimentation basin.
a. Volume of sedimentation basin:

$$V = (2 \text{ d})(3800 \text{ m}^3/\text{d}) = 7{,}600 \text{ m}^3$$

b. Surface area of sedimentation basin:

$$A_s = \frac{(7600 \text{ m}^3)}{1.25 \text{ m}} = 6080 \text{ m}^2$$

The aspect ratio for the surface area of the sedimentation basin (ratio of width to length) depends on the geometry of the available site.

4. Determine the depth required for the storage of sludge.
a. Determine the mass of accumulated sludge per square meter.

$$\text{Mass per unit area} = \left(\frac{\text{accumulated sludge}}{\text{surface area of lagoon}} \right)$$

$$= \frac{979{,}388 \text{ kg}}{6080 \text{ m}^2} = 161.1 \text{ kg/m}^2$$

b. Determine the required depth d of the sludge blanket, assuming that the deposited solids are distributed evenly in the lagoon and will compact to an average final value of 15 percent of the initial solids volume and that the density of the accumulated solids is equal to 1.06.

$$d = \frac{(161.1 \text{ kg/m}^2)}{(1.06)(0.15)(1000 \text{ kg/m}^3)}$$

$$d = 1.0 \text{ m}$$

c. Total depth of basin excluding freeboard is

Total depth = 1.25 m + 1.0 m = 2.25 m

Comment Where it is difficult to provide the total required depth, it may be necessary to increase the detention time or clean the sedimentation basins more frequently.

Dual-Powered Flow-Through Lagoon System

The dual-powered flow-through lagoon system is a modification of the conventional flow-through lagoon pioneered by Professor Rich (1982, 1996, and 1999). The dual-powered flow-through lagoon system is comprised of a complete-mix lagoon followed by two or three facultative lagoons that serve as settling basins (see Fig. 8–48). The operating principle of the dual-powered flow-through lagoon system may be outlined as follows: From basic considerations, it can be reasoned that secondary treatment involves the following functions: (1) bioconversion of the influent substrate to biomass and end products, (2) flocculation of the biomass, (3) separation of the solids, (4) stabilization of the solids, and (5) storage of the solids until they are reused or disposed of. These functions are optimized in the dual-powered flow-through lagoon system. Bioconversion and flocculation occur in the first complete-mix lagoon. The solids separation, stabilization, and storage are accomplished in the facultative lagoons following the complete-mix lagoon. To minimize the growth of algae, the detention time should be limited and the facultative lagoons should be divided into a series of cells (Rich, 1999).

Figure 8–48

Dual-power flow-through lagoon system: (a) with separate facultative lagoon and (b) with contiguous facultative lagoon.

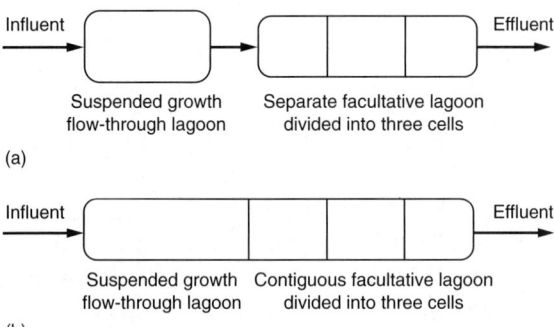

Influent → Effluent

Suspended growth flow-through lagoon

Separate facultative lagoon divided into three cells

(a)

Influent → Effluent

Suspended growth flow-through lagoon

Contiguous facultative lagoon divided into three cells

(b)

The hydraulic detention time for the complete-mix lagoon will typically vary from 1.5 to 3 d. The energy input in the complete-mix lagoon is on the order of 6 kW/10^3 m^3 (30 hp/Mgal). The total retention time for the facultative lagoons is on the order of 3 d. The energy input into the facultative lagoons is below the threshold value for the suspension of solids, but is sufficient to oxidize soluble organic constituents released from the anaerobic degradation of the solids accumulated on the bottom of the facultative lagoons. The corresponding energy input of the facultative lagoons is on the order of 1.0 to 1.25 kW/10^3 m^3 (5 to 6.25 hp/Mgal). The overall detention time for the two lagoons is typically 4.5 to 6 d. The same procedure used for the design of the flow-through suspended growth lagoon, as detailed in Example 8–14, is used for the design of the complete-mix lagoon of the dual-powered lagoon system. The design procedure for the facultative lagoons is similar to the procedure used in Example 8–15 for the design of a large earthen sedimentation basin. Additional details on dual-powered flow-through lagoon systems may be found in Rich (1999).

8–9 BIOLOGICAL TREATMENT WITH MEMBRANE SEPARATION

Membrane biological reactors (MBRs), consisting of a biological reactor (bioreactor) with suspended biomass and solids separation by microfiltration membranes with nominal pore sizes ranging from 0.1 to 0.4 μm, are finding many applications in wastewater treatment. Membrane biological reactor systems may be used with aerobic or anaerobic suspended growth bioreactors to separate treated wastewater from the active biomass (see Fig. 8–49). The membrane systems described in this section produce an effluent quality equal to the combination of secondary clarification and effluent microfiltration. Membrane biological reactors have been used for treatment of both municipal and industrial wastewater (Brindle and Stephenson, 1996; Van Dijk and Roncken, 1997; Trussell et al., 2000) and for water-reuse applications (Cicek et al., 1998). Membrane systems used in advanced wastewater applications are discussed in Chap. 11. In Sec. 11–6 of Chap. 11, various types of membrane materials, membrane designs, and operating conditions are also described. An overview of membrane biological reactors, process description, operational issues, and process capabilities is described in this section.

Overview of Membrane Biological Reactors

The concept of MBR systems consists of utilizing a bioreactor and microfiltration as one unit process for wastewater treatment thereby replacing, and in some cases supplementing, the solids separation function of secondary clarification and effluent filtration. The ability to eliminate secondary clarification and operate at higher MLSS concentrations provides the following advantages: (1) higher volumetric loading rates and thus shorter reactor hydraulic retention times; (2) longer SRTs resulting in less sludge production; (3) operation at low DO concentrations with potential for simultaneous nitrification-denitrification in long SRT designs; (4) high-quality effluent in terms of low turbidity, bacteria, TSS, and BOD; and (5) less space required for wastewater treatment. Disadvantages of MBRs include high capital costs, limited data on membrane life, potential high cost of periodic membrane replacement, higher energy costs, and the need to control membrane fouling.

Figure 8–49

Schematic diagram of membrane bioreactors: (a) integrated MBR with an immersed membrane module, and (b) bioreactor with an external membrane separation unit.

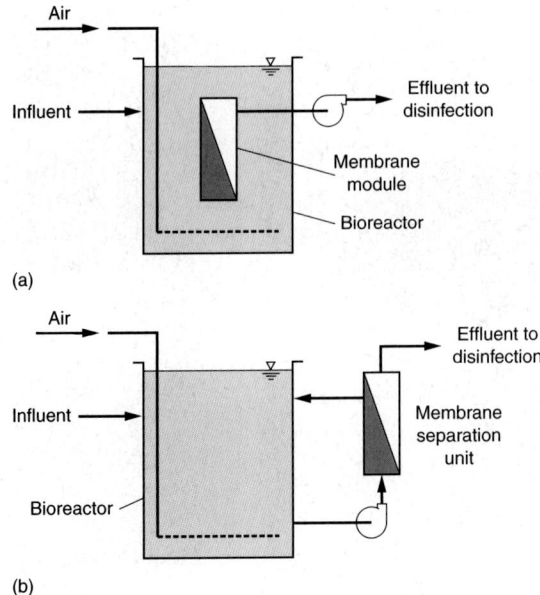

(a)

(b)

Process Description

Membrane bioreactor systems have two basic configurations: (1) the integrated bioreactor that uses membranes immersed in the bioreactor and (2) the recirculated MBR in which the mixed liquor circulates through a membrane module situated outside the bioreactor.

In the integrated MBR system shown on Fig. 8–49a, the key component is the microfiltration membrane that is immersed directly into the activated-sludge reactor. The membranes are mounted in modules (sometimes called cassettes) that can be lowered into the bioreactor. The modules are comprised of the membranes, support structure for the membranes, feed inlet and outlet connections, and an overall support structure. The membranes are subjected to a vacuum (less than 50 kPa) that draws water (permeate) through the membrane while retaining solids in the reactor. To maintain TSS within the bioreactor and to clean the exterior of the membranes, compressed air is introduced through a distribution manifold at the base of the membrane module. As the air bubbles rise to the surface, scouring of the membrane surface occurs; the air also provides oxygen to maintain aerobic conditions.

A proprietary integrated MBR system is shown on Fig. 8–50, the Zenogem® process manufactured by Zenon Environmental, Inc. The Kubota submerged membrane bioreactor marketed by Enviroquip is similar. The Zenogem® microfiltration cassette shown on Fig. 8–50 is composed of hollow fiber membranes and has overall dimensions of 0.91 m wide by 2.13 m long, and approximately 2.44 m high (3 ft by 7 ft by 8 ft). Support facilities required include permeate pumps, chemical storage tanks, chemical feed pumps and all process controls for the membranes, including the motor control center for the membrane process. The membrane support equipment also includes an air-scour system and a back-pulse water-flushing system. The air-scour system consists of coarse bubble diffusers located in the aeration basin and provides continuous agitation on the

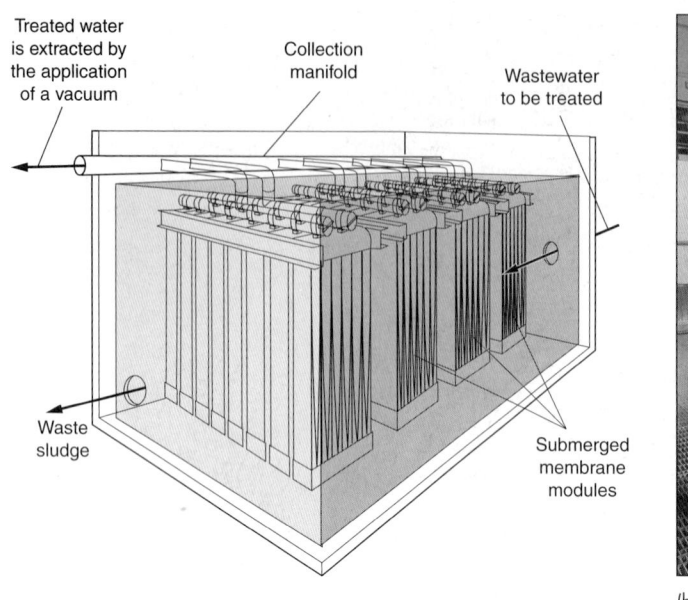

(b)

Figure 8–50

Typical membrane bioreactors: (a) schematic of placement of membrane bundles in an activated sludge reactor and (b) membrane bundle in position to be placed in a membrane bioreactor. *(Courtesy Zenon Environmental, Inc.)*

outside of the membranes to minimize solids deposition. The air supply for the air-scour system is typically provided in addition to the activated-sludge process air.

The membrane unit manufactured by Kubota consists of a series of membrane cartridges composed of fine porous membranes mounted on both sides of a supporting plate. The cartridges, which have a thickness of 6 m, can be removed individually for cleaning and replacement. Each of the cassettes or modules has a series of tubes that extract effluent from the membranes and connect to a collection manifold. The permeate support facilities are similar to those for the Zenon system.

As of 2001, in addition to Zenon and Kubota, submerged MBR systems in the United States are also marketed by Mitsubishi Rayon Corporation (Mitsubishi) from Japan (Merlo et al., 2000).

Besides submerged membranes, an in-line, recirculated MBR design is being marketed by Lyonnaise-des-Eaux Degremont of France. For this device, activated sludge from the bioreactor is pumped to a pressure-driven tubular membrane where solids are retained inside the membrane and water passes through to the outside. The driving force is the pressure created by high cross velocity through the membrane. The solids are recycled to the activated-sludge basin. The membranes are backwashed systematically to remove solids and cleaned chemically to control pressure buildup.

By replacing solids separation by gravity settling in secondary clarifiers, membranes avoid issues of filamentous sludge bulking and other floc settling and clarification problems, and the aeration tank MLSS concentration is no longer controlled by secondary clarifier solids loading limitations. The MBR systems can operate at much higher MLSS concentrations (15,000 to 25,000 mg/L) than conventional activated-sludge processes (Cote et al., 1998). Although high concentrations of MLSS as noted above have been reported, MLSS concentrations in the range of 8000 to 10,000 mg/L appear to be most cost-effective when all factors are considered.

The membrane flux rate, defined as the mass or volume rate of transfer through the membrane surface [in terms of $L/m^2 \cdot h$ $(gal/ft^2 \cdot d)$] is an important design and operating parameter that affects the process economics. Lower flux rates are expected at higher MLSS concentrations. As more applications occur for MBR with activated-sludge treatment, more information will be available for long-term membrane flux rates and membrane life.

Membrane Fouling Control

In the activated-sludge reactor, biomass coats the outer layer of the membranes used in an integrated MBR during effluent withdrawal. Finer particles may penetrate the inner pores of the membrane, causing an increase in pressure loss. Continuous membrane fouling control methods are used during the operation of the MBR, with periodic more aggressive cleaning to maintain the filtration capacity of the membrane.

A method developed by Zenon Environmental to control fouling on the outside surfaces of the membrane fibers involves a three-step process. First, coarse bubble aeration is provided at the bottom of the membrane tank directly below the membrane fibers. The air bubbles flow upward between the vertically oriented fibers, causing the fibers to agitate against one another to provide mechanical scouring. Second, filtration is interrupted about every 15 to 30 min and the membrane fibers are backwashed with permeate for 30 to 45 s (Giese and Larsen, 2000). The system remains on-line during backwashing, and the total time for backflushing is about 45 min per day. Typically, a low concentration of chlorine (<5 mg/L) is maintained in the backflush water to inactivate and remove microbes that colonize the outer membrane surface. Third, about three times per week a strong sodium hypochlorite solution (about 100 mg/L) or citric acid is used in the backflush mode for 45 min in a procedure called a "maintenance clean." After the 45-min *in situ* cleaning, the system is flushed with permeate for 15 min. An additional permeate flush-to-drain operation is performed for 10 to 15 min to purge the system of free chlorine once the vacuum is initiated. The total system downtime during a maintenance clean is about 75 min.

The back-pulse system is similar to a typical flushing or cleaning system. A chlorine solution is pumped periodically through the membranes in the reverse direction, similar to backwashing a filter. The cassettes can be removed easily from the aeration basin by an overhead hoist system for a periodic chemical-bath cleaning. When removed for cleaning, the cassettes are submerged in a high-concentration chlorine solution bath in a separate small tank or basin located adjacent to the aeration basin. External cleaning occurs about every 3 to 6 months.

The combination of air scour, backflushing, and maintenance cleaning is not completely effective in controlling membrane fouling, and the pressure drop across the membrane increases with time. The pressure drop across the membranes is monitored to indicate fouling problems and cleaning needs. At a maximum operating pressure drop of ~60 kPa, the membranes are removed from the aeration basin for a recovery cleaning (Fernandez et al., 2000). During recovery cleaning, a membrane cassette is soaked in a tank containing a 1500 to 2000 mg/L sodium hypochlorite solution for

Table 8–30
Typical operational and performance data for a membrane bioreactor[a]

Parameter	Unit	Range
Operational data		
COD loading	kg/m³·d	1.2–3.2
MLSS	mg/L	5000–20,000
MLVSS	mg/L	4000–16,000
F/M	g COD/g MLVSS·d	0.1–0.4
SRT	d	5–20
τ	h	4–6
Flux	L/m²·d	600–1100
Applied vacuum	kPa	4–35
DO	mg/L	0.5–1.0
Performance data		
Effluent BOD	mg/L	<5
Effluent COD	mg/L	<30
Effluent NH₃	mg/L	<1
Effluent TN	mg/L	<10
Effluent turbidity	NTU	<1

[a]From Stephenson et al. (2000).

about 24 h. Spare membranes are typically installed in the aeration tank during recovery cleaning so that there is no reduction in treatment capacity. A similar fouling control procedure has been reported for the Mitsubishi MBR membrane (Merlo et al., 2000). In the Kubota MBR process, the flat-plate membranes are not removed for cleaning, and an infrequent backflush with a 0.5 percent solution of hypochlorite has been shown to be effective (Stephenson et al., 2000).

Process Capabilities

The treatment capability of MBR is evaluated in terms of BOD, TSS, coliform, and nitrogen removal based on laboratory, pilot-plant, and full-scale plant studies. Because the activated-sludge effluent from MBRs is treated by filtration through a nominal 0.40-μm membrane, very low concentrations of effluent suspended solids, turbidity, and BOD are produced that provide an effluent suitable for water reuse following disinfection. Reported operating performance characteristics for MBR systems are summarized in Table 8–30. Low effluent BOD and turbidity concentrations are possible for MBR systems with MLSS concentrations in the range of 6000 to 16,000 mg/L.

Full-scale and pilot-plant systems have been operated with the anoxic/aerobic MLE biological nitrogen-removal process with the result that effluent total nitrogen concentrations of <10 mg/L have been achieved (Mourato et al., 1999; ReVoir et al., 2000; and Giese et al., 2000). Influent recycle flowrate ratios of 4.0 to 6.0 have been used in those studies to feed nitrate to a separate preanoxic tank.

Nitrogen removal in MBR systems has also been done in bench- and pilot-scale tests by using intermittent aeration or operation at low DO concentration to achieve SNdN. An intermittent aeration study was conducted with an MLSS concentration of 6000 mg/L and on/off aeration intervals of 15 min (Fernandez et al., 2000). An effluent total nitrogen concentration ranging from 7 to 10 mg/L was achieved. Full nitrification was not possible at high MLSS concentrations (up to 13,000 mg/L), probably due to DO limitations in the activated sludge. In a laboratory study, an MBR was operated successfully at low DO concentrations (<1.0 mg/L) to accomplish SNdN (Choo and Stensel, 2000).

8–10 SIMULATION DESIGN MODELS

Activated-sludge processes used today are much more complex with their applications in single-sludge systems expanding from BOD removal to nitrification, denitrification, and phosphorus removal. The reactions involved in these processes are carried out by different types of bacteria that include a mixture of heterotrophic bacteria that are phosphorus-storing and non-phosphorus-storing, and among these, bacteria that can and cannot use nitrate as an electron acceptor, as well as autotrophic nitrifying bacteria. Competition exists between various types of heterotrophic bacteria for carbonaceous substrates. Furthermore, the importance of the wastewater components with regard to nonsoluble, soluble, biodegradable, and nonbiodegradable substrates on reaction rates, oxygen consumption, and sludge production is better understood. Computer modeling provides the tool to incorporate the large number of components and reactions to evaluate activated-sludge performance under both dynamic and steady-state conditions, and to easily design multiple-staged reactors as well as a single-stage complete-mix reactor.

A long list of complex equations would be needed to describe the various reactions in an activated-sludge process involving numerous components such as organic substrates (soluble and particulate), inorganic substrates (ammonia, nitrate, and phosphorus), dissolved oxygen, and various heterotrophic and autotrophic bacteria. Instead of showing the model in terms of numerous equations, a universal approach is to use a matrix model format that simply shows the process reactions and stoichiometric factors that link the components to the various reactions. The advantage is a simple, relatively short format is developed that describes the process. The purpose of this section is to provide a basic introduction to the matrix model approach showing the components and reactions for the model in its simplest form for the activated-sludge process and how the matrix format describes the process. Further, the approach also shows how the matrix model can be interpreted to describe a complete set of equations for a given process component. For example, the Activated Sludge Model No. 1 (ASM1) (Henze et al., 1987*a*) is described, though a more comprehensive version that incorporates biological phosphorus-removal mechanisms is available (Henze et al., 1995; Barker and Dold, 1997). Various software programs are available that apply models for different reactor configurations.

The ASM1 model only describes reactions by heterotrophic bacteria under aerobic and anoxic conditions consuming carbonaceous substrates and autotrophic nitrifying bacteria oxidizing ammonia to nitrate. A more complex model that includes phosphorus-storing bacteria with appropriate anaerobic, anoxic, and aerobic reactions has been developed and is termed the ASM2 model (Henze et al., 1995). The ASM2 model is incorporated into the model results presented by Barker and Dold (1997).

The models are based on *growth* as opposed to *substrate utilization.* Monod specific growth rate kinetics are used in the model to describe the growth of autotrophic or heterotrophic bacteria; and the substrate, oxygen, and nutrient utilization rates are related to the growth rates by stoichiometric factors. Another important feature of the model is that it uses COD as the common measure of organic substrate and biomass, so that a COD balance exists for substrate utilization, biomass growth, and oxygen consumption. The models also use the lysis-regrowth model for endogenous respiration instead of a net endogenous decay coefficient as presented in this chapter for activated-sludge design. In the lysis-regrowth model, endogenous decay results in the release of biomass particulate material, some of which is biodegradable and is hydrolyzed to provide a source of rbCOD. Another portion remains as cell debris, similar to what has been incorporated for endogenous respiration in the activated-sludge design model presented in this chapter. In order to compute the same amount of sludge production as the endogenous respiration model, the decay coefficient value in the lysis-regrowth model is higher.

Model Matrix Format, Components, and Reactions

A convenient matrix format is used to describe the model without having to present the large number of equations involved. The ASM1 model with components and reaction terms is shown in Table 8–31. The model follows 13 wastewater characterization components, which are defined in Table 8–32. These components have been presented previously in Sec. 8–2. Complete wastewater characterization is an essential feature for using ASM1 and ASM2 models. Components 1 to 5 and 12 are particulate organic material, represented as COD, and are components of the activated-sludge MLVSS concentration. Components 6 to 11 are all soluble substrates and include electron acceptors (oxygen, S_o and nitrate, S_{NO}). The soluble nonbiodegradable COD is represented by S_i, and as shown, no stoichiometric factors are given as it is not involved in any reactions.

Various reaction terms that are described as processes are shown in the last column of Table 8–31. For example, r_j (where $j = 2$) is the growth reaction for heterotrophic bacteria under anoxic conditions. The heterotrophic bacteria growth rate is affected by the readily degradable substrate concentration (S_s), the DO concentration (S_o), and the nitrate concentration (S_{NO}). The term η represents the fraction of heterotrophic bacteria that can use nitrate in place of DO. For cell decay, the death-lysis model is used, which is shown by processes 4 and 5 for heterotrophic and autotrophic bacteria, respectively. The cell debris material produced during biomass decay is indicated by X_o and the production of degradable particulate substrate is indicated by a stoichiometric factor for X_s. The conversion of soluble organic nitrogen (S_{NS}) to ammonia (S_{NH}) is described by process 6. The hydrolysis of particulate organics and particulate organic nitrogen into soluble substrates (S_s and S_{NS}, respectively) under either aerobic or anoxic conditions is described by processes 7 and 8, respectively.

Within the matrix are stoichiometric factors that relate the rate of change in a particular component to a process reaction rate. For example, the process rate j_1 is the growth rate of aerobic heterotrophic bacteria in mg/L·d. Stoichiometric factors relate changes in model components to the growth rate. For component 3, the stoichiometric factor is 1 as

$$R_{B,H} = \hat{\mu}_H \left(\frac{S_s}{K_s + S_s} \right) \left(\frac{S_o}{K_{O,H} + S_o} \right) X_{B,H} \qquad (8\text{–}77)$$

where $R_{B,H}$ = growth rate of heterotrophic biomass, g/m³·d
S_s = readily biodegradable substrate concentration, g COD/ m³
$\hat{\mu}_H$ = maximum specific growth rate, g VSS/g VSS·d
K_s = half-velocity coefficient for readily biodegradable substrate, g COD/ m³
S_o = dissolved oxygen concentration, g/m³
$K_{O,H}$ = half velocity coefficient for dissolved oxygen concentration, g/m³
$X_{B,H}$ = heterotrophic biomass concentration g/m³

Component 7, the readily degradable substrate S_s (see Table 8–32), is used for growth and $-(1/Y_H)$ relates the rate of change in S_s to the rate of heterotrophic growth:

$$R_{s_s} = -\frac{1}{Y_H} (RX_{B,H}) \tag{8-78}$$

The total rate of change for S_s is equal to the sum of the stoichiometric factors in the column below S_s times the respective process rate.

The stoichiometric terms for oxygen consumption are described as follows: For heterotrophic growth ($\Psi_{j,1}$), the term $(1-Y_H)$ is the g O_2 used/g COD removed. The term $(1-Y_H)$ is divided by Y_H (g cell COD/g COD used) to obtain the stoichiometric factor as g O_2/g cell COD produced, to fit to the matrix format. The stoichiometric term for autotrophic growth contains the factor 4.57. The term is required because ammonia, the substrate for the nitrifying bacteria, is expressed as nitrogen in the matrix S_{NH}, and oxygen is expressed as COD. The oxygen equivalent for ammonia is 4.57 g O_2/g NH_4-N. The amount in the numerator is lower by Y_A, which accounts for the ammonia used in cell synthesis.

Model Applications

The models may be used for a number of purposes: (1) as a research tool to evaluate biological processes and to better understand important parameters that affect a certain type of performance, (2) for wastewater treatment-plant design, and (3) for evaluating the treatment capacity of a given facility. For items 2 and 3, accurate and representative wastewater characterization is critical and dynamic simulations are often used to include the effect of variable flow and concentrations with time.

To evaluate the capacity of an existing plant, the model is calibrated using wastewater characterization and plant performance data. Calibrations based only on comparing the model predictions to the plant effluent concentration are not necessarily valid. Low effluent concentrations normally exist for all soluble degradable components, and thus the ability of the model to predict the plant performance is clouded by analytical accuracy and practical considerations. Intermediate soluble substrate concentrations from aerobic stages and/or anoxic and anaerobic stages provide a more reliable indication of the ability to describe the kinetics for the site. Oxygen uptake rate data are more meaningful for model calibration, since the data reflect different factors in the model including kinetic rates for different reactions and stoichiometric ratios for cell yield and decay. Where oxygen uptake rate data can be obtained for staged systems, it provides a valuable set of data for model calibration.

Typical values for model parameters have been selected and are summarized in Table 8–33. These are referred to as default values, but do not ensure that the model can then predict the performance of an activated sludge accurately, as some of the coefficient values can be different at different sites. One of the parameters found to vary the most at different sites and often adjusted during model calibration is the maximum specific growth rate of nitrifying bacteria μ_A. Variations in nitrification kinetics may be due to differences in wastewater characteristics and inhibitors on nitrification

Table 8-31
Process kinetics and stoichiometry for multiple events in suspended growth cultures as presented by the IAWQ Task Group for mathematical modeling

Component	1	2	3	4	5	6	7	8	9	10
Process	X_i	X_s	$X_{B,H}$	$X_{B,A}$	X_D	S_i	S_s	S_o	S_{NO}	S_{NH}
1. Aerobic growth of heterotrophs			1				$-\dfrac{1}{Y_H}$	$\dfrac{1-Y_H}{Y_H}$		$-i_{N/XB}$
2. Anoxic growth of heterotrophs			1				$-\dfrac{1}{Y_H}$		$-\dfrac{1-Y_H}{2.86\,Y_H}$	$-i_{N/XB}$
3. Aerobic growth of autotrophs				1				$\dfrac{4.57-Y_A}{Y_A}$	$\dfrac{1}{Y_A}$	$-i_{N/XB}-\dfrac{1}{Y_A}$
4. Death and lysis of heterotrophs		$1-f'_D$	-1		f'_D					
5. Death and lysis of autotrophs		$1-f'_D$		-1	f'_D					
6. Ammonification of soluble organic nitrogen										
7. Hydrolysis of particulate organics		-1					1			1
8. Hydrolysis of particulate organic nitrogen										
Observed conversion rates, $ML^{-1}T^{-1}$						$r_i = \displaystyle\sum_{j=1}^{n} \psi_{ij} r_j$				

Note: All organic compounds (cols. 1–7) and oxygen (8) are expressed as COD; all nitrogenous components (9–12) are expressed as nitrogen. Coefficients must be multiplied by -1 to be expressed as oxygen.

or may also reflect other kinetic changes for which μ_A adjustments improve the overall fit.

The example on page 865 is intended to illustrate how the model matrix is interpreted to describe the rate of concentration change of a model component. Additional information on application of the model may be found in U.S. EPA (1993).

| **Table 8-31** (Continued)

11	12	13	Process rate r, $ML^{-3}T^{-1}$
S_{NS}	X_{NS}	S_{Alk}	
		$\dfrac{-i_{N/XB}}{14}$	$\hat{\mu}_H\left(\dfrac{S_s}{K_s + S_s}\right)\left(\dfrac{S_o}{K_{O,H} + S_o}\right)X_{B,H}$
		$\dfrac{1 - Y_H}{14(2.86)Y_H}$ $\dfrac{-i_{N/XB}}{14}$	$\hat{\mu}_H\left(\dfrac{S_s}{K_s + S_s}\right)\left(\dfrac{K_{O,H}}{K_{O,H} + S_o}\right)\left(\dfrac{S_{NO}}{K_{NO} + S_{NO}}\right)\eta_g X_{B,H}$
		$\dfrac{-i_{N/XB}}{14} - \dfrac{1}{7Y_A}$	$\hat{\mu}_A\left(\dfrac{S_{NH}}{K_{NH} + S_{NH}}\right)\left(\dfrac{S_o}{K_{O,A} + S_o}\right)X_{B,A}$
	$-i_{N/X_H} - f_{D,N/X_D}$		$b_{LH}X_{B,H}$
	$-i_{N/X_A} - f_{D,N/X_O}$		$b_{LA}X_{BA}$
-1		$\dfrac{1}{14}$	$k_a S_{NS}X_{B,H}$
			$k_h\left[\dfrac{(X_s/X_{B,H})}{K_h + (X_s/X_{B,H})}\right]\left[\left(\dfrac{S_o}{K_{O,H} + S_o}\right) + \eta_h\left(\dfrac{K_{O,H}}{K_{O,H} + S_o}\right)\left(\dfrac{S_{NO}}{K_{NO} + S_{NO}}\right)\right]X_{B,H}$
1	-1		$r_7(X_{NS}/X_S)$

Table 8–32
Definition of ASM1 model components in Table 8–31

Component		Definition
Number	**Symbol**	
1	X_i	Inert particulate organic matter, mg/L
2	X_s	Slowly biodegradable substrate, mg/L as COD
3	$X_{O,H}$	Heterotrophic biomass, mg/L as COD
4	$X_{O,A}$	Autotrophic biomass, mg/L as COD
5	X_o	Debris from biomass death and lysis, mg/L as COD
6	S_i	Inert soluble organic matter, mg/L as COD
7	S_s	Readily biodegradable substrate, mg/L as COD
8	S_o	Dissolved oxygen, mg/L as O_2 (COD)
9	S_{NO}	Nitrate–nitrogen, mg/L as N
10	S_{NH}	Ammonia–nitrogen, mg/L as N
11	S_{NS}	Soluble biodegradable organic nitrogen, mg/L as N
12	X_{NS}	Particulate degradable organic nitrogen, mg/L as N
13	S_{ALK}	Alkalinity, molar units

Table 8–33
Typical stoichiometric parameter values for the ASM1 model[a]

Symbol	Description	Units	Value
Y_H		g biomass COD/g COD used	0.60
f_D		g cell debris/g biomass COD	0.08
i_n/xb		g N/g active biomass COD	0.086
i_n/xd		g N/g biomass debris COD	0.06
Y_A		g biomass COD/g N oxidized	0.24
μ_H	Maximum specific growth rate of heterotrophic bacteria	d^{-1}	6.0
K_s	Half-velocity constant for heterotrophic bacteria	mg/L	20.0
$K_{O,H}$	DO half-velocity constant	mg/L	0.10
K_{NO}	Nitrate half-velocity constant	mg/L	0.20
$b_{L,H}$	Decay/lysis coefficient, heterotrophic	g/g·d	0.40
η_g	Fraction of heterotrophs using nitrate under anoxic growth	g/g	0.80
η_h	Anoxic/aerobic hydrolysis rate fraction	g/g	0.40
K_a	Ammonification rate constant	L/mg COD·d	0.16
K_h	Particulate hydrolysis maximum specific rate constant	g/g·d	2.21
K_X	Hydrolysis half-velocity constant	g/g·d	0.15
μ_A	Autotrophic maximum specific growth rate	g/g·d	0.76
K_{NH}	Autotrophic half-velocity constant	mg/L	1.0
$K_{O,A}$	Half-velocity DO constant for autotrophs	mg/L	0.75
$b_{L,A}$	Decay constant for autotrophs	g/g·d	0.07

[a] Adapted from Grady et al. (1999).

EXAMPLE 8–16 **Apply ASM1 Model Matrix** Use the ASM1 model matrix to describe the reaction rates for (1) readily degradable COD (S_s) and (2) particulate organic substrates (X_s).

Solution For S_s, $i = 7$, from matrix there are reaction terms where $j = 1, 2, 7$. These stoichiometric factors are $j = 1$, factor $= -1/Y_H$, $j = 2$, factor $= -1/Y_H$, $j = 7$, factor $= 1$. Multiplying the stoichiometric factors by the process rate terms (r_j) yields the following for the rate of change of S_s:

$$R_{S_s} = -\frac{1}{Y_H} \mu_H \left(\frac{S_s}{K_s + S_s}\right)\left(\frac{S_o}{K_{O,H} + S_o}\right)X_{B,H} \tag{1}$$

$$-\frac{1}{Y_H} \mu_H \left(\frac{S_s}{K_s + S_s}\right)\left(\frac{K_{O,H}}{K_{O,H} + S_o}\right)\left(\frac{S_{NO}}{K_{NO} + S_{NO}}\right)\eta X_{B,H} \tag{2}$$

$$+ k_h \left[\frac{(X_s/X_{B,H})}{K_X + (X_s/X_{B,H})}\right]\left(\frac{S_o}{K_{O,H} + S_o}\right) + \eta_h\left(\frac{K_{O,H}}{K_{O,H} + S_o}\right)\left(\frac{S_{NO}}{K_{NO} + S_{NO}}\right)X_{B,H} \tag{7}$$

For X_s, $i = 2$, from matrix there are reaction terms where $j = 4, 5,$ and 7.

$$R_{X_s} = (1 - f_d') b_{L,H} X_{B,H} + (1 - f_d') b_{L,A} X_{B,A}$$
$$- k_h \left[\frac{(X_s/X_{B,H})}{K_X + (X_s/X_{B,H})}\right]\left(\frac{S_o}{K_{O,H} + S_o}\right)$$
$$+ \eta_b\left(\frac{K_{O,H}}{K_{O,H} + S_o}\right)\left(\frac{S_{NO}}{K_{NO} + S_{NO}}\right)X_{B,H}$$

PROBLEMS AND DISCUSSION TOPICS

8–1 Given the following laboratory test BOD and UBOD test results for one of the following wastewaters (to be selected by the instructor), determine the biodegradable COD (bCOD) concentration. Assume that values for f_d and Y_H are 0.15 and 0.40, respectively.

Test Parameter	Unit	Wastewater		
		A	B	C
BOD	mg/L	120	200	200
UBOD	mg/L	180	300	340

8–2 An influent wastewater sample is evaluated in a laboratory respirometer test to determine its readily biodegradable COD (rbCOD) concentration. The respirometer bottle is prepared by adding 500 mL of the influent sample with 500 mL of an activated-sludge mixed liquor. The respirometer records the accumulative oxygen consumption with time. The oxygen consumption occurs at a relatively constant rate initially (Phase A), and then the rate declines within 3 minutes to another relatively constant rate (Phase B), which continues for a number of hours. Finally, the oxygen consumption rate declines dramatically again and continues at a relatively constant rate (Phase C). The respirometer data is summarized in the following table for three different samples (to be selected by instructor). For the selected sample determine the rbCOD concentration of the wastewater in mg/L, assuming the biomass yield for the heterotropic bacteria is 0.4 g VSS/g COD and the oxygen equivalent of the biomass is 1.42 g COD/g VSS.

| Phase | Duration of phase, h | Respirometer accumulative oxygen consumption for each phase, mg | | |
| | | Sample | | |
		A	B	C
A	0.8	64	80	58
B	3.2	192	288	192
C	2.0	40	50	46

8-3 For one of the following wastewater samples (as selected by instructor) shown in the following table, with values for conventional wastewater characterization parameters, determine (a) the non-biodegradable volatile suspended solids (nbVSS) concentration, (b) the inert total suspended solids (iTSS) concentration, and (c) the average COD/VSS ratio. Assume that the bCOD/BOD ratio equals 1.6.

| Parameter | Unit | Wastewater | | |
		A	B	C
TSS	mg/L	220	220	225
VSS	mg/L	200	200	210
BOD	mg/L	200	155	270
sBOD	mg/L	80	80	170
COD	mg/L	544	400	700
sCOD	mg/L	160	160	300

8-4 Using one of the following wastewater characteristics below (as selected by instructor), determine: (a) biodegradable COD (bCOD), (b) slowly biodegradable COD (sbCOD), (c) non-biodegradable COD (nbCOD), and (d) nonbiodegradable soluble and particulate COD (nbsCOD and nbpCOD) concentrations. Assume that the bCOD/BOD and biodegradable soluble COD/sBOD ratios are both 1.6.

| Parameter | Unit | Wastewater | | |
		A	B	C
COD	mg/L	450	550	420
sCOD	mg/L	130	170	110
rbCOD	mg/L	40	60	30
BOD	mg/L	200	220	160
sBOD	mg/L	60	80	50

8-5 Given the following wastewater characteristics (as selected by instructor), determine (a) the organic nitrogen, (b) the nonbiodegradable particulate organic nitrogen (nbpON), and (c) the biodegradable organic nitrogen (bON) concentrations.

Parameter	Unit	Wastewater		
		A	B	C
TKN	mg/L	40	45	50
NH$_4$-N	mg/L	25	30	35
Soluble nonbiodegradable organic nitrogen	mg/L	1.5	1.5	1.5
VSS	mg/L	180	180	190
Nonbiodegradable VSS fraction	Percent	40	40	40

8-6 Using an observed yield value from Fig. 8–7, and Eqs. (8–14) and (7–54) (see Table 8–5), determine (a) the aeration tank volume (m^3) and (b) the amount of sludge wasted daily in kg TSS/d for an activated-sludge system designed to treat a 6000 m^3/d wastewater flow with an influent BOD concentration of 120, 140, or 160 mg/L (as selected by instructor). The SRT is 6 d, the mixed-liquor temperature is 10°C, and primary treatment is used. What is the aeration tank volume and amount of daily sludge production if the SRT is increased to 12 d? Assume that the MLVSS and MLSS concentrations are 2500 mg/L and 3000 mg/L, respectively.

8-7 The following information is given for an activated-sludge system design:

Parameter	Unit	Value
Flow	m^3/d	10,000
Influent BOD	mg/L	150
Effluent BOD	mg/L	2
τ	h	4
SRT	d	6
Synthesis yield, Y	g VSS/g bCOD	
Wastewater A		0.40
Wastewater B		0.50
Wastewater C		0.30
Cell debris yield, f_d	g VSS/g VSS	0.15
Endogenous decay, k_d	g VSS/g VSS·d	0.08
nbVSS	mg/L	40
Temperature	°C	10

Note: Value A, B, or C to be selected by instructor.

Assume no nitrification occurs due to the SRT selected and low temperature. Determine (a) the aeration tank oxygen requirements in kg/d, (b) the aeration tank oxygen uptake rate in mg/L·hr, and (c) the aeration tank biomass concentration (mg/L).

8-8 The following information is given for an activated-sludge system designed with a long enough SRT to provide complete nitrification:

Parameter	Unit	Value
Flow	m³/d	10,000
Influent BOD	mg/L	150
Effluent BOD	mg/L	2
Influent TKN	mg/L	35
Effluent NH$_4$-N	mg/L	0.5
τ	h	8.0
SRT	d	15
Temperature	°C	10
Cell debris yield, f_d	g VSS/g VSS	0.10
Synthesis yield, Y	g VSS/g bCOD	
Wastewater A		0.40
Wastewater B		0.50
Wastewater C		0.30
Endogenous decay, k_d	g VSS/g VSS·d	0.08
Nitrifier yield, Y_n	g VSS/g NH$_4$-N	0.12
Nitrifier decay, k_{dn}	g VSS/g VSS·d	0.06

Note: Value A, B, or C to be selected by instructor.

Determine (a) the aeration tank oxygen requirements in kg/d, (b) the aeration tank oxygen uptake rate in mg/L·hr, (c) the aeration tank biomass concentration (mg/L), and (d) the portion of the total oxygen required that is needed for nitrification.

8-9 Using Eqs. (8–15) and (8–16), compare the amount of sludge wasted daily as (a) VSS, (b) TSS, and (c) biomass for operation at an SRT of 10 and 20 days for the following wastewater and design conditions. Assume complete nitrification. Repeat the calculation without accounting for cell debris. How much error is introduced?

Parameter	Unit	Value
Flow	m³/d	15,000
Influent BOD	mg/L	200
Effluent BOD	mg/L	2
Influent TKN	mg/L	35
Effluent NH$_4$-N	mg/L	0.5
Heterotrophic yield, Y	g VSS/g bCOD	0.4
Heterotrophic decay, k_d	g VSS /g VSS·d	0.10
Cell debris yield, f_d	g VSS /g VSS	0.15
Nitrifier yield, Y_n	g VSS /g NH$_4$-N	0.12

(continued)

(*Continued*)

Parameter	Unit	Value
Nitrifier decay, k_{dn}	gVSS/gVSS·d	0.08
nbVSS	mg/L	
Wastewater A		100
Wastewater B		120
Wastewater C		80
Temperature	°C	15

Note: Value A, B, or C to be selected by instructor.

8-10 An industrial wastewater from food processing is to be treated in an activated-sludge process. The wastewater consists of soluble carbonaceous BOD with low nitrogen and phosphorus concentrations. For the summary of wastewater characteristics and design assumptions given in the following table, determine the amount of nitrogen and phosphorus that must be added to the influent flow as mg/L and in kg/d. Assume that residual NH_4-N and the soluble phosphorus concentrations of 0.10 mg/L are needed to prevent nutrient limitations, and no nitrification.

Parameter	Unit	Value
Flow	m³/d	3000
Influent soluble bCOD	mg/L	
Wastewater A		2000
Wastewater B		3000
Wastewater C		2500
Effluent soluble bCOD	mg/L	5
Influent NH_4-N	mg/L	20
Influent phosphorus as P	mg/L	5
SRT	d	10
Synthesis yield, Y	g VSS/g COD	0.4
Heterotrophic decay, k_d	g VSS/g VSS·d	0.10
Cell debris yield, f_d	g VSS/g VSS	0.10

Note: Value A, B, or C to be selected by instructor.

8-11 A 2-L settleometer is used to perform an SVI test. The MLSS concentration for the test is 3500 mg/L and the settled sludge volume after 30 min is 840 mL. What is the sludge volume index?

8-12 According to Eq. (8–20) the secondary clarifier surface overflow rate may be approximated from an estimate of the mixed-liquor initial interfacial settling velocity (V_i), and Eq. (8–22) may be used to estimate V_i from the expected SVI value and design MLSS concentration. Using these equations, determine the surface overflow rate for a clarifier designed for an expected SVI of up to 180, 200, or 220 mL/g (as selected by instructor). Assume a safety factor value of 2.0. How does the calculated overflow rate compare to the range of design values in Table 8–7 for an air activated-sludge system without a selector? Assume the MLSS is equal to 3000 mg/L.

8-13 Two 20-m-diameter circular clarifiers are used for liquid-solids separation for an air activated-sludge system. The MLSS concentration is 3000 mg/L and the return activated-sludge recycle ratio is 50 percent. Assuming average solids loading rate values, within the range shown in Table 8–7, of 4, 5, or 6 kg/m²·h (as selected by instructor), determine the allowable average influent flowrate in m³/d and the return MLSS concentration.

8-14 A completely mixed activated-sludge system treating domestic wastewater is operated at an SRT of 15 days at 12°C, such that complete nitrification occurs. The clarifier surface overflow rate is 1 m/h at average flow conditions, but the clarifier has a high sludge blanket with significant solids loss in the effluent. Describe (a) the specific steps you recommend to investigate the cause of the bulking sludge condition, (b) possible short-term immediate action to reduce the effluent TSS concentration, and (c) the selector alternatives that can be considered for bulking sludge control. Which one would you use and why?

8-15 Using the kinetic coefficients in Table 8–10 and 8–11, plot the following as a function of SRT (ranging from 3.0 to 20.0 days) for the municipal wastewater described in the table below. Assume that the MLSS concentration is 2500 mg/L. The parameters to plot are (a) solids wasted as kg TSS/d, (b) aeration tank volume (m³) and τ (h), (c) observed yield, as g TSS/g BOD$_r$, and g TSS/g bCOD$_r$, (d) effluent soluble bCOD concentration, (e) effluent NH$_4$-N concentration, and (f) oxygen requirements, kg/d.

Parameter	Unit	Value
Flow	m³/d	20,000
BOD	mg/L	
Wastewater A		220
Wastewater B		250
Wastewater C		180
bCOD/BOD	g/g	1.6
TSS	mg/L	220
VSS	mg/L	200
nbVSS	mg/L	
Wastewater A		100
Wastewater B		120
Wastewater C		80
TKN	mg/L	40
Temperature	°C	15
Aeration tank DO	mg/L	2.0

Note: Value A, B, or C to be selected by instructor.

8-16 Solve Problem 8–15 using primary clarification, assuming 35 percent BOD removal, 65 percent TSS and VSS removal, 10 percent TKN removal, and 80 percent nbVSS removal.

8-17 An activated-sludge system is to be designed for nitrification to achieve an effluent NH$_4$-N concentration of 0.50, 0.80, or 1.0 mg/L (as selected by the instructor) at a reactor DO concentration of 2.0 mg/L and 10°C temperature. The peak/average TKN load is 1.8. Determine the design SRT. Use coefficients from Table 8–11.

8-18 A completely mixed activated-sludge system has a hydraulic retention time (τ) of 8.3, 10.8, or 13.1 hours (as selected by the instructor), an aeration tank DO concentration of 2.0 mg/L, and an MLSS concentration of 3000 mg/L. For the municipal wastewater characteristics given below determine (a) the aeration tank average SRT and (b) the nitrification safety factor for a desired average effluent NH_4-N concentration of 0.5 mg/L.

Parameter	Unit	Value
Temperature	°C	10
Flow	m³/d	15,000
BOD removal	mg/L	130
nbVSS	mg/L	30
TSS	mg/L	70
VSS	mg/L	60
TKN	mg/L	40

8-19 A 3000 m³/d industrial wastewater with a soluble COD concentration of 1800 mg/L is to be treated in a completely mixed activated-sludge process at 15°C and 2500 mg/L MLSS concentration. Using the kinetic coefficients and assumptions provided below determine (a) the aeration tank volume (m³) and τ (h), (b) the oxygen required (kg/d), (c) the sludge production (kg TSS/d), (d) the effluent sBOD concentration from the secondary clarifiers, (e) the clarifier diameter (m), assuming two clarifiers, and (f) the air flowrate for fine bubble diffused air aeration. Assume very little excess NH_4-N exists after cell synthesis needs are met so that nitrification is not significant.

Parameter	Unit	Value
bCOD/BOD	g/g	1.6
μ_m	g VSS/g VSS·d	3.0
K_s	mg bCOD/L	60.0
Y	g VSS/g bCOD	0.40
k_d	g VSS/gVSS·d	0.08
f_d	g VSS/g VSS	0.15
SRT	d	
Wastewater A		8.0
Wastewater B		12.0
Wastewater C		16.0
alpha	Unitless	0.45
F (diffuser fouling factor)	Unitless	0.90
beta	Unitless	1.0

(continued)

(Continued)

Parameter	Unit	Value
Elevation	m	300
Effective DO saturation depth	m	2.5
Aeration tank liquid depth	m	5.0
Clean H_2O oxygen transfer efficiency	%	30

Note: Value A, B, or C to be selected by instructor.

8-20 An oxidation ditch process is designed for the following wastewater using the conventional design approach of no primary treatment, a 24-hour aeration tank detention time, and 3500 mg/L MLSS concentration. The lowest expected mixed-liquor temperature is 10°C and an average effluent NH_4-N concentration of 1.0 mg/L is required with a 1.5 safety factor for peak loads. Mechanical surface aerators are used to provide a DO concentration in the aeration zone of 2.0 mg/L. Determine (a) the SRT (d), (b) the sludge production (kg TSS/d), (c) the MLVSS concentration, (d) the oxygen required (kg/d), (e) the total required aeration horsepower, and (f) the ratio of the total volume provided to the necessary nitrification volume.

Parameter	Unit	Values		
		Wastewater A	Wastewater B	Wastewater C
Flow	m³/d	4000	4000	4000
BOD	mg/L	270	250	200
nbVSS	mg/L	130	120	100
TSS	mg/L	250	230	200
VSS	mg/L	240	215	180
TKN	mg/L	40	40	40
Clean H_2O oxygen transfer efficiency	kg O_2/kW·hr	0.9	0.9	0.9
α	Unitless	0.90	0.90	0.90
β	Unitless	0.98	0.98	0.98
Elevation	m	500	500	500

Note: Value A, B, or C to be selected by instructor.

8-21 For the same wastewater characteristics and mixed-liquor temperature as in Problem 8–20 and assuming uniform continuous flow, two sequencing batch reactor tanks are operated under the following conditions:

Fraction of tank liquid depth used for decant depth = 0.20

Aeration time = 2 h

Settle time = 1 h

Decant time = 0.5 h

Determine (a) the fill time (hr), (b) the total time/cycle (hr), (c) the total full volume of each tank (m^3), (d) the SBR SRT assuming an MLSS concentration of 3500 mg/L, and (e) the decant pumping rate (m^3/min).

8–22   A sequencing batch reactor treating the following wastewater is operated under the following conditions:

Temperature = 10°C

Aeration time/cycle = 2.5 h

Fill volume/total liquid volume per cycle = 0.30

SRT = 20 d

DO = 2.0 mg/L

Two SBR reactor tanks are used and the full volume for each SBR tank = 3000 m^3.

Parameter	Unit	Value		
		Wastewater A	Wastewater B	Wastewater C
Flow	m³/d	6000	6000	6000
BOD	mg/L	250	250	250
TKN	mg/L	45	40	30

Note: Value A, B or C to be selected by instructor.

Determine the nitrification safety factor, defined as the actual aeration time/required aeration time per cycle to achieve an effluent NH$_4$-N concentration of 0.50 mg/L. Use kinetic coefficients values from Table 8–11.

8–23   An SBR tank has a total liquid depth of 5.5 m when full. The desired operating MLSS concentration is 3500 mg/L. If the SVI is 150, 180, or 200 mL/g (as selected by instructor), determine the possible fill volume to total liquid volume ratio, assuming a 0.6 m clear liquid depth above the settled sludge blanket.

8–24   Repeat computations for Example 8–5 using the same design conditions and assumptions, but for an effluent NH$_4$-N concentration of 1.0, 2.0, and 4.0 mg/L (as selected by instructor). How do the advantages of the staged system compare to the single-stage nitrification system as the desired effluent NH$_4$-N concentration is increased?

8–25   A four-stage activated-sludge system with equal tank volumes of 240 m^3/stage is used to treat an industrial wastewater with a soluble BOD concentration of 300 mg/L. The influent flow is 4000 m^3/day and the RAS recycle ratio of 0.5. The active biomass concentration is 1600 mg/L. For the following biokinetic information determine: (a) the substrate concentration in each stage (mg/L as soluble bCOD), (b) the oxygen required per stage (kg/d), and (c) the percent of the total oxygen required per stage. Hint: to start the solution assume a 4th stage soluble bCOD concentration of 1.0 mg/L.

Parameter	Unit	Value
k, maximum specific substrate utilization rate	g COD/g VSS·d	1.2
K_s, half velocity coefficient	mg COD/L	
Wastewater A		50
Wastewater B		75
Wastewater C		100
Y, yield	g VSS/g COD removed	0.35
k_d, endogenous decay	g VSS/g VSS·d	0.10

Note: Value A, B, or C to be selected by instructor.

8-26 Compare the following activated-sludge processes in terms of effluent quality, space requirements, complexity, energy requirements, operational requirements, and ability to handle variable flows and loads: completely mixed, pure oxygen, contact stabilization, and oxidation ditch.

8-27 A four-pass step-feed activated-sludge system, shown in the figure below, has equal aeration tank volumes in each pass of 240 m³. Using the design parameters given below, determine the MLVSS concentration in each tank.

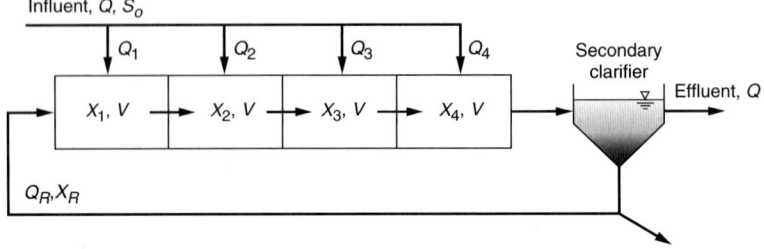

Parameter	Unit	Values Wastewater A	Wastewater B	Wastewater C
X_R	mg VSS/L	10,000	10,000	10,000
Q_R	m³/d	2,000	4,000	6,000
Q	m³/d	4,000	4,000	4,000
Q_1	m³/d	800	800	800
Q_2	m³/d	1,200	1,200	1,200
Q_3	m³/d	1,000	1,000	1,000
Q_4	m³/d	1,000	1,000	1,000

Note: Value A, B, or C to be selected by instructor.

8-28 The following describes the operating conditions of an anoxic/aerobic process as shown in Example 8–5 for 85 percent nitrogen removal.

Parameter	Unit	Value
Flow	m³/d	1,000
BOD	mg/L	200
rbCOD	mg/L	
Wastewater A		60
Wastewater B		95
Wastewater C		120
Alkalinity	mg/L as C$_a$CO$_3$	200
TKN	mg/L	35
Temperature	°C	15
MLSS	mg/L	3,500
Biomass (VSS)	mg/L	1,620
RAS (TSS)	mg/L	10,000
Aerobic volume	m³	460
Aerobic SRT	days	10.0
Biomass nitrogen content	g N/g VSS	0.12
Effluent NH$_4$-N	mg/L	1.0

Note: Value A, B, or C to be selected by instructor.

Determine (a) the internal recycle ratio and flowrate (m³/d), (b) the anoxic tank volume and τ for a single-stage anoxic reactor, and (c) the anoxic tank volume and τ for each stage of a three-stage anoxic tank with equal volumes per stage, (d) the final alkalinity concentration, and (e) the oxygen required (kg/d) for the anoxic/aerobic process versus the aerobic system without the pre-anoxic tank.

8-29 An existing activated-sludge system is operated at a minimal temperature of 10°C. The system is to be modified to an anoxic/aerobic process with the anoxic tank accounting for 10 percent of its total volume. For the following design wastewater conditions and total tank volume, determine (a) the effluent NH$_4$-N and NO$_3$-N concentrations, assuming a 1.5 safety factor for nitrification, and (b) the minimal internal recycle ratio needed to match the nitrate removal capacity of the anoxic tank. Use necessary coefficients from Tables 8–10 and 8–11.

Parameter	Unit	Value
Flow	m³/d	8000
bCOD	mg/L	240
rbCOD	mg/L	
Wastewater A		25
Wastewater B		50
Wastewater C		75
nbVSS	mg/L	60

(continued)

(*Continued*)

Parameter	Unit	Value
TSS	mg/L	80
VSS	mg/L	70
TKN	mg/L	40
MLSS	mg/L	3500
Tank volume	m^3	3600
RAS ratio	Unitless	0.50
Biomass nitrogen content	g N/g VSS	0.12
Aeration tank DO	mg/L	2.0

Note: Value A, B, or C to be selected by instructor.

8-30 An oxidation ditch system is operated with a single mechanical surface aerator such that one-half of the tank volume is aerobic with the DO concentration varying from 0 to 2.0 mg/L. For the information given below, determine the effluent NH_4-N (assuming 1.5 safety factor) and NO_3-N concentrations.

Parameter	Unit	Value
Ditch volume	m^3	4600
MLSS	mg/L	3500
Temperature	°C	10
BOD	mg/L	
Wastewater A		250
Wastewater B		220
Wastewater C		200
nbVSS	mg/L	80
TKN	mg/L	40
TSS	mg/L	220
VSS	mg/L	210

Note: Value A, B, or C to be selected by instructor.

8-31 For the following wastewater characteristics design an SBR system, using two tanks, to produce an effluent NO_3-N and NH_4-N concentration of 6 mg/L and 1.0 mg/L, respectively. Assume that the aeration, settle, and decant times per cycle are 2.0, 1.0, and 0.5 hours, respectively, and the MLSS concentration is 4000 mg/L. Determine (a) the fill volume fraction, (b) the volume of each SBR tank, (c) the decant pumping rate, and (d) the nitrification safety factor. Assume the aeration DO concentration is 2.0 mg/L and only anoxic mixing occurs during the fill period.

Parameter	Unit	Value
Flow	m³/d	5000
BOD	mg/L	250
rbCOD	mg/L	50
nbVSS	mg/L	120
TKN	mg/L	
Wastewater A		45
Wastewater B		40
Wastewater C		35
TSS	mg/L	220
VSS	mg/L	210
Temperature	°C	12

Note: Value A, B, or C to be selected by
instructor.

8-32   For the suspended-growth postanoxic denitrification process design example shown in Example
8–9 a design SRT of 5 d was used. Determine (a) the change in the methanol dose requirement
(kg/d), (b) the amount of sludge produced (kg/d), and (c) the anoxic tank volume if a design SRT
of 15 d is used instead. If methanol costs $0.05/kg, would the longer SRT design be more eco-
nomical?

8-33   An A^2O system for biological nitrogen and phosphorus removal is designed with a 15-d SRT. An
observed yield of 0.60 g TSS/g BOD removed is estimated and the following wastewater characteristics following primary treatment have been obtained:

Parameter	Unit	Value
BOD	mg/L	
Wastewater A		150
Wastewater B		140
Wastewater C		120
P	mg/L	8
TKN	mg/L	35
TSS	mg/L	82
VSS	mg/L	72
pH	units	7.2

Note: Value A, B, or C to be selected by
instructor.

Determine (a) the estimated effluent soluble phosphorus concentration for the biological
phosphorus-removal process and (b) the phosphorus content (in percent on a dry weight basis) of

the waste sludge. What additional wastewater characterization parameter would you recommend obtaining to better estimate the effluent phosphorus concentration?

8–34 An A²O process is operated with a 12 d SRT and achieves the following effluent nutrient concentrations with internal and return activated sludge (RAS) recycle ratios of 3.0 and 0.5, respectively.

Effluent soluble P concentration = 0.50 mg/L

Effluent NO_3-N concentration = 5.0 mg/L

Assuming that the influent wastewater characteristics and SRT remain the same, determine (a) the change in effluent soluble phosphorus and NO_3-N concentrations if the internal recycle ratio is changed to 2.0, 2.5, or 2.8 (as selected by instructor) and the RAS ratio is increased to 1.0 and (b) how much additional influent rbCOD is consumed by nitrate fed to the anaerobic zone.

8–35 State whether the effluent phosphorus concentration for an A²O system will increase, decrease, or remain the same for each of the following changes in the wastewater characteristics or process operating conditions. Refer to basic process fundamentals to explain the basis for your answer.

1. The SRT is increased.
2. The influent rbCOD concentration increases.
3. The clarifier effluent suspended solids concentration increases.
4. A higher NO_3-N concentration exists in the return activated-sludge recycle.
5. The influent particulate BOD concentration increases.

8–36 Given the settling data in the following table from tests with mixed liquor from an activated-sludge plant, determine the percent RAS recycle rate if the clarifier overflow rate is 0.82, 1.0, or 1.2 m/h (as selected by instructor) and the RAS MLSS concentration is 10,500 mg/L. What will the recycle rate be if the RAS MLSS concentration is 15,000 mg/L?

| | Sludge liquid/solids interface depth in column test as a function of time and initial mixed-liquor concentration | | | | | |
| | MLSS concentration, mg/L | | | | | |
Time, min	1000	2000	3000	5000	10,000	15,000
0	0	0	0	0	0	0
10	117.1	90.5	41.2	17.1	4.9	3.0
20	189.0	167.1	84.1	34.1	10.1	6.1
30	192.1	182.9	127.7	50.9	14.9	9.1
40	193.0	188.1	156.1	68.0	20.1	11.9
50	193.0	189.0	166.2	85.1	25.9	14.0
60	193.9	189.9	172.0	102.1	31.1	15.9

Note: Data in table corresponds to the distance from the top of the settling column to the sludge interface at indicated time, cm.

8–37 Two secondary clarifiers are operating with an overflow rate of 1 m/h and the activated-sludge tank MLSS concentration is 4000 mg/L. Based on sludge thickening tests, the interfacial settling velocity can be described by the following relationship:

$$V_i = V_o(e^{-kX})$$

where V_i = interfacial settling velocity, m/d
$\qquad X$ = MLSS concentration, g/L
$\qquad V_o$ = 172 m/d
$\qquad k$ = 0.4004 L/g

(a) Plot the solids flux due to thickening as a function of the MLSS concentration in g/L.
(b) On the same curve draw the overflow rate operating flux line and show the operating state point.
(c) Determine the solids flux rate and recycle ratio for operating the clarifiers at an underflow concentration of 10, 11, or 12 g/L (as selected by instructor).
(d) Determine the MLSS concentration possible with only one clarifier in operation using the above underflow concentration. Indicate the new overflow rate operating flux line and operating state point.

8-38 Design an aerated lagoon to treat 10,000 m³/d of wastewater under the following conditions:

1. Influent soluble BOD and suspended solids = 150 mg/L
2. Overall first-order BOD removal-rate constant = 2.0 d⁻¹ at 20°C
3. Summer temperature = 27°C
4. Winter temperature = 7°C
5. Wastewater temperature = 15°C
6. Temperature coefficient = 1.07
7. α = 0.85; β = 1.0
8. Elevation = 1250 m
9. Oxygen concentration to be maintained = 2.0 mg/L
10. Lagoon depth = 2 m
11. Hydraulic residence time = 10 d
12. Temperature proportionality constant f = 12 × 10⁶

Determine the surface area, summer and winter temperatures in the lagoon, and the effluent BOD in summer and winter. If the growth yield is approximately 0.5 (BOD basis), determine the biological solids concentration in the lagoon, the oxygen requirements, and the power requirements for summer and winter conditions. Use surface aerators rated at 1.5 kg O_2/kW·h.

8-39 A conventional activated-sludge process with an aeration tank volume of 4600 m³ is operated with a 6-d SRT and 2500 mg/L MLSS concentration, when treating the following wastewater after primary treatment:

Parameter	Unit	Value
Flow	m³/d	15,000
BOD	mg/L	150
nbVSS	mg/L	35
TSS	mg/L	80
VSS	mg/L	68
TKN	mg/L	35

The system must be upgraded to treat additional flow and to provide nitrification to produce an effluent NH_4-N concentration of 1.0 mg/L. A 12-d SRT is selected and membrane separation will replace the secondary clarifier operation, due to limited space for additional aeration tanks. Assuming the same wastewater characteristics, and using the same aeration tank volume with an MLSS concentration of 10,000, 12,000, or 15,000 mg/L (as selected by instructor), determine (a) the new wastewater treatment flowrate possible, (b) the previous and new volumetric BOD loadings (kg/m^3·d) and F/M ratios, and (c) the membrane surface area needed assuming a flux rate of 900 L/m^2·d.

8–40 Using the ASM1 Model matrix described in Table 8–31, write the sum of all the rate equations that describe the rate of change of the following model components:

Heterotrophic biomass concentration
Autotrophic biomass concentration
Readily biodegradable COD concentration
Slowly biodegradable COD concentration

REFERENCES

Adams, C. E., and W. W. Eckenfelder, Jr. (eds.) (1974) *Process Design Techniques for Industrial Waste Treatment,* Enviro, Nashville.

Albertson, O. E. (1987) "The Control of Bulking Sludges: From the Early Innovators to Current Practice," *Journal Water Pollution Control Federation,* vol. 59, p. 172.

Anderson, N. E. (1945) "Design of Final Settling Tanks for Activated Sludge," *Sewage Works Journal,* vol. 17, no. 1.

Arceivala, S. J. (1998) *Wastewater Treatment for Pollution Control,* 2d ed., Tata, McGraw-Hill Publishing Company Limited, New Delhi.

Ardern, E., and W. T. Lockett (1914) "Experiments on the Oxidation of Sewage without the Aid of Filters," *Journal Society of Chemical Industries,* vol. 33, p. 523.

ASCE (1998) *Design of Municipal Wastewater Treatment Plants,* 4th ed., ASCE Manual and Report on Engineering Practice, No. 76, American Society of Civil Engineers, Reston, VA.

Barker, P. L., and P. L. Dold (1997) "General Model for Biological Nutrient Removal in Activated Sludge Systems: Model Presentation," *Water Environment Research,* vol. 69, no. 5, pp. 969–984.

Barnard, J. L. (1973) "Biological Denitrification," *Water Pollution Control* (G.B.), vol. 72, no. 6, pp. 705–720.

Barnard, J. L. (1974) "Cut P and N without Chemicals," *Water and Wastes Engineering,* vol. 11, pp. 41–44.

Barnard, J. L. (1975) "Biological Nutrient Removal without the Addition of Chemicals," *Water Research,* vol. 9, pp. 485–490.

Barnard, J. L. (1984) "Activated Primary Tanks for Phosphate Removal," *Water S.A.,* vol. 10, pp. 121–126.

Bradstreet, K. A., and G. R. Johnson (1994) "Study of Critical Operational Parameters for Biological Nitrogen Reduction at a Municipal Wastewater Treatment Plant," *Proceedings,* Water Environment Federation 67th Annual Conference & Exposition, pp. 669–680.

Brindle, K., and T. Stephenson (1996) "The Application of Membrane Biological Reactors for the Treatment of Wastewaters," *Biotechnology and Bioengineering,* vol. 49, p. 601.

Burdick, C. R., D. R. Refling, and H. D. Stensel (1982) "Advanced Biological Treatment to Achieve Nutrient Removal," *Journal of Water Pollution Control Federation,* vol. 54, no. 7.

Carrio, L., F. Streett, K. Mahoney, C. deBarbadillo, J. Anderson, K. Abraham, and N. Passerelli (2000) "Practical Considerations for Design of a Step-Feed Biological Nutrient Removal System," *Proceedings of the Municipal Wastewater Treatment Symposium: Nitrogen and Phosphorus Removal,* 73d Annual Conference, Water Environment Federation, Anaheim, CA.

Choo, K.-H., and H. D. Stensel (2000) "Sequencing Batch Membrane Reactor Treatment: Nitrogen Removal and Membrane Fouling Evaluation," *Water Environment Research,* vol. 72, pp. 490–498.

Chudoba, J., P. Grau, and V. Ottova (1973) "Control of Activated Sludge Filamentous Bulking II. Selection of Microorganisms by Means of a Selector," *Water Research,* vol. 7, pp. 1389–1406.

Cicek, N., J. P. Franco, M. T. Suidan, and V. Urbain (1998) "Using a Membrane Bioreactor to Reclaim Wastewater," *Journal of American Water Works Association,* vol. 76, p. 356.

Clark, H. W., and G. O. Adams (1914) "Sewage Treatment by Aeration and Contact in Tanks Containing Layers of Slate," *Engineering Record* 69, 158.

Cote, P., H. Buisson, and P. Matthieu (1998) "Immersed Membranes Activated Sludge process Applied to the Treatment of Municipal Wastewater," *Water Science Technology,* vol. 38, pp. 437–482.

Crites, R., and G. Tchobanoglous (1998) *Small and Decentralized Wastewater Management Systems,* McGraw-Hill, New York.

Crosby, R. M., and J. H. Bender (1980) "Hydraulic Considerations That Affect Clarifier Performance," EPA Technology Transfer.

Crown Zellerbach Corp. (1970) "Aerated Lagoon Treatment of Sulfite Pulping Effluents," U.S. Environmental Protection Agency, Water Pollution Control Series, Report No. 12040 ELW.

Daigger, G. T. (1995) "Development of Refined Clarifier Operating Diagrams Using an Updated Settling Characteristics Database," *Water Environment Research,* vol. 67, pp. 95–100.

Daigger, G. T., G. D. Waltrip, E. D. Rumm, and L. A. Morales (1988) "Enhanced Secondary Treatment Incorporating Biological Nutrient Removal," *Journal of Water Pollution Control Federation,* vol. 60, pp. 1833–1842.

Dold, P. L., W. K. Bagg, and G. v. R. Marais (1986) "Measurement of the Readily Biodegradable COD Fraction (S_{bs}) in Municipal Wastewater by Ultrafiltration," *Research Report* No. WS7, University of Cape Town, Cape Town, South Africa.

Dold, P. L., G. A. Ekama, and G. v. R. Marais (1986) "Evaluation of the General Activated Sludge Model," Proposed by the IAWPRC Task Group, *Water Science and Technology,* vol. 18, pp. 63–69.

Eikelboom, D. H. (1975) "Filamentous Organism in Activated Sludge," *Water Research,* vol. 9, p. 365.

Eikelboom, D. H. (2000) *Process Control of Activated Sludge Plants by Microscopic Investigation,* IWA Publishing, London.

Ekama, G. A., P. L. Dold, and G. v. R. Marais (1986) "Procedures for Determining Influent COD Fractions and the Maximum Specific Growth Rate of Heterotrophs in Activated Sludge Systems," *Water Science Technology,* vol. 18, no. 6.

Ekama, G. A., I. P. Srebritz, and G. v. R. Marais (1983) "Considerations in the Process Design of Nutrient Removal Activated Sludge Processes," *Water Science and Technology,* vol. 15, pp. 283–318.

Fernandez, A., J. Lozier, G. Daigger (2000) "Investigating Membrane Bioreactor Operation for Domestic Wastewater Treatment: A Case Study," Municipal Wastewater Treatment

Symposium: Membrane Treatment Systems, *Proceedings,* 73d Annual Conference, Water Environment Federation, Anaheim, CA.

Fillos, J., D. Katehis, K. Ramalingam, L. A. Carrio, and K. Gopalakrishan (2000) "Determination of Nitrifier Growth Rates in New York City Water Pollution Control Plants," Research Symposium: Nitrogen Removal, *Proceedings,* 73d Annual Conference, Water Environment Federation, Anaheim, CA.

Fleckseder, H. R., and J. F. Malina (1970) "Performance of the Aerated Lagoon Process," Technical Report CRWR-71, Center for Research in Water Resources, University of Texas, Austin, TX.

Giese, T. P., and M. D. Larsen (2000) "Pilot Testing New Technology at the Kitsap County Sewer District No. 5/City of Port Orchard, WA. Joint Wastewater Treatment Facility," Municipal Wastewater Treatment Symposium: Membrane Treatment Systems, *Proceedings,* 73d Annual Conference, Water Environment Federation, Anaheim, CA.

GLUMRBSS (1978) *Recommended Standards for Sewage Works* (Ten State Standards), Great Lakes–Upper Mississippi River Board of State Sanitary Engineers.

Goronsky, M. (1979) "Intermittent Operation of the Extended Aeration Process for Small Systems," *Journal of Water Pollution Control Federation,* vol. 51, p. 272.

Grady, C. P. L., Jr., G. T. Daigger, and H. C. Lim (1999) *Biological Wastewater Treatment,* 2d ed., Marcel Dekker, Inc., New York.

Grady, C. P. L. Jr., W. Gujer, G. v. R. Marais, and T. Matsuo (1986) "A Model for Single-Sludge Wastewater Treatment Systems," *Water Science Technology,* vol. 18, no. 6, p. 47.

Hellinga, C., A. A. J . C. Schellen, J. W. Mulder, M. C. M. van Loosdrecht, and J. J. Heijnen (1998) "The Sharon Process: An Innovative Method for Nitrogen Removal from Ammonium-Rich Waste Water," *Water Science Technology,* vol. 37, pp. 135–142.

Henze, M. (1991) "Capabilities of Biological Nitrogen Removal Processes from Wastewater," *Water Science Technology,* vol. 23, pp. 669–679.

Henze, M., C. P. L. Grady, W. Gujer, G. v. R. Marais, and T. Matsuo (1987*a*) *Activated Sludge Model No. 1,* IAWPRC Scientific and Technical Reports, no. 1, IAWPRC, London.

Henze, M., C. P. L. Grady, Jr., W. Gujer, G. v. R. Marais, and T. Matsuo (1987*b*) "A General Model for Single Sludge Wastewater Treatment Systems," *Water Research* (G.B.), vol. 21, p. 545.

Henze, M., W. Gujer, T. Mino, T. Matsuo, M. C. Wentzel, and G. v. R. Marais (1995) *Activated Sludge Model No. 2,* IAWQ Scientific and Technical Reports, no. 3.

Hong, S. N., Y. D. Feng, and R. D. Holbrook (1997) "Enhancing Denitrification in the Secondary Anoxic Zone by RAS Addition: A Full Scale Evaluation," *Proceedings,* Water Environment Federation 70th Annual Conference & Exposition, pp. 411–417.

Huang, J .Y. C., and M. D. Cheng (1984) "Measurements and New Applications of Oxygen Uptake Rates in Activated Sludge Process," *Journal Water Pollution Control Federation,* vol. 56, p. 787.

Irvine, R. L., L. H. Ketchum, R. Breyfogle, and E. F. Barth (1983) "Municipal Applications of Sequencing Batch Treatment," *Journal Water Pollution Control Federation,* vol. 59, no. 5.

Jenkins, D., M. G. Richards, and G. T. Daigger (1993) *The Causes and Cures of Activated Sludge Bulking and Foaming,* 2d ed., Lewis Publishers, Ann Arbor, MI.

Kang, S. J., et al. (1987) *Handbook, Retrofitting POTWs for Phosphorus Removal in the Chesapeake Bay Drainage Basin,* EPA 625/6-87-017, U.S. Environmental Protection Agency, Cincinnati, OH.

Keinath, T. M. (1985) "Operational Dynamics and Control of Secondary Clarifiers," *Journal Water Pollution Control Federation,* vol. 57, no. 7, pp. 770–776.

Keinath, T. M., M. D. Ryckman, C. H. Dana, and D. A. Hofer (1977) "Activated Sludge-Unified System Design and Operation," *Journal Environmental Engineering,* vol. 103, p. 829.

Levin, G. V., G. J. Topol, and A. G. Tarnay (1975) "Operation of Full Scale Biological Phosphorus Removal Plant," *Journal of Water Pollution Control Federation,* vol. 47, pp. 577–590.

Ludzack, F. T., and M. B. Ettinger (1962) "Controlling Operation to Minimize Activated Sludge Effluent Nitrogen," *Journal of Water Pollution Control Federation,* vol. 34, no. 9.

Mamais, D., D. Jenkins, and P. Pitt (1993) "A Rapid Physical-Chemical Method for the Determination of Readily Biodegradable Soluble COD in Municipal Wastewater," *Water Research,* vol. 27, pp. 195–197.

Mancini J. L., and E. L. Barnhart (1968) "Industrial Waste Treatment in Aerated Lagoons," in E. F. Gloyna and W. W. Eckenfelder, Jr. (eds.), *Advances in Water Quality Improvement,* University of Texas Press, Austin, TX.

McKinney, R. E. (1962) "Mathematics of Complete Mixing Activated Sludge," *Journal Sanitary Engineering Division,* American Society of Civil Engineers, Eng., 88, SA3.

McKinney, R. E., and H. H. Benjes, Jr. (1965) "Evaluation of Two Aerated Lagoons," *Journal of Sanitary Engineering,* American Society of Civil Engineers, vol. 91, pp. 43–55.

McKinney, R. E., J. M. Symons, W. G. Shifron, and M. Vezina (1958) "Design and Operation of a Complete Mixing Activated Sludge System," *Sewage and Industrial Wastes,* vol. 30, pp. 287–295.

McWhirter, J. R. (1978) *The Use of High-Purity Oxygen in the Activated Sludge Process,* vol. I and III, CRC Press, West Palm Beach, FL.

Merlo, R. R., S. Adham, P. Gagliardo, R. S. Trussell, and R. R. Trussell (2000) "Application of Membrane Bioreactor Technology for Water Reclamation," *Proceedings,* Water Environment Federation, 73th Annual Conference, Anaheim, CA.

Metcalf & Eddy (1930) *Sewerage and Sewage Disposal, A Textbook,* McGraw-Hill Book Co., New York.

Metcalf & Eddy (1991) *Wastewater Engineering: Treatment Disposal, and Reuse,* G. Tchobanoglous and F. L. Burton (eds.), McGraw-Hill, New York.

Mourato, D., D. Thompson, C. Schneider, N. Wright, M. Devol, and S. Rogers (1999) "Upgrade of a Sequential Batch Reactor into a Zenogem® Membrane Bioreactor," *Proceedings,* Water Environment Federation 72nd Annual Conference, New Orleans, LA.

Murakami, C., and R. Babcock, Jr. (1998) "Effect of Anoxic Selector Detention Time and Mixing Rate on Denitrification Rate and Control of *Sphaerotilus Natans* Bulking," *Proceedings,* Water Environment Federation 71st Annual Conference & Exposition, pp. 89–98.

Murphy, K., and A. Wilson (1974) "Characterization of Mixing in Aerated Lagoons," *Journal Environmental Engineering Division,* American Society of Civil Engineers, vol. 100, p. 1105.

Oldham, W. K., and G. M. Stevens (1985) "Operating Experiences with the Kelowna Pollution Control Centre," *Proceedings of the Seminar on Biological Phosphorus Removal in Municipal Wastewater Treatment,* Penticton, British Columbia, Canada.

O'Neill, M., and N. J. Huren (1995) "Achieving Simultaneous Nitrification and Denitrification of Wastewater at Reduced Costs," *Water Science Technology,* vol. 32, p. 303.

Orhon, D., and N. Arton (1994) Modelling of Activated Sludge Systems, Technomic Publishing Co., Inc., Lancaster, PA.

Parker, D. S., D. J. Kinnear, and E. J. Wahlberg (2001) "Review of Folklore in Design and Operation of Secondary Clarifiers," *Journal of Environmental Engineering,* vol. 127, pp. 476–484.

Parkin, G. F., and P. L. McCarty (1981) "Sources of Soluble Organic Nitrogen in Activated Sludge Effluents," *Journal Water Pollution Control Federation,* vol. 53, pp. 89–98.

Rabinowitz, B., and W. K. Oldham (1985) "The Use of Primary Sludge Fermentation in the Enhanced Biological Phosphorus Removal Process," *Proceedings of University of British Columbia Conference on New Directions and Research in Waste Treatment and Management,* Vancouver, Canada.

Randall, C. W., J. L. Barnard, and H. D. Stensel (1992) *Design and Retrofit of Wastewater Treatment Plants for Biological Nutrient Removal,* Technomics Publishing, Lancaster, PA.

Reardon, R. D., T. Kolby, and M. Odo (1996) "The LOTT Nitrogen Removal Facilities: A First Year Evaluation," *Proceedings,* Water Environment Federation 69th Annual Conference & Exposition, pp. 695–705.

Reed, S. C., R. W. Crites, and E. J. Middlebrooks (1995) *Natural Systems for Waste Management and Treatment,* 2nd ed., McGraw-Hill, New York.

Rensink, J. H. (1974) "New Approach to Preventing Bulking Sludge," *Journal Water Pollution Control Federation,* vol. 46, no. 8, pp. 1888–1894.

ReVoir, G. J., II, D. R. Refling, and H. J. Losch. (2000) "Wastewater Process Enhancements Utilizing Submerged Membrane Technology," *Proceedings of the Municipal Wastewater Treatment Symposium: Membrane Treatment Systems*; 73d Annual Conference, Water Environment Federation, Anaheim, CA.

Rich, L. G. (1982) "Design Approach to Dual-Powered Aerated Lagoons," *Journal Environmental Engineering Division,* American Society of Civil Engineers, vol. 108, no. 3, pp. 532–548.

Rich, L. G. (1996) "Modification of Design Approach to Aerated Lagoons," *Journal Environmental Engineering Division,* American Society of Civil Engineers, vol. 122, no. 2, pp. 149–153.

Rich, L. G. (1999) *High Performance Aerated Lagoon Systems,* American Academy of Environmental Engineers, Annapolis, MD.

Rittman, B. E., and W. E. Langeland (1985) "Simultaneous Denitrification with Nitrification in Single-Channel Oxidation Ditches," *Journal Water Pollution Control Federation,* vol. 57, p. 300.

Shao, Y. J., M. Starr, K. Kaporis, H. S. Kim, and D. Jenkins (1997) "Polymer Addition as a Solution to *Nocardia* Foaming Problems," *Water Environment Research,* vol. 69, no. 1.

Slater, N. J., and Y. Bian (1995) "Application of the Cyclic Activated Sludge System to Municipal and Industrial Wastewater Treatment," Seminar Proceedings, Wastewater Treatment Technology—Looking into the 21st Century, *The Chartered Institution of Water and Environmental Management.*

Stensel, H. D. (1981) "Biological Nitrogen Removal System Design," *Water,* American Institute of Chemical Engineers, p. 237.

Stensel, H. D., and T. E. Coleman (2000) "Technology Assessments: Nitrogen Removal Using Oxidation Ditches," *Water Environment Research Foundation.*

Stensel, H. D., T. E. Coleman, W. B. Denham, and D. Fleishman (1995) "Innovative Processes Used to Upgrade Oxidation Ditch for Nitrogen Removal and SVI Control," *Proceedings of the 68th Annual Conference and Exposition,* Water Environment Federation, Miami Beach, FL.

Stensel, H. D., and G. Horne (2000) "Evaluation of Denitrification Kinetics at Wastewater Treatment Facilities," *Proceedings, Research Symposium,* Water Environment Federation 73d Annual Conference & Exposition, Anaheim, CA.

Stensel, H. D., R. C. Loehr, and A. W. Lawrence (1973) "Biological Kinetics of Suspended-Growth Denitrification," *Journal Water Pollution Control Federation,* vol. 45, p. 249.

Stenstrom, M. K., and S. S. Song (1991) "Effects of Oxygen Transport Limitations on Nitrification in the Activated Sludge Process," *Research Journal Water Pollution Control Federation,* vol. 63, p. 208.

Stephens, H. L., and H. D. Stensel (1998) "Effect of Operating Conditions on Biological Phosphorus Removal," *Water Environment Research,* vol. 70, pp. 360–369.

Stephenson, T., J. Simon, B. Jefferson, and K. Brindle (2000) *Membrane Bioreactors for Wastewater Treatment,* IWA Publishing, London.

Strom, P. F., and D. Jenkins (1984) "Identification and Significance of Filamentous Microorganisms in Activated Sludge," *Journal Water Pollution Control Federation,* vol. 56, no. 5.

Trivedi, H., and N. Heinen (2000) "Simultaneous Nitrification/Denitrification by Monitoring NAOH Fluorescence in Activated Sludge," *Proceedings of the Facility Operations II: Innovative Technology Forum;* 73d Annual Conference, Water Environment Federation, Anaheim, CA.

Trussell, R. S., S. Adham, P. Gagliarado, R. Merlo, and R. R. Trussell (2000) "WERF: Application of MBR Technology for Wastewater Treatment," *Proceedings of the 73d Annual Water Environment Federation Conference,* Anaheim, CA, Oct. 14–18, 2000.

U.S. EPA (1992) *Wastewater Treatment/Disposal for Small Communities,* EPA/625/R-92/005, U.S. Environmental Protection Agency, Washington, DC.

U.S. EPA (1993) *Nitrogen Control Manual,* EPA/625/R-93/010, Office of Research and Development, U.S. Environmental Protection Agency, Washington, DC.

Van Dijk, L., and G. C. G. Roncken (1997) "Membrane Bioreactor for Wastewater Treatment: The State of the Art and New Developments," *Water Science and Technology,* vol. 35, p. 53.

Van Huyssteen, J. A., J. L. Barnard, and J. Hendriksz (1990) "The Olifantsfontein Nutrient Removal Plant," *Water Science and Technology,* vol. 22, pp. 1–8.

Wahlberg, E. J. (1995) "Update on Secondary Clarifiers; Design, Operation, and Performance," *Proceedings,* Fourth National Wastewater Treatment Technology Transfer Workshop, Kansas City, MO.

Wahlberg, E. J., and T. M. Keinath (1988) "Development of Settling Flux Curves Using SVI," *Journal Water Pollution Control Federation,* vol. 60, pp. 2095–2100.

Wanner, J. (1994) *Activated Sludge Bulking and Foaming Control,* Technomic Publishing, Lancaster, PA.

WEF (1995) *Standard Methods for the Examination of Water and Wastewater,* 19th Edition, Water Environment Federation, Alexandria, VA.

WEF (1996) *Operation of Municipal Wastewater Treatment Plants,* 5th ed., Manual of Practice no. 11, vol. 2, Water Environment Federation, Alexandria, VA.

WEF (1998) *Design of Wastewater Treatment Plants,* 4th ed., Manual of Practice no. 8, Water Environment Federation, Alexandria, VA.

Wentzel, M. C., P. L. Dold, G. A. Ekama, and G. v. R. Marais (1990) "Enhanced Polyphosphate Organism Cultures in Activated Sludge Systems, Part III, Kinetic Model," *Water Science,* vol. 15, pp. 89–102.

Wentzel, M. C., G. A. Ekama, R. E. Lowenthal, P. L. Dold, and G. v. R. Marais (1989) "Enhanced Polyphosphate Organism Cultures in Activated Sludge Systems, Part II, Experimental Behavior," *Water Science,* vol. 15, pp. 71–88.

Wilson, T. E. (1996) "A New Approach to Interpret Settling Data," *Proceedings of Water Environment Federation 69th Annual Conference and Exposition,* vol. 1, Part 1. *Wastewater Treatment Research,* Alexandria, VA.

Wilson, T. E., and J. S. Lee (1982) "Comparison of Final Clarifier Design Techniques," *Journal of Water Pollution Control Federation,* vol. 54, p. 1376.

Wood, D. K., and G. Tchobanoglous (1975) "Trace Element in Biological Waste Treatment," *Journal Water Pollution Control Federation,* vol. 47, no. 7.

WPCF (1985) *Clarifier Design,* Manual of Practice FD-8, Water Environment Federation, Alexandria, VA.

WPCF (1987) *Activated Sludge,* Manual of Practice OM-9, Water Environment Federation, Alexandria, Va.

WRC (1984). *Theory, Design and Operation of Nutrient Removal Activated Sludge Processes,* Water Research Commission, Pretoria, South Africa.

Wuhrmann, K. (1964) "Nitrogen Removal in Sewage Treatment Processes," *Proceedings International Association of Theoretical and Applied Limnology,* vol. 15, part 2, pp. 580–596.

9 Attached Growth and Combined Biological Treatment Processes

The concept of attached growth processes was introduced in Chap. 7, along with the fundamental mechanisms of mass transfer of substrate and electron acceptors into the biofilms that develop in attached growth systems. In this chapter, various attached growth and combined biological processes used for wastewater treatment are discussed. The attached growth processes considered in detail in this chapter include (1) trickling filters, (2) rotating biological contactors, and (3) combined attached and suspended growth processes. In addition, the activated-sludge processes with fixed-film packing, submerged attached growth processes, and attached growth denitrification processes are introduced and discussed. The use of attached growth biofilters for odor control is discussed in Sec. 15–3 in Chap. 15.

9–1   BACKGROUND

To introduce the discussion and analysis of attached growth processes, it will be helpful to briefly review the evolution of these processes and to consider the impact of mass transfer on the performance of these processes.

Evolution of Attached Growth Processes

Attached growth processes can be grouped into three general classes: (1) nonsubmerged attached growth processes, (2) suspended growth processes with fixed-film packing, and (3) submerged attached growth aerobic processes. The evolution of each of the processes in these groupings is discussed below.

Nonsubmerged Attached Growth Processes.   Trickling filters with rock packing have been a common, simple, and low-energy process used for secondary treatment since the early 1900s. A trickling filter is a nonsubmerged fixed-film biological reactor using rock or plastic packing over which wastewater is distributed continuously. Treatment occurs as the liquid flows over the attached biofilm. The concept of a trickling filter grew from the use of contact filters in England in the late 1890s. Originally they were watertight basins filled with broken stones and were operated in a cyclic mode. The bed was filled with wastewater from the top, and the wastewater was allowed to contact the packing for a short time. The bed was then drained and allowed to rest before the cycle was repeated. A typical cycle required 12 h (6 h for operation and 6 h of resting). The limitations of the contact filter included a relatively high incidence of clogging, the long rest period required, headloss, and the relatively low loading that could be used. Because of the clogging problems, larger packing was used until a rock size of 50 to 100 mm (2 to 4 in) was reached (Crites and Tchobanoglous 1998).

In the 1950s, plastic packing began to replace rock in the United States. The use of plastic packing allowed the use of higher loading rates and taller filters (also known as biotowers) with less land area, improved process efficiency, and reduced clogging. In the 1960s, practical designs were developed for rotating biological contactors (RBCs), which provided an alternative attached growth process where the packing is rotated in the wastewater treatment tank, versus pumping and applying the wastewater over a static packing. Both trickling filters and RBCs have been used as aerobic attached growth processes for BOD removal only, combined BOD removal and nitrification, and for tertiary nitrification after secondary treatment by suspended growth or attached growth processes. The principal advantages claimed for these aerobic attached growth processes over the activated-sludge process are as follows:

- Less energy required
- Simpler operation with no issues of mixed liquor inventory control and sludge wasting
- No problems of bulking sludge in secondary clarifiers
- Better sludge thickening properties
- Less equipment maintenance needs
- Better recovery from shock toxic loads

In comparison to the activated-sludge process, many disadvantages often cited for trickling filters, such as a poorer effluent quality in terms of BOD and TSS concentrations, greater sensitivity to lower temperatures, odor production, and uncontrolled solids sloughing events, are related more to the specific process and final clarifier designs than to the actual process capabilities (WEF, 2000). In general, the actual limitations of the processes (1) make it difficult to accomplish biological nitrogen and phosphorus removal compared to single-sludge biological nutrient removal suspended growth designs, and (2) result in an effluent with a higher turbidity than activated-sludge treatment. Trickling filters and RBCs have also been used in combined processes with activated sludge to utilize the benefits of both processes, in terms of energy savings and effluent quality.

Suspended Growth Processes with Fixed-Film Packing. The placement of packing material in the aeration tank of the activated-sludge process dates back to the 1940s with the Hays and Griffith processes (WEF, 2000). Present-day designs use more engineered packings and include the use of packing materials that are suspended in the aeration tank with the mixed liquor, fixed packing material placed in portions of the aeration tank, as well as submerged RBCs. The advantages claimed for these activated-sludge process enhancements are as follows:

- Increased treatment capacity
- Greater process stability
- Reduced sludge production
- Enhanced sludge settleability
- Reduced solids loadings on the secondary clarifier
- No increase in operation and maintenance costs

Submerged Attached Growth Processes. Beginning in the 1970s and extending into the 1980s, a new class of aerobic attached growth processes became established alternatives for biological wastewater treatment. These are upflow and downflow packed-bed reactors and fluidized-bed reactors that do not use secondary clarification. Their unique advantage is the small footprint with an area requirement that is a fraction (one-fifth to one-third) of that needed for activated-sludge treatment. Though they are more compact, their capital costs are generally higher than that for activated-sludge treatment. In addition to BOD removal, submerged attached growth processes have also been used for tertiary nitrification and denitrification following suspended or attached growth nitrification.

Downflow and upflow packed-bed reactors, fluidized-bed reactors, and submerged RBCs can be used for postanoxic denitrification. Trickling filters and upflow packed-bed reactors are also used for preanoxic denitrification.

Mass Transfer Limitations

A significant process feature of attached growth processes in contrast to activated-sludge treatment is the fact that the performance of biofilm processes is often diffusion-limited. Substrate removal and electron donor utilization occur within the depth of the attached growth biofilm and subsequently the overall removal rates are a function of diffusion rates and the electron donor and electron acceptor concentrations at various locations in the biofilm. By comparison, the process kinetics for the activated-sludge process are generally characterized by the bulk liquid concentrations.

The diffusion-limited concept is especially important when considering the measurable bulk liquid DO concentrations on attached growth process biological reaction rates. Where a DO concentration of 2 to 3 mg/L is generally considered satisfactory for most suspended growth aerobic processes, such low DO concentrations can be limiting for attached growth processes. For uninhibited nitrification in the biofilm a much higher DO concentration may be required, as shown in Chap. 7, depending on the ammonia concentration.

The concept of diffusion limitations on nitrification rates and the ability to develop anaerobic layers within the biofilm may be exploited to accomplish both nitrification and denitrification in attached growth processes with positive bulk liquid DO concentrations. Investigators have shown how aerobic and anaerobic layers can be developed in the biofilm to accomplish nitrogen removal by nitrification and denitrification (Chui et al., 1996; Richter and Kruner, 1994; and Meaney and Strickland, 1994).

9-2 TRICKLING FILTERS

Trickling filters have been used to provide biological wastewater treatment of municipal and industrial wastewaters for nearly 100 years. As noted above, the trickling filter is a nonsubmerged fixed-film biological reactor using rock or plastic packing over which wastewater is distributed continuously. Treatment occurs as the liquid flows over the attached biofilm. The depth of the rock packing ranges from 0.9 to 2.5 m (3 to 8 ft) and averages 1.8 m (6 ft). Rock filter beds are usually circular, and the liquid wastewater is distributed over the top of the bed by a rotary distributor (see Fig. 9–1). Many conventional trickling filters using rock as the packing material have been converted to plastic packing to increase treatment capacity. Virtually all new trickling filters are now constructed with plastic packing.

Trickling filters that use plastic packing have been built in round, square, and other shapes with depths varying from 4 to 12 m (14 to 40 ft). In addition to the packing, other components of the trickling filter include a wastewater dosing or application system, an underdrain, and a structure to contain the packing. The underdrain system is important both for collecting the trickling filter effluent liquid and as a porous structure through which air can circulate. The collected liquid is passed to a sedimentation tank where the solids are separated from the treated wastewater. In practice, a portion of the liquid col-

(a)

(b)

(c)

(d)

Figure 9–1

Typical examples of trickling filters: (a) conventional shallow-depth rock trickling filter, (b) seven-sided trickling filter (older design) with rock replaced by random plastic packing, (c) intermediate-depth trickling filter converted to tower trickling filter, and (d) one of four tower trickling filters 10 m high and 50 m in diameter with plastic packing. Blowers, used to provide air for biological treatment, are located in the enclosures shown at the bottom left- and right-hand side of the tower filter. See Fig. 9–4 for examples of the rotary distributors used to apply wastewater to the top of the filter packing.

lected in the underdrain system or the settled effluent is recycled to the trickling filter feed flow, usually to dilute the strength of the incoming wastewater and to maintain enough wetting to keep the biological slime layer moist.

Influent wastewater is normally applied at the top of the packing through distributor arms that extend across the trickling filter inner diameter and have variable openings to provide a uniform application rate per unit area. The distributor arms are rotated by the force of the water exiting through their opening or by the use of electric drives. The electric drive designs provide more control flexibility and a wider range of distributor rotational speeds than possible by the simple hydraulic designs. In some cases, especially for square or rectangular filters, fixed flat-spray nozzles have been used.

Primary clarification is necessary before rock trickling filters, and generally used also before trickling filters with plastic packing, though fine screens (smaller than 3-mm openings) have been used successfully with plastic packing (WEF, 1998). With increases in plastic and rubber floatable materials in wastewater, screening of these materials is important to reduce fouling of the packing. In some installations a wire-mesh screen is placed over the top of plastic packing to collect debris that can be vacuumed off periodically.

A slime layer develops on the rock or plastic packing in the trickling filters and contains the microorganisms for biodegradation of the substrates to be removed from the liquid flowing over the packing. The biological community in the filter includes aerobic and facultative bacteria, fungi, algae, and protozoans. Higher animals, such as worms, insect larvae, and snails, are also present.

Facultative bacteria are the predominating organisms in trickling filters, and decompose the organic material in the wastewater along with aerobic and anaerobic bacteria. *Achromobacter, Flavobacterium, Pseudomonas,* and *Alcaligenes* are among the bacterial species commonly associated with the trickling filter. Within the slime layer, where adverse conditions prevail with respect to growth, the filamentous forms *Sphaerotilus natans* and *Beggiatoa* will be found. In the lower reaches of the filter, the nitrifying bacteria will be present. The fungi present are also responsible for waste stabilization, but their role is usually important only under low-pH conditions or with certain industrial wastes. At times, fungi growth can be so rapid that the filter clogs and ventilation becomes restricted. Among the fungi species that have been identified are *Fusazium, Mucor, Penicillium, Geotrichum, Sporatichum,* and various yeasts (Hawkes, 1963; Higgins and Burns, 1975).

Algae can grow only in the upper reaches of the filter where sunlight is available. *Phormidiun, Chlorella.* and *Ulothrix* are among the algae species commonly found in trickling filters (Hawkes, 1963; Higgins and Burns, 1975). Generally, algae do not take a direct part in waste degradation, but during the daylight hours they add oxygen to the percolating wastewater. From an operational standpoint, the algae may be troublesome because they can cause clogging of the filter surface, which may cause odors.

The protozoa in the filter are predominantly of the ciliate group, including *Vorticella, Opercularia,* and *Epistylis* (Hawkes, 1963; Higgins and Burns, 1975). Their function is to feed on the biological films and, as a result, effluent turbidity decreases and the biofilm is maintained in a higher growth state. The higher animals, such as worms, snails, and insects, feed on the biological film. Snails are especially troublesome in trickling filters used mainly for nitrification, where they have been known to consume enough of the nitrifying bacteria to significantly reduce treatment efficiency (Timpany and Harrison, 1989).

The slime layer thickness can reach depths as much as 10 mm. Organic material from the liquid is adsorbed onto the biological film or slime layer. In the outer portions of the biological slime layer (0.1 to 0.2 mm), the organic material is degraded by aerobic microorganisms. As the microorganisms grow and the slime layer thickness increases, oxygen is consumed before it can penetrate the full depth, and an anaerobic environment is established near the surface of the packing. As the slime layer increases in thickness, the substrate in the wastewater is used before it can penetrate the inner depths of the biofilm. Bacteria in the slime layer enter an endogenous respiration state and lose their ability to cling to the packing surface. The liquid then washes the slime off the packing, and a new slime layer starts to grow. The phenomenon of losing the slime layer is called *sloughing* and is primarily a function of the organic and hydraulic loading on the filter. The hydraulic loading accounts for shear velocities, and the organic loading accounts for the rate of metabolism in the slime layer. Hydraulic loading and trickling filter sloughing can be controlled by using a wastewater distributor with an electric motor drive to vary rotational speed (Albertson, 1989).

The mechanisms of biological film loss in plastic and rock packing are different. Continuous, small-scale sloughing of the film occurs in high-rate plastic filters due to hydraulic shear, while large-scale, spring-time sloughing occurs in rock filters located in temperate zones. Sloughing is due to the activity of insect larvae, which become active in the warmer spring temperatures and consume and mechanically dislodge thick biofilms that accumulate over the winter. When a rock filter sloughs, the effluent before settling will contain higher amounts of BOD and TSS than the applied wastewater (Hawkes, 1963).

Trickling Filter Classification and Applications

Trickling filter applications and loadings, based on historical terminology developed originally for rock filter designs, are summarized in Table 9–1. Trickling filter designs

Table 9–1
Historical classification of trickling filters applications[a]

Design characteristics	Low or standard rate	Intermediate rate	High rate	High rate	Roughing
Type of packing	Rock	Rock	Rock	Plastic	Rock/plastic
Hydraulic loading, $m^3/m^2 \cdot d$	1–4	4–10	10–40	10–75	40–200
Organic loading, kg BOD/$m^3 \cdot d$	0.07–0.22	0.24–0.48	0.4–2.4	0.6–3.2	>1.5
Recirculation ratio	0	0–1	1–2	1–2	0–2
Filter flies	Many	Varies	Few	Few	Few
Sloughing	Intermittent	Intermittent	Continuous	Continuous	Continuous
Depth, m	1.8–2.4	1.8–2.4	1.8–2.4	3.0–12.2	0.9–6
BOD removal efficiency, %	80–90	50–80	50–90	60–90	40–70
Effluent quality	Well nitrified	Some nitrification	No nitrification	No nitrification	No nitrification
Power, kW/10^3 m^3	2–4	2–8	6–10	6–10	10–20

[a] Adapted from Metcalf & Eddy, Inc. (1979) and WEF (2000).
Note: $m^3/m^2 \cdot d \times 24.5424 = gal/ft^2 \cdot d$.
$kg/m^3 \cdot d \times 62.4280 = lb/10^3 \ ft^3 \cdot d$.
$kW/10^3 \ m^3 \times 5.0763 = hp/10^3 \ gal$.

are classified by hydraulic or organic loading rates. Rock filter designs have been classified as low- or standard-rate, intermediate-rate, and high-rate. Plastic packing is used typically for high-rate designs; however, plastic packing has also been used at lower organic loadings, near the high end of those used for intermediate-rate rock filters. Much higher organic loadings have been used for rock or plastic packing designs in "roughing" applications where only partial BOD removal occurs.

Low-Rate Filters. A low-rate filter is a relatively simple, highly dependable device that produces an effluent of consistent quality with an influent of varying strength. The filters may be circular or rectangular in shape. Generally, feed flow from a dosing tank is maintained by suction level controlled pumps or a dosing siphon. Dosing tanks are small, usually with only a 2-min detention time based on twice the average design flow, so that intermittent dosing is minimized. Even so, at small plants, low nighttime flows may result in intermittent dosing and recirculation may be necessary to keep the packing moist (U.S. EPA, 1974). If the interval between dosing is longer than 1 or 2 h, the efficiency of the process deteriorates because the character of the biological slime is altered by a lack of moisture.

In most low-rate filters, only the top 0.6 to 1.2 m (2 to 4 ft) of the filter packing will have appreciable biological slime. As a result, the lower portions of the filter may be populated by autotrophic nitrifying bacteria, which oxidize ammonia nitrogen to nitrite and nitrate forms. If the nitrifying population is sufficiently well established, and if climatic conditions and wastewater characteristics are favorable, a well-operated low-rate filter can provide good BOD removal and a highly nitrified effluent.

With a favorable hydraulic gradient, the ability to use gravity flow is a distinct advantage. If the site is too flat to permit gravity flow, pumping will be required. Odors are a common problem, especially if the wastewater is stale or septic, or if the weather is warm. Filters should not be located where the odors would create a nuisance. Filter flies (*Psychoda*) may breed in the filters unless effective control measures are used.

Intermediate- and High-Rate Filters. High-rate filters use either a rock or plastic packing. The filters are usually circular and flow is usually continuous. Recirculation of the filter effluent or final effluent permits higher organic loadings, provides higher dosing rates on the filter to improve the liquid distribution and better control of the slime layer thickness, provides more oxygen in the influent wastewater flow, and returns viable organisms. Recirculation also helps to prevent ponding in the filter and to reduce the nuisance from odors and flies (U.S. EPA, 1974). Intermediate- and high-rate trickling filters may be designed as single- or two-stage processes. Flow diagrams for various trickling filter configurations are shown on Fig. 9–2. Two filters in series operating at the same hydraulic application rate ($m^3/m^2 \cdot h$) will typically perform as if they were one unit with the same total depth (WEF, 2000).

Roughing Filters. Roughing filters are high-rate-type filters that treat an organic load of more than 1.6 kg/$m^3 \cdot$d (100 lb BOD/10^3 ft$^3 \cdot$d) and hydraulic loadings up to 190 $m^3/m^2 \cdot$d (3.2 gal/ft$^2 \cdot$min). In most cases, roughing filters are used to treat wastewater prior to secondary treatment. Most roughing filters are designed using plastic packing (WPCF, 1988). One of the advantages of roughing filters is the low energy

Figure 9–2

Typical trickling filter process flow diagrams: (a) single stage and (b) two-stage. Where used, the most common flow diagrams are the first two of each series.

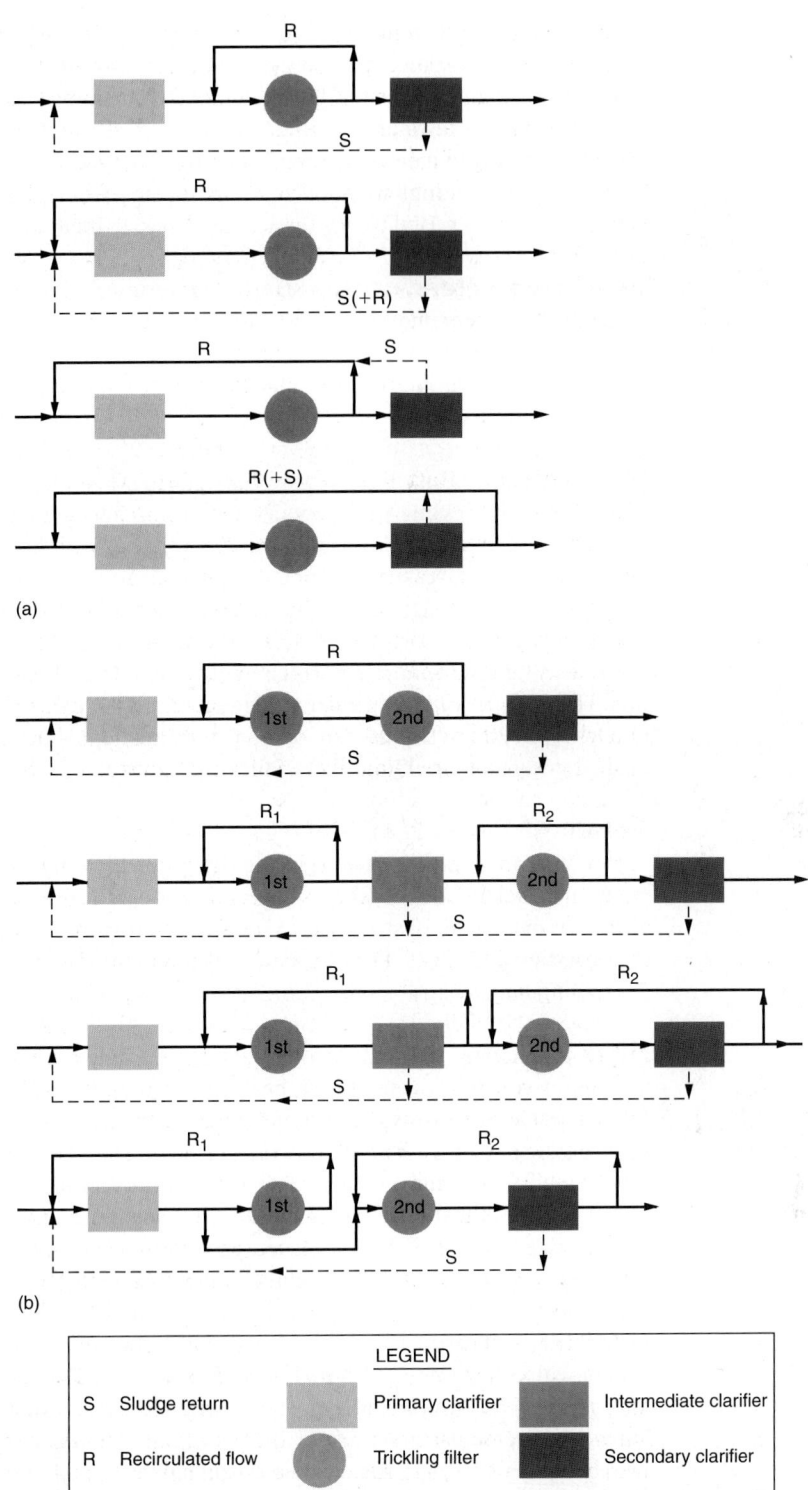

(a)

(b)

LEGEND

S   Sludge return

R   Recirculated flow

Primary clarifier

Trickling filter

Intermediate clarifier

Secondary clarifier

requirement for BOD removal of higher strength wastewaters as compared to activated-sludge aeration. Because the energy required is only for pumping the influent wastewater and recirculation flows, the amount of BOD removal per unit of energy input can increase as the wastewater strength increases until more recirculation is needed to dilute the influent wastewater concentration or to increase wetting efficiency. The energy requirement for a roughing application may range from 2 to 4 kg BOD applied/kWh versus 1.2 to 2.4 kg BOD/kWh for activated-sludge treatment.

Two-Stage Filters.   A two-stage filter system, with an intermediate clarifier to remove solids generated by the first filter, is most often used with high-strength wastewater (Fig. 9–2b). Two-stage systems are also used where nitrification is required. The first-stage filter and intermediate clarifier reduce carbonaceous BOD, and nitrification takes place in the second stage.

Nitrification.   Both BOD removal and nitrification can be accomplished in rock or plastic packing trickling filters operated at low organic loadings (Stenquist et al., 1974; Parker and Richards, 1986). Heterotrophic bacteria, with higher yield coefficients and faster growth rates, are more competitive than nitrifying bacteria for space on the fixed-film packing. Thus, significant nitrification occurs only after the BOD concentration is appreciably reduced. Bruce et al. (1975) demonstrated that the effluent BOD had to be less than 30 mg/L to initiate nitrification and less than 15 mg/L for complete nitrification. Harremöes (1982) considered the soluble BOD, and concluded that a concentration less than 20 mg/L is needed to initiate nitrification. Nitrification can also be accomplished in separate trickling filters following secondary treatment.

Design of Physical Facilities

Factors that must be considered in the design of trickling filters include (1) type and physical characteristics of filter packing to be used; (2) dosing rate; (3) type and dosing characteristics of the distribution system; (4) configuration of the underdrain system; (5) provision for adequate airflow (i.e., ventilation), either natural or forced air; and (6) settling tank design.

Filter Packing.   The ideal filter packing is a material that has a high surface area per unit of volume, is low in cost, has a high durability, and has a high enough porosity so that clogging is minimized and good air circulation can occur. Typical trickling filter packing materials are shown on Fig. 9–3. The physical characteristics of commonly used filter packings, including those shown on Fig. 9–3, are reported in Table 9–2. Until the mid-1960s, the material used was either high-quality granite or blast-furnace slag. Since the 1960s, plastic packing material, either cross-flow or vertical-flow, has become the packing of choice in the United States.

Where locally available, rock has the advantage of low cost. The most suitable material is rounded river rock or crushed stone, graded to a uniform size so that 95 percent is within the range of 75 to 100 mm (3 to 4 in). The specification of size uniformity is a way of ensuring adequate pore space for wastewater flow and air circulation. Other important characteristics of filter packing materials are strength and durability. Durability may be determined by the sodium sulfate test, which is used to test the soundness of concrete aggregates (U.S. EPA, 1974). Because of the weight of the pack-

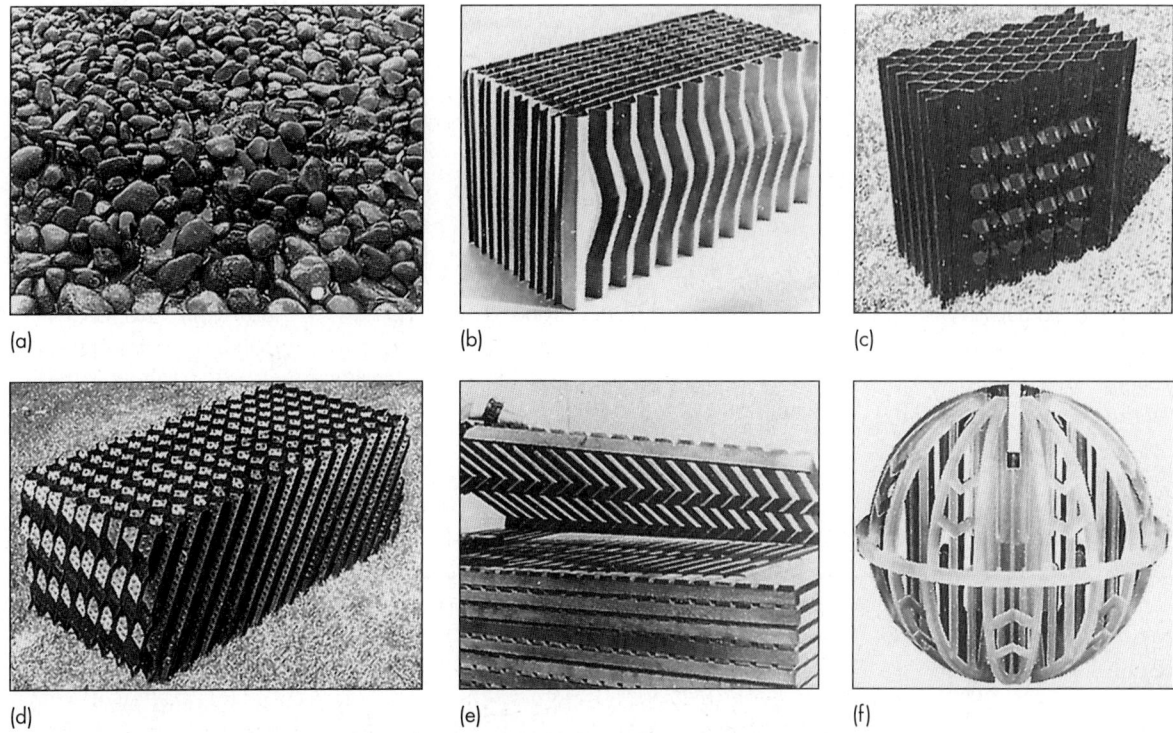

(a) (b) (c)

(d) (e) (f)

Figure 9–3

Typical packing material for trickling filters: (a) rock, (b) and (c) plastic vertical-flow, (d) plastic cross-flow, (e) redwood horizontal, and (f) random pack. *[Figs. (c) and (d) from American Surfpac Corp., (e) from Neptune Microfloc, and (f) from Jaeger Products, Inc.]* Note: the random pack material is often used in air stripping towers.

ing, the depth of rock filters is usually on the order of 2 m (6 ft). The low void volume of rock limits the space available for airflow and increases the potential for plugging and flow short circuiting. Because of plugging, the organic loadings to rock filters are more commonly in the range of 0.3 to 1.0 kg BOD/m³·d.

Various forms of plastic packings are shown on Fig. 9–3. Molded plastic packing materials have the appearance of a honeycomb. Flat and corrugated sheets of polyvinyl chloride are bonded together in rectangular modules. The sheets usually have a corrugated surface for enhancing slime growth and retention time. Each layer of modules is turned at right angles to the previous layer to further improve wastewater distribution. The two basic types of corrugated plastic sheet packing are vertical and cross flow (see Fig. 9–3b, c, and d). Both types of packing are reported to be effective in BOD and TSS removal over a wide range of loadings (Harrison and Daigger, 1987; Aryan and Johnson, 1987). Biotowers as deep as 12 m (40 ft) have been constructed using plastic packing, with depths in the range of 6 m (20 ft) being more common. In biotowers with vertical plastic packing, cross-flow packing can be used for the uppermost layers to enhance

Table 9-2

Physical properties of trickling filter packing materials

Packing material	Nominal size, cm	Approx. unit weight, kg/m³	Approx. specific surface area, m²/m³	Void space, %	Application[a]
River rock (small)	2.5–7.5	1250–1450	60	50	N
River rock (large)	10–13	800–1000	45	60	C, CN, N
Plastic—conventional	61 × 61 × 122	30–80	90	>95	C, CN, N
Plastic—high specific surface area	61 × 61 × 122	65–95	140	>94	N
Plastic random packing— conventional	Varies	30–60	98	80	C, CN, N
Plastic random packing—high specific surface area	Varies	50–80	150	70	N

[a] C = BOD removal; N = tertiary nitrification; CN = combined BOD and nitrification.

Note: kg/m³ × 0.0624 = lb/ft³.

m²/m³ × 0.0305 = ft²/ft³.

the distribution across the top of the filter. The high hydraulic capacity, high void ratio, and resistance to plugging offered by these types of packing can best be used in a high-rate-type filter. Redwood or other wood packings have been used in the past, but with the limited availability of redwood, wood packing is seldom used currently.

Plastic packing has the advantage of requiring less land area for the filter structure than rock due to the ability to use higher loading rates and taller trickling filters. Grady et al. (1999) noted that when loaded at the similar low organic loadings rates (less than 1.0 kg BOD/m³·d), the performance of rock filters compared to filters with plastic packing is similar. At higher organic loading rates, however, the performance of filters with plastic packing is superior. The higher porosity, which provides for better air circulation and biofilm sloughing, is a likely explanation for the improved performance.

Dosing Rate. The dosing rate on a trickling filter is the depth of liquid discharged on top of the packing for each pass of the distributor. For higher distributor rotational speeds, the dosing rate is lower. In the past, typical rotational speeds for distributors were about 0.5 to 2 min per revolution (WEF, 2000). With two to four arms, the trickling filter is dosed every 10 to 60 s. Results from various investigators have indicated that reducing the distributor speed results in better filter performance. Hawkes (1963) showed that rock trickling filters dosed every 30 to 55 min/rev outperformed a more conventional operation of 1 to 5 min/rev. Besides improved BOD removal, there were dramatic reductions in the *Psychoda* and *Anisopus* fly population, biofilm thickness, and odors. Albertson and Davies (1984) showed similar advantages from an investigation of reduced distributor speed. At a higher dosing rate, the larger water volume applied per revolution (1) provides greater wetting efficiency, (2) results in greater agitation, which causes more solids to flush out of the packing, (3) results in a thinner biofilm, and (4) helps to wash away fly eggs. The thinner biofilm creates more surface area and results in a more aerobic biofilm.

If the high dosing rate is sustained to control the biofilm thickness, the treatment efficiency may be decreased because the liquid contact time in the filter is less. A daily intermittent high dose, referred to as a *flushing dose,* is used to control the biofilm thickness and solids inventory. A combination of a once-per-day high flushing rate and a lower daily sustained dosing rate is recommended as a function of the BOD loading as shown in Table 9–3 (WEF, 2000). The data in Table 9–3 are guidelines to establish a dosing range. Optimization of the dosing rate and flushing rate and frequency is best determined from field operation. Flexibility in the distributor design is needed to provide a range of dosing rates to optimize the trickling filter performance.

The rotational speed for a rotary distributor can be determined using the following relationship (Albertson, 1989):

$$n = \frac{(1 + R)(q)(10^3 \text{ mm/m})}{(A)(DR)(60 \text{ min/h})} \tag{9–1}$$

where n = rotational speed, rev/min
   q = influent applied hydraulic loading rate, m^3/m^2·h
   R = recycle ratio
   A = number of arms in rotary distributor assembly
   DR = dosing rate, mm/pass of distributor arm

The dosing rate, DR, has also been referred to as the SK value, which stands for *Spülkraft,* a term used in the German regulations to define dosing in the early 1980s. To achieve the suggested dosing rates, the speed of the rotary distributor can be controlled by (1) reversing the location of some of the existing orifices to the front of the distributor arm, (2) adding reversed deflectors to the existing orifice discharges, or (3) converting the rotary distributor to a variable-speed electric drive (Albertson, 1995).

Table 9–3
A guideline for trickling filter dosing rate as a function of BOD loading[a]

BOD loading kg/m^3·d	Operating dose, mm/pass[b]	Flushing dose, mm/pass[b]
0.25	10–30	≥200
0.50	15–45	≥200
1.00	30–90	≥300
2.00	40–120	≥400
3.00	60–180	≥600
4.00	80–240	≥800

[a] From WEF (2000).
[b] mm/pass represents the amount of liquid applied for each pass of each distributor arm.
Note: kg/m^3·d × 62.4280 = lb/10^3 ft^3·d.

Distribution Systems. A distributor consists of two or more arms that are mounted on a pivot in the center of the filter and revolve in a horizontal plane (see Fig. 9–4). The arms are hollow and contain nozzles through which the wastewater is discharged over the filter bed. The distributor assembly may be driven either by the dynamic reaction of the wastewater discharging from the nozzles or by an electric motor. The flow-driven rotary distributor for trickling filtration has been used traditionally for the process because it is reliable and easy to maintain. Motor drives are used in more recent designs. The speed of rotation, which varies with the flowrate and the organic loading rate, can be determined using Eq. (9–1). Clearance of 150 to 225 mm (6 to 9 in) should be allowed between the bottom of the distributor arm and the top of the bed. The clearance permits the wastewater streams from the nozzles to spread out and cover the bed uniformly, and it prevents ice accumulations from interfering with the distributor motion during freezing weather.

Distributors are manufactured for trickling filters with diameters up to 60 m (200 ft). Distributor arms may be of constant cross section for small units, or they may be tapered to maintain minimum transport velocity. Nozzles are spaced unevenly so that greater flow per unit of length is achieved near the periphery of the filter than at the center. For uniform distribution over the area of the filter, the flowrate per unit of length should be proportional to the radius from the center. Headloss through the distributor is in the range of 0.6 to 1.5 m (2 to 5 ft). Important features that should be considered in selecting a distributor are the ruggedness of construction, ease of cleaning, ability to handle large variations in flowrate while maintaining adequate rotational speed, and corrosion resistance of the material and its coating system.

Figure 9–4

Typical distributors used to apply wastewater to trickling filter packing:
(a) view of conventional rock filter with two-arm rotary distributor,
(b) view of early (circa 1920) rock filter with a fixed distribution system *(Library of Congress)*, and (c) view of top of tower trickling filter with four-arm rotary distributor.

(a)

(b)

(c)

In the past, fixed nozzle distribution systems were used for shallow rock filters (see Fig. 9–4b). Fixed nozzle distribution systems consist of a series of spray nozzles located at the points of equilateral triangles covering the filter bed. A system of pipes placed in the filter is used to distribute the wastewater uniformly to the nozzles. Special nozzles having a flat spray pattern are used, and the pressure is varied systematically so that the spray falls first at a maximum distance from the nozzle and then at a decreasing distance as the head slowly drops. In this way, a uniform dose is applied over the whole area of the bed. Half-spray nozzles are used along the sides of the filter. In current practice, fixed nozzle systems are seldom used.

Underdrains. The wastewater collection system in a trickling filter consists of underdrains that catch the filtered wastewater and solids discharged from the filter packing for conveyance to the final sedimentation tank. The underdrain system for a rock filter usually has precast blocks of vitrified clay or fiberglass grating laid on a reinforced-concrete subfloor (see Fig. 9–5). The floor and underdrains must have sufficient strength to support the packing, slime growth, and the wastewater. The floor and underdrain block slope to a central or peripheral drainage channel at a 1 to 5 percent grade. The effluent channels are sized to produce a minimum velocity of 0.6 m/s (2 ft/s) at the average daily flowrate (WPCF, 1988). Underdrains may be open at both ends, so that they may be inspected easily and flushed out if they become plugged. The underdrains also allow ventilation of the filter, providing the air for the microorganisms that live in the filter slime. The underdrains should be open to a circumferential channel for ventilation at the wall as well as to the central collection channel.

The underdrain and support system for plastic packing consists of either a beam and column or a grating. A typical underdrain system for a tower filter is shown on Fig. 9–6. The beam and column system typically has precast-concrete beams supported by columns or posts. The plastic packing is placed over the beams, which have channels in their tops to ensure free flow of wastewater and air. All underdrain systems should be designed so that forced-air ventilation can be added at a later date if filter operating conditions should change.

Airflow. An adequate flow of air is of fundamental importance to the successful operation of a trickling filter to provide efficient treatment and to prevent odors. Natural draft has historically been the primary means of providing airflow, but it is not always adequate and forced ventilation using low-pressure fans provides more reliable and controlled airflow.

Figure 9–5

Typical underdrain for rock filter: (a) fiberglass grating and (b) vitrified clay block.

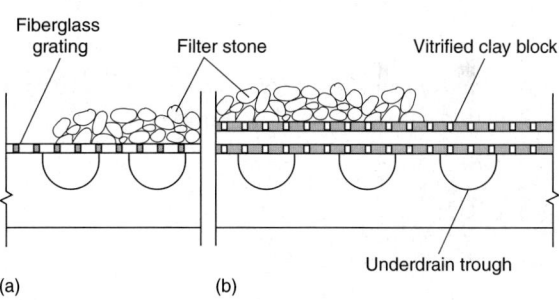

Figure 9-6

Typical underdrain system for tower filter.

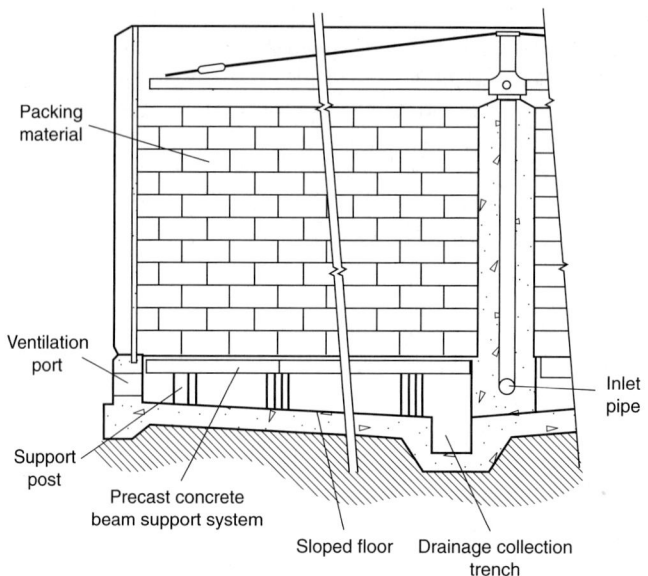

In the case of natural draft, the driving force for airflow is the temperature difference between the ambient air and the air inside the pores. If the wastewater is colder than the ambient air, the pore air will be cold and the direction of flow will be downward. If the ambient air is colder than the wastewater, the flow will be upward. The latter is less desirable from a mass transfer point of view because the partial pressure of oxygen (and thus the oxygen transfer rate) is lowest in the region of highest oxygen demand. In many areas of the country, there are periods, especially during the summer, when essentially no airflow occurs through the trickling filter because temperature differentials are negligible.

Draft, which is the pressure head resulting from the temperature and moisture differences, may be determined from Eq. (9–2) (Schroeder and Tchobanoglous, 1976):

$$D_{air} = 353\left(\frac{1}{T_c} - \frac{1}{T_h}\right)Z \qquad (9\text{–}2)$$

where D_{air} = natural air draft, mm of water
T_c = cold temperature, K
T_h = hot temperature, K
Z = height of the filter, m

A more conservative estimate of the average pore air temperature is obtained by using the log mean temperature T_m for T_n in Eq. (9–2).

$$T_m = \frac{T_2 - T_1}{\ln\ (T_2/T_1)} \qquad (9\text{–}3)$$

where T_2 = colder temperature, K
T_1 = warmer temperature, K

The volumetric air flowrate may be estimated by setting the draft equal to the sum of the head losses that result from the passage of air through the filter and underdrain system (Albertson and Okey, 1988).

Where natural draft is used, the following needs to be included in the design:

1. Underdrains and collecting channels should be designed to flow no more than half full to provide a passageway for the air.
2. Ventilating access ports with open grating types of covers should be installed at both ends of the central collection channel.
3. Large-diameter filters should have branch collecting channels with ventilating manholes or vent stacks installed at the filter periphery.
4. The open area of the slots in the top of the underdrain blocks should not be less than 15 percent of the area of the filter.
5. One square meter gross area of open grating in ventilating manholes and vent stacks should be provided for each 23 m² of filter area (10 ft²/250 ft²).

The use of forced- or induced-draft fans is recommended for trickling filter designs to provide a reliable supply of oxygen. The costs for a forced-draft air supply are minimal compared to the benefits. For a 3800 m³/d (1.0 Mgal/d) wastewater treatment flow the estimated power requirement is only about 0.15 kW (0.2 hp) (WEF, 2000). As an approximation, an airflow of 0.3 m³/m²·min (1 ft³/ft²·min) of filter area in either direction is recommended. A downflow direction has some advantage by providing contact time for treating odorous compounds released at the top of the filter and by providing a richer air supply where the oxygen demand is highest. Forced-air designs should provide multiple air distribution points by the use of fans around the periphery of the tower or the use of air headers below the packing material, as there is very little headloss through the filter packing to promote air distribution. For applications with extremely low air temperature, it may be necessary to restrict the flow of air through the filter to keep it from freezing.

Little has been done to quantify the amount of oxygen used in trickling filters and the actual oxygen transfer efficiency. The following formulations are based on earlier work by Dow Chemical during the development of plastic packing material for trickling filter applications. In developing these formulations, it was assumed that the oxygen transfer efficiency was about 5 percent. The required oxygen supply is given as follows:

BOD removal only:

$$R_o = (20 \text{ kg/kg})[0.80e^{-9L_B} + 1.2e^{-0.17L_B}](\text{PF}) \tag{9–4}$$

BOD removal and nitrification:

$$R_o = (40 \text{ kg/kg})[0.80e^{-9L_B} + 1.2e^{-0.17L_B} + 4.6N_{ox}/\text{BOD}](\text{PF}) \tag{9–5}$$

where R_o = oxygen supply, kg O_2/kg BOD applied
L_B = BOD loading to filter, kg BOD/m³·d
N_{ox}/BOD = ratio of influent nitrogen oxidized to influent BOD, mg/mg
PF = peaking factor, maximum to average load

The air application rate at 20°C and 1.0 atm is computed as follows: From Appendix B, the density of air at 20°C and 1.0 atm is 1.204 kg/m³, and the percent of oxygen by weight in air is 23.18 percent. Thus, the volume of oxygen per kg of air is 3.58 m³/kg [1/(1.204 kg/m³)(0.2318)] and the required airflow is given by

$$AR_{20} = \frac{(R_o)(Q)(S_o)(3.58 \text{ m}^3/\text{kg O}_2)}{(10^3 \text{ g/kg})(1440 \text{ min/d})} \tag{9-6}$$

where AR_{20} = airflow rate at 20°C and 1.0 atm, m³/min
Q = wastewater flowrate, m³/d
S_o = primary effluent BOD, g/m³

The airflow rate is corrected for temperature and pressure according to the ideal gas law:

$$AR_{T_A} = AR_{20}\left(\frac{273.15 + T_A}{273.15}\right)\left(\frac{760}{P_o}\right) \tag{9-7}$$

where AR_{T_A} = airflow rate at ambient air temperature, °C
T_A = ambient air temperature, °C
P_o = pressure at treatment plant site, mm Hg

A further correction to the calculated airflow is recommended for temperatures above 20°C to account for the lower oxygen saturation concentration at higher temperatures and the higher biological uptake rates in the filter. For each degree Celsius above 20°C the airflow rate is increased by 1 percent.

$$AR_{T>20°C} = AR_T\left(1 + \frac{T_A - 20}{100}\right) \tag{9-8}$$

The pressure drop through the packing is related to the superficial air velocity as follows:

$$\Delta P = N_p\left(\frac{v^2}{2g}\right) \tag{9-9}$$

where ΔP = total headlosses, kPa
g = acceleration of gravity, 9.81 m/s²
v = superficial velocity, Q/A, m/s
N_p = tower resistance as number of velocity heads lost

The tower resistance term N_p is the sum of all the individual headlosses related to the airflow. Headloss occurs as air moves through the inlet, underdrain, and packing material. The packing loss in terms of number of velocity headlosses was developed by Dow Chemical for the original vertical packing:

$$N_p = 10.33(D)e^{(1.36 \times 10^{-5})(L/A)} \tag{9-10}$$

where N_p = packing headloss in terms of velocity heads
D = packing depth, m
L = liquid loading rate, kg/h
A = tower cross-section area, m²

Although similar correlations have not been developed for other packing materials, recommended correction factors that can be used to obtain N_p values for other packing materials based on the value determined using Eq. (9–10) are given in Table 9–4. To estimate the total headloss, the value of N_p, computed using Eq. (9–10), is often multiplied by a factor of 1.3 to 1.5 to include inlet, underdrain, and other minor losses.

Table 9–4

Correction factors for computing headloss in nonvertical trickling filter packings based on Eq. (9–10)[a]

Packing	Specific surface area, m²/m³	Correction factor
Rock	45	2.0
Plastic cross flow	100	1.3
Plastic cross flow	140	1.6
Plastic random	100	1.6

[a] Adapted from WEF (2000).
Note: m²/m³ × 0.3048 = ft²/ft³.

Settling Tanks. The function of settling tanks that follow trickling filters is to produce a clarified effluent. They differ from activated-sludge settling tanks in that the clarifier has a much lower suspended solids content and sludge recirculation is not necessary. All the sludge from trickling filter settling tanks is sent to sludge-processing facilities or returned to the primary clarifiers to be settled with primary solids. Trickling filter performance has historically suffered from poor clarifier designs. The use of shallow clarifiers for trickling filter applications, with relatively high overflow rates, was recommended in previous versions of the "Ten States Standards" (GLUMRB1997). Unfortunately, the use of shallow clarifiers typically resulted in poor clarification efficiency. Clarifier overflow rates recommended currently in the Ten States Standards are more in line with those used for the activated-sludge process. Clarifier designs for trickling filters should be similar to designs used for activated-sludge process clarifiers (see Sec. 8–7 in Chap. 8), with appropriate feedwell size and depth, increased sidewater depth, and similar hydraulic overflow rates. Recommended overflow rates as a function of the clarifier sidewater depth are given on Fig. 9–7. With proper clarification designs, single-stage trickling filters can achieve a less than 20 mg/L concentration of BOD and TSS.

Figure 9–7

Recommended trickling filter clarifier overflow rates as a function of the clarifier sidewall depth. (Adapted from WEF, 2000.)

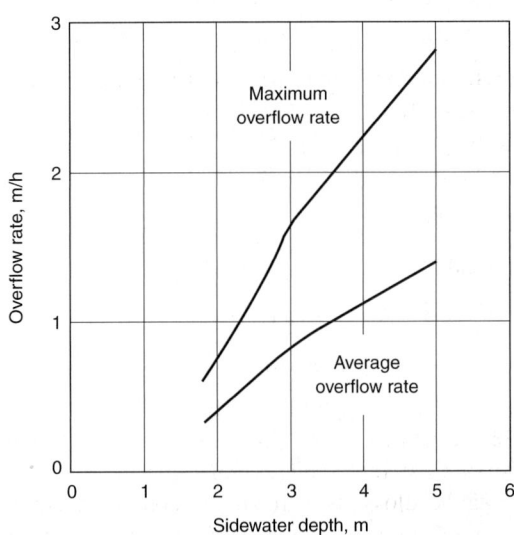

EXAMPLE 9–1 **Determine Trickling Filter Airflow Rate Requirements and Pressure Drop for Forced Ventilation** Determine the forced-ventilation air requirement and pressure drop in a trickling filter with cross-flow plastic packing, given the following design and operating information.

Wastewater characteristics:

1. Wastewater flowrate = 19,000 m^3/d (220 L/s)
2. Primary effluent BOD = 140 mg/L (g/m^3)
3. Temperature = 20°C

Design assumptions:

1. BOD loading = 0.56 kg BOD/m^3·d
2. Organic loading peaking factor = 1.4
3. Tower diameter = 22 m
4. Number of towers = 2
5. Depth of packing = 6.1 m
6. Headloss correction factor for inlet and other minor losses = 1.5
7. Headloss correction factor for cross-flow packing = 1.3 (see Table 9–4)
8. Air temperature = −7 to 28°C

Solution

1. Determine the required oxygen supply rate using Eq. (9–4).

$$R_o = (20 \text{ kg/kg})[0.80e^{-9L_B} + 1.2e^{-0.17L_B}](PF)$$

$$L_B = 0.56 \text{ kg BOD/m}^3\text{·d}$$

$$PF = 1.4$$

$$R_o = 20[0.80e^{-9(0.56)} + 1.2e^{-0.17(0.56)}](1.4) = 30.69 \text{ kg O}_2/\text{kg BOD applied}$$

2. Determine airflow rate for the given conditions.
   a. Using Eq. (9–6), determine the airflow rate at standard conditions.

$$AR_{STD} = \frac{R_o(Q)(S_o)(3.58 \text{ m}^3/\text{kg O}_2)}{(10^3 \text{ g/kg})(1440 \text{ min/d})}$$

$$= \frac{(30.69 \text{ kg/kg})(19,000 \text{ m}^3/\text{d})(140 \text{ g/m}^3)(3.58 \text{ m}^3/\text{kg O}_2)}{(10^3 \text{ g/kg})(1440 \text{ min/d})}$$

$$= 203 \text{ m}^3/\text{min}$$

   b. Correct the air flowrate for temperature and pressure using Eq. (9–7).

$$AR_{TA} > AR_{STD}\left(\frac{273.15 + T_A}{273.15}\right)$$

$$T_A = 28°C$$

$$AR_{28} = (203 \ \text{m}^3/\text{min}) \left(\frac{273.15 + 28}{273.15} \right) = 223.8 \ \text{m}^3/\text{min}$$

    c. Correct the air flowrate for lower oxygen saturation using Eq. (9–8).

$$AR = AR_{TA} \left(1 + \frac{T_A - 20}{100} \right)$$

$$T_A = 28°C$$

$$AR = (223.8) \left(1 + \frac{28 - 20}{100} \right) = 241.7 \ \text{m}^3/\text{min}$$

3. Determine pressure drop in the cross-flow packing.
    a. Determine the value of N_p using Eq. (9–10). The value of N_p will then be used to determine the pressure drop using Eq. (9–9).

$$N_P = 10.33 \ (D)e^{(1.36 \times 10^{-5})(L/A)}$$

Solve for L/A, kg/m²·h

Hydraulic application rate $= q = Q/A$

$Q = (19{,}000 \ \text{m}^3/\text{d})(1 \ \text{d}/24 \ \text{h}) = 792 \ \text{m}^3/\text{h}$

Area of single tower $A = \dfrac{\pi D^2}{4} = \dfrac{\pi (22\text{m})^2}{4} = 380 \ \text{m}^2$

$q = \dfrac{792}{380} = 2.08 \ \text{m}^3/\text{m}^2 \cdot \text{h}$

$\dfrac{L}{A} = (2.08 \ \text{m}^3/\text{m}^2\text{·h})(10^3 \ \text{L/m}^3)(1 \ \text{kg/L})$

     $= 2080 \ \text{kg/m}^2\text{·h}$

Packing depth $= 6.1 \ \text{m}$

$N_P = 10.33 \ (6.1)e^{(1.36 \times 10^{-5})(2080)}$

     $= 64.8$

Headloss correction factor for cross-flow packing $= 1.3$

Headloss correction factor for inlet and other losses $= 1.5$

$N_P = (1.5)(1.3)(64.8) = 126.3$

    b. Solve for the pressure drop using Eq. (9–9)

$$\Delta P = N_P \left(\frac{v^2}{2g} \right)$$

Superficial velocity $v = \dfrac{\text{air flowrate/tower}}{\text{area of tower}}$

$$v = \frac{(241.7/2)}{380.0} = 0.32 \text{ m/min} = 0.005 \text{ m/s}$$

$$\Delta P = 126.3\left(\frac{v^2}{2g}\right)$$

$$= \frac{(126.3)(0.005 \text{ m/s})^2}{2(9.8 \text{ m/s}^2)} = 0.00016 \text{ m}$$

Air density at 28°C = 1.175 kg/m³ (see Appendix B)

$$\Delta P = (0.00016 \text{ m})(1.175 \text{ kg/m}^3)(9.8 \text{ m/s}^2)$$

$$= 0.0018 \text{ N/m}^2 = 0.0018 \text{ Pa} = 1.8 \times 10^{-6} \text{ kPa}$$

4. Compare to natural draft pressure per Eqs. (9–2) and (9–3).
   a. Determine log mean temperature using Eq. (9–3).

   Wastewater temperature = 20°C

   Air temperature = 28°C

   $$T_m = \frac{T_2 - T_1}{\ln(T_2/T_1)} = \frac{28 - 20}{\ln(28/20)} = 23.8°C$$

   b. Determine draft using Eq. (9–2).

   $$D_{air} = 353\left(\frac{1}{T_c} - \frac{1}{T_m}\right)Z$$

   $$T_c = 273.15 + 20 = 293.15 \text{ K}$$

   $$T_m = 273.15 + 23.8 = 296.95 \text{ K}$$

   $$D_{air} = 353\left(\frac{1}{293.15} - \frac{1}{296.95}\right)6.1 = 0.094 \text{ mm}$$

   c. Compare draft to estimated headloss.

   Convert mm of water to pressure expressed in Pa.

   $$\text{Draft} = (0.094 \text{ mm H}_2\text{O})\left(\frac{9.797 \text{ Pa}}{\text{mm H}_2\text{O}}\right) = 0.921 \text{ Pa}$$

   Thus, draft (0.921 Pa) > headloss (0.0018 Pa)

Comment More draft is available for these temperature differences than is needed; but for periods where the wastewater and air temperatures are very close, sufficient oxygen will not be available, which may result in the formation of malodorous gases such as hydrogen sulfide. Note that the pressure drop for the necessary air flowrate is very low and multiple air feed points are needed to assure uniform air distribution.

Process Design Considerations

The trickling filter process appears simple, consisting of a bed of packing material through which wastewater flows and an external clarifier. In reality, a trickling filter is a very complex system in terms of the characteristics of the attached growth and internal hydrodynamics. In view of these complexities, trickling filter designs are based mainly on empirical relationships derived from pilot-plant and full-scale plant experience. In this section trickling filter performance for BOD removal and nitrification, features that affect performance, and commonly used process design approaches are reviewed.

Effluent Characteristics. Historically, trickling filters have been considered to have major advantages of using less energy than activated-sludge treatment and being easier to operate, but have disadvantages of more potential for odors and lower-quality effluent. Some of these shortcomings, however, have been due more to inadequate ventilation, poor clarifier design, inadequate protection from cold temperatures, and the dosing operation. With proper design, trickling filters have been used successfully in a number of applications. Typical applications, process loadings, and effluent quality are summarized in Table 9–5.

Loading Criteria. In the activated-sludge process, biodegradation efficiency was shown to be related to the average SRT for the biomass or the F/M ratio. For both of these parameters, the solids or biomass can be sampled and reasonably well quantified. However, for trickling filters quantifying the biomass in the system is not possible, and only recently has progress been made to control the solids inventory to some degree by the dosing operation (Albertson, 1989). The attached growth is not uniformly distributed in the trickling filter (Hinton and Stensel, 1994), the biofilm thickness can vary, the biofilm solids concentration may range from 40 to 100 g/L, and the liquid does not uniformly flow over the entire packing surface area, which is referred to as the wetting efficiency. With the inability to quantify the biological and hydrodynamic properties of field trickling filter systems, broader parameters such as volumetric organic loading, unit area loadings, and hydraulic application rates have been used as design and operating parameters to relate to treatment efficiency.

Table 9–5
Trickling filter applications, loadings, and effluent quality

| Application | Loading | | Effluent quality | |
	Unit	Range	Unit	Range
Secondary treatment	kg BOD/m³·d[a]	0.3–1.0	BOD, mg/L	15–30
			TSS, mg/L	15–30
Combined BOD removal and nitrification	kg BOD/m³·d g TKN/m²·d[b]	0.1–0.3 0.2–1.0	BOD, mg/L NH_4-N, mg/L	<10 <3
Tertiary nitrification	g NH_4-N/m²·d	0.5–2.5	NH_4-N, mg/L	0.5–3
Partial BOD removal	kg BOD/m³·d	1.5–4.0	% BOD removal	40–70

[a] Volumetric loading.
[b] Loading based on packing surface area.
Note: kg/m³·d $\times$ 62.4280 = lb/10³ ft³·d.
g/m²·d $\times$ 0.204 = lb/10³ ft²·d.

Figure 9–8

Example of trickling filter performance at 20°C. Effect of BOD loading removal efficiency for plastic media filter.

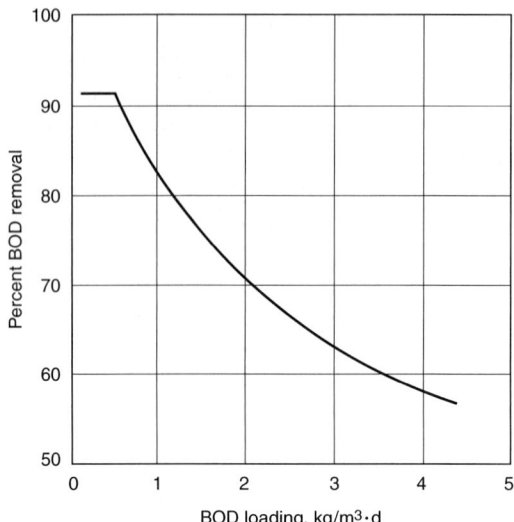

For BOD removal, the volumetric BOD loading has been correlated well with treatment performance for both BOD removal and nitrification in combined BOD and nitrification trickling filter designs. The original design model for rock trickling filters was developed by the National Research Council (NRC) in the early 1940s at military installations (Mohlman et al., 1946). The NRC formulations were based on field data for BOD removal efficiency and the organic loading rate. The NRC design model was used even though there was a significant amount of data scatter. Bruce and Merkens (1970 and 1973) found that the organic loading rate controlled trickling filter performance and not the hydraulic application rate. For combined BOD removal and nitrification systems, nitrification efficiency has been related to the volumetric BOD loading (Stenquist and Kelly, 1980; U.S. EPA, 1975; and Daigger et al., 1993). Examples of the effect of BOD loading on BOD removal are illustrated on Fig. 9–8. At low BOD loadings, BOD removal efficiency reaches a maximum plateau level of about 90 percent. In actual plant operations considerable scatter will exist around the curve due to variations in solids sloughing, wastewater characteristics (sBOD fraction), and clarification efficiency.

For tertiary nitrification applications, very little BOD is applied to the trickling filter and a thin biofilm develops on the packing that consists of a high proportion of nitrifying bacteria. The nitrification removal efficiency is related to the packing surface area and correlated with the specific nitrogen loading rate in terms of g NH_4-N removed/m² packing surface area·d (Duddles et al., 1974; Okey and Albertson, 1989; Parker et al., 1990; U.S. EPA, 1993; and Anderson et al., 1994). Surface nitrification rates for tertiary nitrification towers with plastic packing are shown by Fig. 9–9, along with the effect of temperature on nitrification rates. The specific nitrification rates for the Central Valley facility are much higher than reported previously for trickling filters. The higher rate was credited to the use of cross flow instead of vertical plastic packing material and better biofilm control procedures. Higher NH_4-N concentrations were also present. The lower rates for the Zurich facility were attributed to a poorer wetting efficiency and a less uniform biofilm coverage. The analysis of trickling filter loading is illustrated in Example 9–2.

Figure 9–9

Effect of temperature on nitrification rates for tower trickling filters. (Adapted from Parker et al., 1990.)

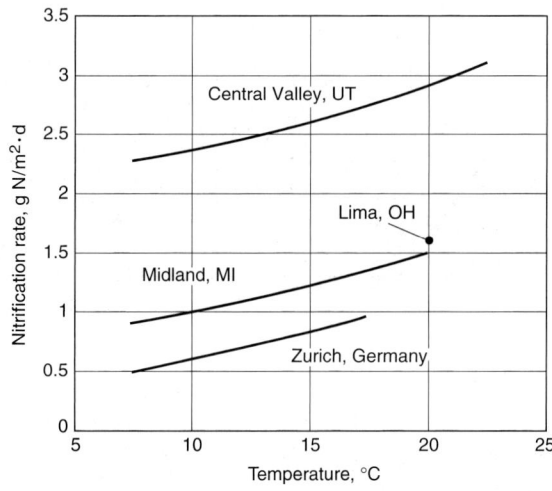

EXAMPLE 9–2   **Trickling Filter Loading**   A 10-m-diameter single-stage trickling filter contains conventional cross-flow plastic packing at a depth of 6.1 m. Primary effluent with the characteristics given below is applied to the filter. What is the volumetric BOD and TKN loading? What is the specific TKN loading? What is the approximate BOD removal efficiency at 20°C? Can nitrification be expected?

Primary effluent wastewater characteristics

Item	Unit	Value
Flowrate	m^3/d	4000
BOD	g/m^3	120
TSS	g/m^3	80
TKN	g/m^3	25

Note: g/m^3 = mg/L

Solution

1. Determine the volume of the trickling filter packing material.

    Volume $V = (A)(D)$

    $$A = \frac{\pi(10\ m^2)}{4} = 78.5\ m^2$$

    $V = (78.5\ m^2)(6.1\ m) = 479\ m^3$

2. Determine the BOD loading.

   BOD loading rate $= QS_o/V$

   $$= \frac{(4000 \text{ m}^3/\text{d})(120 \text{ g/m}^3)(1 \text{ kg}/10^3 \text{ g})}{479 \text{ m}^3} = 1.0 \text{ kg/m}^3\cdot\text{d}$$

3. Determine the TKN loading rate.

   $$\text{TKN} = \frac{(4000 \text{ m}^3/\text{d})(25 \text{ g/m}^3)(1 \text{ kg}/10^3\text{g})}{479 \text{ m}^3} = 0.21 \text{ kg TKN/m}^3\cdot\text{d}$$

4. Estimate the approximate BOD removal efficiency.

   From Fig. 9–8, at a loading of 1.0 kg BOD/m³·d, the BOD removal efficiency is about 82 percent.

5. Can nitrification be expected?

   No. Based on the data given in Table 9–5, the BOD loading for combined BOD removal and nitrification is too high (1.0 kg/m³·d), even though the TKN loading is appropriate. At the higher BOD loading, the heterotrophic bacteria outcompete the nitrifying bacteria for sites on the packing surface and thus control the slime biomass population.

6. What is the specific TKN loading?

   From Table 9–2, the specific surface area of the packing is $\approx 90 \text{ m}^2/\text{m}^3$

   Total surface area $= (90 \text{ m}^2/\text{m}^3)(479 \text{ m}^3) = 43,110 \text{ m}^2$

   $$\text{Specific TKN loading} = \frac{QN_o}{A} = \frac{(4000 \text{ m}^3/\text{d})(25 \text{ g/m}^3)}{43,110 \text{ m}^2}$$

   $$= 2.3 \text{ g N/m}^2\cdot\text{d}$$

Comment In summary, it is important to note that wide variations will be observed in the nitrification rates depending on site-specific conditions.

BOD Removal Design. The first empirical design equations for BOD removal were developed for rock trickling filters from an analysis of trickling filter performance at 34 plants at military installations treating domestic wastewater (Mohlman et al., 1946). The effect of volumetric BOD loading and recirculation ratio on treatment performance was accounted for in the equations. The equations given below should only be used as an estimate of performance as they are based on a limited data base and the influent BOD values at the installations sampled were relatively high compared to most municipal primary effluents. The BOD removal includes the effect of the secondary clarifier, so that if the equation overpredicts treatment performance, improved and deeper secondary clarifier designs used today may help in meeting expected treatment performance.

For a single-stage or first-stage rock filter, the NRC equation is

$$E_1 = \frac{100}{1 + 0.4432\sqrt{\dfrac{W_1}{VF}}} \tag{9–11}$$

where E_1 = BOD removal efficiency for first-stage filter at 20°C, including recirculation, percent
W_1 = BOD loading to filter, kg/d
V = volume of filter packing, m^3
F = recirculation factor

The recirculation factor is calculated using Eq. (9–12):

$$F = \frac{1 + R}{(1 + R/10)^2} \tag{9–12}$$

where F = recirculation factor
R = recycle ratio, unitless

The recirculation factor represents the average number of passes of the influent organic matter through the filter. The factor $R/10$ accounts for the fact that the benefits of recirculation decrease as the number of passes increase. Recycle ratios used generally vary from 0 to 2.0. For a two-stage trickling filter system (see Fig. 9–2b) the BOD removal efficiency of the second stage is given as follows:

$$E_2 = \frac{100}{1 + \dfrac{0.4432}{1 - E_1}\sqrt{\dfrac{W_2}{VF}}} \tag{9–13}$$

where E_2 = BOD removal efficiency for the second-stage filter at 20°C, percent
E_1 = fraction of BOD removal in the first-stage filter
W_2 = BOD loading applied to the second-stage filter, kg/d

The effect of wastewater temperature on the BOD removal efficiency is calculated as follows:

$$E_T = E_{20}(1.035)^{T-20} \tag{9–14}$$

where E_T = BOD removal efficiency at temperature T in °C, percent
E_{20} = BOD removal efficiency at 20°C, percent

EXAMPLE 9–3   Trickling Filter Sizing Using NRC Equations   A municipal wastewater having a BOD of 200 g/m^3 is to be treated by a two-stage trickling filter. The desired effluent quality is 25 g/m^3 of BOD. If both of the filter depths are to be 1.83 m and the recirculation ratio is 2:1, find the required filter diameters. Assume the following design assumptions apply. (Note: g/m^3 = mg/L.)

Design assumptions:

1. Flowrate = 7570 m³/d
2. Wastewater temperature = 20°C
3. $E_1 = E_2$

Solution

1. Compute E_1 and E_2.

$$\text{Overall efficiency} = \left\{ \frac{[(200 - 25)\ \text{g/m}^3]}{(200\ \text{g/m}^3)} \right\} (100) = 87.5\%$$

$$E_1 + E_2 (1 - E_1) = 0.875$$

$$E_1 = E_2 = 0.646$$

2. Compute the recirculation factor using Eq. (9–12).

$$F = \frac{1 + R}{(1 + R/10)^2} = \frac{1 + 2}{(1.2)^2} = 2.08$$

3. Compute the BOD loading for the first filter.

$$W_1 = (200\ \text{g/m}^3)(7570\ \text{m}^3/\text{d})(1\ \text{kg}/10^3\ \text{g}) = 1514\ \text{kg BOD/d}$$

4. Compute the volume for the first stage using Eq. (9–11).

$$E_1 = \frac{100}{1 + 0.4432 \sqrt{\dfrac{W_1}{VF}}}$$

$$64.6 = \frac{100}{1 + 0.4432 \sqrt{\dfrac{1514}{V(2.08)}}}$$

$$V = 476\ \text{m}^3$$

5. Compute the diameter of the first filter.

$$A = \frac{V}{D} = \frac{476\ \text{m}^3}{1.83\ \text{m}} = 260\ \text{m}^2$$

$$D = 18.2\ \text{m}$$

6. Compute the BOD loading for the second-stage filter.

$$W_2 = (1 - E_1)\ W_1 = (1 - 0.646)(1514\ \text{kg BOD/d}) = 536\ \text{kg BOD/d}$$

7. Compute the volume of the second-stage filter using Eq. (9–13).

$$E_2 = \cfrac{100}{1 + \cfrac{0.4432}{1 - E_1}\sqrt{\cfrac{W_2}{VF}}}$$

$$64.6 = \cfrac{100}{1 + \cfrac{0.4432}{1 - 0.646}\sqrt{\cfrac{536}{V(2.08)}}}$$

$V = 1345 \text{ m}^3$

8. Compute the diameter of the second filter.

$$A = \frac{V}{D} = \frac{1345 \text{ m}^3}{1.83 \text{ m}} = 735 \text{ m}^2$$

$D = 30.6 \text{ m}$

9. Compute the BOD loading to each filter.
   a. First-stage filter:

   $$\text{BOD loading} = \frac{(1514 \text{ kg/d})}{476 \text{ m}^3} = 3.18 \text{ kg/m}^3\text{·d}$$

   b. Second-stage filter:

   $$\text{BOD loading} = \frac{(536 \text{ kg/d})}{1345 \text{ m}^3} = 0.40 \text{ kg/m}^3\text{·d}$$

10. Compute the hydraulic loading to each filter.
    a. First-stage filter:

    $$\text{Hydraulic loading} = \frac{(1 + 2)(7570 \text{ m}^3/\text{d})}{(1440 \text{ min/d})(260 \text{ m}^2)}$$
    $$= 0.061 \text{ m}^3/\text{m}^2\text{·min}$$

    b. Second-stage filter:

    $$\text{Hydraulic loading} = \frac{(1 + 2)(7570 \text{ m}^3/\text{d})}{(1440 \text{ m/d})(735 \text{ m}^2)}$$
    $$= 0.022 \text{ m}^3/\text{m}^2\text{·min}$$

Comment To accommodate standard rotary distributor mechanisms, the diameters of the two filters should be rounded to the nearest 1.0 m. To reduce construction costs, the two trickling filters are often made the same size. Where two filters of equal diameter are used, the removal efficiencies will be unequal. In many cases, the hydraulic loading rate will be limited by state standards.

Formulations for Plastic Packing. In the general equations developed for trickling filters with plastic packing, BOD removal was related to the hydraulic application rate. The formulations are based on the early work of Velz (1948), who observed that the BOD remaining with depth in a trickling filter could be modeled as a first-order relationship, and Howland (1958) and Schulze (1960), who described the hydraulic detention time. Howland (1958) proposed that the liquid contact time with the biofilm is proportional to the filter depth and inversely proportional to the hydraulic application rate as follows:

$$t = \frac{CD}{q^n} \tag{9-15}$$

$$q = \frac{Q}{A} \tag{9-16}$$

where t = liquid contact time, min
$\quad C$ = constant for packing used
$\quad D$ = depth of packing, m
$\quad q$ = hydraulic loading rate, L/m²·min
$\quad n$ = hydraulic constant for the packing material used, unitless
$\quad Q$ = influent flowrate, L/min
$\quad A$ = filter cross section area, m²

According to Eq. (9–15), as the influent flowrate to the trickling filter increases, the detention time does not decrease in direct proportion to the flow, because the liquid film thickness increases.

By considering the change in BOD concentration in the filter with time for depth as a first-order reaction, the following equation is derived:

$$\frac{ds}{dt} = -kS \tag{9-17}$$

where k = an experimentally determined rate constant
$\quad S$ = BOD concentration at time t

and using Eq. (9–15) to determine the time in the filter, Schulze derived the following equation:

$$\frac{S_e}{S_o} = e^{-kD/Q^n} \tag{9-18}$$

where S_e = BOD concentration of settled filter effluent, mg/L (g/m³)
$\quad S_o$ = influent BOD concentration, mg/L (g/m³)
$\quad k$ = an experimentally determined rate constant
$\quad D$ = packing depth, m
$\quad Q$ = hydraulic application rate, m³/m²·d
$\quad n$ = constant, characteristic of packing used

The values of k and n determined by Schulze at 20°C were 0.69/d and 0.67.

Germain applied the Schulze equation in 1966 to trickling filters with plastic packing (WEF, 2000) as follows:

$$\frac{S_e}{S_o} = e^{-kD/q^n} \qquad (9\text{-}19)$$

where S_e = BOD concentration of settled filter effluent, mg/L (g/m^3)
  S_o = influent BOD concentration to the filter, mg/L (g/m^3)
  k = wastewater treatability and packing coefficient, $(L/s)^{0.5}/m^2$ (based on $n = 0.5$)
  D = packing depth, m
  q = hydraulic application rate of primary effluent, excluding recirculation, $L/m^2 \cdot s$
  n = constant characteristic of packing used

The value for n is normally assumed to be 0.50 and pilot-plant or full-scale plant influent and effluent BOD concentration data are used to solve for k. Values for k were developed from more than 140 pilot-plant studies by Dow Chemical Company with vertical plastic packing with a specific surface area of about 90 m^2/m^3. Similar tests have been done by other suppliers for a variety of packings. Most of these tests have been done with packing depths of 6.1 to 6.7 m (20 to 22 ft).

It should be noted that the clarification design and dosing cycle and method can affect pilot-plant results used to calculate a value for k (Daigger and Harrison, 1987). In summary, the value for k is affected by many factors, including the wastewater characteristics, filter and clarifier design, and operating conditions.

The commonly accepted temperature correction for k is as follows:

$$k_T = k_{20}(1.035)^{T-20} \qquad (9\text{-}20)$$

Other formulations have been proposed (WEF, 2000) to describe the performance of plastic packing filters, including models by Eckenfelder (1963) and Eckenfelder and Barnhart (1963). One of the modified equations termed the modified Velz equation, as given below, includes a factor for the specific surface area of the packing and recirculation flow.

$$S_e = \frac{S_o}{(R+1)\exp\left\{\dfrac{k_{20}A_sD\theta^{T-20}}{[q(R+1)]^n}\right\} - R} \qquad (9\text{-}21)$$

where S_o = influent BOD, mg/L
  S_e = effluent BOD, mg/L
  R = recirculation ratio, recycle flowrate divided by influent flowrate
  k_{20} = filter treatability constant at 20°C, $(L/s)^{0.5}/m^2$
  A_s = clean packing specific surface area, m^2/m^3
  D = depth of packing, m
  θ = temperature correction coefficient, 1.035
  q = hydraulic application rate, $L/m^2 \cdot s$
  n = constant characteristic of packing used

Because the BOD removal is determined as a function of hydraulic application rate, application of Eqs. (9–19) and (9–21) without regard to the fundamental effect of organic loading can lead to erroneous designs. For example, to achieve the same BOD removal efficiency, Eq. (9–19) would predict a smaller packing volume requirement by increasing the packing depth greater than 6.1 m. However, as the volume is reduced, the organic loading increases and thus the treatment efficiency should decline. By assuming that the BOD removal efficiency is equal at the same organic loading, the value for k had to be adjusted for depth and influent BOD concentration. The k value is normalized to a specified depth and influent BOD concentration, as follows (WEF, 2000):

$$k_2 = k_1 \left(\frac{D_1}{D_2} \right)^{0.5} \left(\frac{S_1}{S_2} \right)^{0.5} \tag{9–22}$$

where k_2 = normalized value of k for the site-specific packing depth and influent BOD concentration
k_1 = k value at depth of 6.1 m (20 ft) and influent BOD of 150 mg/L (g/m^3)
S_1 = 150 g BOD/m^3
S_2 = site-specific influent BOD concentration, g BOD/m^3
D_1 = 6.1 m (20 ft) packing depth, m
D_2 = site-specific packing depth, m

Normalized values of k_1 at 20°C, determined from Dow Chemical Company pilot-plant studies, are summarized in Table 9–6. The k_1 values reported in Table 9–6 can be used as a rough approximation of treatability differences for different wastewaters. Wastewaters that would have the lowest treatment efficiencies are from refineries, Kraft mills, and textile mills.

Recirculation. The minimum hydraulic application rate recommended by Dow Chemical (WEF, 2000) is 0.5 L/m^2·s (0.75 gal/ft^2·min) to provide maximum efficiency. Shallow tower designs require recirculation to provide minimum wetting rates. When above the minimum hydraulic application rate, recirculation was reported to have little

Table 9–6
Normalized Germain equation k_1 values from pilot-plant studies for different wastewaters

Type of wastewater	k_1 value, (L/s)$^{0.5}$/m^2
Domestic	0.210
Fruit canning	0.181
Kraft mill	0.108
Meat packing	0.216
Pharmaceutical	0.221
Potato processing	0.351
Refinery	0.059
Sugar processing	0.165
Synthetic dairy	0.170
Textile mill	0.107

Note: (L/s)$^{0.5}$/m^2 × 0.3704 = (gal/min)$^{0.5}$/ft^2.

benefit (Germain, 1966). For filters with low hydraulic application rates and higher organic loadings, recirculation may improve efficiency. For design systems such as rock filters with low hydraulic application rates, recirculation provides a higher flow to improve wetting and flushing of the filter packing.

Solids Production.   Solids production from trickling filter processes will depend on the wastewater characteristics and the trickling filter loading. At lower organic loading rates, a greater amount of the particulate BOD is degraded, the biomass has a longer SRT, and, as a result, less biomass is produced. A procedure that can be used to evaluate the solids production for trickling filters is presented later in this chapter in Sec. 9–4, which deals with combined trickling filter activated-sludge processes.

Mass Transfer Limitations.   One of the concerns in the process design for trickling filters is at what organic loading the filter performance becomes limited by oxygen transfer. When this condition occurs, treatment efficiency at the higher organic load is limited and odors may be produced due to anaerobic activity in the biofilm. Based on an evaluation of the data in the literature, for influent BOD concentrations in the range of 400 to 500 mg/L, oxygen transfer may become limiting (Schroeder and Tchobanoglous, 1976). Hinton and Stensel (1994) reported that oxygen availability controlled organic substrate removal rates at soluble biodegradable COD loadings above 3.3 $kg/m^3 \cdot d$.

EXAMPLE 9–4   **Design of Trickling Filter with Plastic Packing**   Given the following design flowrates and primary effluent wastewater characteristics, determine the following design parameters for a trickling filter design assuming 2 towers at 6.1 m depth, cross-flow plastic packing with a specific surface area of 90 m^2/m^3, a packing coefficient n value of 0.5, and a 2-arm distributor system. The required minimum wetting rate = 0.5 $L/m^2 \cdot s$. Assume a secondary clarifier depth of 4.2 m.

Design Conditions

Item	Unit	Primary effluent	Target effluent
Flow	m^3/d	15,140	
BOD	g/m^3	125	20
TSS	g/m^3	65	20
Minimum temp.	°C	14	

Note: g/m^3 = mg/L

Using the above information, determine:

1. Diameter of tower trickling filter, m
2. Volume of packing required, m^3
3. Recirculation rate required, if any

4. Total pumping rate, m^3/h
5. Flushing and normal dose rate, mm/pass
6. Flushing and normal distributor speeds, min/rev
7. Clarifier diameter, m (assume the ratio of the peak to average flowrate is 1.5)

Solution

1. Determine k_{20} for the design conditions using Eq. (9–22).

$$k_2 = k_1 \left(\frac{D_1}{D_2}\right)^{0.5} \left(\frac{S_1}{S_2}\right)^{0.5}$$

a. Solve for k_2.

From Table 9–6, $k_1 = 0.210$ $(L/s)^{0.5}/m^2$

Trickling filter depth = 6.1 m

$$= 0.210 \left(\frac{6.1\ m}{6.1\ m}\right)^{0.5} \left[\frac{(150\ g/m^3)}{(125\ g/m^2)}\right]^{0.5} = 0.230 (L/s)^{0.5}/m^2$$

b. Correct k_2 for temperature effect using Eq. (9–20).

  i. $k_T = k_{20} (1.035)^{T-20}$

  ii. $k_{14} = 0.230(1.035)^{14-20} = 0.187$

2. Determine the hydraulic application rate and the filter area, volume, and diameter.
   a. Using Eq. (9–19), determine the hydraulic application rate.

$$\frac{S_e}{S_o} = e^{-kD/q^n}$$

$q = [kD/\ln (S_o/S_e)]^{1/n}$

$q = [0.187(6.1)/\ln (125/20)]^{1/0.5}$

$q = 0.3875$ $L/m^2 \cdot s$

   b. Determine the tower area.

$Q = 15,140$ $m^3/d = 175.2$ L/s

Filter area = Q/q = (175.2 L/s)/(0.3875 $L/m^2 \cdot s$) = 452.2 m^2

   c. Determine the packing volume.

Packing volume = (452.2 m^2) (6.1 m) = 2758 m^3

   d. Determine the tower diameter.

Area/tower = 452.2 m^2/2 = 226.1 m^2

Diameter = 17 m each for two filters

3. Determine the recirculation rate and the recirculation ratio.
   a. Determine the recirculation rate.

   The minimum wetting rate = 0.5 L/m²·s

   $q + q_r = 0.5$ L/m²·s

   $q_r = 0.5 - 0.39 = 0.11$ L/m²·s

   b. Determine the recirculation ratio.

   $R = q_r/q = 0.11/0.39 = 0.28$

4. Determine the pumping rate.

   $q + q_r = 0.5$ L/m²·s

   Total pumping rate = (0.5 L/m²·s)(452.2 m²)

   = 226 L/s = 814 m³/h

5. Determine flushing and normal dose rate using the data given in Table 9–3.
   a. Determine BOD loading.

   BOD loading = $Q\, S_o/V$

   $$= \frac{(15{,}140 \text{ m}^3/\text{d})(125 \text{ mg/L})(1 \text{ kg}/10^3 \text{ g})}{2758 \text{ m}^3}$$

   = 0.69 kg/m³·d

   b. Determine the dosing rates.

   From Table 9–3, the estimated flushing and operation dose rates are:

   i. Flushing dose = 300 mm/pass

   ii. Operating dose = 50 mm/pass

6. Determine the distributor speed using Eq. (9–1).
   a. For flushing:

   $$n = \frac{(1 + R)\, q(10^3 \text{ mm/min})}{(A)(DR)(60 \text{ min/h})}$$

   $A = 2$

   $$q = \frac{(15{,}140 \text{ m}^3/\text{d})}{(452.2 \text{ m}^2)(24 \text{ hr/d})} = 1.4 \text{ m}^3/\text{m}^2\cdot\text{h}$$

   $R = 0.28$

   $$n = \frac{(1 + 0.28)(1.4)(1000)}{(2)(300)(60)} = 0.0498 \text{ rev/min (i.e., 20 min/rev)}$$

b. For normal operation:

$$n = \frac{(1 + 0.28)(1.4)(1000)}{(2)(50)(60)} = 0.30 \text{ rev/min (i.e., 3.33 min/rev)}$$

Note: Because of the different speed requirements for normal and flushing operation, a distributor drive with variable speed capability should be used.

7. Determine clarifier diameter.

Clarifier depth = 4.2 m

From Fig. 9–7, the recommended overflow rate for peak and average flowrates is 1.1 and 2.4 m/h, respectively. Because the ratio of the peak to average flowrate is 1.5, the average overflow rate controls the design.

Flowrate = (15,140 m³/d)/(24 h/d) = 630.8 m³/h

Clarifier area = (630.8 m³/h)/(1.1 m/h) = 573.5 m²

Use 2 clarifiers.

Area for each = 573.5 m²/2 = 286.7 m²

Diameter of each = 14.1 m

8. Design summary:

Parameter	Unit	Value
Number of filters	number	2
Diameter	m	17
Depth	m	6.1
Total packing volume	m³	2758
BOD loading	kg/m³·d	0.69
Hydraulic application rate	L/m²·s	0.39
Total pumping rate	m³/h	814
Recirculation ratio	unitless	0.28
Distributor arms	number	2
Normal distributor speed	min/rev	3.33
Flushing distributor speed	min/rev	20
Clarifiers	number	2
Clarifier depth	m	4.2
Clarifier diameter	m	14.1

Nitrification Design

Two types of process design approaches have been used to accomplish biological nitrification in trickling filters, either in a combined system along with BOD removal or in a tertiary application following secondary treatment and clarification for BOD removal.

The secondary treatment process may be a suspended growth or fixed-film process. Empirical design approaches based on pilot-plant and full-scale plant results are again used to guide nitrification designs in view of the difficulty in predicting the actual biofilm coverage area, wetting efficiency, and biofilm thickness and density.

Major impacts on nitrification performance are the influent BOD concentration and dissolved oxygen concentration within the trickling filter bulk liquid. As the BOD to TKN ratio of the influent wastewater increases, a greater proportion of the trickling filter packing area is covered by heterotrophic bacteria and the apparent nitrification rate ($kg/m^3 \cdot d$) based on the total trickling filter volume is decreased. A number of investigations have shown that BOD, if at high enough concentration, inhibits nitrification. Studies by Harremöes (1982) showed that nitrification (1) could occur at a maximum rate at soluble BOD (sBOD) concentrations below 5 mg/L, (2) was inhibited in proportion to the sBOD concentration above 5 mg/L, and (3) was insignificant, in proportion to the sBOD concentration of 30 mg/L or more. In a study with a flat plat experimental design, Huang and Hopson (1974) demonstrated a steady inhibition of nitrification rates occurred as the sBOD concentration was increased from 1.0 to 8.0 mg/L. Figueroa and Silverstein (1991) found that nitrification rates in fixed-film processes are inhibited at BOD concentrations above 10 mg/L, which finding is in agreement with observations by others (Parker and Richards, 1986).

Design Basis for Combined BOD Removal and Nitrification.   Nitrification efficiency has been correlated with the volumetric BOD loading for rock trickling filters (U.S. EPA, 1975). For 90 percent nitrification efficiency, a BOD loading of less than 0.08 kg $BOD/m^3 \cdot d$ (5 lb $BOD/10^3$ $ft^3 \cdot d$) is recommended. At a loading of about 0.22 kg $BOD/m^3 \cdot d$ (14 lb $BOD/10^3$ $ft^3 \cdot d$), about 50 percent nitrification efficiency could be expected. It was noted that increased recirculation rates improved nitrification performance.

Instead of volumetric BOD loading values, the nitrification efficiency has been related to the BOD loading based on the packing surface area. In comparing nitrification performance for both rock and plastic packing, Parker and Richards (1986) found that the nitrification efficiency was similar at similar BOD surface loading rates (g $BOD/m^2 \cdot d$) for both packings. A surface loading rate as low as 2.4 g $BOD/m^2 \cdot d$ is necessary for ≥ 90 percent NH_4-N removal.

Daigger et al. (1994) found that the oxidation of BOD and NH_4-N in trickling filters with plastic packing could be characterized by a volumetric oxidation rate defined as follows:

$$\text{VOR} = \frac{[S_o + 4.6(NO_x)]Q}{(V)(10^3 \text{ g/kg})} \tag{9–23}$$

where VOR = volumetric oxidation rate, $kg/m^3 \cdot d$
 S_o = influent BOD concentration, g/m^3
 NO_x = amount of ammonia-nitrogen oxidized, g/m^3
 Q = influent flowrate, m^3/d
 V = packing volume, m^3

Using Eq. (9–23), the volumetric oxidation rate for three trickling filter plants with nitrification was determined and the 90th percentile value varied from 0.75 to 1.0 $kg/m^3 \cdot d$.

Figure 9–10

Effect of influent wastewater BOD/TKN ratio on nitrification rate in trickling filters with plastic packing used for both BOD removal and nitrification. *[Adapted from Okey and Albertson, WEF (2000).]*

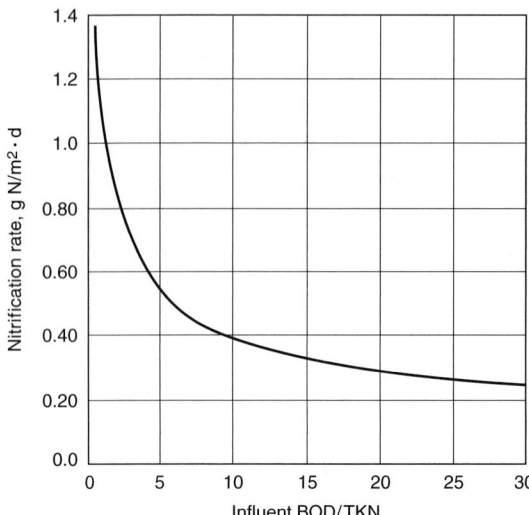

The amount of nitrification can then be estimated by using the VOR value and the influent BOD concentration.

Using data from four different studies, Okey and Albertson (WEF, 2000) found a linear relationship between the specific nitrification rate (g/m²·d) and the influent BOD/TKN ratio for combined systems. The range of results and the relationship is illustrated by Fig. 9–10. The data included in this correlation were for operation at temperatures ranging from 9 to 20°C. The following equation, represented on Fig. 9–10, is 25 percent below the mean of the observations reported on Fig. 9–10 to allow a design that provides nitrification rates consistent with most of the data.

$$R_n = 0.82 \left(\frac{\text{BOD}}{\text{TKN}} \right)^{-0.44} \tag{9-24}$$

where R_n = nitrification rate, g N/m²·d
$\dfrac{\text{BOD}}{\text{TKN}}$ = influent BOD to TKN ratio, g/g

The authors concluded that the DO concentration had a greater effect on the nitrification rates than temperature. The effect of DO concentration is supported by fundamental mass transfer considerations in which it can be shown that a bulk liquid DO concentration of 2.8 mg/L is required for nitrification without oxygen diffusion limitations, at a liquid NH_4-N concentration of 1.0 mg/L.

EXAMPLE 9–5 **Combined BOD Removal and Nitrification in a Trickling Filter with Plastic Packing** Determine the volume of plastic packing required for 90 percent TKN removal in trickling filter with a depth of 6.1 m for the wastewater characteristics given. How does the computed volume compare to the volumetric oxidation rate prediction?

Assume the specific surface area of plastic packing material is 100 m²/m³. Also, determine the hydraulic application rate.

Wastewater characteristics

Parameter	Unit	Value
Flow	L/s	100
BOD	g/m³	160
TKN	g/m³	25
TSS	g/m³	70

Note: g/m³ = mg/L

Solution

1. Determine the specific TKN removal rate using Eq. (9–24).

$$R_n = 0.82\left(\frac{BOD}{TKN}\right)^{-0.44}$$

BOD/TKN = 160/25 = 6.4

$R_n = 0.82(6.4)^{-0.44}$

$= 0.36 \text{ g/m}^2 \cdot \text{d}$

2. Determine the TKN removal.

Q = 100 L/s = 8640 m³/d

TKN removal = 0.90(8640 m³/d)(25 g/m³)

$= 194,400 \text{ g/d}$

3. Determine the required surface area of packing.

$$A_s = \frac{(194,400 \text{ g/d})}{R_n} = \frac{(194,400 \text{ g/d})}{(0.36 \text{ g/m}^2 \cdot \text{d})} = 540,000 \text{ m}^2$$

4. Determine the volume of packing material.

$$\text{Vol} = \frac{540,000 \text{ m}^2}{(100 \text{ m}^2/\text{m}^3)} = 5400 \text{ m}^3$$

5. Determine the hydraulic application rate.

$$\text{Filter area} = \frac{\text{volume}}{\text{depth}} = \frac{5400 \text{ m}^3}{6.1 \text{ m}} = 885 \text{ m}^2$$

Hydraulic application rate q

$$q = \frac{Q}{A} = \frac{(100 \text{ L/s})}{885 \text{ m}^2} = 0.11 \text{ L/m}^2 \cdot \text{s}$$

To meet the minimum hydraulic application rate given previously as 0.5 L/m²·s, recirculation will be required.

6. Determine the BOD loading based on volume and surface area.

a. Loading based on volume

$$BOD \ loading = \frac{(8640 \ m^3/d)(160 \ g/m^3)(1 \ kg/10^3 \ g)}{5400 \ m^3} = 0.26 \ kg/m^3 \cdot d$$

b. Loading based on area

$$BOD \ loading = (0.26 \ kg/m^3 \cdot d)[1/(100 \ m^2/m^3)](10^3 \ g/kg)$$

$$= 2.6 \ g/m^2 \cdot d$$

7. Determine the volumetric oxidation rate using Eq. (9–23).

$$VOR = \frac{[S_o + 4.6(NO_x)]Q}{V(10^3 \ g/kg)}$$

$$S_o = 160 \ g/m^3$$

$$NO_x = 0.90(25) = 22.5 \ g/m^3$$

$$VOR = \frac{[(160 \ g/m^3) + 4.6(22.5 \ g/m^3)](8640 \ m^3/d)}{(5400 \ m^3)(10^3 \ g/kg)}$$

$$= 0.42 \ kg/m^3 \cdot d$$

Comments The computed value for the BOD loading, based on the packing material surface area, is within the range reported by Parker and Richards (1986). The computed value for the volumetric oxidation rate is a little lower than the volumetric oxidation rates reported by Daigger et al. (1994).

Tertiary Nitrification. A number of facilities exist where trickling filters with plastic packing are used after secondary treatment for nitrification. The influent BOD concentration is relatively low at <10 mg/L and in some cases less than 5 mg/L. Nitrification trickling filter performance will depend on the ammonia loading rate, oxygen availability, temperature, and packing design. Effluent NH_4-N concentrations will vary with summer and winter operation and can range from <1.0 mg/L at warm temperatures and from <1 to 4 mg/L at cold temperatures. Hydraulic application rates may range from 0.40 to 1.0 L/m²·s. The process performance has been described as relative to a packing surface nitrification rate. Surface nitrification rates, as given on Fig. 9–9, have ranged from 1.0 to 3.0 g/m²·d with about a 20 to 50 percent decline in the observed rate as the temperature decreases from 20 to 10°C (Parker et al., 1990). The higher nitrification rates, reported on Fig. 9–9 for Central Valley, represent the results obtained at higher hydraulic loading rates, with effluent NH_4-N concentrations above 5 mg/L. Other investigators have observed minimal temperature effects for tertiary nitrification, and have attributed the minimal observed rate change more to the effect of dissolved oxygen concentration and hydraulics (Okey and Albertson, 1989).

It is generally well accepted that in the upper portion of the trickling filter the nitrification rate is limited by oxygen availability and diffusion into the biofilm. To overcome the oxygen limitation, forced-draft air is generally used to assure maximum oxygen availability. Higher hydraulic rates that provide better wetting efficiency and agitation of the biofilm surface generally produce better performance. Because plugging is less of an issue with the exception of snails (see subsequent section), a medium-density packing material is preferred (i.e., specific surface area of about 100 m²/m³) to provide more area as a function of the percent of the reactor volume.

In a large fraction of the nitrification tower, the NH_4-N concentration is high enough so that the nitrification rate is oxygen-limited and thus zero-order with respect to nitrogen. Farther down in the packing as the NH_4-N concentration decreases, the nitrification rate is limited by the NH_4-N concentration and thus decreases. The decline in nitrification rate is further affected by a lower growth of nitrifying bacteria due to the low amount of NH_4-N available. The use of nitrification trickling filters in series with operational modifications has been shown to compensate for this limitation (Boller and Gujer, 1986). The order of operation of the towers is reversed every few days so that a higher nitrifying bacteria population can be developed and be available where the NH_4-N concentration is low. Anderson et al. (1994) showed a 20 percent improvement in nitrification efficiency with this method.

The generally accepted design approach to determine the volume of packing required and hydraulic application rates for nitrification trickling filters is based on an empirical equation that can be used to describe the NH_4-N surface removal flux (WEF, 2000).

$$r_n(Z, T) = r_{n,\max}\left(\frac{N}{K_n + N}\right)e^{-rZ} \qquad (9\text{–}25)$$

where r_n (Z, T) = NH_4-N surface removal rate or nitrification rate at depth z and temperature T, g/m²·d
$r_{n,\max}$ = maximum NH_4-N surface removal rate at temperature $T(°C)$, g/m²·d
N = bulk liquid NH_4-N concentration, mg/L
K_n = half-velocity constant, mg/L
r = empirical constant describing the decrease in rates as a function of depth
Z = depth in tower, m

The decline in nitrification rate with depth, accounted for by the term rZ, could be due to (1) a decrease in nitrifying biofilm growth where the NH_4-N concentration is lower, (2) predator organism activity, and (3) a change in wetting efficiency. The value for $r_{n,\max}$ is assumed to be controlled by the oxygen transfer characteristics of the tower and biofilm. The value of $r_{n,\max}$ has been determined by observations on zero-order nitrification rates in tertiary trickling filter operations where N is significantly greater than K_n. An accepted value for K_n is 1.5 mg/L (Grady et al., 1999). No temperature correction is recommended between 10 and 25°C. Below 10°C the following temperature correction coefficient is recommended (WEF, 2000):

$$r_{n,\max(T)} = r_{n,\max(10)}(1.045)^{T-10} \qquad (9\text{–}26)$$

Specific values for $r_{n,\max}$, K_n, and r will be site-specific and can be obtained by pilot-plant studies or possibly an analysis of full-scale plants with similar wastewater characteristics.

Table 9–7
Reported maximum
surface specific
nitrification rates for
tertiary nitrification
trickling filters

Location	Packing[a]	Range of $r_{n,max}$ values, g N/m²·d	Reference
Central Valley, UT	XF 138	2.1–2.9	Parker, 1990
Malmö, Sweden	XF 138	1.6–2.8	Parker, 1995
Littleton/Englewood, CO	XF 138	1.2–2.3	Parker, 1997
Midland, MI	VF 89	1.1–1.8	WEF, 2000
Lima, OH	VF 89	1.2–1.8	WEF, 2000
Zurich, Switzerland	VF 92	1.6	WEF, 2000
Zurich, Switzerland	XF 223	1.2	WEF, 2000

[a] XF = cross flow, VF = vertical, 89, 92, 138, 223 m²/m³.
Note: g/m²·d × 0.204 = lb/10³ ft²·d.

Without specific data, conservative design values must be used. Zero-order specific nitrification rates [$r_{n,max(T)}$] vary widely as reported in Table 9–7. Cross-flow plastic packing appears to produce higher nitrification rates. The higher nitrification rates may be due to better flow distribution and greater wetting of the packing to develop a higher biofilm surface area. However, the higher-density cross-flow packing material did not appear to do as well.

The design procedure to determine volume of packing required for nitrifying towers requires a stepwise design procedure using zero-order and first-order nitrification rates based on Eq. (9–25). In the proposed procedure, the depth effect is ignored because of a lack of data, and, instead, the amount of ammonia oxidized is calculated at incremental depths. First, the zero-order rate [$r_{n,max(T)}$] is used to calculate the volume of packing needed to reduce the NH_4-N concentration to a value that is high enough to approximate the zero-order rate. An NH_4-N concentration of 5 to 7 mg/L is used typically. Below a concentration value of 5 to 7 mg/L, the volume of packing required can be calculated in increments of decreasing NH_4-N concentration, to determine the surface removal rate for each increment and the corresponding packing volume requirements.

EXAMPLE 9–6 Tertiary Nitrification Trickling Filter Design   A two-stage nitrification filter is to treat a flow of 6000 m³/d after secondary treatment. The influent NH_4-N concentration is 25 g/m³ and an effluent NH_4-N concentration of 1.0 g/m³ is desired. From pilot-plant studies the $r_{n,max}$ value is 1.5 g/m²·d and the K_n value is 1.5 g/m³, at a temperature of 15°C. A medium-density cross-flow plastic packing material with a specific surface area of 138 m²/m³ will be used, as little BOD or TSS exists in the influent wastewater to pose plugging problems. Use a hydraulic application rate of 1.0 L/m²·s to provide good packing wetting. Assume $r = 0$ and that the effluent NH_4-N concentration from the zero-order nitrification rate portion of the tower is 6 g/m³. (Note g/m³ = mg/L.) For the purpose of this example, synthesis has been neglected, which represents a conservative approach.

Solution

1. Determine the volume of packing material at a zero-order nitrification rate to reduce the NH_4-N concentration from 25 to 6.0 g/m^3. Because no recirculation is used, the NH_4-N concentration at the top of the tower is 25 g/m^3.

$$r_{n,\,max} = 1.5 \text{ g/m}^2\text{·d}$$

NH_4-N removed at zero order $= [(25 - 6) \text{ g/m}^3](6000 \text{ m}^3/\text{d}) = 114{,}000 \text{ g/d}$

$$\text{Packing surface area} = \frac{(114{,}000 \text{ g/d})}{(1.5 \text{ g/m}^2\text{·d})} = 76{,}000 \text{ m}^2$$

$$\text{Volume of packing material} = \frac{76{,}000 \text{ m}^2}{(138 \text{ m}^2/\text{m}^3)} = 551 \text{ m}^3$$

2. Solve for first-order rates in concentration increments of 1 g/m^3.

The required computations for the concentration increment from $6 \rightarrow 5$ g/m^3 are illustrated below.

Average NH_4-N $= 5.5$ g/m^3

Determine the rate of surface removal using Eq. (9–25).

$$r_n = r_{n,\text{max}}\left(\frac{N}{K_n + N}\right)$$

$$r_n = (1.5 \text{ g/m}^2\text{·d})\left(\frac{5.5}{1.5 + 5.5}\right) = 1.18 \text{ g/m}^2\text{·d}$$

NH_4-N removed $= (1.0 \text{ g/m}^3)(6000 \text{ m}^3/\text{d}) = 6000 \text{ g/d}$

$$\text{Required packing surface area} = \frac{(6000 \text{ g/d})}{(1.18 \text{ g/m}^2\text{·d})} = 5085 \text{ m}^2$$

Volume $= 5085 \text{ m}^2/(138 \text{ m}^2/\text{m}^3) = 36.8 \text{ m}^3$

3. Calculations for other increments are summarized in the following table.

NH_4-N increment, g/m^3	Avg. NH_4-N, g/m^3	r_n, $g/m^2\text{·d}$	NH_4-N removed, g/d	Volume, m^3
25–6	15.5	1.50	114,000	551
6–5	5.5	1.18	6000	37
5–4	4.5	1.13	6000	38
4–3	3.5	1.05	6000	41
3–2	2.5	0.94	6000	46
2–1	1.5	0.75	6000	58
			Total	771

4. Determine cross-section area of trickling filter.

$$\text{Area} = \frac{\text{flow}}{\text{hydraulic application rate}}$$

$$\text{Area} = \frac{(6000 \text{ m}^3/\text{d})(1 \text{ d}/86{,}400 \text{ s})(10^3 \text{ L/m}^3)}{(1.0 \text{ L/m}^2 \cdot \text{s})} = 69 \text{ m}^2$$

$$\text{Diameter} = \sqrt{\frac{4(69 \text{ m}^2)}{\pi}} = 9.4 \text{ m}$$

5. Determine the depth of the trickling filter packing.

$$\text{Volume} = (A)(\text{depth})$$

$$\text{Depth} = \frac{771 \text{ m}^3}{69 \text{ m}^2} = 11 \text{ m}$$

Depth per trickling filter in series

$$\text{Depth/tower} = \frac{11 \text{ m}}{2} = 5.5 \text{ m}$$

Predator Problems. A significant problem for nitrifying filters is the development of a snail population, which may graze on the biofilm to reduce the nitrifying bacteria population and nitrification performance. In addition, snails can cause problems with plugging of channels and pumps, accumulating in digesters, and causing wear and tear on equipment. A sump can be provided in an effluent collection chamber upstream of the secondary clarifiers to facilitate removal of snails from the effluent. Methods proposed to control snails are periodic flooding of the trickling filter, reducing the distributor speed to create higher flushing rates, high pH dosing, chlorination, saline water dosing, and dosing with copper sulfate at 0.4 g/L (WEF, 2000). Periodic flooding does not appear to control snails but does eliminate filter flies. Some success has been claimed for reduced distributor speed and high pH treatment. An alkaline backwash to control the snail population and increase nitrification rates was demonstrated successfully at the Littleton/Englewood, CO, wastewater treatment plant (Parker et al., 1997). The nitrification tower trickling filters were flooded on three separate occasions, from October to November 1993, after nitrification efficiency had declined due to predator growth. For the first backwash, the pH was 10, but a pH of 9.0 was used for the last two backwashes to minimize loss of nitrification activity. After the high-pH backwash, nitrification efficiency improved and was maintained long-term.

9-3 ROTATING BIOLOGICAL CONTACTORS

Rotating biological contactors (RBCs) were first installed in West Germany in 1960 and later introduced in the United States. Hundreds of RBC installations were installed in the 1970s and the process has been reviewed in a number of reports (U.S. EPA, 1984,

1985, and 1993; WEF, 1998 and 2000). An RBC consists of a series of closely spaced circular disks of polystyrene or polyvinyl chloride that are submerged in wastewater and rotated through it (see Fig. 9–11). The cylindrical plastic disks are attached to a horizontal shaft and are provided at standard unit sizes of approximately 3.5 m (12 ft) in diameter and 7.5 m (25 ft) in length. The surface area of the disks for a standard unit is about 9300 m² (100,000 ft²), and a unit with a higher density of disks is also available with approximately 13,900 m² (150,000 ft²) of surface area. The RBC unit is partially

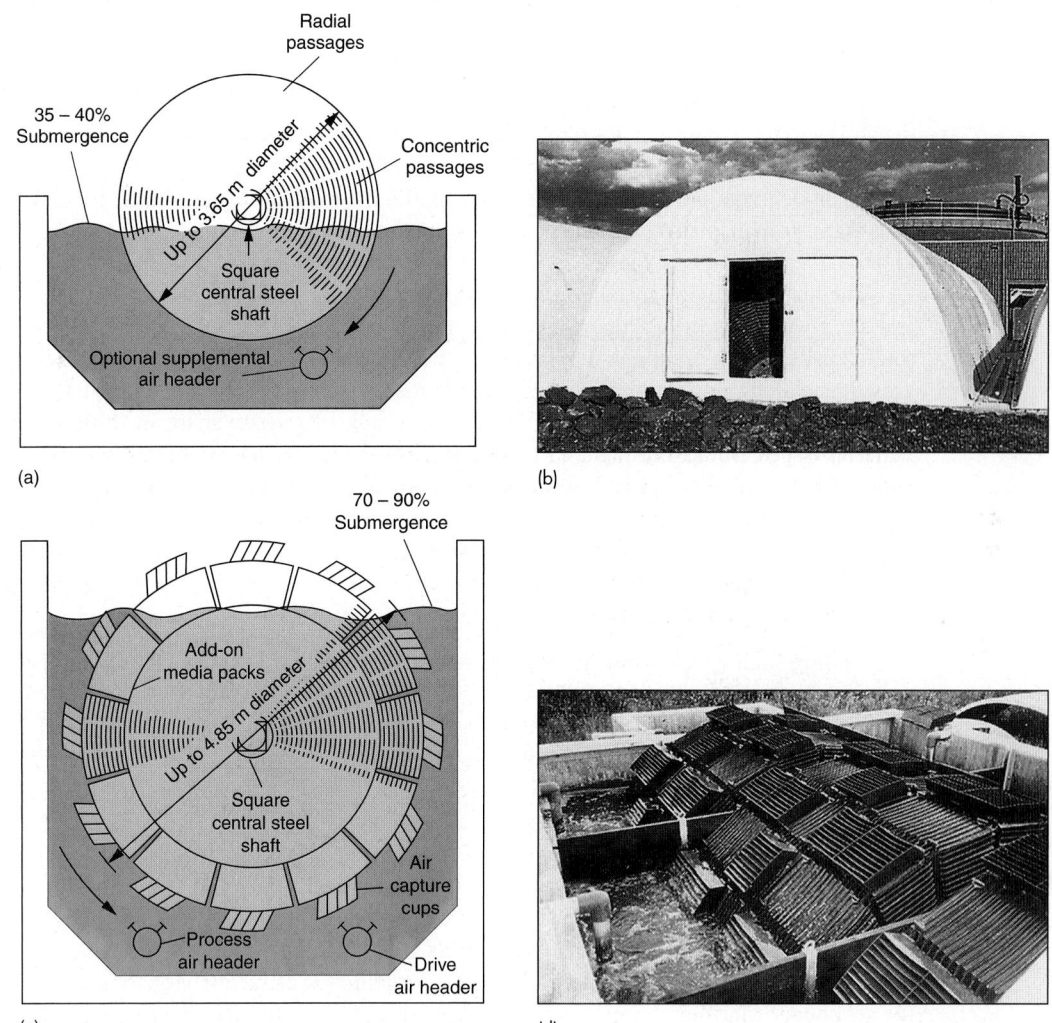

Figure 9–11

Typical RBC units: (a) conventional RBC with mechanical drive and optional air input, (b) conventional RBC in enclosed reactor, (c) submerged-type RBC equipped with air capture cups (air is used both to rotate and to aerate the biodisks), and (d) typical submerged RBC equipped with air capture cups. *(From Envirex Inc.)*

submerged (typically 40 percent) in a tank containing the wastewater, and the disks rotate slowly at about 1.0 to 1.6 revolutions per minute (see Fig. 9–11a). Mechanical drives are normally used to rotate the units, but air-driven units have also been installed. In the air-driven units, an array of cups (see Fig. 9–11c) is fixed to the periphery of the disks and diffused aeration is used to direct air to the cups to cause rotation. As the RBC disks rotate out of the wastewater, aeration is accomplished by exposure to the atmosphere. Wastewater flows down through the disks, and solids sloughing occurs. Similar to a trickling filter, RBC systems require pretreatment of primary clarification or fine screens and secondary clarification for liquid/solids separation.

A submerged RBC design was also introduced in the early 1980s but has seen limited applications. The submergence is 70 to 90 percent and air-drive units are used to provide oxygen and rotation. The advantages claimed for the submerged unit are reduced loadings on the shaft and bearings, improved biomass control by air agitation, the ability to use larger bundles of disks, and ease of retrofit into existing aeration tanks. However, because of the comparatively low levels of dissolved oxygen in the liquid, biological degradation activity by the submerged units may be oxygen-limited. To prevent algae growth, protect the plastic disks from the effects of ultraviolet exposure, and to prevent excessive heat loss in cold weather, RBC units are covered (see Fig. 9–11b).

The history of RBC installations has been troublesome due to inadequate mechanical design and lack of full understanding of the biological process. Structural failure of shafts, disks, and disk support systems has occurred. Development of excessive biofilm growth and sloughing problems has also led to mechanical shaft, bearing, and disk failures. Many of these problems were related to a lack of conservatism in design and scale-up issues from pilot-plant to full-scale units. Many of the problems associated with earlier installations have been solved and numerous RBC installations are operating successfully.

Process Design Considerations

There are many similarities between RBC design considerations and those described for trickling filters. Both systems develop a large biofilm surface area and rely on mass transfer of oxygen and substrates from the bulk liquid to the biofilm. The complexity in the physical and hydrodynamic characteristics requires that the design of the RBC process be based on fundamental information from pilot-plant and field installations. As for trickling filters, the organic loading affects BOD removal efficiency and the nitrogen loading after a minimal BOD concentration is reached affects the nitrification efficiency. In contrast to the trickling filter where the wastewater flow approaches a plug flow hydraulic regime, the RBC units are rotated in a basin containing the wastewater, so that separate baffled basins are needed to develop the benefits of a staged biological reactor design. The design of an RBC system must include the following considerations: (1) staging of the RBC units, (2) loading criteria, (3) effluent characteristics, and (4) secondary clarifier design. Typical design information for RBCs is presented in Table 9–8.

Staging of RBC Units. Staging is the compartmentalization of the RBC disks to form a series of independent cells. Based on mass transfer and biological kinetic fundamentals, higher specific substrate removal rates will occur in RBC biofilms at higher

Table 9–8
Typical design information for rotating biological contactors

Parameter	Unit	Treatment level[a]		
		BOD removal	**BOD removal and nitrification**	**Separate nitrification**
Hydraulic loading	m³/m²·d	0.08–0.16	0.03–0.08	0.04–0.10
Organic loading	g sBOD/m²·d	4–10	2.5–8	0.5–1.0
	g BOD/m²·d	8–20	5–16	1–2
Maximum 1st-stage organic loading	g sBOD/m²·d	12–15	12–15	
	g BOD/m²·d	24–30	24–30	
NH₃ loading	g N/m²·d		0.75–1.5	
Hydraulic retention time	h	0.7–1.5	1.5–4	1.2–3
Effluent BOD	mg/L	15–30	7–15	7–15
Effluent NH₄-N	mg/L		<2	1–2

[a] Wastewater temperature above 13°C (55°F).
Note: g/m²·d × 0.204 = lb/10³ ft²·d.
m³/m²·d × 24.5424 = gal/ft²·d.

bulk liquid substrate concentrations. Because a low effluent substrate concentration and high specific substrate removal rates are generally the ultimate treatment goal, reduced disk area requirements can be realized only by using staged-RBC units.

The RBC process application typically consists of a number of units operated in series. The number of stages depends on the treatment goals, with two to four stages for BOD removal and six or more stages for nitrification. Stages can be accomplished by using baffles in a single tank or by use of separate tanks in series. Staging promotes a variety of conditions where different organisms can flourish in varying degrees from stage to stage. The degree of development in any stage depends primarily on the soluble organic concentration in the stage bulk liquid. As the wastewater flows through the system, each subsequent stage receives an influent with a lower organic concentration than the previous stage. Typical RBC staging arrangements are illustrated on Fig. 9–12.

For small plants, RBC drive shafts are oriented parallel to the direction of flow with disk clusters separated by baffles (see Fig. 9–12a). In larger installations, shafts are mounted perpendicular to flow with several stages in series to form a process train (see Fig. 9–12b). To handle the loading on the initial units, step feed (see Fig. 9–12d) or a tapered system (see Fig. 9–12e) may be used. Two or more parallel flow trains should be installed so the units can be isolated for turndown or repairs. Tank construction may be reinforced concrete or steel, with steel preferred at smaller plants. Treatment systems employing RBCs have been used for BOD removal, pretreatment of industrial wastewater, combined BOD removal and nitrification, tertiary nitrification, and denitrification. The principal advantages of the RBC process are simplicity of operation and relatively low energy costs.

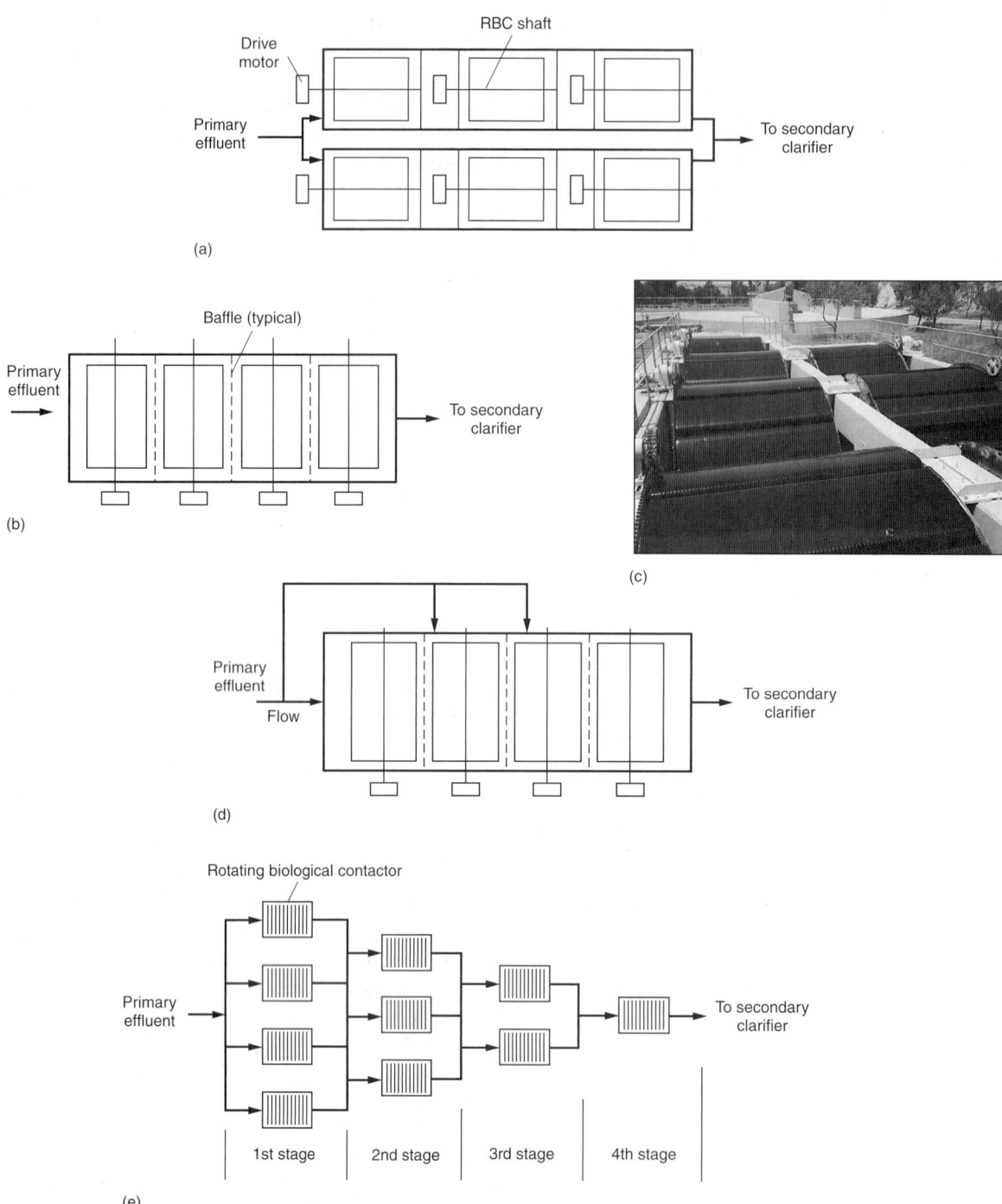

Figure 9–12

Typical RBC staging arrangements: (a) flow parallel to shaft, (b) flow perpendicular to shaft, (c) view of RBCs with flow perpendicular to shaft, (d) step feed flow, and (e) tapered feed flow parallel to shaft.

The History of RBC Loading Criteria. Based on experience, the performance of an RBC system is related to the specific surface loading rate of total and soluble BOD for BOD removal and NH_4-N for nitrification. For successful treatment, the loading rates must be within the oxygen transfer capability of the system. Poor performance, odors, and biofilm sloughing problems have occurred when the oxygen demand due to the BOD loading has exceeded the oxygen transfer capability. A characteristic of this problem is the development of *Beggiatoa,* a reduced-sulfur oxidizing bacteria, on the outer portion of the biofilm, which prevents sloughing (U.S. EPA, 1984). A thick biofilm can develop to create enough weight to stress the structural strength of the plastic disks and shaft.

Under overloaded conditions, anaerobic conditions develop deep in the attached film. Sulfate is reduced to H_2S, which diffuses to the outer layer of the biofilm, where oxygen is available. *Beggiatoa,* a filamentous bacteria, which is able to oxidize the H_2S and other reduced sulfur compounds, forms a tenacious whitish biofilm that does not slough under the normal RBC rotational sheer conditions. In designing RBC units, it is important to select a low enough BOD loading for the initial units in the staged design to prevent overloading. Odor problems are most frequently caused by excessive organic loadings, particularly in the first stage.

Because the soluble BOD is used more rapidly in the first stage of an RBC system, most manufacturers of RBC equipment specify a specific soluble BOD loading in the range of 12 to 20 g $sBOD/m^2{\cdot}d$ (2.5 to 4.1 lb $sBOD/10^3\ ft^2{\cdot}d$) for the first stage. Assuming a 50 percent soluble BOD fraction, the total BOD loading ranges from 24 to 30 g $BOD/m^2{\cdot}d$. For some designs that involve higher-strength wastewaters, the loading criteria are met by splitting the flow to multiple RBC units in the first stage or using a step feeding approach as shown on Fig. 9–12d.

For nitrification, the design approach for RBC systems can be very similar to that shown for tertiary nitrification trickling filters after the sBOD concentration is depleted in RBC units preceding nitrification. An sBOD concentration of less than 15 mg/L must be met before a significant nitrifying population can be developed on the RBC disks (Pano and Middlebrooks, 1983). The maximum nitrogen surface removal rate has been observed to be about 1.5 g $N/m^2{\cdot}d$ (U.S. EPA, 1985), which is quite similar to values observed for trickling filters.

Effluent Characteristics. Treatment systems with RBCs can be designed to provide secondary or advanced levels of treatment. Effluent BOD characteristics for secondary treatment are comparable to well-operated activated-sludge processes. Where a nitrified effluent is required, RBCs can be used to provide combined treatment for BOD and ammonia nitrogen, or to provide separate nitrification of secondary effluent. Typical ranges of effluent characteristics are indicated in Table 9–8. An RBC process modification in which the disk support shaft is totally submerged has been used for denitrification of wastewater (see Sec. 9–7).

Physical Facilities for RBC Process

The principal elements of an RBC unit and their importance in the process are described in this section. The suppliers of RBC equipment differ in their disk designs, shafts, and packing support, and configuration designs. The principal elements of an RBC system design are the shaft, disk materials and configuration, drive system, enclosures, and settling tanks.

Shafts. The RBC shafts are used to support and rotate the plastic disks. Maximum shaft length is presently limited to 8.23 m (27 ft) with 7.62 m (25 ft) occupied by disks. Shorter shaft lengths ranging from 1.52 to 7.62 m (5 to 25 ft) are also available. Shaft shapes include square, round, and octagonal, depending on the manufacturer. Steel shafts are coated to protect against corrosion and thickness ranges from 13 to 30 mm (0.5 to 1.25 in) (WEF, 1998). Structural details and the life expectancy of the disk shaft are important design considerations.

Disk Materials. High-density polyethylene is the material used most commonly for the manufacture of RBC disks, which are available in different configurations or corrugation patterns. Corrugations increase the available surface area and enhance structural stability. The types of RBC disks, classified based on the total area of disks on the shaft, are commonly termed low- (or standard) density, medium-density, and high-density. Standard-density disks, defined as disks with a surface area of 9300 m^2 (100,000 ft^2) per 8.23 m (27 ft) shaft, have larger spaces between disks and are normally used in the lead stages of an RBC process flow diagram. Medium- and high-density disk assemblies have surface areas of 11,000 to 16,700 m^2 (120,000 to 180,000 ft^2) per 8.23-m (27-ft) shaft, and are used typically in the middle and final stages of an RBC system where thinner biological growths occur.

Drive Systems. Most RBC units are rotated by direct mechanical drive units attached directly to the central shaft. Motors are typically rated at 3.7 or 5.6 kW (5 or 7.5 hp) per shaft. Air-drive units are also available. The air-drive assembly consists of deep plastic cups attached to the perimeter of the disks, an air header located beneath the disks, and an air compressor. Airflows necessary to achieve design rotational speeds are about 5.3 m^3/min (190 scfm) for a standard-density shaft and 7.6 m^3/min (270 scfm) for a high-density shaft. The release of air into the cups creates a buoyant force that causes the shaft to turn. Both systems have proved to be mechanically reliable. Variable-speed features can be provided to regulate the speed of rotation of the shaft.

Tankage. Tankage for RBC systems has been optimized at 0.0049 m^3/m^2 (0.12 gal/ft^2) of disk area, resulting in a stage volume of 45 m^3 (12,000 gal) for a shaft with a disk area of 9300 m^2. Based on this volume, a detention time of 1.44 h is provided for a hydraulic loading of 0.08 m^3/m^2·d (2 gal/ft^2·d). A typical sidewater depth is 1.5 m (5 ft) to accommodate a 40 percent submergence of the disks.

Enclosures. Segmented fiberglass reinforced plastic covers are usually provided over each shaft. In some cases, units have been housed in a building for protection against cold weather, to improve access, or for aesthetic reasons. RBCs are enclosed to (1) protect the plastic disks from deterioration due to ultraviolet light, (2) protect the process from low temperatures, (3) protect the disks and equipment from damage, and (4) control the buildup of algae in the process.

Settling Tanks. Settling tanks for RBCs are similar to trickling filter settling tanks in that all of the sludge from the settling tanks is removed to the sludge processing facilities. Typical design overflow rates for settling tanks used with RBCs are similar to that described for trickling filters with plastic packing in Sec. 9–2.

RBC Process Design

Empirical design approaches have been developed for RBC systems based on pilot-plant and full-scale plant data and that consider such fundamental factors as the disk surface area and specific loadings in terms of g/m^2 disk area·d. Approaches for designing staged RBC systems for BOD removal and nitrification are presented in this section.

BOD Removal. Design models for BOD removal in RBC systems are reviewed in WEF (2000). In a design comparison, the models generally resulted in lower recommended BOD loadings than that determined from manufacturer's literature and were, in some cases, similar for BOD removals below 90 percent. Of these, a second-order model by Opatken (U.S. EPA 1985) is selected to estimate RBC surface area requirements, as the model was developed with data from nine full-scale plants and includes staged reactor designs.

The second-order model was converted to SI units by Grady et al. (1999), and terms were converted to account for disk surface area. The model can be used to estimate the soluble BOD concentration in each stage.

$$S_n = \frac{-1 + \sqrt{1 + (4)(0.00974)(A_s/Q)S_{n-1}}}{(2)(0.00974)(A_s/Q)} \tag{9–27}$$

where S_n = sBOD concentration in stage n, mg/L
A_s = disk surface area on stage n, m^2
Q = flowrate, m^3/d

Because Eq. (9–27) applies only to sBOD concentrations, a secondary clarifier effluent sBOD/BOD ratio of 0.50 is assumed to design for an effluent BOD concentration. Similarly, without sBOD concentration data for the primary effluent fed to the RBC system, an sBOD/BOD ratio of 0.50 to 0.75 can be assumed. Because the design is based on sBOD, the first-stage RBC soluble unit organic loading rate should be equal to or less than 12 to 15 g sBOD/m^2·d to determine the first-stage disk area and effluent sBOD concentration from Eq. (9–27). The computational procedure used to size an RBC system for BOD removal is summarized in Table 9–9 and illustrated in Example 9–7.

Table 9–9
Computation procedure for the design of a rotating biological contactor (RBC) process

Item	Description
1	Determine influent and effluent sBOD concentrations and wastewater flowrate
2	Determine the RBC disk area for the first stage based on a maximum sBOD of 12 to 15 g sBOD/m^3·d
3	Determine the number of RBC shafts using a standard disk density of 9300 m^2/shaft
4	Select the number of trains for the design, flow per train, number of stages, and disk area/shaft in each stage. For the lower loaded stages a higher disk density may be used
5	Based on the design assumptions made in Step 4, calculate the sBOD concentration in each stage. Determine if the effluent sBOD concentration will be achieved. If not, modify the number of stages, number of shafts per stage, and/or disk area per stage. If the effluent sBOD concentration is met, evaluate alternatives to further optimize the design. Note that the procedure lends itself to spreadsheet calculations
6	Develop the secondary clarifier design

Note: g/m^3·d × 0.0624 = lb/10^3 ft^3·d.

EXAMPLE 9–7 **Staged RBC Design for BOD Removal** Given the following design conditions, develop a process design for a staged RBC system.

Parameter	Unit	Primary effluent	Target effluent
Flowrate	m³/d	4000	
BOD	g/m³	140	20
sBOD	g/m³	90	10
TSS	g/m³	70	20

Note: g/m³ = mg/L

Solution

1. Determine number of RBC shafts for the first stage.

    Assume 1st-stage sBOD = 15 g/m²·d

    sBOD loading = (90 g/m³) 4000 m³/d = 360,000 g/d

    $$\text{Disk area required} = \frac{(360,000 \text{ g/d})}{(15 \text{ g/m}^2\text{·d})} = 24,000 \text{ m}^2$$

    Use 9300 m²/shaft

    $$\text{Number of shafts} = \frac{24,000 \text{ m}^2}{(9300 \text{ m}^2/\text{shaft})} = 2.6$$

    Use 3 shafts for first stage at 9300 m²/shaft.

2. Select number of trains and number of stages.

    Assume: 3 trains with 3 stages/train

    $$\text{Flowrate/train} = \frac{(4000 \text{ m}^3/\text{d})}{3 \text{ trains}} = 1333.3 \text{ m}^3/\text{d}$$

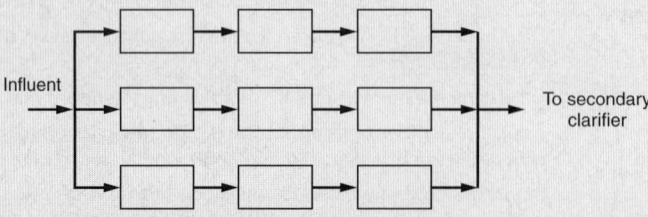

3. Calculate sBOD concentration in each stage using the shaft area and flow to each train. Use Eq. (9–27).

a. Stage 1

$$S_1 = \frac{-1 + \sqrt{1 + (4)(0.00974)(A_s/Q)S_o}}{(2)(0.00974)(A_s/Q)}$$

$S_o = 90 \text{ g/m}^3$

$A_s/Q = 9300 \text{ m}^2/(1333.3 \text{ m}^3/\text{d}) = 6.97 \text{ d/m}$

$$S_1 = \frac{-1 + \sqrt{1 + 24.6}}{0.136} = 29.8 \text{ g/m}^3$$

b. Repeat calculation similar to (*a*) above. Solving for S_2 and S_3 yields

$S_2 = 14.8 \text{ g/m}^3$

$S_3 = 9.1 \text{ g/m}^3$

Because the goal was 10 g/m^3 for S_3, the proposed design is satisfactory.

4. Determine the organic and hydraulic loadings.

a. First stage organic loading

$$L_{org} = \frac{(4000 \text{ m}^3/\text{d})(90 \text{ g sBOD/m}^3)}{(3)(9300 \text{ m}^2)} = 12.9 \text{ g sBOD/m}^2 \cdot \text{d}$$

b. Overall organic loading

$$L_{org} = \frac{(4000 \text{ m}^3/\text{d})(140 \text{ g/m}^3)}{(3 \text{ stage})(3 \text{ shaft/stage})(9300 \text{ m}^2/\text{shaft})} = 6.7 \text{ g BOD/m}^2 \cdot \text{d}$$

c. Hydraulic loading

$$\text{HLR} = \frac{(4000 \text{ m}^3/\text{d})}{(3 \text{ stage})(3 \text{ shafts/stage})(9300 \text{ m}^2/\text{shaft})} = 0.05 \text{ m}^3/\text{m}^2 \cdot \text{d}$$

5. Summary:

Parameter	Unit	Value
No. of trains	number	3
Flowrate/train	m³/d	1333.3
No. of stages	number	3
Total disk area/stage	m²	9300
First-stage sBOD loading	g BOD/m²·d	12.9
Total number of shafts	number	3
Overall organic loading	g BOD/m²·d	6.7
Hydraulic loading/shaft	m³/m²·d	0.05

Comment At the lower sBOD concentrations in stages 2 and 3, some nitrification is likely.

Nitrification. Treatment systems employing RBC units can be used to develop nitrifying biofilms for nitrification of secondary effluents or at low sBOD loadings where nitrification can occur in BOD removal systems. For tertiary nitrification the same procedure used for the design of trickling filters (Sec. 9–2) can be followed. A $r_{n,\max}$ value of 1.5 g N/m^2·d is recommended based on field test results (U.S. EPA, 1984). For combined BOD removal and nitrification, nitrification will be prevented or inhibited by the addition of sBOD to the RBC unit. The nitrifying bacteria can compete for space on the RBC disk once the sBOD concentration is reduced to 10 to 15 mg/L. The sBOD concentration remaining in an RBC tank will be related to the sBOD loading. Pano and Middlebrooks (1983) provide a relationship to show the effect of the sBOD loading on the nitrification rates:

$$F_{r_n} = 1.00 - 0.1\ \text{sBOD} \tag{9–28}$$

where F_{r_n} = fraction of nitrification rate possible without sBOD effect
    sBOD = soluble BOD loading, g/m^2·d

At an sBOD loading rate of 10 g sBOD/m^2·d, the nitrification rate is predicted to be zero.

9–4 COMBINED AEROBIC TREATMENT PROCESSES

Several treatment process combinations have been developed that couple trickling filters with the activated-sludge process. The combined biological processes are known as dual processes or coupled trickling filter/activated-sludge systems. Combined processes have resulted as part of plant upgrading where either a trickling filter or activated-sludge process is added; they have also been incorporated into new treatment plant designs (Parker et al., 1994). Combined processes have the advantages of the two individual processes, which can include (1) the stability and resistance to shock loads of the attached growth process, (2) the volumetric efficiency and low energy requirement of attached growth process for partial BOD removal, (3) the role of attached growth pre-treatment as a biological selector to improve activated-sludge settling characteristics, and (4) the high-quality effluent possible with activated-sludge treatment. The three principal types of combined processes are described below.

Trickling Filter/Solids Contact and Trickling Filter/ Activated-Sludge Processes

The first group of the combined treatment processes, as illustrated on Fig. 9–13, is commonly referred to as the *trickling filter/solids contact* (TF/SC) or *trickling filter/activated-sludge* (TF/AS) process. The principal difference between these processes is the shorter aeration period in the TF/SC process of minutes versus hours for the TF/AS process. Both processes use a trickling filter (with either rock or plastic packing), an activated-sludge aeration tank, and a final clarifier. In both processes, the trickling filter effluent is fed directly to the activated-sludge process without clarification and the return activated sludge from the secondary clarifier is fed to the activated-sludge aeration basin. A process modification that has been incorporated only into the TF/SC process (see Fig. 9–13a, c, and d) is a return-sludge aeration tank and flocculating center-feed well for the clarifier. The TF/AS process is illustrated on Fig. 9–13b.

The most common application for the TF/AS process is where the trickling filter is designed as a roughing filter for 40 to 70 percent BOD removal and may be referred to

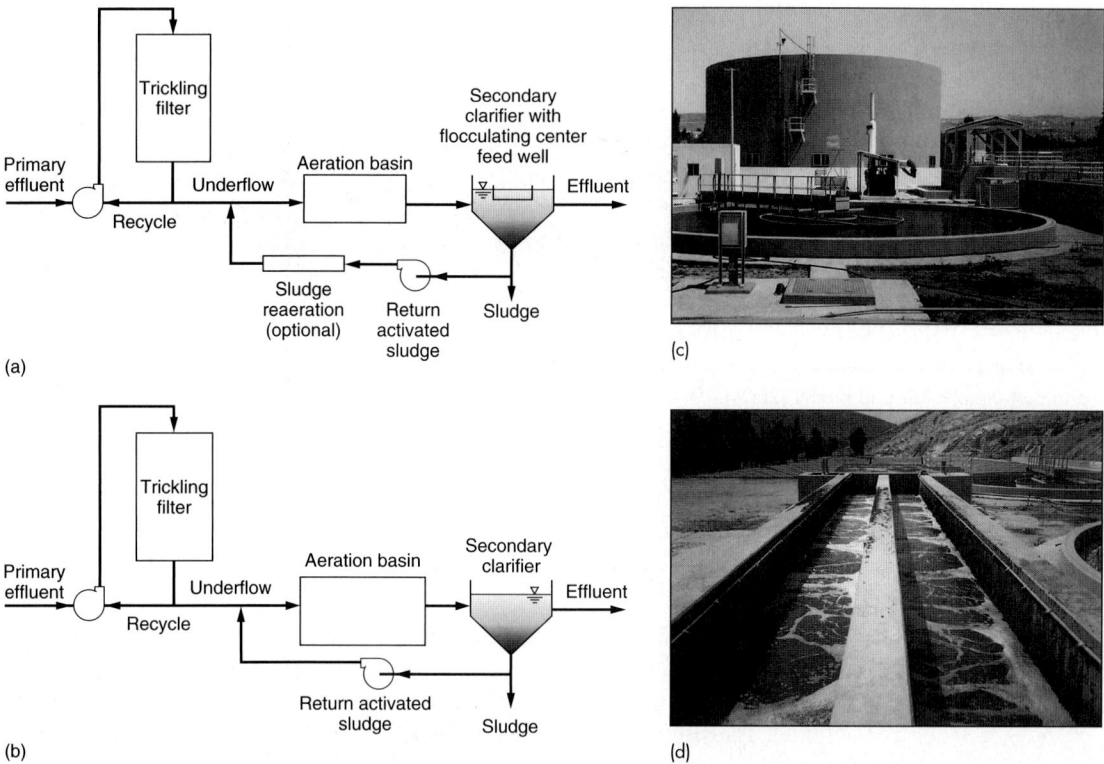

(a)

(b)

(c)

(d)

Figure 9–13

Combined trickling filter/activated-sludge processes: (a) schematic flow diagram of trickling filter/solids contact (TF/SC) process, (b) schematic flow diagram of trickling filter/activated-sludge (TF/AS) process, (c) view of trickling filter/solids contact (TF/SC) process, and (d) view of solids contact unit shown on the right-hand side of photo in (c).

as a *roughing filter/activated-sludge* (RF/AS) process. The trickling filter loading is about 4 times that used for the TF/SC process (see Table 9–10). The aeration basin hydraulic retention time may be 50 to 70 percent of that used in the conventional activated-sludge process. The RF/AS process is attractive for treating higher-strength industrial waste-water because of the relatively low energy use per quantity of BOD removed on the trickling filter. The use of the trickling filter also results in good SVI values for the activated-sludge mixed liquor, as it acts as a biological selector in removing soluble BOD (Biesinger et al., 1980).

The main differences between the TF/SC and RF/AS processes are in their trickling filter loadings and activated-sludge SRT values, which are summarized in Table 9–10. A relatively low organic load for the trickling filter is used for the TF/SC process, and the purpose of the aeration tank is to remove remaining soluble BOD and to develop a flocculent activated-sludge mass that incorporates dispersed solids from trickling filter sloughing. The process is able to produce an advanced treatment effluent quality, which is low in TSS and BOD concentrations, near 10 mg/L and typically less

Table 9–10

Process parameters for trickling filter/solids contact (TF/SC) and roughing filter/activated sludge (RF/AS) processes[a]

Process	Trickling filter loading,[b] kg BOD/m³·d	Activated sludge			Clarifier peak overflow rate, m/h
		τ, min	SRT, d	MLSS, mg/L	
TF/SC	0.3[c]–1.2	10–60	0.3–2.0	1000–3000	1.8–3.0
RF/AS	1.2–4.8	10–60	2.0–7.0	2500–4000	2.0–3.5

[a] Adapted in part from Parker and Bratby (2001).
[b] Loading rates are for cross-flow plastic packing.
[c] Lower value for combined oxidation-nitrification.
Note: kg/m³·d × 62.4280 = lb/10³ ft³·d.
m/h × 0.4090 = gal/ft²·min.

than 15 mg/L (Norris et al., 1982; Harrison et al., 1984). The detention time and SRT of the TF/SC aeration contact basin is quite low, ranging from 0.15 to 1.0 h and 0.5 to 2.0 d, respectively. In some designs an aerated channel is used to provide the aerobic contact time. The sBOD concentration has been estimated by the following plug-flow reactor model, based on first-order kinetics (Matasci et al., 1986).

$$\ln \frac{S}{S_o} = - k_{20}\theta^{(T-20)}Xt \qquad (9\text{–}29)$$

where S_o = mixed-liquor soluble BOD at the contact tank inlet, mg/L
S = soluble carbonaceous BOD after time t, mg/L
k_{20} = first-order reaction rate coefficient at 20°C, L/mg·min
θ = temperature correction coefficient (assume $\theta = 1.035$)
T = wastewater temperature, °C
X = MLVSS, mg/L
t = contact time based on total flow in the tank, min

The value of S_o at the beginning of the contact tank can be related to the soluble effluent BOD in the trickling filter effluent by the following mass balance:

$$(1 + R)S_o = RS_e + S_1 \qquad (9\text{–}30)$$

where R = return-sludge recycle ratio
S_e = trickling filter soluble BOD
S_1 = aerobic solids-contact tank effluent soluble BOD

Reported values for k_{20} range from 2.0 to 3.3 × 10⁻⁵ L/mg·min. Newbry et al. (1988) observed values of 1.4 × 10⁻⁵ and 3.5 × 10⁻⁵ L/mg·min for different operating periods that had different amounts of inert material in the wastewater. Newbry et al. (1988) further showed that at an SRT of 1.0 d or more for the aeration basin, the effluent sBOD and TSS concentration could be minimized. They also showed that the trickling filter organic loading affected the TF/SC effluent TSS and BOD concentrations. Based on the results reported by Newbry et al. (1988), the trickling filter BOD loading should be less than 7.0 kg BOD/m³·d with an aeration tank SRT of 1.0 d to maximize performance.

The TF/SC process has also been demonstrated for nitrification applications. At a low enough organic loading, nitrification can occur in the trickling filter. At BOD loadings ranging from 0.40 to 0.70 kg/m³·d, nitrification occurred in a high-density trickling filter tower at Windsor, Ontario, at temperatures ranging from 11 to 20°C (Parker et al., 1998). The trickling filter effluent NH_4-N concentration averaged 2.8 mg/L, and the final effluent concentration averaged 1.9 mg/L with an aeration contact zone detention time of 15 min and 2 to 4 d SRT. Chemical treatment was used with primary clarification to result in low enough BOD (30 to 70 g/m³) and TSS concentrations to allow use of the high-density plastic packing without plugging problems. Daigger et al. (1994) also showed similar performance characteristics for low-loaded trickling filters in the TF/SC process and noted that nitrifying bacteria sloughed from the trickling filter result in nitrification in the contact activated-sludge process at lower than theoretical SRTs due to the seed population.

Activated Biofilter and Biofilter Activated-Sludge Processes

The second group of combined processes is similar to the first, described above, with the exception that the RAS is returned directly to the trickling filter as illustrated on Fig. 9–14 and an aeration basin may or may not be used. The two combined processes are termed

Figure 9–14

Combined trickling filter/activated-sludge process with return sludge recycle to trickling filter: (a) schematic flow diagram of activated biofilter (ABF) and (b) schematic flow diagram of biofilter/activated sludge (BF/AS).

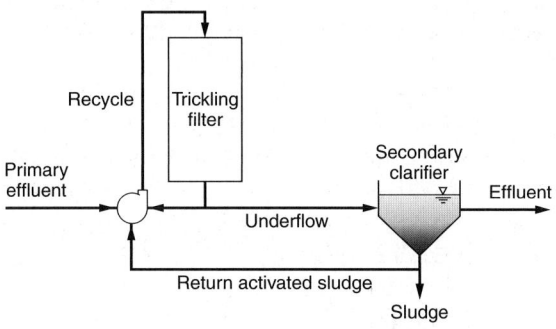

(a)

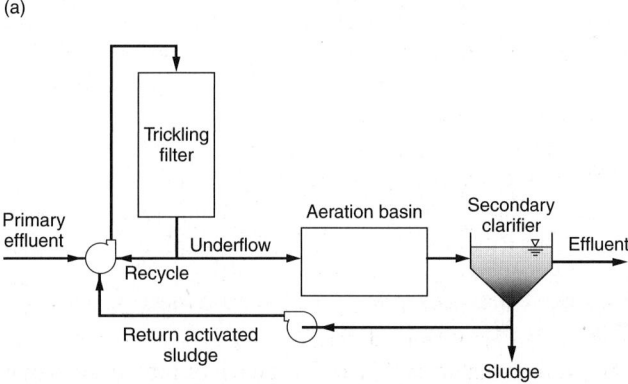

(b)

Table 9-11

Process parameters for activated biofilter (ABF) and biofilter/activated sludge (BF/AS)

| Process | Trickling filter loading, kg BOD/m³·d | τ, h | Activated sludge | | Clarifier peak overflow rate, m/h |
			SRT, d	MLSS, mg/L	
ABF	0.36–1.2	—	0.5–2.0	1500–4000	1.8–3.0
BF/AS	1.2–4.8	2–4	2.0–7.0	1500–4000	2.0–3.5

Note: kg/m³·d × 62.4280 = lb/10³ ft³·d.
m/h × 0.4090 = gal/ft²·min.

the *activated biofilter* (ABF) and the *biofilter/activated-sludge* (BF/AS) process. The ABF and BF/AS processes are not used much today, in part because the original designs relied on redwood filter packing, which is available only at prohibitive cost.

In both cases, rock cannot be used because of the potential plugging and oxygen availability problems created by feeding return activated sludge to the trickling filter. High-rate plastic packing can be used in lieu of redwood packing. The ABF process can produce a secondary effluent quality at low organic loads (see Table 9–11) to the trickling filter. At higher organic loading rates, an acceptable secondary effluent quality with the ABF process has been difficult to produce. To improve effluent quality, the ABF process is followed by a short HRT aeration basin, which fits the description of the BF/AS process (see Fig. 9–14b) (Harrison and Timpany, 1988). In essence the BF/AS process is very similar to the RF/AS process with the exception of RAS return to the trickling filter instead of to the aeration basin. Harrison et al. (1984) showed similar performance for BF/AS and RF/AS processes for full-scale plants surveyed. Typical parameters for the ABF and the BF/AS processes are given in Table 9–11.

Series Trickling Filter-Activated-Sludge Process

In the third approach employing combined processes, a trickling filter and an activated-sludge process are operated in series, with an intermediate clarifier between the trickling filter and activated-sludge process (see Fig. 9–15). The combination of a trickling filter process followed by an activated-sludge process is often used (1) to upgrade an existing activated-sludge system, (2) to reduce the strength of wastewater where industrial and domestic wastewater is treated in common treatment facilities, and (3) to protect a nitrification activated-sludge process from toxic and inhibitory substances. In systems treating high-strength wastes, intermediate clarifiers are used between the trickling filters and the activated-sludge units to reduce the solids load to the activated-sludge system and to minimize the aeration volume required. Typical process design parameters for the trickling filter activated-sludge process are given in Table 9–12.

Design Considerations for Combined Trickling Filter Activated-Sludge Systems

For the combined processes without intermediate clarification, the activated-sludge design is affected by the trickling filter design loading and performance. The amount of

Figure 9–15

Schematic flow diagram of combined trickling filter activated-sludge process with intermediate clarifier.

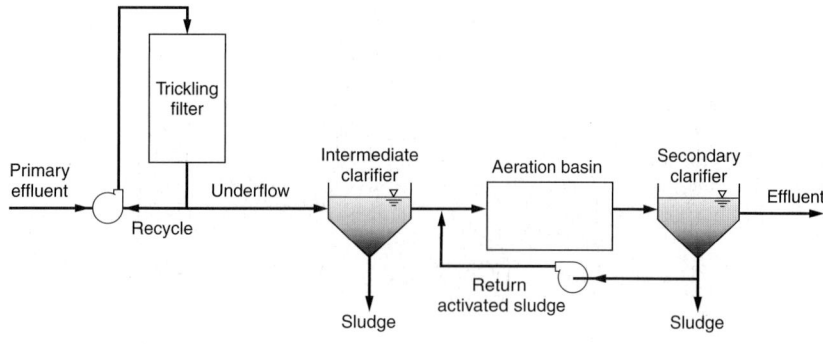

Table 9–12

Process parameters for series trickling filter–activated-sludge process with intermediate clarifier

Parameter	Unit	Value
Trickling filter loading	kg BOD/m³·d	1.0–4.8
Intermediate clarifier	m/h	2.0–3.5
Activated sludge:		
τ	h	a
SRT	d	2.0–7.0
F/M	unitless	0.2–0.5
MLSS	mg/L	1500–4000
Final clarifier	m/h	2.0–3.5

[a] Function of the wastewater temperature.
Note: kg/m³·d × 62.4280 = lb/10³ ft³·d.
m/h × 0.4090 = gal/ft²·min.

oxygen required in the activated-sludge aeration basin depends on how much BOD is removed in the trickling filter. All of the solids produced, which are the biological solids plus the influent nonbiodegradable solids, will end up in the activated-sludge aeration tank, but the amount of biomass is affected by how much BOD is removed in the trickling filter. An estimate of the SRT in the trickling filter as a function of the BOD loading is presented on Fig. 9–16. At higher BOD loadings the concentration of biomass produced in the trickling filter can be determined based on the estimated SRT and amount of BOD removed in the trickling filter.

The amount of BOD removed in the trickling filter is difficult to predict. Both particulate and soluble BOD are removed by biomass in the trickling filter, and current empirical design models (Sec. 9–2) are generally based on influent and final settled BOD, and thus do not distinguish between particulate (pBOD) and soluble BOD (sBOD) removal rates. These models may be used to estimate the sBOD removal as the final suspended solids concentration after settling and their BOD contribution can be estimated. However, pBOD not degraded in the trickling filter will most likely be degraded in the activated-sludge process, therefore affecting the oxygen demand. Thus,

Figure 9–16

Equivalent SRT for biosolids in a trickling filter as a function of the BOD loading. *(Adapted from WEF, 2000.)*

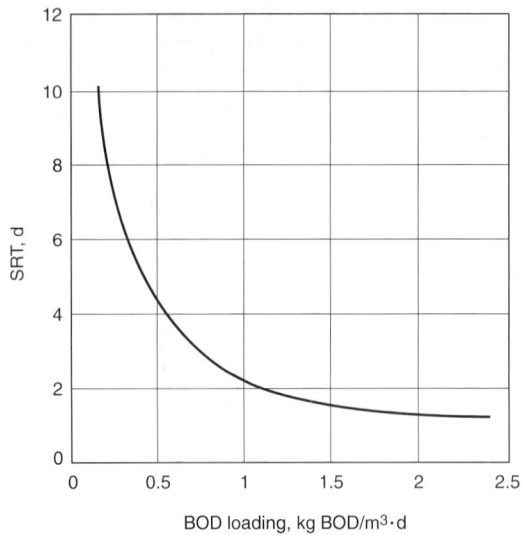

Figure 9–17

Approximate amount of particulate BOD degraded in a trickling filter as a function of organic loading. *(From Bogus, 1989.)*

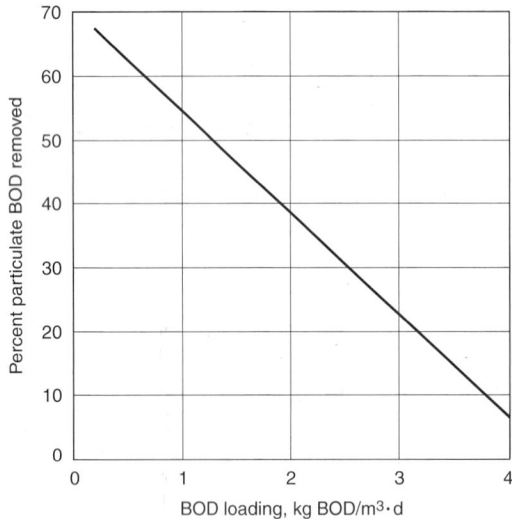

to determine the oxygen required for the activated-sludge process, the amount of pBOD degraded in the trickling filter is critical.

The removal of pBOD was studied in a combined trickling filter activated-sludge pilot plant over a wide range of trickling filter BOD loadings. Intensive sampling with COD and BOD solids balances on the trickling filter were used to determine the amount of pBOD degradation (Bogus, 1989). The amount of pBOD degraded increased as the BOD loading to the trickling filter was decreased. An estimate of the percent of influent pBOD degraded as a function of the BOD loading is provided on Fig. 9–17. With an estimate of the amount of pBOD and sBOD removal in the trickling filter and the trickling filter SRT, the amount of biomass produced can be calculated. With that information the amount of oxygen demand satisfied in the trickling filter can be estimated. The trickling filter biomass and nondegraded pBOD and sBOD concentrations can then be used to estimate the activated-sludge aeration basin oxygen demand. The sBOD/BOD ratio in the effluent from the trickling filter will typically vary from 0.25 to 0.50. A solids balance is also done to determine the basin volume as a function of the design SRT and MLSS concentration. The biomass yield and endogenous decay coefficients and basic equations are the same as that presented in Sec. 7–4 and 7–5 in Chap. 7.

The design procedure to determine the oxygen requirements, sludge production, and aeration volume of the activated-sludge basin for a combined trickling filter activated-sludge process including RF/AS and TF/SC processes is summarized in Table 9–13. An example of a TF/SC design is provided in Example 9–8 using the procedure outlined in Table 9–13.

Table 9–13
Computation procedure for the design of a combined trickling filter activated-sludge process

Item	Description
1	Select a BOD loading for the trickling filter that is compatible with the combined process selection
2	Select the SRT of the activated-sludge process. For highly loaded trickling filters longer SRTs will be used in the activated-sludge process
3	Determine the trickling filter tower size and BOD removal. Estimate the soluble BOD removal in the trickling filter
4	Based on the BOD loading to the trickling filter, use Fig. 9–17 to determine fraction of particulate BOD removed in the trickling filter
5	Use Fig. 9–16 to estimate the biomass SRT in the trickling filter. Use this value to estimate the biomass production
6	From the biomass production in the trickling filter and the amount of BOD removed, perform a mass balance on the ultimate BOD, including biomass production, to determine the amount of oxygen demand satisfied in the trickling filter
7	Assume that the portion of influent BOD not degraded in the trickling filter is degraded in the activated-sludge basin, if the SRT is ≥ 10 d. Otherwise, base the soluble BOD removal on Fig. 9–17. From the BOD removal calculate the biomass production in the activated-sludge tank. Adjust the biomass produced from the trickling filter by the loss due to endogenous decay in the activated-sludge tank
8	Based on the total biomass produced, perform an ultimate BOD mass balance to determine the total oxygen demand for the entire system. Subtract the amount of oxygen demand satisfied in the trickling filter to obtain the oxygen requirements for the activated-sludge tank. Add a sufficient peaking factor to account for variable loadings to short detention time basins
9	Sum the sludge production from biomass, nonbiodegradable VSS (nbVSS) in the influent wastewater, and inorganic TSS (TSS—VSS) in the influent wastewater
10	Using the net sludge production SRT and assumed MLSS concentration, calculate the volume of the activated-sludge aeration tank
11	Evaluate solids and hydraulic loadings for the secondary clarifier design

EXAMPLE 9–8 **Trickling Filter/Solids Contact (TF/SC) Process Design** For the following wastewater characteristics, determine the following design elements for a trickling filter/solids contact (TF/SC) process:

1. Trickling filter diameter and hydraulic application rate
2. The amount of oxygen demand satisfied in the trickling filter, kg/d
3. The amount of oxygen required in the activated-sludge aeration tank, kg/d
4. The amount of solids wasted per day, kg/d
5. The volume and hydraulic retention time of the aeration tank

Wastewater characteristics

Item	Unit	Value
Flow	m³/d	5000
BOD	g/m³	130
sBOD	g/m³	90
TSS	g/m³	60
VSS	g/m³	52
nbVSS	g/m³	20
Temperature	°C	15

Note: g/m³ = mg/L

For TF/SC design assume the following conditions apply:

1. Number of trickling filters = 2
2. Plastic packing coefficient $k_{20} = 0.21(L/s)^{0.5}/m^2$ (see Table 9–6)
3. Depth of packing = 6.1 m
4. Trickling filter BOD loading = 0.70 kg BOD/m³·d
5. Biomass yield Y = 0.6 g VSS/g BOD
6. Endogenous decay k_d = 0.08 g/g·d
7. UBOD/BOD = 1.6
8. Trickling filter effluent sBOD = 0.5 BOD
9. Activated-sludge SRT = 1.0 d
10. MLSS = 2500 g/m³
11. Secondary clarifier sBOD/BOD = 0.5
12. Biomass VSS/TSS ratio = 0.85

Solution

1. Determine the hydraulic application rate and the trickling filter diameter.
   a. Determine the hydraulic application rate.
      The organic volumetric loading rate, as given by Eq. (7–67), is

$$L_{org} = \frac{QS_o}{(V)(10^3 \text{ g/kg})} = \frac{QS_o}{(A)(D)(10^3 \text{ g/kg})} = \frac{qS_o}{D(10^3 \text{ g/kg})}$$

where A = cross-sectional area, m^2
D = depth of packing, m
q = hydraulic application rate, $m^3/m^2{\cdot}d$

Solving for q yields

$$q = \frac{(L_{org})(D)(10^3 \text{ g/kg})}{S_o}$$

Substitute known quantities and solve for q

$$q = \frac{(0.70 \text{ kg/m}^3{\cdot}d)(6.1 \text{ m})(10^3 \text{ g/kg})}{(130 \text{ g/m}^3)}$$

$$= 32.85 \text{ m}^3/\text{ m}^2{\cdot}d = 0.38 \text{ L/m}^2{\cdot}s$$

b. Determine the trickling filter tower diameter.

$$A = \frac{Q}{q} = \frac{(5000 \text{ m}^3/d)}{(32.85 \text{ m}^3/\text{m}^2{\cdot}d)} = 152.2 \text{ m}^2$$

For 2 towers, area/tower = 76.1 m^2

Diameter/tower = 9.8 m

2. Determine the amount of soluble and particulate BOD removed in the trickling filter and the BOD in the effluent from the trickling filter using Eq. (9–19).
   a. Determine the soluble BOD removed using Eq. (9–19) and the fact that sBOD = 0.5 BOD.

$$\frac{S_e}{S_o} = e^{-kD/q^n}$$

Correct the removal rate coefficient for temperature using Eq. (9–20).

$$k_{15°C} = k_{20°C}(1.035)^{15-20}$$

$$= 0.21(1.035)^{-5} = 0.177 \text{ (L/s)}^{0.5}/\text{m}^2$$

Normalize the removal rate coefficient using Eq. (9–22).

$$k_{15°C} = k_1\left(\frac{D_1}{D_2}\right)^{0.5}\left(\frac{S_1}{S_2}\right)^{0.5} = 0.177\left(\frac{6.1}{6.1}\right)^{0.5}\left(\frac{150}{130}\right)^{0.5}$$

$$= 0.177(1.074) = 0.19 \text{(L/s)}^{0.5}/\text{m}^2$$

$$\frac{S_e}{S_o} = e^{-0.19(6.1)/0.38^{0.5}}$$

$$\frac{S_e}{S_o} = 0.152$$

$$S_e = 0.152(130) = 19.7 \text{ g/m}^3$$

The sBOD removed in the trickling filter effluent is:

sBOD removed = 90 g/m^3 − 0.5(19.7 g/m^3) = 80.1 g/m^3

b. Determine the particulate BOD removed.

Particulate BOD in influent = BOD − sBOD

pBOD = (130 − 90) g/m^3 = 40 g/m^3

From Fig. 9–17, amount of pBOD degraded at 0.70 kg BOD/m^3·d = 56%

pBOD in trickling filter effluent = (1 − 0.56)(40 g/m^3) = 17.6 g/m^3

The pBOD degraded in the trickling filter effluent is:

pBOD removed = 40 g/m^3 − 17.6 g/m^3 = 22.4 g/m^3

c. Determine the BOD in the trickling filter effluent.

Trickling filter effluent BOD = sBOD + pBOD = (9.9 + 17.6) g/m^3

$$= 27.5 \text{ g/m}^3$$

3. Determine oxygen demand satisfied in the trickling filter.

From Fig. 9–16, the trickling filter SRT = 2.5 d at an organic rate of 0.70 kg BOD/m^3·d

Determine biomass produced using Eq. (7–43) (Note, SRT = τ):

$$X = \frac{Y(S_o - S)}{1 + (k_d)\text{SRT}}$$

Substrate removal (S_o − S) in trickling filter = (130 − 27.5) = 102.5 g/m^3

$$X_{TF} = \frac{0.6(102.5 \text{ g/m}^3)}{[1 + (0.08 \text{ g/g·d})(2.5 \text{ d})]} = 51.2 \text{ g/m}^3$$

For short SRT values, cell debris [Eq. (7–50)] is negligible and is not included here.

Determine the oxygen satisfied in trickling filter with a COD balance.

O$_2$ used = bCOD$_{IN}$ − bCOD$_{OUT}$ − 1.42X

$$= 1.6(130 \text{ g/m}^3 - 27.5 \text{ g/m}^3) - 1.42(51.2 \text{ g/m}^3)$$

Oxygen used in trickling filter = 91.1 g/m^3

4. Determine biomass produced in solids contact (SC) aeration tank.

Approximate BOD removed = 27.5 g/m^3, SRT = 1 d

Biomass due to oxidation of organic matter

$$X_{SC} = \frac{0.6(27.5 \text{ g/m}^3)}{[1 + (0.08 \text{ g/g·d})(1 \text{ d})]} = 15.3 \text{ g/m}^3$$

Trickling filter biomass remaining after endogenous decay of filter biomass

$$X_{TF, decay} = \frac{(51.2 \text{ g/m}^3)}{[1 + (0.08 \text{ g/g·d})(1 \text{ d})]} = 47.4 \text{ g/m}^3$$

Total biomass produced or fed to the SC aeration tank = 47.4 g/m^3 + 15.3 g/m^3 = 62.7 g/m^3

5. Determine oxygen demand in SC aeration tank in g/m^3 and kg O$_2$/d.

Total oxygen consumed = 1.6(130 g/m^3) − 1.42(62.7 g/m^3) = 119 g/m^3

Activated-sludge oxygen demand = total demand − TF demand

$$= 119 \text{ g/m}^3 - 91.1 \text{ g/m}^3 = 27.9 \text{ g/m}^3$$

kg O$_2$/d = (27.9 g/m^3) (5000 m^3/d) (1 kg/10^3 g) = 140 kg/d

6. Determine the amount of solids wasted per day (TSS).
   a. Determine the inert inorganic solids

   Inert solids = TSS − VSS = 60 g/m^3 − 52 g/m^3 = 8 g/m^3

   b. Determine the total solids wasted per day expressed in g/m^3.

   Total solids = biomass + inert inorganics + nbVSS

   $$= \frac{(62.7 \text{ g VSS/m}^3)}{(0.85 \text{ g VSS/g TSS})} + 8 \text{ g/m}^3 + 20 \text{ g/m}^3 = 101.8 \text{ g/m}^3$$

   c. Determine the total solids wasted per day expressed in kg/d.

   $$P_X = (101.8 \text{ g/m}^3) (5000 \text{ m}^3/\text{d}) (1 \text{ kg/10}^3 \text{ g})$$

   $$= 509 \text{ kg/d solids wasted}$$

7. Determine the solids contact aeration tank volume and the corresponding hydraulic detention time.
   a. Determine the solids contact aeration tank volume using Eq. (7–55).

   $$(X_{TSS})(V) = P_{X, TSS}(SRT)$$

   $$V = \frac{P_{X,TSS}(SRT)}{X_{TSS}} = \frac{(509 \text{ kg/d})(1.0 \text{ d})(10^3 \text{ g/kg})}{(2500 \text{ g/m}^3)} = 203.6 \text{ m}^3$$

   b. Determine the hydraulic detention time in the solids contact aeration tank.

   $$\tau = \frac{(203.6 \text{ m}^3)(24 \text{ h/d})}{(5000 \text{ m}^3/\text{d})} = 0.98 \text{ h}$$

Comments Most of the BOD removal and oxygen demand is satisfied in the trickling filter. The g O$_2$ supplied in the activated sludge per g BOD removed for the system is 0.21 [(27.9 mg/L)/(130 mg/L)] which agrees well with published information. The clarifier recycle flowrate is determined in a similar manner to that shown in Chap. 8 for the suspended growth processes.

9–5 **ACTIVATED SLUDGE WITH FIXED-FILM PACKING**

Several types of synthetic packing materials have been developed for use in activated-sludge processes. These packing materials may be suspended in the activated-sludge mixed liquor or fixed in the aeration tank. A term used to describe these types of processes is an integrated fixed-film activated-sludge process (Sen et al., 1994*a*, 1994*b*). These processes are intended to enhance the activated-sludge process by providing a greater biomass concentration in the aeration tank and thus offer the potential to reduce the basin size requirements. They have also been used to improve volumetric nitrification rates and to accomplish denitrification in aeration tanks by having anoxic zones within the biofilm depth. Because of the complexity of the process and issues related to understanding the biofilm area and activity, the process designs are empirical and based on prior pilot-plant or limited full-scale results. In this section these processes are introduced and described and some design considerations and parameters are presented.

Processes with Internal Suspended Packing for Attached Growth

There are now more than 10, and counting, different variations of processes in which a packing material of various types is suspended in the aeration tank of the activated-sludge process (see Fig. 9–18). Typical examples of activated-sludge treatment processes with suspended packing include the Captor®, Linpor®, and Kaldnes®.

Captor® and Linpor®.   In the Captor® and Linpor® processes foam pads with a specific density of about 0.95 g/cm³ are placed in the bioreactor in a free-floating fashion and retained by an effluent screen (see Fig. 9–19). The pad volume can account for 20 to 30 percent of the reactor volume. Dimensions of the various packing materials are presented in Table 9–14. Mixing from the diffused aeration system circulates the foam pads in the system, but without additional mixing methods, they may tend to accumulate at the effluent end of the aeration basin and float at the surface (Brink et al., 1996; Golla et al., 1994). An air knife has been installed to continuously clean the screen and a pump is used to return the packing material to the influent end of the reac-

Figure 9–18

View of suspended packing in activated-sludge aeration tank with proper mixing. Without proper mixing the suspended packing tends to accumulate on the surface of the reactor.

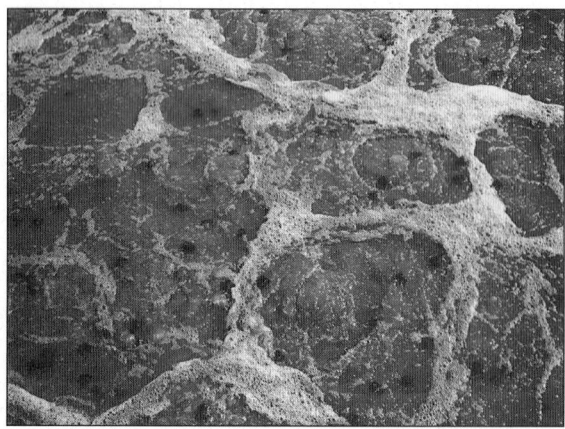

Figure 9–19

Typical flow diagram for Captor® and Limpor® processes.

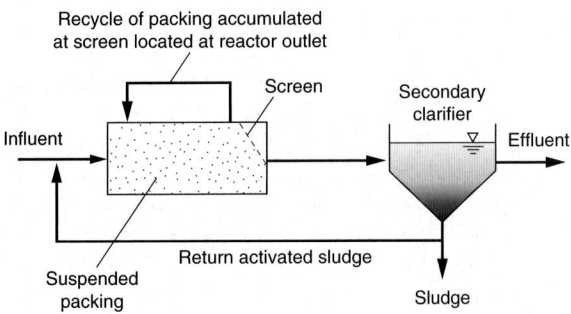

Table 9–14
Packing material for suspended fixed-film activated-sludge process

Name	Packing specifications	
	Material	**Dimensions, mm**
Captor®	Polyurethane	30 × 25 × 25
Linpor®	Polyurethane	10 to 13 cubes

tor. Solids are removed from a conventional secondary clarifier and wasting is from the return line as in the activated-sludge process.

The principal advantage for the sponge packing systems is the ability to increase the loading on an existing plant without increasing the solids load on existing secondary clarifiers, as most of the biomass is retained in the aeration basin. Loading rates for BOD of 1.5 to 4.0 kg/m³·d with equivalent MLSS concentrations of 5000 to 9000 mg/L have been achieved with these processes (WEF, 2000). Based on the results with full-scale and pilot-scale tests with the sponge packing installed it appears that nitrification can occur at apparent lower SRT values, based on the suspended growth mixed liquor, than those for activated sludge without internal packing.

Kaldnes®. A *moving-bed biofilm reactor* (MBBR) has been developed by a Norwegian company, Kaldnes Miljøteknologi. The process consists of adding small cylindrical-shaped polyethylene carrier elements (specific density of 0.96 g/cm³) in aerated or non-aerated basins to support biofilm growth (see Fig. 9–20). The small cylinders are about 10 mm in diameter and 7 mm in height with a cross inside the cylinder and longitudinal fins on the outside. The biofilm carriers are maintained in the reactor by the use of a perforated plate (5 × 25 mm slots) at the tank outlet. Air agitation or mixers are applied in a manner to continuously circulate the packing. The packing may fill 25 to 50 percent of the tank volume. The specific surface area of the packing is about 500 m²/m³ of bulk packing volume. The MBBR does not require any return activated-sludge flow or backwashing. A final clarifier is used to settle sloughed solids. The MBBR process provides an advantage for plant upgrading by reducing the solids loading on existing clarifiers (Rusten et al., 1998 and 2000). The presence of packing material discourages the use of more efficient fine bubble aeration equipment, which would require periodic drainage of the aeration and removal of the packing for diffuser cleaning.

Two different applications in which the MBBR has been applied are illustrated on Fig. 9–21. The first is a more common design application for BOD removal, nitrification, and denitrification (Rusten et al., 1995). For the anoxic-aerobic treatment mode (see Fig. 9–21a), a 6-stage reactor design is used. Chemical addition for phosphorus removal is done after the MBBR. The Kaldnes packing is added to provide a specific packing surface area of 200 to 400 m²/m³ in the reactors. In the second application the MBBR is used in place of the trickling filter in the solids contact process (see

Figure 9–20

Typical reactors used with suspended packing materials: (a) aerobic and (b) anaerobic/anoxic with submerged internal mixer.

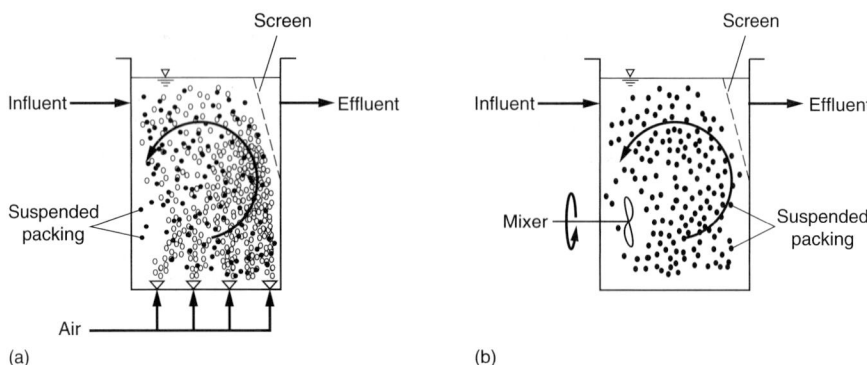

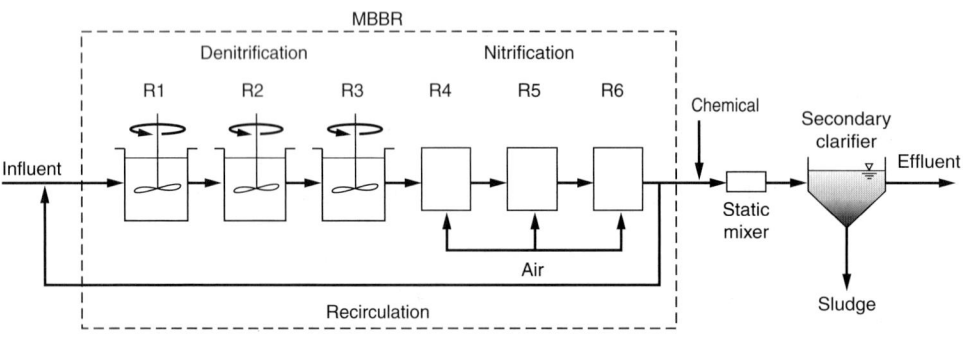

(a)

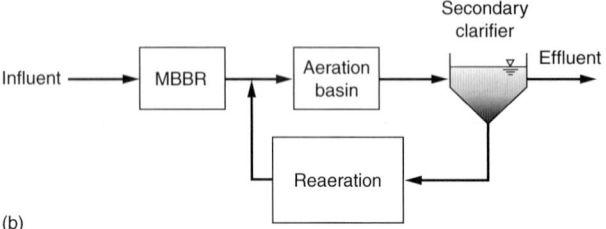

(b)

Figure 9–21

Schematic flow diagram for the application of the moving-bed bioreactor treatment (MBBR) process: (a) removal of BOD and nutrients and (b) solids contact process in which the MBBR process replaces the trickling filter as shown in Fig. 9–13a. *(Adapted from Kaldnes Miljøteknologi.)*

Table 9–15
Typical process design parameters for a moving-bed biofilm reactor (MBBR)[a]

Parameter	Unit	Range of values
MBBR:		
Anoxic detention time	h	1.0–1.2
Aerobic detention time	h	3.5–4.5
Biofilm area	m^2/m^3	200–250
BOD loading	$kg/m^3{\cdot}d$	1.0–1.4
Secondary clarifier hydraulic application rate	m/h	0.5–0.8

[a] Adapted in part from Rusten et al. (2000).
Note: $m^2/m^3 \times 0.3048 = ft^2/ft^3$.
$kg/m^3{\cdot}d \times 62.4280 = lb/10^3 \ ft^3{\cdot}d$.
$m/h \times 0.4090 = gal/ft^2{\cdot}min$.

Table 9–16
Typical operating parameters for a moving-bed biofilm reactor/solids contact (MBBR/SC) process[a]

Parameter	Unit	Range of values
MBBR:		
Biofilm surface area	m^2/m^3	300–350
Organic load	$kg \ BOD/m^3{\cdot}d$	4.0–7.0
MLSS concentration	mg/L	2500–4500
SC:		
SRT	d	2–3
MLSS concentration	mg/L	1500–2500
SVI	mL/g	90–120
Detention time	h	0.6–0.8
Reaeration tank	h	0.6–0.8
MLSS	mg/L	6000–8500

[a] Adapted in part from Rusten et al. (1998).
Note: $m^2/m^3 \times 0.3048 = ft^2/ft^3$.
$kg/m^3{\cdot}d \times 62.4280 = lb/10^3 \ ft^3{\cdot}d$.

Fig. 9–21b). Typical process design parameters for the moving-bed biofilm reactor and the moving-bed biofilm reactor/solids contact (MBBR/SC) processes are reported in Tables 9–15 and 16, respectively.

Processes with Internal Fixed Packing for Attached Growth

There are now more than half a dozen, and counting, different variations of processes in which a fixed packing material is placed in the aeration tank of the activated-sludge process. Three typical examples of fixed packing processes include the Ringlace® and BioMatrix® processes, Bio-2-Sludge® process, and submerged RBCs.

Figure 9–22

Placement of Ringlace® packing in an activated-sludge reactor: (a) schematic of placement of packing in activated-sludge reactor and (b) isometric view of packing placed in activated-sludge reactor. (Adapted from Ringlace, Inc.)

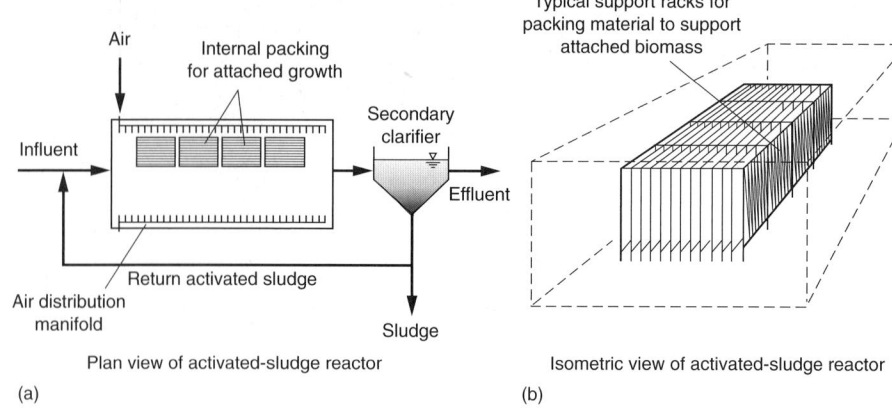

Plan view of activated-sludge reactor

(a)

Isometric view of activated-sludge reactor

(b)

Ringlace®. Ringlace® packing is a looped polyvinyl chloride material that is about 5 mm in diameter. It is placed in about 25 to 35 percent of the activated-sludge basin volume in modules with individual strands at 40 to 100 mm apart (see Fig. 9–22). The specific surface area provided ranges from 120 to 500 m²/m³ of tank volume. The packing placement location in the aeration tank is important. To provide efficient contact with the wastewater the packing should be placed along one side of the aeration vessel with the aeration equipment providing a spiral roll pattern for flow through the packing (Sen et al., 1993). Spiral roll aeration is usually less efficient than full floor coverage aeration with fine bubble diffusers.

The location along the length of the tank is also important for nitrification and denitrification system operations. Randall and Sen (1996) recommend a location where sufficient BOD remains to develop a biofilm growth, but where the BOD demand is low enough so that ammonia oxidation can occur in the film. However, they noted that the optimal rate can be difficult to achieve as variations in BOD loading can vary the biofilm growth on the packing and the competition between heterotrophic and autotrophic bacteria for surface area. In some applications, the advantages of using a fixed internal packing were negated due to the growth of bristle worms in the biofilm.

Bio-2-Sludge® Process. A schematic of the Bio-2-Sludge® process is illustrated on Fig. 9–23. The PVC packing bundle with a surface area of 90 to 165 m²/m³ and a minimum opening of 20 × 20 mm to prevent clogging is positioned along the walls of the aeration tank. The air diffusion system is designed to create a mixed liquid recirculation flow through the packing (WEF, 2000).

Submerged Rotating Biological Contactors. Rotating biological contactor units have been installed in activated sludge. The *submerged rotating biological contactor* (SRBC) is operated at approximately 85 percent submergence (see Fig. 9–11c). The SRBC units can be as large as 5.5 m (18 ft) diameter with a surface area of 28,800 m². The rotation is driven by aeration and may be mechanically assisted. The submerged operation reduces the load on the packing shaft.

Figure 9–23

Schematic of Bio-2-Sludge® process.

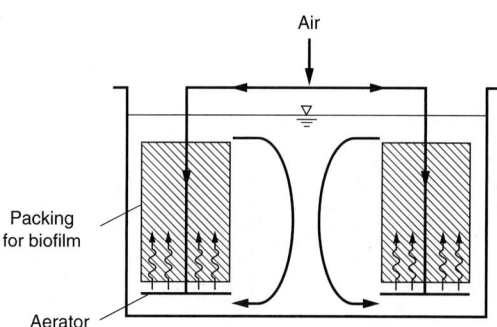

9–6 SUBMERGED ATTACHED GROWTH PROCESSES

Aerobic submerged fixed-film processes consist of three phases: a packing, biofilm, and liquid. The BOD and/or NH_4-N removed from the liquid flowing past the biofilm is oxidized. Oxygen is supplied by diffused aeration into the packing or by being predissolved into the influent wastewater. Aerobic fixed-film processes include downflow packed-bed reactors, upflow packed-bed reactors, and upflow fluidized-bed reactors. The type and size of packing is a major factor that affects the performance and operating characteristics of submerged attached growth processes. Designs differ by their packing configuration and inlet and outlet flow distribution and collection. No clarification is used with aerobic submerged attached growth processes, and excess solids from biomass growth and influent suspended solids are trapped in the system and must be periodically removed. Most designs require a backwashing system much like that used in a water filtration plant to flush out accumulated solids, usually on a daily basis.

The major advantages of submerged attached growth processes are their relatively small space requirement, the ability to effectively treat dilute wastewaters, no sludge settling issues as in activated-sludge process, and aesthetics. Also for many processes solids filtration occurs to produce a high-quality effluent. Such fixed-film systems have equivalent hydraulic retention times of less than 1 to 1.5 h, based on their empty tank volumes. Their disadvantages include a more complex system in terms of instrumentation and controls, limitations of economies of scale for application to larger facilities, and generally a higher capital cost than activated-sludge treatment.

A wide variety of submerged attached growth processes have been used. The purpose of this section is to describe the more common processes used and their design loadings and performance capability. The processes described are the downflow Biocarbone® process, the upflow Biofor® and Biostyr® process, and the aerobic upflow fluidized-bed reactor.

Downflow Submerged Attached Growth Processes

The Biocarbone® process is a typical example of a downflow submerged attached growth process. Over 100 facilities have been constructed worldwide since the development of the process in France in the early 1980s (WEF, 1998). The process has also been termed the biological aerated filter (BAF) (Stensel et al., 1988). Although activated carbon packing was used in the original design, a 3- to 5-mm fired clay material is used in

Figure 9–24

Schematic of Biocarbone® downflow biological process.

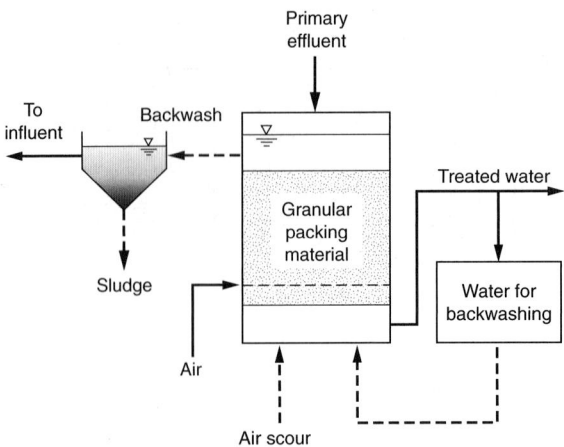

Table 9–17
Typical design loadings for Biocarbone® aerobic submerged attached growth process[a]

Application	Unit	Loading range
BOD removal	kg BOD/m³·d	3.5–4.5
BOD removal and nitrification	kg BOD/m³·d	2.0–2.75
Tertiary nitrification	kg N/m³·d	1.2–1.5

[a] Adapted from Mendoza and Stephenson (1999) and WEF (1998).
Note: kg/m³·d × 62.4280 = lb/10³ ft³·d.

current designs. Plant installations range in size from 2000 to 80,000 m³/d. A schematic of the Biocarbone® process is shown on Fig. 9–24. The system is designed much like a water filter with the addition of an air header about 300 mm above the underdrain nozzles to sparge air through the packing. The air header is uniformly arranged across the bed to assure that oxygen is provided throughout the entire bed of packing. The actual oxygen transfer efficiency is in the range of 5 to 6 percent, which is comparable to fine bubble diffused aeration performance for that depth (Stensel et al., 1988).

Backwashing is normally done once per day or when the headloss increases to about 1.8 m. The design must consider both organic loading and hydraulic application rate. Hydraulic application rates in the range of 2.4 to 4.8 m³/m²·h are recommended to prevent excessive headloss. The Biocarbone® process has been used in aerobic applications for BOD removal only, combined BOD removal and nitrification, and tertiary nitrification. The process has also been used for denitrification as have all the attached growth processes shown here, and these applications are presented in Chap. 10. Typical design loadings for Biocarbone applications are presented in Table 9–17. In combined BOD removal and nitrification systems, the nitrification rate is about 0.45 kg/m³·d (Stensel et al., 1988). Higher DO concentrations in the range of 3 to 5 mg/L are recommended for efficient nitrification. For BOD removal applications, effluent BOD and TSS concentrations are generally <10 mg/L and for nitrification, effluent NH₄-N concentrations may range from 1 to 4 mg/L. A similar process, but using an expanded shale

Figure 9–25

Schematic of Biofor® upflow biological reactor.

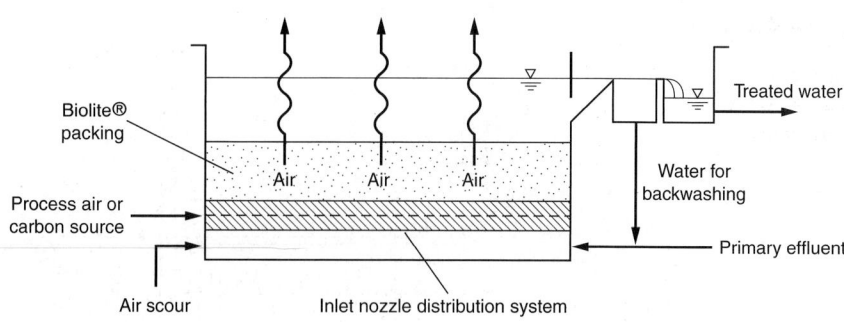

Biolite® packing

Process air or carbon source

Air scour

Inlet nozzle distribution system

Air Air Air

Treated water

Water for backwashing

Primary effluent

Table 9–18

Recommended loading limits for the Biofor® process[a]

Item	Unit	COD removal	Tertiary nitrification
Installed packing porosity	%		~40
Loading	kg COD/m³·d	10–12	
	kg N/m³·d		1.5–1.8
Hydraulic application	m³/m²·h	5.0–6.0	10–12

[a]Adapted from Pujol et al. (1994).
Note: kg/m³·d × 62.4280 = lb/10³ ft³·d.
m³/m²·h × 3.2808 = ft³/ft²·h.

packing, has been developed and installed in England by the Thames Water Utilities (Smith and Edwards, 1994).

Upflow Submerged Attached Growth Processes

Two upflow submerged attached growth processes that have been applied with success include the Bifor® and Biostyr® processes. The key features of these processes are described below.

Biofor® Process. The Biofor® process, an upflow submerged aerobic attached growth process (see Fig. 9–25), is being used at more than 100 installations in Europe and North America (Ninassi et al., 1998). The upflow reactor has a typical bed depth of 3 m but designs in the range of 2- to 4-m bed depth have been used (Pujol et al., 1994). The packing, termed Biolite®, is an expanded clay material with a density greater than 1.0 and a 2- to 4-mm size range. Inlet nozzles distribute the influent wastewater up through the bed, and an air header (Oxazur® system) provides process air across the bed area. Backwashing is typically done once per day with a water flush rate of 10 to 30 m/h to expand the bed (Lazarova et al., 2000). Fine screening of the wastewater is needed to protect the inlet nozzles. The Biofor® process has been applied for BOD removal and nitrification, tertiary nitrification, and denitrification. Recommended operating limits for the Biofor® process are shown in Table 9–18. The loadings are similar to that used for the Biocarbone® process. A study of 12 plants by Canler and Perret

Figure 9–26

Schematic of Biostyr®
process showing
arrangement for
nitrification-denitrification.

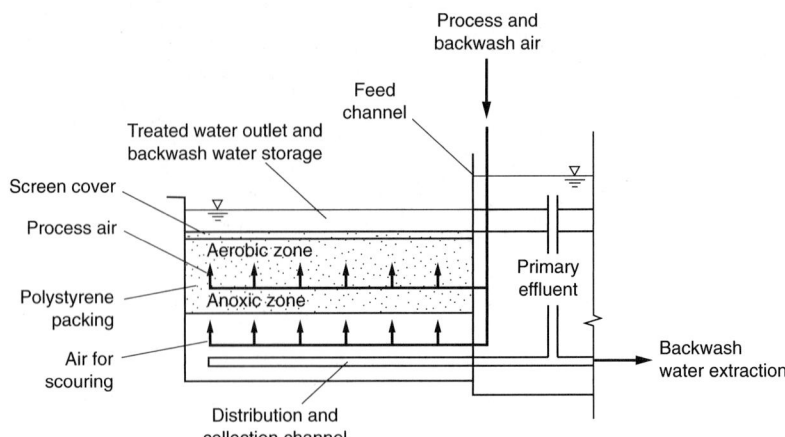

(1994) showed similar treatment performance results for COD removal as a function of the hydraulic application rate and COD loading for both the Biocarbone® and Biofor® systems.

Biostyr® Process. The Biostyr® process is an upflow process, developed in Denmark, and uses 2- to 4-mm (specific area about 1000 m^2/m^3) polystyrene beads that have a specific density less than water. The installed packing porosity is about 40 percent, providing an effective area of about 400 m^2/m^3 for biofilm growth. Packing depths range from 1.5 to 3 m. A schematic of the Biostyr® process is shown on Fig. 9–26. The bed can be operated entirely aerobic by providing air at the bottom or as an anoxic/aerobic bed by providing air at an intermediate level. Nitrified effluent is recycled for the anoxic/aerobic operation. The floating packing is retained by nozzle plates and is compressed as the wastewater flows upward to provide filtration. Backwash water is stored above the treatment bed. During the backwash cycle treated water flows down through the packing at a very high rate, which results in a downward expansion of the originally compressed packing. The solids retained in the lower portion of the reactor and excess biomass produced on the packing are flushed out into a backwash collection tank. The normal backwash procedure consists of repeated rinse (water flushing) and air scouring steps. Typically four water phases and three air phases are used (Holbrook et al., 1998; Borregaard, 1997).

The Biostyr® process has been used for BOD removal only, combined BOD removal and nitrification, tertiary nitrification, and postdenitrification. Typical loadings possible for the various types of treatment with the Biostyr® process are presented in Table 9–19 (Le Tallec et al, 1997; Borregaard, 1997; Payraudeau et al., 2000). The organic loadings are in the same range as that for the Biocarbone® and Biofor® processes. Nitrification testing work by Payraudeau et al. (2000) suggests that a higher nitrification load is possible for tertiary nitrification. At loadings of 1.5 to 1.8 kg N/m^3·d, 85 to 90 percent nitrogen oxidation was shown. Effluent performance for a number of full-scale facilities was given by Borregaard (1997), but the process loadings were not defined. Average effluent BOD, TSS, and NH$_4$-N concentrations of 7, 11, and 1.8 mg/L, respectively, were shown for long-term operation.

Table 9–19
Typical design loadings for the Biostyr® process

Application	Unit	Value
BOD only	kg COD/m³·d	8–10
BOD removal and nitrification	kg COD/m³·d	4–5
Tertiary nitrification	kg N/m³·d	1.0–1.7

Note: kg/m³·d × 62.4280 = lb/10³ ft³·d.

Figure 9–27

Schematic of fluidized-bed biological reactor (FBBR).

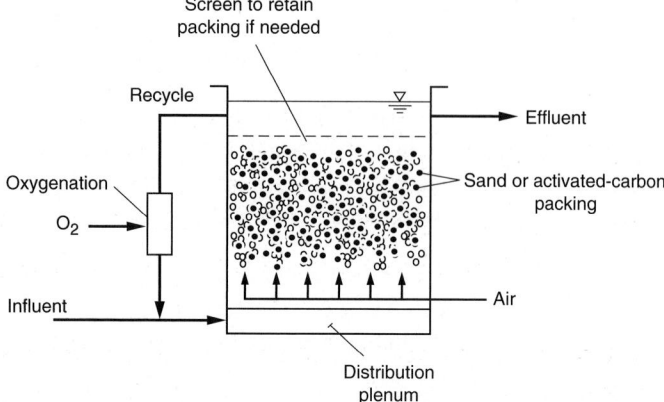

Fluidized-Bed Bioreactors (FBBR)

In fluidized-bed bioreactors the wastewater is fed upward to a bed of 0.4- to 0.5-mm sand or activated carbon. Bed depths are in the range of 3 to 4 m. The specific surface area is about 1000 m²/m³ of reactor volume, which is greater than any of the other fixed-film packing. Upflow velocities are 30 to 36 m/h. Effluent recirculation is necessary to provide the fluid velocity within the necessary treatment detention times. Hydraulic retention times in FBBRs range from 5 to 20 min. As the biofilm increases in size, the packing becomes lighter and accumulates at the top of the bed where it can be removed and agitated periodically to remove excess solids.

A schematic of an FBBR system is shown on Fig. 9–27. For aerobic applications recirculated effluent is passed through an oxygenation tank to predissolve oxygen. Adding air to the fluidized-bed reactor would discharge packing to the effluent. For municipal wastewater treatment, FBBRs have been used mainly for postdenitrification (see Sec. 9–7). Anaerobic applications of the FBBRs are described in Chap. 10. Aerobic FBBRs are frequently used to treat groundwater contaminated with hazardous substances. In these applications activated carbon is used for the packing to provide both carbon adsorption and biological degradation (Sutton and Mishra, 1994). The main advantages for the FBBR technology in this application are (1) it provides an extraordinarily long SRT for microorganisms necessary to degrade the xenobiotic and toxic compounds; (2) shock loads or nonbiodegradable toxic compounds can be absorbed onto the activated carbon; (3) high-quality effluent is produced low in TSS and COD concentration; (4) the oxygenation method prevents stripping and emission of toxic organic compounds to the atmosphere; and (5) the system operation is simple and reliable.

9-7 ATTACHED GROWTH DENITRIFICATION PROCESSES

Biological denitrification processes have been applied following secondary treatment nitrification processes to reduce the nitrate/nitrite produced. Typically, an exogenous carbon source is added to provide an electron donor that can be oxidized biologically using nitrate or nitrite. Both attached growth and suspended growth processes have been used for postanoxic denitrification with success. Suspended growth processes used for denitrification are discussed in Sec. 8–5 in Chap. 8. The different types of attached growth postanoxic processes, illustrated on Fig. 9–28, include (1) downflow and upflow packed-bed reactors, (2) upflow fluidized-bed reactors, and (3) submerged rotating biological contactors. All of the processes require carbon addition in the form of methanol to the influent. Methanol to NO_3-N dose ratios are in the range of 3.0 to 3.5 kg methanol/kg NO_3-N. The general physical designs of the downflow Biocarbone® and upflow Biofor® and Biostyr® systems have been described in Sec. 9–6. Specific design loadings used for the various systems for attached growth postanoxic denitrification are presented and discussed in this section. Attached growth preanoxic denitrification processes are also considered.

Downflow Packed-Bed Postanoxic Denitrification Processes

Downflow packed-bed denitrification filters (see Fig. 9–28a) are deep bed filters [1.2 to 2.0 m (4 to 6.5 ft)] that have been used in many installations (U.S. EPA, 1993) for

Figure 9–28

Postanoxic attached growth denitrification processes: (a) downflow packed-bed reactor, (b) upflow packed-bed reactor, (c) fluidized-bed denitrification reactor, and (d) submerged rotating biological contactor.

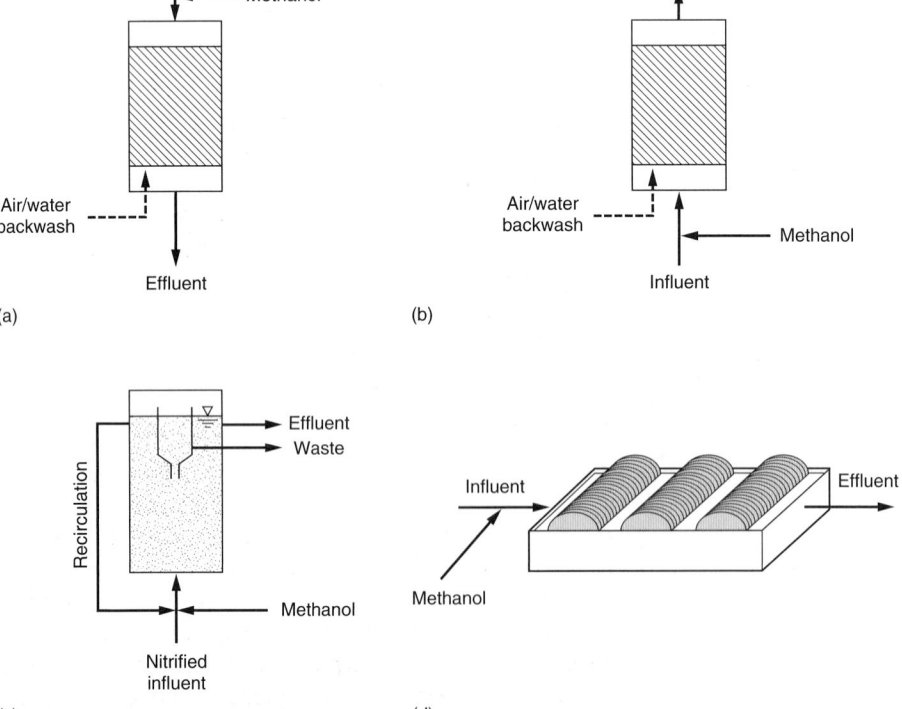

Table 9–20

Typical design criteria for downflow denitrification filters[a]

Parameter	Unit	Value Range	Value Typical
Packing:			
Type		Sand	Sand
Effective size	mm	1.8–6.0	4.0
Sphericity	unitless	0.8–0.9	0.82
Specific gravity	unitless	2.5–2.7	2.6
Depth	m	1.2–1.8	1.6
Hydraulic application rate:			
20°C	$m^3/m^2 \cdot d$	60–120	100
10°C	$m^3/m^2 \cdot d$	30–90	80
NO_3-N loading:			
20°C	$kg/m^3 \cdot d$	1.4–1.8	1.6
10°C	$kg/m^3 \cdot d$	0.8–1.2	1.0
Empty-bed contact time	min	20–30	20
Methanol to NO_3-N ratio	unitless	3.0–3.5	3.2
Air scour before backwash:			
Flowrate	$m^3/m^2 \cdot min$	1.2–1.6	1.5
Duration	s	20–40	30
Backwash:			
Air flowrate	$m^3/m^2 \cdot h$	90–120	110
Water flowrate	$m^3/m^2 \cdot h$	15–25	20
Duration	min	10–20	15
Frequency	times/d	1	1
Water flush (bump):			
Flowrate	$m^3/m^2 \cdot d$	10–14	12
Duration	min	3–5	4
Frequency	times/d	12	12

[a] Adapted from U.S. EPA (1993) and Bailey et al. (1998).
Note: $m^3/m^2 \cdot d \times 24.5424 = gal/ft^2 \cdot d$.
$kg/m^3 \cdot d \times 62.4280 = lb/10^3 \ ft^3 \cdot d$.

postanoxic nitrate removal. The denitrification filter used most commonly is a proprietary process of TETRA Technologies. The filters provide both suspended solids removal and denitrification by microbial growth on the filter packing. Typical design criteria are reported in Table 9–20. Sand is the filter packing and the size selected is small enough to provide effective filtration and sufficient surface area for microbial growth but large enough to accommodate solids capture and microbial growth without excessive headloss. At hydraulic application rates of 80 to 100 $m^3/m^2 \cdot d$ (Table 9–20), effluent TSS concentrations of < 5.0 mg/L are commonly achieved. With proper control of the methanol dose, total effluent nitrogen concentrations of 1 to 3 mg/L are possible. Overdosing of the methanol can lead to odor production, due to biological sulfate reduction in the filter.

During operation of the denitrification filter, headloss gradually increases because of solids accumulation (filtration), biomass growth, and accumulation of nitrogen gas due to denitrification. The filter is "bumped" periodically by a hydraulic surge to remove nitrogen gas and backwashed for solids removal. A water-only flush is used at a rate of 12 m/h for about 3 to 5 min for a "bump" to release the accumulated nitrogen gas. The bump frequency may vary from once every 2 to 4 h. An air and water backwash is required every 24 to 48 h, depending on the solids accumulation and headloss. The solids storage capacity is estimated to be about 4.0 kg TSS/m³ before high headloss occurs. The backwash typically consists of an air scour followed by an air and water backwash.

EXAMPLE 9–9 **Denitrification Filter Design**   Given the following flow and secondary effluent characteristics for feed to postanoxic denitrification filters, determine the following design parameters to achieve effluent TSS and NO_3-N concentrations of less than 5.0 and 2.0 mg/L, respectively. Use the typical values given in Table 9–20 for the process design.

Design parameters to be determined:

1. Filter dimensions
2. Backwash water rate
3. Backwash air rate
4. Nitrogen release backwash water bump rate
5. Methanol requirements
6. Solids production

Wastewater characteristics

Item	Unit	Value
Flowrate	m³/d	8000
TSS	g/m³	20
NO_3-N	g/m³	25
Temperature	°C	15

Note: g/m³ = mg/L

Operational parameters and assumptions

1. Filter backwash frequency = 1/d
2. Water backwash flowrate = 20 m³/m²·h (see Table 9–20)
3. Air backwash rate = 110 m³/m²·h (see Table 9–20)
4. Water flush (bump) frequency = once/2 h

5. One filter required for standby
6. Synthesis yield with methanol = 0.18 g VSS/g COD_r
7. VSS/TSS = 0.85

Solution

1. Determine the dimensions of the denitrification filter.
   a. Determine filter size based on nitrogen loading.

   NO_3-N removal = $[(25.0 - 2.0) \text{ g/m}^3]$ (8000 m^3/d) (1 kg/10^3 g)

   $$= 184 \text{ kg/d}$$

   From Table 9–20, select NO_3-N loading at 15°C of 1.3 kg/m^3·d and a depth D of 1.6 m.

   $$\text{Volume} = \frac{(184 \text{ kg/d})}{(1.3 \text{ kg/m}^3\text{·d})} = 141.5 \text{ m}^3$$

   Area = V/D = 141.5 m^3/1.6 m

   $$= 88.4 \text{ m}^2$$

   b. Determine filter size based on filtration hydraulic loading bases.

   From Table 9–20, select hydraulic application rate of 90 m^3/m^2·d

   Filter area = (8000 m^3/d)/(90 m^3/m^2·d)

   $$= 88.9 \text{ m}^2$$

   Thus, the filter size is controlled by the hydraulic application rate.

   Assume 5 filters installed with 1 filter used as standby. Use of the filters should be rotated so that active biofilm is maintained in all of the filters.

   $$\text{Area/filter} = \frac{88.9 \text{ m}^2}{4} = 22.2 \text{ m}^2$$

   Use a square configuration.

   Filter dimensions = 4.7 × 4.7 × 1.6 m

2. Determine air and water backwash flowrates and percent of product water used as backwash water.
   a. Determine backwash air flowrate at an air backwash application rate of 110 m^3/m^2·h.

   Air flowrate = (110 m^3/m^2·h) (22.2 m^2/filter) = 2442 m^3/h

   b. Determine backwash water flowrate at a water backwash application rate of 20 m^3/m^2·h.

   Water flowrate = (20 m^3/m^2·h) (22.2 m^2/filter) = 444 m^3/h

   $$= 7.4 \text{ m}^3/\text{min}$$

c. Determine volume of backwash water at 1 backwash/24 h for each filter with a duration of 15 min (Table 9–20)

Backwash water volume = (7.4 m³/min)(15 min/filter)(4 filters)

$$= 444 \text{ m}^3/\text{d}$$

d. Determine percent of product water used as backwash water.

$$(444 \text{ m}^3/\text{d})/(8000 \text{ m}^3/\text{d}) = 0.0555$$

$$= 5.6\%$$

3. Determine volume of water used to flush (bump) filters at 1 bump/2 h with a duration of 3 min/bump and a flowrate of 12 m/d.

Water volume = (88.9 m²) (12 m/d) (3 min/bump) (1 bump/2 h·filter) (24 h/d)

(1 h/60 min)

$$= 640 \text{ m}^3/\text{d}$$

Percent of product water used for nitrogen release bumping:

$$= (640 \text{ m}^3/\text{d}) / (8000 \text{ m}^3/\text{d}) = 0.08$$

$$= 8.0\%$$

4. Total product water used = 8.0 + 5.6 = 13.6%

Actual filtration rate required including product water used for backwashing and bumping:

$$(90 \text{ m}^3/\text{m}^2\text{·d})(1.136) = 102 \text{ m}^3/\text{m}^2\text{·d}$$

Note: An iterative calculation can be performed to develop an actual filtration rate approaching 90 m³/m²·d by using a lower rate in step 1b.

5. Methanol (CH₃OH) requirements.

Nitrate removal = 184 kg/d

Use 3.2 kg methanol/kg NO₃-N (see Table 9–20)

Methanol = (3.2 kg/kg)(184 kg/d) = 588.8 kg/d

6. Determine the solids production.

Solids = filtered solids + biomass production

Use effluent TSS = 5 mg/L (g/m³)(given value)

Filter solids = [(20 − 5) g/m³] (8000 m³/d) (1 kg/10³ g)

$$= 120 \text{ kg/d}$$

Biomass production:

Based on the stoichiometric balance for methanol oxidation, there are 1.5 g COD/g CH$_3$OH (CH$_3$OH + 1.5O$_2$ → CO$_2$ + 2H$_2$O)

$$\text{Biomass} = \left(\frac{0.18 \text{ g VSS}}{\text{g COD}}\right)\left(\frac{1.5 \text{ g COD}}{\text{g CH}_3\text{OH}}\right)\left(\frac{588.8 \text{ kg}}{\text{d}}\right)$$

$$= 159 \text{ kg VSS/d}$$

TSS = (159 kg VSS/d)/(0.85 g VSS/g TSS) = 187.1 kg TSS/d

Total solids = 120 kg/d + 187.1 kg/d = 307.1 kg/966d

Solids storage assuming 1 backwash/d

Filter volume = (22.2 m^2)(1.6 m)(1 backwash/d) = 36 m^3/filter·d

$$\text{Solids storage} = \left(\frac{307.1 \text{ kg}}{\text{d}}\right)\left(\frac{1}{4 \text{ filters}}\right)\left[\frac{\text{filter}}{(36 \text{ m}^3/\text{d})}\right] = 2.1 \text{ kg TSS/m}^3$$

Comment   Because the solids storage value is well below 4.0 kg TSS/m^3, backwashing may be required only once every 2 days. However, to maintain the health of the filter, backwashing once per day is recommended (see Sec. 11–4 in Chap. 11).

The downflow packed-bed Biocarbone® process, as described in Sec. 9–6, has been demonstrated at pilot scale for nitrogen removal. An effluent total nitrogen concentration of less than 8 mg/L was achieved with an NO$_3$-N loading of 4 kg NO$_3$-N/m^3·d at 10 to 12°C in a postanoxic application (Jepsen and Jansen, 1993). The effective size of the packing was 4 to 6 mm and the packing depth was 2.0 m. Nitrogen gas bumping was not advocated for this system.

Upflow Packed-Bed Postanoxic Denitrification Reactors

As illustrated on Fig. 9–28b, upflow packed-bed reactors have been demonstrated for postanoxic denitrification for both the Biostyr® and Biofor® processes. As described previously, the packing used in these processes is a 2.0-m depth of 2- to 5-mm polystyrene beads for the Biostyr® (Jepsen and Jansen, 1993) and a 2-m depth of a 2- to 4-mm expanded clay (Biolite) for the Biofor® process (Aesoy et al., 1998). Reported NO$_3$-N loading rates for these processes are in the range from 3.0 to 4.0 kg NO$_3$-N/m^3·d to achieve effluent NO$_3$-N concentrations below 5.0 mg/L (Pujol et al., 1994; Borregaard, 1997), assuming a sufficient amount of electron donor is supplied. Higher hydraulic application rates of up to 330 m^3/m^2·d have been claimed for these processes. No special backwash "bumping" is provided for nitrogen gas release during backwash as compared to the proprietary Tetra® filter.

Fluidized-Bed Reactors for Postanoxic Denitrification

In the fluidized-bed reactor (FBR), the upward flow to the reactor containing sand or other suitable packing material is at a sufficient velocity to expand and fluidize the sand

Table 9–21

Design criteria for upflow fluidized-bed reactors for postanoxic denitrification[a]

Parameter	Unit	Value Range	Typical
Packing:			
Type		Sand	Sand
Effective size	mm	0.3–0.5	0.4
Sphericity	unitless	0.8–0.9	0.8–0.85
Uniformity coefficient	unitless	1.25–1.50	≤1.4
Specific gravity	unitless	2.4–2.6	2.6
Depth	m	1.5–2.0	2.0
Bed expansion	%	75–150	100
Upflow velocity	m/min	0.60–0.70	0.6
Hydraulic application rate	$m^3/m^2{\cdot}d$	400–600	500
Recirculation ratio	unitless	2:1–5:1	2:1–5:1
NO_3-N loading:			
13°C	$kg/m^3{\cdot}d$	2.0–4.0	3.0
20°C	$kg/m^3{\cdot}d$	3.0–6.0	5.0
Empty-bed contact time	min	10–20	15
Methanol to NO_3-N ratio	unitless	3.0–3.5	3.2

[a] Adapted from U.S. EPA, 1993; Sadick et al., 1994; and Sadick et al., 1996.
Note: $m^3/m^2{\cdot}d \times 24.5424 = gal/ft^2{\cdot}d$.
$kg/m^3{\cdot}d \times 62.4280 = lb/10^3\ ft^3{\cdot}d$.

particles coated with biofilms (see Fig. 9–28c). The intense mixing provides good mass transfer and the dense biofilm that develops results in an equivalent reactor biomass concentration of 20 to 30 g/L for postdenitrification applications (U.S. EPA, 1993).

Design criteria for fluidized-bed reactors for postanoxic denitrification are summarized in Table 9–21. Effluent recirculation ratios of 2:1 to 5:1 are used to maintain the high upflow liquid velocity to fluidize the sand. Volumetric NO_3-N loadings of 2.0 to 6.0 $kg/m^3{\cdot}d$ have been shown to be feasible for denitrification applications, depending on the temperature. Empty-bed liquid retention times are only 10 to 20 min, and effluent NO_3-N concentrations of 2 to 4 mg/L have been achieved when treating municipal nitrified wastewaters (Sadick et al. 1996).

The FBR systems are operated without the need of effluent filtration or clarification. The sand packing is removed at the top of the reactor and passed through a high shear pump to separate biomass from the sand. The cleaned packing material is returned to the FBR. Packing with thicker biofilms migrates to the top of the bed due to the lower overall particle density caused by the biofilm growth. The excess biomass is removed from the top of the bed to control the bed packing depth. Various separation techniques including screens with spray washers, external cyclones, and hydraulic separators (U.S. EPA, 1993) are used to separate the sheared biomass from the packing material. In a properly designed and operated system the FBR effluent TSS concentration can be in

Table 9–22
Postanoxic
denitrification rates
in submerged RBCs[a]

Effluent NO$_3$-N concentration, mg/L	Specific denitrification rate relative to packing area, kg NO$_3$-N/10^3 m^2·d
1.0	0.40
6.0	3.25

[a] Adapted from WEF (1998).
Note: kg/10^3 m^2·d × 0.2048 = lb/10^3 ft^2·d.

the range of 15 to 20 mg/L. The influent manifold system is also a critical physical design element of the FBR. The manifold must be designed to distribute the influent flow uniformly across the fluidized bed. The Reno Sparks, NV, wastewater-treatment plant was the first full-scale installation of a FBR for postanoxic denitrification, with operation beginning in the early 1980s.

Submerged Rotating Biological Contactors

Submerged RBCs (see Fig. 9–28d) have been proposed for postanoxic denitrification applications and have shown good performance at pilot-plant scale. Denitrification rates for submerged RBCs are reported in Table 9–22.

Attached Growth Preanoxic Denitrification Processes

Preanoxic denitrification, as illustrated on Fig. 9–29a, can be used with attached growth processes. In the first case, an attached growth process is used for denitrification using organic material in the influent wastewater for the electron donor to reduce NO$_3$-N. Nitrate is provided in the recycle stream, which is pumped at a rate that is 3 to 4 times the influent flowrate. Attached growth preanoxic processes that have been used for pre-anoxic denitrification include: (1) trickling filters (Nasr et al., 2000, and Dorias and Baumenn, 1994), (2) the Biofor® process at pilot scale (Ninassi et al., 1998), and (3) a moving bed fixed film reactor (Lazarova et al., 1998). An important advantage of the preanoxic denitrification treatment scheme is that influent BOD is used for nitrate reduction, and thus the cost of methanol addition is eliminated. A disadvantage is the effect of the nitrate recycle flows on the design and operating cost. Where trickling filters are used for nitrification, the energy required for pumping the recycle and influent flow to the trickling filter increases significantly. The preanoxic loading conditions for 80 percent removal of NO$_3$-N fed to the preanoxic Biofor® reactor are summarized in Table 9–23.

An alternative approach for preanoxic denitrification is shown on Fig. 9–29b for a system with an attached growth process for nitrification. The nitrified trickling filter effluent is recycled to provide nitrate for a suspended growth preanoxic reactor (Melhart, 1994). An intermediate clarifier is used to separate the denitrifying mixed liquor and to provide return activated sludge to the anoxic tank. The recycle flow needed to provide nitrate to the preanoxic zone has a significant effect on the clarifier size and pumping requirements to the trickling filter, and thus the overall system economics.

Part of a full-scale trickling filter facility at Salisbury, MD, was modified to evaluate biological nitrogen removal by converting an existing trickling filter to a preanoxic

Figure 9–29

Preanoxic biological
nitrogen removal
processes: (a) attached
growth denitrification
process and
(b) suspended growth
denitrification process
with attached growth
nitrification.

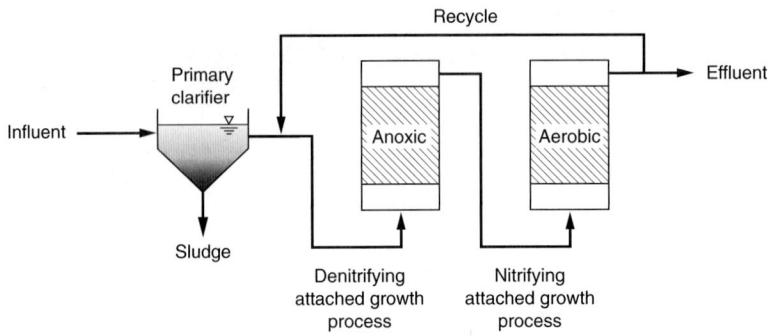

(a)

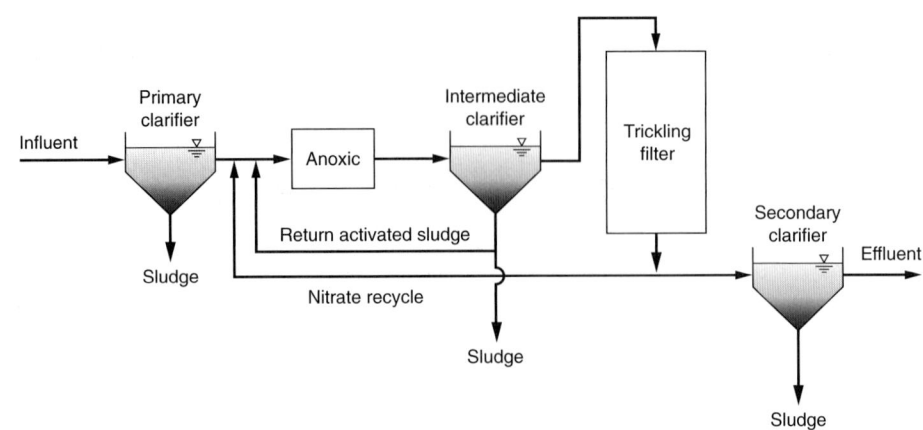

(b)

Table 9–23

Preanoxic reactor
loading conditions
for Biofor® process
in attached growth
biological nitrogen
removal treatment
system[a]

Parameter	Unit	Value	
		Range	Typical
Nitrogen loading	kg NO$_3$-N/m^3·d	1.0–1.5	1.2
Hydraulic application rate including recycle	m^3/m^2·d	480–750	600

[a] Adapted from Ninassi et al. (1998).
Note: kg/m^3·d × 62.4280 = lb/10^3 ft^3·d.
m^3/m^2·d × 24.5424 = gal/ft^2·d.

filter followed by a nitrification trickling filter (Nasr et al., 2000). Both filters were converted from rock to plastic packing with an increase in height and packing depth. The
preanoxic filter was operated as a submerged reactor, and the internal recycle ratio from
the effluent of the nitrification trickling filter to the preanoxic filter was 5:1 based on
the plant influent flowrate. With most of the BOD removal occurring in the preanoxic
filter, a higher DO concentration (3 to 6 mg/L) was obtained in the nitrification filter
effluent. The trickling filter nitrification rates were on the order of 1.5 to 2.0 g N/m^2·d

and controlled the treatment process flowrate. An effluent total nitrogen concentration of less than 8 mg/L was achieved at temperatures as low as 13°C. Long-term operating data were not sufficient to assess solids inventory and control issues in the preanoxic trickling filter. The high DO concentration in the nitrification trickling filter effluent and its effect on nitrate removal in the preanoxic filter was of concern. In another design using trickling filters, Dorias and Baumenn (1994) used a nonsubmerged preanoxic trickling filter with the recycle and influent feed flows supplied by the trickling filter distributor. A gastight cover over the trickling filter was necessary to develop sustained anoxic conditions, and the preanoxic trickling filter effluent NO_3-N concentrations ranged from 2.0 to 4.0 mg/L with specific NO_3-N removal rates of 70 to 100 $g/m^3 \cdot d$.

The Biofor® attached growth process for preanoxic denitrification and nitrification in series was demonstrated in a pilot plant with municipal wastewater (Ninassi et al, 1998). An internal recycle ratio of 2.5, based on the effluent flowrate, was used and nitrogen removal efficiency was affected by the preanoxic reactor feed bCOD/NO_3-N ratio and internal recycle ratio. A COD/N value greater than 10 was needed to maximize performance.

PROBLEMS AND DISCUSSION TOPICS

9-1 A 20-m-diameter plastic packing trickling filter, containing cross-flow plastic packing at a 6.1-m depth with a specific surface area of 90 m^2/m^3, receives domestic wastewater after primary treatment. The average flowrate is 400, 450, or 480 m^3/h (value to be selected by instructor) and the BOD concentration is 130 mg/L. Determine and compare the effluent BOD concentration and percent BOD removal at 20°C and 15°C. Assume an n value of 0.5.

9-2 Two 15-m-diameter trickling filters containing conventional cross-flow plastic packing at a 6-m depth are used to treat a pharmaceutical wastewater at an average flowrate of 2120 m^3/d. The influent BOD concentration is 500, 800, or 1000 mg/L (value to be selected by instructor) and the temperature is 20°C. Each tower has a 2-arm distributor. Determine: (a) the operating dose and flushing dose from Table 9–3 and distributor speeds in revolutions/min for each case; and (b) the recirculation ratio and total pumping rate in m^3/h to each filter.

9-3 For Example 9–4, using the same design criteria given for the trickling towers, influent BOD concentration and temperature, and effluent BOD concentration, determine the following using a plastic packing depth of 4.0, 5.0, or 6.0 m (value to be selected by the instructor):

1. Volume of plastic packing, m^3
2. Hydraulic application rate in $L/m^2 \cdot s$
3. Volumetric BOD loading in $kg/m^3 \cdot d$

Compare the results to the values determined for the same design parameters in the example.

9-4 The following data were obtained for a pilot-plant study involving the treatment of a combined domestic-industrial wastewater with a tower trickling filter filled with plastic packing at a depth of 6.1 m. The diameter of the pilot-plant tower was 1 m and the specific surface area of the plastic packing was 90 m^2/m^3. The wastewater temperature at the time of the test was 15, 18, or 22°C (value to be determined by the instructor). During the testing the influent BOD concentration was 350 mg/L. The following table summarizes the average BOD removal efficiency at different flowrates. Using these data, determine the value of the wastewater treatability coefficient, k (assume the value of n is 0.50), at the test temperature and at 20°C.

Pilot-plant test results:

Flowrate, m³/d	BOD removal efficiency, %
6	88
12	82
18	67
24	63
48	54

9-5 The following design information is given for a domestic wastewater (wastewater A, B, or C to be selected by instructor):

Design information	Wastewater		
	A	B	C
Flow, m³/d	10,000	10,000	15,000
BOD, mg/L	260	300	220
Minimum temp., °C	15	15	15
TSS, mg/L	240	280	200

Using the above information, and assuming 30 percent BOD removal in primary clarification, a packing depth of 6.1 m, and a value for n of 0.50, determine the following design parameters for a trickling filter treatment system:

1. Primary and secondary clarifier diameters (m)
2. Trickling filter tower diameter (m)
3. Packing volume (m³)
4. Recirculation ratio, if required
5. Total pumping rate (m³/h)
6. Flushing and normal dose rates (mm/pass)

9-6 Determine the diameters and packing volume of two 1.5-m deep rock trickling filters, operated in series, for a domestic wastewater treatment application with the following characteristics and requirements using the NRC model.

Parameter	Unit	Value
Flow	m³/d	5000
Influent BOD	mg/L	(A) 220
		(B) 200
		(C) 180
Temperature	°C	14

(continued)

(Continued)

Parameter	Unit	Value
Primary clarifier BOD removal efficiency	%	35
Trickling filter average effluent BOD	mg/L	20

Note: Value A, B, or C to be selected by instructor

9–7 An 18-m-diameter trickling filter tower containing conventional plastic packing at a 6.1-m depth treats a primary effluent flowrate of 7570 m³/d. The primary effluent BOD is 120, 150, or 160 mg/L (value to be selected by instructor) and the wastewater temperature is 18°C. The air temperature varies from 2 to 23°C. The BOD loading peaking factor is 1.5. Assume a factor of 1.5 for the inlet and outlet pressure losses for the tower. Determine the following:

1. The required oxygen supply rate (kg/h)
2. The required air flowrate at the warmest temperature (m³/min)
3. The airflow pressure drop across the packing (Pa).

9–8 Two 20-m-diameter plastic tower trickling filters containing 6.1 m of conventional plastic media (90 m² area/m³ volume) receive a primary clarifier effluent at an average flowrate of 11,200 m³/d. The TKN concentration is 24 mg/L and the BOD concentration is 150, 130, or 120 mg/L (value to be selected by the instructor). The temperature is 18°C. Evaluate the trickling filter BOD loading and determine the nitrogen removal efficiency due to nitrification.

9–9 Prepare a table to compare the advantages and limitations of a tower trickling filter with plastic packing versus activated-sludge treatment in terms of space, ease of process operation, sludge settling characteristics, energy requirements, maintenance, treatment flexibility, nitrification reliability, potential odors, and potential for future nitrogen or phosphorus removal.

9–10 A 20-m-diameter trickling filter tower containing a 5-m depth of high-density plastic packing (138 m² area/m³ packing volume) is used in a tertiary nitrification application. The influent flowrate is 37,000, 39,000, or 40,000 m³/d (value to be selected by the instructor) and the NH_4-N concentration is 20 mg/L. Assuming that the nitrification $r_{n,max}$ value is 1.5 g/m²·d and the K_n value is 1.5 g/m³, determine the trickling filter effluent NH_4-N concentration (mg/L). Why is no gravity settling tank used for the trickling filter effluent?

9–11 A TF/AS process is used after primary clarification, and the primary effluent wastewater characteristics are given below. For the following design parameters, compare the effect of designing the plastic tower trickling filter step for 40 percent versus 75 percent BOD removal:

1. Trickling filter diameter (m) and hydraulic application rate (L/m²·s)
2. Oxygen required in the activated-sludge aeration tank (kg/d)
3. The amount of solids wasted per day (kg/d)
4. The volume (m³) and hydraulic retention time (h) of the aeration tank.

Use the following assumptions for the trickling filter and activated-sludge designs.

Trickling filter:

Plastic packing treatability coefficient, $k_{20} = 0.18(L/s)^{0.5}/m^2$
Packing depth = 6.1 m
Number of towers = 2
50 percent of theoretical effluent BOD is soluble.

Activated sludge:

SRT = 5.0 d (no nitrification)
MLSS = 3000 mg/L
Biomass yield, Y, = 0.6 g VSS/g BOD removed
Endogenous decay, k_d=0.12 g VSS/g VSS·d
UBOD/BOD = 1.6

Wastewater characteristics:

Item	Unit	Value
Flow	m³/d	8000
BOD	mg/L	A (150)
		B (140)
		C (110)
sBOD	mg/L	80
TSS	mg/L	65
VSS	mg/L	55
nbVSS	mg/L	22
Temperature	°C	12

Note: Value A, B, or C to be selected by instructor

Which design is preferred? State your reasons.

9-12 An existing activated-sludge facility with primary treatment is operated at an SRT of 12 d and minimal temperature of 12°C so that complete nitrification is maintained. The system is operated with an MLSS concentration of 2200 mg/L, and the SVI is generally in the range of 180 to 200 mL/g. At these conditions, the treatment flow is 8,000 m³/d. The city engineer requests that your firm design a plastic tower pretreatment system for about 60 percent BOD removal before the activated-sludge unit. You are to list for the city engineer the potential impacts of converting the existing activated-sludge process to a TF/AS process, including flow capacity, solids production, oxygen requirements, energy demand, sludge settling characteristics, and effluent NH_4-N and BOD concentrations. Provide an explanation of the basis of your opinion on the potential impacts.

9-13 Prepare a plot of soluble BOD removal efficiency in percent versus contact time in minutes for a contact tank to be used in a TF/SC process. The operating conditions are:

1. Temperature = 12, 15, or 18°C (value to be selected by instructor)
2. MLVSS = 2000 mg/L
3. $k_{20} = 3.0 \times 10^{-5}$ L/mg·min

9–14 An industrial waste is to be treated with a tower trickling filter followed by an activated-sludge process (TF/AS process). The wastewater flowrate is 20,000 m³/d and is equalized. Primary settling is not used because the wastewater contains mainly soluble organic substances. The tower trickling filter contains conventional plastic packing medium and the operational SRT for the activated-sludge process is to be 5 d during the critical summer period and vary from 5 to 15 d during the winter. The lowest average sustained winter temperature (at least two weeks) is 5°C and the highest average sustained summer temperature is 26°C. The characteristics of the industrial waste, data derived from pilot-plant studies, and related design data are presented below. Using these data, size the units and determine the following:

1. Concentration of mixed liquor suspended solids to be maintained during summer and winter operation
2. Recycle rates around the filter and activated-sludge process
3. Quantity of sludge to be disposed
4. Effluent BOD concentration from the trickling and activated-sludge processes
5. Quantity of nutrients that must be added.

Wastewater characteristics:

BOD = 1200, 1500, or 1800 mg/L (value to be selected by instructor)
TSS = 100 mg/L
VSS = 0 mg/L
Total nitrogen as N = 10 mg/L
Total phosphorus as P = 4 mg/L

Trickling filter pilot-plant results:

$k_{20°C}$ = 0.075 (L/s)$^{0.5}$/m²
Net solids yield = 0.5 g VSS/g BOD removal
Temperature correction value, θ = 1.06

Activated-sludge pilot-plant results:

Solids synthesis yield, Y = 0.6 g VSS/g BOD removal
Endogenous decay, k_d = 0.12 g VSS/g VSS·d
k = 6.0 g BOD/g VSS·d
K_s = 90 mg BOD/L
θ = 1.035

Design parameters:

Trickling filter hydraulic application rate = 0.10 m³/m²·min

Activated-sludge SRT: summer = 5 d; winter = 5–15 d

9–15 The following wastewater is to be treated with an upflow attached growth process for BOD removal only (assume the Biostyr® process) to achieve an effluent BOD concentration of 20 mg/L or less. Assume a media depth of 2 m and maximum surface area of 100 m² for the Biostyr® process. Determine (a) the reactor volume and number of Biostyr® units, and (b) the area requirements of a plastic tower trickling filter and secondary clarifier to treat the same wastewater. Compare the area requirements for the two processes. Assume a plastic packing depth of 6.1 m, a packing treatability coefficient of 0.21 (L/s)$^{0.5}$/m², and 20°C.

Design wastewater characteristics:

Parameter	Unit	Value
COD	mg/L	400
Flow	m³/d	A (15,000)
		B (18,000)
		C (20,000)
BOD	mg/L	160
TSS	mg/L	80
Temperature	°C	18

Note: Value A, B, or C to be selected by instructor

9-16 A secondary effluent has the following characteristics:

Parameter	Unit	Value
Flowrate	m³/d	5000
TSS	mg/L	15
NO_3-N	mg/L	30
Temperature	°C	18

Design a downflow denitrification filter with a sand media depth of 1.6 m to reduce the NO_3-N to an effluent concentration of 4.0, 2.0, or 1.0 mg/L (value to be determined by instructor). Determine the following:

1. Denitrification filter media volume; m³
2. Filter hydraulic application rate, m³/m²·d
3. Number of filters and filter dimensions, assuming square tanks with a maximum dimension of 10 m by 10 m
4. Methanol dose in mg/L and kg/d
5. Amount of solids produced, kg/d

REFERENCES

Aesoy, A., H. Odegaard, K. Bach, R. Pujol, and M. Hamon (1998) "Denitrification in a Packed Bed Biofilm Reactor (Biofor) Experiments with Different Carbon Sources," *Water Research,* vol. 32, no. 5, pp. 1463–1470.

Albertson, O. E. (1989) "Optimizing Rotary Distributor Speed for Trickling Filters," *Water Pollution Control Federation Operators Forum,* vol. 2, no. 1.

Albertson, O. E. (1995) "Excess Biofilm Control by Distributor Speed Modulation," *Journal Environmental Engineering,* vol. 121, p. 330.

Albertson, O. E., and G. Davies (1984) "Analysis of Process Factors Controlling Performance of Plastic Bio-Packing," *Proceedings of the 57th Annual Water Pollution Control Federation Conference,* New Orleans.

Albertson, O. E., and R. N. Okey (1988) "Trickling Filters Need to Breathe Too," Presented at the Iowa Water Pollution Control Federation, Des Moines, Iowa, June 1988.

Anderson, B., H. Aspegren, D. S. Parker, and M. P. Lutz (1994) "High Rate Nitrifying Trickling Filters," *Water Science and Technology,* vol. 29, pp. 47–60.

Aryan, A. F., and S. H. Johnson (1987) "Discussion of a Comparison of Trickling Filter Packing," *Journal Water Pollution Control Federation,* vol. 59, p. 915.

Bailey, W., A. Tesfaye, T. Dakita, A. Benjamin, M. McGrath, T. Sadick, E. Daigger, and M. Tucker (1998) "Demonstration of Deep Bed Denitrification at the Blue Plains Wastewater Treatment Plant," *Proceedings of the 71st Water Environment Federation Annual Conference.*

Biesinger, M. G., H. D. Stensel, and D. Jenkins (1980) "Brewery Wastewater Treatment without Activated Sludge Bulking Problems," *Proceedings of the 34th Annual Purdue Industrial Wastewater Conference,* West Lafayette, IN.

Bogus, B. J. (1989) *A Spreadsheet Design Model for the Trickling Filter/Suspended Growth Process,* Master's Thesis, University of Washington, Department of Civil Engineering, Seattle, WA.

Boller, M., and W. Gujer (1986) "Nitrification in Tertiary Trickling Filters Followed by Deep-Bed Filters," *Water Research,* vol. 20, p. 1363.

Borregaard, V. R. (1997) "Experience with Nutrient Removal in a Fixed-Film System at Full-Scale Wastewater Treatment Plants," *Water Science and Technology,* vol. 36, pp. 129–137.

Brink, W. P., R. R. Copithorn, D. E. Schwinn, D. Sen, R. J. White, and A. Daniel (1996) "Full-Scale Experience of IFAS Processes Using Sponge Packing for BOD, Ammonia, and Nitrogen Removal," *Proceedings, 69th Water Environment Federation Annual Conference and Exposition.*

Bruce, A. M., J. C. Merkins, and B. A. O. Haynes (1975) "Pilot Studies on the Treatment of Domestic Sewage by Two-Stage Biological Nitrification—With Special Reference to Nitrification," *Water Pollution Control* (G.B.), vol. 74, pp. 80–100.

Bruce, A. M., and J. C. Merkens (1970) "Recent Studies of High Rate Biological Filtration," *Water Pollution Control,* vol. 72, pp. 499–523.

Bruce, A. M., and J. C. Merkens (1973) "Further Studies of Partial Treatment of Sewage by High-Rate Biological Filtration," *Journal Institute Water Pollution Control* (G.B.), vol. 72, no. 5, p. 449.

Canler, J. P., and J. H. Perret (1994) "Biological Aerated Filters: Assessment of the Process Based on 12 Sewage Treatment Plants," *Water Science and Technology,* vol. 29, p. 13.

Chui, P. C., Y. Terashima, J. H. Tay, and H. Ozaki (1996) "Performance of a Partly Aerated Biofilter in the Removal of Nitrogen," *Water Science and Technology,* vol. 34, no. 1–2, pp. 187–194.

Crites, R., and G. Tchobanoglous (1998) *Small and Decentralized Wastewater Management Systems,* McGraw-Hill, New York.

Daigger, G. T., and J. R. Harrison (1987) "A Comparison of Trickling Filter Packing Performance," *Journal Water Pollution Control Federation,* vol. 59, pp. 679–685.

Daigger, G. T., T. A. Heineman, G. Land, and R. S. Watson (1994) "Practical Experience with Combined Carbon Oxidation and Nitrification in Plastic Packing Trickling Filters," *Water Science and Technology,* vol. 29, p. 189.

Daigger, G. T., L. E. Norton, R. S. Watson, D. Crawford, and R. B. Sieger (1993) "Process and Kinetic Analysis of Nitrification in Coupled Trickling Filter/Activated Sludge Processes," *Water Environment Research,* vol. 65, pp. 750–758.

Dorias, B., and P. Baumenn (1994) "Denitrification in Trickling Filters," *Water Science and Technology,* vol. 30, no. 6, pp. 101–184.

Duddles, G. A., S. E. Richardson, and E. Barth (1974) "Plastic Packing Trickling Filters for Biological Nitrogen Control," *Journal Water Pollution Control Federation,* vol. 46, pp. 937–946.

Eckenfelder, W. W., Jr. (1963) "Trickling Filter Design and Performance," *Transactions. ASCE,* vol. 128.

Eckenfelder, W. W., Jr., and E. L. Barnhart (1963) "Performance of a High Rate Trickling Filter Using Selected Media," *Journal Water Pollution Control Federation,* vol. 35, no. 12.

Figueroa, L., and J. Silverstein (1991) "Pilot-Scale Trickling Filter Nitrification at the Longmont WWTP," *Proceedings Environmental Engineering Specialty Conference,* American Society of Civil Engineers, Reno, NV.

Germain, J. E. (1966) "Economical Treatment of Domestic Waste by Plastic-Medium Trickling Filters," *Journal Water Pollution Control Federation,* vol. 38, p. 192.

GLUMRB (1997) *Recommended Standards for Wastewater Facilities* (Ten States Standards), Great Lakes—Upper Mississippi River Board of State Sanitary Engineering Health Education Services Inc., Albany, NY.

Golla, P. S., M. P. Reedy, M. K. Simms, and T. J. Laken (1994) "Three Years of Full-Scale Captor® Process Operations at Moundsville WWTP," *Water Science and Technology,* vol. 29, p. 175.

Grady, C. P. L., Jr., G. T. Daigger, and H. D. Lim (1999) *Biological Wastewater Treatment,* 2d ed., revised and expanded, Marcel Dekker, New York.

Harremöes, P. (1982) "Criteria for Nitrification in Fixed-Film Reactors," *Water Science and Technology,* vol. 14, p. 167.

Harrison, J. R., and G. T. Daigger (1987) "A Comparison of Trickling Filter Packing," *Journal Water Pollution Control Federation,* vol. 59, p. 679.

Harrison, J. R., and P. L. Timpany (1988) "Design Considerations with the Trickling Filter Solids Contact Process," *Proceedings, Joint Canadian Society for Civil Engineers and American Society of Civil Engineers Environmental Engineering Conference,* Vancouver, Canada, July 1988.

Harrison, J. R., G. T. Daigger, and J. W. Filtert (1984) "A Survey of Combined Trickling Filter and Activated Sludge Processes," *Journal Water Pollution Control Federation,* vol. 56, p. 1073.

Hawkes, H. A. (1963) *The Ecology of Waste Water Treatment,* The Macmillan Co., New York, pp. 82–99.

Higgins. I. J., and R. G. Burns (1975) *The Chemistry and Microbiology of Pollution,* Academic Press, London.

Hinton, S. W., and H. D. Stensel (1994) "Oxygen Utilization of Trickling Filter Biofilms," *Journal Environmental Engineering,* vol. 120, p. 1284.

Holbrook, R. D., S.-N. Hong, S. M. Heise, and V. R. Andersen (1998) "Pilot and Full-Scale Experience with Nutrient Removal in a Fixed Film System," *Proceedings 70th Annual Water Environment Federation Conference and Exposition,* October 1998.

Howland, W. E. (1958) "Flow Over Porous Media as in a Trickling Filter," *Proceedings of the 12th Industrial Waste Conference,* p. 435, Purdue University, Lafayette, IN.

Huang, C. S., and N. E. Hopson (1974) "Nitrification Rate in Biological Processes," *Journal of Environmental Engineering,* EE2, American Society of Civil Engineers, p. 409.

Jepsen, S. E., and J. L. C. Jansen (1993) "Biological Filters for Post-Denitrification," *Water Science and Technology,* vol. 27, no. 5–6, pp. 369–379.

Jones, R. M., D. Sen, and R. Lambert (1998) "Full Scale Evaluation of Nitrification Performance in an Integrated Fixed Film Activated Sludge Process," *Water Science and Technology,* vol. 38, p. 7.

Lazarova, V., F. Ferre, and L. Duval (1998) "Innovative Approach for High Rate Nitrogen Removal: Real Time Control of the Coupling of Turbo N DN and Turbo N Biofilm Reactors," *Proceedings of the 71st Annual Water Environment Federation Conference,* New Orleans, LA.

Lazarova, V., J. Perera, M. Bowen, and P. Shields (2000) "Application of Aerated Biofilters for Production of High Quality Water for Industrial Reuse In West Berlin," *Water Science and Technology,* vol. 41, p. 417.

Le Tallec, X., S. Zeghal, A. Vidal, and A. LeSouëf (1997) "Effect of Influent Quality Variability on Biofilter Operation," *Water Science and Technology,* vol. 26, p. 111.

Matasci, R. N., C. Kaempfer, and J. A. Heidman (1986) "Full-Scale Studies of the Trickling Filter/Solids Contact Process," *Journal Water Pollution Control Federation,* vol. 58, p. 68.

Meaney, B. J., and J. E. T. Strickland (1994) "Operating Experiences with Submerged Filters for Nitrification and Denitrification," *Water Science and Technology,* vol. 29, no. 10–11, pp. 119–125.

Melhart, G. F. (1994) "Upgrading Existing Trickling Filter Plants for Denitrification," *Water Science and Technology,* vol. 30, no. 6, pp. 173–179.

Mendoza, L., and T. Stephenson (1999) "A Review of Biological Aerated Filters (BAFs) for Wastewater Treatment," *Environmental Engineering Science,* vol. 16, p. 201.

Metcalf & Eddy, Inc. (1979) *Wastewater Engineering: Treatment, Disposal, and Reuse,* 2d ed., McGraw-Hill, New York.

Mohlman, F. W. et al. (1946) "Sewage Treatment at Military Installations," National Research Council, Subcommittee Report, *Sewage Works Journal,* vol. 18, no. 5, pp. 787–1028.

Nasr, S. M., W. D. Hankins, C. Messick, and D. Winslow (2000) "Full Scale Demonstration of an Innovative Trickling Filter BNR Process," *Proceedings of 73rd Annual Water Environment Federation Conference,* Anaheim, CA.

Newbry, B. W., G. T. Daigger, and D. Taniguchi-Dennis (1988) "Unit Process Tradeoffs for Combined Trickling Filter and Activated Sludge Processes," *Journal Water Pollution Control Federation,* vol. 60, no. 10, pp. 1813–1821.

Ninassi, M. V., J. G. Peladan, and R. Pujol (1998) "Pre-denitrification of Municipal Wastewater: The Interest of Up-flow Biofiltration," *Proceedings of the 70th Annual Water Environment Federation Conference and Exposition,* October 1998.

Norris, D. P., D. S. Parker, M. L. Daniels, and E. L. Owens (1982) "Production of High Quality Trickling Filter Effluent without Tertiary Treatment," *Journal Water Pollution Control Federation,* vol. 54, p. 1087.

Okey, R. W., and O. E. Albertson (1989) "Diffusion's Role in Regulating and Masking Temperature Effects in Fixed Film Nitrification," *Journal Water Pollution Control Federation,* vol. 61, p. 500.

Pano, A., and J. E. Middlebrooks (1983) "Kinetics of Carbon and Ammonia Nitrogen Removal in RBCs," *Journal Water Pollution Control Federation,* vol. 55, p. 956.

Parker, D. (1999) "Trickling Filter Mythology," *Journal of Environmental Engineering,* vol. 105, p. 618.

Parker, D., M. Lutz, B. Anderson, and H. Aspegren (1995) "Effect of Operating Variables on Nitrification Rates in Trickling Filters," *Water Environment Research,* vol. 67, p. 1111.

Parker, D. S., and H. J. R. Bratby (2001) "Review of Two Decades of Experience with TF/SC Process," *Journal of Environmental Engineering,* vol. 127, no. 5, pp. 380–387.

Parker, D. S., and T. Richards (1986) "Nitrification in Trickling Filters," *Journal Water Pollution Control Federation,* vol. 58, p. 896.

Parker, D. S., T. Jacobs, E. Bower, D. Stowe, and G. Farmer (1997) "Maximizing Trickling Filter Nitrification Rates Through Biofilm Control: Research Review and Full-Scale Application," *Water Science Technology (G.B.),* vol. 36, no. 1, p. 255.

Parker, D. S., D. J. Kinnear, and E. J. Wahlberg (2001) "Review of Folklore in Design and Operation of Secondary Clarifiers," *Journal of Environmental Engineering,* vol. 127, pp. 476–484.

Parker, D. S., S. Krugel, and H. McConnell (1994) "Critical Process Design Issues in the Selection of the TF/SC Process for a Large Secondary Treatment Plant," *Water Science and Technology,* vol. 29, p. 209.

Parker, D. S., M. P. Lutz, and A. M. Pratt (1990) "New Trickling Filter Applications in the U.S.A.," *Water Science Technology,* vol. 22, p. 215.

Parker, D. S., L. S. Romano, and H. S. Horneck (1998) "Making a Trickling Filter/Solids Contact Process Work for Cold Weather Nitrification and Phosphorus Removal," *Water Environment Research,* vol. 70, p. 181.

Payraudeau, M., C. Paffoni, and M. Gousailles (2000) "Tertiary Nitrification in an Upflow Biofilter on Floating Packing: Influence of Temperature and COD Load," *Water Science and Technology,* vol. 41, p. 21.

Pujol, R., M. Hamon, X. Kendel, and H. Lemmel (1994) "Biofilters: Flexible, Reliable Biological Reactors," *Water Science and Technology,* vol. 29. no. 10/11, pp. 33–38.

Randall, C. W., and D. Sen (1996) "Full-Scale Evaluation of an Integrated Fixed-Film Activated Sludge (IFAS) Process for Enhanced Nitrogen Removal," *Water Science and Technology,* vol. 33, p. 155.

Richter, K.-U., and G. Kruner (1994) "Elimination of Nitrogen in Two Flooded and Statically Packed Bed Biofilters with Aerobic and Anaerobic Microsites," *Water Research,* vol. 28, no. 3, pp. 709–716.

Rogalla, F., A. LaMouche, W. Specht, and B. Kleiher (1994) "High Rate Aerated Biofilters for Plant Upgrading," *Water Science and Technology,* vol. 24, p. 81.

Rusten, B., B. G. Hellström, F. Hellström, O. Sehested, E. Skjelfoss, and B. Svendsen (2000) "Pilot Testing and Preliminary Design of Moving Bed Biofilm Reactors for Nitrogen Removal at the FREVAR Wastewater Treatment Plant," *Water Science and Technology,* vol. 41, p. 13.

Rusten, B., L. J. Horn, and H. Odegaard (1995) "Nitrification of Municipal Wastewater NO Moving-Bed Biofilm Reactors," *Water Environment Research,* vol. 67, p. 75.

Rusten, B., M. McCoy, R. Proctor, and J. G. Siljuden (1998) "The Innovative Moving Bed Biofilm Reactor/Solids Contact Reaeration Process for Secondary Treatment of Municipal Wastewater," *Water Environment Research,* vol. 70, p. 1083.

Sadick, T. E., J. P. Semon, D. Palumbo, G. T. Daigger, and P. Keenan (1994) "Demonstration of Fluidized-Bed Denitrification in the Long Island Sound Region," *Proceedings of the 67th Water Environment Federation Annual Conference.*

Sadick, T. E., J. P. Semon, D. Palumbo, P. Keenan, and G. Daigger (1996) "Fluidized-Bed Denitrification," *Water Environment and Technology,* vol. 8, no. 8, p. 81.

Schroeder, E. D., and G. Tchobanoglous (1976) "Mass Transfer Limitations in Trickling Filter Designs," *Journal Water Pollution Control Federation,* vol. 48, pp. 771–775.

Schulze, K. L. (1960) "Load and Efficiency in Trickling Filters," *Journal Water Pollution Control Federation,* vol. 33, no. 3, pp. 245–260.

Sen, D., G. D. Favren, R. R. Copithorn, and C. W. Randall (1993) "Full Scale Evaluation of Nitrification and Denitrification on Fixed Film Packing (Ringlace) for Design of Single-Sludge Nitrogen Removal System," *Proceedings of the 66th Annual Water Environment Federation Conference and Exposition.*

Sen, D., M. Pramod, and C. W. Randall (1994b) "Performance of Fixed Film Packing Integrated in Activated Sludge Reactors to Enhance Nitrogen Removal," *Water Science and Technology,* vol. 30, p. 13.

Sen, D., C. W. Randall, K. Jensen, G. Favren, R. R. Copithorn, T. A. Young, and W. P. Brink (1994a) "Design Parameters for Integrated Fixed Film Activated Sludge (IFAS) Processes to Enhance Biological Nitrogen Removal," *Proceedings 67th Water Environment Federation Annual Conference and Exposition,* October 1994.

Smith, A. J., and W. Edwards (1994) "Operating Experiences with Submerged Aerated Filters in the U.K," *Proceedings of the 66th Annual Water Environmental Federation Conference and Exposition.*

Stenquist, R. J., and R. A. Kelly (1980) *Converting Rock Trickling Filters to Plastic Packing— Design and Performance,* EPA-600/2-80-120, U.S. Environmental Protection Agency, Washington, DC.

Stenquist, R. J., D. S. Parker, and J. J. Dosh (1974) "Carbon Oxidation—Nitrification in Synthetic Packing Trickling Filters," *Journal Water Pollution Control Federation,* vol. 46, no. 10, pp. 2327–2339.

Stensel, H. D., R. C. Brenner, K. M. Lee, H. Melcer, and K. Rackness (1988) "Biological Aerated Filter Evaluation," *Journal Environmental Engineering,* vol. 14, p. 65T.

Sutton, P. M., and P. N. Mishra (1994) "Activated Carbon Based Biological Fluidized Beds for Contaminated Water and Wastewater Treatment: A State-of-the-Art Review," *Water Science and Technology,* vol. 29, p. 309.

Timpany, P. L., and J. R. Harrison (1989) "Trickling Filter Solids Contact Performance from the Operator's Perspective," *Proceedings of the 62nd Annual Water Pollution Control Federation Conference,* San Francisco.

U.S. EPA (1974) *Process Design Manual for Upgrading Existing Wastewater Treatment Plants,* Office of Technology Transfer, U.S. Environmental Protection Agency, Washington, DC.

U.S. EPA (1975) *Process Design Manual for Nitrogen Control,* U.S. EPA Technology Transfer, EPA-625/1-77-007, U.S. Environmental Protection Agency, Washington, DC.

U.S. EPA (1984) *Design Information on Rotating Biological Contactors,* EPA/600/2-84/106, U.S. Environmental Protection Agency, Cincinnati, OH.

U.S. EPA (1985) *Review of Current RBC Performance and Design Procedures,* EPA/600/2-85/033, U.S. Environmental Protection Agency, Cincinnati, OH.

U.S. EPA (1993) *Manual Nitrogen Control,* EPA/625/R-93/010, Office of Research and Development, U.S. Environmental Protection Agency, Washington, DC.

Velz, C. J. (1948) "A Basic Law for the Performance of Biological Beds," *Sewage Works Journal,* vol. 20, no. 4.

Wanner, O., and W. Gujer (1985) "Competition in Biofilms," *Water Science and Technology,* vol. 17, p. 27.

WEF (1998) *Design of Municipal Wastewater Treatment Plants, Manual of Practice 8,* Water Environment Federation, Alexandria, VA.

WEF (2000) *Aerobic Fixed-Growth Reactors; A Special Publication,* Water Environment Federation, Alexandria, VA.

WPCF (1988) *O&M of Trickling Filters, RBCs, and Related Processes, Manual of Practice OM-10,* Water Pollution Control Federation, Alexandria, VA.

10

Anaerobic Suspended and Attached Growth Biological Treatment Processes

Anaerobic treatment processes include anaerobic suspended growth, upflow and downflow anaerobic attached growth, fluidized-bed attached growth, upflow anaerobic sludge blanket (UASB), anaerobic lagoons, and membrane separation anaerobic processes. Anaerobic suspended growth processes associated with biological phosphorus removal are discussed in Chap. 8. The anaerobic digestion of sludge is presented in Chap. 14. The purpose of this chapter is to present process designs for other anaerobic treatment processes used to remove organic material from liquid streams. The various types of processes are described along with their typical design loadings and treatment process capabilities. Before considering the individual anaerobic treatment processes, it will be helpful to consider the rationale for the use of anaerobic treatment processes.

10-1 THE RATIONALE FOR ANAEROBIC TREATMENT

The rationale for and interest in the use of anaerobic treatment processes can be explained by considering the advantages and disadvantages of these processes. The principal advantages and disadvantages of anaerobic treatment are listed in Table 10–1 and are discussed below.

Advantages of Anaerobic Treatment Processes

Of the advantages cited in Table 10–1, energy considerations, lower biomass yield, fewer nutrients required, and higher volumetric loadings are examined further in the following discussion.

Energy Considerations. Anaerobic processes may be net energy producers instead of energy users, as is the case for aerobic processes. An energy balance comparison for a high-strength wastewater at 20°C is presented in Table 10–2. For the con-

Table 10-1
Advantages and disadvantages of anaerobic processes compared to aerobic processes

Advantages	• Less energy required
	• Less biological sludge production
	• Fewer nutrients required
	• Methane production, a potential energy source
	• Smaller reactor volume required
	• With acclimation most organic compounds can be transformed
	• Rapid response to substrate addition after long periods without feeding
Disadvantages	• Longer start-up time to develop necessary biomass inventory
	• May require alkalinity and/or specific ion addition
	• May require further treatment with an aerobic treatment process to meet discharge requirements
	• Biological nitrogen and phosphorus removal is not possible
	• Much more sensitive to the adverse effect of lower temperatures on reaction rates
	• May be more susceptible to upsets due to toxic substances
	• Potential for production of odors and corrosive gases

Table 10–2
Comparison of energy balance for aerobic and anaerobic processes for the treatment of a wastewater with the following characteristics: wastewater flowrate = 100 m³/d; wastewater strength = 10 kg/m³; and temperature = 20°C

Energy	Value, kJ/d	
	Anaerobic	**Aerobic**
Aeration[a,b]		-1.9×10^6
Methane produced[c,d]	12.5×10^6	
Increase wastewater temperature to 30°C	-4.2×10^6	
Net energy, kJ/d	8.3×10^6	-1.9×10^6

[a] Oxygen required = 0.8 kg/kg COD removed.
[b] Aeration efficiency = 1.52 kg O_2/kWh and 3600 kJ = 1 kWh.
[c] Methane production = 0.35 m³/kg COD removed.
[d] Energy content of methane = 35,846 kJ/m³ (at 0°C and 1 atm).

ditions given in Table 10–2, the aerobic process requires 1.9×10^6 kJ/d. On the other hand, the anaerobic process produces a total of 12.5×10^6 kJ/d. Of the total energy produced anaerobically, about 4.2×10^6 kJ/d is required to raise the temperature of the wastewater from 20 to 30°C, the low end of the mesophilic temperature range, a more desirable temperature for anaerobic treatment. Thus, the potential net energy production that can be achieved with anaerobic treatment is on the order of 8.3×10^6 kJ/d, or about 4 times the energy required for aerobic treatment.

The wastewater strength is important for comparing energy balances for aerobic and anaerobic processes, where the wastewater temperature must be increased. With the same assumptions used to generate the energy balance presented in Table 10–2, both the aerobic and anaerobic processes would require the same amount of energy input if the wastewater biodegradable COD concentration is 1270 mg/L. At lower COD concentrations, the aerobic process requires less energy. However, heat recovery from the anaerobic effluent stream can modify these values. Further, the lower biomass yield discussed below is still a major advantage offered by anaerobic treatment.

Lower Biomass Yield. Because the energetics of anaerobic processes result in lower biomass production by a factor of about 6 to 8 times, sludge processing and disposal costs are reduced greatly. The major environmental and economic issues associated with the reuse and disposal of biomass produced from aerobic processes are discussed in Chap. 14. The fact that less sludge is produced in anaerobic treatment is a significant advantage over aerobic treatment.

Fewer Nutrients Required. Many industrial wastewaters lack sufficient nutrients to support aerobic growth. The cost for nutrient addition is much less for anaerobic processes because less biomass is produced.

Higher Volumetric Loadings. Anaerobic processes generally have higher volumetric organic loads than aerobic processes, so smaller reactor volumes and less space may be required for treatment. Organic loading rates of 3.2 to 32 kg COD/m³·d may be used for anaerobic processes, compared to 0.5 to 3.2 kg $COD/m^3 \cdot d$ for aerobic processes (Speece, 1996).

Disadvantages of Anaerobic Treatment Processes

Potential disadvantages also exist for anaerobic processes as reported in Table 10–1. Operational considerations, the need for alkalinity addition, and the need for further treatment are highlighted further in the following discussion.

Operational Considerations. The major concerns with anaerobic processes are their longer start-up time (months for anaerobic versus days for aerobic processes), their sensitivity to possible toxic compounds, operational stability, the potential for odor production, and corrosiveness of the digester gas. However, with proper wastewater characterization and process design these problems can be avoided and/or managed.

Need for Alkalinity Addition. The most significant negative factor that can affect the economics of anaerobic versus aerobic treatment is the possible need to add alkalinity. Alkalinity concentrations of 2000 to 3000 mg/L as $CaCO_3$ may be needed in anaerobic processes to maintain an acceptable pH with the high gas phase CO_2 concentration. If this amount of alkalinity is not available in the influent wastewater or cannot be produced by the degradation of proteins and amino acid, a significant cost may be incurred to purchase alkalinity, which can affect the overall economics of the process.

Need for Further Treatment. Anaerobic processes can also be followed by aerobic processes for effluent polishing to utilize the benefits of both processes. Series reactors of anaerobic-aerobic processes have been shown feasible for treating municipal wastewaters in warmer climates resulting in lower energy requirements and less sludge production (Goncalves and Avaujo, 1999; Garuti et al., 1992).

Summary Assessment

In general, for municipal wastewaters with lower concentrations of biodegradable COD, lower temperatures, higher effluent quality needs, and nutrient removal requirements, aerobic processes are favored at present. For industrial wastewaters with much higher biodegradable COD concentrations and elevated temperatures, anaerobic processes may be more economical. In the future, as more is learned about anaerobic treatment processes, it is anticipated that their use will become more widespread in a variety of applications.

10–2 GENERAL DESIGN CONSIDERATIONS FOR ANAEROBIC TREATMENT PROCESSES

The type of wastewater and its characteristics are important in the evaluation and design of anaerobic processes. The characteristics presented here apply to the suspended growth, sludge blanket, attached growth, and membrane separation anaerobic processes presented in subsequent sections. Important factors and wastewater characteristics that need to be considered in the evaluation of anaerobic processes for wastewater treatment are discussed below.

Table 10–3
Examples of types of wastewater treated by anaerobic processes

Alcohol distillation	Landfill leachate
Breweries	Pharmaceuticals
Chemical manufacturing	Pulp and paper
Dairy and cheese processing	Slaughterhouse and meatpacking
Domestic wastewater	Soft drink beverages
Fish and seafood processing	Sugar processing

Characteristics of the Wastewater

A wide variety of wastewaters have been treated by anaerobic processes including those reported in Table 10–3. Anaerobic processes are attractive, especially for high strength and warm temperature wastewaters because: (1) aeration is not required, thus saving energy cost, and (2) the low amount of solids generated. Food processing and distillery wastewaters, for example, can have COD concentrations ranging from 3000 to 30,000 mg/L. Other considerations that may apply to different wastewater sources are the presence of potential toxic streams, flow variations, inorganic concentrations, and seasonal load variations. Anaerobic processes are capable of responding quickly to wastewater feed after long periods without substrate addition. In some cases with warmer climates, anaerobic treatment has also been considered for municipal wastewater treatment.

Flow and Loading Variations. Wide variations in influent flow and organic loads can upset the balance between acid fermentation and methanogenesis in anaerobic processes. For soluble, easily degradable substrates, such as sugars and soluble starches, the acidogenic reactions can be much faster at high loadings and may increase the reactor volatile fatty acids (VFA) and hydrogen concentrations and depress the pH. Higher hydrogen concentrations can inhibit propionic and butyric acid conversion. The lower pH can inhibit methanogenesis. Flow equalization or additional capacity must be provided to meet peak flow and loading conditions.

Organic Concentration and Temperature. As discussed in Sec. 10–1, the wastewater strength and temperature greatly affect the economics and feasibility of anaerobic treatment. Reactor temperatures of 25 to 35°C are generally preferred to support more optimal biological reaction rates and to provide more stable treatment. Generally, COD concentrations greater than 1500 to 2000 mg/L are needed to produce sufficient quantities of methane to heat the wastewater without an external fuel source. At 1300 mg/L COD or less, aerobic treatment may be the preferred selection.

Anaerobic treatment can be applied at lower temperatures and has been sustained at 10 to 20°C in suspended and attached growth reactors. At the lower temperatures, slower reaction rates occur and longer SRTs, larger reactor volumes, and lower organic COD loadings are needed (Banik and Dague, 1996; Collins et al., 1998). Further, at temperatures in the range from 10 to 20°C, the degradation of long chain fatty acids is often rate limiting. If long chain fatty acids accumulate, foaming may occur in the reactor. When higher SRTs are needed, the solids loss in an anaerobic reactor can become a critical limiting factor. Anaerobic reactors generally produce more dispersed, less flocculent solids than aerobic systems, with effluent TSS concentrations for suspended growth processes in the 100 to 200 mg/L range. For dilute wastewaters, the effluent TSS concentration will limit the possible SRT of the process and treatment potential. Either a lower treatment performance occurs or it is necessary to operate the reactor at a higher temperature. Thus, the method used to retain solids in the anaerobic reactor is important in the overall process design and performance.

Fraction of Nondissolved Organic Material. The composition of the wastewater in terms of its particulate and soluble fractions affects the type of anaerobic reactor selected and its design. Wastewaters with high solids concentrations are treated more appropriately in suspended growth reactors than by upflow or downflow attached growth processes. Where greater conversion of particulate organic matter is required, longer SRT values may be needed if solids hydrolysis is the rate-limiting step as compared to acid fermentation or methanogenesis in anaerobic treatment.

Wastewater Alkalinity. With the high CO_2 content (typically in the range from 30 to 50 percent) in the gas produced in anaerobic treatment, alkalinity concentrations in the range from 2000 to 4000 mg/L as $CaCO_3$ are typically required to maintain the pH at or near neutral. The level of alkalinity needed is seldom available in the influent wastewater, but may be generated in some cases by the degradation of protein and amino acids (e.g., meatpacking wastewaters). The requirement to purchase chemicals for pH control can have a significant impact on the economics of anaerobic treatment.

The relationship between pH and alkalinity as outlined in Appendix F is controlled by the bicarbonate chemistry as follows:

$$\frac{[HCO_3^-][H^+]}{[H_2CO_3]} = K_{a1} \tag{10-1}$$

where K_{a1} = first acid dissociation constant, which is a function of ionic strength and temperature

The carbonic acid (H_2CO_3) concentration is determined using Henry's law [Eq. (2–46)] and the partial pressure of the CO_2 in the atmosphere above the water.

$$x_g = \frac{P_T}{H} p_g \tag{2-46}$$

where x_g = mole fraction of gas in water, mole gas/mole water

$$= \frac{\text{mole gas } (n_g)}{\text{mole gas } (n_g) + \text{mole water } (n_w)}$$

P_T = total pressure, usually 1.0 atm

H = Henry's law constant, $\dfrac{\text{atm (mole gas/mole air)}}{\text{(mole gas/mole water)}}$

p_g = mole fraction of gas in air, mole gas /mole of air

Once the carbonic acid concentration is known, the bicarbonate (HCO_3^-) alkalinity needed to maintain the required pH can be estimated. The use of the above equations is illustrated in Example 10–1.

EXAMPLE 10–1 **pH and Alkalinity in an Anaerobic Process** Determine the alkalinity required in kg $CaCO_3$/d to maintain a pH value of 7.0 in an anaerobic suspended growth process at 35°C, with a 30 percent CO_2 content in the gas above the water. The influent wastewater flowrate is 2000 m^3/d, the alkalinity is 400 mg/L as $CaCO_3$, and no alkalinity producing substances are present. At 35°C, Henry's constant for CO_2, computed using Eq. (2–48) and the data given in Table 2–8, is 2092 atm and the value of K_{a1} is 4.85×10^{-7} mole/L (see Table F–2 in Appendix F).

Solution

1. Determine the concentration of HCO_3^- required to maintain the pH at or near a value of 7.0.

   a. Determine the concentration of H_2CO_3 using Eq. (2–47).

   $$x_{H_2CO_3} = \frac{P_g}{H} = \frac{(1\ atm)(0.30)}{2092\ atm} = 1.434 \times 10^{-4}$$

   Because 1 liter of water contains 55.6 mole [1000 g/(18 g/mole)], the mole fraction of H_2CO_3 is equal to

   $$x_{H_2CO_3} = \frac{mole\ gas\ (n_g)}{mole\ gas\ (n_g) + mole\ water\ (n_w)}$$

   $$1.434 \times 10^{-4} = \frac{[H_2CO_3]}{[H_2CO_3] + [55.6\ mole/L]}$$

   Because the number of moles of dissolved gas in a liter of water is much less than the number of moles of water,

   $$[H_2CO_3] \approx (1.434 \times 10^{-4})[55.6\ mole/L] \approx 7.97 \times 10^{-3}\ mole/L$$

   b. Determine the concentration of HCO_3^- required to maintain the pH at or near a value of 7.0 using Eq. (10–1).

   $$[HCO_3^-] = \frac{[4.85 \times 10^{-7}\ mole/L][7.97 \times 10^{-3}\ mole/L]}{[10^{-7}\ mole/L]}$$

   $$= 0.0387\ mole/L$$

   $$HCO_3^- = (0.0387\ mole/L)(61\ g/mole)(10^3\ mg/g) = 2361\ mg/L$$

2. Determine the amount of alkalinity required per day.

   $$Equivalents\ of\ HCO_3^- = \frac{(2.361\ g/L)}{(61\ g/eq)} = 0.0387\ eq/L$$

   $$1\ eq\ CaCO_3 = \frac{mw}{2} = \frac{(100\ g/mole)}{2} = 50\ g\ CaCO_3/eq$$

   $$Alkalinity\ as\ CaCO_3 = (0.0387\ eq/L)\ (50\ g/eq)(10^3\ mg/g)$$

   $$= 1925\ mg/L\ as\ CaCO_3$$

$$\text{Alkalinity needed} = (1935 - 400) \text{ mg/L}$$
$$= 1535 \text{ mg/L as CaCO}_3$$
$$\text{Daily alkalinity addition} = (1535 \text{ g/m}^3)(2000 \text{ m}^3/\text{d})(1 \text{ kg}/10^3 \text{ g})$$
$$= 3070 \text{ kg/d}$$

Comment Based on the results of the above analysis, it is clear that a large quantity of alkalinity may be required and, as a consequence, a significant cost can be incurred. Because the addition of lime for alkalinity can lead to the formation of precipitates, the preferred form of alkalinity to be added is sodium bicarbonate.

The results of similar calculations to those presented in Example 10–1 for different temperatures and gas phase CO_2 concentrations are reported in Table 10–4. The data presented in Table 10–4 were derived using the carbonate equilibrium constants given in Table 6–16 in Chap. 6 and Henry's constants derived from the data given in Table 2–8 in Chap. 2. The values presented in Table 10–4 can be used to estimate the alkalinity requirements. For wastewaters with a higher total dissolved solids concentration and ionic strength, the alkalinity requirements will generally be much greater.

Nutrients. Though anaerobic processes produce less sludge and thus require less nitrogen and phosphorus for biomass growth, many industrial wastewaters may lack sufficient nutrients. Thus, the addition of nitrogen and/or phosphorus may be needed. Depending on the characteristics of the substrate and the SRT value, typical nutrient requirements for nitrogen, phosphorus, and sulfur are in the range from 10 to 13, 2 to 2.6, and 1 to 2 mg per 100 mg of biomass, respectively. The values for nitrogen and phosphorus are consistent with the values for these constituents estimated on the basis of the composition of the cell biomass (see Sec. 8–3 in Chap. 8). Further, to maintain maximum methanogenic activity, liquid phase concentrations of nitrogen, phosphorus, and sulfur on the order of 50, 10, and 5 mg/L, respectively are desirable (Speece, 1996).

Macronutrients. The importance of trace metals to stimulate methanogenic activity has been noted and discussed by Speece (1996). The recommended requirements for iron, cobalt, nickel, and zinc are 0.02, 0.004, 0.003, and 0.02 mg/g acetate produced,

Table 10–4
Estimated minimum alkalinity as $CaCO_3$ required to maintain a pH of 7.0 as a function of temperature and percent carbon dioxide during anaerobic digestion

Temperature, °C	Gas phase CO_2, %			
	25	30	35	40
20	900	1050	1200	1400
25	1100	1300	1500	1700
30	1300	1600	1800	2100
35	1500	1800	2100	2400
40	1700	2100	2400	2800

respectively. Examples of increased anaerobic activity were noted after trace additions of iron, nickel, or cobalt. The exact amounts of trace nutrients needed can vary for different wastewaters, and thus trial approaches are used to assess their benefit for anaerobic processes with high VFA concentrations. A recommended dose of trace metals per liter of reactor volume is 1.0 mg $FeCl_2$, 0.1 mg $CoCl_2$, 0.1 mg $NiCl_2$, and 0.1 $ZnCl_2$.

Inorganic and Organic Toxic Compounds. Proper analysis and treatability studies are needed to assure that a chronic toxicity does not exist for wastewater treated by anaerobic processes. At the same time, the presence of a toxic substance does not mean the process cannot function. Some toxic compounds inhibit anaerobic methanogenic reaction rates, but with a high biomass inventory and low enough loading, the process can be sustained. Toxic and inhibitory inorganic and organic compounds of concern for anaerobic processes are presented in Tables 10–5 and 10–6, respectively.

Acclimatization to toxic concentrations has also been shown (Speece, 1996). Pretreatment steps may be used to remove the toxic constituents, and, in some cases, phase separation can prevent toxicity problems by providing for degradation of the toxic constituents in the acid phase, before exposure of the more sensitive methanogenic bacteria to the toxic constituents (Lettinga and Hulshoff Pol, 1991).

Solids Retention Time

The solids retention time is a fundamental design and operating parameter for all anaerobic processes. In general, SRT values greater than 20 d are needed for anaerobic processes at 30°C for effective treatment performance, with much higher SRT values at lower temperatures.

Table 10–5
Toxic and inhibitory inorganic compounds of concern for anaerobic processes[a]

Substance	Moderately inhibitory concentration, mg/L	Strongly inhibitory concentration, mg/L
Na^+	3500–5500	8,000
K^+	2500–4500	12,000
Ca^{2+}	2500–4500	8,000
Mg^{2+}	1000–1500	3,000
Ammonia-nitrogen NH_4^+	1500–3000	3,000
Sulfide, S^{2-}	200	200
Copper, Cu^{2+}		0.5 (soluble)
		50–70 (total)
Chromium, Cr(VI)		3.0 (soluble)
		200–250 (total)
Chromium, Cr(III)		180–420 (total)
		2.0 (soluble)
Nickel, Ni^{2+}		30.0 (total)
Zinc, Zn^{2+}		1.0 (soluble)

[a] From Parkin and Owen (1986).

Table 10–6
Toxic and inhibitory organic compounds of concern for anaerobic processes[a]

Compound	Concentration resulting in 50 percent reduction in activity, mM[b]
1-Chloropropene	0.1
Nitrobenzene	0.1
Acrolein	0.2
1-Chloropropane	1.9
Formaldehyde	2.4
Lauric acid	2.6
Ethyl benzene	3.2
Acrylonitrile	4
3-Chlorol-1,2-propanediol	6
Crotonaldehyde	6.5
2-Chloropropionic acid	8
Vinyl acetate	8
Acetaldehyde	10
Ethyl acetate	11
Acrylic acid	12
Catechol	24
Phenol	26
Aniline	26
Resorcinol	29
Propanol	90

[a] From Parkin and Owen (1986).
[b] mM = millimole.

Expected Methane Gas Production

Higher-strength wastewaters will produce a greater amount of methane per volume of liquid treated to provide a relatively higher amount of energy to raise the liquid temperature, if needed. As derived in Sec. 7–12 in Chap. 7, the amount of methane (CH_4) produced per unit of COD converted under anaerobic conditions is equal to 0.35 L CH_4/g COD at standard conditions (0°C and 1 atm). The quantity of methane at other than standard conditions is determined by using the universal gas law [Eq. (2–44)] to determine the volume of gas occupied by one mole of CH_4 at the temperature in question.

$$V = \frac{nRT}{P} \qquad (2\text{–}44)$$

where V = volume occupied by the gas, L
$\quad n$ = moles of gas, mole

R = universal gas law constant, 0.082057 atm·L/mole·K
T = temperature, K (273.15 + °C)
P = absolute pressure, atm

Thus, at 35°C, the volume occupied by one mole of CH_4 is

$$V = \frac{(1\ \text{mole})(0.082057\ \text{atm·L/mole·K})[(273.15 + 35)\text{K}]}{1.0\ \text{atm}} = 25.29\ \text{L}$$

Because the COD of one mole of CH_4 is equal to 64 g, the amount of CH_4 produced per unit of COD converted under anaerobic conditions at 35°C is equal to 0.40 L as determined below.

(25.29 L)/(64 g COD/mole CH_4) = 0.40 L CH_4/g COD

If the composition of the waste is known, and neglecting the amount of the constituent used for cell synthesis, the following relationship, first proposed by Buswell and Boruff (1932) and subsequently extended by Sykes (2000), can be used to estimate the amount of methane (CH_4), carbon dioxide (CO_2), ammonia (NH_3), and hydrogen sulfide (H_2S) that will be produced under anaerobic conditions.

$$C_v H_w O_x N_y S_z + \left(v - \frac{w}{4} + \frac{x}{2} + \frac{3y}{4} + \frac{z}{2}\right) H_2O \rightarrow$$

$$\left(\frac{v}{2} + \frac{w}{8} + \frac{x}{4} + \frac{3y}{8} + \frac{z}{4}\right) CH_4 + \left(\frac{v}{2} - \frac{w}{8} + \frac{x}{4} + \frac{3y}{8} + \frac{z}{4}\right) CO_2 + yNH_3 + zH_2S$$

$$(10\text{–}2)$$

The gaseous ammonia (NH_3) that is formed will react with the carbon dioxide to form the ammonium ion and bicarbonate according to the following relationship.

$$NH_3 + H_2O + CO_2 \rightarrow NH_4^+ + HCO_3^- \tag{10–3}$$

The reaction given by Eq. (10–3) is representative of the formation of alkalinity under anaerobic conditions, due to the conversion of organic compounds containing proteins (i.e., nitrogen). The expected mole fractions of methane, carbon dioxide, and hydrogen sulfide are given by the following three expressions, respectively. In general, the mole fraction of hydrogen sulfide will be somewhat less because of metal complexation/precipitation.

$$f_{CO_2} = \frac{4v - w + 2x - 5y + 2z}{8(v - y + z)} \tag{10–4}$$

$$f_{CH_4} = \frac{4v + w - 2x - 5y - 2z}{8(v - y + z)} \tag{10–5}$$

$$f_{H_2S} = \frac{z}{8(v - y + z)} \tag{10–6}$$

As noted previously, the percentage of carbon dioxide in the gas can be as high as 50 percent. For carbohydrate and starch wastes, alkalinity will be a problem.

Treatment Efficiency Needed

Anaerobic treatment processes are capable of high COD conversion efficiency to methane with minimal biomass production. At SRT values greater than 20 to 50 d, maximum conversion of solids may occur at temperatures above 25°C. However, high-effluent suspended solids (50 to 200 mg/L) are common for anaerobic processes. Without pilot-plant studies and extreme measures to control effluent suspended solids concentrations, such as chemical flocculation or membrane separation, anaerobic processes alone cannot be depended on to achieve secondary treatment levels. Some form of aerobic treatment would be necessary to provide effluent polishing, either attached growth or suspended growth processes. For high-strength wastewaters the combination of anaerobic and aerobic treatment can be economical (Obayashi et al., 1981).

Sulfide Production

Oxidized sulfur compounds, such as sulfate, sulfite, and thiosulfate, may be present in significant concentrations in various industrial wastewaters and to some degree in municipal wastewaters. These compounds can serve as electron acceptors for sulfate-reducing bacteria, which consume organic compounds in the anaerobic reactor and produce hydrogen sulfide (H_2S). For example, using methanol as the electron donor and an f_s value of 0.05 (see Sec. 7–4 in Chap. 7), the overall reaction for the reduction of sulfate to H_2S can be illustrated by the following expression:

$$0.119SO_4^{2-} + 0.167CH_3OH + 0.010CO_2 + 0.003NH_4^+ + 0.003HCO_3^- +$$

$$0.178H^+ \rightarrow 0.003C_5H_7NO_2 + 0.060H_2S + 0.060HS^- + 0.331H_2O \qquad (10\text{–}7)$$

From Eq. (10–7), the amount of COD used for sulfate reduction is 0.89 g COD/g sulfate, which is in the range of 0.67 g COD/g sulfate reduced as reported by Arceivala (1998). The higher value is due to the lower biomass yield coefficient associated with methanol oxidation. Based on the following stoichiometry for H_2S oxidation, 2 moles of oxygen are required per mole of H_2S, as was the case for methane oxidation,

$$H_2S + 2O_2 \rightarrow H_2SO_4 \qquad (10\text{–}8)$$

Thus, the amount of H_2S produced per unit COD is the same as that for methane (0.40 L H_2S/g COD used at 35°C).

Hydrogen sulfide is malodorous and corrosive to metals. Combustion products formed from sulfur oxidation are considered air pollutants. In contrast to methane, H_2S is highly soluble in water, with a solubility of 2650 mg/L at 35°C, for example.

The concentration of oxidized sulfur compounds in the influent wastewater to an anaerobic treatment process is important, as high concentrations can have a negative effect on anaerobic treatment. Sulfate-reducing bacteria compete with the methanogenic bacteria for COD and thus can decrease the amount of methane gas production. While low concentrations of sulfide (less than 20 mg/L) are needed for optimal methanogenic activity, higher concentrations can be toxic (Speece, 1996). Methanogenic activity has been decreased by 50 percent or more at H_2S concentrations ranging from 50 to 250 mg/L (Arceivala, 1998). A comprehensive evaluation of the dynamics of competition between sulfate-reducing and methanogenic bacteria and toxicity effects is given in Maillacheruvu et al. (1993).

Because un-ionized H_2S is considered more toxic than ionized sulfide, pH is important in determining H_2S toxicity. The degree of H_2S toxicity is also complicated by the type of anaerobic biomass present (granular versus dispersed), the particular methanogenic population, and the feed COD/SO_4 ratio. With higher COD concentrations, more methane gas is produced to dilute the H_2S and transfer more H_2S to the gas phase. Hydrogen sulfide exists in aqueous solution as either the hydrogen sulfide gas (H_2S), the ion (HS^-), or the sulfide ion (S^{2-}), depending on the pH of the solution, in accordance with the following equilibrium reactions:

$$H_2S \leftrightarrow HS^- + H^+ \tag{10–9}$$

$$HS^- \leftrightarrow S^{2-} + H^+ \tag{10–10}$$

Applying the law of mass action to Eqs. (10–9) to (10–10) yields

$$\frac{[HS^-][H^+]}{[H_2S]} = K_{a1} \quad \text{and} \quad \frac{[S^{2-}][H^+]}{[HS^-]} = K_{a2} \tag{10–11}$$

where K_{a1} = first acid dissociation constant at 25°C, 1×10^{-7} mole/L
K_{a2} = second acid dissociation constant 25°C, $\sim 10^{-19}$ mole/L (value uncertain)

The percentage H_2S as a function of the pH can be determined using the following relationship:

$$H_2S, \% = \frac{[H_2S] \times 100}{[H_2S] + [HS^-]} = \frac{100}{1 + [HS^-]/[H_2S]} = \frac{100}{1 + K_{a1}/[H^+]} \tag{10–12}$$

Dissociation constants for hydrogen sulfide as a function of temperature are presented in Table 10–7, along with values for ammonia. As illustrated on Fig. 10–1, at a pH value of 7, at 30°C, about 60 percent of the total H_2S is present as gaseous H_2S.

Ammonia Toxicity

Ammonia toxicity may be of concern for anaerobic treatment of wastewaters containing high concentrations of ammonium or proteins and/or amino acids, which can be degraded to produce ammonium. Free ammonia (NH_3), at high enough concentrations, is considered toxic to methanogenic bacteria. As described in Chap. 2, ammonia is a weak base and dissociates in water to form ammonium (NH_4^+) and hydroxyl ions. The

Table 10–7 Acid equilibrium constants for hydrogen sulfide (H_2S) and ammonia (NH_3)	Temperature, °C	$K_{a1,H_2S} \times 10^7$, mole/L	$K_{a,NH_3} \times 10^{10}$, mole/L
	0	0.262	7.28
	10	0.485	6.37
	20	0.862	5.84
	25	1.000	5.62
	30	1.48	5.49
	40	2.44	5.37

Figure 10–1

Percent of hydrogen
sulfide present as H₂S
and HS⁻ as a function
of pH.

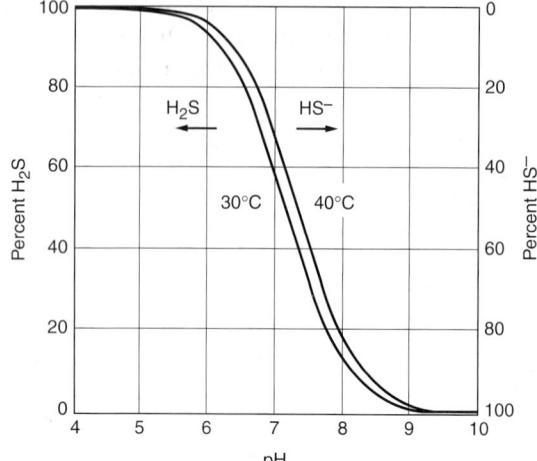

Figure 10–1

Percent of hydrogen
sulfide present as H_2S
and HS^- as a function
of pH.

amount of free ammonia is a function of temperature and pH. Dissociation constants for NH_3 as a function of temperature are given in Table 10–7. At a pH of 7.5 and at 30 to 35°C, 2 to 4 percent of the ammonium present will be as free ammonia (Speece, 1996). The toxicity threshold for ammonia has been reported to be 100 mg/L as NH_3-N (McCarty and McKinney, 1961), but with acclimatization time, higher concentrations may be tolerated. In batch tests, Lay et al. (1998) found a steady inhibition of methanogenic activity as the NH_3-N concentration was increased from 50 to 500 mg/L, with 500 mg/L being the apparent toxicity threshold level.

In stating the toxicity as total ammonium concentration, McCarty (1964) reported a toxicity concentration range of 1500 to 3000 mg/L as NH_4-N at pH above 7.4, with 3000 mg/L being toxic at any pH. However, after long-term acclimatization, much higher NH_4-N concentrations without toxicity have been observed. Moen (2000) found no inhibition effects for both thermophilic and mesophilic digestion of municipal sludge with NH_4-N concentrations ranging from 1900 to 2400 mg/L. Others (van Velsen, 1977; Parkin and Miller, 1982) have reported no effect of ammonia toxicity with long-term acclimatized cultures at NH_4-N concentrations in the range of 5000 to 8000 mg/L.

Liquid-Solids Separation

Efficient liquid-solids separation can enhance the performance of anaerobic treatment processes. Because of the low solids synthesis yield coefficient associated with anaerobic treatment, most of the solids wasting is via the treated effluent flow, and thus the degree of solids capture affects the SRT value that can be maintained. Good solids capture improves the effluent quality in terms of TSS concentration, and can also result in a longer SRT in the anaerobic reactor to increase the level of COD conversion.

10–3 ANAEROBIC SUSPENDED GROWTH PROCESSES

Early applications of anaerobic treatment of industrial wastewaters were suspended growth processes, which were initially designed in a similar manner as anaerobic sludge digesters. Three types of anaerobic suspended growth treatment processes are illustrated on Fig. 10–2: (1) the complete-mix suspended growth anaerobic digester, (2) the anaerobic contact process, and (3) the anaerobic sequencing batch reactor, a more recent suspended growth reactor design development. Each of these processes is described below along with general design considerations for suspended growth processes.

Complete-Mix Process

For the complete-mix anaerobic digester (see Fig. 10–2a) the hydraulic retention and solids retention times are equal (τ = SRT). The reactor τ may be in the range of 15 to 30 d to provide sufficient safety factors for operation and process stability (Parkin and Owen, 1986). The complete-mix digester without sludge recycle is more suitable for wastes with high concentrations of solids or extremely high dissolved organic concentrations, where thickening the effluent solids is difficult so that it is more practical to operate with τ equal to the SRT. As in anaerobic digestion (see Chap. 14), various methods of mixing may be used to utilize the reactor volume more fully. Typical organic loading rates for the complete-mix process are presented in Table 10–8, along with comparative values for the anaerobic contact and anaerobic sequence reactor processes. Examples of typical pilot plants used to assess the feasibility of anaerobic digestion for specific organic wastes are shown on Fig. 10–3.

Anaerobic Contact Process

The anaerobic contact process (see Fig. 10–2b) overcomes the disadvantages of a complete-mix process without recycle. Biomass is separated and returned to the

Figure 10–2

Anaerobic suspended growth treatment processes: (a) complete-mix suspended growth anaerobic digester, (b) anaerobic contact process, and (c) anaerobic sequencing batch reactor (ASBR).

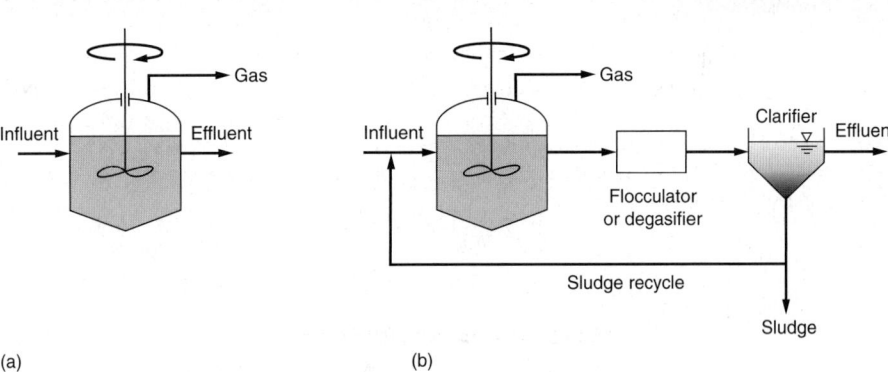

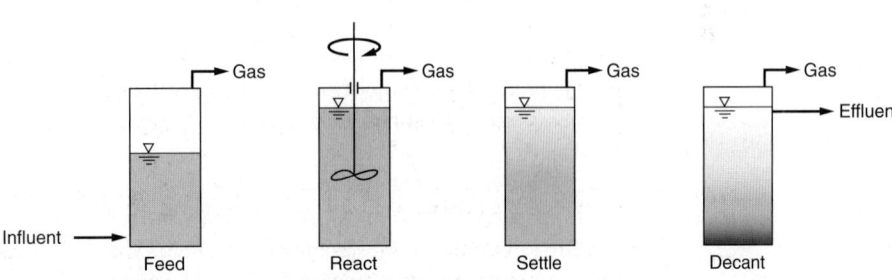

Table 10–8

Typical organic loading rates for anaerobic suspended growth processes at 30°C

Process	Volumetric organic loading, kg COD/m³·d	Hydraulic retention time τ, d
Complete-mix	1.0–5.0	15–30
Anaerobic contact	1.0–8.0	0.5–5
Anaerobic sequence batch reactor	1.2–2.4	0.25–0.50

Note: kg/m³·d × 62.4280 = lb/10³ ft³·d.

(a)

(b)

Figure 10–3

View of typical pilot plants used to assess the feasibility of the anaerobic complete-mix treatment process: (*a*) pilot-scale anaerobic reactors at a food-processing facility and (*b*) pilot-scale anaerobic reactors (under construction) to be used to assess the treatability of animal wastes.

complete-mix or contact reactor so that the process SRT is longer than τ. By separating τ and SRT values, the anaerobic reactor volume can be reduced. Gravity separation is the most common approach for solids separation and thickening prior to sludge recycle; however, a sludge with poor settling properties is commonly produced and alternative separation processes must be used or other methods must be employed to improve solids capture. Separation of solids using gas flotation by dissolving the process off-gas under pressure has been used in place of gravity separation.

Because the reactor sludge contains gas produced in the anaerobic process and gas production can continue in the separation process, solids-liquid separation can be inefficient and unpredictable. Various methods have been used to minimize the effect of trapped gas bubbles in the sludge-settling step. These include gas stripping by agitation or vacuum degasification, inclined-plate separators, and the use of coagulant chemicals

(Malina and Pohland, 1992). Clarifier hydraulic application rates range from 0.5 to 1.0 m/h. Practical reactor MLVSS concentrations are 4000 to 8000 mg/L (Malina and Pohland, 1992). Volumetric organic loadings that are used for anaerobic contact process are presented in Table 10–8.

Anaerobic Sequencing Batch Reactor

The anaerobic sequencing batch reactor (ASBR) process (see Fig. 10–2c) can be considered a suspended growth process with reaction and solids-liquid separation in the same vessel, much like that for aerobic sequencing batch reactors (SBRs) (see Chap. 8). The success of the ASBR depends on the development of a good settling granulated sludge as found in the upflow anaerobic sludge blanket (UASB) processes (Speece, 1996). The operation of ASBRs consists of four steps: (1) feed, (2) react, (3) settle, and (4) decant/effluent withdrawal. During the react period, intermittent mixing for a few minutes each hour is done to provide uniform distribution of substrates and solids (Sung and Dague, 1995). The feasibility of the process was demonstrated in laboratory reactors at temperatures from 5 to 25°C treating a nonfat dry milk synthetic substrate with a 600 mg/L feed COD concentration (Banik and Dague, 1996). The organic loading of the process was changed by selecting hydraulic retention times from 6.0 to 24 h. At 25°C, 92 to 98 percent COD removal was achieved at volumetric organic loadings of 1.2 to 2.4 kg COD/m^3·d. At 5°C, COD removal ranged from 85 to 75 percent for COD loadings from 0.9 to 2.4 kg/m^3·d, respectively.

A critical feature of the ASBR process is the settling velocity of the sludge during the settle period before decanting the effluent. Settling times used are about 30 min. Just before settling, at the end of the react period, organic removal and gas production rates are lower to provide better conditions for solids settling. After sufficient operating time, a dense granulated sludge was developed that improved liquid-solids separation rates. Effluent TSS concentrations ranged from 50 to 100 mg/L over the range of test conditions. Effluent TSS concentrations were higher at lower temperatures. At τ values from 6 to 24 h, the SRT ranged from 50 to 200 d, respectively.

Design of Anaerobic Suspended Growth Processes

Anaerobic suspended growth processes may be designed in a manner similar to completely mixed aerobic activated-sludge processes, because the hydraulic regime and biomass concentration can be reasonably defined. The design procedure is summarized in Table 10–9.

These same design considerations must be used for other anaerobic treatment processes, with the main difference being the need to rely on observed organic loadings to size the reactor volume.

A summary of kinetic coefficients and design values from Chap. 7 is provided in Table 10–10. For temperatures below 25°C, the anaerobic process design is best developed from laboratory treatability studies or pilot-plant test results. A temperature correction value of 1.08 as proposed by Banik and Dague (1996) may be used to develop a preliminary sizing for the treatability studies. The design of an anaerobic suspended growth contact process is presented in Example 10–2.

Table 10-9
Summary of design procedure for anaerobic suspended growth process

Item	Description
1.	Select an SRT to achieve a given effluent concentration and percent COD removal
2.	Determine the daily solids production and mass of solids in the system to maintain the desired SRT
3.	Select the expected solids concentration in the reactor and determine the reactor volume
4.	Determine the gas production rate
5.	Determine the amount of excess sludge wasted and the nutrient needs
6.	Check the volumetric organic loading rate
7.	Determine alkalinity needs

Table 10-10
Summary of design parameters for completely mixed suspended growth reactors treating soluble COD

Parameter	Unit	Value Range	Value Typical
Solids yield, Y			
Fermentation	g VSS/g COD	0.06–0.12	0.10
Methanogenesis	g VSS/g COD	0.02–0.06	0.04
Overall combined	g VSS/g COD	0.05–0.10	0.08
Decay coefficient, k_d			
Fermentation	g/g·d	0.02–0.06	0.04
Methanogenesis	g/g·d	0.01–0.04	0.02
Overall combined	g/g·d	0.02–0.04	0.03
Maximum specific growth rate, μ_m			
35°C	g/g·d	0.30–0.38	0.35
30°C	g/g·d	0.22–0.28	0.25
25°C	g/g·d	0.18–0.24	0.20
Half-velocity constant, K_S			
35°C	mg/L	60–200	160
30°C	mg/L	300–500	360
25°C	mg/L	800–1100	900
Methane			
Production at 35°C	m³/kg COD	0.4	0.4
Density at 35°C	kg/m³	0.6346	0.6346
Content of gas	%	60–70	65
Energy content	kJ/g	50.1	50.1

Note: m³/kg × 16.0185 = ft³/lb.
kg/m³ × 62.4280 = lb/10³ ft³.

EXAMPLE 10–2 **Suspended Growth Anaerobic Contact Reactor Process** Determine the reactor sizing, the gas production rate, the energy available, the solids production rate, and the alkalinity and nutrient requirements for an anaerobic contact process (see Fig. 10–2b) treating the following wastewater to achieve 90 percent COD removal.

Wastewater characteristics

Item	Unit	Value
Flowrate	m³/d	300
COD	g/m³	6000
Soluble COD	g/m³	4000
COD/TSS ratio	g/g	1.8
Degradable fraction of TSS	%	80
Nitrogen	g/m³	10
Phosphorus	g/m³	20
Alkalinity	g CaCO₃/m³	500
Temperature	°C	25

Design parameters and assumptions:

1. Effluent TSS concentration = 150 g/m³
2. Factor of safety for design SRT = 1.5
3. VSS/TSS = 0.85 (from Chap. 7)
4. f_d = 0.15 g VSS cell debris /g VSS biomass decay
5. Use kinetic coefficients from Table 10–10
6. Nutrient requirements based on VSS: N = 12% and P = 2.4%
7. At SRT > 40 d, degradable TSS is transformed
8. MLSS = 6000 g/m³
9. Settling rate = 24 m/d
10. Gas composition = 65% CH_4 and 35% CO_2

Solution

1. Determine design SRT at 25°C.
   At 90 percent COD removal the effluent COD is

   = (1.0 − 0.90) (6000 g/m³) = 600 g/m³

   The assumed effluent TSS concentration equals 150 g/m³.

   Effluent COD from TSS = (150 mg/L) 1.8 g COD/g TSS = 270 g/m³

   Allowable effluent soluble COD = (600 − 270) g/m³ = 330 g/m³

   Rearrange Eq. (7–39) and substitute μ_m for Yk to determine the SRT:

   $$SRT = \left(\frac{\mu_m S_e}{K_s + S_e} - k_d \right)^{-1}$$

From Table 10–10 the kinetic coefficients are:

$\mu_m = 0.20$ g/g·d

$K_S = 900$ mg/L

$k_d = 0.03$ g/g·d

$$SRT = \left[\frac{(0.20 \text{ g/g·d})(330 \text{ g/m}^3)}{[(900 + 330) \text{ g/m}^3]} - 0.03 \text{ g/g·d} \right]^{-1}$$

SRT = 42.3 d

The design SRT with a factor of safety of 1.5 is:

Design SRT = 1.5(42.3) = 63.4 d ≈ 63 d

2.   Determine sludge production.

Calculate nondegraded TSS concentration.

At SRT values greater than 40 d, degradable TSS is transformed.

Nonsoluble COD = (6000 − 4000) g/m³

$\qquad\qquad\qquad$ = 2000 g/m³

Determine the nonsoluble COD as TSS using 1.8 g COD/g TSS (given).

Nonsoluble COD as TSS = (2000 g/m³ COD)/(1.8 g COD/g TSS)

$\qquad\qquad\qquad\qquad\qquad$ = 1111 g/m³ TSS

Degradable fraction of TSS = 0.8(given)

Nondegraded TSS = 0.20(1111) = 222 g/m³ TSS

Use Eq. (8–16) to determine solids production:

$$P_{X,TSS} = \frac{QY(S_o - S)}{[1 + (k_d)SRT](0.85)} + \frac{f_d(k_d)QY(S_o - S)SRT}{[1 + (k_d)SRT](0.85)}$$

$$+ \, Q \text{ (nondegradable TSS)}$$

$S_o - S$ = COD degraded = Influent COD − nondegradable TSS COD
$\qquad\qquad\qquad\qquad\quad$ − effluent soluble degradable COD

$\qquad$ = 6000 g COD/m³ − [(222 g/m³ TSS)(1.8 g COD/g TSS)] − 330 g/m³

$\qquad$ = (6000 − 400 − 330) g/m³

$\qquad$ = 5270 g/m³ COD

Using the following coefficients from Table 10–10 and the given value for f_d, solve for $P_{X,TSS}$

$Y = 0.08$ g VSS/g COD

$k_d = 0.03$ g/g·d

$$P_{X,TSS} = \frac{(300 \text{ m}^3/\text{d})(0.08 \text{ g VSS/g COD})(5270 \text{ g COD/m}^3)}{[1 + (0.03 \text{ g/g·d})(63 \text{ d})](0.85)}$$

$$+ \frac{0.15 \text{ g/g}(0.03 \text{ g/g·d})(300 \text{ m}^3/\text{d})(0.08 \text{ g/g})(5270 \text{ g/m}^3)(63 \text{ d})}{[1 + (0.03 \text{ g/g·d})(63 \text{ d})](0.85)}$$

$$+ 300 \text{ m}^3/\text{d} (222 \text{ g/m}^3)$$

$$= 51{,}488 \text{ g/d} + 14{,}597 \text{ g/d} + 66{,}600 \text{ g/d}$$

$$P_{X,TSS} = 132{,}685 \text{ g/d}$$

3. Determine reactor volume and τ.
   a. Determine the volume using Eq. (7–55).

   $$\text{Volume} = \frac{(P_{X,TSS})(\text{SRT})}{X_{TSS}}$$

   For $X_{TSS} = 6000$ g/m^3 (given), the volume is:

   $$\text{Volume} = \frac{(132{,}685 \text{ g/d})(63 \text{ d})}{(6000 \text{ g/m}^3)} = 1393 \text{ m}^3$$

   b. Determine the hydraulic detention time τ.

   $$\tau = \frac{V}{Q} = \frac{1393 \text{ m}^3}{(300 \text{ m}^3/\text{d})} = 4.64 \text{ d}$$

4. Determine the methane and total gas production rate and energy content.
   a. Methane gas production rate
      From Table 10–10 methane volume at 35°C = 0.4 m^3 CH$_4$/kg COD

      At 25°C, CH$_4$ the volume of methane is

      $$= (0.4)\frac{(273.15 + 25)}{(273.15 + 35)} = 0.39 \text{ m}^3/\text{kg COD}$$

      Total CH$_4$ production rate = (0.39 m^3/kg)(5270 g COD/m^3)(300 m^3/d)/

      $$(1 \text{ kg}/10^3 \text{ g})$$

      $$= 616.6 \text{ m}^3/\text{d}$$

   b. Total gas production rate

      $$\text{Gas production} = \frac{(616.6 \text{ m}^3 \text{ CH}_4/\text{d})}{(0.65 \text{ m}^3 \text{ CH}_4/\text{m}^3 \text{ gas})} = 948.6 \text{ m}^3/\text{d}$$

c. Energy content of gas

At 25°C, the volume occupied by one mole of gas is

$$V = \frac{(1 \text{ mole})(0.082057 \text{ atm·L/mole·K})[(273.15 + 25)\text{K}]}{1.0 \text{ atm}} = 24.5 \text{ L}$$

The total moles of methane produced per day

$$\text{Mole CH}_4/\text{d} = \frac{(616.6 \text{ m}^3/\text{d})}{(24.5 \text{ L/mole})(1 \text{ m}^3/10^3 \text{ L})} = 25{,}167 \text{ mole/d}$$

Mass of methane = $(25{,}167 \text{ mole CH}_4/\text{d})(16 \text{ g CH}_4/\text{mole}) = 4.03 \times 10^5$ g/d

Energy content = $(4.03 \times 10^5 \text{ g/d})(50.1 \text{ kJ/g}) = 20.2 \times 10^6$ kJ/d

5. Determine nutrient requirements.

Biomass production = first two terms in $P_{X,TSS}$ calculation given in step 2.

$$= 51{,}488 \text{ g/d} + 14{,}597 \text{ g/d}$$

$$= 66{,}085 \text{ g/d}$$

For N = 12% and P = 2.4% of VSS, the nutrient requirements are:

N required = $(66{,}085 \text{ g/d})(0.12 \text{ g/g})(0.85) = 6741$ g/d

P required = $(66{,}085 \text{ g/g})(0.024 \text{ g/g})(0.85) = 1348$ g/d

Influent nutrients:

N = $(10 \text{ g/m}^3)(300 \text{ m}^3/\text{d}) = 3000$ g/d

P = $(20 \text{ g/m}^3)(300 \text{ m}^3/\text{d}) = 6000$ g/d

There is sufficient phosphorus in the influent, but nitrogen must be added.

N addition = $(6741 - 3000)$ g/d

$$= 3741 \text{ g/d}$$

$$= 3.74 \text{ kg/d}$$

6. Determine alkalinity requirement.

From Table 10–4, the minimum alkalinity needed at 35% CO_2 = 1500 mg/L as $CaCO_3$

Alkalinity addition = 1500 mg/L − 500 mg/L = 1000 mg/L as $CaCO_3$

$$\text{As NaHCO}_3 = \frac{(84 \text{ g NaHCO}_3/\text{eq})(10^3 \text{ g/m}^3)}{(50 \text{ g/eq CaCO}_3)}$$

$$= 1680 \text{ g NaHCO}_3/\text{m}^3$$

$\text{NaHCO}_3/\text{d} = (1680 \text{ g/m}^3)(300 \text{ m}^3/\text{d})(1 \text{ kg}/10^3 \text{ g}) = 504$ kg/d

7. Determine clarifier diameter.
   (Assume degasifier used before clarifier)

$$\text{Area} = \frac{(Q, \text{m}^3/\text{d})}{(\text{settling rate, m/d})} = \frac{300 \text{ m}^3}{(24 \text{ m/d})} = 12.5 \text{ m}^2$$

Diameter $= 4$ m

Comments Considerable energy (i.e., 20.2×10^6 kJ/d) is generated by the production of methane (CH_4). The methane could be used to heat the anaerobic reactor, which would provide more rapid degradation and thus reduce the anaerobic bioreactor size.

10–4 ANAEROBIC SLUDGE BLANKET PROCESSES

One of the most notable developments in anaerobic treatment process technology was the upflow anaerobic sludge blanket (UASB) reactor in the late 1970s in the Netherlands by Lettinga and his coworkers (Lettinga and Vinken, 1980; Lettinga et al., 1980). The principal types of anaerobic sludge blanket processes include (1) the original UASB process and modification of the original design, (2) the anaerobic baffled reactor (ABR), and (3) the anaerobic migrating blanket reactor (AMBR®). The ABR process was developed by McCarty and coworkers at Stanford University in the early 1980s (Bachmann et al., 1985). Work in the late 1990s at Iowa State University has led to the development of the AMBR process (Angenent et al., 2000). Of these sludge blanket processes, the UASB process is used most commonly, with over 500 installations treating a wide range of industrial wastewaters. A number of pilot studies have been done with the ABR, with a limited number of full-scale installations (Orozco, 1988). In this section all three types of anaerobic sludge blanket processes are described along with their performance, demonstrated process loadings, and key design considerations. The major emphasis is, however, on the UASB process.

Upflow Sludge Blanket Reactor Process

The basic UASB reactor is illustrated on Fig. 10–4a. As shown on Fig. 10–4a, influent wastewater is distributed at the bottom of the UASB reactor and travels in an upflow mode through the sludge blanket. Critical elements of the UASB reactor design are the influent distribution system, the gas-solids separator, and the effluent withdrawal design. Modifications to the basic UASB design include adding a settling tank (see Fig. 10–4b) or the use of packing material at the top of the reactor (see Fig. 10–4c). Both modifications are intended to provide better solids capture in the system and to prevent the loss of large amounts of the UASB reactor solids due to process upsets or changes in the UASB sludge blanket characteristics and density. The use of an external solids capture system to prevent major losses of the system biomass is recommended strongly by Speece (1996). A view of a sludge blanket fixed-film reactor installation is shown on Fig. 10–5.

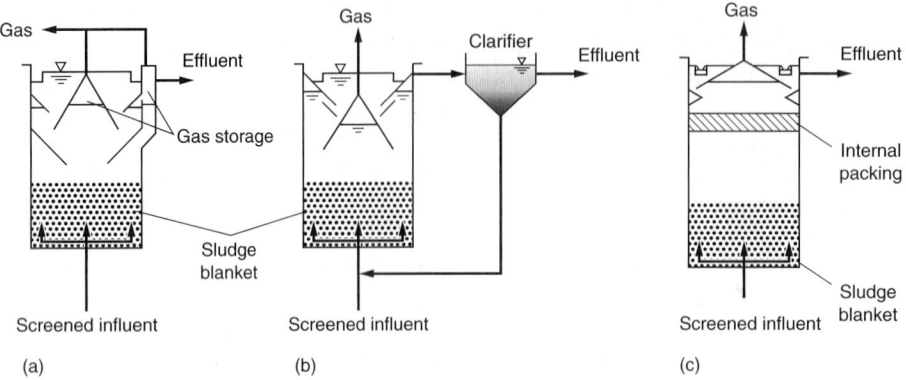

Figure 10–4

Schematic of the UASB process and some modifications: (a) original UASB process, (b) UASB reactor with sedimentation tank and sludge recycle, and (c) UASB reactor with internal packing for fixed-film attached growth, placed above the sludge blanket.

Figure 10–5

View of UASB reactor equipped with internal packing above the sludge blanket. The exterior physical appearance of a UASB reactor without and with internal packing is the same (see Fig. 10–4c for location of internal packing).

The key feature of the UASB process that allows the use of high volumetric COD loadings compared to other anaerobic processes is the development of a dense granulated sludge. Because of the granulated sludge floc formation, the solids concentration can range from 50 to 100 g/L at the bottom of the reactor and 5 to 40 g/L in a more diffuse zone at the top of the UASB sludge blanket. The granulated sludge particles have a size range of 1.0 to 3.0 mm and result in excellent sludge-thickening properties with SVI values less than 20 mL/g. Several months may be required to develop the granulated sludge, and seed is often supplied from other facilities to accelerate the system startup. Variations in morphology were observed for anaerobic granulated sludge developed at 30 and 20°C, but both exhibited similar floc size and settling properties (Soto et al., 1997).

The development of granulated sludge solids is affected by the wastewater characteristics. Granulation is very successful with high carbohydrate or sugar wastewaters, but less so with wastewaters high in protein, resulting in a more fluffy floc instead (Thaveesri et al., 1994). Other factors affecting the development of granulated solids are pH, upflow velocity, and nutrient addition (Annachhatre, 1996). The pH should be maintained near 7.0, and a recommended COD:N:P ratio during startup is 300:5:1, while a lower ratio can be used during steady-state operation at 600:5:1. Control of the upflow velocity is recommended during startup by having it high enough to wash out nonflocculent sludge.

The presence of other suspended solids in the sludge blanket can also inhibit the density and formation of granulated sludge (Lettinga and Hulshoff Pol, 1991). An explanation of the fundamental metabolic conditions associated with granular sludge formation is provided by Speece (1996) based on work by Palns et al. (1987, 1990). The explanation is as follows. The formation of dense granulated sludge floc particles is favored under conditions of near neutral pH, a plug-flow hydraulic regime, a zone of high hydrogen partial pressure, a nonlimiting supply of NH_4-N, and a limited amount of the amino acid cysteine. With a high hydrogen concentration and sufficient NH_4-N, the bacteria responsible for granulation may produce other amino acids, but their synthesis is limited by the cysteine supply. Some of the excess amino acids that are produced are thought to be secreted to form extracellular polypeptides which, in turn, will bind organisms together to form the dense pellets or floc granules.

Design Considerations for UASB Process

A comprehensive review of design considerations for UASB reactors has been provided by Lettinga and Hulshoff Pol (1991). Important design considerations are (1) wastewater characteristics in terms of composition and solids content, (2) volumetric organic load, (3) upflow velocity, (4) reactor volume, (5) physical features including the influent distribution system, and (6) gas collection system.

Wastewater Characteristics. Wastewaters that contain substances that can adversely affect the sludge granulation, cause foaming, or cause scum formation are of concern. Wastewaters with higher concentrations of proteins and/or fats tend to create more of the above problems. The fraction of particulate versus soluble COD is important in determining the design loadings for UASB reactors as well as determining the applicability of the process. As the fraction of solids in the wastewater increases, the ability to form a dense granulated sludge decreases. At a certain solids concentration (greater than 6 g TSS/L) anaerobic digestion and anaerobic contact processes may be more appropriate.

Volumetric Organic Loadings. Typical COD loadings as a function of the wastewater strength, fraction of particulate COD in the wastewater, and TSS concentrations in the effluent are summarized in Table 10–11. Removal efficiencies of 90 to 95 percent for COD have been achieved at COD loadings ranging from 12 to 20 kg COD/$m^3 \cdot$d on a variety of wastes at 30 to 35°C with UASB reactors. Values for τ for high-strength wastewater have been as low as 4 to 8 h at these loadings. Where less than 90 percent COD removal and higher-effluent TSS concentrations are acceptable, higher upflow velocities can be used, which will develop a more dense granulated sludge by flushing out other solids. Thus, the higher volumetric COD loadings are shown for this condition.

Table 10–11

Recommended volumetric COD loading for UASB reactors at 30°C to achieve 85 to 95 percent COD removal[a]

Wastewater COD, mg/L	Fraction as particulate COD	Volumetric loading, kg COD/m³·d		
		Flocculent sludge	Granular sludge with high TSS removal	Granular sludge with little TSS removal
1000–2000	0.10–0.30	2–4	2–4	8–12
	0.30–0.60	2–4	2–4	8–14
	0.60–1.00	na	na	na
2000–6000	0.10–0.30	3–5	3–5	12–18
	0.30–0.60	4–8	2–6	12–24
	0.60–1.00	4–8	2–6	na
6000–9000	0.10–0.30	4–6	4–6	15–20
	0.30–0.60	5–7	3–7	15–24
	0.60–1.00	6–8	3–8	na
9000–18,000	0.10–0.30	5–8	4–6	15–24
	0.30–0.60	na	3–7	na
	0.60–1.00	na	3–7	na

[a] Adapted from Lettinga and Hulshoff Pol (1991).

Note: kg/m³·d × 62.4280 = lb/10³ ft³·d.

Recommended loadings as a function of temperature for wastewaters with mainly soluble COD are presented in Table 10–12. These loadings apply to the sludge blanket volume, and a reactor effectiveness factor of 0.8 to 0.9 as discussed below is used to determine the reactor liquid below the gas collector. The higher loading recommendation for the wastewater containing mainly volatile fatty acids (VFA) is based on the potential of obtaining a more dense granulated sludge. Design τ values are also given for the treatment of domestic wastewater in Table 10–13 based on pilot-plant experience. The τ value needed is longer than that used in aerobic processes for secondary treatment for BOD removal. In addition, an aerobic polishing step would likely be needed. The economic benefits of energy savings and lower sludge production would have to be sufficient to justify the higher capital costs for liquid treatment with a UASB process.

Upflow Velocity. The upflow velocity, based on the flowrate and reactor area, is a critical design parameter. Recommended design velocities are shown in Table 10–14. Temporary peak superficial velocities of 6 m/h and 2 m/h can be allowed for soluble and partially soluble wastewaters, respectively. For weaker wastewaters the allowable velocity and reactor height will determine the UASB reactor volume, and for stronger

Table 10–12
Recommended volumetric organic loadings as a function of temperature for soluble COD substrates for 85 to 95 percent COD removal. Average sludge concentration is 25 g/L[a]

| | Volumetric loading, kg sCOD/m³·d | | | |
| | VFA wastewater | | Non-VFA wastewater | |
Temperature, °C	Range	Typical	Range	Typical
15	2–4	3	2–3	2
20	4–6	5	2–4	3
25	6–12	6	4–8	4
30	10–18	12	8–12	10
35	15–24	18	12–18	14
40	20–32	25	15–24	18

[a]Adapted from Lettinga and Hulshoff Pol (1991).
Note: kg/m³·d × 62.4280 = lb/10³ ft³·d.

Table 10–13
Applicable hydraulic retention times τ for treatment of raw domestic wastewater in a 4-m-high UASB reactor[a]

Temperature, °C	Average τ, h	Maximum τ, for 4- to 6-h peak, h
16–19	10–14	7–9
22–26	7–9	5–7
>26	6–8	4–5

[a]Adapted from Lettinga and Hulshoff Pol (1991).

Table 10–14
Upflow velocities and reactor heights recommended for UASB reactors[a]

| | Upflow velocity, m/h | | Reactor height, m | |
Wastewater type	Range	Typical	Range	Typical
COD nearly 100% soluble	1.0–3.0	1.5	6–10	8
COD partially soluble	1.0–1.25	1.0	3–7	6
Domestic wastewater	0.8–1.0	0.7	3–5	5

[a]Adapted from Lettinga and Hulshoff Pol (1991).
Note: m × 3.2808 = ft.
m/h × 3.2808 = ft/h.

wastewaters it will be determined by the volumetric COD loading. The upflow velocity is equal to the feed rate divided by the reactor cross-section area:

$$v = \frac{Q}{A}$$

(10–13)

where v = design upflow superficial velocity, m/h
A = reactor cross-section area, m²
Q = influent flowrate, m³/h

Reactor Volume and Dimensions. To determine the required reactor volume and dimensions, the organic loading, superficial velocity, and effective treatment volume must all be considered. The effective treatment volume is that volume occupied by the sludge blanket and active biomass. An additional volume exists between the effective volume and the gas collection unit where some additional solids separation occurs and the biomass is dilute. The nominal liquid volume of the reactor based on using an acceptable organic loading is given by

$$V_n = \frac{QS_o}{L_{org}} \qquad (10\text{-}14)$$

where V_n = nominal (effective) liquid volume of reactor, m³
$\quad Q$ = influent flowrate, m³/h
$\quad S_o$ = influent COD, kg COD/m³
$\quad L_{org}$ = organic loading rate, kg COD/m³·d

To determine the total liquid volume below the gas collectors, an effectiveness factor is used, which is the fraction occupied by the sludge blanket. Taking into account the effectiveness factor, which may vary from 0.8 to 0.9, the required total liquid volume of the reactor exclusive of the gas storage area is given by

$$V_L = \frac{V_n}{E} \qquad (10\text{-}15)$$

where V_L = total liquid volume of reactor, m³
$\quad V_n$ = nominal liquid volume of reactor, m³
$\quad E$ = effectiveness factor, unitless

Rearranging Eq. (10–13), the area of the reactor is

$$A = \frac{Q}{v} \qquad (10\text{-}16)$$

The liquid height of the reactor is determined using the following relationship:

$$H_L = \frac{V_L}{A} \qquad (10\text{-}17)$$

where H_L = reactor height based on liquid volume, m
$\quad V_L$ = total liquid reactor volume, m³
$\quad A$ = cross-sectional area, m²

The gas collection volume is in addition to the reactor volume and adds an additional height of 2.5 to 3 m. Thus, the total height of the reactor is

$$H_T = H_L + H_G \qquad (10\text{-}18)$$

where H_T = total reactor height, m
$\quad H_L$ = reactor height based on liquid volume, m
$\quad H_G$ = reactor height to accommodate gas collection and storage, m

Physical Features. The main physical features requiring careful consideration are the feed inlet, gas separation, gas collection, and effluent withdrawal. The inlet and gas separation designs are unique to the UASB reactor. The feed inlet must be designed to provide uniform distribution and to avoid channeling or the formation of dead zones. The avoidance of channeling is more critical for weaker wastewaters, as there would be less gas production to help mix the sludge blanket. A number of inlet feed pipes are used to direct flow to different areas of the bottom of the UASB reactor from a common feed source. Access must be provided to clean the pipes in the event of clogging. Guidelines for determining the area served by the individual inlet feed pipes as a function of the sludge characteristics and organic loading are provided in Table 10–15.

Gas Collection and Solid Separation. The gas solids separator (GSS) is designed to collect the biogas, prevent washout of solids, encourage separation of gas and solid particles, allow for solids to slide back into the sludge blanket zone, and help improve effluent solids removal. A series of upside-down V-shaped baffles is used next to effluent weirs to accomplish the above objectives. Guidelines for the GSS design are summarized in Table 10–16.

Table 10–15
Guidelines for sizing the area served by inlet feed pipes for UASB reactor[a]

Sludge type	COD loading, kg/m³·d	Area per feed inlet, m²
Dense flocculent sludge, >40 kg TSS/m³	<1.0	0.5–1
	1–2	1–2
	>2	2–3
Medium flocculent sludge, 20–40 kg TSS/m³	<1–2	1–2
	>3	2–5
Granular sludge	1–2	0.5–1
	2–4	0.5–2
	>4	>2

[a] Adapted from Lettinga and Hulshoff Pol (1991).
Note: kg/m³·d × 62.4280 = lb/10³ ft³·d.

Table 10–16
Recommended design considerations for the gas solids separator for UASB reactors[a]

- The slope of the settler bottom, i.e., the inclined wall of the gas collector, should be between 45 and 60°
- The surface area of the apertures between the gas collectors should not be smaller than 15 to 20 percent of the total reactor surface area
- The height of the gas collector should be between 1.5 and 2 m at reactor heights of 5–7 m
- A liquid-gas interface should be maintained in the gas collector to facilitate the release and collection of gas bubbles and to control scum layer formation
- The overlap of the baffles installed beneath the apertures should be 100 to 200 mm to avoid upward-flowing gas bubbles entering the settler compartment
- Generally scum layer baffles should be installed in front of the effluent weirs
- The diameter of the gas exhaust pipes should be sufficient to guarantee the easy removal of the biogas from the gas collection cap, particularly in the case where foaming occurs
- In the upper part of the gas cap, antifoam spray nozzles should be installed in the case where the treatment of the wastewater is accompanied by heavy foaming

[a] Adapted from Malina and Pohland (1992).

Advantages for the UASB process are the high loadings and relatively low detention times possible for anaerobic treatment and the elimination of the cost of packing material. Another major advantage is that the UASB process is, as noted previously, a proven process with more than 500 full-scale facilities in operation. Limitations of the process are related to those wastewaters that are high in solids content or where their nature prevents the development of the dense granulated sludge. The process design for the UASB process is illustrated in Example 10–3.

EXAMPLE 10–3 **UASB Treatment Process Design** For a UASB treatment process treating an industrial wastewater, determine the (1) size and dimensions of the reactor, (2) detention time, (3) reactor SRT, (4) average VSS concentration in biomass zone of the reactor, (5) methane gas production, (6) energy available from methane production, and (7) alkalinity requirements for a wastewater with the characteristics given below to achieve greater than 90 percent soluble COD removal. The wastewater is mainly soluble, containing carbohydrate compounds, and a granular sludge is expected. Assume 50 percent of the influent pCOD and VSS is degraded, 90 percent of the influent sulfate is reduced biologically, and the effluent VSS concentration is 150 g/m³. Assume the design parameters given below and the typical values given in Tables 10–10 and 10–12 are applicable.

Wastewater characteristics

Item	Unit	Value
Flowrate	m^3/d	1000
COD	g/m^3	2300
sCOD	g/m^3	2000
TSS	g/m^3	200
VSS	g/m^3	150
Alkalinity	g/m^3 as $CaCO_3$	500
SO_4	g/m^3	200
Temperature	°C	30

Design parameters and assumptions:

1. From Table 10–10,

   $Y = 0.08$ g VSS/g COD

   $k_d = 0.03$ g VSS/g VSS·d

   $\mu_m = 0.25$ g VSS/g VSS·d

2. $f_d = 0.15$ g VSS cell debris/g VSS biomass decay
3. Methane production at 35°C = 0.40 L CH_4/g COD
4. Reactor volume effectiveness factor = 85 percent
5. Height for gas collection = 2.5 m

Solution

1. Determine the reactor volume based on the design organic loading and the use of Eq. (10–14).

   a. From Table 10–12 select the average organic loading of 10 kg sCOD/m^3·d.

   $$V_n = \frac{QS_o}{L_{org}} = \frac{(1000 \text{ m}^3/\text{d})(2 \text{ kg sCOD/m}^3)}{(10 \text{ kg sCOD/m}^3\cdot\text{d})}$$

   $$V_n = 200 \text{ m}^3$$

   b. Determine the total reactor liquid volume using Eq. (10–15).

   $$V_L = \frac{V_n}{E} = \frac{200 \text{ m}^3}{0.85} = 235 \text{ m}^3$$

2. Determine the reactor dimensions.

   a. First determine the reactor cross-sectional area using Eq. (10–16) based on the design superficial velocity. Use the upflow velocity data given in Table 10–14. Because the wastewater is highly soluble, select a velocity of 1.5 m/h.

   $$A = \frac{Q}{v} = \frac{(1000 \text{ m}^3/\text{d})}{(1.5 \text{ m/h})(24 \text{ h/d})} = 27.8 \text{ m}^2$$

   $$A = \frac{\pi D^2}{4} = 27.8 \text{ m}^2 \qquad D = 6 \text{ m}$$

   b. Determine the reactor liquid height using Eq. (10–17).

   $$H_L = \frac{V_L}{A} = \frac{235 \text{ m}^3}{27.8 \text{ m}^2} = 8.4 \text{ m}$$

   c. Determine the total height of the reactor using Eq. (10–18).

   $$H_T = H_L + H_G = 8.4 \text{ m} + 2.5 \text{ m} = 10.9 \text{ m}$$

   d. Reactor dimensions.

   Diameter = 6 m

   Height = 10.9 m

3. Determine the reactor hydraulic detention time τ.

   $$\tau = \frac{V_L}{Q} = \frac{(235 \text{ m}^3)(24 \text{ h/d})}{(1000 \text{ m}^3/\text{d})} = 5.64 \text{ h}$$

4. Determine the reactor SRT.

   a. The value of the SRT can be estimated by assuming that all the wasted biological solids are in the effluent flow. A conservative design approach is to assume that the given effluent VSS concentration consists of biomass. Thus, the following relationship applies:

   $$QX_e = P_{X,VSS} = \text{solids wasted per day}$$

Both Q and X_e are known. The value of $P_{X,VSS}$ is given by Eq. (7–52):

$$P_{X,VSS} = \frac{Q(Y)(S_o - S)}{1 + (k_d)SRT} + \frac{f_d(k_d)Q(Y)(S_o - S)SRT}{1 + (k_d)SRT} + Q(nbVSS) - QX_e$$

b. Develop the data needed to solve the above equation.

   i. The effluent soluble COD concentration at 90% COD removal is

$$S = (1.0 - 0.9)(2000\ g/m^3) = 200\ g/m^3$$

   ii. The effluent nbVSS concentration given that 50 percent of the influent VSS is degraded is:

$$nbVSS = 0.50(150\ g/m^3) = 75\ g/m^3$$

   iii. The pCOD degraded is

$$pCOD\ degraded = 0.50(2300 - 2000)\ g/m^3 = 150\ g/m^3$$

   iv. Total degradable influent COD, S_o

$$S_o = (2000 + 150)\ g/m^3 = 2150\ g/m^3$$

c. Substitute the given parameter values and solve the expression given above for SRT.

$$QX_e = (1000\ m^3/d)(150\ g/m^3)$$

$$= \frac{(1000\ m^3/d)(0.08\ g\ VSS/g\ COD)[(2150 - 200)g/m^3]}{[1 + (0.03\ g\ VSS/g\ VSS\cdot d)SRT]}$$

$$+ \frac{(0.15\ g\ VSS/g\ VSS)(0.03\ g\ VSS/g\ VSS\cdot d)(1000\ m^3/d)(0.08\ g\ VSS/gCOD)[(2150 - 200)g/m^3]SRT}{[1 + (0.03\ g\ VSS/g\ VSS\cdot d)SRT]}$$

$$+ (1000\ m^3/d)(75\ g/m^3)$$

$$150,000\ g/d = 156,144,000/[1 + (0.03)\ SRT] + 702,648\ SRT/$$

$$[1 + (0.03)\ SRT] + 75,000\ g/d$$

$$SRT = 52\ d$$

5. Estimate the effluent soluble COD at an SRT of 52 d at 30°C using Eq. (7–40) and the given coefficients.

$$S = \frac{K_s[1 + (k_d)SRT]}{SRT(Yk - k_d) - 1}$$

$$k = \frac{\mu_m}{Y} = \frac{(0.25\ g\ VSS/g\ VSS\cdot d)}{(0.08\ g\ VSS/g\ COD)} = 3.125\ g\ COD/g\ VSS\cdot d$$

$$S = \frac{(360\ mg/L)[1 + (0.03\ g/g\cdot d)\ 52\ d]}{[(52\ d)[(0.08\ g/g)(3.125\ g/g\cdot d) - (0.03\ g/g\cdot d)] - 1]}$$

$$S = 88.3\ mg/L$$

6. Determine if the computed SRT value is adequate.

$$\text{The fraction of the influent sCOD in effluent} = \frac{(88.3 \text{ mg/L})}{(2000 \text{ mg/L})} = 0.044 = 4.4\%\text{S}$$

Because 4.4 percent is less than 10 percent (specified in problem statement), the process SRT is adequate.

7. Determine the average X_{TSS} concentration in biomass zone of the reactor.
   a. The value of the X_{TSS} can be estimated by using Eq. (7–35) developed previously in Chap. 7 for the SRT.

   $$SRT = \frac{V(X_{TSS})}{(Q - Q_w)X_e + Q_w X_R}$$

   Because it was assumed that all the wasted solids are in the effluent flow, the term $Q_w = 0$ and the value of X_{TSS} can be estimated as follows:

   $$SRT \approx \frac{V X_{TSS}}{Q X_e} \quad \text{and} \quad X_{TSS} \approx \frac{Q X_e SRT}{V}$$

   b. Solve for the value of X_{TSS}, with the volume V equal to the effective volume, V_n.

   $$X_{TSS} \approx \frac{(1000 \text{ m}^3/\text{d})(150 \text{ g/m}^3)(52 \text{ d})(1 \text{ kg}/10^3 \text{ g})}{200 \text{ m}^3} = 39.0 \text{ kg/m}^3$$

   The computed value is within the range of solids concentration values given earlier for the UASB process.

8. Determine the methane gas production and energy produced.
   a. Determine the COD degraded.

   $$COD = (2150 - 200) \text{ g/m}^3 = 1950 \text{ g/m}^3$$

   b. Determine the COD removed with sulfate as the electron acceptor. From Sec. 10–3, 0.67 g COD removed/g SO_4 reduced

   $$COD_{SR} = 0.90(200 \text{ g } SO_4/\text{m}^3)(0.67 \text{ g COD/g } SO_4) = 120.6 \text{ g/m}^3$$

   c. Determine the COD used by methanogenic bacteria.

   $$COD_{MB} = (1950 - 120.6) \text{ g/m}^3(1000 \text{ m}^3/\text{d}) = 1,829,400 \text{ g/d}$$

   d. Determine the methane production rate.

   $$\text{Methane production at 30°C} = (0.40 \text{ L/g})\left(\frac{273.15 + 30}{273.15 + 35}\right) = 0.3935 \text{ L/g}$$

   Amount of CH_4 produced/d $= 0.3935$ L/g $(1,829,400 \text{ g COD/d})$

   $$= 719,869 \text{ L/d}$$

   $$= 719.9 \text{ m}^3/\text{d}$$

   Total gas volume produced (use 65% methane per Table 10–10)
   $= (719.9 \text{ m}^3/\text{d})/(0.65) = 1107.5 \text{ m}^3/\text{d}$

9. Determine energy produced from methane.
   To determine the energy produced, determine the density of methane at 30°C and use the factor of 50.1 kJ/g methane (Table 10–10).
   a. Determine density.

   Density at 35°C = 0.6346 g/L (Table 10–10)

   $$\text{Methane density at 30°C} = (0.6346 \text{ g/L})\left(\frac{273.15 + 35}{273.15 + 30}\right) = 0.6451 \text{ g/L}$$

   b. Determine energy produced.

   Energy produced = (719,869 L CH_4/d)(0.6451 g/L)(50.1 kJ/g)

   $$= 23.3 \times 10^6 \text{ kJ/d}$$

10. Determine the alkalinity requirements.
    From Table 10–4, the estimated alkalinity concentration required at 30°C and 35% CO_2 in the gas phase is 1800 mg/L. Because the alkalinity in the influent is 500 mg/L, the amount of alkalinity that has to be added is

    Alkalinity required = (1800 − 500) mg/L as $CaCO_3$ = 1300 mg/L as $CaCO_3$

    Daily addition = (1300 g/m³)(1000 m³/d)(1 kg/10³ g) = 1300 kg/d as $CaCO_3$

Comments A significant amount of methane gas is produced daily. If the methane can be used for the production of energy by the industrial facility, it could help to offset the cost of adding a considerable amount of alkalinity to maintain the anaerobic reactor pH near 7. Also it is important to note the significance of the effluent solids concentration in determining the system SRT. Anaerobic processes generally produce higher-effluent VSS concentrations compared to aerobic processes. For weak wastewater for which solids production is lower, it may be difficult to maintain long SRT values for high treatment efficiency due to effluent solids loss.

Also, in contrast to the situation presented in this example, manual wasting of sludge may be necessary. If the wastewater had a higher-influent COD concentration, and the effluent VSS concentration remained the same, the concentration of solids (X_{TSS}) in the sludge blanket would have to increase and the blanket level would have to be higher. To avoid a rising sludge blanket, manual wasting of sludge would have to be initiated and the SRT value would have to be less than the computed value of 52 d. For design, it is best to assume that the average VSS concentration of the sludge blanket would be less than 25 to 35 kg/m³.

Anaerobic Baffled Reactor

In the anaerobic baffled reactor (ABR) process, as shown on Fig. 10–6a, baffles are used to direct the flow of wastewater in an upflow mode through a series of sludge blanket reactors. The sludge in the reactor rises and falls with gas production and flow, but moves through the reactor at a slow rate. Various modifications have been made to the ABR to improve performance. The modifications include (1) changes to the baffle

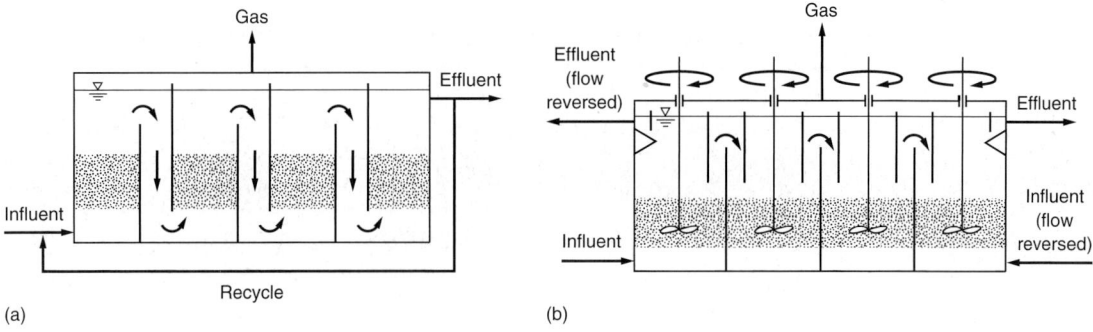

Figure 10-6

Schematic views of alternative sludge blanket processes: (a) anaerobic baffled reactor (ABR) and (b) anaerobic migrating blanket reactor (AMBR).

design, (2) hybrid reactors where a settler has been used to capture and return solids, or (3) packing has been used in the upper portion of each chamber to capture solids (Barber and Stuckey, 1999). Though granulated sludge is not considered essential for the operation and performance of the ABR process, it has been observed in the ABR process (Boopathy and Tilche, 1992).

A number of studies have been done with the ABR process at bench and pilot scale for a wide range of wastewaters and temperatures as low as 13°C. An excellent summary of these studies including organic loading rates, temperature, and percent COD removal is provided by Barber and Stuckey (1999). Typical design loadings for the ABR process are presented in Table 10–17. Many of these studies were operated at τ values in the range of 6 to 24 h. The reactor volatile solids concentrations varied from 4 to 20 g/L.

Advantages claimed for the ABR process include the following:

1. Simplicity, i.e., no packing material, no special gas separation method, no moving parts, no mechanical mixing, and little plugging potential
2. Long SRT possible with low hydraulic retention time τ
3. No special biomass characteristic required
4. Wastewaters with a wide variety of constituent characteristics can be treated
5. Staged operation to improve kinetics
6. Stable to shock loads

The main limitations at this time (2001) with the ABR process are the lack of information based on the operation of full-scale facilities and a full understanding of the system hydraulics.

Anaerobic Migrating Blanket Reactor

The anaerobic migrating blanket reactor (AMBR®) process is similar to the ABR with the added features of mechanical mixing in each stage and an operating approach to maintain the sludge in the system without resorting to packing or settlers for additional

Table 10–17

Design and performance results from bench- and pilot-scale studies on anaerobic treatment of various wastewaters with the ABR process[a]

Wastewater	Temperature, °C	Number of chambers	Influent COD, mg/L	COD loading, kg/m³·d	Percent COD removal
Carbohydrate/protein	35	5	7100–7600	2–10	79–82
Distilling	35	5	51,600	2.2–3.5	90
Carbohydrate/protein	35	5	4000	1–2	94
Molasses	35	3	115,000–900,000	4.3–28	49–88
Swine manure	35	3	58,500	4.0	62–69
Municipal wastewater	18–28	3	264–906	2.2	90
Slaughterhouse	25–30	4	450–550	0.9–4.7	75–90
Pharmaceutical	35	5	20,000	20	36–68
Domestic/industrial	15	8	315	0.9	70
Glucose	35	5	1000–10,000	2–20	72–99

[a] Adapted from Barber and Stuckey (1999).
Note: kg/m³·d × 62.4280 = lb/10³ ft³·d.

solids capture (see Fig. 10–6b). In the AMBR® process, the influent feed point is changed periodically to the effluent side and the effluent withdrawal point is also changed. In this way the sludge blanket remains more uniform in the anaerobic reactor. The flow is reversed when a significant quantity of solids accumulates in the last stage.

The AMBR® process has been shown to be feasible from bench-scale testing treating nonfat dry milk (Angenent et al., 2000) at 15 and 20°C. The organic loading rate was varied from 1.0 to 3.0 kg COD/m³·d with hydraulic retention times ranging from 4 to 12 h. At the higher COD loading, the COD removal efficiency was 59 percent at 15°C. At 20°C, COD removals ranged from 80 to 95 percent at COD loadings of 1 to 2.0 kg/m³·d.

10–5 ATTACHED GROWTH ANAEROBIC PROCESSES

Upflow attached growth anaerobic treatment reactors differ by the type of packing used and the degree of bed expansion. Three types of upflow attached growth processes are illustrated on Fig. 10–7. In the upflow packed-bed reactor (see Fig. 10–7a) the packing is fixed and the wastewater flows up through the interstitial spaces between the packing and biogrowth. Effluent recycle is generally not used for the packed-bed reactor except for high-strength wastewaters. While the first upflow anaerobic packed-bed processes contained rock, a variety of designs employing synthetic plastic packing are used currently. The anaerobic expanded-bed reactor (see Fig. 10–7b) uses a fine-grain sand to support biofilm growth. Recycle is used to provide upflow velocities, resulting in 20 percent bed expansion. Higher upflow velocities are used for fluidized-bed anaer-

Figure 10–7

Upflow anaerobic attached growth treatment reactors: (a) anaerobic upflow packed-bed reactor, (b) anaerobic expanded-bed reactor, and (c) anaerobic fluidized-bed reactor.

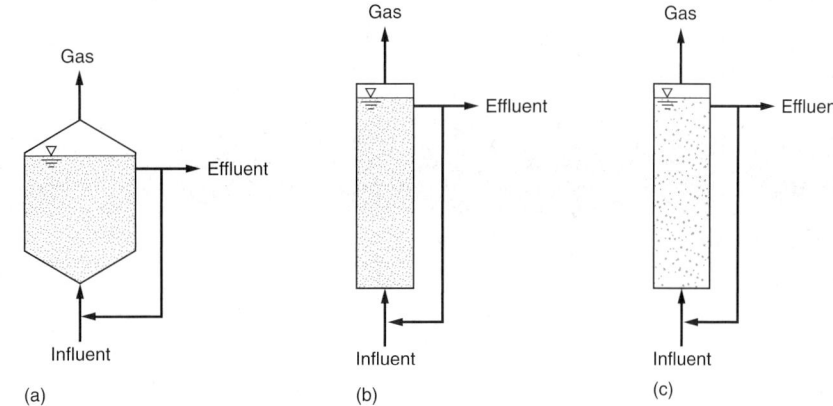

obic reactors (see Fig. 10–7c), which also contain a fine-grain packing. In fluidized-bed systems, both fluidization and mixing of the packing material occurs. The expanded and fluidized-bed reactors have more surface area per reactor volume for biomass growth and better mass transfer than the upflow packed-bed reactor, but have lower solids capture. These different systems are described in more detail in the following paragraphs, along with their typical design conditions and COD loadings.

Upflow Packed-Bed Attached Growth Reactor

Full-scale upflow packed-bed anaerobic filters are used in cylindrical or rectangular tanks at widths and diameters ranging from 2 to 8 m and heights from 3 to 13 m (see Fig. 10–7a). Packing material placement may be in the entire depth or, for hybrid designs, only in the upper 50 to 70 percent. The most common packing materials are corrugated plastic cross-flow or tubular modules, similar to that described in Chap. 9 for aerobic attached growth processes, and plastic pall rings. The specific surface area of the packing averages about 100 m^2/m^3 and, based on research results, no performance improvements were observed at higher packing densities (Song and Young, 1986). Cross-flow packing appeared to provide a higher process performance efficiency over randomly packed medium (Young and Yang, 1989). Typical COD loadings used, hydraulic retention times, and COD removal efficiencies for upflow packed-bed anaerobic reactors are reported in Table 10–18. At loadings of 1.0 to 6.0 kg COD/m^3·d, process efficiencies up to 90 percent are shown for high-strength wastewaters.

A large portion of the biomass responsible for treatment in the upflow attached growth anaerobic processes is loosely held in the packing void spaces and not just attached to the packing material (Young and Dehab, 1983). Low upflow velocities are generally used to prevent washing out the biomass. Over time, solids and biomass will accumulate in the packing to cause plugging and flow short circuiting. At this point, solids must be removed by flushing and draining the packing.

Advantages of upflow attached growth anaerobic reactors are high COD loadings, relatively small reactor volumes, and operational simplicity. The main limitations are the cost of the packing material and operational problems and maintenance associated with solids accumulation and possible packing plugging. The process is best suited for wastewaters with low suspended solids concentrations.

Table 10–18

Examples of process operating conditions and performance for upflow attached growth anaerobic reactors[a]

Wastewater	Packing type	Temp., °C	COD loading kg/m³·d	τ, d	Recycle ratio, R/Q	COD removed, %
Guar gum	Pall rings	37	7.7	1.2	5.0	61
Chemical processing	Pall rings	37	12–15	0.9–1.3	5.0	80–90
	Pall rings	15–25	0.1–1.2	0.5–0.75	0	50–70
Domestic	Tubular	37	0.2–0.7	25–37	0	90–96
Landfill leachate	Cross-flow	35	1.5–2.5	2.0–3.0	0.25	89
Food canning	Cross-flow	30	4–6	1.8–2.5	0	90
Soft drink	2-stage					

[a] Adapted from Young (1991).

Note: kg/m³·d × 62.4280 = lb/10³ ft³·d.

Upflow Attached Growth Anaerobic Expanded-Bed Reactor

In the upflow attached growth anaerobic expanded-bed reactor (AEBR) process (see Fig. 10–7b), the packing material is generally silica sand with a diameter in the range of 0.2 to 0.5 mm and specific gravity of 2.65. For operation with about 20 percent bed expansion, an upflow velocity of about 2 m/h is used. The smaller packing provides a greater surface area per unit volume, theoretically supporting a greater amount of bio-mass growth. The packing void fraction is about 50 percent when expanded and the specific surface area is in the range of 10,000 m²/m³ (Malina and Pohland, 1992). With such a small packing and void volume, the expanded-bed operation is necessary to prevent plugging. Because the expanded-bed system is not fully fluidized, some solids are trapped and some degree of solids degradation occurs (Morris and Jewell, 1981).

Most applications for the AEBR treatment process have been for the treatment of domestic wastewater (Jewell, 1987). More recently, Collins et al. (1998) have investigated the application of the AEBR for treating domestic wastewater at low temperatures. The results of these studies are summarized in Table 10–19, where the data for COD loading and removal efficiency at 20°C are within the same order of magnitude as shown for upflow packed anaerobic bed reactors treating industrial wastewater. The advantages and limitations are similar to those for the fluidized-bed reactor discussed next.

Attached Growth Anaerobic Fluidized-Bed Reactor

The attached growth anaerobic fluidized-bed reactor (FBR) (see Fig. 10–7c) is similar in physical design to the upflow expanded-bed reactor. The packing size (~0.3 mm sand) is similar to the expanded-bed reactor, but the FBR is operated at higher upflow liquid velocities of about 20 m/h to provide about 100 percent bed expansion. Effluent recycle is used to provide sufficient upflow velocity. Reactor depth ranges from 4 to 6 m.

Table 10–19
Performance of a bench-scale upflow anaerobic expanded-bed reactor treating domestic wastewater at low temperatures[a]

Temperature, °C	Organic loading, kg COD/m³·d	Percent COD removal
20	4.4	89
15	4.0	80
10	0.4	71
5	0.3	35

[a] Adapted from Alderman et al. (1998).
Note: kg/m³·d × 62.4280 = lb/10³ ft³·d.

Table 10–20
Benefits of using granular activated carbon (GAC) versus sand as the packing material in anaerobic FBRs

- Higher biomass concentration maintained due to porous structure of GAC
- Adsorption properties help prevent toxic and inhibitory substances from decreasing biological treatment performance
   Concentration of degradable substrate can be reduced to below toxic levels
   Other toxic substances can be removed to protect bacteria
- Adsorption properties may minimize shock loads by sorption of increased organics
- Adsorption properties may help acclimate and enhance biomass degradation of toxic compounds by providing more time of exposure

Besides sand, other packing materials have been considered for use in FBRs including diatomaceous earth, anion and cation exchange resins, and activated carbon (Wang et al., 1984; Kindzierski et al., 1992). Though a greater biomass concentration could be developed with the more porous diatomaceous earth packing, better performance did not result when compared to sand. Diffusion limitations into the biomass within the porous packing structure may explain the lack of improved treatment efficiency.

Activated carbon has been used in many anaerobic FBRs for treating industrial and hazardous waste streams (Hickey et al., 1991; Iza, 1991). The mean diameter of the granular activated carbon particles is 0.6 to 0.8 mm and upflow velocities of 20 to 24 m/h are used. Many benefits associated with using activated carbon in the anaerobic FBR (Wang et al., 1986; Fox et al., 1988) are listed in Table 10–20. The main limitation with activated carbon is the higher cost, but for certain types of industrial and hazardous waste streams the use of activated carbon is a necessity.

Solids capture is minimal in the anaerobic FBR due to the high turbulence and thin biofilms developed. With little solids capture, the process is better suited for wastewaters with mainly soluble COD. Solids discharged in the effluent from sloughed biofilm are minimized by controlling the biofilm inventory in the reactor. As biomass accumulates on the FBR packing, the net particle density decreases and the particle migrates to the top of the reactor. Periodic removal of these solids can control biofilm sloughing and minimize effluent TSS concentrations. The removed particles are mechanically processed to separate biomass from the sand, which is returned to the FBR.

Startup of anaerobic FBRs must be done with more care than the other types of high-rate anaerobic reactors. A higher hydraulic application rate is recommended at first to select for bacteria that more readily attach to the reactor packing under the highly turbulent conditions (Sutton and Huss, 1984; Denac and Dunn, 1988). The startup time can take 3 to 6 months. In a laboratory study by Tay and Zhang (2000) the startup and performance of an aerobic FBR was compared to that for an anaerobic UASB reactor and upflow packed-bed reactor. All three could achieve COD loading of 10 kg $COD/m^3 \cdot d$ at 35°C in 3 months with an influent COD concentration of 5000 mg/L (primarily glucose) and a τ of 12 h. The COD removal efficiency was best for the FBR and UASB reactors, about 96 percent compared to 90 percent for the upflow packed-bed reactor.

Process COD loading values of 10 to 20 kg $COD/m^3 \cdot d$ are feasible for anaerobic FBRs with greater than 90 percent COD removal, depending on the type of wastewater. Treatment performance is higher for FBRs than upflow packed-bed reactors at higher loadings, most likely due to a greater mass transfer rate due to the turbulent mixing. Reactor biomass concentrations of 15 to 20 g/L can be established in anaerobic FBRs (Malina and Pohland, 1992). Anaerobic FBR loadings and performance data are presented in Table 10–21. Bench-scale or pilot-plant studies are normally done before establishing full-scale design loadings.

The advantages for the anaerobic FBR process include the ability to provide high biomass concentrations and relatively high organic loadings, high mass transfer characteristics, the ability to handle shock loads due to its mixing and dilution with recycle, and minimal space requirements. The process is best suited for soluble wastewaters due to its inability to capture solids. Care must also be taken in the inlet and outlet designs to assure good flow distribution. Other disadvantages include the pumping power required to operate the fluidized bed, the cost of reactor packing, the need to control the packing level and wasting with biogrowth, and the length of startup time.

Downflow Attached Growth Processes

Downflow attached growth anaerobic processes, as illustrated on Fig. 10–8, have been applied for treatment of high-strength wastewaters using a variety of packing materials including cinder block, random plastic, and tubular plastic. Packing heights are in the range of 2 to 4 m, and systems are designed to allow recirculation of the reactor efflu-

Table 10–21
Examples of process operating conditions and performance for anaerobic FBRs[a]

Wastewater	Temperature, °C	COD loading, kg/m³·d	τ, h	COD removed, %
Citric acid	35	42	24	70
Starch, whey	35	8.2	105	99
Milk	37	3–5	18–12	71–85
Molasses	36	12–30	3–8	50–95
Glucose	35	10	12	95
Sulfite, pulp	35	3–18	3–62	60–80

[a] Adapted from Denac and Dunn (1988).
Note: kg/m³·d × 62.4280 = lb/10³ ft³·d.

Figure 10–8

Downflow attached growth anaerobic treatment reactor with plastic packing.

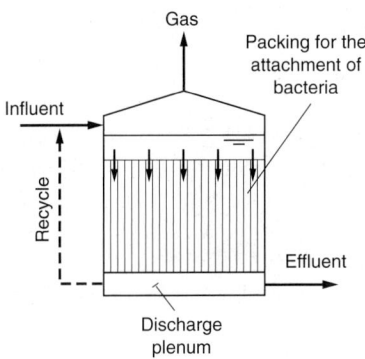

Table 10–22

Examples of process operating conditions and performance for anaerobic downflow attached growth applications[a]

Wastewater	Temperature, °C	COD loading, kg/m³·d	τ, h	COD removed, %
Citrus	38	1–6	24–144	40–80
Cheese whey	35	5–22	2–8	92–97
Sludge heat-treatment liquor	40	20–30		58
Brewery	35	20	1–2	76
Molasses	35	2–13	14–112	56–80
Piggery slurry	35	5–25	0.9–6.0	40–60

[a] Adapted in part from Speece (1996); Lomas et al. (1999); Fredericks et al. (1994); and Jhung and Choi (1995).
Note: kg/m³·d × 62.4280 = lb/10³ ft³·d.

ent. Anaerobic downflow attached growth reactors can be operated at loadings in the range of 5 to 10 kg COD/m³·d for easily degradable wastewaters. Operating performance varies for different wastewaters as shown in Table 10–22. Because plugging of the packing can be of concern, the use of a packing material with a high void volume, such as the vertical plastic packing used in tower trickling filters, is recommended.

The major advantages for the downflow attached growth process, where a higher void space packing material is used, are a simpler inlet flow distribution design, no plugging problems, and a simple operation. For systems with aerobic treatment following anaerobic treatment, the solids are captured in the aerobic process and thus do not accumulate in the attached growth process. Similar to the other anaerobic processes used for high-strength industrial wastewater treatment, benefits include the ability to treat high COD loadings with relatively small reactor volume sizes. Disadvantages include the cost of the packing material, and the somewhat lower organic loading rates to achieve the same treatment efficiency as the UASB and FBR processes.

10-6 OTHER ANAEROBIC TREATMENT PROCESSES

In addition to the anaerobic processes considered in Secs. 10–3 through 10–5, numerous other anaerobic processes have been developed and new processes are being developed continuously. Two such processes, the covered anaerobic earthen lagoon process and the membrane separation anaerobic treatment process, are considered briefly in the following discussion.

Covered Anaerobic Lagoon Process

Anaerobic lagoons have been used for high-strength industrial wastewaters, such as meat-processing wastewaters. Detention times ranged from 20 to 50 d with lagoon depths of 5 to 10 m (Crites and Tchobanoglous, 1998). A schematic of a simple nonproprietary covered anaerobic lagoon process is shown on Fig. 10–9a. A schematic of a proprietary covered anaerobic lagoon process known as the ADI-BVF® reactor is shown on Fig. 10–9b (McMullin et al., 1994). One of the main advantages for the covered lagoon processes is their ability to handle a wide range of waste characteristics including solids and oils and greases. Other advantages include simple and relatively economical construction, the large volume that can provide equalization of loads, the use of a low loading, and a high effluent quality. Disadvantages include the large land area required, potential feed flow distribution inefficiencies, and maintenance of the geomembrane cover.

For the plug-flow covered lagoon process shown on Fig. 10–10a, typical τ and SRT values for treating cow manure are on the order of 20 to 24 d, when treating a wastewater with a total solids concentration of about 8 to 10 percent. In general, process performance is best with fresh manure, as opposed to reconstituted manure that has dried for some period of time. The degree of waste conversion achieved will depend on the concentration of biodegradable COD in the manure. Typically, the VSS/TSS ratio of fresh manure will vary from 0.80 to 0.85. Of the VSS, about 30 to 40 percent is

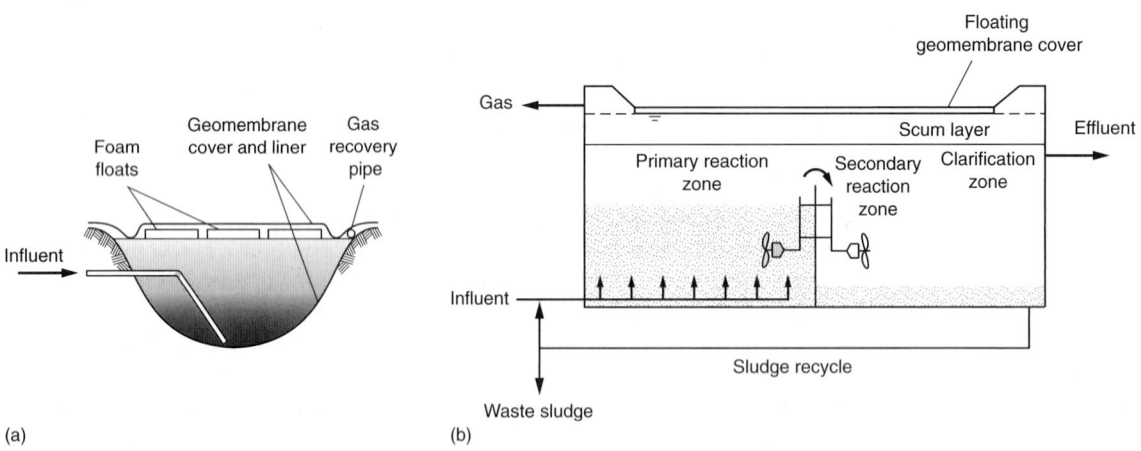

(a)

(b)

Figure 10–9

Schematic of covered anaerobic lagoon process: (a) simple nonproprietary design and (b) proprietary ADI-BVF® lagoon process.

(a)

(b)

Figure 10–10

Views of covered lagoon treatment process: (a) installation used for the treatment of cow manure and (b) installation for the treatment of piggery waste.

biodegradable. Detention times for the covered lagoon process shown on Figs. 10–9a and 10–10b are in the range of 30 to 50 d with solids concentrations varying from 0.5 to 5 percent, although higher values have been used. The SRT value for the covered lagoon process will be higher than the τ value because a large fraction of the influent solids will settle and undergo long-term degradation. It is estimated that the SRT values can vary from about 50 to 100 d. The performance of the covered lagoon process can be improved by building a series of ponds with the same total volume as a single large lagoon. Another consideration for lagoon processes is to separate the acid phase from the methane phase to reduce the total volume required and improve performance.

The ADI-BVF® reactor (see Fig. 10–9b) is typically of earthen and concrete construction with about 3- to 4-m vertical concrete sidewalls. A sloping interior floor and exterior earthen berms are used. The maximum liquid depth ranges from 7 to 9 m. The reactor has a floating geomembrane cover with a layer of closed-cell polyethylene insulation attached to its underside. The cover is sealed to the reactor perimeter, thereby allowing for the collection of biogas, temperature control, and positive odor control. The cover material is arranged in panels, with folds or "troughs" designed to allow rainwater collection in controlled locations and subsequent removal from the cover. Influent piping is situated along the bottom of the influent zone of the reactor in a design which provides distribution of influent throughout the sludge blanket. Sludge recirculation is employed and interior baffles are provided to promote retention of sludge within the influent zone. The wastewater flows through a series of gas-liquid-solids separators which act as internal clarifiers to inhibit the movement of solids to the reactor surface. The reactor also contains a low-speed mixer which operates on an intermittent basis for short periods of time. The ADI-BVF® process is normally operated as a low-loaded anaerobic process, with loadings in the range of 1.0 to 2.0 kg COD/m³·d. The

Table 10–23
Operation and treatment performance for the ADI-BVF® anaerobic covered lagoon process[a]

Wastewater	Temperature, °C	COD loading, kg/m³·d	τ, h	COD removed, %
Potato/pea processing	20–25	0.7	8.0	85
Apple juice	15–30	0.2	17	94

[a]Adapted from McMullin et al., 1994; Malina and Pohland, 1992.
Note: kg/m³·d × 62.4280 = lb/10³ ft³·d.

lagoon is generally not heated so that the temperatures may be 15 to 25°C, depending on the location and time of year. Typical performance data for the ADI-BVF® process are presented in Table 10–23. At very low loads, 80 to 90 percent COD removal is possible.

Membrane Separation Anaerobic Treatment Process

Basic concepts of membrane separation technology are described in detail in Chap. 11. The successful use of membrane separation in activated-sludge treatment is discussed in Chap. 8. Membrane separation has been considered for anaerobic reactors, but the technology is still in a developmental stage. The potential advantages for using membrane separation technology are (1) using higher biomass concentrations in the anaerobic reactor to further reduce its size and increase volumetric COD loadings, (2) allowing much higher SRTs in the anaerobic reactor by almost complete capture of solids that could result in maximum removal of VFAs and degradable soluble COD substances to provide a higher-quality effluent, and (3) maximizing capture of effluent suspended solids to greatly improve the effluent quality from anaerobic treatment. The latter two potential advantages of a longer SRT and low effluent suspended solids concentration should allow anaerobic reactors to produce an effluent quality equal to aerobic secondary treatment processes.

The major design considerations for the application of membrane separation in biological reactors (see Sec. 8–9 in Chap. 8) are (1) the membrane flux rate and (2) the ability to prevent fouling of the membrane to sustain acceptable flux rates. In the submerged membrane reactors used for aerobic suspended growth treatment the membranes are placed in the activated sludge reactor and coarse bubble aeration is used to minimize fouling. For anaerobic processes a similar type of gas agitation system might be possible. In current designs, as shown on Fig. 10–11, an external cross-flow membrane separation unit is used.

To control fouling, high liquid velocities are maintained across the membrane. The solids in the reject stream are recycled back to the anaerobic reactor, while clear permeate is removed for effluent discharge. Fouling problems and loss of active cells have been reported (Choo and Lee, 1998). Organic fouling problems are typically caused by the accumulation of colloidal material and bacteria on the membrane surface. High pumping flowrates across the membrane may lead to the loss of viable bacteria due to cell lysis. Inorganic fouling is due to the formation of struvite ($MgNH_4PO_4$), which precipitates because of a rise in the pH as the flow passes through the membranes and CO_2 escapes from the liquid. Observed performance data derived from a limited number of membrane reactor studies are summarized in Table 10–24. In general, the process

Figure 10–11

Anaerobic bioreactor with external membrane separation of solids in effluent stream.

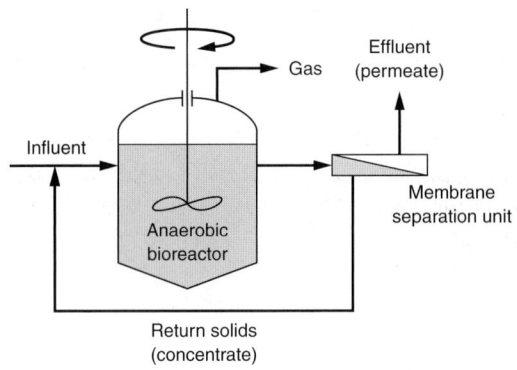

Table 10–24

Summary of operating conditions from a limited number of membrane separation anaerobic reactor studies

Parameter	Unit	Value	
		Range	Typical
COD loading	kg COD/m³·d	2–22	8
τ	d	0.5–15.0	8
SRT	d	30–160	50
MLSS	g/L	8–50	20–25
Solids yield	g VSS/g COD	0.04–0.12	0.08
Final flux	L/m²·h	5–26	18
Specific flux	L/m²·h·bar	2.5–21	14

Note: kg/m³·d × 62.4280 = lb/10³ ft³·d.
L/m²·h × 0.0245 = gal/ft²·h.

organic loading rates are in the range of those used with the sludge blanket and attached growth processes. Further developments in membrane design and fouling control measures could make membrane separation anaerobic reactors a viable technology in the future.

PROBLEMS AND DISCUSSION TOPICS

10-1   The alkalinity and pH are measured in an anaerobic suspended growth reactor operated at 30°C. The alkalinity values are 1800, 2000, or 2200 mg as $CaCO_3$/L (value to be selected by instructor) and the pH is 7.2. Assuming equilibrium between the liquid and gas phases, determine the approximate CO_2 content of the gas phase in percent.

10-2   An industrial wastewater with a flow of 4000 m³/d has a soluble degradable COD concentration of 1800, 2200, or 3000 mg/L (value to be selected by instructor), a temperature of 20°C and an alkalinity concentration of 200 mg/L as $CaCO_3$. Determine and compare the net operating costs or revenue for anaerobic versus aerobic treatment based on the following key parameters and assumptions (labor and maintenance costs are omitted here) for each:

Anaerobic process

Anaerobic operating cost items are related to raising the liquid temperature, and adding alkalinity, versus the revenue from methane production. The following assumptions apply:

1. Reactor temperature = 35°C
2. Heat exchanger efficiency for raising liquid temperature = 80%
3. COD removal efficiency = 95 percent
4. Percent CO_2 of gas phase = 35 percent and pH = 7.0
5. Value of methane = $5/10^6$ kJ
6. Alkalinity is provided as $NaHCO_3$ at $0.90/kg

Aerobic

Major aerobic treatment operating cost items are energy for aeration and sludge processing and disposal. The following assumptions apply:

1. COD removal efficiency = 99 percent
2. g O_2/g COD removal = 1.2
3. Actual aeration efficiency = 1.2 kg O_2/kw-hr
4. Electricity costs = $0.08/kWh
5. Net sludge production = 0.3 g TSS/g COD removed
6. Sludge processing/disposal cost = $0.10/kg dry solids

10-3 A wastewater has a daily average flowrate of 1000, 2000, or 3000 m³/d (value to be selected by instructor) and 4000 mg/L of an organic substance with the following approximate composition: $C_{50}H_{75}O_{20}N_5S$. For anaerobic treatment at 95 percent degradation determine: (a) the alkalinity production in mg/L as $CaCO_3$; and (b) the approximate mole fraction of CO_2, CH_4, and H_2S in the gas phase.

10-4 An industrial wastewater has an average flowrate of 2000 m³/d, an influent COD concentration of 4000, 6000, or 8000 mg/L (value to be selected by instructor), and influent sulfate concentration of 500 mg/L. The percent of COD degraded in an anaerobic treatment process at 35°C is 95 percent, and 98 percent of the sulfate reduced. Determine: (a) the amount of methane produced in m³/d; (b) the amount of methane produced in m³/d, if the sulfate reduction is not accounted for; and (c) the amount of H_2S in the gas phase at a reactor pH value of 7.0.

10-5 A suspended growth anaerobic reactor is operated at an SRT of 30 days at a temperature of 30°C. On a given day, the methane gas production rate (m³/d) decreases by 30 percent. List at least four possible causes that should be investigated and briefly explain the mechanism behind each one.

10-6 A 100 percent soluble industrial wastewater is to be treated by an anaerobic suspended growth process consisting of a mixed covered reactor, a degasifier, and gravity settling. The effluent TSS concentration from the clarifier is 120 mg/L. For the following wastewater characteristics and design assumptions determine and compare the following design parameters for treatment at 25°C and 35°C:

1. The design SRT, d
2. The reactor volume, m³
3. The reactor hydraulic detention time τ, d
4. The methane gas production rate, m³/d
5. The total gas production rate, m³/d

6. The amount of solids to be manually wasted daily, kg/d
7. The nitrogen and phosphorus requirements, kg/d

Wastewater characteristics:

Parameter	Unit	Value
Flow	m³/d	2000
Degradable COD	mg/L	A (4000)
		B (6000)
		C (8000)
Percent sCOD	%	100
Alkalinity	mg/L as CaCO	500

Note: value A, B, or C to be selected by instructor

Other design assumptions:

1. Reactor MLSS concentration = 5000 mg/L
2. Factor of safety for SRT = 1.5
3. VSS/TSS ratio = 0.85
4. f_d = 0.15 g VSS cell debris/g VSS biomass decay
5. Gas phase percent methane = 65 percent
6. Nitrogen content of biomass = 0.12 g N/g VSS5
7. Phosphorus content of biomass = 0.02 g P/g VSS
8. Use the appropriate kinetic coefficients and design information provided in Table 10–10

10-7 An anaerobic process is being considered for the treatment of a soluble industrial wastewater at 30°C. A design SRT of 30 d is required to provide the desired level of 95 percent soluble COD degradation. An effluent VSS concentration of 100, 150, or 200 mg/L (value to be selected by instructor) from biomass growth is assumed. Using the appropriate kinetic coefficient values from Table 10–10, determine the influent sCOD concentration that must be present to allow operation at a 30-d SRT, if all the biomass wasted is via the effluent solids losses.

10-8 Design a single UASB reactor to treat an industrial wastewater at 30°C with the following wastewater characteristics and using the assumptions given below. Design for 97 percent degradation of the soluble COD, and assume 50 percent particulate COD degradation and an effluent VSS concentration of 200 mg/L. Using the given information, determine:

1. The reactor liquid volume, m³
2. The reactor area (assume a circular reactor will be used), m²
3. The reactor area diameter, liquid depth, and total height, m
4. The hydraulic retention time, d
5. The average SRT, d
6. The amount of solids to be manually wasted daily, kg/d
7. The average MLVSS and MLSS concentration in the UASB reactor, mg/L
8. The methane gas production rate, m³/d
9. The energy value of the gas, kJ/d
10. The alkalinity requirement, kg as CaCO₃/d

Design Assumptions:

1. Kinetic coefficients from Table 10–10.
2. Reactor effectiveness factor (E) = 0.85
3. f_d = 0.15 g VSS/g VSS biomass decayed
4. pH = 7.0
5. Percent CO_2 in gas phase = 35 percent
6. Maximum reactor liquid height = 8 m

Wastewater characteristics:

Parameter	Unit	Value
Flow	m³/d	500
Total bCOD	mg/L	A (6000)
		B (7000)
		C (8000)
Particulate COD	Percent	40
Particulate COD/VSS ratio	g/g	1.8
Particulate VSS/TSS ratio	g/g	0.85
Alkalinity	mg/L as $CaCO_3$	300

Note: value A, B, or C to be selected by instructor

10-9 A domestic wastewater is to be treated using the UASB process at 25°C. The wastewater characteristics are given in the following table. Determine: (a) the reactor hydraulic retention time (hours); (b) the COD loading rate (kg COD/m³·d); and (c) the reactor liquid height (m) and diameter (m). What effluent BOD and TSS concentration may be expected from the UASB reactor? Describe an aerobic secondary treatment process you would select to add after the UASB process to meet an effluent BOD concentration of 20 mg/L or less. Would alkalinity have to be added to the UASB reactor to maintain the pH near 7.0? Explain the basis for your answer.

Wastewater characteristics:

Parameter	Unit	Value
Flow	m³/d	A (3000)
		B (4000)
		C (5000)
COD	mg/L	450
BOD	mg/L	180
TSS	mg/L	180
Alkalinity	mg/L as $CaCO_3$	150

Note: value A, B, or C to be selected by instructor

10-10   A brewery wastewater with a flow of 1000 m³/d and COD (mainly soluble) of 4000 mg/L is to be treated at 35°C in an upflow attached growth anaerobic reactor, which contains cross-flow plastic packing, with the aim of 90 percent COD removal. Assume that the attached growth SRT is 30 days. Determine: (a) the reactor volume (m³) and dimensions; (b) the methane gas production rate (m³/d); and (c) the effluent TSS concentration (mg/L).

10-11   An industrial wastewater has a degradable COD concentration of 8000 mg/L and 4000 mg/L VSS concentration with 50 percent of the VSS degradable. Briefly critique the compatibility of the following processes for treatment of this wastewater and describe the potential impact of the influent solids on the reactor operation and performance.

Processes:

UASB
Anaerobic fluidized-bed reactor
Anaerobic baffled reactor
Upflow packed-bed reactor
Downflow attached growth reactor
Anaerobic covered lagoon

10-12   From the literature within the past three years, identify and summarize an application of a membrane separation anaerobic treatment process. Include a description of the wastewater treated, the reactor design, the organic loading rate, the temperature, the membrane fouling control strategy, the membrane flux rate over time, the reactor solids concentration, the membrane cleaning method or restoration method, and any significant operating and performance issues.

REFERENCES

Alderman, B. J., T. L. Theis, and A. G. Collins (1998) "Optimal Design for Anaerobic Pretreatment of Municipal Wastewater," *Journal of Environmental Engineering,* vol. 124, no. 1, pp. 4–10.

Angenent, L. T., G. C. Banik, and S. Sung (2000) "Psychrophilic Anaerobic Pretreatment of Low-Strength Wastewater Using the Anaerobic Migrating Blanket Reactor," *Proceedings of the 73rd Annual Water Environment Federation Conference,* New Orleans, LA.

Annachhatre, A. P. (1996) "Anaerobic Treatment of Industrial Wastewaters," *Resources, Conservation, and Recycling,* vol. 16, pp. 161–166.

Arceivala, S. J. (1998) *Wastewater Treatment for Pollution Control,* 2d ed., McGraw-Hill, New Delhi.

Bachmann, A., V. L. Beard, and P. L. McCarty (1985) "Performance Characteristics of the Anaerobic Baffled Reactor," *Water Research,* vol. 19, no. 1, pp. 99–106.

Banik, G. C., and R. R. Dague (1996) "ASBR Treatment of Dilute Wastewater at Psychrophilic Temperatures," *Proceedings of the 69th Annual Water Environment Federation Conference,* Chicago.

Barber, W. P., and D. C. Stuckey (1999) "The Use of the Anaerobic Baffled Reactor (ABR) for Wastewater Treatment: A Review," Water Research, vol. 33, no. 7, pp. 1559–1578.

Boopathy, R., and A. Tilche (1992) "Pelletization of Biomass in a Hybrid Anaerobic Baffled Reactor (HABR) Treating Acidified Wastewater," *Bioresource Technology,* vol. 40, no. 2, pp. 101–107.

Buswell, A. W., and C. B. Boruff (1932) "The Relationship Between Chemical Composition of Organic Matter and the Quality and Quantity of Gas Production During Digestion," *Sewage Works Journal,* vol. 4, no. 3, p. 454.

Choo, K. H., and C. H. Lee (1998) "Hydrodynamic Behavior of Anaerobic Biosolids During Crossflow Filtration in the Membrane Anaerobic Bioreactor," *Water Research,* vol. 32, pp. 3387–3397.

Collins, A. G., T. L. Theis, S. Kilambi, L. He, and S. G. Paulostathis (1998) "Anaerobic Treatment of Low-Strength Domestic Wastewater Using an Anaerobic Expanded Bed Reactor," *Journal Environmental Engineering,* vol. 124, no. 7, pp. 652–655.

Crites, R., and G. Tchobanoglous (1998) *Small and Decentralized Wastewater Management Systems,* McGraw-Hill, New York.

Denac, M., and I. J. Dunn (1988) "Packed- and Fluidized-Bed Biofilm Reactor Performance for Anaerobic Wastewater Treatment," *Biotechnology and Bioengineering,* vol. 32, pp. 159–173.

Elmaleh, S., and L. Abdelmoumni (1998) "Experimental Test to Evaluate Performance of an Anaerobic Reactor Provided with an External Membrane Unit," *Water Science and Technology,* vol. 38, no. 8, pp. 385–392.

Fox, P., M. T. Suidan, and J. T. Pfeffer (1988) "Anaerobic Treatment of Biologically Inhibitory Wastewater," *Journal Water Pollution Control Federation,* vol. 60, pp. 86–92.

Fredericks, D. W., T. P. Hovanec, D. E. Foster, and L. M. Szendry (1994) "Start-up and Operational Performance of a Downflow Anaerobic Filter on Citrus Processing Wastewater," *Proceedings of the 67th Annual Water Environment Federation Conference.*

Fullen, W. J. (1953) "Anaerobic Digestion of Packing Plant Wastes," *Sewage and Industrial Wastes,* vol. 25, p. 577.

Garuti, G., M. Dohanyos, and A. Tilche (1992) "Anaerobic-Aerobic Combined Process for the Treatment of Sewage with Nutrient Removal: The Ananox® Process," *Water Research,* vol. 25, no. 7, pp. 383–394.

Goncalves, R. F., and V. L. de Avaujo (1999) "Combining Upflow Anaerobic Sludge Blanket (UASB) Reactors and Submerged Aerated Biofilters for Secondary Domestic Wastewater Treatment," *Water Science and Technology,* vol. 40, no. 8, p. 79.

Hickey, R. F., W. M. Wu, M. C. Viega, and R. June (1991) "Start-up Operation, Monitoring, and Control of High-Rate Anaerobic Treatment Systems," *Water Science and Technology,* vol. 24, no. 8, pp. 207–256.

Iza, J. (1991) "Fluidized Bed Reactors for Anaerobic Wastewater Treatment," *Water Science and Technology,* vol. 24, no. 8, pp. 109–132.

Jewell, W. J. (1987) "Anaerobic Sewage Treatment," *Environmental Science and Technology,* vol. 21, no. 1, pp. 14–21.

Jhung, J. K., and F. Choi (1995) "A Comparative Study of UASB and Anaerobic Fixed Film Reactors with Development of Sludge Granulation," *Water Research,* vol. 29, no. 1, pp. 271–277.

Kindzierski, W. B., M. R. Gray, P. M. Fedorak, and S. E. Hrudey (1992) "Activated Carbon and Synthetic Resins as Support Material for Methanogenic Phenol-Degrading Consortia—Comparison of Surface Characteristics and Initial Colonization," *Water Environment Research,* vol. 64, pp. 786–795.

Lay, J. J., Y. Y. Li, and T. Noike (1998) "The Influence of pH and Ammonia Concentration on the Methane Production in High-Solids Digestion Processes," *Water Environment Research,* vol. 70, pp. 1075–1082.

Lettinga, G., and L. W. Hulshoff Pol (1991) "UASB-Process Designs for Various Types of Wastewaters," *Water Science and Technology,* vol. 24, no. 8, pp. 87–107.

Lettinga, G., and J. N. Vinken (1980) "Feasibility of the Upflow Anaerobic-Sludge Blanket (UASB) Process for the Treatment of Low-Strength Wastes," *Proceedings of the 35th Industrial Waste Conference,* Purdue University, 1980.

Lettinga, G., A. F. M. van Velsen, S. W. Hobma, W. J. de Zeeuw, and A. Klapwijk (1980) "Use of the Upflow Sludge Blanket (USB) Reactor Concept for Biological Wastewater Treatment," *Biotechnology and Bioengineering,* vol. 22, pp. 699–734.

Lomas, J. M., C. Urbano, and L. M. Camarero (1999) "Evaluation of a Pilot Scale Downflow Stationary Fixed Film Anaerobic Reactor Treating Piggery Slurry in the Mesophilic Range," *Biomass & Bioenergy,* vol. 17, pp. 49–58.

McCarty, P. L. (1964) "Anaerobic Waste Treatment Fundamentals: I. Chemistry and Microbiology; II. Environmental Requirements and Control; III. Toxic Materials and Their Control; IV. Process Design," *Public Works,* nos. 9–12.

McCarty, P. L., and R. E. McKinney (1961) "Salt Toxicity in Anaerobic Digestion," *Journal of the Water Pollution Control Federation,* vol. 33, p. 399.

McMullin, M. J., R. C. Landine, and R. C. Landine (1994) "Anaerobic Pretreatment of Beverage Industry Wastewater-Case History," *Proceedings of the 67th Annual Water Environment Federation Conference,* Chicago, IL.

Maillacheruvu, K. Y., G. F. Parkin, C. Y. Peng, W. C. Kuo, Z. I. Oonge, and V. Lebduschka (1993) "Sulfide Toxicity in Anaerobic Systems Fed Sulfate and Various Organics," *Water Environment Research,* vol. 65, pp. 100–109.

Malina, J. F., and F. G. Pohland (1992) *Design of Anaerobic Processes for the Treatment of Industrial and Municipal Wastes,* Water Quality Management Library, vol. 7, Technomic Publishing Co., Lancaster, PA.

Moen, G. M. (2000) *A Comparison of the Performance and Kinetics of Thermophilic and Mesophilic Anaerobic Digestion,* Master of Science Thesis, Department of Civil and Environmental Engineering, University of Washington, Seattle, WA.

Morris, J. W., and W. J. Jewell (1981) "Organic Particulate Removal with the Anaerobic Attached-Film-Expanded-Bed Process," *Proceedings of 36th Industrial Waste Conference,* Purdue University, pp. 621–630.

Obayashi, A., H. D. Stensel, and E. G. Kominek (1981) "Anaerobic Treatment of High Strength Industrial Wastewater," *Chemical Engineering Progress,* vol. 77, no. 4, p. 79.

Orozco, A. (1988) "Pilot and Full-Scale Anaerobic Treatment of Low-Strength Wastewater at Sub-Optimal Temperature (13°) with a Hybrid Plug Flow Reactor," *Proceedings of the 8th International Conference on Anaerobic Digestion,* vol. 2, pp. 183–191, Sendai, Japan.

Palns, S. S., R. E. Lowenthal, P. L. Dold, and G. R. Marais (1987) "Hypothesis for Pelletisation in the Upflow Anaerobic Sludge Bed Reactor," *Water* (South African Journal), vol. 13, pp. 69–80.

Palns, S. S., R. E. Lowenthal, M. C. Wentzel, and G. R. Marais (1990) "Effect of Nitrogen Limitation on Pelletisation in Upflow Anaerobic Sludge Bed (UASB) Systems," *Water* (South African Journal), vol. 16, pp. 165–170.

Parkin, G. F., and S. W. Miller (1982) "Response of Methane Fermentation to Continuous Addition of Selected Industrial Toxicants," *Proceedings of the 37th Purdue Industrial Waste Conference,* Lafayette, IN.

Parkin, G. F., and W. E. Owen (1986) "Fundamentals of Anaerobic Digestion of Wastewater Sludges," *Journal Environmental Engineering,* vol. 112, no. 5, pp. 867–920.

Rajeshwari, K. V., M. Balakrishnan, A. Kansal, K. Lata, and V. V. N. Kishore (2000) "State-of-the-Art of Anaerobic Digestion Technology for Industrial Wastewater Treatment," *Renewable and Sustainable Energy Reviews,* vol. 4, pp. 135–156.

Randall, C. W., J. L. Barnard, and H. D. Stensel (1992) *Design and Retrofit of Wastewater Treatment Plants for Biological Nutrient Removal,* vol. 5, Water Quality Management Library, Technomic Publishing Co., Lancaster, PA.

Schroepfer, G. J., W. J. Fullen, A. S. Johnson, N. R. Ziemke, and J. J. Anderson (1955) "The Anaerobic Contact Process as Applied to Packinghouse Wastes," *Sewage and Industrial Wastes,* vol. 27, p. 61.

Song, K. H., and J. C. Young (1986) "Media Design Factors for Fixed-Bed Anaerobic Filters," *Journal Water Pollution Control Federation,* vol. 58, pp. 115–121.

Soto, M., P. Ligero, V. I. Ruiz, M. C. Veiga, and R. Blazquez (1997) "Sludge Granulation in UASB Digesters Treating Low Strength Wastewaters of Mesophilic and Psychrophilic Temperatures," *Environmental Technology,* vol. 18, pp. 1133–1141.

Speece, R. E. (1996) *Anaerobic Biotechnology for Industrial Wastewaters,* Archae Press, Nashville, TN.

Steffen, A. J., and M. Bedker (1961) "Operation of Full-Scale Anaerobic Contact Treatment Plant for Meat Packing Wastes," *Proceedings of the 16th Industrial Waste Conference,* Purdue University, Lafayette, IN, p. 423.

Stephenson, T., S. Judd, B. Jefferson, and K. Brindle (2000) *Membrane Bioreactors for Wastewater Treatment,* IWA Publishing, Cornwall, U.K.

Sung, S., and R. R. Dague (1995) "Laboratory Studies on the Anaerobic Sequencing Batch Reactor," *Water Environment Research,* vol. 67, pp. 294–301.

Sutton, P. M., and D. A. Huss (1984) "Anaerobic Fluidized Biological Treatment: Pilot to Full-Scale Demonstration," *Proceedings of the 57th Annual Water Pollution Control Federation Conference,* October, New Orleans, LA.

Sykes, R. M. (2001) Reviewers comments and additions.

Tay, J. H., and X. Zhang (2000) "Stability of High-Rate Anaerobic Systems. I. Performance under Shocks," *Journal Environmental Engineering,* vol. 126, no. 8, pp. 713–725.

Thaveesri, J., K. Gernaey, B. Kaonga, G. Boucneau, and W. Verstraete (1994) "Organic and Ammonium Nitrogen and Oxygen in Relation to Granular Sludge Growth in Lab-Scale UASB Reactors," *Water Science and Technology,* vol. 30, no. 12, pp. 43–53.

van Velsen, A. F. M. (1977) "Anaerobic Digestion of Piggery Waste," *Netherlands Journal of Agricultural Science,* vol. 25, p. 151.

Wang, Y. T., M. T. Suidan, and J. T. Pfeffer (1984) "Anaerobic Activated Carbon Filter for the Degradation of Polycyclic N. Aromatic Compounds," *Journal Water Pollution Control Federation,* vol. 56, pp. 1247–1253.

Wang, Y. T., M. T. Suidan, and B. E. Rittman (1986) "Anaerobic Treatment of Phenol by an Expanded Bed Reactor," *Journal Water Pollution Control Federation,* vol. 58, pp. 227–233.

Young, J. C. (1991) "Factors Affecting the Design and Performance of Upflow Anaerobic Filters," *Water Science and Technology,* vol. 24, no. 8, pp. 133–155.

Young, J. C., and M. F. Dehab (1983) "Effect of Media Design on the Performance of Fixed-Bed Anaerobic Reactors," *Water Science and Technology,* vol. 15, pp. 369–383.

Young, J. C., and B. S. Yang (1989) "Design Considerations for Full-Scale Anaerobic Filters," *Journal Water Pollution Control Federation,* vol. 61, no. 9, pp. 1570–1587.

11

Advanced Wastewater Treatment

Advanced wastewater treatment is defined as the additional treatment needed to remove suspended, colloidal, and dissolved constituents remaining after conventional secondary treatment. Dissolved constituents may range from relatively simple inorganic ions, such as calcium, potassium, sulfate, nitrate, and phosphate, to an ever-increasing number of highly complex synthetic organic compounds. In recent years, the effects of many of these substances on the environment have become understood more clearly. Research is ongoing to determine (1) the environmental effects of potential toxic and biologically active substances found in wastewater and (2) how these substances can be removed by both conventional and advanced wastewater-treatment processes. As a result, wastewater-treatment requirements are becoming more stringent in terms of both limiting concentrations of many of these substances in the treatment plant effluent and establishing whole effluent toxicity limits, as outlined in Chap. 2. To meet these new requirements, many of the existing secondary treatment facilities will have to be retrofitted and new advanced wastewater-treatment facilities will have to be constructed. Therefore, the purpose of this chapter is to present an introduction to the subject of advanced wastewater treatment. The chapter contains an expanded discussion of the need for advanced wastewater treatment, an overview of the available technologies used for the removal of the constituents of concern, identified previously in Chap. 2, and an introduction to the more important of these technologies as applied to the removal of specific constituents found in wastewater. The ultimate disposal of residuals from advanced wastewater treatment is considered in Chap. 14.

11–1   NEED FOR ADVANCED WASTEWATER TREATMENT

The need for advanced wastewater treatment is based on a consideration of one or more of the following factors.

1. The need to remove organic matter and total suspended solids beyond what can be accomplished by conventional secondary treatment processes to meet more stringent discharge and reuse requirements.
2. The need to remove residual total suspended solids to condition the treated wastewater for more effective disinfection.
3. The need to remove nutrients beyond what can be accomplished by conventional secondary treatment processes to limit eutrophication of sensitive water bodies.
4. The need to remove specific inorganic (e.g., heavy metals) and organic constituents (e.g., MBTE and NDMA) to meet more stringent discharge and reuse requirements for both surface water and land-based effluent dispersal and for indirect potable reuse applications (e.g., groundwater recharge).
5. The need to remove specific inorganic (e.g., heavy metals, silica) and organic constituents for industrial reuse (e.g., cooling water, process water, low-pressure boiler makeup water, and high-pressure boiler water).

With increased scientific knowledge derived from laboratory studies and environmental monitoring concerning the impacts of the residual constituents found in secondary effluent, it is anticipated that many of the methods now classified as advanced will become conventional within the next 5 to 10 years.

Compounds containing available nitrogen and phosphorus have received considerable attention since the mid-1960s. Initially, nitrogen and phosphorus in wastewater discharges became important because of their effects in accelerating eutrophication of lakes and promoting aquatic growths. More recently, nutrient control has become a routine part of treating wastewaters used for the recharge of groundwater supplies. Nitrification of wastewater discharges is also required in many cases to reduce ammonia toxicity or to lessen the impact on the oxygen resources in flowing streams or estuaries. As a result of the many concerns over nutrients, nutrient removal has become, for all practical purposes, an integral part of conventional wastewater treatment. Thus, the removal of nitrogen and phosphorus by biological processes is considered in Chaps. 8 and 9. The removal of phosphorus by chemical methods is discussed in Chap. 6.

11-2 TECHNOLOGIES USED FOR ADVANCED TREATMENT

Over the past 20 years, a wide variety of treatment technologies have been studied, developed, and applied for the removal of the residual constituents found in treated effluent. The residual constituents found in treated wastewater are reviewed and the technologies used for the advanced treatment of wastewater are introduced in this section.

Residual Constituents in Treated Wastewater

The typical composition of domestic wastewater was reported previously in Table 3–15. In addition to the constituents reported in Table 3–15, most domestic wastewaters also contain a wide variety of trace compounds and elements, although they are not measured routinely. The potential effects of residual constituents contained in treated effluents may vary considerably. Some of the effects of specific constituents are listed in Table 11–1. The list presented in Table 11–1 is not meant to be exhaustive; rather, it is meant to highlight that a wide variety of substances must be considered in establishing and meeting discharge requirements.

Classification of Technologies

Advanced wastewater-treatment systems may be classified by the type of unit operation or process or by the principal removal function performed. To facilitate a general comparison of the various operations and processes, information on (1) the principal residual constituent removal function and (2) the types of operations or processes that can be used to perform this function are presented in Table 11–2. As will be noted in reviewing Table 11–2, many of the operations and processes can be used for the removal of a number of constituents. The individual constituents listed in Table 11–2 can be grouped into four broad categories requiring removal: (1) residual organic and inorganic colloidal and suspended solids, (2) dissolved organic constituents, (3) dissolved inorganic constituents, and (4) biological constituents. Typical advanced treatment process flow diagrams incorporating many of the technologies listed in Table 11–2 are illustrated on Fig. 11–1.

Removal of Organic and Inorganic Colloidal and Suspended Solids

Removal of organic and inorganic colloidal and suspended solids is typically accomplished by filtration. A general classification of the filtration processes commonly used in wastewater engineering is presented on Fig. 11–2. As illustrated on Fig. 11–2, the fil-

Table 11–1

Typical residual constituents found in treated wastewater effluents and their impacts

Residual constituent	Effect
Inorganic and organic colloidal and suspended solids	
Suspended solids	• May cause sludge deposits or interfere with receiving water clarity
	• Can impact disinfection by shielding organisms
Colloidal solids	• May affect effluent turbidity
Organic matter (particulate)	• May shield bacteria during disinfection, may deplete oxygen resources
Dissolved organic matter	
Total organic carbon	• May deplete oxygen resources
Refractory organics	• Toxic to humans; carcinogenic
Volatile organic compounds	• Toxic to humans; carcinogenic; form photochemical oxidants
Pharmaceutical compounds	• Impact aquatic species (e.g., endocrine disruption, sex reversal)
Surfactants	• Cause foaming and may interfere with coagulation
Dissolved inorganic matter	
Ammonia	• Increases chlorine demand
	• Can be converted to nitrates and, in the process, can deplete oxygen resources
	• With phosphorus, can lead to the development of undesirable aquatic growths
	• Toxic to fish
Nitrate	• Stimulates algal and aquatic growth
	• Can cause methemoglobinemia in infants (blue babies)
Phosphorus	• Stimulates algal and aquatic growth
	• Interferes with coagulation
	• Interferes with lime-soda softening
Calcium and magnesium	• Increase hardness and total dissolved solids
Chloride	• Imparts salty taste
Total dissolved solids	• Interfere with agricultural and industrial processes
Biological	
Bacteria	• May cause disease
Protozoan cysts and oocysts	• May cause disease
Viruses	• May cause disease

ters used for wastewater fall into three general classifications: (1) depth filtration, (2) surface filtration, and (3) membrane filtration. In depth filtration, the removal of suspended material occurs within and on the surface of the filter bed (see Fig. 11–3a). In surface and membrane filtration, the suspended material is removed by straining through a straining surface (e.g., filter cloth) or a thin supported membrane (see Fig. 11–3b). An automobile oil filter and a kitchen colander are diverse examples of surface filters.

Table 11-2

Unit operations and processes for the removal of residual constituents found in treated wastewater effluents[a]

	Unit operation or process (section discussed)				
Residual constituent	**Depth filtration (11–4)**	**Surface filtration (11–5)**	**Micro- and ultra- filtration (11–6)**	**Reverse osmosis (11–6)**	**Electro- dialysis (11–6)**
Inorganic and organic colloidal and suspended solids					
Suspended solids	✔	✔	✔	✔	✔
Colloidal solids	✔	✔	✔	✔	✔
Organic matter (particulate)				✔	✔
Dissolved organic matter					
Total organic carbon				✔	✔
Refractory organics				✔	✔
Volatile organic compounds				✔	✔
Dissolved inorganic matter					
Ammonia[a]				✔	✔
Nitrate[a]				✔	✔
Phosphorus[a]	✔[b]			✔	✔
Total dissolved solids				✔	✔
Biological					
Bacteria			✔	✔	✔
Protozoan cysts and oocysts	✔		✔	✔	✔
Viruses				✔	✔

[a] The biological removal of nitrogen and phosphorus is considered in Chaps. 7 through 10.

[b] Phosphorous removal is accomplished in a two-stage filtration process.

[c] Some carryover can occur.

Removal of Dissolved Organic Constituents

Many treatment methods can be used for the removal of dissolved organic constituents. Because of the complex nature of the dissolved organic constituents, the treatment methods must consider the specific characteristics of the wastewater and the nature of the constituents. Treatment processes used to remove some of the specific dissolved organic constituents include (1) carbon adsorption, (2) reverse osmosis, (3) chemical precipitation, (4) chemical oxidation, (5) advanced chemical oxidation, (6) electrodialysis, and (7) distillation. Chemical precipitation and chemical oxidation are discussed in Chap. 6. Biological treatment for nutrient removal is discussed in Chaps. 7 through 10.

| **Table 11-2** (*Continued*)

Unit operation or process (section discussed)						
Adsorption (11–7)	Air stripping (11–8)	Ion exchange (11–9)	Advanced oxidation processes (11–10)	Distillation (11–11)	Chemical precipitation (6–3, 4, 5)	Chemical oxidation (6–7)
✔		✔		✔	✔	
✔		✔		✔	✔	
				✔		✔
✔		✔	✔	✔	✔	✔
✔			✔	✔		
✔	✔		✔	✔ᶜ		
	✔	✔		✔		
		✔		✔		
				✔	✔	
		✔		✔		
				✔		
✔		✔		✔	✔	
				✔		

Removal of Dissolved Inorganic Constituents

As reported in Table 11–2, a number of different unit operations and processes have been investigated in various advanced wastewater-treatment applications. Although many of them have proved to be technically feasible, other factors, such as cost, operational requirements, and aesthetic considerations, have not been favorable in some cases. Nevertheless, it is important that environmental engineers be familiar with the more important operations and processes so that in any given situation they can consider all treatment possibilities. Removal of dissolved inorganic constituents is accomplished by chemical processes or membrane filtration. The principal unit operations and processes are (1) chemical precipitation (discussed in Chap. 6), (2) ion exchange, (3) ultrafiltration, (4) reverse osmosis, (5) electrodialysis, and (6) distillation. As noted previously, chemical precipitation is discussed in Chap. 6.

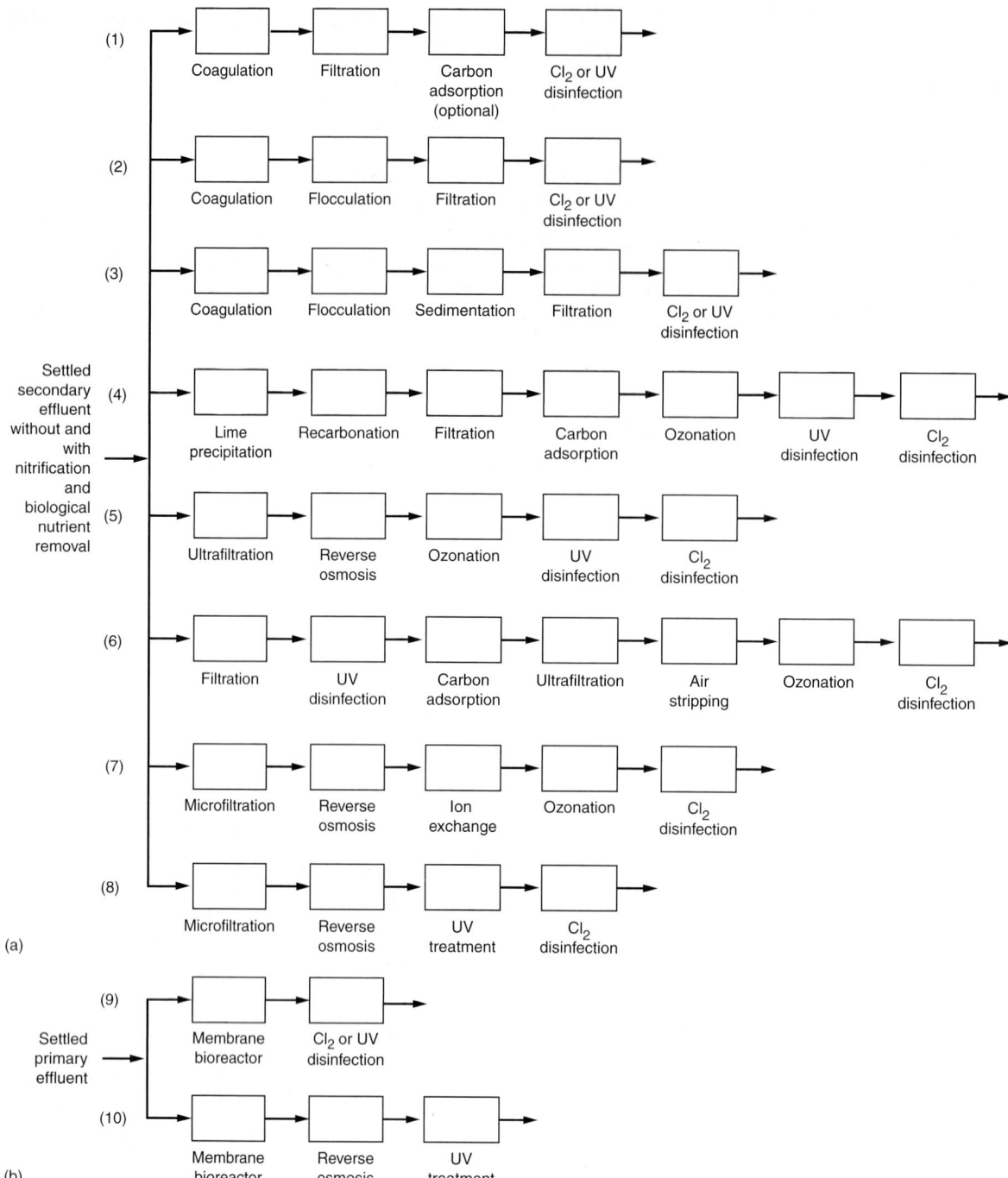

(1) Coagulation → Filtration → Carbon adsorption (optional) → Cl₂ or UV disinfection

(2) Coagulation → Flocculation → Filtration → Cl₂ or UV disinfection

(3) Coagulation → Flocculation → Sedimentation → Filtration → Cl₂ or UV disinfection

(4) Lime precipitation → Recarbonation → Filtration → Carbon adsorption → Ozonation → UV disinfection → Cl₂ disinfection

(5) Ultrafiltration → Reverse osmosis → Ozonation → UV disinfection → Cl₂ disinfection

(6) Filtration → UV disinfection → Carbon adsorption → Ultrafiltration → Air stripping → Ozonation → Cl₂ disinfection

(7) Microfiltration → Reverse osmosis → Ion exchange → Ozonation → Cl₂ disinfection

(8) Microfiltration → Reverse osmosis → UV treatment → Cl₂ disinfection

(a)

Settled secondary effluent without and with nitrification and biological nutrient removal

(9) Membrane bioreactor → Cl₂ or UV disinfection

(10) Membrane bioreactor → Reverse osmosis → UV treatment

Settled primary effluent

(b)

Figure 11–1

Typical process flow diagrams for wastewater treatment employing advanced treatment processes with (a) settled secondary effluent and (b) settled primary effluent. All of the flow diagrams have been used at one time or another. For example, in flow diagram 7, ion exchange is used for the removal of nitrate. In flow diagram 10, UV can be used for both disinfection and the destruction of NDMA.

Figure 11–2

Classification of filtration processes used in wastewater management. Note: Intermittent and recirculating porous medium filters are used for small systems and are not considered in this text.

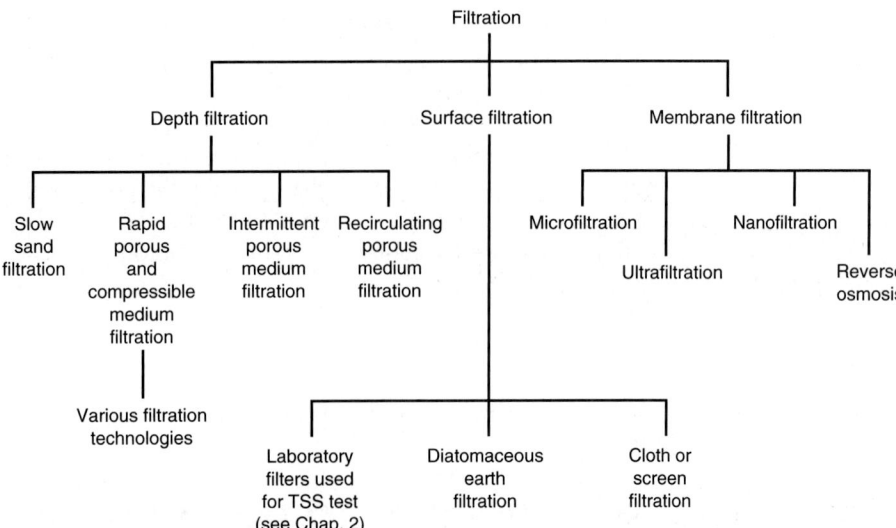

Figure 11–3

Definition sketch for filtration processes: (a) depth filtration and (b) surface filtration.

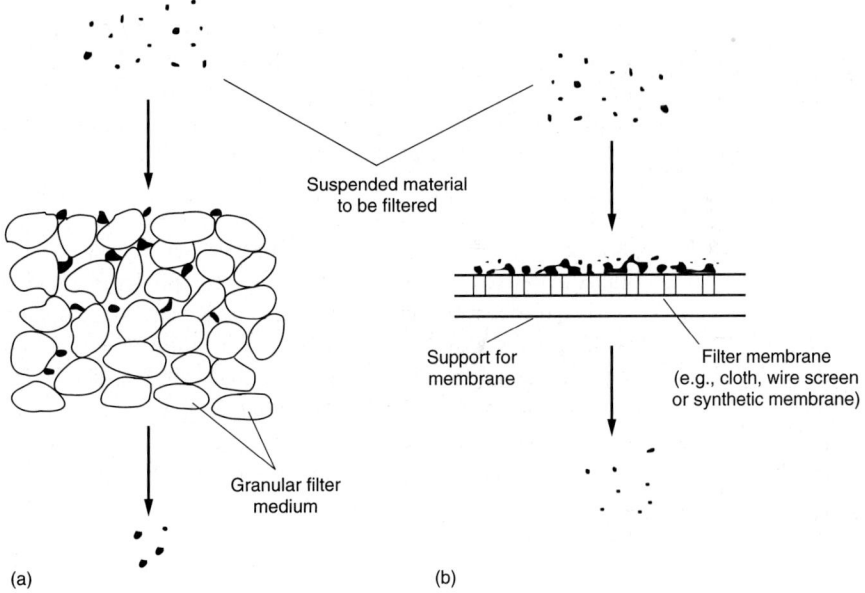

Removal of Biological Constituents

In addition to the constituents discussed above, the removal of biological constituents is also of interest. The unit operations and processes that are useful for the removal of biological constituents including bacteria, protozoan cysts and oocysts, and viruses are reported in Table 11–2. Because the effectiveness of the unit operations and processes listed in Table 11–2 is variable, disinfection of the treated effluent is required for most applications.

Process Selection and Performance Data

Selection of a given operation, process, or combination thereof depends on (1) the use to be made of the treated effluent, (2) the nature of the wastewater, (3) the compatibility of the various operations and processes, (4) the available means to dispose of the ultimate contaminants, and (5) the environmental and economic feasibility of the various systems. Specific factors that should be considered in the selection of treatment processes may be found in Table 4–11 in Chap. 4. It should be noted that in some situations economic feasibility may not be a controlling factor in the design of advanced wastewater-treatment systems, especially where specific constituents must be removed to protect the environment. Based on the variations in performance observed in the field, pilot-plant testing is recommended for the development of treatment performance data and design criteria. Representative performance data are presented in the discussion of the individual technologies.

11–3 INTRODUCTION TO DEPTH FILTRATION

Depth filtration involves the removal of particulate material suspended in a liquid by passing the liquid through a filter bed comprised of a granular or compressible filter medium (see Fig. 11–3a). Although depth filtration is one of the principal unit operations used in the treatment of potable water, the filtration of effluents from wastewater-treatment processes is becoming more common. Depth filtration is now used to achieve supplemental removals of suspended solids (including particulate BOD) from wastewater effluents of biological and chemical treatment processes to reduce the mass discharge of solids and, perhaps more importantly, as a conditioning step that will allow for the effective disinfection of the filtered effluent. Depth filtration is also used as a pretreatment step for membrane filtration (see Sec. 11–6). Single- and two-stage filtration is also used to remove chemically precipitated phosphorus.

Historically, the first depth filtration process developed for the treatment of wastewater was the slow sand filter [typical filtration rates of 30 to 60 L/m²·d (0.75 to 1.5 gal/ft²·d), Frankland (1870), and Dunbar (1908)]. The rapid sand filter (typical filtration rates of 80 to 200 L/m²·min (2.0 to 5.0 gal/ft²·min), the subject of this section, was developed to treat larger volumes of water in a facility with a smaller footprint. To introduce the subject of depth filtration, the purpose of this section is to present (1) a general introduction to the depth filtration process, (2) an introduction to filter clean-water hydraulics, and (3) an analysis of the filtration process. The types of filters that are available and issues associated with their selection and design, including a discussion of the need for pilot-plant studies, are considered in the following section.

Description of the Filtration Process

Before discussing the available filter technologies, it will be useful to first describe the basics of the depth filtration including (1) the physical features of a conventional granular medium-depth filter, (2) filter-medium characteristics, (3) the *filtration process* in which suspended material is removed from the liquid, (4) the operative particle-removal mechanisms that bring about the removal of suspended material within the filter, and (5) the *backwash process*, in which the material that has been retained within the filter is removed.

Physical Features of a Depth Filter. The general features of a conventional rapid granular medium-depth filter are illustrated on Fig. 11–4. As shown, the filtering medium (sand in this case) is supported on a gravel layer, which, in turn, rests on the filter underdrain system. The water to be filtered enters the filter from an inlet channel. Filtered water is collected in the underdrain system, which is also used to reverse the flow to backwash the filter. Filtered water typically is disinfected before being discharged to the environment. If the filtered water is to be reused, it can be discharged to a storage reservoir or to the reclaimed water distribution system. The hydraulic control of the filter is described in a subsequent section.

Filter-Medium Characteristics. Grain size is the principal filter-medium characteristic that affects the filtration operation. Grain size affects both the clear-water headloss and the buildup of headloss during the filter run. If too small a filtering medium is

Figure 11–4

General features and operation of a conventional rapid granular medium-depth filter: (a) flow during filtration cycle, and (b) flow during backwash cycle. (From Tchobanoglous and Schroeder, 1985.)

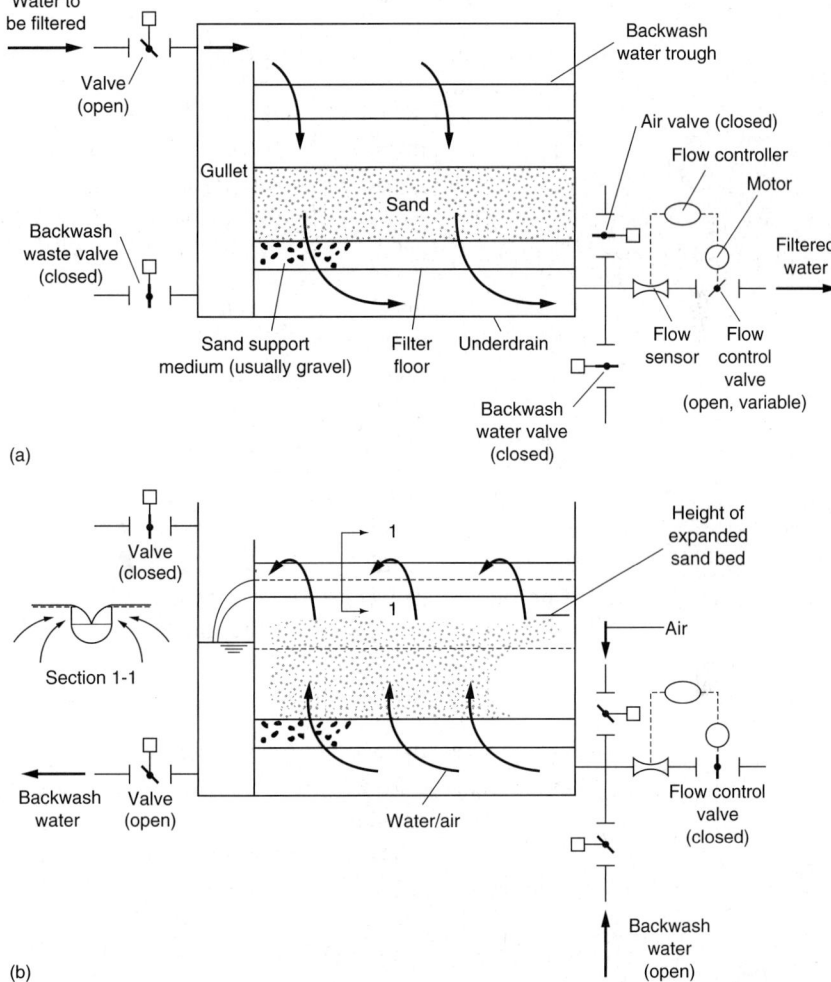

selected, much of the driving force will be wasted in overcoming the frictional resistance of the filter bed. On the other hand, if the size of the medium is too large, many of the small particles in the influent will pass directly through the bed. The size distribution of the filter material is usually determined by sieve analysis using a series of decreasing sieve sizes. The designation and size of opening for U.S. sieve sizes are given in Table 11–3. The results of a sieve analysis are usually analyzed by plotting the cumulative percent passing a given sieve size on arithmetic-log or probability-log paper (see Example 11–1 and Fig. 11–23).

The effective size of a filtering medium is defined as the 10 percent size based on mass and is designated as d_{10}. For sand, it has been found that the 10 percent size by weight corresponds to the 50 percent size by count. The uniformity coefficient (UC) is defined as the ratio of the 60 percent size to the 10 percent size (UC $= d_{60}/d_{10}$). Sometimes it is advantageous to specify the 99 percent passing size and the 1 percent passing size to define the gradation curve for each filter medium more accurately. Addi-

Table 11–3
Designation and size of opening of U.S. sieve sizes

Sieve size or number	Size of opening	
	in	mm
3/8 in	0.375[a]	9.51[a]
1/4 in	0.250[a]	6.35[a]
4	0.187	4.76
6	0.132	3.36
8	0.0937	2.38
10	0.0787[a]	2.00[a]
12	0.0661	1.68
14	0.0555[a]	1.41[a]
16	0.0469	1.19
18	0.0394[a]	1.00[a]
20	0.0331	0.841
25	0.0280[a]	0.710[a]
30	0.0234	0.595
35	0.0197[a]	0.500[a]
40	0.0165	0.420
45	0.0138[a]	0.350[a]
50	0.0117	0.297
60	0.0098[a]	0.250[a]
70	0.0083	0.210
80	0.0070[a]	0.177[a]
100	0.0059	0.149

[a] Size does not follow the ratio $(2)^{0.5}$.

Figure 11–5

Definition sketch for
length of filter run
based on headloss
and effluent turbidity.

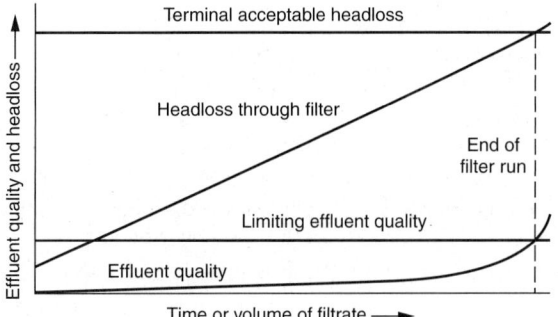

tional information on filter-medium characteristics is presented in the following section
dealing with the design of filters.

The Filtration Process.   During filtration in a conventional downflow depth
filter, wastewater containing suspended matter is applied to the top of the filter bed
(Fig. 11–4a). As the water passes through the filter bed, the suspended matter in the
wastewater is removed by a variety of removal mechanisms as described below. With
the passage of time, as material accumulates within the interstices of the granular
medium, the headloss through the filter starts to build up beyond the initial value, as
shown on Fig. 11–5. After some period of time, the operating headloss or effluent turbidity reaches a predetermined headloss or turbidity value, and the filter must be
cleaned. Under ideal conditions, the time required for the headloss buildup to reach the
preselected terminal value should correspond to the time when the suspended solids in
the effluent reach the preselected terminal value for acceptable quality. In actual practice, one or the other event will govern the backwash cycle.

Particle Removal Mechanisms.   The principal particle removal mechanisms,
believed to contribute to the removal of material within a granular-medium filter, are
identified and described in Table 11–4. The major removal mechanisms (the first six
listed in Table 11–4) are illustrated pictorially on Fig. 11–6. Straining has been identified as the principal mechanism that is operative in the removal of suspended solids during the filtration of settled secondary effluent from biological treatment processes
(Tchobanoglous and Eliassen, 1970). Other mechanisms including impaction, interception, and adhesion are also operative even though their effects are small and, for the
most part, masked by the straining action.

    The removal of the smaller particles found in wastewater (see Fig. 11–6) must be
accomplished in two steps involving (1) the transport of the particles to or near the surface where they will be removed and (2) the removal of particles by one or more of the
operative removal mechanisms. This two-step process has been identified as transport
and attachment (O'Melia and Stumm, 1967).

Backwash Process.   The end of the filter run (filtration phase) is reached when
the suspended solids in the effluent start to increase (break through) beyond an acceptable level, or when a limiting headloss occurs across the filter bed (see Fig. 11–5). Once

Table 11–4
Principal mechanisms and phenomena contributing to removal of material within a granular medium-depth filter

Mechanism/ phenomenon	Description
1. Straining	
a. Mechanical	Particles larger than the pore space of the filtering medium are strained out mechanically
b. Chance contact	Particles smaller than the pore space are trapped within the filter by chance contact
2. Sedimentation	Particles settle on the filtering medium within the filter
3. Impaction	Heavy particles will not follow the flow streamlines
4. Interception	Many particles that move along in the streamline are removed when they come in contact with the surface of the filtering medium
5. Adhesion	Particles become attached to the surface of the filtering medium as they pass by. Because of the force of the flowing water, some material is sheared away before it becomes firmly attached and is pushed deeper into the filter bed. As the bed becomes clogged, the surface shear force increases to a point at which no additional material can be removed. Some material may break through the bottom of the filter, causing the sudden appearance of turbidity in the effluent
6. Flocculation	Flocculation can occur within the interstices of the filter medium. The larger particles formed by the velocity gradients within the filter are then removed by one or more of the above removal mechanisms
7. Chemical adsorption	
a. Bonding	
b. Chemical interaction	Once a particle has been brought in contact with the surface of the filtering medium or with other particles, either one of these mechanisms, chemical or physical adsorption or both, may be responsible for holding it there
8. Physical adsorption	
a. Electrostatic forces	
b. Electrokinetic forces	
c. van der Waals forces	
9. Biological growth	Biological growth within the filter will reduce the pore volume and may enhance the removal of particles with any of the above removal mechanisms (1 through 5)

either of these conditions is reached, the filtration phase is terminated, and the filter must be cleaned (backwashed) to remove the material (suspended solids) that has accumulated within the granular filter bed. Backwashing is accomplished by reversing the flow through the filter (see Fig. 11–4b). A sufficient flow of washwater is applied until the granular filtering medium is fluidized (expanded), causing the particles of the filtering medium to abrade against each other.

The suspended matter arrested within the filter is removed by the shear forces created by backwash water as it moves up through the expanded bed. The material that has accumulated within the bed is then washed away. Surface washing with water and air

Figure 11-6

Removal of suspended particulate matter within a granular filter (a) by straining, (b) by sedimentation or inertial impaction, (c) by interception, (d) by adhesion, (e) by flocculation. (*Adapted from Tchobanoglous and Schroeder, 1985.*)

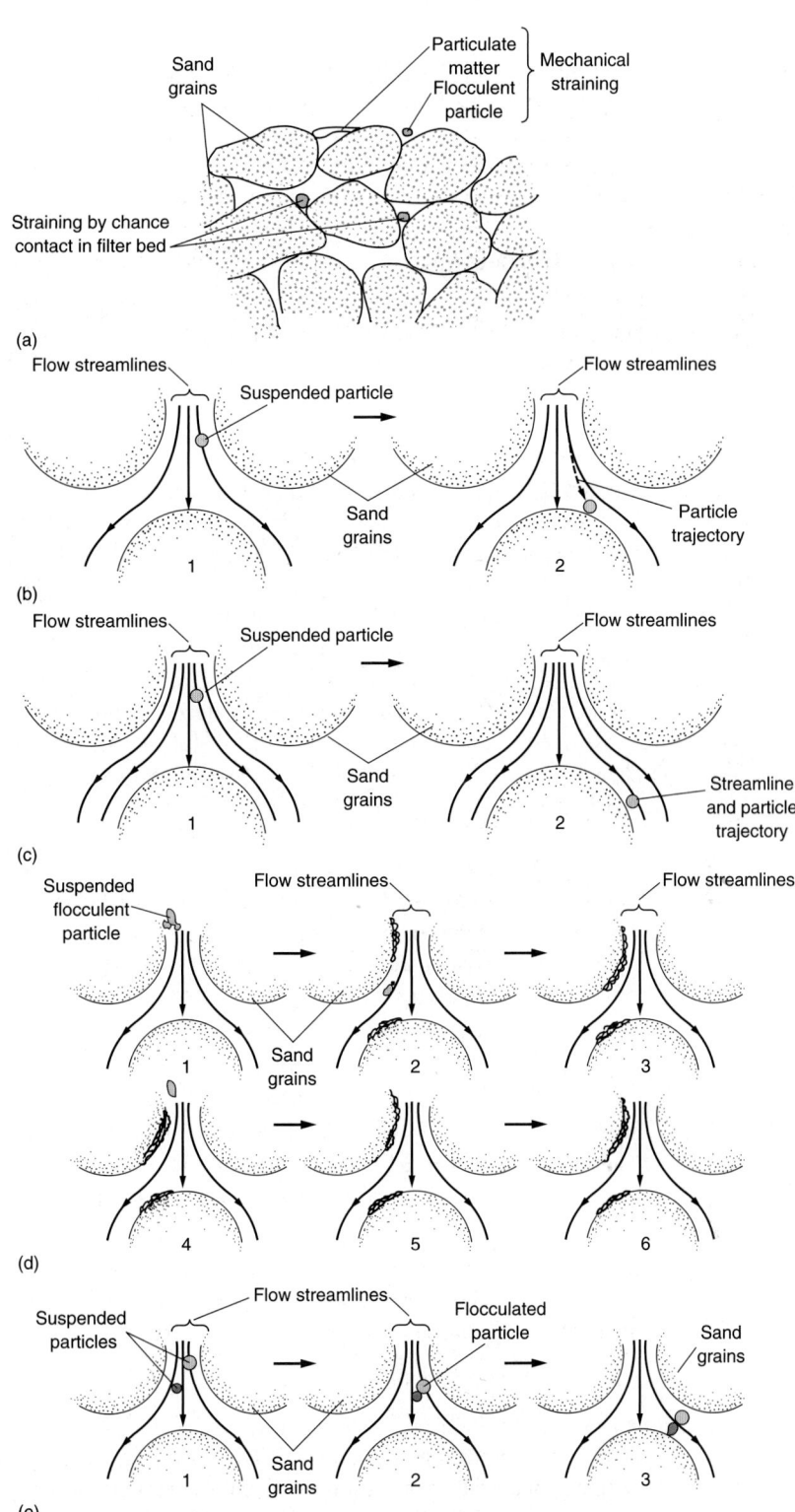

scouring are often used in conjunction with the water backwash to enhance the cleaning of the filter bed. In most wastewater-treatment plant flow diagrams, the washwater containing the suspended solids that are removed from the filter is returned either to the primary settling facilities or to the biological treatment process. Backwash hydraulics are considered in the section dealing with filter hydraulics.

Filter Hydraulics

During the past 60 years considerable effort has been devoted to the modeling of the filtration process. The models fall into two general categories: those models used to predict the clean-water headloss through a filter bed and the filter backwash expansion and those models used to predict the performance of filters for the removal of suspended solids. Filter hydraulics are considered in the following discussion.

Clean-Water Headloss. Over the years several equations have been developed to describe the flow of clean water through a porous medium (Carman, 1937; Fair and Hatch, 1933; Hazen, 1905; Kozeny, 1927; Rose, 1945); these equations are summarized in Table 11–5. In most cases, the equations for the flow of clean water through a porous medium are derived from a consideration of the Darcy-Weisbach equation for flow in a closed conduit and dimensional analysis. The summation term in Eqs. (11–2), (11–6), and (11–8) is included to account for the stratification that occurs in filters. To account for stratification, the mean size of the material retained between successive sieve sizes is assumed to correspond to the mean size of the successive sieves (see Table 11–9 in the following section for sieve sizes), assuming that the particles retained between sieve sizes are substantially uniform (Fair and Hatch, 1933).

In applying the equations given in Table 11–5, some confusion exists over the definition of the shape factor. The definition of the shape factor will depend on whether particle surface, volume, or a linear dimension is of importance in the application (Trussell and Chang, 1999). In Chap. 5, the shape factor for a particle ϕ_p is defined as the ratio of the surface area of an equivalent sphere to the surface area of the particle, for particles of the same volume. For spherical-shaped particles the specific surface area is

$$\frac{A_p}{v_p} = \frac{\pi d^2}{(\pi d^3/6)} = \frac{6}{d} \tag{11-11}$$

where A_p = area of filter-medium particle, m², mm²
v_p = volume of filter-medium particle, m³, mm³
d = diameter of filter-medium particle, m, mm

For irregularly shaped particles of the same volume the specific surface area is:

$$\frac{A_p}{v_p} = \frac{6}{\phi d} = \frac{S}{d} \tag{11-12}$$

where ϕ = approximate shape factor
S = approximate area to volume shape factor

In the literature S has been identified as a shape factor (Fair et al., 1968). The area to volume shape factor is used in the Fair-Hatch equation [(Eq. 11–5)]. Computation of the clean-water headloss through a filter is illustrated in Example 11–1.

Table 11–5
Formulas used to compute the clean-water headloss through a granular porous medium

Equation	No.	Definition of terms
Carman-Kozeny (Carman, 1937)		C = coefficient of compactness (varies from 600 for very closely packed sands that are not quite clean to 1200 for very uniform clean sand)
$$h = \frac{f}{\phi}\frac{1-\alpha}{\alpha^3}\frac{L}{d}\frac{v_s^2}{g}$$	(11–1)	
$$h = \frac{1}{\phi}\frac{1-\alpha}{\alpha^3}\frac{Lv_s^2}{g}\sum f\frac{p}{d_g}$$	(11–2)	C_d = coefficient of drag
		d = grain size diameter, m (ft)
$$f = 150\frac{1-\alpha}{N_R} + 1.75$$	(11–3)	d_g = geometric mean diameter between sieve sizes d_1 and d_2, $\sqrt{d_1 d_2}$, mm (in)
$$N_R = \frac{\phi d v_s \rho}{\mu}$$	(11–4)	d_{10} = effective grain size diameter, mm
		f = friction factor
Fair-Hatch (Fair and Hatch, 1933)		g = acceleration due to gravity, 9.81 m/s² (32.2 ft/s²)
$$h = k\nu S^2\frac{(1-\alpha)^2}{\alpha^3}\frac{L}{d^2}\frac{v_s}{g}$$	(11–5)	h = headloss, m (ft)
		k = filtration constant, 5 based on sieve openings, 6 based on size of separation
$$h = k\nu\frac{(1-\alpha)^2}{\alpha^3}\frac{Lv_s}{g}\left(\frac{6}{\phi}\right)^2\sum\frac{p}{d_g^2}$$	(11–6)	L = depth of filter bed or layer, m (ft)
		N_R = Reynolds number
Rose (Rose, 1945)		p = fraction of particles (based on mass) within adjacent sieve sizes
$$h = \frac{1.067}{\phi}C_d\frac{1}{\alpha^4}\frac{L}{d}\frac{v_s^2}{g}$$	(11–7)	S = shape factor (varies between 6.0 for spherical particles and 8.5 for crushed materials)
$$h = \frac{1.067}{\phi}\frac{Lv_s^2}{\alpha^4 g}\sum C_d\frac{p}{d_g}$$	(11–8)	T = temperature, °C (°F)
		v_h = superficial (approach) filtration velocity, m/d (ft/d)
$$C_d = \frac{24}{N_R} + \frac{3}{\sqrt{N_R}} + 0.34$$	(11–9)	v_s = superficial (approach) filtration velocity, m/s (ft/d)
		α = porosity
Hazen (Hazen, 1905)		μ = viscosity, N·s/m² (lb·s/ft²)
$$h = \frac{1}{C}\frac{60}{T+10}\frac{L}{(d_{10})^2}v_h$$	(11–10)	ν = kinematic viscosity, m²/s (ft²/s)
		ρ = density = kg/m³ (lb·s²/ft⁴)
		ϕ = particle shape factor (1.0 for spheres, 0.82 for rounded sand, 0.75 for average sand, 0.73 for crushed coal and angular sand)

EXAMPLE 11–1   Determination of Clean-Water Headloss in a Granular-Medium Filter   Determine the effective size, the uniformity coefficient, and the clean-water headloss in a filter bed composed of 0.75 m of uniform sand with the size distribution given below for a filtration rate of 160 L/m²·min. Assume that the operating temperature is 20°C. Use the Rose equation [Eq. (11–8)] given in Table 11–5 for computing the headloss. Assume the porosity of the sand in the various layers is 0.40 and use a value of 0.85 for the shape factor for sand.

Sieve size or number	Percent of sand retained	Cumulative percent passing	Geometric mean size[a], mm
6–8	0	100	
8–10	1	99	2.18
10–12	3	96	1.83
12–18	16	80	1.30
18–20	16	64	0.92
20–30	30	34	0.71
30–40	22	12	0.50
40–50	12		0.35

[a] Using sieve size data from Table 11–3, the geometric mean size $d_g = \sqrt{d_1 d_2}$.

Solution

1. Determine the effective size and the uniformity coefficient of the sand. Plot the cumulative percent passing versus the corresponding sieve size. Two different methods of plotting the data are presented below.

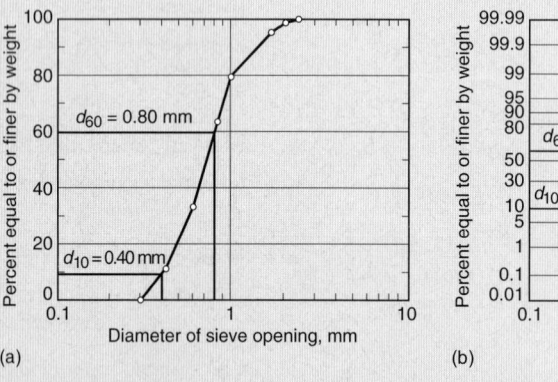
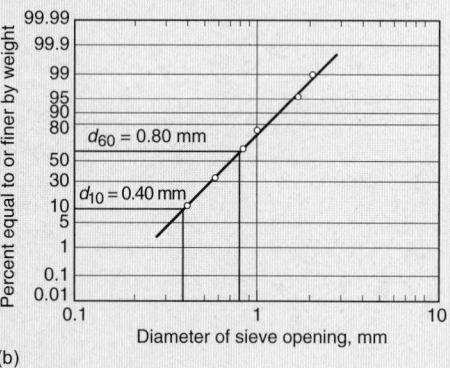

a. The effective size d_{10} read from the graphs is 0.40 mm.
b. The uniformity coefficient is

$$UC = \frac{d_{60}}{d_{10}} = \frac{0.80 \text{ mm}}{0.40 \text{ mm}} = 2.0$$

2. Determine the clean-water headloss using Eq. (11–8).

$$h = \frac{1.067}{\phi} \frac{L v_s^2}{\alpha^4 g} \Sigma C_d \frac{p}{d_g}$$

a. Set up computation table to determine the summation term in Eq. (11–8).

Sieve size or number	Fraction of sand retained	Geometric mean size, $d_g \times 10^3$ m	Reynolds number, N_R	C_d	$C_d(p/d)$, m^{-1}
8–10	0.01	2.18	4.93	6.56	30
10–12	0.03	1.83	4.15	7.60	124
12–18	0.16	1.30	2.93	10.28	1,268
18–20	0.16	0.92	2.08	13.99	2,441
20–30	0.30	0.71	1.60	17.71	7,509
30–40	0.22	0.50	1.13	24.38	10,729
40–50	0.12	0.35	0.80	33.73	11,459
Sum					33,560

    i. Determine the C_d value for each geometric mean as illustrated below.

$$C_d = \frac{24}{N_R} + \frac{3}{\sqrt{N_R}} + 0.34$$

$$N_R = \frac{\phi d v_s \rho}{\mu} = \frac{\phi d v_s}{\nu}$$

$$d = 2.18 \text{ mm}$$

$$v_s = \frac{(160 \text{ L/m}^2 \cdot \text{min})}{(10^3 \text{ L/m}^3)(60 \text{ s/min})} = 0.00267 \text{ m/s}$$

$$\nu = 1.003 \times 10^{-6} \text{ m}^2/\text{s} \text{ (see Appendix C)}$$

$$N_R = \frac{(0.85)(0.00218 \text{ m})(0.00267 \text{ m/s})}{(1.003 \times 10^{-6} \text{ m}^2/\text{s})}$$

$$N_R = 4.93$$

    ii. Determine C_d.

$$C_d = \frac{24}{4.93} + \frac{3}{\sqrt{4.93}} + 0.34 = 6.56$$

b. Determine the headloss through the stratified filter bed using Eq. (11–8).

$$L = 0.75 \text{ m}$$

$$v_s = 0.00267 \text{ m/s}$$

$$\phi = 0.85$$

$$\alpha = 0.40$$

$$g = 9.81 \text{ m/s}^2$$

$$h = \frac{1.067}{(0.85)} \frac{(0.75 \text{ m})(0.00267 \text{ m/s})^2}{(0.40)^4 (9.81 \text{ m/s}^2)} (33,500/\text{m})$$

$$h = 0.90 \text{ m}$$

Backwash Hydraulics. To understand what happens during the backwash operation it will be helpful to refer to Fig. 11–7 in which the pressure drop across a packed bed is illustrated as the backwash velocity through it increases. Between points A and B, the bed is stable, and the pressure drop and Reynolds number N_R are related linearly. At point B, the pressure drop essentially balances the weight of the filter. Between points B and C the filter bed is unstable, and the particles comprising the filter medium adjust their position to present as little resistance to flow as possible. At point C, the loosest possible arrangement is obtained in which the particles are still in contact. Beyond point C, the particles begin to move freely but collide frequently so that the motion is similar to that of particles in hindered settling. Point C is referred to as the "point of fluidization." By the time point D is reached, the particles are all in motion, and, beyond this point, increases in N_R result in very small increases in ΔP as the bed continues to expand and the particles move in more rapid and more independent motion. Ultimately, the particles will stream with the fluid, and the bed will cease to exist at point E.

To expand a filter bed comprised of a uniform filter medium hydraulically, the headloss must equal the buoyant mass of the granular medium in the fluid. Mathematically this relationship can be expressed as

$$h = L_e (1 - \alpha_e)\left(\frac{\rho_s - \rho_w}{\rho_w}\right) \tag{11-13}$$

where h = headloss required to expand the bed
L_e = depth of the expanded bed
α_e = expanded porosity

Figure 11–7

Schematic diagram illustrating the fluidization of a filter bed. (Adapted from Foust et al. 1960.)

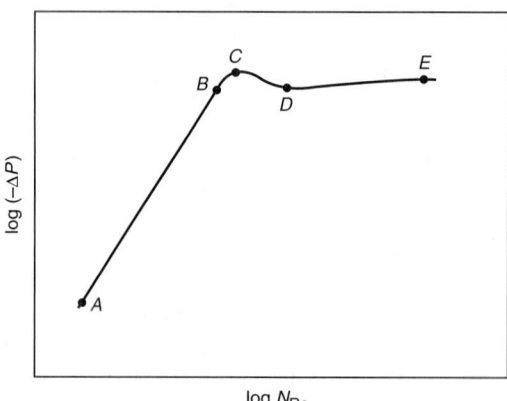

ρ_s = density of the medium

ρ_w = density of water

Because the individual particles are kept in suspension by the drag force exerted by the rising fluid, it can be shown from settling theory (see Sec. 5–5) that

$$C_D A_p \rho_w \frac{v^2}{2} \phi(\alpha_e) = (\rho_s - \rho_w) g v_p \tag{11–14}$$

where v = face velocity of backwash water, m/s

$\phi(\alpha_e)$ = correction factor to account for the fact that v is the velocity of the backwash water and not the particle-settling velocity v_p

and other terms are as defined previously.

From experimental studies (Fair, 1951; Richardson and Zaki, 1954) it has been found that the expanded-bed porosity can be approximated using the following relationships, assuming the Reynolds number is approximately 1.

$$\phi(\alpha_e) = \left(\frac{v_s}{v} \right)^2 = \left(\frac{1}{\alpha_e} \right)^9 \tag{11–15}$$

Thus

$$\alpha_e = \left(\frac{v}{v_s} \right)^{0.22} \tag{11–16}$$

or

$$v = v_s \alpha_e^{4.5} \tag{11–17}$$

where v_s = settling velocity of particle, m/s

However, because the volume of the filtering medium per unit area remains constant, $(1 - \alpha)L$ must be equal to $(1 - \alpha_e)L_e$ so that

$$\frac{L_e}{L} = \frac{1 - \alpha}{1 - \alpha_e} = \frac{1 - \alpha}{1 - (v/v_s)^{0.22}} \tag{11–18}$$

Where the filter medium is stratified, the smaller particles in the upper layers expand first. To expand the entire bed, the backwash velocity must be sufficient to lift the largest particle. To account for filter-bed stratification, Eq. (11–18) is modified assuming that particles retained between sieve sizes are substantially uniform (Fair and Hatch, 1933).

$$\frac{L_e}{L} = (1 - \alpha) \Sigma \frac{p}{(1 - \alpha_e)} \tag{11–19}$$

where p = fraction of filter medium retained between sieve sizes.

Thus, the required backwash velocity and expanded depth can be estimated using Eqs. (11–18) and (11–19), respectively, as illustrated in Example 11–2. Additional details on filter-bed expansion may be found in Amirtharajah (1978), Cleasby and Fan (1982), and Dharmarajah and Cleasby (1986); Kawamura (2000), Leva (1959); and Richardson and Zaki (1954).

EXAMPLE 11-2   **Determination of Required Backwash Velocities for Filter Cleaning**   A stratified sand bed with the size distribution given below is to be backwashed at a rate of 0.75 m/min. Determine the degree of expansion and whether the proposed backwash rate will expand all of the bed. Assume the following data are applicable:

Sieve size or number	Percent of sand retained	Geometric mean size[a], mm
6–8	0	2.83
8–10	1	2.18[b]
10–12	3	1.83
12–18	16	1.30
18–20	16	0.92
20–30	30	0.71
30–40	22	0.50
40–50	12	0.35

[a] Based on sieve sizes given in Table 11–3.
[b] $2.18 = \sqrt{2.38 \times 2.0}$

1. Granular medium = sand
2. Specific gravity of sand = 2.65
3. Depth of filter bed = 0.90 m
4. Temperature = 20°C

Solution

1. Set up computation table to determine the summation term in Eq. (11–19).

$$\frac{L_e}{L} = (1 - \alpha)\Sigma \frac{p}{(1 - \alpha_e)}$$

Sieve size or number	Fraction of sand retained	Geometric mean size d_g, mm	$v_{s\prime}$ m/s	v/v_s	α_e	$p/(1 - \alpha_e)$
8–10	0.01	2.18	0.304	0.041	0.496	0.020
10–12	0.03	1.83	0.270	0.046	0.509	0.061
12–18	0.16	1.30	0.210	0.060	0.538	0.346
18–20	0.16	0.92	0.157	0.080	0.573	0.351
20–30	0.30	0.71	0.123	0.102	0.605	0.760
30–40	0.22	0.50	0.085	0.146	0.655	0.638
40–50	0.12	0.35	0.055	0.227	0.722	0.432
Summation						2.632

a. Determine the particle settling velocity as illustrated in Example 5–5. Alternatively, the settling velocities can be estimated using Fig. 5–21 in Chap. 5. The computed settling velocity values are entered in the computation table.

b. Determine the values of v/v_s and enter the computed values in the computation table. The backwash velocity is $v = 0.75$ m/min $= 0.0125$ m/s.

c. Determine the values of α_e and enter the computed values in the computation table.

$$\alpha_e = \left(\frac{v}{v_s}\right)^{0.22} = \left(\frac{0.0125}{0.304}\right)^{0.22} = 0.496$$

d. Determine the values for column 7 and enter the computed values in the computation table.

$$\frac{p}{1 - \alpha_e} = \frac{1}{1 - 0.496} = 1.98$$

2. Determine the expanded-bed depth using Eq. (11–19).

$$\frac{L_e}{L} = (1 - \alpha)\Sigma\,\frac{p}{(1 - \alpha_e)}$$

$$L_e = (0.9 \text{ m}) (1 - 0.4) (2.63) = 1.42 \text{ m}$$

3. Because the expanded porosity of the largest size fraction (0.496) is greater than the normal porosity of the filter material, the entire filter bed will be expanded.

Comment The expanded depth needs to be known to establish the minimum height of the wash-water troughs above the surface of the filter bed. In practice, the bottom of the backwash-water troughs is set from 50 to 150 mm (2 to 6 in) above the expanded filter bed. The width and depth of the troughs should be sufficient to handle the volume of backwash water used to clean the bed, with a minimum freeboard of 600 mm (24 in) at the upper end of the trough.

Analysis of the Filtration Process

In general, the mathematical characterization of the time-space removal of particulate matter within the filter is based on a consideration of the equation of continuity, together with an auxiliary rate equation.

Equation of Continuity. The equation of continuity for the filtration operation may be developed by considering a suspended-solids mass balance for a section of filter of cross-sectional area A and of thickness Δz, measured in the direction of flow (see Fig. 11–8). Following the approach delineated in Chap. 4, the mass balance would be as follows:

Figure 11-8

Definition sketch for the analysis of the filtration process. (From Tchobanoglous and Schroeder, 1985.)

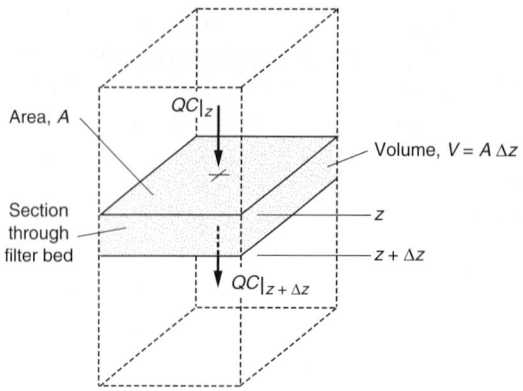

1. General word statement:

$$
\begin{matrix}
\text{Rate of accumulation} & & \text{rate of} & & \text{rate of} \\
\text{of solids within the} & = & \text{flow into the} & - & \text{flow out of the} \\
\text{volume element} & & \text{volume element} & & \text{volume element}
\end{matrix}
\qquad (11\text{--}20)
$$

2. Simplified word statement:

   Accumulation = inflow − outflow

3. Symbolic representation:

$$
\left(\frac{\partial q}{\partial t} + \alpha(t)\frac{\partial \overline{C}}{\partial t} \right)\Delta V = QC + Q\left(C + \frac{\Delta C}{\Delta z}\Delta z \right)
\qquad (11\text{--}21)
$$

where $\partial q/\partial t$ = change in quantity of solids deposited within the filter time, $g/m^3 \cdot min$

   $\alpha(t)$ = average porosity as a function of time

   $\partial \overline{C}/\partial t$ = change in average concentration of solids in pore space with time, $g/m^3 \cdot min$

   ΔV = differential volume, m^3

   Q = volumetric flowrate, L/min (the unit L/min is used for convenience)

   C = concentration of suspended solids, g/m^3

   $\Delta C/\Delta z$ = change in concentration of suspended solids in fluid stream with distance, $g/m^3 \cdot m$

Substituting $A\,\Delta z$ for ΔV and Av for Q where v is the filtration velocity, simplifying, and taking the limit as Δz approaches zero, Eq. (11–21) becomes

$$
-v\frac{\partial C}{\partial z} = \frac{\partial q}{\partial t} + \alpha(t)\frac{\partial \overline{C}}{\partial t}
\qquad (11\text{--}22)
$$

In Eq. (11–22), the first term represents the difference between the mass of suspended solids entering and leaving the section; the second term represents the time rate of change in the mass of suspended solids accumulated within the interstices of the filter medium; and the third term represents the time rate of change in the suspended-solids concentration in the pore space within the filter volume.

In a flowing process, the quantity of fluid contained within the bed is usually small compared with the volume of liquid passing through the bed. In this case, the materials balance equation can be written as

$$-v\frac{\partial C}{\partial z} = \frac{\partial q}{\partial t}$$ (11–23)

Equation (11–23) is the one most commonly found in the literature dealing with filtration theory.

If the shape of the removal curve within the filter does not vary with time, the equation of continuity [Eq. (11–23)] may be written as an ordinary differential equation:

$$-v\frac{dC}{dz} = \frac{dq}{dt}$$ (11–24)

Rate Equation. To solve Eq. (11–24), an additional independent equation is required. The most direct approach is to derive a relationship that can be used to describe the change in concentration of suspended matter with distance, such as

$$\frac{\partial C}{\partial z} = \phi(V_1, V_2, V_3, \ldots)$$ (11–25)

in which V_1, V_2, and V_3 are the variables governing the removal of suspended matter from solution.

An alternative approach is to develop a complementary equation in which the pertinent process variables are related to the amount of material retained (accumulated) within the filter at various depths. In equation form, this may be written as

$$\frac{\partial q}{\partial t} = \phi(V_1, V_2, V_3, \ldots)$$ (11–26)

As noted previously, the principal removal mechanism for the residual suspended solids found in settled secondary effluent from biological treatment processes is straining. The removal of suspended material by straining can be identified by noting (1) the variation in the normalized concentration-removal curves through the filter as a function of time, and (2) the shape of the headloss curve for the entire filter or an individual layer within the filter. If straining is the principal removal mechanism, the shape of the normalized removal curve will not vary significantly with time (see Fig. 11–9a), and the headloss curves will be curvilinear (Tchobanoglous and Eliassen, 1970). If the floc particles that are being filtered are easily broken apart, the normalized removal curves will vary with time as shown on Fig. 11–9b.

From the size and distribution of the influent particles (see Fig. 2–9, Chap. 2) and the shape of the normalized curves (see Fig. 11–9a), it can be concluded that the rate of change of concentration with distance must be proportional to some removal coefficient that is changing with the degree of treatment or removal achieved in the filter. For example, the entire particle-size distribution in the influent is passed through the first layer. The probability of removing particles from the waste stream is p_1. In the second layer, the probability of removing particles is p_2; p_2 is less than p_1, assuming that some of the larger particles will be removed by the first layer. Continuing this argument, it

Figure 11-9

Normalized suspended-solids removal curves as a function of depth and time as observed during (a) the filtration of settled effluent from an activated-sludge treatment process and (b) the filtration of pretreated wastewater (coagulation/flocculation/sedimentation) with a weak floc. When the removal curves do not vary significantly with time as shown on Fig. 11-9a, straining is the predominant removal mechanism.

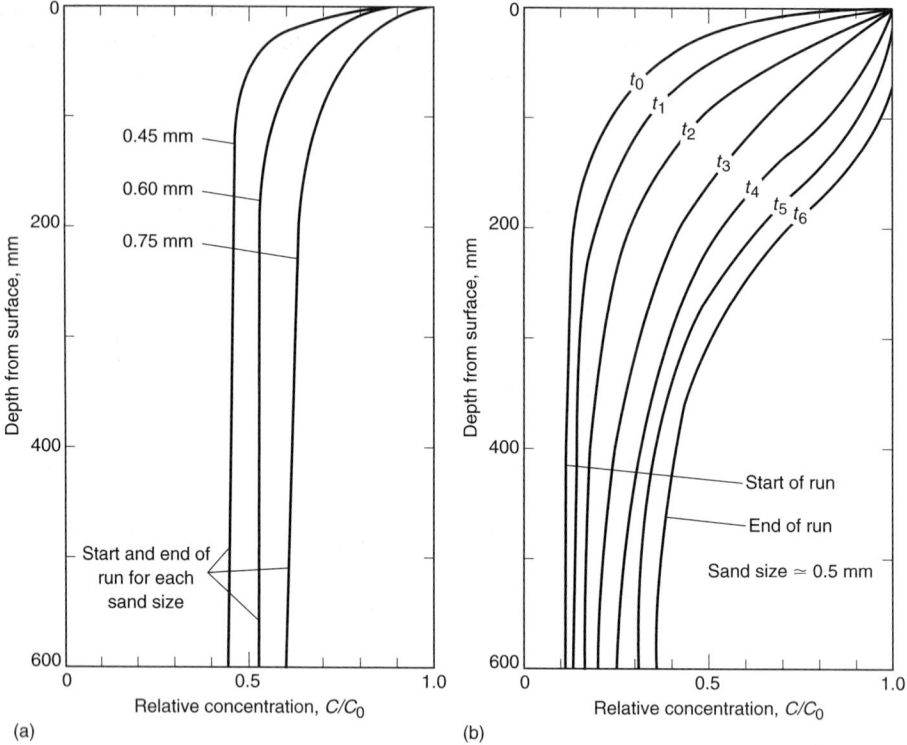

(a)

(b)

can be reasoned that the rate of removal must always be changing as a function of the degree of treatment. This phenomenon can be expressed mathematically using the following equation:

$$\frac{dC}{dz} = \left[\frac{1}{(1 + az)^n}\right] r_o C \qquad (11\text{-}27)$$

where C = concentration, mg/L
$\quad z$ = distance from top of filter bed, cm
$\quad r_o$ = initial removal rate, cm^{-1}
$\quad a, n$ = constants

In Eq. (11-27), the term within brackets is a retardation factor as described in Chap. 4. When the exponent n is equal to zero, the term within the brackets is equal to 1; under these conditions, Eq. (11-27) represents a logarithmic curve. When n equals 1, the value of the term within brackets drops off rapidly in the first 125 mm (5 in) and then more gradually as a function of distance. Therefore, it appears that the exponent n may be related to the distribution of particle sizes in the influent. For example, when dealing with a uniform filter medium and filtering particles of one size, it would be expected that the value of the exponent n would be equal to zero and that the initial removal could be described as a first-order removal function. It should be noted that this equation was verified only for filtration rates up to 400 L/m^2·min (10 gal/ft^2·min).

The value of r_o, the initial removal rate constant, is obtained by computing the slope of the removal curve at near zero depth, because $[1/(1 + az)^n] \approx 1$. The easiest way to determine the slope is to rewrite Eq. (11–27) as follows:

$$\left(\frac{Cr_o}{dC/dz} \right)^{1/n} = 1 + az \tag{11–28}$$

If Eq. (11–28) is plotted functionally, the valve of n is equal to the value that results in a straight-line plot. The slope of the line describing the experimental data will be equal to the constant a.

Generalized Rate Equation. On the basis of experimental results derived from pilot testing and data reported in the literature, there appear to be five major factors that affect the time-space removal of the residual suspended matter from a biological treatment process within a granular filter for a given temperature. These factors are the size of the filter medium, the rate of filtration, the influent particle size and size distribution, the floc strength, and the amount of material removed within the filter. Although a number of different formulations are possible, a generalized rate equation in which all five factors are considered can be developed by multiplying Eq. (11–27) by a factor that takes into account the effect of the material accumulated in the filter. The proposed equation is

$$\frac{dC}{dz} = \frac{1}{(1 + az)^n} r_o C \left(1 - \frac{q}{q_u} \right)^m \tag{11–29}$$

where q = quantity of suspended solids deposited in the filter
q_u = ultimate quantity of solids that can be deposited in the filter
m = a constant related to floc strength

Initially, when the amount of material removed by the filter is low, $q = 0$; $(1 - q/q_u)^m = 1$, and Eq. (11–29) is equivalent to Eq. (11–27). As the upper layers begin to clog, the term $(1 - q/q_u)^m$ approaches zero, and the rate of change in concentration with distance is equal to zero. At the lower depths, the amount of material removed is essentially zero, and the previous analysis applies.

Headloss Development. In the past, the most commonly used approach to determine headloss in a clogged filter was to compute it with a modified form of the equations used to evaluate the clear-water headloss (see Table 11–5). In all cases, the difficulty encountered in using these equations is that the porosity must be estimated for various degrees of clogging. Unfortunately, the complexity of this approach renders most of these formulations useless or, at best, extremely difficult to use.

An alternative approach is to relate the development of headloss to the amount of material removed by the filter. The headloss would then be computed using the expression

$$H_t = H_o + \sum_{i=i}^{n} (h_i)_t \tag{11–30}$$

where H_t = total headloss at time t, m (ft)
H_o = total initial clear-water headloss, m (ft)
$(h_i)_t$ = headloss in the ith layer of the filter at time t, m (ft)

Figure 11-10

Headloss versus
suspended solids
removed for various
sizes of uniform sand
and anthracite. *(Adapted
from Tchobanoglous
and Eliassen, 1970.)*

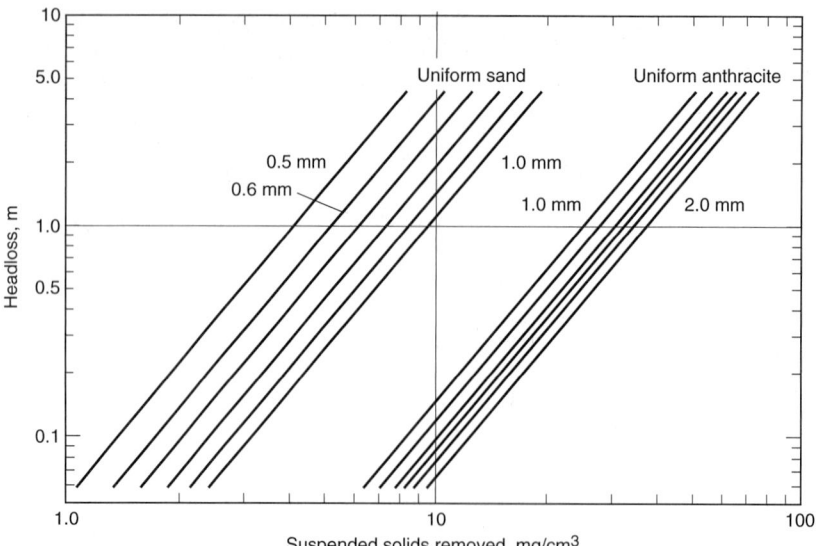

From an evaluation of the incremental headloss curves for uniform sand and
anthracite, the buildup of headloss in an individual layer of the filter was found to be
related to the amount of material contained within the layer (see Fig. 11–10). The form
of the resulting equation for headloss in the ith layer is

$$(h_i)_t = a(q_i)_t^b \tag{11-31}$$

where $(q_i)_t$ = amount of material deposited in the ith layer at time t, mg/cm^3
a, b = constants

In Eq. (11–31), it is assumed that the buildup of headloss is only a function of the
amount of material removed. Determination of the buildup of headloss during the filtration process using the data presented on Fig. 11–10 is illustrated in Example 11–3.

EXAMPLE 11-3 **Analysis of Filtration Data from a Pilot Plant used to Filter Settled Effluent
from an Activated-Sludge Process** The normalized suspended-solids-removal-
ratio curves shown below were derived from a filtration pilot-plant study conducted at
an activated-sludge wastewater-treatment plant. Using these curves and the following
data, develop curves that can be used to estimate (1) the headloss buildup as a function
of the length of run and (2) the length of run to a terminal headloss of 3.0 m (10 ft) as
a function of the filtration rate.

Biological
treatment process

1. Solids retention time SRT = 15 d
2. Average suspended-solids concentration in effluent from secondary settling tank
   = 20 mg/L

Pilot plant

1. Type of filter bed = dual-medium
2. Filter media = graded anthracite and sand
3. Filter-medium characteristics
   a. Anthracite, d_{10} = 1.4 mm, $UC \approx 1.4$
   b. Sand, d_{10} = 0.6 mm, $UC \approx 1.4$
4. Filter-bed depth = 0.3 m
   a. Anthracite = 0.3 m
   b. Sand = 0.3 m
5. Filtration rates = 80, 160, and 240 L/m²·min
6. Temperature = 20°C
7. General observation: average concentration-ratio curves given in the following figure did not vary significantly with time

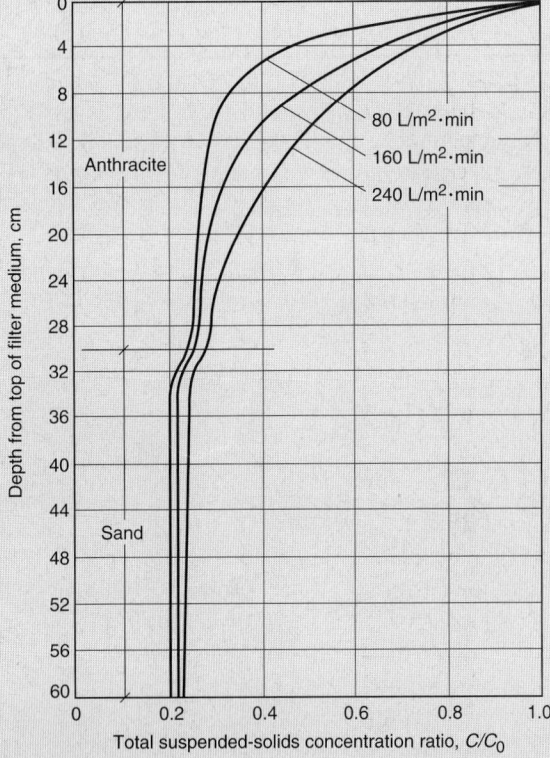

Average suspended-solids concentration-ratio curves versus filter depth for various filtration rates.

Solution

1. To analyze the concentration-ratio curves, rewrite Eq. (11–24) in a form suitable for numerical analysis:

$$-v\frac{\Delta C}{\Delta z} = \frac{\Delta q}{\Delta t}$$

$$-v\frac{C_{z-1} - C_z}{z_{z-1} - z_z} = \frac{q_2 - q_1}{t_2 - t_1}$$

2. Set up a computation table and determine the values of ΔC ($C_{z-1} - C_z$) and the mean grain size for various depth intervals throughout the filter. The required computations are summarized in the following table. The entries in the table are explained below the table.

Depth, cm (1)	Geometric mean grain size, d_g, mm (2)	Filtration rate, L/m²·min								
		80			160			240		
		C_z/C_o (3)	C_z (4)	ΔC (5)	C_z/C_o (6)	C_z (7)	ΔC (8)	C_z/C_o (9)	C_z (10)	ΔC (11)
0		1.0	20.0		1.0	20.0		1.0	20.0	
	1.18			11.4			7.4			5.6
4		0.43	8.6		0.63	12.6		0.72	14.4	
	1.42			2.2			3.2			2.8
8		0.32	6.4		0.47	9.4		0.58	11.6	
	1.56			0.8			2.2			2.0
12		0.28	5.6		0.36	7.2		0.48	9.6	
	1.69			0.2			1.2			1.6
16		0.27	5.4		0.30	6.0		0.40	8.0	
	1.82			0.2			0.4			1.0
20		0.26	5.2		0.28	5.2		0.35	7.0	
	1.98			0.2			0.2			0.8
24		0.25	5.0		0.27	5.4		0.31	6.2	
	2.23			0.2			0.2			0.6
28		0.24	4.8		0.26	5.2		0.28	5.6	
	2.62			0.0			0.0			0.2
30		0.24	4.8		0.26	5.2		0.27	5.4	
	0.50			0.6			0.6			0.5
32		0.215	4.3		0.23	4.6		0.245	4.9	
	0.60			0.1			0.1			0.1
34		0.205	4.1		0.23	4.5		0.24	4.8	
	0.65			0.1			0.1			0.0
36		0.20	4.0		0.22	4.4		0.24	4.8	
	0.69			0.0			0.0			0.0
38		0.20	4.0		0.22	4.4		0.24	4.8	
	0.72			0.0			0.0			0.0
40		0.20	4.0		0.22	4.4		0.24	4.8	

a. As shown in the computation table, the filter depth intervals are entered in col. 1.

b. The mean anthracite and sand particle size associated with each depth interval is entered in col. 2. The mean particle size for each interval can be estimated using the effective size and uniformity coefficient for each medium. For example, at a depth of 4 cm, 4/30ths of the total depth of 30 cm has been passed. Therefore, the particle size associated with a depth of 4 cm corresponds to the sieve size that 13.3 percent [(4/30) × 100] of the anthracite will pass. By plotting the particle size distributions on probability-log paper, and obtaining the equation of the line passing through the plotted points, the 13.3 percent passing anthracite particle size is estimated to be 1.29 mm (see figure below). Assuming the 1 percent size as the cutoff size (1.08 mm), the geometric mean particle size for the interval from zero and 4 cm is, as shown below, 1.18 mm. Following the same procedure, the remaining particle sizes for the anthracite and sand are entered in col. 2.

$$d_g = \sqrt{d_1 d_2} = \sqrt{1.08 \text{ mm} \times 1.29 \text{ mm}} = 1.18 \text{ mm}$$

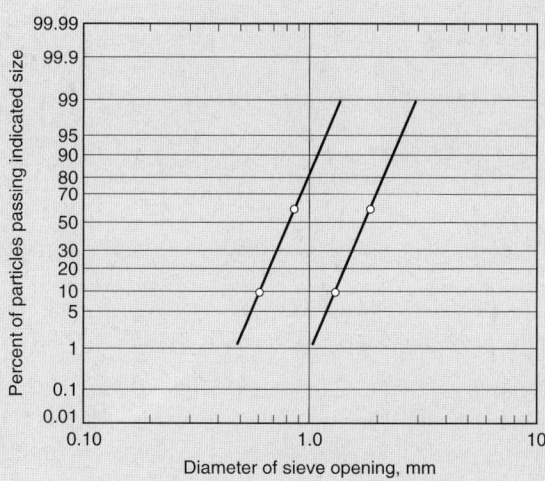

c. The values of C/C_o taken from the normalized removal ratio curves corresponding to the depths given in col. 1 are entered in cols. 3, 6, and 9 for each filtration rate.

d. The value of the concentration at each depth is entered in cols. 4, 7, and 10 for each filtration rate.

e. The concentration difference $\Delta C\,(C_{z-1} - C_z)$ between the depths given in col. 1 is entered in cols. 5, 8, and 11 for each filtration rate.

3. Set up a computation table and determine the buildup of suspended solids and headloss within each layer of the filter for filter runs of various lengths. The necessary computations for a filtration rate of 160 L/m²·min (4 gal/ft²·min) are summarized in the following table.

Depth, cm (1)	ΔC, mg/L (2)	Run length, h					
		10		**15**		**20**	
		Δq, mg/cm³ (3)	Δh, m (4)	Δq, mg/cm³ (5)	Δh, m (6)	Δq, mg/cm³ (7)	Δh, m (8)
0							
	7.4	17.76	0.47	26.64	1.15	35.52	2.17
4							
	3.2	7.68	0.06	11.52	0.14	15.36	0.26
8							
	2.2	5.28	0.02	7.92	0.05	10.56	0.10
12							
	1.2	2.88	—	4.32	0.01	5.76	0.02
16							
	0.4	0.96	—	1.44	—	1.92	—
20							
	0.2	0.48	—	0.72	—	0.96	—
24							
	0.2	0.48	—	0.72	—	0.96	—
28							
	0.0	0.00	—	0.00	—	0.00	—
30							
	0.6	2.88	0.43	4.32	1.04	5.76	1.95
32							
	0.1	0.48	0.01	0.72	0.01	0.96	0.03
36							
	0.1	0.48	0.01	0.72	0.01	0.96	0.02
38							
	0.0	0.0	—	0.0	—	0.0	—
40							
Σ Δh, m			1.00		2.41		4.55

Although the required computations for the other filtration rates are not shown, they are the same. The values of ΔC given in col. 2 are taken from col. 8 of the table prepared in Step 2. The values of q shown in cols. 3, 5, and 7 are determined by using the difference equation given in Step 1. To illustrate, for the anthracite layer between 4 and 8 cm from the top of the filter bed, the value of Δq after 20 h is as follows:

$$-v\frac{\Delta C}{\Delta z} = \frac{\Delta q}{\Delta t}$$

where $v = 160$ L/m²·min $= 0.016$ L/cm²·min

$\Delta C = 3.2$ mg/L

$\Delta z = 4$ cm $- 8$ cm $= -4$ cm

$\Delta t = (20$ h $\times 60$ min/h $- 0) = 1200$ min

$$\Delta q = \left(-\frac{0.016 \text{ L}}{\text{cm}^2 \cdot \text{min}}\right)\left(\frac{3.2 \text{ mg/L}}{-4 \text{ cm}}\right) 1200 \text{ min}$$

$$= 15.36 \text{ mg/cm}^3$$

The value of incremental headloss buildup Δh (col. 8) for the anthracite layer between 4 and 8 cm is obtained from Fig. 11–10 by entering with the value of Δq for this layer. The value of the headloss in the sand layers is determined in a similar manner. To simplify the computations, it is assumed that no intermixing occurs between the anthracite and sand. Once all of the Δh headloss values are entered, the entire column is summed to obtain the total headloss in the filter bed. The total headloss for other time periods and filtration rates is determined in exactly the same manner. Summary data for the other flowrates are as follows:

Run length, h	Headloss, m		
	80 L/m²·min	160 L/m²·min	240 L/m²·min
10	0.34	1.00	1.51
15	—	2.41	3.70
20	1.56	4.55	6.96
30	3.82		

4. The clean-water headloss at the three filtration rates, computed as illustrated previously in Example 11–1, is

Filtration rate, L/m²·min	Clean-water headloss, m
80	0.15
160	0.31
240	0.49

5. The total headloss including the clean-water headloss and the headloss due to the accumulation of material within the filter for each of the filtration rates is summarized below.

Run length, h	Headloss, m		
	80 L/m²·min	160 L/m²·min	240 L/m²·min
0	0.15	0.31	0.49
10	0.49	1.31	2.00
15	—	2.72	4.18
20	1.71		
30	3.97		

6. Plot curves of headloss versus run length for the three flowrates. The required curves, which are plotted using the data given in Step 5, are shown in the following figure.

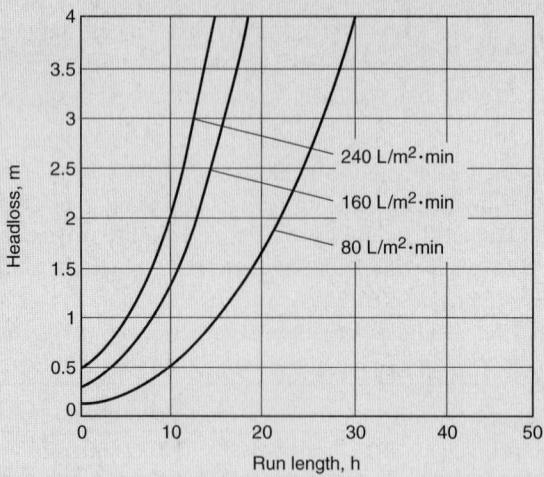

7. Plot the curve of run length to reach a headloss of 3 m versus filtration rate. The required curve is shown below. The data needed to plot this curve are obtained from the headloss curve developed in Step 6 by finding the time required to reach a headloss of 3 m for each filtration rate.

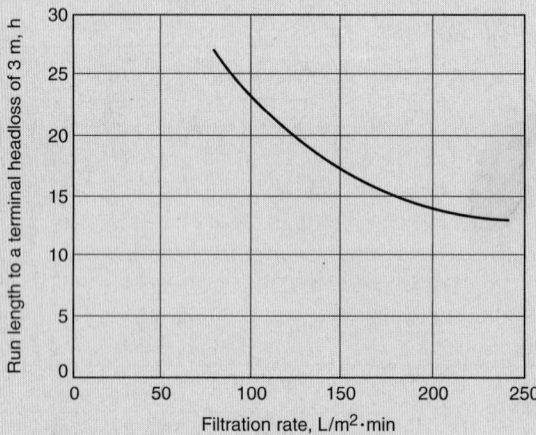

Comment The use of unstratified filter beds for the filtration of treated effluents has been studied by Dahab and Young (1977). They found that unstratified filter beds having the same effective size as that used in the top layer of a dual-medium filter were essentially equivalent to dual-medium filters in terms of effluent quality and length of run.

11–4 SELECTION AND DESIGN CONSIDERATIONS FOR DEPTH FILTERS

The ability to select and design filter technologies must be based on (1) knowledge of the types of filters that are available, (2) a general understanding of their performance characteristics, and (3) an appreciation of the process variables controlling depth filtration. Important design considerations for effluent filtration systems include (1) influent wastewater characteristics, (2) design and operation of the biological treatment process, (3) type of filtration technology to be used, (4) available flow-control options, (5) type of filter backwashing system to be employed, (6) necessary filter appurtenances, and (7) filter control systems and instrumentation. An understanding of issues related to effluent filtration with chemical addition, the type of filter problems encountered in the field, and the importance of pilot-plant studies is also necessary. These subjects are presented and discussed in this section.

Available Filtration Technologies

Historically, the granular-medium filter in which the complete filtration cycle (filtration and backwashing) occurs sequentially, as described in Sec. 11–3, has been used most commonly. However, during the past 20 years, a number of new types of filter technologies have been developed and are now available for the filtration of treated wastewater. The principal types of depth filters that have been used for the filtration of wastewater are identified in Table 11–6 and illustrated on Fig. 11–11. As shown in Table 11–6, the filters can be classified in terms of their operation as semicontinuous or continuous. Filters that must be taken off-line periodically to be backwashed are classified operationally as semicontinuous. Filters in which the filtration and backwash operations occur simultaneously are classified as continuous. Within each of these two classifications there are a number of different types of filters depending on bed depth (e.g., shallow, conventional, and deep bed), the type of filtering medium used (mono-, dual-, and multimedium), whether the filtering medium is stratified or unstratified, the type of operation (downflow or upflow), and the method used for the management of solids (i.e., surface or internal storage). For the mono- and dual-medium semicontinuous filters, a further classification can be made based on the driving force (e.g., gravity or pressure). Another important distinction that must be noted for the filters identified in Table 11–6 is whether they are proprietary or individually designed.

The five types of depth filters used most commonly for wastewater filtration are (1) conventional downflow filters, (2) deep-bed downflow filters, (3) deep-bed upflow continuous-backwash filters, (4) the pulsed-bed filter, and (5) traveling-bridge filters. These filter technologies are described briefly in this section. In addition a new filter employing a synthetic filter medium and a two-stage filtration system which incorporates phosphorus removal are also discussed. Intermittent and recirculating sand and textile filters, usually limited to small systems, are not considered in this text. Information on these filters may be found in Crites and Tchobanoglous (1998). It should be noted that many of the filters to be considered in the following discussion are proprietary and are supplied by the manufacturer as a complete unit. Thus, many of the design details presented in the following section apply only to filters that are designed individually.

Table 11–6
Comparison of principal types of granular-medium filters

| Type of filter (common name) | Type of filter operation | Filter-bed details | | | Typical direction of fluid flow |
		Type of filter bed	Filtering medium	Typical bed depth, mm	
Conventional	Semicontinuous	Mono-medium (stratified or unstratified)	Sand or anthracite	760	Downward
Conventional	Semicontinuous	Dual-medium (stratified)	Sand and anthracite	920	Downward
Conventional	Semicontinuous	Multimedium (stratified)	Sand, anthracite, and garnet	920	Downward
Deep-bed	Semicontinuous	Mono-medium (stratified or unstratified)	Sand or anthracite	1830	Downward
Deep-bed	Semicontinuous	Mono-medium (stratified)	Sand or anthracite	1830	Upward
Deep-bed	Continuous	Mono-medium (unstratified)	Sand	1830	Upward
Pulsed-bed	Semicontinuous	Mono-medium (stratified)	Sand	280	Downward
Fuzzy filter	Semicontinuous	Mono-medium (unstratified)	Synthetic fiber	610[a]	Upward
Traveling-bridge	Continuous	Mono-medium (stratified)	Sand	610	Downward
Traveling-bridge	Continuous	Dual-medium (stratified)	Sand and anthracite	410	Downward

[a] Compressed depth.

Conventional Downflow Filters. Single-, dual-, or multimedium filter materials are utilized in conventional downflow depth filters. Typically sand or anthracite is used as the filtering material in single-medium filters (see Fig. 11–11a). Dual-medium filters usually consist of a layer of anthracite over a layer of sand (see Fig. 11–11b). Other combinations include (1) activated carbon and sand, (2) resin beads and sand, and (3) resin beads and anthracite. Multimedium filters typically consist of a layer of anthracite over a layer of sand over a layer of garnet or ilmenite. Other combinations include (1) activated carbon, anthracite, and sand, (2) weighted, spherical resin beads, anthracite, and sand, and (3) activated carbon, sand, and garnet.

Dual- and multimedium and deep-bed mono-medium depth filters were developed to allow the suspended solids in the liquid to be filtered to penetrate farther into the filter bed, and thus use more of the solids-storage capacity available within the filter bed (see Fig. 11–12). The deeper penetration of the solids into the filter bed also permits longer filter runs because the buildup of headloss is reduced. By comparison, in shal-

| **Table 11–6** (*Continued*)

Backwash operation	Flowrate through filter	Solids storage location	Remarks	Type of design
Batch	Constant/variable	Surface and upper bed	Rapid headloss buildup	Individual
Batch	Constant/variable	Internal	Dual-medium design used to extend filter run length	Individual
Batch	Constant/variable	Internal	Multimedium design used to extend filter run length	Individual
Batch	Constant/variable	Internal	Deep bed used to store solids and extend filter run length	Individual
Batch	Constant	Internal	Deep bed used to store solids and extend filter run length	Proprietary
Continuous	Constant	Internal	Sand bed moves in countercurrent direction to fluid flow	Proprietary
Batch	Constant	Surface and upper bed	Air pulses used to break up surface mat and increase run length	Proprietary
Batch	Constant	Internal		Proprietary
Semicontinuous	Constant	Surface and upper bed	Individual filter cells backwashed sequentially	Proprietary
Semicontinuous	Constant	Surface and upper bed	Individual filter cells backwashed sequentially	Proprietary

low mono-medium beds, most of the removal occurs in the upper few millimeters of the bed (Tchobanoglous and Eliassen, 1970). The operation of conventional downflow filters has been described previously (see Fig. 11–4). Typically, water wash plus surface water and water wash plus air scour are the principal methods used in backwashing mono-, dual-, and multimedium filter beds. With either of these backwashing methods, the filter mediums are fluidized.

Deep-Bed Downflow Filters. The deep-bed downflow filter (see Fig. 11–11*c*) is similar to the conventional downflow filter with the exception that the depth of the filter bed and the size of the filtering medium (usually anthracite) are greater than the corresponding values in a conventional filter. Because of the greater depth and larger medium size (i.e., sand or anthracite), more solids can be stored within the filter bed and the run length can be extended. The maximum size of the filter medium used in these filters will depend on the ability to backwash the filter. In general, deep-bed filters are not fluidized completely during backwashing. To achieve effective cleaning, air scour plus water is used in the backwash operation.

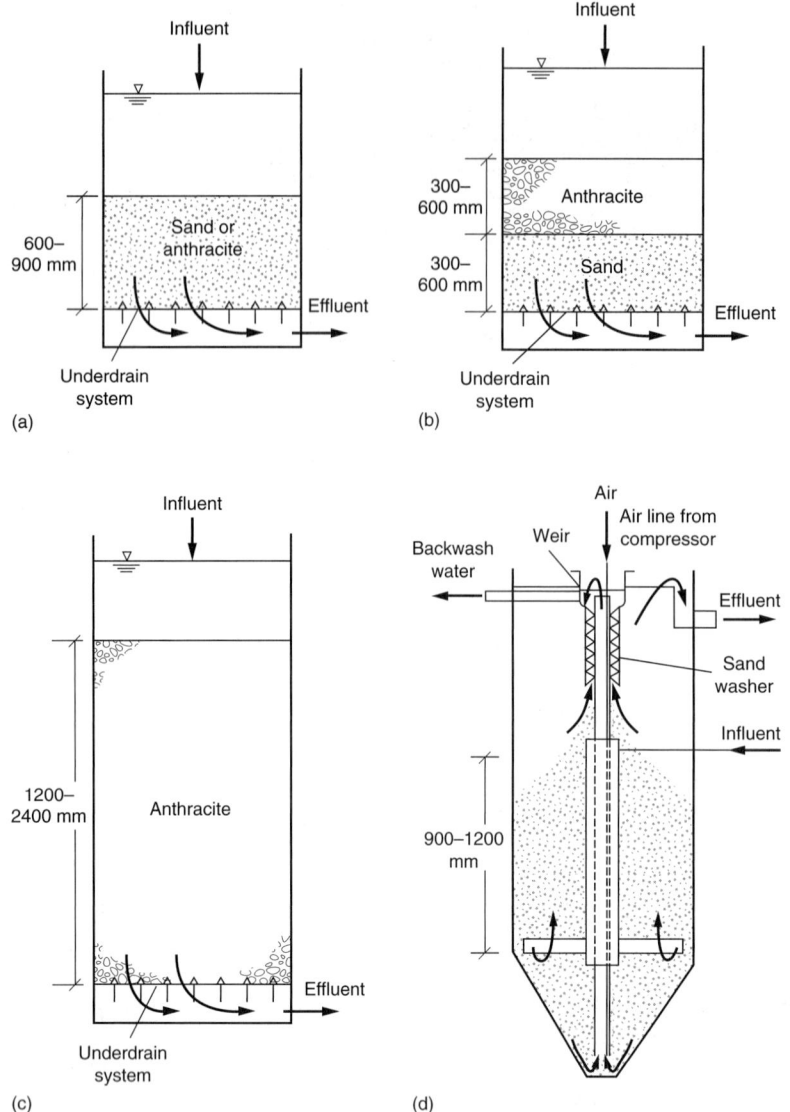

Figure 11–11

Definition sketches for the principal types of granular medium filters: (a) conventional mono-medium downflow filter, (b) conventional dual-medium downflow filter, (c) conventional mono-medium deep-bed downflow filter, (d) continuous backwash deep-bed upflow filter, (e) pulsed-bed filter, (f) traveling-bridge filter, (g) synthetic-medium filter, (h) pressure filter, and (i) slow sand filter.

Deep-Bed Upflow Continuous Backwash Filters. In this filter, shown schematically (see Fig. 11–11d) and pictorially on Fig. 11–13, the wastewater to be filtered is introduced into the bottom of the filter where it flows upward through a series of riser tubes and is distributed evenly into the sand bed through the open bottom of an inlet distribution hood. The water then flows upward through the downward-moving

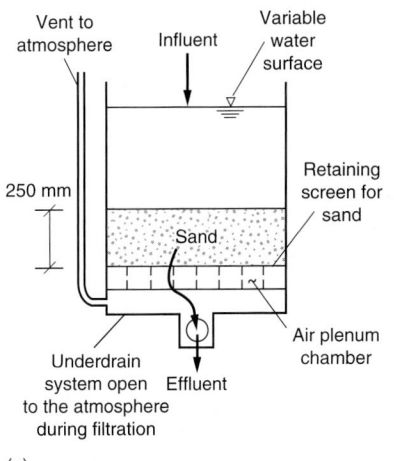

(e)

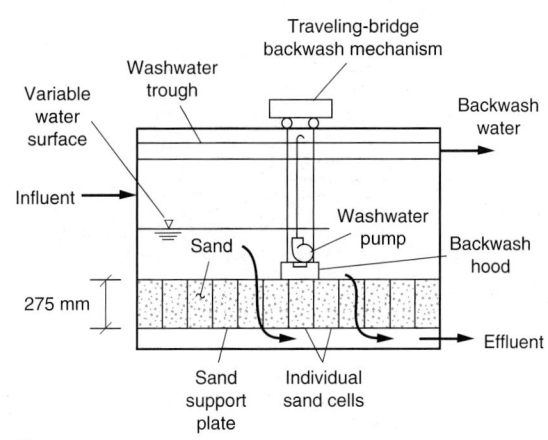

(f)

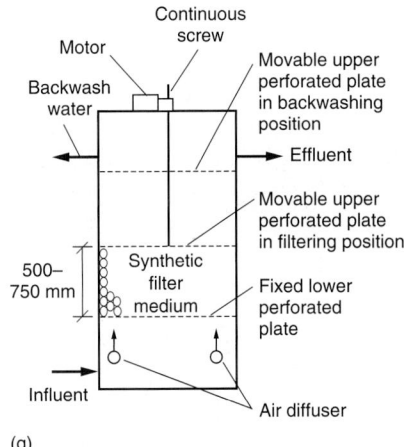

(g)

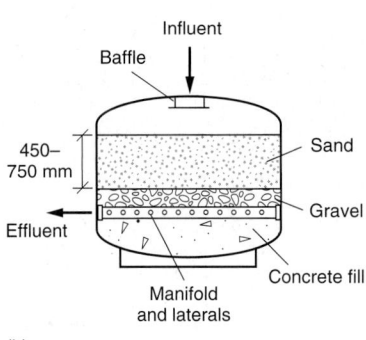

(h)

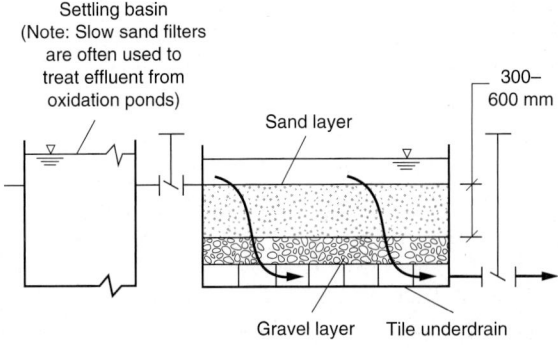

(i)

Figure 11-12

Schematic diagram of filter beds illustrating potential increase in storage capacity: (a) single medium, (b) dual medium, and (c) multimedium.

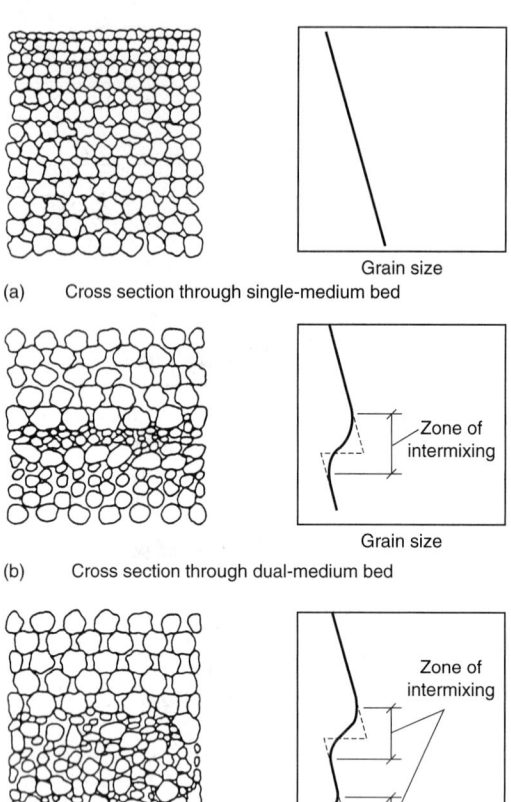

(a)   Cross section through single-medium bed

(b)   Cross section through dual-medium bed

(c)   Cross section through multimedium bed

sand. The clean filtrate exits from the sand bed, overflows a weir, and is discharged from the filter. At the same time sand particles, along with trapped solids, are drawn downward into the suction of an airlift pipe that is positioned in the center of the filter. A small volume of compressed air, introduced into the bottom of the airlift, draws sand, solids, and water upward through the pipe by creating a fluid with a density less than 1.

Impurities are scoured (abraded) from the sand particles during the turbulent upward flow. Upon reaching the top of the airlift, the dirty slurry spills over into the central reject compartment. By setting the filtrate weir above the reject weir, a steady stream of clean filtrate flows upward, countercurrent to the movement of sand, through the washer section. The upflow liquid carries away the solids and reject water. Because the sand has a higher settling velocity than the removed solids, the sand is not carried out of the filter. The sand is cleaned further as it moves down through the washer. The cleaned sand is redistributed onto the top of the sand bed, allowing for a continuous uninterrupted flow of filtrate and reject water.

Pulsed-Bed Filters. The pulsed-bed filter (see Figs. 11–11e and 11–14) is a proprietary downflow gravity filter with an unstratified shallow layer of fine sand as the fil-

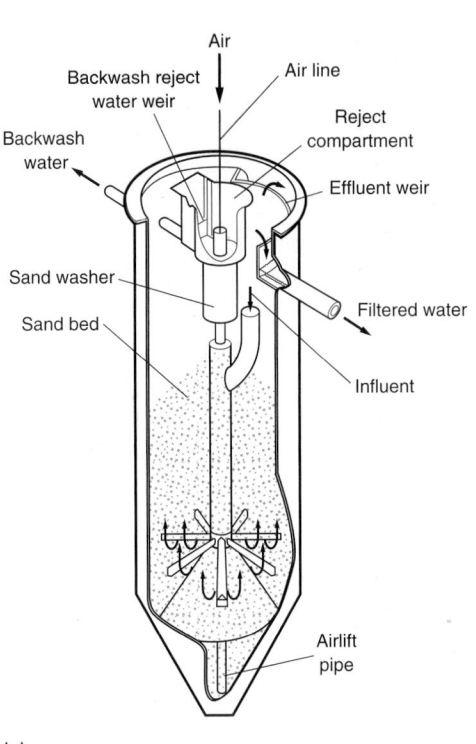

(a)   (b)

Figure 11–13

Deep-bed upflow continuous backwash filter: (a) schematic view and (b) view of test unit with older style influent feed system. *Note:* The influent feed assembly will vary depending on the filter configuration (see Fig. 11–22). *(Courtesy: Parkson Corporation.)*

tering medium. The shallow bed is used for solids storage, as opposed to other shallow-bed filters where solids are principally stored on the sand surface. An unusual feature of this filter is the use of an air pulse to disrupt the sand surface and thus allow penetration of suspended solids into the bed. The air pulse process involves forcing a volume of air, trapped in the underdrain system, up through the shallow filter bed to break up the surface mat of solids and renew the sand surface. When the solids mat is disturbed, some of the trapped material is suspended into the admixture over the sand, but the majority of solids are entrapped within the filter bed. The intermittent air pulse causes a folding over the sand surface, burying solids within the medium and regenerating the filter bed surface. The filter continues to operate with intermittent pulsing until a terminal headloss limit is reached. The filter then operates in a conventional backwash cycle to remove solids from the sand. It should be noted that during normal operation the filter underdrain is not flooded as it is in a conventional filter.

Traveling-Bridge Filters. The traveling-bridge filter (see Figs. 11–11f and 11–15) is a proprietary continuous downflow, automatic backwash, low-head, granular medium-depth filter. The bed of the filter is divided horizontally into long independent

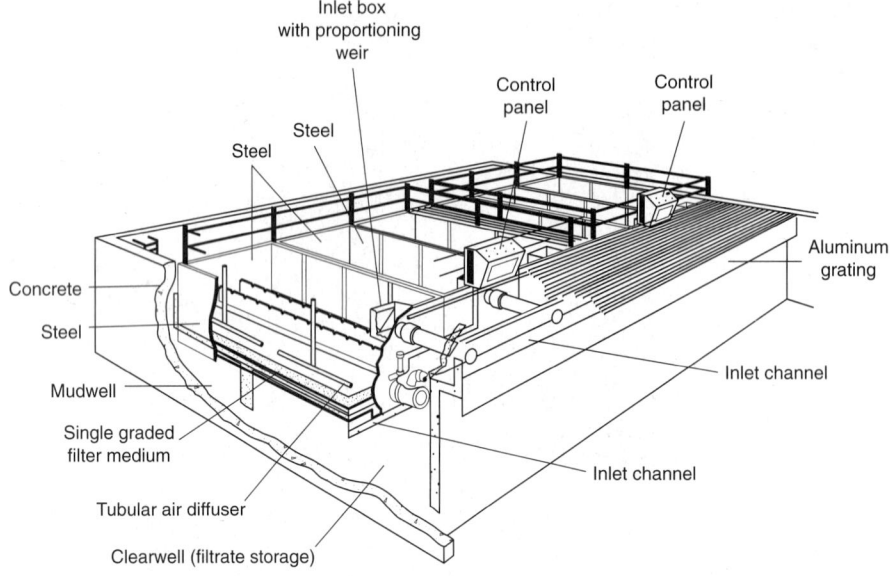

Figure 11–14

Schematic view of pulsed-bed filter. *(Courtesy of U.S. Filter.)*

Figure 11–15

Traveling-bridge filter.
*(Courtesy: Aqua
Aerobic, Inc.)*

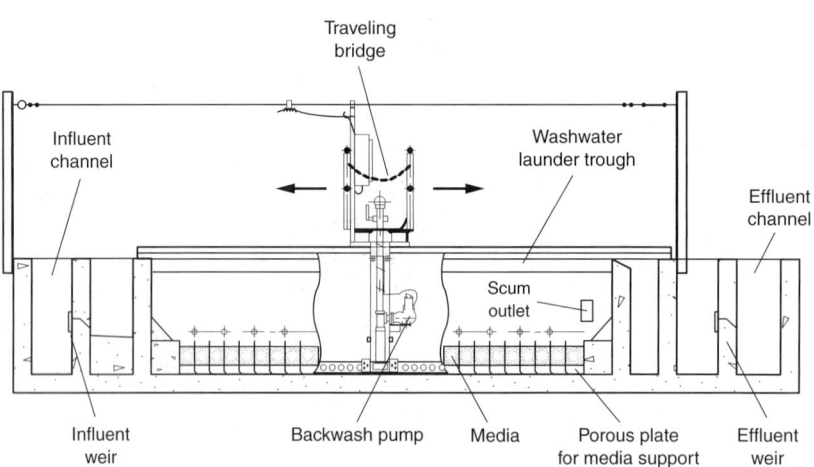

filter cells. Each filter cell contains approximately 280 mm (11 in) of medium. Treated wastewater flows through the medium by gravity and exits to the clearwell plenum via a porous-plate, polyethylene underdrain. Each cell is backwashed individually by an overhead, traveling-bridge assembly, while all other cells remain in service. Water used for backwashing is pumped directly from the clearwell plenum up through the medium and deposited in a backwash trough. During the backwash cycle, wastewater is filtered continuously through the cells that are not being backwashed. The backwash mechanism includes a surface wash pump to assist in breaking up of the surface matting and "mudballing" in the medium. Because the backwashing operation is performed on an "as needed" basis, the backwash cycle is termed semicontinuous (see Table 11–6).

Synthetic-Medium Filters. A synthetic-medium filter, developed originally in Japan, is now being used for the filtration wastewater. The highly porous synthetic filter medium, made of polyvaniladene, has a diameter of approximately 30 mm (1.25 in). Based on displacement tests, the porosity of the uncompacted quasi-spherical filter medium itself is estimated to be about 88 to 90 percent, and the porosity of the filter bed made up of the filter medium is approximately 94 percent. Unusual features of the filter are: (1) the porosity of the filter bed can be modified by compressing the filter medium and (2) the size of the filter bed is increased mechanically to backwash the filter (see Figs. 11–11g and 11–16). The filter medium also represents a departure from conventional filter media in that the fluid to be filtered flows through the medium as opposed to flowing around the filtering medium, as in sand and anthracite filters. As a result of the high porosity, filtration rates of 400 to 1200 L/m²·min (10 to 30 gal/ft²·min) have been pilot-tested (Caliskaner and Tchobanoglous, 2000).

In the filtering mode, secondary effluent is introduced in the bottom of the filter. The influent wastewater flows upward through the filter medium, retained by two porous plates, and is discharged from the top of the filter. To backwash the filter, the upper porous plate is raised mechanically. While flow to the filter continues, air is introduced sequentially from the left and right sides of the filter below the lower porous plate, causing the filter medium to move in a rolling motion. The filter medium is cleaned by the shearing forces as the wastewater moves past the filter and by abrasion as the filter medium rubs against itself. Wastewater containing the solids removed from

Figure 11–16

Schematic diagram of the Fuzzy filter. *(Courtesy: Schreiber Corporation.)*

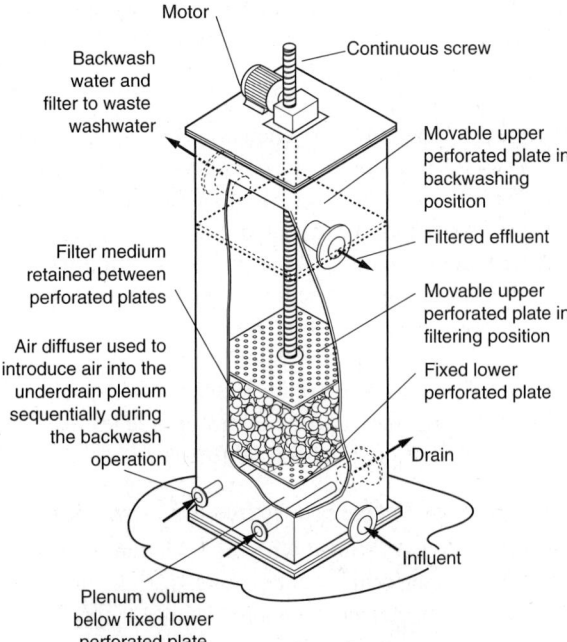

- Motor
- Backwash water and filter to waste washwater
- Continuous screw
- Movable upper perforated plate in backwashing position
- Filtered effluent
- Filter medium retained between perforated plates
- Movable upper perforated plate in filtering position
- Air diffuser used to introduce air into the underdrain plenum sequentially during the backwash operation
- Fixed lower perforated plate
- Drain
- Influent
- Plenum volume below fixed lower perforated plate

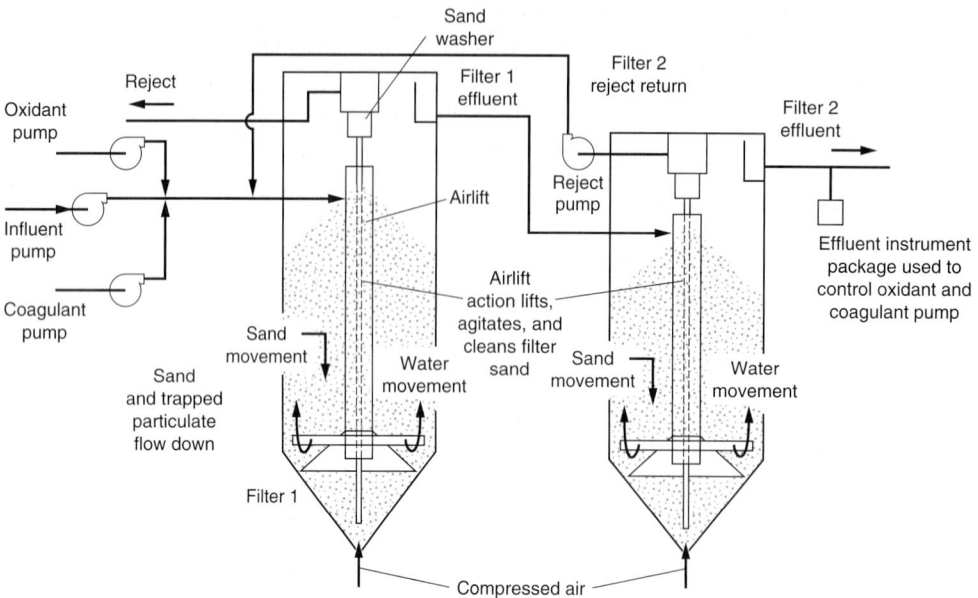

Figure 11–17

Flow diagram for a two-stage filtration process for the removal of turbidity, total suspended solids, and phosphorus. (*Courtesy DSS Environmental, Inc.*)

the filter is diverted for subsequent processing. To put the filter back into operation after the backwash cycle has been completed, the raised porous plate is returned to its original position. After a short flushing cycle, the filtered effluent valve is opened, and filtered effluent is discharged.

Two-Stage Filtration. A schematic flow diagram for a proprietary two-stage filtration process for the removal of turbidity, total suspended solids, and phosphorus is shown on Fig. 11–17. As shown, two deep-bed upflow continuous backwash filters are used in series to produce a high-quality effluent. A large-size sand diameter is used in the first filter to increase the contact time and to minimize clogging. A smaller sand size is used in the second filter to remove residual particles from the first-stage filter. The backwash water from the second filter which contains small particles and residual coagulant is recycled to the first filter to improve floc formation within the first-stage filter and the influent to waste ratio. Based on full-scale installations the reject rate has been found to be less than 5 percent. Phosphorus levels equal to or less than 0.02 mg/L have been achieved in the final filter effluent.

Performance of Different Types of Filter Technologies

The critical question associated with the selection of any granular-medium depth filter is whether it will perform as anticipated. Insight into the performance of granular-medium filters can be gained from a review of the turbidity, total suspended solids, and particle-removal data for depth filters.

Figure 11–18

Performance data for seven different types of depth filters used for wastewater applications tested using the effluent from the same activated-sludge plant at filtration rate 160 L/m²·min (4 gal/ft²·min) with the exception of the Fuzzy filter, which was operated at 800 L/m²·min (20 gal/ft²·min).

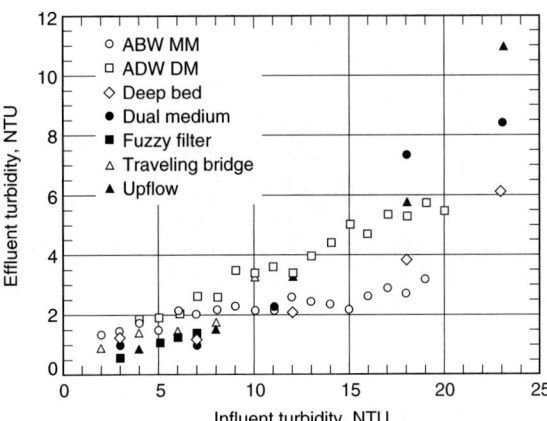

Removal of Turbidity. The results of long-term testing of seven different types of pilot-scale filters on the effluent from the same activated-sludge process (SRT > 10 d), without chemical addition, are shown on Fig. 11–18. Long-term data from other large-scale wastewater reclamation plants are also shown. The principal conclusions to be reached from an analysis of the data presented on Fig. 11–18 are that (1) given a high-quality filter influent (turbidity less than 5 to 7 NTU) all of the filters tested including the large-scale plant are capable of producing an effluent with an average turbidity of 2 NTU or less, and (2) when the influent turbidity is greater than about 5 to 7 NTU, chemical addition will be required with all of the filters to achieve an effluent turbidity of 2 NTU or less.

Removal of Total Suspended Solids. Using the following two relationships between turbidity and suspended solids as given in Chap. 2 [see Eq. (2–16)], influent turbidity values of 5 to 7 NTU correspond to a total suspended solids concentration varying from about 10 to 17 mg/L, and an effluent turbidity of 2 NTU corresponds to a total suspended solids concentration varying from 2.8 to 3.2 mg/L.

Settled secondary effluent

$$\text{TSS, mg/L} = (2.0 \text{ to } 2.4) \times (\text{turbidity, NTU}) \tag{11–32}$$

Filter effluent

$$\text{TSS, mg/L} = (1.3 \text{ to } 1.5) \times (\text{turbidity, NTU}) \tag{11–33}$$

Long-term total suspended-solids performance data for the depth filters at the Donald C. Tillman Wastewater Reclamation Plant in Los Angeles, CA, are shown on Fig. 11–19. For the data shown on Fig. 11–19, the ratio of TSS to turbidity is about 1.33.

Removal of Particle Size. Typical data on the removal of particle sizes found in settled wastewater are shown on Fig. 11–20. As shown, the particle-removal rate is essentially independent of the filtration rate up to about 240 L/m²·min (6 gal/ft²·min).

Figure 11–19

Probability distribution of product water turbidity and TSS for 1998 for the Donald C. Tillman Wastewater Reclamation Plant, Los Angeles, CA.

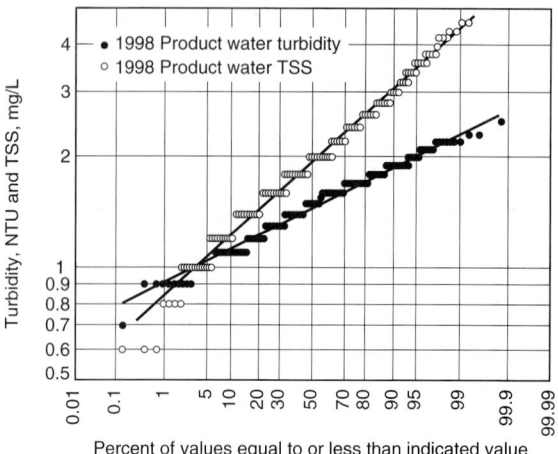

Figure 11–20

Particle-size removal efficiency for a depth filter for effluent from an activated-sludge plant.

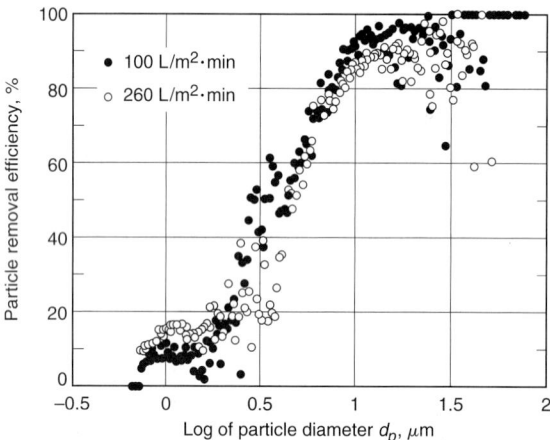

It is significant that most depth filters will pass some particles with diameters up to 20 μm. Particles in the size range from 10 to 20 μm are of importance with respect to disinfection because they are of sufficient size to shield microorganisms (see Chap. 12).

Issues Related to Design and Operation of Treatment Facilities

For new wastewater-treatment plants, extra care should be devoted to the design of the secondary settling facilities. With properly designed settling facilities resulting in an effluent with low TSS (typically 5 mg/L), the decision on what type of filtration system is to be used is often based on plant-related variables, such as the space available, duration of filtration period (seasonal versus year-round), the time available for construction, and costs. For existing plants that have variable suspended solids concentrations in the treated effluent and must be retrofitted with effluent filtration, it may be appropriate to consider the type of a filter that can continue to function even when heavily loaded. The pulsed-bed filter and both downflow and upflow deep-bed coarse-medium filters have been used in such applications.

Importance of Influent Wastewater Characteristics

The most important influent characteristics in the filtration of treated secondary effluents are the suspended-solids concentration, particle size and distribution, and floc strength.

Suspended Solids/Turbidity. Typically, the TSS concentration in the effluent from activated-sludge and trickling-filter plants varies between 6 and 30 mg/L. Because this concentration usually is the principal parameter of concern, turbidity is often used as a means of monitoring the filtration process. Corresponding turbidity values can vary from 3 to 15 NTU.

Particle Size and Distribution. Typical data on the particle size and distribution in the effluent from two activated-sludge plants were presented previously on Fig. 2–9 in Chap. 2. As illustrated, the particles fell into two distinct size ranges, small particles varying in areal size (equivalent circular diameter) from 0.8 to 1.2 μm and large particles varying in size from about 5 to 100 μm. In addition, a few particles larger than about 500 μm are almost always found in settled treated effluent. These particles are light and amorphous and do not settle readily (see discussion of hindered settling in Sec. 5–5). The weight fraction of the smaller particles was estimated to be approximately 40 to 60 percent of the total. This percentage will vary, however, depending on the operating conditions of the biological process and the degree of flocculation achieved in the secondary settling facilities.

The most significant observation related to particle size is that the distribution of sizes is bimodal. This observation is important because it will influence the removal mechanisms that may be operative during the filtration process. For example, it seems reasonable to assume that the removal mechanism for particles 1.0 μm in size would be different from that for particles in the size range from 10 to 100 μm. The bimodal particle-size distribution has also been observed in water-treatment plants.

Floc Strength. Floc strength, which will vary with the type of process and the mode of operation, is also important. For example, the residual floc from the chemical precipitation of biologically processed wastewater may be considerably weaker than the residual biological floc before precipitation. Further, the strength of the biological floc will vary with the mean cell-residence time, increasing with longer mean cell-residence time. The increased strength derives in part from the production of extracellular polymers as the mean cell-residence time is lengthened. At extremely long mean cell-residence times (15 days and longer), it has been observed that the floc strength will decrease due to floc breakup.

Selection of Filtration Technology

In selecting a filter technology, important issues that must be considered include (1) type of filter to be used: proprietary or individually designed, (2) the filtration rate, (3) filtration driving force, (4) number and size of filter units, and (5) the backwash-water requirements.

Type of Filter: Proprietary or Individually Designed. As noted in Table 11–6, the currently available filter technologies are either proprietary or individually designed. With proprietary filters, the manufacturer is responsible for providing the complete filter unit and its controls, based on basic design criteria and performance specifications. In individually designed filters, the designer is responsible for working with several suppliers in developing the design of the system components. Contractors and suppliers then furnish the materials and equipment in accordance with the engineer's design.

Filtration Rate. The rate of filtration is important because it affects the real size of the filters that will be required. For a given filter application, the rate of filtration will depend primarily on the strength of the floc and the size of the filtering medium. For example, if the strength of the floc is weak, high filtration rates will tend to shear the floc particles and carry much of the material through the filter. It has been observed that filtration rates in the range of 80 to 320 L/m²·min (2 to 8 gal/ft²·min) will not affect the effluent quality when filtering settled activated-sludge effluents.

Filtration Driving Force. Either the force of gravity or an applied pressure force can be used to overcome the frictional resistance to flow offered by the filter bed. Gravity filters of the type discussed above are used most commonly for the filtration of treated effluent at large plants. Pressure filters of the type shown on Fig. 11–21 operate in the same manner as gravity filters and are used at smaller plants. The only difference is that, in pressure filters, the filtration operation is carried out in a closed vessel under

Figure 11–21

Typical pressure filter with multimedium filter bed and surface wash used for wastewater filtration.

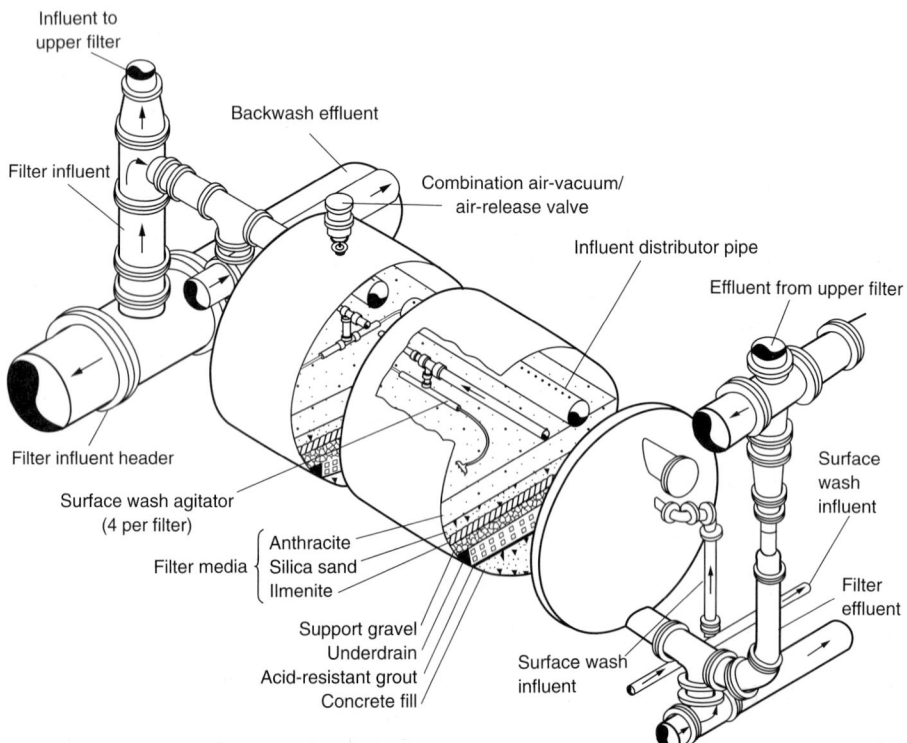

pressurized conditions achieved by pumping. Pressure filters normally are operated at higher terminal headlosses, resulting in longer filter runs and reduced backwash requirements.

Number and Size of Filter Units.   One of the early decisions to be made in the design of a depth filtration system is determining the number and size of filter units that will be required. The surface area required is based on the peak filtration and peak plant flowrates. The allowable peak filtration rate is usually established on the basis of regulatory requirements. Operating ranges for a given filter type are based on past experience, the results of pilot-plant studies, manufacturers' recommendations, and regulatory constraints. The number of units generally should be kept to a minimum to reduce the cost of piping and construction, but it should be sufficient to assure (1) that backwash flowrates do not become excessively large and (2) that when one filter unit is taken out of service for backwashing, the transient loading on the remaining units will not be so high that material contained in the filters will be dislodged. Transient loadings due to backwashing are not an issue with filters that backwash continuously. To meet redundancy requirements, a minimum of two filters will generally be required.

The sizes of the individual units should be consistent with the sizes of equipment available for use as underdrains, washwater troughs, and surface washers. Typically, width-to-length ratios for individually designed gravity filters vary from 1:1 to 1:4. A practical limit for the surface area on an individual depth filter (or filter cell) is about 100 m² (1100 ft²), although larger filters units have been built. The layout of a bank of deep-bed upflow continuous-backwash filters is shown on Fig. 11–22. For proprietary

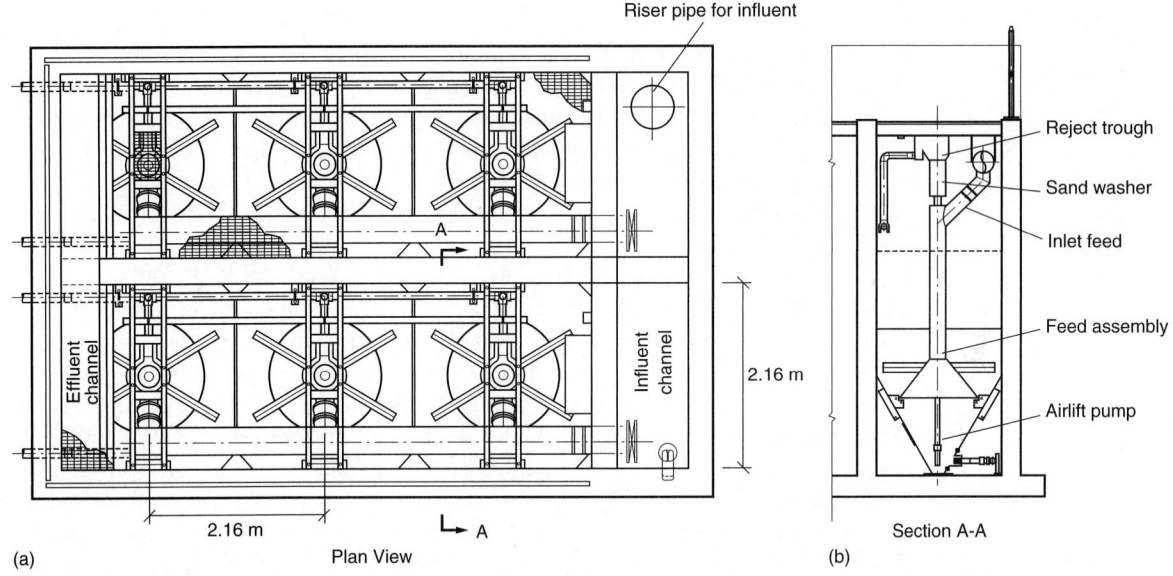

(a)   (b)

Figure 11–22

Layout of a bank of six deep-bed upflow continuous backwash filters. *(Courtesy Parkson Corp.)*

and pressure filters, it is common practice to use standard sizes that are available from manufacturers. Depending on the manufacturing method, shipping constraints will also limit the size of pressure filters. The maximum diameter of vertical pressure filters is about 3.7 m (12 ft). The maximum diameter and length of horizontal pressure filters is about 3.7 m (12 ft) and 12 m (40 ft), respectively.

Backwash-Water Requirements. As noted in Table 11–6, depth filters operate in either a semicontinuous or continuous mode. In semicontinuous operation, the filter is operated until the effluent quality starts to deteriorate or the headloss becomes excessive, at which point the filter is taken out of service and backwashed to remove the accumulated solids. With filters operated in the semicontinuous mode, provision must be made for the backwash water needed to clean the filters. Typically, the backwash water is pumped from a filtered water clearwell or obtained by gravity from an elevated storage tank. For filters that operate continuously, such as the upflow filter (see Figs. 11–11d and 11–13) and the traveling-bridge filter (see Figs 11–11f and 11–15), the filtering and cleaning (backwashing) phases take place simultaneously. In the traveling-bridge filter, the backwash operation can be either continuous or semicontinuous as required. It should be noted that with filters that operate continuously, there is no turbidity breakthrough or terminal headloss.

Filter-Bed Characteristics

The principal variables that must be considered in the design of filters are identified in Table 11–7. In the application of filtration to the removal of residual suspended solids, it has been found that the nature of the particulate matter in the influent to be filtered, the filter-bed configuration, the size of the filter material or materials, and the filtration flowrate are perhaps the most important of the process variables.

Filter-Bed Configuration. The principal types of nonproprietary filter-bed configurations now used for wastewater filtration may be classified according to the number of filtering mediums that are used as mono-medium, dual-medium, or multimedium beds (see Fig. 11–12). In conventional downflow filters, the distribution of grain sizes for each medium after backwashing is from small to large. Typical design data for monomedium filters and dual- and multimedium filters are presented in Tables 11–8 and 11–9, respectively.

Selection of Filter Medium. Once the type of filter to be used has been selected, the next step is to specify the characteristics of the filtering medium, or media if more than one is used. Typically, this process involves the selection of the grain size as specified by the effective size d_{10}, uniformity coefficient UC, the 90 percent size, the specific gravity, solubility, hardness, and depth of the various materials used in the filter bed. Typical particle-size distribution ranges for sand and anthracite filtering material are shown on Fig. 11–23. The 90 percent size designated d_{90}, as read from a grain-size analysis, is commonly used to determine the required backwash rate for depth filters. Typical physical properties of filter materials used in depth filters are summarized in Table 11–10. The sizes of filter materials were given previously in Tables 11–7 and 11–8.

Table 11–7
Principal variables in the design of granular-medium filters[a]

Variable	Significance
1. Required effluent quality	Usually fixed regulatory requirement
2. Influent wastewater characteristics	Affect the removal characteristics of a given filter-bed configuration. To a limited extent the listed influent characteristics can be controlled by the designer
a. Suspended-solids concentration	
b. Floc or particle size and distribution	
c. Floc strength	
d. Floc or particle charge	
e. Fluid properties	
3. Filter-medium characteristics	Affect particle-removal efficiency and headloss buildup
a. Effective size, d_{10}	
b. Uniformity coefficient, UC	
c. Type, grain shape, density, and composition	
4. Filter-bed characteristics	Porosity affects the amount of solids that can be stored within the filter. Bed depth affects initial headloss, length of run. Degree of intermixing will affect performance of filter bed
a. Bed depth	
b. Stratification	
c. Degree of medium intermixing	
d. Porosity	
5. Filtration rate	Used in conjunction with variables 2, 3, and 4 to compute clear-water headloss
6. Chemical usage	
7. Allowable headloss	Design variable
8. Backwash requirements	Affects size of filter piping and pipe gallery

[a] Adapted in part from Tchobanoglous and Eliassen (1970) and Tchobanoglous and Schroeder (1985).

The degree of intermixing in the dual-medium and multimedium beds depends on the density and size differences of the various media. Further, the size of the medium and the specific gravity of each layer are important with respect to filter backwashing and the potential for the loss of filter material. To avoid extensive intermixing the settling rate of the filter mediums comprising the dual- and multimedium filters must have essentially the same settling velocity. The following relationship can be used to establish the appropriate sizes (Kawamura, 2000).

$$\frac{d_1}{d_2} = \left(\frac{\rho_2 - \rho_w}{\rho_1 - \rho_w}\right)^{0.667} \tag{11–34}$$

where d_1, d_2 = effective size of filter mediums
ρ_1, ρ_2 = density of filter mediums
ρ_w = density of water

The application of Eq. (11–34) is illustrated in Example 11–4.

Table 11–8
Typical design data
for depth filters with
mono-medium[a]

Characteristic	Unit	Value	
		Range	**Typical**
Shallow-bed (stratified)			
Anthracite			
Depth	mm	300–500	400
Effective size	mm	0.8–1.5	1.3
Uniformity coefficient	unitless	1.3–1.8	≤1.5
Filtration rate	L/m²·min	80–240	120
Sand			
Depth	mm	300–360	330
Effective size	mm	0.45–0.65	0.45
Uniformity coefficient	unitless	1.2–1.6	≤1.5
Filtration rate	L/m²·min	80–240	120
Conventional (stratified)			
Anthracite			
Depth	mm	600–900	750
Effective size	mm	0.8–2.0	1.3
Uniformity coefficient	unitless	1.3–1.8	≤1.5
Filtration rate	L/m²·min	80–400	160
Sand			
Depth	mm	500–750	600
Effective size	mm	0.4–0.8	0.65
Uniformity coefficient	unitless	1.2–1.6	≤1.5
Filtration rate	L/m²·min	80–240	120
Deep-bed (unstratified)			
Anthracite			
Depth	mm	900–2100	1500
Effective size	mm	2–4	2.7
Uniformity coefficient	unitless	1.3–1.8	≤1.5
Filtration rate	L/m²·min	80–400	200
Sand			
Depth	mm	900–1800	1200
Effective size	mm	2–3	2.5
Uniformity coefficient	unitless	1.2–1.6	≤1.5
Filtration rate	L/m²·min	80–400	200
Fuzzy filter			
Depth	mm	600–1080	800
Effective size	mm	25–30	28
Uniformity coefficient	unitless	1.1–1.2	1.1
Filtration rate	L/m²·min	600–1000	800

[a] Adapted from Tchobanoglous (1988).

Table 11-9
Typical design data for dual- and multimedium depth filters[a]

Characteristic	Unit	Value[b]	
		Range	Typical
Dual-medium			
Anthracite ($\rho = 1.60$)			
Depth	mm	360–900	720
Effective size	mm	0.8–2.0	1.3
Uniformity coefficient	unitless	1.3–1.6	≤1.5
Sand ($\rho = 2.65$)			
Depth	mm	180–360	360
Effective size	mm	0.4–0.8	0.65
Uniformity coefficient	unitless	1.2–1.6	≤1.5
Filtration rate	L/m²·min	80–400	200
Multimedium			
Anthracite (top layer of quad-media filter, $\rho = 1.60$)			
Depth	mm	240–600	480
Effective size	mm	1.3–2.0	1.6
Uniformity coefficient	unitless	1.3–1.6	≤1.5
Anthracite (second layer of quad-media filter, $\rho = 1.60$)			
Depth	mm	120–480	240
Effective size	mm	1.0–1.6	1.1
Uniformity coefficient	unitless	1.5–1.8	1.5
Anthracite (top layer of tri-media filter, $\rho = 1.60$)			
Depth	mm	240–600	480
Effective size	mm	1.0–2.0	1.4
Uniformity coefficient	unitless	1.4–1.8	≤1.5
Sand ($\rho = 2.65$)			
Depth	mm	240–480	300
Effective size	mm	0.4–0.8	0.5
Uniformity coefficient	unitless	1.3–1.8	≤1.5
Garnet ($\rho = 4.2$)			
Depth	m	50–150	100
Effective size	mm	0.2–0.6	0.35
Uniformity coefficient	unitless	1.5–1.8	≤1.5
Filtration rate	L/m²·min	80–400	200

[a] Adapted from Tchobanoglous (1988).
[b] Anthracite, sand, and garnet sizes selected to limit the degree of intermixing. Use Eq. (11–34) for other values of density ρ.

Figure 11-23

Typical particle-size distribution ranges for sand and anthracite used in dual-medium depth filters. Note that for sand the 10 percent size by weight corresponds to the 50 percent size by count.

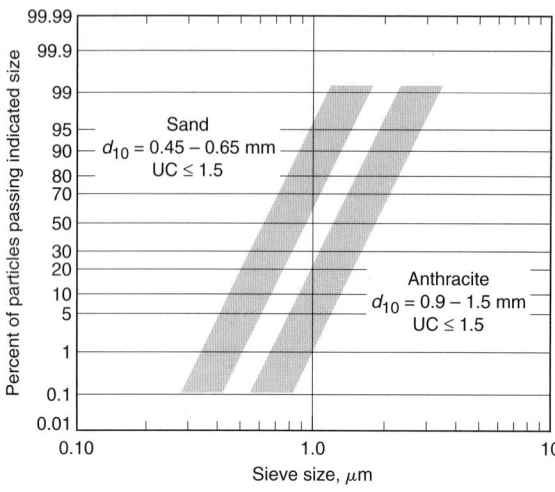

Table 11-10

Typical properties of filter materials used in depth filtration[a]

Filter material	Specific gravity	Porosity, α	Sphericity[b]
Anthracite	1.4–1.75	0.56–0.60	0.40–0.60
Sand	2.55–2.65	0.40–0.46	0.75–0.85
Garnet	3.8–4.3	0.42–0.55	0.60–0.80
Ilmenite	4.5	0.40–0.55	
Fuzzy filter medium		0.87–0.89	

[a] Adapted in part from Cleasby and Logsdon (1999).
[b] Sphericity is defined as ratio of the surface area of an equal volume sphere to the surface area of the filter medium particle.

EXAMPLE 11-4 **Determination of Filter-Medium Sizes** A dual-medium filter bed composed of sand and anthracite is to be used for the filtration of settled secondary effluent. If the effective size of the sand in the dual-medium filter is to be 0.55 mm, determine the effective size of the anthracite to avoid significant intermixing.

Solution

1. Summarize the properties of the filter mediums
   a. For sand
      i. Effective size = 0.55 mm
      ii. Specific gravity = 2.65 (see Table 11–10)
   b. For anthracite
      i. Effective size = to be determined, mm
      ii. Specific gravity = 1.7 (see Table 11–10)

2. Compute the effective size of the anthracite, beyond which extensive intermixing can be expected, using Eq. (11–34)

$$d_1 = d_2 \left(\frac{\rho_2 - \rho_w}{\rho_1 - \rho_w} \right)^{0.667}$$

$$d_1 = 0.55 \text{ mm} \left(\frac{2.65 - 1}{1.7 - 1} \right)^{0.667}$$

$$d_1 = 0.97 \text{ mm}$$

Comment Another approach that can be used to assess whether intermixing will occur is to compare the fluidized bulk densities of the two adjacent layers (e.g., upper 50 mm of sand and lower 100 mm of anthracite).

Filter Flowrate Control

The principal methods now used to control the rate of flow through downflow gravity filters may be classified as (1) constant-rate filtration with fixed head, (2) constant-rate filtration with variable head, and (3) variable-declining-rate filtration. A variety of other control methods are also in use (Cleasby and Logsdon, 1999; Kawumura, 2000).

Constant-Rate Filtration with Fixed Head. In constant-rate filtration with fixed head (see Fig. 11–24a), the flow through the filter is maintained at a constant rate. Constant-rate filtration systems are either influent controlled or effluent controlled. Pumps or weirs are used for influent control whereas an effluent modulating valve that can be operated manually or mechanically is used for effluent control. In effluent control systems, at the beginning of the run, a large portion of the available driving force is dissipated at the valve, which is almost closed. The valve is opened as the headloss builds up within the filter during the run. Because the required control valves are expensive and because they have malfunctioned on a number of occasions, alternative methods of flowrate control involving pumps and weirs have been developed and are coming into wider use.

Constant-Rate Filtration with Variable Head. In constant-rate variable filtration head (see Fig. 11–24b), the flow through the filter is maintained at a constant rate. Pumps or weirs are used for influent control. When the head or effluent turbidity reaches a preset value, the filter is backwashed.

Variable-Rate Filtration with Fixed or Variable Head. In variable-declining-rate filtration (see Fig. 11–24c), the rate of flow through the filter is allowed to decline as the rate of headloss builds up with time. Declining-rate filtration systems are either influent controlled or effluent controlled. When the rate of flow is reduced to the minimum design rate, the filter is removed from service and backwashed.

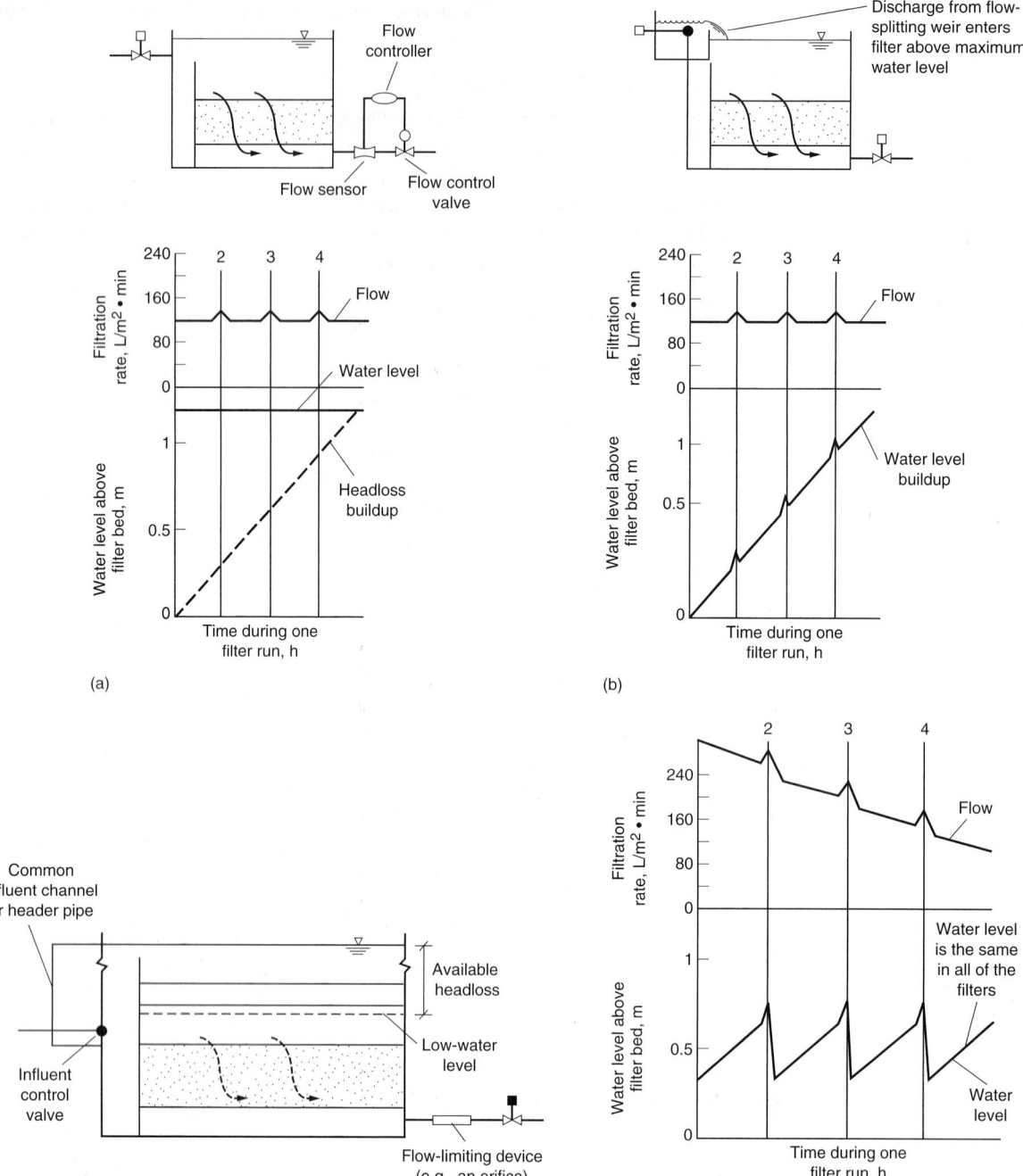

Figure 11–24

Definition sketch for filter operation: (*a*) fixed head, (*b*) variable head, and62 (*c*) variable-flow variable-head. Curves for filters in (*a*), (*b*), and (*c*) are for the operation of one filter in a bank of four filters. The numbers represent the filter that is backwashing during the filter run. In practice, the time before backwashing will not be the same for all of the filters. (*Adapted from Tchobanoglous and Schroeder, 1985.*)

Filter Backwashing Systems

A filter bed can function properly only if the backwashing system cleans the material removed within the filter effectively. The methods commonly used for backwashing granular-medium filter beds operated in the semicontinuous mode include (1) water backwash with auxiliary surface water-wash agitation, (2) water backwash with auxiliary air scour, and (3) combined air-water backwashing. With the first two methods, fluidization of the granular medium is necessary to achieve effective cleaning of the filter bed at the end of the run. With the third method, fluidization is not necessary. Typical backwash flowrates required to fluidize various filter beds are reported in Table 11–11.

Water Backwash with Auxiliary Surface Wash. Surface washers (see Fig. 11–25) are often used to provide the shearing force required to clean the grains of the filtering medium used for wastewater filtration. Operationally, the surface washing cycle is started about 1 or 2 min before the water backwashing cycle is started. Both cycles are continued for about 2 min, at which time the surface wash is terminated. Water usage is as follows: for a single-sweep surface backwashing system, from 20 to 40 L/m²·min (0.5 to 1.0 gal/ft²·min); for a dual-sweep surface backwashing system, from 60 to 80 L/m²·min (1.5 to 2.0 gal/ft²·min).

Water Backwash with Auxiliary Air Scour. For dual- and multimedium depth filters the use of air to scour the filter provides a more vigorous washing action than water alone. Operationally, air is usually applied for 3 to 4 min before the water backwashing cycle begins. In some systems, air is also injected during the first part of the water-washing cycle. Typical air flowrates range from 0.9 to 1.5 m³/m²·min (3 to 5 ft³/ft²·min).

Combined Air-Water Backwash. The combined air-water backwash system is used in conjunction with the single-medium unstratified filter bed. Operationally, air and water are applied simultaneously for several minutes. The specific duration of the combined backwash varies with the design of the filter bed. Ideally, during the backwash operation, the filter bed should be agitated sufficiently so that the grains of the filter medium move in a circular pattern from the top to the bottom of the filter as the air and

Table 11–11
Typical backwash flowrates required to fluidize various filter beds

Type of filter	Size of critical granular medium	Minimum backwash velocity needed to fluidize bed[a]	
		m³/m²·min	m/h
Single medium (sand)	2 mm	1.8–2.0	110–120
Dual-medium (anthracite and sand)	See Table 11–9	0.8–1.2	48–72
Multimedium (anthracite, sand, and garnet or ilmenite)	See Table 11–9	0.8–1.2	48–72
Fuzzy filter	30 mm	0.4–0.6	24–36

[a]Varies with size, shape, and specific gravity of the filter medium and the temperature of the backwash water.
Note: m³/m²·min × 24.5424 = gal/ft²·min.

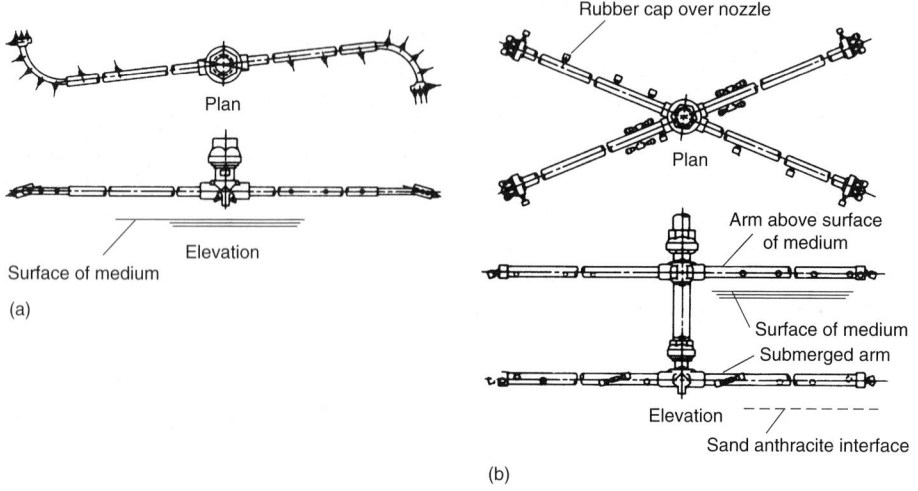

Rubber cap over nozzle

Plan

Plan

Arm above surface of medium

Surface of medium

Submerged arm

Elevation

Sand anthracite interface

(b)

Plan

Elevation

Surface of medium

(a)

Figure 11–25

Surface washing facilities used to clean conventional granular-medium filters: (a) single arm agitator and (b) dual arm agitator.

Table 11–12
Typical air and water backwash rates used with sand and anthracite filters[a]

| Medium | Medium characteristics | | Backwash sequence | Backwash rates, m³/m²·min | |
	Effective size, mm	Uniformity coefficient		Air	Water
Sand and anthracite[b]	0.65(s) 1.2(a)	1.4 1.4	1st − air 2nd − air + water 3rd − water	0.9–1.5 0.9–1.5	0.3–0.5 0.6–1.0
Sand	1	1.4	1st − air + water 2nd − water	0.9–1.5	0.25–0.3 0.5–0.6
Sand	2	1.4	1st − air + water 2nd − water	1.8–2.4	0.4–0.6 0.8–1.2
Anthracite	1.7	1.4	1st − air + water 2nd − water	1.0–1.5	0.35–0.5 0.6–0.8

[a] Adapted from Dehab and Young (1977) and Cleasby and Logsdon (2000).
[b] Dual medium filter bed is fluidized.
Note: m³/m²·min × 24.5424 = gal/ft²·min
m³/m²·min × 3.2808 = ft³/ft²·min

water rise up through the bed. Some typical data on the quantity of water and air required are reported in Table 11–12. The reduced washwater requirements for the air-water backwash system can be appreciated by comparing the values given in Table 11–11 with those given in Table 11–12. At the end of the combined air-water backwash, a 2- to 3-min water backwash at subfluidization velocities is used to remove any air bubbles that may remain in the filter bed. This last step is required to eliminate the possibility of air binding within the filter.

Filter Appurtenances

The principal filter appurtenances are as follows: (1) the underdrain system used to support the filtering materials, collect the filtered effluent, and distribute the backwash water and air (where used); (2) the washwater troughs used to remove the spent backwash water from the filter; and (3) the surface washing systems used to help remove attached material from the filter medium.

Underdrain Systems. The type of underdrain system to be used depends on the type of backwash system. In conventional water-backwashed filters without air scour, it is common practice to place the filtering medium on a support consisting of several layers of graded gravel. The design of a gravel support for a granular medium is delineated in the AWWA Standard for Filtering Material B100-96. Typical underdrain systems are shown on Fig. 11–26.

Washwater Troughs. Washwater troughs are constructed of fiberglass, plastic, or sheet metal or of concrete with adjustable weir plates. The particular design of the trough will depend to some extent on the other equipment to be used in the design and construction of the filter. Loss of filter material during backwashing is a common operating problem. To reduce this problem, baffles can be placed on the underside of the washwater troughs as shown on Fig. 11–27.

Surface Washers. Surface washers for filters can be fixed or mounted on rotary sweeps, as shown on Fig. 11–25. According to data on a number of systems, rotary-sweep washers appear to be the most effective.

Filter Instrumentation and Control Systems

The supervisory control facilities for wastewater filtration include instrumentation systems for the control and monitoring of the filters. The control systems are similar to those used for water treatment; however, full automation of gravity wastewater filters is not required. Although full automation is not required, fully automatic control systems using programmed logic controllers (PLCs) are used routinely.

Flow through the filters may be controlled from a water level upstream of the filters or from the water level in each filter. These water levels are used in conjunction with rate-of-flow controllers or a control valve to limit or regulate the flowrate through a filter. Filter hydraulic operating parameters requiring monitoring include filtered water flowrate, total headloss across each filter, surface wash and backwash water flowrates, and air flowrate if an air/water backwash system is employed. Water-quality parameters in filtered water that are monitored usually include BOD, TSS, phosphorus, and nitrogen. Turbidity may also be monitored in systems where chemical addition is used. Signals from effluent turbidity monitors and effluent flowrate are often used to pace the chemical-feed system.

The sequencing of the backwash cycle for a conventional gravity filter should be semiautomatic preferably, incorporating manual start, followed by automatic operation to carry the backwash cycle through its various steps. The design of backwash systems

Figure 11–26

Typical underdrain systems used for wastewater filters: (a) underdrain system used with gravel or porous plastic cap support layer for filter media *(Courtesy F. B. Leopold Co., Inc.)*, (b) underdrain system used without gravel support layer, and (c, d) air-water nozzle used in filters without gravel support layer. *(Courtesy Walker Process and Infilco-Degremont, Inc., respectively.)* Underdrain systems without a gravel support layer for the filter medium are known as *direct retention underdrains.*

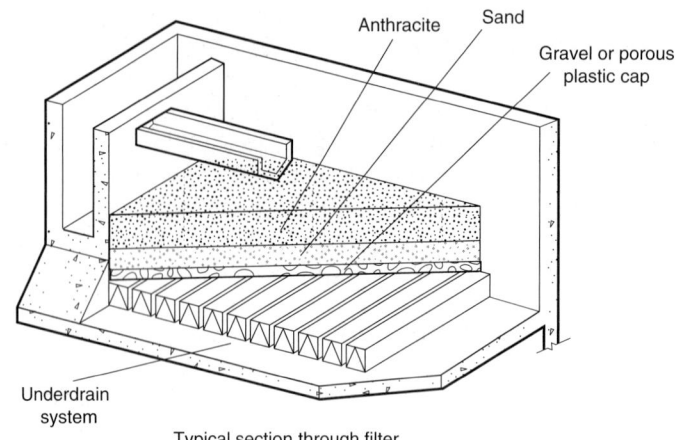

Anthracite Sand

Gravel or porous plastic cap

Underdrain system

Typical section through filter

(a)

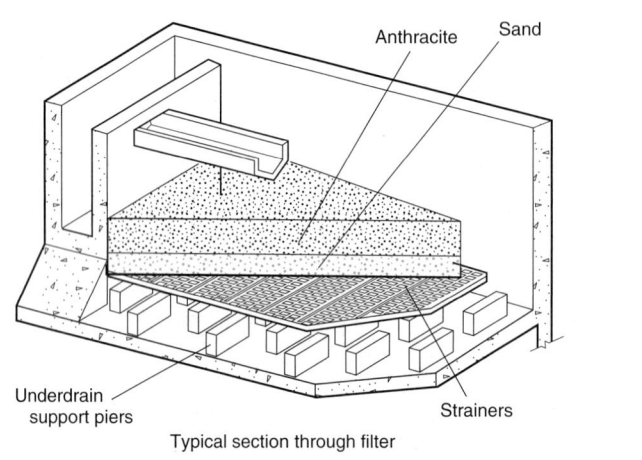

Anthracite Sand

Underdrain support piers

Strainers

Typical section through filter

(b)

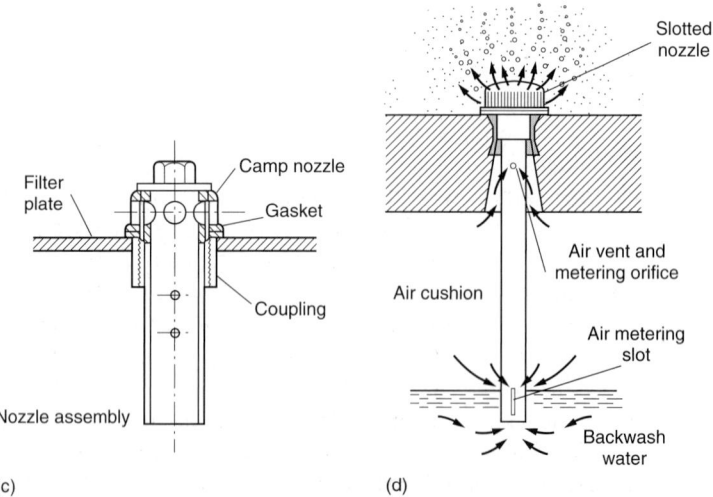

Filter plate

Camp nozzle

Gasket

Coupling

Nozzle assembly

(c)

Slotted nozzle

Air cushion

Air vent and metering orifice

Air metering slot

Backwash water

(d)

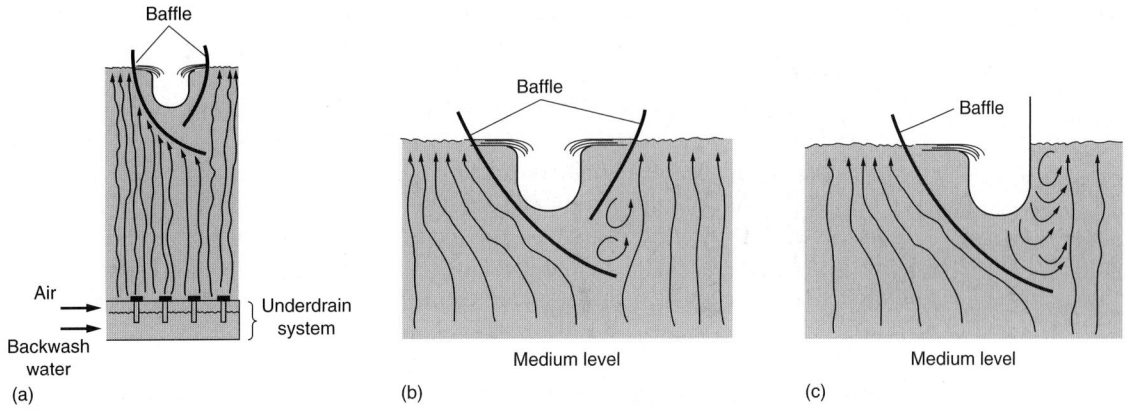

Figure 11-27

Details of baffle system developed for use with single-medium unstratified filter bed to minimize loss of filtering medium during backwash operation. (*a*) Section through filter showing flow patterns during air-water backwash operation. (*b*) Dual-baffle detail. (*c*) Single-baffle detail.

must recognize the impact of maximum wastewater temperatures experienced at treatment plants. Local control units should be provided at the filters to allow for local operation and backwashing by plant operators.

Effluent Filtration with Chemical Addition

Depending on the quality of the settled secondary effluent, chemical addition has been used to improve the performance of effluent filters. Chemical addition has also been used to achieve specific treatment objectives including the removal of specific contaminants such as phosphorus, metal ions, and humic substances. The removal of phosphorus by chemical addition is considered in Chap. 6. To control eutrophication, the contact filtration process is used in many parts of the country to remove phosphorus from wastewater-treatment plant effluents which are discharged to sensitive water bodies. The two-stage filtration process discussed previously (see Fig.11–17) has proved to be very effective, achieving phosphorus levels of 0.2 mg/L in the filtered effluent. Chemicals commonly used in effluent filtration include a variety of organic polymers, alum, and ferric chloride. Use of organic polymers and the effects of the chemical characteristics of the wastewater on alum addition are considered in the following discussion.

Use of Organic Polymers. Organic polymers are typically classified as long-chain organic molecules with molecular weights varying from 10^4 to 10^6. With respect to charge, organic polymers can be cationic (positively charged), anionic (negatively charged), or nonionic (no charge). Polymers are added to settled effluent to bring about the formation of larger particles by bridging as described in Chap. 6. Because the chemistry of the wastewater has a significant effect on the performance of a polymer, the selection of a given type of polymer for use as a filter aid generally requires experimental testing. Common test procedures for polymers involve adding an initial dosage (usually 1.0 mg/L) of a given polymer and observing the effects. Depending upon the effects observed, the dosage should be increased by 0.5 mg/L increments or decreased by 0.25 mg/L increments (with accompanying observation of effects) to obtain an operating range. After the operating range is established, additional testing can be done to establish the optimum dosage.

A recent development is the use of lower-molecular-weight polymers that are intended to serve as alum substitutes. When these polymers are used, the dosage is considerably higher ($\geq$10 mg/L) than with higher-molecular-weight polymers (0.25 to 1.25 mg/L). As with the mixing of alum, the initial mixing step is critical in achieving maximum effectiveness of a given polymer. In general, mixing times of less than 1 s with G values of $>$3500 s^{-1} are recommended (see Table 5–10 in Chap. 5). It should be noted that, as a practical matter, as treatment plants get larger it is difficult to achieve mixing times less than 1 s unless multiple mixing devices are used.

Effects of Chemical Characteristics of Wastewater on Alum Addition. As with polymers, the chemical characteristics of the treated wastewater effluent can have a significant impact on the effectiveness of aluminum sulfate (alum) when it is used as an aid to filtration. For example, the effectiveness of alum is dependent on pH (see Fig. 6–9 in Chap. 6). Although Fig. 6–9 was developed for water-treatment applications, it has been found to apply to most wastewater effluent filtration uses with minor variations. As shown on Fig. 6–9, the approximate regions in which the different phenomena associated with particle removal in conventional sedimentation and filtration processes are operative are plotted as a function of the alum dose and the pH of the treated effluent after alum has been added. For example, optimum particle removal by sweep floc occurs in the pH range of 7 to 8 with an alum dose of 20 to 60 mg/L. Generally, for many wastewater effluents that have high pH values (e.g., 7.3 to 8.5), low alum dosages in the range of 5 to 10 mg/L will not be effective. To operate with low alum dosages, pH control will generally be required.

Filter Problems

The principal problems encountered in wastewater filtration and the control measures that have proved to be effective are reported in Table 11–13. Because these problems can affect both the performance and operation of a filter system, care should be taken in the design phase to provide the necessary facilities that will minimize their impact. When filtering secondary effluent containing residual biological floc, semicontinuous filters should be backwashed at least once every 24 h to avoid the formation of mudballs and the buildup of grease. In most cases, the frequency of backwashing will be more often.

Need for Pilot-Plant Studies

Although the information presented earlier in this section and previously in Sec. 11–3 will help the reader understand the nature of the filtration operation as it is applied to the filtration of treated wastewater, it must be stressed that there is no generalized approach to the design of full-scale filters. The principal reason is the inherent variability in the characteristics of the influent suspended solids to be filtered. For example, changes in the degree of flocculation of the suspended solids in the secondary settling facilities will significantly affect the particle sizes and their distribution in the effluent. This in turn will affect the performance of the filter. Further, because the characteristics of the effluent suspended solids will also vary with the organic loading on the process as well as with the time of day, filters must be designed to function under a rather wide range of oper-

Table 11–13
Summary of commonly encountered problems in depth filtration of wastewater and control measures for those problems

Problem	Description/control
Turbidity breakthrough[a]	Unacceptable levels of turbidity are recorded in the effluent from the filter, even though the terminal headloss has not been reached. To control the buildup of effluent turbidity levels, chemicals and polymers have been added to the filter. The point of chemical or polymer addition must be determined by testing
Mudball formation	Mudballs are an agglomeration of biological floc, dirt, and the filtering medium or media. If the mudballs are not removed, they will grow into large masses that often sink into the filter bed and ultimately reduce the effectiveness of the filtering and backwashing operations. The formation of mudballs can be controlled by auxiliary washing processes such as air scour or water surface wash concurrent with, or followed by, water wash.
Buildup of emulsified grease	The buildup of emulsified grease within the filter bed increases the headloss and thus reduces the length of filter run. Both air scour and water surface wash systems help control the buildup of grease. In extreme cases, it may be necessary to steam clean the bed or to install a special washing system
Development of cracks and contraction of filter bed	If the filter bed is not cleaned properly, the grains of the filter-bed filtering medium become coated. As the filter compresses, cracks develop, especially at the sidewalls of the filter. Ultimately, mudballs may develop. This problem can be controlled by adequately backwashing and scouring
Loss of filter medium or media (mechanical)	In time, some of the filter material may be lost during backwashing and through the underdrain system (where the gravel support has been upset or the underdrain system has been installed improperly). The loss of the filter material can be minimized through the proper placement of washwater troughs and underdrain systems. Special baffles have also proved effective
Loss of filter medium or media (operational)	Depending on the characteristics of the biological floc, grains of the filter material can become attached to it, forming aggregates light enough to be floated away during the backwashing operations. The problem can be minimized by the addition of an auxiliary air and/or water scouring system
Gravel mounding	Gravel mounding occurs when the various layers of the support gravel are disrupted by the application of excessive rates of flow or by uncontrolled air and hydraulic shock loadings during the backwashing operation. To avoid catastrophic failure of the underdrain, remove and replace the filter material and support gravel. The use of a support layer of high-density material such as ilmenite or garnet should also be investigated.

[a] Turbidity breakthrough does not occur with filters that operate continuously.

ating conditions. The best way to ensure that the filter configuration selected for a given application will function properly is to conduct pilot-plant studies (see Fig. 11–28).

Because of the many variables that can be analyzed, care must be taken not to change more than one variable at a time so as to confound the results in a statistical sense. Testing should be carried out at several intervals, ideally throughout a full year, to assess seasonal variations in the characteristics of the effluent to be filtered. All test

(a)

(b)

(c)

Figure 11–28

Typical pilot-plant facilities used to conduct filtration studies: (a) pulsed-bed filter, (b) continuous backwash upflow filter and dual-medium and deep bed monomedium filter columns, and (c) shallow bed monomedium.

results should be summarized and evaluated in several different ways to ensure their proper analysis. Because the specific details of each test program will be different, no generalization on the best method of analysis can be given.

11–5 SURFACE FILTRATION

Surface filtration, as shown on Fig. 11–3b, involves the removal of particulate material suspended in a liquid by mechanical sieving by passing the liquid through a thin septum (i.e., filter material). The mechanical sieving action is similar to a kitchen colander. Materials that have been used as filter septums include woven metal fabrics, cloth fabrics of different weaves, and a variety of synthetic materials. Membrane filters, discussed in the following section, are also surface filtration devices but are differentiated on the basis of the size of the pores in the filter medium. Cloth-medium surface filters have openings in the size range from 10 to 30 μm or larger. In membrane filters the pore size can vary from 0.0001 to 1.0 μm. Surface filtration has been used to remove the residual suspended solids from secondary effluents and from stabilization pond effluents. The focus of this section is on the removal of residual suspended solids from secondary and pond effluents discussed in the previous chapters. In this application, surface filtration is used as a replacement for depth filtration. The two principal types of cloth-medium surface filtration devices are described below.

Discfilter®

The Discfilter® (DF) consists of a series of disks composed of two vertically mounted parallel disks that are used to support the filter cloth. Each disk is connected to a cen-

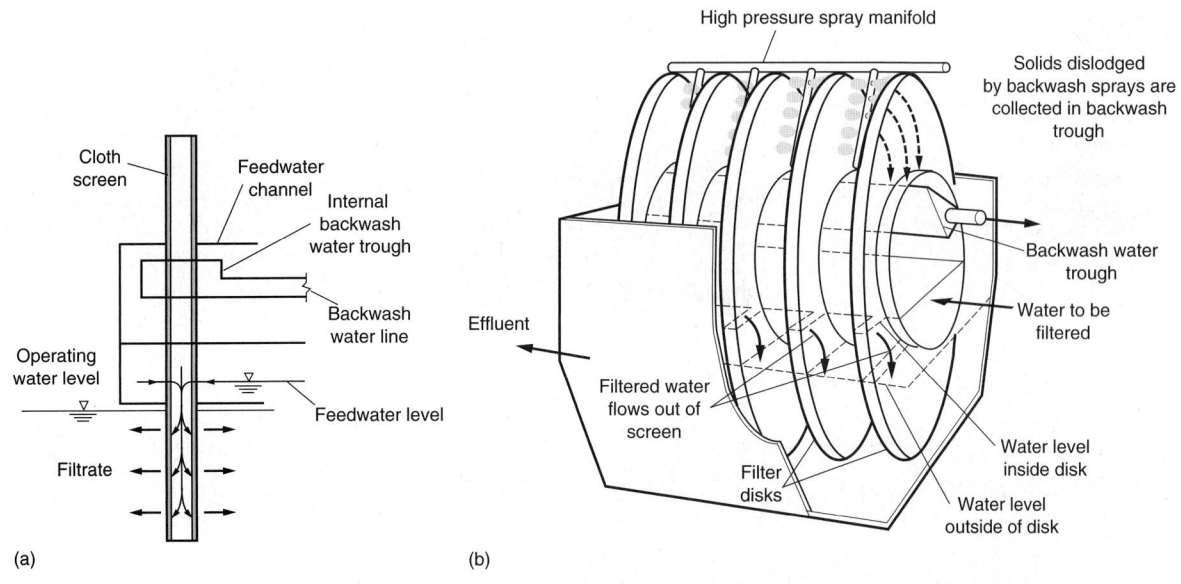

Figure 11–29

Discfilter® surface filter: (a) definition sketch for operation, (b) pictorial schematic, and (c) photo of pilot test unit (*Note:* disk is full size). *(Part b adapted from US Filter.)*

tral feed tube (see Fig. 11–29). The two-dimensional cloth screen material used with the DF can be of either polyester or Type 316 stainless steel. Typical design information for a DF can be found in Table 11–14.

Operation of DF. Operationally, as shown on Fig. 11–29a, water enters through a central channel and flows outward through the filter cloth. During normal operation 60 to 70 percent of the surface area of the DF is submerged and the disk rotates, depending on the headloss, from 1 to 8.5 r/min. The DF has the ability to operate in an intermittent or a continuous-backwash mode. When operating in a continuous-backwash mode, the disks of the DF will both produce filtered water and be backwashed simultaneously. The

Table 11–14
Typical design information for surface filtration of secondary settled effluent using a Discfilter®

Item	Unit	Typical value	Remarks
Size of opening in screen material	μm	20–35	Two-dimensional stainless-steel or polyester screen cloths are available in sizes ranging from 10 to 60 μm
Hydraulic loading rate	m³/m²·min	0.25–0.83	Depends on characteristics of suspended solids that must be removed
Headloss through screen	mm	75–150	Based on submerged surface area of drum
Drum submergence	% height % area	70–75 60–70	Bypass should be provided when headloss exceeds 200 mm
Drum diameter	m	0.75–1.50	Varies depending on screen design; 3 m (10 ft) is most commonly used size; smaller sizes increase backwash requirements
Drum speed	m/min	4.5 at 50 mm headloss 30–40 at 150 mm headloss	Maximum rotational speed is limited
Backwash requirements	% throughput	2 at 350 kPa 5 at 100 kPa	

different stages the disk filter passes through during each rotation are identified on Fig. 11–29*b*. At the start of a rotation feedwater enters the central feed tube from where it is distributed into the disks. While the disk filter is submerged, water and particles smaller than openings in the cloth screen pass through the filter to the effluent collection channel. Those particles larger than the screen are retained.

Backwashing the DF. As the disk continues to rotate past the effluent water level, the remaining feedwater passes through the cloth screen until no water remains in the disk. The screen containing the captured wastewater solids continues to rotate until it passes the backwash spray jets. As the disk rotates past the backwash spray jets, the particles that have been retained on the cloth screen are flushed from the surface of the screen. The mixture of backwash water and solids falls from the disk and is collected in the backwash water trough. After passing by the backwash nozzles, the cleaned screens are ready to begin filtering water again. When the DF is operating in an intermittent backwash mode, the backwash spray jets are activated only when headloss through the filter reaches a preset level.

Cloth-Media Disk Filter®

The Cloth-Media Disk Filter (CMDF), as illustrated on Fig. 11–30, also consists of several disks mounted vertically in a tank. Two different types of cloths can be used in the CMDF: (1) a needle felt cloth, made of polyester or (2) a synthetic pile fabric cloth. The

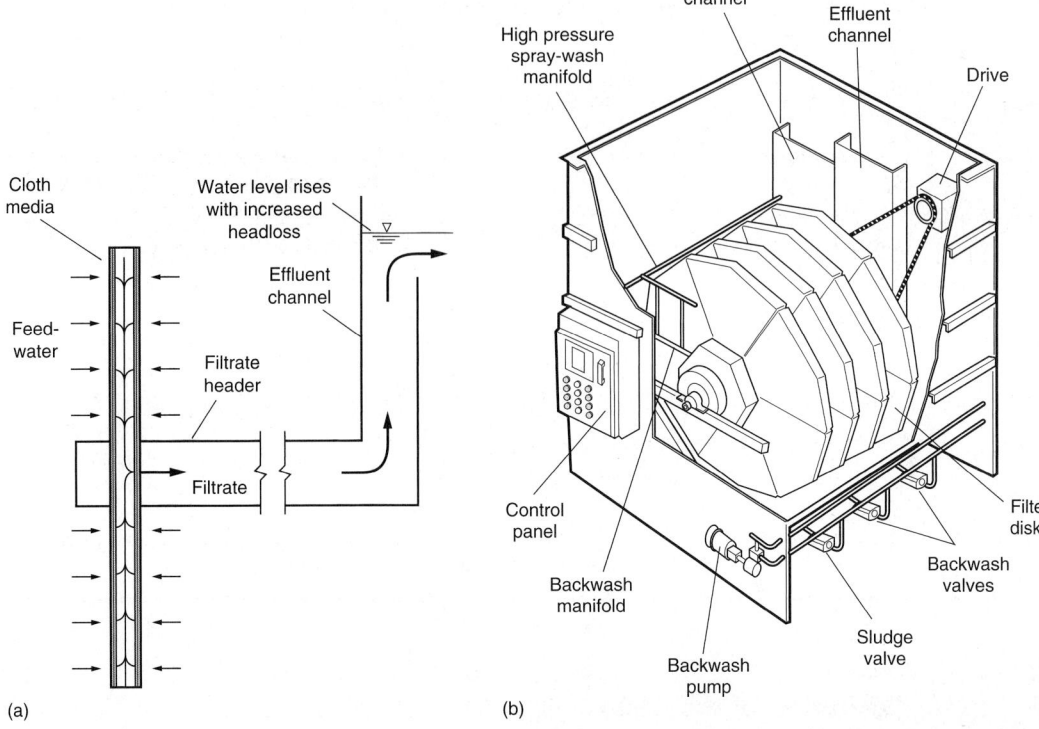

(a)

(b)

(c)

Figure 11–30

Cloth-Media Disk Filter® surface filter: (a) definition sketch for operation, (b) pictorial schematic, and (c) photo of pilot test unit looking down into the filter housing (*Note:* disk is full size). (*Part b adapted from Aqua Aerobic Systems, Inc.*)

needle felt cloth has a random three-dimensional weave to facilitate particle removal. In addition to the normal backwashing, the felt cloth must be cleaned periodically with a high-pressure spray. The pile fabric does not require the use of a high-pressure spray wash and can be cleaned completely by use of a backwash alone. Typical design information for the CMDF can be found in Table 11–15.

Table 11–15
Typical design information for surface filtration of secondary settled effluent using a Cloth-Media Disk Filter®

Item	Unit	Typical value	Remarks
Nominal pore size	μm	10	Polyester three-dimensional needle felt cloths are employed as the filter material
Hydraulic loading rate	m³/m²·min	0.1–0.27	Depends on characteristics of suspended solids that must be removed
Headloss through screen	mm	50–300	Based on solids accumulation on or within the cloth
Disk submergence	% height % area	100 100	
Disk diameter	m	0.90 or 1.80	Two sizes are available
Disk speed	r/min	Stationary during normal operation 1 during backwash	
Backwash and sludge wasting requirements	% throughput	4.5 at 0.1 m³/m²·min 7.2 at 0.27 m³/m²·min	Is a function of hydraulic loading rate and feedwater quality

Operation of the CMDF.   Operationally, as shown on Fig. 11–30a, water enters the feed tank and flows through the filter cloth into a central collection tube or header. The resulting CMDF filtrate is collected in a central tube or a filtrate header (see Fig. 11–30a), where it flows to final discharge over an overflow weir in the effluent channel. As solids accumulate on and in the cloth medium, resistance to flow or headloss increases. When the headloss through the cloth medium reaches a predetermined level, the disks are backwashed. After filtering to waste after the backwash cycle, the filter is put back into operation.

Backwashing the CMDF.   When a backwash cycle is initiated, the disks remain submerged and rotate at 1 r/min. Vacuum suction heads, located on either side of the CMDF, draw filtrate water from the filtrate header back through the cloth media into the vacuum heads while the disk is rotating. This reversal of flow removes particles that have become entrapped on the surface and within the cloth medium. The normal duration of the backwash cycle for the CMDF is 1 min.

Over time, particles will accumulate in the cloth medium that cannot be removed by a typical backwash. This accumulation of particles leads to increased headloss across the filter, an increase in the backwash suction pressure, and shorter run times between backwashes. When the backwash suction pressure reaches 124 kPa (18 lb/in²), or the time exceeds the desired preset time interval, a high-pressure spray wash is initiated automatically. Prior to the high-pressure spray wash, the tank inlet valve is closed and the influent flow is stopped. A standard backwash is initiated to remove the outer layer of solids from the cloth. The filter-to-waste valve is opened, the water level is lowered to below the midline of the disks, and then a high-pressure spray wash is initiated.

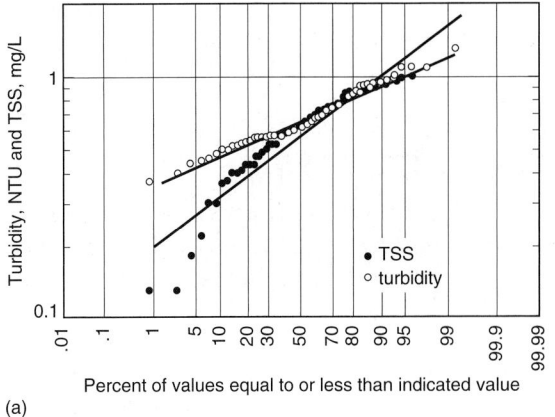
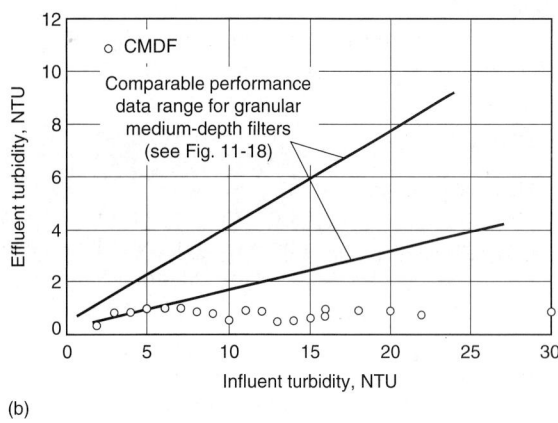

(a)

(b)

Figure 11–31

Performance data for the Cloth-Media Disk Filter® for secondary effluent: (a) effluent probability distributions for turbidity and TSS and (b) effluent turbidity as function of the influent turbidity at a filtration rate of 176 L/m²·min (4.4 gal/ft²·min).

During the high-pressure spray wash, the disks rotate slowly at 1 r/min, while filtrate water is sprayed at a high pressure from the outside of the filter cloth. The high-pressure spray wash flushes the particles that have become lodged inside the cloth filter medium effectively in two revolutions of the disk. After the end of the high-pressure spray wash, the inlet valve to the CMDF opens to allow the wastewater to flow to the filter. The disks continue to rotate and the filter-to-waste valve remains open until the solids that were flushed from the filter cloth to the filtrate side of the CMDF are removed from the filtrate header and the effluent line. The time interval between high-pressure spray washes is a function of the feedwater quality.

Performance Characteristics

To evaluate the performance capabilities of CMDF, a pilot unit employing a full-size disk (0.9 m, 3 ft, see Table 11–15) was used to treat secondary effluent from an activated-sludge process that had TSS and turbidity values ranging from 3.9 to 30 mg/L and 2 to 30 NTU, respectively. Based on a long-term study, it was found, as shown on Fig. 11–31a, that both the TSS and turbidity values of the filtered effluent were less than 1, 92 percent of the time (Riess et al., 2001). The performance of the CMDF as compared to depth filters all tested with effluent from a previous activated-sludge plant treating the same wastewater is shown on Fig. 11–31b. As shown, the effluent from the CMDF remained constant over all of the influent turbidity values tested up to 30 NTU. The DF has not been tested. Based on observations of the cloth surface, it appears that the material removed on the filter also acts as a filter (autofiltration). The percentage of backwash water used ranged between 4 and 10.

In reviewing the data presented on Fig. 11–31b, it is important to note the activated-sludge treatment process is operated as an extended aeration treatment process (SRT > 15 d). If either of the cloth filters is to be used to filter effluent from activated-sludge processes operated with extremely low SRT values (i.e., 1 to 2 *d*), pilot testing should be undertaken, as the characteristics of the residual solids to be filtered are significantly different.

11–6 MEMBRANE FILTRATION PROCESSES

Filtration, as defined in Secs. 11–3 and 11–5, involves the separation (removal) of particulate and colloidal matter from a liquid. In membrane filtration the range of particle sizes is extended to include dissolved constituents (typically 0.0001 to 1.0 μm). The role of the membrane, as shown on Fig. 11–32, is to serve as a selective barrier that will allow the passage of certain constituents and will retain other constituents found in the liquid (Cheryan, 1998). To introduce membrane technologies and their application, the following subjects are considered in this section: (1) membrane process terminology, (2) membrane classification, (3) membrane configurations, (4) application of membrane technologies, (5) electrodialysis, (6) the need for pilot-plant studies, and (7) the disposal of concentrated waste streams, which is considered at the end of this section.

Membrane Process Terminology

Terms commonly encountered when considering the application of membrane processes are summarized in Table 11–16. Referring to Fig. 11–32 and Table 11–16, the influent to the membrane module is known as the *feed stream* (also known as feedwater). The liquid that passes through the semipermeable membrane is known as *permeate* (also known as the product stream or permeating stream) and the liquid containing the retained constituents is known as the *concentrate* (also known as the retentate, reject, retained phase, or waste stream). The rate at which the permeate flows through the membrane is known as the rate of *flux,* typically expressed as kg/m²·d (gal/ft²·d).

Membrane Process Classification

Membrane processes include microfiltration (MF), ultrafiltration (UF), nanofiltration (NF), reverse osmosis (RO), dialysis, and electrodialysis (ED). Membrane processes can be classified in a number of different ways including (1) the type of material from which the membrane is made, (2) the nature of the driving force, (3) the separation mechanism, and (4) the nominal size of the separation achieved. Each of these methods

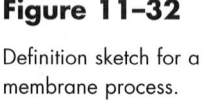

Figure 11–32

Definition sketch for a membrane process.

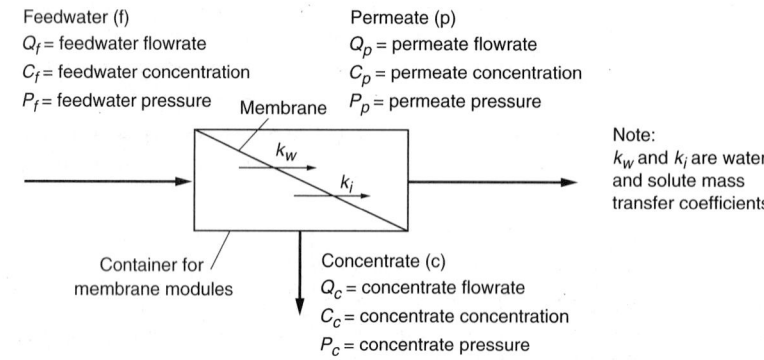

Table 11–16
Terminology used to describe membrane processes[a]

Term	Description
Array or train	Multiple interconnected stages in series
Brine	Concentrate stream containing total dissolved solids greater than 36,000 mg/L.
Concentrate, retentate, retained phase, reject, residual stream	The portion of the feed stream that does not pass through the membrane that contains higher TDS than the feed stream
Feed stream, feedwater	Input stream to the membrane array
Flux	Mass or volume rate of transfer through the membrane surface
Fouling	Deposition of existing solid material in the element on the feed stream of the membrane. Fouling can be either reversible or irreversible
Lumen	The interior of a hollow fiber membrane
Mass transfer coefficient (MTC)	Mass or volume unit transfer through membrane based on driving force
Membrane element	A single membrane unit containing a bound group of spiral-wound or hollow fine-fiber membranes to provide a nominal surface area
Module	A complete unit comprised of the membranes, the pressure support structure for the membranes, the feed inlet and outlet permeate and retentate ports, and an overall support structure
Molecular weight cutoff (MWCO)	The molecular weight of the smallest material rejected by the membrane, usually expressed in Daltons (D)
Permeate, product, permeating stream	The portion of the feed stream that passes through the membrane that contains lower TDS than the feed stream
Reject ion	Percent solute concentration reduction of permeate stream relative to feed stream
Pressure vessel	A single tube that contains several membrane elements in series
Scaling	Precipitation of solids in the element due to solute concentration on the feed stream of the membrane
Size exclusion	Removal of particles by sieving
Solvent	Liquid containing dissolved constituents (TDS), usually water
Solute	Dissolved constituents (TDS) in raw, feed, permeate, and concentrate streams
Stage or bank	Pressure vessels arranged in parallel
Submerged membrane vessel or reactor	Membrane elements are submerged (or immersed) in an open reactor
System arrays	Number of arrays needed to produce the required plant flow
Train or array	Multiple interconnected stages in series

[a] Adapted in part from AWWA (1996), Cheryan (1998), and Taylor and Wiesner (1999).

of classifying membrane processes is considered in the following discussion. The general characteristics of membrane processes including typical operating ranges are reported in Table 11–17. The focus of the following discussion is on pressure-driven membrane processes. Electrodialysis is considered separately following the discussion of the application of pressure-driven membranes.

Table 11–17

General characteristics of membrane processes

Membrane process	Membrane driving force	Typical separation mechanism	Operating structure (pore size)	Typical operating range, μm	Permeate description	Typical constituents removed
Microfiltration	Hydrostatic pressure difference or vacuum in open vessels	Sieve	Macropores (>50 nm)	0.08–2.0	Water + dissolved solutes	TSS, turbidity, protozoan oocysts and cysts, some bacteria and viruses
Ultrafiltration	Hydrostatic pressure difference	Sieve	Mesopores (2–50 nm)	0.005–0.2	Water + small molecules	Macromolecules, colloids, most bacteria, some viruses, proteins
Nanofiltration	Hydrostatic pressure difference	Sieve + solution/ diffusion + exclusion	Micropores (<2 nm)	0.001–0.01	Water + very small molecules, ionic solutes	Small molecules, some hardness, viruses
Reverse osmosis	Hydrostatic pressure difference	Solution/ diffusion + exclusion	Dense (<2 nm)	0.0001– 0.001	Water, very small molecules, ionic solutes	Very small molecules, color, hardness, sulfates, nitrate, sodium, other ions
Dialysis	Concentration difference	Diffusion	Mesopores (2–50 nm)	—	Water + small molecules	Macromolecules, colloids, most bacteria, some viruses, proteins
Electrodialysis	Electromotive force	Ion exchange with selective membranes	Micropores (<2 nm)	—	Water + ionic solutes	Ionized salt ions

Membrane Materials. Membranes used for the treatment of water and wastewater typically consist of a thin skin having a thickness of about 0.20 to 0.25 μm supported by a more porous structure of about 100 μm in thickness. Most commercial membranes are produced as flat sheets, fine hollow fibers, or in tubular form. The flat sheets are of two types, asymmetric and composite. Asymmetric membranes are cast in one process and consist of a very thin (less than 1 μm) layer and a thicker (up to 100 μm) porous layer that adds support and is capable of high water flux. Thin-film composite (TFC) membranes are made by bonding a thin cellulose acetate, polyamide, or other active layer (typically 0.15 to 0.25 μm thick) to a thicker porous substrate, which provides stability. Membranes can be made from a number of different organic and inorganic materials. The membranes used for wastewater treatment are typically organic. The principal types of membranes used include polypropylene, cellulose acetate, aromatic polyamides, and thin-film composite (TFC). The choice of membrane and system configuration is based on minimizing membrane clogging and deterioration, typically based on pilot-plant studies.

Driving Force. The distinguishing characteristic of the first four membrane processes considered in Table 11–17 (MF, UF, NF, and RO) is the application of hydraulic pressure to bring about the desired separation. Where MF membrane elements are submerged in open vessels (see Fig. 8–50), vacuum is used instead of pressure. Dialysis involves the transport of constituents through a semipermeable membrane on the basis of concentration differences. Electrodialysis involves the use of an electromotive force and ion-selective membranes to accomplish the separation of charged ionic species.

Removal Mechanisms. The separation of particles in MF and UF is accomplished primarily by straining (sieving), as shown on Fig. 11–33a. In NF and RO, small particles are rejected by the water layer adsorbed on the surface of the membrane which is known as a *dense* membrane (see Fig. 11–33b). Ionic species are transported across the membrane by diffusion through the pores of the macromolecule comprising the membrane. Typically NF can be used to reject constituents as small as 0.001 μm whereas RO can reject particles as small as 0.0001 μm. Straining is also important in NF membranes, especially at the larger pore size openings.

Size of Separation. The pore sizes in membranes are identified as macropores (>50 nm), mesopores (2 to 50 nm), and micropores (< 2 nm). Because the pore sizes in RO membranes are so small, the membranes are defined as dense. The classification of membrane processes on the basis of the size of separation is shown on Fig. 11–34 and in Table 11–17. Referring to Fig. 11–34, it can be seen that there is considerable overlap in the sizes of particles removed, especially between NF and RO. Nanofiltration is used most commonly in water-softening operations in place of chemical precipitation.

Figure 11–33

Definition sketch for the removal of wastewater constituents: (a) removal of large molecules and particles by sieving (size exclusion) mechanism and (b) rejection of ions by adsorbed water layer. (Adapted from Tchobanoglous and Schroeder, 1985.)

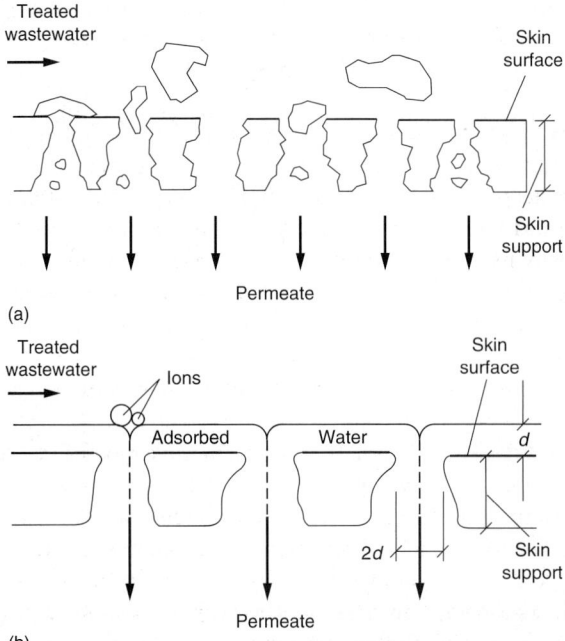

Figure 11–34

Comparison of the size
of the constituents found
in wastewater and the
operating size ranges for
membrane technologies.
The operating size range
for conventional depth
filtration is also shown.

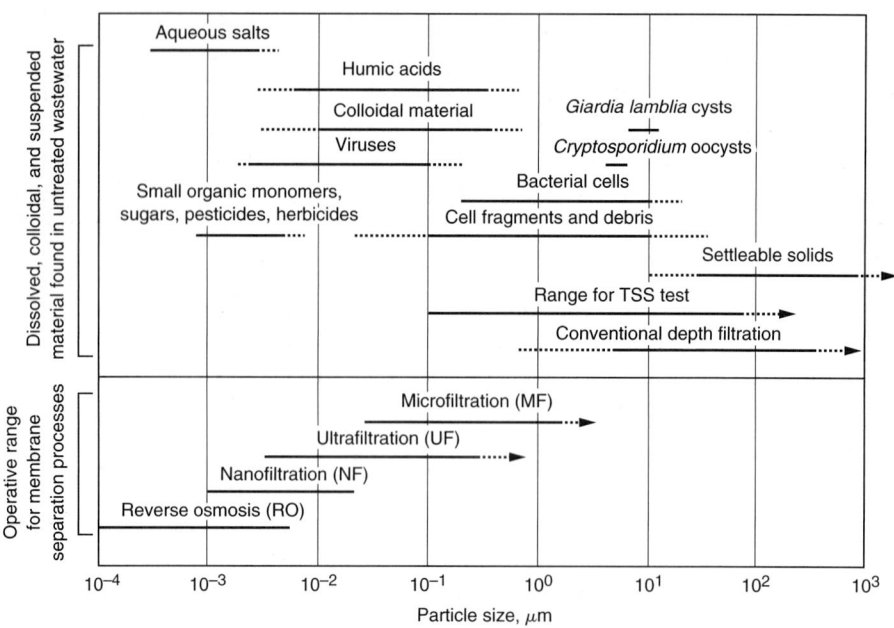

Figure 11–34

Comparison of the size
of the constituents found
in wastewater and the
operating size ranges for
membrane technologies.
The operating size range
for conventional depth
filtration is also shown.

Membrane Configurations

In the membrane field, the term *module* is used to describe a complete unit comprised of the membranes, the pressure support structure for the membranes, the feed inlet and outlet permeate and retentate ports, and an overall support structure. The principal types of membrane modules used for wastewater treatment are (1) tubular, (2) hollow fiber, and (3) spiral wound (AWWA, 1996). Plate and frame and pleated cartridge filters are also available but are used more commonly in industrial applications. The various membranes, described below, are illustrated on Fig. 11–35.

Tubular Modules.   In the tubular configuration the membrane is cast on the inside of a support tube. A number of tubes (either singly or in a bundle) are then placed in an appropriate pressure vessel. The feedwater is pumped through the feed tube and the product water is collected on the outside of the tubes (see Fig. 11–35a). The concentrate continues to flow through the feed tube. These units are generally used for water with high suspended solids or plugging potential. Tubular units are the easiest to clean, which is accomplished by circulating chemicals and pumping a "foamball" or "spongeball" through to mechanically wipe the membrane. Tubular units produce at a low product rate relative to their volume, and the membranes are generally expensive.

Hollow Fiber.   The hollow-fiber membrane module, as shown on Fig. 11–35b, consists of a bundle of hundreds to thousands of hollow fibers. The entire assembly is inserted into a pressure vessel. The feed can be applied to the inside of the fiber (inside-out flow) or the outside of the fiber (outside-in flow).

Spiral Wound.   In the spiral-wound membrane, a flexible permeate spacer is placed between two flat membrane sheets. The membranes are sealed on three sides.

Figure 11–35

Typical membrane elements and modules used in membrane applications: (a) single tubular hollow fiber membrane, (b) bundle of tubular hollow fiber membranes, (c) bundle of hollow fine fiber membranes with flow from the outside to the inside of the fiber, (d) view of an exposed bundle of hollow fine fiber membranes, (e) bundle of hollow fine fiber membranes with flow from the inside to the outside of the fiber, (f) bundles of hollow fine fiber membranes placed in a pressure vessel. *(Figs. c, e, g, h, and k from Crites and Tchobanoglous, 1998.)* *(Figure continues on next page.)*

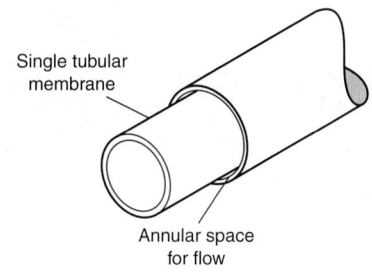

Single tubular membrane

Annular space for flow

(a)

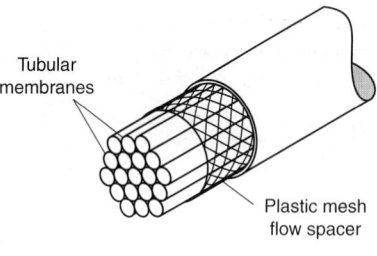

Tubular membranes

Plastic mesh flow spacer

(b)

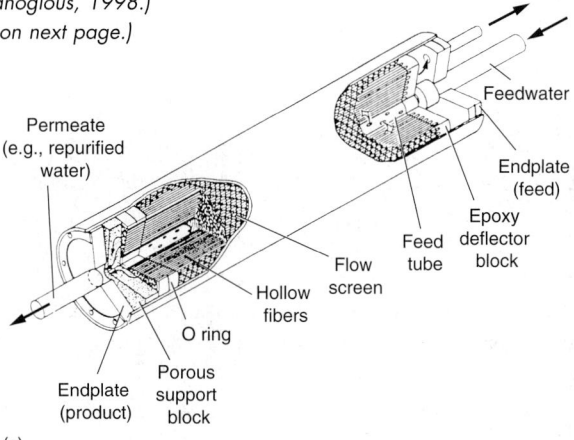

Concentrate

Feedwater

Endplate (feed)

Epoxy deflector block

Feed tube

Flow screen

Hollow fibers

O ring

Porous support block

Endplate (product)

Permeate (e.g., repurified water)

(c)

(d)

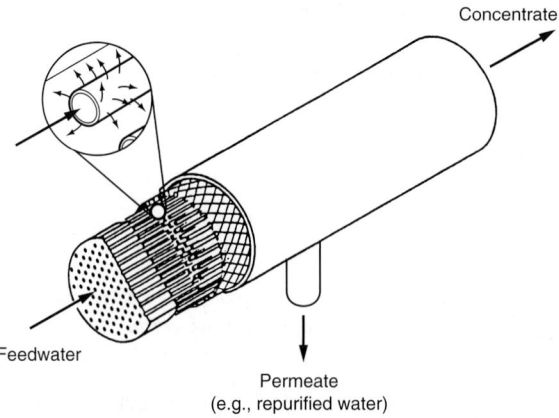

Concentrate

Feedwater

Permeate (e.g., repurified water)

(e)

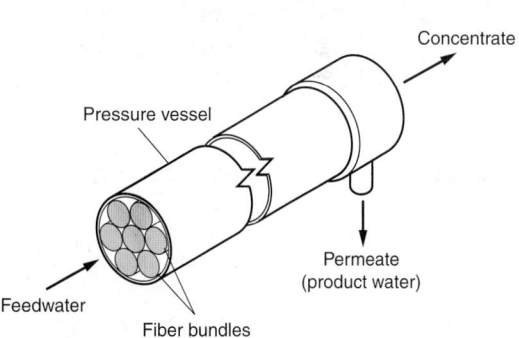

Concentrate

Pressure vessel

Permeate (product water)

Feedwater

Fiber bundles

(f)

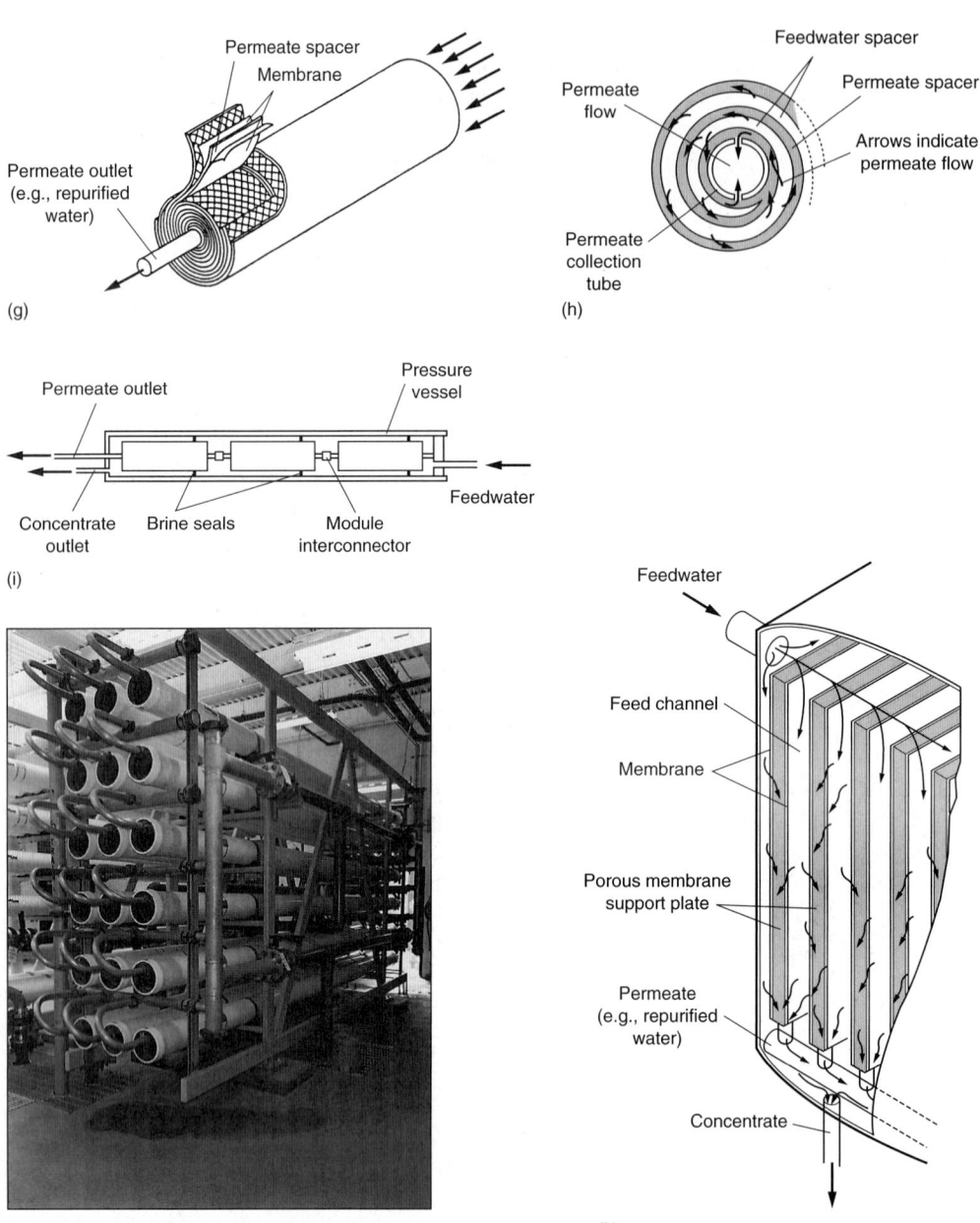

Figure 11–35 (Continued)

Typical membrane elements and modules used in membrane applications: (g) cutaway of spiral-wound thin-film composite membrane module, (h) section through spiral-wound thin-film composite membrane module, (i) pressure vessel containing 3 spiral-wound thin-film composite membrane modules in series, (j) typical installation of reverse osmosis pressure vessels containing 6 spiral-wound thin-film composite membrane modules in series, and (k) parallel plate and frame membrane. (Figs. c, e, g, h, and k from Crites and Tchobanoglous, 1998.)

The open side is attached to a perforated pipe. A flexible feed spacer is added and the flat sheets are rolled into a tight circular configuration (see Figs. 11–35*g* and *h*). Thin-film composites are used most commonly in spiral-wound membrane modules. The term spiral derives from the fact that the flow in the rolled-up arrangement of membranes and support sheets follows a spiral flow pattern.

Plate and Frame.   Plate and frame membrane modules (see Fig. 11–35*k*) are comprised of a series of flat membrane sheets and support plates. The water to be treated passes between the membranes of two adjacent membrane assemblies. The plate supports the membranes and provides a channel for the permeate to flow out of the unit. The plate and frame configuration is used most commonly for electrodialysis modules (sometimes identified as stacks).

Pleated-Cartridge Filters.   Pleated-cartridge filters are used most commonly in microfiltration applications and usually are designed as disposable units. Pleated-cartridge filters are used almost exclusively to concentrate virus from treated wastewater for analysis (see Chap. 2).

Pressure Vessels.   With the exception of the pressure vessels, which need to be manufactured specially, most of the equipment used for membrane filtration is standard to the chemical and process industry. The primary purpose of the pressure vessel (or tube) is to support the membrane and keep the feedwater and product streams isolated. The vessel must also be designed to prevent leaks and pressure losses to the outside, minimize the buildup of salt or fouling, and permit easy replacement of the membranes. Depending on the operating pressure and the characteristic of the feedwater, a variety of materials have been used, including plastic and fiberglass tubes and plumbing components. Steel pressure tubes are required for some reverse-osmosis applications, and stainless steel is required for seawater and brackish water having high TDS. Centrifugal pumps can be used for MF, UF, and NF; positive-displacement pumps or high-pressure turbine pumps are necessary for RO.

Membrane Operation

The operation of membrane processes is quite simple. A pump is used to pressurize the feed solution and to circulate it through the module. A valve is used to maintain the pressure of retentate. The permeate is withdrawn, typically at atmospheric pressure. As constituents in the feedwater accumulate on the membranes (often termed membrane fouling), the pressure builds up on the feed side, the membrane flux (i.e., flow through membrane) starts to decrease, and the percent rejection also starts to decrease (see Fig. 11–36). When the performance has deteriorated to a given level, the membrane modules are taken out of service and backwashed and/or cleaned chemically. The operational configurations and parameters for the various membrane processes are considered in the following discussion.

Microfiltration and Ultrafiltration.   Three different process configurations are used with microfiltration and ultrafiltration units as illustrated on Fig. 11–37. In the first configuration known as *cross flow* (Fig. 11–37*a*) the feedwater is pumped with cross flow tangential to the membrane. Water that does not pass through the membrane

Figure 11–36

Definition sketch for the performance of a membrane filtration system.

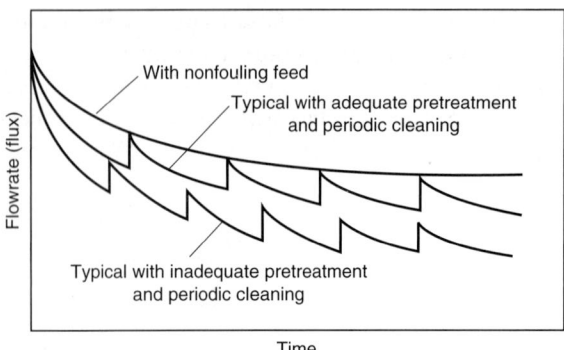

Figure 11–37

Typical operational modes for MF and UF membrane processes: (a) cross flow, (b) cross flow with reservoir, and (c) direct feed. *(Adapted from Jacangelo and Buckley, 1996.)*

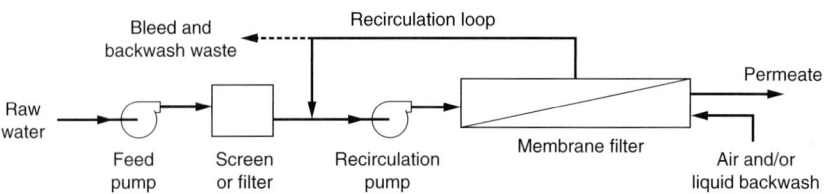

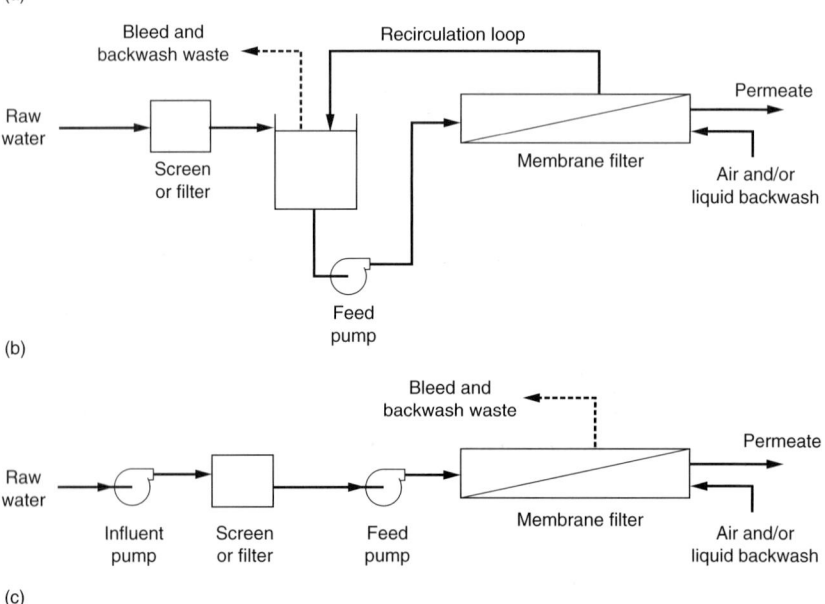

is recirculated through the membrane after blending with additional feedwater. The second configuration, also known as cross flow (Fig. 11–37b), is similar to the first with the exception that the water that does not pass through the membrane is recirculated to a storage reservoir. The third configuration is known as direct feed (also *dead-end*) (Fig. 11–37c) in that there is no cross flow. All of the water applied to the membrane passes through the membrane. Raw feedwater is used periodically to flush the accumulated material from the membrane surface.

For the cross-flow mode of operation (see Fig. 11–37a and b), the transmembrane pressure is given by the following expression:

$$P_{tm} = \left[\frac{P_f + P_c}{2}\right] - P_p \tag{11-35}$$

where P_{tm} = transmembrane pressure gradient, kPa
P_f = inlet pressure of feed stream, kPa
P_c = pressure of concentrate stream, kPa
P_p = pressure of permeate stream, kPa

The overall pressure drop across the filter module for the cross-flow mode of operation is given by

$$P = P_f - P_p \tag{11-36}$$

where P = pressure drop across module, kPa
P_f and P_p as defined above

For the direct-feed mode of operation (see Fig. 11–37c) the transmembrane pressure is given by the following expression:

$$P_{tm} = P_f - P_p \tag{11-37}$$

where P_{tm} = transmembrane pressure gradient, kPa
P_f and P_p as defined above

The total permeate flow from a membrane system is given by

$$Q_p = F_w A \tag{11-38}$$

where Q_p = permeate stream flowrate, kg/s
F_w = transmembrane water flux rate, kg/m²·s
A = membrane area, m²

As would be expected, the transmembrane water flux rate is a function of the quality of the feed stream, the degree of pretreatment, the characteristics of the membrane, and the system operating parameters.

The recovery rate r is defined as

$$r, \% = \frac{Q_p}{Q_f} \times 100 \tag{11-39}$$

where Q_p = permeate stream flow, kg/s
Q_f = feed stream flow, kg/s

It should be noted that there is a difference in the recovery rate (which refers to the water) and the rate of rejection (which refers to the solute) as given below:

$$R, \% = \frac{C_f - C_p}{C_f} \times 100 = 1 - \frac{C_p}{C_f} \times 100 \tag{11-40}$$

The corresponding mass balance equations are

$$Q_f = Q_p + Q_c \tag{11-41}$$

$$Q_f C_f = Q_p C_p + Q_c C_c \tag{11-42}$$

Three different operating modes can be used to control the operation of a membrane process with respect to flux and the transmembrane pressure (TMP). The three modes, illustrated on Fig. 11–38, are (1) constant flux in which the flux rate is fixed and the TMP is allowed to vary (increase) with time; (2) constant TMP in which the TMP is fixed and the flux rate is allowed to vary (decrease) with time, and (3) both the flux rate and the TMP are allowed to vary with time. Traditionally, the constant-flux mode of operation has been used. However, based on the results of a recent study with various wastewater effluents (Bourgeous et al., 1999), it appears the mode in which both the flux rate and the TMP are allowed to vary with time may be the most effective mode of operation.

Reverse Osmosis.   When two solutions having different solute concentrations are separated by a semipermeable membrane, a difference in chemical potential will exist across the membrane (Fig. 11–39). Water will tend to diffuse through the membrane from the lower-concentration (higher-potential) side to the higher-concentration (lower-potential) side. In a system having a finite volume, flow continues until the pressure difference balances the chemical potential difference. This balancing pressure difference is

Figure 11–38

Three modes of membrane operation: (a) constant flux, (b) constant pressure, and (c) nonrestricted flux and pressure. (Adapted from Bourgeous et al., 1999.)

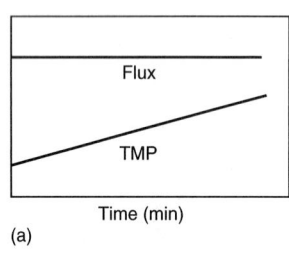
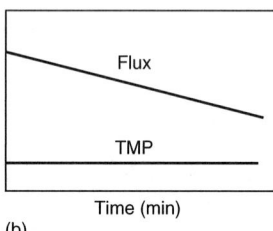
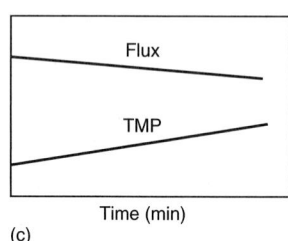

Figure 11–39

Definition sketch of osmotic flow: (a) osmotic flow, (b) osmotic equilibrium, and (c) reverse osmosis.

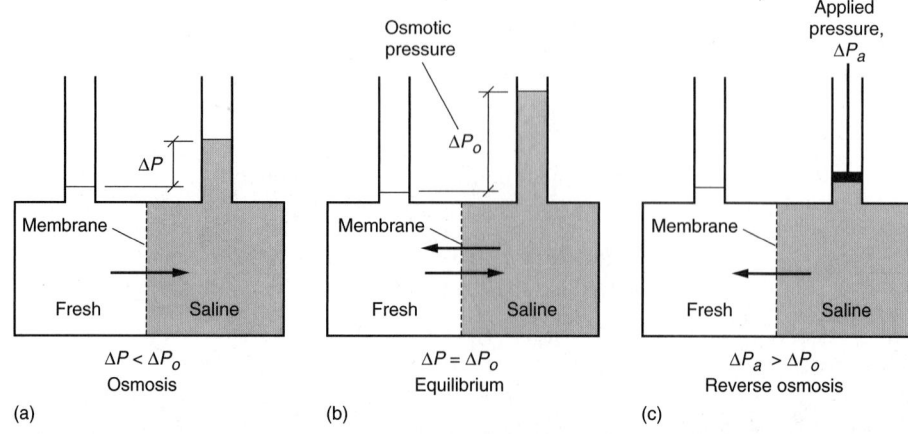

termed the *osmotic pressure* and is a function of the solute characteristics and concentration and temperature. If a pressure gradient opposite in direction and greater than the osmotic pressure is imposed across the membrane, flow from the more concentrated to the less concentrated region will occur and is termed *reverse osmosis* (see Fig. 11–39c).

A number of different models have been developed to determine the surface area of membrane and the number of arrays required (see Fig. 11–40). The basic equations used to develop the various models are as follows (Taylor and Wiesner, 1999). Referring to Fig. 11–32, the flux of water through the membrane is a function of the pressure gradient:

$$F_w = k_w(\Delta P_a - \Delta\Pi) = \frac{Q_p}{A} \qquad (11\text{–}43)$$

where F_w = water flux rate, kg/m²·s, m/s
k_w = water mass transfer coefficient involving temperature, membrane characteristics, and solute characteristics, s/m, m/s·bar
ΔP_a = average imposed pressure gradient, kg/m·s², bar
$$= \left[\frac{P_f + P_c}{2}\right] - P_p$$
$\Delta\Pi$ = osmotic pressure gradient, kg/m·s², bar
$$= \left[\frac{\Pi_f + \Pi_c}{2}\right] - \Pi_p$$
Q_p = permeate stream flow, kg/s, m³/s
A = membrane area, m²

Some solute passes through the membrane in all cases. Solute flux can be described adequately by an expression of the form

$$F_i = k_i\,\Delta C_i = \frac{Q_p C_p}{A} \qquad (11\text{–}44)$$

where F_i = flux of solute species i, kg/m²·s
k_i = solute mass transfer coefficient, m/s

Figure 11–40

Typical flow diagram for reverse osmosis membrane process with pre- and posttreatment.

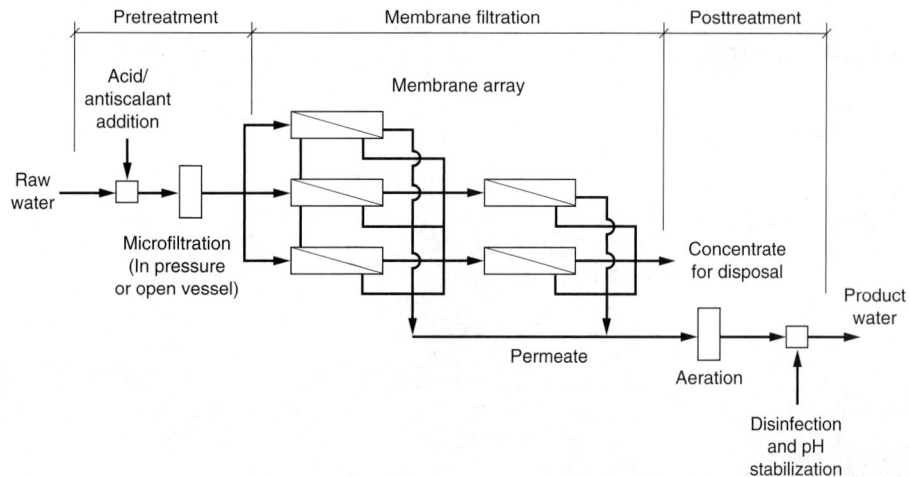

ΔC_i = solute concentration gradient, kg/m^3

$$= \left[\frac{C_f + C_c}{2}\right] - C_p$$

C_f = solute concentration in feed stream, kg/m^3

C_c = solute concentration in concentrate stream, kg/m^3

C_p = solute concentration in permeate stream, kg/m^3

The recovery rate and the rate of rejection are as described above for the micro- and ultrafiltration membranes. Use of the above equations to estimate the required surface area for desalination is illustrated in Example 11–5.

EXAMPLE 11–5 **Determination of Membrane Area Required for Demineralization** A brackish water having a TDS concentration of 3000 g/m^3 is to be desalinized using a thin-film composite membrane having a solvent (water) mass transfer coefficient k_w of 9 × 10^{-9} s/m (9 × 10^{-7} m/s·bar) and a solute (i.e., TDS) mass transfer coefficient k_i of 6 × 10^{-8} m/s. The product water is to have a TDS of no more than 200 g/m^3. The feedwater flowrate is 0.010 m^3/s. The net operating pressure ($\Delta P_a - \Delta\Pi$) will be 2500 kPa (2.5 × 10^6 kg/m·s^2). Assume the recovery rate r will be 90 percent and that all of the water is to be processed through the membrane unit to remove other constituents in addition to TDS. Estimate the rejection rate R and the concentration of the concentrate stream.

Solution

1. The problem involves determination of the membrane area required to produce 0.009 m^3/s (0.9 × 0.010 m^3/s) of water with a TDS concentration equal to or less than 200 g/m^3. If the estimated permeate TDS concentration is well below 200 g/m^3 and TDS is the only constituent of concern, blending of feedwater and permeate can be used to reduce the required membrane area.

2. Estimate membrane area using Eq. (11–43).

$$F_w = k_w(\Delta P_a - \Delta\Pi)$$

$$= (9 \times 10^{-9} \text{ s/m})(2.5 \times 10^6 \text{ kg/m·s}^2) = 2.25 \times 10^{-2} \text{ kg/m}^2\text{·s}$$

$$Q_p = F_w \times A, \; Q_p = rQ_f = 0.9 \; Q_f$$

$$A = \frac{(0.9 \times 0.010 \text{ m}^3/\text{s})(10^3 \text{ kg/m}^3)}{(2.25 \times 10^{-2} \text{ kg/m}^2\text{·s})} = 400 \text{ m}^2$$

3. Estimate permeate TDS concentration using Eq. (11–44) and the estimated area.

$$F_i = k_i \, \Delta C_i = \frac{Q_p C_p}{A}$$

Substituting for ΔC_i and solving for C_p yields

$$C_p = \frac{k_i[(C_f + C_c)/2]A}{Q_p + k_i A}$$

Assume $C_c \approx 10 \; C_f$ and solve for C_p (Note: If the estimated C_c value and the computed value of C_c, as determined below, are significantly different, the value of C_p must be recomputed)

Assume $Q_p = r Q_f$

$$C_p = \frac{(6 \times 10^{-8} \text{ m/s})[(3.0 \text{ kg/m}^3 + 30 \text{ kg/m}^3)/2](400 \text{ m}^2)}{(0.9)(0.01 \text{ m}^3/\text{s}) + (6 \times 10^{-8} \text{ m/s})(400 \text{ m}^2)} = 0.044 \text{ kg/m}^3$$

The permeate solute concentration is lower than necessary. The area could be reduced, if blending were allowed.

4. Estimate the rejection rate using Eq. (11–40)

$$R, \% = \frac{C_f - C_p}{C_f} \times 100$$

$$R = \frac{(3.0 \text{ kg/m}^3 - 0.044 \text{ kg/m}^3)}{(3.0 \text{ kg/m}^3)} \times 100 = 98.5\%$$

5. Estimate the concentrate stream TDS using Eq. (11–42)

$$C_c = \frac{Q_f C_f - Q_p C_p}{Q_c}$$

$$C_c = \frac{(1.0 \text{ L})(3.0 \text{ kg/m}^3) - (0.9 \text{ L})(0.044 \text{ kg/m}^3)}{0.1 \text{ L}} = 29.6 \text{ kg/m}^3$$

The estimated value of C_c used in Step 3 (30 kg/m^3) is ok.

Membrane Fouling

The term *fouling* is used to describe the potential deposition and accumulation of constituents in the feed stream on the membrane. Membrane fouling is an important consideration in the design and operation of membrane systems as it affects pretreatment needs, cleaning requirements, operating conditions, cost, and performance. Constituents in wastewater that can bring about membrane fouling are identified in Table 11–18. Fouling of the membrane, as reported in Table 11–18, can occur in three general forms: (1) a buildup of the constituents in the feedwater on the membrane surface, (2) the formation of chemical precipitates due to the chemistry of the feedwater, and (3) damage to the membrane due to the presence of chemical substances that can react with the membrane or biological agents that can colonize the membrane.

Membrane Fouling Caused by Buildup of Solids. Three accepted mechanisms resulting in resistance to flow due to the accumulation of material within a lumen (see Fig. 11–41) are (1) pore narrowing, (2) pore plugging, and (3) gel/cake formation caused by concentration polarization (Ahn et al., 1998). Gel/cake formation, caused by concentration polarization, occurs when the majority of the solid matter in the feed is larger than the pore sizes or molecular weight cutoff of the membrane. Concentration polarization can be described as the buildup of matter close to or on the membrane surface that causes an increase in resistance to solvent transport across the membrane. Some degree of concentration polarization will always occur in the operation of a membrane system. The formation of a gel or cake layer, however, is an extreme

Table 11–18
Constituents in wastewater that can affect the performance of membranes through the mechanism of fouling

Type of membrane fouling	Responsible constituents	Remarks
Fouling (cake formation sometimes identified as biofilm formation)	Metal oxides Organic and inorganic colloids Bacteria Microorganisms Concentration polarization	Damage to membranes can be limited by controlling these substances (for example, by the use of microfiltration before reverse osmosis)
Scaling (precipitation)	Calcium sulfate Calcium carbonate Calcium fluoride Barium sulfate Metal oxide formation Silica	Scaling can be reduced by limiting salt content, by adding acid to limit the formation of calcium carbonate, and by other chemical treatments (e.g., the addition of antiscalants)
Damage to membrane	Acids Bases pH extremes Free chlorine Bacteria Free oxygen	Damage to membranes can be limited by controlling these substances. Extent of damage depends on the nature of the membrane

Figure 11–41

Modes of membrane fouling: (a) pore narrowing, (b) pore plugging, and (c) gel/cake formation caused by concentration polarization. (Adapted from Bourgeous et al., 1999.)

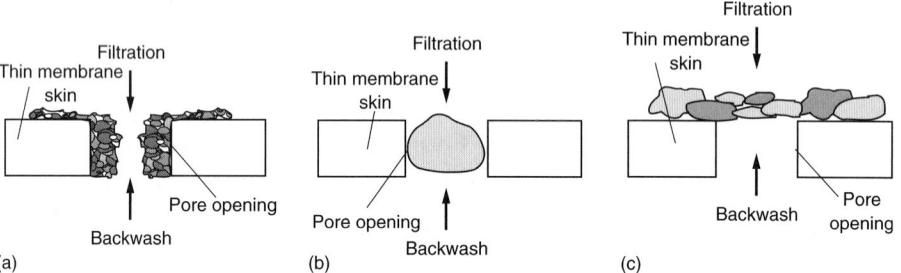

case of concentration polarization where a large amount of matter has actually accumulated on the membrane surface, forming a gel or cake layer. The mechanisms of pore plugging and pore narrowing will occur only when the solid matter in the feedwater is smaller than the pore size or the molecular weight cutoff. As the name describes, pore plugging occurs when particles the size of the pores become stuck in the pores of the membrane. Pore narrowing consists of solid material attaching to the interior surface of the pores, which results in a narrowing of the pores. It has been hypothesized that once the pore size is reduced, concentration polarization is amplified further, causing an increase in fouling (Crozes et al.,1997).

Control of Membrane Fouling. Typically, three approaches are used to control membrane fouling: (1) pretreatment of the feedwater, (2) membrane backflushing, and (3) chemical cleaning of the membranes. Pretreatment is used to reduce the TSS and bacterial content of the feedwater. Often the feedwater will be conditioned chemically to limit chemical precipitation within the units. The most commonly used method of eliminating the accumulated material from the membrane surface is backflushing with water and/or air. Chemical treatment is used to remove constituents that are not removed during conventional backwashing. Chemical precipitates can be removed by altering the chemistry of the feedwater and by chemical treatment. Damage of the membrane due to deleterious constituents typically cannot be reversed.

Pretreatment for Nanofiltration and Reverse Osmosis.   A very high-quality feed is required for efficient operation of a nanofiltration or reverse osmosis unit. Membrane elements in the reverse osmosis unit can be fouled by colloidal matter and constituents in the feed stream. The following pretreatment options have been used singly and/or in combination.

1.   Pretreatment of a secondary effluent by complete treatment, direct filtration, or contact filtration (see Fig. 13–11); by microfiltration; or by ultrafiltration to remove residual suspended solids and colloidal material.
2.   Cartridge filters with a pore size of 5- to 10-μm have also been used to reduce residual suspended solids.
3.   To limit bacterial activity it may be necessary to disinfect the feedwater using either chlorine, ozone, or UV radiation.
4.   The exclusion of oxygen may be necessary to prevent oxidation of iron, manganese, and hydrogen sulfide.
5.   Depending on the type of membrane, removal of chlorine (with sodium bisulfite) and ozone may be necessary.
6.   The removal of iron and manganese may also be necessary to decrease scaling potential.
7.   To inhibit scale formation, the pH of the feed should be adjusted (usually with sulfuric acid) within the range from 4.0 to 7.5.

Regular chemical cleaning of the membrane elements (about once a month) is necessary to restore the membrane flux.

Assessing Need for Pretreatment for NF and RO.   To assess the treatability of a given wastewater with NF and RO membranes, a variety of fouling indexes have been developed over the years. The three principal indexes are the silt density index (SDI), the modified fouling index (MFI), and the mini plugging factor index (MPFI). Fouling indexes are determined from simple membrane tests. The sample must be passed through a 0.45 μm Millipore filter with a 47 mm internal diameter at 210 kPa (30 lb/in^2) gage to determine any of the indexes. The time to complete data collection for these tests varies from 15 min to 2 h, depending on the fouling nature of the water.

The most widely used index is the SDI. The SDI is defined as follows:

$$\text{SDI} = \frac{100[1 - (t_i/t_f)]}{t} \qquad (11\text{–}45)$$

where t_i = time to collect initial sample of 500 mL
t_f = time to collect final sample of 500 mL
t = total time for running the test

The silt density index is a static measurement of resistance that is determined by samples taken at the beginning and end of the test. The SDI does not measure the rate of change of resistance during the test. Recommended SDI values are reported in Table 11–19. The calculation of the SDI is demonstrated in Example 11–6.

The modified fouling index (MFI) is determined using the same equipment and procedure used for the SDI, but the volume is recorded every 30 s over a 15-min filtration period. Derived from a consideration of cake filtration, the MFI is defined as follows:

$$\frac{1}{Q} = a + \text{MFI} \times V \tag{11–46}$$

where Q = average flow, L/s
a = constant
MFI = modified fouling index, s/L^2
V = volume, L

The value of the MFI is obtained as the slope of the straight-line portion of the curve obtained by plotting the inverse flow versus the cumulative volume (see Fig. 11–42). Recommended MFI values are reported in Table 11–19.

Table 11–19

Approximate values for fouling indexes[a]

Membrane process	Fouling index	
	SDI	MFI, s/L^2
Nanofiltration	0–2	0–10
Reverse osmosis hollow fiber	0–2	0–2
Reverse osmosis spiral wound	0–3	0–2

[a] Adapted in part from Taylor and Wiesner (1999) and AWWA (1996).

Figure 11–42

Typical plot to determine the modified fouling factor.

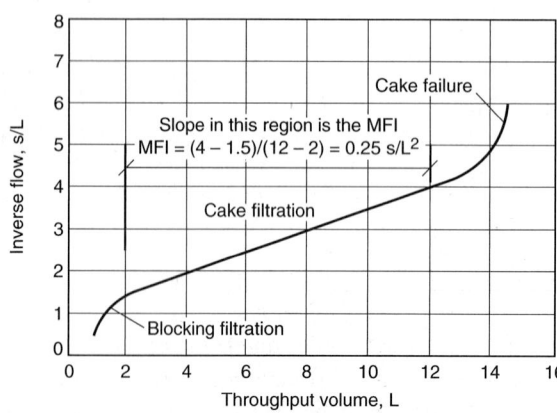

EXAMPLE 11–6 **Silt Density Index for Reverse Osmosis** Determine the silt density index for a proposed feedwater from the following test data. If a spiral-wound RO membrane is to be used, will pretreatment be required?

Test run time = 30 min

Initial 500 mL = 2 min

Final 500 mL = 10 min

Solution

1. Calculate the SDI using Eq. (11–45).

$$SDI = \frac{100[1 - (t_i/t_f)]}{t}$$

$$SDI = \frac{100[1 - (2/10)]}{30} = 2.67$$

2. Compare the SDI to the acceptable criteria.

Comment Calculated SDI value of 2.67 is less than 3 (see Table 11–19); therefore, no further pretreatment would be needed normally. As a practical matter, because the SDI value is close to 3.0 it may be prudent to consider some form of pretreatment to prolong the filtration cycle.

Application of Membranes

With evolving health concerns (see Chap. 2) and the development of new and lower-cost membranes, the application of membrane technologies in the field of environmental engineering has increased dramatically within the past 5 years. The increased use of membranes is expected to continue well into the future. In fact, the use of conventional filtration technology, such as described in Secs. 11–3 and 11–4, may be a thing of the past within 10 to 15 years, especially in light of the need to remove resistant organic constituents of concern (see Chap. 13). Typical applications of membrane technologies for wastewater treatment are reported in Table 11–20. The principal applications of the various membrane technologies for the removal of specific constituents found in wastewater are summarized in Table 11–21. A comparison of the constituents found in wastewater and the operating ranges for the various membrane technologies was presented previously on Fig. 11–34. Typical operating ranges in terms of operating pressures and flux rates, along with the types of membranes used, are reported in Table 11–22.

Typical energy consumption and product recovery values for various membrane systems are presented in Table 11–23. In reviewing the information presented in Table 11–23, it is important to note that the reported operating pressure values for all of the membrane processes, excluding electrodialysis, are considerably lower than comparable values of 5 years ago. It is anticipated that operating pressures will continue to go down as new membranes are developed. At the present time where the use of membranes is being considered, special attention must be devoted to the characteristics of the wastewater to be

Table 11–20

Typical applications for membrane technologies in wastewater treatment[a]

Applications	Description
Microfiltration and ultrafiltration	
Aerobic biological treatment	Membrane is used to separate the treated wastewater from the active biomass in an activated-sludge process. The membrane separation unit can be internal immersed in the bioreactor or external to the bioreactor (see Fig. 11–45). Such processes are known as membrane bioreactor (MBR) processes
Anaerobic biological treatment	Membrane is used to separate the treated wastewater from the active biomass in an anaerobic complete-mix reactor
Membrane aeration biological treatment	Plate and frame, tubular, and hollow membranes are used to transfer pure oxygen to the biomass attached to the outside of the membrane (see Fig. 11–47a). Such processes are known as membrane aeration bioreactor (MABR) processes
Membrane extraction biological treatment	Membranes are used to extract degradable organic molecules from inorganic constituents such as acids, bases, and salts from the waste stream for subsequent biological treatment (see Fig. 11–47b). Such processes are known as extractive membrane bioreactor (EMBR) processes
Pretreatment for effective disinfection	Used to remove residual suspended solids from settled secondary effluent or from the effluent from depth or surface filters to achieve effective disinfection with either chlorine or UV radiation for reuse applications
Pretreatment for nanofiltration and reverse osmosis	Microfilters are used to remove residual colloidal and suspended solids as a pretreatment step for additional processing
Nanofiltration	
Effluent reuse	Used to treat prefiltered effluent (typically with microfiltration) for indirect potable reuse applications such as groundwater injection. Credit is also given for disinfection when using nanofiltration
Wastewater softening	Used to reduce the concentration of multivalent ion contributing to hardness for specific reuse applications
Reverse osmosis	
Effluent reuse	Used to treat prefiltered effluent (typically with microfiltration) for indirect potable reuse applications such as groundwater injection. Credit is also given for disinfection when using reverse osmosis
Effluent dispersal	Reverse osmosis processes have proved capable of removing sizable amounts of selected compounds such as NDMA
Two-stage treatment for boiler use	Two stages of reverse osmosis are used to produce water suitable for high-pressure boilers

[a] Adapted in part from Stephenson et al. (2000).

Table 11–21
Application of membrane technologies for the removal of specific constituents found in wastewater[a]

Constituent	Membrane technology				Comments
	MF	**UF**	**NF**	**RO**	
Biodegradable organics		✔	✔	✔	
Hardness			✔	✔	
Heavy metals			✔	✔	
Nitrate			✔	✔	
Priority organic pollutants		✔	✔	✔	
Synthetic organic compounds			✔	✔	
TDS			✔	✔	
TSS	✔	✔			TSS removed during pretreatment for NF and RO
Bacteria	✔[b]	✔	✔	✔	Used for membrane disinfection. Removed as pretreatment for NF and RO with MF and UF
Protozoan cysts and oocysts and helminth ova	✔	✔	✔	✔	
Viruses			✔	✔	Used for membrane disinfection

[a] Specific removal rates will depend on the composition and constituent concentrations in the treated wastewater.
[b] Variable performance.

processed. Advantages and disadvantages of MF and UF and RO membrane technology are compared in Table 11–24.

Microfiltration. Microfiltration membranes are the most numerous on the market, are the least expensive, and as reported in Table 11–22, are commonly made of polypropylene, acrylonitrile, nylon, and polytetrafluoroethylene. Typical operating information for microfiltration and ultrafiltration membrane technologies used for wastewater including operating pressures and flux rates is also presented in Table 11–22. Microfiltration technologies, as discussed below, can be used in a variety of ways in wastewater-treatment and water reuse systems.

In advanced treatment applications, microfiltration has been used, most commonly as a replacement for depth filtration to reduce turbidity, remove residual suspended solids, and reduce bacteria to condition the water for effective disinfection (see Fig. 11–43a) and as a pretreatment step for reverse osmosis (see Figs. 11–40 and 11–43a). The MF units at Dublin San Ramon Sanitary District, used to produce treated water for specific reuse applications or as a pretreatment step for RO, are shown on Fig. 11–44. Corresponding performance data are presented in Table 11–25. Typical

Table 11-22

Typical characteristics of membrane technologies used in wastewater-treatment applications[a]

Membrane technology	Typical operating range, μm	Operating pressure		Rate of flux		Membrane details	
		lb/in^2	kPa	gal/ft$^2 \cdot$d	L/m$^2 \cdot$d	Type	Configuration
Microfiltration	0.08–2.0	1–15	7–100	10–40	405–1600	Polypropylene, acrylonitrile, nylon, and polytetrafluoroethylene	Spiral wound, hollow fiber, plate and frame
Ultrafiltration	0.005–0.2	10–100	70–700	10–20	405–815	Cellulose acetate, aromatic polyamides	Spiral wound, hollow fiber, plate and frame
Nanofiltration	0.001–0.01	75–150	500–1000	5–20	200–815	Cellulose acetate, aromatic polyamides	Spiral wound, hollow fiber
Reverse osmosis	0.0001–0.001	125–1000	850–7000	8–12	320–490	Cellulose acetate, aromatic polyamides	Spiral wound, hollow fiber, thin-film composite

[a] Adapted from Crites and Tchobanoglous (1998).

Note: kPa $\times$ 0.1450 = lb/in^2

L/m$^2 \cdot$d $\times$ 0.024542 = gal/ft$^2 \cdot$d

Table 11–23
Typical energy consumption and product recovery values for various membrane systems

Membrane process	Operating pressure		Energy consumption, kWh per		Product recovery, %
	lb/in²	kPa	1000 gal	m³	
Microfiltration	15	100	0.1	0.4	94–98
Ultrafiltration	75	525	0.8	3.0	70–80
Nanofiltration	125	875	1.4	5.3	80–85
Reverse osmosis	225	1575	2.7	10.2	70–85
Reverse osmosis	400	2800	4.8	18.2	70–85
Electrodialysis			2.5	9.5	75–85

Table 11–24
Advantages and disadvantages of membrane treatment technologies

Advantages	Disadvantages
Microfiltration and ultrafiltration	
• Can reduce the amount of treatment chemicals	• Uses more electricity; high-pressure systems can be energy-intensive
• Smaller space requirements (footprint); membrane equipment requires 50 to 80 percent less space than conventional plants	• May need pretreatment to prevent fouling; pretreatment facilities increase space needs and overall costs
• Reduced labor requirements; can be automated easily	• May require residuals handling and disposal of concentrate
• New membrane design allows use of lower pressures; system cost may be competitive with conventional wastewater-treatment processes	• Requires replacement of membranes about every 3 to 5 years
	• Scale formation can be a serious problem. Scale-forming potential difficult to predict without field testing
• Removes protozoan cysts, oocysts, and helminth ova; may also remove limited amounts of bacteria and viruses	• Flux rate (the rate of feedwater flow through the membrane) gradually declines over time. Recovery rates may be considerably less than 100 percent
	• Lack of a reliable low-cost method of monitoring performance
Reverse osmosis	
• Can remove dissolved constituents	• Works best on groundwater or low solids surface water or pretreated wastewater effluent
• Can disinfect treated water	
• Can remove NDMA and other related organic compounds	• Lack of a reliable low-cost method of monitoring performance
• Can remove natural organic matter (a disinfection by-product precursor) and inorganic matter	• May require residuals handling and disposal of concentrate
	• Expensive compared to conventional treatment

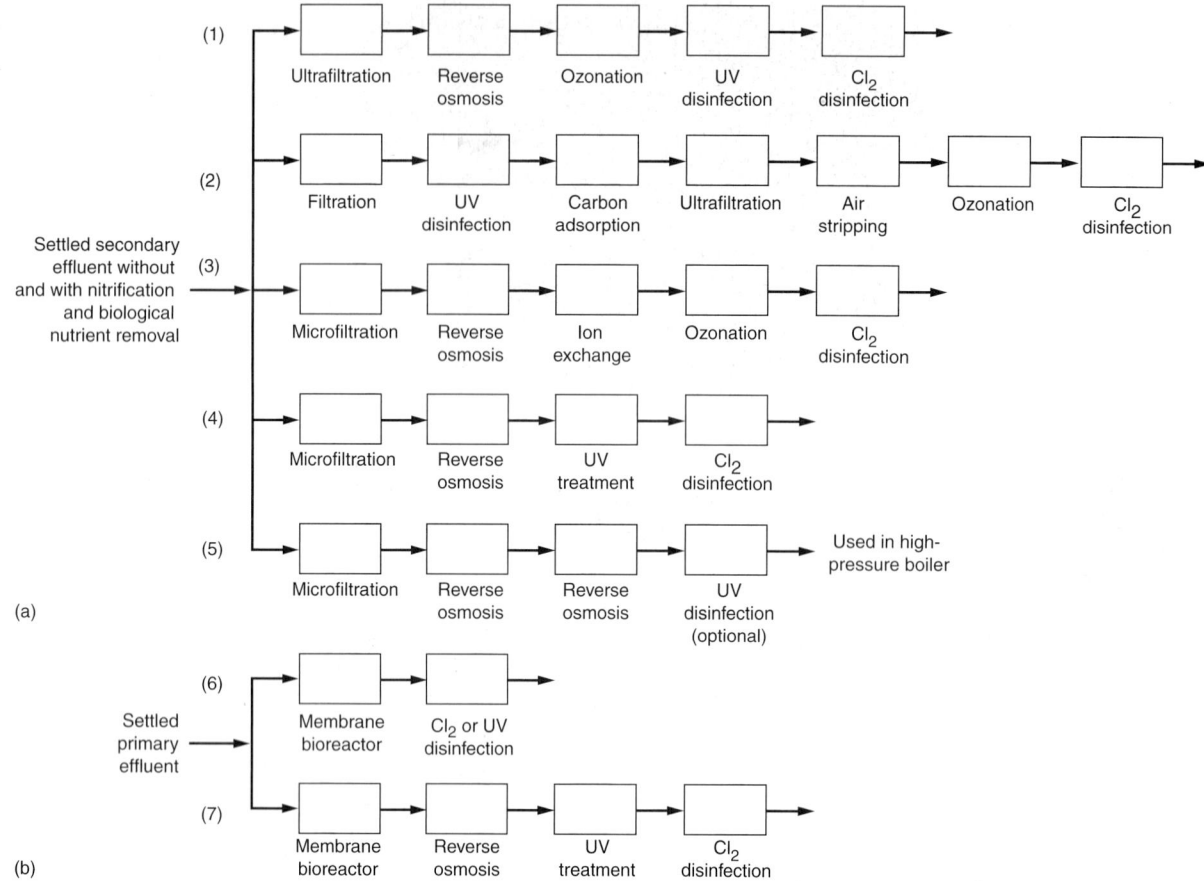

Figure 11–43

Typical flow diagrams employing membranes for advanced wastewater treatment with (a) settled secondary effluent and (b) settled primary effluent. Within the past 5 years, the use of membranes has essentially replaced the use of chemical precipitation in advanced wastewater-treatment applications. In flow diagram 5, two stages of reverse osmosis are used to produce an extremely low TDS water suitable for use in high-pressure boilers. In flow diagram 7, UV can be used for both disinfection and the destruction of NDMA. Chlorine is added in flow diagrams 1 through 4 and 7 to limit the growth of microorganisms in the distribution or conveyance system.

performance data reported in the literature are also included in Table 11–25 for the purpose of comparison. Care should be used in applying the performance data reported in Table 11–25 as it has been found that the performance of MF is to a large extent site-specific, especially with respect to fouling.

A relatively recent development, the use of membranes for biological treatment, promises to be one of the most important uses of membranes in wastewater treatment. As reported in Table 11–20, membranes have been used for both aerobic and anaerobic treatment of wastewaters. Typical membrane bioreactors (MBRs) are illustrated on Fig. 11–45. On Fig. 11–45a, the membrane separation unit is internal, immersed in the bioreactor. Treated effluent is withdrawn from the bioreactor with the application of a

Figure 11–44

Microfiltration (MF) units at Dublin San Ramon Sanitary District used as a pretreatment step for RO.

Table 11–25
Performance summary for the Dublin San Ramon Sanitary District MF for the period from 4/00 through 12/00[a,b]

Constituent	MF influent, mg/L	MF effluent, mg/L	Average reduction, %	Reduction reported in literature, %
TOC	10–31	9–16	57	45–65
BOD	11–32	<2–9.9	86	75–90
COD	24–150	16–53	76	70–85
TSS	8–46	<0.5	97	95–98
TDS	498–622	498–622	0	0–2
NH_3–N	21–42	20–35	7	5–15
NO_3–N	<1–5	<1–5	0	0–2
PO_4^-	6–8	6–8	0	0–2
SO_4^{2-}	90–120	90–120	0	0–1
Cl^-	93–115	93–115	0	0–1
Turbidity	2–50 NTU	0.03–0.08 NTU	>99	

[a] From Whitley Burchett & Associates (1999).
[b] Typical flux rate during test period was 1600 L/m²·d

Figure 11–45

Schematic flow diagrams for the membrane bioreactor activated-sludge process: (a) with internal membrane biosolids separation unit and (b) with external biosolids separation unit.

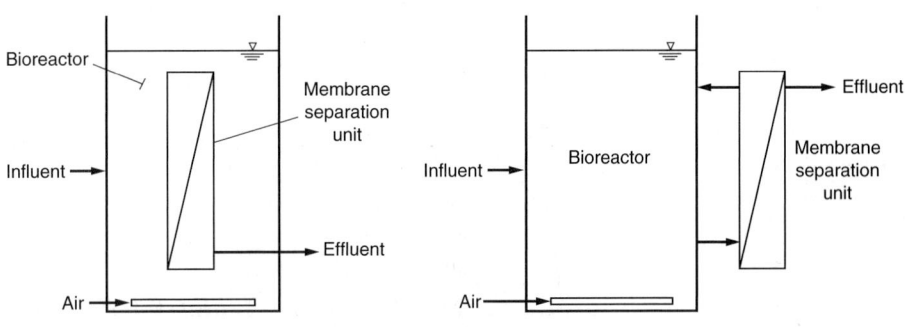

vacuum (on the order of 4 to 50 kPa). On Fig. 11–45*b*, the membrane separation unit is external to the bioreactor. A photograph of a typical membrane separation unit installed in a test reactor is shown on Fig. 11–46. The aeration system installed below the membrane separation unit is used to clean the hollow fibers continuously by the shearing action of the air bubbles as they rise through the liquid (Stephenson et al., 2000).

Typical performance data for the MBR activated-sludge process are presented in Table 11–26. As reported in Table 11–26, the quality of the effluent is ideal for a number of reuse application or for further processing by nanofiltration or reverse osmosis. In addition to offering excellent performance, MBRs require a considerably smaller aerial footprint. It is also interesting to note that MBRs have also been used for the anaerobic treatment of wastewater. The current literature abounds with articles delineating the use of membranes in a variety of wastewater-treatment applications. The biological characteristics of membrane bioreactors are discussed in greater detail in Chap. 8.

In addition to the MBR process, plate and frame, tubular, and hollow membranes have been used to transfer pure oxygen to the biomass attached to the outside of the membrane (see Fig. 11–47*a*). Such processes are known as membrane aeration bioreactor (MABR) processes. In another application, membranes are used to extract degradable soluble organic molecules from inorganic constituents in a waste stream (see Fig. 11–47*b*). The extracted organic constituents are treated in an external bioreactor. Such processes are known as extractive membrane bioreactor (EMBR) processes.

Figure 11–46

Photograph of horizontal type membrane bundle used in membrane bioreactors installed in pilot-scale reactor for testing with clear water.

Table 11–26
Typical performance data for membrane bioreactor used to treat domestic wastewater[a]

Parameter	Unit	Typical
Effluent BOD	mg/L	<5
Effluent COD	mg/L	<30
Effluent NH_3	mg/L	<1
Effluent TN	mg/L	<10
Effluent turbidity	NTU	<1

[a]For additional details on membrane bioreactors, see Sec. 8–9 in Chap. 8.

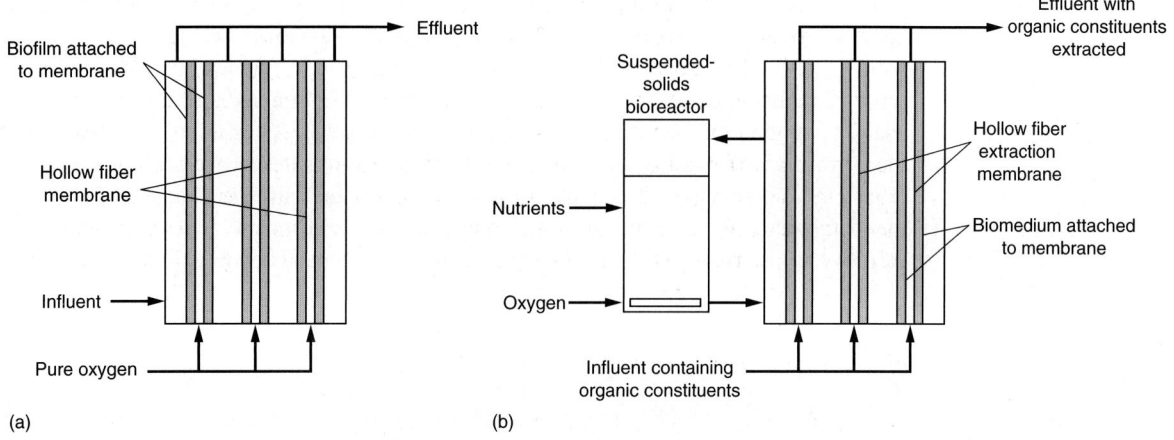

Figure 11–47

Schematic flow diagrams for alternative biological treatment processes employing membranes: (a) membrane aeration bioreactor, (b) extractive membrane bioreactor with external treatment unit.

Additional details on aeration and extractive membrane bioreactors applications may be found in Stephenson et al. (2000).

Ultrafiltration.   Ultrafiltration (UF) membranes are used for many of the same applications as described above for microfiltration. Some UF membranes with small pore sizes have also been used to remove dissolved compounds with high molecular weight, such as colloids, proteins, and carbohydrates. The membranes do not remove sugar or salt. Ultrafiltration is used typically in industrial applications for the production of high-purity process rinse water. Typical operating data are presented in Table 11–22.

Nanofiltration.   Nanofiltration, also known as "loose" RO, can reject particles as small as 0.001 μm. Nanofiltration is used for the removal of selected dissolved constituents from wastewater such as the multivalent metallic ions responsible for hardness. The advantages of nanofiltration over lime softening include the production of a product water that meets the most stringent reuse water quality requirements. Because both inorganic and organic constituents and bacteria and viruses are removed, disinfection requirements are minimized. Although most NF facilities use polyamide TFC membranes in a spiral-wound configuration, more than ten different types of membranes are available. Other membranes include polyamide hollow fiber, polyvinyl acetate spiral wound, and asymmetric cellulose acetate in a tubular configuration.

Reverse Osmosis.   Worldwide, reverse osmosis (RO) is used primarily for desalination. In wastewater treatment, RO is used for the removal of dissolved constituents from wastewater remaining after advanced treatment with depth filtration or microfiltration. The membranes exclude ions, but require high pressures to produce the deionized water. Typical flow diagrams involving the use of reverse osmosis are shown on Figs. 11–40 and 11–43. Typical operating information for reverse osmosis used for wastewater including operating pressures and flux rate rates is reported in Table 11–22. Typical RO units used for the repurification of pretreated water (with MF) are shown on Fig. 11–48. Corresponding performance data are presented in Table 11–27. Typical performance data reported in the literature are also included in Table 11–27. As noted above, care should be used in applying the performance data reported in Table 11–27 as it has been found that the performance of RO is also site-specific, especially with respect to fouling. Disinfection of the RO feedwater is usually practiced to minimize or limit the bacterial growth on the membrane. Care must be taken with polyamide and TFC membranes because they are sensitive to chemical oxidants. The need for cleaning depends on how long the flux rate is maintained.

Figure 11–48

Reverse osmosis (RO) units at Dublin San Ramon Sanitary District used for the repurification of water pretreated with MF (see Fig. 11–44).

Table 11–27
Performance summary for the Dublin San Ramon Sanitary District RO for the period from 4/1999 through 12/1999[a,b]

Constituent	RO influent, mg/L	RO effluent, mg/L	Average reduction, %	Reduction reported in literature, %
TOC	9–16	<0.5	>94	85–95
BOD	<2–9.9	<2	>40	30–60
COD	16–53	<2	>91	85–95
TSS	<0.5	~0	>99	95–100
TDS	498–622	10–31	>97	90–98
NH_3–N	20–35	1–3	96	90–98
NO_3–N	<1–5	0.08–3.2	96	65–85
PO_4^-	6–8	0.1–1	~99	95–99
SO_4^{2-}	90–120	<0.5–0.7	99	95–99
Cl^-	93–115	0.9–5.0	97	90–98
Turbidity	0.03–0.08 NTU	0.03 NTU	50	40–80

[a] From Whitley Burchett & Associates (1999).
[b] Typical flux rate during test period was 348 L/m²·d.

Electrodialysis

In the electrodialysis process, ionic components of a solution are separated through the use of semipermeable ion-selective membranes. Because the electrodialysis process is being evaluated for a variety of reuse applications, the purpose of the following discussion is (1) to introduce the theory of electrodialysis and (2) to consider some applications.

Theory of Electrodialysis. Application of an electrical potential between the two electrodes causes an electric current to pass through the solution, which in turn causes a migration of cations toward the negative electrode and a migration of anions toward the positive electrode (see Fig. 11–49). Because of the alternate spacing of cation- and anion-permeable membranes, cells of concentrated and dilute salts are formed. Wastewater is pumped through the membranes, which are separated by spacers and assembled into stacks. The wastewater is usually retained for about 10 to 20 days in a single stack or stage. Dissolved solids removals vary with the (1) wastewater temperature, (2) amounts of electrical current passed, (3) type and amount of ions, (4) permselectivity of the membrane, (5) fouling and scaling potential of the wastewater, (6) wastewater flowrates, and (7) number and configuration of stages.

The current required for electrodialysis can be estimated using Faraday's laws of electrolysis. Because one Faraday of electricity will cause one gram equivalent of a substance to migrate from one electrode to another, the number of gram equivalents removed per unit time is given by

$$\text{Gram eq/unit time} = QN\eta \qquad (11\text{--}47)$$

Figure 11–49

Schematic of electrodialysis unit.

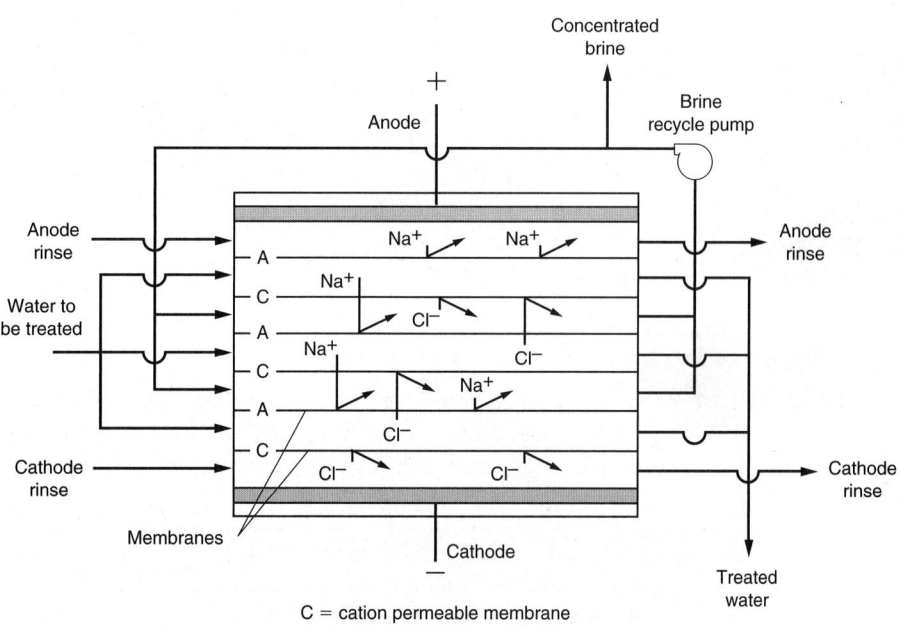

C = cation permeable membrane
A = anion permeable membrane

where Q = flowrate, L/s

N = normality of solution, eq/L

η = electrolyte removal as a fraction

The corresponding expression for the current for a stack of membranes is given by

$$I = \frac{FQN\eta}{nE_c} \tag{11-48}$$

where I = current, amp

F = Faraday's constant

= 96,485 amp·s/gram equivalent = 96,485 A·s/eq

n = number of cells in the stack

E_c = current efficiency expressed as a fraction

In the analysis of the electrodialysis process, it has been found that the capacity of the membrane to pass an electrical current is related to the current density (CD) and the normality (N) of the feed solution. Current density is defined as the current in milliamperes that flows through a square centimeter of membrane perpendicular to the current direction. The relationship between current density and the solution normality is known as the *current density to normality* (CD/N) ratio. High values of the ratio are indicative that there is insufficient charge to carry the current. When high ratios exist, a localized deficiency of ions may occur on the surface of the membrane, causing a condition called *polarization*. Polarization should be avoided, as it results in high electrical resistance leading to excessive power consumption. In practice, CD/N ratios will vary from 500 to 800 when the current density is expressed as mA/cm^2. The resistance of an electrodialysis unit used to treat a particular water must be determined experimentally. Once the resistance R and the current flow I are known, the power required can be computed using Ohm's law as follows:

$$P = E \times I = R(I)^2 \tag{11-49}$$

where P = power, W

E = voltage, V

= $R \times I$

R = resistance, Ω

I = current, A

The application of these relationships is considered in Example 11–7.

EXAMPLE 11–7   Area and Power Requirements for Electrodialysis   Determine the area and power required to demineralize 4000 m^3/d of treated wastewater to be used for industrial cooling water using an electrodialysis unit comprised of 240 cells. Assume the following data apply:

1.   TDS concentration = 2500 mg/L
2.   Cation and anion concentration = 0.010 g-eq/L

3. Efficiency of salt removal = 50 percent
4. The current efficiency = 90 percent
5. The CD/N ratio = 500 mA/cm^2
6. Resistance = 5.0 Ω

Solution

1. Calculate the current using Eq. (11–48).

$$I = \frac{FQN\eta}{nE_c}$$

$Q = (4000 \text{ m}^3/\text{d})(10^3 \text{ L/m}^3)/(86,400 \text{ s/d}) = 46.3 \text{ L/s}$

$$I = \frac{(96,485 \text{ A·s/g eq})(46.3 \text{ L/s})(0.010 \text{ g eq/L})(0.50)}{240 \times 0.90}$$

$I = 103.4 \text{ A}$

2. Determine the power required using Eq. (11–49).

$P = R(I)^2$

$P = (5.0 \text{ }\Omega)(103.4 \text{ A})^2 = 53,477 \text{ W} = 53.5 \text{ kW}$

3. Determine the required surface area.
   a. Determine the current density:

   $CD = (500)(\text{normality}) = 500 \text{ mA/cm}^2 \times 0.010 = 50 \text{ mA/cm}^2$

   b. The required area is

   $$\text{Area} = \frac{(103.4 \text{ A})(10^3 \text{ mA/A})}{(50 \text{ mA/cm}^2)} = 2068 \text{ cm}^2$$

   c. Assuming a square configuration will be used, determine the dimension of each side of the membrane.

   $$\text{Side dimension} = \sqrt{2068 \text{ cm}^2} \approx 45 \text{ cm}$$

Comment   The actual performance will have to be determined from pilot tests.

Application.   The electrodialysis process may be operated in either a continuous or a batch mode. The units can be arranged either in parallel to provide the necessary hydraulic capacity or in series to effect the desired degree of demineralization. A typical flow diagram for electrodialysis membrane process with pretreatment is shown on Fig. 11–50. Makeup water, usually about 10 percent of the feed volume, is required to wash the membranes continuously. A portion of the concentrate stream is recycled to maintain nearly

Figure 11–50

Typical flow diagram for electrodialysis membrane process with pretreatment.

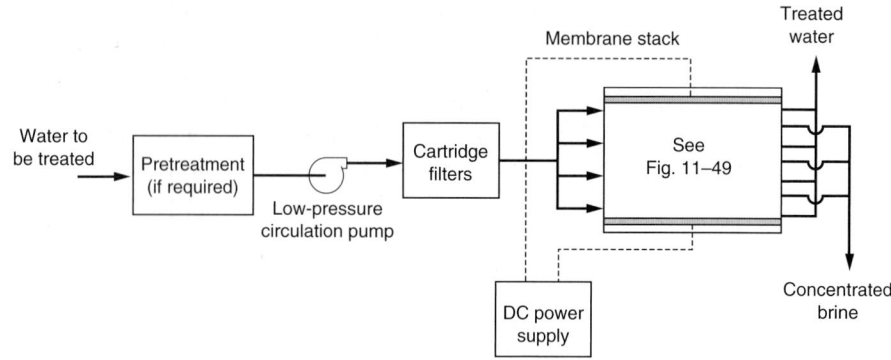

Table 11–28
Typical operating parameters for electrodialysis units

Parameter	Unit	Range
Detention time in stack	d	10–20
CD/N ratio	mA/cm^2	500–800
Membrane resistance, Ω	ohms	4–8
Salt-removal efficiency	%	40–60
Current efficiency	%	85–95
Concentrate stream flow	% of feed	10–20

equal flowrates and pressures on both sides of each membrane. Sulfuric acid is fed to the concentrate stream to maintain a low pH and thus minimize scaling. Typical operating parameters for the electrodialysis process are reported in Table 11–28.

Problems associated with the electrodialysis process for wastewater renovation include chemical precipitation of salts with low solubility on the membrane surface and clogging of the membrane by the residual colloidal organic matter in wastewater-treatment plant effluents. To reduce membrane fouling, activated carbon pretreatment, possibly preceded by chemical precipitation and some form of multimedia filtration, may be necessary.

Pilot Studies for Membrane Applications

Because every wastewater is unique with respect to its chemistry, it is difficult to predict a priori how a given membrane process will perform. As a result, the selection of the best membrane for a given application is usually based on the results of pilot studies. Membrane fouling indexes can be used to assess the need for pretreatment. In some situations, manufacturers of membranes will provide a testing service to identify the most appropriate membrane for a specific water or wastewater.

The elements that comprise a pilot plant include (1) the pretreatment system, (2) tankage for flow equalization and cleaning, (3) pumps for pressurizing the membrane, recirculation, and backflushing with appropriate controls (e.g., variable-frequency drives), (4) the membrane test module, (5) adequate facilities for monitoring the performance of the test module, and (6) an appropriate system for backflushing the membranes. The information collected should be sufficient to allow for the design of the full-scale system and should include as a minimum the following items:

Membrane operating parameters
    Pretreatment requirements including chemical dosages
    Transmembrane flux rate correlated to operating time
    Transmembrane pressure
    Washwater requirements
    Recirculation ratio
    Cleaning frequency including protocol and chemical requirements
    Posttreatment requirements

Typical water quality measurements may include:

Turbidity	Temperature
Particle counts	Heterotrophic plate count (see
Total organic carbon	Chap. 2)
Nutrients	Other bacterial indicators
Heavy metals	The specific constituents that can
Organic priority pollutants	limit recovery such as silica, bar-
Total dissolved solids	ium, calcium, and sulfate
pH	Biotoxicity

Additional specific parameters selected for evaluation will depend on the final use to be made of the repurified water. Typical pilot scale test facilities used to evaluate various membrane treatment options in connection with the production of repurified water for indirect potable reuse are shown on Fig. 11–51.

Disposal of Concentrated Waste Streams

Disposal of the concentrated waste streams produced by membrane processes represents the major problem that must be dealt with in their applications. The principal methods now used for the disposal of the concentrated waste streams are reported in Table 11–29. While small facilities can dispose of small concentrated waste streams by blending with other wastewater flows, this approach is not suitable for large facilities. The concentrate from NF and RO facilities will contain hardness, heavy metals, high-molecular-weight organics, microorganisms, and often hydrogen sulfide gas. The pH is usually high due to the concentration of alkalinity, which increases the likelihood of metal precipitation in disposal wells. As a result, most of the large-scale desalination facilities are located along coastal regions, both in the United States and in other parts of the world. For inland locations, long transmission lines to coastal regions are being considered. While controlled evaporation is technically feasible, because of high operating and maintenance costs this approach is used where no other alternatives are available, and the value of product water is high. The quality and quantity of the concentrated waste stream produced from nanofiltration, reverse osmosis, and electrodialysis can be estimated using simplified recovery and rejection computations as illustrated in Example 11–8.

Figure 11–51

Typical pilot-plant facilities used to test membranes:
(a) nanofiltration and
(b) reverse osmosis.

(a)

(b)

Table 11–29
Disposal options for concentrated brine solutions from membrane processes

Disposal option	Description
Ocean discharge	The disposal option of choice for facilities located in the coastal regions of the United States. Typically, a brine line, with a deep ocean discharge, is used by a number of dischargers. Combined discharge with power-plant cooling water has been used in Florida. For inland locations truck, rail hauling, or pipeline is needed for transportation
Surface water discharge	Discharge of brines to surface waters is the most common method of disposal for concentrated brine solutions
Land application	Land application has been used for some low-concentration brine solutions
Discharge to wastewater-collection system	This option is suitable only for very small discharges such that the increase in TDS is not significant (e.g., less than 20 mg/L)
Deep-well injection	Depends on whether subsurface aquifer is brackish water or is otherwise unsuitable for domestic uses
Evaporation ponds	Large surface area required in most areas with the exception of some southern and western states
Controlled thermal evaporation	Although energy-intensive, thermal evaporation may be the only option available in many areas

EXAMPLE 11–8 **Estimate Quantity and Quality of Waste Streams from a Reverse Osmosis Facility** Estimate quantity and quality of the waste stream, and the total quantity of water that must be processed, from a reverse osmosis facility that is to produce 4000 m^3/d of water to be used for industrial cooling operations. Assume that both the recovery and rejection rates are equal to 90 percent and that the concentration of the feed stream is 400 g/m^3.

Solution

1. Determine the flowrate of the concentrated waste stream and the total amount of water that must be processed.
   a. Combining Eqs. (11–41) and (11–39) results in the following expression for the concentrate stream flowrate:

   $$Q_c = \frac{Q_p(1 - r)}{r}$$

   b. Determine the concentrate stream flowrate.

   $$Q_c = \frac{(4000 \ m^3/d)(1 - 0.9)}{0.9} = 444 \ m^3/d$$

   c. Determine the total amount of water that must be processed to produce 4000 m^3/d of RO water. Using Eq. (11–41), the required amount of water is

   $$Q_f = Q_p + Q_c = 4000 \ m^3/d + 444 \ m^3/d = 4444 \ m^3/d$$

2. Determine the concentration of the permeate stream. The permeate concentration is obtained by writing Eq. (11–40) as follows:

   $$C_p = C_f(1 - R) = 400 \ g/m^3 \ (1 - 0.9) = 40 \ g/m^3$$

3. Determine the concentration of the concentrated waste stream. The required value is obtained by solving Eq. (11–42):

   $$C_c = \frac{Q_f C_f - Q_p C_p}{Q_c}$$

   $$C_c = \frac{(4444 \ m^3/d)(400 \ g/m^3) - (4000 \ m^3/d)(40 \ g/m^3)}{(444 \ m^3/d)}$$

   $$C_c = 3643 \ g/m^3$$

Comment To reduce the volume of the waste stream that must be treated, a variety of concentration methods are currently under investigation.

11–7 ADSORPTION

Adsorption is the process of accumulating substances that are in solution on a suitable interface. Adsorption, as noted in Chap. 4, is a mass transfer operation in that a constituent in the liquid phase is transferred to the solid phase. The *adsorbate* is the substance that is being removed from the liquid phase at the interface. The *adsorbent* is the solid, liquid, or gas phase onto which the adsorbate accumulates. Although adsorption is used at the air-liquid interface in the flotation process, only the case of adsorption at the liquid-solid interface will be considered in this discussion. The adsorption process has not been used extensively in wastewater treatment, but demands for a better quality of treated wastewater effluent, including toxicity reduction, have led to an intensive examination and use of the process of adsorption on activated carbon. Activated carbon treatment of wastewater is usually thought of as a polishing process for water that has already received normal biological treatment. The carbon in this case is used to remove a portion of the remaining dissolved organic matter. The purpose of this section is to introduce the basic concepts of adsorption and to consider carbon adsorption.

Types of Adsorbents

The principal types of adsorbents include activated carbon, synthetic polymeric, and silica-based adsorbents, although synthetic polymeric and silica-based adsorbents are seldom used for wastewater adsorption because of their high cost. Because activated carbon is used most commonly in advanced wastewater-treatment applications, the focus of the following discussion is on activated carbon. The nature of activated carbon, the use of granular carbon and powdered carbon for wastewater treatment, and carbon regeneration and reactivation are discussed below.

Activated Carbon.   Activated carbon is prepared by first making a char from organic materials such as almond, coconut, and walnut hulls; other materials including woods, bone, and coal have also been used. The char is produced by heating the base material to a red heat (less than about 700°C) in a retort to drive off the hydrocarbons, but with an insufficient supply of oxygen to sustain combustion. The carbonization or char-producing process is essentially a pyrolysis process. The char particle is then *activated* by exposure to oxidizing gases such as steam and CO_2 at high temperatures, in the range from 800 to 900°C. These gases develop a porous structure, as illustrated previously on Fig. 4–32 in Chap. 4, in the char and thus creates a large internal surface area. The resulting pore sizes are defined as follows:

Macropores	>25 nm
Mesopores	>1 nm and <25 nm
Micropores	<1 nm

The surface properties that result are a function of both the initial material used and the preparation procedure, so that many variations are possible. The type of base material from which the activated carbon is derived may also affect the pore-size distribution and the regeneration characteristics. After activation, the carbon can be separated into, or prepared in, different sizes with different adsorption capacity. The two size classifications are *powdered activated carbon* (PAC), which typically has a diameter of less than 0.074 mm (200 sieve), and *granular activated carbon* (GAC), which has a diame-

Table 11–30
Comparison of granular and powdered activated carbon

Parameter	Unit	Type of activated carbon[a]	
		GAC	**PAC**
Total surface area	m²/g	700–1300	800–1800
Bulk density	kg/m³	400–500	360–740
Particle density, wetted in water	kg/L	1.0–1.5	1.3–1.4
Particle size range	mm (μm)	0.1–2.36	(5–50)
Effective size	mm	0.6–0.9	na
Uniformity coefficient	UC	≤1.9	na
Mean pore radius	Å	16–30	20–40
Iodine number		600–1100	800–1200
Abrasion number	minimum	75–85	70–80
Ash	%	≤8	≤6
Moisture as packed	%	2–8	3–10

[a] Specific values will depend on the source material used for the production of the activated carbon.

ter greater than 0.1 mm (~140 sieve). The characteristics of granular and powdered activated carbon are summarized in Table 11–30.

Carbon Regeneration and Reactivation. Economical application of activated carbon depends on an efficient means of regenerating and reactivating the carbon after its adsorptive capacity has been reached. *Regeneration* is the term used to describe all of the processes that are used to recover the adsorptive capacity of the spent carbon, exclusive of reactivation, including: (1) chemicals to oxidize the adsorbed material, (2) steam to drive off the adsorbed material, (3) solvents, and (4) biological conversion processes. Typically some of the adsorptive capacity of the carbon (about 4 to 10 percent) is also lost in the regeneration process, depending on the compounds being adsorbed and the regeneration method used (Crittenden, 2000). In some applications, the capacity of the carbon following regeneration has remained essentially the same for years. A major problem with the use of powdered activated carbon is that the methodology for its regeneration is not well defined. The use of powdered activated carbon produced from recycled solid wastes may obviate the need to regenerate the spent carbon, and may be more economical. Additional details on carbon reactivation and regeneration may be found in Sontheimer et al. (1988).

Reactivation of granular carbon involves essentially the same process used to create the activated carbon from virgin material. Spent carbon is reactivated in a furnace by oxidizing the adsorbed organic material and, thus, removing it from the carbon surface. The following series of events occurs in the reactivation of spent activated carbon: (1) the carbon is heated to drive off the absorbed organic material (i.e., absorbate), (2) in the process of driving off the absorbed material some new compounds are formed that remain on the surface of the carbon, and (3) the final step in the reactivation process is to burn off the new compounds that were formed when the absorbed material was burned off. With effective process control, the adsorptive capacity of reactivated carbon will be essentially the same as that of the virgin carbon (Crittenden, 2000). For planning purposes, it is often assumed that a loss of 2 to 5 percent will occur in the reactivation process. It is important to note that most other losses of carbon occur through attrition due to mishandling. For example, right angle bends in piping cause attrition through abrasion and impact. The type of pumping facilities used will also affect the amount of attrition. In general, a 4 to 8 percent loss of carbon is assumed, due to handling. Replacement carbon must be available to make up the loss.

Fundamentals of Adsorption

The adsorption process, as illustrated on Fig. 11–52, takes place in four more or less definable steps: (1) bulk solution transport, (2) film diffusion transport, (3) pore transport, and (4) adsorption (or sorption). *Bulk solution transport* involves the movement of the organic material to be adsorbed through the bulk liquid to the boundary layer of fixed film of liquid surrounding the adsorbent, typically by advection and dispersion in carbon contactors. *Film diffusion transport* involves the transport by diffusion of the organic material through the stagnant liquid film to the entrance of the pores of the adsorbent. *Pore transport* involves the transport of the material to be adsorbed through the pores by a combination of molecular diffusion through the pore liquid and/or by diffusion along the surface of the adsorbent. *Adsorption* involves the attachment of the material to be adsorbed to adsorbent at an available adsorption site (Snoeyink and Summers, 1999). Adsorption can occur on the outer surface of the adsorbent and in the macropores, mesopores, micropores, and submicropores, but the surface area of the macro- and mesopores is small compared with the surface area of the micropores and submicropores and the amount of material adsorbed there is usually considered negligible. Adsorption forces include (Crittenden, 1999):

- Coulombic-unlike charges
- Point charge and a dipole
- Dipole-dipole interactions

Figure 11–52

Definition sketch for the adsorption of an organic constituent with activated carbon.

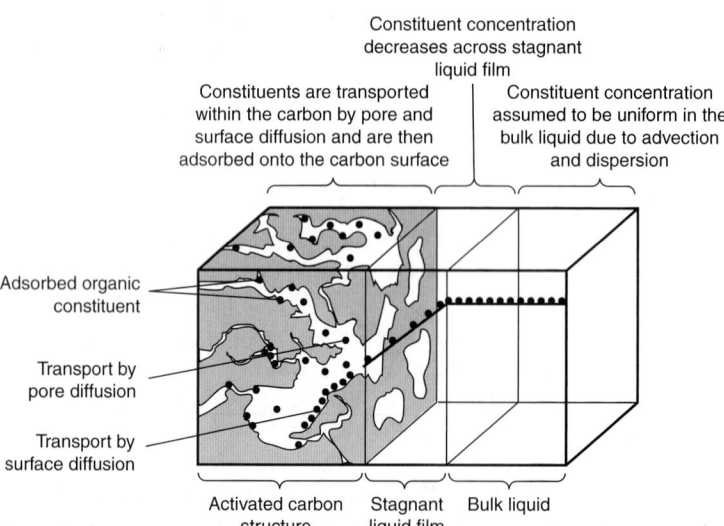

Constituent concentration decreases across stagnant liquid film

Constituents are transported within the carbon by pore and surface diffusion and are then adsorbed onto the carbon surface

Constituent concentration assumed to be uniform in the bulk liquid due to advection and dispersion

Adsorbed organic constituent

Transport by pore diffusion

Transport by surface diffusion

Activated carbon structure

Stagnant liquid film

Bulk liquid

- Point charge neutral species
- London or van der Waals forces
- Covalent bonding with reaction
- Hydrogen bonding

Additional details on the above adsorption forces may be found in Crittenden (1992). Because it is difficult to differentiate between chemical and physical adsorption, the term *sorption* is often used to describe the attachment of the organic material to the activated carbon.

Because the adsorption process occurs in a series of steps, the slowest step in the series is identified as the *rate limiting step.* In general, if physical adsorption is the principal method of adsorption, one of the diffusion transport steps will be the rate limiting, because the rate of physical adsorption is rapid. Where chemical adsorption is the principal method of adsorption, the adsorption step has often been observed to be rate limiting. When the rate of sorption equals the rate of desorption, equilibrium has been achieved and the capacity of the carbon has been reached. The theoretical adsorption capacity of the carbon for a particular contaminant can be determined by developing its adsorption isotherm as described below.

Development of Adsorption Isotherms.   The quantity of adsorbate that can be taken up by an adsorbent is a function of both the characteristics and concentration of adsorbate and the temperature. The characteristics of the adsorbate that are of importance include: solubility, molecular structure, molecular weight, polarity, and hydrocarbon saturation (see Table 11–31). Generally, the amount of material adsorbed is determined as a function of the concentration at a constant temperature, and the resulting function is called an adsorption isotherm. Adsorption isotherms are developed by exposing a given amount of absorbate in a fixed volume of liquid to varying amounts of activated carbon. Typically, more than ten containers are used, and the minimum time allowed for the samples to equilibrate where powdered activated carbon is used is seven days. If granular activated carbon is used, it is usually powdered to minimize adsorption times. The test is illustrated previously on Fig. 4–34 in Chap. 4. At the end of the test period, the amount of absorbate remaining in solution is measured. The absorbent phase concentration after equilibrium is computed using Eq. (11–50). The absorbent phase concentration data computed using Eq. (11–50) are then used to develop adsorption isotherms as described below.

$$q_e = \frac{(C_o - C_e)V}{m} \tag{11–50}$$

where q_e = adsorbent (i.e., solid) phase concentration after equilibrium, mg adsorbate/g adsorbent
C_o = initial concentration of adsorbate, mg/L
C_e = final equilibrium concentration of adsorbate after absorption has occurred, mg/L
V = volume of liquid in the reactor, L
m = mass of adsorbent, g

Freundlich Isotherm.   Equations that are often used to describe the experimental isotherm data were developed by Freundlich, Langmuir, and Brunauer, Emmet, and

Table 11–31
Readily and poorly
adsorbed organics[a]

Readily adsorbed organics	Poorly adsorbed organics

Aromatic solvents
  Benzene
  Toluene
  Nitrobenzenes
	Low-MW ketones, acids, and aldehydes
	Sugars and starches
	Very-high-MW or colloidal organics
	Low-MW aliphatics

Chlorinated aromatics
  PCBs
  Chlorophenols	

Polynuclear aromatics
  Acenaphthene
  Benzopyrenes	

Pesticides and herbicides
  DDT
  Aldrin
  Chlordane
  Atrazine	

Chlorinated nonaromatics
  Carbon tetrachloride
  Chloroalkyl ethers
  Trichloroethene
  Chloroform
  Bromoform	

High-MW hydrocarbons
  Dyes
  Gasoline
  Amines
  Humics	

[a]From Froelich (1978).

Teller (BET isotherm) (Shaw, 1966). Of the three, the Freundlich isotherm is used most commonly to describe the adsorption characteristics of the activated carbon used in water and wastewater treatment. Derived empirically in 1912, the Freundlich isotherm is defined as follows:

$$\frac{x}{m} = K_f C_e^{1/n} \tag{11-51}$$

where x/m = mass of adsorbate adsorbed per unit mass of adsorbent, mg adsorbate/g activated carbon

K_f = Freundlich capacity factor, (mg absorbate/g activated carbon)(L water/ mg adsorbate)$^{1/n}$

C_e = equilibrium concentration of adsorbate in solution after adsorption, mg/L

$1/n$ = Freundlich intensity parameter

The constants in the Freundlich isotherm can be determined by plotting log (x/m) versus log C_e and making use of Eq. (11–51) rewritten as:

$$\log\left(\frac{x}{m}\right) = \log K_f + \frac{1}{n} \log C_e \tag{11-52}$$

The U.S. EPA has developed adsorption isotherms for a variety of toxic compounds, some of which are presented in Table 11–32 for reference (Dobbs and Cohen, 1980). As shown in Table 11–32, the variation in the Freundlich capacity factor for the various compounds is extremely wide (e.g., 14,000 for PCB to 6.8×10^{-5} for N-Dimethylnitrosamine). Because of the wide variation, the Freundlich capacity factor must be determined for each new compound. Application of the Freundlich adsorption isotherm is illustrated in Example 11–9.

Table 11–32
Freundlich adsorption isotherm constants for selected organic compounds[a,b]

Compound	pH	K_f(mg/g)(L/mg)$^{1/n}$	1/n
Benzene	5.3	1.0	1.6–2.9
Bromoform	5.3	19.6	0.52
Carbon tetrachloride	5.3	11	0.83
Chlorobenzene	7.4	91	0.99
Chloroethane	5.3	0.59	0.95
Chloroform	5.3	2.6	0.73
DDT	5.3	322	0.50
Dibromochloromethane	5.3	4.8	0.34
Dichlorobromomethane	5.3	7.9	0.61
1,2-Dichloroethane	5.3	3.6	0.83
Ethylbenzene	7.3	53	0.79
Heptachlor	5.3	1,220	0.95
Hexachloroethane	5.3	96.5	0.38
Methylene chloride	5.3	1.3	1.16
N-Dimethylnitrosamine	na	6.8×10^{-5}	6.60
N-Nitrosodi-n-propylamine	na	24	0.26
N-Nitrosodiphenylamine	3–9	220	0.37
PCB	5.3	14,100	1.03
PCB 1221	5.3	242	0.70
PCB 1232	5.3	630	0.73
Phenol	3–9	21	0.54
Tetrachloroethylene	5.3	51	0.56
Toluene	5.3	26.1	0.44
1,1,1-Trichloroethane	5.3	2–2.48	0.34
Trichloroethylene	5.3	28	0.62

[a] Adapted from Dobbs and Cohen (1980) and LaGrega et al. (2001).
[b] The adsorption isotherm constants reported in this table are meant to be illustrative of the wide range of values that will be encountered for various organic compounds. It is important to note that the characteristics of the activated carbon used as well as the analytical technique used for the analysis of the residual concentrations of the individual compounds will have a significant effect on the coefficient values obtained for specific organic compounds.

Langmuir Isotherm.   Derived from rational considerations, the Langmuir adsorption isotherm is defined as:

$$\frac{x}{m} = \frac{abC_e}{1 + bC_e}$$ (11–53)

where x/m = mass of adsorbate adsorbed per unit mass of adsorbent, mg adsorbate/g activated carbon

a, b = empirical constants

C_e = equilibrium concentration of adsorbate in solution after adsorption, mg/L

The Langmuir adsorption isotherm was developed by assuming: (1) a fixed number of accessible sites are available on the adsorbent surface, all of which have the same energy, and (2) adsorption is reversible. Equilibrium is reached when the rate of adsorption of molecules onto the surface is the same as the rate of desorption of molecules from the surface. The rate at which adsorption proceeds is proportional to the driving force, which is the difference between the amount adsorbed at a particular concentration and the amount that can be adsorbed at that concentration. At the equilibrium concentration, this difference is zero.

Correspondence of experimental data to the Langmuir equation does not mean that the stated assumptions are valid for the particular system being studied, because departures from the assumptions can have a canceling effect. The constants in the Langmuir isotherm can be determined by plotting $C_e/(x/m)$ versus C_e and making use of Eq. (11–53) rewritten as:

$$\frac{C_e}{(x/m)} = \frac{1}{ab} + \frac{1}{a}C_e$$ (11–54)

Application of the Langmuir adsorption isotherm is illustrated in Example 11–9.

EXAMPLE 11–9   **Analysis of Activated-Carbon Adsorption Data**   Determine the Freundlich and Langmuir isotherm coefficients for the following GAC adsorption test data. The liquid volume used in the batch adsorption tests was 1 L. The initial concentration of the adsorbate in solution was 3.37 mg/L. Equilibrium was obtained after 7 days.

Mass of GAC, m, g	Equilibrium concentration of adsorbate in solution, C_e, mg/L
0.0	3.37
0.001	3.27
0.010	2.77
0.100	1.86
0.500	1.33

Solution

1. Derive the values needed to plot the Freundlich and Langmuir adsorption isotherms using the batch adsorption test data.

Adsorbate concentration, mg/L				x/m,[a]	
C_o	C_e	$C_o - C_e$	m, g	mg/g	$C_e/(x/m)$
3.37	3.37	0.00	0.000	—	—
3.37	3.27	0.10	0.001	100	0.0327
3.37	2.77	0.60	0.010	60	0.0462
3.37	1.86	1.51	0.100	15.1	0.1232
3.37	1.33	2.04	0.500	4.08	0.3260

[a] $q_e = \dfrac{x}{m} = \dfrac{(C_o - C_e)V}{m}$

2. Plot the Freundlich and Langmuir adsorption isotherms using the data developed in Step 1 (see following figures).

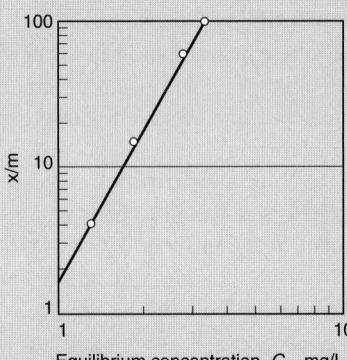

(a)

Equilibrium concentration, C_e, mg/L

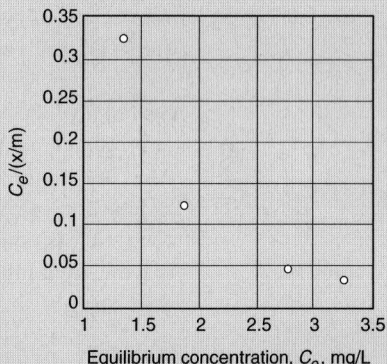

(b)

Equilibrium concentration, C_e, mg/L

3. Determine the adsorption isotherm coefficients.
   a. Freundlich

   When x/m versus C_e is plotted on log-log paper, the intercept when $C_e = 1.0$ is the value of (x/m) and the slope of the line is equal to $1/n$. Thus, $x/m = 1.55$, and $K_f = 1.55$. When $x/m = 1.0$, $C_e = 0.9$, and $1/n = 0.26$. Thus,

   $$\frac{x}{m} = 1.5\, C_e^{0.26}$$

   b. Langmuir

   Because the plot for the Langmuir isotherm is curvilinear, use of the Langmuir adsorption isotherm is inappropriate.

Adsorption of Mixtures. In the application of adsorption to wastewater treatment, mixtures of organic compounds are always encountered. Typically, there is a depression of the adsorptive capacity of any individual compound in a solution of many compounds, but the total adsorptive capacity of the adsorbent may be larger than the adsorptive capacity with a single compound. The amount of inhibition due to competing compounds is related to the size of the molecules being adsorbed, their adsorptive affinities, and their relative concentrations. It is important to note that adsorption isotherms can be determined for a heterogeneous mixture of compounds including total organic carbon (TOC), dissolved organic carbon (DOC), chemical oxygen demand (COD), dissolved organic halogen (DOH), UV absorbance, and fluorescence (Snoeyink and Summers, 1999). The adsorption from mixtures is considered further in Crittenden et al. (1987b, 1987c, 1985) and Sontheimer (1988).

Activated Carbon Adsorption Kinetics

As noted previously, both granular carbon (in downflow and upflow columns) and powdered activated carbon are used for wastewater treatment. The analysis procedures for both types are described briefly in the following discussion.

Mass Transfer Zone. The area of the GAC bed in which sorption is occurring is called the mass transfer zone (MTZ) (see Fig. 11–53). After the water containing the constituent to be removed passes through a region of the bed whose depth is equal to the MTZ, the concentration of the contaminant in the water will have been reduced to its minimum value. No further adsorption will occur within the bed below the MTZ. As the top layers of carbon granules become saturated with organic material, the MTZ will move down in the bed until breakthrough occurs. Typically, breakthrough is said to have occurred when the effluent concentration reaches 5 percent of the influent value. Exhaustion of the adsorption bed is assumed to have occurred when the effluent concentration is equal to 95 percent of the influent concentration. The length of the MTZ is typically a function of the hydraulic loading rate applied to the column and the characteristics of the activated carbon. In the extreme, if the loading rate is too great the

Figure 11–53

Typical breakthrough curve for activated carbon showing movement of mass transfer zone (MTZ) with throughput volume.

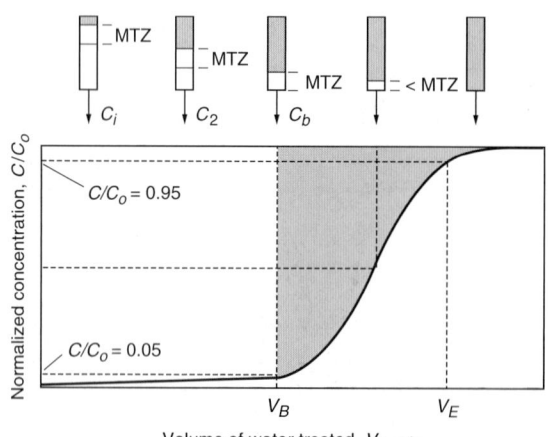

height of the MTZ will be larger than the GAC bed depth, and the adsorbable constituent will not be removed completely by the carbon. At complete exhaustion, the effluent concentration is equal to the influent concentration.

In addition to the applied hydraulic loading rate, the shape of the breakthrough curve will also depend on whether the applied liquid contains nonadsorbable and biodegradable constituents. The impact of the presence of nonadsorbable and biodegradable organic constituents on the shape of the breakthrough curve is illustrated on Fig. 11–54. As shown on Fig. 11–54, if the liquid contains nonadsorbable constituents, the nonadsorbable constituents will appear in the effluent as soon as the carbon column is put into operation. If adsorbable and biodegradable constituents are present in the applied liquid, the breakthrough curve will not reach a C/C_o value of 1.0, but will be depressed, and the observed C/C_o value will depend on the biodegradability of the influent constituents, because biological activity continues even though the adsorption capacity has been utilized. If the liquid contains nonadsorbable and biodegradable constituents, the observed breakthrough curve will not start at zero and will not terminate at a value of 1.0 (Snoeyink and Summers, 1999). The above effects are observed quite commonly in wastewater adsorption applications, especially with respect to the removal of COD.

The height of the MTZ (see Fig. 11–53) varies with the flowrate because dispersion, diffusion, and channeling in a granular medium are directly related to the flowrate. For a symmetrical breakthrough curve it has been shown that the height of the mass transfer zone, H_{MTZ}, is related to the column height Z and the throughput volumes V_B and V_E, as follows (Michaels, 1952; Weber, 1972).

$$H_{MTZ} = Z \left[\frac{V_E - V_B}{V_E - 0.5(V_E - V_B)} \right]$$
(11–55)

where H_{MTZ} = length of mass transfer zone, m
Z = height of the adsorption column, m
V_E = throughput volume to exhaustion, L, m^3
V_B = throughput volume to breakthrough L, m^3

Figure 11–54

Impact of the presence of adsorbable, nonadsorbable, and biodegradable organic constituents on the shape of the activated carbon breakthrough curve. (Adapted from Snoeyink and Summers, 1999.)

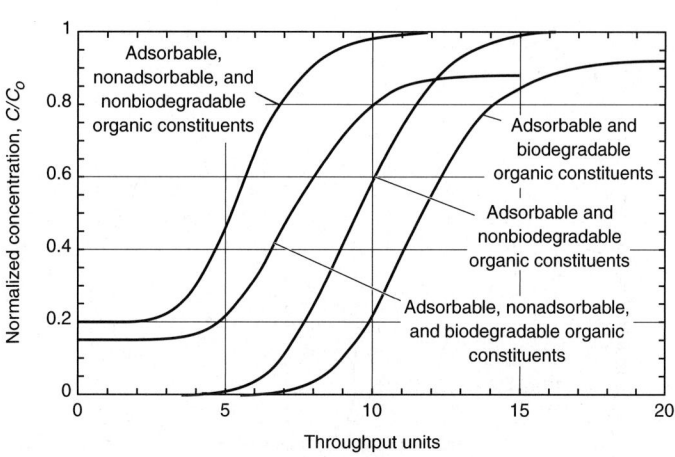

The area above the breakthrough curve represents the mass of adsorbate adsorbed within the column and is equal to:

$$\int (C_o - C)dV$$

integrated from $V = 0$ to $V = V$, the throughput volume at breakthrough. At complete exhaustion, $C = C_o$.

In practice, the only way to use the capacity at the bottom of the carbon adsorption column is to have two or more columns in series and switch them as they are exhausted, or to use multiple columns in parallel so that breakthrough in one column does not affect effluent quality. The optimum flowrate and bed depth, as well as the operating capacity of the carbon, must be established to determine the dimensions and the number of columns necessary for continuous treatment. These parameters can be determined from dynamic column tests, as discussed below.

Carbon Adsorption Capacity. The adsorptive capacity of a given carbon is estimated from isotherm data as follows. If isotherm data are plotted, the resulting isotherm will be as shown on Fig. 11–55. Referring to Fig. 11–55, the adsorptive capacity of the carbon can be estimated by extending a vertical line from the point on the horizontal axis corresponding to the initial concentration C_o, and extrapolating the isotherm to intersect this line. The $q_e = (x/m)_{C_o}$ value at the point of intersection can be read from the vertical axis. The $(q_e)_{C_o}$ value represents the amount of constituent adsorbed per unit weight of carbon when the carbon is at equilibrium with the initial concentration of constituent C_o. The equilibrium condition generally exists in the upper section of a carbon bed during column treatment, and it therefore represents the ultimate capacity of the carbon for a particular waste.

Breakthrough Adsorption Capacity. In the field, the breakthrough adsorption capacity $(x/m)_b$ of the GAC in a full-scale column is some percentage of the theoretical adsorption capacity found from the isotherm. The $(x/m)_b$ of a single column can be assumed to be approximately 25 to 50 percent of the theoretical capacity, $(x/m)_o$. The value of $(x/m)_b$ can be determined using the small-scale column test described later in

Figure 11–55

Typical activated carbon adsorption isotherm.

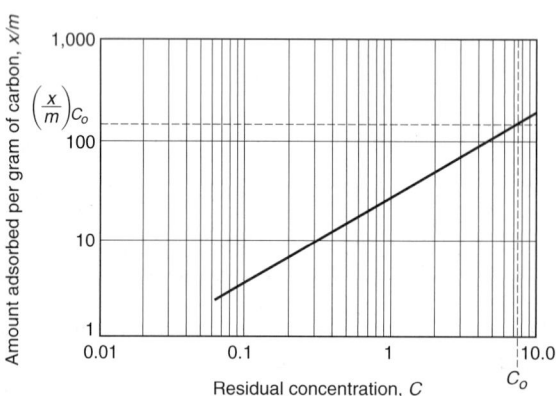

this section. Once $(x/m)_b$ is known, the time to breakthrough can be approximated by solving the following equation for t_b.

$$\left(\frac{x}{m}\right)_b = \frac{x_b}{m_{\mathrm{GAC}}} = Q\left(C_o - \frac{C_b}{2}\right)\frac{t_b}{m_{\mathrm{GAC}}} \qquad (11\text{–}56)$$

Where $(x/m)_b$ = field breakthrough adsorption capacity, g/g

x_b = mass of organic material adsorbed in the GAC column at breakthrough, g

m_{GAC} = mass of carbon in the column, g

Q = flowrate, m³/d

C_o = influent organic concentration, g/m³

C_b = breakthrough organic concentration, g/m³

t_b = time to breakthrough, d

Equation (11–56) was developed assuming that C_o is constant and that the effluent concentration increases linearly with time from 0 to C_b (see Fig. 11–53). The term $(C_o - C_b/2)$ represents the average concentration of the organic matter adsorbed up to the breakthrough point. Rearranging Eq. (11–56), the time to breakthrough can be calculated using the following relationship:

$$t_b = \frac{(x/m)_b m_{\mathrm{GAC}}}{Q(C_o - C_b/2)} \qquad (11\text{–}57)$$

Over the years, a number of equations have been developed to describe the breakthrough curve, including those by Bohart and Adams (1920) and Crittenden et al. (1987*a*).

However, as noted previously, because of the breakthrough phenomenon (see Fig. 11–55), the usual practice is either to use *two or more columns in series and rotate them as they become exhausted, or to use multiple columns in parallel so that breakthrough in a single column will not significantly affect the effluent quality.* With proper sampling from points within the column, constituent (e.g., TOC) breakthrough can be anticipated.

Activated Carbon Treatment Process Applications

Carbon adsorption is used principally for the removal of refractory organic compounds, as well as residual amounts of inorganic compounds such as nitrogen, sulfides, and heavy metals. The removal of taste and odor compounds from wastewater is another important application, especially in reuse applications. Both powdered and granular activated carbon are used and appear to have a low adsorption affinity for low molecular weight polar organic species. If biological activity is low in the carbon contactor or in other biological unit processes, these species are difficult to remove with activated carbon. Typical compounds that can be removed by carbon adsorption are reported in Table 11–31. Under normal conditions, after treatment with carbon, the effluent BOD ranges from 2 to 7 mg/L, and the effluent COD ranges from 10 to 20 mg/L. Under optimum conditions, it appears that the effluent COD can be reduced to less than 10 mg/L.

Treatment with Granular Activated Carbon (GAC). Treatment with GAC involves passing a liquid to be treated through a bed of activated carbon held in a reactor (sometimes called a contactor). Several types of activated carbon contactors are used for advanced wastewater treatment. Typical systems may be either pressure or gravity type, and may be downflow or upflow fixed-bed units having two or three columns in series, or expanded bed upflow-countercurrent type. Typical schematic diagrams of carbon contactors are shown on Fig. 11–56.

Fixed-Bed A fixed-bed column is used most commonly for contacting wastewater with GAC. Fixed-bed columns can be operated singly, in series, or in parallel. A typical granular activated carbon contactor is shown on Fig. 11–57a. Granular-medium filters are commonly used upstream of the activated carbon contactors to remove the organics associated with the suspended solids present in secondary effluent. The water to be treated is applied to the top of the column and withdrawn at the bottom. The carbon is held in place with an underdrain system at the bottom of the column. Provision for backwashing and surface washing is often provided in wastewater applications to limit the headloss buildup due to the removal of particulate suspended solids within the carbon column. Unfortunately, backwashing has the effect of destroying the adsorption front as discussed later.

The advantage of a downflow design is that adsorption of organics and filtration of suspended solids are accomplished in a single step. Although upflow fixed-bed reactors have been used, downflow beds are used more commonly to lessen the chance of accumulating particulate material in the bottom of the bed, where the particulate material would be difficult to remove by backwashing. If soluble organic removal is not maintained at a high level, more frequent regeneration of the carbon may be required. Lack of consistency in pH, temperature, and flowrate may also affect performance of carbon contactors. A typical column installation used for the treatment of wastewater is shown on Fig. 11–57b.

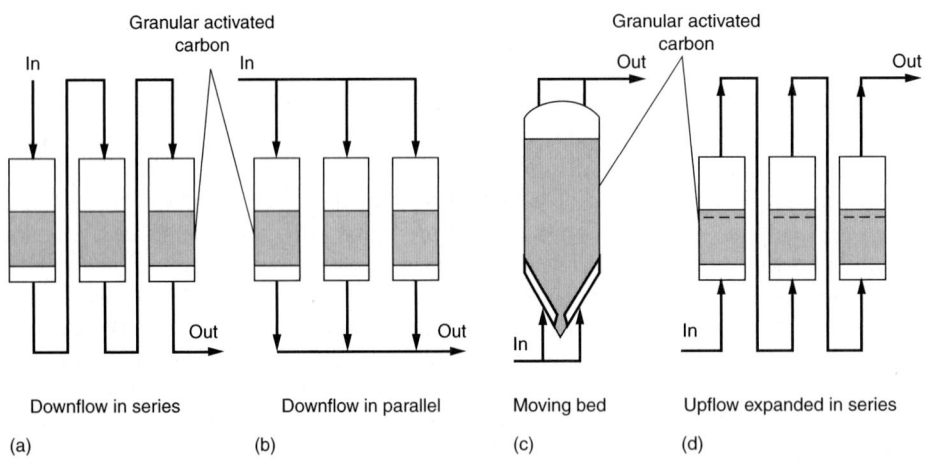

Downflow in series	Downflow in parallel	Moving bed	Upflow expanded in series
(a)	(b)	(c)	(d)

Figure 11–56

Types of activated carbon contactors: (a) downflow in series, (b) downflow in parallel, (c) moving bed, and (d) upflow expanded in series. (Adapted from Calgon Carbon Corp.)

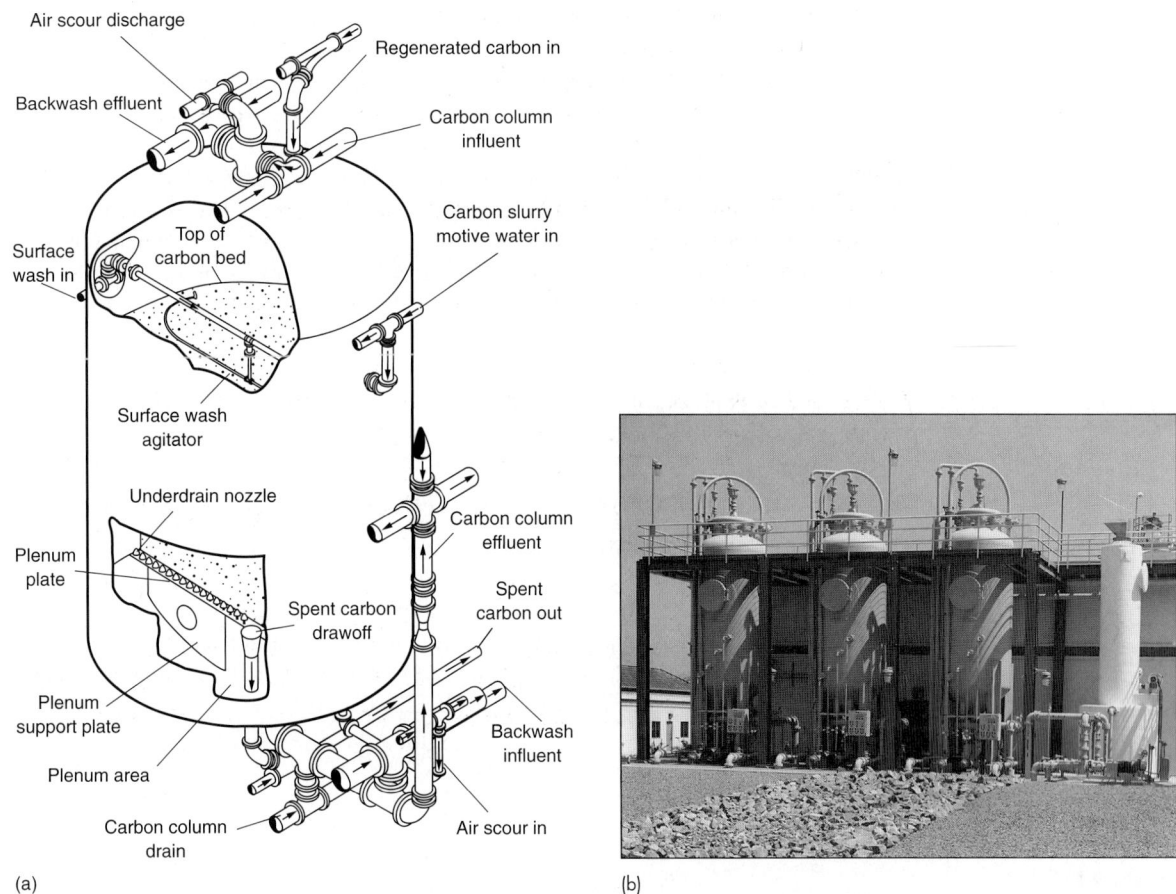

(a)

(b)

Figure 11–57

Activated carbon adsorbers: (a) Typical activated-carbon contactor in a pressure vessel and (b) view of granular activated carbon contactors operated in parallel, used for the treatment of filtered secondary effluent.

Expanded-Bed Expanded-bed, moving-bed, and pulsed-bed carbon contactors have also been developed to overcome the problems associated with headloss buildup. In the expanded-bed system, the influent is introduced at the bottom of the column and the activated carbon is allowed to expand, much as a filter bed expands during backwash. When the adsorptive capacity of the carbon at the bottom of the column is exhausted, the bottom portion of carbon is removed, and an equivalent amount of regenerated or virgin carbon is added to the top of the column. In such a system, headloss does not build up with time after the operating point has been reached. In general, expanded-bed upflow contactors may have more carbon fines in the effluent than downflow contactors because bed expansion leads to the creation of fines as the carbon particles collide and abrade, and allows the fines to escape through passageways created by the expanded bed. At present, few, if any, expanded bed contactors are used for the treatment of wastewater.

Treatment with Powdered Activated Carbon (PAC). An alternative means of achieving adsoption is through the application of powdered activated carbon (PAC). Powdered activated carbon can be applied to the effluent from biological treatment processes, directly to the various biological treatment processes, and in physical-chemical treatment process flow diagrams. In the case of biological treatment plant effluent, PAC is added to the effluent in a contacting basin. After a certain amount of time for contact, the carbon is allowed to settle to the bottom of the tank, and the treated water is then removed from the tank. Because carbon is very fine, a coagulant, such as a polyelectrolyte, may be needed to aid the removal of the carbon particles, or filtration through rapid sand filters may be required. The addition of PAC directly to the aeration basin of an activated-sludge treatment process has proved to be effective in the removal of a number of soluble refractory organics. In physical-chemical treatment processes, PAC is used in conjunction with chemicals used for the precipitation of specific constituents (see discussion in Chap. 6).

Analysis and Design of Granular Activated Carbon Contactor

The sizing of carbon contactors is based on four factors: contact time, hydraulic loading rate, carbon depth, and number of contactors. Typical design information for the first three factors is presented in Table 11–33. Typical specifications for granular activated carbon are given in Table 11–34. A minimum of two parallel carbon contactors is recommended for design. Multiple units permit one or more units to remain in operation while one unit is taken out of service for removal and regeneration of spent carbon, or for maintenance.

Table 11–33
Typical design values for GAC contactors[a]

Parameter	Symbol	Unit	Value
Volumetric flowrate	V	m^3/h	50–400
Bed volume	V_b	m^3	10–50
Cross-sectional area	A_b	m^2	5–30
Length	D	m	1.8–4
Void fraction	α	m^3/m^3	0.38–0.42
GAC density	ρ	kg/m^3	350–550
Approach velocity	V_f	m/h	5–15
Effective contact time	t	min	2–10
Empty bed contact time	EBCT	min	5–30
Operation time	t	d	100–600
Throughput volume	V_L	m^3	10–100
Specific throughput	V_{sp}	m^3/kg	50–200
Bed volumes	BV	m^3/m^3	2,000–20,000

[a] Adapted from Sontheimer et al. (1998).

For the case where the mass transfer rate is fast and the mass transfer zone is a sharp wave front, a steady-state mass balance around a carbon contactor (see Fig. 11–58) reactor yields:

Accumulation = inflow − outflow − amount absorbed

$$0 = QC_o t - QC_e t - m_{GAC} q_e \tag{11–58}$$

where Q = volumetric flowrate, L/h
   C_o = initial concentration of adsorbate, mg/L
   t = time, h
   C_e = final equilibrium concentration of adsorbate, mg/L
   m_{GAC} = mass of adsorbent, g
   q_e = adsorbent phase concentration after equilibrium, mg adsorbate/g adsorbent

Table 11–34
Typical specifications for GAC

Parameter	Unit	Value
Total surface area	m²/g	700–1300[a]
Bulk density	kg/m³	400–440[a]
Particle density, wetted in water	kg/L	1.3–1.5[a]
Effective size	mm	0.8–0.9[a]
Uniformity coefficient	UC	≤1.9
Mean particle diameter	mm	1.5–1.7
Iodine number		850 min
Abrasion number		70 min
Ash	%	8 max
Moisture	%	4–6 max

[a] Depends on the source material for the carbon.

Figure 11–58

Definition sketch for the analysis of activated-carbon adsorption.

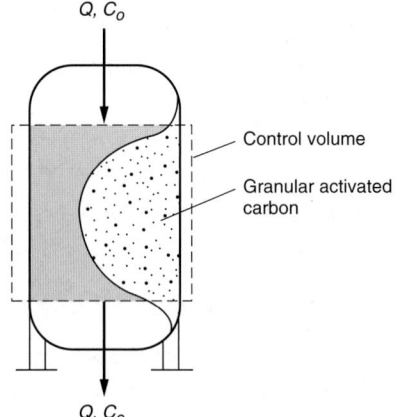

Q, C_o

Control volume

Granular activated carbon

Q, C_e

From Eq. (11–58), the adsorbent usage rate is defined as

$$\frac{m_{GAC}}{Qt} = \frac{C_o - C_e}{q_e} \tag{11–59}$$

If it is assumed that the mass of the adsorbate in the pore space is small compared to the amount adsorbed, then the term $QC_e t$ in Eq. (11–58) can be neglected without serious error and the absorbent usage rate is given by:

$$\frac{m_{GAC}}{Qt} \approx \frac{C_o}{q_e} \tag{11–60}$$

To quantify the operational performance of GAC contactors, the following terms have been developed and are used commonly.

1. Empty-bed contact time (EBCT):

$$EBCT = \frac{V_b}{Q} = \frac{A_b D}{v_f A_b} = \frac{D}{v_f} \tag{11–61}$$

where EBCT = empty bed contact time, h
 V_b = volume of GAC in contactor, m³
 Q = volumetric flowrate, m³/h
 A_b = cross-sectional area of GAC filter bed, m²
 D = length of GAC in contactor, m
 v_f = linear approach velocity, m/h

2. Activated carbon density:
 The density of the activated carbon is defined as

$$\rho_{GAC} = \frac{m_{GAC}}{V_b} \tag{11–62}$$

where ρ_{GAC} = density of granular activated carbon, g/L
 m_{GAC} = mass of granular activated carbon, g
 V_b = volume of GAC filter bed, L

3. Specific throughput, expressed as m³ of water treated per gram of carbon:

$$\text{Specific throughput, m}^3/\text{g} = \frac{Qt}{m_{GAC}} = \frac{V_b t}{EBCT \times m_{GAC}} \tag{11–63}$$

Using Eq. (11–62), Eq. (11–63) can be written as

$$\text{Specific throughput} = \frac{V_b t}{EBCT(\rho_{GAC} \times V_b)} = \frac{t}{EBCT \rho_{GAC}} \tag{11–64}$$

4. Carbon usage rate (CUR) expressed as gram of carbon per m³ of water treated:

$$\text{CUR, g/m}^3 = \frac{m_{GAC}}{Qt} = \frac{1}{\text{Specific throughput}} \tag{11–65}$$

5. Volume of water treated for a given EBCT, expressed in liters, L:

$$\text{Volume of water treated, L} = \frac{\text{Mass of GAC for given EBCT}}{\text{GAC usage rate}} \qquad (11\text{–}66)$$

6. Bed life, expressed in days, d:

$$\text{Bed life, d} = \frac{\text{Volume of water treated for given EBCT}}{Q} \qquad (11\text{–}67)$$

The application of these terms is illustrated in Example 11–10.

EXAMPLE 11–10 **Estimation of Activated Carbon Adsorption Breakthrough Time** A fixed-bed activated carbon adsorber has a fast mass transfer rate, and the mass transfer zone is essentially a sharp wavefront. Assuming the following data apply, determine the carbon requirements to treat a flow of 1000 L/min, and the bed life.

1. Compound = trichloroethylene (TCE)
2. Initial concentration, C_o = 1.0 mg/L
3. Final concentration C_e = 0.005 mg/L
4. GAC density = 450 g/L
5. Freundlich capacity factor K_f = 28 (mg/g)(L/mg)$^{1/n}$ (see Table 11–32)
6. Freundlich intensity parameter, $1/n$ = 0.62 (see Table 11–32)
7. EBCT = 10 min

Solution

1. Estimate the GAC usage rate for toluene. The GAC usage rate is estimated using Eq. (11–59).

$$\frac{m_{GAC}}{Qt} = \frac{C_o - C_e}{q_e} = \frac{C_o - C_e}{K_f C_o^{1/n}}$$

$$= \frac{(1.0 \text{ mg/L})}{28\,(\text{mg/g})\,(\text{L/mg})^{0.62}(1.0 \text{ mg/L})^{0.62}}$$

$$= 0.036 \text{ g GAC/L}$$

2. Determine the mass of carbon required for a 10 min EBCT.

The mass of GAC in the bed = $V_b \rho_{GAC}$ = EBCT × Q × ρ_{GAC}

Carbon required = 10 min (1000 L/min) (450 g/L) = 4.5 × 10^6 g

3. Determine the volume of water treated using a 10 min EBCT.

$$\text{Volume of water treated} = \frac{\text{mass of GAC for given EBCT}}{\text{GAC usage rate}}$$

$$\text{Volume of water treated} = \frac{4.5 \times 10^6 \text{ g}}{(0.036 \text{ g GAC/L})} = 1.26 \times 10^8 \text{ L}$$

4. Determine the bed life.

$$\text{Bed life} = \frac{\text{volume of water treated for given EBCT}}{Q}$$

$$\text{Bed life} = \frac{1.26 \times 10^8 \text{ L}}{(1000 \text{ L/min})(1440 \text{ min/d})} = 87.5 \text{ d}$$

Comment In this example, the full capacity of the carbon in the contactor was utilized based on the assumption that two columns in series will be used. If a single column is to be used, a breakthough curve must be used to arrive at the bed life. Biological activity within the column was not considered.

Small-Scale Column Tests

Over the years, a number of small-scale column tests have been developed to simulate the results obtained with full-scale reactors. One of the early column tests was the *high-pressure minicolumn* (HPMC) technique developed by Rosene et al. (1983), and later modified by Bilello and Beaudet (1983). In the HPMC test procedure, a high-pressure liquid chromatography column loaded with activated carbon is used. Typically the HPMC test procedure is used to determine the capacity of activated carbon for the adsorption of volatile organic compounds. The principal advantage of the HPMC test procedure is that it allows for the rapid determination of the GAC adsorptive capacity under conditions similar to those encountered in the field.

An alternative procedure known as the *rapid small-scale column test* (RSSCT) has been developed by Crittenden et al. (1991). The test procedure allows for the scaling of data obtained from small columns (see Fig. 11-59) to predict the performance of pilot or full-scale carbon columns. In developing the procedure, mathematical models were used to define the relationships between the breakthrough curve for small and large columns. In adsorption columns, the mass transfer mechanisms that are responsible for the spreading of the mass transfer zone are (1) dispersion, (2) film diffusion, and (3) interparticle diffusion. Two different design relationships were developed, one for constant diffusivity and one for proportional diffusivity. In the constant diffusivity model, it is assumed that dispersion is negligible because the hydraulic loading rate is high in the RSSCT, and that mass transfer occurs as a result of film diffusion. Further, it is assumed that the interparticle diffusivity is the same for both the small and large columns. In the proportional diffusivity model, it is assumed that dispersion is negligible because the hydraulic loading rate is high in the RSSCT, and that mass transfer occurs as a result of interparticle diffusion. The relationships for the two cases can be generalized as follows:

$$\frac{\text{EBCT}_{SC}}{\text{EBCT}_{LC}} = \left[\frac{d_{SC}}{d_{LC}}\right]^{2-x} = \frac{t_{SC}}{t_{LC}} \tag{11–68}$$

$$\frac{V_{SC}}{V_{LC}} = \frac{d_{LC}}{d_{SC}} \tag{11–69}$$

Figure 11–59

Small-scale column
used to develop data
for pilot- or full-scale
carbon columns.
*(Adapted from Crittenden
et al., 1991.)*

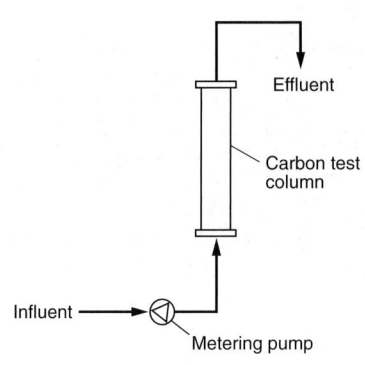

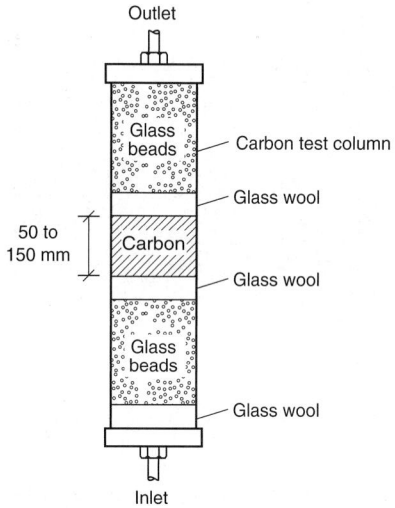

Particle to column diameter = 20:1 (or greater)
Column diameter = 20 to 40 mm
Column length = 300 mm

where d_{SC} = diameter of particle in short column, mm
d_{LC} = diameter of particle in long column, mm
t_{SC} = time in short column, min
t_{LC} = time in long column, min
V_{SC} = superficial velocity in short column, m/h
V_{LC} = superficial velocity in long column, m/h

For constant and proportional diffusivity, the value of x in the exponent in Eq. (11–68) is 0 and 1, respectively. The application of the above equations is illustrated in Example 11–11.

EXAMPLE 11–11

Comparison of Rapid Small-Scale Column Test Parameters to Pilot-Scale Parameters Determine the corresponding parameters for an RSSCT based on the following data proposed for a pilot-scale column. Assume that film diffusion is the controlling mechanism.

Parameter	Unit	Pilot column (LC)	RSSCT (SC)
Particle diameter	mm	0.5	0.1
Carbon density	g/L	450	450
EBCT	min	10	
Loading rate	m/h	5.0	
Flowrate	mL/min	200	

(continued)

(Continued)

Parameter	Unit	Pilot column (LC)	RSSCT (SC)
Column diameter	mm	75	10[a]
Column length	mm	1000	
Mass adsorbent	g		
Time of operation	d	100	
Water volume	L	28,800	

[a] Assumed value for small column.

Solution

1. Estimate the EBCT for the RSSCT.

$$EBCT_{SC} = EBCT_{LC} \left[\frac{d_{SC}}{d_{LC}} \right]^2$$

$$EBCT_{SC} = 10 \text{ min} \left[\frac{0.1}{0.5} \right]^2 = 0.4 \text{ min}$$

2. Estimate the loading rate for the RSSCT.

$$V_{SC} = V_{LC} \frac{d_{LC}}{d_{SC}}$$

$$V_{SC} = 5 \text{ m/h} \frac{0.5}{0.1} = 25 \text{ m/h}$$

3. Estimate flowrate for the RSSCT.

$$A = \frac{\pi}{4} d_{SC}^2 = \frac{\pi}{4} (10 \text{ mm})^2 = 78.5 \text{ mm}^2$$

$$Q_{SC} = (V_{SC})(A)$$

$$Q_{SC} = \frac{(25 \text{ m/h})(10^3 \text{ mm/m})(78.5 \text{ mm}^2)}{(60 \text{ min/h})(10^3 \text{ mm}^3/\text{mL})} = 32.7 \text{ mL/min}$$

4. Estimate column length for the RSSCT.

$$L_{SC} = \frac{Q_{SC} \times EBCT_{SC}}{A} = \frac{(32.7 \text{ mL/min})(0.4 \text{ min})}{(1 \text{ mL}/10^3 \text{ mm}^3)(78.5 \text{ mm}^2)} = 166.7 \text{ mm}$$

5. Estimate mass of adsorbent required for the RSSCT.

$$M_{SC} = EBCT_{LC} \left[\frac{d_{SC}}{d_{LC}} \right]^2 (Q_{SC})(\rho_{SC})$$

$$M_{SC} = 10 \text{ min} \left[\frac{0.1 \text{ mm}}{0.5 \text{ mm}} \right]^2 \left[\frac{(32.7 \text{ mL/min})(450 \text{ g/L})}{10^3 \text{ mL/L}} \right] = 5.9 \text{ g}$$

6. Estimate time of operation for the RSSCT.

$$t_{SC} = t_{LC} \frac{\text{EBCT}_{SC}}{\text{EBCT}_{LC}}$$

$$t_{SC} = 100 \text{ d} \frac{0.4 \text{ min}}{10 \text{ min}} = 4 \text{ d}$$

7. Estimate volume of water required for the RSSCT.

$$V_W = Q_{SC} \times t_{SC}$$

$$V_W = \frac{(32.7 \text{ mL/min})(4 \text{ d})(1440 \text{ min/d})}{(10^3 \text{ mL/L})} = 188.4 \text{ L}$$

8. Summarize the findings for the RSSCT.

Parameter	Unit	Pilot column	RSSCT
Particle radius	mm	0.5	0.1
Carbon density	g/L	450	450
EBCT	min	10	0.4
Loading rate	m/h	5.0	25.0
Flowrate	mL/min	200	32.7
Column diameter	mm	75	10[a]
Column length	mm	1000	166.7
Mass adsorbent	g		5.9
Time of operation	d	100	4
Water volume	L	28,800	188.4

[a] Assumed value for small column.

Comment The time savings in conducting the RSSCT versus the pilot column is apparent. Furthermore, many more tests can be conducted to test alternative configurations and carbon types.

Analysis and Design of Powdered Activated Carbon Contactor

For a powdered activated carbon (PAC) application, the isotherm adsorption data can be used in conjunction with a materials mass balance analysis to obtain an approximate estimate of the amount of carbon that must be added as illustrated below. Here again, because of the many unknown factors involved, column and bench scale tests are recommended to develop the necessary design data.

If a mass balance is written around the contactor (i.e., batch reactor) after equilibrium has been reached, the resulting expression is (Note: the following derivation was presented previously in Chap. 4, but is repeated here for continuity):

$$
\begin{array}{ccc}
\text{Amount} & \text{initial amount} & \text{final amount} \\
\text{adsorbed} = & \text{of absorbate} & - \text{of absorbate} \\
& \text{present} & \text{present}
\end{array}
\tag{11–70}
$$

$$
q_e m = VC_o - VC_e
\tag{11–71}
$$

where q_e = adsorbent phase concentration after equilibrium, mg adsorbate/g adsorbent

m = mass of adsorbent, g
V = volume of liquid in the reactor, L
C_o = initial concentration of adsorbate, mg/L
C_e = final equilibrium concentration of adsorbate after absorption has occurred, mg/L

It should be noted that q_e is in equilibrium with C. If Eq. (11–71) is solved for q_e, the following expression is obtained.

$$
q_e = \frac{V(C_o - C_e)}{m}
\tag{11–72}
$$

If Eq. (11–72) is rewritten as follows:

$$
\frac{V}{m} = \frac{q_e}{C_o - C_e}
\tag{11–73}
$$

The term V/m is defined as the *specific volume* and represents the volume of liquid that can be treated with a given amount of carbon. The graphical representation of the specific volume was presented previously on Fig. 4–33 in Chap. 4. The reciprocal value of the specific volume corresponds to the dose of adsorbent that must be used.

EXAMPLE 11–12 **Estimation of Powdered Activated Carbon (PAC) Adsorption Dose and Cost**
A treated wastewater with a flowrate of 1000 L/min is to be treated with PAC to reduce the concentration of residual organics measured as TOC from 5 to 1 mg/L. The Freundlich adsorption isotherm parameters were developed as discussed previously. Assuming the following data apply, determine the PAC requirements to treat the wastewater flow. If PAC costs $0.50/kg, estimate the annual cost for treatment, assuming the PAC will not be regenerated.

1. Compound = mixed organics
2. Initial concentration, C_o = 5.0 mg/L
3. Final concentration, C_e = 1.0 mg/L
4. GAC density = 450 g/L
5. Freundlich capacity factor, K_f = 150 $(\text{mg/g})(\text{L/mg})^{1/n}$
6. Freundlich intensity parameter, $1/n$ = 0.5

Solution

1. Estimate the PAC dose based on the isotherm data. The PAC dose can be estimated by writing Eq. (11–71) as follows:

$$\frac{m}{V} = \frac{(C_o - C_e)}{q_e}$$

Substituting Eq. (11–15) for q_e and using the given values in the above expression yields

$$\frac{m}{V} = \frac{[(5 - 1)\ \text{mg/L}]}{(150\ \text{mg/g})(\text{L/mg})^{0.5}(1.0\ \text{mg/L})^{0.5}} = 0.0267\ \text{g/L}$$

2. Estimate the annual cost for the PAC treatment.

Annual cost

$$= \frac{(0.0267\ \text{g/L})(1000\ \text{L/min})(1440\ \text{min/d})(365\ \text{d/yr})(\$0.50/\text{kg})}{(10^3\ \text{g/kg})}$$

Annual cost = $7008/year

Comment   For small wastewater flows, it is not usually cost effective to plan for carbon regeneration.

Activated Sludge with Powdered Activated Carbon Treatment

A proprietary process, "PACT," combines the use of powdered activated carbon with the activated-sludge process (see Fig. 11–60). In this process, when the activated carbon is added directly to the aeration tank, biological oxidation and physical adsorption occur simultaneously. A feature of this process is that it can be integrated into existing activated sludge systems at nominal capital cost. The addition of powdered activated carbon has several process advantages, including (1) system stability during shock loads, (2) reduction of refractory priority pollutants, (3) color and ammonia removal, and (4) improved sludge settleability. In some industrial waste applications, where nitrification is inhibited by toxic organics, the application of powdered activated carbon may reduce or limit this inhibition.

The dosage of powdered activated carbon and the mixed liquor–powdered activated carbon suspended solids concentration are related to the SRT as follows:

$$X_p = \frac{X_i \text{SRT}}{\tau} \tag{11–74}$$

where X_p = equilibrium powdered activated carbon-MLSS content, mg/L
   X_i = powdered activated carbon dosage, mg/L
   SRT = solids retention time, d
   τ = hydraulic retention time, d

Carbon dosages typically range from 20 to 200 mg/L. With higher SRT values, the organic removal per unit of carbon is enhanced, thereby improving the process efficiency.

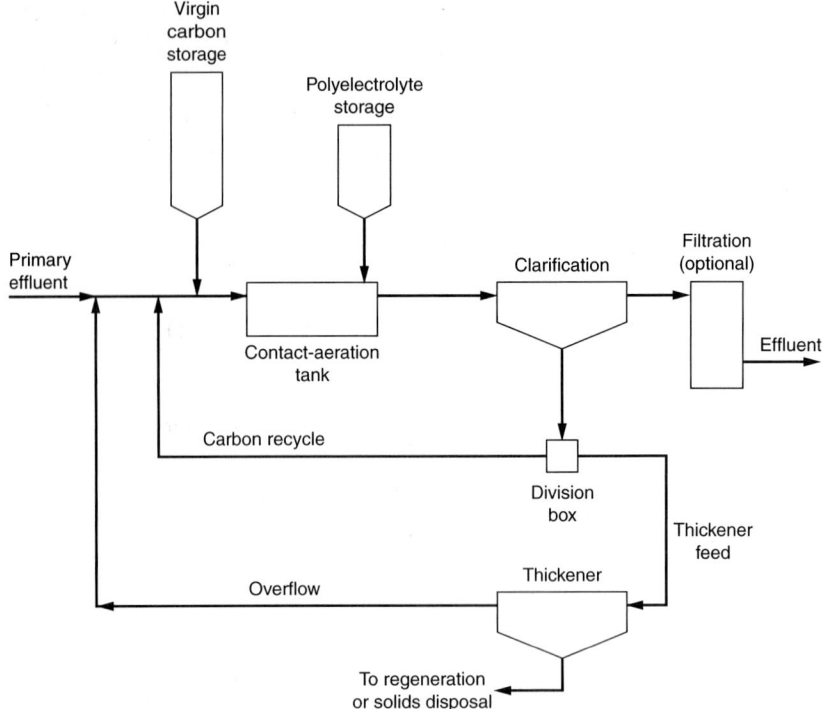

Figure 11-60

Definition sketch for the
application of powdered
activated carbon.

Reasons cited for this phenomenon include (1) additional biodegradation due to decreased toxicity, (2) degradation of normally nondegradable substances due to increased exposure time to the biomass through adsorption on the carbon, and (3) replacement of low-molecular-weight compounds with high-molecular-weight compounds, resulting in improved adsorption efficiency and lower toxicity.

11-8 GAS STRIPPING

Gas stripping involves the mass transfer of a gas from the liquid phase to the gas phase. The transfer is accomplished by contacting the liquid containing the gas that is to be stripped with a gas (usually air) that does not contain the gas initially. The removal of dissolved gases from wastewaters by gas (usually air) stripping has received considerable attention, especially for the removal of ammonia and odorous gases and volatile organic compounds (VOCs). Early work on the air stripping of ammonia from wastewater was conducted at Lake Tahoe, California (Culp and Slechta, 1966; Slechta and Culp, 1967). The removal of VOCs by aeration was considered earlier in Sec. 5–13.

The purpose of this section is to introduce the fundamental principles involved in gas stripping and to illustrate the general application of these principles. A design procedure is also presented. The material presented in this section is applicable to the removal of ammonia (NH_3), carbon dioxide (CO_2), oxygen (O_2), hydrogen sulfide (H_2S), and a variety of VOCs. The focus of the discussion in this section is on the analysis of facilities designed specifically for the removal of gaseous constituents as opposed to the removal of odorous gases (see Sec. 15–3, Chap. 15) and VOCs in aeration systems designed for the biological treatment of wastewater (see Sec. 5–13, Chap. 5).

Analysis of Gas Stripping

Important factors that must be considered in the analysis of gas stripping include (1) the characteristics of the compound(s) to be stripped, (2) the type of contactor to be used and the required number of stages, (3) the materials mass balance analysis of the stripping tower, and (4) the required physical features and dimensions of the required stripping tower.

Characteristics of the Compound(s) to Be Stripped. As noted above, the removal of volatile dissolved compounds by stripping involves contacting the liquid with a gas that does not contain the compound initially. The compound that is to be stripped will come out of solution and enter the gas phase to satisfy the Henry's law equilibrium as discussed in Chap. 2. Compounds such as benzene, toluene, and vinyl chloride that have Henry's law constants greater than 500 atm (mol H_2O/mol air) are readily strippable, compounds such as ammonia 0.75 atm (mol H_2O/mol air) and sulfur dioxide 38 atm (mol H_2O/mol air) are marginally strippable, and compounds such as acetone and methyl ethyl ketone with Henry's law constants less than 0.1 atm (mol H_2O/mol air) are essentially not strippable.

The air stripping of ammonia from wastewater requires that the ammonia be present as a gas. Ammonium ions in wastewater exist in equilibrium with gaseous ammonia, as shown in Eq. (2–38):

$$NH_4^+ \leftrightarrow NH_3 + H^+ \tag{2-38}$$

As the pH of the wastewater is increased above 7, the equilibrium is shifted to the left and the ammonium ion is converted to ammonia, which may be removed by gas stripping. The amount of lime required to raise the pH of wastewater as a function of the alkalinity is given on Fig. 6–11 in Chap. 6.

Methods Used to Contact Phases. In practice, two methods are used to achieve contact between phases so that mass transfer can occur: (1) continuous contact and (2) staged contact. As shown on Fig. 11–61, three flow patterns are used in practice: (1) countercurrent, (2) cocurrent, and (3) cross-flow. In addition, the contact medium may be fixed or mobile (Crittenden, 1999). The most common flow pattern in mass transfer operations is the countercurrent mode. A schematic and photograph of a typical gas stripping tower is shown on Fig. 11–62.

Mass-Balance Analysis for a Continuous Stripping Tower. A steady-state materials balance for the lower portion of a countercurrent continuous stripping tower used for the removal of a dissolved gas from wastewater (see Fig. 11–63) is given by

1. General word statement:

| Moles of solute entering in liquid stream | + | Moles of solute entering in gas stream | = | Moles of solute leaving in liquid stream | + | Moles of solute leaving in gas stream | (11-75) |

Figure 11–61

Typical water and airflow patterns for gas stripping towers: (a) countercurrent flow, (b) cocurrent flow, and (c) cross flow.

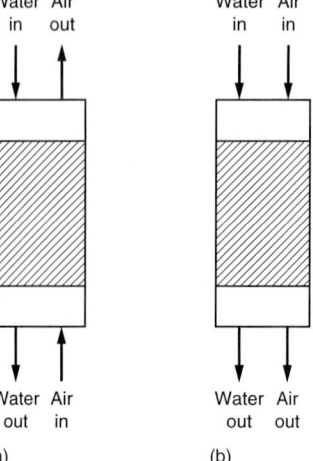

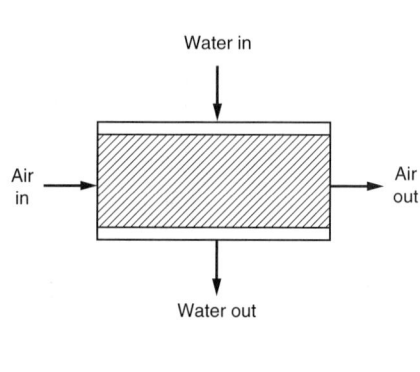

(a)

(b)

(c)

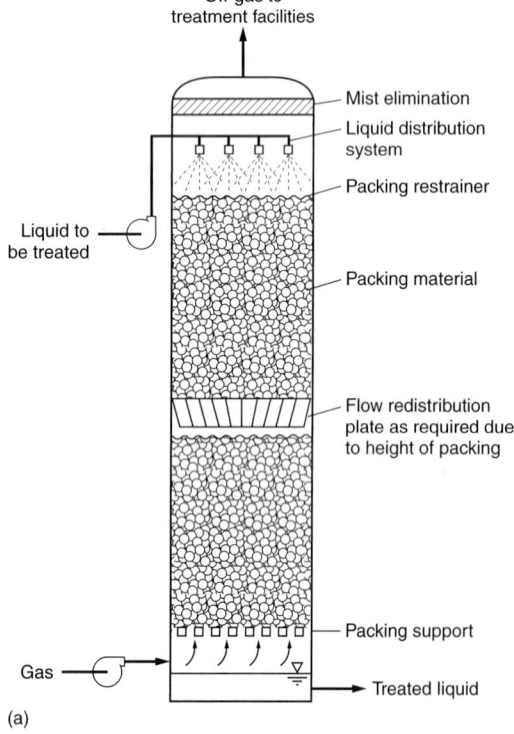

Off-gas to treatment facilities

Mist elimination

Liquid distribution system

Packing restrainer

Liquid to be treated

Packing material

Flow redistribution plate as required due to height of packing

Packing support

Gas

Treated liquid

(a)

(b)

Figure 11–62

Typical stripping towers for the removal of volatile gases from water: (a) schematic and (b) typical operating facility.

Figure 11-63

Definition sketch for the analysis of a continuous countercurrent flow gas stripping tower.

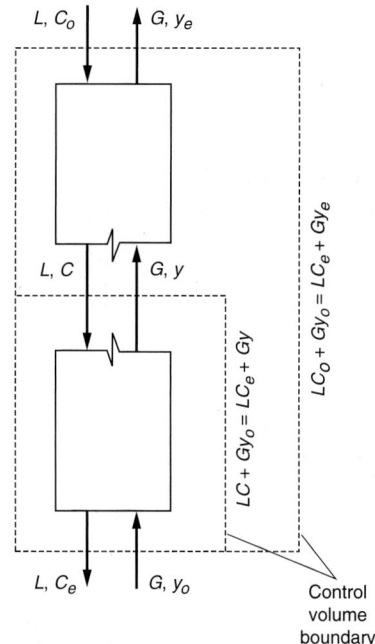

2. Simplified word statement:

Inflow = outflow   (11–76)

3. Symbolic representation (refer to Fig. 11–63):

$$LC + Gy_o = LC_e + Gy$$ (11–77)

where L = moles of incoming liquid (i.e., wastewater) per unit time

C = concentration of solute in liquid at point within the tower, moles of solute per mole of liquid

G = moles of incoming gas per unit time

y_o = concentration of solute in gas entering the bottom of the tower, moles of solute per mole of solute-free gas

C_e = concentration of solute in liquid leaving the bottom of the tower, moles of solute per mole of liquid

y = concentration of solute at a point within the tower, moles of solute per mole of solute-free gas

Combining terms, Eq. (11–77) can be written as

$$(y_o - y) = L/G(C_e - C)$$ (11–78)

If the overall tower is considered, Eq. (11–77) can be written as

$$LC_o + Gy_o = LC_e + Gy_e$$ (11–79)

Combining terms, Eq. (11–79) can be written as

$$(y_o - y_e) = L/G(C_e - C_o)$$ (11–80)

Figure 11–64

Operating lines for various gas stripping conditions: (a) general case, (b) condition when $y_o = 0$, (c) condition when $y_o = 0$ and y_e is in equilibrium with C_o, the constituent concentration in the incoming water, and (d) condition when $y_o = 0$, $C_e = 0$, and y_e is in equilibrium with C_o.

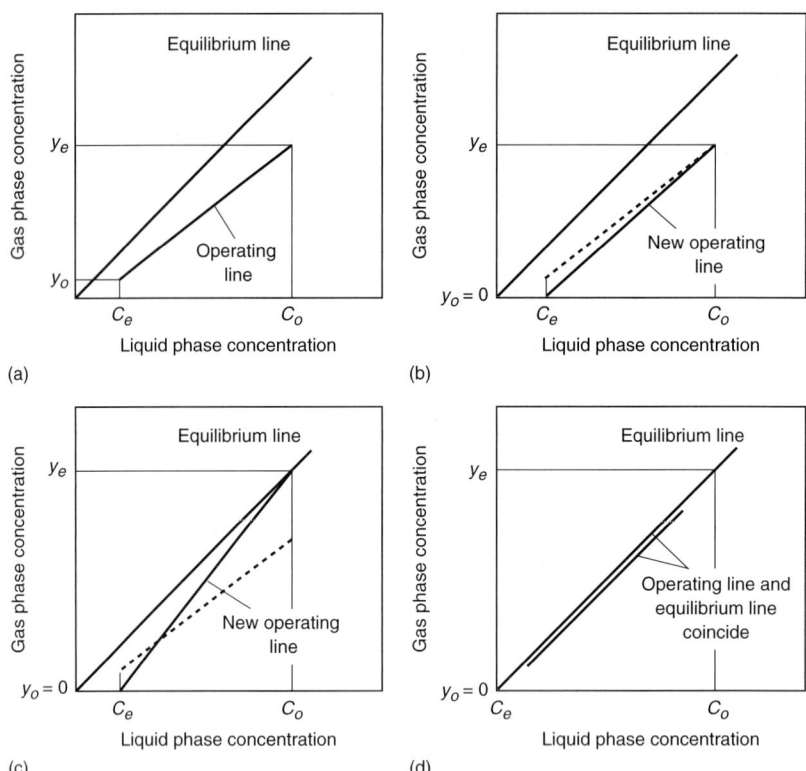

where C_o = concentration of solute in liquid entering at the top of the tower, moles of solute per mole of liquid

y_e = concentration of solute in gas leaving the top of the tower, moles of solute per mole of air

Because Eq. (11–80) is derived solely from a consideration of the equality of input and output, it holds regardless of the internal equilibria that may control the mass transfer. Equation (11–80) represents the equation of a straight line with slope L/G, which passes through the point (C_o, y_e) and point (C_e, y_o). The line passed through these two points (see Fig. 11–64a) is known as the *operating line* and represents the conditions at any point within the column. The equilibrium line is based on Henry's law. For example, equilibrium lines defined by Henry's law for ammonia as a function of temperature are presented on Fig. 11–65. It should be noted that when a gas is being stripped from solution, the operating line will lie below the equilibrium line. If a gas is being absorbed into solution the operating line will lie above the equilibrium line.

If it is assumed that the air entering the bottom of the tower contains no solute (i.e, $y_o = 0$), then Eq. (11–80) can be written as

$$y_e = L/G(C_o - C_e) \tag{11–81}$$

The operating line for the condition defined by Eq. (11–81) is shown on Fig. 11–64b.

Figure 11-65

Equilibrium curves for ammonia in water as a function of temperature based on Henry's law (see also Table 2–8 in Chap. 2).

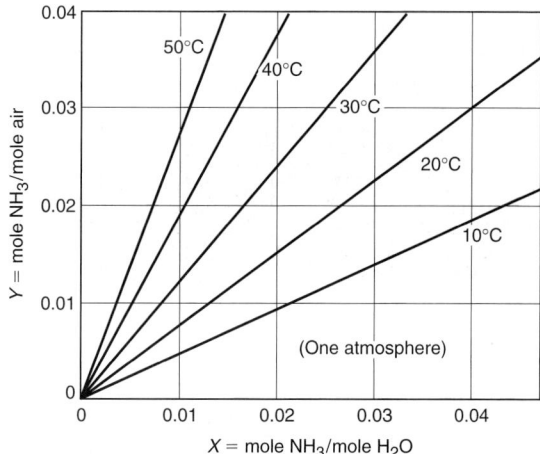

Using Henry's law [see Eq. (2–46)], y_e is defined as follows:

$$y_e = \frac{H}{P_T} C_o'$$

(11–82)

where y_e = concentration of solute in gas leaving the top of the tower, moles of solute per mole of air

H = Henry's law constant, $\dfrac{\text{atm (mole gas/mole air)}}{\text{(mole gas/mole water)}}$

P_T = total pressure, usually 1.0 atm

C_o' = concentration of solute in liquid that is in equilibrium with the gas leaving the tower, moles of solute per mole of liquid

Using Eq. (11–82), Eq. (11–81) can be written as follows:

$$C_o' = \frac{L}{G} \times \frac{P_T}{H} (C_o - C_e)$$

(11–83)

If it is assumed that the concentration of solute in the liquid entering the tower is in equilibrium with the gas leaving the tower, Eq. (11–83) can be written as

$$\frac{G}{L} = \frac{P_T}{H} \times \frac{C_o - C_e}{C_o}$$

(11–84)

The operating line for the condition defined by Eq. (11–84) is shown on Fig. 11–64c. The value of G/L (air to liquid ratio) defined by Eq. (11–84) represents the minimum amount of air that can be used for stripping for the given conditions (i.e., $y_o = 0$ and $y_e = HC_o/P_T$). In practice, from one and a half to three times the theoretical minimum air-to-liquid ratio is used to achieve effective stripping of most constituents. The application of this relationship is illustrated in Example 11–13.

Figure 11–66

Air requirements for ammonia stripping as a function of temperature.

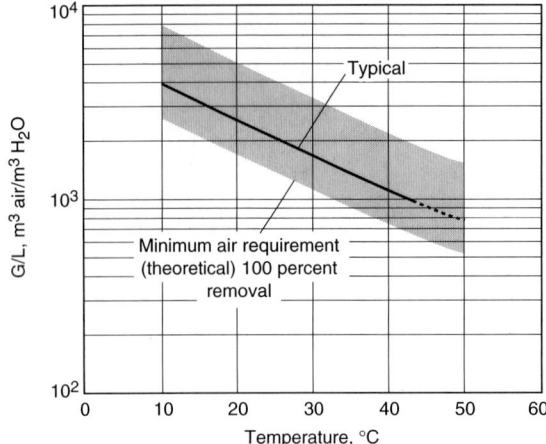

If it is assumed further that the liquid leaving and the air entering the bottom of the tower contains no solute, then Eq. (11–84) can be written as

$$\frac{G}{L} = \frac{P_T \times C_o}{H \times C_o} = \frac{P_T}{H} \tag{11-85}$$

The value of G/L for this condition corresponds to the equilibrium line defined by Henry's law (see Fig. 11–64d), and represents the theoretical minimum amount of air that can be used for stripping for the given conditions (i.e., $y_o = 0$, $C_e = 0$, and $y_e = HC_o/P_T$). The theoretical air-to-liquid ratio for stripping ammonia from wastewater for various temperatures is plotted on Fig. 11–66. The theoretical ratio is derived by assuming the process to be 100 percent efficient with a stripping tower of infinite height—obviously unachievable in practice.

EXAMPLE 11–13   **Air Requirements for Ammonia Stripping**   Determine the theoretical amount of air required at 20°C to reduce the ammonia concentration from 40 to 1 mg/L in a treated wastewater with a flowrate of 4000 m³/d. Assume the pH of the wastewater has been increased to a value of 11 (see Fig. 2–13 in Chap. 2), the Henry's constant for ammonia at 20°C is 0.75 atm, and the air entering the bottom of the tower does not contain any ammonia.

Solution

1. Determine the influent and effluent mole fractions of ammonia in the liquid using Eq. (2–2).

$$x_B = \frac{n_B}{n_A + n_B}$$

where x_B = mole fraction of solute B
n_B = number of moles of solute B
n_A = number of moles of solute A

$$C_o = \frac{[(40 \times 10^{-3})/17]}{[55.5 + (40 \times 10^{-3})/17]} = 4.24 \times 10^{-5} \text{ mole } NH_3/\text{mole } H_2O$$

$$C_e = \frac{[(1 \times 10^{-3})/17]}{[55.5 + (1 \times 10^{-3})/17]} = 1.06 \times 10^{-6} \text{ mole } NH_3/\text{mole } H_2O$$

2. Determine the effluent mole fraction of ammonia in the air leaving the tower using Eq. (11–82).

$$y_e = \frac{H}{P_T} C_o$$

$$H = \frac{(0.75 \text{ atm})(\text{mole } NH_3/\text{mole air})}{(\text{mole } NH_3/\text{mole } H_2O)} = (0.75 \text{ atm})\left(\frac{\text{mole } H_2O}{\text{mole air}}\right)$$

$$y_e = \frac{H}{P_T} \times C_o = \frac{0.75 \text{ atm}}{1.0 \text{ atm}}\left(\frac{\text{mole } H_2O}{\text{mole air}}\right) \times (4.24 \times 10^{-5} \text{ mole } NH_3/\text{mole } H_2O)$$

$$= 3.18 \times 10^{-5} \frac{\text{mole } NH_3}{\text{mole air}}$$

3. Determine the gas-to-liquid ratio using Eq. (11–84) rearranged as follows:

$$\frac{G}{L} = \frac{P_T}{H} \times \frac{(C_o - C_e)}{C_o} = \frac{(C_o - C_e)}{y_e}$$

$$\frac{G}{L} = \frac{[(4.24 \times 10^{-5} - 0.106 \times 10^{-5}) \text{ mole } NH_3/\text{mole } H_2O]}{(3.18 \times 10^{-5} \text{ mole } NH_3/\text{mole air})} = 1.3 \frac{\text{mole air}}{\text{mole } H_2O}$$

4. Convert the moles of air and water to liters of air and water.

For air at 20°C:

1.3 mole $\times$ 24.1 L/mole = 31.33 L

For water:

$$\frac{(1.0 \text{ mole } H_2O)(18 \text{ g/mole } H_2O)}{(10^3 \text{ g/L})} = 0.018 \text{ L}$$

$$\frac{G}{L} = \frac{31.33 \text{ L}}{0.018 \text{ L}} = 1741 \text{ L/L} = 1741 \text{ m}^3/\text{m}^3$$

5. Determine the total quantity of air required based on ideal conditions.

$$\text{Air required} = \frac{(1741 \text{ m}^3/\text{m}^3)(4000 \text{ m}^3/\text{d})}{(1440 \text{ min/d})} = 4835 \text{ m}^3/\text{min}$$

Comment The approximate amount of lime required to raise the pH of wastewater to a value of 11 is given on Fig. 6–11 in Chap. 6. The procedure followed to determine the height of the stripping tower is illustrated in Example 11–14, presented later in this section.

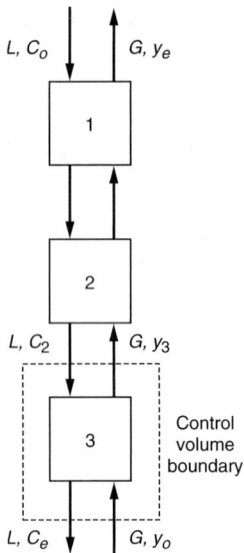

Figure 11–67

Definition sketch for the analysis of a three-stage countercurrent flow gas stripping tower.

Mass-Balance Analysis of a Multistage Stripping Tower. In the analysis of stripping towers, reference is often made to the number of ideal stages required for stripping. The analysis for the number of stages is analogous to the simulation of plug flow with a series of complete-mix reactors as detailed in Chap. 4. Separate stages are used to improve performance of stripping towers. Equilibrium conditions are assumed in each stage of the tower. A steady-state materials balance for the lower portion of a countercurrent staged stripping tower (see Fig. 11–67) is given by

Inflow $=$ outflow

$$LC_2 + Gy_o = LC_e + Gy_3 \qquad (11\text{–}86)$$

or

$$(y_3 - y_o) = L/G(C_2 - C_e) \qquad (11\text{–}87)$$

If an overall mass balance is performed around the tower, the resulting equation is the same as that derived above for the continuous stripping tower [see Eq. (11–79)].

In the 1920s, McCabe and Thiele (1925) developed a graphical procedure for determining the required number of ideal stages. The method is illustrated on Fig. 11–68 for a stripping tower comprised of three stages. The operating line for the three stages is also shown on Fig. 11–68. The number of ideal stages required for stripping of a constituent is obtained as follows. The point C_o, y_e represents the airstream leaving and the water stream entering the top of the stripping column. The composition of the liquid in equilibrium with the constituent concentration in the airstream is found by extending a horizontal line from the point C_o, y_e to the point C_1, y_e. From the point C_1, y_e the value of y_2, the air entering stage 1 from stage 2, is obtained from the equation of the operating line. If a materials mass balance is performed between stages 1 and 2, the resulting expression for y_2 is

$$y_2 = \frac{L}{G}C_1 + \frac{Gy_e - LC_o}{G} \qquad (11\text{–}88)$$

Figure 11–68

Operating line for three-stage countercurrent flow gas stripping tower.

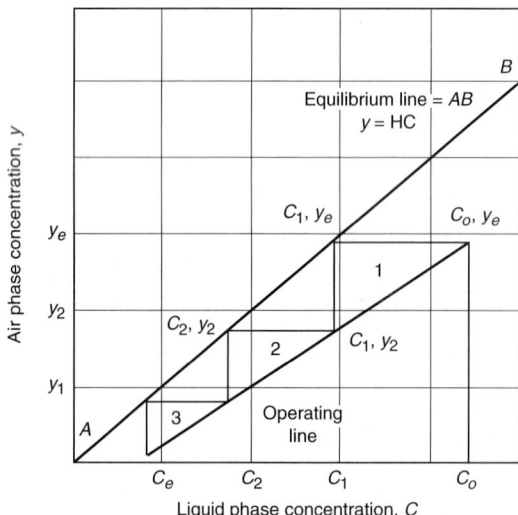

The value of y_2 is obtained by drawing a vertical line from point C_1, y_e to the operating line at point C_1, y_2 as shown on Fig. 11–68. In a similar manner the value of C_2 is obtained by drawing a horizontal line from the point C_1, y_2 to the equilibrium line. This procedure is repeated until the point C_n, y_{n+1} is reached. The number of ideal stages is typically a fractional number (e.g., 4.2, 5.6). In practice, the number of stages is rounded to the next whole number.

Determination of Height of Stripping Tower. The purpose of the following analysis is to illustrate how the height of the packing in the stripping tower is determined, based on an analysis of the mass transfer occurring within the tower. A mass balance performed on the liquid phase within the stripping tower shown on Fig. 11–69 is as follows.

Simplified word statement:

Accumulation = inflow − outflow + volatilization

$$\frac{\partial C}{\partial t} \Delta V = LC|_{z+\Delta z} - LC|_z + r_v \Delta V \tag{11-89}$$

where $\partial C / \partial t$ = change in concentration with time, g/m³·s
ΔV = differential volume, m³
Δz = differential height, m
L = liquid volumetric flowrate, m³/s
C = concentration of constituent C, g/m³
r_v = rate of mass transfer per unit volume per unit time C, g/m³·s

Figure 11–69

Definition sketch for the analysis of mass transfer from the liquid phase to the gas phase within a stripping tower. *Note:* the packing material is not shown. *(Adapted from Hand et al., 1999.)*

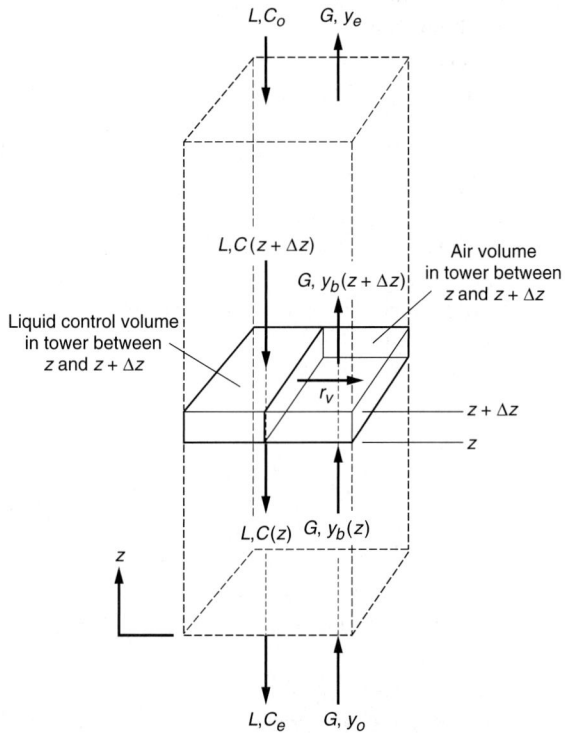

Substituting the area times the differential height ($A\Delta z$) for the differential volume (ΔV) and writing the differential form for the term $LC|_{z+\Delta z}$ in Eq. (11–89) results in the following expression:

$$\frac{\partial C}{\partial t} A\Delta z = L\left(C + \frac{\Delta C}{\Delta z}\Delta z\right) - LC + r_v A\Delta z \tag{11–90}$$

Simplifying Eq. (11–90) and taking the limit as Δz approaches zero yields

$$\frac{\partial C}{\partial t} = \frac{L}{A}\frac{\partial C}{\partial z} + r_v \tag{11–91}$$

The rate of mass transfer due to volatilization as described in Chap. 4 [see Eq. (4–127)] is given by:

$$r_v = -K_L a(C_b - C_s) \tag{11–92}$$

where r_v = rate of mass transfer per unit volume per unit time, g/m³·s
$K_L a$ = volumetric mass transfer coefficient which depends on water-quality characteristics and temperature, 1/s
C_b = concentration in liquid bulk phase at time t, g/m³
C_s = concentration in liquid in equilibrium with gas as given by Henry's law, g/m³

Assuming steady-state conditions within the tower ($\partial C/\partial t = 0$) and substituting Eq. (11–92) for r_v, Eq. (11–91) can now be written as

$$\frac{dC_b}{dz} = \frac{K_L aA}{L}(C_b - C_s) \tag{11–93}$$

The height of the tower packing can be obtained by integrating the above expression:

$$\int_0^Z dz = \frac{L}{K_L aA}\int_{C_e}^{C_o}\frac{dC_b}{(C_b - C_s)} \tag{11–94}$$

To integrate the right-hand side of the above equation, a relationship must be found between C_b and C_s because C_s is changing continuously throughout the height of the tower. From Henry's law, the value of C_s is given by:

$$C_s = \frac{P_T}{H}y \tag{11–95}$$

Substituting a modified form of Eq. (11–81) for y in Eq. (11–95) yields:

$$C_s = \frac{L}{G} \times \frac{P_T}{H}(C_b - C_e) \tag{11–96}$$

If Eq. (11–96) is substituted into Eq. (11–94) and the resulting expression is integrated, the following expression is obtained (Hand et al., 1999):

$$Z = \frac{L}{K_L aA}\left[\frac{C_o - C_e}{C_o - C_e - C_o'}\right]\ln\left[\frac{C_o - C_o'}{C_e}\right] \tag{11–97}$$

where

$$C'_o = \frac{L}{G} \times \frac{P_T}{H} (C_o - C_e) \tag{11-98}$$

It should be noted that if $C'_o = C_o$, Eq. (11–98) is the same as Eq. (11–84).

Design Equations for Stripping Towers. Utilizing the above equations, a number of process models and design equations have been developed. Equations that can be used to determine the height of a stripping tower are as follows:

$$Z = \text{HTU} \times \text{NTU} \tag{11-99}$$

where Z = height of stripping tower packing, m
 HTU = height of a transfer unit, m
 NTU = number of transfer units

The height of a transfer unit is defined as

$$\text{HTU} = \frac{L}{K_L a A} \tag{11-100}$$

where L = liquid volumetric flowrate, m³/s
 $K_L a$ = volumetric mass transfer coefficient, 1/s
 A = cross-sectional area of tower, m²

The HTU is a measure of the mass transfer characteristics of the packing medium.
 The number of transfer units is defined as

$$\text{NTU} = \left[\frac{C_o - C_e}{C_o - C_e - C'_o} \right] \ln \left[\frac{C_o - C'_o}{C_e} \right] \tag{11-101}$$

Substituting Eq. (11–98) in Eq. (11–101) yields

$$\text{NTU} = \left[\frac{S}{S - 1} \right] \ln \left[\frac{(C_o/C_e)(S - 1) + 1}{S} \right] \tag{11-102}$$

where S is known as the stripping factor and is defined as

$$S = \frac{G}{L} \times \frac{H}{P_T} \tag{11-103}$$

A value of $S = 1$ corresponds to the minimum amount of air required for stripping. When $S > 1$ the amount of air is in excess and complete stripping is possible given a tower of infinite height. When $S < 1$, there is insufficient air for stripping. In practice, stripping factors vary from 1.5 to 5.0.
 Values for $K_L a$ for specific compounds are best obtained from pilot-plant studies or by using empirical correlations such as given in Eq. (5–62) in Chap. 5, and repeated here for convenience as Eq. (11–104). It should be noted that many other relationships have been proposed in the literature (Sherwood and Hollaway, 1940, and Onda et al., 1968).

$$K_L a_{\text{VOC}} = K_L a_{O_2} \left(\frac{D_{\text{VOC}}}{D_{O_2}} \right)^n \tag{11-104}$$

where $K_L a_{VOC}$ = system mass transfer coefficient, 1/h

$\qquad K_L a_{O_2}$ = system oxygen mass transfer coefficient, 1/h

$\qquad D_{VOC}$ = diffusion coefficient of VOC in water, cm²/s

$\qquad D_{O_2}$ = diffusion coefficient of oxygen in water, cm²/s

$\qquad n$ = coefficient (0.5 for stripping towers)

Air and water temperature are significant factors in the design of stripping towers because of their effect on air and water viscosities, Henry's law constants, and volumetric mass transfer coefficients. The effect of temperature on the Henry's law constant is illustrated on Fig. 11–65. The value of $K_L a$ can be adjusted for temperature effects using Eq. (2–25) with a theta value of 1.024.

Design of Stripping Towers

In its simplest form a stripping tower consists of a tower (usually circular), a support plate for the packing material, a distribution system for the liquid to be stripped, located above the packing material, and an air supply located at the bottom of the stripping tower (see Fig. 11–62). The process design variables include (1) the type of packing material, (2) the stripping factor, (3) the cross-sectional area of the tower, and (4) the height of the stripping tower. The cross-sectional area will depend on the pressure drop through the packing. Representative design values for stripping of VOCs and ammonia are presented in Table 11–35. The significant difference in the amount of air required for stripping is a clear illustration of the importance of the Henry's constant.

The headloss through the packing is determined using a generalized gas pressure-drop relationship such as that plotted on Fig. 11–70 (Eckert, 1975). The pressure drop is expressed in newtons per square meter per meter of depth (N/m²)/m. The upper line on Fig. 11–70 labeled *approximate flooding* represents the condition that occurs when the amount of water and air applied is so great that the pore spaces fill to the point where water starts to flood within the tower. The units for the x and y axes are as follows:

x axis:

$$x = \frac{L'}{G'}\left(\frac{\rho_G}{\rho_L - \rho_G}\right)^{1/2} \approx \frac{L'}{G'}\left(\frac{\rho_G}{\rho_L}\right)^{1/2} \qquad (11\text{-}105)$$

y axis:

$$y = \frac{(G')^2(C_f)(\mu_L)^{0.1}}{(\rho_G)(\rho_L - \rho_G)} \qquad (11\text{-}106)$$

which can be rewritten as follows:

$$G' = \left[\frac{(\text{value from } y \text{ axis})(\rho_G)(\rho_L - \rho_G)}{(C_f)(\mu_L)^{0.1}}\right]^{1/2} \qquad (11\text{-}107)$$

where L' = liquid loading rate, kg/m²·s

$\qquad G'$ = gas loading rate, kg/ m²·s

$\qquad \rho_G$ = density of gas, kg/m³

$\qquad \rho_L$ = density of liquid, kg/m³

$\qquad C_f$ = packing factor for packing material, unitless

$\qquad \mu_L$ = viscosity of liquid, N/m²·s

Table 11–35
Typical design parameters for stripping towers for the removal of VOC and ammonia[a]

Item	Symbol	Unit	VOC removal[b]	Ammonia removal[c]
Liquid loading rate		L/m²·min	600–1800	40–80
Air-to-liquid ratio[d]	G/L	m³/m³	20–60:1	2000–6000:1
Stripping factor	S	unitless	1.5–5.0	1.5–5.0
Allowable air pressure drop	ΔP	(N/m²)/m	100–400	100–400
Height-to-diameter ratio	Z/D	m/m	≤10:1	≤10:1
Packing depth[e]	Z	m	1–6	2–6
Factor of safety	SF	%D, %Z	20–50	20–50
Wastewater pH	pH	Unitless	5.5–8.5	10.8–11.5
Approximate packing factors				
Pall rings, Intalox saddles	C_f	12.5 mm[f]	180–240	180–240
	C_f	25 mm[f]	30–60	30–60
	C_f	50 mm[f]	20–25	20–25
Berl saddles, Raschig rings	C_f	12.5 mm[f]	300–600	300–600
	C_f	25 mm[f]	120–160	120–160
	C_f	50 mm[f]	45–60	45–60

[a] Adapted in part from Eckert (1970, 1975), Kavanaugh and Trussell (1980), and Hand et al. (1999).
[b] Typical data for VOCs with Henry's constants greater than 500 atm (mole H_2O/mole air).
[c] Ammonia with a Henry's constant of 0.75 atm (mole H_2O/mole air) is considered only marginally strippable, which accounts for the low loading rate and high air-to-liquid ratio.
[d] Ratio is highly temperature-dependent.
[e] For packing depths greater than 5 to 6 m, redistribution of the liquid flow is recommended (see Fig. 11–62a).
[f] Size of packing material.

The packing factor C_f depends on the type and size of the packing. Typical ranges for packing factors that can be used for preliminary assessments are reported in Table 11–35. For more detailed design calculations, current values should be obtained from manufacturers.

To use Fig. 11–70, select a value for L'/G' and compute the corresponding x value. Enter the plot at the computed value of x and move vertically upward to a preselected pressure-drop line. Move horizontally from the point of intersection to the y axis and note the value on the y axis. Using the y-axis value, determine the gas loading rate G' using Eq. (11–107) and the corresponding liquid loading rate L'. To determine the required cross-sectional area the liquid flowrate is divided by the liquid loading rate.

A generalized analysis procedure is as follows:

1. Select a packing material and its corresponding packing factor for use in Eq. (11–105).
2. Select several stripping factors for successive trials (e.g., 2.5, 3.0, 4.0)

Figure 11–70

Generalized pressure-drop curves for packed-bed stripping towers. This chart should only be used with dissolved compounds with Henry's constants ranging from 2 to 500 atm, or 0.002 to 0.4 for unitless values of Henry's constant. (From Eckert, 1975.)

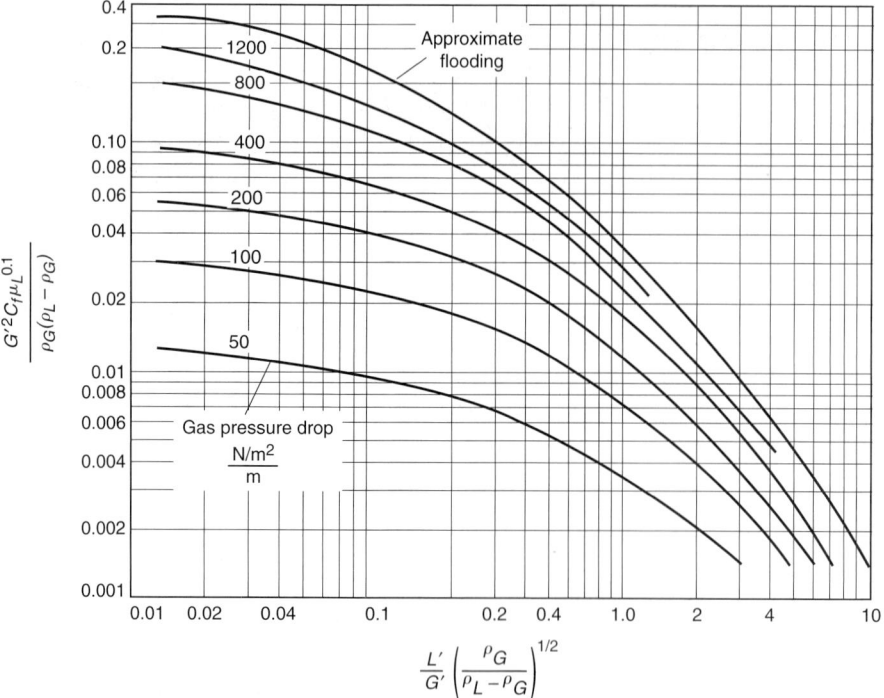

3. Select acceptable pressure drop ΔP (typically a function of the packing material selected).
4. Determine the cross-sectional area of the tower, based on the allowable pressure drop using the data presented on Fig. 11–70 or other appropriate relationships.
5. Determine the height of the transfer units using Eq. (11–100). To apply Eq. (11–100), the value of K_La must be known or estimated using Eq. (11–104).
6. Determine the number of transfer units using Eq. (11–102).
7. Determine the height of the stripping tower using Eq. (11–99).

The design procedure outlined above is illustrated in Example 11–14. Representative design values for stripping towers are given in Table 11–35. To evaluate the stripping process more thoroughly, any of the commercially available software packages can be used.

EXAMPLE 11–14   Determination of Height of Stripping Tower for the Removal of Chloroform
Determine the diameter and height of the stripping tower required to remove chloroform (trichloromethane, $CHCl_3$) from a disinfected effluent. Assume 50 mm Pall rings will be used as the packing material in the stripping tower. The chloroform concentration in a treated wastewater flow of 4000 m^3/d is to be reduced from 150 to 20 $\mu g/L$. Assume that the Henry's constant for chloroform at 20°C is 172 atm (see trichloromethane, Table 5-36), and the air entering the bottom of the tower does not contain any chloroform. Assume the K_La value for chloroform is 0.0120 s^{-1}.

Solution

1. Select a packing material and corresponding packing factor. For 50 mm Pall rings (specified), assume a packing factor of 20 (see Table 11–35).
2. Select a stripping factor. Assume a stripping factor of 3 (see Table 11–35).
3. Select an acceptable pressure drop. Assume a pressure drop of 100 (N/m²)/m (see Table 11–35).
4. Determine the cross-sectional area of the stripping tower using the pressure drop plot given on Fig. 11–70.

   a. Determine the value of the x axis for a stripping factor of 3.

   $$S = \frac{G}{L} \times \frac{H}{P_T} = \frac{G}{L} \times \frac{1}{1.0 \text{ atm}} \left[\frac{172 \text{ atm (mole gas/mole air)}}{(\text{mole gas/mole water})} \right]$$

   $$S = \frac{G}{L} \times \frac{1}{1.0 \text{ atm}} \left[\frac{172 \text{ atm (mole water)(18 g/mole water)}}{(\text{mole air})(28.8 \text{ g/mole air})} \right]$$

   $$= 107.6 \frac{G \text{ g}}{L \text{ g}} = 107.6 \frac{G' \text{ kg}}{L' \text{ kg}}$$

   $$\frac{L'}{G'} = \frac{(107.6 \text{ kg/kg})}{3} = 35.9$$

   $$\frac{L'}{G'} \left(\frac{\rho_G}{\rho_L - \rho_G} \right)^{1/2} \approx \frac{L'}{G'} \left(\frac{\rho_G}{\rho_L} \right)^{1/2} = (35.9 \text{ kg/kg}) \left[\frac{(1.204 \text{ kg/m}^3)}{(998.2 \text{ kg/m}^3)} \right]^{1/2} = 1.25$$

   b. Determine the corresponding value on the y axis. For an x axis value of 1.25 and a pressure drop of 100 (N/m²)/m, the y axis value is 0.006.

   c. Determine the loading rate using Eq. (11–107).

   $$G' = \left[\frac{(\text{value from } y \text{ axis})(\rho_G)(\rho_L - \rho_G)}{(C_f)(\mu_L)^{0.1}} \right]^{1/2}$$

   $$G' = \left[\frac{(0.006)(1.204)(998.2 - 1.204)}{(20)(0.001)^{0.1}} \right]^{1/2} = 0.85 \text{ kg/m}^2 \cdot \text{s}$$

   $$L' = 35.9 \, G' = 35.9 \times 0.85 \text{ kg/m}^2 \cdot \text{s} = 30.1 \text{ kg/m}^2 \cdot \text{s}$$

   d. Substitute known values and solve for the area of the tower.

   $$A = \frac{L \rho_L}{L'} = \frac{(4000 \text{ m}^3/\text{d})(998.2 \text{ kg/m}^3)}{(30.1 \text{ kg/m}^2 \cdot \text{s})} = 1.53 \text{ m}^2$$

5. Determine the height of the transfer unit using Eq. (11–100).

   $$\text{HTU} = \frac{L}{K_L a \, A}$$

   $$\text{HTU} = \frac{L}{K_L a \, A} = \frac{(4000 \text{ m}^3/\text{d})(1 \text{ d}/86,400 \text{ s})}{(0.0120/\text{s})1.53 \text{ m}^2} = 2.52 \text{ m}$$

6. Determine the number of transfer units using Eq. (11–102).

$$\text{NTU} = \left[\frac{S}{S-1} \right] \ln \left[\frac{(C_o/C_e)(S-1)+1}{S} \right]$$

$$\text{NTU} = \left[\frac{3}{3-1} \right] \ln \left[\frac{(150/20)(3-1)+1}{3} \right] = 2.51$$

7. Determine the height of the stripping tower packing using Eq. (11–99).

$$Z = \text{HTU} \times \text{NTU} = 2.52 \times 2.51 = 6.3 \text{ m}$$

Comment In this example, the value of $K_L a$ for chloroform was known. Quite often the required $K_L a$ value must be determined in the field, using pilot-scale facilities. Alternatively Eq. (11–104) can be used to estimate a value for $K_L a$. In some cases, data from the literature or from manufacturers may be used to obtain preliminary sizing. To optimize the design, various stripping ratios must be evaluated. Optimization is best accomplished using one of the commercially available software packages.

Application

As noted previously, air stripping is used to remove ammonia (NH_3), carbon dioxide (CO_2), oxygen (O_2), hydrogen sulfide (H_2S), and a variety of VOCs. The application of air stripping for the removal of ammonia is considered briefly in the following discussion. The removal and treatment of VOCs in aeration systems is considered in Sec. 5–13 in Chap. 5. The VOC management alternatives considered in Chap. 5 are also applicable for the treatment of VOCs removed with a stripping tower. The removal and treatment of odorous compounds are considered in Sec. 15–3 in Chap. 15.

Removal of Ammonia from Wastewater. In general, removal efficiency depends on the temperature, size, and proportions of the facility, and the efficiency of the air-water contact. If the removal of ammonia is unsatisfactory, the tower has not been designed correctly or is overloaded. In this case, additional air volume may improve operation. As the temperature decreases, the amount of air required increases significantly for the same degree of removal (see Fig. 11–66). A typical flow diagram for the removal of ammonia from wastewater by air stripping is shown on Fig. 11–71. The ammonia stripped from the wastewater is converted to the ammonium ion using an acid scrubber.

In most cases where ammonia stripping has been applied, a number of operating problems have developed, the most serious being (1) maintaining the required pH for effective stripping, (2) calcium carbonate scaling within the tower and feed lines, and (3) poor performance during cold weather operation. Maintaining the required pH is a control problem that can be managed with multiple sensors. The amount and nature (soft to extremely hard) of the calcium carbonate scale formed varies with the characteristics of the wastewater and local environmental conditions, and cannot be predicted a priori. Under conditions of icing, the liquid-air contact geometry in the tower is altered, which further reduces the overall efficiency. The best solution for cold weather conditions is to enclose the stripper.

Removal of Ammonia from Digester Supernatant by Steam Stripping. Steam stripping of ammonia is similar to air stripping, with the exception that the process requires temperatures in excess of 95°C (200°F). Although costly, steam stripping has been proposed for the removal of ammonia from digester supernatant in large treatment plants, where separate facilities are used for the treatment of return flows (see discussion in Chap. 14). The flow diagram for the removal of ammonia from dewatering centrate is shown on Fig. 11–72. The ammonia stripped from the supernatant is converted to the ammonium ion by passing the off-gas through an acid bath. Acid scrubbers are also used.

In a large-scale test, two processes were evaluated for the removal of ammonia from centrate: steam stripping and hot air stripping. Steam stripping proved to be the most cost-effective. Based on the results of recent full-scale testing, it has been found that pretreatment of the centrate is essential for effective steam stripping of ammonia. Pretreatment consisting of settling followed by straining proved to be effective. Using a gas-to-liquid ratio of 300 to 1, it was possible to achieve ammonia removals of about 88 percent, at pH values of 7 to 7.5. The influent ammonia concentrations values were on the order of 500 to 600 mg/L (Gopalakrishnan et al., 2000).

Operating problems that have been encountered with steam stripping include (1) extensive fouling (iron deposits, for example) within the heat exchanger and in the stripper due to the presence of waste constituents at elevated temperatures, (2) maintaining the required pH for effective stripping, (3) controlling the steam flow, and

Figure 11–71

Typical flow diagram for the air stripping of ammonia from wastewater.

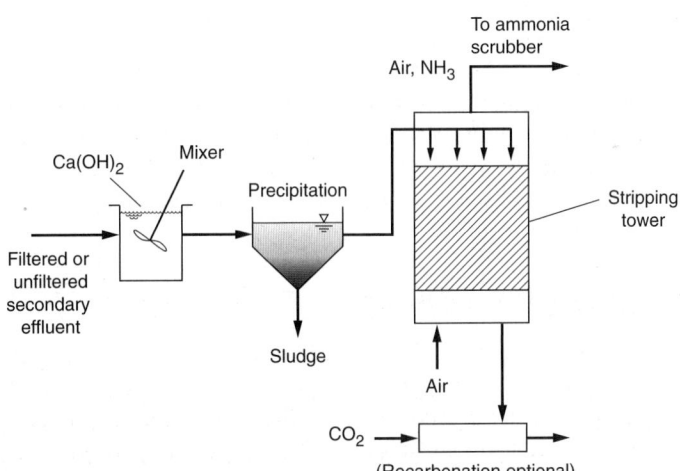

(a)   (b)

Figure 11-72

Stream stripping of ammonia from centrate from dewatering operations: (a) typical flow diagram, and (b) view of pilot-plant facility. *(Courtesy of The New York City Department of Environmental Protection.)*

(4) maintaining the stripping tower temperature. Because of the importance of temperature, steam stripping should be carried out in enclosed facilities. Spiral-wound heat exchangers have proved to be effective. In Europe, an acid wash is used to clean the piping, the heat exchanger, and the stripping column.

11-9   ION EXCHANGE

Ion exchange is a unit process in which ions of a given species are displaced from an insoluble exchange material by ions of a different species in solution. The most widespread use of this process is in domestic water softening, where sodium ions from a cationic-exchange resin replace the calcium and magnesium ions in the treated water, thus reducing the hardness. Ion exchange has been used in wastewater applications for the removal of nitrogen, heavy metals, and total dissolved solids.

Ion-exchange processes can be operated in a batch or continuous mode. In a batch process, the resin is stirred with the water to be treated in a reactor until the reaction is complete. The spent resin is removed by settling and subsequently is regenerated and reused. In a continuous process, the exchange material is placed in a bed or a packed column, and the water to be treated is passed through it. Continuous ion exchangers are usually of the downflow, packed-bed column type. Wastewater enters the top of the column under pressure, passes downward through the resin bed, and is removed at the bottom. When the resin capacity is exhausted, the column is backwashed to remove trapped solids and is then regenerated. Two examples of commercial ion-exchange reactors are shown on Fig. 11–73.

Ion-Exchange Materials

Naturally occurring ion-exchange materials, known as zeolites, are used for water softening and ammonium ion removal. Zeolites used for water softening are complex aluminosilicates with sodium as the mobile ion. Ammonium exchange is accomplished using a naturally occurring zeolite clinoptilolite. Synthetic aluminosilicates are manufactured, but most synthetic ion-exchange materials are resins or phenolic polymers.

Figure 11–73

Two examples of full-scale ion-exchange installations: (a) large downflow packed-bed columns, and (b) ion-exchange canisters on a rotating platform. The canisters are rotated so that one canister can be regenerated while the others are in operation.

(a)

(b)

Table 11–36
Classification of ion-exchange resins[a]

Type of resin	Characteristics
Strong-acid cation resins	Strong-acid resins behave in a manner similar to a strong acid, and are highly ionized in both the acid (R—SO_3H) and salt (R—SO_3Na) form, over the entire pH range
Weak-acid cation resins	Weak-acid cation exchangers have a weak-acid functional group (—COOH), typically a carboxylic group. These resins behave like weak organic acids that are weakly dissociated
Strong-base anion resins	Strong-base resins are highly ionized, having strong-base functional groups such as (OH), and can be used over the entire pH range. These resins are used in the hydroxide (OH) form for water deionization
Weak-base anion resins	Weak-base resins have weak-base functional groups in which the degree of ionization is dependent on pH
Heavy-metal selective chelating resins	Chelating resins behave like weak-acid cation resins but exhibit a high degree of selectivity for heavy-metal cations. The functional group in most of these resins is EDTA, and the resin structure in the sodium form is R—EDTA—Na

[a]Adapted in part from Eckenfelder (2000).

Five types of synthetic ion-exchange resins are in use: (1) strong-acid cation, (2) weak-acid cation, (3) strong-base anion, (4) weak-base anion, and (5) heavy-metal selective chelating resins. The properties of these resins are summarized in Table 11–36.

Most synthetic ion-exchange resins are manufactured by a process in which styrene and divinylbenzene are copolymerized. The styrene serves as the basic matrix of the resin, and divinylbenzene is used to cross-link the polymers to produce an insoluble tough resin. Important properties of ion-exchange resins include exchange capacity, particle size, and stability. The exchange capacity of a resin is defined as the quantity of an exchangeable ion that can be taken up. The exchange capacity of resins is expressed as eq/L or eq/kg (meq/L or meq/g). The particle size of a resin is important with respect to the hydraulics of the ion-exchange column and the kinetics of ion exchange. In general, the rate of exchange is proportional to the inverse of the square of the particle diameter. The stability of a resin is important to the long-term performance of the resin. Excessive osmotic swelling and shrinking, chemical degradation, and structural changes in the resin caused by physical stresses are important factors that may limit the useful life of a resin.

Typical Ion-Exchange Reactions

Typical ion-exchange reactions for natural and synthetic ion-exchange materials are given below.

For natural zeolites (Z):

$$ZNa_2 + \begin{bmatrix} Ca^{2+} \\ Mg^{2+} \\ Fe^{2+} \end{bmatrix} \leftrightarrow Z \begin{bmatrix} Ca^{2+} \\ Mg^{2+} \\ Fe^{2+} \end{bmatrix} + 2Na^+ \qquad (11\text{--}108)$$

For synthetic resins (R):

Strong-acid cation exchange:

$$RSO_3H + Na^+ \leftrightarrow RSO_3Na + H^+ \tag{11-109}$$

$$2RSO_3Na + Ca^{2+} \leftrightarrow (RSO_3)_2Ca + 2Na^+ \tag{11-110}$$

Weak-acid cation exchange:

$$RCOOH + Na^+ \leftrightarrow RCOONa + H^+ \tag{11-111}$$

$$2RCOONa + Ca^{2+} \leftrightarrow (RCOO)_2Ca + 2Na^+ \tag{11-112}$$

Strong-base anion exchange:

$$RR_3'NOH + Cl^- \leftrightarrow RR_3'NCl + OH^- \tag{11-113}$$

Weak-base anion exchange:

$$RNH_3OH + Cl^- \leftrightarrow RNH_3Cl + OH^- \tag{11-114}$$

$$2RNH_3Cl + SO_4^{2-} \leftrightarrow (RNH_3)_2\,SO_4 + 2Cl^- \tag{11-115}$$

Exchange Capacity of Ion-Exchange Resins

Reported exchange capacities vary with the type and concentration of regenerant used to restore the resin. Typical synthetic resin exchange capacities are in the range of 2 to 10 eq/kg of resin; zeolite cation exchangers have exchange capacities of 0.05 to 0.1 eq/kg. Exchange capacity is measured by placing the resin in a known form. A cationic resin would be washed with a strong acid to place all of the exchange sites on the resin in the H^+ form or washed with a strong NaCl brine to place all of the exchange sites in the Na^+ form. A solution of known concentration of an exchangeable ion (e.g., Ca^{2+}) can then be added until exchange is complete and the amount of exchange capacity can be measured, or in the acid case, the resin is titrated with a strong base. Determination of the capacity of an ion resin based on titration is illustrated in Example 11–15.

Exchange capacities for resins often are expressed in terms of grams $CaCO_3$ per cubic meter of resin (g/m^3) or gram equivalents per cubic meter (g eq/m^3). Conversion between these two units is accomplished using the following expression:

$$\frac{1 \text{ eq}}{m^3} = \frac{(1 \text{ eq})(50 \text{ g } CaCO_3/\text{eq})}{m^3} = 50 \text{ g } CaCO_3/m^3 \tag{11-116}$$

Calculation of the required resin volume for an ion-exchange process is also illustrated in Example 11–15.

EXAMPLE 11–15   **Determination of Ion-Exchange Capacity for a New Resin**   A column study was conducted to determine the capacity of a cation-exchange resin. In conducting the study, 0.1 kg of resin was washed with NaCl until the resin was in the R-Na form. The column was then washed with distilled water to remove the chloride ion (Cl^-) from the interstices of the resin. The resin was then titrated with a solution of calcium chloride ($CaCl_2$), and the concentrations of chloride and calcium were measured at various throughput volumes. The measured concentrations of Cl^- and Ca^{2+} and the corresponding throughput volumes are as given below. Using the data given below, determine the exchange capacity of the resin and the mass and volume of a resin required to treat 4000 m³ of water containing 18 mg/L of ammonium ion NH_4^+. Assume the density of the resin is 700 kg/m³.

Throughput volume, L	Constituent, mg/L	
	Cl^-	Ca^{2+}
2	0	0
3	Trace	0
5	7	0
6	18	0
10	65	0
12	71	Trace
20	71	13
26	71	32
28	71	38
32	$C_o = 71$	$C_o = 40$

Solution

1. Prepare a plot of the normalized concentrations of Cl^- and Ca^{2+} as a function of the throughput volume. The required plot is given below.

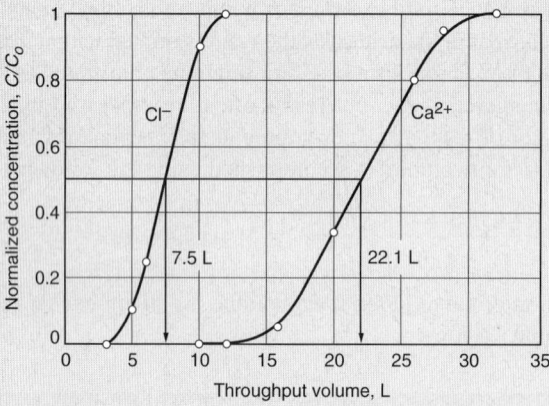

2. Determine the exchange capacity. The exchange capacity (EC) of the resin in meq/kg is

$$EC = \frac{VC_o}{R}$$

where V = throughput volume between the Cl^- and Ca^{2+} breakthrough curves
at $C/C_o = 0.5$

C_o = calcium concentration in meq/L

R = amount of resin in kg

$$EC = \frac{(22.1\ L - 7.5\ L)\left[\dfrac{(40\ mg/L)}{(20\ mg/meq)}\right]}{0.1\ kg\ of\ resin} = 292\ meq/kg\ of\ resin$$

3. Determine the mass and volume of resin required to treat 4000 m³ of water containing 18 mg/L of ammonium ion NH_4^+.

   a. Determine the meq of NH_4^+.

   $$NH_4^+,\ meq/L = \frac{(18\ mg/L\ as\ NH_4^+)}{(18\ mg/meq)} = 1\ meq/L$$

   b. The required exchange capacity is equal to

   $$(1.0\ meq/L)(4000\ m^3)(10^3\ L/m^3) = 4 \times 10^6\ meq$$

   c. The required mass of resin is

   $$R_{mass},\ kg = \frac{4 \times 10^6\ meq}{(292\ meq/kg\ of\ resin)} = 13{,}700\ kg\ of\ resin$$

   d. The required volume of resin is

   $$R_{vol},\ m^3 = \frac{13{,}700\ kg\ of\ resin}{(700\ kg/m^3)} = 19.6\ m^3\ of\ resin$$

Comment In practice, because of leakage and other operational and design limitations, the required volume of resin will usually be about 1.1 to 1.4 times that computed on the basis of exchange capacity. Also, the above computation is based on the assumption that the entire capacity of the resin is utilized.

Ion-Exchange Chemistry

The chemistry of the ion-exchange process may be represented by the following equilibrium expression for the reaction of constituent A on a cation-exchange resin and constituent B in solution.

$$nR^-A^+ + B^{+n} \leftrightarrow R_n^-B^{+n} + nA^+ \qquad (11\text{-}117)$$

where R^- is the anionic group attached to an ion-exchange resin and A and B are cations in solution. The generalized form of the equilibrium expression for the above reaction is

$$\frac{[A^+]_S^n[R_n^-B^{+n}]_R}{[R^-A^+]_R^n[B^{+n}]_S} = K_{A^+ \to B^{+n}} \qquad (11\text{-}118)$$

where $K_{A^+ \to B^{+n}}$ = selectivity coefficient

$\quad [A^+]_S$ = concentration of A in solution

$\quad [R^- A^+]_R$ = concentration A on the exchange resin

The reactions for the removal of sodium (Na^+) and calcium (Ca^{2+}) ions from water using a strong-acid synthetic cationic-exchange resin R, and the regeneration of the exhausted resins with hydrochloric acid (HCl) and sodium chloride (NaCl) are as follows:

Reaction:

$$R^- H^+ + Na^+ \to R^- Na^+ + H^+ \tag{11-119}$$

$$2 R^- Na^+ + Ca^{2+} \to R_2^- Ca^{2+} + 2Na^+ \tag{11-120}$$

Regeneration:

$$R^- Na^+ + HCl \to R^- H^+ + NaCl \tag{11-121}$$

$$R_2^- Ca^{2+} + 2NaCl \to 2R^- Na^+ + CaCl_2 \tag{11-122}$$

The corresponding equilibrium expressions for sodium and calcium are as follows:

For sodium:

$$\frac{[H^+][R^- Na^+]}{[R^- H^+][Na^+]} = K_{H \to Na} \tag{11-123}$$

For calcium:

$$\frac{[Na^+]^2 [R^- Ca^{2+}]}{[R^- Na^+]^2 [Ca^{2+}]} = K_{Na \to Ca} \tag{11-124}$$

The selectivity coefficient depends primarily on the nature and valence of the ion, the type of resin and its saturation, and the ion concentration in wastewater and typically is valid over a narrow pH range. In fact, for a given series of similar ions, exchange resins have been found to exhibit an order of selectivity or affinity for the ions. Approximate selectivity coefficients for cationic and anionic resins are given in Tables 11–37 and 11–38, respectively. The use of the selectivity coefficients given in these tables is illustrated in Example 11–16.

For synthetic cationic and anionic exchange resins, typical series are

$$Li^+ < H^+ < Na^+ < NH_4^+ < K^+ < Rb^+ < Ag^+ \tag{11-125}$$

$$Mg^{2+} < Zn^{2+} < Co^{2+} < Cu^{2+} < Ca^{2+} < Sr^{2+} < Ba^{2+} \tag{11-126}$$

$$OH^- < F^- < HCO^- < Cl^- < Br^- < NO_3^- < ClO_4^- \tag{11-127}$$

In practice, the selectivity coefficients are determined by measurement in the laboratory and are valid only for the conditions under which they were measured. At low concentrations, the value of the selectivity coefficient for the exchange of monovalent ions by divalent ions is, in general, larger than the exchange of monovalent ions by monovalent ions. This fact has, in many cases, limited the use of synthetic resins for the removal of certain substances in wastewater, such as ammonia in the form of the ammonium ion. There are, however, certain natural zeolites that favor NH_4^+ or Cu^{2+}.

Table 11–37
Approximate selectivity scale for cations on 8 percent cross-linked strong-acid ion-exchange resins[a]

Cation	Selectivity	Cation	Selectivity
Li^+	1.0	Co^{2+}	3.7
H^+	1.3	Cu^{2+}	3.8
Na^+	2.0	Cd^{2+}	3.9
NH_4^+	2.6	Be^{2+}	4.0
K^+	2.9	Mn^{2+}	4.1
Rb^+	3.2	Ni^{2+}	3.9
Cs^+	3.3	Ca^{2+}	5.2
Ag^+	8.5	Sr^{2+}	6.5
Mg^{2+}	3.3	Pb^{2+}	9.9
Zn^{2+}	3.5	Ba^{2+}	11.5

[a] Adapted from Bonner and Smith (1957); see also Slater (1991).

Table 11–38
Approximate selectivity scale for anions on strong-base ion-exchange resins[a]

Anion	Selectivity	Anion	Selectivity
HPO_4^{2-}	0.01	BrO_3^-	1.0
CO_3^{2-}	0.03	Cl^-	1.0
OH^- (Type I)	0.06	CN^-	1.3
F^-	0.1	NO_2^-	1.3
SO_4^{2-}	0.15	HSO_4^-	1.6
CH_3COO^-	0.2	Br^-	3.0
HCO_3^-	0.4	NO_3^-	3.0–4.0
OH^- (Type II)	0.5–0.65	I^-	18.0

[a] Adapted from Peterson (1953) and Bard (1966).

Anderson (1975) in a classic paper developed a method that can be used to evaluate the effectiveness of a proposed ion-exchange process using strong ionic resins. In the development proposed by Anderson, it is assumed that at 100 percent leakage, the effluent concentration of a constituent is equal to the influent concentration (i.e., equilibrium has been reached). The equilibrium condition can be assumed to be either the limiting operating exchange capacity of the resin or the capacity corresponding to the maximum regeneration level that can be attained. Using this assumption, Eq. (11–118) is converted from concentration units to units of equivalent fractions by making the following substitutions:

$$X_{A^+} = \frac{[A^+]_S}{C} \quad \text{and} \quad X_{B^+} = \frac{[B^+]_S}{C} \tag{11–128}$$

$$X_{A^+} + X_{B^+} = 1 \tag{11–129}$$

where X_{A^+} and X_{B^+} are the equivalent fractions of A and B in solution and C is the total cationic or anionic concentration in solution.

$$\bar{X}_{A^+} = \frac{[R^-A^+]_R}{\bar{C}} \qquad \text{and} \qquad \bar{X}_{B^+} = \frac{[R^-B^+]_R}{\bar{C}} \tag{11-130}$$

$$\bar{X}_{A^+} + \bar{X}_{B^+} = 1 \tag{11-131}$$

where $\bar{X}_A$ and $\bar{X}_B$ are the equivalent fractions of A and B in the resin and $\bar{C}$ is the total ionic concentration in the resin (i.e., the total resin capacity in eq/L). Substituting the above terms into Eq. (11–118) and simplifying results in the following expression:

$$\frac{\bar{X}_{B^+}X_{A^+}}{\bar{X}_{A^+}X_{B^+}} = K_{A^+\to B^+} \tag{11-132}$$

Substituting for X_A and $\bar{X}_A$ in Eq. (11–132) results in

$$\frac{\bar{X}_{B^+}}{1 - \bar{X}_{B^+}} = [K_{A^+\to B^+}]\left[\frac{X_{B^+}}{1 - X_{B^+}}\right] \tag{11-133}$$

It should be noted that Eq. (11–133) is only valid for exchanges between monovalent ions on fully ionized exchange resins. The distribution of a single monovalent ion A between the solution and the resin for different values of the selectivity coefficient is presented on Fig. 11–74. The distribution curves can be used to assess the effectiveness of a resin for the removal of a given ion, based on the selectivity coefficient.

The following three attributes of Eq. (11–133) were identified by Anderson (1975).

1. The term $\bar{X}_B/(1 - \bar{X}_B)$ corresponds to the state of the resin in an exchange column when the influent and effluent concentrations are the same.
2. The term $\bar{X}_B$ corresponds to the extent to which the resin can be converted to the B^+ form when the resin is in equilibrium with a solution of composition X_B.
3. The term $\bar{X}_B$ also corresponds to the maximum extent of regeneration that can be achieved with a regenerant composition of X_B.

Figure 11–74

Distribution curves for a single monovalent ion A between the solution and the resin for different values of the selectivity coefficient.

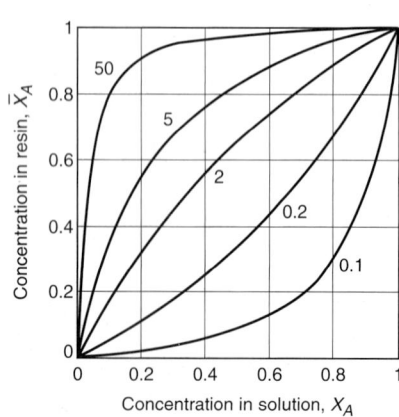

The corresponding equation for exchanges between monovalent and divalent ions on a fully ionized exchange resin is

$$\frac{\bar{X}_{B^{2+}}}{(1 - \bar{X}_{B^{2+}})^2} = [K_{A^+ \to B^{2+}}]\left[\frac{\bar{C}}{C}\right]\frac{X_{B^{2+}}}{(1 - X_{B^{2+}})^2} \tag{11–134}$$

The application of these equations is illustrated in Example 11–16.

Application of Ion Exchange

As noted previously, ion exchange has been used in wastewater applications for the removal of nitrogen, heavy metals, and total dissolved solids.

For Nitrogen Control. For nitrogen control, the ions typically removed from the waste stream are ammonium, NH_4^+, and nitrate, NO_3^-. The ion that the ammonium displaces varies with the nature of the solution used to regenerate the bed. Although both natural and synthetic ion-exchange resins are available, synthetic resins are used more widely because of their durability. Some natural resins (zeolites) have found application in the removal of ammonia from wastewater. Clinoptilolite, a naturally occurring zeolite, has proved to be one of the best natural exchange resins. In addition to having a greater affinity for ammonium ions than other ion-exchange materials, it is relatively inexpensive when compared to synthetic media. One of the novel features of this zeolite is the regeneration system employed. Upon exhaustion, the zeolite is regenerated with lime $Ca(OH)_2$, and the ammonium ion removed from the zeolite is converted to ammonia because of the high pH. A flow diagram for this process is shown on Fig. 11–75. The stripped liquid is collected in a storage tank for subsequent reuse. A problem that must be solved is the formation of calcium carbonate precipitates within the zeolite exchange bed and in the stripping tower and piping appurtenances. As indicated on Fig. 11–76, the zeolite bed is equipped with backwash facilities to remove the carbonate deposits that form within the filter.

Figure 11–75

Typical flow diagram for the removal of ammonia by zeolite exchange. *Note:* The ammonia removed is removed in ammonia scrubber.

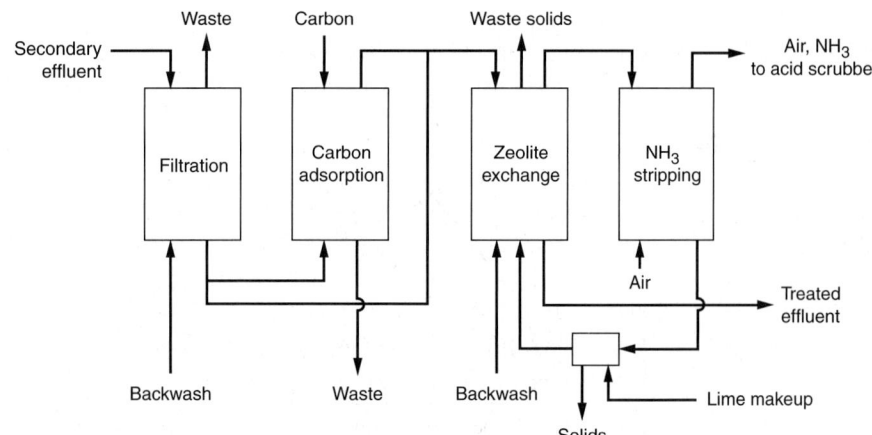

Figure 11-76

Typical ion-exchange test columns used to study the removal of nitrate from water that has been processed with reverse osmosis.

When using conventional synthetic ion-exchange resins for the removal of nitrate, two problems are encountered. First, while most resins have a greater affinity for nitrate over chloride or bicarbonate, they have a significantly lower affinity for nitrate as compared to sulfate, which limits the useful capacity of the resin for the removal of nitrate. The impact of the presence of sulfate on the nitrate removal capacity of conventional resins is illustrated in Example 11–16. Second, because of the lower affinity for nitrate over sulfate, a phenomenon known as *nitrate dumping* can occur. Nitrate dumping occurs when an ion-exchange column is operated past the nitrate breakthrough, at which point the sulfate in the feedwater will displace the nitrate on the resin, causing a release of nitrate. To overcome the problems associated with low affinity and nitrate breakthrough, new types of resins have been developed in which the affinities for nitrate and sulfate have been exchanged. When significant amounts of sulfate are present (i.e., typically greater than 25 percent of the total of the sum of the sulfate and nitrate expressed in meq/L), the use of nitrate-selective resins is advantageous. Because the performance of nitrate-selective resins will vary with the composition of the treated wastewater, pilot testing will usually be required (McGarvey et al., 1989; Dimotsis and McGarvey, 1995). Typical ion-exchange test columns used to study the removal of nitrate from water which has been processed with reverse osmosis are shown on Fig. 11–76.

EXAMPLE 11–16   **Ion Exchange for the Removal of Nitrate without and with Sulfate Present in the Water**   Nitrate is to be removed from two different treated wastewaters with the compositions given below. For the purpose of illustration, assume a conventional ion-exchange resin will be used.

Wastewater A

Cation	Conc., mg/L	mg/meq	meq/L	Anion	Conc., mg/L	mg/meq	meq/L
Ca^{2+}	82.2	20.04	4.10	HCO_3^-	305.1	61.02	5.00
Mg^{2+}	17.9	12.15	1.47	SO_4^{2-}	0.00	48.03	0.00
Na^+	46.4	23.00	2.02	Cl^-	78.0	35.45	2.20
K^+	15.5	39.10	0.40	NO_3^-	50.0	62.01	0.81
		Σ cations	7.99			Σ anions	8.01

Wastewater B

Cation	Conc., mg/L	mg/meq	meq/L	Anion	Conc., mg/L	mg/meq	meq/L
Ca^{2+}	82.2	20.04	4.10	HCO_3^-	220	61.02	3.61
Mg^{2+}	17.9	12.15	1.47	SO_4^{2-}	79.2	48.03	1.65
Na^+	46.4	23.00	2.02	Cl^-	78.0	35.45	2.20
K^+	15.5	39.10	0.40	NO_3^-	50.0	62.01	0.81
		Σ cations	7.99			Σ anions	8.27

Determine the maximum amount of water that can be processed per liter of a strong-base anion-exchange resin with an exchange capacity of 2.0 eq/L.

Solution:
Wastewater A

1. Estimate the selectivity coefficient (see Table 11–38). To apply Eq. (11–133) the system must be reduced to two components. For this purpose, HCO_3^- and Cl^- are combined into a single component. Using a selectivity value of 4 for nitrate, the selectivity coefficient is estimated as follows:

$$K_{HCO_3^- \to NO_3^-} = \frac{4.0}{0.4} = 10.0$$

$$K_{Cl^- \to NO_3^-} = \frac{4.0}{1.0} = 4.0$$

$$K_{[(HCO_3^-)(Cl^-)] \to NO_3^-} = 7.0 \text{ (estimated)}$$

2. For the equilibrium condition ($C_e/C_o = 1.0$), estimate the nitrate equivalent fraction in solution.

$$X_{NO_3^-} = \frac{0.81}{8.01} = 0.101$$

3. Compute the equilibrium resin composition using Eq. (11–133).

$$\frac{\bar{X}_{B^+}}{1 - \bar{X}_{B^+}} = [K_{A^+ \rightarrow B^+}]\left[\frac{X_{B^+}}{1 - X_{B^+}}\right]$$

$$\frac{\bar{X}_{NO_3^-}}{1 - \bar{X}_{NO_3^-}} = 7.0\left[\frac{0.101}{1 - 0.101}\right]$$

$$\bar{X}_{NO_3^-} = 0.44$$

Thus, 44 percent of the exchange sites on the resin can be used for the removal of nitrate.

4. Determine the limiting operating capacity of the resin for the removal of nitrate.

Limiting operating capacity = (2 eq/L of resin)(0.44) = 0.88 eq/L of resin

5. Determine the volume of water that can be treated during a service cycle.

$$\text{Vol} = \frac{(\text{nitrate removal capacity of resin, eq/L of resin})}{(\text{nitrate in solution, eq/L of water})}$$

$$= \frac{(0.88 \text{ eq/L of resin})}{(0.81 \times 10^{-3} \text{ eq/L of water})} = 1090 \frac{\text{L of water}}{\text{L of resin}}$$

Solution:
Wastewater B

1. Estimate the selectivity coefficient (see Table 11–38). To apply Eq. (11–134) the system must be reduced to two components. For this purpose, HCO_3^-, Cl^-, and NO_3^- are combined into a single monovalent component. The selectivity coefficient is estimated as follows:

$$K_{HCO_3^- \rightarrow SO_4^{2-}} = \frac{0.15}{0.4} = 0.4$$

$$K_{Cl^- \rightarrow SO_4^{2-}} = \frac{0.15}{1.0} = 0.15$$

$$K_{NO_3^- \rightarrow SO_4^{2-}} = \frac{0.15}{4.0} = 0.04$$

$$K_{[(NO_3^-)(HCO_3^-)(Cl^-)] \rightarrow SO_4^{2-}} = 0.2 \text{ (estimated)}$$

2. For the equilibrium condition ($C_e/C_o = 1.0$), estimate the sulfate equivalent fraction in solution.

$$X_{SO_4^{2-}} = \frac{1.65}{8.27} = 0.2$$

3. Compute the equilibrium resin composition using Eq. (11–134).

$$\frac{\bar{X}_{B^{2-}}}{(1 - \bar{X}_{B^{2-}})^2} = [K_{A^- \to B^{2-}}] \frac{\bar{C}}{C} \left[\frac{X_{B^{2-}}}{(1 - X_{B^{2-}})^2} \right]$$

$$\frac{\bar{X}_{SO_4^{2-}}}{(1 - \bar{X}_{SO_4^{2-}})^2} = 0.2 \frac{2}{0.00827} \left[\frac{0.2}{(1 - 0.2)^2} \right]$$

$\bar{X}_{SO_4^{2-}} = 0.77$ (determined by successive trials)

Thus, 77 percent of the exchange sites on the resin will be in the divalent form at equilibrium. The relative amount of NO_3^- can be estimated by assuming that the remaining 23 percent of the resin sites are in equilibrium with a solution of NO_3^-, HCO_3^-, and Cl^- with the same relative concentration as the feed.

The equivalent fraction of nitrate in the solution will then be

$$X_{NO_3^-} = \frac{0.81}{6.62} = 0.12$$

The selectivity coefficient for the monovalent system is estimated:

$$K_{HCO_3^- \to NO_3^-} = \frac{4.0}{0.4} = 10.0$$

$$K_{Cl^- \to NO_3^-} = \frac{4.0}{1.0} = 4.0$$

$$K_{[(HCO_3^-)(Cl^-)] \to NO_3^-} = 7.0 \text{ (estimated)}$$

Compute the equilibrium resin composition using Eq. (11–133).

$$\frac{\bar{X}'_{B^+}}{1 - \bar{X}'_{B^+}} = [K_{A^+ \to B^+}] \left[\frac{X'_{B^+}}{1 - X'_{B^+}} \right]$$

$$\frac{\bar{X}'_{NO_3^-}}{1 - \bar{X}'_{NO_3^-}} = 7.0 \left[\frac{0.12}{1 - 0.12} \right]$$

$\bar{X}'_{NO_3^-} = 0.5$

The fraction of the total resin capacity in the nitrate form is then computed.

$$\bar{X}_{NO_3^-} = (1 - \bar{X}_{SO_4^{2-}})(\bar{X}'_{NO_3^-}) = (0.23)(0.5) = 0.115$$

4. Determine the limiting operating capacity of the resin for the removal of nitrate.

Limiting operating capacity = (2 eq/L of resin)(0.115) = 0.23 eq/L of resin

5. Determine the volume of water that can be treated during a service cycle.

$$\text{Vol} = \frac{(\text{nitrate removal capacity of resin, eq/L of resin})}{(\text{nitrate in solution, eq/L of water})}$$

$$= \frac{(0.23 \text{ eq/L of resin})}{(0.81 \times 10^{-3} \text{ eq/L of water})} = 284 \frac{\text{L of water}}{\text{L of resin}}$$

Comment As illustrated in this example, the ionic composition of the wastewater can have a significant effect on the amount of water that can be treated per unit volume of resin, especially where nitrate is to be removed. Because the sulfate is more than 25 percent of the sum of the sulfate and nitrate, the use of a nitrate-selective resin would be advantageous in this application. The approximate nature of these calculations also demonstrates the importance of conducting pilot-plant tests to establish actual throughput volumes.

Removal of Heavy Metals. Metal removal may be required as a pretreatment before discharge to a municipal sewer system. Because of the potential accumulation and toxicity of these metals, it is desirable to remove them from wastewater effluents before release to the environment. Ion exchange is one of the most common forms of treatment used for the removal of metals. Facilities and activities that may discharge wastewater containing high concentrations of metals include metal processing, electronics industries (semiconductors, printed circuit boards), metal plating and finishing, pharmaceuticals and laboratories, and vehicle service shops. High metal concentrations can also be found in leachate from landfills, and stormwater runoff.

Where industries produce effluents with widely fluctuating metal concentrations, flow equalization may be required to make ion exchange feasible, making process design difficult. The economic feasibility of using ion-exchange processes for metal removal greatly improves when the process is used for the removal and recovery of valuable metals. Because it is now possible to manufacture resins for specific applications, the use of resins that have a high selectivity for the desired metal(s) also improves the economics of ion exchange.

Materials used for the exchange of metals include zeolites, weak and strong anion and cation resins, chelating resins, and microbial and plant biomass. Biomass materials are generally more abundant and therefore less expensive when compared to other commercially available resins. Natural zeolites, clinoptilolite (selective for Cs) and chabazite (mixed metals background Cr, Ni, Cu, Zn, Cd, Pb) (Ouki and Kavannagh 1999), have been used to treat wastewater with mixed metal backgrounds. Chelating resins, such as aminophosphonic and iminodiacetic resins, have been manufactured to have a high selectivity for specific metals, such as Cu, Ni, Cd, and Zn.

Ion-exchange processes are highly pH-dependent. Solution pH has a significant impact on the metal species present and the interaction between exchanging ions and the resin. Most metals bind better at higher pH, due to less competition from protons for adsorption sites. Operating and wastewater conditions determine selectivity of resin, pH, temperature, other ionic species, and chemical background. The presence of oxidants, particles, solvents, and polymers may affect the performance of ion-exchange resins. The quantity and quality of regenerate produced and subsequently requiring management must also be considered.

Removal of Total Dissolved Solids. For the reduction of the total dissolved solids, both anionic- and cationic-exchange resins must be used (see Fig. 11–77). The wastewater is first passed through a cation exchanger where the positively charged ions are replaced by hydrogen ions. The cation-exchanger effluent is then passed over an anionic-exchange resin where the anions are replaced by hydroxide ions. Thus, the dissolved solids are replaced by hydrogen and hydroxide ions that react to form water molecules.

Total dissolved solids removal can take place in separate exchange columns arranged in series, or both resins can be mixed in a single reactor. Wastewater application rates range from 0.20 to 0.40 $m^3/m^2 \cdot min$ (5 to 10 $gal/ft^2 \cdot min$). Typical bed depths are 0.75 to 2.0 m (2.5 to 6.5 ft). In reuse applications, treatment of a portion of the wastewater by ion exchange, followed by blending with wastewater not treated by ion exchange, would possibly reduce the dissolved solids to acceptable levels. In some situations, it appears that ion exchange may be as competitive, if not more so, with reverse osmosis.

Figure 11–77

Typical flow diagram for the removal of hardness and for the complete demineralization of water using ion-exchange resins.

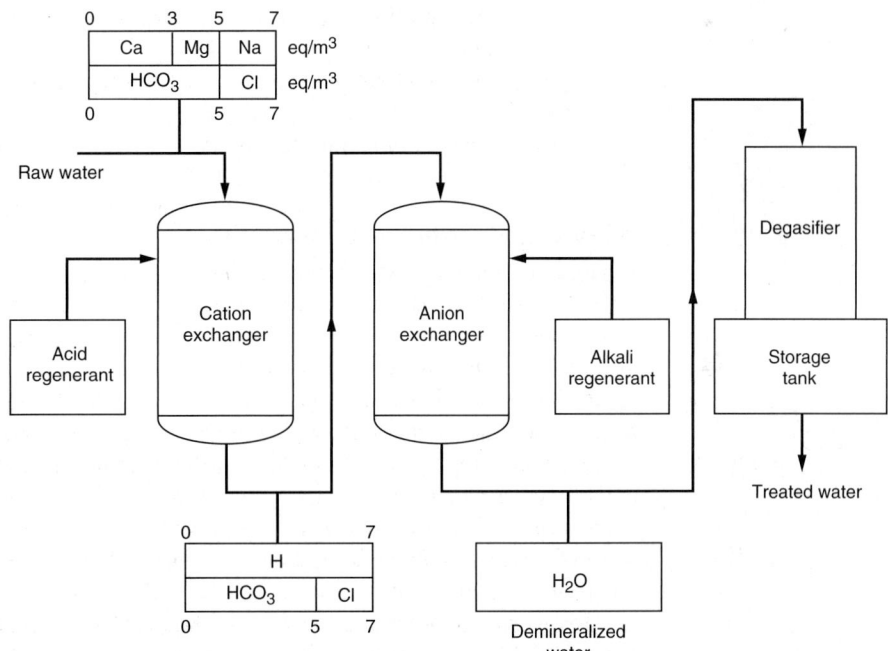

Operational Considerations

To make ion exchange economical for advanced wastewater treatment, it would be desirable to use regenerants and restorants that would remove both the inorganic anions and the organic material from the spent resin. Chemical and physical restorants found to be successful in the removal of organic material from resins include sodium hydroxide, hydrochloric acid, methanol, and bentonite. To date, ion exchange has had limited application because of the extensive pretreatment required, concerns about the life of the ion-exchange resins, and the complex regeneration system required.

High concentrations of influent TSS can plug the ion-exchange beds, causing high headlosses and inefficient operation. Resin binding can be caused by residual organics found in biological treatment effluents. Some form of chemical treatment and clarification is required before ion-exchange demineralization. This problem has been solved partially by prefiltering the wastewater or by using scavenger exchange resins before application to the exchange column.

11–10 ADVANCED OXIDATION PROCESSES

Advanced oxidation processes (AOPs) are used to oxidize complex organic constituents found in wastewater that are difficult to degrade biologically into simpler end products. When chemical oxidation is used, it may not be necessary to oxidize completely a given compound or group of compounds. In many cases, partial oxidation is sufficient to render specific compounds more amenable to subsequent biological treatment or to reduce their toxicity. The oxidation of specific compounds may be characterized by the extent of degradation of the final oxidation products as follows (Rice, 1996):

1. *Primary degradation.* A structural change in the parent compound.
2. *Acceptable degradation (defusing).* A structural change in the parent compound to the extent that toxicity is reduced.
3. *Ultimate degradation (mineralization).* Conversion of organic carbon to inorganic CO_2.
4. *Unacceptable degradation (fusing).* A structural change in the parent compound resulting in increased toxicity.

Theory of Advanced Oxidation

Advanced oxidation processes typically involve the generation and use of the hydroxyl free radical ($HO^\bullet$) as a strong oxidant to destroy compounds that cannot be oxidized by conventional oxidants such as oxygen, ozone, and chlorine. The relative oxidizing power of the hydroxyl radical, along with other common oxidants, is summarized in Table 11–39. As shown, with the exception of fluorine, the hydroxyl radical is one of the most active oxidants known. The hydroxyl radical reacts with the dissolved constituents, initiating a series of oxidation reactions until the constituents are completely mineralized. Nonselective in their mode of attack and able to operate at normal temperature and pressures, hydroxyl radicals are capable of oxidizing almost all reduced materials present without restriction to specific classes or groups of compounds, as compared to other oxidants.

Advanced oxidation processes differ from the other treatment processes discussed (such as ion exchange or stripping) because wastewater compounds are degraded rather

Table 11–39
Comparison of oxidizing potential of various oxidizing agents[a]

Oxidizing agent	Electrochemical oxidation potential (EOP), V	EOP relative to chlorine
Fluorine	3.06	2.25
Hydroxyl radical	2.80	2.05
Oxygen (atomic)	2.42	1.78
Ozone	2.08	1.52
Hydrogen peroxide	1.78	1.30
Hypochlorite	1.49	1.10
Chlorine	1.36	1.00
Chlorine dioxide	1.27	0.93
Oxygen (molecular)	1.23	0.90

[a]From Ozonia (1977).

Table 11–40
Examples of technologies used to produce the reactive hydroxyl free radical, $HO^{\bullet}$[a]

Ozone-based processes	Non-ozone-based processes
Ozone at elevated pH (8 to >10)	H_2O_2 + UV[b]
Ozone + UV_{254} (also applicable in the gas phase)[b]	H_2O_2 + UV + ferrous salts (Fenton's reagent)
Ozone + H_2O_2[b]	Electron-beam irradiation
Ozone + UV_{254} + H_2O_2[b]	Electrohydraulic cavitation
Ozone + TiO_2	Ultrasonics
Ozone + TiO_2 + H_2O_2	Nonthermal plasmas
Ozone + electron-beam irradiation	Pulsed corona discharges
Ozone + ultrasonics	Photocatalysis (UV + TiO_2)
	Gamma radiolysis
	Catalytic oxidation
	Supercritical water oxidation

[a]Adapted in part from Rice (1996).
[b]Processes currently (2001) being used on a commercial scale.

than concentrated or transferred into a different phase. Because secondary waste materials are not generated, there is no need to dispose of or regenerate materials. Additional details on AOPs may be found in Singer and Reckhow (1999).

Technologies Used to Produce Hydroxyl Radicals (HO•)

At the present time, a variety of technologies are available to produce $HO^{\bullet}$ in the aqueous phase. The various technologies are summarized in Table 11–40, according to whether ozone is used in the reaction. Of the technologies reported in Table 11–40, only ozone/UV, ozone/hydrogen peroxide, ozone/UV/hydrogen peroxide, and hydrogen peroxide/UV are being used on a commercial scale (Rice, 1996).

Ozone/UV.   Production of the free radical HO˙ with UV light can be illustrated by the following reactions for the photolysis of ozone (Glaze et al., 1987; Glaze and Kang, 1990):

$$O_3 + UV \text{ (or } h\nu, \lambda < 310 \text{ nm)} \rightarrow O_2 + O(^1D) \tag{11–135}$$

$$O(^1D) + H_2O \rightarrow HO^\bullet + HO^\bullet \text{ (in wet air)} \tag{11–136}$$

$$O(^1D) + H_2O \rightarrow HO^\bullet + HO^\bullet \rightarrow H_2O_2 \text{ (in water)} \tag{11–137}$$

where O_3 = ozone
    UV = ultraviolet radiation (or $h\nu$ = energy)
    O_2 = oxygen
    $O(^1D)$ = excited oxygen atom. The symbol (^1D) is a spectroscopic notation used to specify the atomic and molecular configuration (also known as a singlet oxygen)
    $HO^\bullet$ = hydroxyl radical. The dot ($^\bullet$) that appears next to the hydroxyl and other radicals is used to denote the fact that these species have an unpaired electron.

As shown in Eq. (11–136), the photolysis of ozone in wet air results in the formation of hydroxyl radicals. In water, the photolysis of ozone leads to the formation of hydrogen peroxide [see Eq. (11–137)]. Because the photolysis of ozone in water leads to the formation of hydrogen peroxide, which is subsequently photolyzed to form hydroxyl radicals, the use of ozone in this application is generally not cost-effective. In air, the ozone/UV process can degrade compounds through direct ozonation, photolysis, or reaction with the hydroxyl radical. The ozone/UV process is more effective when the compounds of interest can be degraded through the absorption of the UV irradiation as well as through the reaction with the hydroxyl radicals. A schematic flow diagram of the processes is illustrated on Fig. 11–78.

Figure 11–78

Schematic representation of advanced oxidation process involving the use of ozone and UV radiation.

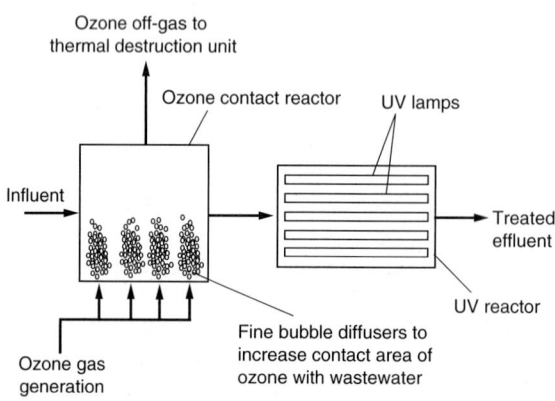

Figure 11–79

Schematic representation of advanced oxidation process involving the use of ozone and hydrogen peroxide.

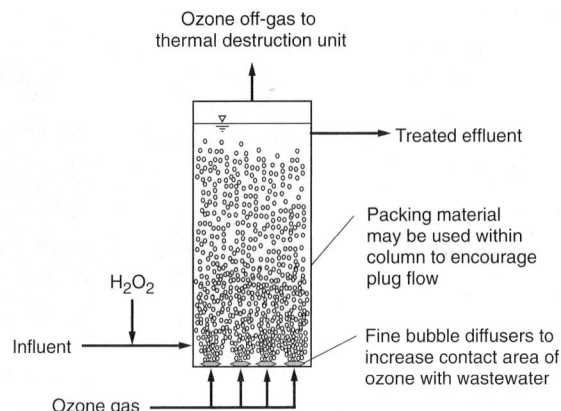

Ozone/Hydrogen Peroxide. For compounds that do not adsorb UV, AOPs involving ozone/H_2O_2 may be more effective. A schematic flow diagram of the processes is illustrated on Fig. 11–79. Compounds in water such as trichloroethylene (TCE) and perchloroethylene (PCE) have been reduced significantly with AOPs using hydrogen peroxide and ozone to generate $HO^\bullet$ (Karimi et al., 1997). The overall reaction for the production of hydroxyl radicals using hydrogen peroxide and ozone is as follows:

$$H_2O_2 + 2O_3 \rightarrow HO^\bullet + HO^\bullet + 3O_2 \qquad (11\text{--}138)$$

Hydrogen Peroxide/UV. Hydroxyl radicals are also formed when water containing H_2O_2 is exposed to UV light (200 to 280 nm). The following reaction can be used to describe the photolysis of H_2O_2:

$$H_2O_2 + UV \text{ (or } h\nu, \lambda \approx 200\text{--}280 \text{ nm)} \rightarrow HO^\bullet + HO^\bullet \qquad (11\text{--}139)$$

In some cases the use of the hydrogen peroxide/UV process has not been feasible because H_2O_2 has a small molar extinction coefficient, requiring high concentrations of H_2O_2 and not using the UV energy efficiently. A schematic flow diagram and a typical installation of the hydrogen peroxide/UV process is shown on Fig. 11–80.

Most recently, the hydrogen peroxide/UV process has been applied to the oxidation of trace constituents found in treated water. The process has been studied for the removal of N-Nitrosodimethylamine (NDMA) and other compounds of concern in treated wastewater including (1) sex and steroidal hormones, (2) human prescription and nonprescription drugs, (3) veterinary and human antibiotics, and (4) industrial and household wastewater products (see Table 2–20 in Chap. 2). At the relatively low concentration range of these compounds encountered in treated water (typically in the μg/L range), their oxidation appears to follow first-order kinetics. The electrical energy required for their oxidation is expressed in EE/O units, defined as the electrical energy input per unit volume per log order of reduction (Bolton et al., 1999). In equation form, the EE/O is given by

$$EE/O = \frac{EE_i}{V[\log(C_i/C_f)]} \qquad (11\text{--}140)$$

Figure 11–80

Hydrogen peroxide and UV radiation advanced oxidation process: (a) schematic flow diagram. The hydrogen peroxide feed system and storage container are placed in a contained area and (b) photograph of typical vertical-flow UV reactor used to treat a flowrate of about 4000 m³/d. Fifteen high-energy UV lamps (20 kW each) are employed in the reactor.

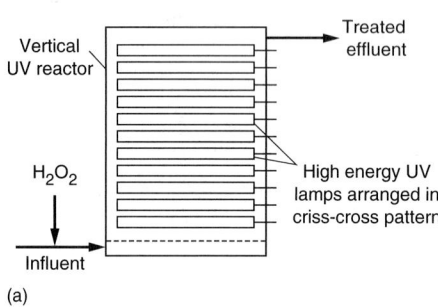

where EE/O = electrical energy input per log reduction, kWh/m³·log order of reduction

EE_i = electrical energy input, kWh

V = volume of liquid treated, m³

C_i = initial concentration, ng

C_f = final concentration, ng

Based on currently available technology (2001), the required EE/O value for a one-log order of reduction (i.e., 100 to 10) of NDMA is on the order of 21 to 265 kWh/ 10^3 m³·log order (0.08 to 1.0 kWh/10^3 gal·log order) with a 5 to 6 mg/L dose of H_2O_2, although in some cases it did not appear that the peroxide was necessary (Soroushian et al., 2001). The required EE/O value will vary significantly with the characteristics of the treated wastewater.

Other Processes. Other reactions that yield HO• include the reactions of H_2O_2 and UV with Fenton's reagent and the adsorption of UV by semiconductor metal oxides such as TiO_2 suspended in water, which acts as a catalyst. Still others are currently under development.

Applications

Based on numerous studies, it has been found that combined AOPs are more effective than any of the individual agents (e.g., ozone, UV, hydrogen peroxide). AOPs are usually applied to low COD wastewaters because of the cost of ozone and/or H_2O_2 required to generate the hydroxyl radicals. Material that was previously resistant to degradation may be transformed into compounds that will require further biological treatment. The application of AOPs for the disinfection of treated wastewater and for the treatment of refractory organic compounds is considered below. Some operational problems are also identified.

Disinfection. Because it was recognized that free radicals generated from ozone were more powerful oxidants than ozone alone, it was reasoned that the hydroxyl free radicals could be used effectively to oxidize microorganisms and refractory organic materials in water and wastewater. Unfortunately, because the half-life of the hydroxyl free radicals is short, on the order of microseconds, it is not possible to develop high concentrations. With extremely low concentrations, the required detention times for microorganism disinfection, based on the CT concept (see Chap 12), are prohibitive.

Oxidation of Refractory Organic Compounds. For the reasons cited above hydroxyl radicals are not used for conventional disinfection; instead they are used more commonly for the oxidation of trace amounts of refractory organic compounds found in highly treated effluents. The hydroxyl radicals, once generated, can attack organic molecules by radical addition, hydrogen abstraction, electron transfer, and radical combination (SES, 1994).

1. *By radical addition.* The addition of the hydroxyl radical to an unsaturated aliphatic or aromatic organic compound (e.g., C_6H_6) results in the production of a radical organic compound that can be oxidized further by compounds such as oxygen or ferrous iron to produce stable oxidized end products. In the following reactions the abbreviation R is used to denote the reacting organic compound.

$$R + HO^{\bullet} \rightarrow ROH \qquad (11\text{–}141)$$

2. *By hydrogen abstraction.* The hydroxyl radical can be used to remove a hydrogen atom from organic compounds. The removal of a hydrogen atom results in the formation of a radical organic compound, initiating a chain reaction where the radical organic compound reacts with oxygen, producing a peroxyl radical, which can react with another organic compound, and so on.

$$R + HO^{\bullet} \rightarrow R^{\bullet} + H_2O \qquad (11\text{–}142)$$

3. *By electron transfer.* Electron transfer results in the formation of ions of a higher valence. Oxidation of a monovalent negative ion will result in the formation of an atom or a free radical.

$$R^n + HO^{\bullet} \rightarrow R^{n-1} + OH^- \qquad (11\text{–}143)$$

4. *By radical combination.* Two radicals can combine to form a stable product.

$$HO^{\bullet} + HO^{\bullet} \rightarrow H_2O_2 \qquad (11\text{–}144)$$

In general, the reaction of hydroxyl radicals with organic compounds, at completion, will produce water, carbon dioxide, and salts; this process is also known as minera.

Operational Problems

High concentrations of carbonate and bicarbonate in some wastewater can react with $HO^{\bullet}$ and reduce the efficiency of advanced oxidation treatment processes. Other factors can also affect the treatment process, such as suspended material, pH, type and nature of the residual TOC, and other wastewater constituents. *Because the chemistry of the wastewater matrix is different for each wastewater, pilot testing is almost always required to test the technical feasibility, to obtain usable design data and information, and to obtain operating experience with a specific AOP.*

11–11   DISTILLATION

Distillation is a unit operation in which the components of a liquid solution are separated by vaporization and condensation. Along with reverse osmosis, distillation can be used to control the buildup of salts in critical reuse applications. Because distillation is expensive, its application is generally limited to applications such as (1) a high degree of treatment is required, (2) contaminants cannot be removed by other methods, and (3) inexpensive heat is available. The purpose of this section is to introduce the basic concepts involved in distillation. As the use of distillation for wastewater reclamation is a recent development, the current literature must be consulted for the results of ongoing studies and more recent applications.

Distillation Processes

Over the past 20 years, a variety of distillation processes, employing a variety of evaporator types and methods of using and transferring heat energy, have been evaluated or used. The principal distillation processes are (1) boiling with submerged-tube heating surface, (2) boiling with long-tube vertical evaporator, (3) flash evaporation, (4) forced circulation with vapor compression, (5) solar evaporation, (6) rotating-surface evaporation, (7) wiped-surface evaporation, (8) vapor reheating process, (9) direct heat transfer using an immiscible liquid, and (10) condensing-vapor-heat transfer by vapor other than steam. Of these types of distillation processes, multistage flash evaporation, multiple-effect evaporation, and the vapor-compression distillation appear most feasible for the reclamation of municipal wastewater.

Multiple-Effect Evaporation.   In multiple-effect evaporation several evaporators (boilers) are arranged in series, each operating at a lower pressure than the preceding one. In a three-stage, vertical-tube evaporator (see Fig. 11–81), after the influent water is pretreated, it enters the heat exchanger in the last stage (No. 2) and progressively warms as it goes through the heat exchangers in the other effects (i.e., stages). As the water moves through the heat exchangers, it condenses the water vapor emanating from the various effects. When the progressively warmed influent water reaches the first stage, it flows down the internal periphery of vertical tubes in a thin film, which is heated by steam. The wastewater feed to the second effect comes from the bottom of the first effect. If entrainment is kept low, almost all of the nonvolatile contaminants can be removed in a single evaporation step. Volatile contaminants, such as ammonia gas and low-molecular-weight organic acids, may be removed in a preliminary evaporation step, but if their concentration is so small that their presence in the final product is not objectionable, this step with its added cost can be eliminated.

Figure 11–81

Schematic of multiple-effect evaporation distillation process. (Adapted from Liu and Liptak, 2000.)

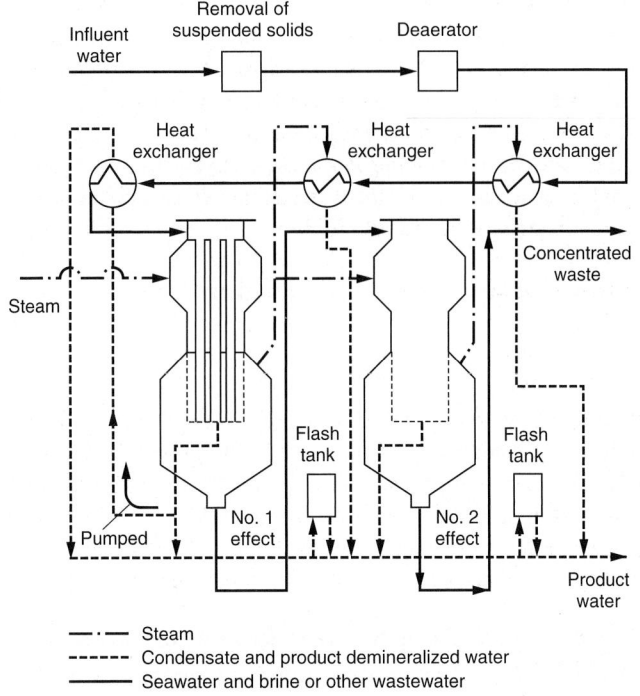

— · — Steam
- - - - - Condensate and product demineralized water
———— Seawater and brine or other wastewater

Figure 11–82

Schematic of multistage flash evaporation distillation process.

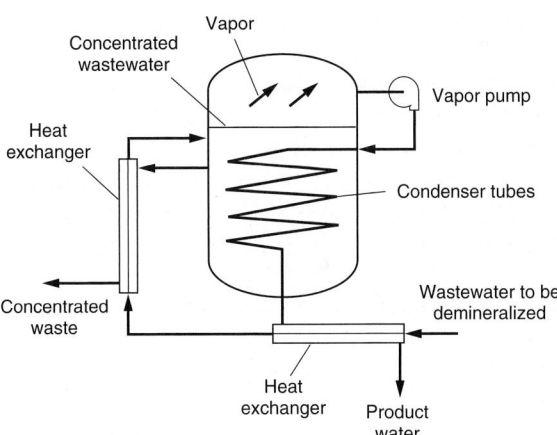

Multistage Flash Evaporation. Multistage flash evaporation systems have been used commercially in desalination for many years. In the multistage flash process (see Fig. 11–82), the influent wastewater is first treated to remove TSS and deaerated before being pumped through heat transfer units in the several stages of the distillation system, each of which is maintained at a lower pressure. Vapor generation or boiling caused by reduction in pressure is known as *flashing*. As the water enters each stage through a pressure-reducing nozzle, a portion of the water is flashed to form a vapor. In turn, the flashed water vapor condenses on the outside of the condenser tubes and is collected in trays (see Fig. 11–82). As the vapor condenses, its latent heat is used to preheat the wastewater that is being returned to the main heater, where it will receive additional heat before being introduced to the first flashing stage. When the concentrated wastewater reaches the lowest-pressure stage, it is pumped out. Thermodynamically, multistage flash evaporation is less efficient than ordinary evaporation. However, by combining a number of stages in a single reactor, external piping is eliminated and construction costs are reduced.

Vapor-Compression Distillation. In the vapor-compression process an increase in pressure of the vapor is used to establish the temperature difference for the transfer of heat. The basic schematic of a vapor-compression distillation unit is shown on Fig. 11–83. After initial heating of the wastewater, the vapor pump is operated so that the vapor under higher pressure can condense in the condenser tubes, at the same time causing the release of an equivalent amount of vapor from the concentrated solution. Heat exchangers can conserve heat from both the condensate and the waste brine. The only energy input required during operation is the mechanical energy for the vapor pump. Hot concentrated wastewater must be discharged at intervals to prevent the buildup of excessive concentrations of salt in the boiler.

Performance Expectations in Reclamation Applications

The theoretical thermodynamic minimum energy required to raise the temperature of wastewater and to provide the latent heat of vaporization is about 2280 kJ/kg. Unfortunately, because of the many irreversibilities in an actual distillation process, the thermodynamic minimum energy requirements are of little practical relevance in the practical evaluation of distillation processes. Typically, about 1.25 to 1.35 times the latent heat of vaporization will be required.

Figure 11–83

Schematic of vapor-compression distillation process. (Adapted from Liu and Liptak, 2000.)

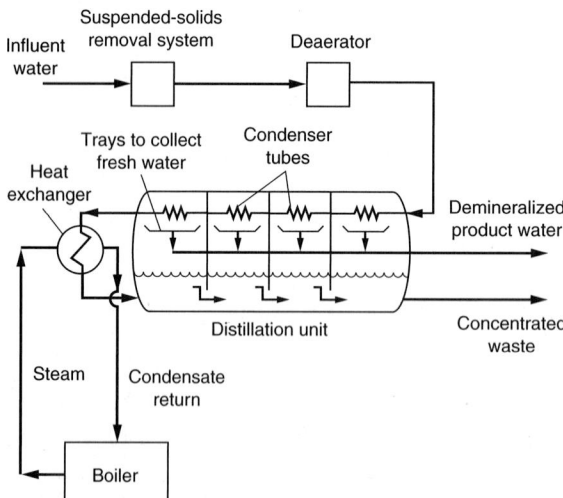

The principal issues with the application of the distillation processes for wastewater reclamation are the carry-over of volatile constituents found in treated wastewater and the degree of subsequent cooling and treatment that may be required to renovate the distilled water. Typical water-quality performance data for a multiple-effect distillation process have been reported for a pilot-scale unit by Rose et al. (1999).

Operating Problems

The most common operating problems encountered include scaling and corrosion. Due to temperature increases, inorganic salts come out of solution and precipitate on the inside walls of pipes and equipment. The control of scaling due to calcium carbonate, calcium sulfate, and magnesium hydroxide is one of the most important design and operational considerations in distillation desalination processes. Controlling the pH minimizes carbonate and hydroxide scales. Most inorganic solutions are corrosive. Cupronickel alloys are used most commonly in seawater desalination. Other metals that are used include aluminum, titanium, and monel.

Disposal of Concentrated Waste

All distillation processes reject part of the influent wastewater. Hence, all of these processes have concentrated wastewater disposal problems. The permissible maximum concentration in the wastewater depends on the solubility, corrosion, and vapor pressure characteristics of the wastewater. Therefore, the waste concentration is an important consideration in process optimization. Disposal of concentrated wastewater brines is essentially the same problem encountered with the membrane process discussed in Section 11–6.

PROBLEMS AND DISCUSSION TOPICS

11-1 The following sieve analysis results were obtained for four different stock sands.

U.S. sieve size designation[a]	Size of opening, mm	Cumulative weight passing, %			
		A	B	C	D
140	0.105	0.4	1.5	0.1	5.0
100	0.149	1.5	4.1	0.8	11.1
70	0.210	4.0	10.0	2.5	20.0
50	0.297	9.5	21.0	8.2	32.0
40	0.420	18.5	40.6	18.5	49.5
30	0.590	31.0	61.0	32.0	62.3
20	0.840	49.0	78.3	58.1	78.3
16	1.190	63.2	90.0	76.3	88.5
12	1.680	82.8	96.0	90.0	94.4
8	2.380	89.0	99.0	96.7	97.8
6	3.360	98.0	99.9	99.0	99.0
4	4.760	100.0	100.0	100.0	100.0

[a]Note: sieve size number 18 has an opening size of 1.0 mm.

a. For sand sample A, B, C, or D (to be selected by instructor) determine the geometric mean size, the geometric standard deviation, the effective size, and the uniformity coefficient for the stock sand.

b. It is desired to produce from the stock sand a filter sand with an effective size of 0.45 mm and a uniformity coefficient of 1.6. Estimate the amount of stock sand needed to obtain one ton of filter sand.

c. What U.S. standard sieve size should be used to eliminate the excess coarse material?

d. If the sand remaining after sieving in part (c) above is placed in a filter, what backwash rise rate would be needed to eliminate the excess fine material?

e. What depth of sieved material would have to be placed in the filter to produce 600 mm of usable filter sand?

f. On log-probability paper, plot the size distribution of the modified sand. Check against the required distribution and sizes.

g. Determine the headloss through 600 mm of the filter sand specified in part (b) for a filtration rate of 160 L/m²·min. Assume the sand is stratified and that the maximum and minimum sand sizes are 1.68 (sieve size 12) and 0.297 mm (sieve size 50), respectively. Assume also that T = 20°C, α for all layers = 0.4, and ϕ = 0.75.

An excellent discussion of the procedures involved in developing a usable filter sand from a stock filter sand may be found in Fair et al. (1968).

11-2 Using the equations developed by Fair and Hatch, Kozeny, and Rose, compare the headloss through a 600-mm sand bed. Assume that the sand bed is composed of spherical unsized sand with a diameter of 0.55 mm, the porosity of the sand is 0.40, and the filtration rate is 240 L/m²·min. The temperature is 18°C.

11-3 Using the equation developed by Rose, determine the headloss through a 750-mm sand bed for a filtration rate of 240 L/m²·min. Assume that the sand bed is composed of spherical unsized sand with a diameter of 0.40, 0.45, 0.5, and 0.6 mm (size to be selected by instructor) and a porosity of 0.40. The kinematic viscosity is equal to 1.306×10^{-6} m²/s.

11-4 Solve Problem 11–3 assuming the bed is stratified. Assume that the given sand sizes correspond to the effective size (d_{10}) and that the uniformity coefficient for all of the sand sizes is equal to 1.5.

11-5 If a 0.3-m layer of uniform anthracite is placed on top of the sand bed in Prob. 11–3, determine the ratio of the headloss through the anthracite to that of the sand. Assume that the grain-size diameter of the anthracite is 2.0 mm and porosity is 0.50. Will intermixing occur?

11-6 Given the particle size distribution A, B, C, or D (to be selected by instructor), determine the effective size (d_{10}) and uniformity coefficient UC, and clean water headloss through a stratified bed 600 mm deep for a filtration rate of 0.00267 m/s. If a layer of anthracite is to be added over 600 mm of sand, determine the effective size required to minimize intermixing. Assume ϕ = 0.85 and α = 0.4.

	Percent of sand retained			
Sieve number	A	B	C	D
6–8	2	0	1	0.1
8–10	8	0.1	2	0.7

(continued)

(Continued)

Sieve number	Percent of sand retained			
	A	**B**	**C**	**D**
10–14	10	0.5	4	1.2
14–20	30	7.4	13	10
20–30	26	32	20	24
30–40	14	30	20	29
40–60	8	25	23	25
Pan	2	5	17	10

11–7 For a given filtration operation it has been found that straining is the operative particulate-matter-removal mechanism and that the change in concentration with distance can be approximated with a first-order equation $(dC/dx = -rC)$. If the initial concentration of particulate matter is 10 mg/L, the removal-rate constant is equal to 2.0 cm^{-1}, and the filtration rate is equal to 100 L/m^2·min, determine the amount of material arrested within the filter in the layer between 3 and 6 cm over a 1.0-h period. Express your answer in mg/cm·m^2. Estimate the headloss in the layer at the end of 1.0 h. Assume a uniform sand size of 0.4, 0.6, 0.8, or 1.0 mm (to be selected by instructor).

11–8 The data in the following table were obtained from a pilot-plant study on the filtration of settled secondary effluent from an activated-sludge treatment process. Using these data, estimate the length of run that is possible with and without the addition of polymer if the maximum allowable headloss is 3 m, the filtration rate is 150 L/m^2·min, and the influent suspended-solids concentration is 15 mg/L. Uniform sand with a diameter of 0.55 mm and a depth of 600 mm was used in the pilot filters. Assume the clean water headloss is 0.4 m.

Depth, cm	Concentration ratio C/C_o	
	With polymer addition	**Without polymer addition**
0	1.00	1.00
4	0.65	0.70
8	0.26	0.52
12	0.20	0.46
16	0.16	0.42
20	0.15	0.40
24	0.13	0.39
28	0.12	0.38
32	0.12	0.38
36	0.12	0.38
40	0.12	0.38
50	0.12	0.38
60	0.12	0.38

11–9 Gravity filters are to be used to treat 24,000 m^3/d of settled effluent at a filtration rate of 200 L/m^2·min (5 gal/ft^2·min). The filtration rate with one filter taken out of service for backwashing is not to exceed 240 L/m²·min (6 gal/ft²·min). Determine the number of units and the area of each unit to satisfy these conditions. If each filter is backwashed for 30 min every 24 h at a wash rate of 960 L/m²·min (24 gal/ft²·min), determine the percentage of filter output used for washing if the filter is out of operation for a total of 30 min/d. What would be the total percentage of filter output used for backwashing if a surface washing system that requires 40 L/m²·min (1 gal/ft²·min) of filtered effluent is to be installed?

11-10 Four different waters are to be desalinized by reverse osmosis using a thin-film composite membrane. For water A, B, C, or D (water to be selected by instructor), determine the required membrane area, the rejection rate, and the concentration of the concentrate stream.

Item	Unit	Water			
		A	**B**	**C**	**D**
Feed flowrate	m³/d	4000	5500	20,000	10,000
Influent TDS	g/m³	2850	3200	2000	2700
Permeate TDS	g/m³	200	500	400	225
Flux rate coefficient k_w	m/s·bar	1.0×10^{-6}	1.0×10^{-6}	1.0×10^{-6}	1.0×10^{-6}
Mass transfer rate coefficient, k_i	m/s	6×10^{-8}	6×10^{-8}	6×10^{-8}	6×10^{-8}
Net operating pressure	kPa	2750	2500	2800	3000
Recovery	%	88	90	89	86

11-11 Using the data given below, determine the recovery and rejection rates for one of the following reverse osmosis units (unit to be selected by instructor).

Item	Unit	Reverse osmosis unit			
		A	**B**	**C**	**D**
Feed flowrate	m³/d	25,000	50,000	20,000	10,000
Concentrate flowrate	m³/d	2500	4800	2200	850
Permeate TDS	g/m³	65	88	125	175
Concentrate TDS	g/m³	1500	2500	1850	2850

11-12 Using the data given below, determine the flux rate coefficient and the mass transfer rate coefficient.

Item	Unit	Reverse osmosis unit			
		A	**B**	**C**	**D**
Feed flowrate	m³/d	4000	5500	20,000	10,000
Influent TDS	g/m³	2500	3300	5300	2700
Permeate TDS	g/m³	20	50	40	23
Net operating pressure	kPa	2750	2500	2800	3000
Area	m²	1600	1700	9600	5500
Recovery	%	88	90	89	86

11-13 Estimate the SDI for the following filtered wastewater samples. If the water is to be treated with reverse osmosis will additional treatment be required?

Test run time, min	Volume filtered, mL			
	A	**B**	**C**	**D**
2		600	400	500
5		300	350	500
10	100	150	0	
20	0	100	0	

11-14 Determine the cost (based on the current price of electricity) to treat a flow of 2500 m^3/d with a TDS concentration of 1300 g/m^3 and a cation and anion concentration of 0.13 g-eq/L using an electrodialysis unit with 300 cells. Assume typical values of operation for the electrodialysis unit.

11-15 A wastewater is to be treated with activated carbon to remove residual COD. The following data were obtained from a laboratory adsorption study in which 1 g of activated carbon was added to a beaker containing 1 L of wastewater at selected COD values. Using these data, determine the more suitable isotherm (Langmuir or Fruendlich) to describe the data (sample to be selected by instructor).

Initial COD, mg/L	Equilibrium COD, mg/L			
	A	**B**	**C**	**D**
140	5	10	0.4	5
250	12	30	0.9	18
300	17	50	2	28
340	23	70	4	36
370	29	90	6	42
400	36	110	10	50
450	50	150	35	63

11-16 Using the results from Prob. 11–15, determine the amount of activated carbon that would be required to treat a flow of 5000 m^3/d to a final COD concentration of 20 mg/L if the COD concentration after secondary treatment is equal to 120 mg/L.

11-17 Using the following carbon adsorption data (data set to be selected by instructor), determine the Freundlich capacity factor (mg absorbate/g activated carbon) and Freundlich intensity parameter, 1/n.

Carbon dose, mg/L	Residual concentration, mg/L					
	A	**B**	**C**	**D**	**E**	**F**
0	25.9	9.20	9.89	27.5	20.4	9.88
5	17.4	7.36		24.8	19.3	7.95
10	13.2	6.86	8.96	24.2	18.6	7.02

(continued)

(*Continued*)

Carbon dose, mg/L	Residual concentration, mg/L					
	A	**B**	**C**	**D**	**E**	**F**
25	10.2	3.86	7.83	18.9	16.1	3.66
50	3.6	1.13	5.81	11.8	12.2	0.98
100	2.5	0.22	4.45	2.3	6.7	0.25
150	2.1	0.18	2.98	1.1	3.1	0.09
200	1.4	0.11	2.01	0.9	1.1	0.04

11-18 Determine the theoretical air flowrate required to remove the following compounds in a stripping tower at the indicated concentrations (compound and water to be selected by instructor). Also estimate the height of the stripping tower for a water flowrate of 3000 m³/d. Values of the Henry's constant may be found in Table 5–36 in Chap. 5. Henry's constant for TCE = 5.53 × 10^{-3} m³·atm/mole.

Compound	K_La, s^{-1}	Water A		Water B	
		Influent	**Effluent**	**Influent**	**Effluent**
Chlorobenzene	0.0163	100	5	120	7
Chloroethene	0.0141	100	5	150	5
TCE	0.0176	100	5	180	10
Toluene	0.0206	100	5	200	15

11-19 A quantity of sodium-form ion-exchange resin (5 g) is added to a water containing 2 meq of potassium chloride and 0.5 meq of sodium chloride. Calculate the residual concentration of potassium if the exchange capacity of the resin is 4.0 meq/g of dry weight and the selectivity coefficient is equal to 1.46.

11-20 Determine the exchange capacity for one of the following resins (resin to be selected by instructor). How much resin would be required to treat a flowrate of 4000 m³/d to reduce the concentration of calcium Ca^{2+} from 125 to 45 mg/L? Assume the mass of resin is 0.1 kg.

Throughput volume, L	Resin A		Resin B	
	Cl^-	Ca^{2+}	Cl^-	Ca^{2+}
0	0	0	0	0
5	2	0	2	0
10	8	0	13	0
15	44	0	29	0
20	65	0	45	0
25	70	0	60	1
30	71	0	69	8
35	71	6	71	17

(*continued*)

(Continued)

Throughput volume, L	Resin A		Resin B	
	Cl⁻	**Ca²⁺**	**Cl⁻**	**Ca²⁺**
40	71	20	71	27
45		34	71	35
50		39		39
55		40		40
60		40		40

11-21 Determine the exchange capacity for one of the resins given in Problem 11–20 (resin to be selected by instructor). How much resin would be required to treat a flowrate of 5500 m³/d to reduce the concentration of magnesium Mg^{2+} from 115 to 15 mg/L?

11-22 Four different wastewaters have been reported to have the following ionic composition data. Estimate the selectivity coefficient and determine the amount of wastewater (A, B, C, D, to be selected by instructor) that can be treated by a strong-base ion-exchange resin, per service cycle, for the removal of nitrate. Assume the resin has an ion-exchange capacity of 1.8 eq/L.

Cation	Conc., mg/L	Anion	Concentration, mg/L			
			A	**B**	**C**	**D**
Ca^{2+}	82.2	HCO_3^-	304.8	152	254	348
Mg^{2+}	17.9	SO_4^{2-}	0	0	0	0
Na^+	46.4	Cl^-	58.1	146.3	124	60
K^+	15.5	NO_3^-	82.5	90	21.5	42

11-23 Four different wastewaters have been reported to have the following ionic composition data. Estimate the selectivity coefficient and determine the amount of wastewater (A, B, C, D, to be selected by instructor) that can be treated by a strong-base ion-exchange resin, per service cycle, for the removal of nitrate. Assume the resin has an ion-exchange capacity of 2.5 eq/L.

Cation	Conc., mg/L	Anion	Concentration, mg/L			
			A	**B**	**C**	**D**
Ca^{2+}	82.2	HCO_3^-	321	180	198.5	69
Mg^{2+}	17.9	SO_4^{2-}	65	36.5	124	136
Na^+	46.4	Cl^-	22	95	56	87
K^+	15.5	NO_3^-	46	93	34.5	97

11-24 For the following list of compounds, determine the most suitable method of treatment to reduce the concentration from 100 to 10 μg/L using the advanced wastewater-treatment processes discussed in this chapter.

Benzene	Heptachlor	Vinyl chloride
Chloroform	N-Nitrosodimethylamine	
Dieldrin	Trichloroethylene (TCE)	

REFERENCES

Ahn, K. H., J. H. Y. Song Cha, K. G. Song, and H. Yoo (1998) "Application of Tubular Ceramic Membranes for Building Wastewater Reuse," *Proceedings IAWQ 19th International Conference,* Vancouver, p. 137.

Amirtharajah, A. (1978) "Optimum Backwashing of Sand Filters," *Journal Environmental Engineering Division, American Society of Civil Engineers,* vol. 104, EE5, pp. 917–932.

Amirtharajah, A., and K. M. Mills (1962) "Rapid Mix Design for Mechanisms of Alum Coagulation," *Journal American Water Works Association,* vol. 74, p. 210.

Anderson, R. E. (1975) "Estimation of Ion Exchange Process Limits by Selectivity Calculations," in I. Zwiebel and N. H. Sneed (eds.), *Adsorption and Ion Exchange,* Symposium Series, American Institute of Chemical Engineers, no. 152, vol. 71.

Anderson, R. E. (1979) "Ion Exchange Separations," in P. A. Scheitzer (ed.), *Handbook of Separation Techniques For Chemical Engineers,* McGraw-Hill, New York, NY.

AWWA (1996) *Water Treatment Membrane Processes,* American Water Works Association, McGraw-Hill, New York.

Ball, W. P., M. D. Jones, and M. C. Kavanaugh (1984) "Mass Transfer of Volatile Organic Compounds in Packed Tower Aeration," *Journal Water Pollution Control Federation,* vol. 56, no. 2, pp. 127–136.

Bard, A. J. (1966) *Chemical Equilibrium,* Harper & Row, Publishers, New York.

Bilello, L. J., and B. A. Beaudet (1983) "Evaluation of Activated Carbon by the Dynamic Minicolumn Adsorption Technique," in M. J. McGuire and I. H. Suffet (eds.), *Treatment of Water by Granular Activated Carbon,* American Chemical Society, Washington, DC.

Bohart, G. S., and E. Q. Adams (1920) "Some Aspects of the Behavior of Charcoal with Respect to Chlorine," *Journal American Chemical Society,* vol. 42, p. 523.

Bolton, J. R., M. I. Stephan, and S. R. Carter (1999) "UV Light Driven Degradation of N-nitrosodimethylamine," In *Proceeding of the 5th International Conference on Advanced Oxidation Technologies for Water Remediation,* Albuquerque, NM.

Bonner, O. D., and L. L. Smith (1957) "A Selectivity Scale for Some Divalent Cations on Dowex 50," *Journal of Physical Chemistry,* vol. 61, no. 3, pp. 326–329.

Bourgeous, K., G. Tchobanoglous, and J. Darby (1999) "Performance Evaluation of the Koch Ultrafiltration (UF) Membrane System for Wastewater Reclamation," Center for Environmental and Water Resources Engineering, Report No. 99-2, Department of Civil and Environmental Engineering, University of California, Davis, CA.

Caliskaner, O., and G. Tchobanoglous (2000) "Modeling Depth Filtration of Activated Sludge Effluent Using a Synthetic Compressible Filter Medium," Presented at the 73rd Annual Conference and Exposition on Water Quality and Wastewater Treatment, Water Environment Federation, Anaheim, CA.

Carman, P. C., (1937) "Fluid Flow Through Granular Beds," *Transactions of Institute of Chemical Engineers,* London, vol. 15, p. 150.

Cheryan, M. (1998) *Ultrafiltration and Microfiltration Handbook,* Technomic Publishing Co., Inc., Lancaster, PA.

Cleasby, J. L., and K. Fan (1982) "Predicting Fluidization and Expansion of Filter Media," *Journal of the Environmental Engineering Division, American Society of Civil Engineers,* vol. 107, EE3, pp. 455–472.

Cleasby, J. L., and G. S. Logsdon (1999) "Granular Bed and Precoat Filtration," Chapter 8, in R. D. Letterman (ed.), *Water Quality and Treatment: A Handbook of Community Water Supplies,* 5th ed., American Water Works Association, McGraw-Hill, New York.

Cornwell, D. A. (1990) "Air Stripping and Aeration," in F. W. Pontius (ed.), *Water Quality and Treatment: A Handbook of Community Water Supplies,* 4th ed., American Water Works Association, McGraw-Hill, Inc., New York.

Coulson, J. M., and J. F. Richardson (1968) *Chemical Engineering,* vol. 2, *Unit Operations,* Pergamon Press, Oxford, England.

Crites, R., and G. Tchobanoglous (1998) *Small and Decentralized Wastewater Management Systems,* McGraw-Hill Book Company, New York.

Crittenden, J. C., P. Luft, D. W. Hand, J. L. Oravitz, S. W. Loper, and M. Art (1985) "Prediction of Multicomponent Adsorption Equilibria Using Ideal Adsorption Solution Theory," *Environmental Science & Technology,* vol. 19, no. 11, pp. 1037–1043.

Crittenden, J. C., D. W. Hand, H. Arora, B. W. Lykins, Jr. (1987a) "Design Considerations for GAC Treatment of Organic Chemicals," *Journal American Water Works Association,* vol. 79, no. 1, pp. 74–82.

Crittenden, J. C., T. F. Speth, D. W. Hand, P. J. Luft, and B. W. Lykins, Jr. (1987b) "Multicomponent Competition in Fixed Beds," *Journal Environmental Engineering Division,* American Society of Civil Engineers, vol. 113, no. EE6, pp. 1364–1375.

Crittenden, J. C., P. J. Luft, and D. W. Hand (1987c) "Prediction of Fixed-Bed Adsorber Removal of Organics in Unknown Mixtures," *Journal Environmental Engineering Division,* American Society of Civil Engineers, vol. 113, no. EE3, pp. 486–498.

Crittenden, J. C., P. S. Reddy, H. Arora, J. Trynoski, D. W. Hand, D. L. Perram, and R. S. Summers (1991) "Predicting GAC Performance with Rapid Small-Scale Column Tests," *Journal American Water Works Association,* vol. 83, no. 1, pp. 77–87.

Crittenden, J. C., K. Vaitheeswaran, D. W. Hand, E. W. Howe, E. M. Aieta, C. H. Tate, M. J. McGuire, and M. K. Davis (1993) "Removal of Dissolved Organic Carbon Using Granular Activated Carbon," *Water Research,* vol. 27, no. 4, pp. 715–721.

Crittenden, J. C. (1999) *Class Notes,* Michigan Technological University, Houghton, MI.

Crittenden, J. C. (2000) Personal communication.

Crozes, G. F., J. G. Jacangelo, C. Anselme, and J. M. Laine (1997) "Impact of Ultrafiltration Operating Conditions on Membrane Irreversible Fouling," *Journal of Membrane Science,* vol. 124, p. 63.

Culp, G. L., and A. Slechta (1966) *Nitrogen Removal from Sewage,* Final Progress Report, U.S. Public Health Service Demonstration Grant 29-01.

Dahab, M. F., and J. C. Young (1977) "Unstratified-Bed Filtration of Wastewater," *Journal Environmental Engineering Division, American Society of Civil Engineers,* vol. 103, EE 12714.

Dharmarajah, A. H., and J. L. Cleasby (1986) "Predicting the Expansion of Filter Media," *Journal American Water Works Association,* vol. 798, no. 12, pp. 66–76.

Dimotsis, G. L., and F. McGarvey (1995) "A Comparison of a Selective Resin with a Conventional Resin for Nitrate Removal," IWC, No. 2.

Dobbs, R. A., and J. M. Cohen (1980) "Carbon Adsorption Isotherms for Toxic Organics," EPA-600/8-80-023, U.S. Environmental Protection Agency, Washington, DC.

Dunbar, Professor, Dr. (1908) *Principles of Sewage Treatment,* Charles Griffen & Company, Ltd., London, England.

Eckenfelder, W. W., Jr. (2000) *Industrial Water Pollution Control,* 3rd ed., McGraw Hill, Boston, MA.

Eckert, J. S. (1970) "Selecting the Proper Distillation Column Packing," *Chemical Engineering Progress,* vol. 66, no. 3, pp. 39–44.

Eckert, J. S. (1975) "How Tower Packings Behave," *Chemical Engineering,* vol. 82, no. 4, pp. 70–76.

Fair, G. M. (1951) "The Hydraulics of Rapid Sand Filters," *Journal Institute of Water Engineers,* vol. 5, p. 171.

Fair, G. M., J. C. Geyer, and D. A. Okun (1968) *Water and Wastewater Engineering,* vol. 2, Wiley, New York.

Fair, G. M., and L. P. Hatch (1933) "Fundamental Factors Governing the Streamline Flow of Water Through Sand," *Journal American Water Works Association,* vol. 25, no. 11, p. 1551.

Foust, A. S., L. A. Wenzel, C. W. Clump, L. Maus, and L. B. Andersen (1960) *Principles of Unit Operations,* John Wiley & Sons, Inc., New York, NY.

Frankland, Sir. E. (1870) *River Pollution Commission of Great Britain, First Report,* London, England.

Froelich, E. M. (1978) "Control of Synthetic Organic Chemicals by Granular Activated Carbon: Theory, Application and Reactivation Alternatives," Presented at the Seminar on Control of Organic Chemical Contaminants in Drinking Water, Cincinnati, OH.

Glaze, W. H., and J.-W. Kang (1990) "Chemical Models of Advanced Oxidation Processes," *Proceedings Symposium on Advanced Oxidation Processes,* Wastewater Technology Centre of Environment Canada, Burlington, Ontario, Canada.

Glaze, W. H., J.-W. Kang, and D. H. Chapin (1987) "The Chemistry of Water Treatment Processes Involving Ozone, Hydrogen Peroxide, and Ultraviolet Radiation," *Ozone Science and Engineering,* vol. 9, no. 4, pp. 335–342.

Gopalakrishnan, K., J. Anderson, L. Carrio, K. Abraham, and B. Stinson (2000) "Design and Operational Considerations for Ammonia Removal from Centrate by Steam Stripping," Presented at the 73rd Annual Conference and Exposition, Water Environment Federation, Los Angeles, CA.

Hand, D. W., J. C. Crittenden, D. R. Hokanson, and J. L. Bulloch (1997) "Predicting the Performance of Fixed-Bed Granular Activated Carbon Adsorbers," *Water Science and Technology,* vol. 35, no. 7, pp. 235–241.

Hand, D. W., D. R. Hokanson, and J. C. Crittenden (1999) "Air Stripping and Aeration," chapter 6, in R. D. Letterman (ed.), *Water Quality and Treatment: A Handbook of Community Water Supplies,* 5th ed., American Water Works Association, McGraw-Hill, New York.

Hazen, A. (1905) *The Filtration of Public Water Supplies,* 3rd ed., John Wiley & Sons, New York.

Jacangelo, J. G., and C. A. Buckley (1996) "Chapter 11 Microfiltration," in J. Mallevialle, P. E. Odendaal, and M. R. Wiesner (editorial group) *Water Treatment Membrane Processes,* McGraw-Hill, New York.

Karimi, A. A., J. A. Redman, W. H. Glaze, and G. F. Stolarik (1997) "Evaluating an AOP for TCE and OPCE Removal," *Journal American Water Works Association,* vol. 89, no. 8, pp. 41–53.

Kavanaugh, M. C., and R. R. Trussell (1980) "Design of Stripping Towers to Strip Volatile Contaminants from Drinking Water," *Journal American Water Works Association,* vol. 72., no. 12, p. 684.

Kawamura, S. (2000) *Integrated Design and Operation of Water Treatment Facilities,* 2nd ed., John Wiley & Sons, In., New York.

Kozeny, J. (1927) Uber Grundwasserbewegung, *Wasserkraft and Wasserwirtschaft,* vol. 22, nos. 4, 6, 7, 8, 10.

LaGrega, M. D., P. L. Buckingham, and J. C. Evans (2001) *Hazardous Waste Management,* McGraw-Hill Book Company, Boston.

Leva, M. (1959) *Fluidization,* McGraw-Hill Book Company, Inc., New York.

Liu, D. H. F., and B. G. Liptak (edited by) (2000) *Wastewater Treatment,* Lewis Publishers, Boca Raton, FL.

McCabe, W. L., and E. W. Thiele (1925) "Graphical Design of Fractionating Columns," *Industrial Engineering Chemistry,* vol. 17, p. 605.

McGarvey, F., B. Bachs, and S. Ziarkowski (1989) "Removal of Nitrates from Natural Water Supplies," Presented at the American Chemical Society Meeting, Dallas TX.

Michaels, A. S. (1952) "Simplified Method of Interpreting Kinetic Data in Fixed Bed Ion Exchange," *Industrial and Engineering Chemistry,* vol. 44, no. 8, pp. 1922–1930.

Miltner, R. J., T. F. Speth, D. D. Endicott, and J. M. Reinhold (1987) Final Internal Report on Carbon Use Rate Data, U.S. Environmental Protection Agency, Washington, DC.

O'Melia, C. R., and W. Stumm (1967) "Theory of Water Filtration," *Journal American Water Works Association,* vol. 59, no. 11.

Onda, K., H. Takeuchi, and Y. Okumoto (1968) "Mass Transfer Coefficients Between Gas and Liquid Phases in Packed Columns," *Japanese Journal Chemical Engineering,* vol. 1, no. 1.

Ouki, S. K., and M. Kavannagh (1999) "Treatment of Metals-Contaminated Wastewaters by Use of Natural Zeolites," *Water Science Technology,* vol. 39, no. 10–11, pp. 115–122.

Ozonia, Ltd. (1977) The Ozat™ Compact Ozone Generation Units, Ozonia North America, Elmwood Park, NJ.

Peterson, S. (1953) *Annual New York Academy of Science,* vol. 57, p. 144.

Rice, R. G. (1996) *Ozone Reference Guide,* Prepared for the Electric Power Research Institute, Community Environment Center, St. Louis, MO.

Richardson, J. F., and W. N. Zaki (1954) *Transactions of the Institute of Chemical Engineers,* vol. 32, pp. 35–53.

Riess, J., K. Bourgeous, G. Tchobanoglous, and J. Darby (2001) "Evaluation of the Aqua-Aerobics Cloth Medium Disk Filter (CMDF) for Wastewater Recycling in California," Center for Environmental and Water Resources Engineering, Report 01-2, Department of Civil and Environmental Engineering, University of California, Davis, CA.

Rose, H. E. (1945) "An Investigation of the Laws of Flow of Fluids through Beds of Granular Materials," *Proceedings Institute of Mechanical Engineers, British,* vol. 153, p. 141.

Rose, J., P. Hauch, D. Friedman, and T. Whalen (1999) "The Boiling Effect: Innovation for Achieving Sustainable Clean Water," *Water 21,* 9/10.

Rosene, M. R. et al. (1983) "High Pressure Technique for Rapid Screening of Activated Carbons for Use in Potable Water," in M. J. McGuire and I. H. Suffet (eds.) *Treatment of Water by Granular Activated Carbon,* American Chemical Society, Washington, DC.

SES (1994) *The UV/Oxidation Handbook,* Solarchem Environmental Systems, Markham, Ontario, Canada.

Shaw, D. J. (1966) *Introduction to Colloid and Surface Chemistry,* Butterworth, London, England.

Sherwood, T. K., and F. A. Hollaway (1940) "Performance of Packed Towers-Liquid Film Data for Several Packings," *Transaction American Institute of Industrial Chemical Engineers,* vol. 36, p. 39.

Singer, P. C., and D. A. Reckhow (1999) "Chemical Oxidation," chapter 12, in R. D. Letterman (ed.), *Water Quality and Treatment: A Handbook of Community Water Supplies,* 5th ed., American Water Works Association, McGraw-Hill, New York.

Slater, M. J. (1991) *Principles of Ion Exchange Technology,* Butterworth Heinemann, New York.

Slechta, A., and G. L. Culp (1967) "Water Reclamation Studies at the South Lake Tahoe Public Utility District," *Journal Water Pollution Control Federation,* vol. 39, no. 5.

Snoeyink, V. L., and R. S. Summers (1999) "Adsorption of Organic Compounds," in R. D. Letterman (ed.), *Water Quality and Treatment: A Handbook of Community Water Supplies,* 5th ed., American Water Works Association, McGraw-Hill, New York.

Sontheimer, H., J. C. Crittenden, and R. S. Summers (1988) *Activated Carbon for Water Treatment,* 2nd ed., in English, DVGW-Forschungsstelle, Engler-Bunte-Institut, Universitat Karlsruhe, Germany.

Soroushian, F., Y. Shen, M. Patel, and M. Wehner (2001) "Evaluation and Pilot Testing of Advanced Treatment Processes for NDMA Removal and Reformation," Proceedings of the AWWA Annual Conference, *American Water Works Association,* Washington, D.C.

Stephenson, T., S. Judd, B. Jefferson, and K. Brindle (2000) *Membrane Bioreactors for Wastewater Treatment,* IWA Publishing, London.

Taylor, J. S., and M. Wiesner (1999) "Membranes," chapter 11, in R. D. Letterman (ed.), *Water Quality and Treatment: A Handbook of Community Water Supplies,* 5th ed., American Water Works Association, McGraw-Hill, New York.

Tchobanoglous, G. (1970) "Physical and Chemical Processes for Nitrogen Removal—Theory and Application," *Proceedings Twelfth Sanitary Engineering Conference,* University of Illinois, Urbana, IL.

Tchobanoglous, G. (1988) "Filtration of Secondary Effluent for Reuse Applications," Presented at the 61st Annual Conference of the WPCF, Dallas, TX.

Tchobanoglous, G., and R. Eliassen (1970) "Filtration of Treated Sewage Effluent," *Journal Sanitary Engineering Division,* ASCE, vol. 96, no. SA2, p. 243.

Tchobanoglous, G., and E. D. Schroeder (1985) *Water Quality: Characteristics, Modeling, Modification,* Addison-Wesley Publishing Company, Reading, MA.

Treybal, R. E. (1980) *Mass Transfer Operations,* 3rd ed., Chemical Engineering Series, McGraw-Hill, New York.

Trussell, R. R., and M. Chang (1999) "Review of Flow Through Porous Media as Applied to Head Loss in Water Filters," *Journal of Environmental Engineering, American Society of Civil Engineers,* vol. 125, no. 11.

Weber, W. J., Jr. (1972) *Physicochemical Processes for Water Quality Control,* Wiley-Interscience, New York.

Whitley Burchett & Associates (1999) *Clean Water Revival Groundwater Replenishment System: Performance and Reliability Summary Report,* Report prepared for Dublin San Ramon Services District, Walnut Creek, CA.

12

Disinfection Processes

Disinfection refers to the partial destruction of disease-causing organisms. All the organisms are not destroyed during the process. The fact that all of the organisms are not destroyed differentiates disinfection from sterilization, which is the destruction of all organisms. In the field of wastewater treatment, the four categories of human enteric organisms of the greatest consequence in producing disease are bacteria, protozoan oocysts and cysts, helminths, and viruses. Diseases caused by these waterborne microorganisms have been discussed previously in Chap. 2 and are reported in Table

2–24. The purpose of this chapter is to introduce the reader to the general concepts involved in the disinfection of wastewater and the design of disinfection systems. Topics to be considered include (1) the regulatory requirements for wastewater disinfection, (2) a general introduction to disinfection theory, (3) disinfection with chlorine and related compounds, (4) disinfection with chlorine dioxide, (5) dechlorination, (6) the design of chlorination and dechlorination facilities, (7) disinfection with ozone, (8) other chemical disinfection methods, (9) disinfection with UV radiation, and (10) a comparison of the disinfection methods. Because of the concern over the use of chlorine, the use of UV radiation as an alternative disinfectant is emphasized in this chapter.

12-1 REGULATORY REQUIREMENTS FOR WASTEWATER DISINFECTION

From the early 1900s to the early 1970s, treatment objectives were concerned primarily with (1) the removal of colloidal, suspended, and floatable material, (2) the treatment of biodegradable organics, and (3) the elimination of pathogenic organisms, but generally with unspecified treatment goals. Passage of the Federal Water Pollution Control Act Amendments of 1972 (Public Law 92-500), as amended in 1977 and 1978 (Clean Water Act, Public Laws 95-217 and 97-117, respectively), stimulated substantial changes in wastewater treatment to achieve the principal objective of these acts, which was "to restore and maintain the chemical, physical, and biological integrity of the Nations Waters." In effect, the goal of the Clean Water Act was to make the nation's waters "fishable and swimmable."

To implement the Clean Water Act, the U.S. EPA in 1973 adopted regulations for secondary treatment. These regulations were contained in Part 133, Title 40, of the Code of Federal Regulations (40 CFR 133). Secondary treatment was defined in terms of biochemical oxygen demand (BOD), total suspended solids (TSS), pH, and fecal coliform (FC) bacteria (see Table 1–3 in Chap. 1). The fecal coliform bacteria requirement was deleted in 1976 and replaced with a provision that allowed individual states to develop site-specific water-quality standards including disinfection standards for wastewater-treatment plants (WWTP). As a consequence, today (2001) some states have fecal coliform standards varying from ≤2.2 to 5000 MPN/100 mL and total coliform standards varying from ≤2.2 to 10,000 MPN/100 mL. Further, a number of states have seasonal standards. More than 40 states have variable standards, depending on the quality of the receiving water and, if the treated wastewater is to be reused, on the reuse application. The most common standard for receiving waters is 200 MPN FC/100 mL. Because the regulations for disinfection are site-specific and are under continual review, the governing regulatory authority must be consulted to determine current requirements.

12-2 DISINFECTION THEORY

The purpose of this section is to introduce the basic concepts involved in disinfection. The background material presented in this section is intended to serve as a basis for the discussion of disinfectants used for wastewater. Topics to be discussed include a review of the requirements for an ideal disinfectant, current disinfection methods and means, mechanisms of disinfectants, and a general discussion of the factors influencing the action of disinfectants.

Characteristics of an Ideal Disinfectant

The characteristics for an ideal disinfectant are reported in Table 12–1. As shown, an ideal disinfectant would have to possess a wide range of characteristics. Although such a compound may not exist, the characteristics set forth in Table 12–1 should be considered in evaluating proposed or recommended disinfectants. It is also important that the disinfectant be safe to handle and apply, and that its strength or concentration in treated waters be measurable. The latter consideration is an issue with the use of ozone and UV disinfection where no residual is measured.

Disinfection Methods and Means

Disinfection is most commonly accomplished by the use of (1) chemical agents, (2) physical agents, (3) mechanical means, and (4) radiation. Each of these techniques is considered briefly in the following discussion.

Chemical Agents. Chemical agents that have been used as disinfectants include (1) chlorine and its compounds, (2) bromine, (3) iodine, (4) ozone, (5) phenol and phenolic compounds, (6) alcohols, (7) heavy metals and related compounds, (8) dyes, (9) soaps and synthetic detergents, (10) quaternary ammonium compounds, (11) hydrogen peroxide, (12) peracetic acid, (13) various alkalies, and (14) various acids. Of these, the most common disinfectants are the oxidizing chemicals, and chlorine is the one used most universally. Although not discussed in this chapter, bromine and iodine have also been used for wastewater disinfection. Ozone is a highly effective disinfectant, and its use is increasing even though it leaves no residual (see Sec. 12–7). Highly acid or alkaline water can also be used to destroy pathogenic bacteria, because water with a pH greater than 11 or less than 3 is relatively toxic to most bacteria.

Table 12–1
Characteristics of an ideal disinfectant

Characteristic	Properties/response
Availability	Should be available in large quantities and reasonably priced
Deodorizing ability	Should deodorize while disinfecting
Homogeneity	Solution must be uniform in composition
Interaction with extraneous material	Should not be absorbed by organic matter other than bacterial cells
Noncorrosive and nonstaining	Should not disfigure metals or stain clothing
Nontoxic to higher forms of life	Should be toxic to microorganisms and nontoxic to humans and other animals
Penetration	Should have the capacity to penetrate through surfaces
Safety	Should be safe to transport, store, handle, and use
Solubility	Must be soluble in water or cell tissue
Stability	Should have low loss of germicidal action with time on standing
Toxicity to microorganisms	Should be effective at high dilutions
Toxicity at ambient temperatures	Should be effective in ambient temperature range

Physical Agents. Physical disinfectants that can be used are heat, light, and sound waves. Heating water to the boiling point, for example, will destroy the major disease-producing non-spore-forming bacteria. Heat is commonly used in the beverage and dairy industry, but it is not a feasible means of disinfecting large quantities of wastewater because of the high cost. However, pasteurization of sludge is used extensively in Europe.

Sunlight is also a good disinfectant, due primarily to the ultraviolet (UV) radiation portion of the electromagnetic spectrum. The decay of microorganisms observed in oxidation ponds is due, in part, to their exposure to the UV component of sunlight. Special lamps developed to emit ultraviolet rays have been used successfully to disinfect water and wastewater. The efficiency of the process depends on the penetration of the rays into water. The contact geometry between the ultraviolet radiation source and the water is extremely important because suspended matter, dissolved organic molecules, and water itself will absorb the radiation, in addition to the microorganisms.

Mechanical Means. Bacteria and other organisms are also removed by mechanical means during wastewater treatment. Typical removal efficiencies for various treatment operations and processes are reported in Table 12–2. The first four operations listed may be considered to be physical. The removals accomplished are byproducts of the primary function of the process.

Radiation. The major types of radiation are electromagnetic, acoustic, and particle. Gamma rays are emitted from radioisotopes, such as cobalt-60. Because of their penetration power, gamma rays have been used to disinfect (sterilize) both water and wastewater. Although the use of a high-energy electron-beam device for the irradiation of wastewater or sludge has been studied extensively, there are no commercial devices or full-scale installations in operation.

Comparison of Wastewater Disinfectants. Using the criteria defined in Table 12–1, disinfectants that have been used in wastewater are compared in Table 12–3.

Table 12–2
Removal or destruction of bacteria by different treatment processes

Process	Percent removal
Coarse screens	0–5
Fine screens	10–20
Grit chambers	10–25
Plain sedimentation	25–75
Chemical precipitation	40–80
Trickling filters	90–95
Activated sludge	90–98
Chlorination of treated wastewater	98–99.999

Table 12-3

Comparison of ideal and actual characteristics of commonly used disinfectants[a,b]

Characteristic[a]	Chlorine	Sodium hypochlorite	Calcium hypochlorite	Chlorine dioxide	Ozone	UV radiation
Availability/cost	Low cost	Moderately low cost	Moderately low cost	Moderately low cost	Moderately high cost	Moderately high cost
Deodorizing ability	High	Moderate	Moderate	High	High	na
Homogeneity	Homogeneous	Homogeneous	Homogeneous	Homogeneous	Homogeneous	na
Interaction with extraneous material	Oxidizes organic matter	Active oxidizer	Active oxidizer	High	Oxidizes organic matter	Absorbance of UV radiation
Noncorrosive and nonstaining	Highly corrosive	Corrosive	Corrosive	Highly corrosive	Highly corrosive	na
Nontoxic to higher forms of life	Highly toxic to higher life forms	Toxic	Toxic	Toxic	Toxic	Toxic
Penetration	High	High	High	High	High	Moderate
Safety concern	High	Moderate	Moderate	High	Moderate	Low
Solubility	Moderately	High	High	High	High	na
Stability	Stable	Slightly unstable	Relatively stable	Unstable, must be generated as used	Unstable, must be generated as used	na
Toxicity to microorganisms	High	High	High	High	High	High
Toxicity at ambient temperatures	High	High	High	High	High	High

[a] See Table 12–1 for a description of each characteristic.

[b] na = not applicable.

Mechanisms of Disinfectants

The five principal mechanisms that have been proposed to explain the action of disinfectants are (1) damage to the cell wall, (2) alteration of cell permeability, (3) alteration of the colloidal nature of the protoplasm, (4) alteration of the organism DNA or RNA, and (5) inhibition of enzyme activity.

Damage or destruction of the cell wall will result in cell lysis and death. Some agents, such as penicillin, inhibit the synthesis of the bacterial cell wall.

Agents such as phenolic compounds and detergents alter the permeability of the cytoplasmic membrane. These substances destroy the selective permeability of the membrane and allow vital nutrients, such as nitrogen and phosphorus, to escape.

Heat, radiation, and highly acid or alkaline agents alter the colloidal nature of the protoplasm. Heat will coagulate the cell protein and acids or bases will denature proteins, producing a lethal effect.

UV radiation can cause the formation of double bonds in microorganisms as well as rupturing some DNA strands. When UV photons are absorbed by the DNA in bacteria and protozoa and the DNA and RNA in viruses, covalent dimers can be formed from adjacent thymines in DNA or uracils in RNA. The formation of double bonds disrupts the replication process so that the organism can no longer reproduce and is thus inactivated.

Another mode of disinfection is the inhibition of enzyme activity. Oxidizing agents, such as chlorine, can alter the chemical arrangement of enzymes and inactivate the enzymes.

A comparison of the mechanism of disinfection using chlorine, ozone, and UV radiation, commonly used disinfectants for wastewater, is presented in Table 12–4. To a large extent, observed performance differences for the various disinfectants can be explained on the basis of the operative removal mechanisms.

Factors Influencing the Action of Disinfectants

In applying the disinfection agents or means that have been described, the following factors must be considered: (1) contact time, (2) concentration of the disinfectant,

Table 12–4
Mechanisms of disinfection using chlorine, UV, and ozone

Chlorine	Ozone	UV radiation
1. Oxidation	1. Direct oxidation/destruction of cell wall with leakage of cellular constituents outside of cell	1. Photochemical damage to RNA and DNA (e.g., formation of double bonds) within the cells of an organism
2. Reactions with available chlorine		
3. Protein precipitation	2. Reactions with radical byproducts of ozone decomposition	2. The nucleic acids in microorganisms are the most important absorbers of the energy of light in the wavelength range of 240–280 nm
4. Modification of cell wall permeability	3. Damage to the constituents of the nucleic acids (purines and pyrimidines)	
5. Hydrolysis and mechanical disruption	4. Breakage of carbon-nitrogen bonds leading to depolymerization	3. Because DNA and RNA carry genetic information for reproduction, damage of these substances can effectively inactivate the cell

(3) intensity and nature of physical agent or means, (4) temperature, (5) types of organisms, and (6) nature of suspending liquid. The subjects introduced in this section are considered further in the subsequent sections. It is important to note that in the discussion that follows, the various formulations used to explain the action of disinfectants are based on an analysis of data derived using discrete organisms in solution. The presence of suspended material, which may provide shielding of microorganisms, is considered in the subsequent section dealing with chlorine and in the section dealing with UV disinfection.

Contact Time. Perhaps one of the most important variables in the disinfection process is contact time. Working in England in the early 1900s, Harriet Chick observed that for a given concentration of disinfectant, the longer the contact time, the greater the kill (see Fig. 12–1). This observation was first reported in the literature in 1908 (Chick, 1908). In differential form, Chick's law is

$$\frac{dN_t}{dt} = -kN_t \tag{12–1}$$

where dN_t/dt = rate of change in the concentration of organisms with time
k = inactivation rate constant, T^{-1}
N_t = number of organisms at time t
t = time

If N_o is the number of organisms when t equals 0, Eq. (12–1) can be integrated to

$$\frac{N_t}{N_o} = e^{-kt} \tag{12–2}$$

Figure 12–1

Effect of time and concentration on survival of dispersed *E. coli* in a batch reactor using chlorine as disinfectant at 20°C.

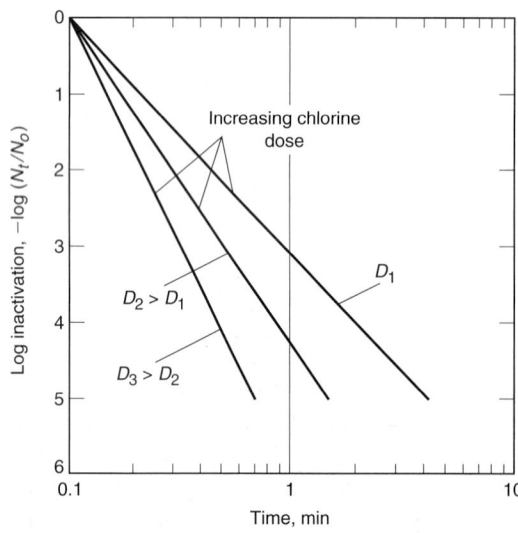

or

$$\ln \frac{N_t}{N_o} = -kt \qquad (12\text{–}3)$$

The value of the inactivation rate constant k in Eq. (12–3) can be obtained by plotting $-\ln (N_t/N_o)$ versus the contact time t.

Concentration of Disinfectant. Also working in England in the early 1900s, Herbert Watson reported that the inactivation rate constant was related to the concentration as follows (Watson, 1908):

$$k = k'C^n \qquad (12\text{–}4)$$

where k = inactivation rate constant
$\qquad k'$ = die-off constant
$\qquad C$ = concentration of disinfectant
$\qquad n$ = coefficient of dilution

Combining the expressions proposed by Chick and Watson in differential form yields (Hass and Kara, 1984)

$$\frac{dN_t}{dt} = -k'C^nN_t \qquad (12\text{–}5)$$

The integrated form of Eq. (12–5) is

$$\frac{N_t}{N_o} = e^{-k'C^nt} \qquad (12\text{–}6)$$

or

$$\ln \frac{N_t}{N_o} = -k'C^nt \qquad (12\text{–}7)$$

The linearized form of Eq (12–7) is:

$$\ln C = -\frac{1}{n} \ln t + \frac{1}{n} \ln \left[\frac{1}{k'}\left(-\ln \frac{N_t}{N_o}\right) \right]$$

The value of n can be obtained by plotting C versus t on log-log paper for a given level of inactivation (see Example 12–1). The following explanation has been offered for various values of n:

$n = 1$ both the concentration and time are equally important

$n > 1$ concentration is more important than time

$n < 1$ time is more important than concentration

Departures from this rate law are common, as shown on Fig. 12–2. As shown on Fig. 12–2a, there can be a lag or shoulder effect in which constituents in the suspending liquid react initially with the disinfectant, rendering the disinfectant ineffective. The tailing effect in which large particles shield the organisms to be disinfected is shown on

Figure 12–2

Departures observed from Chick's law: (a) lag or shoulder effect in which constituents in the suspending liquid react initially with the disinfectant, (b) tailing effect in which large particles shield the organisms to be disinfected, and (c) combined lag and tailing effects.

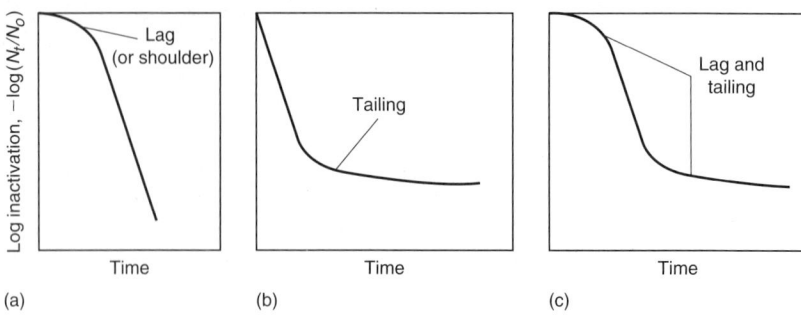

Fig. 12–2b. The combined effects of lag and tailing are illustrated on Fig. 12–2c. These observed phenomena are discussed in greater detail in the following section dealing with the disinfection of wastewater. Further, Eq. (12–6) as applied to wastewater fails to account for the variable, heterogeneous characteristics of wastewater.

Using an n value of 1 and assuming the C value corresponds to the chlorine residual C_R, the relationship given by Eq. (12–6) is used by a number of regulatory agencies to control the wastewater disinfection process. This subject is considered in greater detail in the discussion dealing with chlorine disinfection (Sec. 12–3).

EXAMPLE 12–1 Application of the Modified Chick/Watson Equation Given the following chlorination test survival data for *E. coli*, expressed as a percentage, determine the values of the constants in the Chick and Watson equation [Eq. (12–6)] for 99 percent reduction. The results were obtained using a batch reactor.

	Percent survival				
	Contact time, min[a]				
Free Cl$_2$,[b] mg/L	1	3	5	10	20
0.05	97	82	63	21	0.3
0.07	93	60	28	0.5	—
0.14	67	11	0.7	—	—

[a] Test conditions pH = 8.5; temp = 5°C.
[b] HOCl (hypochlorous acid) and OCl$^-$ (hypochlorite ion). See also discussion in Sec. 12–3.

Solution

1. To determine the constants in the Chick and Watson equation form of Chick's law, convert the survival data to log removal values and then plot on arithmetic and log-log paper to determine the time required for a given degree of kill.

a. Convert the given percentage removal data to log removal values.

	$-\ln (N_t/N_o)$				
	Contact time, min				
Free Cl$_2$, mg/L	**1**	**3**	**5**	**10**	**20**
0.05	0.030	0.198	0.462	1.561	5.809
0.07	0.073	0.511	1.273	5.298	
0.14	0.400	2.207	4.962		

b. Plot the converted values on both arithmetic and log-log paper to determine the best fit of the data. After plotting the data, it was found that the data could be represented best by a straight line as shown on the following log-log plot.

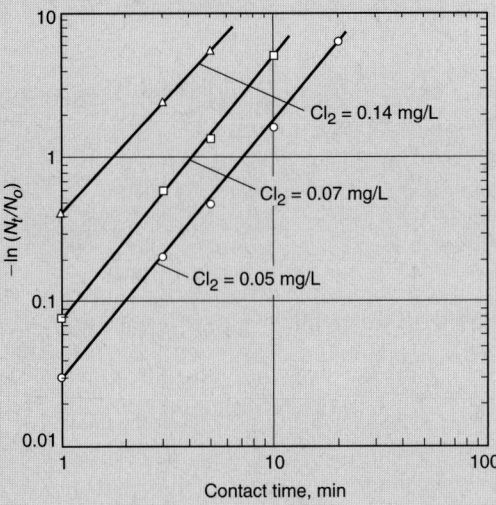

c. Determine the required contact time for a log inactivation value of 99 percent $[-\ln (N_t/N_o) = 4.61]$ for the various concentrations. The required values are:

Concentration, mg/L	Time, min
0.05	18.0
0.07	10.0
0.14	4.7

2. Determine the constants in the Chick and Watson equation using the data from Step 1.

a. Plot the concentration versus time values determined in Step 1c.

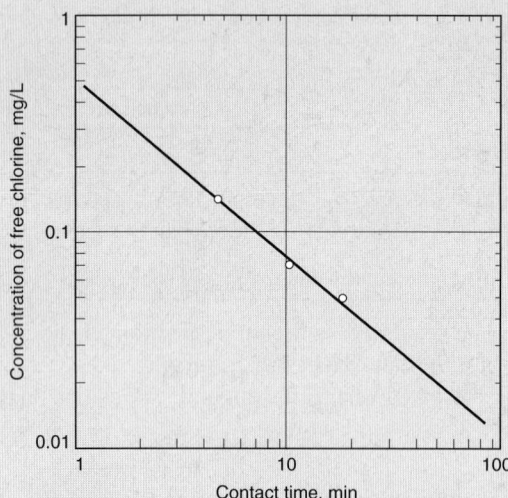

b. Use the linear form of Eq. (12–7) to determine the constants

$$\ln C = -\frac{1}{n} \ln t + \frac{1}{n} \ln \left[\frac{1}{k'} \left(-\ln \frac{N_t}{N_o} \right) \right]$$

$$\ln C = -\frac{1}{n} \ln t + \frac{1}{n} (\text{constant})$$

From the above plot, the value of n equals

$$\text{Slope} = -\frac{1}{n} = -\frac{\log 0.5 - \log 0.011}{\log 100 - \log 1} = -\frac{[-0.30 - (-1.96)]}{2 - 0}$$

$$= -\frac{1.66}{2} = -0.83$$

$$n = 1.20$$

When $t = 1$, the Y-intercept $= \ln 0.5 = \frac{1}{n} \ln \left[\frac{1}{k'} \left(-\ln \frac{N_t}{N_o} \right) \right]$

$$n(\ln 0.5) = \ln \left[\frac{1}{k'} \left(-\ln \frac{N_t}{N_o} \right) \right]$$

$$1.20(-0.69) = \ln \left[\frac{1}{k'} (-\ln 0.01) \right]$$

$$-0.44 = \left[\frac{1}{k'} (4.61) \right]$$

$$k' = -\frac{1}{0.44}(4.61) = -10.48$$

$$\ln\frac{N_t}{N_o} = -10.48C^{1.20}t$$

Comment   The equation developed in this example can also be used to determine the degree of inactivation at different temperatures (see Example 12–2).

Intensity and Nature of Physical Agent.   As noted earlier, heat and light are physical agents that have been used from time to time in the disinfection of wastewater. It has been found that their effectiveness is a function of intensity. For example, if the decay of organisms can be described with a first-order reaction [see Eq. (12–1)], then the effect of the intensity of the physical disinfectant is reflected in the constant k through some functional relationship.

Temperature.   The effect of temperature on rate of kill with chemical disinfectants can be represented by a form of the van't Hoff–Arrhenius relationship. Increasing the temperature results in a more rapid kill. In terms of the time t required to effect a given percentage kill, the relationship is

$$\ln\frac{t_1}{t_2} = \frac{E(T_2 - T_1)}{RT_1T_2} \tag{12-8}$$

where t_1, t_2 = time for given percentage kill at temperatures T_1 and T_2, K, respectively
E = activation energy, J/mole (cal/mole)
R = gas constant, 8.3144 J/mole·K = (1.99 cal/mole·K)

Some typical values for the activation energy for various chlorine compounds at different pH values are reported in Table 12–5. The effect of temperature is considered in Example 12–2.

Table 12–5
Activation energies for aqueous chlorine and chloramines at normal temperatures (~20°C)[a]

Compound	pH	E, Cal/mole	E, J/mole
Aqueous chlorine	7.0	8,200	34,340
	8.5	6,400	26,800
	9.8	12,000	50,250
	10.7	15,000	62,810
Chloramines	7.0	12,000	50,250
	8.5	14,000	58,630
	9.5	20,000	83,750

[a]From Fair et al. (1948).

EXAMPLE 12–2 **Effect of Temperature on Disinfection Times** Estimate the time required for a 99.9 percent kill for a chlorine dosage of 0.05 mg/L at a temperature of 20°C and pH of 8.5, using the data and relationship developed in Example 12–1. The test data given in Example 12–1 were obtained at a temperature of 5°C using a batch reactor.

Solution

1. Estimate the time required for a 99.9 percent kill using the equation developed in Example 12–1.

$$\ln \frac{N_t}{N_o} = -10.48C^{1.20}t$$

$$\ln \frac{0.10}{100} = -(10.48)(0.05)^{1.20}t$$

$$t = \frac{-6.91}{(-10.48)(0.027)} = 24.4 \text{ min at } 5°C$$

2. Estimate the time required at 20°C using the modified form of the van't Hoff–Arrhenius equation [Eq. (12–8)] and the data given in Table 12–5.

$$\ln \frac{t_1}{t_2} = \frac{E(T_2 - T_1)}{RT_1T_2}$$

$$\ln \frac{t_1}{20.4 \text{ min}} = \frac{(26,800 \text{ J/mole})[(278.15 - 293.15)\text{K}]}{(8.3144 \text{ J/mole·K})[(293.15 \text{ K})(278.15 \text{ K})]}$$

$$\ln \frac{t_1}{24.4 \text{ min}} = -0.60$$

$$\frac{t_1}{24.4 \text{ min}} = e^{-0.60} = 0.549$$

$$t = 13.4 \text{ min at } 20°C$$

Types of Organisms. The effectiveness of various disinfectants will be influenced by the type, nature, and condition of the microorganisms. For example, viable, growing bacteria cells are often killed more easily than older cells that developed a slime coating. In contrast, bacterial spores are extremely resistant, and many of the chemical disinfectants normally used will have little or no effect. Similarly, many of the viruses and protozoa of concern respond differently to each of the chemical disinfectants. In some cases, other disinfecting agents, such as heat or UV radiation, may have to be used. The inactivation of different microorganism groups is considered further in the following sections.

Nature of Suspending Liquid. In reviewing the development of the various relationships proposed for the inactivation of microorganisms, it is important to note that most of the tests were conducted in batch reactors using distilled or buffered water, under laboratory conditions. In practice, the nature of the suspending liquid must be evaluated carefully. For example, extraneous organic material will react with most oxidizing disinfectants and reduce their effectiveness. The presence of suspended matter will reduce the effectiveness of disinfectants by absorption of the disinfectant and by shielding the entrapped bacteria. The characteristics of the suspending liquid are examined in more detail in the following sections and in the discussion of UV disinfection (Sec. 12–9).

12–3 DISINFECTION WITH CHLORINE

As noted earlier, of all the chemical disinfectants, chlorine is the one used most commonly throughout the world. The reason is that chlorine satisfies most of the requirements specified in Table 12–1. The first use of chlorine for the disinfection of wastewater in the United States was in 1914 at Altoona, PA and Milwaukee, WI (Enslow, 1938). Specific topics considered in this section include a brief description of the characteristics of the various chlorine compounds, a review of chlorine chemistry and breakpoint chlorination, an analysis of the performance of chlorine as a disinfectant and the factors that may influence the effectiveness of the chlorination process, a discussion of the formation of disinfection byproducts (DBPs), and a consideration of the potential impacts of the discharge of DBPs to the environment. Disinfection with chlorine dioxide, chlorination and dechlorination, and the design of chlorination and dechlorination facilities are considered in the following three sections, respectively.

Characteristics of Chlorine Compounds

The principal chlorine compounds used at wastewater-treatment plants are chlorine (Cl_2), sodium hypochlorite (NaOCl), calcium hypochlorite [$Ca(OCl)_2$], and chlorine dioxide (ClO_2). Many large cities have switched from chlorine gas to sodium hypochlorite because of the safety concerns related to handling and storage of liquid chlorine. The characteristics of Cl_2, NaOCl, and $Ca(OCl)_2$ are considered in the following discussion. The characteristics of chlorine dioxide and its use as a disinfectant are discussed in the following section.

Chlorine.   Chlorine (Cl_2) can be present as a gas or a liquid. Chlorine gas is greenish yellow in color and about 2.48 times as heavy as air. Liquid chlorine is amber colored and about 1.44 times as heavy as water. Unconfined liquid chlorine vaporizes rapidly to a gas at standard temperature and pressure with 1 liter of liquid yielding about 450 liters of gas. Chlorine is moderately soluble in water, with a maximum solubility of about 1 percent at 10°C (50°F). Chlorine is supplied as a liquefied gas under high pressure in containers varying in size from 45 kg (100 lb) and 68 kg (150 lb) cylinders, 908 kg (1 ton) containers, multiunit railcars containing fifteen 908 kg (1 ton) containers, and railcars with capacities of 14.5, 27.2, and 49.9 Mg (16, 30, and 55 tons) (see Fig. 12–3). Selection of the size of chlorine pressure vessel depends on an analysis of the rate of chlorine usage, cost of chlorine, facilities requirements, and dependability of supply. The properties of chlorine are summarized in Table 12–6.

Figure 12–3

Typical supply and storage containers for chlorine: (a) 908 kg (1.0 ton) containers and (b) railcars of varying capacity.

(a)

(b)

Table 12–6
Properties of chlorine, chlorine dioxide, and sulfur dioxide[a]

Property	Unit	Chlorine (Cl_2)	Chlorine dioxide (ClO_2)	Sulfur dioxide (SO_2)
Molecular weight	g	70.91	67.45	64.06
Boiling point (liquid)	°C	−33.97	11	
Melting point	°C	−100.98	−59	
Latent heat of vaporization at 0°C	kJ/kg	253.6	27.28	376.0
Liquid density at 15.5°C	kg/m³	1422.4	1640[b]	1396.8
Solubility in water at 15.5°C	g/L	7.0	70.0[b]	120
Specific gravity of liquid at 0°C (water = 1)	s.g.	1.468		1.486
Vapor density at 0°C and 1 atm.	kg/m³	3.213	2.4	2.927
Vapor density compared to dry air at 0°C and 1 atm	unitless	2.486	1.856	2.927
Specific volume of vapor at 0°C and 1 atm	m³/kg	0.3112	0.417	0.342
Critical temperature	°C	143.9	153	157.0
Critical pressure	kPa	7811.8		7973.1

[a] Adapted in part from U.S. EPA (1986); White (1999).
[b] At 20°C.

Although the use of chlorine for the disinfection of both potable water supplies and treated wastewater has been of great significance from a public health perspective, serious concerns have been raised of its continued use. Important concerns include:

1. Chlorine is a highly toxic substance that is transported by rail and truck, both of which are prone to accidents.

2. Chlorine is a highly toxic substance that potentially poses health risks to treatment-plant operators and the general public if released by accident.

3. Because chlorine is a highly toxic substance, stringent requirements for containment and neutralization must be implemented as specified in the Uniform Fire Code (UFC).

4. Chlorine reacts with the organic constituents in wastewater to produce odorous compounds.

5. Chlorine reacts with the organic constituents in wastewater to produce byproducts, many of which are known to be carcinogenic and/or mutagenic.

6. Residual chlorine in treated wastewater effluent is toxic to aquatic life.

7. Concern exists over the discharge of chloro-organic compounds to the environment whose long-term effects are not known.

Sodium Hypochlorite. Many of the safety concerns related to the transport, storage, and feeding of liquid-gaseous chlorine are eliminated by the use of either sodium or calcium hypochlorite. Sodium hypochlorite (NaOCl) (i.e., liquid bleach), is only available as liquid and usually contains 12.5 to 17 percent available chlorine at the time it is manufactured. Sodium hypochlorite can be purchased in bulk lots of 12 to 15 percent of available chlorine or manufactured onsite. The solution decomposes more readily at high concentrations and is affected by exposure to light and heat. A 16.7 percent solution stored at 26.7°C (80°F) will lose 10 percent of its strength in 10 days, 20 percent in 25 days, and 30 percent in 43 days. It must therefore be stored in a cool location in a corrosion-resistant tank (see Fig. 12–4). Another disadvantage of sodium hypochlorite is the chemical cost. The purchase price may range from 150 to 200 percent of the cost of liquid chlorine. The handling of sodium hypochlorite requires special design considerations because of its corrosiveness and the presence of chlorine fumes.

Several proprietary systems are available for the generation of sodium hypochlorite from sodium chloride (NaCl) or seawater. These systems are electric power intensive and, in the case of generation from seawater, result in a very dilute solution, a maximum of 0.8 percent hypochlorite. The onsite generation systems have been used only on a limited basis due to their complexity and high power cost.

Figure 12–4

Typical onsite storage and feeding equipment for sodium hypochlorite. The facility shown is typical of those used at smaller treatment plants; in larger facilities multiple storage tanks are used.

Calcium Hypochlorite. Calcium hypochlorite [Ca(OCl)$_2$] is available commercially in either a dry or a wet form. High-test calcium hypochlorite contains at least 70 percent available chlorine. In dry form it is available as an off-white powder or as granules, compressed tablets, or pellets. A wide variety of container sizes are available depending on the source. Calcium hypochlorite granules or pellets are readily soluble in water, varying from about 21.5 g/100 mL at 0°C (32°F) to 23.4 g/100 mL at 40°C (104°F). Because of its oxidizing potential, calcium hypochlorite should be stored in a cool, dry location away from other chemicals in corrosion-resistant containers. With proper storage conditions, the granules are relatively stable. Hypochlorite is more expensive than liquid chlorine, loses its available strength on storage, and may be difficult to handle. Because it tends to crystallize, calcium hypochlorite may clog metering pumps, piping, and valves. Calcium hypochlorite is used most commonly at small installations.

Chemistry of Chlorine Compounds

The following discussion deals with the reactions of chlorine in water and the reaction of chlorine with ammonia.

Chlorine Reactions in Water. When chlorine in the form of Cl$_2$ gas is added to water, two reactions take place: hydrolysis and ionization.

Hydrolysis may be defined as the reaction in which chlorine gas combines with water to form hypochlorous acid (HOCl).

$$Cl_2 + H_2O \leftrightarrow HOCl + H^+ + Cl^- \tag{12-9}$$

The equilibrium constant K_H for this reaction is

$$K_H = \frac{[HOCl][H^+][Cl^-]}{[Cl_2]} = 4.5 \times 10^{-4} \text{ (mole/L)}^2 \text{ at } 25°C \tag{12-10}$$

Because of the magnitude of the equilibrium constant, large quantities of chlorine can be dissolved in water.

Ionization of hypochlorous acid to hypochlorite ion (OCl$^-$) may be defined as

$$HOCl \leftrightarrow H^+ + OCl^- \tag{12-11}$$

The ionization constant K_i for this reaction is

$$K_i = \frac{[H^+][OCl^-]}{[HOCl]} = 3 \times 10^{-8} \text{ mole/L at } 25°C \tag{12-12}$$

The variation in the value of K_i with temperature is reported in Table 12–7.

The total quantity of HOCl and OCl$^-$ present in water is called the "free available chlorine." The relative distribution of these two species (see Fig. 12–5) is very important because the killing efficiency of HOCl is about 40 to 80 times that of OCl$^-$. The percentage distribution of HOCl at various temperatures can be computed using Eq. (12–13) and the data in Table 12–7.

$$\frac{[HOCl]}{[HOCl] + [OCl^-]} = \frac{1}{1 + [OCl^-]/[HOCl]} = \frac{1}{1 + K_i[H^+]} = \frac{1}{1 + K_i 10^{pH}} \tag{12-13}$$

Table 12–7
Values of the ionization constant of hypochlorous acid at different temperatures[a]

Temperature, °C	$K_i \times 10^8$, mole/L
0	1.5
5	1.7
10	2.0
15	2.3
20	2.6
25	2.9

[a]Computed using equation from Morris (1966).

Figure 12–5

Percentage distribution of hypochlorous acid and hypochlorite in water as a function of pH at 0 and 20°C.

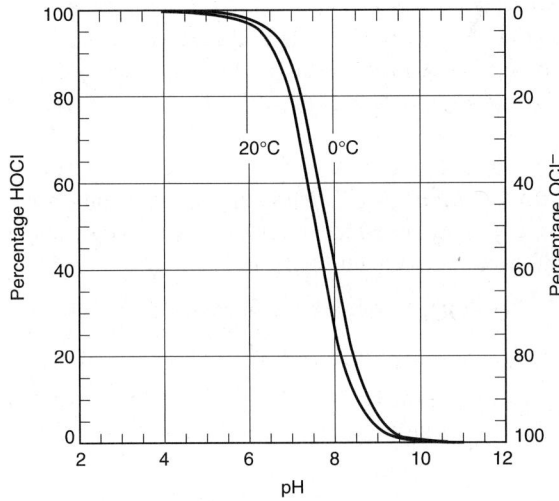

Hypochlorite Reactions in Water. Free available chlorine can also be added to water in the form of hypochlorite salts. Both calcium and sodium hypochlorite hydrolyze to form hypochlorous acid (HOCl) as follows:

$$Ca(OCl)_2 + 2H_2O \rightarrow 2HOCl + Ca(OH)_2 \qquad (12\text{–}14)$$

$$NaOCl + H_2O \rightarrow HOCl + NaOH \qquad (12\text{–}15)$$

The ionization of hypochlorous acid was discussed previously [see Eq. (12–11)].

Chlorine Reactions with Ammonia. As noted in Chaps. 2 and 3, untreated wastewater contains nitrogen in the form of ammonia and various combined organic forms. The effluent from most treatment plants also contains significant amounts of nitrogen, usually in the form of ammonia, or nitrate if the plant is designed to achieve nitrification (see Chaps. 8 and 9). Because hypochlorous acid is a very active oxidizing agent, it will react readily with ammonia in the wastewater to form three types of chloramines in successive reactions:

$$NH_3 + HOCl \rightarrow NH_2Cl \text{ (monochloramine)} + H_2O \qquad (12\text{--}16)$$

$$NH_2Cl + HOCl \rightarrow NHCl_2 \text{ (dichloramine)} + H_2O \qquad (12\text{--}17)$$

$$NHCl_2 + HOCl \rightarrow NCl_3 \text{ (nitrogen trichloride)} + H_2O \qquad (12\text{--}18)$$

These reactions are dependent on the pH, temperature, and contact time, and on the ratio of chlorine to ammonia (White, 1999). The two species that predominate, in most cases, are monochloramine (NH_2Cl) and dichloramine ($NHCl_2$). The ratio of dichloramine to monochloramine as a function of the ratio of chlorine to ammonia at various pH values is presented in Table 12–8. The amount of nitrogen trichloride present is negligible up to chlorine-to-nitrogen ratios of 2.0. The chlorine in these compounds is called "combined available chlorine." As will be discussed subsequently, chloramines also serve as disinfectants, although they are slow-reacting. When chloramines are the only disinfectants, the measured residual chlorine is defined as "combined chlorine residual" as opposed to free chlorine in the form of hypochlorous acid and hypochlorite ion.

Actual and Available Chlorine. The percent actual and available chlorine can be used to compare the effectiveness of compounds containing chlorine. The percent actual chlorine is determined as follows:

$$(Cl_2)_{\text{actual}}, \% = \frac{(\text{weight of chlorine in compound})}{(\text{molecular weight of compound})} \times 100 \qquad (12\text{--}19)$$

Available chlorine is the term used to compare the "oxidizing power" of chlorine compounds. The oxidizing power of chlorine is based on the value of the valence of the

Table 12–8
Ratio of dichloramine to monochloramine under equilibrium conditions as a function of pH and applied molar dose ratio of chlorine to ammonia[a]

Molar ratio Cl_2:NH_4	pH			
	6	7	8	9
0.1	0.13	0.014	1E-03	0.000
0.3	0.389	0.053	5E-03	0.000
0.5	0.668	0.114	0.013	1E-03
0.7	0.992	0.213	0.029	3E-03
0.9	1.392	0.386	0.082	0.011
1.1	1.924	0.694	0.323	0.236
1.3	2.700	1.254	0.911	0.862
1.5	4.006	2.343	2.039	2.004
1.7	6.875	4.972	4.698	4.669
1.9	20.485	18.287	18.028	18.002

[a]From U.S. EPA (1986).

chloride in the compound that is reduced to a valence value of -1. For example, the half reaction for hypochlorous acid is as given below (see Table 6–11, Chap. 6)

$$HOCl + H^+ + 2e^- \rightarrow Cl^- + H_2O \tag{12–20}$$

As shown in Eq. (12–20), the electron change was 2. The percent available chlorine is given by the following relationship:

$$(Cl_2)_{available} = (Cl \text{ equivalent}) [(Cl_2)_{actual}, \%)] \tag{12–21}$$

Thus, for HOCl the actual percentage of chlorine is 67.7 percent $[(35.5/52.5) \times 100]$ and the available chlorine is 135.4 percent (2×67.7). Values for the actual and available percent chlorine are given in Table 12–9 for chlorine and various compounds containing chlorine that have been used as disinfectants.

Breakpoint Reaction with Chlorine

The maintenance of a residual (combined or free) for the purpose of wastewater disinfection is complicated because free chlorine not only reacts with ammonia, as noted previously, but also is a strong oxidizing agent. The term "breakpoint chlorination" is the term applied to the process whereby enough chlorine is added to react with all oxidizable substances such that if additional chlorine is added it will remain as free chlorine. The main reason for adding enough chlorine to obtain a free chlorine residual is that effective disinfection can usually then be assured. Breakpoint chlorination chemistry, acid generation, and the buildup of dissolved solids are considered in the following discussion.

Breakpoint Chlorination Chemistry. The stepwise phenomena that result when chlorine is added to wastewater containing oxidizable substances and ammonia can be explained by referring to Fig. 12–6. As chlorine is added, readily oxidizable substances, such as Fe^{2+}, Mn^{2+}, H_2S, and organic matter, react with the chlorine and

Table 12–9
Actual and available chlorine in compounds containing chlorine[a]

Compound	Molecular weight	Chlorine equivalent[a]	Actual chlorine, %	Available chlorine, %
Cl_2	71	1	100	100
Cl_2O	87	2	81.7	163.4
ClO_2	67.5	5	52.5	260
$CaClOCl$	127	1	56	56
$Ca(OCl)_2$	143	2	49.6	99.2
$HOCl$	52.5	2	67.7	135.4
$NaClO_2$	90.5	4	39.2	157
$NaOCl$	74.5	2	47.7	95.4
$NHCl_2$	86	2	82.5	165
NH_2Cl	51.5	2	69	138

[a] Valence change to obtain reduced form of chloride (Cl^{-1}).

Figure 12-6

Generalized curve obtained during breakpoint chlorination of wastewater.

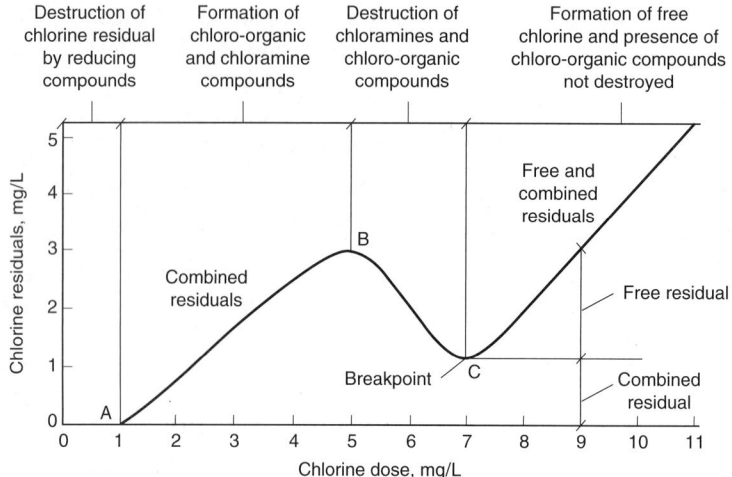

reduce most of it to the chloride ion (point A on Fig. 12–6). After meeting this immediate demand, the added chlorine continues to react with the ammonia to form chloramines between points A and B, as discussed above. For mole ratios of chlorine to ammonia less than 1, monochloramine and dichloramine will be formed. The distribution of these two forms is governed by their rates of formation, which are dependent on the pH and temperature. Between point B and the breakpoint, point C, some chloramines will be converted to nitrogen trichloride [see Eq. (12–18)], the remaining chloramines will be oxidized to nitrous oxide (N_2O) and nitrogen (N_2), and the chlorine will be reduced to the chloride ion. With continued addition of chlorine, most of the chloramines will be oxidized at the breakpoint. Continued addition of chlorine past the breakpoint, as shown on Fig. 12–6, will result in a directly proportional increase in the free available chlorine (unreacted hypochlorite). Theoretically, the weight ratio of chlorine to ammonia nitrogen at the breakpoint is 7.6 to 1 (see Example 12–3). The weight ratio to reach point B is about 5.0 to 1.

Possible reactions to account for the appearance of N_2 and N_2O and the disappearance of chloramines during breakpoint chlorination are as follows (Saunier, 1976; Saunier and Selleck, 1976):

$$NH_4^+ + HOCl \rightarrow NH_2Cl + H_2O + H^+ \tag{12-22}$$

$$NH_2Cl + HOCl \rightarrow NHCl_2 + H_2O \tag{12-23}$$

$$0.5NHCl_2 + 0.5H_2O \rightarrow 0.5NOH + H^+ + Cl^- \tag{12-24}$$

$$0.5NHCl_2 + 0.5NOH \rightarrow 0.5N_2 + 0.5HOCl + 0.5H^+ + 0.5Cl^- \tag{12-25}$$

The overall reaction, obtained by summing Eq. (12–22) through Eq. (12–25), is given as

$$NH_4^+ + 1.5HOCl \rightarrow 0.5N_2 + 1.5H_2O + 2.5H^+ + 1.5Cl^- \tag{12-26}$$

Occasionally, serious odor problems have developed during breakpoint-chlorination operations because of the formation of nitrogen trichloride and related compounds. The presence of additional compounds that will react with chlorine, such as organic nitrogen, may greatly alter the shape of the breakpoint curve, as shown on Fig. 12–7. The

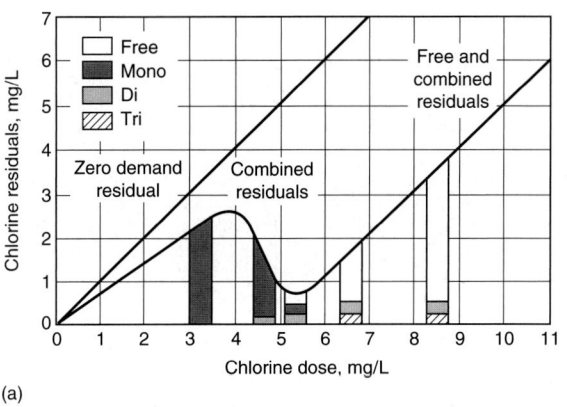
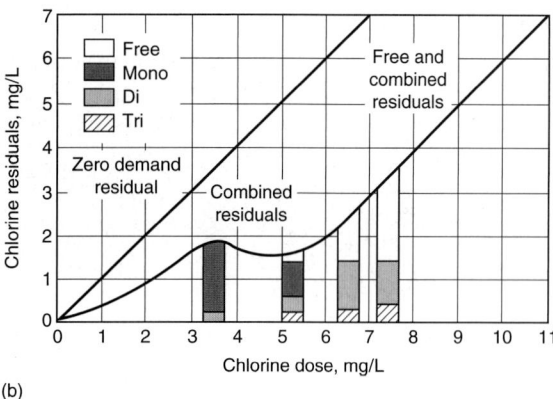

(a)

(b)

Figure 12–7

Curves of chlorine residual versus chlorine dosage for wastewater: (a) for wastewater containing ammonia nitrogen and (b) for wastewater containing nitrogen in the form of ammonia and organic nitrogen. (Adapted from White, 1999).
Note: $1.8(°C) + 32 = °F$.

corresponding drop in pH is usually slight. The formation of disinfection byproducts is considered later in this section. The amount of chlorine that must be added to reach a desired level of residual is called the "chlorine demand."

Acid Generation. In practice, the hydrochloric acid formed during chlorination [see Eq. (12–9)] will react with the alkalinity of the wastewater, and under most circumstances, the pH drop will be slight. Stoichiometrically, 14.3 mg/L of alkalinity, expressed as $CaCO_3$, will be required for each 1.0 mg/L of ammonia nitrogen that is oxidized in the breakpoint-chlorination process as illustrated in Example 12–3.

EXAMPLE 12–3   Analysis of Chlorine Breakpoint Stoichiometry   Determine the stoichiometric weight ratio of chlorine to ammonia nitrogen at the breakpoint and the amount of alkalinity required for each mg/L of ammonia nitrogen oxidized at the breakpoint.

Solution

1. Determine the molecular weight ratio of hypochlorous acid (HOCl), expressed as Cl_2, to ammonia (NH_4^+), expressed as N, using the overall reaction for the breakpoint phenomenon given by Eq. (12–26).

    $$NH_4^+ + 1.5HOCl \rightarrow 0.5N_2 + 1.5H_2O + 2.5H^+ + 1.5Cl^-$$

    (17)   1.5(52.45)

    (14)   1.5(2 × 35.45)

    $$\text{Molecular ratio} = \frac{Cl_2}{N} = \frac{1.5(2 \times 35.45)}{14} = 7.60$$

2.   Determine the alkalinity required per 1.0 mg/L of ammonia nitrogen oxidized at the breakpoint. Assuming the pH of the solution is alkaline, the following expression can be written to describe the oxidation of ammonia:

$$NH_4^+ + Cl_2 \rightarrow 0.5N_2 + 4H^+ + 2Cl^-$$

Assuming lime will be used to neutralize the acidity, the required alkalinity ratio is

$$2CaO + 2H_2O \rightarrow 2Ca^{2+} + 4OH^-$$

$$\text{Alkalinity required} = \frac{2(100 \text{ mg/millimole of } CaCO_3)}{(14 \text{ mg/millimole of } NH_4^+ \text{ as N})} = 14.3$$

Thus, stoichiometrically, 14.3 mg/L of alkalinity are required per mg/L of ammonia as N.

Comment   The ratio computed in Step 1 will vary somewhat, depending on the actual reactions involved. In practice, the actual ratio typically has been found to vary from 8:1 to 10:1. Similarly, in Step 2, the stoichiometric coefficients will also depend on the actual reactions involved. In practice, it has been found that about 15 mg/L of alkalinity are required because of the hydrolysis of chlorine.

Buildup of Total Dissolved Solids.   In addition to the formation of hydrochloric acid, the chemicals added to achieve the breakpoint reaction will also contribute an incremental increase to the total dissolved solids of the wastewater. In situations where the level of total dissolved solids may be critical with respect to reuse applications, this incremental buildup from breakpoint chlorination should always be checked. The total dissolved solids contribution for each of several chemicals that may be used in the breakpoint reaction is summarized in Table 12–10. The magnitude of the possible buildup of total dissolved solids is illustrated in Example 12–4 in which the use of breakpoint chlorination is considered for the seasonal control of nitrogen.

Table 12–10
Effects of chemical addition on total dissolved solids in breakpoint chlorination

Chemical addition	Increase in total dissolved solids per unit of NH_4^+-N consumed
Breakpoint with chlorine gas	6.2:1
Breakpoint with sodium hypochlorite	7.1:1
Breakpoint with chlorine gas—neutralization of all acidity with lime (CaO)	12.2:1
Breakpoint with chlorine gas—neutralization of all acidity with sodium hydroxide (NaOH)	14.8:1

EXAMPLE 12–4 **Analysis of Breakpoint-Chlorination Process Used for Seasonal Control of Nitrogen** Estimate the daily required chlorine dosage and the resulting buildup of total dissolved solids when breakpoint chlorination is used for the seasonal control of nitrogen. Assume that the following data apply to this problem:

1. Plant flowrate = 3800 m³/d (1.0 Mgal/d)
2. Effluent characteristics
   a. BOD = 20 mg/L
   b. Total suspended solids = 25 mg/L
   c. NH_3–N = 23 mg/L
3. Required effluent NH_3–N concentration = 1.0 mg/L

Solution

1. Estimate the required Cl_2 dosage. Assume that the required mass ratio of chlorine to ammonia is 9:1.

   kg Cl_2/d = (3800 m³/d)[(23 − 1) g/m³](9.0)(10³ g/kg) = 752.4 kg/d

2. Determine the increment of total dissolved solids added to the wastewater. Using the data reported in Table 12–10, the total dissolved solids increase per mg/L of ammonia consumed is equal to 6.2.

   Total dissolved solids increment = 6.2(23 − 1) mg/L = 136.4 mg/L

Comment In this example, it was assumed that the acid produced from the breakpoint reaction would not require the addition of a neutralizing agent such as NaOH (sodium hydroxide). If the addition of NaOH were required, the total dissolved solids would have increased significantly. It is noted that although breakpoint chlorination can be used to control nitrogen, it may be counterproductive if in the process the treated effluent is rendered unusable for other applications because of the buildup of total dissolved solids and the formation of disinfection byproducts.

Measurement and Reporting of Disinfection Process Variables

To provide a framework in which to consider the effectiveness of disinfection and the factors that affect the disinfection of wastewater, it will be appropriate to consider how the effectiveness of the chlorination process is now assessed and how the results are analyzed. When using chlorine for the disinfection of wastewater, the principal parameters that can be measured, apart from environmental variables such as pH and temperature, are the number of organisms and the chlorine residual remaining after a specified period of time.

Number of Organisms Remaining. The coliform group of organisms can be determined using the most probable number (MPN) procedure as discussed in Chap. 2. The organisms remaining can also be determined by the plate-count procedure using an agar mixture as the plating medium. Either the standard "pour-plate" method or the "spread-plate" method can be used. The plates should be incubated at 37°C (98.6°F), because this temperature results in the optimum growth of *E. coli,* and the colonies should be counted after a 24-h incubation period.

Measurement of Chlorine Residual. The chlorine residual (free and combined) should be measured using the amperometric method, which has proved to be the most consistently reliable method now available. Also, because almost all the commercial analyzers of residual chlorine use it, the adoption of this method will allow the results of independent studies to be compared directly.

Reporting of Results. Disinfection process laboratory results are reported in terms of the number of organisms and the chlorine residual remaining after a specified period of time. When the results are plotted, it is common practice to plot the logs of removal versus the corresponding $C_R t$ value. A typical plot for the inactivation of *Giardia* cysts with free chlorine is illustrated on Fig. 12–8.

Germicidal Efficiency of Chlorine and Various Chlorine Compounds

Numerous tests have shown that when all the physical parameters controlling the chlorination process are held constant, the germicidal efficiency of disinfection, as measured by the survival of "discrete bacteria," depends primarily on the residual bactericidal chlorine present C_R and the contact time t. A comparison of the relative germicidal efficiency of hypochlorous acid (HOCl), hypochlorite ion (OCl$^-$), and monochloramine (NH$_2$Cl) is presented on Fig. 12–9. For a given contact time or chlorine residual, the germicidal efficiency of hypochlorous acid, in terms of either time or residual, is significantly greater than that of either the hypochlorite ion or monochloramine. It should

Figure 12–8

Plot of laboratory disinfection data for the inactivation of *Giardia* cysts with free chlorine at 10°C. *(Adapted from U.S. EPA, 1999b.)*

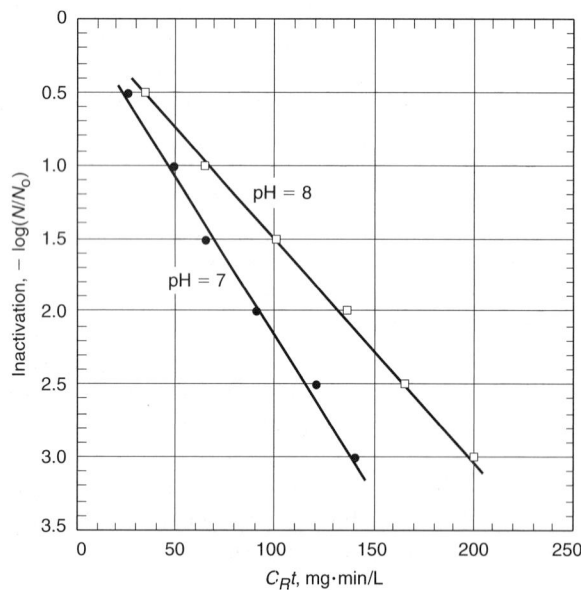

Figure 12-9

Comparison of the germicidal efficiency of hypochlorous acid, hypochlorite ion, and monochloramine for 99 percent destruction of *E. coli* at 2 to 6°C with $C_R t$ values added for the purpose of comparison. (From Butterfield et al., 1943.)

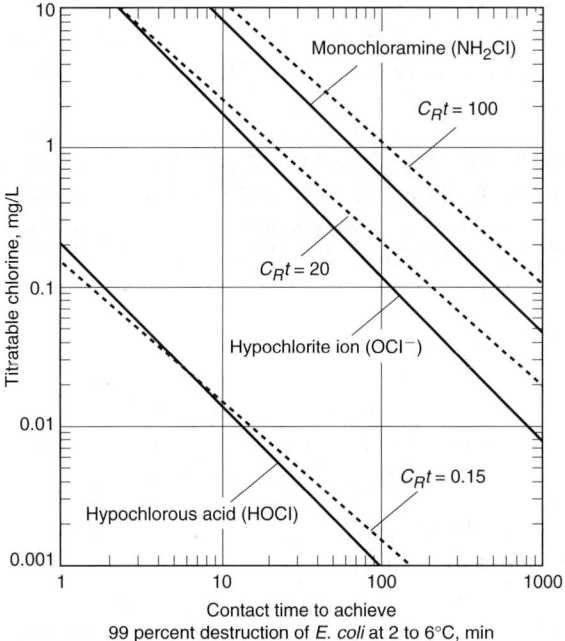

be noted, however, that given an adequate contact time, monochloramine is nearly as effective as free chlorine in achieving disinfection. In addition to the data for the chlorine compounds, $C_R t$ values have been added for the purpose of comparison. As shown, the disinfection data presented on Fig. 12–9 can be represented quite well with the $C_R t$ relationship.

Referring to Fig. 12–9, it is clear that hypochlorous acid offers the most positive way of achieving disinfection. For this reason, with proper mixing, the formation of hypochlorous acid following the breakpoint is most effective in achieving wastewater chlorination. However, when free chlorine is present, the formation of disinfection byproducts is enhanced, as discussed later in this section. If sufficient chlorine cannot be added to achieve the breakpoint reaction, great care must be taken to ensure that the proper contact time is maintained to ensure effective disinfection. Because of the equilibrium between hypochlorous acid and the hypochloric ion, maintenance of the proper pH is also important if effective disinfection is to be achieved.

In view of the renewed interest in public health, environmental water quality, and wastewater reclamation, the effectiveness of the chlorination process is of great concern. Generalized data on the relative germicidal effectiveness of chlorine and other disinfectants for the disinfection of different microorganisms are presented in Table 12–11. It is very important to note that the data presented in Table 12–11 were derived primarily using batch reactors operated under controlled conditions and, as such, are of limited use other than for the purpose of illustrating the relative differences in the effectiveness of the different disinfectants for different organism groups. As shown, there are significant differences in the effectiveness of the various disinfectants for each organism group.

Table 12-11
Estimated range of C_Rt values for various levels of inactivation of dispersed bacteria, viruses, and protozoan oocysts and cysts in filtered secondary effluent (pH $\sim$ 7 and T $\sim$ 20°C)[a]

Disinfectant	Unit	Inactivation			
		1-log	2-log	3-log	4-log
Bacteria					
Chlorine (free)	mg·min/L	0.1–0.2	0.4–0.8	1.5–3	10–12
Chloramine	mg·min/L	4–6	12–20	30–75	200–250
Chlorine dioxide	mg·min/L	2–4	8–10	20–30	50–70
Ozone	mg·min/L		3–4		
UV radiation[b]	mJ/cm²		30–60	60–80	80–100
Virus					
Chlorine (free)	mg·min/L		2.5–3.5	4–5	6–7
Chloramine	mg·min/L		300–400	500–800	200–1200
Chlorine dioxide	mg·min/L		2–4	6–12	12–20
Ozone	mg·min/L		0.3–0.5	0.5–0.9	0.6–1.0
UV radiation[b]	mJ/cm²		20–30	50–60	70–90
Protozoan cysts					
Chlorine (free)	mg·min/L	20–30	35–45	70–80	
Chloramine	mg·min/L	400–650	700–1000	1100–2000	
Chlorine dioxide	mg·min/L	7–9	14–16	20–25	
Ozone	mg·min/L	0.2–0.4	0.5–0.9	0.7–1.4	
UV radiation[b,c]	mJ/cm²	5–10	10–15	15–25	

[a] Adapted in part from Montgomery (1985), U.S. EPA (1986), WEF (1996), U.S. EPA (1999b).
[b] UV dose = UV intensity $\times$ time [see Eq. (12–68)].
[c] Based on the results of infectivity studies.

Factors That Affect Disinfection Efficiency of Chlorine

The purpose of the following discussion is to explore the important factors that affect the disinfection efficiency of chlorine compounds in wastewater applications. These include (1) the importance of initial mixing, (2) the chemical characteristics of the wastewater, (3) the impact of particles found in wastewater, (4) particles with coliform organisms, and (5) the characteristics of the microorganisms.

Initial Mixing. The importance of initial mixing on the disinfection process cannot be overstressed. It has been shown that the application of chlorine in a highly turbulent regime ($N_R \geq 10^4$) will result in kills two orders of magnitude greater than when chlorine is added separately to a conventional rapid-mix reactor under similar conditions. Although the importance of initial mixing is well delineated, the optimum level of turbulence is not known. Based on recent findings, questions have now been raised about the form in which the chlorine compounds are added.

In some plants where chlorine injectors are used, there is concern over the practice of using chlorinated wastewater for the chlorine injection water. The concern is that if nitrogenous compounds are present in the wastewater, a portion of the chlorine that is added will react with these compounds, and by the time chlorine solution is injected, it will be in the form of monochloramine or dichloramine. The formation of chloramines can be a problem if adequate retention time is not available in the chlorine contact basin. Again, it should be remembered that hypochlorous acid (HOCl) and monochloramine (NH_2Cl) are equally effective as disinfecting compounds; only the contact time required is different (see Fig. 12–9).

The formation of disinfection byproducts (DBPs) is another major concern with the use of chlorine inductors, in which molecular chlorine is added directly to the wastewater (see Sec. 12–6). When wastewater is exposed to free chlorine, competing reactions such as the formation of chloramines (free chlorine and ammonia), the formation of DBPs, and the formation of N-nitrosodimethylamine (free chlorine, nitrite, and amines) can occur. The predominant reaction will depend on the applicable kinetic rates for the various reactions. The formation and control of DBPs is discussed later in this section.

Chemical Characteristics of the Wastewater. It has often been observed that, for treatment plants of similar design with exactly the same effluent characteristics measured in terms of BOD, COD, and nitrogen, the effectiveness of the chlorination process varies significantly from plant to plant. To investigate the reasons for this observed phenomenon, and to assess the effects of the compounds present in the chlorination process, Sung (1974) studied the characteristics of the compounds in untreated and treated wastewater. Among the more important conclusions derived from Sung's study are the following:

1. In the presence of interfering organic compounds, the total chlorine residual cannot be used as a reliable measure for assessing the bactericidal efficiency of chlorine.
2. The degree of interference of the compounds studied depended on their functional groups and their chemical structure.
3. Saturated compounds and carbohydrates exert little or no chlorine demand and do not appear to interfere with the chlorination process.
4. Organic compounds with unsaturated bonds may exert an immediate chlorine demand, depending on their functional groups. In some cases, the resulting compounds may titrate as chlorine residual and yet may possess little or no disinfection potential.
5. Compounds with polycyclic rings containing hydroxyl groups and compounds containing sulfur groups react readily with chlorine to form compounds which have little or no bactericidal potential, but which still titrate as chlorine residual.
6. To achieve low bacterial counts in the presence of interfering organic compounds, additional chlorine and longer contact times will be required.

From the results of this work, it is easy to see why the efficiency of chlorination at plants with the same general effluent characteristics can be quite different. Clearly, it is not the value of the BOD or COD that is significant, but the nature of the compounds that make up the measured values. Thus, the nature of the treatment process used in any plant will also have an effect on the chlorination process. The impact of wastewater characteristics on chlorine disinfection is presented in Table 12–12. The presence of oxidizable compounds such as humics and iron will cause the inactivation curve to have a lag or shoulder effect as shown on Fig. 12–10. In effect, the added chlorine is being utilized in the oxidization of these substances and is not available for the inactivation of microorganisms.

Because more wastewater-treatment plants are now removing nitrogen, operational problems with chlorine disinfection are now reported more frequently. In treatment plants where the effluent is nitrified completely, the chlorine added to the wastewater will be present as free chlorine, after satisfying any immediate chlorine demand. In general, the presence of free chlorine will reduce significantly the required chlorine dosage. However, the presence of free chlorine may lead to the formation of *N*-nitrosodimethylamine (NDMA), an undesirable disinfection byproduct (see subsequent discussion under the heading Formation and Control of Disinfection By-products). In treatment plants that do not nitrify completely, or partially nitrify, control of the chlorination process is especially difficult because of the variation in the effectiveness of the chlorine compounds. Some of the chlorine is used to satisfy the demand of the residual nitrite and/or ammo-

Table 12–12

Impact of wastewater constituents on the use of chlorine for wastewater disinfection

Constituent	Effect
BOD, COD, TOC, etc.	Organic compounds that comprise the BOD and COD can exert a chlorine demand. The degree of interference depends on their functional groups and their chemical structure
Humic materials	Reduce effectiveness of chlorine by forming chlorinated organic compounds that are measured as chlorine residual but are not effective for disinfection
Oil and grease	Can exert a chlorine demand
TSS	Shield embedded bacteria
Alkalinity	No or minor effect
Hardness	No or minor effect
Ammonia	Combines with chlorine to form chloramines
Nitrite	Oxidized by chlorine, formation of *N*-nitrosodimethylamine (NDMA)
Nitrate	Chlorine dose is reduced because chloramines are not formed. Complete nitrification may lead to the formation of NDMA due the presence of free chlorine. Partial nitrification may lead to difficulties in establishing the proper chlorine dose
Iron	Oxidized by chlorine
Manganese	Oxidized by chlorine
pH	Affects distribution between hypochlorous acid and hypochlorite ion
Industrial discharges	Depending on the constituents, may lead to diurnal and seasonal variations in the chlorine demand

Figure 12–10

Typical disinfection curve obtained with wastewater containing oxidizable constituents and suspended solids.

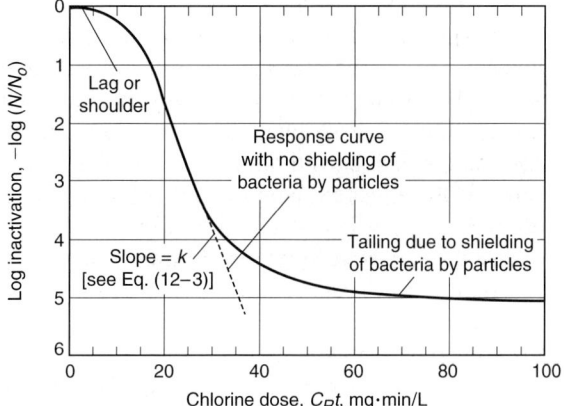

nia. Because of the uncertainties involved in knowing to what degree the plant is nitrifying at any point in time, the chlorine dosage that is added is based on the dosage required if the disinfection of the wastewater is to be accomplished by combined chlorine compounds resulting in excessive chlorine use.

Impact of Particles Found in Wastewater. Another factor that must be considered is the presence of suspended solids in the wastewater to be disinfected. As shown on Fig. 12–10, when suspended solids are present, the disinfection process is controlled by two different mechanisms. The large bacterial inactivation that is observed initially, after the shoulder effect, is of individual free-swimming bacteria and bacteria in small clumps. The straight-line portion of the bacterial inactivation can be described using Eq. (12–3). In the second portion of the curve the bacterial kill is controlled by the presence of suspended solids. Further the slope of the second portion of the curve is a function of the particle size distribution.

Particles with Coliform Organisms. In addition to the particle size distribution, the number of particles with associated coliform organisms is another factor that will impact the performance of both chlorine and UV disinfection for unfiltered effluents. Based on recent studies dealing with UV disinfection, as discussed in Sec. 12–9, it has been found that for activated-sludge plants, the number of particles with associated coliform organisms is a function of the solids retention time (Darby et al., 1999). If particles contain significant numbers of organisms, the organisms can provide protection to other organisms embedded within the particle by limiting the penetration of chlorine through diffusion. Thus, when suspended solids are present, a single equation based on a single mechanism cannot be used to describe the observed results (see Fig. 12–10). Unfortunately, the observed variability caused by the presence of particles usually is masked by the addition of excess chlorine to overcome both chemical and particle effects.

Characteristics of the Microorganisms. Other important variables in the chlorination process are the type, characteristics, and age of the microorganisms. For a young bacterial culture (1 day old or less) with a chlorine dosage of 2 mg/L, only 1 min was needed to reach a low bacterial number. When the bacterial culture was 10 days old or more, approximately 30 min was required to achieve a comparable reduction for the same applied chlorine dosage. It is likely that the resistance offered by the polysaccharide sheath, which the microorganisms develop as they age, accounts for this observation. In the activated-sludge treatment process, the operating mean cell residence time, which to some extent is related to the age of the bacterial cells in the system, will, as discussed previously, affect the performance of the chlorination process. The relative effectiveness of chlorine and other disinfectants was reported previously in Table 12–11. Because the data were not derived using wastewater, they are only provided to illustrate broad differences between organism groups.

Some representative data on the effectiveness of chlorine in killing *E. coli* and three enteric viruses are reported on Fig. 12–11. Again, because newer analytical techniques have been developed, the data presented on Fig. 12–11 are meant only to illustrate the differences in the resistances of different organisms. From the available evidence on the viricidal effectiveness of the chlorination process, it appears that chlorination beyond the breakpoint to obtain free chlorine will be required to kill many of the viruses of concern. Where breakpoint chlorination is used, it will be necessary to dechlorinate the treated wastewater before reuse to reduce any residual toxicity that may remain after chlorination. Recently, based on the use of integrated cell culture PCR techniques, it has been reported that the inactivation of poliovirus may require five times more chlorine than thought previously (Blackmer et al., 2000).

Modeling the Chlorine Disinfection Process

As discussed in Sec. 12–1, a number of models have been developed to describe the disinfection process. The use of these conventional models for predicting microorganism inactivation has, for the most part, proved to be of limited value. Two types of effluent are considered in the following discussion: (1) secondary and filtered secondary effluent and (2) microfiltration and reverse osmosis effluent.

Figure 12–11

Concentration of chlorine as HOCl required for 99 percent kill of *E. coli* and three enteric viruses at 0 to 6°C.

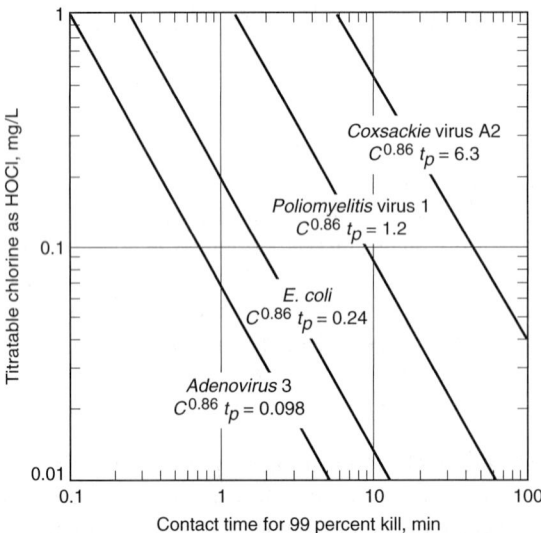

Coxsackie virus A2
$C^{0.86} t_p = 6.3$

Poliomyelitis virus 1
$C^{0.86} t_p = 1.2$

E. coli
$C^{0.86} t_p = 0.24$

Adenovirus 3
$C^{0.86} t_p = 0.098$

Titratable chlorine as HOCl, mg/L

Contact time for 99 percent kill, min

Secondary and Filtered Secondary Effluent. When considering the disinfection of both secondary and filtered secondary effluent, both the lag or shoulder effect and the effect of the residual particles (see Fig. 12–10) must be considered. As noted previously, depending on the constituents in the treated wastewater, a shoulder region may be observed in which there is no reduction in the number of organisms as the result of the addition of disinfectant. As additional chlorine is added beyond some limiting value, a log-linear reduction in the number of organisms will be observed with increased chlorine dosages. If particles (typically greater than 20 μm) are present, the disinfection curve will start to diverge from the log-linear form and a tailing region will be observed due to particle shielding of the microorganisms. The tailing region is of importance as more restrictive standards are to be achieved (e.g., 23 MPN/100 mL). It is interesting to note that the first attempt to determine the effect of suspended solids on disinfection dates back to 1918 to work conducted by H. P. Eddy at Cleveland, OH (Pearse et al., 1934). The tailing region was also identified in an early paper dealing with the disinfection of wastewater effluents (Tiedeman, 1927).

Because the lag and tailing regions, identified on Fig. 12–10, are not considered, most classic disinfection models are not applicable. Nevertheless, it is appropriate to review some of the models that have been developed more recently. In 1957, Gard noted that the rate of inactivation was not constant, but decreased with time, resulting in a tailing effect (Gard, 1957). Gard proposed the following relationship to define the decreasing rate constant:

$$\frac{dN}{dt} = -\frac{kN}{1 + a(Ct)} \tag{12-27}$$

The integrated form of Eq. (12–27) is

$$\frac{N}{N_o} = \frac{1}{[1 + a(Ct)]^{k/a}} \tag{12-28}$$

where N = number of organisms remaining after disinfection at time t
  N_o = number of organisms present before disinfection
  k = first-order inactivation rate at time $t = 0$
  a = rate coefficient
  C = concentration of chemical agent held constant over time, mg/L
  t = contact time, min

It is interesting to note that the form of Eq. (12–28) corresponds to the retarded rate expression given by Eq. (4–102) in Chap. 4.

In the early 1970s, Collins conducted extensive experiments on the disinfection of various wastewaters (Collins, 1970; Collins and Selleck, 1972). Using the batch reactor whose contents were well stirred, Collins and Selleck found that the reduction of coliform organisms in a chlorinated primary treated effluent followed a linear relationship when plotted on log-log paper. The original equation developed to describe the observed results is

$$\frac{N}{N_o} = \frac{1}{(1 + 0.23C_R t)^3} \tag{12-29}$$

It will be noted that the original form of the equation developed by Collins did not account for the lag effect, but did account for the tailing effect (see Fig. 12–12).

Figure 12–12

Coliform survival in a batch reactor as a function of amperometric chlorine residual and contact time (temperature range 11.5 to 18°C). It should be noted that if the $C_R t$ values had been plotted on an arithmetic scale, a tailing effect would be observed in the plotted values. (From Collins, 1970; Collins and Selleck, 1972; Sellect et al., 1978.)

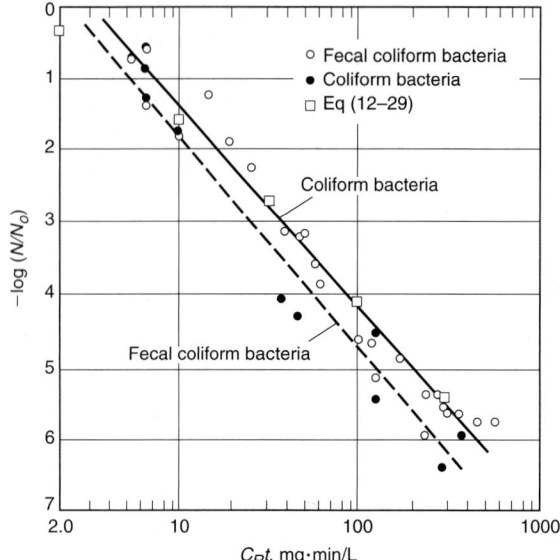

A refinement of the original Collins model for the disinfection of secondary effluent in which both lag and tailing effects are observed (see Fig. 12–12), as proposed by Selleck et al. (1978) and later modified by White (1999), is

$$\frac{N}{N_o} = 1 \qquad \text{for } C_R t < b \tag{12–30}$$

$$\frac{N}{N_o} = [(C_R t)/b]^{-n} \qquad \text{for } C_R t > b \tag{12–31}$$

where C_R = chlorine residual remaining at the end of time t
t = contact time
n = slope of inactivation curve
b = value of x-intercept when $N/N_o = 1$ or log $N/N_o = 0$ (see Fig. 12–13)

Typical values for the coefficients n and b for secondary effluent for coliform and fecal coliform organisms are 2.8 and 4.0 and 2.8 and 3.0, respectively (Roberts et al., 1980; White, 1999). However, because of the variability of the chemical composition of the secondary effluent and the variable particle size distribution, it is recommended that the constants be determined for each wastewater. When the $C_R t$ values become large (>100 mg·min/L), Eq. (12–29) is essentially the same as Eq. (12–31). It should be noted that numerous other models have been proposed for wastewater disinfection including an empirical model proposed by Hom (1972), which was subsequently rationalized by Haas and Joffe (1994).

An alternative approach that can be used to model the disinfection process in secondary and filtered secondary effluent is to model the inactivation of the free-swimming microorganisms and the inactivation of the microorganisms associated with particles

Figure 12–13

Definition sketch for the application of Eq. (12–31).

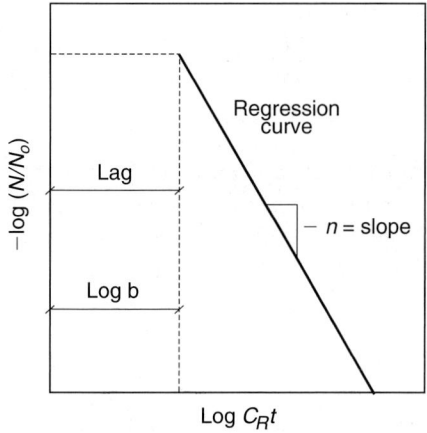

Figure 12–14

Disinfection curves for chlorine and UV radiation for dispersed coliform organisms, particle-associated coliform organisms, and the combined curve for both the free-swimming and associated coliform organisms.

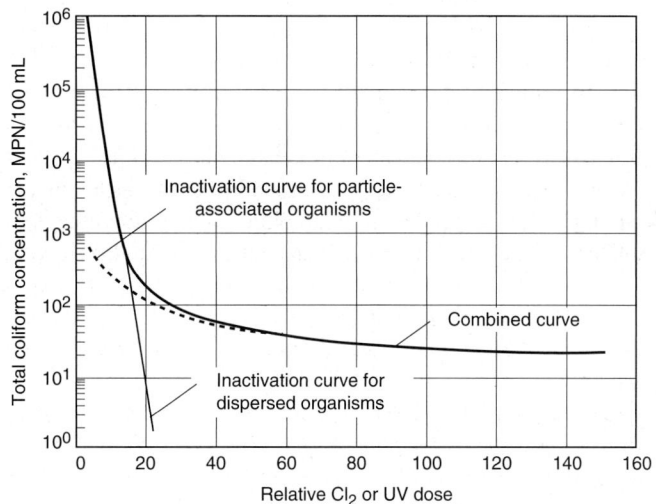

separately as has been proposed for UV disinfection (see Fig. 12–14). The resulting formulation takes the following form:

$$N(t) = N_D(0)e^{-k(C_R t)} + \frac{N_P(0)}{k(C_R t)}\left(1 - e^{-k(C_R t)}\right) \qquad (12\text{–}32)$$

where $N(t)$ = total number of surviving coliform bacteria at time t

$N_D(0)$ = total number of disperse coliform bacteria prior to application of disinfectant at time $t = 0$

$N_P(0)$ = total number of particles containing at least one coliform bacterium at time $t = 0$

k = inactivation rate coefficient

C_R = residual chlorine at time t

t = time

It will be recognized that the first term in Eq. (12–32) corresponds to the Chick and Watson formulation as given by Eq. (12–6). The inactivation rate coefficient in Eq. (12–32) could also be modified to account for a lag effect and the fact that the inactivation may not be constant but decreases with time.

Microfiltration and Reverse Osmosis Effluent. The most important characteristic of these effluents is that they do not contain particles that can shield microorganisms. For these effluents, the Chick-Watson model, as given by Eq. (12–7), can be used to model the disinfection process with chlorine. A lag term can also be added, if needed. Typically, the lag effect is reduced considerably, especially with reverse osmosis effluent.

Review of the $C_R t$ Concept

Use of the $C_R t$ concept to control the disinfection process is now becoming more common in the wastewater field. In some states, the $C_R t$ value and the chlorine contact time are specified in regulatory requirements. For example, the state of California requires a $C_R t$ value of 450 mg·min/L and a modal contact time of 90 min at peak flow for certain reclamation applications. It is assumed, based on past testing, that a $C_R t$ value of 450 mg·min/L will produce a four-log inactivation of poliovirus. As the use of the $C_R t$ concept becomes more common in the wastewater field, there are a number of limitations that must be considered in the application of this concept for regulatory purposes. Most of the $C_R t$ values reported in the literature are obtained: (1) using complete mixed batch reactors (i.e., ideal flow conditions) in a laboratory setting under controlled conditions, (2) using discrete organisms grown in the laboratory in pure culture, and (3) using a buffered fluid for the suspension of the discrete organisms.

Further, many of the $C_R t$ values reported in the literature were based on older analytical techniques. As a consequence, $C_R t$ values used for regulatory purposes often do match what is observed in the field. Referring to Fig. 12–10, it can be seen that in the tailing region, the residual concentration of microorganisms is essentially independent of the $C_R t$ value. As shown on Fig. 12–15, it is clear that a $C_R t$ value of 450 mg·min/L, as used by the state of California, will not result in a four-log reduction of virus, when the measured residual chlorine is combined chlorine (i.e., mono- and dichloramine). Clearly, site-specific testing is required to establish the appropriate chlorine dose.

Required Chlorine Dosage for Disinfection

The required chemical dosage for disinfection can be estimated by considering (1) the initial chlorine demand of the wastewater, (2) the allowance needed for decay during the chlorine contact time, and (3) the required chlorine residual concentration determined using Eq. (12–31) for the organism under consideration (e.g., bacteria, virus, or protozoan oocysts and cysts). The chlorine dosage required to meet the initial demand will depend on the wastewater constituents. It is important to remember that the chlorine added to meet the initial demand due to inorganic compounds will be reduced to the chloride ion and will not be measured as chlorine residual. On the other hand, chlorine that combines with humic materials will not be effective as a disinfectant but will nevertheless be measured as a chlorine residual contributing to the lag term b in Eq. (12–31). Typical decay values for chlorine residual are on the order of 2 to 4 mg/L for contact time of about 1 h. Typical chlorine dosage values for various wastewaters

Figure 12–15

Inactivation of MS2 coliphage and poliovirus with combined chlorine. (From BioVir Laboratories, 2001.)

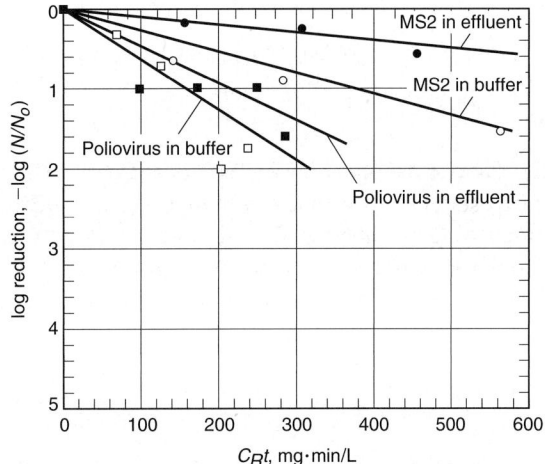

Table 12–13

Typical chlorine dosages, based on combined chlorine unless otherwise indicated, required to achieve different effluent total coliform disinfection standards for various wastewaters based on a 30-min contact time[a]

Type of wastewater	Initial coliform count, MPN/100 mL	Chlorine dose, mg/L Effluent standard, MPN/100 mL			
		1000	**200**	**23**	**≤2.2**
Raw wastewater	10^7–10^9	15–40			
Primary effluent	10^7–10^9	10–30	20–40		
Trickling filter effluent	10^5–10^6	3–10	5–20	10–40	
Activated-sludge effluent	10^5–10^6	2–10	5–15	10–30	
Filtered activated-sludge effluent	10^4–10^6	4–8	5–15	6–20	8–30
Nitrified effluent	10^4–10^6	4–12	6–16	8–18	8–20
Filtered nitrified effluent	10^4–10^6	4–10	6–12	8–14	8–16
Microfiltration effluent	10^1–10^3	1–3	2–4	2–6	4–10
Reverse osmosis[b]	~0	0	0	0	0–2
Septic tank effluent	10^7–10^9	20–40	40–60		
Intermittent sand filter effluent	10^2–10^4	1–5	2–8	5–10	8–18

[a] Adapted in part from U.S. EPA (1986); White (1999).
[b] Based on free chlorine.

for total coliform, based on a contact time of 30 min, are reported in Table 12–13. It should be noted that the dosage values given in Table 12–13 are only meant to serve as a guide for the initial estimation of the required chlorine dose. As noted above, site-specific testing is required to establish the appropriate chlorine dose. Estimation of the required chlorine dose is illustrated in Example 12–5.

EXAMPLE 12–5 **Estimate the Required Chlorine Dose for a Typical Secondary Effluent**   Estimate the chlorine dose needed to disinfect a filtered secondary effluent assuming a shoulder effect exists and that the following conditions apply:

1. Effluent total coliform count before disinfection = $10^7/100$ mL
2. Required summer effluent total coliform count = 23/100 mL
3. Required winter effluent total coliform count = 240/100 mL
4. Initial effluent chlorine demand = 4 mg/L
5. Demand due to decay during chlorine contact = 2.5 mg/L
6. Required chlorine contact time = 60 min

Solution

1. Estimate the required chlorine residual using Eq. (12–31).

$$N/N_o = (C_R t/b)^{-n}$$

Use the typical values given above for the coefficients.

$$b = 4.0$$

$$n = 2.8$$

   a. Summer

$$23/10^7 = (C_R t/4.0)^{-2.8}$$

$$(23/10^7)^{-1/2.8} = (C_R t/4.0)$$

$$(234.3)4 = C_R (60)$$

$$C_R = 15.6 \text{ mg/L}$$

   b. Winter

$$240/10^7 = (C_R t/4.0)^{-2.8}$$

$$C_R = 3.0 \text{ mg/L}$$

2. The required chlorine dosage is
   a. Summer

   Chlorine dosage = 4.0 mg/L + 2.5 mg/L + 15.6 mg/L = 22.1 mg/L

   b. Winter

   Chlorine dosage = 4.0 mg/L + 2.5 mg/L + 3.0 mg/L = 9.5 mg/L

Comment   The chlorine dosage increases significantly as the effluent standards become more stringent. In the above computation, it was assumed that the wastewater to be disinfected remained in the chlorine contact tank for the full 60 min. Thus, it is clear that the proper design of a plug-flow chlorine contact basin is critical to the effective use of chlorine as disinfectant. The design of chlorine contact basins is discussed in Sec. 12–6 and in Chap. 4.

Formation and Control of Disinfection Byproducts

In the early 1970s, it was found that the use of oxidants, such as chlorine and ozone, in water-treatment plants for disinfection; for taste, odor, and color removal; and other in-plant uses resulted in the production of undesirable disinfection byproducts (DBPs) (U.S. EPA, 1999*b*). The DBPs occurring most frequently and with the highest concentration are trihalomethanes (THMs) and haloacetic acids (HAAs). In addition to trihalomethanes and haloacetic acids, a variety of other DBPs is also produced. The principal DBPs that have been identified are reported in Table 12–14. Many of the compounds identified in Table 12–14 have also been identified in treated effluent that has been disinfected using chlorine.

Formation of DBPs is of great concern because of the potential impact of these compounds on public health and the environment. Chloroform, for example, is a well-known animal carcinogen, and many of the haloforms are also thought to be animal carcinogens. In addition, many of these compounds have been classified as probable human carcinogens. Still others of these compounds are known to cause chromosomal aberrations and sperm abnormalities. Recognizing the many unknowns and the potential public health and environmental risks associated with these compounds, the U.S. EPA has moved aggressively to control their formation in drinking water.

Formation of DBPs Due to the Addition of Chlorine for Disinfection.

Trihalomethanes (THMs) and other DBPs are formed as a result of a series of complex reactions between free chlorine and a group of organic acids known collectively as humic acids. The reactions lead to the formation of single carbon molecules that are often designated as HCX_3, where X is either a chlorine (Cl^-) or bromine (Br^-) atom. For example, the chemical formula for chloroform is $HCCl_3$.

The rate of formation of DBPs is dependent on a number of factors including:

- Presence of organic precursors
- Free chlorine concentration
- Bromide concentration
- pH
- Temperature

The type and concentration of the organic precursor will affect both the rate of the reaction and the extent to which the reaction is completed. The presence of free chlorine was thought to be necessary for the THM formation reaction to proceed, but it appears that THMs can form in the presence of combined chlorine, but at a very reduced rate. It is important to note that initial mixing can affect the formation of THMs because of the competing reactions between chlorine and ammonia and chlorine and humic acids. If bromide is present, it can be oxidized to bromine by free chlorine. In turn the bromine ion can combine with the organic precursors to form THMs, including bromodichloromethane, dibromochloromethane, and bromoform. The rate of formation of THMs has been observed to increase with both pH and temperature. Additional details on the formation of THMs may be found in U.S. EPA (1999*b*).

Another DBP that has recently surfaced in disinfected treated wastewater effluents is *N*-nitrosodimethylamine (NDMA), which is a member of a class of compounds known as nitrosamines. As a class of compounds nitrosamines are among the most

Table 12–14

Representative disinfection byproducts resulting from the chlorination of wastewater containing organic and selected inorganic constituents[a]

Disinfectant residuals	**Halogenated organic byproducts**
Free chlorine	Trihalomethanes (THMs)
Hypochlorous acid	Chloroform
Hypochlorite ion	Bromodichloromethane (BDCM)
Chloramines	Dibromochloromethane (DBCM)
Monochloramine	Bromoform
Dichloramine	Total trihalomethanes
Trichloramine	Haloacetic acids (HAAs)
Chlorine dioxide	Monochloroacetic acid
	Dichloroacetic acid (DCA)
Inorganic byproducts	Trichloroacetic acid (TCA)
Chlorate ion	Monobromoacetic acid
Chlorite ion	Dibromoacetic acid
Bromate ion	Total haloacetic acids
Iodate ion	Haloacetonitriles
Hydrogen peroxide	Chloroacetonitrile (CAN)
Ammonia	Dichloroacetonitrile (DCAN)
	Trichloroacetonitrile (TCAN)
Organic oxidation byproducts	Bromochloroacetonitrile (BCAN)
Aldehydes	Dibromoacetonitrile (DCAN)
Formaldehyde	Total haloacetonitriles
Acetaldehyde	Haloketones
Chloroacetaldehyde	1,1-Dichloropropanone
Dichloroacetaldehyde	1,1,1-Trichloropropanone
Trichloroacetaldehyde (chloral hydrate)	Total haloketones
Glyoxal (also methyl glyoxal)	Chlorophenols
Hexanal	2-Chlorophenol
Heptanal	2,4-Dichlorophenol
Carboxylic acids	2,4,6-Trichlorophenol
Hexanoic acid	Chloropicrin
Heptaoic acid	Chloral hydrate
Oxalic acid	Cyanogen chloride
Assimilable organic carbon	N-organochloramines
Nitrosoamines	(MX) 3-chloro-4-(dichloromethyl)-5Hydroxy-2(5H)-furanone
N-nitrosodimethylamine (NDMA)	

[a]Adapted, in part, from U.S. EPA (1999b).

powerful carcinogens known (Snyder, 1995). The compounds in this class have been found to produce cancer in every species of laboratory animal tested. Formation of NDMA can be illustrated with the following two reactions:

$$NO_2^- + HCl \rightarrow HNO_2 + Cl^- \qquad (12\text{--}33)$$

nitrite   hydrochloric   nitrous   chloride
anion   acid   acid   ion

$$\overset{\displaystyle NO}{\overset{\displaystyle |}{HNO_2 + CH_3-NH-CH_3 \rightarrow CH_3-N-CH_3}} \qquad (12\text{--}34)$$

nitrous acid   dimethylamine   *N*-nitrosodimethylamine

The concern in biological wastewater-treatment facilities is that some nitrite may leak through the process. While the concentration of nitrite may be too low to measure by conventional means, concentrations of NDMA as low as 5 ng/L (ppt) are being measured and the action level for NDMA in drinking water in California has been set temporarily (2001) at 20 ng/L. Based on a limited number of test locations, it has been observed that the concentrations of NDMA in the incoming wastewater can be quite variable, with concentrations varying from below 10 to greater than 10,000 ng/L being measured. In addition to the formation of NDMA as outlined above, it appears the addition of chlorine serves to amplify the concentration of any NDMA that may be present in the treated effluent before chlorination.

Control of DBP Formation Due to the Addition of Chlorine for Disinfection. The principal means of controlling the formation of THMs and other related DBPs in wastewater is to avoid the direct addition of free chlorine. Based on the evidence to date, it appears that the use of chloramines will not lead to the formation of THMs and other related DBPs in amounts that would be of concern relative to current standards. It is important to note that if chloramines are to be used for disinfection, the chloramine solution must be prepared with a potable water supply containing little or no ammonia; treated plant water cannot be used. If the formation of DBPs is of concern due to the presence of specific organic precursors (i.e., humic materials), the practice of breakpoint chlorination cannot be used. Further, if humic materials are present consistently, it may be appropriate to investigate alternative means of disinfection such as UV radiation.

With respect to NDMA it appears that with proper control and operation of the biological treatment process, the potential for the formation or amplification of this compound can be reduced. Removals of 50 to 70 percent have been reported for NDMA when using reverse osmosis employing thin-film composite membranes. The use of UV disinfection has also proved to be effective in the control of NDMA. At present, a variety of studies are under way to develop means for controlling the discharge of effluents with high concentrations of NDMA.

Environmental Impacts

The environmental impacts associated with the use of chlorine and chlorine compounds as a wastewater disinfectant include the regrowth of organisms and the discharge of DBPs.

Regrowth of Microorganisms.   In many locations, a regrowth of microorganisms has been observed in receiving water bodies and in long transmission pipelines following dechlorination of secondary effluent disinfected with chlorine. It has been hypothesized that regrowth (also known as aftergrowth), which occurs in films on the pipe surface exposed to treated wastewater, results because the amount of organic matter in secondary effluent is sufficient to sustain the limited number of organisms remaining after disinfection, and because predators such as protozoa are absent. Regrowth is an especially important issue in transmission lines used for the transport of reclaimed water. In many reuse applications, a chlorine residual is maintained (a common practice in water distribution systems) in the pipeline to control regrowth. In very long pipelines, it may be necessary to add additional chlorine at intermediate points along the length of the pipeline. Additional details on regrowth may be found in the vast literature dealing with regrowth in water supply distribution systems, as the controlling factors are similar, with the exception that the residual organic content in treated wastewater is higher and not all treated wastewaters are disinfected completely.

Discharge of DBPs.   It has been shown that many of the DBPs can cause environmental impacts at very low concentrations. The occurrence of DBPs and compounds such as NDMA raises serious questions about the continued use of free chlorine for wastewater disinfection.

12–4   DISINFECTION WITH CHLORINE DIOXIDE

Chlorine dioxide (ClO_2) is another bactericide, equal to or greater than chlorine in disinfecting power. Chlorine dioxide has proved to be an effective virocide, being more effective in achieving inactivation of viruses than chlorine. A possible explanation is that because chlorine dioxide is adsorbed by peptone (a protein), and that viruses have a protein coat, adsorption of ClO_2 onto this coating could cause inactivation of the virus. In the past, ClO_2 did not receive much consideration as a wastewater disinfectant due to its high costs.

Characteristics of Chlorine Dioxide

Chlorine dioxide (ClO_2) is, under atmospheric conditions, a yellow to red unpleasant-smelling irritating unstable gas with a high specific gravity. Physical properties of chlorine dioxide are reported in Table 12–6. Because chlorine dioxide is unstable and decomposes rapidly, it is usually generated onsite before its application. Chlorine dioxide is generated by mixing and reacting a chlorine solution in water with a solution of sodium chlorite ($NaClO_2$) according to the following reaction:

$$2NaClO_2 + Cl_2 \rightarrow 2ClO_2 + 2NaCl \tag{12–35}$$

Based on Eq. (12–35), 1.34 mg sodium chlorite reacts with 0.5 mg chlorine to yield 1.0 mg chlorine dioxide. Because technical-grade sodium chlorite is only about 80 percent pure, about 1.68 mg of the technical-grade sodium chlorite would be required to produce 1.0 mg of chlorine dioxide. Sodium chlorite may be purchased and stored as a liquid (generally a 25 percent solution) in refrigerated storage facilities.

The chlorine and liquid sodium chlorite solutions are brought together at the base of a porcelain ring filled reaction tower (see Fig. 12–16). As this combined solution

Figure 12–16

Schematic flow diagram for the generation of chlorine dioxide. (Adapted from Wallace and & Tiernan.)

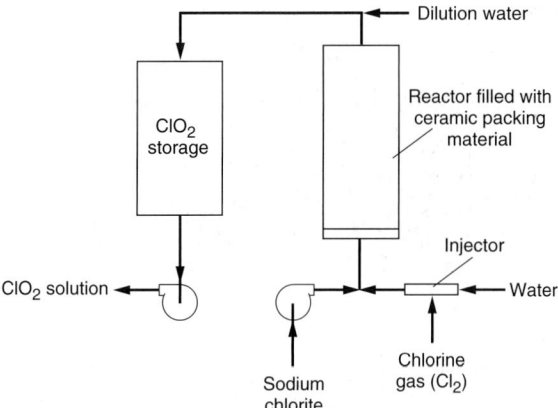

flows upward, chlorine dioxide is generated. A contact time of about 1 min is generally adequate. To increase the reaction rate and obtain the highest yield of chlorine dioxide, a slight excess of chlorine is recommended. Because sodium chlorite is about ten times as expensive as chlorine on a weight basis, economical considerations must be taken into account. The solution discharged from the tower is only partly chlorine dioxide, with the remaining portion being chlorine in solution as hypochlorous acid.

Chlorine Dioxide Chemistry

The active disinfecting agent in a chlorine dioxide system is free dissolved chlorine dioxide (ClO_2). At the present time, the complete chemistry of chlorine dioxide in an aqueous environment is not clearly understood. Because ClO_2 does not hydrolyze in a manner similar to the chlorine compounds discussed in the previous section, the oxidizing power of ClO_2 is often referred to as "equivalent available chlorine." The definition of the term equivalent available chlorine is based on a consideration of the following oxidation half reaction for ClO_2:

$$ClO_2 + 5e^- + 4H^+ \rightarrow Cl^- + 2H_2O \tag{12–36}$$

As shown in Eq. (12–36), the chlorine atom undergoes a 5-electron change in its conversion from chlorine dioxide to the chloride ion. Because the weight of chlorine in ClO_2 is 52.6 percent and there is a 5-electron change, the equivalent available chlorine content is equal to 263 percent as compared to chlorine. Thus, ClO_2 has 2.63 times the oxidizing power of chlorine. The concentration of ClO_2 is usually expressed in g/m^3. On a molar basis, 1 mole of ClO_2 is equal to 67.45 g, which is equivalent to 177.5 g (5×35.45) of chlorine. Thus, 1 g/m^3 of ClO_2 is equivalent to 2.63 g/m^3 of chlorine.

Effectiveness of Chlorine Dioxide as a Disinfectant

Chlorine dioxide has an extremely high oxidation potential, which probably accounts for its potent germicidal powers. Because of its extremely high oxidizing potential, possible bactericidal mechanisms may include inactivation of critical enzyme systems or disruption of protein synthesis. It should be noted, however, that when ClO_2 is added to wastewater it is often reduced to chlorite according to the following reaction:

$$ClO_2 + e^- \rightarrow ClO_2^- \tag{12–37}$$

Equation (12–37) may help to explain the variability that is sometimes observed in the performance of ClO_2 as a disinfectant.

Modeling the Chlorine Dioxide Disinfection Process. As discussed previously in Sec. 12–3, the models that have been developed to describe the disinfection process with chlorine can also be used with appropriate caution, for chlorine dioxide. As with chlorine, the shoulder effect and the effect of the residual particles must be considered. Further, the differences between (1) secondary and filtered secondary effluent and (2) microfiltration and reverse osmosis effluent must also be considered.

Required Chlorine Dioxide Dosages for Disinfection. The required chlorine dioxide dosage will depend on the pH and the specific organism under investigation. Relative $C_R t$ values for chlorine dioxide are given in Table 12–11, presented in Sec. 12–3. In general, the effectiveness of chlorine dioxide is similar to that of combined chlorine for bacteria. The values given in Table 12–11 can be used as a starting point. However, there is a significant difference in the effectiveness of chlorine dioxide for the disinfection of virus, which is essentially the same as that for free chlorine. Chlorine dioxide appears to be more effective than free chlorine in the inactivation of protozoan cysts. Because the data on chlorine dioxide in the literature are limited, site-specific testing is recommended to establish appropriate dosage ranges.

Byproduct Formation and Control

The formation of DBPs is of great concern with the use of chlorine dioxide. The formation and control of DBPs with chlorine dioxide is considered in the following discussion.

Formation of DBPs Due to the Addition of Chlorine Dioxide for Disinfection. The principal DBPs formed when chlorine dioxide is used as a disinfectant are chlorite (ClO_2^-) and chlorate (Cl_2O_2), both of which are potentially toxic. The principal sources of the chlorite ion are from the process used to generate the chlorine dioxide and from the reduction of chlorine dioxide. As given by Eq. (12–35), all of the sodium chlorite reacts with chlorine to form chlorine dioxide. Unfortunately, on occasion some unreacted chlorite ion can escape from the reactor where the chlorine dioxide is being generated and find its way into the wastewater that is being treated. The second source of chlorite is from the reduction of chlorine dioxide as discussed above [see Eq. (12–37)]. The chlorate ion can be derived from the oxidation of chlorine dioxide, from the impurities in the sodium chlorite feedstock, and from the photolytic decomposition of chlorine dioxide.

The chlorine dioxide residuals and other end products are believed to degrade more quickly than chlorine residuals and therefore may not pose as serious a threat to aquatic life as the chlorine residuals. An advantage in using chlorine dioxide is that it does not react with ammonia to form the potentially toxic chlorinated DBPs. It has also been reported that halogenated organic compounds are not produced to any appreciable extent. This finding is specifically true with respect to the formation of chloroform, which is a suspected carcinogenic substance.

Control of DBP Formation Due to the Addition of Chlorine Dioxide for Disinfection. The formation of chlorite can be controlled by careful management of the feedstock or increasing the chlorine dose beyond the stoichiometric amount. Treatment methods for the removal of the chlorite ion involve reducing the chlorite ion to the chloride ion using either ferrous iron or sulfite. At the present time there are no cost-effective methods for the removal of the chlorate ion. The control of the chlorate ion depends primarily on the effective management of the facilities used for the production of chlorine dioxide (White, 1999).

Environmental Impacts

The environmental impacts associated with the use of chlorine dioxide as a wastewater disinfectant are not well known. It has been reported that the impacts are less adverse than those associated with chlorination. Chlorine dioxide does not dissociate or react with water as does chlorine. However, because chlorine dioxide is normally produced from chlorine and sodium chlorite, free chlorine may remain in the resultant chlorine dioxide solution (depending on the process) and impact the receiving aquatic environment, as do chlorine residuals. A free chlorine dioxide residual will also remain, but it has been found to be less harmful to aquatic life than chlorine.

12–5 DECHLORINATION

In cases where low-level chlorine residuals may have potential toxic effects on aquatic organisms, dechlorination of treated effluent is practiced. Dechlorination may be accomplished by reacting the residual chlorine with a reducing agent such as sulfur dioxide or sodium metabisulfite or by adsorption on activated carbon. The design of dechlorination systems is considered in the following section.

Need for Dechlorination

Chlorination is one of the most commonly used methods for the destruction of pathogenic and other harmful organisms that may endanger human health. As noted in the previous sections, however, certain organic constituents in wastewater interfere with the chlorination process. Many of these organic compounds may react with the chlorine to form toxic compounds that can have long-term adverse effects on the beneficial uses of the waters to which they are discharged. To minimize the effects of these potentially toxic chlorine residuals on the environment, dechlorination of wastewater is necessary.

Dechlorination of Wastewater Treated with Chlorine and Chlorine Compounds

Where effluent toxicity requirements are applicable, or where dechlorination is used as a polishing step following the breakpoint-chlorination process for the removal of ammonia nitrogen, sulfur dioxide is used most commonly for dechlorination. Other chemicals that have been used are sodium sulfite (Na_2SO_3), sodium bisulfite ($NaHSO_3$), sodium metabisulfite ($Na_2S_2O_5$), and sodium thiosulfate ($Na_2S_2O_3$). Activated carbon has also been used for dechlorination. The use of these chemicals for dechlorination is discussed below.

Dechlorination with Sulfur Dioxide. Sulfur dioxide (SO_2) is available commercially as a liquefied gas under pressure in steel containers with capacities of 45, 68, and 908 kg (100, 150, and 2000 lb). Properties of sulfur dioxide are reported in Table 12–6. Sulfur dioxide is handled in equipment very similar to standard chlorine systems. When added to water, sulfur dioxide reacts to form sulfurous acid ($H_2SO_3^-$), a strong reducing agent. In turn, the sulfurous acid dissociates to form HSO_3^- that will react with free and combined chlorine, resulting in formation of chloride and sulfate ions. Sulfur dioxide gas successively removes free chlorine, monochloramine, dichloramine, nitrogen trichloride, and poly-*n*-chlor compounds as illustrated in Eqs. (12–38) through (12–42).

Reactions between sulfur dioxide and free chlorine:

$$SO_2 + H_2O \rightarrow HSO_3^- + H^+ \tag{12–38}$$

$$\underline{HOCl + HSO_3^- \rightarrow Cl^- + SO_4^{2-} + 2H^+} \tag{12–39}$$

$$SO_2 + HOCl + H_2O \rightarrow Cl^- + SO_4^{2-} + 3H^+ \tag{12–40}$$

Reactions between sulfur dioxide and monochloramine, dichloramine, and nitrogen trichloride are

$$SO_2 + NH_2Cl + 2H_2O \rightarrow Cl^- + SO_4^{2-} + NH_4^+ + 2H^+ \tag{12–41}$$

$$SO_2 + NHCl_2 + 2H_2O \rightarrow 2Cl^- + SO_4^{2-} + NH_3 + 2H^+ \tag{12–42}$$

$$SO_2 + NCl_3 + 3H_2O \rightarrow 3Cl^- + SO_4^{2-} + NH_4^+ + 2H^+ \tag{12–43}$$

For the overall reaction between sulfur dioxide and chlorine [Eq. (12–40)], the stoichiometric weight ratio of sulfur dioxide to chlorine is 0.903:1 (see Table 12–15). In practice, it has been found that about 1.0 to 1.2 mg/L of sulfur dioxide will be required for the dechlorination of 1.0 mg/L of chlorine residue (expressed as Cl_2). Because the reactions of sulfur dioxide with chlorine and chloramines are nearly instantaneous, contact time is not usually a factor and contact chambers are not used, but rapid and positive mixing at the point of application is an absolute requirement.

The ratio of free chlorine to the total combined chlorine residual before dechlorination determines whether the dechlorination process is partial or proceeds to completion. If the ratio is less than 85 percent, it can be assumed that significant organic nitrogen is present and that it will interfere with the free residual chlorine process.

In most situations, sulfur dioxide dechlorination is a very reliable unit process in wastewater treatment, provided that the precision of the combined chlorine residual monitoring service is adequate. Excess sulfur dioxide dosages should be avoided, not only because of the chemical wastage but also because of the oxygen demand exerted by the excess sulfur dioxide. The relatively slow reaction between excess sulfur dioxide and dissolved oxygen is given by the following expression:

$$HSO_3^- + 0.5O_2 \rightarrow SO_4^{2-} + H^+ \tag{12–44}$$

The result of this reaction is a reduction in the dissolved oxygen contained in the wastewater, a corresponding increase in the measured BOD and COD, and a possible drop in the pH. All these effects can be eliminated by proper control of the dechlorination system.

Dechlorination with Sulfite Compounds.   When sodium sulfite (Na_2SO_3) and sodium bisulfite ($NaHSO_3$) and sodium metabisulfite ($Na_2S_2O_5$) are used for dechlorination, the following reactions occur. The stoichiometric weight ratios of these compounds needed per mg/L of residual chlorine are given in Table 12–15.

Reaction between sodium sulfite and free chlorine residual and combined chlorine residual, as represented by monochloramine:

$$Na_2SO_3 + Cl_2 + H_2O \rightarrow Na_2SO_4 + 2HCl \tag{12–45}$$

Table 12–15
Typical information on the quantity of dechlorinating compound required for each mg/L of residual chlorine

	Dechlorinating compound		Quantity, mg/(mg/L) residual	
Name	**Formula**	**Molecular weight**	**Stoichiometric amount**	**Range in use**
Sulfur dioxide	SO_2	64.09	0.903	1.0–1.2
Sodium sulfite	Na_2SO_3	126.04	1.775	1.8–2.0
Sodium bisulfite	$NaHSO_3$	104.06	1.465	1.5–1.7
Sodium metabisulfite	$Na_2S_2O_5$	190.10	1.338	1.4–1.6
Sodium thiosulfate	$Na_2S_2O_3$	112.12	0.556	0.6–0.9

$$Na_2SO_3 + NH_2Cl + H_2O \rightarrow Na_2SO_4 + Cl^- + NH_4^+ \qquad (12\text{–}46)$$

Reaction between sodium bisulfite and free chlorine residual and combined chlorine residual, as represented by monochloramine:

$$NaHSO_3 + Cl_2 + H_2O \rightarrow NaHSO_4 + 2HCl \qquad (12\text{–}47)$$

$$NaHSO_3 + NH_2Cl + H_2O \rightarrow NaHSO_4 + Cl^- + NH_4^+ \qquad (12\text{–}48)$$

Reactions between sodium metabisulfite and free chlorine residual and combined chlorine residual, as represented by monochloramine:

$$Na_2S_2O_5 + Cl_2 + 3H_2O \rightarrow 2NaHSO_4 + 4HCl \qquad (12\text{–}49)$$

$$Na_2S_2O_5 + 2NH_2Cl + 3H_2O \rightarrow Na_2SO_4 + H_2SO_4 + 2Cl^- + 2NH_4^+ \qquad (12\text{–}50)$$

Dechlorination with Sodium Thiosulfate. Although sodium thiosulfate $(Na_2S_2O_3)$ is often used as a dechlorinating agent in analytical laboratories, its use in full-scale wastewater-treatment plants is limited for the following reasons. It appears the reaction of sodium thiosulfate with residual chlorine is stepwise, creating a problem with uniform mixing. The ability of sodium thiosulfate to remove residual chlorine is a function of the pH (White, 1999). The reaction with residual chlorine is only stoichiometric at pH = 2, making prediction of the required dose impossible in wastewater applications. As reported in Table 12–15, the stoichiometric weight ratio of sodium thiosulfate per mg/L of residual chlorine is 0.556.

Dechlorination with Activated Carbon. Carbon adsorption for dechlorination provides complete removal of both combined and free residual chlorine. When activated carbon is used for dechlorination, the following reactions occur:

Reactions with free chlorine residual:

$$C + 2Cl_2 + 2H_2O \rightarrow 4HCl + CO_2 \qquad (12\text{–}51)$$

Reactions with combined residual as represented by mono- and dichloramine:

$$C + 2NH_2Cl + 2H_2O \rightarrow CO_2 + 2NH_4^+ + 2Cl^- \qquad (12\text{–}52)$$

$$C + 4NHCl_2 + 2H_2O \rightarrow CO_2 + 2N_2 + 8H^+ + 2Cl^- \qquad (12\text{–}53)$$

Granular activated carbon is used in either a gravity or pressure filter bed. If carbon is to be used solely for dechlorination, it must be preceded by an activated-carbon process for the removal of other constituents susceptible to removal by activated carbon. In treatment plants where granular activated carbon is used to remove organics, either the same or separate beds can also be used for dechlorination.

Because granular carbon in column applications has proved to be very effective and reliable, activated carbon should be considered where dechlorination is required. However, this method is quite expensive. It is expected that the primary application of activated carbon for dechlorination will be in situations where high levels of organic removal are also required.

Dechlorination of Chlorine Dioxide with Sulfur Dioxide

Dechlorinating wastewater disinfected with chlorine dioxide can be achieved using sulfur dioxide. The reaction that takes place in the chlorine dioxide solution can be expressed as

$$SO_2 + H_2O \rightarrow H_2SO_3 \tag{12-54}$$

$$5H_2SO_3 + 2ClO_2 + H_2O \rightarrow 5H_2SO_4 + 2HCl \tag{12-55}$$

Based on Eq. (12-55), it can be seen that 2.5 mg of sulfur dioxide will be required for each mg of chlorine dioxide residual (expressed as ClO_2). In practice, 2.7 mg SO_2/mg ClO_2 would normally be used.

12-6 DESIGN OF CHLORINATION AND DECHLORINATION FACILITIES

The chemistry of chlorine in water and wastewater has been discussed in the previous sections, along with an analysis of how chlorine functions as a disinfectant. However, chlorine has been applied for a wide variety of objectives other than disinfection in the wastewater-treatment field including prechlorination for hydrogen sulfide control, activated-sludge bulking control, and odor control. Therefore, the purpose of this section is to discuss briefly the design of chlorination and dechlorination facilities for a variety of purposes.

Important practical design factors that must be considered in the design of chlorine mixing and contact facilities include (1) estimation of the chlorine dosage, (2) application flow diagrams, (3) dosage control, (4) injection and initial mixing, (5) chlorine contact basin design, (6) maintenance of solids transport velocity, (7) outlet control and chlorine residual measurement, (8) chlorine storage facilities, (9) chemical containment and neutralization facilities, and (10) dechlorination facilities. These topics are considered briefly in the following discussion. Additional details on the design of chlorination and dechlorination facilities may be found in U.S. EPA (1986), WEF (1996), and White (1999).

Sizing Chlorination Facilities

To aid in the design and selection of the required chlorination facilities and equipment, it is important to know the uses, including dosage ranges, to which chlorine and its compounds have been applied. Chlorination capacities for disinfection are generally selected to meet the specific design criteria of the state or other regulatory agencies con-

Table 12-16
Typical dosages for various chlorination applications in wastewater collection and treatment

Application	Dosage range, mg/L
Collection:	
Corrosion control (H₂S)	2–9[a]
Odor control	2–9[a]
Slime growth control	1–10
Treatment:	
BOD reduction	0.5–2[b]
Digester and Imhoff tank foaming control	2–15
Digester supernatant oxidation	20–140
Ferrous sulfate oxidation	—[c]
Filter fly control	0.1–0.5
Filter ponding control	1–10
Grease removal	2–10
Sludge bulking control	1–10

[a] Per mg/L of H_2S.
[b] Per mg/L of BOD_5 destroyed.
[c] $6FeSO_4 \cdot 7H_2O + 3Cl_2 \rightarrow 2Fe_2(SO_4)_3 + 42H_2O$.

trolling the receiving body of water. In any case, where the residual in the effluent is specified or the final number of coliform bacteria is limited, onsite testing is preferred to determine the dosage of chlorine required. Typical chlorine dosages for disinfection have been given previously in Table 12–13. Chlorine dosages can also be estimated using the equations given previously in Sec. 12–2. Typical chlorine dosages for applications other than disinfection are given in Table 12–16. A range of dosage values is given because they will vary depending on the characteristics of the wastewater. In the absence of more specific data, the maximum values given in Tables 12–13 and 12–16 can be used as a guide in sizing chlorination equipment.

EXAMPLE 12–6 **Sizing of Chlorination Facilities** Determine the capacity of a chlorinator for a treatment plant with an average wastewater flow of 1000 m³/d (0.26 Mgal/d). The peak daily factor for the treatment plant is 3.0 and the maximum required chlorine dosage (set by state regulations) is to be 20 mg/L.

Solution

1. Determine the capacity of the chlorinator at peak flow.

$$Cl_2, \text{ kg/d} = (20 \text{ g/m}^3)(1000 \text{ m}^3/\text{d})(3)(1 \text{ kg}/10^3 \text{ g})$$

$$= 60 \text{ kg/d}$$

Use the next largest standard-size chlorinator: two 90 kg/d (200 lb/d) units with one unit serving as a spare. Although the peak capacity will not be required during most of the day, it must be available to meet the chlorine requirements at peak flow. Best design practice calls for the availability of a standby chlorinator.

2. Estimate the daily consumption of chlorine. Assume an average dosage of 10 mg/L.

$$Cl_2, kg/d = (10 \text{ g/m}^3)(1000 \text{ m}^3/\text{d})(1 \text{ kg}/10^3 \text{ g})$$

$$= 10 \text{ kg/d}$$

Comment In sizing and designing chlorination systems, it is also important to consider the low flow/dosage requirements. The chlorination system should have sufficient turndown capability for these conditions so that excessive chlorine is not applied. In some cases, consideration should be given to the use of two units to cover high and low flow dosage requirements, when the turndown required gets to be too high.

Application Flow Diagrams

In this section, the flow diagrams and equipment used to inject (feed) chlorine and its related compounds into the wastewater are discussed.

Flow Diagram for Chlorine. Chlorine may be applied directly as a gas or in an aqueous solution. Typical chlorine/sulfur dioxide chlorination/dechlorination process flow diagrams are shown on Fig. 12–17. Chlorine can be withdrawn from storage containers in either liquid or gas form. If withdrawn as a gas, the evaporation of the liquid in the container results in frost formation that restricts gas withdrawal rates to 18 kg/d (40 lb/d) for 68 kg (150 lb) cylinders and 205 kg/d (450 lb/d) for 908 kg (1 ton) containers at 21°C (70°F). Evaporators are used normally where the maximum rate of chlorine gas withdrawal from a 908 kg (1 ton) container must exceed approximately 180 kg/d (400 lb/d). Although multiple-ton cylinders can be connected to provide more than 180 kg/d (400 lb/d), the use of an evaporator conserves space. Evaporators are almost always used when the total dosage exceeds 680 kg/d (1500 lb/d). Chlorine evaporators are available in sizes ranging from 1816 to 4540 kg/d (4000 to 10,000 lb/d) capacities; chlorinators are available normally in sizes ranging from 227 to 4540 kg/d (500 to 10,000 lb/d).

Flow Diagram for Liquid Hypochlorite Solutions. A typical sodium hypochlorite/sodium bisulfite chlorination/dechlorination process flow diagram is shown on Fig. 12–18. For small treatment plants, the most satisfactory means of feeding sodium or calcium hypochlorite is through the use of low-capacity proportioning pumps (see Fig. 12–4). Generally, the pumps are available in capacities up to 450 L/d (120 gal/d), with adjustable stroke for any value below this. Large capacities or multiple units are available from some of the manufacturers. The pumps can be arranged to

Figure 12–17

Schematic flow diagrams for chlorination/dechlorination: (a) using a chlorine injector system and (b) using a molecular chlorine vapor induction system.

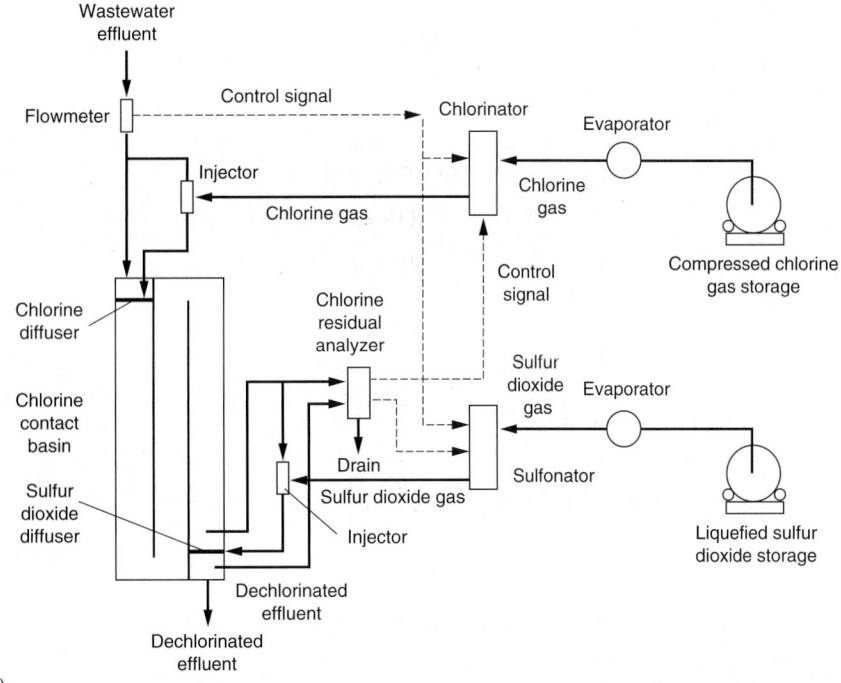

(a)

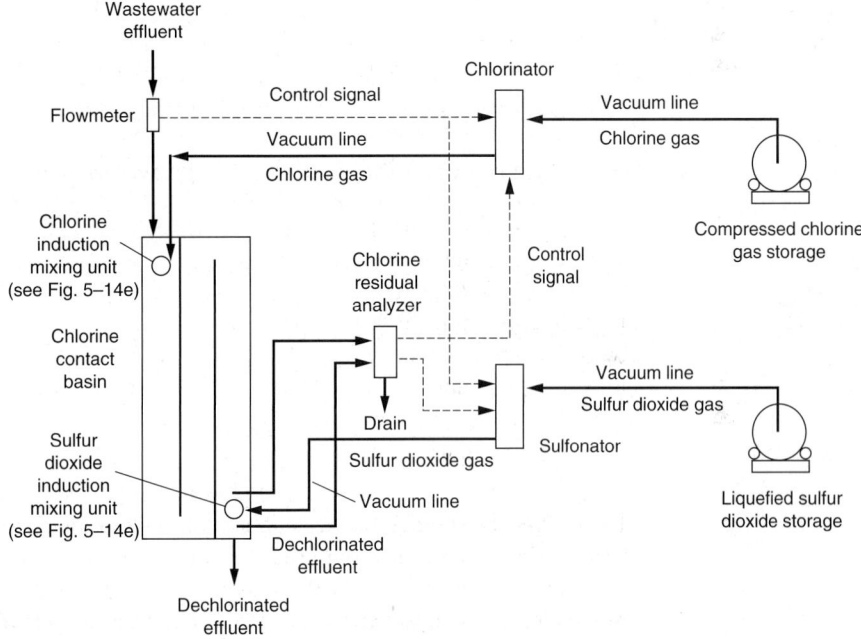

(b)

Figure 12–18

Schematic flow diagrams for sodium hypochlorite chlorination with sulfur dioxide dechlorination.

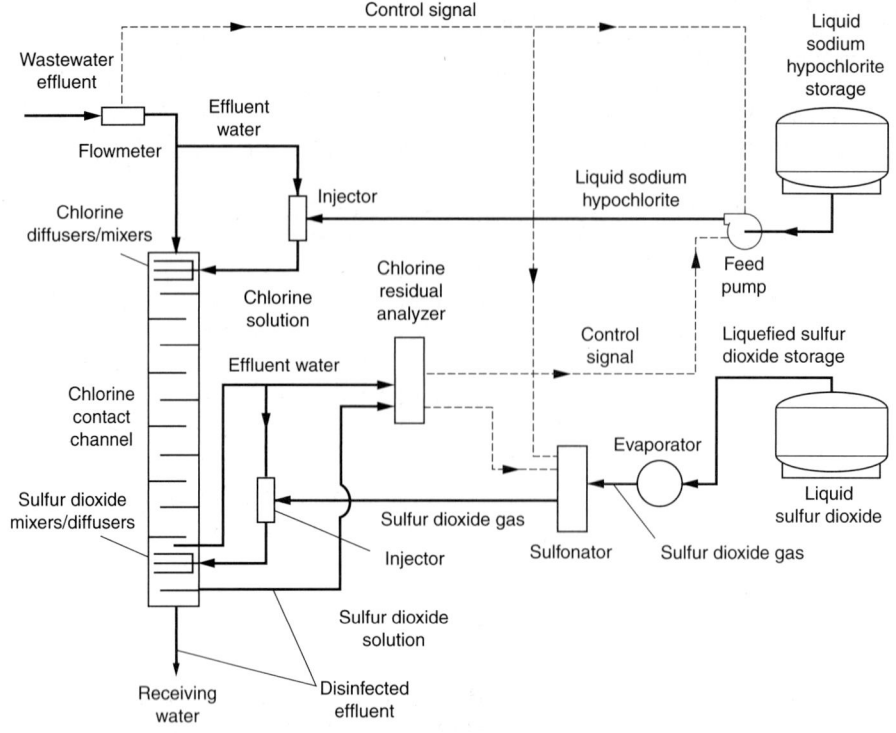

feed at a constant rate, or they can be provided with variable speed and with analog signals for varying the feed rate. The stroke length can also be controlled.

Flow Diagram for Dry Calcium Hypochlorite Feed System.   For small wastewater flowrates up to about 400 m³/d (10^5 gal/d) chlorine in the form of dry calcium hypochlorite tablets is used for disinfection. Two of the most common types of tablet chlorinators (nonpressurized and pressurized) are shown on Fig. 12–19. The schematic flow diagrams for the two tablet chlorinators are essentially the same, a side stream of water is diverted from the main discharge line, chlorine at relatively high concentrations is added to the side stream, and the chlorinated side stream is discharged back into the main flow by means of a pump (see Fig. 12–19a) or by reducing the pressure in the main discharge line (see Fig. 12–19b). As shown on Fig. 12–19a, in the nonpressurized tablet chlorinator the side stream contacts the bottom surface of the chlorine tablets that rest on a screen. The chlorine tablets have been designed to erode at a more-or-less constant rate, releasing a controlled amount of chlorine. The amount of chlorine added is dependent on the flowrate through the tablet chlorinator. The tablet chlorinator shown on Fig. 12–19b is pressurized and hypochlorite is released as water flows over the hypochlorite tablets. Dry calcium hypochlorite tablets, typically 75 mm (3 in) in diameter, contain about 65 to 70 percent available chlorine. For small treatment plants, the use of chlorine tablets eliminates the hazards associated with handling chlo-

Figure 12-19

Schematic flow diagrams for calcium hypochlorite tablet chlorinators: (a) nonpressurized (Adapted from PPG Industries, Inc.) and (b) pressurized. (Adapted from PPG Industries, Inc.)

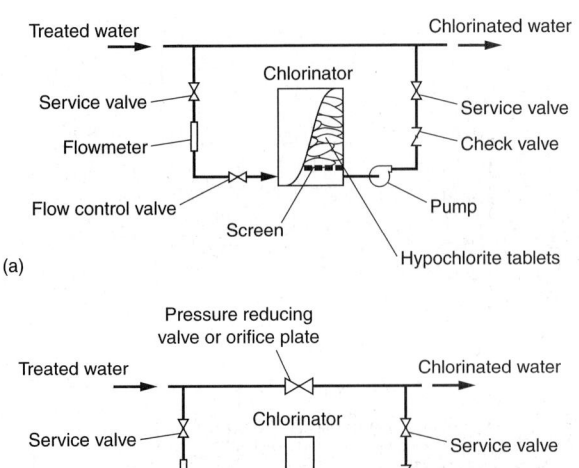

rine cylinders. Further, because there are no moving parts, tablet chlorinators are simple to operate and maintain.

Flow Diagram for Chlorine Dioxide. The chlorine dioxide, generated onsite (see Fig. 12–16), is present in an aqueous solution that is applied in the same manner as that used for typical chlorination systems. A schematic process flow diagram of a typical chlorine dioxide installation is shown on Fig. 12–20.

Dosage Control

Dosage may be controlled in several ways. The simplest method is manual control, where the operator changes the feed rate to suit conditions. The required dosage is usually determined by measuring the chlorine residual after 15 min of contact time and adjusting the dosage to obtain a residual of 0.5 mg/L. A second method is to pace the chlorine flowrate to the wastewater flowrate as measured by a primary meter such as a magnetic meter, Parshall flume, or flow tube. The flowmeter may be placed before the inlet or after the outlet of the chlorine contact basin. A third method is to control the chlorine dosage by automatic measurement of the chlorine residual. An automatic analyzer with signal transmitter and recorder is required. When using on-line analyzers a sample should be taken and analyzed within 2 min so the control system will have time to react. Finally, a compound system that incorporates both the second and third methods may be used. In a compound system, the control signals obtained from the wastewater flowmeter and from the residual recorder and fed to a programmable logic controller (PLC) provide more precise control of chlorine dosage and residual.

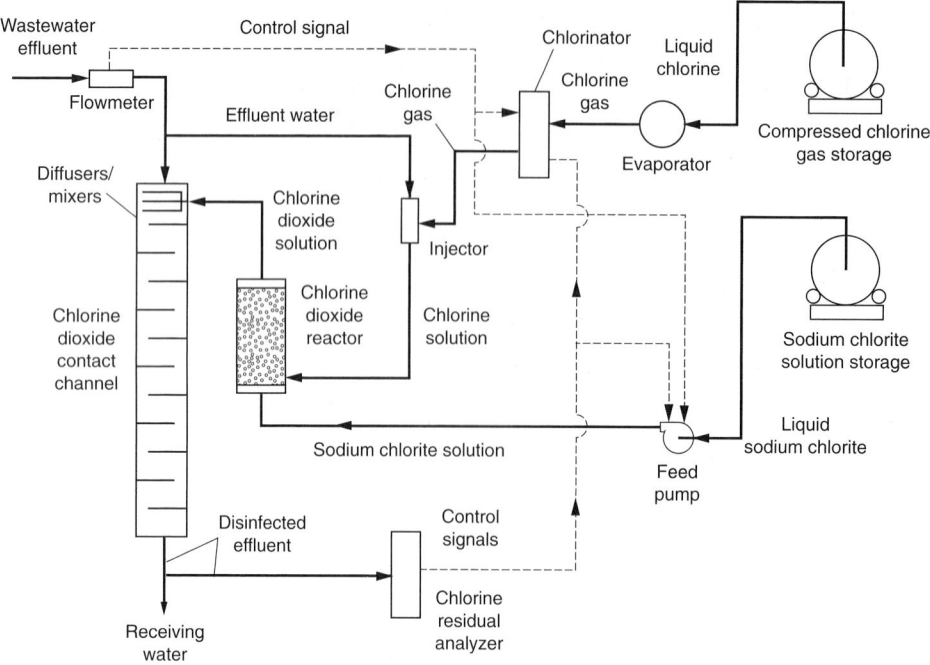

Figure 12–20

Typical flow diagram for the addition of chlorine dioxide.

Injection and Initial Mixing

As pointed out in Sec. 12–3, other things being equal, effective mixing of the chlorine solution with the wastewater, the contact time, and the chlorine residual are the principal factors involved in achieving effective bacterial kill. The addition of chlorine solution is often accomplished with a diffuser, which may be a plastic pipe with drilled holes through which the chlorine solution can be distributed into the path of wastewater flow (see Fig. 12–21). Unfortunately, the use of diffusers for adding chlorine is not very effective. To optimize the performance of disinfection systems, the chlorine should be introduced and mixed as rapidly as possible (ideally in less than a second). Techniques that can be used to mix chlorine in a fraction of a second were introduced and discussed in Chap. 5. Effective methods of mixing chlorine with the wastewater include the use of (1) in-line static mixers, (2) in-line mixers, (3) high-speed induction mixers, (4) pressurized water jets, and (5) pumps. Two typical mixing devices are shown on Fig. 12–22. Additional mixing devices are illustrated on Fig. 5–14 in Chap. 5. A complete analysis of mixing may be found in Chap. 5.

Chlorine Contact Basin Design

The principal design objective for chlorine contact basins is to ensure that some defined percentage of the flow remains in the chlorine contact basin for the design contact time to ensure effective disinfection. The contact time is usually specified by the regulatory

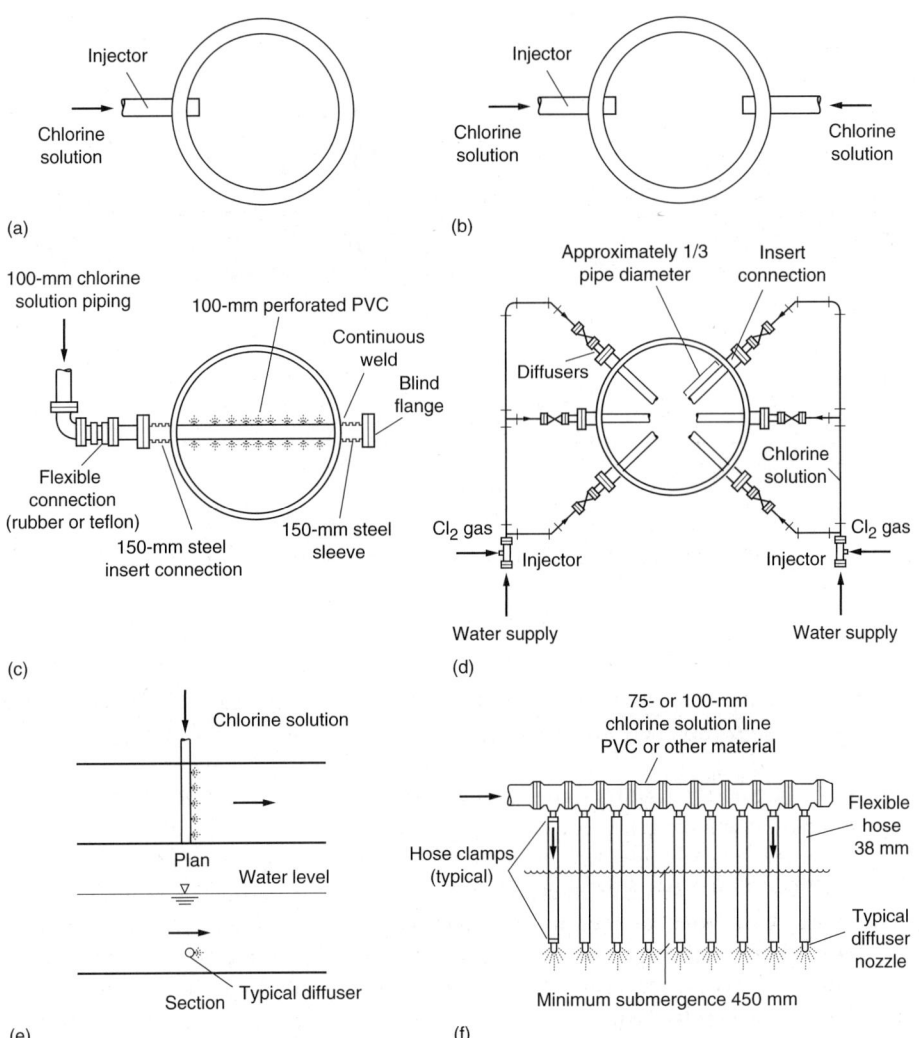

Figure 12–21

Typical diffusers used to inject chlorine solution: (a) single injector for small pipe, (b) dual injector for small pipe, (c) across the pipe diffuser for pipes larger than 0.9 m (3 ft) in diameter, (d) diffuser system for large conduits, (e) single across-the-channel diffuser, and (f) typical hanging nozzle type chlorine diffuser for open channels. (Adapted from White, 1999.)

agency and may range from 30 to 120 min; periods of 15 to 90 min at peak flow are common. For example, for reuse applications the Department of Health Services of the State of California requires a $C_R t$ value of 450 mg·min/L with a modal contact time of 90 min at peak flow. Issues related to the design and analysis of chlorine contact basins considered in the following discussion include (1) basin configuration, (2) the use of baffles and guide vanes, (3) number of chlorine contact basins, (4) precipitation of

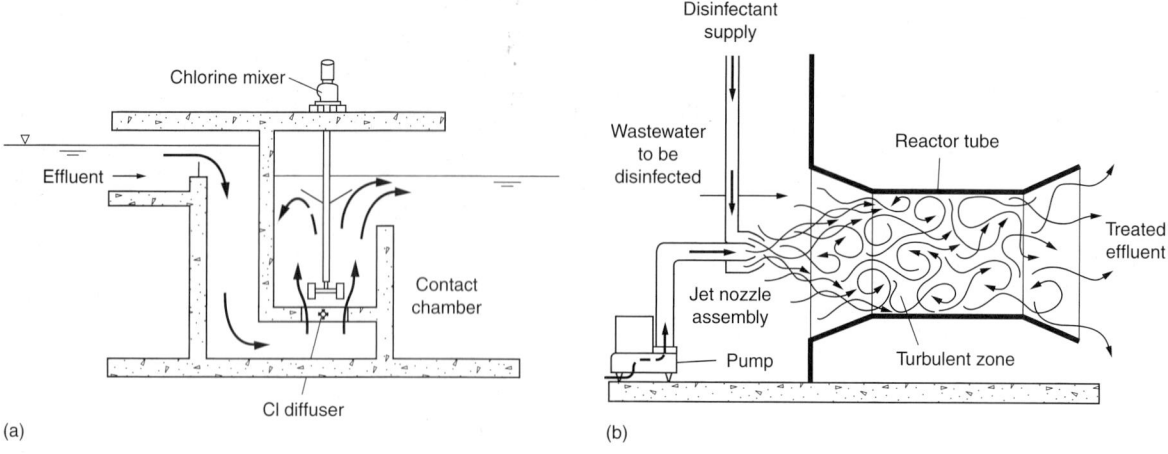

Figure 12-22

Typical mixers for the addition of chlorine: (a) in-line turbine mixer and (b) injector pump type. *(From Pentech-Houdaille.)* For additional types of chlorine mixers see Fig. 5–14 in Chap. 5.

solids in chlorine contact basins, (5) solids transport velocity, and (6) a procedure for predicting disinfection performance.

Chlorine Contact Basin Configuration.   To be assured that a given percentage of the flow will remain in the chlorine contact basin for a given period of time, the most common approach is to use long plug-flow, around-the-end type of contact basins (see Fig. 12–23) or a series of interconnected basins or compartments. Plug-flow chlorine contact basins that are built in a serpentine fashion (e.g., folded back and forth) to conserve space require special attention in their design to eliminate the formation of hydraulic dead zones that will reduce the hydraulic detention times. Length-to-width ratios (L/W) of at least 20 to 1 [preferably 40 to 1, Marske and Boyle (1973)] and the use of baffles and guide vanes, as described below, will help to minimize short circuiting. In some small plants, chlorine contact basins have been constructed of large-diameter sewer pipe. The importance of dispersion in the design of chlorine contact basins was considered in early papers by Marske and Boyle (1973), Trussell and Chao (1977), and Hart (1979). The design of a chlorine contact basin based on dispersion is considered in Example 12–7.

Use of Baffles and Deflection Guide Vanes.   To improve the hydraulic performance of chlorine contact basins, it has become common practice to use either submerged baffles, deflection guide vanes, or combinations of the two. Submerged baffles are used to break up density currents caused by temperature gradients, to limit short circuiting, and to minimize the effect of hydraulic dead spaces. The location of the baffles is critical in improving the performance of chlorine contact basins. The recommended location for baffles, at the beginning of each pass, and the effect on the corresponding tracer response curves, is illustrated on Fig. 12–24. As shown on Fig. 12–24, the addi-

Figure 12–23

Typical examples of chlorine contact basins: (a) spiral wraparound design, (b) serpentine design with short channels, (c) serpentine design with long channels, and (d) back-and-forth straight long-channel design.

(a) (b)

(c) (d)

tion of baffles improves the hydraulic performance of the chlorine contact basin significantly. The open area in submerged baffles will typically vary from 6 to 10 percent of the cross-sectional area of flow. The headloss through each baffle can be estimated using the following expression.

$$h = \frac{1}{2g}\left(\frac{Q}{Cna}\right)^2 \tag{12–56}$$

where h = headloss, m

g = acceleration due to gravity, 9.81 m/s^2

Q = discharge through chlorine contact basin channel, m^3/s

C = discharge coefficient, unitless (typically about 0.8)

n = number of openings

a = area of individual opening, m^2

A typical serpentine chlorine contact basin equipped with removable baffles is shown on Fig. 12–25. The spacing of the baffles can be adjusted to improve performance.

An alternative approach that has been used to improve the performance of chlorine contact basins is through the addition of deflection guide vanes, as shown on Fig. 12–26. The placement and number of vanes will depend on the layout of the chlorine contact basin. Two or three guide vanes are used most commonly. The beneficial effect of adding guide vanes is illustrated on Fig. 12–27.

Figure 12–24

Effect of baffle addition on hydraulic performance of chlorine contact basin. The recommended location for baffles is at the beginning of each pass. *(Adapted from Hart, 1979 and Montgomery, 1985.)*

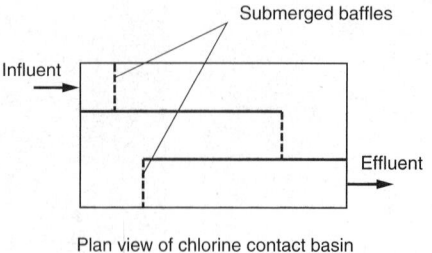

Plan view of chlorine contact basin

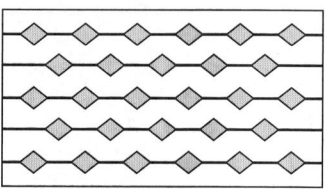

Typical baffle detail
(total area of openings is equal to
6 to 10 percent of cross-sectional area)

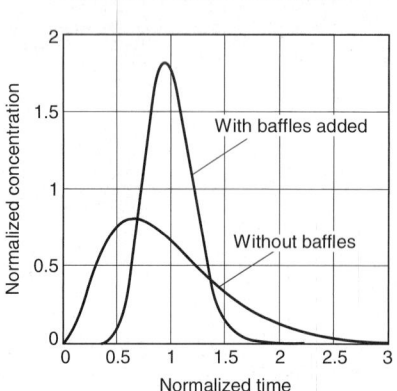

Figure 12–25

Typical chlorine contact basin with removable baffles.

Number of Chlorine Contact Basins. For most treatment plants, two or more contact basins should be used to meet reliability and redundancy requirements to facilitate maintenance and cleaning. Provisions should also be included for draining and scum removal. Vacuum-type cleaning equipment may be used as an alternative to draining the basin for removal of accumulated solids. Bypassing the contact basin for maintenance should only be practiced on rare occasions, with the approval of regulatory agencies. If the time of travel in the outfall sewer at the maximum design flow is suffi-

Figure 12–26

Chlorine contact basin with flow-deflection vanes; (a) schematic and (b) photograph of empty chlorine contact basin designed with guide vanes.

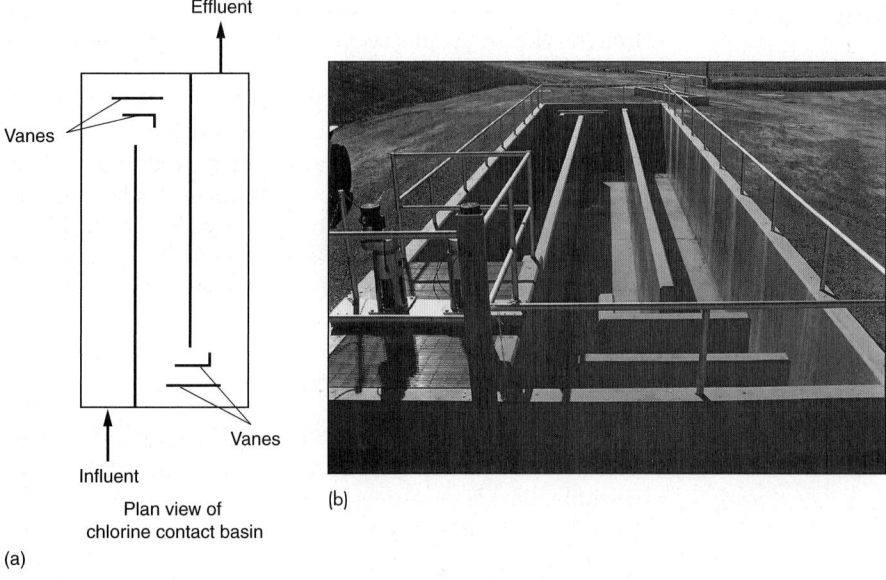

Effluent

Vanes

Vanes

Influent

Plan view of chlorine contact basin

(a)

(b)

Figure 12–27

Effect of chlorine contact basin design on hydraulic performance. (Adapted from Louie and Fohrman, 1968.)

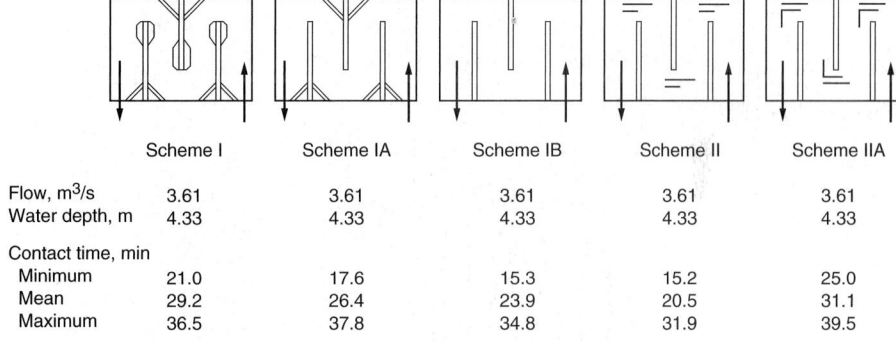

	Scheme I	Scheme IA	Scheme IB	Scheme II	Scheme IIA
Flow, m³/s	3.61	3.61	3.61	3.61	3.61
Water depth, m	4.33	4.33	4.33	4.33	4.33
Contact time, min					
Minimum	21.0	17.6	15.3	15.2	25.0
Mean	29.2	26.4	23.9	20.5	31.1
Maximum	36.5	37.8	34.8	31.9	39.5

cient to equal or exceed the required contact time, it may be possible to eliminate the chlorine contact chambers, provided regulatory authorities agree.

Precipitation in Chlorine Contact Basins. A problem often encountered in the operation of chlorine contact basins is the formation and precipitation of a light flocculent material. The principal cause of the formation and precipitation of floc is the lowering of the pH that results from the addition of chlorine. The problem occurs most frequently where alum is used for phosphorus removal in separate precipitation facilities or is added before the effluent filters. For a variety of reasons including high pH and inadequate initial mixing, not all of the alum added will react completely to form a floc that can be removed by precipitation or filtration. However, when the pH is lowered in the chlorine contact basin due to the addition of chlorine, some of the unreacted alum may form a floc. Thus, in addition to meeting reliability and redundancy requirements, a minimum of two chlorine contact basins is necessary to allow one basin to be removed from service so that the accumulated solids can be removed from the basins.

Solids Transport Velocity.   The horizontal velocity at minimum flow in a chlorine contact basin should, in theory, be sufficient to scour the bottom or to limit the deposition of sludge solids that may have passed through the settling tank. To limit excessive deposition, horizontal velocities should be at least 2 to 4.5 m/min (6.5 to 15 ft/min). In general, it will be difficult to achieve such velocities and simultaneously meet stringent dispersion requirements (see Example 12–7). If floc particles form, it will generally be impossible to avoid the accumulation of a sludge layer in the chlorine contact basins, another reason at least two chlorine contact basins should be used.

EXAMPLE 12–7   **Design of a Chlorine Contact Basin Based on Dispersion**   Design a chlorine contact basin for an average flow of 4000 m^3/d. The estimated peaking factor is 2.0. The detention time at peak flow is to be 90 min. A minimum of two parallel channels must be used for redundancy requirements. The dimensions of the chlorine contact basin should be such as to achieve a dispersion number of about 0.015 at peak flow. Also check the dispersion number at average flow. What will happen if the low flow drops to 33 percent of the average flow in the early morning hours? Based on the resulting calculations, will solids deposition occur, requiring periodic scouring or cleaning of the chlorine contact basin?

Solution

1.   Assume some trial cross-sectional dimensions for the chlorine contact basin and determine the corresponding length and flow velocity.

   a.   Assumed dimensions

   Width = 2 m

   Depth = 3 m

   Number of parallel channels = 2

   b.   Determine required length.

   $$L = \frac{(2 \times 4000 \text{ m}^3/\text{d})}{(2)(1440 \text{ min/d})} \times (90 \text{ min}) \times \frac{1}{(2 \text{ m} \times 3 \text{ m})} = 41.7 \text{ m}$$

   c.   Check velocity at peak flow.

   $$V = \frac{(2 \times 4000 \text{ m}^3/\text{d})}{(2)(1440 \text{ min/d})(60 \text{ s/min})} \times \frac{1}{(2 \text{ m} \times 3 \text{ m})} = 0.0077 \text{ m/s}$$

2.   Check the dispersion number for the chlorine contact basin using Eq. (4–48).

   $$D = 1.01\nu(N_R)^{0.875}$$

   where D = coefficient of dispersion, m^2/s
      ν = kinematic viscosity, m^2/s (see Appendix C)
      N_R = Reynolds number, unitless

a. Compute the Reynolds number.

$N_R = 4vR/v$

v = velocity in open channel, m/s

R = hydraulic radius = area/wetted perimeter m

$v = 0.0077$ m/s

$v = 1.003 \times 10^{-6}$ m²/s

$$N_R = \frac{(4)(0.0077 \text{ m/s})[(2.0 \text{ m} \times 3.0 \text{ m})/(2 \times 3.0 \text{ m} + 2.0 \text{ m})]}{(1.003 \times 10^{-6} \text{ m}^2/\text{s})} = 23{,}031$$

b. Compute the coefficient of dispersion.

$D = 1.01v(N_R)^{0.875}$

$D = (1.01)(1.003 \times 10^{-6} \text{ m}^2/\text{s})(23{,}031)^{0.875} = 6.648 \times 10^{-3} \text{ m}^2/\text{s}$

c. Determine the dispersion number using Eq. (4–43).

$$d = \frac{D}{uL} = \frac{Dt}{L^2} = \frac{(0.006648 \text{ m}^2/\text{s})(90 \text{ min} \times 60 \text{ s/min})}{(41.7 \text{ m})^2} = 0.0206$$

Because the computed dispersion number (0.0206) is greater than the desired value (0.015), an alternative design must be evaluated. For the alternative design, assume three parallel channels will be used.

3. Assume new trial cross-sectional dimensions for the chlorine contact basin and determine the new length and flow velocity.

a. Assumed dimensions

Width = 1.25 m

Depth = 3 m

Number of parallel channels = 3

b. Determine required length.

$$L = \frac{(2 \times 4000 \text{ m}^3/\text{d})}{(3)(1440 \text{ min/d})} \times (90 \text{ min}) \times \frac{1}{(1.25 \text{ m} \times 3 \text{ m})} = 44.4 \text{ m}$$

c. Check velocity at peak flow.

$$V = \frac{(2 \times 4000 \text{ m}^3/\text{d})}{(3)(1440 \text{ min/d})(60 \text{ s/min})} \times \frac{1}{(1.25 \text{ m} \times 3 \text{ m})} = 0.0082 \text{ m/s}$$

4. Check the dispersion number for the chlorine contact basin.

a. Compute the Reynolds number.

$N_R = 4vR/v$

$v = 0.0082$ m/s

$$\nu = 1.003 \times 10^{-6} \text{ m}^2/\text{s}$$

$$N_R = \frac{(4)(0.0082 \text{ m/s})[(1.25 \text{ m} \times 3.0 \text{ m})/(2 \times 3.0 \text{ m} + 1.25 \text{ m})]}{(1.003 \times 10^{-6} \text{ m}^3/\text{s})} = 16{,}915$$

b. Compute the coefficient of dispersion.

$$D = 1.01 \nu N_R^{0.875}$$

$$D = (1.01)(1.003 \times 10^{-6} \text{ m}^2/\text{s})(16{,}915)^{0.875} = 5.07 \times 10^{-3} \text{ m}^2/\text{s}$$

c. Determine the dispersion number.

$$d = \frac{D}{uL} = \frac{Dt}{L^2} = \frac{(0.00507 \text{ m}^2/\text{s})(90 \text{ min} \times 60 \text{ s/min})}{(44.4 \text{ m})^2} = 0.0139$$

Because the computed dispersion number (0.0139) is smaller than the desired value (0.015), the proposed design is acceptable.

5. Check the dispersion number for the chlorine contact basin at average flow.
    a. Compute the Reynolds number.

$$N_R = 4vR/\nu$$

$$v = 0.0082/2 = 0.0041 \text{ m/s}$$

$$\nu = 1.003 \times 10^{-6} \text{ m}^2/\text{s}$$

$$N_R = \frac{(4)(0.0041 \text{ m/s})[(1.25 \text{ m} \times 3.0 \text{ m})/(2 \times 3.0 \text{ m} + 1.25 \text{ m})]}{(1.003 \times 10^{-6} \text{ m}^2/\text{s})} = 8457$$

b. Determine the coefficient of dispersion.

$$D = 1.01 \nu N_R^{0.875}$$

$$D = 1.01 \times 1.003 \times 10^{-6} \text{ m}^2/\text{s } (8457)^{0.875} = 2.77 \times 10^{-3} \text{ m}^2/\text{s}$$

c. Determine the dispersion number using Eq. (4–43).

$$d = \frac{D}{uL} = \frac{Dt}{L^2} = \frac{(0.00277 \text{ m}^2/\text{s})(90 \text{ min} \times 60 \text{ s/min})}{(44.4 \text{ m})^2} = 0.0076$$

d. Because the velocity is reduced at average flow, the computed dispersion number is equivalent to about 66 complete-mix reactors in series.

Comment Under all flow conditions, deposition of residual suspended solids would be expected in the chlorine contact basin, especially so at low flow.

Predicting Disinfection Performance. An extremely important issue in the design of chlorine contact basins is being able to predict the performance of the proposed design. To predict performance, the actual residence time that a given molecule of the fluid spends in the reactor must be known. The residence time in the reactor can be determined using some of the analytical techniques developed in Chap. 4. From Chap. 4, it was noted that the Peclet number divided by 2 is equal to the number of reactors in series. The relationship of the Peclet number to the dispersion number is

$$P_e = \frac{uL}{D} = \frac{1}{d} \tag{4–47}$$

For complete-mix reactors in series, the normalized residence time distribution curve for n reactors in series is given by

$$E(\theta) = \frac{n}{(n-1)!} (n\theta)^{n-1} e^{-n\theta} \tag{4–63}$$

Further, the fraction of tracer that has been in the reactor for less than time t is defined as follows:

$$F(\theta) = \int_0^t E(\theta) d\theta \approx \sum_0^n E(\theta) \Delta\theta \tag{4–32}$$

Thus, for a given dispersion number, the Peclet number can be used to determine the corresponding number of complete-mix reactors in series needed to achieve that dispersion number. Knowing the number of reactors in series, the value of $E(\theta)$ can be computed for various values of θ, the normalized detention time. The value of $F(\theta)$ can then be determined by summing the area under the $E(\theta)$ curve. The amount of flow that has been in the reactor for less than time θ can now be determined. Coupling normalized microorganism inactivation-dose response data, obtained using a batch reactor, with the normalized detention-time data, the actual performance for the chlorine contact basin can be estimated using what is known as a segregated flow model (SFM). In the SFM approach, it is assumed that each block of fluid that enters a chlorine contact basin does not interact with other blocks of water. Thus, each block of water corresponds to an ideal plug flow reactor, each having a different residence time as defined by the value of $E(\theta)$, as given above. The reduction in organisms that would occur in each block of water can then be estimated for the period of time the block of water has remained in the chlorine contact basin. The overall performance is obtained by summing the results for each block of water. The SFM approach can be described as follows (Fogler, 1999).

Word statement

Mean reduction in number of microorganisms spending between time t and $t + dt$ in the chlorine contact basin	=	number of organisms remaining after spending a time t in the chlorine contact basin based on batch test results	×	fraction of flow that remained in the chlorine contact basin between time t and $t + dt$	(12–57)

In equation form

$$d\overline{N} = N(\theta) \times E(\theta) dt \tag{12–58}$$

The values of $N(\theta)$ and $E(\theta)$ are obtained from batch disinfection and tracer or dispersion prediction studies. Application of the above equations for predicting the hydraulic performance and the effluent microorganism concentration using the SFM are illustrated in Example 12–8 using the data from Example 12–7.

EXAMPLE 12–8 **Estimation of Performance of a Chlorine Contact Basin** Using the design information from Example 12–7, determine the fraction of flow that has not remained in the chlorine contact basin for the full hydraulic detention time. Determine how much larger the chlorine contact basin must be to be assured that 90 percent of the flow remains in the chlorine contact basin for the full design hydraulic detention time. Using the following normalized-dose response data for an enteric virus, estimate the performance of the chlorine contact basin in terms of the residual number of organisms remaining in the effluent.

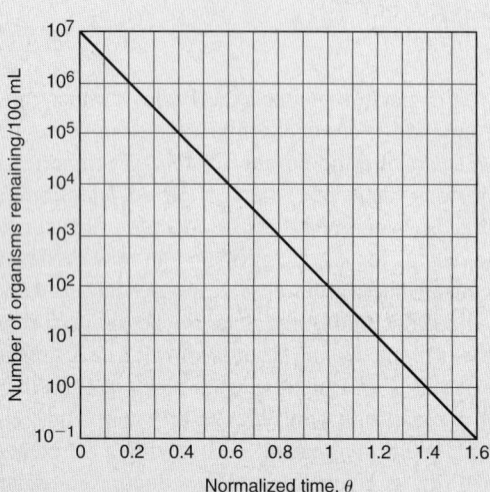

Number of organisms remaining/100 mL vs. Normalized time, θ

Solution

1. Determine the number of complete-mix reactors in series.
   a. From Example 12–7, the dispersion number at peak flow is

   $$d = 0.0139$$

   b. The number of complete-mix reactors in series is

   $$\text{Number of reactors in series} = \frac{P_e}{2} = \frac{1}{2d} = \frac{1}{(2)\,0.0139} = 36$$

2. Determine the percentage of the flow that has been in the chlorine contact basin for less than the hydraulic detention time.
   a. Set up a computation table and compute $E(\theta)$ using Eq. (4–63), as given above.

   $$E(\theta) = \frac{n}{(n-1)!}\,(n\theta)^{n-1}e^{-n\theta}$$

Normalized time, θ	$E(\theta)$	$E(\theta) \times \Delta\theta \times 100$	Cumulative percent, $F(\theta)$
0.30	0.0000	0.000	0.000
0.40	0.0000	0.000	0.000
0.50	0.0046	0.046	0.046
0.60	0.0737	0.737	0.783
0.70	0.4435	4.435	5.218
0.80	1.2976	12.976	18.193
0.90	2.1878	21.878	40.071
1.00	2.3881	23.881	63.952
1.10	1.8337	18.337	82.290
1.20	1.0531	10.531	92.821
1.30	0.4739	4.739	97.560
1.40	0.1733	1.733	99.293
1.50	0.0530	0.530	99.822
1.60	0.0139	0.139	99.961
1.70	0.0031	0.032	99.992
1.80	0.0006	0.006	99.999
1.90	0.0001	0.001	100.00
2.00	0.0000	0.000	100.00

b. Plot the cumulative percent values from the above table.

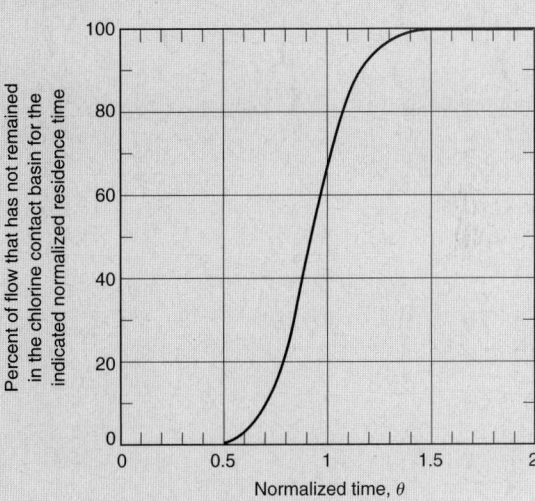

   c. From the computation table and graph given above, the percentage of the flow that has been in the chlorine contact basin for less than the hydraulic residence time is 64 percent. In fact, about 18 percent of the flow has left the chlorine contact basin before 80 percent of the actual hydraulic detention time has elapsed.

3. Estimate how much larger the chlorine contact basin must be to be assured that 90 percent of the flow remains in the chlorine contact basin for the full hydraulic detention time. From the above graph, the size of the chlorine contact basin would have to be increased by a factor of 1.2.

4. Estimate the performance of the chlorine contact basin.

   a. Set up a computation table to determine the number of organisms remaining in the effluent from the chlorine contact basin. The SFM approach, described above, will be used for this analysis. In effect, flow in each time period is treated as a batch reactor for the time interval it has remained in the reactor. The corresponding concentration of microorganisms leaving in any given volume of liquid is taken from the normalized-dose response curve. The computation table for the application of the SFM is given below. The data in columns (1) and (3) are from the computation table prepared in Step 2 above, except that the data in column (3) are divided by 100. The data in column (2) are from the normalized-dose response curve obtained as part of the process analysis for the design of the chlorine contact basin.

Normalized time, θ (1)	Number of organisms remaining, $N(\theta)$ MPN/100 mL (2)	$E(\theta) \times \Delta\theta$ (3)	Number of organisms remaining in effluent, ΔN MPN/100 mL (4)
0.30	300,000	0.00000	0.000
0.40	100,000	0.00000	0.00
0.50	30,000	0.00046	13.80
0.60	10,000	0.00737	73.70
0.70	3,000	0.04435	133.05
0.80	1,000	0.12976	129.76
0.90	300	0.21878	65.63
1.00	100	0.23881	23.88
1.10	30	0.18337	5.50
1.20	10	0.10531	1.05
1.30	3	0.04739	0.14
1.40	1	0.01733	0.02
1.50	0.3	0.00530	—
1.60	0.1	0.00139	—

(continued)

(Continued)

Normalized time, θ (1)	Number of organisms remaining, $N(\theta)$ MPN/100 mL (2)	$E(\theta) \times \Delta\theta$ (3)	Number of organisms remaining in effluent, ΔN MPN/100 mL (4)
1.70	0.03	0.00032	—
1.80	0.01	0.00006	—
1.90	0.003	0.00001	—
2.00	0.001	0.00000	
Total		1.00000	446.53

b. The number of organisms in the effluent leaving the chlorine contact basin is

Organisms in effluent $N = \Sigma[N(\theta) \times E(\theta)\Delta\theta] = 447$ MPN/100 mL

c. By comparison, if it was assumed that the basin had performed as an ideal plug-flow reactor, then the organism concentration in the effluent would have been estimated to be 100 MPN/100 mL.

Comment The SFM method of analysis used to determine the number of organisms in the effluent is useful for estimating the performance of reactors with varying amounts of dispersion such as chlorine contact basins.

Chlorine Residual Measurement

The flow at the inlet or outlet of the chlorine contact basin may be metered, as noted previously, by means of a magnetic meter, Parshall flume, or flow tube. Control devices for chlorination in direct proportion to the flowrate may be operated from these meters or from the main plant flowmeter using an analog flow signal transmitter. Final determination of the success of a chlorine contact basin must be based on samples taken and analyzed to correlate chlorine residual and the MPN of coliform or other indicator organisms. When the chlorine residual is used for chlorinator control, chlorine residual sample pumps should be located at the front end of the first pass of the contact basin immediately after rapid mixing. More precise control of the chlorine feed can be maintained as compared to monitoring the chlorine residual at the chlorine contact basin effluent. Chlorine residual measurements should also be taken at the contact basin outlet to ensure compliance with the regulatory agency requirements. In the event that no chlorine contact basin is provided and the effluent pipeline is used for contact, the sample can be obtained at the point of chlorination, held for the theoretical detention time, and the residual determined. The sample is then dechlorinated and subsequently analyzed for bacteria using standard laboratory procedures.

Chlorine Storage Facilities

Storage and handling facilities for chlorine can be designed with the aid of information developed by the Chlorine Institute. Although all the safety devices and precautions that must be designed into the chlorine handling facilities are too numerous to mention, the following are fundamental:

1. Chlorine gas is toxic and very corrosive. Adequate exhaust ventilation at floor level should be provided because chlorine gas is heavier than air. The ventilation system should be capable of at least 60 air changes per hour.
2. Chlorine storage and chlorinator equipment rooms should be walled off from the rest of the plant and should be accessible only from the outdoors. A fixed glass viewing window should be included in an inside wall, to check for leaks before entering the equipment rooms. Fan controls should be located at the room entrance. Air masks should also be located nearby in protected but readily accessible locations.
3. Temperatures in the scale and chlorinator areas should be controlled to avoid freezing.
4. Dry chlorine liquid and gas can be handled in black steel piping, but chlorine solution is highly corrosive and should be handled in schedule 80 polyvinylchloride (PVC) piping.
5. Adequate storage of standby cylinders should be provided. The amount of storage should be based on the availability and dependability of the supply and the quantities used. Cylinders in use are set on scales and the loss of weight is used as a positive record of chlorine dosage.
6. Chlorine cylinders should be protected from direct sunlight in warm climates to prevent overheating of the full cylinders.
7. In larger systems, chlorine residual analyzers should be provided for monitoring and control purposes to prevent the under- or overdosing of chlorine.
8. The chlorine storage and feed facilities should be protected from fire hazards. In addition, chlorine leak detection equipment should be provided and connected to an alarm system.

Chlorine Containment Facilities

In 1991, the International Conference of Building Officials revised Article 80: Hazardous Material of the Uniform Fire Code (UFC). The revisions were extensive and covered a variety of issues. The provisions of the new code apply to new facilities and to old facilities, if it is determined that they constitute a distinct hazard to life or property. The new code provisions contained in the following divisions apply to the chemicals used for disinfection: I General Provisions, II Classification by Hazard, III Storage Requirements, and IV Dispensing, Use, and Handling. The classification of hazardous materials used for wastewater disinfection is summarized in Table 12–17. Storage requirements include provisions for spill control and containment, ventilation, treatment, and storage. Emergency caustic scrubbing systems are also required to neutralize leaking chlorine and sulfur dioxide gas. Many of the same topics contained in the storage requirements section also apply to the dispensing, use, and handling. Hazardous material management, provision for standby power, security, and alarms are among the

Table 12–17
Classification of hazardous materials used in wastewater disinfection

Category	Typical chemicals
Physical hazards:	
Compressed gases	Oxygen, ozone, chlorine, ammonia, sulfur dioxide
Oxidizers	Oxygen, ozone, chlorine, hydrogen peroxide, acids, chlorine
Health hazards:	
Highly toxic and toxic material	Chlorine, chlorine dioxide, ozone, acids, bases
Corrosives	Acids, bases, chlorine, sulfur dioxide, ammonia, hypochlorite, sodium bisulfite
Other health hazards— irritants, suffocating, etc.	Chlorine, sulfur dioxide, ammonia

(a)

(b)

Figure 12–28

Containment system for chlorine storage containers. Chlorine cylinders are sealed totally within the containment vessel. Vessels are available for one or two 68 kg (150 lb) cylinders and 908 kg (1 ton) chlorine storage containers: (a) containment vessel for one 908 kg chlorine container and (b) view of multiple 908 kg chlorine container containment vessels. (*Courtesy TGO Technologies, Inc.*)

additional topics covered. It is extremely important to review current UFC regulations in the design of new facilities and in the refurbishing of existing facilities.

A relatively recent development (1996) for the storage of chlorine and sulfur dioxide in either 68 kg (150 lb) or 908 kg (1 ton) cylinders is the total containment vessel (see Fig. 12–28) manufactured by TGO Technologies, Inc. Known as the *ChlorTainer*™, the total enclosure vessel is a passive device with no need for pumps, fans or scrubbers, or caustic circulation systems, which meets all of the requirements of the UFC. Should a chlorine and sulfur gas cylinder fail while in the containment vessel, the gas contained within the vessel is processed at the normal rate through the existing chlorination facilities. The containment vessels do require a specially designed truck unloading platform so that the gas cylinders can be moved to and placed in the containment vessel (see Fig. 12–28a). In addition to eliminating the need for specialized chemical handling and storage facilities, the total containment vessel will shut down automatically during a seismic event and does not require mechanical devices to function.

Dechlorination Facilities

Dechlorination of chlorinated effluents is accomplished most commonly using sulfur dioxide. Where granular activated carbon is used for the removal of residual organic material, the carbon can also be used for the dechlorination of chlorinated effluents.

Sulfur Dioxide.   The principal elements of a sulfur dioxide dechlorination system include the sulfur dioxide containers, scales, sulfur dioxide feeders (sulfonators), solution injectors, diffuser, mixing chamber, and interconnecting piping. For facilities requiring large withdrawal rates of SO_2, evaporators are used because of the low vaporization pressure of 240 kN/m^2 at 21°C (35 lb_f/in^2 at 70°F). Common sulfonator sizes are 216, 864, and 3409 kg/d (475, 1900, and 7500 lb/d). The key control parameters of this process are (1) proper dosage based on precise (amperometric) monitoring of the combined chlorine residual and (2) adequate mixing at the point of application of sulfur dioxide.

Sodium Bisulfite.   Sodium bisulfite is available as white powder, a granular material, or a liquid. The liquid form is used most commonly for dechlorination at wastewater-treatment facilities. Although available in solution strengths up to 44 percent, a 38 percent solution is most typical. In most applications, a diaphragm-type pump is used to meter the sodium bisulfite. The reaction between sodium bisulfite and chlorine residual was presented previously [see Eqs. (12–47 and 12–48)]. Based on Eq. (12–47), each mg/L of chlorine residual requires about 1.46 mg/L of sodium bisulfite and 1.38 mg/L of alkalinity as $CaCO_3$ will be consumed.

Granular Activated Carbon.   The common method of activated-carbon treatment used for dechlorination is downflow through either an open or an enclosed vessel. The activated-carbon system, while significantly more costly than other dechlorination approaches, may be appropriate when activated carbon is being used as an advanced wastewater-treatment process. Typical hydraulic loading rates and contact times for activated-carbon columns used for dechlorination are 3000 to 4000 $L/m^2\cdot d$ and 15 to 25 min, respectively.

12–7   DISINFECTION WITH OZONE

Ozone was first used to disinfect water supplies in France in the early 1900s. Its use increased and spread into several western European countries and eventually to North America. There are well over 1000 ozone disinfection installations worldwide, almost entirely for treating water supplies. Approximately 200 installations are in operation in North America. A common use for ozone at these installations is for the control of taste-, odor-, and color-producing agents. Although historically used primarily for the disinfection of water, recent advances in ozone generation and solution technology have made the use of ozone economically more competitive for wastewater disinfection.

Ozone can also be used in wastewater treatment for odor control and in advanced wastewater treatment for the removal of soluble refractory organics, in lieu of the carbon-adsorption process. The characteristics of ozone, the chemistry of ozone, the generation of ozone, an analysis of the performance of ozone as a disinfectant, and the application of the ozonation process are considered in the following discussion.

Ozone Properties

Ozone is an unstable gas produced when oxygen molecules dissociate into atomic oxygen. Ozone can be produced by electrolysis, photochemical reaction, or radiochemical reaction by electrical discharge. Ozone is often produced by ultraviolet light and lightning during a thunderstorm. The electrical discharge method is used for the generation of ozone in water and wastewater disinfection applications. Ozone is a blue gas at normal room temperatures, and has a distinct odor. Ozone can be detected at concentrations of 2×10^{-5} to 1×10^{-4} g/m^3 (0.01 to 0.05 ppm$_v$). Because it has an odor, ozone can usually be detected before health concerns develop. The stability of ozone in air is greater than it is in water, but in both cases is on the order of minutes. Gaseous ozone is explosive when the concentration reaches about 240 g/m^3 (20 percent weight in air). The properties of ozone are summarized in Table 12–18. The solubility of ozone in water is governed by Henry's law. Typical values of Henry's constant for ozone are presented in Table 12–19.

Ozone Chemistry

Some of the chemical properties displayed by ozone may be described by its decomposition reactions, which are thought to proceed as follows:

$$O_3 + H_2O \rightarrow HO_3^+ + OH^- \tag{12–59}$$

$$HO_3^+ + OH^- \rightarrow 2HO_2 \tag{12–60}$$

Table 12–18
Properties of ozone[a]

Property	Unit	Value
Molecular weight	g	48.0
Boiling point	°C	-111.9 ± 0.3
Melting point	°C	-192.5 ± 0.4
Latent heat of vaporization at 111.9°C	kJ/kg	14.90
Liquid density at -183°C	kg/m^3	1574
Vapor density at 0°C and 1 atm	g/mL	2.154
Solubility in water at 20.0°C	mg/L	12.07
Vapor pressure at -183°C	kPa	11.0
Vapor density compared to dry air at 0°C and 1 atm	unitless	1.666
Specific volume of vapor at 0°C and 1 atm	m^3/kg	0.464
Critical temperature	°C	-12.1
Critical pressure	kPa	5532.3

[a] Adapted in part from Rice (1996); U.S. EPA (1986); White (1999).

Table 12-19
Values of Henry's constant for ozone[a]

Temperature, °C	Henry's constant, atm/mole fraction
0	1940
5	2180
10	2480
15	2880
20	3760
25	4570
30	5980

[a]U.S. EPA (1986).

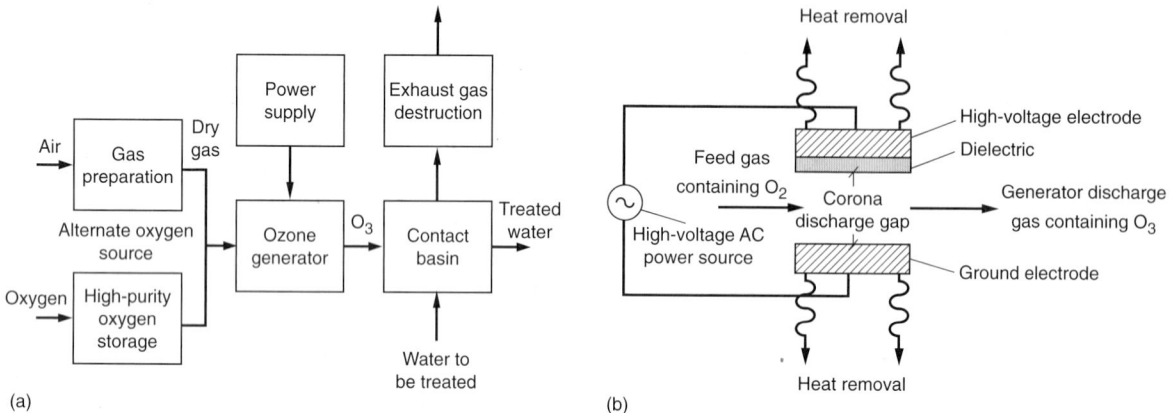

Figure 12-29

Elements of ozone disinfection system: (a) schematic flow diagram for complete ozone disinfection system and (b) schematic detail of the generation of ozone. (Adapted from U.S. EPA, 1986.)

$$O_3 + HO_2 \rightarrow HO + 2O_2 \qquad (12\text{-}61)$$

$$HO + HO_2 \rightarrow H_2O + O_2 \qquad (12\text{-}62)$$

The free radicals formed, HO_2 and HO, have great oxidizing powers, and are probably the active form in the disinfection process. These free radicals also possess the oxidizing power to react with other impurities in aqueous solutions.

Ozone Disinfection Systems Components

A complete ozone disinfection system, as illustrated on Fig. 12–29a, is comprised of the following components: (1) power supply, (2) facilities for the preparation of the feed gas, (3) the ozone generation facilities, (4) facilities for contacting the ozone with the liquid to be disinfected, and (5) facilities for the destruction of the off-gas (Rice, 1996).

Table 12–20
Typical energy requirements for the application of ozone

Component	kWh/lb ozone	kWh/kg ozone
Air preparation (compressor and dryers)	2–3	4.4–6.6
Ozone generation:		
Air feed	6–9	13.2–19.8
Pure oxygen	3–6	6.6–13.2
Ozone contacting	1–3	2.2–6.6
All other uses	0.5–1	1.2–2.2

Energy Requirements. The major requirement for power is the conversion of oxygen into ozone. Additional power is required for preparation of the feed gas, contacting the ozone, destroying the residual ozone, and for the controls, instrumentation, and monitoring facilities. The energy requirements for the major components are reported in Table 12–20.

Preparation of Feed Gas. Ozone can be generated using air, oxygen-enriched air, or high-purity oxygen. If air is used for ozone generation, it must be conditioned by removing the moisture and particulate matter before being introduced into the ozone generator. The following steps are involved in conditioning the air: (1) gas compression, (2) air cooling and drying, and (3) air filtration. If high-purity oxygen is used, the conditioning steps are not required.

Ozone Generation. Because ozone is chemically unstable, it decomposes to oxygen very rapidly after generation, and thus must be generated onsite. The most efficient method of producing ozone today is by electrical discharge. Ozone is generated from either air or high-purity oxygen when a high voltage is applied across the gap of narrowly spaced electrodes (see Fig. 12–29b). The high-energy corona created by this arrangement dissociates one oxygen molecule, which re-forms with two other oxygen molecules to create two ozone molecules. The gas stream generated by this process from air will contain about 1 to 3 percent ozone by weight, and from pure oxygen about three times that amount, or 3 to 10 percent ozone with the latest ozone generators.

Ozone Contact Reactors. The concentration of ozone generated from either air or pure oxygen is so low that the transfer efficiency to the liquid phase is an extremely important economic consideration. For this reason, deep and covered contact chambers are normally used. A schematic process flow diagram of a typical ozone disinfection system is shown on Fig. 12–30. As shown on Fig. 12–30, the ozone is contacted in a three-chamber covered contactor. Ozone is introduced by means of porous diffusers or injectors into the bottom of the first chamber. Fast ozone reactions occur in the first chamber. The combined wastewater-ozone mixture then enters the second chamber, where slower reactions occur. Disinfection generally occurs in the second chamber. The third chamber is used to complete the slow reactions and to allow the ozone to decompose. Specially designed in-line static mixers have also been developed for the dissolution of ozone. A properly designed diffuser system should normally achieve a 90 percent transfer of ozone.

Figure 12–30

Typical flow diagram for the application of ozone for disinfection.

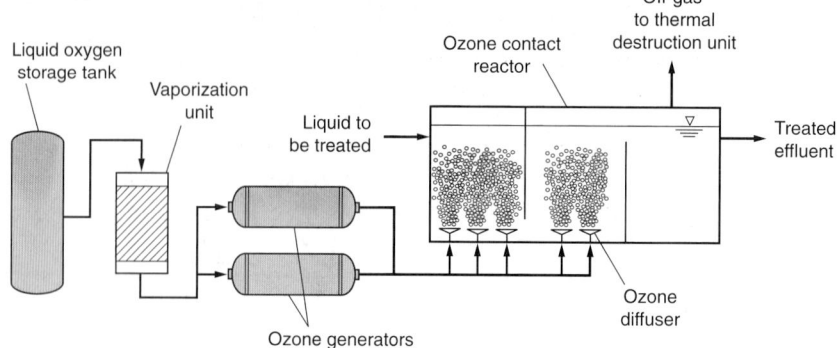

Destruction of Off-Gases. The off-gases from the contact chamber must be treated to destroy any remaining ozone, as it is an extremely irritating and toxic gas. The product formed by destruction of the remaining ozone is pure oxygen, which can be recycled if pure oxygen is being used to generate the ozone. In some designs ozone is introduced in both the first and second chambers.

Effectiveness of Ozone as a Disinfectant

Ozone is an extremely reactive oxidant, and it is generally believed that bacterial kill through ozonation occurs directly because of cell wall disintegration (cell lysis). The impact of the wastewater characteristics on ozone disinfection is reported in Table 12–21. The presence of oxidizable compounds will cause the ozone inactivation curve to have a shoulder effect, as discussed previously for chlorine (see Fig. 12–10).

Ozone is also a very effective viricide and is generally believed to be more effective than chlorine. Ozonation does not produce dissolved solids and is not affected by the ammonium ion or pH influent to the process. For these reasons, ozonation is considered as an alternative to either chlorination or hypochlorination, especially where dechlorination may be required and high-purity oxygen facilities are available at the treatment plant. Data on the relative germicidal effectiveness of ozone for the disinfection of different microorganisms were presented previously in Table 12–11.

Modeling the Ozone Disinfection Process

As discussed previously in Sec. 12–2, the models that have been developed to describe the disinfection process with chlorine have also been adapted for ozone. Equation (12–63) has been modified as follows (Finch and Smith, 1989, 1990; U.S. EPA, 1986):

$$N/N_o = 1 \quad \text{for } U < q \tag{12–63}$$

$$N/N_o = [(U)/q]^{-n} \quad \text{for } U > q \tag{12–64}$$

where N = number of organisms remaining after disinfection
N_o = number of organisms present before disinfection
U = utilized (or transferred) ozone dose, mg/L

Table 12–21
Impact of wastewater constituents on the use of ozone for wastewater disinfection

Constituent	Effect
BOD, COD, TOC, etc.	Organic compounds that comprise the BOD and COD can exert an ozone demand. The degree of interference depends on their functional groups and their chemical structure
Humic materials	Affects the rate of ozone decomposition and the ozone demand
Oil and grease	Can exert an ozone demand
TSS	Increase ozone demand and shielding of embedded bacteria
Alkalinity	No or minor effect
Hardness	No or minor effect
Ammonia	No or minor effect, can react at high pH
Nitrite	Oxidized by ozone
Nitrate	Can reduce effectiveness of ozone
Iron	Oxidized by ozone
Manganese	Oxidized by ozone
pH	Affects the rate of ozone decomposition
Industrial discharges	Depending on the constituents, may lead to diurnal and seasonal variations in the ozone demand

n = slope of dose response curve

q = value of x intercept when $N/N_o = 1$ or log $N/N_o = 0$ (assumed to be equal to the initial ozone demand)

The required ozone dosage must be increased to account for the transfer of the applied ozone to the liquid. The required dosage can be computed with the following expression:

$$D = U\left(\frac{100}{TE}\right)$$ (12–65)

where D = total required ozone dosage, mg/L

U = utilized (or transferred) ozone dose, mg/L

TE = ozone transfer efficiency, %

Typical ozone transfer efficiencies vary from about 80 to 90 percent.

EXAMPLE 12–9 Estimate the Required Ozone Dose for a Typical Secondary Effluent Estimate the ozone dose needed to disinfect a filtered secondary effluent to an MPN value of 240/100 mL using the following disinfection data obtained from pilot-scale installation. Assume the starting coliform concentration will be 1×10^6/100 mL and that the ozone transfer efficiency is 80 percent.

Test number	Initial coliform count N_o, MPN/100 mL	Ozone transferred, mg/L	Final coliform count N, MPN/100 mL	$-\log(N/N_o)$
1	95,000	1	1500	1.80
2	470,000	2	1200	2.59
3	3,500,000	5	730	3.68
4	820,000	7	77	4.03
5	9,200,000	14	92	5.00

Solution

1.   Determine the coefficients in Eq. (12–64) using the pilot-plant data.

  a.   Linearize Eq. (12–64) and plot the log inactivation data versus the ozone dose on loglog paper to determine the constants in

$$N/N_o = [(U)/q]^{-n}$$

$$\log(N/N_o) = -n(\log U - \log q)$$

  b.   The required log-log plot is given below.

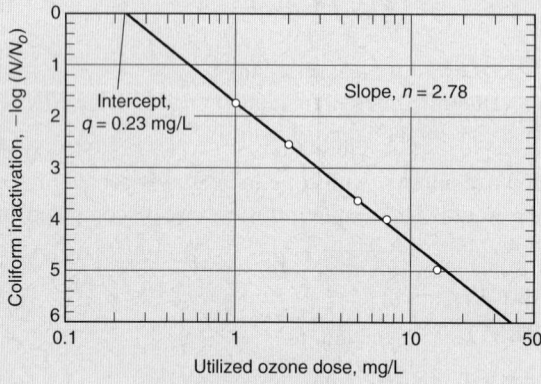

  c.   The required coefficients are:

$$q = 0.23 \text{ mg/L}$$

$$n = 2.72$$

2.   Determine the ozone dose required to achieve an effluent coliform concentration of 240 MPN/100 mL.

  a.   Rearrange Eq. (12–64) to solve for U.

$$U - q\,(N/N_o)^{-1/n}$$

  b.   Solve for U.

$$U = q\,(N/N_o)^{-1/n} = (0.23 \text{ mg/L})(240/10^6)^{-1/2.72} = 4.93 \text{ mg/L}$$

3. Determine the dose that must be applied using Eq. (12–65), assuming a transfer efficiency of 80 percent.

$$D = U\left(\frac{100}{TE}\right) = (4.93 \text{ mg/L})\left(\frac{100}{80}\right) = 6.16 \text{ mg/L}$$

Required Ozone Dosages for Disinfection

The required ozone dosage for disinfection can be estimated by considering (1) the initial ozone demand of the wastewater and (2) the required ozone dose using Eqs. (12–64) and (12–65). The ozone dosages required to meet the initial demand will depend on the wastewater constituents. Typical values for the ozone demand for various wastewaters based on a contact time of 15 min are reported in Table 12–22. It should be noted that the dosage values given in Table 12–22 are only meant to serve as a guide for the initial estimation of the required ozone dose. In most cases, pilot-scale studies (see Fig. 12–31) will need to be conducted to establish the required dosage ranges.

Byproduct Formation and Control

As with chlorine, the formation of unwanted byproducts is one of the problems associated with the use of ozone as a disinfectant. The formation and control of DBPs when using ozone are considered in the following discussion.

Table 12–22
Typical ozone dosages required to achieve different effluent coliform disinfection standards for various wastewaters based on a 15-min contact time[a]

Type of wastewater	Initial coliform count, MPN/100 mL	Ozone dose, mg/L Effluent standard, MPN/100 mL			
		1000	200	23	<2.2
Raw wastewater	10^7–10^9	15–40			
Primary effluent	10^7–10^9	10–40			
Trickling filter effluent	10^5–10^6	4–10			
Activated-sludge effluent	10^5–10^6	4–8	4–10	16–30	30–40
Filtered activated-sludge effluent	10^4–10^6	6–8	4–10	16–25	30–40
Nitrified effluent	10^4–10^6	3–6	4–6	8–20	18–24
Filtered nitrified effluent	10^4–10^6	3–6	3–8	4–15	15–20
Microfiltration effluent	10^1–10^3	2–6	2–6	3–8	4–8
Reverse osmosis	Nil				1–2
Septic tank effluent	10^7–10^9	15–40			
Intermittent sand filter effluent	10^2–10^4	4–8	10–15	12–20	16–25

[a] Adapted in part from White (1999).

Figure 12–31

Typical pilot-scale contactor used to test the effectiveness of ozone for disinfection and oxidation of organics.

Table 12-23

Representative disinfection byproducts resulting from the ozonation of wastewater containing organic and selected inorganic constituents[a]

Aldehydes	Brominated byproducts[b]
Formaldehyde	Bromate ion
Acetaldehyde	Bromoform
Glyoxal	Brominated acetic acids
Methyl glyoxal	Bromopicrin
	Brominated acetonitriles
Acids	Cyanogen bromide
Acetic acids	Other
Formic acid	Hydrogen peroxide
Oxalic acid	
Succinic acid	
Aldo- and ketoacids	
Pyruvic acid	

[a] Adapted in part from U.S. EPA (1999b).
[b] The bromide ion must be present to form brominated byproducts.

Formation of DBPs Due to the Addition of Ozone for Disinfection.

One advantage of ozone is that it does not form chlorinated DBPs such as THMs and HAAs (see Table 12–14). Ozone does, however, form other DBPs (see Table 12–23) including aldehydes, various acids, and aldo- and ketoacids when significant amounts of bromide are not present. In the presence of bromide, the following DBPs may also be produced: inorganic bromate ion, bromoform, brominated acetic acid, bromopicrin, brominated acetonitriles, and cyanogen bromide (see Table 12–14). On occasion, hydrogen peroxide can also be generated. The specific amounts and the relative distribution of compounds will depend on the nature of the precursor compounds that are present. Because the chemical characteristics of wastewater will vary from location to location, pilot testing will be required to assess the effectiveness of ozone as a disinfectant.

Control of DBP Formation Due to the Addition of Ozone for Disinfection. Because the nonbrominated compounds appear to be readily biodegradable, they can be removed by passage through a biologically active filter or carbon column or other biologically active process. The nonbrominated compounds can also be removed by soil application. The removal of DBPs formed when bromine is present is more complex. If brominated DBPs are going to be a problem, it may be appropriate to investigate alternative means of disinfection such as UV radiation.

Environmental Impacts of Using Ozone

It has been reported that ozone residuals can be acutely toxic to aquatic life (Ward et al., 1976). However, because ozone dissipates rapidly, ozone residuals will normally not be found by the time the effluent is discharged into the receiving water. Several investigators have reported that ozonation can produce some toxic mutagenic and/or carcinogenic compounds. These compounds are usually unstable, however, and are present only for a matter of minutes in the ozonated water. White (1999) has reported that ozone destroys certain harmful refractory organic substances such as humic acid (precursor of trihalomethane formation) and malathion. Whether toxic intermediates are formed during ozonation depends on the ozone dose, the contact time, and the nature of the precursor compounds. White (1999) have reported that ozone treatment ahead of chlorination for disinfection purposes reduces the likelihood for the formation of THMs.

Other Benefits of Using Ozone

An additional benefit associated with the use of ozone for disinfection is that the dissolved oxygen concentration of the effluent will be elevated to near saturation levels as ozone rapidly decomposes to oxygen after application. The increase in oxygen concentration may eliminate the need for reaeration of the effluent to meet required dissolved oxygen water-quality standards. Further, because ozone decomposes rapidly, no chemical residual persists in the treated effluent that may require removal, as is the case with chlorine residuals.

12–8 OTHER CHEMICAL DISINFECTION METHODS

Because of the concerns over the effectiveness of disinfection processes and concern over the formation of DBPs, ongoing research is continuing into the evaluation of alternative disinfection methods. The use of peracetic acid, peroxone, and combined disinfection processes is introduced and considered briefly in this section. Because research on these and other disinfection methods is ongoing, current conference proceedings and literature must be consulted for the latest findings.

Peracetic Acid

In the late 1980s, the use of peracetic acid (PAA) was proposed as a wastewater disinfectant. Peracetic acid, made up of acetic acid and hydrogen peroxide, has been used for many years as a disinfectant and sterilizing agent in hospitals. Peracetic acid is also used as a bactericide and fungicide, especially in food processing. Interest in the use of PAA as a wastewater disinfectant arises from considerations of safety and the possibility that its use will not result in the formation of DBPs. The use of PAA is considered briefly in this section as an example of the continuing search for alternative wastewater disinfectants to chlorine.

Peracetic Acid Chemistry and Properties. Commercially available PAA, also known as ethaneperoxide acid, peroxyacetic acid, or acetyl hydroxide, is only available as quaternary equilibrium solution containing acetic acid, hydrogen peroxide, peracetic acid, and water. The pertinent reaction is as follows:

$$CH_3CO_2H + H_2O_2 \leftrightarrow CH_3CO_3H + H_2O \qquad (12\text{--}66)$$

$$\underset{\substack{\text{Acetic} \\ \text{acid}}}{} \quad \underset{\substack{\text{Hydrogen} \\ \text{peroxide}}}{} \quad \underset{\substack{\text{Peracetic} \\ \text{acid}}}{}$$

The undissociated PAA (CH_3CO_3H) is considered to be the biocidal form in the equilibrium mixture; however, the hydrogen peroxide may also contribute to the disinfection process. Hydrogen peroxide is also more stable than PAA. The properties of PAA are summarized in Table 12–24.

Effectiveness of Peracetic Acid as a Disinfectant. The effectiveness of PAA has been studied by Lefevre et al. (1992), Lazarova et al. (1998), Liberti et al. (1999), Gehr (2000), and Colgan and Gehr (2001), among others. The findings to date are mixed concerning the bactericidal effectiveness of PAA, as well as the impact of wastewater characteristics on the effectiveness of PAA. The current literature must be consulted for more information on the application of PAA.

In a recent report by the U.S. EPA (1999a), PAA was included among a total of five possible disinfectants for use on combined sewer overflows (CSOs). Based on data for disinfection of secondary treatment plant effluents, it was suggested that PAA be strongly considered for CSO disinfection. Among the desirable attributes listed are absence of persistent residuals and byproducts, not affected by pH, short contact time, and high effectiveness as a bactericide and virucide.

Table 12-24
Properties of various peracetic acid (PAA) formulations[a]

Property	Unit	PAA, %		
		1.0	5	15
Weight PAA	%	0.8–1.5	4.5–5.4	14–17
Weight hydrogen peroxide	%	min 6	19–22	13.5–16
Weight acetic acid	%	9	10	28
Weight available oxygen	%	3–3.1	9.9–11.5	9.3–11.1
Stabilizers	Yes/no	Yes	Yes	Yes
Specific gravity		1.10	1.10	1.12

[a] Adapted from Solvay Interox (1997).

Formation of Disinfection Byproducts. Based on the limited data available, the principal end products identified were CH_3COOH (acetic acid or vinegar), O_2, CH_4, CO_2, and H_2O, none of which is considered toxic in the concentrations typically encountered.

Ozone/Hydrogen Peroxide (Peroxone)

Peroxone is an advanced oxidation process (AOP) involving the use of ozone and hydrogen peroxide (see Sec. 11–10 in Chap. 11). The peroxone process is comprised of two steps: ozone dissolution and hydrogen peroxide addition. The addition of hydrogen peroxide is used to accelerate the decomposition of ozone leading to the production of relatively high concentrations of hydroxyl radicals according to the following reaction:

$$H_2O_2 + 2O_3 \rightarrow HO^{\bullet} + HO^{\bullet} + 3O_2 \tag{12–67}$$

The disinfection characteristics of peroxone are currently under investigation along with other AOPs. In general, the peroxone process and other AOPs have been found to be equal to or more effective than ozone for pathogen inactivation. In wastewater disinfection, the peroxone process would most likely be used primarily with highly treated effluents (e.g., effluent from microfiltration and reverse osmosis processes). It should be noted that a variety of other combination processes, many of which are classified as advanced oxidation processes, are currently under investigation for the disinfection of water and wastewater (U.S. EPA, 1999b).

Combined Chemical Disinfection Processes

Interest in the sequential or simultaneous use of two or more disinfectants has increased within a few years, especially in the water supply field. Reasons for the increased interest in the use of multiple disinfectants include (U.S. EPA, 1999b):

- The use of less reactive disinfectants, such as chloramines, has proved to be quite effective in reducing the formation of DBPs, and more effective for controlling biofilms in the distribution system.

- Regulatory and consumer pressure to produce water that has been disinfected to achieve high levels of inactivation for various pathogens has forced both the water and wastewater industry to search for more effective disinfectants. To meet more stringent disinfection standards, higher disinfectant doses have been used, which unfortunately has resulted in the production of increased levels of DBPs.

- Based on the results of recent research, it has been shown that the application of sequential disinfectants is more effective than the added effect of the individual disinfectants. When two (or more) disinfectants are used to produce a synergistic effect by either simultaneous or sequential application to achieve more effective pathogen inactivation, the process is referred to as *interactive disinfection* (U.S. EPA, 1999b).

Currently, extensive research is being conducted on these processes. The application of sequential or simultaneous use of multiple disinfectants is, at present, site-specific, depending on the treatment technologies employed. The current literature must be reviewed to assess the suitability and effectiveness of these new disinfection technologies.

12–9 ULTRAVIOLET (UV) RADIATION DISINFECTION

The germicidal properties of the radiation emitted from ultraviolet (UV) light sources have been used in a wide variety of applications since its use was pioneered in the early 1900s. First used on high-quality water supplies, ultraviolet light as a wastewater disinfectant has evolved during the 1990s with the development on new lamps, ballasts, and ancillary equipment. With the proper dosage, ultraviolet radiation has proved to be an effective bactericide and virucide for wastewater, while not contributing to the formation of toxic byproducts. To develop an understanding of the application of UV for the disinfection of wastewater, the following topics are considered in this section: (1) source of UV radiation, (2) UV system components and configurations, (3) the germicidal effectiveness of UV radiation, (4) modeling the UV disinfection process, (5) estimating the UV dose, (6) validation of UV system performance, (7) design of UV systems, (8) troubleshooting UV systems, and (9) the environmental impacts of disinfection with UV radiation.

Source of UV Radiation

The portion of the electromagnetic spectrum in which UV radiation occurs, as shown on Fig. 12–32a, is between 100 and 400 nm. The UV radiation range is characterized further according to wavelength as long-wave (UV-A), also known as near-ultraviolet radiation, middle-wave (UV-B), and short-wave (UV-C), also known as far UV. The germicidal portion of the UV radiation band is between about 220 and 320 nm, principally in the UV-C range. To produce UV radiation, lamps that contain mercury vapor are charged by striking an electric arc. The energy generated by the excitation of the mercury vapor contained in the lamp results in the emission of UV light. In general, UV disinfection systems fall into three categories based on the internal operating parameters of the UV lamp: *low-pressure low-intensity, low-pressure high-intensity,* and *medium-pressure high-intensity* systems. As will be discussed subsequently, UV disinfection systems may also be classified as open-channel or closed-pipe based on their hydraulic characteristics. In the brief discussion of UV lamps presented below, it is important to note that UV lamp technology is changing rapidly. It is therefore imperative that current manufacturers' literature be consulted when designing a UV disinfection facility.

Low-Pressure Low-Intensity UV Lamps. Low-pressure low-intensity UV lamps generate essentially monochromatic radiation at a wavelength of 254 nm, which is close to the 260 nm (255 to 265 nm) wavelength that is considered to be most effective for microbial inactivation (see Fig. 12–32c). In all cases, mercury-argon lamps are used to generate the UV-C region wavelengths. Low-pressure low-intensity UV lamps are of a slimline design with an overall length of 0.75 to 1.5 m and diameters varying from 15 to 20 mm. These lamps operate optimally at a lamp wall temperature of 40°C and an internal pressure of 0.007 mm Hg (Note: 1 mm Hg = 133.322 N/m^2). The output of low-pressure low-intensity lamps is about 25 to 27 W at 254 nm for a power input of 70 to 80 W. Approximately 85 to 88 percent of the lamp output is monochromatic at 254 nm, making it an efficient choice for disinfection processes. Comparative information on the operational characteristics of UV lamps is presented in Table 12–25.

Quartz sleeves are used to isolate the UV lamps from direct water contact and to control the lamp wall temperature by buffering the effluent temperature extremes to

Figure 12–32

Definition sketch for ultraviolet (UV) radiation disinfection: (a) identification of the ultraviolet radiation portion of the electromagnetic spectrum, (b) identification of the germicidal portion of the UV radiation spectrum, and (c) UV radiation spectra for both low-pressure low-intensity and medium-pressure high-intensity UV lamps and the relative UV adsorption for DNA superimposed over spectra of the UV lamps.

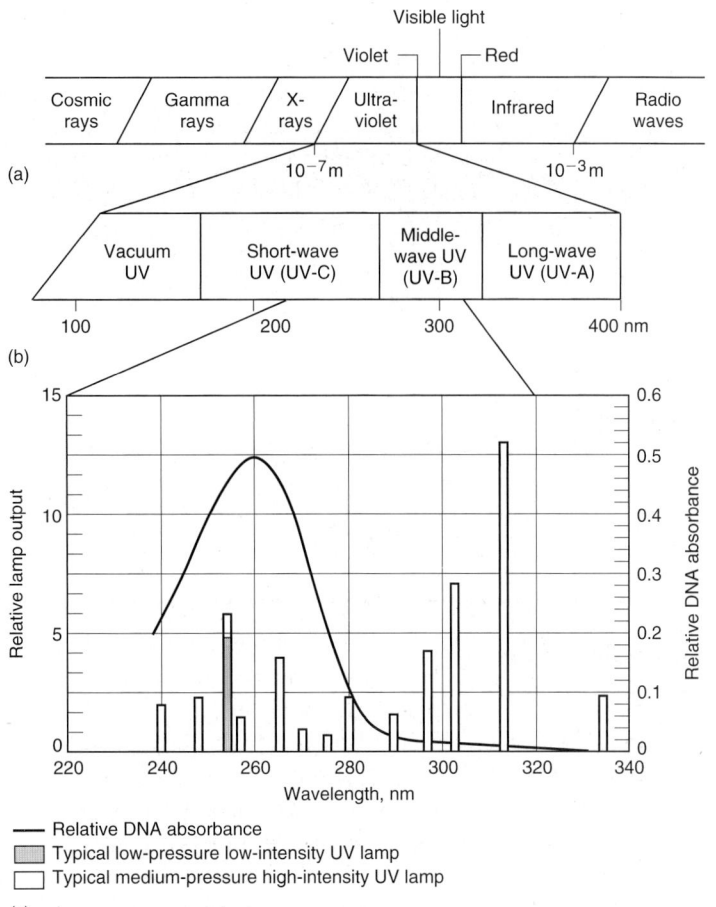

which the UV lamps are exposed, thereby maintaining a fairly uniform UV lamp output. Because there is an excess of liquid mercury in the low-pressure low-intensity UV lamp, the mercury vapor pressure is controlled by the coolest part of the lamp wall. If the lamp wall does not remain at its optimum temperature of 40°C, some of the mercury in the lamp condenses back to its liquid state, thereby decreasing the number of mercury atoms available to release photons of UV; hence UV output declines. The output of UV disinfection systems also decreases with time due to a reduction in the electron pool within the UV lamp, deterioration of the electrodes, and the aging of the quartz sleeve. The useful life of a low-pressure low-intensity UV lamp will vary from 9000 to 13,000 h depending on the number of on-off cycles per day. The useful life of the quartz sleeve is about 4 to 8 years.

Low-Pressure High-Intensity UV Lamps. Low-pressure high-intensity UV lamps are similar to the low-pressure low-intensity lamps with the exception that a mercury-indium amalgam is used in place of mercury. Low-pressure high-intensity UV lamps operate at a higher current discharge and pressures between 0.001 and

Table 12–25
Typical operational characteristics for UV lamps

Item	Unit	Type of lamp		
		Low-pressure low-intensity	Low-pressure high-intensity	Medium-pressure high-intensity
Power consumption	W	70–100	200–500	
	kW		1.2[a]	2–5
Lamp current	mA	350–550	Variable	Variable
Lamp voltage	V	220	Variable	Variable
Efficiency	%	30–40	25–35	10–12[b]
Lamp output at 254 nm	W	25–27	60–400	
Temperature	°C	35–45	90–150	600–800
Pressure	mmHg	0.007	0.001–0.01	
Lamp length	m	0.75–1.5	Variable	Variable
Lamp diameter	mm	15–20	Variable	Variable

[a] Very high output lamp.
[b] Output in the germicidal range (~250–260 μm).

0.01 mm Hg. Use of the mercury amalgam allows greater UV-C output, typically from 2 to 4 times the output of conventional low-intensity lamps. However, one manufacturer offers a lamp that is said to have 20 times the output at 254 nm. The amalgam in the low-pressure high-intensity UV lamps is used to maintain a constant level of mercury atoms, and thus provides greater stability over a broad temperature range, and greater lamp life (25 percent greater than other low-pressure lamps). Current manufacturers' literature must be reviewed for lamp specifications, as new low-pressure high-intensity lamps are being developed continuously.

Medium-Pressure High-Intensity UV Lamps. Medium-pressure high-intensity UV lamps have been developed over the last decade. Medium-pressure high-intensity UV lamps, which operate at temperatures of 600 to 800°C and pressures of 10^2 to 10^4 mm Hg, generate polychromatic radiation (see Fig. 12–32c). About 27 to 44 percent of the total energy of a medium-pressure high-intensity lamp is in the germicidal UV-C wavelength range. Only about 7 to 15 percent of the output is near 254 nm. However, medium-pressure high-intensity UV lamps generate approximately 50 to 100 times the total UV-C output of the conventional low-pressure low-intensity UV lamp. Their use is limited primarily to higher wastewater flows, stormwater overflows, or on space-limited sites because fewer lamps are required and the footprint of the disinfection system is greatly reduced (i.e., contact time is reduced).

Because the high-intensity UV lamp operates at temperatures at which all the mercury is vaporized, the UV output can be modulated across a range of power settings (typically 60 to 100 percent) without significantly changing the spectral distribution of the lamp. The ability to modulate the power is significant with respect to total power usage. Further, because of the high operating temperature, mechanical wiping of the quartz sleeve is essential to avoid the formation of an opaque film on the surface of the sleeve. Although there are a number of manufacturers of high-intensity UV lamps, most of the lamp manufacturers do not market complete UV disinfection systems. The particular UV lamp selected by UV system manufacturers is chosen on the basis of an integrated design approach in which the UV lamp, ballast, and reactor design are interdependent.

Emerging UV Lamp Technologies. New technologies are being developed that may have applications for wastewater disinfection. Two types of lamps that are being developed and applied to wastewater include (1) the pulsed energy broad-band xenon lamp (pulsed UV) and (2) the narrow-band excimer UV lamp. The operating principle of the pulsed UV lamp involves the conversion of alternating current (AC) to direct current (DC) which is stored in a capacitor. The energy stored in the capacitor is released through a high-speed switch and pulsed to produce an intense UV radiation field. Each time the switch closes, the lamp is fired, producing ionized gas (a plasma). The plasma expands until it reaches the wall of the lamp. The electric current carried by the plasma results in heating, which in turn raises the temperature of the plasma to about 10,000 K. At this elevated temperature an intense broad-band spectrum including UV, visible, and infrared wavelengths (see Fig. 12–32a) is emitted by the plasma. It is estimated that the radiation produced by the pulsed UV lamp is 20,000 times as intense as sunlight at sea level (EPRI, 1996; O'Brien et al., 1996).

Narrow-band excimer lamps produce essentially monochromatic radiation. A corona discharge is used to form excited dimers (two monomers joined together, e.g., two gas molecules joined together). Once in the excited state, the two molecules are termed *excimers.* Current flow in the discharge gap is comprised of a large number of microdischarges lasting on the order of nanoseconds. During these short discharges, electrons collide with gas molecules, causing collisions between excited gas molecules, resulting in the formation of dimers. Because the dimers are unstable, they collapse into their normal state, and in the process release energy in the form of photons. Gases that have been used for the purpose include xenon (Xe), xenon chloride (XeCl), krypton (Kr), and krypton chloride (KrCl). The specific wavelength emitted by excimer UV lamps will depend on the gas used in the lamp. Commercial excimer lamps are available that emit essentially monochromatic light in three wavelengths: 172, 222, and 308 nm. Numerous other lamps are under development (EPRI, 1996; O'Brien et al., 1996).

Because the developments in UV technology are occurring at such a rapid pace, it is essential that the current literature be consulted when designing UV disinfection systems. Note that in most cases, emerging technologies do not have a proven track record of cost-effective, reliable performance.

UV Disinfection System Components and Configurations
The principal components of a UV disinfection system consist of (1) the UV lamps, (2) the quartz sleeves in which the UV lamp is placed, (3) the supporting structure for the UV lamps and the quartz sleeves, (4) the ballasts used to supply regulated power to the UV lamps, and (5) the power supply which is used to power the ballasts. Three types of ballasts are used: (1) standard (core coil), (2) energy-efficient (core coil), and

(3) electronic (solid-state). Ballasts are used to limit the current to a lamp. Because UV lamps are arc-discharge devices, the more current in the arc, the lower the resistance becomes. Without a ballast to limit current, the lamp would destroy itself. Thus, matching the lamp and ballast is of critical importance in the design of UV disinfection systems. In addition to the type of lamp used, UV systems for the disinfection of wastewater can also be classified according to whether the flow occurs in open or closed channels. Each of these system configurations is described below.

Open-Channel Disinfection Systems. The principal components of low-pressure low- and high-intensity open-channel UV systems used for the disinfection of wastewater are illustrated on Fig. 12–33. As shown, lamp placement can be either horizontal and parallel to the flow (see Fig. 12–33a) or vertical and perpendicular to the flow (see Fig. 12–33b). Examples of typical low-pressure low-intensity UV disinfection systems for wastewater are shown on Figs. 12–34 and 12–35. The design flowrate is usually divided equally among a number of open channels. Each channel typically contains two or more banks of UV lamps in series, and each bank is comprised of a specified number of modules (or racks of UV lamps). It is important to note that a standby bank or channel should be provided for system reliability. Each module contains a specified number of UV lamps encased in quartz sleeves. The number of UV lamps per module is 2, 4, 8, 12, or 16. A spacing of 75 mm (3 in) between the centers of UV lamps is currently the most frequently used lamp configuration by UV manufacturers. A weighted flap gate, an extended sharp-crested weir, or an automatic level controller is used to control the depth of flow through each disinfection channel. To overcome the effect of fouling, which reduces the intensity of light in the liquid medium, the lamps must be removed occasionally from the flow channel and cleaned. Mechanical cleaning systems are used with low-pressure high-intensity systems to avoid fouling of the quartz sleeves.

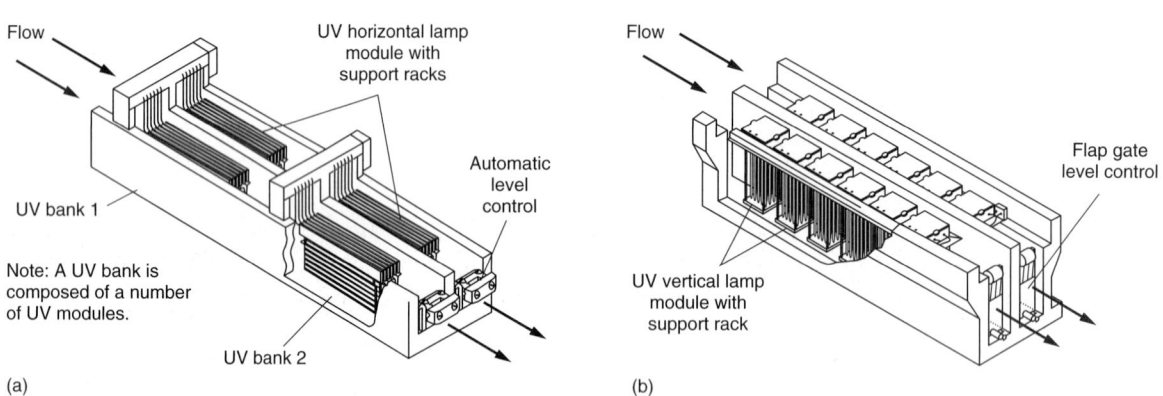

(a)

(b)

Figure 12–33

Isometric cutaway views of typical open-channel UV disinfection systems with cover grating removed: (a) horizontal lamp system parallel to flow (Adapted from Trojan Technologies, Inc.) and (b) vertical lamp system perpendicular to flow. (Adapted from Infilco Degremont, Inc.)

Figure 12–34

Typical example of open-channel UV disinfection system with horizontal lamp placement:
(a) view of two-channel UV disinfection system with four UV banks per channel. Overhead crane is used to remove lamp modules and (b) typical lamp module with 8 UV lamps.

(a)

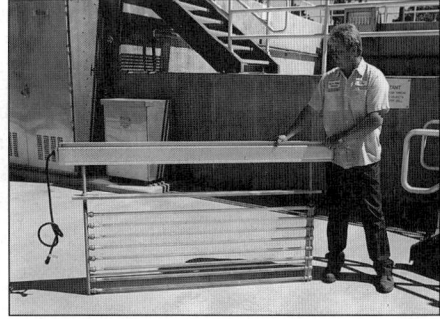

(b)

Figure 12–35

Typical example of open-channel UV disinfection system with vertical lamp placement: (a) view of single-channel UV disinfection system comprised of seven UV banks per channel with one of the UV banks removed for cleaning and (b) overhead crane used to remove modules for cleaning. Modules to be cleaned are placed in a tank shown on the base of the left-hand crane support.

(a)

(b)

Closed-Channel Disinfection Systems. A number of low- and medium-pressure high-intensity UV disinfection systems are designed to operate in closed channels. In most design configurations, the direction of flow is perpendicular to the placement of the lamps, as shown on Fig. 12–36. There are, however, design configurations in which the direction of flow is parallel to the UV lamps (see Fig. 12–37). In the medium-pressure UV disinfection system shown on Fig. 12–38, the lamps are arranged in modules and are positioned in a reactor with a fixed geometry. Because high-intensity UV lamps (see Fig. 12–38c) operate at a lamp wall temperature of between 600 and 800°C, the UV output of these lamps is unaffected by the effluent temperature. Essentially all of the closed or fixed geometry systems used for the disinfection of wastewater incorporate some form of mechanical wiping of the quartz sleeves to maintain performance.

Germicidal Effectiveness of UV Radiation

Ultraviolet light is a physical rather than a chemical disinfecting agent. Radiation penetrates the cell wall of the microorganism and is absorbed by the nucleic acids (see Fig. 12–39), which either prevents replication or causes death of the cell to occur. The effectiveness of the UV disinfection process depends on a number of variables including the characteristics of the UV disinfection system, the overall system hydraulics, the presence of particles, the characteristics of the microorganisms, and the chemical charac-

Figure 12–36

Typical closed-channel UV disinfection system with lamps placed perpendicular to the flow: (a) Isometric cutaway view of disinfection reactor and (b) view of typical installation. (*Courtesy Aquionics Inc.*)

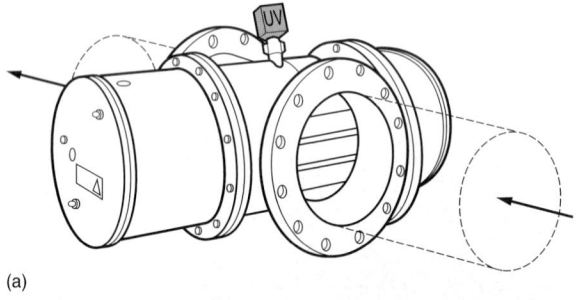
(a)

(b)

Figure 12-37

Typical closed-channel
UV disinfection system
with one or more lamps
placed parallel to the
flow: (a) Isometric
cutaway view of
disinfection reactor and
(b) view of typical full-
scale unit used in a pilot
test program.

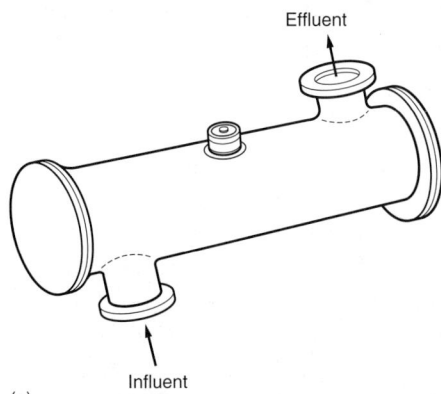

Effluent

Influent

(a)

(b)

teristics of the wastewater. Before considering these subjects, it is appropriate to consider the definition of UV dose to provide a frame of reference.

Definition of UV Dose. The effectiveness of UV disinfection is based on the UV dose to which the microorganisms are exposed. The UV dose D is defined as follows:

$$D = I \times t \tag{12–68}$$

where $D =$ UV dose, mJ/cm^2 (note mJ/cm^2 = mW·s/cm^2)

$I =$ UV intensity, mW/cm^2

$t =$ exposure time, s

It will be noted that the UV dose term is analogous to the dose term used for chlorine disinfection (i.e., $C_R t$). As given by Eq. (12–68), the UV dose can be varied by changing either the intensity or the exposure time. Because the UV intensity is attenuated with distance from the quartz sleeve (modeled using Beer's law), the average UV intensity within a UV disinfection system is often computed mathematically. As will be discussed later, the use of mathematical modeling has yet not proved to be satisfactory for the design of UV disinfection systems given the many variables that can affect performance.

Impact of System Characteristics. Problems with the application of Eq. (12–68) for use in the design of UV disinfection reactors are associated with (1) inaccurate knowledge of the UV intensity and (2) the exposure time associated with all of the pathogens passing through a UV disinfection system. In practice, field-scale UV

Figure 12–38

Views of medium-pressure high-intensity UV disinfection system: (a) schematic view through UV reactor *(Courtesy of Trojan Technologies, Inc.)*, (b) view of UV system with one lamp module removed from the UV reactor, and (c) close-up of medium-pressure high-intensity UV lamps. The UV lamps can be seen inside the quartz sleeve.

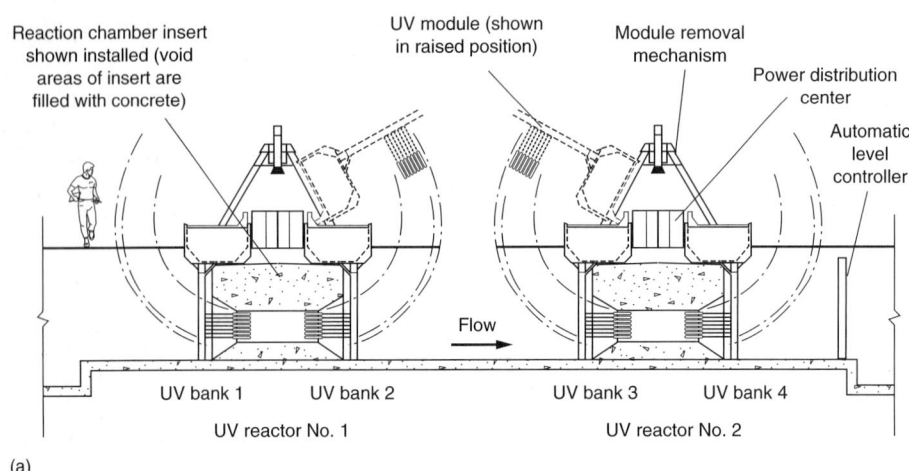

(a)

(b)

(c)

Figure 12–39

Formation of double bonds in microorganisms exposed to ultraviolet radiation.

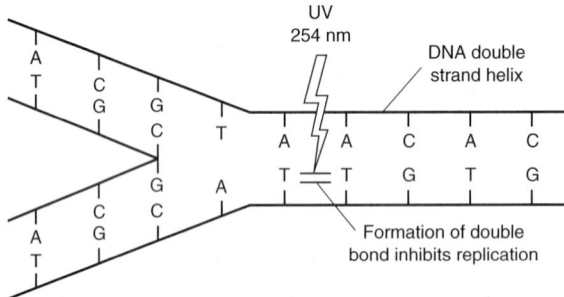

disinfection reactors have dose distributions resulting from both the internal intensity profiles and exposure time distribution. The internal intensity profiles are a reflection of the nonhomogeneous placement of lamps within the system, lack of ideal radial mixing within the system, the scattering/absorbing effects of particulate material, and the absorbance of the liquid medium. The distribution associated with exposure time is a reflection of nonideal hydraulics leading to longitudinal mixing.

Nonideal Approach Hydraulics. One of the most serious problems encountered with UV disinfection systems in open channels is achieving a uniform velocity field in the approach and exit channel. Achieving a uniform velocity field is especially difficult when UV systems are retrofitted into existing open channels, such as converted chlorine contact basins, a practice that is not recommended if the performance of the UV disinfection system is to be optimized. Possible remedies are discussed in the section titled "Troubleshooting UV Systems" following the discussion of the design of UV systems.

Impact of Particles. Aside from the hydraulic impacts, the presence of particles within wastewater also affects the distribution of applied intensity to embedded organisms. Many organisms of interest in wastewater (e.g., coliform bacteria) occur in both a disperse state (i.e., not bound to other objects) and a particle-associated state (i.e., bound to other objects such as other bacteria or cellular debris). Coliform bacteria are of particular importance because of the central role they play in discharge permits (i.e., coliform bacteria are used as indicators for the presence of other pathogenic organisms and their inactivation is assumed to correlate with the inactivation of other pathogenic organisms). Disperse coliform bacteria are readily inactivated because they are exposed fully to UV light. Treatment process related disinfection problems most always result from the influence of particle-associated organisms (Oliver and Cosgrove, 1975; Severin, 1980; Ho and Bohm, 1981; Qualls et al., 1983; Qualls and Johnson, 1985; Scheible, 1987; Cairns, 1993; Emerick and Darby, 1993; Darby et al., 1995; Emerick et al., 1999; Emerick et al., 2000). In fact, coliform bacteria can associate with particles to such a degree that they are completely shielded from UV light resulting in a residual coliform bacteria concentration (Scheible, 1987; Parker and Darby, 1995).

It has been hypothesized (Emerick et al., 2000) that a minimum particle size (wastewater specific, but on the order of 10 μm) governs the ability to shield coliform bacteria from UV light. Due to the inherent porous nature of wastewater particles, particles smaller than that critical size are unable to reduce the applied intensity and thus embedded organisms are inactivated in a manner similar to dispersed organisms. Particles greater than the critical size can shield coliform bacteria similarly. Particle size does not appear to be a governing factor once the critical size is exceeded because coliform bacteria are located randomly within particles and are not typically located in the most shielded regions within particles.

Characteristics of the Microorganisms. The effectiveness of the UV disinfection process depends on the characteristics of the microorganisms. The relative effectiveness of UV radiation for disinfection of various discrete organisms is reported in Table 12–26. Additional data on the relative effectiveness of UV radiation for the disinfection of representative microorganisms of concern in wastewater are presented in Table 12–26. The values given in Tables 12–11 and 12–26 are only meant to serve as a guide in assessing the required UV dose. Knowledge concerning the required UV dose for specific pathogen inactivation is changing continuously as improved methods of analysis are applied. For example, before infectivity studies were conducted, it was thought that UV radiation at reasonable dosage values (i.e., less than 200 mJ/cm^2) was not effective for the inactivation of *Cryptosporidium parvum* and *Giardia lamblia*. However, based on infectivity studies, it has been found that both of these protozoan are inactivated with extremely

Table 12–26
Estimated relative effectiveness of UV radiation for the disinfection of representative microorganisms of concern in wastewater

Organism	Dosage relative to total coliform dosage[a]
Bacteria:	
Fecal coliform	0.5–0.9
Pseudomonas aeruginosa	1.5–2.0
Salmonella typhosa	0.7–0.9
Staphylococcus aureus	1.0–1.5
Total coliform	1.0
Viruses:	
Adenovirus	0.7–0.9
Coxsackie A2	1.0–1.5
F specific bacteriophage	0.4–0.8
Polio type 1	0.9–1.1
MS-2 bacteriophage	0.9–1.0
Protozoa:	
Acanthamoeba castellanii	10–12 (?)
Acanthamoeba culbertsoni	10–12 (?)
Cryptosporidium parvum oocysts[b]	0.2–0.4
Giardia lamblia cysts[b]	0.2–0.6
Other:	
Clostridum spores	

[a] Relative doses based on discrete nonclumped single organisms in suspension. If the organisms are clumped or particle associated, the relative dosages have no meaning.
[b] Based on infectivity studies.

low UV dosage values (typically in the range of 5 to 15 mJ/cm^2). The current literature should be consulted to obtain the most contemporary information regarding required UV dosages for the inactivation of specific microorganisms.

Effect of Wastewater Chemical Constituents. The effect of wastewater constituents on UV disinfection is presented in Table 12–27. Dissolved contaminants impact UV disinfection either directly via absorbance impacts (increasing absorbance serves to attenuate UV light to a larger degree) or via fouling of UV lamps such that a reduced intensity is applied to the bulk liquid medium. One of the most perplexing problems encountered in the application of UV disinfection for wastewater disinfection is the variation typically observed in the absorbance (or transmittance) of wastewater within a treatment plant. Often, the variations in transmittance are caused by industrial discharges, which can lead to diurnal as well as seasonal variations. Common industrial impacts are related to the discharge of inorganic and organic dyes, wastes containing metals, and complex organic compounds. However, stormwater inflows can also cause

Table 12–27

Impact of wastewater constituents on the use of UV radiation for wastewater disinfection

Constituent[a]	Effect
BOD, COD, TOC, etc.	No or minor effect, unless humic materials comprise a large portion of the BOD
Humic materials	Strong adsorbers of UV radiation
Oil and grease	Can accumulate on quartz sleeves of UV lamps, can absorb UV radiation
TSS	Absorption of UV radiation, can shield embedded bacteria
Alkalinity	Can impact scaling potential. Also affects solubility of metals that may absorb UV light
Hardness	Calcium, magnesium, and other salts can form mineral deposits on quartz tubes, especially at elevated temperatures
Ammonia	No or minor effect
Nitrite	No or minor effect
Nitrate	No or minor effect
Iron	Strong adsorber of UV radiation, can precipitate on quartz tubes, can adsorb on suspended solids and shield bacteria by adsorption
Manganese	Strong adsorber of UV radiation
pH	Can affect solubility of metals and carbonates
TDS	Can impact scaling potential and the formation of mineral deposits
Industrial discharges	Depending on the constituents (e.g., dyes), may lead to diurnal and seasonal variations in the transmittance
Stormwater inflow	Depending on the constituents, may lead to short-term as well as seasonal variations in the transmittance

[a] Inorganic constituents including bicarbonate, chloride ion, and nitrate can affect the direct UV photolysis of constituents such as NDMA.

wide variations, especially when humic materials are present. In either case, the solution to the problem of varying transmittance will require monitoring of industrial discharges, the implementation of source control programs, and correcting sources of infiltration. In some cases, biological treatment will mitigate the influent variations. In some extreme situations, it may be concluded that UV disinfection will simply not work.

Where the use of UV disinfection is being assessed, it is useful to install on-line transmittance monitoring equipment to document the variations that occur in the transmittance with time. The scaling potential of wastewater, as defined by the Langelier saturation index (see Sec. 6–7 in Chap. 6), should also be checked to assess whether scaling may be a problem. This recommendation is especially important where the feasibility of using high-intensity UV lamps is being assessed.

Modeling the UV Disinfection Process

Although, in general, it is believed that the concentration of suspended solids has a deleterious impact on UV disinfection performance, Emerick et al. (1999) reported that among different treatment processes there is no correlation between the total suspended solids concentration and the number of particles containing coliform bacteria. This lack of correlation underlies the need for inactivation models based on more fundamental water-quality parameters.

The use of series-event or multihit kinetics has been suggested to describe the initial resistance that homogeneous populations of organisms tend to exhibit to UV light in addition to the subsequent loglinear inactivation behavior (Severin et al., 1983). However, the measurement of the overall response of a mixed population of bacteria (e.g., coliform bacteria) in wastewater tends to mask the initial resistances of specific bacterial species/strains. For UV doses greater than 10 mJ/cm^2 (i.e., as is typically applied for wastewater disinfection), the following equation can be used for modeling the log-linear inactivation of disperse coliform bacteria in a batch system (Jagger, 1967; Oliver and Cosgrove, 1975; Qualls and Johnson, 1985):

$$N_D(t) = N_D(0)e^{-kIt} \tag{12-69}$$

where $N_D(t)$ = total number of surviving disperse coliform bacteria at time t
$N_D(0)$ = total number of disperse coliform bacteria prior to UV light application (at time $t = 0$)
k = inactivation rate coefficient, cm^2/mW·s
I = average intensity of UV light in bulk solution, mW/cm^2
t = exposure time, s

The fundamental difference between disperse coliform bacteria and particle-associated coliform bacteria is the amount of UV intensity reaching the organism. The above equation is applicable only to disperse organisms because all members of that group receive the same intensity of UV light (assuming perfectly mixed conditions). An organism embedded within a particle will receive a reduced UV light intensity relative to that applied to the bulk solution. Knowledge of the distribution of applied intensities would allow a model, analogous to that presented above, to be developed to describe the inactivation of both disperse and particle-associated coliform bacteria. Emerick et al. (2000) demonstrated the applicability of the following modeling equation for describing the inactivation of both disperse and particle-associated coliform bacteria (see Fig. 12–14, presented previously) when knowledge of the applied intensity to the bulk liquid medium is known:

$$N(t) = N_D(0)e^{-kd} + \frac{N_P(0)}{k\,d}(1 - e^{-kd}) \tag{12-70}$$

where $N(t)$ = total number of surviving coliform bacteria at time t
$N_D(0)$ = total number of disperse coliform bacteria prior to application of disinfectant at time $t = 0$
$N_P(0)$ = total number of particles containing at least one coliform bacterium at time $t = 0$
k = inactivation rate coefficient, cm^2/mJ
d = UV dose, mJ/cm^2

Equation (12–70) is best used to describe the underlying constraints to UV disinfection performance. The numbers of particles containing coliform bacteria, the inactivation rate coefficient, and the applied UV dose (product of intensity and exposure time) have fundamental impacts on UV disinfection performance. From experience it has been found that it is more convenient to design disinfection systems using collimated-beam inactivation data and validated UV disinfection equipment, as will be discussed subsequently.

Estimating UV Dose

Three methods have been used to estimate the UV dose. In the first method, an average UV dose is determined by assuming an average system UV intensity and exposure time. The average UV intensity is estimated using a computational procedure known as the point source summation (PSS) method (U.S. EPA, 1992). The PSS method is currently less used by designers due to its dependence on system-specific hydraulics (i.e., pilot-study results are a function of the pilot unit used during the course of study). The second method involves the use of computational fluid dynamics (CFD) to integrate both the distribution of UV intensities and velocity profiles within the reactor to obtain a distribution of UV doses within a system (Blatchley et al., 1995). Although the CFD method is promising, its use is limited at the present time (2001) because (1) the methodology is not standardized, (2) the methodology has not been validated thoroughly over a range of disinfection systems, and (3) the reporting of a distribution of UV doses, though accurate, is problematic for UV disinfection system specification. In the third, and most widely used, method the UV dose is determined by bioassays. Use of the bioassay approach in designing UV disinfection systems is discussed below.

Determination of UV Dose by Collimated-Beam Bioassay.   The most common procedure for determining the required UV dose for the inactivation of microorganisms involves the use of a collimated beam and a small reactor to which a known UV dose is applied. A typical collimated-beam device is shown on Fig. 12–40. Use of a monochromatic low-pressure low-intensity lamp in the collimated-beam apparatus allows for accurate characterization of the applied UV intensity. Use of a batch reactor allows for accurate determination of exposure time. The applied UV dose, as defined by Eq. (12–68), can be controlled by varying either the UV intensity or the exposure time. Because the geometry is fixed, the depth-averaged UV intensity within the petri dish sample (i.e., the batch reactor) can be computed using the following relationship based on Beer's law.

$$I_{\text{avg}} = I_o \times \frac{(1 - e^{-kd})}{kd} \qquad (12\text{–}71)$$

where I_{avg} = average UV intensity, mW/cm^2
   I_o = average incident UV intensity at the surface of the sample, mW/cm^2
   k = absorbance coefficient
   = 2.303 × (a.u./cm)
   a.u. = absorbance units, cm^{-1}
   d = depth of sample, cm

Typical absorbance values for various wastewaters at 254 nm are:

1.   Primary effluent = 0.55 to 0.30/cm
2.   Secondary effluent = 0.35 to 0.15/cm
3.   Nitrified secondary effluent = 0.25 to 0.10/cm

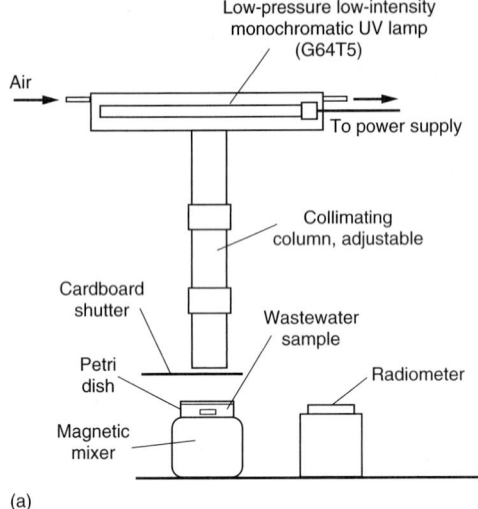

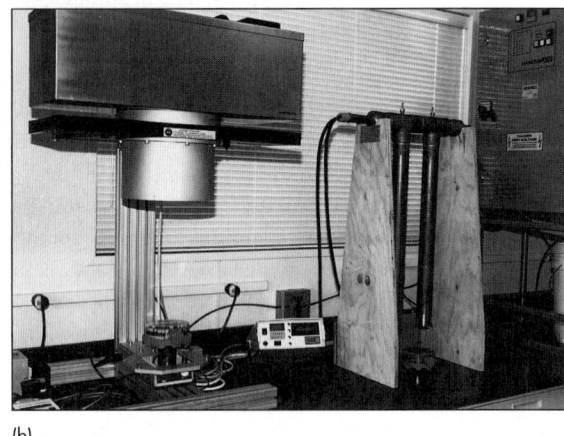

(a)
(b)

Figure 12–40

Collimated-beam device used to develop dose-response curves for UV disinfection (a) schematic and (b) view of two different types of collimated-beam devices. The collimated beam on the left is of European design; the collimated beam on the right is of the type shown schematically in (a).

4. Filtered secondary effluent = 0.25 to 0.10/cm
5. Microfiltered secondary effluent = 0.10 to 0.04/cm
6. Reverse osmosis effluent = 0.05 to 0.01/cm

Knowledge of the average UV intensity and exposure time allows calculation of applied UV dose using Eq. (12–68). The UV dose is then correlated to the microorganism inactivation results, typically measured using an MPN procedure for bacteria, a plaque count procedure for viruses (see Chap. 2), or an animal infectivity procedure for protozoans.

The transmittance T can be derived from absorbance measurements using the following relationship:

$$T, \% = 10^{-(a.u./cm)} \times 100 \tag{12–72}$$

Typical transmittance values over the entire UV spectrum for seven activated-sludge processes and two lagoon systems are shown on Fig. 12–41.

Reporting and Using Bioassay Collimated-Beam Test Results.
The results of collimated-beam bioassays are reported in the form of dose-response curves (see Fig. 12–42). The inactivation curve shown on Fig. 12–42a is for discrete organisms exposed to UV light, whereas the curve shown on Fig. 12–42b is for wastewater containing particulate material. To assess both the inherent and process variability, as discussed in Sec. 15–2 in Chap. 15, the collimated-beam test must be repeated to obtain statistical significance. Because the radiation output of the medium-pressure high-intensity UV lamps is polychromatic (see Fig. 12–32), the inactivation performance of these lamps is often correlated to the performance of low-pressure (i.e., monochro-

Figure 12–41

Percent transmittance versus wavelength for seven activated-sludge plants and two wastewater lagoons.

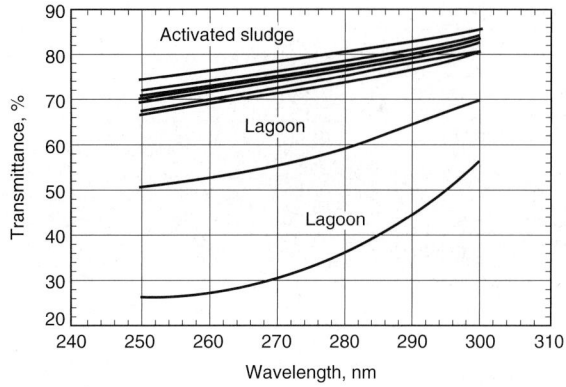

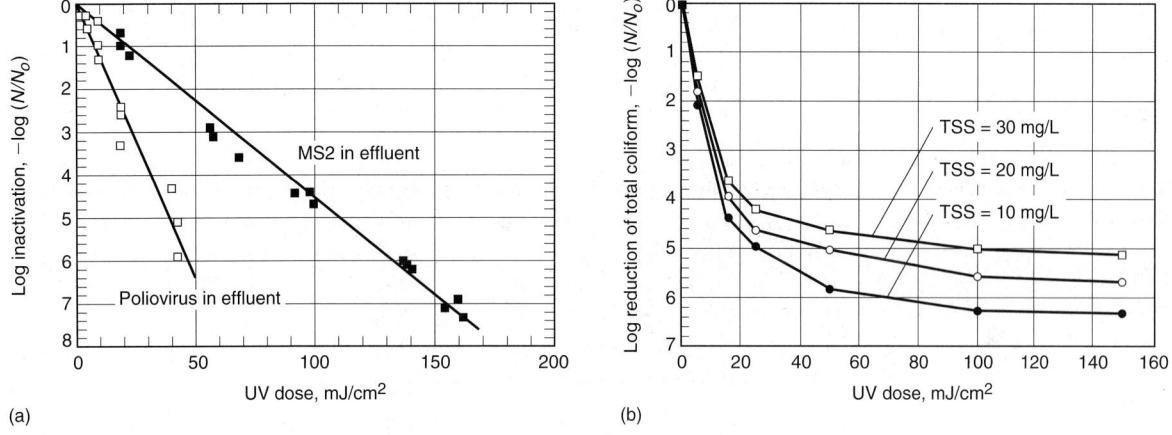

(a) (b)

Figure 12–42

Typical dose-response curves for UV disinfection developed from data obtained using a collimated-beam device: (a) for dispersed microorganisms (Cooper et al., 2000) and (b) wastewater containing varying concentrations of TSS.

matic wavelength) lamps for comparative purposes. An example of the process used to determine the required UV dose using collimated-beam test results is provided in Example 12–10.

EXAMPLE 12–10   **Determination of Dose Requirement Using Results from Collimated-Beam Testing**   The following dose-response data were obtained by conducting collimated-beam tests once per month over a 12-month period for a given wastewater. Using these data, determine (1) the mean, standard deviation, and 75 percent confidence interval associated with the surviving number of total coliform bacteria at each UV dose investigated and (2) the dose required (site-specific) to comply with a permit limitation of 23 total coliform bacteria per 100 mL (30 d median) at a confidence level of 75 percent.

Test	Survival at applied UV dose, mJ/cm²						
number	0	20	40	60	80	100	120
1	3,500,000	280	43	6.8	5.5	5.4	6.0
2	79,000	920	23	6.8	5.5	36	22
3	920,000	58	17	13	10	1.8	1.8
4	430,000	540	110	24	430	14	8.1
5	9,200,000	2,800	540	24	46	1.8	21
6	210,000	54,000	9,200	920	110	2.0	5.5
7	16,000,000	36	23	13	5.5	17	5.5
8	1,700,000	180	46	4.0	4.0	69	4.5
9	920,000	540	49	21	1.8	3.6	5.5
10	5,600,000	2,400	31	69	19	24	1.8
11	79,000	920	280	280	81	12	1.8
12	4,400,000	110	9.1	84	22	54	95

Solution

1. Determine the mean, standard deviation, and confidence interval for the monthly dose-response data. Because biological UV dose-response data are generally log-normally distributed, log-transform the observed survival data to enable use of student-t statistics (student-t statistics must be used because there are not enough data to apply normal statistics; typically 30 samples are required).

   a. Log-transform the observed number of surviving total coliform bacteria. For example, for test 1 the log-transform data point associated with a UV dose of 40 is

   $$\log (43) = 1.63$$

   b. Determine the average and standard deviation for the log-transformed data for each investigated UV dose.
   For the UV dose of 20, the average is 2.75.
   For the UV dose of 20, the standard deviation is 0.86.
   The observed mean and standard deviation for each UV dose is provided in the following table.

Test	Log survival at applied UV dose, mJ/cm²						
number	0	20	40	60	80	100	120
1	6.54	2.45	1.63	0.83	0.74	0.73	0.78
2	4.90	2.96	1.36	0.83	0.74	1.56	1.34
3	5.96	1.76	1.23	1.11	1.00	0.26	0.26
4	5.63	2.73	2.04	1.38	2.63	1.15	0.91
5	6.96	3.45	2.73	1.38	1.66	0.26	1.32

(continued)

(*Continued*)

Test number	Log survival at applied UV dose, mJ/cm²						
	0	**20**	**40**	**60**	**80**	**100**	**120**
6	5.32	4.73	3.96	2.96	2.04	0.30	0.74
7	7.20	1.56	1.36	1.11	0.74	1.23	0.74
8	6.23	2.26	1.66	0.60	0.60	1.84	0.65
9	5.96	2.73	1.69	1.32	0.26	0.56	0.74
10	6.75	3.38	1.49	1.84	1.28	1.38	0.26
11	4.90	2.96	2.45	2.45	1.91	1.08	0.26
12	6.64	2.04	0.96	1.92	1.34	1.73	1.98
Average	6.08	2.75	1.88	1.48	1.25	1.01	0.83
Standard deviation	0.78	0.86	0.83	0.70	0.70	0.58	0.51

   c. Determine the 75 percent confidence interval. Because the permit is based on a 30-day median value, designing based on the mean survival risks occasional permit violations. The 75 percent confidence interval is often used to ensure compliance with a median permit limit.

      i. For a dose of 60 mJ/cm², the 75 percent confidence interval is calculated using the following expression (Larson and Faber, 2000):

$$75\% \text{ confidence limit} = \bar{x} \pm t_{0.125}\left(\frac{s}{\sqrt{n}}\right)$$

      where $\bar{x}$ = mean survival at a specific UV dose = 1.48

          $t_{0.125}$ = student-t value associated with a 75 percent level of confidence
            = 1.214 (obtained from statistical tables, Larson and Faber, 2000). Note that the degrees of freedom are
            $n - 1 = 12 - 1 = 11$
        n = number of replicates = 12
        s = sample standard deviation = 0.70

$$75\% \text{ confidence limit} = 1.48 \pm 1.214\left(\frac{0.70}{\sqrt{12}}\right) = 1.48 \pm 0.245$$

      ii. Transform the mean and confidence interval back to base 10. The mean and confidence interval associated with each investigated UV dose is provided in the table given below.

UV dose, mJ/cm²	Surviving total coliform per 100 mL		
	Average	**Lower 75% C.I.**	**Upper 75% C.I.**
0	1,200,000	623,000	2,320,000
20	560	280	1,200
40	76	38	150

(continued)

(*Continued*)

UV dose, mJ/cm²	Surviving total coliform per 100 mL		
	Average	**Lower 75% C.I.**	**Upper 75% C.I.**
60	30	17	54
80	18	10	32
100	10	6	17
120	7	4	10

2. Estimate the required UV dose.

An inspection of the upper 75 percent confidence intervals in the above table leads to the conclusion that a design UV dose of 100 mJ/cm² is adequate to obtain a 30 d median survival of 23 total coliform bacteria per 100 mL.

Comment The variability in the data reported in Step 1 is representative of what will be observed in practice based on limited testing. To gain a better understanding of the variability associated with the wastewater of interest, it is recommended that replicate tests be conducted (a minimum of three tests is recommended).

Ultraviolet Disinfection Guidelines

Recently, the National Water Research Institute and the American Water Works Association Research Foundation published "Ultraviolet Disinfection Guidelines for Drinking Water and Wastewater Reclamation" (NWRI and AWWARF, 2000). The following elements are considered in the UV guidelines: (1) reactor design, (2) reliability design, (3) monitoring and alarm design, (4) the field commissioning test, (5) performance monitoring, and (6) an engineering report for unrestricted effluent reuse applications. Some of the items may not be applicable when utilizing UV disinfection for less demanding uses.

Application of UV Guidelines. The guidelines that cover reclaimed water are similar to those that cover drinking water systems. The primary difference is that recommended doses are provided for reclaimed water systems whereas there is no mention of recommended doses for drinking water systems. For reclaimed water systems, the recommended design UV doses are 100 mJ/cm² for granular medium filtration effluent, 80 mJ/cm² for membrane filtration effluent, and 50 mJ/cm² for reverse osmosis effluent. The different dose requirements reflect the different virus density concentrations expected within each type of process effluent. The dosages selected are intended to provide four logs of poliovirus inactivation with a factor of safety of about 2. In addition to differing dose recommendations as a function of effluent quality, there are differing design transmittance recommendations. For granular medium, microfiltration, and reverse osmosis effluents, the design transmittances are 55, 65, and 90 percent, respectively. The differing transmittance values are based on field observations made to date. All UV disinfection systems installed for either drinking water or unrestricted reuse applications must undergo validation testing prior to their installation. Although the guidelines do not apply to the disinfection of secondary effluent, the general design issues addressed are applicable.

Relationship of UV Guidelines to UV System Design. The design of a UV disinfection system requires three general steps: (1) determination of the UV dose required, based on bioassay testing, for adequate inactivation of the target microorganism(s), (2) validation of manufacturer-specific UV disinfection system performance, and (3) determination of an optimal UV system configuration (e.g., the number of lamps per module, modules per bank, banks per channel, and the overall number of channels). The first two issues are addressed directly in the guidelines, and general guidance is provided on design aspects. Determination of the UV dose required to comply with a permit limitation was discussed previously and illustrated in Example 12–10. Specific details on the culture of the microorganisms and the conduct of the test are given in the guidelines. The determination of an optimal UV system configuration is considered following the discussion of the approach used for the validation of manufacturer-specific UV disinfection system performance presented below.

UV System Performance Validation. Validation testing consists of quantifying the inactivation of a virus surrogate (e.g., MS2 bacteriophage) as a function of flowrate through the disinfection system. The flowrate is normalized to the number of lamps within the system, such that the design of the UV disinfection systems can be based on the hydraulic loading rate (e.g., L/min·lamp). Validation testing is important because the test results can be used to compare competing UV disinfection technologies and it eliminates the need to make choices based on manufacturers' claims, often not verified by an independent third party. Validation testing of UV disinfection equipment consists of the following steps:

1. Selection of a representative test water for use in the validation testing of the disinfection system.
2. Selection of the configuration of the UV disinfection system to be tested (for low-pressure low-intensity UV systems, a minimum of two banks must be tested but typically more are used; see Fig. 12–43a). If the power to the UV lamps cannot

Figure 12–43

Typical pilot-scale UV disinfection test reactors used for performance validation: (a) low-pressure low-intensity, and (b) medium-pressure high-intensity.

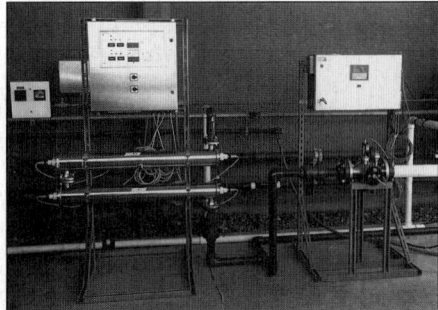

(a)   (b)

be turned down to simulate the end of life lamp performance, then aged UV lamps must be used in the test.

3. Testing of the hydraulic performance of the UV disinfection system. Hydraulic testing is done to verify the uniformity of the approach and exit velocities.

4. Quantification of the inactivation of the viral indicator as a function of hydraulic loading rate through the UV test reactor. Two typical dosing arrangements are illustrated on Fig. 12–44.

5. Simultaneously conduct a collimated-beam test on the test water to determine the inactivation of the viral indicator as a function of applied UV dose.

6. Verify the accuracy of the laboratory collimated-beam dose-response test data. The laboratory test data must fall within the area bound by the following two equations:

$$-\log_{10}(N/N_o) = 0.040 \times [\text{UV dose, mJ/cm}^2] + 0.64 \qquad (12\text{–}73)$$

$$-\log_{10}(N/N_o) = 0.033 \times [\text{UV dose, mJ/cm}^2] + 0.20 \qquad (12\text{–}74)$$

7. Assign UV doses to the pilot reactor based on the measured inactivation observed during the conduct of the collimated-beam test as a function of applied UV dose.

The steps required in conducting a validation test are illustrated in Example 12–11.

Figure 12–44

Schematic flow diagram for the conduct of field evaluation of UV systems using seeded microorganisms and UV transmittance adjustment: (a) seeding with premixed diluted solution from batch tank and (b) seeding with concentrated solutions.

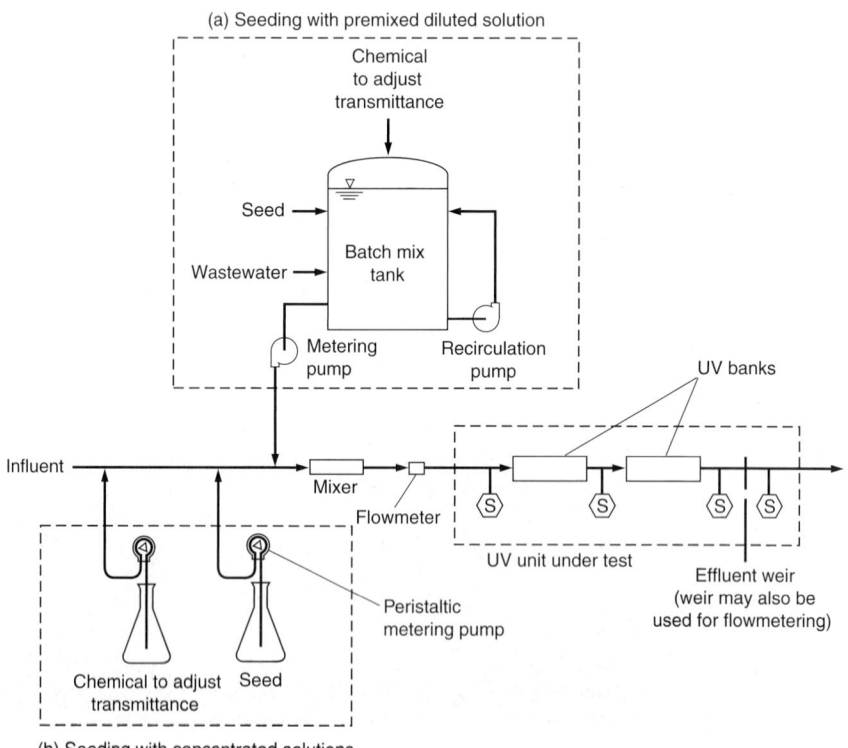

EXAMPLE 12–11   **Conduct a Pilot Test to Validate UV Disinfection System Performance**   The manufacturer has supplied a pilot-scale UV disinfection system to be tested for the assignment of UV doses as a function of lamp hydraulic loading rate. The steps involved are delineated below. Because the UV disinfection system will be used for filtered secondary effluent, determine the range of flows expressed as L/min·lamp over which the UV disinfection system will deliver a dose of 100 mJ/cm^2 at a confidence level of 75 percent.

Solution

1. Set up a pilot testing facility.
   a. The manufacturer chooses to make use of a four lamp per bank pilot facility with three banks in series for the total applied dose. Each bank of lamps is hydraulically independent of subsequent banks. Therefore, the results can be applied to full-scale reactors up to 40 lamps per bank (i.e., full-scale facility can utilize up to 10 times as many lamps per bank as the pilot facility tested).
   b. Aged lamps were placed in the pilot facility to simulate the performance of the UV lamps at the end of their warranted life.
   c. Testing was conducted on tertiary effluent from a local wastewater reclamation facility. Normal transmittance of the tertiary effluent is 75 percent. Transmittance reducing agent (e.g., instant coffee) was injected into the process stream until the transmittance was lowered to 55 percent.
   d. The manufacturer has specified that the UV disinfection system should be tested for hydraulic loading rates ranging from 20 to 80 L/min·lamp.
   e. The titer of the virus indicator (i.e., MS2 coliphage) to be used for performance testing was approximately 1×10^{11} phage/mL. Therefore, it was decided to test the system under the conditions outlined in the following table:

Hydraulic loading rate, L/min·lamp (1)	Process flow, L/min (2)	Virus titer concentration, phage/mL (3)	Virus titer injection flowrate, L/min (4)	Approximate resulting virus concentration in process flow, phage/mL (5)
20	240	1×10^{11}	0.024	1×10^{7}
40	480	1×10^{11}	0.048	1×10^{7}
60	720	1×10^{11}	0.072	1×10^{7}
80	960	1×10^{11}	0.096	1×10^{7}

Notes on column entries:

(1) Desired range to be tested as specified by the manufacturer.

(2) The pilot system contained three banks with 12 lamps total. Therefore, at a hydraulic loading rate of 20 L/min·lamp, the process flowrate needs to be (12 lamps)(20 L/min·lamp) = 240 L/min.

(3) Provided by the laboratory.

(4) It was desired to obtain a virus titer in the process flow of about 1×10^7 phage/mL. Therefore, at 240 L/min, the solution containing the virus had to be injected at a rate of 0.024 L/min to obtain the desired initial titer.

2. Test the system.
   a. Each flowrate was tested randomly with respect to order. Three distinct replicate samples were collected per flowrate. An inlet sample (i.e., that containing the concentration of phage prior to any inactivation) was collected with each process replicate.
   b. Determine the inlet viral indicator concentration for each test. The results are provided in the following table.

Flowrate, L/min	Replicate	Inlet concentration, phage/mL	Log-transformed inlet conc. log (phage/mL)	Average log-transformed inlet conc. log (phage/mL)
240	1	5.25×10^6	6.72	
240	2	1.00×10^7	7.00	6.93
240	3	1.15×10^7	7.06	
480	1	1.00×10^7	7.00	
480	2	1.23×10^7	7.09	7.07
480	3	1.29×10^7	7.11	
720	1	1.23×10^7	7.09	
720	2	1.05×10^7	7.02	7.03
720	3	9.55×10^6	6.98	
960	1	1.23×10^7	7.09	
960	2	1.20×10^7	7.08	7.02
960	3	7.94×10^6	6.90	

   c. Determine the outlet concentration for each trial. The results are provided in the following table. Notice that at the low flowrate investigated (240 L/min), only two operational banks were investigated rather than three. Only two banks were tested because three operational banks resulted in no detectable viruses in the effluent. Because the banks were hydraulically independent, it is allowed under the UV Guidelines to investigate the inactivation for only two banks and extrapolate to performance expected for additional banks of lamps.

Flowrate, L/min	Replicate	Number of operational banks	Outlet concentration, phage/mL	Log-transformed outlet conc., log (phage/mL)
240	1	2	2.09×10^2	2.32
240	2	2	1.44×10^2	2.16
240	3	2	1.66×10^2	2.22

(continued)

(*Continued*)

Flowrate, L/min	Replicate	Number of operational banks	Outlet concentration, phage/mL	Log-transformed outlet conc., log (phage/mL)
480	1	3	3.80×10^2	2.58
480	2	3	3.31×10^2	2.52
480	3	3	3.09×10^2	2.49
720	1	3	1.32×10^4	4.12
720	2	3	6.03×10^3	3.78
720	3	3	4.27×10^3	3.63
960	1	3	4.79×10^4	4.68
960	2	3	1.86×10^5	5.27
960	3	3	6.61×10^4	4.82

    d. Determine the 75 percent level of confidence for each investigated flowrate. The results are provided in the following table. Note that the 75 percent level of confidence is determined using the student-t distribution (a minimum of 30 samples are required for use of the normal distribution).

Flowrate, L/min	Replicate	Log inactivation	Average log inactivation	Sample standard deviation	75% confidence log inactivation
240	1	4.61[a]			
240	2	4.77	4.69	0.08	4.63
240	3	4.71			
480	1	4.49			
480	2	4.55	4.54	0.04	4.50[b]
480	3	4.58			
720	1	2.91			
720	2	3.25	3.19	0.25	2.95
720	3	3.40			
960	1	2.34			
960	2	1.75	2.10	0.31	1.81
960	3	2.20			

[a] Sample calculation. From the previous table, the average inlet log concentration was observed to be 6.93. Therefore, the log inactivation for replicate 1 is $6.93 - 2.32 = 4.61$.

[b] Sample calculation. For the flowrate of 480 L/min, the 75 percent level of confidence occurs at 4.50 as shown below.

$$75\% \text{ confidence limit} = \bar{x} \pm t_{0.125}\left(\frac{s}{\sqrt{n}}\right) = 4.54 - 1.214\frac{0.05}{\sqrt{3}} = 4.54 - 0.04 = 4.50$$

3. Conduct the required collimated-beam test. The results are provided in the following table:

Dose	Surviving concentration, phage/mL	Log survival, log (phage/mL)	Log inactivation
0	1.00×10^7	7.0	0.0
20	1.12×10^6	6.05	0.95[a]
40	6.76×10^4	4.83	2.17
60	1.95×10^4	4.29	2.71
80	4.37×10^3	3.64	3.36
100	1.20×10^3	3.08	3.92
120	7.08×10^1	1.85	5.15
140	1.48×10^1	1.17	5.83

[a]Sample calculation. Log inactivation = 7.00 − 6.05 = 0.95.

a. Plot the collimated-beam test results. Compare to the quality-control range provided in the UV Guidelines. A plot of the results is provided in the figure given below. All of the data points fall within the acceptable range. The equation of the linear regression used to determine the required dose as a function of log inactivation is given on the figure.

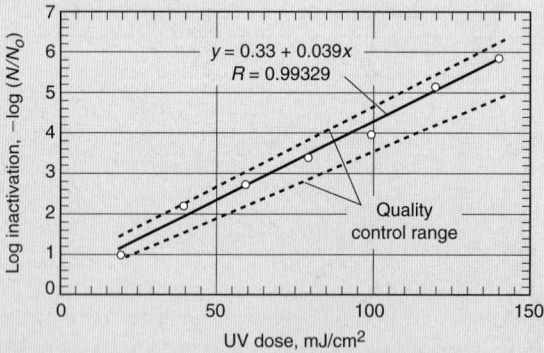

b. Assign doses to investigated hydraulic loading rates, and present results graphically. The calculated doses are provided in the table given below.

Flowrate, L/min	Hydraulic loading rate, L/min·lamp	75% confidence log-inactivation	Equivalent UV dose[b], mJ/cm²
240	20	(1.5)(4.63) = 6.95[a]	170
480	40	4.50	107

(continued)

(*Continued*)

Flowrate, L/min	Hydraulic loading rate, L/min·lamp	75% confidence log-inactivation	Equivalent UV dose[b], mJ/cm^2
720	60	2.95	67.2
960	80	1.81	37.9

[a]The inactivation for this flowrate was extrapolated from the two-bank results. Because the system is a three-bank system, the inactivation for three banks is 150 percent greater than the inactivation observed with two operational banks.

[b]Sample calculation. Using the linear regression expression derived from the collimated-beam test, the equivalent UV dose at a flowrate of 480 L/min is

$$x(\text{UV dose}) = \frac{y\,(\text{log inactivation}) - 0.33}{0.039}$$

$$\text{Dose, mJ/cm}^2 = \frac{\text{log inactivation} - 0.33}{0.039} = \frac{4.50 - 0.33}{0.039} = 107 \text{ mJ/cm}^2$$

The results are plotted in the following figure.

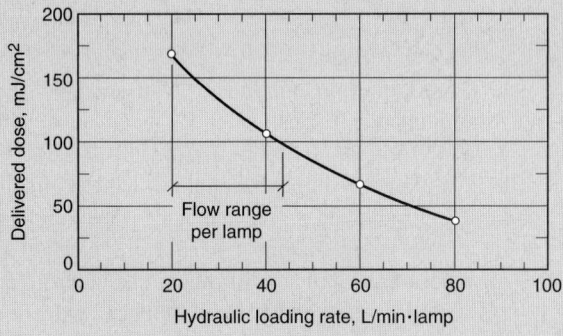

4. Determine the flow per lamp over which the system will deliver 100 mJ/cm^2. From the plot given above, the system is capable of applying a dose of 100 mJ/cm^2 within the range of 20 to 43 L/min·lamp.

Comment Because aged lamps were used and the transmittance value was adjusted to 55 percent, the test results represent the performance of the UV disinfection system under the worst possible conditions, which provides a factor of safety under typical operating conditions. When the lamps are new, it may not be necessary to operate all three banks, depending on the test results. The curve given above in Step 3 will be used to determine the optimal configuration of the full-scale UV disinfection system, as will be discussed and illustrated below.

Selection and Sizing of a UV Disinfection System

Factors that affect the minimum number of UV lamps necessary for disinfection are (1) the hydraulic loading rate determined in the equipment validation test as outlined in the previous example, (2) the aging and fouling characteristics of the UV lamp/quartz sleeve assembly, (3) wastewater quality and its variability, and (4) the nature of the discharge permit itself and the level of confidence desired in meeting that permit. Hydraulic behavior has a significant impact on field reactor performance. The flow per lamp determined using the collimated-beam bioassay has a corresponding velocity that maintains that inactivation performance. The process configuration must maintain adequate system velocity to ensure that the bioassay results are applicable to the field installation. Although beyond the scope of this presentation, the UV Guidelines cited above should be reviewed carefully before undertaking the design of a UV system. The selection and sizing procedure is illustrated in Example 12–12.

EXAMPLE 12–12 **Design of a UV Disinfection System** Design a UV disinfection system that will deliver a minimum design dose of 100 mJ/cm². Assume for the purpose of this example that the following data apply:

1. Wastewater characteristics
   a. Minimum design flow = 6000 m³/d = 4167 L/min (diurnal low flow)
   b. Maximum design flow = 21,000 m³/d = 14,584 L/min (peak hour flow with recycle streams)
   c. Minimum transmittance = 55 percent
2. System characteristics
   a. Horizontal lamp configuration
   b. Validated system performance as determined in Example 12–11
   c. System headloss coefficient = 1.8 (manufacturer-specific)
   d. Lamp/sleeve diameter = 23 mm
   e. Cross-sectional area of quartz sleeve = 4.15 × 10⁻⁴ m²
   f. Lamp spacing = 75 mm (center to center)
   g. One standby UV bank will be required per channel

Solution

1. Determine the number of UV channels.
   a. The manufacturer has provided validated information for a three-bank system at a flow range of 20 to 80 L/min·lamp. The system is capable of applying a dose of 100 mJ/cm² within the range of 20 to 43 L/min·lamp (see Example 12–11). Therefore, the system has an approximate 2:1 flow variation capacity for the design dose under consideration.
   b. Use two channels.
      i. From 4167 to 8000 L/min use one channel.
      ii. From 8000 to 14,584 L/min split the flow between two channels such that each channel is loaded between 4000 and 7300 L/min.
2. Determine the number of lamps required per channel.

At 8000 L/min, the total number of required lamps is

$$\text{Lamps required at 8000 L/min} = \frac{(8000 \text{ L/min})}{(43 \text{ L/lamp·min})} = 186 \text{ lamps}$$

3. Determine the minimum number of lamps per bank.

$$\frac{186 \text{ lamps}}{3 \text{ banks}} = 62 \frac{\text{lamps}}{\text{bank}}$$

4. Configure the UV disinfection system.
   Typically, 2, 4, 8, or 16 lamps per module are available. For an 8 lamp module, eight modules are required per bank for a total of 64 lamps per bank.

5. Check that the design falls within the manufacturer recommended range.
   a. At low flow:

   $$\frac{(4167 \text{ L/min})}{192 \text{ lamps}} = 21.7 \text{ L/lamp·min}$$

   b. At high flow:

   $$\frac{(14,584 \text{ L/min})}{384 \text{ lamps}} = 38.0 \text{ L/lamp·min}$$

   Both of these hydraulic loading rates fall within the acceptable range for the UV disinfection system provided by the manufacturer.

6. Check whether the headloss for the selected configuration is acceptable.
   a. Determine the channel cross-sectional area.

   $$\text{Cross-sectional area of channel} = (8 \times 0.075 \text{ m}) \times (8 \times 0.075 \text{ m})$$
   $$= 0.36 \text{ m}^2$$

   b. Determine the net channel cross-sectional area by subtracting the cross-sectional area of the quartz sleeves (4.15×10^{-4} m²/lamp).

   $$A_{\text{channel}} = 0.36 \text{ m}^2 - [(8 \times 8) \text{ lamps/bank}] \times (4.15 \times 10^{-4} \text{ m}^2/\text{lamp})$$
   $$= 0.33 \text{ m}^2$$

   c. Determine the maximum velocity in each channel.

   $$v_{\text{channel}} = \frac{(14,584 \text{ L/min})}{(2 \text{ channel})(0.33 \text{ m}^2)(10^3 \text{ L/m}^3)(60 \text{ s/min})} = 0.37 \text{ m/s}$$

   d. Determine the headloss per UV channel.

   $$h_{\text{channel}} = 1.8 \frac{v^2}{2g}$$

   $$h_{\text{channel}} = 1.8 \frac{(0.37 \text{ m/s})^2(10^3 \text{ mm/m})}{2(9.81 \text{ m/s}^2)} (4 \text{ banks}) = 50 \text{ mm}$$

Note that four banks were used to determine system headloss. Use of four banks includes a redundant bank of lamps in each channel. Because the clear spacing between quartz sleeves is 52 mm (75 mm − 23 mm), the headloss cannot exceed 26 mm total (one-half the clear spacing between the quartz sleeves) without exposing the uppermost row of lamps to the air. To allow for the calculated 50 mm of total headloss, each UV channel will require a stepped channel bottom. A 24-mm step between the second and third bank of lamps is required to allow for the expected headloss and to allow the third and fourth banks of lamps to be set lower.

7. Summarize the system configuration.

   Minimum required system utilizes two channels, each channel containing four banks of lamps in series, three operational banks and one redundant bank. Each bank contains eight modules, each of which contains eight lamps.

Troubleshooting UV Disinfection Systems

Problems associated with UV disinfection are related primarily to the inability to achieve permit limitations. Some issues that must be considered when diagnosing problems associated with UV disinfection systems are discussed below.

UV Disinfection System Hydraulics. Perhaps one of the most serious problems encountered in the field is erratic or reduced inactivation performance due to poor system hydraulics. The most common hydraulic problems are related to (1) the creation of density currents which can cause the incoming wastewater to move along the bottom or top of the UV lamp banks, resulting in short circuiting; (2) inappropriate entry and exit conditions which can lead to the formation of eddy currents, which ultimately create uneven velocity profiles that induce short circuiting; (3) the creation of dead spaces or zones within the reactor, resulting in short circuiting. The occurrence of short-circuiting or dead zones reduces the average contact time, leading to ineffective use of the UV system.

The principal hydraulic design features that can be used to improve system hydraulics in open channels include the use of (1) submerged perforated diffusers, (2) corner fillets in rectangular open-channel systems with horizontal lamp placement, (3) flow deflectors in open-channel systems with vertical lamp placement, and (4) serpentine effluent overflow weirs used in conjunction with perforated diffusers. In some cases, power input to mix the incoming flow may be necessary. Some of these corrective measures for open-channel UV disinfection systems are illustrated on Fig. 12–45. Submerged perforated baffles should have an open area of about 4 to 6 percent of the cross-sectional area of the flow channel. The headloss dissipated through a submerged baffle can be estimated using Eq. (12–56), presented previously in the discussion of the use of submerged baffles for chlorine contact basins (see Sec. 12–6). In closed UV disinfection systems, the use of perforated plates (see Fig. 12–46) is typically not required when the units are plumbed correctly. The use of computational fluid dynamics is of great value in studying the effect of various physical interventions in bringing about a more uniform approach velocity flow field.

Figure 12–45

Typical examples of physical features that can be used to improve the hydraulic performance of open-channel UV disinfection systems: (a) horizontal lamp placement and (b) vertical lamp placement.

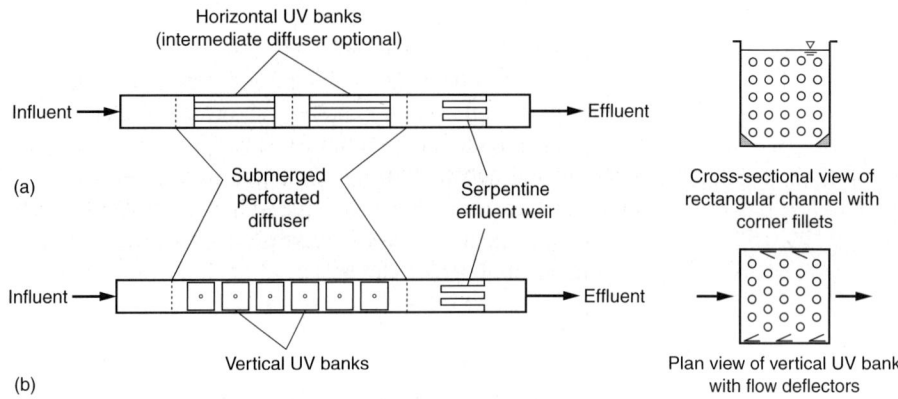

Figure 12–46

Use of perforated diffusers to improve the hydraulic performance of closed-channel UV disinfection system: (a) potential dead spaces without diffuser, (b) potential dead spaces with diffuser, and (c) relative improvement in hydraulic performance.

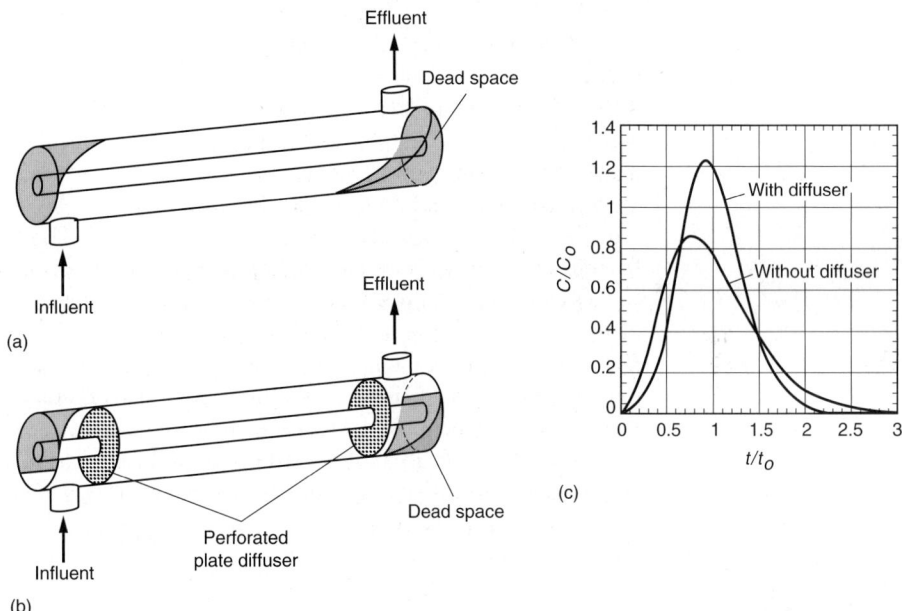

Biofilms on Walls of UV Channels and on UV Equipment. Another serious problem encountered with UV disinfection systems is the development of biofilms on the exposed surfaces of the UV reactor. The problem is especially serious in open-channel systems covered with standard grating. It has been found that if the UV channels are exposed to any light, even very dim light, biofilms (typically fungal and filamentous bacteria) will develop on the exposed surfaces. The problem with biofilms is that they can harbor and effectively shield bacteria. When the clumped biofilms break away from the attachment surface, bacteria can be shielded as the clumps pass through the disinfection system. The best control measure is to completely cover the UV channels. In addition, the channels can be occasionally cleaned and disinfected using hypochlorite, paracetic acid (see Sec. 12–8), or another suitable cleaning agent/disinfectant.

It should be noted that biofilm development can also occur in closed UV systems, but the severity is usually less, with the exception of UV systems in which medium-pressure high-intensity UV lamps are employed. Because medium-pressure high-intensity UV lamps emit some light in or near the visible light range (see Fig. 12–32), they can stimulate the growth of microorganisms on exposed surfaces. In some cases, growths approaching 300 mm in length have been found attached to the lamp support structure. The amount of light emitted in the visible light range will vary with each type of lamp (i.e., manufacturer). Removal of these growths with a suitable disinfectant must be conducted on a periodic basis.

Overcoming the Impact of Particles by Increasing the UV Intensity. It was thought at one time that the impact of particles on the performance of UV disinfection systems could be overcome by increasing the UV intensity. Unfortunately, it is not possible to increase the UV intensity enough to overcome the shading effect imparted by particles. An example is illustrated on Fig. 12–47. As shown, increasing the UV intensity tenfold has little effect on reducing the number of surviving particle-associated coliform bacteria because the absorption of UV radiation by wastewater particles is typically 10,000 times or more greater than the bulk liquid medium. Particles essentially block the transmission of UV light. Particles larger than some critical size (a function of the size of the target organism) will effectively shield the embedded microorganisms (Emerick et al., 1999; Emerick et al., 2000). Further, because large particles have little effect on turbidity, effluents with low measured turbidity values (i.e., ≤2 NTU) can still be difficult to disinfect due to the presence of large undetected particles (Ekster, 2001).

Because the effectiveness of UV disinfection is governed primarily by the number of particles containing coliform bacteria, to improve the performance of a UV disinfection system either the number of particles with associated coliform bacteria must be reduced (e.g., by selecting an appropriate upstream treatment process), or the particles themselves must be removed (e.g., by improved clarifier design or use of some form of filtration). Currently, to meet the stringent coliform bacteria requirements for body contact wastewater reuse applications (i.e., equal to or less than 2.2 MPN/100 mL), effluent filtration is required.

Figure 12–47

Impact of UV intensity on the effectiveness of UV disinfection of effluent from an air-activated sludge wastewater-treatment plant.

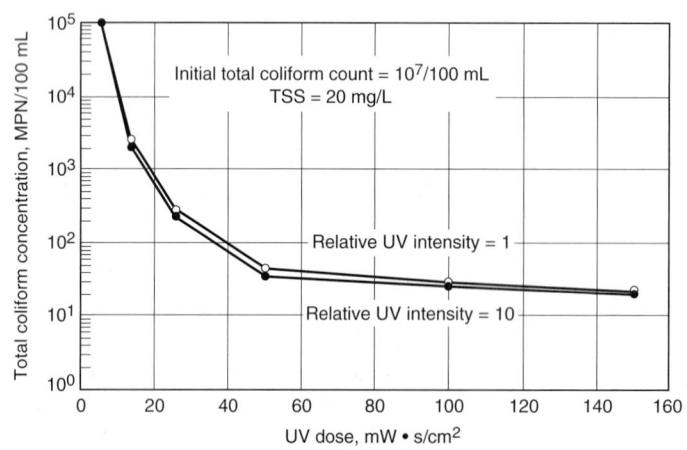

Figure 12–48

Fraction of particles in settled wastewater with one or more associated coliform organisms as function of the solids retention time (SRT).

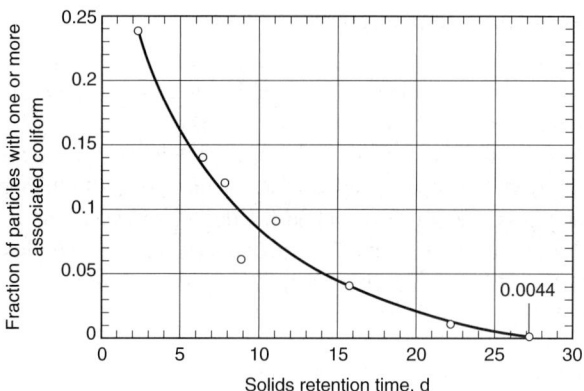

Effect of Upstream Treatment Processes on UV Performance. The number of particles with associated coliform bacteria is another factor that will impact the performance of both chlorine and UV disinfection for unfiltered effluents (when coliform bacteria are used as the regulatory indicator). It has been observed that for activated-sludge plants the number of particles with associated coliform organisms is a function of the solids retention time (SRT). The relationship between the fraction of wastewater particles with one or more associated coliform organisms and the SRT is illustrated on Fig. 12–48. As illustrated, longer SRTs result in a decrease in the fraction of particles containing coliform bacteria. Without effluent filtration, the UV performance achieved will depend on the operation of the upstream processes (Darby et al., 1999). The use of deep final clarifiers will reduce the number of large particles that may shield bacteria. In general, without some form of filtration, it will be difficult to achieve extremely low coliform concentrations in the settled effluent from an activated-sludge plant operated at low SRT values (e.g., 1 to 2 d).

Environmental Impacts of UV Radiation Disinfection

On the basis of the evidence to date, it appears that the compounds formed at the UV dosages used for the disinfection (50 to 140 mJ/cm^2) of wastewater are harmless or are broken down into more innocuous forms. Thus, the disinfection of wastewater with ultraviolet light is considered to have no adverse environmental impacts. The environmental impacts and types of compounds formed when UV radiation is used for the destruction of trace organics such as NDMA and endocrine disruptors, typically at UV dosages greater than about 400 mJ/cm^2, is not known at present (2001). The use of hydrogen peroxide and UV for the destruction of selected organic compounds is considered in Sec. 11–10 in Chap. 11, which deals with *Advanced Oxidation Processes.*

12–10 COMPARISON OF ALTERNATIVE DISINFECTION TECHNOLOGIES

The advantages and disadvantages of the four technologies considered in this chapter for disinfection of wastewater (i.e., chlorine, chlorine dioxide, ozone, and UV) are reviewed briefly in the following discussion.

Germicidal Effectiveness

A general comparison of the germicidal effectiveness of the disinfection technologies by classes of organisms was presented previously in Table 12–11. Additional information was presented in Table 12–13 for chlorine, Table 12–22 for ozone, and Table 12–26 for UV. It is important to note that the values given in these tables are only meant to serve as a guide in assessing the effectiveness of these technologies. Because the characteristics of each wastewater and the degree of treatment will significantly impact the effectiveness of the various disinfection technologies, site-specific testing must be conducted to evaluate the effectiveness of alternative disinfection technologies and to establish appropriate dosing ranges.

Advantages and Disadvantages

The general advantages and disadvantages of using chlorine, chlorine dioxide, ozone, and UV for wastewater disinfection are summarized in Table 12–28.

Table 12–28

Advantages and disadvantages of chlorine, chlorine dioxide, ozone, and UV for the disinfection of wastewater[a]

Advantages	Disadvantages
Chlorine	
1. Well-established technology	1. Hazardous chemical that can be a threat to plant workers and the public; thus strict safety measures must be employed
2. Effective disinfectant	2. Relatively long contact time required as compared to other disinfectants
3. Chlorine residual can be monitored and maintained	3. Combined chlorine is less effective in inactivating some viruses, spores, cysts at low dosages used for coliform organisms
4. Combined chlorine residual can also be provided by adding ammonia	4. Residual toxicity of treated effluent must be reduced through dechlorination
5. Germicidal chlorine residual can be maintained in long transmission lines	5. Formation of trihalomethanes and other DBPs[b] (see Table 12–14)
6. Availability of chemical system for auxiliary uses such as odor control, dosing RAS lines, and disinfection plant water systems	6. Release of volatile organic compounds from chlorine contact basins
7. Oxidizes sulfides	7. Oxidizes iron, magnesium, and other inorganic compounds (consumes disinfectant)
8. Relatively inexpensive (cost is increasing with implementation of Uniform Fire Code regulations)	8. Oxidizes a variety of organic compounds (consumes disinfectant)
	9. TDS level of treated effluent is increased
9. Available as calcium and sodium hypochlorite considered to be safer than chlorine gas	10. Chloride content of the wastewater is increased
	11. Acid generation; pH of the wastewater can be reduced if alkalinity is insufficient
	12. Increased safety regulations, especially in light of the Uniform Fire Code
	13. Chemical scrubbing facilities may be required to meet Uniform Fire Code regulations

(continued)

| **Table 12–28** (*Continued*)

Advantages	Disadvantages
Chlorine dioxide	
1. Effective disinfectant	1. Unstable, must be produced onsite
2. More effective than chlorine in inactivating most viruses, spores, cysts, and oocysts	2. Oxidizes iron, magnesium, and other inorganic compounds (consumes disinfectant)
3. Biocidal properties not influenced by pH	3. Oxidizes a variety of organic compounds
4. Under proper generation conditions, halogen-substituted DBPs are not formed	4. Formation of DBPs (i.e., chlorite and chlorate)
	5. Potential for the formation of halogen-substituted DBPs
5. Oxidizes sulfides	6. Decomposes in sunlight
6. Provides residuals	7. Can lead to the formation of odors
	8. TDS level of treated effluent is increased
	9. Operating costs can be high (e.g., must test for chlorite and chlorate)
Ozone	
1. Effective disinfectant	1. No immediate measure of whether disinfection was successful
2. More effective than chlorine in inactivating most viruses, spores, cysts, and oocysts	2. No residual effect
3. Biocidal properties not influenced by pH	3. Less effective in inactivating some viruses, spores, cysts at low dosages used for coliform organisms
4. Shorter contact time than chlorine	4 Formation of DBPs (see Table 12–23)
5. Oxidizes sulfides	5. Oxidizes iron, magnesium, and other inorganic compounds (consumes disinfectant)
6. Requires less space	6. Oxidizes a variety of organic compounds (consumes disinfectant)
7. Contributes dissolved oxygen	7. Off-gas requires treatment
	8. Safety concerns
	9. Highly corrosive and toxic
	10. Energy-intensive
	11. Relatively expensive
	12. Highly operational and maintenance-sensitive
	13. Lack of chemical system that can be used for auxiliary uses such as dosing RAS lines
	14. May be limited to plant where generation of high-purity oxygen already exists

(*continued*)

| **Table 12-28** (Continued)

Advantages	Disadvantages
UV	
1. Effective disinfectant	1. No immediate measure of whether disinfection was successful
2. No residual toxicity	2. No residual effect
3. More effective than chlorine in inactivating most viruses, spores, cysts	3. Less effective in inactivating some viruses, spores, cysts at low dosages used for coliform organisms
4. No formation of DBPs at dosages used for disinfection	4. Energy-intensive
5. Does not increase TDS level of treated effluent	5. Hydraulic design of UV system is critical
6. Effective in the destruction of resistant organic constituents such as NDMA	6. Relatively expensive (price is coming down as new and improved technology is brought to the market)
7. Improved safety compared to the use of chemical disinfectants	7. Large number of UV lamps required where low-pressure low-intensity systems are used
8. Requires less space than chlorine disinfection	8. Low-pressure low-intensity lamps require acid washing to remove scale
9. At higher UV dosages than required for disinfection, UV radiation can be used to reduce the concentration of trace organic constituents of concern such as NDMA (see Sec. 11–10 in Chap. 11)	9. Lack of chemical system that can be used for auxiliary uses such as odor control, dosing RAS lines, and disinfection of plant water systems

[a] Adapted in part from Crites and Tchobanoglous (1998) and U.S. EPA (1999b).
[b] DBPs = disinfection byproducts.

PROBLEMS AND DISCUSSION TOPICS

12-1 Determine the inactivation rate constant for one of the following four waters (water to be selected by instructor) assuming Chick's law applies.

Log of organisms remaining	Time, min Water			
	A	**B**	**C**	**D**
7	0.0	0.0	0.0	0.0
6	2.0	6	4	3.0
5	4.0	12	8.1	5.9
4	6.5	18	12.6	8.6
3	8.5	23.9	17.5	12.5
2	10.0	30.1	20	15.2
Combined chlorine residual, mg/L	10	4	6	8

12-2 The following data were obtained for a 99 percent inactivation of a given bacterial strain for four different waters at 10°C. Using these data develop a general relationship that can be used to model the observed data for one of the waters (to be selected by instructor). Using the relationship developed, determine the chlorine dose required to achieve a 99.99 percent inactivation in 60 min at 25°C.

Combined chlorine residual, mg/L	Contact time, min			
	Water			
	A	B	C	D
2	12	240	90	38
3	4.0	92	30	10
4	1.9	47	12	4.2
5	1.0	28	7	2.1
6		18	4.1	1.2
7		12	2.8	
8		9	1.9	
9		6.8	1.2	
10		5.1	1.0	

12-3 Disinfection data were obtained for four different waters. Using these data, determine the coefficients for Eq. (12–31) (water to be selected by instructor). Using the equation developed, determine the time required to achieve a 4.5-log reduction.

$C_R t$, mg·min /L	$-\log N/N_o$			
	Water			
	A	B	C	D
0	0.0	0.0	0.0	0.0
5	0.0	0.4	0.0	0.8
10	0.0	1.2	0.0	1.8
20	0.0	2.2	1.0	2.8
50	1.0	3.0	2.0	3.8
100	2.0	3.9	2.8	4.8
200	3.0	4.8	3.6	5.6
500	4.0	6.0	4.6	—
Combined chlorine residual, mg/L	4.3	5.1	3.8	4.7

12-4 The data in the following table were obtained in a series of laboratory tests performed on an effluent from a secondary wastewater treatment process with a fecal coliform count of $10^6/100$ mL.

Chlorine dosage, mg/L	Residual fecal coliform count, no./100 mL; contact time, min.		
	15	30	60
1	10,000	2000	500
2	3,000	350	90
4	400	65	20
6	110	30	12
8	54	19	6
10	30	10	1

a. Plot the data on log-log paper. Using these data, determine the value of the exponent n and
the constant in Eq. (12–7) for residual coliform counts of 200/100 mL and 1000/100 mL.

b. The following data apply to the wastewater treatment plant.

Item	Unit	May–Oct.	Nov.–Apr.
Average flow	m³/d	20,000	26,000
Peak daily flow	m³/d	40,000	52,000
Maximum permissible fecal coliform count in effluent	MPN/100 mL	200	1000

Determine the required volume in m³ of a chlorine contact chamber designed to provide 60-min
contact at the average winter flow. Using the equations developed in part a, determine the minimum dosage required in mg/L to give the required kill under each of the four flow conditions
given above. Assuming that the yearly chlorine requirement can be computed on the basis of the
average flow for each of the two 6-month periods, determine the minimum yearly chlorine
requirement in kilograms. (Courtesy E. Foree.)

12-5 The chlorine residuals measured when various dosages of chlorine were added to four different
wastewaters are given below. Determine (wastewater to be selected by instructor) (a) the breakpoint dosage and (b) the design dosage to obtain a residual of 1.0, 1.5, 2.0, and 2.5 mg/L (residual to be selected by instructor) free available chlorine.

Dosage, mg/L	Residual, mg/L			
	A	B	C	D
0	0.0	0.0	0.0	0.0
1	0.6	1.0	0.95	1.0
2	0.2	2.0	1.7	1.98
3	1.0	2.98	2.3	2.9
4	2.0	3.95	1.9	3.4
5	3.0	4.3	1.0	2.8

(continued)

(*Continued*)

Dosage, mg/L	A	B	C	D
		Residual, mg/L		
6		3.7	1.7	1.8
7		2.7	2.7	1.2
8		1.6	3.7	2.1
9		0.8		3.1
10		1.7		4.1
11		2.8		
12				

12–6 Compare the volume required for a complete-mix reactor to that for a plug-flow reactor to achieve a 10^4 reduction in the coliform count of a treated effluent. Assume that in both cases the chlorine residual to be maintained is 5 mg/L.

12–7 Using the following dose-response data for an enteric virus and the tracer data for four different chlorine contact basins, determine for one of the basins (to be selected by instructor) the expected effluent microorganism concentration using a segregated flow model. Also estimate the chlorine residual that would be required to achieve 4 logs of removal with the existing basins.

Dose-response data for enteric viruses

$C_R t$, mg/L·min[a]	Log number of organisms remaining
0	10^7
100	$10^{6.2}$
200	$10^{5.4}$
400	$10^{3.8}$
600	$10^{2.1}$
800	$10^{0.6}$
1000	10^{-1}

[a]Combined chlorine residual = 5.0.

Tracer data for chlorine contact basins

Time, min	A	B	C	D
		Tracer concentration, mg/L		
		Chlorine contact basin		
0	0.0	0.0	0.0	0.0
10	0.0	0.0	0.0	0.0
20	0.0	0.0	0.0	0.0

(continued)

(*Continued*)

Time, min	Tracer concentration, mg/L Chlorine contact basin			
	A	**B**	**C**	**D**
30	0.1	0.0	0.0	0.0
40	2.0	0.0	0.0	0.0
50	7.3	1.1	0.1	0.0
60	7.0	7.0	1.3	0.1
70	5.2	7.3	8.0	1.5
80	3.3	5.7	8.5	7.5
90	1.7	4.2	6.2	8.0
100	0.7	2.9	2.9	5.5
110	0.2	1.7	1.3	3.5
120	0.0	0.9	0.4	1.8
130		0.3	0.0	0.9
140		0.1		0.3
150		0.0		0.1
160				0.0
τ, min	80	85	90	100

12–8 Determine the amount of sulfur dioxide (SO_2), sodium sulfite (Na_2SO_3), sodium bisulfite ($NaHSO_3$), sodium metabisulfite ($Na_2S_2O_5$), and activated carbon (C) that would be required per year to dechlorinate treated effluent containing a chlorine residual of 4.0, 3.5, 6.0, or 5.2 mg/L as Cl_2 (residual to be selected by instructor) from a plant with an average flowrate of 1000, 3800 5000, or 8250 m^3/d (flowrate to be selected by instructor).

12–9 Review the current literature and prepare an assessment of the use of chlorine gas versus sodium hypochlorite for the disinfection of wastewater. A minimum of 4 articles and/or reports dating back to 1995 should be cited in your assessment.

12–10 Estimate the ozone dose needed to disinfect a filtered secondary effluent to an MPN value of 240/100 mL using the following disinfection data obtained from pilot-scale installation. Assume the starting coliform concentration will be 1×10^6/100 mL and that the ozone transfer efficiency is 80 percent.

Test number	Initial coliform count, N_o MPN/100 mL	Ozone transferred, mg/L	Final coliform count, MPN/100 mL	$-\log (N/N_o)$
1	95,000	3.1	1500	1.80
2	470,000	4.0	1200	2.59
3	3,500,000	4.5	730	3.68
4	820,000	5.0	77	4.03
5	9,200,000	6.5	92	5.00

12-11 Review the current literature and prepare an assessment of the use of ozone for the disinfection of wastewater. A minimum of 4 articles and/or reports dating back to 1995 should be cited in your assessment.

12-12 Determine the mean, standard deviation, and confidence interval for the following MS2 inactivation data, obtained using a collimated-beam device. What UV dose would be required to achieve a 4-log inactivation of MS2 with a confidence interval of 75 percent?

Log reduction, $-\log N/N_o$	Applied UV dose, mJ/cm^2				
	Test				
	1	2	3	4	5
1	17	21	26	24	20
2	37	43	51	47	40
3	56	66	80	70	60
4	75	89	105	94	80
5	94	110	131	120	100
6	114	133	160	143	121
7	131	155	185	170	142

12-13 If the intensity of the UV radiation measured at the water surface in a petri dish is 10 mW/cm^2, determine the average UV intensity to which a sample will be exposed if the depth of water in the petri dish is 10, 22, 14, 15, or 16 mm (water depth to be selected by instructor).

12-14 Assume the intensity of UV radiation measured at the water surface in a petri dish in Problem 12–12 is 5 mW/cm^2, and that the computed UV dose was based on a water depth of 10 mm. What would be the effect if the actual water depth in the petri dish were 20 mm? For one of the test runs (to be selected by instructor) plot the actual test results reported in Problem 12–12 and the corrected values using a water depth of 20 mm.

12-15 The following data were obtained for filtered wastewater with a transmittance of 65 percent at 254 nm with a UV pilot-test unit comprised of 4 UV lamps in each of two UV banks. Determine the mean, standard deviation, and 75 percent confidence interval for one of the tests (to be selected by instructor). For the test to be analyzed determine the range of flows expressed as L/min·lamp over which the UV disinfection system will deliver a dose of 90 mJ/cm^2.

Test	Pilot flowrate, L/min	Log$_{10}$ MS2 bacteriophage inactivation		
		Replicate 1	Replicate 2	Replicate 3
1	200	4.51	4.47	4.48
1	400	3.48	3.50	3.53
1	700	2.50	2.57	2.45
2	300	5.55	5.50	5.60
2	500	4.51	4.44	4.53
2	800	3.05	2.90	3.20

(continued)

(*Continued*)

Test	Pilot flowrate, L/min	Log$_{10}$ MS2 bacteriophage inactivation		
		Replicate 1	Replicate 2	Replicate 3
3	400	4.30	4.25	4.32
3	550	3.55	3.60	3.50
3	750	2.60	2.55	2.63
4	280	4.40	4.37	4.46
4	480	3.43	3.40	3.39
4	620	2.73	2.66	2.76

12-16 The following data were obtained for filtered wastewater with a transmittance of 55 percent at 254 nm with a UV pilot-test unit comprised of 6 UV lamps in a single UV bank operated at various ballast settings. Determine (1) the mean, standard deviation, and 75 percent confidence interval for one of the given ballast output settings (to be selected by instructor) and (2) the range of flows expressed as L/min·lamp over which the UV disinfection system will deliver a dose of 120 mJ/cm^2.

Pilot flowrate, L/min	Ballast output, %	Log$_{10}$ MS2 bacteriophage inactivation		
		Replicate 1	Replicate 2	Replicate 3
400	100	6.58	6.62	6.53
	80	5.43	5.42	5.41
	60	4.94	4.98	4.93
	50	4.67	4.74	4.77
560	100	5.82	5.69	5.61
	80	4.62	4.85	4.61
	60	4.32	4.38	4.06
	50	4.11	4.07	4.14
732	100	5.54	5.5	5.44
	80	4.49	4.58	4.62
	60	4.22	3.96	3.76
	50	4.06	3.97	3.94

12-17 The following data were obtained in a pilot test of a UV disinfection system. If a UV dose of 30 mJ/cm^2 is to be maintained, determine the flowrate per lamp for each of the ballast settings. What is the lowest ballast setting that can be used to meet the required dosage?

Flowrate, L/min	Flowrate per lamp, L/lamp	Ballast output, %	Log inactivation	UV dose based on collimated beam, mJ/cm^2
100	25	60	1.15	26.2
140	35	60	1.01	21.8

(*continued*)

(*Continued*)

Flowrate, L/min	Flowrate per lamp, L/lamp	Ballast output, %	Log inactivation	UV dose based on collimated beam, mJ/cm²
188	47	60	0.93	20.5
100	25	80	1.66	34.6
140	35	80	1.45	29.4
188	47	80	1.51	28.5
100	25	100	1.85	40.7
140	35	100	1.59	31.7
188	47	100	1.5	29.2

12–18 Tests were conducted at the extreme hydraulic conditions to evaluate the independence of three different UV disinfection systems comprised of three UV banks in series. It is assumed that demonstration of hydraulic independence at the hydraulic extremes is adequate to allow testing to proceed on a single bank to evaluate UV performance versus ballast output. Using an analysis of variance, determine whether all three UV banks were operating similarly or differently. The test flowrate is to be selected by the instructor.

Flowrate, L/min	UV bank	Replicate 1	2	3	Average
380	1	1.26	1.16	1.10	1.17
	2	0.99	0.86	1.29	1.05
	3	1.31	1.01	1.09	1.14
				Average =	1.12
710	1	0.77	0.56	0.62	0.65
	2	0.67	0.69	0.61	0.66
	3	0.82	0.75	0.83	0.80
				Average =	0.70
1000	1	0.55	0.56	0.50	0.54
	2	0.57	0.65	0.61	0.61
	3	0.60	0.51	0.58	0.56
				Average =	0.57

12–19 Review the current literature and prepare an assessment of the use of low-pressure low-intensity versus low-pressure high-intensity UV disinfection systems for the disinfection of wastewater. A minimum of 4 articles and/or reports dating back to 1995 should be cited in your assessment.

12–20 Review the current literature and prepare an assessment of the use of medium-pressure high-intensity UV disinfection systems for the disinfection of wastewater. A minimum of 4 articles and/or reports dating back to 1995 should be cited in your assessment.

REFERENCES

Blackmer, F., K. A. Reynolds, C. P. Gerba, and I. L. Pepper (2000) "Use of Integrated Cell Culture-PCR to Evaluate the Effectiveness of Poliovirus Inactivation by Chlorine," *Applied and Environmental Microbiology,* vol. 66, no. 5, pp. 2267–2268.

Blatchley, E. R., et al. (1995) "UV Pilot Testing: Intensity Distributions and Hydrodynamics," *Journal of Environmental Engineering, American Society of Civil Engineers,* vol. 121, p. 258.

Butterfield, C. T., E. Wattie, S. Megregian, and C. W. Chambers (1943) "Influence of pH and Temperature on the Survival of Coliforms and Enteric Pathogens When Exposed to Free Chlorine," *U.S. Public Health Service Report* 58.

Cairns, W. L. (1993) "Comparing Disinfection by Ultraviolet Light and Chlorination—The Implications of Mechanism for Practice," in the *Proceedings of the Planning, Design, and Operation Effluent Disinfection Systems Specialty Conference, Water Environment Federation,* Whippany, NJ, p. 555.

Chick, H. (1908) "Investigation of the Laws of Disinfection," *Journal of Hygiene* (British), vol. 8, p. 92.

Colgan S., and R. Gehr (2001) "Disinfection," *Water Environment and Technology,* vol. 13, no. 11.

Collins, H. F. (1970) "Effects of Initial Mixing and Residence Time Distribution on the Efficiency of the Wastewater Chlorination Process," paper presented at the California State Department of Health Annual Symposium, Berkeley and Los Angeles, CA.

Collins, H. F., and R. E. Selleck (1972) "Process Kinetics of Wastewater Chlorination," *SERL Report* 72-5, Sanitary Engineering Research Laboratory, University of California, Berkeley, CA.

Cooper, R. C., A. T. Salveson, R. Sakaji, G. Tchobanoglous, D. A. Requa, and R. Whitley (2000) "Comparison of the Resistance of MS-2 and Poliovirus to UV and Chlorine Disinfection," Presented at the California Water Reclamation Meeting, Santa Rosa, CA.

Crites, R., and G. Tchobanoglous (1998) *Small and Decentralized Wastewater Management Systems,* McGraw-Hill, New York.

Darby, J., R. Emerick, F. Loge, and G. Tchobanoglous (1999) "The Effect of Upstream Treatment Processes on UV Disinfection Performance," Project 96-CTS-3, *Water Environment Research Foundation,* Washington DC.

Darby, J. L., M. Heath, J. Jacangelo, F. Loge, P. Swain, and G. Tchobanoglous (1995) *Comparative Efficiencies of UV Irradiation to Chlorination: Guidance for Achieving Optimal UV Performance,* Project 91-WWD-1, Water Environment Research Foundation, Alexandria, VA.

Darby, J. L., K. E. Snider, and G. Tchobanoglous (1993) "Ultraviolet Disinfection for Wastewater Reclamation and Reuse Subject to Restrictive Standards," *Water Environment Research,* vol. 65, no. 2, Reuse Applications in California.

Ekster, A. (2001) Personal communication.

Emerick, R. W., and J. L. Darby (1993) "Ultraviolet Light Disinfection of Secondary Effluents: Predicting Performance Based on Water Quality Parameters," in the *Proceedings of the Planning, Design, and Operation Effluent Disinfection Systems, Specialty Conference, Water Environment Federation,* Whippany, NJ.

Emerick, R. W., F. Loge, T. Ginn, and J. L. Darby (2000) "Modeling the Inactivation of Particle-Associated Coliform Bacteria," *Water Environment Research,* vol. 72, no. 4.

Emerick, R. W., F. J. Loge, D. Thompson, and J. L. Darby (1999) "Factors Influencing Ultraviolet Disinfection Performance Part II: Association of Coliform Bacteria with Wastewater Particles," *Water Environment Research,* vol. 71, no. 6.

Enslow, L. H. (1938) "Chapter VII Chlorine in Sewage Treatment," in L. Pearse, ed., *Modern Sewage Disposal,* Federation of Sewage Works Associations, New York.

EPRI (1996) "UV Disinfection for Water and Wastewater Treatment," Report CR-105252, Electric Power Research Institute, Inc., Report prepared by Black & Veatch, Kansas City, MO.

Fair, G. M., J. C. Morris, S. L. Chang, I. Weil, and R. P. Burden (1948) "The Behavior of Chlorine as a Water Disinfectant," *Journal American Water Works Association,* vol. 40, p. 1051.

Finch, G. R., and D. W. Smith (1989) "Ozone Dose-Response of *Escherichia coli* in Activated Sludge Effluent," *Water Research,* vol. 23, p. 1017.

Finch, G. R., and D. W. Smith (1990) "Evaluation of Empirical Process Design Relationships for Ozone Disinfection of Water and Wastewater," *Ozone Science Engineering,* vol. 12, p. 157.

Fogler, H. S. (1999) *Elements of Chemical Reaction Engineering,* 3d ed., Prentice-Hall, Englewood Cliffs, NJ.

Gard, S. (1957) "Chemical Inactivation of Viruses," in *CIBA Foundation Symposium on the Nature of Viruses,* Little, Brown and Company, Boston, MA.

Gehr, R. (2000) Seminar lecture notes, Mexico City.

Glaze, W. H., et al. (1987) "The Chemistry of Water Treatment Process Involving Ozone, Hydrogen Peroxide, and Ultraviolet Radiation," *Ozone Science Engineering,* vol. 9, no. 4, p. 335.

Hart, F. L. (1979) "Improved Hydraulic Performance of Chlorine Contact Chambers," *Journal Water Pollution Control Federation,* vol. 51, no. 12, pp. 2868–2875.

Haas, C. N., and J. Joffe (1994) "Disinfection under Dynamic Conditions: Modification of Hom's Model for Decay," *Environmental Science and Technology,* vol. 28, no. 7, pp. 1367–1369.

Haas, C. N., and S. B. Kara (1984) "Kinetics of Microbial Inactivation by Chlorine: I. Review of Results in Demand-Free System," *Water Research* (G.B.) vol. 18, p. 1443.

Ho, K. W. A., and P. Bohm (1981) "UV Disinfection of Tertiary and Secondary Effluents," *Journal Water Pollution Research* (Canadian), vol. 16, no. 33.

Hom, L. W. (1972) "Kinetics of Chlorine Disinfection in an Eco-System," *Journal of Environmental Engineering Division,* American Society of Civil Engineers, vol. 98, SA1, pp. 183–194.

Jacob, S. M., and J. S. Dranoff (1970) "Light Intensity Profiles in a Perfectly Mixed Photoreactor," *Journal American Institute of Chemical Engineering,* vol. 16, p. 359.

Jagger, J. H. (1967) *Introduction to Research in UV Photobiology,* Prentice-Hall, Englewood Cliffs, NJ.

Larson, R., and B. Faber (2000) *Elementary Statistics,* Prentice-Hall, Upper Saddle River, NJ.

Lazarova, V., M. L. Janex, L. Fiksdal, C. Oberg, I. Barcina, and M. Ponimepuy (1998) "Advanced Wastewater Disinfection Technologies: Short and Long Term Efficiency," *Water Science Technology,* vol. 38, no. 12, pp. 109–117.

Lefevre, F., J. M. Audic, and F. Ferrand (1992) "Peracetic Acid Disinfection of Secondary Effluents Discharged Off Coastal Seawater," *Water Science Technology,* vol. 25, no. 12, pp. 155–164.

Liberti, L., A. Lopez, and M. Notarnicola (1999) "Disinfection with Peracetic Acid for Domestic Sewage Re-use in Agriculture," *Journal Water Environment Management* (Canadian) vol. 13, no. 8, pp. 262–269.

Loge, F., J. L. Darby, and G. Tchobanoglous (1996*b*) "UV Disinfection of Wastewater: A Probablistic Approach to Design," *Journal of Environmental Engineering Division,* American Society of Civil Engineers, vol. 122, no. 12, pp. 1078–1084.

Loge, F. J., R. W. Emerick, M. Heath, J. Jacangelo, G. Tchobanoglous, and J. L. Darby (1996*a*) "Ultraviolet Disinfection of Wastewater Secondary Effluents: Prediction of Performance and Design," *Water Environment Research,* vol. 68, no. 5, pp. 900–916.

Louie, D., and M. Fohrman (1968) "Hydraulic Model Studies of Chlorine Mixing and Contact Chambers," *Journal Water Pollution Control Federation,* vol. 40, no. 2.

Malcolm Pirnie and HDR Engineering (1991) *Guidance Manual for Compliance with the Filtration and Disinfection Requirements for Public Water Systems Using Surface Water Sources,* American Water Works Association.

Manglik, P. K., J. R. Johnston, T. Asano, and G. Tchobanoglous (1988) "Effect of Particles on Chlorine Disinfection of Wastewater," *Proceedings of Water Reuse Symposium IV Implementing Water Reuse, American Water Works Association Research Foundation,* Denver, CO.

Marske, D. M., and J. D. Boyle (1973) "Chlorine Contact Chamber Design—A Field Evaluation," *Water & Sewage Works,* vol. 120, no. 1, pp. 70–77.

Montgomery, J. M., Consulting Engineers, Inc. (1985) *Water Treatment Principles and Design,* Wiley, New York.

Morris, J. C. (1966) "The Acid Ionization Constant of HOCL from 5°C to 35°C," *Journal Physical Chemistry,* vol. 70, p. 3798.

NWRI (1993) *UV Disinfection Guidelines for Wastewater Reclamation in California and UV Disinfection Research Needs Identification,* National Water Research Institute Prepared for the California Department of Health Services, Sacramento, CA.

NWRI and AWWARF (2000) Ultraviolet Disinfection Guidelines for Drinking Water and Wastewater Reclamation, NWRI-00-03, National Water Research Institute and American Water Works Association Research Foundation, Fountain Valley, CA.

O'Brien, W. J., G. L. Hunter, J. J. Rosson, R. A. Hulsey, and K. E. Carns (1996) "Ultraviolet System Design: Past, Present, and Future, Proceedings Disinfecting Wastewater for Discharge & Reuse," *Water Environment Federation,* Alexandria, VA.

Oliver, B. G., and E. G. Cosgrove (1975) "The Disinfection of Sewage Treatment Plant Effluents Using Ultraviolet Light," *Journal of Chemical Engineering* (Canadian), vol. 53, p. 170.

Parker, J. A., and J. L. Darby (1995) "Particle-Associated Coliform in Secondary Effluents: Shielding from Ultraviolet Light Disinfection," *Water Environment Research,* vol. 67, p. 1065.

Pearse, L., et al. (1934) *Chlorination in Sewage Treatment,* Report of Committee on Sewage Treatment, American Public Health Association (also in *Sewage Work Journal,* vol. 7, pp. 997–1108, 1935).

Qualls, R. G., and J. D. Johnson (1985) "Modeling and Efficiency of Ultraviolet Disinfection Systems," *Water Research* (British), vol. 8, p. 1039.

Qualls, R. G., M. P. Flynn, and J. D. Johnson (1983) "The Role of Suspended Particles in Ultraviolet Disinfection," *Journal Water Pollution Control Federation,* vol. 57, pp. 1006–1011.

Rice, R. G. (1996) *Ozone Reference Guide,* Prepared for the Electric Power Research Institute, Community Environment Center, St. Louis, MO.

Roberts, P. V., E. M. Aieta, J. D. Berg, and B. M. Chow (1980) "Chlorine Dioxide for Wastewater Disinfection: A Feasibility Evaluation," *Technical Report* 21, Civil Engineering Department, Stanford University, Stanford, CA.

Saunier, B. M. (1976) Kinetics of Breakpoint Chlorination and Disinfection, Ph.D. Thesis, University of California, Berkeley, CA.

Saunier, B. M., and R. E. Selleck (1976) "The Kinetics of Breakpoint Chlorination in Continuous Flow Systems," paper presented at the American Water Works Association Annual Conference, New Orleans, LA.

Scheible, O. K. (1987) "Development of a Rationally Based Design Protocol for the Ultraviolet Light Disinfection Process," *Journal Water Pollution Control Federation,* vol. 59, p. 25.

Selleck, R. E., B. M. Saunier, and H. F. Collins (1978) "Kinetics of Bacterial Deactivation with Chlorine," *Journal of Environmental Engineering Division,* American Society of Civil Engineers, vol. 104, EE6, pp. 1197–1212.

Severin, B. F. (1980) "Disinfection of Municipal Effluents with Ultraviolet Light," *Journal Water Pollution Control Federation,* vol. 52, p. 2007.

Severin, B. F, M. T. Suidan, and R. S. Engelbrecht (1983) "Kinetic Modeling of UV Disinfection of Water," *Water Research* (British), vol. 17, p. 1669.

Snyder, C. H. (1995) *The Extraordinary Chemistry of Ordinary Things,* 2d ed., Wiley, New York.

Solvay Interox (1997) Proxitane™ Peracetic Acid, Company Brochure.

Sung, R. D. (1974) Effects of Organic Constituents in Wastewater on the Chlorination Process, Ph.D. Thesis, Department of Civil Engineering, University of California, Davis, CA.

Tiedeman, W. D. (1927) "Efficiency of Chlorinating Sewage Tank Effluents," Engineering News Record, vol. 98, pp. 944–948.

Trussell, R. R., and J. L. Chao (1977) "Rational Design of Chlorine Contact Facilities," *Journal Water Pollution Control Federation,* vol. 49, no. 4, pp. 659–667.

U.S. EPA (1986) *Design Manual, Municipal Wastewater Disinfection,* U.S. Environmental Protection Agency, EPA/625/1-86/021, Cincinnati, OH.

U.S. EPA (1992) *User's Manual for UVDIS, Version 3.1, UV Disinfection Process Design Manual,* U.S. Environmental Protection Agency, EPA G0703, Risk Reduction Engineering Laboratory, Cincinnati, OH.

U.S. EPA (1999a) *Combined Sewer Overflow Technology Fact Sheet, Alternative Disinfection Methods,* U.S. Environmental Protection Agency, EPA832-F-99-033, Cincinnati, OH.

U.S. EPA (1999b) *Alternative Disinfectants and Oxidants Guidance Manual,* U.S. Environmental Protection Agency, EPA 815-R-99-014, Cincinnati, OH.

Venosa, A. D., et al. (1978) "Comparative Efficiencies of Ozone Utilization and Microorganism Reduction in Different Ozone Contactors," in A. D. Venosa (ed.), *Progress in Wastewater Disinfection Technology,* U.S. Environmental Protection Agency, EPA-600/9-79-018, Municipal Environment Research Laboratory, Cincinnati, OH.

Ward, R. W. et al. (1976) *Disinfection Efficiency and Residual Toxicity of Several Wastewater Disinfectants,* EPA-600/2-76-156, U.S. Environmental Protection Agency, Cincinnati, OH.

Watson, H. E. (1908) "A Note on the Variation of the Rate of Disinfection with Change in the Concentration of the Disinfectant," *Journal of Hygiene* (British), vol. 8.

WEF (1996) *Wastewater Disinfection,* Manual of Practice FD-10, Water Environment Federation, Alexandria, VA.

White, G. C. (1999) *Handbook of Chlorination and Alternative Disinfectants,* 4th ed., A Wiley-Interscience Publication, John Wiley & Sons, Inc., New York.

13 Water Reuse

Continued population growth, contamination of both surface water and groundwater, uneven distributions of water resources, and periodic droughts have forced water agencies to search for new sources of water supply. Use of highly treated wastewater effluent, now discharged to the environment from municipal wastewater treatment plants, is receiving more attention as a reliable water resource. In many parts of the country, water reuse is already an important element in water resources planning and implementation. While water reuse is a viable option, water conservation, efficient use of existing water supplies, and new water resources development and management are other alternatives that must be evaluated.

The purpose of this chapter is to introduce the basic concepts and issues involved in water reclamation and reuse. Because water reuse is a rapidly evolving field, especially with respect to water quality, emerging issues are considered in the epilogue to this chapter. The chapter is organized in 10 sections dealing with: (1) a general introduction to the water reuse, including the definition of terms commonly used, (2) public health and environmental considerations in water reuse, (3) risk assessment, (4) a brief review of the principal treatment technologies used for water reclamation, (5) storage of reclaimed water, (6) agricultural and landscape irrigation, (7) industrial water reuse,

(8) groundwater recharge with reclaimed water, (9) potable water reuse, and (10) planning considerations for water reclamation and reuse. Separate sections are devoted to agricultural and landscape irrigation, industrial water reuse, and groundwater recharge because these are the largest and most important users of reclaimed water. Although the quantities of water used for potable reuse are limited, potable reuse is discussed separately because it is a hotly debated subject that encompasses important issues related to advanced wastewater treatment, public health, and public acceptance.

13-1 WASTEWATER RECLAMATION AND REUSE: AN INTRODUCTION

The purpose of this section is (1) to introduce the technical terms used in water reuse practices, (2) to present the role of water reclamation and reuse in the context of the hydrologic cycle, (3) to provide a brief historical perspective on the subject of water reclamation and reuse, (4) to provide an introduction to the principal water reuse applications, and (5) to review briefly the need and status of water reuse.

Definition of Terms

Terms used frequently in the field of water reclamation, recycling, and reuse are summarized in Table 13–1. It should be noted that the definitions given in Table 13–1 have evolved from several water reuse regulations and in response to questions raised by consumers and the public at large.

The Role of Water Recycling in the Hydrologic Cycle

The inclusion of planned water reclamation, recycling, and reuse in water resource systems reflects the increasing scarcity of water sources to meet societal demands, technological advancements, increased public acceptance, and improved understanding of public health risks. As the link between wastewater, reclaimed water, and water reuse has become delineated more clearly, increasingly smaller recycle loops are possible (Asano and Levine, 1996). Traditionally, the hydrologic cycle has been used to represent the continuous transport of water in the environment. The water cycle consists of freshwater and saltwater surface resources, subsurface groundwater, water associated with various land use functions, and atmospheric water vapor. Many subcycles to the hydrologic cycle exist, including the engineered transport of water. Water reclamation, recycling, and reuse represent significant components of the hydrologic cycle in urban, industrial, and agricultural areas. A conceptual overview of the cycling of water from surface and groundwater resources to water treatment facilities, irrigation, municipal, and industrial applications, and to water reclamation and reuse facilities is shown on Fig. 13–1.

The major pathways of water reuse include irrigation, industrial use, surface water replenishment, and groundwater recharge. Surface water replenishment and groundwater recharge also occur through natural drainage and through infiltration of irrigation and stormwater runoff. The potential use of reclaimed water for a potable water source is also shown on Fig. 13–1. The quantity of water transferred via each pathway depends on the watershed characteristics, climatic and geohydrologic factors, the degree of water utilization for various purposes, and the degree of direct or indirect water reuse.

The water used or reused for agricultural and landscape irrigation includes agricultural, residential, commercial, and municipal applications. Industrial reuse is a general

Table 13-1
Definition of terms used in water reuse applications

Term	Description
Beneficial uses	The many ways water can be used, either directly by people, or for their overall benefit. Examples include municipal water supply, agricultural and industrial applications, navigation, fish and wildlife, and water contact recreation
Direct potable reuse	A form of reuse by the incorporation of reclaimed water directly into a potable water supply system, often implying the blending of reclaimed water with potable water
Direct reuse	The use of reclaimed water which has been transported from the wastewater reclamation plant to the water reuse site without intervening discharge to a natural body of water, including such uses as agricultural and landscape irrigation
Dual distribution system	Two sets of pipelines for water delivery, one for potable water and another for reclaimed water
Indirect potable reuse	Potable reuse by incorporation of reclaimed water into a raw water supply, allowing mixing and assimilation by discharge into an impoundment or natural body of water, such as in a domestic water supply reservoir or groundwater
Indirect reuse	Use of reclaimed water indirectly by passing through a natural body of water or use of groundwater that has been recharged with reclaimed water
Nonpotable water reuse	All reuse applications that do not involve either indirect or direct potable use
Planned reuse	Deliberate direct or indirect use of reclaimed water, without relinquishing control over the water during its delivery
Potable water reuse	An augmentation of drinking water supplies directly or indirectly by reclaimed water that is highly treated to protect public health
Reclaimed water	Water that, as a result of wastewater treatment, is suitable for a direct beneficial use or a controlled use that would not otherwise occur
Recycled water	Reclaimed water that has been used beneficially. The term recycled water is used synonymously with reclaimed water
Unplanned reuse	Reuse of treated wastewater following discharge (control relinquished), such as in the diversion of water from a river downstream of a discharge of treated wastewater
Water reclamation	Treatment or processing of wastewater to make it reusable. Also, this term is often used to include delivery of reclaimed water to place of use and its actual use
Water recycling	The use of wastewater that is captured and redirected back into the same water use scheme. Recycling is practiced predominantly in industries, such as manufacturing, and normally involves only one industrial plant or one user
Water reuse	The use of treated wastewater for a beneficial use, such as agricultural irrigation and industrial cooling

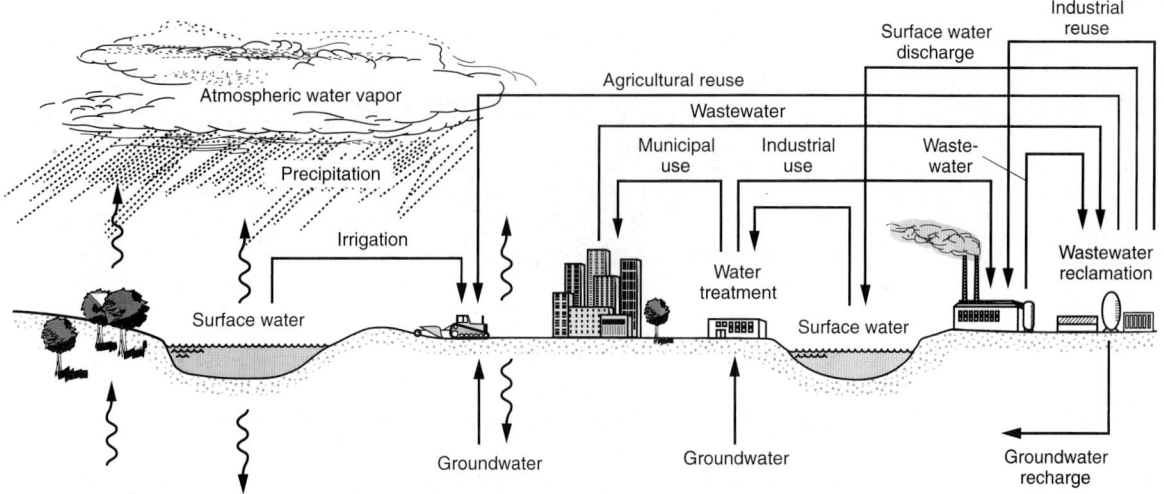

Figure 13–1

The role of engineered treatment, reclamation, and reuse facilities in the cycling of water through the hydrologic cycle. *(Adapted from Asano and Levine, 1996.)*

category encompassing water use for a diversity of industries that include power plants, pulp and paper, and other industries with high rates of water utilization. In some cases, closed-loop recycle systems have been developed that treat water from a single process stream and recycle the water back to the same process with some makeup water. In other cases, reclaimed municipal water is used for industrial purposes such as in cooling towers. The National Aeronautical and Space Administration has studied closed-loop systems for long-duration space missions and space stations.

How small this water-recycling loop should be will depend on public health, engineering, economics, aesthetics, and more importantly, public acceptance. Planned water reclamation and reuse have gained considerable attention worldwide in recent decades as an alternative and new water resource in the context of integrated water resources management.

Historical Perspective

Early developments in the field of water reuse are synonymous with the historical practice of land application for the disposal of wastewater. With the advent of sewerage systems in the nineteenth century, domestic wastewater was used at *sewage farms* and by 1900 there were numerous sewage farms in Europe and in the United States. While these sewage farms were used primarily for waste disposal, incidental use was made of the water for crop production or other beneficial uses (Veatch, 1938).

During the past century, a number of water reclamation and reuse projects have been developed as a matter of necessity to meet growing need for reliable water. Selected examples of historical development of water reuse in the United States are presented in Table 13–2.

Table 13-2

Selected examples of historic development of water reuse in the United States and other parts of the world

Year	Location	Water reuse application
United States		
1912–1985	Golden Gate Park, San Francisco, CA	Watering lawns and supplying ornamental lakes
1926	Grand Canyon National Park, AZ	Toilet flushing, lawn sprinkling, cooling water, and boiler feed water
1929	City of Pomona, CA	Irrigation of lawns and gardens
1942	City of Baltimore, MD	Metals cooling and steel processing at the Bethlehem Steel Company
1960	City of Colorado Springs, CO	Landscape irrigation for golf courses, parks, cemeteries, and freeways
1961	Irvine Ranch Water District, CA	Irrigation and domestic water including toilet flushing in high-rise buildings
1962	County Sanitation Districts of Los Angeles County, CA	Groundwater recharge using spreading basins at the Montebello Forebay
1976	Orange County Water District, CA	Groundwater recharge by direct injection into the aquifers at the Water Factory 21
1977	City of St. Petersburg, FL	Irrigation of parks, golf courses, school yards, residential lawns, and cooling tower makeup water
1985	City of El Paso, TX	Groundwater recharge by direct injection into the Hueco Bolson aquifers, and power plant cooling water
1987	Monterey Regional Water Pollution Control Agency, Monterey, CA	Agricultural irrigation of food crops eaten uncooked including artichoke, celery, broccoli, lettuce, and cauliflower
Outside of the United States		
1890	Mexico City, Mexico	Drainage canals were built to take untreated wastewater to irrigate an important agricultural area north of the city, a practice that still continues today. Untreated or minimally treated wastewater from Mexico City is delivered to the Valley of Mexico where it is used to irrigate about 90,000 ha of agricultural lands
1962	La Soukra, Tunisia	Irrigation with reclaimed water for citrus plants and groundwater recharge to reduce saltwater intrusion into groundwater
1968	City of Windhoek, Namibia	Advanced wastewater reclamation system to augment potable water supplies
1969	City of Wagga Wagga, Australia	Landscape irrigation of sporting fields, lawns, and cemeteries
1977	Dan Region Project, Tel-Aviv, Israel	Groundwater recharge via basins. Pumped groundwater is transferred via a 100-km-long conveyance system to southern Israel for unrestricted crop irrigation
1984	Tokyo Metropolitan Government, Japan	Water recycling project in Shinjuku District of Tokyo providing reclaimed water for toilet flushing in 19 high-rise buildings in highly congested metropolitan area
1989	Consorci de la Costa Brava, Girona, Spain	Golf course irrigation

Table 13–3
Estimates of water reclamation in the United States in 1995[a]

Category	Quantity	
	Mgal/d	**Mm³/d**
Public supply	57	0.2157
Commercial	19	0.0719
Industrial	110	0.4164
Thermoelectric	100	0.3785
Mining	14	0.0530
Irrigation	718	2.7176
Total	1018	3.8531

[a]Adapted from Solley et al. (1998).

According to the most recent survey on estimated use of water in the United States, the use of reclaimed water in 1995 was 3.85×10^6 m³/d (1018 Mgal/d), which was a 36 percent increase in 5 years (Solley et al., 1998). Estimates of water reclamation in the United States in 1995 by water use category are shown in Table 13–3. Most of the water reuse sites are located in the arid and semiarid western and southwestern states, including Arizona, California, Colorado, Nevada, Texas, and Utah. However, an increasing number of water reuse projects are being implemented in the humid regions of the United States, which include Florida, Maryland, and Missouri, for water pollution abatement as well as for water supply benefits. Because of costs of treatment, health issues, and safety concerns, water reuse applications have been limited primarily to nonpotable uses such as agricultural and landscape irrigation. However, where no other possibilities exist for expanding freshwater supplies, some communities are continuing to investigate and evaluate the potential for indirect and direct potable reuse options. While the quantities of reclaimed water involved in these potable water reuse projects are small, the technological, public health, aesthetic, and public acceptance issues are of fundamental importance and are a greater challenge in water reuse than in drinking water supply.

Wastewater Reuse Applications

In the planning and implementation of water reclamation and reuse, the reclaimed water application will usually govern the wastewater treatment needed to protect public health and the environment, and the degree of reliability required for the treatment processes and operations. The seven principal categories of municipal wastewater reuse are listed in Table 13–4 in descending order of projected volume of use along with potential constraints for their application.

- The first category, agricultural irrigation, is the largest current use of reclaimed water (see Fig. 13–2). This reuse category offers significant future opportunities for water reuse. Reviewing the base data for agricultural irrigation, California is by far the largest user of irrigation water, withdrawing about 110 million m³/d (29 Bgal/d), 22 percent of the national total. California, together with Idaho, Colorado, Texas, and Montana, accounts for 54 percent of the national total (Solley et al., 1998).

Table 13–4
Categories of municipal wastewater reuse and potential issues/constraints

Wastewater reuse categories	Issues/constraints
1. Agricultural irrigation 　Crop irrigation 　Commercial nurseries	• Surface and groundwater contamination if not managed properly • Marketability of crops and public acceptance
2. Landscape irrigation 　Parks 　School yards 　Freeway medians 　Golf courses 　Cemeteries 　Greenbelts 　Residential	• Effect of water quality, particularly salts, on soils and crops • Public health concerns related to pathogens (e.g., bacteria, viruses, and parasites) • Use area control including buffer zone may result in high user costs
3. Industrial recycling and reuse 　Cooling water 　Boiler feed 　Process water 　Heavy construction	• Constituents in reclaimed water related to scaling, corrosion, biological growth, and fouling • Public health concerns, particularly aerosol transmission of pathogens in cooling water • Cross connection of potable and reclaimed water lines
4. Groundwater recharge 　Groundwater replenishment 　Saltwater intrusion control 　Subsidence control	• Possible contamination of groundwater aquifer used as a source of potable water • Organic chemicals in reclaimed water and their toxicological effects • Total dissolved solids, nitrates, and pathogens in reclaimed water
5. Recreational/environmental uses 　Lakes and ponds 　Marsh enhancement 　Stream-flow augmentation 　Fisheries 　Snowmaking	• Health concerns related to presence of bacteria and viruses (e.g., enteric infections and ear, eye, and nose infections) • Eutrophication due to nitrogen and phosphorus in receiving water • Toxicity to aquatic life
6. Nonpotable urban uses 　Fire protection 　Air conditioning 　Toilet flushing	• Public health concerns about pathogens transmitted by aerosols • Effects of water quality on scaling, corrosion, biological growth, and fouling • Cross connection of potable and reclaimed water lines
7. Potable reuse 　Blending in water supply reservoirs 　Pipe-to-pipe water supply	• Constituents in reclaimed water, especially trace organic chemicals and their toxicological effects • Aesthetics and public acceptance • Health concerns about pathogen transmission, particularly enteric viruses

Figure 13–2

Typical irrigation of agricultural crops using reclaimed water: (a) artichokes using moveable impact-type sprinkler system, (b) orchard using fixed impact-type sprinkler system, (c) animal fodder using moveable water gun sprinkler system, and (d) experimental eucalyptus grove using impact-type sprinkler system (Australia). The experimental eucalyptus grove (d) is instrumented to assess the effect of nitrogen and other nutrients in reclaimed water.

- The second category, landscape irrigation, includes the irrigation of parks; playgrounds; golf courses (see Fig. 13–3); freeway medians; landscaped areas around commercial, office, and industrial developments; and landscaped areas around residences. Many landscape irrigation projects involve dual distribution systems—one distribution network for potable water and another for reclaimed water.

- The third major use of reclaimed water is in industrial activities, primarily for cooling and process needs. Cooling water is the predominant industrial water reuse and, for either cooling towers or cooling ponds, creates the single largest demand for water in many industries (see Fig. 13–22 in Sec. 13–7). Industrial uses vary greatly, and to provide adequate water quality, additional treatment is often required beyond conventional secondary wastewater treatment.

Figure 13–3

Typical golf course irrigated with reclaimed water.

- The fourth reuse application for reclaimed water is groundwater recharge, via either spreading basins (see Fig. 13–4) or direct injection to groundwater aquifers. Groundwater recharge involves assimilation of reclaimed water for replenishment, storage in groundwater aquifers, or establishing hydraulic barriers against saltwater intrusion in coastal areas.

- A fifth use of reclaimed water, recreational/environmental uses, involves a number of nonpotable uses related to land-based water features such as the development of recreational lakes, marsh enhancement, and stream-flow augmentation (see Fig. 13–5). Reclaimed water impoundments can be incorporated into urban landscape developments. Man-made lakes, golf course storage ponds, and water traps can be supplied with reclaimed water. Storage of reclaimed water is discussed in Sec. 13–5. Examples of recreational impoundments include Las Colinas, TX; Santee, CA; and Lubbock, TX. The use of reclaimed water in an environmental water feature is illustrated on Fig. 13–6. Reclaimed water has been applied to wetlands for a variety of reasons including creation, restoration, and enhancement of habitat, provision for additional treatment prior to discharge to receiving water, and provision for a wet weather disposal alternative for reclaimed water.

- The sixth reuse category, nonpotable urban uses, includes such uses as fire protection, air conditioning, toilet flushing, construction water, and flushing of sanitary sewers. Typically, for economic reasons, these uses are incidental depending on the location of the wastewater reclamation plant and whether these applications can be coupled with other ongoing reuse applications such as landscape irrigation.

- The seventh reuse opportunity involves potable reuse, which could occur by blending in water supply storage reservoirs or, in the extreme, by direct input into the water distribution system (i.e., so called "pipe-to-pipe" potable reuse).

While potentially large quantities of reclaimed municipal wastewater can be used in the first five categories, the quantities associated with the sixth and seventh reuse categories are minor at present; particularly potable water reuse.

Need for Water Reuse

To understand the potential and significance of water reuse, it is helpful to compare reclaimed water use with freshwater withdrawals on a national scale. The estimates of

Figure 13–4

Rio Hondo surface spreading basins in southern California. The spreading basins cover an area of 231 ha (570 ac). Water to be recharged is applied to the 20 shallow basins over a period of 7 days. The basins are then drained for 7 days and dried for an additional 7 days before being put back into service. *(Photo courtesy of County Sanitation Districts of Los Angeles County.)*

Figure 13–5

Stream flow augmentation with reclaimed water. During the summer months, the flow in the river shown in the photo is comprised essentially of reclaimed water. Streams in which a large fraction of the flow is reclaimed water are known as *effluent dominated waterways*.

Figure 13–6

View of the stilling pond maintained with reclaimed wastewater at the Japanese Garden located adjacent to the Tillman Water Reclamation Plant in Los Angeles, CA.

Table 13-5
Estimates of freshwater withdrawals and water reclamation in the United States in years 1975, 1985, and 1995[a]

Category	Quantity, Mm³/d (Bgal/d)		
	1975	**1985**	**1995**
Total freshwater withdrawals:			
Ground	310.37 (82)	277.06 (73.2)	289.17 (76.4)
Surface	984.10 (260)	1003.03 (265)	999.24 (264)
Reclaimed water	1.93 (0.5)	2.27 (0.6)	3.86 (1.02)

[a]Adapted from Solley et al. (1998).

freshwater withdrawals and quantity of reclaimed water used are reported in Table 13–5. The ratio of municipal wastewater reclamation and reuse to total freshwater withdrawals is expected to remain low in the near future. However, because the reclaimed water is a locally controllable water resource, the actual quantity of water reuse will increase significantly (see Table 13–5).

In recent years, the desirability and benefits of water reuse have been well recognized by several states. For example, in the California State Water Code it is clearly noted that "it is the intention of the Legislature that the State undertake all possible steps to encourage development of water recycling facilities so that recycled water may be made available to help meet the growing water requirements of the State." Today, technically proven wastewater treatment processes exist that can be used to produce water of almost any quality for water reuse. Thus, water reuse has a rightful place and has an important role in the planning and efficient use of water resources. However, public concerns about public health and environmental issues will need to be addressed in a more rigorous and transparent manner than has occurred in the past.

13-2 PUBLIC HEALTH AND ENVIRONMENTAL ISSUES IN WATER REUSE

Despite the existence of technically proven advanced wastewater treatment processes, as discussed in Chap. 11, long-term *safety* of reclaimed water and the *impact* on the environment are still difficult to quantify. Given that it is possible to produce water of almost any quality, public health and environmental issues that now must be addressed are: what constituents must be removed and to what extent must they be removed? These questions are of even greater significance in developing countries that are experiencing extreme water shortages and where high levels of treatment are not economically feasible. The purpose of this section is (1) to evaluate the impact of new scientific findings and knowledge concerning the constituents in water, (2) to examine the public health significance and environmental impact of the constituents in treated wastewater that is to be reused, and (3) to review existing water reuse criteria and guidelines.

Constituents in Reclaimed Water

The constituents in municipal wastewater subject to treatment may be classified as conventional, nonconventional, and emerging. Typical constituents included under each

Table 13–6
Classification of typical constituents found in wastewater

Classification	Constituent
Conventional	Total suspended solids
	Colloidal solids
	Biochemical oxygen demand
	Chemical oxygen demand
	Total organic carbon
	Ammonia
	Nitrate
	Nitrite
	Total nitrogen
	Phosphorus
	Bacteria
	Protozoan cysts and oocysts
	Viruses
Nonconventional	Refractory organics
	Volatile organic compounds
	Surfactants
	Metals
	Total dissolved solids
Emerging	Prescription and nonprescription drugs[a]
	Home care products
	Veterinary and human antibiotics
	Industrial and household products
	Sex and steroidal hormones
	Other endocrine disrupters

[a] Pharmaceutically active substances.

category are described in Table 13–6. Expected concentrations in the reclaimed water after various levels of treatment are reported later in this chapter in Tables 13–16 through 13–18. The term *conventional* is used to define those constituents measured in mg/L that have served as the basis for the design of most conventional wastewater treatment plants. *Nonconventional* applies to those constituents that may have to be removed or reduced using advanced wastewater treatment processes before the water can be used beneficially. The term *emerging* is applied to those classes of compounds measured in the micro- or nanogram/L range that may pose long-term health concerns and environmental problems as more is known about the compounds. In some cases, these compounds cannot be removed effectively, even with advanced treatment processes.

Public Health Issues

Over the past 10 years, a variety of new analytical techniques have been developed (see Chap. 2) that can be used to measure extremely low concentrations (e.g., 10^{-12} g/m^3 or ng/L) of an ever-increasing variety of inorganic and organic constituents in water. As a consequence, hundreds of organic compounds have been detected in untreated and treated wastewater at relatively low concentrations. Further, it has been found that the effectiveness of conventional and advanced wastewater treatment processes varies considerably with respect to removal of these individual compounds.

For most of the *emerging* compounds listed in Table 13–6, there is little or no information concerning health or environmental effects. Unfortunately, some of the compounds that have been identified in reclaimed water are known to have both acute and chronic health effects, depending on their concentrations and exposure pathways. The dilemma posed by the presence of these trace amounts of organic compounds in reclaimed water is that the perceived or potential adverse health effects that may be caused by these compounds have not been quantified fully in current water reuse criteria and guidelines or drinking water standards. Where it is desired to conserve water resources via conservation and water reuse rather than to develop environmentally costly new water resources, the emphasis in the evaluation of water reuse options will be on public health protection, treatment process reliability, value of water, and public policy.

Environmental Issues

Environmental risks associated with previously unknown or unrecognized chemicals in treated effluents have long been a concern for environmental scientists and engineers. Ternes (1998), for example, investigated the occurrence of 32 drug residues in German municipal wastewater treatment plants. In general, more than 60 percent of the drug residues that were detected at parts per billion (ppb) or parts per trillion (ppt) levels in the wastewater influent were removed in the activated-sludge process. The long-term environmental effects of these compounds found in treated effluent are, for the most part, not well understood, and the true significance of these trace quantities of constituents in the aquatic environment and in water supplies is still an issue of concern. The most troubling finding is that the concentrations of these compounds that cause environmental effects to aquatic organisms are considerably lower than the concentrations that cause any measurable effects in humans.

The Evolution of Water Reuse Guidelines in the United States

To protect public health, considerable efforts have been made to establish conditions and regulations that would allow for safe use of reclaimed water for a variety of water reuse applications (see Table 13–4). Early guidelines and, to a large extent, current water reuse criteria and guidelines are based on the control of conventional parameters such as BOD, TSS, and coliforms used for the design and monitoring of wastewater treatment plants (Ongerth and Ongerth, 1982). Thus, most current criteria and guidelines do not include pathogenic bacteria such as *Salmonella* and enteric viruses such as *Hepatitis,* emerging pathogens, or the trace organic compounds discussed above.

Federal Guidelines. The U.S. Environmental Protection Agency (U.S. EPA, 1992) has suggested reclaimed water quality guidelines for the following reuse categories:

• Urban reuse

• Restricted-access-area irrigation

• Agricultural reuse—food crops

• Agricultural reuse—nonfood crops

• Recreational impoundments

• Landscape impoundments

• Construction uses

• Industrial reuse

• Groundwater recharge

• Indirect potable reuse

For each reuse category, levels of treatment, minimum reclaimed water quality, reclaimed water monitoring, and setback distances are suggested (U.S. EPA, 1992). The guidelines are summarized in Table 13–7 for the two principal levels of treatment—disinfected tertiary (filtered secondary effluent) and disinfected secondary effluents.

State Guidelines and Regulations. Regulations or guidelines adopted by individual states are summarized in Table 13–8. The listing includes 18 states with regulations and 18 states with guidelines on water reuse (U.S. EPA, 1992*a*). Both the regulations and the guidelines vary considerably from state to state. States such as California, Arizona, Florida, and Texas have developed comprehensive regulations that strongly encourage water reuse as an alternative water resource and a water conservation strategy. In these four states regulations have been developed for water quality requirements, treatment processes, and monitoring and reporting for the full spectrum of water reuse applications. California water reuse regulations have been developed progressively since 1918, and are the most comprehensive with regard to public health. In December 2000, the state of California revised the *Water Recycling Criteria* (Title 22 regulations), which are summarized in Table 13–9 (State of California, 2000). The representative uses and application methods are listed in Table 13–10.

Assessment of Water Reuse Guidelines. The goal of essentially virus-free reclaimed water (i.e., disinfected tertiary) contained in the *California Water Recycling Criteria* (see Tables 13–9 and 13–10) should not be interpreted to mean that the practice of using such water is risk-free. There is always a statistical probability of infection due to the exposure to reclaimed water. Because of the paucity of microbiological and infectious disease data, difficulties exist in correlating causative agents and the disease outcome and its epidemiological significance. However, the practice of water reclamation and reuse cannot be construed as *unsafe* compared to other sources of available water, such as polluted river and irrigation water. The *safety* of water reuse practice is defined by the acceptable level of risks developed by the regulatory agencies responsible for

Table 13–7
Summary of EPA suggested guidelines for water reuse[a]

Level of treatment	Types of reuse	Reclaimed water quality	Reclaimed water monitoring	Setback distances
1. Disinfected tertiary[b]	Urban reuse[c] Food crop irrigation Recreational impoundments	pH = 6–9 $BOD_5 \leq 10$ mg/L Turb. ≤ 2 NTU E. coli = none Res. $Cl_2 \geq 1$ mg/L	pH = weekly BOD = weekly Turb. = cont. E. coli = daily Res. Cl_2 = cont.	15 m (50 ft) to potable water supply wells[d]
2. Disinfected secondary	Restricted access area irrigation Food crop irrigation (commercially processed) Nonfood crop irrigation Landscape impoundments (restricted access) Construction Wetlands habitat	pH = 6–9 $BOD_5 = 30$ mg/L TSS = 30 mg/L E. coli = 200/100 mL Res. $Cl_2 \geq 1$ mg/L	pH = weekly BOD = weekly TSS = daily E. coli = daily Res. Cl_2 = cont.	30 m (100 ft) to areas accessible to the public (if spray irrigation) 90 m (300 ft) to potable water supply well

[a] From U.S. EPA (1992a).
[b] Filtration of secondary effluent.
[c] Uses include landscape irrigation, vehicle washing, toilet flushing, use in fire protection, and commercial air conditioners.
[d] Setback increases to 150 m (500 ft) if impoundment bottom is not sealed.

health and environmental risk management and endorsed by the public for the necessity of such undertaking in the integrated water resources management.

Although the *California Water Recycling Criteria* reported in Table 13–9 lack explicit epidemiological evidence to assess fully the associated health risks in water reuse practices, the criteria have been in effect since 1978 as attainable and enforceable regulations. Studies conducted using the available enteric virus monitoring data in California and quantitative microbiological risk assessment methodology appear to prove the validity of the *California Water Recycling Criteria* (Asano et al., 1992; Tanaka et al., 1998). Additional safety measures that have been used for nonpotable water reuse applications include (1) installation of separate storage and distribution systems for potable water, (2) use of color-coded pipes (normally purple) and labels to distinguish reclaimed water from potable water distribution piping, (3) cross-connection and back-flow prevention devices, (4) periodic use of tracer dyes to detect the occurrence of cross contamination in potable supply lines, and (5) irrigation during off hours to further minimize the potential for human contact.

Table 13–8
Summary of state reuse regulations and guidelines[a]

State	Regulations or guidelines	Regulations or guidelines for each type			
		Unrestricted urban reuse	Restricted urban reuse	Nonfood crops	Food crops
Alabama	G			X	
Arizona	R	X	X	X	X
Arkansas	G	X	X	X	X
California	R	X	X	X	X
Colorado	G	X	X	X	X
Delaware	G	X	X	X	
Florida	R	X	X	X	X
Georgia	G	X	X	X	
Hawaii	G	X	X	X	X
Idaho	R	X	X	X	X
Illinois	R[b]	X	X	X	
Indiana	R			X	X
Kansas	G	X	X	X	X
Maryland	G		X	X	
Michigan	R			X	X
Missouri	R		X	X	
Montana	G	X	X	X	X
Nebraska	G		X	X	
Nevada	G[a]	X	X	X	X
New Jersey	R			X	
New Mexico	G	X	X	X	X
New York	G			X	
North Carolina	R		X		
North Dakota	G			X	
Oklahoma	G		X	X	
Oregon	R	X	X	X	X
South Carolina	G	X	X	X	
South Dakota	G	X	X	X	
Tennessee	R	X	X	X	
Texas	R	X	X	X	X
Utah	R	X	X	X	X
Vermont	R			X	
Washington	G	X	X	X	X
West Virginia	R			X	X
Wisconsin	R			X	
Wyoming	R	X	X	X	X

[a] From U.S. EPA (1992).
[b] Draft or proposed.

Water Reclamation Criteria in Other Countries

Reclaimed water quality criteria for protecting health in developing countries are often established in relation to the limited resources available for public works and other health delivery systems that may yield greater health benefits for the funds spent. Confined wastewater collection systems and wastewater treatment facilities are often non-existent, and reclaimed water often provides an essential water supply and fertilizer source. For most developing countries, the greatest concern with the use of wastewater for irrigation is that untreated or inadequately treated wastewater contains numerous intestinal nematodes (e.g., *Ascaris* and *Trichuris* species and hookworms) and bacterial pathogens that are difficult to control. These infectious agents, as well as enteric viruses, can damage the health of both the general public consuming the crops contaminated with wastewater and farmworkers and their families.

The World Health Organization (WHO, 1989) has recommended that irrigation of crops likely to be eaten uncooked, sports fields, and public parks should be irrigated with wastewater treated by a series of stabilization ponds. The ponds are designed to achieve a microbiological quality of less than or equal to 1 intestinal nematode per liter and fecal coliforms of less than or equal to 1000 per 100 mL (geometric mean). In comparison, California's criteria are significantly more stringent (see Table 13–9). The criteria recommended by WHO for irrigation with reclaimed water have been accepted as

Table 13–9
California water recycling criteria[a]

Category of reclaimed water	Criteria for:		Suitable uses
	Total coliform, MPN/100 mL	Turbidity, NTU	
Disinfected tertiary[b]	<2.2	2 average 5 maximum	All uses shown in Table 13–4
Disinfected secondary—2.2	<2.2	na	All uses shown in Table 13–4 except irrigation of parks and playgrounds,[c] food crops contacted by reclaimed water, nonrestricted impoundments
Disinfected secondary—23	<23	na	Same restrictions as disinfected secondary—2.2, except no food crop irrigation, no nonrestricted impoundment, and no watering of yards
Undisinfected secondary[d]	na	na	Drip or surface irrigation of fodder, fiber, seed orchard, and tree crops and sugar beets (commercially processed food crops)

[a] From California Code of Regulations, Title 22, Division 4, Chapter 3 Water Recycling Criteria, Sections 60301 et seq., Dec. 2, 2000.
[b] Filtered through natural undisturbed soils or filter media, such as sand or diatomaceous earth.
[c] Urban areas such as parks, playgrounds, school yards, residential yards, and golf courses associated with residences.
[d] Undisinfected wastewater means wastewater in which the organic matter has been stabilized, is nonputrescible, and contains dissolved oxygen.
na = not applicable.

Table 13–10

Representative uses and application methods for reclaimed water in California

General use	Conditions in which use is allowed[a]			
	Disinfected tertiary reclaimed water	**Disinfected secondary— 2.2 reclaimed water**	**Disinfected secondary— 23 reclaimed water**	**Undisinfected secondary reclaimed water**
All water uses other than potable use or food preparation;[b] and other than groundwater recharge (governed by other regulations)	Allowed[b]	Not allowed	Not allowed	Not allowed
Irrigation of:				
Parks, playgrounds, school yards, residential yards, and golf courses associated with residences	Spray, drip, or surface	Not allowed	Not allowed	Not allowed
Restricted-access golf courses, cemeteries, freeway landscapes	Spray, drip, or surface	Spray, drip, or surface	Spray, drip, or surface	Not allowed
Nonedible vegetation at other areas with limited public exposure[c]	Spray, drip, or surface	Spray, drip, or surface[c]	Spray, drip, or surface[c]	Not allowed
Sod farms	Spray, drip, or surface	Spray, drip, or surface	Spray, drip, or surface	Not allowed
Ornamental plants for commercial use	Spray, drip, or surface	Spray, drip, or surface	Spray, drip, or surface	Not allowed
All food crops	Spray, drip, or surface	Not allowed	Not allowed	Not allowed
Food crops that are above ground and not contacted by reclaimed water	Spray, drip, or surface	Drip or surface	Not allowed	Not allowed
Pasture for milking animals and other animals	Spray, drip, or surface	Spray, drip, or surface	Spray, drip, or surface	Not allowed
Fodder (e.g., alfalfa), fiber (e.g., cotton), and seed crops not eaten by humans	Spray, drip, or surface	Spray, drip, or surface	Spray, drip, or surface	Drip or surface
Orchards and vineyards bearing food crops	Spray, drip, or surface	Drip or surface	Drip or surface	Drip or surface
Orchards and vineyards not bearing food crops during irrigation	Spray, drip, or surface	Spray, drip, or surface	Spray, drip, or surface	Drip or surface
Christmas trees and other trees not bearing food crops	Spray, drip, or surface	Spray, drip, or surface	Spray, drip, or surface	Drip or surface
Food crop which must undergo commercial pathogen-destroying processing before consumption (e.g., sugar beets)	Spray, drip, or surface	Spray, drip, or surface	Spray, drip, or surface	Drip or surface

(continued)

| **Table 13-10** (Continued)

	Conditions in which use is allowed[a]			
General use	**Disinfected tertiary reclaimed water**	**Disinfected secondary— 2.2 reclaimed water**	**Disinfected secondary— 23 reclaimed water**	**Undisinfected secondary reclaimed water**
Impoundments:				
Supply for nonrestricted recreational impoundment	Allowed	Not allowed	Not allowed	Not allowed
Supply for restricted recreational impoundment	Allowed	Allowed	Not allowed	Not allowed
Supply for basins at fish hatcheries	Allowed	Allowed	Not allowed	Not allowed
Landscape impoundment without decorative fountain	Allowed	Allowed	Allowed	Not allowed
Other uses:				
Flushing toilets and urinals and priming drain	Allowed	Not allowed	Not allowed	Not allowed
Supply for commercial and public laundries for clothing and other linens	Allowed	Not allowed	Not allowed	Not allowed
Air conditioning and industrial cooling utilizing cooling towers[d]	Allowed	Not allowed	Not allowed	Not allowed
Industrial process with exposure of workers[e]	Allowed	Not allowed	Not allowed	Not allowed
Industrial cooling not utilizing cooling towers, spraying, or feature that creates aerosols or other mist	Allowed	Allowed	Allowed	Not allowed
Industrial process without exposure of workers	Allowed	Allowed	Allowed	Not allowed
Industrial boiler feed	Allowed	Allowed	Allowed	Not allowed
Fire fighting by dumping from aircraft	Allowed	Allowed	Allowed	Not allowed
Fire fighting other than by dumping from aircraft	Allowed	Not allowed	Not allowed	Not allowed
Water jetting for consolidation of backfill material around potable water pipelines during water shortages	Allowed	Not allowed	Not allowed	Not allowed
Water jetting for consolidation of backfill material around pipelines for reclaimed water, sewage, storm drainage, and gas and conduits for electricity	Allowed	Allowed	Allowed	Not allowed
Dampening soil for compaction at construction sites, landfills, and elsewhere	Allowed	Allowed	Allowed	Not allowed

(continued)

| **Table 13-10** (Continued)

General use	Conditions in which use is allowed[a]			
	Disinfected tertiary reclaimed water	**Disinfected secondary— 2.2 reclaimed water**	**Disinfected secondary— 23 reclaimed water**	**Undisinfected secondary reclaimed water**
Other uses:				
Washing aggregate and making concrete	Allowed	Allowed	Allowed	Not allowed
Dampening roads and other surfaces for dust control	Allowed	Allowed	Allowed	Not allowed
Dampening brushes and street surfaces in street sweeping	Allowed	Allowed	Allowed	Not allowed
Flushing sanitary sewers[f]	Allowed	Allowed	Allowed	Not allowed
Washing yards, lots, and sidewalks	Allowed	Allowed	Not allowed	Not allowed
Supply for decorative fountain	Allowed	Not allowed	Not allowed	Not allowed

[a] See California Code of Regulations, Title 22, Division 4, Chapter 3 Water Recycling Criteria, Sections 60301 et seq., Dec. 2, 2000.

[b] Disinfected tertiary effluent is suitable for all water uses that are not for potable use or food preparation, do not involve incorporation of reclaimed water into drink or food for humans, and do not conflict with provisions of the California Code of Regulations, federal regulations, statute, or other law.

[c] Disinfected secondary—2.2 reclaimed water and disinfected secondary—23 reclaimed water are suitable for irrigation of landscape vegetation and nonedible plants where (a) the public would have access and exposure to irrigation water similar to that which would occur at a golf course or cemetery, and (b) children do not have direct access and exposure to irrigation water. There is no concern regarding access and exposure when disinfected tertiary reclaimed water is used.

[d] The industrial process generates mist or could involve facial contact with reclaimed water.

[e] The industrial process does not generate mist or involve facial contact with reclaimed water.

[f] The regulatory agency may approve, for flushing sanitary sewers, the use of disinfected wastewater, notwithstanding the fact that the median concentration of total coliform bacteria is higher than that of disinfected secondary—23 reclaimed water. For example, suitable for such use is wastewater that is always disinfected so that the median concentration of total coliform bacteria does not exceed 240/100 mL.

an attainable goal for stabilization ponds in several Mediterranean and Latin American countries. In some Gulf Coast countries that have recently developed wastewater treatment facilities for water reuse, such as Abu Dhabi in the United Arab Emirates (whose population has grown from about 15,000 in the mid-1950s to almost one million in the late 1990s), more stringent water reuse criteria, similar to the California regulations, have been adopted. More stringent regulations were adopted to maintain an already high standard of health, and to transform the desert into a green environment, by planting numerous trees and shrubs for public parks and gardens.

What Level of Treatment Is Necessary?

Although water reuse guidelines were developed for protection of health, it is evident that they were based on the control of conventional parameters such as pH, BOD, TSS, and pathogenic organisms (see Tables 13–7 and 13–9). Protozoan pathogens, helminths, or the trace organic compounds, as discussed above, are not included in current water reuse regulations. It is clear, as delineated in Chap. 11 and in the following section, that water of high quality can be produced using a combination of advanced wastewater treatment processes. Thus, important questions that must be asked are (1) should the technology-based highest level of treatment that can be achieved be specified for water reuse applications, regardless of different exposure to reclaimed water in various applications, or (2) should the decision to use reclaimed water be based on risk assessment related to the exposure to reclaimed water for a specific application? Is the science sufficiently advanced to undertake such risk-based decisions in water reuse practice? Although there are no definite answers to these questions, the issues surrounding reclaimed water quality and the corresponding exposure risks must nevertheless be addressed if the *safety* of water reuse practice is to be debated on a rational basis, especially with respect to water quality, public health, liability, and public acceptance. In fact, the same questions must be addressed with regard to the treated wastewater now discharged to the environment.

13–3 INTRODUCTION TO RISK ASSESSMENT

When health effects can occur as the result of an environmental action, risk analysis is used to quantify the corresponding risks. Typically, a complete risk analysis is divided into two parts: (1) risk assessment and (2) risk management. Risk assessment involves the study and analysis of the potential effect of certain hazards to human health. Using statistical information on cause and effect, risk assessment is intended to be a tool for making informed decisions. Risk management is the process of reducing risks that are determined to be unacceptable.

Both risk assessment and risk management are introduced in the following discussion to provide a perspective on the approaches used in risk analysis. Noncarcinogenic effects and ecological risk assessment are also considered briefly. Additional details may be found in Pepper et al. (1996) and U.S. EPA (1990). Finally, risk assessment for water reuse is discussed with respect to enteric viruses, which are a major concern in industrialized countries.

Risk Assessment

Environmental risk analysis takes place in four discrete steps, as diagrammed on Fig. 13–7 (Neely, 1994).

1. Hazard identification
2. Exposure assessment
3. Dose response assessment
4. Risk characterization

These four steps are described below, along with a brief discussion of the use of risk assessment in setting standards.

Hazard Identification. This step involves weighing the available evidence and determining whether a substance or constituent exhibits a particular adverse health hazard. As part of hazard identification, evidence is gathered on the potential for a sub-

Figure 13–7

Definition sketch for the conduct of health effects risk assessment.

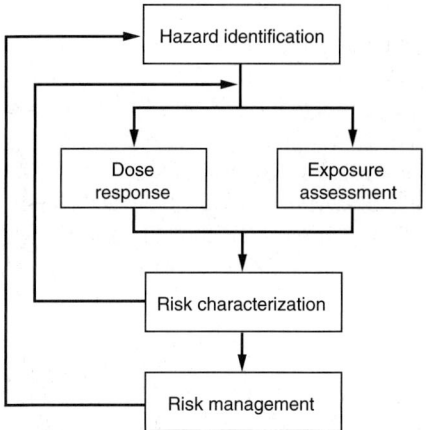

stance to cause adverse health effects in humans or unacceptable environmental impacts. For humans, the principal sources for this information are clinical studies, controlled epidemiological studies, experimental animal studies, and from evidence gathered from accidents and natural disasters.

Exposure Assessment. Exposure is the process by which an organism comes into contact with a hazard; exposure or access is what bridges the gap between a hazard and a risk (Kolluru et al., 1996). For humans, exposure can occur through different pathways including inhalation of air, ingestion of water or food, absorption through the skin via dermal contact, or absorption through the skin via radiation. The key steps in exposure assessment are identification of a potential receptor population, evaluation of exposure pathways and routes, and quantification of exposures. For example, an exposure scenario to assess the impacts of drinking groundwater that contains a known amount of trichloroethylene (TCE) would be as follows: an adult weighing 70 kg drinks 2 L of groundwater containing 50 μg/L of TCE every day for 70 years. In this case, the adult is the receptor, drinking groundwater is the pathway, and drinking 2.0 L/d containing 50 μg/L of TCE for 70 years is the quantification of exposure.

Dose-Response Assessment. The fundamental goal of a dose-response assessment is to define a relationship (typically mathematical) between the amount of a toxic constituent to which a human is exposed and the risk that there will be an unhealthy response to that dose in humans. The relative sensitivity of epidemiological studies in defining excess risk is illustrated on Fig. 13–8a. Typical dose-response relationships for carcinogenic and noncarcinogenic constituents are illustrated on Fig. 13–8b. It should be noted that it is assumed that there is no threshold for potentially carcinogenic constituents. Although the dose-response curve for a carcinogenic constituent is shown passing through the origin, data are not available at extremely low doses (see Fig. 13–9a). Therefore, mathematical models have been developed to define the dose response at low concentrations. Typical dose-response models that have been proposed and used for human exposure include (1) the single-hit model, (2) the multistage model, (3) the linear multistage model, (4) the multihit model, and (5) the probit model. The characteristics of these models are summarized in Table 13–11.

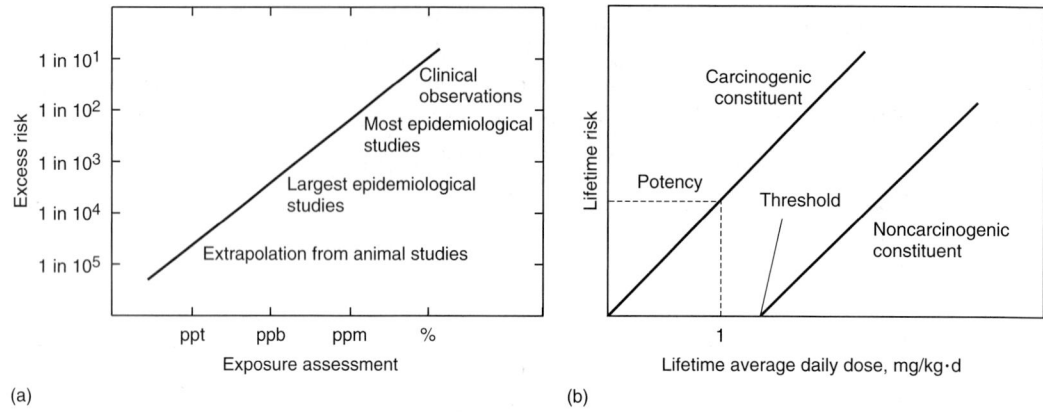

Figure 13–8

Definition sketches for risk assessment: (a) relative sensitivity of epidemiological studies in defining excess risk (adapted from NRC, 1993) and (b) dose-response curves for carcinogenic and noncarcinogenic constituents. As shown, it is assumed that the dose-response curve for a carcinogenic constituent has no threshold value.

Figure 13–9

Definition sketch for dose-response curves: (a) illustration of where data are available and where data are required, and (b) two different models used to define the dose-response relationship.

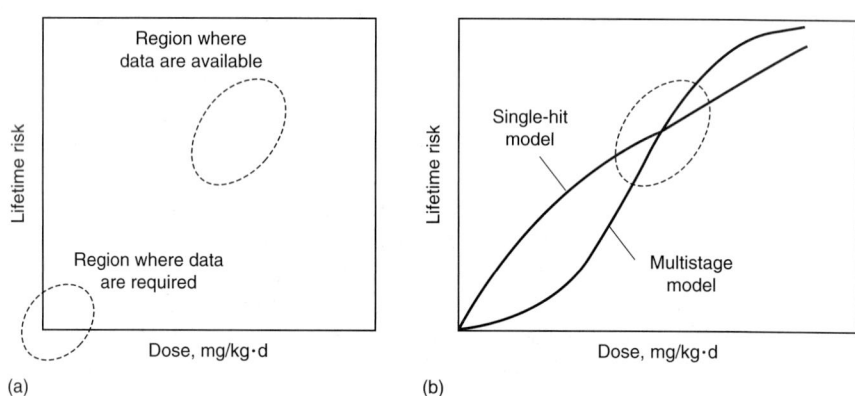

The mathematical function used to describe the relationship between risk and dose for the single-hit model is

$$P(d) = 1 - \exp\left[-(q_0 - q_1 d)\right] \tag{13-1}$$

where $P(d)$ = lifetime risk (probability) of developing cancer
q_0 and q_1 = empirical parameters picked to fit the data
d = dose

The mathematical formulation used to describe the relationship between risk and dose for the multistage model is

$$P(d) = 1 - \exp\left[-\sum_{i=0}^{n} q_i d^i\right] \tag{13-2}$$

Table 13-11
Models used to
assess nonthreshold
effects of toxic
constituents[a]

Model[b]	Description
One-hit	A single exposure can lead to the development of a tumor
Multistage	The formation of a tumor is the result of a sequence of biological events
Linear multistage	Modification of the multistage model. The model is linear at low doses with a constant of proportionality that statistically will produce less than 5 percent chance of underestimating risk
Multihit	Several interactions are required before cell becomes transformed
Probit	Tolerance of exposed population is assumed to follow a log-normal (probit) distribution

[a] Adapted from Cockerham and Shane (1994), Pepper et al. (1996).
[b] In all of the models cited above, it is assumed that exposure to the toxic constituent will always produce an effect regardless of the dose.

where $P(d)$ = lifetime risk (probability) of developing cancer
  q_i = positive empirical parameters picked to fit the data
  d = dose

The relationship between these two models and the risk data given on Fig. 13–9a is shown on Fig. 13–9b. A variety of models have also been proposed to define the risk associated with low levels of microbial pathogens (Haas, 1983).

The U.S. EPA has defined lifetime risk as follows:

$$\text{Lifetime risk} = \text{CDI} \times \text{PF} \tag{13-3}$$

where CDI = chronic daily intake over a 70-year lifetime, mg/kg·d
  PF = potency factor, $(\text{mg/kg·d})^{-1}$

The chronic daily intake (CDI) is computed as follows:

$$\text{CDI} = \frac{\text{total dose, mg}}{(\text{body weight, kg})(\text{lifetime, d})} \tag{13-4}$$

In its most general form, the total dose is defined as

$$\text{Total dose} = \left(\begin{array}{c}\text{constituent}\\\text{concentration}\end{array}\right)\left(\begin{array}{c}\text{intake}\\\text{rate}\end{array}\right)\left(\begin{array}{c}\text{exposure}\\\text{duration}\end{array}\right)\left(\begin{array}{c}\text{absorption}\\\text{factor}\end{array}\right) \tag{13-5}$$

Recommended standard values for daily intake calculations have also been developed by EPA. The average body weights used for an adult and child are 70 and 10 kg, respectively, and the corresponding rates of water ingestion are 2 and 1 liter(s) per day (U.S. EPA, 1986b).

The potency factor PF, often identified as the *slope factor,* is the slope of the dose-response curve, at very low doses (see Fig. 13–8b). The U.S. EPA has selected the linear multistage model as the basis for assessing risk. In effect, the PF corresponds to the risk resulting from a lifetime average dose of 1.0 mg/kg·d. The U.S. EPA maintains a database on toxic substances known as the Integrated Risk Information System (IRIS) (U.S. EPA, 1996). Typical toxicity data for several chemical constituents are reported in

Table 13-12
Toxicity data for selected potential carcinogenic chemical constituents[a,b]

Chemical constituent	CASRN[c]	Potency factor, PF	
		Oral route, $(mg/kg \cdot d)^{-1}$	Inhalation route, $(\mu g/kg \cdot d)^{-1}$
Arsenic, inorganic	7440-38-2	1.5 E+0	3.0 E−2
Benzene	71-43-2	1.5 to 5.5 E−2	1.54 to 5.45 E−5
Bromate	15541-45-4	7 E−1	na
Chloroform	67-66-3	6.1 E−3	1.6 E−4
Dieldrin	60-57-1	1.6 E+1	3.2 E−2
Heptachlor	76-44-8	4.5 E+0	9.1 E−3
N-Nitrosodiethylamine	55-18-5	1.2 E+2	3.0 E−1
N-Nitrosodimethylamine	62-75-9	5.1 E+1	9.8 E−2
Vinyl chloride[d]	75-01-4	7.2 E−1	3.1 to 6.2 E−5

[a] U.S. EPA IRIS database (1996) (http://www.epa.gov/iris).
[b] Because the data in the IRIS database is being revised continuously, it is important to check the database for the most current values.
[c] Chemical Abstracts Service Registry Number.
[d] Continuous lifetime exposure during adulthood.

Table 13–12. Comparing the magnitude of the given values listed in Table 13–12, the relative potency of the chemical constituents can be assessed (e.g., for the oral route, the potency of arsenic is about 245 times that of chloroform). Because of the numerous uncertainties involved in the development of the database, it is important to remember that the values given in the IRIS database cannot be used to predict the incidence of human disease or the type of effects a given chemical constituent will have on an individual. The data included extrapolations of animal data to humans and from high experimental dosages to the low environmental dosages encountered in real life. Use of the data given in Table 13–12 is illustrated in Example 13–1.

EXAMPLE 13–1   **Risk Assessment for Drinking Groundwater Containing Trace Amounts of N-Nitrosodimethylamine (NDMA)**   Estimate the incremental cancer risk for an adult associated with drinking 2 L per day of groundwater containing 2.0 $\mu g/L$ of N-nitrosodimethylamine (NDMA) using data from the Integrated Risk Information System (IRIS) (U.S. EPA, 1996). To limit NDMA exposure to acceptable cancer risk of 1 in 100,000, determine the concentration of NDMA that can be allowed in extracted groundwater.

Solution

1.   Compute the CDI using Eq. 13–4.

$$CDI = \frac{average\ daily\ dose,\ mg/d}{body\ weight,\ kg}$$

$$CDI = \frac{(2.0 \ \mu g/L)(2 \ L/d)(1 \ mg/10^3 \ \mu g)}{70 \ kg} = 0.57 \times 10^{-4} \ mg/kg \cdot d$$

2. Compute the lifetime risk using Eq. 13–3 and data from Table 13–12.

Incremental lifetime risk = CDI × PF

The potency factor from Table 13–12 for the oral route for NDMA is $5.1 \times 10 \ (mg/kg \cdot d)^{-1.}$ Thus,

Incremental lifetime risk = $(0.57 \times 10^{-4} \ mg/kg \cdot d) [5.1 \times 10 \ (mg/kg \cdot d)^{-1}]$

$$= 2.9 \times 10^{-3}$$

From the results of this analysis, the estimated probability of developing elevated cancer risks as a result of drinking the groundwater containing 2.0 $\mu g/L$ of NDMA is 2.9 per 1000 persons.

3. Determine the concentration of NDMA to limit the acceptable cancer risk to 1 in 100,000
   a. Estimate the CDI

   $$10^{-5} = (CDI) [5.1 \times 10 \ (mg/kg \cdot d)^{-1}]$$

   $$CDI = 1.96 \times 10^{-7} \ mg/kg \cdot d$$

   b. Estimate the concentration of NDMA

   $$\frac{(C \ \mu g/L)(2 \ L/d)(1 \ mg/10^3 \ \mu g)}{70 \ kg} = 1.96 \times 10^{-7} \ mg/kg \cdot d$$

   $$C = 0.0069 \ \mu g/L = 6.9 \ ng/L$$

Comment The U.S. EPA–proposed drinking water standard for NDMA is 2.0 ng/L. Because of concern over the carcinogenocity of NDMA, the Ontario, Canada, Drinking Water Objective has been set at 9 ng/L, based on a cancer risk to 1 in 100,000 estimated by the U.S. EPA (Andrews and Taguchi, 2001).

In addition to the carcinogenic dose-response information, the U.S. EPA has developed reference doses (RfD) for a number of constituents based on the assumption that thresholds exist for certain toxic effects (see Fig. 13–8), such as cellular *necrosis* (localized death of living tissue), but may not exist for other toxic effects, such as carcinogenicity. In general, RfDs are established based on reported results from human epidemiological data, long-term animal studies, and other available toxicological information. The RfD values represent an estimate (with uncertainty spanning perhaps an order of magnitude) of a daily exposure to the human population (including sensitive subgroups) that is likely to be without an appreciable risk of deleterious effects during a lifetime (U.S. EPA, 1989a). RfD values are available in the IRIS database (U.S. EPA, 1996) and in the Health Effects Assessment Summary Tables (U.S. EPA, 1991).

The RfD is used as a reference point for assessing the potential effects of other doses. Usually, doses that are less than the RfD are not likely to be associated with health risks. As the frequency of exposure exceeds the RfD and the size of excess increases, the probability increases that adverse health effects may be observed in a human population. The RfD is derived using the following formula:

$$RfD = \frac{NOAEL\ or\ LOAEL}{(UF_1 \times UF_2...) \times MF} \tag{13–6}$$

where NOAEL = no observable adverse effect level
LOAEL = lowest observable adverse effect level
UF_1, UF_2 = uncertainty factors
MF = modifying factor

In the above equation, uncertainty factors are based on experimental species, effects, and duration of the study, while modifying factors represent professional assessments reflecting the confidence in the study. The LOAEL is used only when a suitable NOAEL is unavailable.

It must be recognized that the present state of knowledge concerning the impacts of specific constituents is incomplete. Thus, each step in risk assessment involves uncertainty. In hazard identification, most assessments depend on animal tests and yet animal biological systems are different from human ones. In dose response, it is often unknown whether safe levels or thresholds exist for any toxic chemical. Exposure assessment usually involves modeling, with the attendant uncertainty as to substance releases, release characteristics, meteorology, and hydrology. Because of the uncertainties associated with any risk assessment, the results of such an analysis should only be used as a guide in decision making (Haas et al., 1999).

Risk Characterization. The final step in risk assessment is risk characterization, in which the question of who is affected and what are the likely effects are defined to the extent they are known. Risk characterization involves the integration of exposure and dose-response assessments to arrive at the quantitative probabilities that effects will occur in humans for a given set of exposure conditions.

Risk Assessment in Standard Setting. Examples of risk assessment in wastewater management include health effects from consuming highly treated reclaimed water and health and environmental effects from land application of biosolids (see Sec. 14–17 in Chap. 14). An acceptable risk of 1 in 10,000 is often used in environmental risk assessment (U.S. EPA, 1986a). Risks of less than 1 in 10,000 are considered minimal.

Risk Management

Risk management involves the development of standards and guidelines and management strategies for specific constituents including both toxic constituents and infectious agents. For example, if a toxic constituent or infectious microorganisms are present at higher than the maximum allowable concentration based on the risk assessment, risk management involves the determination of what management and/or technology is necessary to limit the risk to an acceptable level. Thus, the development and screening of alternatives; selection, design, and implementation; and monitoring and review are important elements of health risk management.

Ecological Risk Assessment

Ecological risk assessment is similar to risk assessment for humans in that the ecological effects of exposure to one or more stressors are assessed. A stressor is defined as a substance, circumstance, or energy field that can cause an adverse effect on a biological system. It should be noted that ecological risk assessments are undertaken for a variety of reasons such as to assess the potential impacts of the discharge of treated effluent to an existing wetland or applying biosolids on land. The framework for ecological risk assessment is illustrated on Fig. 13–10 involving (1) problem formulation, in which the characteristics of the stressor are identified, (2) identification and characterization of the ecosystem at risk and the exposure modes, (3) identification of likely ecological risks, and (4) risk characterization, in which all of the information and data are integrated along with input from the risk manager (U.S. EPA, 1992*b*). Because the field of ecological risk assessment is continually undergoing change, the latest reports and publications should be consulted.

Risk Assessment for Water Reuse

In less-developed countries where advanced levels of wastewater treatment are not possible or are economically out of reach, a number of investigators have sought to assess the risk of using reclaimed water of varying quality in different reuse applications by controlling possible transmission routes of excreta-related infections. The WHO *Health Guidelines for the Use of Wastewater in Agriculture and Aquaculture* (1989) is such an example where high concentrations of pathogens exist in wastewater and partially treated effluents.

In the United States, the constituents in reclaimed water that have received the most attention are enteric viruses because of their low-dose infectivity, long-term survival in the environment, difficulties in monitoring them, and their low removal and inactivation

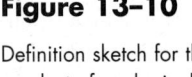

Figure 13–10

Definition sketch for the conduct of ecological risk assessment.

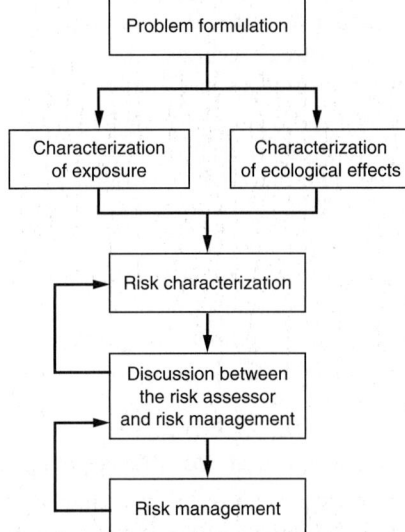

efficacy in conventional wastewater treatment. However, more recently, other inorganic and organic constituents that may be present in treated wastewater such as arsenic and NDMA have received considerable attention because of their significant potency factors (see Table 13–12).

Health risks associated with enteric viruses in reclaimed water that are typically encountered in the California water reuse conditions were analyzed by a quantitative microbial risk assessment approach (Tanaka et al., 1998). Past monitoring data from four wastewater treatment facilities in California on enteric virus concentrations in unchlorinated secondary effluents were used. To assess potential health risks associated with the use of reclaimed water in various reuse applications, four exposure scenarios were tested: (1) golf course irrigation, (2) food crop irrigation, (3) recreational impoundments, and (4) groundwater recharge. Because enteric virus concentrations in unchlorinated secondary effluents were found to vary over a wide range, characterizing their variability was found to be extremely important in this study.

Two concepts related to safety of water reuse were used: (1) the *reliability*, defined as the probability that the risk of infection does not exceed an acceptable risk, and (2) the *expectation*, defined by acceptable annual risk in which exposure to the enteric viruses may be estimated stochastically by numerical simulation such as the Monte Carlo methods. In the U.S. EPA Surface Water Treatment Rule (SWTR) (U.S. EPA, 1989*b*), it is assumed that one infection per 10,000 population per year due to pathogens is acceptable in the public water supply. Therefore, if 10^{-4} annual risk of infection (less than or equal to 1 infection per 10,000 population per year) is set as an acceptable risk for water reclamation and reuse, the reliability (percent of time that infection risk due to exposure to enteric viruses in reclaimed water is less than the acceptable risk) is presented in Table 13–13.

From the results of the analysis presented in Table 13–13, the reliability or relative safety of water reuse can be assessed in comparison to the domestic water supply meeting the SWTR. When the effluent from the full treatment with high chlorine dose of about 10 mg/L is used, there is virtually no difference in terms of a probability of enteric virus infection whether reclaimed water or domestic water is used for golf course irrigation, crop irrigation, and groundwater recharge. However, depending on the water quality of the secondary effluent, there is a considerable difference in water reuse for recreational impoundment where body contact sports and swimming may take place. Similar observations can be made for the use of chlorinated secondary effluent and the reclaimed water from contact filtration with low chlorine dose of less than 5 mg/L.

The utility of the study conducted by Tanaka et al. (1998), as described above, was that their findings were able to provide a basis for conservative assessment of microbiological requirements promulgated in the *California Wastewater Reclamation Criteria* (State of CA, 1978) for variety of water reuse applications. Based on the professional judgment and the research findings, California Department of Health Services was able to ensure safe and reliable water reuse practices.

Limitations in Risk Assessment for Water Reuse

Although risk assessment is used in a variety of settings as an aid to decision making, a number of serious limitations exist with the application of risk assessment to water

Table 13–13

Reliability of various water reuse applications meeting the criterion of one enteric virus infection per 10,000 populations per year[a]

Treatment process	Secondary effluent from plant	Reliability, %			
		Golf course irrigation	Crop irrigation	Recreational impoundment	Groundwater recharge
Full treatment or contact filtration with high chlorine dose removing 5.2 logs of viruses	A	100	100	77	100
	B	100	100	99	100
	C	100	100	98	100
	D	99	100	62	100
Chlorination of secondary effluent removing 3.9 logs of viruses	A	95	100	10	100
	B	100	100	81	100
	C	99	100	93	100
	D	84	100	11	100
Contact filtration with low chlorine removing 4.7 logs of viruses	A	100	100	48	100
	B	100	100	96	100
	C	100	100	97	100
	D	97	100	39	100

[a] Adapted from Tanaka et al. (1998).

reuse. The principal limitations of risk assessment as applied to water reuse are: (1) the relative nature of risk assessment as practiced currently, (2) inadequate consideration of secondary infections in microbial risk assessment, and (3) most importantly, the limited availability of valid dose-response data.

Because it is impossible to determine absolute risk, with the present state of knowledge, relative health risk as opposed to absolute risk, as illustrated above, must be used to assess the safety of water reuse practices and to evaluate alternative reuse applications. In a recently completed study, the relative human and ecological risks involved in three different effluent discharge options were examined using a Bayesian analysis, based on available information and expert opinion. Arsenic, *Cryptosporidium parvum,* rotavirus, and n-nitrosodimethylamine (NDMA) were used as indicators of human health risks. The three alternative discharge options evaluated were: (1) injection to a deep subsurface aquifer located below a drinking water aquifer, (2) discharge to surface canals where the treated effluent can infiltrate into and become mixed with the existing natural groundwater, and (3) discharge to the ocean where body contact sports including swimming and beach activities may be involved (Englehardt et al., 2001, 2002). The Bayesian approach used in this study may be applicable to reuse projects although it has not been attempted. Regardless of the approach used, risk assessment is not well understood by the public; thus, if project evaluation risk is influenced by risk assessment, it is imperative that the process be made as transparent as possible.

In chemical risk assessment, the basis for the analysis is the individual who is ingesting the constituent of concern. Although the same approach can be used for microbial risk assessment, as discussed above, more rigorous quantitative methodologies are required because enteric viruses can be transmitted from person to person. Because of the possibility of disease transmission from person to person, secondary infection must be considered in assessing microbial risks, especially where large populations are exposed (e.g., swimming in a pond containing reclaimed water). To describe the transmission of enteric virus infection and disease within an exposed population, it has been proposed that consideration must be given to the quantification of persons who are: (1) infectious and symptomatic, (2) infectious and not symptomatic, (3) not infectious and not symptomatic, and (4) not infectious and not symptomatic with short-term or partial immunity and their movement between these states (Olivieri et al., 1999). Such an approach has been applied in assessing the risk associated with the land application of digested sludge in a study conducted for the WEF (2001). Because of the complexity of disease transmission mechanisms involved in person-to-person contact, additional research must be conducted before microbial risk assessment can be considered routine undertaking in water reuse applications.

Finally, the most serious limitation in the application of risk assessment to water reuse is the limited availability of dose-response data for most of the constituents of concern in water reuse. As noted previously, numerous uncertainties are involved in the development of the dose-response data including extrapolations of animal data to humans and mathematical extrapolations from high experimental dosages to the extremely low dosages encountered in most water reuse risk assessments. Unfortunately, the current situation is often described as being *model rich* (i.e., mathematical models) and *data poor*. Because dose-response data serve as the basis for most mathematical modeling, the results obtained from modeling efforts must be used judiciously. At this time, what is critically needed is a national dialogue to develop a consensus on acceptable health and environmental risk.

13–4 WATER RECLAMATION TECHNOLOGIES

The required water quality for reclaimed water varies with each reuse application (see Tables 13–9 and 13–10). The focus of this section is to consider treatment concepts and technologies that are of particular importance to water reuse. The topics considered include (1) a review of constituent removal technologies, (2) a review of selected treatment process flow diagrams used in water reclamation applications, (3) predicting the performance of the combinations of treatment operations and processes, and (4) a discussion of treatment process reliability.

Constituent Removal Technologies

As noted in the previous section, the constituents in wastewater subject to treatment may be classified as conventional, nonconventional, and emerging (see Table 13–6). Conventional constituents are removed by the conventional treatment technologies discussed in Chaps. 5 through 10. Advanced treatment technologies, as discussed in Chap. 11, are used most commonly for the removal of nonconventional constituents. The removal of emerging constituents occurs in both conventional and advanced treatment processes, but the

Table 13–14
Treatment levels achievable with various combinations of unit operations and processes used for wastewater reclamation

Treatment process	Typical effluent quality, mg/L except turbidity, NTU						
	TSS	BOD$_5$	COD	Total N	NH$_3$-N	PO$_4$-P	Turbidity
Activated sludge + granular medium filtration	4–6	<5–10	30–70	15–35	15–25	4–10	0.3–5
Activated sludge + granular medium filtration + carbon adsorption	<5	<5	5–20	15–30	15–25	4–10	0.3–3
Activated sludge/nitrification, single stage	10–25	5–15	20–45	20–30	1–5	6–10	5–15
Activated sludge/nitrification-denitrification separate stages	10–25	5–15	20–35	5–10	1–2	6–10	5–15
Metal salt addition to activated sludge + nitrification/denitrification + filtration	≤5–10	≤5–10	20–30	3–5	1–2	≤1	0.3–2
Biological phosphorus removal[a]	10–20	5–15	20–35	15–25	5–10	≤2	5–10
Biological nitrogen and phosphorus removal + filtration	≤10	<5	20–30	≤5	≤2	≤2	0.3–2
Activated sludge + granular medium filtration + carbon adsorption + reverse osmosis	≤1	≤1	5–10	<2	<2	≤1	0.01–1
Activated sludge/nitrification-denitrification and phosphorus removal + granular medium filtration + carbon adsorption + reverse osmosis	≤1	≤1	2–8	≤1	≤0.1	≤0.5	0.01–1
Activated sludge/nitrification-denitrification and phosphorus removal + microfiltration + reverse osmosis	≤1	≤1	2–8	≤0.1	≤0.1	≤0.5	0.01–1

[a] Removal process occurs in the main flowstream as opposed to sidestream treatment.

levels to which individual constituents are removed are not well defined. Typical performance data for selected treatment process combinations are presented in Table 13–14.

Conventional Wastewater Treatment Process Flow Diagrams for Water Reclamation

Unrestricted urban water reuse involves the use of reclaimed water where frequent public exposure is likely in the water reuse application. Disinfected, filtered effluent (see Tables 13–7, 13–9, and 13–10) is required for unrestricted urban water reuse. The three principal treatment process flow diagrams used to produce the required effluent quality are (1) full treatment, (2) direct filtration, and (3) contact filtration.

Figure 13-11

Comparison of tertiary treatment systems for wastewater reclamation: (a) complete treatment, (b) direct filtration, (c) contact filtration with optional granular activated-carbon adsorption.

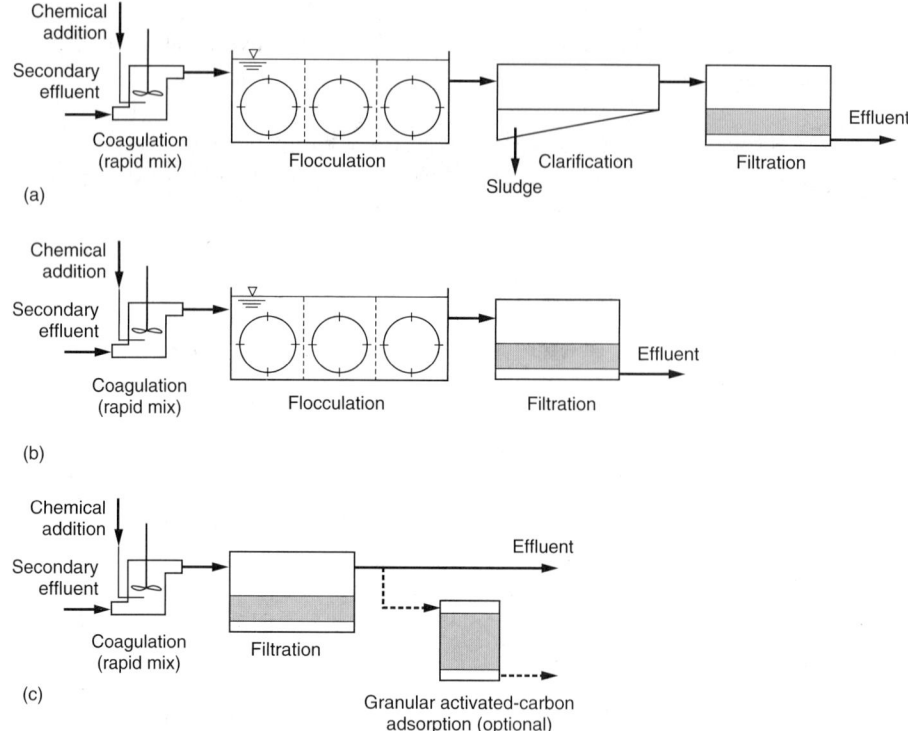

Full Treatment. The full or complete treatment flow diagram, as shown on Fig. 13–11a, is essentially a water treatment flow diagram that involves coagulation, flocculation, clarification, filtration, and disinfection. Using secondary effluent as the feed source, the treatment efficacy for this flow diagram in terms of solids, bacteria, and enteric viruses is substantial, and essentially virus-free water can be attained following disinfection (Asano et al., 1992). The disadvantage of the full treatment flow diagram is the added cost of clarification and solids handling (Richards et al., 1993).

Direct Filtration. Direct filtration is the full treatment process flow diagram without the clarifier (see Fig. 13–11b). The flocculation basin is used to develop a floc with the addition of alum and polymer for coagulation. A turbidity value of 10 NTU in the conventional secondary effluent is typically used as the economic dividing line when deciding between the use of full treatment versus contact or direct filtration (Tchobanoglous and Burton, 1991). When the effluent from a waste-stabilization lagoon is being filtered, higher influent turbidities are expected; thus, air flotation and the use of chemicals are often required.

If the turbidity of the conventional secondary effluent is less than 10 NTU, contact or direct filtration can be used to meet the stringent turbidity requirement of 2 NTU imposed by the *California Water Recycling Criteria* without the need for chemicals. When the turbidity of the secondary effluent is above 7 to 9 NTU, coagulant chemical addition will normally be required to obtain effluent turbidity value of 2 NTU and the filters may need to be operated at lower than a conventional loading rate of 200 L/m²·min (5 gal/ft²·min).

Contact Filtration. In contact filtration, flocculation and clarification facilities are omitted and the system relies on in-line coagulation prior to filtration (see Fig. 13–11c). In the "Pomona Virus Study" (CSDLAC, 1977), it was demonstrated that, with adequate disinfection contact time, the equivalent of full treatment virus attenuation/disinfection could be achieved.

Advanced Wastewater Treatment Process Flow Diagrams

Several examples of existing advanced wastewater treatment process flow diagrams for water reuse are illustrated on Figs. 13–12 through 13–14. A 38,000 m³/d (10 Mgal/d) advanced wastewater treatment system in operation since 1985 at El Paso, TX, is shown on Fig. 13–12, where reclaimed water is injected directly into a potable water aquifer. At Orange County, CA, groundwater recharge with direct injection of reclaimed water has been in operation since 1976. A schematic flow diagram of the original 57,000 m³/d (15 Mgal/d) Water Factory 21 water reclamation facility is shown on Fig. 13–13a. As shown, the advanced wastewater treatment process includes lime clarification for metals and nutrients removal, recarbonation, granular medium filtration, activated-carbon adsorption, demineralization by reverse osmosis, and chlorination (NRC, 1998). The replacement flow diagram currently (2001) under design is shown on Fig. 13–13b. The 3,800 m³/d (1 Mgal/d) Denver Potable Water Reuse Demonstration Plant is illustrated on Fig. 13–14.

Performance Expectations for Water Reclamation Processes

Over the past 10 years, a considerable amount of operating data and information has become available concerning the performance and reliability of conventional and advanced wastewater treatment processes for the removal of conventional constituents. Limited information is available for the removal of nonconventional and emerging constituents. A summary of unit operations and processes used for the removal of conventional and nonconventional constituents is presented in Table 13–15.

Removal of Conventional Constituents. Typical performance data for conventional and advanced treatment processes for the removal of constituents are reported in Table 13–16. As reported in Table 13–16, it is clear that for conventional constituents a very high treatment efficacy can be attained using a variety of treatment process flow diagrams. The typical variability in the performance of treatment processes for conventional constituents is illustrated on Fig. 13–15 with respect to the effluent turbidity and TSS, and on Fig. 13–16 with respect to the removal of MS2 *bacteriophage (coliphage)*.

Removal of Nonconventional Emerging Constituents and Compounds. Corresponding data for the removal of nonconventional constituents are reported in Table 13–17 for complete treatment (as shown on Fig. 13–11a) plus reverse osmosis.

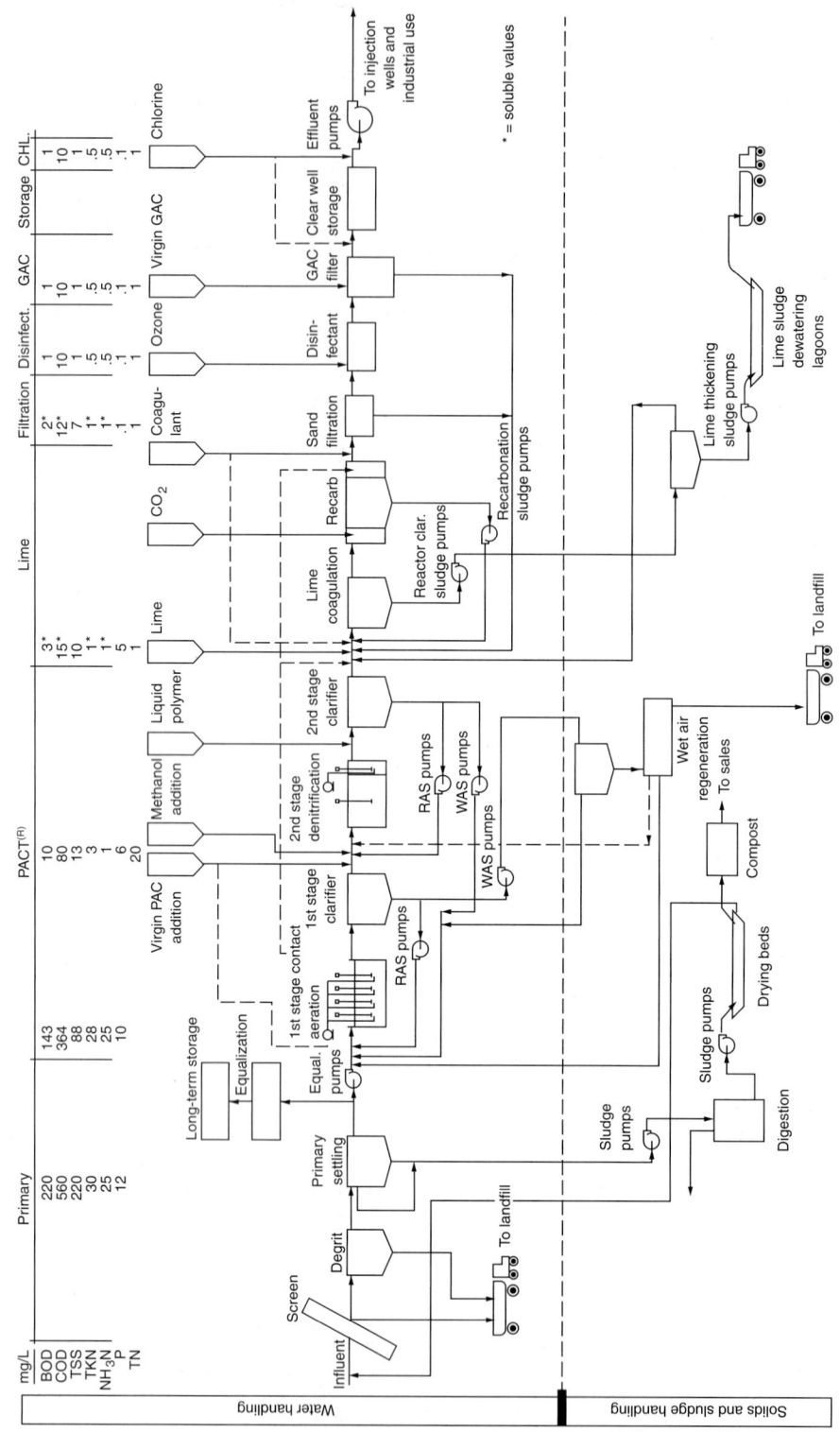

Figure 13–12

Multistage wastewater reclamation processes used in El Paso, TX, for direct injection of reclaimed municipal wastewater.

Figure 13–13

Schematic flow diagrams of treatment processes used at Water Factory 21, Orange County Water District, CA (a) original process flow diagram and (b) replacement flow diagram currently (2001) under design.

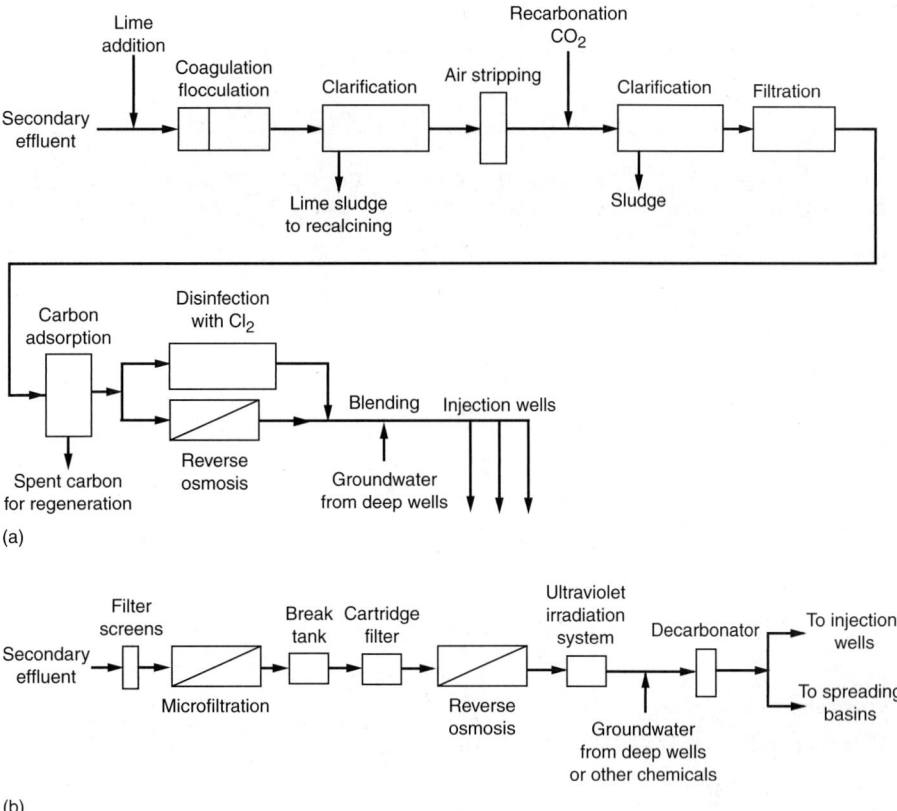

Figure 13–14

Flow diagram for the Denver Potable Water Reuse Demonstration Plant for the Health Effects Study.

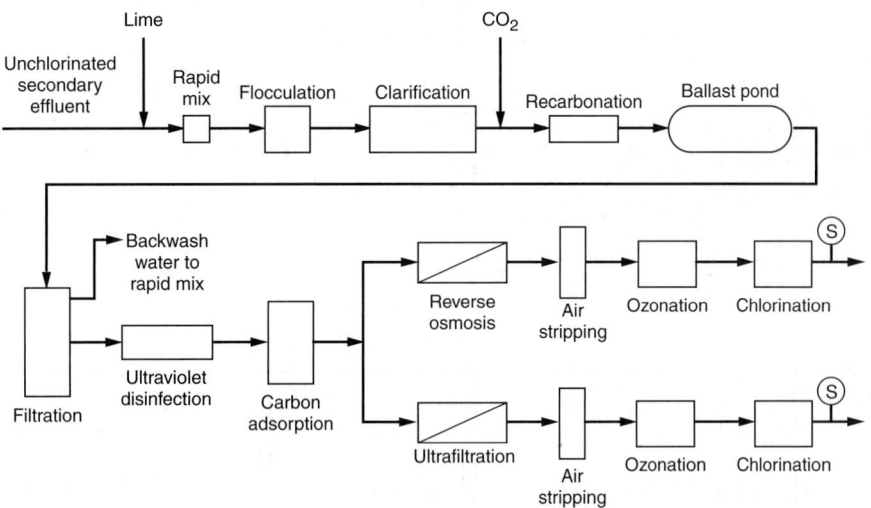

Table 13–15
Unit processes and operations used in wastewater reclamation and potential for contaminant removal

Constituent	Unit process or operation								
	Primary treatment	Activated sludge	Nitrification	Denitrification	Trickling filter	RBC	Coag-Floc.-Sed.	Filtration after A/S	Carbon adsorption
BOD	X	+	+	O	+	+	+	X	+
COD	X	+	+	O	+		+	X	X
TSS	+	+	+	O	+	+	+	+	+
NH_3-N	O	+	+	X		+	O	X	X
NO_3-N				+				X	O
Phosphorus	O	X	+	+			+	+	+
Alkalinity		X					X	+	
Oil and grease	+	+	+				X		X
Total coliform		+	+		O		+		+
TDS									
Arsenic	X	X	X				X	+	O
Barium		X	O				X	O	
Cadmium	X	+	+		O	X	+	X	O
Chromium	X	+	+		O	+	+	X	X
Copper	X	+	+		+	+	+	O	X
Fluoride							X		O
Iron	X	+	+		X	+	+	+	+
Lead	+	+	+		X	+	+	O	X
Manganese	O	X	X		O		X	+	X
Mercury	O	O	O		O	+	O	X	O
Selenium	O	O	O				O	+	O
Silver	+	+	+		X		+		X
Zinc	X	X	+		+	+	+		+
Color	O	X	X		O		+	X	+
Foaming agents	X	+	+		+		X		+
Turbidity	X	+	+	O	X		+	+	+
TOC	X	+	+	O	X		+	X	+

Symbols: O = 25% removal of influent concentration.

X = 25–50%.

+ = >50%.

Blank denotes no data, inconclusive results, or an increase.

| Table 13–15 (Continued)

Ammonia stripping	Selective ion exchange	Breakpoint chlorination	Reverse osmosis	Overland flow	Irrigation	Infiltration-percolation	Chlori-nation	Ozone
	X		+	+	+	+		O
O	X		+	+	+	+		+
	+		+	+	+	+		
+	+	+	+	+	+	+		
				X				
			+	+	+	+		
						X		
				+	+	+		
		+		+	+	+	+	+
			+					
						O		
						+		
						X		
						X		
			+					
						+		
			+	+	+	+		+
			+	+	+	+		O
			+	+	+	+		
O	O		+	+	+	+		+

Table 13–16
Removal of wastewater constituents in a water reclamation facility (all units are mg/L unless otherwise noted)[a]

	Raw conc.	Primary effluent Conc.	%R	Secondary effluent Conc.	%R	Tertiary effluent Conc.	%R	AWT effluent Conc.	%R	Overall %R
Conventional:										
CBOD[b]	185	149	19	13	74	4.3	5	NA		98
TSS	219	131	40	9.8	55	1.3	4	NA		99+
TOC	91	72	21	14	64	7.1	8	0.6	7	99+
TS	1452	1322	9	1183	10	1090	6	43	72	97
Turb. (NTU)	100	88	12	14	74	0.5	14	0.27	0	99+
Ammonia-N	22	21	5	9.5	52	9.3	1	0.8	39	96
Nitrate-N	0.1	0.1	0	1.4	0	1.7	0	0.7	0	0
TKN	31.5	30.6	3	13.9	53	14.2	0	0.9	41	97
Phosphate-P	6.1	5.1	16	3.4	28	0.1	54	0.1	0	98
Nonconventional:										
Arsenic	0.0032	0.0031	3	0.0025	19	0.0015	30	0.0003	40	92
Boron	0.35	0.38	0	0.42	0	0.31	13	0.29	3	17
Cadmium	0.0006	0.0005	17	0.0012	0	0.0001	67	0.0001	0	83
Calcium	74.4	72.2	3	66.7	7	70.1	0	1.0	88	99
Chloride	240	232	3	238	0	284	0	15	90	94
Chromium	0.003	0.004	0	0.002	32	0.001	24	0.001	28	83
Copper	0.063	0.070	0	0.043	33	0.009	52	0.011	0	83
Iron	0.60	0.53	11	0.18	59	0.05	22	0.04	2	94
Lead	0.008	0.008	0	0.008	0	0.001	93	0.001	0	91
Magnesium	38.5	38.1	1	39.3	0	6.4	82	1.5	13	96
Manganese	0.065	0.062	4	0.039	37	0.002	57	0.002	0	97
Mercury	0.0003	0.0002	33	0.0001	33	0.0001	0	0.0001	0	67
Nickel	0.007	0.010	0	0.004	33	0.004	11	0.001	45	89
Selenium	0.003	0.003	0	0.002	16	0.002	0	0.001	64	80
Silver	0.002	0.003	0	0.001	75	0.001	0	0.001	0	75
Sodium	198	192	3	198	0	211	0	11.9	91	94
Sulfate	312	283	9	309	0	368	0	0.1	91	99+
Zinc	0.081	0.076	6	0.024	64	0.002	27	0.002	0	97

[a] Adapted from WCPH (1996a). Primary treatment consisted of a rotary-drum screen followed by disk screens (see Chap. 5), secondary treatment was with water hyacinths, tertiary treatment involved lime precipitation and depth filtration, and AWT (advanced wastewater treatment) was comprised of reverse osmosis, air stripping, and carbon adsorption.

[b] Raw and primary effluent results are BOD, not CBOD. Note: Conc. = concentration; %R = % removal.

Figure 13-15

Probability distribution of measured effluent turbidity and TSS in the filtered secondary effluent from the Tillman Wastewater Reclamation Plant, Los Angeles, CA.

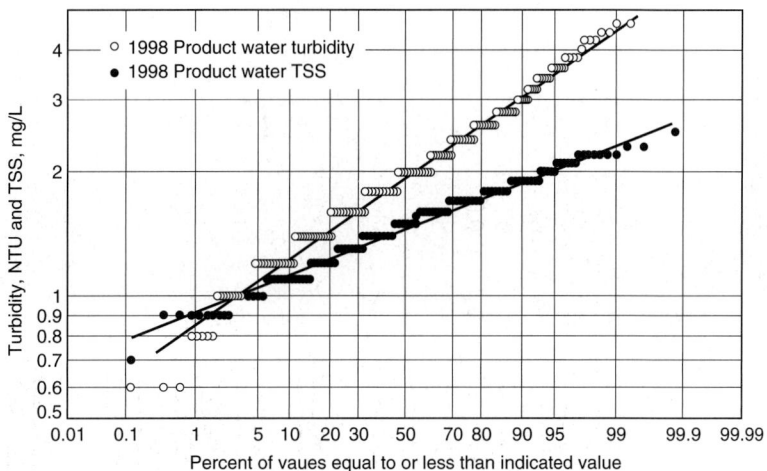

Figure 13-16

Probability distribution of measured indigenous bacteriophage concentrations in primary influent, secondary effluent, filter effluent, and product water at the Tillman Wastewater Reclamation Plant, Los Angeles, CA.

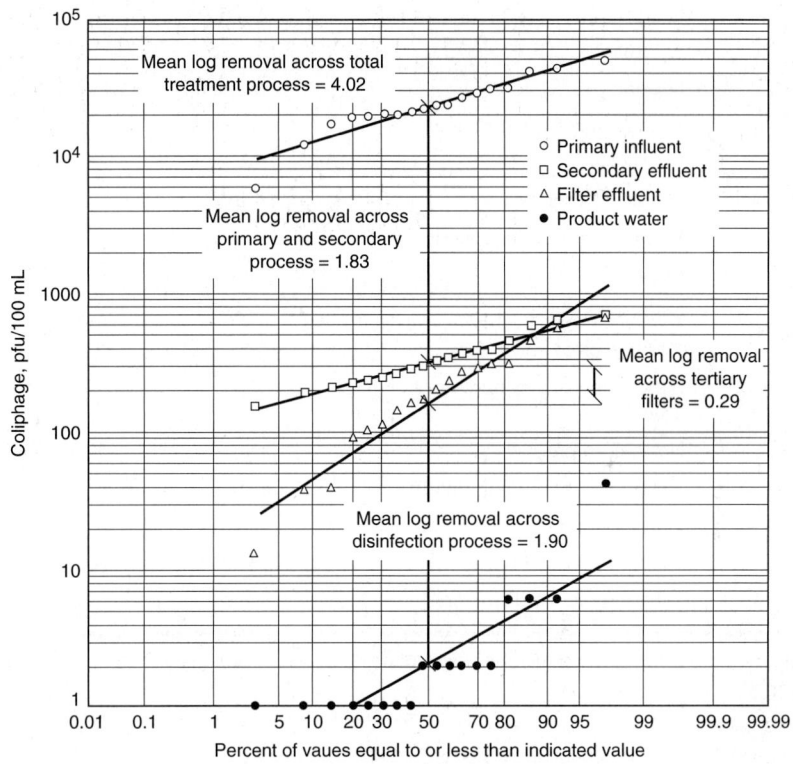

Table 13-17
Typical range of effluent quality after various levels of treatment

Constituent classification	Constituent	Range of effluent quality, mg/L		
		After secondary with BNR plus disinfection	After secondary with BNR plus depth filtration plus disinfection	After secondary with BNR plus microfiltration plus reverse osmosis plus disinfection
Conventional	Total suspended solids	5–20	1–4	≤1
	Colloidal solids	5–10	1–5	≤1
	Bichemical oxygen demand	5–20	1–5	≤1
	Total organic carbon	10–20	0–5	0–2
	Ammonia	0.1–1	0.1–1	≤0.1
	Nitrate	1–10	1–10	≤1
	Nitrite	0.001–0.1	0.001–0.1	≤0.001
	Total nitrogen	2–12	2–12	≤1
	Phosphorus	0.1–0.5	0.1–0.5	≤0.5
	Turbidity	2–6	≤2	0.1–1
	Bacteria	2.2–240	≤2.2	≈0
	Protozoan cysts and oocysts[a]	5–10	≤1	≈0
	Viruses[b]	10^1–10^4	≤10^{-4}	≈0
Nonconventional	Refractory organics	1–5	1–5	1–5
	Volatile organic compounds	1–2	1–2	≤1
	Metals	—[c]	—[c]	≤1
	Surfactants	1–2	1–1.5	≤1
	Total dissolved solids	500–700	500–700	10–50
Emerging	Prescription and nonprescription drugs[d]	Unknown	Unknown	Unknown
	Home-care products	Unknown	Unknown	Unknown
	Veterinary and human antibiotics	Unknown	Unknown	Unknown
	Industrial and household products	Unknown	Unknown	Unknown
	Sex and steroidal hormones	Unknown	Unknown	Unknown
	Other endocrine disrupters	Unknown	Unknown	Unknown

Notes: It is assumed the secondary treatment includes biological nutrient removal (BNR).
[a] Value per 100 mL.
[b] Plaque-forming units/100 mL.
[c] Depends on specific metal and operation of biological treatment process.
[d] Pharmaceutically active substances.

Table 13-18
Reported removal ranges for selected emerging constituents of concern

	Removal range, percent		
Constituent	**Secondary treatment**	**Microfiltration**	**Reverse osmosis**
N-nitrosodimethylamine (NDMA)	50–75	50–75	50–75
17β-Estradiol			50–100
Alkylphenols ethoxylates (APEOs)	40–80	40–80	40–80

Note: Significant variations have been observed in the concentrations of these constituents in the influent wastewater.

In reviewing the data presented in Table 13–17, it can be concluded that the treatment performance for nonconventional constituents is quite variable and not as well defined. Limited data for emerging constituents are reported in Table 13–18. The impact of the constituents that remain after various treatment processes is of profound importance with respect to the long-term protection of public health and the environment. The emerging constituents were unable to be measured in the ppb or ppt range until just a few years ago, and their impacts are mostly unknown at present. The typical variability in the performance of treatment processes for nonconventional and emerging constituents and compounds is illustrated on Fig. 13–17 with respect to the removal of total dissolved solids (TDS) and seeded MS2 bacteriophage used as a surrogate parameter.

Predicting the Performance of Treatment Process Combinations

By logically combining the various unit processes and operations described in Table 13–15, and based on the experience obtained with the advanced wastewater treatment plants for water reuse shown on Figs. 13–12 through 13–14, it is now possible to produce nearly any quality of water from municipal wastewater. One of the problems, however, is that it is not possible to define directly the performance of the combined processes for selected constituents by field testing. For example, the combined performance of a treatment process for removing enteric viruses through advanced treatment processes, consisting of secondary treatment and advanced treatment—microfiltration and reverse osmosis followed by UV disinfection, cannot directly be tested because it is not possible to grow enteric viruses at high enough concentrations to test the removal in combined treatment process and operation.

It is possible, however, to test the processes individually and to synthesize the results stochastically to predict the performance of the combined process. The procedure for combining performance data from two processes may be outlined as follows. First sufficient data must be gathered to define the performance probability density function for each of the treatment process (see Fig. 13–18). The combined density function is obtained by randomly sampling both distributions and combining the results. The application of this approach is illustrated in Example 13–2.

Figure 13–17

Probability distribution of measured values in effluent from a reverse osmosis unit at Dublin San Ramon Services District: (a) TDS and (b) seeded MS2 bacteriophage. (From Whitley Burchett & Associates, 1999, CA.)

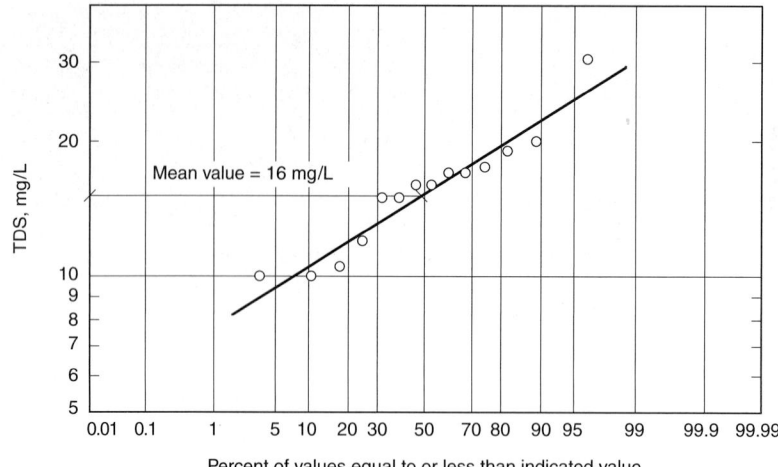

Mean value = 16 mg/L

(a)

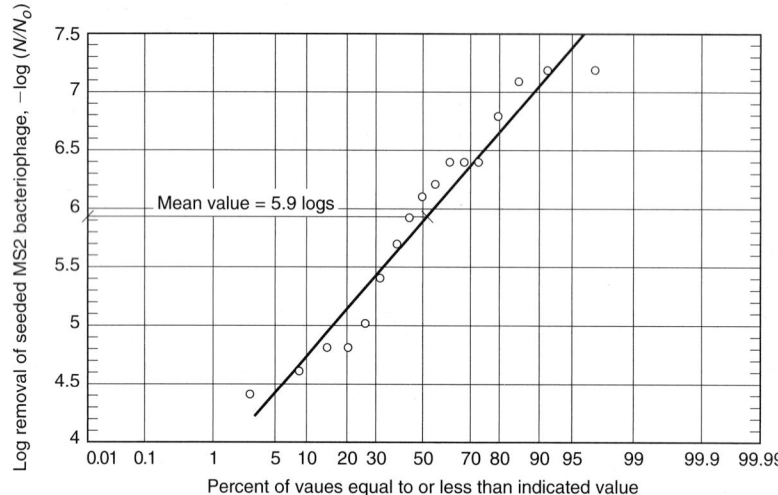

Mean value = 5.9 logs

(b)

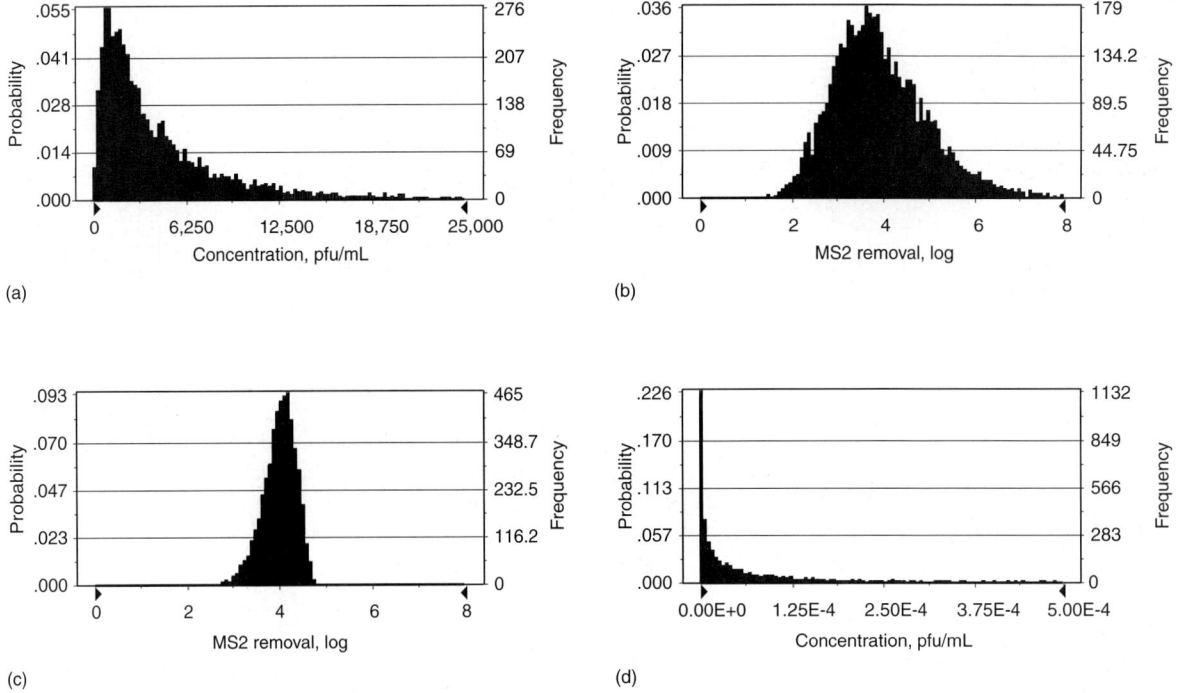

Figure 13–18

Prediction of effluent quality for combined processes by means of simulation. Probability frequency functions generated by 5000 Monte Carlo simulations based on measured performance data for: (a) influent MS2 coliphage concentration (log-normal distribution), (b) MS2 coliphage removal by ultrafiltration (log-normal distribution), and (c) MS2 coliphage removal by reverse osmosis (Weibull distribution). The predicted effluent quality frequency function, given in plot (d), is obtained by randomly sampling 5000 times from the distributions given in plots (a), (b), and (c). (Adapted from Olivieri, et al., 1999.)

EXAMPLE 13–2   Performance Estimation of Combined Treatment Processes Estimate the combined performance of the following two treatment processes. The performance of the individual processes for a given constituent has been measured and is reported below in unitless performance numbers.

Process A	Process B
1.4	3.5
2.2	5.0
3.0	6.0
4.0	6.7
5.0	8.0

(continued)

(*Continued*)

Process A	Process B
6.0	9.0
7.0	11.0
8.0	12.0
10.0	16.0
18.0	20.0

Solution

1. The first step is to determine the nature of the performance distributions. Plotting the performance distributions on both arithmetic and log-probability paper, it was found that the performance distributions were log-normal. The plots for the two processes are shown in the following plot.

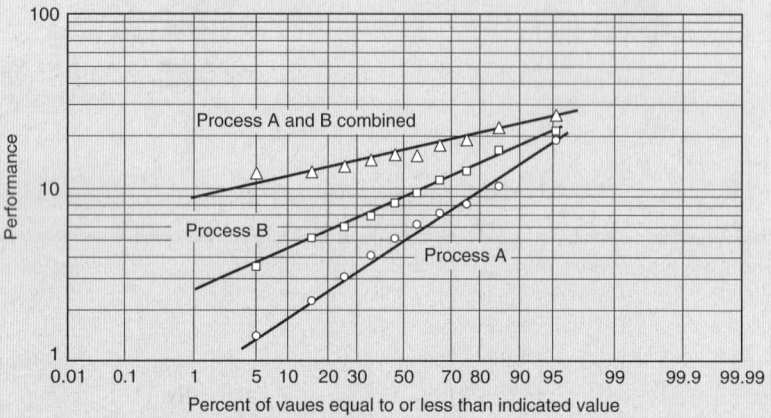

2. The expected performance of the combined processes can be estimated by combining randomized performance values for the individual processes.
   a. The individual performance values are randomized using a random-number table or a random-number generator. The procedure is as follows. Starting at any random location within the table, proceed systematically to find the required numbers. The random sequence numbers correspond to the order in which the required numbers are found in the table. For example, starting in row five, the random numbers for Process A are as summarized in the following table. The random numbers for Process B were obtained by starting at row 16.
   b. The next step is to add the corresponding performance values for the two treatment processes as shown in column 5. For example, for the random number 1, the performance value for Process A is 5.0 and the value for Process B is 9.0 for a total of 14.0.
   c. The combined performance values are arranged in ascending order for plotting as shown in column 6.

d. The combined performance is plotted on the graph developed in Step 1.

Process A (1)	Random number (2)	Process B (3)	Random number (4)	Combined processes A and B (5)	Combined processes A and B[a] (6)
1.4	8	3.5	9	14.0	11.5
2.2	4	5.0	2	15.0	12.0
3.0	7	6.0	5	24.7	13.0
4.0	10	6.7	3	18.2	14.0
5.0	1	8.0	10	13.0	15.0
6.0	6	9.0	1	17.0	15.0
7.0	5	11.0	6	15.0	17.0
8.0	9	12.0	7	21.4	18.2
10.0	2	16.0	4	11.5	21.4
18.0	3	20.0	8	12.0	24.7

[a] Performance of combined processes based on random numbers arranged in ascending order.

Comment The above procedure is used when it is not possible to field-test a series of processes that are to be used in combination.

Treatment Process Reliability

The reliability of a water reclamation plant can be assessed in terms of its ability to produce acceptable reclaimed water consistently. There are two categories of problems that can affect the performance and reliability of a water reclamation plant: (1) problems caused by the influent wastewater variability, even though the water reclamation plant is designed, operated, and maintained properly, and (2) problems caused by mechanical breakdown, design deficiencies, and operational failures. For the first category of problems, evaluation of the influent water quality variability and the corresponding operational reliability is of particular importance in the design of water reclamation and reuse systems. With respect to the second category of problems, operation and maintenance is cited most frequently as the leading cause of poor plant performance. Thus, reliability is a function of both the inherent process reliability as well as the mechanical process reliability. Each of these topics is considered further in Sec. 15–2 in Chap. 15.

13–5 STORAGE OF RECLAIMED WATER

The storage of reclaimed water has become an important issue in a number of reuse applications and especially where communities are moving toward zero discharge of treated effluent. Reservoirs used for the storage of reclaimed water can be either open or enclosed, with open reservoirs being the most common. To introduce some of the issues involved in the storage of reclaimed water the following topics are considered in this section: (1) the need for storage, (2) meeting water quality discharge requirements, (3) operation of open storage reservoirs, (4) problems involved with storage of reclaimed water in open and enclosed reservoirs, and (5) management strategies that can be used to overcome the problems encountered in the storage of reclaimed water in open and enclosed reservoirs. The following discussion deals with the storage of reclaimed water for nonpotable applications, more specifically for agricultural, landscape, and industrial applications.

Need for Storage

In many reclaimed water reuse applications, storage facilities for reclaimed water have become an integral part of the water recycling facilities. In golf course watering and landscape irrigation, reclaimed water is applied in the evening or early morning hours. Because of low flow during these time periods, reclaimed water produced during the daytime hours must be stored to meet the irrigation demand. A typical storage reservoir and golf course application site is shown on Fig. 13–19. A large enclosed reservoir is also shown on Fig. 13–20. Because of increasingly restrictive discharge standards to streams and rivers, many communities are moving toward applying treated effluent to land, with the ultimate objective of zero discharge to water bodies. In some locations, treated effluent must be stored during the summer months, because of low-flow conditions in streams, and discharged during the winter high-flow months.

Meeting Water Quality Discharge Requirements

An issue involving water reuse applications where one or more storage reservoirs are involved is where the applicable discharge requirements from the treatment process

(a)   (b)

Figure 13–19

Storage and irrigation facilities for reclaimed water: (a) example of large storage reservoir for reclaimed water. The top of the dam used to create the reservoir is shown on the left-hand side, and (b) view from the top of the dam. The battery of agricultural-type filters shown at the base of the dam (just left of center) are used to remove algae and other suspended matter before the reclaimed water is disinfected and applied to the golf course shown in the background behind the clubhouse.

Figure 13-20

Aerial view of a large enclosed storage reservoir for reclaimed water located adjacent to a residential area. The top of the reservoir has been landscaped and is used as a soccer field. *(Photo courtesy Metropolitan Wastewater Department, City of San Diego, aerial photography by Jeran-Aero Graphics.)*

should be met. Historically, water quality discharge requirements were monitored at the point of discharge to the environment (i.e., the point of reuse if all of the water is recycled). In more recent times, it has become common to meet the applicable discharge requirements at the end of the treatment plant. However, where storage is used, it will still be necessary to meet the applicable requirements for reuse when the water is withdrawn. In many cases it will be necessary to disinfect again to meet applicable reuse requirements. To avoid the need to disinfect twice (i.e., at the treatment plant and at the point of reuse), it has been proposed that the reclaimed water storage reservoir be considered part of the treatment process. If the storage reservoir is considered to be part of the treatment process, the water would only have to be disinfected to meet the applicable discharge standards when it is withdrawn for use. If the effluent were not disinfected before being placed in storage, the number of organisms to be disinfected would be reduced by sedimentation and natural decay during storage, depending on the storage conditions. It should be noted, however, that many operating agencies choose to disinfect at the treatment plant to limit the growth of microorganisms on the walls of the transmission pipelines and other locations in the distribution systems.

Operation of Storage Reservoirs

Given the above applications, a number of operating schemes are in use for both open and enclosed reservoirs including (1) off-line storage of peak flows during the daytime hours for nighttime use, (2) flow-through in-line storage, (3) long-term storage of winter flows for summer discharge, and (4) long-term storage of summer flows for winter discharge. Off-line storage is common in water reuse applications to (1) meet the off-hour irrigation water requirements, and (2) minimize pumping costs associated with meeting peak water use demands. Long-term storage of winter flows for reuse during the summer irrigation period is quite common. Estimation of the irrigation area and reservoir capacity required for irrigation and storage requirements is illustrated in Example 13-3. Long-term storage of summer flows for winter discharge occurs where there are limited local water reuse opportunities and low stream flow limits the amount of treated effluent that can be discharged during the summer months. The type of storage facility required will depend on the amount of storage required to fulfill the rate and duration of application as well as the schedule for reservoir filling.

EXAMPLE 13–3 Determination of Irrigation Area and Reservoir Storage Capacity Requirements for a No Discharge Reclaimed Water Reuse System Determine the irrigation area required for a wastewater land application system. Calculate the reservoir storage volume, assuming that effluent will be stored in the reservoir until the end of the irrigation season. Find the reservoir area, assuming that the reservoir depth may not exceed 6.0 m. Monthly average wastewater flowrates, rainfall, and crop evapotranspiration (ET) values for the system are provided below in Computation Table 1. Assume that water surface evaporation rates are approximately equal to crop ET rates. Based on percolation tests conducted in the land application area, the soil saturated hydraulic conductivity is 37 mm/d. Use a practical application rate of 10 percent of the saturated hydraulic conductivity when calculating maximum allowable percolation rate. The allowable application rate will depend on local soil conditions but will vary from about 10 to 15 percent.

Solution

1. Determine the monthly depth of wastewater that can be applied to the land application area. Follow the procedure shown in Computation Table 1.
   a. From Computation Table 1, *natural percolation* is defined as the portion of total percolation attributable to excess rainfall (i.e., rainfall that exceeds the evaporation rate). Calculate natural percolation by subtracting the crop ET value from the rainfall value for each month. Where crop ET exceeds rainfall, natural percolation equals zero.
   b. Determine the design total percolation depth for each month by multiplying the saturated hydraulic conductivity times the application factor (10 percent), times the number of days per month. For January:

$$\text{January design total percolation} = \left(\frac{37 \text{ mm}}{\text{d}}\right) \times (0.10) \times (31 \text{ d}) = 115 \text{ mm}$$

   Where natural percolation exceeds the design total percolation value for a given month (e.g., in January and February), total percolation is assumed to equal natural percolation.
   c. Calculate the applied percolation depth for each month by subtracting the natural percolation from the total percolation. Applied percolation refers to the portion of percolation attributable to applied wastewater.
   d. Net ET is defined as the crop ET requirement that is not met by rainfall (see Sec. 13–5). Calculate net ET for each month by subtracting rainfall from crop ET. Where rainfall exceeds crop ET, net ET equals zero.

e. Find the depth of applied wastewater for each month by adding applied percolation to net ET. As shown in col. 10 of Computation Table 1, the annual depth of applied wastewater is 1871 mm.

2. Follow the procedure shown in Computation Table 2 to determine the required size of the land application area and storage reservoir.

    a. Determine the wastewater inflow volume to the storage reservoir by multiplying the average monthly wastewater flowrate times the number of days for that month. For January:

$$\text{January wastewater inflow} = \left(\frac{8000 \text{ m}^3}{\text{d}}\right) \times (31 \text{ d}) = 248{,}000 \text{ m}^3$$

    b. Determine the monthly rainfall inflow volume to the storage reservoir by multiplying the monthly rainfall in col. 4 of Computation Table 1 times an assumed area for the storage reservoir. (The assumed area will be modified later to provide the desired depth. Changing assumed values to achieve a given objective is done most easily by using a spreadsheet program, such as the Microsoft Excel.)

    c. Determine the monthly evaporation outflow volume from the storage reservoir by multiplying the monthly crop ET depth in col. 5 of Computation Table 1 times the assumed area for the storage reservoir.

    d. Determine the monthly irrigation outflow volume from the storage reservoir by multiplying the applied wastewater depth in col. 10 of Computation Table 1 times an assumed area for the land application area. Like the assumed area for the storage reservoir, the assumed value for the land application area will be modified later.

    e. Calculate the monthly change in storage reservoir volume by adding the inflow values (wastewater + rainfall) and subtracting the outflow values (evaporation and irrigation) for each month.

3. In a water balance, the annual sum of inflow and outflow values to the storage reservoir should equal zero. Change the assumed value for the land application area until the sum is zero. If the sum is positive, the land application area should be made larger. Conversely, if the sum is negative, the land application area should be made smaller.

4. The annual sum of monthly volume changes for the storage reservoir (now zero) consists of both positive (net inflow) and negative (net outflow) values. Calculate the net volume of wastewater stored in the reservoir, beginning with the first positive volume change after summer. The running volume sum should terminate with zero volume (see the October value in col. 7 of Computation Table 2).

5. Determine the approximate monthly depth of wastewater in the storage reservoir by dividing the monthly storage volume by the assumed storage reservoir area. If the depth exceeds 6.0 m, increase the reservoir area. Conversely, if the depth is much less than 6.0 m, decrease the reservoir area. Due to changes in the reservoir size, the land application area may also have to be adjusted slightly to maintain a zero sum of volume changes.

Computation Table 1
Water balance for determination of land area required for irrigation

Month (1)	Days (2)	Daily WW flow, m³/d (3)	Rainfall, mm (4)	Crop ET, mm (5)	Land application area			Net ET, mm (9)	Applied WW, mm (10)
					Percolation, mm				
					Natural (6)	Applied (7)	Total (8)		
Jan	31	8000	183	28	155	0	155	0	0
Feb	28	7500	178	51	127	0	127	0	0
Mar	31	6000	114	69	45	70	115	0	70
Apr	30	5000	76	99	0	111	111	23	134
May	31	4000	10	142	0	115	115	132	247
Jun	30	4000	3	178	0	111	111	175	286
Jul	31	4000	3	218	0	115	115	215	330
Aug	31	4000	5	188	0	115	115	183	298
Sep	30	4000	15	150	0	111	111	135	246
Oct	31	5000	30	94	0	115	115	64	179
Nov	30	6000	102	51	51	60	111	0	60
Dec	31	7000	122	30	92	23	115	0	23
Total	365	—	841	1298	470	944	1414	927	1871

Computation Table 2
Water balance for determination of volume of storage reservoir

Month (1)	Inflow, m³		Outflow, m³		Volume, m³		Depth, m (8)
	Wastewater (2)	Rainfall (3)	Evaporation (4)	Irrigation (5)	Change (6)	Net (7)	
Jan	248,000	29,738	4,550	0	273,188	610,171	3.75
Feb	210,000	28,925	8,288	0	230,638	840,808	5.17
Mar	186,000	18,525	11,213	70,167	123,145	963,954	5.93
Apr	150,000	12,350	16,088	134,898	11,365	975,319	6.00
May	124,000	1,625	23,075	248,353	-145,803	829,516	5.10
Jun	120,000	488	28,925	287,916	-196,354	633,162	3.90
Jul	124,000	488	35,425	331,909	-242,847	390,315	2.40
Aug	124,000	813	30,550	299,695	-205,432	184,883	1.14
Sep	120,000	2,438	24,375	247,648	-149,586	35,297	0.22
Oct	155,000	4,875	15,275	179,897	-35,297	0	0.00
Nov	180,000	16,575	8,288	60,402	127,885	127,885	0.79
Dec	217,000	19,825	4,875	22,852	209,098	336,983	2.07
Total	1,958,000	136,663	210,925	1,883,738	0	—	—

6. Determine the final storage reservoir and land application area requirements, without construction and buffer allowances.

   a. Determine the storage area. From Computation Table 2, the critical month for storage is April. Thus,

   $$\text{Storage reservoir area} = \left(\frac{975{,}319 \text{ m}^3}{6 \text{ m}}\right)\left(\frac{1 \text{ ha}}{10^5 \text{ m}^2}\right) = 1.63 \text{ ha}$$

   b. Determine the land application area. From Computation Table 2, the total amount of water applied is 1,883,738 m³ and the depth of application from Computation Table 1 is 1.871 m. Thus,

   $$\text{Land application area} = \left(\frac{1{,}883{,}738 \text{ m}^3}{1.871 \text{ m}}\right)\left(\frac{1 \text{ ha}}{10^5 \text{ m}^2}\right) = 10.1 \text{ ha}$$

Problems Involved with Storage of Reclaimed Water

The problems encountered with the storage of reclaimed water in open and enclosed reservoirs derive from the residual constituents remaining in the water after treatment and the chemical characteristics of the potable water supply.

Problems with Open Reservoirs. Changes in water quality during storage or impoundment can be significant. The most common problems encountered with the storage of reclaimed water in open reservoirs are listed in Table 13–19 under the categories of physical, chemical, and biological. As noted in Table 13–19, most of the physical parameters are also of aesthetic concerns. In many cases, aesthetic concerns will be the controlling factor in deciding whether the community will accept a proposed water-recycling program. The principal problems with the storage of reclaimed water in open reservoirs are:

- Release of odors, principally hydrogen sulfide
- Temperature stratification
- Loss of chlorine residual
- Low dissolved oxygen resulting in odors and fish kills
- Excessive growth of algae and phytoplankton
- High levels of turbidity and color
- Regrowth of microorganisms
- Water quality deterioration due to bird and rodent populations

Many of these problems are interrelated. For example, the production of hydrogen sulfide odors is related to reservoir stratification caused by temperature differences. The production of hydrogen sulfide is also related to the sulfate concentration in the potable water supply and the amount added during usage. If the sulfate levels in the reclaimed water are greater than about 50 mg/L, hydrogen sulfide production is likely to occur in stratified reclaimed water storage reservoirs. Unfortunately, the amount of residual

Table 13-19
Problems
encountered in the
operation of open
reservoirs used for
the storage of
reclaimed water

Reservoir problem	Description
Physical/aesthetic:	
Color	The presence of color can affect the aesthetic acceptance of the water. Often caused by the presence of humic materials and fine silts and clays in runoff and the presence of color in the reclaimed water
Odors (primarily H$_2$S)	One of the most common problems encountered with the storage of reclaimed water. In addition to causing odors, H$_2$S has a chlorine demand
Temperature	Water may be unusable during certain times of the year
Temperature stratification	Usually occurs once or twice a year depending on the latitude
Turbidity	The presence of turbidity can affect the aesthetic acceptance of the water. Turbidity can be caused by runoff containing silt and clay and by algal growth
Chemical:	
Chlorine	Chlorine and compounds containing chlorine may be toxic to aquatic life in open reservoirs
Dissolved oxygen	Low DO can cause fish kills and allow the release of odors in open reservoirs
Nitrogen	Nutrient capable of stimulating phytoplankton
Phosphorus	Nutrient capable of stimulating phytoplankton
Biological:	
Algae	Presence of excess algae can cause odors, increase turbidity, and clog filters
Aquatic foul	The presence of excessive numbers of aquatic birds can degrade the water quality of the stored water
Bacteria	Regrowth is a common occurrence in open storage reservoirs. May affect possible applications
Chlorophyll	Presence of excess algae and plant matter
Helminths	May affect possible reuse applications
Insects (mosquitoes)	May require spraying of insecticides
Phytoplankton	Presence of excess algae can cause odors, increase turbidity, and clog filters
Protozoa	May affect possible reuse applications
Viruses	May affect possible reuse applications

nitrogen and phosphorus in most reclaimed waters is more than adequate for the growth of algae and other phytoplankton. Materials carried into the storage reservoir by local runoff often cause turbidity and color. Fine clays and silts are difficult to settle and color caused by humic substances is extremely difficult to remove. In some cases the reclaimed water may also have a high color content due to the presence of untreated humic substances that can pass through a treatment process.

Problems with Enclosed Reservoirs. The most common problems encountered with the storage of reclaimed water in enclosed reservoirs are listed in Table 13–20 under the categories of physical, chemical, and biological. The principal problems with the storage of reclaimed water in enclosed reservoirs are:

- Stagnation
- Release of odors, principally hydrogen sulfide
- Loss of chlorine residual (much slower compared to open reservoirs)
- Regrowth of microorganisms

Management Strategies for Open and Enclosed Reservoirs

Management strategies that have been used to overcome the problems cited in Tables 13–19 and 13–20, and discussed above, are summarized in Table 13–21. The strategies for open and enclosed reservoirs are considered in the following discussion.

Strategies for Open Reservoirs. Although all of the strategies listed in Table 13–21 have been used, the most effective strategy for open reclaimed water storage reservoirs has been the use of aeration to provide both oxygen and destratification. A variety of aeration systems have been used to provide oxygen and to eliminate stratification including surface aerators with high pumping capacity, brush aerators, static tube aerators, diffused aeration systems, and the Speece cone (see Secs. 5–4 and 5–12 in Chap. 5). Of these devices the Speece cone is superior with respect to the transfer of

Table 13–20
Problems encountered in the operation of enclosed reservoirs used for the storage of reclaimed water

Reservoir problem	Description
Physical/aesthetic:	
Color	Often caused by the presence of humic materials in reclaimed water
Odors (primarily H_2S)	One of the most common problems encountered with the storage of reclaimed water. In addition to causing odors, H_2S has a chlorine demand
Turbidity	The presence of turbidity can affect the aesthetic acceptance of the water
Chemical:	
Chlorine	Chlorine and compounds containing chlorine may cause odors. Chlorine is used commonly to control biological growths
Dissolved oxygen	Lack of oxygen can lead to the release of odors in enclosed reservoirs
Biological:	
Bacteria	Regrowth has occurred in enclosed storage reservoirs. May affect possible applications
Insects (mosquitoes)	Insects can enter improperly sealed reservoirs. May require spraying of insecticides
Viruses	May affect possible reuse applications

Table 13–21

Management strategies for open and enclosed reservoirs used for the storage of reclaimed water

Management strategies	Comments
Open storage reservoirs	
Aeration/destratification	Installation of aeration facilities can be used to maintain aerobic conditions and eliminate thermal stratification. May result in release of phosphorus from bottom sediments
Alum precipitation	Alum precipitation has been used to remove suspended solids and phosphorus. Can be used to stop release of phosphorus from sediments
Biomanipulation	Control of microorganism growth rates
Copper sulfate addition	Copper sulfate is applied to control the growth of algae. The use of copper may be eliminated because of toxicity concerns over accumulation of copper
Destratification (including recirculation)	Submerged or aspirating mixers can be used to eliminate thermal stratification. Recirculating pumps can also be used. May result in release of phosphorus from bottom sediments
Dilution	Water from other sources can be blended with water from the storage reservoir to manage the water quality
Dredging	Accumulated sediment can be removed annually to limit the formation of deposits and the generation of hydrogen sulfide
Filtration	Water from the storage reservoir can be filtered through a rock filter, a slow sand filter, or a disk-type filter to remove algae and to improve the clarity of the water
Natural microorganism decay	The effectiveness of natural decay will depend on the operation of the reservoir and the detention time
Nutrient removal	Removal of nutrients (e.g., nitrogen and phosphorus) to control aquatic growths
Photooxidation	With proper mixing, advantage can be taken of the beneficial effects of exposing the water to sunlight
Wetlands treatment	Water from the storage reservoir can be passed through a constructed wetland to improve the clarity of the effluent and to remove algae
Withdrawal from selected depths	Varying water quality can be obtained by drawing off water at selected depths within the reservoir
Enclosed storage reservoirs	
Aeration	Maintain residual level of DO to eliminate the formation of odors
Chlorination	Used to control the growth of microorganisms
Recirculation	Adequate recirculation can limit the growth of microorganisms and the formation of odors

oxygen and for mixing the contents of the reservoir. The power input required for aeration and destratification is on the order of 0.30 to 0.50 kW/1000 m^3. Obviously, the actual power requirement will vary with the physical characteristics of the reservoir including surface area, aspect ratio, depth of the reservoir, and temperature.

The growth of plankton in reservoirs can be controlled using copper sulfate or more selective algicides. The use of chlorine in open reservoirs is not recommended as a control measure. Under certain circumstances chlorine can combine with odor-causing compounds present in the reservoir to form odors with greater intensity. The effective use of these chemicals, such as copper sulfate, requires microscopic examination of the water to determine the number and type of organisms involved. Ideally, the control chemicals should be applied just at the time when the number of organisms starts to increase rapidly.

Treatment is usually needed when the number of organisms exceeds 500 to 1000 units per milliliter. The required dosage of copper sulfate is influenced by the temperature, alkalinity, and carbon dioxide content of the water. Based on operating experience with water supply storage reservoirs, a copper sulfate dose of 0.3 mg/L can be used without laboratory control. Safe dosages for most fish are 0.5 mg/L, with the exception of trout, for which the safe dosage is on the order of 0.10 mg/L. Because many of the organisms can be controlled with lower dosages, savings are possible with proper control. The dosage of copper sulfate is usually based on the volume of shallow reservoirs or on the upper circulating portion of the reservoir, where the plankton are typically found, in stratified reservoirs. If the depth of the circulation pattern is unknown, a rule of thumb is to use an active depth of about 4.5 m. The frequency of application should be based on microscopic examination.

Several methods are used to apply copper sulfate to reservoir waters. The most common methods are (1) the burlap bag method, (2) the spray method, (3) the blower method, and (4) the continuous dosing method. In the burlap bag method, copper sulfate is placed in a bag in a dose equivalent to 3.0 mg/L. The bag is then dragged through the water from the stern of a rowboat or a motorboat. In the spray and blower methods, copper sulfate is either sprayed or blown over the water surface from a boat, using tree spraying equipment or air blowers. In many reservoirs, copper sulfate is applied continuously using commercial or locally made dosing equipment (Tchobanoglous and Schroeder, 1985).

Strategies for Enclosed Reservoirs. In enclosed reservoirs, effective management strategies include providing facilities to (1) recirculate the contents of the storage basin and (2) add additional chlorine to maintain a residual. In addition to using aeration devices and pumps to promote circulation, the inlet and withdrawal piping can be configured to promote circulation. Generally, the addition of chlorine will be limited to small off-line storage reservoirs, typically used for landscape irrigation and some industrial reuse applications.

13–6 AGRICULTURAL AND LANDSCAPE IRRIGATION

Irrigation of crops was developed in the United States, along with the settlement of the arid West, because irrigation was needed to raise crops and livestock. In the humid eastern United States, irrigation was used to supplement natural rainfall, to increase the number of plantings per year and yield of crops, and to reduce the risk of crop failures during drought periods. Irrigation is also used to maintain recreational lands such as parks and golf courses. An important use of reclaimed water in recent years is for the irrigation of landscaped areas and golf courses in the urban environment.

Evaluation of Irrigation Water Quality

Although irrigation has been practiced throughout the world (see Fig. 13–21) for several millennia, it is only in the last century that the importance of the quality of the irrigation

(a)

(b)

Figure 13–21

Typical examples of worldwide use of reclaimed water for agricultural crops (a) date palms using drip irrigation in Acaba Jordan and (b) fodder crop (recently harvested) using flood irrigation in Australia near Brisbane.

water has been recognized. The design approach to irrigation with reclaimed municipal wastewater depends upon whether emphasis is placed on water supply for irrigation or on wastewater treatment via land treatment and disposal.

Physical and Chemical Characteristics. The physical and chemical characteristics of irrigation water are of particular importance in arid zones where extremes of temperature and low humidity result in high rates of evapotranspiration (ET). Evapotranspiration refers to the water lost through evaporation from the soil and surface water bodies and transpiration from plants. Water used for irrigation can vary greatly in quality depending upon type and quantity of dissolved salts. The consequence of evapotranspiration is salt deposition from the applied water, which tends to accumulate in the soil profile. The physical and mechanical properties of the soil, such as degree of dispersion of the soil particles, stability of aggregates, soil structure, and permeability, are sensitive to the types of exchangeable ions present in irrigation water. Thus, when irrigation with reclaimed water is being planned, not only is the crop yield important, but also soil properties must be taken into consideration. The problems, however, are no different from those caused by salinity or trace elements in any water supply and are of concern only if they restrict the use of the water or require special management to maintain acceptable crop yields.

Numerous irrigation water quality guidelines have been proposed. The guidelines presented in Table 13–22 were developed by the University of California Committee of

Table 13–22

Guidelines for interpretations of water quality for irrigation[a]

Potential irrigation problem	Units	Degree of restriction on use		
		None	Slight to moderate	Severe
Salinity (affects crop water availability):				
EC_w	dS/m or mmho/cm	<0.7	0.7–3.0	>3.0
TDS	mg/L	<450	450–2000	>2000
Permeability (affects infiltration rate of water into the soil. Evaluate using EC_w and SAR or adj R_{Na} together)[b]				
adj R_{Na} = 0–3		and EC_w ≥0.7	0.7–0.2	<0.2
3–6		≥1.2	1.2–0.3	<0.3
6–12		≥1.9	1.9–0.5	<0.5
12–20		≥2.9	2.9–1.3	<1.3
20–40		≥5.0	5.0–2.9	<2.9
Specific ion toxicity (affects sensitive crops):[c]				
Sodium (Na)				
Surface irrigation	SAR	<3	3–9	>9
Sprinkler irrigation	mg/L	<70	>70	
Chloride (Cl)				
Surface irrigation	mg/L	<140	140–350	>350
Sprinkler irrigation	mg/L	<100	>100	
Boron (B)	mg/L	<0.7	0.7–3.0	>3.0
Trace elements (see Table 13–23)				
Miscellaneous effects (affects susceptible crops):				
Nitrogen (total-N)	mg/L	<5	5–30	>30
Bicarbonate (HCO₃) (overhead sprinkling only)	mg/L	<90	90–500	>500
pH	unit	Normal range 6.5–8.4		
Residual chlorine (overhead sprinkling only)	mg/L	<1.0	1.0–5.0	>5.0

[a] Adapted from Ayers and Westcot (1985) and Pettygrove and Asano (1985).

[b] For wastewater irrigation, it is recommended that SAR be adjusted to include a more correct estimate of calcium in the soil water. A procedure is given in Eq. (13–11) and Table 13–24. The adjusted sodium adsorption ratio (adj R_{Na}) calculated by this procedure is to be substituted for the SAR value in this table.

[c] See also Table 13–23.

Consultants and were expanded subsequently by Ayers and Westcot (1985). The long-term influence of water quality on crop production, soil conditions, and farm management is emphasized, and the guidelines are applicable to both freshwater and reclaimed water. Four categories of potential management problems associated with water quality in irrigation are (1) salinity, (2) specific ion toxicity, (3) water infiltration rate, and (4) other problems.

Salinity. Salinity of an irrigation water is determined by measuring its electrical conductivity and is the most important parameter in determining the suitability of a water for irrigation. The electrical conductivity (EC) of a water is used as a surrogate measure of total dissolved solids (TDS) concentration. The electrical conductivity is expressed as decisiemens per meter (dS/m) or mmho/cm. It should be noted that one dS/m is equivalent to one mmho/cm. Values for salinity are also reported as TDS in mg/L. For most agricultural irrigation purposes, the values for EC and TDS are related to each other and can be converted within an accuracy of about 10 percent using Eq. (13–7) (Pettygrove and Asano, 1985).

$$\text{TDS (mg/L)} \approx \text{EC (dS/m or mmho/cm)} \times 640 \tag{13–7}$$

The presence of salts affects plant growth in three ways: (1) osmotic effects, caused by the total dissolved salt concentration in the soil water; (2) specific ion toxicity, caused by the concentration of individual ions; and (3) soil particle dispersion, caused by high sodium and low salinity. With increasing soil salinity in the root zone, plants expend more of their available energy on adjusting the salt concentration within the tissue (osmotic adjustment) to obtain needed water from the soil. Consequently, less energy is available for plant growth.

In irrigated areas, salts originate from the local groundwater or from salts in the applied water. Salts tend to concentrate in the root zone due to evapotranspiration, and plant damage is tied closely to an increase in soil salinity. Establishing a net downward flux of water and salt through the root zone is the only practical way to manage a salinity problem. Under such conditions, good drainage is essential to allow a continuous movement of water and salt below the root zone. Long-term use of reclaimed water for irrigation in which only the conventional constituents have been removed is not possible without adequate drainage.

If more water is applied than the plant uses, the excess water will percolate below the root zone, carrying with it a portion of the accumulated salts. Consequently, the soil salinity will reach some constant value dependent on the leaching fraction. The fraction of applied water that passes through the entire rooting depth and percolates below is called the leaching fraction (LF):

$$\text{LF} = \frac{D_d}{D_i} = \frac{(D_i - \text{ET}_c)}{D_i} \tag{13–8}$$

where LF = leaching fraction
D_d = depth of water leached below the root zone, mm (in)
D_i = depth of water applied at the surface, mm (in)
ET_c = crop evapotranspiration, mm (in)

A high leaching fraction results in less salt accumulation in the root zone. If the salinity of irrigation water (EC_w) and the leaching fraction are known, salinity of the drainage water that percolates below the rooting depth can be estimated by using Eq. (13–9):

$$EC_{dw} = \frac{EC_w}{LF} \tag{13-9}$$

where EC_{dw} = salinity of the drainage water percolating below the root zone
EC_w = salinity of irrigation water

The EC_{dw} value is used to assess the potential effects on crop yield and on groundwater. For salinity management, it is often assumed that EC_{dw} is equal to the salinity of the saturation extract of the soil sample, EC_e. This assumption is, however, conservative in that EC_{dw} occurs at the soil-water potential of field capacity and EC_e occurs at a potential of zero, by definition, at laboratory condition. For a quick check, the value of EC_{dw} can be estimated as twice the value of EC_e for most soils.

EXAMPLE 13-4 **Calculation of Drainage Water Quality** A crop is irrigated with reclaimed water whose salinity (EC_w), measured by electrical conductivity, is 1.0 dS/m. If the crop is irrigated to achieve a leaching fraction of 0.15 (i.e., 85 percent of the applied water is used by the crop or lost through evapotranspiration): (1) Calculate the salinity of the deep percolate water, and (2) Determine the appropriate leaching fraction to maintain crop yield. The crop is known to suffer significant loss in yield when TDS of the soil water exceeds 5000 mg/L.

Solution

1. After many successive irrigations, the salt accumulation in the soil will approach an equilibrium concentration based on the salinity of the applied irrigation water and the leaching fraction. Thus, the salinity of the water (drainage water) that percolates below the root zone can be estimated using Eq. (13–9).

$$EC_{dw} = \frac{EC_w}{LF} = \frac{1.0}{0.15} = 6.7 \text{ dS/m}$$

2. Estimate the TDS of the drainage water using Eq. (13–7).

TDS (mg/L) $\approx$ EC (mmho/cm or dS/m) $\times$ 640

TDS (mg/L) $\approx$ 6.7 $\times$ 640 = 4290 mg/L (ok less than 5000 mg/L)

3. Determine the leaching fraction using Eq. (13–9).

$$LF = \frac{EC_w}{EC_{dw}} = \frac{1.0 \times 640}{4290} = 0.15$$

Thus, to prevent loss in yield, a minimum of 15 percent of the applied water will be needed to carry salts below root zone and 85 percent will be consumed by evapotranspiration.

Specific Ion Toxicity. If the decline of crop growth is due to excessive concentrations of specific ions, rather than osmotic effects alone, it is referred to as "specific ion toxicity." As shown in Table 13–22, the ions of most concern in wastewater are sodium, chloride, and boron. The most prevalent toxicity from the use of reclaimed water is from boron. The source of boron is usually household detergents or discharges from industrial plants. The quantities of chloride and sodium also increase as a result of domestic usage, especially where water softeners are used.

For sensitive crops, specific ion toxicity is difficult to correct short of changing the crop or the water source. The problem is also accentuated by hot and dry climatic conditions due to high evapotranspiration rates. The suggested maximum trace element concentrations for irrigation waters are reported in Table 13–23. In severe cases, these elements tend to accumulate in plants and soils, which could result in human and animal health hazards or cause phytotoxicity in plants (Ayers and Westcot, 1985).

Water Infiltration Rate. In addition to the sodium toxicity discussed above, another indirect effect of high sodium content is the deterioration of the physical condition of a soil (formation of crusts, waterlogging, reduced soil permeability). If the infiltration rate is greatly reduced, it may be impossible to supply the crop or landscape plant with enough water for good growth. In addition, reclaimed water irrigation systems are often located on less desirable soils or soils already having permeability and management problems. It may be necessary, in these cases, to modify soil profiles by excavating and rearranging the affected land.

The water infiltration problem occurs within the top few centimeters of the soil and is mainly related to the structural stability of the surface soil. To predict a potential infiltration problem, the sodium adsorption ratio (SAR) is often used (Ayers and Westcot, 1985).

$$ SAR = \frac{Na^+}{\sqrt{\dfrac{Ca^{2+} + Mg^{2+}}{2}}} \tag{13-10} $$

where the cation concentrations are expressed in meq/L.

The adjusted sodium adsorption ratio (adj R_{Na}) is a modification of Eq. (13–10), which takes into account changes in calcium solubility in the soil water (Ayers and Westcot, 1985; and Suarez, 1981).

$$ adj\ R_{Na} = \frac{Na^+}{\sqrt{\dfrac{Ca_x^{2+} + Mg^{2+}}{2}}} \tag{13-11} $$

where Na^+ and Mg^{2+} concentrations are expressed in meq/L, and the value of Ca_x^{2+}, also expressed in meq/L, is obtained from Table 13–24.

Use of the adj R_{Na} value is preferred in irrigation applications with reclaimed water because it reflects the changes in calcium in the soil water more accurately. At a given sodium adsorption ratio, the infiltration rate increases as salinity increases or decreases as salinity decreases. Therefore, SAR, or adj R_{Na} and electrical conductivity (EC_w) of

Table 13–23
Recommended maximum concentrations of trace elements in irrigation waters[a]

Element	Recommended maximum concentration,[b] mg/L	Remarks
Al (aluminum)	5.0	Can cause nonproductivity in acid soils (pH < 5.5), but more alkaline soils at pH > 5.5 will precipitate the ion and eliminate any toxicity
As (arsenic)	0.10	Toxicity to plants varies widely, ranging from 12 mg/L for Sudan grass to less than 0.05 mg/L for rice
Be (beryllium)	0.10	Toxicity to plants varies widely, ranging from 5 mg/L for kale to 0.5 mg/L for bush beans
Cd (cadmium)	0.010	Toxic to beans, beets, and turnips at concentrations as low as 0.1 mg/L in nutrient solutions. Conservative limits recommended because of its potential for accumulation in plants and soils to concentrations that may be harmful to humans
Co (cobalt)	0.050	Toxic to tomato plants at 0.1 mg/L in nutrient solution. Tends to be inactivated by neutral and alkaline soils
Cr (chromium)	0.10	Not generally recognized as an essential growth element. Conservative limits recommended because of lack of knowledge on toxicity to plants
Cu (copper)	0.20	Toxic to a number of plants at 0.1 to 1.0 mg/L in nutrient solutions
F (fluoride)	1.0	Inactivated by neutral and alkaline soils
Fe (iron)	5.0	Not toxic to plants in aerated soils but can contribute to soil acidification and loss of reduced availability of essential phosphorus and molybdenum. Overhead sprinkling may result in unsightly deposits on plants, equipment, and buildings
Li (lithium)	2.5	Tolerated by most crops up to 5 mg/L; mobile in soil. Toxic to citrus at low levels (>0.075 mg/L). Acts similar to boron
Mn (manganese)	0.20	Toxic to a number of crops at a few tenths mg to a few mg/L, but usually only in acid soils
Mo (molybdenum)	0.010	Not toxic to plants at normal concentrations in soil and water. Can be toxic to livestock if forage is grown in soils with high levels of available molybdenum
Ni (nickel)	0.20	Toxic to a number of plants at 0.5 to 1.0 mg/L; reduced toxicity at neutral or alkaline pH
Pb (lead)	5.00	Can inhibit plant cell growth at very high concentrations
Se (selenium)	0.020	Toxic to plants at concentrations as low as 0.025 mg/L and toxic to livestock if forage is grown in soils with relatively high levels of added selenium. An essential element for animals but in very low concentrations
Sn (tin)	—	Effectively excluded by plants; specific tolerance unknown
Ti (titanium)	—	(See remark for tin)
W (tungsten)	—	(See remark for tin)
V (vanadium)	0.10	Toxic to many plants at relatively low concentrations
Zn (zinc)	2.0	Toxic to many plants at widely varying concentrations; reduced toxicity at pH > 6.0 and in fine-textured or organic soils

[a] Adapted from Ayers and Westcot (1985) and NAS (1972).
[b] The maximum concentration is based on a water application rate of 1.25 m/yr (4 ft/yr) that is consistent with good agricultural practice.

Table 13–24
Values of Ca_x^{2+} used in Eq. (13–11) as a function of the HCO_3^-/Ca^{2+} ratio and solubility[a]

Ratio of HCO_3^-/Ca^{2+} in meq/L	Values of Ca_x^{2+}, meq/L											
	Salinity of applied water (EC$_w$), dS/m or mmhos/cm											
	0.1	0.2	0.3	0.5	0.7	1.0	1.5	2.0	3.0	4.0	6.0	8.0
0.05	13.20	13.61	13.92	14.40	14.79	15.26	15.91	16.43	17.28	17.97	19.07	19.94
0.10	8.31	8.57	8.77	9.07	9.31	9.62	10.02	10.35	10.89	11.32	12.01	12.56
0.15	6.34	6.54	6.69	6.92	7.11	7.34	7.65	7.90	8.31	8.64	9.17	9.58
0.20	5.24	5.40	5.52	5.71	5.87	6.06	6.31	6.52	6.86	7.13	7.57	7.91
0.25	4.51	4.65	4.76	4.92	5.06	5.22	5.44	5.62	5.91	6.15	6.52	6.82
0.30	4.00	4.12	4.21	4.36	4.48	4.62	4.82	4.98	5.24	5.44	5.77	6.04
0.35	3.61	3.72	3.80	3.94	4.04	4.17	4.35	4.49	4.72	4.91	5.21	5.45
0.40	3.30	3.40	3.48	3.60	3.70	3.82	3.98	4.11	4.32	4.49	4.77	4.98
0.45	3.05	3.14	3.22	3.33	3.42	3.53	3.68	3.80	4.00	4.15	4.41	4.61
0.50	2.84	2.93	3.00	3.10	3.19	3.29	3.43	3.54	3.72	3.87	4.11	4.30
0.75	2.17	2.24	2.29	2.37	2.43	2.51	2.62	2.70	2.84	2.95	3.14	3.28
1.00	1.79	1.85	1.89	1.96	2.01	2.09	2.16	2.23	2.35	2.44	2.59	2.71
1.25	1.54	1.59	1.63	1.68	1.73	1.78	1.86	1.92	2.02	2.10	2.23	2.33
1.50	1.37	1.41	1.44	1.49	1.53	1.58	1.65	1.70	1.79	1.86	1.97	2.07
1.75	1.23	1.27	1.30	1.35	1.38	1.43	1.49	1.54	1.62	1.68	1.78	1.86
2.00	1.13	1.16	1.19	1.23	1.26	1.31	1.36	1.40	1.48	1.54	1.63	1.70
2.25	1.04	1.08	1.10	1.14	1.17	1.21	1.26	1.30	1.37	1.42	1.51	1.58
2.50	0.97	1.00	1.02	1.06	1.09	1.12	1.17	1.21	1.27	1.32	1.40	1.47
3.00	0.85	0.89	0.91	0.94	0.96	1.00	1.04	1.07	1.13	1.17	1.24	1.30
3.50	0.78	0.80	0.82	0.85	0.87	0.90	0.94	0.97	1.02	1.06	1.12	1.17
4.00	0.71	0.73	0.75	0.78	0.80	0.82	0.86	0.88	0.93	0.97	1.03	1.07
4.50	0.66	0.68	0.69	0.72	0.74	0.76	0.79	0.82	0.86	0.90	0.95	0.99
5.00	0.61	0.63	0.65	0.67	0.69	0.71	0.74	0.76	0.80	0.83	0.88	0.93
7.00	0.49	0.50	0.52	0.53	0.55	0.57	0.59	0.61	0.64	0.67	0.71	0.74
10.00	0.39	0.40	0.41	0.42	0.43	0.45	0.47	0.48	0.51	0.53	0.56	0.58
20.00	0.24	0.25	0.26	0.26	0.27	0.28	0.29	0.30	0.32	0.33	0.35	0.37

[a] Adapted from Pettygrove and Asano (1985).

irrigation water should be used in combination to evaluate the potential permeability problem as shown in Table 13–22.

Reclaimed water is normally high in calcium, and there is little concern for the water dissolving and leaching too much calcium from the surface soil (see Example 13–5). However, reclaimed water is sometimes high in sodium; the resulting high SAR is a major concern in planning irrigation projects with reclaimed water.

EXAMPLE 13–5 **Calculation of Adjusted Sodium Adsorption Ratio and Evaluation of Potential Water Infiltration Problems** The following water quality analyses were reported on an aerated lagoon effluent which will be used for irrigating agricultural land. Using the reported water quality data: (1) calculate adj R_{Na}, and (2) determine whether an infiltration problem may develop by using this effluent for irrigation.

Water quality parameter	Concentration, mg/L
BOD	39
TSS	160
Total N	4.4
Total P	5.5
pH[a]	7.7
Cations:	
$\quad Ca^{2+}$	37
$\quad Mg^{2+}$	46
$\quad Na^+$	410
$\quad K^+$	27
Anions:	
$\quad HCO_3^-$	295
$\quad SO_4^{2-}$	66
$\quad Cl^-$	526
Electrical conductivity (dS/m)	2.4
TDS	1536
Boron	1.2
Alkalinity (total, as $CaCO_3$)	242
Hardness (total, as $CaCO_3$)	281

[a] Unitless.

Solution

1. Calculate the adj R_{Na} using Eq. (13–11) and the values given in Table 13–24.
   a. Convert the concentrations of the related water quality parameters to meq/L

   $Ca^{2+} = 37/20.04 = 1.85$ meq/L

   $Mg^{2+} = 46/12.15 = 3.79$ meq/L

   $Na^+ = 410/23 = 17.83$ meq/L

   $HCO_3^- = 295/61 = 4.84$ meq/L

  b. Determine the value of Ca_x^{2+} in Eq. (13–11) using the given water quality data.
    i. Salinity of applied water $(EC_w) = 2.4$ dS/m
    ii. Ratio of $HCO_3^-/Ca^{2+} = 4.84/1.85 = 2.62$
    iii. From Table 13–24, the value of $Ca_x^{2+} = 1.20$ meq/L
  c. The adj R_{Na} is

$$ \text{adj } R_{Na} = \cfrac{Na^+}{\sqrt{\cfrac{Ca_x^{2+} + Mg^{2+}}{2}}} = \cfrac{17.83}{\sqrt{\cfrac{1.20 + 3.79}{2}}} = 11.29 $$

2. Determine whether an infiltration problem will develop.

  Entering Table 13–22 with adj $R_{Na} = 11.29$ and $EC_w = 2.4$ dS/m, no restrictions are indicated for use of this reclaimed wastewater.

Nutrients. The nutrients in reclaimed water provide fertilizer value for crop or landscape production. However, in certain instances, when nutrients are in excess of plant needs they can cause problems, as described below. Nutrients that are important to agriculture and landscape management include N, P, and occasionally K, Zn, B, and S. The most beneficial and the most frequently excessive nutrient in reclaimed water is nitrogen.

  The nitrogen in reclaimed water can replace equal amounts of commercial fertilizer during the early to midseason crop-growing period. Excessive nitrogen in the latter part of the growing period may be detrimental to many crops, causing excessive vegetative growth, delayed or uneven maturity, or reduced crop quality. If alternate low-nitrogen water is available, a switch in water supplies or blending of reclaimed water with other water supplies has been used to keep nitrogen under control, otherwise expensive seasonal denitrification would be required.

Other Problems

Clogging problems with sprinkler and drip irrigation systems have been reported, particularly with oxidation pond effluent. Biological growth (slimes) in the sprinkler head, emitter orifice, or supply line causes plugging, as do heavy concentrations of algae and suspended solids. The most frequent clogging problems occur with drip irrigation systems. From the standpoint of public health, such systems are often considered ideal, as they are totally enclosed, minimizing the problems of worker exposure to reclaimed water or spray drift.

  In reclaimed water that is chlorinated, chlorine residuals of less than 1 mg/L do not affect plant foliage, but chlorine residuals in excess of 5 mg/L can cause severe plant damage when reclaimed water is sprayed directly on foliage.

EXAMPLE 13–6 **Suitability and Effects of Various Irrigation Waters** Analyses of four representative waters in California are presented in the following data table. The waters are: (1) the relatively unpolluted Sacramento River, (2) a moderately saline groundwater in San Joaquin County, and (3) two reclaimed municipal wastewaters from the cities of Fresno and Bakersfield. Assuming the following conditions are applicable, determine the suitability of these waters for irrigation.

1. Daily crop water demand varies during the growing season and among crop types. The range of water demand may be from a low of 2 mm (0.08 in/d) to a high of 7.6 to 10 mm/d (0.3 to 0.4 in/d).
2. On-farm management of reclaimed wastewater must take crop water demand into account, and the irrigation objective should be to use the reclaimed wastewater efficiently to produce a crop.

Constituents[a]	Sacramento River	San Joaquin County groundwater	Fresno wastewater effluent	Bakersfield wastewater effluent
EC (dS/m)	0.11	1.25	0.69	0.77
pH	7.1	7.7	8.6	7.0
Ca^{2+}	10	100	24.0	47
Mg^{2+}	5	33	12.8	5
Na^+	6	92	80	109
K^+	1.5	3.9	13.8	26
SAR	0.4	2.0	3.3	4.1
HCO_3^-	42	190	236	218
SO_4^{2-}	7.3	110	—	62
Cl^-	2.2	200	70	107
NO_3^--N + NH_3-N	0.08	5.9	14[b]	0.5[c]
B	—	1.4	0.43	0.38
TDS[d]	72	800	442	477
As			<0.002	
Cd			<0.002	<0.01
Cr			<0.02	
Pb			<0.05	

[a] All concentrations are expressed in mg/L except electrical conductivity (EC) and pH.
[b] Total Kjeldahl-N.
[c] NO_3^--N. A total N of 20 to 25 mg/L was reported.
[d] TDS values estimated using Eq. (13–7).

Solution

1. Evaluation for the suitability of various irrigation waters is made based on the information given in Table 13–22.

	Degree of problem[a]			
Problem area	**Sacramento River**	**San Joaquin County groundwater**	**Fresno wastewater effluent**	**Bakersfield wastewater effluent**
Salinity	N	S–M	N	S–M
Infiltration	SV	N	S–M	S–M
Toxicity (sensitive crops only)				
Na+				
Surface irrigation	N	N	S–M	S–M
Sprinkler irrigation	N	S–M	S–M	S–M
Cl−				
Surface irrigation	N	S–M	N	N
Sprinkler irrigation	N	S–M	N	S–M
B	—	S–M	N	N
Miscellaneous (susceptible crops only)				
N	N	S–M	S–M	N
HCO_3^-	N	S–M	S–M	S–M

[a] N = no problem, S–M = slight to moderate problem, and SV = severe problem expected when using respective water for a long period of time.

2. Although the water quality of the two reclaimed waters is such that minor water infiltration, toxicity, and miscellaneous problems may be anticipated from use, normal agronomic practice used in the area has proved to be adequate to allow full production of adapted crops. The cities of Fresno and Bakersfield have a long history of using reclaimed water for irrigation.

13–7 INDUSTRIAL WATER REUSE

Industrial water use includes water for such purposes as processing, washing, and cooling in facilities that manufacture products. Major water-using industries include steel, chemical and allied products, paper and allied products, and petroleum refining. Industrial water reuse represents a significant potential application for reclaimed water in the United States and other industrialized countries. Industrial uses of reclaimed water from external sources such as reclaimed municipal wastewater include evaporative cooling water, boiler feedwater, process water, and landscaping and maintenance of industrial grounds. Water recycling within an industrial plant is normally an integral part of the industrial process, and treated and reclaimed water is recycled to conserve water and avoid stringent discharge requirements.

Industrial Water Use

Total industrial water use during 1995 was estimated to be 103×10^6 m³/d (27,100 Mgal/d). Industrial freshwater use was an estimated 97×10^6 m³/d (25,500 Mgal/d) during 1995, and represents about 7 percent of freshwater use for all off-stream categories. Self-supplied industrial withdrawals were an estimated 78×10^6 m³/d (20,700 Mgal/d) of freshwater and 6×10^6 m³/d (1660 Mgal/d) of saline water. Surface water was the source for 82 percent of self-supplied industrial withdrawals; groundwater, 18 percent; and reclaimed wastewater less than 1 percent. Public-supply deliveries to industries were about 18×10^6 m³/d (4750 Mgal/d) and accounted for 12 percent total public-supply withdrawals. The Great Lakes and Ohio water resources regions had the largest fresh- and saline-water withdrawals for industrial purposes in 1995 (Solley et al., 1998).

Cooling Tower Makeup Water

Among several industrial water reuse applications, cooling tower makeup water (see Fig. 13–22) represents a significant water use for many industries and is currently the predominant industrial water reuse application. For industries such as electric power generating stations, oil refining, and many other types of manufacturing plants, one-quarter to more than one-half of the total water use may be cooling tower makeup water. Because a cooling tower normally operates as a closed-loop system, it can be viewed as a separate water system with its own specific set of water quality requirements, largely independent of the particular industry involved. Thus, using reclaimed water for cooling tower makeup water is relatively easy and is practiced in many locations in the United States.

The onsite water reclamation plant for the cooling tower operations at the Palo Verde Nuclear Generating Station in Arizona is shown on Fig. 13–23. Secondary effluent from the cities of Tolleson and Phoenix is pumped 61 km (38 miles) to this site. Before using the effluent, it is subjected to advanced treatment consisting of (1) biological nitrification, (2) lime and soda ash addition for softening and phosphorus

Figure 13–22

Typical industrial cooling towers operated with reclaimed wastewater.

Figure 13–23

Onsite wastewater reclamation plant for the cooling tower operations at the Palo Verde Nuclear Generating Station. The secondary treated effluents from cities of Tolleson and Phoenix, AZ, are pumped 61 km (38 mi) to this site and undergo advanced wastewater treatment. *(Photo courtesy of Arizona Nuclear Power Project.)*

removal, (3) filtration, and (4) chlorination. The purpose of the advanced treatment is to reduce corrosion and scaling in the cooling tower systems.

Water and Salt Balances in Cooling Tower

The basic principle of cooling tower operation is that of evaporative condensation and exchange of sensible heat. The air and water mixture releases latent heat of vaporization. Water exposed to the atmosphere evaporates and as the water changes to vapor, heat is consumed, which amounts to approximately 2.3 kJ/g (1000 Btu/lb) of water evaporated.

Under normal operating conditions, the loss of water discharged from the cooling tower to the atmosphere as hot, moist vapor amounts to approximately 1.2 percent for each 5.5°C (10°F) of cooling range. Drift, or water lost from the top of the tower to the wind, is the second mechanism by which water is lost from the cooling system. About 0.005 percent of the recirculating water is lost in this way. While evaporation results in a loss of water from the system, the salt concentration is increased because salts are not removed by evaporation. To prevent the formation of precipitates in the resulting more-concentrated tower water, a portion of the concentrated cooling water is bled off and replaced with low-salt-makeup water to maintain a proper salt balance. This highly saline water bled off from the cooling tower system is called *blowdown*.

The total makeup water flow for the cooling tower system includes all three of the water losses described above (i.e., vapor loss, drift, and blowdown). The definition sketch for a recirculating evaporative cooling tower is shown on Fig. 13–24.

Figure 13–24

Definition sketch for salt balance in the recirculating, evaporative cooling tower.

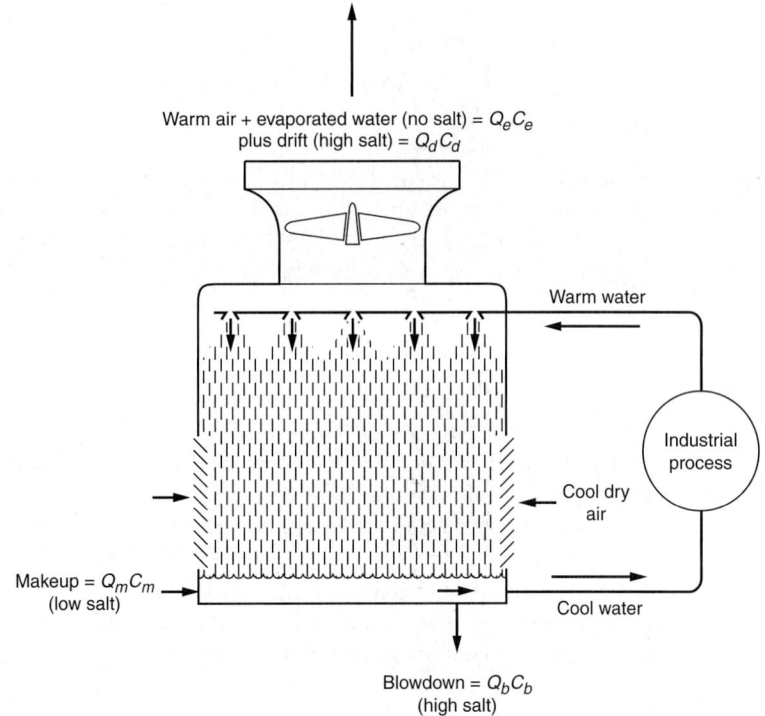

Warm air + evaporated water (no salt) = $Q_e C_e$
plus drift (high salt) = $Q_d C_d$

Warm water

Industrial process

Cool dry air

Makeup = $Q_m C_m$
(low salt)

Cool water

Blowdown = $Q_b C_b$
(high salt)

Salt in makeup = salt out in blowdown and drift

The water balance around the cooling tower is

Accumulation = in − out

At steady state the accumulation term is equal to zero and

$$Q_m = Q_b + Q_d + Q_e \qquad (13\text{--}12)$$

where Q_m = makeup water flow, L/min
Q_b = blowdown flow, L/min
Q_d = drift flow, L/min
Q_e = evaporation loss, L/min

Drift flow Q_d is normally small enough to be ignored, as noted previously (<0.005 percent).

In a similar way, the salt balance in the cooling tower is

$$Q_m C_m = Q_b C_b + Q_d C_d + Q_e C_e \qquad (13\text{--}13)$$

where C_m = salt concentration in makeup water flow, mg/L
C_b = salt concentration in blowdown flow, mg/L
C_d = salt concentration in drift flow, mg/L
C_e = salt concentration in evaporation loss, mg/L

Because Q_d is negligible, the term $Q_d C_d$ can be omitted without serious error. Further, because the concentration of salt in the evaporation water is also negligible under normal operating conditions Eq. (13–13) can be reduced to Eq. (13–14).

$$Q_m C_m = Q_b C_b \tag{13–14}$$

The magnitude of the blowdown flow (and thus the makeup flow) is dependent upon the concentration of potential precipitants in the makeup water. The ratio of the concentration of the salt C_b in the blowdown to its concentration C_m in the makeup water is known as the cycles of concentration.

$$\text{Cycles of concentration} = C_b / C_m \tag{13–15}$$

Combining Eqs. (13–14) and (13–15) yields

$$\text{Cycles of concentration} = Q_m / Q_b \tag{13–16}$$

It can be seen in Eq. (13–16) that the cycles of concentration also equal the ratio of the makeup flow to the blowdown flow.

When the cycles of concentration are on the order of 3 to 7, some of the dissolved solids in the circulating water can exceed their solubility limits and precipitate, causing scale formation in pipes and coolers. To avoid scale formation, sulfuric acid is often used to convert calcium and magnesium carbonates into more soluble sulfate compounds. The amount of acid used must be limited to maintain some residual alkalinity in the system. If the pH of the system is reduced too far below 7, accelerated corrosion can occur (see Example 13–6).

Common Water Quality Problems in Cooling Tower Systems

Four general water quality problems are encountered in industrial cooling tower operations: (1) scaling, (2) metallic corrosion, (3) biological growths, and (4) fouling in heat exchanger and condensers. Both freshwater and reclaimed water contain constituents that can cause these problems, but their concentrations in reclaimed water are generally higher.

Scaling. Scaling refers to the formation of hard deposits, usually on hot surfaces, which reduce the efficiency of heat exchange. Calcium scales (calcium carbonate, calcium sulfate, and calcium phosphate) are the principal cause of cooling tower scaling problems. Magnesium scales (magnesium carbonate and phosphate) can also be a problem. Silica deposits are particularly difficult to remove from heat exchanger surfaces; however, most waters contain relatively small quantities of silica.

Reducing the potential for scaling in wastewater is achieved by controlling the formation of calcium phosphate, which is the first calcium salt to precipitate if phosphate is present. Treatment is usually accomplished by removing phosphates by precipitation (see Chap. 11). Other treatment methods, such as ion exchange, reduce scale formation by the removal of calcium and magnesium; however, these techniques are comparatively expensive and their use is limited.

Metallic Corrosion. In cooling systems, corrosion can occur when an electrical potential between dissimilar metal surfaces is created. The corrosion cell consists of an anode, where oxidation of one metal occurs, and a cathode, where reduction of another metal takes place. Water quality greatly affects metallic corrosion. Contaminants such as TDS increase the electrical conductivity of the solution and thereby accelerate the corrosion reaction. Dissolved oxygen and certain metals (manganese, iron, and aluminum) promote corrosion because of their relatively high oxidation potential.

The corrosion potential of cooling water can be controlled by the addition of chemical corrosion inhibitors. The chemical requirements to control corrosion in reclaimed water are usually much higher than for freshwater because the concentration of TDS is often two to five times higher in wastewater.

Biological Growth. The warm, moist environment inside the cooling tower makes an ideal environment for promoting biological growth. Nutrients, particularly N and P, and available organics further encourage the growth of microorganisms that can attach and deposit on heat-exchanger surfaces, inhibiting heat transfer and water flow. Biological growths may also settle and bind other debris present in the cooling water, which may further inhibit effective heat transfer. Certain microorganisms also create corrosive byproducts during their growth. Biological growths are usually controlled by the addition of biocides as part of the internal chemical treatment process that may include the addition of acid for pH control, the use of biocides, and scale and biofoul inhibitors. Because reclaimed water contains a greater concentration of organic matter, it may require larger dosages of biocides. It is possible, however, that most of the nutrients and available organic matter are removed from the reclaimed water during biological and chemical treatment.

When reclaimed water is used for cooling, the assurance of adequate disinfection is a primary concern to protect the health of workers. The disinfection requirements for use of reclaimed water in industrial processes are made on a case-by-case basis. The most stringent requirement, similar to unrestricted reclaimed water use in food crop irrigation, would be appropriate if exposure to spray were possible. Protection of the neighboring public as well as plant operators is of prime importance.

Fouling. Fouling refers to the process of attachment and growth of deposits of various kinds in cooling tower recirculation systems. The deposits consist of biological growths, suspended solids, silt, corrosion products, and inorganic scales. Inhibition of heat transfer in the heat exchangers can result. Control of fouling is achieved by the addition of chemical dispersants that prevent particles from aggregating and subsequently settling. Dispersants are also added at the point of use, as is the usual case for freshwater cooling systems. Also, the chemical coagulation and filtration processes required for phosphorus removal are effective in reducing the concentration of contaminants that contribute to fouling.

In most cases, disinfected secondary effluent is supplied for noncritical, once-through cooling. For recirculating cooling tower operation, most wastewaters contain constituents which, if not removed, would limit industries to very low cycles of concentration in their cooling towers. Additional treatment includes lime clarification, alum precipitation, or ion exchange. Treatment processes used for both external and internal treatment of cooling or boiler makeup water are summarized in Table 13–25. In many cases, the water quality requirements for the use of reclaimed water are the same as those for freshwater. Water quality requirements at the point of use for cooling waters for both once-through and makeup for recirculation are reported in Table 13–26.

Table 13–25
Processes used in treating water for cooling or boiler makeup

Processes	Cooling		Boiler makeup
	Once-through	Recirculated	
Suspended solids and colloids removal:			
Straining	X	X	X
Sedimentation	X	X	X
Coagulation		X	X
Filtration		X	X
Aeration		X	X
Microfiltration		X	X
Dissolved-solids modification softening:			
Cold lime		X	X
Hot lime soda			X
Hot lime zeolite			X
Cation-exchange sodium		X	X
Nanofiltration			X
Alkalinity reduction cation exchange:			
Hydrogen		X	X
Cation-exchange hydrogen and sodium		X	X
Anion exchange			X
Dissolved-solids removal:			
Evaporation			X
Demineralization		X	X
Reverse osmosis/nanofiltration		X	X
Ion exchange		X	X
Dissolved-gases removal:			
Degasification			
Mechanical		X	X
Vacuum	X		X
Heat			X
Internal conditioning:			
pH adjustment	X	X	X
Hardness sequestering	X	X	X
Hardness precipitation			X
Corrosion inhibition general		X	X
Embrittlement			X
Oxygen reduction			X
Sludge dispersal	X	X	X
Biological control			
Chemicals	X	X	
Ozone		X	
Ultraviolet light		X	

Table 13-26

Water quality requirements at point of use for steam generation and cooling in heat exchangers[a]

| Characteristic | Boiler feedwater — Quality of water prior to the addition of chemicals used for internal conditioning | | | | Cooling water | | | |
| | Industrial | | | Electrical utilities | Once-through | | Makeup for recirculation | |
	Low pressure	Intermediate pressure	High pressure		Fresh	Brackish[b]	Fresh	Brackish[b]
Silica (SiO$_2$)	30	10	0.7	0.01	50	25	50	25
Aluminum (Al)	5	0.1	0.01	0.01	c	c	0.1	0.1
Iron (Fe)	1	0.3	0.05	0.01	c	c	0.5	0.5
Manganese (Mn)	0.3	0.1	0.01	0.01	c	c	0.5	0.02
Calcium (Ca)	c	0.4	0.01	0.01	200	420	50	420
Magnesium (Mg)	c	0.25	0.01	0.01	c	c	c	c
Ammonia (NH$_4$)	0.1	0.1	0.1	0.07	c	c	c	c
Bicarbonate (HCO$_3$)	170	120	48	0.5	600	140	24	140
Sulfate (SO$_4$)	c	c	c	c	680	2,700	200	2,700
Chloride (Cl)	c	c	c	c,e	600	19,000	500	19,000
Dissolved solids	700	500	200	0.5	1000	35,000	500	35,000
Copper (Cu)	0.5	0.05	0.05	0.01	c	c	c	c
Zinc (Zn)	c	0.01	0.01	0.01	c	c	c	c
Hardness (CaCO$_3$)	350	1.0	0.07	0.07	850	6,250	650	6,250
Alkalinity (CaCO$_3$)	350	100	40	1	500	115	350	115

(continued)

| Table 13-26 (Continued)

	Boiler feedwater				Cooling water			
	Quality of water prior to the addition of chemicals used for internal conditioning							
	Industrial			Electrical utilities	Once-through		Makeup for recirculation	
Characteristic	Low pressure	Intermediate pressure	High pressure		Fresh	Brackish[b]	Fresh	Brackish[b]
pH, units	7.0–10.0	8.2–10.0	8.2–9.0	8.8–9.4	5.0–8.3	6.0–8.3	c	c
Organics:								
Methylene blue active substances	1	1	0.5	0.1	c	c	1	1
Carbon tetrachloride extract	1	1	0.5	c,d	f	f	1	2
Chemical oxygen demand (COD)	5	5	1.0	1.0	75	75	75	75
Hydrogen sulfide (H$_2$S)	c	c	c	c	—	c	c	c
Dissolved oxygen (O$_2$)	2.5	0.007	0.007	0.007	Present	Present	c	c
Temperature	c	c	c	c	c	c	c	c
Suspended solids	10	5	0.5	0.05	5000	2500	100	100

Note: Unless otherwise indicated, units are mg/L and values that normally should not be exceeded. No one water will have all the maximum values shown.

[a] From NAS (1972).

[b] Brackish water—dissolved solids more than 1000 mg/L.

[c] Accepted as received (if meeting other limiting values); has never been a problem at concentrations encountered.

[d] Zero, not detectable by test.

[e] Controlled by treatment for other constituents.

[f] No floating oil.

EXAMPLE 13–7 **Estimation of Blowdown Water Composition** Reclaimed water with the chemical characteristics given below is being considered for use as makeup water for a cooling tower. Calculate the composition of the blowdown flow if 5 cycles of concentration are to be used. Assume that the temperature of the hot water entering the cooling tower is 50°C (120°F) and the solubility of $CaSO_4$ is about 2200 mg/L as $CaCO_3$ at this temperature.

Constituent	Concentration, mg/L
Total hardness (as $CaCO_3$)	118
Ca²⁺ (as $CaCO_3$)	85
Mg²⁺ (as $CaCO_3$)	33
Total alkalinity (as $CaCO_3$)	90
SO_4^{2-}	20
Cl^-	19
SiO_2	2

Where mw of $CaCO_3$ = 100, $CaSO_4$ = 136, H_2SO_4 = 98, and SO_4 = 96.

Solution

1. Determine the total hardness in the circulating water.
   a. When the total alkalinity is less than total hardness, Ca and Mg are also present in forms other than carbonate hardness.
   b. Setting the cycles of concentration equal to 5 and using Eq. (13–15), the total hardness in circulating water is

   C_b = (cycles of concentration) (C_m)

   = 5 × 118 = 590 mg/L $CaCO_3$

2. Determine the total amount of H_2SO_4 that must be added to convert the $CaCO_3$ to $CaSO_4$.
   a. To convert from $CaCO_3$ to $CaSO_4$, sulfuric acid is injected into the circulating water and the following reaction occurs:

   $$CaCO_3 + H_2SO_4 \rightarrow CaSO_4 + H_2O + CO_2$$
   $$\quad 100 \qquad 98 \qquad\quad 136$$

   b. The alkalinity in the circulating water, if not converted into sulfates, is 5 × 90 = 450 mg/L as $CaCO_3$.
   c. If 10 percent of the alkalinity is left unconverted to avoid corrosion, the amount of alkalinity remaining is 0.1 × 450 = 45 mg/L as $CaCO_3$.
   d. The amount of alkalinity that must be converted is 450 − (0.1 × 450) = 405 mg/L as $CaCO_3$.
   e. The amount of sulfate that must be added for the conversion is

   $$SO_4^{2-} = (405 \text{ mg/L})\left(\frac{96}{100}\right) = 389 \text{ mg/L}$$

f. Converting to mg/L $CaSO_4$ yields

$$CaSO_4 = (389 \text{ mg/L})\left(\frac{136}{96}\right) = 551 \text{ mg/L}$$

3. Determine the required sulfuric acid concentration in the circulating water.

$$H_2SO_4 = (389 \text{ mg/L } SO_4)\left(\frac{98}{96}\right) = 397 \text{ mg/L}$$

4. Determine the sulfate concentration in the circulating water contributed by the makeup water.
    a. Sulfate from makeup water is $5 \times 20 = 100$ mg/L as SO_4.
    b. If combined with Ca^{2+}, the concentration is $CaSO_4 = (100 \text{ mg/L})\left(\frac{136}{96}\right) = 142$ mg/L.

5. The solubility of $CaSO_4$ at 50°C (120°F) is about 2200 mg/L. In the circulating water, 142 mg/L $CaSO_4$ was originally present after 5 cycles of concentration and 551 mg/L was formed by the addition of sulfuric acid. Therefore, an additional 1507 mg/L of $CaSO_4$ formation is theoretically permissible before the solubility limit is exceeded. The cycles of concentration could have been carried out much higher before $CaSO_4$ would precipitate in the system.

6. Determine the concentrations of Cl and SiO_2 in the circulating water.
    a. Chloride

$$Cl^- = 5 \times 19 = 95 \text{ mg/L}$$

    b. Silica

$$SiO_2 = 5 \times 2 = 10 \text{ mg/L}$$

7. Summarize the composition of the blowdown flow after 5 cycles of concentration.

Parameter	Concentration, mg/L	
	Initial	Final
Total hardness (as $CaCO_3$)	118	590
Total alkalinity (as $CaCO_3$)	90	45
SO_4^{2-}	20	489
Cl^-	19	95
SiO_2	2	10

13–8 GROUNDWATER RECHARGE WITH RECLAIMED WATER

Groundwater recharge has been used to (1) reduce, stop, or even reverse declines of groundwater levels, (2) protect underground freshwater in coastal aquifers against saltwater intrusion from the ocean, and (3) store surface water, including flood or other surplus water, and reclaimed water for future use. Groundwater recharge is also achieved incidentally in land treatment and disposal systems where municipal and industrial wastewater is disposed of via percolation and infiltration.

Groundwater recharge with reclaimed water is an approach to water reuse that results in the planned augmentation of groundwater supplies. There are several advantages to storing water underground: (1) the cost of artificial recharge may be less than the cost of equivalent surface reservoirs; (2) the aquifer serves as an eventual distribution system and may eliminate the need for surface pipelines or canals; (3) water stored in surface reservoirs is subject to evaporation, to potential taste and odor problems due to algae and other aquatic productivity, and to pollution; (4) suitable sites for surface reservoirs may not be available or environmentally acceptable; and (5) the inclusion of groundwater recharge in an indirect potable reuse project may also provide health, psychological, and aesthetic secondary benefits as a result of the transition between reclaimed water and groundwater.

Groundwater Recharge Methods

Two methods of groundwater recharge are used commonly with reclaimed water: (1) surface spreading in basins, and (2) direct injection into groundwater aquifers.

Groundwater Recharge by Surface Spreading. Surface spreading (see Figs. 13–4, 13–25, and 13–26) is the simplest, oldest, and most widely used method of groundwater recharge. In surface spreading, recharge waters percolate from the spreading basins through an unsaturated groundwater (vadose) zone. Infiltration basins are the most favored methods of recharge because they allow efficient use of space and require relatively little maintenance. Based on recent findings, the assumption that the vadose zone can be counted for treatment of trace organics has come into question. It appears, based on large-scale tracer studies, that much of the flow moves down through the soil column in preferential flow paths and receives little or no treatment as it passes through the soil.

Where hydrogeological conditions are favorable for groundwater recharge with spreading basins, water reclamation can be implemented relatively simply by the rapid infiltration (also known as the soil-aquifer treatment (SAT) system or geopurification). Typical SAT system schematics are shown on Fig. 13–27, where the necessary treatment for conventional constituents can be obtained by filtration as the treated effluent (normally the secondary treated effluent) percolates through the soil and the vadose zone, down to the groundwater, and then some distance through the aquifer. The extracted groundwater can be used for irrigation for a variety of food crops.

Because recharged groundwater may be an eventual source of potable water supply, groundwater recharge with reclaimed water often involves treatment beyond the conventional secondary treatment. For surface spreading operations practiced in California, common wastewater reclamation processes prior to recharge include primary and secondary wastewater treatment, and tertiary granular-medium filtration followed by disinfection with chlorine or UV radiation. The RIX (rapid infiltration/extraction) spreading ground in San Bernardino County, CA, is shown on Fig. 13–28. The water extracted from the area surrounding the spreading basin, which has been filtered by passage through the soil, is disinfected by UV radiation before being discharged to the Santa Ana River.

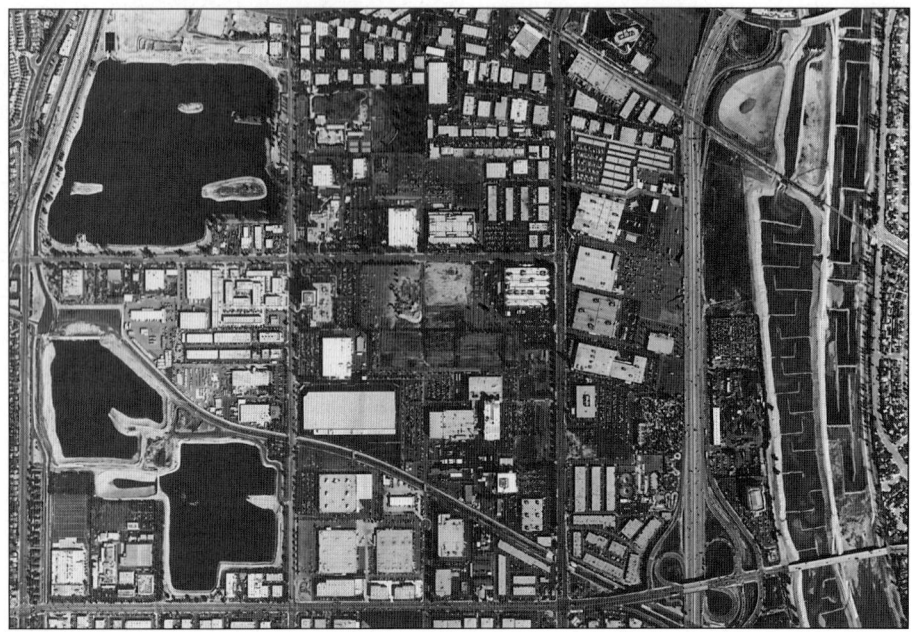

Figure 13–25

Groundwater replenishment facilities operated by the Orange County Water District in southern CA. Anaheim Lake, Kraemer Basin, and Miller Basin, shown top to bottom on the left-hand side of the photo, are spreading basins where Santa Ana River water and imported water is recharged to replenish the Orange County Groundwater Basin. The three basins can recharge up to 1.11×10^5 m³/yr (90,000 ac-ft/yr). In the future, the Groundwater Replenishment System, a joint project of the Orange County Water District and the Orange County Sanitation District, will be used to recharge reclaimed water, treated by secondary treatment, microfiltration, reverse osmosis, and ultraviolet light, at Kraemer Basin. Infiltration facilities used to recharge about 8.6×10^4 m³/yr (70,000 ac-ft/yr) of Santa Ana River water are shown on the right hand side of the photo. *(Photo courtesy of the Orange County Water District, aerial photography by Robert J. Lung, Associates.)*

Figure 13–26

Inflatable rubber dam used to divert flow from the Santa Ana River in southern California to groundwater spreading basins. In the summer the flow in the river is comprised almost entirely of treated effluent from upstream treatment plants.

Figure 13–27

Typical schematics of soil-aquifer treatment (SAT) systems with recovery of renovated water by: (a) subsurface drains, (b) wells surrounding the spreading basins, and (c) wells midway between two parallel strips of basins.

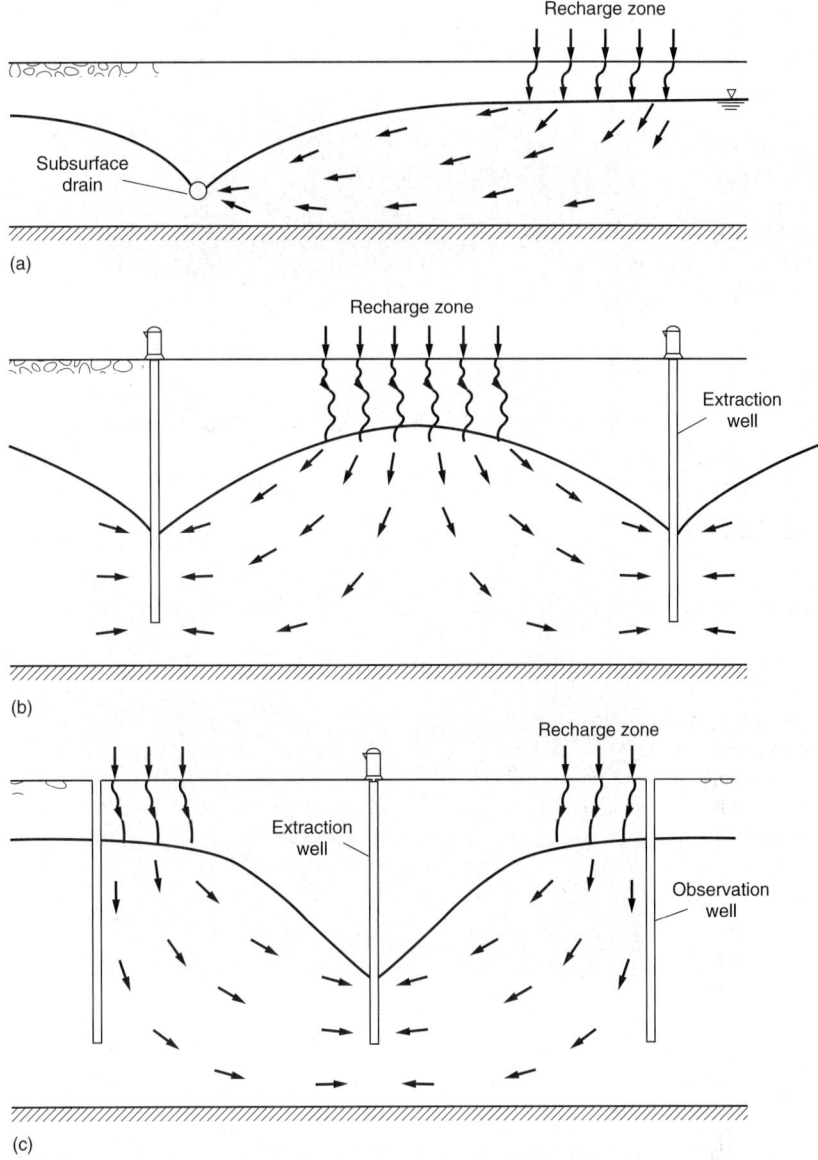

Groundwater Recharge by Direct Injection. Direct subsurface recharge is achieved when water is conveyed and injected directly into groundwater aquifer. In direct injection, generally, highly treated reclaimed water is injected directly into the saturated groundwater zone, usually into a well-confined aquifer. Groundwater recharge by direct injection is practiced, in most cases, where groundwater is deep or where the topography or existing land use makes surface spreading impractical or too expensive. This method of groundwater recharge is particularly effective in creating freshwater barriers in coastal aquifers against intrusion of saltwater from the sea.

Figure 13-28

Rapid infiltration and extraction surface spreading basins. Reclaimed water that has infiltrated is pumped out from a series of wells surrounding the spreading basins, disinfected with ultraviolet irradiation, and released to the Santa Ana River.

Figure 13-29

View (from effluent end) of lime clarification facilities used for the pretreatment of secondary effluent for reverse osmosis filtration. The flocculation basins are located at the head end of the clarification facilities. Lime storage and slaking facilities are located in the building just behind the flocculation facilities. In the future, lime clarification will be replaced by microfiltration (see Fig. 13-13b).

Pretreatment Requirements for Groundwater Recharge

Pretreatment requirements for groundwater recharge vary considerably, depending on the purpose of groundwater recharge, sources of reclaimed water, recharge methods, and location (see Fig. 13–29). For example, pretreatment for municipal wastewater for application to a SAT system may include only primary treatment or treatment in a stabilization pond. However, pretreatment processes that leave high algal concentrations in the recharge water should be avoided. Algae can severely clog the soil of infiltration basins. While the water recovered from the SAT system has much better water quality than the influent water, it could still be lower quality than the native groundwater. Thus, the SAT system should be designed and managed to avoid encroachment into the native groundwater and to use only a portion of the aquifer as shown on Fig. 13–27. The distance between infiltration basins and wells or drains should be as large as possible, usually at least 45 to 106 m (150 to 350 ft) to allow for adequate soil-aquifer treatment.

More detailed discussions regarding the capability of various advanced wastewater treatment process combinations are found in Sec. 13–4. To avoid potential health risks, careful attention must be paid to groundwater recharge operations when a possibility exists to augment substantial portions of potable groundwater supplies. Both in surface spreading and in direct injection, locating the extraction wells as great a distance as possible from the spreading basins or the injection wells increases the flow path length and residence time of the recharged water. These separations in space and in time contribute to the assimilation of the recharged reclaimed water with the other aquifer contents, and the loss of identity of the recharged water originated from municipal wastewater.

Fate of Contaminants in Groundwater

Understanding of the behavior of stable organic constituents and bacterial and viral pathogens is crucial in evaluating the feasibility of groundwater recharge using reclaimed water. Treated effluents contain trace quantities of nonconventional and emerging constituents and compounds even when the most advanced treatment technology is used. The transport and fate of these substances in the subsurface environment are governed by various mechanisms, which include biodegradation by microorganisms, chemical oxidation and reduction, sorption and ion exchange, filtration, chemical precipitation, dilution, volatilization, and photochemical reactions (in spreading basins) (see Table 4–5 in Chap. 4).

Particulate Constituents. Particulate constituents including microorganisms in reclaimed water are removed by filtration and retained effectively by the soil matrix. Factors affecting the movement, removal, and inactivation of viruses in groundwater are discussed in Bitton and Gerba (1984) and Ward et al. (1985).

Dissolved Inorganic and Organic Constituents. In addition to the common dissolved mineral constituents, reclaimed water contains many dissolved trace elements. The physical action of filtration does not, however, accomplish the removal of these dissolved inorganic contaminants. For trace metals to be retained in the soil matrix, physical, chemical, or microbiological reactions are required to immobilize the dissolved constituents. In a groundwater recharge system, the impact of microbial activities on the attenuation of inorganic constituents is small. Physical and chemical reactions in the soil that are important with respect to trace metal elements include cation exchange, precipitation, surface adsorption, and chelation and complexation. Although soils do not possess unlimited capability in attenuating inorganic constituents, it has been found in experimental studies that soils do have capacities for retaining large amounts of trace metal elements. Therefore, it is conceivable that a site used for groundwater recharge may be effective in retaining trace metals for extended periods of time.

Removal of dissolved organic constituents is affected primarily by biodegradation and adsorption during groundwater recharge operations. Biodegradation offers the potential of permanent conversion of toxic organic substances into harmless products. The rate and extent of biodegradation are strongly influenced by the nature of the organic substances, as well as by the presence of electron acceptors such as dissolved oxygen, nitrate, and sulfate. Biodegradation of easily degradable substances takes place

Figure 13–30

Effect of dispersion, sorption, and biodecomposition on the time change in concentration of an organic compound. Expected responses to a step change in concentration are shown. (From McCarty et al., 1981.)

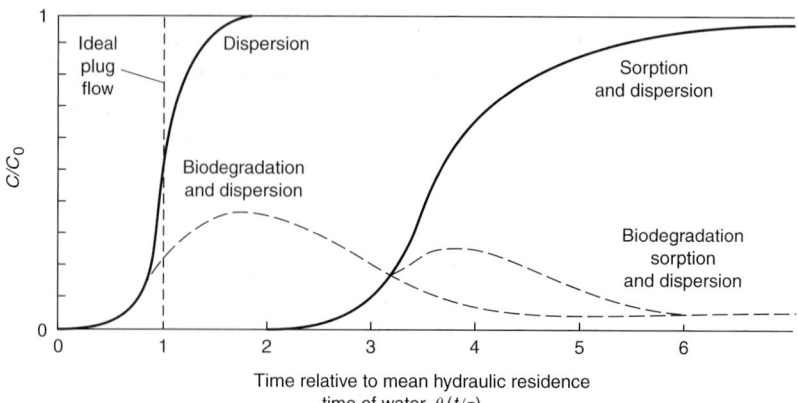

almost exclusively in the first one or 0.6 m (2 ft) of travel. There is increasing evidence that a significant portion of the observed degradation occurs in the biological mat (*Schmutzdecke*) that forms on the infiltrating surface in a spreading basin. The fate of some of the more resistant organic compounds found in recharge water is still poorly understood. The effects of dispersion, sorption, and biodecomposition on the time change in concentration of an organic compound at an aquifer observation well are illustrated on Fig. 13–30. The observed concentration is C, and C_o represents the concentration in the injection water (McCarty et al., 1981).

Among the end products of complete biodegradation of dissolved organic constituents are carbon dioxide and water under aerobic conditions, or carbon dioxide, nitrogen, hydrogen sulfide, and methane under anaerobic conditions. However, the degradation process does not necessarily proceed to completion. Degradation may terminate at an intermediate stage, leaving a residual organic product that, under the particular conditions, cannot be degraded further at an appreciable rate.

Pathogen Removal.   Groundwater contamination by pathogenic microorganisms has not received as much attention as surface water. It has been generally assumed that groundwater is free of pathogenic microorganisms. However, a number of well-documented disease outbreaks have been traced to contaminated groundwater. The fate of bacterial pathogens and viruses in the subsurface environment is determined by their survival characteristics and their retention in the soil matrix. Both survival and retention are largely determined by (1) climate, (2) nature of the soil, and (3) nature of microorganisms.

Temperature and rainfall are two important climatic factors that will affect viral and bacterial survival and movement. At higher temperatures, inactivation or natural die-off is fairly rapid. For bacteria, and probably viruses, the die-off rate is approximately doubled with each 10°C rise in temperature between 5 and 30°C. Rainwater, because of its lower pH value, can elute adsorbed virus particles, which may then move with the groundwater. The physical and chemical characteristics of the soil will also play a major role in determining survival and retention of microorganisms. Soil properties influence moisture-holding capacity, pH, and organic matter. All of these factors control the survival of bacteria and viruses in the soil.

Resistance of microorganisms to environmental factors varies among different species as well as strains. Bacteria are removed largely by filtration processes in the soil while adsorption appears to be the major factor controlling virus retention.

Groundwater Recharge Guidelines

The major concern with groundwater recharge with reclaimed water is the potential adverse health effects that may be caused by the introduction of pathogens and trace amounts of toxic constituents. Because of the increasing concern for long-term health effects, every effort should be made to reduce the number of chemical species and concentration of specific organic constituents in the recharge water.

A source control program to limit the quantities of potentially harmful constituents entering the collection system must be an integral part of any groundwater recharge project. Extreme caution is necessary in controlling the quality of the treated effluent to be recharged, because restoring a groundwater basin once it is contaminated is difficult and expensive. Additional cost would be incurred if groundwater quality changes resulting from recharge necessitated the treatment of extracted groundwater or the development of additional water sources. The level of municipal wastewater treatment necessary to produce a suitable reclaimed water for groundwater recharge depends upon the groundwater quality objectives, hydrogeologic characteristics of the groundwater basin, and the amount of reclaimed water recharged in relation to other waters recharged. Factors to be considered in the formulation of the groundwater recharge guidelines are summarized in Table 13–27.

In the United States, federal requirements for groundwater recharge in the context of water reclamation and reuse have not been established. As a consequence, water reuse requirements for groundwater recharge are presently regulated by state agencies on a case-by-case determination. The state of California initially considered developing statewide groundwater recharge regulations with reclaimed water in the mid-1970s (Nellor et al., 1985). As interest in indirect potable reuse via groundwater recharge grew, the state of California agencies (State Water Resources Control Board, Department of Water Resources, and Department of Health Services) formed a Scientific Advisory Panel on Groundwater Recharge with Reclaimed Water, and its report was issued in 1987 (State of California, 1987). Proposed regulations by the Department of Health Services have undergone several iterations and, when adopted, will replace the more general groundwater recharge regulations that are included in the current *Water Recycling Criteria* (State of California, 2000). A conceptual framework for the proposed regulations on groundwater recharge is summarized in Table 13–28. Some issues remain to be resolved, for example, nitrogen limits, criteria to identify impairment of groundwater suitable for use as drinking water source, and unregulated trace organics.

13–9 PLANNED INDIRECT AND DIRECT POTABLE WATER REUSE

With the increased attention being paid to drinking water quality, much of the research on drinking water is of relevance to planned indirect and direct potable reuse. Three classes of constituents of special concern in water reclamation and reuse where potable water supply may be affected are (1) enteric viruses, (2) organic constituents including both industrial chemicals and home products and medicines, and (3) heavy metals. The

Table 13–27
Factors to be considered in the formulation of groundwater recharge guidelines in the United States

Surface spreading	
Treatment	Source control of toxic chemicals
	Primary sedimentation and secondary biological treatment
	Tertiary granular-medium filtration (possibly activated-carbon adsorption for organics removal)
	Disinfection
Depth to groundwater	Percolation through an unsaturated zone of undisturbed soil
	Depth to groundwater in the range of 3 to 15 m (10 to 50 ft) depending on percolation rate of the soils
Retention time in ground	6 to 12 months depending on the type of pretreatment
Maximum percent reclaimed wastewater	20 to 50 percent on the annual basis at extraction wells, depending on organics removal
Horizontal distance	150 to 300 m (500 to 1000 ft) depending on pretreatment
Monitoring	Extensive, including the contaminants in the drinking water regulations
Direct injection	
Treatment	Source control of toxic chemicals
	Primary sedimentation and secondary biological treatment
	Chemical coagulation, clarification, and granular-medium filtration
	Activated-carbon adsorption
	Volatile organics removal
	Reverse osmosis or other membrane process
	Disinfection
Depth to groundwater	Not applicable (direct injection to groundwater aquifers)
Retention time in ground	12 months
Maximum percent reclaimed wastewater	20 percent on the annual basis at extraction wells
Horizontal distance	300 to 600 m (1000 to 2000 ft)
Monitoring	Quite extensive, including the contaminants in the drinking water regulations

ramification of many of these constituents with respect to health effects is not well understood, and, as a result, regulatory agencies are proceeding with extreme caution in permitting water reuse applications that affect potable water supplies. Where little possibility exists for developing additional freshwater supplies, some communities are developing plans for indirect potable reuse (Crook et al., 1999). Although the quantities involved in potable reuse are small, the technological and public health interests are the greatest, and hence considerable research has been directed toward potable water reuse in recent years. Both planned indirect and direct reuse are considered in this section.

Table 13–28
Conceptual frameworks for the proposed California groundwater recharge criteria[a]

Contaminant type	Type of recharge	
	Surface spreading	**Subsurface injection**
Pathogenic microorganisms:		
Secondary treatment	TSS ≤30 mg/L	
Filtration	≤2 NTU	
Disinfection	4-log virus inactivation, ≤2.2 total coliform per 100 mL	
Retention time underground	6 months	12 months
Horizontal separation	152 m (500 ft)	61 m (200 ft)
Regulated contaminants	Meet all drinking water MCLs	
Unregulated contaminants:		
Secondary treatment	BOD ≤30 mg/L, TOC ≤16 mg/L	
Reverse osmosis	Four options available depending on meeting 1 mg/L TOC requirement	100 percent treatment to: $TOC \leq \dfrac{1 \text{ mg/L}}{RWC^b}$
Spreading criteria for SAT 50 percent TOC removal credit	Depth to groundwater at initial percolation rates of: <5 mm/min = 0.9 m (<0.2 in/min = 10 ft) <7.5 mm/min = 3.1 m (<0.3 in/min = 20 ft)	NA
Mound monitoring option	Demonstrate feasibility of mound compliance point	NA
Reclaimed water contribution	≤50 percent	

[a] Adapted from California Department of Health Services Working Paper (2000).
[b] RWC = percent reclaimed water contribution in groundwater extracted by drinking water wells.

Planned Indirect Potable Water Reuse

While direct potable reuse is not practiced in the United States, indirect potable reuse is implemented to augment several drinking water systems through groundwater recharge, and several pilot facilities have been constructed to evaluate the potential for indirect and direct potable reuse. Several examples are shown below (Asano, 1998; NRC, 1998):

- The Denver Water Department's Direct Potable Water Reuse Demonstration Project, 1979–1990
- The Potomac Estuary Experimental Water Treatment Plant, 1980–1983
- The County Sanitation Districts of Los Angeles County Ground Water Recharge Projects, 1962–present
- The Orange County Water District, California, Water Factory 21, 1972–present

- The Fred Hervey Water Reclamation Plant, El Paso, TX, Groundwater Recharge, 1985–present
- The city of San Diego's Total Resource Recovery Project, 1984–1999
- The city of Tampa, FL, Water Resources Recovery Project, 1993

Planned indirect potable reuse systems in operation today include such groundwater recharge operations as the Whittier Narrows Groundwater Recharge Project in Los Angeles County, CA; Orange County Water District in Orange County, CA, and in El Paso, TX. Another example of indirect potable reuse is the Occoquan Reservoir in northern Virginia. Highly treated effluent from the 57,000 m³/d (15 Mgal/d) Manassas, VA, advanced wastewater treatment plant is discharged directly into the Occoquan Reservoir, a principal drinking water reservoir for more than 660,000 people.

Planned Direct Potable Water Reuse

The only well-studied case of a planned direct potable reuse system in operation is in Windhoek, Namibia. For nearly 30 years, this facility has been used intermittently to forestall water emergencies during drought conditions. The current treatment process sequences involve primary, secondary treatment, and maturation ponds at the Gammans wastewater treatment plant. The secondary effluent is then directed to the Goreangab water reclamation plant, which includes alum addition, dissolved air flotation, chlorination, and lime addition, followed by settling, sand filtration, chlorination, carbon filtration, and final chlorination. The final effluent is then blended with other drinking water sources before distribution (Harhoff and van der Merwe, 1996).

In almost all the cases where potable water reuse has been implemented, alternative sources of water have been developed in the ensuing years and the need to adopt direct (e.g., pipe-to-pipe) potable water reuse has been avoided. In proposing direct potable water reuse, serious consideration must be given to whether the need is for a short-term emergency or for normal use over a prolonged period. The major emphasis placed upon the potable water reuse today concerns the chronic health effects that might result from ingesting the mixture of unregulated inorganic and organic contaminants in trace amounts that remain in water, even after subjecting it to the most advanced treatment methods such as microfiltration and reverse osmosis.

Planned Potable Water Reuse Criteria

It has been argued that there should be a single water quality standard for potable water and that if reclaimed water can meet this standard, it should be acceptable. It must be recognized, however, that current drinking water standards have evolved on the presumption of water supplies derived from relatively unpolluted freshwater sources. While great advances have been made in analytical methods to identify chemical contaminants in water, only a small fraction of the contaminants present in surface water and groundwater can be identified (see discussion in Sec. 2–7 in Chap. 2). This analytical limitation has frustrated attempts to develop comprehensive potable water reuse criteria for various sources of water. In assessing water being considered for possible potable reuse, comparison should be made with the highest-quality water available locally to rationalize the selection.

While the implementation of direct potable use of reclaimed water is obviously limited to extreme situations, research relating to potable reuse has continued in several locations as identified in the previous section. As the proportional quantities of treated waste-water discharged into the nation's waters increase, as experienced in New Orleans, LA, on the Mississippi River, much of the research that addressed only potable water reuse is becoming of equal relevance to *unplanned indirect potable reuse* such as municipal water intakes downstream from wastewater discharges or from increasingly polluted rivers.

What Is the Ultimate Water Reuse Goal?

While many technological advances have been made in producing high-quality treated wastewater, a number of obstacles remain to be overcome before potable reuse is accepted. Much of the objection to planned potable reuse in the United States arises from concerns in the scientific community whether drinking water standards are adequate to ensure the safety of all waters "regardless of source." Some argue that drinking water standards apply only to—and were designed only for—waters derived from a relatively pristine source. Although this argument has a long-standing basis in public health and normal sanitation practice, it is becoming more difficult to determine what is the best available water source (NRC, 1998). Some lessons learned from implementing indirect potable reuse projects were reviewed by Hartling (2001).

Support from the public sector is also lacking; some of the reasons may be (1) the immediate need for potable reuse has not been demonstrated convincingly, (2) few *champions* for water reuse have rallied support, (3) political leaders prefer not to be advocates for a potentially controversial issue, (4) few media presentations make an appealing case for supporting potable reuse, and (5) few forums are available for public education and informed discussions of the issues surrounding water reuse.

Beyond engineering feasibility for indirect potable reuse, future research by social and political scientists may involve an effort to answer the following questions:

- Why were planned indirect potable water reuse projects not approved and implemented? What were the main influences on public opinion in each case?

- What should be the roles and strategies of the media, government officials, and other interested parties with respect to the indirect potable reuse projects?

- To what extent do public concerns about health effects, growth inducement, and cost influence community acceptance of potable water reuse projects as compared to the prospects of diminished water resources?

13–10 PLANNING FOR WASTEWATER RECLAMATION AND REUSE

In effective planning for water reclamation and reuse, the objectives and basis for conducting the planning should be defined clearly. The optimum water reclamation and reuse project is best achieved by integrating both wastewater treatment and water supply needs into one plan. This integrated approach is somewhat different from planning for conventional wastewater treatment facilities where planning is done only for conveyance, treatment, and disposal of municipal wastewater.

Planning Basis

Effective water reclamation and reuse facilities plans should include the following elements: (1) assessment of wastewater treatment and disposal needs, (2) assessment of water supply and demand, (3) assessment of water supply benefits based on water reuse potential, (4) analysis of reclaimed water market, (5) engineering and economic analyses of alternatives, (6) implementation plan with financial analysis, and (7) a public information program. Important factors that should be considered in the planning process are briefly introduced in this section.

Project Objectives.   Water reclamation and reuse can serve the functions of both water pollution control and water supply. It has been only in the last three decades that increased attention has been given to the water supply benefits in the facilities planning process for the municipal wastewater treatment. Ignoring the water supply potential of municipal wastewater has often resulted in facilities that now hinder the development of this alternative water resource. For example, in some locations optimum water reuse would have been better achieved if smaller satellite wastewater treatment plants had been constructed with reuse in mind, instead of a large regional wastewater treatment facility.

Because many water and wastewater agencies are established with a single-purpose function, planning by these agencies tends to be single-purpose as well. Optimum water reclamation and reuse is best achieved in a framework of multiple-purpose planning and with cooperative efforts of both wastewater management and water supply agencies. Once these multiple benefits and beneficiaries of water reuse are recognized, additional options may be available for sharing project responsibilities and costs among project sponsors.

Project Study Area.   The project study area is another critical planning issue. There are two study area horizons to consider in project planning. The first planning horizon is established based on the direct service area of the project facilities plan. The second horizon extends to the area that accrues less direct costs or benefits from a project, which should be accounted for in evaluating the project. Thus, the study area for facilities design includes (1) the collection system area to be served by the wastewater treatment facilities and (2) the area that can potentially be served by reclaimed water. To evaluate project benefits and costs, the project study area must include (1) the area impacted by the environmental effects of the wastewater, and (2) the area benefiting from the supplemental or alternative water supply of the reclaimed water.

The traditional approach to planning is to equate the study area with the project sponsor's jurisdictional boundaries. Such boundaries, however, may not suit the optimum design of a water reuse project or include the areas of benefit. Because water supply is typically dependent on regional water resources outside of the facilities study area, it is essential to look beyond this study area to obtain an understanding of the water resources situation. For instance, overdrafted groundwater basins may be having their most serious impacts on communities at great distances beyond the project area. Thus, implementing a water reuse project in one community that reduces groundwater overdraft could result in benefits to a water supply in another community.

Market Assessment

In planning a water reclamation project, it is essential to find potential customers who are capable and willing to use reclaimed water. The success of water reuse projects is

Table 13–29
Reclaimed wastewater market assessment: background information and survey[a]

1. Inventory potential users and uses of reclaimed wastewater
2. Determine health-related requirements regarding water quality and application requirements (e.g., treatment reliability, backflow prevention, use area controls, irrigation methods) for each type of application of reclaimed wastewater
3. Determine regulatory requirements to prevent nuisance and water quality problems, such as restrictions to protect groundwater
4. Develop assumptions regarding probable water quality that would be available in the future with various levels of treatment and compare those to regulatory and user requirements
5. Develop an estimate of future freshwater supply costs to potential users of reclaimed wastewater
6. Survey potential reclaimed wastewater users, obtaining the following information:
   a. Specific potential uses of reclaimed wastewater
   b. Present and future quantity needs
   c. Timing and reliability of needs
   d. Water quality needs
   e. Onsite facilities modifications to convert to reclaimed wastewater and meet regulatory requirements for protection of public health and prevention of pollution problems from reclaimed wastewater
   f. Capital investment of the user for onsite facilities modifications, changes in operational costs, desired pay-back period or rate of return, and desired water cost savings
   g. Plans for changing use of site in future
7. Inform potential users of applicable regulatory restrictions, probable water quality available with different levels of treatment, reliability of the reclaimed water, future costs, and quality of freshwater compared to reclaimed water
8. Determine the willingness of potential users to use reclaimed wastewater now or in the future

[a]Developed from Asano and Mills (1990).

largely dependent on securing of markets for the reclaimed water. A market assessment consists of three parts: (1) background information including potential uses of reclaimed water, (2) a survey of potential reclaimed water users and their needs, and (3) community support. Background and survey information necessary for a market assessment for reclaimed water is listed in Table 13–29. The results of this assessment form the basis for developing alternatives and determining financial feasibility of a project.

Monetary Analyses

In many cases, monetary factors tend to be the overriding concern in determining whether and how to implement a water reuse project, even though the *make or break* for most projects will be based on technical, environmental, and social factors. Greater emphasis must be placed on environmental considerations, public acceptance, and public policy issues rather than mere cost-effectiveness as a measure of the feasibility of a water reuse project.

Economic and Financial Analyses.   Monetary analyses, based on established water resources economics, fall into two categories: economic analysis and financial analysis. The economic analysis is focused on the value of the resources invested in a project to construct and operate it, measured in monetary terms. The financial analysis is focused on the perceived costs and benefits of a project from the viewpoints of the project sponsor, participants, and others impacted by a project. These perceived costs and benefits may not reflect the actual value of resources invested due to subsidies or monetary transfers.

Whereas economic analysis evaluates water reclamation and reuse projects in the context of impacts on society, financial analysis focuses, instead, on the local ability to raise money from project revenues, government grants, loans, and bonds to pay for the project. The basic result of the economic analysis is to answer the question, *should* a water reuse project be constructed? Equally important, however, is the question, *can* a water reuse project be constructed? Both orientations, therefore, are necessary. However, only water reuse projects that are viable in the economic context are given further consideration for a financial analysis (Asano and Mills, 1990).

Cost and Price of Water.   An important factor in the monetary analysis of water reuse projects is to distinguish between the cost and the price of water. In an economic analysis, only the future flow of resources invested in or derived from a project is considered. Past resource investments are considered sunk costs that are irrelevant to future investment decisions. Thus, debt service on past investments is not included in an economic analysis. The price of water is the purchase price paid to a water wholesaler or retailer to purchase water. Water price usually reflects a melding of current and past expenditures for a combination of projects, as well as water system administration costs, which are generally fixed costs. The costs of relevance to an economic analysis are only costs for future construction, operations, and maintenance.

To determine the water supply benefit of a water reuse project in an economic analysis, the project is usually compared to the development of a new freshwater supply. In performing such an analysis, the relevant costs for comparison are the future stream of costs (1) to construct new freshwater facilities, and (2) to operate and maintain all of the facilities needed to treat and deliver the new increment of water supply developed. In this case, the present and future price charged for freshwater would not provide a valid basis of comparison to judge the water supply benefit of a water reuse project.

On the other hand, consideration of prices charged for freshwater and reclaimed water are important to determine financial feasibility. The perceived cost of water to customers is the price charged, and thus prices will be evaluated by potential reclaimed water users in determining willingness to participate in a water reuse project.

Other Planning Factors

In addition to the monetary analyses, a number of factors have to be evaluated during the planning for a water reclamation and reuse project. Factors of particular significance in project development are related to (1) water demand characteristics, (2) supplemental water supply and emergency backup systems, (3) water quality requirements, and (4) determination of optimum project size. In essence, a water reuse project is a relatively small-scale water supply project with considerations of matching supply and

Table 13–30
Outline for wastewater reclamation and reuse facilities plan[a]

1. Study area characteristics: geography, geology, climate, groundwater basins, surface waters, land use, population growth

2. Water supply characteristics and facilities: agency jurisdictions, sources and qualities of supply, description of major facilities, water use trends, future facilities needs, groundwater management and problems, present and future freshwater costs, subsidies and customer prices

3. Wastewater characteristics and facilities: agency jurisdictions, description of major facilities, quantity and quality of treated effluent, seasonal and hourly flow and quality variations, future facilities needs, need for source control of constituents affecting reuse, description of existing reuse (users, quantities, contractual and pricing agreements)

4. Treatment requirements for discharge and reuse and other restrictions: health and water quality–related requirements, user-specific water quality requirements, use area controls

5. Potential water reuse customers: description of market analysis procedures, inventory of potential reclaimed water users, and results of user survey

6. Project alternative analysis: capital and operation and maintenance costs, engineering feasibility, economic analysis, financial analysis, energy analysis, water quality impacts, public and market acceptance, water rights impacts, environmental and social impacts, comparison of alternatives and selection

   a. Treatment alternatives

   b. Alternative markets: based on different levels of treatment and service areas

   c. Pipeline route alternatives

   d. Alternative reclaimed water storage locations and options

   e. Freshwater alternatives

   f. Water quality management alternatives

   g. No project alternative

7. Recommended plan: description of proposed facilities, preliminary design criteria, projected cost, list of potential users and commitments, quantity and variation of reclaimed water demand in relation to supply, reliability of supply and need for supplemental or backup water supply, implementation plan, operational plan

8. Construction financing plan and revenue program: sources and timing of funds for design and construction; pricing policy of reclaimed water; cost allocation between water supply benefits and pollution control purposes; projection of future reclaimed water use, freshwater prices, reclamation project costs, unit costs, unit prices, total revenue, subsidies; sunk costs and indebtedness; analysis of sensitivity to changed conditions

[a]Developed from Asano and Mills (1990).

demand, appropriate levels of wastewater treatment, reclaimed water storage, and supplemental or backup freshwater supply.

Planning Report

The results of the completed planning effort are documented in a facilities planning report on water reclamation and reuse. An outline is shown in Table 13–30, which also serves as a checklist for planning considerations. All of the items listed in Table 13–30 have been found at one time or another to affect the evaluation of water reclamation and reuse projects. Thus, while all of the factors shown do not deserve an in-depth analysis, they should at least be considered in the planning process. While the emphasis on the wastewater treatment or water supply aspects will vary depending on whether a project is single- or multiple-purpose, the nature of water reclamation and reuse is such that both aspects must be considered.

13-11 EPILOGUE ON WATER REUSE ISSUES

Inadequate water supply and water quality deterioration represent serious contemporary concerns for many municipalities, industries, agriculture, and the environment in various parts of the world. Several factors have contributed to these problems such as continued population growth in urban areas, contamination of surface water and groundwater, uneven distribution of water resources, and frequent droughts caused by the extreme global weather patterns. For more than a quarter century, a recurring thesis in environmental and water resources engineering has been that improved wastewater treatment provides a treated effluent of such quality that it should not be wasted but put to beneficial use. This conviction in responsible engineering, coupled with the vexing problem of increasing water shortage and environmental pollution, provides a realistic framework for considering reclaimed water as a water resource in many parts of the world. Water reuse has been dubbed as the *greatest challenge of the 21st century* as water supplies remain practically the same and water demands increase because of increasing population and per capita consumption. Water reuse accomplishes two fundamental functions: (1) the treated effluent is used as a water resource for beneficial purposes, and (2) the effluent is kept out of streams, lakes, and beaches; thus reducing pollution of surface water and groundwater.

The main concerns in water reuse are: (1) reliable treatment of wastewater to meet strict water quality requirements for the intended reuse, (2) protection of public health, and (3) gaining public acceptance. Water reuse also requires close examinations of infrastructure and facilities planning, and water utility management involving effective integration of water and reclaimed water supply functions. Whether water reclamation and reuse will be appropriate in a specific locale depends upon careful economic considerations, potential uses for the reclaimed water, and stringency of waste discharge requirements. Public policy wherein the desire to conserve and reuse rather than develop additional water resources with considerable environmental expenditures may be an important consideration. Through integrated water resources planning, the use of reclaimed water may provide sufficient flexibility to allow a water agency to respond to short-term needs as well as increase long-term water supply reliability.

Groundwater recharge with reclaimed water and direct potable water reuse share many of the public health concerns encountered in drinking water withdrawn from polluted rivers and reservoirs. Three classes of constituents are of special concern where reclaimed water is used in such applications: (1) enteric viruses and other emerging pathogens; (2) organic constituents including industrial chemicals, home products, and medicines; and (3) heavy metals. The impact of many of these constituents in trace quantity are not well understood with respect to long-term health effects, and, as a result, regulatory agencies are proceeding with extreme caution in permitting water reuse applications that affect potable water supplies. In all the cases in the United States where potable water reuse has been contemplated, alternative sources of water have been developed in the ensuing years and the need to adopt direct potable water reuse has been avoided. As the proportional quantities of treated wastewater discharged into the nation's waters increase, as experienced in New Orleans, Louisiana, on the Mississippi River, much of the research that addresses groundwater recharge and potable water reuse is becoming of equal relevance to *unplanned indirect potable reuse* such as municipal water intakes located downstream from wastewater discharges or from increasingly polluted rivers and reservoirs. At this time, what is critically needed is a national dialogue to develop a consensus on acceptable health and environmental risk.

Reclaimed water is a locally controllable water resource that exists right at the doorstep of the urban environment, where water is needed the most and priced the highest. Closing the water cycle not only is technically feasible in industries and municipalities but also makes economic sense. While potable reuse is still a distant possibility and may never be implemented in the United States, groundwater recharge with advanced wastewater treatment technologies is a viable option backed by the decades of experiences in Arizona, California, New York, and Texas as well as in Australia, Israel, Germany, the Netherlands, and the United Kingdom. Water reuse has become an essential element of future water resources development and management but with more challenges.

PROBLEMS AND DISCUSSION TOPICS

13-1 Using the data given in Table 13–12, estimate the incremental cancer risk for an adult associated with drinking groundwater containing 2.0 μg/L of one of the following constituents (to be selected by the instructor): inorganic arsenic, benzene, bromate, chloroform, dieldrin, heptachlor, n-nitrosodiethylamine, or vinyl chloride. To limit the constituent exposure to an acceptable cancer risk of 1 in 100,000, determine the concentration of the constituent that can be allowed in extracted groundwater. Compare your computed value to the value given in the U.S. EPA IRIS database (http://www.epa.gov/iris).

13-2 The removal performance for 5 individual processes for a given constituent has been measured and is reported below. If process A is to be combined with one of the other processes, estimate the performance of the combined processes A and B, A and C, A and D, or A and E (process combination to be selected by the instructor). Using the performance for combined processes, estimate the combined removal at the 90 and 99 percent probability levels.

Process	Geometric removal, M_g	Geometric standard deviation, s_g
A	2.0	1.40
B	5.0	1.40
C	2.5	1.26
D	4.2	1.31
E	3.4	1.29

13-3   The performance of 5 individual treatment processes for a given constituent has been measured and is reported below. For the processes studied, it was observed that the percentage removal was independent of the initial concentration. Estimate the performance for one of the following process combinations A + B + C, B + C + D, C + D + E, or D + E + A (to be selected by the instructor) and plot the resulting probability distribution. Using the performance for combined processes, estimate the combined removal at the 88 and 95 percent probability levels.

Process	Arithmetic mean removal, %	Arithmetic standard deviation
A	60	12
B	30	17
C	55	5
D	40	10
E	50	5

13-4   The following average monthly performance data were obtained for a wastewater reclamation plant before effluent filtration. Compare and contrast the reported performance data to similar data in the literature. Cite a minimum of three references in your review.

Month	Concentration, mg/L CBOD	COD
Jan	8.1	47.0
Feb	10.0	58.5
Mar	9.0	49.2
Apr	6.5	42.0
May	5.0	44.0
Jun	5.0	43.3
Jul	4.5	54.5
Aug	5.2	53.8
Sep	6.7	73.6
Oct	7.1	62.1
Nov	7.5	50.5
Dec	7.7	44.6

13-5   The following wastewater flowrate data are from four different activated-sludge plants. Assuming the information and data given in Example 13-3 are applicable, use the procedure outlined in Example 13-3 to determine the land area required for storage and land application. The treatment plant will be selected by the instructor.

		Daily wastewater flowrate, m³/d			
		Treatment plant			
Month	**Days**	**A**	**B**	**C**	**D**
Jan	31	5000	7000	4000	7400
Feb	28	4750	6000	3750	6850
Mar	31	4500	5400	3470	6400
Apr	30	4200	5000	3200	6100
May	31	4000	4700	3000	5800
Jun	30	4000	4600	3000	5600
Jul	31	4000	4550	3000	5450
Aug	31	4000	4570	3000	5400
Sep	30	4000	4700	3200	5500
Oct	31	4000	4950	3700	5600
Nov	30	4300	5400	4400	5900
Dec	31	4600	6000	5400	6300
Total	365				—

13-6 The following effluent water quality data were obtained from 4 different activated-sludge plants.

	Concentration, mg/L			
Water quality constituent	**Treatment plant**			
	A	**B**	**C**	**D**
BOD	5	7	6	11
COD	55	45	60	42
TSS	4	8	7	10
NH_3-N	5	1.0	0.5	1.4
NO_3^--N	9	10	27.5	5
Total − P	6.3	7.3	6.9	6
pH	?	?	?	?
Cations				
Ca^{2+}	60	140	175	85
Mg^{2+}	40	37	60	33
Na^+	25	33	45	151
K^+	16	12	19	15
Anions				
HCO_3^-	320	440	550	272
SO_4^{2-}	68	132	240	168
Cl^-	35	65	74	210

(continued)

(*Continued*)

Water quality constituent	Concentration, mg/L			
	Treatment plant			
	A	**B**	**C**	**D**
EC (mmhos/cm)	1.0	1.4	1.9	1.6
TDS	640	930	1265	1000
B	0.55	0.8	1.0	0.7
Temperature °C	20	28	24	26

For one of these effluents (to be selected by instructor): (1) determine the SAR and adj R_{Na}, (2) estimate the pH, (3) determine whether infiltration problems will develop in using these effluents for irrigation, (4) estimate the salinity (expressed as EC) of drainage water if the crop is irrigated to achieve a leaching fraction of 0.1, and (5) estimate also the TDS concentration of the drainage water from the crop irrigation.

13-7   For the waters given in Problem 13–6, determine the appropriate leaching fraction to maintain crop yield, if the crop is known to suffer significant loss in yield when TDS of the soil water exceeds 3000, 4000, 5000, or 6000 mg/L (TDS value to be selected by instructor).

13-8   A common chemical precipitation reaction in water softening is shown in the following equation:

$$Ca(HCO_3)_2 + Ca(OH)_2 \rightarrow 2CaCO_3 + 2H_2O$$

Given a treated wastewater flowrate of 3500, 4000, 4200, and 5500 m^3/d (to be selected by instructor), determine how much $Ca(OH)_2$ is required each day to reduce the Ca^{2+} concentration from 175 mg/L as $CaCO_3$ to 35, 50, 65, and 80 mg/L as $CaCO_3$ (to be selected by instructor) so that the treated wastewater can be used for industrial cooling. Determine also the kilograms of $CaCO_3$ and the sludge volume produced each day if the solids are settled to 1 percent by weight and the specific gravity of $CaCO_3$ is 2.8.

13-9   Reclaimed water with the chemical characteristics given below is being considered for use as makeup water for a cooling tower. Calculate the composition of the blowdown flow if 3, 4, or 5 cycles of concentration are to be used (cycles of concentration and water to selected by instructor). Assume that the temperature of the hot water entering the cooling tower is 55°C. Estimate the solubility of $CaCO_3$ using the data in Table 6–16.

Constituent	Concentration, mg/L			
	A	**B**	**C**	**D**
Ca^{2+} (as $CaCO_3$)	120	90	88	65
Mg^{2+} (as $CaCO_3$)	25	30	12	25
Alkalinity (as $CaCO_3$)	120	100	150	145
SO_4^{2-}	20	15	25	18
Cl^-	16	14	20	18
SiO_2	2.2	2.5	3.1	5.0

13–10 In your state which federal, state, or local agencies are involved in setting water reclamation regulations? Discuss the pros and cons of the federal government establishing national water reuse criteria.

13–11 Review the current literature (a minimum of five articles should be reviewed and cited) and list health and regulatory factors affecting implementation of water reuse projects. What is the rationale for setting less stringent microbiological standards in developing countries where enteric diseases are rampant among the population?

13–12 Discuss the feasibility of wastewater reclamation and reuse by means of decentralized wastewater treatment facilities.

REFERENCES

Andrews, S. A., and V. Y. Taguchi (2001), "NDMA—Canadian Issues," *Water Quality Technical Conference Proceedings,* American Water Works Assn., Denver, CO.

Asano, T. (ed.) (1998) *Wastewater Reclamation and Reuse,* Water Quality Management Library, vol. 10, Technomic Publishing Co., Inc., Lancaster, PA.

Asano, T., and A. D. Levine (1996) "Wastewater Reclamation, Recycling and Reuse: Past, Present, and Future," *Water Science & Technology,* vol. 33. no. 10–11, pp. 1–14.

Asano, T., and R. A. Mills (1990) "Planning and Analysis for Water Reuse Projects," *Journal American Water Works Association,* vol. 82, no. 1, pp. 38–47.

Asano, T., L. Y. C. Leong, N. G. Rigby, and R. H. Sakaji (1992) "Evaluation of the California Wastewater Reclamation Criteria Using Enteric Virus Monitoring Data," *Water Science and Technology,* vol. 26, no. 7–8, pp. 1513–1524.

Asano, T., D. Richard, R. W. Crites, and G. Tchobanoglous (1992) "Evolution of Tertiary Treatment Requirements in California," *Water Environment & Technology,* vol. 4, no., 2, pp. 36–41.

Ayers, R. S., and D. W. Westcot (1985) *Water Quality for Agriculture,* FAO Irrigation and Drainage Paper 29, Rev. 1, Food and Agriculture Organization of the United Nations, Rome, Italy.

Bitton, G., and C. P. Gerba (1984) *Groundwater Pollution Microbiology,* Wiley, New York.

Cockerham, L. G., and B. S. Shane (1994) *Basic Environmental Toxicology,* CRC Press, Boca Raton, FL.

Crook, J., A. A. MacDonald, and R. R. Trussell (1999) "Potable Use of Reclaimed Water," *Journal American Water Works Association,* vol. 9, no. 9.

CSDLAC (1977) *Pomona Virus Study Final Report,* County Sanitation Districts of Los Angeles County, Whittier, CA.

Englehardt, J. D., V. P. Amy, F. Bloetscher, D. A. Chin, L. E. Fleming, S. Gokgoz, J. B. Rose, H. Solo-Gabriele, and G. Tchobanoglous (2002) "Comparative Assessment of Microbial and NDMA Risks Among Wastewater Disposal Methods in Southeast Florida," Presented at the WEF Disinfection 2002 Specialty Conference, St. Petersburg, FL.

Englehardt, J. D., V. P. Amy, F. Bloetscher, D. A. Chin, L. E. Fleming, S. Gokgoz, J. R. Proni, J. B. Rose, H. Solo-Gabriele, and G. Tchobanoglous (2001) *Comparative Assessment of Human and Ecological Impacts from Municipal Wastewater Disposal Methods in Southeast Florida.* Report prepared for Florida Water Environment Association Utility Council, University of Miami, Coral Gables, FL.

Haas, C. N. (1983) "Estimation of Risk Due to Low Levels of Microorganisms: A Comparison of Alternative Methodologies," *American Journal of Epidemiology,* vol. 118, pp. 573–582.

Haas, C. N., J. B. Rose, and C. P. Gerba (1999) *Quantitative Microbial Risk Assessment,* Wiley, New York.

Harhoff, J., and B. van der Merwe (1996) "Twenty-five years of Wastewater Reclamation in Windhoek, Namibia," *Water Science & Technology,* vol. 33, no. 10–11, pp. 25–35.

Hartling, C. E. (2001) "Laymanization—An Engineer's Guide to Public Relations," *Water Environment Technology,* vol. 13, no. 4, pp. 45–48.

Kolluru, R. V., S. M. Bartell, R. M. Pitblado, and R. S. Stricoff (1996) *Environmental Assessment and Management Handbook,* McGraw-Hill, New York.

McCarty, P. L., B. E. Rittmann, and M. Reinhard (1981) "Processes Affecting the Movement and Fate of Trace Organics in the Subsurface Environment," *Environmental Science & Technology,* vol. 15, no. 1.

NAS (1972) *Water Quality Criteria, A Report of the Committee on Water Quality Criteria,* National Academy of Science, National Academy of Engineering, Superintendent of Documents, Government Printing Office, Washington, DC.

Neely, W. B. (1994) *Introduction to Chemical Exposure and Risk Assessment,* Lewis Publishers, Boca Raton, FL.

Nellor, M. H., R. B. Baird, and J. R. Smyth (1985) "Health Effects of Indirect Potable Water Reuse," *Journal American Water Works Association,* vol. 77, no. 7, pp. 88–96.

NRC (1993) *Managing Wastewater in Coastal Urban Areas,* National Research Council, National Academy Press, Washington, DC.

NRC (1998) *Issues in Potable Reuse: The Viability of Augmenting Drinking Water Supplies with Reclaimed Water,* National Research Council, National Academy Press, Washington, DC.

Olivieri, A., D. Eisenberg, J. Soller, J. Eisenberg, R. C. Cooper, G. Tchobanoglous, R. R. Trussell, and P. Gagliardo (1999) "Estimation of Pathogen Removal in an Advanced Water Treatment Facility Using Monte Carlo Simulation," *Water Science and Technology,* vol. 40. no. 4–5, pp. 223–233.

Ongerth, H. J., and J. E. Ongerth (1982) "Health Consequences of Wastewater Reuse," *Annual Review of Public Health,* vol. 3, p. 419.

Pepper, I. L., C. P. Gerba, and M. L. Brusseau (eds.) (1996) *Pollution Science,* Academic Press, San Diego, CA.

Pettygrove, G. S., and T. Asano (eds.) (1985) *Irrigation with Reclaimed Municipal Wastewater—A Guidance Manual,* Lewis Publishers, Inc., Chelsea, MI.

Richards, D., R. W. Crites, T. Asano, and G. Tchobanoglous (1993) "A Systematic Approach to Estimating Wastewater Reclamation Costs in California," *Proceedings of the 66th Annual Conference & Exposition,* Water Environment Federation, Anaheim, CA, pp. 235–245.

Soller, J. A., J. N. Eisenberg, and A. W. Olivieri (1999) "Evaluation of Pathogen Risk Assessment Framework," Prepared by EOA Inc. and U.C. Berkeley School of Public Health for ILSI Risk Science Institute, Oakland, CA.

Solley, W. B., R. R. Pierce, and H. A. Perlman (1998) *Estimated Use of Water in the United States,* U.S. Geological Survey Circular 1200, U.S. Department of the Interior, U.S. Geological Survey, Denver, CO.

State of California (1978) *Wastewater Reclamation Criteria, An Excerpt from the California Code of Regulations, Title 22, Division 4,* Environmental Health, Department of Health Services, Berkeley, CA.

State of California (1987) *Report of the Scientific Advisory Panel on Groundwater Recharge with Reclaimed Wastewater,* Prepared for State of California, State Water Resources Control Board, Department of Water Resources, and Department of Health Services, November 1987.

State of California (2000) Code of Regulations, Title 22, Division 4, Chapter 3. *Water Recycling Criteria,* Sections 60301 et seq., Dec. 2, 2000.

Suarez, D. L. (1981) "Relation between pH_c and Sodium Adsorption Ratio (SAR) and Alternative Method of Estimating SAR of Soil or Drainage Waters," *Journal Soil Science Society of America,* vol. 45, p. 469.

Tanaka, H., T. Asano, E. D. Schroeder, and G. Tchobanoglous (1998) "Estimating the Safety of Wastewater Reclamation and Reuse Using Enteric Virus Monitoring Data," *Water Environment Research,* vol. 70, no. 1, pp. 39–51.

Tchobanoglous, G., and F. L. Burton (Metcalf & Eddy, Inc.) (1991) *Wastewater Engineering: Treatment, Disposal, and Reuse,* McGraw-Hill, New York.

Tchobanoglous, G., and E. D. Schroeder (1985) *Water Quality: Characteristics, Modeling, and Modification,* Addison-Wesley, Reading, MA.

Ternes, T. (1998) "Occurrence of Drugs in German Sewage Treatment Plants and Rivers," *Water Research,* vol. 32, no. 11, pp. 3245–3260.

U.S. EPA (1982) *Evaluation and Documentation of Mechanical Reliability of Conventional Wastewater Treatment Plant Components,* EPA 600/2-82-044, U.S. Environmental Protection Agency, Washington, DC.

U.S. EPA (1986a) *Risk Assessment and Management: Framework for Decision Making,* EPA 600/9-85-002, U.S. Environmental Protection Agency, Washington, DC.

U.S. EPA (1986b) *Guidelines for Carcinogen Risk Assessment,* Federal Register, Sept. 24, 1986.

U.S. EPA (1989a) *Risk Assessment, Guidance for Superfund, Volume I, Human Health Evaluation Manual (Part A),* Office of Emergency and Remedial Response, EPA 540/1-89-002, U.S. Environmental Protection Agency, Washington, DC.

U.S. EPA (1989b) "National Primary Drinking Water Regulations: Filtration, Disinfection; Turbidity, *Giardia Lamblia,* viruses, *Legionella* and Heterotrophic Bacteria; Final Rule," *Federal Register,* vol. 54, no. 124, p. 27486.

U.S. EPA (1990) *Risk Assessment, Management and Communication of Drinking Water Contamination,* EPA 625/4-89/024, U.S. Environmental Protection Agency, Washington, DC.

U.S. EPA (1991) *Health Effects Assessment Summary Tables,* Annual FY 1991, publication 9200.6-303 (91-1) Office of Solid Waste and Emergency Response, U.S. Environmental Protection Agency, Washington, DC.

U.S. EPA (1992a) *Manual—Guidelines for Water Reuse,* EPA/625/R-92/004, U.S. Environmental Protection Agency and U.S. Agency for International Development, Washington, DC.

U.S. EPA (1992b) *Framework for Ecologic Risk Assessment,* EPA 630/R-92/001, U.S. Environmental Protection Agency, Washington, DC.

U.S. EPA (1996) *Integrated Risk Information System (IRIS), Electronic On-Line Database of Summary Health Risk Assessment and Regulatory Information on Chemical Substances Assessment,* EPA 1630/R-92/001, U.S. Environmental Protection Agency, Washington, DC.

Veatch, N. T., Jr. (1938) "Chapter XVI The Use of Sewage Effluents in Agriculture," in L. Pearse, ed., *Modern Sewage Disposal,* Federation of Sewage Works Associations, New York.

Ward, C. H., W. Giger, and P. L. McCarty (ed.) (1985) *Groundwater Quality,* Wiley, New York.

WCPH (1996a) *Total Resource Recovery Project, Final Report,* Prepared for City of San Diego, Western Consortium for Public Health, Water Utilities Department, San Diego, CA.

WCPH (1996b) *Total Resource Recovery Project Aqua III San Pasqual Health Effects Study Final Summary Report,* Prepared for City of San Diego, Western Consortium for Public Health, Water Utilities Department, San Diego, CA.

WEF (2001) "A Dynamic Model to Assess Microbial Health Risks Associated with Beneficial Uses of Biosolids," Project 98-REM-1, Currently in progress, Water Environment Federation, Alexandria, VA.

Whitley Burchett & Associates (1999) *Clean Water Revival Groundwater Replenishment System: Performance and Reliability Summary Report,* Report prepared for Dublin San Ramon Services District, Walnut Creek, CA.

WHO (1989) *Health Guidelines for the Use of Wastewater in Agriculture and Aquaculture,* World Health Organization, Geneva, Switzerland.

14

Treatment, Reuse, and Disposal of Solids and Biosolids

The constituents removed in wastewater-treatment plants include screenings, grit, scum, solids, and biosolids. The solids and biosolids (formerly collectively called sludge) resulting from wastewater-treatment operations and processes are usually in the form of a liquid or semisolid liquid, which typically contains from 0.25 to 12 percent solids by weight, depending on the operations and processes used. The term *biosolids,* as defined by the Water Environment Federation (WEF 1998), reflects the fact that wastewater solids are organic products that can be used beneficially after treatment with processes such as stabilization and composting. The term *sludge* is used only before beneficial use criteria (discussed in Sec. 14–2) have been achieved. The term *sludge* is generally used in conjunction with a process descriptor, such as *primary sludge, waste-activated sludge,* and *secondary sludge.* In cases where it is uncertain whether beneficial use criteria have been met, the term *solids* is used.

Of the constituents removed by treatment, solids and biosolids are by far the largest in volume, and their processing, reuse, and disposal present perhaps the most complex problem facing the engineer in the field of wastewater treatment. For this reason, a separate chapter has been devoted to this subject. The disposal of grit and screenings is discussed in Chap. 5.

The problems of dealing with solids and biosolids are complex because (1) they are composed largely of the substances responsible for the offensive character of untreated wastewater; (2) the portion of biosolids produced from biological treatment requiring disposal is composed of the organic matter contained in the wastewater but in another form, and it, too, will decompose and become offensive; and (3) only a small part is solid matter. The principal purpose of this chapter is to describe the solids operations and the processes that are used (1) to reduce the water and organic content and (2) to render the processed solids suitable for reuse or final disposal.

The principal methods used for solids processing are listed in Table 14–1. Thickening (concentration), conditioning, dewatering, and drying are used primarily to remove

Table 14–1
Solids processing methods

Unit operation, unit process, or treatment method	Function	See Sec.
Pumping	Transport of sludge and liquid biosolids	14–4
Preliminary operations:		
Grinding	Particle size reduction	14–5
Screening	Removal of fibrous materials	14–5
Degritting	Grit removal	14–5
Blending	Homogenization of solids streams	14–5
Storage	Flow equalization	14–5
Thickening:		
Gravity thickening	Volume reduction	14–6
Flotation thickening	Volume reduction	14–6
Centrifugation	Volume reduction	14–6
Gravity-belt thickening	Volume reduction	14–6
Rotary-drum thickening	Volume reduction	14–6
Stabilization:		
Alkaline stabilization	Stabilization	14–8
Anaerobic digestion	Stabilization, mass reduction	14–9
Aerobic digestion	Stabilization, mass reduction	14–10
Autothermal aerobic digestion (ATAD)	Stabilization, mass reduction	14–10
Composting	Stabilization, product recovery	14–11
Conditioning:		
Chemical conditioning	Improve dewaterability	14–12
Other conditioning methods	Improve dewaterability	14–12
Dewatering:		
Centrifuge	Volume reduction	14–13
Belt-filter press	Volume reduction	14–13
Filter press	Volume reduction	14–13
Sludge drying beds	Volume reduction	14–13
Reed beds	Storage, volume reduction	14–13
Lagoons	Storage, volume reduction	14–13
Heat drying		
Direct dryers	Weight and volume reduction	14–14
Indirect dryers	Weight and volume reduction	14–14
Incineration:		
Multiple-hearth incineration	Volume reduction, resource recovery	14–15
Fluidized-bed incineration	Volume reduction	14–15
Coincineration with solid waste	Volume reduction	14–15
Application of biosolids to land:		
Land application	Beneficial use, disposal	14–17
Dedicated land disposal	Disposal, land reclamation	14–17
Landfilling	Disposal	
Conveyance and storage	Solids transport and storage	14–18

moisture from solids; digestion, composting, and incineration are used primarily to treat or stabilize the organic material in the solids. To make the study of these operations and processes more meaningful, the first two sections of this chapter are devoted to a discussion of the sources, characteristics, and quantities of solids; the current regulatory environment; and a presentation of representative solids-treatment process flow diagrams. Because the pumping of sludge is a fundamental part of wastewater-treatment plant design, a separate discussion (Sec. 14–4) is devoted to sludge and scum pumping. The various methods used in the processing of solids are discussed in Secs. 14–5 through 14–15. Stabilization of solids and biosolids is introduced in Sec. 14–7 and is divided into four subsequent sections for more detailed discussion: alkaline stabilization, anaerobic digestion, aerobic digestion, and composting (see Secs. 14–8 through 14–11). The preparation of solids balances for treatment facilities is described in Sec. 14–16. The application of biosolids to land and conveyance and storage of biosolids after processing is discussed in Secs. 14–17 and 14–18.

14–1 SOLIDS SOURCES, CHARACTERISTICS, AND QUANTITIES

To design solids processing, treatment, and disposal facilities properly, the sources, characteristics, and quantities of the solids to be handled must be known. Therefore, the purpose of this section is to present background data and information on these topics that will serve as a basis for the material to be presented in the subsequent sections of this chapter.

Sources

The sources of solids in a treatment plant vary according to the type of plant and its method of operation. The principal sources of solids and the types generated are reported in Table 14–2. For example, in a complete mix activated-sludge process, if the wasting of solids is accomplished from the mixed liquor line or aeration chamber, the activated-sludge settling tank is not a source of solids. On the other hand, if wasting is accomplished from the activated-sludge return line, the activated-sludge settling tank constitutes a solids source. Processes used for thickening, digesting, conditioning, and dewatering of solids produced from primary and secondary settling tanks also constitute sources.

Characteristics

To treat and dispose of the solids produced from wastewater-treatment plants in the most effective manner, it is important to know the characteristics of the solids that will be processed. The characteristics vary depending on the origin of the solids, the amount of aging that has taken place, and the type of processing to which they have been subjected (see Table 14–3).

General Composition. Typical data on the chemical composition of untreated sludge and digested biosolids are reported in Table 14–4. Many of the chemical constituents, including nutrients, are important in considering the ultimate disposal of the processed solids and the liquid removed during processing. The measurement of pH,

Table 14-2

Sources of solids from conventional wastewater-treatment plants

Unit operation or process	Types of solids	Remarks
Screening	Coarse solids	Coarse solids are removed by mechanical and hand-cleaned bar screens. In small plants, screenings are often comminuted for removal in subsequent treatment units
Grit removal	Grit and scum	Scum-removal facilities are often omitted in grit-removal facilities
Preaeration	Grit and scum	In some plants, scum-removal facilities are not provided in preaeration tanks. If the preaeration tanks are not preceded by grit-removal facilities, grit deposition may occur in preaeration tanks
Primary sedimentation	Primary solids and scum	Quantities of solids and scum depend upon the nature of the collection system and whether industrial wastes are discharged to the system
Biological treatment	Suspended solids	Suspended solids are produced by the biological conversion of BOD. Some form of thickening may be required to concentrate the waste sludge stream from the biological treatment system
Secondary sedimentation	Secondary biosolids and scum	Provision for scum removal from secondary settling tanks is a requirement of the U.S. EPA
Solids processing facilities	Solids, compost, and ashes	The characteristics of the end products depend on the characteristics of the solids treated and the operations and processes used. Regulations for the disposal of residuals are stringent

alkalinity, and organic acid content is important in process control of anaerobic digestion. The content of heavy metals, pesticides, and hydrocarbons has to be determined when incineration and land application methods are contemplated. The thermal content of solids is important where a thermal reduction process such as incineration is considered.

Specific Constituents. Solids characteristics that affect their suitability for application to land and for beneficial use include organic content (usually measured as volatile solids), nutrients, pathogens, metals, and toxic organics. The fertilizer value of the sludge and solids, which should be evaluated where they are to be used as a soil conditioner, is based primarily on the content of nitrogen, phosphorus, and potassium (potash). Typical nutrient values of wastewater biosolids as compared to commercial fertilizers are reported in Table 14–5. In most land application systems, biosolids provide sufficient nutrients for good plant growth. In some applications, the phosphorus and potassium content may be low and require augmentation.

Trace elements are those inorganic chemical elements that, in very small quantities, can be essential or detrimental to plants and animals. The term "heavy metals" is used to denote several of the trace elements present in sludge and biosolids. Concentrations of heavy metals may vary widely, as indicated in Table 14–6. For the application of biosolids to land, concentrations of heavy metals may limit the application rate and the useful life of the application site (see Sec. 14–17).

Table 14–3
Characteristics of solids and sludge produced during wastewater treatment

Solids or sludge	Description
Screenings	Screenings include all types of organic and inorganic materials large enough to be removed on bar racks. The organic content varies, depending on the nature of the system and the season of the year
Grit	Grit is usually made up of the heavier inorganic solids that settle with relatively high velocities. Depending on the operating conditions, grit may also contain significant amounts of organic matter, especially fats and grease
Scum/grease	Scum consists of the floatable materials skimmed from the surface of primary and secondary settling tanks and from grit chambers and chlorine contact tanks, if so equipped. Scum may contain grease, vegetable and mineral oils, animal fats, waxes, soaps, food wastes, vegetable and fruit skins, hair, paper and cotton, cigarette tips, plastic materials, condoms, grit particles, and similar materials. The specific gravity of scum is less than 1.0 and usually around 0.95
Primary sludge	Sludge from primary settling tanks is usually gray and slimy and, in most cases, has an extremely offensive odor. Primary sludge can be readily digested under suitable conditions of operation
Sludge from chemical precipitation	Sludge from chemical precipitation with metal salts is usually dark in color, though its surface may be red if it contains much iron. Lime sludge is grayish-brown. The odor of chemical sludge may be objectionable, but is not as objectionable as the odor of primary sludge. While chemical sludge is somewhat slimy, the hydrate of iron or aluminum in it makes it gelatinous. If the sludge is left in the tank, it undergoes decomposition similar to primary sludge, but at a slower rate. Substantial quantities of gas may be given off and the sludge density increased by long residence times in storage
Activated sludge	Activated sludge generally has a brown flocculent appearance. If the color is dark, the sludge may be approaching a septic condition. If the color is lighter than usual, there may have been underaeration with a tendency for the solids to settle slowly. Sludge in good condition has an inoffensive "earthy" odor. The sludge tends to become septic rapidly and then has a disagreeable odor of putrefaction. Activated sludge will digest readily alone or when mixed with primary sludge
Trickling-filter sludge	Humus sludge from trickling filters is brownish, flocculent, and relatively inoffensive when fresh. It generally undergoes decomposition more slowly than other undigested sludges. When trickling-filter sludge contains many worms, it may become inoffensive quickly. Trickling-filter sludge digests readily
Aerobically digested biosolids	Aerobically digested biosolids are brown to dark brown and have a flocculent appearance. The odor of aerobically digested sludge is not offensive; it is often characterized as musty. Well-digested aerobic sludge dewaters easily on drying beds
Anaerobically digested biosolids	Anaerobically digested biosolids are dark brown to black and contain an exceptionally large quantity of gas. When thoroughly digested, they are not offensive, the odor being relatively faint and like that of hot tar, burnt rubber, or sealing wax. Primary sludge, when anaerobically digested, produces about twice as much methane gas as does waste activated sludge. When drawn off onto porous beds in thin layers, the solids first are carried to the surface by the entrained gases, leaving a sheet of comparatively clear water. The water drains off rapidly and allows the solids to sink down slowly onto the bed. As the solids dry, the gases escape, leaving a well-cracked surface with an odor resembling that of garden loam
Compost	Composted solids are usually dark brown to black, but the color may vary if bulking agents such as recycled compost or wood chips have been used in the composting process. The odor of well-composted solids is inoffensive and resembles that of commercial garden-type soil conditioners

Table 14–4
Typical chemical composition of untreated sludge and digested biosolids[a]

Item	Untreated primary sludge		Digested primary sludge		Untreated activated sludge
	Range	Typical	Range	Typical	Range
Total dry solids (TS),%	5–9	6	2–5	4	0.8–1.2
Volatile solids (% of TS)	60–80	65	30–60	40	59–88
Grease and fats (% of TS):					
Ether soluble	6–30	—	5–20	18	—
Ether extract	7–35	—	—	—	5–12
Protein (% of TS)	20–30	25	15–20	18	32–41
Nitrogen (N, % of TS)	1.5–4	2.5	1.6–3.0	3.0	2.4–5.0
Phosphorus (P_2O_5, % of TS)	0.8–2.8	1.6	1.5–4.0	2.5	2.8–11
Potash (K_2O, % of TS)	0–1	0.4	0–3.0	1.0	0.5–0.7
Cellulose (% of TS)	8–15	10	8–15	10	—
Iron (not as sulfide)	2.0–4.0	2.5	3.0–8.0	4.0	—
Silica (SiO_2, % of TS)	15–20	—	10–20	—	—
pH	5.0–8.0	6.0	6.5–7.5	7.0	6.5–8.0
Alkalinity (mg/L as $CaCO_3$)	500–1500	600	2500–3500	3000	580–1100
Organic acids (mg/L as HAc)	200–2000	500	100–600	200	1100–1700
Energy content, kJ/kg TSS	23,000–29,000	25,000	9000–14,000	12,000	19,000–23,000

[a] Adapted, in part, from U.S. EPA (1979).
Note: kJ/kg × 0.4303 = Btu/lb.

Table 14–5
Comparison of nutrient levels in commercial fertilizers and wastewater biosolids[a]

Product	Nutrients, %		
	Nitrogen	Phosphorus	Potassium
Fertilizers for typical agricultural use[a]	5	10	10
Typical values for stabilized wastewater biosolids (based on TS)	3.3	2.3	0.3

[a] The concentrations of nutrients may vary widely depending upon the soil and crop needs.

Quantities

Data on the quantities of solids produced from various processes and operations are presented in Table 14–7. Although the data in Table 14–7 are useful as presented, it should be noted that the quantity of solids produced would vary widely. Corresponding data on the solids concentrations to be expected from various processes are given in Table 14–8.

Table 14–6

Typical metal content in wastewater solids[a]

Metal	Dry solids, mg/kg	
	Range	Median
Arsenic	1.1–230	10
Cadmium	1–3410	10
Chromium	10–99,000	500
Cobalt	11.3–2490	30
Copper	84–17,000	800
Iron	1000–154,000	17,000
Lead	13–26,000	500
Manganese	32–9870	260
Mercury	0.6–56	6
Molybdenum	0.1–214	4
Nickel	2–5300	80
Selenium	1.7–17.2	5
Tin	2.6–329	14
Zinc	101–49,000	1700

[a]U.S. EPA (1984).

Quantity Variations. The quantity of solids entering the wastewater-treatment plant daily may be expected to fluctuate over a wide range. To ensure capacity capable of handling these variations, the designer of solids processing and disposal facilities should consider (1) the average and maximum rates of solids production, and (2) the potential storage capacity of the treatment units within the plant. The variation in daily quantity that may be expected in large cities is shown on Fig. 14–1. The curve is characteristic of large cities having a number of large collection lines laid on flat slopes; even greater variations may be expected at small plants.

A limited quantity of solids may be stored temporarily in the sedimentation and aeration tanks. Where digestion tanks with varying levels are used, their large storage capacity provides a substantial dampening effect on peak digested solids loads. In solids-treatment systems where digestion is used, the design is usually based on maximum monthly loadings; however, digesters that have a 15 d solids residence time frequently use maximum 15 d loadings. Where digestion is not used, the solids-treatment process should be designed based on the inherent storage capacity available in the solids handling system. For example, the mechanical dewatering system following gravity thickening could be based on the maximum 1 or 3 d solids production. Certain components of the solids operations system, such as sludge pumping and thickening, may need to be sized to handle the maximum-day conditions.

Volume-Mass Relationships. The volume of sludge depends mainly on its water content and only slightly on the character of the solid matter. A 10 percent sludge, for example, contains 90 percent water by weight. If the solid matter is composed of

Table 14–7

Typical data for the physical characteristics and quantities of sludge produced from various wastewater-treatment operations and processes

Treatment operation or process	Specific gravity of solids	Specific gravity of sludge	Dry solids, lb/10³ gal		Dry solids, kg/10³ m³	
			Range	Typical	Range	Typical
Primary sedimentation	1.4	1.02	0.9–1.4	1.25	110–170	150
Activated sludge (waste biosolids)	1.25	1.005	0.6–0.8	0.7	70–100	80
Trickling filter (waste biosolids)	1.45	1.025	0.5–0.8	0.6	60–100	70
Extended aeration (waste biosolids)	1.30	1.015	0.7–1.0	0.8[a]	80–120	100[a]
Aerated lagoon (waste biosolids)	1.30	1.01	0.7–1.0	0.8[a]	80–120	100[a]
Filtration	1.20	1.005	0.1–0.2	0.15	12–24	20
Algae removal	1.20	1.005	0.1–0.2	0.15	12–24	20
Chemical addition to primary tanks for phosphorus removal						
Low lime (350–500 mg/L)	1.9	1.04	2.0–3.3	2.5[b]	240–400	300[b]
High lime (800–1600 mg/L)	2.2	1.05	5.0–11.0	6.6[b]	600–1300	800[b]
Suspended growth nitrification	—	—	—	—	—	—[c]
Suspended growth denitrification	1.20	1.005	0.1–0.25	0.15	12–30	18
Roughing filters	1.28	1.02	—	—[d]	—	—[d]

[a] Assuming no primary treatment.
[b] Solids in addition to that normally removed by primary sedimentation.
[c] Negligible.
[d] Included in biosolids production from secondary treatment processes.

fixed (mineral) solids and volatile (organic) solids, the specific gravity of all of the solid matter can be computed using Eq. (14–1).

$$\frac{W_s}{S_s \rho_w} = \frac{W_f}{S_f \rho_w} + \frac{W_v}{S_v \rho_w} \qquad (14\text{–}1)$$

Table 14-8
Expected solids
concentrations from
various treatment
operations and
processes

Operation or process application	Solids concentration % dry solids	
	Range	Typical
Primary settling tank:		
Primary sludge	5–9	6
Primary sludge to a cyclone degritter	0.5–3	1.5
Primary sludge and waste activated sludge	3–8	4
Primary sludge and trickling-filter humus	4–10	5
Primary sludge with iron salt addition for phosphorus removal	0.5–3	2
Primary sludge with low lime addition for phosphorus removal	2–8	4
Primary sludge with high lime addition for phosphorus removal	4–16	10
Scum	3–10	5
Secondary settling tank:		
Waste activated sludge with primary settling	0.5–1.5	0.8
Waste activated sludge without primary settling	0.8–2.5	1.3
High-purity oxygen with primary settling	1.3–3	2
High-purity oxygen without primary settling	1.4–4	2.5
Trickling-filter humus	1–3	1.5
Rotating biological contactor waste sludge	1–3	1.5
Gravity thickener:		
Primary sludge	5–10	8
Primary sludge and waste activated sludge	2–8	4
Primary sludge and trickling-filter humus	4–9	5
Dissolved air flotation thickener:		
Waste activated sludge with polymer addition	4–6	5
Waste activated sludge without polymer addition	3–5	4
Centrifuge thickener (waste activated sludge only)	4–8	5
Gravity-belt thickener (waste activated sludge with polymer addition)	4–8	5
Anaerobic digester:		
Primary sludge	2–5	4
Primary sludge and waste activated sludge	1.5–4	2.5
Primary sludge and trickling-filter humus	2–4	3
Aerobic digester:		
Primary sludge	2.5–7	3.5
Primary sludge and waste activated sludge	1.5–4	2.5
Primary sludge and trickling-filter humus	0.8–2.5	1.3

Figure 14-1

Peak sludge loading as a percentage of the average daily load.

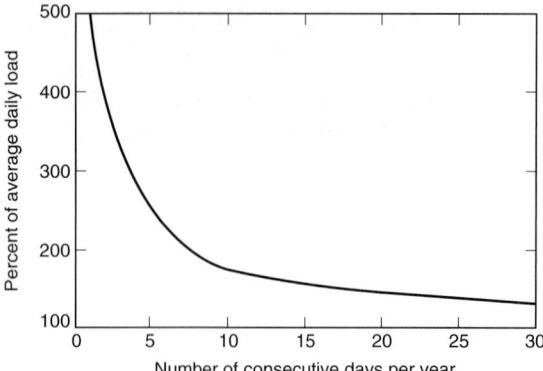

where W_s = weight of solids
S_s = specific gravity of solids
ρ_w = density of water
W_f = weight of fixed solids (mineral matter)
S_f = specific gravity of fixed solids
W_v = weight of volatile solids
S_v = specific gravity of volatile solids

Therefore, if one-third of the solid matter in a sludge containing 90 percent water is composed of fixed mineral solids with a specific gravity of 2.5, and two-thirds is composed of volatile solids with a specific gravity of 1.0, then the specific gravity of all solids S_s would be equal to 1.25, as follows:

$$\frac{1}{S_s} = \frac{0.33}{2.5} + \frac{0.67}{1} = 0.802$$

$$S_s = \frac{1}{0.802} = 1.25$$

If the specific gravity of the water is taken to be 1.0, the specific gravity of the sludge S_{sl} is 1.02, as follows:

$$\frac{1}{S_{sl}} = \frac{0.1}{1.25} + \frac{0.9}{1.0} = 0.98$$

$$S_{sl} = \frac{1}{0.98} = 1.02$$

The volume of sludge may be computed with the following expression:

$$V = \frac{M_s}{\rho_w S_{sl} P_s} \tag{14-2}$$

where V = volume, m³
M_s = mass of dry solids, kg
ρ_w = specific weight of water, 10^3 kg/m³

S_{sl} = specific gravity of the sludge
P_s = percent solids expressed as a decimal

For approximate calculations for a given solids content, it is simple to remember that the volume varies inversely with the percent of solid matter contained in the sludge as given by

$$\frac{V_1}{V_2} = \frac{P_2}{P_1} \quad \text{(approximate)}$$

where V_1, V_2 = sludge volumes
P_1, P_2 = percent of solid matter

The application of these volume and weight relationships is illustrated in Example 14–1.

EXAMPLE 14–1 **Volume of Untreated and Digested Sludge** Determine the liquid volume before and after digestion and the percent reduction for 500 kg (dry basis) of primary sludge with the following characteristics:

	Primary	Digested
Solids, %	5	10
Volatile matter, %	60	60 (destroyed)
Specific gravity of fixed solids	2.5	2.5
Specific gravity of volatile solids	≈1.0	≈1.0

Solution

1. Compute the average specific gravity of all the solids in the primary sludge using Eq. (14–1).

$$\frac{1}{S_s} = \frac{0.4}{2.5} + \frac{0.6}{1.0} = 0.76$$

$$S_s = \frac{1}{0.76} = 1.32 \quad \text{(primary solids)}$$

2. Compute the specific gravity of the primary sludge.

$$\frac{1}{S_{sl}} = \frac{0.05}{1.32} + \frac{0.95}{1}$$

$$S_{sl} = \frac{1}{0.99} = 1.01$$

3. Compute the volume of the primary sludge using Eq. (14–2).

$$V = \frac{500 \text{ kg}}{(1000 \text{ kg/m}^3)(1.01)(0.05)}$$

$$= 9.9 \text{ m}^3$$

4. Compute the percentage of volatile matter after digestion.

$$\text{Volatile matter, \%} = \frac{\text{total volatile solids after digestion}}{\text{total solids after digestion}} \times 100$$

$$= \frac{0.4(0.6 \times 1000)}{400 + 0.4(600)} \times 100 = 37.5\%$$

5. Compute the average specific gravity of all the solids in the digested sludge using Eq. (14–1).

$$\frac{1}{S_s} = \frac{0.625}{2.5} + \frac{0.375}{1} = 0.625$$

$$S_s = \frac{1}{0.625} = 1.6 \qquad \text{(digested solids)}$$

6. Compute the specific gravity of the digested sludge (S_{ds}).

$$\frac{1}{S_{ds}} = \frac{0.1}{1.6} + \frac{0.90}{1} = 0.96$$

$$S_{ds} = \frac{1}{0.96} = 1.04$$

7. Compute the volume of digested sludge using Eq. (14–2).

$$V = \frac{200 \text{ kg} + 0.4(300 \text{ kg})}{(1000 \text{ kg/m}^3)(1.04)(0.10)}$$

$$= 3.1 \text{ m}^3$$

8. Determine the percentage reduction in the sludge volume after digestion.

$$\text{\% reduction} = \frac{(9.9 - 3.1) \text{ m}^3}{9.9 \text{ m}^3} \times 100 = 68.7$$

14–2 REGULATIONS FOR THE REUSE AND DISPOSAL OF SOLIDS IN THE UNITED STATES

In selecting the appropriate methods of solids processing, reuse, and disposal, consideration must be given to the appropriate regulations. In the United States, regulations (40 CFR Part 503) were promulgated in 1993 by the U.S. Environmental Protection Agency that established pollutant numerical limits and management practices for the reuse and disposal of solids generated from the processing of municipal wastewater and septage (Federal Register, 1993). The regulations were designed to protect public health and the environment from any reasonably anticipated adverse effects of pollutants contained in the biosolids.

The regulations addressed by 40 CFR Part 503 cover specifically (1) land application of biosolids, (2) surface disposal of biosolids, (3) pathogen and vector reduction in treated biosolids, and (4) incineration. Each of these subjects is discussed below. The regulations directly affect selection of many of the processes used in solids treatment, especially for sludge stabilization, i.e., alkaline stabilization, anaerobic digestion, aerobic digestion, and composting. In some cases, to achieve compliance, appropriate treatment requirements or methods are stipulated by the regulations. Additional discussion regarding regulations for applying biosolids on land is provided in Sec. 14–17.

Land Application

Land application relates to biosolids reuse and includes all forms of applying bulk or bagged biosolids to land for beneficial uses at agronomic rates, i.e., rates designed to provide the amount of nitrogen needed by crop or vegetation while minimizing the amount that passes below the root zone. The regulations establish two levels of biosolids quality with respect to heavy metals concentrations—pollutant ceiling and pollutant concentrations ("high" quality biosolids); two levels of quality with respect to pathogen densities—Class A and Class B; and two types of approaches for meeting vector attraction—biosolids processing or use of physical barriers. Vector attraction reduction decreases the potential for spreading infectious disease by vectors such as rodents, insects, and birds.

Surface Disposal

The surface disposal part of the Part 503 regulations apply to (1) dedicated surface disposal sites; (2) monofills, i.e., solids-only landfills; (3) piles or mounds; and (4) impoundments or lagoons. Disposal sites and solids placed on those sites for final disposal are addressed in the surface disposal rules. Surface disposal does not include placement of solids for storage or treatment purposes. Where surface disposal sites do not have a liner or leachate collection system, limits are established for pollutants such as arsenic and nickel and vary based on the distance of the active surface disposal site boundary from the site property line (see Federal Register, 1993).

Pathogen and Vector Attraction Reduction

The 40 CFR Part 503 regulations divide the quality of biosolids into two categories, referred to as Class A and Class B (see definitions in Table 1–1). Class A biosolids must meet specific criteria to ensure they are safe to be used by the general public and for nurseries, gardens, and golf courses. Class B biosolids have lesser treatment requirements than Class A, and typically are used for application to agricultural land or disposed of in a landfill.

When biosolids are prepared for sale or given away for land application to lawns and home gardens or are marketed in containers, they must meet one of the following Class A biosolids criteria:

- A fecal coliform density of less than 1000 MPN/g total dry solids
- *Salmonella* sp. density of less that 3 MPN per 4 g total dry solids (3 MPN/4 g TS)

Bulk biosolids applied to lawns and home gardens or sold or given away in bags or other containers must meet the Class A criteria for pathogen reduction (see Table 14–9)

Table 14–9

Pathogen reduction alternatives[a]

Class A:	
In addition to meeting the requirements in one of the six alternatives listed below, fecal coliform or *Salmonella* sp. bacterial levels must meet specific densities at the time of biosolids use or disposal, when prepared for sale or giveaway in a bag or other container for application to the land, or when prepared to meet the requirements in 503.10(b), (c), (e), or (f)	
Alternative 1	Thermally treated biosolids: use one of four time-temperature regimes
Alternative 2	Biosolids treated in a high pH–high temperature process: specifies pH, temperature, and air-drying requirements
Alternative 3	For biosolids treated in other processes: demonstrate that the process can reduce enteric viruses and viable helminth ova. Maintain operating conditions used in the demonstration
Alternative 4	Biosolids treated in unknown processes: demonstration of the process is unnecessary. Instead, test for pathogens—*Salmonella* sp. bacteria, enteric viruses, and viable helminth ova—at the time the biosolids are used or disposed of or are prepared for sale or giveaway in a bag or other container for application to the land, or when prepared to meet the requirements in 503.10(b), (c), (e), or (f)
Alternative 5	Use of PFRP: Biosolids are treated in one of the processes to further reduce pathogens (PFRP)
Alternative 6	Use of a process equivalent to PFRP: biosolids are treated in a process equivalent to one of the PFRPs, as determined by the permitting authority
Class B:	
The requirements in one of the three alternatives below must be met in addition to Class B site restrictions for the application of biosolids to land	
Alternative 1	Monitoring of indicator organisms: test for fecal coliform density as an indicator for all pathogens at the time of biosolids use or disposal
Alternative 2	Use of PSRP: biosolids are treated in one of the processes to significantly reduce pathogens (PSRP)
Alternative 3	Use of processes equivalent to PSRP: biosolids are treated in a process equivalent to one of the PSRPs, as determined by the permitting authority

[a]From U.S. EPA (1992).

and one of several vector attraction reduction processing options (see Table 14–10). In addition, other requirements that must be met that are related to either (1) time/temperature for heat drying, (2) alkaline treatment, or (3) testing for virus. Alternatively, biosolids can be treated by a prescribed process that reduces pathogens beyond detectable levels (see Process to Further Reduce Pathogens discussed below).

Class B pathogen requirements are the minimum level of pathogen reduction for land application and surface disposal. The only exception to achieving at least Class B level occurs when the solids are placed in a surface disposal facility that is covered daily. Biosolids that do not qualify as Class B cannot be land applied. To meet Class B requirements, biosolids must be treated by a process that reduces but does not eliminate pathogens (see PSRP, also discussed below), or that must be tested to meet fecal coliform limits of less than 2.0×10^6 MPN/g TS or less than 2.0×10^6 CFU/g TS.

Table 14–10
Vector attraction reduction[a]

Requirement	What is required?	Most appropriate for
Option 1 503.33(b)(1)	At least 38% reduction in volatile solids during biosolids treatment	Biosolids processed by Anaerobic biological treatment Aerobic biological treatment Chemical oxidation
Option 2 503.33(b)(2)	Less than 17% additional volatile solids loss during bench-scale anaerobic batch digestion of the biosolids for 40 additional days at 30–37°C (86–99°F)	Only for anaerobically digested biosolids
Option 3 503.33(b)(3)	Less than 15% additional volatile solids reduction during bench-scale aerobic batch digestion for 30 additional days at 20°C (68°F)	Only for aerobically digested biosolids with 2% or less solids—e.g., biosolids treated in extended aeration plants
Option 4 503.33(b)(4)	SOUR at 20°C (68°F) is 1.5 mg O_2/h·g total solids	Biosolids from aerobic processes (should not be used for composted sludges). Also for biosolids that have been deprived of oxygen for longer than 1–2 h
Option 5 503.33(b)(5)	Aerobic treatment of the biosolids for at least 14 d at over 40°C (104°F) with an average temperature of over 45°C (113°F)	Composted biosolids (Options 3 and 4 are likely to be easier to meet for biosolids from other aerobic processes)
Option 6 503.33(b)(6)	Addition of sufficient alkali to raise the pH to at least 12 at 25°C (77°F) and maintain a pH of 12 for 2 h and a pH of 11.5 for 22 more hours	Alkali-treated biosolids (alkalies include lime, fly ash, kiln dust, and wood ash)
Option 7 503.33(b)(7)	Percent solids of 75% prior to mixing with other materials	Biosolids treated by an aerobic or anaerobic process (i.e., biosolids that do not contain unstabilized solids generated in primary wastewater treatment)
Option 8 503.33(b)(8)	Percent solids of 90% prior to mixing with other materials	Biosolids that contain unstabilized solids generated in primary wastewater treatment (e.g., any heat-dried sludges)
Option 9 503.33(b)(9)	Biosolids are injected into soil so that no significant amount of biosolids is present on the land surface 1 hour after injection, except Class A biosolids, which must be injected within 8 h after the pathogen reduction process	Liquid biosolids applied to the land. Domestic septage applied to agricultural land, a forest, or a reclamation site
Option 10 503.33(b)(10)	Biosolids are incorporated into the soil within 6 h after application to land. Class A biosolids must be applied to the land surface within 8 h after the pathogen reduction process, and must be incorporated within 6 h after application.	Biosolids applied to the land. Domestic septage applied to agricultural land, forest, or a reclamation site

[a]From U.S. EPA (1992).

To meet pathogen and vector reduction requirements, two levels of preapplication treatment are required, and have been defined by the U.S. EPA as Processes to Further Reduce Pathogens (PFRP) and Processes to Significantly Reduce Pathogens (PSRP). These processes are defined in Tables 14–11 and 14–12. Because PSRPs reduce but do not eliminate pathogens, PSRP-treated biosolids still have the potential to transmit disease. Because PFRPs reduce pathogens below detectable levels, there are no pathogen-related restrictions for land application. Minimum frequency of monitoring, record-keeping, and reporting requirements must be met, however.

Incineration

The Part 503 regulations establish requirements for wastewater biosolids-only incinerators. The regulations cover incinerator feed solids, the furnace itself, operation of the furnace, and exhaust gases from the stack. The rule indirectly limits emissions of heavy metals and directly limits total hydrocarbon emissions from incinerator stacks. Pollutant limits for wastewater solids fired in an incinerator are established for beryllium, mercury, lead, arsenic, cadmium, chromium, and nickel. Incinerators must also meet a monthly average limit of 100 ppm for total hydrocarbons (THCs), corrected for moisture level (0 percent) and oxygen content (to 7 percent) to control toxic organic compound emissions (Federal Register, 1993). Monitoring and reporting are also required.

Table 14–11
Regulatory definition of processes to further reduce pathogens (PFRP)[a]

Process	Definition
Composting	Using either within-vessel or static aerated pile composting, the temperature of the biosolids is maintained at 55°C or higher for 3 d. Using windrow composting, the temperature of the wastewater sludge is maintained at 55°C or higher for 15 d or longer. During this period, a minimum of five windrow turnings is required
Heat drying	Dewatered biosolids are dried by direct or indirect contact with hot gases to reduce the moisture content to 10 percent or lower. Either the temperature of solids particles exceeds 80°C or the wet-bulb temperature of the gas stream in contact with the biosolids as the biosolids leave the dryer exceeds 80°C
Heat treatment	Liquid biosolids are heated to a temperature of 180°C or higher for 30 min
Thermophilic aerobic digestion	Liquid biosolids are agitated with air or oxygen to maintain aerobic conditions, and the MCRT is 10 d at 55 to 60°C
Beta-ray irradiation	Biosolids are irradiated with beta rays from an accelerator at dosages of at least 1.0 megarad (Mrad) at room temperature (approximately 20°C)
Gamma-ray irradiation	Biosolids are irradiated with gamma rays from certain isotopes such as 60-cobalt or 135-cesium at dosages of at least 1.0 Mrad at room temperature (approximately 20°C)
Pasteurization	The temperature of the biosolids is maintained at 70°C or higher for at least 30 min

[a]Federal Register (1993).

Table 14–12
Regulatory definition of processes to significantly reduce pathogens (PSRP)ᵃ

Process	Definition
Aerobic digestion	Biosolids are agitated with air or oxygen to maintain aerobic conditions for an SRT and temperature between 40 d at 20°C and 60 d at 15°C
Air drying	Biosolids are dried on sand beds or on paved or unpaved basins for a minimum of 3 months. During 2 of the 3 months, the ambient average daily temperature exceeds 0°C
Anaerobic digestion	Biosolids are treated in the absence of air between an SRT of 15 d at temperatures of 35 to 55°C and an SRT of 60 d at a temperature of 20°C. Times and temperatures between these endpoints may be calculated by linear interpolation
Composting	Using either within-vessel, static aerated pile, or windrow composting, the temperature of the biosolids is raised to 40°C or higher for 5 d. For 4 h during the 5 d, the temperature in the compost pile exceeds 55°C
Lime stabilization	Sufficient lime is added to raise the pH of the biosolids to pH 12 and maintained for 2 h of contact

ᵃFederal Register (1993).

14–3 SOLIDS PROCESSING FLOW DIAGRAMS

A generalized flow diagram incorporating the unit operations and processes to be discussed in this chapter is presented on Fig. 14–2. As shown, an almost infinite number of combinations are possible. In practice, the most commonly used process flow diagram for biosolids processing involves biological treatment. Typical flow diagrams incorporating biological processing are presented on Fig. 14–3. Thickeners may be used depending upon the source of sludge and the method of sludge stabilization, dewatering, and disposal. Following biological digestion, any of the several methods shown may be used to dewater the sludge; the choice depends on economic evaluation, beneficial use requirements, and local conditions. In instances where biological stabilization is not used, dewatered biosolids undergo thermal decomposition in either multiple-hearth or fluidized-bed incinerators.

14–4 SLUDGE AND SCUM PUMPING

Sludge produced in wastewater-treatment plants must be conveyed from point to point in the plant in conditions ranging from a watery sludge or scum to a thick sludge. Sludge may also be pumped off-site for long distances for treatment and disposal. For each type of sludge and pumping application, a different type of pump may be needed (see Table 14–13).

Pumps

Pumps used most frequently to convey sludge include the plunger, progressive cavity, centrifugal, torque flow, diaphragm, high-pressure piston, and rotary-lobe types. Other types of pumps such as peristaltic (hose or rotor) pumps and concrete slurry pumps

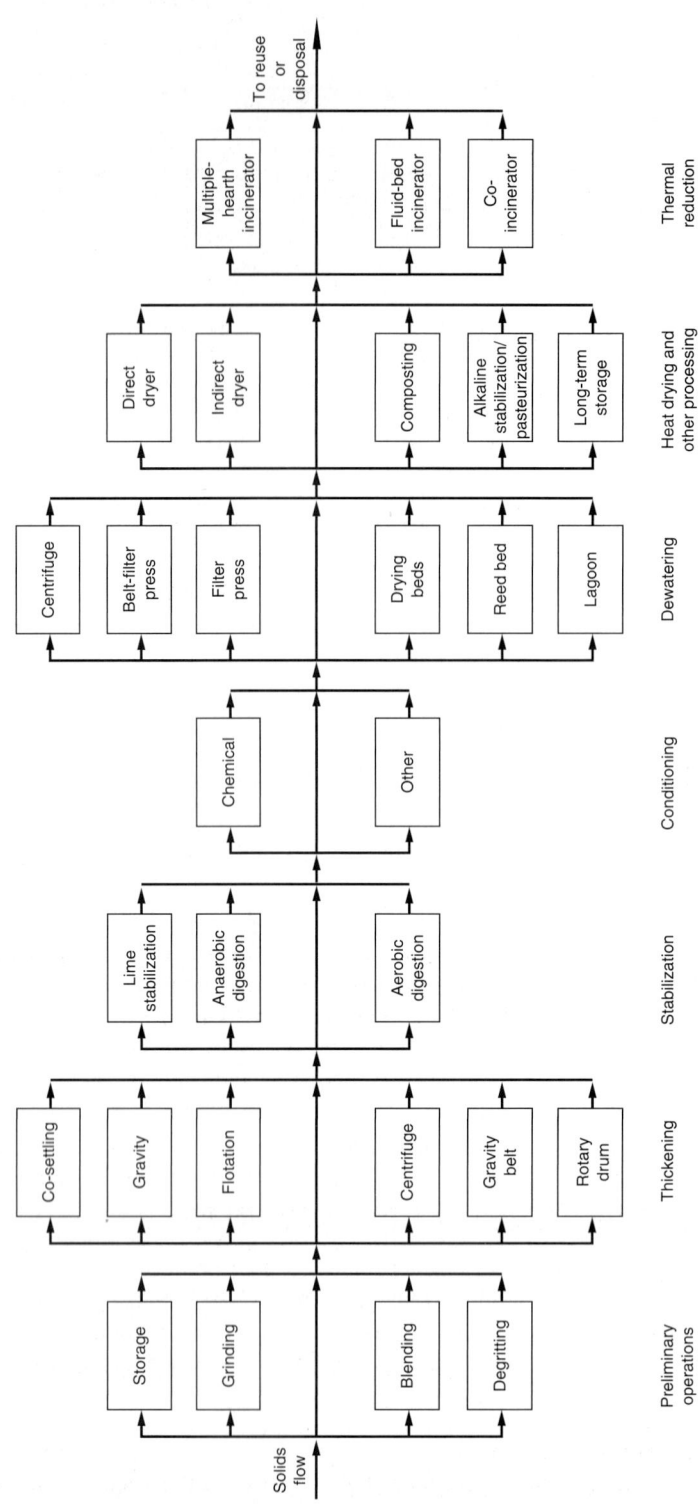

Figure 14-2

Generalized sludge-processing flow diagram.

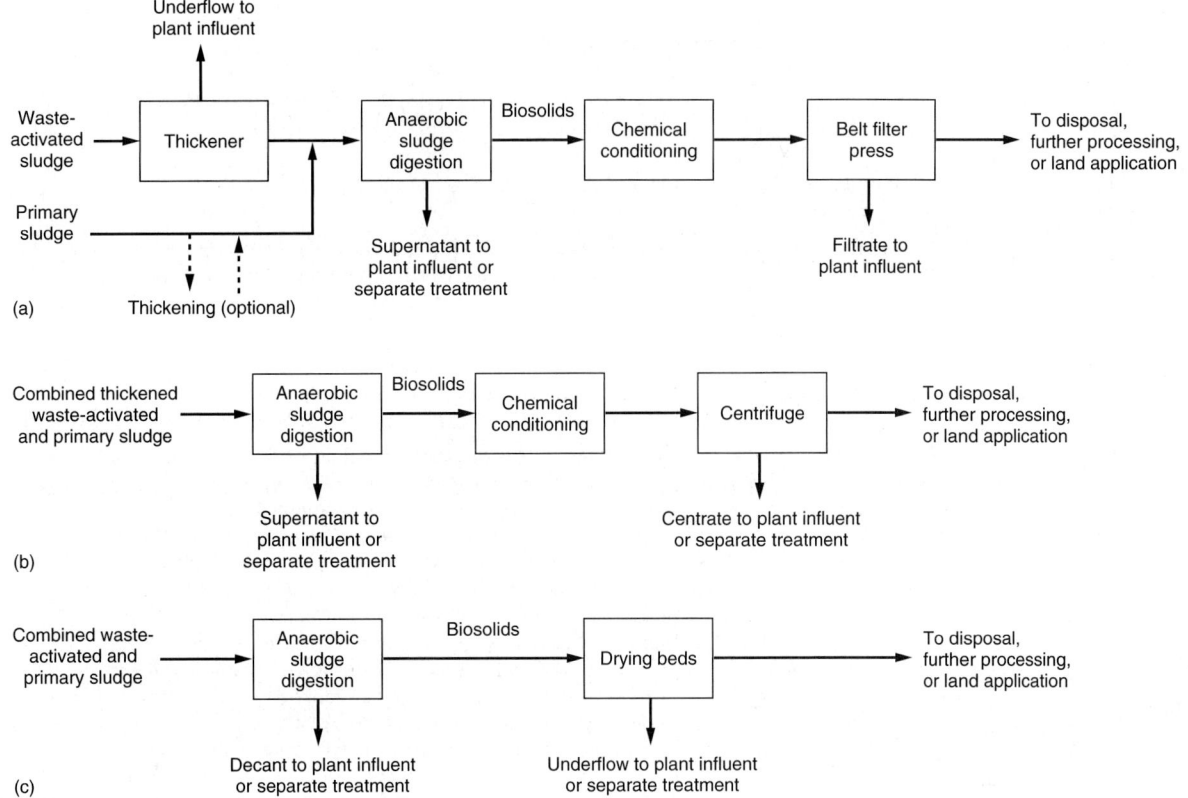

Figure 14–3

Typical sludge treatment flow diagrams with biological digestion and three different sludge dewatering processes: (a) belt-filter press, (b) centrifuge, (c) drying beds. In some plants, flows that are to be returned to the headworks are stored in equalization basins for return to the treatment process during the early morning hours when the plant load is reduced.

have also been used to pump sludge. Diaphragm and centrifugal pumps also are used extensively for pumping scum. "Chopper" type centrifugal pumps are also used for pumping sludge and scum containing rags, plastics, and other fibrous materials that require shredding. The advantages and disadvantages of each type of pump are summarized in Table 14–14.

Plunger Pumps. Plunger pumps (see Fig. 14–4a) have been used frequently, and if rugged enough for the service, have proved to be quite satisfactory. The advantages of plunger pumps are as follows:

1. Pulsating action of simplex and also duplex pumps tends to concentrate the sludge in the hoppers ahead of the pumps and resuspend solids in pipelines when pumping at low velocities.
2. They are suitable for suction lifts up to 3 m (10 ft) and are self-priming.

Table 14–13

Application of pumps to types of sludge and biosolids[a]

Type of sludge or solids	Applicable pump	Comment
Ground screenings	Pumping screenings should be avoided	Pneumatic ejectors may be used
Grit	Torque flow centrifugal	The abrasive character of grit and the presence of rags make grit difficult to handle. Hardened casings and impellers should be used for torque flow pumps. Pneumatic ejectors may also be used
Scum	Plunger; progressive cavity; diaphragm; centrifugal; chopper	Scum is often pumped by the sludge pumps; valves are manipulated in the scum and sludge lines to permit this. In larger plants separate scum pumps are used. Scum mixers are often used to ensure homogeneity prior to pumping. Pneumatic ejectors may also be used
Primary sludge	Plunger; centrifugal torque flow; diaphragm progressive cavity; rotary lobe; chopper; hose	In most cases, it is desirable to obtain as concentrated a sludge as practicable from primary sedimentation tanks, usually by collecting the sludge in hoppers and pumping intermittently, allowing the solids to collect and consolidate between pumping periods. The character of untreated primary solids will vary considerably, depending on the characteristics of the solids in the wastewater, and the types of treatment units and their efficiency. Where biological treatment follows, the quantity of solids from (1) waste-activated sludge, (2) humus sludge from settling tanks following trickling filters, (3) overflow liquors from digestion tanks, (4) and centrate or filtrate return from dewatering operations will also affect the sludge characteristics. In many cases, the character of the sludge is not suitable for the use of conventional nonclog centrifugal pumps. Where sludge contains rags, chopper pumps may be used
Sludge from chemical precipitation	Same as primary sludge	May contain large amounts of inorganic constituents depending on the type and amount of chemicals used
Trickling-filter humus	Nonclog and torque flow centrifugal; progressive cavity; plunger; diaphragm	Humus is usually of homogeneous character and can be easily pumped
Return or waste-activated sludge	Nonclog and torque flow centrifugal; progressive cavity; plunger; diaphragm	Sludge is dilute and contains only fine solids so that nonclog pumps may be used. For nonclog pumps, slow speeds are recommended to minimize the breakup of flocculent particles
Thickened or concentrated sludge	Plunger; progressive cavity; diaphragm; high-pressure piston; rotary lobe; hose	Positive-displacement pumps are most applicable for concentrated sludge because of their ability to generate movement of the sludge mass. Torque flow pumps may be used but may require the addition of flushing or dilution facilities
Digested biosolids	Plunger; torque flow centrifugal; progressive cavity; diaphragm; high-pressure piston; rotary lobe	Well-digested biosolids are homogeneous, containing 5 to 8% solids and a quantity of gas bubbles, but may contain up to 12% solids. Poorly digested biosolids may be difficult to handle. If good screening and grit removal are provided, nonclog centrifugal pumps may be considered

[a] Adapted, in part, from U.S. EPA (1979).

Table 14–14
Advantages and disadvantages of various types of sludge pumps[a]

Type of pump	Advantages	Disadvantages
Plunger	Can pump heavy solids concentrations (up to 15%)	Low efficiency
	Self-priming and can handle suction lifts up to 3 m (10 ft)	High maintenance if operated continuously
	Constant but adjustable capacity regardless of variations in head	Depending on downstream processes, pulsating flow may not be acceptable
	Cost-effective choice for flowrates up to 30 L/s (500 gal/min) and heads up to 60 m (200 ft)	
	Pulsating action of simplex and duplex pumps sometimes helps to concentrate sludge in hoppers ahead of pumps and resuspend solids in pipelines when pumping at low velocities	
	High-pressure capability	
Progressing cavity	Provides a relatively smooth flow	Stator will burn out if pump is operated dry; needs a run dry protection system
	Pumps greater than 3 L/s (50 gal/min) capacity can pass solids of about 20 mm (0.8 in) in size	Smaller pumps usually require grinders to prevent clogging
	Easily controlled flowrates	Power cost escalates when pumping heavy sludge
	Minimal pulsation	Grit in sludge may cause excessive stator wear
	Relatively simple operation	
	Stator/rotor tends to act as a check valve, thus preventing backflow through pump. An external check valve may not be required	Seals and seal water required
Diaphragm	Pulsating action may help to concentrate sludge in hoppers ahead of pumps and resuspend solids in pipelines when pumping at low velocities	Depending on downstream processes, pulsating flow may not be acceptable
	Self-priming with suction lifts up to 3 m (10 ft)	Requires a source of compressed air
	Can pump grit with relatively minimum wear	Operation may be excessively noisy
	Relatively simple operation	Low head and efficiency
		High maintenance if operated continuously
Centrifugal Nonclog (mixed flow)	Has high volume and excellent efficiency for activated-sludge pumping applications	Not recommended for other sludge pumping applications because of potential clogging due to rags and other debris
	Relatively low cost	
Recessed impeller	Because of recessed impeller design, pump can pass large solids and grit	Low efficiency—about 5 to 20% lower than standard nonclog pump
	Can pump digested sludges up to approximately 4%	Limited to raw sludges with solids concentrations of 2.5% and less
		Abrasion-resistant impellers cannot be trimmed to modify pumping characteristics

(continued)

| **Table 14–14** (Continued) | | |

Type of Pump	**Advantages**	**Disadvantages**
Chopper	Reduces clogging at pump suction May eliminate need for grinder or comminutor Can handle higher solids concentrations than nonclog pumps	Relatively low efficiency—efficiency ranges from about 40 to 60% Requires a level of maintenance similar to grinders
Rotary lobe	Provides a relatively smooth flow Does not require a check valve in most applications with low to moderate discharge static heads Able to run dry for short periods of time without significant damage Low speed and low maintenance	Because of close tolerances between rotating lobes, grit will cause excessive wear, thus reducing pumping efficiency Fluid pumped must act as a lubricant Cost of pumping increases with volume
Peristaltic hose	Has self-priming capabilities Because it is a positive-displacement pump, it is capable of metering flow Relatively simple to maintain Can pump sludge with abrasive grit	Depending on downstream processes, pulsating flow may not be acceptable High starting torque (two to three times running torque) Replacement hoses may be expensive
High-pressure piston	Can be used to pump thickened sludge long distances Can pump at rates of 30 L/s (500 gal/min) at pressures up to 13,800 kPa (2000 lb_f/in^2) Can run dry without major damage Unobstructed internal flow path; can pass large solids	High capital cost Requires skilled maintenance personnel

[a]Adapted, in part, from WEF (1998).

3. Low pumping rates can be used with large port openings.
4. Positive delivery is provided unless some object prevents the ball check valves from seating.
5. They have constant but adjustable capacity, regardless of large variations in pumping head.
6. High discharge heads may be provided for.
7. Heavy solids concentrations may be pumped if the equipment is designed for the load conditions.

Plunger pumps come with one, two, or three plungers (called simplex, duplex, or triplex units) with capacities of 2.5 to 3.8 L/s (40 to 60 gal/min) per plunger, and larger models are available. Pump speeds should be between 40 and 50 r/min. The pumps should be designed for a minimum head of 24 m (80 ft) in small plants and 35 m (115 ft) or more in large plants, because grease accumulations in sludge lines cause a progres-

Figure 14–4

Typical sludge and scum pumps used in wastewater-treatment plants: (a) plunger, (b) progressive cavity. *(Figure continues on next page)*

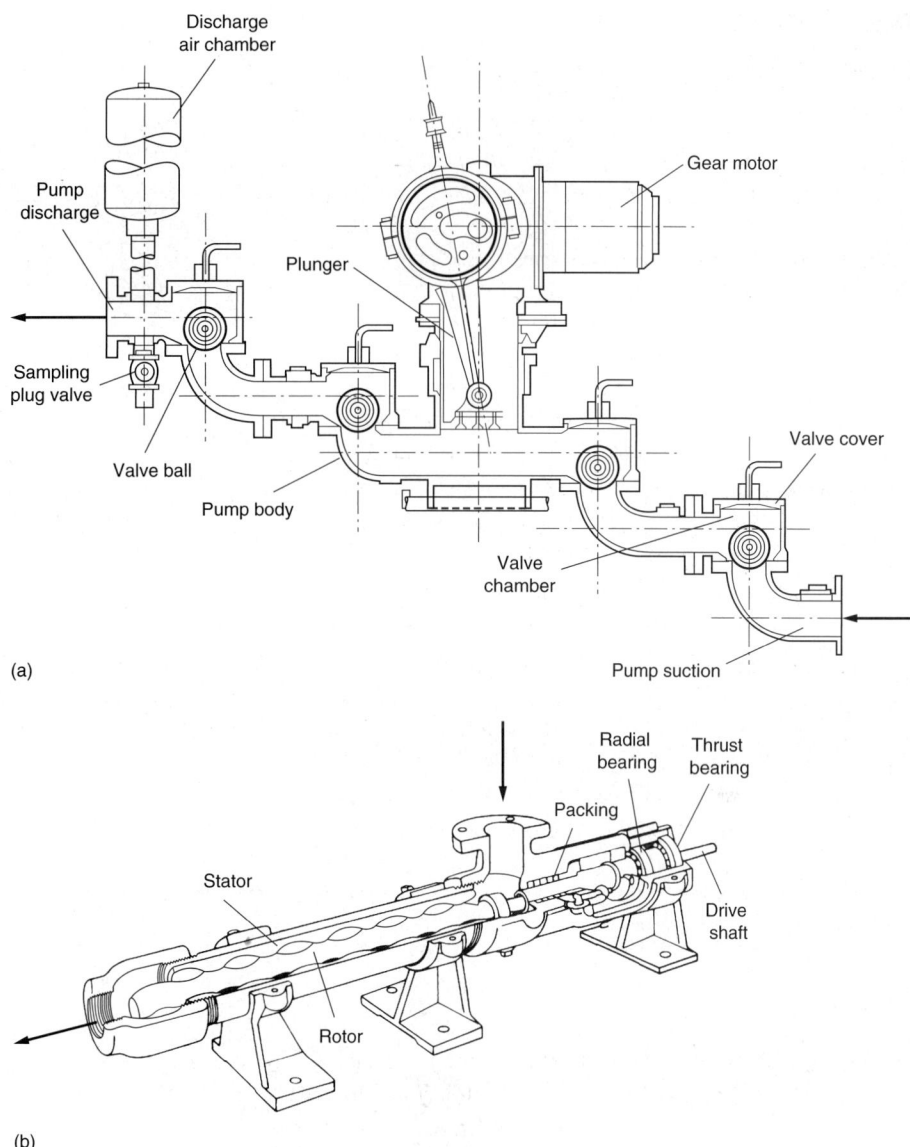

(a)

(b)

sive increase in head with use. Heavy-duty pumps are rated at 450 kPa (100 lb$_f$/in^2) up to 730 kPa (165 lb$_f$/in^2). Capacity is decreased by shortening the stroke of the plunger; however, the pumps seem to operate more satisfactorily at or near full stroke. For this reason, many pumps are provided with variable-speed drives for speed control of capacity. A plunger pump differs from a centrifugal or torque-flow pump in that its discharge is pulsing due to the action of a piston; consequently, the actual flow while sludge is moving in the pipeline is greater than average pumping capacity. The headloss calculations,

Figure 14–4

(Continued)

Typical sludge and scum pumps used in wastewater-treatment plants: (c) nonclog centrifugal, (d) torque flow, (e) diaphragm, (f) high-pressure piston. *(Figure continues on next page)*

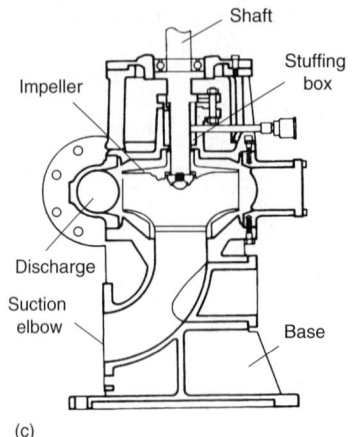

(c)

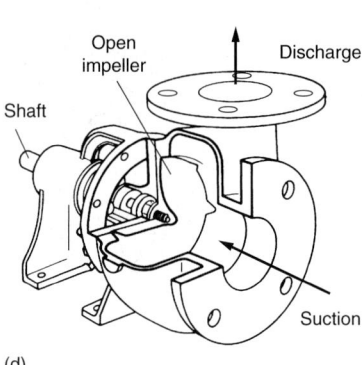

(d)

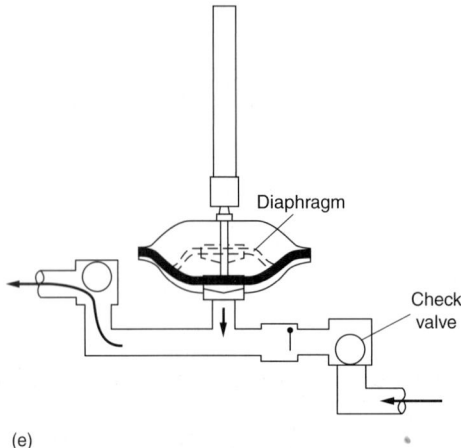

(e)

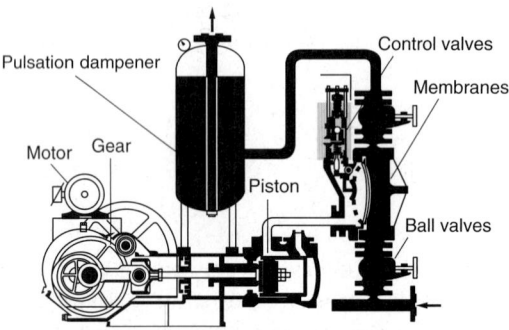

(f)

Figure 14–4
(*Continued*)

Typical sludge and scum pumps used in wastewater-treatment plants: (*g*) rotary lobe, (*h*) hose.

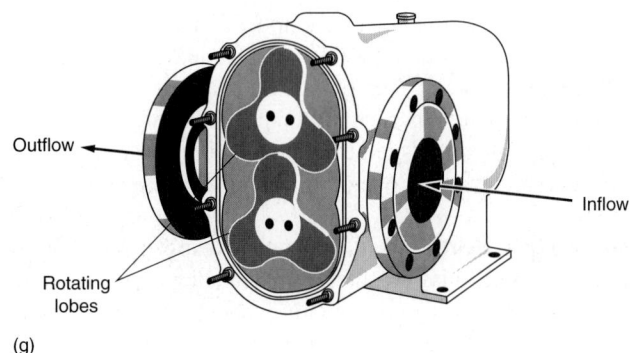

(g)

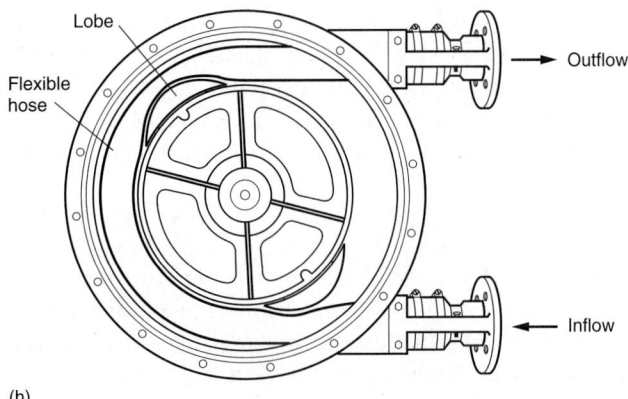

(h)

Table 14–15
Factors for computing peak pulsating flow when using plunger pumps

Type of plunger pump	Actual pulsating peak flow
Simplex	3.1 × design flow
Duplex	1.55 × design flow
Triplex	1.2 × design flow

therefore, must be based on the peak pulsating flow rather than the design flow. The factors given in Table 14–15 illustrate how the actual peak pulsating flow can be determined.

Progressive Cavity Pumps.   The progressive cavity pump (see Fig. 14–4*b*) has been used successfully on almost all types of sludges. The pump is composed of a single-threaded rotor that operates with a minimum of clearance in a double-threaded helix stator made of rubber. A volume or "cavity" moves progressively from suction to discharge when the rotor turns. The pump is self-priming at suction lifts up to 8.5 m (28 ft), but it must not be operated dry or it will burn out the rubber stator. Progressive cavity pumps are available in capacities up to 75 L/s (1200 gal/min) and may be operated at discharge heads of 135 m (450 ft) with sludge. For primary sludges, a grinder normally precedes these pumps. The pumps are expensive to maintain because of wear of the rotors and the stators, particularly in primary sludge pumping applications where grit is present. Advantages of the pumps are (1) the flowrates are controlled easily, (2) pulsation is minimal, and (3) operation is relatively simple.

Centrifugal Pumps. Centrifugal pumps of nonclog design (see Fig. 14–4c) are commonly used to pump sludge. In centrifugal pumping applications, the problem is choosing the proper size. At any given speed, centrifugal pumps operate well only if the pumping head is within a relatively narrow range; the variable nature of sludge, however, causes pumping heads to change. The selected pumps must have sufficient clearance to pass the solids without clogging and have a small enough capacity to avoid pumping a sludge diluted by large quantities of wastewater overlying the sludge blanket. Throttling the discharge to reduce the capacity is impractical because of frequent stoppages; hence it is absolutely essential that these pumps be equipped with variable-speed drives. Centrifugal pumps of special design: torque-flow and "chopper" type have been used for pumping primary sludge.

Torque-flow pumps (see Fig. 14–4d) have impellers that are fully recessed and are very effective in conveying sludge. The size of particles that can be handled is limited only by the diameter of the suction or discharge openings. The rotating impeller develops a vortex in the sludge so that the main propulsive force is the liquid itself. Most of the fluid does not actually pass through the vanes of the impeller, thereby minimizing abrasive contact; however, pumps used in sludge service are recommended to have nickel or chrome abrasion-resistant volute and impellers. The pumps can operate only over a narrow head range at a given speed, so the system operating conditions must be carefully evaluated. Variable speed control is recommended where the pumps are expected to operate over a wide range of head conditions. For high-pressure applications, multiple pumps may be used, connected together in series.

Chopper-type pumps have a cutter knife attached to a nonclog impeller that agitates and breaks up large solids that tend to block the pump suction. Incoming solids are chopped by sharpened impeller blades that turn across the cutter bar. Chopper pumps are manufactured in sizes up to 150 L/s (2400 gal/min) in both horizontal and vertical configurations.

Slow-speed centrifugal and mixed-flow pumps are commonly used for returning activated sludge to the aeration tanks. Screw pumps are also being used for this service.

Diaphragm Pumps. Diaphragm pumps use a flexible membrane that is pushed and pulled to contract and enlarge an enclosed cavity (see Fig. 14–4e). Flow is directed through this cavity by check valves, which may be either ball or flap type. The capacity of a diaphragm pump is altered by changing either the length of the diaphragm stroke or the number of strokes per minute. Pump capacity can be increased and flow pulsations smoothed out by providing two pump chambers and using both strokes of the diaphragm for pumping. Diaphragm pumps are relatively low capacity and low head; the largest available air-diaphragm pump delivers 14 L/s (220 gal/min) against 15 m (50 ft) of head.

High-Pressure Piston Pumps. High-pressure piston pumps are used in high-pressure applications such as pumping sludge long distances. Several types of piston pumps have been developed for high-pressure applications, and are similar in action to plunger pumps. The high-pressure piston pumps use separate power pistons or membranes to separate the drive mechanisms from contacting the sludge. A piston membrane type of pump is shown on Fig. 14–4f. Advantages of these types of pumps are (1) they can pump relatively small flowrates at high pressures, up to 13,800 kN/m² (2000 lb$_f$/in²), (2) large solids up to the discharge pipe diameter can be passed, (3) a range of solids concentrations can be handled, and (4) the pumping can be accomplished in a single stage. The pumps, however, are very expensive.

Rotary-Lobe Pumps. Rotary-lobe pumps (see Fig. 14–4g) are positive-displacement pumps in which two rotating synchronous lobes push the fluid through the pump. Rotational speed and shearing stresses are low. For sludge pumping, lobes are made of hard metal or hard rubber. An advantage cited for the rotary-lobe pump is that lobe replacement is less costly than rotor and stator replacement for progressive cavity pumps. Rotary-lobe pumps, like other positive-displacement pumps, must be protected against pipeline obstructions.

Hose Pumps. Peristaltic hose pumps (see Fig. 14–4h) have also been used to a limited extent for pumping sludge. The pump works by alternately compressing and relaxing a specially designed resilient hose. The hose is compressed between the inner wall of the pump housing and the compression shoes on the rotor. A lubricant is used to reduce heat and wear on the hose. The pumped sludge only comes in contact with the inner wall of the hose, which cushions entrained abrasives during compression. The pumps are available in capacities ranging from 36 to 1250 L/min (10 to 330 gal/min). As a positive-displacement pump, the pump output is directly proportional to speed at either high or low discharge pressures. The primary disadvantages of the hose pump are the pulsating flow, hose wear, and the relatively high cost of hose replacement.

Headloss Determination

The headloss encountered in the pumping of sludge depends on the flow properties (rheology) of sludge, the pipe diameter, and the flow velocity. It has been observed that headlosses increase with increased solids content, increased volatile content, and lower temperatures. When the percent volatile matter multiplied by the percent solids exceeds 600, difficulties may be encountered in pumping sludge.

Water, oil, and most other fluids are "Newtonian," which means that the pressure drop is proportional to the velocity and viscosity under laminar flow conditions. As the velocity increases past a critical value, the flow becomes turbulent. Dilute sludges such as unconcentrated activated and trickling-filter sludges behave similar to water. Concentrated wastewater sludges, however, are non-Newtonian fluids. The pressure drop under laminar conditions for non-Newtonian fluids is not proportional to flow, so the viscosity is not a constant. Special procedures may be used to determine headloss under laminar-flow conditions, and the velocity at which turbulent flow begins. In this section both the simplified approach of calculating headloss and a method using the sludge rheology will be discussed.

The headloss in pumping unconcentrated activated and trickling-filter sludges may be from 10 to 25 percent greater than for water. Primary, digested, and concentrated sludges at low velocities may exhibit a plastic-flow phenomenon in which a definite pressure is required to overcome resistance and start flow. The resistance then increases approximately with the first power of the velocity throughout the laminar range of flow, which extends to about 1.1 m/s (3.5 ft/s), the lower critical velocity. Above the higher critical velocity at about 1.4 m/s (4.5 ft/s), the flow may be considered turbulent. In the turbulent range, the losses for well-digested sludge may be more than two to three times the losses for water. The losses for primary and concentrated sludges, especially those conditioned with polymer, and scum may be considerably greater. The risk of under-estimating the headloss also increases as the piping distance and solids concentrations increase. Where possible, particularly in long-distance sludge pumping, hydraulic stud-ies should be conducted to confirm the ranges of headloss characteristics.

Simplified Headloss Computations. Relatively simple procedures are used to compute headloss for short sludge pipelines. The accuracy of these procedures may be adequate, especially at solids concentrations less than 3 percent by weight. To determine the headloss, the factor k is obtained from Fig. 14–5a for a given solids con-tent and type of sludge. The headloss when pumping sludge is computed by multiply-ing the headloss of water, determined by using the Darcy-Weisbach, Hazen-Williams, or Manning equations, by k.

Figure 14–5

Headloss multiplication factors: (a) for different sludge types and concentrations, (b) for different pipeline velocities and sludge concentrations.

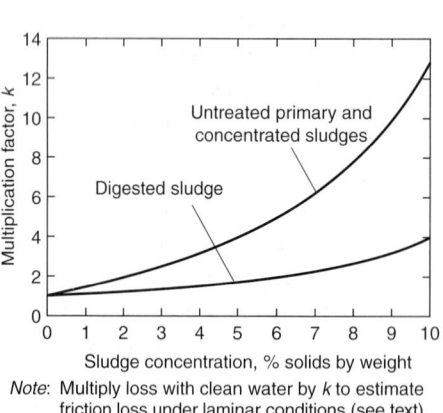

(a)

Note: Multiply loss with clean water by k to estimate friction loss under laminar conditions (see text).

(b)

The values given on Fig. 14–5a should be used only when (1) velocities are at least 0.8 m/s (2.5 ft/s), (2) velocities do not exceed 2.4 m/s (8 ft/s), (3) thixotropic behavior is not considered, and (4) the pipe is not obstructed by grease or other materials.

Another approximate method makes use of empirical multiplication factor charts (see Fig. 14–5b). The approximate method involves only velocity and percent solids consideration.

Usually, the consistency of untreated primary sludge changes during pumping. At first, the most concentrated sludge is pumped. When most of the sludge has been pumped, the pump must handle a dilute sludge that has essentially the same hydraulic characteristics as water. The change in characteristics causes a centrifugal pump to operate farther out on its head-capacity curve, beyond the areas of best efficiency. The pump motor should be sized for the additional load, and a variable-speed drive should be considered to reduce the flow under changing sludge characteristics. If the pump motor is not sized for the maximum load when pumping water at top speed, it is likely to be overloaded or damaged if the overload devices do not function or are set too high.

To determine the operating speeds and motor power required for a centrifugal pump handling sludge, system curves should be computed (1) for the most dense sludge anticipated, (2) for average conditions, and (3) for water. The curves should be plotted on a graph of the pump curves for a range of available speeds. The maximum and minimum speeds required of a particular pump are obtained from the intersection of the pump head-capacity curves with the system curves at the desired capacity. Where the maximum speed head-capacity curve intersects the system curve for water determines the power required. In constructing the system curves for sludge for velocities from 0 to 1.1 m/s (3.5 ft/s), the headloss can be considered constant at the figure computed for 1.1 m/s. The intersection of the pump curves with the system curve for average conditions can be used to estimate hours of operations, average speed, and power costs.

Because the usual flow formulas cannot be used in the plastic and laminar range, the engineer must rely on judgment and experience. In this range, capacities will be small, and plunger, progressive cavity, or rotary-lobe pumps should be used with ample head and capacity as recommended previously.

Application of Rheology to Headloss Computations. For pumping sludge over long distances, an alternative method of computing headloss characteristics has been developed based on the flow properties of the sludge. A method of computing headloss for laminar-flow conditions was derived originally by Babbitt and Caldwell (1939), based on the results of experimental and theoretical studies. Additional studies have been performed for the transition from laminar to turbulent flow (Mulbarger et al., 1981; U.S. EPA, 1979) and are summarized in Sanks et al. (1998). Long-distance pumping of mixtures of untreated (raw) primary and secondary sludge is discussed by Carthew et al. (1983). The approach used in those studies for turbulent flow, which is of critical importance for long pipelines, is described below. For laminar and transitional flow, computational procedures described in Sanks et al. (1998) are recommended.

As stated previously, water, oil, and most other common fluids are "Newtonian," which means the pressure drop is directly proportional to the velocity and viscosity under laminar-flow conditions. As the velocity increases past a critical value, the flow becomes turbulent. The transition from laminar to turbulent flow depends on the Reynolds number, which is inversely proportional to the fluid viscosity. Wastewater sludge, however, is a non-Newtonian fluid. The pressure drop under laminar conditions is not proportional to flow, so the viscosity is not a constant. The precise Reynolds number at which turbulent-flow characteristics are encountered is uncertain for sludges.

Sludge has been found to behave much like a Bingham plastic, a substance with a straight-line relationship between shear stress and flow only after flow begins. A Bingham plastic is described by two constants: the yield stress s_y and the coefficient of rigidity η. Typical ranges of values for yield stress and coefficient of rigidity are shown on Fig. 14–6a and b. If the two constants can be determined, the pressure drop over a wide range of velocities can be obtained using ordinary equations for water and the use of Fig. 14–6c. As observed on Fig. 14–6a and b, published data quantifying yield stress and the coefficient of rigidity values for wastewater sludges are highly variable. Pilot studies should be conducted to determine the rheological data for specific applications. Procedures for developing yield stress and the coefficient of rigidity using a pipeline viscometer and rotational viscometer are also given by Carthew et al. (1983).

Two dimensionless numbers can be used to determine the pressure drop due to friction for sludge: Reynolds number and Hedstrom number. Reynolds number is calculated by using the following expression:

$$N_R = \frac{\rho VD}{\eta} \quad \text{SI units} \tag{14–3a}$$

$$N_R = \frac{\gamma VD}{\eta} \quad \text{U.S. customary units} \tag{14–3b}$$

where N_R = Reynolds number, dimensionless
   ρ = density of sludge, kg/m³
   γ = specific weight of sludge, lb/ft³
   V = average velocity, m/s (f/s)
   D = diameter of pipe, m (ft)
   η = coefficient of rigidity, kg/m·s (lb/ft·s)

Hedstrom number, which is reviewed by Hill et al. (1986), is calculated as follows:

$$He = \frac{D^2 s_y \rho}{\eta^2} \quad \text{SI units} \tag{14–4a}$$

$$He = \frac{D^2 s_y g_c \gamma}{\eta^2} \quad \text{U.S. customary units} \tag{14–4b}$$

where He = Hedstrom number, dimensionless
   s_y = yield stress, N/m² (lb$_f$/ft²)
   g_c = 32.2 lb$_m$·ft/lb$_f$·s²

Other terms are as defined previously.

Using the calculated Reynolds number and the Hedstrom number, the friction factor f can be determined from Fig. 14–6c. The pressure drop for turbulent conditions can then be calculated from the following relationship:

$$\Delta p = \frac{2f\rho LV^2}{D} \quad \text{SI units} \tag{14–5a}$$

Figure 14–6

Curves for computing pipeline headloss by the sludge rheology method: (a) yield stress vs. percent sludge solids, (b) coefficient of rigidity vs. percent sludge solids, and (c) friction factor for sludge analyzed as a Bingham plastic. (Adapted from Carthew et al., 1983.)

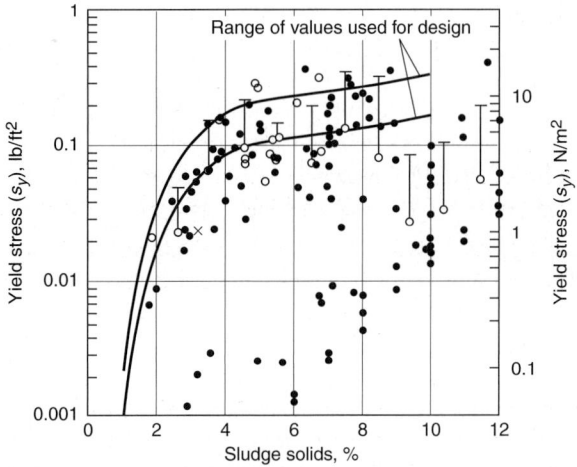

LEGEND
o - Raw primary sludge
× - Secondary sludge
• - Digested sludge
⊥ - Median + standard deviation

Note: $lb/ft^2 \times 47.8803 = N/m^2$

(a)

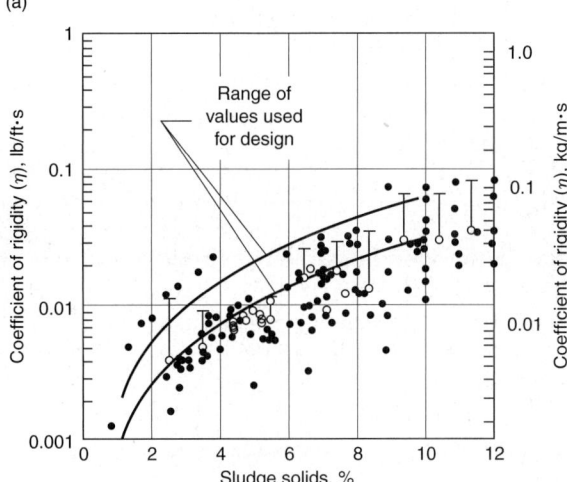

Note: $lb/ft^2 \cdot s \times 1.488 = kg/m \cdot s$

(b)

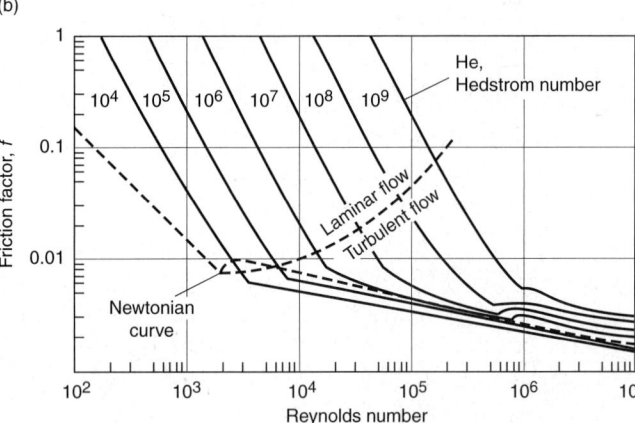

(c)

$$\Delta p = \frac{2 f \rho L V^2}{g_c D} \qquad \text{U.S. customary units} \qquad (14\text{--}5b)$$

where Δp = pressure drop due to friction, N/m² (lb$_f$/ft²)
$\quad\quad\; f$ = friction factor (from Fig. 14–6c)
$\quad\quad L$ = length of pipeline, m (ft)
Other terms as defined previously.

In using Eqs. (14–3), (14–4), and (14–5), it should be noted that the Reynolds number is not the same as the Reynolds number based on viscosity. In plastic flow, an effective viscosity may be defined, but it is variable and can be much greater than the coefficient of rigidity. Consequently, the two Reynolds numbers can differ greatly. The friction factor f will usually differ significantly from the f values reported in standard hydraulic texts for clear water, which may be four times the values used on Fig. 14–6c. These equations apply to the entire range of laminar and turbulent flows, except that Fig. 14–6c does not allow for pipe roughness. To allow for pipe roughness, if customary water formulas for headloss give a higher pressure drop than Eq. (14–5), then roughness is dominant, the flow is fully turbulent, and the pressure drop given by the water headloss formula will be reasonably accurate. A safety factor on the order of 1.5 is recommended for worst-case design conditions (Mulbarger et al., 1981). The use of Eqs. (14–3), (14–4), and (14–5) is illustrated in Example 14–2.

EXAMPLE 14–2   Computation of Headloss Using Sludge Rheology   Calculate the headloss in a 200-mm-diameter pipeline 10,000 m long conveying untreated (raw) sludge at an average flowrate 0.04 m³/s. Determine also if the flow is turbulent. By testing, the following sludge rheology data were found:

Yield stress s_y = 1.3 N/m²

Coefficient of rigidity η = 0.035 kg/m·s

Specific gravity = 1.01

Solution

1. Calculate the pipeflow velocity
    a. Determine the pipe cross-sectional area.

    $$A = \pi \frac{D^2}{4} = 3.14 \frac{(0.2 \text{ m})^2}{4} = 0.031 \text{ m}^2$$

    b. Determine velocity.

    $$V = \frac{Q}{A} = \frac{(0.04 \text{ m}^3/\text{s})}{0.031 \text{ m}^2} = 1.29 \text{ m/s}$$

2. Compute sludge specific weight.

    $$\rho = 1000 \text{ kg/m}^3 \times 1.01 = 1010 \text{ kg/m}^3$$

3. Compute Reynolds number using Eq. (14–3).

$$N_R = \frac{\rho V D}{\eta} = \frac{(1010 \text{ kg/m}^3)(1.29 \text{ m/s})(0.2 \text{ m})}{(0.035 \text{ kg/m·s})} = 7.4 \times 10^3$$

4. Compute Hedstrom number using Eq. (14–4).

$$He = \frac{D^2 s_y \rho}{\eta^2} = \frac{(0.2 \text{ m})^2(1.3 \text{ N/m}^2)(1010 \text{ kg/m}^3)}{(0.035 \text{ kg/m·s})^2} = 4.29 \times 10^4$$

5. Determine friction factor f from Fig. 14–6c using the computed Reynolds and Hedstrom numbers.

$f = 0.007$

Note, on Fig. 14–6c, that the flow is in the turbulent zone.

6. Compute pressure drop using Eq. (14–5).

$$\Delta p = \frac{2 f \rho L V^2}{D} = \frac{2(0.007)(1010 \text{ kg/m}^3)(10,000 \text{ m})(1.29 \text{ m/s})^2}{0.2 \text{ m}}$$

$$= 1,176,519 \text{ kg·m/s}^2 \ (\text{N/m}^2)$$

Convert to meters of water.

$$\Delta p = \frac{1,176,159 \text{ kg·m/s}^2}{(1000 \text{ kg/m}^3)(9.81 \text{ m/s}^2)} = 120 \text{ m}$$

Comment In this example, only one set of rheology data was used. In actual design, test data should be used for a range of probable conditions so that a family of headloss curves can be developed for the range of operating conditions. In addition, appropriate safety factors should be used for worst-case conditions. Comparison of the headloss to the headloss for water using the Hazen-Williams formula is left as a homework problem.

Sludge Piping

In wastewater treatment plants, conventional sludge piping should not be smaller than 150 mm (6 in) in diameter although smaller-diameter glass-lined pipe has been used successfully. Sludge piping may not need to be larger than 200 mm (8 in), unless the velocity exceeds 1.5 to 1.8 m/s (5 to 6 ft/s), in which case the pipe is sized to maintain that velocity. Gravity sludge withdrawal lines should not be less than 200 mm (8 in) in diameter. It is common practice to install a number of cleanouts in the form of plugged tees or crosses instead of elbows so that the lines can be rodded if necessary. Pump connections should not be smaller than 100 mm (4 in) in diameter.

A liberal number of hose gates should be installed in the piping, and an ample supply of high-pressure flushing water should be available for clearing stoppages. The flushing water should be plant effluent. The flushing water system should have a capacity of not less than 0.010 m³/s (150 gal/min) at 500 kN/m² (~70 lb_f/in²). In large plants with larger piping, a greater capacity should be available, and the available pressure should be increased to 700 kN/m^2 (100 lb$_f$/in^2).

Grease has a tendency to coat the inside of piping used for transporting primary sludge and scum. Grease accumulation is more of a problem in large plants than in small ones. The coating results in a decrease in effective diameter and a large increase in pumping head. For this reason, low capacity positive-displacement pumps are designed for heads greatly in excess of the theoretical head. Centrifugal pumps, with their larger capacity, usually pump a more dilute sludge, often containing some waste-water, and head buildup due to grease accumulations appears to occur more slowly. In some plants, provisions have been made for melting the grease by circulating hot water, steam, or digester supernatant through the main sludge lines.

In treatment plants, friction losses are low because the pipe runs are short; consequently, there is little difficulty in providing an ample safety factor. In the design of long sludge lines, however, special design features should be considered including (1) providing two pipes unless a single pipe can be shut down for several days without causing problems; (2) providing for external corrosion and pipe loads; (3) adding facilities for applying dilution water for flushing the line; (4) providing means to insert a pipe cleaner; (5) including provisions for steam injection, especially in cold climates and where excessive grease accumulation occurs; (6) providing air relief and blowoff valves for the high and low points, respectively, and (7) considering the potential effects of waterhammer. A discussion of waterhammer in force mains is provided in the companion volume to this text (Metcalf & Eddy, 1981).

14–5 PRELIMINARY OPERATIONS

Grinding, degritting, blending, and storage of solids and biosolids are necessary to provide a relatively constant, homogeneous feed to subsequent processing facilities. Blending and storage can be accomplished either in a single unit designed to do both or separately in other plant components. Screening of raw sludge or digested biosolids is sometimes required in reuse applications for the removal of plastics, rags, and other material. Each of these preliminary operations is discussed in this section.

Grinding

Sludge grinding is a process in which large and stringy material contained in sludge is cut or sheared into small particles to prevent clogging or wrapping around rotating equipment. A typical sludge grinder is shown on Fig. 14–7. Some of the processes that must be preceded by sludge grinders and the purposes of grinding are reported in Table 14–16. Grinders historically have required high maintenance, but newer designs of slow-speed grinders have been more durable and reliable. These designs include improved bearings and seals, hardened steel cutters, overload sensors, and mechanisms that reverse the cutter rotation to clear obstructions or shut down the unit if the obstruction cannot be cleared.

Screening

Because raw wastewater screens can allow significant quantities of solids to pass through, sludge screening is an alternative to grinding. Screening is advantageous in that

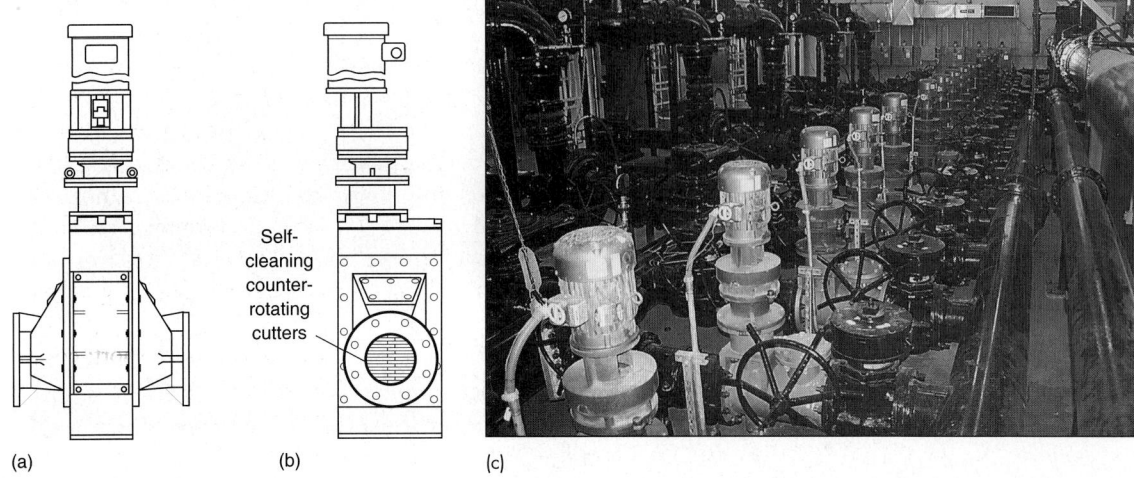

Figure 14–7

Typical in-line sludge grinder: (a) side view, (b) end view, and (c) view of typical installation with multiple grinders. (Courtesy Franklin Miller.)

Table 14–16
Operations and processes requiring the grinding of sludge

Operation or process	Purpose of grinding
Pumping with progressive cavity pumps	Prevent clogging and reduce wear
Solid-bowl centrifuges	Prevent clogging. Large solid bowl units generally can handle larger particles and may not require sludge grinding
Belt-filter press	Prevent clogging of the sludge distribution system, prevent warping of rollers, reduce wear on belts, and provide more uniform dewatering

nuisance material is removed from the solids stream. Step screens, shown on Fig. 5–4c in Chap. 5, can be used for the removal of fine solids from septage, primary sludge, or biosolids. Screen openings normally range from 3 to 6 mm (0.12 to 0.24 in), although openings up to 10 mm (0.4 in) can be used.

Another type of sludge screen is an in-line screen that can be installed in a pipeline (see Fig. 14–8). The screen removes material by passing the flow stream through a screen with 5-mm (0.2-in) openings. Material captured by the screen moves by a screw conveyor onto a finer screen [2-mm (0.08-in) -diameter perforations] where it is dewatered and compacted. Material is ejected from the screen when sufficient solids build up to overcome the force on the unit's discharge cone. Screenings solids concentrations range from 30 to 50 percent. Allowable operating pressure is reported to be 100 kPa (14 lb_f/in^2) (Arakaki et al., 1998). The screened sludge is dilute and may require thickening.

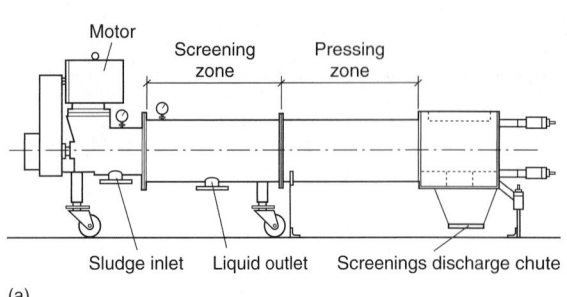

(a)   (b)

Figure 14-8

Sludge screenings press: (a) schematic and (b) view of a large installation. *(Courtesy City of San Diego.)*

Degritting

In some plants where separate grit-removal facilities are not used ahead of the primary sedimentation tanks, or where the grit-removal facilities are not adequate to handle peak flows and peak grit loads, it may be necessary to remove the grit before further processing of the sludge. Where further thickening of the primary sludge is desired, a practical consideration is sludge degritting. The most effective method of degritting sludge is through the application of centrifugal forces in a flowing system to achieve separation of the grit particles from the organic sludge. Such separation is achieved through the use of cyclone degritters, which have no moving parts. The sludge is applied tangential to a cylindrical feed section, thus imparting a centrifugal force. The heavier grit particles move to the outside of the cylinder section and are discharged through a conical feed section. The organic sludge is discharged through a separate outlet.

The efficiency of the cyclone degritter is affected by pressure and by the concentration of the organics in the sludge. To obtain effective grit separation, the sludge must be relatively dilute 1 to 2 percent TS. As the sludge concentration increases, the particle size that can be removed decreases. The general relationship between sludge concentration and effectiveness of removal for primary sludges is shown in Table 14–17.

Blending

Sludge is generated in primary, secondary, and advanced wastewater-treatment processes. Primary sludge consists of settleable solids carried in the raw wastewater. Secondary sludge consists of biological solids as well as additional settleable solids. Sludge produced in the advanced wastewater may consist of biological and chemical solids. Sludge is blended to produce a uniform mixture to downstream operations and processes. Uniform mixtures are most important in short-detention-time systems, such as sludge dewatering, heat treatment, and incineration. Provision of well-blended sludge

Table 14–17
Grit-removal efficiency using cyclone degritters for primary sludge[a]

Primary sludge concentration, % total solids	Mesh of removal[b]
1	150
2	100
3	65
4	28–35

[a] For a 300-mm (12-in) hydroclone at 42 kN/m² (6 lb$_f$/in² gage) at 13 L/s (200 gal/min).

[b] About 95 percent or more of indicated particle size is removed.

Note: Normal design range is for 1 to 1.5 percent feed solids.

with consistent characteristics to these treatment units will enhance greatly plant operability and performance.

Sludge from primary, secondary, and advanced processes can be blended in several ways:

1. *In primary settling tanks.* Secondary or tertiary sludges can be returned to the primary settling tanks where they will mix and co-settle with the primary sludge.
2. *In pipes.* Blending in pipes requires careful control of sludge sources and feed rates to ensure the proper blend. Without careful control, wide variations in sludge consistency may be expected.
3. *In sludge-processing facilities with long detention times.* Aerobic and anaerobic digesters (complete-mix type) can blend the feed sludges uniformly.
4. *In a separate blending tank.* This practice provides the best opportunity to control the quality of the blended sludges.

In treatment plants of less than 0.05 m³/s (~1 Mgal/d) capacity, blending is accomplished usually in the primary settling tanks. In large facilities, optimum efficiency is achieved by separately thickening sludges before blending.

Storage

Storage should be provided to smooth out fluctuations in the rate of solids and biosolids production and to allow solids to accumulate during periods when subsequent processing facilities are not operating, e.g., night shifts, weekends, and periods of unscheduled equipment downtime. Storage is particularly important in providing a uniform feed rate ahead of the following processes: mechanical dewatering, lime stabilization, heat drying, and thermal reduction.

Short-term solids storage may be accomplished in wastewater settling tanks or in thickening tanks. Long-term solids storage may be accomplished in stabilization processes with long detention times, e.g., aerobic and anaerobic digestion, or in specially designed separate tanks. In small installations, sludge is usually stored in the settling tanks and digesters. In large installations that do not use aerobic and anaerobic digestion, sludge is often stored in separate blending and storage tanks. Such tanks may be sized to retain the sludge for a period of several hours to a few days. If sludge is stored longer than 2 to 3 days, it will deteriorate, become odorous, and be more difficult to dewater.

The determination of the required storage volume is illustrated in Example 14–3. Sludge is often aerated to prevent septicity and to promote mixing. Mechanical mixing may be necessary to assure complete blending of the sludge. Chlorine, iron salts, potassium permanganate, and hydrogen peroxide have been used with limited success to arrest septicity and to control the odors from sludge storage and blending tanks. Sodium hydroxide or lime may also be used for odor control by raising the pH and keeping the hydrogen sulfide in solution. In cases where sludge storage occurs in enclosed tanks, ventilation should be provided along with appropriate odor-control technologies such as chemical scrubbers or biofilters (see Chap. 15).

EXAMPLE 14–3   **Determination of Volume Required for Sludge Storage**   Assume that the yearly average rate of sludge production from an activated-sludge treatment plant is 12,000 kg/d. Develop a curve of sustained sludge mass loading rates that can be used to determine the size of sludge-storage facilities required with various downstream sludge-processing units. Then, using the developed curve, determine the volume required for sludge storage, assuming that sludge accumulated for 7 d is to be processed in 5 working days, and that sludge accumulated for 14 d is to be processed in 10 working days. Note that the 5- and 10-d work periods correspond to 1 and 2 weeks, respectively, assuming that certain sludge-processing facilities, such as belt-filter presses, will not be operated on the weekends.

Solution

1. Develop a curve of sustained sludge mass loadings.
    a. Because no information is specified, it will be assumed that the sustained sludge production will mirror the sustained BOD plant loadings given on Fig. 5–6a and used in Example 5–3.
    b. Set up an appropriate computation table and compute the values necessary to plot the curve.

Length of sustained peak, d (1)	Peaking factor[a] (2)	Peak solids mass loading, kg/d (3)	Total sustained loading, kg[b] (4)
1	2.4	28,800	28,800
2	2.1	25,200	50,400
3	1.9	22,800	68,400
4	1.8	21,600	86,400
5	1.7	20,400	102,000
10	1.4	16,800	168,000
15	1.3	15,600	234,000
365	1.0	12,000	

[a] From Fig. 3–8a.
[b] Total mass produced for the corresponding sustained period given in col. 1.

    c. Plot the sustained solids loading curve (see following figure).

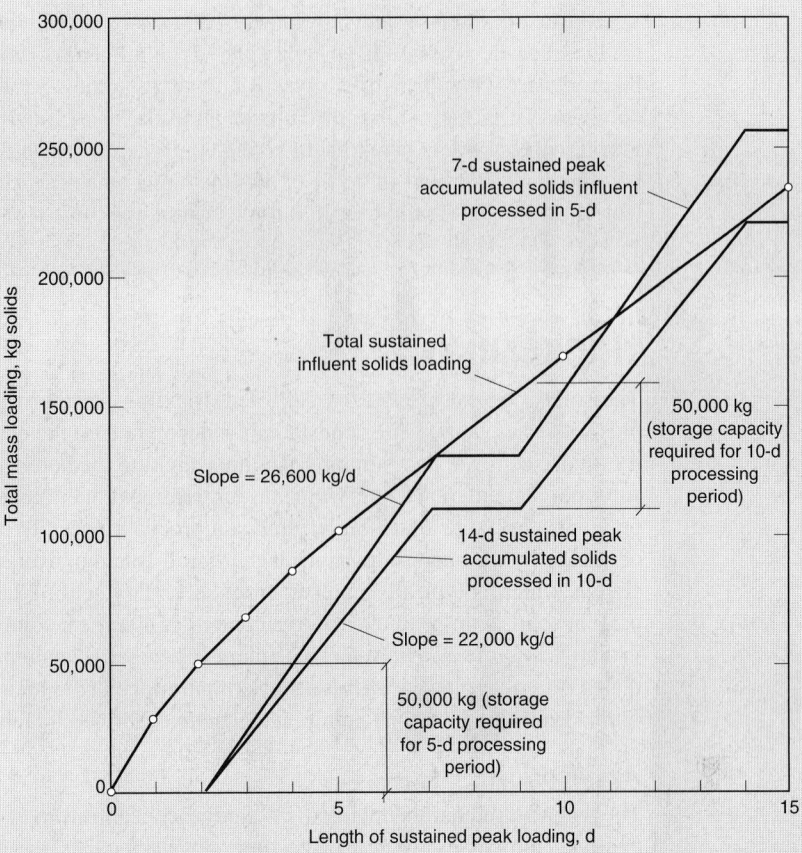

2. Determine the sludge storage volume required for the stated operating conditions.
   a. Determine the daily rate at which sludge must be processed to handle the 7-d sustained peak (from figure) in 5 working days.

   $$kg/d = \frac{133,000}{5\ d} = 26,600\ kg/d$$

   b. Determine the daily rate at which sludge must be processed to handle the 14–d sustained peak (from figure) in 10 working days.

   $$kg/d = \frac{220,000}{10\ d} = 22,000\ kg/d$$

   c. Assuming that the sludge storage facilities are empty on Friday just before the weekend, plot on the figure the average daily rate at which sludge must be processed during the 5- and 10-d periods.
   d. From the figure, the required storage capacity in pounds of solids is
      i. Capacity based on 5 working days = 50,000 kg
      ii. Capacity based on 10 working days = 50,000 kg

Comment   The downstream processing equipment can now be sized using the daily rate at which sludge must be processed. For example, if the number of kilograms per hour that can be processed with a belt-filter press is known, then the size and number of units can be computed from the number of shifts to be used per day and the assumed value of the actual working hours per shift. In sizing equipment, a trade-off analysis should always be performed between the cost of storage and processing facilities versus labor costs (for both one shift and two shifts) to determine the most cost-effective combination.

14-6 THICKENING

The solids content of primary, activated, trickling filter, or mixed sludge (i.e., primary plus waste activated) varies considerably, depending on the characteristics of the sludge, the sludge removal and pumping facilities, and the method of operation. Representative values of percent total solids from various treatment operations or processes were shown previously in Table 14–8. Thickening is a procedure used to increase the solids content of sludge by removing a portion of the liquid fraction. To illustrate, if waste-activated sludge, which is typically pumped from secondary settling tanks with a content of 0.8 percent solids, can be thickened to a content of 4 percent solids, then a fivefold decrease in sludge volume is achieved. Thickening is generally accomplished by physical means, including co-settling, gravity settling, flotation, centrifugation, gravity belt, and rotary drum. Typical sludge-thickening methods are described in Table 14–18.

Application

The volume reduction obtained by sludge concentration is beneficial to subsequent treatment processes, such as digestion, dewatering, drying, and combustion from the following standpoints: (1) capacity of tanks and equipment required, (2) quantity of chemicals required for sludge conditioning, and (3) amount of heat required by digesters and amount of auxiliary fuel required for heat drying or incineration, or both.

For large facilities where sludge must be transported a significant distance, such as to a separate plant for processing, a reduction in sludge volume may result in a reduction of pipe size and pumping costs. For smaller facilities, the requirements of a minimum practicable pipe size and minimum velocity may necessitate pumping of significant volumes of wastewater in addition to sludge, thereby diminishing the value of volume reduction. Volume reduction is very desirable when liquid sludge is transported by tank trucks for direct application to land as a soil conditioner.

Sludge thickening is achieved at all wastewater treatment plants in some manner— in the primary clarifiers, in sludge-digestion facilities, or in specially designed separate units. If separate units are used, the recycled flows are returned normally to the wastewater treatment facilities. In treatment plants of less than 4000 m³/d (~1 Mgal/d) capacity, separate sludge thickening is seldom practiced. In small plants, gravity thickening is accomplished in the primary settling tank or in the sludge-digestion units, or both. In larger treatment facilities, the additional costs of separate sludge thickening are often justified by the improved control over the thickening process and the higher concentrations attainable.

Table 14–18
Occurrence of thickening methods in solids processing

Method	Type of solids	Frequency of use and relative success
Gravity, co-settling in clarifier	Primary and waste activated	Occasional use; may negatively impact the effectiveness of primary clarifier
Gravity, thickening in separate tank	Untreated primary sludge	Commonly used with excellent results; sometimes used with hydroclone degritting of sludge. Can be odorous
	Untreated primary and waste-activated sludge	Often used. For small plants, generally satisfactory results with solids concentrations in the range of 4 to 6 percent. For large plants, results are marginal. Can be odorous in warm weather
	Waste-activated sludge	Seldom used; poor solids concentration (2 to 3 percent)
Dissolved air flotation	Untreated primary and waste-activated sludge	Limited use; results similar to gravity thickeners
	Waste-activated sludge	Commonly used, but use is decreasing because of high operating cost; good results (3.5 to 5 percent solids concentration)
Solid-bowl centrifuge	Waste-activated sludge	Often used in medium to large plants; good results (4 to 6 percent solids concentration)
Gravity-belt thickener	Waste-activated sludge	Often used; good results (3 to 6 percent solids concentration)
Rotary-drum thickener	Waste-activated sludge	Limited use; good results (3 to 6+ percent solids concentration)

Description and Design of Thickeners

The following discussion is intended to introduce the reader to the operations used for the thickening of sludges. Most of the equipment is mechanical; therefore, the design engineer usually is more concerned with its proper application to meet a given treatment objective than with the theory of mechanical design.

In designing thickening facilities, it is important to (1) provide adequate capacity to meet peak demands and (2) prevent septicity, with its attendant odor problems, during the thickening process. The six methods of thickening discussed in this section are co-settling thickening, gravity, dissolved air flotation, centrifugal, gravity belt, and rotary drum.

Co-settling Thickening. Primary clarifiers are often used to thicken solids for downstream processing. To thicken solids, a sludge blanket must be created to consolidate the solids without allowing the clarified water to be pulled through. Oftentimes solids retention times of 12 to 24 h or more are maintained in clarifiers to achieve thickened solids concentration levels in the clarifier underflow. Excessive retention of solids in the clarifier can cause septic conditions and gasification, and reduce the levels of TSS and BOD removal. Typical effects of sludge blanket retention on TSS removal are illustrated on Fig. 14–9.

Figure 14-9

Effect of sludge blanket retention time on TSS removal for cothickening of primary sludge. (Albertson and Walz, 1997.)

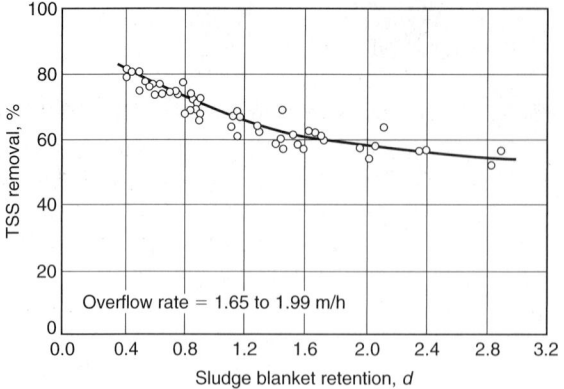

Figure 14-10

Schematic diagram of a sludge co-settling thickening system.

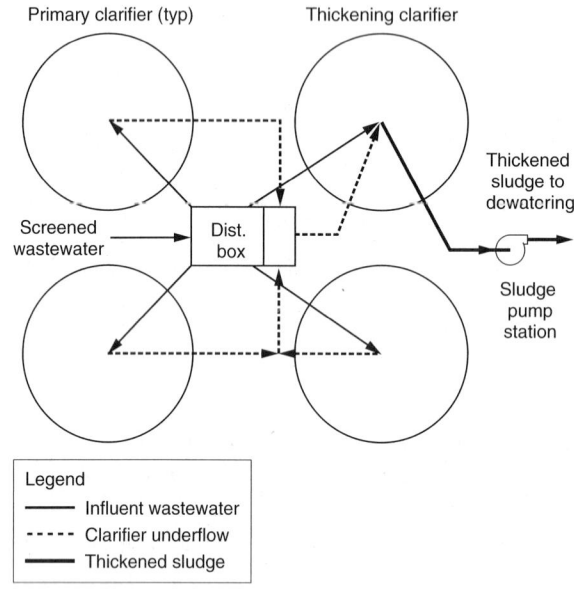

Successful thickening of solids in primary clarifiers has been achieved by a combination of the following: (1) using one clarifier in a bank of clarifiers for co-settling thickening; dilute solids underflow (less that 1 percent solids) from the other clarifers is discharged to the thickening clarifier, (2) maintaining the solids inventory for about 6 to 12 h, and (3) providing for the addition of coagulating chemicals such as polymer and ferric chloride to condition the solids to enhance settling. The need for chemical addition depends upon the clarifer overflow rates. Underflow solids concentrations on the order of 3 to over 5 percent have been reported (Albertson and Walz, 1997). By controlling the sludge blanket within the above solids retention parameters, clarifier removal rates are enhanced and solids thickening is achieved. A schematic diagram of the co-settling thickening system is shown on Fig. 14–10.

Gravity Thickening. Gravity thickening is one of the most common methods used and is accomplished in a tank similar in design to a conventional sedimentation tank. Normally, a circular tank is used, and dilute sludge is fed to a center feed well. The feed sludge is allowed to settle and compact, and the thickened sludge is withdrawn from the conical tank bottom. Conventional sludge-collecting mechanisms with deep trusses (see Fig. 14–11) or vertical pickets stir the sludge gently, thereby opening up channels for water to escape and promoting densification. The supernatant flow that results is drawn off and returned to either the primary settling tank, the influent of the treatment plant, or a return-flow treatment process. The thickened sludge is pumped to the digesters or dewatering equipment as required; thus, storage space must be provided for the sludge. As indicated in Table 14–18, gravity thickening is most effective on primary sludge. Gravity thickeners are designed on the basis of solids loading and thickener overflow rate. Typical solids loadings based on existing data are listed in Table 14–19. Recommended maximum hydraulic overflow rates range from 15.5 to 31 $m^3/m^2 \cdot d$

Figure 14–11

Schematic diagram of a gravity thickener: (a) plan and (b) section.

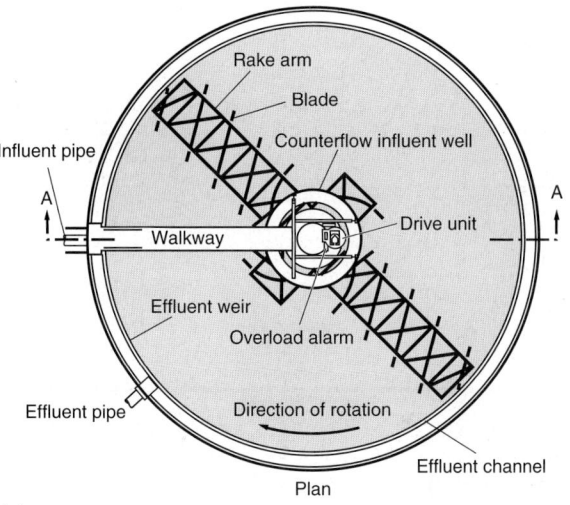

(a)

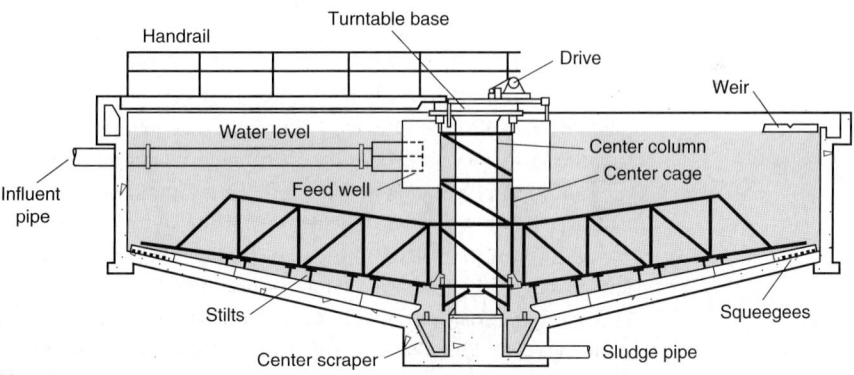

(b)

Table 14-19
Typical concentrations of unthickened and thickened sludges and solids loadings for gravity thickeners[a]

Type of sludge or biosolids	Solids concentration, %		Solids loading	
	Unthickened	Thickened	lb/ft²·d	kg/m²·d
Separate:				
Primary sludge	2–6	5–10	20–30	100–150
Trickling-filter humus sludge	1–4	3–6	8–10	40–50
Rotating biological contactor	1–3.5	2–5	7–10	35–50
Air-activated sludge	0.5–1.5	2–3	4–8	20–40
High-purity oxygen-activated sludge	0.5–1.5	2–3	4–8	20–40
Extended aeration-activated sludge	0.2–1.0	2–3	5–8	25–40
Anaerobically digested primary sludge from primary digester	8	12	25	120
Combined:				
Primary and trickling-filter humus sludge	2–6	5–9	12–20	60–100
Primary and rotating biological contactor	2–6	5–8	10–18	50–90
Primary and waste-activated sludge	0.5–1.5	4–6	5–14	25–70
	2.5–4.0	4–7	8–16	40–80
Waste-activated sludge and trickling-filter humus sludge	0.5–2.5	2–4	4–8	20–40
Chemical (tertiary) sludge:				
High lime	3–4.5	12–15	24–61	120–300
Low lime	3–4.5	10–12	10–30	50–150
Iron	0.5–1.5	3–4	2–10	10–50

[a] Adapted from WEF (1996).

(380 to 760 gal/ft²·d) for primary sludges, 4 to 8 m³/m²·d (100 to 200 gal/ft²·d) for waste-activated sludge, and 6 to 12 m³/m²·d (150 to 300 gal/ft²·d) for combined primary and waste-activated sludge (WEF, 1980). High hydraulic loadings can cause excessive solids carryover. Conversely, low hydraulic loadings can cause septic conditions and odors and floating sludge can result.

Provisions for dilution water and occasional chlorine addition are frequently included to improve process performance by maintaining the hydraulic loading. Polymer addition is frequently provided. To maintain aerobic conditions in gravity thickeners, especially when wastewater is warm (22 to 28°C), provisions should be included for adding up to 24 to 30 m³/m²·d (600 to 750 gal/ft²·d) of dilution water (final effluent) to the thickening tank. The dilution water may also remove certain soluble organic and inorganic compounds that consume large amounts of conditioning chemicals used in dewatering. Dilution water that is part of supernatant returned and recycled to the liquid process must be considered in process design.

Because the thickening characteristics of wastewater solids can vary considerably, it is desirable to design a thickening facility using criteria based on a testing program. Testing programs that can be used include batch settling tests, bench-scale settling tests, and pilot-scale testing. The latter method is recommended wherever possible because data can be obtained from a variety of operating parameters. Test methods are described in WEF (1998).

In operation, a sludge blanket is maintained on the bottom of the thickener to aid in concentrating the sludge. An operating variable is the sludge volume ratio, which is the volume of the sludge blanket held in the thickener divided by the volume of the thickened sludge removed daily. Values of the sludge volume ratio normally range between 0.5 and 20 d; the lower values are required during warm weather. Alternatively, sludge blanket depth should be measured. Blanket depths may range from 0.5 to 2.5 m (2 to 8 ft); shallower depths are maintained in the warmer months.

EXAMPLE 14–4 **Design a Gravity Thickener for Combined Primary and Waste-Activated Sludge** Design a gravity thickener for a wastewater treatment plant having primary and waste-activated sludge with the following characteristics:

Type of sludge	Specific gravity	Solids, %	Flowrate, m³/d
Average design conditions:			
Primary sludge	1.03	3.3	400
Waste activated	1.005	0.2	2250
Peak design conditions:			
Primary sludge	1.03	3.4	420
Waste activated	1.005	0.23	2500

Solution

1. Compute the dry solids at peak design conditions.
   a. Primary sludge

   $$\text{kg/dry solids} = (420 \text{ m}^3/\text{d})(1.03)(0.034 \text{ g/g})(10^3 \text{ kg/m}^3)$$
   $$= 14{,}708 \text{ kg/d}$$

   b. Waste-activated sludge

   $$\text{kg/dry solids} = (2500 \text{ m}^3/\text{d})(1.005)(0.0023 \text{ g/g})(10^3 \text{ kg/m}^3)$$
   $$= 5779 \text{ kg/d}$$

   c. Combined sludge mass = $14{,}708 + 5779 = 20{,}487$ kg/d
   d. Combined sludge flowrate = $2{,}500 + 420 = 2{,}920$ m³/d
2. Compute solids concentration of the combined sludge, assuming the specific gravity of the combined sludge is 1.02.

   $$\% \text{ solids} = \frac{(20{,}487 \text{ kg/d})}{(2920 \text{ m}^3/\text{d})(1.02)(10^3 \text{ kg/m}^3)} \times 100\% = 0.69\%$$

3. Compute surface area based on solids loading rate. Because the solids concentration is between 0.5 and 1.5%, select a solids loading rate of 50 kg/m²·d from Table 14–19.

$$\text{Area} = \frac{(20{,}487 \text{ kg/d})}{(50 \text{ kg/m}^2\text{·d})} = 409.7 \text{ m}^2$$

4. Compute hydraulic loading rate.

$$\text{Hydraulic loading} = \frac{(2920 \text{ m}^3/\text{d})}{409.7 \text{ m}^2} = 7.13 \text{ m}^3/\text{m}^2\text{·d}$$

5. Compute diameter of thickener; assume two thickeners.

$$\text{Diameter} = \sqrt{\frac{4 \times 409.7 \text{ m}^2}{2 \times \pi}} = 16.15 \text{ m}$$

Comment The hydraulic loading rate of 7.13 m³/m²·d at peak design flow is at the lower end of the recommended rate. To prevent septicity and odors, dilution water should be provided. Calculation of the dilution water requirements for average design flow is a homework problem. The thickener size of 16.15 m is within the maximum size of 20 m customarily recommended by thickener equipment manufacturers for use in municipal wastewater treatment. In actual design, round the thickener diameter to the nearest 0.5 m, or, in this case, 16 m.

Flotation Thickening. In dissolved-air flotation, air is introduced into a solution that is being held at an elevated pressure. A typical unit used for thickening waste-activated sludge is shown on Fig. 14–12. When the solution is depressurized, the dissolved air is released as finely divided bubbles carrying the sludge to the top, where it is removed. In locations where freezing is a problem or where odor control is of concern, flotation thickeners are normally enclosed in a building.

Flotation thickening is used most efficiently for waste sludges from suspended-growth biological treatment processes, such as the activated-sludge process or the suspended-growth nitrification process. Other sludges such as primary sludge, trickling-filter humus, aerobically digested sludge, and sludges containing metal salts from chemical treatment have been flotation thickened.

The float solids concentration that can be obtained by flotation thickening of waste-activated sludge is influenced primarily by the air-to-solids ratio, sludge characteristics (in particular the sludge volume index, SVI), solids loading rate, and polymer application. Although float solids concentrations have ranged historically between 3 and 6 percent by weight, float solids concentration is difficult to predict during the design stage without bench- or pilot-plant testing. The air-to-solids ratio is probably the most important factor affecting performance of the flotation thickener, and is defined as the weight ratio of air available for flotation to the solids to be floated in the feed

Figure 14–12

Typical dissolved air flotation unit used for thickening waste-activated sludge.

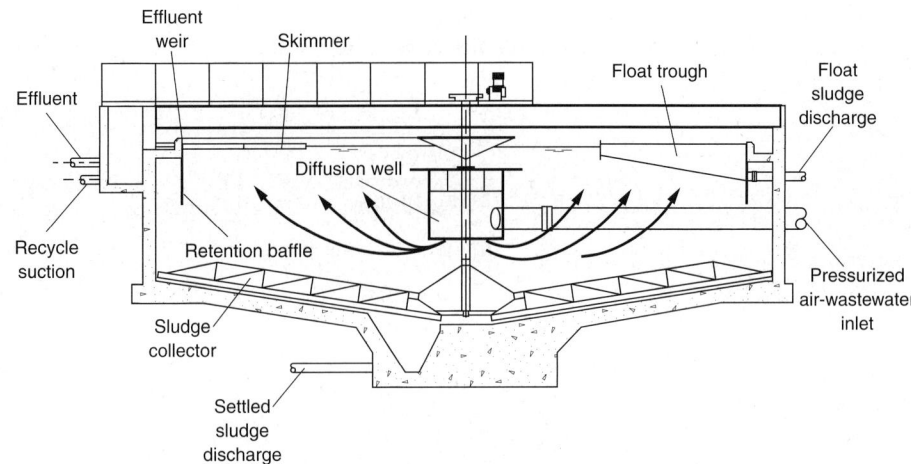

stream. The air-to-solids ratio at which float solids are maximized varies from 2 to 4 percent. The SVI is also important because better thickening performance has been reported when the SVI is less than 200, using nominal polymer dosages. At high SVIs, the float concentration deteriorates and high polymer dosages are required.

Higher loadings can be used with dissolved-air flotation thickeners than are permissible with gravity thickeners, because of the rapid separation of solids from the wastewater. Flotation thickeners typically are designed for the solids loadings given in Table 14–20. For design without the benefit of pilot studies, the minimum loadings should be used. The higher solids loadings generally result in lower concentrations of thickened sludge. Operational difficulties may arise when the solids loading rate exceeds approximately 10 kg/m^2·h (2.0 lb/ft^2·h). The increased amount of float created at high solids loading necessitates continuous skimming, often at high skimming speeds.

Primary tank effluent or plant effluent is recommended as the source of air-charged water rather than flotation tank effluent, except when chemical aids are used, because of the possibility of fouling the air-pressure system with solids. The use of polymers as flotation aids is effective in increasing the solids recovery in the floated sludge from 85 to 98 or 99 percent, and in reducing the recycle loads. Polymer dosages for thickening waste-activated sludge are 2 to 5 kg of dry polymer per Mg of dry solids (4 to 10 lb/ton).

Centrifugal Thickening. Centrifuges are used both to thicken and to dewater sludges. As indicated in Table 14–18, their application in thickening is limited normally to waste-activated sludge. Thickening by centrifugation involves the settling of sludge particles under the influence of centrifugal forces. The basic type of centrifuge used for sludge thickening is the solid-bowl centrifuge (see Fig. 14–13).

The solid-bowl centrifuge consists of a long bowl, normally mounted horizontally and tapered at one end. Sludge is introduced into the unit continuously, and the solids concentrate on the periphery. An internal helical scroll, spinning at a slightly different

Table 14–20

Typical solids loadings for dissolved air flotation units[a,b]

Type of sludge or biosolids	Loading, lb/ft²·h		Loading, kg/m²·h	
	Without chemical addition	With chemicals	Without chemical addition	With chemicals
Air-activated sludge:				
Mixed liquor	0.25–0.6	Up to 2.0	1.2–3.0	Up to 10
Settled	0.5–0.8	Up to 2.0	2.4–4.0	Up to 10
High-purity oxygen-activated sludge	0.6–0.8	Up to 2.0	3.0–4.0	Up to 10
Trickling-filter humus sludge	0.6–0.8	Up to 2.0	3.0–4.0	Up to 10
Primary + air-activated sludge	0.6–1.25	Up to 2.0	3.0–6.0	Up to 10
Primary + trickling-filter humus sludge	0.83–1.25	Up to 2.0	4.0–6.0	Up to 10
Primary sludge only	0.83–1.25	Up to 2.5	4.0–6.0	Up to 12.5

[a] Adapted, in part, from U.S. EPA (1979) and WEF (1998).
[b] Loading rates necessary to produce a minimum 4 percent solids concentration in the float.

Figure 14–13

Schematic diagram of a centrifuge used for sludge thickening.

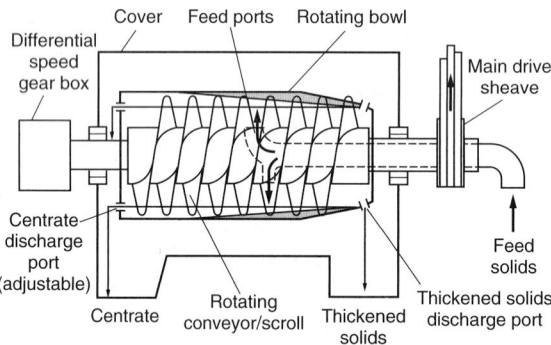

speed, moves the accumulated sludge toward the tapered end where additional solids concentration occurs and the thickened sludge is discharged.

Under normal conditions, thickening can be accomplished by centrifugal thickening without polymer addition. Maintenance and power costs for the centrifugal thickening process, however, can be substantial. Therefore, the process is usually attractive only at facilities larger than 0.2 m³/s (5 Mgal/d), where space is limited and skilled operators are available, or for sludges that are difficult to thicken by more conventional means. Many systems are designed with standby polymer systems for use to improve system performance. Polymer dosages for thickening waste-activated sludge range from 0 to 4 kg of dry polymer per Mg of dry solids (0 to 8 lb/ton).

The performance of a centrifuge is often quantified by the concentration achieved in the thickened solids product and the TSS recovery (sometimes termed "capture").

The recovery is calculated as the thickened dry solids as a percentage of the feed solids. Using the commonly measured solids concentrations, the recovery is calculated by the following expression (WEF, 1998):

$$R = \frac{\text{TSS}_P(\text{TSS}_F - \text{TSS}_C)}{\text{TSS}_F(\text{TSS}_P - \text{TSS}_C)} \times 100 \qquad (14\text{–}6)$$

where R = percent recovery
TSS_P = total suspended solids concentration in thickened solids product, percent by weight
TSS_F = total suspended solids concentration in feed, percent by weight
TSS_C = total suspended solids concentration in centrate, percent by weight

For a constant feed concentration, the percent recovery increases as the concentration of solids in the centrate decreases. In concentrating sludge solids, recovery is important because with a higher recovery lesser amounts of biodegradable solids are returned to the treatment process for further treatment. In developing a mass balance for the treatment plant, return flows (also termed sidestream flows) from thickening, stabilization, and dewatering processes must be taken into account (see Sec. 14–16).

The principal operational variables include the following: (1) characteristics of the feed sludge (its water-holding structure and the sludge volume index); (2) rotational speed; (3) hydraulic loading rate; (4) depth of the liquid pool in the bowl; (5) differential speed of the screw conveyor; and (6) the need for polymers to improve the performance. Because the interrelationships of these variables will be different in each location, specific design recommendations are not available; in fact, bench-scale or pilot-plant tests are recommended.

Gravity-Belt Thickening. The development of gravity-belt thickeners stemmed from the application of belt presses for sludge dewatering. In belt-press dewatering, particularly for sludges having solids concentrations less than 2 percent, effective thickening occurred in the gravity drainage section of the press. The equipment developed for thickening consists of a gravity belt that moves over rollers driven by a variable-speed drive unit (see Fig. 14–14). The sludge is conditioned with polymer and fed into a feed/distribution box at one end, where the sludge is distributed evenly across the width of the moving belt. The water drains through the belt as the concentrating sludge is carried toward the discharge end of the thickener. The sludge is ridged and furrowed by a series of plow blades placed along the travel of the belt, allowing the water released from the sludge to pass through the belt. After the thickened sludge is removed, the belt travels through a wash cycle. The gravity-belt thickener has been used for thickening waste-activated sludge, anaerobically and aerobically digested sludge, and some industrial sludges. Polymer addition is required. Testing is recommended to verify that the solids can be thickened at typical polymer dosages.

Typical hydraulic loading rates for gravity-belt thickeners are given in Table 14–21. In lieu of pilot-plant data, a value of 800 L/m·min (200 gal/m·min) is suggested as a design value; the higher the feed rate, the greater the operator attention required to maintain stable operation. Solids loading rates range on the order of 200 to 600 kg/m·h. Systems are often designed for a maximum of 5 to 7 percent thickened solids. Solids capture typically ranges between 90 and 98 percent (WEF, 1998). Polymer dosages for

Figure 14–14

Gravity-belt thickener: (a) schematic diagram, and (b) pictorial view showing the inlet to the thickener on the top left-hand side. The sludge moves along the belt to the far end while water drains through the porous belt. (Courtesy Ashbrook Corporation.)

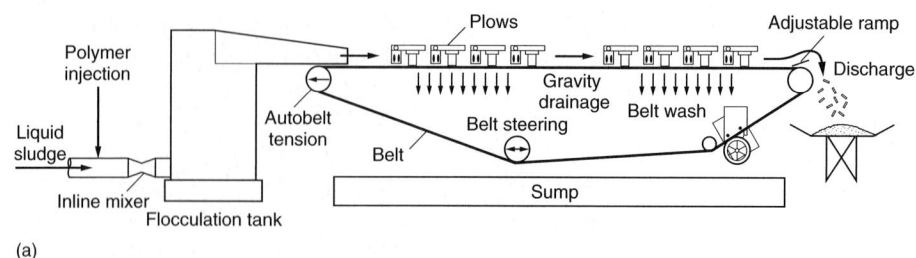

(a)

(b)

Table 14–21
Typical hydraulic loading rates for gravity belt thickeners[a,b]

Belt size (effective dewatering width), m	Hydraulic loading range	
	gal/min	L/s
1.0	100–250	6.7–16
1.5	150–375	9.5–24
2.0	200–500	12.7–32
3.0	300–750	18–47

[a] Assumes 0.5 to 1.0 percent feed solids for municipal sludges. Variations in sludge density, belt porosity, polymer reaction rate, and belt speed will act to increase or decrease the rates of flow for any given size belt.
[b] Adapted from WEF (1998).

thickening waste-activated sludge range from 3 to 7 kg of dry polymer per Mg of dry solids (6 to 14 lb/ton).

Rotary-Drum Thickening. Rotary media-covered drums are also used to thicken sludges. A rotary-drum thickening system consists of a conditioning system (including a polymer feed system) and rotating cylindrical screens (see Fig. 14–15). Polymer is mixed with dilute sludge in the mixing and conditioning drum. The conditioned sludge is

Figure 14–15

Rotary-drum thickener.
*(Courtesy Parkson
Corporation.)*

Table 14–22

Typical performance ranges for rotary-drum thickeners[a]

Type of solids or biosolids	Feed, % TS	Water removed, %	Thickened solids, %	Solids recovery, %
Primary	3.0–6.0	40–75	7–9	93–98
Waste activated	0.5–1.0	70–90	4–9	93–99
Primary + waste activated	2.0–4.0	50	5–9	93–98
Aerobically digested	0.8–2.0	70–80	4–6	90–98
Anaerobically digested	2.5–5.0	50	5–9	90–98

[a]WEF (1996).

then passed to rotating-screen drums, which separate the flocculated solids from the water. Thickened sludge rolls out the end of the drums, while separated water decants through the screens. Some designs also allow coupling of the rotary-drum unit to a belt-filter press for combination thickening and dewatering.

Rotary-drum thickeners can be used as a prethickening step before belt-press dewatering and are typically used in small- to medium-sized plants for waste-activated sludge thickening. The addition of large amounts of polymer for conditioning can be of concern because of floc sensitivity and shear potential in the rotating drum (WEF, 1998). Rotary-drum thickeners are available in capacities up to 24 L/s (400 gal/min). Typical performance data for rotary drum thickeners are given in Table 14–22.

14–7 INTRODUCTION TO STABILIZATION

Solids and biosolids are stabilized to (1) reduce pathogens, (2) eliminate offensive odors, and (3) inhibit, reduce, or eliminate the potential for putrefaction. The success in achieving these objectives is related to the effects of the stabilization operation or process on the volatile or organic fraction of the solids and biosolids. Survival of pathogens, release of odors, and putrefaction occur when microorganisms are allowed to flourish in the organic fraction of the sludge. The means to eliminate these nuisance conditions is mainly related to the biological reduction of the volatile content and the addition of chemicals to the solids or biosolids to render them unsuitable for the survival of microorganisms.

Stabilization is not practiced at all wastewater-treatment plants, but it is used by an overwhelming majority of plants ranging in size from small to very large. In addition to the health and aesthetic reasons cited above, stabilization is used for volume reduction, production of usable gas (methane), and improving the dewaterability of sludge.

The principal methods used for stabilization of sludge are (1) alkaline stabilization, usually with lime; (2) anaerobic digestion; (3) aerobic digestion, including autothermal thermophilic aerobic digestion; and (4) composting. These processes are generally defined in Table 14–23. Each of the processes is discussed in more detail in the following sections, and their ability to mitigate or stabilize the effects related to pathogens, putrefaction, and odors is given in Table 14–24. Heat treatment and the addition of oxidizing chemicals, processes that seldom are used in the United States for stabilization, are not included in this text. For information about these processes, however, Metcalf & Eddy (1991) and WEF (1998) are recommended.

When designing a stabilization process, it is important to consider the sludge quantity to be treated, the integration of the stabilization process with the other treatment units, and the objectives of the stabilization process. The objectives of the stabilization process are often affected by existing or pending regulations. If sludge is to be applied on land, pathogen reduction has to be considered. The effect of regulations on application of biosolids to land is discussed in Sec. 14–17.

14–8 ALKALINE STABILIZATION

A method to eliminate nuisance conditions in sludge is through the use of alkaline material to render the sludge unsuitable for the survival of microorganisms. In the lime stabilization process, lime is added to untreated sludge in sufficient quantity to raise the pH to 12 or higher. The high pH creates an environment that halts or substantially retards the microbial reactions that can otherwise lead to odor production and vector attraction. The sludge will not putrefy, create odors, or pose a health hazard so long as the pH is maintained at this level. The process can also inactivate virus, bacteria, and other microorganisms present.

Chemical Reactions in Lime Stabilization

The lime stabilization process involves a variety of chemical reactions that alter the chemical composition of the sludge. The following simplified equations are illustrative of the types of reactions that may occur (WEF, 1998):

Calcium: $Ca^{2+} + 2HCO_3^- + CaO \rightarrow 2CaCO_3 + H_2O$ (14-7)

Phosphorus: $2PO_4^{3-} + 6H^+ + 3CaO \rightarrow Ca_3(PO_4)_2 + 3H_2O$ (14-8)

Carbon dioxide: $CO_2 + CaO \rightarrow CaCO_3$ (14-9)

Table 14-23
Description of stabilization processes

Process	Description	Comments
Alkaline stabilization	Addition of an alkaline material, usually lime, to maintain a high pH level to effect the destruction of pathogenic organisms	An advantage of alkaline stabilization is that a rich soil-like product results with substantially reduced pathogens. A disadvantage is that the product mass is increased by the addition of the alkaline material. Some alkaline stabilization processes are capable of producing a Class A sludge
Anaerobic digestion	The biological conversion of organic matter by fermentation in a heated reactor to produce methane gas and carbon dioxide. Fermentation occurs in the absence of oxygen	Methane gas can be used beneficially for the generation of heat or electricity. The resulting biosolids may be suitable for land application. The process requires skilled operation as it may be susceptible to upsets and recovery is slow
Aerobic digestion	The biological conversion of organic matter in the presence of air (or oxygen), usually in an open-top tank	Process is much simpler to operate than an anaerobic digester, but no usable gas is produced. The process is energy-intensive because of the power requirements necessary for mixing and oxygen transfer
Autothermal thermophilic digestion	Process is similar to aerobic digestion except higher amounts of oxygen are added to accelerate the conversion of organic matter. Process operates at temperatures of 40 to 80°C, autothermally in an insulated tank	Process is capable of producing a Class A sludge. Skilled operators are required and the process is a high-energy user (to produce air or oxygen)
Composting	The biological conversion of solid organic matter in an enclosed reactor or in windrows or piles	A variety of solids or biosolids can be composted. Composting requires the addition of a bulking agent to provide an environment suitable for biological activity. Volume of compost produced is usually greater than the volume of wastewater solids being composted. Class A or Class B sludge can be produced. Odor control is very important, as process is odorous

Table 14-24
Relative degree of attenuation achieved with various sludge stabilization processes[a]

Process	Degree of attenuation		
	Pathogens	Putrefaction	Odor potential
Alkaline stabilization	Good	Fair	Fair
Anaerobic digestion	Fair	Good	Good
Aerobic digestion	Fair	Good	Good
Autothermal thermophilic digestion (ATAD)	Excellent	Good	Good
Composting	Fair	Good	Poor to fair
Composting (thermophilic)	Excellent	Good	Poor to fair

[a]Adapted, in part, from WEF (1998).

Reactions with organic contaminants:

Acids: $RCOOH + CaO \rightarrow RCOOCaOH$ (14-10)

Fats: $Fat + Ca(OH)_2 \rightarrow glycerol + fatty\ acids$ (14-11)

Other reactions also occur, such as the hydrolysis of polymers, especially polymeric carbohydrates and proteins, and the hydrolysis of ammonia from amino acids.

Initially, lime addition raises the pH of the sludge. Then, reactions occur such as those in the above equations. If insufficient lime is added, the pH decreases as the reactions take place. Therefore, excess lime is required.

Biological activity produces compounds, such as carbon dioxide and organic acids, that react with lime. If biological activity in the sludge being stabilized is not sufficiently inhibited, these compounds will be produced, reducing the pH and resulting in inadequate stabilization. Many odorous, volatile off-gases are also produced, especially ammonia, which require collection and treatment in odor-control systems such as chemical scrubbers or biofilters (see Chap. 15).

Heat Generation

If quicklime, CaO (or any compound high in quicklime), is added to sludge, it initially reacts with water to form hydrated lime. This reaction is exothermic and releases approximately 64 kJ/g·mole (2.75×10^4 Btu/lb·mole) (U.S. EPA, 1982). The reaction between quicklime and carbon dioxide is also exothermic, releasing approximately 180 kJ/g·mole (7.8×10^4 Btu/lb·mole). These reactions can result in substantial temperature rise (see discussion of lime posttreatment).

Application of Alkaline Stabilization Processes

Three methods of alkaline stabilization are commonly used: (1) addition of lime to sludge prior to dewatering, termed "lime pretreatment," (2) the addition of lime to sludge after dewatering, or "lime posttreatment," and (3) advanced alkaline stabilization technologies (WEF 1998). Either hydrated lime, $Ca(OH)_2$, or quicklime is used most commonly for lime stabilization. Fly ash, cement kiln dust, and carbide lime have also been used as a substitute for lime in some cases.

Lime Pretreatment. Pretreatment (before dewatering) of liquid sludge with lime has been used for either (1) the direct application of liquid sludge to land, or (2) combining benefits of sludge conditioning and stabilization prior to dewatering. In the former case, large quantities of liquid sludge have to be transported to land disposal sites, which limits utilization of lime pretreatment of sludge to small treatment plants. When pretreatment is used prior to dewatering, dewatering has been accomplished by a pressure-type filter press. Lime pretreatment is seldom used with centrifuges or belt-filter presses because of abrasive wear and scaling problems.

Lime pretreatment of liquid sludge requires more lime per unit weight of sludge processed than that necessary for dewatering. The higher lime dose is needed to attain the required pH because of the chemical demand of the liquid. In addition, sufficient contact time must be provided before dewatering so as to effect a high level of pathogen kill. The recommended design objective is to maintain the pH above 12 for about 2 h to ensure pathogen destruction (the minimum U.S. EPA criterion for lime stabilization),

Table 14–25

Typical lime dosages for pretreatment sludge stabilization[a]

| Type of sludge | Solids concentration, % | | Lime dosage[b] | | | |
| | | | lb Ca(OH)$_2$/ton dry solids | | g Ca(OH)$_2$/kg dry solids | |
	Range	Average	Range	Average	Range	Average
Primary	3–6	4.3	120–340	240	60–170	120
Waste activated	1–1.5	1.3	420–860	600	210–430	300
Anaerobically digested mixed	6–7	5.5	280–500	380	140–250	190
Septage	1–4.5	2.7	180–1020	400	90–510	200

[a] Adapted from WEF (1995a).
[b] Amount of Ca(OH)$_2$ required to maintain a pH of 12 for 30 min.

and to provide enough residual alkalinity so that the pH does not drop below 11 for several days. The lime dosage required varies with the type of sludge and solids concentration. Typical dosages are reported in Table 14–25. Generally, as the percent solids concentration increases, the required lime dose increases. Testing should be performed for specific applications to determine the actual dosage requirements.

Because lime stabilization does not destroy the organics necessary for bacterial growth, the sludge must be treated with an excess of lime or disposed of before the pH drops significantly. An excess dosage of lime may range up to 1.5 times the amount needed to maintain the initial pH of 12. For additional details about pH decay following lime stabilization, WEF (1995a) is recommended.

Lime Posttreatment.   In lime posttreatment, hydrated lime or quicklime is mixed with dewatered sludge in a pugmill, paddle mixer, or screw conveyor to raise the pH of the mixture. Quicklime is preferred because the exothermic reaction of quicklime and water can raise the temperature of the mixture above 50°C, sufficient to inactivate worm eggs. The theoretical temperature increase by the addition of quicklime is illustrated on Fig. 14–16.

Lime posttreatment has several significant advantages when compared to lime pretreatment: (1) dry lime can be used; therefore, no additional water is added to the dewatered sludge; (2) there are no special requirements for dewatering; and (3) scaling problems and associated maintenance problems of lime-sludge dewatering equipment are eliminated. Adequate mixing is critical for a posttreatment stabilization system so as to avoid pockets of putrescible material. A lime posttreatment stabilization system consists typically of a dry lime feed system, dewatered sludge cake conveyor, and a lime-sludge mixer (see Fig. 14–17). Good mixing is especially important to ensure contact between lime and small particles of sludge. When the lime and sludge are well mixed, the resulting mixture has a crumbly texture, which allows it to be stored for long periods or easily distributed on land by a conventional manure spreader. A potential disadvantage of

Figure 14–16

Theoretical temperature increase in postlime stabilized sludge using quicklime. (Roediger, 1987.)

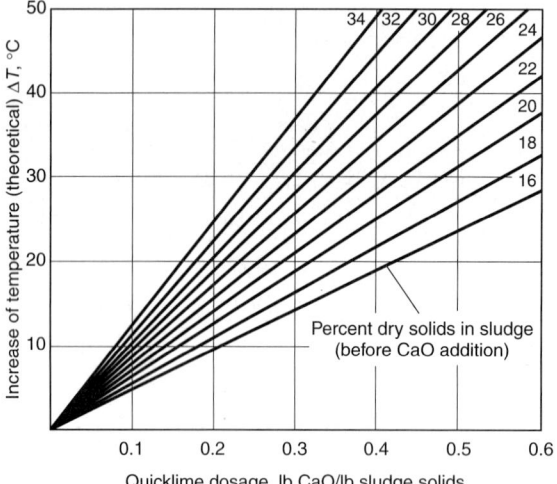

Note : In practice, higher temperature values are expected. 1.8 (°C) + 32 = °F

Figure 14–17

Typical lime posttreatment system. (From Roediger Pittsburgh.)

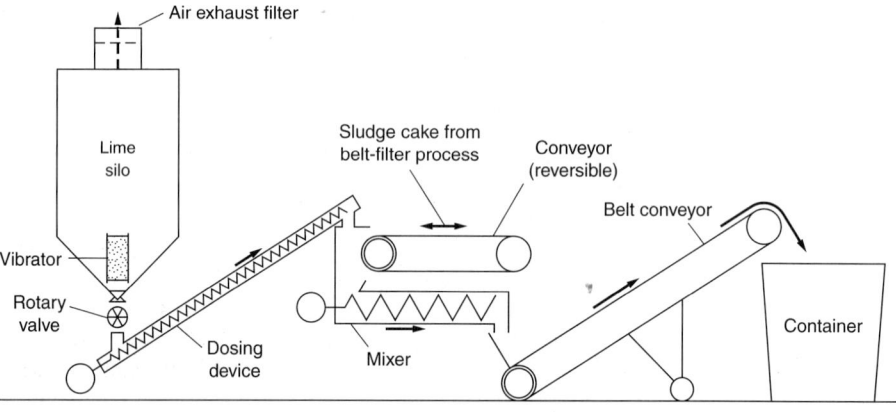

lime posttreatment of anaerobically digested sludge may be the release of odorous gases, specifically trimethyl amine (Novak, 2001).

Advanced Alkaline Stabilization Technologies. Alkaline stabilization using materials other than lime is used by a number of municipalities. Most of the technologies that rely on additives, such as cement kiln dust, lime kiln dust, or fly ash, are modifications of conventional dry lime stabilization. The most common modifications include the addition of other chemicals, a higher chemical dose, and supplemental drying. These processes alter the characteristics of the feed material and, depending on the process, may increase product stability, decrease odor potential, and provide product enhancement. To utilize these technologies, dewatered sludge is required.

Pasteurization may be accomplished by the exothermic reaction of quicklime with water to achieve a process temperature of 70°C and maintain it for more than 30 min.

Figure 14–18

Schematic diagram of a chemical fixation system. (U.S. EPA, 1989.)

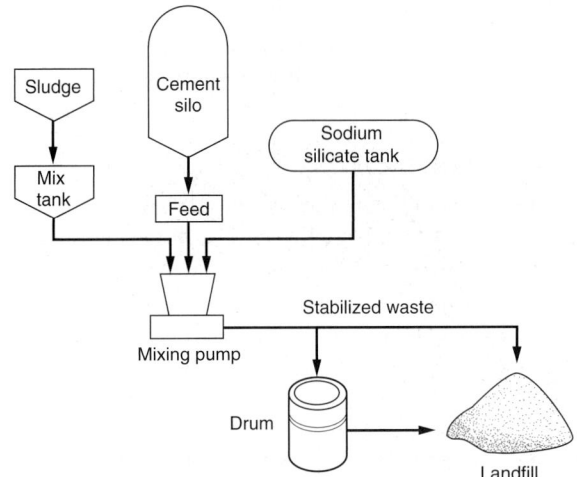

To meet Class A sludge criteria, the pasteurization reaction must be carried out under carefully controlled and monitored mixing and temperature conditions to ensure uniform treatment and inactivation of pathogens by the heat generated during the reaction. The process produces a soil-like material that is not subject to liquefaction under mechanical stress.

A chemical stabilization/fixation technology typically involves the addition of pozzolanic materials to dewatered sludge cake. Chemical fixation/solidification processes have also been applied for the treatment of industrial sludge and hazardous wastes to immobilize the undesirable constituents. The addition of pozzolanic materials causes cementitious reactions and produces, after drying, a soil-like material of approximately 35 to 50 percent solids content. The principal use of this material is landfill cover. A typical process flow diagram is shown on Fig. 14–18.

Several other process variations of advanced alkaline stabilization are available, some of which are proprietary. Additional information may be found in WEF (1998).

14–9 ANAEROBIC DIGESTION

Anaerobic digestion is among the oldest processes used for the stabilization of solids and biosolids. As described in Chap. 10, anaerobic digestion involves the decomposition of organic matter and inorganic matter (principally sulfate) in the absence of molecular oxygen. The major applications of anaerobic digestion are in the stabilization of concentrated sludges produced from the treatment of municipal and industrial wastewater. Great progress has been made in the fundamental understanding and control of the process, the sizing of tanks, and the design and application of equipment. Because of the emphasis on energy conservation and recovery and the desirability of obtaining beneficial use of wastewater biosolids, anaerobic digestion continues to be the dominant process for stabilizing sludge. Furthermore, anaerobic digestion of municipal wastewater sludge can, in many cases, produce sufficient digester gas to meet most of the energy needs for plant operation.

In this section, a brief review is provided of process fundamentals followed by discussions of mesophilic anaerobic digestion, the most common basic process used; thermophilic digestion; and phased digestion. Phased digestion covers many of the new developments in anaerobic digestion.

Process Fundamentals

As described in Chap. 7, the three types of chemical and biochemical reactions that occur in anaerobic digestion are hydrolysis; fermentation, also called acidogenesis (the formation of soluble organic compounds and short-chain organic acids); and methanogenesis (the bacterial conversion of organic acids into methane and carbon dioxide). Important environmental factors in the anaerobic digestion process are (1) solids retention time, (2) hydraulic retention time, (3) temperature, (4) alkalinity, (5) pH, (6) the presence of inhibitory substances, i.e., toxic materials, and (7) the bioavailability of nutrients and trace metals. The first three factors are important in process selection and are discussed in this section. Alkalinity is a function of feed solids and is important in controlling the digestion process. The effects of pH and inhibitory substances are discussed in Chaps. 7 and 10. The presence of nutrients and trace metals necessary for biological growth is described in Sec. 10–2 in Chap. 10.

Solids and Hydraulic Retention Times.   Anaerobic digester sizing is based on providing sufficient residence time in well-mixed reactors to allow significant destruction of volatile suspended solids (VSS) to occur. Sizing criteria that have been used are (1) solids retention time SRT, the average time the solids are held in the digestion process, and (2) the hydraulic retention time τ, the average time the liquid is held in the digestion process. For soluble substrates, the SRT can be determined by dividing the mass of solids in the reactor (M) by the mass of solids removed daily (M/d). The hydraulic retention time τ is equal to the volume of liquid in the reactor (L^3) divided by the quantity of biosolids removed (L^3/d). For digestion systems without recycle, SRT = τ.

The three reactions (hydrolysis, fermentation, and methanogenesis) are directly related to SRT (or τ). An increase or decrease in SRT results in an increase or decrease in the extent of each reaction. There is a minimum SRT for each reaction. If the SRT is less than the minimum SRT, bacteria cannot grow rapidly enough and the digestion process will fail eventually (WEF, 1998).

Temperature.   As discussed in Sec. 7–5, temperature not only influences the metabolic activities of the microbial population but also has a profound effect on such factors as gas transfer rates and the settling characteristics of biological solids. In anaerobic digestion, temperature is important in determining the rate of digestion, particularly the rates of hydrolysis and methane formation. The design operating temperature establishes the minimum SRT required to achieve a given amount of VSS destruction. Most anaerobic digestion systems are designed to operate in the mesophilic temperature range, between 30 and 38°C (85 and 100°F). Other systems are designed for operation in the thermophilic temperature range of 50 to 57°C (122 to 135°F). Newly developed systems, as discussed in a latter part of this section, use a combination of mesophilic and thermophilic digestion in separate stages.

While selection of the design operating temperatures is important, maintaining a stable operating temperature is more important because the bacteria, especially the methane formers, are sensitive to temperature changes. Generally, temperature changes greater than 1°C/d affect process performance, and thus changes less than 0.5°C/d are recommended (WEF, 1998).

Alkalinity. Calcium, magnesium, and ammonium bicarbonates are examples of buffering substances found in a digester. The digestion process produces ammonium bicarbonate from the breakdown of protein in the raw sludge feed: the others are found in the feed sludge. The concentration of alkalinity in a digester is, to a great extent, proportional to the solids feed concentration. A well-established digester has a total alkalinity of 2000 to 5000 mg/L (WEF, 1996).

The principal consumer of alkalinity in a digester is carbon dioxide, and not volatile fatty acids as is commonly believed (Speece, 2001). Carbon dioxide is produced in the fermentation and methanogenesis phases of the digestion process (see Sec. 7–12 in Chap. 7). Due to the partial pressure of gas in a digester, the carbon dioxide solubilizes and forms carbonic acid, which consumes alkalinity. The carbon dioxide concentration in the digester gas is therefore reflective of the alkalinity requirements. Supplemental alkalinity can be supplied by the addition of sodium bicarbonate, lime, or sodium carbonate.

Description of Mesophilic Anaerobic Digestion Processes

The operation and physical facilities for mesophilic anaerobic digestion in single-stage high-rate, two-stage, and separate digesters for primary sludge and waste-activated sludge are described in this section. Standard-rate, sometimes called low-rate, digestion is seldom used for digester design (because of the large tank volume required and the lack of adequate mixing) and is not covered in this text. For information about standard-rate digestion, the reader is referred to the previous edition of this text (Metcalf & Eddy, 1991) and WEF (1998). The processes described below normally operate in the mesophilic range; high-rate digesters also operate in the thermophilic range. Thermophilic digestion is discussed at the end of the section. An aerial view of a large digester installation is shown on Fig. 14–19.

Figure 14–19

Aerial view of several large anaerobic digesters at Boston, MA. *(Courtesy Massachusetts Water Resources Authority, Kevin Kerwin, photographer, April 22, 1998.)*

Single-Stage High-Rate Digestion. Heating, auxiliary mixing, uniform feeding, and thickening of the feed stream characterize the single-stage high-rate digestion process. The sludge is mixed by gas recirculation, pumping, or draft-tube mixers (separation of scum and supernatant does not take place), and sludge is heated to achieve optimum digestion rates (see Fig. 14–20a).

Uniform feeding is very important, and sludge should be pumped to the digester continuously or on a 30-min to 2-h time cycle to help maintain constant conditions in the reactor. In digesters fed on a daily cycle of 8 or 24 h, it is important to withdraw digested sludge from the digester *before* adding the feed sludge, because the pathogen kill is significantly greater when compared to using the feed sludge to displace the waste sludge (Speece, 2001). Because there is no supernatant separation in the high-rate digester, and the total solids are reduced by 45 to 50 percent and given off as gas, the digested sludge is about half as concentrated as the untreated sludge feed. Digestion tanks may have fixed roofs or floating covers. Any or all of the floating covers may be of the gas-holder type, which provides excess gas storage capacity. Alternatively, gas may be stored in a separate low-pressure gas holder or compressed and stored under pressure.

Figure 14–20

Schematic diagram of typical anaerobic digesters: (a) single-stage high-rate, and (b) two-stage.

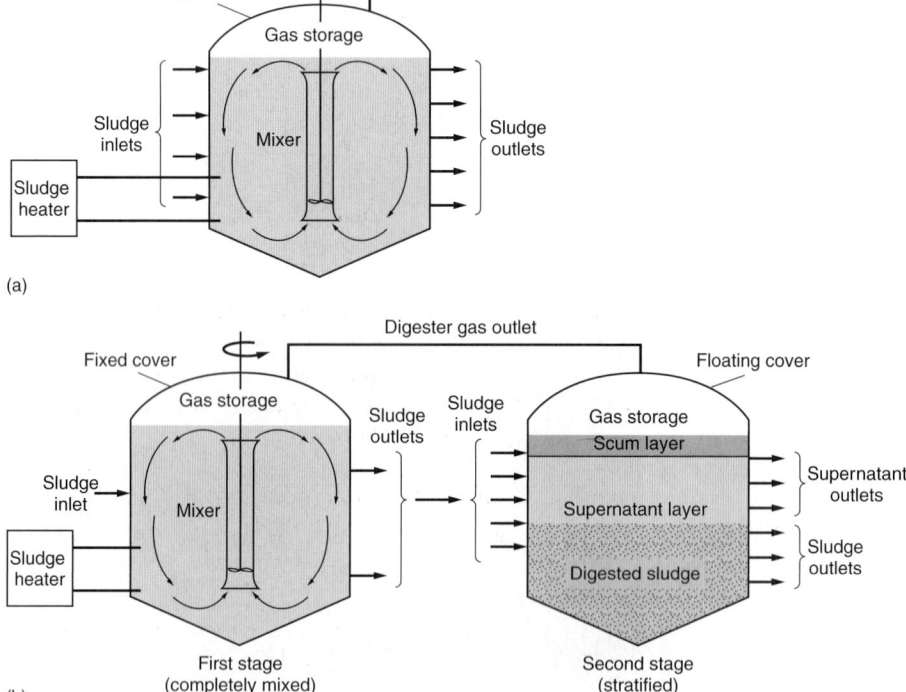

Two-Stage Digestion. Two-stage digestion, which was frequently used in the past, is seldom used in modern digester design. In two-stage digestion, a high-rate digester is coupled in series with a second tank (see Fig. 14–20b). The first tank is used for digestion and is heated and equipped with mixing facilities. The second tank is usually unheated and used principally for storage. The tanks may be identical, in which case either one may be the primary. Tanks may have fixed roofs or floating covers, the same as single-stage digestion. In other cases, the second tank may be an open tank or a sludge lagoon. Two-stage digestion of the type described above is seldom used, mainly because of the expense of building a large tank that is not fully utilized and because the second tank was of negligible benefit, operationally.

Because anaerobically digested solids may not settle well, the supernatant withdrawn from the second-stage tank may contain high concentrations of suspended solids. Reasons for poor settling characteristics include incomplete digestion in the primary digester (which generates gases in the secondary digester and causes floating solids) and fine-sized solids that have poor settling characteristics. Supernatant returned to the liquid processing system could cause upset conditions and might require separate treatment. Where two-stage digestion is used, return flows from the second tank must be accounted for in the solids mass balance. Less than 10 percent of the gas generated comes from the second stage.

In some installations, the second stage is a heated and mixed reactor to achieve further stabilization prior to dewatering or other subsequent processing. Additional discussion is provided later in this section on two-phase mesophilic digestion that provides more effective utilization of tank capacity.

Separate Sludge Digestion. Most wastewater-treatment plants employing anaerobic digestion use common tanks for the digestion of a mixture of primary and biological sludge. The solid-liquid separation of digested primary sludge, however, is downgraded by even small additions of biological sludge, particularly activated sludge. The rate of reaction under anaerobic conditions is also slowed slightly. In separate sludge digestion, the digestion of primary and biological sludges is accomplished in separate tanks. Reasons cited for separate digestion include (1) the excellent dewatering characteristics of the digested primary sludge are maintained, (2) the digestion process is specifically tailored to the sludge being treated, and (3) optimum process control conditions can be maintained. Design criteria and performance data for the separate anaerobic digestion of biological sludges, however, are very limited. In some cases, especially where biological phosphorus removal is practiced, biological sludge is digested aerobically instead of anaerobically to prevent resolubilization of the phosphorus under anaerobic conditions.

Process Design for Mesophilic Anaerobic Digestion

Ideally, the design of anaerobic sludge digestion processes should be based on an understanding of the fundamental principles of biochemistry and microbiology discussed in Chap. 7 in Sec. 7–12. Because these principles have not been appreciated fully in the past, a number of empirical methods have also been used in the design of digesters. The purpose of the following discussion is to illustrate the various methods that have been used to design single-stage, high-rate digesters in terms of size. These methods are based on (1) solids retention time, (2) the use of volumetric loading factors, (3) volatile solids destruction, (4) observed volume reduction, and (5) loading factors based on population.

Solids Retention Time. Digester design based on SRT involves application of the principles discussed in Chap. 10. To review briefly, the respiration and oxidation end products of anaerobic digestion are methane gas and carbon dioxide. The quantity of methane gas can be calculated using Eq. (14–12):

$$V_{CH_4} = (0.35)[(S_o - S)(Q)(10^3 g/kg)^{-1} - 1.42\, P_x]$$ (14-12)

where V_{CH_4} = volume of methane produced at standard conditions (0°C and 1 atm), m³/d

0.35 = theoretical conversion factor for the amount of methane produced, m³, from the conversion of 1 kg of bCOD at 0°C (conversion factor at 35°C = 0.40, see Example 7–9 in Chap. 7)

Q = flowrate, m³/d
S_o = bCOD in influent, mg/L
S = bCOD in effluent, mg/L
P_x = net mass of cell tissue produced per day, kg/d

The theoretical conversion factor for the amount of methane produced from the conversion of 1 g of bCOD is derived in Sec. 7–12 in Chap. 7.

For a complete-mix high-rate digester without recycle, the mass of biological solids synthesized daily, P_x, can be estimated using Eq. (14–13).

$$P_x = \frac{YQ(S_o - S) \times (10^3\, g/kg)^{-1}}{1 + k_d\,(SRT)}$$ (14-13)

where Y = yield coefficient, g VSS/g bCOD
k_d = endogenous coefficient, d⁻¹ (typical values range from 0.02 to 0.04)
SRT = solids retention time, d
other terms as defined previously

For a complete-mix digester, the SRT is the same as the hydraulic retention time τ.

Typical anaerobic reaction values for Y and k_d are given in Table 10–10 and range from 0.05 to 0.10 and 0.02 to 0.04, respectively. Typical values for SRT at various temperatures are reported in Table 14–26. In practice for high-rate digestion, however, values for SRTs range from 10 to 20 days. Grady, Daigger, and Lim (1999) observed that (1) a lower SRT limit of 10 days at a temperature of 35°C is sufficient to ensure an adequate safey factor against a washout of the menthanogenic population, and (2) incremental changes in volatile solids destruction are relatively small for SRT values above 15 days at 35°C. In selecting the design SRT for anaerobic digestion, peak hydraulic loading must be considered. The peak loading can be estimated by combining poor thickener performance with the maximum sustained plant loading expected during seven continuous days during the design period (U.S. EPA, 1979). The application of Eqs. (14–12) and (14–13) in the process design of a high-rate digester is illustrated in Example 14–5.

Table 14–26

Suggested solids retention times for use in the design of complete-mix anaerobic digesters[a]

Operating temperature, °C	SRT (minimum)	SRT$_{des}$
18	11	28
24	8	20
30	6	14
35	4	10
40	4	10

[a]McCarty (1964) and (1968).

Note: 1.8 (°C) + 32 = °F.

EXAMPLE 14–5 **Estimating Single-Stage, High-Rate Digester Volume and Performance** Estimate the size of digester required to treat the sludge from a primary treatment plant designed to treat 38,000 m³/d (10 Mgal/d) of wastewater. Check the volumetric loading, and estimate the percent stabilization and the amount of gas produced per capita. For the wastewater to be treated, it has been found that the quantity of dry volatile solids and biodegradable COD removed is 0.15 kg/m³ and 0.14 kg/m³, respectively. Assume that the sludge contains about 95 percent moisture and has a specific gravity of 1.02. Other pertinent design assumptions are as follows:

1. The hydraulic regime of the reactor is complete-mix.
2. τ = SRT = 10 days at 35°C (see Table 14–26).
3. Efficiency of waste utilization (solids conversion) E = 0.70.
4. The sludge contains adequate nitrogen and phosphorus for biological growth.
5. Y = 0.08 kg VSS/kg bCOD utilized and k_d = 0.03 d⁻¹.
6. Constants are for a temperature of 35°C.
7. Digester gas is 65 percent methane.

Solution

1. Determine the daily sludge volume using Eq. (14–2).

$$\text{Sludge volume} = \frac{(0.15 \text{ kg/m}^3)(38,000 \text{ m}^3/\text{d})}{1.02(10^3 \text{ kg/m}^3)(0.05)} = 111.8 \text{ m}^3/\text{d}$$

2. Determine the bCOD loading.

bCOD loading = $(0.14 \text{ kg/m}^3)(38,000 \text{ m}^3/\text{d})$ = 5320 kg/d

3. Compute the digester volume.

$$\tau = \frac{V}{Q}$$

$V = Q\tau = (111.8 \text{ m}^3/\text{d})(10 \text{ d})$ = 1118 m³

4. Compute the volumetric loading.

$$kg\ bCOD/m^3 \cdot d = \frac{(5320\ kg/d)}{1118\ m^3} = 4.76\ kg/m^3 \cdot d$$

5. Compute the quantity of volatile solids produced per day using Eq. (14–13).

$$P_x = \frac{YQ(S_o - S)(10^3\ g/kg)}{1 + (k_d)SRT}$$

$S_o = 5320\ kg/d$

$S = 5320(1 - 0.70) = 1596\ kg/d$

$S_o - S = 5320 - 1596 = 3724\ kg/d$

$$P_x = \frac{(0.08)[(5320 - 1596)\ kg/d]}{1 + (0.03\ d^{-1})(10\ d)} = 229.2\ kg/d$$

6. Compute the percent stabilization.

$$Stabilization,\ \% = \frac{[3724\ kg/d - 1.42\ (229.2\ kg/d)]}{(5320\ kg/d)} \times 100 = 63.9\%$$

7. Compute the volume of methane produced per day at 35°C using Eq. (14–12) (conversion factor at 35°C = 0.40).

$$V_{CH_4} = (0.40)[(S_o - S)(Q)/(10^3\ g/kg) - 1.42\ P_x]$$

$$V_{CH_4} = \{(0.40\ m^3/kg)[(5320 - 1596)\ kg/d] - 1.42(229.2\ kg/d)\}$$

$$= 1359\ m^3/d$$

8. Estimate the total gas production.

$$Total\ gas\ volume = \frac{1359}{0.65} = 2091\ m^3/d$$

Loading Factors. One of the most common methods used to size digesters is to determine the required volume based on a loading factor. Although a number of different factors have been proposed, the two most favored are based on (1) the mass of volatile solids added per day per unit volume of digester capacity, and (2) the mass of volatile solids added to the digester each day per mass of volatile solids in the digester. Of the two, the first method is preferred. Loading criteria are based generally on sustained loading conditions (see Chap. 3), typically peak 2-week or peak month solids production with provisions for avoiding excessive loadings during shorter periods. Typical design criteria for sizing mesophilic high-rate anaerobic digesters are given in Table 14–27. The upper limit of volatile solids loading rates is typically determined by the rate of accumulation of toxic materials, particularly ammonia, or washout of methane formers (WEF, 1998).

Table 14–27
Typical design criteria for sizing mesophilic high-rate complete-mix anaerobic sludge digesters[a]

Parameter	U.S. customary units		SI units	
	Units	Value	Units	Value
Volume criteria:				
Primary sludge	ft³/capita	1.3–2.0	m³/capita	0.03–0.06
Primary sludge + trickling-filter humus sludge	ft³/capita	2.6–3.3	m³/capita	0.07–0.09
Primary sludge + activated sludge	ft³/capita	2.6–4	m³/capita	0.07–0.11
Solids loading rate	lb VSS/ 10³ ft³·d	100–300	kg VSS/m³·d	1.6–4.8
Solids retention time	d	15–20	d	15–20

[a] Adapted, in part, from U.S. EPA (1979).

Excessively low volatile solids loading rates can result in designs that are costly to build and are troublesome to operate. In a survey conducted by Speece (1988) of 30 digester installations in the United States, one of the most significant observations was the relatively low solids content in the sludge feed to the digesters. The average TSS in the sludge feed was 4.7 ± 1.6 percent and the average volatile solids content was 70 percent. The average VSS value in the digesters was a dilute 1.6 percent. Dilute sludge results in the following adverse effects in digester operation: (1) reduced τ, (2) reduced VS destruction, (3) reduced methane generation, (4) reduced alkalinity, (5) increased volumes of digested biosolids and supernatant, (6) increased heating requirements, (7) increased dewatering capacity, and (8) increased hauling cost for liquid biosolids. One cautionary measure should be noted, however. A potential problem with ammonia toxicity could occur if the waste-activated sludge is thickened too much. Thus, in planning the design and operation of anaerobic digesters, consideration should be given to optimizing volatile solids loading to effectively utilize digester capacity. The effect of solids concentration and hydraulic detention time on volatile solids loading is reported in Table 14–28.

Estimating Volatile Solids Destruction. The degree of stabilization obtained is often measured by the percent reduction in volatile solids. The reduction in volatile solids can be related either to the SRT or to the detention time based on the untreated sludge feed. The amount of volatile solids destroyed in a high-rate complete-mix digester can be estimated by the following empirical equation (Liptak, 1974):

$$V_d = 13.7 \ln (\text{SRT}_{des}) + 18.9 \qquad (14\text{–}14)$$

where V_d = volatile solids destruction, %
    SRT = time of digestion, d (range 15 to 20 d)

Volatile solids destruction estimates can also be made using Table 14–29.

Because the untreated sludge feed can be measured easily, this method is also used commonly. In plant operation, calculation of volatile solids reduction should be made

Table 14–28
Effect of sludge concentration and hydraulic detention time on volatile solids loading factors[a]

Sludge concentration, %	Volatile solids loading factor							
	lb/ft³·d				kg/m³·d			
	10 d[b]	12 d	15 d	20 d	10 d	12 d	15 d	20 d
2	0.09	0.07	0.06	0.04	1.4	1.2	0.95	0.70
3	0.13	0.11	0.09	0.07	2.1	1.8	1.4	1.1
4	0.18	0.15	0.12	0.09	2.9	2.4	1.9	1.4
5	0.22	0.19	0.15	0.11	3.6	3.0	2.4	1.8
6	0.27	0.22	0.18	0.13	4.3	3.6	2.9	2.1
7	0.31	0.26	0.21	0.16	5.0	4.2	3.3	2.5
8	0.36	0.30	0.24	0.18	5.7	4.8	3.8	2.9

[a] Based on 70 percent volatile content of sludge, and a sludge specific gravity of 1.02 (concentration effects neglected).
[b] Hydraulic detention time, d.

Table 14–29
Estimated volatile solids destruction in high-rate complete-mix mesophilic anaerobic digestion[a]

Digestion time, d	Volatile solids destruction, %
30	65.5
20	60.0
15	56.0

[a] Adapted from WEF (1998).

routinely as a matter of record whenever sludge is drawn to processing equipment or drying beds. Alkalinity and volatile acids content should also be checked daily as a measure of the stability of the digestion process.

In calculating the volatile solids reduction, the ash content of the sludge is assumed to be conservative; that is, the number of pounds of ash going into the digester is equal to that being removed. A typical calculation of volatile solids reduction is presented in Example 14–6.

EXAMPLE 14–6   **Determination of Volatile Solids Reduction**   From the following analysis of untreated and digested biosolids, determine the total volatile solids reduction achieved during digestion. It is assumed that (1) the weight of fixed solids in the digested biosolids equals the weight of fixed solids in the untreated sludge and (2) the volatile solids are the only constituents of the untreated sludge lost during digestion.

	Volatile solids, %	**Fixed solids, %**
Untreated sludge	70	30
Digested sludge	50	50

Solution

1. Determine the weight of the digested solids. Because the quantity of fixed solids remains the same, the weight of the digested solids based on 1.0 kg of dry untreated sludge, as computed below, is 0.6 kg.

$$\text{Fixed solids in untreated sludge, } 30\% = \frac{(0.3 \text{ kg})100}{0.3 \text{ kg} + 0.7 \text{ kg}}$$

Let X equal the weight of volatile solids after digestion. Then

$$\text{Fixed solids in untreated sludge, } 50\% = \frac{(0.3 \text{ kg}) 100}{0.3 \text{ kg} + X \text{ kg}}$$

$$X = \frac{(0.3 \text{ kg})100}{50} - 0.3 \text{ kg} = 0.3 \text{ kg to volatile solids}$$

Weight of digested solids = 0.3 kg + X = 0.6 kg

2. Determine the percent reduction in total and volatile suspended solids.
   a. Percent reduction of total suspended solids

   $$R_{\text{TSS}} = \frac{(1.0 - 0.6) \text{ kg}}{1.0 \text{ kg}} \times 100 = 40\%$$

   b. Percent reduction in volatile suspended solids

   $$R_{\text{VSS}} = \frac{(0.7 - 0.3) \text{ kg}}{0.7 \text{ kg}} \times 100 = 57.1\%$$

Population Basis. Digestion tanks are also designed on a volumetric basis by allowing a certain number of cubic meters per capita (cubic feet per capita). Detention times range from 10 to 20 d for high-rate digesters (U.S. EPA, 1979). These detention times are recommended for design based on total tank volume, plus additional storage volume if sludge is dried on beds and weekly sludge withdrawals are curtailed because of inclement weather.

Typical design criteria for heated anaerobic digesters based on population are shown in Table 14–27. The criteria are applied only where analyses and volumes of sludge to be digested are not available. The capacities shown in Table 14–27 should be increased 60 percent in a municipality where the use of garbage grinders is universal and should be increased on a population-equivalent basis to allow for the effect of industrial wastes.

Figure 14–21

Typical shapes of anaerobic digesters: (a) cylindrical with reinforced concrete construction, (b) conventional German design with reinforced concrete construction, and (c) egg-shaped with steel shell.

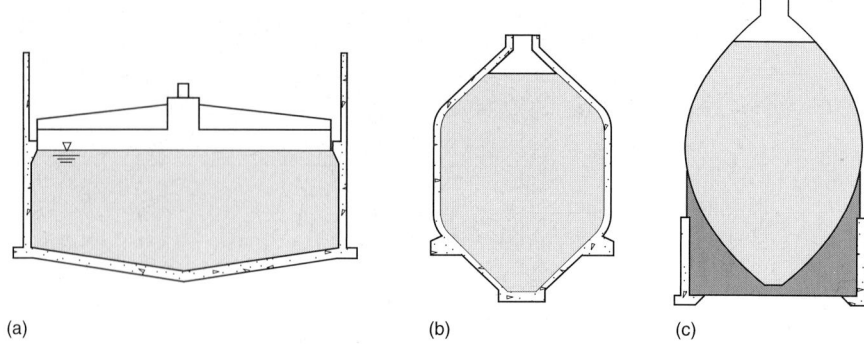

(a)   (b)   (c)

Selection of Tank Design and Mixing System

Most anaerobic digestion tanks are either cylindrical, conventional German design, or egg-shaped (see Fig. 14–21). The most common shape used in the United States is a shallow, vertical cylinder shown on Fig. 14–21a. Rectangular tanks were used in the past, but they experienced great difficulty in mixing the tank contents uniformly. German designers have worked on optimizing the shape of digesters and two basic types have emerged: the conventional German digester and the egg-shaped digester. The conventional German digester (see Fig. 14–21b) is a deep cylindrical vessel with steeply sloped top and bottom cones (Stukenberg, et al, 1992). The egg-shaped digester, shown on Fig. 14–21c, is similar in appearance to an upright egg, and the design is sometimes modified to a sphere-cone shape. Egg-shaped tanks have been used extensively in Europe, especially in Germany, and are growing in popularity in the United States. Essentially all of the modern digester designs in the United States are of either the cylindrical or egg-shaped type. Cylindrical and egg-shaped digesters and the mixing systems used for each type of tank are discussed in the following paragraphs. Advantages and disadvantages of each type of digester are summarized in Table 14–30.

Proper mixing is one of the most important considerations in achieving optimum process performance. Various systems for mixing the contents of the digesters have been used; the most common types involve the use of (1) gas injection, (2) mechanical stirring, and (3) mechanical pumping. Some digester installations use a combination of gas mixing and recirculation by pumping. The advantages and disadvantages of the various mixing systems are summarized in Table 14–31; typical design parameters are shown in Table 14–32.

Cylindrical Tanks. Cylindrical sludge digestion tanks are seldom less than 6 m (20 ft) or more than 38 m (125 ft) in diameter. The water depth should not be less than 7.5 m (25 ft) at the sidewall because of the difficulty in mixing shallow tanks, and the depth may be as much as 15 m (50 ft). The floor of the digester is usually conical with the bottom sloping to the center, with a minimum slope of 1 vertical to 6 horizontal where the sludge is drawn off (see Fig. 14–22). An alternative design uses a "waffle" bottom to minimize grit accumulation and to reduce the need for frequent digester cleaning (see Fig. 14–23).

Table 14–30

Comparison of cylindrical and egg-shaped anaerobic digesters[a]

Type of digester	Advantages	Disadvantages
Cylindrical	Reactor shape allows a relatively large volume for gas storage	Reactor shape results in inefficient mixing and dead spaces
	Reactor can be equipped with gas-holder covers	Poor mixing results in grit and silt accumulation
	Low profile	Large surface area provides space for scum accumulation and foam formation
	Conventional construction techniques can be applied; construction costs can be competitive	Cleaning is required for removal of grit and scum accumulation; digester may be required to be taken out of service
Egg-shaped	Minimum grit accumulation	Very little gas storage volume; external gas storage is required if gas is recovered
	Reduced scum formation	High-profile structures; may be aesthetically objectionable
	Higher mixing efficiency	
	More homogeneous biomass is obtained	Difficult access to top-mounted equipment; installation requires a high stair tower or an elevator
	Lower operating and maintenance costs; cleaning frequency significantly reduced	
	Smaller footprint; less land area is required	Greater foundation design requirements and seismic considerations
	Foaming is minimized (except for gas mixing)	Foaming of gas-mixed digester may be a problem in collecting gas
		Higher construction costs
		Construction limited to specialty contractors

[a] Adapted, in part, from Brinkman and Voss (1998).

Gas-injection systems used in cylindrical tanks are classified as unconfined or confined (Fig. 14–24a and b). Unconfined gas systems collect gas at the top of the digesters, compress the gas, and then discharge the gas through a pattern of bottom diffusers or through a series of radially placed top-mounted lances. Unconfined gas systems mix the digester contents by releasing gas bubbles that rise to the surface, carrying and moving the sludge. These systems are suitable for digesters with fixed, floating, or gas-holder covers. In confined gas systems, gas is collected at the top of the digesters, compressed, and discharged through confined tubes. Two major types of confined systems are the gas lifter and the gas piston. The gas lifter system consists of submerged gas pipes or lances inserted into an eductor tube or gas lifter. Compressed gas is released from the lances or pipes, and the gas bubbles rise, creating an air-lift effect. In the gas piston system, gas bubbles are released intermittently at the bottom of a cylindrical tube or piston. The bubbles rise and act like a piston, pushing the sludge to the surface. These systems are suitable for fixed, floating, or gas-holder covers.

Table 14–31
Summary of advantages and disadvantages of various anaerobic digester mixing systems[a]

Type of mixer	Advantages	Disadvantages
All systems	Increased rate of biosolids stabilization	Corrosion and tear of ferrous-metal piping and supports. Equipment wear by grit. Equipment plugging and operational interference by rags
Gas injection:		
Unconfined		
Cover-mounted lances	Lower maintenance and less hindrance to cleaning than bottom-mounted diffusers. Effective against scum buildup	Corrosion of gas piping and equipment. High maintenance for compressor. Potential gas-seal problem. Compressor problems if foam gets inside. Solids deposition. Plugging of gas lances. Entire tank contents are not mixed
Bottom-mounted diffusers	Better movement of bottom deposits than cover-mounted lances	Corrosion of gas piping and equipment. High compressor maintenance. Potential gas-seal problem. Foaming. Incomplete mixing. Scum formation. Diffuser plugging. Bottom deposits can alter mixing patterns. Requires digester dewatering for maintenance
Confined:		
Gas lifters	Better mixing and gas production and better movement of bottom deposits than cover-mounted lances. Lower power requirements	Corrosion of gas piping and equipment. High maintenance for compressor. Potential gas-seal problem. Corrosion of gas lifter. Lifter interferes with digester cleaning. Scum buildup. Does not provide good top mixing. Requires digester dewatering for maintenance if bottom mounted
Gas pistons	Good mixing efficiency	Corrosion of gas piping and equipment. High maintenance for compressor. Potential gas-seal problem. Equipment internally mounted. Pistons interfere with digester cleaning. Requires digester dewatering for maintenance
Mechanical stirring:		
Low-speed turbines	Good mixing efficiency	Wear of impellers and shafts. Bearing failures. Interference of impellers with rags. Requires oversized gear boxes. Gas leaks at shaft seal. Long overhung loads
Low-speed mixers	Breaks up scum layers	Not designed to mix entire tank contents. Bearing and gear box failures. Impeller wear. Interference of impellers by rags
Mechanical pumping:		
Internal draft tubes	Good top-to-bottom mixing	Sensitive to liquid level. Corrosion and wear of impeller. Bearing and gear box failures. Requires oversized gear box
External draft tubes	Same as internal draft tube. Draft-tube maintenance easier than internal type	Same as internal draft tube
Pumps	Better mixing control. Scum layer and sludge deposits can be recirculated. Pumps easier to maintain than compressors	Impeller wear. Plugging of pumps by rags. Bearing failures

[a] Adapted from WEF (1987b) and Metcalf & Eddy (1984a).

Table 14–32
Typical design parameters for anaerobic digester mixing systems[a]

Parameter	Type of mixing system	Typical values[b]	
		U.S. customary units	**SI units**
Unit power	Mechanical systems	0.025–0.04 hp/10³ gal of digester volume	0.005–0.008 kW/m³ of digester volume
Unit gas flow[c]	Gas mixing		
	Unconfined	4.5–5 ft³/10³ ft³·min	0.0045–0.005 m³/m³·min
	Confined	5–7 ft³/10³ ft³·min	0.005–0.007 m³/m³·min
Velocity gradient G	All	50–80 s⁻¹	50–80 s⁻¹
Turnover time of tank contents	Confined gas mixing and mechanical systems	20–30 min	20–30 min

[a] Adapted from WEF (1987b).
[b] Actual design values may differ depending on the type of mixing system, manufacturer, and digestion process or function.
[c] Quantity of gas delivered by the gas-injection system divided by the digester gas volume.

Figure 14–22

Typical cross section through a high-rate, gas-mixed cylindrical digester.

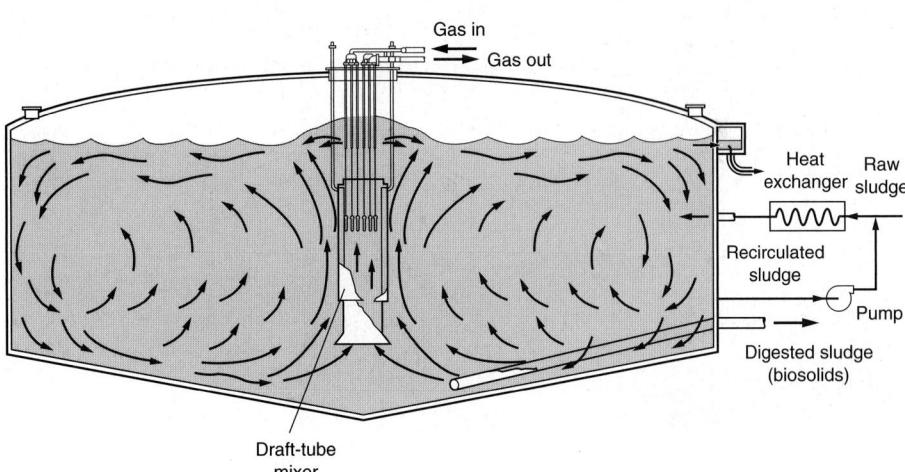

Mechanical stirring systems commonly use low-speed turbines or mixers (see Fig. 14–24c). In both systems, the rotating impeller(s) displaces the sludge, mixing the digester contents. Low-speed turbine systems usually have one cover-mounted motor with two turbine impellers located at different sludge depths. A low-speed mixer system usually has one cover-mounted mixer. Mechanical stirring systems are most suitable for digesters with fixed or floating covers.

Most mechanical pumping systems consist of propeller-type pumps mounted in internal or external draft tubes, or axial-flow or centrifugal pumps and piping installed

Figure 14–23

Typical waffle-bottom
anaerobic digester:
(a) plan view, (b) section.

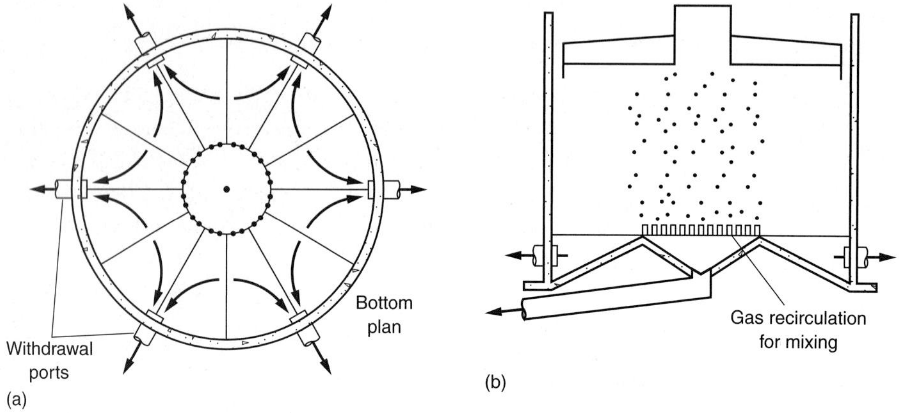

Withdrawal
ports

Bottom
plan

(a)

Gas recirculation
for mixing

(b)

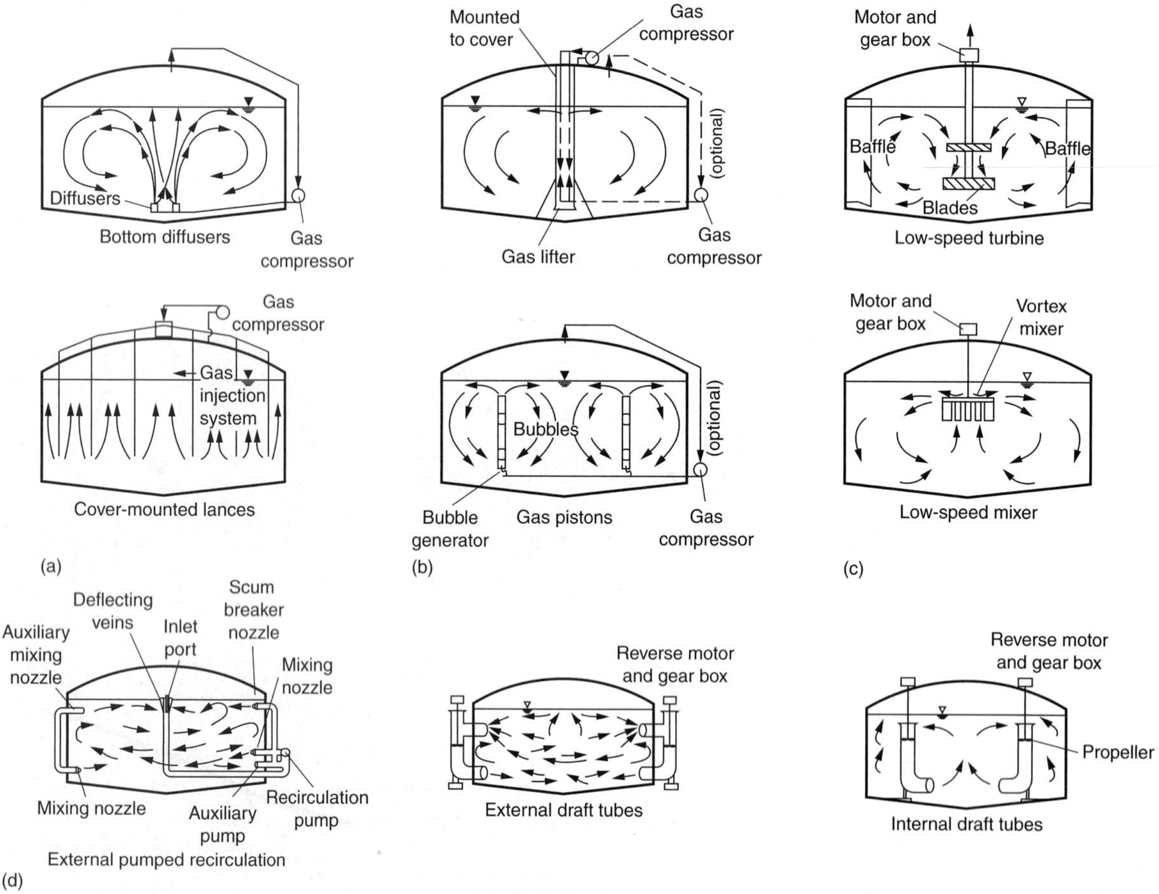

Diffusers

Bottom diffusers

Gas
compressor

Mounted
to cover

Gas
compressor

Gas lifter

Gas
compressor

(optional)

Motor and
gear box

Baffle

Baffle

Blades

Low-speed turbine

Gas
compressor

Gas
injection
system

Cover-mounted lances

(a)

Bubble
generator

Bubbles

Gas pistons

Gas
compressor

(optional)

(b)

Motor and
gear box

Vortex
mixer

Low-speed mixer

(c)

Deflecting
veins

Scum
breaker
nozzle

Auxiliary
mixing
nozzle

Inlet
port

Mixing
nozzle

Mixing nozzle

Auxiliary
pump

Recirculation
pump

External pumped recirculation

(d)

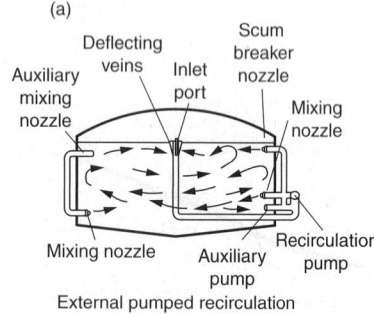

Reverse motor
and gear box

External draft tubes

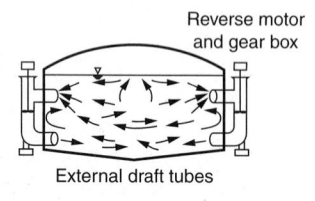

Reverse motor
and gear box

Propeller

Internal draft tubes

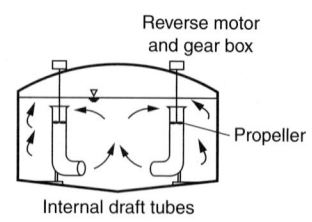

Figure 14–24

Devices used for mixing contents of anaerobic high-rate digesters: (a) unconfined gas-injection systems, (b) confined gas-injection systems, (c) mechanical stirring systems, (d) mechanical pumping systems. (Metcalf & Eddy, 1984a.)

externally (see Fig. 14–24*d*). Mixing is promoted by the circulation of sludge. Mechanical pumping systems are suitable for digesters with fixed covers.

Egg-Shaped Tanks.   The purpose of the egg-shaped design is to enhance mixing and to eliminate the need for cleaning. The digester sides form a steep cone at the bottom so that grit accumulation is minimized (see Fig. 14–25*a*). Other advantages cited for the egg-shaped design include better mixing, better control of the scum layer, and smaller land-area requirements. Steel construction is more common for egg-shaped tanks in the United States; reinforced concrete construction requires complex formwork and special construction techniques. The structures are relatively high as compared to other treatment plant structures (see Fig. 14–25*b*), and may require an elevator for access to the top of the structure. In recent designs in Boston, MA (see Fig. 14–19), and Baltimore, MD, the heights of the digesters were over 40 m (130 ft).

Egg-shaped digester mixing systems are similar to those for cylindrical tanks and consist of unconfined gas mixing, mechanical draft-tube mixing, or pumped recirculation mixing (see Fig. 14–26). Gas mixing is considered by some to be relatively ineffective in mixing the digester contents below the level of the injection nozzles. The mechanical draft tube and pumped recirculation mixing systems, however, are considered able to provide sufficient energy to mix even the sludge in the bottom cone of the digester. The mechanical draft-tube mixer, which can be operated in either an up- or

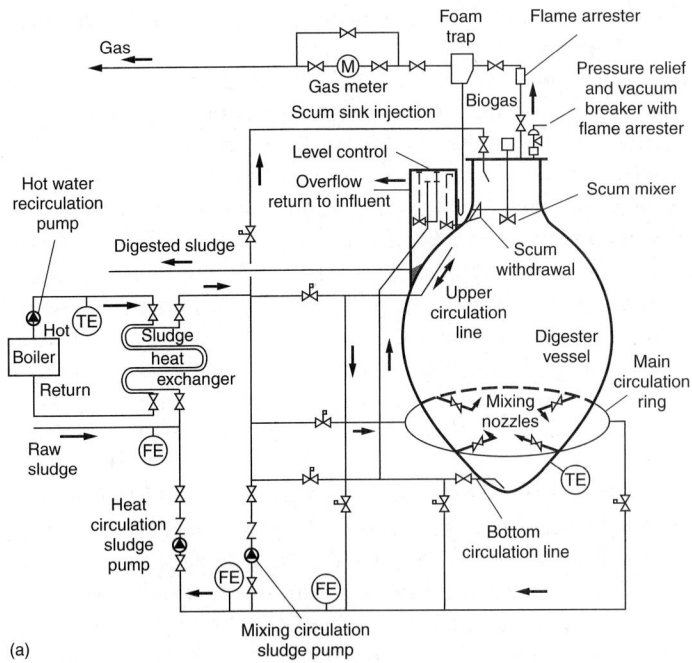

(a)

(b)

Figure 14–25

Egg-shaped anaerobic digester: (*a*) schematic diagram from Walker Process catalog, (*b*) pictorial view.

Figure 14–26

Mixing systems for egg-shaped anaerobic digesters. *(From Stukenberg et al., 1992.)*

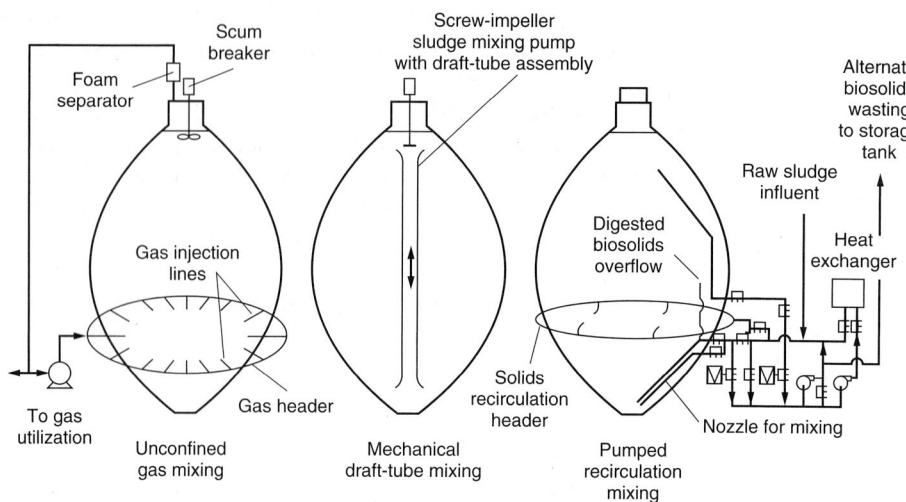

down-pumping mode, also provides a positive means of mixing at the surface to control scum and foam (Stukenberg et al., 1992).

Recirculation mixing is generally more effective when the sludge is taken from the bottom and discharged near the gas-liquid interface or above the gas-liquid interface to break up scum that may have accumulated. Recirculation mixing is also effective for foam control in gas-mixed digesters.

Any or all of the mixing systems may be used, and all may be operated during any one day, although gas and mechanical draft-tube mixing are seldom used at the same time. Most digesters are fitted with a gas lance or hydraulic jet near the bottom of the cone to stir any accumulated grit.

A combination jet-pump draft-tube mixing system is also used that permits mixing in three zones of the digester. One jet pump is attached to the bottom of the centrally located vertical draft tube, and a second jet pump is attached to the top. In this configuration, the draft tube can function for pumping sludge upward or downward for periodic blending of bottom sludge and scum with the tank contents. A third pump is located at the vessel perimeter to create a swirling action. External recirculation pumps are also provided for sludge heating and additional circulation of the tank contents. The system is designed to circulate the tank volume 10 times per day (Clark and Ruehrwein, 1992).

Methods for Enhancing Solids Loading and Digester Performance

Opportunities for enhancing the performance of anaerobic digesters include thickening the digester feed sludge or thickening a portion of the digesting sludge to increase the SRT. Recirculating a portion of the digested sludge and cothickening with untreated primary and waste sludge was reported originally by Torpey and Melbinger (1967). The solids concentration in the feed sludge improved and the performance of the digester, as measured by volatile solids destruction, increased significantly. The thickening system was installed at wastewater plants in New York City. In a recent study by Maco et al. (1998), the effects of thickening digested solids, either thickened separately or combined with prethickening of untreated sludge, increased the SRT of the digestion process and the production of biogas and decreased the hydraulic retention time.

The value of thickening the feed sludge to the digester is indicated by data presented in Table 14–28. For example, for a 15-d hydraulic retention time and an average TSS of 3 percent, the volatile solids loading factor in Table 14–28 is 1.4 kg/m³·d. By improving the feed solids TSS to 6 percent, the VSS loading can be increased to 2.9 kg/m³·d, near the middle of the solids loading range given in Table 14–27. In this hypothetical example, a doubling in digester capacity is achieved. In evaluating digested solids recycling to reduce the size and number of digesters or increase the solids processing capacity of existing digesters, sludge rheology and sludge handling equipment require evaluation. While most digesters can accommodate increases in solids concentrations, the limits imposed by pumping and mixing systems require careful evaluation (Maco et al., 1998).

Gas Production, Collection, and Use

Gas from anaerobic digestion contains about 65 to 70 percent CH_4 by volume, 25 to 30 percent CO_2, and small amounts of N_2, H_2, H_2S, water vapor, and other gases. Digester gas has a specific gravity of approximately 0.86 relative to air. Because production of gas is one of the best measures of the progress of digestion and because digester gas can be used as fuel, the designer should be familiar with its production, collection, and use.

Gas Production. The volume of methane gas produced during the digestion process can be estimated using Eq. (14–12), which has been discussed previously. Total gas production is usually estimated from the percentage of volatile solids reduction. Typical values vary from 0.75 to 1.12 m³/kg (12 to 18 ft³/lb) of volatile solids destroyed. Gas production can fluctuate over a wide range, depending on the volatile solids content of the sludge feed and the biological activity in the digester. Excessive gas production rates sometimes occur during startup and may cause foaming and escape of foam and gas from around the edges of floating digester covers. In egg-shaped and shallow cylindrical digesters, foaming can clog the gas outlet unless foam control is provided. If stable operating conditions have been achieved and the foregoing gas production rates are being maintained, a well-digested sludge can be obtained.

Gas production can also be estimated crudely on a per capita basis. The normal yield is 15 to 22 m³/10³ persons·d (0.6 to 0.8 ft³/person·d) in primary plants treating normal domestic wastewater. In secondary treatment plants, the gas production is increased to about 28 m³/10³ persons·d (1.0 ft³/person·d).

Gas Collection. In cylindrical digesters, gas is collected under the cover of the digester. Three principal types of covers are used: (1) floating, (2) fixed, and (3) membrane. Floating covers fit on the surface of the digester contents and allow the volume of the digester to change without allowing air to enter the digester (see Fig. 14–27a). Gas and air must not be allowed to mix, or an explosive mixture may result. Explosions have occurred in wastewater-treatment plants. Gas piping and pressure-relief valves must include adequate flame traps. The covers may also be installed to act as gas holders for a limited storage of gas. High-rate digesters produce about two volumes of gas per volume of digester capacity/d (Speece 2001). Floating covers can be used for single-stage digesters or in the second stage of two-stage digesters.

Fixed covers provide a free space between the roof of the digester and the liquid surface (see Fig. 14–27b). Gas storage must be provided so that (1) when the liquid volume is changed, gas, and not air, will be drawn into the digester; otherwise an overflow weir with a U-shaped trap needs to be provided to maintain a liquid seal, and (2) gas will not be lost by displacement. Gas can be stored either at low pressure in external gas holders that use floating covers or at high pressure in pressure vessels if gas compressors are used. Gas not used should be burned in a flare. Gas meters should be installed to measure gas produced and gas used or wasted.

A recent development in gas-holder covers for cylindrical tanks is the membrane cover (see Fig. 14–27c). This cover consists of a support structure for a small center gas dome and flexible air and gas membranes. An air-blower system is provided to pressurize the air space between the two membranes and vary the air space volume. Only

Figure 14–27

Types of anaerobic digester covers: (a) floating, (b) fixed, and (c) schematic and view of membrane gas cover.

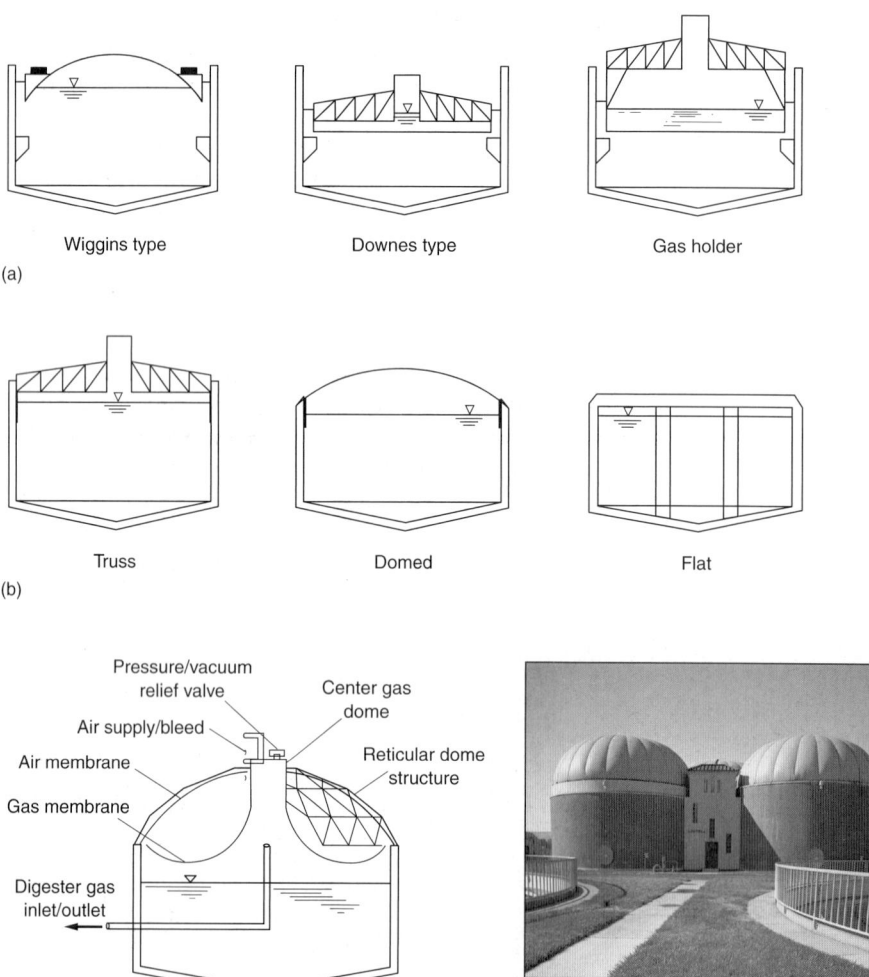

Wiggins type   Downes type   Gas holder

(a)

Truss   Domed   Flat

(b)

Pressure/vacuum relief valve
Air supply/bleed
Air membrane
Gas membrane
Center gas dome
Reticular dome structure
Digester gas inlet/outlet

(c)

the gas membrane and the center gas dome are in contact with the digester contents. The gas membrane is made from a flexible polyester fabric (WEF, 1998).

In egg-shaped digesters, the volume available for gas storage is small. For efficient utilization of digester gas, supplemental external storage may be required.

Use of Digester Gas. Methane gas at standard temperature and pressure (20°C and 1 atm) has a lower heating value of 35,800 kJ/m³ (960 Btu/ft³). Lower heating value is the heat of combustion less the heat of vaporization of any water vapor present. Because digester gas is only 65 percent methane, the lower heating value of digester gas is approximately 22,400 kJ/m³ (600 Btu/ft³). By comparison, natural gas, which is a mixture of methane, propane, and butane, has a heating value of 37,300 kJ/m³ (1000 Btu/ft³).

In large plants, digester gas may be used as fuel for boiler and internal-combustion engines which are, in turn, used for pumping wastewater, operating blowers, and generating electricity. Hot water from heating boilers or from engine jackets and exhaust-heat boilers may be used for sludge heating and for building heating, or gas-fired sludge-heating boilers may be used. Energy recovery is more efficient if prime movers are designed to run hot because heat rejected at high temperatures can be put to a greater variety of uses than heat rejected at low temperatures.

Digester gas can be used in cogeneration. Cogeneration is generally defined as a system for generating electricity and producing another form of energy (usually steam or hot water). Digester gas can be used to power an engine-generator to generate electricity, and the jacket water from the internal-combustion engine can then be used for digester or building heating. Surplus power, over and above that required for plant purposes, can sometimes be sold to local electric utilities.

Because digester gas contains hydrogen sulfide, nitrogen, particulates, and water vapor, the gas frequently has to be cleaned in dry or wet scrubbers before it is used in internal-combustion engines. Hydrogen sulfide concentrations in excess of approximately 100 ppm by volume may require the installation of hydrogen sulfide removal equipment (WEF, 1998).

Digester Heating

The heat requirements of digesters consist of the amount needed (1) to raise the incoming sludge to digestion tank temperatures, (2) to compensate for the heat losses through walls, floor, and roof of the digester, and (3) to make up the losses that might occur in the piping between the source of heat and the tank. The sludge in digestion tanks is heated by pumping the sludge and supernatant through external heat exchangers and back to the tank.

Analysis of Heat Requirements. In computing the energy required to heat the incoming sludge to the temperature of the digester, it is assumed that the specific heat of most sludges is essentially the same as that of water. The assumption that the specific heats of sludge and water are essentially the same has proved to be acceptable for engineering computations. The heat loss through the digester sides, top, and bottom is computed using the following expression:

$$q = UA\Delta T \tag{14–15}$$

where q = heat loss, J/s, Btu/h

$\quad$ U = overall coefficient of heat transfer, J/m²·s·°C (Btu/ft²·h·°F)

$\quad$ A = cross-sectional area through which the heat loss is occurring, m² (ft²)

$\quad$ ΔT = temperature drop across the surface in question, °C (°F)

In computing the heat losses from a digester using Eq. (14–15), it is common practice to consider the characteristics of the various heat transfer surfaces separately and to develop transfer coefficients for each one. The application of Eq. (14–15) in the computation of digester heating requirements is illustrated in Example 14–7.

Heat-Transfer Coefficients. Typical overall heat-transfer coefficients are reported in Table 14–33. As shown, separate entries are included for the walls, bottom, and top of the digester.

Digestion tank walls may be surrounded by earth embankments that serve as insulation, or they may be of compound construction consisting of approximately 300 mm (12 in) of concrete, insulation, or an insulating air space, plus brick facing or corrugated aluminum facing over rigid insulation. The heat transfer from plain concrete walls

Table 14–33

Typical values of heat transfer coefficients used for computing digester heat losses[a]

Item	U.S. customary, Btu/ft²·°F·h	SI units, W/m²·°C
Plain concrete walls (above ground):		
$\quad$ 300 mm (12 in) thick, not insulated	0.83–0.90	4.7–5.1
$\quad$ 300 mm (12 in) thick with air space plus brick facing	0.32–0.42	1.8–2.4
$\quad$ 300 mm (12 in) thick wall with insulation	0.11–0.14	0.6–0.8
Plain concrete walls (below ground):		
$\quad$ Surrounded by dry earth	0.10–0.12	0.57–0.68
$\quad$ Surrounded by moist earth	0.19–0.25	1.1–1.4
Plain concrete floors:		
$\quad$ 300 mm (12 in) thick in contact with moist earth	0.5	2.85
$\quad$ 300 mm (12 in) thick in contact with dry earth	0.3	1.7
Floating covers:		
$\quad$ With 35-mm (1.5-in) wood deck, built-up roofing, and no insulation	0.32–0.35	1.8–2.0
$\quad$ With 25-mm (1-in) insulating board installed under roofing	0.16–0.18	0.9–1.0
Fixed concrete covers:		
$\quad$ 100 mm (4 in) thick and covered with built-up roofing, not insulated	0.70–0.88	4.0–5.0
$\quad$ 100 mm (4 in) thick and covered, but insulated with 25-mm (1-in) insulating board	0.21–0.28	1.2–1.6
$\quad$ 225 mm (9 in) thick, not insulated	0.53–0.63	3.0–3.6
Fixed steel covers 6 mm (0.25 in) thick	0.70–0.95	4.0–5.4

[a] Adapted, in part, from U.S. EPA (1979).

below ground level and from floors depends on whether they are below the groundwater level. If the groundwater level is not known, it may be assumed that the sides of the tank are surrounded by dry earth and that the bottom is saturated earth. Because the heat losses from the tank warm up the adjacent earth, it is assumed that the earth forms an insulating blanket 1.5 to 3 m (5 to 10 ft) thick before stable ambient earth temperatures are reached. In northern climates, frost may penetrate to a depth of 1.2 m (4 ft). Therefore, the ground temperature can be assumed to be 0°C (32°F) at this depth and to vary uniformly above this depth to the design air temperatures at the surface. Below the frost depth, normal winter ground temperatures can be assumed, which are 5 to 10°C (10 to 20°F) higher at the base of the wall. Alternatively, an average temperature may be assumed for the entire wall below grade.

The loss through the roof depends on the type of construction, the absence or presence of insulation and its thickness, the presence of air space (as with floating covers between the skin plate and the roofing), and whether the underside of the roof is in contact with sludge liquor or gas.

Radiation from roofs and aboveground walls also contributes to heat losses. At the temperatures involved, the effect is small and is included in the coefficients normally used, such as those given in the foregoing discussion. For the theory of radiant-heat transmission, the reader is referred to McAdams (1954). Heat requirements for a digester are determined in Example 14–7.

When external heaters are installed, the sludge is pumped at high velocity through the tubes while water circulates at high velocity around the outside of the tubes. The circulation promotes high turbulence on both sides of the heat transfer surface and results in higher heat transfer coefficients and better heat transfer. Another advantage of external heaters is that untreated cold sludge on its way into the digesters can be warmed, intimately blended, and seeded with sludge liquor before entering the tank. Heat exchangers require cleaning periodically to maintain heat transfer efficiency.

Digestion tanks have also been heated using internal heating systems. Some arrangements have included pipes mounted to the interior face of the digester wall and mixing tubes equipped with hot-water jackets. Because of inherent operating and maintenance problems with this type of heating system, internal heating is not recommended. Reported problems include caking of sludge on the heating surface and the inability to inspect or service the equipment unless the tank is dewatered (WEF, 1987a).

EXAMPLE 14–7   **Estimation of Digester Heating Requirements**   A digester with a capacity of 45,000 kg/d (100,000 lb/d) of sludge is to be treated by circulation of sludge through an external hot water heat exchanger. Assuming that the following conditions apply, find the heat required to maintain the required digester temperature. If all heat were shut off for 24 h, what would be the average drop in temperature of the tank contents?

1. Concrete digester dimensions:

    Diameter = 18 m

    Side depth = 6 m

    Middepth = 9 m

2. Heat-transfer coefficients:

Dry earth embanked for entire depth, $U = 0.68$ W/m²·°C

Floor of digester in groundwater, $U = 0.85$ W/m²·°C

Roof exposed to air, $U = 0.91$ W/m²·°C

3. Temperatures:

Air $= -5°C$

Earth next to wall $= 0°C$

Incoming sludge $= 10°C$

Earth below floor $= 5°C$

Sludge contents in digester $= 32°C$

4. Specific heat of sludge $= 4200$ J/kg·°C

Solution

1. Compute the heat requirement for the sludge.

$q = (45{,}000$ kg/d)$[(32 - 10)°C](4200$ J/kg·°C)

$\quad = 41.6 \times 10^8$ J/d

2. Compute the area of the walls, roof, and floor.

Wall area $= \pi\,(18)\,(6) = 339.2$ m²

Floor area $= \pi\,(9)\,(9^2 + 3^2)^{1/2} = 268.2$ m²

Roof area $= \pi\,(9^2) = 254.5$ m²

3. Compute the heat loss by conduction using Eq. (14–15).

$q = UA\Delta T$

a. Walls:

$q = 0.68$ W/m²·°C $(339.3$ m²)$(32 - 0°C)(86{,}400$ s/d) $= 6.38 \times 10^8$ J/d

b. Floor:

$q = 0.85$ W/m²·°C $(268.2$ m²) $(32 - 5°C)(86{,}400$ s/d) $= 5.32 \times 10^8$ J/d

c. Roof:

$q = 0.91$ W/m²·°C $(254.5$ m²) $(32 + 5°C)(86{,}400$ s/d) $= 7.40 \times 10^8$ J/d

d. Total losses:

$q_t = (6.38 + 5.32 + 7.40) \times 10^8$ J/d $= 19.1 \times 10^8$ J/d

4. Compute the required heat-exchanger capacity.

Capacity $=$ heat required for sludge and heat required for digester

$\quad\quad = (41.6 + 19.1) \times 10^8$ J/d $= 60.7 \times 10^8$ J/d

5. Determine the effect of heat shutoff.

a. Digester volume $= \pi\left(\dfrac{D^2}{4}\right)h_s + \pi\left(\dfrac{D^2}{12}\right)h_c$

$$= \pi\left(\dfrac{18^2}{4}\right)(6) + \pi\left(\dfrac{18^2}{12}\right)(3) = 1526.8 + 254.5$$

$$= 1781.3 \text{ m}^3$$

b. Weight of sludge $= (1781.3 \text{ m}^3)(10^3 \text{ kg/m}^3)$

$$= 1.78 \times 10^6 \text{ kg}$$

c. Drop in temperature $= \dfrac{(60.7 \times 10^8 \text{ J/d})(1.0)}{(1.78 \times 10^6 \text{ kg})(4200 \text{ J/kg·°C})} = 0.81\text{°C/d}$

Heating Equipment. The contents of the digester can be heated by tube-in-tube, spiral-plate, or water-bath external heat exchangers. The tube-in-tube and spiral-heat exchangers are similar in design. A tube-in-tube exchanger consists of two concentric pipes, one containing the circulating sludge and the other containing hot water. Flow through the pipes is countercurrent. Spiral-plate heat exchangers (see Fig. 14–28a and b) are composed of two long strips of plate that are wrapped to form a pair of concentric passages. The flow regime is also countercurrent. Water temperatures are kept generally below 68°C (154°F) to prevent caking of the sludge. Heat-transfer coefficients for external heat exchangers range from 0.9 to 1.6 W/m²·°C (WEF 1998).

Operation of a water-bath heat exchanger involves circulation of the sludge through a heated water bath (see Fig. 14–28c). The heat transfer rate is increased by pumping hot water in and out of the bath. Recirculation pumps allow the sludge feed to be heated before introduction to the digester.

Boilers and cogeneration systems are used typically to supply heat to the circulating water in the heat exchangers. Boilers can be fueled by digester gas; however, natural gas or fuel oil may be used as auxiliary fuel for times when sufficient digester gas is not available, such as for digester startup. If a cogeneration system is provided that uses digester gas to fuel an internal-combustion engine for generating electricity or powering pumps or blowers, heat from the engine jacket water can be used in the heat exchanger.

Thermophilic Anaerobic Digestion

Thermophilic digestion occurs at temperatures between 50 and 57°C (120 and 135°F), conditions suitable for thermophilic bacteria. Because biochemical reaction rates increase with temperature, doubling with every 10°C (18°F) rise in temperature until a limiting temperature is reached, thermophilic digestion is much faster than mesophilic digestion. Advantages cited for thermophilic digestion include increased solids destruction capability, improved dewatering, and increased bacterial destruction. Disadvantages of thermophilic digestion are higher energy requirements for heating, poorer-quality supernatant containing larger quantities of dissolved solids, odors, and less

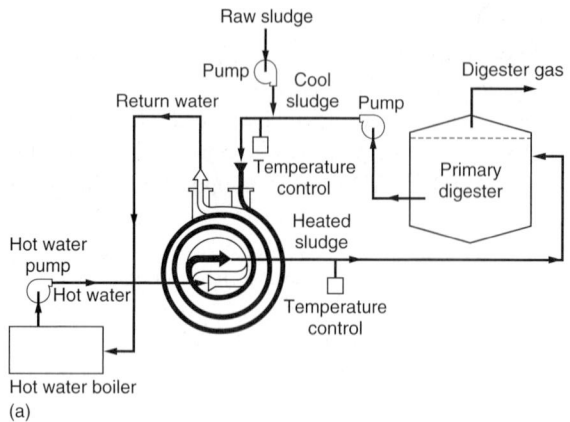

(b)

Figure 14–28

Heat exchangers used for heating digesting sludge: (a) schematic diagram of a spiral type, (b) pictorial view of a spiral type, and (c) schematic of a water bath type heat exchanger.

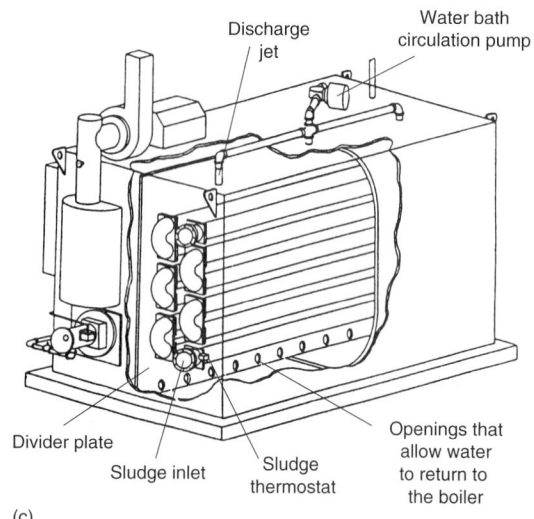

process stability (WEF, 1987a). Single-stage thermophilic digesters have been used only in limited applications; for municipal sludge treatment, they have been mainly used as the first stage of a temperature-phased anaerobic digestion process (Moen, 2000).

Although there may be greater reductions in pathogens in thermophilic digestion than in mesophilic digestion, U.S. federal regulations controlling land application of biosolids do not classify thermophilic digestion as a process to significantly reduce pathogens (PSRP). Both mesophilic and thermophilic digestion are classified as processes to further reduce pathogens (PFRP). Therefore, single-stage thermophilic digestion has significant limitations, as cited above.

Two-Phased Anaerobic Digestion

In search for improved anaerobic digestion performance, several options for phasing or staging the digestion process in multiple reactors have been investigated. Four basic phasing methods have been evaluated: (1) staged mesophilic digestion, (2) temperature-phased digestion, (3) acid/gas (A/G) phased digestion, and (4) staged thermophilic digestion (Schafer and Farrell, 2000*a* and 2000*b*). These phasing options are shown on Fig. 14–29 and are discussed below. Typical SRTs are also noted on Fig. 14–29.

Staged Mesophilic Digestion. Although digestion performed in two tanks coupled in series has been done in the past, little information is available about the operation of two-stage heated and mixed high-rate digesters. Researchers Torpey and Garber found that there were few benefits in volatile solids reduction and gas production in two

Figure 14–29

Options for staged anaerobic digestion: (*a*) staged mesophilic digestion, (*b*) temperature-phased thermophilic-mesophilic digestion, (*c*) temperature-phased mesophilic-thermophilic digestion, (*d*) acid/gas phased digestion with mesophilic acid-phase, (*e*) acid/gas phased digestion with thermophilic acid/gas phase, and (*f*) staged thermophilic digestion. (Adapted from Schafer and Farrell (2000a) and Moen (2000).)

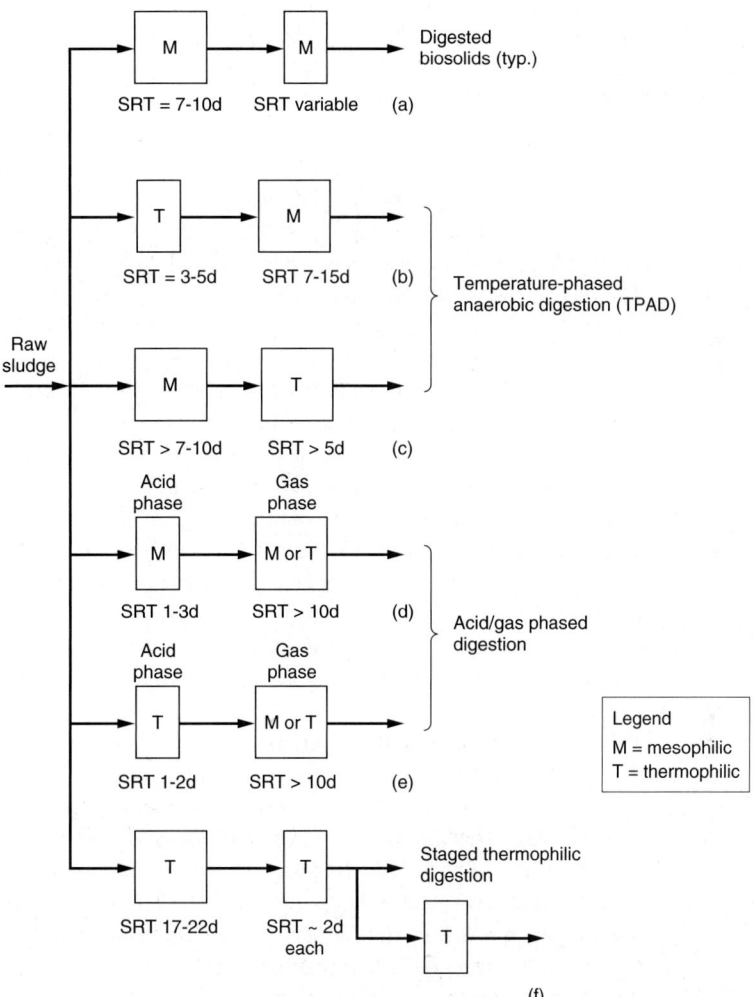

series tanks as compared to a single-stage high-rate process. More recent testing indicates that two-stage mesophilic digestion may produce more stable, less odorous biosolids that are easier to dewater (Schafer and Farrell, 2000a). Staged mesophilic digestion is shown on Fig. 14–29a.

Temperature-Phased Digestion. Temperature-phased anaerobic digestion (TPAD), shown on Fig. 14–27b, was developed in Germany and is an approach that incorporates the advantages of thermophilic digestion and mitigates the disadvantages through the addition of a mesophilic phase that enhances stabilization. The design of the temperature-phased process utilizes the advantage of the greater thermophilic digestion rate, which generally is four times faster than mesophilic digestion. The TPAD process has shown the capability for absorbing shock loadings better, as compared to single-stage mesophilic or thermophilic digestion. The process can operate in either of two modes, thermophilic-mesophilic or mesophilic-thermophilic. In the thermophilic-mesophilic mode (shown on Fig. 14–29b), the thermophilic phase is designed to operate at 55°C (130°F) with a 3 to 5 d detention time. The mesophilic phase is designed to operate at 35°C (95°F) with a 10 d or greater detention. The total average detention time of 15 d compares to the typical 10 to 20 d range of the single-stage high-rate mesophilic digestion process. The volatile suspended solids (VSS) destruction efficiencies of the TPAD process are on the order of 15 to 25 percent greater than single-stage mesophilic digestion (Schafer and Farrell, 2000b).

Through greater hydrolysis and biological activity in the thermophilic phase, the system tends to have greater VSS destruction and gas production. Foaming is also reduced. The mesophilic phase provides additional VSS destruction and conditions the sludge for further processing. The main advantages of the mesophilic phase are (1) the destruction of odorous compounds (mostly fatty acids) that are common to the thermophilic digestion process and (2) the improved stability of the digestion operation. The process is also reported to be capable of meeting Class A sludge requirements (WEF, 1998).

A second temperature-phased digestion process shown on Fig. 14–29c has a mesophilic stage that precedes the thermophilic stage. Limited results from full-scale and pilot testing show that the volatile solids reduction is greater than that from single-stage mesophilic digestion (Schafer and Farrell, 2000b).

Design considerations for the temperature-phased anaerobic digestion process include selection of the heating and mixing systems to ensure proper temperature control of each stage, sizing of the gas-handling equipment to meet the greater gas production rates, and control of the pumping systems for digester feed and heating (WEF, 1998).

Acid/Gas Phased Digestion. In the acid/gas digestion process, anaerobic digestion proceeds through the three distinct phases of digestion described earlier—hydrolysis, fermentation (acidogenesis), and methanogenesis. In the first stage, solubilization of particulate matter occurs and volatile acids are formed. The first phase is conducted at a pH of 6 or less and at a short SRT conducive to the production of high concentrations of volatile acids (> 6000 mg/L). The second phase is conducted at a neutral pH and a longer SRT, to suit the environmental conditions for the methane-generating bacteria and maximize gas production. Advantages of this method of digestion are (1) greater volatile solids reduction can be achieved, (2) digester foaming can be controlled, and (3) either stage can be operated at mesophilic or thermophilic temperatures (see Fig. 14–29d and e). Three full-scale plants using the acid/gas process are in operation at the time of writing of this text (2001) (Wilson and Dichtl, 2000). Volatile solids reductions range from 50 to 60 percent. At the Belmont Wastewater Plant in Indianapolis, IN, where a thermophilic acid phase and mesophilic gas phase system was pilot tested, it was found that the process was effective in meeting Class A requirements for pathogen reduction (Schafer and Farrell, 2000a).

Staged Thermophilic Digestion.   A staged thermophilic digestion process (see Fig. 14–29f) uses a large reactor followed by one or more smaller reactors to reduce pathogen short circuiting and achieve a Class A sludge. At the Annacis Island Wastewater Treatment Plant in Vancouver, BC, the first stage is followed by three subsequent stages. Volatile solids reductions for the digestion system are reported to be on the order of 63 percent (Schafer and Farrell, 2000b).

14–10   AEROBIC DIGESTION

Aerobic digestion may be used to treat (1) waste-activated sludge only, (2) mixtures of waste-activated sludge or trickling-filter sludge and primary sludge, or (3) waste sludge from extended aeration plants. Aerobic digestion has been used primarily in plants of a size less than $0.2 \text{ m}^3/\text{s}$ (5 Mgal/d), but in recent years the process has been employed in larger wastewater-treatment plants with capacities up to $2 \text{ m}^3/\text{s}$ (50 Mgal/d) (WEF, 1998). Advantages of aerobic digestion as compared to anaerobic digestion are: (1) volatile solids reduction in a well-operated aerobic digester is approximately equal to that obtained anaerobically; (2) lower BOD concentrations in supernatant liquor; (3) production of an odorless, humuslike, biologically stable end product; (4) recovery of more of the basic fertilizer values in the sludge; (5) operation is relatively easy; (6) lower capital cost; and (7) suitability for digesting nutrient-rich biosolids. In cases where separate sludge digestion is considered, aerobic digestion of biological sludge may be an attractive application. The major disadvantages of the aerobic digestion process are that (1) high power cost is associated with supplying the required oxygen; (2) digested biosolids produced have poorer mechanical dewatering characteristics; and (3) the process is affected significantly by temperature, location, tank geometry, concentration of feed solids, type of mixing/aeration device, and type of tank material. An additional disadvantage is that a useful byproduct such as methane is not recovered.

As discussed in Sec. 14–2 and Table 14–12, aerobic digestion is one of the processes defined to meet PSRP requirements. To meet Class B requirements for pathogen reduction, the regulations state the solids retention times must be at least 40 d at 20°C and 60 d at 15°C. In many instances, plants that have facilities designed for SRTs less than 40 d and wish to meet the Class B requirements for pathogen reduction have had to add additional storage capacity or thickeners. If the design engineer uses aerobic digestion for stabilization and does not meet the above SRTs, it will be necessary to monitor the performance of the process to demonstrate that the pathogen reduction criterion has been met. Monitoring is also required to demonstrate that the volatile solids reduction requirements are met for compliance to the vector attraction criterion (U.S. EPA, 1999).

Process Description

Aerobic digestion is similar to the activated-sludge process. As the supply of available substrate (food) is depleted, the microorganisms begin to consume their own protoplasm to obtain energy for cell maintenance reactions. When energy is obtained from cell tissue the microorganisms are said to be in the endogenous phase. Cell tissue is oxidized aerobically to carbon dioxide, water, and ammonia. In actuality, only about 75 to 80 percent of the cell tissue can be oxidized; the remaining 20 to 25 percent is composed of inert components and organic compounds that are not biodegradable. The ammonia is subsequently oxidized to nitrate as digestion proceeds. Nonbiodegradable volatile suspended solids will remain in final product from aerobic digestion. Considering the biomass wasted to a digester and the formula $C_5H_7NO_2$ is representative for cell mass of a microorganism, the biochemical changes in an aerobic digester can be described by the following equations:

Biomass destruction:

$$C_5H_7NO_2 + 5O_2 \rightarrow 4CO_2 + H_2O + NH_4HCO_3 \tag{14-16}$$

Nitrification of released ammonia nitrogen:

$$NH_4^+ + 2O_2 \rightarrow NO_3 + 2H^+ + H_2O \tag{14-17}$$

Overall equation with complete nitrification:

$$C_5H_7NO_2 + 7O_2 \rightarrow 5CO_2 + 3H_2O + HNO_3 \tag{14-18}$$

Using nitrate nitrogen as electron acceptor (denitrification):

$$C_5H_7NO_2 + 4NO_3^- + H_2O \rightarrow NH_4^+ + 5HCO_3^- + 2NO_2 \tag{14-19}$$

With complete nitrification/denitrification:

$$2C_5H_7NO_2 + 11.5O_2 \rightarrow 10CO_2 + 7H_2O + 2N_2 \tag{14-20}$$

As given by Eqs. (14–16) through (14–18), the conversion of organic nitrogen to nitrate results in an increase in the concentration of hydrogen ions and subsequently a decrease in pH if sufficient buffering capacity is not available in the sludge. Approximately 7 kg of alkalinity, expressed as $CaCO_3$, are destroyed per each kg of ammonia oxidized. Theoretically, approximately 50 percent of the alkalinity consumed by nitrification can be recovered by denitrification. If the dissolved oxygen is kept very low (less than 1 mg/L), however, nitrification will not occur. In practice, cycling of the aerobic digester between aeration and mixing has been found to be effective in maximizing denitrification while maintaining pH control. In situations where the buffering capacity is insufficient resulting in pH depressions below 5.5, it may be necessary to install alkalinity feed equipment to maintain the desired pH.

Where activated or trickling-filter sludge is mixed with primary sludge and the combination is to be digested aerobically, direct oxidation of the organic matter in the primary sludge and oxidation of the cell tissue will both occur. Aerobic digesters can be operated as batch or continuous flow reactors (see Fig. 14–30).

Three proven variations of the process are most commonly used: (1) conventional aerobic digestion, (2) high-purity oxygen aerobic digestion, and (3) autothermal aerobic digestion (ATAD). Aerobic digestion accomplished with air is the most commonly used process, so it is considered in greater detail in the following discussion.

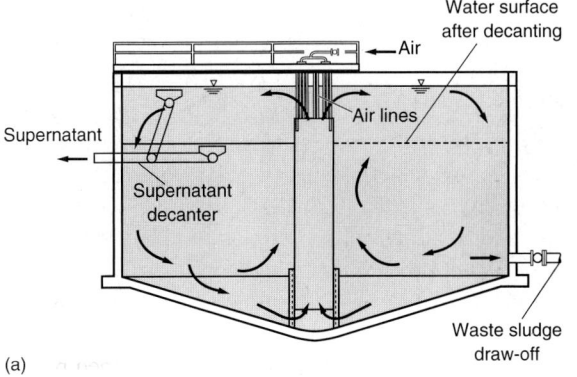

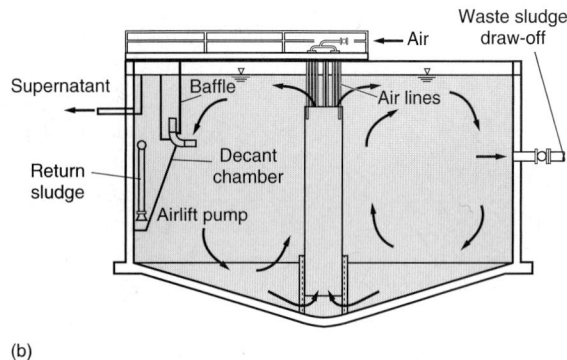

Figure 14–30

Examples of aerobic digesters: (a) batch operation type with air addition, (b) continuous operation type with air addition, and (c) view of empty aerobic digester with mechanical surface aerator.

Conventional Air Aerobic Digestion

Factors that must be considered in designing conventional aerobic digesters include temperature, solids reduction, tank volume, feed solids concentration, oxygen requirements, energy requirements for mixing, and process operation. Typical design criteria for aerobic digestion are presented in Table 14–34.

Temperature. Because the majority of aerobic digesters are open tanks, digester liquid temperatures are dependent on weather conditions and can fluctuate extensively. As with all biological systems, lower temperatures retard the process while higher temperatures accelerate it. In considering the temperature effects, heat losses should be minimized by using concrete instead of steel tanks, placing the tanks below grade instead of above grade or providing insulation for above-grade tanks, and using subsurface instead of surface aeration. In extremely cold climates, consideration should be given to heating the sludge or the air supply, covering the tanks, or both. The design should provide for the necessary degree of sludge stabilization at the lowest expected liquid operating temperature, and should provide the maximum oxygen requirements at the maximum expected liquid operating temperature.

Table 14-34
Design criteria for aerobic digesters[a]

Parameter	U.S. customary units		SI units	
	Units	Value	Units	Value
SRT[b]	d		d	
At 20°C		40		40
At 15°C		60		60
Volatile solids loading	lb/ft³·d	0.1–0.3	kg/m³·d	1.6–4.8
Oxygen requirements:				
Cell tissue[c]	lb O₂/lb VSS	~2.3	kg O₂/kg VSS	~2.3
BOD in primary sludge	Destroyed	1.6–1.9	Destroyed	1.6–1.9
Energy requirements for mixing:				
Mechanical aerators	hp/10³ ft³	0.75–1.5	kW/10³ m³	20–40
Diffused air mixing	ft³/10³ ft³·min	20–40	m³/m³·min	0.02–0.040
Dissolved oxygen residual in liquid	mg/L	1–2	mg/L	1–2
Reduction of volatile suspended solids	%	38–50	%	38–50

[a] Adapted, in part, from WEF (1995a); Federal Register (1993).
[b] To meet pathogen reduction requirements (PSRP) of 40 CFR Part 503 regulations.
[c] Ammonia produced during carbonaceous oxidation oxidized to nitrate.

Volatile Solids Reduction. A major objective of aerobic digestion is to reduce the mass of the solids for disposal. This reduction is assumed to take place only with the biodegradable content of the sludge, although there may be some destruction of the nonorganics as well. Volatile solids reductions ranging from 35 to 50 percent are achievable by aerobic digestion. Optional criteria for meeting vector attraction requirements of 40 CFR Part 503 are (1) a minimum of 38 percent reduction in volatile solids during biosolids treatment or (2) less than a specific oxygen uptake rate (SOUR) of (1.5 mg O₂/h)/g of total sludge solids at 20°C (U.S. EPA, 1999).

The change in biodegradable volatile solids in a completely mixed digester can be represented by a first-order biochemical reaction at constant-volume conditions:

$$\frac{dM}{dt} = -k_d M \tag{14-21}$$

where $\dfrac{dM}{dt}$ = rate of change of biodegradable volatile solids (M) per unit of time (Δ mass/time), MT^{-1}

k_d = reaction rate constant, T^{-1}

M = mass of biodegradable volatile solids remaining at time t in the aerobic digester, M

The time factor in Eq. (14–17) is the solids retention time (SRT) in the aerobic digester. Depending on how the aerobic digester is being operated, time t can be equal to or considerably greater than the theoretical hydraulic residence time (t). Use of the biodegradable portion of the volatile solids in the equation recognizes that approximately 20 to 35 percent of the waste-activated sludge from wastewater-treatment plants with primary treatment is not biodegradable. The percentage of nonbiodegradable volatile solids in waste-activated sludge from contact stabilization processes (no primary tanks) ranges from 25 to 35 percent (WEF, 1998).

The reaction rate term k_d is a function of the sludge type, temperature, and solids concentration. Representative values for k_d may range from 0.05 d^{-1} at 15°C to 0.14 d^{-1} at 25°C for waste-activated sludge. Because the reaction rate is influenced by several factors, it may be necessary to confirm decay coefficient values by bench-scale or pilot-scale studies.

Solids destruction is primarily a direct function of both basin liquid temperature and the SRT (sometimes referred to as sludge age), as indicated on Fig. 14–31. The data were derived from both pilot- and full-scale studies. The plot on Fig. 14–31 relates volatile solids reduction to degree-days (temperature times sludge age). Initially, as the degree-days increase, the rate of volatile solids reduction increases rapidly. As the degree-days approach 500, the curve begins to flatten. To produce well-stabilized biosolids, at least 550 degree-days are recommended for the aerobic digestion system (Enviroquip, 2000). The use of Fig. 14–31 is demonstrated in Example 14–8, the design of an aerobic digester.

Tank Volume and Detention Time Requirements. The tank volume is governed by the detention time necessary to achieve the desired volatile solids reduction. In the past, SRTs of 10 to 20 d were the norm for the design of aerobic digestion systems (Metcalf & Eddy, 1991). To meet the pathogen reduction requirements of 40 CFR Part 503 regulations, the SRT criteria (see Table 14–34) in conventional aerobic digesters take precedence over the vector attraction criteria of 38 percent solids reduction for sizing the tank volume.

The digester tank volume can be calculated by Eq. (14–22) (WEF, 1998)

$$V = \frac{Q_i(X_i + YS_i)}{X(k_dP_v + 1/\text{SRT})} \qquad (14\text{–}22)$$

Figure 14–31

Volatile solids reduction in an aerobic digester as a function of digester liquid temperature and digester sludge age. *(From WEF, 1995a.)*

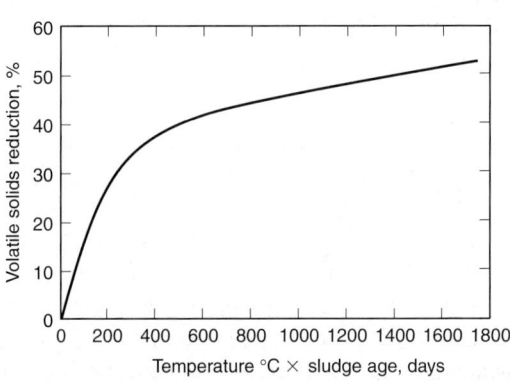

where V = volume of aerobic digester, m^3 (ft^3)

Q_i = influent average flowrate to digester, m^3/d (ft^3/d)

X_i = influent suspended solids, mg/L

Y = fraction of the influent BOD consisting of raw primary solids, (expressed as a decimal)

S_i = influent BOD, mg/L

X = digester suspended solids, mg/L

k_d = reaction rate constant, d^{-1}

P_v = volatile fraction of digester suspended solids (expressed as a decimal)

SRT = solids retention time, d

The term YS_i can be neglected if primary sludge is not included in the sludge load to the aerobic digester.

If the aerobic digestion process is operated in a complete-mix, staged configuration (two or three stages), the total SRT should be divided approximately equally among the stages. For more information on staging aerobic digestion, Enviroquip (2000) should be consulted.

Feed Solids Concentration. The concentration of the feed solids digester is important in the design and operation of the aerobic digester. If thickening precedes aerobic digestion, higher feed solids concentrations will result in higher oxygen input levels per digester volume, longer SRTs, smaller digester volume requirements, easier process control (less decanting in batch-operated systems), and subsequently increased levels of volatile solids destruction (WEF 1998). However, feed solids concentrations greater than 3.5 to 4 percent may affect the ability of the mixing and aeration system in maintaining well-mixed tank contents with adequate dissolved oxygen levels necessary to support the biological process. At feed solids concentrations greater than 4 percent, the aeration equipment must be evaluated carefully to ensure adequate mixing and aeration are achieved.

Oxygen Requirements. The oxygen requirements that must be satisfied during aerobic digestion are those of the cell tissue and, with mixed sludges, the BOD in the primary sludge. The oxygen requirement for the complete oxidation of cell tissue (including nitrification), computed using Eqs. (14–16) and (14–17), is equal to 7 mole/mole of cells, or about 2.3 kg/kg of cells. The oxygen requirement for the complete oxidation of the BOD contained in primary sludge varies from about 1.6 to 1.9 kg/kg destroyed. The oxygen residual should be maintained at 1 mg/L or above under all operating conditions.

Energy Requirements for Mixing. To ensure proper operation, the contents of the aerobic digester should be well mixed. In large tanks, multiple mixing devices should be installed to ensure good distribution of the mixing energy. Typical energy requirements for mixing are given in Table 14–34. In general, because of the large amount of air that must be supplied to meet the oxygen requirement, adequate mixing should be achieved; nevertheless, mixing power requirements should be checked, particularly when feed solids concentrations are greater than 3.5 percent. If polymers are used in the thickening process, especially for centrifuge thickening, a greater amount of unit energy may be required for mixing.

If fine-pore diffused air mixing is used, considerations for selecting the aeration system should include limitations of feed solids concentration on achieving good mixing. Recommendations on feed solids limitations should be obtained from manufacturers of aeration equipment. In addition, the potential for diffuser fouling should be evaluated, especially if the process operation requires decanting.

Process Operation. Depending on the buffering capacity of the system, the pH may drop to a low value of about 5.5 at long hydraulic detention times. The potential drop in pH is due to the increased presence of nitrate ions in solution and the lowering of the buffering capacity due to air stripping. Filamentous growths may also develop at low pH values. The pH should be checked periodically and adjusted if found to be excessively low. Dissolved oxygen levels and respiration rates should also be checked to ensure proper process performance.

Aerobic digesters that do not include prethickening should be equipped with decanting facilities for thickening the digested solids before discharge to subsequent operations. Operator control and visibility of the decanting operation are important design considerations. If the digester is operated so that the incoming sludge is used to displace supernatant and the solids are allowed to build up, the solids retention time will not be equal to the hydraulic retention time.

EXAMPLE 14–8   Aerobic Digester Design   Design an aerobic digester to treat the waste sludge produced by the activated-sludge treatment plant. Assume that the following conditions apply:

1. The amount of waste sludge to be digested is 2057 kg TSS/d.
2. The minimum and maximum liquid temperatures are 15°C for winter operation and 25°C for summer operation.
3. The system must achieve 40 percent volatile solids reduction in the winter.
4. The minimum SRT for winter conditions is 60 d.
5. Waste sludge is concentrated to 3 percent, using a gravity-belt thickener.
6. The specific gravity of the waste sludge is 1.03.
7. Sludge concentration in the digester is 70 percent of the incoming thickened sludge concentration.
8. The reaction rate coefficient k_d is 0.06 d^{-1} at 15°C.
9. Volatile fraction of digester TSS is 0.80.
10. No primary solids are included in the influent to the digester.
11. Diffused-air mixing is used.
12. Air temperature in diffused air system = 20°C.

Solution

1. Compute the volatile solids reduction for winter conditions using Fig. 14–31 and compute the percent volatile solids reduction under summer (maximum) conditions.

a. For winter conditions, the degree-days from Fig. 14–31 are 15°C × 60 d = 900 degree-days. From Fig. 14–31, the volatile solids reduction is 45 percent, which exceeds the winter requirements of 40 percent.

In order to meet the pathogen reduction requirements, the SRT must be 60 d; therefore, the required volume is 66.6 m³/d × 60 d = 3996 m³.

b. During the summer, the liquid temperature will be 25°C, and the degree-days will be 25 × 60 = 1500. From Fig. 14–31, the volatile solids reduction in the summer will be 50 percent.

2. Compute the winter and summer volatile solids reduction based on a total mass of volatile suspended solids

Total mass of VSS (VSS_M) = (0.8)(2057 kg/d) = 1646 kg/d

a. Winter: 1646 × 0.45 = 741 kg VSS_M reduced/d
b. Summer: 1646 × 0.50 = 823 kg VSS_M reduced/d

3. Determine oxygen requirements (see Table 14–34 for oxygen requirements).
a. Winter: 741 × 2.3 = 1704 kg O_2/d
b. Summer: 724 × 2.3 = 1893 kg O_2/d

4. Compute the volume of air required per d at standard conditions. For the density of air, see Appendix B-1.

a. Winter: $V = \dfrac{1704\ \text{kg}}{(1.204\ \text{kg/m}^3)(0.232)} = 6100\ \text{m}^3/\text{d}$

b. Summer: $V = \dfrac{1893\ \text{kg}}{(1.204\ \text{kg/m}^3)(0.232)} = 6777\ \text{m}^3/\text{d}$

Assuming an oxygen transfer efficiency of 10 percent, the air flowrates are

Winter: $q = \dfrac{(6100\ \text{m}^3/\text{d})}{(0.10)(1440\ \text{min/d})} = 42.4\ \text{m}^3/\text{min}$

Summer: $q = {} = \dfrac{(6777\ \text{m}^3/\text{d})}{(0.10)(1440\ \text{min/d})} = 47.1\ \text{m}^3/\text{min}$

5. Compute the volume of sludge to be disposed of per day using Eq. (14–2).

$V = \dfrac{2057\ \text{kg}}{(10^3\ \text{kg/m}^3)(1.03)(0.03)} = 66.6\ \text{m}^3/\text{d}$

6. Determine the volume of the aerobic digester (winter conditions govern) using Eq. (14–22).

$V = \dfrac{(66.6\ \text{m}^3/\text{d})(30{,}000\ \text{g/m}^3)}{(30{,}000\ \text{g/m}^3)[(0.7)(0.06/\text{d})(0.80) + (1/60\ \text{d})]}$

$= 1318\ \text{m}^3$

7. Compute the air requirement per m³ of digester volume.

$q = \dfrac{(47.1\ \text{m}^3/\text{min})}{1318\ \text{m}^3} = 0.036\ \text{m}^3/\text{min·m}^3$

8. Check the mixing requirements. Because the air requirement computed in Step 7 is within the range of values given in Table 14–34, adequate mixing should prevail.

Comment The above example is based on a single-stage aerobic digester. If a two-stage or more digester were used, a significant reduction in tank volume is possible. In a multistage arrangement, the air distribution between tanks would vary based on the expected demand as most of the volatile solids reduction will occur in the first stage where the biomass is most active.

Dual Digestion

Aerobic thermophilic digestion has also been used extensively in Europe as a first stage in the dual digestion process. The second stage is mesophilic anaerobic digestion. Dual digestion has also been tried in the United States using high-purity oxygen in the first stage. Residence times in the aerobic digester range typically from 18 to 24 h, and the reactor temperature ranges from 55 to 65°C. Typical residence time in the anaerobic digester is 10 d. The advantages of using aerobic thermophilic digestion in dual digestion are (1) increased levels of pathogen reduction, (2) improved overall volatile solids reduction, (3) increased methane gas generation in the anaerobic digester, (4) less organic material in and fewer odors produced by the stabilized sludge, and (5) equivalent volatile solids reductions can be achieved in one-third less tankage than a single-stage anaerobic digester. Prior hydrolysis in the aerobic reactor results in increased degradation during subsequent anaerobic digestion and gas production. Approximately 10 to 20 percent of the volatile solids is liquefied in the aerobic digester, while COD reduction is less than 5 percent. Provisions for foam suppression and odor control are required (Roediger and Vivona, 1998).

Autothermal Thermophilic Aerobic Digestion (ATAD)

Autothermal thermophilic aerobic digestion (ATAD), illustrated on Fig. 14–32, represents a variation of both conventional and high-purity oxygen aerobic digestion. In the ATAD process, the feed sludge is generally prethickened and the reactors are insulated to conserve the heat produced from the oxidation of volatile solids during the digestion process. Thermophilic operating temperatures (generally in the range of 55 to 70°C) can be achieved without external heat input by using the heat released by the exothermic microbial oxidation process. Approximately 20,000 kJ of heat is produced per kg of volatile solids destroyed. Because supplemental heat is not provided (other than the heat introduced by aeration and mixing), the process is termed autothermal.

Within the ATAD reactor, sufficient levels of oxygen, volatile solids, and mixing allow aerobic microorganisms to degrade organic matter to carbon dioxide, water, and nitrogen byproducts. The major advantages of ATAD are (1) retention times required to achieve a given suspended solids reduction are decreased significantly (to about 5 to 6 d) to achieve volatile solids reductions of 30 to 50 percent, similar to conventional aerobic digestion; (2) simplicity of operation; (3) greater reduction of bacteria and viruses are achieved as compared to mesophilic anaerobic digestion; and (4) when the reactor is well mixed and maintained at 55°C and above, pathogenic viruses, bacteria,

Figure 14–32

Autothermal thermophilic
aerobic digester (ATAD)
system: (a) system
schematic, and
(b) reactor schematic.

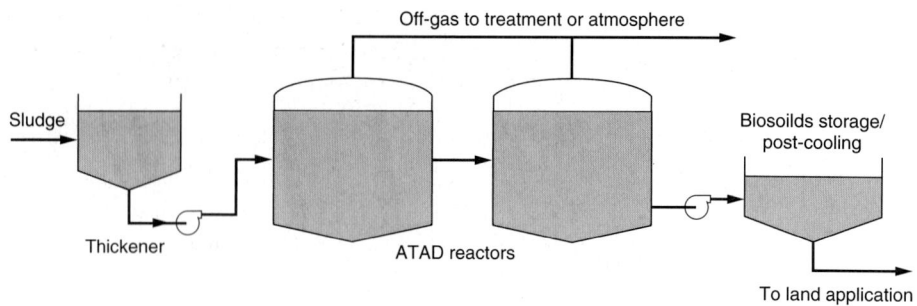

(a)

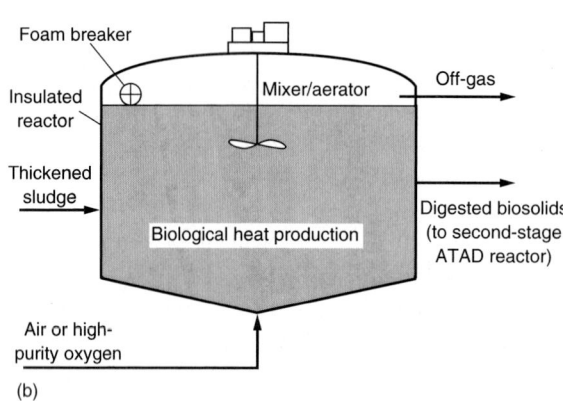

(b)

viable helminth ova, and other parasites can be reduced to below detectable levels, thus meeting the pathogen reduction requirements of Class A biosolids. Disadvantages cited are (1) objectionable odors are formed, (2) poor dewatering characteristics of ATAD biosolids, and (3) lack of nitrification. Because the ATAD system is capable of producing Class A biosolids, it is growing in popularity. As of June 2000, there are approximately 35 ATAD systems operating in North America (25 in the United States and 10 in Canada). In excess of 40 plants are operating in Europe (Stensel and Coleman, 2000).

Process Theory. The biochemical conditions in thermophilic aerobic digesters differ significantly from conventional aerobic digesters. Because of the high operating temperatures, nitrification is inhibited and aerobic destruction of volatile solids occurs as described by Eq. (14–18) without the subsequent reactions described by Eqs. (14–19) through (14–22). Additionally, most, if not all, ATAD systems may be operating under microaerobic conditions where oxygen demand exceeds oxygen supply (Stensel and Coleman, 2000). Under microaerobic conditions, proteinaceous cellular material will undergo fermentation where protein is represented as peptone as described by Eq. (14–23) (Chu and Mavinic, 1998):

$$4CH_2NH_2COOH + 4H_2O \rightarrow 3CH_3COOH + 2(NH_4)_2CO_3 \qquad (14\text{--}23)$$

Both Eqs. (14–16) and (14–23) result in the production of ammonia that reacts with water and carbon dioxide to form ammonium bicarbonate and ammonium carbonate to increase alkalinity. Because nitrification does not occur, the pH in the ATAD system will typically range from 8 to 9, higher than in conventional aerobic digesters. Ammonia-nitrogen produced will be present in the off-gas and in solution with concentrations of several hundred mg/L in each. Most of the ammonia nitrogen will be returned to the liquid process in sidestreams from the odor-control and dewatering facilities. The acetic acid (or acetate) produced by the fermentation of proteins is one of the volatile fatty acids. Acetic acid will be oxidized subsequently in the presence of sufficient dissolved oxygen as described by Eq. (14–24):

$$CH_3COOH + 2O_2 \rightarrow 2CO_2 + 2H_2O \tag{14–24}$$

Anaerobic conditions will occur at times in ATAD systems and will most likely take place in the pre-ATAD sludge holding facilities and in the first-stage ATAD reactors during and immediately after sludge transfers and batch feeding. Under anaerobic conditions, reduced sulfur compounds can be formed that can affect the design and performance of odor-control systems.

Process Design. ATAD systems are designed to have short hydraulic retention times within insulated reactors (see Fig. 14–32b). As long as the ATAD system is well mixed and sufficient oxygen is provided, the temperature in the reactor will rise until a balance occurs; i.e., the heat lost equals the heat input from the exothermic reaction and mechanical energy input. The temperature will continue to rise until the process becomes oxygen mass-transfer-limited.

Factors that must be considered in designing an ATAD system include prethickening, number and type of reactors, postcooling/thickening, feed characteristics, detention time, feed cycle, aeration and mixing, temperature and pH, and foam and odor control. Nearly all of the ATAD systems currently installed in the United States utilize two or more reactors operated in series (see Fig. 14–32a). Design considerations for ATAD systems are presented in Table 14–35; typical design criteria are summarized in Table 14–36.

ATAD systems must be designed to (1) transfer sufficient oxygen to meet the high demand of the reactors and (2) supply the required oxygen while minimizing the latent heat loss in the exhaust air. It is difficult to define the oxygen transfer rate in an ATAD system while using typical design procedures used for selecting and sizing aeration equipment for wastewater treatment processes. The oxygen transfer coefficient α (alpha) and the oxygen saturation coefficient β (beta) have not been quantified under the environmental conditions present in an ATAD reactor (Stensel and Coleman, 2000). Factors affecting oxygen transfer are the high temperatures (that would reduce α values) and the foam layer and low dissolved oxygen levels (that might increase oxygen transfer). Nearly all ATAD systems utilize a type of aspirating aerator to introduce oxygen into the reactors. The types include hollow-shaft propeller or turbine aerators, pumped venturi aspirators, and jet aspirators. With all air aspirating systems, the equipment provides both mixing and oxygen transfer. Typical energy requirements for mixing and aeration are given in Table 14–36.

Substantial amounts of foam are generated in the ATAD process as cellular proteins, lipids, and oil and grease materials are broken down and released into solution. The foam layer contains high concentrations of biologically active solids that provide

Table 14–35
Typical design
considerations for
an autothermal
aerobic digester
(ATAD) system[a]

System component	Design consideration
Prethickening system	Thickening or blending facilities may be required to maintain an influent COD to the ATAD reactor greater than 40 g/L
Reactors	Number of reactors; a minimum of two enclosed insulated reactors in series should be provided and equipped with mixing, aeration, and foam control equipment
Screening	Fine screening, 6- to 12-mm (0.25- to 0.5-in.) clear openings, of raw wastewater or solids feed stream should be provided for the removal of inert materials, plastics, and rags
Feed cycle	Continuous or batch processing is acceptable, except batch processing provides greater assurance in meeting Class A pathogen reduction requirements
Foam control	Foam suppression is required to ensure effective oxygen transfer and enhanced biological activity. Freeboard of 0.5 to 1.0 m (1.65 to 3.3 ft) is recommended
Post-ATAD storage/dewatering	Postprocess cooling is necessary to achieve solids consolidation and to enhance dewaterability. A minimum of 20 d detention may be necessary unless heat exchangers are used for cooling the processed biosolids
Odor control	Because of high temperatures in the ATAD system, relatively high concentrations of ammonia are released. Reduced sulfur compounds also result, which can include hydrogen sulfide, carbonyl sulfide, methyl mercaptan, ethyl mercaptan, dimethyl sulfide, and dimethyl disulfide. Odor-control systems may include wet scrubbers, biofilters, or a combination of both (see Chap. 15)
Sidestreams	Liquid sidestreams from odor-control and dewatering systems, when returned to the liquid processing system, may contain constituents that could affect process performance unless accounted for or treated separately

[a] Adapted in part from WEF (1998) and Stensel and Coleman (2000).

insulation of the reactor and improved oxygen utilization. It is important, therefore, that the foam layer be effectively managed and controlled. Mechanical foam cutters are used most commonly for foam control, but other methods such as spray systems have been employed. A freeboard of 0.5 to 1.0 m is generally recommended for controlling the foam layer (Stensel and Coleman, 2000).

Where ATAD systems are followed by mechanical dewatering, post-ATAD storage is recommended to allow for cooling of the biosolids to improve belt-filter press and centrifuge performance. Post-ATAD storage coupled with long detention times in the final-stage ATAD reactors may further increase the reduction of volatile solids.

Process Control. The provisions of the 40 CFR Part 503 regulations applicable for meeting the Class A biosolids requirements with the ATAD process are complex because several alternative pathogen-reduction requirements are given. The basic

Table 14–36
Typical design criteria for autothermal aerobic digestion (ATAD)[a]

Parameter	U.S. customary units			SI units		
	Units	Range	Typical	Units	Range	Typical
Reactor:						
HRT	d	4–30	6–8	d	4–30	6–8
Volumetric loading						
TSS, 40 to 60 g/L	lb/10³ ft³·d	320–520		kg/m³·d	5–8.3	
VSS, 25 g/L min	lb/10³ ft³·d	200–260		kg/m³·d	3.2–4.2	
Temperature	°C			°C		
Stage 1		35–50	40		35–50	40
Stage 2		50–70	55		50–70	55
Aeration and mixing:						
Mixer type			Aspirating			Aspirating
Oxygen transfer efficiency	lb O$_2$/kWh		4.4	kg O$_2$/kWh		2
Energy requirement	hp/10³ ft³	5–6.4		W/m³	130–170	

[a] Adapted, in part, from Stensel and Coleman (2000).

requirement that needs to be demonstrated is (1) fecal coliform densities are less than 1000 MPN/g of total solids (dry weight basis), or (2) *Salmonella* sp. bacteria concentrations are below detection limits of 3 MPN/4 g of total solids (dry weight basis). For compliance with these pathogen regulations for Class A biosolids, the withdrawal and feeding of the sludge to the reactors is performed on a batch basis. (In flow-through systems, it is possible that some pathogens might pass through.) Two or more reactors in a series configuration are used typically to ensure that all particles in the reactor are subjected to the time and temperature requirements and that no insufficiently treated biosolids are released to the environment. The ATAD pumping system is designed to withdraw and feed the daily amount of sludge in 1 h or less. The reactor is then isolated for the remaining 23 h each day at a minimum temperature of 55°C.

High-Purity Oxygen Digestion

High-purity oxygen aerobic digestion is a modification of the aerobic digestion process in which high-purity oxygen is used in lieu of air. The resultant biosolids are similar to biosolids from conventional aerobic digestion. Influent sludge concentrations vary from 2 to 4 percent. Recycle flows are similar to those achieved by conventional aerobic digestion. High-purity oxygen aerobic digestion is particularly applicable in cold weather climates because of its relative insensitivity to changes in ambient air temperatures due to the increased rate of biological activity and the exothermal nature of the process.

While one variation of the high-purity aerobic digestion process uses open tanks, aerobic digestion is usually done in closed tanks similar to those used in the high-purity oxygen activated-sludge process. Using closed tanks for high-purity oxygen aerobic digestion will generally result in higher operating temperatures because of the exothermic nature of the digestion process. Maintenance of these higher temperatures in the digester results in a significant increase in the rate of volatile suspended solids destruction. Where covered tanks are used, a high-purity oxygen atmosphere is maintained above the liquid surface, and oxygen is transferred into the sludge via mechanical aerators. Where an open aeration tank is used, oxygen is introduced to the liquid sludge by a special diffuser that produces minute oxygen bubbles. The bubbles dissolve before reaching the air-liquid interface.

The major disadvantage of high-purity oxygen aerobic digestion is the increased cost associated with oxygen generation. As a result, high-purity oxygen aerobic digestion is cost-effective generally only when used in conjunction with the high-purity oxygen activated-sludge system. Also, neutralization may be required to offset the reduced buffering capacity of the system.

14–11 COMPOSTING

Composting is a cost-effective and environmentally sound alternate for the stabilization of wastewater biosolids. Increasingly stringent air pollution regulations and biosolids disposal requirements coupled with the anticipated shortage of available landfills have accelerated the development of composting as a viable sludge-management option.

Composting is a process in which organic material undergoes biological degradation to a stable end product. Sludge that has been composted properly is a nuisance-free, humuslike material. Approximately 20 to 30 percent of the volatile solids are converted to carbon dioxide and water. As the organic material in the sludge decomposes, the compost heats to temperatures in the pasteurization range of 50 to 70°C (120 to 160°F), and enteric pathogenic organisms are destroyed. Properly composted biosolids may be used as soil conditioners in agricultural or horticultural applications, subject to any limitations based on the constituents in the composed biosolids (WEF, 1995b).

Although composting may be accomplished under anaerobic or aerobic conditions, essentially all municipal wastewater biosolids composting applications are under mostly aerobic conditions (composting is never completely aerobic). Aerobic composting accelerates material decomposition and results in the higher rise in temperature necessary for pathogen destruction. Aerobic composting also minimizes the potential for nuisance odors.

The anticipated daily production of biosolids from a wastewater-treatment facility will have a pronounced effect on the alternate composting systems available for use, as will the availability of land for the construction of the composting facility. Other factors affecting the type of composting system are the nature of the biosolids produced; stabilization, if any, of the biosolids prior to composting; and the type of dewatering equipment and chemicals used. Biosolids that are stabilized by aerobic or anaerobic digestion prior to composting may result in reducing the size of the composting facilities by up to 40 percent.

Process Microbiology

The composting process involves the complex destruction of organic material coupled with the production of humic acid to produce a stabilized end product. The microorganisms involved fall into three major categories: bacteria, actinomycetes, and fungi. Although the interrelationship of these microbial populations is not fully understood, bacterial activity appears to be responsible for the decomposition of proteins, lipids, and fats at thermophilic temperatures, as well as for much of the heat energy produced. Fungi and actinomycetes are also present at varying levels during the mesophilic and thermophilic stages of composting and appear to be responsible for the destruction of complex organics and the cellulose supplied in the form of amendments or bulking agents.

During the composting process, three separate stages of activity and associated temperatures are observed: mesophilic, thermophilic, and cooling (see Fig. 14–33). In the initial mesophilic stage, the temperature in the compost pile increases from ambient to approximately 40°C (104°F) with the appearance of fungi and acid-producing bacteria. As the temperature in the composting mass increases to the thermophilic range of 40 to 70°C (104 to 160°F), these microorganisms are replaced by thermophilic bacteria, actinomycetes, and thermophilic fungi. It is in the thermophilic temperature range that the maximum degradation and stabilization of organic material occur. The cooling stage is characterized by a reduction in microbial activity, and replacement of the thermophilic organisms with mesophilic bacteria and fungi. During the cooling period, further evaporative release of water from the composted material will occur, as well as stabilization of pH and completion of humic acid formation.

Process Description

Most composting operations consist of the following basic steps (see Fig. 14–34): (1) preprocessing, the mixing of dewatered sludge with an amendment and/or a bulking agent; (2) high-rate decomposition, aerating the compost pile either by the addition of air, by mechanical turning, or by both; (3) recovery of the bulking agent (at the end of either the high-rate decomposition or curing phase, if practicable); (4) further curing and storage, which allows further stabilization and cooling of the compost; (5) postprocessing, screening for the removal of nonbiogradable material such as metals and plastics or grinding for size reduction, and (6) final disposition. A portion of the final product is sometimes recycled to the preprocessing step to aid in conditioning the compost mixture.

Figure 14–33

Phases during composting as related to carbon dioxide respiration and temperature.
(Epstein, 1997.)

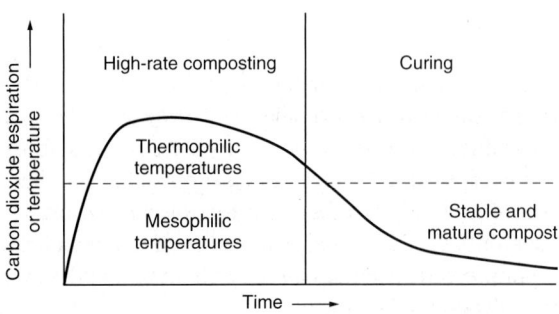

Figure 14–34

Generalized process diagram for composting showing inputs of sludge (feed substrate), amendments, and bulking agents. *(Haug, 1993.)*

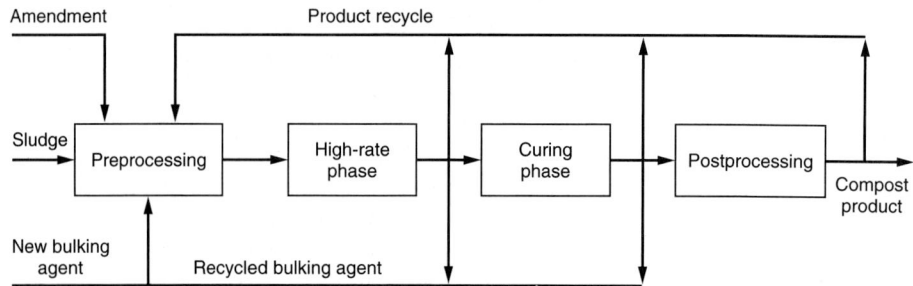

The high-rate decomposition stage of composting has been more engineered and controlled due to the need to reduce odors, supply high aeration rates, and maintain process control. The curing stage is often less engineered, less controlled, and given only small consideration in some designs. The curing stage is an integral part of the system design and operation, and both stages need to be designed and operated properly to produce a mature compost product (Haug, 1993).

The two principal methods of composting now in use in the United States may be classified as *agitated* or *static*. In the agitated method the material to be composted is agitated periodically to introduce oxygen, to control the temperature, and to mix the material to obtain a uniform product. In the static method, the material to be composted remains static and air is blown through the composting material. The most common agitated and static methods of composting are known as the windrow and static pile methods, respectively. Proprietary composting systems in which the composting operation is carried out in a reactor of some type are known as in-vessel composting systems (Tchobanoglous et al., 1993).

An amendment is an organic material added to the feed substrate primarily to reduce the bulk weight, reduce moisture content, and increase the air voids for proper aeration. Amendments can also be used to increase the quantity of degradable organics in the mixture. Commonly used amendments are sawdust, straw, recycled compost, and rice hulls. A bulking agent is an organic or inorganic material that is used to provide structural support and to increase the porosity of the mixture for effective aeration. Wood chips are the most commonly used bulking agents and can be recovered and reused.

Aerated Static Pile. The aerated static pile system consists of a grid of aeration or exhaust piping over which a mixture of dewatered sludge and bulking agent is placed (see Fig. 14–35a). In a typical static pile system, the bulking agent consists of wood chips, which are mixed with the dewatered sludge by a pug-mill type or rotating-drum mixer or by movable equipment such as a front-end loader. Material is composted for 21 to 28 days and is typically followed by a curing period of 30 days or longer. Typical pile heights are generally about 2 to 2.5 m (6 to 8 ft). A layer of screened compost is often placed on top of the pile for insulation. Disposable corrugated plastic drainage pipe is commonly used for air supply and each individual pile is recommended to have an individual blower for more effective aeration control. Screening of the cured compost usually is done to reduce the quantity of the end product requiring ultimate dis-

Figure 14–35

Composting systems:
(a) aerated static pile,
(b) view of compost
windrows, and
(c) equipment for turning
and reforming compost
windrows.

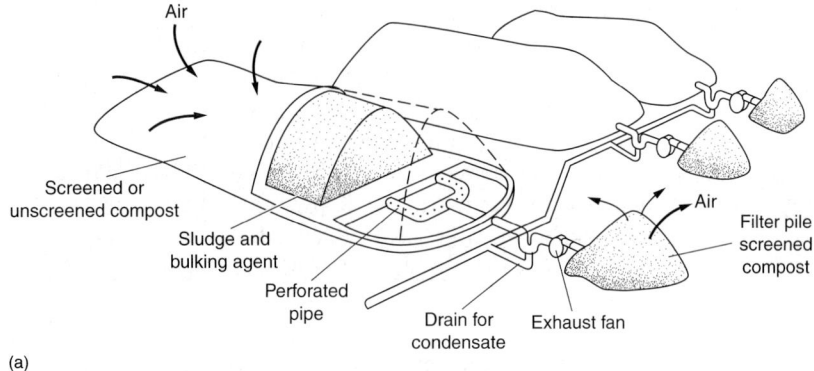

Air

Screened or
unscreened compost

Sludge and
bulking agent

Perforated
pipe

Drain for
condensate

Exhaust fan

Air

Filter pile
screened
compost

(a)

(b)

(c)

posal and to recover the bulking agent. For improved process and odor control, many facilities cover or enclose all or significant portions of the system.

Windrow. In a windrow system, the mixing and screening operations are similar to those for the aerated static pile operation. Windrows are constructed from 1 to 2 m (3 to 6 ft) high and 2 to 4.5 m (6 to 14 ft) at the base (see Fig. 14–35b). The rows are turned and mixed periodically during the composting period. Supplemental mechanical aeration is used in some applications. The composting period is about 21 to 28 d. Under typical operating conditions, the windrows are turned a minimum of five times while the temperature is maintained at or above 55°C. In windrow composting, aerobic conditions are difficult to maintain throughout the cross-sectional area of the windrow. Thus, the microbial activity within the pile may be aerobic, facultative, anaerobic, or various combinations thereof, depending on when and how often the pile is turned. Turning of the windrows is often accompanied by the release of offensive odors. The release of odors occurs typically when anaerobic conditions develop within the windrow. Specialized equipment is available to mix the sludge and bulking agent and to turn the composting windrows (see Fig. 14–35c). Some windrow operations are covered or enclosed, similar to aerated static piles.

In-Vessel Composting Systems. In-vessel composting is accomplished inside an enclosed container or vessel. Mechanical systems are designed to minimize odors and process time by controlling environmental conditions such as air flow, temperature, and oxygen concentration. The advantages of in-vessel composting systems are better process and odor control, faster throughput, lower labor costs, and smaller area requirements.

In-vessel composting systems can be divided into two major categories: plug flow and dynamic (agitated bed). In plug-flow systems, the relationship between particles in the composting mass stays the same throughout the process, and the system operates on the basis of a first-in, first-out principle. In a dynamic system, the composting material is mechanically mixed during the processing. In-vessel systems can be further categorized based on the geometric shape of the vessels or containers used. Examples of plug-flow reactors are shown on Fig. 14–36 and examples of dynamic-type systems are illustrated on Fig. 14–37.

Design Considerations

Many factors must be considered in the design of a composting system (see Table 14–37). Each of the factors has to be considered carefully to meet the special requirements of each system. A design approach using a materials balance is particularly important because the amount of each component (sludge or biosolids, bulking agent, and amendment) used during each phase of the process is determined. In a materials balance, the following parameters must be measured or calculated for each component: (1) total volume, (2) total wet weight, (3) total solids content (dry weight), (4) volatile solids content (dry weight), (5) water content (weight), (6) bulk density (wet weight/unit volume), (7) percent water content, and (8) percent volatile solids of the compost mix. An important output of the materials balance is to determine the composition of the compost mix. The compost mix should be about 40 percent dry solids to ensure ade-

Figure 14–36

Plug-flow in-vessel composting reactors: (a) unmixed horizontal plug-flow reactor and (b) unmixed tunnel (horizontal plug-flow) reactor.

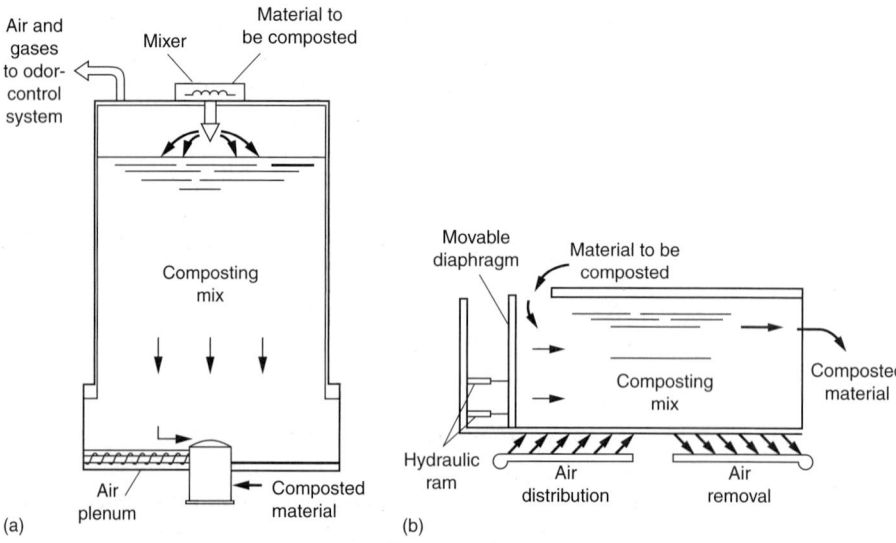

Figure 14-37

Dynamic (mixed) in-vessel
composting units:
(a) vertical reactor and
(b) horizontal reactor.

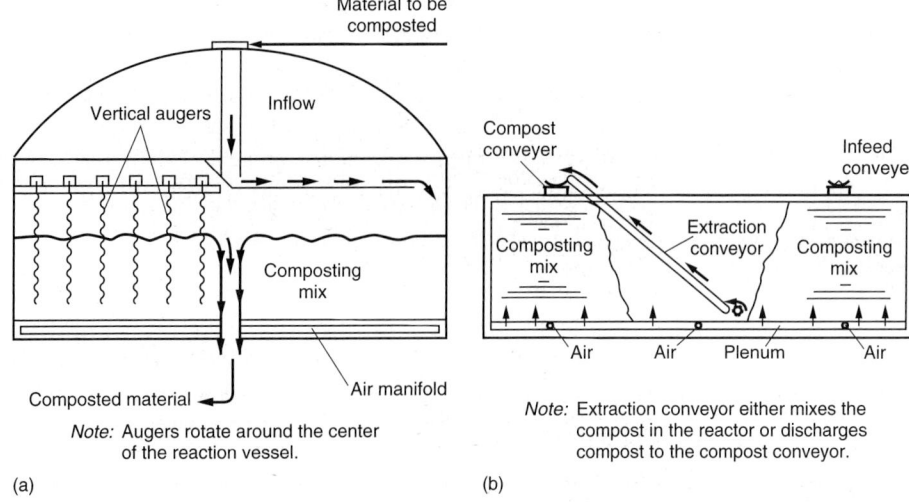

Note: Augers rotate around the center
of the reaction vessel.

(a)

Note: Extraction conveyor either mixes the
compost in the reactor or discharges
compost to the compost conveyor.

(b)

quate composting in windrow and static-pile composting. In-vessel systems require similar solids requirements, but slightly lower values may be used, depending on the aeration system. Starting the compost process with solids higher than 40 percent soon results in a "dry mix." A dry mix is dusty, and sufficient biological activity and temperature levels are difficult to maintain. The addition of water will likely be required for the duration of the process; thus, provisions for a water supply should be made.

The effect of moisture content in the dewatered sludge on the compost mix is illustrated on Fig. 14–38. The moisture content of the sludge affects the wet weight of the mixture and the amount of amendment that has to be used. Using Fig. 14–38a, for example, if the sludge cake contains 24 percent solids, the wet weight of the mix is about 6.7 Mg (tonne) per dry Mg of sludge. If the sludge solids content decreases to 16 percent, the wet weight increases to about 11 Mg per dry Mg of sludge. The additional moisture content would require larger materials-handling systems and larger reactors. The amendment requirements, as indicated on Fig. 14–38b, would triple over the same range of sludge solids. In compost-system design, the types of sludge-dewatering system and the consistency of resulting product have to be evaluated carefully.

Postprocessing is often used to prepare the finished compost for marketing. Preparation includes conveying the finished compost from the active composting area to the curing, screening, and preparation areas. Trommel screens and belt shredders are used frequently; shredding can precede or follow curing. In some cases, double screening is preferable, especially for the horticultural market to meet product quality requirements. Particle size of the finished product for general use ranges typically from 6 to 25 mm (1/4 to 1 in).

Cocomposting with Municipal Solid Wastes

Cocomposting of sludge and municipal solid wastes is a possible alternative where integrated waste-disposal facilities are considered. Mixing the sludge with the organic fraction of municipal solid waste or source separated yard wastes is beneficial because

Table 14–37
Design considerations for aerobic sludge composting processes[a]

Item	Comment
Type of sludge	Both untreated sludge and digested biosolids can be composted successfully. Untreated sludge has a greater potential for odors, particularly for windrow systems. Untreated sludge has more energy available, will degrade more readily, and has higher oxygen demand
Amendments and bulking agents	Amendment and bulking agent characteristics (i.e., moisture content, particle size, and available carbon) affect the process and quality of product. Bulking agents should be readily available. For characteristics of typical bulking agents, see Table 14–38
Carbon-nitrogen ratio	The initial C/N ratio should be in the range of 20:1 to 35:1 by weight. At lower ratios, ammonia is given off. Carbon should be checked to ensure it is readily biodegradable
Volatile solids	The volatile solids of the composting mix should be greater than 30 percent of the total solids content. Dewatered sludge will usually require an amendment or bulking agent to adjust the solids content
Air requirements	Air with at least 50 percent of the oxygen remaining should reach all parts of the composting material for optimum results, especially in mechanical systems
Moisture content	Moisture content of the composting mixture should be not greater than 60 percent for static pile and windrow composting and not greater than 65 percent for in-vessel composting
pH control	pH of the composting mixture should generally be in the range of 6 to 9. To achieve optimum aerobic decomposition, pH should remain in the 7 to 7.5 range
Temperature	For best results, temperature should be maintained between 50 and 55°C for the first few days and between 55 and 60°C for the remainder of the active composting period. If the temperature is allowed to increase beyond 65°C for a significant period of time, biological activity will be reduced
Control of pathogens	If properly conducted, it is possible to kill all pathogens, weeds, and seeds during the composting process. To achieve this level of control, the temperature must be maintained between 60 and 70°C for 24 h. For temperatures and times of exposure required for the destruction of common pathogens, see Table 14–39
Mixing and turning	To prevent drying, caking, and air channeling, material in the process of being composted should be mixed or turned on a regular schedule or as required. Frequency of mixing or turning will depend on the type of composting operation
Heavy metals and trace organics	Heavy metals and trace organics in the sludge and finished compost should be monitored to ensure that the concentrations do not exceed the applicable regulations for end use of the product
Site constraints	Factors to be considered in selecting a site include available area, access, proximity to treatment plant and other land uses, climatic conditions, and availability of buffer zone

[a] Adapted in part from U.S. EPA (1985); Tchobanoglous et al. (1993).

(1) sludge dewatering may not be required, and (2) the overall metals content of the composted material will be less than that of the composted sludge alone. Liquid treatment plant sludges typically have a solids content ranging from 3 to 8 percent. A 2 to 1 mixture of compostable municipal solid or yard wastes to sludge is recommended as a minimum. Both static and agitated compost systems have been tried (Tchobanoglous et al., 1993).

Table 14–38
Characteristics of compost bulking agents used in the aerobic composting of sludge from wastewater treatment[a]

Bulking agent	Comments
Wood chips	May have to be purchased
	High recovery rate by screening
	Provides supplemental carbon source
Chipped brush	Possibly available as a waste material
	Low recovery rate by screening
	Provides supplemental carbon source
	Longer curing time of compost
Leaves and yard waste	Must be shredded
	Wide range of moisture content
	Readily available source of carbon
	Relatively low porosity
	Nonrecoverable
Shredded tires	Often mixed with other bulking agents
	Supplemental carbon is not available
	Nearly 100 percent recoverable
	May contain metals
Ground waste lumber	Possibly available as waste material
	Often a poor source of supplemental carbon

[a] Adapted, in part, from WEF (1998).

Table 14–39
Temperature and time of exposure required for destruction of some common pathogens and parasites[a]

Organism	Observations
Salmonella typhosa	No growth beyond 46°C; death within 30 min at 55–60°C and within 20 min at 60°C; destroyed in a short time in compost environment
Salmonella sp.	Death within 1 h at 55°C and within 15–20 min at 60°C
Shigella sp.	Death within 1 h at 55°C
Escherichia coli	Most die within 1 h at 55°C and within 15–20 min at 60°C
Entamoeba histolytica cysts	Death within a few minutes at 45°C and within a few seconds at 55°C
Taenia saginata	Death within a few minutes at 55°C
Trichinella spiralis larvae	Quickly killed at 55°C; instantly killed at 60°C
Brucella abortus or Br. suis	Death within 3 min at 62–63°C and within 1 h at 55°C
Micrococcus pyogenes var. aureus	Death within 10 min at 50°C
Streptococcus pyogenes	Death within 10 min at 54°C
Mycobacterium tuberculosis var. hominis	Death within 15–20 min at 66°C or after momentary heating at 67°C
Corynebacterium diphtheria	Death within 45 min at 55°C
Necator americanus	Death within 50 min at 45°C
Ascaris lumbricoides eggs	Death in less than 1 h at temperatures over 50°C

[a] From Tchobanoglous et al. (1993).
Note: $1.8 \times (°C) + 32 = °F$.

Figure 14-38

Effect of sludge solids content on compost mix and amendment quantities: (a) effect of sludge solids on wet weight of compost mix, (b) amendment requirements vs sludge solids content. (U.S. EPA, 1989.)

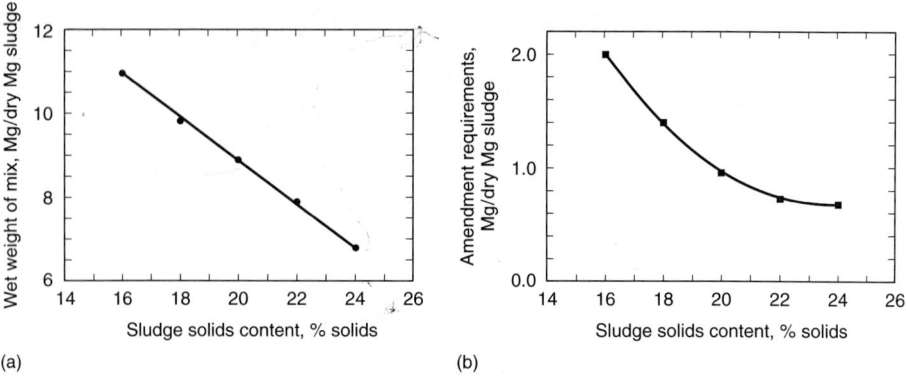

(a)

(b)

Public Health and Environmental Issues

The principal public health and environmental issues concerning compost operations relate to exposure to pathogens and bioaerosols. Exposure to pathogens can occur during the composting process or through the use of the product if the composting process is not executed properly and the resulting product is not disinfected. The potential modes of infection for workers are (1) inhalation of aerosols containing airborne microorganisms, (2) dermal contact, or (3) oral through inadvertent contact of dust or contaminated food or through hand-to-mouth contact such as cigarette smoking. Ingestion of contaminated product or contamination by the product to cigarettes or food is the greatest potential source of pathogen invasion by workers or users (Epstein, 1997).

Compost bioaerosols are organisms or biological agents that can be dispersed through the air and affect human health. Bioaerosols can contain living organisms including bacteria, fungi, actinomycetes, arthropods, and protozoa and microbial products such as endotoxin and microbial enzymes. During composting, bioaerosols are not only present in waste materials but also can be generated during the process. The level and type of bioaerosols are a function of feedstock. The two bioaerosols of greatest interest to worker health and the environment surrounding composting facilities are *Aspergillus fumigatus* and endotoxin. *A. fumigatus,* a common fungus, is of concern to both worker health and populations surrounding composting facilities, as it can cause lung disease. Endotoxin, part of the cell wall of gram-negative bacteria that is released to the environment during the composting process, is of primary concern to workers in composting, recycling, and other solid-waste processing facilities. There is little evidence that exposure to airborne endotoxin causes toxic conditions. Most of the data, however, concerning worker illness is associated with composting municipal solid waste, principally in Europe. Proper ventilation, dust control, and use of dust masks reduce worker exposure to bioaerosols (Epstein, 1997).

14-12   CONDITIONING

Sludge and biosolids are chemically conditioned expressly to improve their dewatering characteristics. Other conditioning methods, heat treatment and freeze-thaw, have also been used to a limited extent or experimentally and are discussed briefly in this section.

Chemical Conditioning

The use of chemicals to condition sludge and biosolids for dewatering is economical because of the increased yields and greater flexibility obtained. Chemical conditioning can reduce the 90 to 99 percent incoming moisture content to 65 to 85 percent, depending on the nature of the solids to be treated. Chemical conditioning results in coagulation of the solids and release of the absorbed water. Conditioning is used in advance of mechanical dewatering systems such as centrifugation, belt-filter presses, and pressure-filter presses. Chemicals used include ferric chloride, lime, alum, and organic polymers.

Adding conditioning chemicals to sludges and biosolids may increase the dry solids. Polymers do not increase the dry solids while iron salts and lime can increase the dry solids by 20 to 30 percent.

Chemicals are most easily applied and metered in the liquid form. Dissolving tanks are needed if the chemicals are received as dry powder. In most plants, these tanks should be large enough for at least one day's supply of chemicals and should be furnished in duplicate. In large plants, tankage sufficient for one shift is usually adequate. The tanks must be fabricated or lined with corrosion-resistant material. Polyvinyl chloride, polyethylene, and rubber are suitable materials for tank and pipe linings for acid solutions. Metering pumps must be corrosion-resistant. These pumps are generally of the positive-displacement type with variable-speed or variable-stroke drives to control the flowrate.

Factors Affecting Chemical Conditioning.
Factors that affect the selection of the type and dosage of the conditioning agents are the properties of the solids and the type of mixing and dewatering devices to be used. Important solids properties include source, solids concentration, age, pH, and alkalinity. Sources such as primary sludge, waste-activated sludge, and digested biosolids are good indicators of the range of probable conditioner doses required. Solids concentrations will affect the dosage and the dispersal of the conditioning agent. The pH and alkalinity may affect the performance of the conditioning agents, in particular the inorganic conditioners. When lime is used to maintain a high pH for dewatering, strong ammonia odor and lime scaling problems may occur. The method of dewatering may also affect the selection of the conditioning chemical because of the differences in mixing equipment used by various vendors and the characteristics of particular methods of dewatering. For example, polymers are used commonly in centrifuge and belt-press dewatering but are used less frequently for pressure filtration. Laboratory- or pilot-scale testing is recommended to determine the types of chemical conditioning agents required, particularly for solids and biosolids that may be difficult to dewater.

Dosage.
The chemical dosage required for any sludge is determined in the laboratory. Tests used for selecting chemical dosage include the Buchner funnel test for the determination of specific resistance of sludge, capillary suction time test (CST), and the standard jar test. The Buchner funnel test is a method of testing sludge drainability or dewatering characteristics using various conditioning agents. The capillary suction test relies on gravity and the capillary suction of a piece of thick filter paper to draw out water from a small sample of conditioned sludge. The standard jar test, the easiest method to use, consists of testing standard volumes of sludge samples (usually 1 L)

with different conditioner concentrations, followed by rapid mixing, flocculation, and settling using standard jar test apparatus. Detailed descriptions of testing procedures are provided in WEF (1988).

In general, it has been observed that the type of sludge has the greatest impact on the quantity of chemical required. Difficult-to-dewater sludges that require larger doses of chemicals, generally do not yield as dry a cake, and have poorer quality of filtrate or centrate. Sludge types, listed in the approximate order of increasing conditioning chemical requirements, are as follows:

1. Untreated (raw) primary sludge
2. Untreated mixed primary and trickling-filter sludge
3. Untreated mixed primary and waste-activated sludge
4. Anaerobically digested primary sludge
5. Anaerobically digested mixed primary and waste-activated sludge
6. Untreated waste-activated sludge
7. Aerobically digested sludge

Typical levels of polymer addition for various types of sludge using belt-filter press and centrifuge dewatering are shown in Table 14–40. Actual dosages in any given case may vary considerably from the indicated values. Polymer dosages will also vary greatly depending on the molecular weight, ionic strength, and activity levels of the polymers used. Manufacturers should be consulted for applicability and dosage information. Dosages of ferric chloride and lime also vary widely.

Mixing. Intimate admixing of sludge and coagulant is essential for proper conditioning. The mixing must not break the floc after it has formed, and the detention should

Table 14–40

Typical levels of polymer addition for belt-filter press and solid-bowl centrifuge sludge dewatering[a]

Type of sludge	U.S. customary units, lb/ton dry solids		SI units, kg/Mg dry solids	
	Belt-filter press	Solid-bowl centrifuge	Belt-filter press	Solid-bowl centrifuge
Primary	2–8	1–5	1–4	1–2.5
Primary and waste-activated	4–16	4–10	2–8	2–5
Primary and trickling-filter	4–16	—	2–8	—
Waste-activated	8–20	10–16	4–10	5–8
Anaerobically digested primary	4–10	6–10	2–5	3–5
Anaerobically digested primary and air waste-activated	3–17	4–10	1.5–8.5	2–5
Aerobically digested primary and air waste-activated	4–16	—	2–8	—

[a] Adapted from WEF (1983); WEF (1988).

be kept to a minimum so that sludge reaches the dewatering unit as soon after conditioning as possible. Mixing requirements vary depending on the dewatering method used. A separate mixing and flocculation tank is usually provided ahead of pressure filters; a separate flocculation tank may be provided for a belt-filter press or the conditioner may be added directly to the sludge feed line of the belt-press unit; and in-line mixers are usually used with a centrifuge. It is generally desirable in design to provide at least two locations for the addition of conditioning chemicals.

Other Conditioning Methods

Other conditioning methods that have been used or evaluated include heat-treatment, sludge preheating, and freeze-thaw conditioning.

Heat Treatment. Heat treatment is a process that has been used for the conditioning and stabilization of sludge, but it is seldom used in new installations. Heat treatment involves heating the sludge for short periods of time under pressure. The treatment coagulates the solids, breaks down the gel structure, and reduces the water affinity of sludge solids. As a result, the sludge is sterilized and is dewatered readily. Supernatant from the heat-treatment unit is high in BOD and may require special sidestream treatment before it is introduced into the mainstream wastewater treatment process.

Advantages cited for heat treatment are (1) the solids content of the dewatered sludge can range from 30 to 50 percent, depending on the degree of oxidation achieved, (2) the processed sludge does not normally require chemical conditioning, (3) the process stabilizes sludge and will destroy most pathogenic organisms, (4) the processed sludge will have a heating value of 28 to 30 kJ/g (12,000 to 13,000 Btu/lb) of volatile solids, and (5) the process is relatively insensitive to changes in sludge composition.

The major disadvantages associated with heat treatment are (1) high capital cost due to its mechanical complexity and the use of corrosion-resistant materials; (2) close supervision, skilled operators, and a strong preventive maintenance program are required; (3) the process produces sidestreams with high concentrations of organics, ammonia nitrogen, and color; (4) significant odorous gases are produced that require extensive containment, treatment, and/or destruction; and (5) scale formation in the heat exchangers, pipes, and reactor requires acid washing or high-pressure water jets. Because of these disadvantages, interest in heat treatment has declined significantly and few new facilities have been constructed.

Sludge Preheating. In pilot testing of centrifuges used in tandem with heat dryers it has been shown that as much as six percentage points incremental increase in cake solids concentration can be achieved by conditioning the sludge with heat to 60°C. The downside of preheating is the increase in soluble BOD in the centrate that could lead to an increase in treatment of the recycle loads (Garelli et al., 1992). Additional factors in evaluating preheating are the source of the heat and the cost of producing the heat. If heat is obtained from a source that otherwise would be wasted, sludge preheating may be an effective and economical method of conditioning to improve dewatering performance.

Freeze-Thaw Conditioning. It is a well-known fact that natural freezing of water and wastewater-treatment plant residuals in cold climates enhances their dewatering characteristics. Freezing and thawing convert the jellylike consistency of the residuals to a granular-type material that drains readily. Similar results have been achieved by the use of mechanical freeze/thaw equipment.

Residuals can be described as being composed of several types of water: free water, interstitial water, surface water, and bound water. Free water refers to water that surrounds the sludge floc but does not move with the solids. Interstitial water is defined as the water that "is trapped within the floc structure and travels with the floc or is held by capillary forces between the particles." Surface water is held on the surface of the floc and cannot be removed by mechanical means. Finally, bound water is that which is "bound to the particles and can only be released by thermochemical destruction of the particles" (Vesilind and Martel, 1990).

When sludge freezes, the free water begins to freeze first. As the free water binds by crystallization, it seeks more free water to bind and grow with while pushing the floc particles to the ice front. Once free water is frozen, the interstitial water is extracted by diffusion and added to the growing crystalline structure.

The initial concentration of solids in the sludge, the rate of freezing, and the duration of freezing conditions (curing time) are cited as important variables to consider in optimizing the freeze/thaw process (Parker et al., 2000). The concentration of sludge solids in the residual matter is important in sizing the freezing equipment; thickening before freezing can decrease the size of the conditioning facility and reduce energy consumption. Freeze rate and curing time are closely related. Curing time is defined as the time at which the ice block is kept at subfreezing temperatures. Because of the design of the freezing vessel, areas close to the freezing surface will freeze before the more remote areas. Curing time allows extra freezing time to ensure the portion of the ice that was frozen last had enough time to completely dehydrate.

Pilot- and demonstration-scale testing has proved that the mechanical freeze-thaw system technology works on difficult-to-dewater water and wastewater sludges. The core temperature of the frozen sludge is critical to achieving separation of the bound water. To be effective, sludge has to be frozen for at least 30 min at temperatures of minus 10 to 20°C. The time to reach freezing temperature varies depending upon the size of the ice block and solids concentration. After thawing and dewatering, dewatered sludge cake can range from 25 to 40 percent solids with a filtrate that is very low in TSS (EPRI, 1998). Natural freeze-thaw dewatering using sludge-drying beds is discussed in the following section.

14–13 DEWATERING

Dewatering is a physical unit operation used to reduce the moisture content of sludge and biosolids for one or more of the following reasons:

1. The costs for trucking sludge and biosolids to the ultimate disposal site become substantially lower when the volume is reduced by dewatering.
2. Dewatered sludge and biosolids are generally easier to handle than thickened or liquid sludge. In most cases, dewatered sludge may be shoveled, moved about with tractors fitted with buckets and blades, and transported by belt conveyors.

3. Dewatering is required normally prior to the incineration of the sludge to increase the calorific value by removal of excess moisture.

4. Dewatering is required before composting to reduce the requirements for supplemental bulking agents or amendments.

5. In some cases, removal of the excess moisture may be required to render biosolids odorless and nonputrescible.

6. Dewatering is required prior to landfilling sludge and biosolids in monofills to reduce leachate production at the landfill site.

Several techniques are used in dewatering devices for removing moisture. Some of these techniques rely on natural evaporation and percolation to dewater the solids. In mechanical dewatering devices, mechanically assisted physical means are used to dewater the sludge more quickly. The physical means include filtration, squeezing, capillary action, and centrifugal separation and compaction. In the past, vacuum filtration was used to a significant extent for municipal sludge dewatering, but it has largely been replaced by alternative mechanical dewatering equipment. Vacuum filters may sometimes be used in industrial applications. Imperforate basket centrifuges have also been used for dewatering; the most notable application was at the County Sanitation Districts of Los Angeles. Because of the improved design and performance of solid-bowl centrifuges, imperforate basket centrifuges are seldom used in new dewatering installations. Vacuum filters and imperforate basket centrifuges are not discussed in this text; information for vacuum filters and imperforate basket centrifuges may be found in the preceding edition of this text (Metcalf & Eddy, 1991) or WEF (1998).

The selection of the dewatering device is determined by the type of sludge to be dewatered, characteristics of the dewatered product, and the space available. For smaller plants where land availability is not a problem, drying beds or lagoons are generally used. Conversely, for facilities situated on constricted sites, mechanical dewatering devices are often chosen. Odor control is an important design consideration as the level of odor release varies based on the type of sludge and the mechanical equipment selected. High shear dewatering and conveyance equipment can increase odor release, especially from anaerobically digested sludge (Murthy, 2001; Novak, 2001) (see also Sec. 15–3 in Chap. 15).

Some sludges, particularly those that are aerobically digested, are not readily amenable to mechanical dewatering. These sludges can be dewatered on sand beds with good results. When a particular sludge must be dewatered mechanically, it is often difficult or impossible to select the optimum dewatering device without conducting bench-scale or pilot studies. Trailer-mounted, full-size equipment is available from several manufacturers for field-testing purposes.

The dewatering processes that are commonly used include centrifuges, belt-filter presses, recessed-plate filter presses, drying beds, and lagoons. The advantages and disadvantages of the various methods of sludge dewatering are summarized in Table 14–41.

Centrifugation

For separating liquids of different density, thickening slurries, or removing solids, the centrifugation process is widely used in industry. The process is applicable to the dewatering

Table 14-41

Comparison of alternative methods for dewatering various types of sludge and biosolids[a]

Dewatering method	Advantages	Disadvantages
Solid-bowl centrifuge	Clean appearance, good odor containment, fast startup and shutdown capabilities	Scroll wear potentially a high maintenance problem
	Produces a relatively dry sludge cake	Requires grit removal and possibly a sludge grinder in the feed stream
	Low capital cost-to-capacity ratio	Skilled maintenance personnel required
	High installed capacity to building area ratio	Moderately high suspended solids content in centrate
Belt-filter press	Low energy requirements	High odor potential
	Relatively low capital and operating costs	Requires sludge grinder in feed stream
	Less complex mechanically and is easier to maintain	Very sensitive to incoming sludge feed characteristics
	High-pressure machines are capable of producing very dry cake	Automatic operation generally not advised
	Minimal effort required for system shutdown	
Recessed-plate filter press	Highest cake solids concentration	Batch operation
	Low suspended solids in filtrate	High equipment cost
		High labor cost
		Special support structure requirements
		Large floor area required for equipment
		Skilled maintenance personnel required
		Additional solids due to large chemical addition require disposal
Sludge drying beds	Lowest capital cost method where land is readily available	Requires large area of land
	Small amount of operator attention and skill required	Requires stabilized sludge
	Low energy consumption	Design requires consideration of climatic effects
	Low to no chemical consumption	Sludge removal is labor-intensive
	Less sensitive to sludge variability	
	Higher solids content than mechanical methods	
Sludge lagoons	Low energy consumption	Potential for odor and vector problems
	No chemical consumption	Potential for groundwater pollution
	Organic matter is further stabilized	More land-intensive than mechanical methods
	Low capital cost where land is available	Appearance may be unsightly
	Least amount of skill required for operation	Design requires consideration of climatic effects

[a] Adapted, in part, from U.S. EPA (1979) and WEF (1998).

Figure 14–39

Typical solid-bowl centrifuge dewatering installation (see also Fig. 14–40.

of wastewater sludges and has been widely used in both the United States and Europe. Solid-bowl centrifugal devices used for thickening sludge (see Sec. 14–6) may also be used for sludge and biosolids dewatering. In this section, standard solid-bowl and "high-solids" centrifuges are discussed. The high-solids centrifuge is a modification of the standard centrifuge.

Solid-Bowl Centrifuge. In the solid-bowl machine (see Figs. 14–39 and 14–40), sludge is fed at a constant flowrate into the rotating bowl, where it separates into a dense cake containing the solids and a dilute stream called "centrate." The centrate contains fine, low-density solids and is returned to the wastewater treatment system. The sludge cake, which contains approximately 70 to 80 percent moisture, is discharged from the bowl by a screw feeder into a hopper or onto a conveyor belt. Depending on the type of sludge, solids concentration in the cake varies generally from 10 to 30 percent range. Sludge-cake concentrations above 25 percent are desirable for disposal by incineration or by hauling to a sanitary landfill.

Solid-bowl centrifuges are suitable generally for a variety of dewatering applications. The units can be used to dewater sludge and biosolids with no prior chemical conditioning, but the solids capture and centrate quality are improved considerably when solids are conditioned with polymers. Chemicals for conditioning are added to the sludge feed line or to the sludge within the bowl of the centrifuge. Dosage rates for conditioning with polymers vary from 1.0 to 7.5 kg/10^3 kg (2 to 15 lb/ton) of sludge (dry solids basis). Typical performance data for solid-bowl centrifuges are reported in Table 14–42.

High-Solids Centrifuge. High-solids (also called "high-torque") centrifuges are modified solid-bowl centrifuges that are designed to produce a dryer solids cake. These units have a slightly longer bowl length to accommodate a longer "beach" section, a lower differential bowl speed to increase residence time, and a modified scroll to provide a pressing action within the beach end of the unit. The high-solids units are capable of achieving solids contents in excess of 30 percent in dewatering municipal wastewater sludges, although a higher polymer usage may be required.

Table 14–42

Typical dewatering performance data for solid-bowl centrifuges for various types of sludge and biosolids[a]

Type of sludge	Cake solids, %	Solids capture, % Without chemicals	With chemicals
Untreated:			
Primary	25–35	75–90	95+
Primary and trickling filter	20–25	60–80	95+
Primary and air activated	12–20	55–65	92+
Waste sludge:			
Trickling filter	10–20	60–80	92+
Air activated	5–15	60–80	92+
Oxygen activated	10–20	60–80	92+
Anaerobically digested:			
Primary	25–35	65–80	92+
Primary and trickling filter	18–25	60–75	90+
Primary and air activated	15–20	50–65	90+
Aerobically digested:			
Waste activated	8–10	60–75	90+

[a] Adapted, in part, from U.S. EPA (1979).

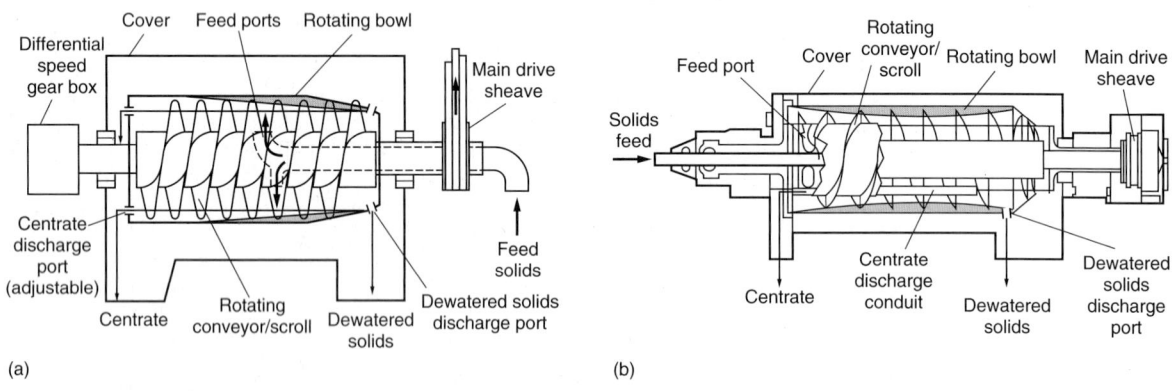

(a)   (b)

Figure 14–40

Schematic diagrams of two solid-bowl centrifuge configurations for dewatering sludge: (a) countercurrent and (b) cocurrent.

Design Considerations. Two basic designs of centrifuges are used: countercurrent flow and cocurrent flow (see Fig. 14–40a and b). The main difference in the designs is the location of the sludge feed ports, removal of centrate, and internal flow patterns of the liquid and solid phases. In the countercurrent design, the feed slurry enters at the junction of the cylindrical conical section; solids travel to the conical end while the liquid phase moves in the opposite direction. In the cocurrent design, the solid phase travels the full length of the bowl as does the liquid phase (WEF, 1998).

Process variables affecting centrifuge performance, as measured by the sludge cake solids and TSS recovery, include feed flowrate, rotational speed, differential speed of the scroll, depth of the settling zone, chemical use, and the physicochemical properties of the suspended solids and suspending liquid. Important properties are particle size and shape, particle density, temperature, and liquid viscosity.

The major difficulty encountered in the operation of centrifuges has been the disposal of the centrate, which can be relatively high in suspended, nonsettling solids. The return of centrate to the wastewater treatment units has resulted in the passage of fine solids through the treatment system, thereby reducing effluent quality. Two methods can be used to control the fine solids discharge and to increase the capture—increased residence time or chemical conditioning. Longer residence of the liquid stream is accomplished by reducing the feed rate or by using a centrifuge with a larger bowl volume. Particle size can be increased by coagulating the sludge prior to centrifugation. Solids capture (measured in percent of influent solids) may be increased from a range of 50 to 80 percent to a range of 80 to 95 percent by longer residence time and chemical conditioning.

The addition of lime will also aid in the control of odors that may develop when centrifuging untreated sludge. Untreated primary sludge can usually be dewatered to a lower moisture content than digested sludge, because it has not been subjected to the liquefying action of the digestion process, which reduces particle size. Chemical conditioning is usually desirable when dewatering combined primary and waste-activated sludge, regardless of whether it has been digested.

Selection of units for plant design is dependent on manufacturer's rating and performance data. Several manufacturers have portable pilot-plant units, which can be used for field testing if sludge is available. Wastewater sludges from supposedly similar treatment processes but in different localities may differ markedly from each other. For this reason, pilot-plant tests should be run whenever possible before final design decisions are made.

The area required for a centrifuge installation is less than that required for other dewatering devices of equal capacity, and the initial cost is lower. Higher power costs will partially offset the lower initial cost. Special consideration must also be given in providing sturdy foundations and soundproofing because of the vibration and noise that result from centrifuge operation. An adequate electric power source is required because large motors may be used.

Because centrifuges are enclosed, odor generation may be better contained as compared to other types of dewatering systems. Ventilation of the centrifuge facility to control potential odors and moisture accumulation should be provided, however.

Belt-Filter Press

Belt-filter presses are continuous-feed dewatering devices that use the principles of chemical conditioning, gravity drainage, and mechanically applied pressure to dewater sludge (see Fig. 14–41). The belt-filter press was introduced in the United States in the early 1970s and has become one of the predominant sludge-dewatering devices. It has proved to be effective for almost all types of municipal wastewater sludge and biosolids.

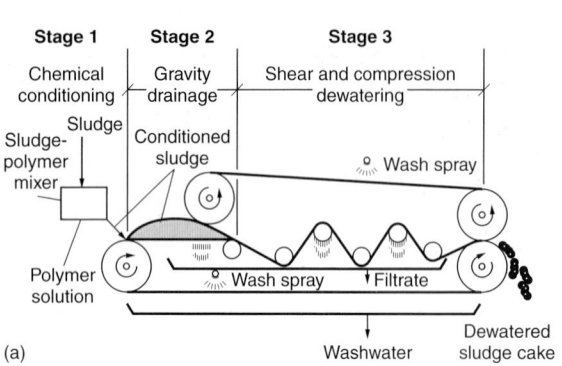

Figure 14–41

Belt-press dewatering: (a) three basic stages of belt-press dewatering, (b) pictorial view of a typical installation.

Description. In most types of belt-filter presses, conditioned sludge is first introduced on a gravity drainage section where it is allowed to thicken. In this section, a majority of the free water is removed from the sludge by gravity. On some units, this section is provided with vacuum assistance, which enhances drainage and may help to reduce odors. Following gravity drainage, pressure is applied in a low-pressure section, where the sludge is squeezed between opposing porous cloth belts. On some units, the low-pressure section is followed by a high-pressure section where the sludge is subjected to shearing forces as the belts pass through a series of rollers. The squeezing and shearing forces thus induce release of additional quantities of water from the sludge. The final dewatered sludge cake is removed from the belts by scraper blades.

System Operation and Performance. A typical belt-filter press system consists of sludge feed pumps, polymer feed equipment, sludge-conditioning tank (flocculator), belt-filter press, sludge cake conveyor, and support systems (sludge feed pumps, washwater pumps, and compressed air). Some units do not use a sludge-conditioning tank. A schematic diagram of a typical belt-filter press installation is shown on Fig. 14–42.

Many variables affect the performance of the belt-filter press: sludge characteristics, method and type of chemical conditioning, pressures developed, machine configuration (including gravity drainage), belt porosity, belt speed, and belt width. The belt-filter press is sensitive to wide variations in sludge characteristics, resulting in improper conditioning and reduced dewatering efficiency. Sludge blending facilities should be included in the system design where the sludge characteristics are likely to vary widely. Based on actual operating experience, it has been found that the solids throughput is greater and the cake dryness is improved with higher solids concentrations in the feed sludge. Typical belt-filter press performance data for various types of sludge are reported in Table 14–43.

Figure 14–42

Schematic diagram of
a belt-press dewatering
system. *(Metcalf & Eddy,
1984b.)*

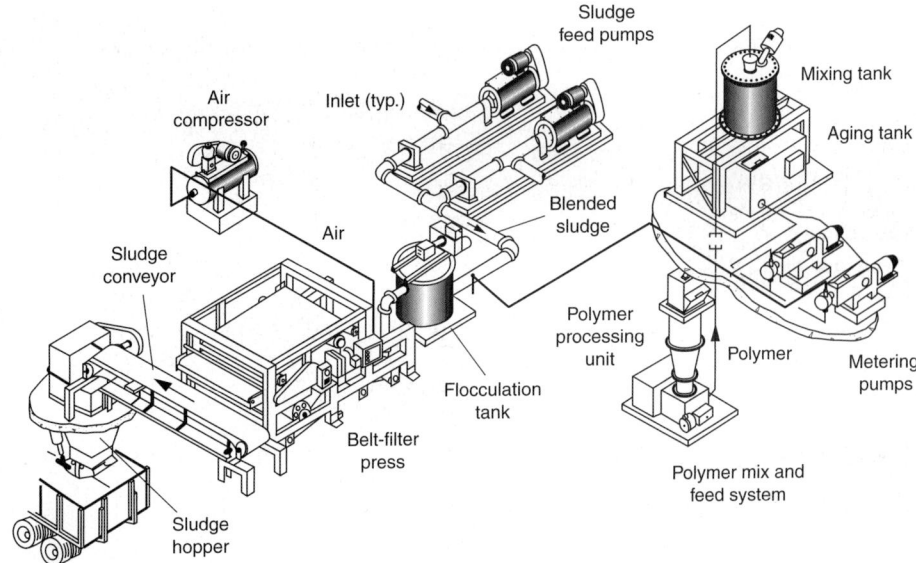

Design Considerations. Belt-filter presses are available in metric sizes from 0.5 to 3.5 m in belt width. The most common size used for municipal sludge applications is 2.0 m. Sludge loading rates vary from 90 to 680 kg/m·h (200 to 1500 lb/m·h) depending on the sludge type and feed concentrations. Hydraulic throughput based on belt width ranges from 1.6 to 6.3 L/m·s (25 to 100 gal/m·min). Design of a belt-filter press is illustrated in Example 14–9.

Safety considerations in design should include adequate ventilation to remove hydrogen sulfide or other gases, and equipment guards to prevent loose clothing from being caught between the rollers.

Filter Presses

In a filter press, dewatering is achieved by forcing the water from the sludge under high pressure. Advantages cited for the filter press include high concentrations of cake solids achieved, good filtrate clarity, and high solids capture. Disadvantages include mechanical complexity, high chemical costs, high labor costs, and limitations on filter cloth life.

Various types of filter presses have been used to dewater sludge. The two types used most commonly are the fixed-volume and variable-volume recessed-plate filter presses.

Fixed-Volume, Recessed-Plate Filter Press. The fixed-volume, recessed-plate filter press consists of a series of rectangular plates, recessed on both sides, that are supported face to face in a vertical position on a frame with a fixed and movable head (see Fig. 14–43). A filter cloth is hung or fitted over each plate. The plates are held together with sufficient force to seal them to withstand the pressure applied during the filtration process. Hydraulic rams or powered screws are used to hold the plates together.

Table 14–43
Typical dewatering performance data for belt-filter presses for various types of sludge and biosolids[a]

Type of sludge	Dry feed solids, %	Loading per meter of belt length		Dry polymer,[b] g/kg dry solids	Cake solids, %	
		L/s	kg/h		Typical	Range
Raw primary	3–7	1.8–3.2	360–550	1–4	28	26–32
Waste-activated (WAS)	1–4	0.7–2.5	45–180	3–10	15	12–20
Primary + WAS (50:50)[c]	3–6	1.3–3.2	180–320	2–8	23	20–28
Primary + WAS (40:60)[c]	3–6	1.3–3.2	180–320	2–10	20	18–25
Primary + trickling filter	3–6	1.3–3.2	180–320	2–8	25	23–30
Anaerobically digested:						
Primary	3–7	1.3–3.2	360–550	2–5	28	24–30
WAS	3–4	0.7–2.5	45–135	4–10	15	12–20
Primary + WAS	3–6	1.3–3.2	180–320	3–8	22	20–25
Aerobically digested:						
Primary + WAS, unthickened	1–3	0.7–3.2	135–225	2–8	16	12–20
Primary + WAS (50:50), thickened	4–8	0.7–3.2	135–225	2–8	18	12–25
Oxygen-activated WAS	1–3	0.7–2.5	90–180	4–10	18	15–23

WAS = waste-activated sludge.
[a] WEF (1998).
[b] Polymer needs based on high-molecular-weight polymer (100 percent strength, dry basis).
[c] Ratio is based on dry solids or the primary and WAS.

EXAMPLE 14–9 Belt-Filter Press Design A wastewater treatment plant produces 72,000 L/d of thickened biosolids containing 3 percent solids. A belt-filter press installation is to be designed based on a normal operation of 8 h/d and 5 d/wk, a belt-filter press loading rate of 275 kg/m·h, and the following data. Compute the number and size of belt-filter presses required and the expected solids capture, in percent. Determine the daily hours of operation required if a sustained 3-d peak solids load occurs.

1. Total solids in dewatered sludge = 25 percent.
2. Total suspended solids concentration in filtrate = 900 mg/L = 0.09 percent.
3. Washwater flowrate = 90 L/min per m of belt width.
4. Specific gravities of sludge feed, dewatered cake, and filtrate are 1.02, 1.07, and 1.01, respectively.

Solution

1. Compute average weekly sludge production rate.

Wet biosolids $= (72,000 \text{ L/d})(7 \text{ d/wk})(10^3 \text{ g/L})(1 \text{ kg/}10^3 \text{ g})(1.02)$

$$= 514,080 \text{ kg/wk}$$

Dry solids $= 514,080 \times 0.03 = 15,422 \text{ kg/wk}$

2. Compute daily and hourly dry solids-processing requirements.

Daily rate $= 15,442 \text{ kg/wk/5 operating d/wk}$

$$= 3084 \text{ kg/d}$$

Hourly rate $= 3084/8 = 385.5 \text{ kg/h (per 8 h operating d)}$

3. Compute belt-filter press size

$$\text{Belt width} = \frac{(385.5 \text{ kg/h})}{(275 \text{ kg/m·h})} = 1.40 \text{ m}$$

Use one 1.5-m belt-filter press and provide one identical size for standby.

4. Compute filtrate flowrate by developing solids balance and flow balance equations.

   a. Develop daily solids balance equation.

   Solids in sludge feed $=$ solids in sludge cake $+$ solids in filtrate

   $$3084 = (S \text{ L/d})(1.07)(0.25) + (F \text{ L/d})(1.01) \times (0.0009)$$

   $$3084 = 0.2675S + 0.00091F$$

   where S = sludge cake flowrate, L/d
   F = filtrate flowrate, L/d

   b. Develop flowrate equation.

   Sludge flowrate $+$ washwater flowrate $=$ filtrate flowrate $+$ cake flowrate

   Daily sludge flowrate $= (72,000 \text{ L/d})(7/5) = 100,800 \text{ L/d}$

   Washwater flowrate $= (90 \text{ L/min·m})(1.5 \text{ m})(60 \text{ min/h})(8 \text{ h/d})$

   $$= 64,800 \text{ L/d}$$

   $$100,800 + 64,800 = 165,600 = F + S$$

   c. Solve the mass balance and flowrate equations simultaneously.

   $$F = 154,600 \text{ L/d}$$

5. Determine solids capture.

$$\text{Solids capture} = \frac{\text{solids in feed} - \text{solids in filtrate}}{\text{solids in feed}} \times 100\%$$

$$= \frac{(3084 \text{ kg/d}) - [(154{,}600 \text{ L/d})(1.01)(0.0009)(10^3 \text{ g/L})(1 \text{ kg/}10^3 \text{ g})]}{(3084 \text{ kg/d})} \times 100\%$$

$$= 95.4 \text{ percent}$$

6. Determine operating requirements for sustained peak biosolids load.
   a. Determine peak 3-d load.
      From Fig. 3–9b, the ratio of peak to average mass loading for 3 consecutive days is 2. The peak load is 72,000 (2) = 144,000 L/d.
   b. Determine daily operating time requirements, neglecting sludge in storage.

   $$\text{Dry solids/d} = 144{,}000/\text{d} (1.02)(0.03)$$

   $$= 4406 \text{ kg/d}$$

   $$\text{Operating time} = \frac{(4406 \text{ kg/d})}{(275 \text{ kg/m·h})(1.5 \text{ m})} = 10.7 \text{ h}$$

   The operating time can be accomplished by running the standby belt-filter press in addition to the duty press, or operating the duty press for an extended shift.

Comment   The value of sludge storage is important in dewatering applications because of the ability to schedule operations to suit labor availability most efficiently. Scheduling sludge dewatering operations during the day shift is also desirable if sludge has to be hauled off-site.

In operation, chemically conditioned sludge is pumped into the space between the plates, and pressure of 690 to 1550 kN/m² (100 to 225 lb$_f$/in²) is applied and maintained for 1 to 3 h, forcing the liquid through the filter cloth and plate outlet ports. The plates are then separated and the sludge is removed. The filtrate normally is returned to the influent of the treatment plant. The sludge cake thickness varies from about 25 to 38 mm (1 to 1.5 in), and the moisture content varies from 48 to 70 percent. The filtration cycle time varies from 2 to 5 h and includes the time required to (1) fill the press, (2) maintain the press under pressure, (3) open the press, (4) wash and discharge the cake, and (5) close the press. Depending on the degree of automation incorporated into the machine, operator attention must be devoted to the filter press during feed, discharge, and wash intervals.

Variable-Volume, Recessed-Plate Filter Press.   Another type of filter press used for wastewater sludge dewatering is the variable-volume recessed-plate filter press, commonly called the "diaphragm press." This type of filter press is similar to the fixed-volume press except that a rubber diaphragm is placed behind the filter media, as shown on Fig. 14–44. The rubber diaphragm expands to achieve the final squeeze pres-

Figure 14–43

Typical fixed-volume, recessed-plate filter press used for dewatering sludge: (a) schematic, (b) pictorial view of a typical installation.

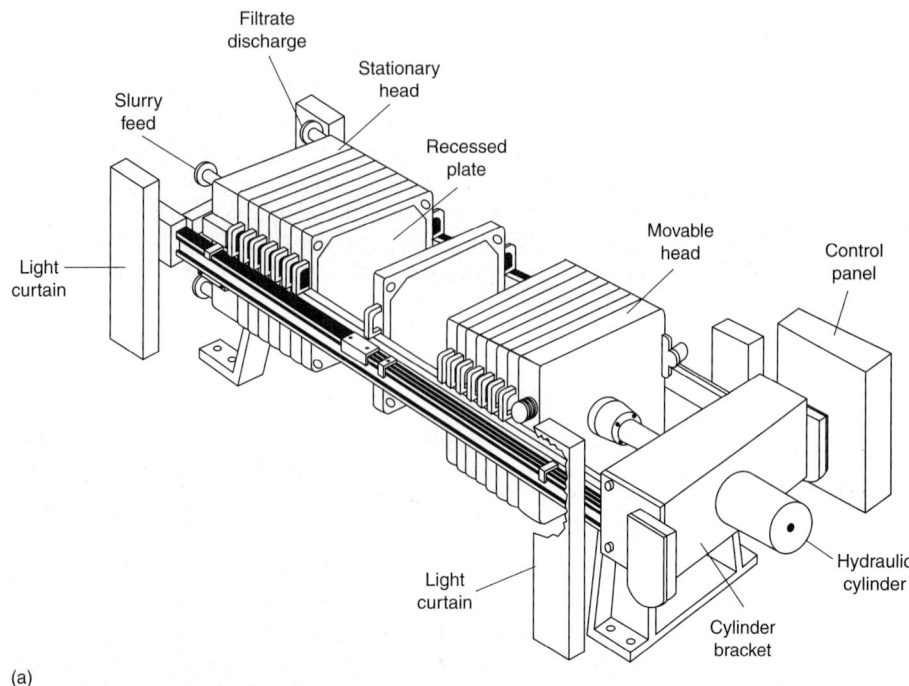

(a)

(b)

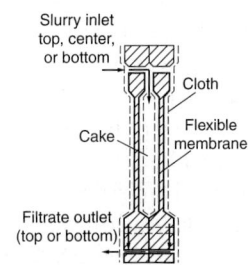

Figure 14–44

Cross section of a variable-volume recessed-plate filter press.

sure, thus reducing the cake volume during the compression step. Generally about 10 to 20 min are required to fill the press and 15 to 30 min of constant pressure are required to dewater the cake to the desired solids content. Variable-volume presses are generally designed for 690 to 860 kN/m² (100 to 125 lb$_f$/in²) for the initial stage of dewatering followed by 1380 to 2070 kN/m² (200 to 300 lb$_f$/in²) for final compression. Variable-volume presses can handle a wide variety of sludges with good performance results but require considerable maintenance (Metcalf & Eddy, 1984c).

Design Considerations. Several operating and maintenance problems have been identified for recessed-plate filter presses ranging from difficulties in the chemical feed and sludge-conditioning system to excessive downtime for equipment maintenance. Features that should be considered in the design of a filter press installation include (1) adequate ventilation in the dewatering room (6 to 12 air changes per hour are recommended depending on the ambient temperature), (2) high-pressure washing systems, (3) an acid wash circulation system to remove calcium scale when lime is used, (4) a sludge grinder ahead of the conditioning tank, (5) cake breakers or shredders following the filter press (particularly if the dewatered sludge is incinerated), and (6) equipment to facilitate removal and maintenance of the plates (Metcalf & Eddy, 1984c).

Sludge Drying Beds

Drying beds are the most widely used method of sludge dewatering in the United States. Sludge drying beds are typically used to dewater digested biosolids and settled sludge from plants using the extended aeration activated-sludge treatment process without prethickening. After drying, the solids are removed and either disposed of in a landfill or used as a soil conditioner. The principal advantages of drying beds are low cost, infrequent attention required, and high-solids content in the dried product. The principal disadvantages are the large space required, effects of climatic changes on drying characteristics, labor-intensive sludge removal, insects, and potential odors. Five types of drying beds are used: (1) conventional sand, (2) paved, (3) artificial media, (4) vacuum-assisted, and (5) solar. Because conventional sand and paved drying beds are used most extensively, more detailed discussion is provided for these types. For additional information on the other types of drying beds, the reader is referred to U.S. EPA (1979) and U.S. EPA (1987c).

Conventional Sand Drying Beds. Conventional sand drying beds are generally used for small- and medium-sized communities. For cities with populations over 20,000, consideration should be given to alternative means of sludge dewatering. In larger municipalities, the initial cost, the cost of removing the sludge and replacing sand, and the large area requirements generally preclude the use of sand drying beds.

In a typical sand drying bed, sludge is placed on the bed in a 200 to 300 mm (8 to 12 in) layer and allowed to dry. Sludge dewaters by drainage through the sludge mass and supporting sand and by evaporation from the surface exposed to the air (see Fig. 14–45). Most of the water leaves the sludge by drainage; thus the provision of an adequate underdrainage system is essential. Drying beds are equipped with lateral drainage lines (perforated plastic pipe or vitrified clay pipe laid with open joints), sloped at a minimum of 1 percent and spaced 2.5 to 6 m (8 to 20 ft) apart. The drainage lines should be adequately supported and covered with coarse gravel or crushed stone. The sand layer should be from 230 to 300 mm (9 to 12 in) deep with an allowance for some loss from cleaning operations. Deeper sand layers generally retard the draining process. Sand should have a uniformity coefficient of not over 4.0 and an effective size of 0.3 to 0.75 mm.

The drying area is partitioned into individual beds 6 m wide by 6 to 30 m long (approximately 20 ft wide by 20 to 100 ft long), or a convenient size so that one or two beds will be filled in a normal loading cycle. The interior partitions commonly consist

Figure 14–45

Typical conventional sand drying bed: (a) plan and pictorial views and (b) cross-sectional view. Insert—view of sludge drying beds with sludge in various stages of dryness.

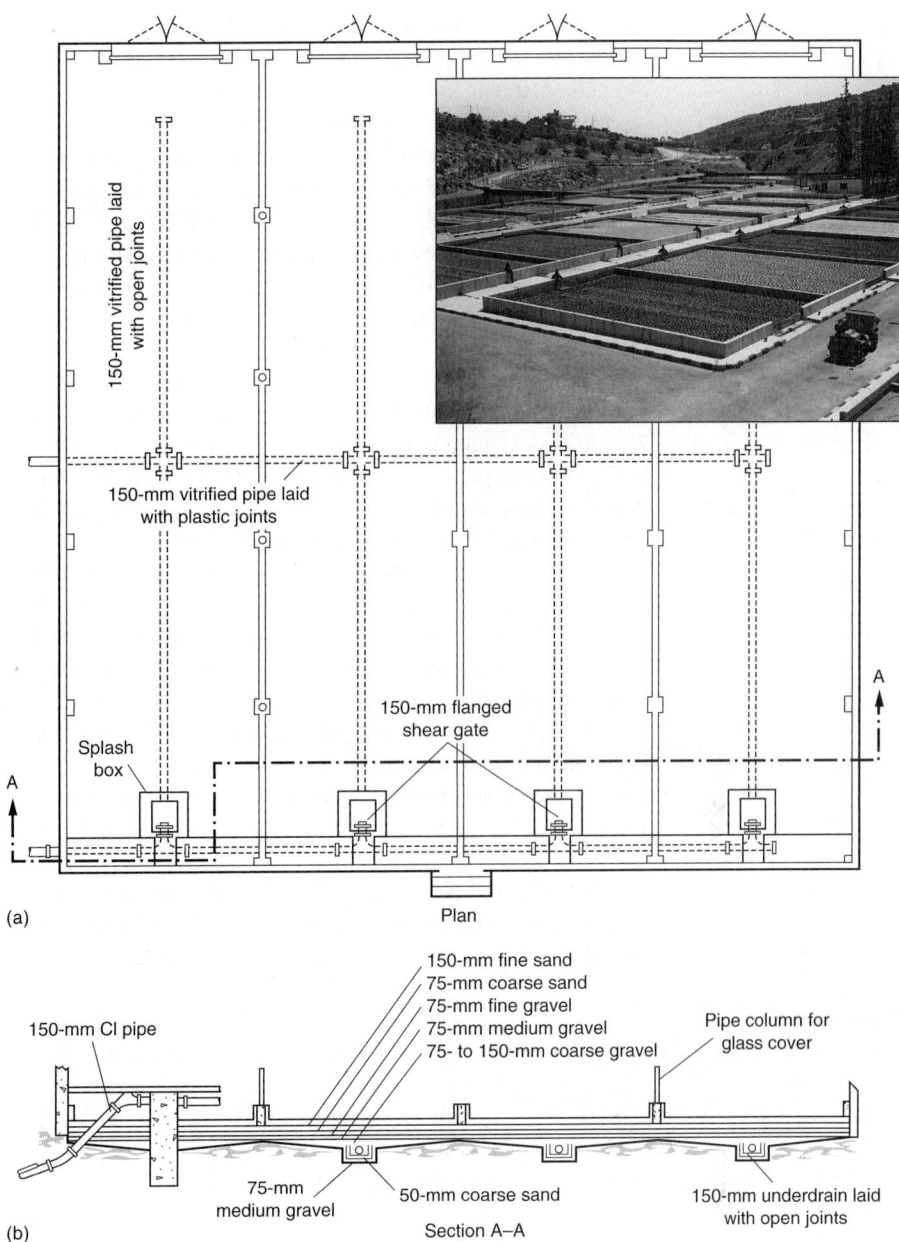

(a)

Plan

(b)

Section A–A

of two or three planks, one on top of the other, to a height of 380 to 460 mm (15 to 18 in), stretching between slots in precast-concrete posts. The outer boundaries may be of similar construction or may be earthen embankments for open drying beds. Concrete foundation walls are required if the beds are to be covered.

Piping to the sludge beds should be designed for a velocity of at least 0.75 m/s (2.5 ft/s) and preferably should drain to the beds. Cast iron or plastic pipe is commonly used. Provisions should be included to flush the lines, if necessary, and to prevent their freezing in cold climates. Distribution boxes are required to divert the sludge flow into the bed selected. Splash plates are placed in front of the sludge outlets to spread the sludge over the bed and to prevent erosion of the sand.

Sludge can be removed from the drying bed after it has drained and dried sufficiently. Dried sludge has a coarse, cracked surface and is black or dark brown. The moisture content is approximately 60 percent after 10 to 15 d under favorable conditions. Sludge removal is accomplished by manual shoveling into wheelbarrows or trucks, or by a scraper, front-end loader, or special mechanical sludge removal equipment. Provisions should be made for driving a truck onto or alongside of the bed to facilitate loading.

Open beds are used where adequate area is available and sufficiently isolated to avoid complaints caused by occasional odors. Open sludge beds should be located at least 100 m (about 300 ft) from dwellings to avoid odor nuisance. Covered beds with greenhouse types of enclosures are used where it is necessary to dewater sludge continuously throughout the year regardless of the weather, and where sufficient isolation does not exist for the installation of open beds (see Solar Drying Beds).

Sludge bed loadings are computed on a per capita basis or on a unit loading of pounds of dry solids per square foot per year (kilograms of dry solids per square meter per year). Typical data for various types of biosolids are shown in Table 14–44. With covered drying beds, more biosolids can be applied per year because of the protection from rain and snow.

As discussed previously, the effects of freezing and thawing in cold climates have been observed to improve the dewatering characteristics of sludge. After freezing, solids concentrations exceeding 20 percent may occur when the material thaws and may increase to over 50 percent with additional drying time. An 75-mm (3-in) layer of sludge has been found to be practical for most locations in moderately cold climates that experience alternating freeze-thaw periods. In colder climates, greater depths [up to 225 mm (9 in)] may be feasible. The type of facility used for natural freeze-thaw

Table 14–44

Typical area requirements for open sludge drying beds for various types of biosolids[a]

| Type of biosolids | Area[a] | | Sludge loading rate | |
	ft²/person	m²/person	lb dry solids/ft²·yr	kg dry solids/m²·yr
Primary digested	1.0–1.5	0.1	25–30	120–150
Primary and trickling-filter humus digested	1.25–1.75	0.12–0.16	18–25	90–120
Primary and waste-activated digested	1.75–2.5	0.16–0.23	12–20	60–100
Primary and chemically precipitated digested	2.0–2.5	0.19–0.23	20–33	100–160

[a]Corresponding area requirements for covered beds vary from about 70 to 75 percent of those for the open beds.

dewatering is similar to conventional sand drying beds, or deep lined and underdrained trenches can be used. Dewatered biosolids are recommended to be removed once each year; biosolids stored in deep trenches should be removed before the onset of warm weather (Reed et al., 1995).

Paved Drying Beds. Two types of paved drying beds have been used as an alternate to sand drying beds: a drainage type and a decanting type. The drainage type functions similarly to a conventional bed in that underdrainage is collected, but sludge removal is improved by using a front-end loader. Sludge drying may also be improved by frequent agitation with mobile equipment. With this design, the beds are normally rectangular in shape and are 6 to 15 m (20 to 50 ft) wide by 20 to 45 m (~70 to 150 ft) long with vertical sidewalls. Concrete or bituminous concrete linings are used, overlaying a 200 to 300 mm (8 to 12 in) sand or gravel base. The lining should have a minimum 1.5 percent slope to a center unpaved drainage area. For a given amount of sludge, this type of paved drying bed requires more area than conventional sand beds.

The decanting type of paved drying bed is advantageous for warm, arid, and semi-arid climates. This type of drying bed uses low-cost impermeable paved beds that depend on decanting of the supernatant and mixing of the drying sludge for enhanced evaporation. Features of this design include (1) a soil-cement mixture paving material, (2) drawoff pipes for decanting the supernatant, and (3) a sludge feed pipe at the center of the bed (see Fig. 14–46). Decanting may remove about 20 to 30 percent of the water with a good settling sludge. Solids concentration may range from 40 to 50 percent for a 30 to 40 d drying time in an arid climate for a 300 mm (12 in) sludge layer.

The water losses and bottom area of the drying bed may be determined by the following equations (U.S. EPA, 1987a, 1987b):

The water content in the applied sludge can be determined by

$$W_o = 1.04(S)\left[\frac{1 - s_o}{s_o}\right] \tag{14–25}$$

where W_o = total water content in the applied sludge, kg/yr
 1.04 = assumed specific gravity of biosolids
 S = annual sludge production, dry solids, kg/yr
 s_o = percent dry solids in applied sludge, as a decimal

Figure 14–46

Paved sludge drying bed for decantation and evaporation: (a) isometric view, (b) cross section.

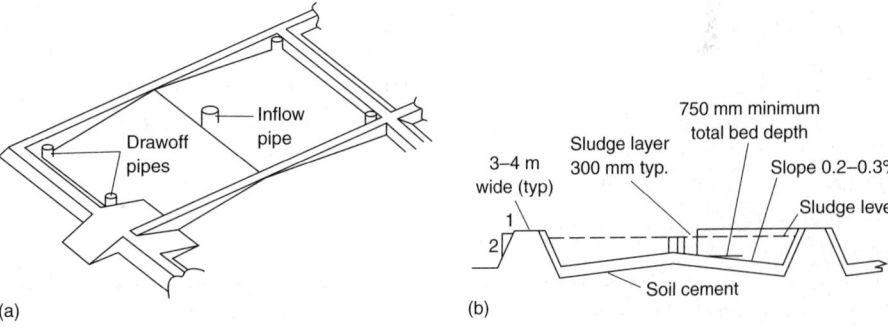

The water content after decanting can be determined by

$$W_d = 1.04(S)\left[\frac{1 - s_d}{s_d}\right]$$

(14–26)

where W_d = total water content after decanting, kg/yr
s_d = percent dry solids in decanted sludge, as a decimal

The water content to be removed by evaporation is given by

$$W_e = W_d - 1.04(S)\left[\frac{1 - s_e}{s_e}\right] + (P)(A)(10^3 \text{ kg/m}^3)$$

(14–27)

where W_e = water to be evaporated after decanting, kg/yr
s_e = percent dry solids required after evaporation, as a decimal
P = annual precipitation, m/yr
A = bottom area of paved bed, m^2 (ft^2)

The evaporation rate for a given location is given by

$$R_e = (10)(k_e)(E_p)$$

(14–28)

where R_e = annual evaporation rate, kg/m^2·yr
10 = factor used for converting cm/yr to kg/m^2·yr
k_e = reduction factor for evaporation from sludge versus a free water surface. Use 0.6 for preliminary estimate; pilot test to determine factor for final design
E_p = free water pan evaporation rate, cm/yr

The area required can be estimated by combining Eqs. (14–27) and (14–28), which yields

$$A = \frac{1.04 \, S\left[(1 - s_d)/s_d - (1 - s_e)/s_e\right] + (10^3 \text{ kg/m}^3)(P)(A)}{(10)(k_e)(E_p)}$$

(14–29)

Application of the equations is shown in Example 14–10.

EXAMPLE 14–10 **Paved Drying Bed Design** Design a paved drying bed for a wastewater treatment plant that produces 180,000 kg/yr of dry solids. Assume the following conditions:

1. Percent of dry solids in applied biosolids (before decanting) = 6 percent
2. Percent of dry solids in biosolids after decanting = 8 percent
3. Percent dry solids required for final disposal = 30 percent
4. Free water pan evaporation rate = 120 cm/yr
5. k_e = 0.6.

Solution

Determine evaporation rate by Eq. (14–28).

$$R_e = (10)(0.6)(120) = 720 \text{ kg/m}^2\text{·yr}$$

Determine bed area by Eq. (14–29).

$$A = \frac{(1.04)(180{,}000)[(1 - 0.08)/0.08 - (1 - 0.35)/0.35] + (0.6)(A)(10^3)}{720}$$

$A = 14{,}741 \ m^2 \ (1.47 \ ha \ or \ 3.63 \ ac)$

Divide area into beds of approximately 0.25 ha each for operating flexibility. The number of beds required is 6.

Comment   A detailed month-by month analysis of weather records and sludge production rates should be made when sizing the bed area. Testing should also be done to determine the solids concentrations reasonably achievable after settling and decanting the biosolids.

Artificial-Media Drying Beds.   Artificial-media can be used such as stainless-steel wedgewire or high-density polyurethane formed into panels. In a wedgewire drying bed, liquid sludge is introduced onto a horizontal, relatively open drainage medium (see Fig. 14–47). The medium consists of small stainless-steel wedge-shaped bars with the flat part of the wedge on top. The slotted openings between the bars are 0.25 mm (0.01 in) wide. The wedgewire is formed into panels and installed in a false floor. An outlet valve is used to control the drainage. Advantages cited for this method of dewatering are (1) no clogging of the wedgewire, (2) drainage is constant and rapid, (3) throughput is higher than sand beds, (4) aerobically digested sludges can be dried, and (5) beds are relatively easy to maintain. The principal disadvantage is that the capital costs are higher than for conventional drying beds. Typical solids loading rates are 2 to 5 kg/m²·cycle. The number of operating cycles varies depending on local climatic conditions. Typically the biosolids will dry to 8 to 12 percent solids after 24 h of drying.

In the high-density polyurethane media system, special 300 mm- (12 in-) square interlocking panels are formed for installation on a sloped slab or in prefabricated steel self-dumping trays. Each panel has an 8 percent open area for dewatering and contains

Figure 14–47

Cross section of a wedgewire sludge drying bed.

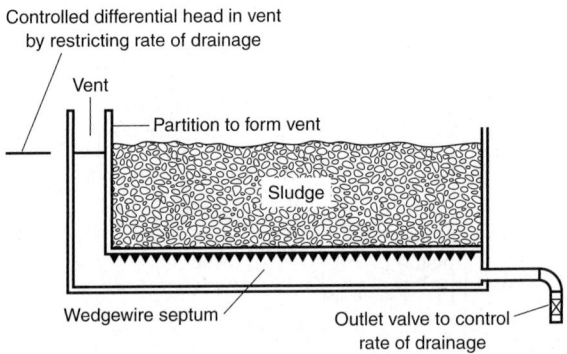

Controlled differential head in vent by restricting rate of drainage

Vent

Partition to form vent

Sludge

Wedgewire septum

Outlet valve to control rate of drainage

a built-in underdrain system. The system can be designed for installation in open or covered beds. Advantages of this method of dewatering are (1) dilute sludges can be dewatered including aerobically digested waste-activated sludge, (2) filtrate contains low suspended solids, and (3) fixed units can be cleaned easily with a front-end loader. Drying rates are comparable to those of wedgewire media.

Vacuum-Assisted Drying Beds.   A method used to accelerate dewatering and drying is the vacuum-assisted sludge drying bed (see Fig. 14–48). Dewatering and drying are assisted by the application of a vacuum to the underside of porous filter plates. Operation of this method consists usually of the following steps: (1) preconditioning the biosolids with a polymer, (2) filling the beds with biosolids, (3) dewatering the biosolids initially by gravity drainage followed by applying a vacuum, (4) allowing the biosolids to air-dry for approximately 24 to 48 h, (5) removal of the dewatered biosolids by a front-end loader, and (6) washing of the surface of the porous plates with a high-pressure hose to remove the remaining biosolids residue. Typical solids concentrations produced by vacuum-assisted drying beds are reported in Table 14–45 (WEF, 1998). The principal advantages cited for this method are the reduced cycle time required for biosolids dewatering, thereby reducing the effects of weather on biosolids drying, and the smaller area required as compared to other types of drying beds. The principal disadvantage is that adequate polymer conditioning must be used for successful operation or further processing may be required for additional moisture reduction (U.S. EPA 1987*a*).

Solar Drying Beds.   A method used to enhance the dewatering and drying of liquid, thickened, or dewatered biosolids is solar drying in covered drying beds. The solar drying system, which is a sophisticated "greenhouse," consists of a rectangular base structure, translucent chamber, sensors to measure atmospheric drying conditions, air louvers, ventilation fans, a mobile electromechanical device (called a "mole") that agitates and moves the drying biosolids, and a microprocessor that controls the drying environment (see Fig. 14–49). The system's main source of drying energy is solar radiation.

Figure 14–48

Isometric view of a vacuum-assisted sludge drying bed.

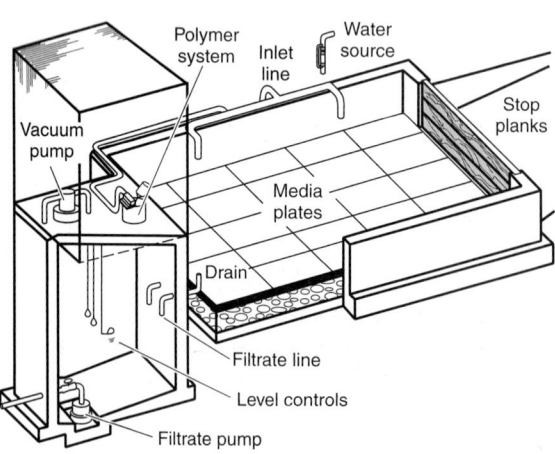

Table 14-45

Typical performance data for vacuum-assisted sludge drying beds for various types of biosolids[a]

Type of biosolids	Dry feed solids %	Dry solids loading		Cycle time h	Polymer dose		Cake solids range %
		lb/ft²	kg/m²		lb/ton	g/kg	
Anaerobically digested:							
Primary	1–7	2–4	10–20	8–24	4–40	2–20	12–26
Primary + WAS	1–4	1–4	5–20	18–24	30–40	15–20	15–20
Primary + TF	3–10	3–6	15–30	18–24	40–52	20–26	20–26
Aerobically digested:							
Conventional WAS	1–4	1–3	5–15	8–24	2–34	1–17	10–23
Oxidation ditch	1–2	1–2	5–10	8–24	4–14	2–7	10–20

[a] Adapted from WEF (1998).
WAS = waste-activated sludge.
TF = trickling-filter humus sludge.

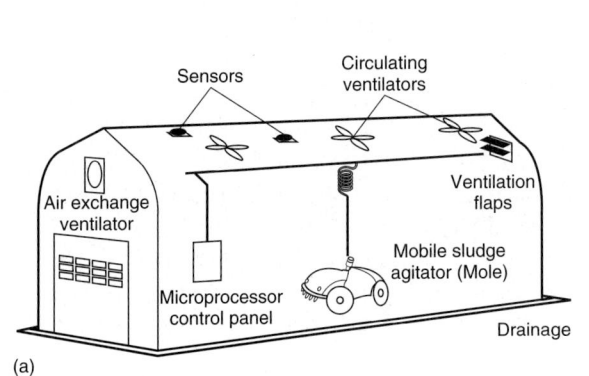

(a)   (b)

Figure 14-49

Solar sludge drying bed system employing mobile sludge agitator: (a) schematic and (b) interior view of a typical installation. (Courtesy Parkson, Corp.)

The microprocessor evaluates the number of climatic variables and then initiates one or more operations that optimize the moisture-absorbing potential of the ambient air. Ventilating fans provide circulating air movement to exhaust the moisture-laden air. The mole tills the sludge bed, bringing moist biosolids to the surface to accelerate drying. Ultimately, a dry pelletized material is produced with a solids content as high as 90 percent.

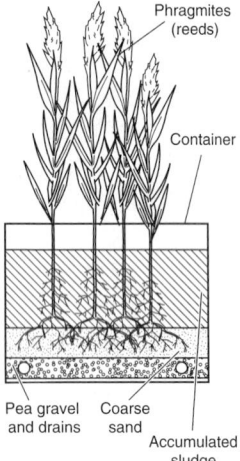

Phragmites (reeds)

Container

Pea gravel and drains

Coarse sand

Accumulated sludge

Figure 14–50

Cross section of a reed bed for dewatering and storage of biosolids.

Reed Beds

Reed beds can be used for biosolids dewatering at treatment plants with capacities up to $0.2 \text{ m}^3/\text{s}$ (5 Mgal/d). Reed beds are similar in appearance to subsurface flow constructed wetlands, which consist of channels or trenches filled with sand or rock to support emergent vegetation. The difference between reed beds used for biosolids application and subsurface flow wetlands is that the liquid biosolids are applied to the surface of the beds (as compared to subsurface application) and the filtrate flows through the gravel to underdrains. A typical reed bed for biosolids dewatering, treatment, and storage is shown on Fig. 14–50.

Typically, reed beds are constructed of washed river-run gravel in the following layers: (1) a 250 mm (10 in) deep drainage layer composed of 20 mm (0.8 in) washed gravel, (2) a 250 mm (10 in) deep layer composed of 4 to 6 mm (0.16 to 0.24 in) washed gravel, and (3) a 100 to 150 mm (4 to 6 in) layer of sand (0.4 to 0.6 mm). Sometimes an even coarser bottom layer is used. At least 1 m (3 ft) of freeboard above the sand layer is provided for a 10-year accumulation of sludge. Phragmites (reeds) are planted on 300 mm (12 in) centers in the gravel layer just below the sand. Other wetland vegetation can be used, although reeds are the most popular. The first sludge application is made after the reeds are well established. Harvesting of the reeds is practiced typically in the winter by cutting the tops back to a level above the sludge blanket. Harvesting is necessary whenever the plant growth becomes too thick and restricts the even flow of sludge. The harvested material can be composted, burned, or landfilled (Crites and Tchobanoglous, 1998).

The purpose of the plants is to provide a pathway for continuous drainage of water from the sludge layer. Movement and growth of the plants create pathways for water to drain from the biosolids into the underdrains. The plants also absorb water from the sludge. Oxygen transfer to the plant roots assists in the biological stabilization and mineralization of the sludge. The reed bed system is a form of passive composting. The design loading rates for reed beds range from 30 to 60 $\text{kg/m}^2\cdot\text{yr}$ (6 to 12 $\text{lb/ft}^2\cdot\text{yr}$). Loading rates as high as 100 $\text{kg/m}^2\cdot\text{yr}$ (20 $\text{lb/ft}^2\cdot\text{yr}$) have been used, depending on the nature of the sludge and climatic conditions. The liquid sludge is applied intermittently, as in sand drying beds. The typical sludge depth applied is 75 to 100 mm (3 to 4 in) every week to 10 d (Crites and Tchobanoglous, 1998; Cooper et al., 1996).

Lagoons

Drying lagoons may be used as a substitute for drying beds for the dewatering of digested sludge. Lagoons are not suitable for dewatering untreated sludges, limed sludges, or sludges with a high-strength supernatant because of their odor and nuisance potential. The performance of lagoons, like that of drying beds, is affected by climate; precipitation and low temperatures inhibit dewatering. Lagoons are most applicable in areas with high evaporation rates. Dewatering by subsurface drainage and percolation is limited by increasingly stringent environmental and groundwater regulations. If a groundwater aquifer used for a potable water supply underlies the lagoon site, it may be necessary to line the lagoon or otherwise restrict significant percolation.

Unconditioned digested biosolids are discharged to the lagoon in a manner suitable to accomplish an even distribution. Biosolids depths usually range from 0.75 to 1.25 m (2.5 to 4 ft). Evaporation is the prime mechanism for dewatering. Facilities for decanting of supernatant are usually provided, and the liquid is recycled to the treatment facility. Biosolids are removed mechanically, usually at a solids content of 25 to 30 percent. The cycle time for lagoons varies from several months to several years. Typically, biosolids are pumped to the lagoon for 18 months, and then the lagoon is rested for 6 months. Solids loading criteria range from 36 to 39 kg/m^3·yr (2.2 to 2.4 lb/ft^3·yr) of lagoon capacity (U.S. EPA, 1987c). A minimum of two cells is essential, even in very small plants, to ensure availability of storage space during cleaning, maintenance, or emergency conditions.

14-14 **HEAT DRYING**

Heat drying involves the application of heat to evaporate water and to reduce the moisture content of biosolids below that achievable by conventional dewatering methods. The advantages of heat drying include reduced product transportation costs, further pathogen reduction, improved storage capability, and marketability (WEF, 1998).

Heat-Transfer Methods

The classification of dryers is based on the predominant method of transferring heat to wet solids. These methods are convection, conduction, radiation, or a combination of both. Dryers using infrared radiation are relatively new and have been used mostly for demonstration testing.

Convection.   In convection (direct drying) systems the wet sludge directly contacts the heat-transfer mechanism, usually hot gases. Under equilibrium conditions of constant rate drying, mass transfer is proportional to (1) the area of wetted surface exposed, (2) the difference between water content of the drying air and saturation humidity at the wet-bulb temperature of the sludge-air interface, and (3) other factors, such as velocity and turbulence of drying air expressed as a mass transfer coefficient. The heat-transfer rate for evaporation is determined by the following equation (U.S. EPA, 1979; WEF, 1998):

$$q_{\text{conv}} = h_c A(T_g - T_s) \tag{14-30}$$

where q_{conv} = convective heat-transfer rate, kJ/h (Btu/h)
$\quad\quad h_c$ = convection heat-transfer coefficient, kJ/m^2·h·°C (Btu/ft^2·h·°F)
$\quad\quad A$ = area of the heated surface, m^2 (ft^2)
$\quad\quad T_g$ = gas temperature, °C (°F)
$\quad\quad T_s$ = temperature at sludge/gas interface, °C (°F)

The convection heat-transfer coefficient can be obtained from dryer manufacturers or from pilot studies.

Conduction.   In conduction (indirect) drying systems, a solid retaining wall separates the wet sludge from the heat-transfer medium, usually steam or another hot fluid. Heat transfer for conduction is determined by the following equation (U.S. EPA, 1979; WEF, 1998):

$$q_{\text{cond}} = h_{\text{cond}} A(T_m - T_s) \tag{14-31}$$

where q_{cond} = conductive heat-transfer rate, kJ/h (Btu/h)

$\quad h_{cond}$ = conductive heat-transfer coefficient, kJ/m²·h·°C (Btu/ft²·h·°F)

$\quad A$ = area of wetted surface exposed to gas, m² (ft²)

$\quad T_m$ = temperature of heating medium, °C (°F)

$\quad T_s$ = temperature sludge at drying surface, °C (°F)

The conductive heat-transfer coefficient, a composite term, includes the effects of the heat-transfer surface films of the sludge and the medium. The conduction heat-transfer coefficient can be obtained from dryer manufacturers or from pilot studies.

Radiation. In radiation (infrared) drying systems, infrared lamps, electric resistance elements, or gas-fired incandescent refractories supply radiant energy that transfers to the wet sludge and evaporates moisture. Radiation heat transfer is expressed as follows (U.S. EPA, 1979; WEF, 1998):

$$q_{rad} = C_s A\sigma (T_r^4 - T_s^4) \tag{14-32}$$

where q_{rad} = radiation heat-transfer rate, kJ/h (Btu/h)

$\quad C_s$ = emissivity of the drying surface, dimensionless

$\quad A$ = sludge surface area exposed to radiant source, m² (ft²)

$\quad \sigma$ = Stefan-Boltzmann constant, 20.41 × 10⁻⁸ kJ/m²·h·K⁴ (1.713 × 10⁻⁹ Btu/ft²·h·°R⁴)

$\quad T_r$ = absolute temperature of radiant source, K (°R)

$\quad T_s$ = absolute temperature of the sludge drying surface, K (°R)

Process Description

Heat dryers are classified as follows: direct, indirect, combined direct-indirect, and infrared. Direct and indirect dryers are described below as they are the types most commonly used for municipal biosolids drying. Coal, oil, gas, infrared radiation, or dried sludge may be used as the means of supplying the energy for heat drying.

Direct Dryers. Direct (convection) dryers that have been used for drying municipal wastewater sludge are the flash dryer, rotary dryer, and fluidized-bed dryer. Although approximately 50 municipal flash dryers have been installed in the United States since 1940, only five or six currently remain in operation (WEF, 1998). Rotary dryers are used for a variety of industrial applications, and to a lesser extent for municipal wastewater sludge. Fluidized-bed drying of wastewater sludge is a relatively new application in the United States.

Flash Dryer Flash drying involves pulverizing the sludge in a cage mill or by an atomized suspension technique in the presence of hot gases. The equipment is designed so that the particles remain in contact with the turbulent hot gases long enough to accomplish mass transfer of moisture from sludge to the gases.

One operation involves a cage mill that receives a mixture of wet sludge or sludge cake and recycled dried sludge. The mixture contains approximately 40 to 50 percent moisture. The hot gases and sludge are forced up a duct in which most of the drying takes place and to a cyclone, which separates the vapor and solids. It is possible to

---

achieve a moisture content of 8 to 10 percent in this operation. The dried sludge may be used or sold as soil conditioner or it may be incinerated in a furnace in any proportion up to 100 percent of production. Flash-dried cake is extremely dusty, which creates the potential for fires and explosions (see following discussion on fire and explosion hazards).

Rotary Dryer Rotary dryers have been used for the drying of raw primary sludge, waste-activated sludge, and digested primary biosolids. A rotary dryer consists of a cylindrical steel shell that is rotated on bearings and usually is mounted with its axis at a slight slope from the horizontal (see Fig. 14–51a). The feed sludge is mixed with previously dried sludge cake in a blender located ahead of the dryer (see Fig. 14–51b). The blended feed sludge has a moisture content of approximately 65 percent that improves its ability to move through the dryer without sticking. The mixture and hot gases are conveyed to the discharge end of the dryer. During conveyance, axial flights along the

Figure 14–51

Rotary sludge dryer:
(a) isometric view,
(b) system schematic.

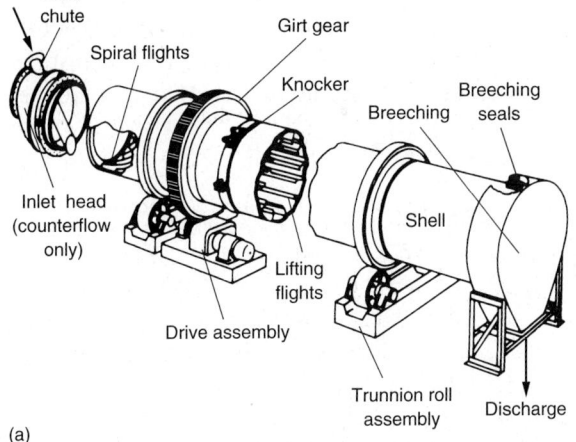

(a)

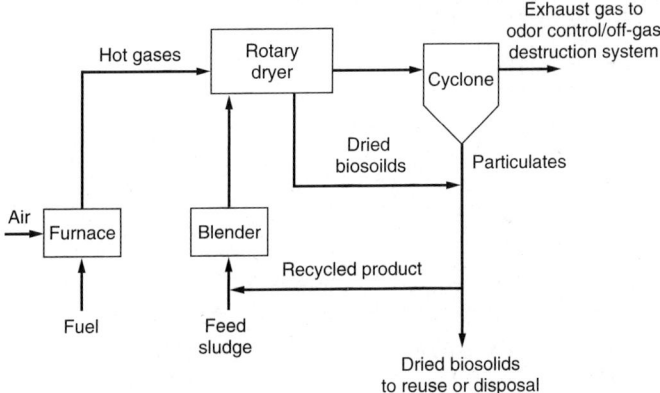

(b)

rotating interior wall pick up and cascade the sludge through the dryer. The product, which has a dry solids content of 90 to 95 percent, is screened and the oversize material passes through a crusher and then is transported to a recycle bin. The dried product is amenable to handling, storage, and marketing as a fertilizer or soil conditioner. Recently, a large rotary dryer facility for drying dewatered bisolids was installed in the Boston area for the Massachusetts Water Resources Authority.

Fluidized-Bed Dryer A new type of fluidized-bed dryer, developed in Europe, has been introduced recently into the United States (see Fig. 14–52). The dryer has the capability of producing a pellet product, similar to that obtained from rotary drying systems (Holcomb et al., 2000). The dryer system has certain components that are similar to rotary dryers: product classification and cooling of the product before storage and loading. The main differences are that a steam boiler and fluid-bed reactor are used in lieu of a hot gas furnace and rotary drum. The heat required for evaporation is supplied by steam via an in-bed heat exchanger. A uniform temperature of $\sim 120°C$ ($\sim 230°F$) is maintained in the bed through intimate contact between the sand granules and the fluidizing air.

Advantages cited for the dryer are (1) the product has a granular consistency suitable for use as a fuel or in soil blending to augment landfill cover, (2) the product has a low odor, (3) the system can be equipped for automatic operation, and (4) it has a smaller footprint than a rotary dryer. Disadvantages are that there is only one unit in

Figure 14–52

View of a fluidized-bed reactor. *(From Andritz.)*

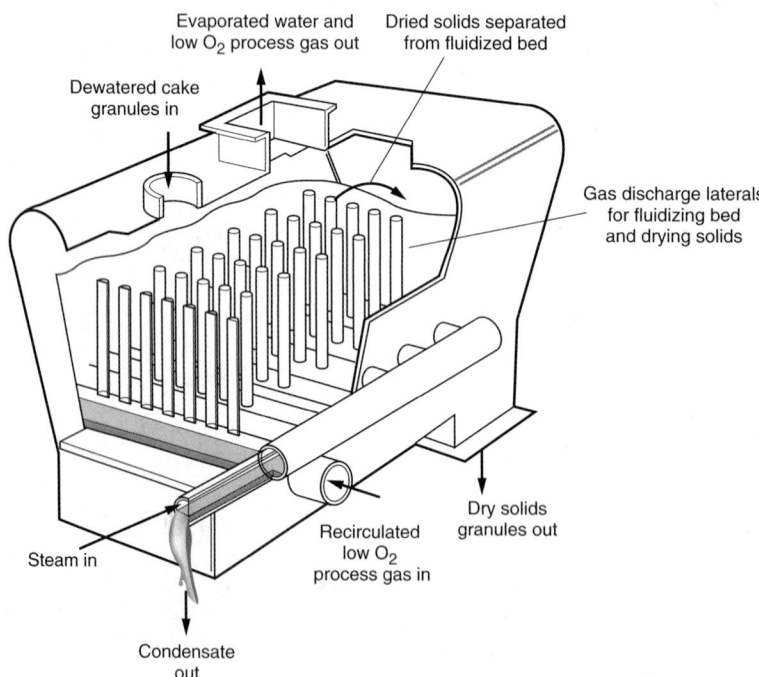

Evaporated water and low O_2 process gas out

Dried solids separated from fluidized bed

Dewatered cake granules in

Gas discharge laterals for fluidizing bed and drying solids

Steam in

Recirculated low O_2 process gas in

Dry solids granules out

Condensate out

operation in the United States (although several are in operation in Europe), and it uses a larger amount of electric power than a rotary-drum dryer because it requires a large induced-draft fan (Holcomb et al, 2000).

Indirect Dryers. Indirect dryers are designed in either a horizontal or vertical configuration. Horizontal dryers employ paddles, hollow flights, or disks mounted on one or more rotating shafts to convey biosolids through the dryer (see Fig. 14–53a). A heated medium, usually steam or oil, is circulated through the jacketed shell of the dryer and the hollow core of the rotating assembly. Dewatered biosolids are fed perpendicular to the dryer shaft and pass horizontally in a helical pattern through the dryer. The dryer performs the dual function of heat transfer and solids conveying. Drying occurs as the biosolids particles are broken up through agitation and come into contact with the heated metal surfaces in the dryer. The granular particles are both hot and abrasive. Moving parts, as in most dryers, will become abraded, and corrosion will accelerate the deterioration of the metal. The design of the agitator has to allow for efficient heat transfer, mixing of the sludge mass, and minimum fouling of the agitator. A weir at the discharge end ensures complete submergence of the heat-transfer surface in the biosolids being dried. The water vapor derived from the drying operation may be drawn off under a slightly negative pressure by an induced-draft fan located in the off-gas duct.

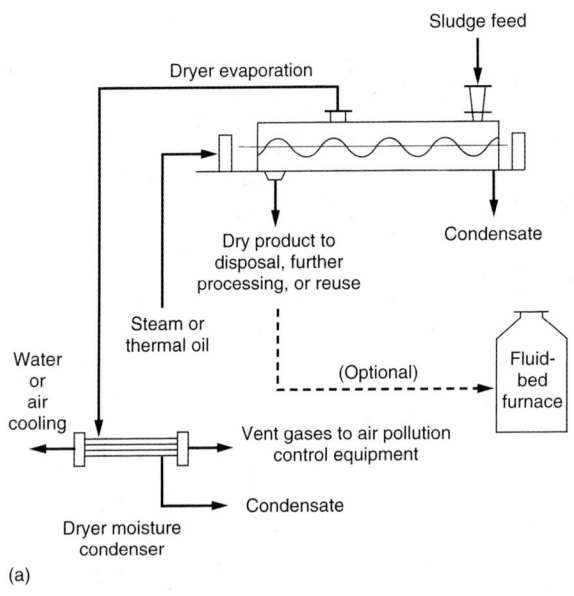

Figure 14–53

Indirect dryer: (a) system schematic, (b) exterior view of a horizontal dryer, and (c) interior view of paddle flights. (Courtesy Komline-Sanderson.)

Figure 14–54

Cross section of a vertical indirect dryer. (From Pelletech.)

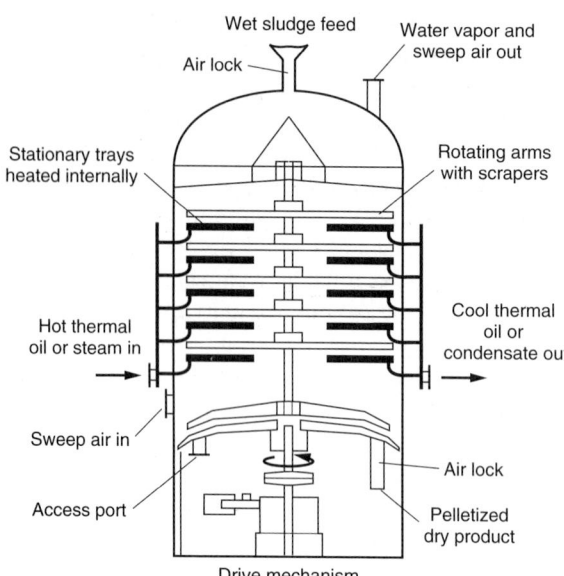

In vertical indirect dryers, sludge contacts a metal surface heated by a medium, such as steam or oil, and the heat is conductively transferred to the sludge (see Fig. 14–54). The sludge does not come in contact with the heating medium. Dewatered product (approximately 20 percent solids) mixed with recycled product is fed through the top inlet of a multistage dryer. Rotating arms move the sludge from the heated stationary tray to another in a rotating zigzag motion until it exits at the bottom as a dried, pelletized product. The rotating arms are equipped with adjustable scrapers that move and tumble the sludge in thin layers (20 to 30 mm) over the heated stationary trays. Dried product exits the dryer at a moisture content ranging from 5 to 8 percent and at a size of 2 to 4 mm (O'Brien and Schweizer, 1990).

In indirect dryers, sludge is dried to a specific level of dryness and discharged to a product conveyor for transfer to storage. Solids concentrations in the dryer product can range from 65 to over 95 percent depending on the ultimate use or disposal of the product.

Product Characteristics

The characteristics of the finished biosolids product depend on the type of sludge, the type of preprocessing, and the physical configuration of the drying surfaces. Raw primary sludge produces a fibrous, dusty material that is difficult to manage. Digested sludge can produce an amorphous particle that can be handled and transported. Particles will vary in size from 6 to 8 mm. The best size range for marketing may be from 3 to 5 mm. To maximize marketing potential, screening for sizes smaller and larger than the selected size range will be required. The fines and oversize particles can be returned to blend with the incoming sludge (see Fig. 14–51b), hence increasing the solids content entering the dryer but not changing the amount of moisture that must be evaporated.

In the range of 45 to 65 percent dry solids, the sludge mass passes through a very plastic and sticky phase that can impede the rotation of drying paddles or disks, particularly in horizontal dryers. Care should be taken in selecting equipment that is capable of being restarted if the dryer should be stopped while the sludge mass is in this phase.

Product Transport and Storage

Although the granular product from heat drying is reasonably durable, long conveyors that create an abrasive action such as screw conveyors and drag conveyors should be avoided. Even the use of pneumatic and dense-phase pneumatic conveyors may be too abrasive and cause crumbling and dust formation. Open or folded belt conveyors may be the preferred choices.

Upon exiting the dryer, the hot biosolids should be cooled to below 50°C (120°F) before placing in silos or storage vessels. The combination of initial heat plus heat that could be introduced by biological activity in the silo can cause smoldering or an open flame. This condition can occur where the drying operation is expected to pass through frequent start-and-stop cycles and drying may not be complete. In general, the product entering storage should be from 92 to 98 percent dry solids.

Fire and Explosion Hazards

With the fine particles and high levels of dryness in heat-dried sludge, hazards due to fire and explosion may exist when dried sludge is conveyed or stored. An organic dust suspended in air can rapidly combust if exposed to an ignition source. The heat of combustion can rapidly increase the volume and/or pressure of hot combustion products. If the pressure exceeds the rupture strength of the containing vessel, an explosion occurs. The phenomenon is called "deflageration," and deflageration explosions are the most serious concern when handling dried biosolids (Haug et al., 1993). Design considerations that are recommended for preventing dust explosions are given in Table 14–46.

Air Pollution and Odor Control

Two important control measures associated with heat drying of sludge are fly ash collection and odor control. Cyclone separators having efficiencies of 75 to 80 percent are suitable for vent gas temperatures up to 340 or 370°C (650 or 700°F). Wet scrubbers have higher efficiencies and will condense some of the organic matter in the vent gas but may carry over water droplets.

Sludge drying occurs in direct dryers at temperatures of approximately 370°C (700°F), whereas temperatures ranging from 650 to 760°C (1200 to 1400°F) are required for complete incineration. To achieve destruction of odors, the exhaust gases must reach approximately 730°C (1350°F). Thus, if the gases evolved in the drying process are reheated in an incinerator to a minimum of 730°C, odors will be diminished greatly. At lower temperatures, partial oxidation of odor-producing compounds may occur, resulting in an increase in the intensity or disagreeable character of odor produced. In the absence of an incinerator, thermal oxidizers have proved most effective in odor destruction. Thermally treated sludge vapors, aldehydes, and various species of sulfides and disulfides (methyl, dimethyl, and carbonyl) are difficult to remove without the use of a thermal oxidizer.

Table 14–46
Prevention measures to avoid dust hazards in the heat drying of sludge and biosolids[a]

Item	Prevention measure
Venting system (for processing, conveyance, and storage components)	Provide explosion relief vents Size explosion vents for "worst case" explosion in an air atmosphere
Nitrogen padding	Provide a nitrogen inerting atmosphere for all dried biosolids conveyance and processing facilities. Maintain oxygen levels below 5 percent by volume to reduce potential for self-heating and ignition of hot biosolids
Electrical equipment	Design in accordance with appropriate National Fire Protection Association criteria. If dust is present, all equipment must be dusttight and electronic cabinets nitrogen purged. Motor control centers that contain sparking devices, such as starters and relays, must be located outside classified areas
Ducts and vessels	Electrically bond and ground all conductive elements of the system that contact dried biosolids
Maintenance	Keep areas clean to prevent accumulation of dust. Any vessel containing powder must have powder removed before opening or the powder must be cooled to ambient temperatures before safe entry clearance is given
Miscellaneous	Eliminate or move outside all heat sources from classified areas. Equip electric motors located in a Class II, Division 2 area with Class F insulation to reduce "skin" temperatures

[a]Adapted from Haug et al. (1993).

14–15 INCINERATION

Incineration of sludge involves the total conversion of organic solids to oxidized end products, primarily carbon dioxide, water, and ash. The major advantages of incineration are (1) maximum volume reduction thereby lessening disposal requirements, (2) destruction of pathogens and toxic compounds, and (3) energy recovery potential (U.S. EPA, 1985a). Disadvantages include (1) high capital and operating cost, (2) highly skilled operating and maintenance staffs are required, (3) the residuals produced (air emissions and ash) may have adverse environmental effects, and (4) disposal of residuals, which may be classified as hazardous wastes, if they exceed prescribed maximum pollutant concentrations. Incineration is used most commonly by medium- to large-sized plants with limited disposal options.

Sludges processed by incineration are usually dewatered, untreated sludges. It is normally unnecessary to stabilize sludge before incineration. In fact, such practice may be detrimental because stabilization, specifically aerobic and anaerobic digestion, decreases the volatile content of the sludge and consequently increases the requirement for an auxiliary fuel. Sludges may be incinerated separately or in combination with municipal solid wastes.

The incineration processes considered in the following discussion include multiple-hearth incineration, fluidized-bed incineration, and coincineration with municipal solid waste. Before discussing these processes, it will be helpful to review some fundamental aspects of complete combustion.

Fundamental Aspects of Complete Combustion

Combustion is the rapid exothermic oxidation of combustible elements in fuel. Incineration is complete combustion. The predominant elements in the carbohydrates, fats, and proteins composing the volatile matter of sludge are carbon, oxygen, hydrogen, and nitrogen (C-O-H-N). The approximate percentages of these may be determined in the laboratory by a technique known as *ultimate analysis.*

Oxygen requirements for complete combustion of a material may be determined from a knowledge of its constituents, assuming that carbon and hydrogen are oxidized to the ultimate end products CO_2 and H_2O. The formula becomes

$$C_aO_bH_cN_d + (a + 0.25c - 0.5b)O_2 \rightarrow aCO_2 + 0.5cH_2O + 0.5dN_2 \qquad (14\text{--}33)$$

The theoretical quantity of air required will be 4.35 times the calculated quantity of oxygen, because air is composed of 23 percent oxygen on a mass basis. To ensure complete combustion, excess air amounting to about 50 percent of the theoretical amount will be required. A materials balance must be made to include the above compounds and the inorganic substances in the sludge, such as the inert material and moisture, and the moisture in the air. The specific heat of each of these substances and of the products of combustion must be taken into account in determining the heat required for the incineration process.

Heat requirements will include the sensible heat Q_s in the ash, plus the sensible heat required to raise the temperature of the flue gases to 760°C (1400°F) or whatever higher temperature of operation is selected for complete oxidation and elimination of odors, less the heat recovered in preheaters or recuperators. Latent heat Q_e must also be furnished to evaporate all of the moisture in the sludge. The total heat required may be expressed as

$$Q = \Sigma Q_s + Q_e = \Sigma C_pW_s (T_2 - T_1) + W_w\lambda \qquad (14\text{--}34)$$

where Q = total heat, kJ (Btu)

$\quad Q_s$ = sensible heat in the ash, kJ (Btu)

$\quad Q_e$ = latent heat, kJ (Btu)

$\quad C_p$ = specific heat for each category of substance in ash and flue gases, kJ/kg·°C (Btu/lb·°F)

$\quad W_s$ = mass of each substance, kg (lb)

$\quad W_w$ = mass of water, kg (lb)

$\quad T_1, T_2$ = initial and final temperatures

$\quad \lambda$ = latent heat of evaporation, kJ/kg (Btu/lb)

Reduction of moisture content of the sludge is the principal way to lower heat requirements, and the moisture content may determine whether additional fuel will be needed to support combustion.

The heating value of a sludge may be determined by (U.S. EPA, 1979; WEF, 1998):

$$Q = 33,829C + 144,696(H - O/8) + 9420S \qquad (14\text{--}35)$$

where Q = total value, kJ/kg dry solids (Btu/lb = kJ/kg $\times$ 0.43)
     C = carbon, percent by weight expressed as a decimal
     H = hydrogen, percent by weight expressed as a decimal
     O = oxygen, percent by weight expressed as a decimal
     S = sulfur, percent by weight expressed as a decimal

The fuel value of sludge ranges widely depending on the type of sludge and the volatile solids content. The fuel value of untreated primary sludge is the highest, especially if it contains appreciable amounts of grease and skimmings. Where kitchen food grinders are used, the volatile and thermal content of the sludge will also be high. Digested biosolids have significantly lower heating values than raw sludge. Typical heating values for various types of sludge are reported in Table 14–47. The heating value for sludge is equivalent to that of some of the lower grades of coal.

To design an incinerator for solids volume reduction, a detailed heat balance must be prepared. Such a balance must include heat losses through the walls and pertinent equipment of the incinerator, as well as losses in the stack gases and ash. Approximately 4.0 to 5 MJ (4000 to 5500 Btu) are required to evaporate each kg (2.2 lb) of water in the sludge. Heat is obtained from the combustion of the volatile matter in the sludge and from the burning of auxiliary fuels. For untreated primary sludge incineration, the auxiliary fuel is needed only for warming up the incinerator and maintaining the desired temperature when the volatile content of the sludge is low. The design should include provisions for auxiliary heat for startup and for assuring complete oxidation at the desired temperature under all conditions. Fuels such as oil, natural gas, or excess digester gas are suitable.

Multiple-Hearth Incineration

Multiple-hearth incineration is used to convert dewatered sludge cake to an inert ash. Because the process is complex and requires specially trained operators, multiple-hearth furnaces are normally used only in large plants. Multiple-hearth incinerators have been

Table 14–47
Typical heating values for various types of sludge and biosolids[a]

Type of sludge/biosolids	Btu/lb of total solids[b]		kJ/kg of total solids[b]	
	Range	Typical	Range	Typical
Raw primary	10,000–12,500	11,000	23,000–29,000	25,000
Activated	8,500–10,000	9,000	20,000–23,000	21,000
Anaerobically digested primary	4,000–6,000	5,000	9,000–14,000	12,000
Raw chemically precipitated primary	6,000–8,000	7,000	14,000–18,000	16,000
Biological filter	7,000–10,000	8,500	16,000–23,000	20,000

[a]Adapted, in part, from WEF (1988).
[b]Lower value applies to plants with long solids retention times.

Figure 14–55

Typical multiple-hearth incinerator.

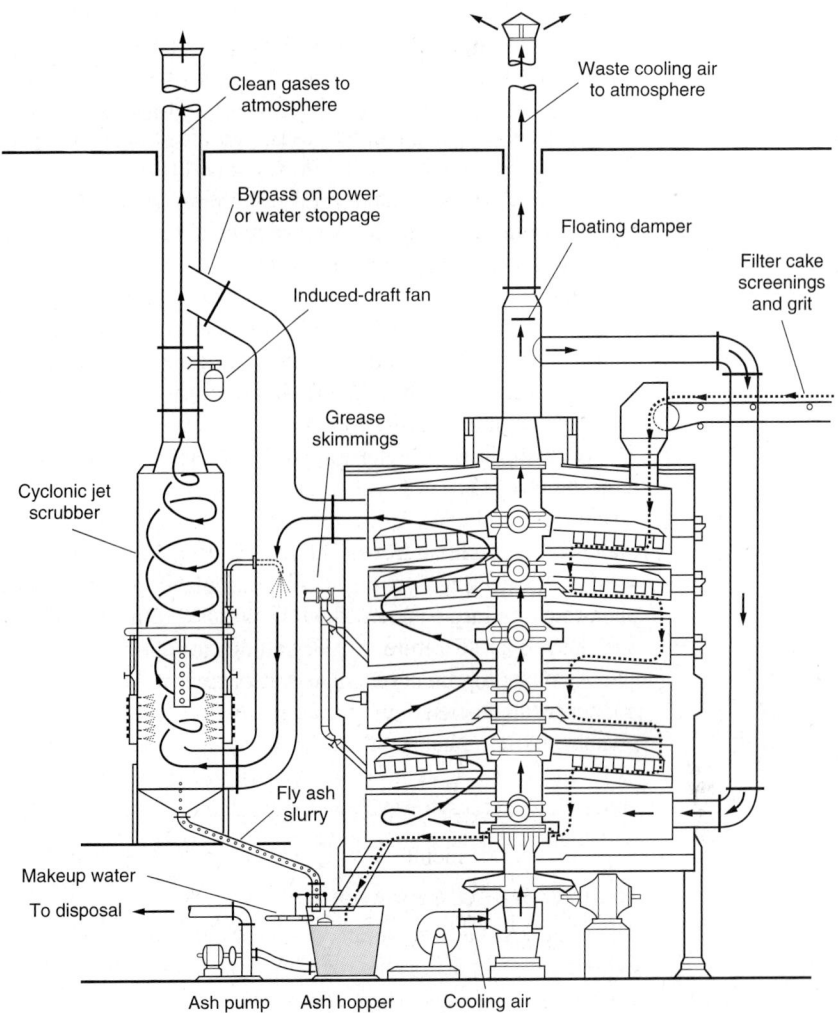

used at smaller facilities where land for the disposal of sludge is limited and at chemical treatment plants for the recalcining of lime sludges.

As shown on Fig. 14–55, sludge cake is fed onto the top hearth and is slowly raked to the center. From the center, sludge cake drops to the second hearth, where the rakes move it to the periphery. The sludge cake then drops to the third hearth and is again raked to the center. The hottest temperatures are on the middle hearths, where the sludge burns and where auxiliary fuel is also burned as necessary to warm up the furnace and to sustain combustion. Preheated air is admitted to the lowest hearth and is further heated by the sludge as the air rises past the middle hearths where combustion is occurring. The air then cools as it gives up its heat to dry the incoming sludge on the top hearths.

The highest moisture content of the flue gas is found on the top hearths where sludge with the highest moisture content is heated and some water is vaporized. Cooling air is initially blown into the central column and hollow rabble arms to keep them from overheating. A large portion of this air, after passing out of the central column at the top, is recirculated to the lowest hearth as preheated combustion air.

A multiple-hearth furnace may also be designed as a dryer only. In this case, a furnace is needed to provide hot gases, and the sludge and gases both proceed downward through the furnace in parallel flow. Parallel flow of product and hot gases is frequently used in drying operations to prevent burning or scorching heat-sensitive materials.

Feed sludge must contain more than 15 percent solids because of limitations on the maximum evaporating capacity of the furnace. Auxiliary fuel is required usually when the feed sludge contains between 15 and 30 percent solids. Feed sludge containing more than 50 percent solids may create temperatures in excess of the refractory and metallurgical limits of standard furnaces. Average loading rates of wet cake are approximately 40 kg/m²·h (8 lb/ft²·h) of effective hearth area but may range from 25 to 75 kg/m²·h (5 to 15 lb/ft²·h).

In addition to dewatering, required ancillary processes include ash handling systems and some type of wet or dry scrubber to meet air pollution requirements. In wet scrubbers, scrubber water comes in contact with and removes most of the particulate matter in the exhaust gases. The recycle BOD and COD is nil, and the total suspended solids content is a function of the particulates captured in the scrubber. Under proper operating conditions, particulate discharges to the air from wet scrubbers are less than 0.65 kg/10^3 kg (1.3 lb/ton) of dry sludge input.

Ash handling may be either wet or dry. In the wet system, the ash falls into an ash hopper located beneath the furnace, where it is slurried with water from the exhaust gas scrubber. After agitation, the ash slurry is pumped to a lagoon or is dewatered mechanically. In the dry system, the ash is conveyed mechanically to a storage hopper for discharge into a truck for eventual disposal as fill material. The ash is usually conditioned with water. Ash density is about 5.6 kg/m³ (0.35 lb/ft³) dry and 880 kg/m³ (55 lb/ft³) wet.

Fluidized-Bed Incineration

The fluidized-bed incinerator used commonly for sludge incineration is a vertical, cylindrically shaped, refractory-lined steel shell that contains a sand bed (media) and fluidizing air orifices to produce and sustain combustion (see Fig. 14–56). The fluidized-bed incinerator ranges in size from 2.7 to 7.6 m (9 to 25 ft) in diameter. The sand bed, when quiescent, is approximately 0.8 m (2.5 ft) thick, and rests on a brick dome or refractory-lined grid. The sand bed support area contains orifices, called "tuyeres," through which air is injected into the incinerator at a pressure of 20 to 35 kN/m² (3 to 5 lb$_f$/in²) to fluidize the bed. At low velocities, combustion gas "bubbles" appear within the fluidized bed. The main bed of suspended particles remains at a certain elevation in the combustion chamber and "boils" in place. Units that function in this manner are called "bubbling-bed" incinerators. The mass of suspended solids and gas, when active and at operating temperature, expands to about double the at-rest volume. Sludge is mixed quickly within the fluidized bed by the turbulent action of the bed. The minimum temperature needed in the sand bed prior to injection of sludge is approximately 700°C (1300°F). The temperature of the sand bed is controlled between 760 and 820°C (1400 and 1500°F). Evaporation of the water and combustion of the sludge solids takes place rapidly. Combustion gases and ash leave the bed and are transported through the freeboard area to the gas outlet through the top of the incinerator. No ash exits from the bed

Figure 14–56

Typical fluidized-bed
incinerator.

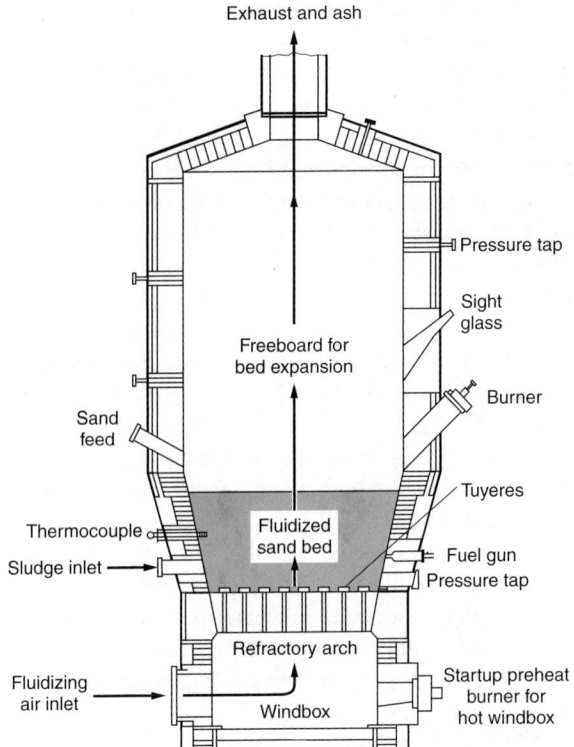

at the bottom of the incinerator. Combustion gases and entrained ash are scrubbed, normally with a venturi scrubber.

Recycle flows consist of scrubber water produced at a rate of approximately 25 to 40 L/kg (3 to 5 gal/lb) of dry solids feed to the fluidized bed. Most of the ash (99 percent) is captured in the scrubber water, and the total suspended solids content is approximately 20 to 30 percent of the dry solids feed. The recycle flow is normally directed to an ash lagoon. BOD and COD are nil. Particulates and other air emissions are comparable to those from the multiple-hearth incinerator.

The combustion process is controlled by varying the sludge feed rate and the air flow to the reactor to oxidize completely all the organic material. If the process is operated continuously or with shutdowns of short duration, there is no need for auxiliary fuel after startup.

A modification of the fluidized-bed incineration technology is the "circulating-bed" incinerator. In the circulating-bed unit, the reactor gas passes through the combustion chamber at much higher velocities, ranging from 3 to 8 m/s (10 to 25 ft/s). At these velocities, the bubbles in the fluidized bed disappear and streamers of solids and gas prevail. The entire mass of entrained particles flow up the reactor shaft to a particle separator, are deposited in storage momentarily, and are recirculated back to the primary combustion zone in the bottom of the reactor. Ash is removed continuously from the bottom of the bed. On turndown, the circulating bed becomes a bubbling bed.

Like the multiple hearth, the fluidized bed, though reliable, is complex and requires the use of trained personnel. Because fluidized-bed incinerators are complex, they are normally used in medium to large plants, but may be used in plants with lower flow ranges where land for the disposal of sludge is limited.

Coincineration with Municipal Solid Waste

Coincineration is the process of incinerating wastewater sludges with municipal solid wastes. The major objective is to reduce the combined costs of incinerating sludge and solid wastes. Coincineration is not practiced widely. The process has the advantages of producing the heat energy necessary to evaporate water from sludges, supporting combustion of solid wastes and sludge, and providing an excess of heat for steam generation, if desired, without the use of auxiliary fossil fuels. In properly designed systems, the hot gases from the process can be used to remove moisture from sludges to a content of 10 to 15 percent. Direct feeding of sludge cake containing 70 to 80 percent moisture over solid wastes on traveling or reciprocating grates has been found to be ineffective.

For systems operating without heat recovery, a disposal ratio of 1 kg (2.2 lb) of dry wastewater solids to 4.6 kg (11 lb) of solid wastes is fired in normal operation. In the case of the water-walled boiler with heat recovery, the ratio is approximately 1 kg of dry (industrial plant) solids to 7 kg (17 lb) of solid wastes.

Based on past experience in municipal solid-waste disposal, the application of coincineration will likely continue to proceed very slowly, despite the advantages to the community in combining the two waste-disposal functions.

Air-Pollution Control

Incineration methods for wastewater sludge have the potential to be significant contributors to air pollution. Air contaminants associated with incineration can be divided into two categories: (1) odors and (2) combustion emissions. Odors are particularly offensive to the human senses and special attention is required to minimize nuisance odor emissions. Combustion emissions vary depending upon the type of thermal reduction technology employed and the nature of the sludge and auxiliary fuel used in the combustion process. Combustion emissions of particular concern are particulates, oxides of nitrogen, acid gases, and specific constituents such as hydrocarbons and heavy metals (mercury, beryllium, etc.). Regulations promulgated by the U.S. EPA concerning incinerator emissions are discussed in Sec. 14–2.

14–16 SOLIDS MASS BALANCES

Solids-processing facilities, such as thickening, digestion, and dewatering, produce waste streams that must be recycled to the treatment process or to treatment facilities designed specifically for the purpose. The recycled flows impose an incremental solids, hydraulic, and organic and nutrient load on the wastewater-treatment facilities that must be considered in the plant design. When the flows are recycled to the treatment process, they should be directed to the head of the plant and blended with the plant flow following preliminary treatment. Equalization facilities can be provided for the recycled flows so that their reintroduction into the plant flow will not cause a shock loading on the subsequent treatment processes. To predict the incremental loads imposed by the recycled flows, it is necessary to perform a materials mass balance for the treatment system.

Preparation of Solids Mass Balances

Typically, a materials mass balance is computed on the basis of average flow and average BOD and total suspended solids concentrations. To size certain facilities properly, such as sludge storage tanks and plant piping, it is also important to perform a materials mass balance for the maximum expected concentration of BOD and TSS in the untreated wastewater. However, the maximum concentrations will not usually result in a proportional increase in the recycled BOD and TSS. The principal reason is that the storage capacity in the wastewater and sludge-handling facilities tends to dampen peak solids loads. For example, for a maximum TSS load equal to twice the average value, the resulting peak solids loading to a dewatering unit may be only 1.5 times the average loading. Further, it has been shown that periods of maximum hydraulic loading typically do not correlate with periods of maximum BOD and TSS. Therefore, coincident maximum hydraulic loadings should not be used in the preparation of a materials mass balance for maximum organic loadings (see Chap. 5). The preparation of a mass balance is illustrated in Example 14–11.

Performance Data for Solids-Processing Facilities

To prepare a materials mass balance, it is necessary to have information on the operational performance and efficiency of the various unit operations and processes that are used for the processing of waste solids. Representative data on the solids capture and expected solids concentrations for the most commonly used operations are reported in Tables 14–48 and 14–49. These data were derived from an analysis of the records from

Table 14–48
Typical solids concentrations and capture values for various solids-processing methods

Operation	Solids concentration, %		Solids capture, %	
	Range	Typical	Range	Typical
Gravity thickeners:				
Primary sludge only	4–10	6	85–92	90
Primary and waste-activated	2–6	4	80–90	85
Flotation thickeners:				
With chemicals	4–6	5	90–98	95
Without chemicals	3–5	4	80–95	90
Centrifuge thickeners:				
With chemicals	4–8	5	90–98	95
Without chemicals	3–6	4	80–90	85
Belt-filter press:				
With chemicals	15–30	22	85–98	93
Filter press:				
With chemicals	20–50	36	90–98	95
Centrifuge dewatering:				
With chemicals	10–35	22	85–98	92
Without chemicals	10–30	18	55–90	80

Table 14–49
Typical BOD and total suspended-solids (TSS) concentrations in the return flows from various processes[a]

Operation	BOD, mg/L		Suspended solids, mg/L	
	Range	Typical	Range	Typical
Gravity thickening supernatant:				
Primary sludge	100–400	250	80–350	200
Primary sludge + waste-activated sludge	60–400	300	100–350	250
Flotation thickening subnatant	50–1200	250	100–2500	300
Centrifuge thickening centrate	170–3000	1000	500–3000	1000
Aerobic digestion supernatant	100–1700	500	100–10,000	3400
Anaerobic digestion (two-stage, high-rate) supernatant	500–5000	1000	1000–11,500	4500
Centrifuge dewatering centrate	100–2000	1000	200–20,000	5000
Belt-filter press filtrate	50–500	300	100–2000	1000
Recessed-plate-filter press filtrate	50–250		50–1000	
Sludge lagoon supernatant	100–200		5–200	
Sludge drying bed underdrainage	20–500		20–500	
Composting leachate		2000		500
Incinerator scrubber water	30–80		600–8000	
Depth filter washwater	50–500		100–1000	
Microscreen washwater	100–500		240–1000	
Carbon adsorber washwater	50–400		100–1000	

[a] Adapted, in part, from U.S. EPA (1987b) and WEF (1998).

a number of installations throughout the United States. The wide variation that can occur in the reported values is apparent; thus, the values in Tables 14–48 and 14–49 should be used *only* if no other information is available. Wherever possible, local conditions and data should be used in performing the mass balance.

Impact of Return Flows and Loads

In addition to performance data for expected solids capture and constituent concentrations for the various process components, data for the expected concentrations of BOD and TSS in the return flows must also be included in preparing of mass balances. If the quantities and characteristics of recycled flows and loads are not accounted for properly, the facilities that receive them may be underdesigned significantly. The major impacts of return flows and measures that can mitigate these impacts are summarized in Table 14–50.

Table 14–50

Major impacts and potential mitigation measures for return flows from sludge and biosolids-processing facilities[a]

Source of return flow	Impact	Process impacted	Mitigation measure
Sludge thickening	Effluent degradation by colloidal SS	Sedimentation	Add flocculent aid ahead of sedimentation tank
			Separately thicken primary and biological sludges
			Optimize gravity thickener dilution water
	Floating sludge	Sedimentation	Minimize gravity thickener detention time
			Remove sludge continuously and uniformly
	Odor release and septicity	Recycle point	Reduce gravity thickener detention time
			Return flows ahead of aerated grit chamber
			Provide odor containment, ventilation, and treatment (scrubber or biofilter)
		Biological	Return odorous flows to aeration tank
			Remove sludge continuously and uniformly
			Provide separate return flow treatment (with other recycle streams)
	Solids buildup	Sedimentation	Increase dewatering unit operation time
		Biological	Remove sludge continuously and uniformly
			Include recycle loads in mass balance analysis
Sludge dewatering	Effluent degradation by colloidal suspended solids	Sedimentation	Optimize dewatering unit solids capture by improved sludge conditioning
			Add flocculent aid ahead of sedimentation tank
			Return centrate/filtrate to thickener
			Provide separate return flow treatment (with other recycle streams)
	Solids buildup	Sedimentation	Increase dewatering unit operation time
		Biological	Remove sludge continuously and uniformly
			Reduce trickling-filter recycle rate
			Include recycle loads in mass balance analysis
Sludge stabilization	Effluent degradation by excessive BOD load	Biological	Optimize supernant/decant removal, i.e., remove smaller amounts over a longer period of time, or reschedule removal to off-peak periods
			Provide separate return flow treatment
			Increase RBC speed
			Increase MLVSS in activated-sludge system (decrease F:M ratio)
			Increase dissolved oxygen level in activated-sludge process
	Effluent degradation by nutrients	Biological	Regulate digester supernatant/decant removal
			Thicken sludge before stabilization
			Provide separate return flow treatment

(continued)

| Table 14-50 (Continued)

Source of return flow	Impact	Process impacted	Mitigation measure
Washwater from depth filters	Hydraulic surges	Sedimentation	Provide backwash storage for flow equalization
			Schedule filter backwashing for off-peak periods

^aAdapted, in part, from U.S. EPA (1987b).

EXAMPLE 14–11   Preparation of Solids Mass Balance for a Secondary Treatment Facility   Prepare a solids balance for the treatment flow diagram shown in the following figure, using an iterative computational procedure.

1.   Definition of terms
   BOD_C = biochemical oxygen demand expressed as a concentration, g/m^3
   BOD_M = biochemical oxygen demand expressed as a mass, kg/d
   TSS_C = total suspended solids expressed as a concentration, g/m^3
   TSS_M = total suspended solids expressed as a mass, kg/d
   Assume for the purpose of this example that the following data apply:
2.   Wastewater flowrates
   a.   Average dry weather flow = 21,600 m^3/d
   b.   Peak dry weather flow = 2.5(21,600 m^3/d) = 53,900 m^3/d
3.   Influent characteristics
   a. BOD_C = 375 g/m^3
   b. TSS_C = 400 g/m^3
   c. TSS_C after grit removal = 360 g/m^3
4.   Solids characteristics
   a.   Concentration of primary solids = 6%
   b.   Concentration of thickened waste-activated sludge = 4%
   c.   Total suspended solids in digested sludge = 5%
   d.   For the purposes of this example, assume that the specific gravity of the solids from the primary sedimentation tank and the flotation thickener is equal to 1.0
   e.   Fraction of the biological solids that are biodegradable = 65%
   f.   The value of BOD_C can be obtained by multiplying the value of UBOD by a factor of 0.68 (corresponds to a k value of 0.23 d^{-1} in the BOD equation, see Chap. 2)
5.   Effluent characteristics
   a. BOD_C = 20 g/m^3
   b. TSS_C = 22 g/m^3

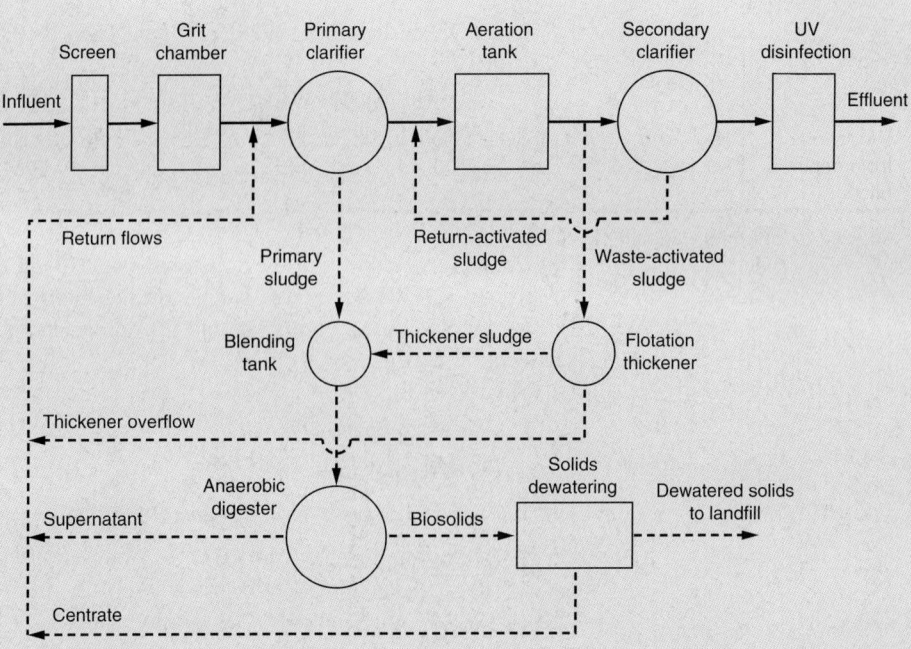

Solution

1. Convert the given constituent quantities to daily mass values.
   a. BOD_M in influent:

   $$BOD_M = (21{,}600 \text{ m}^3/\text{d})(375 \text{ g/m}^3)/(10^3 \text{ g/kg})$$

   $$= 8100 \text{ kg/d}$$

   b. TSS_M in influent:

   $$TSS_M = (21{,}600 \text{ m}^3/\text{d})(400 \text{ g/m}^3)/(10^3 \text{ g/kg})$$

   $$= 8640 \text{ kg/d}$$

   c. TSS_M after grit removal (influent to primary settling tanks):

   $$TSS_M = (21{,}600 \text{ m}^3/\text{d})(360 \text{ g/m}^3)/(10^3 \text{ g/kg})$$

   $$= 7776 \text{ kg/d}$$

2. Estimate the concentration of soluble BOD_C in the effluent using the following relationship:

   Effluent BOD_C = influent soluble BOD_C escaping treatment + BOD_C of effluent TSS_C

   a. Determine the BOD_C of the effluent TSS_C.
      i. Biodegradable portion of effluent TSS_C is 0.65 (22 g/m^3) = 14.3 g/m^3

        ii. UBOD of the biodegradable effluent TSS_C is $[0.65(22 \text{ g/m}^3)](1.42 \text{ g/g}) = 20.3 \text{ g/m}^3$

        iii. BOD_C of effluent suspended solids $= 20.3 \text{ g/m}^3 (0.68) = 13.8 \text{ g/m}^3$

    b. Solve for the influent soluble BOD_C escaping treatment.

$$20 \text{ g/m}^3 = S + 13.8 \text{ g/m}^3$$

$$S = 6.2 \text{ g/m}^3$$

3. Prepare the first iteration of the solids balance. (In the first iteration, the effluent wastewater TOTAL suspended solids and the biological solids generated in the process are distributed among the unit operations and processes that make up the treatment system.)

    a. Primary setting

        i. Operating parameters:

            BOD_C removed $= 33\%$

            TSS_C removed $= 70\%$ (see also Fig. 5–46)

        ii. BOD_M removed $= 0.33(8100 \text{ kg/d}) = 2700 \text{ kg/d}$

        iii. BOD_M to secondary $= (8100 - 2700) \text{ kg/d} = 5400 \text{ kg/d}$

        iv. TSS_M removed $= 0.7(7776 \text{ kg/d}) = 5443 \text{ kg/d}$

        v. TSS_M to secondary $= (7776 - 5443) \text{ kg/d} = 2333 \text{ kg/d}$

    b. Determine the volatile fraction of primary sludge.

        i. Operating parameters:

            Volatile fraction of TSS_C in influent $= 67\%$
            Volatile fraction of grit $= 10\%$
            Volatile fraction of incoming TSS_C discharged to the secondary process $= 85\%$

        ii. Volatile suspended solids (VSS_M) in influent prior to grit removal $= 0.67$ $(8640 \text{ kg/d}) = 5789 \text{ kg/d}$

        iii. VSS_M removed in grit chamber $= 0.10(8640 - 7776) \text{ kg/d} = 86 \text{ kg/d}$

        iv. VSS_M in secondary influent, kg/d $= 0.85(2333 \text{ kg/d}) = 1983 \text{ kg/d}$

        v. VSS_M in primary sludge, kg/d $= (5789 - 86 - 1983) \text{ kg/d} = 3710 \text{ kg/d}$

        vi. Volatile fraction in primary sludge $= [(3710 \text{ kg/d})/(5443 \text{ kg/d})](100\%)$ $= 68.2\%$

    c. Secondary process

        i. Operating parameters:

            Mixed-liquor $TSS_C = 4375 \text{ g/m}^3$

            Volatile fraction of mixed-liquor suspended solids $= 0.80 \ TSS_C$

            $Y_{obs} = 0.3125$

        ii. Determine the effluent mass quantities.

            $BOD_M = (21,600 \text{ m}^3/\text{d})(20 \text{ g/m}^3)/(10^3 \text{ g/kg}) = 432 \text{ kg/d}$

            $TSS_M = (21,600 \text{ m}^3/\text{d})(22 \text{ g/m}^3)/(10^3 \text{ g/kg}) = 475 \text{ kg/d}$

iii. Estimate the mass of volatile solids produced in the activated-sludge process that must be wasted. [The required value is computed using Eq. (8–14)].

$$P_{x,\text{VSS}} = Y_{\text{obs}} \, Q(S_o - S)/(10^3 \text{ g/kg})$$

$$= \frac{0.3125\,(21{,}600 \text{ m}^3/\text{d})[\,(250 - 6.2) \text{ g/m}^3\,]}{(10^3 \text{ g/kg})} = 1646 \text{ kg/d}$$

Note: The actual flowrate will be the primary influent less the flowrate of the primary underflow. However, the primary underflow is normally small and can be neglected. If the underflow is significant, the actual flowrate should be used to determine the volatile solids production.

iv. Estimate the TSS_M that must be wasted assuming the volatile fraction represents 0.80 of the total solids.

$$\text{TSS}_M = 1646/0.80 = 2057 \text{ kg/d}$$

Note: If it is assumed that the fixed solids portion of the influent suspended solids equals 0.15, the mass of fixed solids in the input from the primary settling facilities is equal to $0.15 \times 2333 = 350$ kg/d. This value can then be compared with the fixed solids determined in the above computations, which is equal to $2057 - 1646 = 411$ kg/d. The ratio of these values is 1.18[(411 kg/d)/(350 kg/d)]. Values that have been observed for this ratio vary from about 1.0 to 1.3; a value of 1.15 is considered to be the most representative.

v. Estimate the waste quantities discharged to the thickener. (It is assumed in this example that wasting is from the biological reactor.)

$$\text{TSS}_M = (2057 - 475) \text{ kg/d} = 1582 \text{ kg/d}$$

$$\text{Flowrate} = \frac{(1582 \text{ kg/d})(10^3 \text{ g/kg})}{(4375 \text{ g/m}^3)} = 362 \text{ m}^3/\text{d}$$

The assumed concentration value of MLSS of 4375 g/m^3 in the aeration tank will increase when the recycled BOD_C and TSS_C are taken into consideration in the second and subsequent iterations of the mass balance.

d. Flotation thickeners
   i. Operating parameters:

Concentration of thickened sludge = 4%

Assumed solids recovery = 90%

Assumed specific gravity of feed and thickened sludge = 1.0

   ii. Determine the flowrate of the thickened sludge.

$$\text{Flowrate} = \frac{(1582 \text{ kg/d})(0.9)}{(10^3 \text{ kg/m}^3)(0.04)} = 35.6 \text{ m}^3/\text{d}$$

iii. Determine the flowrate recycled to the plant influent.

Recycled flowrate $= (362 - 35.6)$ m³/d $= 326.4$ m³/d

iv. Determine the TSS_M to the digester.

$TSS_M = (1582$ kg/d$)(0.9) = 1424$ kg/d

v. Determine the TSS_M recycled to the plant influent.

$TSS_M = (1582 - 1424)$ kg/d $= 158$ kg/d

vi. Determine the BOD_C of the TSS_C in the recycled flow.

$$TSS_C \text{ in recycled flow} = \frac{(158 \text{ kg/d})(10^3 \text{ g/kg})}{(326 \text{ m}^3/\text{d})} = 485 \text{ g/m}^3$$

BOD_C of the $TSS_C = (485$ g/m³$)(0.65)(1.42)(0.68)$
$$= 304.6 \text{ g/m}^3$$

$BOD_M = (304.6$ g/m³$)(326$ m³/d$)/(10^3$ g/kg$) = 99$ kg/d

e. Sludge digestion
   i. Operating parameters:

   SRT $= 10$ d

   VSS destruction during digestion $= 50\%$

   Gas production $= 1.12$ m³/kg of VSS destroyed

   BOD_C in digester supernatant $= 1000$ g/m³ (0.1%)

   TSS_C in digester supernatant $= 5000$ g/m³ (0.5%)

   TSS_C in digested sludge $= 5\%$

   ii. Determine the total solids fed to the digester and the corresponding flowrate.

   $TSS_M =$ solids from primary settling plus waste solids from thickener

   $TSS_M = 5443$ kg/d $+ 1424$ kg/d $= 6867$ kg/d

   $$\text{Total flowrate} = \frac{(5443 \text{ kg/d})}{0.06(10^3 \text{ kg/m}^3)} + \frac{(1424 \text{ kg/d})}{0.04(10^3 \text{ kg/m}^3)}$$
   $$= (90.7 + 35.6) \text{ m}^3/\text{d} = 126.3 \text{ m}^3/\text{d}$$

   iii. Determine the VSS_M fed to the digester.

   $VSS_M = 0.682(5443$ kg/d$) + 0.80(1424$ kg/d$)$
   $$= (3712 + 1139) \text{ kg/d} = 4851 \text{ kg/d}$$

$$\text{Percent VSS}_M \text{ in mixture fed to digester} = \frac{(4851 \text{ kg/d})}{(6867 \text{ kg/d})} (100)$$

$$= 70.6\%$$

iv. Determine the VSS_M destroyed.

$$\text{VSS}_M = 0.5(4851 \text{ kg/d}) = 2426 \text{ kg/d}$$

v. Determine the mass flow to the digester.
Primary sludge at 6% solids:

$$\text{Mass flow} = \frac{(5443 \text{ kg/d})}{0.06} = 90{,}717 \text{ kg/d}$$

Thickened waste-activated sludge at 4% solids:

$$\text{Mass flow} = \frac{(1424 \text{ kg/d})}{0.04} = 35{,}600 \text{ kg/d}$$

Total mass flow $= (90{,}717 + 35{,}600) \text{ kg/d} = 126{,}317 \text{ kg/d}$

Note: The total mass flow can also be computed by multiplying the total flowrate to the digester by the density of the combined primary sludge and the thickened biosolids, if known.

vi. Determine the mass quantities of gas and sludge after digestion. Assume that the total mass of fixed solids does not change during digestion and that 50% of the volatile solids is destroyed.

Fixed solids $= \text{TSS}_M - \text{VSS}_M = (6867 - 4851) \text{ kg/d} = 2016 \text{ kg/d}$

TSS_M in digested sludge $= 2016 \text{ kg/d} + 0.5(4851 \text{ kg/d}) = 4441 \text{ kg/d}$

Gas production assuming that the density of digester gas is equal to 0.86 times that of air (1.204 kg/m^3, see Appendix B):

Gas $= (1.12 \text{ m}^3/\text{kg})(0.5)(4851 \text{ kg/d})(0.86)(1.204 \text{ kg/m}^3) = 2813 \text{ kg/d}$

Mass balance of digester output:

Mass input $= 126{,}317 \text{ kg/d}$

Less gas $= -2813 \text{ kg/d}$

Mass output $= 123{,}504 \text{ kg/d}$ (solids and liquid)

vii. Determine the flowrate distribution between the supernatant at 5000 mg/L and digested sludge at 5% solids. Let S = kg/d of supernatant suspended solids.

$$\frac{S}{0.005} + \frac{4441 - S}{0.05} = 123{,}504 \text{ kg/d}$$

$$S + 444.1 - 0.1S = 617.5 \text{ kg/d}$$

$$0.9S = 173$$

$$S = 192 \text{ kg/d}$$

$$\text{Digested solids} = (4441 - 192) \text{ kg/d} = 4249 \text{ kg/d}$$

$$\text{Supernatant flowrate} = \frac{(192 \text{ kg/d})}{0.005(10^3 \text{ kg/m}^3)} = 38.4 \text{ m}^3/\text{d}$$

$$\text{Digested sludge flowrate} = \frac{(4249 \text{ kg/d})}{0.05(10^3 \text{ kg/m}^3)} = 85 \text{ m}^3/\text{d}$$

viii. Establish the characteristics of the recycled flow.

$$\text{Flowrate} = 38.4 \text{ m}^3/\text{d}$$

$$\text{BOD}_C = (38.4 \text{ m}^3/\text{d})(1000 \text{ g/m}^3)/(10^3 \text{ g/kg}) = 38 \text{ kg/d}$$

$$\text{TSS}_M = (38.4 \text{ m}^3/\text{d})(5000 \text{ g/m}^3)/(10^3 \text{ g/kg}) = 192 \text{ kg/d}$$

f. Sludge dewatering. (*Note:* In the analysis that follows, the weight of the polymer or other sludge-conditioning chemicals that may be added was not considered. In some cases, their contribution can be significant and must be considered.)

  i. Operating parameters for centrifuge:

  Sludge cake = 22% solids

  Specific gravity of sludge = 1.06

  Solids capture = 93%

  Centrate BOD_C = 2000 mg/L

  ii. Determine the sludge-cake characteristics.

  $$\text{Solids} = (4249 \text{ kg/d})(0.93) = 3952 \text{ kg/d}$$

  $$\text{Volume} = \frac{(3952 \text{ kg/d})}{1.06(0.22)(10^3 \text{ kg/m}^3)} = 16.9 \text{ m}^3/\text{d}$$

  iii. Determine the centrate characteristics.

  $$\text{Flow} = (85 - 16.9) \text{ m}^3/\text{d} = 68.1 \text{ m}^3/\text{d}$$

  $$\text{BOD}_M \text{ (at 2000 g/m}^3) = (2000 \text{ g/m}^3)(68.1 \text{ m}^3/\text{d})/(10^3 \text{ g/kg})$$

  $$= 136 \text{ kg/d}$$

  $$\text{TSS}_M = (4249 \text{ kg/d})(0.07) = 297 \text{ kg/d}$$

g. Prepare a summary table of the recycle flows and waste characteristics for the first iteration.

Operation	Flow, m³/d	BOD$_M$, kg/d	TSS$_M$, kg/d
Flotation thickener	326.0	99	158
Digester supernatant	38.4	38	192
Centrate	68.1	136	297
Totals	432.5	273	647[a]

[a] The volatile fraction of the returned suspended solids will typically vary from 50 to 75 percent. A value of 60 percent will be used for the computation in the second iteration.

4. Prepare the second iteration of the solids balance.
   a. Primary settling
      i. Operating parameters = same as those in the first iteration
      ii. TSS$_M$ and BOD$_M$ entering the primary tanks

   $$TSS_M = \text{influent } TSS_M + \text{recycled } TSS_M$$

   $$= 7776 \text{ kg/d} + 647 \text{ kg/d} = 8423 \text{ kg/d}$$

   $$\text{Total } BOD_M = \text{influent } BOD_M + \text{recycled } BOD_M$$

   $$= 8100 \text{ kg/d} + 273 \text{ kg/d} = 8373 \text{ kg/d}$$

      iii. BOD$_M$ removed = 0.33(8373 kg/d) = 2763 kg/d
      iv. BOD$_M$ to secondary = (8,373 − 2763) kg/d = 5610 kg/d
      v. TSS$_M$ removed = 0.7(8423 kg/d) = 5896 kg/d
      vi. TSS$_M$ to secondary = (8423 − 5896) kg/d = 2527 kg/d
   b. Determine the volatile fraction of the primary sludge and effluent suspended solids.
      i. Operating parameters:

      Incoming wastewater = same as those for the first iteration

      Volatile fraction of solids in recycle returned to headworks = 60%

      ii. Although the computations are not shown, the computed change in the volatile fractions determined in the first iteration is slight and, therefore, the values determined previously are used for the second iteration. If the volatile fraction of the return is less than about 50%, the volatile fractions should be recomputed.
   c. Secondary process
      i. Operating parameters = same as those for the first iteration and as follows:

      Aeration tank volume = 4700 m³

      SRT = 10 d

      $Y = 0.50$ kg/kg

      $k_d = 0.06$ d^{-1}

ii. Determine the BOD_C in the influent to the aeration tank.

Flowrate to aeration tank = influent flowrate + recycled flowrate

$$= (21,600 + 432.5) \text{ m}^3/\text{d} = 22,033 \text{ m}^3/\text{d}$$

$$BOD_C = \frac{(5610 \text{ kg/d})(10^3 \text{ g/kg})}{(22,032.5 \text{ m}^3/\text{d})} = 255 \text{ g/m}^3$$

iii. Determine the new concentration of mixed liquor VSS.

$$X_{vss} = \frac{(Q)(Y)(S_o - S)\text{SRT}}{(\text{SRT})(Q)(Y)(S_o - S)[(1 + (k_d)\text{SRT}]}$$

$$X_{vss} = \frac{(22,032.5 \text{ m}^3/\text{d})(0.5)[(255 - 6.2) \text{ g/m}^3](10 \text{ d})}{4700 \text{ m}^3[1 + (0.06 \text{ d}^{-1})(10 \text{ d})]} = 3648 \text{ g/m}^3$$

iv. Determine the mixed liquor suspended solids.

$$X_{SS} = \frac{X_{VSS}}{0.8}$$

$$X_{SS} = 3648/0.8 = 4560 \text{ g/m}^3$$

v. Determine the cell growth.

$$P_{x,VSS} = Y_{obs} \, Q(S_o - S)/(10^3 \text{ g/kg})$$

$$= \frac{0.3125(22,032.5 \text{ m}^3/\text{d})[(255 - 6.2) \text{ g/m}^3]}{(10^3 \text{ g/kg})} = 1714 \text{ kg/d}$$

$$P_{x,TSS} = 1714/0.8 = 2143 \text{ kg/d}$$

vi. Determine the waste quantities discharged to the thickener.

Effluent TSS_M = 432 kg/d (specified in the first iteration)

Total TSS_M to be wasted to the thickener = (2143 − 432) kg/d

$$= 1711 \text{ kg/d}$$

$$\text{Flowrate} = \frac{(1711 \text{ kg/d})(10^3 \text{ g/kg})}{(4560 \text{ g/m}^3)} = 375 \text{ m}^3/\text{d}$$

d. Flotation thickeners
    i. Operating parameters:

Concentration of thickened sludge = 4%

Assumed solids recovery = 90%

Assumed specific gravity of feed and thickened sludge = 1.0

ii. Determine the flowrate of the thickened sludge.

$$\text{Flowrate} = \frac{(1711 \text{ kg/d})(0.9)}{(10^3 \text{ kg/m}^3)(0.04)} = 38.5 \text{ m}^3/\text{d}$$

iii. Determine the flowrate recycled to the plant influent.

Recycled flowrate = $(375 - 38.5) \text{ m}^3/\text{d} = 336.5 \text{ m}^3/\text{d}$

iv. Determine the TSS_M to the digester.

$\text{TSS}_M = (1711 \text{ kg/d})(0.9) = 1540 \text{ kg/d}$

v. Determine the TSS_M recycled to the plant influent.

$\text{TSS}_M = (1711 - 1540) \text{ kg/d} = 171 \text{ kg/d}$

vi. Determine the BOD_C of the TSS_C in the recycled flow.

$$\text{TSS}_C \text{ in recycled flow} = \frac{(171 \text{ kg/d})(10^3 \text{ g/kg})}{(336.5 \text{ m}^3/\text{d})} = 508 \text{ g/m}^3$$

BOD_C of $\text{TSS}_C = (508 \text{ g/m}^3)(0.65)(1.42)(0.68) = 319 \text{ g/m}^3$

$\text{BOD}_M = (319 \text{ g/m}^3)(336.5 \text{ m}^3/\text{d})(10^3 \text{ g/kg})^{-1} = 107 \text{ kg/d}$

e. Sludge digestion
   i. Operating parameters = same as those in the first iteration
   ii. Determine the total solids fed to the digester and the corresponding flowrate.

$\text{TSS}_M = \text{TSS}_M$ from primary settling plus waste TSS_M from thickener

$\text{TSS}_M = 5443 \text{ kg/d} + 1540 \text{ kg/d} = 6983 \text{ kg/d}$

$$\text{Total flowrate} = \frac{(5443 \text{ kg/d})}{0.06(10^3 \text{ kg/m}^3)} + \frac{(1540 \text{ kg/d})}{0.04(10^3 \text{ kg/m}^3)}$$

$$= (90.7 + 38.5) \text{ m}^3/\text{d} = 129.2 \text{ m}^3/\text{d}$$

iii. Determine the total VSS_M fed to the digester.

$\text{VSS}_M = 0.682(5443 \text{ kg/d}) + 0.80(1540 \text{ kg/d})$

$= (3712 + 1232) \text{ kg/d} = 4944 \text{ kg/d}$

$$\text{Percent VSS in mixture fed to digester} = \frac{(4944 \text{ kg/d})}{(6983 \text{ kg/d})}(100)$$

$$= 71.3\%$$

iv. Determine the VSS destroyed.

VSS destroyed = 0.5(4944 kg/d) = 2472 kg/d

v. Determine the mass flow to the digester.

Primary sludge at 6% solids:

$$\text{Mass flow} = \frac{(5443 \text{ kg/d})}{0.06} = 90{,}717 \text{ kg/d}$$

Thickened waste-activated sludge at 4% solids:

$$\text{Mass flow} = \frac{(1540 \text{ kg/d})}{0.04} = 38{,}500 \text{ kg/d}$$

Total mass flow = (90,717 + 38,500) kg/d = 129,217 kg/d

vi. Determine the mass quantities of gas and sludge after digestion. Assume that the total mass of fixed solids does not change during digestion and that 50% of the volatile solids is destroyed.

Fixed solids = TSS_M − VSS_M = (6983 − 4944) kg/d = 2039 kg/d

TSS in digested sludge = 2039 kg/d + 0.5(4944) kg/d = 4511 kg/d

Gas production assuming that the density of digester gas is equal to 0.86 times that of air (1.204 kg/m^3):

Gas = (1.12 m^3/kg)(0.5)(4944 kg/d)(0.86)(1.204 kg/m^3) = 2867 kg/d

Mass balance of digester output:

Mass input = 129,217 kg/d

Less gas = − 2867 kg/d

Mass output = 126,350 kg/d (solids and liquid)

vii. Determine the flowrate distribution between the supernatant at 5000 mg/L and digested sludge at 5 percent solids. Let S = kg/d of supernatant suspended solids.

$$\frac{S}{0.005} + \frac{4511 - S}{0.05} = 126{,}350 \text{ kg/d}$$

S + 451.1 − 0.1S = 631.8 kg/d

0.9S = 180.7

S = 201 kg/d

Digested TSS_M = (4511 − 201) kg/d = 4310 kg/d

$$\text{Supernatant flowrate} = \frac{(201 \text{ kg/d})}{0.005(10^3 \text{ kg/m}^3)} = 40.2 \text{ m}^3/\text{d}$$

$$\text{Digested sludge flowrate} = \frac{(4310 \text{ kg/d})}{0.05(10^3 \text{ kg/m}^3)} = 86.2 \text{ m}^3/\text{d}$$

viii. Establish the characteristics of the recycled flow.

Flowrate $= 40.2 \text{ m}^3/\text{d}$

$\text{BOD}_M = (40.2 \text{ m}^3/\text{d})(1000 \text{ g/m}^3)/(10^3 \text{ g/kg}) = 40 \text{ kg/d}$

$\text{TBB}_M = (40.2 \text{ m}^3/\text{d})(5000 \text{ g/m}^3)/(10^3 \text{ g/kg}) = 201 \text{ kg/d}$

f. Sludge dewatering
  i. Operating parameters for centrifuge = same as those in the first iteration
  ii. Determine the sludge-cake characteristics.

$\text{TSS}_M = (4310 \text{ kg/d})(0.93) = 4008 \text{ kg/d}$

$$\text{Volume} = \frac{(4008 \text{ kg/d})}{1.06(0.22)(10^3 \text{ kg/m}^3)} = 17.2 \text{ m}^3/\text{d}$$

iii. Determine the centrate characteristics.

Flow $= (86.2 - 17.2) \text{ m}^3/\text{d} = 69 \text{ m}^3/\text{d}$

BOD_M (at 2000 g/m^3) $= (2000 \text{ g/m}^3)(69 \text{ g/m}^3)/(10^3 \text{ g/kg}) = 138 \text{ kg/d}$

$\text{TSS}_M = (4310 \text{ kg/d})(0.07) = 302 \text{ kg/d}$

g. Prepare a summary table of the recycle flows and waste characteristics for the second iteration.

| Operation/process | Flow, m³/d | BOD$_M$, kg/d | TSS$_M$, kg/d | Incremental change from previous iteration | | |
				Flow, m³/d	BOD$_M$, kg/d	TSS$_M$, kg/d
Flotation thickener	336.5	107	171	10.5	8	13
Digester supernatant	40.2	40	201	1.8	2	9
Belt-press filtrate	69.0	138	302	0.9	2	5
Totals	445.7	285	674	13.2	12	27

5. Because the incremental change in the return quantities is less than 5 percent, the values summarized in the above table are acceptable for design. Given that the above computations would be done on a spreadsheet program, additional iterations could be made to obtain an incremental change of less than 1 percent. The flow, TSS$_M$, and BOD$_M$ values for the various processes from the second iteration are presented in following figure.

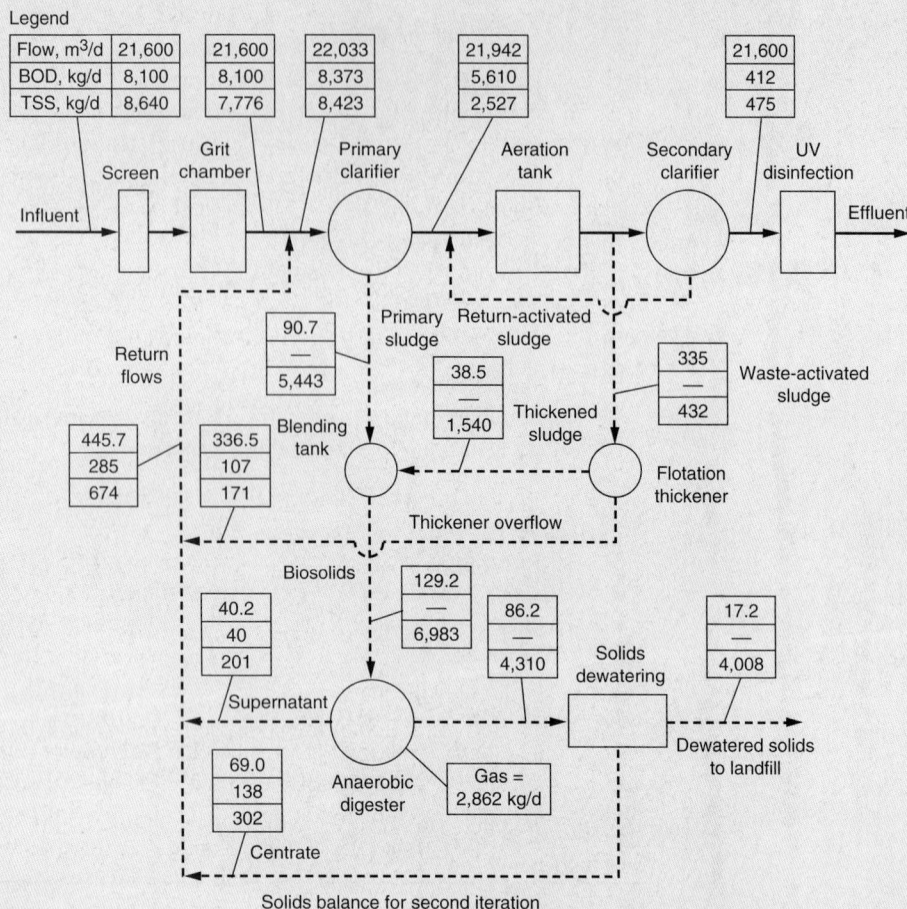

Solids balance for second iteration

Comment   In this example, an iterative approach was used to illustrate the computational steps in preparing a solids mass balance. Solids balances can be prepared using a specially designed spreadsheet or a proprietary solids balance software program. In general, if the iterative computational procedure is used, similar to the method used in this example, it should be carried out until the incremental change in all of the return quantities from the previous iteration is equal to or less than 5 percent.

14–17 APPLICATION OF BIOSOLIDS TO LAND

Land application of biosolids is defined as the spreading of biosolids on or just below the soil surface. Biosolids may be applied to (1) agricultural land, (2) forest land, (3) disturbed land, and (4) dedicated land disposal sites. In all four cases, the land application is designed with the objective of providing further biosolids treatment. Sunlight, soil microorganisms, and desiccation combine to destroy pathogens and many toxic organic substances. Trace metals are trapped in the soil matrix and nutrients are taken up by plants and converted to useful biomass. In some cases, a geomembrane liner is installed below a dedicated land disposal area.

To qualify for application to agricultural and nonagricultural land, biosolids or material derived from biosolids must meet at least the pollutant ceiling concentrations, Class B requirements for pathogens, and vector attraction requirements. Bulk biosolids applied to lawns and home gardens and biosolids that are sold or given away in bags or containers must meet the Class A criteria and one of several available vector-attraction reduction processes.

The application of biosolids to land for agricultural purposes is beneficial because organic matter improves soil structure, tilth, water holding capacity, water infiltration, and soil aeration; and macronutrients (nitrogen, phosphorus, potassium) and micronutrients (iron, manganese, copper, chromium, selenium, and zinc) aid plant growth. Organic matter also contributes to the cation-exchange capacity (CEC) of the soil which allows the soil to retain potassium, calcium, and magnesium. The presence of organic matter improves the biological diversity in soil and improves the availability of nutrients to the plants (Wegner, 1992). Nutrients in the biosolids also serve as a partial replacement for expensive chemical fertilizers.

Land application can also be of great value in silviculture and site reclamation. Forest utilization has been practiced extensively in the northwest, and biosolids application has been recognized as being beneficial to forest growth (WEF, 1998). Reclamation of disturbed land such as superfund sites has also been successful (Henry and Brown, 1997).

Site Evaluation and Selection

A critical step in land application of biosolids is finding a suitable site. The characteristics of the site will determine the actual design and will influence the overall effectiveness of the land-application concept. The sites considered potentially suitable will depend on the land-application option or options being considered, such as application to agricultural lands, forest lands, etc. The site-selection process should include an initial screening on the basis of the factors and criteria described in the following discussion. For screening purposes, it is necessary to have at least a rough estimate of land-area requirements for each feasible option.

Ideal sites for land application of biosolids have deep silty loam to sandy loam soils, groundwater deeper than 3 m (10 ft), slopes at 0 to 3 percent, no wells, wetlands, or streams, and few neighbors. Site characteristics of importance are topography, soil characteristics, soil depth to groundwater, and accessibility and proximity to critical areas.

Topography. Topography is important as it affects the potential for erosion and runoff. Suitability of site topographies also depends on the type of biosolids and the method of application. As shown in Table 14–51, liquid biosolids can be spread, sprayed, or injected onto sites with rolling terrain up to 15 percent in slope. Dewatered sludge is usually spread on agricultural land that requires a tractor and spreader. Forested sites can accommodate slopes up to 30 percent if adequate setbacks from streams are provided.

Soil Characteristics. In general, desirable soil characteristics include (1) loamy soil, (2) slow to moderate permeability, (3) soil depth of 0.6 m (2 ft) or more, (4) alkaline

Table 14–51
Typical slope
limitations for
land application
of biosolids

Slope, %	Comment
0–3	Ideal; no concern for runoff or erosion of liquid or dewatered biosolids
3–6	Acceptable; slight risk of erosion; surface application of liquid or dewatered biosolids is acceptable
6–12	Injection of liquid biosolids required for general cases, except in closed drainage basin and/or when extensive runoff control is provided; surface application of dewatered biosolids is generally acceptable
12–15	No application of liquid biosolids should be made without extensive runoff control; surface application of dewatered biosolids is acceptable, but immediate incorporation into the soil is recommended
Over 15	Slopes greater than 15 percent are suitable only for sites with good permeability where the length of slope is short and where the area with steep slope is a minor part of the total application area

or neutral soil pH (pH > 6.5), and (5) well drained to moderately well drained. Practically any soil can be adapted to a well-designed and well-operated system.

Soil Depth to Groundwater. A basic philosophy inherent in federal and state regulations is to design biosolids application systems that are based on sound agronomic principles, so that biosolids pose no greater threat to groundwater than current agricultural practices. Because the groundwater fluctuates on a seasonal basis in many soils, difficulties are encountered in establishing an acceptable minimum depth to groundwater. The quality of the underlying groundwater and the biosolids application option have to be considered carefully, especially where groundwater nondegradation restrictions apply. Generally, the greater the depth to the water table, the more desirable a site is for biosolids application. At least 1 m (3 ft) to groundwater is preferred for land-application sites. Seasonal water-table fluctuations to within 0.5 m (1.5 ft) of the surface can be tolerated. If the shallow groundwater is excluded as a drinking-water aquifer, the groundwater depth can be as shallow as 0.5 m before problems with trafficability of the soil arise. The presence of faults, solution channels, and other similar connections between soil and groundwater is undesirable unless the depth of overlying soil is adequate. When a specific site or sites are being considered for biosolids application, a detailed field investigation may be necessary to obtain the required groundwater information.

Accessibility and Proximity to Critical Areas. Buffer zones or setbacks are needed to separate the active application area from sensitive areas such as residences, wells, roads, surface waters, and property boundaries. Local and state regulations often include minimum distances for setbacks depending on the method of application; example minimum setback distances used in California are listed in Table 14–52.

U.S. EPA Regulations for Beneficial Use and Disposal of Biosolids

As discussed in Sec. 14–2, the U.S. EPA published regulations for biosolids (sewage sludge is the term used in the regulations) use and disposal under the code of Federal

Table 14–52
Typical setback distances for land application of biosolids[a]

Setback from	Minimum distance	
	ft	m
Property boundaries	10	3
Domestic water supply wells	500	150
Nondomestic water supply wells	100	30
Public roads and onsite occupied residences	50	15
Surface waters (wetlands, creeks, ponds, lakes, underground aqueducts, and marshes)	100	30
Primary agricultural drainangeways	33	10
Occupied nonagricultural buildings and offsite residences	500	150
Domestic water supply reservoir	400	120
Primary tributary to a domestic water supply	200	60
Domestic surface water supply intake	2500	750

[a]California State Water Resources Control Board (2000).

Table 14–53
U.S. EPA sludge regulations for land application of biosolids

Classification	Class A: no restrictions[a] Class B: site restrictions
Management practices	See Table 14–54
Pathogen-reduction alternatives	See Table 14–9
Vector attraction reduction	See Table 14–10
Site restrictions for Class B biosolids	See Table 14–55
Metal limits and loading rates	See Table 14–57

[a]Other than bag labeling (similar to fertilizer).

Regulations (CFR), 40 CFR Part 503. For land application, the regulations provide numerical limits on 10 metals, management practice guidance, and requirements for monitoring, record keeping, and reporting. The regulations are summarized in Table 14–53 and discussed in the following paragraphs.

Management Practices. Management practices that must be followed when biosolids are applied on land are specified in the Part 503 rule (see Table 14–54). The practices vary depending on whether the material that is applied is hauled in bulk or in individual bags.

Pathogen-Reduction Alternatives. As discussed in Chap. 1 and in Sec. 14–2, the Part 503 pathogen-reduction requirements for biosolids are divided into Class A and Class B categories (see Table 14–9). The goal of the Class A requirements is to reduce the pathogens in the biosolids (including *Salmonella* sp. bacteria, enteric viruses, and

Table 14–54
Land-application
management
practices under U.S.
EPA Part 503 rule[a]

For bulk biosolids[b]

Bulk biosolids cannot be applied to flooded, frozen, or snow-covered agricultural land, forests, public contact sites, or reclamation sites in such a way that the biosolids enter a wetland or other waters of the United States (as defined in 40 CFR Part 122.2), except as provided in a permit issued pursuant to Section 402 (NPDES permit) or Section 404 (Dredge and Fill Permit) of the Clean Water Act, as amended

Bulk biosolids cannot be applied to agricultural land, forests, or reclamation sites that are 10 m or less from U.S. waters, unless otherwise specified by the permitting authority

If applied to agricultural lands, forests, or public contact sites, bulk biosolids must be applied at a rate that is equal to or less than the agronomic rate for the site. Biosolids applied to reclamation sites may exceed the agronomic rate if allowed by the permitting authority

Bulk biosolids must not harm or contribute to the harm of a threatened or endangered species or result in the destruction or adverse modification of the species' critical habitat when applied to the land. Threatened or endangered species and their critical habitats are listed in Section 4 of the Endangered Species Act. Critical habitat is defined as any place where a threatened or endangered species lives and grows during any stage of its life cycle. Any direct or indirect action (or the result of any direct or indirect action) in a critical habitat that diminishes the likelihood of survival and recovery of a listed species is considered destruction or adverse modification of a critical habitat

For biosolids sold or given away in a bag or other container for application to the land

A label must be affixed to the bag or other container, or an information sheet must be provided to the person who receives this type of biosolids in another container. At a minimum, the label or information sheet must contain the following information:

- The name and address of the person who prepared the biosolids for sale or giveaway in a bag or other container

- A statement that prohibits application of the biosolids to the land except in accordance with the instructions on the label or information sheet

- An AWSAR (annual whole sludge application rate) for the biosolids that does not cause the annual pollutant loading rate limits to be exceeded

---

[a] Adapted from U.S. EPA (1995).
[b] These management practices do not apply if the biosolids are of "exceptional quality."

viable helminth ova) to below detectable levels. When this goal is achieved, Class A biosolids can be land applied without any pathogen-related restrictions on the site (U.S. EPA, 1995).

The goal of the Class B requirements is to ensure that pathogens have been reduced to levels that are unlikely to pose a threat to public health and the environment under specific use conditions. Site restrictions on land application of Class B biosolids minimize the potential for human and animal contact with the biosolids until environmental factors have reduced pathogens to below detectable levels.

Vector Attraction Reduction. There are 10 potential vector attraction reduction measures that can be combined with pathogen-reduction alternatives for an acceptable

Table 14–55
Site restrictions for
Class B biosolids[a]

Restrictions for the harvesting of crops and turf

1. Food crops with harvested parts that touch the biosolids/soil mixture and are totally above ground shall not be harvested for 14 months after application of biosolids

2. Food crops with harvested parts below the land surface where biosolids remains on the land surface for 4 months or longer prior to incorporation into the soil shall not be harvested for 20 months after biosolids application

3. Food crops with harvested parts below the land surface where biosolids remains on the land surface for less than 4 months prior to incorporation shall not be harvested for 38 months after biosolids application

4. Food crops, feed crops, and fiber crops whose edible parts do not touch the surface of the soil shall not be harvested for 30 d after biosolids application

5. Turf grown on land where biosolids are applied shall not be harvested for 1 year after application of the biosolids when the harvested turf is placed on either land with a high potential for public exposure or a lawn, unless otherwise specified by the permitting authority

Restriction for the grazing of animals

1. Animals shall not be grazed on land for 30 d after application of biosolids to the land

Restrictions for public contact

1. Access to land with a high potential for public exposure, such as a park or ball field, is restricted for 1 year after biosolids application. Examples of restricted access include posting with no trespassing signs or fencing

2. Access to land with a low potential for public exposure (e.g., private farmland) is restricted for 30 d after biosolids application. An example of restricted access is remoteness

[a]From U.S. EPA (1995).

land-application project using Class B biosolids (see Table 14–10). The list in Table 14–10 also includes some stabilization processes that reduce pathogens.

Site Restrictions for Class B Biosolids.   Site restrictions, listed in Table 14–55, depend on the crops to be used and the contact control for animals and the public. Food crops and turf grass are given the longest time restrictions because of the potential for public exposure (U.S. EPA, 1995).

Exceptional Quality Biosolids.   The category of "exceptional quality" biosolids has been defined as those biosolids that meet metal standards, Class A pathogen reduction standards, and vector reduction standards as defined in the Part 503 regulations.

Design Loading Rates

Design loading rates for land application of biosolids can be limited by pollutants (heavy metals) or by nitrogen. The long-term loadings of heavy metals are based on U.S. EPA 503 regulations. The annual loading rate is usually limited by the nitrogen loading rate.

Nitrogen Loading Rates. Nitrogen loading rates are set typically to match the available nitrogen provided by commercial fertilizers (Chang et al., 1995). Because municipal biosolids represent a slow-release organic fertilizer, a combination of ammonia and organic nitrogen must be made according to Eq. (14–36).

$$L_N = [(NO_3) + k_v(NH_4) + f_n(N_o)]F \qquad (14\text{--}36)$$

where L_N = plant available nitrogen in the application year, g N/kg (lb N/ton)
NO_3 = percent nitrate nitrogen in biosolids, decimal
k_v = volatilization factor for ammonia loss
= 0.5 for surface-applied liquid sludge
= 0.75 for surface-applied dewatered sludge
= 1.0 for injected liquid or dewatered sludge
NH_4 = percent ammonia nitrogen in sludge, decimal
f_n = mineralization factor for organic nitrogen
= 0.5 for warm climates and digested sludge
= 0.4 for cool climates and digested sludge
= 0.3 for cold climates or composted sludge
N_o = percent organic nitrogen in sludge, decimal
F = conversion factor, 1000 g/kg of dry solids (lb/ton)

To use Eq. (14–36) requires knowledge of the method of application, the nitrogen content of the biosolids (nitrate, ammonia, and organic), the type of stabilization, and the type of climate. The use of the mineralization factors simplifies the previously used method of calculating the amount of organic nitrogen mineralized each year and adding up the total for an annual equivalent. The use of Eq. (14–36) is also appropriate if biosolids are applied to a single site once every 2 to 3 years.

The loading rate based on nitrogen loadings is then calculated from Eq. (14–37).

$$L_{SN} = U/N_p \qquad (14\text{--}37)$$

where L_{SN} = biosolids loading rate based on N, Mg/ha·yr (tons/ac·yr)
U = crop uptake of nitrogen, kg/ha (lb/ac) (see Table 14–56)
N_p = plant available nitrogen in sludge, g/kg (lb/ton)

Loading Rates Based on Pollutant Loading. The pollutants of concern are those listed in Table 14–57. To calculate the biosolids loading rate based on pollutant loading use Eq. (14–38).

$$L_S = \frac{L_C}{CF} \qquad (14\text{--}38)$$

where L_S = maximum amount of biosolids that can be applied per year, Mg/ha·yr (tons/ac·yr)
L_C = maximum amount of constituent that can be applied per year, kg/ha·yr (lb/ac·yr)
C = pollutant concentration in biosolids, decimal (mg/kg)
F = conversion factor, 0.001 kg/Mg (2000 lb/ton)

Table 14–56
Typical nitrogen uptake values for selected crops[a]

Crop	Nitrogen uptake lb/ac·yr	kg/ha·yr	Crop	Nitrogen uptake lb/ac·yr	kg/ha·yr
Forage crops			**Tree crops**		
Alfalfa	200–600	220–670	Eastern forests:		
Brome grass	115–200	130–220	Mixed hardwoods	200	225
Coastal Bermuda grass	350–600	390–670	Red pine	100	110
Kentucky bluegrass	175–240	195–270	White spruce	200	225
Quack grass	210–250	235–280	Pioneer succession	200	225
Orchard grass	220–310	250–350	Aspen sprouts	100	110
Reed canary grass	300–400	335–450	Southern forests:		
Ryegrass	160–250	180–280	Mixed hardwoods	250	280
Sweet clover[b]	155	175	Loblolly pine	200–250	225–280
Tall fescue	130–290	145–325	Lake states forest:		
Field crops			Mixed hardwoods	100	110
Barley	110	120	Hybrid poplar	140	155
Corn	155–180	175–200	Western forest:		
Cotton	65–100	70–110	Hybrid poplar	270	300
Grain sorghum	120	135	Douglas fir	200	225
Potatoes	200	225			
Soybeans	220	245			
Wheat	140	155			

[a] Adapted from U.S. EPA (1981).
[b] Legume crops can fix nitrogen from the air but will take up most of their nitrogen from applied wastewater.

Land Requirements. Once the minimum biosolids loading rate is determined [by comparing the values from Eqs. (14–37) and (14–38)], the field area can be calculated using Eq. (14–39).

$$A = \frac{B}{L_S} \tag{14–39}$$

where A = application area required, ha
B = biosolids production, Mg of dry solids/yr
L_S = design loading rate, Mg of dry solids/ha·yr

Table 14-57
Metals concentrations and loading rates for land application of biosolids[a]

Pollutant	Ceiling concentration[b]		Cumulative pollutant loading rate[c]		Pollutant concentration for exceptional quality[d]		Annual pollutant loading rate[e]	
	lb/ton	mg/kg	lb/ac	kg/ha	lb/ton	mg/kg	lb/ac	kg/ha
Arsenic	0.15	75	37	41	0.08	41	1.78	2.0
Cadmium	0.17	85	35	39	0.08	39	1.70	1.9
Chromium[f]	—	—	—	—	—	—	—	—
Copper	8.60	4300	1338	1500	3.00	1500	66.91	75
Lead	1.68	840	268	300	0.60	300	13.38	15
Mercury	0.11	57	15	17	0.03	17	0.76	0.85
Molybdenum[f]	0.15	75	—	—	—	—	—	—
Nickel	0.84	420	374	420	0.84	420	18.74	21
Selenium	0.20	100	89	100	0.20	100	4.46	5.0
Zinc	15.00	7500	2498	2800	15.00	2800	124.91	140

[a] Adapted from Federal Register (1993).
[b] Dry weight basis, Table 1 from Part 503 regulations, instantaneous maximum.
[c] Dry weight basis, Table 2 from 503 regulations.
[d] Dry weight basis, Table 3 from 503 regulations, monthly average.
[e] Table 4 from 503 regulations.
[f] A Feb. 25, 1994, Federal Register notice deleted chromium; deleted the molybdenum values for Tables 2, 3, and 4; and raised the selenium value in Table 3 from 36 to 100.

EXAMPLE 14-12 Metals Loadings in Land Application   A community has stockpiled biosolids in a storage lagoon. The lagoon needs to be cleaned and the biosolids disposed of to make room for a plant expansion. The metals concentrations (mg/kg) in the lagoon are as follows:

As = 40	Hg = 15
Cd = 56	Ni = 510
Cu = 2500	Se = 10
Pb = 750	Zn = 3400

Determine if the biosolids are acceptable for land application.

Solution

1. Compare the concentrations for the above metals to the ceiling concentration (column 2) and the pollutant concentration for exceptional quality (column 4) in Table 14-57.

a. All metals concentrations are under the ceiling limits in column 2. The biosolids are suitable for land application.
b. Cadmium, copper, lead, nickel, and zinc exceed the values for exceptional quality. Calculations of annual loadings are necessary.

2. Calculate the allowable annual biosolids loading rates, using Eq. (14–38), for the four metals using the annual pollutant loading rates in Table 14–57.

a. Cadmium-based loading rate.

$$L_S = \frac{L_{Cd}}{C_{Cd}(0.001)} = \frac{(1.9 \text{ kg/ha·yr})}{[(56 \text{ mg/kg})(0.001 \text{ kg/Mg})/(\text{mg/kg})]}$$
$$= 33.9 \text{ Mg/ha·yr}$$

b. Copper-based loading rate.

$$L_S = \frac{75}{2500(0.001)} = 30 \text{ Mg/ha·yr}$$

c. Lead-based loading rate.

$$L_S = \frac{15}{750(0.001)} = 20 \text{ Mg/ha·yr}$$

d. Zinc-based loading rate.

$$L_S = \frac{140}{3400(0.001)} = 41.1 \text{ Mg/ha·yr}$$

3. Compare the whole biosolids loading rates to determine the limiting rate. The 20 Mg/ha·yr biosolids loading based on lead is limiting.

Comment Nitrogen loadings typically are more limiting than metals loadings. If the nitrogen loading rate exceeds 20 Mg/ha·yr, then the lead loading rate will determine the whole biosolids loading rate.

Application Methods

Application methods for biosolids range from direct injection of liquid biosolids to surface spreading of dewatered biosolids. The method of application selected will depend on the physical characteristics of the biosolids (liquid or dewatered), site topography, and the type of vegetation present (annual field crops, existing forage crops, trees, or preplanted land).

Liquid Biosolids Application. Application of biosolids in the liquid state is attractive because of its simplicity. Dewatering processes are not required, and the liquid biosolids can be transferred by pumping. Typical solids concentrations of liquid biosolids applied to land range from 1 to 10 percent. Liquid biosolids may be applied to land by vehicular application or by irrigation methods similar to those used for wastewater distribution.

Vehicular application may be by surface distribution or by subsurface injection or incorporation. Limitations to vehicular application include limited tractability on wet soil and potential reduction in crop yields due to soil compaction from truck traffic. Use of high-flotation tires can minimize these problems.

Surface distribution may be accomplished by tank truck or tank wagon equipped with rear-mounted spreading manifolds or by tank trucks mounted with high-capacity spray nozzles or guns. Specially designed, all-terrain biosolids application vehicles with spray guns are ideally suited for biosolids application on forest. Vehicular surface application is the most common method used for field and forage croplands. The procedure used commonly for annual crops is to (1) spread biosolids prior to planting, (2) allow the biosolids to dry partially, and (3) incorporate the biosolids by disking or plowing. The process is repeated then after harvest.

Liquid biosolids can be injected below the soil surface by using tank wagon or tank trucks with injection shanks or incorporated immediately after surface application by using plows or disks equipped with biosolids distribution manifolds and covering spoons (see Fig. 14–57). Advantages of injection or immediate incorporation methods include minimization of potential odors and vector attraction, minimization of ammonia loss due to volatilization, elimination of surface runoff, and minimum visibility leading to better public acceptance. Injection shanks and plows are very disruptive to perennial forage crops or pastures. To minimize such effects, special grassland biosolids injectors have been developed (Crites and Tchobanoglous, 1998).

Irrigation methods include sprinkling and furrow irrigation. Typically, large-diameter, high-capacity sprinkler guns are used to avoid clogging problems. Sprinkling has been used mainly for application to forested lands and occasionally for application

(a)

(b)

Figure 14–57

Land application of liquid sludge: (a) self-contained vehicle used to haul and to inject the liquid sludge into the ground. Self-contained vehicles of the type shown are used for relatively small amounts of liquid sludge and (b) tractor equipped with subsurface liquid sludge injection tines. The liquid sludge to be injected is supplied by a hose connected to the injection device. The tethered sludge supply hose is dragged along by the tractor. The injected liquid sludge is disked in the ground using the tractor and disk such as shown in Fig. 14–58b.

to dedicated disposal sites that are relatively isolated from public view and access. Sprinklers can operate satisfactorily on land too rough or wet for tank trucks or injection equipment and can be used throughout the growing season. Disadvantages to sprinkling include power costs of high-pressure pumps, contact of biosolids with all parts of the crop, possible foliage damage to sensitive crops, potential odors and vector attraction problems, and potentially high visibility to the public.

Furrow irrigation can be used to apply biosolids to row crops during the growing season. Disadvantages associated with furrow irrigation are localized settling of solids and the potential for ponding of biosolids in the furrows, both of which can result in odor problems.

Dewatered Biosolids Application. Application of dewatered biosolids to the land is similar to an application of semisolid animal manure. The use of conventional manure spreaders is an important advantage because farmers can apply biosolids on their lands with their own equipment. Typical solids concentrations of dewatered biosolids applied to land range from 15 to 20 percent. Dewatered biosolids are spread most commonly using tractor-mounted box spreaders or manure spreaders followed by plowing or disking into the soil (see Fig. 14–58). For high application rates bulldozers, loaders, or graders may be used. For forest application, a side-slinging vehicle has been tested that can apply dewatered biosolids up to 60 m (200 ft) (Leonard et al., 1992).

Application to Dedicated Lands

Disturbed land reclamation and dedicated land disposal are two types of high-rate land application. Disturbed land reclamation consists of a one-time application of 110 to 220 Mg/ha (50 to 100 dry tons/ac) to correct adverse soil conditions. Lack of soil fertility and poor physical properties can be corrected by biosolids application to allow revegetation programs to proceed. For disturbed land reclamation to be the sole avenue

(a)

(b)

Figure 14–58

Land application of dewatered sludge: (a) typical example of vehicle used to apply dewatered sludge on the surface of the soil and (b) typical tractor and rotating two-way disk used to disk dewatered and liquid sludge into the ground.

for biosolids reuse, a large area of disturbed land must be available on an ongoing basis. Dedicated land disposal requires a site where high rates of biosolids application are acceptable environmentally on a continuing basis. Biosolids for a dedicated land disposal (DLD) operation should meet at least Class B requirements.

Site Selection. Siting criteria for a dedicated land disposal site are presented in Table 14–58. Major issues in DLD siting are nitrogen control and the avoidance of groundwater contamination. Groundwater contamination can be avoided by (1) locating sites remote from useful aquifers, (2) intercepting of leachate, and (3) constructing an impervious geological barrier. Low percolation rates and deep aquifers will substantially reduce or eliminate potential contamination effects.

Where groundwater nondegradation restrictions apply, it has been found that for most DLD sites it is less costly to excavate the site entirely, install a geomembrane liner, and replace the excavated material, than to dispose of the sludge by some other means (e.g., dewatering and landfilling). The limited amount, if any, of leachate collected from the liner is returned to the treatment plant for processing.

Loading Rates. Annual biosolids loading rates have ranged from 12 to 2250 Mg/ha (5 to 1000 tons/ac). The higher rates have been associated with sites that:

- Receive dewatered biosolids
- Mechanically incorporate the biosolids into the soil
- Have relatively low precipitation
- Have no leachate problems because of site conditions or project design

Table 14–58
Criteria for dedicated land disposal (DLD) sites for biosolids[a]

Parameter	Unacceptable condition	Ideal condition
Slope	Deep gullies, slope >12%	<3%
Soil permeability	>1 × 10⁵ cm/s[b]	≤10⁻⁷ cm/s[c]
Soil depth	<0.6 m (2 ft)	>3 m (10 ft)
Distance to surface water	<90 m (300 ft) to any pond or lake used for recreational or livestock purposes, or any surface water body officially classified under state law	>300 m (1000 ft) from any surface water
Depth to groundwater	<3m (10 ft) to groundwater table (wells tapping shallow aquifers)[d]	>15 m (50 ft)
Supply wells	Within 300 m (1000 ft) radius	No wells within 600 m (2000 ft)

[a] Adapted from U.S. EPA (1983).
[b] Permeable soil can be used for DLD if appropriate engineering design preventing DLD leachate from reaching the groundwater is feasible.
[c] When low-permeability soils are at or too close to the surface, liquid disposal operations can be hindered due to water ponding.
[d] If an exempted aquifer underlies the site, poor-quality leachate may be permitted to enter groundwater.

Design loading rates for DLD can be estimated using Eq. (14–40).

$$L_S = \frac{E(TS)F}{100} - TS \tag{14–40}$$

where L_S = annual biosolids loading rate, Mg/ha (tons/ac)
E = net evaporation rate from soil, mm/yr (in/yr)
TS = total solids content, percent by weight
F = conversion factor, 10 Mg/mm (113.3 tons/in)

The net soil evaporation can be estimated from Eq. (14–41).

$$E = (f)E_L - P \tag{14–41}$$

where E = net evaporation rate from soil, mm/yr (in/yr)
f = 0.7
E_L = pan evaporation rate, mm/yr (in/yr)
P = annual precipitation, mm/yr (in/yr)

It should be noted that infiltration into the soil is not considered in Eq. (14–41). If infiltration is allowed, the term E should be increased by the annual infiltration rate in mm/yr.

Once the annual loading rate is calculated, the field area can be determined using Eq. (14–39) (dividing the biosolids production by the loading rate). Other area requirements include buffer zones, surface runoff control, roads, and supporting facilities.

Landfilling

Landfilling of biosolids in a monofill is covered under 40 CFR Part 503. Landfilling of biosolids in a sanitary landfill with municipal solid waste is regulated by the U.S. EPA under 40 CFR 258. If an acceptable site is convenient, landfilling can be used for disposal of biosolids, grit, screenings, and other solids. Stabilization may be required depending on state or local regulations. Dewatering of biosolids is usually required to reduce the volume to be transported and to control the generation of leachate from the landfill. In many cases, solids concentration is an important factor in determining the acceptability of biosolids in landfills. The sanitary landfill method is most suitable if it is also used for disposal of the other types of solid wastes. In a true sanitary landfill, the wastes are deposited in a designated area, compacted in place with a tractor or roller, and covered with a 350 mm (14 in) layer of clean soil. With daily coverage of the newly deposited wastes, nuisance conditions, such as odors and flies, are minimized.

14–18 BIOSOLIDS CONVEYANCE AND STORAGE

The solids removed as biosolids from preliminary and biological treatment processes are concentrated and stabilized by biological and thermal means and are reduced in volume in preparation for final disposal. Because the methods of conveyance and final disposal often determine the type of stabilization required and the amount of volume reduction that is needed, they are considered briefly in the following discussion.

Conveyance Methods

Biosolids may be transported long distances by (1) pipeline, (2) truck, (3) barge, (4) rail, or any combination of these four modes. To minimize the danger of spills, odors, and dissemination of pathogens to the air, liquid biosolids should be transported in closed vessels, such as tank trucks, railroad tank cars, or covered or tank barges. Stabilized, dewatered biosolids can be transferred in open vessels, such as dump trucks, or in railroad gondolas. If biosolids are hauled long distances, the vessels should be covered.

The method of transportation chosen and its costs are dependent on a number of factors, including (1) the nature, consistency, and quantity of biosolids to be transported; (2) the distance from origin to destination; (3) the availability and proximity of the transit modes to both origin and destination; (4) the degree of flexibility required in the transportation method chosen; and (5) the estimated useful life of the ultimate disposal facility.

Each transportation method contributes a minor air pollutant load, either directly or indirectly. A certain amount of air pollution is produced from the facility that generates electricity necessary for sludge pumping. The engines that move trucks, barges, and railroad cars also produce some air pollutants. On a mass (tonnage) basis, the transportation mode that contributes the lowest pollutant load is piping. Next, in sequence, are barging and unit train rail transportation. The highest pollutant load is from trucking. Other factors of environmental concern include traffic, noise, and construction disturbance.

Storage

It is often necessary to store biosolids that have been digested anaerobically before they are disposed of or used beneficially. Storage of liquid biosolids can be accomplished in storage basins, and storage of dewatered biosolids can be done on storage pads.

Storage Basins and Lagoons.   Biosolids stored in basins become more concentrated and are further stabilized by continued anaerobic biological activity. Long-term storage is effective in pathogen destruction.

Depth of the biosolids storage basins may vary from 3 to 5 m (10 to 16 ft). Solids loading rates vary from about 0.1 to 0.25 kg VSS/m²·d (20 to 50 lb VSS/10³ ft²·d) of surface area. If the basins are not loaded too heavily ($\leq$0.1 kg VSS/m²·d), it is possible to maintain an aerobic surface layer through the growth of algae and by atmospheric reaeration. Alternatively, surface aerators can be used to maintain aerobic conditions in the upper layers.

The number of basins to be used should be sufficient to allow each basin to be out of service for a period of about 6 months. Stabilized and thickened biosolids can be removed from the basins using a mud pump mounted on a floating platform or by mobile crane using a drag line. Biosolids concentrations as high as 35 percent solids have been achieved in the bottom layers of these basins.

Long-term storage of solids in lagoons is simple and economical if the treatment plant is in a remote location. A lagoon is an earthen basin into which untreated solids or digested biosolids are deposited. In lagoons with untreated solids, the organic matter is stabilized by anaerobic and aerobic decomposition, which may give rise to objectionable odors. The stabilized solids settle to the bottom of the lagoon and accumulate, and excess liquid from the lagoon, if there is any, is returned to the plant for treatment. Lagoons should be located away from highways and dwellings to minimize possible nuisance conditions and should be fenced to keep out unauthorized persons. Lagoons should be relatively shallow, 1.25 to 1.5 m (4 to 5 ft), if they are to be cleaned by scraping. If the lagoon is used only for digested biosolids, the nuisances mentioned should not be a problem. If subsurface drainage and percolation are potential problems, the lagoon should be lined. Solids may be stored indefinitely in a lagoon and may be removed periodically after draining and drying. The accumulation of solids can be estimated using the procedure illustrated in Example 8–15 in Chap. 8.

Storage Pads.   Where dewatered biosolids have to be stored prior to land application, sufficient storage area should be provided based on the number of consecutive days that biosolids hauling could occur without applying biosolids to land. Allowances also have to be made for paved access and for area to maneuver the biosolids hauling trucks, loaders, and application vehicles. The storage pads should be constructed of concrete or bituminous concrete and designed to withstand the truck loadings and biosolids piles. Provisions for leachate and stormwater collection and disposal also have to be included in the design of sludge storage pads.

PROBLEMS AND DISCUSSION TOPICS

14-1 The water content of a solids slurry is reduced from 98 to 95 percent. What is the percent reduction in volume by the approximate method and by the more exact method, assuming that the solids contain 70 percent organic matter of specific gravity 1.00 and 30 percent mineral matter of specific gravity 2.00? What is the specific gravity of the 98 and the 95 percent slurry?

14-2 Consider an activated-sludge treatment plant with a capacity of 40,000 m^3/d. The untreated wastewater contains 200 mg/L suspended solids. The plant provides 60 percent removal of the suspended solids in the primary settling tank. If the primary sludge alone is pumped, it will contain 5 percent solids. Assume that 400 m^3/d of waste-activated sludge containing 0.5 percent solids is to be wasted to the digester. If the waste-activated sludge is thickened in the primary settling tank, the resulting mixture will contain 3.5 percent solids. Calculate the reduction in daily volume of biosolids pumped to the digester that can be achieved by thickening the waste-activated sludge in the primary settling tank as compared with discharging the primary and waste-activated sludge directly to the digester. Assume complete capture of the waste-activated sludge in the primary settling tank.

14-3 For Example 14–4 for gravity thickening, calculate the amount of dilution water required at average design flow using the data provided to maintain a hydraulic loading rate of 12 $m^3/m^2 \cdot d$ for the thickener size computed in the example.

14-4 Determine the required digester volume for the treatment of the sludge quantities specified in Example 14–5 using the (a) volatile solids loading factor, and (b) volumetric per capita allowance methods. Set up a comparison table to display the results obtained using the three different procedures for sizing digesters (two in this problem and one in Example 14–5). Assume the following data apply:

1. Volatile solids loading method
    a. Solids concentration = 5%
    b. Detention time = 10 d
    c. Loading factor = 3.8 kg VSS/$m^3 \cdot d$

2. Volumetric loading method
   a. Per capita contribution = 0.72 g/capita·d
   b. Volume required = 50 m³/10³ capita·d

14-5 A wastewater-treatment plant is planning to provide for separate anaerobic sludge digestion for its primary sludge. The plant receives an influent wastewater with the following characteristics:

Average flow = 8000 m³/d

Suspended solids removed by primary sedimentation = 200 mg/L

Volatile matter in settled solids = 75%

Water in untreated sludge = 96%

Specific gravity of mineral solids = 2.60

Specific gravity of organic solids = 1.30

(a) Determine the required digester volume using an SRT of 12 d.
(b) Determine the minimum digester capacity using the recommended loading parameters of kilograms of volatile matter per cubic meter per day and cubic meters per 1000 persons.

14-6 Consider an industrial waste consisting mainly of carbohydrates in solution. Pilot-plant experiments using a complete-mix anaerobic digester without recycle yielded the following data:

Run	COD influent kg/d	X_T reactor, kg	P_x effluent kg/d
1	1000	428	85.7
2	500	115	46

Assuming a waste-utilization efficiency of 80 percent, estimate the percentage of added COD that can be stabilized when treating a waste load of 5000 kg/d. Assume that the design solids retention time (SRT) is 10 d.

14-7 A digester is loaded at a rate of 300 kg COD/d. Using a waste-utilization efficiency of 75 percent, what is the volume of gas produced when SRT = 40 d? $Y = 0.10$ and $k_d = 0.02$ d^{-1}.

14-8 Volatile acid concentration, pH, or alkalinity should not be used alone to control a digester. How should they be correlated to predict most effectively how close to failure a digester is at any time?

14-9 A digester is to be heated by circulation of sludge through an external hot water heat exchanger. Using the following data, find the heat required to maintain the required digester temperature:

(a) U_x = overall heat-transfer coefficient, W/m²·°C
(b) U_{air} = 0.85, U_{ground} = 1.2, U_{cover} = 1.0
(c) Digester is a concrete tank with floating steel cover; diameter = 11 m and sidewall depth = 8 m, 4 m of which is above the ground surface. The tank walls and floor are 300 mm thick.
(d) Sludge fed to digester = 15 m³/d at 14°C
(e) Outside temperature = −15°C

(*f*) Average ground temperature = 5°C

(*g*) Sludge in tank is to be maintained at 35°C

(*h*) Assume a specific heat of the sludge = 4200 J/kg·°C

(*i*) Sludge contains 4% solids

(*j*) Assume a cone-shaped cover with center 0.6 m above digester top, and a cone-shaped bottom with center 1.2 m below bottom edge.

14–10   The ultimate elemental analysis of a dried sludge yields the following data:

Element	Percent
Carbon	52.1
Oxygen	38.3
Hydrogen	2.7
Nitrogen	6.9
Total	100.0

How many kg of air will be required per kg of sludge for its complete oxidation?

14–11   Compute the fuel value of the sludge from a primary settling tank having a composition (by weight) of 64.5 percent carbon, 8.5 percent hydrogen, 21.0 percent oxygen, and 4 percent sulfur.

14–12   A community of 25,000 persons has asked you to serve as a consultant on their sludge disposal problems. Specifically, you have been asked to determine if it is feasible to compost the sludge from the primary clarifier with the community's solid waste. If this plan is not feasible, you have been asked to recommend a feasible solution. Currently the waste solids from the community's biological process are thickened in the primary clarifier. Assume the following data are applicable:

Solid waste data:

   Waste production = 2 kg/person·d

   Compostable fraction = 55%

   Moisture content of compostable fraction = 22%

Sludge production:

   Net sludge production = 0.12 kg/person·d

   Concentration of sludge in underflow from primary clarifier = 5%

   Specific gravity of underflow solids = 1.08

Compost:

   Final moisture content of composted biosolids/solid waste mixture = 55%

14–13   Prepare a solids balance for the peak loading condition for the treatment plant used in Example 14–11 for one of the parameter series as selected by the instructor. Enter your final values on the solids balance figure in Example 14–11.

	Parameter series			
	A	**B**	**C**	**D**
Peak flow rate, m³/d	54000	60000	50000	54000
Average BOD at peak flow, mg/L	340	300	350	300
Average TSS at peak flow, mg/L	350	320	330	320
TSS after grit removal, mg/L	325	300	310	300

Use data given in Example 14–11 for other parameters.

14–14 Discuss the advantages and disadvantages of dissolved air flotation, centrifugation, and gravity belt thickeners for thickening waste-activated sludge.

14–15 Using the Internet and a reference buyer's guide from the Water Environment Federation or similar organization to select candidate equipment manufacturers, obtain descriptive information for egg-shaped digesters and autothermal aerobic digestion (ATAD) processes from at least one manufacturer of each process. Compare the advantages and disadvantages of each process and submit the material obtained from the Internet web sites.

14–16 Prepare a solids balance, using the iterative technique delineated in Example 14–11, for the following flow diagram and one of the following parameter series as selected by the instructor. Also determine the effluent flowrate and suspended solids concentration.

	Parameter series			
	A	**B**	**C**	**D**
Influent characteristics				
Flowrate, m³/d	4000	10,000	20,000	40,000
Suspended solids, mg/L	1000	350	400	300
Sedimentation tank				
TSS removal efficiency, %	75	60	65	60
Underflow TSS concentration, %	7	6.5	6	5.5
Specific gravity of sludge	1.1	1.1	1.1	1.1
Alum addition				
Dosage, mg/L	10	10	20	15
Chemical solution, kg alum/L of solution	0.5	0.5	0.5	0.5
Filters				
TSS removal efficiency, %	90	90	95	92
Washwater solids concentration, %	6	6	6.8	6.5
Specific gravity of backwash	1.08	1.08	1.089	1.085
Thickener				
Supernatant TSS, mg/L	400	300	200	250
Concentration of solids in underflow, %	12	8	9	8

(continued)

(*Continued*)

	Parameter series			
	A	**B**	**C**	**D**
Chemical addition				
Dosage, % of underflow solids from thickener	0.8	1.0	1.0	1.0
Chemical solution, kg/L of solution	2.0	2.0	2.0	2.0
Filter press				
TSS concentration in filtrate, mg/L	200	300	250	200
Concentration of dewatered solids, %	40	38	42	42
Specific gravity of sludge cake	1.6	1.5	1.65	1.65

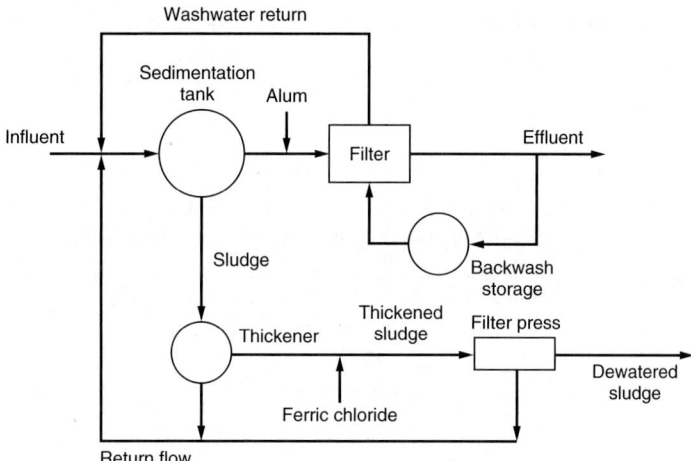

In preparing the solids balance assume that all of the unit operations respond linearly such that the removal efficiency for recycled solids is the same as that for the solids in the influent wastewater. Also assume that the distribution of the chemicals added to improve the performance of the filter and filter press is proportional to the total solids in the return flows and the effluent solids.

14-17 A municipality with a population of 200,000 has hired you as a consultant to investigate alternative mechanical sludge dewatering options. The three alternatives to be investigated are belt-press dewatering, centrifugation, and pressure filter-press dewatering. The biosolids to be dewatered are stabilized by anaerobic digestion, are a mixture of primary sludge and waste-activated sludge, and have a solids concentration of 5 percent. Ultimate disposal is by landfilling at a site located 50 kilometers from the treatment plant. Compare the various dewatering alternatives and recommend one. State the reasons for your recommendation.

14-18 Determine the dry sludge application rate for Reed canary grass on the basis of satisfying crop nitrogen uptake. Assume that biosolids containing 3 percent organic nitrogen by weight are applied to a soil that has an initial nitrogen content of zero. Use a mineralization rate of 30 percent for the first year, 15 percent for the second year, and 5 percent for the third and subsequent years.

14-19   Biosolids containing 50 ppm of cadmium on a dry basis are to be applied to land. If the limiting mass loading to the soil is set at 10 kg/ha, what would be the safe loading rate for 50 yr of application?

14-20   Compare the advantages and disadvantages of land application of liquid biosolids to dewatered biosolids. Assume the land application site is located 15 kilometers from the treatment plant in primarily an agricultural area and the biosolids are transported by truck. The biosolids are stabilized by anaerobic digestion and the liquid biosolids concentration is 6 percent and the dewatered solids concentration is 25 percent.

14-21   In problem 14–20, what would be the advantages and limitations of conveying the liquid biosolids by pipeline to the land application site? What types of facilities, i.e., structures, equipment, and vehicles, would be required and what are the operating and maintenance considerations?

REFERENCES

Albertson, O. E., and T. Walz (1997) "Optimizing Primary Clarification and Thickening," *Water Environment & Technology,* vol. 9, no. 12, pp. 41–45.

Arakaki, G., R. Vander Schaaf, S. Lewis, and G. Himaka (1998) "Design of Sludge Screening Facilities," *Proceedings of the 71st Annual Conference & Exposition,* vol. 2, Water Environment Federation, Alexandria, VA.

ASCE (1988) "Belt Filter Press Dewatering of Wastewater Sludge," American Society of Civil Engineers Task Force on Belt Press Filters, *Journal Environmental Engineering Division,* vol. 114, no. 5, pp. 991–1006.

ASCE (2000) *Conveyance of Residuals from Water and Wastewater Treatment,* ASCE Manuals and Reports on Engineering Practice No. 98, American Society of Civil Engineers, Reston, VA.

Babbitt, H. E., and D. H. Caldwell (1939) "Laminar Flow of Sludge in Pipes," *University of Illinois Bulletin* 319.

Brinkman, D., and D. Voss (1998) "Egg-Shaped Digesters—Are They All They're Cracked Up to Be?" *Proceedings of the 71st Annual Conference & Exposition,* vol. 2, Water Environment Federation, Alexandria, VA.

California State Water Resources Control Board (2000) "General Waste Discharge Requirements for the Discharge of Biosolids to Land for Use as a Soil Amendment in Agricultural, Silvicultural, Horticultural, and Land Reclamation Activities," Water Quality Order No. 2000-10-DWQ.

Carthew, G. A., C. A. Goehring, and J. E. van Teylingen (1983) "Development of Dynamic Head Loss Criteria for Raw Sludge Pumping," *Journal Water Pollution Control Federation,* vol. 55, p. 472.

Chang, A. C., A. L. Page, and T. Asano (1995) "Developing Human Health-Related Chemicals Guidelines for Reclaimed Wastewater and Sewage Sludge Applications in Agriculture," World Health Organization, Geneva, Switzerland.

Chu, A., and D. S. Mavinic (1998) "The Effects of Macromolecular Substrates and a Metabolic Inhibitor on Volatile Fatty Acid Metabolism in Thermophilic Aerobic Digestion," *Water Sci. Tech.* (G.B.), vol. 38, no. 55.

Clark, S. E., and D. N. Ruehrwein (1992) "Egg-Shaped Digester Mixing Improvements," *Water Environment & Technology,* vol. 4, no. 1.

Cooper, P. F., G. D. Job, M. B. Green, and R. B. E. Shutes (1996) *Reed Beds and Constructed Wetland for Wastewater Treatment,* WRc Swindon, ISBN 1.898920-27-3, Swindon, Wiltshire, England.

Crites, R. W., and G. Tchobanoglous (1998) *Small and Decentralized Wastewater Management Systems,* McGraw-Hill, New York.

Eckenfelder, W. W., Jr. (2000) *Industrial Water Pollution Control,* 3d ed., McGraw-Hill, New York.

Enviroquip (2000) *Aerobic Digestion Workshop,* vol. III, Enviroquip, Inc., Austin, TX.

EPRI (1998) *Mechanical Freeze/Thaw of Water and Wastewater Residuals,* Status Report, TR-112063, Electric Power Research Institute, Palo Alto, CA.

Epstein, E. (1997) *The Science of Composting,* Technomic Publishing Co., Lancaster, PA.

Farrell, J. (1999) "Summary of Designs," *Aerobic Digestion Workshop,* vol. III, sponsored by Enviroquip, Inc., Austin, TX.

Federal Register (1993) 40 CFR Part 503, Standards for the Disposal of Sewage Sludge.

Garelli, B. A., B. J. Swartz, and R. W. Dring (1992) "Improved Centrifuge Dewatering by Steam and Carbon Dioxide Injection," *Water Environment & Technology,* vol. 4, no. 6.

Grady, C. P. L., Jr., G. T. Daigger, and H. C. Lim (1999) *Biological Wastewater Treatment,* 2d ed., Marcel Dekker, New York.

Haug, R. T. (1993) *The Practical Handbook of Compost Engineering,* Lewis Publishers, Boca Raton, FL.

Haug, R. T., F. M. Lewis, G. Petino, W. J. Harnett (1993) "Explosion Protection and Fire Prevention at a Biosolids Drying Facility," *Proceedings of the 66th Annual Conference & Exposition,* Water Environment Federation, Alexandria, VA.

Henry, C., and S. Brown (1997) "Restoring a Superfund Site with Biosolids and Fly Ash," *Biocycle,* vol. 38, no. 11.

Hill, R. A., P. E. Snoek, and R. L. Gandhi (1986) "Hydraulic Transport of Solids," in *Pump Handbook,* by I. J. Karassik, W. C. Krutzsch, W. H. Fraser, and J. P. Medina (eds.), McGraw-Hill, New York.

Holcomb, S. P., B. Dahl, T. A. Cummings, and G. P. Shimp (2000) "Fluidized Bed Drying Replaces Incineration at Pensacola, Florida," *Proceedings of the 73rd Annual Conference & Exposition on Water Quality and Wastewater Treatment,* Water Environment Federation, Alexandria, VA.

Leonard, P., R. King, and M. Lucas (1992) "Fertilizing Forests with Biosolids; How to Plan, Operate, and Maintain a Long-Term Program," *Proceedings, The Future of Municipal Sludge (Biosolids) Management,* WEF Specialty Conference, pp. 233–250, Water Environment Federation, Alexandria, VA.

Liptak, B. G. (1974) *Environmental Engineers' Handbook,* Chilton Book Co., Radnor, PA.

Maco, R. S., H. D. Stensel, and J. F. Ferguson (1998) "Impacts of Solids Recycling Strategies on Anaerobic Digester Performance," *Proceedings of the 71st Annual Conference & Exposition,* Water Environment Federation, Alexandria, VA.

McAdams, W. H. (1954) *Heat Transmission,* 2d ed., McGraw-Hill, New York.

McCarty, P. L. (1964) "Anaerobic Waste Treatment Fundamentals," *Public Works,* vol. 95, nos. 9–12.

McCarty, P. L. (1968) "Anaerobic Treatment of Soluble Wastes," in E. F. Gloyna and W. W. Eckenfelder, Jr. (eds.), *Advances in Water Quality Improvement,* University of Texas Press, Austin, TX.

McFarland, M. J. (2000) *Biosolids Engineering,* McGraw-Hill, New York.

Metcalf & Eddy, Inc. (1981) *Wastewater Engineering—Collection and Pumping of Wastewater,* McGraw-Hill, New York.

Metcalf & Eddy, Inc. (1984a) "Improved Design and Operation of Anaerobic Digester Mixing Systems," Draft Report to U.S. Environmental Protection Agency, Contract 68-03-3208.

Metcalf & Eddy, Inc. (1984b) "Improved Design and Operation of Belt Filter Presses," Draft Report to U.S. Environmental Protection Agency, Contract 68-03-3208.

Metcalf & Eddy, Inc. (1984c) "Improved Design and Operation of Recessed Plate Filter Presses," Draft Report to U.S. Environmental Protection Agency, Contract 68-03-3208.

Metcalf & Eddy, Inc. (1991) *Wastewater Engineering: Treatment, Disposal, Reuse,* 3d ed., McGraw-Hill, New York.

Moen, G. (2000) "Comparison of Thermophilic and Mesophilic Digestion," Master's Thesis, Department of Civil and Environmental Engineering, University of Washington, Seattle, WA.

Mulbarger, M. C., S. R. Copas, J. R. Kordic, and F. M. Cash (1981) "Pipeline Friction Losses for Wastewater Sludges," *Journal Water Pollution Control Federation,* vol. 51, no. 8.

Murthy, S. (2001) Personal communication.

Nayyar, M. L. (1992) *Piping Handbook,* 6th ed., McGraw-Hill, New York.

Novak, J. (2001) Personal communication.

O'Brien, E. M. and E. Schweizer (1990) "Belgian Sludge Processing Holds Promise in US," *Water Environment & Technology,* vol. 2, no. 4.

Parker, P. J., A. G. Collins, and J. R. DeWolfe (2000) "Freeze-Thaw Residuals Conditioning," *Journal American Water Works Association,* vol. 92, no. 4.

Reed, S. C., R. W. Crites, and E. J. Middlebrooks (1995) *Natural Systems for Waste Management and Treatment,* 2d ed., McGraw-Hill, New York.

Roediger, H. (1987) "Using Quicklime—Hygienization and Solidification of Dewatered Sludge, Water Environment Federation," *Operations Forum,* pp. 18–21.

Roediger, M., and M. A. Vivona (1998) "Processes for Pathogen Reduction to Produce Class A Solids," *Proceedings of the 71st Annual Conference & Exposition Water Environment Federation,* pp. 137–148, Alexandria, VA.

Sanks, R. L., G. Tchobanoglous, D. Newton, B. E. Bosserman, and G. M. Jones (1998) *Pumping Station Design,* 2d ed., Butterworths, Stoneham, MA.

Schafer, P. L., and J. B. Farrell (2000a) "Turn Up the Heat," *Water Environment & Technology,* vol. 12, no. 11.

Schafer, P. L., and J. B. Farrell (2000b) "Performance Comparisons for Staged and High-Temperature Anaerobic Digestion Systems," *Proceedings of WEFTEC 2000,* Water Environment Federation, Alexandria, VA.

Sparr, A. E. (1971) "Pumping Sludge Long Distances," *Journal Water Pollution Control Federation,* vol. 43, no. 8.

Speece, R. E. (1988) "A Survey of Municipal Anaerobic Sludge Digesters and Diagnostic Activity Assays," *Wat. Res,* vol. 22, no. 3, pp. 365–372.

Speece, R. E. (2001) Personal communication.

Stensel, H. D., and T. E. Coleman (2000) "Assessment of Innovative Technologies for Wastewater Treatment: Autothermal Aerobic Digestion (ATAD)," *Preliminary Report,* Project 96-CTS-1.

Stukenberg, J. R., J. H. Clark, J. Sandine, and W. Naydo (1992) "Egg-Shaped Digesters: from Germany to the U.S.," *Water Environment & Technology,* vol. 4, no. 4, pp. 42–51.

Tchobanoglous, G., H. Theisen, and S. Vigil (1993) *Integrated Solid Waste Management,* McGraw-Hill, New York.

Torpey, W. N., and N. R. Melbinger (1967) "Reduction of Digested Sludge Volume by Controlled Recirculation," *Journal Water Pollution Control Federation,* vol. 39, p. 1464.

U.S. EPA (1974) *Process Design Manual for Upgrading Existing Wastewater Treatment Plants,* U.S. Environmental Protection Agency Office of Technology Transfer, Washington, D.C.

U.S. EPA (1979) *Process Design Manual Sludge Treatment and Disposal,* U.S. Environmental Protection Agency.

U.S. EPA (1981) *Process Design Manual for Land Treatment of Municipal Wastewater,* EPA/625/1-81-013, U.S. Environmental Protection Agency.

U.S. EPA (1983) *Process Design Manual for Land Application of Municipal Sludge,* EPA 625/1-83-016.

15

Issues Related to Treatment-Plant Performance

In the previous chapters of this book, special attention has been devoted to the unit operations and chemical and biological unit processes used for the treatment of wastewater. While the importance of these subjects cannot be overstated, a number of other issues related to treatment-plant performance are of equal, if not of greater, importance in the planning and design of wastewater management facilities. The two most important issues are the need to treat a broader range of constituents and the fact that discharge permit limits are becoming more stringent. In response to these issues, the following topics are addressed in this chapter: (1) the need for upgrading treatment-plant performance, (2) treatment process reliability and the selection of design values, (3) odor management, (4) process control strategies, (5) upgrading wastewater treatment plants, (6) energy efficiency in wastewater treatment, and (7) considerations in the design of new wastewater treatment plants.

15–1 NEED FOR UPGRADING TREATMENT-PLANT PERFORMANCE

To meet both current and future wastewater discharge permit requirements and to meet more stringent requirements for discharge and reuse, most wastewater treatment facilities will need to undergo significant modifications and improvements.

Meeting Current and Future Needs

In the 20th century the primary focus of wastewater treatment was on the removal and treatment of settleable and floatable solids, organic matter expressed as biochemical oxygen demand (BOD), total suspended solids (TSS), and pathogenic microorganisms (see Chap. 1). Late in the century, nutrient removal and odors also became an issue. In the 21st century, wastewater treatment will still include the aforementioned constituents, but will also include a wide variety of human and veterinary antibiotics, human prescription and nonprescription drugs, industrial and household wastewater products, and sex and steroidal hormones (see Table 2–20 in Chap. 2). To treat many of these compounds, it will be necessary to optimize and upgrade existing treatment facilities and operations. In some cases, additional treatment beyond secondary treatment, such as the advanced treatment processes described in Chap. 11, will be required. For example, a considerable amount of upgrading will be required to meet new odor requirements at existing wastewater treatment plants. In some situations, new facilities will have to be designed and built.

Meeting More Stringent Discharge Requirements

Based on past events, it is clear that the discharge requirements specified in NPDES permits will become more stringent in the 21st century. For example, the following constituent values were specified in a recently issued (2001) NPDES permit in California to be complied with by the year 2006.

Constituent	Unit	Limit Type	Value	Type of sample
BOD	mg/L	Monthly average	7	Daily composite
	mg/L	Weekly average	8.5	Daily composite
	mg/L	Daily maximum	10	Daily composite
TSS	mg/L	Monthly average	10	Daily composite
	mg/L	Weekly average	15	Daily composite
Trihalomethanes (THMs)				
Bromodichloromethane	μg/L	Daily maximum	1.14	Monthly grab
	μg/L	Monthly average	0.56	Monthly grab
Chloroform	μg/L	Daily average	9.3	Monthly grab
	μg/L	Monthly average	5.7	Monthly grab
Dibromochloromethane	μg/L	Daily average	0.92	Monthly grab
	μg/L	Monthly average	0.41	Monthly grab

Comparing the BOD and TSS values given above and the secondary standards given in Table 1–3 in Chap. 1, it is clear that treatment processes will have to be designed and operated differently to meet these standards. The total daily average for the three THMs is astounding, given that the current (2001) limit for total THMs in drinking water is 100 μg/L. Further, the constituent values given above are specified as *not-to-exceed* values, subject to fines and legal action for any exceedances. The State of California has enacted legislation (AB 307) by which fines can be levied for an individual exceedance, depending on the constituents and circumstances, and a mandatory fine is levied if more than four exceedances occur during any six-month period.

Discharge Limits for Wastewater Treatment Plants

Because not-to-exceed values are impossible to achieve with current wastewater treatment-plant facilities without massive expenditures, it is appropriate to define what is an acceptable level of performance. For example, 6 and 3 exceedances per year correspond to a 98.3 and 99.2 percent level of compliance, respectively. The U.S. EPA recommends an average frequency for excursions of both acute and chronic criteria not to exceed once in 3 years or 99.9 percent (U.S. EPA, 1994). It is interesting to note that not-to-exceed standards have been applied in England and Europe for the past 10 years; however, the not-to-exceed limit is usually set at the 95 percent level. For values that exceed 95 percent, fines can be levied. With nondegradation groundwater requirements that are making their way into new NPDES permits, even 99.9 percent compliance may not be acceptable.

To meet permit limits, such as those given above, at the 98.3, 99.2, or 99.9 percent compliance level, it will be necessary to consider (1) the variability of the influent wastewater characteristics, (2) the inherent variability of biological treatment processes, and (3) the reliability of the mechanical treatment-plant components. Even with effective design and operation, it may not be possible to meet very stringent discharge requirements at the 99.9 percent compliance level for a number of constituents without the use of additional treatment processes. Management of the additional residuals that would result from enhanced treatment is another issue that must be addressed. The need for process optimization, additional treatment processes, or even the implementation of new treatment technologies will have to be addressed on a case-by-case basis, depending on the constituent or constituents that must be removed and the levels to which those constituents must be removed. The question of what NPDES permit limits are set and whether these limits are appropriate is left to the courts and to another book.

15-2 TREATMENT PROCESS RELIABILITY AND SELECTION OF DESIGN VALUES

Reliability of a treatment plant or a treatment process may be defined as the probability of adequate performance for a specified period of time under specified conditions or, in terms of treatment-plant performance, the percent of the time that effluent concentrations meet specified permit requirements. For example, a treatment process with a reliability of 99 percent is expected to meet the performance requirements 99 percent of the time. For 1 percent of the time, or three to four times per year, the not-to-exceed daily permit limits would be expected to be exceeded. Such a level of performance may, or may not, be acceptable depending on the permit requirements. For each specific case where the reliability concept is to be employed, the levels of reliability must be evaluated including the cost of the facilities required to achieve the specified levels of reliability and the associated operating and maintenance costs. Thus, the purpose of this section is to examine how treatment process variability is assessed and how the performance of combined processes can be evaluated. The specific topics to be discussed are (1) variability in wastewater treatment, (2) selection of process design parameters, (3) the performance of combined processes, and (4) the development of input-output relationships.

Variability in Wastewater Treatment

Three categories of variability that can affect the design, performance, and reliability of a wastewater treatment plant are (1) variability of the influent wastewater flowrate and characteristics, (2) inherent variability in wastewater treatment processes, and (3) variability caused by mechanical breakdown, design deficiencies, and operational failures. Each of these categories of variability is considered in the following discussion. Following the discussion of these topics, an approach to the selection of design values in which the observed variability can be taken into account is introduced and illustrated subsequently in Example 15-1.

Variability of Influent Wastewater Flowrate and Characteristics.

Because of the way modern human life is organized, there is variability in the influent wastewater flowrates and characteristics observed at wastewater treatment plants (see Chap. 3). A typical example of the variability that can be observed in influent wastewater flowrates and characteristics is illustrated on Fig. 15–1. As shown on Fig. 15–1a, the summer flowrates are very stable and follow a log-normal distribution, whereas the entire daily set of flowrate data is influenced significantly by the high winter flowrates and is therefore extremely variable. In fact, the complete daily flowrate data set cannot be modeled with either an arithmetic or a log-normal distribution. As will be discussed subsequently, such variability, which is not uncommon, is of concern where stringent discharge requirements must be met. In some cases, it will be necessary to reduce the amount of infiltration/inflow in the collection system and/or install equalization facilities to improve treatment process performance. The use of equalization facilities has the added benefit of reducing the size of the individual treatment units that may be needed. In general, as the capacity of the treatment facility increases, the observed variability in flowrates will tend to decrease.

Inherent Variability in Wastewater Treatment Processes.

Because of the variability of the influent wastewater flowrate and characteristics, the variability of all treatment processes due to design limitations, the inherent variability of biological

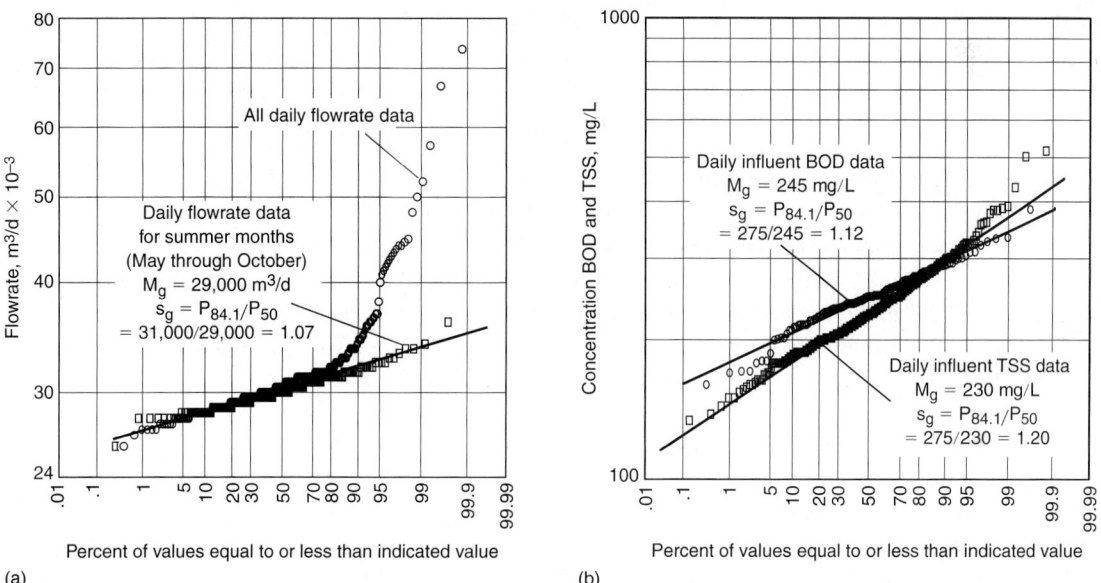

Figure 15–1

Probability distributions for daily influent wastewater characteristics collected over one year: (a) flowrate and (b) biochemical oxygen demand (BOD) and total suspended solids (TSS).

treatment processes due to the presence of living microorganisms and the laws of chance, all physical, chemical, and biological treatment processes exhibit some measure of variability with respect to the performance that can be achieved. Based on an analysis of effluent performance data from a number of wastewater treatment processes, it has been observed that the performance of most processes for most constituents can be modeled with a log-normal distribution. In some cases, when dealing with advanced wastewater treatment processes, many of the effluent constituent values will be reported as nondetect (ND), method detection limit (MDL), or level of quantification (LOQ) (see Sec. 2–1 in Chap. 2). When such values occur in a data set, the data set is said to be *censored* in a statistical sense. Methods that can be used to estimate summary statistics for censored data sets are discussed in Helsel and Cohn (1998) and Helsel (1990). Typical performance summaries for various processes for a variety of constituents that have been presented in previous chapters of this textbook are summarized in Table 15–1. The variability in these distributions, measured in terms of the geometric standard deviation, s_g, is also presented in Table 15–1 for the purpose of discussion and comparison. The greater the numerical value of s_g, the greater the observed range in the measured values.

Table 15–1
Variability observed in selected treatment processes taken from various figures presented in previous chapters

Constituent	Figure	Unit	Statistic[a]		
			Mg = P_{50}	$P_{84.1}$	s_g[b]
Flowrate	3–11	m³/d	8000	9100	1.14
Primary influent BOD	3–11	mg/L	175	220	1.26
Primary influent TSS	3–11	mg/L	150	195	1.27
Effluent turbidity from granular-medium filtration following biological treatment	11–19	NTU	1.4	1.8	1.29
Effluent TSS from granular-medium filtration following biological treatment	11–19	mg/L	1.95	2.8	1.44
Effluent turbidity from disk filter following biological treatment	11–31a	NTU	0.65	0.85	1.31
Effluent TSS from disk filter following biological treatment	11–31a	mg/L	0.58	0.90	1.55
Primary influent coliphage	13–16	pfu/100 mL	2.1×10^4	3.5×10^4	1.67
Secondary effluent coliphage following activated-sludge treatment	13–16	pfu/100 mL	300	500	1.67
Filter effluent coliphage following secondary sedimentation	13–16	pfu/100 mL	140	450	3.21
Coliphage following disinfection with chlorine	13–16	pfu/100 mL	2	5	2.50
TDS in reverse osmosis effluent following microfiltration	13–17a	mg/L	16	21	1.31
Seeded coliphage removal with reverse osmosis	13–17b	log removal	5.9	6.8	1.15

[a] All of the reported distributions are log-normal.
[b] $s_g = P_{84.1}/P_{50}$.

In general, it has been found that the performance of conventional biological treatment processes for wastewater constituents that can be modified by treatment, such as BOD, TOC, etc., can be described most frequently with a log-normal distribution. For constituents that are not modified significantly such as inorganic TDS, both arithmetic and log-normal distributions can be used to model process performance. Where the variability in performance is not great, both the arithmetic and log-normal distributions can be used to model the observed performance. The Weibull distribution (Kokoska and Zwillinger, 2000) has also proven to be useful in the analysis of the performance of advanced wastewater treatment processes (WCPH, 1996, 1997). It should also be noted that censored effluent data sets are encountered often in assessing the performance of advanced wastewater treatment processes (see above).

Mechanical Process Reliability. In addition to the variability in the influent wastewater flowrate and characteristics and the inherent variability in the response of wastewater treatment processes, the variability associated with the mechanical equipment used at wastewater treatment facilities must also be considered in analyzing what design values and how much standby equipment must be available to meet stringent standards at some specified reliability value (e.g., 99 or 99.9 percent). Several approaches are available for analyzing mechanical reliability of a treatment plant including the following (WCPH, 1996).

1. Critical component analysis (CCA)
2. Failure modes and effects analysis
3. Event tree analysis
4. Fault tree analysis

All four of these approaches are cited frequently in the literature and are used by a variety of industries. The critical component analysis (CCA) approach was developed by the U.S. EPA to determine the in-service reliability, maintainability, and operational availability of selected critical wastewater treatment components (U.S. EPA, 1982). The objective of the CCA is to determine which mechanical components in the wastewater treatment plant will have the most immediate impact on effluent quality should failure occur (WCPH, 1996, 1997). The statistical parameters used most commonly in applying the CCA method, as defined in Table 15–2, are *mean time before failure* (MTBF),

Table 15–2
Statistical measures used to assess equipment reliability[a]

Statistical measure	Description
Mean time before failure (MTBF)	A measure of the mechanical reliability of equipment, determined by the number of failures. The usual approach is to divide the operating hours by the number of failures
Expected time before failure (ETBF)	Similar to the MTBF, but the actual elapsed time is used as the total time in service
Inherent availability (AVI)	Fraction of calendar time that the component or unit was operating
Operating availability (AVO)	Fraction of time the component or unit can be expected to be operational excluding preventive maintenance

[a]Adapted from U.S. EPA (1982), WCPH (1996, 1997).

Table 15–3
Summary statistics
on the mechanical
reliability for
Aqua III[a,b]

	Statistical measure[c]			
	MTBF, year	90% CL[d] MTBF, year	AVO	AVI
Preliminary (headworks)	0.35	0.57	0.9953	0.9998
Primary	0.82	0.65	0.9967	0.9981
Secondary	2.12	1.75	0.9757	0.9953
Tertiary	2.24	1.78	0.9994	0.9995
UV disinfection	0.58	0.25	0.9991	0.9984
Reverse osmosis	1.22	0.99	0.9900	0.9903
Aeration tower	1.16	0.50	0.7835	0.9995
Carbon tower	1.86	1.02	0.9963	0.9999
Product water	0.56	0.45	0.9771	0.9964

[a] Adapted from WCPH (1997).
[b] Aqua III data 10/9/94 through 9/30/95.
[c] See Table 15–11 for definitions of statistical measures.
[d] 90 percent confidence level.

expected time before failure (ETBF), *inherent availability* (AVI), *and operating availability* (AVO).

In a study of the San Pasqual Aqua III treatment system in San Diego, CA, a complete process reliability analysis was performed (WCPH, 1996, 1997). The Aqua III treatment system was designed to produce 3800 m³/d (1.0 Mgal/d) of reclaimed water. The treatment facility includes preliminary treatment (coarse screening and grit removal), primary treatment (rotary drum and disk screens), secondary treatment (water hyacinth ponds), and tertiary treatment (a package plant consisting of coagulation, softening, sedimentation, and filtration). Advanced wastewater treatment consists of ultraviolet disinfection, reverse osmosis, air stripping, and granular activated carbon adsorption. Sodium hypochlorite is used for plant effluent disinfection, with the required contact time taking place within the distribution system. The results of the process reliability analysis are presented in Table 15–3. As shown, the preliminary treatment process has the lowest MTBF. Typical problems experienced with the preliminary treatment facilities included tripped breaker, packing leak, and gearbox failure. With the exception of three treatment processes, the AVO for the remaining processes was greater than 0.99. The AVI was greater than 0.99 for all of the treatment process (WCPH, 1996, 1997). Information such as that presented in Table 15–3 is used to determine maintenance schedules and the requirements for standby parts and backup components.

Selection of Process Design Parameters to Meet Discharge Permit Limits

Because of the variations in effluent quality, treatment plants must now be designed to produce an average effluent concentration below the permit requirements. The question is: What mean value should be used for process design to be assured that constituent concentrations in the effluent will be equal to or less than a specified limit with a specified degree of reliability? Two approaches can be used to estimate the design mean value needed to meet prescribed standards: (1) a statistical approach involving the coefficient of reliability and (2) a graphical approach. Both of these approaches are described and illustrated in the following discussion. Whether the mean value arrived at can be designed for is a question that is addressed subsequently.

Statistical Approach to Selection of Mean Design Value. One approach that can be used to determine the mean design value involves the use of the coefficient of reliability (COR) approach developed by Niku et al. (1979, 1981). In the COR method, the mean constituent values (design values) are related to the standards that must be achieved on a probability basis. The mean value, m_d, may be obtained by the relationship

$$m_d = (COR)X_s \qquad (15–1)$$

where m_d = mean constituent design value (e.g., g/m³, mg/L)
  X_s = a fixed standard that must be met at a specified reliability level (e.g., g/m³, mg/L)
  COR = coefficient of reliability, unitless

The coefficient of reliability is determined using the following expression.

$$COR = [(V_x^2 + 1)^{1/2}] \exp\{-Z_{1-\alpha}[\ln (V_x^2 + 1)]^{1/2}\} \qquad (15–2)$$

where V_x = coefficient of variation of the ratio of the existing distribution = σ_x/m_x
   (see also Table 3–10 in chap. 3)
  σ_x = standard deviation of performance values (i.e., daily, weekly, or monthly) from an existing treatment process
  m_x = mean of performance values (i.e., daily, weekly, or monthly) from an existing treatment process
  $Z_{1-\alpha}$ = number of standard deviations away from mean of a normal distribution
  $1 - \alpha$ = cumulative probability of occurrence (reliability level)

Values of $Z_{1-\alpha}$ for various cumulative probability levels, $1 - \alpha$, are given in Table 15–4. Values of COR for determining effluent concentrations for different coefficients of variation at different levels of reliability are reported in Table 15–5. Selection of an appropriate value of V_x must be based on data from operating facilities. The use of the reliability concept is illustrated in Example 15–1, following the discussion of the graphical approach to the selection of process design values.

Graphical Approach to Selection of Mean Design Value. Another method of determining appropriate mean design values to meet the specified effluent standards is the graphical probability method. If it is assumed that the geometric standard deviation can be used as a measure of reliability and that the value remains approximately constant at other design values, then the required effluent value can be set at the specified reliability level (e.g., 10 mg/L at 99 percent) and a line is passed through the value with the same geometric standard deviation as the measured data. The value at a probability value of 50 percent is the new mean design value.

Table 15-4
Values of standardized normal distribution[a]

Cumulative probability $1 - \alpha$	Percentile $Z_{1-\alpha}$
99.9	3.090
99	2.326
98	2.054
95	1.645
92	1.405
90	1.282
80	0.842
70	0.525
60	0.253
50	0

[a]From Niku et al. (1979).

Table 15-5
Coefficient of reliability as a function of V_x and reliability[a]

V_x	Reliability, %							
	50	80	90	92	95	98	99	99.9
0.3	1.04	0.81	0.71	0.69	0.64	0.57	0.53	0.42
0.4	1.08	0.78	0.66	0.63	0.57	0.49	0.44	0.33
0.5	1.12	0.75	0.61	0.58	0.51	0.42	0.37	0.26
0.6	1.17	0.73	0.57	0.54	0.47	0.37	0.32	0.21
0.7	1.22	0.72	0.54	0.50	0.43	0.33	0.28	0.17
0.8	1.28	0.71	0.52	0.48	0.40	0.30	0.25	0.15
0.9	1.35	0.70	0.50	0.46	0.38	0.28	0.22	0.12
1.0	1.41	0.70	0.49	0.44	0.36	0.26	0.20	0.11
1.2	1.56	0.70	0.46	0.41	0.33	0.22	0.17	0.08
1.5	1.80	0.70	0.45	0.39	0.30	0.19	0.14	0.06

[a]From Niku et al. (1979).

The graphical approach is illustrated on Fig. 15–2. The plotted data correspond to the monthly total copper concentration values in the effluent from a wastewater treatment plant. If the permit limit is to be 10 μg/L at the 99.9 percent reliability value, that value is plotted and a line with the same geometric standard deviation is drawn through the point. The value at a probability of 50 percent on the line drawn through permit limit corresponds to the required mean design value, in this case 2.1 μg/L. For many constituents, it will be found that the required mean design value cannot be met with an existing process. Where the required mean value cannot be met with a single process,

Figure 15–2

Probability distributions: (a) monthly effluent total copper (Cu) concentration values collected over a period of 2 years and (b) corresponding distribution to achieve 99.9 percent compliance with a copper permit limit of 10 μg/L, drawn with the same geometric standard deviation as the original distribution.

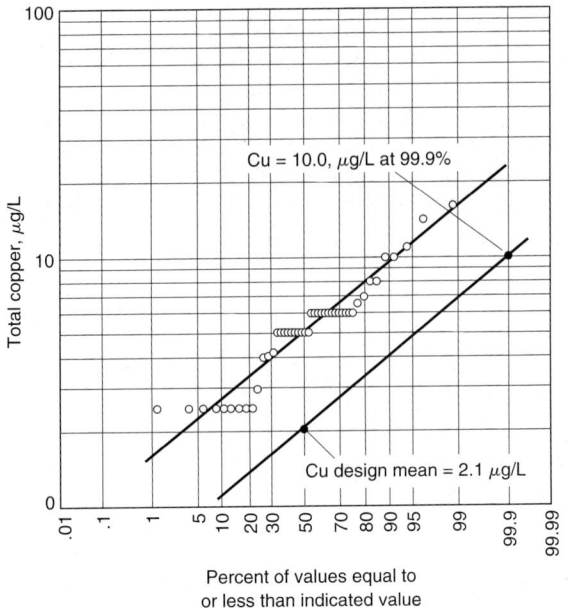

it may be necessary to use two or more treatment processes in series. The typical geometric standard deviation values given in Table 15–1 can also be used for the design of new treatment plants if the discharge permit limits are known.

EXAMPLE 15–1 **Estimating Effluent Design BOD and TSS Concentrations Based on Reliability Considerations** An existing activated sludge plant is required to be expanded and upgraded to meet new permit requirements. The new effluent requirements are as given below. Determine the mean design effluent BOD and TSS concentrations required to meet the 99 percent reliability level for the monthly standard and the 99.9 percent reliability level for the weekly standard using the COR method and the log-probability graphical method. Average monthly effluent BOD and TSS data for the existing facility for a period of 1 year are also given below.

Effluent requirements

Constituent	Monthly mean	Weekly mean
BOD, mg/L	15	20
TSS, mg/L	15	20

Monthly performance data

Month	BOD, mg/L	TSS, mg/L
Jan	34.0	15.0
Feb	27.1	18.0
Mar	29.0	17.5
Apr	25.0	22.5
May	25.1	22.0
Jun	22.0	24.9
Jul	21.7	28.0
Aug	20.5	25.1
Sep	17.0	19.5
Oct	18.5	20.0
Nov	23.1	20.1
Dec	24.0	21.5

Solution Part 1 COR Method

1. Determine the statistics for the given data using a standard statistical software package.

Parameter	Value BOD	Value TSS
Minimum	17	15
Maximum	34	28
Sum	287.0	254.1
Points	12	12
Mean	23.92	21.18
Median	23.55	20.80
RMS	24.33	21.46
Std deviation	4.65	3.64
Variance	21.67	13.22
Std error	1.34	1.05
Skewness	0.61	0.21
Kurtosis	0.05	−0.53

2. Determine the coefficient of reliability using Eq. (15–2) at a cumulative probability of 99%.

$$\text{COR} = [(V_x^2 + 1)^{1/2}] \exp\{-Z_{1-a}[\ln (V_x^2 + 1)]^{1/2}\}$$

    a.   Determine the value of V_x using the results of the statistical analysis.

       i. For BOD

$$V_x = \frac{\sigma_x}{m_x} = \frac{4.65}{23.92} = 0.194$$

      ii. For TSS

$$V_x = \frac{\sigma_x}{m_x} = \frac{3.64}{21.18} = 0.172$$

    b.   The value of $Z_{1-\alpha}$ for a cumulative probability of 99% from Table 15–4 is 2.326.

    c.   Determine the coefficient of reliability.

       i. For BOD

$$COR = [(0.194^2 + 1)^2]\exp\{-2.326[\ln(0.194^2 + 1)]^{1/2}\} = 0.69$$

      ii. For TSS

$$COR = [(0.172^2 + 1)^2]\exp\{-2.326[\ln(0.172^2 + 1)]^{1/2}\} = 0.71$$

3.   Determine the coefficient of reliability using Eq. (15–2) at a cumulative probability of 99.9%.

    a.   Determine the value of V_x from Step 2a.

       i. For BOD

$$V_x = 0.194$$

      ii. For TSS

$$V_x = 0.172$$

    b.   The value of $Z_{1-\alpha}$ for a cumulative probability of 99.9% from Table 15–4 is 3.090.

    c.   Determine the coefficient of reliability.

       i. For BOD

$$COR = [(0.194^2 + 1)^2]\exp\{-3.090[\ln(0.194^2 + 1)]^{1/2}\} = 0.59$$

      ii. For TSS

$$COR = [(0.172^2 + 1)^2]\exp\{-3.090[\ln(0.172^2 + 1)]^{1/2}\} = 0.63$$

4.   Determine the design effluent concentrations for 99% reliability for the monthly standard.

    a.   Mean design BOD = (COR) X_s

$$= 0.69 \times 15 \text{ mg/L} = 10.4 \text{ mg/L}$$

    b.   Mean design TSS = 0.71 × 15 mg/L = 10.7 mg/L

5.   Determine the design effluent concentrations for 99.9% reliability for the weekly standard.

    a.   Mean design BOD = (COR) X_s

$$= 0.59 \times 20 \text{ mg/L} = 11.8 \text{ mg/L}$$

    b.   Mean design TSS = 0.63 × 20 mg/L = 12.6 mg/L

6. Use the most conservative values for design.

$$BOD_{design} = 10.4 \text{ mg/L}$$

$$TSS_{design} = 10.7 \text{ mg/L}$$

Solution   Part 2 Log-Probability Graphical Method

1. Plot the monthly data for BOD and TSS on log-probability paper. The required plots for BOD and TSS are shown below.

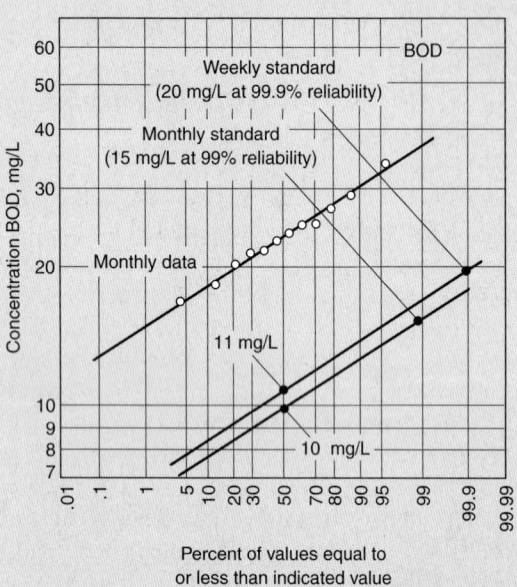
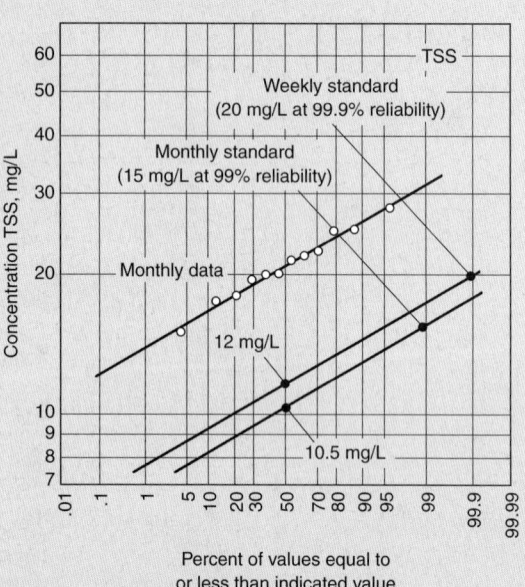

2. Estimate the design effluent concentrations for BOD and TSS for (a) 99% reliability for the monthly standard and (b) 99.9% reliability for the weekly standard.

a. Determine the design effluent concentrations for BOD. The BOD concentrations are determined by passing lines with the same slope as the measured data through the points at 15 mg/L and 99% and 20 mg/L and 99.9% and noting the corresponding values at 50%. The values so determined are:

$$BOD_{design} \text{ at } 15 \text{ mg/L and } 99\% = 10.0 \text{ mg/L}$$

$$BOD_{design} \text{ at } 20 \text{ mg/L and } 99.9\% = 11.0 \text{ mg/L}$$

b. Determine the design effluent concentrations for TSS. The TSS concentrations are determined by passing lines with the same slope as the measured data through the points at 15 mg/L and 99% and 20 mg/L and 99.9% and noting the corresponding values at 50%. The values so determined are:

$$\text{TSS}_{\text{design}} \text{ at 15 mg/L and 99\%} = 10.5 \text{ mg/L}$$

$$\text{TSS}_{\text{design}} \text{ at 20 mg/L and 99.9\%} = 11 \text{ mg/L}$$

It is interesting to note that the values determined graphically are essentially the same as those determined analytically using the COR method.

Comment　When the concept of reliability is used, the mean effluent values selected for design will typically be significantly lower than permit requirements. Loge et al. (2001) demonstrated how the coefficient of variation was reduced (i.e. process variability) when processes are used in series with respect to the disinfection of treated effluent with UV radiation. Based on numerous past designs in England and Europe, it has been found that to achieve the prescribed limit at the 95 percent level the average design value should be about 50 percent of the prescribed limit.

Performance of Combined Processes

Applying the statistical and graphical procedures as illustrated in Example 15–1, it will often be found that the resulting mean design value for a given process is well below the range where any factual knowledge exists on how to design the process. For example, assume that to meet NPDES permit requirements, an activated sludge process must be designed to meet an average effluent suspended solids concentration of 4 mg/L. Realistically, it is now not possible to design a secondary clarifier to meet a specific design value. What is typically assumed is that with good design and effective operation of the secondary process, an average TSS value of 4 or 5 mg/L can be achieved. Unfortunately, such assumptions are not acceptable when not-to-exceed permit limits must be met. In such a situation, it will usually be necessary to add an additional process, such as depth or surface filtration, to meet the permit requirements consistently. In the following example the basis for determining the performance of the combined treatment processes is addressed.

EXAMPLE 15–2　**Estimating the Performance of Combined Treatment Processes Based on Reliability Considerations**　Estimate the combined performance at the 98.3, 99.2, and 99.9 percent levels for an activated sludge process followed by a granular-medium depth filter with respect to the removal of TSS and turbidity. Assume the following data apply to the activated sludge process. Also, assume that no chemicals are to be used with the depth filter.

1. Distribution for effluent TSS is log-normal
2. Geometric mean for effluent TSS, M_g = 15 mg/L
3. Geometric standard deviation for TSS, s_g = 1.25
4. Also $s_g = P_{84.1}/P_{50}$ (see Table 3–10 in Chap. 3)

Solution

1. From Fig. 11–18, develop a relationship between influent turbidity for a typical granular-medium depth filter. From Fig. 11–18, find, for a best fit line by eye, that

when the effluent turbidity is 2 NTU, the influent turbidity is 7.5 NTU, and the intercept of the line of best fit passes through 0.5 NTU. For these conditions the corresponding relationship is:

Filter effluent turbidity, NTU = 0.5 NTU + 0.2(filter influent turbidity, NTU)

2. To determine the TSS values after filtration, use the following two relationships given previously in Chap. 11.
   a. Secondary effluent TSS, mg/L = (2.3 mg/L/NTU) (effluent turbidity, NTU) [Eq. (11–32)]
   b. Filter effluent TSS = (1.4 mg/L/NTU) (filter effluent turbidity, NTU) [Eq. (11–33)]

3. Using the above relationships, determine the values corresponding to 50% and 84.1% (one standard deviation) after filtration
   a. At 50%
      i. Secondary effluent turbidity = (15 mg/L)/[(2.3 mg/L)/NTU] = 6.52 NTU
      ii. Filter effluent turbidity = 0.5 NTU + 0.2(6.52 NTU) = 1.8 NTU
      iii Filter effluent TSS = [(1.4 mg/L)/NTU] (1.8 NTU) = 2.52 mg/L
   b. At 84.1% [use Eq. (3–9) from Chap. 3]
      i. $P_{84.1} = P_{50} \times s_g$ = 15 mg/L × 1.25 = 18.75 mg/L
      ii. Secondary effluent turbidity = (18.75 mg/L)/[(2.3 mg/L)/NTU]
          = 8.15 NTU
      iii. Filter effluent turbidity = 0.5 NTU + 0.2(8.15 NTU) = 2.13 NTU
      iv. Filter effluent TSS = [(1.4 mg/L)/NTU] (2.13 NTU) = 2.98 mg/L

4. Plot the secondary effluent TSS and the filter effluent turbidity and TSS and prepare a summary compliance table.
   a. Log-probability plot

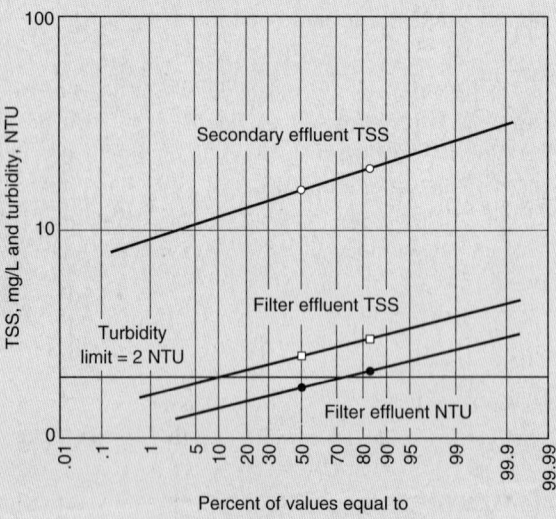

b. Summary table

Item	Unit	Value at indicated probability		
		98.3%	**99.2%**	**99.9%**
TSS	mg/L	2.6	2.9	4.2
Turbidity	NTU	2.5	2.8	3.0

Comment   As shown in the summary table, 6 TSS exceedances per year (98.3 percent probability) would be equal to or greater than 2.6 mg/L, 3 TSS exceedances per year (99.2 percent probability) would be equal to or greater than 2.9 mg/L, and 1 TSS exceedance in 3 years (99.9 percent probability) would be equal to or greater than 4.2 mg/L. If a turbidity level of 2 NTU had to be met for reuse applications, the 2 NTU limit would be exceeded 25 percent of the time, without the use of additional treatment.

Development of Input-Output Data

To predict the performance of different combined processes, input-output relationships such as presented in Example 15–2 will have to be developed. In general, input-output relationships are needed for processes with short treatment retention times such as granular and surface filtration, microfiltration, ultrafiltration, and reverse osmosis. Input-output relationships for processes where the treatment process performance is decoupled from the hydraulic retention time, such as the activated sludge process, are not very useful because of the buffering provided by the process. Further, as illustrated in Sec. 7–5 in Chap. 7, the value of the effluent substrate concentration for the activated sludge process is independent of the influent concentration. The variability observed in the effluent BOD, TSS, and other constituent values from activated sludge processes reflects (1) the inherent variability of the biological process (notwithstanding conclusions reached from the theoretical equations); (2) the diurnal, daily, monthly, and seasonal variation in the influent flowrates and wastewater characteristics; (3) the impact of process adjustments made by operators; and (4) the configuration of the physical facilities. For most processes with short retention times, the output values will be affected by the influent concentrations. It is interesting to note that the performance of the cloth media disk filter, as illustrated on Fig. 11–31*b* in Chap. 11, is essentially independent of the influent turbidity for turbidity values up to 30 NTU.

A further and very important consideration in the development of input-output relationships for processes following the activated sludge process is the solids retention time (SRT) at which the process is operated. As illustrated on Fig. 12–48 in Chap. 12, the number of particles with one or more associated coliform organisms will vary with the SRT. It has also been observed that the performance of downstream processes employing some form of filtration will also vary with the SRT value. Thus, the SRT value must be taken into account when developing input-output relationships for downstream processes. In general, it will not be possible to use input-output relationships developed for processes following activated sludge processes with long SRT values

(> 10 d) to predict the combined performance of downstream processes following activated sludge processes with short SRT values (1 to 2 d).

15–3 ODOR MANAGEMENT

The potential release of odors is a major concern of the public relative to modifying existing wastewater treatment facilities and constructing new facilities. Thus, the control of odors has become a major consideration in the design and operation of wastewater collection, treatment, and disposal facilities, especially with respect to the public acceptance of these facilities. In many instances, projects have been rejected because of the fear of potential odors. In several states, wastewater management agencies are now subject to fines and other legal action over odor violations. In view of the importance of odors in the field of wastewater management, the following discussion will deal with (1) the types of odors encountered, (2) the sources of odors, (3) the movement of odorous gases, (4) strategies for odor control, (5) odor control methods, and (6) the design of odor-control facilities.

Types of Odors

For humans, the importance of odors at low concentrations is related primarily to the psychological stress the odors cause, rather than to the harm they do to the body. The principal types of odors encountered in wastewater management facilities are reported in Table 15–6. With few exceptions, odorous compounds typically contain either sulfur or nitrogen. The characteristic odor of organic compounds containing sulfur is that of decayed organic material. Of the odorous compounds reported in Table 15–6, the rotten egg smell of hydrogen sulfide is the odor encountered most commonly in wastewater management facilities. As noted in Chap. 2, gas chromatography has been used successfully for the identification of specific compounds responsible for odors. Unfortunately, this technique has not proved as successful in the detection and quantification of odors derived from wastewater collection, treatment, and disposal facilities, because of the many compounds that may be involved. It should be noted that at higher concentrations, many of the odorous gases (e.g., hydrogen sulfide) can, depending on exposure, be lethal.

Sources of Odors

The principal sources of odors in wastewater management facilities and the relative potential for release of odor are presented in Table 15–7. Minimization of odors from these sources is the concern of odor management.

Wastewater Collection Systems. The potential for odor release from collection systems is high. The principal sources of odorous compounds in collection systems are from (1) the biological conversion, under anaerobic conditions, of organic matter containing nitrogen and sulfur, and (2) the discharge of industrial wastewater that may contain odorous compounds or compounds that may react with compounds in the wastewater to produce odorous compounds. Odorous gases released to the sewer atmosphere can accumulate and be released at air release valves, cleanouts, access ports (i.e., manholes), and house vents.

Table 15–6

Odor thresholds of odorous compounds and their characteristics associated with wastewater management[a]

Odorous compound	Chemical formula	Molecular weight	Odor threshold, ppm_v[b]	Characteristic odor
Ammonia	NH_3	17.0	46.8	Pungent, irritating
Chlorine	Cl_2	71.0	0.314	Pungent, suffocating
Chlorophenol	ClC_6H_4OH	128.56	0.00018	Medicinal odor
Crotyl mercaptan	$CH_3-CH=CH-CH_2-SH$	90.19	0.000029	Skunklike
Dimethyl sulfide	CH_3-S-CH_3	62	0.0001	Decayed cabbage
Diphenyl sulfide	$(C_6H_5)_2S$	186	0.0047	Unpleasant
Ethyl mercaptan	CH_3CH_2-SH	62	0.00019	Decayed cabbage
Ethyl sulfide	$(C_2H_5)_2SH$	90.21	0.000025	Nauseating odor
Hydrogen sulfide	H_2S	34	0.00047	Rotten eggs
Indole	C_8H_6NH	117	0.0001	Fecal, nauseating
Methyl amine	CH_3NH_2	31	21.0	Putrid, fishy
Methyl mercaptan	CH_3SH	48	0.0021	Decayed cabbage
Skatole	C_9H_9N	131	0.019	Fecal odor, nauseating
Sulfur dioxide	SO_2	64.07	0.009	Pungent, irritating
Thiocresol	$CH_3-C_6H_4-SH$	124	0.000062	Skunklike, rancid
Trimethyl amine	$(CH_3)_3N$	59	0.0004	Pungent, fishy

[a] Adapted in part from Patterson et al. (1984), U.S. EPA (1985), and WEF (1998).
[b] Parts per million by volume.

Wastewater Treatment Facilities. In considering the potential for the generation and release of odors from treatment plants, it is common practice to consider the liquid and solids processing facilities separately. The headworks and preliminary treatment operations have the highest potential for release of odor, especially for treatment plants that have long collection systems where anaerobic conditions can be created (see Fig. 15–3). Sidestream discharges including return flows from filter backwashing and from sludge and biosolids processing facilities are often a major source of odors, especially where these flows are allowed to discharge freely into a control structure or mixing chamber.

Sludge and Biosolids Handling Facilities. Typically, the most significant sources of odors at wastewater treatment plants are sludge-thickening facilities, anaerobic digesters, and sludge-loadout facilities. The highest potential for odor release occurs when unstabilized sludge is handled (e.g., turned, spread, or stored).

One of the major contributors to odor in solids processing and an important item to consider in treatment plant design is shear. Shear is the cutting or tearing of solids by shear stress. When solids undergo mixing by either high shear dewatering or conveyance equipment, particle size reduction occurs and odor production increases.

Table 15–7

Sources of odor in wastewater management systems[a]

Location	Source/cause	Odor potential
Wastewater-collection system		
Air release valves	Accumulation of odorous gases released from wastewater	High
Cleanouts	Accumulation of odorous gases released from wastewater	High
Access ports (manholes)	Accumulation of odorous gases released from wastewater	High
Industrial wastewater discharges	Odorous compounds may be discharged to wastewater collection system	Variable
Raw wastewater pumping station	Wetwell/septic raw wastewater, solids, and scum deposits	High
Wastewater treatment facilities		
Headworks	Release odorous gases generated in the wastewater collection system due to turbulence in hydraulic channels and transfer points	High
Screening facilities	Putrescible matter removed by screening	High
Preaeration	Release of odorous compounds generated in wastewater collection system	High
Grit removal	Organic matter removed with grit	High
Flow-equalization basins	Basin surfaces/septic conditions due to accumulation of scum and solids deposits	High
Septage receiving and handling facilities	Odorous compounds can be released at septage receiving stations, especially when septage is being transferred	High
Sidestream returns[b]	Return flows from biosolids processing facilities	High
Primary clarifiers	Effluent weirs and troughs/turbulence that releases odorous gases. Scum—either floating or accumulated on weirs and baffles/putrescible matter. Floating sludge/septic conditions	High/moderate
Fixed-film processes (trickling filters or RBCs)	Biological film/septicity due to insufficient oxygen, high organic loading, or plugging of trickling filter medium; turbulence causing release of odorous material	Moderate/high
Aeration basins	Mixed liquor/septic return sludge, odorous sidestream flows, high organic loading, poor mixing, inadequate DO, solids deposits	Low/moderate
Secondary clarifiers	Floating solids/excessive solids retention	Low/moderate
Sludge and biosolids facilities		
Thickeners, solids holding tanks	Floating solids; weirs and troughs/scum and solids septicity due to long holding periods, solids deposits, and temperature increases; odor release by turbulence	High/moderate
Aerobic digestion	Incomplete mixing in reactor	Low/moderate
Anaerobic digestion	Leaking hydrogen sulfide gas/upset conditions, high sulfate content in solids	Moderate/high
Sludge storage basins	Lack of mixing, formation of scum layer	Moderate/high
Mechanical dewatering by belt-filter press, recessed plate-filter press, or centrifuge	Cake solids/putrescible matter; chemical addition, ammonia release	Moderate/high

(continued)

| **Table 15–7** (*Continued*) |

Location	Source/cause	Odor potential
Sludge and biosolids facilities (*continued*)		
Sludge loadout facilities	Release of odors during the transfer of biosolids from storage to transfer facilities	High
Composting facilities	Composting solids/insufficient aeration, inadequate ventilation	High
Alkaline stabilization	Stabilized solids/ammonia generation resulting from reaction with lime	Moderate
Incineration	Air emissions/combustion temperature is not high enough to destroy all organic substances	Low
Sludge drying beds	Drying solids/excess putrescible matter due to insufficient stabilization	Moderate/high

[a] Adapted in part from WEF (1995).
[b] Sidestreams could include digester decant, dewatering return flows, or backwash water.

Figure 15–3

Typical examples of headworks facilities: (*a*) uncovered influent channel leading to in-line comminutor and (*b*) coarse bar screens housed in an enclosed building with odor-control facilities.

(a)　　(b)

Solids that exit a dewatering facility can be sheared enough to release odors. The major mechanism appears to be the release of proteinaceous biopolymer. Once released, these proteins are degraded, liberating a number of odorous compounds, but mostly mercaptans. The increase in solution protein also makes dewatering poorer. The solution proteins can be "coagulated" by addition of polymer, but the synthetic polymers are degraded and the protein becomes degradable. The synthetic polymer can also generate methylamines when degraded (Novak, 2001; Murthy, 2001).

Trimethylamine (TMA) is present in the liquid phase in many anaerobically digested sludges. Trimethylamine is like ammonia in that it is soluble below pH 9, but above this pH level it is a gas and can be released into the air. Adding lime to digested sludge for odor control may, in fact, liberate odor by converting TMA to a gas (Novak, 2001; Murthy, 2001). Some plants may be unable to land apply dewatered sludge because of the increased odor production. Thus, in evaluating processing and disposal options, the ramifications of odor generation and control have to be evaluated carefully.

Movement of Odors from Wastewater Treatment Facilities

Under quiescent meteorological conditions, odorous gases that develop at treatment facilities tend to hover over the point of generation (e.g., sludge thickening facilities, sludge storage lagoons), because the odorous gases are more dense than air. Depending on the local meteorological conditions, it has been observed that odors may be measured at undiluted concentrations at great distances from the point of generation. The following events appear to happen: (1) in the evening or early morning hours, under quiescent meteorological conditions, a cloud of odors will develop over the wastewater treatment unit prone to the release of odors; and (2) the concentrated cloud of odors can then be transported (i.e., pushed along), without breaking up, over great distances by the weak evening or early morning breezes, as they develop. In some cases, odors have been detected at distances of up to 25 km from their source. This transport phenomenon has been termed the *puff movement* of odors (Tchobanoglous and Schroeder, 1985). The *puff movement* of odors was first described by Wilson (1975). The most common method used to mitigate the effects of the odor puff is to install barriers to induce turbulence, thus breaking up and dispersing the cloud of concentrated odors, and/or to use wind generators to maintain a minimum velocity across the source.

Strategies for Odor Management

Strategies for the management and control of odors are presented and discussed below. An overview of some of the methods used to control and treat odorous gases is presented in the following section. Where there are chronic odor problems at treatment facilities, approaches to solving these problems may include (1) control of odor-causing wastewaters discharged to the collection system and treatment plant that creates odor problems, (2) control of odors generated in the wastewater-collection system, (3) control of odors generated in wastewater treatment facilities, (4) installation of odor containment and treatment facilities, (5) application of chemicals to the liquid (wastewater) phase, (6) use of odor masking and neutralizing agents, (7) use of gas-phase turbulence-inducing structures and facilities, and (8) establishment of buffer zones.

Control of Discharges to Wastewater-Collection System.   The elimination and/or control of wastewater discharges containing odorous compounds to the collection system can be accomplished by (1) the adoption of more stringent waste discharge ordinances and enforcement of their requirements, (2) requiring pretreatment of industrial wastewater, and (3) providing flow equalization at the source to eliminate slug discharges of wastewater.

Odor Control in Wastewater-Collection Systems.   The release of odors from the liquid phase in wastewater-collection systems can be limited by (1) maintain-

Figure 15–4

Typical uses of commercial oxygen in wastewater-collection systems for odor control: (a) sidestream oxygenation and reinjection of wastewater into a gravity sewer, (b) injection of oxygen into a hydraulic fall, and (c) injection of oxygen into two-phase flow in force main. (From Speece et al., 1990.)

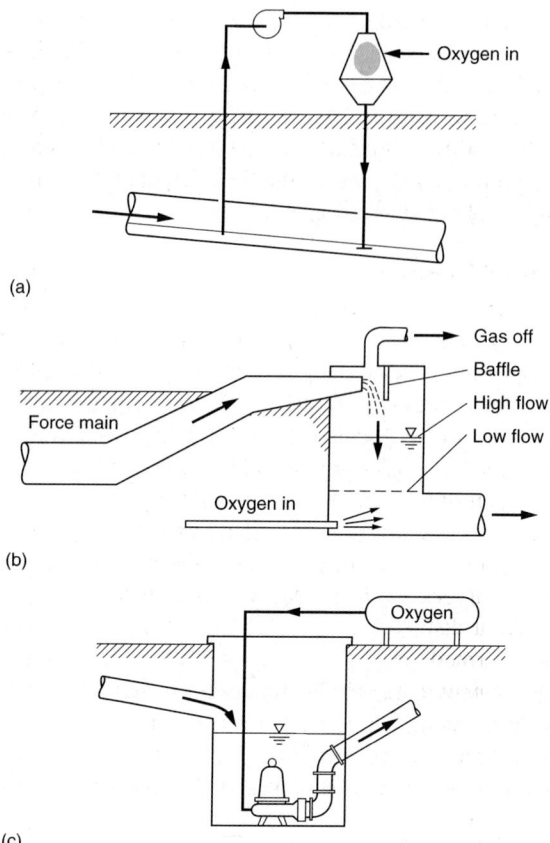

(a)

(b)

Force main

Gas off
Baffle
High flow
Low flow

Oxygen in

(c)

Oxygen

ing aerobic conditions through the addition of hydrogen peroxide, pure oxygen, or air at critical locations in the collection system and to long force mains (see Fig. 15–4), (2) controlling anaerobic microbial growth by disinfection or pH control, (3) oxidizing or precipitating odorous compounds by chemical addition, (4) design of the wastewater collection system to minimize the release of odors due to turbulence, and (5) off-gas treatment at selected locations.

Odor Control in Wastewater Treatment Facilities. With the proper attention to design details, such as the use of submerged inlets and weirs, the elimination of hydraulic jumps in influent piping and channels, the elimination of physical conditions leading to the formation of turbulence, proper process loadings, containment of odor sources, off-gas treatment, and good housekeeping, the routine release of odors at treatment plants can be minimized. It must also be recognized, however, that odors will develop occasionally. When they do, it is important that immediate steps be taken to control them. Often, this will involve operational changes or the addition of chemicals, such as chlorine, hydrogen peroxide, lime, or ozone.

Design and operational changes that can be instituted can include (1) minimization of free-fall turbulence by controlling water levels, (2) reduction of overloading of plant processes, (3) increasing the aeration rate in biological treatment processes, (4) increasing the plant treatment capacity by operating standby process units, (5) reducing solids inventory and sludge backlog, (6) increasing the frequency of pumping of sludge and scum, (7) adding chlorinated dilution water to sludge thickeners, (8) controlling the release of aerosols, (9) increasing the frequency of disposal of grit and screenings, (10) cleaning odorous accumulations more frequently, and (11) containment, ventilation, and treatment of odorous gases.

Odor Containment and Treatment. Odor containment includes the installation of covers, collection hoods, and air handling equipment for containing and directing odorous gases to disposal or treatment systems. In cases where the treatment facilities are close to developed areas, it has become common practice to cover treatment units such as the headworks (see Fig. 15–3b), primary clarifiers (see Figs. 15–5a and b), trickling filters (see Fig. 15–5c and d), sludge thickeners (see Fig. 15–5e), sludge-processing facilities, and sludge-loadout facilities (see Fig. 15–5f). Where covers are used, the trapped gases must be collected and treated. The specific method of treatment will depend on the characteristics of the odorous compounds. Typical containment alternatives for the control of odor emissions from wastewater management facilities are reported in Table 15–8. Methods used to treat odorous gases are summarized in Table 15–9.

Chemical Additions to Wastewater for Odor Control. Odors can be eliminated in the liquid phase through the addition of a variety of chemicals to achieve (1) chemical oxidation, (2) chemical precipitation, and (3) pH control. The most common oxidizing chemicals that can be added to wastewater include oxygen, air, chlorine, sodium hypochlorite, potassium permanganate, hydrogen peroxide, and ozone. While all of these compounds will oxidize hydrogen sulfide (H_2S) and other odorous compounds, their use is complicated by the chemical matrix in which the odorous gases exist. The only way to establish the required chemical dosages for the removal of chemical compounds is through bench- or pilot-scale testing. Odorous compounds can also be reduced by precipitation. For example, ferrous chloride and ferrous sulfate can be used for the control of H_2S odors by precipitation of the sulfide ion as ferrous sulfide. As with the oxidation reactions, the required chemical dosage can be determined only through bench- or pilot-scale testing. The release of H_2S can also be controlled by increasing the pH value of the wastewater. Increasing the pH of the wastewater results in reduced bacterial activity and also shifts the equilibrium so that the sulfide ion is present as HS^-. With most of the odor-control methods involving the addition of chemicals to wastewater, some residual product is formed that must ultimately be dealt with. Shock treatment involving the addition of sodium hydroxide (NaOH) can be used to reduce microbial slimes in sewers. The high pH also reduces sulfide (S^{2-}) formation. Additional details on chemical addition may be found in Rafson (1998).

Use of Odor Masking and Neutralization. On occasion, chemicals have been added to wastewater or offgases to mask an offensive odor with a less offensive odor. Masking chemicals are based on essential oils with the most common aromas being vanilla, citrus, pine, or floral (Williams, 1996). Typically, enough masking chem-

Figure 15–5

Typical odor containment facilities at wastewater treatment plants:
(a) covered primary sedimentation tanks,
(b) covered primary sedimentation tanks with odor-control facilities,
(c) covered trickling filter,
(d) view inside covered trickling filter; the large pipe shown overhead is used to remove odorous gases for treatment in a chemical scrubber,
(e) covered sludge thickener, and
(f) enclosed sludge-loadout facility.

ical is added to wastewater to overpower the offensive odor. Masking chemicals, however, do not modify or neutralize the offensive odors. Neutralization involves finding chemical compounds that can be combined with the odorous gases in the vapor state so that the combined gases cancel each other's odor, produce an odor of lower intensity, or eliminate the odorous compounds. Although odor masking and neutralization are viable options for short-term management of odor problems, the key to long-term odor management is to identify the source of the odors and implement corrective measures.

Table 15–8
Odor containment alternatives for the control of odor emissions from wastewater management facilities

Facility (sources)	Suggested control strategies
Wastewater sewers	Seal existing access ports (i.e., manholes). Eliminate the use of structures that create turbulence and enhance volatilization. Ventilate sewer reaches that develop positive pressure to odor control facilities
Sewer appurtenances	Isolate and cover existing appurtenances
Pump stations	Vent odorous gases from wetwell to treatment unit. Use variable-speed pumps to reduce the size of the wetwell
Bar racks	Cover existing units. Reduce headloss through bar racks
Comminutors	Cover existing units. Use in-line enclosed comminutors
Parshall flume	Cover existing units. Use alternative measuring device
Grit chamber	Cover existing aerated grit chambers. Reduce turbulence in conventional horizontal-flow grit chambers; cover if necessary. Avoid the use of aerated grit chambers
Equalization basins	Cover existing units. Use submerged mixers, and reduce airflow
Primary and secondary sedimentation tanks	Cover existing units. Replace conventional overflow weirs with submerged weirs
Biological treatment	Cover existing units. Use submerged mixers and reduce aeration rate
Sludge thickeners	Cover existing units
Transfer channels	Use enclosed transfer channels

Use of Turbulence-Inducing Structures and Facilities for Odor Dispersion. In a number of wastewater treatment plants, physical facilities used to induce atmospheric turbulence have been constructed specifically for the purpose of gas-phase odor reduction. The high barrier fence [3.7 m (12 ft)] shown on Fig. 15–6 surrounds sludge-storage lagoons. Operationally, any odorous gases that develop under quiescent conditions over the lagoons are diluted as they move away from the storage lagoons, due to the local turbulence induced by the barrier. Trees are also used commonly to dilute odorous gases by inducing turbulence (i.e., the formation of eddies) and mixing. Trees are also known to help purify the air as a result of respirometric activity.

Use of Buffer Zones. The use of buffer zones can also help in reducing the impact of odors on developed areas. Typical buffer zone distances used by regulatory agencies are presented in Table 15–10. If buffer zones are used, odor studies should be conducted that identify the type and magnitude of the odor source, meteorological conditions, dispersion characteristics, and type of adjacent development. Trees that grow rapidly are often planted at the periphery of the buffer zones to further reduce the impact of odors.

Odor-Treatment Methods

The general classification of odor-treatment methods is presented in Table 15–9, along with typical applications in wastewater management. Odor-treatment methods are designed either to treat the odor-producing compounds in the wastewater stream or to

Table 15–9
Methods used to treat odorous gases found in wastewater management facilities[a]

Method	Description and/or application
Physical methods	
Adsorption on activated carbon	Odorous gases can be passed through beds of activated carbon to remove odors. Carbon regeneration can be used to reduce costs. Additional details may be found in Chap. 11
Adsorption on sand, soil, or compost beds	Odorous gases can be passed through sand, soil, or compost beds. Odorous gases from pumping stations may be vented to the surrounding soils or to specially designed beds containing sand or soils. Odorous gases collected from treatment units may be passed through compost beds
Dilution with odor-free air	Gases can be mixed with fresh air sources to reduce the odor unit values. Alternatively, gases can be discharged through tall stacks to achieve atmospheric dilution and dispersion
Masking agents	Perfume scents can be sprayed in fine mists near offending process units to overpower or mask objectionable odors. In some cases, the odor of the masking agent is worse than the original odor. Masking agents should not be confused with neutralizing agents described below
Oxygen injection	The injection of oxygen (either air or pure oxygen) into the wastewater to control the development of anaerobic conditions has proved to be effective
Scrubbing towers	Odorous gases can be passed through specially designed gas scrubbing towers to remove odors
Turbulence-inducing facilities	Use of wind breaks, such as high fences and trees, and propeller fans
Chemical methods	
Chemical oxidation	Oxidizing the odor compounds in wastewater is one of the most common methods used to achieve odor control. Chlorine, ozone, hydrogen peroxide, and potassium permanganate are among the oxidants that have been used. Chlorine also limits the development of a slime layer
Chemical precipitation	Chemical precipitation refers to the precipitation of sulfide with metallic salts, especially iron
Neutralizing agents	Compounds that can be sprayed or atomized in fine mists to chemically react with, neutralize, and/or dissolve odorous compounds
Scrubbing with various alkalies	Odorous gases can be passed through specially designed chemical scrubbing towers to remove odors. If the level of carbon dioxide is high, costs may be prohibitive
Thermal oxidation	Combustion of off-gases at temperatures from 800 to 1400°C will eliminate odors. Lower temperatures (400 to 800°C) are used with catalytical incineration
Biological methods	
Activated-sludge aeration tanks	Odorous gases can be combined with the process air for activated-sludge aeration tanks to remove odorous compounds
Biotrickling filters	Specially designed biotrickling filters can be used to remove odorous compounds biologically. Typically, the filters are filled with plastic packing of various types on which biological growths can be maintained
Compost filters	Gases can be passed through biologically active beds of compost to remove odors
Sand and soil filters	Gases can be passed through biologically active beds of compost to remove odors
Trickling filters	Odorous gases can be passed through existing trickling filters to remove odorous compounds

[a] Adapted in part from U.S. EPA (1985).

Figure 15–6

High barrier fence placed around sludge-holding lagoons to induce air turbulence and mixing, and thus limit the release of concentrated odors off the treatment-plant site.

Table 15–10
Suggested minimum buffer distances from treatment units for odor containment[a,b]

Treatment process unit	Buffer distance	
	ft	m
Sedimentation tank	400	125
Trickling filter	400	125
Aeration tank	500	150
Aerated lagoon	1000	300
Sludge digester (aerobic or anaerobic)	500	150
Sludge-handling units	1000	300
Open drying beds	500	150
Covered drying beds	400	125
Sludge-holding tank	1000	300
Sludge-thickening tank	1000	300
Vacuum filter	500	150
Wet air oxidation	1500	450
Effluent recharge bed	800	250
Secondary effluent filters		
Open	500	150
Enclosed	200	75
Advanced wastewater treatment		
Tertiary effluent filters		
Open	300	100
Enclosed	200	75
Denitrification	300	100
Polishing lagoon	500	150
Land disposal	500	150

[a]Source: New York State Department of Environmental Conservation.
[b]Actual buffer distance requirements will depend on local conditions.

treat the foul air. Most of the methods in Table 15–9 are meant to be used to treat the foul air (i.e., the gas phase). As noted above, to control the release of odorous gases from treatment facilities, it has become more common to cover wastewater treatment processes (see Fig. 15–5).

The principal methods used to treat odorous gases include the use of (1) chemical scrubbers, (2) activated-carbon absorbers, (3) vapor-phase biological treatment processes (i.e., compost filters), (4) treatment in conventional biological treatment processes, and (5) thermal processes. Each of these methods is discussed below. The specific method of odor control and treatment that should be applied will vary with local conditions. However, because odor-control measures are expensive, the cost of making process changes or modifications to the facilities to eliminate odor development should always be evaluated and compared to the cost of various alternative odor-control measures before their adoption is suggested.

Chemical Scrubbers. The basic design objective of a chemical scrubber is to provide contact between air, water, and chemicals (if used) to provide oxidation or entrainment of the odorous compounds. The principal wet-scrubber types, as shown on Fig. 15–7, include single-stage countercurrent packed towers, countercurrent spray

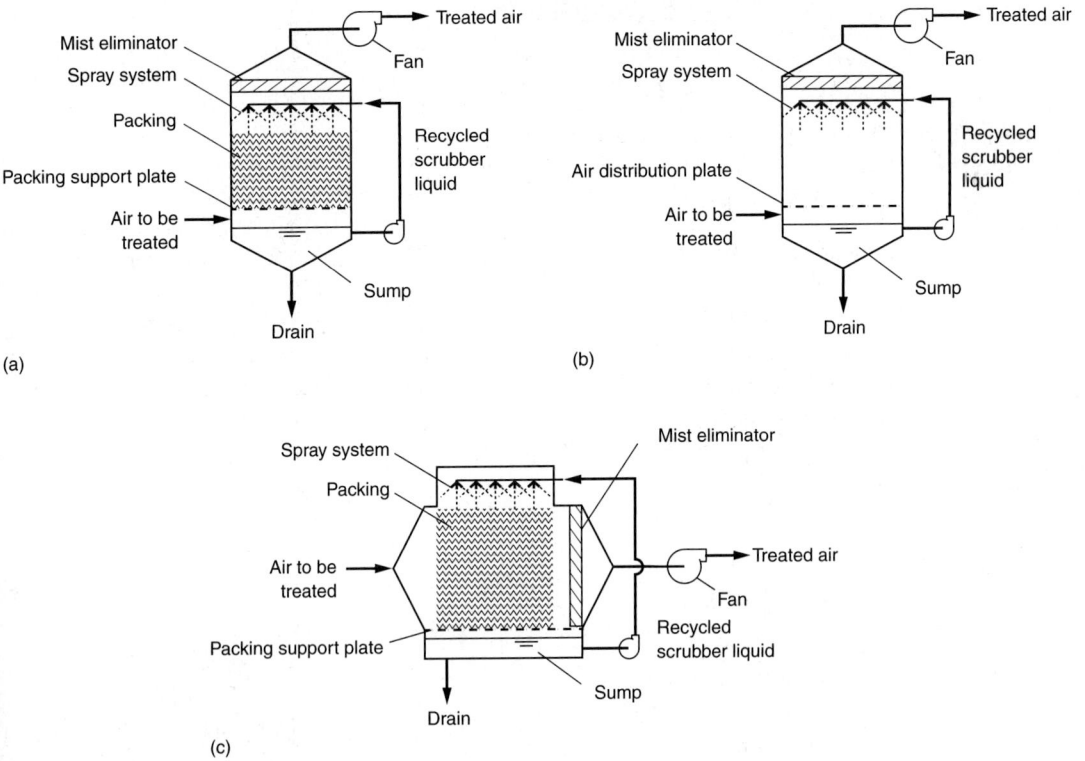

Figure 15–7

Typical wet-scrubber systems for odor control: (a) countercurrent packed tower, (b) spray chamber absorber, and (c) cross-flow scrubber.

Figure 15-8

Typical sodium hypochlorite scrubber used to treat the odors from the trickling filters shown in Fig. 15-5c and d.

chamber absorbers, and cross-flow scrubbers. In most single-stage scrubbers, such as shown on Fig. 15–8, the scrubbing fluid (usually sodium hypochlorite) is recirculated. The commonly used oxidizing scrubbing liquids are sodium hypochlorite, potassium permanganate, and hydrogen peroxide solutions. Because of safety and handling issues, chlorine gas is not used commonly in scrubbing applications at wastewater treatment facilities. Sodium hydroxide is also used in scrubbers where H_2S concentrations in the gas phase are high. Typical simplified scrubbing reactions with sodium hypochlorite, potassium permanganate, and hydrogen peroxide are as follows:

With sodium hypochlorite

$$H_2S + 4NaOCl + 2NaOH \rightarrow Na_2SO_4 + 2H_2O + 4NaCl \qquad (15\text{–}3)$$
(34.06)   (4 × 74.45)

$$H_2S + NaOCl \rightarrow S^{\circ}\downarrow + NaCl + H_2O \qquad (15\text{–}4)$$
(34.06)   (74.45)

With potassium permanganate

$$3H_2S + 2KMnO_4 \rightarrow 3S + 2KOH + 2MnO_2 + 2H_2O \text{ (acidic pH)} \qquad (15\text{–}5)$$
(3 × 34.06) (2 × 142.04)

$$3H_2S + 8KMnO_4 \rightarrow 3K_2SO_4 + 2KOH + 8MnO_2 + 2H_2O \text{ (basic pH)} \qquad (15\text{–}6)$$
(3 × 34.06) (8 × 142.04)

With hydrogen peroxide

$$H_2S + H_2O_2 \rightarrow S^\circ\downarrow + 2H_2O \qquad pH < 8.5 \tag{15-7}$$
$$\text{(34.06)} \quad \text{(34.0)}$$

In the reaction given by Eq. (15–3), 8.74 mg/L of sodium hypochlorite is required per mg/L of hydrogen sulfide, or 9.29 mg/L if the hydrogen sulfide is expressed as sulfide. In addition, in the reaction given by Eq. (15–3), 2.35 mg/L of sodium hydroxide (caustic) will be required per mg/L of hydrogen sulfide to make up for alkalinity consumed in the reaction. In practice, the required sodium hypochlorite dosage for the reaction given by Eq. (15–3) will vary from 8 to 10 mg/L per mg/L of H_2S. For the reaction given by Eq. (15–4), 2.19 mg/L of sodium hypochlorite is required per mg/L of hydrogen sulfide.

When potassium permanganate is used, the reactions that occur are typically various combinations of the reactions given by Eqs. (15–5) and 15–6). Reaction products that can occur, depending on the local wastewater chemistry, include elemental sulfur, sulfate, thionates, dithionates, and manganese sulfide. Stoichiometrically, about 2.8 and 11.1 mg/L of $KMnO_4$ are required for each mg/L of H_2S oxidized as given by Eqs. (15–5) and (15–6), respectively. However, based on operating data from actual field installations, about 6 to 7 mg/L $KMnO_4$ are required for each mg/L of H_2S oxidized. Potassium permanganate is generally used in smaller installations because of the cost (U.S. EPA, 1985; WEF, 1995).

In the reaction given in Eq. (15–7), 1.0 mg/L of hydrogen peroxide is required for each mg/L of sulfide expressed as hydrogen sulfide. In practice, the required dosage can vary from 1 to 4 mg/L per mg/L of H_2S. Because the systems used to carry out the reactions defined by Eqs. (15–3) through (15–7) are complex, especially where competing reactions may occur, some experimentation will be required to establish the correct dosage.

Hypochlorite scrubbers can be expected to remove oxidizable odorous gases when other gas concentrations are minimal. Typical removal efficiencies for single-stage scrubbers are reported in Table 15–11. In cases where the concentrations of odorous components in the exhaust gas from the scrubbers are above desirable levels, multistage

Table 15-11
Effectiveness of hypochlorite wet scrubbers for removal of several odorous gases[a]

Gas	Expected removal efficiency, %
Hydrogen sulfide	98
Ammonia	98
Sulfur dioxide	95
Mercaptans	90
Other oxidizable compounds	70–90

[a] Adapted in part from U.S. EPA (1985).

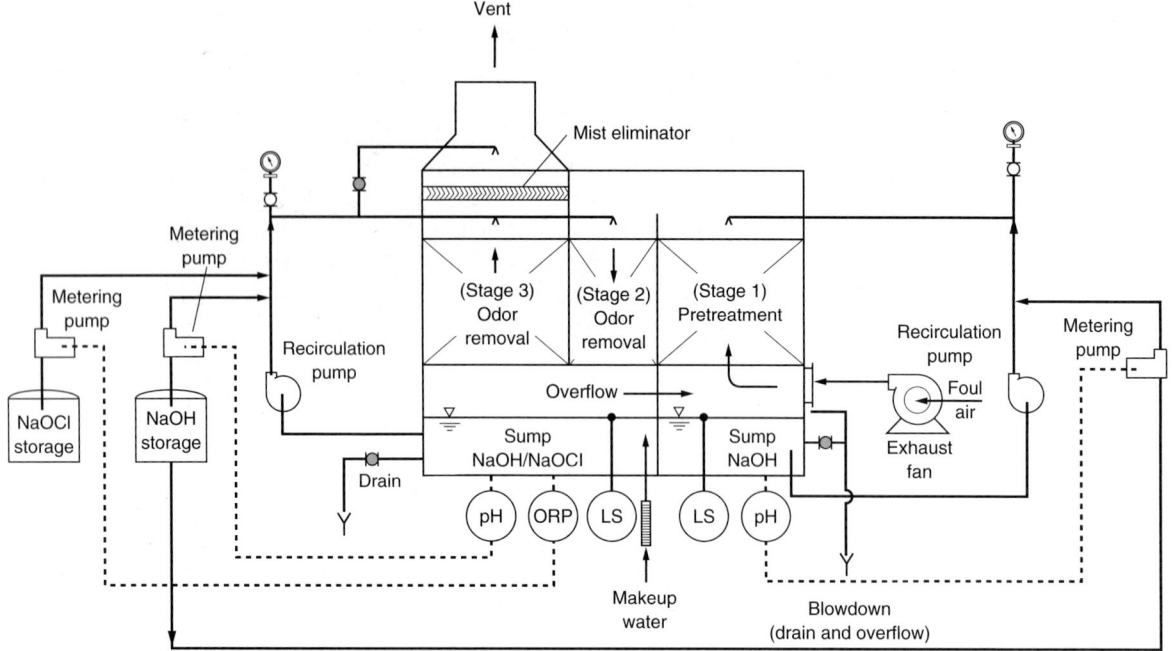

Figure 15–9

Three-stage odor-control process flow diagram. *(From Lo/Pro Systems, Inc.)*

scrubbers (see Fig. 15–9) are often used. In the three-stage scrubber shown on Fig. 15–9, the first stage is a pretreatment stage used to raise the pH so that a portion of the odorous gases (e.g., hydrogen sulfide) is reduced before treatment with chlorine in the second and third stages. The reaction that occurs in the first stage of a three-stage unit can be represented as follows.

$$H_2S + 2NaOH \rightarrow Na_2S + 2H_2O \tag{15–8}$$

To reduce maintenance problems due to precipitation, it is recommended that a low hardness (less than 50 mg/L as $CaCO_3$) be used for the makeup water.

Activated-Carbon Adsorbers. Activated-carbon adsorbers are used commonly for odor control (see Fig. 15–10). The rate of adsorption for different constituents or compounds will depend on the nature of the constituents or compounds being adsorbed (nonpolar versus polar). It has also been found that the removal of odors depends on the concentration of the hydrocarbons in the odorous gas. Typically, hydrocarbons are adsorbed preferentially before polar compounds such as H_2S are removed (note activated carbon is nonpolar). Thus, the composition of the odorous gases to be treated must be known if activated carbon is to be used effectively. Because the life of a carbon bed is limited, carbon must be regenerated (see Sec. 11–7 in Chap. 11) or replaced regularly for continued odor removal. To prolong the life of the carbon, two-stage systems

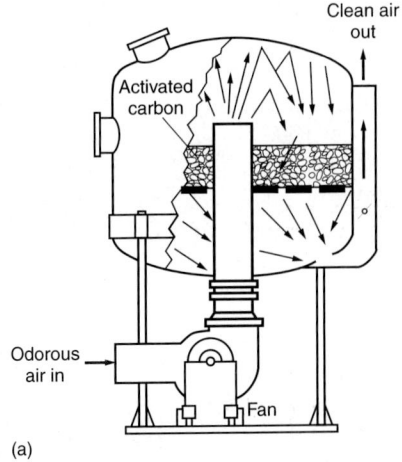

Clean air out

Activated carbon

Odorous air in

Fan

(a)

(b)

Figure 15–10

Typical activated-carbon systems for odor control: (a) schematic of a single-bed adsorber. In dual-bed adsorbers the foul air is introduced between the beds and is withdrawn from above the upper bed and below the lower bed, and (b) view of multiple dual-bed activated-carbon odor-control units. Because of operating problems, the activated carbon adsorbers have been replaced by the chemical scrubber shown in the background.

have been used, with the first stage being a wet scrubber followed by activated-carbon adsorption.

Vapor-Phase Biological Treatment Processes. The two principal biological processes used for the treatment of odorous gases present in the vapor phase are (1) biofilters and (2) biotrickling filters (Eweis et al., 1998). The use of microbial growths for the treatment of odors was the subject of an early patent by Pomeroy (1957), one of the important early researchers in the area of odor management in wastewater collection and treatment facilities.

Biofilters Biofilters are packed-bed filters. In open biofilters (see Fig. 15–11a), the gases to be treated move upward through the filter bed. In closed biofilters (see Fig. 15–11b), the gases to be treated are either blown or drawn through the packing material. As the odorous gases move through the packing in the biofilter, two processes occur simultaneously: sorption (i.e., absorption/adsorption) and bioconversion. Odorous gases are absorbed into the moist surface biofilm layer and the surfaces of the biofilter packing material. Microorganisms, principally bacteria, actinomycetes, and fungi, attached to the packing material, oxidize the absorbed/adsorbed gases and renew the treatment capacity of the packing material. Moisture content and temperature are important environmental conditions that must be maintained to optimize microorganism activity (Williams and Miller, 1992a; Yang and Allen, 1994; Eweis et al., 1998). Although compost biofilters are used commonly, one drawback is the large surface area (footprint) required for these units.

Figure 15–11

Typical packed-bed biofilters: (a) open-bed type and (b) enclosed type.

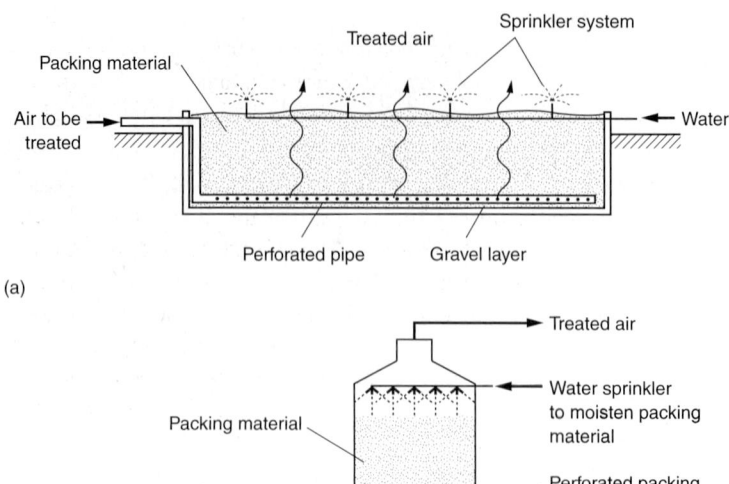

(a)

(b)

Figure 15–12

Typical covered biological stripping tower used for odor control.

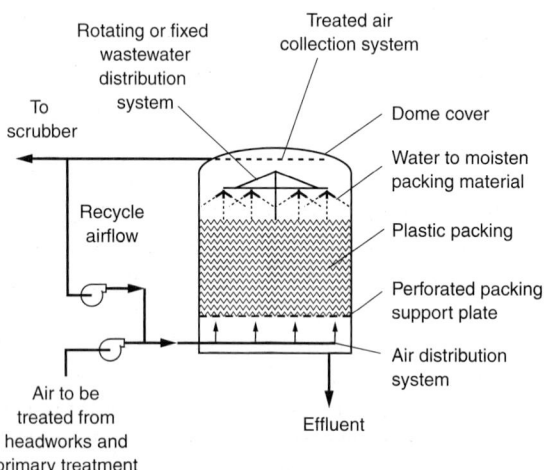

Biotrickling Filters Biotrickling filters are essentially the same as biofilters with the exception that moisture is provided continuously or intermittently by applying (typically spraying) a liquid (e.g., treated effluent) over the packing (see Fig. 15–12). The liquid is recirculated and nutrients are often added. Because water is lost in the gas leaving the filter, makeup water must be provided. Similarly, because of the accumulation of salts in the recycled water, a blowdown stream is required. Compost is not a suitable packing material for biotrickling filters because water will accumulate within the compost, thereby limiting the free movement of air within the filter. Typical packing materials include Pall rings, Raschig rings, lava rock, and granular activated carbon (Eweis et al., 1998; see also Sec. 11–8, Gas Stripping, in Chap. 11).

Conventional Biological Treatment Processes. The ability of microorganisms to oxidize hydrogen sulfide and other similar odorous compounds dissolved in the liquid under aerobic conditions is the basic concept used for the treatment of odors in liquid-based systems. The two principal types of conventional liquid-based systems used in wastewater treatment plants are the activated sludge process and the trickling filter process. In the activated sludge process, the odorous compounds are introduced into the aeration basin either with the existing air supply or injected separately through a manifold system. A major concern with this method of odor management is the high rate of corrosion in the air piping and blowers that occurs due to the presence of moist air containing hydrogen sulfide. The ability to transfer the odorous gaseous compounds to the liquid phase is also of concern.

With conventional uncovered trickling filters the major issues are how to transfer the air containing the odorous compounds to the trickling filter and how to avoid the release of untreated odorous compounds to the atmosphere. To control the release of odorous compounds, existing trickling filters that are to be used for odor control must be covered (see Fig. 15–5c).

Thermal Processing. Three thermal processing techniques have been used: (1) thermal oxidation, (2) catalytic oxidation, and (3) recuperative and regenerative thermal oxidation. The oxidation of methane (CH_4) can be used to illustrate the basic principle of all three thermal processes.

$$CH_4(gas) + 2O_2(gas) \rightarrow CO_2(gas) + 2H_2O(vapor) + heat \qquad (15-9)$$

If the gas to be combusted does not liberate enough heat to sustain the combustion process, it is usually necessary to use an external fuel source such as fuel oil, natural gas, or propane. Unfortunately, because of the low concentrations of odorous combustible gases in most waste streams, sustainable thermal oxidation is seldom possible, and large amounts of fuel are typically required to maintain the combustion temperatures needed to eliminate odors.

Thermal oxidation involves preheating the odorous gases before passing them into the combustion chamber so that complete oxidation can be achieved. Combustion occurs at temperatures in the range from 425 to 760°C (800 to 1400°F). Thermal oxidation is used, more commonly, for concentrated waste streams. The flaring of odorous gases is a relatively crude form of thermal combustion. Depending on the design of the combustion facility, incomplete combustion can occur due to variations in gas flow. For this method of odor control to be sustainable, the waste gas must typically contain 50 percent of the fuel value of the gas stream to be combusted.

Catalytic oxidation is a flameless oxidation process that occurs at 310 to 425°C (600 to 800°F) in the presence of a catalyst. Common catalysts include platinum, palladium, and rubidium. The decrease in temperature as compared to complete thermal

Figure 15–13

Typical self-contained chemical odor-stripping unit.

oxidation reduces the energy requirements significantly. However, because the catalysts can become fouled, the gas to be oxidized must not contain particulate material or constituents that will result in a residue. Additional information on the physical facilities used for thermal processing of VOCs may be found in Sec. 5–13 in Chap. 5.

Recuperative and regenerative thermal oxidation processes are used to reduce fuel consumption by preheating the incoming air, especially in large installations. In recuperative oxidizers, thin wall tubes are used to transfer heat recovered from exhaust air to the incoming air. In regenerative oxidizers, ceramic packing material is used to capture the heat from the hot exhaust gas and subsequently to release it to the incoming air. To maintain optimal heat recovery, the exhaust and incoming air are cycled through the packing material so that the incoming air is always passed through the hottest packing material. Typically, three beds of packing material are used in regenerative oxidizers.

Selection and Design of Odor-Control Facilities

The following steps are involved in the selection and design of odor-control and treatment facilities:

1. Determine the characteristics and volumes of the gas to be treated.
2. Define the exhaust requirements for the treated gas.
3. Evaluate climatic and atmospheric conditions.
4. Select one or more odor-control and treatment technologies to be evaluated.
5. Conduct pilot tests to determine design criteria and performance.
6. Perform life cycle economic analysis.

Many of the chemical odor-control technologies are supplied as complete packages, designed to meet a given performance specification. The analysis of chemical scrubbers and the design of biofilters is considered in the following discussion.

Design Considerations for Chemical Scrubbers

Most chemical scrubbers are supplied as a complete unit (see Fig. 15–13). Typical design factors for chemical scrubbers are presented in Table 15–12. Determination of the chemical requirements for odor scrubbing is illustrated in Example 15–3.

Table 15–12
Typical design factors for chemical scrubbers[a]

Item	Units	Value
Packing depth	m	1.8–3
Gas residence time in packing	s	1.3–2.0
Scrubbing liquid flowrate	kg H_2O/kg airflow	1.5–2.5
	L/s per m^3/s airflow	2–3
Makeup water flow	L/s per kg sulfide at pH 11	0.075
	L/s per kg sulfide at pH 12.5	0.004
pH	Unitless	11–12.5
Temperature	°C	15–40
Caustic usage	kg NaOH/kg sulfide	2–3

[a] Adapted in part from WEF (1995), Devinny et al. (1999).

EXAMPLE 15–3   **Chemical Requirements for Odor Scrubbing**   Hydrogen sulfide is to be scrubbed from a waste airstream using sodium hypochlorite. Determine the chemical (i.e., sodium hypochlorite and caustic) and water requirements for the following conditions.

1. Waste airstream flowrate = 1000 m^3/min
2. H_2S concentration in waste stream = 20 ppm_v at 20°C
3. Specific weight of air = 0.0118 kN/m^3 at 20°C
4. Density of air = 1.204 kg/m^3 at 20°C (see Appendix B-1)
5. Assume liquid to gas ratio for scrubber = 1.75
6. Density of 50 percent NaOH solution = 1.52 kg/L

Solution

1. Determine the volume occupied by one mole of a gas at a temperature of 20°C and a pressure of 1.0 atm using Eq. (2–44).

$$V = \frac{nRT}{P}$$

$$V = \frac{(1 \text{ mole})(0.082057 \text{ atm·L/mole·K})\left[(273.15 + 20) \text{ K}\right]}{1.0 \text{ atm}}$$

$$= 24.055 \text{ L} \qquad \text{use } 24.1 \text{ L}$$

2. Estimate the sodium hypochlorite requirement.
   a. Using Eq. (2–45), convert the H_2S concentration from ppm_v to g/m^3.

$$20 \text{ ppm}_v = \left(\frac{20 \text{ m}^3}{10^6 \text{ m}^3}\right)\left[\frac{(34.08 \text{ g/mole } H_2S)}{(24.1 \times 10^{-3} m^3/\text{mole of } H_2S)}\right]$$

$$H_2S \text{ concentration} = 28.3 \times 10^{-3} \text{ g/m}^3$$

b. Determine the amount of H_2S that must be treated per day.

$$(1000 \text{ m}^3/\text{min}) \times (28.3 \times 10^{-3} \text{ g/m}^3)(1440 \text{ min/d})(1 \text{ kg}/10^3\text{g})$$

$$= 40.8 \text{ kg/d}$$

c. Estimate the sodium hypochlorite dose. From Eq. (15–3), 8.74 mg/L of sodium hypochlorite are required per mg/L of sulfide, expressed as hydrogen sulfide.

$NaOCl_2$ required per day $= (40.8 \text{ kg/d}) \times (8.74) = 356.6 \text{ kg/d}$

3. Estimate the water requirement for the scrubbing tower.
   a. Determine the mass air flowrate.

$$(1000 \text{ m}^3/\text{min})(1.204 \text{ kg/m}^3) = 1204 \text{ kg/min}$$

   b. Determine the water flowrate.

$$(1204 \text{ kg/min})(1.75) = 2107 \text{ kg/min} = 2.1 \text{ m}^3/\text{min}$$

4. Determine the amount of sodium hydroxide (caustic) that must be added to replace the alkalinity consumed in the reaction.
   a. From the reaction given by Eq. (15–3), 2.35 mg/L of NaOH is required for each mg/L of H_2S removed.
   b. Determine the amount of NaOH required.

$NaOH = 40.8 \text{ kg/d} \times 2.35 = 95.9 \text{ kg/d}$

   c. Determine the volume of NaOH required. The amount of caustic per liter at 50% NaOH is

$NaOH = (1.52 \text{ kg/L})(0.50) = 0.76 \text{ kg/L}$

$$\text{Volume of 50\% NaOH} = \frac{(95.9 \text{ kg/d})}{(0.76 \text{ kg/L})} = 126.2 \text{ L/d}$$

Comment   The water requirement for the scrubbing tower will be specified initially by the scrubber supplier and field adjusted based on the results of pilot-plant studies and past operating experience. If the wastewater has sufficient alkalinity, it may not be necessary to add sodium hydroxide.

Design Considerations for Odor-Control Biofilters

Important design considerations for biofilters include (1) the type and composition of the packing material, (2) facilities for gas distribution, (3) maintenance of moisture within the biofilter, and (4) temperature control. Each of these topics is considered below. Design and operational parameters are presented and discussed following the discussion of the above topics. Additional details on biofilters may be found in van Lith (1989), Allen and Yang (1991, 1992), WEF (1995), Eweis et al., (1998), and Devinny et al. (1999).

Packing Material.   The requirements for the packing used in biofilters are (1) sufficient porosity and near-uniform particle size, (2) particles with large surface areas and significant pH-buffering capacities, and (3) the ability to support a large population of microflora (WEF, 1995). Packing materials used in biofilters include compost, peat, and a variety of synthetic mediums. Although soil and sand have been used in the past, they are less used today because of excessive headloss and clogging problems (Bohn and Bohn, 1988). Bulking materials used to maintain the porosity of compost and peat biofilters include perlite, Styrofoam™ pellets, wood chips, bark, and a variety of ceramic and plastic materials. A typical recipe for a compost biofilter is as follows (Schroeder, 2001):

> Compost = 50 percent by volume
> Bulking agent = 50 percent by volume
> 1 meq $CaCO_3$/g of packing material by weight

Optimal physical characteristics of a packing material include a pH between 7 and 8, air-filled pore space between 40 and 80 percent, and organic matter content of 35 to 55 percent (Williams and Miller, 1992a). When compost is used, additional compost must be added periodically to account for the loss due to biological conversion. Bed depths of up to 1.8 m (6.0 ft) have been used. However, because most of the removal takes place in the first 20 percent of the bed, the use of deeper beds is not recommended.

Gas Distribution.   An important design feature of a biofilter is the method used to introduce the gas to be treated. The most commonly used gas distribution systems include (1) perforated pipes, (2) prefabricated underdrain systems, and (3) plenums. Perforated pipes are usually placed in a gravel layer below the compost (see Fig. 15–14). Where perforated pipes are used, it is important to size the pipe so that it

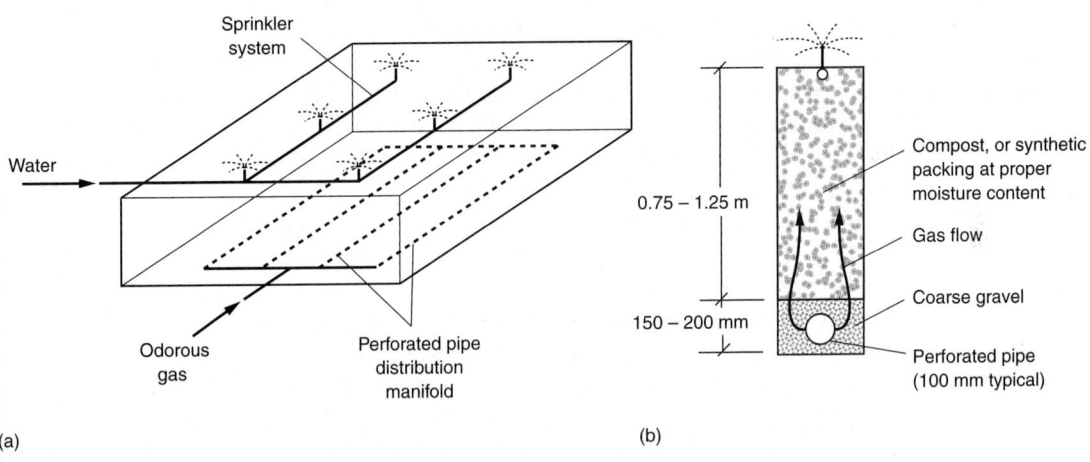

(a)

(b)

Figure 15–14

Definition sketch for open biofilters: (a) open bed and (b) trench type.

performs as a reservoir and not a manifold to assure uniform distribution (Crites and Tchobanoglous, 1998). A variety of prefabricated underdrain systems are available which allow for the movement of gas upward through the compost bed and allow for the collection of drainage. Air plenums are used to equalize the air pressure to achieve uniform flow upward through the compost bed. The height of air plenums will typically vary from 200 to 500 mm.

Moisture Control.   Perhaps the most critical item in the successful operation of a biofilter is to maintain the proper moisture within the filter bed. If the moisture content is too low, biological activity will be reduced. If the moisture content is too high, the flow of air will be restricted and anaerobic conditions may develop within the bed. Also, biofilters tend to dry out unless moisture or humidity is added. The optimal moisture content is between about 50 and 65 percent, defined as follows:

$$\text{Moisture content, \%} = \left(\frac{\text{mass of water}}{\text{mass of water} + \text{mass of dry packing}} \right) \times 100 \qquad (15\text{--}10)$$

Moisture can be supplied by adding water to the top of the bed (usually by spraying) or by humidifying the incoming gas in a humidification chamber. The relative humidity of the gas entering the biofilter should be 100 percent at the operating temperature of the biofilter (Eweis et al., 1998). In biotrickling filters the liquid application rate is typically about 0.75 to 1.25 $\text{m}^3/\text{m}^2 \cdot \text{d}$.

Temperature Control.   The operating temperature range for biofilters is between 15 and 45°C, with the optimal range being between 25 and 35°C. In cold climates, biofilters must be insulated and the incoming gas must be heated. Where the incoming gas is warmer, it may have to be cooled before being introduced to the biofilter. Operation at higher temperatures (e.g., 45 to 60°C) is often possible, as long as the temperature remains relatively constant.

Design and Operating Parameters for Biofilters.   The sizing of biofilters is typically based on a consideration of the gas residence time in the bed, the unit air loading rate, and the constituent elimination capacity. Terms that will be encountered in the literature and the relationships commonly used to describe the performance of bulk media filters are summarized in Table 15–13. The empty bed residence time (EBRT) [see Eq. (15–11)], used to define the relationship between the volume of the contactor and the volumetric gas flowrate, is similar to Eq. (11–61) used for the analysis of activated-carbon systems. The true residence time is determined by incorporating the porosity, α [see Eq. (15–12)]. Surface and volumetric mass loading rates are often used to define the operation of bulk media filters. The elimination capacity, as given by Eq. (15–18), is used to compare the performance of different odor-control systems.

The residence time for foul air from wastewater treatment facilities is typically between 15 and 60 s and surface loading rates have ranged up to 120 $\text{m}^3/\text{m}^2 \cdot \text{min}$ for H_2S concentrations up to 20 mg/L. Constituent elimination rates are determined experimentally and are usually reported as a function of the constituent loading rate (e.g., mg $H_2S/\text{m}^3 \cdot \text{h}$). An essentially linear, 1 to 1, constituent elimination rate up to a critical load-

Table 15-13

Parameters used for the design and analysis of bulk media filters[a]

Parameter		Definition
Empty bed residence time		EBRT = empty bed residence time, h
$$EBRT = \frac{V_f}{Q} \quad (15\text{-}11)$$		V_f = total volume of filter bed contactor, m^3
		Q = volumetric flowrate, m^3/h
Actual residence time in filter		RT = residence time, h, min, s
$$RT = \frac{V_f \times \alpha}{Q} \quad (15\text{-}12)$$		α = porosity of filter bed contactor
		SLR = surface loading rate, $m^3/m^2{\cdot}h$
Surface loading rate		A_f = surface area of filter bed contactor, m^2
$$SLR = \frac{Q}{A_f} \quad (15\text{-}13)$$		VLR = volumetric loading rate, $m^3/m^3{\cdot}h$
		RE = removal efficiency, %
Surface mass loading rate		C_o = influent gas concentration, g/m^3
$$SLR_m = \frac{Q \times C_o}{A_f} \quad (15\text{-}14)$$		EC = elimination rate (capacity), $g/m^3{\cdot}h$
		C_e = effluent gas concentration, g/m^3
Volume loading rate		
$$VLR = \frac{Q}{V_f} \quad (15\text{-}15)$$		
Volume mass loading rate		
$$VLR_m = \frac{Q \times C_o}{V_f} \quad (15\text{-}16)$$		
Removal efficiency		
$$RE = \frac{(C_o - C_e)}{C_o} \times 100 \quad (15\text{-}17)$$		
Elimination rate (capacity)		
$$EC = \frac{Q(C_o - C_e)}{V_f} \quad (15\text{-}18)$$		

[a] Adapted in part from from Eweis et al. (1998); Devinny et al. (1999).

ing rate has been observed for hydrogen sulfide and other odorous compounds (see Fig. 15–15). Yang and Allen (1994) have reported a linear 1 to 1 elimination rate for H$_2$S, with loading rates for compost filters up to a maximum value of about 130 g S/m³·h, beyond which the elimination rate becomes essentially constant at a rate of 130 g S/m³·h with increased loading. It should be noted that H$_2$S is eliminated easily as it passes through a biofilter.

Typical design criteria for biofilters are presented in Table 15–14. Typical biofilters are shown on Fig. 15–16. Some states regulate the design of compost biofilters including loading rates, biofilter emission rates, odor-sampling procedures, and setbacks from property lines. A typical odor-emission limit at the surface of the biofilter is 50 dilutions to threshold [see Eq. (2–52) in Chap. 2] (Finn and Spencer, 1997). The design of a compost biofilter for the elimination of hydrogen sulfide is illustrated in Example 15–4.

Figure 15–15

Typical elimination capacity versus applied load. The relationship is linear up to a critical value, after which the elimination capacity approaches a maximum value asymptotically.

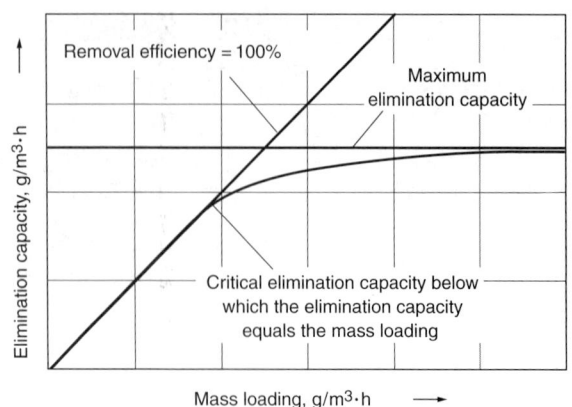

Table 15–14
Typical design factors for biofilters[a]

Item	Units	Type of biofilter	
		Biofilter	**Biotrickling filter**
Oxygen concentration	Parts oxygen/ parts oxidizable gas	100	100
Moisture			
Compost filter	%	50–65	50–65
Synthetic media	%	55–65	55–65
Temperature, optimum	°C	15–35	15–35
pH	Unitless	6–8	6–8
Porosity	%	35–50	35–50
Gas residence time	s	30–60	30–60
Depth of medium	m	1–1.25	1–1.25
Inlet odorous gas concentration	g/m^3	0.01–0.5	0.01–0.5
Surface loading rate	$m^3/m^2{\cdot}h$	10–100[b]	10–100[b]
Volume loading rate	$m^3/m^3{\cdot}h$	10–100	10–100
Liquid application rate	$m^3/m^2{\cdot}d$		0.75–1.25
Elimination rate (capacity)			
H_2S (in compost filter)	$g/m^3{\cdot}h$	80–130	80–130
Other odorous gases	$g/m^3{\cdot}h$	20–100	20–100
Back pressure, maximum	mm of water	50–100	50–100

[a] Adapted in part from van Lith (1989), Yang and Allen (1994), WEF (1995), and Devinny et al. (1999).

[b] Loading rates as high as 500 $m^3/m^2{\cdot}h$ have been reported, depending on the compound and its concentration.

Figure 15–16

Typical soil/compost bulk biofilters filter for odor control: (a) compost biofilter used for odor control at an influent pump station at a wastewater treatment plant (Florida) and (b) soil-type biofilter with grass covering used for odor control at a wastewater treatment plant (Okinawa, Japan).

(a)

(b)

EXAMPLE 15–4 **Design of a Compost Biofilter for Odor Control** Determine the size of compost biofilter needed to scrub the air from a 100 m³ enclosed volume using the design criteria given in Table 15–14. Also estimate the mass of the buffer compound needed to neutralize the acid formed as a result of treatment within the filter. Assume 12 air changes per hour are needed. Assume a bed porosity of 40 percent. Will the volume selected be adequate if the air contains 40 ppm$_v$ of H$_2$S in addition to other odorous constituents? Assume an elimination rate of 65 g S/m³·h, which incorporates a factor of safety of 2 as compared to the maximum rate given in Table 15–14. The temperature of the air is 20°C.

Solution

1. Estimate the airflow to be scrubbed.

   Flow = volume/time

   Flow = (100 m³)(12 changes per hour) = 1200 m³/h

2. Select a loading rate from Table 15–14; use 90 m³/m²·h
3. Select a filter-bed depth from Table 15–14; use 1.0 m
4. Calculate the area needed for the filter bed.

   Area = gas flow/ loading rate

   Area = (1200 m³/h)/(90 m³/m²·h)

   Area = 13.3 m²

5. Check the empty bed residence time using Eq. (15–11)

$$\text{EBRT} = \frac{V_f}{Q} = \frac{(13.3 \text{ m}^2)(1 \text{ m})}{(1200 \text{ m}^3/\text{h})}$$

   = 0.011 h = 39.9 s   (OK 39.9 s > 30 s)

6. Determine if the volume of the biofilter determined in Step 5 is adequate to treat the H_2S.

   a. Determine the concentration of H_2S in g/m^3 using Eq. (2–45). From Example 15–3 the volume of gas occupied by one mole of a gas at a temperature of 20°C and a pressure of 1.0 atm is 24.1 L. Thus, the concentration of H_2S is

   $$g/m^3 = \left(\frac{40\ L^3}{10^6\ L^3}\right)\left[\frac{(34.08\ g/mole\ H_2S)}{(24.1 \times 10^{-3}\ m^3/mole\ of\ H_2S)}\right]$$

   $$= 0.057\ g/m^3$$

   b. Determine the mass loading rate of S^{2-} in g S/h.

   $$M_s = \left(\frac{1200\ m^3}{h}\right)\left(\frac{0.057\ g\ H_2S}{m^3}\right)\left(\frac{32\ g\ S^{2-}}{34.08\ g\ H_2S}\right)$$

   $$= 64.2\ g\ S^{2-}/h$$

   c. Determine the required volume assuming an elimination rate of 65 g $S/m^3{\cdot}h$.

   $$V = \frac{(64.2\ g\ S/h)}{(65\ g\ S/m^3{\cdot}h)} = 0.99\ m^3$$

   Because the volume of the bed ($13.3\ m^3$) is significantly greater than the required volume, the removal of H_2S will not be an issue.

7. Determine the mass of the buffer compound needed to neutralize the acid formed as a result of treatment within the filter.

   a. Determine the mass of H_2S in kg applied per year.

   $$H_2S,\ kg/year = \frac{(1200\ m^3/h)(0.057\ g/m^3)(24\ h/d)(365\ d/yr)}{(10^3\ g/kg)}$$

   $$= 599.2\ kg/year$$

   b. Determine the mass of buffer compound required. Assume the following equation applies:

   $$H_2S + Ca(OH)_2 + 2O_2 \rightarrow CaSO_4 + 2H_2O$$

   $$34.06 \qquad 74.08$$

   Thus, about 2.05 kg of $Ca(OH)_2$ will be required per kg of H_2S. If the compost biofilter has a useful life of 2 years, then a total of 2457 kg of $Ca(OH)_2$ equivalent will be required to be added to the bed. Typically, 1.25 to 1.5 times as much are added. The buffer compound is mixed in with the compost and the bulking agent.

Comment   Based on the results of the computation carried out in Step 6, it is clear why compost and soil filters are so effective in the elimination of H_2S.

15–4 INTRODUCTION TO AUTOMATIC PROCESS CONTROL

The purpose of most wastewater automatic control systems is to maintain one or several process parameters such as SRT, DO, or clarifier sludge depth at a fixed value. If there were no changes in external conditions, such as flowrate variation, process control would be a simple task. However, external conditions are changing constantly, and, as a result, some dynamic control over treatment processes is needed. The purpose of this section is to introduce the basic concepts and design considerations involved in automatic process control. Terms commonly encountered when discussing process control are summarized in Table 15–15.

Table 15–15
Definition of terms used in wastewater process control applications

Term	Description
Bias gain coefficient	Constant value usually equal to 50 percent of the maximum value of a manipulated parameter
Cascade control mode	One feedback controller, identified as the primary controller, is used to calculate a set point of a second feedback controller
Control loop	Structure of a control system
Controlled parameter	Process parameter that is to be kept equal to the target value
Disturbance	An external cause of a measurable change in the operation or performance of a treatment process
Error	Difference between measured value of a controlled parameter and set point under non-steady-state operating conditions
Feedback control	A control method in which the value of a manipulated parameter is based on errors of a controlled parameter
Feed forward control	A control method in which the value of a manipulated parameter is based on a disturbance value
Integral control mode (I)	The value of a manipulated parameter is calculated based on a linear relationship with the accumulated error of a controlled variable
Manipulated parameters	Process parameters that can be changed directly to control the process
Offset	Difference between measured value of a controlled parameter and set point under steady-state operating conditions
On-off control	Simplest and the least expensive type of feedback control
Proportional control mode	The value of a manipulated parameter is calculated based on a linear relationship with the current error of a controllable variable
Proportional gain coefficient (P)	Coefficient relating manipulated parameter with the current error of controllable parameter
Proportional-integral control	Combination of two independent control modes: proportional (P) and integral (I)
Set point	Target operating value for a controlled variable. In advanced control systems the set point is not a fixed value

Process Disturbances

Factors that cause a change in the operation or performance of a treatment process are termed *disturbances*. There are two major types of disturbances: external, such as variations in incoming flow and wastewater characteristics; and internal, such as the actions of an operator or interaction among process units within a treatment plant. Both internal and external disturbances are discussed below.

External Disturbances.   Continuous changes of influent flow, wastewater composition, and temperature introduce external disturbances to practically every process at a wastewater treatment plant. For example, the ratio of maximum to minimum oxygen demand, as well as the solids loading, will vary usually from 3:1 to 5:1 between the daytime hours (peak hours) and nighttime hours (off-peak hours). For small treatment plants this ratio is sometimes as high as 16:1. Wide variations in loads can lead to (1) inconsistent performance, (2) underloaded or overloaded unit processes, and (3) over- or underdesigned treatment facilities. Process performance deteriorates even further during storm events when hydraulic loadings increase as much as an order of magnitude, causing a loss of biological solids, and resulting in potential violations of the discharge permit.

To control a disturbance caused by widely fluctuating flows and loads, an equalization basin can be constructed to distribute the incoming mass load more evenly over the day and reduce loads during storm or peak flow events. The basin fills with wastewater during the peak hours, when flow and oxygen demand are high, and empties during the night when flow and oxygen demand are low (see Sec. 5–3 in Chap. 5). A goal of equalization basin operation is to dampen the peak incoming hydraulic load to the plant. With the introduction of on-line water-quality analyzers, more beneficial oxygen demand equalization is sometimes used. Use of diurnal load equalization (1) simplifies and reduces the cost of plant operation, (2) reduces capital costs for aerators, clarifiers, blowers, pumps, and other equipment, and (3) can reduce the variability of the constituent concentrations in the effluent. The same equalization basin can be used to reduce flow to a plant during storms.

Another method to reduce peak flow coming to a treatment plant during storm events is utilization of spare volume of the collection system. Sophisticated flow control systems are used to maximize collection-system storage capacity. Such systems utilize extensive flow monitoring and control devices including hydraulic brake regulators (Tchobanoglous, 1981) and often rely on a dynamic computer model of wastewater collection systems.

Internal Disturbances.   In addition to external disturbances, there are also internal disturbances that are sometimes have a greater effect than the external disturbances. An example of an internal process disturbance is a change in the growth of filamentous organisms in the mixed liquor that reduces the settling performance of the final clarifiers. For successful design and operation of a wastewater treatment plant it must be recognized that every treatment process at the plant is not an isolated unit, but an element of a complex system. Change in operation of one element affects not only the particular process unit, but also the entire system. The actions of an operator, such as a changing the number of process units in operation, changing the *set points* (target operating

Table 15–16

Typical examples of internal disturbances at wastewater treatment plants

Source	Parameter affected	Process directly affected
Intermittent operation of pumps and other mechanical equipment	Flow	All downstream processes
Unequal flow distribution among parallel process units	Flow, TSS, oxygen demand	Parallel process units
Change in effluent ammonia and nitrite concentration	Chlorine demand	Disinfection
Change in mass of wasted solids	Solids loading	Sludge processing
Change in blanket depth of primary and secondary clarifiers	Effluent TSS, BOD, sludge TSS concentration	Downstream liquid processes, thickeners
Change in recycle flowrate and composition	Solids loading and oxygen demand	All processes located downstream of junction of the recycled flow and the main flow

values for controlled variables), and changing the flow routing, if not implemented carefully, may cause significant internal disturbances.

Another source of internal disturbances is the mode of operation of an individual unit process. Each of the process units has to not only reliably and efficiently perform its function, but also should minimize disturbances to other process units or, better yet, compensate these units for external disturbances. For example, the ammonia load in sludge supernatant from dewatering facilities that is recycled back for biological treatment can increase oxygen demand by as much as 30 percent. If this load is introduced during the nighttime hours, it will help to spread oxygen demand evenly and utilize less expensive nighttime energy; recycling supernatant during peak hours may cause problems with control, effluent quality, and cost of aeration. A list of some disturbances caused by changes in the operating modes of treatment processes is presented in Table 15–16.

Control Systems for Wastewater Treatment Plants

Even with the best efforts it is impossible, however, to eliminate disturbances completely. Thus, to maintain optimum process efficiency and reliability, it is necessary to compensate for disturbances by changing some of the process parameters. Process parameters that are to be maintained at target values are called *controlled parameters*. Process parameters that can be changed directly to maintain controlled parameters at target values are called *manipulated parameters*. Examples of controlled and manipulated parameters for wastewater treatment processes are given in Table 15–17.

Control System Architecture. The typical structure of a control system is termed a *control loop* and is shown on Fig. 15–17. As illustrated on Fig. 15–17, a control loop is comprised of (1) an instrument for input measurement, (2) a controller for calculating the action of a final control element, (3) a final control element that is used to implement control action, and (4) signal transmitters for communication between the other

Table 15-17

Typical examples of controlled and manipulated parameters

Process	Controlled parameter	Manipulated parameter
Primary treatment and gravity thickening	• Sludge depth	• Rate of sludge removal
Activated sludge	• Dissolved oxygen concentration in an aerator	• Oxygen supply rate
	• Sludge depth in a clarifier	• Return sludge flowrate
	• Solids-retention time	• Rate of sludge removal
Chlorination	• Chlorine residual	• Chlorine dosage
Dissolved air flotation	• Air to solids ratio	• Airflow
	• TSS in the supernatant	• Polymer flow
Gravity belt thickening	• Percent solids	• Belt solids loading
	• Capture percent	• Polymer flowrate

Figure 15-17

General structure and components of a control loop used to control dissolved oxygen (DO) in an activated-sludge biological treatment process.

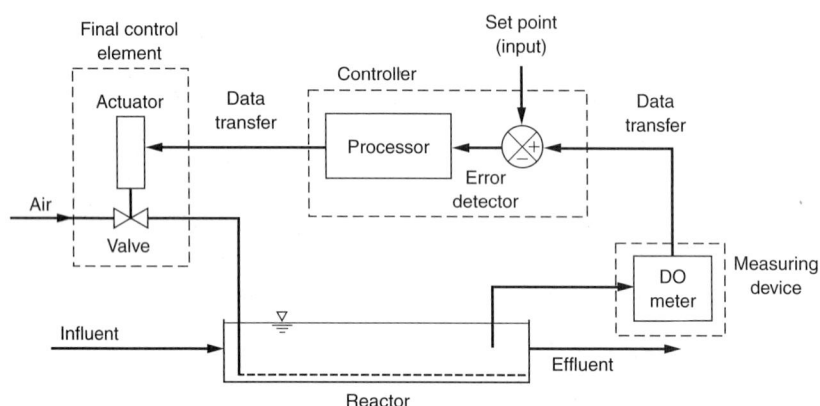

elements. In wastewater treatment plants, control-loop systems are used to control flow, pressure, liquid level, constituent concentrations, temperature, and other operating variables. The elements of the control loop are described in greater detail at the end of this section following the discussion of control algorithms.

Manual versus Automatic Control. The general control-loop structure described above and illustrated on Fig. 15–17 exists in any type of control, whether manual or automatic. For instance, manual control of solids-retention time (SRT) in an activated-sludge process consists of the following steps:

1. An operator samples both mixed liquor and return sludge.
2. Samples are brought to laboratory for analysis.
3. TSS analysis is performed during a period of several hours.

4. The results from the laboratory analysis are communicated to a person responsible for waste flow calculation.
5. The person responsible for operation determines if the sludge wasting flowrate should be changed and, if so, how much.
6. Instructions on waste sludge flow change are given to an operator.
7. The operator changes position of a valve on waste flow line using flowmeter readings.

The same structure exists in an automatic SRT control system as illustrated on Fig. 15–18. The difference, however, is that TSS measurements are conducted on-line, calculations are performed continuously by a computer, the valve position is changed by an actuator, and communication between computer and other elements is conducted without human intervention through cables and fiber optics, or by wireless. Comparison of activated-sludge performance with manual versus automatic control is illustrated on Fig. 15–19.

Advantages of Automatic Control. Based on the performance records of numerous wastewater treatment plants, automatic control systems not only reduce workload, they also allow for more precise control of process parameters. Improved control of process parameters stabilizes and improves operation of not only a process under control, but also processes located downstream. For example, automation of SRT improves operation not only of the activated-sludge process, but also of sludge-thickening processes due to more stable sludge mass load on the thickening facility.

Failure of any element (an instrument, most often) of an automatic control system may lead to serious problems if a control algorithm does not automatically recognize such a failure has occurred and compensate for it. A control algorithm is the set of instruction or logic used by the controller to bring about a change in a controllable

Figure 15–18

Automatic control system for an activated-sludge process based on solids-retention time (SRT).

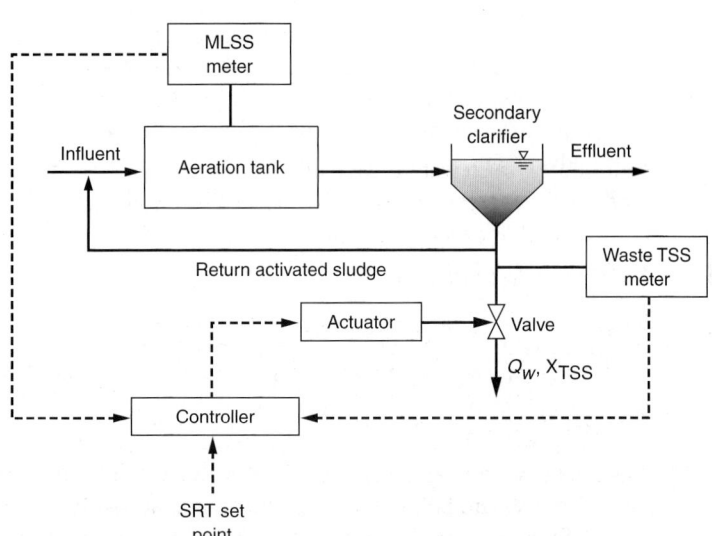

Figure 15–19

Performance comparison for a wastewater treatment plant with manual and automatic solids-retention time (SRT) process control.

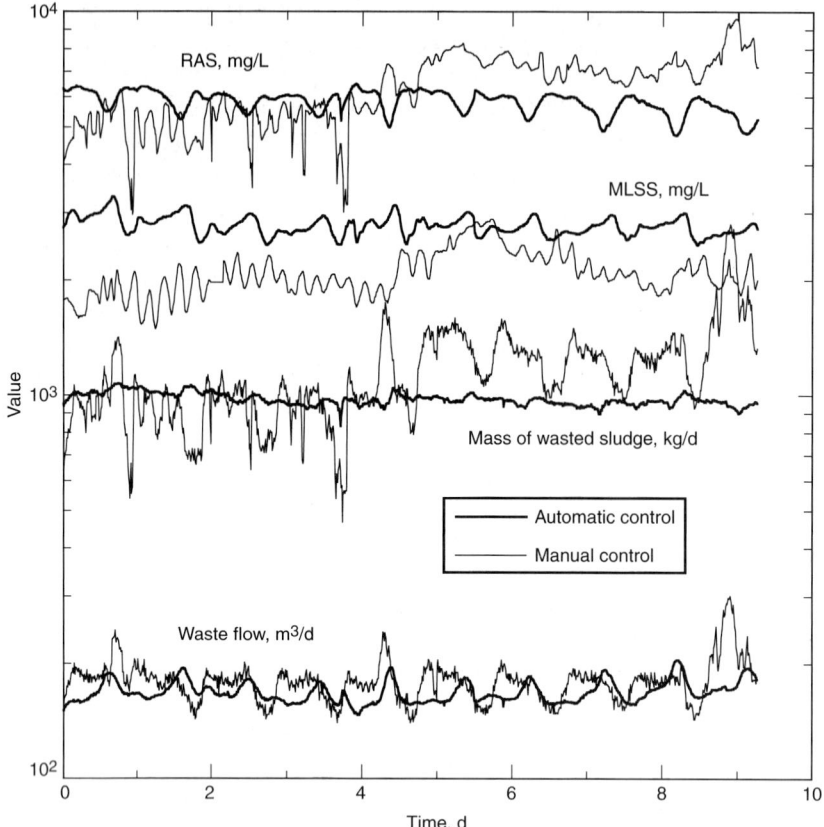

parameter. In the past, poor reliability was a major reason for the limited utilization of automatic control in the wastewater industry. Progress in computer hardware, and especially software, has made it possible for modern control systems to function reliably, even if one of the components fails (Colby and Ekster, 1997; Olson and Newell, 1992). Other advantages accrue through automatic control such as reduced operating staff, unattended operation for one or more shifts, and automatic data logging and archiving.

Control Algorithms

To compensate for process disturbances, an appropriate change of a manipulated variable must be made. The challenge of process control is to determine how much change is required. Inadequate change will not completely compensate for a disturbance; a large change may cause more problems than the disturbance itself. For example, if the airflow supply is reduced too much in the evening hours, when oxygen demand is low, effluent quality will deteriorate significantly. A new value of a manipulated parameter can be determined either by human (manual control) or automatically (automatic control). Several methodologies that can be used for the calculation of a manipulated variable that will be used in an automatic control are discussed below. Additional details on control systems may be found in Shinskey (1996), Smith and Corripo (1997), Coughanowr (1991), and WEF (1993).

There are three general types of control: (1) feedback, (2) feed-forward, and (3) combination feed-forward and feedback. Feedback control involves measurement of a process variable to make adjustments to an input variable. The input variable then becomes the manipulated variable. A feedback controller cannot achieve perfect control continuously because it can respond only when an error is detected in the output. Feed-forward control, on the other hand, is used to detect disturbances at the input before the system responds and to make adjustments to one or more manipulated variables to reduce or eliminate the effect of the disturbance. Thus, the essential difference between feedback and feed-forward control is that the former acts to compensate for observed changes, whereas the latter tries to compensate for disturbances (WEF, 1993). In situations where control cannot be achieved accurately or completely with feedback control, a combination of feed-forward and feedback control can be used. Each of the three general methods of control is discussed in the following paragraphs. Cascade and adaptive control, two other methods used for process control, are also discussed.

Feed-Forward Control. The feed-forward control method consists of measuring disturbances and changing manipulated parameters so that the controlled parameters will be maintained within a desired range. For example, if the influent characteristics and flowrate (disturbances) are measured and it was possible to calculate the required change in airflow (manipulated parameter) supplied to an activated-sludge system to maintain constant dissolved oxygen (DO) concentration in the aeration tank (controlled parameter), it would be possible to implement feed-forward control.

In a very basic, so-called *static feed-forward control* algorithm, a manipulated parameter is calculated based on a linear relationship with a controlled variable. For DO, the magnitude of the required airflow signal is calculated using the following formula:

$$Q_A = (K_{ff})Q_L \tag{15–19}$$

where Q_A = airflow, m^3/min
K_{ff} = feed-forward *gain* coefficient, $(m^3/min)/(m^3/d)$
Q_L = liquid flowrate, m^3/d

In more sophisticated dynamic feed-forward control algorithms, it is common practice to take into account process dynamics.

Unfortunately, most of the time measurements of disturbances as well as calculation of a required change in a manipulated parameter are very difficult tasks. As a result, feed-forward control has limited application in the automation of wastewater treatment processes. The current-limited applications of feed-forward control include control of chemical addition and control of return activated-sludge flow from clarifiers to the aeration tanks. In both cases, only one disturbance (influent flow change) is used for calculation of change in manipulated parameters. Because the effect of other disturbances (water-quality characteristics, for example) is not considered, a controlled parameter cannot be maintained within a narrow range. Development of advanced on-line water

Figure 15–20

Typical on-line analyzer for total suspended solids (TSS). *(Courtesy of Royce Instruments.)*

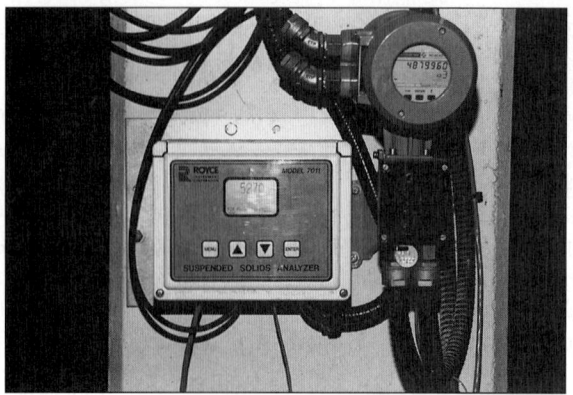

analyzers (see Fig. 15–20) and computer modeling of wastewater treatment processes may help to make this type of control more feasible in the future.

Feedback Control. The principle of feedback control is one of the most intuitive concepts for process control. An action is taken to correct a less than desirable situation; the result of corrective action is then measured and a new corrective action is applied. So instead of measuring a disturbance, a change in a controlled parameter (such as DO discussed above) is measured, and based on the measured change a manipulated parameter (airflow) is calculated. There are several types of feedback control, including on-off, proportional, integral, proportional-derivative, proportional-integral, and proportional-integral-derivative. Of these control modes, the most common are *on-off* and *proportional-integral*. As in the feed-forward discussion, DO will be used to illustrate the various control methods.

On-Off Control On-off control is the simplest and the least expensive type of feedback control. Control action is intermittent and initiated when a controlled parameter is approaching a predetermined value. Turning the aeration system on and off as the oxygen level falls below or rises above a given DO level in an aeration tank is an important example in wastewater treatment. Another example is the control of the water level in the wetwell of a pump station. Pumping is initiated when water elevation reaches a high mark and is stopped when water level reaches a low mark. Frequency of start-stop cycles of mechanical equipment is usually limited due to an extremely high electrical load that a motor experiences during startup. Thus, in wetwell level control, the high and low operating levels should be selected far enough apart to limit the frequency of pump starts (such as one start every 20 min) to avoid overheating the electric motor. As a result, on-off control cannot maintain a controlled parameter within a narrow range. On-off control actions also cause a significant disturbance to be transmitted to downstream processes. Variable-speed pumps are often used to limit disturbances to downstream processes.

Proportional Control In a proportional control mode, a manipulated parameter is calculated based on a linear relationship with the error for the controlled variable. For purposes of this discussion, error is defined as the difference between the target value of the controlled variable and the current value. For DO, the magnitude of the required airflow signal is calculated using the following formula:

$$Q_A = (K_p)(DO_{sp} - DO) + M = K_p(\varepsilon) + M \qquad (15-20)$$

where Q_A = airflow, m³/min
$\quad\quad K_p$ = controller *proportional gain* coefficient, (m³/min)/(mg/L)
$\quad\quad DO$ = dissolved oxygen concentration, mg/L
$\quad\quad DO_{sp}$ = targeted dissolved oxygen concentration at the set point, mg/L
$\quad\quad M$ = constant, often identified as the controller *bias coefficient* or *base value,* which corresponds to the output from the controller when the error is zero. The bias coefficient is often initially set equal to 50 percent of maximum value of the manipulated variable, m³/min
$\quad\quad \varepsilon$ = difference between measured value and set point defined as the *error* under non-steady-state operating conditions and as the *offset* under steady-state conditions

The controller relationship represented by Eq. (15–20) corresponds to line 1 on Fig. 15–21. Assume, for example, when the set point (i.e., target value) DO_{sp} is reached in the aeration basin at point *A,* the line that characterizes the airflow demand-DO relationship (line 2) intercepts the controller line 1. When the BOD load is increased, the airflow demand-DO relationship changes (line 3). A new air demand will be satisfied at point *B,* and the system will reach a new equilibrium (or so-called steady state). As noted above, the difference between the new value of the controlled variable and set point under steady-state conditions is called *offset.* Under non-steady-state conditions, the offset is called an *error.* Increasing the absolute value of the gain can decrease the offset. However, an offset cannot be eliminated completely because, to achieve zero offset, proportional gain needs to be increased indefinitely (i.e., line 1 has to become parallel to the *Y*

Figure 15–21

Schematic representation of proportional control for dissolved oxygen (DO).

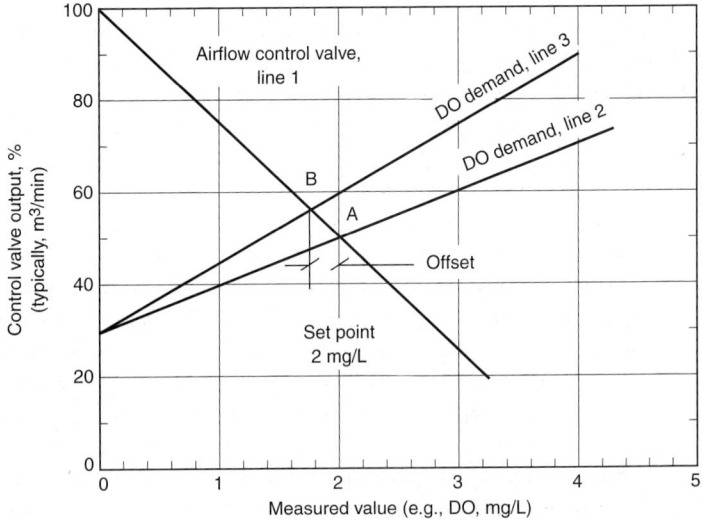

axis). The dynamics of the process and accuracy of measurements impose considerable limitations on the value K_p and, as a result, K_p cannot be increased indefinitely.

Integral Control In the integral control mode, the area under the curve of a plot of the offset versus time is used to determine the corrective action. The change in the manipulated parameter is calculated based on a proportional relationship. The integral mode is the best control mode to eliminate offset. If any offset is present, an action based on integral control will continue to make changes until the offset is eliminated. Integral control action for DO control can be described by the following equation:

$$Q_A = (K_p/t_i)\int(DO_{sp} - DO)\, dt = (K_i)\int(\varepsilon)\, dt \qquad (15\text{--}21)$$

where Q_A = airflow, m³/min
$\qquad t_i$ = *integral time,* usually measured in minutes, min
$\qquad K_i$ = controller integration gain factor
$\qquad$ = K_p/t_i, (m³/min²)/(mg/L)
$\qquad t$ = time, min
$\qquad$ Other terms as defined above.

If the parameter versus time is defined by a series of discrete time step measurements, the theoretical airflow is typically approximated as

$$Q_A = (K_p/t_i)\Sigma(DO_{sp} - DO)\, \Delta t = (K_i)\Sigma(\varepsilon)\Delta t \qquad (15\text{--}22)$$

where Δt = differential time, min

Theoretically, using integral control, the set point can always be maintained precisely without offset. However, if the dynamics of the control loop are slow and the K_i value is high, the system becomes unstable because of overly aggressive control action. Reduction of the K_i value will make the control system too slow and make it impossible to compensate for disturbances within a reasonable period of time.

Proportional-Integral (PI) Control Proportional-integral (PI) control is by far the most popular type of feedback control. A PI algorithm is a combination of two independent modes: proportional (P) and integral (I). A combination of the proportional and integral modes (PI) provides reliable and accurate control. Mathematically the proportional-integral control can be represented by the following equation:

$$Q_A = K_p(\varepsilon) + M + (K_i)\int(\varepsilon)\, dt \qquad (15\text{--}23)$$

where the terms are as defined previously.

The basic concept of PI control is to respond rapidly to an error by a proportional action and slowly eliminate an offset by an integral action. The ratio between K_p and K_i will depend on the dynamics of the control loop. For fast loops, such as flow control, integral control action should be increased; for slow processes, such as DO and level control, proportional action should be used to avoid process oscillation. The process of finding the proper values of the coefficients K_p and K_i is called *control-loop tuning.* Incorrect tuning in many cases is a cause of ineffective automatic control and sometimes leads to complete abandonment of automatic control. There are several recommended methods of tuning that use a successive trial approach (Corripo, 1990). Computer modeling is sometimes also necessary to determine tuning coefficients for slow control loops such as SRT control of an activated-sludge system (Ekster et al., 1997).

Proportional-Derivative (PD) Control Improved process control can be achieved by adding a derivative term, obtained from a plot of the offset versus time, to the proportional control term. The resulting proportional-derivative control equation for DO is:

$$Q_A = K_p(\varepsilon) + M + K_p t_d \frac{d(\varepsilon)}{dt} \tag{15-24}$$

where t_d = derivative time, min

The purpose of the derivative term is to anticipate the change in a system by considering the time rate of change of the error term. The PD method of control is used in control loops where the dynamics of the process or processes are fast. Because the process dynamics of wastewater treatment processes are relatively slow, the PD method of control rarely is used in wastewater applications.

Proportional-Integral-Derivative (PID) Control Combining the three modes discussed above results in an even more sophisticated and effective control system, as given by the following equation.

$$Q_A = K_p(\varepsilon) + M + (K_i)\int(\varepsilon)dt + K_p t_d \frac{d(\varepsilon)}{dt} \tag{15-25}$$

The difficulty with using the PID control method is that three variables must be adjusted simultaneously to tune the control loop.

Feed-Forward and Feedback Control. Because of the limitations of feedback control, i.e., a manipulated variable is changed only when an error is detected or reaction time is slow, process response can be improved by adding feed-forward control. Under a combined control strategy, a manipulated variable responds initially to a disturbance (e.g., change of flow) using a feed-forward control algorithm. The error caused by inaccurate disturbance compensation is corrected by applying feedback PI control action. A formula for a combination of static feed-forward DO control with PI feedback control is given below.

$$Q_A = (K_{ff})Q_L + [K_p(\varepsilon) + M + (K_i)\Sigma(\varepsilon)\Delta t] \tag{15-26}$$

where the terms are as defined previously.

With proper control-loop tuning, an error caused by imperfect disturbance compensation by feed-forward control action is much less than an error when only feedback control is used. A combination of feed-forward and feedback control makes controller tuning more difficult and time-consuming because an additional tuning coefficient is required for feed-forward control action. As a result, the combination of feed-forward and feedback controls is not used very often. Advances in computer modeling of wastewater treatment processes may simplify control-loop tuning in the future, and this advanced control mode will be used more frequently. Application of combined feedback and feed-forward control is illustrated in Example 15–5.

EXAMPLE 15–5 **Maintaining Proper Flow of Air for Oxygen Control in a Complete-Mix Activated-Sludge Process** Changes in the DO and flowrate for the last 7 min in the aeration tank are shown in the following table. Using these data and the operating values given below, determine the airflow control signal at the end of the seventh minute to maintain dissolved oxygen concentration at 2 mg/L using a combined feed-back and feed-forward control mode.

Time, min	DO, mg/L	Wastewater flowrate, m^3/d
1	1.9	18,200
2	1.9	18,600
3	2.0	18,900
4	2.1	19,300
5	2.2	19,700
6	2.2	20,000
7	2.3	21,000

Applicable operating values:

1. The feed-forward control gain factor K_{ff} is $(0.0075\ m^3/min)/(m^3/d)$.
2. Proportional gain K_p is $(15\ m^3/min)/(g/m^3)$.
3. Integration time t_i is 10 min.
4. Bias is equal to 0.

Solution

1. Determine the value of the integral gain.

   $$K_i = K_p/t_i = [(15\ m^3/min)/(g/m^3)]/10\ min = (1.5\ m^3/min^2)/(g/m^3)$$

2. Determine the magnitude of the airflow signal using Eq. (15–26).

   $$Q_A = (K_{ff})Q_L + [K_p(\varepsilon) + M + (K_i)\Sigma(\varepsilon)\Delta t]$$

   $$Q_A = [(0.0075\ m^3/min)/(m^3/d)](21,000\ m^3/d) + [(15\ m^3/min)/(g/m^3)(2.3$$
   $$- 2.0)\ g/m^3] + [(1.5\ m^3/min^2)/(g/m^3)] \times \Sigma[(2.0 - 1.9) + (2.0 - 1.9)$$
   $$+ (2.0 - 2.0) + (2.0 - 2.1) + (2.0 - 2.2) + (2.0 - 2.2)$$
   $$+ (2.0 - 2.3)\ g/m^3] \times 1.0\ min$$

   $$= 152.1\ m^3/min$$

   Thus, at the end of seventh minute the airflow should be equal to 152.1 m^3/min.

Comment The computations illustrated above are performed by the controller on a continuous basis, subject to the input. The gain and bias factors are characteristic of the controller.

Figure 15–22

Typical example of a cascade DO control for the activated-sludge process.

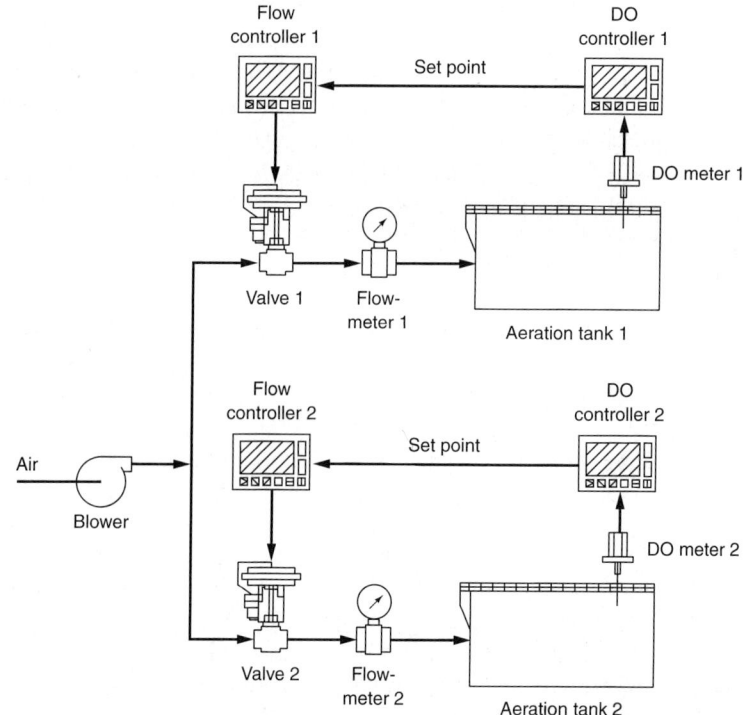

Cascade Control. In a cascade control mode, one feedback controller, identified as the *primary controller,* is used to calculate a set point of another feedback controller that is called the *secondary controller.* Interaction among control loops is the most pressing reason for utilization of cascade control mode. For example, when a blower is used to supply air to two aeration tanks as shown on Fig. 15–22, an increase of airflow to one tank (Tank 1, for example) will cause a decrease of airflow to another tank (Tank 2), which will eventually lead to a decrease of DO in Tank 2. In a standard PI control system, it will take a long time before the DO error in Tank 2 will be compensated by the DO controller.

In a cascade control system, in addition to DO, airflow is also controlled. The required set point for airflow is calculated by the DO controller, so this set point is a manipulated variable for the DO controller. As soon as airflow to Tank 2 starts to decrease because of the increased airflow to Tank 1, the airflow controller, in response to an error, will trigger a fast opening of corresponding control valve 2. Manifold air pressure (P) will decrease and, in response to an error, a pressure controller will make a blower produce more air, so air demand of both tanks is satisfied. Everything will happen much faster than in the previous scenario and the DO in Tank 2 will not be affected by the increased air demand of Tank 1.

Availability of an intermediate parameter that can be controlled to compensate for interaction among control loops is a criterion for application of cascade control. In the aeration example, the airflow to each tank is such a parameter. Cascade control allows faster compensation of disturbances caused by interacting controllers, in this example, the DO controllers. As in the case of combined feed-forward feedback control, cascade control often requires additional measurements and increases the complexity of tuning.

Distribution of flow among parallel process units is another popular application of a cascade control. The distributed method of control is called *the most opened valve* algorithm. According to this control method, a primary controller maintains one valve almost fully open (90 percent) and flowrate through this valve serves as a set point for flowrate through other parallel control valves. If opening of any valve exceeds 90 percent, this valve becomes *fully open,* its opening is reduced to 90 percent, and flowrate through this valve becomes a set point for other control loops. The described control algorithm guarantees that pressure in the distribution system will be just enough to provide the required flow through each valve. By always maintaining an optimum pressure in the system, energy use by pumps or blowers is minimized.

Adaptive Control. Most wastewater treatment processes cannot be described by simple linear equations and, as a result, a regular PI control algorithm cannot control these processes precisely. In some cases, adaptive PI control algorithms utilizing variable gains provide better results. Gain values can be programmed initially or calculated on-line using an iteration technique. Another method of adaptive control is to use self-calibrating computerized models of the wastewater treatment process to calculate control actions. Such models can be mechanistic or statistical. Adaptive algorithms use much more processing power than regular control algorithms and require more sophisticated programming techniques. To maintain robustness of adaptive algorithms, boundary conditions often have to be used. Despite numerous benefits, adaptive control is rarely used due to its complexity.

Selection of Set Points. Optimization of set points is a necessary part of any control strategy. Such optimization is a tedious and complex task. For process control parameters, set points should be selected based on information provided in Table 15–18. Before final selection, it is recommended that several set points are evaluated.

Process Control Diagrams

Wastewater treatment plants are designed by a team of engineers. The process engineer, who takes the lead in the conceptual design, develops the project documentation that allows the engineers of the various other disciplines, (i.e. mechanical, structural, instrumentation, and electrical) to complete the detail of the project design. The required documentation is normally developed in three stages: (1) block flow diagrams (BFDs), (2) process flow diagrams (PFDs), and (3) piping and instrument diagrams (P&IDs), with each succeeding document adding to the detailed definition of the process flow and control.

Block Flow Diagrams (BFDs). The BFDs, typically developed in conjunction with the client, are used to identify the major process treatment elements that will be used to treat the wastewater prior to discharge or reuse. In some cases, more than one BFD may be developed.

Table 15–18

Considerations in the selection of set points for various treatment-plant processes

Process	Parameter	Potential problems if value is	
		Lower than optimum	**Higher than optimum**
On-off pumping	Wetwell levels	• Damage to the pump motor • Vortexing of pump; cavitation • Increased pumping head (increased energy cost)	• Possible overflow and flooding of downstream processes • Reduced pump efficiency
Primary treatment	Sludge depth	• Reduction of efficiency of subsequent sludge treatment processes due to excess water in thin sludge discharged from primary clarifiers • Increased return flow from sludge processing facilities	• Increased effluent TSS, soluble and particulate BOD and H_2S generation • Increased floating solids due to gasification • Increased mechanical stress on sludge and scum collection mechanism
Conventional activated sludge	Solids-retention time	• Poor constituent (BOD, NH_3, etc.) removal in aeration basin • Poor sludge settleability due to dispersed growth • Deterioration of waste sludge thickening due to thin sludge discharged from secondary clarifiers	• Increased clarifier solids loading that can cause an increase in effluent TSS • Low F/M foaming and bulking • Increased energy demand to sustain endogenous respiration
	Dissolved oxygen (DO) concentration	• Poor constituent removal in aeration basin • Poor settleability due to low DO bulking • High effluent turbidity due to deterioration of sludge floc • Possible development of filamentous organisms	• High energy cost due to excessive DO • Break-up of floc and, as a result, increased effluent TSS • Nitrification encouraged
	Return flowrate from the clarifier	• Increased effluent TSS due to (a) increased sludge depth in clarifier and (b) solids floating due to denitrification	• Deterioration of waste sludge thickening due to thin sludge discharged from secondary clarifiers • Increased filamentous growth causing poor settleability and foaming • Increased energy demand because of excessive pumping

(continued)

| **Table 15–18** (Continued)

Process	Parameter	Potential problems if value is	
		Lower than optimum	**Higher than optimum**
Step-feed activated sludge	Flow to first compartment	• Increased clarifier solids loading	• Poor constituent removal in aeration basin • Low DO in first stage (compartment)
Biological nutrient removal	Internal recycle	• Incomplete denitrification	• Decreased denitrification rate due to increased oxygen concentration in the anoxic compartment
	Methanol feed rate	• Incomplete denitrification	• Increased cost of methanol, aeration, and sludge processing • Increased energy demand for recycle pumping
Dissolved air flotation thickening	Air/solids ratio	• Increased TSS in recycle flow	• Increased energy demand • Decreased sludge concentration
Gravity thickening	Sludge depth	• Decreased sludge concentration	• Increased TSS in the supernatant
Gravity belt thickening	Belt speed	• Sludge spills at inlet to the belt	• Decreased solids concentration
Chemical addition for water-solids separation	Chemical dosage rate	• Poor water-solids separation due to inadequate flocculation	• Increased cost of chemicals and sludge processing • Increased turbidity and TSS in settled effluent
Chlorination	Chlorine dosage rate	• Poor disinfection • Permit violation	• Poor disinfection due to destruction of chloramines • Increased cost of chlorine • Excessive SO_2 demand and cost for dechlorination • Formation of potential carcinogenic constituents • Increased toxic to aquatic organisms

Process Flow Diagrams (PFDs). After the block flow diagrams have been agreed to, the next step is to develop PFDs in which the flow paths for all of the liquid and solids are identified. The PFDs are used to define the treatment process in greater detail including the major equipment involved in each of the various processes and material balances for the anticipated operating conditions (see Sec. 14–16 in Chap. 14). In the PFDs, the overall process control scheme for the entire facility is identified, along with the major control loops for the individual unit processes. It is important to note that the PFDs should include the critical process conditions that may be encountered in the operation of the facility (i.e., peak flows, minimum flows, etc.). Many treatment processes have been hampered by systems designed in strict accordance with "normal" or "average" steady-state conditions, with little or no allowance for the flow or load extremes under which the unit processes may be expected to operate (see discussion in Sec. 15–2).

Piping and Instrument Diagrams (P&IDs). Once the PFDs are adopted, the P&IDs are then developed. The P&IDs are used as a pictorial representation of the treatment plant including the process equipment, with installed spares, the associated piping, valves, and instrumentation necessary for the control and safe operation of the facility. Because they encompass the complete process design, the P&IDs usually form the basis for the project scope and project control, as well as a guide for the detailed design by the discipline engineers. The P&IDs also become the principal form of communication between the designers and the owner/operator of the facility; they are also important in the training of operators. To ensure proper definition of the treatment process facilities, the capacities of equipment, piping, and instrumentation along with data for sizes, and materials of construction are normally included on the P&IDs. While the process engineer will, typically, prepare an initial issue of the P&ID, incorporating the equipment, piping, and basic instrumentation scheme, it then becomes the responsibility of the mechanical and instrumentation engineers to develop and expand the details and bring the P&IDs to the level of detail needed to proceed with the detailed design and construction drawings. One of most important tasks is to define the exact details of how the individual process control loops will function, using one or more of the control methods described in the remainder of this section.

Description of Automatic Control System Elements

The elements of an automatic control system include (1) on-line instrumentation, (2) signal transmitters, (3) controllers, and (4) final control elements. Each of these elements is described and discussed below.

On-line Instrumentation. Frequency of measurements and adjustments depends on how long it takes for a parameter to change after a disturbance affecting the parameter in question is introduced. The time required for a parameter to achieve 63.2 percent of the difference between the initial and final conditions after a disturbance is introduced is called the *time constant* (Smith and Corripo, 1997). The magnitude of time constants for some important parameters for different wastewater treatment processes is given in Table 15–19. To provide reliable control, these parameters need to be measured from ten to thirty times more frequently than the corresponding time constant. So, it is clear that for control of most parameters on-line instrumentation is necessary.

The wastewater treatment industry used on-line instrumentation for manual control long before automation was introduced. Every plant is equipped with flow, pressure, and liquid-level meters. These instruments are necessary to operate a plant reliably and efficiently, whether process adjustments are conducted manually or automatically. Methods used for flowrate and liquid-level measurement are presented in Tables 15–20 and 15–21, respectively. More recently, instruments for the on-line measurement of turbidity, TSS, DO, and residual chlorine concentration have become an integral part of

Parameter	Time constant	
	Unit	Typical value
Pressure	s	<1
Flowrate in pressure pipe	s	1–5
Flowrate in open channels and clarifiers	min	5–30
Dissolved oxygen	min	10–20
Sludge blanket depth in secondary clarifier	min	30–120
Ammonia concentration in the aeration basin	h	>1
Ammonia concentration in the effluent	h	>1
Solids-retention time	d	>1
	d	$3 \times SRT^a$

[a] Approximate time to reach steady-state operation after the SRT value has been changed.

the control systems used at medium- and large-sized treatment plants. Summary information on the methods used for the measurement of these parameters of treatment process performance variables is given in Table 15–22.

Nutrient removal has made the measurement of NH_3, NO_2^-, NO_3^-, and PO_4 a necessity for proper process control. The three methods currently used for the measurement of nutrients are colorimetric, UV absorption, and ion selectivity. On-line instruments utilizing these methods are complex and expensive devices. Most of the time these analyzers can measure nutrients only in streams with low TSS concentrations, so some manufacturers supply sample filtration equipment with the instruments. At the same time, a new generation of instruments with ion-selective electrodes that do not require sample preconditioning can measure nutrients *in situ*. Most of the older nutrient analyzers provide only batch-type measurements and require sample pumping to an analyzer.

Signal Transmitters. Information from instruments used to measure process parameters must be converted into electrical signals and transmitted to a controller. Controller output has to be transmitted to a receiver of a final control element. Until recently, analog direct current was the most popular type of electrical signal used for data transmission. Speed of data transfer is almost instantaneous and resistance of the transmission wire does not affect the quality of transmission. Transmitters and receivers are adjusted so that values of measured and controlled parameters correspond to a 4 to 20 mA (milliampere) signal.

Modern computer-based control systems do not require separate wiring from each transmitter and valve to the controllers. Data links or *data highways* can be used to transmit many encoded digital signals back and forth at very high speeds. Rules of data transmission through networks are called *protocols*. Signals can be transmitted using either ethernet or special protocols used for field buses (proprietary local area networks). Transmitting media can be copper wires, fiber optics, or air (wireless transmission).

Table 15–20

Typical methods used for flow measurements at wastewater treatment plants

Measurement principle	Application	Advantages	Limitations
Measurement of headloss in venturi tubes, orifice plates, flow tubes	Closed conduits	• Simplicity • Low cost	• Maximum to minimum measuring range should not exceed 4:1 • Prevention of sensing lines clogging is often required • Cannot be used for thick sludges • Periodic maintenance and calibration is required • The method of flow measurement is intrusive and causes pressure loss • Minimum of straight runs of piping required before and after meters
Water-level measurement in Parshall flumes and upstream of the weirs	Open channels	• Simplicity • Wide flow range • Flumes come in prefabricated sizes and configurations	• The method is flow-intrusive and causes extra headloss • Periodic cleaning of weirs and Parshall flumes is required for some applications • Slow response time • High quality of manufacturing and installation of Parshall flumes and weirs is required • Most of the time, cannot be installed without modifications to an existing channel
Magnetic	Closed conduits	• Low pressure drop • High accuracy • Low maintenance	• High cost especially for large diameters • Poor accuracy for velocities less than 1 m/s • Meter size may be less than line size
Combination of reflective sonic (Doppler) velocity and level sensors installed inside the conduit	Open conduits	• No pressure drop • Low maintenance • Low cost • Good accuracy • No construction	• Conduit size should not be less than an equivalent diameter of 600 mm (24 in)
Reflective sonic (Doppler) installed outside the pipe	Closed conduits	• No pressure drop • Low maintenance • Low cost • Noninvasive installation	• Potential problem for clean fluids • Inadequate accuracy for some applications, especially for velocities less than 0.9 m/s (3 ft/s) • Pipe should have less than 50 mm (2 in) thickness and cannot be constructed, or lined with, an aggregate material • Sensitive to pipe vibration
Transmissive sonic (through beam)	Closed and open conduits	• No pressure drop • Ease of installation • Low and convenient maintenance • Noninvasive installation • Good accuracy for large conduits	• Cannot be used for thick sludges • Accuracy is reduced for small-diameter conduits and pipes • Extremely long runs of straight pipe are required

Table 15–21
Methods used for level measurements

Methods	Application	Advantages	Limitations
Hydrostatic head	Open tanks, wetwell, open channel	• Simplicity	• Dirt and grease may interfere with measurements • May be affected by temperature change
Bubbler	Head created by flumes and weirs	• Simplicity	• Compressed air and constant air rate-of-flow regulator are required • Automatic purging is often required
Float	Wetwells of pump stations	• Simplicity • Low cost	• Grease and solids may interfere with measurements
Ultrasonic	Water level in open channels, wetwell, reservoirs	• No contact with fluid • No maintenance • Reasonable cost	• Presence of foam may create problems • Sometimes difficult to calibrate
Ultrasonic	Sludge level in clarifiers	• No contact with fluid • No maintenance	• Bubble interference • Poor accuracy in secondary clarifier with high SVI • Certain flow conditions can cause sudden decrease in accuracy • Inability to measure depth less than 75 mm (3 in)
Light adsorption with the lifting mechanism	Sludge level in clarifiers	• Ability to measure solids profile • Insensitivity to bubbles	• High cost • Presence of moving parts

Conventional transmitters and receivers use a different type of electrical signal than controllers. To translate from one type of electrical signal to another, special devices, called input/output (I/O) devices, are used. The I/O device can be located in either the same box as a main controller or the same box as a field secondary controller, or can reside separately. Historically, I/O devices were located in close proximity to a main controller; however, most recently, there is a tendency to locate I/O devices in the field and connect them with the main controller through a data highway or through wireless communication. Data transmitted through the network are not only used for control, but are also supplied to a data management system for plantwide availability and storage.

Controllers.   In the simplest automatic control loop, an on-off relay is used as a controller. The next step in control sophistication is the programmable logic controller (PLC). These controllers were developed originally for manufacturing industries. By

Table 15–22
Typical analyzers used for measuring wastewater treatment process performance variables

Measured parameter	Operating principle	Operating range, mg/L	Advantages	Disadvantages
Total suspended solids	Light scatter or adsorption	12–80,000[a]	• Low cost • Good accuracy • Infrequent maintenance	• Requires periodic calibration due to changing particle size distribution • Requires color compensation and prevention of fouling • Reduced accuracy for dark sludges
	Ultrasonic	10,000–100,000	• Infrequent maintenance • Insensitivity to color	• Requires periodic calibration due to changing particle size distribution • Bubble interference
	Microwave	10,000–500,000	• Insensitivity to color, bubbles, fouling, and particle size distribution	• High cost
Dissolved oxygen in aeration channels	Electrochemical with oxygen-permeable membrane	0–20	• Low cost • Good accuracy	• Requires replacing membrane periodically • Requires periodic inspection, cleaning, and recalibration • For values below 1 mg/L galvanic system is desirable
	Electrochemical	0–15	• Low maintenance • Acceptable accuracy	• High initial cost and high cost of replacement parts • Interference with Cl_2, Fe, H_2S, and some other chemicals • Requires prevention of fouling and cleaning
	Fluorescence	0–10	• Very low maintenance	• Delayed response • Interference with fluorescent chemicals
Chlorine in the chlorine contact tank	Amperometric with membrane	0.1–10	• Low maintenance • Low cost	• Requires prevention of fouling • Interference with turbidity and conductivity
	Amperometric	0.1–50	• Insensitivity to turbidity and conductivity	• High initial and maintenance costs • Requires frequent calibration
Turbidity of effluent	Light scatter	0.1–1000	• Direct measurement	• Measurements are affected by particle size distribution, color, and flow • Results depend on type of instrument utilized • Poor correlation with suspended solids

[a] Range is covered with different sensors.

allowing programmers to input electrical ladder diagrams into a PLC, the programming requirements were simplified greatly. Only discrete logic (i.e., on-off functions, etc.) could be controlled initially with these devices. The next level in complexity is a controller that is capable to accomplish only simple mathematical operations (e.g., PI calculations) for one loop only. Such control elements are called single loop PI controllers and have processors similar to a simple calculator. A multi loop PI controller has an ability to implement cascade and other types of more sophisticated control algorithms. Modern PLC controllers combine functionality of discrete controllers and sophisticated multi loop controllers. Operators can still use ladder logic diagrams to program these devices, although more complex programming languages may be available.

A single processor failure in multi loop PI controller may lead to problems with all of the processes controlled by this device. Attempts to improve control reliability have led to development of distributed control systems (DCS) that uses duplicating I/O devices, processors, and operator's terminals. Such systems, despite their high cost, have become very popular for large treatment plants. In the past, however, DCS could utilize only simple control algorithms, and, as a result, it was not always possible to implement algorithms that could be used to address instrument failures or adaptive algorithms. Distributed control systems could not provide the required reliability for complex control loops. Current DCS are very sophisticated and are able to accomplish most of the tasks necessary for reliable automation of a wastewater treatment plant.

Personal and industrial computers are becoming increasingly part of a control system. Systems composed of a net of PLCs used for control and computers used to provide an interface between a human (operator) and a control system often are used as affordable and expandable alternatives to DCS. Computers are now being used not only for human-controller interfaces but also as controllers, and, in some cases, as a net of personal computers to replace the more expensive DCS. To improve reliability of control systems utilizing personal computers, such computer systems are used for only complex calculations, while single loop or multi loop controllers located in the field are used to implement simpler control algorithms. Failure of a main computer in such cases will not cause process interruptions, because the controllers are located in the field. Another method of improving reliability is to use *smart I/O* devices that can control the system when the main computer fails. Data transfer through either conventional point-to-point wiring, wireless communication devices, or through a data highway can be used for communication between computers and field located controllers or distributed I/O devices.

The uniqueness of wastewater treatment processes has motivated some manufacturers to supply control systems for automation of just one process or even for automatic control of one parameter within processes such as shown on Fig. 15–23. Such control systems usually utilize sophisticated proprietary control algorithms and operate either separately or as part of larger nonproprietary control systems. Savings on design, commissioning, and operation sometimes compensate for an increased initial cost of such systems.

Final Control Elements.   A final control element is a necessary part of any control loop, manual or automatic. In process control systems, the final control element is the device used to bring about the desired change. At wastewater treatment plants, the final control element used most commonly is a valve. Use of these devices for auto-

Figure 15–23

View of typical activated-sludge solids-retention time (SRT) process controller.

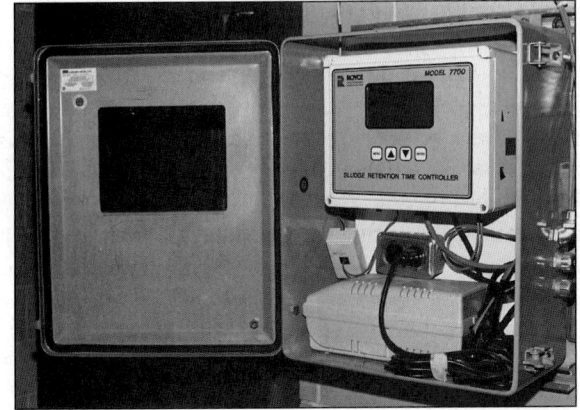

matic control introduces additional requirements for their selection. Control valves used to change the flowrate by modulating the valve setting (i.e., changing valve opening) are the most common types of final control elements. Proper selection of a control valve is very important for performance of automatic control. Selection should be based on required valve type, valve sizing, and valve linearity. Based on the type of throttling element movement, all valves are divided into two groups: those with linear motions and those with rotary motions. The first group includes globe, gate, and diaphragm valves; the second group includes ball, butterfly, and plug valves. Each of these valves has its application depending on the type of physical and chemical properties of controlled flow, the range between required minimum and maximum flow, and other parameters.

Independent of the type of valve, the following empirical formula is used for valve sizing:

$$K_v = \frac{Q}{\sqrt{\Delta P / \text{sg}}} \qquad (15\text{–}27)$$

where K_v = metric valve capacity coefficient
Q = maximum design flow through the valve, m³/h
ΔP = headloss through the valve at maximum design flow, kg/cm²
(Note: $\Delta P = 1$ m $= 0.1$ kg/cm²)
sg = specific gravity relative to water

The above units must be used when calculating the value of K_v. In U.S. customary units, the valve capacity coefficient is given as C_v. The relationship between the K_v and C_v is (Coughanowr, 1991):

$$K_v = 0.856 \, C_v \qquad (15\text{–}28)$$

A decrease in ΔP can cause an increase in the valve size and thus the cost. An increase in size may cause a decrease of the valve operating range. Conversely, an increase in ΔP usually results in a greater headloss and increased energy use for pumping system applications. For optimum system operation some authors (Bauman, 1998) recommend use of $\Delta P = 0.35$ kg/cm² (3.5 m) for rotary control valves and $\Delta P = 0.70$ kg/cm² (7 m) for ball valves. For control of airflow to activated-sludge aeration tanks using butterfly valves, a much lower ΔP value is often used.

Valves are selected on the basis of their K_v values. Valves are typically supplied only in discrete sizes, so the next highest size to K_v [calculated by using Eq. (15–27)] should be chosen. If the valve size chosen has a considerably higher capacity than required, the valve may not be able to handle low-flow conditions. For stable operation of a control loop, it is critical that valve opening at the lowest design flow should not be less than the minimum flowrate value recommended by the valve manufacturer (usually 5 to 10 percent of maximum flow). If the range of flowrates is too wide and one valve cannot satisfy that range, two parallel valves of different sizes should be installed. The larger valve should be selected based on the maximum flow and a small valve should be sized for a maximum flowrate corresponding to the flowrate when the large valve is 20 percent open.

In some applications, sluice gates and adjustable weirs are used as final control elements. However, the use of these devices for automatic control is challenging for two reasons. First, such devices usually operate very slowly because of their weight and, as a result, are poorly suited for rapid control. Second, the relationship between flow and the positions of these control elements is complex. For example, weir characteristics depend on whether flow over the weir is a free fall or the weir is submerged. If both conditions exist, use of the weir as the final control element may present a problem for developing a control algorithm.

A standard valve or gate can be automated by adding an actuator, usually a pneumatic or electrical device that positions the throttling element inside the valve body. In a manually operated valve, a person plays the role of an actuator. Pneumatic actuators function by using a flexible diaphragm or a piston cylinder. Pneumatic actuators usually are fast and well suited for automatic control. Electrical actuators use an AC reversing motor or a stepping motor and are usually more expensive than pneumatic actuators. Electric actuators are used mostly when instrument air to power the activator is not available or when high valve stem forces are required. Because an electric reversing motor usually is very slow and can have a built-in dead zone (an operating zone in which the motor does not respond to the control signal), modification of control logic is also required. In addition, a backlash can occur in the gear train between the motor and the valve shaft.

Sizing of an actuator depends on the force it has to produce to overcome the fluid pressure acting on the valve plug, including packing friction (0.5 to 5 m), and plug weight.

EXAMPLE 15–6   Determine the Size of Valve and Piston Actuator   Determine whether a 150-mm (6-in) valve can be used in the discharge line from a submersible pump for control of return activated-sludge flow from a wetwell to an aeration tank. Determine the size of the piston of a pneumatic actuator for this valve, assuming that packing friction is equal to 5 percent of fluid pressure, an allowance for moving the valve is 5 percent of the hydrostatic pressure, and weight of the valve plug is negligible. The maximum and minimum flowrates are 3.1 and 0.6 m³/min, respectively.

The valve has the following characteristics:

Percent opening	K_v
10	20
20	40
30	60
40	82
50	110
60	140
70	180
80	225
90	280
100	350

The characteristics of the submersible pump are:

Flowrate, m³/min	Pressure, m
0.0	14.2
0.6	14.1
1.24	13.8
1.87	13.3
2.74	12.0
3.1	11.2

Design conditions:

1. The difference in elevation between the water level in the wetwell and the aeration tank is 2 m.
2. The pipeline leading from the wetwell to the aeration tank contains 4 elbows.
3. Assume that most of the friction losses in the piping system are associated with elbows and exit loss. The headloss coefficients for the elbows and pipe exit are 1.2 and 1.0, respectively.
4. Air pressure for an actuator is 55 m.

Solution

1. Determine the dynamic headloss using the following relationship.

$$H_{total} = H_{static} + H_{dynamic}$$

$$= H_{static} + (K_{elbow} + K_{exit})[(v)^2/2\,g]$$

a. At the maximum flowrate (3.1 m³/min) the dynamic headloss is:
    The velocity in the pipeline is

$$v = Q/A = Q/(\pi D^2/4)$$

$$v = (3.1 \text{ m}^3/\text{min})(1.0 \text{ min}/60 \text{ s})/(3.14/4)(0.15 \text{ m})^2 = 2.92 \text{ m/s}$$

$$H_{\text{dynamic}} = (1.2 \times 4 + 1.0)(2.92 \text{ m/s})^2/(2 \times 9.81 \text{ m/s}^2)$$

$$= 2.53 \text{ m}$$

b. At the minimum flowrate (0.6 m³/min) the dynamic headloss is

$$v = (0.6 \text{ m}^3/\text{min})(1.0 \text{ min}/60 \text{ s})/(3.14/4)(0.15 \text{ m})^2 = 0.0.57 \text{ m/s}$$

$$H_{\text{dynamic}} = (1.2 \times 4 + 1.0)(0.76 \text{ m/s})^2/(2 \times 9.81 \text{ m/s}^2)$$

$$= 0.09 \text{ m}$$

2. Determine the pressure loss across the valve using the following relationship:

$$\Delta P = \text{pump head} - \text{static head} - \text{dynamic headloss}$$

a. At the maximum flowrate:

$$\Delta P = (11.2 - 2 - 2.53) \text{ m} = 6.67 \text{ m} = 0.67 \text{ kg/cm}^2$$

b. At the minimum flowrate:

$$\Delta P = (14.1 - 2 - 0.09) \text{ m} = 12.01 \text{ m} = 1.2 \text{ kg/cm}^2$$

3. Determine the valve capacity coefficient using Eq. (15–27) and the corresponding valve opening.
    a. At the maximum flowrate:

$$K_v = \frac{Q}{\sqrt{\Delta P/\text{sg}}} = \frac{(3.1 \text{ m}^3/\text{min})(60 \text{ min/h})}{\sqrt{(0.67 \text{ kg/cm}^2)/1.0}} = 228$$

Referring to the table given above, the valve will be about 81% open at the maximum flowrate.

b. At the minimum flowrate:

$$K_v = \frac{Q}{\sqrt{\Delta P/\text{sg}}} = \frac{(0.6 \text{ m}^3/\text{min})(60 \text{ min/h})}{\sqrt{(1.2 \text{ kg/cm}^2)/1.0}} = 32.9$$

Referring to the table given above, the valve will be about 16% open at the minimum flowrate.

c. Conclusion: Because the valve opening range is within 10 to 100% range, the valve is acceptable for the control of the return activated-sludge flowrate.

4. Determine the required diameter of the piston actuator at shutoff head. The piston actuator must have a diameter such that, with the available air pressure, it is possible to overcome the hydrostatic pressure on the valve when fully shut, plus

the frictional resistance of the valve assembly plus an allowance for moving the valve.

a. Determine area of the piston using the following relationship:

$$A_{\text{piston}} \times \text{pressure} = A_{\text{pipe/valve}} \times \Sigma \text{ pressure}$$

b. Substitute known values and solve for the piston diameter. Assume the area of pipe and valve are the same.

$$(3.14/4)(D_{\text{piston}})^2 \times 55 \text{ m}$$

$$= (3.14/4)(0.15 \text{ m})^2 \times 14.2 \text{ m } (1 + 0.05 + 0.05)$$

$$D_{\text{piston}} = 0.15 \text{ m} \sqrt{\frac{1.1 \times 14.2 \text{ m}}{55 \text{ m}}} = 0.080 \text{ m} = 80 \text{ mm}$$

Comment As with the previous example, the computations illustrated in this example are typical of those that must be made to size each control element properly. In this example the maximum velocity in the sludge return line was 2.92 m/s, which is in the range that is used normally (2 to 3 m/s), although flowrates as high as 4.0 m/s have been used.

Instead of using a control valve that reduces available pressure developed by a pump or a blower, a pump or blower may be equipped with a variable-speed drive that changes the pump or blower output without wasting energy by inducing increased headloss. Pump operation at a lower speed may be somewhat less efficient than at a design speed. In most wastewater applications where water and air volumes vary widely, use of a variable-speed drive is a preferable way for flow control because the incremental increase in equipment cost is small compared to energy savings. A life-cycle cost analysis should be performed to evaluate the economic benefits. In small-plant applications, the savings in power consumption may not be enough to justify the expense of a variable-speed drive.

15–5 ENERGY EFFICIENCY IN WASTEWATER TREATMENT

Water and wastewater utilities in the United States consume about 2 percent of the total amount of electricity produced (Batts et al., 1993). Typically, 30 percent of the operating cost of a wastewater treatment plant is budgeted for energy use. During the next 20 to 30 years, the electricity requirements for wastewater treatment in the United States are expected to increase an additional 30 to 40 percent. In an era where there are concerns about the adequacy of fuel supplies, cost of energy, and the increasingly higher levels of treatment that result in increased energy consumption, the design and operation of wastewater treatment plants are focused increasingly on improving the efficiency of electric energy use and reducing the cost of treatment. An overview of the use of electricity in wastewater treatment plant operation and measures that can be employed for improving energy efficiency and the cost of operation are presented in this section.

Overview of the Use of Electricity in Wastewater Treatment

The operational requirements for wastewater collection and treatment systems vary directly as the wastewater load (see Fig. 3–6). If a diurnal electricity demand curve were developed for the treatment facilities, it would be of a similar shape to the flowrate and loading curves shown on Fig. 3–6. The peak energy demand would likely occur from midday to the early evening hours when other peak demands for electricity occur in the community. As the wastewater load changes during the course of a day, the requirements for pumping, aeration, and solids processing change accordingly. Some plants modify schedules for equipment operation to meet load conditions; others operate their system components (such as aeration blowers) continuously at full capacity, regardless of the load.

As described in Table 1–7, approximately 85 percent of the wastewater treatment plants in the United States provide secondary or higher levels of treatment. In conventional secondary treatment, most of the electricity is used for (1) biological treatment by either the activated-sludge process that requires energy for aeration or trickling filters that require energy for influent pumping and effluent recirculation; (2) pumping systems for the transfer of wastewater, liquid sludge, biosolids, and process water; and (3) equipment for the processing, dewatering, and drying of solids and biosolids. A typical distribution of energy use in a conventional activated-sludge treatment plant, the most common type of plant used in wastewater treatment, is illustrated on Fig. 15–24. In activated-sludge treatment, approximately 1100 to 2400 MJ of electricity are required to process each 1000 m³ (1200 to 2500 kWh per Mgal) of wastewater. The amount of elec-

Figure 15–24

Distribution of energy usage in a typical wastewater treatment plant employing the activated-sludge process. (From EPRI, 1994.)

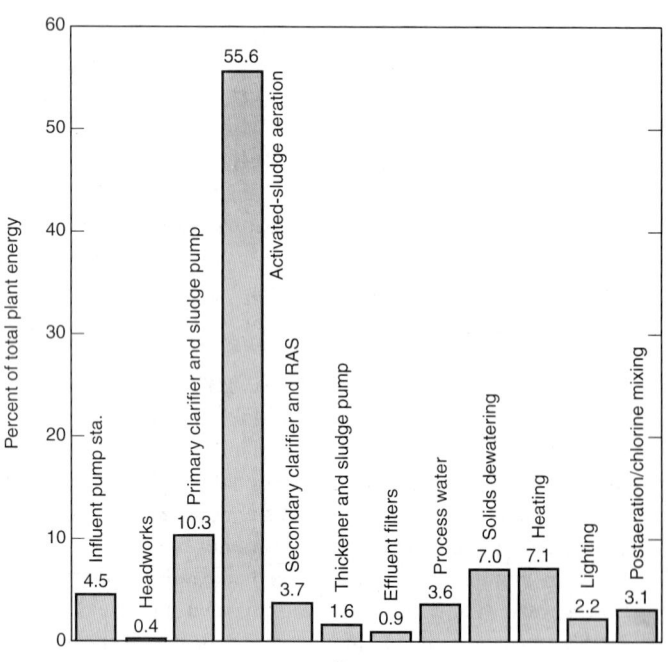

Figure 15–25

Comparison of electrical energy used for different types of treatment processes as a function of flowrate (Burton, 1996).

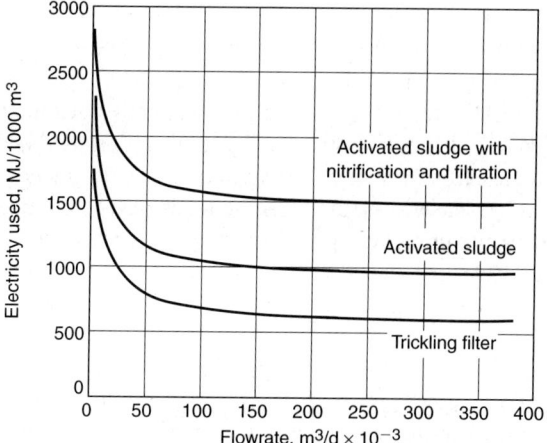

tricity consumed varies generally according to plant size and type of treatment system employed (see Fig. 15–25).

Combinations of physical operations (such as pumping, screening, settling, and filtration) and chemical and biological processes are used for the removal of constituents, as described in the preceding chapters. Various types of electric motor-driven equipment are involved in these operations and processes including pumps, blowers, mixers, sludge collectors, and centrifuges. Treatment processes and equipment requiring electric energy in a municipal wastewater treatment plant are presented in Table 15–23.

Advanced wastewater treatment plants require even greater amounts of electric energy. Plants that have biological treatment for nutrient removal and filtration use on the order of 30 to 50 percent more electricity for aeration, pumping, and solids processing than conventional activated-sludge treatment (see Fig. 15–25) (Burton, 1996).

With the introduction of new technologies for wastewater treatment, the energy requirements will change; the magnitude of potential impacts is summarized in Table 15–24. The impacts can be either reductions in energy use due to more efficient equipment or systems, or increases due to changes in treatment capacity requirements, higher levels of treatment, or new technologies whose operation is based on electric energy, i.e., membrane treatment, UV disinfection, and autothermal aerobic digestion (ATAD).

Measures for Improving Energy Efficiency

Opportunities exist for managing energy use in wastewater treatment by employing the concept of demand-side management. The electric power industry has long recognized the importance of integrating traditional supply-side planning with demand-side management to reduce peak demand. The goal of demand-side management is to change the electrical load characteristics (the amount of energy used at different times of the day) by improving energy efficiency and managing equipment operation. In demand-side management, it is also recognized that continued load growth will occur as the systems expand to meet new domestic and industrial wastewater collection and treatment requirements.

Table 15–23

Equipment commonly used in wastewater treatment facilities requiring electrical energy

Process or operation	Commonly used equipment	Process or operation	Commonly used equipment
Pumping and preliminary treatment	Chemical feeders for prechlorination, influent pumps, screens, screenings press, grinders and macerators, blowers for preaeration and aerated grit chambers, grit collectors, grit pumps, air lift pumps	Advanced wastewater treatment	Blowers for nitrification aeration, mechanical aerators, mixers, trickling filter pumps, pumps for depth filters, blowers for air backwash, pumps for membrane filtration
Primary treatment	Flocculators, clarifier drives, sludge and scum pumps, blowers for channel aeration	Solids processing	Pumps, grinders, thickener drives, chemical feeders, mixers for anaerobic digesters and blending tanks, aerators for aerobic digesters, centrifuges, belt presses, heat dryer drives, incinerator drives, conveyors
Secondary (biological) treatment	Blowers for channel and activated-sludge aeration, mechanical aerators, trickling filter pumps, trickling filter distributors, clarifier drives, return and waste activated-sludge pumps	Ancillary systems	
		Odor control	Odor control fans, chemical feeders
		Process water	Pumps
		Plant air	Compressors
Disinfection	Chemical feeders, evaporators, exhaust fans, neutralization facilities, mixers, injector water pumps, UV lamps		

Measures for improving energy efficiency require knowledge of how much energy is used in the wastewater treatment plant by the various elements of the treatment process. One of the best ways to understand energy use and the potential for effecting improved efficiency and energy management is to conduct an energy audit of an existing facility. Different levels of an energy audit can be performed including a preliminary "walk-through" to obtain an overview of principal energy-using equipment or a detailed process audit in which the unit energy use by process components is evaluated (EPRI, 1994). In general, one of the initial tasks in an audit is the evaluation of the largest energy-consuming operations or processes, such as aeration at activated-sludge plants or pumping systems at trickling-filter plants. As illustrated on Fig. 15–24, about 50 percent of the total plant energy at an activated-sludge plant is used for aeration, and, thus, aeration is a prime candidate for considering alternative energy-management measures that can be used to achieve reductions in power use and cost. Other recommendations for energy savings, based on audits of existing wastewater treatment plants, are summarized in Table 15–25.

Table 15–24
Energy impacts of new technologies on wastewater treatment[a]

Technology	Energy impact	
	kWh/Mgal	MJ/1000 m³
Fine pore diffusers (for aeration)	−125 to −150	−120 to −140
Ultrafine pore diffusers	−180 to −220	−170 to −210
Dissolved oxygen control systems (as compared to manual control)	−50 to −100	−48 to −95
Energy-efficient blower control systems, i.e., inlet guide vanes, inlet butterfly valves, or adjustable-speed drives	−50 to −150	−48 to −140
Energy-efficient aeration blowers (as compared to blowers with inlet guide vanes)	−100 to −150	−95 to −140
UV (ultraviolet) disinfection	+50 to +200	+48 to +190
Membranes		
Microfiltration	+200 to +400	+190 to +380
Reverse osmosis	+1000 to +2000	+950 to +1900

[a]Adapted in part from Burton (1998).

Table 15–25
Summary of common audit recommendations for energy savings at several U.S. wastewater treatment plants[a]

- Install adjustable-speed drives on pumps and blowers for variable flow operations
- Install DO monitoring and control in aeration tanks
- Conduct periodic pumps tests and repair or replace inefficient pumps
- Operate emergency generators during peak periods to reduce power demand
- Install electric load monitoring devices
- Install capacitors to improve power factor
- Change or reduce pumping operations
- Replace oversized motors
- Change selected operations to off-peak periods

[a] Burton, 1996.

Implementing operational changes or retrofitting processes and equipment can reduce energy use and cost. Operational changes can be made to a process or system usually with little or no increase in cost. Retrofitting processes or equipment, however, often requires significant expenditures for modifying or replacing existing equipment, or for constructing new physical facilities and installing equipment. A detailed cost analysis has to be made of each change considered to determine if the savings generated by making the change justify the capital cost of the change.

15–6 UPGRADING WASTEWATER TREATMENT-PLANT PERFORMANCE

Upgrading of wastewater treatment plants is defined as (1) modifying existing operations and facilities to improve operation, maintenance, and performance and (2) adding new facilities that enhance performance, improve efficiency of treatment, and provide higher levels of treatment in conformance with new regulations for reuse, discharge, or disposal.

Establishing proper control of treatment processes was always considered a primary duty of treatment-plant operating personnel. In the past, the criterion of good control and operation was producing an effluent with water-quality indicators that on the average did not exceed limits established by a regulatory authority, typically prescribed in an NPDES permit. Operational parameters and procedures usually were chosen based on intuitive approach and anecdotal evidence.

In recent years, many secondary treatment plants were converted to biological nutrient-removal and water-reclamation facilities. Some plants are even required to produce an effluent of a quality approaching that of drinking water. In a number of cases, because of funding limitations for plant expansion, some plants operate in excess of their design capacity. At the same time, more emphasis is being placed on operational reliability, and, in some cases, plants are cited and operating agencies are fined for even the slightest violation of water-quality standards. Finally, because of the need to improve plant efficiency and reduce capital and operating cost, privatization of facilities construction and operation has entered the wastewater field, resulting in cost competition between public and private operating entities.

As a result of these social, economic, and technological changes, the wastewater treatment-plant staff is faced with many challenges. Often, these challenges can be addressed by optimizing process parameters and operational procedures, modernizing facilities, and retrofitting existing equipment and processes. Development of such improvements usually requires the use of sophisticated tools and protocols rather than just intuitive approaches. The purpose of this section is to introduce the concepts of treatment process optimization and approaches used to upgrade the physical, liquid treatment, and solids processing facilities at existing wastewater treatment plants. The material presented in this section is intended to serve as a guide to topics of specialized interest related to plant upgrading. In the previous chapters and sections, many of the elements of the design of a wastewater treatment plant are discussed. Issues related to plant upgrading are reviewed, applicable theory or practical considerations are considered, and references to sources of information contained in this text are included in this section.

Process Optimization

Analysis of plant operating data is a first step in facility evaluation and optimization. Utilization of several methods can simplify this analysis including the use of (1) histograms, (2) linear correlation, (3) on-line process monitoring, (4) computer models, and (5) pilot-scale testing. Each of these methods is described in Table 15–26.

Histograms.   Histograms are particularly useful if the frequency of occurrence of undesirable values, such as effluent TSS, is high. Depending on the frequency of occurrence and critical nature of the parameter, significant changes may have to be made to

Table 15–26
Methods used to
evaluate process
performance

Method	Description
Histograms	Graphs used to display frequency of occurrence of parameters such as wastewater characteristics, flow, and cost of chemicals and electricity
Linear correlation	A statistical method used to evaluate data from historical records such as flowrates and water-quality parameters
On-line process monitoring	Instruments are used continuously or intermittently to record and track important operating parameters such as flowrates, dissolved oxygen concentrations, chlorine residual concentrations, and tank levels. Data from such monitoring can be used to identify trends in operation so that process changes can be implemented before a problem occurs
Computer models	Computer modeling is a useful tool to simulate existing process operations and the effects of possible changes such as modifications to operating strategies or the addition of new equipment or processes
Pilot-scale testing	Pilot-scale testing is useful in evaluating the performance of new or alternative technologies and in developing criteria that can be used for the design of full-scale facilities

the process operation or perhaps to the process unit infrastructure to achieve correction. Infrequent occurrences of undesirable values often means that only minor adjustments to operation and maintenance practices may be necessary to achieve the desired results.

Another application of histograms is in establishing the priorities for the tasks that have to be performed to avoid an undesirable event. This type of histogram, called a Pareto chart, represents a graph displaying relative contribution of small problems causing one larger problem. The Pareto chart was named after Italian economist Vilfredo Pareto (1848–1923), who made the famous observation that 80 percent of the world wealth was owned by only 20 percent of people. This principle is valid in many situations, and can be equated to process operations as 80 percent of process problems are due to 20 percent of the causes. Pareto analysis is used to identify the 20 percent that causes the majority of process troubles. Analysis includes following four steps:

1. Identifying the causes of a particular problem
2. Determining the frequency of each cause
3. Calculating the percentage of each cause of the total occurrences
4. Plotting the percentage in descending order

Use of a Pareto chart in analyzing the occurrence of tank overflows due to specific causes, such as instrumentation and electrical instrumentation failures, is illustrated on Fig. 15–26.

Linear Correlation. Another statistical method that is often used to evaluate historical data is an analysis of the linear correlation between a parameter that needs to be optimized and other variables that may have some effect on that parameter. Many models used in process control are linear because they are well understood and are simple to analyze and solve. Usually, linear correlation is written as follows:

$$y = a_0 + a_1 x_1 \qquad (15\text{--}29)$$

Figure 15–26

Pareto plot used to illustrate relative contribution of various causes to tank overflows.

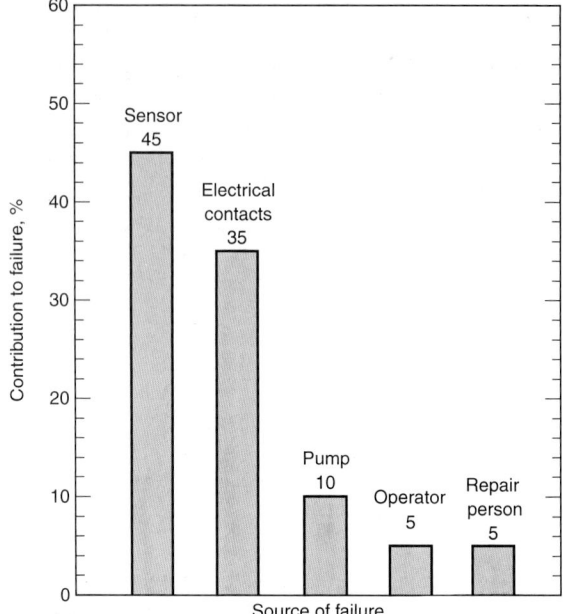

where y = parameter to be optimized
  a_0, a_1 = coefficients
    x_1 = variable

The higher the probability that coefficient a_1 is not zero, the higher is the effect of the variable x_1 on the parameter y. This probability can be calculated using various statistical software packages. Usually, a 95 percent probability that a coefficient a_1 is not equal to zero is considered an indication that a variable x_1 has an effect on the optimized parameter y. The sign of the a_1 coefficient is used to determine in which direction the variable affects the optimized parameter. It is also important to note that lack of correlation between an optimized parameter and a variable does not necessarily mean that they are not related in general. In some cases, an appropriate relationship may be more complex than just a linear relationship, or the effect of their relationship was minimal within examined ranges.

On-line Process Monitoring. Since the 1950s many industries have been using a variety of statistical analyses for making continuous process improvements. In the wastewater industry, statistical process control is not as popular because effluent water-quality characteristics are not distributed normally (i.e., bell-shaped). As a result, the methods required for implementation of statistical control for municipal wastewater treatment processes have to be more sophisticated than ones used in other industries.

Conventional analysis of historical data includes the analysis of data obtained through grab and composite sampling. Such data often do not reflect the full dynamics

Figure 15–27

Comparison of actual dissolved oxygen concentration versus the target value.

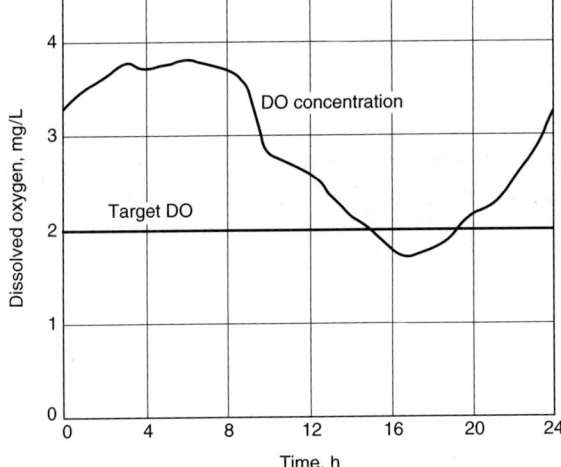

of a treatment process. On-line instrumentation, however, can be used to provide information that is more representative of the process dynamics. An example of on-line monitoring of the dissolved oxygen (DO) concentration is shown on Fig. 15–27. The DO values vary significantly from the target value of 2 mg/L, indicating for much of the time the wastewater is overaerated. Installation of an automatic DO control can prevent overaeration as well as underaeration, resulting in savings in aeration energy and improved process control.

Computer Models. Computer modeling of treatment processes is an effective tool for development of facility improvements because it is now possible to understand the dynamics of treatment processes and to optimize operational parameters. The effect of changes in the operational modes on process performance can also be analyzed. For example, the results of computer modeling of the activated-sludge process and the hydraulic modeling of clarifiers can be used to evaluate whether an existing activated-sludge plant, designed for conventional treatment (i.e., BOD and TSS removal), can be modified to remove nutrients or/and to treat a higher flow than specified in the original design. Computer modeling is also very helpful in the evaluation of retrofit options, such as the installation of baffles in aeration tanks and clarifiers, changing of media size in filters, etc. In many cases, computer modeling can simplify physical modeling. The use of modeling in activated-sludge design is discussed in Sec. 8–10 in Chap. 8.

To provide reliable results, a computer model requires calibration to verify the values that have been chosen for the coefficients included in the model. These coefficients often depend on wastewater characteristics and the design features of the particular process. Calibration involves modifying the coefficient values until model output matches the plant performance data collected using on-line water quality and flow monitors, composite samplers, and other data-collection equipment. Sometimes to improve the quality of the model calibration, the process and equipment need to be operated under extreme conditions. Such experiments are called *stress testing,* and require careful planning and execution to avoid threatening plant reliability. When applying stress testing to clarifier operation, for example, care must be taken to ensure TSS values do not exceed operational or permit limits.

Pilot-Scale Testing.   A full-scale physical model of a treatment process is an ultimate tool for evaluation of proposed improvements. In recent years, more use has been made of pilot-scale testing of proposed improvements before they are implemented at full scale. Pilot-scale testing was discussed previously in Sec. 4–9 in Chap. 4, and a large-scale pilot facility was shown on Fig. 8–3. Pilot-scale units can be operated under a variety of different conditions, and the quality of effluent and other parameters can be compared between old and new modes of operation. Such piloting not only provides the most accurate information about the benefits of proposed process improvements, but also provides an opportunity for training the operating staff in the operation of new processes.

Upgrading Existing Wastewater Treatment Facilities

Upgrading of a wastewater treatment plant can encompass a large number or only a few factors. The scope of this text cannot include identifying and discussing all of the factors that may go into the upgrading of a specific facility, but the text does identify many of the issues confronting a number of facilities. Examples of significant issues have been prepared that include the potential resolution of an operational or physical facility problem and the sources of information available in this text. The various issues related to upgrading are divided into three categories: (1) operational changes that improve plant performance, (2) upgrading of physical facilities for liquid and solids treatment facilities, and (3) potential process modifications for meeting new requirements for constituent removal, and are discussed in the following paragraphs.

Operational Modifications.   Some of the common problems faced in the operation of treatment facilities relate to the changing nature of the wastewater to be treated, variations in flows and loads, utilizing and maintaining each process component to achieve its maximum capability, and maintaining quality control so that treated effluent and plant residuals meet exacting reuse or permit requirements. Examples of operational changes that can be used to resolve plant performance issues are presented in Table 15–27.

Upgrading Physical Facilities.   Most treatment plants contain all of the essential elements necessary to meet treatment goals, but in some instances, the capacity of some of the components is underutilized or overloaded, hydraulic bottlenecks exist that constrain effective and efficient operation, and inadequacies in facilities design affect plant operations and maintenance. Examples of issues relating to the upgrading of physical facilities for the treatment of wastewater (liquid treatment facilities) are summarized in Table 15–28. In some cases, a capacity evaluation is required in which a full-scale test program is required to determine the capability of the critical elements in the treatment process. For example, short circuiting may exist in the primary clarifiers, resulting in the carryover of settleable material to the biological treatment units, thus causing overloading conditions or increasing the cost of treatment. Stress testing of the

Table 15–27

Examples of operational changes that can be used to improve plant performance

Issue	Possible remedial action	See Sec. no.
General plant		
Odors from open tanks and channels	Reduce turbulence by controlling water levels to eliminate free falls and splashing	15–2
	Add chemicals (such as chlorine, ferric chloride, or hydrogen peroxide) to influent wastewater	15–2
	Modify process loading	15–2
Wide influent flow variations	Conduct collection system infiltration/inflow (I/I) investigation to identify sources of extraneous flow	3–2
Short circuiting in clarifiers and chlorine contact tanks	Conduct dye testing	4–4
	Conduct stress tests	
Headworks		
Odorous grit	Adjust air flowrates in aerated grit chambers to obtain cleaner grit	5–6
	Add lime to dewatered grit	
Odors and vectors in headworks	Cover grit and screenings receptacles	15–2
	Add lime to dewatered grit and screenings	
Inadequate grit removal	Analyze channel and grit chamber hydraulics	5–6
	Adjust flow distribution to grit chamber	
	Add temporary baffles to prevent short circuiting	
	Adjust air flowrate in aerated grit chambers	
Grit deposition in channels	Modify/adjust channel flow through velocity	
Primary clarifiers		
Poor solids removal	Check for short circuiting/modify baffling	4–4
	Improve flow distribution	8–7
	Add chemicals to influent	6–3
	Reduce return flows from other processes	14–14
Low solids concentration in primary sludge	Modify sludge pumping rate/install timers	
	Increase sludge blanket depth	
Aeration tanks		
Low DO	Determine DO profile in tank and adjust air flowrate	5–11
	Conduct oxygen transfer test	5–12
	Assess diffuser fouling/clean fouled diffusers	8–4, 8–5
	Change conventional plug-flow operation to step feed (if possible)	
	Check wastewater characteristics (rbCOD and nbVSS)	8–4
High DO	Turn off aeration equipment during low flow and loading periods	
	Install timers to control blower or mechanical aerator operation	(continued)

| **Table 15-27** (Continued) | | |

Issue	Possible remedial action	See Sec. no.
Filamentous organisms in mixed liquor	Examine mixed liquor microscopically to identify types of organisms	8–3
	Increase sludge wasting	
	Chlorinate return sludge	
Process is nitrifying	Check SRT	
Process is not nitrifying	Check SRT and temperature	
Low pH	Check if nitrification is occurring	
	Add alkalinity	
Foaming	Identify nature of foam	8–7
	Change MLSS concentration	8–7
	Add defoaming agent to spray water	
Nocardia foaming	Use dilute chlorine solution spray on foam	8–3, 8–7
	Reduce oil and grease discharges to collection system	
Trickling filters		9–2
Poor BOD removal	Reduce dosing rate	
Solids washout at peak flows	Reduce recirculation rate	
Biological nitrogen-removal process		8–5
Inadequate removal	Check rbCOD, MLSS, and temperature/revise feed	
	Increase methanol feed	
Final clarifiers		8–3
Bulking sludge	Increase or decrease dissolved oxygen concentration	
	Increase F/M ratio	
	Modify return and waste activated-sludge pumping rates	
	Chlorinate influent wastewater	
	Chlorinate return activated sludge	
Rising sludge	Increase return activated-sludge pumping rate to reduce sludge blanket depth in clarifier	
	Conduct state point analysis of final clarifiers	
Poor solids separation	Perform state point analysis	8–7
Disinfection		12–6
High coliform count in effluent	Improve chlorine mixing	12–6
	Conduct dye tests to determine short circuiting	4–4
	Check process for partial nitrification	8–4

(continued)

| Table 15–27 (Continued)

Issue	Possible remedial action	See Sec. no.
Solids processing		
Low solids concentration from dissolved air flotation thickener	Check air-solids ratio Reduce solids loading rate Add/increase polymer feed	14–6
Poor anaerobic digester performance	Change frequency of solids feeding Increase concentration of feed solids Check adequacy of mixing Remove sludge and grit deposits Increase SRT	14–9
Poor aerobic digester performance	Check temperature/adjust SRT Increase concentration of feed solids Check adequacy of mixing Increase DO Check pH/adjust alkalinity	14–10
Odors from composting operations	Increase aeration by air addition or frequency of turning	14–11
Poor compost quality	Perform materials balance/adjust feed composition	14–11
Excessive moisture in compost mixture	Change compost mixture by adding amendment or bulking agent Improve sludge dewatering operations	14–11

process components may be necessary to determine the limits of operating capability of the key components. In other cases, dye testing to document hydraulic flow patterns or measuring dissolved oxygen transfer rates to determine aeration efficiency may be employed.

Although a major focus of treatment-plant design and operation is on the liquid treatment facilities because of the standards for treated effluent reuse and disposal, solids processing facilities often receive less attention. Solids processing is often the most vexing problem for many plants because of operational difficulties, increasingly stringent requirements for reuse, and limited options for disposal. However, some of the new technologies described in this text can be used to upgrade the design and operation of solids processing facilities. Example upgrade options are described in Table 15–29.

Upgrading to Meet New Constituent Removal Requirements. As discussed in Chaps. 1, 13, and 14 and in earlier sections of this chapter, standards for constituent removal have changed in recent years and will continue to change as more scientific information is developed and as the reuse of treated wastewater and biosolids becomes increasingly important. Examples of many of the current and future issues and upgrade options in resolving these issues are summarized in Table 15–30.

Table 15-28
Examples of upgrading of physical liquid treatment facilities to improve plant performance

Issue	Remedial action/upgrade option	See Sec. no.
General plant		
Odors	Cover structures	15-3
	Add odor collection and treatment system	15-3
	Reduce turbulence by eliminating free falls and sharp bends	15-3
	Add chemical feed facilities	6-3
Wide influent flow variations	Add upstream flow equalization	5-3
	Install variable-speed drives on pumps	15-5
	Install small-capacity pumps for low flows	15-5
Flow control/ distribution	Improve flow splitting	8-7
	Add metering	15-4
Return flows from sludge-processing facilities upset biological process	Provide flow storage/equalization	3-2
	Provide sidestream treatment of return flows	14-16
	Modify operations/upgrade solids processing facilities to reduce load	14-16
Headworks		
Inadequate screenings removal	Modify/replace screens to prevent screenings carryover	5-1
	Install fine screens	5-1
	Replace comminutors	5-1, 5-2
Odorous, wet screenings	Install screenings press	5-1
	Replace screens with macerators	5-2
	Enclose and ventilate screening equipment	15-3
Odorous grit	Install grit washer	5-6
	Enclose and ventilate grit equipment	15-3
Inadequate grit removal	Add permanent baffles to prevent short circuiting	5-6
	Replace/upgrade grit-removal equipment	5-6
Primary clarifiers		
Inadequate solids removal in primary clarifiers	Add chemical treatment and flocculation	5-7, 6-3
	Add high-rate clarification	5-8
	Install baffles at effluent weirs	8-7
	Add energy dissipation inlet	5-7
Aeration tanks		
Low DO	Install DO probes for DO monitoring	15-4
	Replace coarse bubble with fine-pore diffusers	5-12
	Change diffuser placement to a grid pattern	5-12
High DO	Install variable-speed drives on centrifugal blowers to provide turndown capability	5-12, 15-5
	Install inlet guides vanes on centrifugal blowers to provide turndown capability	5-12
	Install variable-speed drives on positive-displacement blowers	5-12
	Install timers and two-speed motors on mechanical aerators	5-12
	Install automatic DO control system	15-4

(continued)

| **Table 15–28** (Continued) | | |

Issue	Remedial action/upgrade option	See Sec. no.
Unbalanced DO profile in plug-flow aeration tanks	Change to step-feed process	8–4, 8–5
	Add DO control system	15–4
Solids deposition	Increase mixing capacity	5–12
Nocardia foaming	Add selector	8–3
Trickling filter		
Plugging and ponding of rock filters	Install plastic packing	9–2
Odors and poor BOD removal	Increase airflow by improving natural draft or adding ventilation system	9–2
Biological treatment system		
Insufficient reactor and solids separation capacity	Add chemical treatment	6–3
	Add high-rate clarification to reduce loading on biological treatment system	5–8
	Add membrane bioreactor	8–9
Solids washout from high flowrates	Add flow equalization	5–2
	Add high-rate clarification process for excess flows	5–8
	Use contact stabilization process	8–3
Secondary clarifiers		
Inadequate solids separation in secondary clarifiers	Modify flow distribution	8–7
	Modify circular clarifier center feedwell	8–7
	Add flocculating center feedwell	8–7
	Install baffles at effluent weirs	8–7
	Add tube or plate settlers	5–5
	Modify effluent weir configuration	8–7
	Modify sludge collector to improve solids withdrawal	8–7
Disinfection		
Inadequate chlorine disinfection	Add/replace chlorine mixers	5–4, 12–6
	Add/modify baffles to reduce short circuiting in chlorine contact tank	12–6
	Add chlorine residual control system	12–6
TSS in effluent	Add depth filters before disinfection	11–4
Excessive chlorine residual	Add chlorine residual analyzer and automatic control system	12–6
	Add dechlorination facilities	12–6
	Replace chlorination system with UV	12–9

Table 15–29
Examples of upgrading physical solids treatment facilities to improve plant performance

Issue	Remedial action/upgrade option	See Sec. no.
Thickening		14–6
Low solids concentration in primary sludge	Add gravity thickening	
	Add co-settling thickening	
Insufficient gravity thickening of waste-activated sludge	Use alternative thickeners (dissolved air flotation, or centrifuge)	
Alkaline stabilization		
Odor and vector problems in dewatered sludge	Add postlime stabilization	14–8
Anaerobic digesters		14–9
Excessive hydraulic overloading	Add sludge thickening prior to digestion	14–6
Inadequate mixing	Upgrade digester mixing system	14–9
	Install egg-shaped digester	14–9
Poor digestion of mixed primary and biological sludge	Install separate digesters	14–9
Inadequate solids destruction	Install two-phased anaerobic digestion process	
Aerobic digestion		14–10
Insufficient pathogen removal	Increase SRT by adding thickening or additional aerobic digester capacity	
	Add ATAD process	
Inadequate mixing	Increase mixing energy	
Composting		
Excessive plastics and inert material in product to be reused	Install fine screens in plant influent	5–1
	Install sludge screens	14–5
Dewatering and drying		
Excessive water in dewatered sludge cake	Add sludge thickeners	14–6
	Install high solids centrifuge dewatering	14–13
	Install filter press	14–13
	Add solar drying beds	14–13
	Add heat dryers	14–14
Sludge lagoons and drying beds		
Odors	Construct turbulence-inducing structures	15–3
Land application of biosolids		
Excessive attraction of vectors	Modify preapplication treatment methods or method of biosolids application	14–17
Excessive pathogen levels	For Class A biosolids, use one of the six prescribed alternative treatments	14–2, 14–17
	For Class B biosolids, use one of three prescribed alternatives for monitoring or treatment	14–2, 14–17

Table 15–30

Potential process modifications for meeting new standards for constituent removal

Issue	Remedial action/upgrade option	See Sec. no.
TSS discharge standards	Investigate alternative solids separation facilities	
	Chemical treatment to enhance settling	6–3
	Tube or plate settlers for final clarifiers	5–5
	High rate clarification	5–8
	Depth filtration	11–4
	Surface filtration	11–5
	Membrane separation	8–9, 11–6
BOD/COD standards	Investigate alternative treatment facilities	
	Supplemental chemical treatment	6–6
	Nitrification	8–4, 9–2
	Combined aerobic biological processes	9–4
	Membrane biological reactors	8–9
	Adsorption	11–7
	Advanced oxidation	11–10
Removal of nitrogen and phosphorus	Investigate alternative removal facilities	
	Chemical treatment for phosphorus removal	6–4
	Activated-sludge selector	8–3
	Suspended growth processes	
	Nitrification	8–4
	Nitrogen removal	8–5
	Phosphorus removal	8–6
	Attached growth processes	
	Nitrification	9–2, 9–3
	Nitrogen removal	9–7
	Ammonia stripping of digester supernatant	11–8
	Ion exchange for nitrogen removal	11–9
New disinfection standards	Add depth filtration (prior to disinfection)	11–3, 11–4
	Improve chlorine mixing and dispersion	12–2, 12–6
	Add dechlorination system	12–5, 12–6
	Replace chlorination with UV	12–9

(continued)

| **Table 15–30** (*Continued*)

Issue	Remedial action/upgrade option	See Sec. no.
VOC emission requirements	Investigate alternative advanced treatment systems	
	Adsorption	5–13, 11–7
	Air stripping	5–13, 11–9
	Advanced oxidation	11–10
	Thermal or catalytic incineration	5–13
Removal of residual solids for water reuse	Investigate alternative advanced treatment systems	
	Depth filtration	11–4
	Surface filtration	11–5
	Microfiltration	11–6
	Activated-carbon adsorption	11–7
	Ion exchange	11–9
	Advanced oxidation	11–10
Removal of trace constituents	Investigate alternative treatment systems	
	Chemical precipitation and oxidation	6–5, 6–6
	Microfiltration and/or reverse osmosis	11–6
	Microfiltration and/or reverse osmosis with UV oxidation	11–6, 11–10
Part 503 biosolids regulations for Class A land application	Investigate alternative processes to further reduce pathogens (PFRP) including	
	Thermophilic aerobic digestion	14–10
	Composting	14–11
	Heat drying	14–14
	Heat treatment	14–12
Part 503 biosolids regulations for Class B land application	Investigate alternative processes to significantly reduce pathogens (PSRP) including	
	Lime stabilization	14–8
	Anaerobic digestion	14–9
	Aerobic digestion	14–10
	Composting	14–11
	Air drying	14–13

15–7 **IMPORTANT DESIGN CONSIDERATIONS FOR NEW WASTEWATER TREATMENT PLANTS**

During the late 1980s and continuing onward to the present time (2001), the focus on wastewater engineering has changed from the goals of percent removal of solids and organic matter to the specific requirements for constituent removal for health and environmental protection. During the next decade as additional demands are placed on the quality and quantity of the nation's water supplies, greater emphasis will be placed on maintaining high standards of wastewater treatment plant performance consistently and reliably. In many cases, treated effluent will have to approach or meet drinking water standards or will have to meet not-to-exceed effluent requirements that contain harsh penalties for violation. As scientific research continues in defining wastewater constituents that may cause adverse effects, greater levels of treatment will be needed and, in some cases, will require the use of new technologies. The purpose of this section is to highlight important issues that must be considered in the planning and design of new treatment facilities. Additional information on many of the topics identified in this section may be found in the previous section.

Process Design Considerations for Liquid Streams

One of the most important issues that the designer of new plants in the United States will have to face is the exceedance criterion for effluent discharges described in Sec. 15–1. The focus of treatment-plant design will be on the nature and variability of the wastewater constituents and the processes that will be required to meet the treatment objectives. In some cases, it may be necessary to develop source control programs to limit the effects of both flow and constituent concentrations for discharges to the collection system. The source control program may also include reduction of excessive extraneous flow such as infiltration/inflow and wet weather discharges to the collection system and may even necessitate upstream flow equalization. Facilities to handle in-plant variations of flows and loads including return flows will also have to be considered. Some of the important factors that will have to be considered in the design of new facilities to meet possible future standards and possible technological solutions are listed in Table 15–31.

Process Design Considerations for Solids Processing

In selecting the appropriate methods of solids processing, reuse, and disposal, consideration must be given to the appropriate regulations. In the United States, pollutant numerical limits and management practices for the reuse and disposal of solids generated from the processing of municipal wastewater are defined in the 40 CFR Part 503 regulations (see Chap. 14, Sec. 14–2). The regulations are designed to protect public health and the environment from any reasonably anticipated adverse effects of pollutants contained in the biosolids. As many communities investigate options for solids processing, reuse, and disposal, greater emphasis is being placed on producing a cleaner product and meeting the Class A biosolids requirements. Many of the important issues that have to be considered in solids processing are also summarized in Table 15–31. Control of pathogens and vectors is of special importance for the protection of public health.

Table 15–31
Important considerations in the design of new wastewater treatment plants

Issue	Description	See Sec. or Chap.
Liquid stream processing		
Constituent and treatment process variability	New treatment processes should be designed to meet not-to-exceed permit limits taking into account constituent and treatment process variability	15–2
Flow equalization	Improved performance by eliminating flow surges through treatment facilities	5–3
Organic load equalization	Improved performance by equalizing organic loading to treatment processes throughout the day	5–3
Automatic process control	Provision for automatic process control of dissolved oxygen and solids-retention time (SRT)	15–4
Enhanced disinfection for reuse	Improved and alternative technologies for disinfection and the control of disinfection by-products	12
Advanced treatment processes	Removal of residual constituents and constituents not achieved by conventional treatment	11
Conventional and advanced oxidation processes	Removal of specific constituents may require the use of advanced oxidation processes	6–6, 11–10
Combined processes for specific constituents	To meet stringent not-to-exceed permit limits, two or more processes may have to be used in a series arrangement	13–4
Water reuse	Issues related to risk assessment will have to be addressed	13–3
Energy management	Installation of physical facilities to allow for improved energy efficient operation	15–6
Solids processing		
Improved screening	Removal of extraneous materials that end up in the biosolids	5–1
Grit removal	Removal of grit that can settle in primary sedimentation tanks and digesters	5
Enhanced pathogen control	To meet requirements for Class A biosolids	14–7 to 14–14
Enhanced vector control	To meet requirements for Class A biosolids	14–11
Separate treatment of return flows	Improved performance of liquid waste streams, especially for the removal of nitrogen	14–16
Odor control		
Odor control in collection system	Implementation of source control program or addition of oxidizing agents to control odors in collection system	15–3
Odor formation in treatment facilities	Careful attention to hydraulic design to avoid dead zones with respect to flow	15–3
Odor containment	Cover facilities to eliminate odors	
Odor treatment	Separate or combined treatment of odorous gases from odor containment facilities	15–3
Process operation		
Computer simulation models	Development of improved operational strategies through the use of mathematical simulation models	8–10

Odor Control

Gaining public acceptance for siting new wastewater treatment plants will depend, to a large extent, on meeting community concerns over odors. The prevention, control, and treatment of odors are now mandatory parts of any new treatment-plant design. Proper planning for odor management can mitigate odor-related issues and help restore public confidence in facilities operation. The important considerations in odor control are listed in Tables 15–8 and 15–9.

PROBLEMS AND DISCUSSION TOPICS

15–1 Search the literature and find 3 probability plots of effluent characteristics. How do the reported variabilities, as given by the geometric standard deviation, compare with the values given in Table 15–1?

15–2 A superintendent of a wastewater treatment plant has requested that his staff collect and analyze 6 effluent samples of a specific constituent for submittal to the regulatory agency. If the regulatory agency, in turn, sets the not-to-be-exceeded value (i.e., 99.9 percent) in the effluent discharge permit at the maximum value measured in the 6 samples, has the superintendent potentially shot himself in the foot? Explain and illustrate your analysis.

15–3 Using the data in Table 15–1, estimate the total logs of removal for coliphage with a treatment process composed of activated sludge followed by microfiltration and reverse osmosis for influent coliphage concentrations of 20,000, 40,000, 28,000, 50,000 pfu (plaque-forming units)/100 mL (value to be selected by the instructor). Assume no removal of coliphage through the microfilter and that the logs of removal achieved with the particular reverse osmosis membrane used are 2, 3.2., 3.0, or 3.7 (log removal value to be selected by the instructor). What is the removal achieved at 99 and 99.9 percent reliability?

15–4 The following monthly effluent constituent concentrations have been obtained from 4 different existing activated-sludge treatment plants. Each of these plants is to be replaced with a new plant to meet new and more stringent discharge requirements. For one of these plants (to be selected by the instructor), determine the mean design value assuming the following maximum monthly permit limits must be met: (*a*) BOD and TSS = 15 mg/L at 99 and 99.9 percent reliability, (*b*) BOD and TSS = 10 mg/L at 99 and 99.9 percent reliability, (*c*) BOD = 5 mg/L and TSS = 8 mg/L. On average, what percentage improvement would be required to meet the new discharge permit requirements? In your opinion, is the percentage improvement reasonable?

| | Constituent concentration, mg/L | | | | | | | |
| | Plant A | | Plant B | | Plant C | | Plant D | |
Month	BOD	TSS	BOD	TSS	BOD	TSS	BOD	TSS
Jan	11.0	14.0	4.0	5.0	10.0	40.0	8.4	6.5
Feb	14.0	11.0	5.0	5.0	15.0	39.0	10.2	4.2
Mar	7.0	10.0	6.0	6.0	17.0	23.0	17.9	5.9
Apr	6.0	6.0	7.5	7.0	20.0	30.0	10.3	10.3
May	11.0	13.0	9.0	8.0	25.0	33.0	13.2	10.9
Jun	8.0	12.0	13.0	10.0	29.0	10.0	9.3	8.10

(*continued*)

(Continued)

	Constituent concentration, mg/L							
	Plant A		**Plant B**		**Plant C**		**Plant D**	
Month	**BOD**	**TSS**	**BOD**	**TSS**	**BOD**	**TSS**	**BOD**	**TSS**
Jul	9.0	8.0	16.0	14.0	30.0	18.0	8.6	6.9
Aug	10.0	8.0	18.0	17.0	25.0	50.0	12.0	8.2
Sep	16.0	10.0	6.5	15.0	35.5	60.0	13.7	9.1
Oct	10.2	10.0	10.0	1.5	25.0	70.0	13.8	14.0
Nov	7.5	10.0	12.0	10.0	40.5	77.0	16.3	14.0
Dec	4.8	10.0	5.5	9.0	50.0	82.0	17.0	18.2

15-5 Verify that for each mg/L of H_2S removed with chlorine 14.7 mg/L of alkalinity as $CaCO_3$ will be required.

15-6 Determine the amount of hydrogen peroxide (H_2O_2) required for the oxidation of hydrogen sulfide (H_2S).

15-7 Using the following half reaction for permanganate (MnO_4^-), estimate the amount of permanganate that would be required per day to oxidize 100 ppm_v of H_2S from a foul airstream with a flowrate of 1500, 2000, 1800, or 2200 m^3/min (flowrate to be selected by instructor).

$$MnO_4^- + 4H^+ + 3e^- \rightarrow MnO_2(s) + 2H_2O$$

15-8 Determine the amount of ferrous sulfate ($FeSO_4$) that would be required to remove 150 mg/L of H_2S from digester supernatant. Assume the sulfide ion in H_2S will be converted to ferrous sulfide in an exchange reaction.

15-9 Four different waste airstreams have been sampled and the results are summarized below. For one of these waste airstreams (to be selected by the instructor), determine the chemical requirements. Sodium hypochlorite and sodium hydroxide are to be used in the chemical scrubber.

Item	Unit	Plant A	Plant B	Plant C	Plant D
Waste airstream flowrate	m^3/min	1000	2500	3200	1800
H_2S concentration	ppm_v	75	45	65	35
Liquid to gas ratio	kg/kg	1.85	2.0	2.1	1.9
Temperature	°C	28	33	30	25
Density of 50 percent NaOH solution	kg/L	1.52	1.52	1.52	1.52

15-10 Using the design criteria given in Table 15–14, determine the size of compost filter needed to scrub 65 ppm_v H_2S from foul air at a flowrate of 1500, 1880, 2100, or 2300 m^3/min (value to be selected by the instructor). Also estimate the mass of the buffer compound needed to neutralize the acid formed as a result of treatment within the filter. Assume a packed-bed porosity of 43 percent. The temperature of the foul air is 20°C.

15-11 From an analysis of several incidents of tank overflow it was found that the overflow was caused 9 times by level-sensor failure, 7 times by malfunction of electrical contacts, 2 times by pump failure, 2 times by gasket failure, and 1 time by a repairperson error. Prepare a Pareto plot and develop recommendations for limiting tank overflow. From the literature prepare a brief discussion of the use of Pareto charts for process applications (cite at least two references).

15-12 A poorly sized valve has an unnecessary pressure drop of 4 m. If the flow through the valve is 5000 m³/d, determine the annual cost attributable to the pressure drop across the valve. Use the current cost of electricity in your region (e.g., $0.12 / kWh).

15-13 Given the following valve capacity factors for four different valves A, B, C, or D (to be selected by the instructor), determine whether the valve is adequate for the pump described in the following table. The pipe diameter is 150 mm. Assume the static head is 3.25, 3.5, 3.9, or 4.3 m (to be selected by the instructor) and the dynamic headloss is $4v^2/2g$ (the factor 4 includes allowances for friction, exit, and miscellaneous pipe losses).

Flowrate, m³/min	Pressure, m
0.00	17.75
0.76	17.63
1.70	17.25
2.46	16.63
3.97	14.30
4.73	12.00

The valve has the following characteristics:

Percent opening	Valve capacity factor, K_v			
	Valve A	Valve B	Valve C	Valve D
10	15	35	23	20
20	34	50	40	52
30	68	68	62	95
40	120	95	90	142
50	200	121	115	200
60	280	160	140	270
70	380	202	180	350
80	500	240	240	420
90	650	290	330	510
100	850	340	480	600

15-14 Average monthly performance data have been collected over a period of a year for an activated-sludge process treating primary settled effluent followed by depth filters operated in parallel. Using these data, estimate what overall average monthly removal would be expected with the activated-sludge process followed by one of the depth filters (to be selected by the instructor) assuming the influent constituent concentration is 150, 200, 275, or 300 mg/L (to be selected by the instructor). What would be the expected constituent concentrations that could be achieved 99 and 99.9 percent of the time?

Removal, %			
Biological treatment	Depth filter A	Depth filter B	Depth filter C
80	65	45	41
98	65	50	44
80	65	49	49
84	65	55	45
90	65	58	47
85	65	68	45
78	65	70	43
93	65	40	46
88	65	45	45
92	65	57	43
94	65	61	48
89	65	54	42

15–15 An existing conventional activated-sludge wastewater treatment plant having a design capacity of 15,000 m^3/d is considering upgrading the existing solids-processing facilities to include either aerobic digestion or anaerobic digestion. List the factors the design engineer will have to consider in selecting the process and comment on the energy implications associated with the process selection.

15–16 Review three articles in the literature in which aeration systems have been retrofitted with fine-pore diffusers. Document the physical changes that were made including the type of replacement diffusers and the amount of savings in electric energy that was achieved.

REFERENCES

Allen, E. R., and Y. Yang (1991) "Biofiltration Control of Hydrogen Sulfide Emissions," *Proceedings of the 84th Annual Meeting of the Air and Waste Management Association,* Vancouver, BC, Canada.

Allen, E. R., and Y. Yang (1992) "Operational Parameters for the Control of Hydrogen Sulfide Emissions Using Biofiltration," *Proceedings of the 85th Annual Meeting of the Air and Waste Management Association,* Kansas City, MO.

Batts, C.W., F. L. Burton, and M. Jones (1993) "Demand Side Management Opportunities in the Wastewater Industry," Water Environment Federation, *Proceedings of the 66th Annual Conference & Exposition.*

Bauman, H. D. (1998) *Control Valve Primer,* 3d ed., Instrument Society of America, Research Triangle Park, NC.

Bohn, H. L., and R. K. Bohn (1988) "Soil Beds Weed Out Air Pollutants," *Chemical Engineering,* vol. 95, no. 6, pp. 73–76.

Burton, F. L. (1996) *Water and Wastewater Industries: Characteristics and Energy Management Opportunities,* CR-10691, Electric Power Research Institute, St. Louis, MO.

Burton, F. L. (1998) "Saving on Wastewater Treatment," *Energy Magazine,* vol. XXIII, no. 1, pp. 17–20, Norwalk, CT.

Colby, S., and A. Ekster (1997) "Control-Loop Fault Detection," *Water Environment and Technology,* vol. 9, no. 11, pp. 43–46.

Corripo, A. B. (1990) *Tuning of Industrial Controllers,* Instrument Society of America , Research Triangle Park, NC.

Coughanowr, D. R. (1991) *Process Systems Analysis and Control,* McGraw-Hill, New York.

Crites, R.W., and G. Tchobanoglous (1998) *Small and Decentralized Wastewater Management Systems,* McGraw-Hill, New York.

Devinny, J. S., M. A. Deshusses, and T. S. Webster (1999) *Biofiltration for Air Pollution Control,* Lewis Publishers, Boca Raton, FL.

Ekster, A., et al. (1997) "Automation of Waste Control," *Proceedings of the Water Environment Federation 70th Annual Conference and Exposition,* vol. 5, pp. 575–586.

EPRI (1994) *Energy Audit Manual for Water and Wastewater Facilities,* Electric Power Research Institute, St. Louis, MO.

EPRI (1998) *Quality Energy Retrofits for Wastewater Systems,* CR-109081, Electric Power Research Institute, Palo Alto, CA.

Eweis, J. B., S. J. Ergas, D. P. Y. Chang, and E. D. Schroeder (1998) *Bioremediation Principles,* McGraw-Hill, Boston, MA.

Finn, L., and R. Spencer (1997) "Managing Biofilters for Consistent Odor and VOC Treatment," *BioCycle,* vol. 38, no. 1.

Helsel, D. R. (1990) "Statistical Treatment of Data Below the Detection Limit," *Environmental Science and Technology,* vol. 24, no. 12, pp. 1767–1774.

Helsel, D. R., and T. A. Cohn (1988) "Estimation of Descriptive Statistics for Multiply Censored Water Quality Data," *Water Resources Research,* vol. 24, no. 12, pp. 1997–2004.

Kokoska, S., and D. Zwillinger (2000) *Standard Probability and Statistics Tables and Formulae,* student ed., Chapman & Hall/CRC, Boca Raton, FL.

Loge, F. J., K. Bourgeous, R. W. Emerick, and J. L. Darby (2001) "Variations in Wastewater Quality Parameters Influencing UV Disinfection Performance: Relative Impact of Filtration," *Journal of Environmental Engineering,* vol. 127, no. 9, pp. 832–837.

Murthy, S. (2001) Personal communication.

Niku, S., E. D. Schroeder, and F. J. Samaniego (1979) "Performance of Activated Sludge Processes and Reliability-Based Design," *Journal Water Pollution Control Federation,* vol. 51, p. 2841.

Niku, S., E. D. Schroeder, G. Tchobanoglous, and F. J. Samaniego (1981) "Performance of Activated Sludge Processes: Reliability, Stability and Variability," Environmental Protection Agency, EPA Grant No. R805097-01, pp. 1–124.

Novak, J. T. (2001) Personal communication.

Olson, G., and B. Newell (1992) *Wastewater Treatment Systems. Modeling, Diagnosis and Control,* International Water Association Publishing, London.

Patterson, R. G., R. C. Jain, and S. Robinson (1984) "Odor Controls for Sewage Treatment Facilities," Presented at Annual Meeting of the Air Pollution Control Association, San Francisco, CA.

Pomeroy, R. D. (1957) "Deodorizing Gas Streams by the Use of Microbiological Growths," U.S. Patent No. 2,793,096.

Rafson, H. J. (ed.) (1998) *Odor and VOC Control Handbook,* McGraw-Hill, New York.

Schroeder, E. D. (2001) Personal Communication, Department of Civil and Environmental Engineering, University of California at Davis, Davis, CA.

Shinskey, F. G. (1996) *Process Control Systems: Application, Design, and Tuning,* 4th ed., McGraw-Hill, New York.

Smith, C. A., and A. B. Corripo (1997) *Principles and Practice of Automatic Process Control,* Wiley, New York.

Speece, R. E., N. Nirmalakhandan, and G. Tchobanoglous (1990) "Commercial Oxygen Use in Water-Quality Management," *Water Environment and Technology,* vol. 7, no. 7, pp. 54–61.

Tchobanoglous, G. (1981) *Wastewater Engineering: Collection and Pumping of Wastewater,* McGraw-Hill, New York.

Tchobanoglous, G., and E. D. Schroeder (1985) *Water Quality: Characteristics, Modeling, Modification,* Addison-Wesley, Reading, MA.

U.S. EPA (1982) *Evaluation and Documentation of Mechanical Reliability of Conventional Wastewater Treatment Plant Components,* EPA 600/2-82-044, U.S. Environmental Protection Agency, Washington, DC.

U.S. EPA (1985) *Design Manual, Odor and Corrosion Control in Sanitary Sewerage Systems and Treatment Plants,* EPA/625/1-85/018, U.S. Environmental Protection Agency, Washington, DC.

U.S. EPA (1994) *Water Quality Standards Handbook,* 2d ed., EPA-823-B-94-005a, U.S. Environmental Protection Agency, Washington, DC.

van Lith, C. (1989) "Design Criteria for Biofilters," *Proceedings of the 82nd Annual Meeting of the Air and Waste Management Association,* Anaheim, CA.

WCPH (1996) *Total Resource Recovery Project, Final Report,* prepared for City of San Diego, Water Utilities Department, Western Consortium for Public Health, Oakland, CA.

WCPH (1997) *Total Resource Recovery Project Aqua III San Pasqual Health Effects Study Final Summary Report,* Prepared for City of San Diego Water Utilities Department, Western Consortium for Public Health, Oakland, CA.

WEF (1993) *Instrumentation in Wastewater Treatment Plants,* WEF Manual of Practice No. 21, Water Environment Federation, Alexandria, VA.

WEF (1995) *Odor Control in Wastewater Treatment Plants,* WEF Manual of Practice No. 22, Water Environment Federation, Alexandria, VA.

WEF (1997) *Biofiltration: Controlling Air Emissions through Innovative Technology,* Project 92-VOC-1, Water Environment Federation, Alexandria, VA.

Williams, D. G. (1996) *The Chemistry of Essential Oils,* Michelle Press, Dorset, England.

Williams, T. O., and F. C. Miller (1992*a*) "Odor Control Using Biofilters," *BioCycle,* vol. 33, no. 10.

Williams, T. O., and F. C. Miller (1992*b*) "Biofilters and Facilities Operations," *BioCycle,* vol. 33, no. 11.

Wilson, G. (1975) *Odors: Their Detection and Measurement,* EUTEK Process Development and Engineering, Sacramento, CA.

Yang, Y., and E. R. Allen (1994) "Biofiltration Control of Hydrogen Sulfide: Design and Operational Parameters," *Journal Air and Water Management Association,* vol. 44, pp. 863–868.

Appendix

Table A-1
Metric conversion factors (SI units to U.S. customary units)

Multiply SI unit		by	To obtain U.S. customary unit	
Name	**Symbol**		**Symbol**	**Name**
Acceleration				
meters per second squared	m/s^2	3.2808	ft/s^2	feet per second squared
meters per second squared	m/s^2	39.3701	in/s^2	inches per second squared
Area				
hectare (10,000 m²)	ha	2.4711	ac	acre
square centimeter	cm^2	0.1550	in^2	square inch
square kilometer	km^2	0.3861	mi^2	square mile
square kilometer	km^2	247.1054	ac	acre
square meter	m^2	10.7639	ft^2	square foot
square meter	m^2	1.1960	yd^2	square yard
Energy				
kilojoule	kJ	0.9478	Btu	British thermal unit
joule	J	2.7778×10^{-7}	kW·h	kilowatt-hour
joule	J	0.7376	ft·lb$_f$	foot-pound (force)
joule	J	1.0000	W·s	watt-second
joule	J	0.2388	cal	calorie
kilojoule	kJ	2.7778×10^{-4}	kW·h	kilowatt-hour
kilojoule	kJ	0.2778	W·h	watt-hour
megajoule	kJ	0.3725	hp·h	horsepower-hour
Force				
newton	N	0.2248	lb$_f$	pound force
Flowrate				
cubic meters per day	m^3/d	264.1720	gal/d	gallons per day
cubic meters per day	m^3/d	2.6417×10^{-4}	Mgal/d	million gallons per day
cubic meters per second	m^3/s	35.3147	ft^3/s	cubic feet per second
cubic meters per second	m^3/s	22.8245	Mgal/d	million gallons per day
cubic meters per second	m^3/s	15,850.3	gal/min	gallons per minute
liters per second	L/s	22,824.5	gal/d	gallons per day
liters per second	L/s	2.2825×10^{-2}	Mgal/d	million gallons per day
liters per second	L/s	15.8508	gal/mm	gallons per minute

(continued)

| **Table A-1** (*Continued*)

Multiply SI unit		by	To obtain U.S. customary unit	
Name	**Symbol**		**Symbol**	**Name**
Length				
centimeter	cm	0.3937	in	inch
kilometer	km	0.6214	mi	mile
meter	m	39.3701	in	inch
meter	m	3.2808	ft	foot
meter	m	1.0936	yd	yard
millimeter	mm	3.9370×10^{-2}	in	inch
Mass				
gram	g	3.5274×10^{-2}	oz	ounce
gram	g	2.2046×10^{-3}	lb	pound
kilogram	kg	2.2046	lb	pound
megagram (10^3 kg)	Mg	1.1023	ton	ton (short: 2000 lb)
Power				
kilowatt	kW	0.9478	Btu/s	British thermal units per second
kilowatt	kW	1.3410	hp	horsepower
watt	W	0.7376	ft-lb$_f$/s	foot-pounds (force) per second
Pressure (force/area)				
Pascal (newtons per square meter)	Pa (N/in^2)	1.4504×10^{-4}	lb$_f$/in^2	pounds (force) per square inch
Pascal (newtons per square meter)	Pa (N/in^2)	2.0885×10^{-2}	lb$_f$/ft^2	pounds (force) per square foot
Pascal (newtons per square meter)	Pa (N/in^2)	2.9613×10^{-4}	inHg	inches of mercury (60°F)
Pascal (newtons per square meter)	Pa (N/in^2)	4.0187×10^{-3}	in H$_2$O	inches of water (60°F)
kilopascal (kilonewtons per square meter)	kPa (kN/m^2)	0.1450	lb$_f$/in^2	pounds (force) per square inch
kilopascal (kilonewtons per square meter)	kPa (kN/m^2)	9.8688×10^{-3}	atm	atmosphere (standard)
Temperature				
degree Celsius (centigrade)	°C	1.8(°C) + 32	°F	degree Fahrenheit
degree Kelvin	K	1.8(K) − 459.67	°F	degree Fahrenheit
Velocity				
meters per second	m/s	2.2369	mi/h	miles per hour
meters per second	m/s	3.2808	ft/s	feet per second
Volume				
cubic centimeter	cm^3	6.1024×10^{-2}	in^3	cubic inch
cubic meter	m^3	35.3147	ft^3	cubic foot
cubic meter	m^3	1.3079	yd^3	cubic yard
cubic meter	m^3	264.1720	gal	gallon
cubic meter	m^3	8.1071×10^{-4}	ac·ft	acre·foot
liter	L	0.2642	gal	gallon
liter	L	3.5315×10^{-2}	ft^3	cubic foot
liter	L	33.8150	oz	ounce (U.S. fluid)

Table A–2
U.S. customary conversion factors (U.S. customary units to SI units)

Multiply U.S. customary unit			To obtain SI unit	
Name	**Symbol**	**by**	**Symbol**	**Name**
Acceleration				
feet per second squared	ft/s^2	0.3048^a	m/s^2	meters per second squared
inches per second squared	in/s^2	2.54×10^{-2a}	m/s^2	meters per second squared
Area				
acre	ac	4046.8564	m^2	square meter
acre	ac	0.4047	ha^a	hectare
square foot	ft^2	9.2903×10^{-2a}	m^2	square meter
square inch	in^2	6.4516	cm^2	square centimeter
square mile	mi^2	2.5900	km^2	square kilometer
square yard	yd^2	0.8361	m^2	square meter
Energy				
British thermal unit	Btu	1.0551	kJ	kilojoule
British thermal units per kilowatt-hour	Btu/kWh	1.0551	kJ/kW·h	kilojoules per kilowatt-hour
watt-hour	W·h	3.600^a	kJ	kilojoule
watt-second	W·s	1.000^a	J	Joule
Force				
pound (force)	lb_f	4.4482	N	Newton
Flowrate				
cubic feet per minute	ft^3/min	4.7190×10^{-4}	m^3/s	cubic meters per second
cubic feet per minute	ft^3/min	0.4719	L^a/s	liters per second
cubic feet per second	ft^3/s	2.8317×10^{-2}	m^3/s	cubic meters per second
gallons per day	gal/d	4.3813×10^{-5}	m^3/s	cubic meters per second
gallons per day	gal/d	4.3813×10^{-2}	L/s	liters per second
gallons per minute	gal/min	6.3090×10^{-5}	m^3/s	cubic meters per second
gallons per minute	gal/min	6.3090×10^{-2}	L^a/s	liters per second
million gallons per day	Mgal/d	43.8126	L/s	liters per second
million gallons per day	Mgal/d	3.7854×10^3	m^3/d	cubic meters per day
million gallons per day	Mgal/d	4.3813×10^{-2}	m^3/s	cubic meters per second
Length				
foot	ft	0.3048^a	m	meter
inch	in	2.54^a	cm	centimeter
inch	in	2.54×10^{-2a}	m	meter
inch	in	25.4^a	mm	millimeter
miles	mi	1.6093	km	kilometer
yard	yd	0.9144^a	m	meter

(continued)

| **Table A–2** (Continued)

Multiply U.S. customary unit			To obtain SI unit	
Name	**Symbol**	**by**	**Symbol**	**Name**
Mass				
ounce	oz	28.3495	g	gram
pound (mass)	lb_m	4.5359×10^2	g	gram
ton (short: 2000 lb)	ton	0.9072	Mg	megagram (10^3 kilogram)
ton (long: 2240 lb)	ton	1.0160	Mg	megagram (10^3 kilogram)
Power				
British thermal unit	Btu/s	1.0551	kW	kilowatt
foot pound (force) per second	$ft \cdot lb_f/s$	1.3558	W	watt
horsepower	hp	0.7457	kW	kilowatt
horsepower-hour	hp-h	2.6845	MJ	megajoule
kilowatt-hour	kWh	3.6000	MJ	megajoule
Pressure (force/area)				
atmosphere (standard)	atm	1.0133×10^2	kPa (kN/m^2)	kilopascal (kilonewton per square meter)
inches of mercury (60°F)	in Hg (60°F)	3.3768×10^3	Pa (N/in^2)	Pascal (newtons per square meter)
inches of water (60°F)	in H_2O (60°F)	2.4884×10^2	Pa (N/in^2)	Pascal (newtons per square meter)
pounds per square foot	lb_f/ft^2	47.8803	Pa (N/m^2)	pascal (newtons per square meter)
pounds per square inch	lb_f/in^2	6.8948×10^3	Pa (N/m^2)	Newtons per square meter
pounds per square inch	lb_f/in^2	6.8948	kPa (N/m^2)	kilonewtons per square meter
Temperature				
degree Fahrenheit	°F	0.555 (°F − 32)	°C	degree Celsius (centigrade)
degree Fahrenheit	°F	0.555 (°F + 459.67)	K	degree Kelvin
Velocity				
feet per minute	ft/min	5.0800×10^{-3}	m/s	meters per second
feet per second	ft/s	0.3048ª	m/s	meters per second
miles per hour	mi/h	1.6093	km/h	kilometers per hour
miles per hour	mi/h	0.44704ª	m/s	meters per second
Volume				
acre-foot	ac-ft	1.2335×10^3	m^3	cubic meter
cubic foot	ft^3	2.8317×10^{-2}	m^3	cubic meter
cubic foot	ft^3	28.3168	L	liter
cubic inch	in^3	16.3781	cm^3	cubic centimeter
cubic yard	yd^3	0.7646	m^3	cubic meter
gallon	gal	3.7854×10^{-3}	m^3	cubic meter
gallon	gal	3.7854	L	liter
ounce (U.S. fluid)	oz	2.9573×10^{-2}	L	liter

ªExact conversion.

Table A-3

Conversion factors used commonly in environmental engineering and scientific computations

	To convert, multiply in direction shown by arrows		
SI units	$\rightarrow$	$\leftarrow$	**U.S. units**
cm^2/s^a	0.1550	6.4516	in^2/s
g	3.5274×10^{-2}	28.3495	ounce
g/m^3	8.3454	0.1198	lb/Mgal
ha	2.4711	0.4047	ac
$J/m^2 \cdot {}^\circ C \cdot s$	0.1763	5.6735	$Btu/ft^2 \cdot {}^\circ F \cdot h$
kg	2.2046	0.4536	lb
kg/ha	0.8922	1.1209	lb/ac
$kg/kW \cdot h$	1.6440	0.6083	$lb/hp \cdot h$
kg/m^2	0.2048	4.8824	lb/ft^2
kg/m^3	6.2429×10^{-2}	16.0181	lb/ft^3
kg/m^3	8345.3205	1.1983×10^{-4}	lb/Mgal
$kg/m^3 \cdot d$	62.4280	1.6018×10^{-2}	$lb/10^3 \, ft^3 \cdot d$
$kg/m^3 \cdot h$	6.2428×10^{-2}	16.0185	$lb/ft^3 \cdot h$
kJ	0.9478	1.0551	Btu
kJ/kg	0.4303	2.3241	Btu/lb
kPa (gage)	0.1450	6.8948	lb_f/in^2 (gage)
kPa Hg (15.5°C)	0.2961	3.3768	in Hg (60 °F)
kW/m^3	5.0763	0.1970	$hp/10^3 \, gal$
$kW/10^3 \, m^3$	3.7973×10^{-2}	26.3342	$hp/10^3 \, ft^3$
L	0.2642	3.7854	gal
$L/m \cdot min$	8.0520×10^{-2}	12.4193	$gal/ft \cdot min$
$L/m^2 \cdot d$	2.4542×10^{-2}	40.7465	$gal/ft^2 \cdot d$
$L/m^2 \cdot min$	2.4542×10^{-2}	40.7465	$gal/ft^2 \cdot min$
$L/m^2 \cdot min$	35.3420	2.8295×10^{-2}	$gal/ft^2 \cdot d$
m	3.2808	0.3048	ft
m/h	3.2808	0.3048	ft/h
m/h	5.4681×10^{-2}	18.2880	ft/min
m/h	0.4090	2.4448	$gal/ft^2 \cdot min$
m^2	10.7639	9.2903×10^{-2}	ft^2
$m^2/10^3 m^3 \cdot d$	2.4542×10^{-3}	407.4611	$ft^2/Mgal \cdot d$
m^2/s	10.7639	9.2903×10^{-2}	ft^2/s
m^3	8.1071×10^{-4}	1.2335×10^3	$ac \cdot ft$
m^3	264.1720	3.7854×10^{-3}	gal
m^3	1.3079	0.7646	yd^3
$m^3/capita$	35.3147	2.8317×10^{-2}	$ft^3/capita$
m^3/d	2.6417×10^{-4}	3.7854×10^3	Mgal/d

(continued)

Table A–3 (*Continued*)

	To convert, multiply in direction shown by arrows		
SI units	$\rightarrow$	$\leftarrow$	**U.S. units**
m^3/h	0.5886	1.6990	ft^3/min
m^3/s	35.3147	2.8317×10^{-2}	ft^3/s
m^3/s	22.8245	4.3813×10^{-2}	Mgal/d
$m^3/ha\cdot d$	106.9064	9.3540×10^{-3}	gal/ac·d
m^3/kg	16.0185	6.2428×10^{-2}	ft^3/lb
$m^3/m\cdot d$	80.5196	1.2419×10^{-2}	gal/ft·d
$m^3/m\cdot min$	10.7639	9.2903×10^{-2}	$ft^3/ft\cdot min$
$m^3/m^2\cdot d$	24.5424	4.0746×10^{-2}	$gal/ft^2\cdot d$
$m^3/m^2\cdot d$	1.7043×10^{-2}	58.6740	$gal/ft^2\cdot min$
$m^3/m^2\cdot d$	1.0691	0.9354	Mgal/ac·d
$m^3/m^2\cdot h$	3.2808	0.3048	$ft^3/ft^2\cdot h$
$m^3/m^2\cdot h$	589.0173	1.6977×10^{-3}	$gal/ft^2\cdot d$
$m^3/m^2\cdot min$	24.5425	4.0746×10^{-2}	$gal/ft^2\cdot min$
m^3/m^3	0.1337	7.4805	ft^3/gal
$m^3/10^3\ m^3$	133.6805	7.04805×10^{-3}	$ft^3/Mgal$
$m^3/m^3\cdot min$	133.6805	7.04805×10^{-3}	$ft^3/10^3\ gal\cdot min$
$m^3/m^3\cdot min$	1000.0	1×10^{-3}	$ft^3/10^3\ ft^3\cdot min$
Mg/ha	0.4461	2.2417	ton/ac
mm	3.9370×10^{-2}	25.4	in
$N\cdot s/m^2$	2.0885×10^{-2}	47.8810	$lb\cdot s/ft^2$
$W/m^2\cdot K$	0.1762	5.6745	$Btu/ft^2\cdot h\cdot °F$

[a]Although cm is not an SI unit, the coefficient of diffusion is most often expressed in units of cm^2/s in the literature.

Table A–4
Abbreviations for
SI Units

Abbreviation	SI unit
°C	degree Celsius
cm	centimeter
g	gram
g/m^2	gram per square meter
g/m^3	gram per cubic meter ($= mg/L$)
ha	hectare
J	Joule
K	Kelvin
kg	kilogram
kg/capita·d	kilogram per capita per day
kg/ha	kilogram per hectare
kg/m^3	kilogram per cubic meter
kJ	kilojoule
kJ/kg	kilojoule per kilogram
kJ/kW·h	kilojoule per kilowatt-hour
km	kilometer
km^2	square kilometer
km/h	kilometer per hour
km/L	kilometer per liter
kN/m^2	kiloNewton per square meter
kPa	kiloPascal
ks	kilosecond
kW	kilowatt
L	liter
L/s	liters per second
m	meter
m^2	square meter
m^3	cubic meter
mm	millimeter
m/s	meter per second
mg/L	milligram per liter ($= g/m^3$)
m^3/s	cubic meter per second
MJ	megajoule
N	Newton
N/m^2	Newton per square meter
Pa	Pascal (usually given as kiloPascal)
W	Watt

Table A–5
Abbreviations for
US Customary Units

Abbreviation	U.S. customary unit
ac	acre
ac-ft	acre foot
Btu	British thermal unit
Btu/ft^3	British thermal unit per cubic foot
d	day
ft	foot
ft^2	square foot
ft^3	cubic foot
ft/min	feet per minute
ft/s	feet per second
ft^3/min	cubic feet per minute
ft^3/s	cubic feet per second
°F	degree Fahrenheit
gal	gallon
$gal/ft^2 \cdot d$	gallon per square foot per day
$gal/ft^2 \cdot min$	gallon per square foot per minute
gal/min	gallon per minute
hp	horsepower
hp-h	horsepower-hour
in	inch
kWh	kilowatt-hour
lb_f	pound (force)
lb_m	pound (mass)
lb/ac	pound per acre
$lb/ac \cdot d$	pound per acre per day
$lb/capita \cdot d$	pound per capita per day
lb/ft^2	pound per square foot
lb/ft^3	pound per cubic foot
lb/in^2	pound per square inch
lb/yd^3	pound per cubic yard
Mgal/d	million gallons per day
mi	mile
mi^2	square mile
mi/h	mile per hour
ppb	part per billion
ppm	part per million
ton (2000 lb_m)	ton (2000 lb mass)
yd	yard
yd^2	square yard
yd^3	cubic yard

Physical Properties of Selected Gases and the Composition of Air

Appendix B

Table B–1
Molecular weight, specific weight, and density of gases found in wastewater at standard conditions (0°C, 1 atm)[a]

Gas	Formula	Molecular weight	Specific weight, lb/ft³	Density g/L
Air	—	28.97	0.0808	1.2928
Ammonia	NH_3	17.03	0.0482	0.7708
Carbon dioxide	CO_2	44.00	0.1235	1.9768
Carbon monoxide	CO	28.00	0.0781	1.2501
Hydrogen	H_2	2.016	0.0056	0.0898
Hydrogen sulfide	H_2S	34.08	0.0961	1.5392
Methane	CH_4	16.03	0.0448	0.7167
Nitrogen	N_2	28.02	0.0782	1.2507
Oxygen	O_2	32.00	0.0892	1.4289

[a] Adapted from Perry, R. H., D. W. Green, and J. O. Maloney (eds.) (1984) *Chemical Engineers' Handbook*, 6th ed., McGraw-Hill, New York.

Table B–2
Composition of dry air at 0°C and 1.0 atmosphere[a]

Gas	Formula	Percent by volume[b,c]	Percent by weight
Nitrogen	N_2	78.03	75.47
Oxygen	O_2	20.99	23.18
Argon	Ar	0.94	1.30
Carbon dioxide	CO_2	0.03	0.05
Other[d]	—	0.01	—

[a] Values reported in the literature vary depending on the standard conditions.
[b] Adapted from *North American Combustion Handbook*, 2d ed., North American Mfg. Co., Cleveland, OH.
[c] For ordinary purposes air is assumed to be composed of 79 percent N_2 and 21 percent O_2 by volume.
[d] Hydrogen, neon, helium, krypton, xenon.
Note: $(0.7803 \times 28.02) + (0.2099 \times 32.00) + (0.0094 \times 39.95) + (0.0003 \times 44.00) = 28.97$ (see Table B–1).

B-1   DENSITY OF AIR AT OTHER TEMPERATURES

In SI units

The following relationship can be used to compute the density of air ρ_a.

$$\rho_a = \frac{PM}{RT}$$

where P = atmospheric pressure = $1.01325 \cdot 10^5$ N/m^2
   M = mole of air (see Table B–1) = 28.97 kg/kg-mole
   R = universal gas constant = 8314 N·m/kg-mole·K
   T = temperature, K (Kelvin) = (273.15 + °C)

For example, at 20°C, the density of air is

$$\rho_{a,20°C} = \frac{(1.01325 \times 10^5 \text{ N/m}^2)(28.97 \text{ kg/kg-mole})}{(8314 \text{ N·m/kg-mole·K})[(273.15 + 20)\text{K}]} = 1.204 \text{ kg/m}^3$$

In U.S. customary units

The following relationship can be used to compute the specific weight of air γ_a at other temperatures at atmospheric pressure.

$$\gamma_a = \frac{P\ (144 \text{ in}^2/\text{ft}^2)}{RT}$$

where P = atmospheric pressure = 14.7 lb/in^2
   R = universal gas constant = 53.3 ft·lb/lb-air·°R
   T = temperature, °R (Rankine) = (459.67 + °F)

For example, at 68°F, the specific weight of air is

$$\gamma_{a,68°F} = \frac{(14.7 \text{ lb/in}^2)(144 \text{ in}^2/\text{ft}^2)}{(53.3 \text{ ft·lb/lb-air·°R})[(459.67 + 68)°\text{R}]} = 0.0753 \text{ lb/ft}^3$$

B-2   CHANGE IN ATMOSPHERIC PRESSURE WITH ELEVATION

In SI units

The following relationship can be used to compute the change in atmospheric pressure with elevation:

$$\frac{P_b}{P_a} = \exp\left[-\frac{gM\ (z_b - z_a)}{RT}\right]$$

where P = pressure, 1.01325×10^5 N/m^2
   g = 9.81 m/s^2
   M = mole of air (see Table B–1) = 28.97 kg/kg-mole
   z = elevation, m

$$R = \text{universal gas constant} = 8314 \text{ N·m/kg-mole·K}$$
$$= 8314 \text{ kg·m}^2/\text{s}^2\text{·kg-mole·K}$$
$$T = \text{temperature, K (Kelvin)} = (273.15 + \,°\text{C})$$

In U.S. customary units

The following relationship can be used to compute the change in atmospheric pressure with elevation.

$$\frac{P_b}{P_a} = \exp\left[-\frac{gM(z_b - z_a)}{g_c RT} \right]$$

where P = pressure, lb/in^2
$\quad\quad g = 32.2 \text{ ft/s}^2$
$\quad\quad g_c = 32.2 \text{ ft·lb}_m/\text{lb·s}^2$
$\quad\quad M$ = mole of air (see Table B–1) = 28.97 lb/lb-mole
$\quad\quad z$ = elevation, ft
$\quad\quad R$ = universal gas constant = 53.3 ft·lb/lb-air·°R
$\quad\quad T$ = temperature, °R = (459.67 + °F)

Physical Properties of Water

Appendix C

The principal physical properties of water are summarized in Table C–1 in SI units and in U.S. customary units in Table C–2 . They are described briefly below (Vennard and Street, 1975; Webber, 1971).

SPECIFIC WEIGHT

The specific weight g of a fluid is its weight per unit volume. In SI units, it is expressed in kilonewtons per cubic meter (kN/m^3). The relationship between g, γ, and the acceleration due to gravity g is $\gamma = \rho g$. At normal temperatures γ is 9.81 kN/m^3 (62.4 lb$_f$/ft^3).

DENSITY

The density of ρ of a fluid is its mass per unit volume. In SI units it is expressed in kilograms per cubic meter (kg/m^3). For water, ρ is 1000 kg/m^3 at 4°C. There is a slight decrease in density with increasing temperature.

MODULUS OF ELASTICITY

For most practical purposes, liquids may be regarded as incompressible. The bulk modulus of elasticity E is given by

$$K = \frac{\Delta p}{(\Delta V/V)}$$

where Δp is the increase in pressure, which when applied to a volume V, results in a decrease in volume ΔV. For water K is approximately 2.150 kN/m^2 at normal temperatures and pressures.

DYNAMIC VISCOSITY

The viscosity of a fluid μ is a measure of its resistance to tangential or shear stress. Viscosity in SI units is expressed in Newton seconds per square meter (N·s/m^2).

KINEMATIC VISCOSITY

In many problems concerning fluid motion, the viscosity appears with the density in the form μ/ρ and it is convenient to use a single term ν, known as the kinematic viscosity and expressed in square meters per second (m^2/s) in SI units. The kinematic viscosity of a liquid diminishes with increasing temperature.

SURFACE TENSION

Surface tension is the physical property that enables a drop of water to be held in suspension at a tap, a glass to be filled with liquid slightly above the brim and yet not spill, or a needle to float on the surface of a liquid. The surface-tension force across any imaginary line at a free surface is proportional to the length of the line and acts in a direction perpendicular to it. The surface tension per unit length s is expressed in Newtons per meter (N/m) in SI units. There is a slight decrease in surface tension with increasing temperature.

VAPOR PRESSURE

Liquid molecules that possess sufficient kinetic energy are projected out of the main body of a liquid at its free surface and pass into the vapor. The pressure exerted by this vapor is known as the vapor pressure p_v. In SI units vapor pressure is expressed in kilonewtons per square meter (kN/m²). The vapor pressure of water at 15°C is 1.72 kN/m².

REFERENCES

Vennard, J.K., and R.L. Street (1975) *Elementary Fluid Mechanics,* 5th ed., Wiley, New York.
Webber, N.B. (1971) *Fluid Mechanics for Civil Engineers,* SI ed., Chapman and Hall, London.

Table C-1
Physical Properties of Water (SI Units)[a]

Temp-erature, °C	Specific weight γ, kN/m³	Density[b] ρ, kg/m³	Modulus of elasticity[b] $E/10^6$, kN/m²	Dyamic viscosity, $\mu \times 10^3$, N·s/m²	Kinematic viscosity, $\nu \times 10^6$, m²/s	Surface tension[c] σ, N/m	Vapor pressure p_v, kN/m²
0	9.805	999.8	1.98	1.781	1.785	0.0765	0.61
5	9.807	1000.0	2.05	1.518	1.519	0.0749	0.87
10	9.804	999.7	2.10	1.307	1.306	0.0742	1.23
15	9.798	999.1	2.15	1.139	1.139	0.0735	1.70
20	9.789	998.2	2.17	1.002	1.003	0.0728	2.34
25	9.777	997.0	2.22	0.890	0.893	0.0720	3.17
30	9.764	995.7	2.25	0.798	0.800	0.0712	4.24
40	9.730	992.2	2.28	0.653	0.658	0.0696	7.38
50	9.689	988.0	2.29	0.547	0.553	0.0679	12.33
60	9.642	983.2	2.28	0.466	0.474	0.0662	19.92
70	9.589	977.8	2.25	0.404	0.413	0.0644	31.16
80	9.530	971.8	2.20	0.354	0.364	0.0626	47.34
90	9.466	965.3	2.14	0.315	0.326	0.0608	70.10
100	9.399	958.4	2.07	0.282	0.294	0.0589	101.33

[a] Adapted from Vennard and Street (1975).
[b] At atmospheric pressure.
[c] In contact with air.

Table C–2
Physical properties of water (U.S. customary units)[a]

Temperature, °F	Specific weight γ, lb/ft³	Density[b] ρ, slug/ft³	Modulus of elasticity[b] $E/10^3$, lb$_f$/in²	Dyamics viscosity, $\mu \times 10^5$, lb·s/ft²	Kinematic viscosity, $\nu \times 10^5$, ft²/s	Surface tension[c] σ, lb/ft	Vapor pressure p_{v}, lb$_f$/in²
32	62.42	1.940	287	3.746	1.931	0.00518	0.09
40	62.43	1.940	296	3.229	1.664	0.00614	0.12
50	62.41	1.940	305	2.735	1.410	0.00509	0.18
60	62.37	1.938	313	2.359	1.217	0.00504	0.26
70	62.30	1.936	319	2.050	1.059	0.00498	0.36
80	62.22	1.934	324	1.799	0.930	0.00492	0.51
90	62.11	1.931	328	1.595	0.826	0.00486	0.70
100	62.00	1.927	331	1.424	0.739	0.00480	0.95
110	61.86	1.923	332	1.284	0.667	0.00473	1.27
120	61.71	1.918	332	1.168	0.609	0.00467	1.69
130	61.55	1.913	331	1.069	0.558	0.00460	2.22
140	61.38	1.908	330	0.981	0.514	0.00454	2.89
150	61.20	1.902	328	0.905	0.476	0.00447	3.72
160	61.00	1.896	326	0.838	0.442	0.00441	4.74
170	60.80	1.890	322	0.780	0.413	0.00434	5.99
180	60.58	1.883	318	0.726	0.385	0.00427	7.51
190	60.36	1.876	313	0.678	0.362	0.00420	9.34
200	60.12	1.868	308	0.637	0.341	0.00413	11.52
212	59.83	1.860	300	0.593	0.319	0.00404	14.70

[a] Adapted from Vennard and Street (1975).
[b] At atmospheric pressure.
[c] In contact with the air.

Solubility of Dissolved Oxygen in Water as a Function of Salinity and Barometric Pressure

Table D–1
Dissolved-oxygen concentration in water as a function of temperature and salinity
(barometric pressure = 760 mm Hg[a]

Temp. °C	Dissolved-oxygen concentration, mg/L									
	Salinity, parts per thousand									
	0	5	10	15	20	25	30	35	40	45
0	14.60	14.11	13.64	13.18	12.74	12.31	11.90	11.50	11.11	10.74
1	14.20	13.73	13.27	12.83	12.40	11.98	11.58	11.20	10.83	10.46
2	13.81	13.36	12.91	12.49	12.07	11.67	11.29	10.91	10.55	10.20
3	13.45	13.00	12.58	12.16	11.76	11.38	11.00	10.64	10.29	9.95
4	13.09	12.67	12.25	11.85	11.47	11.09	10.73	10.38	10.04	9.71
5	12.76	12.34	11.94	11.56	11.18	10.82	10.47	10.13	9.80	9.48
6	12.44	12.04	11.65	11.27	10.91	10.56	10.22	9.89	9.57	9.27
7	12.13	11.74	11.37	11.00	1065	10.31	9.98	9.66	9.35	9.06
8	11.83	11.46	11.09	10.74	10.40	10.07	9.75	9.44	9.14	8.85
9	11.55	11.19	10.83	10.49	10.16	9.84	9.53	9.23	8.94	8.66
10	11.28	10.92	10.58	10.25	9.93	9.62	9.32	9.03	8.75	8.47
11	11.02	10.67	10.34	10.02	9.71	9.41	9.12	8.83	8.56	8.30
12	10.77	10.43	10.11	9.80	9.50	9.21	8.92	8.65	8.38	8.12
13	10.53	10.20	9.89	9.59	9.30	9.01	8.74	8.47	8.21	7.96
14	10.29	9.98	9.68	9.38	9.10	8.82	8.55	8.30	8.04	7.80
15	10.07	9.77	9.47	9.19	8.91	8.64	8.38	8.13	7.88	7.65
16	9.86	9.56	9.28	9.00	8.73	8.47	8.21	7.97	7.73	7 50
17	9.65	9.36	9.09	8.82	8.55	8.30	8.05	7.81	7.58	7.36
18	9.45	9.17	8.90	8.64	8.39	8.14	7.90	7.66	7.44	7.22
19	9.26	8.99	8.73	8.47	8.22	7.98	7.75	7.52	7.30	7.09
20	9.08	8.81	8.56	8.31	8.07	7.83	7.60	7.38	7.17	6.96
21	8.90	8.64	8.39	8.15	7.91	7.69	7.46	7.25	7.04	6.84
22	8.73	8.48	8.23	8.00	7.77	7.54	7.33	1.12	6.91	6.72
23	8.56	8.32	8.08	7.85	7.63	7.41	7.20	6.99	6.79	6.60
24	8.40	8.16	7.93	7.71	7.49	7.28	7.07	6.87	6.68	6.49
25	8.24	8.01	7.79	7.57	7.36	7.15	6.95	6.75	6.56	6.38

(continued)

| **Table D–1** (Continued)

Temp. °C	Dissolved-oxygen concentration, mg/L									
	Salinity, parts per thousand									
	0	5	10	15	20	25	30	35	40	45
26	8.09	7.87	7.65	7.44	7.23	7.03	6.83	6.64	6.46	6.28
27	7.95	7.73	7.51	7.31	7.10	6.91	6.72	6.53	6.35	6.17
28	7.81	7.59	7.38	7.18	6.98	6.79	6.61	6.42	6.25	6.08
29	7.67	7.46	7.26	7.06	6.87	6.68	6.50	6.32	6.15	5.98
30	7.54	7.33	7.14	6.94	6.75	6.57	6.39	6.22	6.05	5.89
31	7.41	7.21	7.02	6.83	6.65	6.47	6.29	6.12	5.96	5.80
32	7.29	7.09	6.90	6.72	654	6.36	6.19	6.03	5.87	5.71
33	7.17	6.98	6.79	6.61	6.44	6.26	6.10	5.94	5.78	5.63
34	7.05	6.86	6.68	6.51	6.33	6.17	6.01	5.85	5.69	5 54
35	6.93	6.75	6.58	6.40	6.24	6.07	5.92	5.76	5.61	5 46
36	6.82	6.65	6.47	6.31	6.14	5.98	5.83	5.68	5.53	5.39
37	6.72	6.54	6.37	6.21	6.05	5.89	5.74	5.59	5.45	5.31
38	6.61	6.44	6.28	6.12	5.96	5.81	5.66	5.51	5.37	5 24
39	6.51	6.34	6.18	6.03	5.87	5.72	5.58	5.44	5.30	5.16
40	6.41	6.25	6.09	5.94	5.79	5.64	5.50	5.36	5.22	5.09

[a]From Colt, J. (1984) "Computation of Dissolved Gas Concentrations in Water as Functions of Temperature. Salinity ana Pressure," *American Fisheries Society Special Publication* 14, Bethesda, MD. 1984.

Table D–2

Dissolved-oxygen concentration in water as a function of temperature and barometric pressure (salinity = 0 ppt)[a]

Temp, °C	Dissolved-oxygen concentration, mg/L									
	Barometric pressure, millimeters of mercury									
	735	740	745	750	755	760	765	770	775	780
0	14.12	14.22	14.31	14.41	14.51	14.60	14.70	14.80	14.89	14.99
1	13.73	13.82	13.92	14.01	14.10	14.20	14.29	14.39	14.48	14.57
2	13.36	13.45	13.54	13.63	13.72	13.81	13.90	14.00	14.09	14.18
3	13.00	1109	13.18	13.27	13.36	1145	13.53	13.62	13.71	13.80
4	12.66	12.75	12.83	12.92	13.01	13.09	13.18	13.27	13.35	13.44
5	12.33	12.42	12.50	12.59	12.67	12.76	12.84	12.93	13.01	13.10
6	12.02	12.11	12.19	12.27	12.35	12.44	12.52	12.60	12.68	12.77
7	11.72	11.80	11.89	11.97	12.05	12.13	12.21	12.29	12.37	12.45
8	11.44	11.52	11.60	11.67	11.75	11.83	11.91	11.99	12.07	12.15
9	11.16	11.24	11.32	11.40	11.47	11.55	11.63	11.70	11.78	11.86
10	10.90	10.98	11.05	11.13	11.20	11.28	11.35	11.43	11.50	11.58
11	10.65	10.72	10.80	10.87	10.94	11.02	11.09	11.16	11.24	11.31
12	10.41	10.48	10.55	10.62	10.69	10.77	10.84	10.91	10.98	11.05
13	10.17	10.24	10.31	10.38	10.46	10.53	10.60	10.67	10.74	10.81
14	9.95	10.02	10.09	10.16	10.23	10.29	10.36	10.43	10.50	10.57
15	9.73	9.80	9.87	9.94	10.00	10.07	10.14	10.21	10.27	10.34
16	9.53	9.59	9.66	9.73	9.79	9.86	9.92	9.99	10.06	10.12
17	9.33	9.39	9.46	9.52	9.59	9.65	9.72	9.78	9.85	9.91
18	9.14	9.20	9.26	9.33	9.39	9.45	9.52	9.58	9.64	9.71
19	8.95	9.01	9.07	9.14	9.20	9.26	9.32	9.39	9.45	9.51
20	8.77	8.83	8.89	8.95	9.02	9.08	9.14	9.20	9.26	9.32
21	8.60	8.66	8.72	8.78	8.84	8.90	8.96	9.02	9.08	9.14
22	8.43	8.49	8.55	8.61	8.67	8.73	8.79	8.84	8.90	8.96
23	8.27	8.33	8.39	8.44	8.50	8.56	8.62	8.68	8.73	8.79
24	8.11	8.17	8.23	8.29	8.34	8.40	8.46	8.51	8.57	8.63
25	7.96	8.02	8.08	8.13	8.19	8.24	8.30	8.36	8.41	8.47
26	7.82	7.87	7.93	7.98	8.04	8.09	8.15	8.20	8.26	8.31
27	7.68	7.73	7.79	7.84	7.89	7.95	8.00	8.06	8.11	817
28	7.54	7.59	7.65	7.70	7.75	7.81	7.86	7.91	7.97	8.02
29	7.41	7.46	7.51	7.57	7.62	7.67	7.72	7.78	7.83	7.88
30	7.28	7.33	7.38	7.44	7.49	7.54	7.59	7.64	7.69	7.75

(continued)

| **Table D–2** (Continued)

Temp, °C	Dissolved-oxygen concentration, mg/L									
	Barometric pressure, millimeters of mercury									
	735	740	745	750	755	760	765	770	775	780
31	7.16	7.21	7.26	7.31	7.36	7.41	7.46	7.51	7.46	7.62
32	7.04	7.09	7.14	7.19	7.24	7.29	7.34	7.39	7.44	7.49
33	6.92	6.97	7.02	7.07	7.12	7.17	7.22	7.27	7.31	7.36
34	6.80	6.85	6.90	6.95	7.00	7.05	7.10	7.15	7.20	7.24
35	6.69	6.74	6.79	6.84	6.89	6.93	6.98	7.03	7.08	7.13
36	6.59	6.63	6.68	6.73	6.78	6.82	6.87	6.92	6.97	7.01
37	6.48	6.53	6.57	6.62	6.67	6.72	6.76	6.81	6.86	6.90
38	638	6.43	6.47	6.52	6.56	6.61	6.66	6.70	6.75	6.80
39	6.28	6.33	6.37	6.42	6.46	6.51	6.56	6.60	6.65	6.69
40	6.18	6.23	6.27	6.32	6.36	6.41	6.46	6.50	6.55	6.59

[a]From Colt, J. (1984) "Computation of Dissolved Gas Concentrations in Water as Functions of Temperature, Salinity. and Pressure," *American Fisheries Society Special Publication* 14, Bethesda, MD, 1984.

Note: ppt = parts per thousand.

MPN Tables and Their Use   **Appendix** E

When three serial sample volumes (e.g., dilutions) are used in the bacteriological testing of water, the resulting MPN (most probable number) values per 100 mL can be determined using Table E–1. The MPN values given there are based on serial sample volumes of 10, 1, and 0.1mL. If lower or higher serial sample volumes are used, the MPN values given in Table E–1 must be adjusted accordingly. For example, if sample volumes used are 100, 10, and 1 ml, the MPN values from the table are multiplied by 0.1. Similarly, if the sample volumes are 1, 0.1, and 0.01 ml, the MPN values from the table are multiplied by 10.

In situations where more than three test dilutions have been run, the following rule is applied to select the three dilutions to be used in determining the MPN value (Standard Methods, 1998): Choose the highest dilution that gives positive results in all five portions tested (no lower dilution giving any negative results) and the two next higher dilutions. Use the results at these three volumes in computing the MPN value. In the examples given in the accompanying table, the significant dilution results are shown in boldface. The number in the numerator represents positive tubes, that in the denominator, the total tubes planted.

	1.0 mL	0.1 mL	0.01 mL	0.001 mL	0.0001 mL	Combination of positives	MPN/ 100 mL
a	**5/5**	**5/5**	**2/5**	0/5		**5-2-0**	4900
b	**5/5**	5/5	**4/5**	**2/5**	0/5	**5-4-2**	2200
c	**5/5**	**0/5**	**1/5**	0/5	0/5	**0-1-0**	18
d	**5/5**	5/5	3/5	1/5	1/5		
(d)ᵃ	**5/5**	**5/5**	**3/5**	**2/5**	0/5	**5-3-2**	1400
e	**5/5**	4/5	1/5	1/5	0/5		
(e)ᵃ	**5/5**	**4/5**	**2/5**	0/5	0/5	**5-4-2**	2200

ᵃAdjusted values used to determine the MPN using Table E–1.

In example c, the first three dilutions are used so as to throw the positive result in the middle dilution. Where positive results occur in dilutions higher than the three chosen according to the above rule, they are incorporated into the result of the highest chosen dilution up to a total of five. The result of applying this procedure to the data is illustrated in examples d and e.

REFERENCE

Standard Methods (1998) *Standard Methods for the Examination of Water and Wastewater,* 20th ed., American Public Health Association, New York.

Table E-1
Most probable number (MPN) of coliforms per 100 mL of sample

10 mL	1 mL	0.1 mL	MPN	10 mL	1 mL	0.1 mL	MPN	10 mL	1 mL	0.1 mL	MPN
0	0	0		1	0	0	2.0	2	0	0	4.5
0	0	1	18	1	0	1	4.0	2	0	1	6.8
0	0	2	16	1	0	2	6.0	2	0	2	9.1
0	0	3	5.4	1	0	3	8.0	2	0	3	12
0	0	4	7.2	1	0	4	10	2	0	4	14
0	0	5	9.0	1	0	5	12	2	0	5	16
0	1	0	1.8	1	1	0	4.0	2	1	0	6.8
0	1	1	3.6	1	1	1	6.1	2	1	1	9.2
0	1	2	5.5	1	1	2	8.1	2	1	2	12
0	1	3	7.3	1	1	3	10	2	1	3	14
0	1	4	9.1	1	1	4	12	2	1	4	17
0	1	5	11	1	1	5	14	2	1	5	19
0	2	0	3.7	1	2	0	6.1	2	2	0	9.3
0	2	1	5.5	1	2	1	8.2	2	2	1	12
0	2	2	7.4	1	2	2	10	2	2	2	14
0	2	3	9.2	1	2	3	12	2	2	3	17
0	2	4	11	1	2	4	15	2	2	4	19
0	2	5	13	1	2	5	17	2	2	5	22
0	3	0	5.6	1	3	0	8.3	2	3	0	12
0	3	1	7.4	1	3	1	10	2	3	1	14
0	3	2	9.3	1	3	2	13	2	3	2	17
0	3	3	11	1	3	3	15	2	3	3	20
0	3	4	13	1	3	4	17	2	3	4	22
0	3	5	15	1	3	5	19	2	3	5	25
0	4	0	7.5	1	4	0	11	2	4	0	15
0	4	1	9.4	1	4	1	13	2	4	1	17
0	4	2	11	1	4	2	15	2	4	2	20
0	4	3	13	1	4	3	17	2	4	3	23
0	4	4	15	1	4	4	19	2	4	4	25
0	4	5	17	1	4	5	22	2	4	5	28
0	5	0	9.4	1	5	0	13	2	5	0	17
0	5	1	11	1	5	1	15	2	5	1	20
0	5	2	13	1	5	2	17	2	5	2	23
0	5	3	15	1	5	3	19	2	5	3	26
0	5	4	17	1	5	4	22	2	5	4	29
0	5	5	19	1	5	5	24	2	5	5	32

| **Table E-1** (Continued) |

Number of positive tubes				Number of positive tubes				Number of positive tubes			
10 mL	1 mL	0.1 mL	MPN	10 mL	1 mL	0.1 mL	MPN	10 mL	1 mL	0.1 mL	MPN
3	0	0	7.8	4	0	0	13	5	0	0	23
3	0	1	11	4	0	1	17	5	0	1	31
3	0	2	13	4	0	2	21	5	0	2	43
3	0	3	16	4	0	3	25	5	0	3	58
3	0	4	20	4	0	4	30	5	0	4	76
3	0	5	23	4	0	5	36	5	0	5	95
3	1	0	11	4	1	0	17	5	1	0	33
3	1	1	14	4	1	1	21	5	1	1	46
3	1	2	17	4	1	2	26	5	1	2	64
3	1	3	20	4	1	3	31	5	1	3	84
3	1	4	23	4	1	4	36	5	1	4	110
3	1	5	27	4	1	5	42	5	1	5	130
3	2	0	14	4	2	0	22	5	2	0	49
3	2	1	17	4	2	1	26	5	2	1	70
3	2	2	20	4	2	2	32	5	2	2	95
3	2	3	24	4	2	3	38	5	2	3	120
3	2	4	27	4	2	4	44	5	2	4	150
3	2	5	31	4	2	5	50	5	2	5	180
3	3	0	17	4	3	0	27	5	3	0	79
3	3	1	21	4	3	1	33	5	3	1	110
3	3	2	24	4	3	2	39	5	3	2	140
3	3	3	28	4	3	3	45	5	3	3	180
3	3	4	31	4	3	4	52	5	3	4	210
3	3	5	35	4	3	5	59	5	3	5	250
3	4	0	21	4	4	0	34	5	4	0	130
3	4	1	24	4	4	1	40	5	4	1	170
3	4	2	28	4	4	2	47	5	4	2	220
3	4	3	32	4	4	3	54	5	4	3	280
3	4	4	36	4	4	4	62	5	4	4	350
3	4	5	40	4	4	5	69	5	4	5	430
3	5	0	25	4	5	0	41	5	5	0	240
3	5	1	29	4	5	1	48	5	5	1	350
3	5	2	32	4	5	2	56	5	5	2	540
3	5	3	37	4	5	3	64	5	5	3	920
3	5	4	41	4	5	4	72	5	5	4	1600
3	5	5	45	4	5	5	81				

Carbonate Equilibrium

The chemical species that comprise the carbonate system include gaseous carbon dioxide $[(CO_2)_g]$, aqueous carbon dioxide $[(CO_2)_{aq}]$, carbonic acid $[H_2CO_3]$, bicarbonate $[HCO_3^-]$, carbonate $[CO_3^{2-}]$, and solids containing carbonates. In waters exposed to the atmosphere, the equilibrium concentration of dissolved CO_2 is a function of the liquid phase CO_2 mole fraction and the partial pressure of CO_2 in the atmosphere. Henry's law (see Chap. 2) is applicable to the CO_2 equilibrium between air and water; thus

$$x_g = \frac{P_T}{H} p_g \tag{2-46}$$

where x_g = mole fraction of gas in water, mole gas/mol water

$$= \frac{\text{mole gas } (n_g)}{\text{mole gas } (n_g) + \text{mole water } (n_w)}$$

P_T = total pressure, usually 1.0 atm

H = Henry's law constant, $\dfrac{\text{atm (mole gas/mole air)}}{\text{(mole gas/mole water)}}$

p_g = mole fraction of gas in air, mole gas /mole of air

The concentration of aqueous carbon dioxide is determined using Eq (2-46). At sea level, where the average atmospheric pressure is 1 atm, or 101.4 kPa, carbon dioxide comprises approximately 0.03 percent of the atmosphere. Values of the Henry's law constant for CO_2 as a function of temperature are given in Table F–1. The values in Table F–1 were computed using Eq. (2-48) and the data given in Table 2-8 in Chap. 2.

Aqueous carbon dioxide $[(CO_2)_{aq}]$ reacts reversibly with water to form carbonic acid.

$$(CO_2)_{aq} + H_2O \leftrightarrow H_2CO_3 \tag{F-1}$$

Table F–1
Henry's Law constant for CO_2 as a function of temperature

T, °C	H, atm
0	793
10	1072
20	1420
30	1846
40	2360
50	2971
60	3689

The corresponding equilibrium expression is

$$\frac{[H_2CO_3]}{[CO_2]} = K_m \tag{F-2}$$

The value of K_m at 25°C is 1.58×10^{-3}. Note that K_m is unitless. Because of the difficulty of differentiating between $(CO_2)_{aq}$ and H_2CO_3 in solution and the observation that very little H_2CO_3 is ever present in natural waters, an effective carbonic acid value $(H_2CO_3^*)$ is used which is defined as:

$$H_2CO_3^* \leftrightarrow (CO_2)_{aq} + H_2CO_3 \tag{F-3}$$

Because carbonic acid is a diprotic acid it will dissociate in two steps—first to bicarbonate and then to carbonate. The first dissociation of carbonic acid to bicarbonate can be represented as:

$$H_2CO_3^* \leftrightarrow H^+ + HCO_3^- \tag{F-4}$$

The corresponding equilibrium relationship is defined as:

$$\frac{[H^+][HCO_3^-]}{[H_2CO_3^*]} = K_{a1} \tag{F-5}$$

The value of first acid dissociation constant K_{a1} at 25°C is 4.467×10^{-7} mole/L. Values of K_{a1} at other temperatures are given in Table F–2, which is repeated here from Table 6-16 in Chap. 6.

The second dissociation of carbonic acid is from bicarbonate to carbonate as given below.

$$HCO_3^- \leftrightarrow H^+ + CO_3^{2-} \tag{F-6}$$

The corresponding equilibrium relationship is defined as:

$$\frac{[H^+][CO_3^{2-}]}{[HCO_3^-]} = K_{a2} \tag{F-7}$$

The value second acid dissociation constant K_{a2} at 25°C is 4.477×10^{-11} mole/L. Values of K_{a2} at other temperatures are given in Table F–2.

The distribution of carbonate species as function of pH is illustrated on Fig. F–1.

To illustrate the use of the data presented in Tables F–1 and F–2, it will be helpful to estimate the pH of a water, assuming the atmosphere above the water contains 30 percent CO_2, the bicarbonate (HCO_3^-) concentration in the water is 610 mg/L, and the temperature of the water is 20°C.

1. Use Eq. (2-47) to determine concentration of H_2CO_3 in the water. The value of Henry's constant from Table F–1 is 1420 atm, thus

$$x_{H_2CO_3} = \frac{P_T}{H} = \frac{0.30 \text{ atm}}{1420 \text{ atm}} = 2.113 \times 10^{-4}$$

Because one liter of water contains 55.6 mole [1000 g/(18 g/mole)], the mole fraction of H_2CO_3 is equal to:

Table F–2
Carbonate equilibrium constants as function of temperature[a]

Temperature, °C	Equilibrium constant, mole/L[b]	
	$K_{a1} \times 10^7$	$K_{a2} \times 10^{11}$
5	3.020	2.754
10	3.467	3.236
15	3.802	3.715
20	4.169	4.169
25	4.467	4.477
30	4.677	5.129
40	5.012	6.026

[a] Adapted from Table 6–16 in Chap. 6.
[b] The reported values have been multiplied by the indicated exponents. Thus, the value of K_{a2} at 20°C is equal to 4.169×10^{-11}.

Figure F–1

Log concentration versus pH diagram for a 10^{-3} molar solution of carbonate at 25°C. By sliding the constituent curves up or down, pH values can be obtained at different concentration values.

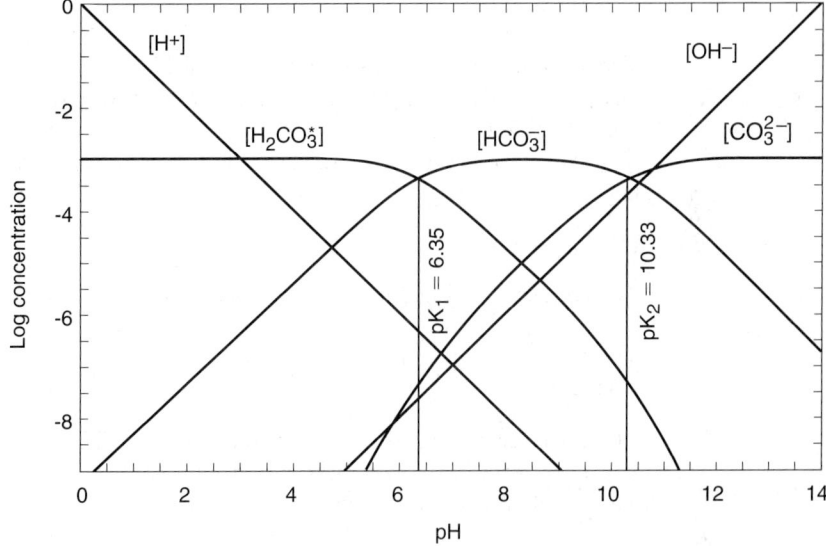

$$x_{H_2CO_3} = \frac{\text{mole gas } (n_g)}{\text{mole gas } (n_g) + \text{mole water } (n_w)}$$

$$2.113 \times 10^{-4} = \frac{[H_2CO_3]}{[H_2CO_3] + [55.6 \text{ mole/L}]}$$

Because the number of moles of dissolved gas in a liter of water is much less than the number of moles of water,

$$[H_2CO_3] \approx (2.113 \times 10^{-4})[55.6 \text{ mole/L}] \approx 11.75 \times 10^{-3} \text{ mole/L}$$

2. Use Eq. (F–5) to determine the pH of the water. The value of K_{a1} at 20°C from Table F–2 is 4.169×10^{-7} mole/L, thus

$$[H^+] = \frac{K_{a1}[H_2CO_3^*]}{[HCO_3^-]}$$

Substitute known values and solve for $[H^+]$

$$[H^+] = \frac{[4.169 \times 10^{-7} \text{ mole/L}][11.75 \times 10^{-3} \text{ mole/L}]}{[(610 \text{ mg/L})/(61{,}000 \text{ mg/mole})]}$$

$$= 4.90 \times 10^{-7} \text{ mole/L}$$

$$pH = -\log[H^+] = -\log [4.90 \times 10^{-7}] = 6.31$$

From Fig. F–1, the amount of carbonate present is essentially non measurable.

Moody Diagrams for the Analysis of Flow in Pipes

Appendix G

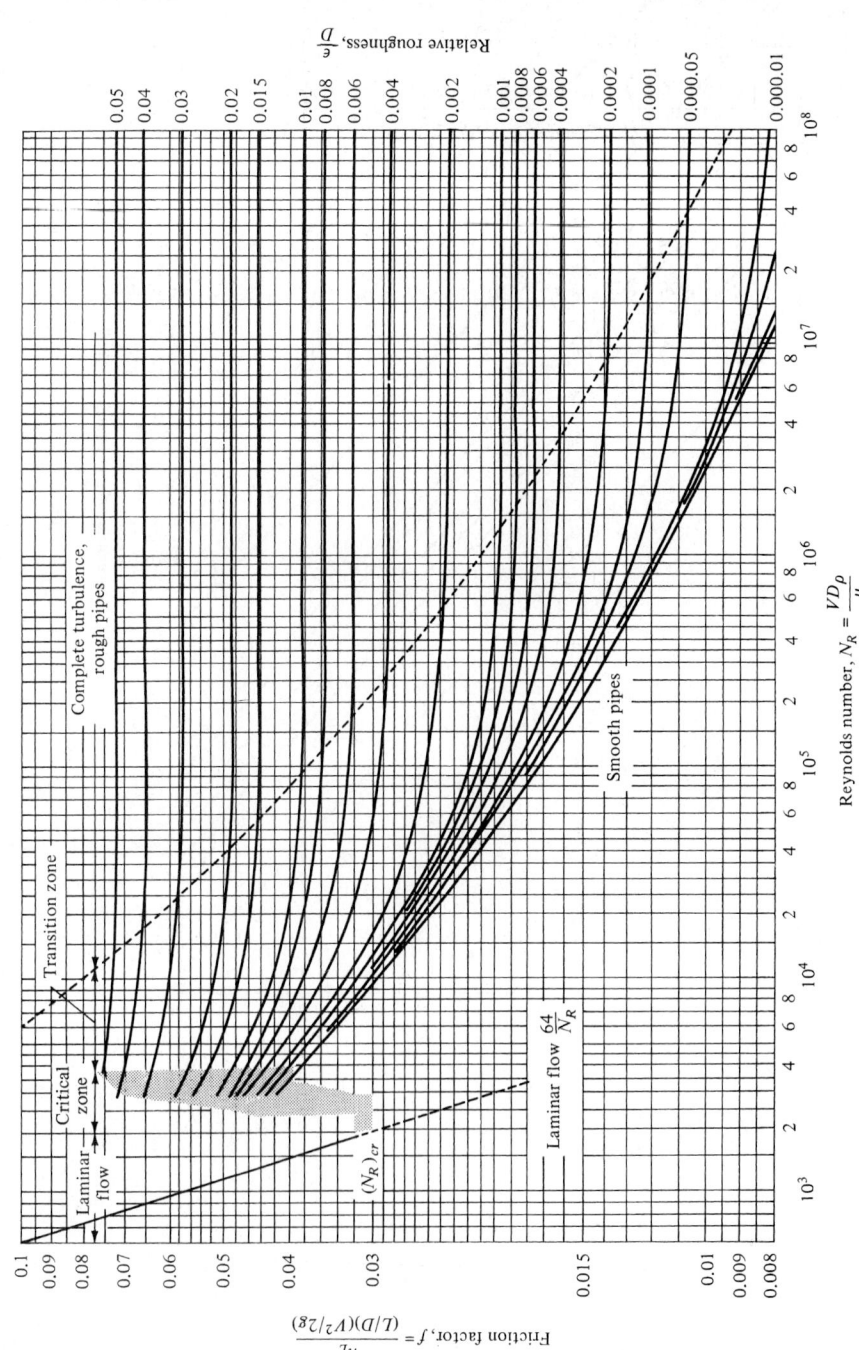

Figure G–1

Moody diagram for friction factor in pipes versus Reynolds number and relative roughness. (From Moody, L.F. (1944) Friction Factors for Pipe Flow, Transactions American Society of Civil Engineers vol. 66, p. 671.)

Figure G–2

Moody diagram for relative roughness as a function of diameter for pipes constructed of various materials. (Adapted from Moody, L.F. (1944) Friction Factors for Pipe Flow, Transactions American Society of Civil Engineers vol. 66, p. 671.)

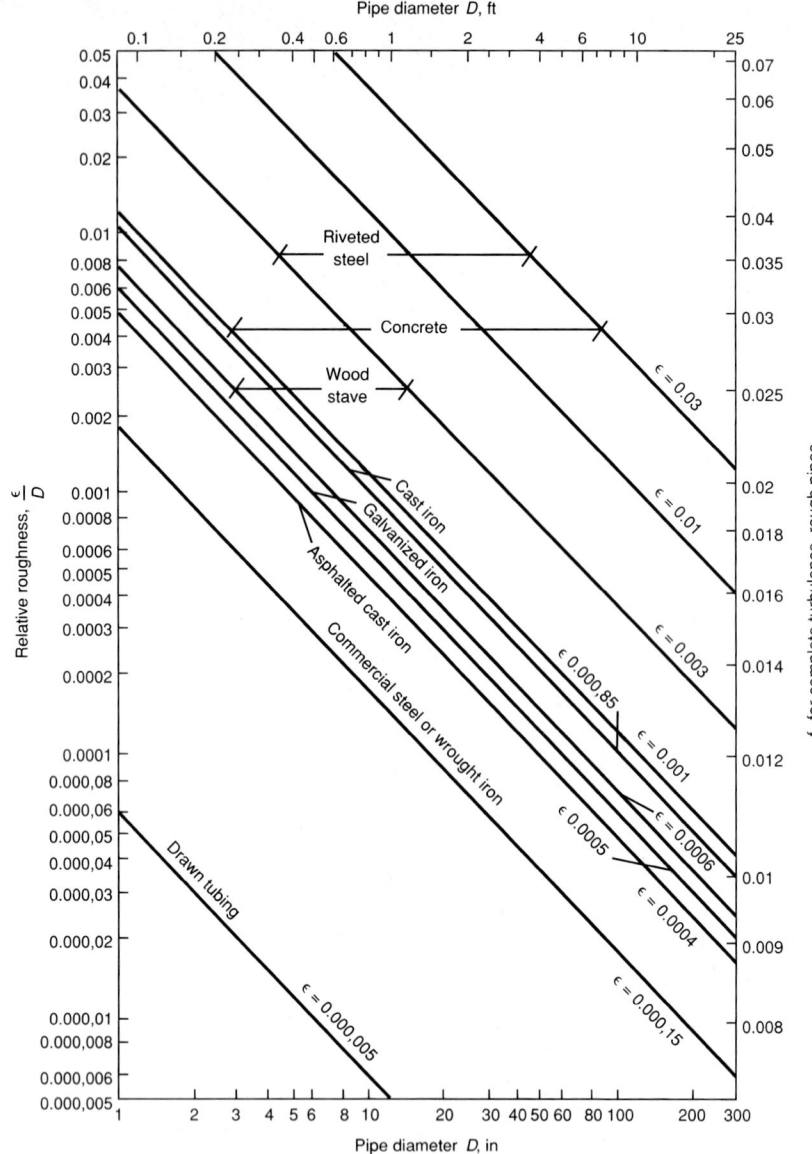

Name Index

Subject Index

Because a number of the subjects covered in this text can be referenced (i.e., indexed) under different alphabetical listings, it has been necessary to develop an approach to limit the degree of duplication, yet not affect the utility of the index. The approach used is as follows. Each subject with multiple subentries is indexed in detail under one letter of the alphabet. Where the same subject is indexed under another letter of the alphabet, inclusive page numbers are given and a *See also* citation is given to the location where the subject is indexed in detail. For example, Activated carbon adsorption, 1138–1162, is followed by (See also Adsorption). Where a subject is indexed in detail (e.g., Biochemical oxygen demand) and a commonly used abbreviation (e.g., BOD) is also cited under the same letter of the alphabet, the abbreviated entry is followed by a *See* citation. For example, the index entry BOD is followed by (See Biochemical oxygen demand). Where an abbreviation occurs under a different letter of the alphabet than where the detailed citation is given, inclusive page numbers are given followed by the *See also* citation. For older or unused terms, the *see* citation is used to direct the reader to the appropriate term used. To access the number of data tables in the textbook more easily, an index entry followed by the capital letter T in parenthesis [e.g., (T)] is used to denote a data table related to the subject matter.

Atomic numbers and atomic masses[a]

Actinium	Ac	89	227.0278	Mercury	Hg	80	200.59
Aluminum	Al	13	26.98154	Molybdenum	Mo	42	95.94
Americium	Am	95	(243)	Neodymium	Nd	60	144.24
Antimony	Sb	51	121.75	Neon	Ne	10	20.179
Argon	Ar	18	39.948	Neptunium	Np	93	237.0482
Arsenic	As	33	74.9216	Nickel	Ni	28	58.70
Astatine	At	85	(210)	Niobium	Nb	41	92.9064
Barium	Ba	56	137.33	Nitrogen	N	7	14.0067
Berkelium	Bk	97	(247)	Nobelium	No	102	(259)
Beryllium	Be	4	9.01218	Osmium	Os	76	190.2
Bismuth	Bi	83	208.9804	Oxygen	O	8	15.9994
Boron	B	5	10.81	Palladium	Pd	46	106.4
Bromine	Br	35	79.904	Phosphorus	P	15	30.97376
Cadmium	Cd	48	112.41	Platinum	Pt	78	195.09
Calcium	Ca	20	40.08	Plutonium	Pu	94	(244)
Californium	Cf	98	(251)	Polonium	Po	84	(209)
Carbon	C	6	12.011	Potassium	K	19	39.0983
Cerium	Ce	58	140.12	Praseodymium	Pr	59	140.9077
Cesium	Cs	55	132.9054	Promethium	Pm	61	(145)
Chlorine	Cl	17	35.453	Protactinium	Pa	91	231.0389
Chromium	Cr	24	51.996	Radium	Ra	88	226.0254
Cobalt	Co	27	58.9332	Radon	Rn	86	(222)
Copper	Cu	29	63.546	Rhenium	Re	75	186.207
Curium	Cm	96	(247)	Rhodium	Rh	45	102.9055
Dysprosium	Dy	66	162.50	Rubidium	Rb	37	85.4678
Einsteinium	Es	99	(254)	Ruthenium	Ru	44	101.07
Erbium	Er	68	167.26	Samarium	Sm	62	150.4
Europium	Eu	63	151.96	Scandium	Sc	21	44.9559
Fermium	Fm	100	(257)	Selenium	Se	34	78.96
Fluorine	F	9	18.99840	Silicon	Si	14	28.0855
Francium	Fr	87	(223)	Silver	Ag	47	107.868
Gadolinium	Gd	64	157.25	Sodium	Na	11	22.98977
Gallium	Ga	31	69.72	Strontium	Sr	38	87.62
Germanium	Ge	32	72.59	Sulfur	S	16	32.06
Gold	Au	79	196.9665	Tantalum	Ta	73	180.9479
Hafnium	Hf	72	178.49	Technetium	Tc	43	(97)
Helium	He	2	4.00260	Tellurium	Te	52	127.60
Holmium	Ho	67	164.9304	Terbium	Tb	65	158.9254
Hydrogen	H	1	1.0079	Thallium	Tl	81	204.37
Indium	In	49	114.82	Thorium	Th	90	232.0381
Iodine	I	53	126.9045	Thulium	Tm	69	168.9342
Iridium	Ir	77	192.22	Tin	Sn	50	118.69
Iron	Fe	26	55.847	Titanium	Ti	22	47.90
Krypton	Kr	36	83.80	Tungsten	W	74	183.85
Lanthanum	La	57	138.9055	Uranium	U	92	238.029
Lawrencium	Lr	103	(260)	Vanadium	V	23	50.9414
Lead	Pb	82	207.2	Xenon	Xe	54	131.30
Lithium	Li	3	6.941	Ytterbium	Yb	70	173.04
Lutetium	Lu	71	174.97	Yttrium	Y	39	88.9059
Magnesium	Mg	12	24.305	Zinc	Zn	30	65.38
Manganese	Mn	25	54.9380	Zirconium	Zr	40	91.22
Mendelevium	Md	101	(258)				

[a]From *Pure Applied Chemistry.*, vol. 47, p. 75 (1976). A value in parentheses is the mass number of the longest lived isotope of the element.